LDA lithium diisopropylamide
LDMAN lithium 1-(dimethylamino)naphthalenide
LHMDS =LiHMDS
LICA lithium isopropylcyclohexylamide
LiHMDS lithium hexamethyldisilazide
LiTMP lithium 2,2,6,6-tetramethylpiperidide
LTMP =LiTMP
LTA lead tetraacetate
lut lutidine

m-CPBA *m*-chloroperbenzoic acid
MA maleic anhydride
MAD methylaluminum bis(2,6-di-*t*-butyl-4-methylphenoxide)
MAT methylaluminum bis(2,4,6-tri-*t*-butylphenoxide)
Me methyl
MEK methyl ethyl ketone
MEM (2-methoxyethoxy)methyl
MIC methyl isocyanate
MMPP magnesium monoperoxyphthalate
MOM methoxymethyl
MoOPH oxodiperoxomolybdenum(pyridine)-(hexamethylphosphoric triamide)
mp melting point
MPM =PMB
Ms mesyl (methanesulfonyl)
MS mass spectrometry; molecular sieves
MTBE methyl *t*-butyl ether
MTM methylthiomethyl
MVK methyl vinyl ketone

n refractive index
NaHDMS sodium hexamethyldisilazide
Naph naphthyl
NBA *N*-bromoacetamide
nbd norbornadiene (bicyclo[2.2.1]hepta-2,5-diene)
NBS *N*-bromosuccinimide
NCS *N*-chlorosuccinimide
NIS *N*-iodosuccinimide
NMO *N*-methylmorpholine *N*-oxide
NMP *N*-methyl-2-pyrrolidinone
NMR nuclear magnetic resonance
NORPHOS bis(diphenylphosphino)bicyclo[2.2.1]-hept-5-ene
Np =Naph

PCC pyridinium chlorochromate
PDC pyridinium dichromate
Pent *n*-pentyl
Ph phenyl
phen 1,10-phenanthroline
Phth phthaloyl
Piv pivaloyl
PMB *p*-methoxybenzyl
PMDTA *N,N,N′,N″,N″*-pentamethyldiethylenetriamine
PPA polyphosphoric acid
PPE polyphosphate ester
PPTS pyridinium *p*-toluenesulfonate
Pr *n*-propyl
PTC phase transfer catalyst/catalysis
PTSA *p*-toluenesulfonic acid
py pyridine

RAMP (R)-1-amino-2-(methoxymethyl)pyrrolidine
rt room temperature

salen bis(salicylidene)ethylenediamine
SAMP (*S*)-1-amino-2-(methoxymethyl)pyrrolidine
SET single electron transfer
Sia siamyl (3-methyl-2-butyl)

TASF tris(diethylamino)sulfonium difluorotrimethylsilicate
TBAB tetrabutylammonium bromide
TBAF tetrabutylammonium fluoride
TBAD =DBAD
TBAI tetrabutylammonium iodide
TBAP tetrabutylammonium perruthenate
TBDMS *t*-butyldimethylsilyl
TBDPS *t*-butyldiphenylsilyl
TBHP *t*-butyl hydroperoxide
TBS =TBDMS
TCNE tetracyanoethylene
TCNQ 7,7,8,8-tetracyanoquinodimethane
TEA triethylamine
TEBA triethylbenzylammonium chloride
TEBAC =TEBA
TEMPO 2,2,6,6-tetramethylpiperidinoxyl
TES triethylsilyl
Tf triflyl (trifluoromethanesulfonyl)
TFA trifluoroacetic acid
TFAA trifluoroacetic anhydride
THF tetrahydrofuran
THP tetrahydropyran; tetrahydropyranyl
Thx thexyl (2,3-dimethyl-2-butyl)
TIPS triisopropylsilyl
TMANO trimethylamine *N*-oxide
TMEDA *N,N,N′,N′*-tetramethylethylenediamine
TMG 1,1,3,3-tetramethylguanidine
TMS trimethylsilyl
Tol *p*-tolyl
TPAP tetrapropylammonium perruthenate
TBHP *t*-butyl hydroperoxide
TPP tetraphenylporphyrin
Tr trityl (triphenylmethyl)
Ts tosyl (*p*-toluenesulfonyl)
TTN thallium(III) nitrate

UHP urea–hydrogen peroxide complex

Z =Cbz

Handbook of Reagents
for Organic Synthesis

Reagents for Organocatalysis

OTHER TITLES IN THIS COLLECTION

Reagents for Heteroarene Functionalization
Edited by André Charette
ISBN 978 1 118 72659 4

Catalytic Oxidation Reagents
Edited by Philip L. Fuchs
ISBN 978 1 119 95327 2

Reagents for Silicon-Mediated Organic Synthesis
Edited by Philip L. Fuchs
ISBN 978 0 470 71023 4

Sulfur-Containing Reagents
Edited by Leo A. Paquette
ISBN 978 0 470 74872 5

Reagents for Radical and Radical Ion Chemistry
Edited by David Crich
ISBN 978 0 470 06536 5

Catalyst Components for Coupling Reactions
Edited by Gary A. Molander
ISBN 978 0 470 51811 3

Fluorine-Containing Reagents
Edited by Leo A. Paquette
ISBN 978 0 470 02177 4

Reagents for Direct Functionalization for C-H Bonds
Edited by Philip L. Fuchs
ISBN 0 470 01022 3

Reagents for Glycoside, Nucleotide, and Peptide Synthesis
Edited by David Crich
ISBN 0 470 02304 X

Reagents for High-Throughput Solid-Phase and Solution-Phase Organic Synthesis
Edited by Peter Wipf
ISBN 0 470 86298 X

Chiral Reagents for Asymmetric Synthesis
Edited by Leo A. Paquette
ISBN 0 470 85625 4

Activating Agents and Protecting Groups
Edited by Anthony J. Pearson and William R. Roush
ISBN 0 471 97927 9

Acidic and Basic Reagent
Edited by Hans J. Reich and James H. Rigby
ISBN 0 471 97925 2

Oxidizing and Reducing Agents
Edited by Steven D. Burke and Rick L. Danheiser
ISBN 0 471 97926 0

Reagents, Auxiliaries, and Catalysts for C–C Bond Formation
Edited by Robert M. Coates and Scott E. Denmark
ISBN 0 471 97924 4

Handbook of Reagents for Organic Synthesis

Reagents for Organocatalysis

Edited by

Tomislav Rovis
Colorado State University, Fort Collins, CO, USA

WILEY

This edition first published 2016

Registered office
John Wiley & Sons Ltd, The Atrium, Southern Gate, Chichester, West Sussex, PO19 8SQ, United Kingdom

For details of our global editorial offices, for customer services and for information about how to apply for permission to reuse the copyright material in this book please see our website at www.wiley.com.

Library of Congress Cataloging-in-Publication Data

Names: Rovis, Tomislav, 1968-
Title: Handbook of reagents for organic synthesis : reagents for organocatalysis / edited by Tomislav Rovis, Colorado State University, Fort Collins, CO, USA.
Description: Chichester, West Sussex : John Wiley & Sons, Inc., 2016. | Includes bibliographical references and index.
Identifiers: LCCN 2016023290 (print) | LCCN 2016023985 (ebook) | ISBN 9781119061007 (cloth) | ISBN 9781119061014 (pdf) | ISBN 9781119061021 (epub)
Subjects: LCSH: Organic compounds–Synthesis–Handbooks, manuals, etc. | Catalysis. | Chemical tests and reagents.
Classification: LCC QD262 .H26745 2016 (print) | LCC QD262 (ebook) | DDC 547/.215–dc23
LC record available at https://lccn.loc.gov/2016023290

A catalogue record for this book is available from the British Library.

ISBN 13: 978-1-119-06100-7

Set in 9½/11½ pt Times Roman by Thomson Press (India) Ltd., New Delhi.
Printed and bound in Singapore by Markono Print Media Pte Ltd.

e-EROS Editorial Board

Contents

Preface

The eight-volume *Encyclopedia of Reagents for Organic Synthesis (EROS)*, authored and edited by experts in the field, first published in 1995, provided mini-reviews describing the properties and reactions of approximately 3000 reagents. In 2002, the entire *EROS* collection with updates and additions was made available on the Internet under the acronym *e-EROS*. The second edition of the encyclopedia, *EROS-II*, was published in March 2009 containing the entire collection of reagents - 4111 at the time of publication in a 14-volume set. While the comprehensive nature of *EROS-II* and the dynamic expansion of *e-EROS* render them invaluable as reference works, their very size limits their practicability in a laboratory environment. For this reason, a series of sharply targeted and inexpensive one-volume *Handbooks of Reagents for Organic Synthesis (HROS)* was introduced by the original editors of EROS in 1999:

Reagents, Auxiliaries and Catalysts for C–C Bond Formation
Edited by Robert M. Coates and Scott E. Denmark

Oxidizing and Reducing Agents
Edited by Steven D. Burke and Rick L. Danheiser

Acidic and Basic Reagents
Edited by Hans J. Reich and James H. Rigby

Activating Agents and Protecting Groups
Edited by Anthony J. Pearson and William R. Roush

This series has continued over the last several years with the publication of a further series of *HROS* volumes, each edited by a current member of the *e-EROS* editorial board:

Chiral Reagents for Asymmetric Synthesis
Edited by Leo A. Paquette

Reagents for High-Throughput Solid-Phase and Solution-Phase Organic Synthesis
Edited by Peter Wipf

Reagents for Glycoside, Nucleotide, and Peptide Synthesis
Edited by David Crich

Reagents for Direct Functionalization of C–H Bonds
Edited by Philip L. Fuchs

Fluorine-Containing Reagents
Edited by Leo A. Paquette

Catalyst Components for Coupling Reactions
Edited by Gary A. Molander

Reagents for Radical and Radical Ion Chemistry
Edited by David Crich

Sulfur-Containing Reagents
Edited by Leo A. Paquette

Reagents for Silicon-Mediated Organic Synthesis
Edited by Philip L. Fuchs

Catalytic Oxidation Reagents
Edited by Philip L. Fuchs

Reagents for Heteroarene Functionalization
Edited by André B. Charette

This series now continues with the present volume entitled Reagents for Organocatalysis, edited by Tomislav Rovis, a member of the online e-EROS Editorial Board. This 16th volume in the HROS series is an affordable and practical compilation of reagents of broad and timely significance. We believe our colleagues will find it useful to own and routinely employ.

Philip L. Fuchs
Purdue University
West Lafayette, IN, USA

Introduction

Organocatalysis, the concept of using small organic molecules as catalysts for organic transformations, has its origins in sporadic and unrelated sprinklings of papers published throughout the course of the last century. The most notable of these is the proline-catalyzed Hajos-Parrish reaction, but there were important contributions by others as well, from Cinchona alkaloid catalysis of ketene aldehyde additions to the use of chiral phase transfer catalysts to solve a problem facing Merck process chemists. The concept was arguably invigorated by near simultaneous demonstrations that catalytic amounts of chiral ketones were capable of epoxidizing olefins in the presence of oxone to the advent and use of hydrogen bond donor catalysts for a number of organic transformations in the late 1990s. The field was christened Organocatalysis by MacMillan in a series of important papers around the turn of the millennium. This event, although not the first appearance of the term, gave the field an identity and a flag to rally around.

More than a decade later, Organocatalysis is now a vibrant field, wide-ranging in impact and broad in coverage. Numerous catalysts have been introduced for a myriad of transformations. Catalytic functional groups range from amines to ketones, phosphoric acids, ureas and thioureas, ammonium salts, diols, and amidines and carbenes, to name but a few. Bond transformations amenable to its control similarly span the entire spectrum: oxidations, reductions, rearrangements, coupling, and carbon-carbon bond-forming reactions of all types. The creativity of chemists is on full display in this field, and it is clear that more is yet to come.

Perhaps one of the most lasting outcomes of this body of work is the (re)definition of concepts and our deep understanding of organic chemistry. Hydrogen-bond-mediated control has led to reevaluation of its strength and nature; divalent carbon is no longer a mere curiosity; electrostatic, ion pairing, and solvent cage effects are not only meaningful but can also be controlled to arrive at a desired outcome.

Complete coverage of this field within the confines of a single volume would be close to impossible. With this volume, we have attempted to select as wide a variety of catalysts, reagents, and occasionally substrates for prototypical organocatalytic reactions as possible. Contributions have been made by many of the field's leading experts, often discussing catalysts developed within their own laboratories. Numerous reagents have been updated to reflect the current state of the art in the field.

As a further service to the field, a list of recent reviews and monographs follows as an introduction to the interested reader.

Tomislav Rovis
Colorado State University
Fort Collins, CO, USA

Catalyst Finder

What follows is a compilation of the catalysts in this volume, grouped by type. We begin with achiral catalysts of all reaction families and then delineate chiral, single enantiomer catalysts by reactant type.

Achiral

Amine

Brønsted Acid

Chiral Ketone

Hydrogen Bond

Lewis Base

A list of General Abbreviations appears on the front Endpapers

N-Heterocyclic Carbene (NHC)

Phase Transfer

Phosphine

Recent Review Articles and Monographs

Recent Reviews

Abbasov, M. E.; Romo, D. The ever-expanding role of asymmetric covalent organocatalysis in scalable, natural product synthesis. *Nat. Prod. Rep.* **2014**, *31*, 1318–1327.

Aitken, L. S.; Arezki, N. R.; Dell'Isola, A.; Cobb, A. J. A. Asymmetric organocatalysis and the nitro group functionality. *Synthesis* **2013**, *45*, 2627–2648.

Akiyama, T.; Itoh, J.; Fuchibe, K. Recent progress in chiral Brønsted acid catalysis. *Adv. Synth. Catal.* **2006**, *348*, 999-1010.

Akiyama, T.; Mori, K. Stronger Brønsted acids: recent progress. *Chem. Rev.* **2015**, *115*, 9277–9306.

Albrecht, L.; Jiang, H.; Jorgensen, K. A. A simple recipe for sophisticated cocktails: organocatalytic one-pot reactions–concept, nomenclature, and future perspectives. *Angew. Chem., Int. Ed.* **2011**, *50*, 8492–8509.

Aleman, J.; Cabrera, S. Applications of asymmetric organocatalysis in medicinal chemistry. *Chem. Soc. Rev.* **2013**, *42*, 774–793.

Angnes, R. A.; Li, Z.; Correia, C. R. D.; Hammond, G. B. Recent synthetic additions to the visible light photoredox catalysis toolbox. *Org. Biomol. Chem.* **2015**, *13*, 9152–9167.

Atodiresei, I.; Vila, C.; Rueping, M. Asymmetric organocatalysis in continuous flow: opportunities for impacting industrial catalysis. *ACS Catal.* **2015**, *5*, 1972–1985.

Bae, H. Y. Malonic acid half thioesters (MAHTs) as efficient enolate precursors in biomimetic catalysis. *Synlett* **2015**, *26*, 705–706.

Beatty, J. W.; Stephenson, C. R. J. Amine functionalization via oxidative photoredox catalysis: methodology development and complex molecule synthesis. *Acc. Chem. Res.* **2015**, *48*, 1474–1484.

Bencivenni, G. Organocatalytic strategies for the synthesis of axially chiral compounds. *Synlett* **2015**, *26*, 1915–1922.

Bertelsen, S.; Nielsen, M.; Joergensen, K. A. Radicals in asymmetric organocatalysis. *Chem., Int. Ed.* **2007**, *46*, 7356–7359.

Biju, A. T.; Kuhl, N.; Glorius, F. Extending NHC-catalysis: coupling aldehydes with unconventional reaction partners. *Acc. Chem. Res.* **2011**, *44*, 1182–1195.

Busschaert, N.; Caltagirone, C.; Van Rossom, W.; Gale, P. A. Applications of supramolecular anion recognition. *Chem. Rev.* **2015**, *115*, 8038–8115.

Chauhan, P.; Mahajan, S.; Enders, D. Organocatalytic carbon-sulfur bond-forming reactions. *Chem. Rev.* **2014**, *114*, 8807–8864.

Cheong, P. H.-Y.; Legault, C. Y.; Um, J. M.; Celebi-Olcum, N.; Houk, K. N. Quantum mechanical investigations of organocatalysis: mechanisms, reactivities, and selectivities. *Chem. Rev.* **2011**, *111*, 5042–5137.

Davis, R. L.; Stiller, J.; Naicker, T.; Jiang, H.; Jorgensen, K. A. Asymmetric organocatalytic epoxidations: reactions, scope, mechanisms, and applications. *Angew. Chem., Int. Ed.* **2014**, *53*, 7406–7426.

Donslund, B. S.; Johansen, T. K.; Poulsen, P. H.; Halskov, K. S.; Jorgensen, K. A. *Angew. Chem., Int. Ed.* **2015**, *54*, 13860–13874.

Douglas, J.; Churchill, G.; Smith, A. D. NHCs in asymmetric organocatalysis: recent advances in azolium enolate generation and reactivity. *Synthesis* **2012**, *44*, 2295–2309.

Enders, D.; Niemeier, O.; Henseler, A. Organocatalysis by N-heterocyclic carbenes. *Chem. Rev.* **2007**, *107*, 5606–5655.

Flanigan, D. M.; Romanov-Michailidis, F.; White, N. A.; Rovis, T. Organocatalytic reactions enabled by *N*-heterocyclic carbenes. *Chem. Rev.* **2015**, *115*, 9307–9387.

Fraile, A.; Aleman, J. Inter- and intramolecular dienamine organocatalytic strategies for the synthesis of tetrahydroisoquinolines and tricyclic derivatives via [3+2] and [4+2] cycloadditions. *Synlett* **2015**, *26*, 1940–1954.

Holland, M. C.; Gilmour, R. Deconstructing covalent organocatalysis. *Angew. Chem., Int. Ed.* **2015**, *54*, 3862–3871.

Jacobsen, E. N.; MacMillan, D. W. C. Organocatalysis. *Proc. Natl. Acad. Sci. USA*, **2010**, *107*, 20618–20619.

James, T.; van Gemmeren, M.; List, B. Development and applications of disulfonimides in enantioselective organocatalysis. *Chem. Rev.* **2015**, *115*, 9388–9409.

Kiesewetter, M. K.; Shin, E. J.; Hedrick, J. L.; Waymouth, R. M. Organocatalysis: opportunities and challenges for polymer synthesis. *Macromolecules* **2010**, *43*, 2093–2107.

Liu, T.-Y.; Xie, M.; Chen, Y.-C. Organocatalytic asymmetric transformations of modified Morita-Baylis-Hillman adducts. *Chem. Soc. Rev.* **2012**, *41*, 4101–4112.

Longbottom, D. A.; Franckevicius, V.; Kumarn, S.; Oelke, A. J.; Wascholowski, V.; Ley, S. V. Practical organocatalysis with *(S)*- and *(R)*-5-pyrrolidin-2-yl-1*H*-tetrazoles. *Aldrichimica Acta* **2008**, *41*, 3–11.

MacMillan, D. W. C. The advent and development of organocatalysis. *Nature* **2008**, *455*, 304–308.

Mahlau, M.; List, B. Asymmetric counteranion-directed catalysis (ACDC): a remarkably general approach to enantioselective synthesis. *Israel J. Chem.* **2012**, *52*, 630–638.

Marcelli, T.; Hiemstra, H. Cinchona alkaloids in asymmetric organocatalysis. *Synthesis* **2010**, 1229–1279.

Marinetti, A.; Voituriez, A. Enantioselective phosphine organocatalysis. *Synlett* **2010**, 174–194.

Moore, J. L.; Rovis, T. Lewis base catalysts 6: carbene catalysts. *Top. Curr. Chem.* **2009**, *290*, 77–144.

Nicewicz, D. A.; Hamilton, D. S. Organic photoredox catalysis as a general strategy for anti-Markovnikov alkene hydrofunctionalization. *Synlett* **2014**, *25*, 1191–1196.

Ouellet, S. G; Walji, A. M.; MacMillan, D. W C. Enantioselective organocatalytic transfer hydrogenation reactions using Hantzsch esters. *Acc. Chem. Res.* **2007**, *40*, 1327–1339.

Prier, C. K.; Rankic, D. A.; MacMillan, D. W. C. Visible light photoredox catalysis with transition metal complexes: applications in organic synthesis. *Chem. Rev.* **2013**, *113*, 5322–5363.

Ryan, S. J.; Candish, L.; Lupton, D. W. Acyl anion free N-heterocyclic carbene organocatalysis. *Chem. Soc. Rev.* **2013**, *42*, 4906–4917.

Selig, P. Guanidine organocatalysis. *Synthesis* **2013**, *45*, 703–718.

van Gemmeren, M.; Lay, F.; List, B. Asymmetric catalysis using chiral, enantiopure disulfonimides. *Aldrichimica Acta* **2014**, *47*, 3–13.

Vetica, F.; de Figueiredo, R. M.; Orsini, M.; Tofani, D.; Gasperi, T. Recent advances in organocatalytic cascade reactions toward the formation of quaternary stereocenters. *Synthesis* **2015**, *47*, 2139–2184.

Volla, C. M. R.; Atodiresei, I.; Rueping, M. Catalytic C–C bond-forming multi-component cascade or domino reactions: pushing the boundaries of complexity in asymmetric organocatalysis. *Chem. Rev.* **2014**, *114*, 2390-2431.

Vora, H. U.; Rovis, T. Asymmetric *N*-heterocyclic carbene (NHC) catalyzed acyl anion reactivity. *Aldrichimica Acta* **2011**, *44*, 3-11

Vora, H. U.; Wheeler, P.; Rovis, T. Exploiting acyl and enol azolium intermediates via *N*-heterocyclic carbene catalyzed reactions of alpha-reducible aldehydes. *Adv. Synth. Catal.* **2012**, *354*, 1617-1639.

Selected Books

Dalko, P. *Comprehensive Enantioselective Organocatalysis: Catalysts, Reactions and Applications, Volume 1: Privileged Catalysts;* Weinheim, 2013; 237 pp.

Dalko, P. *Comprehensive Enantioselective Organocatalysis: Catalysts, Reactions and Applications, Volume 2: Activations;* Weinheim, 2013; 717 pp.

Dalko, P. *Comprehensive Enantioselective Organocatalysis: Catalysts, Reactions and Applications, Volume 3: Reactions and Applications;* Weinheim, 2013; 1438 pp.

List, B. (Ed.) *Asymmetric Organocatalysis*, Topics in Current Chemistry; Springer, 2010; 460 pp.

List, B. *Science of Synthesis: Asymmetric Organocatalysis 1–Lewis Base and Acid Catalysts;* Stuttgart, 2012; 954 pp.

Maruoka, K. *Science of Synthesis: Asymmetric Organocatalysis 2-Brønsted Base and Acid Catalysts, and Additional Topics;* Stuttgart, 2012; 974 pp.

Maruoka, K.; Shibasaki, M. *Comprehensive Chirality, Volume 6: Synthetic Methods V-Organocatalysis;* Amsterdam, 2012; 607 pp.

Torres, R. R. *Stereoselective Organocatalysis: Bond Formation Methodologies and Activation Modes;* Hoboken, 2013; 662 pp.

Waser, M. *Asymmetric Organocatalysis in Natural Product Syntheses;* Berlin, 2012; 197 pp.

Short Note on InChIs and InChIKeys

The IUPAC International Chemical Identifier (InChI^{TM}) and its compressed form, the InChIKey, are strings of letters representing organic chemical structures that allow for structure searching with a wide range of online search engines and databases such as Google and PubChem. While they are obviously an important development for online reference works, such as *Encyclopedia of Reagents for Organic Synthesis* (*e-EROS*), readers of this volume may be surprised to find printed InChI and InChIKey information for each of the reagents.

We introduced InChI and InChIKey to e-EROS in autumn 2009, including the strings in all HTML and PDF files. While we wanted to make sure that all users of *e-EROS*, the second print edition of *EROS*, and all derivative handbooks would find the same information, we appreciate that the strings will be of little use to the readers of the print editions, unless they treat them simply as reminders that *e-EROS* now offers the convenience of InChIs and InChIKeys, allowing the online users to make best use of their browsers and perform searches in a wide range of media.

If you would like to know more about InChIs and InChIKeys, please go to the *e-EROS* website: www.wileyonlinelibrary.com/ref/eros and click on the InChI and InChIKey link.

A

4-Acetamido-2,2,6,6-tetramethyl-1-piperidinyloxyl[1]

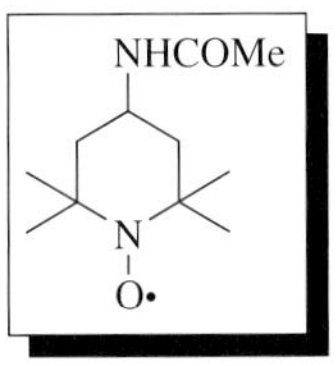

[14691-89-5] $C_{11}H_{21}N_2O_2$ (MW 213.30)
InChI = 1/C11H21N2O2/c1-8(14)12-9-6-10(2,3)13(15)11(4,5)7-9/h9H,6-7H2,1-5H3,(H,12,14)/f/h12H
InChIKey = UXBLSWOMIHTQPH-XWKXFZRBCK

(reagent and catalyst for selective oxidation of alcohols[2])

Alternate Names: 4-acetamido-TEMPO; 4-acetylamino-TEMPO.
Physical Data: mp 146–147 °C,[2] 147.5 °C.[3]
Solubility: nearly insol hexane, ether; sol ethanol, acetone, acetonitrile, methylene chloride; slightly sol water, from which it can be recrystallized.
Form Supplied in: red or pink solid, commercially available.
Preparative Method: can be easily prepared in high yield from 4-amino-2,2,6,6-tetramethylpiperidine.[2]
Handling, Storage, and Precautions: completely stable and non-hygroscopic.

Oxidation of Alcohols. 4-Acetamido-TEMPO is a representative of the nitroxide radicals (**1**) which have been used as reagents or catalysts for the oxidation of organic compounds.[1] Several nitroxide radicals have been converted to the corresponding oxoammonium salts (**2**) and used as stoichiometric oxidizing reagents.[1] Nitroxide radicals have also been used as catalysts in the the presence of a secondary, stoichiometric oxidant.[1]

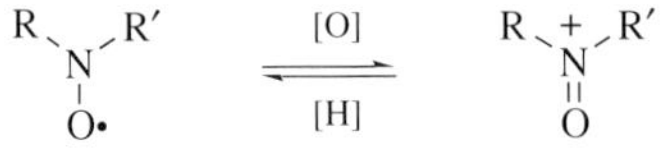

(**1**) (**2**)

4-Acetamido-TEMPO (**3**) is not, itself, a reagent for the oxidation of alcohols. Rather, it serves as a convenient precursor for the in situ preparation of an oxoammonium salt (**4**), which is the true oxidant. The oxoammonium salt (**4**) is the product of the acid-catalyzed disproportionation of (**3**) to (**4**) and (**5**) in the presence of p-Toluenesulfonic Acid (eq 1) and is a highly selective reagent for alcohol oxidation (eq 2).[1] The reactions are carried out in methylene chloride in which $TsOH{\cdot}H_2O$ is essentially insoluble. The only products of the reaction are the desired carbonyl compound product (**7**) and the hydroxylamine salt (**5**), which is completely insoluble in methylene chloride. Product isolation simply involves the filtration of (**5**) and the evaporation of methylene chloride. Compound (**5**) can be converted back to the nitroxide radical (**3**) in quantitative yield.[2]

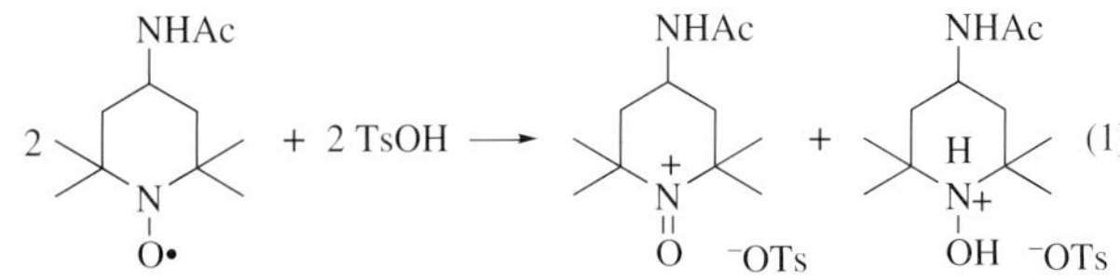

(**3**) (**4**) (**5**)

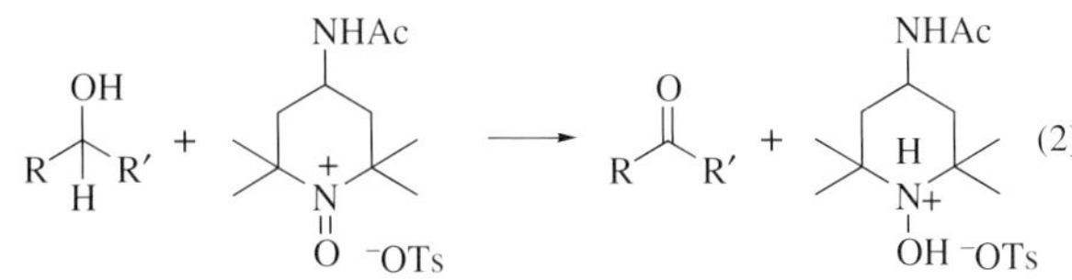

(**6**) (**4**) (**7**) (**5**)

The reaction has been used for the oxidation of a variety of alcohols with excellent yields of isolated products.[2] Primary alcohols are converted to the corresponding aldehydes, and no over-oxidation is observed. Secondary alcohols react as well as primary alcohols and provide the corresponding ketones. The mildness of this reaction has been demonstrated by the oxidations of nerol (**8**) and geraniol (**9**) to the corresponding *cis*-citral (**10**) and *trans*-citral (**11**).

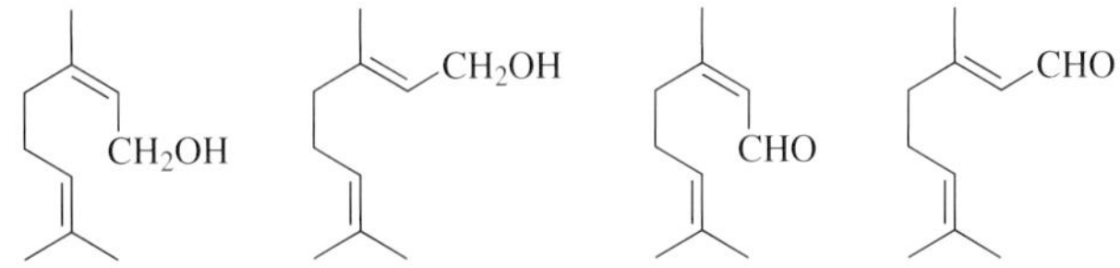

(**8**) (**9**) (**10**) (**11**)

Oxidation with this reagent does not take place when there is an oxygen or nitrogen in the β-position to the alcohol being oxidized and does not take place with 1,2-diols or sugars. Amines,[1,4] thiols,[5] phenols,[6] indoles,[7] benzyl ethers,[1] and ketones (very slow)[1] react with oxoammonium salts and may interfere with alcohol oxidation. However, sulfides, most ethers, amides, esters, and double bonds do not react with oxoammonium salts and should not interfere.[1]

A number of other oxoammonium salts have been described in the literature.[1] In addition, TEMPO-type nitroxide radicals have been used as specific catalysts for the oxidation of alcohols using one or several secondary, stoichiometric oxidants.[1,8] Neither 4-acetamido-TEMPO nor any of its derivatives have been used in this manner, but they should function satisfactorily.

---

1. Bobbitt, J. M.; Flores, M. C. L., *Heterocycles* **1988**, *27*, 509 (general review on oxoammonium salts, but they are called nitrosonium salts in the title).
2. Ma, Z.; Bobbitt, J. M., *J. Org. Chem.* **1991**, *56*, 6110.
3. Rozantsev, E. G.; Kokhanov, Y. V., *Bull. Acad. Sci. USSR, Div. Chem. Sci.* **1966**, *8*, 1422.
4. Hunter, D. H.; Racok, J. S.; Rey, A. W.; Ponce, Y. Z., *J. Org. Chem.* **1988**, *53*, 1278.
5. Kashiwagi, Y.; Ohsawa, A.; Osa, T.; Ma, Z.; Bobbitt, J. M., *Chem. Lett.* **1991**, 581.
6. Bobbitt, J. M.; Ma, Z., *Heterocycles* **1992**, *33*, 641.

7. Bobbitt, J. M.; Guttermuth, M. C. F.; Ma, Z.; Tang, H., *Heterocycles* **1990**, *30*, 1131.

8. (a) Anelli, P. L.; Montanari, F.; Quici, S., *Org. Synth.* **1990**, *69*, 212. (b) Inokuchi, T.; Matsumoto, S.; Torii, S., *J. Org. Chem.* **1991**, *56*, 2416. (c) Leanna, M. R.; Sowin, T. J.; Morton, H. E., *Tetrahedron* **1992**, *33*, 5029.

James M. Bobbitt & Zhenkun Ma
University of Connecticut, Storrs, CT, USA

4-Acetylamino-2,2,6,6-tetramethyl-piperidine-1-oxoammonium Tetrafluoroborate

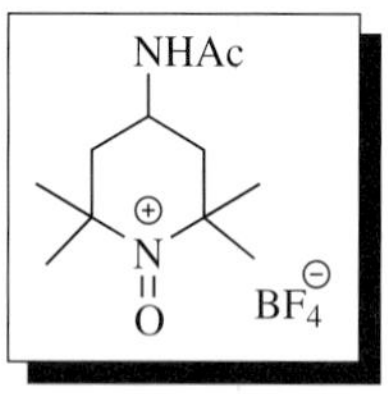

[219543-09-6] $C_{11}H_{21}N_2O_2F_4B$ (MW 300.09)
InChI = 1S/C11H20N2O2.BF4/c1-8(14)12-9-6-10(2,3)13(15)11(4,5)7-9;2-1(3,4)5/h9H,6-7H2,1-5H3;/q;-1/p+1
InChIKey = HTMHEICBCHCWAU-UHFFFAOYSA-O

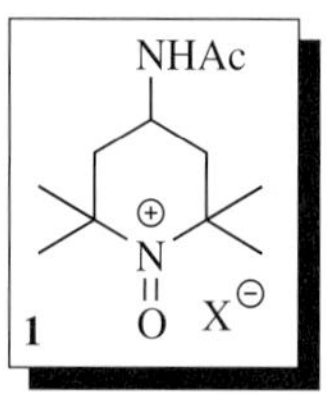

4-Acetylamino-2,2,6,6-tetramethylpiperidine-1-oxoammonium tetrafluoroborat
1a, X = BF$_4^-$
[219543-09-6] $C_{11}H_{21}N_2O_2F_4B$ (MW 300.10)[3]
InChI = 1S/C11H20N2O2.BF4/c1-8(14)12-9-6-10(2,3)13(15)11(4,5)7-9;2-1(3,4)5/h9H,6-7H2,1-5H3;/q;-1/p+1
InChIKey = HTMHEICBCHCWAU-UHFFFAOYSA-O

4-Acetylamino-2,2,6,6-tetramethylpiperidine-1-oxoammonium perchlorate
1b, X = ClO_4^-
[219543-08-5] $C_{11}H_{21}N_2O_6Cl$ (MW 312.72)[3,13]
InChI = 1S/C11H20N2O2.ClHO4/c1-8(14)12-9-6-10(2,3)13(15)11(4,5)7-9;2-1(3,4)5/h9H,6-7H2,1-5H3;(H,2,3,4,5)
InChIKey = QPNXYTPYENYALX-UHFFFAOYSA-N

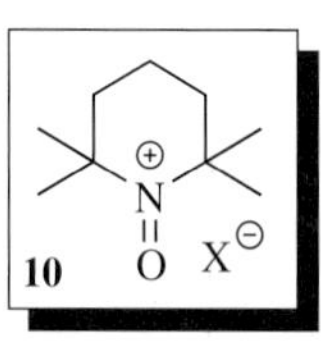

2,2,6,6-Tetramethylpiperidine-1-oxoammonium tetrafluoroborate
10a, X = BF_4^-
[33247-78-8] $C_9H_{18}NOF_4B$ (MW 243.05)[14]
InChI = 1S/C9H18NO.BF4/c1-8(2)6-5-7-9(3,4)10(8)11;2-1(3,4)5/h5-7H2,1-4H3;/q+1;-1
InChIKey = GAJDGWHXJQBONI-UHFFFAOYSA-N

2,2,6,6-Tetramethylpiperidine-1-oxoammonium perchlorate
10b, X = ClO_4^-
[31198-93-3] $C_9H_{18}NO_5Cl$ (MW 255.09)[14,15]
InChI = 1S/C9H18NO.ClHO4/c1-8(2)6-5-7-9(3,4)10(8)11;2-1(3,4)5/h5-7H2,1-4H3;(H,2,3,4,5)/q+1;/p-1
InChIKey = CPLCZRHFOLVQQN-UHFFFAOYSA-M

2,2,6,6-Tetramethylpiperidine-1-oxoammonium chloride
10c, X = Cl^-
[26864-01-7] $C_9H_{18}NOCl$ (MW 191.70)[16]
InChI = 1S/C9H18NO.ClH/c1-8(2)6-5-7-9(3,4)10(8)11;/h5-7H2,1-4H3;1H/q+1;/p-1
InChIKey = UXJJNTSDDNCDLD-UHFFFAOYSA-M

2,2,6,6-Tetramethylpiperidine-1-oxoammonium bromide
10d, X = Br^-
[85917-27-7] $C_9H_{18}NOBr$ (MW 236.15)[17]
InChI = 1S/C9H18NO.BrH/c1-8(2)6-5-7-9(3,4)10(8)11;/h5-7H2,1-4H3;1H/q+1;/p-1
InChIKey = PJAVCFQMWXUNQV-UHFFFAOYSA-M

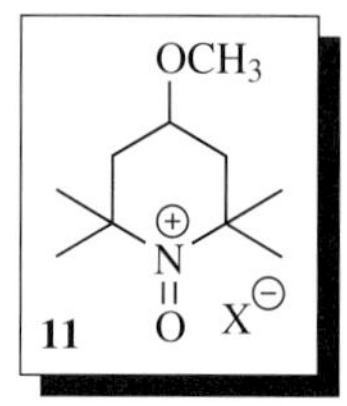

4-Methoxy-2,2,6,6-tetramethylpiperidine-1-oxoammonium-tetrafluoroborate
11a, X = BF_4^-
[135023-09-5] $C_{10}H_{20}NO_2F_4B$ (MW 273.07)[18]
InChI = 1S/C10H20NO2.BF4/c1-9(2)6-8(13-5)7-10(3,4)11(9)12;2-1(3,4)5/h8H,6-7H2,1-5H3;/q+1;-1
InChIKey = YIZRNAPYVZDWEL-UHFFFAOYSA-N

4-Methoxy-2,2,6,6-tetrame-thylpiperidine-1-oxoammonium chloride
11b, X = Cl^-
[95407-70-8] $C_{10}H_{20}NO_2Cl$ (MW 221.72)[19]
InChI = 1S/C10H20NO2.ClH/c1-9(2)6-8(13-5)7-10(3,4)11(9)12;/h8H,6-7H2,1-5H3;1H/q+1;/p-1
InChIKey = XTQPMZGDXUYFLV-UHFFFAOYSA-M

4-Methoxy-2,2,6,6-tetrame-thylpiperidine-1-oxoammonium bromide
11c, X = Br^-
[90246-27-8] $C_{10}H_{20}NO_2Br$ (MW 266.17)[19]
InChI = 1S/C10H20NO2.BrH/c1-9(2)6-8(13-5)7-10(3,4)11(9)12;/h8H,6-7H2,1-5H3;1H/q+1;/p-1
InChIKey = IMQPWXWUDKGRRG-UHFFFAOYSA-M

(reagent used to oxidize alcohols to aldehydes or ketones. Of special interest, allyl alcohols can be oxidized without isomerization and phenolic benzyl alcohols can be oxidized without blocking the phenol group; all of these reagents have been used for the oxidation of primary and secondary alcohols to aldehydes or ketones, respectively. They also have a certain number of additional minor oxidation reactions. Many other salts are known[1])

Physical Data: mp 195–196 °C dec (Kofler). **1a**, mp 193–194 °C (dec.); **1b**, mp 177–178 °C (dec.); **10a**, mp 162–163 °C; **10b**, mp 157–158 °C; **10c**, mp 118–119 °C; **10d**, mp 76–78 °C; **11a**, mp 162–162.5 °C; **11b**, mp 121–123 °C; **11c**, mp 206–207 °C.

Solubility: partially soluble in water (6 g/100 mL at 0 °C, 8 g/100 mL at 20 °C, and about 100 g/100 mL at 100 °C; reacts very slowly with hot water), slightly soluble in methylene chloride (0.04 g/100 mL at 20 °C). **1a** and **1b** are slightly soluble in water and methylene chloride, while insoluble in ether; **10a**–**10d** are insoluble in ether; **11a**–**11c** are insoluble in carbon tetrachloride. Otherwise, solubilities are variable. Most reactions are carried out in methylene chloride, less often in acetonitrile.

Form Supplied in: all of these salts are prepared from the corresponding parent nitroxides, **5**, **12**, and **13**, which are commercially available. None of the salts are commercially available currently.

Preparative Methods: in a slightly modified procedure,[3] 100 g of 4-acetylamino-2,2,6,6-tetramethyl-1-piperidineoxyl (0.469 mol) is stirred with 50 mL of H_2O, and 90 g of 48% aqueous HBF_4 (0.492 mol) is added over about 1 h. The mixture is placed in an ice bath, and 332.2 g of commercial bleach (5.25% NaOCl, 0.235 mol) is added over about 2 h. The mixture is allowed to stir in ice for about 3 h and then filtered. The solid bright yellow salt is pressed as dry as possible and washed with two 50-mL portions of ice water and two 100-mL portions of CH_2Cl_2. The salt is dried at room temperature and in air to constant weight to give 90–95 g (about 65%) of the reagent, mp 195–196 °C. Further drying in a vacuum desiccator reduced the weight only slightly. For recovery of the remainder of the salt, the combined filtrates (water and CH_2Cl_2) are treated with 20 g of $NaHCO_3$ and 20 mL of 95% ethanol to reduce the oxoammonium salt to the starting nitroxide. The layers are separated and the aqueous phase is extracted with three 50-mL portions of CH_2Cl_2. The extracts are combined, dried over $MgSO_4$, and evaporated to give 25–30 g of nitroxide, mp 145–147 °C, suitably pure for the preparation of more oxoammonium salt. Oxoammonium salts are prepared from commercially available, nitroxide free radicals, **5**, **12**, and **13**. There are essentially two methods for the preparation of the oxoammonium salts from nitroxides. The first is based on the acid-catalyzed disproportionation of nitroxide in strong acid as shown for **1a** in eq 1.[14,20] The method has been used both in water and dry ether for the preparation of many oxoammonium salts. The second is based on the halogens as shown for **11b** in eq 2[16,19] In the halide reactions, the elemental halogens act both as oxidizing agents and the source of the anions.

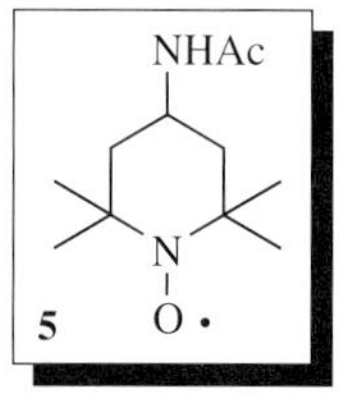

4-acetylamino-2,2,6,6-tetramethyl-1-piperidineoxyl(4-acetylamino-TEMPO)

[14691-89-5]

InChI = 1S/C11H21N2O2/c1-8(14)12-9-6-10(2,3)13(15)11(4,5)7-9/h9H,6-7H2,1-5H3,(H,12,14)

InChIKey = UXBLSWOMIHTQPH-UHFFFAOYSA-N

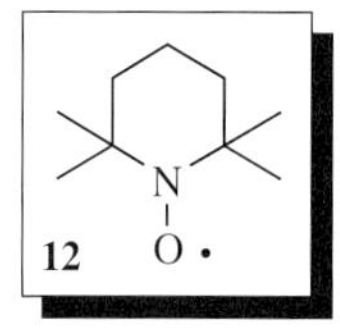

2,2,6,6-tetramethyl-1-piperidineoxyl(TEMPO)

[2564-83-2]

InChI = 1S/C9H18NO/c1-8(2)6-5-7-9(3,4)10(8)11/h5-7H2,1-4H3

InChIKey = QYTDEUPAUMOIOP-UHFFFAOYSA-N

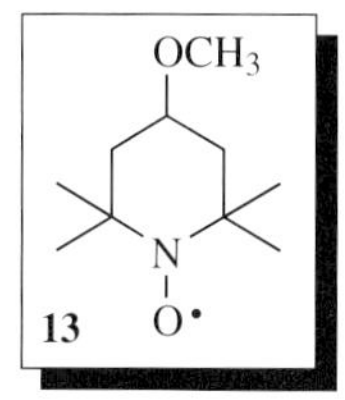

4-methoxy-2,2,6,6-tetramethyl-1-piperidineoxyl(4-methoxy-TEMPO)

[95407-69-5]

InChI = 1S/C10H20NO2/c1-9(2)6-8(13-5)7-10(3,4)11(9)12/h8H,6-7H2,1-5H3

InChIKey = SFXHWRCRQNGVLJ-UHFFFAOYSA-N

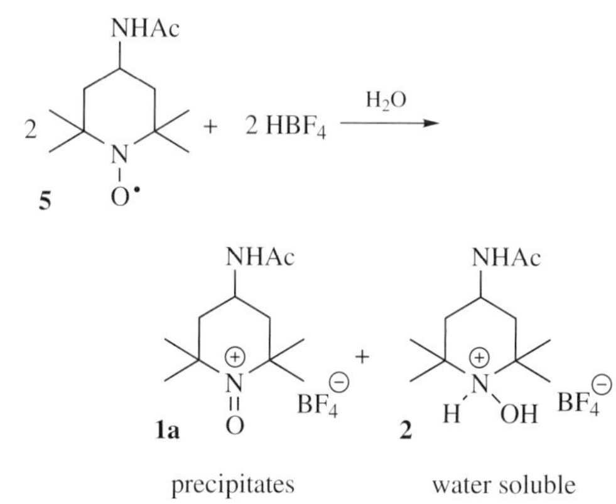

(1)

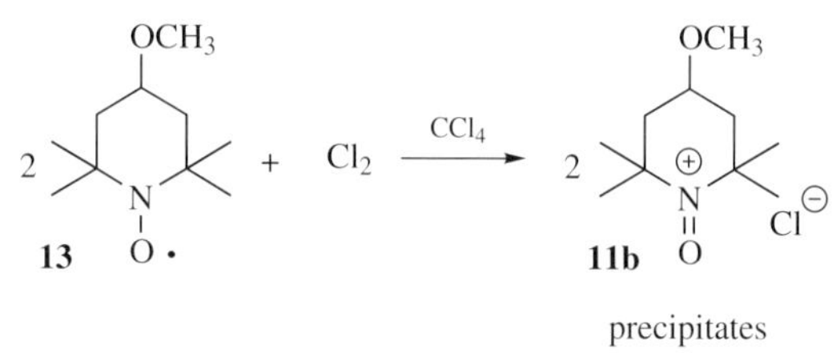

(2)

The preparation of **1a** and its applications are described in a tested procedure in *Organic Syntheses*.[20] The procedure involves the use of a second oxidant, bleach, to increase the yield of salt.

Purity: recrystallized, if necessary, from an equal weight of water. Solution heated and cooled as rapidly as possible. The purity of the salt can be assayed using 1-decanol as substrate. It is normally close to 100%.

Handling, Storage, and Precautions: the salt (tetrafluoroborate) is non-hygroscopic and indefinitely stable. No hazards are known. For anhydrous reactions, the previously dried salt can be dried in a vacuum over P_2O_5 at 35 °C, although the reduction in weight is slight. The perchlorate salts **1b** and **10b** have been used extensively,[3] but since 1**b** detonated on drying under vacuum,[13] they should both be avoided. Compound **1a** appears to be nonhygroscopic, stable indefinitely, and to have ideal solubility properties in methylene chloride. For these reasons, it will be the main theme of this article. The other salts have various degrees of instabilities and some are quite hygroscopic. It should be understood that the other oxoammonium salts have similar properties. Nothing is known about their toxicity.

Original Commentary

James M. Bobbitt & Nabyl Merbouh
University of Connecticut, Storrs, CT, USA

Introduction. The title reagent **1a** has some remarkable oxidizing properties, and is commercially available from several sources. It can also be prepared from 4-amino-2,2,6,6-tetramethylpiperidine.[3] In this preparation, one caution applies. When the nitroxide was prepared, potassium carbonate was used as a base.[3] This should be replaced by sodium carbonate since potassium tetrafluoroborate is quite insoluble in water and causes impurity problems with the oxoammonium salt. The nitroxide has interesting solubility properties. It is only slightly soluble in ether and cold water. It can be extracted from an aqueous layer with several portions of CH_2Cl_2, or from an ether layer with several portions of water. If necessary, the nitroxide can be recrystallized from ethyl acetate.

Reference 3 describes the synthesis and reactions of two oxoammonium salts, the perchlorate and the tetrafluoroborate, but all of the oxidations described were carried out with perchlorate. Unfortunately, the perchlorate proved to be unsafe, and the tetrafluoroborate is now the salt of choice. The oxidizing properties and solubilities of the two salts are essentially identical.

Oxidations in CH_2Cl_2 with Silica Gel. The reaction shown below is actually carried out in a neutral to slightly acidic medium. Silica gel is a potent catalyst, but is not needed for the oxidation of allyl or benzyl alcohols. The oxoammonium salt **1** is bright yellow and its reduction product, the hydroxyamine salt **2**, is white in color. The oxidant is sufficiently soluble in the solvent to allow the reaction to occur, but **2** is almost completely insoluble. Thus, a yellow slurry of the reagent **1** (about 5–10% excess) and the appropriate alcohol is stirred in CH_2Cl_2 (1–2 mmol in 10 mL), with or without silica gel, until the slurry becomes white (eq 3). Filtration through a ca. 0.5-cm pad of silica gel removes the salt **2** and a small trace of **1**. The aldehydes or ketones formed are quite pure and are isolated in good yields.[3]

The hydroxyamine salt **2** can be converted back to nitroxide in the following manner. The precipitated reagent and silica gel are stirred with a small portion of warm water and filtered to remove silica gel. The filtrate is basified with $NaHCO_3$ and treated with an excess of 30% H_2O_2. The nitroxide forms in about 24 h and can be extracted with CH_2Cl_2.

This oxidation method has some advantages as well as disadvantages. Advantages are that the reaction is colorimetric (yellow to white), essentially quantitative, and there is no further oxidation to an acid. If there is a hydroxy group in the product which can form a stable hemiacetal, the product will be a lactone. Since **2** is almost insoluble in CH_2Cl_2, the reaction work-up requires only a filtration. The reaction is ideal for preparing unstable or volatile aldehydes (or ketones) on an 'as needed' basis. Since the products are essentially pure as formed, the oxidation can be combined with various other reactions, such as Wittig or Grignard reactions. Wittig reactions work especially well when the Wittig reagent is added to the solution of aldehyde or ketone, either in CH_2Cl_2 or in diethyl ether. Isolated yields are as high as 90% for the combined reactions.

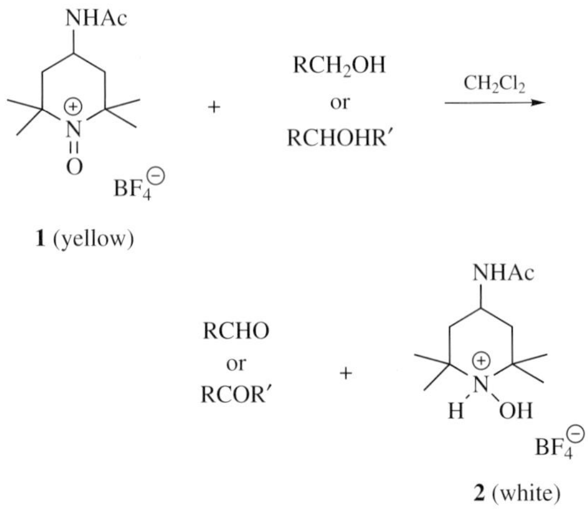

(3)

The oxidation of allyl alcohols when there is a possibility of double bond isomerization is accomplished without isomerization,[3] as in the case of **3**.[4] Benzyl alcohols containing one phenol group can be cleanly oxidized to aldehydes without phenol oxidation,[3] as in the case of **4**.[5] Two phenol groups give decomposition. Yields of aldehydes or ketones are shown in parentheses.

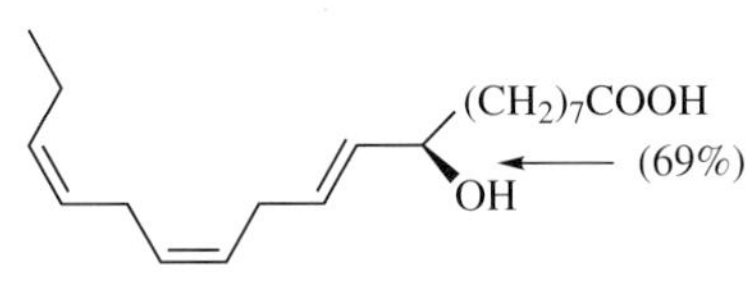

3

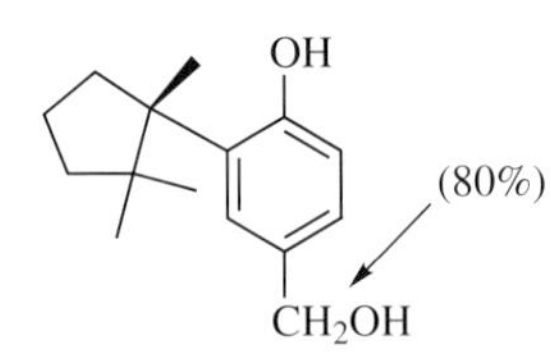

4

There are also some disadvantages for this oxidation method. The oxidation does not take place when there is a β-oxygen or a nitrogen group attached to the alcohol being oxidized (but see below). Since the reaction conditions are slightly acidic, acetals are opened. In addition, there are two slow reactions that interfere with slow oxidations, such as those of primary alcohols. Benzyl groups are slowly oxidized and cleaved.[6] This causes no problem with benzyl alcohols or allyl alcohols, which are oxidized rapidly. In the second reaction, activated double bonds (containing oxygen, nitrogen, or three alkyl groups) are slowly oxidized.[3,7] Again, this is not a problem with allyl or benzyl alcohols.

One more reaction of certain oxoammonium salts is known and may be of major importance. This is the oxidation of ketones to α-diketones,[8,9] and the oxidation of 1,3-diketones to 1,2,3-triketones.[9]

Oxidations Under Basic Conditions. Oxidations can also be carried out in a base, for example pyridine in CH_2Cl_2, but the reaction takes a different course as shown in eq 4.[10] Equal amounts (2 equiv) of salt and the base must be used. The products are the orange-colored nitroxide (**5**) and the base salt of tetrafluoroboric acid.

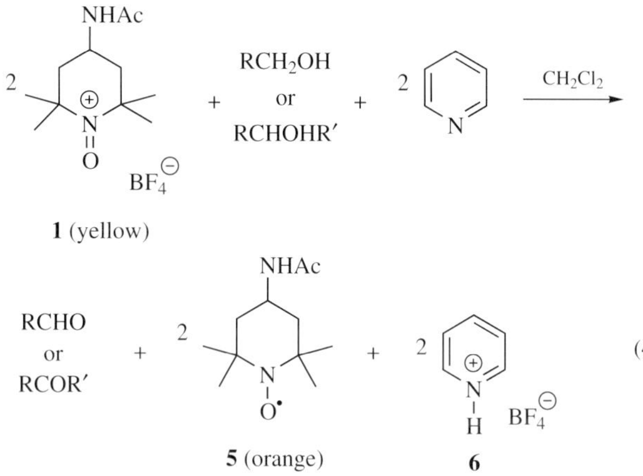

1 (yellow) **5** (orange) **6** (4)

Fortunately, compounds **5** and **6** can both be easily removed for product isolation. The orange-colored nitroxide is only slightly soluble in dry ether, but the pyridine salt is completely insoluble in CH_2Cl_2. Thus, the filtration of the reaction mixture removes **6**, and evaporation of CH_2Cl_2 leaves a residue of product and nitroxide. This residue can be extracted with ether, leaving most, but not all of the nitroxide behind. The last traces of nitroxide can be removed on a short column of silica gel.

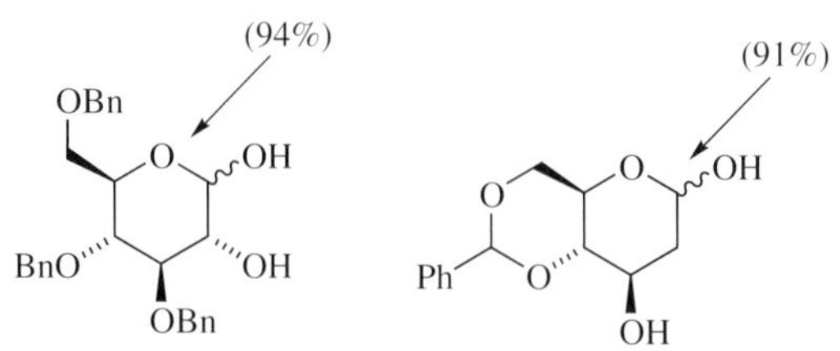

7 **8**

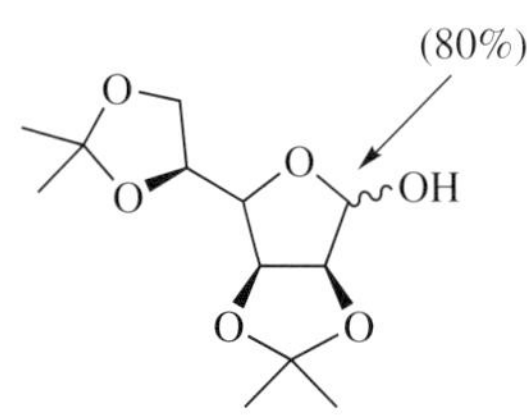

9

The reaction has been successfully applied to several sugar derivatives[10] such as **7**, **8**, and **9**. It is noteworthy that only oxidation of the anomeric carbon takes place, but the secondary alcohol groups in the sugar are not touched. The reaction has also been successfully applied to other acetals and β-oxygen alcohols, thus alleviating some of the objections of the simpler reaction in CH_2Cl_2 alone.

First Update

James M. Bobbitt
University of Connecticut, Storrs, CT

Nabyl Merbouh
Simon Fraser University, Burnaby, Canada

Alcohol Oxidations.

***With* 1a.** A typical oxidation with **1a** is shown eq 5, but it is probably specific for this reagent due to the desirable solubilities of the salt and its reduced product, **2**.[3] In essence, the substrate is dissolved in methylene chloride and treated with the bright yellow **1a**, which is only slightly soluble. The salt is best mixed with an equal weight of chromatographic grade silica gel to give faster reactions. The yellow slurry is stirred until it becomes a white slurry and no starch iodide paper test is observed. The slurry is filtered to recover **2** (insoluble in methylene chloride), and the filtrate contains a high yield of essentially pure carbonyl compound. Compound **2** can be recycled back to nitroxide, **5**.

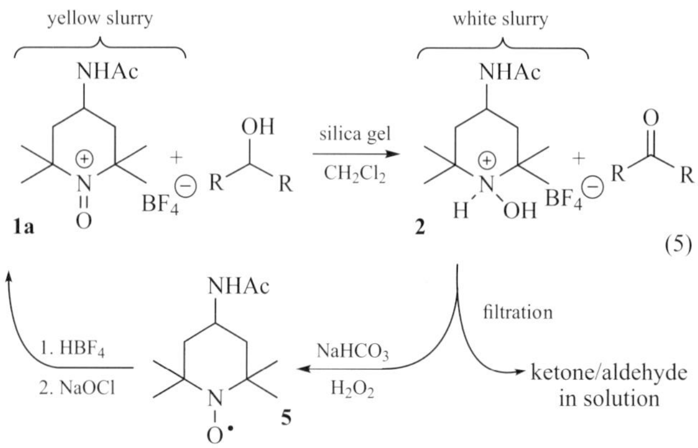

1a **2** **5** (5)

There are some specific advantages to this reaction. First, the reaction is colorometric. Second, product isolation is trivial. Third, the reaction is stereospecific in that a double bond and chiral configurations are maintained even when the oxidation is on adjacent atoms. Finally, the reaction can be carried out in the presence of a phenol group.

There are also disadvantages. Alcohols having a *beta* oxygen or nitrogen function are not oxidized.[3,21] Free amino groups react rapidly (see below). Aliphatic alcohol oxidations (which are slow) can be complicated sometimes by reactions with isolated, activated double bonds (see below) and, in some cases, by benzyl ethers (see below).[6] Frequently, these secondary reactions are quite slow in methylene chloride and faster in acetonitrile.

Allylic and benzylic alcohols react in 2–3 h. Aliphatic secondary alcohols and acetylenic alcohols react slightly faster than primary alcohols. All of the reactions are faster in the presence of silica gel. The reaction is quite specific in that no carboxylic acids are formed from primary alcohols and carbon–carbon bond cleavage is almost unknown. The one exception is that for diols in which five- or six-membered hemiacetal rings can be formed, the products are lactones.[22]

Oxidations with Other Salts. Oxidations with the other salts give comparable reactions, although some salts are not stable or are hygroscopic. The specific anions seem to be important in determining the rates of the various oxidations.[18,19] In general, the halide salts react faster than the tetrafluoroborate or perchlorate mediated oxidations. The reaction conditions are similar to those given above, but the solubilities of the products may not be the same.

Oxidations in Base. Since amine groups react readily with oxoammonium salts (see below), only such organic bases as pyridine and its derivatives can be used to influence these reactions. Only one such reaction has been carefully studied, the oxidative dimerization of primary alcohols containing a beta oxygen and those give good yields of dimeric esters as shown in eq 6.[23]

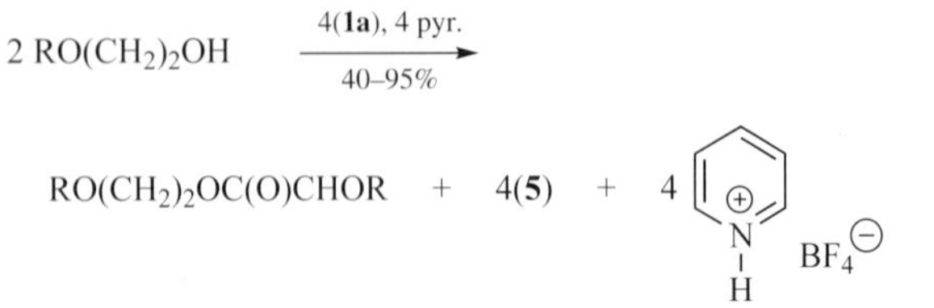

(6)

This reaction is complicated by the fact that the reduced product from the oxoammonium salt, **2**, reacts with pyridine to give a free hydroxyamine base, which in turn reacts with more oxoammonium salt **1a** to give nitroxide **5** as the final product.

Role of Oxoammonium Ions in Nitroxide Catalyzed Oxidations. Oxidations with stoichiometric amounts of oxoammonium salts, although convenient, have not been extensively used, primarily because the salts are not commercially available. The vast majority of oxoammonium oxidations have been catalytic reactions in which a nitroxide is used with a primary oxidant such as NaOCl (or many other similar reagents). In effect, the oxoammonium ion generated from nitroxide in such reactions provides a high degree of specificity for alcohol oxidation, with the convenience of cheaper and more available oxidants. Catalytic reactions have been extensively reviewed[1,2,11,12,24] and will not be considered in this paper.

Oxidations of Activated Methylene Groups to Carbonyl Groups. These reactions are not well known. Some reactions, such as those in eqs 7–9,[9,25–27] probably take place through enol forms. Others, such as those in eq 10[28] do not, and are not easily explained.

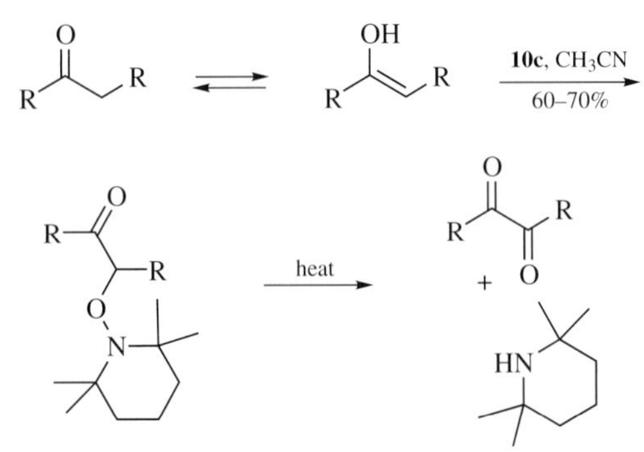

(7)

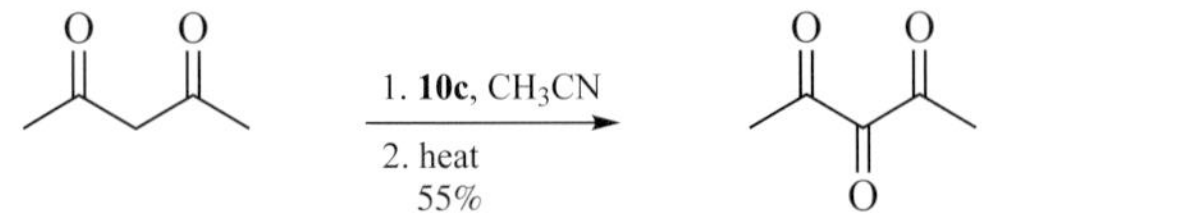

(8)

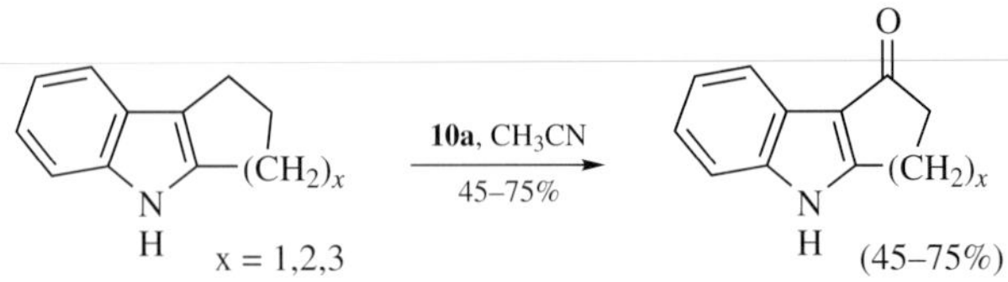

(9)

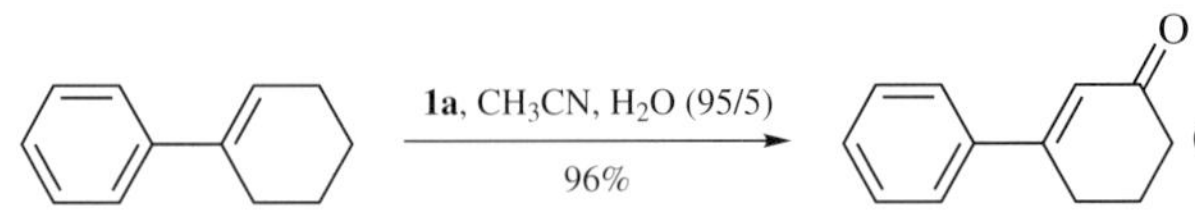

(10)

Reactions with Double Bonds.

Isolated Double Bonds. These reactions take place slowly in methylene chloride and faster in acetonitrile (eq 11). The alkene must contain three or four alkyl substituents[29] or two phenyl groups.[7]

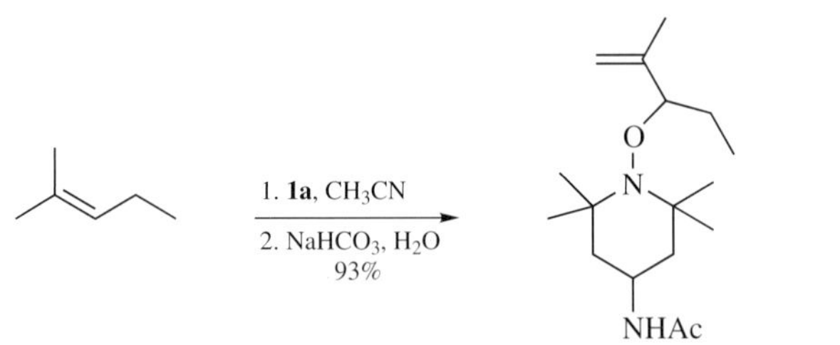

(11)

Enol Ether and Enol Amine Reactions. These reactions are similar to the one discussed above for double bond reactions except that the intermediate is trapped as an ether (eq 12)[7] or as a thymine derivative (eq 13).[30]

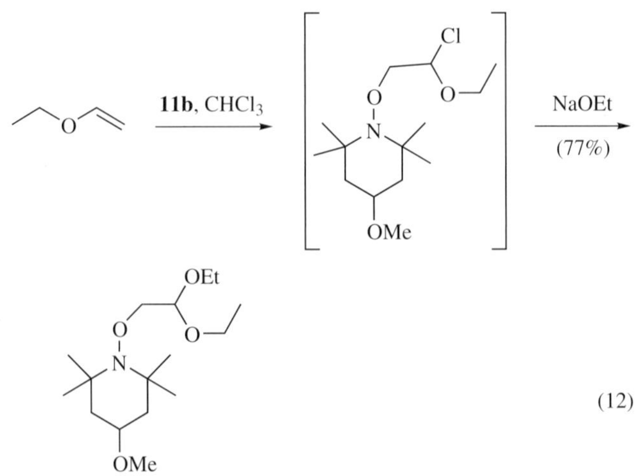

(12)

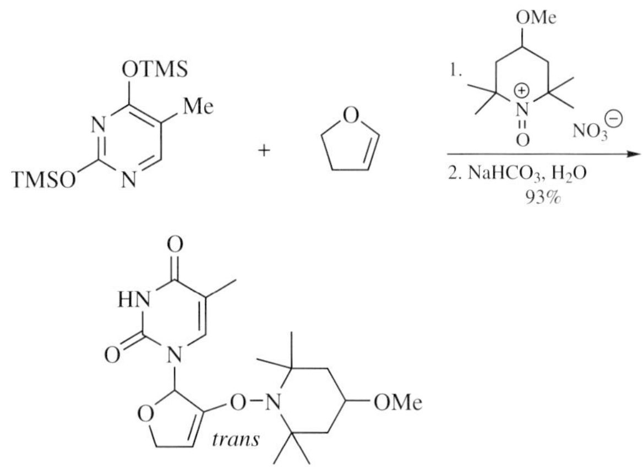

(13)

Reactions with Enolate Ions and Grignard Reagents.[31,32] Equations **14** and **15**.

(14)

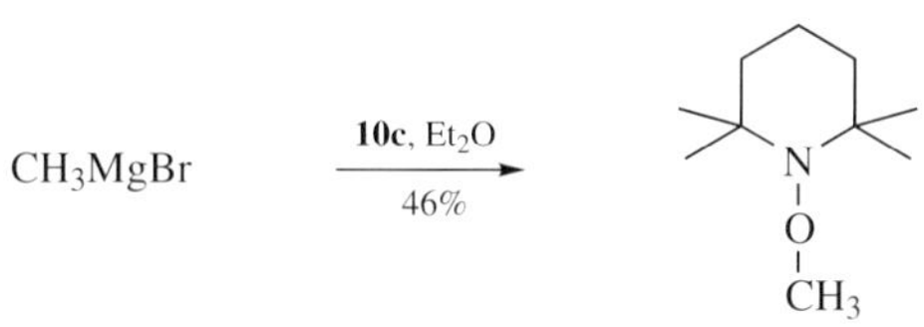

(15)

Reactions with Amines.[33–35] Reactions with amines are complicated by the fact that the starting amine reacts with any reduced oxidant (such as **2**) to give the free hydroxylamine base, which reacts with oxoammonum salt to give nitroxide (reverse of eq 3). The reactions have not been well explored, and yields are low (eqs 16 and 17). The apparent reaction, however, is much like alcohol oxidation in that an imine is formed by the loss of two hydrogens. Amine oxidation has been more successfully explored in nitroxide catalyzed electrochemical oxidations in the presence of 2,6-lutidine.[36]

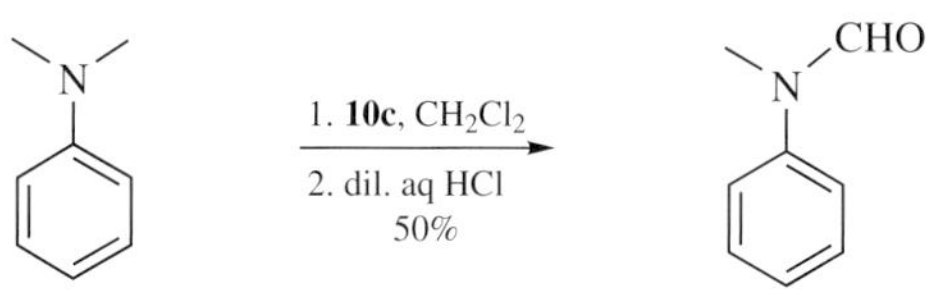

(16)

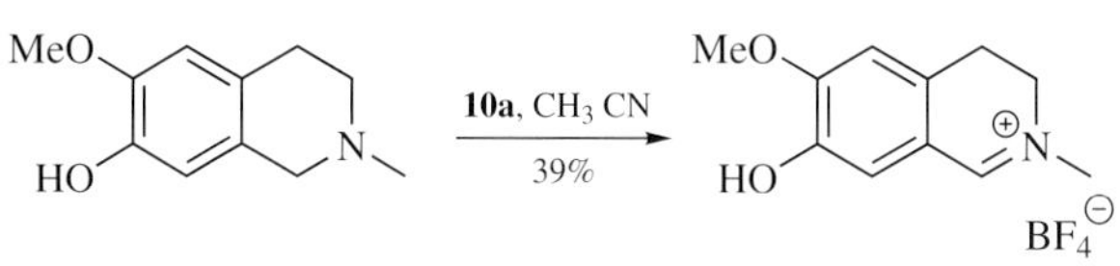

(17)

Reactions with Phenols.[35,37] Equation 18.

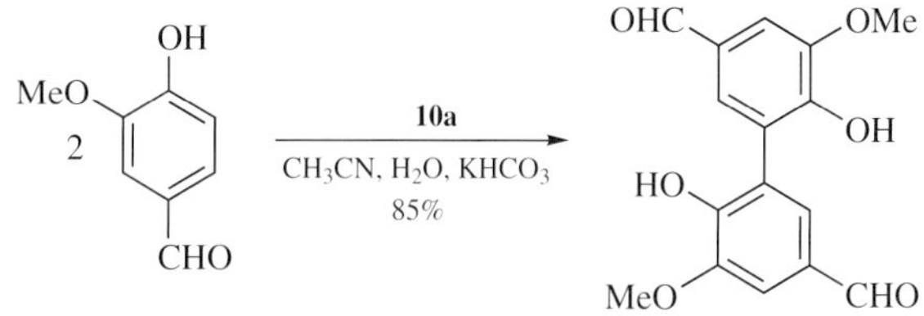

(18)

Reactions with Benzyl Ethers.[6] These reactions only show up only when alcohol oxidation is very slow. For example, benzyl alcohols can be oxidized in the presence of benzyl ethers as in eq 19.[3]

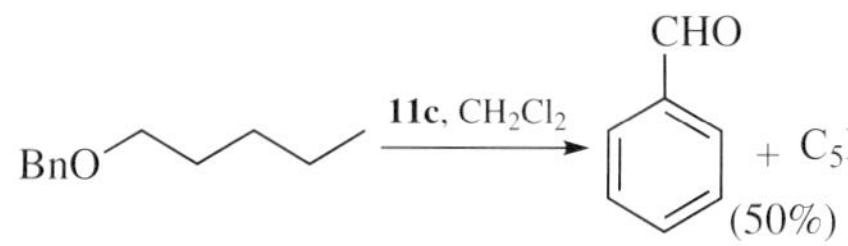

+ $C_5H_{11}Br$ + other products (19)

Reactions with Sulfur Compounds.[38] Thiols are oxidized to disulfides (eq 20). Otherwise sulfur oxidations are controversial. Sulfides and sulfones are reported to be unreactive.[33] Salt **1a** reacts almost violently with DMSO if one attempts to measure an NMR spectrum in that solvent, but the products are not known.[39]

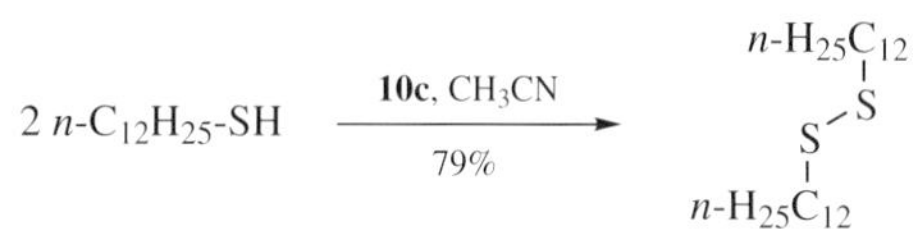

(20)

Second Update

Tsutomu Inokuchi & Li-Jian Ma
Okayama University, Okayama, Japan

Metal-free Aerobic Oxidation of Alcohols.[40,41] Environmentally benign aerobic alcohol oxidations have gained importance recently. TEMPO and its derivatives are used for aerobic oxidation of alcohols in various nonmetallic and transition metal-mediated reactions. A transition metal-free catalytic system consisting of catalytic sodium nitrite[42–44] or PEG (polyethylene glycol)–NO_2[45] with air as the terminal oxidant under acidic conditions was recently developed (eq 21).

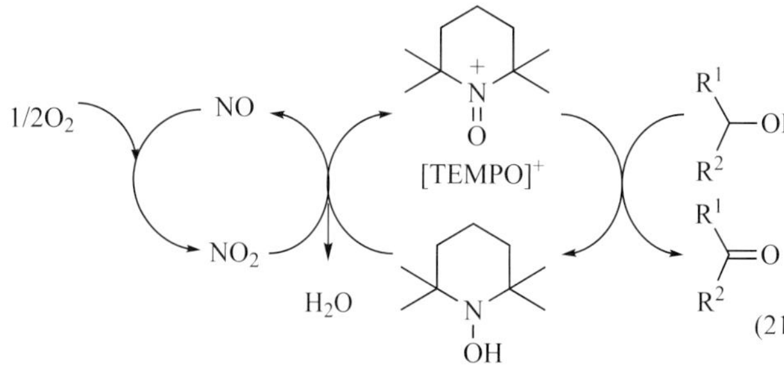

(21)

In place of $NaNO_2$, *tert*-butyl nitrite (TBN) was identified as an efficient NO equivalent for the activation of molecular oxygen.[46,47] Various alcohols including benzylic alcohols substituted with sulfide were converted into their corresponding carbonyl compounds with TEMPO/HBr/TBN as catalyst, which is suitable for the oxidation of solid alcohols with high melting points and/or low solubility using a minimal amount of solvent to form a slurry (eq 22).

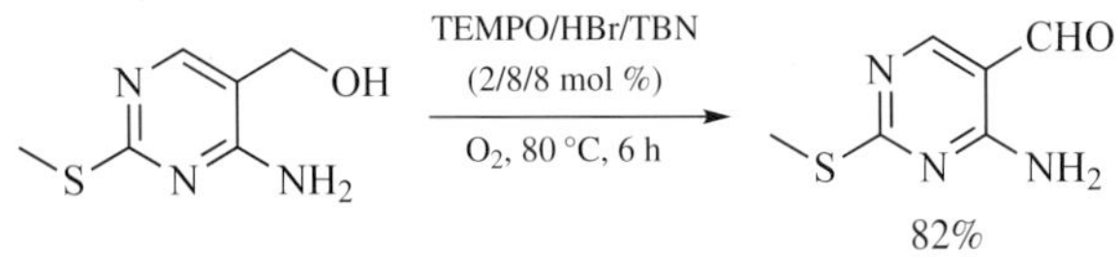

(22)

The cascade systems of O_2–$NaNO_2$–bromine or O_2–Bu_4NBr–TEMPO[48,49] or O_2–$NaNO_2$–$PhI(OAc)_2$ system[50] have also been developed (eqs 23 and 24).

A novel catalyst system that uses 1–4 mol % of TEMPO in combination with 4–6 mol % of aqueous hydroxylamine as a NO source is also effective (eq 25).[51]

Metal-catalyzed Aerobic Oxidation of Alcohols.[52–61] Aerobic metal-mediated TEMPO oxidations of alcohols have been examined. These processes have been mainly conducted with copper catalysts. This catalyst system is also useful for oxidation of benzylic alcohols and the reaction is explained by the mechanism below (eq 26).

Ionic liquid $[bmpy]PF_6$ was used as a reaction medium,[60] and fluorinated bipyridine ligands were successfully used for aerobic oxidation of alcohols under the fluorous biphasic system (FBS).[62] The sequential reaction including alcohol oxidation by TEMPO/Cu system and the asymmetric aldol reaction by peptide catalysis was realized using resin-supported TEMPO catalysts.[63] Simultaneous irradiation with visible light improved the catalytic oxidation with Cu/TEMPO systems, especially for unprotected alkyl glycosides (eq 27).[55]

Other effective metals in place of Cu include a vanadyl complex,[64] polymer incarcerated ruthenium (PIRu),[65] and combined bimetals such as Cu–Mn oxides,[66,67] and Mn–Co.[68,69]

Cascade Aerobic Alcohol Oxidations. Efficient and clean aerobic oxidation using 1 atm of oxygen or air, producing aldehydes/ketones from the widest spectrum of alcohols, including allenols and propargylic alcohols, at room temperature within a few hours was developed by cascade reaction of O_2–NO/NO_2–Fe(III)–TEMPO systems (eqs 28 and 29).[70] These processes have been mainly conducted with combined system involving the 4-acetamido-TEMPO/$FeCl_3$/$NaNO_2$ catalysts[71] or TEMPO/cobaloxime/$Co(NO_3)_2$ catalyst.[72]

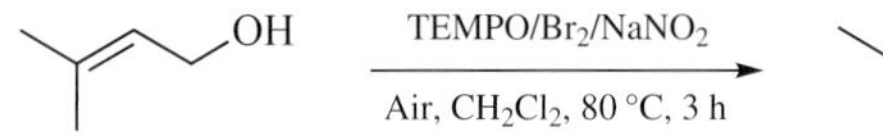

(24)

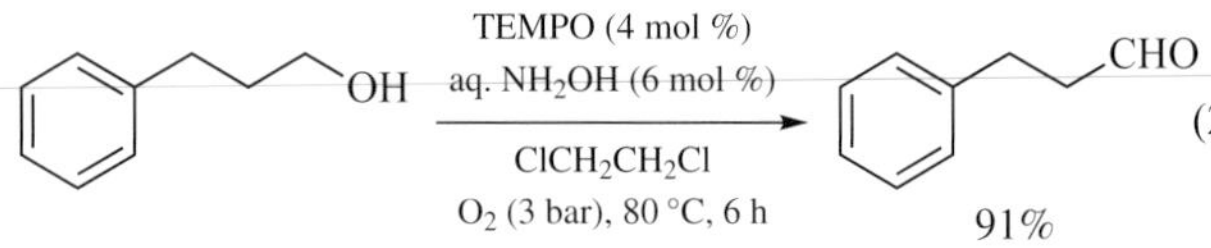

(25)

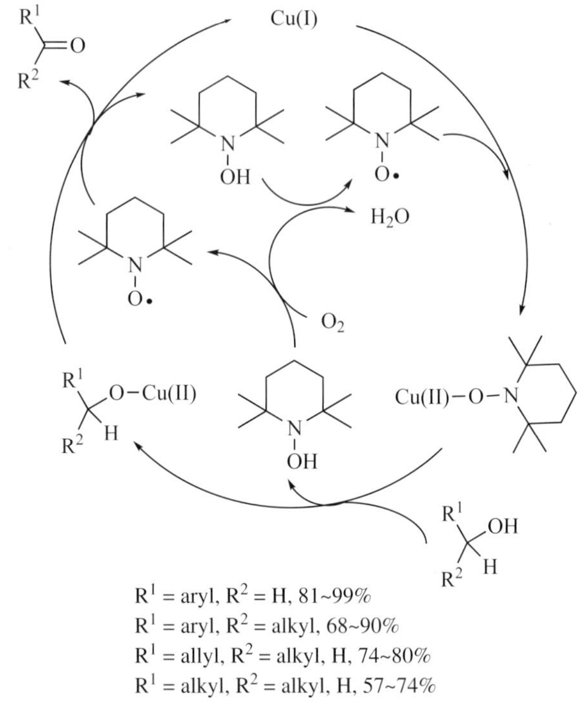

(26)

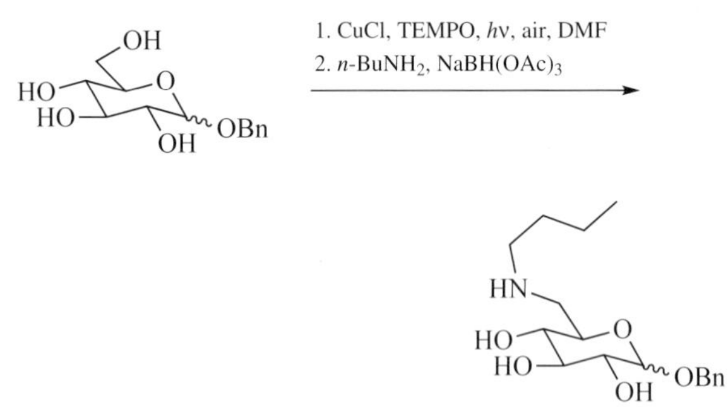

(27)

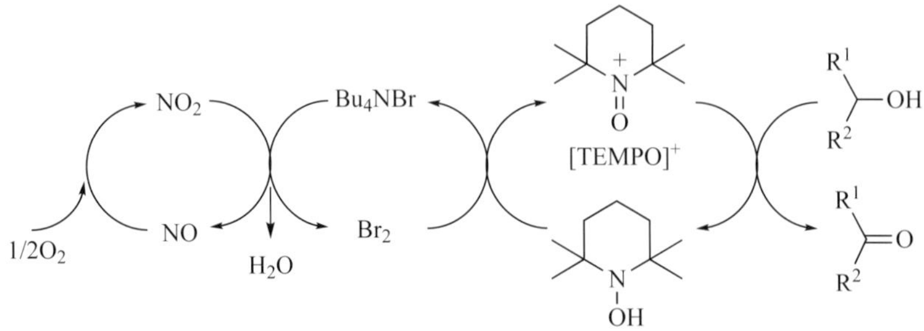

(23)

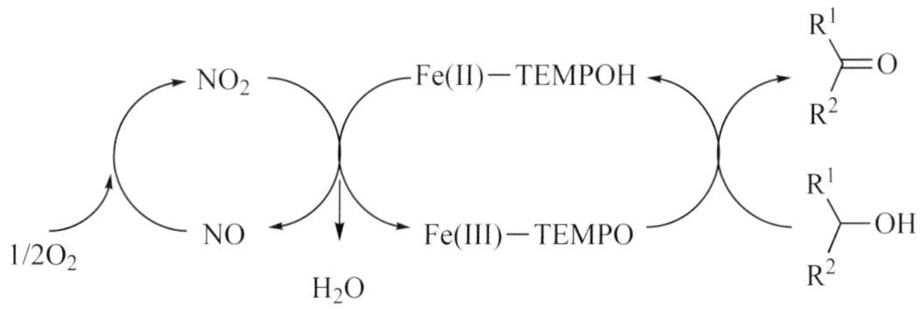

(28)

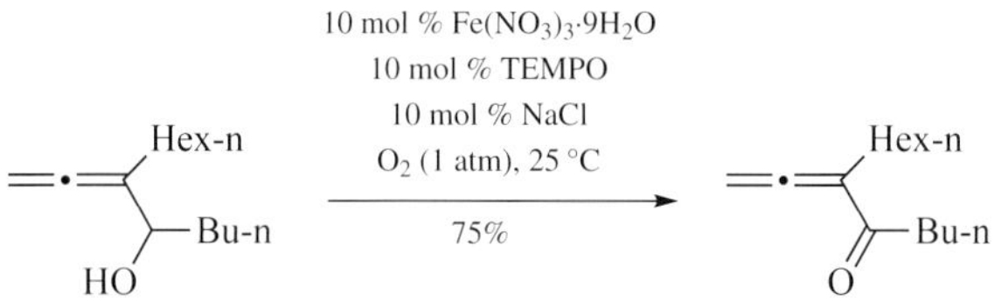

(29)

A combination TiO_2–dye cascade system was developed.[73] Upon irradiation with visible light, excited dye molecules (Alizarin Red) inject electrons into TiO_2 and form dye radicals, which can oxidize TEMPO to $TEMPO^+$, which selectively oxidizes alcohols in the presence of O_2 to the corresponding aldehydes. The selectivity is improved when the irradiation exploits visible light.

Supported TEMPO or Immobilized TEMPO. A combination of IL-supported TEMPO (**14**) with BAIB (bis(acetoxy)-iodobenzene)[74] or IL-supported TEMPO with tetra-*n*-butylammonium peroxymonosulfate (n-Bu_4NHSO_5), available from Oxone and n-Bu_4NHSO_4 in an ionic liquid medium, has been developed.[75]

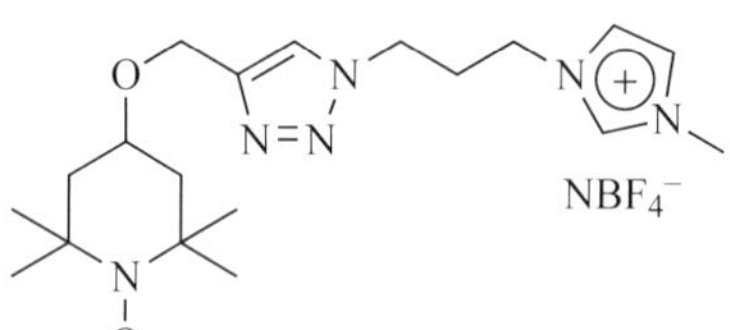

14

Organically modified silica-based SiliaCat TEMPOs (**15**, **16**) are an oxidizing catalyst that can efficiently replace homogeneous stable nitroxyl radicals and effect the oxidation in either organic solvents or water, with or without KBr as cocatalyst.[76,77]

Oxidation–hydrocyanation of γ,δ-unsaturated alcohols using immobilized TEMPO/PhI(OAc)$_2$ in combination with *Hb*HNL (hydroxynitrile lyase from *Hevea brasiliensis*) proceeds smoothly, the resulting cyanohydrin derivatives being obtained in good overall yields and with high ees (eq 30).[78]

Other immobilized TEMPOs shown below include phosphonium-supported TEMPO (**17**),[79] polyurethane- and polystyrene-supported TEMPO (**18**),[80] molecularly imprinted polymeric TEMPO (**19**),[81] a highly active and easily recoverable TEMPO with the attachment of multiple triazole moieties and perfluoroalkyl chains (**20**),[82] readily isolable "fluorous-tagged" TEMPOs,[83a] and TEMPO-linked metalloporphyrins.[83b]

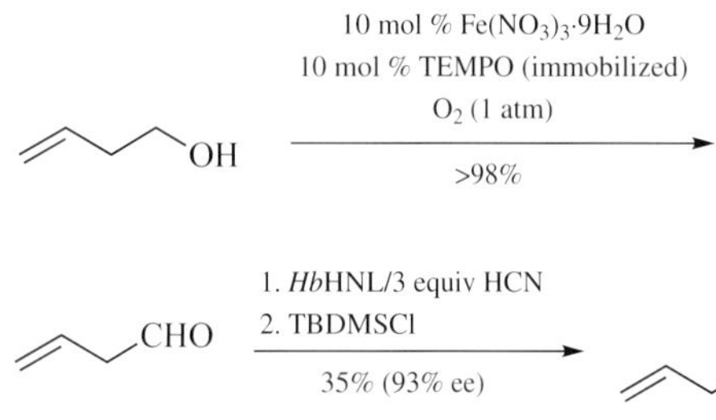

(30)

Site-isolated encapsulated linear polymeric catalysts bound with TEMPO have 2.5 times faster reaction rates than the analogous resin-supported catalysts, in the oxidation of *sec*-phenethyl alcohol to acetophenone (eq 31).[84]

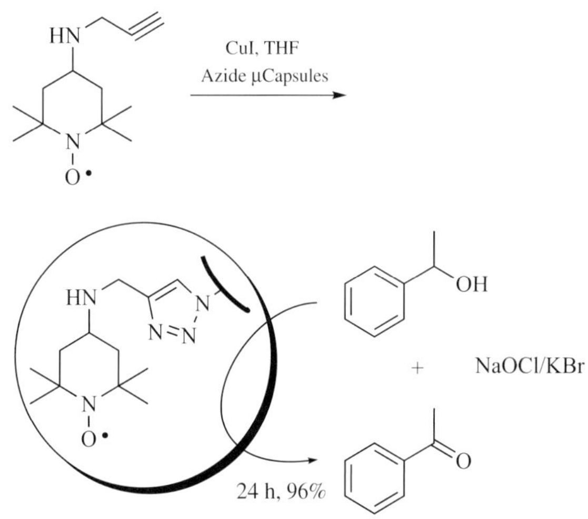

(31)

TEMPO–iodobenzene hybrid catalyst was used in the double catalytic reaction of arylmethyl and alkylmethyl alcohols as a substrate in a 9% acetic acid solution of peracetic acid (PAA) as a cooxidant, giving the corresponding carboxylic acids in good to excellent yields (56~99%) (eq 32).[85]

Industrial Applications. Sustainability of the TEMPO oxidation finds a variety of industrial applications as highly selective oxidation catalysts for the production of pharmaceuticals, flavors and fragrances, agrochemicals, and a variety of other specialty chemicals, and examples are reviewed.[86] In developing the manufacturing route for the cathepsin K inhibitor SB-462795, the TEMPO oxidation of a secondary alcohol for the final chemical stage has been exploited (eq 33).[87]

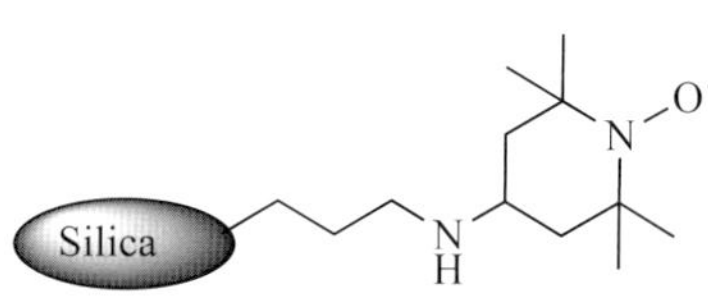

15

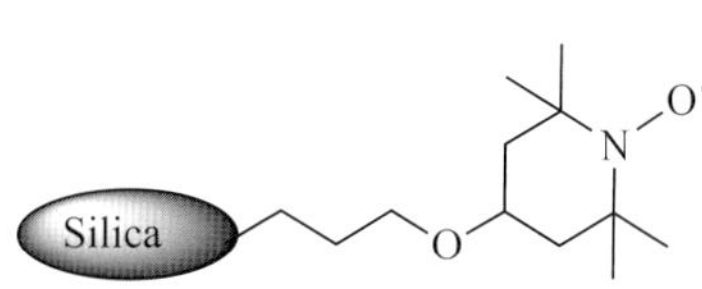

16

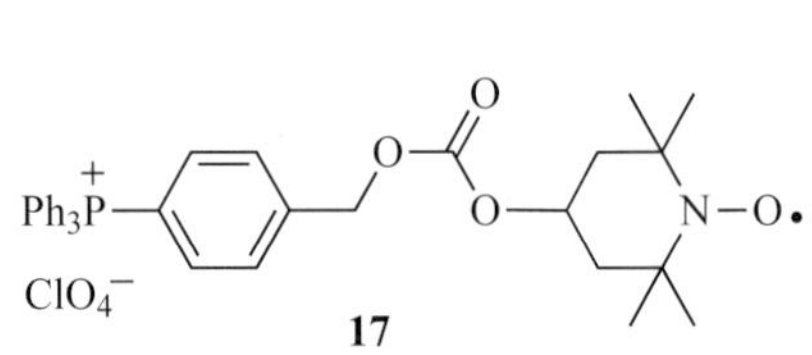

17

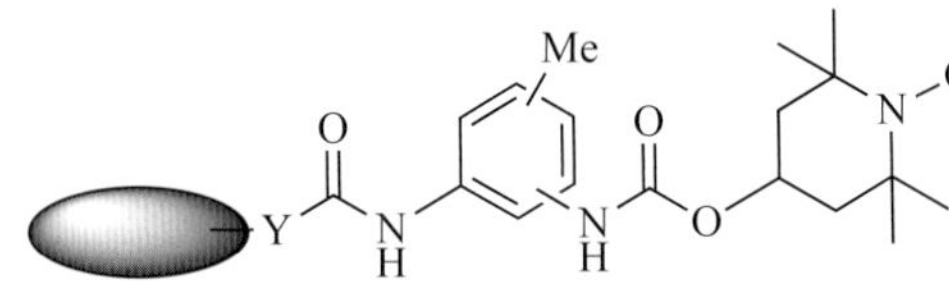

18

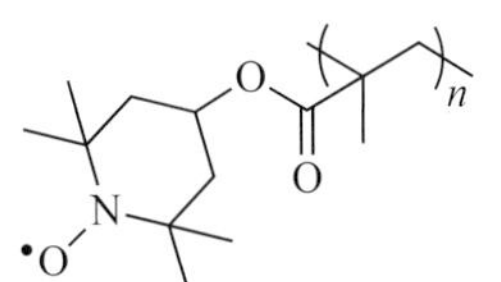

19

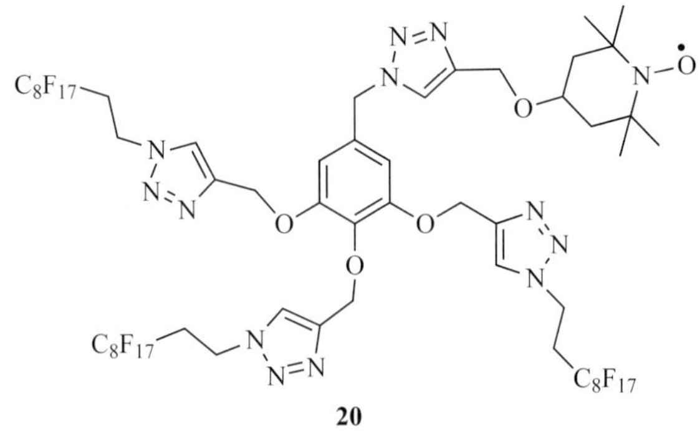

20

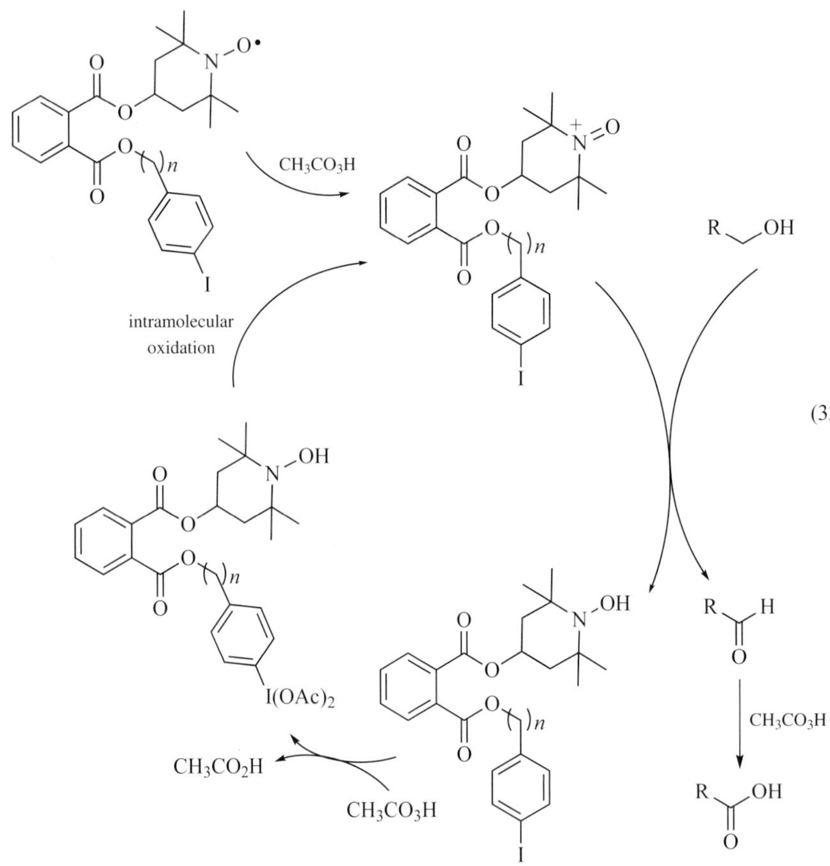

(32)

Conditions: catalyst (0.1 equiv), CH_3CO_3H (10 equiv) in CH_3CO_2H at 30 °C
R = aryl, 56~99%, R = alkyl, 75~99%

Other examples are synthesis of betulinic acid by oxidation with 4-acetamido-TEMPO/$NaClO_2$/NaOCl (86%)[88] and unprotected dialdo-glycosides with TEMPO–TCC (trichloroisocyanuric acid) (65∼100%) (eq 34).[89]

In the pilot plant synthesis of **(2-chlorophenyl)[2-(phenylsulfonyl)pyridin-3-yl]methanone**, reactivities of TEMPO derivatives, such as 4-MeO-TEMPO and 4-AcNH-TEMPO, were compared (eq 35).[90]

Technical production of aldehydes by continuous bleach oxidation of alcohols was achieved in a tube reactor. This system is essential for the synthesis of activated α-substituted aldehydes and also recommended for nonactivated aldehydes (eq 36).[91]

Applications for Chiral Intermediates. An efficient tandem reaction system was developed, in which primary alcohols were oxidized to the corresponding aldehydes followed by an asymmetric α-oxyamination with a resin-supported peptide catalyst (eq 37).[92] A similar transformation has also been achieved using an anodic oxidation (eq 38).[93]

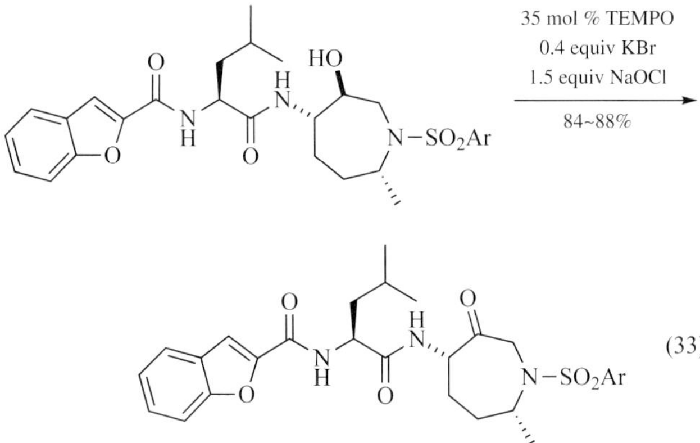

(33)

SB-462795 (Ar = 2-pyridyl)

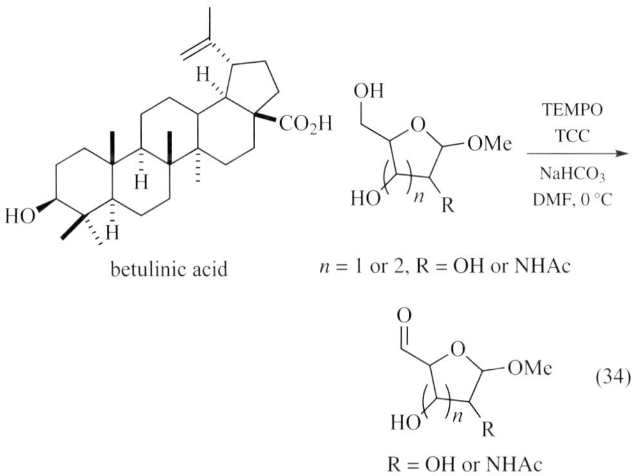

(34)

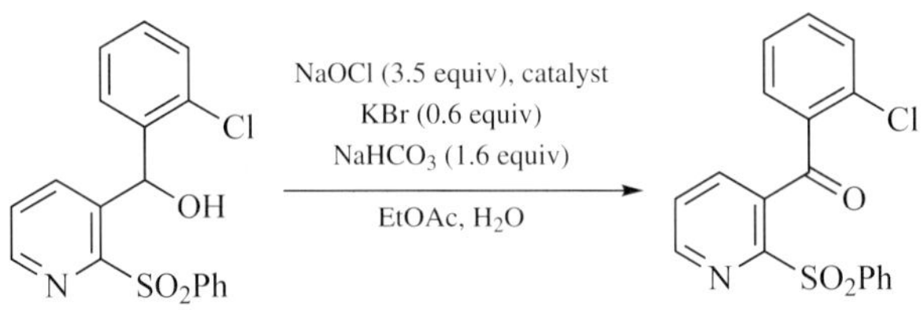

(35)

Catalyst Loading (equiv)	Yield (%) TEMPO	4-MeO-TEMPO	4-AcNH-TEMPO
0.07	93	98	90
0.01	53	62	47
0.005	25	54	46

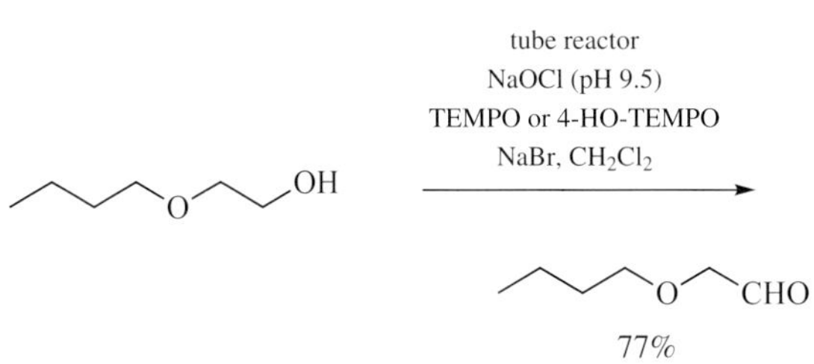

(36)

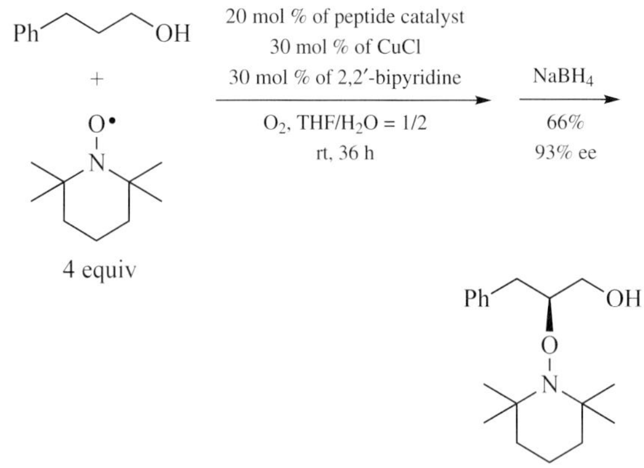

(37)

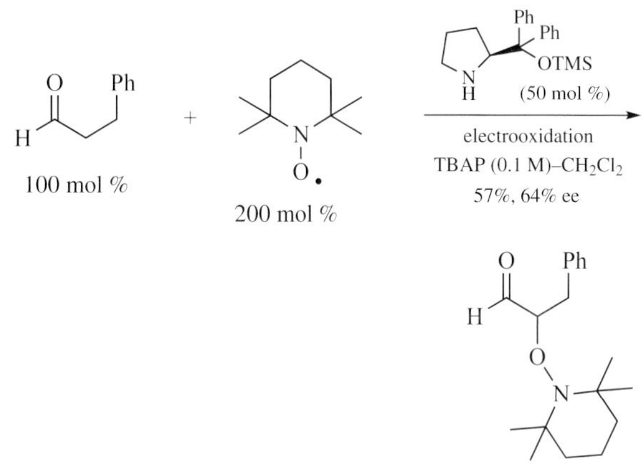

(38)

A chiral cobalt-catalyzed reaction features enantioselective aerobic oxidative kinetic resolution of (+)-α-hydroxy esters, using molecular oxygen as the sole oxidant (eq 39).[94]

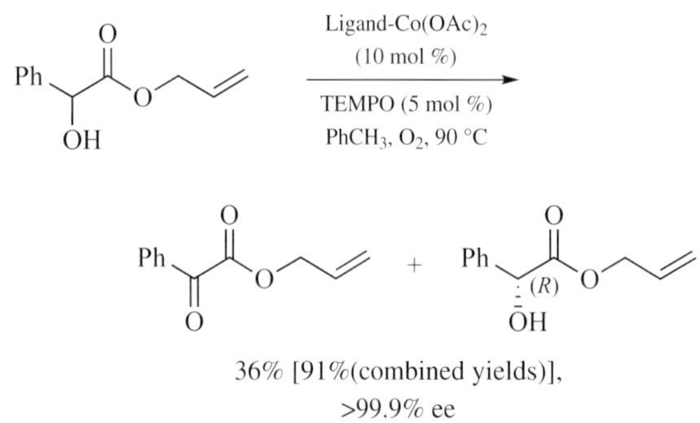

(39)

A convenient oxidation for a variety of electron-rich benzylic alcohols using HIO_3 and I_2O_5 as safer stoichiometric oxidants with catalytic amounts of TEMPO or 4-hydroxy-TEMPO in water was devised.[95]

This article includes some reactions of 4-methoxy-2,2,-6,6-tetramethyl-1-piperidinyloxy (4-methoxy-TEMPO) [95407-69-5], 4-acetylamino-2,2,6,6-tetramethyl-1-piperidinyloxy (4-acetylamino-TEMPO) [14691-89-5], and derivatives of 4-hydroxy-2,2,6,6-tetramethyl-1-piperidinyloxy (4-hydroxy-TEMPO) [2226-96-2] and 1-piperidinyloxy, 2,2,6,6-tetramethyl-4-(2-propyn-1-yloxy)- [147045-24-7].

1. Bobbitt, J. M.; Flores, M. C. L., *Heterocycles* **1988**, *27*, 509.
2. De Nooy, A. E. J.; Besemer, A. C.; Van Bekkum, H., *Synthesis* **1996**, *10*, 1153.
3. Bobbitt, J. M., *J. Org. Chem.* **1998**, *63*, 9367.
4. Koch, T.; Hoskovec, M.; Boland, W., *Tetrahedron* **2002**, *58*, 3271.
5. Abad, A.; Agulló, C.; Cuñat, A. C.; Perni, R. H., *Tetrahedron: Asymmetry* **2000**, *11*, 1607.
6. Miyazawa, T.; Endo, T., *Tetrahedron Lett.* **1986**, *27*, 3395.
7. Takata, T.; Tsujino, Y.; Nakanishi, S.; Nakamura, K.; Yoshida, E.; Endo, T., *Chem. Lett.* **1999**, 937.
8. Liu, Y.-D.; Ren, T.; Guo, Q.-X., *Chin. J. Chem.* **1996**, *14*, 252.
9. Golubev, V. A.; Miklyush, R. V., *Zh. Org. Khim.* **1972**, *8*, 1356 (English translation: **1972**, *8*, 1376).
10. Merbouh, N.; Bobbitt, J. M.; Brückner, C., *Tetrahedron Lett.* **2001**, *42*, 8793.
11. Merbouh, N.; Bobbitt, J. M.; Brückner, C., *Org. Prep. Proced. Int.* **2004**, *36*, 3.
12. Bobbitt, J. M.; Brückner, C.; Merbouh, N. *Oxoammonium Salt and Nitroxide Catalyzed Oxidations of Alcohols*; John Wiley & Sons, Inc.: Hoboken, NJ, **2007**; Vol. *in preparation*.
13. (a) Bobbitt, J. M., *Chem. Eng. News* **1999**, *77* (29), 6. (b) Bobbitt, J. M., *Molecules* **1999**, *4*, M102–U4.
14. Golubev, V. A.; Zhdanov, R. I.; Gida, V. M.; Rozantsev, É. G., *Izv. Akad. Nauk USSR, Ser. Khim.* **1971**, *4*, 768 (English translation).
15. Atovmyan, L. O.; Golubev, V. A.; Golovina, N. I.; Klitskaya, G. A., *Izv. Akad. Nauk USSR, Ser. Khim.* **1975**, *16*, 92 (English translation).
16. Golubev, V. A.; Zhdanov, R. I.; Rozantsev, E. G., *Izv. Akad. Nauk USSR, Ser. Khim.* **1970**, *19*, 186.
17. Kagiya, T.; Komuro, C.; Sakano, K.; Nishimoto, S., *Chem. Lett.* **1983**, 365.
18. Liu, Y.; Guo, H.; Liu, Z., *Huaxue Xuebao* **1991**, *49*, 187 (in Chinese).
19. Miyazawa, T.; Endo, T.; Shiihashi, S.; Okawara, M., *J. Org. Chem.* **1985**, *50*, 1332.
20. Bobbitt, J. M.; Merbouh, N., *Org. Synth.* **2005**, *82*, 80.
21. Yamaguchi, M.; Takata, T.; Endo, T., *Tetrahedron Lett.* **1988**, *29*, 5671.
22. Miyazawa, T.; Endo, T., *J. Am. Chem. Soc.* **1985**, *50*, 3930.
23. Merbouh, N.; Bobbitt, J. M.; Brückner, C., *J. Org. Chem.* **2004**, *69*, 5116.
24. Adam, W.; Saha-Möller, C. R.; Ganeshpure, P. A., *Chem. Rev.* **2001**, *101*, 3499.
25. Golubev, V. A.; Rudyk, T. S.; Sen, V. D.; Aleksandrov, A. L., *Izv. Akad. Nauk USSR, Ser. Khim.* **1976**, *1976*, 763 (English translation).
26. Ren, T.; Liu, Y. C.; Guo, Q. X., *Bull. Chem. Soc. Jpn.* **1996**, *69*, 2935.
27. Bobbitt, J. M.; Guttermuth, M. C. L.; Ma., Z.; Tang, H., *Heterocycles* **1990**, *30*, 1131.
28. Breton, T.; Liaigre, D.; Belgsir, E., *Tetrahedron Lett.* **2005**, *46*, 2487.
29. Pradhan, P. P.; Bobbitt, J. M.; Bailey, W. F., *Org. Lett.* **2006**, *8*, 5485.
30. Church, K. M.; Holloway, L. M.; Matley, R. C.; Brower, R. J., III, *Nucleosides, Nucleotides Nucleic Acids* **2004**, *23*, 1723.
31. Schämann, M.; Schäfer, H. J., *Synlett* **2004**, 1601.
32. Golubev, V. A.; Kobylyanskii, E. V., *Zh. Org. Khim.* **1972**, *8*, 2607.
33. Hunter, D. H.; Barton, D. H. R.; Motherwell, W. J., *Tetrahedron Lett.* **1984**, *25*, 603.
34. Hunter, D. H.; Racok, J. S.; Rey, A. W.; Ponce, Y. Z., *J. Org. Chem.* **1988**, *53*, 1278.
35. Bobbitt, J. M.; Ma, Z., *Heterocycles* **1992**, *33*, 641.
36. Semmelhack, M. F.; Schmid, C. R., *J. Am. Chem. Soc.* **1983**, *105*, 6732.
37. Liu, Y. C.; Wang, W.; Guo, Q. X., *Chin. Chem. Lett.* **1996**, *7*, 790.
38. Ma, Z. Ph. D., University of Connecticut, **1991**.
39. Bobbitt, J. M.; Pradhan, P. P., unpublished.
40. Vogler, T.; Studer, A., *Synthesis* **2008**, *13*, 1979.
41. Lenoir, D., *Angew. Chem., Int. Ed.* **2006**, *45*, 3206.
42. Miao, C.-X.; He, L.-N.; Wang, J.-L.; Wu, F., *J. Org. Chem.* **2010**, *75*, 257.
43. Miao, C.-X.; He, L.-N.; Wang, J.-Q.; Wang, J.-L., *Adv. Synth. Catal.* **2009**, *351*, 2209.
44. Karimi, B.; Biglari, A.; Clark, J. H.; Budarin, V., *Angew. Chem, Int. Ed.* **2007**, *46*, 7210.
45. Tao, J.-W.; Lu, Q.-Q.; Chu, C.-H.; Liu, R.-H.; Liang, X.-M., *Synthesis* **2010**, *23*, 3974.
46. He, X.-J.; Shen, Z.-L.; Mo, W.-M.; Sun, N.; Hu, B.-X.; Hu, X. Q., *Adv. Synth. Catal.* **2009**, *351*, 89.
47. Xie, Y.; Mo, W.; Xu, D.; Shen, Z.; Sun, N.; Hu, B.; Hu, X., *J. Org. Chem.* **2007**, *72*, 4288.
48. Liu, R.; Liang, X.; Dong, C.; Hu, X., *J. Am. Chem. Soc.* **2004**, *126*, 4112.
49. Liu, R.; Dong, C.; Liang, X.; Wang, X.; Hu, X., *J. Org. Chem.* **2005**, *70*, 729.
50. Herrerias, C. I.; Zhang, T. Y.; Li, C.-J., *Tetrahedron Lett.* **2006**, *47*, 13.
51. Wertz, S.; Studer, A., *Adv. Synth. Catal.* **2011**, *353*, 69.
52. Chrobok, A.; Baj, S.; Pudlo, W.; Jarzebski, A., *Appl. Catal. A: Gen.***2010**, *389*, 179.
53. Figiel, P. J.; Sibaouih, A.; Ahmad, J. U.; Nieger, M.; Raeisaenen, M. T.; Leskelae, M.; Repo, T., *Adv. Synth. Catal.* **2009**, *351*, 2625.
54. Ahmad, J. U.; Figiel, P. J.; Raeisaenen, M. T.; Leskelae, M.; Repo, T., *Appl. Catal. A: Gen.* **2009**, *371*, 17.
55. Gassama, A.; Hoffmann, N., *Adv. Synth. Catal.* **2008**, *350*, 35.
56. Lin, L.; Ji, L.; Wei, Y., *Catal. Commun.* **2008**, *9*, 1379.
57. Mannam, S.; Alamsetti, S. K.; Sekar, G., *Adv. Synth. Catal.* **2007**, *349*, 2253.
58. Figiel, P. J.; Leskela, M.; Repo, T., *Adv. Synth. Catal.* **2007**, *349*, 1173.
59. Jiang, N.; Ragauskas, A. J., *J. Org. Chem.* **2006**, *71*, 7087.
60. Jiang, N.; Ragauskas, A. J., *Org. Lett.* **2005**, *7*, 3689.
61. Wu, X.-E.; Ma, Li; Ding, M.-X.; Gao, L.-X., *Chem. Lett.* **2005**, *34*, 312.
62. Lu, N.; Lin, Y.-C., *Tetrahedron Lett.* **2007**, *48*, 8823.
63. Akagawa, K.; Takigawa, S.; Mano, E.; Kudo, K., *Tetrahedron Lett.* **2011**, *52*, 770.
64. Du, Z.-T.; Ma, J.-P.; Ma, H.; Wang, M.; Huang, Y.-Z.; Xu, J., *Catal. Commun.* **2010**, *11*, 732.
65. Matsumoto, T.; Ueno, M.; Kobayashi, J.; Miyamura, H.; Mori, Y.; Kobayashi, S., *Adv. Synth. Catal.* **2007**, *349*, 531.
66. Yang, G.; Ma, J.; Wang, W.; Zhao, J.; Lin, X.; Zhou, L.; Gao, X., *Catal. Lett.* **2006**, *112*, 83.
67. Yang, G.; Zhu, W.; Zhang, P.; Xue, H.; Wang, W.; Tian, J.; Song, M., *Adv. Synth. Catal.* **2008**, *350*, 542.
68. Gilhespy, M.; Lok, M.; Baucherel, X., *Catal. Today* **2006**, *117*, 114.
69. Gilhespy, M.; Lok, M.; Baucherel, X., *Chem. Commun.* **2005**, 1085.
70. Ma, S.-M.; Liu, J.-X.; Li, S.-H.; Chen, B.; Cheng, J.-J.; Kuang, J.-Q.; Liu, Y.; Wan, B.-Q.; Wang, Y.-L.; Ye, J.-T., *Adv. Synth. Catal.* **2011**, *353*, 1005.
71. (a) Yin, W.-L.; Chu, C.-H.; Lu, Q.-Q.; Tao, J.-W.; Liang, X.-M.; Liu, R. H., *Adv. Synth. Catal.* **2010**, *352*, 113. (b) Wang, N.; Liu, R.; Chen, J.; Liang, X., *Chem. Commun.* **2005**, 5322.
72. Jing, Y.-Y.; Jiang, J.; Yan, B.; Lu, S.; Jiao, J.-M.; Xue, H.-Z.; Yang, G.-Y.; Zheng, G.-X., *Adv. Synth. Catal.* **2011**, *353*, 1146.
73. Zhang, M.; Chen, C.-C.; Ma, W.-H.; Zhao, J.-C., *Angew. Chem., Int. Ed.* **2008**, *47*, 9730.
74. Fall, A.; Sene, M.; Gaye, M.; Gomez, G.; Fall, Y., *Tetrahedron Lett.* **2010**, *51*, 4501.
75. Zhu, C.-J.; Ji, L.; Wei, Y.-Y., *Catal. Commun.* **2010**, *11*, 1017.
76. Michaud, A.; Pandarus, V.; Tremblay, L.; Ciriminna, R.; Pagliaro, M.; Beland, F., *Top. Catal.* **2010**, *53*, 1110.
77. Michaud, A.; Gingras, G.; Morin, M.; Beland, F.; Ciriminna, R.; Avnir, D.; Pagliaro, M., *Org. Process Res. Dev.* **2007**, *11*, 766.

78. Vugts, D. J.; Veum, L.; al-Mafraji, K.; Lemmens, R.; Schmitz, R. F.; de Kanter, F. J. J.; Groen, M. B.; Hanefeld, U.; Orru, R. V. A., *Eur. J. Org. Chem.* **2006**, 1672.
79. Roy, M.-N.; Poupon, J.-C.; Charette, A. B., *J. Org. Chem.* **2009**, *74*, 8510.
80. Subhani, M. A.; Beigi, M.; Eilbracht, P., *Adv. Synth. Catal.* **2008**, *350*, 2903.
81. Anderson, C. D.; Shea, K. J.; Rychnovsky, S. D., *Org. Lett.* **2005**, *7*, 4879.
82. Gheorghe, A.; Chinnusamy, T.; Cuevas-Yanez, E.; Hilgers, P.; Reiser, O., *Org. Lett.* **2008**, *10*, 4171.
83. (a) Holczknecht, O.; Cavazzini, M.; Quici, S.; Shepperson, I.; Pozzi, G., *Advanced Synthesis & Catalysis* **2005**, *347*(5), 677. (b) Huang, J.-Y.; Li, S.-J.; Wang, Y.-G., *Tetrahedron Lett.* **2006**, *47*, 5637.
84. Mason, B. P.; Bogdan, A. R.; Goswami, A.; McQuade, D. T., *Organic Lett.* **2007**, *9*, 3449.
85. Yakura, T.; Ozono, A., *Adv. Synth. Catal.* **2011**, *353*, 855.
86. Ciriminna, R.; Pagliaro, M., *Org. Process Res. Dev.* **2010**, *14*, 245.
87. Goodman, S. N.; Dai, Q.-Y.; Wang, J.; Clark, W. M., *Org. Process Res. Dev.* **2011**, *15*, 123.
88. Csuk, R.; Schmuck, K.; Schaefer, R., *Tetrahedron Lett.* **2006**, *47*, 8769.
89. Angelin, M.; Hermansson, M.; Dong, H.; Ramstroem, O., *Eur. J. Org. Chem.* **2006**, 4323.
90. Kopach, M. E.; Kobierski, M. E.; Coffey, D. S.; Alt, C. A.; Zhang, T.; Borghese, A.; Trankle, W. G.; Roberts, D. J.; Moynihan, H.; Lorenz, K., *Org. Process Res. Dev.* **2010**, *14*, 1229.
91. Fritz-Langhals, E., *Org. Process Res. Dev.* **2005**, *9*, 577.
92. Akagawa, K.; Fujiwara, T.; Sakamoto, S.; Kudo, K., *Chem. Commun.* **2010**, *46*, 8040.
93. Bui, N.-N.; Ho, X.-H.; Mho, S.-I.; Jang, H.-Y., *Eur. J. Org. Chem.* **2009**, 5309.
94. Alamsetti, S. K.; Sekar, G., *Chem. Commun.* **2010**, *46*, 7235.
95. Liu, Z.-Q.; Shang, X.; Chai, L.; Sheng, Q., *Catal. Lett.* **2008**, *123*, 317.

2′-Amino-[1,1′-binaphthalen]-2-ol[1]

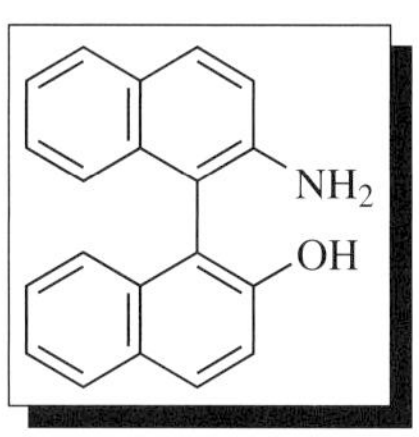

(±)
[134532-03-9] $C_{20}H_{15}NO$ (MW 285.34)
InChI = 1S/C20H15NO/c21-17-11-9-13-5-1-3-7-15(13)19(17)20-16-8-4-2-6-14(16)10-12-18(20)22/h1-12,22H,21H2
InChIKey = HIXQCPGXQVQHJP-UHFFFAOYAR
(*R*)-(+) *[137848-28-3]*
InChI = 1/C20H15NO/c21-17-11-9-13-5-1-3-7-15(13)19(17)20-16-8-4-2-6-14(16)10-12-18(20)22/h1-12,22H,21H2
InChIKey = HIXQCPGXQVQHJP-UHFFFAOYAR
(*S*)-(−) *[137848-29-4]*
InChI = 1/C20H15NO/c21-17-11-9-13-5-1-3-7-15(13)19(17)20-16-8-4-2-6-14(16)10-12-18(20)22/h1-12,22H,21H2
InChIKey = HIXQCPGXQVQHJP-UHFFFAOYAR

(chiral ligand[1,2] and catalyst[3])

Alternate Name: 2-amino-2′-hydroxy-1,1′-binaphthyl, NOBIN.

Physical Data: (±) mp 239–241 °C (benzene). (*S*)-(−) 171–173 °C (benzene), $[\alpha]_D$ −117 (*c* 1.0, THF)[1b] or −120 (*c* 1.5, THF).[1e]

Solubility: (±) soluble very well in THF; well-soluble in CH_2Cl_2, AcOEt; less soluble in toluene and benzene; sparingly soluble in MeOH, EtOH; insoluble in hexane. (*R*)-(+) and (*S*)-(−) are soluble very well in THF; well in CH_2Cl_2, AcOEt, toluene, and benzene; sparingly soluble in MeOH, EtOH, and hexane. Enantiopure NOBIN is much better soluble in heptane than the racemate, which can be used for obtaining a pure enantiomer from an enantiomerically enriched sample.[4]

Preparative Methods: racemic 2′-amino-[1,1′-binaphthalen]-2-ol (NOBIN) is most conveniently prepared in one step by the oxidative cross-coupling of a 1:1 mixture of 2-amino-naphtha-lene with 2-naphthol. The original method, employing the Cu(II)-mediated coupling in solution (eq 1),[1a] has been later improved in form of a Fe(III)-mediated coupling in a quasi-solid-state.[5] When the Cu(II)-coupling is carried out in the presence of a chiral amine, such as α-methylbenzylamine, the resulting NOBIN is obtained in up to 46% ee;[1b,1c] the latter product can be purified by crystallization to ∼100% ee.[1b] Resolution of the racemate can be attained by crystallization with (1*S*)-(+)-10-camphorsulfonic acid from a 10:1 mixture of chlorobenzene and ethanol,[1e] which gives (*S*)-(−)-NOBIN,[1e] or a mixture of toluene and ethanol; kinetic resolution of the corresponding imine with various amino alcohols is rather inefficient. An improved procedure relies on the co-crystallization of (±)-NOBIN with 0.5 equiv of *N*-benzylcin-chonidinium chloride from acetone, which gives (*R*)-(+)- NOBIN.[7] The absolute configuration was determined by CD spectroscopy (namely, by comparison with the spectra of 1,1′-binaphthalene-2,2′-diol and 2,2′-diamino(1,1′-binaphthalene)),[1b] and confirmed by chemical correlation,[5] and X-ray analysis.[7] An alternative synthesis, starting with (*R*)-1,1′-binaphthalene-2,2′-diol (BINOL), employs the Hartwig–Buchwald amination (eq 2) of the protected monotriflate (Ar = 4-MeO-C_6H_4).[8] Although less straightforward, the latter protocol has the advantage of avoiding the use of toxic 2-aminonaphthalene.

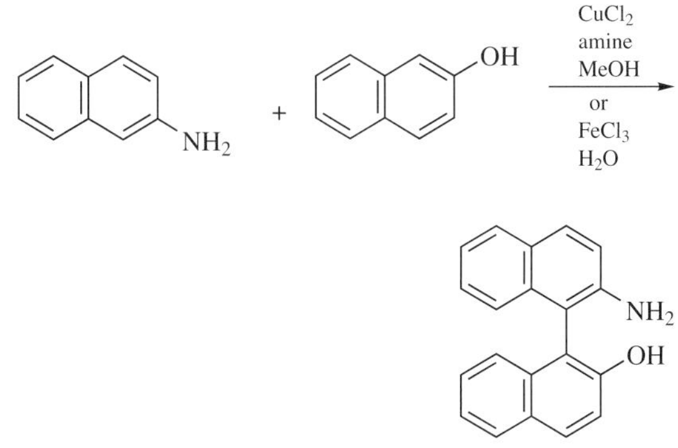

(1)

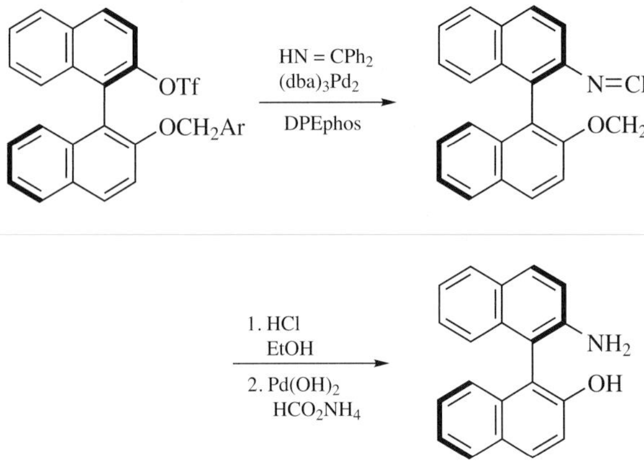

(2)

Drying: standard drying during the work up; not hygroscopic.

Handling, Storage, and Precautions: keep tightly closed, store in a cool dark place; may slowly deteriorate when exposed to direct sunshine and air.

(*R*)-Dimethyl-NOBIN and its *N,N*-dialkyl analogues,[2] obtained from (*R*)-NOBIN on reductive amination of the corresponding aldehydes and ketones, respectively, catalyze the Et_2Zn addition to aromatic aldehydes (eq 3), affording the (*R*)-alcohols with up to 88% ee (for PhCHO and dimethyl-NOBIN).[2]

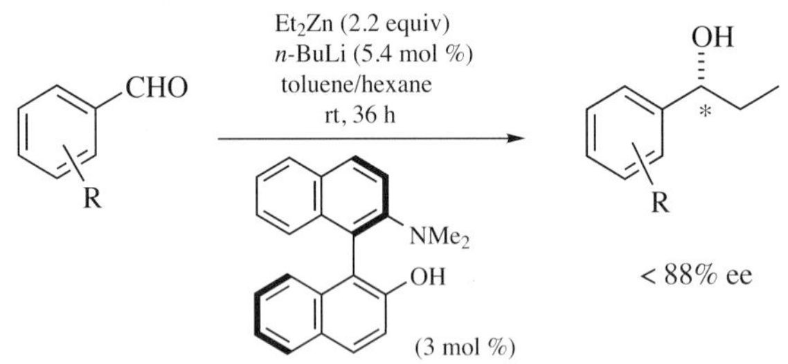

(3)

Imines, derived from methyl esters of amino acids, can be alkylated with $PhCH_2Br$ and other good electrophiles, under the phase transfer conditions in the presence of solid NaOH and (*S*)-NOBIN (10 mol %) in a nonpolar solvent to give unnatural α-alkyl amino acids (eq 4). Best results were obtained for α-methylphenylalanine (62% ee).[3a,3b] Alkylation of the glycine-derived nickel complex under the similar conditions (eq 5), with (*R*)-NOBIN as catalyst, exhibits very high enantioselectivities (98% ee for phenylalanine and 99.5% ee for DOPA).[3c] Strong nonlinear effect has been observed for the latter alkylation (NOBIN of 40% ee exhibits the same asymmetric induction as its enantiopure counterpart).[3c] Furthermore, no second alkylation occurs in this case since the enolization of the primary product is precluded by steric hindrance.

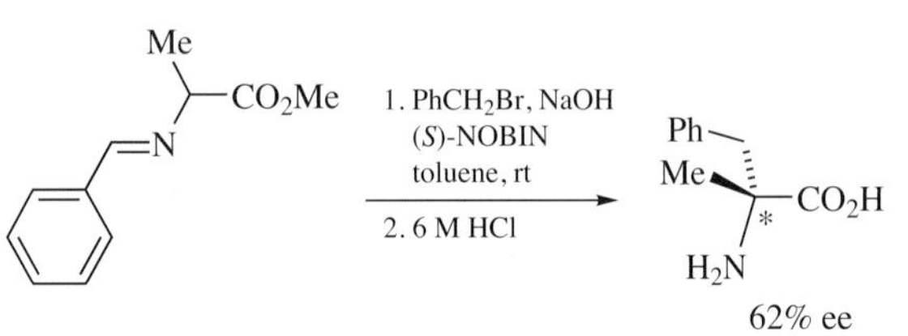

(4)

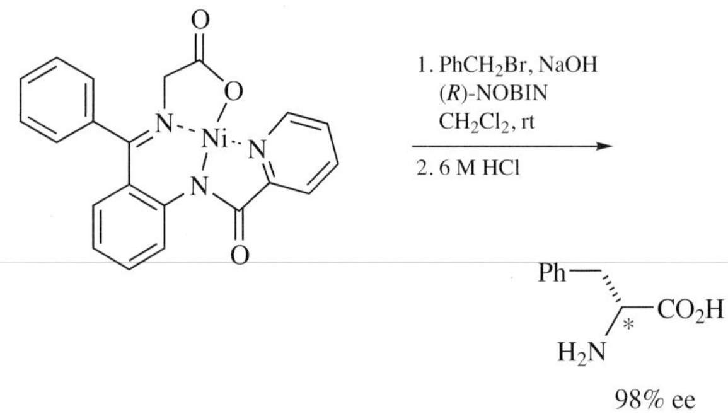

(5)

High asymmetric induction has been reported for aldol reaction of ethyl acetate silyl enol ether or methoxypropene (eq 6) catalyzed by a titanium(IV) complex of the (*R*)-NOBIN-derived imine (Carreira ligand).[9] Methoxy olefins react with a similarly high enantioselectivity as their silylated counterparts.[9]

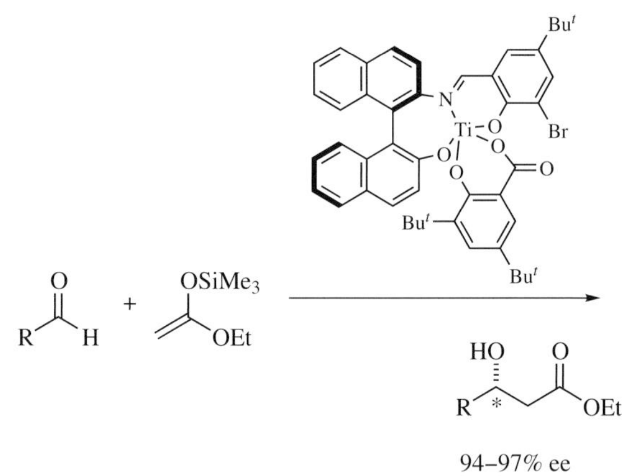

(6)

NOBIN can serve as a starting material for the preparation of various *N*-monoalkyl- and *N,N*-dialkyl-NOBIN analogues via reductive amination. Aldehydes react with NOBIN in the presence of $NaBH_4$ and H_2SO_4 to give dialkyl amino analogues, whereas ketones afford monoalkylated products (eq 7).[2] Hartwig–Buchwald arylation of NOBIN produces *N*-monoaryl derivatives (eq 8).[10] Alternatively, *N*-aryl-NOBIN derivatives can be prepared by amination of the protected monotriflate derived from BINOL with various aromatic and heteroaromatic amines.[8]

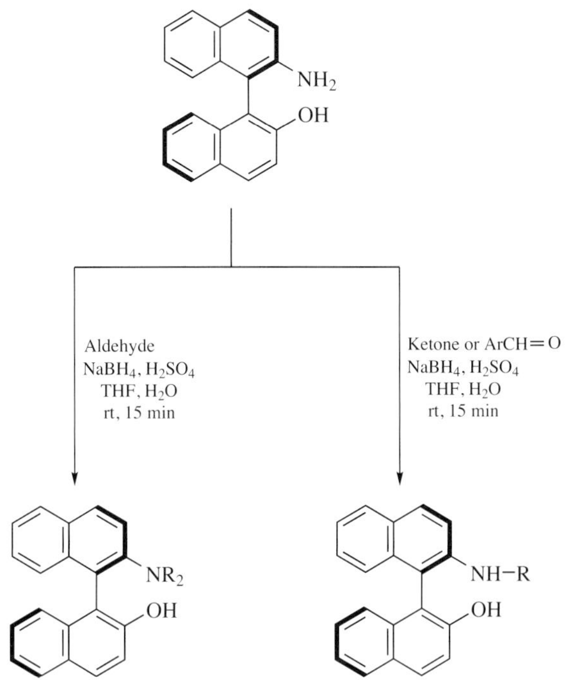

(7)

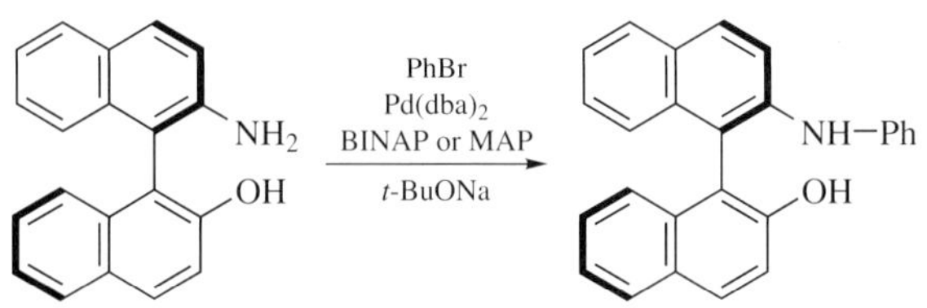

(8)

Employing the Newman–Kwart rearrangement, NOBIN and dimethyl-NOBIN can be converted into the corresponding (*S*)-aminothiol without racemization (eq 9).[11]

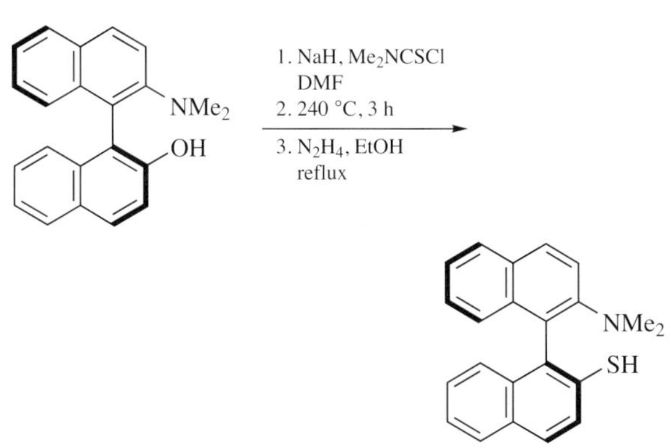

(9)

New aminophosphine ligands (MAP) have been obtained from the triflate derived from dimethyl NOBIN by Pd(0)-catalyzed coupling (eq 10)[12] and applied in asymmetric synthesis.

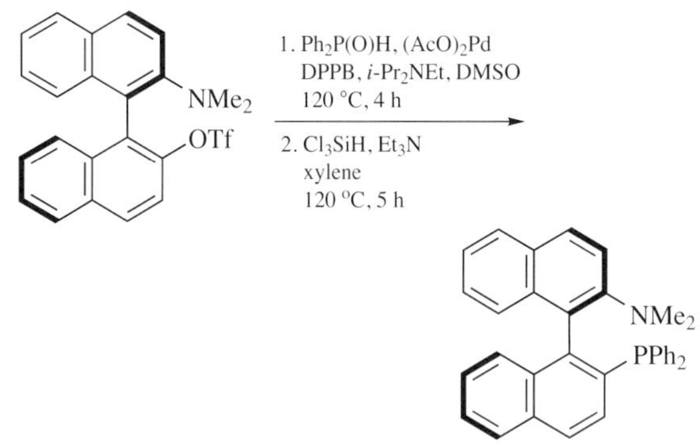

(10)

O-methyl-NOBIN and the corresponding imines have been prepared by Curtius rearrangement (eq 11);[13] the imines were employed in the asymmetric transfer of the $Fe(CO)_3$ unit to form planar-chiral diene complexes.[13c]

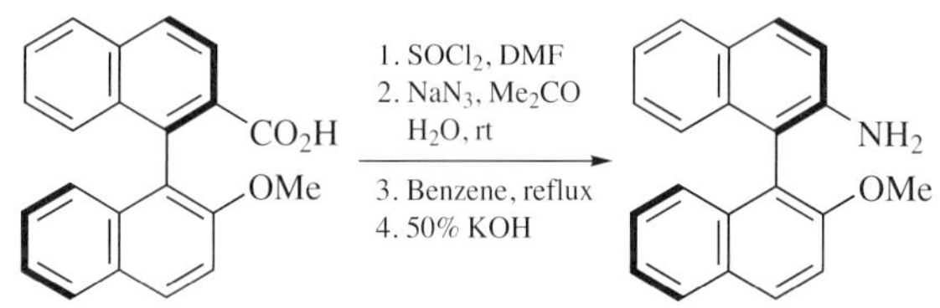

(11)

5,5′,6,6′,7,7′,8,8′-octahydro-NOBIN (H_8-NOBIN)[14] has been prepared by partial reduction of NOBIN with Ni–Al alloy[15] in a dilute aqueous alkaline solution (eq 12) and utilized as starting material for the synthesis of the corresponding aminophosphine[14] and as an *O,N*-ligand for Et_2Zn addition to aldehydes.[16]

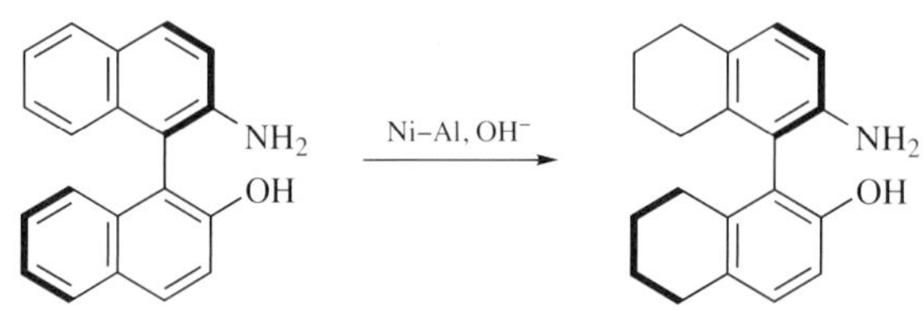

(12)

The NOBIN derivative, prepared by the Cu(II)-mediated coupling of 2-aminonaphthalene with 3-hydroxy-2-naphthoate, exhibits a solid-state $B_{AL}2$ intermolecular migration of the ester CH_3 group to nitrogen (eq 13).[17]

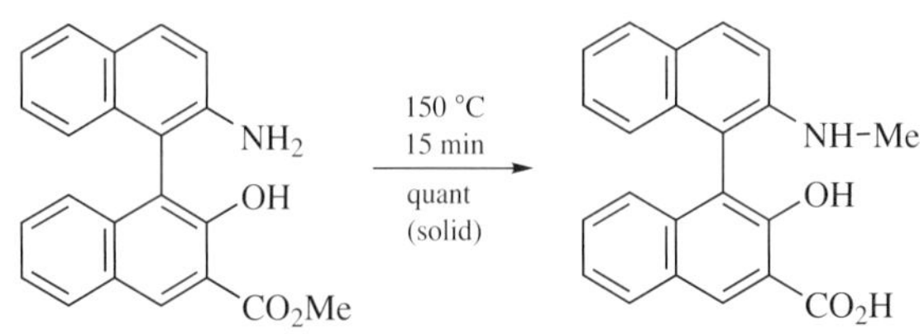

(13)

1. (a) Smrčina, M.; Lorenc, M.; Hanuš, V.; Kočovský, P., *Synlett.* **1991**, 231. (b) Smrčina, M.; Lorenc, M.; Hanuš, V.; Sedmera, P.; Kočovský, P., *J. Org. Chem.* **1992**, *57*, 1917. (c) Smrčina, M.; Poláková, J.; Vyskočil, Š.; Kočovský, P., *J. Org. Chem.* **1993**, *58*, 4534. (d) Smrčina, M.; Vyskočil, Š.; Máca, B.; Polášek, M.; Claxton, T. A.; Abbott, A. P.; Kočovský, P., *J. Org. Chem.* **1994**, *59*, 2156. (e) Smrčina, M.; Vyskočil, Š.; Polívková, J.; Poláková, J.; Kočovský, P., *Collect. Czech. Chem. Commun.* **1996**, *61*, 1520.
2. (a) Vyskočil, Š.; Jaracz, S.; Smrčina, M.; Štícha, M.; Hanuš, V.; Polášek, M.; Kočovský, P., *J. Org. Chem.* **1998**, *63*, 7727. (b) Vyskočil, Š.; Meca, L.; Kubišta, J.; Maloň, P.; Kočovský, P., *Collect. Czech. Chem. Comm.* **2000**, *65*, 539.

3. (a) Belokon, Y. N.; Kochetkov, K. A.; Churkina, T. D.; Ikonnikov, N. S.; Vyskočil, Š.; Kagan, H. B., *Tetrahedron: Asymmetry* **1999**, *10*, 1723. (b) Belokon, Y. N.; Kochetkov, K. A.; Churkina, T. D.; Ikonnikov, N. S.; Chesnokov, A. A.; Larionov, O. V.; Singh, I.; Parmar, V. S.; Vyskočil, Š.; Kagan, H. B., *J. Org. Chem.* **2000**, *65*, 7041. (c) Belokon, Y. N.; Kochetkov, K. A.; Churkina, T. D.; Ikonnikov, N. S.; Larionov, O. V.; Harutjunan, S. R.; Vyskočil, Š.; North, M.; Kagan, H. B., *Angew. Chem., Int. Ed.* **2001**, *40*, 1948.
4. Mahmoud, H.; Han, Y.; Segal, B. M.; Cai, L., *Tetrahedron Asymmetry* **1998**, *9*, 2035.
5. (a) Ding, K.; Xu, Q.; Wang, Y.; Liu, J.; Yu, Z.; Du, B.; Wu, Y.; Koshima, H.; Matsuura, T., *Chem. Commun.* **1997**, 693. (b) Vyskočil, Š.; Smrčina, M.; Lorenc, M.; Hanuš, V.; Polášek, M.; Kočovský, P., *Chem. Commun.* **1998**, 585.
6. Carreira, E. M., personal communication.
7. Ding, K.; Wang, Y.; Yun, H.; Liu, J.; Wu, Y.; Terada, M.; Okubo, Y.; Mikami, K., *Chemistry-Eur. J.* **1999**, *5*, 1734.
8. Singer, R. A.; Buchwald, S. L., *Tetrahedron Lett.* **1999**, *40*, 1095.
9. (a) Carreira, E. M.; Singer, R. A.; Lee, W.; *J. Am. Chem. Soc.* **1994**, *116*, 8837. (b) Carreira, E. M.; Lee, W.; Singer, R. A., *J. Am. Chem. Soc.* **1995**, *117*, 3649. (c) Singer, R. A.; Carreira, E. M., *J. Am. Chem. Soc.* **1995**, *117*, 12360.
10. Vyskočil, Š.; Smrčina, M.; Kočovský, P., *Tetrahedron Lett.* **1998**, *39*, 9289.
11. Smrčina, M.; Vyskočil, Š.; Polívková, J.; Poláková, J.; Sejbal, J.; Hanuš, V.; Polášek, M.; Verrier, H.; Kočovský, P., *Tetrahedron: Asymmetry* **1997**, *8*, 537.
12. Vyskočil, Š.; Smrčina, M.; Hanuš, V.; Polášek, M.; Kočovský, P., *J. Org. Chem.* **1998**, *63*, 7738.
13. (a) Hattori, T.; Shijo, M.; Kumaga, S.; Miyano, S., *Chem. Express* **1991**, *6*, 335. (b) Hattori, T.; Hotta, H.; Suzuki, T.; Miyano, S., *Bull. Chem. Soc. Jpn.* **1993**, *66*, 613. (c) Knölker, H.-J.; Hermann, H., *Angew. Chem., Int. Ed. Engl.* **1996**, *35*, 341. (d) Hungerhoff, B.; Metz, P., *Tetrahedron* **1999**, *55*, 14941.
14. Wang, Y.; Guo, H.; Ding, K., *Tetrahedron: Asymmetry* **2000**, *11*, 4153.
15. Guo, H.; Ding, K. L., *Tetrahedron Lett.* **2000**, *41*, 10061.
16. Guo, H.; Wang, Y.; Ding, K., *Chin. J. Chem.* **2001**, *19*, 52.
17. Smrčina, M.; Vyskočil, Š.; Hanuš, V.; Polášek, M.; Langer, V.; Chew, B. G. M.; Zax, D. B.; Verrier, H.; Harper, K.; Claxton, T. A.; Kočovský, P., *J. Am. Chem. Soc.* **1996**, *118*, 487.

Pavel Kočovský
University of Glasgow, Glasgow, UK

(9*R*)-9-Amino-9-deoxyquinidine

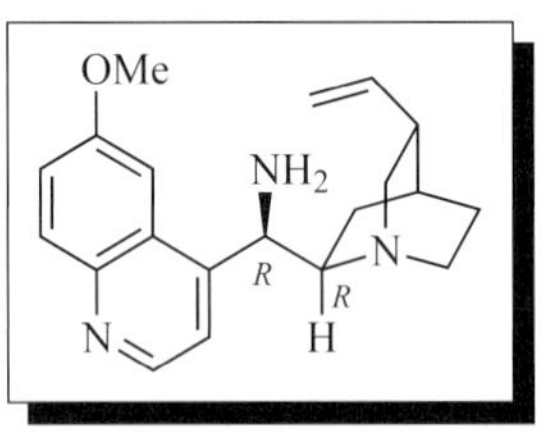

[287979-82-2] $C_{20}H_{25}N_3O$ (MW 323)

InChI = 1S/C20H25N3O/c1-3-13-12-23-9-7-14(13)10-19(23)20(21)16-6-8-22-18-5-4-15(24-2)11-17(16)18/h3-6,8,11,13-14,19-20H,1,7,9-10,12,21H2,2H3/t13?,14-,19+,20+/m0/s1

InChIKey = RIEKOKLABHBCGI-PAWLDSKOSA-N

(used as catalyst and for the preparation of (co)-catalysts in enantioselective reactions)

Alternate Name: 6′-Methoxy-(9*R*)-cinchonan-9-amino; 9-deoxyepiquinidine.

Physical Data: bp 240 °C, 0.01 Torr[1] $[\alpha]_D^{25}$ + 70 (c = 1.5, $CHCl_3$)[2].

Solubility: THF, CH_2Cl_2, and $CHCl_3$.

Form Supplied in: slightly yellow oil.

Purification: distillation under vacuum or chromatography using Et_2O-MeOH-Et_3N (10:1:0.5) as eluent.

Preparative Methods: prepared from quinidine, Hydrazoic acid, and diisopropyl azodicarboxylate (DIAD) by Mitsunobu reaction followed by Staudinger reduction with Triphenylphosphine and hydrolysis (eq 1).

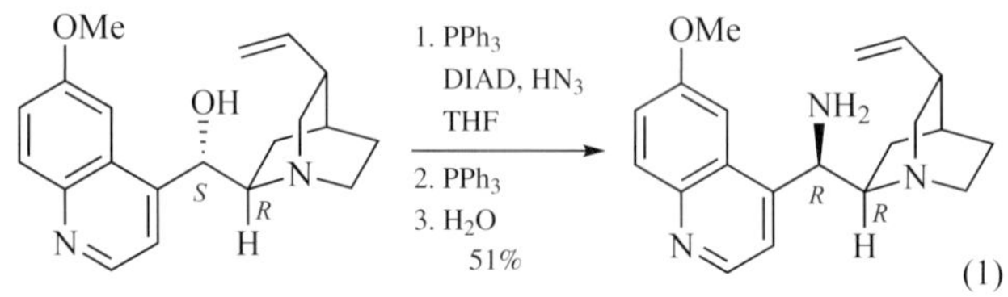

(1)

The synthetic method for (9*R*)-9-amino-9-deoxyquinidine (ADQD) was modified under mild conditions in three steps (eq 2).[18]

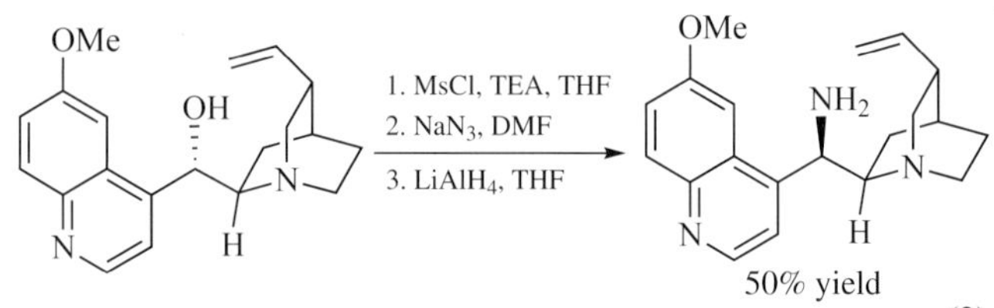

(2)

Original Commentary

Marie-Claire Lasne & Jacques Rouden
Université de Caen, Caen, France

Asymmetric Reactions of α, β Unsaturated Ketones. (9*R*)-9-Amino-9-deoxyquinidine in combination with an acid is a highly efficient catalyst of Michael additions (eq 3),[3] hydroperoxidation (eq 4),[4] and Diels–Alder reaction of pyrones (eq 5).[5] The primary amine group activates the carbonyl group that is inert to secondary amine catalysis. It also promotes the 1,3-cycloaddition of 2-cyclohexen-1-one with azomethine imine with a higher catalytic activity than its quinine analogue.[6]

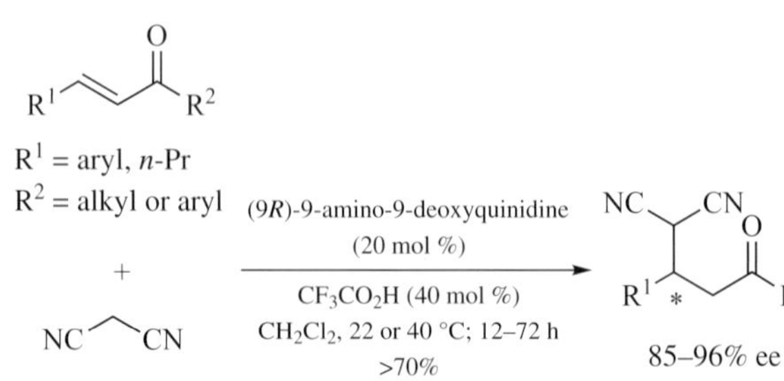

(3)

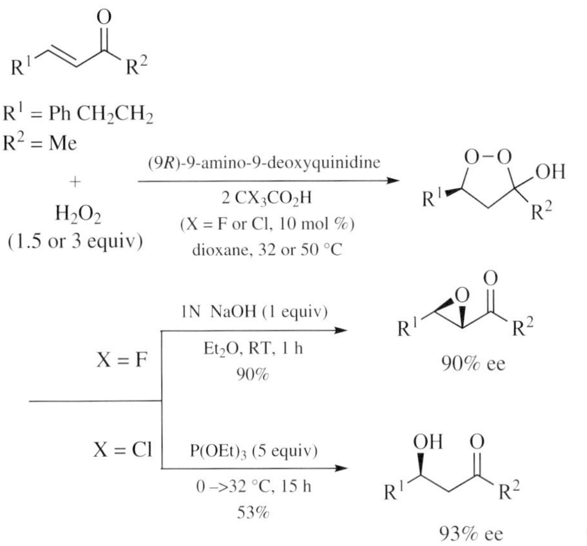

(4)

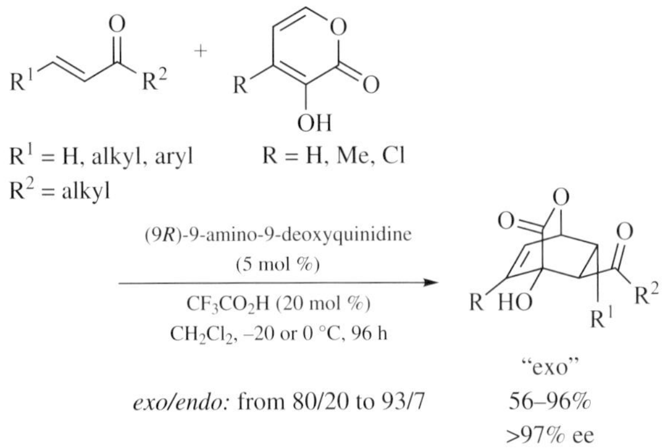

(5)

Desymmetrization.[7] (9*R*)-9-Amino-9-deoxyquinidine was an efficient and highly enantioselective catalyst for the intramolecular aldolization of symmetrical 4-substituted-2,6-heptanediones (eq 6).

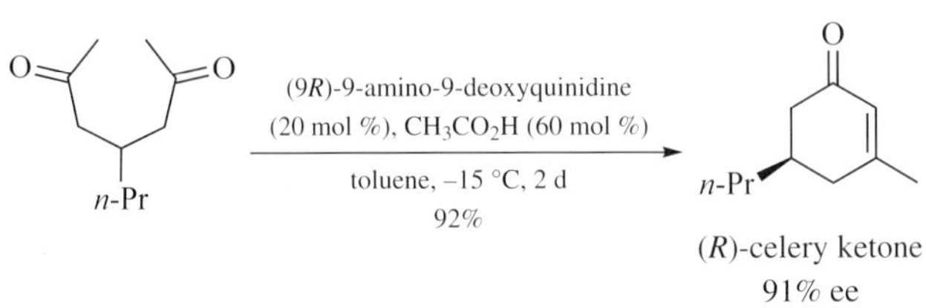

(6)

Asymmetric Hydrogenation.[2] (9*R*)-9-Amino-9-deoxyquinidine was tested as ligand in both iridium- and rhodium- catalyzed asymmetric hydrogenations of aryl ketones (eq 7). It was slightly less efficient than its cinchonine-derivative analogue that led to the same enantiomer.

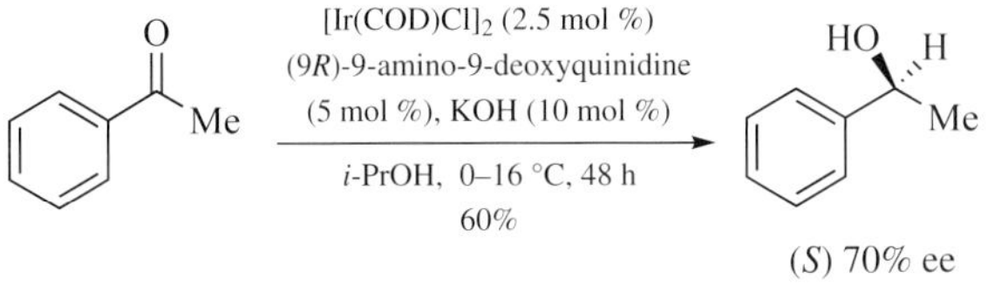

(7)

Preparation of derivatives. Amides (eq 8),[1,8] sulfonamides (eq 9),[9] and thioureas (eq 10)[10,11] derived from (9*R*)-9-amino-9-deoxyquinidine have been prepared according to the standard synthetic methods. These compounds have been used as (co)-organocatalysts in various enantioselective transformations, including aldol reactions,[8] Michael-type additions,[9,10,12] Diels–Alder reactions,[13] decarboxylative protonations,[1,10,14] dynamic kinetic resolutions,[15] hydrophosphinylations,[16] and Mannich reactions.[17]

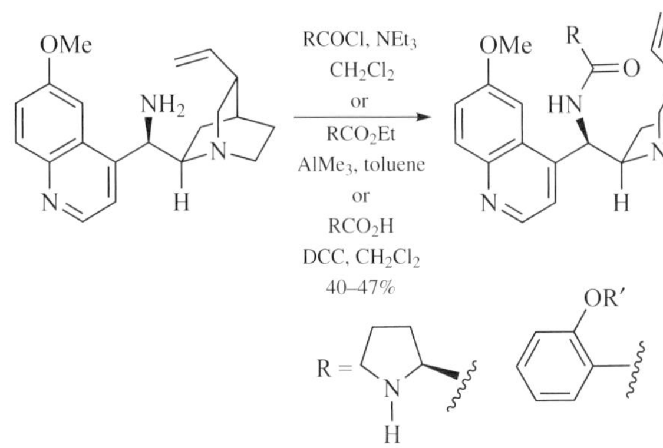

(8)

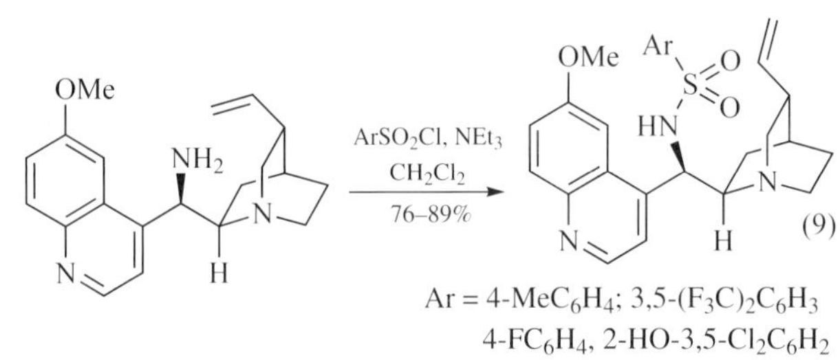

(9)

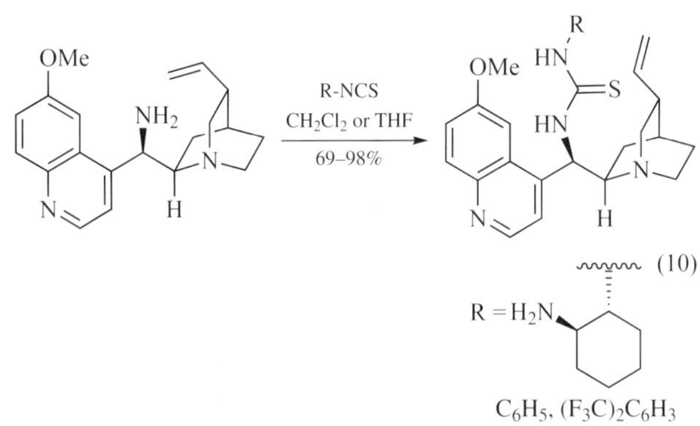

(10)

Related Reagents. (9*S*)-Amino-9-deoxyquinine.

First Update

Wei Du, Zhi Zhou, & Ying-Chun Chen
Sichuan University, Chengdu, China

Asymmetric Enamine Catalysis. Michael addition reactions catalyzed by ADQD through enamine activation are efficient protocols for the α-functionalization of ketones.[19] Heterocyclic spirooxindoles with an all-carbon quaternary stereocenter were constructed through the ADQD/(L)-camphorsulfonic acid

(L-CSA)-mediated Michael addition of simple ketones to isatylidenemalonitrile in high yields and with excellent enantioselectivity (eq 11).[19a]

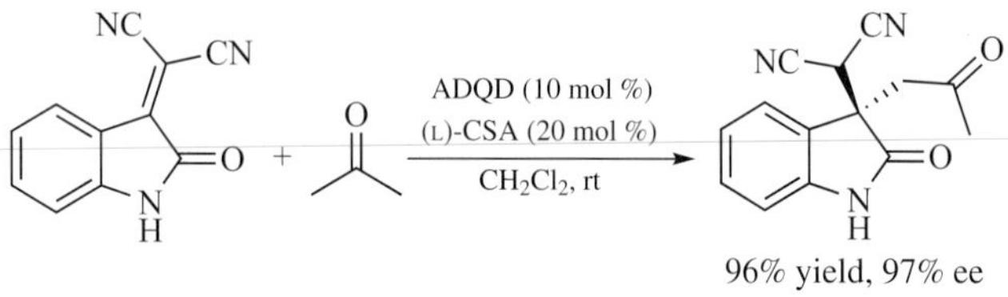

(11)

1,3-Benzodithiolylium ions were proven to be good acceptors for the enamine intermediates of α-substituted aldehyde and ADQD, and the products possessing a chiral quaternary carbon were obtained with good enantioselectivity (eq 12).[20] Similar strategy was applied to cyclopentanone with the identical catalyst under the basic conditions, affording the adduct with high ee.[21]

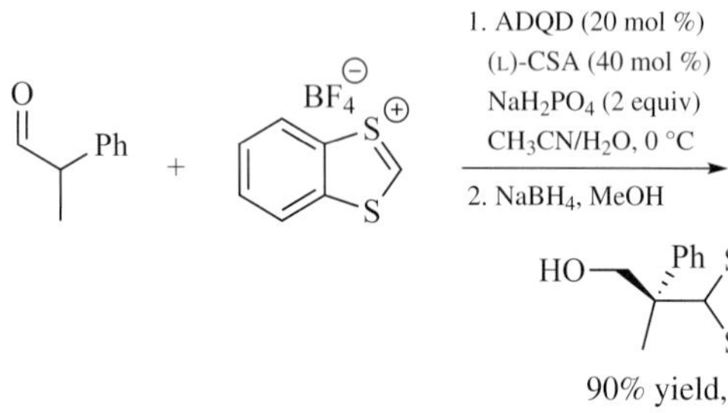

(12)

The asymmetric α-amination catalyzed by ADQD was more challenging, and inferior results were observed with ADQD in comparison with that of (9*S*)-9-amino-9-deoxyquinine (ADQ).[22] On the other hand, the α-benzoyloxylation of cyclic ketones using ADQD as the catalyst furnished the desired products with good enantioselectivity (eq 13).[23]

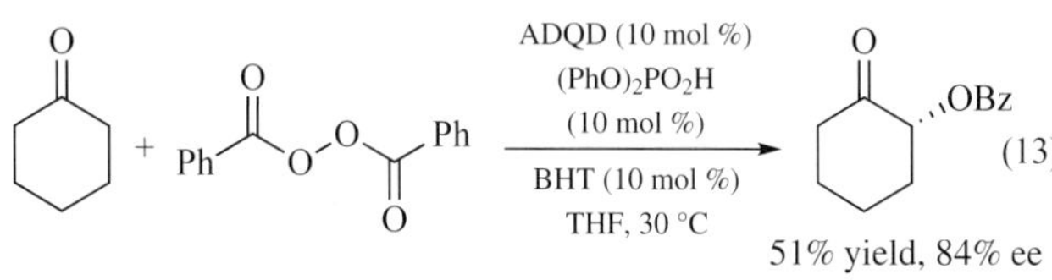

(13)

Asymmetric Dienamine Catalysis. β-Substituted cyclic enones could be activated by dienamine catalysis of ADQD to undergo [4+2] cycloadditions with allylidene- or alkynylidenemalononitrile substrates, and the densely functionalized *endo*-bicycle[2.2.2]octanes were produced with excellent stereoselectivity (eq 14).[24] The similar dienamine catalysis was applied to atroposelective formal [4+2] desymmetrization of *N*-arylmaleimides with α,β-unsaturated ketones, furnishing enantioenriched succinimides with a chiral axis and three stereogenic centers.[25] Remote control of axial chirality was also realized by ADQD-catalyzed desymmetrization of *N*-arymaleimides via vinylogous Michael addition, and high levels of enantioselectivity was achieved although with moderate diastereoselectivity (eq 15).[26] The combination of ADQD and a BINOl-derived phosphoric acid was a highly efficient catalytic system for double Michael addition of isatylidene malononitriles with α,β-unsaturated ketones, providing chiral spirocyclic products in high yields and with excellent diastereo- and enantioselectivity (eq 16).[27] Similar dienamine catalysis was applied to benzylidenepyrazolone (eq 17)[28] and (*E*)-nitroacrylate (eq 18),[29] respectively, affording the desired double Michael adducts with moderate diastereoselectivity and high enantioselectivity. Asymmetric [4+2] formal cycloaddition of cyclic *N*-sulfimines and acyclic enones or ynones proceeding through a stepwise Mannich–Michael addition process was accomplished under the catalysis of ADQD, which stood for a concise route for the synthesis of sulfamidate-fused 2,6-disubstituted piperidin-4-ones with high levels of enantioselectivity (eq 19).[30]

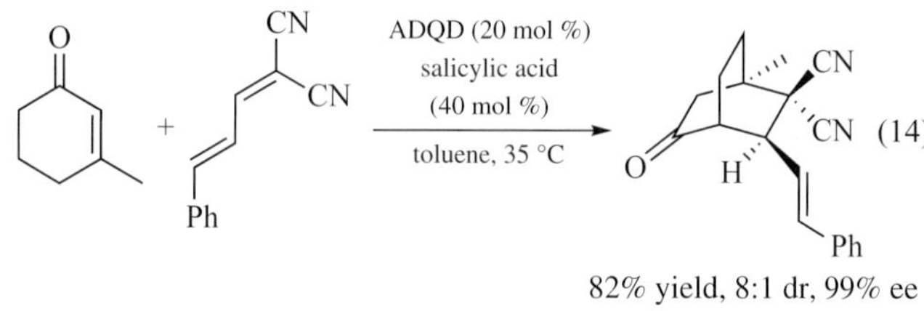

(14)

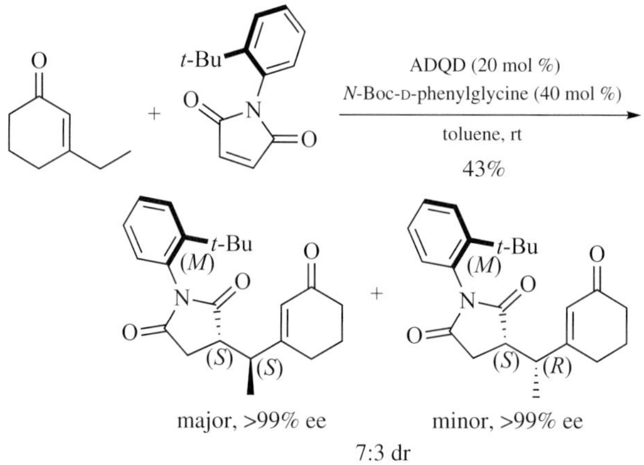

(15)

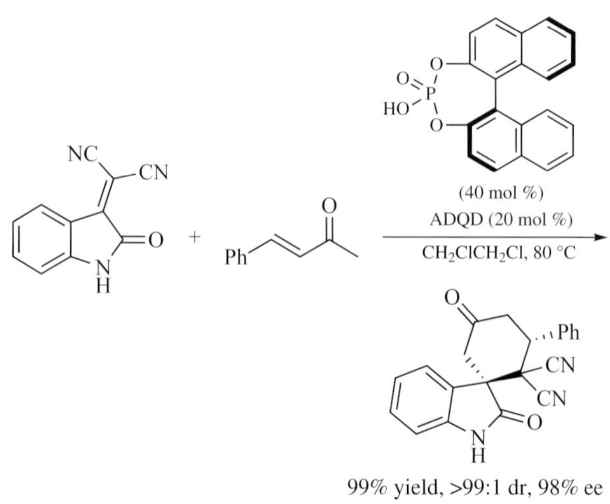

(16)

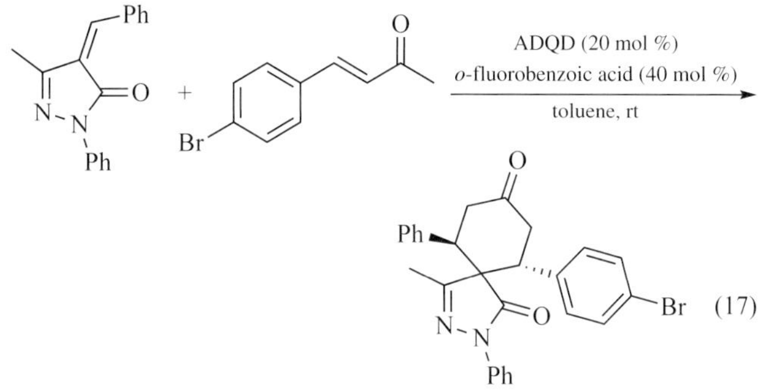

(17)

96% yield, 17:3 dr, 99%/99% ee

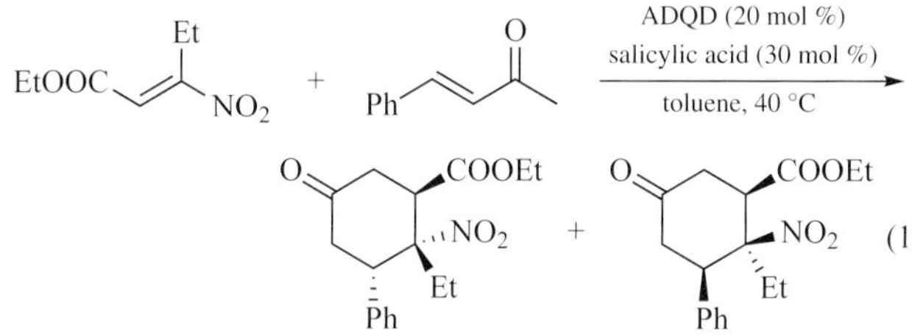

(18)

55% yield, 1:2 dr, 92%/87% ee

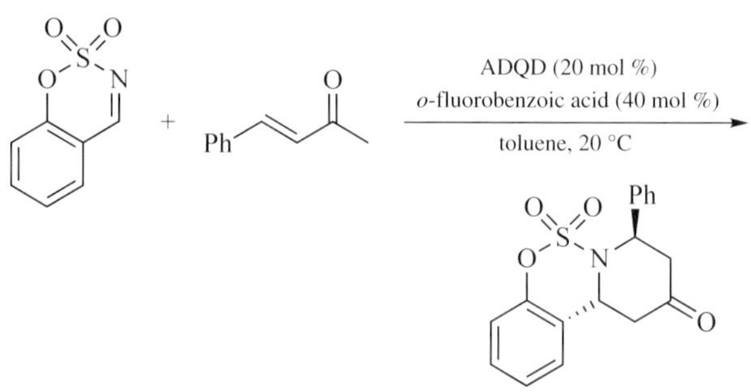

(19)

80% yield, >19:1 dr, 97% ee

Asymmetric Trienamine Catalysis. Asymmetric trienamine catalysis represents a challenging activation mode regarding the reactivity and stereocontrol at the remote sites. Apart from the successful ε-functionalization of 2,4-dienals under the catalysis of secondary amine,[31] ADQD-mediated Diels-Alder reaction via trienamine activation of 2,4-dienones was accomplished with good enantioselectivity (eq 20).[32] However, this strategy was not suitable for those 2,4-dienones with an enolizable α'-CH group owing to the inherent preference for the formation of cross-conjugated trienamines. Accordingly, the interrupted cyclic 2,5-dienone substrates were designed, and the δ, ε–C=C bonds were activated by the ADQD/salicylic acid system, delivering the enantiopure aza-Diels-Alder adducts in excellent yields (eq 21).[33] Asymmetric bisvinylogous 1,4-addition of this type of substrates to nitroalkenes has been discovered as well, affording ε-addition products exclusively with good enantioselectivity in the presence of ADQD (eq 22).[34] The concept of induced trienamine catalysis was further developed with the involvement of the heteroarene π-system, and dearomatizing Diels-Alder reaction products with high enantioselectivity were obtained under the catalysis of ADQD (eq 23).[35] The deconjugated 3,5-dienone substrates have also been successfully applied to the β,ε-regioselective Diels-Alder cycloadditions with 3-olefinic oxindole-based dienophiles under the catalysis of ADDQ, giving spirocyclic oxindoles with high stereoselectivity (eq 24).[36]

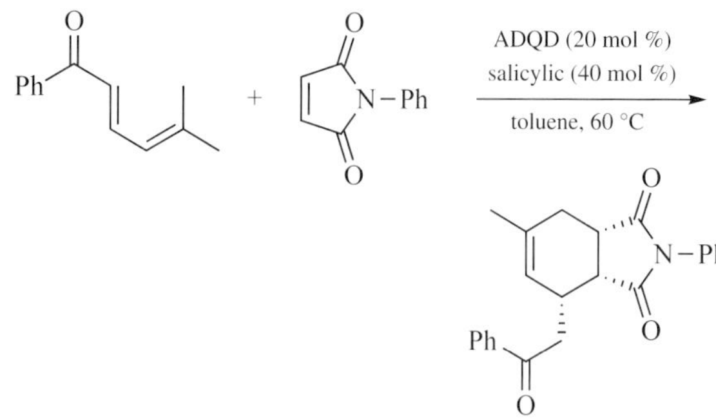

(20)

80% yield, >19:1 dr, 88% ee

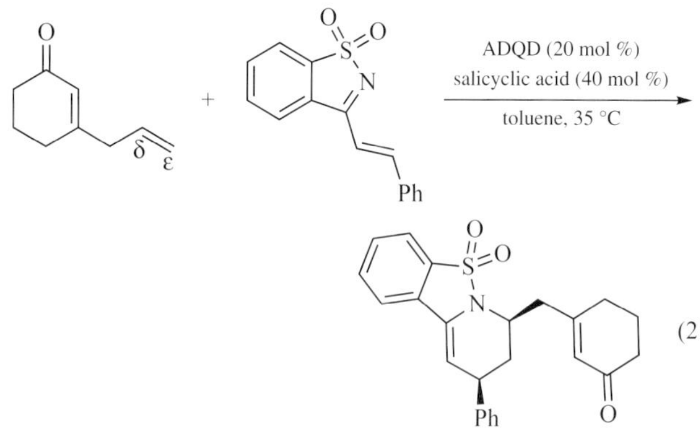

(21)

92% yield, >19:1 dr, >99.5% ee

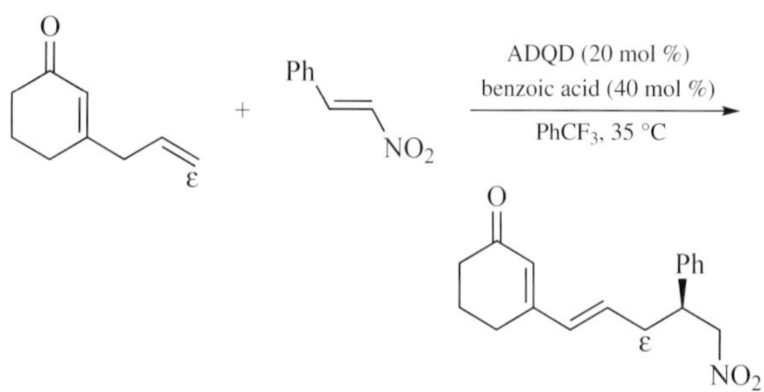

(22)

71% yield, 83% ee

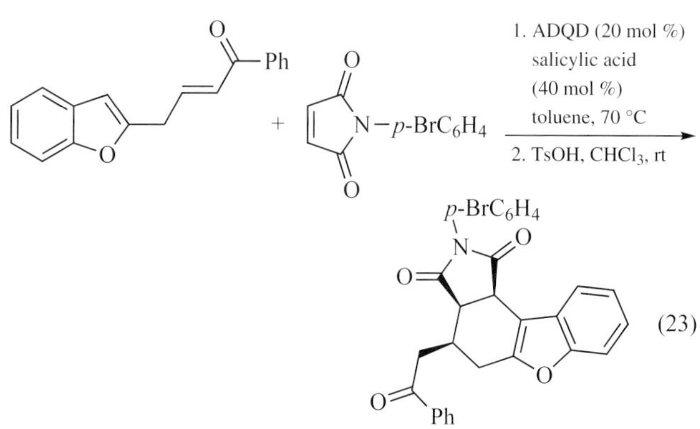

(23)

60% yield, >19:1, 82% ee

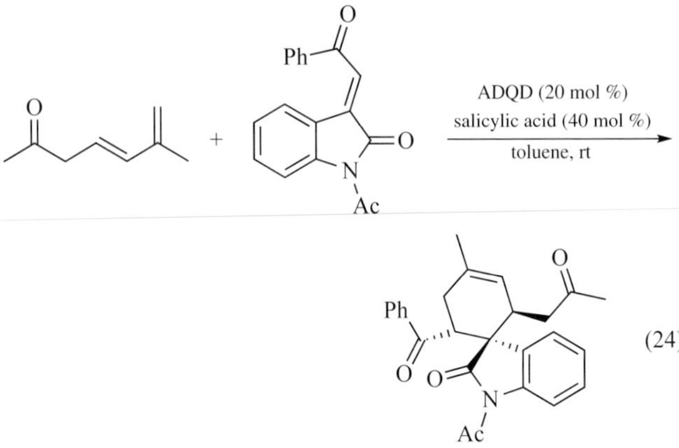

(24)

86% yield, >19:1 dr, 94% ee

Asymmetric Photochemical Catalysis. The enamine species generated from ketones and ADQD can complex with electron-deficient benzyl bromide, which is named electron donor–acceptor (EDA) complex.[37] Under the visible light irradiation, electron transfer occurred between the electron-rich enamine species and the benzyl bromide, followed by fragmentation of bromide anion and radical combination to form the α-alkylated ketones with excellent enantioselectivity (eq 25).

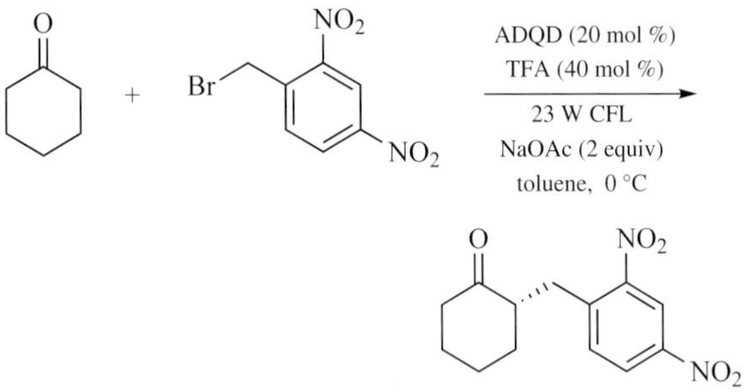

(25)

74% yield, 91% ee

Asymmetric Iminium Catalysis. ADQD can activate enals or enones via decreasing the LUMO energy of the unsaturated system. Enantioselective Diels-Alder reaction of 2-pyrones with simple α,β-unsaturated ketones was catalyzed by ADQD with good to excellent diastereo- and enantioselectivity.[5] The substrates containing an acidic carbon, such as nitro(phenylsulfonyl)methane (eq 26)[38] and oxindoles,[39] could serve as good nucleophiles to attack the β-position of the enones in the presence of ADQD, giving the corresponding Michael adducts in high yields and with excellent enantioselectivity. ADQD contains multifunction groups, including quinuclidine and the primary amine, which could act as Lewis base catalyst or iminium catalyst, respectively, to render the Michael addition of nitroalkenes to enones with high enantioselectivity, although with low diastereoselectivity (eq 27).[40] The protonated quinuclidine moiety could also serve as a Brønsted acid and facilitate the nucleophilic attack of the hydrazides to enones via H-bonding interaction (eq 28).[41] Highly enantioselective peroxidation reaction of enones was developed by utilizing ADQD together with TFA as the catalyst, establishing asymmetric synthesis of biologically interesting peroxides (eq 29).[42] The combination of organo- and metal catalysis has been realized for enantioselective synthesis of bicyclic enones. The process involved the ADQD activation of α,β-unsaturated ketones, followed by Michael addition of propargylated nucleophiles. Then, gold catalyzed *exo-dig* cyclization and double bond migration process ensues finally generating the enantioenriched bicyclic compounds in very good yields (eq 30).[43]

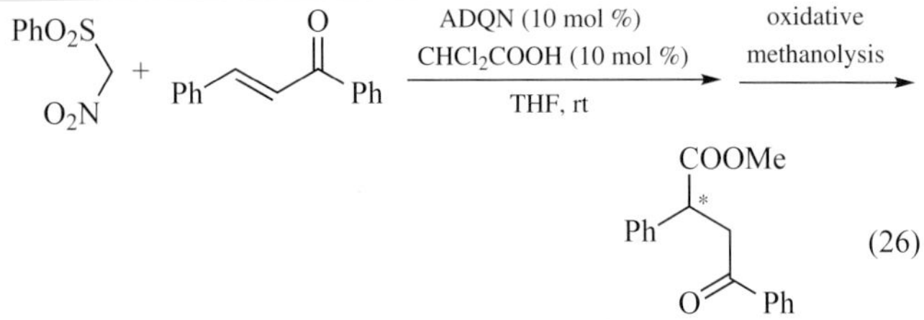

(26)

51% yield, 92% ee

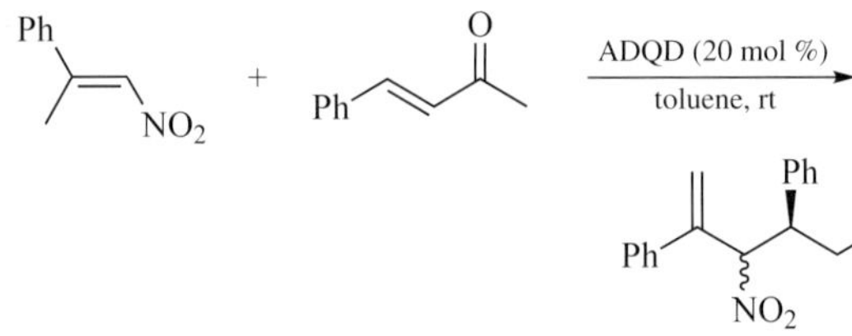

(27)

89% yield, *syn:anti* = 1:1.2
91% (*syn*) / 91% (*anti*) ee

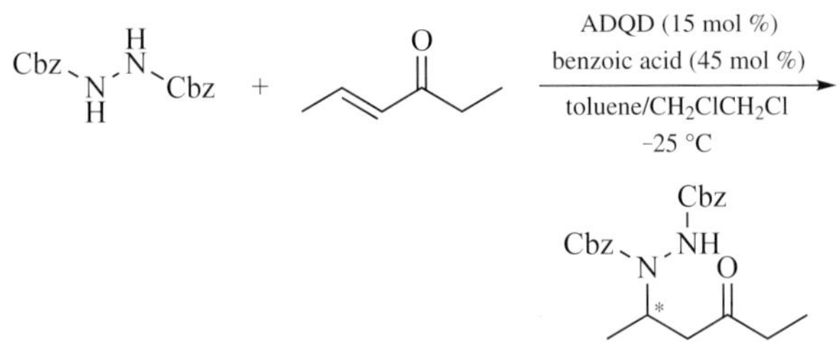

(28)

60% yield, 83% ee

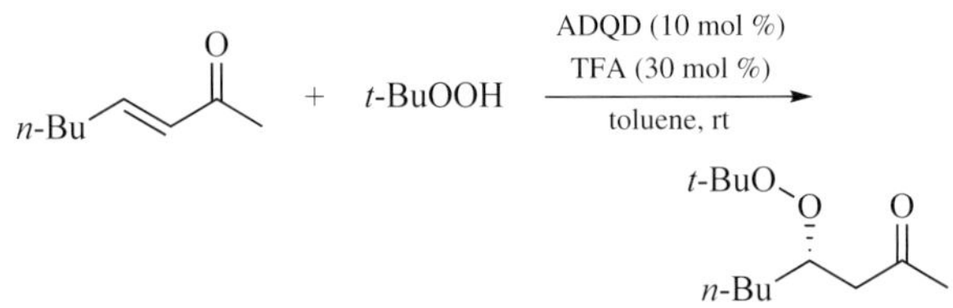

(29)

90% yield, 90% ee

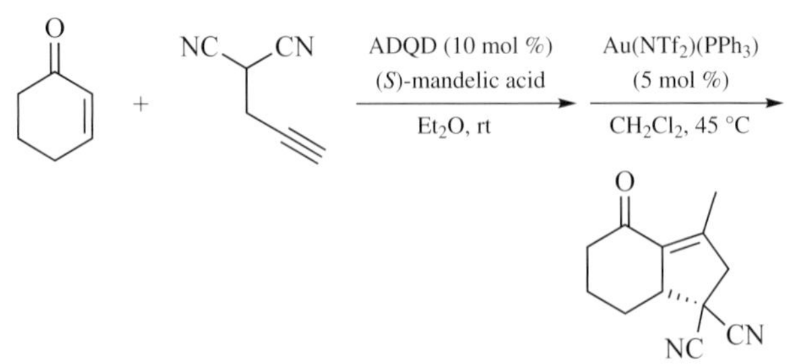

(30)

85% yield, 92% ee

Dienamine/Dienamine Catalysis. ADQD-catalyzed asymmetric cascade reactions always involve two successive activation modes, which could construct more complicated skeletons in one step, representing highly efficient and economical protocols in organic chemistry. The enantioselective [5+3] formal cycloaddition of β-substituted 2-cyclopentenones with bis-electrophilic 1-azadienes was catalyzed by ADQD and 5-nitrosalicylic acid via cascade cross-conjugated dienamine/*endo*-dienamine activation modes, and the enantiopure adducts were obtained in moderate to excellent yields (eq 31).[44]

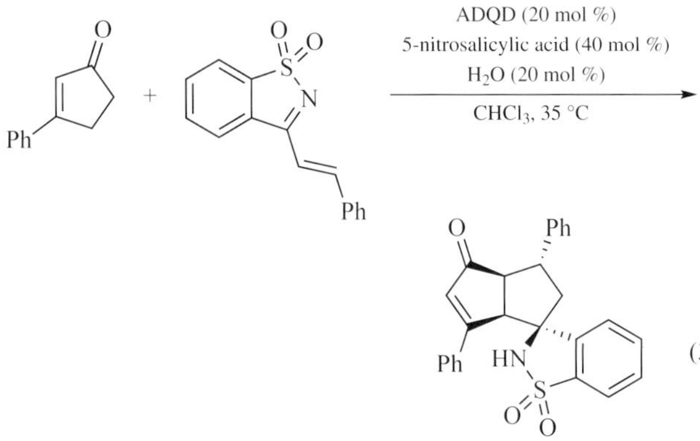

(31)

85% yield, >19:1 dr, >99% ee

Iminium/Enamine Catalysis. A highly stereoselective desymmetrization cascade Michael–Michael addition reaction of cyclic α,α-dicyanoalkenes with α,β-unsaturated ketones was catalyzed by ADQD and TFA via iminium/enamine activation, affording the bicyclic product with four stereogenic carbon centers in good yields and with excellentee values (eq 32).[45] Based on the well-established asymmetric epoxidation reactions catalyzed by cinchona alkaloids,[4,46] cyclopropanation products with good enantioselectivity were assembled through ADQD-catalyzed cascade reactions of bromonitroalkenes and α,β-unsaturated cyclic enones (eq 33). The addition of 1 equiv of *N*-methyl morpholine (NMM) was proven to be crucial for high yields.[47]

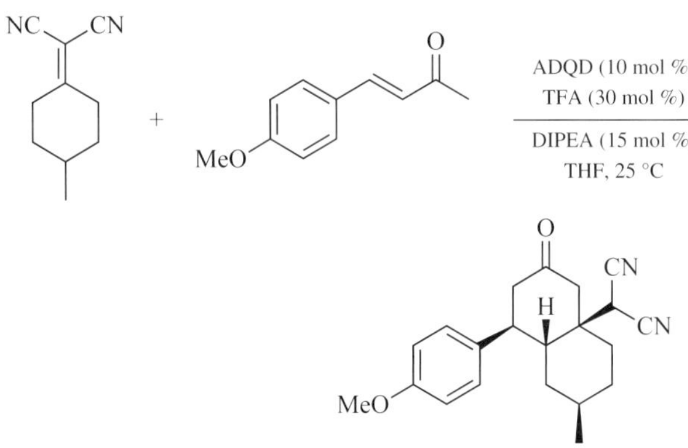

(32)

81% yield, >99:1 dr, 90% ee

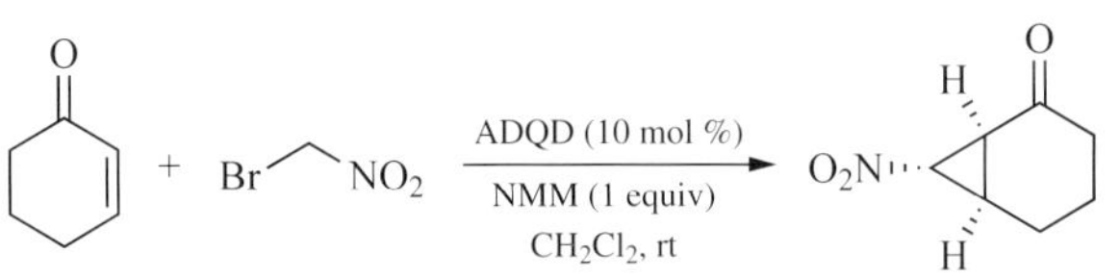

(33)

83% yield, 88% ee

Preparation of Chiral Catalysts. ADQD-derived bifunctional catalysts that contain a Brønsted basic quinuclidine moiety and an H-bonding donor in the same molecule have been demonstrated as powerful catalysts for numerous asymmetric reactions. The aryl sulfonamides from ADQD[9] have been further developed and investigated. Some bulky aromatic sulfonamides were prepared under the reported procedure for the asymmetric decarboxylative aldol reaction,[48] while those possessing heteroaromatic substituents were found to be the optimal catalysts for the addition reaction of ethyl thioglycolate to ketimines derived from isatins (eq 34).[49]

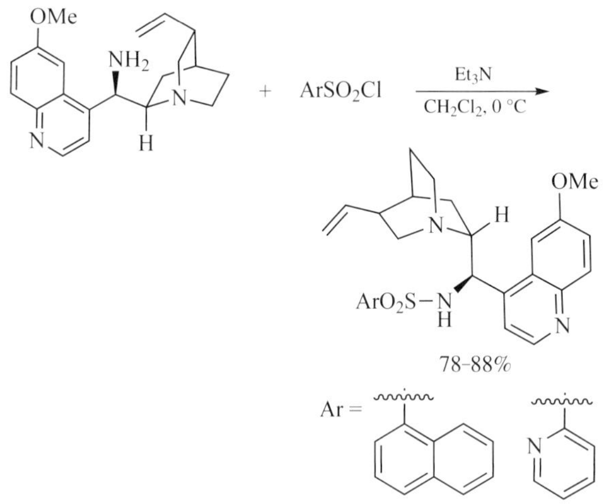

(34)

The synthetic routes to ADQD-derived thioureas had been well established,[10–11] and new types of catalysts were further explored with respect to their application in asymmetric reactions. Some novel ureas (eq 35)[50] and thioureas (eq 36)[51] were synthesized and exhibited moderate to excellent catalytic activity in asymmetric reactions. In addition, a series of ADQD-derived thioureas bearing a primary or a secondary amine group have also been designed and prepared in two steps (eq 37), and they promoted the conjugate addition reactions with high levels of enantioselectivity.[52] Furthermore, multifunctional catalysts have also been developed by incorporating more H-bonding donors with the ADQD–thiourea skeleton (eq 38), as the intramolecular H-bond could provide dramatic improvement in the H-bond donor capacity through the cooperative effects.[53]

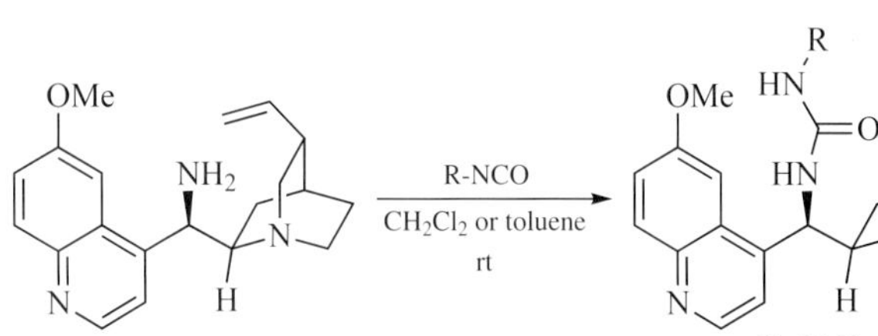

51–81%

R = Ts, (*R*)-SO-*t*-Bu

3,5-$(OMe)_2C_6H_3$

3,5-$(CF_3)_2C_6H_3$

(35)

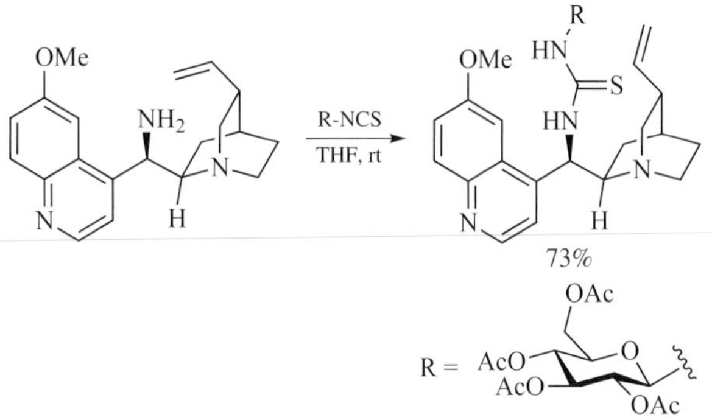

(36)

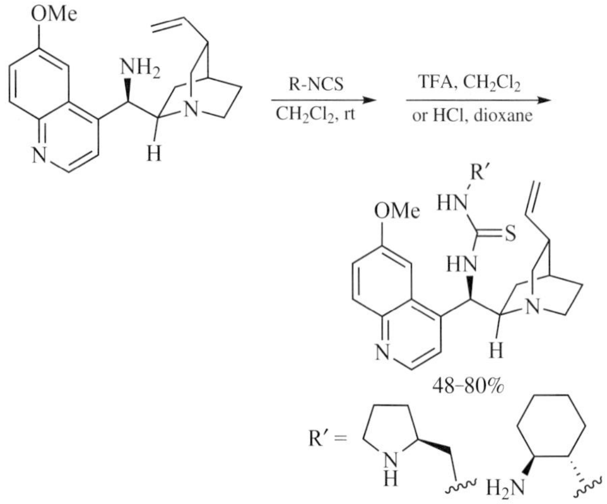

(37)

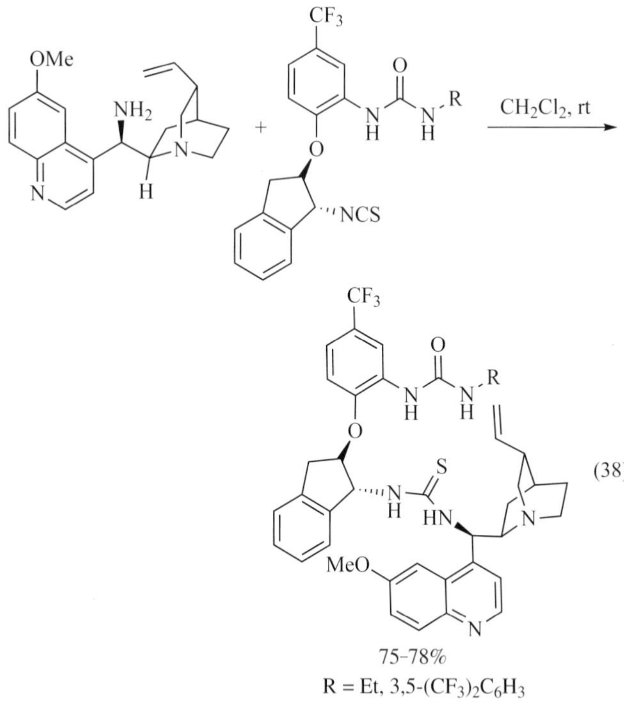

(38)

ADQD-based chiral catalysts with a squaramide unit as the H-bonding donor can be easily prepared through a two-step pathway (eq 39). They were successfully applied in many enantioselective transformations, such as the addition reactions[54] and AgOAc cooperatively catalyzed Mannich/cyclization cascade reactions.[55] ADQD-derived immobilized bifunctional squaramide was also reported (eq 40).[56]

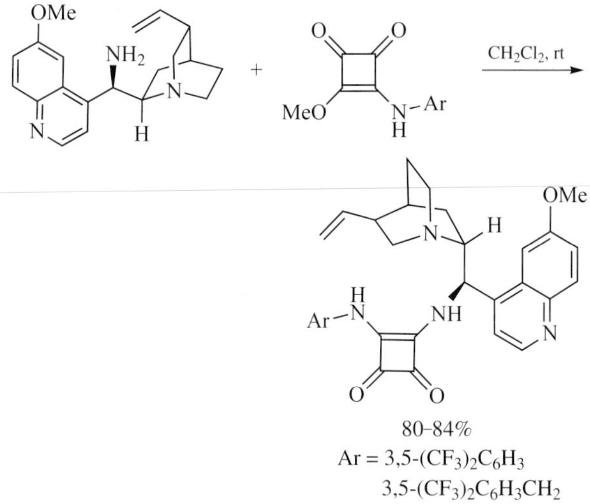

(39)

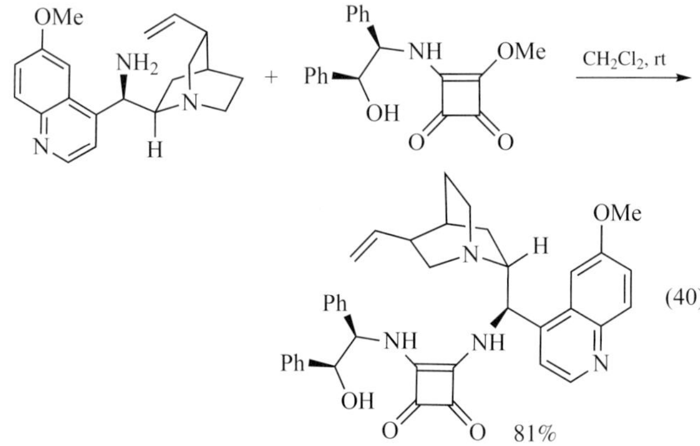

(40)

Guanidine-based organic molecules are versatile catalysts in a wide range of organic reactions, and a new chiral cinchona alkaloid-derived catalyst by introducing the guanidine moiety into ADQD (eq 41) was demonstrated to be effective in the enantioselective amination reactions.[57]

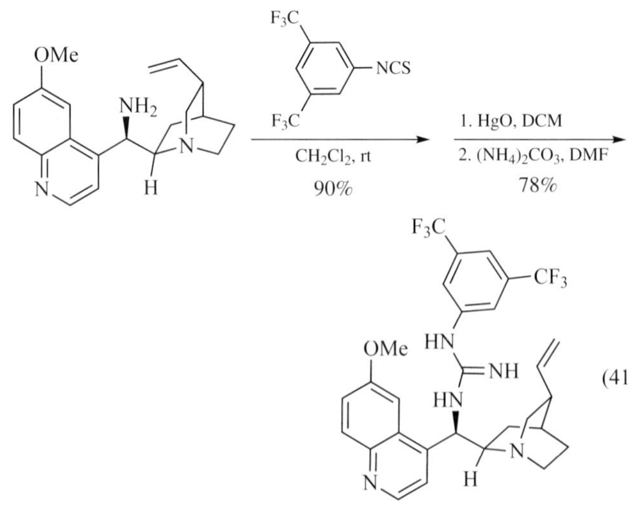

(41)

A series of aniline-containing ADQD catalysts that could be smoothly prepared through the substitution and reduction processes (eq 42) have been developed.[58] Highly enantioselective asymmetric isomerization of β,γ-unsaturated cyclohexenones to their chiral α,β-unsaturated isomers could be realized by using this new catalyst.

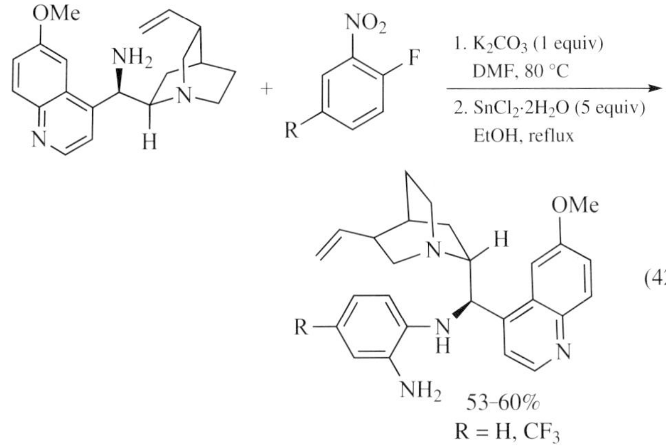

(42)

New Schiff bases derived from cinchona alkaloids served as chiral ligands for the Cu(II)-catalyzed asymmetric Henry reaction. The Schiff base derived from ADQD and salicylaldehyde (eq 43) turned out to be the optimal ligand for the reaction.[59]

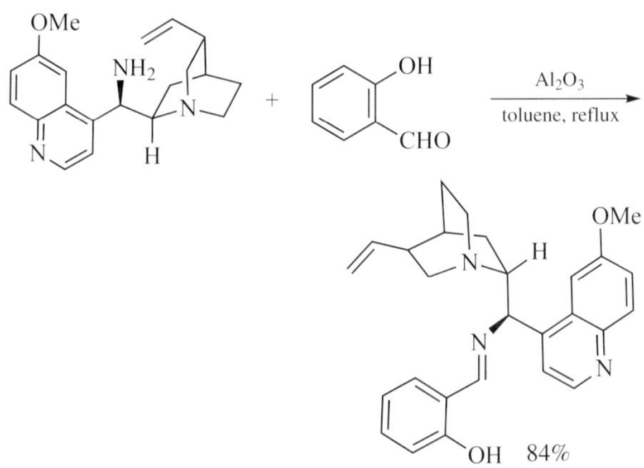

(43)

A new class of chiral phase transfer catalysts bearing amino groups was developed by using 9-amino-(9-deoxy)cinchona alkaloid as the starting materials (eq 44), and they significantly improved the asymmetric α-alkylation of glycine Schiff base.[60]

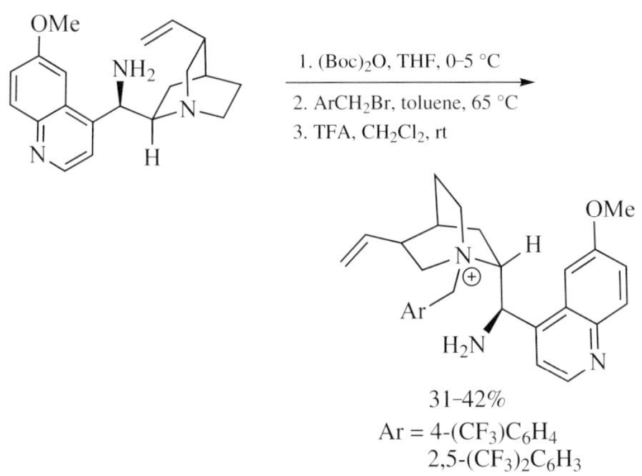

(44)

Related Reagent. (9*S*)-Amino-9-deoxyquinine.

1. Brunner, H.; Schmidt, P., *Eur. J. Org. Chem.* **2000**, 2119.
2. He, W.; Liu, P.; Zhang, B.L.; Sun, X.L.; Zhang, S.Y., *Appl. Organometal. Chem.* **2006**, *20*, 328.
3. Li, X.; Cun, L.; Lian, C.; Zhong, L.; Chen, Y.; Liao, J.; Zhu, J.; Deng, J., *Org. Biomol. Chem.* **2008**, *6*, 349.
4. Reisinger, C. M.; Wang, X.; List, B., *Angew. Chem. Int. Ed.* **2008**, *47*, 8112.
5. Singh, R. P.; Bartelson, K.; Wang, Y.; Su, H.; Lu, X.; Deng, L., *J. Am. Chem. Soc.* **2008**, *130*, 2422.
6. Chen, W.; Du, W.; Duan, Y.-Z.; Wu, Y.; Yang, S.-Y.; Chen, Y.-C., *Angew. Chem. Int. Ed.* **2007**, *46*, 7667.
7. Zhou, J.; Wakchaure, V.; Kraft, P.; List, B., *Angew. Chem. Int. Ed.* **2008**, *47*, 7656.
8. Chen, J.-R.; An, X.-L.; Zhu, X.-Y.; Wang, X.-F.; Xiao, W.-J., *J. Org. Chem.* **2008**, *73*, 6006.
9. Luo, J.; Xu, L.-W.; Hay, R. A. S.; Lu, Y., *Org. Lett.* **2009**, *11*, 437.
10. Vakulya, B.; Varga, S.; Csámpai, A.; Soós, T., *Org. Lett.* **2005**, *7*, 1967. (b) Amere, M.; Lasne, M.-C.; Rouden, J., *Org. Lett.* **2007**, *9*, 2621.
11. Reviews on thioureas: (a) Connon, S. J., *Chem. Commun.* **2008**, 2499. (b) Connon, S. J.; *Chem. Eur. J.* **2006**, *12*, 5418.
12. (a) Li, D. R.; Murugan, A.; Falck, J. R., *J. Am. Chem. Soc.* **2008**, *130*, 46. (b) Vakulya, B.; Varga, S.; Soós, T., *J. Org. Chem.* **2008**, *73*, 3475. (c) Wang, B.; Wu, F.; Wang, Y.; Liu, X.; Deng, L., *J. Am. Chem. Soc.* **2007**, *129*, 768. (d) Li, P.; Wang, Y.; Liang, X.; Ye, J., *Chem. Commun.* **2008**, 3302. (e) Mandal, T.; Zhao, C.-G.;, *Angew. Chem. Int. Ed.* **2008**, *47*, 7714. (f) Elsner, P.; Jiang, H.; Nielsen, J. B.; Pasi, F.; Jørgensen, K. A., *Chem. Commun.* **2008**, 5827.
13. Wang, L.; Li, H.; Wang, Y.-Q.; Liu, Y.; Foxman, B. M.; Deng, L., *J. Am. Chem. Soc.* **2007**, *129*, 6364.
14. Seitz, T.; Baudoux, J.; Bekolo, H.; Cahard, D.; Plaquevent, J.-C.; Lasne, M.-C.; Rouden, J., *Tetrahedron* **2006**, *62*, 6155.
15. Berkessel, A.; Mukherjee, S.; Müller, T. N.; Cleemann, F.; Roland, K.; Brandenburg, M.; Neudörfl, J.-M.; Lex, J., *Org. Biomol. Chem.* **2006**, *4*, 4319.
16. Wang, F.; Liu, X.; Cui, X.; Xiong, Y.; Zhou, X.; Feng, X., *Chem. Eur. J.* **2009**, *15*, 589.
17. Song, J.; Shih, H.-W; Deng, L., *Org. Lett.* **2007**, *9*, 603.
18. Cassani, C.; Martín-Rapún, R.; Arceo, E.; Bravo, F.; Melchiorre, P., *Nat. Protoc.* **2013**, *8*, 325.
19. (a) Zhao, H.; Lan, Y.-B.; Liu, Z.-M.; Wang, Y.; Wang, X.-W.; Tao, J.-C., *Eur. J. Org. Chem.* **2012**, 1935. (b) Lee, Y.-T.; Das, U.; Chen, Y.-R.; Lee, C.-J.; Chen, C.-H.; Yang, M.-C.; Lin, W., *Adv. Synth. Catal.* **2013**, *355*, 3154.
20. Gualandi, A.; Petruzziello, D.; Emer, E.; Cozzi, P. G., *Chem. Commun.* **2012**, *48*, 3614.
21. Capdevila, M. G.; Emer, E.; Gualandi, A.; Petruzziello, D.; Grilli, S.; Cozzi, P. G., *ChemCatChem* **2012**, *4*, 968.
22. (a) Liu, C.; Zhu, Q.; Huang, K.-W.; Lu, Y., *Org. Lett.* **2011**, *13*, 2638. (b) Desmarchelier, A.; Yalgin, H.; Coeffard, V.; Moreau, X.; Greck, C., *Tetrahedron Lett.* **2011**, *52*, 4430.
23. Lifchits, O.; Demoulin, N.; List, B., *Angew. Chem., Int. Ed.* **2011**, *50*, 9680.
24. Feng, X.; Zhou, Z.; Zhou, R.; Zhou, Q.-Q.; Dong, L.; Chen, Y.-C., *J. Am. Chem. Soc.* **2012**, *134*, 19942.
25. Eudier, F.; Righi, P.; Mazzanti, A.; Ciogli, A.; Bencivenni, G., *Org. Lett.* **2015**, *17*, 1728.
26. Di Iorio, N.; Righi, P.; Mazzanti, A.; Mancinelli, M.; Ciogli, A.; Bencivenni, G., *J. Am. Chem. Soc.* **2014**, *136*, 10250.
27. Lan, Y.-B.; Zhao, H.; Liu, Z.-M.; Liu, G.-G.; Tao, J.-C.; Wang, X.-W., *Org. Lett.* **2011**, *13*, 4866.
28. Zhang, J.-X.; Li, N.-K.; Liu, Z.-M.; Huang, X.-F.; Geng, Z.-C.; Wang, X.-W., *Adv. Synth. Catal.* **2013**, *355*, 797.
29. Massolo, E.; Benaglia, M.; Annunziata, R.; Palmieri, A.; Celentano, G.; Forni, A., *Adv. Synth. Catal.* **2014**, *356*, 493.

30. Liu, Y.; Kang, T.-R.; Liu, Q.-Z.; Chen, L.-M.; Wang, Y.-C.; Liu, J.; Xie, Y.-M.; Yang, J.-L.; He, L., *Org. Lett.* **2013**, *15*, 6090.
31. (a) Jia, Z.-J.; Jiang, H.; Li, J.-L.; Gschwend, B.; Li, Q.-Z.; Yin, X.; Grouleff, J.; Chen, Y.-C.; Jørgensen, K. A., *J. Am. Chem. Soc.* **2011**, *133*, 5053. (b) Jia, Z.-J.; Zhou, Q.; Zhou, Q.-Q.; Chen, P.-Q.; Chen, Y.-C., *Angew. Chem., Int. Ed.* **2011**, *50*, 8638.
32. Xiong, X.-F.; Zhou, Q.; Gu, J.; Dong, L.; Liu, T.-Y.; Chen, Y.-C., *Angew. Chem., Int. Ed.* **2012**, *51*, 4401.
33. Feng, X.; Zhou, Z.; Ma, C.; Yin, X.; Li, R.; Dong, L.; Chen, Y.-C., *Angew. Chem., Int. Ed.* **2013**, *52*, 14173.
34. Zhou, Z.; Feng, X.; Yin, X.; Chen, Y.-C., *Org. Lett.* **2014**, *16*, 2370.
35. Xiao, Y.-C.; Yue, C.-Z.; Chen, P.-Q.; Chen, Y.-C., *Org. Lett.* **2014**, *16*, 3208.
36. Chen, P.-Q.; Xiao, Y.-C.; Yue, C.-Z.; Chen, Y.-C., *Org. Chem. Front.* **2014**, *1*, 490.
37. Arceo, E.; Bahamonde, A.; Bergonzini, G.; Melchiorre, P., *Chem. Sci.* **2014**, *5*, 2438.
38. Prakash, G. K. S.; Wang, F.; Zhang, Z.; Ni, C.; Haiges, R.; Olah, G. A., *Org. Lett.* **2012**, *14*, 3260.
39. Sun, W.; Hong, L.; Liu, C.; Wang, R., *Tetrahedron: Asymmetry* **2010**, *21*, 2493.
40. Ling, J.-B.; Wang, W.-P.; Xu, P.-F., *Chem. Cat. Chem.* **2011**, *3*, 302.
41. Campbell, N. R.; Sun, B.; Singh, R. P.; Deng, L., *Adv. Synth. Catal.* **2011**, *353*, 3123.
42. Lu, X.; Liu, Y.; Sun, B.; Cindric, B.; Deng, L., *J. Am. Chem. Soc.* **2008**, *130*, 8134.
43. Zweifel, T.; Hollmann, D.; Prüger, B.; Nielsen, M.; Jørgensen, K. A., *Tetrahedron: Asymmetry* **2010**, *21*, 1624.
44. Yin, X.; Zheng, Y.; Feng, X.; Jiang, K.; Wei, X.-Z.; Gao, N.; Chen, Y.-C., *Angew. Chem., Int. Ed.* **2014**, *53*, 6245.
45. Kang, T.-R.; Xie, J.-W.; Du, W.; Feng, X.; Chen, Y.-C., *Org. Biomol. Chem.* **2008**, *6*, 2673.
46. Wang, X.; Reisinger, C. M.; List, B., *J. Am. Chem. Soc.* **2008**, *130*, 6070.
47. Lv, J.; Zhang, J.; Lin, Z.; Wang, Y., *Chem. Eur. J.* **2009**, *15*, 972.
48. Bae, H. Y.; Sim, J. H.; Lee, J.-W.; List, B.; Song, C. E., *Angew. Chem., Int. Ed.* **2013**, *52*, 12143.
49. Nakamura, S.; Takahashi, S.; Nakane, D.; Masuda, H., *Org. Lett.* **2015**, *17*, 106.
50. (a) Zhang, W.; Zheng, S.; Liu, N.; Werness, J. B.; Guzei, I. A.; Tang, W., *J. Am. Chem. Soc.* **2010**, *132*, 3664. (b) Rong, Z.-Q.; Pan, H.-J.; Yan, H.-L.; Zhao, Y., *Org. Lett.* **2014**, *16*, 208. (c) Zhang, W.; Liu, N.; Schienebeck, C. M.; Decloux, K.; Zheng, S.; Werness, J. B.; Tang, W., *Chem. Eur. J.* **2012**, *18*, 7296. (d) Miyaji, R.; Asano, K.; Matsubara, S., *Org. Lett.* **2013**, *15*, 3658. (e) Phelan, J. P.; Patel, E. J.; Ellman, J. A., *Angew. Chem., Int. Ed.* **2014**, *53*, 11329.
51. Kong, S.; Fan, W.; Wu, G.; Miao, Z., *Angew. Chem., Int. Ed.* **2012**, *51*, 8864.
52. (a) Wen, S.; Li, P.; Wu, H.; Yu, F.; Liang, X.; Ye, J., *Chem. Commun.* **2010**, *46*, 4806. (b) Chen, J.-R.; Cao, Y.-J.; Zou, Y.-Q.; Tan, F.; Fu, L.; Zhu, X.-Y.; Xiao, W.-J., *Org. Biomol. Chem.* **2010**, *8*, 1275.
53. Probst, N.; Madarász, Á.; Valkonen, A.; Pápai, I.; Rissanen, K.; Neuvonen, A.; Pihko, P. M., *Angew. Chem., Int. Ed.* **2012**, *51*, 8495.
54. George, J.; Sridhar, B.; Reddy, B. V. S., *Org. Biomol. Chem.* **2014**, *12*, 1595.
55. Zhao, M.-X.; Bi, H.-L.; Jiang, R.-H.; Xu, X.-W.; Shi, M., *Org. Lett.* **2014**, *16*, 4566.
56. Kardos, G.; Soós, T., *Eur. J. Org. Chem.* **2013**, 4490.
57. Han, X.; Zhong, F.; Lu, Y., *Adv. Synth. Catal.* **2010**, *352*, 2778.
58. Lee, J. H.; Deng, L., *J. Am. Chem. Soc.* **2012**, *134*, 18209.
59. Wei, Y.; Yao, L.; Zhang, B.; He, W.; Zhang, S., *Tetrahedron* **2011**, *67*, 8552.
60. Peng, W.; Wan, J.; Xie, B.; Ma, X., *Org. Biomol. Chem.* **2014**, *12*, 8336.

(9*S*)-Amino-9-deoxyquinine

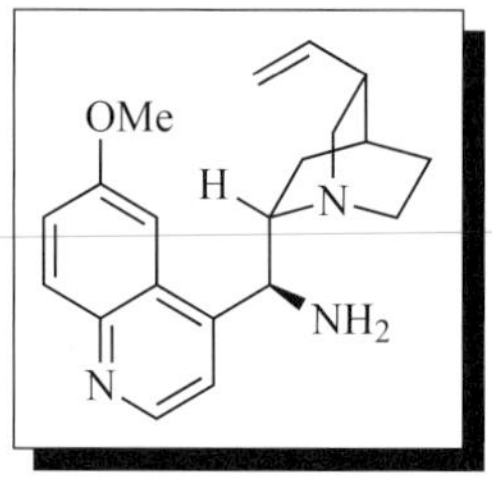

[168960-95-0] $C_{20}H_{25}N_3O$ (MW 323.432)

InChI = 1S/C20H25N3O/c1-3-13-12-23-9-7-14(13)10-19(23)20(21)16-6-8-22-18-5-4-15(24-2)11-17(16)18/h3-6,8,11,13-14,19-20H,1,7,9-10,12,21H2,2H3/t13?,14-,19-,20-/m0/s1

InChIKey = RIEKOKLABHBCGI-HQUBYHCZSA-N

(chiral auxiliary for asymmetric synthesis)

Alternate Names: (9*S*)-amino-9-deoxyepiquinine, ADQ.

Physical Data: slightly yellow oil, $[\alpha]_D = +80$ (c = 1.1, $CHCl_3$, 25 °C).[1a]

Analysis of Reagent Purity: ^{1}H NMR, ^{13}C NMR.

Form Supplied in: slightly yellow oil.

Solubility: sol in CH_2Cl_2, CH_3OH, and most organic solvents.

Preparative Methods: the title compound was prepared from (8*S*,9*R*)-quinine;[1] the analogous compound 9S-Amino-9-deoxyepidihydroquinine (ADHQ, CAS number: 852913-53-2) was prepared via the same method from (8*S*,9*R*)-dihydroquinine and generally has the same utility as ADQ.[2]

A process involving mesylation, S_N2 reaction of sodium azide followed by reduction gave ADQ in 50% yield (eq 1).[3]

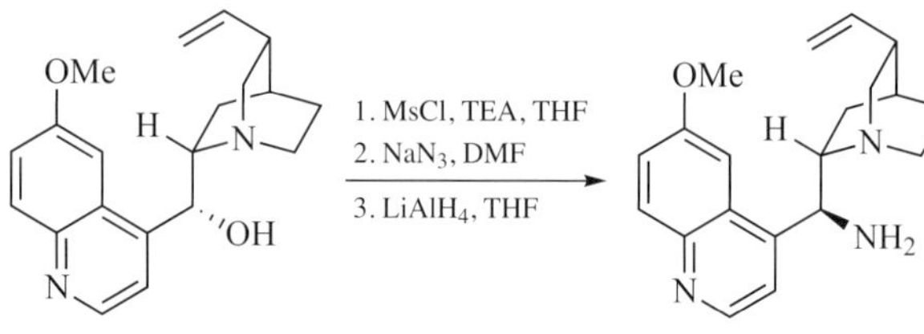

(1)

50% yield

Cinchona alkaloid ADQ has been extensively applied as aminocatalyst for a variety of carbonyl compounds.[4]

Purification: the title compound ADQ is purified by column chromatography on silica gel (EtOAc/MeOH/conc. aq. NH_4OH = 50/50/1 as eluant).[1c]

Handling, Storage, and Precautions: store in a cool, dry place under an Ar atmosphere.

Original Commentary

Tai-Ran Kang & Ying-Chun Chen
Sichuan University, Chengdu, China

Introduction. In the past few years, chiral primary amines have received growing attention as organocatalysts in asymmetric synthesis. (9*S*)-Amino-9-deoxyquinine (ADQ) has been shown to be an effective catalyst for the activation of aldehydes, ketones, and some other substrates, such as nitroolefins, for various asymmetric reactions.[5] In the case of the α-functionalization of aldehydes or ketones, nucleophilic enamines are formed between ADQ and carbonyl compounds, which then attack the electrophiles. For α,β-unsaturated systems, the catalyst activates the substrate by decreasing the energy of the LUMO in the intermediate iminium ions, thereby facilitating the transfer of the chirality by addition of nucleophiles to the β-position.

Asymmetric Michael Reactions. The asymmetric addition of acetophenone to benzylidenemalononitrile can be catalyzed by ADQ in the presence of *p*-TSA in THF to give the corresponding Michael product in 73% yield and 71% ee (eq 2).[6]

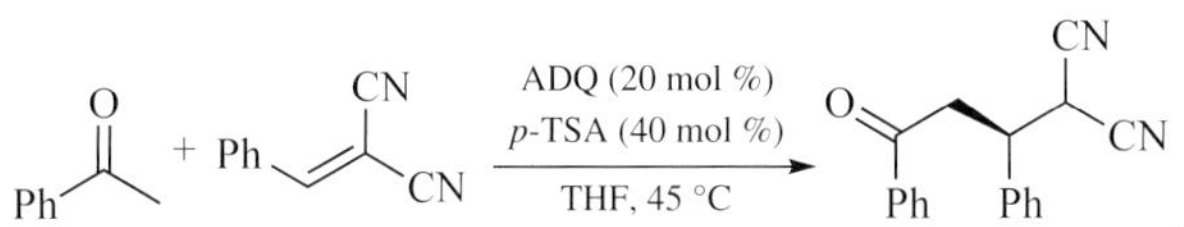

73% yield, 71% ee (2)

The asymmetric Michael addition of aliphatic ketones to vinyl sulfones has also been investigated under catalysis by ADQ and benzoic acid.[7] A novel catalyst (9*S*)-amino-9-deoxyepidihydroquinine, prepared from dihydroquinine, has been found to be capable of promoting the asymmetric Michael addition of enolizable carbonyl compounds to nitroolefins with high enantio- and diastereoselectivity.[2] In the presence of catalyst ADQ or ADHQ and an acidic additive, α,β-unsaturated ketones efficiently undergo asymmetric Michael or aza-Michael addition reactions with various nucleophiles, including indoles,[8a] oximes,[8b] thiols (eq 3),[8c] 1,3-cyclic dicarbonyl compounds (eq 4),[9] α,α-dicyanoalkenes (eq 5),[10] *N*-Boc-protected *N*-benzyloxyamines (eq 6),[11] and malononitrile,[12] forming the β-functionalized carbonyl derivatives with high enantioselectivity.

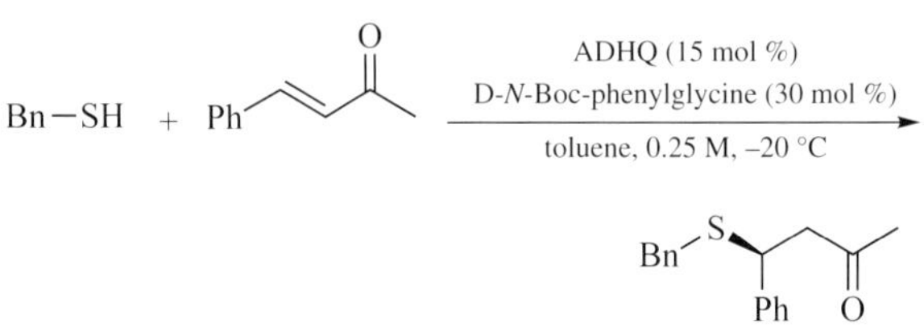

81% yield, 85% ee (3)

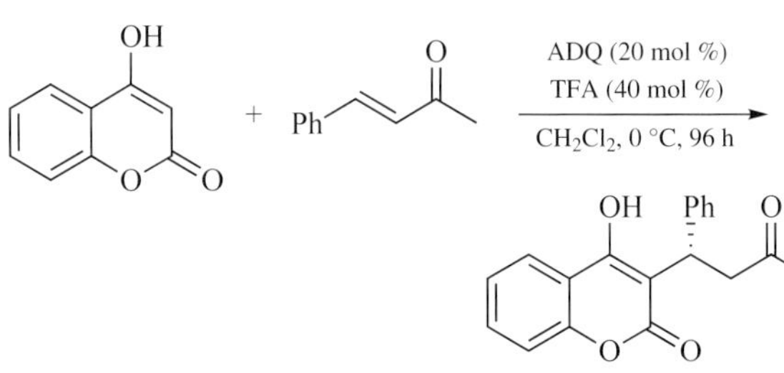

88% yield, 96% ee (4)

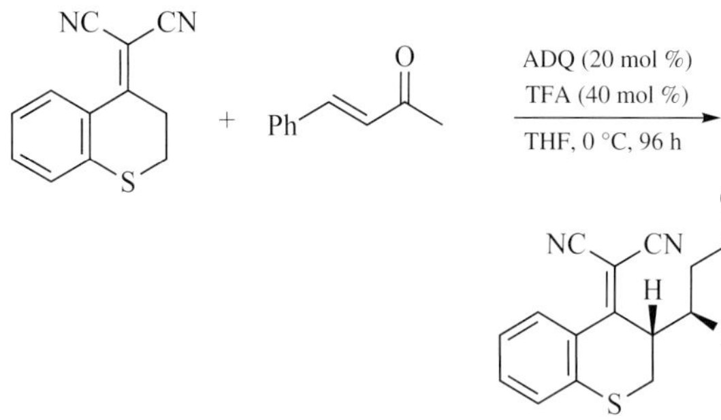

88% yield, 93% ee (5)

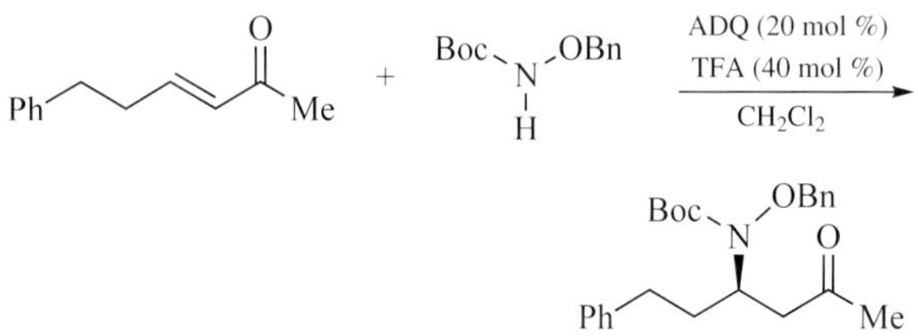

80% yield, 93% ee (6)

Domino Reactions. Employing an ADQ–TFA–DIPEA catalytic system, the reaction of prochiral α,α-dicyanoalkenes and α,β-unsaturated ketones could be carried out successfully to afford the corresponding double Michael addition cycloproducts with excellent enantio- and diastereoselectivity in good yield (eq 7).[13]

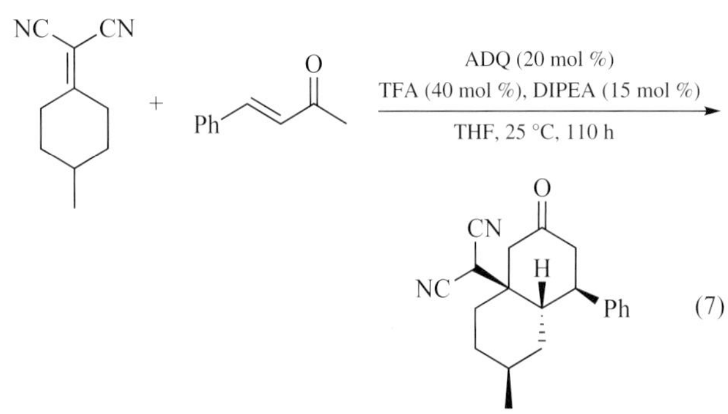

81% yield, 99% ee, >99:1 dr (7)

Except for the classical iminium–enamine strategy, Zhong found that nitroolefins could be activated by the primary amine moiety of ADQ via multiple H-bonds, thus enhancing the electrophilic character of the reacting carbon center and controlling stereochemistry (eq 8).[14] Based on the same mechanism, the tandem asymmetric Michael–Henry reactions of α-substituted β-ketoesters with nitroolefins have also been developed to provide synthetically useful, highly functionalized chiral cycloadducts.[15]

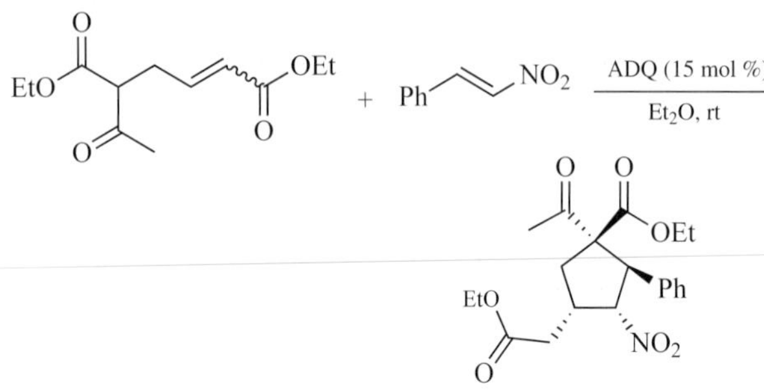

(8)

91% yield, 97% ee, >99:1 dr

Enantioselective Synthesis of Nitrocyclopropanes. The conjugate addition of bromonitromethane to cyclohexenone has been developed and is efficiently catalyzed by ADQ to provide functionalized nitrocyclopropanes in high yields with excellent stereoselectivities (eq 9).[16]

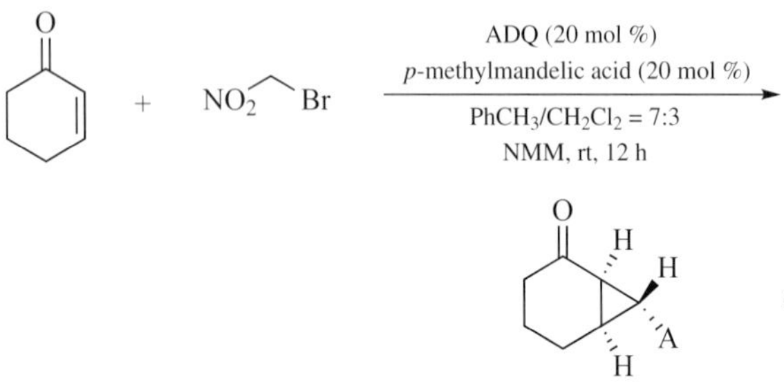

(9)

99% yield, 98% ee

Enantioselective Peroxidation of α,β-Unsaturated Ketones. ADQ served as an effective catalyst in the direct asymmetric peroxidation reaction of α,β-unsaturated ketones with hydroperoxides, such as aqueous hydrogen peroxide or peroxyacids, to afford not only epoxides (eq 10) but also enantioselectivity adducts via a peroxidation pathway.[17] These reactions proceeded generally with high yield and excellent enantioselectivity.

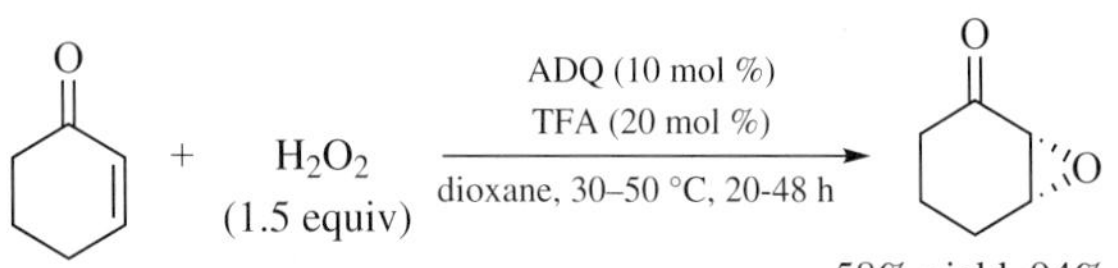

(10)

58% yield, 94% ee

Asymmetric Intramolecular Aldol Reaction. ADQ was used to catalyze the desymmetrizing intramolecular aldol reaction of 4-substituted 2,6-heptandiones leading to chiral 5-substituted-3-methyl-2-cyclohexene-1-ones with high ee values (eq 11).[18]

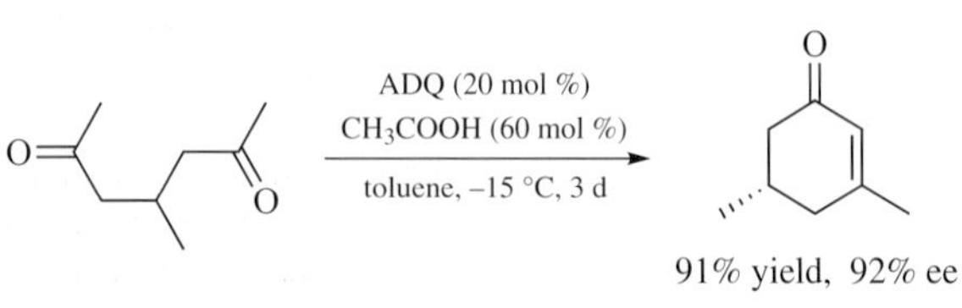

(11)

91% yield, 92% ee

Stereoselective Diels–Alder Reactions. ADQ was applied to promote the asymmetric Diels–Alder reaction of α,β-unsaturated ketones and 2-pyrones (eq 12).[19] When 20 mol % TFA was used as the acid additive, this reaction was significantly improved and afforded the *exo*-Diels–Alder adducts with good diastereoselectivity and enantioselectivity.

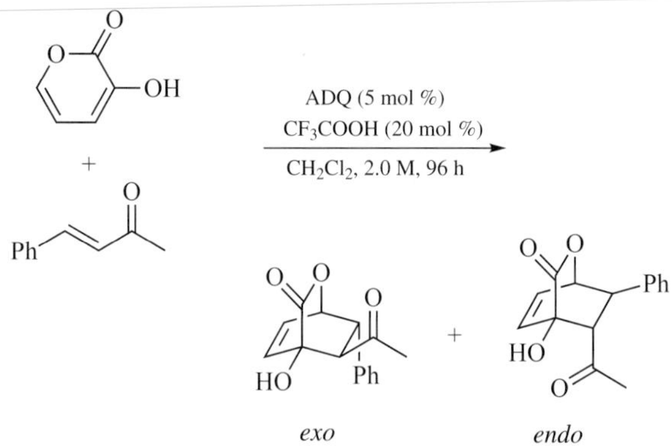

(12)

exo:*endo* = 82:18
for *exo* 74% yield
for *exo* 96% ee

1,3-Dipolar Cycloadditions. Chen et al. found that ADQ in combination with an acid catalyzes the 1,3-dipolar cycloaddition of 2-cyclohexen-1-one and azomethine imines. Unfortunately, only low to moderate ee values could be obtained. However, excellent levels of enantioselectivity could be obtained when a primary amine catalyst DMADQ with a free hydroxyl group, which was prepared from ADQ, was used. A hydrogen bond between DMADQ and the carbonyl group of dipole was proposed to be beneficial for the stereocontrol (eq 13).[20]

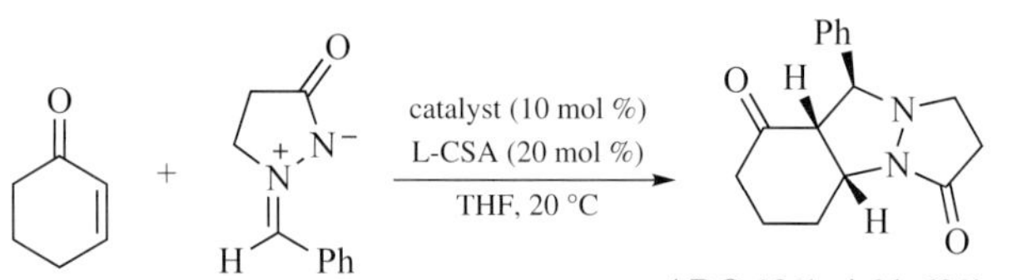

ADQ 19% yield, 60% ee
DMADQ 96% yield, 87% ee

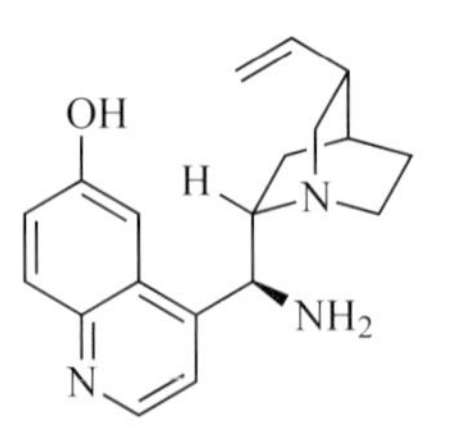

(13)

DMADQ

Enantioselective Direct α-Amination of Aromatic Ketones. The asymmetric direct α-amination of *p*-chloropropiophenone and diethyl azodicarboxylate (DEAD) can be catalyzed by ADQ in the presence of *p*-TSA in 2-PrOH to give the corresponding adduct in 51% yield and 96% ee (eq 14).[21]

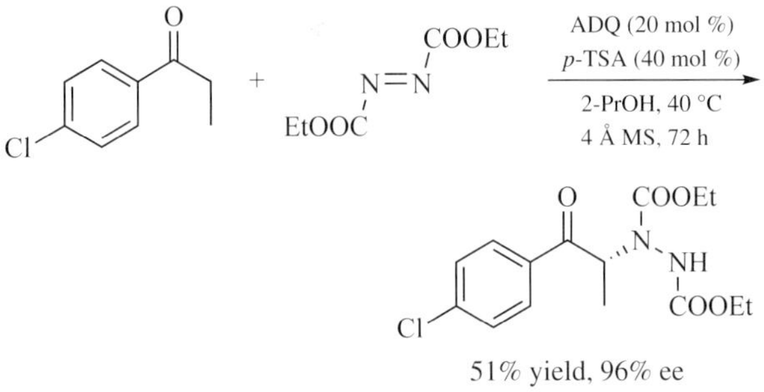

(14)

Asymmetric Transfer Hydrogenation of Aromatic Ketones. ADQ was applied as a chiral ligand in the asymmetric transfer hydrogenation of aromatic ketones with a rhodium catalytic system using *i*-propanol as the hydrogen source (eq 15).[22] Good to excellent conversions and enantioselectivities were observed for a variety of aromatic ketones.

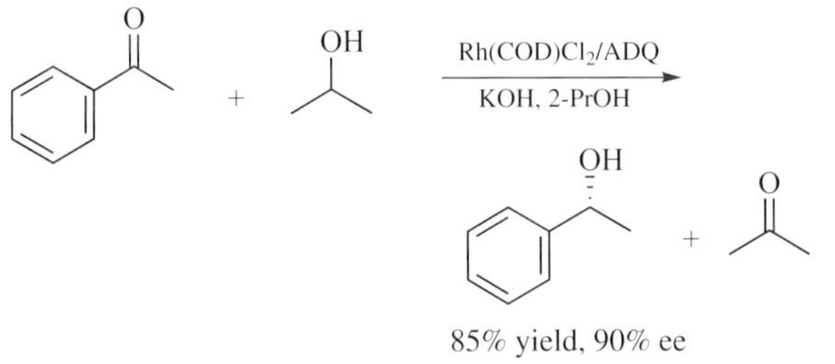

(15)

Preparation of Chiral Catalysts and Ligands. ADQ-derived thioureas are a type of bifunctional organocatalysts that combine a basic bridgehead nitrogen and a readily tunable hydrogen-bonding group, which originates from the 9-amino functionality (eq 16). They have emerged as powerful tools for the asymmetric construction of chiral molecules.[23] Other (9*S*)-amino-9-deoxyquinine functional reagents, such as chiral diamine ligand (eq 17)[24] and chiral bidentate Schiff's base (eq 18),[25] have been reported.

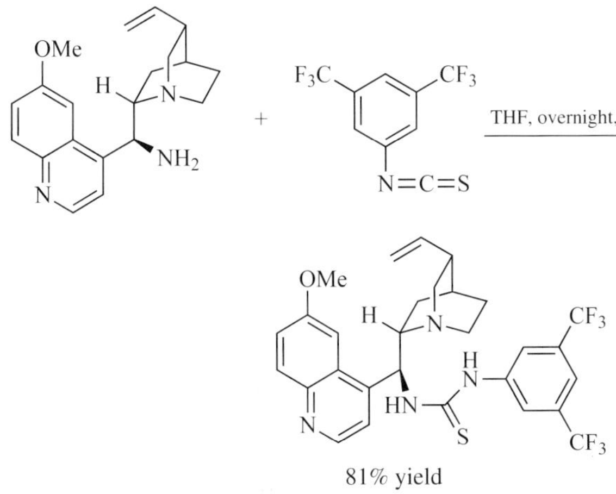

(16)

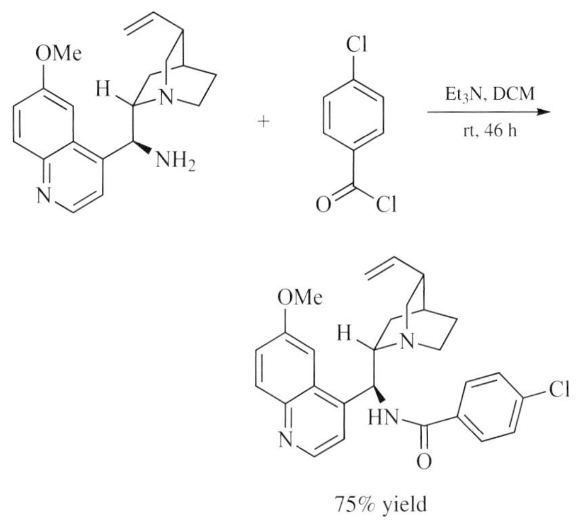

(17)

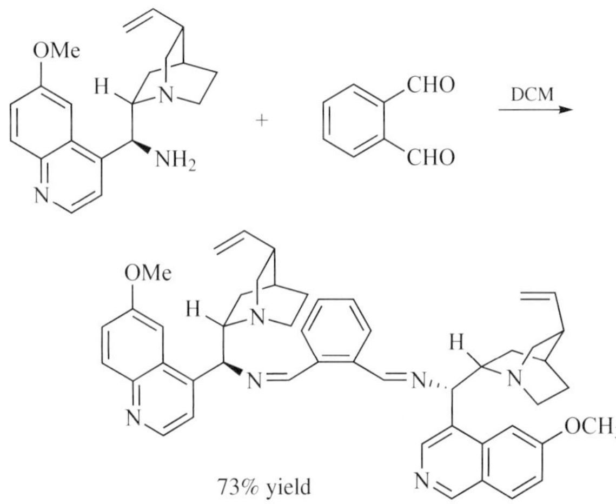

(18)

Related Reagent. (9*R*)-Amino-9-deoxyquinidine.

First Update

Wei Du, Gu Zhan & Ying-Chun Chen
Sichuan University, Chengdu, China

Enamine Catalysis–*C*-Centered Electrophiles. One of the most important catalytic strategies of ADQ is enamine mode, by increasing the HOMO energy of the substrates and facilitating the reactions with the electrophiles. Michael acceptors, such as benzylidenemalononitrile,[4] isatylidene malononitriles,[26] and nitroolefins,[27] are good electrophiles for the α-alkylation of aldehyde, or ketones. α-Branched aldehydes were activated by ADQ and *p*-nitrobenzoic acid, and attacked heteroaryl vinyl sulfones, delivering the Michael adducts having a quaternary stereogenic center with excellent enantioselectivity (eq 19).[28]

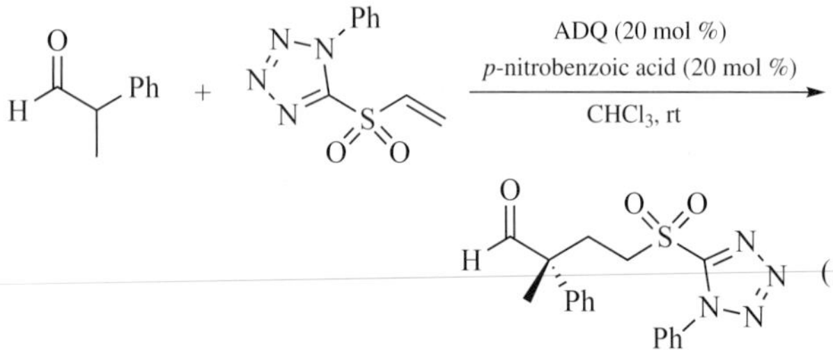

(19)

76% yield, 94% ee

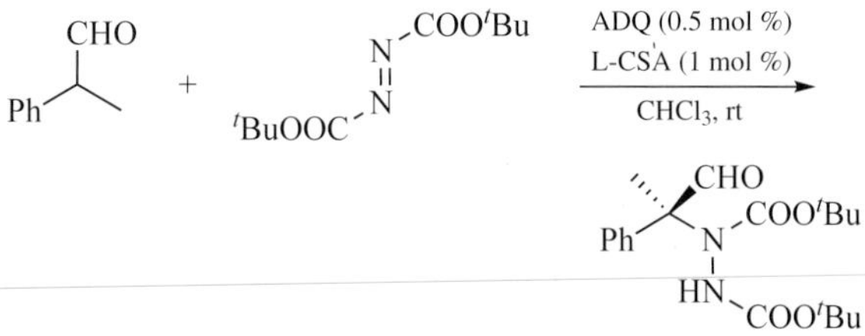

(22)

96% yield, 95% ee

Asymmetric Mannich reactions between imines and ketones were efficiently catalyzed by ADQ.[29] Three-component reaction of aromatic aldehydes, ureas, and β-ketoesters for the asymmetric synthesis of 3,4-dihydropyrimidine-2(1*H*)-ones (DHPMs) was realized by employing ADQ as a catalyst, together with Brønsted acid[30] or Lewis acid additive.[31] For the latter case, ADQ served as both catalyst and ligand: the enamine intermediates derived from ADQ and β-ketoesters would bind to the Lewis acid (Nb salt), further promoting the reaction with imines (eq 20).

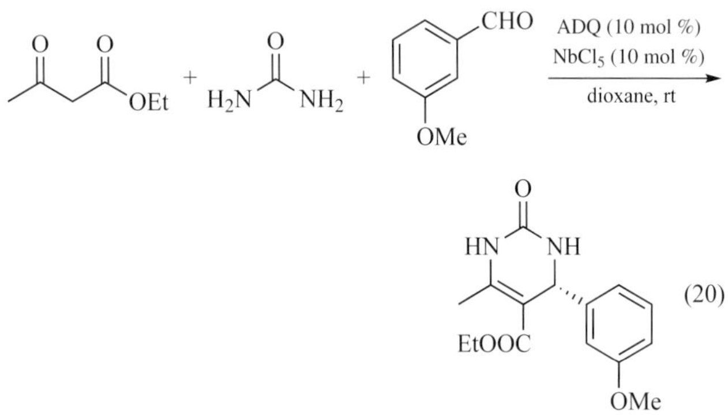

(20)

88% yield, 84% ee

Aldehydes or ketones catalyzed by ADQ could also attack some activated carbonyl groups, including α-ketophosphonates[32] (eq 21) and isatins,[33] to provide enantioenriched cross-aldol products in good to excellent yields.

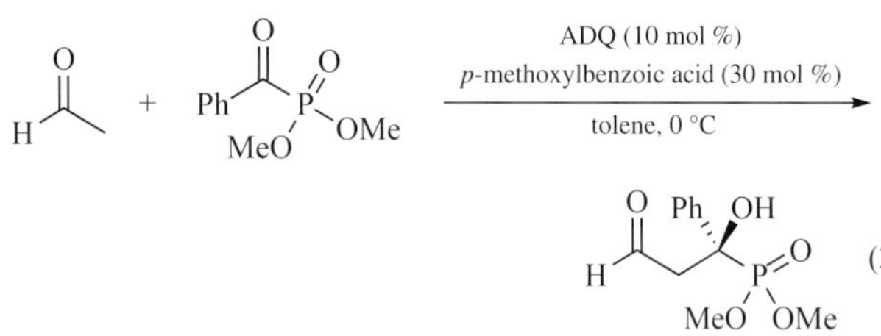

(21)

67% yield, 96% ee

Enamine Catalysis–*N*-Centered Electrophiles. The enantioselective α-amination reaction between aromatic ketones and DEAD catalyzed by ADQ has been reported.[19] Using α-branched aldehydes and dialkyl azodicarboxylates as the substrates, the amination products with a quaternary chiral center were obtained with excellent enantioselectivity.[34] A catalytic system of ADQ/L (−)-camphorsulfonic acid (L-CSA) exhibited excellent efficiency, as the quantitative amination product was formed with 95% ee in the presence of 0.5 mol % of catalyst (eq 22).[34a]

Enamine Catalysis–*O*-Centered Electrophiles. Benzoyl peroxide could be attacked by enamine intermediates of cyclic ketones and ADQ.[35] In addition, the quinuclidine moiety of ADQ was protonated by the acid additive, to assist the leaving of the benzoyloxylate, finally giving the α-benzoyloxylated ketones with very good ee values (eq 23). The substrate scope was expanded to α-branched aldehydes to construct a quaternary oxygenated stereocenter with moderate to good enantioselectivity.[36]

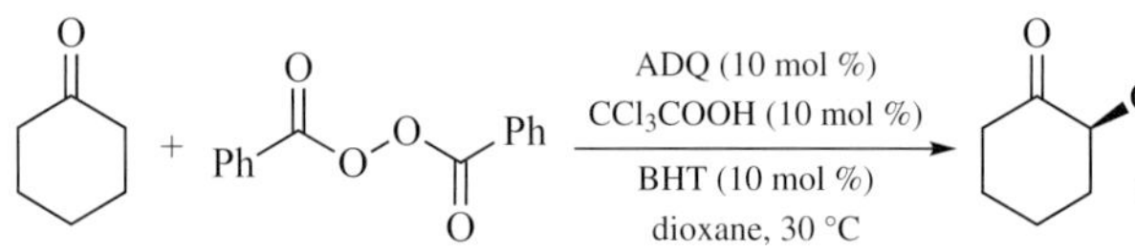

(23)

67% yield, 94% ee

BHT = 2,6-di-*tert*-butyl-4-methylphenol

Dienamine Catalysis. The γ-regioselective vinylogous Michael additions of β-methylcyclohexenone to nitroalkenes have been reported via dienamine catalysis of DMAQD or AQD.[37] The similar strategy was employed with the beta-ethyl cyclic enones and *N*-(2-*tert*-butylphenyl)maleimide, and outstanding atroposelectivity was achieved under the catalysis of ADQ and *N*-Boc-L-phenylglycine (eq 24).[38]

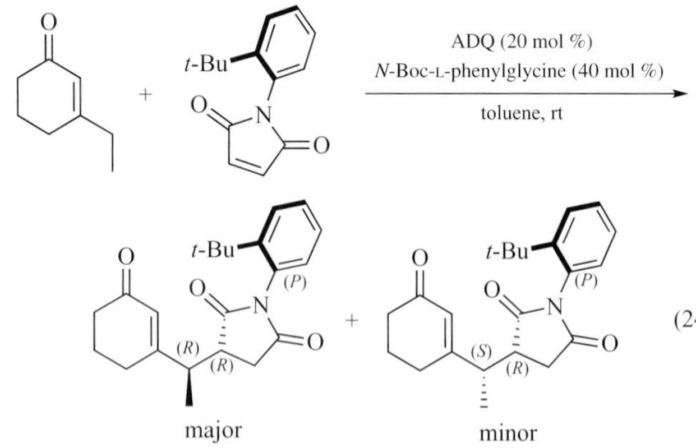

(24)

75% yield, 70:30 dr, >99%/>99% ee

The cyclic enones were also activated by ADQ, serving as electron-rich dienes, which underwent [4 + 2] cycloadditions with allylidenemalononitriles to afford the *endo*-products with excellent ee values (eq 25).[39] Interestingly, *exo*-products were formed by using DMADQ instead of ADQ. Therefore, stereodivergent cycloadditions were realized by delicately tuning the primary amine catalyst.

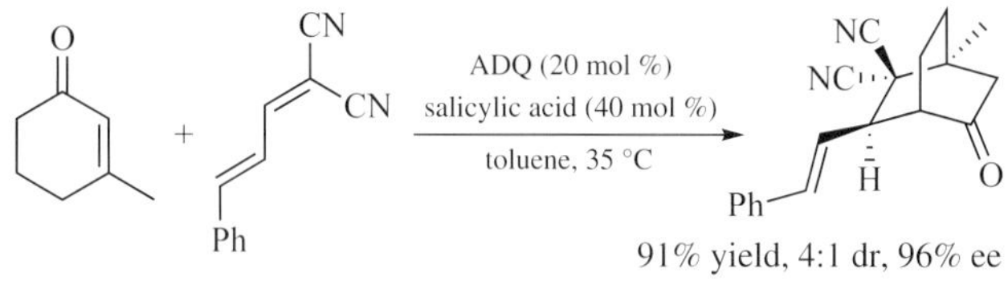

(25)

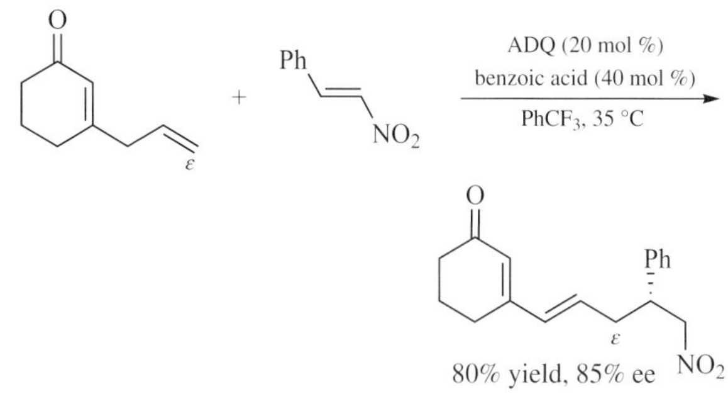

(28)

Trienamine Catalysis. Trienamine catalysis with 2,4-dienals has been developed in 2011, and this protocol is fascinating and challenging due to the regioselectivity and remote stereocontrol.[40] δ,δ-Disubstituted 2,4-dienones without an α'-CH group were designed to facilitate the β,ε-regioselective Diels–Alder reaction with *N*-substituted maleimides, and multifunctional cyclohexene derivatives were constructed with high ee values by the catalysis of ADQ (eq 26).[41] Later, this strategy was further developed by introducing interrupted cyclic 2,5-dienones with an α'-CH group as the substrates, realizing the different δ,ε-regioselective inverse-electron-demand aza-Diels–Alder reaction (eq 27).[42] A remote ε-regioselective 1,4-addition reaction to nitroalkenes was also developed with good enantioselectivity (eq 28).[43] The formal trienamine intermediates were also generated with heteroarene-substituted enones, and the enantioselective dearomatic Diels–Alder reactions proceeded smoothly under the catalysis of ADQ via π-system activation (eq 29).[44]

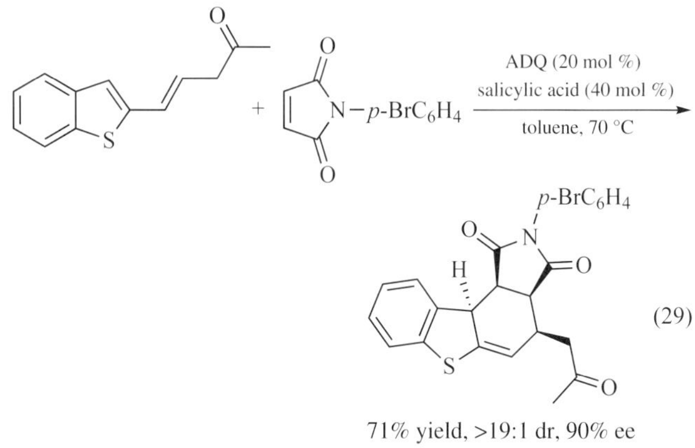

(29)

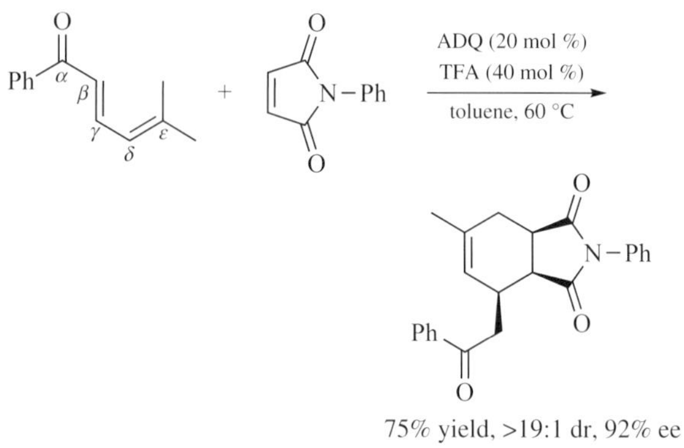

(26)

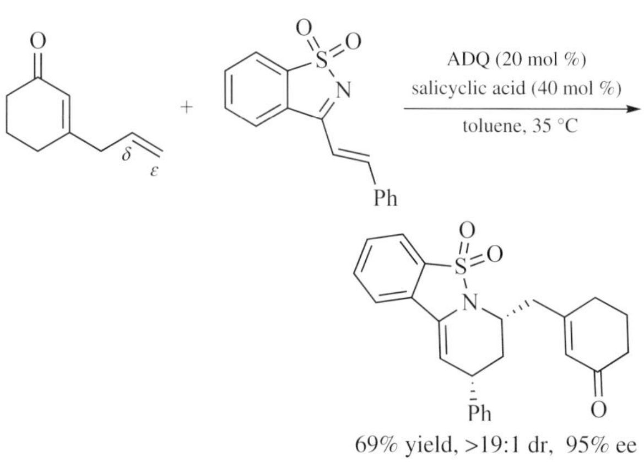

(27)

Iminium Catalysis–*C*-Centered Nucleophiles. Another universal application of ADQ is in iminium catalysis, in which the LUMO energy of α,β-unsaturated carbonyls can be decreased to favor the asymmetric Michael additions with various nucleophiles. The compounds bearing an acidic C-atom, such as malononitrile,[45] bis(phenylsulfonyl)methane (eq 30),[46] and 1,3-dicarbonyl compounds,[7,47] are good nucleophiles for Michael acceptors under the catalysis of ADQ, delivering β-alkylated ketones with good enantioselectivity. Indoles are also highly nucleophilic, and the Friedel–Crafts alkylation of indoles with α,β-unsaturated ketones catalyzed by ADQ was fully investigated with the assistance of computational study.[6a,48] Acrolein was applied to react with 2,3-disubstituted indoles to enantioselective synthesis of highly substituted tricyclic hydrocarbazoles (eq 31).[49]

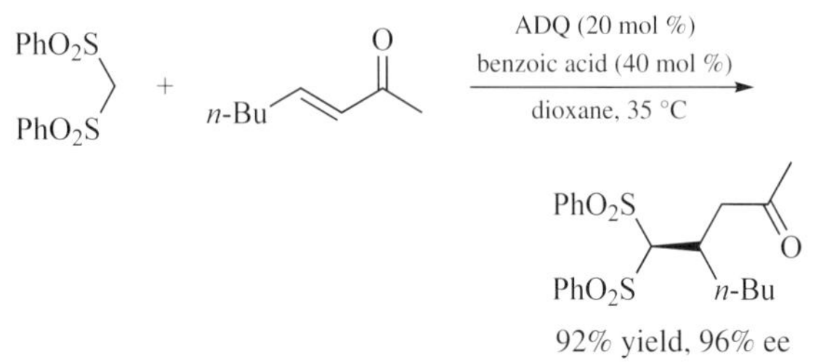

(30)

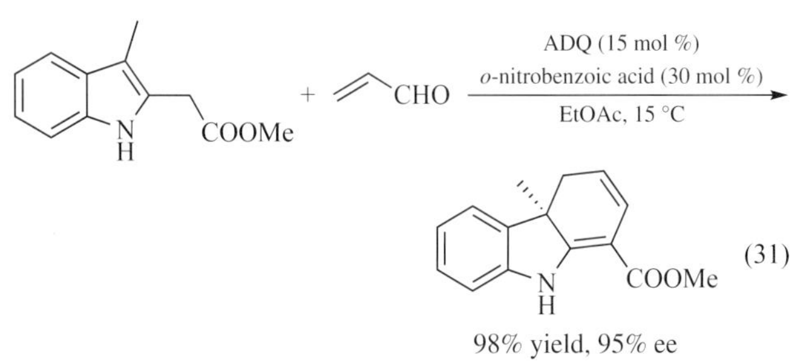

(31)

Stabilized phosphorus ylides could attack the β-position of α,β-unsaturated ketone in the presence of ADQ and *N*-Boc-L-proline, and chiral α-methylene-δ-ketoesters were obtained with good to excellent stereocontrol followed by Wittig reaction with formaldehyde (eq 32).[50]

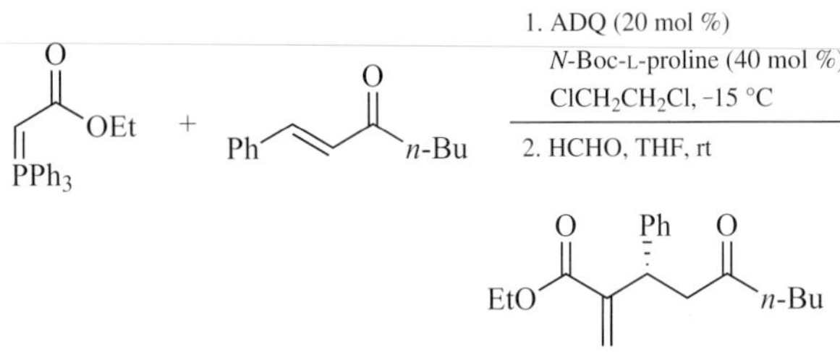

(32)

56% yield, 95% ee

Nitroalkenes has been demonstrated to be nucleophilic though the activation of Lewis bases.[51] Thus, the unique feature of ADQ possessing both primary and tertiary amines promotes the conjugate addition of the nitroalkenes to the enones, generating the γ-nitroketones with excellent enantioselectivity but with poor diastereoselectivity (eq 33).[52]

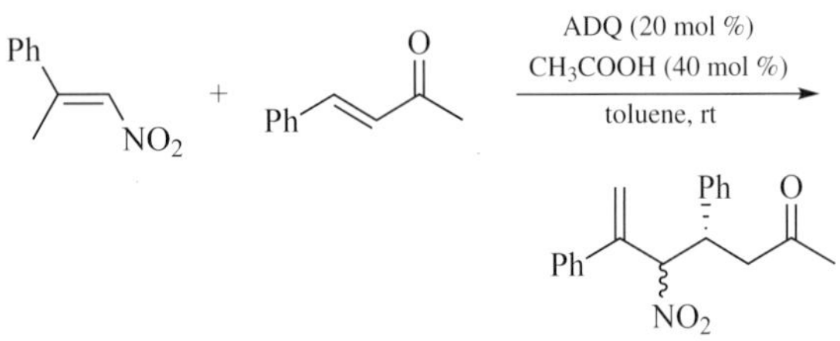

(33)

96% yield, *syn:anti* = 1:1.1
99% (*syn*) / 98% (*anti*) ee

Apart from the β-functionalization of enones, an unexpected δ-regioselective nucleophilic substitution of cyclic Morita-Baylis-Hillman (MBH) alcohols with indoles was achieved with the AQD/TFA catalytic system (eq 34).[53] Good results were obtained in terms of both reactivity and enantioselectivity. Iminium catalysis of the primary amine together with Brønsted acid catalysis of the protonated quinuclidine moiety was proposed to explain the observed results.

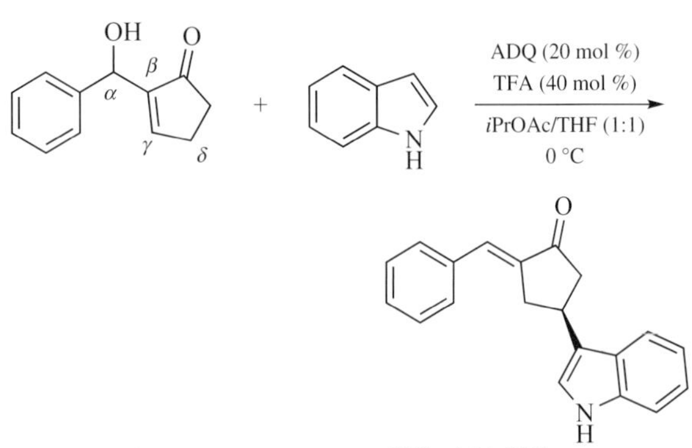

(34)

92% yield, 89% ee

Iminium Catalysis–*N*-Centered Nucleophiles. Great progress has been made in asymmetric aza-Michael additions by means of iminium catalysis. Protected hydroxylamines,[9] hydrazides,[54] and some *N*-heterocycles[55] are ideal nucleophiles to attack the ADQ-activated α,β-unsaturated ketones enantioselectively. An intramolecular aza-Michael addition was reported by employing a novel substrate with the combination of enone and carbamate groups (eq 35).[56] The excellent enantiocontrol was induced by the catalysis of ADQ, and an array of 2-substituted piperidines were synthesized in high yields. This methodology was applied to enantioselective synthesis of phenanthroquinolizidine alkaloids such as cryptopleurine and boehmeriasin.[57]

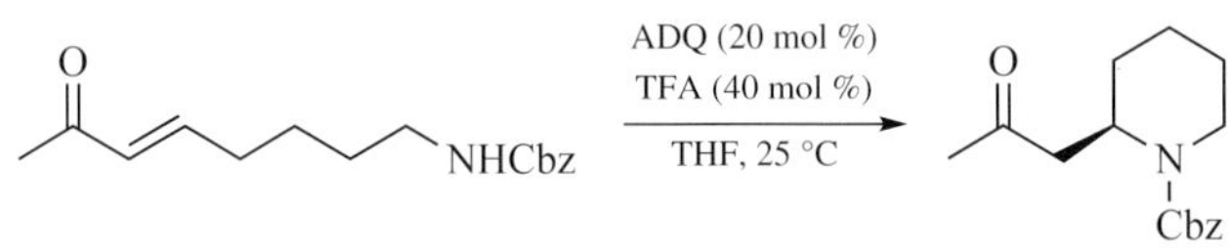

95% yield, 98% ee

(35)

Vinylogous Iminium Catalysis. Vinylogous iminium catalysis of cinchona alkaloids was discovered through δ-functionalization of the extended π system of 2,4-unsaturated ketones. ADQ showed moderate reactivity for this remote functionalization, while the modified ADQ with 2(-phenyl-substituted quinoline ring was more suitable and provided the desired product with excellent ee (eq 36).[58]

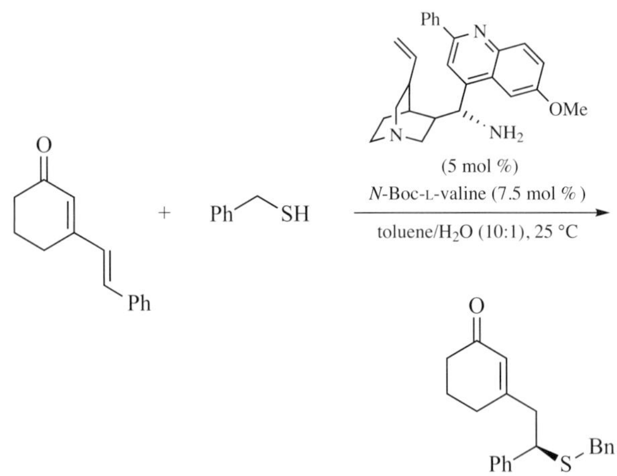

(36)

66% yield, 91% ee

Enamine/Iminium Catalysis. In combination with the well-established enamine catalysis of ADQ, elegant work involving the H-bonding-iminium activation of the MBH alcohols was developed to furnish bicyclic α-alkylidene-ketones with olefinic oxindoles in excellent enantio- and diastereoselectivities (eq 37).[59] Benzylideneacetone could be also activated by ADQ to attack Michael acceptors, giving enantioenriched cyclized products via enamine/iminium catalysis.[60]

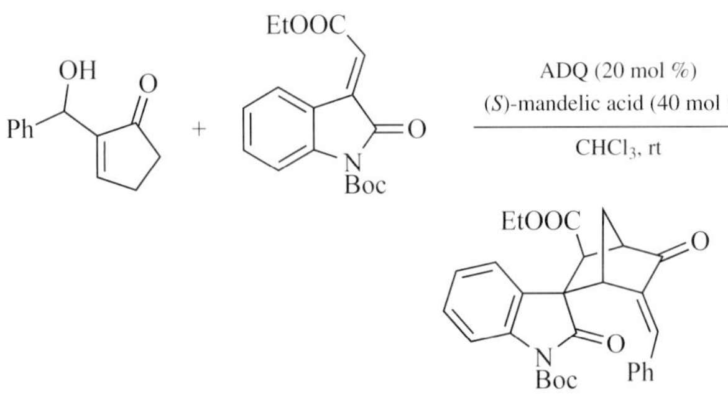

(37)

84% yield, >19:1 dr, 90% ee

Iminium/Enamine Catalysis. The epoxidation of α,β-unsaturated ketones catalyzed by ADQ has been well established in recent years.[15,61] The general mechanism probably involved iminium-mediated conjugate addition of hydrogen (or alkyl) peroxide followed by enamine-promoted ring closure. An aziridination reaction of enones, which proceeded through the similar pathway with BocNHOTs as the nucleophile, was also reported to construct chiral oxazolidinones after sequential double S_N2 reactions (eq 38).[62]

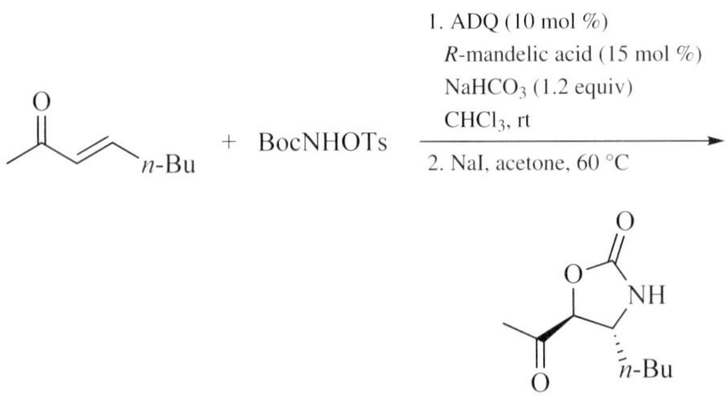

(38)

64% yield, >20:1 dr, 98% ee

Enantioselective Michael–Michael addition reactions between α,α-dicyanoalkenes and enones were accomplished under the catalysis of ADQ.[8,11] Later, both intra- (eq 39) and intermolecular double Michael reactions of indole derivatives via the ADQ-induced iminium/enamine catalysis were reported,[63] generating a series of chiral tetracyclic compounds in satisfying yields and with good stereoselectivity.

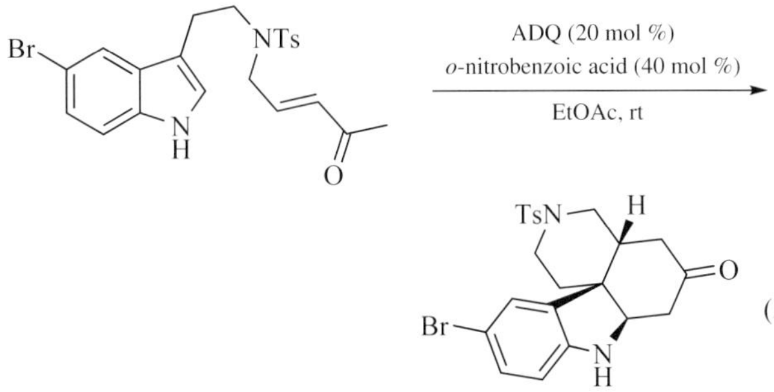

(39)

72% yield, 89:11 dr, 95% ee

The ADQ-catalyzed one-pot reaction of α,α-disubstituted aldehyde, di-*tert*-butylazodicarboxylate (D*t*BAD), and α,β-unsaturated aldehyde was carried out through a sequence of α-amination, aza-Michael addition of hydrazine and aldol condensation.[64] 3-Pyrrolines with a quaternary chiral center were synthesized with high level of enantioselectivity (eq 40).

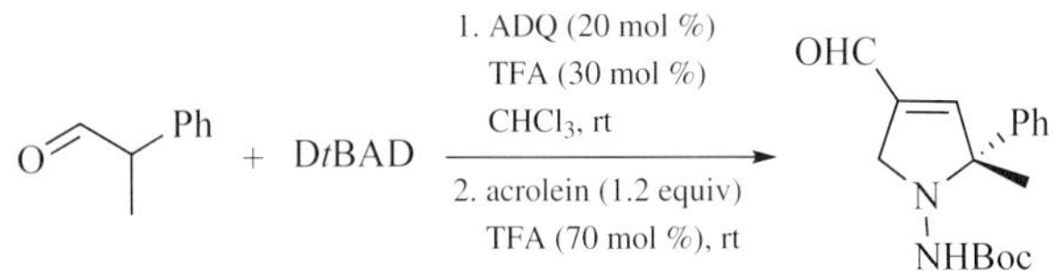

(40)

83% yield, 94% ee

Asymmetric C(sp^3)–H functionalization has been achieved by utilizing ADQ and HOTf as the catalysts.[65] The reaction underwent iminium ion-mediated 1,5-hydride transfer followed by enamine-mediated 6-*endo* cyclization reaction to afford tetrahydroquinolines in good yields with moderate to excellent ee values (eq 41).

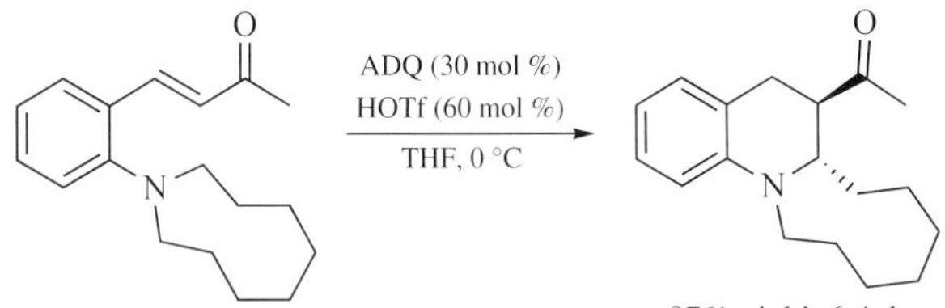

(41)

87% yield, 6:4 dr
81% / 90% ee

Enamine/Enamine Catalysis. [3 + 3] annulations of 3-vinyl-1,2-benzoisothiazole-1,1-dioxides with aliphatic ketones were disclosed via cascade enamine/enamine catalysis of ADQ, through tandem Michael addition–Mannich reaction to construct spirocyclic compounds with high level of enantioselectivity (eq 42).[66]

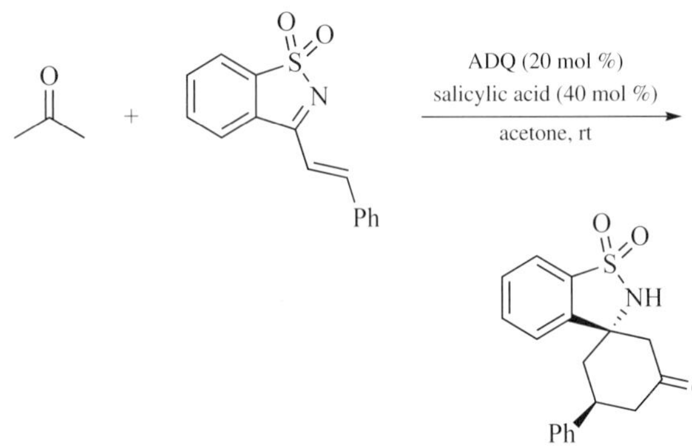

(42)

85% yield, >19:1 dr, 97% ee

Dienamine/Dienamine Catalysis. An asymmetric [5 + 3] formal cycloaddition reaction of 3-substituted 2-cyclopentenones and cyclic 1-azadienes was developed through a similar tandem Michael addition–Mannich reaction sequence. This reaction proceeded via an unusual cascade dienamine/dienamine catalysis, and fused architectures incorporating a spirocyclic motif were produced in excellent stereoselectivity (eq 43).[67]

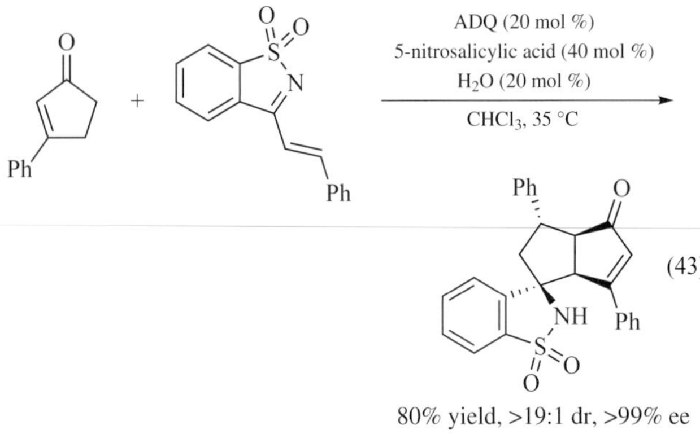

(43)

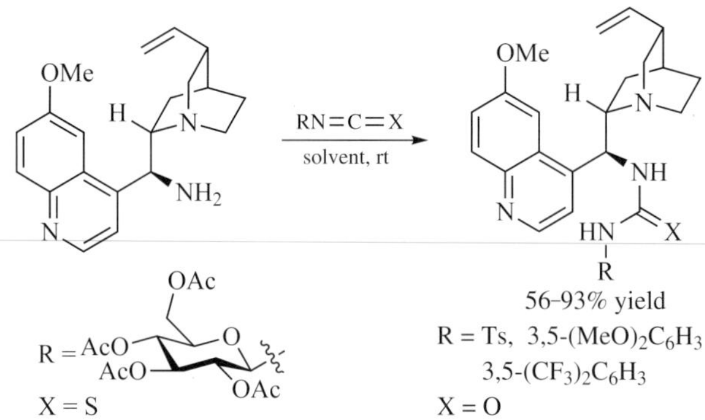

(46)

Iminium/Iminium catalysis. ADQ-promoted iminium/iminium catalysis has been developed with enones[68] and dienones.[69] For the latter case, a number of densely functionalized cyclohexanones with excellent enantioselectivity were obtained by double conjugate addition of malononitrile to dienones (eq 44).

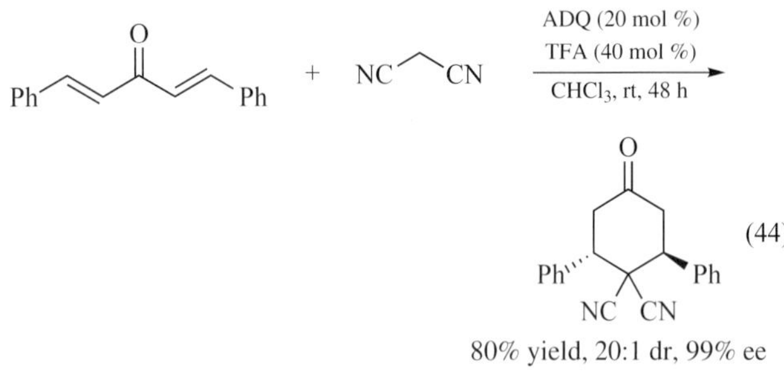

(44)

Metal Ligand. The nitrogen atom of quinuclidine moiety in cinchona alkaloids can coordinate to metal to assist asymmetric transformations.[70] ADQ-ligated Ir or Rh complexes catalyzed the enantioselective transfer hydrogenation of ketones to prepare chiral secondary alcohols (eq 45).[20b]

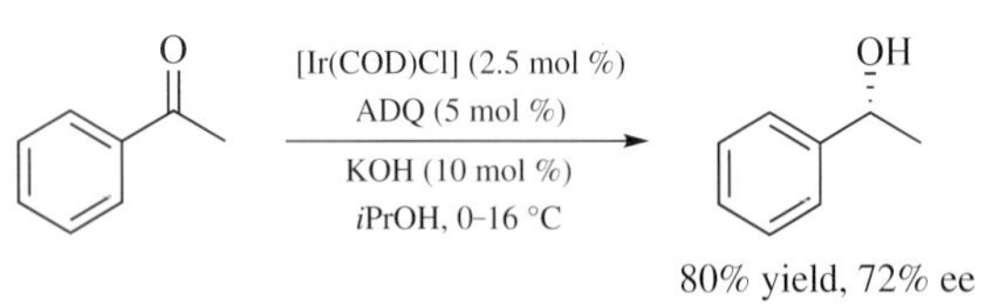

(45)

Preparation of Chiral Catalysts. The ADQ-derived chiral catalysts with a hydrogen-bonding group are widely investigated. Differently substituted ureas or thioureas derived from ADQ have been prepared (eq 46).[71]

In addition, the thiourea–amine bifunctional catalysts have been developed by a combination of chiral amines with cinchona alkaloids (eq 47),[72] which were found to be highly efficient catalysts for the conjugate addition.

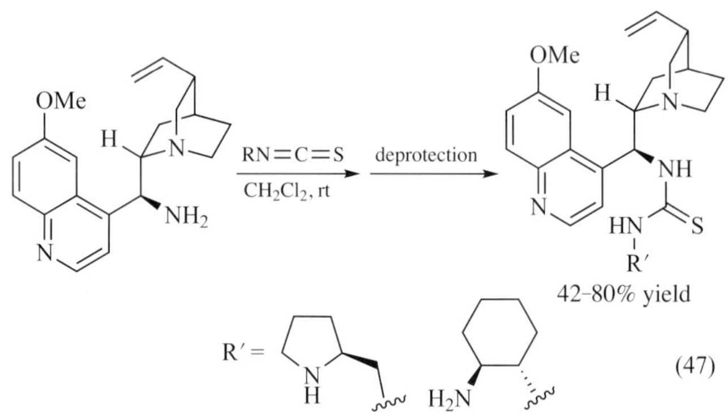

(47)

Multifunctional catalysts were further synthesized with the existing bifunctional catalyst scaffolds, by simple insertion of chiral building blocks, such as amino acids (eq 48).[73] Since the peptides exhibited high efficacy as H-bonding donors, the ureidopeptides-based cinchona alkaloids were designed for asymmetric Michael and Mannich reactions (eq 49).[74]

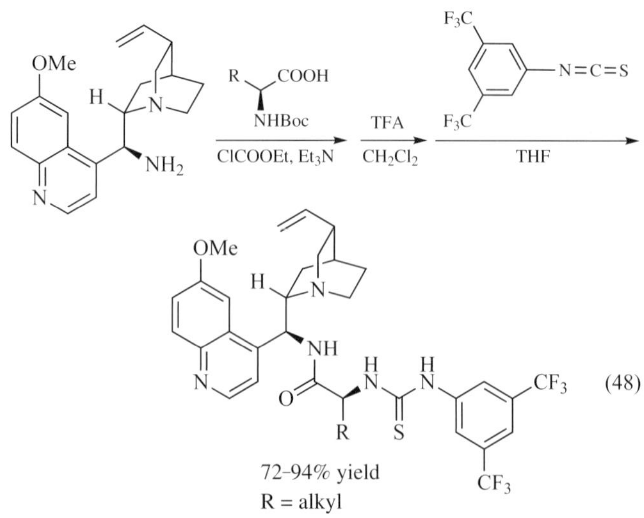

(48)

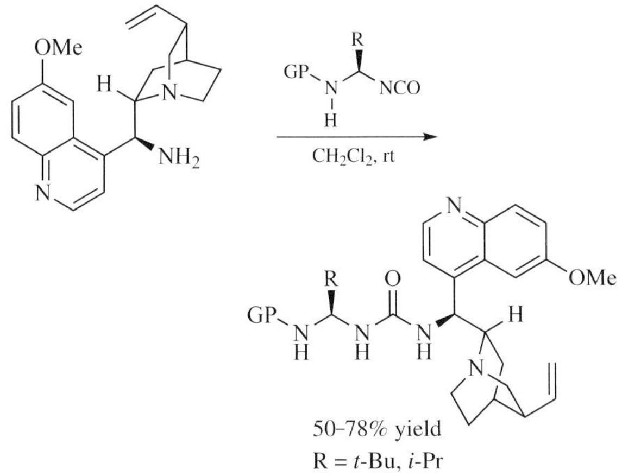

(49)

ADQ-based squaramide[75] (eq 50) and sulfonamide[76] (eq 51) bifunctional catalysts can be prepared through simple and modular pathways.

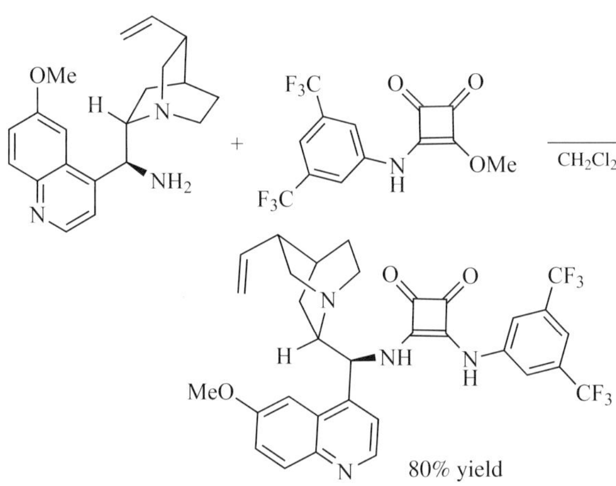

(50)

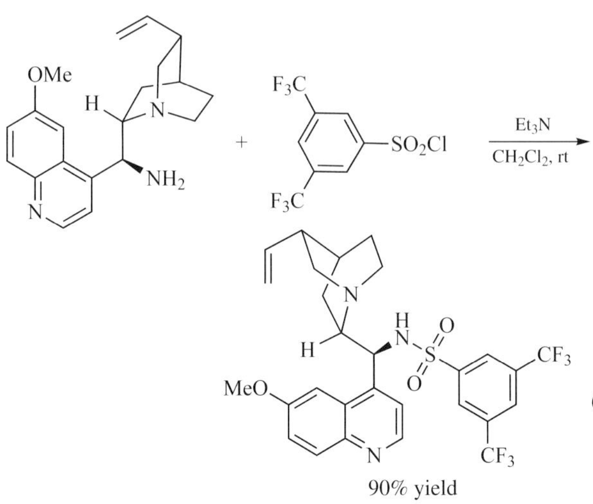

(51)

The ammonium-type of phase-transfer catalysts bearing multiple H-bonding donors derived from ADQ were readily prepared and provided high levels of diastereo- and enantioselectivities in nitro-Mannich reactions of amidosulfones (eq 52).[77]

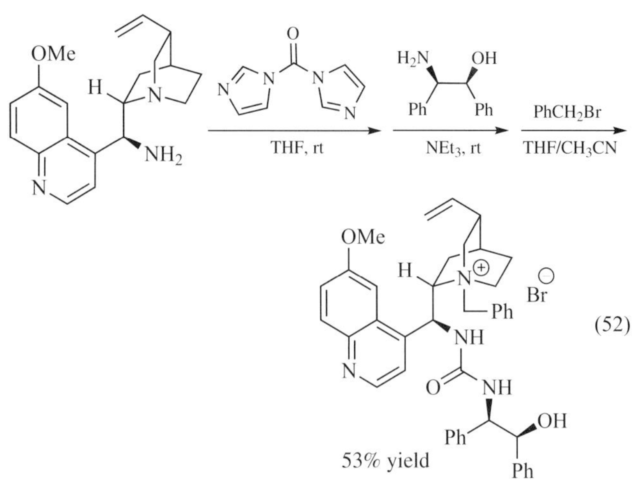

(52)

The incorporation of phosphine moiety into ADQ, which served as a chiral bidentate ligand, showed excellent reactivity in terms of both yield and enantioselectivity in silver-catalyzed cyclization reactions (eq 53).[78]

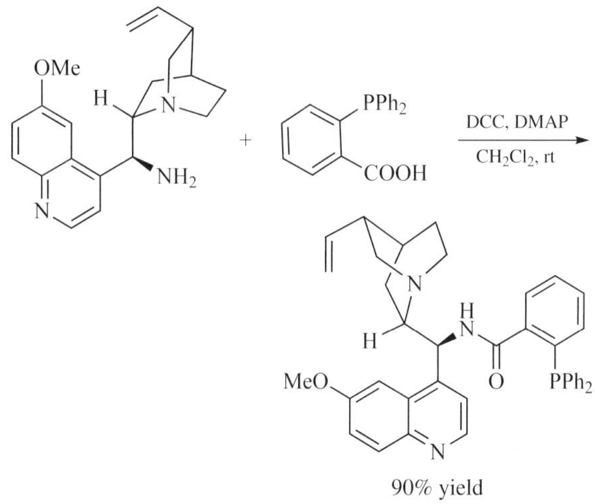

(53)

Beside the functionalization on the 9-amino group of ADQ, the modification on the 2′-position of the quinoline ring has been achieved by organolithium or Grignard reagents-mediated alkylation followed by oxidation (eq 54).[79]

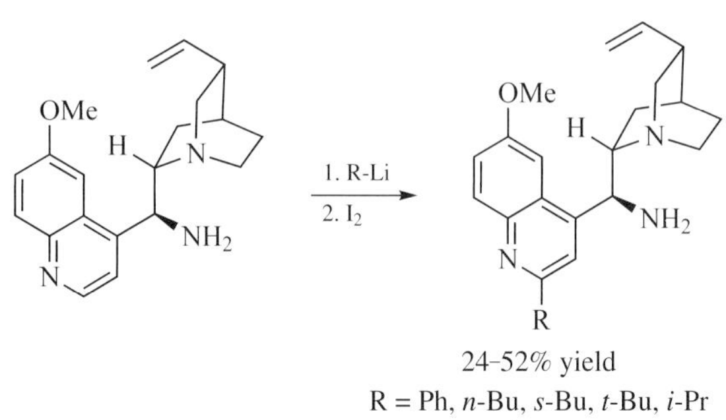

(54)

1. (a) Brunner, H.; Bügler, J.; Nuber, B., *Tetrahedron: Asymmetry* **1995**, *6*, 1699. (b) Brunner, H.; Schmidt, P., *Eur. J. Org. Chem.* **2000**, 2119. (c) Vakulya, B.; Varga, S.; Csámpai, A.; Soós, T., *Org. Lett.* **2005**, *7*, 1967.
2. McCooey, S. H.; Connon, S. J., *Org. Lett.* **2007**, *9*, 599.

3. Cassani, C.; Martín-Rapún, R.; Arceo, E.; Bravo, F.; Melchiorre, P., *Nat. Protoc.* **2013**, *8*, 325.
4. (a) Melchiorre, P., *Angew. Chem., Int. Ed.* **2012**, *51*, 9748; (b) Jiang, L.; Chen, Y.-C., *Catal. Sci. Technol.* **2011**, *1*, 354.
5. For reviews, see (a) Peng, F. Z.; Shao, Z. H., *J. Mol. Catal. A: Chem.* **2008**, *285*, 1. (b) Chen, Y. C., *Synlett* **2008**, 1919. (c) Bartoli, G.; Melchiorre, P., *Synlett* **2008**, 1759. (d) Xu, L. W.; Luo, J.; Lu, Y., *Chem. Commun.* **2009**, 1807.
6. Yue, L.; Du, W.; Liu, Y. K.; Chen, Y. C., *Tetrahedron Lett.* **2008**, *49*, 3881.
7. Zhu, Q.; Cheng, L.; Lu, Y., *Chem. Commun.* **2008**, 6315.
8. (a) Bartoli, G.; Bosco, M.; Carlone, A.; Pesciaioli, F.; Sambri, L.; Melchiorre, P., *Org. Lett.* **2007**, *9*, 1403. (b) Carlone, A.; Bartoli, G.; Bosco, M.; Pesciaioli, F.; Ricci, P.; Sambri, L.; Melchiorre, P., *Eur. J. Org. Chem.* **2007**, 5492. (c) Ricci, P.; Carlone, A.; Bartoli, G.; Bosco, M.; Sambri, L.; Melchiorr, P., *Adv. Synth. Catal.* **2008**, *350*, 49.
9. Xie, J. W.; Yue, L.; Chen, W.; Du, W.; Zhu, J.; Deng, J. G.; Chen, Y. C., *Org. Lett.* **2007**, *9*, 413.
10. Xie, J. W.; Chen, W.; Li, R.; Zeng, M.; Du, W.; Yue, L.; Chen, Y. C.; Wu, Y.; Zhu, J.; Deng, J. G., *Angew. Chem., Int. Ed.* **2007**, *46*, 389.
11. Lu, X. J.; Deng, L., *Angew. Chem., Int. Ed.* **2008**, *47*, 7710.
12. Li, X. F.; Cun, L. F.; Lian, C. X.; Zhong, L.; Chen, Y. C.; Liao, J.; Zhu, J.; Deng, J. G., *Org. Biomol. Chem.* **2008**, *6*, 349.
13. Kang, T. R.; Xie, J. W.; Du, W.; Feng, X.; Chen, Y. C., *Org. Biomol. Chem.* **2008**, *6*, 2673.
14. Tan, B.; Shi, Z. G.; Chua, P. J.; Zhong, G. F. *Org. Lett.* **2008**, *10*, 3425.
15. Tan, B.; Chua, P. J.; Zeng, X. F.; Lu, M.; Zhong, G. F., *Org. Lett.* **2008**, *10*, 3489.
16. Lv, J.; Zhang, J. M.; Lin, Z.; Wang, Y. M., *Chem. Eur. J.* **2009**, *15*, 972.
17. (a) Reisinger, C. M.; Wang, X.; List, B. *Angew. Chem., Int. Ed.* **2008**, *47*, 8112. (b) Wang, X. W.; Reisinger, C. M.; List, B., *J. Am. Chem. Soc.* **2008**, *130*, 6070. (c) Lu, X. J.; Liu, Y.; Sun, B. F.; Cindric, B.; Deng, L., *J. Am .Chem .Soc.* **2008**, *130*, 8134.
18. Zhou, J.; Wakchaure, V.; Kraft, P.; List, B., *Angew. Chem., Int. Ed.* **2008**, *47*, 7656.
19. Singh, R. P.; Bartelson, K.; Wang, Y.; Su, H.; Lu, X. J.; Deng, L., *J. Am. Chem. Soc.* **2008**, *130*, 2422.
20. Chen, W.; Du, W.; Duan, Y. Z.; Wu, Y.; Yang, S. Y.; Chen, Y. C., *Angew. Chem., Int. Ed.* **2007**, *46*, 7667.
21. Liu, T. Y.; Cui, H. L.; Zhang, Y.; Jiang, K.; Du, W.; He, Z. Q.; Chen, Y. C., *Org. Lett.* **2007**, *9*, 3671.
22. (a) He, W.; Zhang, B. L.; Jiang, R.; Liu, P.; Sun, X. L.; Zhang, S. Y., *Tetrahedron Lett.* **2006**, *47*, 5367. (b) He, W.; Liu, P.; Zhang, B. L.; Sun, X. L.; Zhang, S. Y., *Appl. Organomet. Chem.* **2006**, *20*, 328.
23. For a review on 9-thiourea derivatives of cinchona alkaloids as bifunctional catalysts, see Connon, S. J., *Chem. Commun.* **2008**, 2499.
24. He, W.; Zhang, L. B.; Liu, P.; Sun, X. L.; Zhang, S. Y., *Cuihua Xuebao* **2006**, *27*, 527.
25. He, W.; Liu, P.; Nan, P. J.; Jin, Y.; Zhang, S. Y., *Huaxue Shiji* **2006**, *28*, 321.
26. Ramachary, D. B.; Venkaiah, C.; Madhavachary, R., *Org. Lett.* **2013**, *15*, 3042.
27. Ramachary, D. B.; Sakthidevi, R., *Org. Biomol. Chem.* **2010**, *8*, 4259.
28. Rodrigo, E.; Morales, S.; Duce, S.; Ruano, J. L. G.; Cid, M. B., *Chem. Commun.* **2011**, *47*, 11267.
29. Wang, Y.-Q.; Cui, X.-Y.; Ren, Y.-Y.; Zhang, Y., *Org. Biomol. Chem.* **2014**, *12*, 9101.
30. Ding, D.; Zhao, C.-G., *Eur. J. Org. Chem.* **2010**, 3802.
31. Cai, Y.-F.; Yang, H.-M.; Li, L.; Jiang, K.-Z.; Lai, G.-Q.; Jiang, J.-X.; Xu, L.-W., *Eur. J. Org. Chem.* **2010**, 4986.
32. Perera, S.; Naganaboina, V. K.; Wang, L.; Zhang, B.; Guo, Q.; Rout, L.; Zhao, C.-G., *Adv. Synth. Catal.* **2011**, *353*, 1729.
33. (a) Guo, Q.; Zhao, J. C.-G., *Tetrahedron Lett.* **2012**, *53* 1768; (b) Kumar, A.; Chimni, S. S., *Eur. J. Org. Chem.* **2013**, 4780.
34. (a) Liu, C.; Zhu, Q.; Huang, K.-W.; Lu, Y., *Org. Lett.* **2011**, *13*, 2638; (b) Desmarchelier, A.; Yalgin, H.; Coeffard, V.; Moreau, X.; Greck, C., *Tetrahedron Lett.* **2011**, *52*, 4430.
35. Lifchits, O.; Demoulin, N.; List, B., *Angew. Chem., Int. Ed.* **2011**, *50*, 9680.
36. Demoulin, N.; Lifchits, O.; List, B., *Tetrahedron* **2012**, *68*, 7568.
37. Bencivenni, G.; Galzerano, P.; Mazzanti, A.; Bartoli, G.; Melchiorre, P., *Proc. Natl. Acad. Sci. USA* **2010**, *107*, 20642.
38. Di Iorio, N.; Righi, P.; Mazzanti, A.; Mancinelli, M.; Ciogli, A.; Bencivenni, G., *J. Am. Chem. Soc.* **2014**, *136*, 10250.
39. Feng, X.; Zhou, Z.; Zhou, R.; Zhou, Q.-Q.; Dong, L.; Chen, Y.-C., *J. Am. Chem. Soc.* **2012**, *134*, 19942.
40. Jia, Z.-J.; Jiang, H.; Li, J.-L.; Gschwend, B.; Li, Q.-Z.; Yin, X.; Grouleff, J.; Chen, Y.-C.; Jørgensen, K. A., *J. Am. Chem. Soc.* **2011**, *133*, 5053.
41. Xiong, X.-F.; Zhou, Q.; Gu, J.; Dong, L.; Liu, T.-Y.; Chen, Y.-C., *Angew. Chem., Int. Ed.* **2012**, *51*, 4401.
42. Feng, X.; Zhou, Z.; Ma, C.; Yin, X.; Li, R.; Dong, L.; Chen, Y.-C., *Angew. Chem., Int. Ed.* **2013**, *52*, 14173.
43. Zhou, Z.; Feng, X.; Yin, X.; Chen, Y.-C., *Org. Lett.* **2014**, *16*, 2370.
44. Xiao, Y.-C.; Yue, C.-Z.; Chen, P.-Q.; Chen, Y.-C., *Org. Lett.* **2014**, *16*, 3208.
45. (a) Xie, J.-W.; Huang, X.; Fan, L.-P.; Xu, D.-C.; Li, X.-S.; Su, H.; Wen, Y.-H., *Adv. Synth. Catal.* **2009**, *351*, 3077; (b) Huang, X.; Li, P.; Li, X.-S.; Xu, D.-C.; Xie, J.-W., *Org. Biomol. Chem.* **2010**, *8*, 4527.
46. Sun, X.; Yu, F.; Ye, T.; Liang, X.; Ye, J., *Chem. Eur. J.* **2011**, *17*, 430.
47. Hack, D.; Chauhan, P.; Deckers, K.; Hermann, G. N.; Mertens, L.; Raabe, G.; Enders, D., *Org. Lett.* **2014**, *16*, 5188.
48. Moran, A.; Hamilton, A.; Bo, C.; Melchiorre, P., *J. Am. Chem. Soc.* **2013**, *135*, 9091.
49. Huang, J.; Zhao, L.; Liu, Y.; Cao, W.; Wu, X., *Org. Lett.* **2013**, *15*, 4338.
50. Lin, A.; Wang, J.; Mao, H.; Ge, H.; Tan, R.; Zhu, C.; Cheng, Y., *Org. Lett.* **2011**, *13*, 4176.
51. Sun, X.; Sengupta, S.; Petersen, J. L.; Wang, H.; Lewis, J. P.; Shi, X., *Org. Lett.* **2007**, *9*, 4495.
52. Ling, J.-B.; Wang, W.-P.; Xu, P.-F., *ChemCatChem* **2011**, *3*, 302.
53. Qiao, Z.; Shafiq, Z.; Liu, L.; Yu, Z.-B.; Zheng, Q.-Y.; Wang, D.; Chen, Y.-J., *Angew. Chem., Int. Ed.* **2010**, *49*, 7294.
54. Campbell, N. R.; Sun, B.; Singh, R. P.; Deng, L., *Adv. Synth. Catal.* **2011**, *353*, 3123.
55. (a) Luo, G.; Zhang, S.; Duan, W.; Wang, W., *Synthesis* **2009**, *2009*, 1564; (b) Gogoi, S.; Zhao, C.-G.; Ding, D., *Org. Lett.* **2009**, *11* 2249; (c) Ma, S.; Wu, L.; Liu, M.; Xu, X.; Huang, Y.; Wang, Y., *RSC Adv.* **2013**, *3* 11498; (d) Lee, S.-J.; Ahn, J.-G.; Cho, C.-W., *Tetrahedron: Asymmetry* **2014**, *25* 1383; (e) Li, P.; Fang, F.; Chen, J.; Wang, J., *Tetrahedron: Asymmetry* **2014**, *25*, 98.
56. Liu, J.-D.; Chen, Y.-C.; Zhang, G.-B.; Li, Z.-Q.; Chen, P.; Du, J.-Y.; Tu, Y.-Q.; Fan, C.-A., *Adv. Synth. Catal.* **2011**, *353*, 2721.
57. Zeng, C.; Liu, H.; Zhang, M.; Guo, J.; Jiang, S.; Yu, S., *Synlett* **2012**, 2251.
58. Tian, X.; Liu, Y.; Melchiorre, P., *Angew. Chem., Int. Ed.* **2012**, *51*, 6439.
59. Stiller, J.; Kowalczyk, D.; Jiang, H.; Jørgensen, K. A.; Albrecht, Ł., *Chem. Eur. J.* **2014**, *20*, 13108.
60. (a) Liang, J.; Chen, Q.; Liu, L.; Jiang, X.; Wang, R., *Org. Biomol. Chem.* **2013**, *11*, 1441; (b) Massolo, E.; Benaglia, M.; Annunziata, R.; Palmieri, A.; Celentano, G.; Forni, A., *Adv. Synth. Catal.* **2014**, *356*, 493.
61. (a) Li, B.-S.; Zhang, E.; Zhang, Q.-W.; Zhang, F.-M.; Tu, Y.-Q.; Cao, X.-P., *Chem. Asian J.* **2011**, *6*, 2269; (b) Liu, L. L.; Chiu, P., *Chem. Commun.* **2011**, *47* 3416; (c) Lifchits, O.; Mahlau, M.; Reisinger, C. M.; Lee, A.; Farès, C.; Polyak, I.; Gopakumar, G.; Thiel, W.; List, B., *J. Am. Chem. Soc.* **2013**, *135*, 6677.
62. Cruz, D. C.; Sanchez-Murcia, P. A.; Jørgensen, K. A., *Chem. Commun.* **2012**, *48*, 6112.

63. (a) Cai, Q.; Zheng, C.; Zhang, J.-W.; You, S.-L., *Angew. Chem., Int. Ed.* **2011**, *50*, 8665; (b) Cai, Q.; You, S.-L., *Org. Lett.* **2012**, *14*, 3040.
64. Desmarchelier, A.; Coeffard, V.; Moreau, X.; Greck, C., *Chem. Eur. J.* **2012**, *18*, 13222.
65. Kang, Y. K.; Kim, D. Y., *Adv. Synth. Catal.* **2013**, *355*, 3131.
66. He, X.-L.; Xiao, Y.-C.; Du, W.; Chen, Y.-C., *Chem. Eur. J.* **2015**, *21*, 3443.
67. Yin, X.; Zheng, Y.; Feng, X.; Jiang, K.; Wei, X.-Z.; Gao, N.; Chen, Y.-C., *Angew. Chem., Int. Ed.* **2014**, *53*, 6245.
68. Paradisi, E.; Righi, P.; Mazzanti, A.; Ranieri, S.; Bencivenni, G., *Chem. Commun.* **2012**, *48*, 11178.
69. Li, X.-M.; Wang, B.; Zhang, J.-M.; Yan, M., *Org. Lett.* **2011**, *13*, 374.
70. Song, C. E. ed. *Cinchona Alkaloids in Synthesis and Catalysis: Ligands, Immobilization and Organocatalysis*. Wiley-VCH: Weinheim, 2009; p 29.
71. (a) Zhang, W.; Zheng, S.; Liu, N.; Werness, J. B.; Guzei, I. A.; Tang, W., *J. Am. Chem. Soc.* **2010**, *132*, 3664; (b) Rong, Z.-Q.; Pan, H.-J.; Yan, H.-L.; Zhao, Y., *Org. Lett.* **2014**, *16* 208; (c) Phelan, J. P.; Patel, E. J.; Ellman, J. A., *Angew. Chem., Int. Ed.* **2014**, *53* 11329; (d) Oh, J.-S.; Lee, J.-W.; Ryu, T. H.; Lee, J. H.; Song, C. E., *Org. Biomol. Chem.* **2012**, *10* 1052; (e) Kong, S.; Fan, W.; Wu, G.; Miao, Z., *Angew. Chem., Int. Ed.* **2012**, *51* 8864; (f) Zhang, W.; Liu, N.; Schienebeck, C. M.; Decloux, K.; Zheng, S.; Werness, J. B.; Tang, W., *Chem. Eur. J.* **2012**, *18*, 7296.
72. (a) Wen, S.; Li, P.; Wu, H.; Yu, F.; Liang, X.; Ye, J., *Chem. Commun.* **2010**, *46*, 4806; (b) Chen, J.-R.; Cao, Y.-J.; Zou, Y.-Q.; Tan, F.; Fu, L.; Zhu, X.-Y.; Xiao, W.-J., *Org. Biomol. Chem.* **2010**, *8* 1275; (c) Jin, Z.; Xu, J.; Yang, S.; Song, B.-A.; Chi, Y. R., *Angew. Chem., Int. Ed.* **2013**, *52*, 12354.
73. (a) Zhu, Q.; Lu, Y., *Angew. Chem., Int. Ed.* **2010**, *49*, 7753; (b) Dou, X.; Lu, Y., *Chem. Eur. J.* **2012**, *18*, 8315.
74. (a) Diosdado, S.; Etxabe, J.; Izquierdo, J.; Landa, A.; Mielgo, A.; Olaizola, I.; López, R.; Palomo, C., *Angew. Chem., Int. Ed.* **2013**, *52*, 11846; (b) Diosdado, S.; López, R.; Palomo, C., *Chem. Eur. J.* **2014**, *20*, 6526.
75. (a) Yang, W.; Du, D.-M., *Org. Lett.* **2010**, *12*, 5450; (b) Badiola, E.; Fiser, B.; Gómez-Bengoa, E.; Mielgo, A.; Olaizola, I.; Urruzuno, I.; García, J. M.; Odriozola, J. M.; Razkin, J.; Oiarbide, M.; Palomo, C., *J. Am. Chem. Soc.* **2014**, *136* 17869; (c) Manoni, F.; Connon, S. J., *Angew. Chem., Int. Ed.* **2014**, *53* 2628; (d) Zhao, M.-X.; Bi, H.-L.; Jiang, R.-H.; Xu, X.-W.; Shi, M., *Org. Lett.* **2014**, *16* 4566; (e) Bera, K.; Namboothiri, I. N. N., *J. Org. Chem.* **2015**, *80* 1402.
76. (a) Oh, S. H.; Rho, H. S.; Lee, J. W.; Lee, J. E.; Youk, S. H.; Chin, J.; Song, C. E., *Angew. Chem., Int. Ed.* **2008**, *47*, 7872; (b) Youk, S. H.; Oh, S. H.; Rho, H. S.; Lee, J. E.; Lee, J. W.; Song, C. E., *Chem. Commun.* **2009**, 2220; (c) Park, S. E.; Nam, E. H.; Jang, H. B.; Oh, J. S.; Some, S.; Lee, Y. S.; Song, C. E., *Adv. Synth. Catal.* **2010**, *352* 2211; (d) Nakamura, S.; Takahashi, S.; Nakane, D.; Masuda, H., *Org. Lett.* **2015**, *17*, 106.
77. Wang, B.; Liu, Y.; Sun, C.; Wei, Z.; Cao, J.; Liang, D.; Lin, Y.; Duan, H., *Org. Lett.* **2014**, *16*, 6432.
78. (a) Wei, Y.; Yao, L.; Zhang, B.; He, W.; Zhang, S., *Tetrahedron* **2011**, *67*, 8552; (b) Shao, P.-L.; Liao, J.-Y.; Ho, Y. A.; Zhao, Y., *Angew. Chem., Int. Ed.* **2014**, *53*, 5435.
79. (a) Lee, A.; Michrowska, A.; Sulzer-Mosse, S.; List, B., *Angew. Chem., Int. Ed.* **2011**, *50*, 1707; (b) Lee, A.; Reisinger, C. M.; List, B., *Adv. Synth. Catal.* **2012**, *354*, 1701.

(1R,2S)-1-Amino-2,3-dihydro-1H-inden-2-ol[1]

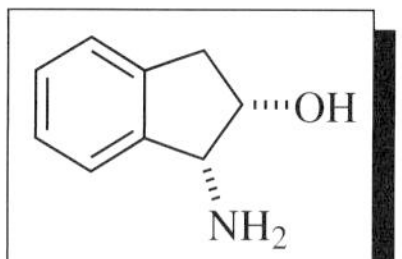

1

[136030-00-7] $C_9H_{11}NO$ (MW 149.19)

InChI = 1S/C9H11NO/c10-9-7-4-2-1-3-6(7)5-8(9)11/h1-4,8-9,11H,5,10H2/t8-,9+/m0/s1

InChIKey = LOPKSXMQWBYUOI-DTWKUNHWSA-N

(synthetic building block used in pharmaceutical compounds, and as an asymmetric control element in chiral auxiliaries and asymmetric catalysts)

Physical Data: mp 122–124 °C.

Solubility: soluble in ethanol, isopropanol, dichloromethane, and toluene (hot).

Form Supplied in: colorless crystalline solid; commercially available in either enantiomeric form.

Analysis of Reagent Purity: NMR/CHN analysis. Chiral HPLC to confirm ee.

Preparative Method: several routes from indene. See main text for details.

Purification: recrystallization from toluene.[2d]

Handling, Storage, and Precautions: relatively air- and moisture-stable, colorless, and odorless crystalline solid, irritating to skin, eyes, and respiratory system.

Original Commentary

Martin Wills & Richard Eaves
University of Warwick, Coventry, UK

Preparation. A number of methods have been reported for both the racemic and asymmetric preparations of 1-amino-2,3-dihydro-1*H*-inden-2-ol **1**, most commonly starting from inexpensive and readily available indene. These methods have been described in detail in recent reviews.[1] The valuable properties of **1** as both a component of a medicinally active compound and as a chirality control element, derive primarily from its rigid and well-defined stereochemical structure. As a result, the compound is most desirable in enantiomerically pure form. One of the most efficient asymmetric syntheses of **1**, which may be employed for the synthesis of either enantiomer of the target molecule, involves an asymmetric epoxidation (89% yield, 88% ee) of indene to give epoxide **2** using the well-established Jacobsen catalyst. This is followed by a Ritter reaction using oleum in acetonitrile resulting in conversion to the oxazoline **3** which is subsequently hydrolysed to the amino alcohol. Fractional crystallization with a homochiral diacid permits purification to > 99% ee (eq 1).[2]

The enantioselective synthesis of **1** has also been achieved by a number of methods including enzymatic resolution of a keto ester precursor to the racemate followed by conversion of the ester to an amino group,[3a] enzymatic resolution of an amino azide precursor followed by reduction,[4] enzymatic resolution through *O*-acylation of a racemic *N*-benzylcarbamate derivative of **1**,[3b] and the resolution via the formation of an amide with a homochiral amino acid.[5] Bioconversion of indene to *trans-2S,1S*-bromoindanol furnishes a key intermediate towards the synthesis of **1**.[6a,6b] Desymmetrization of 2-TBS-protected indanol through an enantioselective oxidation provides access to a ketone precursor of **1** in up to 70% ee.[6c]

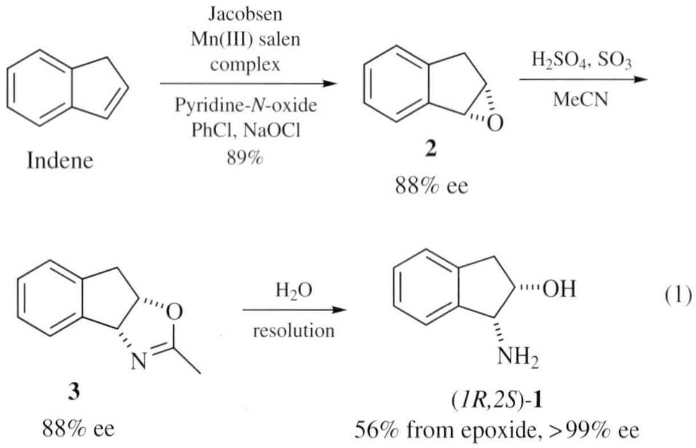

Indene; **2**, 88% ee; **3**, 88% ee; (*1R,2S*)-**1**, 56% from epoxide, >99% ee (1)

Application as a Synthetic Building Block in Pharmaceutical Compounds. The best-known application of the (1*S*,2*R*)-enantiomer of *cis*-aminoindanol is as a component of Indinavir **4**, the primary component of a Crixivan® combination therapy (with other reverse transcriptase inhibitors) for AIDS.[7] An excellent account of the synthetic approach to Indinavir, as well as the use of **1** in other drugs, can be found in a recent review.[1a]

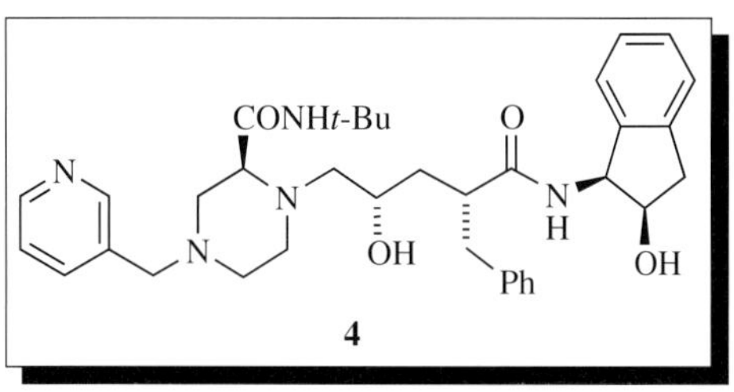

4

Application as a Component of a Chiral Auxiliary. The rigid structure and well-defined conformational rigidity of **1** makes it an ideal building block for a chiral auxiliary. Three different types have been described in some detail. The 'Evans-auxiliary'-type oxazolidinone derivative **5** has given excellent results in aldol reactions with aldehydes (eq 2).[8] The reaction illustrated, proceeding via a boron enolate, is selective for the syn diastereoisomer of product, i.e. **6**, often with > 99% de. Following the reaction, the aldol product can be removed from the auxiliary using lithium hydroxide in a water/THF mixture.

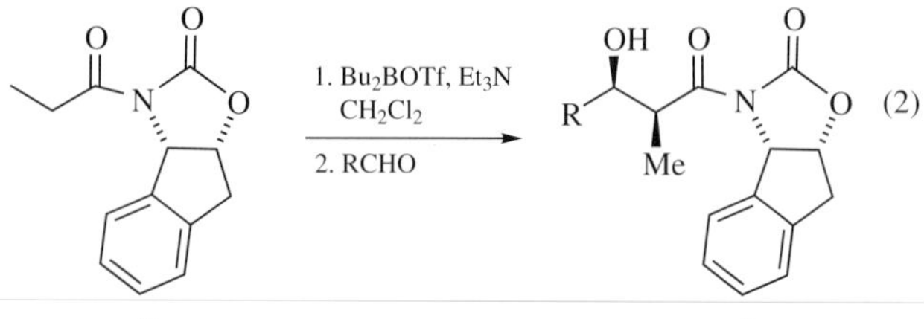

5 **6** (2)

The aldol reaction illustrated in eq 2 has been applied to the targeted synthesis of a number of complex molecules including Tylosin,[8] Hapalosin,[9] the antibiotic Sinefungin,[10] and the HIV protease Saquinavir® inhibitor.[11] Oxazolidinone-type chiral auxiliaries derived from **1** have also been employed for the control of Diels–Alder reactions of attached acryloyl or crotonyl groups.[12]

Asymmetric aldol reactions may also be controlled with high diasteroselectivity, but this time for the anti isomer, in reactions of *N*-tosyl derivatives of esters derived from **7** (eq 3).[13] Diastereoselectivities of up to 99:1 were achieved in the illustrated titanium (IV)-mediated reaction, which has been employed for the synthesis of dipeptide isosteres for incorporation into pharmaceutical building blocks.[14] The selectivity reverses when α- or β- alkoxy aldehydes are employed as electrophiles.

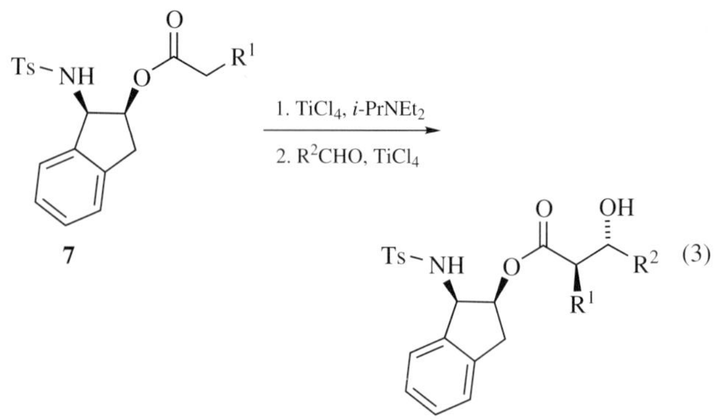

7 (3)

The *N*-tosyl class of auxiliaries derived from **1** have also been successfully applied to the diastereocontrol of Diels–Alder reactions[15,7] and the selective reduction of attached α-keto esters to furnish α-hydroxy ester products.[16]

A third class of chiral auxiliary derived from **1** contains a bridging isopropylidene group between the oxygen and nitrogen atoms, the removable group being appended to nitrogen. This class of auxiliary has been employed in homoaldol reactions via Zn(II) species,[17] to the stereocontrol (in several cases >99% de) of [2,3]-sigmatropic rearrangements[18] and, in the example illustrated in eq 4, the asymmetric synthesis of amino acids through electrophilic amination of attached copper(I) enolates.[19] α-Amino acids may also be prepared through the diastereoselective alkylation of glycine derivatives of the same auxiliary.[20] Addition of organometallic reagents to α-keto amides derived from the same auxiliary provides a means for the asymmetric synthesis of α,α-disubstituted-α-hydroxy acids with excellent enantioselectivity.[21]

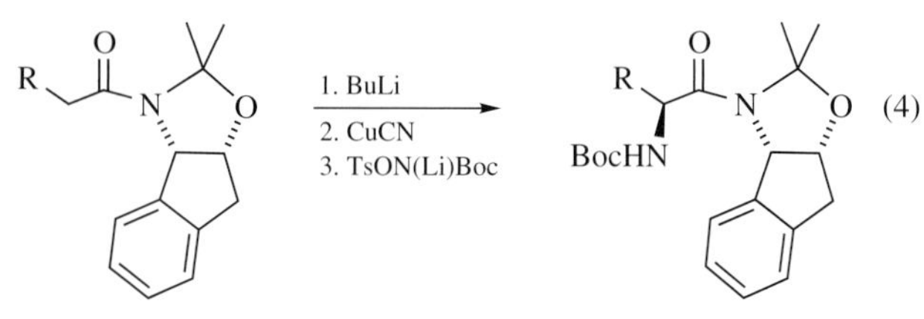

(4)

In a recent application, *cis*-aminoindanol has been employed as a rigid diastereocontrol element in the alkylation of bicyclic lactams and thiolactams of which they are a component.[22] The resulting products form the basis of an enantioselective synthesis of alkaloids.

Application as a Component of an Asymmetric Catalyst. Amino alcohol (**1**) has proven to be a highly versatile ligand for use in asymmetric catalysts for a series of reactions.[1] One of the most comprehensively studied uses it as an oxazaborolidine derivative such as **8** for the asymmetric control of the reduction of ketones by borane. Although its use was first described with stoichiometric levels of **1** being employed for the reduction of both ketones and oximes,[3] development of the system has delivered a catalytic method requiring only 5–10 mol % catalyst.[23] Enantiomeric excesses of over 85% and as high as 96% have been achieved for a range of ketone substrates. α-Chloro and α-bromo ketones are particularly excellent substrates; the reaction in eq 5 is the key step in a highly efficient asymmetric synthesis of the asthma drug (*R,R*)-formoterol[24b,c] and the histamine receptor antagonist Fexofenadine.[24d]

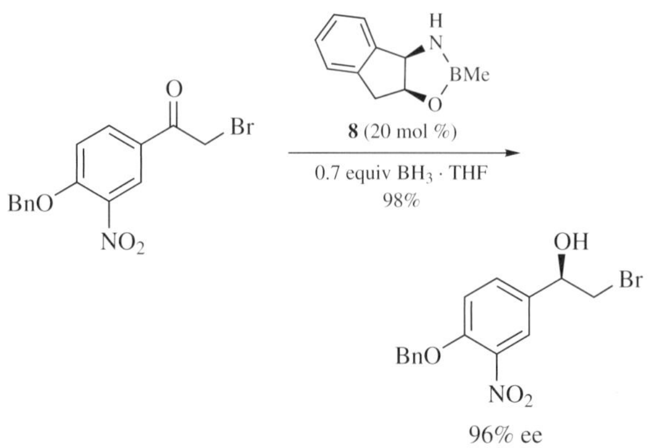

(5)

N,N-Dialkyl derivatives of **1** have been successfully applied to the asymmetric addition of dialkylzinc reagents to aldehydes, giving products of moderate enantiomeric excess.[23] In addition, ruthenium(II) complexes of **1** have been demonstrated to be excellent catalysts for the control of the enantioselective transfer hydrogenation of ketones to alcohols at catalyst loadings as low as 1 mol %.[25] The ruthenium/**1** complex has been applied to a range of ketone substrates, including cyclic enones and α-amino and alkoxy substituted derivatives.

Metal complexes of bis-oxazoline derivatives **9** of **1** have been employed for the asymmetric catalysis of the Diels–Alder reactions of acryloyl-*N*-oxazolines with dienes. Detailed studies have been carried out into the effect of the bite angle of the ligand and the nature of the bridging group, on the efficiency of the reaction.[26] In the reaction shown in eq 6, the copper(II) complex of the six-membered chelate ligand catalyzes the addition reaction to give a product in 94% yield and 98% ee at a loading of only 8 mol %. Use of the same ligand with magnesium(II) in place of copper(II) resulted in a reversal of the enantioselectivity, an effect which has been rationalized by a change in coordination at the metal from square planar to tetrahedral. Hetero Diels–Alder reactions have also been achieved using metal complexes of bis-oxazolines derived from **1**.[27] In addition, magnesium(II) complexes of the bis-oxazolines act as effective asymmetric control elements for the asymmetric conjugate [1,4] addition of free-radicals to oxazolidinone-bound cinnamates.[28] Copper(I) complexes of **9** have been employed for the control of carbenoid insertions into silicon–hydrogen bonds.[29]

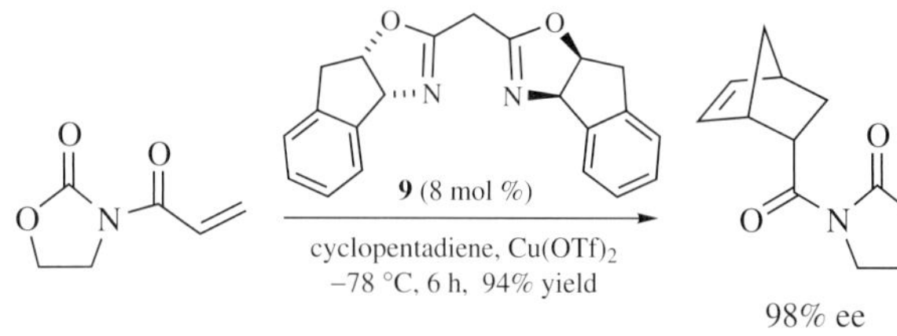

(6)

Bis-oxazoline ligands bearing a bridging 2,6-pyridine group (often referred to as 'pybox' ligands) have been employed for the asymmetric catalysis of alkene cyclopropanations, giving products in good enantioselectivity and diastereoselectivity at very low (0.2 mol % metal) catalyst loadings.[26] However, ligands derived from certain other 1,2-amino alcohols gave superior results.

In a detailed study, bis-oxazoline **9** has been employed for the enantiocontrol of a palladium-catalyzed annulation of allenes with aryl and vinylic iodides. This procedure provided an efficient means for the asymmetric synthesis of several classes of heterocyclic target structures including indoles, cyclic ethers, and lactones.[30] Bis-oxazoline **9**, and certain derivatives bearing different bridging groups, have been employed in the copper-catalyzed allylic acyloxylation reaction of cyclic alkenes. Enantiomeric excesses of up to 78% were achieved using the most efficient catalyst.[31]

Tridentate salen ligands **10** derived from **1** have given excellent results in the enantiocontrol of the hetero Diels–Alder addition reaction of dienes with aldehydes (eq 7)[32] and in the asymmetric additions of TMS-azide to *meso*-epoxide[33] and trimethylsilyl cyanide to benzaldehyde (up to 85% ee).[34] Phosphino-oxazolines derived from **1** have been employed for the asymmetric control of palladium-catalyzed allylic substitution reactions; products of 70–90% ee were obtained.[35] Photolysis of crystalline adducts of enantiomerically pure **1** with prochiral alcohols results in asymmetric inductions of up to 79% in a rare example of a solid-state enantioselective reaction.[36]

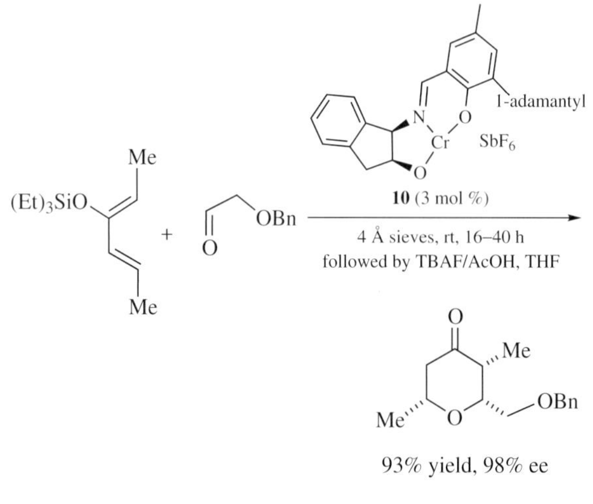

(7)

Related Reagents. (*1S,2R*) enantiomer [26456-43-7].

First Update

David M. Tschaen & Rebecca T. Ruck
Merck & Co., Inc., Rahway, NJ, USA

Since the original EROS report, a comprehensive review of cis-aminoindanol by Gallou and Senanayake appeared in 2006.[37,38]

Preparation of cis-Aminoindanol. Several reports describing alternative methods to prepare *cis*-aminoindanol have appeared in the literature.[39] Biocatalytic approaches starting from indene have been developed and were summarized in a mini review in 2002.[40] One enzymatic approach that converts 2-acetoxyindanone to *cis*-aminoindanol via a tandem lipase/transaminase sequence was reported.[41] A more recent report involving biocatalytic methodology applied a continuous flow kinetic resolution of racemic *cis*-aminoindanol using immobilized Novozyme 435.[42] An efficient classical resolution procedure using 2-phenylpropionic acid as resolving agent was developed as an alternative to the previously described tartrate salt resolution.[43]

Application as a Building Block in Pharmaceutical Compounds. In addition to the well-known application in Crixivan,[7] cis-aminoindanol has been incorporated into other active HIV-1 protease inhibitors such as C_2-symmetric substrate **11**.[44]

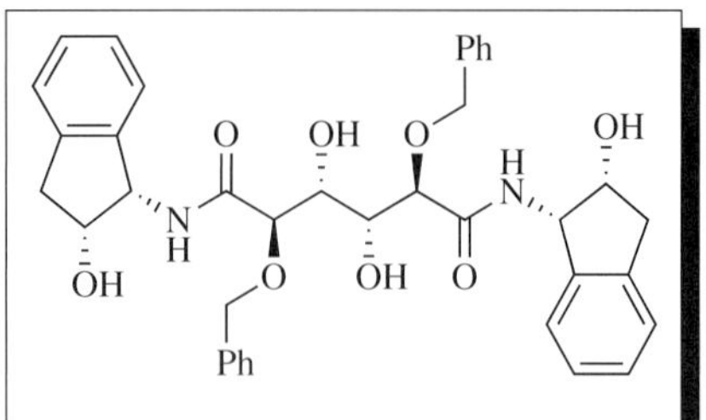

11

Aminoindanol has also been utilized as an important component in other compounds of pharmaceutical interest, including azapurine derivatives such as compound **12** (eq 8) that showed moderate inhibition of BVDV.[45]

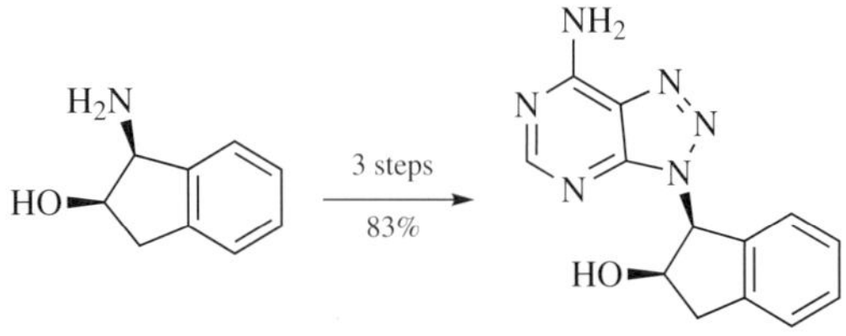

12

(8)

A series of aggrecanase inhibitors similar to the structures **13** and **14** for the potential treatment of OA also contain the cis-aminoindanol pharmacophore.[46,47]

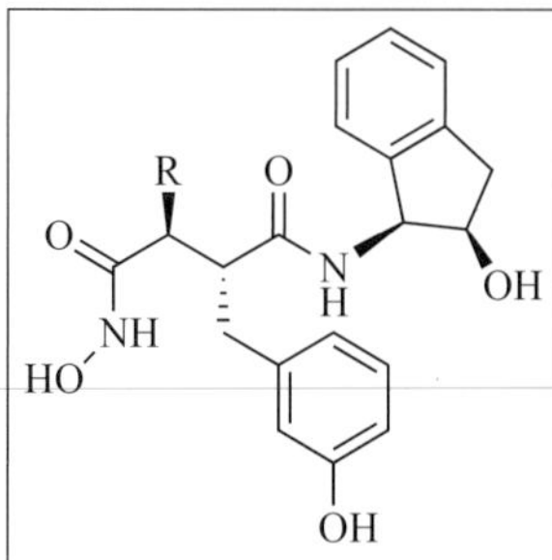

13 R = H

14 R = NH

Application as a Resolving Agent. Derivatization of racemic carboxylic acids using aminoindanol was utilized as an effective method for resolving the carboxylic acids.[48] A number of cyclopropene carboxylic acids were converted to the corresponding oxazolidinones and efficiently separated via simple flash chromatography. Reduction of the *N*-acyloxazolidinones provided the enantiomerically pure hydroxymethylcyclopropene analogs (eq 9).[49]

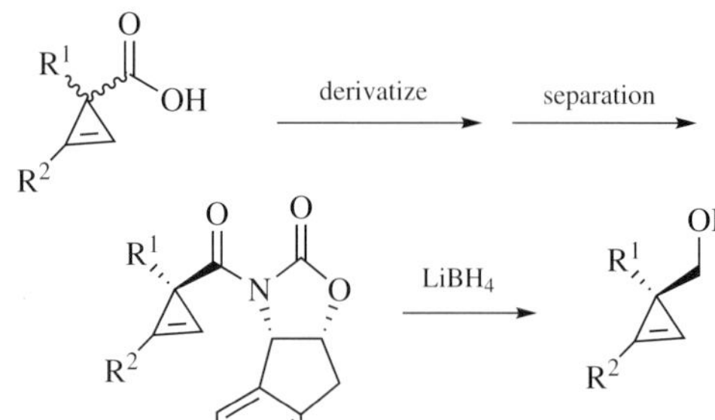

85% yield
>99% ee

(9)

Application as a Chiral Solvating Agent. Investigation of cis-aminoindanol as an effective chiral solvating agent has been reported.[50–52] The enantiomeric recognition of a series of carboxylic acids, including 2-chloromandelic acid, was evaluated by studying the corresponding ^{1}H NMR spectra in the presence of amino alcohols such as cis-aminoindanol.

Application as a Component of a Chiral Auxiliary. Since the original review, new applications of aminoindanol as a chiral auxiliary to effect stereoselective alkylations and Michael additions have been reported. One interesting report described the synthesis of an endothelin receptor antagonist that utilized aminoindanol as a chiral auxiliary to facilitate the stereoselective alkylation of the corresponding amide enolate. Aminoindanol was also used as an auxiliary, in the form of an adjacent *N,O*-acetal, to control the stereoselectivity (96:4 dr) of a conjugate addition (eq 10).[53]

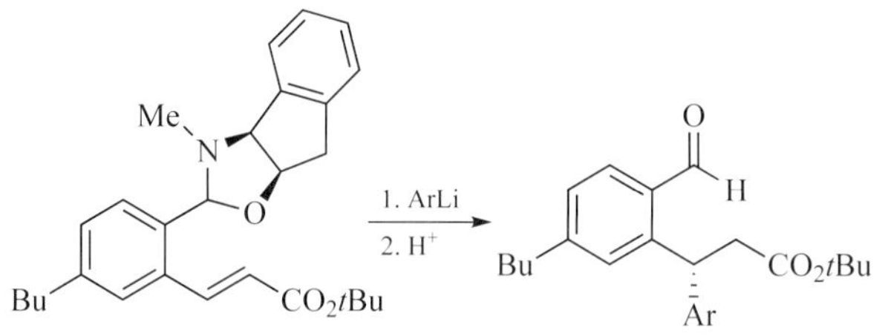

92% yield (10)
96:4 dr

Another example of the use as a chiral auxiliary involved a stereoselective approach to prepare 3-aryl δ-lactones.[54] The *N*-methyl aminoindanol-derived amide enolate underwent Michael addition to an unsaturated ester to afford the product in 94:6 dr (eq 11). Interestingly, the amide derived from the bridging isopropylidine *N,O*-acetal (similar to that shown in eq 4) provided the product with low selectivity (54:46).

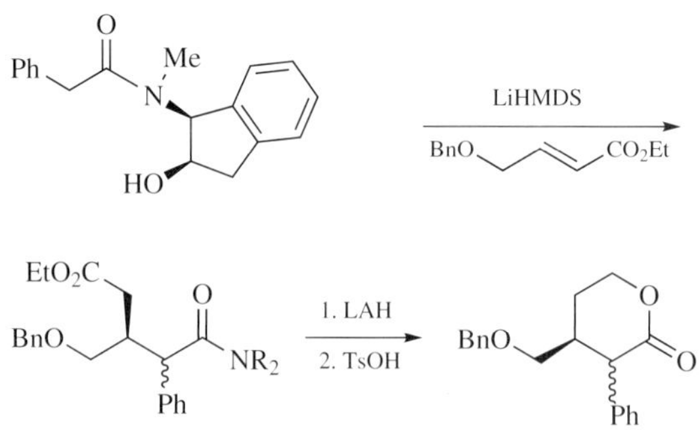

(11)

94:6 dr

Aminoindanol was previously reported as an ester or amide auxiliary on dienophiles to facilitate stereoselective Diels–Alder reactions.[12] More recently, it was described as an auxiliary on the diene component where it provided the product in excellent stereoselectivity.[55] The auxiliary was removed to prepare compounds such as ketone **15** with excellent ee values (eq 12).

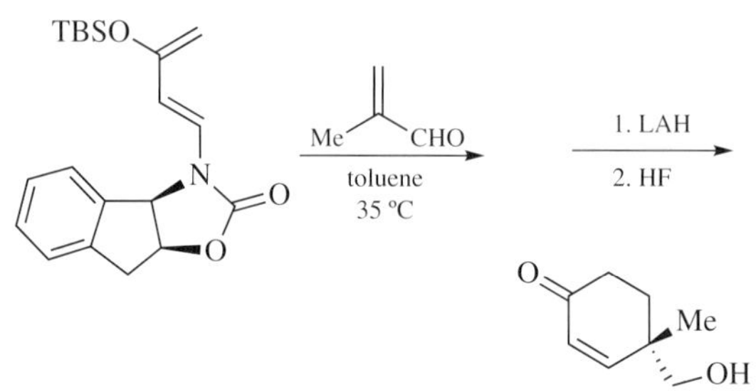

(12)

15
49% yield
96% ee

Aminoindanol was also reported as a chiral auxiliary in a number of stereoselective Claisen rearrangements.[56–58] One investigation described the impact of substituents on the aminoindanol moiety with diastereomeric ratios ranging from moderate (3:1) to excellent (40:1) depending on the substituents X and R (eq 13). This type of rearrangement was also applied as the key transformation in a formal synthesis of epiuleine.

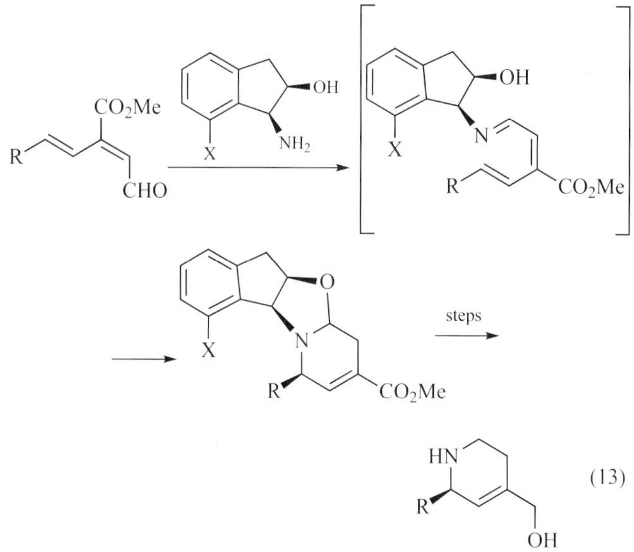

(13)

55–69% yield
Depending on R and X

An interesting example of a Ficini–Claisen rearrangement using aminoindanol as an auxiliary was reported to proceed in moderate stereoselectivity.[59] A recent report described a novel, stereoselective synthesis of dihydropyridoindoles via reaction of a nitrone with an electron-deficient allene that featured cis-aminoindanol as an auxiliary (eq 14).[60]

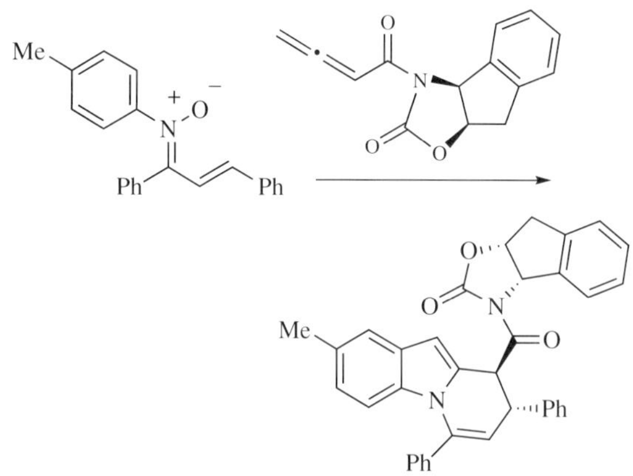

(14)

72% yield
20:1 dr

The preparation of enantiopure sulfonamides using *cis*-aminoindanol as an auxiliary was reported by Senanayake and coworkers.[61] The oxathiazolidine oxide **16** was prepared in excellent *endo*/*exo* selectivity (97:3) via treatment of the *N*-sulfonylated aminoindanol with thionyl chloride and 3,5-lutidine in THF (eq 15). Addition of various organometallic reagents (RM) to compound **16**, followed by treatment with lithium amide in ammonia, led to the S–O bond cleavage and formation of the sulfinamides **17** with excellent ee values.

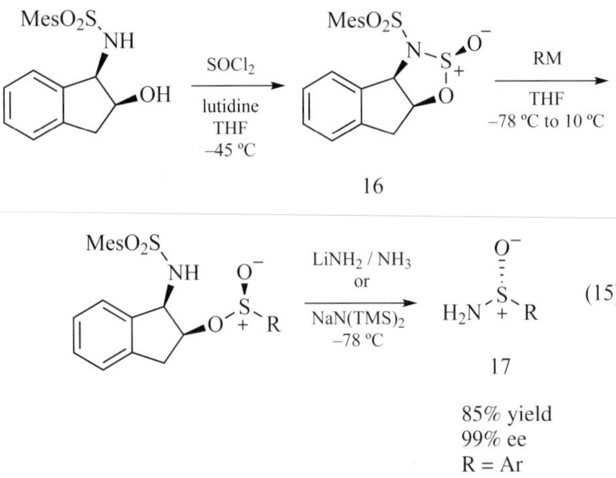

(15)

In a somewhat analogous approach, aminoindanol was utilized as an effective chiral auxiliary for the stereoselective preparation of *P*-stereogenic aminophosphines.[62] The condensation of aminoindanol with *tert*-butyl phosphine derivatives led to the formation of *tert*-butyl-oxazaphospholidine **18**. Subsequent treatment of compound **18** with organometallic reagents produced the ring-opened intermediates **19** with inversion of configuration at the phosphorus center. Subsequent treatment with Li/NH_3 to promote reductive cleavage produced the desired aminophosphines in good yields and high ee values (eq 16).

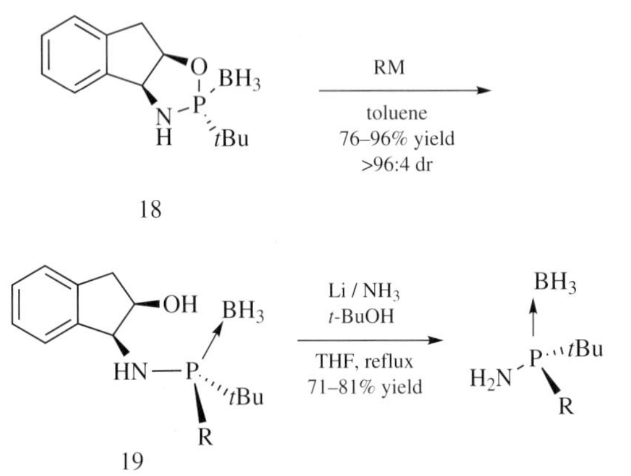

(16)

There are interesting reports of aminoindanol functioning as a chiral auxiliary in solid-state transformations.[63,64] Heating a diastereomeric mixture of compound **20** (83:17) in the solid state at 65 °C for 24 h led to the formation of a diastereomerically pure sample (eq 17). Monitoring this thermal epimerization carefully indicated that no melting was observed during this experiment. This result is in contrast to the result of stirring the mixture in a DMSO solution at room temperature that led to a 1:1 mixture of diastereomers.

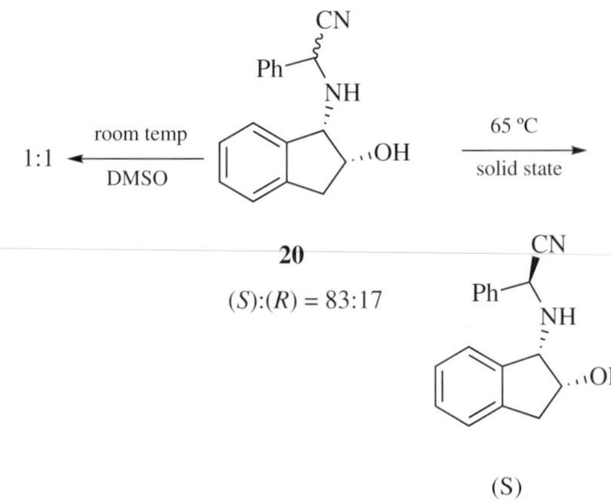

(17)

Application in the Stereoselective Synthesis of Bicyclic Morpholines. A recent report described the use of cis-aminoindanol as a component of a SnAP reagent (Stannyl Amine Protocol). The SnAP reagent **21**, derived from *cis*-aminoindanol, was coupled with aldehydes to provide a series of bicyclic morpholines in moderate to good yields and high dr values (>20:1) (eq 18).[65] Ketones were also found to be viable coupling partners albeit with diminished yields of the corresponding products observed.

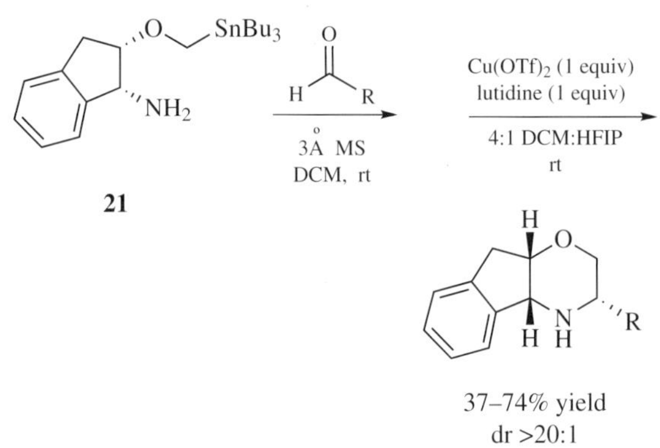

(18)

Organometallic Reactions (with cis-Aminoindanol as All or Part of the Ligand).

Aminoindanol as the Ligand. The use of aminoindanol as a chiral ligand in the titanium-mediated stereoselective oxidation of sulfur was described as a key step in the asymmetric synthesis of (*S*)-tenatoprazole.[66] Utilizing 0.5 equiv of Ti(O*i*Pr)$_4$ and 1 equiv of *cis*-aminoindanol, along with 2.3 equiv of cumene hydroperoxide (CHP) as the stoichiometric oxidant, provided the desired sulfoxide product in high yield (90%) and with excellent ee (99%) (eq 19). Of the various aminoalcohols screened for this reaction, *cis*-aminoindanol provided the best enantio- and chemoselectivities. In comparison, the enantioselectivity using *trans*-aminoindanol was 10% lower, and the conversion was reduced by 20%.

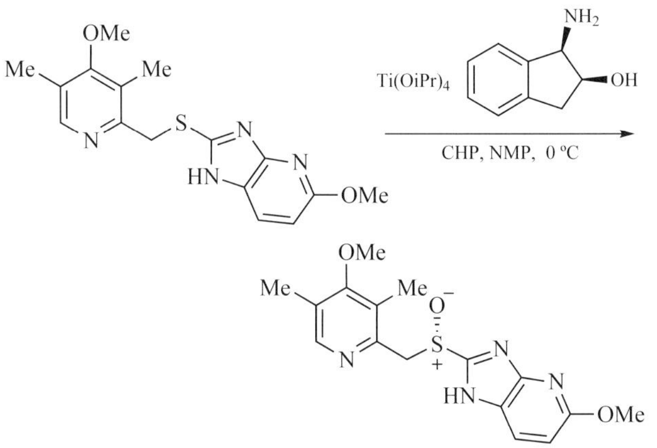

90% yield
99% ee (19)

Aminoindanol has been used as a ligand in asymmetric reductions. Employment of a chiral oxazaborolidine was further studied for utility[67,68] and applications[69] in the asymmetric reduction of ketones and was extended to the desymmetrization of *meso*-imides[70–72] and pyrrolidine-2,5-diones.[73] The parent aminoalcohol has also served as a competent ligand in the asymmetric transfer hydrogenations of carbonyl compounds and imines. The mechanism of catalysis using an aminoindanol–rhodium complex was studied in detail for the transfer hydrogenation of acetophenone.[74] The relevant ruthenium complexes were effective in the kinetic resolution of secondary alcohols[75] due to their ability to oxidize alcohols to the corresponding ketones.[76] Deuterated benzyl alcohols were also accessed via this methodology.[77] The mechanism of this transformation was studied in detail in order to render applicable to industrial applications.[78] This same catalytic system was applied to the asymmetric reduction of imine substrates. High diastereoselectivities were achieved for the transfer hydrogenation of the *tert*-butylsulfinylimine of acetophenone and similar ketimines, with protected products readily hydrolyzed to the corresponding amines (eq 20).[79,80] This methodology was extended to the enantioselective reduction of trifluoromethyl aryl PMP-protected ketimines.[81]

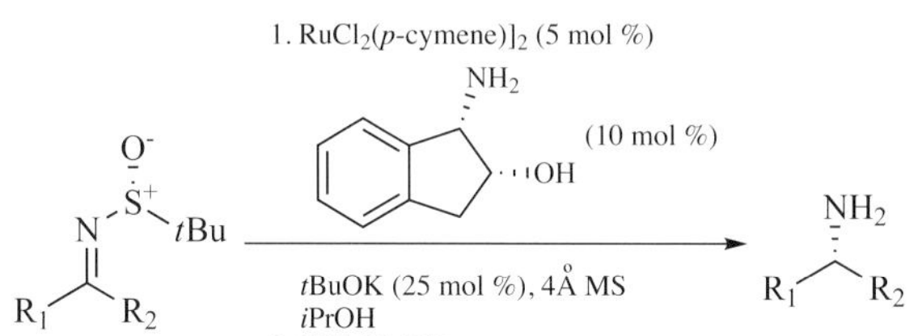

(20)

81–99% yield
69–99% ee

Tridentate Schiff Base Aminoindanol-derived Ligands. Additional publications utilizing tridentate Schiff base ligands derived from aminoindanol have been reported. Chromium(III) complexes[82] were employed in diastereoselective[83] and inverse electron demand[84,85] hetero-Diels–Alder reactions, quinone Diels–Alder reactions,[86,87] and hetero-ene reactions (eq 21).[88–90] Oxidations using vanadium(V) complexes have enabled the commercial scale production of enantiopure *tert*-butylsulfinamide (eq 22)[91] and an asymmetric synthesis of functionalized tetrahydrofurans.[92] Diethylzinc, in the presence of catalytic quantities of this class of ligand, was shown to effect enantio- and diasteroselective carbometallation reactions.[93,94]

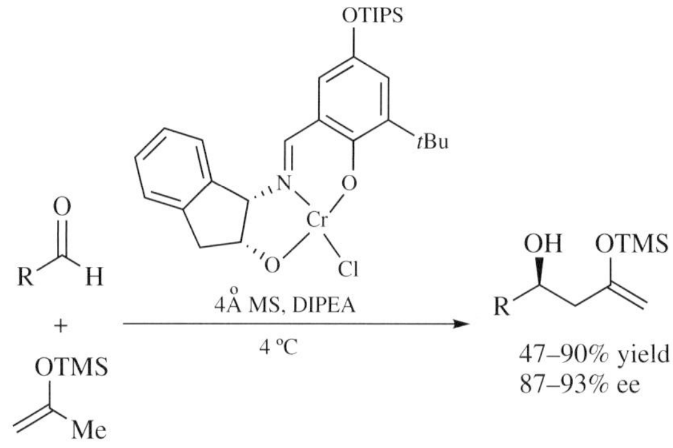

47–90% yield
87–93% ee (21)

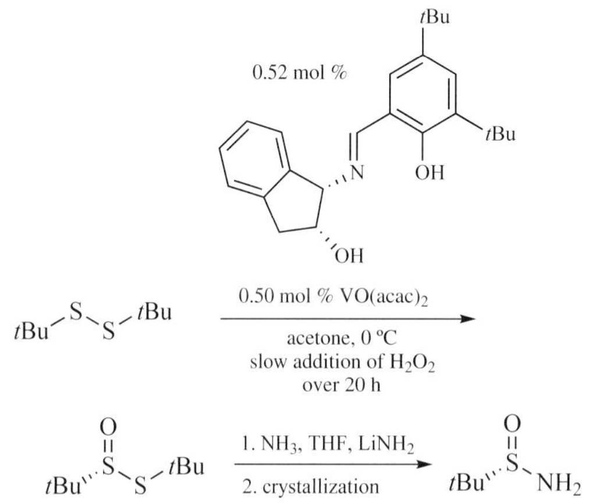

(22)

65% yield
99% ee

Indanyl Bisoxazoline (BOX) Complexes. Oxazoline-based ligands derived from aminoindanol bound to a series of metals have become commonplace in asymmetric catalysis. Bisoxazoline (BOX) ligands represent a workhorse in the field, with the Inda-BOX ligand rising to prominence. The parent system **22a** catalyzed a highly enantio- and diastereoselective Friedel–Crafts-type alkylation of indoles onto β,γ-unsaturated α-ketoesters to generate products that go on to cyclize to potential intermediates in the synthesis of complex natural alkaloids (eq 23).[95] The cyclopropyl-functionalized Inda-Box **22b** has been used in combination with $Cu(OTf)_2$ or MgX_2 to effect Lewis acid-catalyzed asymmetric conjugate addition and cycloaddition reactions. The first such result involved copper-catalyzed conjugate addition of 1,3-dicarbonyl compounds onto nitroalkenes; the second generation version used the magnesium complex to provide critical intermediates toward the synthesis of 4-substituted pyrrolidines such as clinical candidate ABT-546 and rolipram.[96,97] [3 + 2]-cycloaddition was achieved between α,β-unsaturated pyrazolidinone imides and nitrones using the copper–Inda-BOX complex **22b** to afford the corresponding *exo* products in high enantiosselectivities (eq 24).[98] This copper–Inda-BOX complex also catalyzed [3 + 2] cycloadditions between the same pyrazolidinone starting materials and azomethine imines.[99] Asymmetric Nazarov cyclizations using stoichiometric Cu–Inda-BOX provided the architecture of enokipodin B.[100,101] An oxidative cyclization between malonates and indolmethyl C–H bonds, in the presence of

Cu–Inda-BOX **22b** and DDQ, provided entry into complex indole structures.[102]

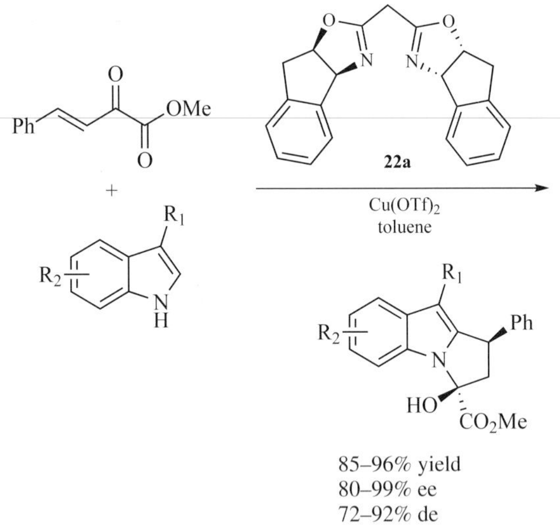

(23)

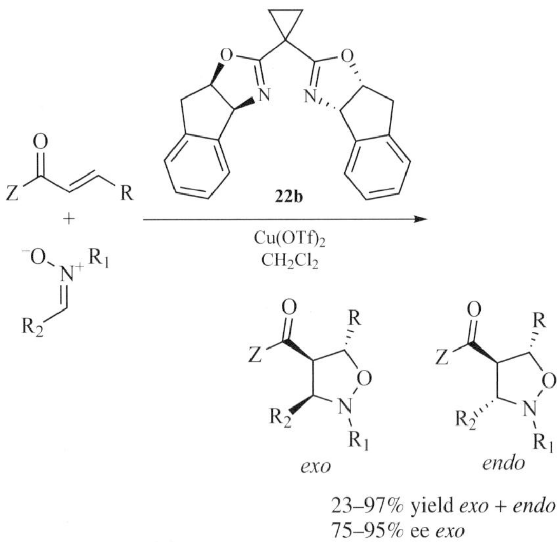

(24)

The corresponding magnesium complex of ligand **22b** has demonstrated similar catalysis for Lewis acid-mediated reactions. In the presence of this catalyst, *N*-benzylhydroxylamine added 1,4 to pyrrolidinone-derived enoates.[103,104] A similar protocol was employed to generate enantioenriched β-amino acids.[105,106] Pyrazolidinone crotonate derivatives underwent [3 + 2]-cycloadditions with nitrile oxides to provide the *exo*-dihydroisoxazoline products in high enantioselectivities.[107] Nitrile imines (masked as hydrazonyl bromides) and oxazolidinone crotonates (eq 25),[108] α,β-unsaturated pyrazolidinone imides[109] or methyleneindolines[110] reacted to afford highly enantioenriched dihydropyrazoles. Mg–Inda-BOX catalyzed the enantioselective [3 + 2]-cycloaddition between pyrazolidinone crotonates and ethyl diazoacetate to afford pyrazolines bearing quaternary *tert*-alkyl amino stereocenters and was applied to the asymmetric syntheses of manzacidins A and C.[111] Diels–Alder reactions, including kinetic resolution of chiral pyrazolidinones, were similarly achieved with this catalyst system.[112–114] α''-Hydroxyenones underwent Mg–Inda-BOX-catalyzed addition of nitromethane to generate precursors to γ-nitro carboxylic acids.[115] Methylenecyclopropanes reacted with *N*-tosyl aromatic aldimines in the presence of Mg–Inda-BOX to provide enantioenriched pyrrolidines (eq 26).[116]

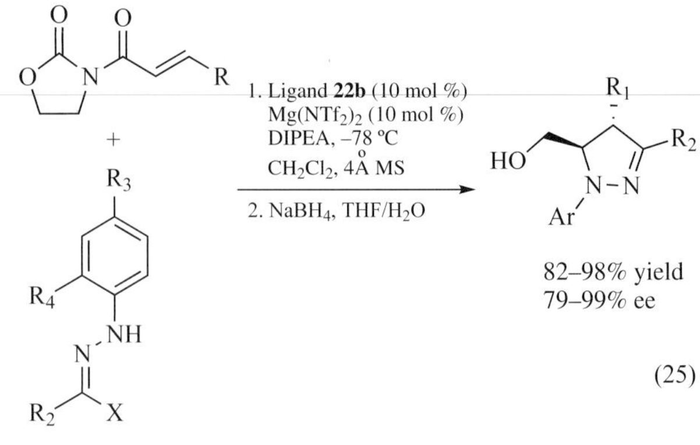

(25)

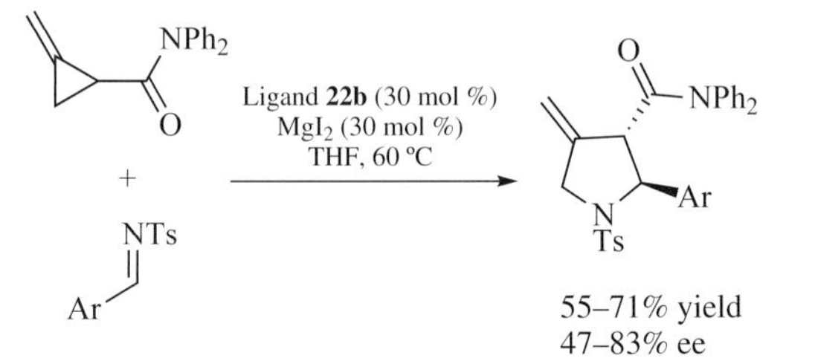

(26)

The same Lewis acid Inda-BOX **22b** catalysts have been used in combination with radical sources in conjugate addition reactions. Alkyl halides underwent radical [1,4]-addition to α,β-unsaturated oxazolidinones[117,118] and imides[119] with concomitant trapping by allyl(tributyl)tin. The system was subsequently optimized to utilize a silane as radical source instead of tin.[120] Use of 2-iodoethanol or 3-iodopropanol as the haloalkane resulted in the synthesis of chiral 6- and 7-membered ring lactones.[121] Chiral γ-lactams were synthesized by a radical addition–cyclization strategy.[122] Additional examples of this Inda-BOX ligand and other metal sources are the iron-Indabox **22b** catalyzed kinetic resolution of *N*-sulfonyl oxaziridines (eq 27)[123] and nickel-Inda-BOX catalyzed reductive cross-coupling between vinyl and benzyl electrophiles.[124]

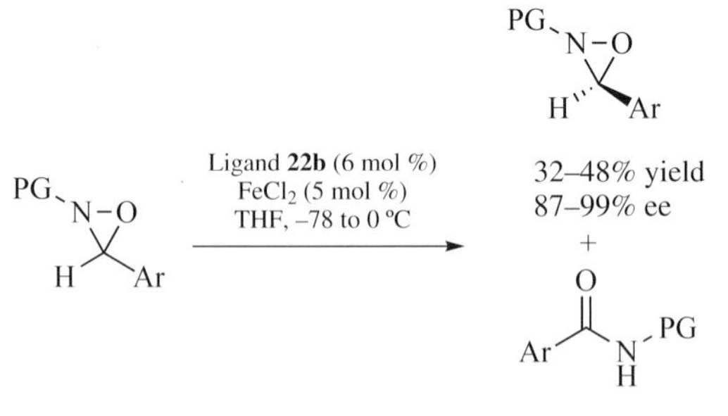

(27)

The aminoindanol pyridine bis(oxazoline) ligand (Inda-pybox **23**)[125] has also seen frequent use in Lewis acid-promoted asymmetric catalysis, enabling reactions with electrophiles featuring two-point binding potential.[126] The Sc(III) complex was employed in Friedel–Crafts alkylations of electron-rich aryl nucleophiles with α,β-unsaturated acyl phosphonates[127] and α,β-unsaturated 2-acyl imidazoles (eq 28).[128,129] A scandium-Inda-pybox **23** catalyzed enantioselective Nazarov cyclization[130] set the stage for employment of this complex in glyoxamide ene reactions.[131] *Anti*-selective allylsilane additions to glyoxamides provided syntheses of D-alloisoleucine and D-leucine,[132] and vinylsilane additions provided entry to β,γ-unsaturated α-hydroxyacids.[133] 2,3-Dihydroquinazolines were generated enantioselectively via intramolecular amidation of imines.[134,135] Isatins have also proven competent electrophiles in asymmetric scandium-Inda-pybox reactions, with indoles[136] and allylsilanes[137–139] as competent nucleophilic partners. The corresponding In(III)-Inda-pybox **23** complex catalyzed allylstannane additions to aldehydes.[140] It proved superior with two-point binding catalysis, such as pyrrole additions to isatins[141] in addition to glyoxylate[142,143] and trifluoropyruvate[144,145] carbonyl-ene reactions, with the methodology extended to Mukaiyama aldol[146] and hetero-Diels-Alder (eq 29)[147] reactions. A La-Inda-pybox **23** complex catalyzed a diastereoselective Diels–Alder reaction to access the core of muironolide A.[148,149]

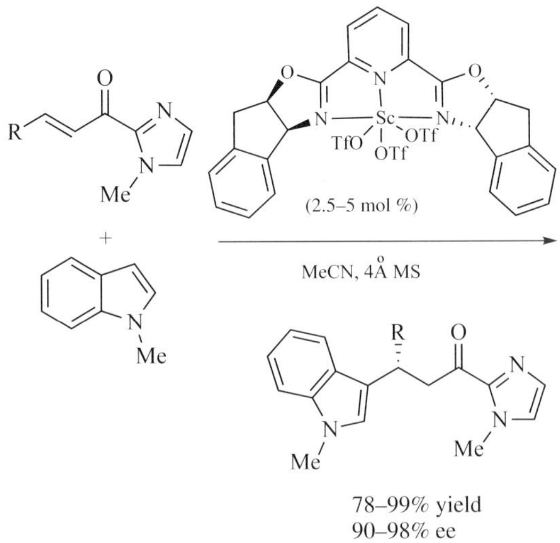

(28)

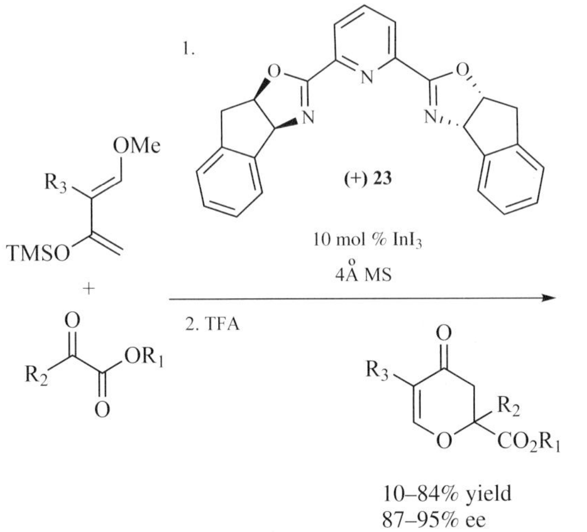

(29)

Calcium triflate complexed to Inda-pybox **23** catalyzed the enantioselective conjugate addition of glycine benzophenone imine to acrylate esters in high ee values,[150] and the calcium chloride complex catalyzed the [1,4]-addition of dimethyl malonate to α,β-unsaturated ketones.[151] An iridium-Inda-pybox catalyst effected the *syn*-selective reductive aldol of aldehydes and methyl acrylate.[152] A nickel-Inda-pybox **23** complex was employed in asymmetric Negishi cross-couplings of racemic propargylic electrophiles and arylzinc reagents (eq 30).[153–155] Ruthenium-Inda-pybox[156] catalyzed the enantioselective amination of benzylic and allylic C–H bonds.[157]

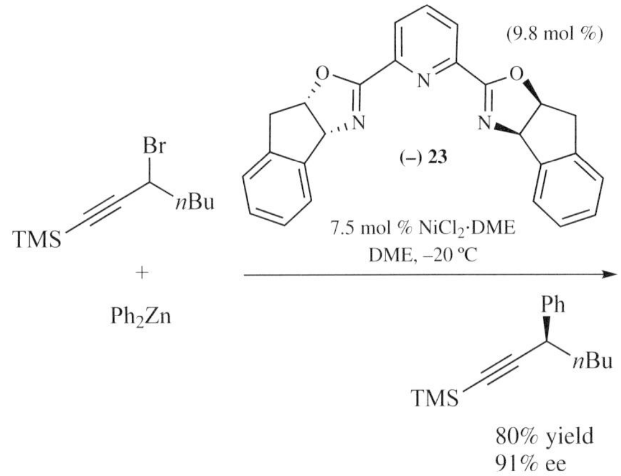

(30)

Aminoindanol-derived tris(oxazoline) (In-TOX **24**) ligands[158] have seen recent use in combination with metals to effect asymmetric Lewis acid-catalyzed reactions. The scandium complex catalyzed α-arylations of cyclic ketones with aryldiazomethane derivatives.[159,160] Enantioselective ring openings of 2-substituted cyclopropane-1,1-dicarboxylates with secondary amines were catalyzed by a nickel(II) In-TOX complex (eq 31).[161] These donor–acceptor cyclopropanes, in the presence of the same nickel catalyst, underwent [3 + 3] cycloaddition with aromatic azomethine imines,[162] and in the presence of the analogous copper catalyst with ligand **24**, reacted with indoles to afford cyclopentannulated products, such as the core of borreverine.[163] This copper complex catalyzed asymmetric tandem Friedel–Crafts–lactonization reactions between phenols and trifluoromethylpyruvate to yield enantioenriched 3-hydroxy-3-trifluoromethyl benzofuran-2-ones.[164]

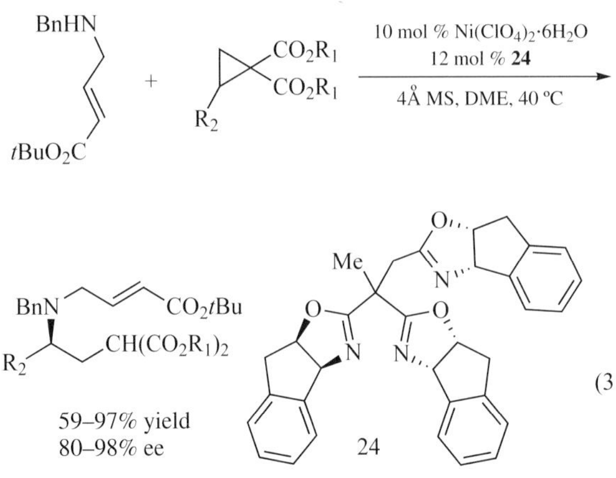

(31)

Additional aminoindanol-based oxazoline ligands have also been reported for asymmetric catalysis. A phosphaferrocene oxazoline copper complex catalyzed the conjugate addition of diethylzinc to acyclic enones.[165] An amino-bridging ligand (indan–ambox) was used in the ruthenium-catalyzed asymmetric hydrogenation of acetophenone derivatives.[166] A semicorrin derivative, in combination with calcium, was also employed in the glycine imine conjugate addition described above.[150]

Cis-Aminoindanol as a Component in Organocatalysis.

Prolinamide and Prolinethioamide Organocatalysts. Aminoindanol has been incorporated into a series of prolinamide organocatalysts. These bifunctional catalysts facilitate asymmetric additions of ketones to β-nitrostyrenes with high *syn* diasteroselectivities (up to 94%) and good enantioselectivities (up to 80% ee) (eq 32).[167,169–170] The addition of additives such as acids, as well as the reaction solvent, were determined to have an impact on reaction rate and stereoselectivity, with polar aprotic solvents such as NMP having an overall beneficial effect.

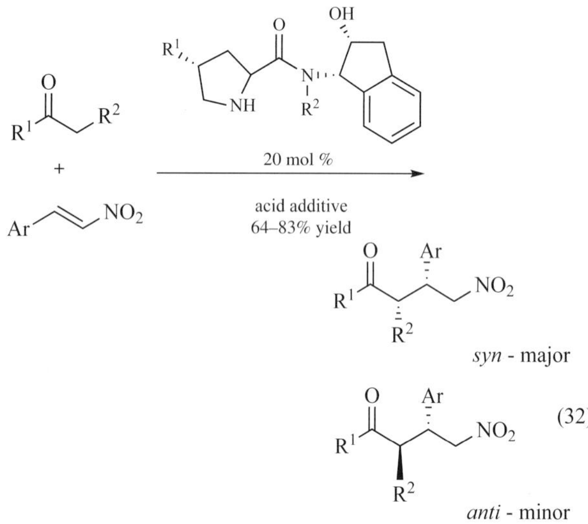

(32)

These prolinamide catalysts, as well as the prolinethioamide analogs, can also catalyze stereoselective aldol reactions. High diastereoselectivity (*anti* aldol) and ee were observed for the reaction between cyclohexanone and 4-nitrobenzaldehyde catalyzed by the aminoindanol-derived prolinamide catalyst or prolinethioamide organocatalyst (eq 33).[171]

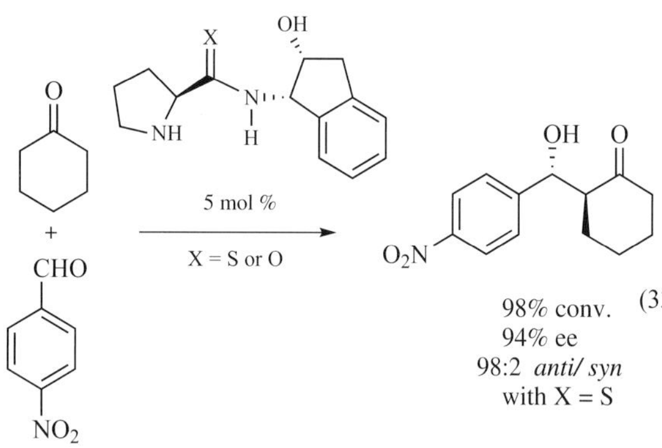

(33)

Urea and Thiourea-derived Organocatalysts. A variety of thiourea catalysts derived from aminoindanol have been reported and used in asymmetric transformations. The conjugate addition of hydrazones in the presence of thioamide **25** provided good stereoselectivity (eq 34).[172] The products resulting from this condensation can be transformed to the corresponding nitriles or carboxylates by cleavage of the hydrazones.

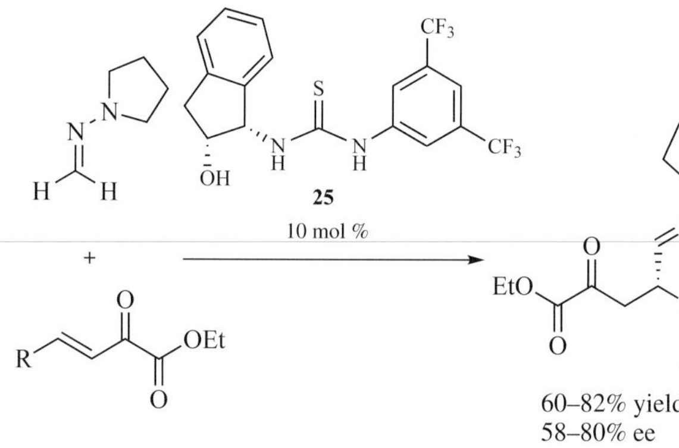

(34)

Conjugate addition at the C3 position of indole with nitrostyrene was catalyzed by aminoindanol-derived thiourea catalyst **26** (eq 35). The addition of exogenous Brønsted acids to the reaction mixture was shown to improve the reactivity and the stereoselectivity.[173,174] Furthermore, by screening a number of alternative thiourea catalysts, it was determined that the reaction rate was increased substantially, and the enantioselectivity improved to 94% using the quinolinium thioamide catalyst.

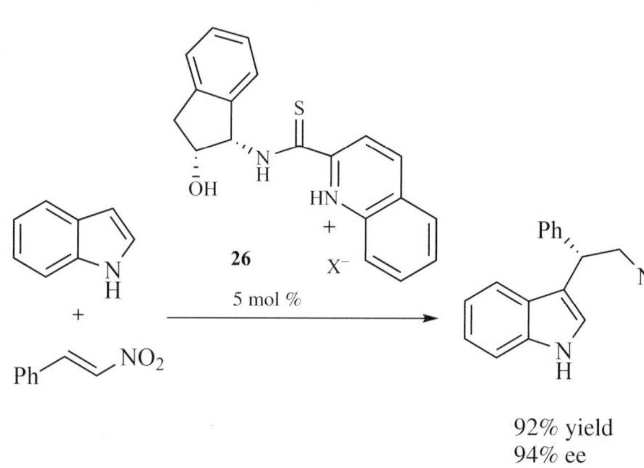

(35)

Aminoindanol-derived *N*-sulfinyl urea organocatalyst **27** was shown to facilitate an aza-Henry reaction with good stereoselectivity (eq 36).[175]

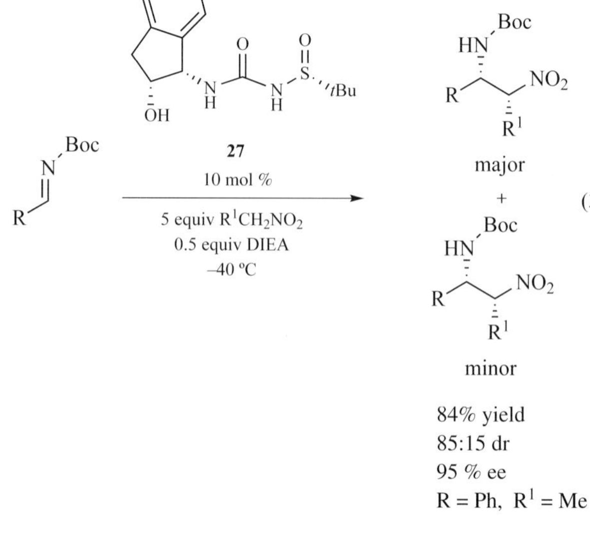

(36)

***N*-Heterocyclic Carbene (NHC) Catalysts Derived from Aminoindanol.** A number of *N*-heterocyclic carbene (NHC) organocatalysts derived from aminoindanol, with the general triazolium core structure shown in Figure 1, have been reported.[176,177] Modifying the aryl substituent on the nitrogen was demonstrated to significantly impact the stability[178] of these catalysts as well as their reactivities and selectivities.[179] Detailed experimental procedures to prepare these aminoindanol-derived catalysts have been described.[180,181] Late-stage functionalization of these aminoindanol-derived triazolium structures has also provided access to an expanding library of these catalysts.[182]

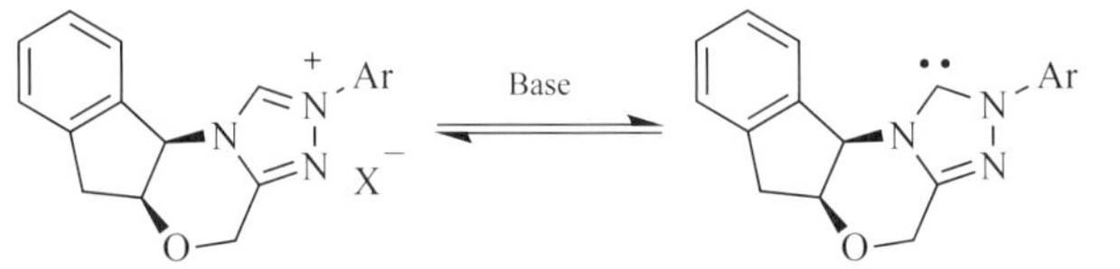

These aminoindanol-derived NHC organocatalysts have been used to facilitate a variety of stereoselective reactions. One investigation of the reactivity and mechanism of these aminoindanol-derived triazolium organocatalysts led to the isolation and characterization of the Breslow-type adducts of these compounds with iminium salts.[183] The use of these aminoindanol-derived NHC organocatalysts in asymmetric Stetter reactions was reported (eq 37).[184–186] The effect of functional groups, such as an ester, ketone, or nitrile, on the Michael acceptor and their impact on the Stetter reaction was also investigated.[187] The scope of the reaction was demonstrated to be broad and included both aromatic and aliphatic aldehydes.[188]

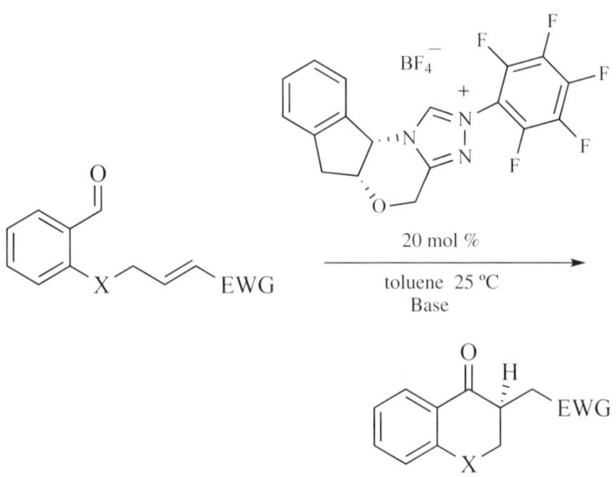

(37)

high yield/high ee

The stereoselective generation of quaternary centers in Stetter reactions catalyzed by an NHC catalyst was reported.[189,190] An interesting application of the aminoindanol-derived NHC organocatalyst to facilitate a stereoselective, one pot tandem asymmetric Michael/Stetter reaction between salicylaldehydes and electron-deficient alkynes was also described.[191] The application of these aminoindanol-based NHC catalysts for the asymmetric synthesis of hydrobenzofuranes via an intramolecular Stetter reaction was published.[192,193] The extension of this methodology for the Stetter reaction onto vinyl phosphates and phosphine oxides was also reported.[194] A catalytic asymmetric Stetter using an aminoindanol-derived organocatalyst was applied as the key transformation in a novel synthesis of the spirofuranone-lactam core of FD-838.[195] This general approach was further expanded, and a novel tandem photoisomerization/asymmetric Stetter was incorporated in order to introduce other key functional groups in the synthesis of the cephalimysin A framework (eq 38).[196]

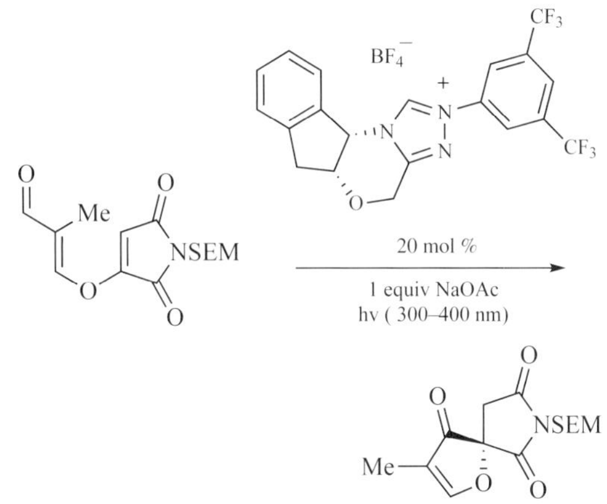

(38)

62% yield
95% ee

An example of a cross aza-benzoin reaction using aminoindanol-derived NHC catalysts was reported as a viable method to stereoselectively prepare aminoketones in good yields and high ee values. The aminoketone products were readily converted to useful amino alcohol building blocks (eq 39).[197]

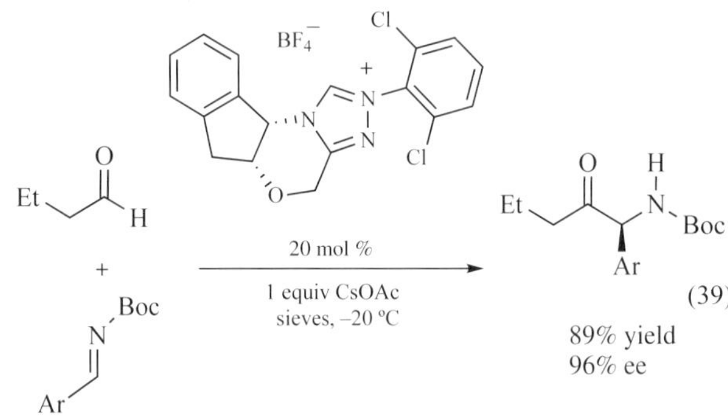

(39)

Aminoindanol-derived NHC catalysts were demonstrated to facilitate various types of Diels–Alder reactions. An all carbon intramolecular [4 + 2] annulation using aminoindanol-derived NHC catalysts was described.[198] An interesting example of an aminoindanol-derived NHC-catalyzed oxodiene Diels–Alder reaction that produced pyrenones in excellent enantioselectivities was reported (eq 40).[199] The *N*-mesityl-substituted azolium salt was discovered to be a very effective catalyst for this reaction.

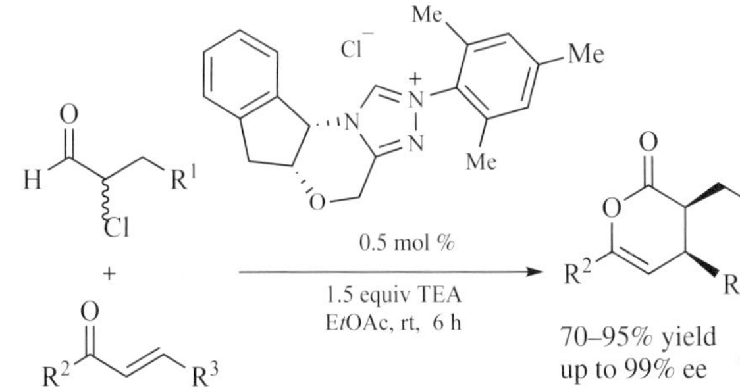

(40)

An enantioselective azadiene Diels–Alder reaction catalyzed by the aminoindanol-derived NHC was published.[200] Excellent dr (>50:1) and high ee (up to 99%) were observed in this reaction (eq 41). A report detailing the computational modeling of the aminoindanol-derived NHC-catalyzed hetero-DA reaction has identified some of the key interactions that influence the stereoselectivity of the reaction.[201]

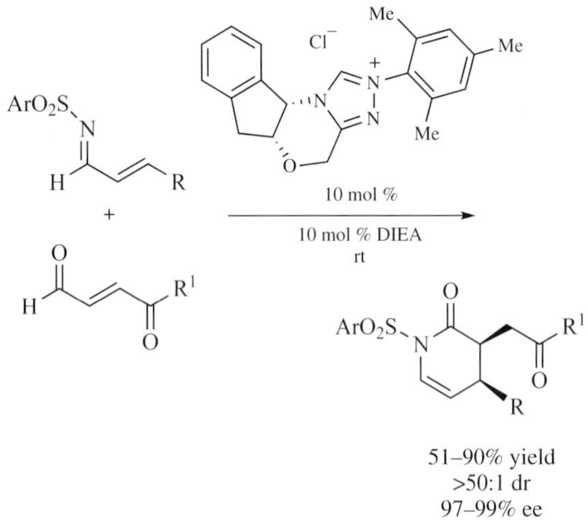

(41)

A recent report detailed the use of aminoindanol-derived NHC catalysts in a redox hetero-Diels–Alder reaction. Azolium enolate precursors generated from aroyloxyaldehydes underwent hetero-DA reactions with α,β-unsaturated trichloromethyl ketones to stereoselectively produce *syn*-dihydropyranones in good yields (eq 42).[202]

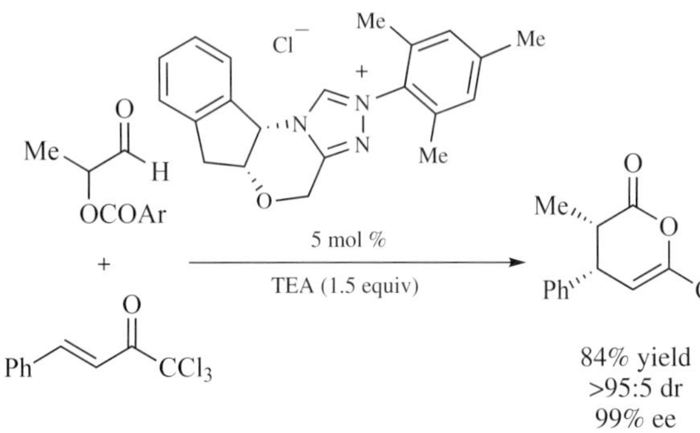

(42)

One report described an aminoindanol-derived NHC-catalyzed asymmetric oxidative hetero-DA reaction, starting with simple aliphatic aldehydes, to provide the products with good to excellent diastereoselectivities and high enantioselectivities (eq 43).[203]

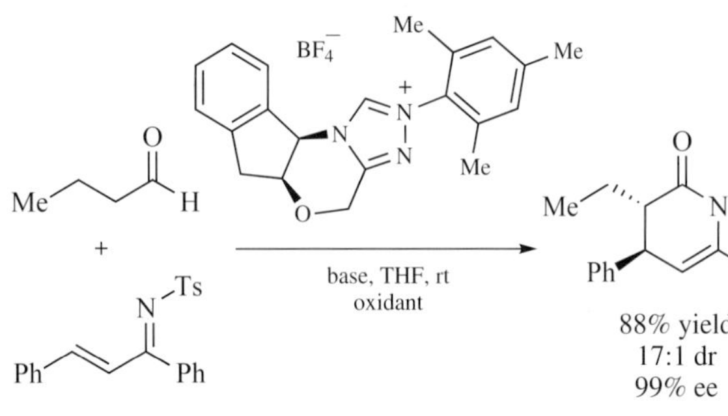

(43)

A novel, catalytic asymmetric α-acylation of tertiary amines with aldehydes, facilitated by an aminoindanol-derived NHC catalyst in combination with a photoredox catalyst, was described (eq 44).[204] This dual-catalysis (NHC/photocatalyst) system provided good to excellent yields and selectivities for a variety of aldehyde starting materials. The aminoindanol-derived NHC with 2,4,6-tribromophenyl substitution was found to be the optimal catalyst for this reaction.

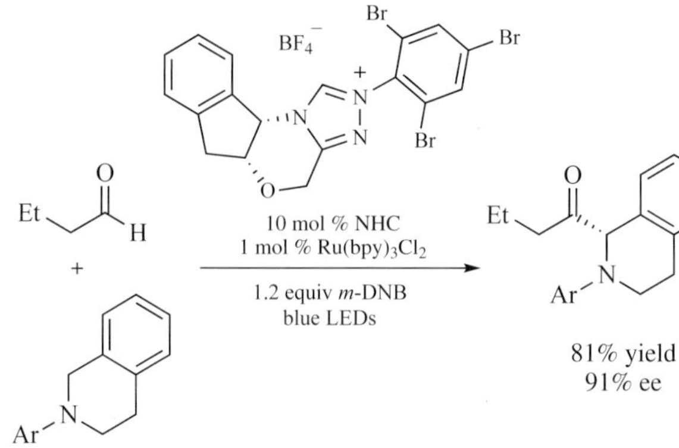

(44)

An NHC-organocatalyzed reaction of enals with α-arylidene pyrazolinones under oxidative conditions to prepare interesting spirofused products was reported. This cascade reaction sequence is a formal [3 + 3]-annulation reaction and proceeds via a vinylogous Michael addition/spiroannulation dehydrogenation to prepare spirocyclohexadienones in good yields and high ee values (eq 45).[205]

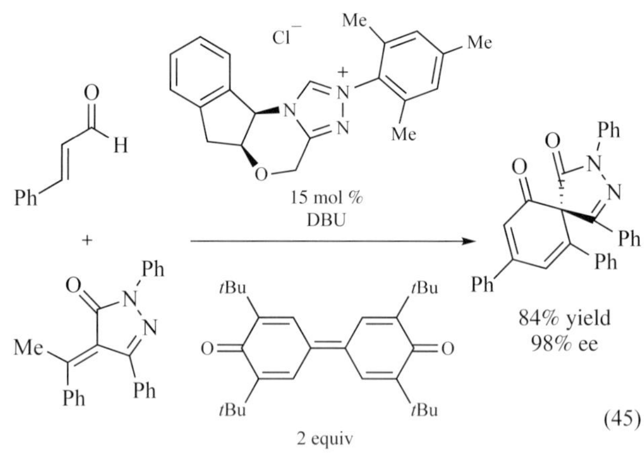

(45)

Enantioselective protonation of enolates, generated from *gem*-dichloroaldehydes, using NHC catalysis to stereoselectively generate α-chloroesters was reported (eq 46).[206] A similar strategy was successfully applied to the conversion of the dichloroaldehydes to the corresponding α-chloro carboxylic acids.[207] This methodology was extended to the enantioselective synthesis of α-deuterio- and α-fluoroacids.

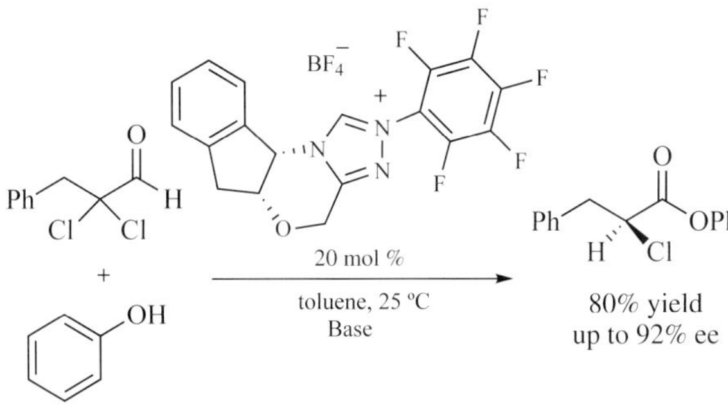

(46)

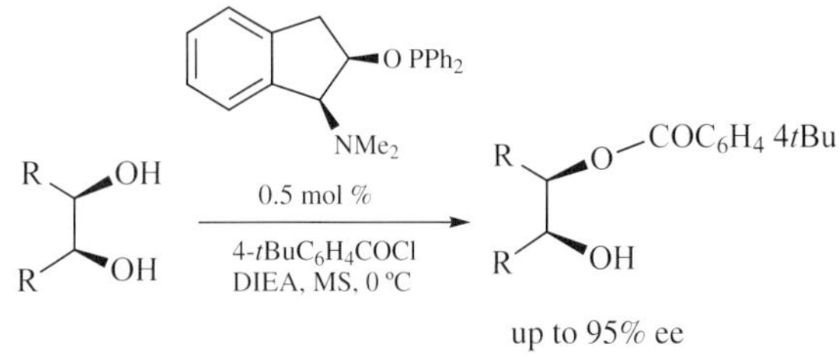

(49)

A novel application of a chiral hydroxamic acid cocatalyst **28** derived from aminoindanol, which works in combination with an achiral NHC catalyst, was demonstrated for the catalytic kinetic resolution of cyclic secondary amines (eq 47).[208] The enantioenriched amines were separated from the corresponding amides via operationally simple aqueous extraction.

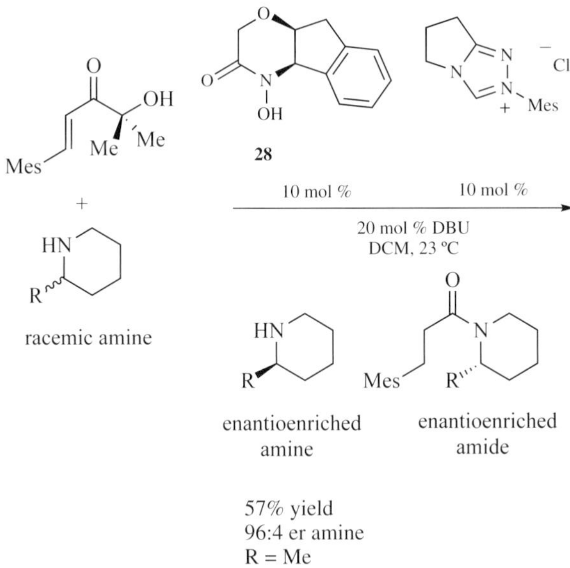

(47)

An alternative series of aminoindanol-derived NHC organocatalysts based on the imidazoline salt core structure was also reported (eq 48).[209]

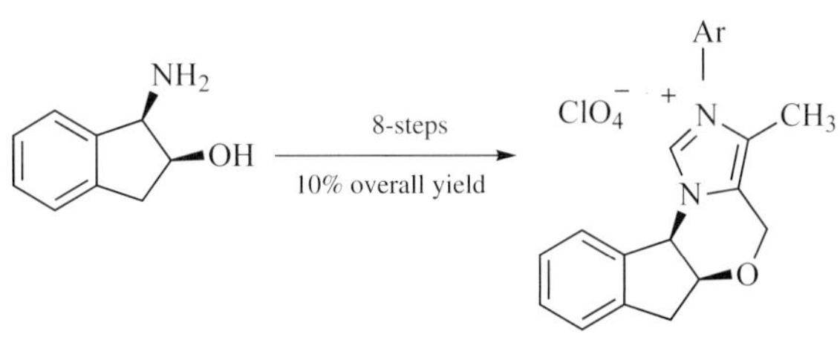

(48)

Other Applications of Aminoindanol as a Ligand for Organocatalysis. An amino phosphinite, derived from aminoindanol, was readily prepared and demonstrated to be an effective catalyst for the asymmetric desymmetrization of *meso*-1,2-diols. The enantioselective acylation of diols produced the monoesters in good yields with high ee values (up to 95%) in the presence of 5 mol % catalyst (eq 49).[210]

1. (a) Ghosh, A. K.; Fidanze, S.; Senanayake, C. H., *Synthesis* **1998**, 937. (b) Senanayake, C. H., *Aldrichimica Acta.* **1998**, *31*, 3.
2. (a) Senanayake, C. H.; Roberts, F. E.; DiMichele, L. M.; Ryan, K. M.; Liu, J.; Fredenburgh, L. E.; Foster, B. S.; Douglas, A. W.; Larsen, R. D.; Verhoeven, T. R.; Reider, P. J., *Tetrahedron Lett.* **1995**, *36*, 3993. (b) Senanayake, C. H.; Smith, G. B.; Ryan, K. M.; Fredenburgh, L. E.; Liu, J.; Roberts, F. E.; Hughes, D. L.; Larsen, R. D.; Verhoeven, T. R.; Reider, P. J., *Tetrahedron Lett.* **1996**, *37*, 3271. (c) Hughes, D. L.; Smith, G. B.; Liu, J.; Dezeny, G. C.; Senanayake, C. H.; Larsen, R. D.; Verhoeven, T. R.; Reider, P. J., *J. Org. Chem.* **1997**, *62*, 2222. (d) Larrow, J. F.; Roberts, E.; Verhoeven, T. R.; Ryan, K. M.; Senanayake, C. H.; Reider, P. J.; Jacobsen, E. N.; Lodise, S. A.; Smith, A. B., Jr, *Org. Synth.* **1999**, *76*, 46.
3. (a) Didier, E.; Loubinoux, B.; Ramos Tombo, G. M.; Rihs, G., *Tetrahedron* **1991**, *47*, 4941. (b) Luna, A.; Maestro, A.; Astorga, C.; Gotor, V., *Tetrahedron: Asymmetry* **1999**, *10*, 1969.
4. (a) Ogasawara, K.; Takahashi, M., *Synthesis* **1996**, 954. (b) Ghosh, A. K.; Kincaid, J. F.; Haske, M. G., *Synthesis* **1997**, 541. (c) Ghosh, A. K.; Cesti, P.; Battistel, E., *J. Org. Chem.* **1988**, *53*, 5531.
5. Thompson, W. J.; Fitzgerald, P. M. D.; Holloway, M. K.; Emini, E. A.; Darke, P. L.; McKeever, B. M.; Schleif, W. A.; Quintero, J. C., Zugay, J. A.; Tucker, T. J.; Schwering, J. E.; Homnick, C. F.; Nunberg, J.; Springer, J. P.; Huff, J. R., *J. Med. Chem.* **1992**, *35*, 1685.
6. (a) Zhang, J.; Roberge, C.; Reddy, J.; Conners, N.; Chartrain, M.; Buckland, B.; Greasham, R., *Enzyme and Microbial Technology* **1999**, *24*, 86. (b) Igarashi, Y.; Otsutomo, S.; Harada, M.; Nakano, S., *Tetrahedron: Asymmetry* **1997**, *8*, 2833. (c) Komiya, N.; Noji, S.; Murahashi, S.-I., *Tetrahedron Lett.* **1998**, *39*, 7921.
7. Vacca, J. P.; Dorsey, B. D.; Schleif, W. A.; Levin, R. B.; McDaniel, S. L.; Darke, P. L.; Zugay, J.; Quintero, J. C.; Blahy, D. M.; Roth, E.; Sardana, V. V.; Schlabach, A. J.; Graham, P. I.; Condra, J. H.; Gotlib, L.; Holloway, M. K.; Lin, J.; Chen, I.-W.; Vastag, K.; Ostovic, D.; Anderson, P. S.; Emini, E. A.; Huff, J. R., *Proc. Natl. Acad. Sci. USA* **1994**, *91*, 4096.
8. Ghosh, A. K.; Duong, T. T.; McKee, S. P., *Chem. Commun.* **1992**, 1673.
9. Ghosh, A. K.; Liu, W.; Xu, Y.; Chen, Z., *Angew. Chem.* **1996**, *108*, 73; *Angew. Chem., Int. Edn.* **1996**, *35*, 74.
10. Ghosh, A. K.; Liu, W., *J. Org. Chem.* **1996**, *61*, 6175.
11. Ghosh A. K.; Hussain, K. A.; Fidanze, S., *J. Org. Chem.* **1997**, *62*, 6080.
12. Davies, I. W.; Senanayake, C. H.; Castonguay, L.; Larsen, R. D.; Verhoeven, T. R.; Reider, P. J., *Tetrahedron Lett.* **1995**, *36*, 7619.
13. (a) Ghosh, A. K.; Onishi, M., *J. Am. Chem.* **1996**, *118*, 2527. (b) Ghosh, A. K.; Fidanze, S.; Onishi, M.; Huissain, K. A., *Tetrahedron Lett.* **1997**, *38*, 7171.
14. (a) Ghosh, A. K.; Fidenze, S., *J. Org. Chem.* **1998**, *63*, 6146. (b) Ghosh, A. K.; Bischoff, A., *Org. Lett.* **2000**, *2*, 1573.
15. Ghosh, A. K.; Mathivanan, P., *Tetrahedron: Asymmetry* **1996**, *7*, 375.
16. Ghosh A. K.; Chen, Y., *Tetrahedron Lett.* **1995**, *36*, 6811.
17. (a) Armstrong, J. D., Jr; Hartner, F. W., Jr; DeCamp, A. E.; Volante, R. P.; Shinkai, I., *Tetrahedron Lett.* **1992**, *33*, 6599. (b) McWilliams, J. C.; Armstrong, J. D., Jr; Zheng, N.; Bhupathy, M.; Volante, R. P.; Reider, P. J., *J. Am. Chem. Soc.* **1996**, *118*, 11970.

18. Kress, M. H.; Yang, C.; Yasuda, N.; Grabowski, E. J. J., *Tetrahedron Lett.* **1997**, *38*, 2633.
19. Zheng N.; Armstrong, J. D., Jr; McWilliams, C.; Volante, R. P., *Tetrahedron Lett.* **1997**, *38*, 2817.
20. Lee, J.; Choi, W.-B.; Lynch, J. E.; Volante, R. P.; Reider, P. J., *Tetrahedron Lett.* **1998**, *39*, 3679.
21. Senanayake, C. H.; Fang, K.; Grover, P.; Bekale, R. P.; Vandenbossche, C. P.; Wald, S. A., *Tetrahedron Lett.* **1999**, *40*, 819.
22. (a) Watson, D. J.; Lawrence, C. M.; Meyers, A. I., *Tetrahedron Lett.* **2000**, *41*, 815. (b) Mechelke, M. F.; Meyers, A. I., *Tetrahedron Lett.* **2000**, *41*, 4339.
23. (a) Hong, Y.; Gao, Y.; Nie, X.; Zepp, C. M., *Tetrahedron Lett.* **1994**, *35*, 6631. (b) DiSimone, B.; Savoia, D.; Tagliavini, E.; Umani-Ronchi, A., *Tetrahedron: Asymmetry* **1995**, *6*, 301. (c) Jones, S.; Atherton, J. C. C., *Tetrahedron: Asymmetry* **2000**, *11*, 4543.
24. (a) Hett, R.; Fang, Q. K.; Gao, Y.; Hong, Y.; Butler, H. T.; Nie, X.; Wald, S. A., *Tetrahedron Lett.* **1997**, *38*, 1125. (b) Hett, R.; Senanayake, C. H.; Wald, S. A., *Tetrahedron Lett.* **1999**, *39*, 1705. (c) Hett, R.; Fang, Q. K.; Gao, Y.; Wald, S. A.; Senanayake, C. H., *Org. Proc. Res and Dev.* **1998**, *2*, 96. (d) Fang, Q. K.; Senanayake, C. H.; Wilkinson, H. S.; Wald, S. A.; Li, H., *Tetrahedron Lett.* **1998**, *39*, 2701.
25. (a) Palmer, M.; Walsgrove, T.; Wills, M., *J. Org. Chem.* **1997**, *62*, 5226. (b) Kenny, J. A.; Palmer, M. J.; Smith, A. R. C.; Walsgrove, T.; Wills, M., *Synlett* **1999**; 1615. (c) Wills, M.; Gamble, M.; Palmer, M.; Smith, A.; Studley, J.; Kenny, J., *J. Mol. Catal A: Chemical* **1999**, *146*, 139. (d) Kawamoto, A.; Wills, M., *Tetrahedron: Asymmetry* **2000**, *11*, 3257. (e) Kenny, J. A.; Versluis, K.; Heck, A. J. R.; Walsgrove, T.; Wills, M., *Chem. Commun.* **2000**; 99. (f) Hennig, M.; Puntener, K.; Scalone, M.; *Tetrahedron: Asymmetry* **2000**, *11*, 1849.
26. (a) Davies, I. W.; Senanayake, C. H.; Larsen, R. D.; Verhoeven, T. R.; Reider, P. J., *Tetrahedron Lett.* **1996**, *37*, 1725. (b) Kanemasa, S.; Oderaotoshi, Y.; Yamamoto, H.; Tanaka, J.; Wada, E.; Curran, D. P., *J. Org. Chem.* **1997**, *62*, 6454. (c) Ghosh, A. K.; Mathivnan, P.; Cappiello, J., *Tetrahedron Lett.* **1996**, *37*, 3815. (d) Davies, I. W.; Garena, L.; Castonguay, L.; Senanayake, C. H.; Larsen, R. D.; Verhoeven, T. R.; Reder, P. J., *Chem. Commun.* **1996**, 1753. (e) Davies, I. W.; Garena, L.; Cai, D.; Larsen, R. D.; Verhoeven, T. R.; Reider, P. J., *Tetrahedron Lett.* **1997**, *38*, 1145. (f) Davies, I. W.; Senenayake, C. H.; Larsen, R. D.; Verhoeven, T. R.; Reider, P. J., *Tetrahedron Lett.* **1996**, *37*, 813.
27. (a) Ghosh, A. K.; Mathivanan, P.; Cappiello, J.; Krishnan, K., *Tetrahedron: Asymmetry* **1996**, *7*, 2165. (b) Ghosh, A. K.; Mathivanan, P.; Cappiello, J., *Tetrahedron Lett.* **1997**, *38*, 2427.
28. (a) Wu, J. H.; Radinov, R.; Porter, N. A., *J. Am. Chem. Soc.* **1995**, *117*, 11029. (b) Sibi, M. P.; Ji, J.; Wu, J.-H.; Gurtler, S.; Porter, N. A., *J. Am. Chem. Soc.* **1996**, *118*, 9200. (c) Sibi, M. P.; Ji, J., *J. Org. Chem.* **1997**, *62*, 3800. (d) Sibi, M. P.; Shay, J. J.; Ji, J., *Tetrahedron Lett.* **1997**, *38*, 5955.
29. Dakin, L. A.; Schaus, S. E.; Jacobsen, E. N.; Panek, J. S., *Tetrahedron Lett.* **1998**, *39*, 8947.
30. Zenner, J. M.; Larock, R. C., *J. Org. Chem.* **1999**, *64*, 7312.
31. Clark, J. S.; Tolhurst, K. F.; Taylor, M.; Swallow, S., *J. Chem. Soc., Perkin Trans 1* **1998**; 1167.
32. Dossetter, A. G.; Jamison, T. F.; Jacobsen, E. N., *Angew. Chem., Int. Edn.* **1999**, *38*, 2398.
33. Li, Z.; Fernandez, M.; Jacobsen, E. N., *Org. Lett.* **1999**, *1*, 1611.
34. Flores-Lopez, L. Z.; Parra-Hake, M.; Somanathan, R.; Walsh, P. J., *Organometallics* **2000**, *19*, 2153.
35. Weise, B.; Helmchen, G., *Tetrahedron Lett.* **1998**, *32*, 5727.
36. Rademacher, K.; Scheffer, J. R.; Trotter, J., *Tetrahedron Lett.* **2000**, *56*, 6739.
37. Gallou, I.; Senanayake, C. H., *Chem. Rev.* **2006**, *106*, 2843.
38. Senanayake, C. H.; Krishnamurthy, D.; Gallou, I. *In Handbook of Chiral Compounds*, ; Ager, D., Ed.; Taylor Francis: New York, 2006; 457.
39. Kim, E. J.; An, K. M.; Ko, S. Y., *Bull. Kor. Chem. Soc.* **2006**, *27*, 2019.
40. O'Brien, X. M.; Parker, J. A.; Lessard, P. A., *Appl. Microbiol. Biotechnol.* **2002**, *59*, 389.
41. Yun, H.; Kim, J.; Kinnera, K.; Kim, B.-G., *Biotechnol. Bioeng.* **2006**, *93*, 391.
42. Kim, Y.-J.; Choi, Y.-S.; Yang, S.; Yang, W. R.; Jeong, J.-H., *Synlett* **2015**, *26*, 1981.
43. Sakurai, R; Sakai, K., *Tetrahedron: Asymmetry* **2003**, *14*, 411.
44. Muhlman, A.; Classon, B.; Hallberg, A.; Samuelson, B., *J. Med. Chem.* **2001**, *44*, 3402.
45. Ugliarolo, E. A.; Gagey, D.; Lantano, B.; Moltrasio, G. Y.; Campos, R. H.; Cavallaro, L. V.; Moglioni, A. G., *Bioorg. Med. Chem.* **2012**, *20*, 5986.
46. Shieh, H. S.; Tomasselli, A. G.; Mathis, K. J.; Schnute, M. E.; Woodard, S. S.; Caspers, N.; Williams, J. M.; Kiefer, J. R.; Munie, G.; Wittwer, A.; Malfait, A.-M.; Tortorella, M. D., *Protein Sci.* **2011**, *20*, 735.
47. Yao, W.; Chao, M.; Wasserman, Z. R.; Liu, R.-Q.; Covington, M. B.; Newton, R.; Christ, D.; Wexler, R. R.; Decicco, C. P., *Bioorg. Med. Chem. Lett.* **2002**, *12*, 101.
48. Fujiwara, T.; Sasaki, M.; Omata, K.; Kabuto, C.; Kabuto, K.; Takeuchi, Y., *Tetrahedron: Asymmetry* **2004**, *15*, 555.
49. Liao, L.; Zhang, F.; Yan, N.; Golen, J. A.; Fox, J. M., *Tetrahedron* **2004**, *60*, 1803.
50. Bozkurt, S.; Durmaz, M.; Naziroglu, H. N.; Yilmaz, M.; Sirit, A., *Tetrahedron: Asymmetry* **2011**, *22*, 541.
51. Estivill, C.; Pomares, M.; Kotev, M.; Ivanov, P.; Virgili, A., *Tetrahedron: Asymmetry* **2005**, *16*, 2993.
52. Palomino-Schatzlein, M.; Virgili, A.; Gil, S.; Jaime, C., *J. Org. Chem.* **2006**, *71*, 8114.
53. Song, Z. J.; Zhao, M.; Desmond, R.; Devine, P.; Tschaen, D. M.; Tillyer, R.; Frey, L.; Heid, R.; Xu, F.; Foster, B.; Li, J.; Reamer, R.; Volante, R.; Grabowski, E. J.; Dolling, U. H.; Reider, P. J.; Okada, S.; Kato, Y.; Mano, E., *J. Org. Chem.* **1999**, *64*, 9658.
54. Smitrovich, J. H.; Boice G. N.; Qu, C.; DiMichele, L.; Nelson, T. D.; Huffman M. A.; Murry, J.; McNamara, J.; Reider, P. J., *Org. Lett.* **2002**, *4*, 1963.
55. Janey, J. M.; Iwama, T.; Kozmin, S. A.; Rawal, V. H., *J. Org. Chem.* **2000**, *65*, 9059.
56. Tanaka, K.; Katsumura, S.; *J. Am. Chem. Soc.* **2002**, *124*, 9660.
57. Kobayashi, T.; Tanaka, K.; Miwa, J.; Katsumura, S., *Tetrahedron: Asymmetry* **2004**, *15*, 185.
58. Tanaka, K; Kobayashi, T.; Mori, H.; Katsumura, S., *J. Org. Chem.* **2004**, *69*, 5906.
59. Mulder, J. A.; Hsung, R. P.; Frederick, M. O.; Tracey, M. R.; Zificsak, C. A., *Org. Lett.* **2002**, *4*, 1383.
60. Mo, D.-L.; Wink, D. J.; Anderson, L. L., *Chem Eur. J.* **2014**, *20*, 13217.
61. Han, Z.; Krishnamurthy, D.; Grover, P. Fang, K.; Senanayake, C., *J. Am. Chem. Soc.* **2002**, *124*, 7880.
62. Leon, T.; Riera, A.; Verdaguer, X., *J. Am. Chem. Soc.* **2011**, *133*, 5740.
63. Sakurai, R.; Suzuki, S.; Hashimoto, J.; Baba, M; Itoh, O.; Uchida, A.; Hattori, T.; Miyano, S.; Yamura, M., *Org. Lett.* **2004**, *6*, 2241.
64. Chen, S.; Patrick, B. O.; Scheffer, J. R., *J. Org. Chem.* **2004**, *69*, 2711.
65. Geoghegan, K.; Bode, J. W., *Org. Lett.* **2015**, *17*, 1934.
66. Delamare, M.; Belot, S.; Caille, J. C.; Martinet, F.; Kagan, H. B.; Henryon, V., *Tetrahedron Lett.* **2009**, *50*, 1702.
67. Gilmore, N. J.; Jones, S., *Tetrahedron: Asymmetry* **2003**, *14*, 2115.
68. Gilmore, N. J.; Jones, S.; Muldowney, M. P., *Org. Lett.* **2004**, *6*, 2805.
69. Adams, H.; Gilmore, N. J.; Jones, S.; Muldowney, M. P.; von Reuss, S. H.; Vemula, R., *Org. Lett.* **2008**, *10*, 1457.
70. Dixon, R. A.; Jones, S., *Tetrahedron: Asymmetry* **2002**, *13*, 1115.
71. Barker, M. D.; Dixon, R. A.; Jones, S.; Marsh, B. J., *Tetrahedron* **2006**, *62*, 11663.

72. Marsh, B. J.; Adams, H.; Barker, M. D.; Kutama, I. U.; Jones, S., *Org. Lett.* **2014**, *16*, 3780.
73. Barker, M. D.; Dixon, R. A.; Jones, S.; Marsh, B. J., *Chem. Commun.* **2008**, 2218.
74. Sun, X.; Manos, G.; Blacker, J.; Martin, J.; Gavriilidis, A., *Org. Proc. Res. Dev.* **2004**, *8*, 909.
75. Faller, J. W.; Lavoie, A. R., *Org. Lett.* **2001**, *3*, 3703.
76. Brownwell, K. R.; McCrory, C. C. L.; Chidsey, C. E. D.; Perry, R. H.; Zare, R. N.; Waymouth, R. M., *J. Am. Chem. Soc.* **2013**, *135*, 14299.
77. Faller, J. W.; Lavoie, A. R., *Organometallics* **2002**, *21*, 3493.
78. Wisman, R. V.; de Vries, J. G.; Deelman, B.-J.; Heeres, H. *J. Org. Proc. Res. Dev.* **2006**, *10*, 423.
79. Guijarro, D.; Pablo, O.; Yus, M., *Tetrahedron Lett.* **2009**, *50*, 5386.
80. Guijarro, D.; Pablo, O.; Yus, M., *J. Org. Chem.* **2010**, *75*, 5265.
81. Dai, X.; Cahard, D., *Adv. Synth. Catal.* **2014**, *356*, 1317.
82. Chavez, D. E.; Jacobsen, E. N., *Org. Synth.* **2005**, *82*, 34.
83. Joly, G. D.; Jacobsen, E. N., *Org. Lett.* **2002**, *4*, 1795.
84. Gademann, K.; Chavez, D. E.; Jacobsen, E. N., *Angew. Chem., Int. Ed.* **2002**, *41*, 3059.
85. Chavez, D. E.; Jacobsen, E. N., *Org. Lett.* **2003**, *5*, 2563.
86. Jarvo, E. R.; Lawrence, B. M.; Jacobsen, E. N., *Angew. Chem., Int. Ed.* **2005**, *44*, 6043.
87. Boezio, A. A.; Jarvo, E. R.; Lawrence, B. M.; Jacobsen, E. N., *Angew. Chem., Int. Ed.* **2005**, *44*, 6046.
88. Ruck, R. T.; Jacobsen, E. N., *J. Am. Chem. Soc.* **2002**, *124*, 2882.
89. Ruck, R. T.; Jacobsen, E. N., *Angew. Chem., Int. Ed.* **2003**, *42*, 4771.
90. Grachan, M. L.; Tudge, M. T.; Jacobsen, E. N., *Angew. Chem., Int. Ed.* **2008**, *47*, 1469.
91. Weix, D. J.; Ellman, J. A., *Org. Lett.* **2003**, *5*, 1317.
92. Hartung, J.; Drees, S.; Greb, M.; Schmidt, P.; Svoboda, I.; Fuess, H.; Murso, A.; Stalke, D., *Eur. J. Org. Chem.* **2003**, 2388.
93. Unger, R.; Weisser, F.; Chinkov, N.; Stanger, A.; Cohen, T.; Marek, I., *Org. Lett.* **2009**, *11*, 1853.
94. Unger, R.; Cohen, T.; Marek, I., *Tetrahedron* **2010**, *66*, 4874.
95. Cheng, H.-G.; Liu, L.-Q.; Wang, T.; Yang, Q.-Q.; Liu, X.-P.; Li, Y.; Deng, Q.-H.; Chen, J.-R.; Xiao, W.-J., *Angew. Chem., Int. Ed.* **2013**, *52*, 3250.
96. Ji, J.; Barnes, D. M.; Zhang, J.; King, S. A.; Wittenberger, S. J.; Morton, H. E., *J. Am. Chem. Soc.* **1999**, *121*, 10215.
97. Barnes, D. M.; Ji, J.; Fickers, M. G.; Fitzgerald, M. A.; King, S. A.; Morton, H. E.; Plagge, F. A.; Preskill, M.; Wagaw, S. H.; Wittenberger, S. J.; Zhang, J., *J. Am. Chem. Soc.* **2002**, *124*, 13098.
98. Sibi, M. P.; Ma, Z.; Jasperse, C. P., *J. Am. Chem. Soc.* **2004**, *126*, 718.
99. Sibi, M. P.; Rane, D.; Stanley, L. M.; Soeta, T., *Org. Lett.* **2008**, *10*, 2971.
100. Huang, J.; Frontier, A. J., *J. Am. Chem. Soc.* **2007**, *129*, 8060.
101. Leboeuf, D.; Wright, C. M.; Frontier, A. J., *Chem. Eur. J.* **2013**, *19*, 4835.
102. Guo, C.; Song, J.; Luo, S.-W.; Gong, L.-Z., *Angew. Chem., Int. Ed.* **2010**, *49*, 5558.
103. Sibi, M. P.; Liu, M., *Org. Lett.* **2000**, *2*, 3393.
104. Sibi, M. P.; Liu, M., *Org. Lett.* **2001**, *3*, 4181.
105. Sibi, M. P.; *Chen. J., Org. Lett.* **2002**, *4*, 2933.
106. Sibi, M. P.; Prabagaran, N.; Ghorpade, S. G.; Jasperse, C. P., *J. Am. Chem. Soc.* **2003**, *125*, 11796.
107. Sibi, M. P.; Itoh, K.; Jasperse, C. P., *J. Am. Chem. Soc.* **2004**, *126*, 5366.
108. Sibi, M. P.; Stanley, L. M.; Jasperse, C. P., *J. Am. Chem. Soc.* **2005**, *127*, 8276.
109. Sibi, M. P.; Stanley, L. M.; Soeta, T., *Adv. Synth. Catal.* **2006**, *348*, 2371.
110. Gerten, A. L.; Slade, K. M.; Stanley, L. M., *Org. Biomol. Chem.* **2013**, *11*, 7834.
111. Sibi, M. P.; Stanley, L. M.; Soeta, T., *Org. Lett.* **2007**, *9*, 1553.
112. Sibi, M. P.; Matsunaga, H., *Tetrahedron Lett.* **2004**, *45*, 5925.
113. Sibi, M. P.; Chen, J.; Stanley, L. M., *Synlett* **2007**, 298.
114. Sibi, M. P.; Kawashima, K.; Stanley, L. M., *Org. Lett.* **2009**, *11*, 3894.
115. Palomo, C.; Pazos, R.; Oiarbide, M.; Garcia, J. M., *Adv. Synth. Catal.* **2006**, *348*, 1161.
116. Taillier, C.; Lautens, M., *Org. Lett.* **2007**, *9*, 591.
117. Sibi, M. P. Chen, J., *J. Am. Chem. Soc.* **2001**, *123*, 9472.
118. Sibi, M. P.; Petrovic, G., *Tetrahedron: Asymmetry* **2003**, *14*, 2879.
119. Sibi, M. P.; Petrovic, G.; Zimmerman, J., *J. Am. Chem. Soc.* **2005**, *127*, 2390.
120. Sibi, M. P.; Yang, Y.-H.; Lee, S., *Org. Lett.* **2008**, *10*, 5349.
121. Sibi, M. P.; Guerrero, M. A., *Synthesis* **2005**, 1528.
122. Miyabe, H.; Asada, R.; Toyoda, A.; Takemoto, Y., *Angew. Chem., Int. Ed.* **2006**, *45*, 5863.
123. Williamson, K. S.; Sawicki, J. W.; Yoon, T. P., *Chem. Sci.* **2014**, *5*, 3524.
124. Cherney, A. H.; Reisman, S. E., *J. Am. Chem. Soc.* **2014**, *136*, 14365.
125. Coulomb, J.; Sibi, M. P. *e-EROS Encyclopedia of Reagents for Organic Synthesis*, **2007**.
126. Evans, D. A.; Fandrick, K. R.; Song, H.-J.; Scheidt, K. A.; Xu, R., *J. Am. Chem. Soc.* **2007**, *129*, 10029.
127. Evans, D. A.; Scheidt, K. A.; Frandrick, K. R.; Lam, H. W.; Wu, J., *J. Am. Chem. Soc.* **2003**, *125*, 10780.
128. Evans, D. A.; Fandrick, K. R.; Song, H.-J., *J. Am. Chem. Soc.* **2005**, *127*, 8942.
129. Evans, D. A.; Fandrick, K. R., *Org. Lett.* **2006**, *8*, 2249.
130. Liang, G.; Trauner, D., *J. Am. Chem. Soc.* **2004**, *126*, 9544.
131. Evans, D. A.; Wu, J., *J. Am. Chem. Soc.* **2005**, *127*, 8006.
132. Evans, D. A.; Aye, Y.; Wu, J., *Org. Lett.* **2006**, *8*, 2071.
133. Evans, D. A.; Aye, Y., *J. Am. Chem. Soc.* **2006**, *128*, 11034.
134. Prakash, M.; Kesavan, V., *Org. Lett.* **2012**, *14*, 1896.
135. Prakash, M.; Jayakumar, S.; Kesavan, V., *Synthesis* **2013**, *45*, 2265.
136. Hanhan, N. V.; Sahin, A. H.; Chang, T. W.; Fettinger, J. C.; Franz, A. K., *Angew. Chem., Int. Ed.* **2010**, *49*, 744.
137. Hanhan, N. V.; Ball-Jones, N. R.; Tran, N. T.; Franz, A. K., *Angew. Chem., Int. Ed.* **2012**, *51*, 989.
138. Hanhan, N. V.; Tang, Y. C.; Tran, N. T.; Franz, A. K., *Org. Lett.* **2012**, *14*, 2218.
139. Ball-Jones, N. R.; Badillo, J. J.; Tran, N. T.; Franz, A. K., *Angew. Chem., Int. Ed.* **2014**, *53*, 9462.
140. Lu, J.; Hong, M.-L.; Ji, S.-J.; Loh, T.-P., *Chem. Commun.* **2005**, 1010.
141. Gituerrez, E. G.; Wong, C. J.; Sahin, A. H.; Franz, A. K., *Org. Lett.* **2011**, *13*, 5754.
142. Zhao, J.-F.; Tsui, H.-Y.; Wu, P.-J.; Lu, J.; Loh, T.-P., *J. Am. Chem. Soc.* **2008**, *130*, 16492.
143. Zhao, Y.-J.; Li, B.; Tan, S.; Shen, Z.-L.; Loh, T.-P., *J. Am. Chem. Soc.* **2010**, *132*, 10242.
144. Zhao, J.-F.; Tjan, T.-B. W.; Tan, B.-H.; Loh, T.-P., *Org. Lett.* **2009**, *11*, 5714.
145. Zhao, J. F.; Tan, B. H.; Zhu, M. K.; Tjan, T. B. W.; Loh, T. P., *Adv. Synth. Catal.* **2010**, *352*, 2085.
146. Zhao, J.-F.; Tan, B.-H.; Loh, T.-P., *Chem. Sci.* **2011**, *2*, 349.
147. Zhao, B.; Loh, T.-P., *Org. Lett.* **2013**, *15*, 2914.
148. Young, K.; Xiao, Q.; Zakarian, A., *Eur. J. Org. Chem.* **2015**, 2337.
149. Xiao, Q.; Young, K.; Zakarian, A., *J. Am. Chem. Soc.* **2015**, *137*, 5907.
150. Hut'ka, M.; Tsubogo, T.; Kobayashi, S., *Adv. Synth. Catal.* **2013**, *355*, 1561.
151. Lippur, K.; Kaabel, S.; Jarving, I.; Rissanen, K.; Kanger, T., *J. Org. Chem.* **2015**, *80*, 6336.
152. Zhao, C.-X.; Duffey, M. O.; Taylor, S. J.; Morken, J. P., *Org. Lett.* **2001**, *3*, 1829.

153. Smith, S. W.; Fu, G. C., *J. Am. Chem. Soc.* **2008**, *130*, 12645.
154. Oelke, A. J.; Sun, J., *J. Am. Chem. Soc.* **2012**, *134*, 2966.
155. Schley, N. D.; Fu, G. C., *J. Am. Chem. Soc.* **2014**, *136*, 16588.
156. Tse, M. K.; Bhor, S.; Klawonn, M.; Anilkumar, G.; Jiao, H.; Dobler, C.; Spannenberg, A.; Magerlein, W.; Hugl, H.; Beller, M., *Chem. Eur. J.* **2006**, *12*, 1855.
157. Milczek, E.; Boudet, N.; Blakey, S., *Angew. Chem., Int. Ed.* **2008**, *47*, 6825.
158. Liao, S.; Sun, X.-L.; Tang, Y., *Acc. Chem. Res.* **2014**, *47*, 2260.
159. Rendina, V. L.; Moebius, D. C.; Kingsbury, J. S., *Org. Lett.* **2011**, *13*, 2004.
160. Rendina, V. L.; Kaplan, H. Z.; Kingsbury, J. S., *Synthesis* **2012**, *44*, 686.
161. Zhou, Y.-Y.; Wang, L.-J.; Li, J.; Sun, X.-L.; Tang, Y., *J. Am. Chem. Soc.* **2012**, *134*, 9066.
162. Zhou, Y.-Y.; Li, J.; Ling, L.; Liao, S.-H.; Sun, X.-L.; Li, Y.-X.; Wang, L.-J.; Tang, Y., *Angew. Chem., Int. Ed.* **2013**, *52*, 1452.
163. Xiong, H.; Xu, H.; Liao, S.; Xie, Z.; Tang, Y., *J. Am. Chem. Soc.* **2013**, *135*, 7851.
164. Ren, H.; Wang, P.; Wang, L.; Tang, Y., *Org. Lett.* **2015**, *17*, 4886.
165. Shintani, R.; Fu, G. C., *Org. Lett.* **2002**, *4*, 3699.
166. Li, W.; Hou, G.; Wang, C.; Jiang, Y.; Zhang, X., *Chem. Commun.* **2010**, *46*, 3979.
167. Almasi, D.; Alonso, D.A.; Gomez-Bengoa, E.; Nagel, Y.; Najera, C., *Eur. J. Org Chem.* **2007**, *2328*.
168. Dohery, S.; Knight, J. G.; McRae, A.; Harrington, R. W.; Clegg, W., *Eur. J. Org. Chem.* **2008**, *1759*.
169. Almasi, D; Alonso, D.; Najera, C., *Adv. Synth. Catal.* **2008**, *350*, 2467.
170. Gruttadauria, M.; Pia Salvo, A. M.; Giacalone, F.; Agreigento, P.; Noto, R., *Eur. J. Org. Chem.* **2009**, 5437.
171. Almasi, D.; Alonso, D. A.; Balaguer, A.-N.; Najera, C., *Adv. Synth. Catal.* **2009**, *352*, 1123.
172. Herrera, R. P.; Monge, D.; Martin-Zamora, E.; Fernandez, R.; Lassaletta, J. M., *Org. Lett.* **2007**, *9*, 3303.
173. Ganesh, M.; Seidel, D., *J. Am. Chem. Soc.* **2008**, *130*, 16464.
174. Marques-Lopez, E.; Alcaine, A.; Tejero, T.; Herrera, R. P., *Eur. J. Org. Chem.* **2011**, *3700*.
175. Robak, M. A.; Trincado, M.; Ellman, J. A., *J. Am. Chem. Soc.* **2007**, *129*, 15110.
176. Mahatthananchai, J.; Bode, J. W., *Acc. Chem. Res.* **2014**, *47*, 696.
177. Vora, H. U.; Wheeler, P.; Rovis, T., *Adv. Synth. Catal.* **2012**, *354*, 1617.
178. Zhao, X.; Glover, G. S.; Oberg, K. M.; Dalton, D. M.; Rovis, T., *Synlett* **2013**, *24*, 1229.
179. Mahatthananchai, J.; Bode, J. W., *Chem. Sci.* **2012**, *3*, 192.
180. Vora, H. U.; Lathrop, S. P.; Reynolds, N. T.; Kerr, M. S.; Read de Alaniz, J.; Rovis, T., *Org. Syn.* **2010**, *87*, 350.
181. Struble, J. R.; Bode, J. W., *Org. Syn.* **2010**, *87*, 362.
182. Ozboya, K. E.; Rovis, T., *Synlet* **2014**, *25*, 2665.
183. DiRocco, D. A.; Oberg, K. M.; Rovis, T., *J. Am. Chem. Soc.* **2012**, *134*, 6143.
184. Read De Alaniz, J.; Rovis, T., *Synlett* **2009**, *1189*.
185. Kerr, M.; Read de Alaniz, J.; Rovis, T., *J. Am. Chem. Soc.* **2002**, *124*, 10298.
186. Read de Alaniz, J.; Rovis, T., *J. Am. Chem. Soc.* **2005**, *127*, 6284.
187. Kerr, M. S.; Rovis, T., *Synlett* **2003**, *12*, 1934.
188. Read de Alaniz, J.; Kerr, M. S.; Moore, J. L.; Rovis, T., *J.Org. Chem.* **2008**, *73*, 2033.
189. Kerr, M. S.; Rovis, T., *J. Am. Chem. Soc.* **2004**, *126*, 8876.
190. Moore, J. L.; Kerr, M. S.; Rovis, T., *Tetrahedron* **2006**, *62*, 11477.
191. Filloux, C. M.; Lathrop, S. P.; Rovis, T., *Proc. Natl. Acad. Sci. USA* **2010**, *107*, 20666.
192. Liu, Q.; Rovis, T., *J. Am. Chem. Soc.* **2006**, *128*, 2552.
193. Liu, Q.; Rovis, T., *Org. Process Res. Dev.* **2007**, *11*, 598.
194. Cullen, S.; Rovis, T., *Org. Lett.* **2008**, *10*, 3141.
195. Orellana, A.; Rovis, T., *Chem. Commun.* **2008**, *730*.
196. Lathrop, S. P.; Rovis, T., *Chem. Sci.* **2013**, *4*, 1668.
197. DiRocco, D. A.; Rovis, T., *Angew. Chem., Int. Ed.* **2012**, *57*, 5904.
198. Candish, L.; Levens, A.; Lupton, D. W., *J. Am. Chem Soc.* **2014**, *136*, 14397.
199. He, M.; Uc, G. J.; Bode, J. W., *J. Am. Chem. Soc.* **2006**, *128*, 15088.
200. He, M.; Struble, J. R.; Bode, J. W., *J. Am. Chem Soc.* **2006**, *128*, 8418.
201. Allen, S. E.; Mahatthananchai, J.; Bode, J. W.; Kozlowski, M., *J. Am. Chem. Soc.* **2012**, *134*, 12098.
202. Attaba, N.; Taylor, J. E.; Slawin, A. M. Z.; Smith, A., *J. Org. Chem.* **2015**, *80*, 9728.
203. Zhao, X.; Ruhl, K. E.; Rovis, T., *Angew. Chem., Int. Ed.* **2012**, *51*, 12330.
204. DiRocco, D.; Rovis, T., *J. Am. Chem. Soc.* **2012**, *134*, 8094.
205. Yetra, S. R.; Mondal, S.; Mukherjee, S.; Gonnade, R. G.; Biju, A. T., *Angew. Chem., Int. Ed.* **2015**, *54*, 1.
206. Reynolds, N.; Rovis, T., *J. Am. Chem. Soc.* **2005**, *127*, 16406.
207. Vora, H. U.; Rovis, T., *J. Am. Chem. Soc.* **2010**, *132*, 2860.
208. Binanzer, M.; Hsieh, S. Y.; Bode, J. W., *J. Am. Chem. Soc.* **2011**, *133*, 19698.
209. Struble, J. R.; Bode, J. W., *Tetrahedron* **2008**, *64*, 6961.
210. Aida, H.; Mori, K.; Yamaguchi, Y.; Mizuta, S.; Moriyama, T.; Yamamoto, I.; Fujimoto, T., *Org. Lett.* **2012**, *14*, 812.

(2*S*,4*S*)-2-(Anilinomethyl)-1-ethyl-4-hydroxypyrrolidine[1]

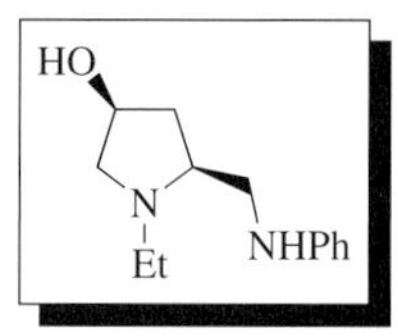

[77937-75-8] $C_{13}H_{20}N_2O$ (MW 220.31)

InChI = 1/C13H20N2O/c1-2-15-10-13(16)8-12(15)9-14-11-6-4-3-5-7-11/h3-7,12-14,16H,2,8-10H2,1H3/t12-,13-/m0/s1

InChIKey = IOSRZSGQMNZZGN-STQMWFEEBA

(chiral catalyst for the enantioselective conjugate addition of thiols to α,β-unsaturated ketones[2] and esters)

Physical Data: bp 150–170 °C/2 mmHg; $[\alpha]_D^{23}$ −58.1° (*c* 1.03, EtOH).

Solubility: sol toluene, benzene, CH_2Cl_2, CCl_4, ethanol.

Preparative Methods: (2*S*,4*S*)-1-benzyloxycarbonyl-4-hydroxyproline (*Z*-allohydroxyproline)[3] is acetylated with Acetic Anhydride and Pyridine to give the acetate (**1**). The acetate (**1**) is treated with aniline–1,3-Dicyclohexylcarbodiimide and the subsequent removal of the benzyloxycarbonyl group affords (**2**). Acetylation of (**2**) followed by reduction with Lithium Aluminum Hydride affords (2*S*,4*S*)-2-(anilinomethyl)-1-ethyl-4-hydroxypyrrolidine (**3**) in 60% overall yield from (**1**) (eq 1).[2]

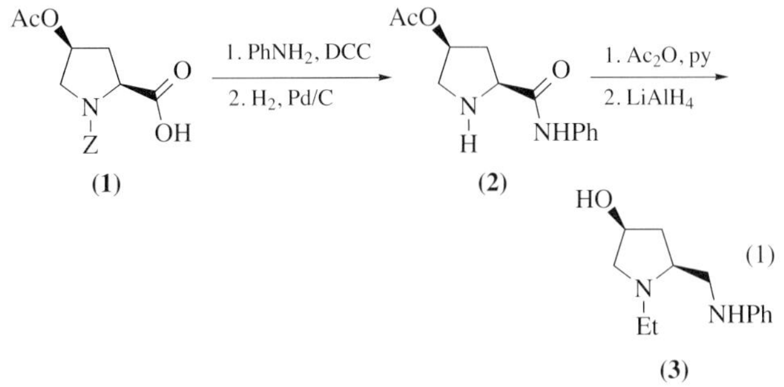

Enantioselective Addition of Thiols to α,β-Unsaturated Ketones and Diisopropyl Maleate. (2*S*,4*S*)-2-(Anilinomethyl)-1-ethyl-4-hydroxypyrrolidine (**3**) is a chiral amino alcohol which acts as a chiral base catalyst for the enantioselective conjugate addition of aromatic thiols to α,β-unsaturated ketones.[2] When an aromatic thiol is treated with a 2-cycloalkenone in the presence of 1 mol % of amino alcohol (**3**) in toluene at −5 °C, 3-arylthio-1-cycloalkanones with 42–88% ee are obtained in 74–85% yields (eq 2).

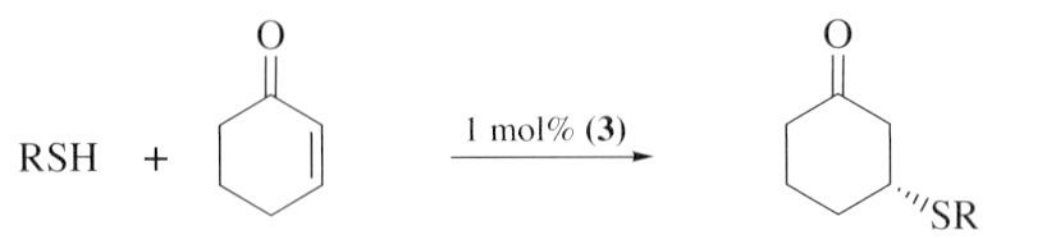 (2)

The results are shown in Table 1. The reaction between various aromatic thiols and 2-cyclohexen-1-one in the presence of (**3**) affords the adduct in good ee. The best selectivity is observed in the addition of 4-*t*-butylbenzenethiol to 2-cyclohexen-1-one using 1 mol % of (**3**). (*R*)-3-(4-*t*-Butylphenylthio)cyclohexan-1-one is obtained in 88% ee. When cinchonine is employed as a chiral catalyst in the same reaction, the corresponding adduct with (*S*) configuration is obtained in 67% ee.[4]

Table 1 Enantioselective addition of thiols to 2-cyclohexenones catalyzed by amino alcohol (**3**)

R	Enone	Yield (%)	ee (%)
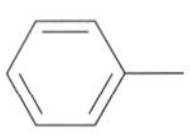	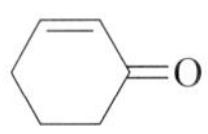	83	77
	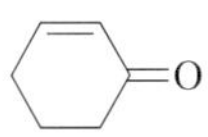	75	73
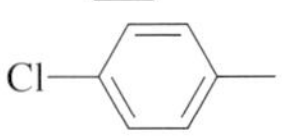	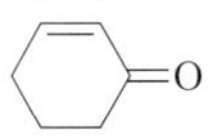	84	47
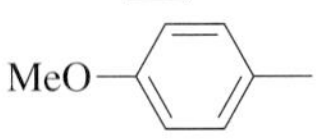	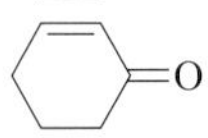	75	83
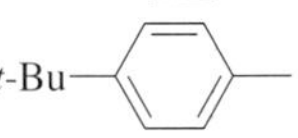	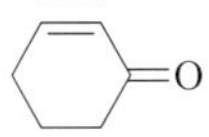	74	88
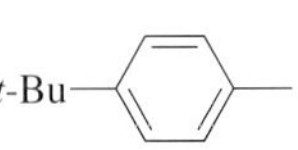	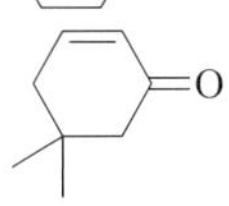	74	42

In the addition of benzenethiol to 2-cyclohexen-1-one, the enantioselectivities using (**3**)[2b] and cinchonidine[4b] are 77 and 54% ee, respectively. Thus (**3**) is more enantioselective than cinchonine and cinchonidine in the addition of RSH to 2-cyclohexen-1-one. However, in the addition of 4-*t*-butylbenzenethiol to a substituted enone (5,5-dimethyl-2-cyclohexen-1-one), the cinchonidine catalyst showed higher enantioselectivity (62% ee)[4] than catalyst (**2**) (42% ee).

Cinchonidine is more enantioselective than (**3**) in the addition to acyclic α,β-unsaturated esters. When benzenethiol is allowed to react with diisopropyl maleate using chiral catalyst (**3**), (*R*)-diisopropyl phenylthiosuccinate with 6% ee is obtained. The same reaction employing cinchonine afforded the (*S*) adduct with 81% ee (eq 3).[5]

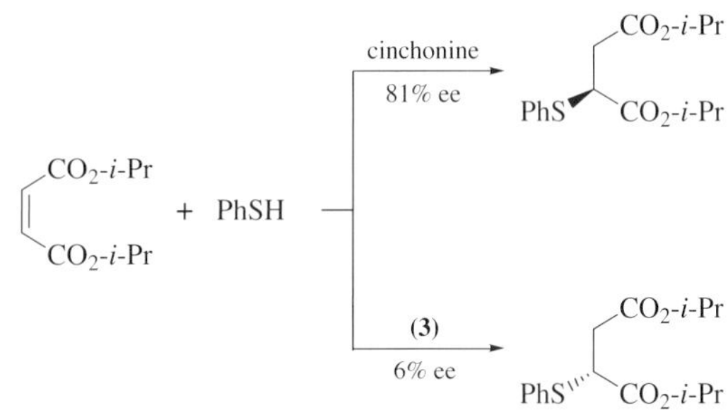 (3)

1. Mukaiyama, T., *Tetrahedron* **1981**, *37*, 4111.
2. (a) Mukaiyama, T.; Ikegawa, A.; Suzuki, K., *Chem. Lett.* **1981**, 165. (b) Suzuki, K.; Ikegawa, A.; Mukaiyama, T., *Bull. Chem. Soc. Jpn.* **1982**, *55*, 3277.
3. Patchett, A. A.; Witkop, B., *J. Am. Chem. Soc.* **1957**, *79*, 185.
4. (a) Helder, R.; Arends, R.; Bolt, W.; Hiemstra, H.; Wynberg, H., *Tetrahedron* **1977**, 2181. (b) Hiemstra, H.; Wynberg, H., *J. Am. Chem. Soc.* **1981**, *103*, 417.
5. Yamashita, H.; Mukaiyama, T., *Chem. Lett.* **1985**, 363.

Kenso Soai
Science University of Tokyo, Tokyo, Japan

(S)-2-(Anilinomethyl)pyrrolidine[1]

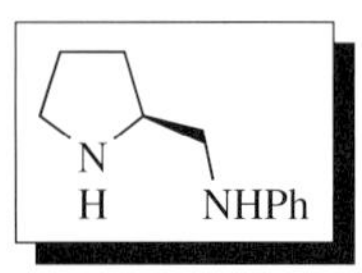

[64030-44-0] $C_{11}H_{16}N_2$ (MW 176.25)
InChI = 1S/C11H16N2/c1-2-5-10(6-3-1)13-9-11-7-4-8-12-11/h1-3,5-6,11-13H,4,7-9H2/t11-/m0/s1
InChIKey = MCHWKJRTMPIHRA-NSHDSACASA-N

(chiral ligand of $LiAlH_4$ for enantioselective reduction of ketones;[2] can form chiral aminals for diastereoselective alkylation,[3,4] 1,2-,[5] and 1,4-additions[6])

Physical Data: bp 111–112 °C/0.55 mmHg; $[\alpha]_D^{24}$ + 19.7° (*c* 1.04, EtOH).
Solubility: sol ether, THF.
Form Supplied in: colorless oil.
Preparative Methods: reaction of (*S*)-*N*-(benzyloxycarbonyl)-proline with (1) ethyl chloroformate and *N*-methylmorpholine, (2) aniline, (3) hydrogenolysis with Pd/C, and (4) reduction with $LiAlH_4$ affords the title compound in 59% overall yield.[2b]

Original Commentary

Kenso Soai
Science University of Tokyo, Tokyo, Japan

Enantioselective Reduction of Ketones. (*S*)-2-(Anilinomethyl)pyrrolidine (**1**) is a chiral ligand which, in combination with Lithium Aluminum Hydride, generates a chiral reagent for the enantioselective reduction of ketones. (*S*)-1-Phenylethanol with 92% ee is obtained from the enantioselective reduction of acetophenone with $LiAlH_4$–(**1**) (eq 1).[2a,b] When (*S*)-2-(2,6-xylidinomethyl)pyrrolidine is used instead of (**1**), the ee of the alcohol obtained reaches 95%.[2c] The ee's of aromatic alcohols obtained using $LiAlH_4$–(**1**) are comparable with other highly enantioselective reductions[1] and with highly enantioselective alkylation of aldehydes.[7]

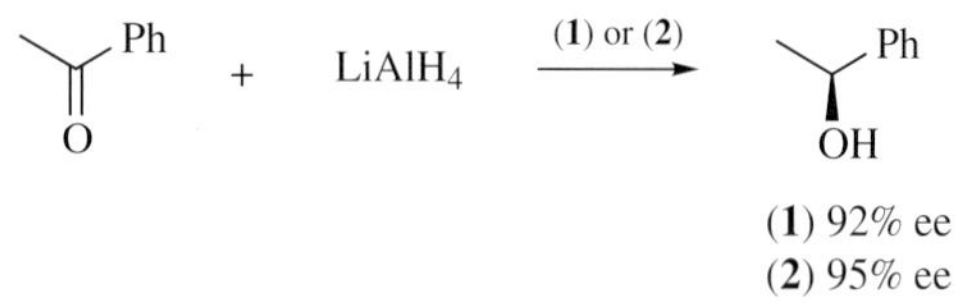

(1)

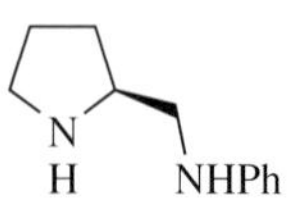
(**1**)

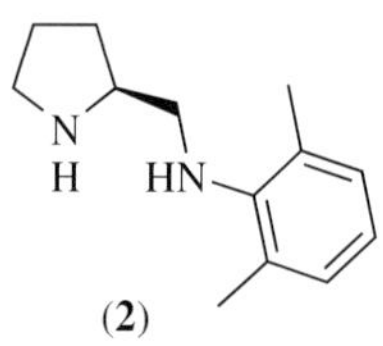
(**2**)

Diastereoselective Alkylation of Chiral Keto- and Formylaminals. Diamine (**1**) forms a chiral ketoaminal by condensation with phenylglyoxal monohydrate. Diastereoselective addition of a Grignard reagent to the ketoaminal and subsequent hydrolysis affords optically active *t*-α-hydroxyaldehydes with >94% ee (eq 2).[3a] Various α-hydroxyaldehydes are synthesized by this method.[3b] Diastereoselectivities are comparable with those of the reactions of chiral 1,3-oxathiane.[8]

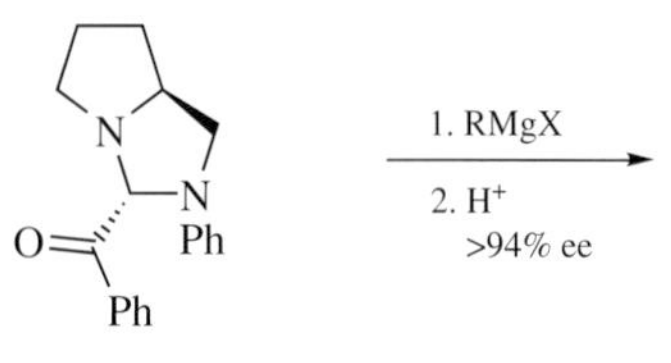

(2)

The method has been applied to the diastereoselective synthesis of naturally occurring compounds such as frontalin (84–100% ee)[3c] and (−)-malyngolide (95% ee).[3d] On the other hand, diastereoselective alkylation of chiral formylaminal with Grignard reagents and the subsequent hydrolysis afford optically active *S*-α-hydroxyaldehydes with moderate stereoselectivity (60% ee).[4]

Diastereoselective 1,2- and 1,4-Additions of Chiral Aminals. The organolithium reagent derived from a chiral bromoaminal and BuLi adds to pentanal to afford an optically active lactol. Subsequent oxidation affords the optically active lactone with 88% ee (eq 3).[5]

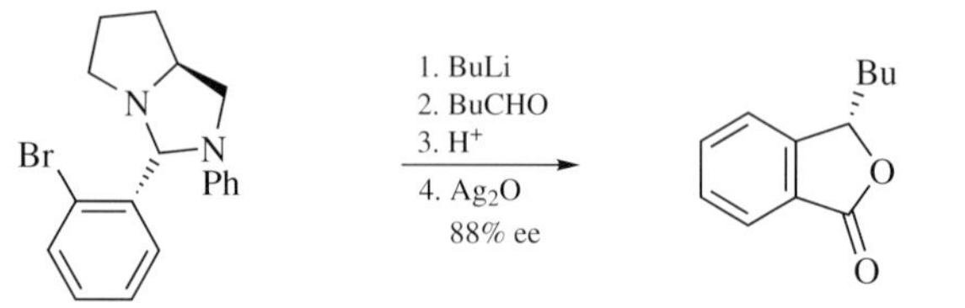

(3)

Diastereoselective 1,4- (conjugate) additions of Grignard reagents to a chiral α,β-unsaturated aminals afford optically active 3-substituted succinaldehydic acid methyl esters with 85–93% ee (eq 4).[6]

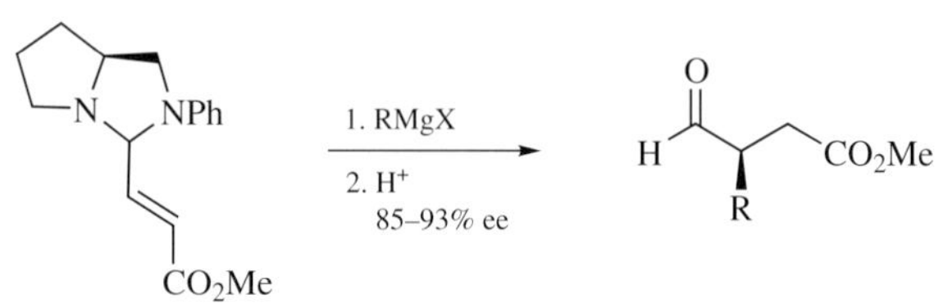

(4)

Related Reagents. (2S,4S)-2-(Anilinomethyl)-1-ethyl-4-hydroxypyrrolidine; (R,R)-1,2-Diphenyl-1,2-diaminoethane N, N″-Bis[3,5-bis(trifluoromethyl)benzenesulfonamide]; (2S,2′S)-2-Hydroxymethyl-1-[(1-methylpyrrolidin-2-yl)methyl]pyrrolidine.

First Update

Andrea Lindsay Frederick
Eli Lilly and Company, Indianapolis, IN, USA

(*S*)-2-(Anilinomethyl)pyrrolidine (AMP) is readily available in high optical purities.[9] Two main routes of preparation have been demonstrated in the literature: a two-step protocol from (*S*)-glutamic acid[10] and a four-step protocol from (*S*)-proline,[2b] which was later optimized.[11] Although (*S*)-AMP has been explored as a possible nAChR ligand for the treatment of Alzheimer's disease[12] and as a peripheral group in dendrimer synthesis,[13] the main utility of this molecule is as a chiral ligand for asymmetric transformations. The use of (*S*)-AMP has been made attractive by the structural rigidity it achieves when chelated with various metals, although poor water solubility has limited its utility in some cases.[14] (*S*)-2-(Anilinomethyl)pyrrolidine has also found use as a derivatizing agent, when modified to contain a chiral phosphorous, for the determination of enantiomeric purity of chiral halohydrins,[15] amines, and alcohols.[16]

Enantioselective Reductions. The synthetic utility of Mukaiyama's pioneering methodology for the asymmetric reduction of prochiral ketones or imines with (*S*)-AMP and $LiAlH_4$[17]

has been investigated in the synthesis of natural products, such as sarcophytol A (92% ee),[18] and pharmaceutical targets, including dilevalol (eq 5)[19] and S18986 (eq 6),[20] with varying success.

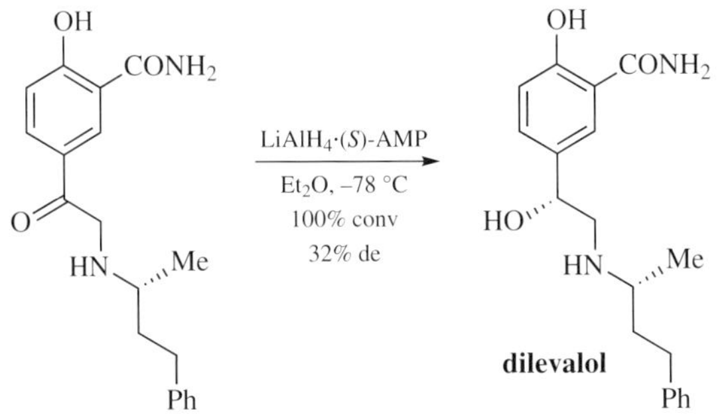

(5)

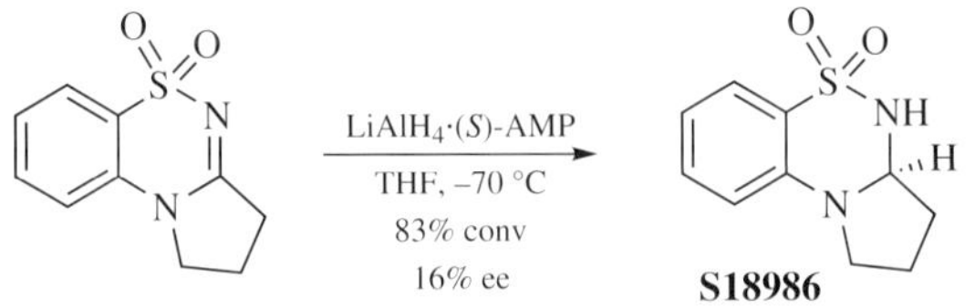

(6)

Expansion of the substrate scope of this reaction to include lactones has led to moderate atropo-enantioselectivity (60%) using the $LiAlH_4$·(*S*)-AMP adduct (eq 7), although very large excesses of $LiAlH_4$ and (*S*)-AMP are required (>10 equiv).[21]

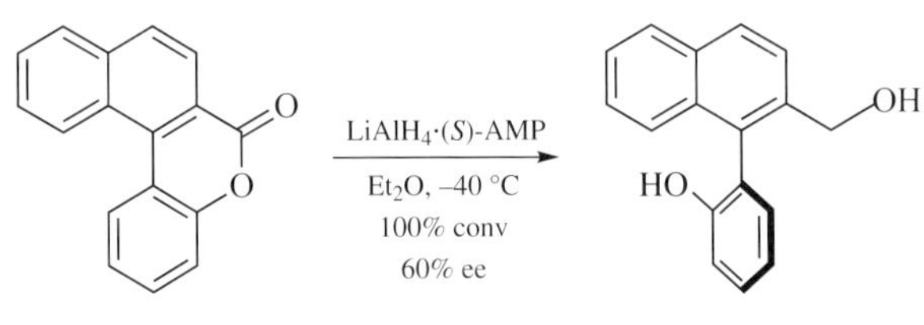

(7)

Several groups have sought to further develop this methodology to include alternative reducing agents. The use of borane·dimethyl sulfide, instead of $LiAlH_4$, has been successful in achieving the reduction of simple ketones (eq 8).[22] Similarly, the use of microwave irradiation effects the reduction of acetophenone in high yield (99%) and enantioselectivity (72%).[23]

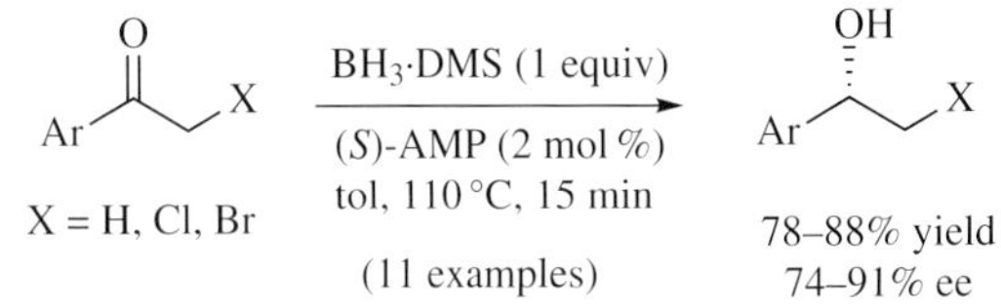

(8)

Modification of the (*S*)-AMP ligand to contain a chiral phosphorous center (**1**) gives high yields and selectivities in the reduction of prochiral ketones in the presence of borane·dimethyl sulfide (eq 9).[24] However, the group on the phosphorous does not play a significant role in the chiral induction process.[25] Therefore, the structure can be simplified to guanidine **2**, without substantial erosion of yield or ee.[26]

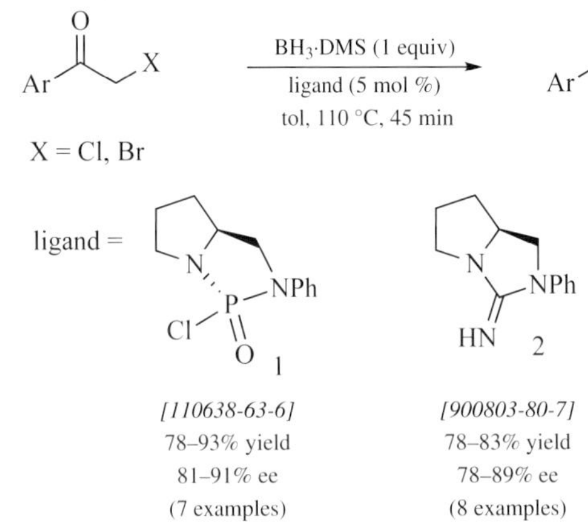

(9)

Chiral reducing agents have also been prepared from $SnCl_2$, diisobutyl aluminum hydride (DIBAL), and (*S*)-AMP-modified ligands (eq 10), providing the chiral alcohols in good yield (60–87%), but poor to moderate enantiomeric excess (3–85%).[27] Additionally, a ruthenium(II) complex prepared from $[Ru(p\text{-cymene})Cl_2]_2$ and (*S*)-AMP selectively reduces acetophenone in the presence of sodium isopropoxide in excellent yield (93%) and high ee (89%).[28] However, poor enantioselectivities (8%) were achieved in the reduction of acetophenone with $ZnEt_2$ and polymethylhydrosiloxane in the presence of (*S*)-AMP.[29]

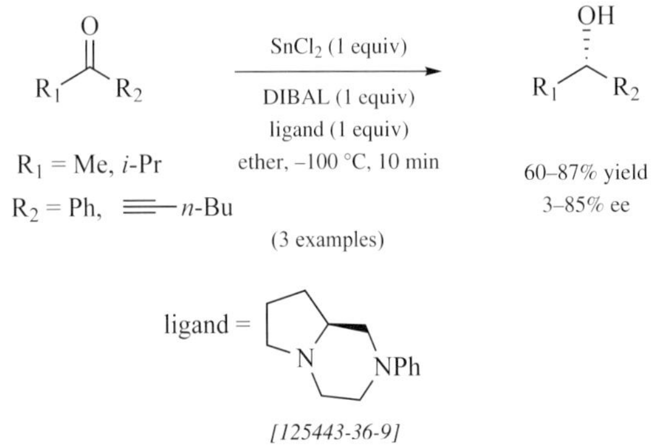

(10)

Rhodium-catalyzed asymmetric hydrogenations of activated ketones using homochiral bis(aminophosphane)s prepared from (*S*)-AMP have been demonstrated in good enantioselectivities, with the selectivity mainly controlled by the pyrrolidinic aminophosphine residue (eq 11).[30]

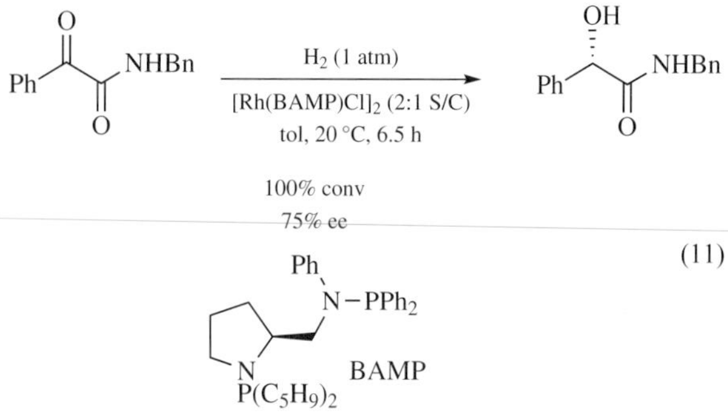

(11)

[176207-36-6]

A promising new class of catalysts for enantioselective hydrogenations is iridium phosphinooxazoline complexes. High enantioselectivities were observed in the reduction of ethyl β-methyl cinnamate using modified (*S*)-AMP ligands for chiral induction (eq 12).[31]

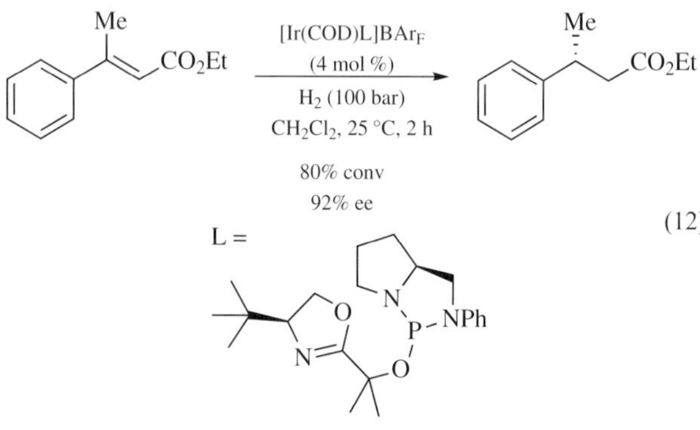

(12)

[866044-26-0]

1,2-Addition Reactions. The synthetic utility of Mukaiyama's bicyclic aminal alkylation methodology[3a] has been demonstrated as the key step in the synthesis of marine natural product (+)-tanikolide, in excellent enantioselectivity (>99%).[32] Extension of this methodology has led to the stereoselective preparation of both diastereomers of diphenylphosphinoyl hydroxy aldehydes and 1,2-diols, depending on the metalation conditions employed (eq 13).[33] Because the diastereoselectivity of the aminals could be improved with crystallization, this method was found to give the diol product, after aminal hydrolysis with HCl and reduction with $LiAlH_4$, in higher enantioselectivity than the corresponding Sharpless asymmetric dihydroxylation reaction (>97 vs. 86% ee).[34]

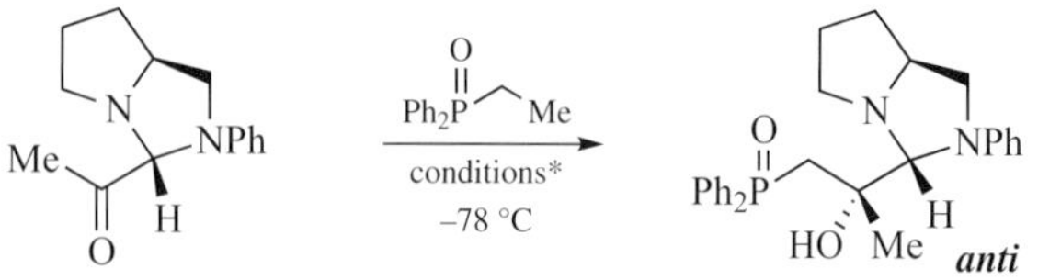

(13)

Conditions*	*anti*:*syn*	Yield (%)
MeLi·LiBr, toluene	86:14	47
BuLi, THF	36:64	54

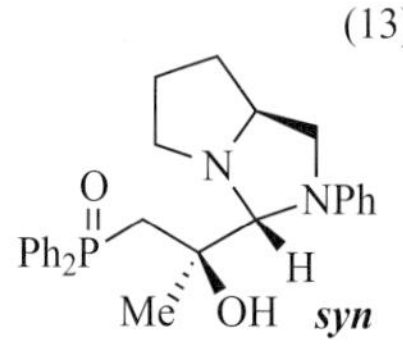

Additionally, modified (*S*)-AMP ligands have been employed in the enantioselective addition of diethylzinc to aldehydes (eq 14), giving moderate to excellent yields (57–95%) and enantioselectivities (14–94%), with aryl aldehydes proceeding in the highest yields and selectivities.[35]

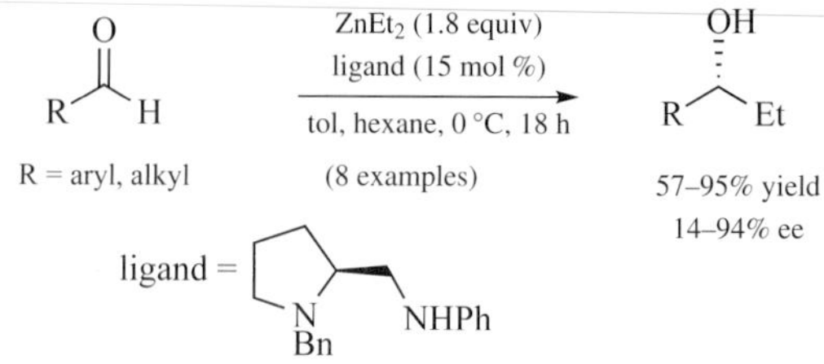

(14)

[1415831-69-4]

Enantioselective Allylic Alkylation. A number of structurally similar phosphine ligands have been developed and utilized for asymmetric alkylations. ESPHOS and SEMI-ESPHOS are examples of 2-(anilinomethyl)pyrrolidine-modified diazaphospholidine ligands that have found utility in the stereocontrol of asymmetric alkylation reactions (eq 15).[36] Further modification of this general structure has led to FerriESPHOS and DiPhenESPHOS, which were found to be superior to ESPHOS in providing chiral induction.[37]

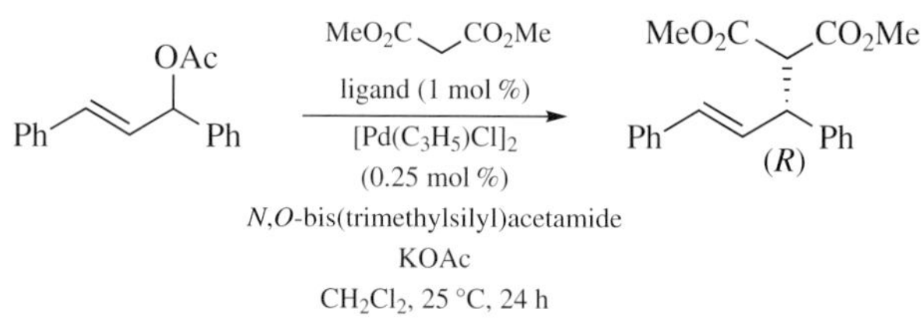

Ligand	Yield (%)	ee (%)
ESPHOS *[254990-08-4]*	77	81 (*R*)
SEMI-ESPHOS *[207129-47-3]*	91	81 (*S*)
FerriESPHOS *[746663-41-2]*	87	92 (*S*)
DiPhenESPHOS *[745820-65-9]*	53	91 (*S*)

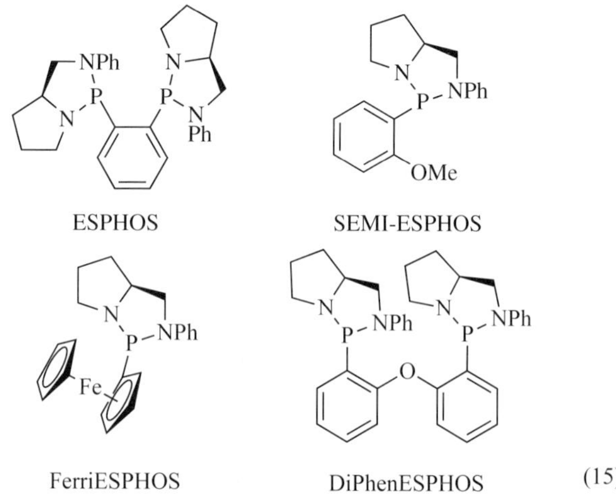

(15)

While other (*S*)-AMP-derived ligands containing chiral pyridine-phosphines,[38] oxazoline-phosphines,[31] and hydrobenzoin amidophosphites[39] have been developed for this transformation, they were generally found to proceed in lower enantiomeric excesses than ESPHOS-derived ligands. Interestingly, dramatic simplification of the frequently used QUIPHOS ligand[40] increased enantioselectivities (eq 16).[41]

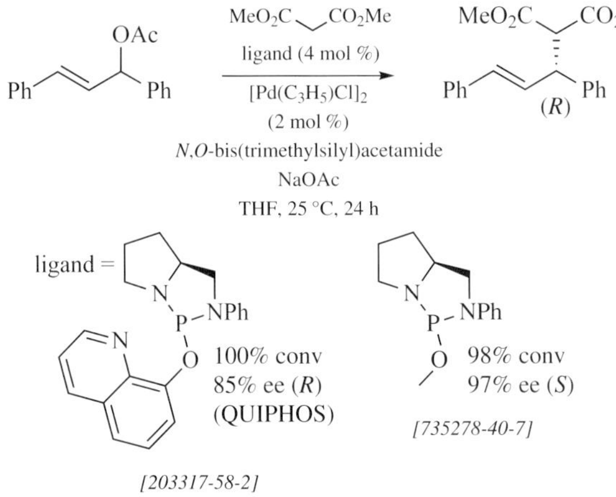

(16)

Enantioselective Allylic Amination Reactions. Allylic aminations have also been reported utilizing conditions similar to those described for allylic alkylations. QUIPHOS was found to be an excellent ligand for this transformation, giving 94% enantiomeric excess (eq 17),[40] although SEMI-ESPHOS and its derivatives,[42] AMP-hydrobenzoin amidophosphites,[39] and AMP-monodentate diamidophosphites[41] were also effective.

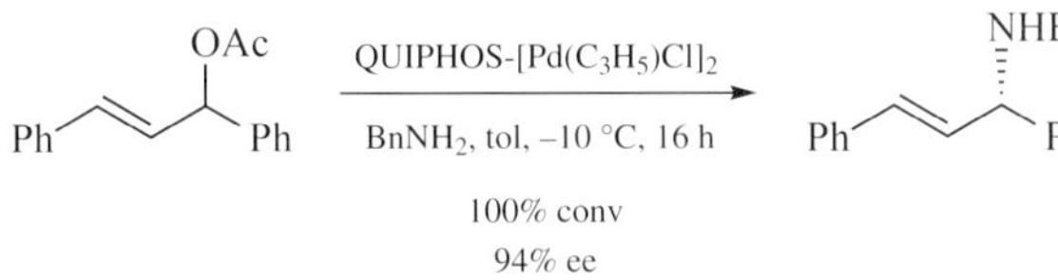

(17)

1,4-Addition Reactions. (*S*)-AMP has also been utilized as an efficient catalyst in enantioselective Michael additions of ketones to nitroalkenes (eq 18), providing high yields (77–99%) and enantiomeric excesses (59–89%).[43] The high enantiomeric excesses are likely achieved due to the hydrogen-bond donating ability of the catalyst, allowing for selective organization of the enamine and nitroalkene. Brine can be used instead of toluene in this transformation, providing a more environment-friendly solvent option.[44]

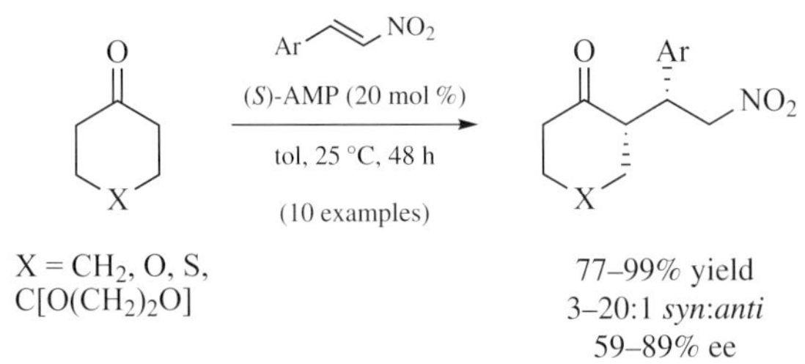

(18)

Although (*S*)-AMP can be utilized as a chiral ligand in zinc-mediated conjugate addition reactions of Grignard reagents to enones, the enantioselectivities were poor (21%).[45] Similarly, the use of phosphorous triamide ligands based on (*S*)-AMP[46] or the QUIPHOS ligand[47] in copper-catalyzed alkylzinc 1,4-addition reactions (eq 19) gave poor to modest enantioselectivities (0–61%).

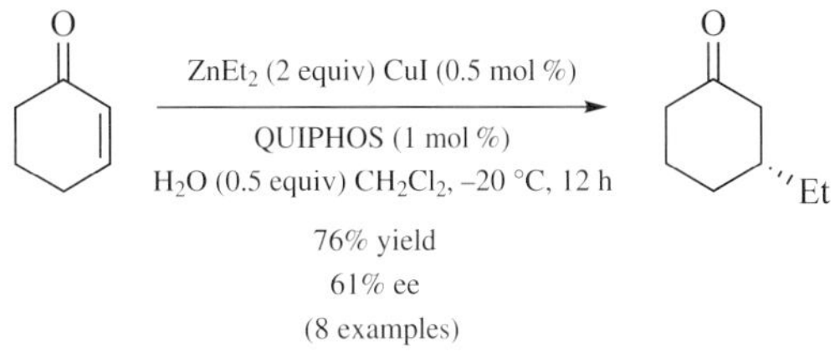

(19)

Asymmetric Allylations and Crotylations. The structural rigidity of chiral phosphoramides derived from (*S*)-AMP[48] has facilitated their use as catalysts in asymmetric allylation and crotylation reactions. Not surprisingly, (*S*)-AMP-derived catalysts gave lower enantioselectivities in the allylation of benzaldehyde than bulkier naphthyl or anthranyl derivatives (eq 20).[49]

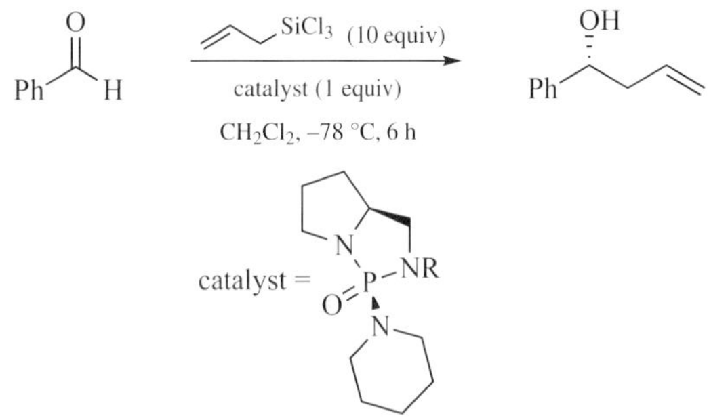

(20)

R =	Yield (%)	ee (%)
Ph *[189107-19-5]*	74	49
1-naphthyl *[181653-69-0]*	74	71
1-anthranyl *[189107-24-2]*	57	81

Enantioselective Hydroformylation. Hydroformylation of vinyl acetate in the presence of rhodium(I) complexes and ESPHOS, an (*S*)-AMP-derived chiral bidentate ligand, led to excellent results (eq 21). Monodentate SEMI-ESPHOS, on the other hand, gave very low yields of racemic product, even at elevated temperatures and pressures.[50] Modification of the ESPHOS[37] and SEMI-ESPHOS[51] ligands for use in the hydroformylation reaction, unfortunately, led to inferior results.

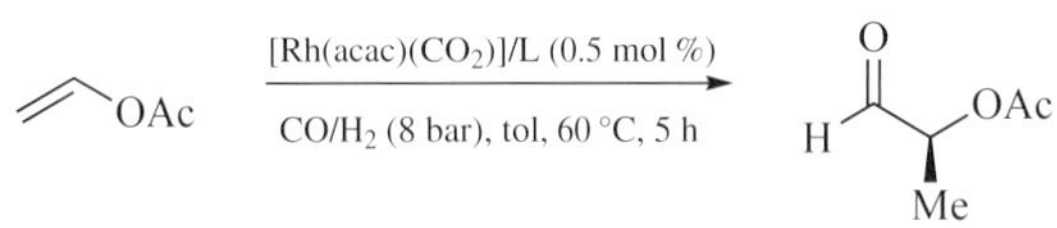

(21)

Ligand	Yield (%)	branched:linear	ee (%)
ESPHOS	90	94.5:5.5	89
SEMI-ESPHOS	<15	70:30	<2

Diels-Alder Reactions. (*S*)-AMP also serves as an effective chiral auxiliary in stereoselective Diels–Alder reactions. Exposure of the corresponding (*S*)-AMP-hydroxamic acid to *n*-Pr_4NIO_4 led to rapid oxidation to the acylnitroso dienophile, which upon exposure to 1,3-cyclohexadiene (eq 22) provided the cycloadduct in high yield (79%) and moderate diastereoselectivity (64%).[52]

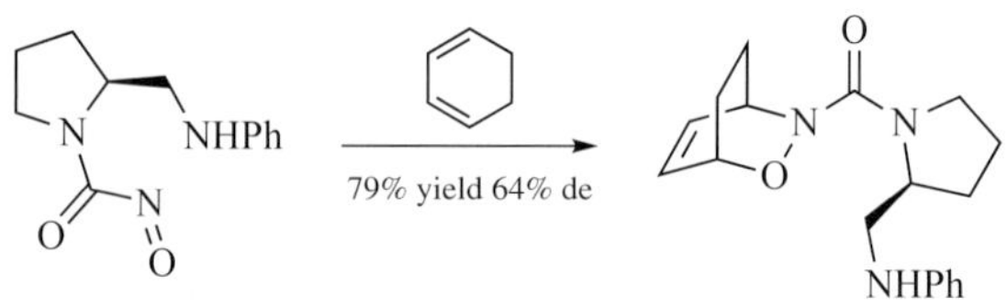

(22)

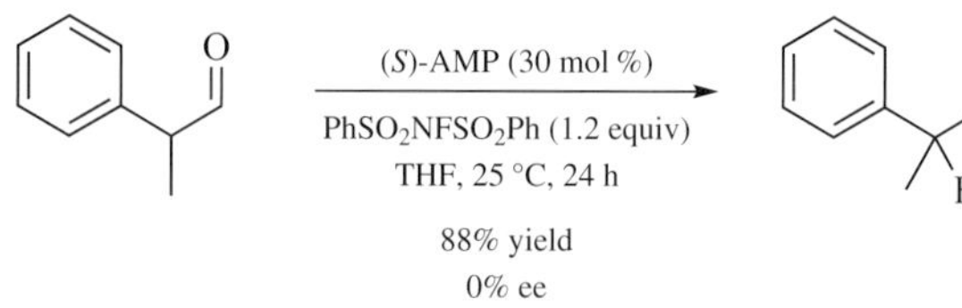

(26)

The use of QUIPHOS as a chiral ligand in the presence of a Cu(II) catalyst has led to a highly selective Diels–Alder reaction between 3-acryloyl-1,3-oxazolidine-2-one and cyclopentadiene (eq 23).[53]

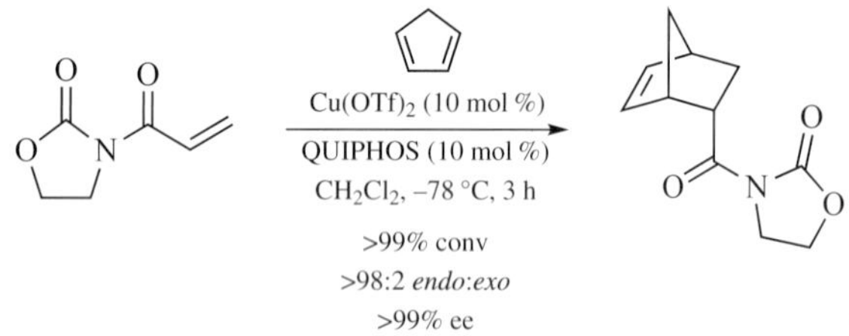

(23)

Asymmetric Cyclopropanation. Chiral iminodiazaphospholidine AMP ligands, with chiral phosphorous centers, have been utilized in copper-catalyzed cyclopropanation reactions (eq 24) with high *trans/cis* selectivity, although with variable enantiomeric excesses (12–95%).[54]

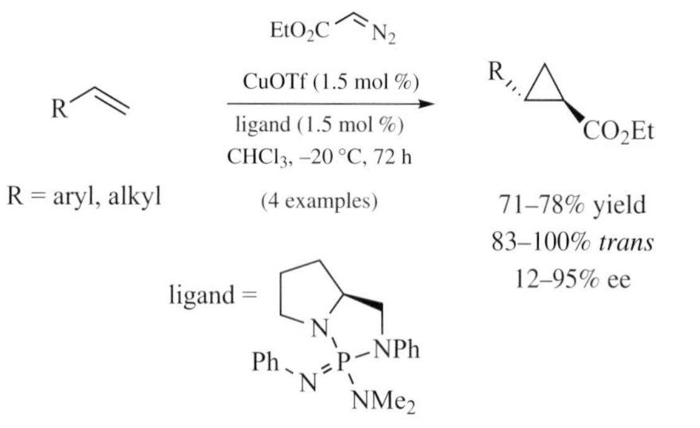

(24)

[232618-90-5]

Enantioselective Carboxylation. Carboxylation of chiral expoxides can be achieved in the presence of Nb(V) oxide with complete retention of configuration. With racemic epoxides, carboxylation in the presence of Nb(IV) and a chiral ligand, such as (*S*)-AMP, led to a small induction in enantiomeric excess (5% for AMP and up to 22% with other ligands, eq 25).[55]

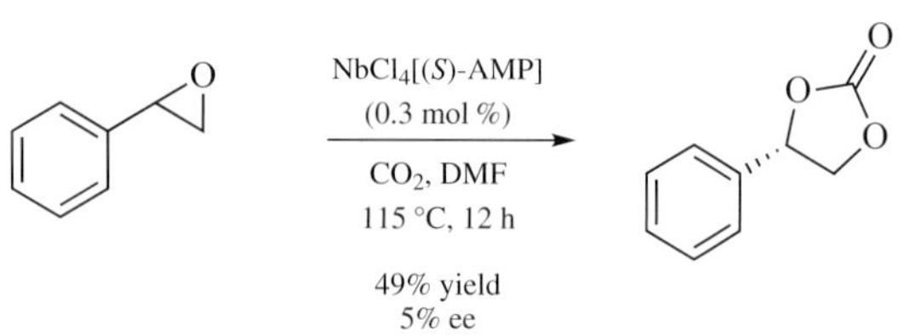

(25)

α-Fluorination of Aldehydes. While other proline-derived ligands provided modest enantioselection in the α-fluorination of 2-phenylpropionaldehyde, (*S*)-AMP (eq 26), unfortunately, led to racemic product, albeit in high yield (88%).[56]

Chiral Resolution. (*S*)-AMP has been utilized to aid in the stereoselective synthesis of chiral auxiliaries. Dynamic thermodynamic resolution of epimeric aminals prepared from aromatic amides and (*S*)-AMP facilitated the formation of atropisomeric amide chiral auxiliaries (eq 27),[57] which have found utility in the formation of substituted alcohols.[58]

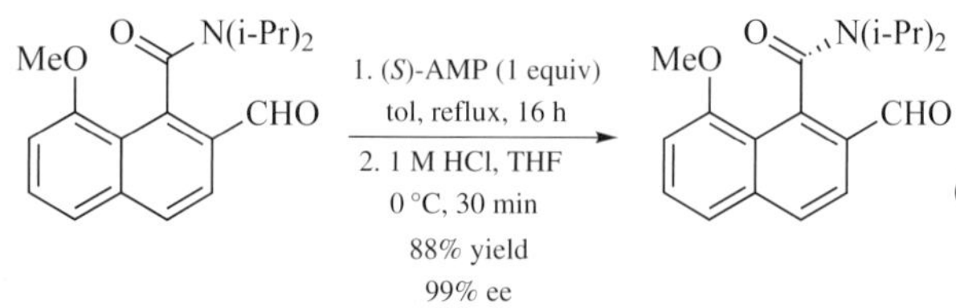

(27)

Enantioselective Oxidative Coupling. In addition to the wide use of (*S*)-AMP as a ligand in asymmetric reductions, it has also found use in oxidation reactions, such as the catalytic, asymmetric aerobic oxidative coupling of naphthols to BINOL derivatives. This can be achieved in the presence of CuCl and oxygen with (*S*)-AMP as a chiral ligand in 30% ee (eq 28), although modification of the substituents on (*S*)-AMP has led to enantiomeric excesses up to 73%.[59]

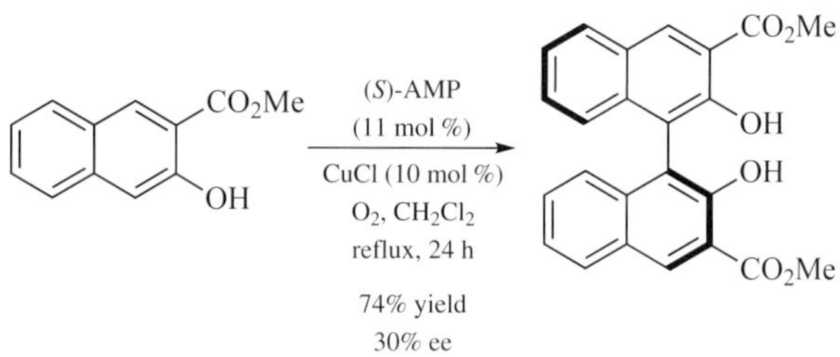

(28)

Second Update

Carina I. C. Crucho & Maria Teresa Barros
Universidade Nova de Lisboa, Caparica, Portugal

Since the discovery of L-proline-catalyzed reactions, several proline derivatives have been extensively studied as organocatalysts.[60] In the realm of proline-based organocatalysts, (*S*)-2-(anilinomethyl)pyrrolidine (AMP) has emerged as an important small organic molecule in the catalysis of organic transformations.

Asymmetric Allylic Alkylation. A series of diamidophosphite ligands were prepared by the treatment of ferrocene-based imino alcohols with chlorophosphites derived from AMP (eq 29).[61] These *P**-chiral diamidophosphites gave up to 97% ee in the Pd-catalyzed allylic substitution of 1,3-diphenylallyl acetate with dimethyl malonate, which was one of the best results for all known chiral *P*,*N*-phosphite ligands.

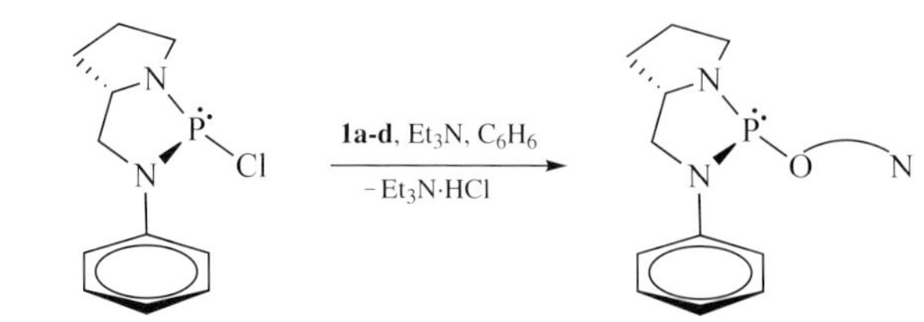

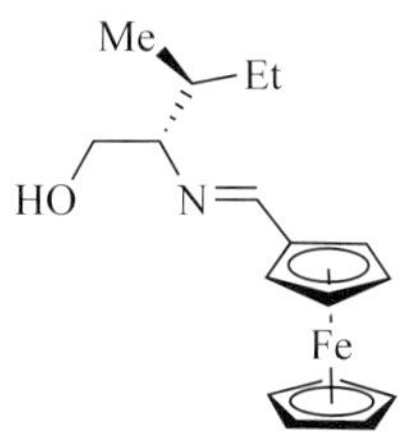

1a

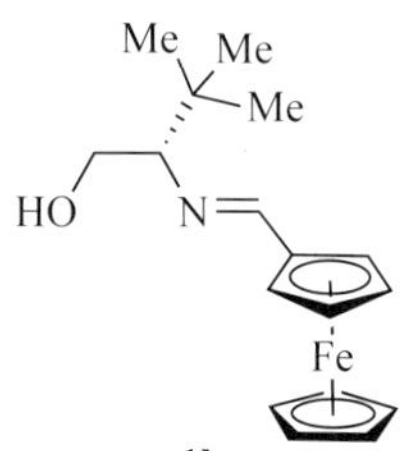

1b

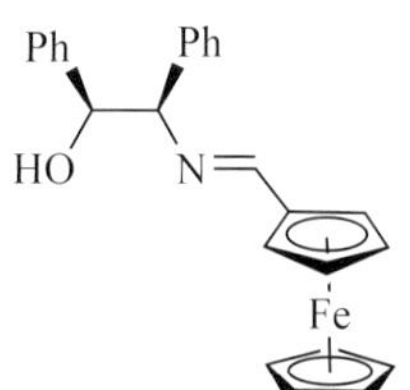

1c

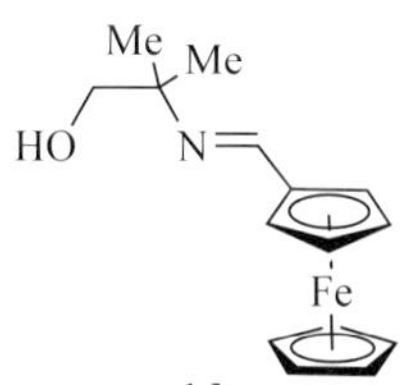

1d

(29)

Aldol Addition Reaction. The aldol reaction is a powerful method for the construction of carbon-carbon bonds with concomitant formation of the β-hydroxycarbonyl structure unit. An eco-friendly polymer-supported catalyst was developed from an optically active pyrrolidine-containing polyaniline, poly(2-anilinomethylpyrrolidine) (PANiMp), for the direct aldol reaction in an aqueous medium (eq 30).[62] The PANiMp-based catalyst was useful for the aldol reaction; however, the enantioselectivity was not observed. These promising results could present a new opportunity for developing a polyaniline-based chiral catalyst.

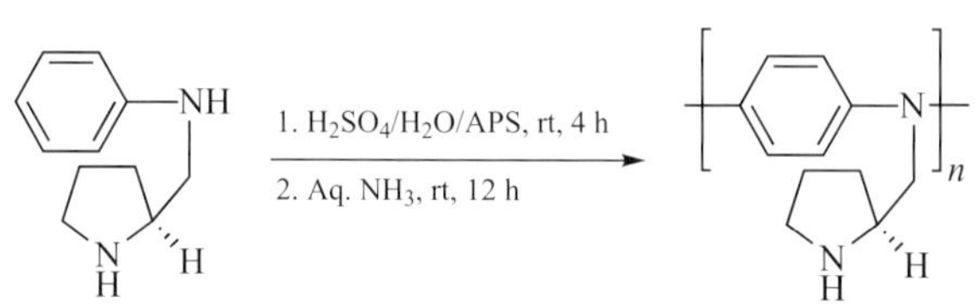

(30)

Diels-Alder Reactions. *P*-Chiral triaminophosphines with an AMP design have been used for many enantioselective metal-catalyzed reactions. Toselli et al. studied structural modifications on the pyrrolidine ring for enantioselective cobalt-catalyzed [6 + 2] cycloaddition (eq 31).[63]

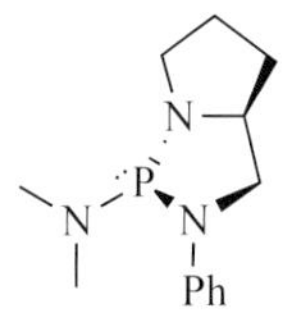

2

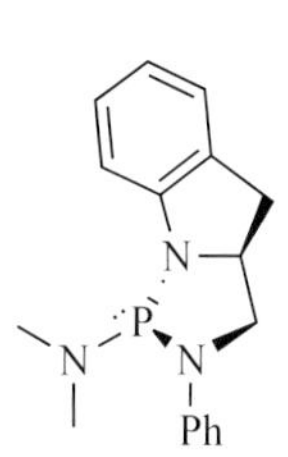

3

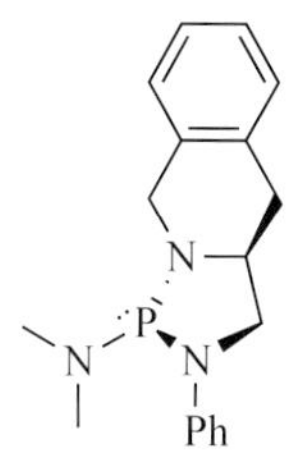

4

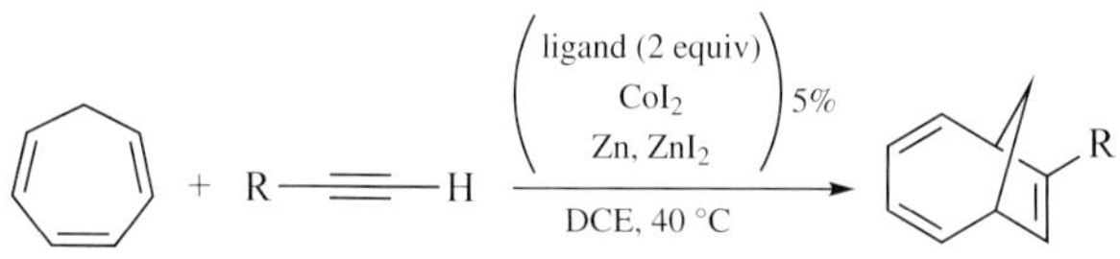

(31)

5a: R = Ph
5b: R = $SiMe_3$

Entry	Chiral ligand	R	Product	Yield (%)	ee (%)
1	**2**	Ph	(−)-**5a**	19	5
2	**3**	Ph	(+)-**5a**	6	41
3	**4**	Ph	–	0	–
4	**2**	SiMe3	(−)-**5b**	13	46
5	**3**	SiMe3	(−)-**5b**	42	29
6	**4**	SiMe3	–	0	–

Enantioselective Reductions. 2-(Anilinophenylmethyl)-pyrrolidine and 2-(anilinodiphenylmethyl)pyrrolidine chiral diamines were efficient chiral catalysts for the enantioselective borane reduction of prochiral ketones (eq 32).[64] Using (*S*)-2-(anilinodiphenylmethyl)pyrrolidine, chiral secondary alcohols were obtained from prochiral ketones with good to excellent enantiomeric excesses (up to 98% ee). Furthermore, the authors demonstrated that the aniline moiety was necessary to achieve easy formation of diazaborolidine under the mild reaction conditions and high enantioselectivity.

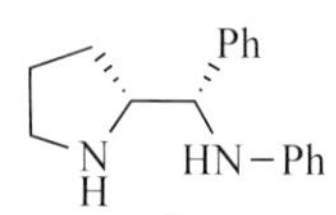

6

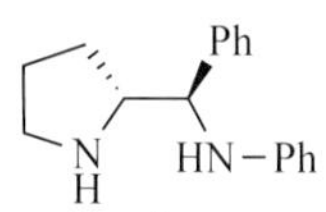

7

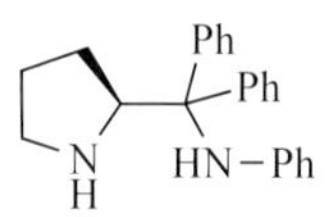

8

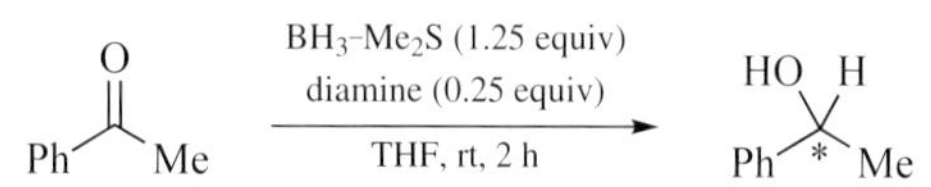

(32)

Diamine	Yield (%)[a]	ee (%) (configuration)[b]
(2*R*,1′*S*)-**6**	82	84 (*S*)
(2*R*,1′*R*)-**7**	86	95 (*S*)
(*S*)-**8**	93	93 (*R*)

[a] Isolated yield.

[b] ee was determined by HPLC analysis, and the absolute configuration was determined by specific rotation.

1. Grandbois, E. R.; Howard, S. I.; Morrison, J. D. In *Asymmetric Synthesis*, Morrison, J. D., Ed.; Academic: New York, 1983; Vol. *2*, Chapter 3.
2. (a) Mukaiyama, T; Asami, M.; Hanna, J.; Kobayashi, S., *Chem. Lett.* **1977**, 783. (b) Asami, M.; Ohno, H.; Kobayashi, S.; Mukaiyama, T., *Bull. Chem. Soc. Jpn.* **1978**, *51*, 1869. (c) Asami, M.; Mukaiyama, T., *Heterocycles* **1979**, *12*, 499.
3. (a) Mukaiyama, T.; Sakito, Y.; Asami, M., *Chem. Lett.* **1978**, 1253. (b) Mukaiyama, T.; Sakito, Y.; Asami, M., *Chem. Lett.* **1979**, 705. (c) Sakito, Y.; Mukaiyama, T., *Chem. Lett.* **1979**, 1027. (d) Sakito, Y.; Tanaka, S.; Asami, M.; Mukaiyama, T., *Chem. Lett.* **1980**, 1223.
4. Asami, M.; Mukaiyama, T., *Chem. Lett.* **1983**, 93.
5. Asami, M.; Mukaiyama, T., *Chem. Lett.* **1980**, 17.
6. Asami, M.; Mukaiyama, T., *Chem. Lett.* **1979**, 569.
7. Soai, K.; Niwa, S., *Chem. Rev.* **1992**, *92*, 833.
8. Eliel, E. L.; Koskimies, J. K.; Lohri, B., *J. Am. Chem. Soc.* **1978**, *100*, 1614.
9. Huang, K.; Breitbach, Z. S.; Armstrong, D. W., *Tetrahedron: Asymmetry* **2006**, *17*, 2821.
10. Iriuchijima, S., *Synthesis* **1978**, 684.
11. Dunina, V. V.; Gorunova, O. N.; Stepanova, V. A.; Zykov, P. A.; Livantsov, M. V.; Grishin, Y. K.; Churakov, A. V.; Kuz'mina, L. G., *Tetrahedron: Asymmetry* **2007**, *18*, 2011.
12. Elliott, R. L.; Kopecka, H.; Gunn, D. E.; Lin, N.-H.; Garvey, D. S.; Ryther, K. B.; Holladay, M. W.; Anderson, D. J.; Campbell, J. E.; Sullivan, J. P.; Buckley, M. J.; Gunther, K. L.; O'Neill, A. B.; Decker, M. W.; Arnerić, S. P., *Bioorg. Med. Chem. Lett.* **1996**, *6*, 2283.
13. Cheng, G.; Slawin, A. M. Z.; Vautravers, N. R.; André, P.; Morris, R. E.; Samuel, I. D. W.; Cole-Hamilton, D., *Org. Biomol. Chem.* **2011**, *9*, 1189.
14. Huang, Y.; Coull, J. M., *J. Am. Chem. Soc.* **2008**, *130*, 3238.
15. Reymond, S.; Brunel, J. M.; Buono, G., *Tetrahedron: Asymmetry* **2000**, *11*, 1273.
16. Oshikawa, T.; Yamashita, M.; Kumagai, S.; Seo, K.; Kobayashi, J., *J. Chem. Soc., Chem. Commun.* **1995**, 435.
17. Mukaiyama, T., *Tetrahedron* **1981**, *37*, 4111.
18. Takayanagi, H.; Kitano, Y.; Morinaka, Y., *J. Org. Chem.* **1994**, *59*, 2700.
19. Cerami, J.; Gala, D.; Hou, D.; Kalliney, S.; Mas, J. L.; Nyce, P.; Peer, L.; Wu, G., *Org. Process Res. Dev.* **2000**, *4*, 107.
20. Desos, P.; Serkiz, B.; Morain, P.; Lepagnol, J.; Cordi, A., *Bioorg. Med. Chem. Lett.* **1996**, *6*, 3003.
21. Bringmann, G.; Hartung, T., *Tetrahedron* **1993**, *49*, 7891.
22. Basavaiah, D.; Rao, K. V.; Reddy, B. S., *Tetrahedron: Asymmetry* **2006**, *17*, 1041.
23. Olivares-Romero, J. L.; Juaristi, E., *Tetrahedron* **2008**, *64*, 9992.
24. Basavaiah, D.; Reddy, G. J.; Chandrashekar, V., *Tetrahedron: Asymmetry* **2002**, *13*, 1125.
25. Basavaiah, D.; Chandrashekar, V.; Das, U.; Reddy, G. J., *Tetrahedron: Asymmetry* **2005**, *16*, 3955.
26. Basavaiah, D.; Rao, K. V.; Reddy, B. S., *Tetrahedron: Asymmetry* **2006**, *17*, 1036.
27. (a) Falorni, M.; Lardicci, L.; Giacomelli, G.; Marchetti, M., *Tetrahedron Lett.* **1989**, *30*, 3551.
    (b) Falorni, M.; Giacomelli, G.; Marchetti, M.; Culeddu, N.; Lardicci, L., *Tetrahedron: Asymmetry* **1991**, *2*, 287.
28. Aitali, M.; Allaoud, S.; Karim, A.; Meliet, C.; Mortreux, A., *Tetrahedron: Asymmetry* **2000**, *11*, 1367.
29. Mimoun, H.; Yves de Saint Laumer, J.; Giannini, L.; Scopelliti, R.; Floriani, C., *J. Am. Chem. Soc.* **1999**, *121*, 6158.
30. Roucoux, A.; Suisse, I.; Devocelle, M.; Carpentier, J.-F.; Agbossou, F.; Mortreux, A., *Tetrahedron: Asymmetry* **1996**, *7*, 379.
31. Hilgraf, R.; Pfaltz, A., *Adv. Synth. Catal.* **2005**, *347*, 61.
32. Zhang, C.; Hosoda, N.; Asami, M., *Tetrahedron: Asymmetry* **2007**, *18*, 2185.
33. O'Brien, P.; Warren, S., *J. Chem. Soc., Perkin Trans. 1* **1996**, 2117.
34. (a) O'Brien, P.; Warren, S., *Tetrahedron Lett.* **1995**, *36*, 2681. (b) O'Brien, P.; Warren, S., *J. Chem. Soc., Perkin Trans. 1* **1996**, 2129.
35. Hosoda, N.; Ito, H.; Takimoto, T.; Asami, M., *Bull. Chem. Soc. Jpn.* **2012**, *85*, 1014.
36. Breeden, S.; Wills, M., *J. Org. Chem.* **1999**, *64*, 9735.
37. Clarkson, G. J.; Ansell, J. R.; Cole-Hamilton, D. J.; Pogorzelec, P. J.; Whittell, J.; Wills, M., *Tetrahedron: Asymmetry* **2004**, *15*, 1787.
38. Brunel, J. M.; Constantieux, T.; Labande, A.; Lubatti, F.; Buono, G., *Tetrahedron Lett.* **1997**, *38*, 5971.
39. Gavrilov, K. N.; Chuchelkin, I. V.; Zheglov, S. V.; Groshkin, N. N.; Novikov, I. M.; Rastorguev, E. A.; Davankov, V. A., *Russ. Chem. Bull. Int. Ed.* **2011**, *60*, 2063.
40. Brunel, J. M.; Constantieux, T.; Buono, G., *J. Org. Chem.* **1999**, *64*, 8940.
41. Tsarev, V. N.; Lyubimov, S. E.; Shiryaev, A. A.; Zheglov, S. V.; Bondarev, O. G.; Davankov, V. A.; Kabro, A. A.; Moiseev, S. K.; Kalinin, V. N.; Gavrilov, K. N., *Eur. J. Org. Chem.* **2004**, 2214.
42. Edwards, C. W.; Shipton, M. R.; Alcock, N. W.; Clase, H.; Wills, M., *Tetrahedron* **2003**, *59*, 6473.
43. Pansare, S. V.; Kirby, R. L., *Tetrahedron* **2009**, *65*, 4557.
44. Singh, S.; Chimni, S. S., *Tetrahedron: Asymmetry* **2012**, *23*, 1068.
45. Jansen, J. F. G. A.; Feringa, B. L., *J. Org. Chem.* **1990**, *55*, 4168.
46. Barta, K.; Hölscher, M.; Franció, G.; Leitner, W., *Eur. J. Org. Chem.* **2009**, 4102.
47. (a) Delapierre, G.; Constantieux, T.; Brunel, J. M.; Buono, G., *Eur. J. Org. Chem.* **2000**, 2507. (b) Delapierre, G.; Brunel, J. M.; Constantieux, T.; Buono, G., *Tetrahedron: Asymmetry* **2001**, *12*, 1345.
48. Peyronel, J.-F.; Samuel, O.; Fiaud, J.-C., *J. Org. Chem.* **1987**, *52*, 5320.
49. Iseki, K.; Kuroki, Y.; Takahashi, M.; Kishimoto, S.; Kobayashi, Y., *Tetrahedron* **1997**, *53*, 3513.
50. Breeden, S.; Cole-Hamilton, D. J.; Foster, D. F.; Schwarz, G. J.; Wills, M., *Angew. Chem., Int. Ed.* **2000**, *39*, 4106.
51. Vautravers, N. R.; Cole-Hamilton, D. J., *Chem. Commun.* **2009**, 92.
52. Brouillard-Poichet, A.; Defoin, A.; Streith, J., *Tetrahedron Lett.* **1989**, *30*, 7061.
53. Brunel, J. M.; Campo, B. D.; Buono, G., *Tetrahedron Lett.* **1998**, *39*, 9663.
54. Brunel, J. M.; Legrand, O.; Reymond, S.; Buono, G., *J. Am. Chem. Soc.* **1999**, *121*, 5807.
55. Aresta, M.; Dibenedetto, A.; Gianfrate, L.; Pastore, C., *Appl. Catal. A: Gen.* **2003**, *255*, 5.
56. Steiner, D. D.; Mase, N.; Barbas, C. F., *Angew. Chem., Int. Ed.* **2005**, *44*, 3706.
57. (a) Clayden, J.; Lai, L. W., *Angew. Chem., Int. Ed.* **1999**, *38*, 2556. (b) Clayden, J.; Lai, L. W.; Helliwell, M., *Tetrahedron* **2004**, *60*, 4399.
58. Clayden, J.; McCarthy, C.; Cumming, J. G., *Tetrahedron Lett.* **2000**, *41*, 3279.
59. (a) Nakajima, M.; Kanayama, K.; Miyoshi, I.; Hashimoto, S., *Tetrahedron Lett.* **1995**, *36*, 9519. (b) Nakajima, M.; Miyoshi, I.; Kanayama, K.; Hashimoto, S., *J. Org. Chem.* **1999**, *64*, 2264.
60. (a) Kotsuki, H.; Ikishima, H.; Okuyama, A., *Heterocycles* **2008**, *75*, 757. (b) Kotsuki, H.; Ikishima, H.; Okuyama, A., *Heterocycles* **2008**, *75*, 493.
61. Tsarev, V. N.; Lyubimov, S. E.; Bondarev, O. G.; Korlyukov, A. A.; Antipin, M. Y.; Petrovskii, P. V.; Davankov, V. A.; Shiryaev, A. A.; Benetsky, E. B.; Vologzhanin, P. A.; Gavrilov, K. N., *Eur. J. Org. Chem.* **2005**, 2097.
62. Saravanan, C.; Palaniappan, S., *J. Appl. Polym. Sci.* **2010**, *116*, 2536.
63. Toselli, N.; Fortrie, R.; Martin, D.; Buono, G., *Tetrahedron: Asymmetry* **2010**, *21*, 1238.
64. Hosoda, N.; Kamito, H.; Takano, M.; Takebe, Y.; Yamaguchi, Y.; Asami, M., *Tetrahedron* **2013**, *69*, 1739.

9-Azabicyclo[3.3.1]nonane-*N*-oxyl

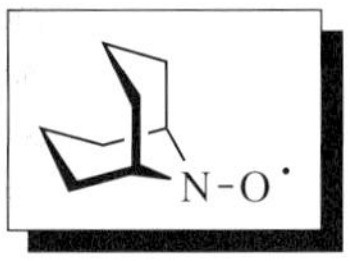

[31785-68-9] $C_8H_{14}NO$ (MW 140.20)
InChI = 1S/C8H14NO/c10-9-7-3-1-4-8(9)6-2-5-7/h7-8H,
1-6H2/t7-,8+
InChIKey = SZAKAAGHPLUEPM-OCAPTIKFSA-N

(stable nitroxyl radical, highly efficient organocatalyst for the oxidation of both primary and secondary alcohols to aldehydes and ketones, respectively; trapping agent for carbon radicals)

Solubility: soluble in a wide variety of common organic solvents such as CH_2Cl_2, $CHCl_3$, CH_3CN, benzene, fluorobenzene, *n*-hexane, and cyclopentane.

Form Supplied in: red solid.

Preparative Method: the title reagent can be prepared in three steps commencing with the decarboxylative double Mannich condensation of acetonedicarboxylic acid with glutaraldehyde and ammonia. Hydrazine reduction of the resultant bicyclic amino ketone using the Huang–Minlon modification of the Wolff–Kishner reduction followed by oxidation of the amine function with cat Na_2WO_4 and urea hydrogen peroxide (UHP) furnished ABNO as a red solid in 42% overall yield after purification by silica gel column chromatography (eq 1).[1]

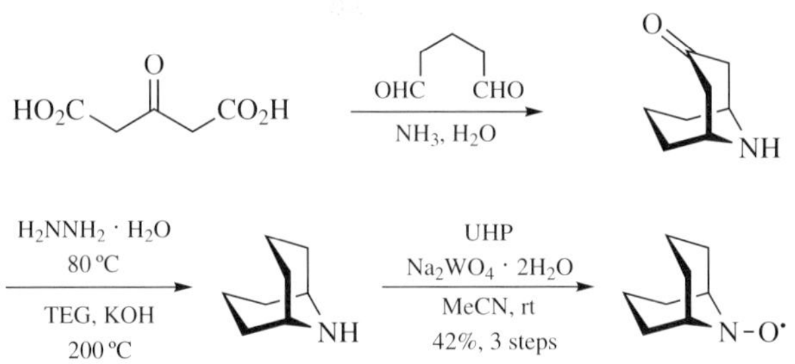

(1)

Purification: checked by TLC; purified by column chromatography on silica as needed; analytical samples can be prepared by sublimation: phase change 50 and ca. 85 °C, mp 126–129 °C.

Handling, Storage, and Precautions: care should be taken in handling this free radical reagent due to potential toxicity; should be stored in dark.

Oxidation of Alcohols to Aldehydes and Ketones. ABNO is an unhindered stable nitroxyl radical and can be used as a highly reactive catalyst for the oxidation of alcohols to the corresponding carbonyl compounds.[1] As shown in eqs 2 and 3, ABNO is significantly more reactive than 2,2,6,6-tetramethylpiperidin-1-oxyl (TEMPO) for the oxidation of primary alcohols to aldehydes when either sodium hypochlorite in combination with KBr (Anelli's conditions[2]) or $PhI(OAc)_2$ (Margarita's conditions[3]) was used as the bulk oxidant. Its activity is comparable with that of 1-methyl-2-azaadamantane *N*-oxyl (1-Me-AZADO); the latter is less readily available and requires a lengthy synthesis starting from 1,3-adamantanediol.[4]

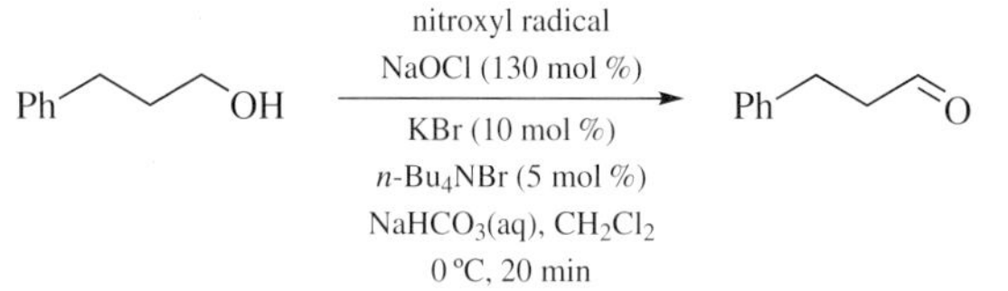

(2)

	Yield (%)		
mol %	TEMPO	1-Me-AZADO	ABNO
1	90	91	90
0.1	88	90	88
0.01	23	91	85

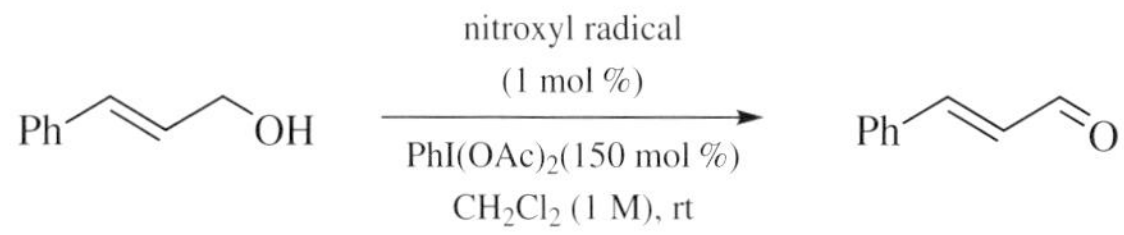

(3)

	Yield/reaction time
TEMPO	43%/360 min
1-Me-AZADO	93%/40 min
ABNO	91%/30 min

Unlike TEMPO, which is known to catalyze the selective oxidation of primary alcohols in the presence of secondary alcohols,[5] ABNO is capable of catalyzing the oxidation of secondary alcohols to ketones with very comparable efficacy to that of 1-Me-AZADO. In the presence of 1 mol % of ABNO, a variety of structurally hindered secondary alcohols were oxidized with NaOCl to the corresponding ketones in high yields (eq 4).

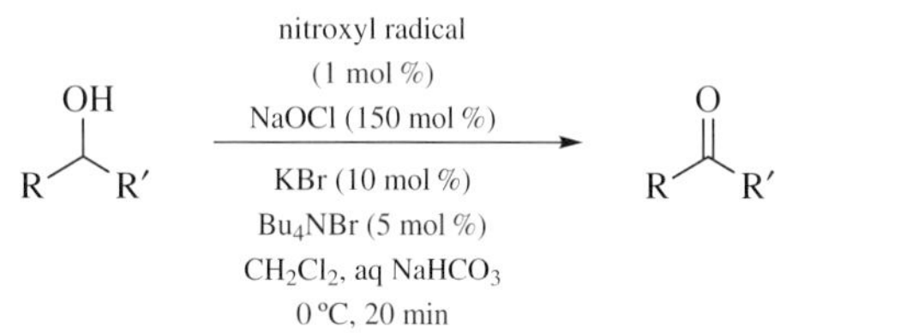

(4)

Alcohols	Ketones	Yield (%)
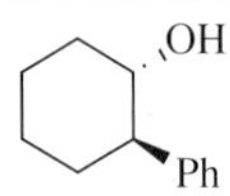	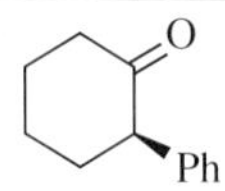	100 with ABNO 99 with 1-Me-AZADO 16 with TEMPO
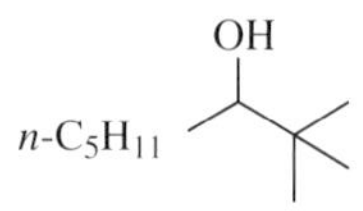	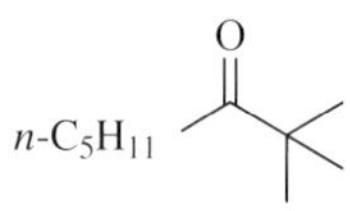	95 with ANBO 93 with 1-Me-AZADO 15 with TEMPO
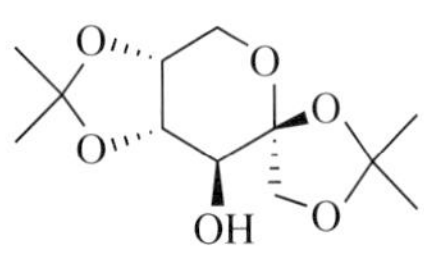	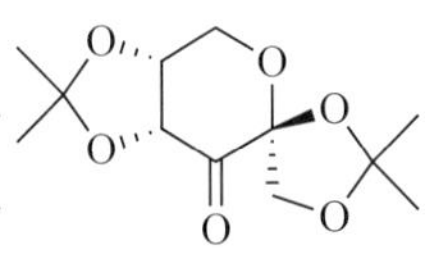	98 with ANBO 99 with 1-Me-AZADO 8 with TEMPO
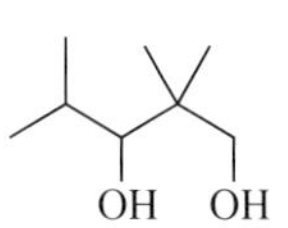	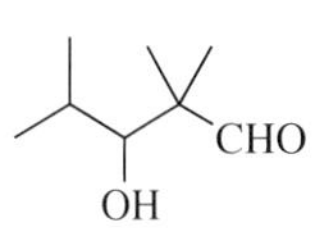	72 with ANBO 96 with 1-Me-AZADO

ABNO exhibits kinetic superiority compared to 1-Me-AZADO in the oxidation of structurally demanding alcohols. For example, ABNO was found to completely oxidize a hindered secondary alcohol, 1-(2-adamantanyl)-3-phenylpropanol, immediately after the addition of the bulk oxidant, while 1-Me-AZADO took 10 min to complete the reaction (eq 5). Notably, ABNO (10 mol %) oxidized a congested bicyclic alcohol substrate after 2 h with 98% yield (eq 6) whereas the use of 1-Me-AZADO as the catalyst gave only 36% yield even after 24 h under otherwise identical conditions.

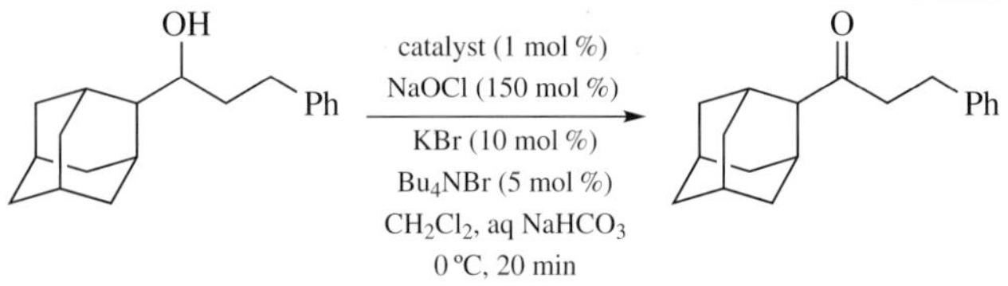

Catalyst	Catalytic efficiency
ABNO	94% after 2 min
1-Me-AZADO	48% after 2 min >90% after 10 min

(5)

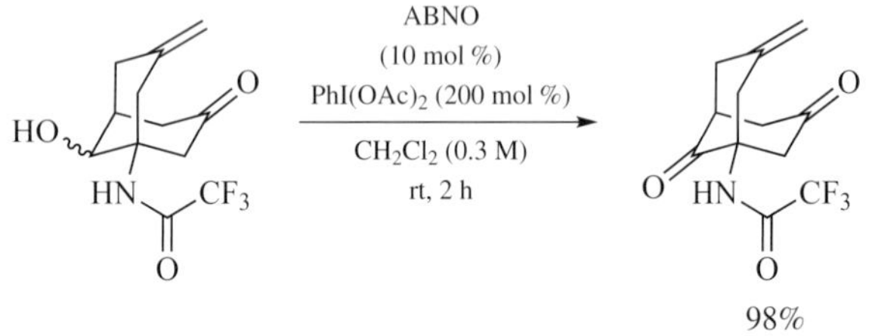

(6)

Trapping of Transient Carbon-centered Radicals. ABNO has been used as an efficient radical trapping agent in a number of kinetic studies involving fast and irreversible rearrangements in free radical chemistry. Thus, cyclopropylcarbinyl radicals, generated either by hydrogen abstraction of the parent hydrocarbons with *tert*-butoxyl radicals (eqs 7 and 8)[6,7] or from photolysis of an azo compound precursor (eq 9),[8] were trapped by ANBO to give the hydroxylamine adducts along with the trapped products of the ring-opened radicals.

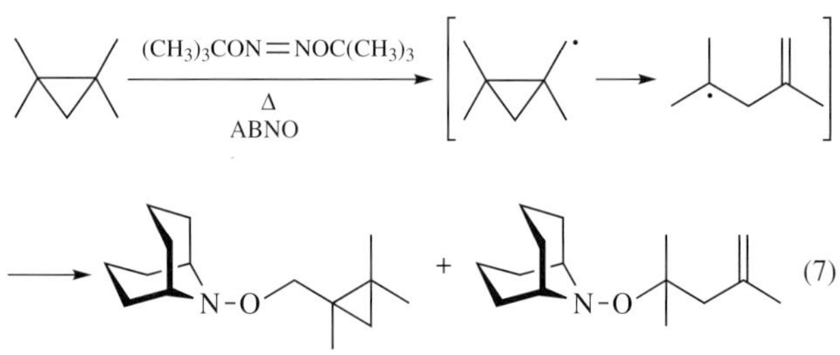

(7)

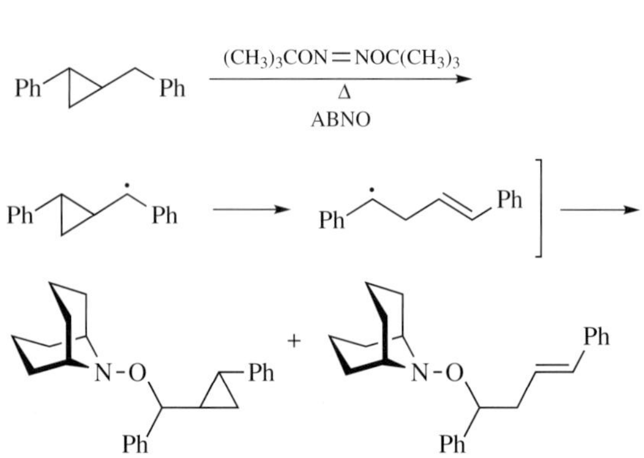

(8)

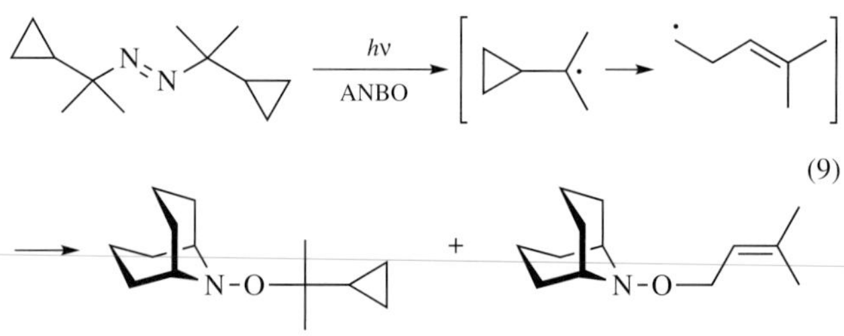

(9)

Photolysis of a trisazoalkane in the presence of ABNO led to the isolation of a stable trishydroxylamine adduct, pointing to the potential utility of the trisazoalkane as a star polymer initiator (eq 10).[9]

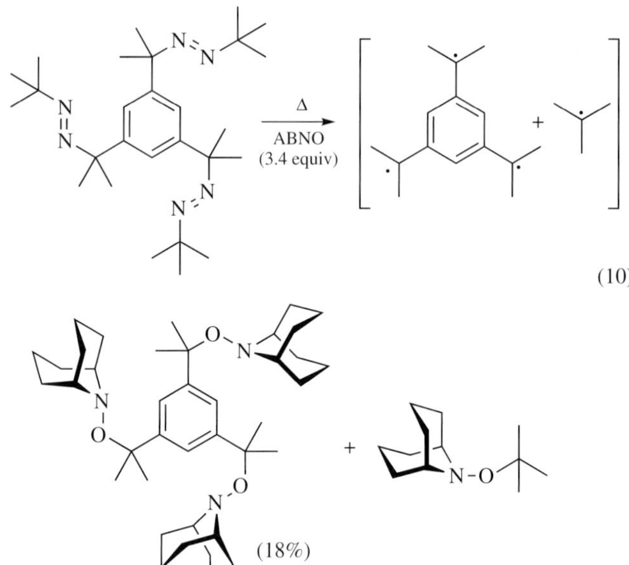

(10)

Related Reagents. 2,2,6,6-Tetramethylpiperidin-1-oxyl (TEMPO); 1-methyl-2-azaadamantane *N*-oxyl (1-Me-AZADO); *N*-hydroxyphthalimide (NHPI).

1. Shibuya, M.; Tomizawa, M.; Sasano, Y. I.; Iwabuchi, Y., *J. Org. Chem.* **2009**, *74*, 4619.
2. (a) Anelli, P. L.; Banfi, C.; Montanari, F.; Quici, S., *J. Org. Chem.* **1987**, *52*, 2559. (b) Anelli, P. L.; Banfi, S.; Montanari, F.; Quici, S., *J. Org. Chem.* **1989**, *54*, 2970.
3. De Mico, A.; Margarita, R.; Parlanti, L.; Vescovi, A.; Piancatelli, G., *J. Org. Chem.* **1997**, *62*, 6974.
4. Shibuya, M.; Tomizawa, M.; Suzuki, I.; Iwabuchi, Y., *J. Am. Chem. Soc.* **2006**, *128*, 8412.
5. (a) Siedlecka, R.; Skarzewski, J.; Michowski, J., *Tetrahedron Lett.* **1990**, *31*, 2177. (b) de Nooy, A. E. J.; Besemer, A. C.; van Bekkum, H., *Tetrahedron* **1995**, *51*, 8023.
6. Bowry, V. W.; Ingold, K. U., *J. Am. Chem. Soc.* **1992**, *114*, 4992.
7. Hollis, G.; Hughes, L.; Bowry, V. W.; Ingold, K. U., *J. Org. Chem.* **1992**, *57*, 4284.
8. Engel, P. S.; He, S.-L.; Banks, J. T.; Ingold, K. U.; Lusztyk, J., *J. Org. Chem.* **1997**, *62*, 1210.
9. Engel, P. S.; Pan, P.; Ying, Y.; Alemany, L. B., *J. Am. Chem. Soc.* **2001**, *123*, 3706.

Qingwei Yao
Northern Illinois University, DeKalb, IL, USA

B

Benzoyl trimethylsilane

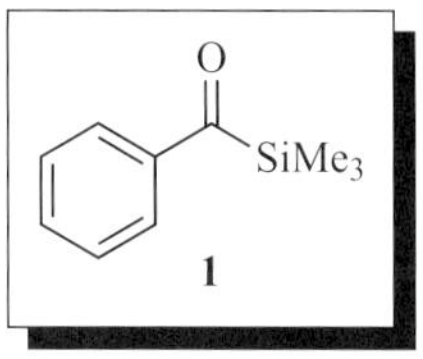

[5908-41-8] $C_{10}H_{14}OSi$ (MW 178.3)
InChI = 1S/C10H14OSi/c1-12(2,3)10(11)9-7-5-4-6-8-9/h4-8H,1-3H3
InChIKey = KJPBNYMAKNNPDA-UHFFFAOYSA-N

(reagent used as an acyl anion equivalent and an electrophile in wide array of reactions)

Physical Data: bp 214 °C; *d* 0.939 g cm^{-3}.
Solubility: soluble in ether, alcohol, toluene, and most organic solvents.
Form Supplied in: this reagent is not commercially available.
Purification: purified using silica gel flash column chromatography, elution with 2.5% diethyl ether in hexanes.[1]
Handling, Storage, and Precautions: because of the light-sensitive nature of the title compound, its exposure to light needs to be minimized.

Synthesis. Compound **1** has been synthesized via dibromination of benzyl-TMS, followed by the AgOAc-mediated hydrolysis of the dibromide (eq 1).[1] Compound **1** has also been synthesized via Pd-catalyzed reactions using hexamethyldisilane (eqs 2 and 3)[2,3] and from the hydrolysis of the dithioacetal **2** (eq 4).[4] Other reagents such as $HgCl_2$,[4] $NaClO_4/H_2O$,[5] and $CuO/CuCl_2$[6] have also been utilized for the hydrolysis of **2**.

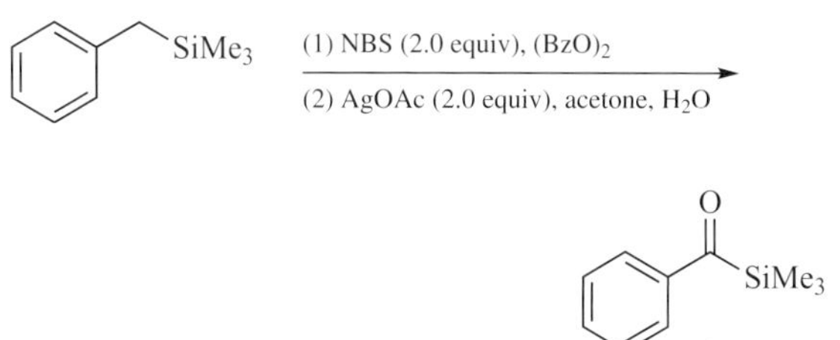

(1)

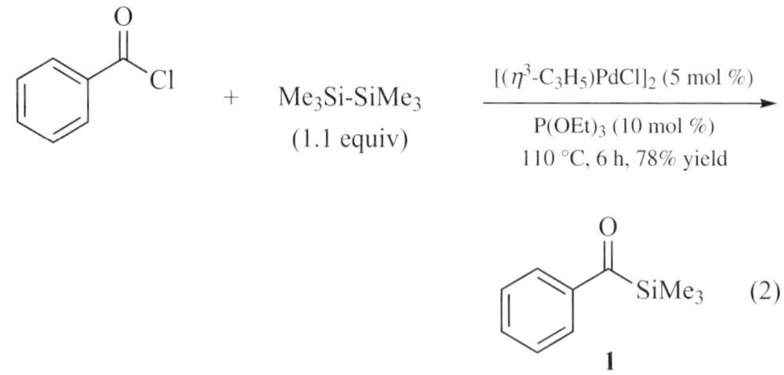

(2)

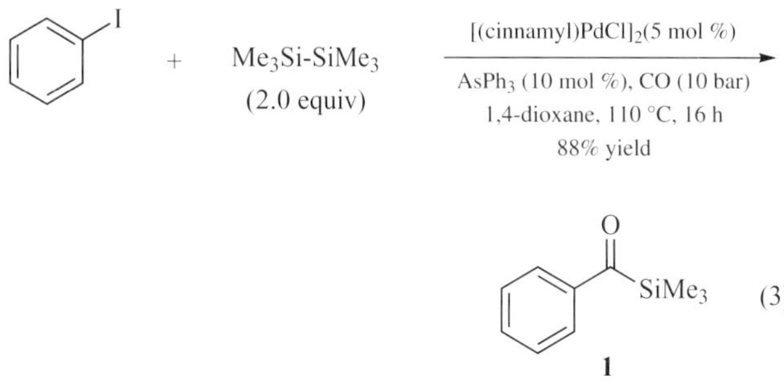

(3)

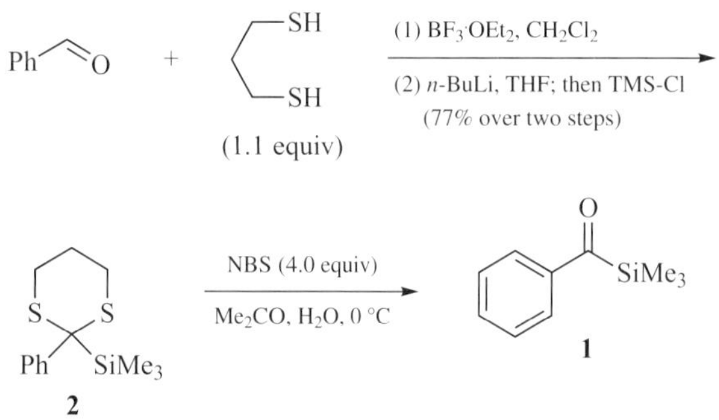

(4)

As Acyl Anion Equivalent. Scheidt and coworkers reported conjugate addition of **1** to unsaturated ketones and esters. *N*-heterocyclic carbene (NHC) **4** as a catalyst for this transformation.[7] The reaction furnished 1,4-dicarbonyl compounds in good yield (Table 1). Other thiazolium salts were also found to catalyze the conjugate addition.[8]

Bharadwaj and Scheidt reported a one-pot catalytic multicomponent synthesis of highly functionalized pyrroles using a tandem sila–Stetter/Paal–Knorr reaction.[9] The scope of this transformation is shown in Table 2. The reaction was tolerant of electron-donating substituents on the aromatic ring at the β-position (Table 2, entries 1 and 4). The placement of electron-withdrawing substituents at the α-position of the enone resulted in a decrease of the yield of the pyrrole (Table 2, entry 2). The anion generated from acyl silanes is reported to undergo nucleophilic addition into imines. The reaction furnished α-aminoketones and was catalyzed by thiazolium salt **5**.[10] The scope of this transformation is described in Table 3. *p*-Substituted phenyl imines were found to be excellent coupling partners for this transformation (Table 3, entries 1–3). Other aromatic imines also underwent the transformation in good yield (Table 3, entries 5 and 6).

$La(CN)_3$-catalyzed intermolecular benzoin condensation of **1** with aldehydes is reported. It resulted in the formation of α-hydroxyketones (Table 4).[11] Aromatic (Table 4, entry 1),

Table 1 Conjugate addition of **1** into conjugated ketones and esters

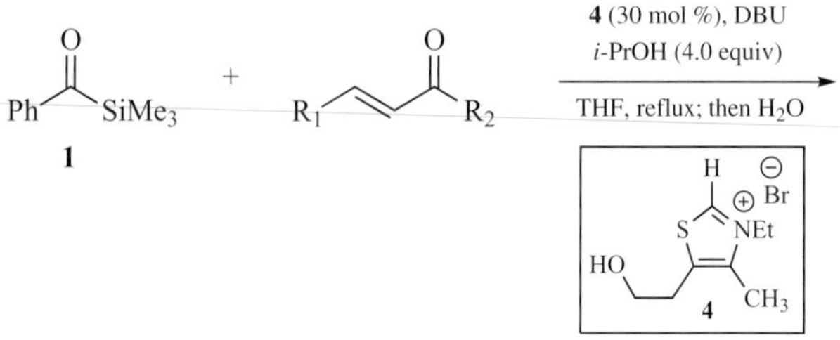

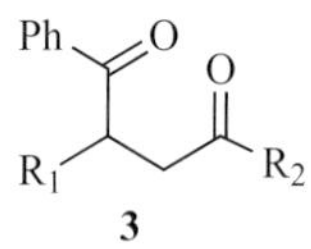

Entry	R_1	R_2	% Yield
1	Ph	Ph	77
2	Ph	4-OMePh	80
3	Ph	4-ClPh	82
4	4-OMePh	Ph	77
5	4-HOPh	Ph	50

Table 2 Sila–Stetter/Paal–Knorr reaction for the construction of substituted pyrroles[9]

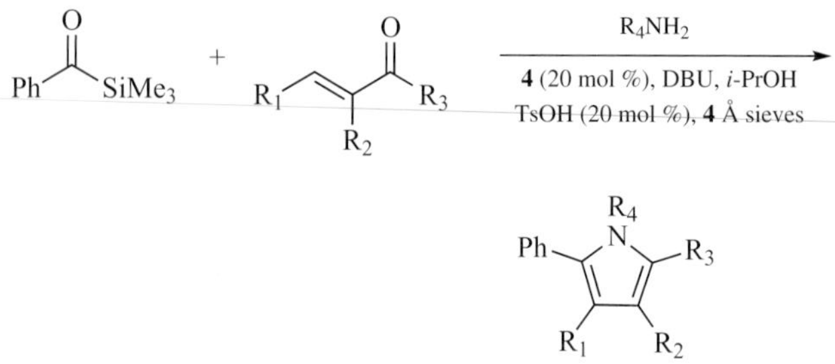

Entry	R_1	R_2	R_3	R_4	Yield (%)
1	4-MePh	H	Ph	Ph	70
2	Ph	COPh	Ph	Ph	35
3	Ph	Ph	Ph	Ph	66
4	3-MePh	H	Ph	Ph	62
5	4-ClPh	H	Ph	Ph	58

conjugated (Table 4, entry 2), and aliphatic (Table 4, entry 3) aldehydes underwent the transformation smoothly. Other cyanide sources, such as KCN/18-crown-6, were found to be competent catalysts for the intermolecular benzoin condensation (eq 5).[12] $La(CN)_3$-catalyzed benzoin condensation with ketones is also reported (eq 6).[13] Acylsilane **1** was found to undergo a nucleophilic addition into α,β-unsaturated amide **10** resulting in the formation of γ-ketoamide **12** (eq 7).[14] Johnson and coworkers reported metallophosphite-catalyzed enantioselective 1,3-silylacylation of nitrones (Table 5).[15]

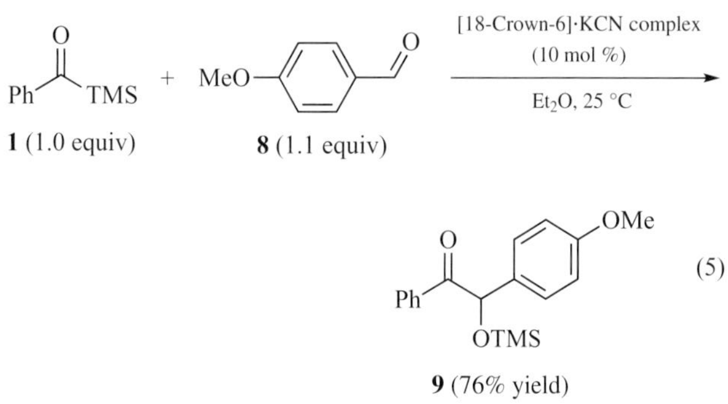

(5)

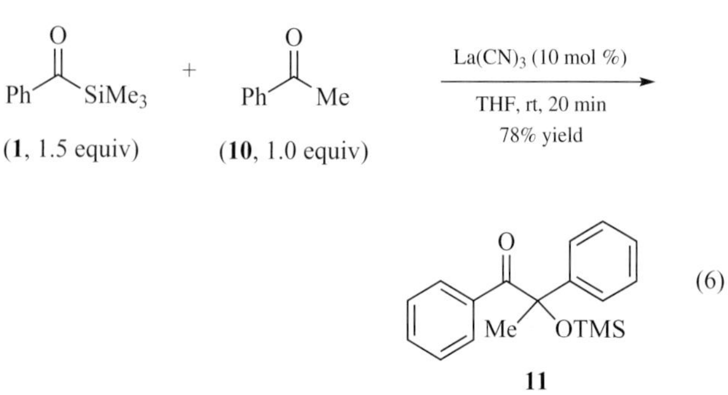

(6)

Reduction Reactions. A variety of reagents are shown to reduce **1** to the benzylic alcohol **16** (Table 6). The (*S*,*S*)-Noyori catalyst furnished the product **16** in high yield and with excellent enantioselectivity (Table 6, entry 1).[1] A combination of diol **17** and 2,4-dinitrobenzenesulfonic acid resulted in the isolation of the product in poor yield (Table 6, entry 2).[16] A mixture of Ti(O-*i*Pr)$_4$/Et_2Zn resulted in the isolation of the racemic **16** in good yield.[17]

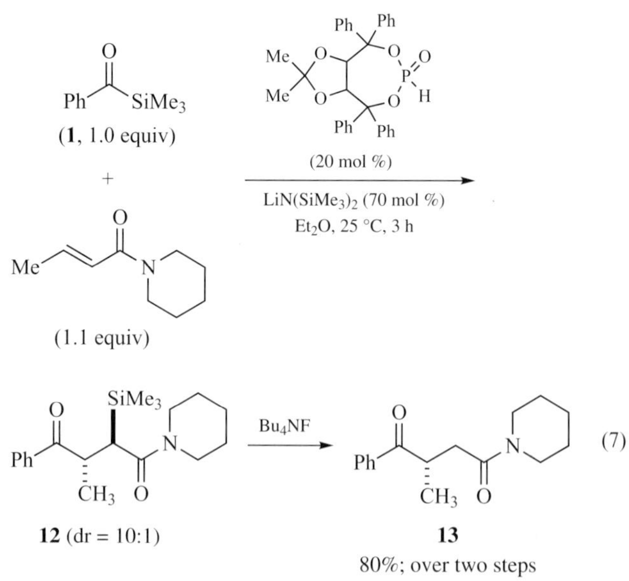

(7)

Reactions with Nucleophiles. The addition of the anion derived from sulfinyl ketimine **18** furnished *cis*-2-aminocyclopropanol **19** as the only diastereomer (eq 8).[18]

Table 3 Synthesis of α-aminoketones via addition of **1** into aryl imines[10]

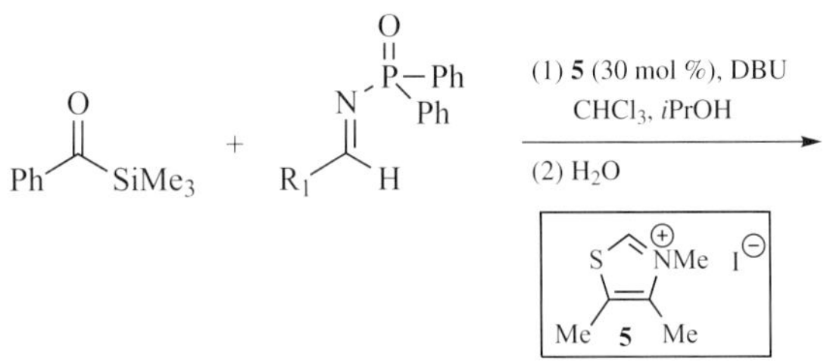

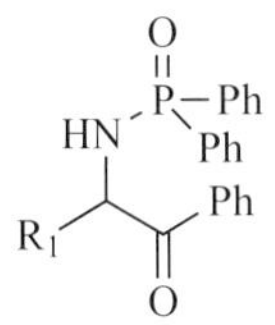

Entry	R_1	Isolated yield (%)
1	4-F-Ph	89
2	4-Cl-Ph	85
3	4-Me-Ph	94
4	3,4-Cl-Ph	67
5	2-naphthyl	80
6	2-thiophene	80

Table 4 Synthesis of α-hydroxyketones using **1** as an acyl anion equivalent

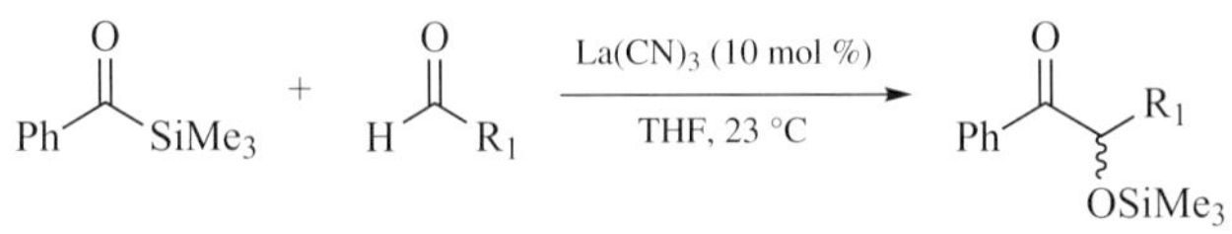

entry	R_1	isolated yield (%)
1	4-ClPh	87
2	(*E*)-CH=CHPh	50
3	CH_2CH_2Ph	64

Addition of the carbanion derived from sulfoxide **20** into acylsilane **1** resulted in the formation of silyl enol ether **21** (eq 9).[19] Lithiated ethyl diazoacetate can be added into **1** to form intermediate **22**. Reaction of the latter with electrophiles led to the formation of substituted β-keto esters (Table 7).[20] Addition of enantiopure lithium enolate derived from *N*-acetyl oxazolidinone **24** to **1** resulted in the formation of homoenolate **25**. The latter was quenched with electrophiles to furnish β-hydroxyketone products in good yield and dr (Table 8).[21] The homoenolate derived from the reaction of **24** and **1** was quenched with aldimine **27** resulting in the formation of carbinol **28**. Amine deprotection and cyclization of the latter produced γ-lactam **29** in high yield and with excellent ee.[22]

Table 5 Enantioselective 1,3-silylacylation of nitrone **13**

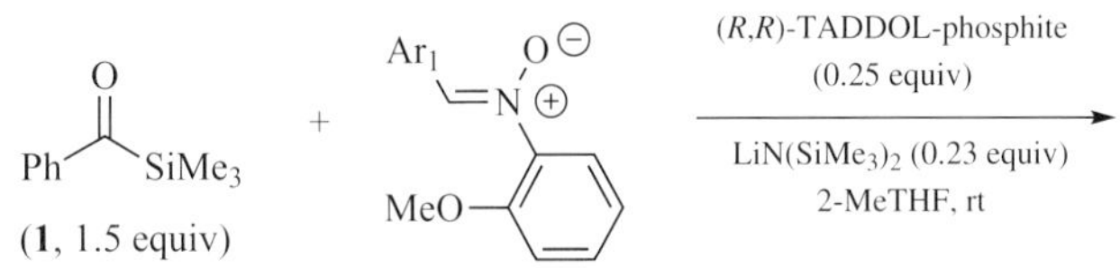

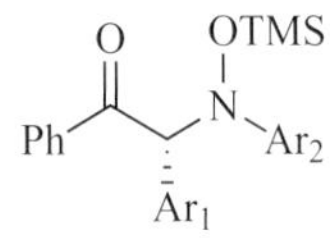

15

Entry	Ar_1	Isolated yield (%)	er
1	Ph	68	97:3
2	4-OMePh	84	97:3
3	4-NMe_2	76	96.5:3.5
4	2-MePh	77	97:3
5	4-ClPh	82	98:2

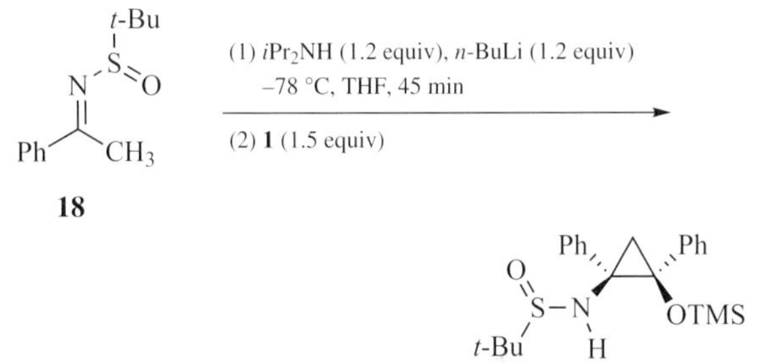

18

19 (74% yield) (8)

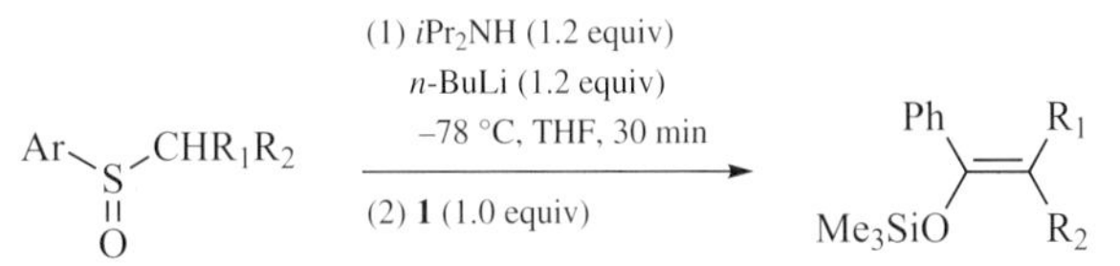

20 (1.2 equiv) **21** (59–92% yield) (9)

Ar = *p*-tol, Ph
R_1 = H, CH_3, CH_2=CH
R_2 = H, CH_3

Wittig reaction of **1** with fluorotribromomethane in presence of PPh_3 and Et_2Zn resulted in the formation of bromofluoroolefin **27** as a mixture of *E*- and *Z*-isomers (eq 10).[23] Shindo et al. reported a *Z*-selective olefination of acylsilanes. Nucleophilic addition of ynolate **28** (generated in situ from the dibromoester **28**) to **1** resulted in the formation of α,β-unsaturated lithium carboxylate. The treatment of the latter with MeI furnished the corresponding methyl ester **30** in excellent stereoselectivity (eq 11).[24]

Table 6 Synthesis of benzylic alcohol **16**

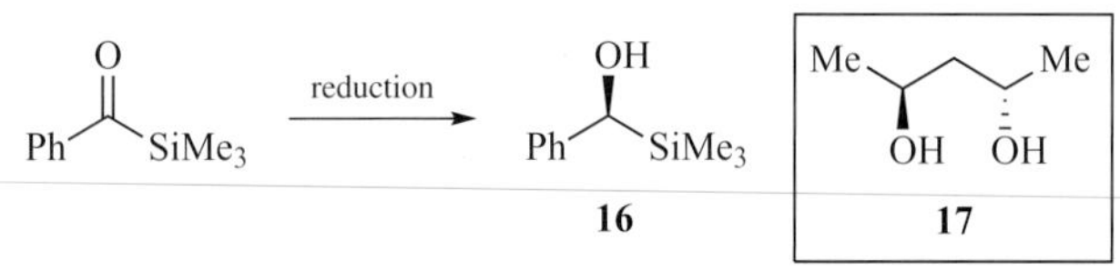

Entry	Catalyst	Conditions	Yield (%), ee (%)
1	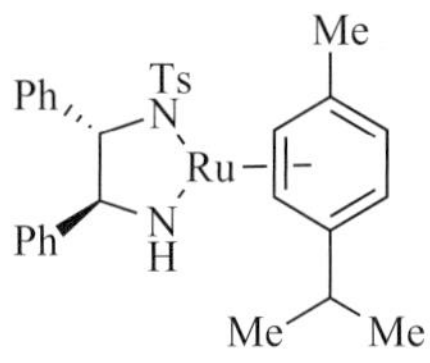 (0.5 mol %)	*i*PrOH, 3.4 h, rt	91, 98
2	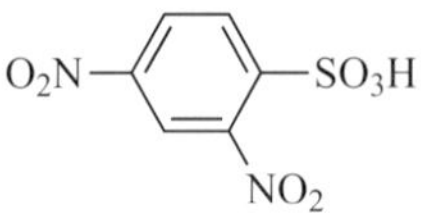 (5 mol %)	diol **17** (1.1 equiv), C_6H_6, reflux, 30 min	17, 74
3	none	Ti(O-*i*Pr)$_4$ (1.0 equiv) Et_2Zn (2.0 equiv) Et_2O, 12 h, rt	78, n.d.

Table 7 Synthesis of substituted *β*-keto esters

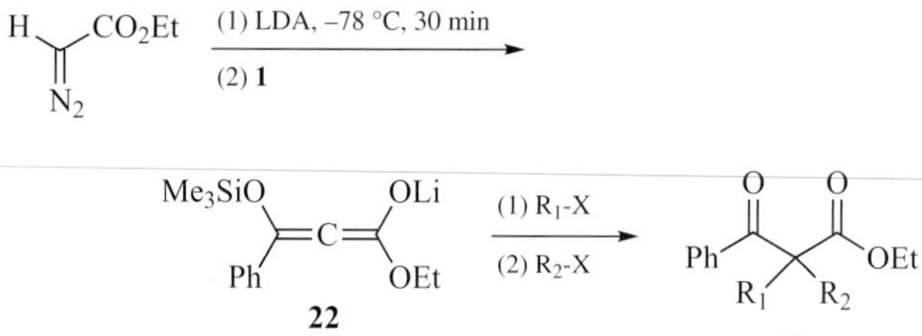

Entry	Electrophile	Product	Isolated yield (%)
1	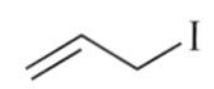	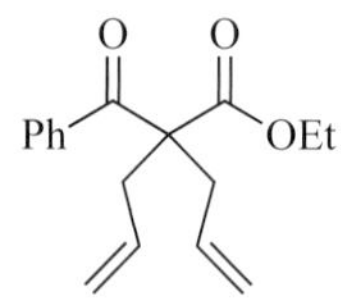	91
2	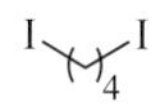	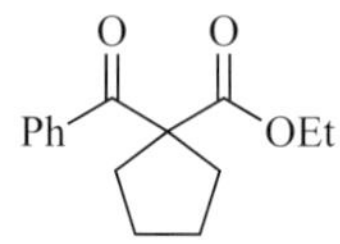	84
3	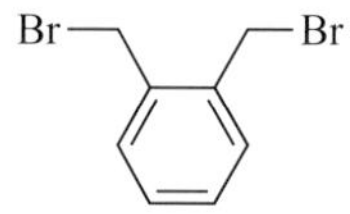	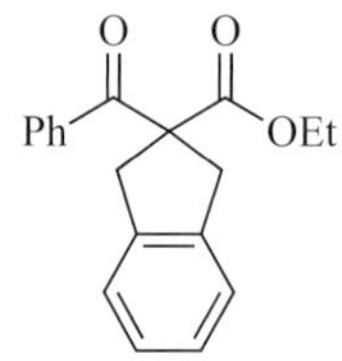	77
4	CH_3I; then allyl iodide	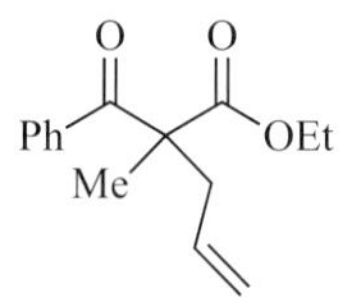	81

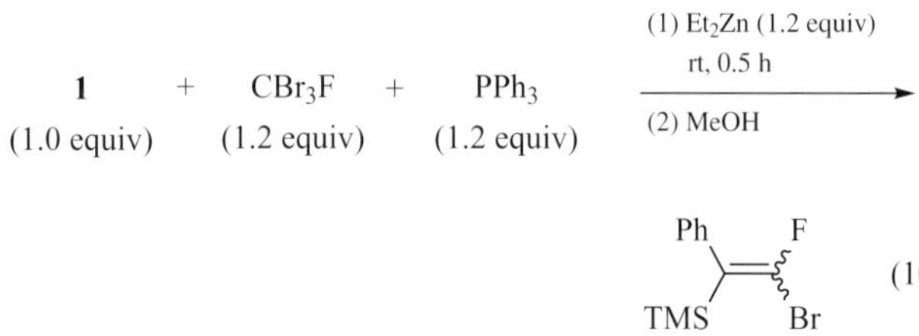

(10)

27 (53% yield)

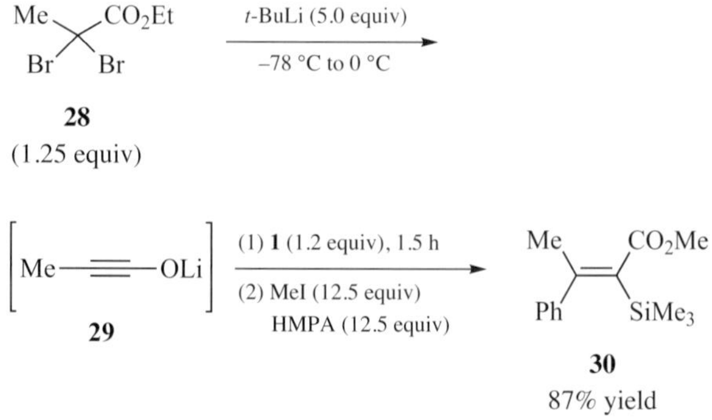

(11)

30 87% yield Z : E > 99 : 1

Miscellaneous Reactions. Irradiation of a mixture of **1** and boronic ester **31** followed by the treatment of the crude reaction mixture with 1 M HCl resulted in the formation of diaryl ketone **32** (eq 12).[25] Photocatalyzed silylacylation of dimethyl acetylenedicarboxylate with **1** resulted in the formation of conjugated enone **33** in good yield (eq 13).[26] Schmink and Krska reported a reversed-polarity synthesis of diarylketones via Pd-catalyzed cross-coupling of **1** with substituted aryl bromides (Table 9).[27]

Table 8 Synthesis of β-hydroxyketones **26**

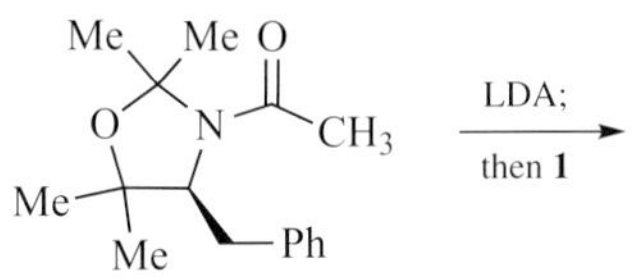

24

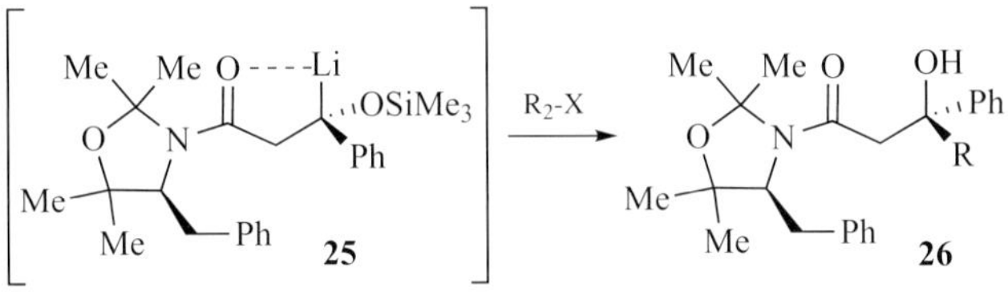

25 **26**

Entry	R_2-X	T (°C)	Isolated yield (%)	dr
1	BnBr	–78	80	3:1
2	BnBr	0	79	10:1
3	AllylBr	0	76	10:1
4	MeI	0	78	15:1

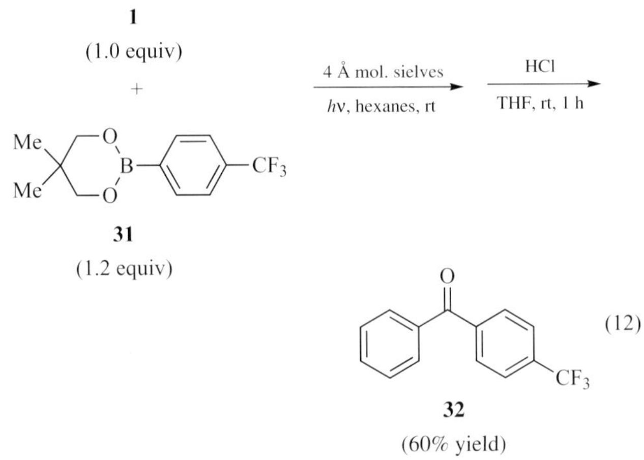

1 (1.0 equiv) + **31** (1.2 equiv); 4 Å mol. sielves, $h\nu$, hexanes, rt; HCl, THF, rt, 1 h → **32** (60% yield) (12)

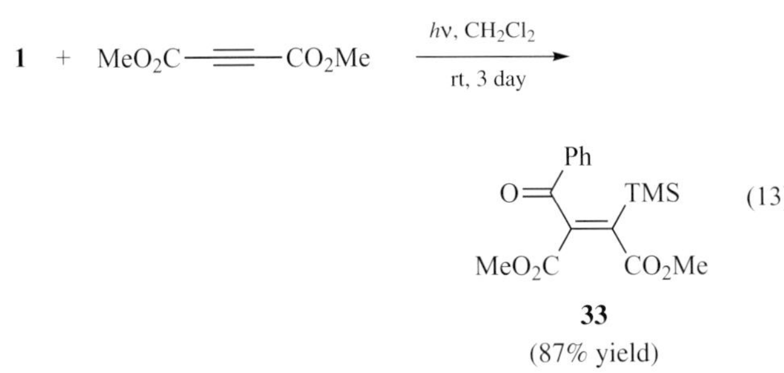

1 + MeO_2C–≡–CO_2Me; $h\nu$, CH_2Cl_2, rt, 3 day → **33** (87% yield) (13)

Table 9 Reversed-polarity synthesis of phenyl ketones

1 + Ar-Br —[Catalyst **34** (2 mol %), H_2O (6 equiv), K_3PO_4 (3.0 equiv); 2-MeTHF, 60 °C, 20–48 h]→

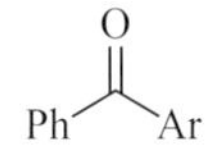

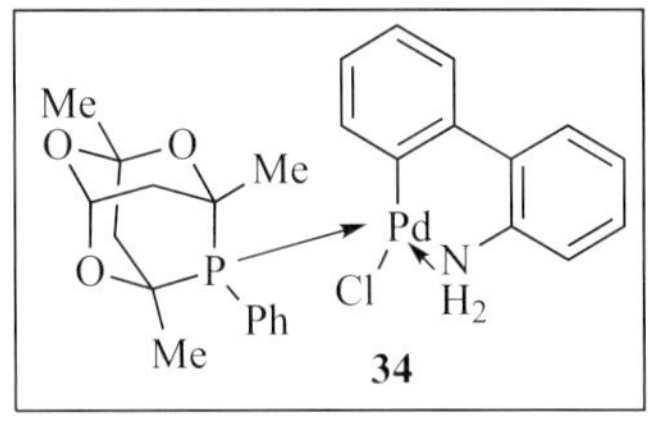

34

Entry	Ar-Br	Yield (%)
1	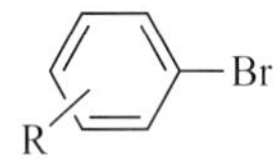	R = 4-OCH_3; 76 R = 3-OCH_3; 69 R = 2-OCH_3; 62 R = 4-Cl; 79 R = 3-Cl; 75 R = 2-CH_3; 85 R = 4-$N(CH_3)_2$; 67 R = 4-CF_3; 56 R = 4-$CH(OCH_3)_2$; 75
2	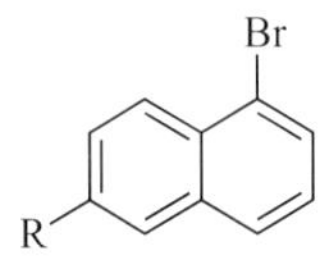	R = H; 63 R = OCH_3; 79
3	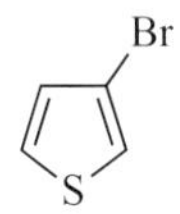	83

1. Cossrow, J.; Rychnovsky, S. D., *Org. Lett.* **2002**, *4*, 147.
2. Yamamoto, K.; Hayashi, A.; Suzuki, S.; Tsuji, J., *Organometallics* **1987**, *6*, 974.
3. Wu, X.-F.; Neumann, H.; Beller, M., *Tetrahedron Lett.* **2012**, *53*, 582.
4. Patrocinio, A. F.; Moran, J. P. S., *J. Organomet. Chem.* **2000**, *603*, 220.
5. Suda, K.; Watanabe, J.-i.; Takanami, T., *Tetrahedron Lett.* **1992**, *33*, 1355.
6. Ito, K.; Tamashima, H.; Iwasawa, N.; Kusama, H., *J. Am. Chem. Soc.* **2011**, *133*, 3716.
7. Mattson, A. E.; Bharadwaj, A. R.; Scheidt, K. A., *J. Am. Chem. Soc.* **2004**, *126*, 2314.
8. Mattson, A. E.; Bharadwaj, A. R.; Zuhl, A. M.; Scheidt, K. A., *J. Org. Chem.* **2006**, *71*, 5715.
9. Bharadwaj, A. R.; Scheidt, K. A., *Org. Lett.* **2004**, *6*, 2465.
10. Mattson, A. E.; Scheidt, K. A., *Org. Lett.* **2004**, *6*, 4363.
11. Bausch, C. C.; Johnson, J. S., *J. Org. Chem.* **2004**, *69*, 4283.
12. Linghu, X.; Bausch, C. C.; Johnson, J. S., *J. Am. Chem. Soc.* **2005**, *127*, 1833.
13. Tarr, J. C.; Johnson, J. S., *Org. Lett.* **2009**, *11*, 3870.
14. Nahm, M. R.; Linghu, X.; Potnick, J. R.; Yates, C. M.; White, P. S.; Johnson, J. S., *Angew. Chem., Int. Ed.* **2005**, *44*, 2377.

15. Garrett, M. R.; Tarr, J. C.; Johnson, J. S., *J. Am. Chem. Soc.* **2007**, *129*, 12944.
16. Matsuo, J.-i.; Hattori, Y.; Hashizume, M.; Ishibashi, H., *Tetrahedron* **2010**, *66*, 6062.
17. Gao, G.; Bai, X.-F.; Li, F.; Zheng, L.-S.; Zheng, Z.-J.; Lai, G.-Q.; Jiang, K.; Li, F.; Xu, L.-W., *Tetrahedron Lett.* **2012**, *53*, 2164.
18. Liu, B.; Lu, C.-D., *J. Org. Chem.* **2011**, *76*, 4205.
19. Honda, M.; Nakajima, T.; Okada, M.; Yamaguchi, K.; Suda, M.; Kunimoto, K.-K.; Segi, M., *Tetrahedron Lett.* **2011**, *52*, 3740.
20. Galliford, C. V.; Scheidt, K. A., *Chem. Commun.* **2008**, 1926.
21. Lettan, II, R. B.; Galliford, C. V.; Woodward, C. C.; Scheidt, K. A., *J. Am. Chem. Soc.* **2009**, *131*, 8805.
22. Lettan, II, R. B.; Woodward, C. C.; Scheidt, K. A., *Angew. Chem., Int. Ed.* **2008**, *47*, 2294.
23. Lei, X.; Dutheuil, G.; Pannecoucke, X.; Quirion, J.-C., *Org. Lett.* **2004**, *6*, 2101.
24. Shindo, M.; Matsumoto, K.; Mori, S.; Shishido, K., *J. Am. Chem. Soc.* **2002**, *124*, 6840.
25. Ito, K.; Tamashima, H.; Iwasawa, N.; Kusama, H., *J. Am. Chem. Soc.* **2011**, *133*, 3716.
26. Zhang, H.-J.; Becker, P.; Huang, H.; Porwerdjan, R.; Pan, F.-F.; Bolm, C., *Adv. Synth. Catal.* **2012**, *354*, 2157.
27. Schmink, J. R.; Krska, S. W., *J. Am. Chem. Soc.* **2011**, *133*, 19574.

Amol A. Kulkarni
Howard University, Washington, DC, USA

Benzyl 1,3-Butadiene-1-carbamate

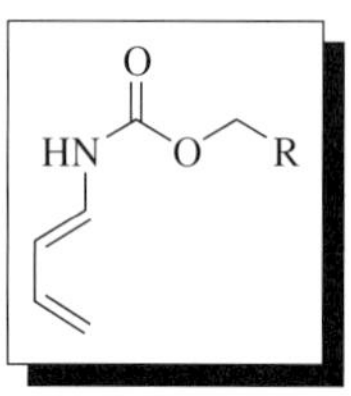

(**1**; R = Ph)
[65899-49-2] $C_{12}H_{13}NO_2$ (MW 203.09)
InChI = 1S/C12H13NO2/c1-2-3-9-13-12(14)15-10-11-7-5-4-6-8-11/h2-9H,1,10H2,(H,13,14)
InChIKey = LSQDRSXFUSBRBP-UHFFFAOYSA-N
(**2**; R = Me)
[61759-61-3] $C_7H_{11}NO_2$ (MW 141.08)
InChI = 1S/C7H11NO2/c1-3-5-6-8-7(9)10-4-2/h3,5-6H,1,4H2,2H3,(H,8,9)
InChIKey = XETZLCYONPVAHN-UHFFFAOYSA-N

(1,3-dienes: useful aminobutadiene equivalents for Diels–Alder reaction,[1] regio- and stereoselective reactions even with poor dienophiles;[2] used in the synthesis of complex heterocyclic natural products[3] and aminoanthraquinones[4])

Physical Data: [5–7] (**1**) mp 74–75 °C; ^{1}H NMR ($CDCl_3$) δ 7.32 (s, Ph), 6.71 (br d, J = 9 Hz, =CHNH), 6.26 (dt, J = 10, 17 Hz, CH=CH$_2$), 5.4–5.8 (m, CH=CHNH and NH), 5.15 (s, CH$_2$Ph), 4.8–5.2 (m, =CH$_2$); ^{13}C NMR ($CDCl_3$) δ 127.2, 112.5, 134.6, 113.5, 128.3, 128.4, 128.7, 136.0, 153.7, 67.5. (**2**) mp 44–45 °C; ^{1}H NMR ($CDCl_3$) δ 7.7 (br d, NH), 5.3–6.9 (m, vinylic), 4.5–5.1 (m, =CH$_2$), 4.13 (q, J = 7 Hz, Me); ^{13}C NMR ($CDCl_3$) δ 127.6, 112.1, 134.8, 113.2, 14.5, 61.7, 154.1.[2,5]

Solubility: (**1**) readily sol dioxane and toluene; (**2**) sol dioxane.

Preparative Method: The synthesis of various *N*-acyl-1-amino-1,3-dienes have been described in detail.[6–10] The following, which uses a Curtius rearrangement (eq 1),[7] illustrates the procedure commonly used in the synthesis of (**1**) and (**2**).

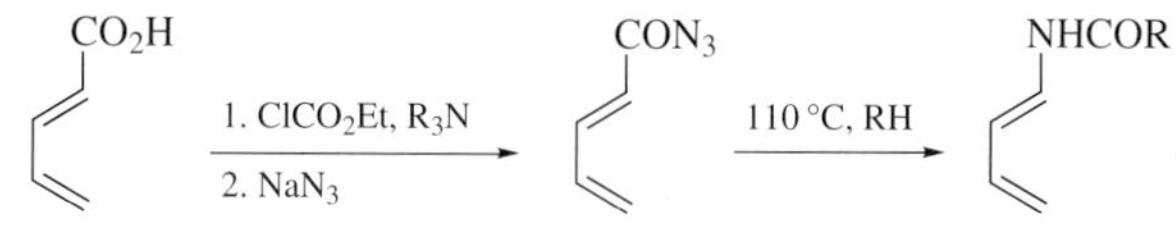

(1)

(**1**) R = OBn, 53%
(**2**) R = OEt, 71%

Handling, Storage, and Precautions: in general, acyl azides are potentially explosive. Therefore solutions containing acyl azides should not be evaporated to dryness.[5] Both (**1**) and (**2**) are stable solids, which can be stored in a freezer for several months. However, they are acid-sensitive and may be decomposed by traces of deuterium chloride present in chloroform-*d*.[5] In rigorously degassed (oxygen-free) solutions, these diene carbamates are stable for extended periods even at 140 °C and this has been a critical factor in their successful reactivity with poor dienophiles.[2]

Original Commentary

Dev P. Arya & David J. Jebaratnam
Northeastern University, Boston, MA, USA

Diels–Alder Reactions. Both (**1**) and (**2**) react with a broad range of dienophiles, providing a variety of amino-functionalized cycloadducts in high yields. Reaction with methyl acrylate yields cyclohexenes with high regio- and stereoselectivity (*endo*) (eq 2).[6] In contrast, with the same dienophile, 2,4-pentanedienoic acid requires longer reaction times and gives a mixture of all the possible stereo- and regioisomers.[1]

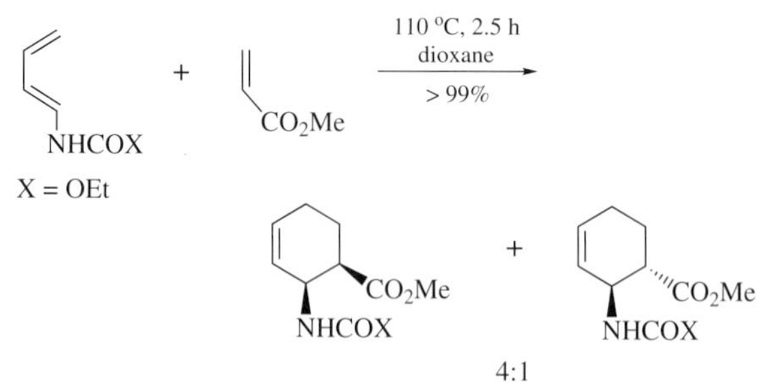

(2)

The ability to give *endo* selectivity, coupled with their higher reactivity, is, in part, responsible for the versatility of (**1**) and (**2**) in natural product synthesis. The construction of three of the chiral centers found in (±)-pumiliotoxin C was achieved in a single step (eq 3).[11,12] While other dienes exhibited lower stereoselectivities with *trans*-crotonaldehyde,[13,14] (**1**) and (**2**) gave the *endo* products in 61 and 67% yields, respectively; these were key intermediates in the synthesis of (±)-pumiliotoxin C.[12] Additionally, use of a chiral dienophile, (5*R*)-5-hydroxy-6,6-dimethylhept-2-en-4-one, permitted the asymmetric synthesis of (+)-pumiliotoxin C.[15]

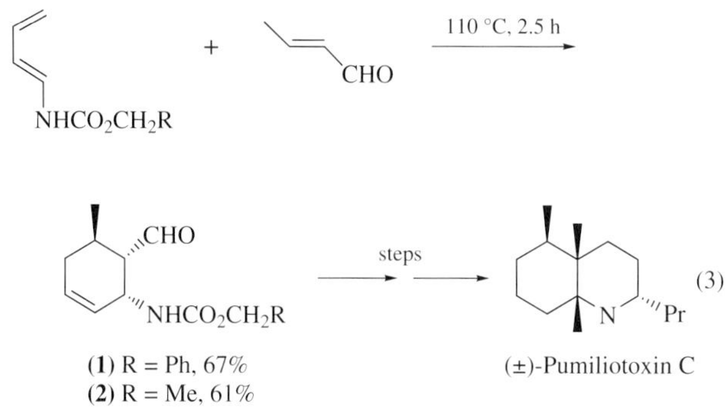

The cycloaddition of benzyl *trans*-1,3-butadiene-1-carbamate (**1**) and ethyl atropate affords the major product *endo*; this was a key intermediate in the synthesis of (±)-tilidine (eq 4).[16] Of significance is the opposite stereoselectivity exhibited by the related *trans*-1-(dialkylamino)-1,3-dienes ($R^1 = R^2 = Me$).[17,18]

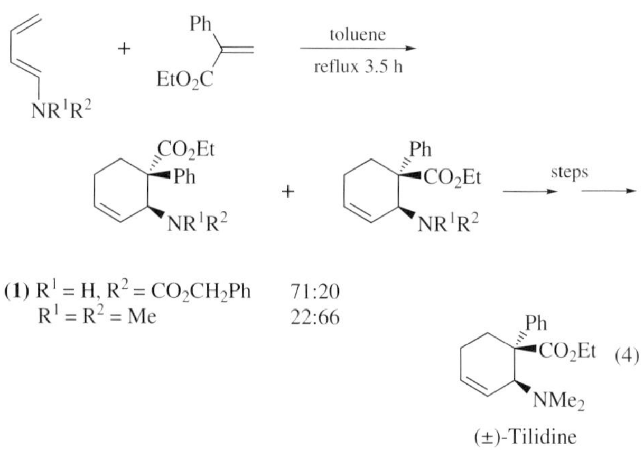

The most remarkable feature of dienes (**1**) and (**2**) is their ability to give *endo* cycloaddition products, in good yields, with extremely poor dienophiles such as methyl *trans*-crotonate, *trans*-crotonaldehyde, 2-cyclohexenone, styrene, and 3,4-methylenedioxystyrene.[2,11,19–24] The reaction of (**1**) with phenylacetylene (140 °C, 114 h),[2] however, led to a mixture of 13 products, suggesting an upper limit to the reactivity of these dienes.

To date, Diels–Alder reactions involving (**1**) and (**2**) have been the key step in syntheses of gephyrotoxin,[26] (±)-perhydrogephyrotoxin,[25] (+)-pumiliotoxin,[11,15] (±)-isogabaculine,[27] and (±)-tilidine.[16] In addition, substituted aminoanthraquinones have been synthesized by the regioselective cycloaddition of (**1**) to naphthoquinones followed by aromatization.[4,28] Also noteworthy is the synthesis of epimeric *cis*-decahydroquinoline-5-carboxylic acids, which involved the reaction of (**1**) with *trans*-3-acetylcrotonaldehyde.[29] Additionally, the presence of an easily modifiable phenyl group in (**1**) has led to the generation of a water-soluble diene (eq 5), which has been used in antibody-catalyzed Diels–Alder reactions.[30] Exclusive formation of either isomer (*endo* or *exo*) has been possible, depending on the transition state analog used for the elicitation of catalytic antibodies.[30]

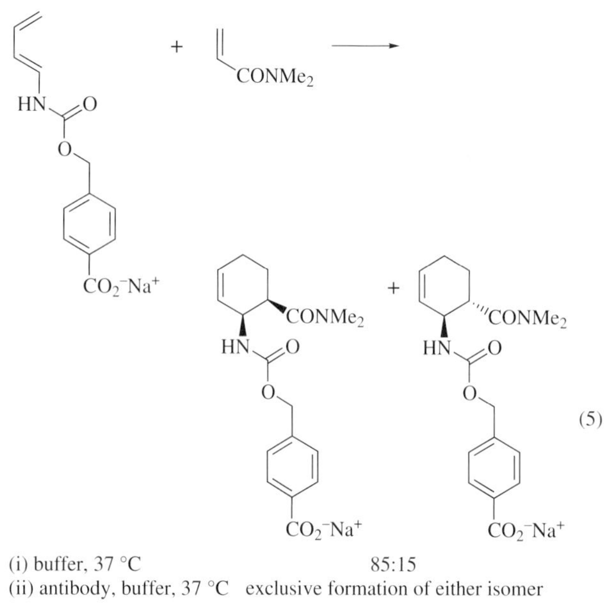

First Update

Peng Jiao
Beijing Normal University, Beijing, China

Benzyl 1,3-butadiene-1-carbamate is a nitrogen-containing electron-rich diene. Overman et al. demonstrated its utility in thermal Diels–Alder reactions with various dienophiles.[2,16] Catalytic asymmetric Diels–Alder reactions and Povarov reactions involving benzyl 1,3-butadiene-1-carbamate have been documented. The chiral catalysts (cat. **3**–**9**) used in these asymmetric reactions are listed below.

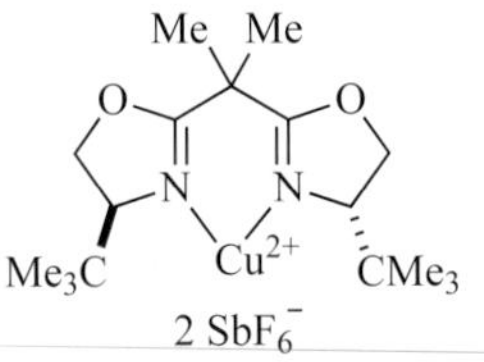

cat. **3**

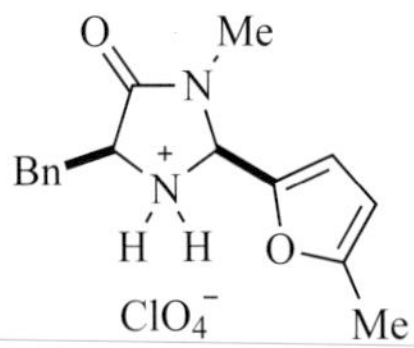

cat. **4a**

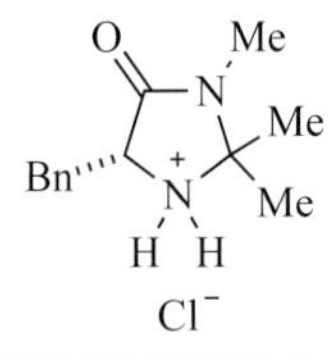

cat. **4b**

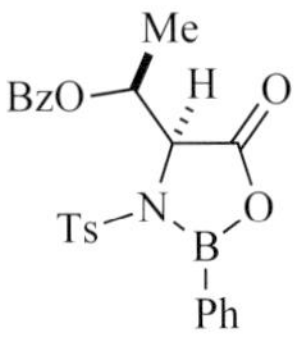

cat. **5**

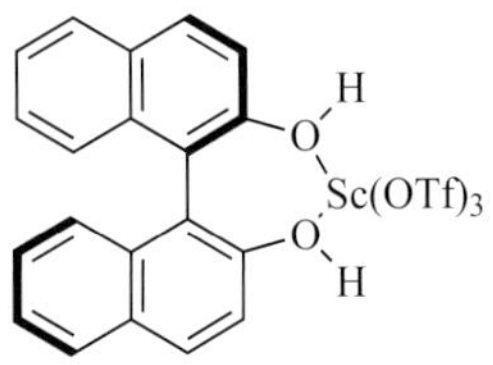

cat. **6**

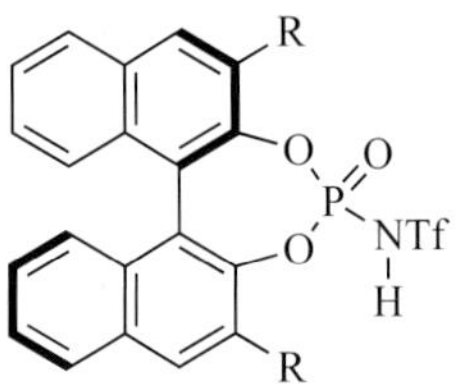

R = 9-anthracenyl

cat. **7**

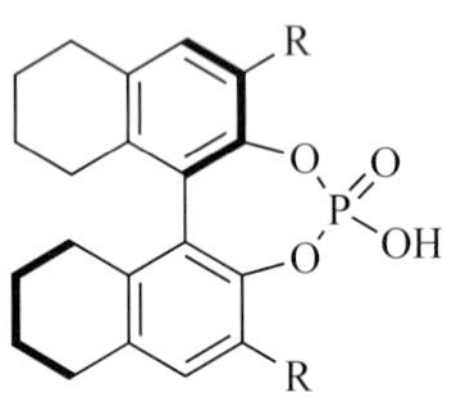

R = 9-anthracenyl

cat. **8**

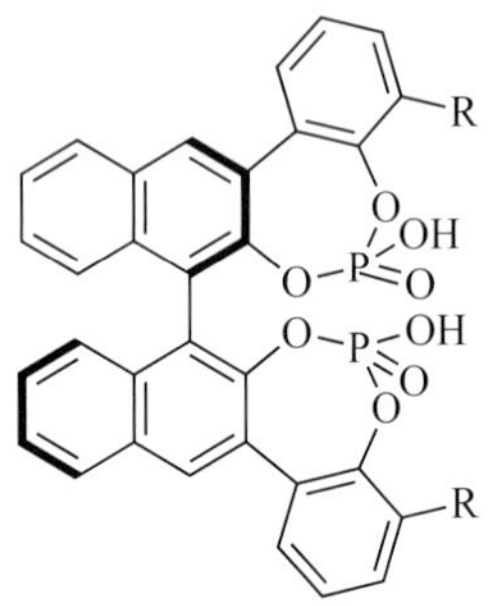

R = 2,4,6-(i-Pr)$_3$C$_6$H$_2$

cat. **9**

Asymmetric Diels–Alder Reactions. Benzyl 1,3-butadiene-1-carbamate reacted with N-acryloyl-2-oxazolidinone under the catalysis of 'Cu(SbF$_6$)$_2$–BOX' complex (cat. **3**) (eq 6).[31] The *cis*-Diels–Alder product was obtained in 90% ee. The same reaction of benzyl 1,3-butadiene-1-carbamate catalyzed by 'Sc(OTf)$_3$–BINOL' complex (cat. **6**) gave 90% ee and 92% yield for the *endo* cycloaddition product (eq 7).[32]

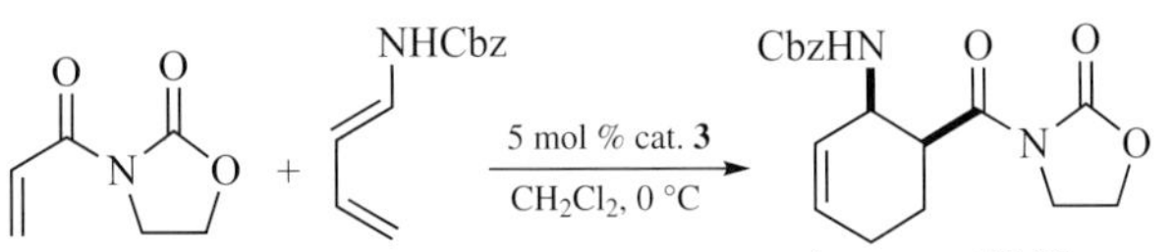

(6)

cis:trans = 72:28
cis: 90% ee, 54% yield

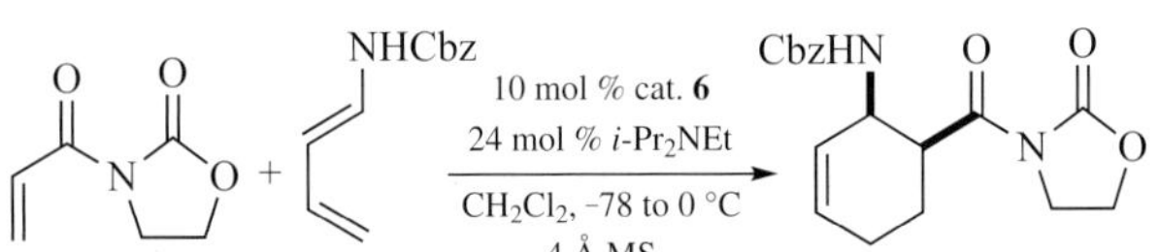

(7)

endo:exo = 94:6
90% ee, 92% yield

The perchloric acid salt of MacMillan's imidazolidinone organocatalyst (cat. **4a**) delivered 98% ee and 91% yield in the asymmetric Diels–Alder reaction of benzyl 1,3-butadiene-1-carbamate with ethyl vinyl ketone (eq 8).[33] Oxazaborolidinone (cat. **5**) gave 86% ee and >98:2 *endo:exo* in the reaction of benzyl 1,3-butadiene-1-carbamate with ethyl vinyl ketone, albeit in only 23% yield (eq 9).[34] The reaction of benzyl 1,3-butadiene-1-carbamate with phenyl vinyl ketone gave the Diels–Alder product in 76% ee and 40% yield under the catalysis of (R)-BINOL-derived N-triflyl chiral phosphoramide (cat. **7**) (eq 10).[35]

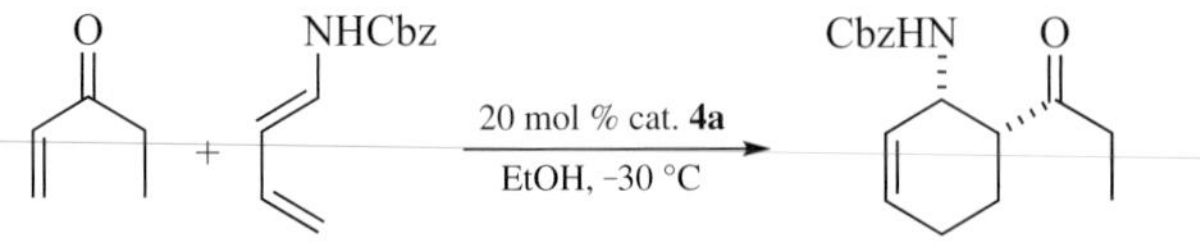

(8)

91% yield
endo:exo > 100:1
endo: 98% ee

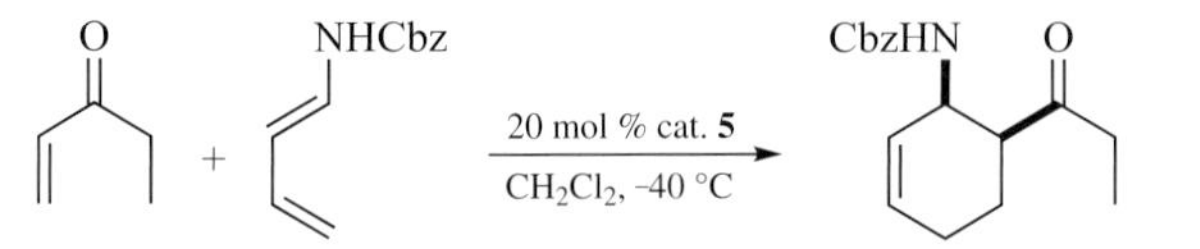

(9)

23% yield
endo:exo > 98:2
endo: 86% ee

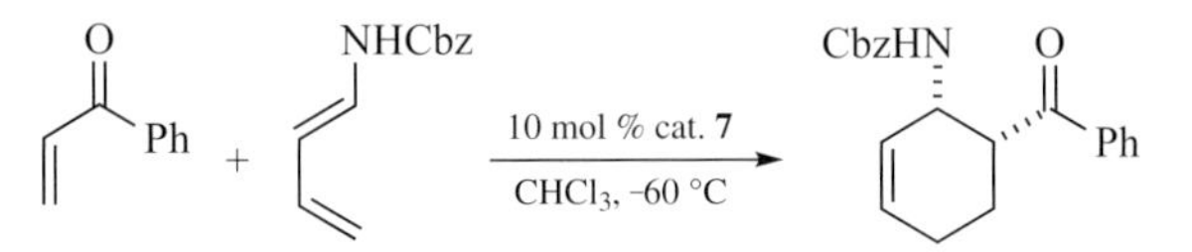

(10)

76% ee
40% yield

A bisphosphoric acid bearing a 1,1′-binaphthyl backbone (cat. **9**) catalyzed the asymmetric Diels–Alder reactions of benzyl 1,3-butadiene-1-carbamate with acrolein and 2-alkylacroleins, delivering the cycloadducts in excellent ee (97–99%) and high yields (86–91%) (eq 11).[36] The hydrochloric acid salt of MacMillan's imidazolidinone organocatalyst (cat. **4b**) gave 89% ee and 68% yield in the reaction of benzyl 1,3-butadiene-1-carbamate with acrolein (eq 12).[32]

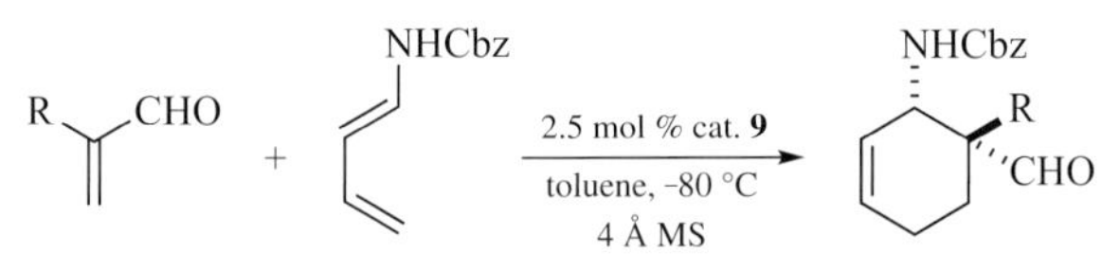

(11)

R = H, Me, Et, Bn
97–99% ee
86–91% yield

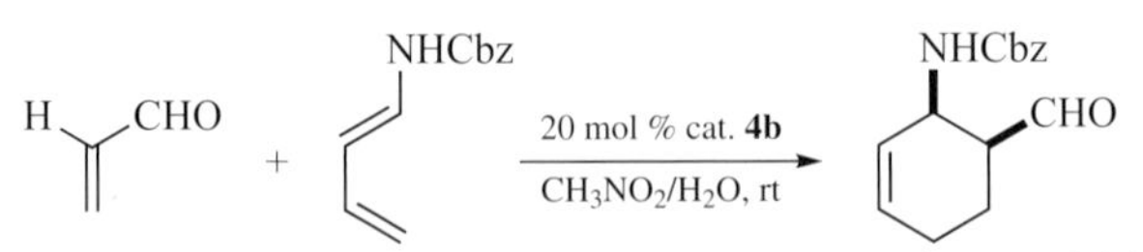

(12)

endo:exo = 14:1
89% ee, 68% yield

Asymmetric Povarov Reactions. Asymmetric vinylogous Povarov reactions of benzyl 1,3-butadiene-1-carbamate with series of imines yielded the products in 84 to >99% ee and 25–94% yields (eq 13).[37] An (R)-H$_8$-BINOL-derived phosphoric acid (cat. **8**) was used as the catalyst.

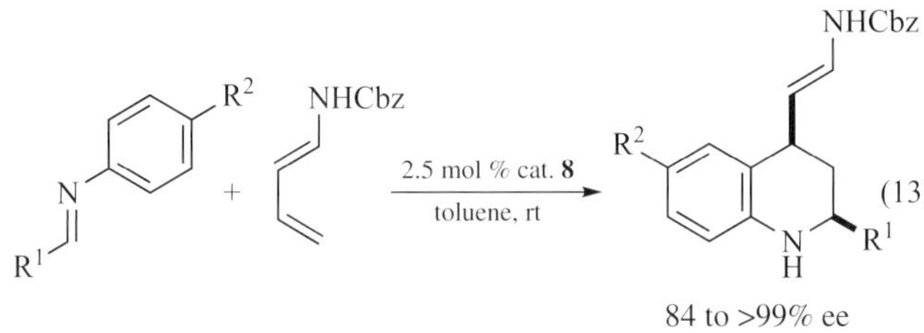

(13)

84 to >99% ee
25–94% yield

R^1 = Ph, 2(4)-BrC_6H_4, 4-$MeOC_6H_4$, 3,4-$(MeO)_2C_6H_3$
2-thienyl, 1(2)-naphthyl, *i*-Pr, $PhCH_2CH_2$
R^2 = OMe, H, Cl, Br

1. Overman, L. E.; Taylor, G. F.; Houk, K. N.; Domelsmith, L. N., *J. Am. Chem. Soc.* **1978**, *100*, 3182.
2. Overman, L. E.; Freerks, R. L.; Petty, C. B.; Clizbe, L. A.; Ono, R. K.; Taylor, G. F.; Jessup, P. J., *J. Am. Chem. Soc.* **1981**, *103*, 2816.
3. Overman, L. E., *Acc. Chem. Res.* **1980**, *13*, 218.
4. Chigr, M.; Fillion, H.; Rougny, A., *Tetrahedron Lett.* **1987**, *28*, 4529.
5. Overman, L. E.; Jessup, P. J.; Petty, C. B.; Roos, J., *Org. Synth.* **1980**, *59*, 1.
6. Overman, L. E.; Taylor, G. F.; Jessup, P. J., *Tetrahedron Lett.* **1976**, *36*, 3089.
7. Overman, L. E.; Taylor, G. F.; Petty, C. B.; Jessup, P. J., *J. Org. Chem.* **1978**, *43*, 2164.
8. Weinstock, J., *J. Org. Chem.* **1961**, *26*, 3511.
9. Overman, L. E.; Clizbe, L. A., *J. Am. Chem. Soc.* **1976**, *98*, 2352.
10. Overman, L. E.; Petty, C. B.; Ban, T.; Huang, G. T., *J. Am. Chem. Soc.* **1983**, *105*, 6335.
11. Overman, L. E.; Jessup, P. J., *Tetrahedron Lett.* **1977**, 1253.
12. Overman, L. E.; Jessup, P. J., *J. Am. Chem. Soc.* **1978**, *100*, 5179.
13. Kobuke, Y.; Fueno, T.; Furukawa, J., *J. Am. Chem. Soc.* **1970**, *92*, 6548.
14. Seguchi, K.; Sera, A.; Otsuki, Y.; Maruyama, K., *Bull. Chem. Soc. Jpn.* **1975**, *48*, 3641.
15. Masamune, S.; Reed, L. A.; Davis, J. T.; Choy, W., *J. Org. Chem.* **1983**, *48*, 4441.
16. Overman, L. E.; Petty, C. B.; Doedens, R. J., *J. Org. Chem.* **1979**, *44*, 4183.
17. Satzinger, G., *Justus Liebigs Ann. Chem./Liebigs Ann. Chem.* **1969**, *728*, 64
(*Chem. Abstr.* **1970**, *72*, 313 33g).
18. Satzinger, G., *Justus Liebigs Ann. Chem./Liebigs Ann. Chem.* **1972**, *756*, 43
(*Chem. Abstr.* **1972**, *77*, 113 681f).
19. (a) Sauer, J., *Angew. Chem., Int. Ed. Engl.* **1966**, *5*, 211.
(b) Sauer, J., *Angew. Chem., Int. Ed. Engl.* **1967**, *6*, 16.
20. Alder, K.; Rickert, H. F., *Ber. Dtsch. Chem. Ges./Chem. Ber.* **1938**, *71*, 379.
21. Danishefsky, S.; Kitahara, T., *J. Org. Chem.* **1975**, *40*, 538.
22. Trost, B. M.; Vladuchik, W. C.; Bridges, A. J., *J. Am. Chem. Soc.* **1980**, *102*, 3554.
23. Berson, J. A.; Hamlet, Z.; Mueller, W. A., *J. Am. Chem. Soc.* **1962**, *84*, 297.
24. Danishefsky, S.; Kitahara, T.; Yan, C. F.; Morris, J., *J. Am. Chem. Soc.* **1979**, *101*, 6996.
25. Overman, L. E.; Fukaya, C., *J. Am. Chem. Soc.* **1980**, *102*, 1454.
26. Overman, L. E.; Lesuisse, D.; Hashimoto, M., *J. Am. Chem. Soc.* **1983**, *105*, 5373.
27. Danishefsky, S.; Hershenson, F. M., *J. Org. Chem.* **1979**, *44*, 1180.
28. Chigr, M.; Fillion, H.; Rougny, A.; Berlion, M.; Riondel, J.; Beriel, H., *Chem. Pharm. Bull.* **1990**, *38*, 688.
29. Witiak, D. T.; Tomita, K.; Patch, R. J.; Enna, S. J., *J. Med. Chem.* **1981**, *24*, 788.
30. Gouverneur, V. E.; Houk, K. N.; Teresa, B. P.; Beno, B.; Janda, K. D.; Lerner, R. A., *Science* **1993**, *262*, 204.
31. Evans, D. A.; Barnes, D. M.; Johnson, J. S.; Lectka, T.; von Matt, P.; Miller, S. J.; Murry, J. A.; Norcross, R. D.; Shaughnessy, E. A.; Campos, K. R., *J. Am. Chem. Soc.* **1999**, *121*, 7582.
32. Wipf, P.; Wang, X., *Tetrahedron Lett.* **2000**, *41*, 8747.
33. Northrup, A. B.; MacMillan, D. W. C., *J. Am. Chem. Soc.* **2002**, *124*, 2458.
34. Singh, R. S.; Adachi, S.; Tanaka, F.; Yamauchi, T.; Inui, C.; Harada, T., *J. Org. Chem.* **2008**, *73*, 212.
35. Kong, L.; Han, X.; Jiao, P., *Chem. Commun.* **2014**, *50*, 14113.
36. Momiyama, N.; Konno, T.; Furiya, Y.; Iwamoto, T.; Terada, M., *J. Am. Chem. Soc.* **2011**, *133*, 19294.
37. Caruana, L.; Fochi, M.; Ranieri, S.; Mazzanti, A.; Bernardi, L., *Chem. Commun.* **2013**, *49*, 880.

3-Benzyl-5-(2-hydroxyethyl)-4-methyl-1,3-thiazolium Chloride

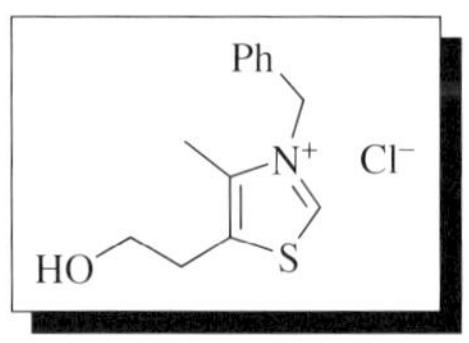

[4568-71-2] $C_{13}H_{16}ClNOS$ (MW 269.79)

InChI = 1S/C13H16NOS.ClH/c1-11-13(7-8-15)16-10-14(11)9-12-5-3-2-4-6-12;/h2-6,10,15H,7-9H2,1H3;1H/q+1;/p-1

InChIKey = IWSVLBKHBJGMAA-UHFFFAOYSA-M

(in the presence of bases, catalyst for the preparation of acyloins and benzoins[2] from aldehydes, and for addition of aldehydes to electrophilic double bonds[3])

Physical Data: mp 141–143 °C.
Solubility: sol H_2O, MeOH, EtOH; slightly sol cold MeCN; insol ether, pentane.
Form Supplied in: colorless crystals, widely available.
Analysis of Reagent Purity: ^{1}H NMR; mp.
Preparative Method: *Organic Syntheses* procedure.[2]
Handling, Storage, and Precautions: store dry.

Original Commentary

Heinrich Kuhlmann
RWTH Aachen, Aachen, Germany

Formation of Benzoins and Acyloins from Aldehydes. Many thiazolium salts catalyze the formation of benzoins from aromatic aldehydes and of acyloins from aliphatic aldehydes in the presence of a base as shown in the preparation of furoin from furan-2-carbaldehyde (eq 1).[2]

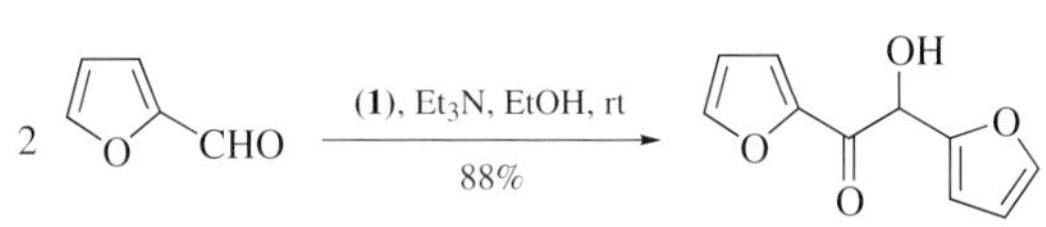

(1)

This catalytic reaction has long been known,[4] the most prominent salt being thiamine (vitamin B_1). Other azolium salts[5] and nucleophilic carbenes, or dimers thereof, also catalyze this reaction.[6] Chiral thiazolium salts have been used.[7] The reaction can be performed as a phase-transfer reaction at preparative scale in water.[8] The title reagent (**1**) has been used in the preparative large-scale formation of acyloins and benzoins in the presence of Triethylamine in ethanol under anhydrous conditions (eq 2).

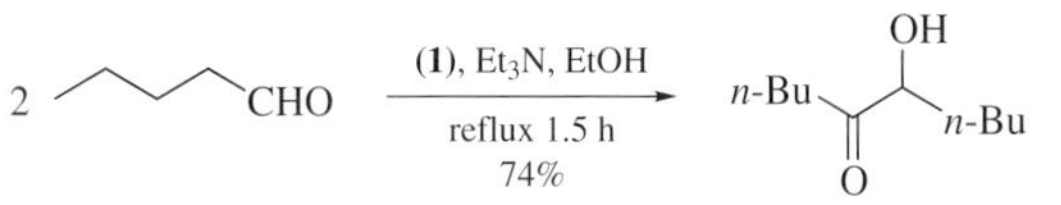

(2)

Simple aliphatic acyloins,[9] mixed aliphatic acyloins,[10] and acyloins containing other functional groups[11] have been prepared.

'Formose' Reaction. A very special case of acyloin formation is observed when formaldehyde is used in that the construction of simple sugars occurs. Only some of the numerous possible products are found in the 'formose' mixture.[12] A similar result has been reported with ethylbenzothiazolium bromide as catalyst.[13]

Addition of Aldehydes to Activated Double Bonds. This reaction has been reviewed.[3] (**1**) in the presence of bases (e.g. Et_3N or NaOAc) catalyzes the addition of aliphatic and heterocyclic aldehydes to α,β-unsaturated ketones (to yield γ-diketones), esters (to yield γ-ketoesters), and nitriles (to yield γ-ketonitriles), for example (eq 3).[14] This further illustrates the 'umpolung' effect of this reagent on the aldehyde carbonyl reactivity.

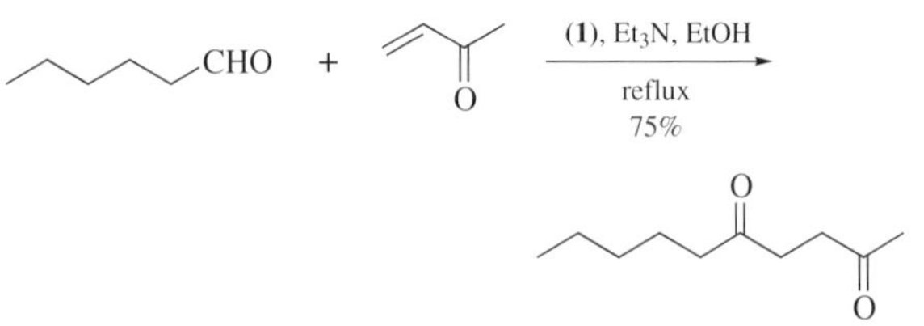

(3)

This reaction is used in the synthesis of tri- and polyketones,[15] cyano ketones,[16] 1,4-diketones,[17] and oxocarboxylic esters.[16b,18] Mannich bases (β-dialkylamino ketones) or their quaternary salts can replace the unsaturated ketones.[3] α-Keto carboxylic acids can replace the aldehydes, because these readily decarboxylate under the reaction conditions (eq 4).[19]

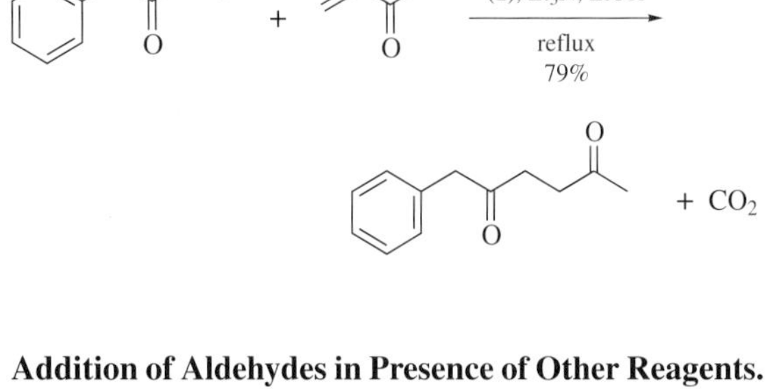

(4)

Addition of Aldehydes in Presence of Other Reagents. The presence of Pentacarbonyliron changes the reaction pathway to afford different products (eq 5).

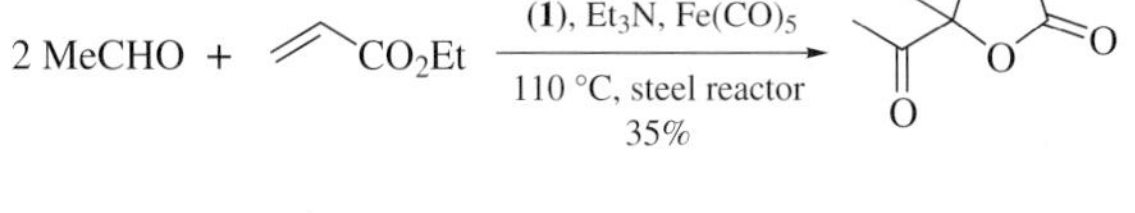

(5)

Table 1 Properties of Some Salts

R	X	CAS No.	MW	Mp (°C)	Empirical formula
Et[a,b]	Br	[54016-70-5]	252.17	84–86	$C_8H_{14}BrNOS$
Me[a,b,c]	I	[16311-69-6]	285.14	85–87	$C_7H_{12}INOS$

[a] Hygroscopic.

[b] Widely available.

[c] Light sensitive.

Michael addition of ethyl acrylate to acetoin and transesterification yields the product shown (35%) together with ethyl levulinate (10%).[20] For details on additions of aldehydes in presence of oxidants.

Related Reagents. Table 1 shows some widely used 1,3-thiazolium salts analogs of (**1**) that are especially well suited with aromatic aldehydes for the above benzoin preparations and for aldehyde additions to double bonds.[3]

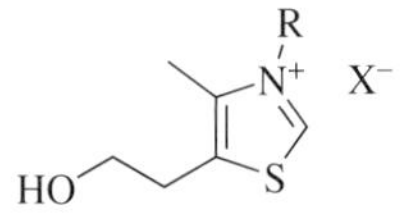

(2)

Other Reagents. Sodium Cyanide or Potassium Cyanide are the classical catalysts for benzoin formation from aromatic aldehydes[21] and they can also be used for the addition of aromatic aldehydes to activated double-bonds.[3] Carboxylic esters and metals such as sodium are used in the classic acyloin condensation[22] and its modern variations.[23] Numerous other acyl anion equivalents have been described.[24]

First Update

Pouyan Haghshenas & Michel Gravel
University of Saskatchewan, Saskatoon, Saskatchewan, Canada

Addition of Aldehydes to Acylimines. The title reagent (**1**) catalyzes the umpolung addition of aldehydes to acylimines (aza-benzoin reaction) (eq 6).[25] The title reagent (**1**) and closely related compounds such as 3,4-dimethyl-5-(2-hydroxyethyl)-thiazolium iodide [16311-69-6] catalyze the aza-benzoin reaction between various aldehydes and acylimines masked as their sulfinic acid adduct under mildly basic conditions. The use of triethylamine is advised for the reaction, but potassium carbonate could also be used. The reaction generally works best with electron-poor aromatic and heteroaromatic aldehydes, but it is also tolerant to aliphatic aldehydes and enals. However, aliphatic imines (as their sulfinic acid adducts) cannot be used in this reaction.

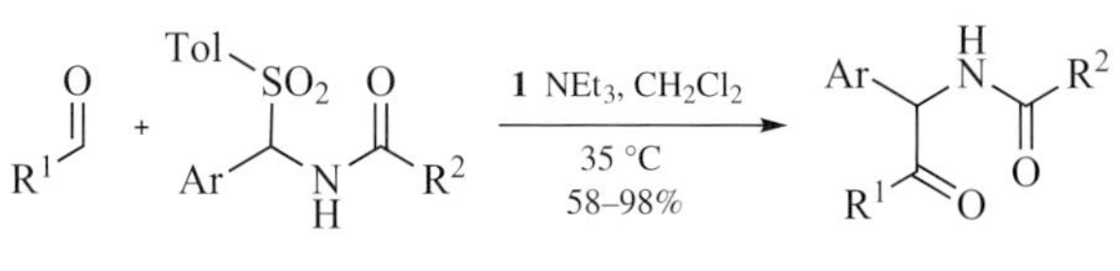

R^1 = Ar, HetAr, Alk, PhCH=CH (6)
R^2 = H, Alk, Ar, O-tBu

Internal Redox Esterification. The title reagent (**1**) may be used to catalyze the internal redox reaction of α-bromo aliphatic aldehydes in which the acyl azolium intermediate is intercepted by alcohols (eq 7).[26] The title reagent (**1**) shows superior reactivity to 1,3-bis(2,4,6-trimethylphenyl)imidazolium chloride [141556-45-8] and is comparable to 2,5,6,7-tetrahydro-2-phenyl-3*H*-pyrrolo[2,1-*c*]-1,2,4-triazol-3-ylidenium chloride [828914-68-7] in this reaction.

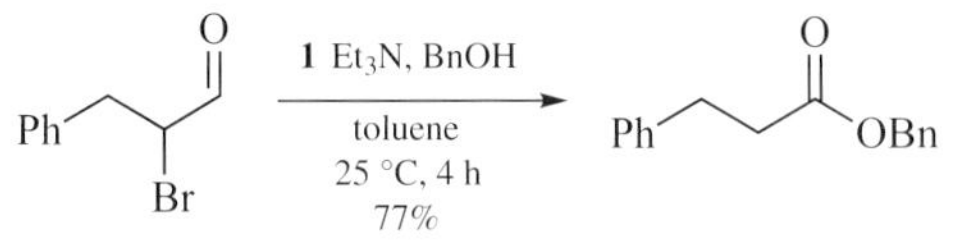

(7)

Stetter Reaction Using α-Keto Acids. The decarboxylative Stetter reaction of aldehydes and α,β-unsaturated 2-acyl *N*-methyl-imidazole may be catalyzed using the title reagent (**1**) (eq 8).[27] This catalyst is reported to be superior to triazolium and imidazolium salts and the reaction is best performed by using the sodium salt of α-keto acids. Reduced catalytic loadings (<20 mol %) are associated with incomplete conversion of the starting material. Elevated temperatures (>63 °C) are advised for obtaining useful yields of the product. The reaction is tolerant to a variety of aromatic, and heteroaromatic, conjugated 2-acyl *N*-methyl-imidazoles. In contrast, electrophiles bearing β-alkyl substituents are associated with low yields of the 1,4-dicarbonyl products. Many α-keto acids are well tolerated with the exception of those bearing an α-substituent. The newly formed stereogenic center readily epimerizes under the reaction conditions.

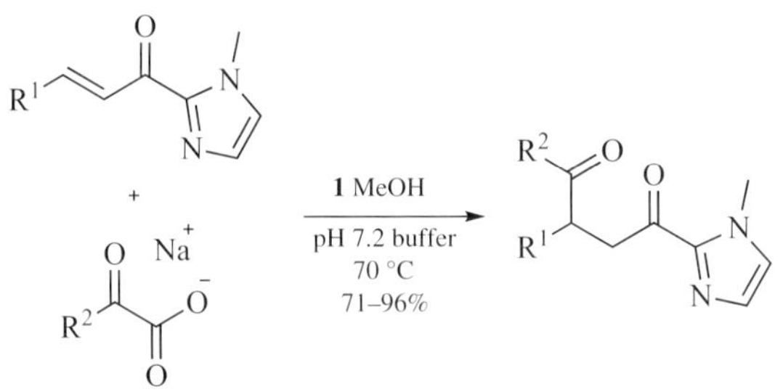

(8)

R^1= Ar, HetAr
R^2= Alk, Ar, HetAr

Intramolecular Cross-benzoin Reaction. The title reagent may be used to catalyze the intramolecular cross-benzoin reaction between an aldehyde and a neighboring ketone (eq 9).[28] For this purpose, this reagent is more effective than sodium cyanide. 3-Ethyl-5-(2-hydroxyethyl)-4-methyl-1,3-thiazolium bromide *[54016-70-5]* shows superior reactivity to the title reagent (**1**) in this reaction.

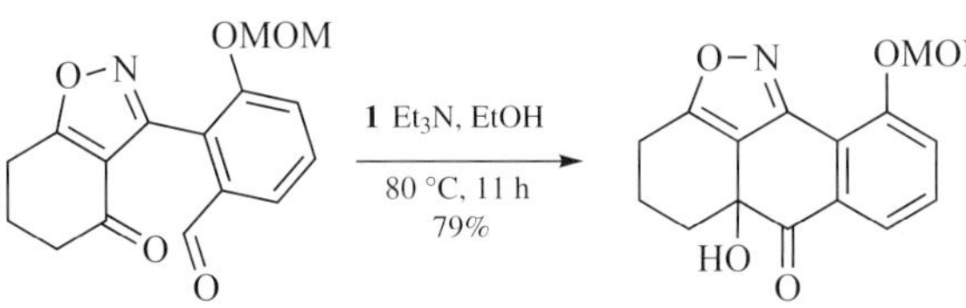

(9)

General Considerations. The pK_a of the title reagent is lower than that of corresponding imidazolium reagents.[29] This provides access to the free carbene using weaker bases.[30] NMR and computational studies suggest that in aqueous media the decomposition pathway consists of a hydrolytic opening of the thiazolium ring.[31,32]

Related Reagents. Other related thiazolium catalysts have been employed for ring opening of epoxyaldehydes,[33] conjugate addition of acylsilanes to unsaturated esters and ketones,[34] and intermolecular cross-benzoin reaction between aromatic aldehydes and acetaldehyde.[35]

1. (a) Fieser, L. F.; Fieser, M., *Fieser & Fieser* **1977**, *6*, 38. *Fieser & Fieser* **1979**, *7*, 16 *Fieser & Fieser* **1982**, *10*, 27. (b) *Beilstein* **1983**, *27(III/IV)*, 1758.
2. Stetter, H.; Kuhlmann, H., *Org. Synth.* **1984**, *62*, 170.
3. (a) Stetter, H.; Kuhlmann, H., *Org. React.* **1991**, *40*, 407. (b) Stetter, H., *Angew. Cheml* **1976**, *88*, 695; *Angew. Chem., Int. Ed. Engl.* **1976**, *15*, 639.
4. (a) Breslow, R., *J. Am. Chem. Soc.* **1958**, *80*, 3719. (b) Castells, J.; Domingo, L.; López-Calahorra, F.; Martí, J., *Tetrahedron Lett.* **1993**, *34*, 517.
5. For example, 1,3,4-thiadiazolium salts: Alemagna, A.; Bachetti, T., *Gazz. Chim. Ital.* **1978**, *108*, 77.
6. (a) Lachmann, B.; Steinmaus, H.; Wanzlik, H.-W., *Tetrahedron* **1971**, *27*, 4085. (b) Bis(1,3-dialkylimidazolidin-2-ylidenes): Lappert, M. F.; Maskell, R. K., *J. Chem. Soc., Chem. Commun.* **1982**, 580.
7. (a) Yano, Y.; Tamura, Y.; Tagaki, W., *Bull. Chem. Soc. Jpn.* **1980**, *53*, 740. (b) Tagaki, W.; Tamura, Y.; Yano, Y., *Bull. Chem. Soc. Jpn.* **1980**, *53*, 478. (c) Leeper, F..; Smith, D. H. C., *J. Chem. Soc., Chem. Commun.* **1990**, 961. (d) Jimenez, L.; Diederich, *Tetrahedron Lett.* **1989**, *30*, 2759.
8. Tagaki, W.; Hara, H., *J. Chem. Soc., Chem. Commun.* **1973**, 891.
9. Stetter, H.; Rämsch, R. Y.; Kuhlmann, H., *Synthesis* **1976**, 733.
10. (a) Stetter, H.; Dämbkes, G., *Synthesis* **1977**, 403. (b) Stetter, H.; Dämbkes, G., *Synthesis* **1980**, 309.
11. Stetter, H.; Rämsch, R. Y., *Synthesis* **1981**, 477.
12. (a) Castells, J.; Geijo, F.; López-Calahorra, F., *Tetrahedron Lett.* **1980**, *21*, 4517. (b) Castells, J.; López-Calahorra, F.; Geijo, F., *Carbohydr. Res.* **1983**, *116*, 197.
13. Matsumoto, T.; Inoue, S., *J. Chem. Soc., Chem. Commun.* **1983**, 171.
14. Stetter, H.; Kuhlmann, H.; Haese, W., *Org. Synth.* **1987**, *65*, 26.
15. (a) Stetter, H.; Basse, W.; Kuhlmann, H.; Landscheidt, A.; Schlenker, W., *Chem. Ber.* **1977**, *110*, 1007. (b) Stetter, H.; Landscheidt, A., *Chem. Ber.* **1979**, *112*, 1410. (c) Stetter, H.; Nienhaus, J., *Chem. Ber.* **1980**, *113*, 979.
16. (a) Stetter, H.; Kuhlmann, H., *Chem. Ber.* **1976**, *109*, 2890. (b) Stetter, H.; Landscheidt, A., *Chem. Ber.* **1979**, *112*, 2419. (c) Stetter, H.; Basse, W.; Nienhaus, J., *Chem. Ber.* **1980**, *113*, 690.
17. (a) Stetter, H.; Hilboll, G.; Kuhlmann, H., *Chem. Ber.* **1979**, *112*, 84. (b) Stetter, H.; Lappe, P., *Chem. Ber.* **1980**, *113*, 1890.
18. Stetter, H.; Basse, W.; Wiemann, K., *Chem. Ber.* **1978**, *111*, 431.
19. Stetter, H.; Lorenz, G., *Chem. Ber.* **1985**, *118*, 1115.
20. Noels, A. F.; Hubert, A. J., *J. Mol. Catal.* **1983**, *22*, 235.
21. Ide, W. S.; Buck, J. S., *Org. React.* **1948**, *4*, 269.
22. (a) McElvain, S. M., *Org. React.* **1948**, *4*, 256. (b) Bloomfield, J. J.; Owsley, D. C.; Nelke, J. M., *Org. React.* **1976**, *23*, 259. (c) Finley, K. T., *Chem. Rev.* **1964**, *64*, 573.
23. Rühlmann, K., *Synthesis* **1971**, 236.
24. (a) Lever, O. W. Jr., *Tetrahedron* **1976**, *32*, 1943. (b) Albright, J. D., *Tetrahedron* **1983**, *39*, 3207.
25. Murry, J. A.; Frantz, D. E.; Soheili, A.; Tillyer, R.; Grabowski, E. J. J.; Reider, P. J., *J. Am. Chem. Soc.* **2001**, *123*, 9696.

26. Reynolds, N. T.; Read de Alaniz, J.; Rovis, T., *J. Am. Chem. Soc.* **2004**, *126*, 9518.
27. Myers, M. C.; Bharadwaj, A. R.; Milgram, B. C.; Scheidt, K. A., *J. Am. Chem. Soc.* **2005**, *127*, 14675.
28. Hachisu, Y.; Bode, J. W.; Suzuki, K., *J. Am. Chem. Soc.* **2003**, *125*, 8432.
29. Massey, R. S.; Collett, C. J.; Lindsay, A. G.; Smith, A. D.; O'Donoghue, A. C., *J. Am. Chem. Soc.* **2012**, *134*, 20421.
30. Amyes, T. L.; Diver, S. T.; Richard, J. P.; Rivas, F. M.; Toth, K., *J. Am. Chem. Soc.* **2004**, *126*, 4366.
31. Hollóczki, O.; Kelemen, Z.; Nyulászi, L., *J. Org. Chem.* **2012**, *77*, 6014.
32. Breslow, R., *J. Am. Chem. Soc.* **1958**, *80*, 3719.
33. Chow, K. Y.-K.; Bode, J. W., *J. Am. Chem. Soc.* **2004**, *126*, 8126.
34. Mattson, A. E.; Bharadwaj, A. R.; Scheidt, K. A., *J. Am. Chem. Soc.* **2004**, *126*, 2314.
35. Jin, M. Y.; Kim, S. M.; Han, H.; Ryu, D. H.; Yang, J. W., *Org. Lett.* **2011**, *13*, 880.

(*S*)-5-Benzyl-2-mesityl-6,6-dimethyl-6,8-dihydro-5*H*-[1,2,4]triazolo[3,4-*C*][1,4]-oxazin-2-ium Tetrafluoroborate

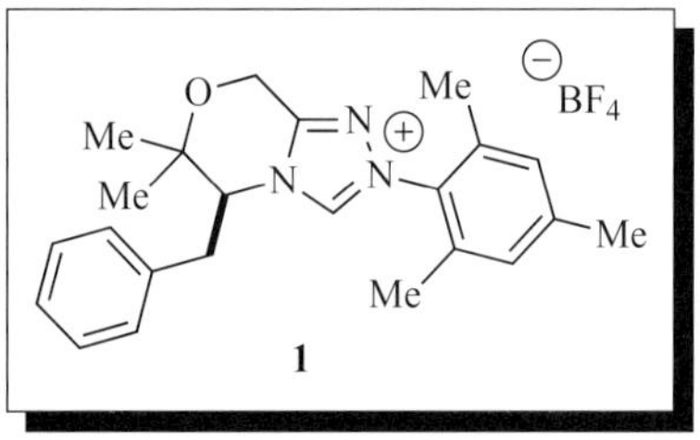

[925706-40-7] $C_{23}H_{28}BF_4N_3O$ (MW 449.29)

InChI = 1S/C23H28N3O.BF4/c1-16-11-17(2)22(18(3)12-16)26-15-25-20(13-19-9-7-6-8-10-19)23(4,5)27-14-21(25)24-26;2-1(3,4)5/h6-12,15,20H,13-14H2,1-5H3;/q+1;-1/t20-;/m0./s1

InChIKey = PVYSZYZMWZCJCT-BDQAORGHSA-N

(precatalyst for *N*-heterocyclic carbene catalysis)

Physical Data: ochre solid, mp 200–206 °C.
Solubility: organic solvents such as THF, methanol, CH_2Cl_2.
Form Supplied in: solid.
Purification: recrystallization from methanol.
Preparative Methods: synthesized in seven steps from the L-phenylalanine methyl ester.

Introduction. Substituted triazolium salts have emerged as powerful precatalysts in carbene-catalyzed reactions. Triazoliums, upon deprotonation with base, form a reactive *N*-heterocyclic carbene (NHC) intermediate (**2**) that can undergo reactions with aldehydes and α,β-unsaturated aldehydes to induce unconventional polarity reversal reactions (eq 1). Addition of **2** to aldehydes produces a "Breslow intermediate" **3**, an overall acyl anion equivalent, allowing a reversal of polarity (umpolung). Similarly, reaction of **2** with an α,β-unsaturated aldehyde yields an extended homoenolate system (**4**) where nucleophilic character resides on the α- and β-positions (eq 2). The subsequent reaction of **3** or **4** with various electrophiles to generate complex molecules has been recently reviewed.[1–3]

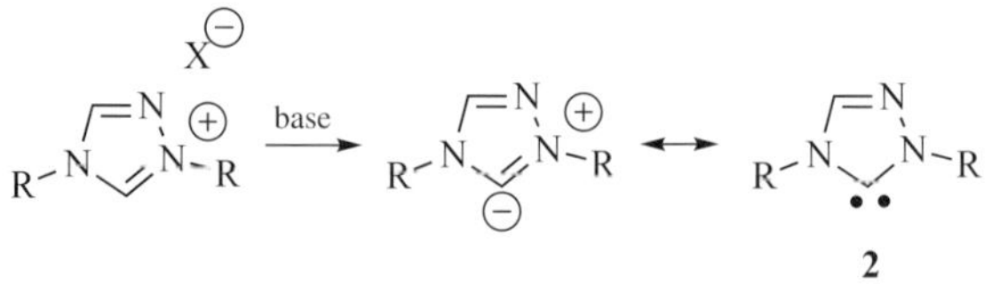

(1)

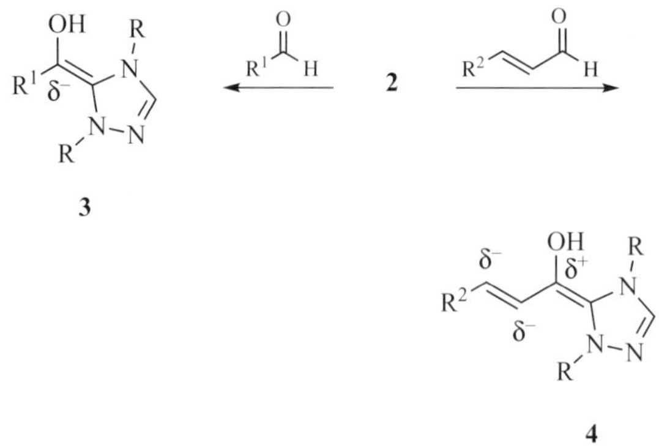

(2)

The title compound (**1**) is a bulkier version of other chiral triazolium salts such as **5**.[4,5] Triazolium salt **5** was found to promote benzoin condensations with up to 82% enantiomeric excess. Precatalyst **1** generates a more sterically encumbered carbene, which has the potential to prevent carbene dimerization and allow more efficient catalytic activity.

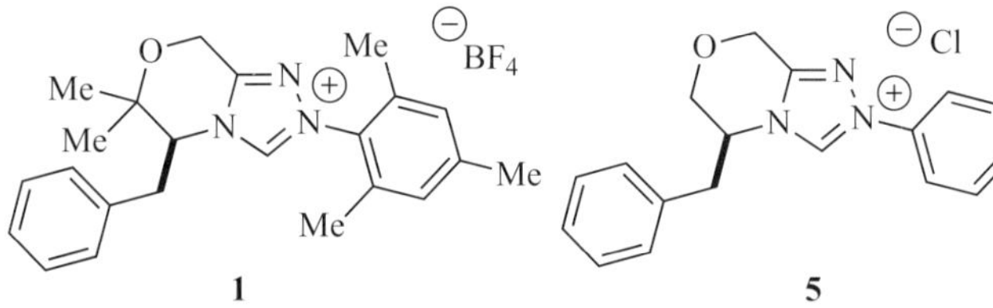

Benzoin Reaction. The first catalytic application of a phenylalanine-derived triazolium salt is the benzoin condensation reaction.[4] Triazolium salt **5** produced the highest enantioselectivity (80% ee) for this reaction, albeit in only 45% yield (eq 3). Since this report, the asymmetric variant of this reaction has seen significant improvement.

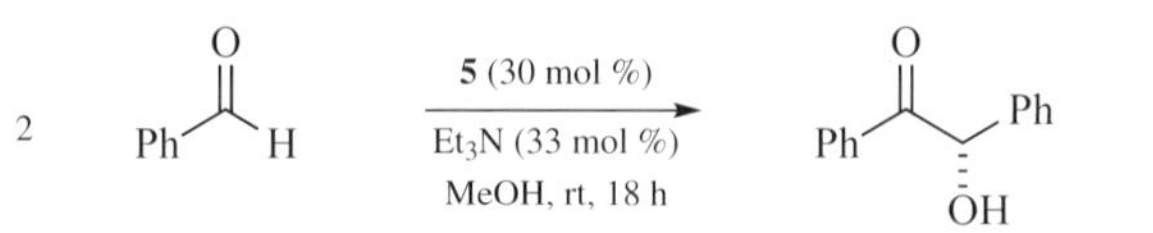

(3)

45% yield
80% ee

Diol Desymmetrization. The enantioselective desymmetrization of 1,2-cyclohexane diol through NHC-catalyzed oxidative acylation was promoted by triazolium **6** in good yield and with moderate enantioselectivity (eq 4).[6] The title compound **1** was also shown to promote this reaction in 57% yield, but with diminished enantioselectivity (20% ee).

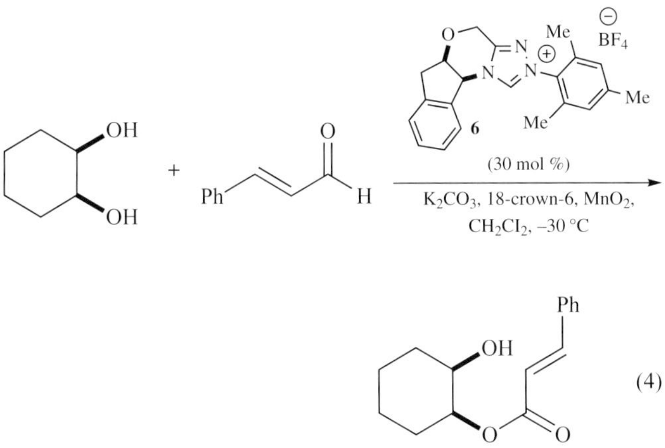

(4)

58% yield, 80% ee

with **1**: 57% yield, 20% ee

Intramolecular Michael Reaction. The Michael reaction typically relies on a stoichiometric amount of base to form the required enol or enolate nucleophile, which undergoes conjugate addition to α,β-unsaturated carbonyl compounds. Carbene catalysis offers a unique opportunity to catalytically generate enol species. Using carbene catalysis, an intramolecular Michael reaction was achieved in high yields with excellent levels of selectivity (eq 5).[7] Employing 20 mol % of **1** furnished the dihydropyranone product in 62% yield with excellent diastereo- and enantioselectivity. Precatalyst **6** was found to be the optimal catalyst for this reaction.

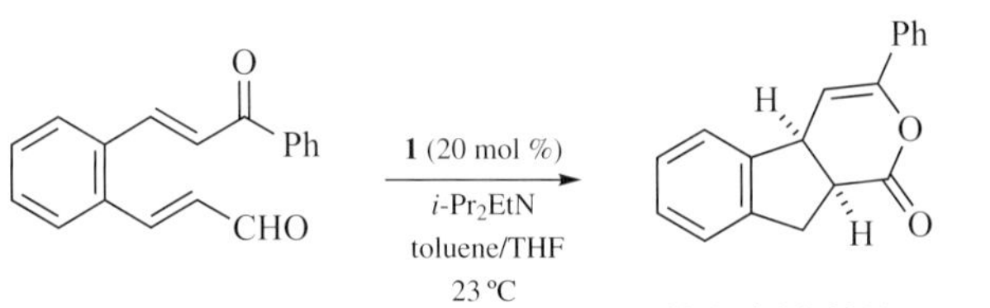

(5)

62% yield, 93% ee
>20:1 dr

Cyclopentene Synthesis. Cyclopentene formation through NHC catalysis was also achieved through an intramolecular cyclization reaction. Initial formation of a β-lactone, followed by decarboxylation, gave substituted cyclopentenes with an all-carbon quaternary center in high yields with excellent diastereo- and enantioselectivity (eq 6).[8,9] Precatalyst **7** was found to be the optimal catalyst for this method, however, triazolium **1** provided the cyclopentene in 38% yield and with 76% ee.

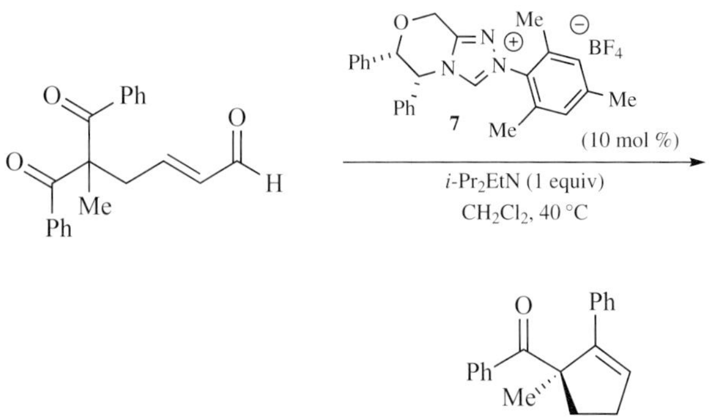

(6)

80% yield, 93% ee

with **1**: 38% yield, 76% ee

The dimerization of enals to provide cyclopentenes with high diastereo- and enantioselectivity was made possible through NHC/Lewis acid cooperative catalysis. A triazolium salt derived from tryptophan was the optimal catalyst for the reaction; however, excellent diastereoselectivity with low yield and enantioselectivity was observed with **1** (eq 7).[10]

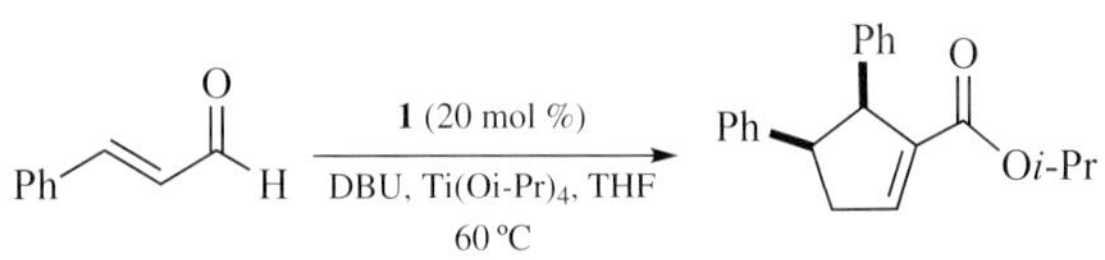

(7)

28% yield, 66% ee
>20:1 dr

Mannich Reaction. Mannich reactions often require a stoichiometric amount of base to promote an efficient addition of an enolate to an activated imine electrophile. New catalytic methods to render this reaction asymmetric are important for the synthesis of β-amino carbonyl compounds. A novel protocol for the catalytic generation of an enolate equivalent through NHC catalysis with α-aryloxyaldehydes was developed. The title compound **1** was found to be the optimal catalyst precursor for an enantioselective Mannich reaction of α-aryloxyaldehydes and *N*-tosyl imines (eq 8).[11] Good yields and high enantioselectivities were achieved through this rebound catalysis approach. The generated carbene reacts with the aryloxyaldehyde to form an aryloxy anion, which later rebounds to turn over the NHC catalyst through an acylation event. Subsequent addition of an alcohol or amine nucleophile provides β-amino carbonyl compounds.

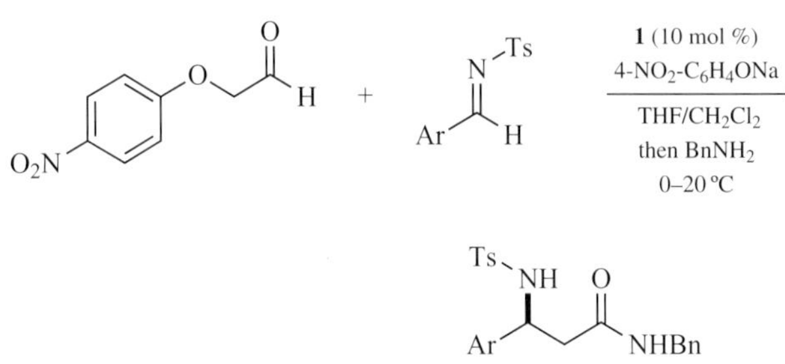

(8)

yields: 56–75%
ee: 88–95%

Intramolecular Lactonization. The use of phenylalanine-derived triazolium **8** in an intramolecular lactonization reaction was disclosed.[12] A good yield was obtained; however, only a 40% enantiomeric excess was achieved (eq 9). A formal synthesis of salinosporamide was completed using this protocol; however, the key carbene-catalyzed cyclization produced a 1:1.5 dr (desired:undesired).

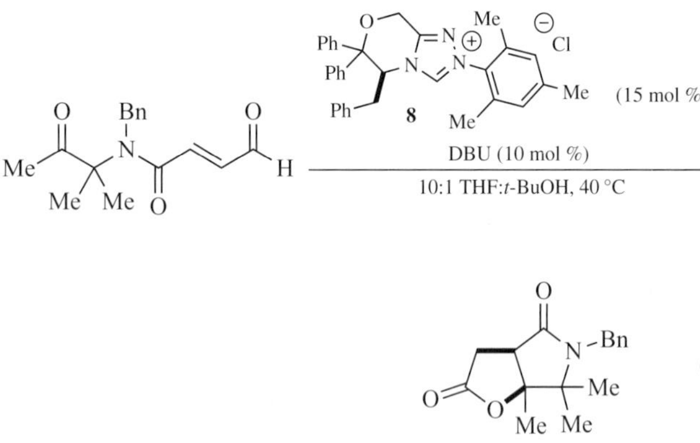

(9)

73% yield, 40% ee

Conjugate Addition of Alcohols. NHCs are well known to act as Lewis bases in catalytic processes; however, their Brønsted basic properties are much less understood. The conjugate addition of alcohols to activated alkenes wherein the NHC acts as a Brønsted base was reported.[13] Enantioselectivity was not observed with catalytic amounts of **1** (eq 10); however, triazolium **6** afforded an 11% ee.

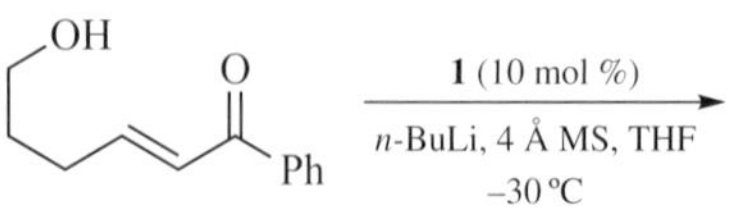

(10)

86% yield, 0% ee

Intermolecular Stetter Reaction. The synthesis of α-amino acid derivatives using an NHC-catalyzed intermolecular Stetter reaction was reported.[14] Employing a phenylalanine-derived precatalyst **9**, γ-keto-α-amino acid derivatives were synthesized in excellent yields with high levels of enantioselectivity (eq 11). The reaction was proven to be amenable to scale-up protocols with an example of a 5 mmol scale.

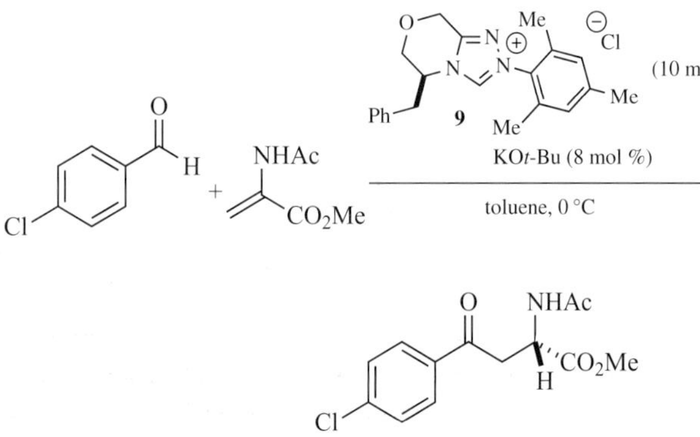

(11)

5 mmol scale: 95%, 94% ee

Cyclopentane Synthesis. Highly functionalized cyclopentane rings are a common motif in organic synthesis and their construction remains an ongoing challenge. A recent report described a cooperative catalysis route to cyclopentanes using a combination of an NHC and $Ti(Oi\text{-}Pr)_4$.[15] Reacting α,β-unsaturated aldehydes and β,γ-unsaturated α-ketoesters with triazolium salts, DBU, and $Ti(Oi\text{-}Pr)_4$ provides highly substituted cyclopentanes in excellent yields with high levels of diastereo- and enantioselectivity (eq 12). The title compound **1** gave the product in 76% yield, 12:1 dr, and with 70% ee. The optimal catalyst for this transformation, however, was an aminoindanol-derived triazolium catalyst.

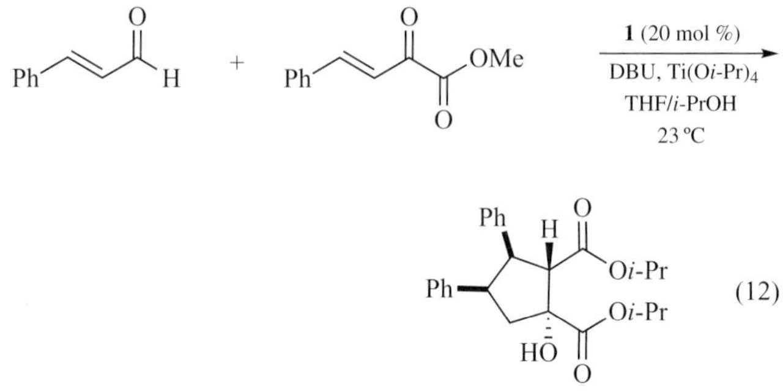

(12)

76% yield, 70% ee
12:1 dr

Hydroacylation of Unactivated Alkenes. Catalytic amounts of **9** was utilized in a highly efficient and selective intramolecular hydroacylation reaction.[16] Various chromanones were synthesized from *O*-allylated *o*-vanillin derivatives in the presence of 10 mol % of **7** and 20 mol % of DBU in dioxane at 80 °C (eq 13). The products were limited to 3-methyl-3-aryl chroman-4-one compounds; however, greater than 99% ee was achieved for most cases.

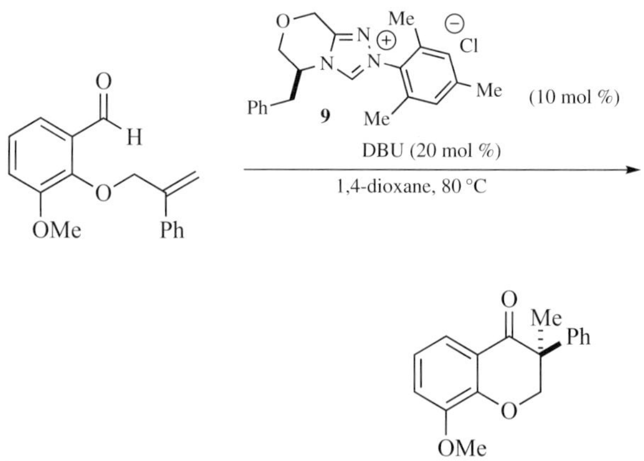

(13)

99% yield, >99% ee

Related Reagents. (4*S*,5*S*)-4-(2,2-Dimethyl-4-phenyl-1,3-dioxan-5-yl)-1-phenyl-4*H*-1,2,4-triazol-1-ium perchlorate; 6,7-dihydro-2-phenyl-5-(phenylmethyl)-5*H*-pyrrolo[2,1-*c*]-1,2,4-triazolium chloride.

1. Enders, D.; Balensiefer, T., *Acc. Chem. Res.* **2004**, *37*, 534.
2. Nair, V.; Vellalath, S.; Babu, B. P., *Chem. Soc. Rev.* **2008**, *37*, 2691.
3. Phillips, E. M.; Chan, A.; Scheidt, K. A., *Aldrichim. Acta* **2009**, *42*, 55.
4. Knight, R. L.; Leeper, F. J., *J. Chem. Soc., Perkin Trans. 1* **1998**, 1891.
5. Enders, D.; Breuer, K.; Teles, J. H., *Helv. Chim. Acta* **1996**, *79*, 1217.
6. Maki, B. E.; Chan, A.; Phillips, E. M.; Scheidt, K. A., *Org. Lett.* **2007**, *9*, 371.
7. Phillips, E. M.; Wadamoto, M.; Chan, A.; Scheidt, K. A., *Angew. Chem., Int. Ed.* **2007**, *46*, 3107.
8. Wadamoto, M.; Phillips, E. M.; Reynolds, T. E.; Scheidt, K. A., *J. Am. Chem. Soc.* **2007**, *129*, 10098.
9. Phillips, E. M.; Wadamoto, M.; Scheidt, K. A., *Synthesis* **2009**, 687.
10. Cohen, D. T.; Cardinal-David, B.; Roberts, J. M.; Sarjeant, A. A.; Scheidt, K. A., *Org. Lett.* **2011**, *13*, 1068.
11. Kawanaka, Y.; Phillips, E. M.; Scheidt, K. A., *J. Am. Chem. Soc.* **2009**, *131*, 18028.
12. Struble, J. R.; Bode, J. W., *Tetrahedron* **2009**, *65*, 4957.
13. Phillips, E. M.; Riedrich, M.; Scheidt, K. A., *J. Am. Chem. Soc.* **2010**, *132*, 13179.
14. Jousseaume, T.; Wurz, N. E.; Glorius, F., *Angew. Chem., Int. Ed.* **2011**, *50*, 1410.
15. Cohen, D. T.; Cardinal- David, B.; Scheidt, K. A., *Angew. Chem., Int. Ed.* **2011**, *50*, 1678.
16. Piel, I.; Steinmetz, M.; Hirano, K.; Fröhlich, R.; Grimme, S.; Glorius, F., *Angew. Chem., Int. Ed.* **2011**, *50*, 4983.

Chad C. Eichman, Daniel T. Cohen & Karl A. Scheidt
Northwestern University, Evanston, IL, USA

3-Benzyl-4-methyl-1,3-thiazolium Chloride

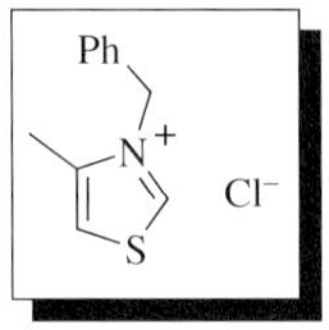

[4209-18-1] $C_{11}H_{12}ClNS$ (MW 225.74)
InChI = 1/C11H12NS.ClH/c1-10-8-13-9-12(10)7-11-5-3-2-4-6-11;/h2-6,8-9H,7H2,1H3;1H/q+1;/p-1/fC11H12NS.Cl/h;1h/qm;-1
InChIKey = ZGBNEPNQIDHABT-WJULMTHWCM

(benzoin condensation catalyst; in the presence of bases, catalyzes the preparation of cyclic acyloins from dialdehydes, and of aromatic esters from aromatic aldehydes)

Physical Data: mp 188 °C; monohydrate 136 °C.[2]
Solubility: sol H_2O, MeOH, EtOH and related solvents; insol ether or pentane.
Analysis of Reagent Purity: ^{1}H NMR, mp.
Preparative Method: colorless crystals prepared from 4-methyl-1,3-thiazole and benzyl chloride, or benzylthioformamide and chloroacetone.[2]
Handling, Storage, and Precautions: store dry.

Formation of Benzoins and Acyloins from Aldehydes. Many thiazolium salts catalyze the formation of benzoins from aromatic aldehydes and of acyloins from aliphatic aldehydes, both in presence of bases (eq 1). This catalytic reaction has long been known,[3] the most prominent salt being thiamine (vitamin B_1). For details on the use of other salts. The title reagent is the only one known to catalyze the formation of five- and six-membered cyclic acyloins in the presence of a base (eq 2).[4]

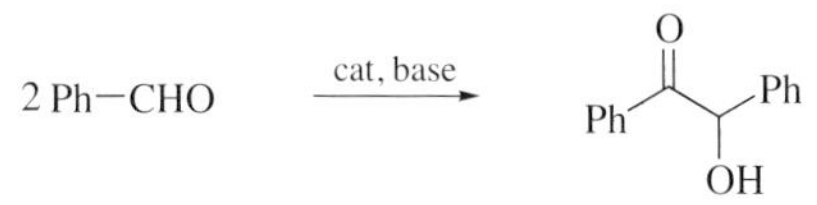
(1)

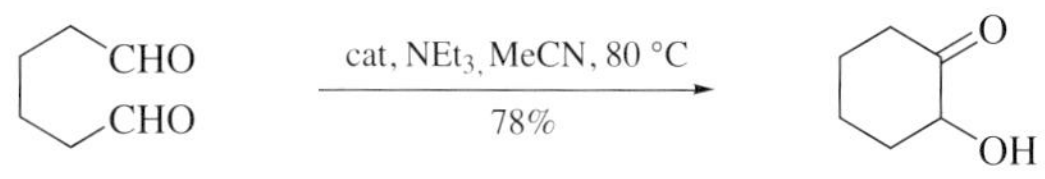
(2)

Oxidative Reactions. Aromatic aldehydes in presence of a nitro compound, an alcohol, and a base react in a different way (eq 3). No formation of benzoins is observed, with only the carboxylic esters and nitrobenzene reduction products being formed.[5] Quinones react analogously to nitro compounds.[6] A test for the catalytic activity of azolium salts uses flavin as oxidant and indicator;[7a] details are given in a related bioorganic model study on the reaction of thiamine.[7b]

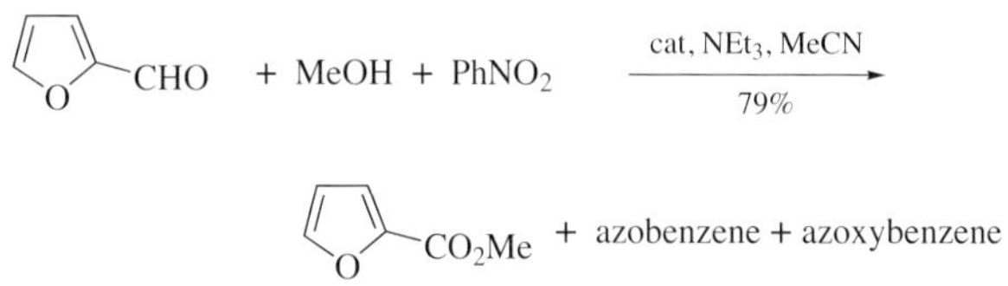
(3)

Addition of Aldehydes to Electrophilic Double Bonds.[8] Full details are given under the entry for 3-Benzyl-5-(2-hydroxyethyl)-4-methyl-1,3-thiazolium Chloride. The reagent in the presence of bases catalyzes addition, but shows a tendency to form byproducts (eq 4). In this case the byproduct results from Michael addition to the benzoin instead of to the activated aldehyde.[9]

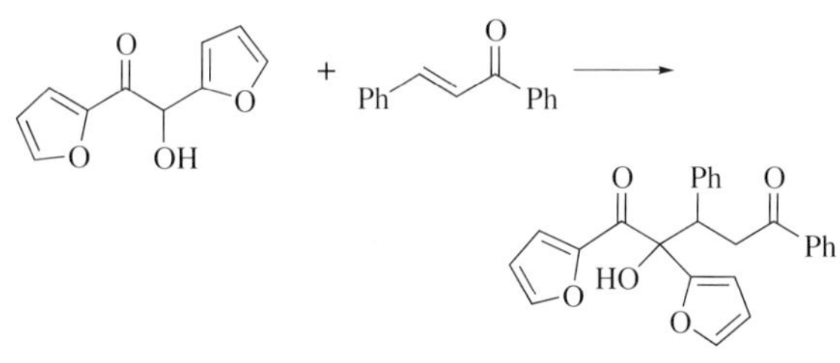
(4)

Reduction and Hydrolysis of the Reagent. The reagent, like other thiazolium salts, is easily reduced by Sodium Borohydride[10] to the thiazoline and the ring is easily cleaved by aqueous bases.[2b]

Other Reagents. Acyloins are produced by the action of metals such as sodium on carboxylic esters;[11] benzoins are produced by the catalytic action of cyanides on aromatic aldehydes.[12]

Related Reagents. 3-Benzyl-5-(2-hydroxyethyl)-4-methyl-1,3-thiazolium Chloride; 3-Benzylthiazolium Bromide.

---

1. (a) Fieser L. F.; Fieser M., *Fieser & Fieser* **1979**, *7*, 18. (b) *Beilstein* **1983**, *27-III/IV*, 971.
2. (a) Karimullah *J. Chem. Soc* **1937**, 961. (b) Sykes P.; Todd A. R., *J. Chem. Soc* **1951**, 534.
3. Breslow R. *J. Am. Chem. Soc.* **1958**, *80*, 3719.
4. Cookson, R. C; Lane R. M., *J. Chem. Soc., Chem. Commun.* **1976**, 804.
5. (a) Castells J.; Llitjós H.; Moreno-Mañas M. *Tetrahedron Lett.* **1977**, 205. (b) Castells J.; Pujol F.; Llitjós H.; Morena-Mañas M. *Tetrahedron* **1982**, *38*, 337. (c) Inoue H.; Higashiura, *J. Chem. Soc., Chem. Commun.* **1980**, 549. (d) Shinkai S.; Yamashita T.; Kusano Y.; Manabe O., *J. Org. Chem.* **1980**, *45*, 4947.
6. Vovk A. I.; Murav'eva I. V.; Yasnikov A. A. *Zh. Obshch. Khim.* **1985**, *55*, 1400 (*Chem. Abstr.* **1985**, *103*, 214 513p).
7. (a) Yano Y.; Tsukagoshi Y. *J. Chem. Res. (S)* **1984**, 406. (b) Rastetter W. H.; Adams J., *J. Org. Chem.* **1981**, *46*, 1882.
8. Stetter H.; Kuhlmann H., *Org. React.* **1991**, *40*, 407.
9. Castells J.; Duñach E.; Geijo F.; López-Calahorra F.; Prats M.; Sanahuja O.; Villanova L. *Tetrahedron Lett.* **1980**, *21*, 2291.
10. Clarke G. M.; Sykes P. *J. Chem. Soc. (C)* **1967**, 1269.
11. (a) McElvain S. M. *Org. React.* **1948**, *4*, 256. (b) Bloomfield J. J.; Owsley D. C.; Nelke J. M. *Org. React.* **1976**, *23*, 259. (c) Finley K. T., *Chem. Rev.* **1964**, *64*, 573. (d) Rühlmann K. *Synthesis* **1971**, 236.
12. Ide W. S.; Buck J. S. *Org. React.* **1948**, *4*, 269.

John Wiley & Sons, Ltd

Heinrich Kuhlmann
RWTH, Aachen, Germany

cis-4-Benzyloxy-α,α-bis-(3,5-dimethylphenyl)-*L*-prolinol (2*S*,4*R*)-[1]

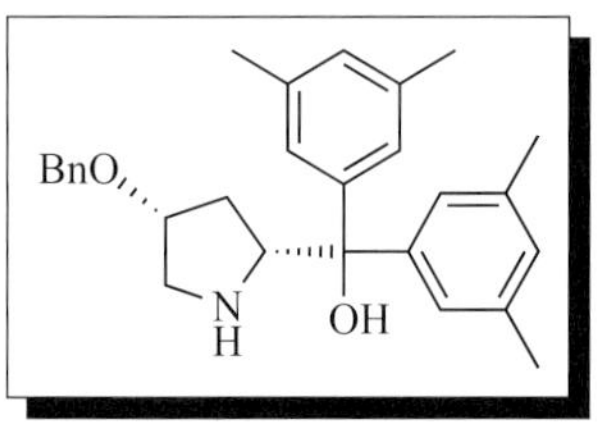

[923985-71-1] $C_{28}H_{33}NO_2$ (MW 415.57)

InChI = 1S/C28H33NO2/c1-19-10-20(2)13-24(12-19)28(30,25-14-21(3)11-22(4)15-25)27-16-26(17-29-27)31-18-23-8-6-5-7-9-23/h5-15,26-27,29-30H,16-18H2,1-4H3/t26-,27-/m1/s1

InChIKey = HSSMHALMJXLUDA-KAYWLYCHSA-N

(organocatalyst used for the asymmetric epoxidation of electron deficient olefins)

Physical Data: white solid, mp 127 °C; $[\alpha]_{589}$ 44.7° (*c* 0.45, $CHCl_3$).[1]

Solubility: sol most organic solvents.

Analysis of Reagent Purity: enantiomeric ratio is determined by HPLC.

Preparative Method: addition of (2*R*, 4*R*)-*N*-ethoxy carbonyl-4-hydroxy-proline methyl ester (**3**) to (3,5-dimethylphenyl)-magnesium bromide followed by etherification of C(4)–OH with benzyl bromide and subsequent alkaline hydrolysis (12% overall from *trans*-4-hydroxyl-*L*-proline **2**, eq 1).[1]

Purification: recrystallization from hexane.

Handling, Storage, and Precautions: no special information available. Care should be exercised to avoid contact of this reagent with the eyes and skin.

Asymmetric Epoxidation of Chalcones. Zhao and coworkers reported a highly enantioselective asymmetric epoxidation of α,β-enones using easily obtained prolinol **1** as a catalyst and TBHP as an oxidant (eq 2).[1] High yields and excellent enantioselectivities have been obtained for a number of substrates with different steric and electronic properties (Table 1).

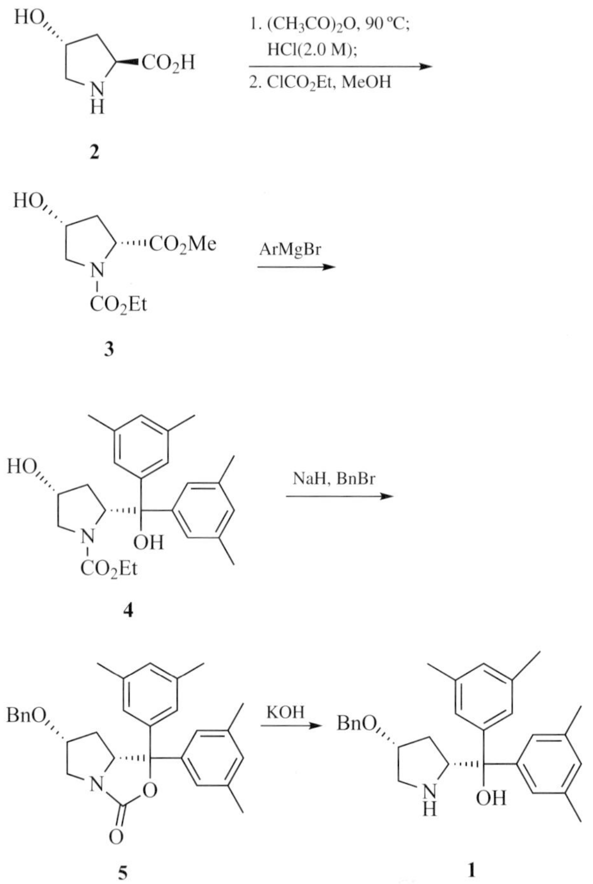

(1)

Although the reaction mechanism is not exactly clear, a possible mechanism for epoxidation of enones was proposed according to the catalytic cycle by Lattanzi (eq 3).[2] First, the catalyst **1** activated TBHP by deprotonation to produce the ammonium salt with a *tert*-butyl hydroperoxide anion. Next, the tertiary hydroxyl activated the enone by forming a hydrogen bond with the carbonyl group. A hypothetical transition state was proposed, parallel to Lattanzi's model. Furthermore, catalyst **1** had a benzyloxy at the *cis*-4-position of the prolinol backbone, which was proximal to the reaction zone. The steric hindrance between benzyloxy and the methyl of the *tert*-butyl hydroperoxide anion induced a frontside nucleophilic attack to the double bond, which produced an increase in enantioselectivity compared to Lattanzi's catalyst.[1] The combined steric effects of the 3,5-dimethylphenyl and benzyloxy presumably established a conformation favoring high enantioselectivity.

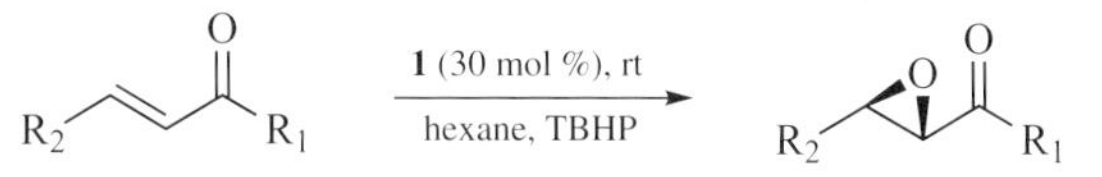

(2)

Table 1 Asymmetric epoxidation of chalcones catalyzed by **1**

Entry	R_1	R_2	Time [h]	Yield [%]	ee [%]
1	Ph	Ph	144	75	94
2[a]	p-$CH_3OC_6H_4$	Ph	144	70	96
3	p-ClC_6H_4	Ph	96	80	90
4	p-BrC_6H_4	Ph	100	82	89
5	p-$NO_2C_6H_4$	Ph	120	86	91
6	Ph	p-$CH_3C_6H_4$	122	72	94
7	Ph	p-ClC_6H_4	105	76	96
8[a]	Ph	p-$NO_2C_6H_4$	96	90	94
9[a]	p-$CH_3OC_6H_4$	p-ClC_6H_4	144	66	94
10	p-ClC_6H_4	p-$CH_3C_6H_4$	144	70	92

[a]Reaction was performed in CCl_4.

Asymmetric Epoxidation of α,β-Unsaturated Trifluoromethyl Ketones. Zhao and coworkers also employed catalyst **1** in the asymmetric epoxidation of α,β-unsaturated trifluoromethyl ketones (eq 4).[3] In the above epoxidation of chalcones, the rate of the reaction was accelerated slightly with an electron-withdrawing substituent, such as the nitrophenyl ring. Changing the nitrophenyl moiety to a trifluoromethyl group greatly increased the substrate reactivity and the corresponding epoxides were obtained in shorter time with lower catalyst loading. The resulting epoxides were quickly transformed to their hydrates during aqueous work-up. A series of *trans*-α,β-unsaturated trifluoromethyl ketones were evaluated in the presence of catalyst **1** (10 mol %) at 0 °C (Table 2). In most examples, *trans*-(2*S*, 3*R*)-epoxides were obtained in 89–96% ee with high yields (Entries 1-5) except in the case of the

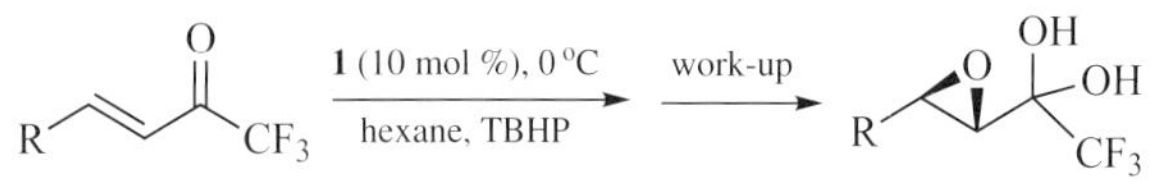

(4)

Table 2 Asymmetric epoxidation of α,β-unsaturated trifluoromethyl ketones using **1**

Entry	R	Time [h]	Yield [%]	ee[%]
1	Ph	12	85	96
2	p-FC_6H_4	12	88	90
3	p-ClC_6H_4	12	84	90
4	p-BrC_6H_4	12	82	95
5	p-$CH_3C_6H_4$	18	80	95
6	p-EtC_6H_4	18	81	78
7	o-ClC_6H_4	28	65	53
8	m-ClC_6H_4	18	79	89
9	o,p-*di*-$ClFC_6H_4$	28	70	53

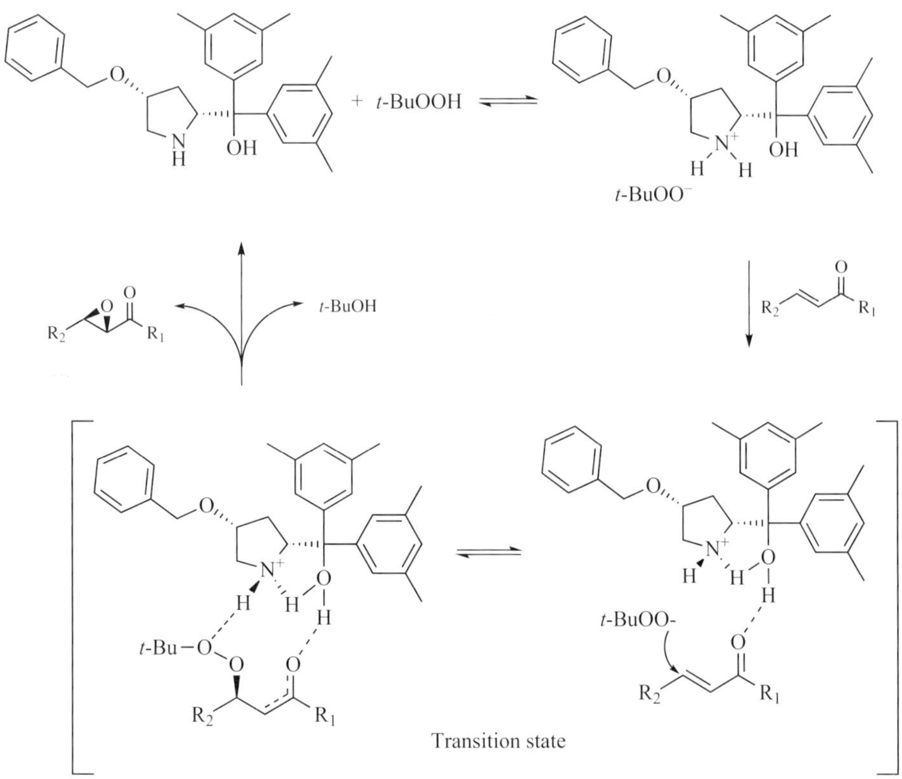

(3)

p-ethyl substituent (Entry 6). Substituents at the *o*-position were deleterious for the epoxidation (Entries 7, 9), while *m*-substitution was well tolerated (Entry 8).

Hydrazinolysis of the trifluoromethyl epoxide **6** with phenyl hydrazine afforded heterocycle **7** bearing a trifluoromethyl substituted quaternary carbon center in 85% yield (eq 5).[4] These types of heterocyclic compounds are useful intermediates in the preparation of biologically interesting compounds.

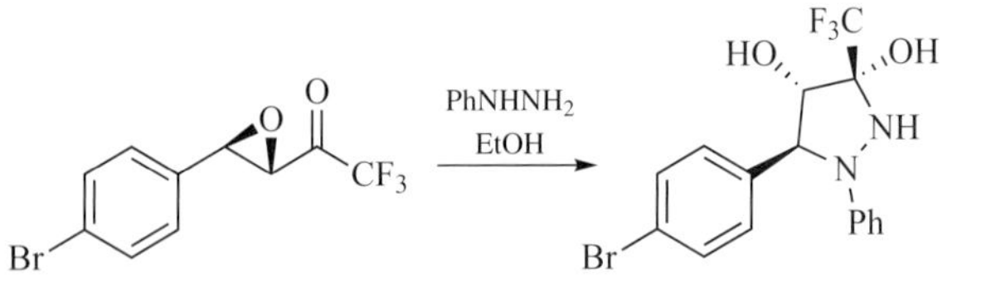

(5)

6 **7**

1. Li, Y. W.; Liu, X. Y.; Yang, Y. Q.; Zhao, G., *J. Org. Chem.* **2007**, *72*, 288.
2. (a) Lattanzi, A., *Org. Lett.* **2005**, *7*, 2579. (b) Lattanzi, A., *Adv. Synth. Catal.* **2006**, *348*, 339.
3. (a) Zheng, C. W.; Li, Y. W.; Yang, Y. Q.; Wang, H. F.; Cui, H. F.; Zhang, J. K.; Zhao, G., *Adv. Synth. Catal.* **2009**, *351*, 1685. (b) Zheng, C. W.; Zhao, G., *Chin. Sci. Bull.* **2010**, *55*, 1712.
4. Zhang, X. Q.; Li, X. J.; Allan, G. F.; Sbriscia, T.; Linton, O.; Lundeen, S. G.; Sui, Z. H., *J. Med. Chem.* **2007**, *50*, 3857.

Gang Zhao
Shanghai Institute of Organic Chemistry, Chinese Academy of Sciences, Peoples' Republic of China

Zheng-Qing Ye
Shanghai Institute of Organic Chemistry, Chinese Academy of Sciences, Peoples' Republic of China

N-Benzylquininium Chloride

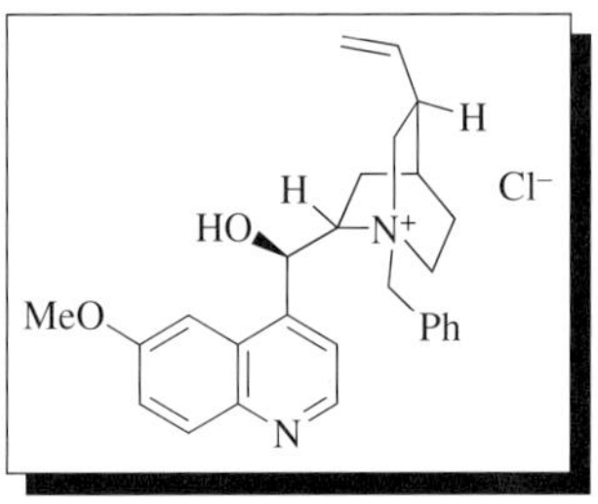

[67174-25-8] $C_{27}H_{31}ClN_2O_2$ (MW 451.01)

InChI = 1/C27H31N2O2.ClH/c1-3-20-18-29(17-19-7-5-4-6-8-19)14-12-21(20)15-26(29)27(30)23-11-13-28-25-10-9-22 (31-2)16-24(23)25;/h3-11,13,16,20-21,26-27,30H,1,12,14- 15,17-18H2,2H3;1H/q+1;/p-1/t20-,21-,26-,27+,29+;/m0./ s1/fC27H31N2O2.Cl/h;1h/qm;-1

InChIKey = JYDIJFKNXHPWBJ-QWAPBUHSDD

(chiral, nonracemic quaternary ammonium salt; catalyst for a variety of phase transfer reactions under basic conditions[1])

Alternate Names: Quibec; BQC.

Physical Data: mp 180.181 °C, $[\alpha]_D^{19}$ −235.5° (c = 1.5, H_2O);[2] monohydrate: mp 169.172 °C (dec), $[\alpha]_D$ .212.5° (c = 0.5, EtOH).[3]

Solubility: freely sol H_2O, alcohols, acetone; slightly sol EtOAc; sparingly sol $CHCl_3$.[4]

Form Supplied in: solid.

Purification: recrystallized from absolute EtOH.[3]

Original Commentary

Mary Ellen Bos
R. W. Johnson Pharmaceutical Research Institute, Raritan, NJ, USA

The combined use of catalytic amounts of *N*-benzylquininium chloride (**1**) with hydroxide bases (NaOH or KOH) has been explored for a variety of phase transfer reactions, including epoxidations, alkylations, and Michael reactions. The degree of stereoselectivity in product formation induced by the reagent can vary widely.[5]

BQC is derived from quinine, which is a member of the *cinchona* family of alkaloids. Ammonium salts derived from quinidine, a diastereomer of (**1**) at the hydroxyl substituent, have been used less frequently in catalysis than BQC. Quinidinium salts often give rise to products with enantioselectivity opposite to that from (**1**). Other related compounds, such as those derived from cinchonine and cinchonidine (which lack the methoxy substituent on the quinoline nucleus), have found application in organic synthesis. The *cinchona* alkaloids, as well as salt derivatives in which the benzyl group bears various substituents, have also been studied.[6] Results from polymer-bound catalysts have not been promising.[7]

Benzylquininium chloride has shown good to excellent selectivity in the epoxidation of α,β-unsaturated ketones.[8] Oxidation of quinone (**2**) in the presence of (**1**) with aqueous t-Butyl Hydroperoxide and Sodium Hydroxide in toluene gave rise to a 95% chemical yield of epoxide **3** in 78% ee (eq).[2] Recrystallization improved the ee to 100% with 63% mass recovery. Aqueous **Hydrogen Peroxide** decreased both the yield (89%) and enantioselectivity (50% ee).

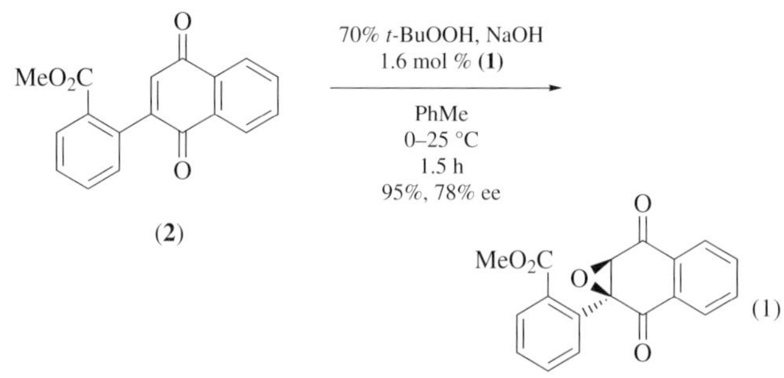

(1)

(**2**) (**3**)

Enantiomerically enriched epoxides have also been generated using (**1**) in the Darzens reaction of α-chloro ketones with aldehydes, as well as through ring closure of racemic halohydrins.[9] The extent of enantioselectivity for both reactions is similar (optical purity 6–8% ± 1), suggesting a moderate degree of kinetic resolution occurring in each case.

Benzylquininium chloride has been studied as a catalyst for the asymmetric Michael reaction. Reaction of amidomalonate (**5**) and chalcone (**4**) with catalytic base and a variety of chiral, nonracemic ammonium salts in the absence of solvent produced **6** in yields of 41–68% and 20–68% ee (eq).[10] The quinine-derived salt (**1**) was of intermediate effectiveness (38% ee, 47% yield) when compared to ephedrine-based catalysts. Although **1** was not specifically tested with regard to solvation effects, it is suggested that increased aggregation of reactive species under solid–liquid PTC conditions leads to enhanced organization and selectivity. Low levels of induction have been observed in other systems.[11] A study comparing various chiral alkaloid salts, bases, and reaction conditions on enantioselectivities in the conjugate addition of thiols and nitroalkanes to enones has been reported.[12] Remarkable results have also been obtained in the cinchonium/cinchonidinium-catalyzed reaction of an indanone with methyl vinyl ketone.[13] A direct comparison with a more efficient lanthanum–binaphthol complex has been reported.[14]

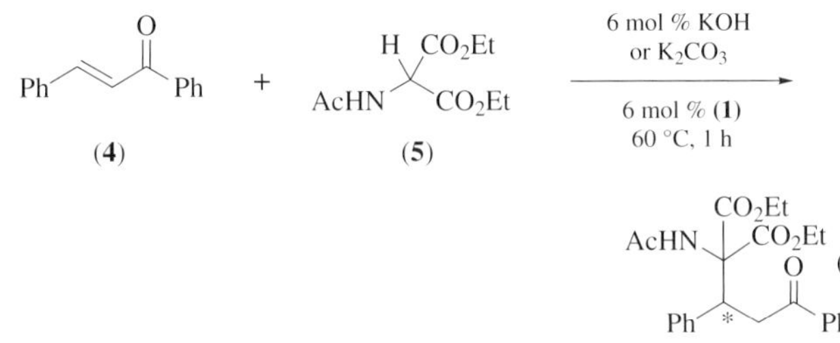
(2)

(**6**)

Phase-transfer alkylations have been studied using *cinchona* alkaloid derivatives; however, results have been more promising with the cinchonium/cinchonidinium series than with the quininium/quinidinium group.[15] For example, catalytic asymmetric alkylation in high yield and selectivity has been achieved with *N*-(*p*-trifluoromethylphenyl)cinchonium bromide (*p*-CF_3BCNB) (eq).[16] A systematic investigation of reaction variables (catalyst type, solvent, concentration, temperature, stirring rate, leaving group, etc.) has produced a general method to prepare α-amino acids in 44–62% ee.[17]

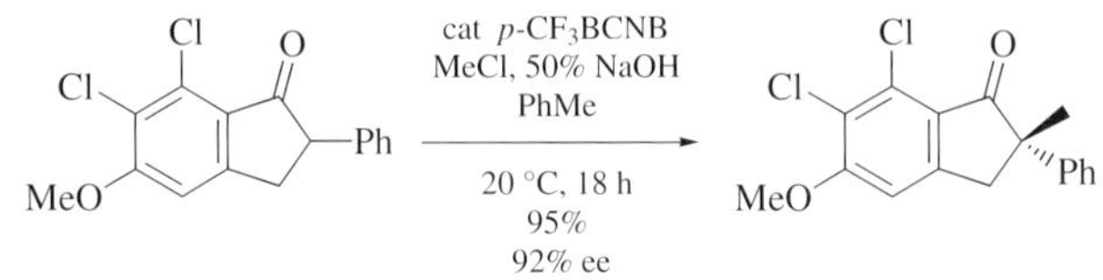
(3)

First Update

Takuya Hashimoto & Keiji Maruoka
Kyoto University, Kyoto, Japan

Phase-transfer-catalyzed Asymmetric Aza-Henry Reaction Utilizing α-Amido Sulfones. Cinchona alkaloid-based quaternary ammonium salts are widely recognized as effective phase-transfer catalysts for the construction of chiral nonracemic compounds. These catalysts have been applied for a variety of asymmetric reactions, such as alkylations, Michael additions, Mannich reactions, aldol reactions, and epoxidations.[18–20] So far, asymmetric phase-transfer catalysis mainly relies on the use of cinchonine or cinchonidine-derived ammonium salts as catalysts, and quininium salts were rarely employed in the highly enantioselective reactions.[21–23] In this context, Palomo et al and Herrera and co-workers reported separately that *N*-benzylquininium chloride **1** acts as a promising catalyst for the asymmetric aza-Henry reaction under solid–liquid phase-transfer conditions (eqs 4–6).[24,25] In both cases, α-amido sulfones are employed to generate reactive *N*-carbamoyl imines in situ, and the authors succeeded in conducting aza-Henry reactions of both aromatic and aliphatic imines, which are prone to isomerize to the undesired ene carbamates. These features bestow additional values on this phase-transfer-catalyzed asymmetric transformation compared to other catalytic asymmetric aza-Henry reactions, which can be applied to only nonenolizable imines.[26–35] Although nitromethane was employed as a representative nucleophile throughout these works, Palomo also demonstrated that the nucleophilic addition of nitroethane led to the *syn*-selective formation of 1,2-nitroamines with high enantioselectivities (eq 5).

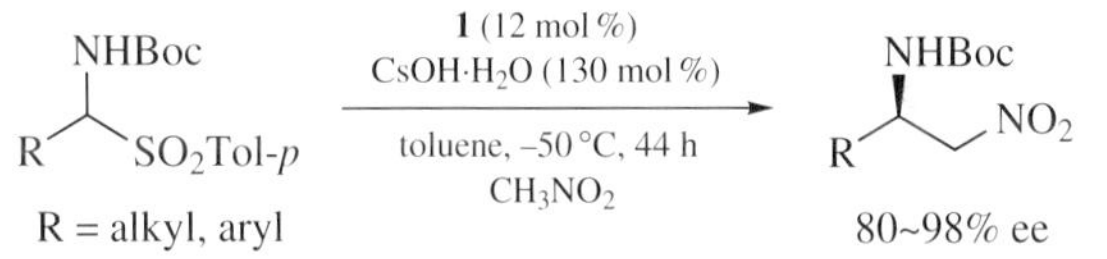
(4)

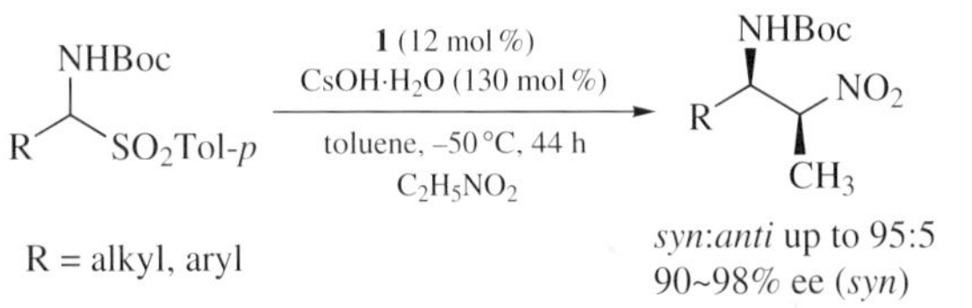
(5)

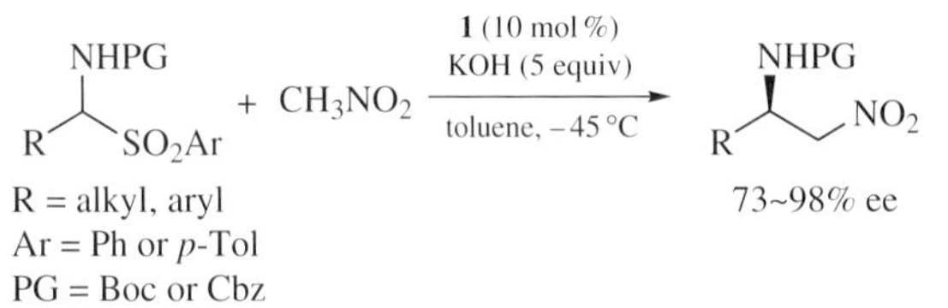
(6)

***N*-Benzylquininium-fluoride-catalyzed Asymmetric Hydrosilylation of Ketones.** Quaternary ammonium fluorides function as unique activators of organosilicon compounds by taking advantage of the strong affinity of fluoride and silicon, and various catalytic asymmetric reactions, such as asymmetric aldol reaction of enol silyl ethers, have been developed employing cinchona alkaloid-derived and synthetic chiral quaternary ammonium fluorides.[36]

N-Benzylquininium chloride can be converted to *N*-benzylquininium fluoride **7** by treatment of **1** with Amberlyst A-26 hydroxide form followed by the neutralization with 1 N HF and evaporation (eq 7). Ammonium fluoride **7** can be utilized as an organocatalyst in the asymmetric hydrosilylation of carbonyl compounds, providing the secondary alcohols in moderate enantioselectivity (eq 8).[37]

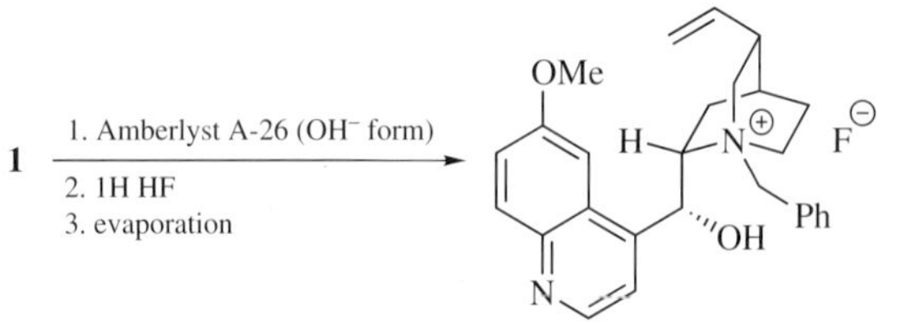

7 (7)

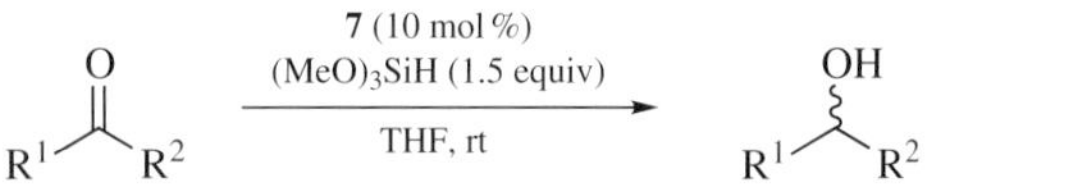

up to 65% ee

(R^1 = Ph, R^2 = Et) (8)

Optical Resolution of the Racemic Compounds by Cocrystallization with *N*-Benzylquininium Chloride. Enantiomers of chiral diols, especially 2,2′-binaphthol derivatives can be easily separated by the formation of inclusion complexes with *N*-alkylcinchonidium halides.[38–43] *N*-Benzylquininium chloride can be utilized in the optical resolution of racemic mixtures of 9,9′-spirobixanthene-1,1′-diol **8**.[44] Thus, (±)-**8** was treated with *N*-benzylcinchonidium chloride in acetonitrile to give the precipitation of the *N*-benzylcinchonidium chloride and (*R*)-**8** in 1:1 molar ratio. (*S*)-**8** can be then obtained from the mother liquor by the formation of the inclusion complex crystal with **1** (eq 9). The usefulness of so-obtained chiral nonracemic diol can be demonstrated by the further elaboration of **8** to the phosphoramidite ligand and application of this ligand to the asymmetric hydrogenation of dehydroamino acid derivatives.

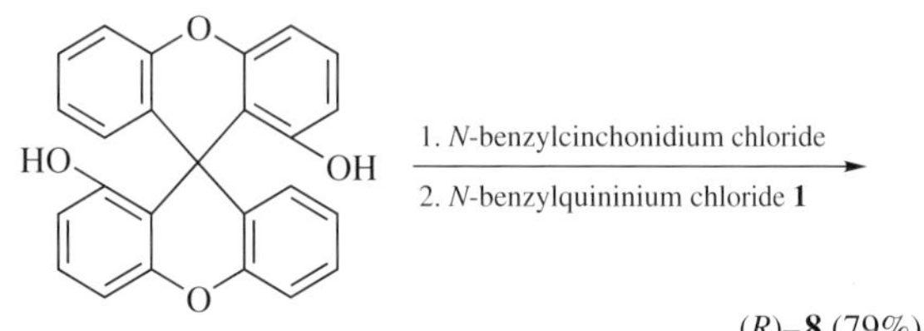

(+/–)-**8**

(*R*)-**8** (79%)
+ (9)
(*S*)-**8** (78%)

Second Update

Martin Smith & Katrina Badiola
University of Oxford, Oxford, UK

Application in Screening for Asymmetric Phase-transfer-catalyzed Reactions. *N*-Benzylquininium chloride (**1**) is a cinchona alkaloid-derived quaternary ammonium salt that is often used in asymmetric transformations, most frequently as a catalyst for reactions under phase-transfer conditions.[18,45–47] Alkylations, epoxidations, conjugate addition reactions, and aza-Henry reactions can be catalyzed by **1** with varying degrees of stereocontrol. As a relatively inexpensive and commercially available catalyst that is also easily accessible from quinine,[2,4,48] **1** is ideal for screening appropriate asymmetric transformations. Catalyst **1** provided a starting point for further optimization in enantioselective solid–liquid alkylation and 1,4-addition reactions to provide precursors for valued monofluoromethylated products (eqs 10 and 11).[49,50]

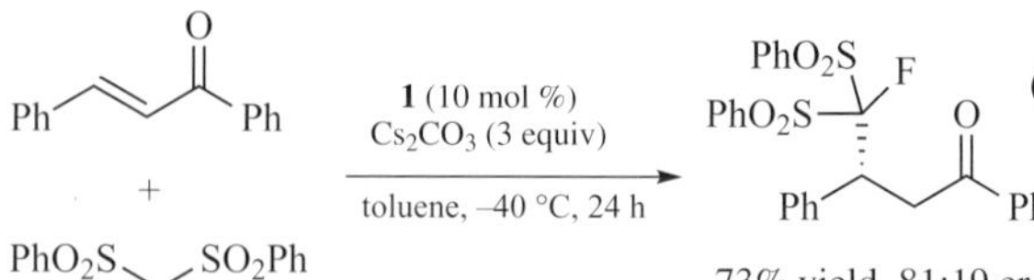

73% yield, 81:19 er (10)

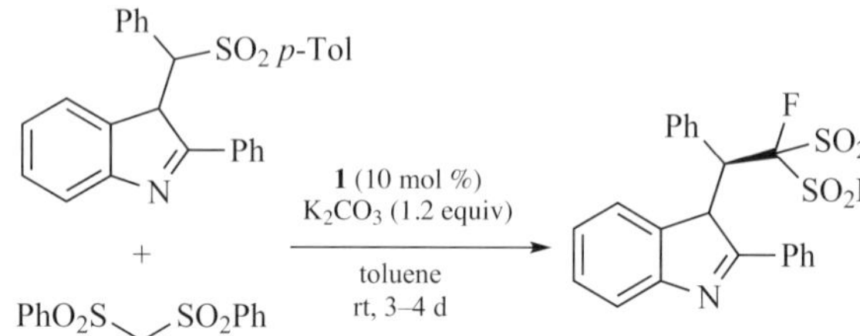

56% yield, 53:47 er (11)

Phase-transfer-catalyzed Reactions Employing α-Amido-sulfones as Proelectrophiles. The scope of phase-transfer-catalyzed enantioselective aza-Henry reaction of azomethines masked as α-amido-sulfones[24,25] has been expanded with the application of extended-chain nitroalkanes as nucleophiles.[51] *N*-Benzylquininium chloride effectively catalyzed the formation of nitroamine adducts with moderate to excellent yields, and enantio- and diastereoselectively (eq 12). Use of the *O*-benzylated derivative of **1** as a catalyst resulted in poor conversion to the desired products, and computational modeling demonstrated a favorable interaction of the nitronate anion of nitromethane to a β-hydroxylated ammonium cation. A catalytic cycle for the reaction was proposed, whereby a complex—consisting of the ammonium cation of **1** and the product anion—is involved in the deprotonation of the nitroalkane substrate. Subsequent functionalization of the nitroamine products provided access to highly enantioenriched 1,2-diamines and vinylogous amino acids. This methodology has also been used as the stereodetermining step in the synthesis of enantioenriched piperidines and diaminopropanoates.[52,53] Non-canonical amino acids can be similarly accessed via an aza-Henry reaction of bromonitromethane with α-amido-sulfones using **1** as a catalyst under solid–liquid phase-transfer reaction conditions (eq 13).[54] These adducts were converted to the corresponding amides with good yields and stereoselectivities over the two steps. Alternative non-cinchona alkaloid-derived catalysts may also be used to effect the transformation.[55]

Other transformations catalyzed by **1** that employ α-amido-sulfones include enantioselective Mannich reactions.[56,57] Moderate to excellent selectivities were obtained when **1** was employed as a catalyst (eqs 14 and 15); however, differently substituted *N*-benzylquininium salts exhibited greater enantiocontrol. Similarly, the addition of phosphinic acids to α-amido-sulfones was catalyzed by **1** with good conversion to the desired product (eq 16), but improved selectivity was observed with *N*-benzylquininium-derived salts.[58] This phase-transfer methodology provides an alternative and mild route to enantioenriched α-amino phosphonic acids.[59–61]

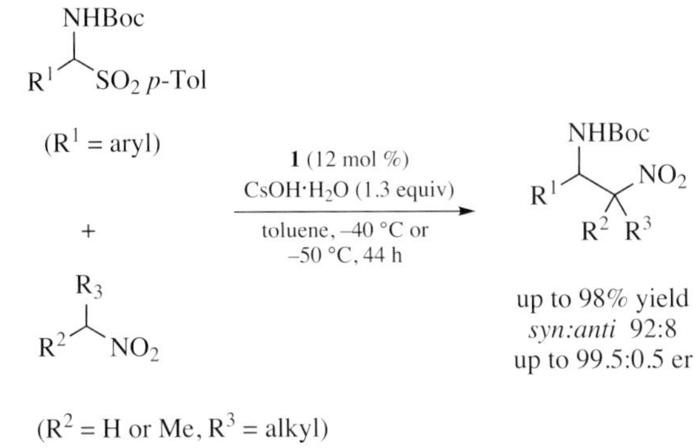

(12)

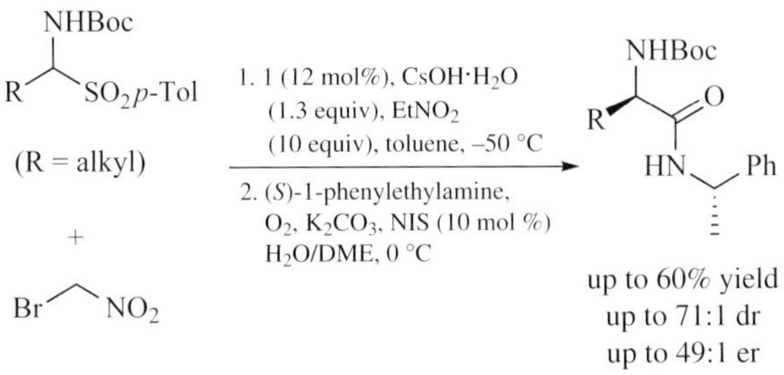

(13)

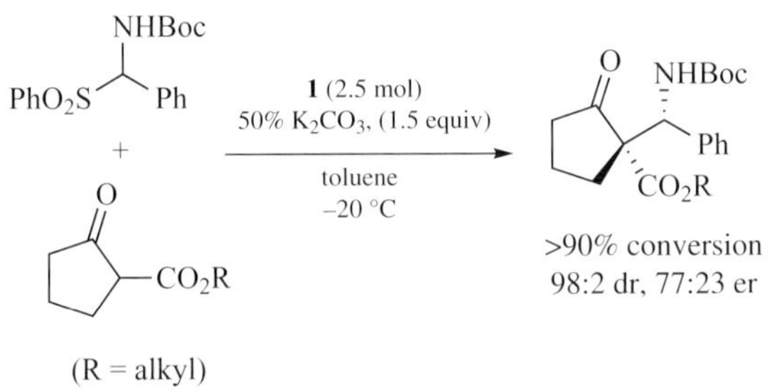

(14)

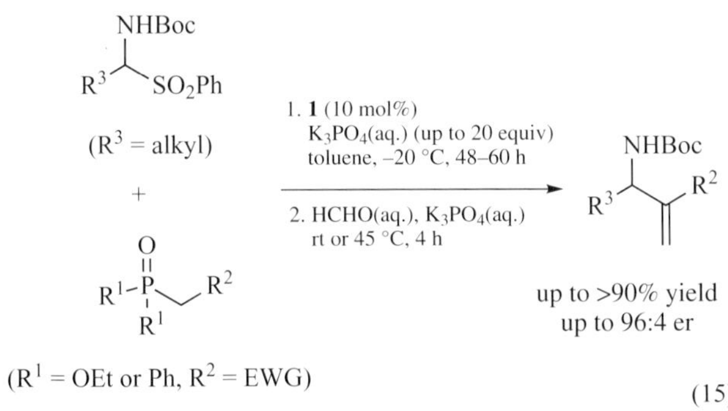

(15)

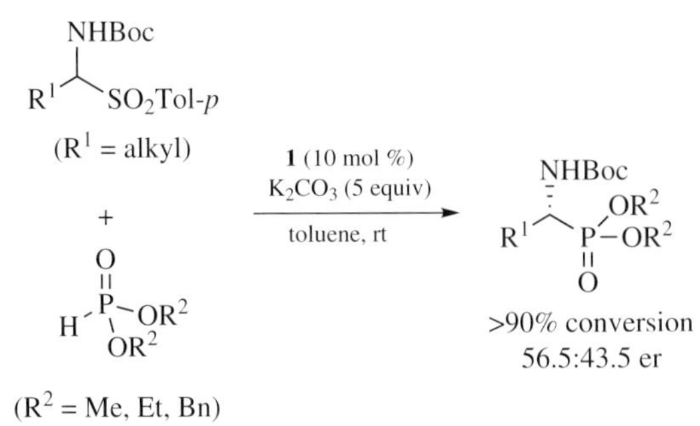

(16)

Enantioselective Syntheses of Nitrogen Heterocycles via Cation-directed Cyclizations. *N*-Benzylquininium chloride can be used as a catalyst in asymmetric cyclization reactions to give small molecule heterocycles with multiple stereogenic centres.[62–65] Highly enantioenriched, functionalized indolines—some containing all-carbon stereogenic centres—can be accessed via cyclization reactions of imine[62,63] and isonitrile[64] precursors (eqs 17–19). The cyclization of the phenylacetonitriles (eq 18) exhibited sigmoidal reaction kinetics. An autoinductive mechanism was proposed involving deprotonation of the hydroxyl group of catalyst **1** to form a zwitterion, which then undergoes a series of proton-exchange events to ultimately give the cyclized products with excellent levels of stereocontrol. The proposed mechanism is not unlike that suggested for the aza-Henry reaction of extended nitroalkanes.[51] The intermediate indolenine produced in the cyclization of the isonitrile substrate was found to be reactive (eq 19), hence the subsequent *in situ* reduction to give enantioenriched indolines.[64] The cyclization to give pyrazolines from electron-poor hydrazine anions and chalcones was catalyzed by **1** under solid–liquid phase-transfer reaction conditions (eq 20).[65] Liquid–liquid phase-transfer conditions were detrimental to the observed enantioselectivity and further optimization found the *N*-(2-*t*-BuO)benzyl-substituted catalyst could produce the cyclized pyrazolines with up to 93% ee.

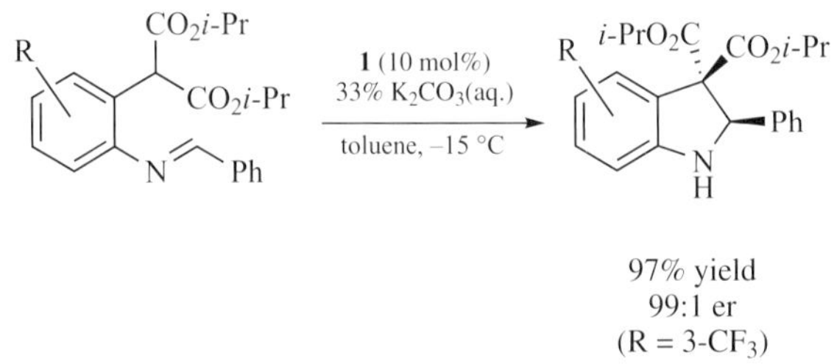

(17)

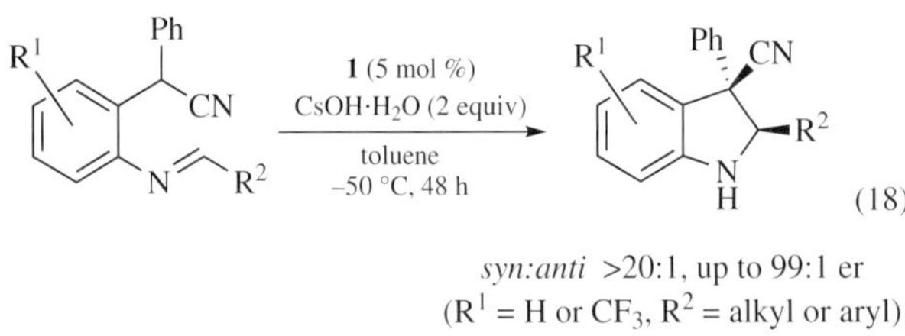

(18)

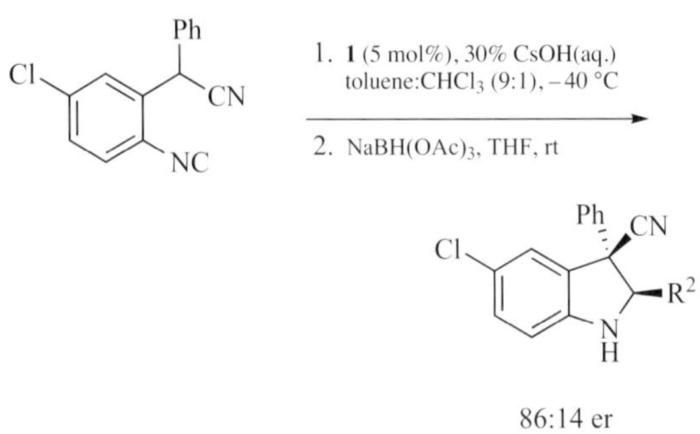

(19)

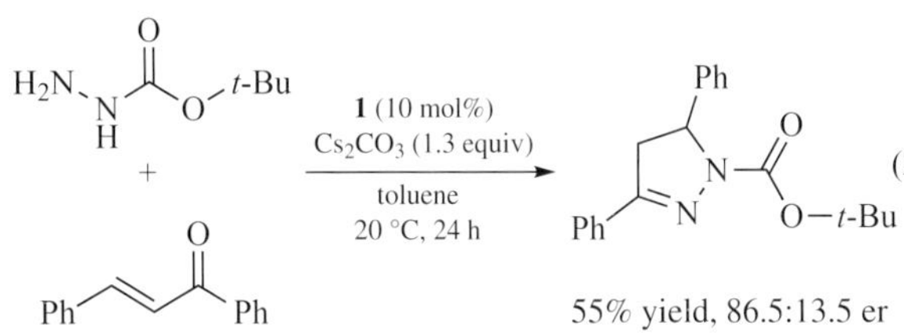

(20)

Enantioselective Synthesis of Planar Chiral Heterocycles. Organic frameworks exhibiting planar chirality are typically cyclic molecules with a rigid topology designated dextral or sinistral.[66] In a phase-transfer-catalyzed cyclization from an achiral diene, the corresponding planar chiral cycle was produced with good yields and moderate enantioselectivity using **1** as a catalyst in the presence of aqueous potassium carbonate (eq 21)[67]; the phase-transfer catalysts (−)sparteine with methyl iodide and a β-naphthyl-BINOL-derived ammonium bromide[68] were ineffective at producing enantioenriched material. The cyclization catalyzed by **1** was limited to 32% ee, but selectivity was improved through the use of the lithium salt of diacetone D-glucose as the chiral catalyst. Alternatives for the synthesis of planar chiral organocycles include palladium[69] and platinium[70] catalysis and cinchona-alkaloid catalysis.[71]

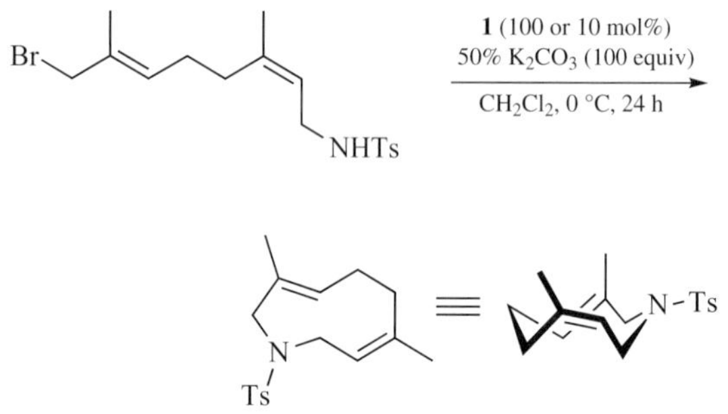

(21)

up to >89% yield
66:34 er

Desymmetrization Reactions Involving *N*-Benzylquininium Chloride. Atropisomers are frameworks that exhibit chirality through restricted rotation about a biaryl σ-bond.[66,72–73] *N*-Benzylquininium chloride can be used as an enantioselective catalyst in reactions to give biaryl atropisomers with excellent yields and enantioselectivities.[74] The biaryls are formed via nucleophilic aromatic substitution reactions of thiophenol to dichloropyrimidines (eq 22). Following the selective desymmetrization of the dichloropyrimidine substrates to give the desired products, a kinetic resolution favoring the minor enantiomer occurs, hence improving the enrichment of the major enantiomer. Phase-transfer reactions for accessing atropisomeric biaryls can also be catalyzed by BINOL-derived ammonium salts.[75,76]

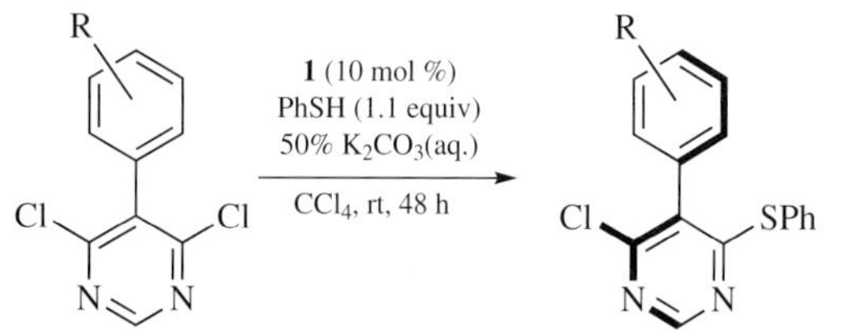

(22)

up to 97% yield
up to 97:4 er
(R = alkyl, halo, or aryl)

Chiral dienes with C_1 or C_2 symmetry can be synthesized via the desymmetrization of norbonene or norbonadiene.[48] In the presence of excess **1** and stoichiometric NaHMDS, reactive cobalt intermediates[77,78] undergo a 1,4-addition reaction with cyclohexanone in good yields and with excellent diastereoselectivity; subsequent demetalation of the diastereomeric intermediates at elevated temperatures produces the enantioenriched, desymmetrized derivatives (eq 23). Use of the *N*-3,5-$(CF_3)_2$-benzylquininium chloride variation of **1** improved the yield and stereoselectivity of the process (up to 73% yield, 83% ee).

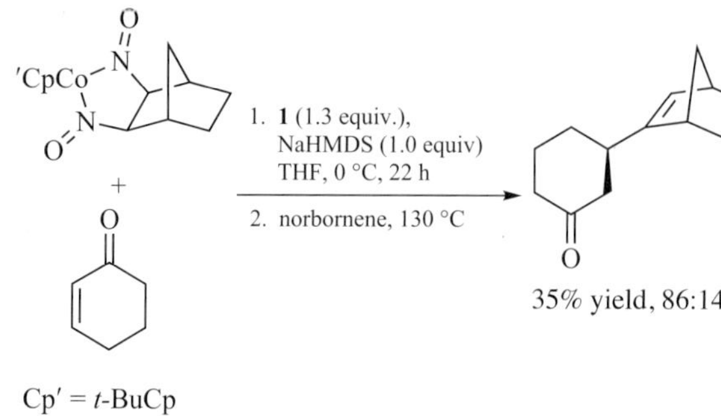

(23)

***N*-Benzylquininium Chloride-catalyzed Enantioselective Protonation via A Retro-Claisen Condensation.** Asymmetric protonation reactions can be used to produce enantioenriched α-substituted carbonyl compounds. Methods include the use of chiral tin, lanthanum, gadolinium, or gold complexes, Brønsted acid catalysis, or cinchona-derived alkaloids in phase-transfer or Brønsted base catalysis.[79] In the presence of **1** as a catalyst under phase-transfer reaction conditions, a retro-Claisen/asymmetric protonation strategy produces the corresponding enantioenriched cyclohexenone with good yield and excellent enantioselectivity (eq 24).[80] Enantioenriched cyclohexenones may be accessed using iminium/enamine, Brønsted acid, NHC, H-bonding, and phase-transfer methods.[81]

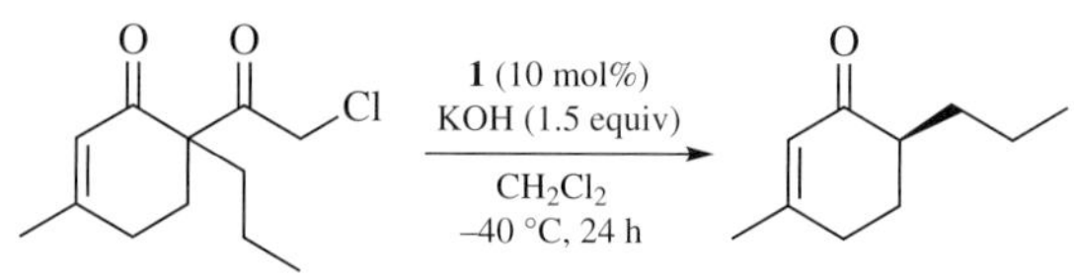

(24)

86% yield
94.5:5.5 er

Counterion Exchange of *N*-Benzylquininium Chloride. Introduction of different counterions to quaternary ammonium salts enables the generation of bifunctional catalysts, where the introduced anion actively participates in the reaction.[82] Aryloxides present a sterically and electronically tuneable alternative to traditional halide counterions that are commonly associated with cinchona-alkaloid-derived phase-transfer catalysts.[83] The aryloxide (**9**) was synthesized from **1** using ion metathesis (eq 25) and evaluated in a vinylogous aldol reaction to give the corresponding furanone with good yield and stereoselectivity (eq 26).[84] Substituents on the *N*-benzyl group of the catalyst were installed to improve the selectivity of a variety of functionalized furanone products (up to 95:5 dr and 94% ee).The aryloxide anion, in the presence of a silylated base, was proposed to promote the formation of the key reaction intermediate: a chiral complex between the catalyst ammonium cation and the substrate enolate.[85,86] The derivative **9** was inefficient as a catalyst in asymmetric protonation reactions of silylenolates.[87]

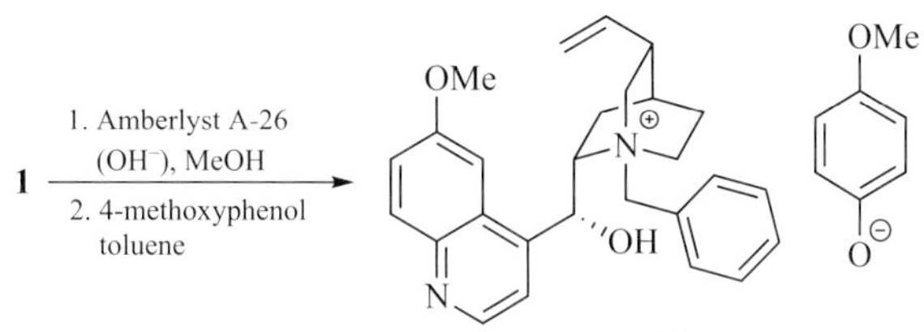

(9) (25)

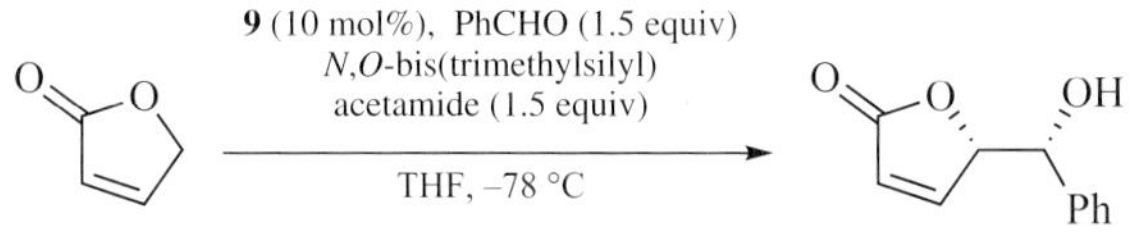

(26)

81% yield
anti:syn 88:12
81.5:18.5 er (*anti*)

Quininium- and cinchonidinium-derived salts in the presence of TMSCN were found to be effective catalysts in the cyanosilylation of nitroalkenes.[88] The counterion of the ammonium cations (e.g., chloride, bromide, and iodide) affected the enantioselectivity of the reaction; the proposed mechanism involved a rate-limiting cyanide ion exchange with the ammonium counterion. The chloride counterion of **1** was easily exchanged for cyanide (eq 27). This new catalyst (**10**) exhibited superior activity to **1** in the cyanosilylation of nitroalkenes (eq 28), but with similarly low enantioselectivity. Bifunctional catalysts bearing thiourea moieties were most effective in this system.

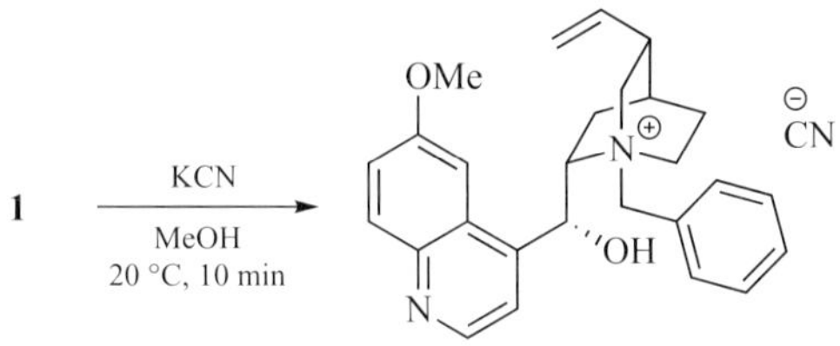

(10) (27)

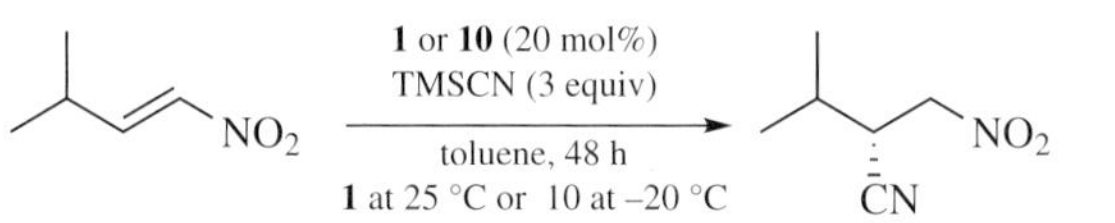

(28)

using **1**, 46% conversion, 55:45 er
using **10**, 97% conversion, 61:39 er

Polymer-supported *N*-Benzylquininium Chloride. Polymer- or gel-immobilized cinchona alkaloid-derived catalysts have been synthesized and applied to enantioselective transformations,[89] including Michael reactions[90–92] and cation-directed alkylations.[93] A polymer-immobilized derivative of **1** (eq 29) was evaluated in the asymmetric Michael reaction of the β-ketoester with *N*-maleimide (eq 30); moderate yield and stereoselectivity was obtained.[90,94–95]

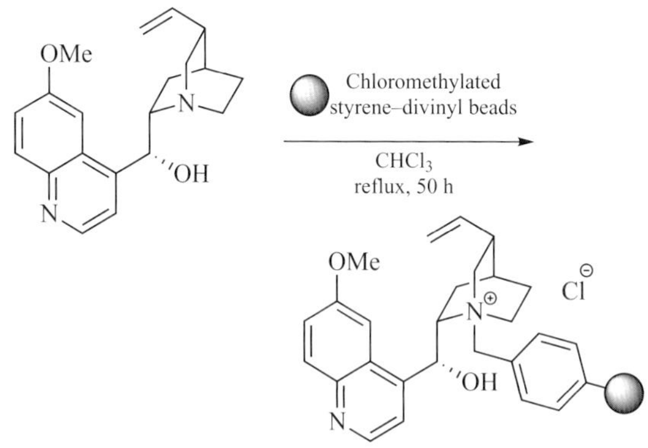

(11) (29)

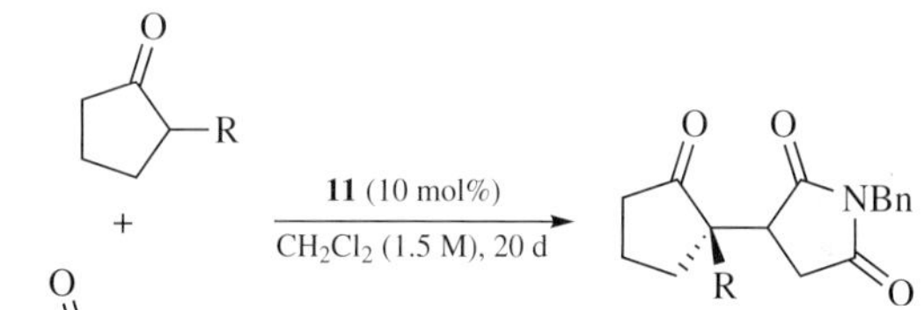

(30)

57% yield
syn:anti 75:25
87.5:12.5 er (syn)

(R = CO_2Et)

1. For reviews of phase transfer reactions, see for example (a) Keller, W. E. *Phase Transfer Reactions. Fluka Compendium*; Thieme: Stuttgart, **1986**; Vols. 1, 2. (b) Dehmlow, E. V. *Phase Transfer Catalysis*; Verlag Chemie: Deerfield Beach, **1980**. (c) Starks, C. M.; Liotta, C., *Phase Transfer Catalysis, Principles and Techniques*; Academic Press: New York, **1978**. (d) Dockx, J. *Synthesis* **1973**, 441. (e) Dehmlow, E. V. *Angew. Chem., Int. Ed. Engl.* **1974**, *13*, 170. For a mechanistic review of hydroxide mediated reactions under PTC conditions, see: (f) Rabinovitz, M.; Cohen, Y.; Halpern, M.; *Angew. Chem., Int. Ed. Engl.* **1986**, *25*, 960.
2. Harigaya, Y.; Yamaguchi, H.; Onda, M., *Chem. Pharm. Bull.* **1981**, *29*, 1321.
3. Colonna, S.; Fornasier, R., *J. Chem. Soc., Perkin Trans. 1* **1978**, 371.
4. Jacobs, W. A.; Heidelberger, M. *J. Am. Chem. Soc.* **1919**, *41*, 2090.
5. Some doubts have been expressed with regard to the accuracy of reported results: Dehmlow, E. V.; Singh, P.; Heider, J. *J. Chem. Res. (S)* **1981**, 292.
6. For example (a) Sera, A.; Takagi, K.; Katayama, H.; Yamada, H.; Matsumoto, K. *J. Org. Chem.* **1988**, *53*, 1157. (b) Wynberg, H.; Helder, R. *Tetrahedron Lett.* **1975**, 4057. (c) Helder, R.; Arends, R.; Bolt, W.; Hiemstra, H.; Wynberg, H. *Tetrahedron Lett.* **1977**, 2181.

7. Kelly, J.; Sherrington, D. C., *Polymer* **1984**, *25*, 1499.
8. (a) Helder, R.; Hummelen, J. C.; Laane, R. W. P. M.; Wiering, J. S.; Wynberg, H., *Tetrahedron Lett.* **1976**, 1831. (b) Wynberg, H.; Marsman, B. *J. Org. Chem.* **1980**, *45*, 158.
9. Hummelen, J. C.; Wynberg, H., *Tetrahedron Lett.* **1978**, 1089.
10. Loupy, A.; Sansoulet, J.; Zaparucha, A.; Merienne, C. *Tetrahedron Lett.* **1989**, *30*, 333.
11. For example: (a) Jianguo, C.; Lingchong, Y. *Synth. Commun.* **1990**, *20*, 2895. (b) Brunner, H.; Zintl, H. *Monatsh. Chem.* **1991**, *122*, 841. (c) Annunziata, R.; Cinquini, M.; Colonna, S. *J. Chem. Soc., Perkin Trans. 1* **1980**, 2422.
12. Colonna, S; Re, A.; Wynberg, H., *J. Chem. Soc., Perkin Trans. 1* **1981**, 547.
13. Conn, R. S. E.; Lovell, A. V.; Karady, S.; Weinstock, L. M., *J. Org. Chem.* **1986**, *51*, 4710.
14. Sasai, H.; Itoh, N.; Suzuki, T.; Shibasaki, M., *Tetrahedron Lett.* **1993**, *34*, 855.
15. (a) Julia, S.; Ginebreda, A.; Guixer, J.; Tomas, A., *Tetrahedron Lett.* **1980**, *21*, 3709. (b) Julia, S.; Ginebreda, A.; Guixer, J.; Masana, J.; Tomas, A.; Colonna, S., *J. Chem. Soc., Perkin Trans. 1* **1981**, 574.
16. (a) Hughes, D. L.; Dolling, U. H.; Ryan, K. M.; Schoenewaldt, E. F.; Grabowski, E. J. J., *J. Org. Chem.* **1987**, *52*, 4745. (b) Bhattacharya, A.; Dolling, U. H.; Grabowski, E. J. J.; Karady, S.; Ryan, K. M.; Weinstock, L. M., *Angew. Chem., Int. Ed. Engl.* **1986**, *25*, 476. (c) Dolling, U. H.; Davis, P.; Grabowski, E. J., J. *J. Am. Chem. Soc.* **1984**, *106*, 446.
17. The ee's can often be improved by recrystallization. O'Donnell, M. J.; Bennett, W. D.; Wu, S., *J. Am. Chem. Soc.* **1989**, *111*, 2353.
18. Maruoka, K.; Ooi, T., *Chem. Rev.* **2003**, *103*, 3013.
19. O'Donnell, M. J., *Acc. Chem. Res.* **2004**, *37*, 506.
20. Lygo, B.; Andrews, P. I., *Acc. Chem. Res.* **2004**, *37*, 518.
21. Lygo, B.; Wainwright, P. G., *Tetrahedron Lett.* **1997**, *38*, 8595.
22. Arai, S.; Ishida, T.; Shioiri, T., *Tetrahedron Lett.* **1998**, *39*, 8299.
23. Perrard, T.; Plaquevent, J. C.; Desmurs, J. R.; Hébrault, D., *Org. Lett.* **2000**, *2*, 2959.
24. Palomo, C.; Oiarbide, M.; Laso, A.; López, R., *J. Am. Chem. Soc.* **2005**, *127*, 17622.
25. Fini, F.; Sgarzani, V.; Pettersen, D.; Herrera, R. P.; Bernardi, L.; Ricci, A., *Angew. Chem., Int. Ed.* **2005**, *44*, 7975.
26. Yamada, K.; Harwood, S. J.; Groger, H.; Shibasaki, M., *Angew. Chem., Int. Ed.* **1999**, *38*, 3504.
27. Yamada, K.; Moll, G.; Shibasaki, M., *Synlett* **2001**, 980.
28. Knudsen, K. R.; Risgaard, T.; Nishiwaki, N.; Gothelf, K. V.; Jørgensen, K. A., *J. Am. Chem. Soc.* **2001**, *123*, 5843.
29. Nishiwaki, N.; Knudsen, K. R.; Gothelf, K. V.; Jørgensen, K. A., *Angew. Chem., Int. Ed.* **2001**, *40*, 2992.
30. Knudsen, K. R.; Jørgensen, K. A., *Org. Biomol. Chem.* **2005**, *3*, 1362.
31. Anderson, J. C.; Howell, G. P.; Lawrence, R. M.; Wilson, C. S., *J. Org. Chem.* **2005**, *70*, 5665.
32. Palomo, C.; Oiarbide, M.; Halder, R.; Laso, A.; Lopez, R., *Angew. Chem., Int. Ed.* **2006**, *45*, 117.
33. Nugent, B. M.; Yoder, R. A.; Johnston, J. N., *J. Am. Chem. Soc.* **2004**, *126*, 3418.
34. Okino, T.; Nakamura, S.; Furukawa, T.; Takemoto, Y., *Org. Lett.* **2004**, *6*, 625.
35. Yoon, T. P.; Jacobsen, E. N., *Angew. Chem., Int. Ed.* **2005**, *44*, 466.
36. Ooi, T.; Maruoka, K., *Acc. Chem. Res.* **2004**, *37*, 526.
37. Drew, M. D.; Lawrence, N. J., Watson, W.; Bowles, S. A., *Tetrahedron Lett.* **1997**, *38*, 5857.
38. Tanaka, K.; Okada, T.; Toda, F., *Angew. Chem., Int. Ed.* **1993**, *32*, 1147.
39. Toda, F.; Tanaka, K., Stein, Z.; Goldberg, I., *J. Org. Chem.* **1994**, *59*, 5748.
40. Cai, D. W.; Huges, D. L.; Verhoeven, T. R.; Reider, P. J., *Tetrahedron Lett.* **1995**, *36*, 7991.
41. Hu, Q. S.; Vitharana, D.; Pu, L., *Tetrahedron: Asymmetry* **1995**, *6*, 2123.
42. Ding, K.; Wang, Y.; Yun, H.; Liu, J.; Wu, Y.; Terada, M.; Okubo, Y.; Mikami, K., *Chem. Eur. J.* **1999**, *5*, 1734.
43. Zhang, J. H.; Liao, J.; Cui, X.; Yu, K. B.; Zhu, J.; Deng, J. G.; Zhu, S. F.; Wang, L. X.; Zhou, Q. L.; Chung, L. W.; Ye, T., *Tetrahedron: Asymmetry* **2002**, *13*, 1363.
44. Wu, S.; Zhang, W.; Zhang, Z.; Zhang, X., *Org. Lett.* **2004**, *06*, 3565.
45. Harigaya, Y.; Yamaguchi, H.; Onda, M., *Chem. Pharm. Bull.* **1981**, *29*, 1321.
46. Ooi, T.; Maruoka, K., *Angew. Chem., Int. Ed.* **2007**, *46*, 4222.
47. Jew, S.-S.; Park, H.G., *Chem. Commun.* **2009**, 7090.
48. Shirakawa, S.; Maruoka, K., *Angew. Chem., Int. Ed.* **2013**, *52*, 4312.
49. Boyd, W. C.; Crimmin, M. R.; Rosebrugh, L. E.; Schomaker, J. M.; Bergman, R. G.; Toste, F. D., *J. Am. Chem. Soc.* **2010**, *132*, 16365.
50. Matsuzaki, K.; Furukawa, T.; Tokunaga, E.; Matsumoto, T.; Shiro, M.; Shibata, N., *Org. Lett.* **2013**, *15*, 3282.
51. Furukawa, T.; Shibata, N.; Mizuta, S.; Nakamura, S.; Toru, T.; Shiro, M., *Angew. Chem., Int. Ed.* **2008**, *47*, 8051.
52. Gomez-Bengoa, E.; Linden, A.; López, R.; Múgica-Mendiola, I.; Oiarbide, M.; Palomo, C., *J. Am. Chem. Soc.* **2008**, *130*, 7955.
53. Kumaraswamy, G.; Pitchaiah, A., *Tetrahedron* **2011**, *67*, 2536.
54. Kumaraswamy, G.; Pitchaiah, A., *Helv. Chim. Acta* **2011**, *94*, 1543.
55. Schwieter, K. E.; Johnston, J. N., *Chem. Sci.* **2015**, *6*, 2590.
56. Schwieter, K. E.; Johnston, J. N., *ACS Catal.* **2015**, *5*, 6559.
57. Mazzotta, S.; Gramigna, L.; Bernardi, L.; Ricci, A., *Org. Process Res. Dev.* **2010**, *14*, 687.
58. Marianacci, O.; Micheletti, G.; Bernardi, L.; Fini, F.; Fochi, M.; Pettersen, D.; Sgarzani, V.; Ricci, A., *Chem. Eur. J.* **2007**, *13*, 8338.
59. Fini, F.; Micheletti, G.; Bernardi, L.; Pettersen, D.; Fochi, M.; Ricci, A., *Chem. Commun.* **2008**, 4345.
60. Enders, D.; Saint-Dizier, A.; Lannou, M.-I.; Lenzen, A., *Eur. J. Org. Chem.* **2006**, 29.
61. Momo, R. D.; Fini, F.; Bernardi, L.; Ricci, A., *Adv. Synth. Catal.* **2009**, *351*, 2283.
62. Kowalczyk, D.; Albrecht, Ł., *Chem. Commun.* **2015**, *51*, 3981.
63. Maciver, E. E.; Thompson, S.; Smith, M. D., *Angew. Chem., Int. Ed.* **2009**, *48*, 9979.
64. Sharma, K.; Wolstenhulme, J. R.; Painter, P. P.; Yeo, D.; Grande-Carmona, F.; Johnston, C. P.; Tantillo, D. J.; Smith, M. D., *J. Am. Chem. Soc.* **2015**, *137*, 13414.
65. Li, M.; Woods, P. A.; Smith, M. D., *Chem. Sci.* **2013**, *4*, 2907.
66. Mahé, O.; Dez, I.; Levacher, V.; Brière, J.-F., *Angew. Chem., Int. Ed.* **2010**, *49*, 7072.
67. Shirakawa, S.; Liu, S.; Kaneko, S., *Chem. Asian J.* **2015**, DOI: 10.1002/asia.201500951.
68. Tomooka, K.; Uehara, K.; Nishikawa, R.; Suzuki, M.; Igawa, K., *J. Am. Chem. Soc.* **2010**, *132*, 9232.
69. Ooi, T.; Kameda, M.; Maruoka, K., *J. Am. Chem. Soc.* **2003**, *125*, 5139.
70. Igawa, K.; Ichikawa, N.; Ano, Y.; Katanoda, K.; Ito, M.; Akiyama, T.; Tomooka, K., *J. Am. Chem. Soc.* **2015**, *137*, 7294.
71. Tomooka, K.; Shimada, M.; Uehara, K.; Ito, M., *Organometallics* **2010**, *29*, 6632.
72. Islas-Gonzalez, G.; Bois-Choussya, M.; Zhu, J., *Org. Biomol. Chem.* **2003**, *1*, 30.
73. Kumarasamy, E.; Raghunathan, R.; Sibi, M. P.; Sivaguru, J., *Chem. Rev.* **2015**, *115*, 11239.
74. Bringmann, G.; Mortimer, A. J. P.; Keller, P. A.; Gresser, M. J.; Garner, J.; Breuning, M., *Angew. Chem., Int. Ed.* **2005**, *44*, 5384.
75. Armstrong, R. J.; Smith, M. D., *Angew. Chem., Int. Ed.* **2014**, *53*, 12822.
76. Shirakawa, S.; Liu, K.; Maruoka, K., *J. Am. Chem. Soc.* **2012**, *134*, 916.
77. Shirakawa, S.; Wu, X.; Liu, S.; Maruoka, K., *Tetrahedron* **2015**, DOI: 10.1016/j.tet.2015.10.074.

78. Schomaker, J. M.; Boyd, W. C.; Stewart, I. C.; Toste, F. D.; Bergman, R. G., *J. Am. Chem. Soc.* **2008**, *130*, 3777.

79. Schomaker, J. M.; Toste, F. D.; Bergman, R. G., *Org. Lett.* **2009**, *11*, 3698.

80. Oudeyer, S.; Brière, J.-F.; Levacher, V., *Eur. J. Org. Chem.* **2014**, 6103.

81. Yamamoto, E.; Gokuden, D.; Nagai, A.; Kamachi, T.; Yoshizawa, K.; Hamasaki, A.; Ishida, T.; Tokunaga, M., *Org. Lett.* **2012**, *14*, 6178.

82. Yang, X.; Wang, J.; Li, P., *Org. Biomol. Chem.* **2014**, *12*, 2499.

83. Brière, J.-F.; Oudeyer, S.; Dalla, V.; Levacher, V., *Chem. Soc. Rev.* **2012**, *41*, 1969.

84. Godemert, J.; Oudeyer, S.; Levacher, V., *ChemCatChem* **2015**, *7*, DOI: 10.1002/cctc.201500616.

85. Claraz, A.; Oudeyer, S.; Levacher, V., *Adv. Synth. Catal.* **2013**, *355*, 841.

86. Yang, Y.; Zheng, K.; Zhao, J.; Shi, J.; Lin, L.; Liu, X.; Feng, X., *J. Org. Chem.* **2010**, *75*, 5382.

87. Pansare, S. V.; Paul, E. K., *Chem. Commun.* **2011**, *47*, 1027.

88. Claraz, A.; Landelle, G.; Oudeyer, S.; Levacher, V., *Eur. J. Org. Chem.* **2013**, 7693.

89. Bernal, P.; Fernández, R.; Lassaletta, J. M., *Chem. Eur. J.* **2010**, *16*, 7714.

90. Jumde, R. P.; Pietro, A. D.; Manariti, A.; Mandoli, A., *Chem. Asian J.* **2014**, *10*, 397.

91. Qin, Y.; Yang, G; Yang, L.; Li, J.; Cui, Y., *Catal. Lett.* **2011**, *141*, 481.

92. Zhao, W.; Zhang, Y.; Qu, C.; Zhang, L.; Wang, J.; Cui, Y., *Catal. Lett.* **2014**, *144*, 1681.

93. Yang, L.; Zhou, D.; Qu, C.; Cui, Y., *Catal. Lett.* **2012**, *142*, 1405.

94. Feng, D.; Xu, J.; Wan, J.; Xie, B.; Ma, X., *Catal. Sci. Technol.* **2015**, *5*, 2141.

95. Manna, M. S.; Mukherjee, S., *Chem. Eur. J.* **2012**, *18*, 15277.

96. Shirakawa, S.; Makino, H.; Yoshidome, T.; Maruoka, K., *Tetrahedron* **2014**, *70*, 7128.

3-Benzylthiazolium Bromide[1]

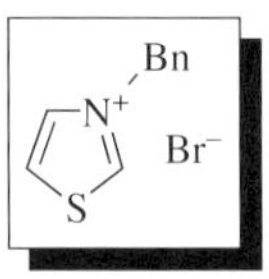

[75066-50-1] $C_{10}H_{10}BrNS$ (MW 256.16)

InChI = 1/C10H10NS.BrH/c1-2-4-10(5-3-1)8-11-6-7-12-9-11;/h1-7,9H,8H2;1H/q+1;/p-1/fC10H10NS.Br/h;1h/qm;-1

InChIKey = BVVXTLWKNXVYDS-XTQWHEQJCJ

(in combination with base is a redox catalyst for the oxidation of aldehydes coupled with the reduction of organic compounds[2])

Physical Data: mp 155–156 °C.

Solubility: freely sol methanol; sol ethanol, DMSO, and DMF; sparingly sol chloroform, dichloromethane, and acetone; insol THF, benzene, and ethyl acetate.

Form Supplied in: white columnar crystals.

Analysis of Reagent Purity: ^{1}H NMR (DMSO-d_6, 270 MHz) δ 5.82 (s, 2H, CH_2), 7.41–7.53 (m, 5H, C_6H_5), 8.37 (dd, 1H, H-5, J = 3.7 and 2.4 Hz), 8.61 (dd, 1H, H-4, J = 3.7 and 1.8 Hz); 10.40 (dd, 1H, H-2, J = 2.4 and 1.8 Hz).

Purification: recrystallized from 5:1 benzene/ethanol.

Handling, Storage, and Precautions: the reagent is somewhat hygroscopic and should be dried before use.

Oxidation of Aldehydes to Methyl Esters. 3-Benzylthiazolium bromide, in methanol containing Triethylamine, catalyzes a redox reaction in which aldehydes are oxidized to methyl esters, and certain organic compounds such as acridine, phenazine, azobenzene, and nitrobenzene are reduced.[2] The redox reactions are carried out as follows. Triethylamine (5*n* mmol) and the substrate (4*n* mmol) are added to a frozen solution of a mixture of aldehyde (6*n* mmol) and 3-benzylthiazolium bromide (*n* mmol) in degassed methanol (10*n* mL). The solution, after melting, is stirred under argon at rt for a 20-h period. To increase the yield in the redox reaction, an excess of aldehyde is added. The presence of air in methanol results in the production of methyl esters by oxidation of aldehydes.

The mechanism of this process is outlined in eq 1. The thiazolium salt is first transformed into an ylide by the action of triethylamine[3] and the ylide then reacts with the aldehyde to form an adduct ('active aldehyde')[4] which undergoes a redox reaction with the organic substrate in methanol to give a methyl ester and the reduced form of the substrate (eq 1).

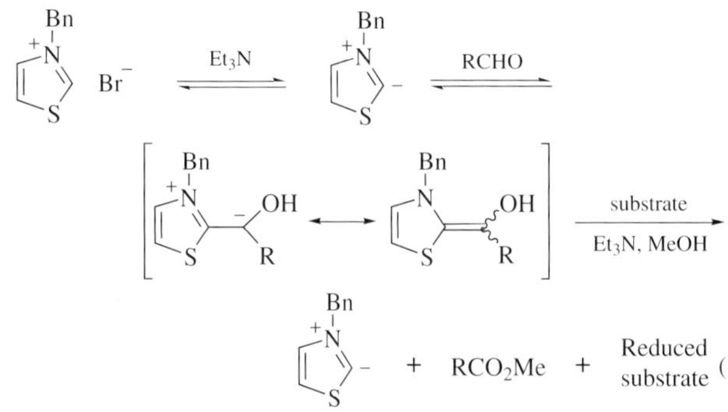

Acridine is reduced by benzaldehyde, 2-furaldehyde, and heptanal in the presence of 3-benzylthiazolium bromide and triethylamine in degassed methanol to give 9,9′-biacridan in satisfactory yields, while aldehydes are oxidized to the corresponding methyl esters (eq 2).[2] The excess aldehydes are converted into α-hydroxy ketones by benzoin condensation.[5] The redox reaction is thought to proceed via the transfer of electrons from the active aldehyde to acridine.

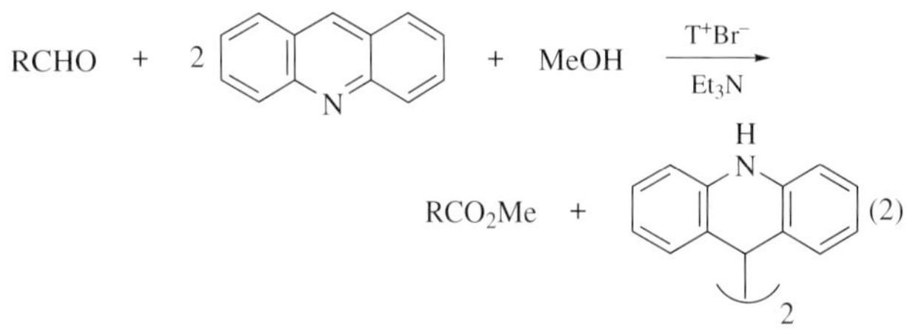

R		
R = Ph	87%	93%
R = 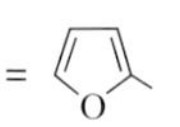	30%	40%
R = $Me(CH_2)_5$	67%	72%

T^+Br^- = 3-benzylthiazolium bromide

Phenazine and azobenzene are reduced, under similar conditions, to 5,10-dihydrophenazine (eq 3) and hydrazobenzene (eq 4), respectively; the latter is converted into benzidine on treatment with hydrochloric acid after the reaction.[2] Nitrobenzene is reduced to aniline and *N*-benzoyl-*N*-phenylhydroxylamine (eq 5).[2,6]

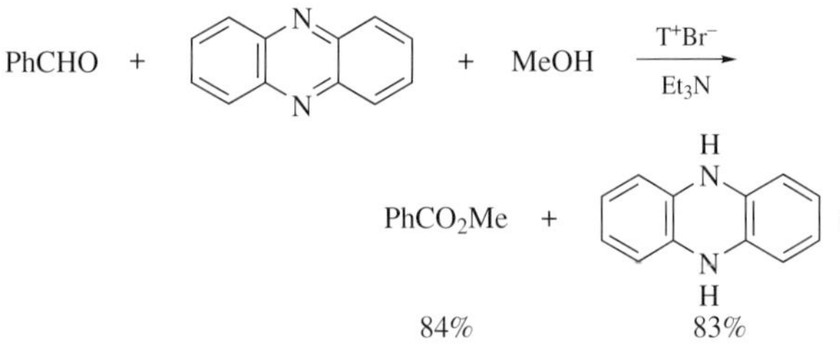

(3)

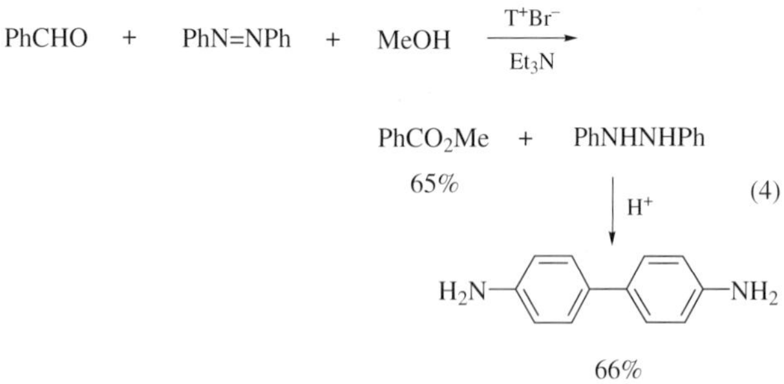

(4)

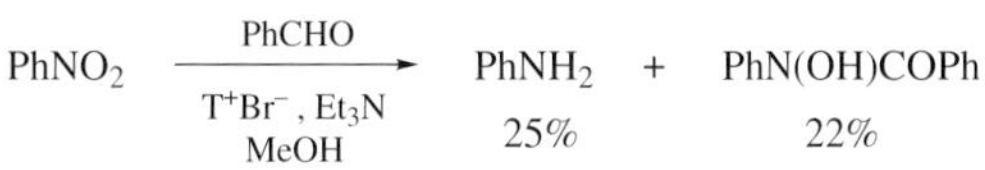

(5)

1. (a) Ugai, T.; Tanaka, S.; Dokawa, S., *J. Pharm. Soc. Jpn.* **1943**, *63*, 269. (b) Metzger, J. V. *Thiazole and its Derivatives*; Interscience: New York, **1979**; Part 1. (c) Metzger, J. V. In *Comprehensive Heterocyclic Chemistry*; Katritzky, A. R.; Rees, C. W., Eds.; Pergamon: Oxford, 1984; *6*, p 235.
2. Inoue, H.; Higashiura, K., *J. Chem. Soc., Chem. Commun.* **1980**, 549.
3. Katritzky, A. R.; Lagowski, J. M. In *Comprehensive Heterocyclic Chemistry*; Katritzky, A. R.; Rees, C. W., Eds.; Pergamon: Oxford, 1984; *5*, p 63.
4. Stetter, H. *Angew. Chem., Int. Ed. Engl.* **1976**, *15*, 639.
5. (a) Hassner, A.; Rai, K. M. L., *Comprehensive Organic Synthesis* **1991**, *1*, 541. (b) Laird, T. In *Comprehensive Organic Chemistry*; Barton, D. H. R.; Ollis, W. D., Eds.; Pergamon: Oxford, 1979; *1*, p 1142. (c) Breslow, R. *J. Am. Chem. Soc.* **1958**, *80*, 3719.
6. Inoue, H.; Tamura, S., *J. Chem. Soc., Chem. Commun.* **1985**, 141.

Hiroo Inoue
University of Osaka Prefecture, Japan

Benzyltriethylammonium Chloride[1]

$BnNEt_3{}^+Cl^-$

[56-37-1] $C_{13}H_{22}ClN$ (MW 227.81)

InChI = 1/C13H22N.ClH/c1-4-14(5-2,6-3)12-13-10-8-7-9-11-13;/h7-11H,4-6,12H2,1-3H3;1H/q+1;/p-1/fC13H22N.Cl/h;1h/qm;-1

InChIKey = HTZCNXWZYVXIMZ-SQHMAPJGCT

(phase transfer catalyst[1])

Alternate Names: TEBA; TEBAC.
Physical Data: mp 185 °C (dec).
Solubility: sol acetone, alcohol, H_2O, 1% aq. NaOH; insol ether, 20% aq. NaOH.
Form Supplied in: anhydrous white solid; monohydrate also available.
Analysis of Reagent Purity: tetraalkylammonium salts can be titrated with potassium 3,5-di-*t*-butyl-2-hydroxybenzenesulfonate and iron(III) chloride.[1a]
Preparative Method: via the reaction of Benzyl Chloride with Triethylamine in absolute ethanol.[2]
Purification: precipitation from acetone with ether.[2]
Handling, Storage, and Precautions: highly hygroscopic.

Introduction. Benzyltriethylammonium chloride is one of many commonly used phase transfer catalysts. Although in specific reactions other phase transfer catalysts have proven superior,[3] TEBA remains the reagent of choice for many reactions.

Catalyst for C–C Bond Formation. TEBA has been used extensively in the formation of carbenes in heterogeneous systems. Dichlorocarbene[2,4] is formed from Chloroform and aqueous Sodium Hydroxide in the presence of TEBA (eq 1).[4a] Dibromocarbenes can also be generated using TEBA.[4b,4d] TEBA has been used as a means of introducing hydroxide or carboxylate ion in alkylations of nitriles,[5] cyanohydrins,[6] Schiff bases,[7] sulfones,[8] malonates,[9] and benzyl ketones.[10] Michael addition reactions[11] have been catalyzed by TEBA, as well as the condensation of *O*-benzoyl cyanohydrins,[12] malonitrile,[13] and sulfones[14] with benzaldehyde. Condensation of 3-methyl-5-pyrazolone with Carbon Disulfide and alkyl bromides was realized under phase transfer conditions using TEBA.[15]

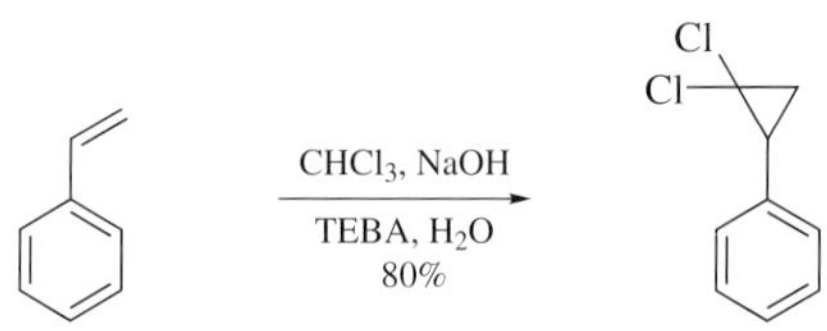

(1)

Catalyst for C–N Bond Formation. TEBA has been used in the alkylation of trifluoroacetamide with α-bromo esters to form, after hydrolysis, α-amino acids (eq 2).[16] TEBA has also been used in the nitration of aromatics,[17] the alkylation of secondary amides,[18] and the dehydration of ureas to form carbodiimides.[19] Reaction of thioamides with dibromoethane and sodium hydroxide catalyzed by TEBA gives thiazolidines.[20]

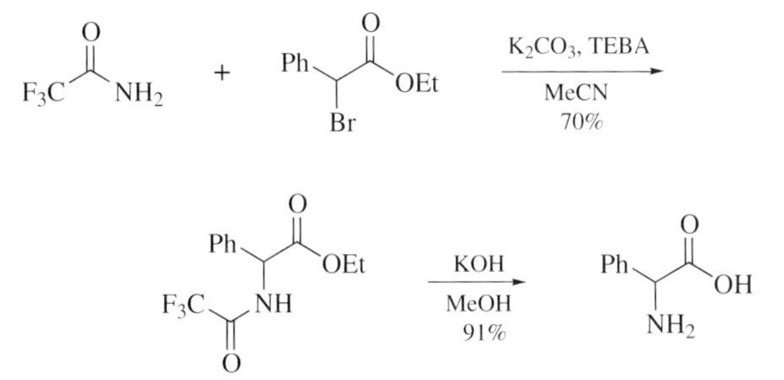

(2)

Catalyst for C–O Bond Formation. Benzyl[21] ethers have been formed using TEBA catalysis. TEBA has been used in a variety of oxidation reactions, including permanganate oxidation of longifolene[22] and benzaldehyde,[23] α-hydroxylation of benzyl nitriles giving phenyl ketones,[24] and Ruthenium(VIII) Oxide catalyzed oxidations of secondary alcohols to ketones with Sodium Periodate.[25]

Catalyst for Carboxylic Acid Functionalization. TEBA facilitates the solubilization of organic acids, permitting formation of a diacid chloride from an insoluble diacid (eq 3).[26] Organic peroxy acids can also be prepared from the corresponding acid using TEBA and potassium persulfate.[27]

(3)

Catalyst for P–X Bond Formation. TEBA has been used in the formation of phosphorodiamidates from bis(dimethylamino) phosphorochloridate[28] and phosphate esters[29] or phosphoramidates[30] from Diethyl Phosphonite. Di-*t*-butyl phosphorohalidates have been prepared from di-*t*-butyl phosphite using TEBA, carbon tetrachloride and aq. NaOH.[31]

Miscellaneous Uses. TEBA has proven very effective in the formation of metal hydrides under phase transfer conditions;[32] for example, $(Ph_3P)_3RuHCl$ and $(i\text{-}Pr_3P)_2IrHCl_2$ are converted to $(Ph_3P)_3RuH_2(H)_2$ and $(i\text{-}Pr_3P)_2IrH_5$,[32a] respectively, upon reaction with hydrogen and aq. NaOH/benzene with TEBA. In metal carbonyl chemistry, TEBA has been used as a phase transfer catalyst.[33] TEBA has also been used in the dibromination of alkenes[34] and azide transfer reactions.[35] Selective tosylation of adenosine has been achieved using Dibutyltin Oxide, p-Toluenesulfonyl Chloride, and TEBA (eq 4).[36]

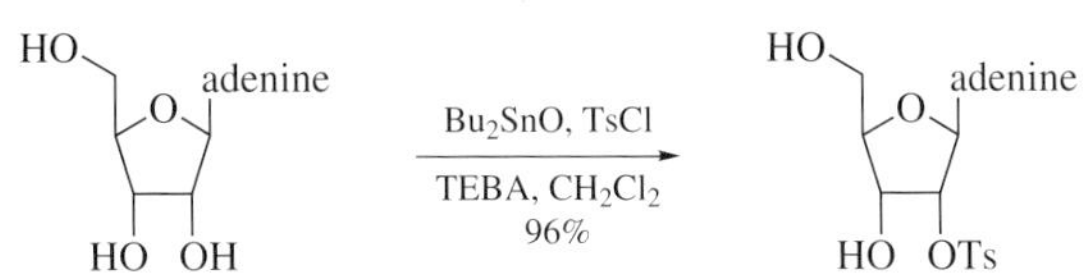

(4)

1. (a) Sjöberg, K. *Aldrichim. Acta* **1980**, *13*, 55. (b) Dehmlow, E. V., *Angew. Chem., Int. Ed. Engl.* **1974**, *13*, 170. (c) Weber, W. P.; Gokel, G. W. *Phase Transfer Catalysis in Organic Synthesis*; Springer: New York, 1977. (d) Starks, C. M.; Liotta, C. *Phase Transfer Catalysis*; Academic: New York, 1978. (e) Dehmlow, E. V., Dehmlow, S. S. *Phase Transfer Catalysis*; Verlag Chemie: Deerfield Beach, FL, **1980**.
2. Gokel, G. W.; Widera, R. P.; Weber, W. P., *Org. Synth., Coll. Vol.* **1976**, *6*, 232.
3. (a) Herriott, A. W., Picker, D., *J. Am. Chem. Soc.* **1975**, *97*, 2345. (b) Xu, C.; Liu, G.; Zhang, Z., *Synth. Commun.* **1987**, *17*, 1839. (c) Dehmlow, E. V., *Tetrahedron Lett.* **1976**, 91.
4. (a) Makosza, M.; Wawrzyniewicz, M. *Tetrahedron Lett.* **1969**, 4659. (b) Dehmlow, E. V.; Schonefeld, J., *Justus Liebigs Ann. Chem.* **1971**, *744*, 42. (c) Sasaki, T.; Kanematsu, K.; Okamura, N., *J. Org. Chem.* **1975**, *40*, 3322. (d) Dunkelblum, E.; Singer, B., *Synthesis* **1975**, 323. (e) de Angelix, F.; Gambacorta, A.; Nicoletti, R., *Synthesis* **1976**, 798. (f) Julia, S.; Ginebreda, A., *Synthesis* **1977**, 682. (g) Baldoli, C.; Luttuada, L.; Licandro, E.; Maiorana, S. Papagni, A., *J. Heterocycl. Chem.* **1989**, *26*, 241.
5. (a) Makosza, M.; Jonczyk, A., *Org. Synth., Coll. Vol.* **1976**, *6*, 897. (b) Makosza, M.; Czyzewski, J.; Jawdosiuk, M., *Org. Synth., Coll. Vol.* **1976**, *6*, 940. (c) Neidlein, R.; Kikelj, D., *Synthesis* **1989**, 612. (d) Chelucci, G.; Giacomelli, G., *J. Heterocycl. Chem.* **1990**, *27*, 307.
6. Makosza, M.; Goetzen, T., *Org. Prep. Proced. Int.* **1973**, *5*, 203.
7. Yaozhong, J.; Zhou, C.; Wu, S.; Chen, D.; Ma, J.; Liu, G., *Tetrahedron* **1988**, *44*, 5343.
8. Jonczyk, A.; Banko, K.; Makosza, M., *J. Org. Chem.* **1976**, *40*, 266.
9. Singh, R.; Danishefsky, S., *J. Org. Chem.* **1975**, *40*, 2969.
10. Jonczyk, A.; Serafin, B.; Makosza, M., *Tetrahedron Lett.* **1971**, 1351.
11. (a) Makosza, M.; Jonczyk, A., *Synthesis* **1976**, 387. (b) Dryanska, V.; Popandova, K.; Ivanov, C., *Synth. Commun.* **1987**, *17*, 211. (c) Yen, C.; Kong, Q.; Lu, W.; Wu, J., *Synth. Commun.* **1992**, *22*, 1651.
12. Rozwadowska, M. D., *Tetrahedron* **1985**, *41*, 3135.
13. Singh, P.; Aggarwal, P., *Indian J. Chem., Sect. B* **1988**, *27B*, 1029.
14. Gokel, G. W.; Gerdes, H. M.; Rebert, N. W., *Tetrahedron Lett.* **1976**, 653.
15. Oliva, A.; Casatro, I.; Castillo, C.; Leon, G., *Synthesis* **1991**, 481.
16. Landini, D.; Penso, M., *J. Org. Chem.* **1991**, *56*, 420.
17. Nishi, B.; Kabra, A.; Narang, C. K.; Mather, N. K., *J. Org. Chem.* **1991**, *56*, 4967.
18. Brehme, R., *Synthesis* **1976**, 113.
19. Jaszay, Z. M.; Petnehazy, I.; Toke, L.; Szajani, B., *Synthesis* **1987**, 520.
20. Zankowska-Jasinska, W.; Mach, K., *Chem. Scr.* **1987**, *27*, 473.
21. Jonczyk, A.; Kmiotek-Skarzynska, I., *J. Org. Chem.* **1989**, *54*, 2756.
22. Nayak, U. R.; Dalavoy, V. S.; Deodar, V. B., *Indian J. Chem., Sect. B* **1989**, *28b*, 1055.
23. Rao, K. H.; Rao, M. B., *J. Indian Chem. Soc.* **1991**, *68*, 132.
24. Donetti, A.; Boniardi, O.; Ezhaya, A., *Synthesis* **1980**, 1009.
25. Morris, P. E., Jr.; Kiely, D. E., *J. Org. Chem.* **1987**, *52*, 1149.
26. Burdett, K., *Synthesis* **1991**, 441.
27. Pande, C. S.; Jain, N., *Synth. Commun.* **1988**, *18*, 2123.
28. Sturtz, G.; Yaouanc, J.-J., *Synthesis* **1980**, 289.
29. Zwierzak, A., *Synthesis* **1976**, 305.
30. Zwierzak, A., *Synthesis* **1975**, 507.
31. Zwierzak, A.; Gajda, T., *Synthesis* **1976**, 243.
32. (a) Grushin, V. V.; Vymenits, A. B.; Volpin, M. E., *J. Organomet. Chem.* **1990**, *382*, 185. (b) Johnstone, R. A.; Wilby, A. H., *Tetrahedron* **1981**, *37*, 3667.
33. des Abbayes, H.; Alper, H., *J. Am. Chem. Soc.* **1977**, *98*, 99.
34. Olah, G. A.; Ho, T.-L.; Gupta, B. G. B., *Synthesis* **1977**, 676.
35. Nakajima, M.; Anselme, J.-P., *Tetrahedron Lett.* **1976**, 4421.
36. Grouiller, A.; Essadiq, H.; Najib, B.; Moliere, P., *Synthesis* **1987**, 1121.

Paul B. Savage

The Ohio State University, Columbus, OH, USA

Bicyclo[2.2.1]heptan-2-one, 7,7-Dimethyl-1-(1*S*,3*S*,4*R*)-2-thiabicyclo-[2.2.1]hept-3-yl-, (1*S*,4*R*)

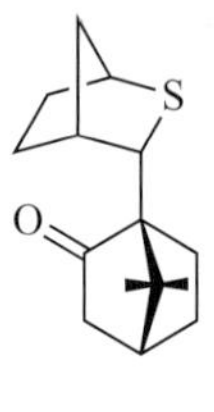

1

[352207-48-8] $C_{15}H_{22}OS$ (MW 250.14)

InChI = 1S/C15H22OS/c1-14(2)10-5-6-15(14,12(16)8-10)13-9-3-4-11(7-9)17-13/h9-11,13H,3-8H2,1-2H3/t9-,10-,11+,13+,15+/m1/s1

InChIKey = OGCMDMCSJMVHFX-SGOHARFASA-N

(used as a chiral reagent and catalyst in a wide array of reactions)

Physical Data: mp 60–62 °C (MeOH); $[\alpha]_D^{20}$ + 35.9° (*c* 19.5, CH_2Cl_2)[1]; $[\alpha]_D^{20}$ + 39.3° (*c* 1.0, MeOH).[2]
Solubility: sol in common organic solvents.
Form Supplied in: white crystalline solid. The sulfide can be made in four steps from camphorsulfonyl chloride in 56% yield.[1]
Purification: recrystallization from pentane or flash chromatography on silica gel using 4% EtOAc in petrol (R_f 0.36).
Handling, Storage, and Precautions: can be stored at room temperature for long periods of time.

Synthesis of Sulfonium Salts. Sulfide **1** is most often used to form ylides. The ylides can be generated in situ in a reaction mixture from the sulfide, or a sulfonium salt can be presynthesized and then reacted to generate the ylide.

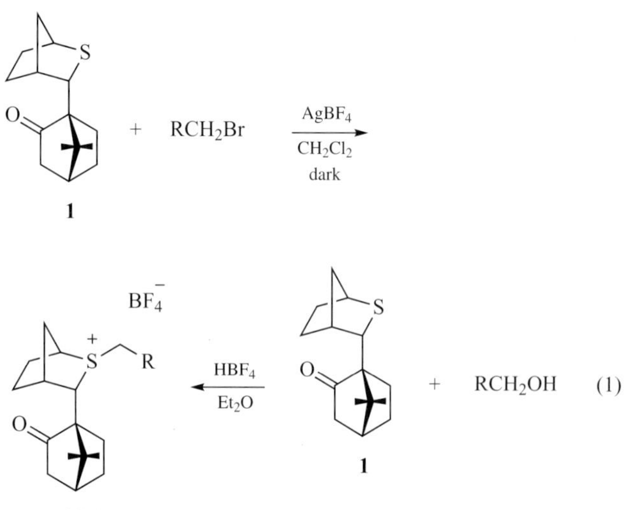

>90%
R = Ar, allyl

(1)

There are two main routes to prepare benzyl or allyl sulfonium salts.[3] The first one consists of reacting sulfide **1** with, for example, benzyl bromide in the presence of silver tetrafluoroborate in the dark. The second one consists of reacting 1 equiv of sulfide with an excess of, for example, benzyl alcohol in diethyl ether in the presence of HBF_4 (eq 1). Both give good yields, but the latter is quicker and has a simple workup.

Sulfonium salts can also be generated by reaction of α-bromocarbonyl compounds with sulfide **1** in either a biphasic mixture of CH_2Cl_2 and aq. $NaBF_4$ (eq 2) or with $NaBF_4$ in acetone.[4,5] This former method has also been used to form benzyl sulfonium salts to good effect.[6]

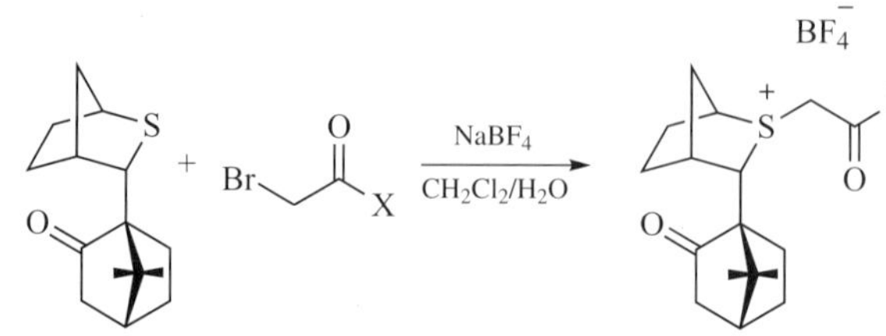

1 X = OR, NR_2 >90% (2)

Vinyl sulfonium salt **2** can be made from sulfide **1** by alkylating with bromoethyl triflate followed by elimination (eq 3).[7]

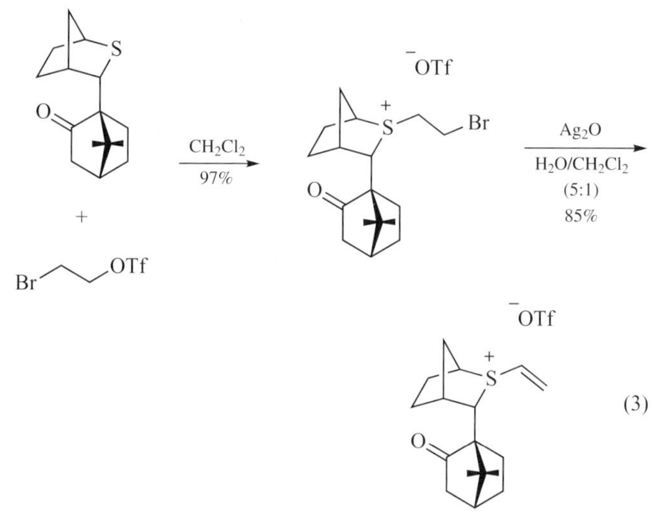

2 (3)

Epoxidation Reactions. Chiral sulfides are very potent reagents for the synthesis of chiral nonracemic epoxides by alkylidenation reactions. Sulfide **1** acts in epoxidation reactions through an ylide intermediate. This ylide can be formed in three main ways: (i) In situ in the reaction pot in the presence of the carbonyl compound, rendering the reaction catalytic in sulfide. (ii) Preforming and isolating a sulfonium salt from the sulfide (see above), which is then deprotonated to give an ylide that can then react with the carbonyl compound (iii) Preformation of a vinyl sulfonium salt followed by conjugate addition to form ylide.

Catalytic Epoxidation via Ylide Formation from an N-Tosylhydrazone Salt.[8] This procedure involves the use of an *N*-tosylhydrazone salt, which is a safer precursor of a diazo compound. The *N*-tosylhydrazone salts are obtained by reaction of an aldehyde with a tosylhydrazine, followed by treatment with base. These salts are then gradually converted into diazo compounds in situ at 40 °C in acetonitrile in the presence of a phase transfer catalyst (PTC), generally benzyltriethylammonium chloride. Then, a metal decomposes the diazo compound to generate a metal carbene that reacts with sulfide **1** to give an ylide. $Rh_2(OAc)_4$ has been found to be the best metal source in most cases. The ylide can then react with a carbonyl group to obtain the epoxide, in a process that regenerates sulfide **1**, which in turn can re-enter the catalytic cycle. Sulfide **1** has been used in this procedure to synthesize *trans*-disubstituted epoxides in moderate to high yields and in general, with good to high diastereomeric and enantiomeric excesses (eq 4). A wide variety of disubstituted epoxides can be prepared using this strategy. The combination of substitution patterns, yields, de, and ee are summarized in Table 1. A Simmons–Smith protocol can be used to generate methylides from sulfide **1**, but the synthesis of terminal epoxides using this ylide does not give high enantioselectivities.[9]

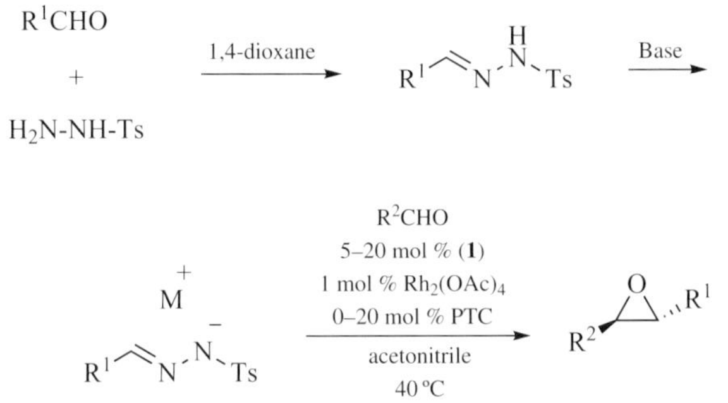

(4)

Table 1 Formation of *trans*-epoxides under the conditions as in eq 4

Entry	R^1	R^2	Yield (%)	de (%)	ee (%)
1	Ph	Ar	68–84	>96	90–94
2	Ph	Heteroaromatic	60–77	>96	89–93
3	Ph	Alkyl	46–58	50–>96	89–94
4	Ph	Vinyl	70	>96	87
5	Ar neutral	Ph	87	>96	94
6	Ar electron-rich	Ph	64–95	60–>96	83–93
7	Ar electron-poor	Ph	70–81	>96	64–93
8	Furyl	Ph	46–53	26–80	61–63
9	α,β-Unsaturated	Ph	12–56	0–>96	20–88

Stoichiometric Epoxidation via Preformation of a Sulfonium Salt.[3] The catalytic process described above is a very powerful method to obtain epoxides, but there are certain limitations: reactions with heteroaromatic aldehydes that bear basic groups, *n*-alkyl aliphatic aldehydes, α,β-unsaturated aldehydes, acetylenic aldehydes, and ketones give low yields, low diastereomeric ratios, or low enantioselectivities. Nevertheless, there is an alternative to get around these limitations—forming the ylide from stoichiometric amounts of sulfonium salt. Although stoichiometric quantities of sulfide are required, the resulting high enantioselectivities, diastereoselectivities, and yields, together with the increase in scope, make this process very attractive. Moreover, once the epoxidation is complete, the sulfide can be recovered by column chromatography in most cases and reused.

There are two main sets of epoxidation conditions that can be used: (A) Use of KOH in a mixture of acetonitrile/water 9:1 at room temperature with 1.1 equiv of carbonyl compound (eq 5). (B) Use of phosphazene base (P_2 base = $[Me_2N]_3P{=}NP({=}NEt)[NMe_2]_2$)) in CH_2Cl_2 at −78 °C mixed with the sulfonium salt for 10 min followed by the addition of the carbonyl compound (eq 6). In some cases, if P_2 base is substituted by KHMDS similar results are obtained.

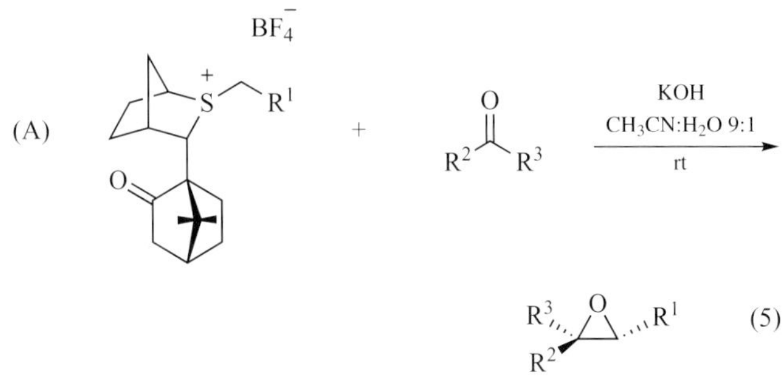

(5)

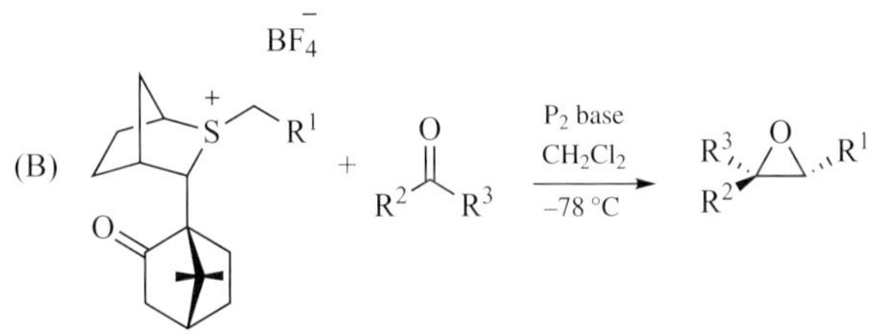

(6)

A wide variety of disubstituted epoxides can be prepared using this strategy. The combination of substitution patterns, yields, de, and ee are summarized in Table 2. In most reactions, yields, diastereoselectivities, and enantioselectivities are better with conditions (B), but conditions (A) have been also reported especially for industrial applications, where P_2 base is not convenient.

Table 2 Formation of *trans*-epoxides from aldehydes under the conditions as eqs 5 and 6

Entry	R^1	R^2	Yield (%)	de (%)	ee (%)
1	Ph	Pyridyl	59–91	82–>98	95–>99
2	Ph	α,β-Unsaturated	30–90	80–>98	90–>99
3	α,β-Unsaturated	Ph	26–96	79–>99	89–>99
4	Ph	Alkynyl	50	20	95–99
5	Ph	Alkyl	87	80	>99

Trisubstituted epoxides can be prepared from ketones using this method. Symmetric ketones performed well to furnish epoxides in high yields (85–93%) and with excellent enantioselectivities (92–>99%). Nonsymmetrical ketones furnish epoxides in good yields and with moderate to excellent de (although in these cases the *cis*-diastereomer is formed preferentially) and moderate to high ee (71–93%). This sulfide and this procedure have been applied to a number of syntheses of complex molecules where the key step is an epoxidation.[6,10–13]

Synthesis of Pyrrolidines and Piperidines Through an Epoxide Intermediate.[14] A specific application of the epoxidation reaction via preformation of a sulfonium salt is the synthesis of pyrrolidines and piperidines. Phenyl-stabilized chiral sulfur ylide, generated by deprotonation of benzyl sulfonium salt with P_2 base, reacts with five-membered ring hemiaminals to give an intermediate *trans*-functionalized epoxide, which upon heating leads to functionalized pyrrolidines with high enantioselectivity and diastereoselectivity and in high yield (eq 7).

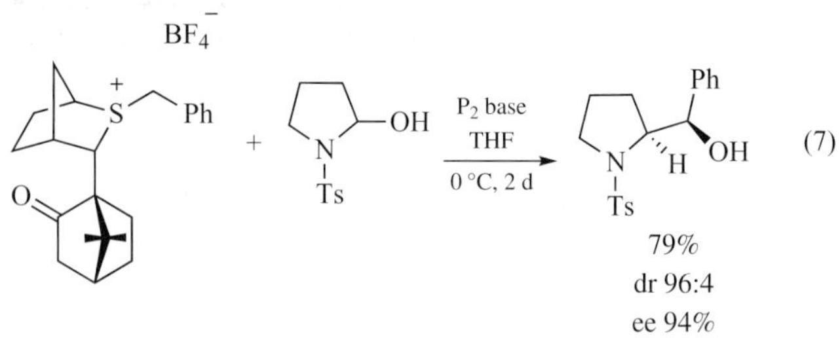

The reaction can be diverted to give piperidines instead in good yields and with excellent enantioselectivity by isolation of the intermediate epoxide and treating it with TMSOTf or $Sc(OTf)_3$ (eq 8).

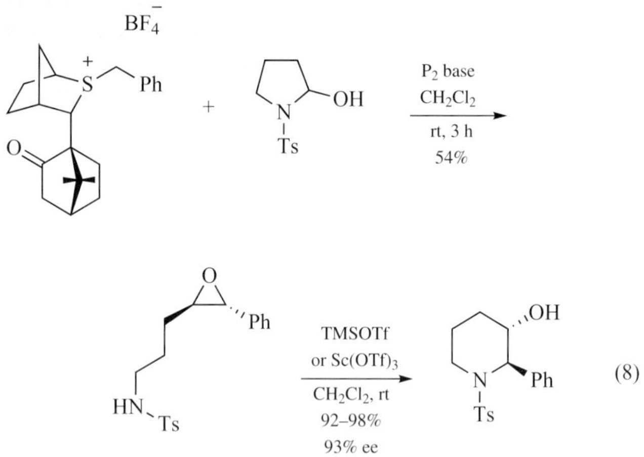

Use of Chiral Aldehyde and Chiral Sulfide.[6,12] The degree of reagent and substrate control in the reaction of chiral sulfur ylides, generated from sulfonium salts, with chiral aldehydes (glyceraldehyde acetonide and meroquinene aldehyde) has also been investigated. In the case of the reaction of phenyl-stabilized ylide with glyceraldehyde acetonide, it was found that the C_1 stereochemistry was largely controlled by the reagent, whereas control at the C_2 center was dependent on the aldehyde used (eq 9).[12,15] The ratio of **4a**:**4b**:**4c** switched from 89:5:6 to 1:21:78 when the

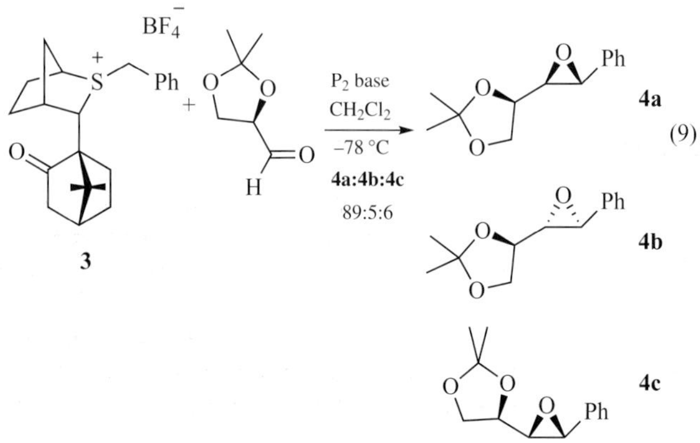

opposite enantiomer of salt **3** was used. In the case of meroquinene aldehyde, where both enantiomers of sulfide **1** were used, no significant matched/mismatched issues arose (eq 10) with the sulfide controlling the outcome (dr 93:7).[6] Thus, the opposite sulfide enantiomer gave essentially the same high selectivity for *trans*-epoxide (>95:5) and diastereoselectivity (89:11), but now in favor of the opposite *trans*-epoxide stereochemistry. The product epoxides were then used in the synthesis of natural products.

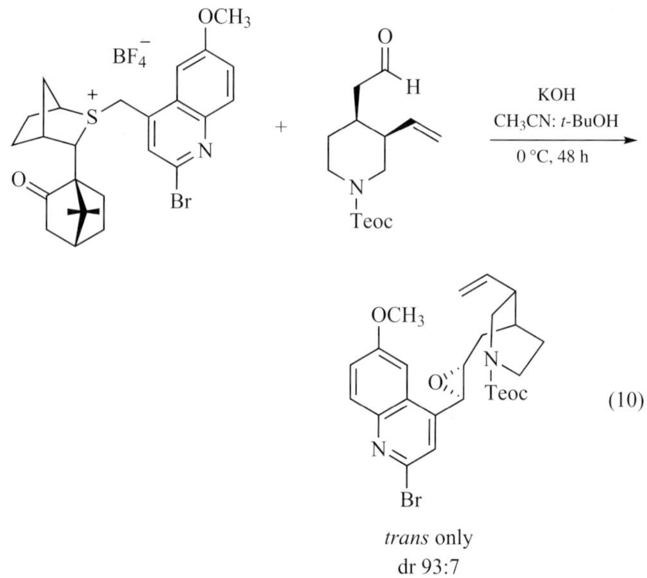

Glycidic Acids.[16] Glycidic acids can be obtained by the epoxidation of aldehydes with carboxylate-stabilized sulfur ylides (thetin salts). Sulfide **1** is conveniently alkylated with α-bromo-*tert*-butyl acetate and the counterion exchanged in situ (see eq 2, X = O-*t*-Bu). The ester can then be hydrolyzed to obtain sulfonium salts that react with carbonyl compounds after deprotonation. Good yields were obtained with dimsyl sodium or LHMDS as bases in DMSO or THF–DMSO mixtures. Diastereoselectivities were high in all cases in favor of the *trans*-isomers. However, the enantioselectivities observed were only moderate (up to 67%) (eq 11). Other chiral sulfides have been reported that can give much better enantioselectivities in the epoxidation of related glycidic amides.[17]

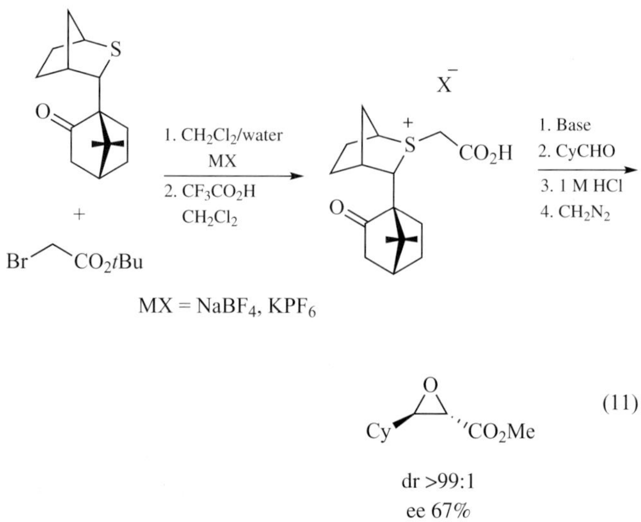

Stoichiometric Epoxidations via Conjugate Addition to Preformed Vinyl Sulfonium Salts.

Fused Epoxides.[7] Vinyl sulfonium salt (**2**) can be made from sulfide **1** (eq 3). This has been used for the preparation of fused epoxides with heterocycles, such as pyrrolidines, piperidines, and azepanes. In these reactions, the vinyl sulfonium salt undergoes conjugate addition with an aminoaldehyde that gives an intermediate ylide that then reacts with the carbonyl group of the aldehyde. The resulting intermediate heterocyclic betaine ring-closes to give a fused epoxide (eq 12). Yields vary from moderate to good (65–80%) and ee values are high (92–99%).

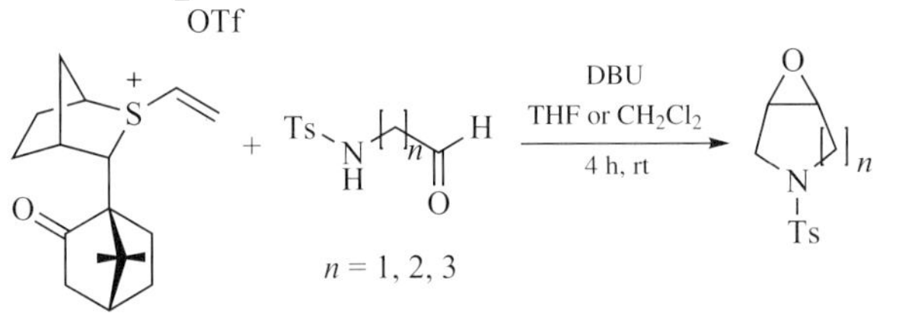
(12)

2

The reaction also works with aromatic aminoaldehydes (such as methyl (2-formylphenyl)carbamate) leading to a tricyclic molecule in a moderate yield (50%) but with high ee (88%). Aminoketones can also be used, although yields are low (30%, probably due to competing reactions) and ee values are good (86%). Racemic and substrate-controlled variants of these reactions can be achieved using ethenyldiphenyl sulfonium trifluoromethanesulfonate.[7,18–20]

Origin of Diastereo and Enantioselectivity in Epoxidation Reactions.[21] For a detailed discussion on the origin of diastereo- and enantioselectivities in these reactions, see the original papers and reviews.[21]

Aziridination Reactions. One of the most important ways of making aziridines involves the reaction of an ylide or a carbene with an imine. Chiral sulfides such as **1** have been used to prepare chiral nonracemic aziridines through two strategies: (i) Catalytic reaction of a sulfide with a metal carbene generated in situ from an *N*-tosylhydrazone salt gives an ylide that then reacts with an imine. (ii) Preformation and isolation of a sulfonium salt from a sulfide, which is then deprotonated to give an ylide that can then react with the imine.

Catalytic Aziridination via Sulfur Ylide Formation from an N-Tosylhydrazone Salt.[22] As described above for epoxidation reactions, this procedure involves the use of an *N*-tosylhydrazone salt to generate a metal carbene, which then reacts with sulfide **1** to form an ylide. This ylide can then react with imines to give aziridines (eq 13). A complication in aziridination is that metal carbenoids can react directly with some imines to yield aziridines. Fortunately, imines bearing electron-withdrawing groups are less reactive to carbenoids and more reactive to ylides than the electron-rich imines. Thus, *N*-tosyl, *N*-diphenylphosphinyl, *N*-[2-(trimethylsilyl)ethanesulfonyl] (SES) and *N*-trichloro-*tert*-butyloxycarbonyl (TcBOC) imines are all suitable substrates.

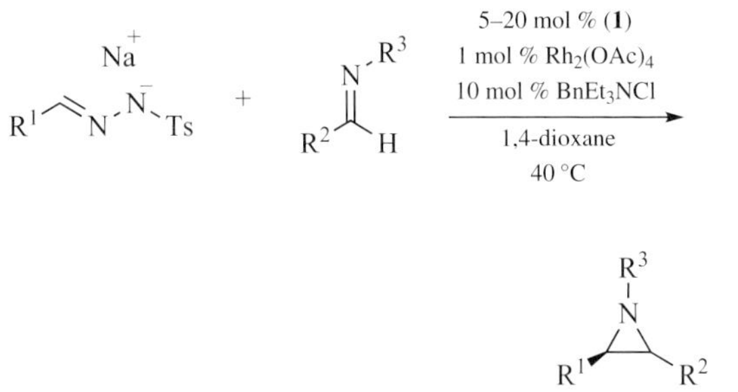
(13)

Sulfide **1** has been used in this procedure to mainly obtain *trans*-disubstituted aziridines in moderate to high yields and, in general, with low to moderate diastereomeric excesses but high enantiomeric excesses (eq 13). 2,3-Diaryl-substituted aziridines have been obtained in moderate to high yields (60–82%) and with moderate dr (3:1) and good ee (92–98%), starting from aromatic *N*-SES-activated imines and from phenyl *N*-tosylhydrazone salt. Other *N*-activating groups on the imine, such as Ts and BOC, have also been tested on the same substrate and their reactions give similar yields, dr, and ee. Disubstituted aziridines bearing a phenyl substituent and an alkyl, heteroaromatic, or cinnamyl imine have also been prepared with similar results. For some selected results, see Table 3. Trisubstituted aziridines can also be prepared from an *N*-activated disubstituted imine, for example, $Ph_2C{=}NSO_2C_8H_7$ reacted with benzyl ylide to give the corresponding aziridine with 84% ee.

Table 3 Asymmetric aziridination of imines under the conditions as in eq 13

Entry	R^1	R^2	R^3	Yield (%)	*trans*: *cis*	ee trans (%)
1	Ph	Ph	SES	75	2.5:1	94
2	Ph	Ar electron-rich	SES	60	2.5:1	92
3	Ph	Ar electron-poor	SES	82	2:1	98
4	Ph	Alkyl	SES	50	2.5:1	98
5	Ph	α,β-Unsaturated	SES	59	8:1	94
6	Ph	Heteroaromatic	Ts	72	8:1	95
7	Ph	Ph	TcBOC	66	6:1	90

Stoichiometric Aziridination via Ylide Formed from a Sulfonium Salt. In some cases, the catalytic method proved unsuccessful and so a stoichiometric method was developed. A sulfonium salt was preformed as described above (eq 1). Deprotonation of the sulfonium salt and reaction with the desired imine gave an aziridine with >90% ee (eq 14).

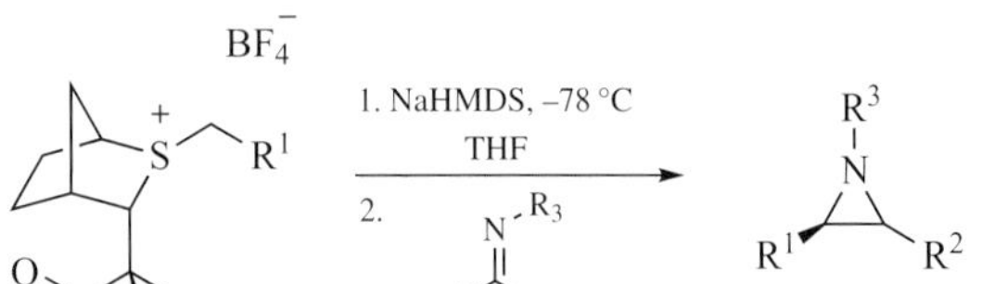
(14)

ee >90%

R^1 = furyl; R^2 = Ph; R^3 = Bz

The enantiomer of sulfide **1** has been employed in both the catalytic and the stoichiometric procedures to produce aziridines with high enantioselectivity, which were used to make the side chain of taxol.[23]

Origin of Diastereo- and Enantioselectivities in Aziridination Reactions.[21] For a detailed discussion of the origin of diastereo- and enantioselectivities, refer to the original papers and reviews.[21]

Cyclopropanation Reactions. Cyclopropanes can be obtained by the reaction of an ylide or a carbene with an olefin. Chiral sulfides such as **1** have been used to prepare chiral nonracemic cyclopropanes through two strategies: (i) catalytic reaction of a sulfide with in situ generation of an ylide that reacts with an electron-deficient alkene. (ii) Use of stoichiometric amounts of a preformed sulfonium salt, which is then deprotonated to give an ylide that can then react with an alkene. Attempts to use sulfide **1** in asymmetric cyclopropanations have been less successful than the corresponding epoxidations and aziridinations.[4,24] Sulfide **1** is not the reagent of choice in these reactions and better results are generally obtained using bicyclic sulfides **5**–**7**.[22,25–30]

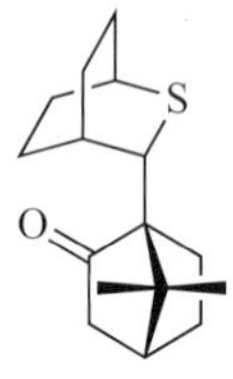

5

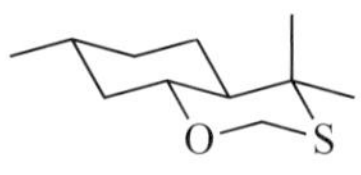

6

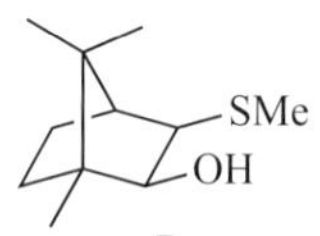

7

Catalytic Cyclopropanations via Ylide Formation from an N-Tosylhydrazone Salt.[22,24] As described before for epoxidation and aziridination reactions, the procedure for catalytic generation of sulfur ylides from sulfides involves the use of, for example, a phenyl-substituted *N*-tosylhydrazone salt to generate a metal carbene that then reacts with sulfide **1** to give the ylide. This ylide reacts in situ with an electron-deficient alkene (Michael acceptor) to give a cyclopropane. Sulfide **1** has been used by this procedure using (*E*)-chalcone to obtain a *trans*-trisubstituted cyclopropane in low yield and with moderate to good diastereomeric and enantiomeric excesses (eq 15).

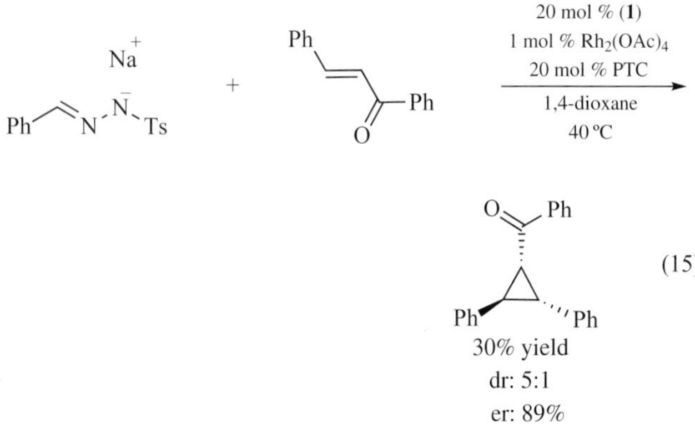

(15)

An ester-stabilized ylide can be generated from ethyl diazoacetate and sulfide using a copper catalyst. Cyclopentenone was reacted under these conditions with the enantiomer of sulfide **1** and a mixture of the *exo*- and *endo*-products was obtained in a 1:1 ratio (eq 16).[24] Slightly higher diastereoselectivities and enantioselectivities were obtained using a preformed salt or ylide (see below).

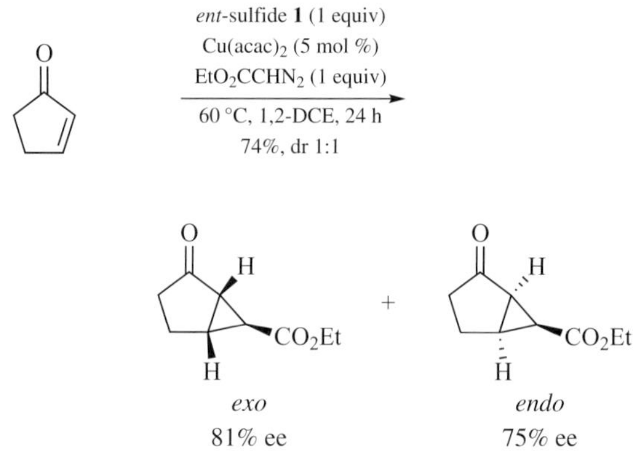

(16)

Stoichiometric Cyclopropanation via Preformation of a Sulfonium Salt or Ylide.[4,24] Sulfonium salts can be preformed as already described (eqs 1 and 2). Deprotonation of the sulfonium salt and reaction with an alkene bearing an electron-withdrawing group leads to cyclopropanes. With cyclopentenone, it was found to be better to isolate a *tert*-butyl ester-stabilized ylide and react it under neutral conditions (eq 17).[24] Less bulky esters gave poorer results.

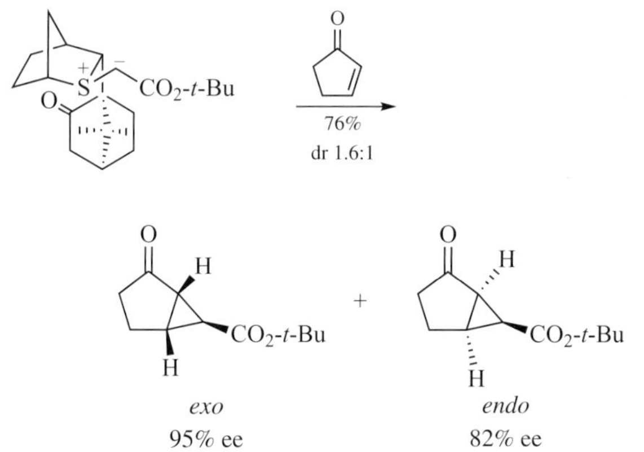

(17)

In the case of acyclic alkenes, stabilized ylides derived from sulfide **1** give quite variable enantioselectivities.[4] Amide-stabilized ylides gave the best results with disubstituted cyclopropanes being obtained in good yields and with high ee (eq 18).

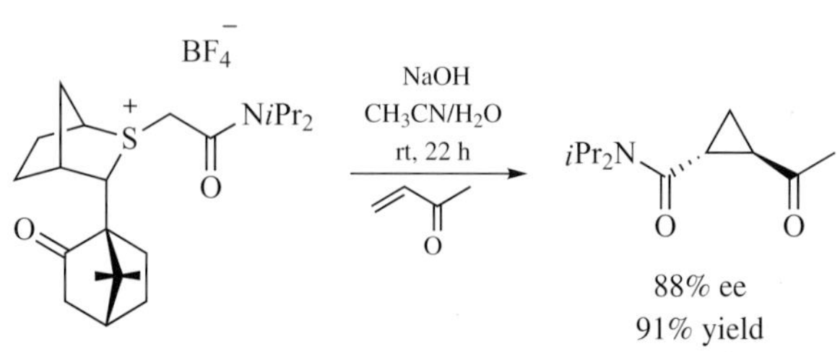

(18)

Reactions of Sulfur Ylides with Organoboranes: Synthesis of Alcohols and Amines.[5,31] Reactions of chiral sulfur ylides with organoboranes lead to new homologated chiral organoboranes, which in turn can be converted to products such as alcohols and primary amines. Benzyl sulfonium salts presynthesized from sulfide **1** can be deprotonated to form ylides that react with trialkylboranes to obtain new organoboranes. When these organoboranes are subjected to a basic oxidative treatment (H_2O_2 and NaOH), alcohols are obtained in good yields (70–98%) and with excellent ee (95–98%). Alternatively, if the organoboranes are treated with hydroxylamine-*O*-sulfonic acid, primary amines are obtained in good yields (68–78%) and with excellent ee (96–99%) (eq 19). For some selected results, see Table 4.[5,31]

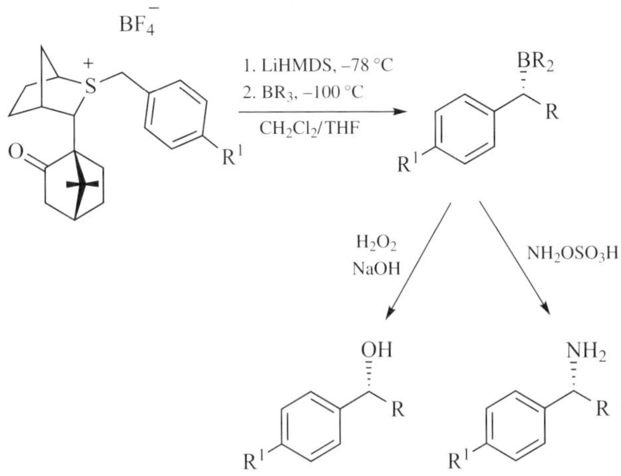

(19)

Table 4 Representative results of the reaction of sulfur ylides with boranes using the conditions as in eq 19

Entry	R[1]	BR_3	Product	Yield (%)	ee (%)
1	H	BEt_3	Alcohol	98	98
2	H	BEt_3	Amine	78	98
3	Me	BEt_3	Alcohol	95	97
4	Me	BEt_3	Amine	74	>99
5	Cl	BEt_3	Alcohol	97	96
6	Cl	BEt_3	Amine	77	>99
7	H	*B*-*i*Pr-9-BBN	Alcohol	97	99

This strategy has been used in the synthesis of some medicinally important compounds that contain a diarylmethylamino or a diarylmethylalkoxy moiety in their structures, such as neobenodine and cetirizine.[5]

When nonsymmetrical R-9-BBN borane derivatives are used, the mixtures of the ring-migration product and the *R*-migration product tend to be obtained; a notable exception is B-*i*Pr-9-BBN where only the *iso*-propyl-migration product is formed (see entry 7 in Table 4).[31] Trimethylsilyl-stabilized ylides derived from sulfide **1** gave good yields, but with low enantioselectivities (20–42%).[32]

The enantioselective synthesis of α-substituted allyl boranes is also possible using sulfonium salts derived from sulfide **1**.[33] These boranes are useful intermediates because they can react with aldehydes to form homoallylic alcohols. In this case, migration of the allylic fragment is observed as the major pathway. The reaction can proceed under two sets of conditions that allow the formation of two different isomeric homoallylic alcohols (eqs 20 and 21). Both routes lead to the formation of products with excellent de (95% in all cases) and ee (>99% in all cases) and in yields that range from good to very good (49–81%). Some selected results are summarized in Table 5. The sulfide **1** could be recovered in >90% yield for reuse. These procedures have been applied to the synthesis of (+)-*iso*-agatharesinol.

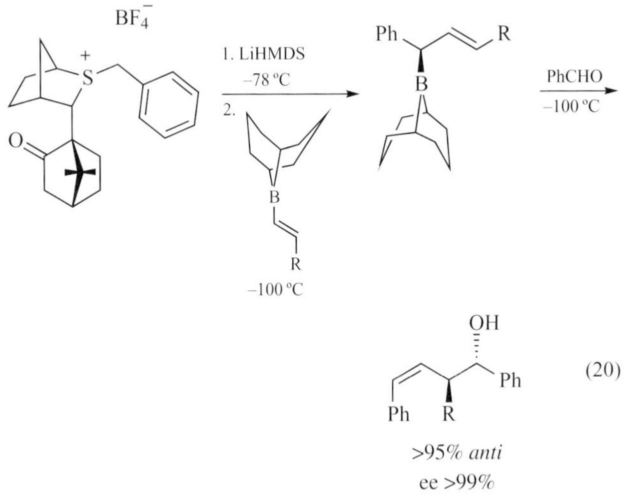

(20)

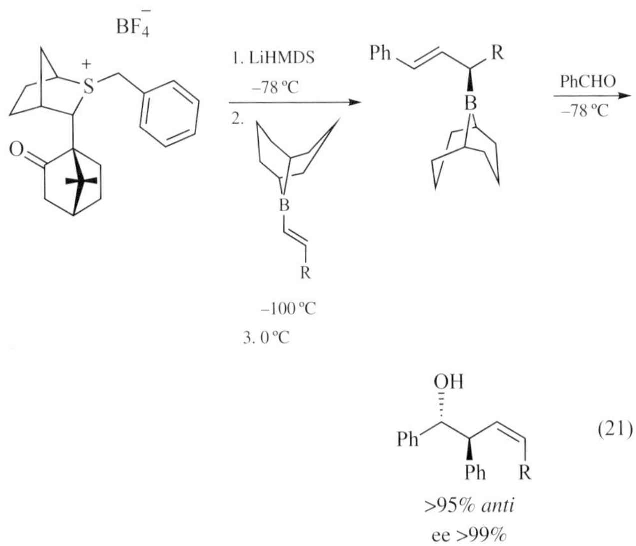

(21)

Table 5 Synthesis of homoallylic alcohols using conditions as in eqs 20–21

Entry	R	Method	Yield (%)	*Z*/*E*
1	*n*Bu	1	79	15:1
2	*n*Bu	2	81	10:1
3	H	1	61[a]	>40:1
4	Me	1	81	40:1
5	Me	2	76	30:1
6	$TMSOCH_2$	1	61[a,b]	>40:1
7	$TMSOCH_2$	2	49[a,b]	>30:1
8	$AcOCH_2CH_2$	1	72[a]	>40:1
9	$AcOCH_2CH_2$	2	56[a]	13:1

[a] The product of boracycle migration was also isolated in yields of 12–20% for entries 3,7, and 8.

[b] The product was isolated as a diol (TMS-deprotected).

Although the reactions of benzyl-stabilized ylides with boranes work well, it has been found that related reactions involving lithiated carbamates have much broader scope and, furthermore, react with boronic esters.[34]

Morita–Baylis–Hillman Reactions.[35] One of the most important ways of functionalizing the α-position of an alkene activated with an electron-withdrawing group is a Morita–Baylis–Hillman reaction (MBH). The reaction typically uses an α,β-unsaturated carbonyl compound with an aldehyde as electrophile and is usually catalyzed by tertiary amines or phosphines (eq 22). The scope of the reaction has been broadened to other Michael acceptors and other electrophiles. Sulfide **1** can be used in one of these modifications, which uses *N*-acyl iminium ions as electrophiles and cyclic enones as activated alkenes, leading to β'-amido-α,β-unsaturated carbonyl compounds in good yield (49–90%) and with excellent ee (80–98%). The reaction involves the use of TMSOTf as a Lewis acid and sulfide **1** as a weak Lewis base (instead of a phosphine or an amine) followed by a basic treatment with DBU (eq 23). For selected results, see Table 6.

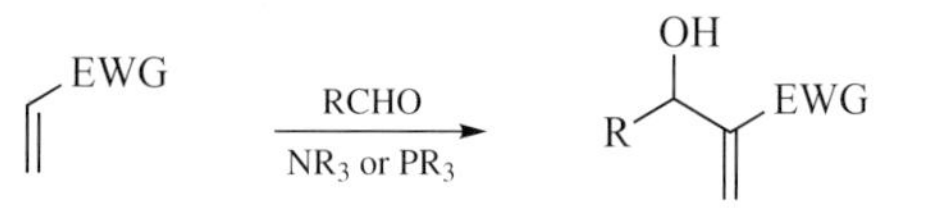

(22)

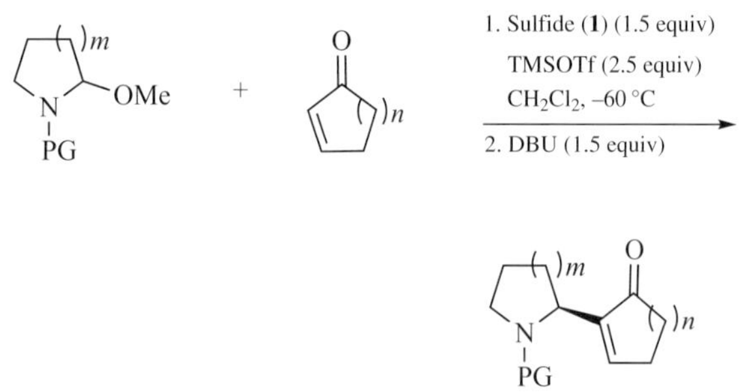

(23)

Table 6 Morita–Baylis–Hillman-type reactions with sulfide **1** according to eq 23

Entry	PG	*m*	*n*	Yield (%)	ee (%)
1	Cbz	1	1	69	82
2	Cbz	1	2	86	80
3	BOC	1	1	75	88
4	BOC	1	2	90	88
5	Cbz	2	1	88	94
6	Cbz	2	2	49	98

Related Reagents. Many chiral sulfides have been described in the literature and their use as catalysts has been reviewed.[21] Recently, Aggarwal and coworkers have developed a new chiral sulfide (**8**), which is very easy and cheap to synthesize in both enantiomeric forms.[36] It has been successfully employed in epoxidation and aziridination reactions via preformation of the sulfonium salt and is probably the most practical sulfide to use for these reactions. Other applications are under investigation.[6,36,37]

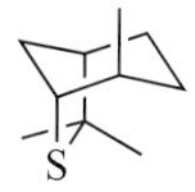

8

1. Aggarwal, V. K.; Fang, G.; Kokotos, C. G.; Richardson, J.; Unthank, M. G., *Tetrahedron* **2006**, *62*, 11297.
2. Aggarwal, V. K.; Alonso, E.; Hynd, G.; Lydon, K. M.; Palmer, M. J.; Porcelloni, M.; Studley, J. R., *Angew. Chem., Int. Ed.* **2001**, *40*, 1430.
3. Aggarwal, V. K.; Bae, I.; Hee-Yoon, L.; Richardson, J.; Williams, D. T., *Angew. Chem., Int. Ed.* **2003**, *42*, 3274.
4. Riches, S. L.; Saha, C.; Fontán Filgueira, N.; Grange, E.; McGarrigle, E. M.; Aggarwal, V. K., *J. Am. Chem. Soc.* **2010**, *132*, 7626.
5. Aggarwal, V. K.; Fang, G. Y.; Schmidt, A. T., *J. Am. Chem. Soc.* **2005**, *127*, 1642.
6. Arshad, M.; Fernández, M. A.; McGarrigle, E. M.; Aggarwal, V. K., *Tetrahedron: Asymmetry* **2010**, *21*, 1771.
7. Unthank, M. G.; Hussain N.; Aggarwal, V. K., *Angew. Chem., Int. Ed.* **2006**, *45*, 7066.
8. Aggarwal, V. K.; Alonso, E.; Bae, I.; Hynd, G.; Lydon, K. M.; Palmer, M. J.; Patel, M.; Porcelloni, M.; Richardson, J.; Stenson, R. A.; Studley, J. R.; Vasse, J. L.; Winn, C. L., *J. Am. Chem. Soc.* **2003**, *125*, 10926.
9. Aggarwal, V. K.; Coogan, M. P.; Stenson, R. A.; Jones, R. V. H.; Fieldhouse, R.; Blacker, J., *Eur. J. Org. Chem.* **2002**, 319.
10. Aggarwal, V. K.; Bae, I.; Hee-Yoon, L., *Tetrahedron* **2004**, *60*, 9725.
11. Morales-Serna, J. A.; Llaveria, J.; Díaz, Y.; Matheu, M. Y.; Castillón, S., *Org. Biomol. Chem.* **2008**, *6*, 4502.
12. Aggarwal, V. K.; Bi, J., *Beilstein J. Org. Chem.* **2005**, *1*, 4.
13. Kotoku, N.; Narumi, F.; Kato, T.; Yamaguchi, M.; Kobayashi, M., *Tetrahedron Lett.* **2007**, *48*, 7147.
14. Kokotos, C. G.; Aggarwal, V. K., *Chem. Commun.* **2006**, 2156.
15. Bi, J.; Aggarwal, V. K., *Chem. Commun.* **2008**, 120.
16. Hebach, C.; Aggarwal, V. K., *Org. Biomol. Chem.* **2005**, *3*, 1419.
17. Aggarwal, V. K.; Charmant, J. P. H.; Fuentes, D.; Harvey, J. N.; Hynd, G.; Ohara, D.; Picoul, W.; Robiette, R.; Smith, C.; Vasse, J.-L.; Winn, C. L., *J. Am. Chem. Soc.* **2006**, *128*, 2105.
18. Unthank, M. G.; Tavassoli, B.; Aggarwal, V. K., *Org. Lett.* **2008**, *10*, 1501.
19. Kokotos, C. G.; McGarrigle, E. M.; Aggarwal, V. K., *Synlett* **2008**, *14*, 2191.
20. Yar, M.; Unthank M. G.; McGarrigle, E. M.; Aggarwal, V. K., *Chem. Asian J.* **2011**, *6*, 372.
21. McGarrigle, E. M.; Myers, E. L.; Illa, O.; Shaw, M. A.; Riches, S. L.; Aggarwal, V. K., *Chem. Rev.* **2007**, *107*, 5841.
22. Aggarwal, V. K.; Alonso, E.; Fang, G.; Ferrara, M.; Hynd, G.; Porcelloni, M., *Angew. Chem., Int. Ed.* **2001**, *40*, 1433.
23. Aggarwal, V. K.; Vasse, J. L., *Org. Lett.* **2003**, *5*, 3987.
24. Aggarwal, V. K.; Grange, E., *Chem. Eur. J.* **2006**, *12*, 568.
25. Solladié-Cavallo, A.; Diep-Vohuule, A.; Isarno, T., *Angew. Chem., Int. Ed.* **1998**, *37*, 1690.
26. Ye, S.; Huang, Z.-Z.; Xia, C.-A.; Tang, Y.; Dai, L.-X., *J. Am. Chem. Soc.* **2002**, *124*, 2432.
27. Zhu, C. Y.; Cao, X. Y.; Zhu, B. H.; Deng, C.; Sun, X.-L.; Wang, B.-Q.; Shen, Q.; Tang, Y., *Chem. Eur. J.* **2009**, *15*, 11465.
28. Deng, X.-M.; Cai, P.; Ye, S.; Sun, X.-L.; Liao, W.-W.; Li, K.; Tang, Y.; Wu, Y.-D.; Dai, L.-X., *J. Am. Chem. Soc.* **2006**, *128*, 9730.
29. Tang, Y.; Ye, S.; Sun, X. L., *Synlett* **2005**, 2720.
30. Huang, K.; Huang, Z.-Z., *Synlett* **2005**, 1621.

31. Fang, G. Y.; Wallner, O. A.; Di Blasio, N.; Ginesta, X.; Harvey, J. N.; Aggarwal, V. K., *J. Am. Chem. Soc.* **2007**, *129*, 14632.

32. Howells, D.; Robiette, R.; Fang, G. Y.; Knowles, L. S.; Woodrow, M. D.; Harvey, J. N.; Aggarwal, V. K., *Org. Biomol. Chem.* **2008**, *6*, 1185.

33. Fang, G. Y.; Aggarwal, V. K., *Angew. Chem., Int. Ed.* **2007**, *46*, 359.

34. Stymiest, J. L.; Dutheuil, G.; Mahmood, A.; Aggarwal, V. K., *Angew. Chem., Int. Ed.* **2007**, *46*, 7491.

35. Myers, E. L.; de Vries, J. G.; Aggarwal, V. K., *Angew. Chem., Int. Ed.* **2007**, *46*, 1893.

36. Illa, O.; Arshad, M.; Ros, A.; McGarrigle, E. M.; Aggarwal, V. K., *J. Am. Chem. Soc.* **2010**, *132*, 1828.

37. Brichacek, M.; Navarro Villalobos, M.; Plichta, A.; Njardarson, J. T., *Org. Lett.* **2011**, *13*, 1110.

Ona Illa
Universitat Autònoma de Barcelona, Barcelona, Spain

Eoghan M. McGarrigle & Varinder K. Aggarwal
University of Bristol, Bristol, UK

[1,1′-Binaphthalene]-2,2′-diol, 3,3′-Bis[3,5-bis(trifluoromethyl)phenyl]-, (1*R*)-

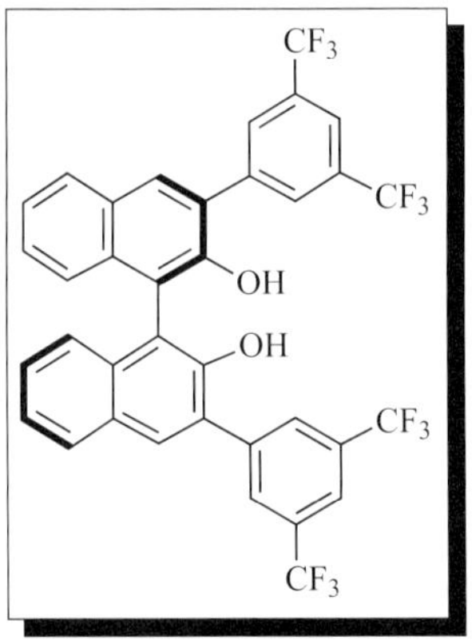

R-isomer
[756491-54-0] $C_{36}H_{18}F_{12}O_2$
InChI = 1S/C36H18F12O2/c37-33(38,39)21-9-19(10-22(15-21)34(40,41)42)27-13-17-5-1-3-7-25(17)29(31(27)49)30-26-8-4-2-6-18(26)14-28(32(30)50)20-11-23(35(43,44)45)16-24(12-20)36(46,47)48/h1-16,49-50H
InChIKey = PGXMYJSTCAQJBY-UHFFFAOYSA-N
S-isomer
[849939-13-5] $C_{36}H_{18}F_{12}O_2$
InChI = 1S/C36H18F12O2/c37-33(38,39)21-9-19(10-22(15-21)34(40,41)42)27-13-17-5-1-3-7-25(17)29(31(27)49)30-26-8-4-2-6-18(26)14-28(32(30)50)20-11-23(35(43,44)45)16-24(12-20)36(46,47)48/h1-16,49-50H
InChIKey = PGXMYJSTCAQJBY-UHFFFAOYSA-N

(chiral diol used as a ligand in asymmetric synthesis)

Alternate Names: 3,3′-bis[3,5-bis(trifluoromethyl)phenyl]-1,1′-bi-2-naphthol; 3,3′-bis[3,5-bis(trifluoromethyl)phenyl]-1,1′-binaphthyl-2,2′-diol; 3,3′-bis[3,5-bis(trifluoromethyl)phenyl]-BINOL.

Physical Data: crystalline solid, mp 97–99 °C, $[\alpha]_{25}^{589}$ −11.9 (*c* 1.1, THF) (*R*-isomer).[1]

Solubility: soluble in methylene chloride and THF; insoluble in hexanes.

Preparative Methods: from [1,1′-binaphthalene]-2,2′-diol (BINOL) via a four-step sequence involving protection with MOMCl, lithiation and iodination, Suzuki coupling, and deprotection (eq 1).[1]

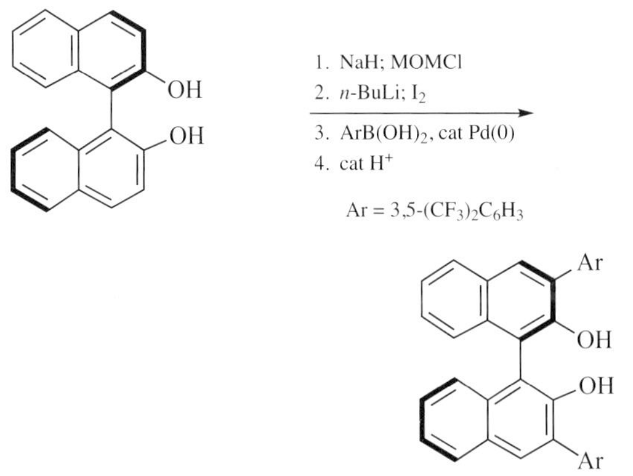

80% overall yield (1)

Alternatively, a route using Suzuki coupling of BINOL dimethyl ether 3,3′-bisboronic acid with the appropriate aryl bromide may be used.[2]

Handling, Storage, and Precautions: stable to air and moisture; while it is expected that this compound should be relatively stereochemically robust, it is possible that racemization may occur under acidic, basic, or thermal conditions.

General. 3,3′-Bis[3,5-bis(trifluoromethyl)phenyl]-1,1′-bi-2-naphthol (**1**) is one of a family of substituted binaphthols that have been developed since the seminal work of Noyori on asymmetric reductions using BINOL-modified aluminum hydrides. Binaphthol **1** may be regarded as a derivative of 3,3′-diphenyl-1,1′-bi-2-naphthol that was shown to be considerably more effective as a chiral ligand than the parent BINOL in certain Diels–Alder reactions.[3] The presence of trifluoromethyl groups often gives higher selectivities than the simpler 3,3′-diphenyl-BINOL. A review on the use of 3,3′-disubstituted BINOL derivatives has been published.[4]

Alkynylation of Imines. Binaphthol **1** may be used as a catalyst for the enantioselective addition of alkynylzinc reagents to *N*-sulfonyl aldimines (eq 2).[5] The selectivity is much higher than that obtained with unsubstituted BINOL (8% ee). Other aryl aldimines also give good selectivities (64 to >99% ee with phenylacetylene) but alkyl aldimines give much lower (18–38% ee) selectivities. Additions of aliphatic alkynes to aryl aldimines generally proceed well with selectivities of 76 to >99% ee.

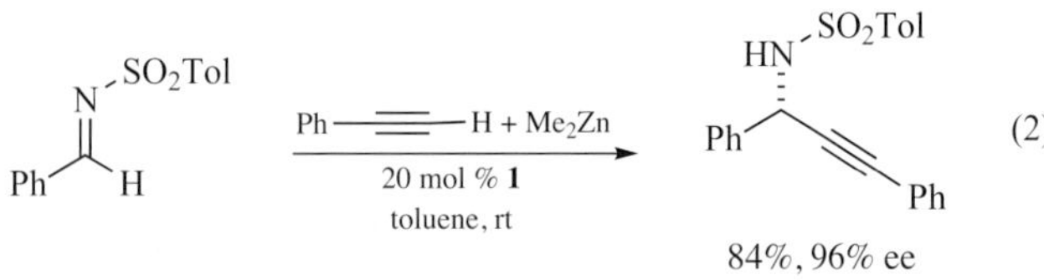

(2)

In an interesting variation of this reaction, use of excess alkyne and heating the reaction mixture permit the formation of protected (*E*)-2-en-3-ynyl amines (eq 3).[6] Use of 3,3′-diphenyl-BINOL rather than ligand **1** in eq 3 affords much lower selectivity (20% ee). This double addition of phenylacetylene to a variety of aryl sulfonimines proceeds with high selectivities (84–99% ee). Sequential addition of two different alkynes also proceeds with high selectivities albeit with slightly lower yields (57–80% compared to 73–92% using an excess of a single alkyne).

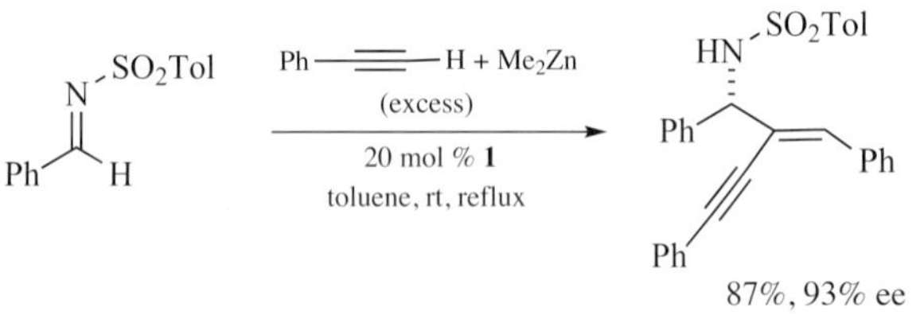

(3)

Addition of Diethylzinc to *N*-Acylimines. Reaction of binaphthol **1** with Et_2Zn and a diimine generates complexes **2** (eq 4).[7] These zinc complexes effectively catalyze the addition of Et_2Zn to *N*-formylimines (formed in situ, eq 5). With a variety of aryl imines, good (75–91%) yields and generally high (71–94% ee, mostly >83% ee) selectivities are obtained. The complex formed from binaphthol **1** is much more selective than that formed with 3,3′-diphenyl-BINOL (66% ee vs. 11% ee under unoptimized conditions). The stereochemistry of the diimine does not affect the selectivity of the additions.[8]

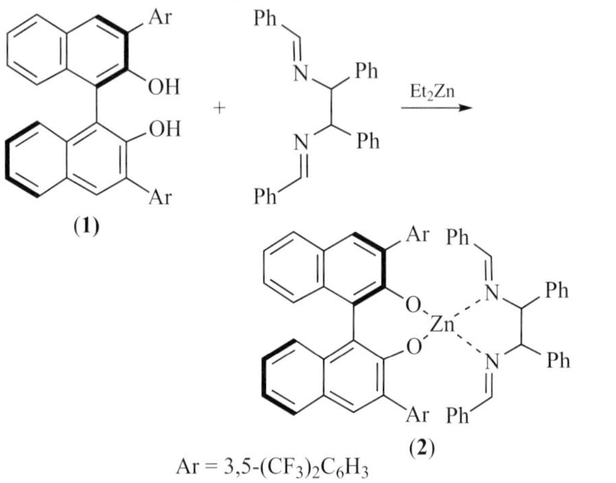

(4)

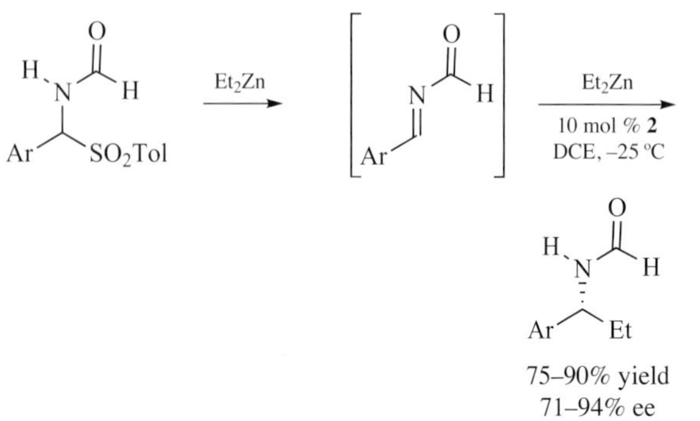

(5)

Conjugate Alkynylation and Alkenylation of Enones. Binaphthols such as **1** can be very effective catalysts for asymmetric conjugate alkynylboration (eq 6)[9] and alkenylboration of enones.[10] Remarkably, conjugate alkynylations of enones using catalytic amounts of binaphthols proceed with selectivities identical to reactions using stoichiometric amounts of preformed alkynylboronates. Addition of an alkenylboronate derived from 1-octyne to chalcone proceeds with 97% ee using binaphthol **1** as catalyst (eq 7). Other 3,3′-disubstituted-BINOLs give comparable selectivities. With 3,3′-diiodo-BINOL, high yields and selectivities (>94% ee) for alkenylboration are observed for a variety of enones, including ones with *β*-alkyl groups.

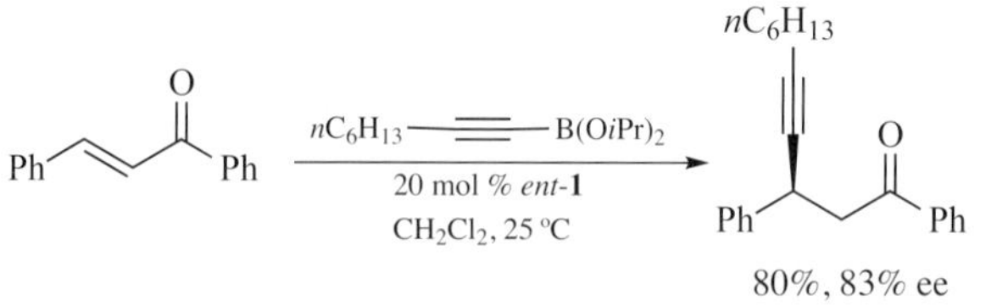

(6)

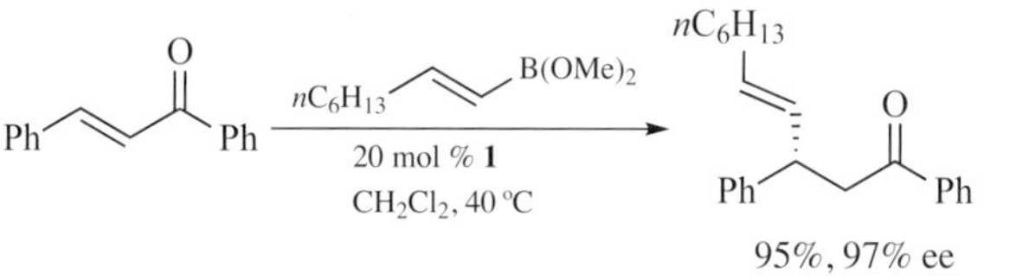

(7)

Hydrophosphonylation of Imines. Binaphthol**1** is readily transformed into phosphate **3** (eq 8), which can serve an as effective chiral Brønsted acid catalyst.[11] For example, phosphate **3** catalyzes the addition of diisopropyl phosphite to a variety of *N*-(4-methoxyphenyl)imines with good selectivities (eq 9).[12] Best selectivities (82–90% ee) are observed with cinnamaldehyde-derived imines. A DFT study suggests that the steric bulk of the bis(3,5-trifluoromethyl)phenyl groups can be used to explain the higher selectivity observed with phosphate **3** compared with other phosphates with less sterically demanding groups.[13]

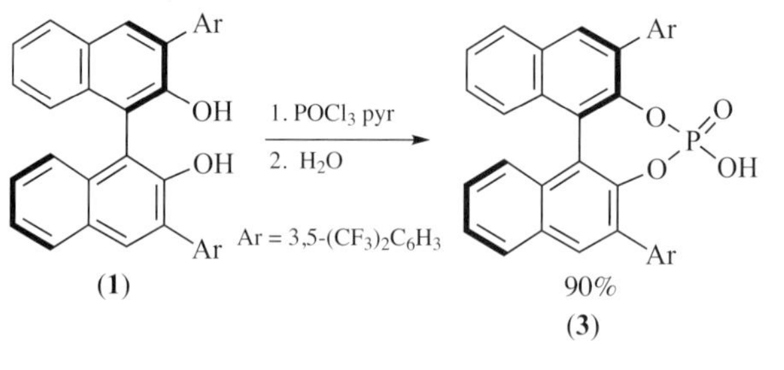

(8)

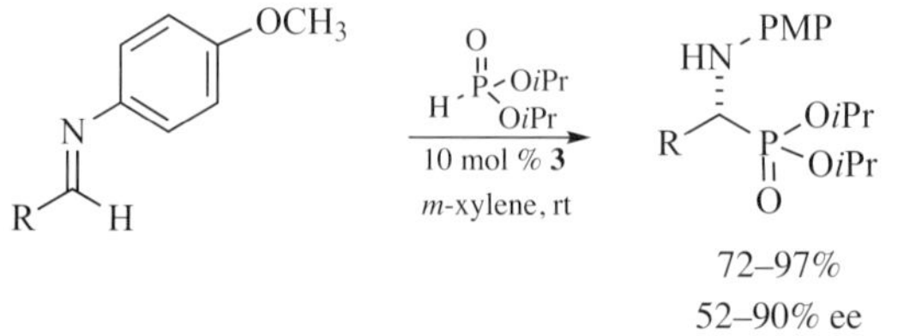

(9)

Hydrophosphonylation of Aldehydes. Binaphthol **1** may be reacted with $Ti(OiPr)_4$ and a cinchona alkaloid such as cinchonidine to generate a bifunctional catalyst that has both a Lewis acidic and a Lewis basic site (eq 10).[14] This in situ generated catalyst is useful for the hydrophosphonylation of aliphatic aldehydes (eq 11). An analogous catalyst prepared from 3,3′-diiodo-BINOL gives high selectivities with aromatic aldehydes (e.g., 99% ee with benzaldehyde) but is less selective than catalyst **4** with aliphatic aldehydes (e.g., 74% ee vs. 92% ee with hydrocinnamaldehyde).

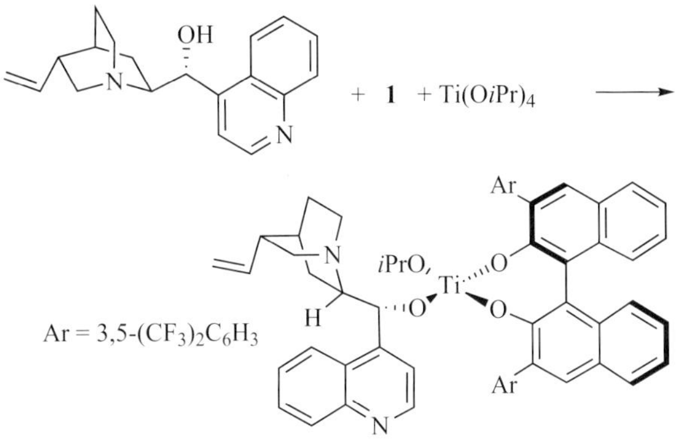

(10)

(**4**)

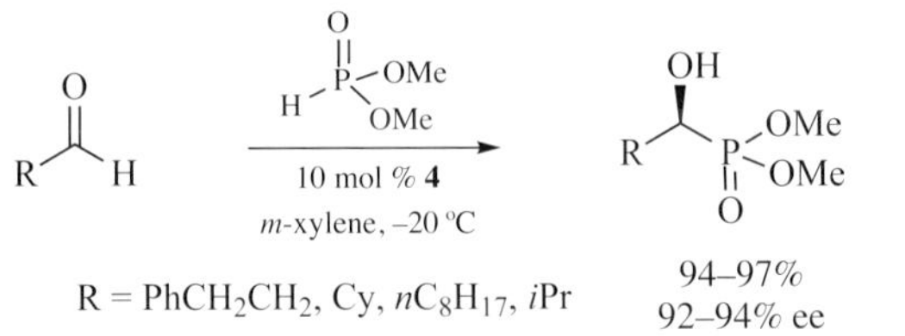

(11)

Asymmetric Desymmetrization of *meso*-Anhydrides. Binaphthol *ent*-**1** can be converted to thiophosphoramide **5**.[15] This bifunctional organocatalyst effectively catalyzes the methanolysis of *meso*-anhydrides (eq 12). Bicyclic anhydrides with four- to Six-membered rings all give high (>92% ee) selectivities while the cyclopropane derivative is slightly less selective (82% ee).

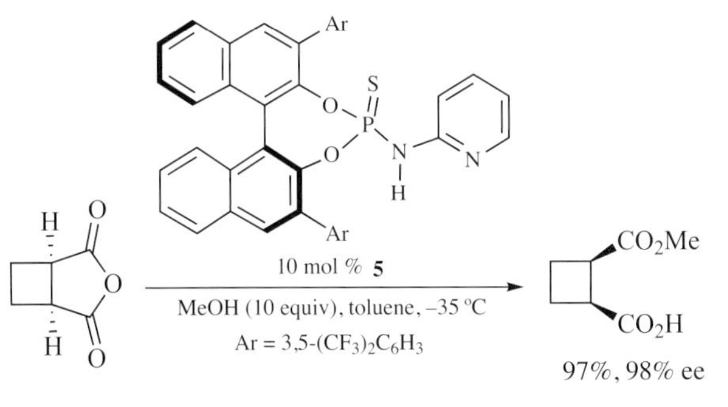

(12)

Allylboration of Aldehydes.[1] Treatment of binaphthol **1** with triallylborane generates allylboronate **6**, which reacts rapidly with benzaldehyde to produce the expected homoallylic alcohol in high yield (eq 13). The observed selectivity (70% ee) is higher than that obtained with 3,3′-diphenyl-BINOL (46% ee) but much higher selectivities (with aryl aldehydes and alkyl aryl ketones) are obtained with 3,3′-bis(trifluoromethyl)-1,1′-bi-2-naphthol (94–99% ee).

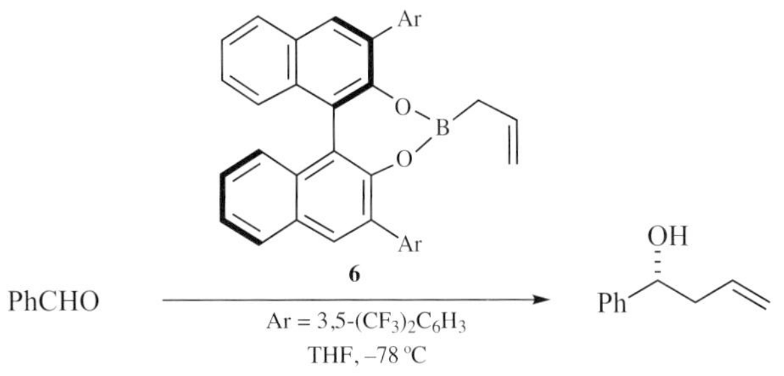

(13)

Diels–Alder Reactions. Substituted binaphthol *ent*-**1** can be used as a ligand for Sn(IV) to make a Lewis acid catalyst.[16] Thus, reaction of binaphthol *ent*-**1** as its dilithium salt with $SnCl_4$ generates complex **7** in situ (eq 14). Complex **7** is an effective Lewis acid for the Diels–Alder reaction of cyclopentadiene and methacrolein (eq 15). Other binaphthols give both lower yields and selectivities.

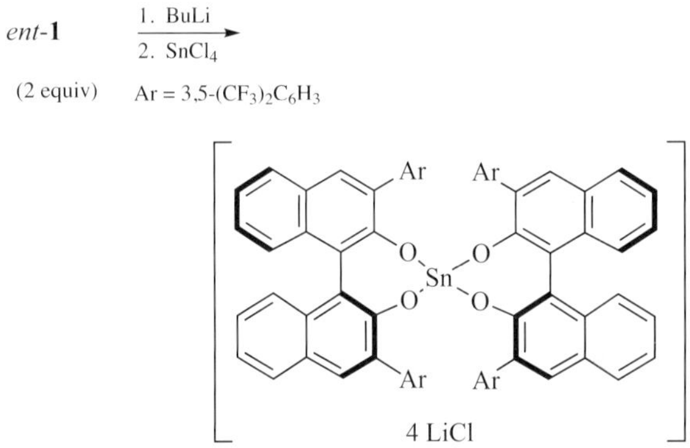

(**7**)

(14)

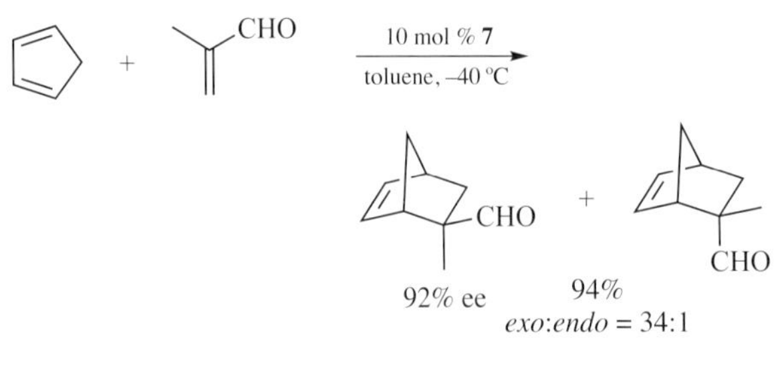

(15)

Asymmetric Hydrosilylation of Styrenes. Conversion of binaphthol *ent*-**1** into phosphoramidite **8** provides a chiral ligand that is effective for the Pd-catalyzed hydrosilylation of styrenes (eq 16).[17] While ligand **8** gives reasonable selectivity (82% ee) with styrene, the corresponding phosphoramidites derived from 3,3′-diphenyl-BINOL (92% ee) and 3,3′-bis(4-methoxyphenyl)-BINOL (96% ee) give higher selectivities.

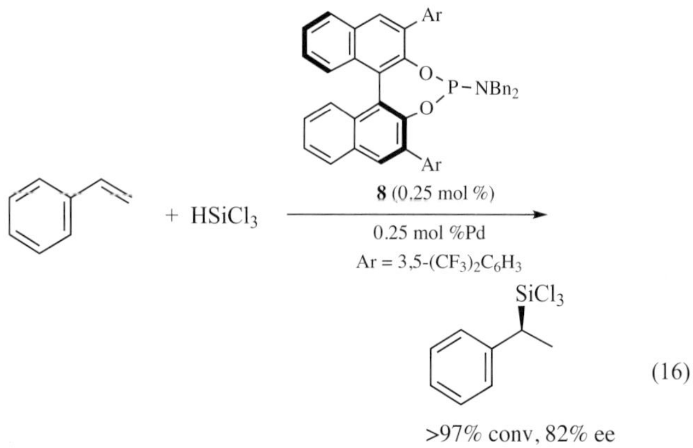

Asymmetric Friedel–Crafts Reactions of β-Trifluoromethyl Enones. Admixture of binaphthol **1** with $Zr(OtBu)_4$ generates a chiral complex that can be used for enantioselective Friedel–Crafts reactions of electron-rich heteroaromatics.[18] For example, indole undergoes reaction with a trifluoromethyl enone to give the expected adduct but conversion is low and selectivity is mediocre (eq 17). Much better reactivity and selectivity is observed using 3,3′-dibromo-BINOL as the ligand (87% yield, 96% ee).

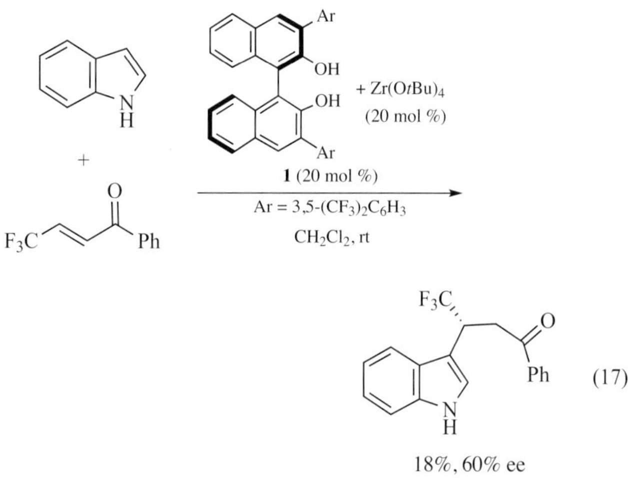

1. Wu, T. R.; Shen, L.; Chong, J. M., *Org. Lett.* **2004**, *6*, 2701.
2. Singh, R.; Czekelius, C.; Schrock, R. R.; Müller, P.; Hoveyda, A. H., *Organometallics* **2007**, *26*, 2528.
3. Kelly, T. R.; Whiting, A.; Chandrakumar, N. S., *J. Am. Chem. Soc.* **1986**, *108*, 3510.
4. Chen, Y.; Yekta, S.; Yudin, A. K., *Chem. Rev.* **2003**, *103*, 3155.
5. Blay, G.; Cardona, L.; Climent, E.; Pedro, J. R., *Angew. Chem., Int. Ed.* **2008**, *47*, 5593.
6. Yin, C.; Hu, X.-Q.; Hui, X.-P.; Xu, P.-F., *Adv. Synth. Catal.* **2009**, *351*, 1512.
7. Zhang, H.-L.; Liu, H.; Cui, X.; Mi, A.-Q.; Jiang, Y.-Z.; Gong, L.-Z., *Synlett* **2005**, 615.
8. Liu, H.; Zhang, H.-L.; Wang, S.-J.; Mi, A.-Q.; Jiang, Y.-Z.; Gong, L.-Z., *Tetrahedron: Asymmetry* **2005**, *16*, 2901.
9. Wu, T. R.; Chong, J. M., *J. Am. Chem. Soc.* **2005**, *127*, 3244.
10. Wu, T. R.; Chong, J. M., *J. Am. Chem. Soc.* **2007**, *129*, 4908.
11. Bhadury, P. S.; Zhang, Y.; Zhang, S.; Song, B.; Yang, S.; Hu, D.; Chen, Z.; Xue, W.; Jin, L., *Chirality* **2009**, *21*, 547.
12. Akiyama, T.; Morita, H.; Itoh, J.; Fuchibe, K., *Org. Lett.* **2005**, *7*, 2583.
13. Akiyama, T.; Morita, H.; Bachu, P.; Mori, K.; Yamanaka, M.; Hirata, T., *Tetrahedron* **2009**, *65*, 4950.
14. Yang, F.; Zhao, D.; Lan, J.; Xi, P.; Yang, L.; Xiang, S.; You, J., *Angew. Chem., Int. Ed.* **2008**, *47*, 5646.
15. Wakchaure, V. N.; List, B., *Angew. Chem., Int. Ed.* **2010**, *49*, 4136.
16. Kano, T.; Konishi, T.; Konishi, S.; Maruoko, K., *Tetrahedron Lett.* **2006**, *47*, 873.
17. Zhang, F.; Fan, Q.-H., *Org. Biomol. Chem.* **2009**, *7*, 4470.
18. Blay, G.; Fernández, I.; Muñoz, M. C.; Pedro, J. R.; Vila, C., *Chem. Eur. J.* **2010**, *16*, 9117.

J. Michael Chong
University of Waterloo, Waterloo, Ontario, Canada

N,N′-(1S)-[1,1′-Binaphthalene]-2,2′-diylbis-(2S,2′S)-pyrrolidine-2-carboxamide

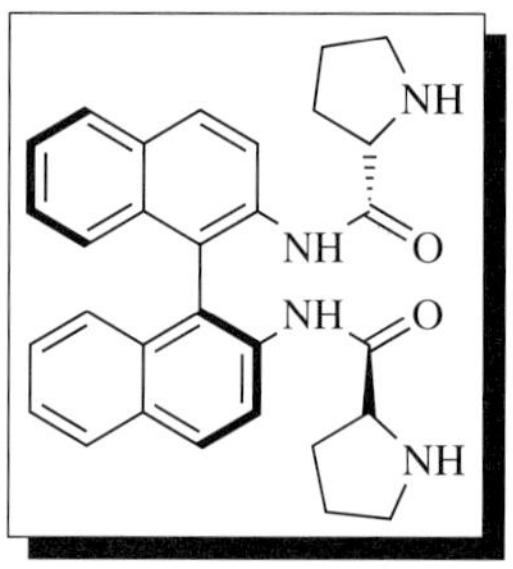

[891496-34-7] $C_{30}H_{30}N_4O_2$ (MW 478.24)

InChI = 1/C30H30N4O2/c35-29(25-11-5-17-31-25)33-23-15-13-19-7-1-3-9-21(19)27(23)28-22-10-4-2-8-20(22)14-16-24(28)34-30(36)26-12-6-18-32-26/h1-4,7-10,13-16,25-26,31-32H,5-6,11-12,17-18H2,(H,33,35)(H,34,36)/f/h33-34H

InChIKey = FYTHJMKJVUTERE-UBXIPSODCL

(used as a catalyst and as a ligand in enantioselective synthesis of aldol products and cyanosilylation of kentones, respectively)

Alternate Names: (S_a)-2,2′-bis[((2*S*)-pyrrolidin-2-ylcarbonyl)amino]-1,1′-binaphthalene, (*S*)-Binam-L-Pro.

Physical Data: mp 230 °C; $[\alpha]^D_{26} = -108.6°$ (c = 1, MeOH).

Solubility: soluble in most polar organic solvents; insoluble in hexane, ether, and water.

Form Supplied in: not commercially available. Prepared from (S_a)-2,2′-diamino-1,1′-binaphthalene and protected L-proline, both commercially available.

Analysis of Reagent Purity: elemental analysis, NMR, HRMS, and X-ray structure.

Preparative Methods: the reagent can be prepared by coupling (S_a)-2,2′-diamino-1,1′-binaphthalene with the mixed anhydride obtained from L-Boc-proline and isobutyl or ethyl chloroformate or by direct coupling of Boc-L-proline with (S_a)-2,2′-diamino-1,1′-binaphthalene (Binam) mediated by 1-ethyl-3-(3-methylaminopropyl)carbodiimide (EDC) and 1-hydroxybenzotriazole (HOBt) or *N,N′*-dicyclohexylcarbodiimide (DCC) and 4-(dimethylamino)pyridine (DMAP). The coupled intermediate was deprotected by using TFA. Alternatively, the reagent can be prepared by coupling (S_a)-2,2′-diamino-1,1′-binaphthalene with the acyl chloride derived from Fmoc-L-proline, and deprotection with piperidine.

Purification: by column chromatography and/or recrystallization from CH_2Cl_2/hexane or $CHCl_3$/ether.

Handling, Storage, and Precaution: the reagent is stable for months when stored in air at room temperature.

Enantioselective Cyanosilylation of Ketones. The generation of chiral quaternary stereocenters is a challenging task, with the addition of trimethylsilyl cyanide (TMSCN) to ketones being one of the most successful examples.[1] (*S*)-Binam-L-Pro was used as a ligand for the enantioselective addition of TMSCN to acetophenone (eq 1). The catalytic complex obtained by reaction with titanium fetra isopropoxide gave the O-TMS cyanohydrin with 47% yield and 33% ee. Better results (up to 90% yield and 94% ee) were obtained under similar reaction conditions using the bisamide derived from 1,2-diphenylethane-1,2-diamine and proline.[2]

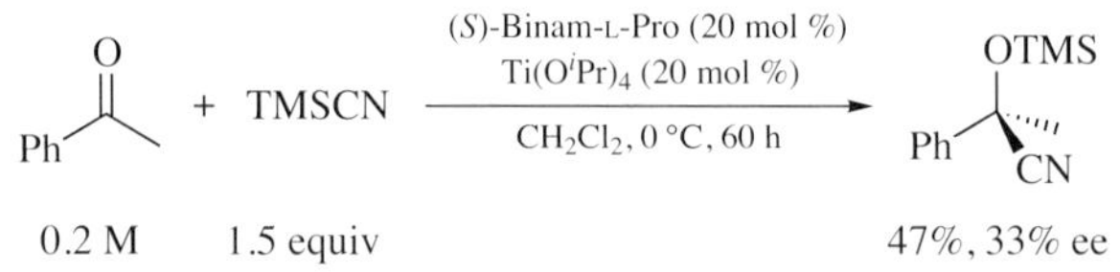

(1)

Enantioselective Direct Aldol Reaction. The use of organocatalytic methods[3] for the enantioselective direct aldol reaction[4] has reached its maturity allowing the synthesis of chiral molecules with high atom efficiency.[5] (*S*)-Binam-L-Pro was first used as an organocatalyst in the intermolecular direct aldol reaction of aromatic aldehydes and alkyl or cyclic ketones using two different protocols.[6] Whereas in one protocol the mixture of 1,4-dioxane/ketone (4:1) at 4 °C were the best reaction conditions, giving the corresponding aldols with 9–79% yield and 50–88% ee's (eq 2),[6a] DMF/water (1:1) at 0 °C or DMF at 25 °C afforded better results in the second instance (52–99% yield, 78–95% ee).[6b] Under the latter reaction conditions, the use of L-proline as catalyst afforded the racemic product.

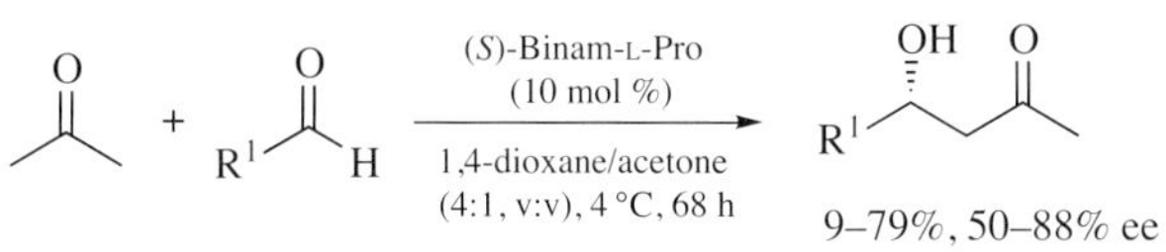

(2)

$R^1 = C_6F_5$, 4-CNC_6H_4, 4-$CF_3C_6H_4$, 2,6-$Cl_2C_6H_3$, 2-ClC_6H_4, 4-ClC_6H_4, 4-BrC_6H_4, Ph, β-naphthyl

When 2-butanone was used as precursor, the reaction took place nearly exclusively at the methyl position, to give the *iso*-isomer preferentially (eq 3). These conditions permitted the recovery by acidic–basic extraction and reuse of catalyst without any detrimental effect on the obtained yields and enantioselectivities during three additional cycles.[6b] Alternatively, $CHCl_3$/ketone (1:1) at −27 °C gave the corresponding aldols with 35–98% yield and 68–95% ee.[7]

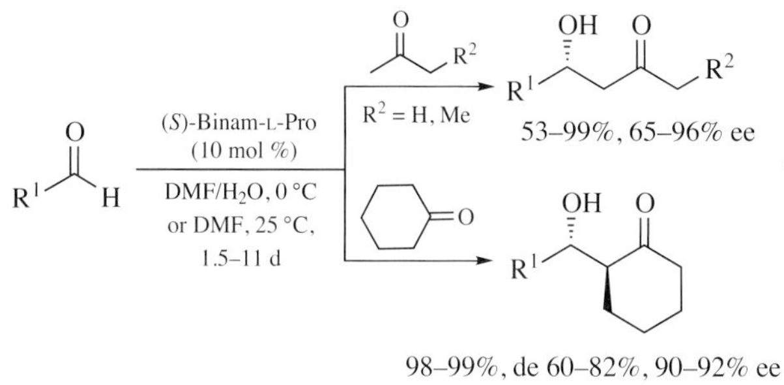

(3)

R^1 = 4-$NO_2C_6H_4$, 2-$NO_2C_6H_4$, 2-ClC_6H_4, Ph, C_6H_{11}, $(CH_3)_2CH$

Under similar conditions, (*S*)-Binam-L-Pro was used as catalyst in the reaction between α-alkoxy ketones (R^1 = OH, OMe, OBn, OTBDMS) and aromatic aldehydes (eq 4) to give mainly the *anti/syn*-regioisomer mixture, with small amounts of corresponding *iso*-regioisomer being formed. The diastereoselectivity was dependent on the nature of R^1 group, with the *anti*-regioisomer obtained as the main product. The enantioselectivity of the process ranged from 73 to 99%. For the case of α-hydroxyacetone (R^1= OH), the best conditions employed DMSO at 25 °C, affording mainly the *anti*-isomer with 85% ee. These results are comparable in terms of regio- and diastereoselectivities to those obtained with L-proline, with the advantage that (*S*)-Binam-L-Pro prolinamide could be recovered when DMF was used as solvent.[8]

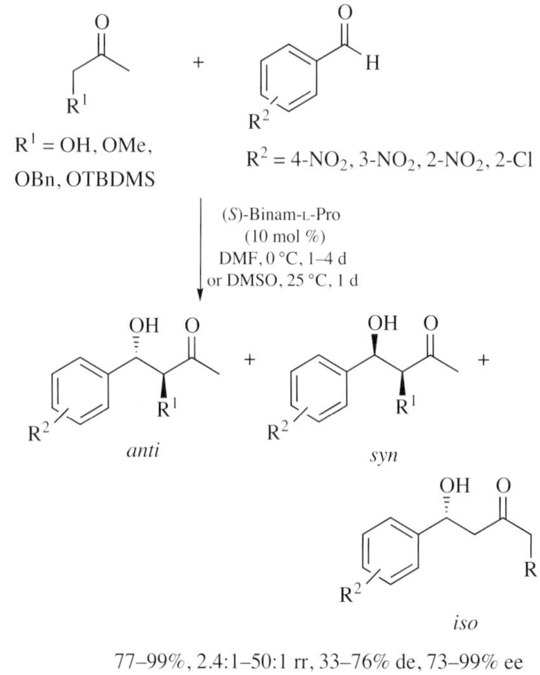

(4)

The reaction rate was highly increased by the addition of catalytic amounts of carboxylic acids, with benzoic acid giving the best results. For instance, using 20 mol % benzoic acid in the reaction between acyclic and cyclic alkyl ketones with 4-nitrobenzaldehyde in DMF:H_2O, the reaction time was reduced from 3 d to only 1.5 h, maintaining the enantioselectivity. This procedure enabled reactions at −20 °C with enhanced enantioselectivity (86–99%). Using benzoic acid as cocatalyst allowed the reaction to be conducted in water without addition of DMF.[9] The combination of (*S*)-Binam-L-Pro (10 mol %) and benzoic acid (20 mol %) in either DMF or pure water permitted the use of less reactive ketones such as α-(methylsulfanyl)acetone, giving mainly the *iso*-isomer with an excellent 93% ee (eq 5).[10] α-Alkoxy ketones give similar yields, regio-, diastereo-, and enantioselectivities to those achieved in the absence of acid, but in shorter reaction times (3–24 h).

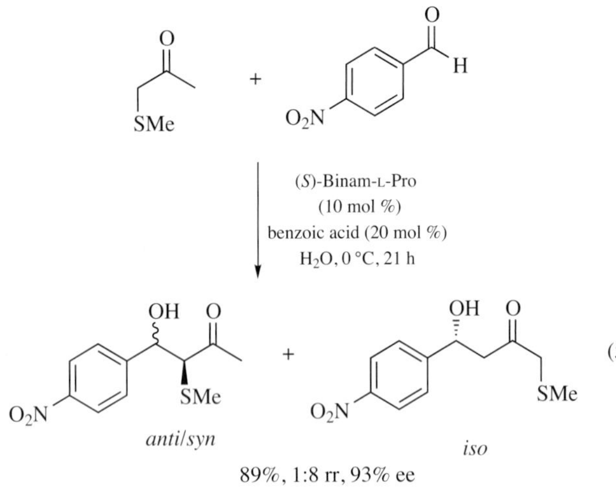

(5)

Other carboxylic acids can be used as cocatalyst. For instance, using acetic acid (10 mol %) as cocatalyst in toluene at −40 °C, the aldol products were obtained in 45–91% yield, 40–96% de, and 67–95% ee but longer reaction times being required (2–3 d).[11] Using a micellar agent stearic acid (20 mol %) as cocatalyst in water at 2 °C permitted reduction of the amount of nucleophilic ketone to 3 equiv, providing 61–99% yield and 58–93% ee in only 12 h.[12]

Using benzoic acid as cocatalyst, α-chloroacetone mainly produces the *anti*-isomer in moderate yields, but high diastereo- and enantioselectivities. These aldol products were easily converted into chiral (3*R*,4*S*)-*trans*-epoxides by treatment with triethylamine (eq 6).[13]

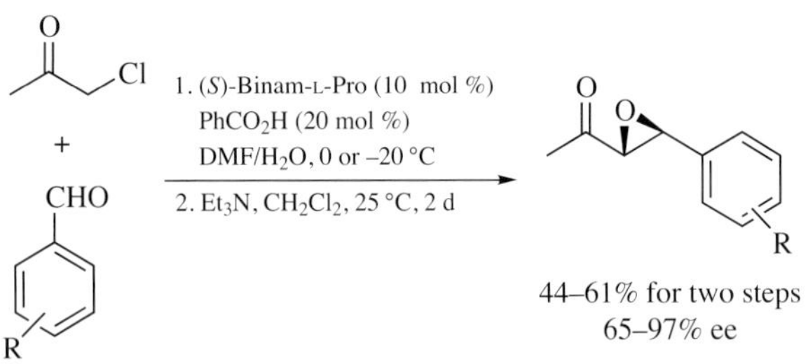

R = 4-NO_2, 2-NO_2, 3-NO_2, 4-NC, 4-MeO_2C, 4-F_3C, 2-Cl, 2-Br, H (6)

A further improvement involved the use of solvent-free conditions. Thus, mixing the reagents by simple magnetic stirring allowed reduction of nucleophile to only 2 equiv, catalyst loading of 5 mol %, and in some cases reduced reaction time. The aldol reaction between cycloalkyl, alkyl, and α-functionalized ketones with aldehydes gave the expected aldol products with similar yields, regio-, diastereo-, and enantioselectivities to those obtained in solution. Furthermore, aldehydes can also be used as nucleophiles providing, after in situ reduction of the aldol products, chiral 1,3-diols with moderate to good enantioselectivities mainly as *anti*-isomers (eq 7).[14]

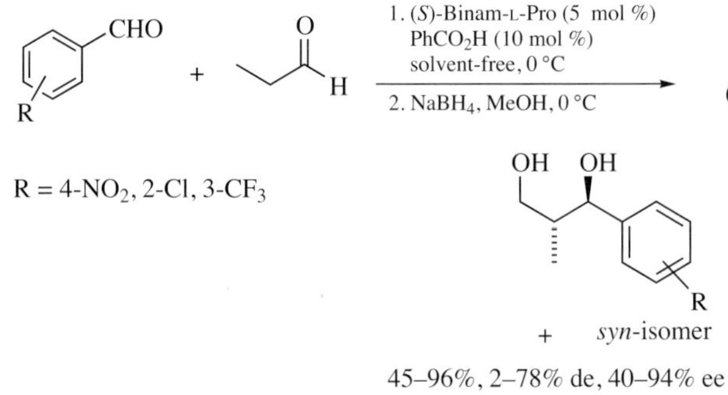

(7)

Enantioselective Mannich Addition. A single example of the use of (*S*)-Binam-L-Pro (10 mol%) as catalyst for this type of transformation has been reported.[7] Thus, the multicomponent[15] Mannich reaction between 4-nitrobenzaldehyde, 4-methoxyaniline, and acetone afforded the corresponding β-amino ketone with 51% ee at 25 °C (eq 8).

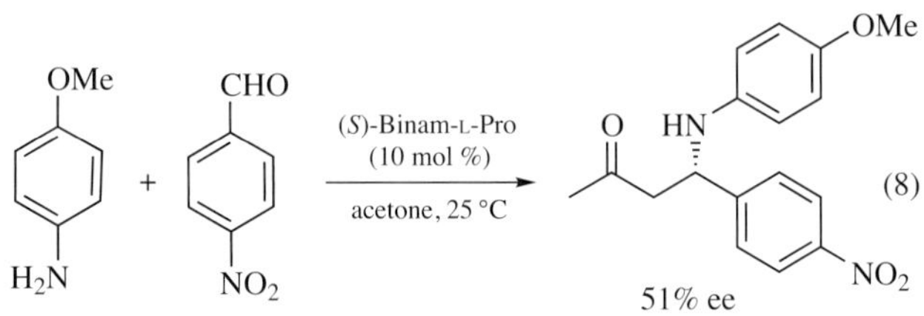

(8)

1. (a) Ramón, D. J.; Yus, M., *Curr. Org. Chem.*, **2004**, *8*, 149. (b) *Quaternary Stereocenters: Challenges and Solutions for Organic Synthesis*; Christoffers, J.; Baro, A., Eds.; Wiley-VCH Verlag: Weinheim, **2005**. (c) Ramón, D. J.; Yus, M., *Chem. Rev.* **2006**, *106*, 2126.
2. Xiong, Y.; Huang, X.; Gou, S.; Huang, J.; Wen, Y.; Feng, X., *Adv. Synth. Catal.* **2006**, *348*, 538.
3. (a) Berkessel, A.; Gröger, H., *Asymmetric Organocatalysis: From Biomimetic Concepts to Application in Asymmetric Synthesis*; Wiley-VCH: Weinheim, **2005**. (b) *Enantioselective Organocatalysis*; Dalko, P. I. Ed.; Wiley-VCH: Weinheim, **2007**. (c) Guillena, G.; Ramón, D. J., *Tetrahedron: Asymmetry* **2006**, *17*, 1465. (d) Guillena, G.; Ramón, D. J.; Yus, M., *Tetrahedron: Asymmetry* **2007**, *18*, 693.
4. Guillena, G.; Nájera, C.; Ramón, D. J., *Tetrahedron: Asymmetry* **2007**, *18*, 2249.
5. (a) Trost, B. M., *Science* **1991**, *254*, 1471. (b) Sheldon, R. A., *Pure Appl. Chem.* **2000**, *72*, 1233. (c) Trost, B. M., *Acc. Chem. Res.* **2002**, *35*, 695. (d) Trost, B. M.; Frederiksen, M. U.; Mathias, U.; Rudd, M. T., *Angew. Chem., Int. Ed.* **2005**, *44*, 6630. (e) Guillena, G.; Ramón, D. J.; Yus, M., *Angew. Chem., Int. Ed.* **2007**, *47*, 2358.
6. (a) Gryko, D.; Kowalczyk, B.; Zawadzki, L., *Synlett* **2006**, 1059. (b) Guillena, G.; Hita, M. C.; Nájera, C., *Tetrahedron: Asymmetry* **2006**, *17*, 729.
7. Guizzetti, S.; Benaglia, M.; Pignataro, L.; Puglisi, A., *Tetrahedron: Asymmetry* **2006**, *17*, 2754.
8. Guillena, G.; Hita, M. C.; Nájera, C., *Tetrahedron: Asymmetry* **2006**, *17*, 1027. (*Corrigendum: Tetrahedron: Asymmetry* **2007**, *18*, 1030).
9. Guillena, G.; Hita, M. C.; Nájera, C., *Tetrahedron: Asymmetry* **2006**, *17*, 1493. (*Corrigendum: Tetrahedron: Asymmetry* **2007**, *18*, 1031).
10. Guillena, G.; Hita, M.C.; Nájera, C.; *ARKIVOC* **2007**, *iv*, 260. (*Corrigendum: ARKIVOC* **2007**, *i*, 146).
11. Ma, G. N.; Zhang, Y. P.; Shi, M., *Synthesis* **2007**, 197.
12. Guizzetti, S.; Benaglia, M.; Raimondi, L.; Celentano, G., *Org. Lett.* **2007**, *9*, 1247.
13. Guillena, G.; Hita, M. C.; Nájera, C.; *Tetrahedron: Asymmetry* **2007**, *18*, 1272.

14. (a) Guillena, G.; Hita, M. C.; Nájera, C.; Viózquez, S. F., *Tetrahedron: Asymmetry* **2007**, *18*, 2300. (b) Guillena, G.; Hita, M. C.; Nájera, C.; Viózquez, S. F., *J. Org. Chem.* **2008**, *73*, 5933.
15. Ramón, D. J.; Yus, M., *Angew. Chem., Int. Ed.* **2005**, *44*, 1602.

Gabriela Guillena & Carmen Nájera
University de Alicante, Alicante, Spain

(*R*)-1,1′-Bi-2,2′-naphthol

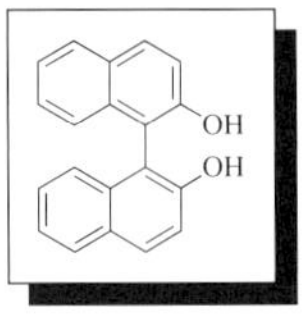

[18531-94-7] $C_{20}H_{14}O_2$ (MW 286.33)
InChI = 1S/C20H14O2/c21-17-11-9-13-5-1-3-7-15(13)19(17)20-16-8-4-2-6-14(16)10-12-18(20)22/h1-12,21-22H
InChIKey = PPTXVXKCQZKFBN-UHFFFAOYSA-N

(chiral ligand and auxiliary[1])

Alternate Name: BINOL.
Physical Data: mp 208–210 °C; $[\alpha]^{21}$ +34° (c = 1, THF).
Solubility: sol toluene, CH_2Cl_2, $EtNO_2$.
Form Supplied in: white solid; widely available.
Preparative Method: racemic 1,1′-bi-2,2′-naphthol (BINOL) is most conveniently prepared by the oxidative coupling reaction of 2-naphthol in the presence of transition metal complexes (eq 1).[2] The resolution of racemic BINOL with cinchonine may be performed via the cyclic phosphate (eq 2).[3] An alternative procedure to provide directly optically active BINOL is the oxidative coupling of 2-naphthol catalyzed by Cu^{II} salt in the presence of chiral amines (eq 3).[4] The best procedure uses (+)-amphetamine as the chiral ligand and provides BINOL in 98% yield and 96% ee. Above 25 °C the Cu^{II}/(+)-amphetamine/(*S*)-BINOL complex precipitates while the more soluble Cu^{II}/(+)-amphetamine/(*R*)-BINOL complex is slowly transformed into the former complex. 9,9′-Biphenanthrene-10,10′-diol has also been prepared in 86% yield and with 98% ee by a similar asymmetric oxidative coupling of 9-phenanthrol in the presence of (*R*)-1,2-diphenylethylamine.[5]

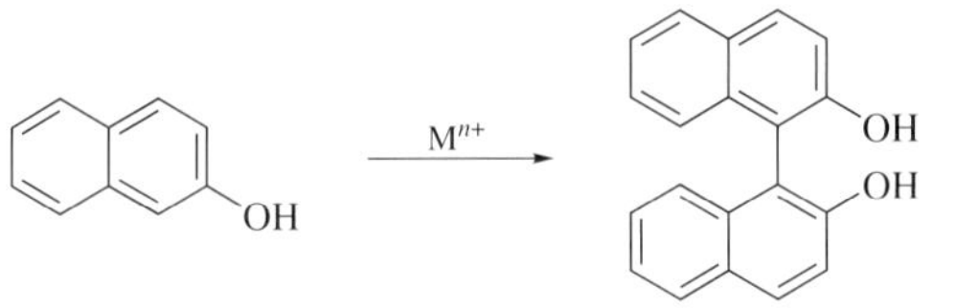

(1)

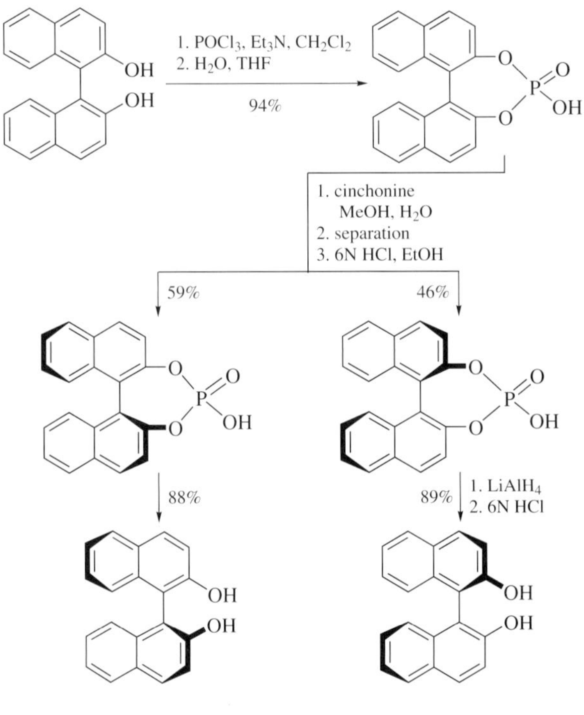

(2)

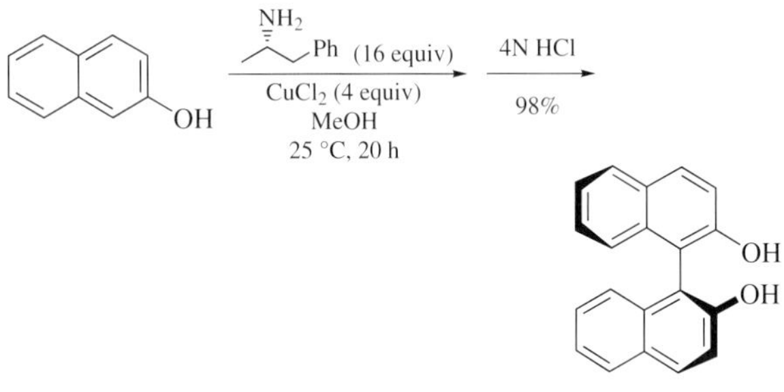

(3)

96% ee

Handling, Storage, and Precautions: keep tightly closed, store in a cool dark place; on heating in butanol at 118 °C for 24 h, BINOL lost ~1% of its optical rotation; at 100 °C for 24 h in dioxane–1.2 N HCl, BINOL lost 56% of its rotation; after 24 h at 118 °C in butanol–0.7 N KOH, BINOL lost 20% of its rotation.

Original Commentary

Koichi Mikami & Yukihiro Motoyama
Tokyo Institute of Technology, Tokyo, Japan

Hydrocarboxylation. The cyclic phosphate resolved according to eq 2 can be used as the chiral ligand in the palladium(II) catalyzed asymmetric hydrocarboxylation of arylethylenes.[6] The 1-arylpropanoic acid is obtained regiospecifically with high enantioselectivity (91% ee, eq 4).

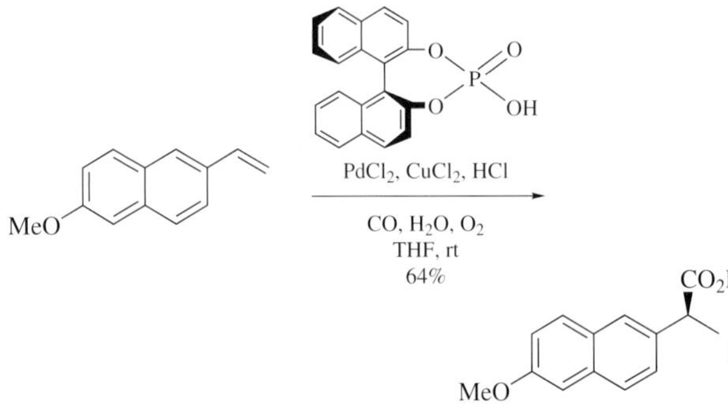

(4)

91% ee

Crown Ethers. BINOL-derived crown ethers have been reported.[7] Crown ethers containing 3,3′-disubstituted BINOL derivatives are particularly effective for asymmetric synthesis. Thus complexes of these crown ethers (e.g. 18-Crown-6) with Potassium Amide or Potassium tert-Butoxide catalyze asymmetric Michael additions. The reaction of methyl 1-oxo-2-indancarboxylate with methyl vinyl ketone with the 3,3′-dimethyl-BINOL–crown ether/KO-*t*-Bu complex gives the Michael product in 48% yield and with 99% ee (eq 5).[8]

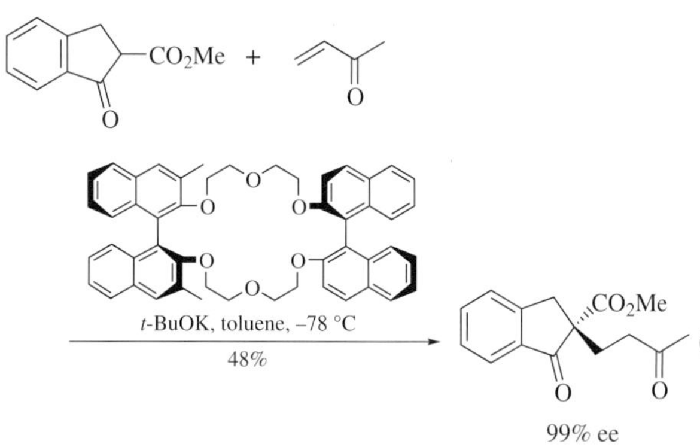

(5)

Polymerization. Complexes of BINOL-derived crown ethers with KO-*t*-Bu or BuLi have been used as initiators in the asymmetric polymerization of methacrylates.[9] Thus optically active polymers are obtained with 80–90% isotacticity. Complexes of BINOL with Diethylzinc or $CdMe_2$ also initiate the asymmetric polymerization of heterocyclic monomers.[10] The chiral initiators selectively polymerize one enantiomer to give an optically active polymer. The unreacted monomer is recovered with 92% ee at 67% conversion in the polymerization of methylthiirane with (*S*)-BINOL/Et_2Zn.

Ullmann Coupling Reaction. Axially dissymmetric biaryls have been synthesized via an intramolecular Ullmann coupling reaction of BINOL-derived aryl diesters (eq 6).[11] In the example shown, the functionalized binaphthyl is obtained with high ee after hydrolysis of the intermediate 12-membered cyclic diester.

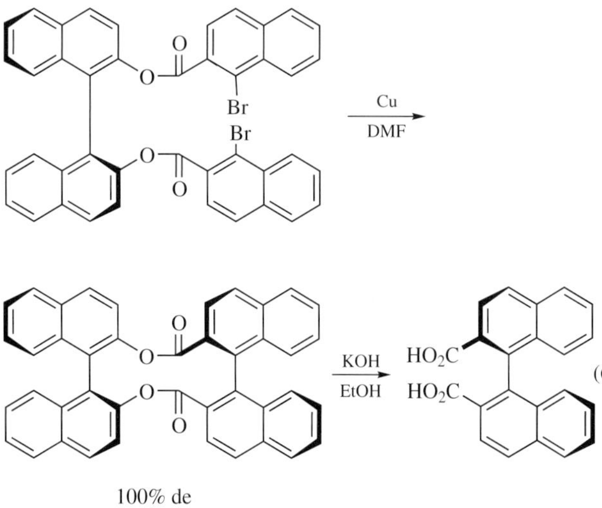

(6)

Reduction of Prochiral Ketones. BINOL has been used as the chiral ligand of the reagent BINAL-H for asymmetric reduction.[12] The reagent reduces prochiral unsaturated ketones to the corresponding secondary alcohols in up to 90% yield and >90% ee (eq 7); (*R*)-BINAL-H leads to the (*R*)-alcohols while (*S*)-BINAL-H gives the (*S*)-alcohols.

Addition Reactions of Chiral Titanium Reagents to Aldehydes. The preparation and use of the BINOL-derived titanium complexes in the enantioselective synthesis of some benzhydrols (>90% ee) have been reported (eq 8).[13]

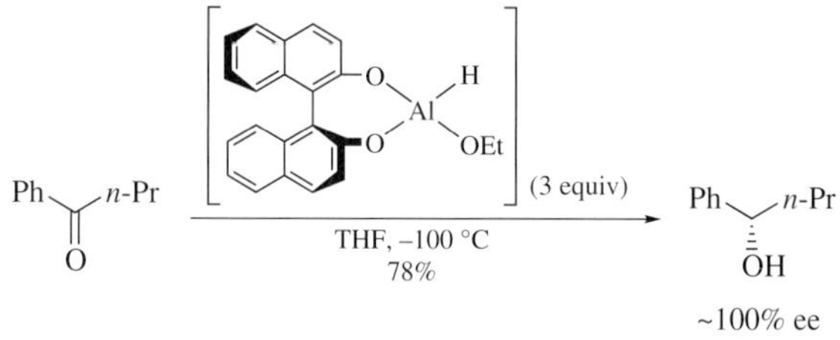

(7)

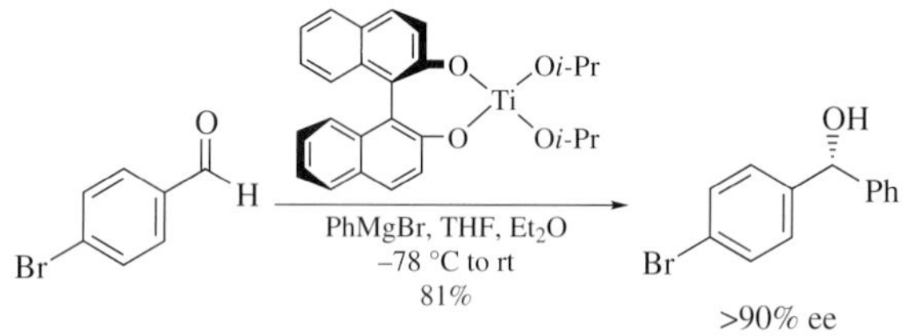

(8)

Cyanosilylation. The chiral titanium reagent, prepared from the lithium salt of BINOL with $TiCl_4$, has been used as a catalyst for the asymmetric addition of cyanotrimethylsilane to aldehydes.[14] In the example shown, the cyanohydrin is obtained with 82% ee (eq 9).

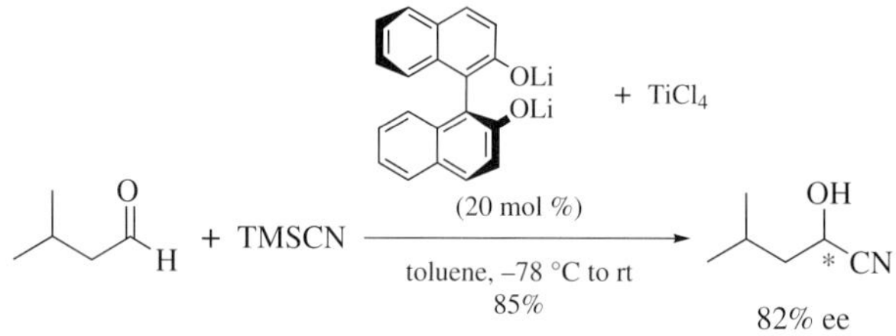

(9)

Diels–Alder Reactions. BINOL and its derivatives are used as the chiral ligand of chiral Lewis acid complexes for enantioselective Diels–Alder cycloadditions. BINOL-$TiCl_2$, prepared from the lithium salt of BINOL with Titanium(IV) Chloride, also catalyzes the enantioselective Diels–Alder reaction of cyclopentadiene with methacrolein (eq 10).[14,15] The *exo* adduct is obtained as the major product (56% yield), but with low enantioselectivity (16% ee). More recently, BINOL-TiX_2 (X = Br or Cl) have been prepared in situ from diisopropoxytitanium dihalides ((*i*-PrO)$_2$$TiX_2$, X = Br[16] or Cl[17]) with BINOL in the presence of molecular sieves (MS 4A).[16] The Diels–Alder reaction of methacrolein with 1,3-dienol derivatives can be catalyzed by BINOL-TiX_2. The *endo* adducts are obtained in high enantioselectivity (eq 11).[18] Asymmetric catalytic Diels–Alder reaction of naphthoquinone derivatives as the dienophile (eq 12)[18] can in principle provide an efficient entry to the asymmetric synthesis of anthracyclinone aglycones. The reaction of the 5-hydroxynaphthoquinone with 1-acetoxy-1,3-diene in the presence of MS-free BINOL-$TiCl_2$ (10 mol%) provides the corresponding Diels–Alder product in high chemical yield and with high enantioselectivity (76–96% ee).[18b] The Diels–Alder product is also obtained by the use of 1 equiv of 3,3′-diphenyl-BINOL/borane complex (eq 13); the structure of the intermediate has been proposed.[19]

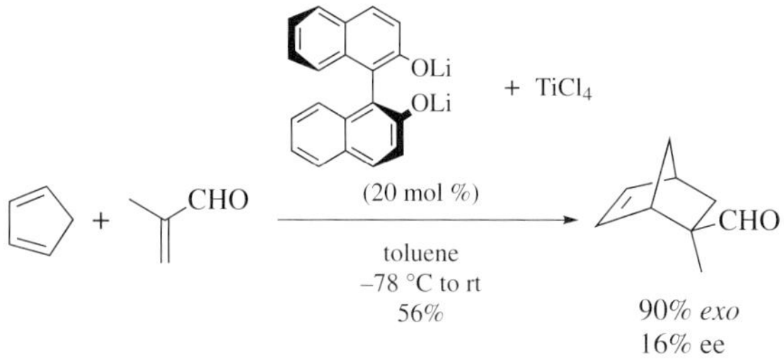

(10)

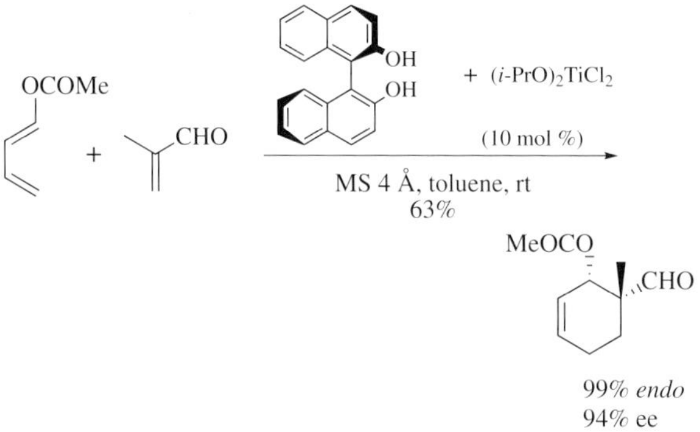

(11)

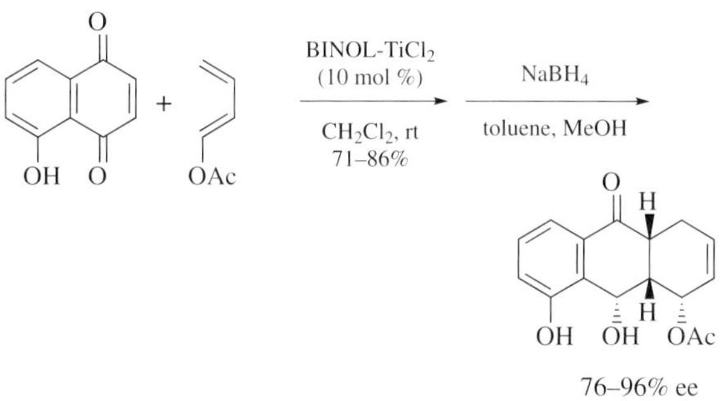

(12)

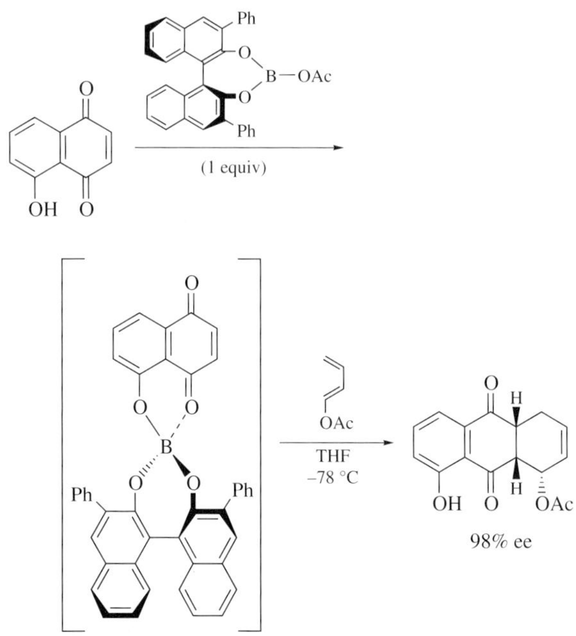

(13)

3,3′-Diphenyl-BINOL-derived chiral aluminum reagents are prepared in situ by addition of Ethylaluminum Dichloride or Diethylaluminum Chloride to 3,3′-diphenyl-BINOL. These chiral aluminum reagents promote the enantioselective Diels–Alder reaction of cyclopentadiene with the oxazolidone dienophile (eq 14).[20] *Endo* products are obtained with a high level of asymmetric induction (>90% ee); however, a stoichiometric amount of the Lewis acid is required. The preparation and use of a C_3 symmetric BINOL-derived boronate has been reported (eq 15).[21] BINOL-B(OAr)$_3$ complexes have recently been developed for the asymmetric Diels–Alder reaction with imines (eq 16).[22]

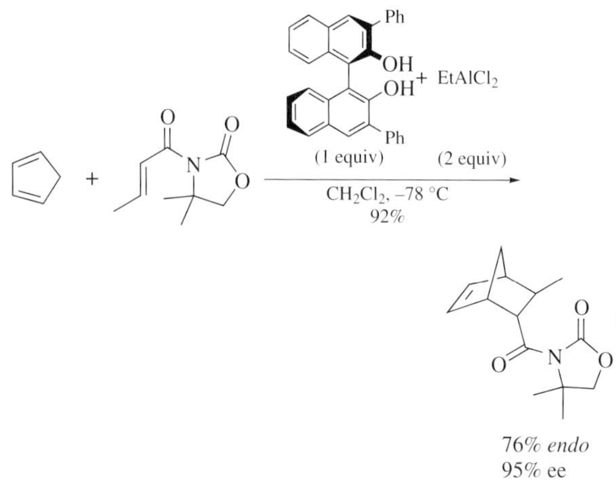

(14)

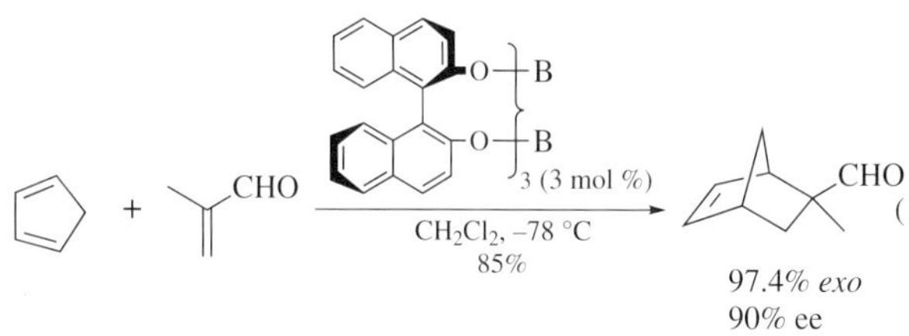

(15)

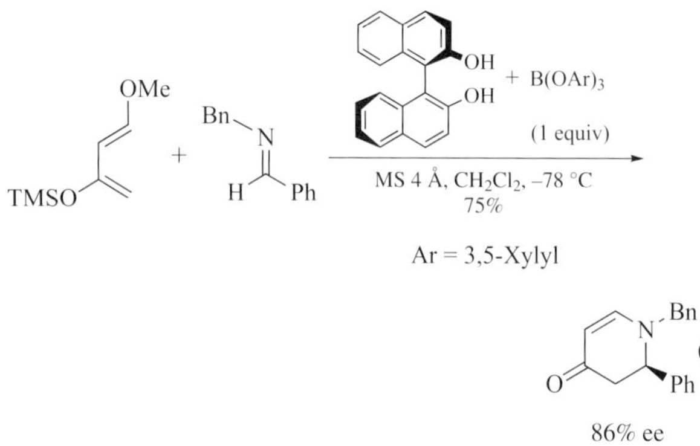

(16)

Hetero Diels–Alder Reaction. Modified BINOL-derived organoaluminum reagents have been used in the asymmetric hetero Diels–Alder reaction of aldehydes (eq 17). The dihydropyrones are obtained with high *cis* diastereoselectivity and enantioselectivity.[23] The hetero Diels–Alder reaction of glyoxylates proceeds smoothly with methoxydienes using BINOL-TiCl$_2$ as a catalyst to give the *cis* product in high enantiomeric excess (eq 18).[18b, 24] The hetero Diels–Alder product thus obtained can be readily converted to the lactone portion of HMG-CoA inhibitors such as mevinolin or compactin.

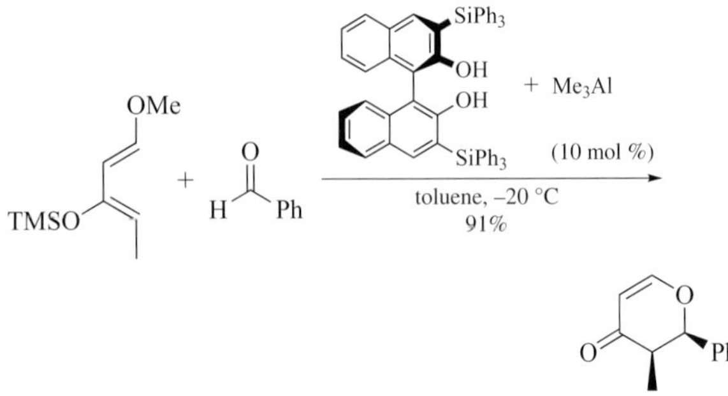

(17)

95% *cis*
95% ee

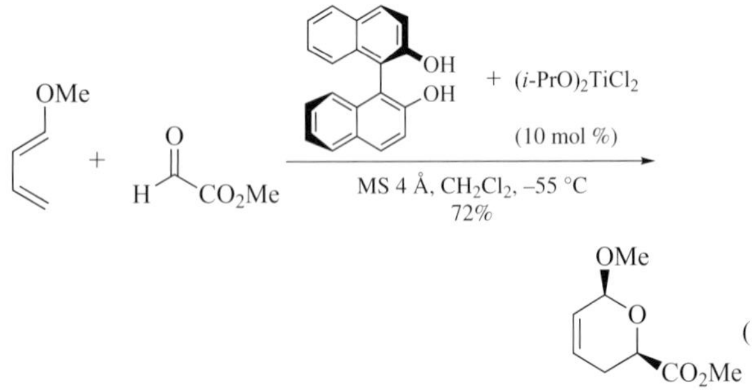

(18)

87% *cis*
96% ee

Carbonyl–Ene Reaction. BINOL-TiX_2 reagent exhibits a remarkable level of asymmetric catalysis in the carbonyl–ene reaction of prochiral glyoxylates, thereby providing practical access to α-hydroxy esters.[16,25] These reactions exhibit a remarkable positive nonlinear effect (asymmetric amplification) that is of practical and mechanistic importance (eq 19).[26] The desymmetrization of prochiral ene substrates with planar symmetry by the enantiofacial selective carbonyl–ene reaction provides an efficient solution to remote internal asymmetric induction (eq 20).[27] The kinetic resolution of a racemic allylic ether by the glyoxylate–ene reaction also provides efficient access to remote but relative asymmetric induction (eq 21).[27] Both the dibromide and dichloride catalysts provide the (2*R*,5*S*)-*syn* product with ≥97% diastereoselectivity and 95% ee.

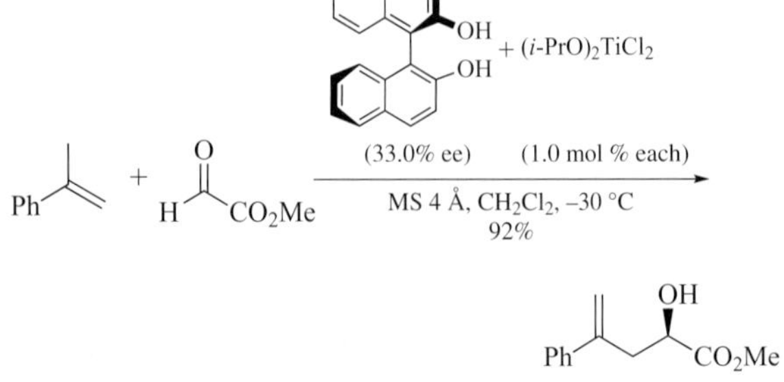

(19)

91.4% ee

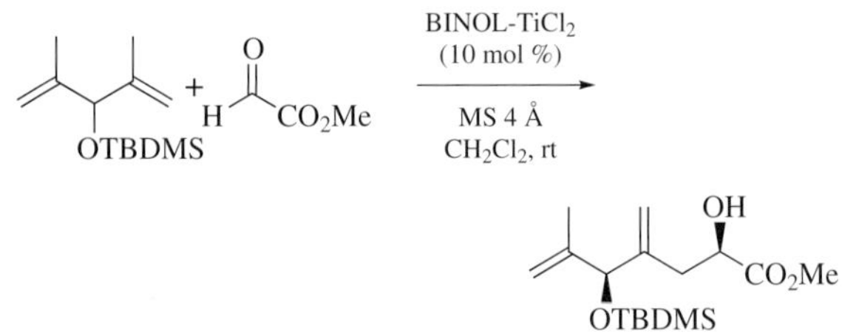

(20)

>99% *syn*
97% ee

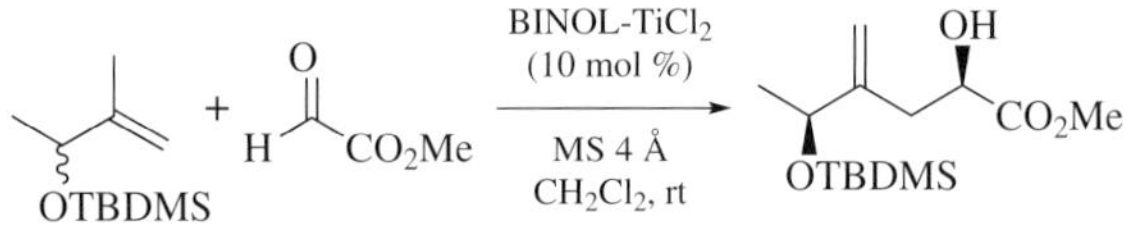

(21)

97% *syn*
>95% ee

Ene Cyclization. An intramolecular (3,4)-ene reaction of unsaturated aldehydes has been accomplished with the BINOL-derived zinc reagent.[28] Cyclization of 3-methylcitronellal with at least 3 equiv of BINOL-Zn reagent afforded the *trans*-cyclohexanol in 86% yield and 88% ee (eq 22). Asymmetric ene cyclizations of type (2,4) are also catalyzed by the BINOL-derived titanium complexes ((*R*)-BINOL-TiX_2, X = ClO_4 or OTf), modified by the perchlorate or trifluoromethanesulfonate ligand. The 7-membered cyclization of type 7-(2,4) gives the oxepane in high ee (eq 23).

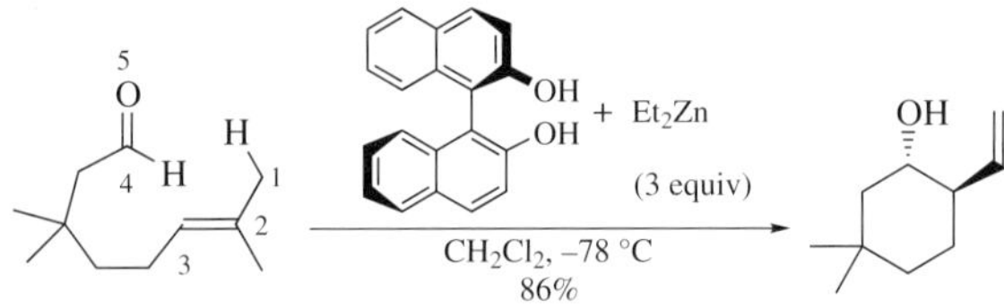

(22)

88% ee

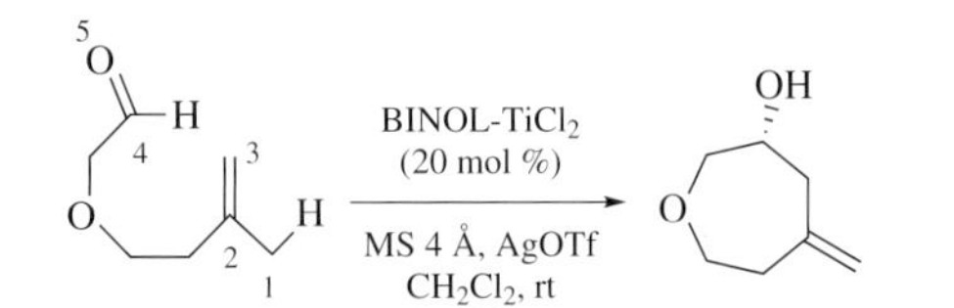

(23)

92% ee

Cationic Cyclization. A cationic cyclization of BINOL-derived neryl ether has been accomplished with an organoaluminum triflate catalyst.[29] Limonene is obtained in 54% yield and 77% ee (eq 24).

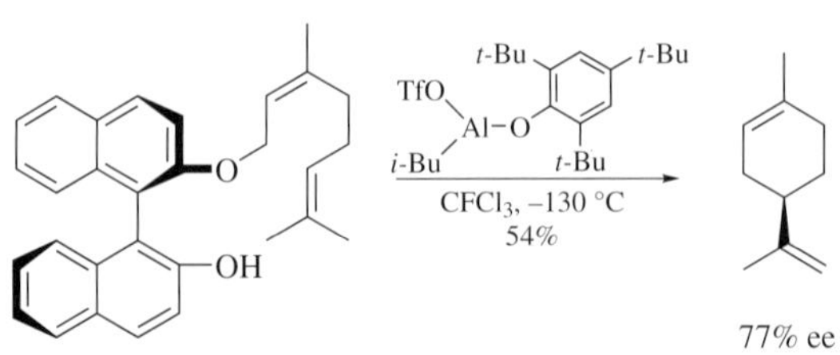

(24)

77% ee

Mukaiyama Aldol Condensation. The BINOL-derived titanium complex BINOL-TiCl_2 is an efficient catalyst for the

Mukaiyama-type aldol reaction. Not only ketone silyl enol ether (eq 25),[30] but also ketene silyl acetals (eq 26)[31] can be used to give the aldol-type products with control of absolute and relative stereochemistry.

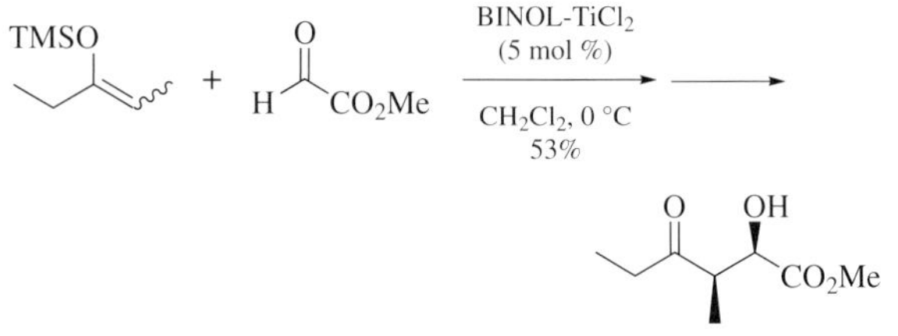

(25)

98% *syn*
99% ee

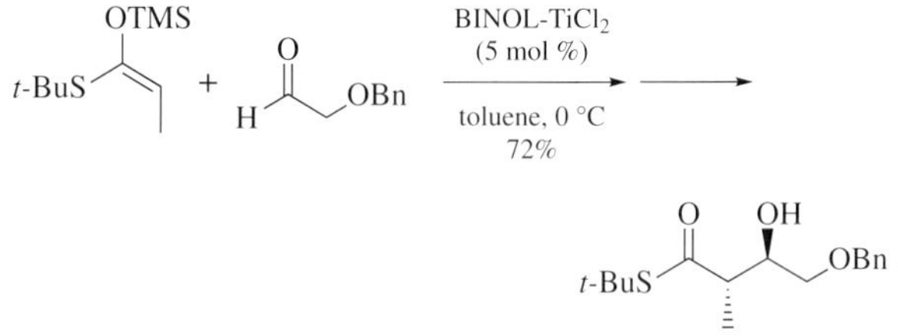

(26)

92% *anti*
90% ee

Nitro-aldol Condensation. A BINOL-derived lanthanide complex has been used as an efficient catalyst for the nitro-aldol reaction (eq 27).[32] Interestingly enough, the presence of water and LiCl in the reaction mixture is essential to obtain the high level of asymmetric induction and chemical yield.

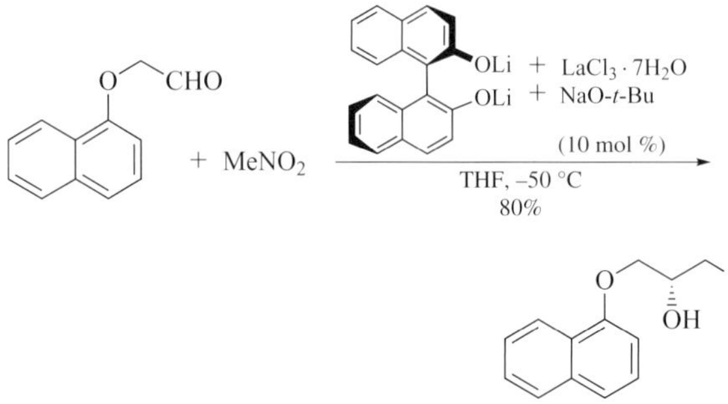

(27)

92% ee

Carbonyl Addition of Allylic Silanes and Stannanes. BINOL-$TiCl_2$ reagent also catalyzes the asymmetric carbonyl addition reaction of allylic silanes and stannanes.[33] Thus the addition reaction of glyoxylate with (*E*)-2-butenylsilane and -stannane proceeds smoothly to give the *syn* product in high enantiomeric excess (eq 28). The reaction of aliphatic and aromatic aldehydes with allylstannane is also catalyzed by BINOL-$TiCl_2$ or BINOL-Ti(O-*i*-Pr)$_2$ to give remarkably high enantioselectivity.[34]

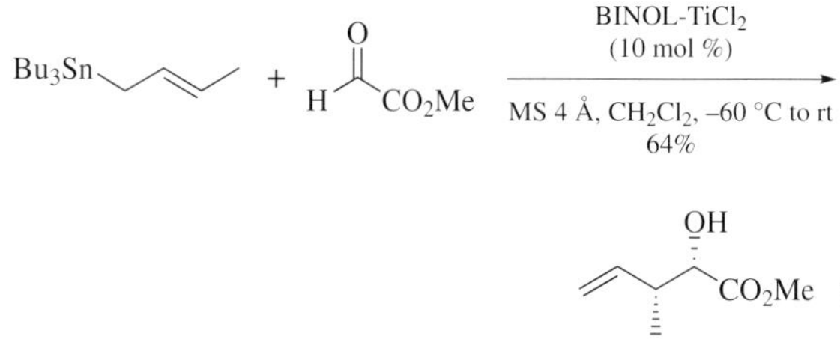

(28)

84% *syn*
86% ee

Claisen Rearrangements. A modified BINOL-derived aluminum reagent is an effective chiral catalyst for asymmetric Claisen rearrangement of allylic vinyl ethers (eq 29).[35] The use of vinyl ethers with sterically demanding C-3 substituents is necessary for the high level of asymmetric induction.

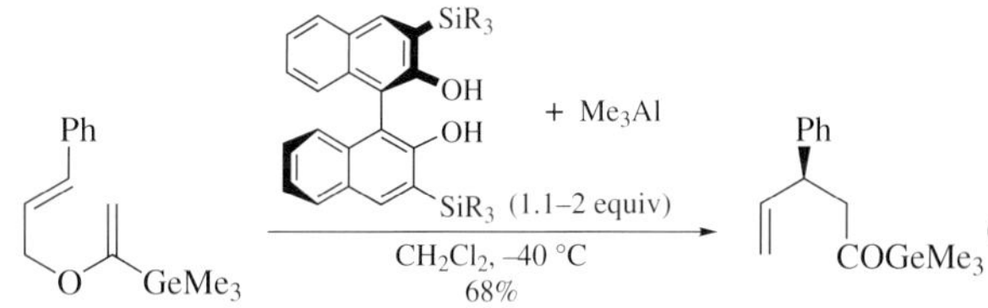

(29)

$R_3 = t\text{-}BuPh_2$

93% ee

Alkylation of BINOL-derived Ester Enolates. The diastereoselective alkylation of BINOL-derived arylacetates affords the optically active 2-arylalkanoic acids (eq 30).[36]

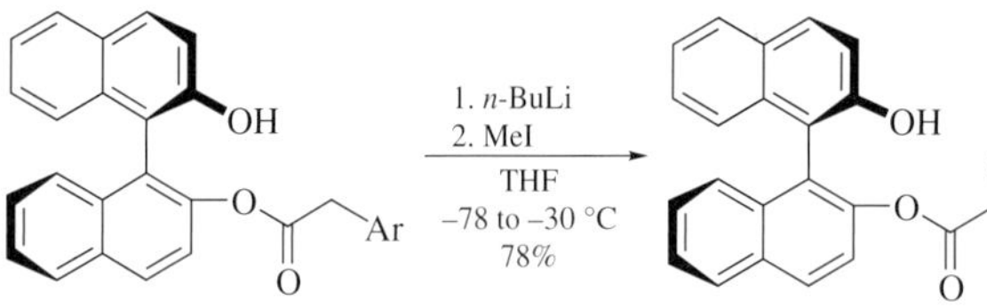

(30)

Ar = 3-pyridyl

90% de

First Update

Jean Michel Brunel
Université Paul Cézanne:, Aix-Marseille, Marseille, France

General. The chemistry of BINOL has been reviewed.[37]

Chemical Resolution of Racemic BINOL. An efficient, practical, and inexpensive chemical resolution method involving the formation of stable chiral organophosphorus diastereomers has been described by Brunel and Tang.[38–40] Complete separation of the two diastereomers, by subsequent oxidation with 30% hydrogen peroxide and reduction with $LiAlH_4$, permit to obtain enantiomerically pure (*R*)- and (*S*)-BINOL in 81% and 85% overall yield, respectively (eq 31).

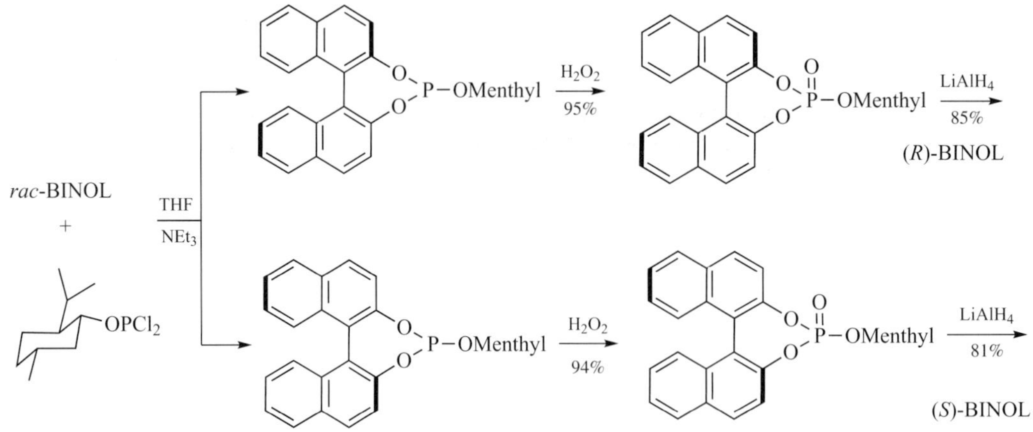

(31)

The preparation of enantiomerically pure BINOL **1** has also been achieved from *rac*-BINOL via a cyclic borate ester formed from the reaction of a racemic 1,1′-binaphtholborane derivative with cinchonine in THF (eq 32).[41–43]

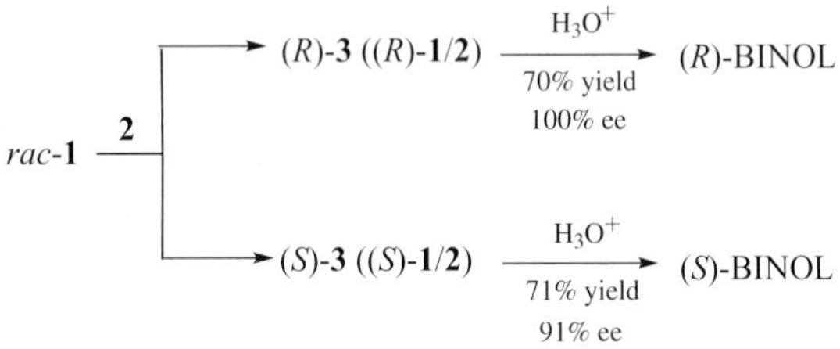

(32)

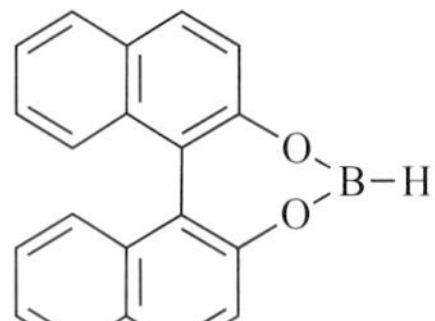

rac-**1**

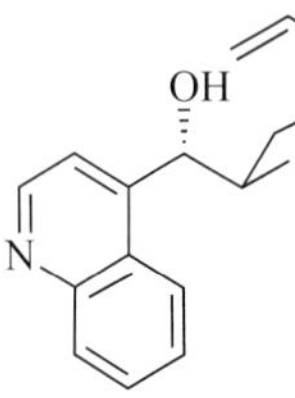

2

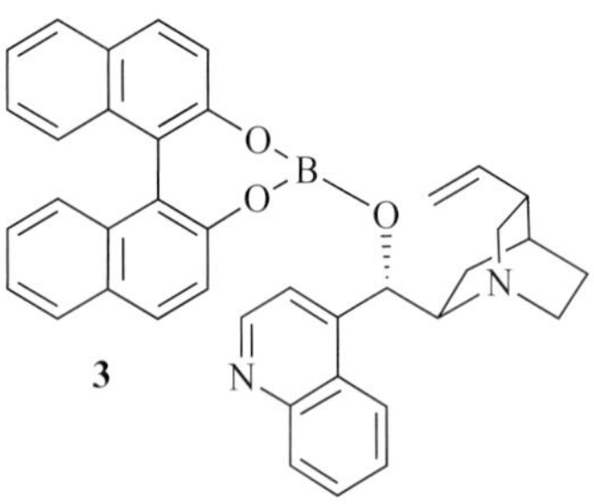

3

Enantioselective Synthesis of BINOL by Oxidative Coupling of 2-Naphthol. Different chiral oxovanadium(IV) complexes have been prepared for the asymmetric catalytic oxidative coupling of 2-naphthols with enantioselectivities of up to 80% ee (eq 33).[44–49]

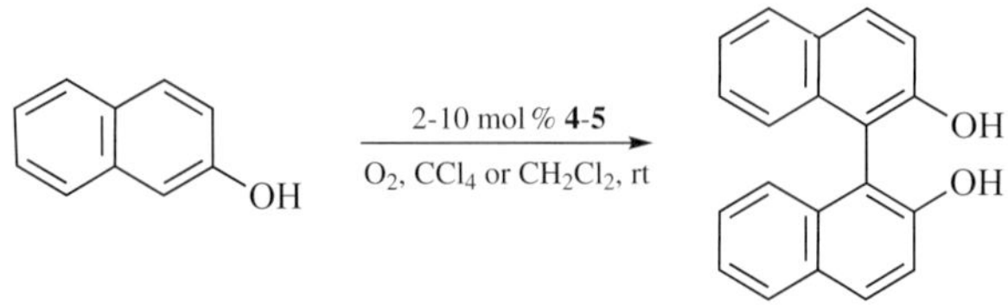

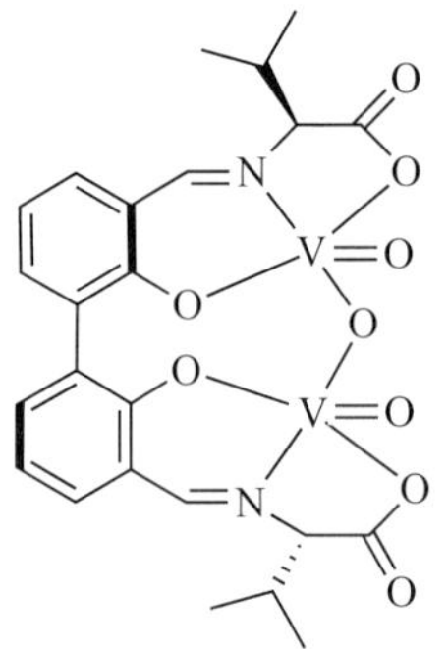

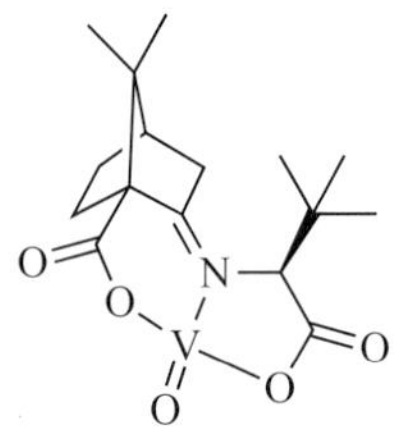

(33)

4 62% yield 90% ee

5 65% yield 87% ee

Reduction of Prochiral Ketones. Synthesis of alcohols can be achieved in high yields and with ee's up to 74% using a chiral aluminum complex generated in situ from aluminum isopropoxide and (*R*)-BINOL (eq 34).[50,51]

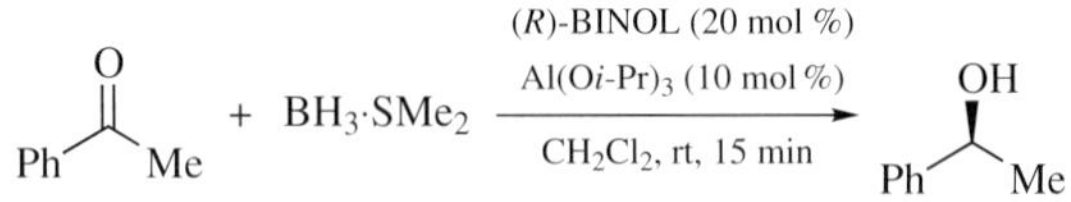

(34)

97% yield
74% ee

Asymmetric catalytic hydrosilylation of prochiral ketones with $HSi(OEt)_3$ using (*R*)-BINOL-Ti(O*i*-Pr)$_2$ complex has been also developed leading to moderate enantioselectivities of up to 55% ee depending on the nature of the ketones (eq 35).[52]

Reduction of Prochiral Imines. An optically active lithium-alkoxide-catalyzed asymmetric reduction of imines with trimethoxyhydrosilane, afforded the expected amines in moderate ee (up to 72%) (eq 36).[53]

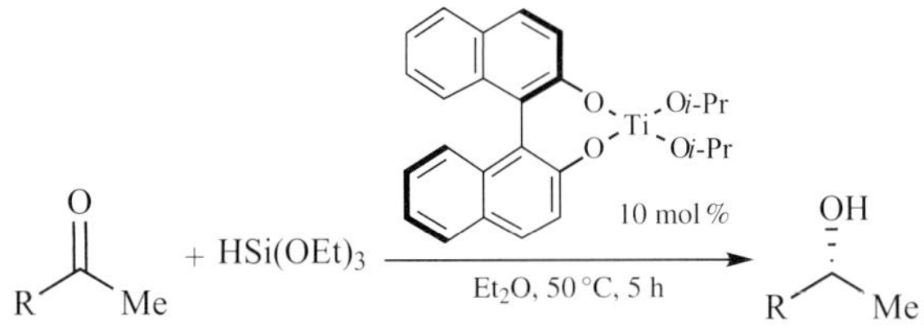

(35)

R =		
n-C_4H_{10}	50% yield	10% ee (N.D.)
m-NO_2Ph	98% yield	39% ee (*R*)
Ph	98% yield	55% ee (*R*)

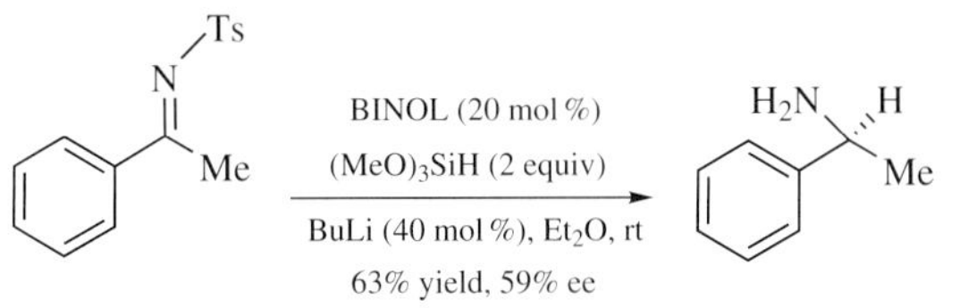

(36)

Epoxidation of Olefins. Asymmetric epoxidation of enones using hydroperoxides such as *tert*-butyl hydroperoxide (TBHP) can be achieved using lanthanum or ytterbium-modified BINOL derivatives as catalysts.[54–57] This catalytic process has been improved using a La-BINOL-triphenyl arsine oxide (Ph_3As=O) complex, when it affords optically active epoxy ketones with broad generality in up to 99% yield and more than 96% ee (eq 37).[58–61]

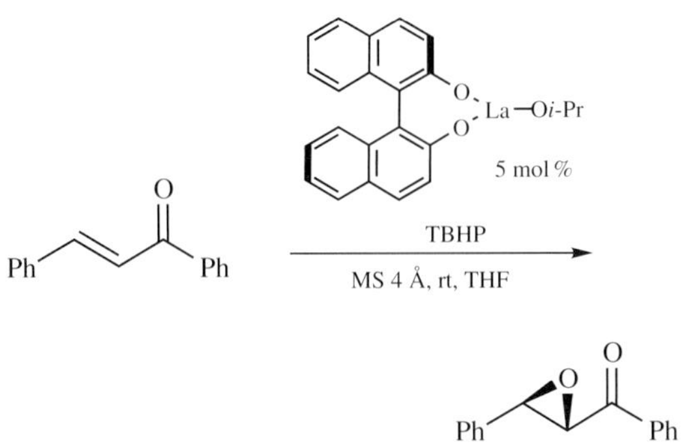

(37)

99% yield
96% ee

Oxidation of Sulfides. The catalytic asymmetric oxidation of sulfides to chiral sulfoxides in moderate yield with *tert*-butyl hydroperoxide was achieved using a titanium complex produced in situ from a titanium alkoxide and (*R*)-BINOL (eq 38).[62–68]

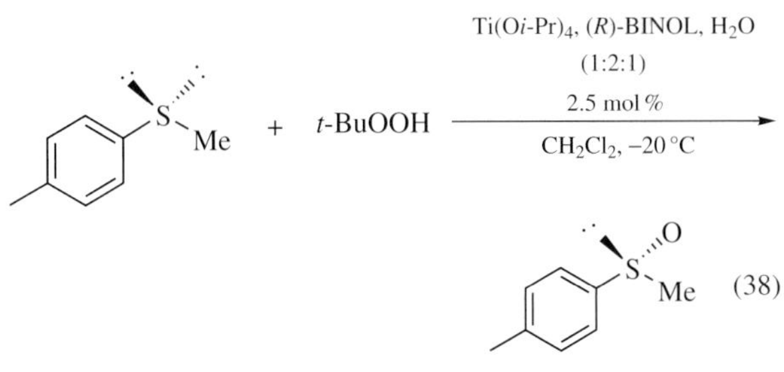

(38)

67% yield
93% ee

The highest enantioselectivities (up to 93% ee) were obtained with 2.5 mol % of catalyst. The presence of more than 1 equiv of water with respect to sulfide was essential for the enantioselective oxidation and to prolong the catalytic activity of the catalyst.

Aldolization. Addition of ketene silyl acetals of esters or thioesters to various aldehydes catalyzed by 10 mol % of a complex prepared from (*R*)-BINOL and Ti(O*i*-Pr)$_4$ has been performed, leading to the silyl ether of the expected aldol product in good yield and enantiomeric excess (eq 39).[69] Introduction of alcohols, phenols, or acids as additives often resulted in an increase in the optical yield of the reaction.[70,71]

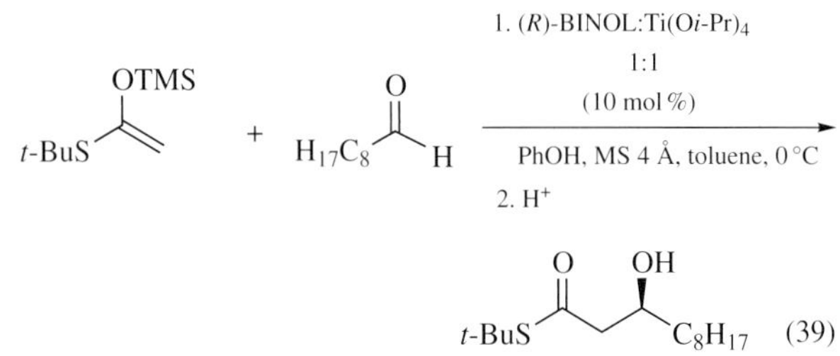

(39)

61% yield
96% ee

Vinylogous and Homoaldol Reaction. BINOL complexes have also been used in various vinylogous aldol reactions involving acyclic[72] or cyclic enol ethers. Thus, in the presence of BINOL-Ti catalyst, siloxydioxine and benzaldehyde led to the corresponding alcohol with very good yield and enantiomeric excess (eq 40).[73–82]

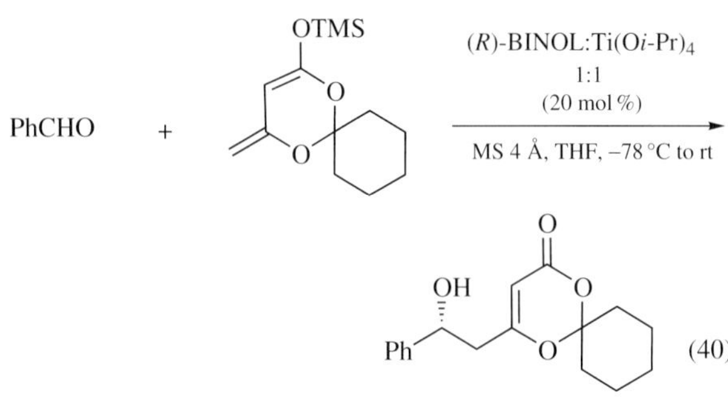

(40)

93% yield
92% ee

2-Trialkylsilyloxyfurans were also added to aldimines in the presence of a BINOL-Ti(O*i*-Pr)$_4$ catalyst: yields were good with diastereomeric (*erythro/threo*) and enantiomeric excesses up to 92% and 54%, respectively (eq 41).[83]

Allylation. Synthesis of allenyl alcohols from 1-trimethylsilyl-4-tri-*n*-butylstannyl-but-2-yne can be achieved in good yields and excellent enantiomeric excesses (eq 42).[58,84–89]

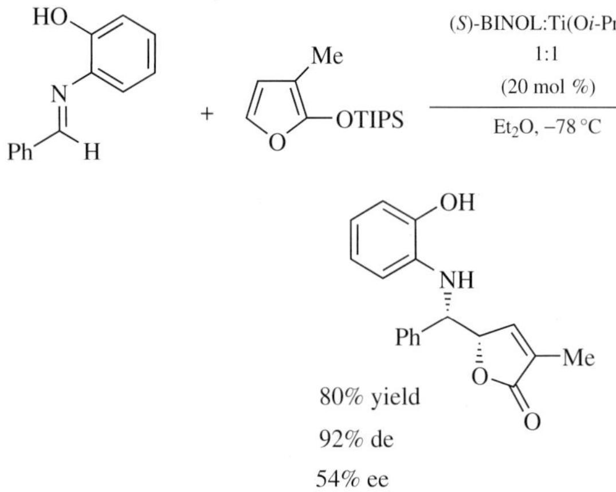

80% yield
92% de
54% ee

(41)

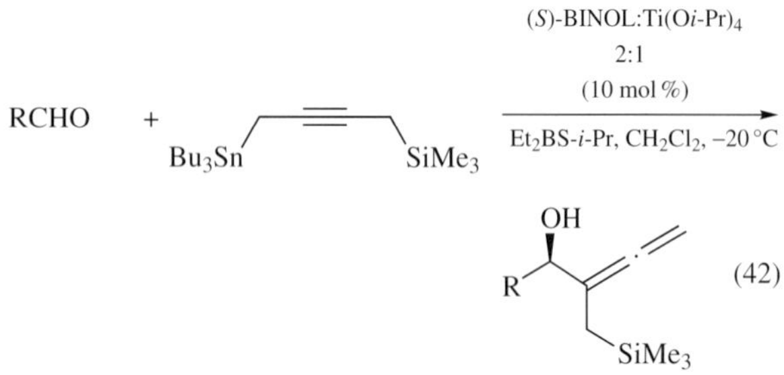

(42)

R = *n*-C$_6$H$_{13}$, 60% yield, 90% ee
R = PhCH$_2$CH$_2$, 62% yield, 92% ee

Bedeschi et al. studied a Lewis acid catalyst obtained from Zr(O*i*-Pr)$_4$-*i*-PrOH and BINOL. The results were very satisfying for aromatic aldehydes when good yields and ee's, along with a dramatic reduction in reaction times, were obtained (eq 43).[90,91]

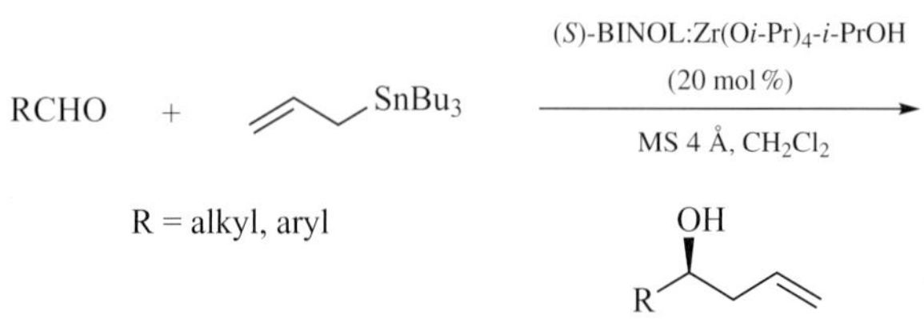

R = alkyl, aryl

(43)

58-84% yield
up to 87% ee

Although 4 Å MS are not necessary for the catalyst preparation, their inclusion is essential to achieve good enantioselectivity and in a reasonable time.[92] It was also demonstrated that the reaction could also be performed with ee's up to 85–87% with a catalyst obtained from ZrCl$_2$(THF)$_2$, BINOL, and calix[4]arenes (eq 44).[93]

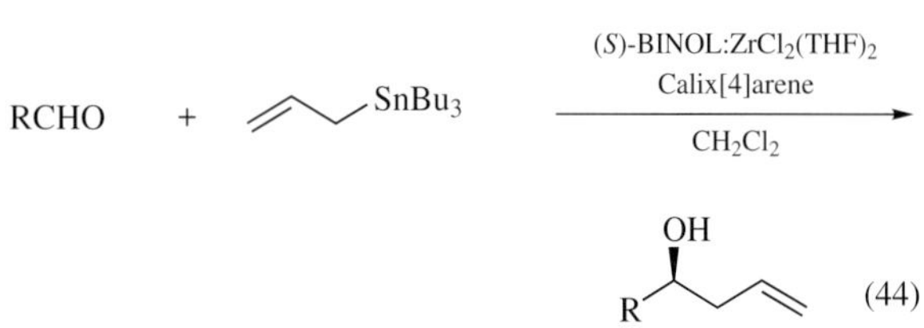

(44)

R = *n*-C$_7$H$_{15}$, 65% yield, 96% ee
R = Ph, 85% yield, 85% ee

Alkynylation. Optically active propargylic alcohols are versatile precursors for the synthesis of many chiral organic compounds. Nevertheless, a single and practical method to make chiral propargylic alcohols from aromatic aldehydes has been developed in the presence of a titanium alkoxide catalyst prepared in situ from Ti(O*i*-Pr)$_4$ and (*R*)-BINOL leading to excellent enantioselectivity (96% ee, eq 45).

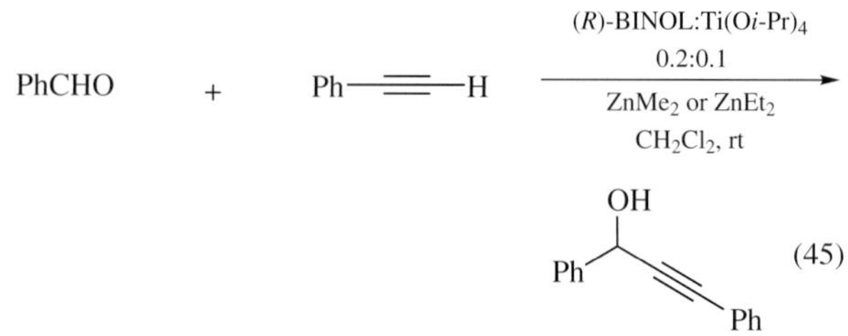

(45)

77% yield
96% ee

The scope of this reaction has been extended by the demonstration that the BINOL-Ti(O*i*-Pr)$_4$ catalyst system is also highly enantioselective for the phenylacetylene addition to aliphatic aldehydes as well as α,β-unsaturated aldehydes (eq 46).[98,99]

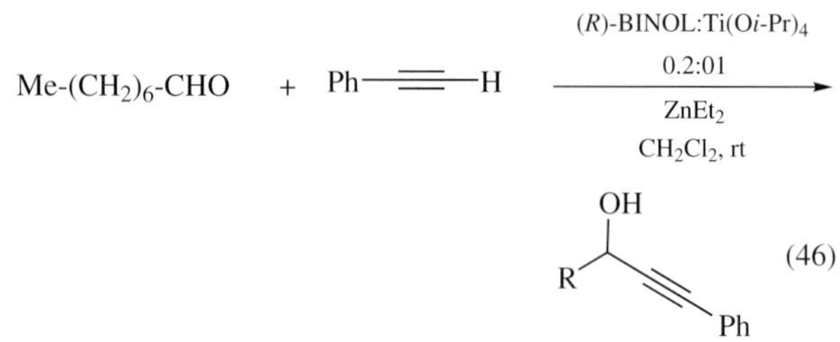

(46)

70% yield
93% ee

This reaction was also successfully applied to phenylacetylene additions to ketones. In this case, most of the chiral tertiary propargyl alcohols generated from aromatic ketones were obtained with up to 85% ee at room temperature (eq 47).[100]

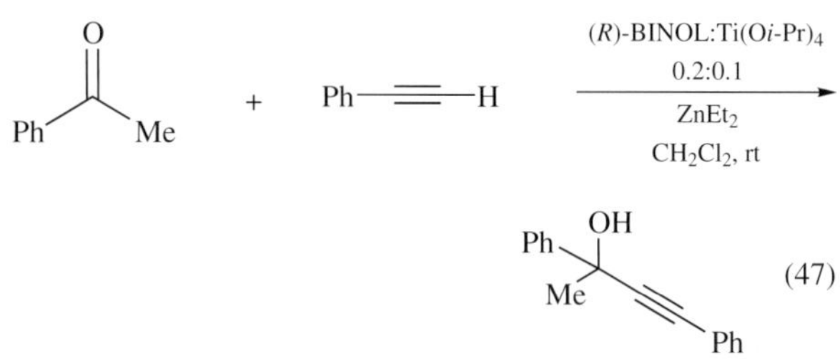

(47)

67% yield
85% ee

Aza-Diels-Alder Reaction. Yamamoto reported the preparation of chiral Lewis acids from (*R*)-BINOL and triarylborates, B(OAr)$_3$.[101] Enantiomeric excesses, up to 90%, were obtained for the aza-Diels-Alder cycloaddition between Danishefsky's dienes and aldimines (eq 48).[102–104]

A significant improvement in the asymmetric aza-Diels-Alder reaction was made with the introduction of chiral lanthanide Lewis acids.[105] The chiral catalyst is prepared from Yb(OTf)$_3$, (*R*)-BINOL, and DBU, and is used in 10–20 mol % in the presence of 1 equiv of an additive (eq 49).

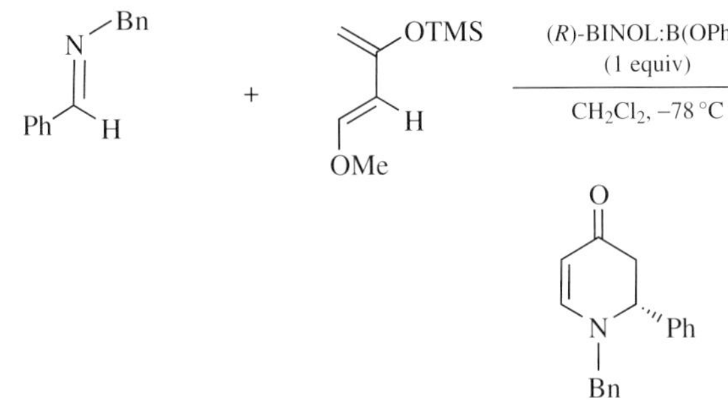

(48)

82% ee
75% yield

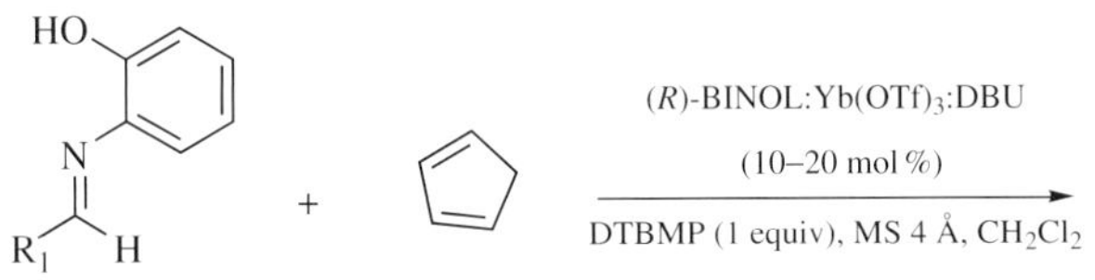

R_1 : α-Naphthyl
DTBMP : 2,6-di-*tert*-butyl-4-methylpyridine

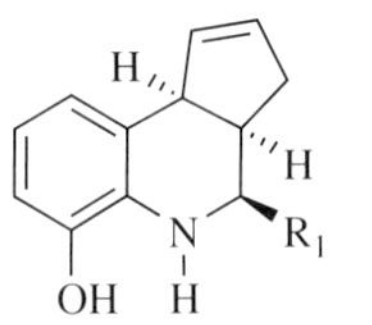

(49)

69% yield
68% ee

Addition of Dialkylzinc and Trialkylaluminium to Aldehydes and Ketones. Nakai[106] and Chan[107] independently reported the asymmetric alkylation of aldehydes with diethylzinc catalyzed by BINOL-Ti complexes (obtained from (*S*)-BINOL and $Ti(Oi\text{-}Pr)_4$). Nakai used a large excess of $Ti(Oi\text{-}Pr)_4$ (vs. BINOL) and 3 equiv of Et_2Zn; thus, the typical proportions were RCHO:Et_2Zn:BINOL:$Ti(Oi\text{-}Pr)_4$ = 1:3:0.2:1.2 (eq 50).

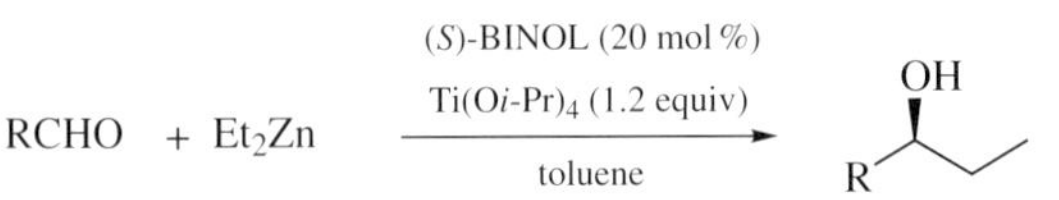

(50)

R = *n*-C_8H_{17}, 94% yield, 86% ee
R = Ph-CH=CH, 97% yield, 82% ee

Trimethylsilylcyanation of Carbonyl Compounds and Imines. Kagan has investigated the addition of TMSCN to aldehydes catalyzed by lithium salts of BINOL converting a range of aldehydes to their corresponding cyanohydrins with ee's of up to 56% (eq 51).[108]

An enantioselective Strecker-type reaction of imines with Et_2AlCN-BINOL has been also examined and α-aminonitriles were obtained in good yields with good enantioselectivities of up to 70% ee in the reaction of *N*-benzylidenebenzhydrylamine (eq 52).[109,110]

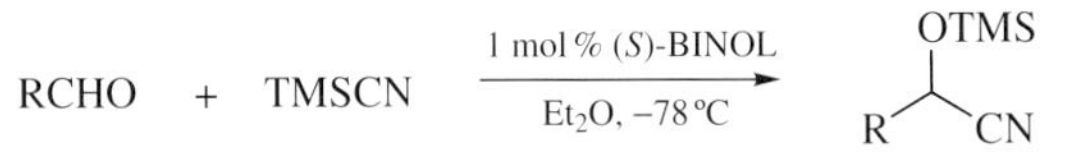

(51)

R = Ph, 96% yield, 56% ee
R = Cyclohexyl, 94% yield, 30% ee

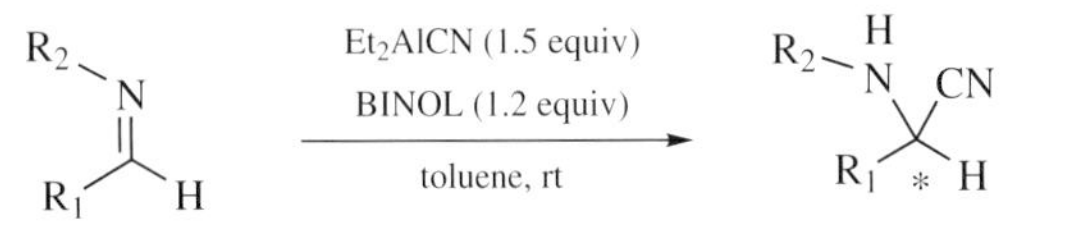

(52)

R_1, R_2 = alkyl, aryl

30–96% yield
0–70% ee

Michael Addition. Albis(binaphthoxide) complex (ALB), an heterobimetallic asymmetric complex, has been successfully used in a catalytic asymmetric Michael addition reaction with ee of up to 99%. The efficiency of the catalyst was improved in presence of 1 equiv of base such as BuLi or KO-*t*-Bu, which accelerates the reaction rate (eq 53).[111–113]

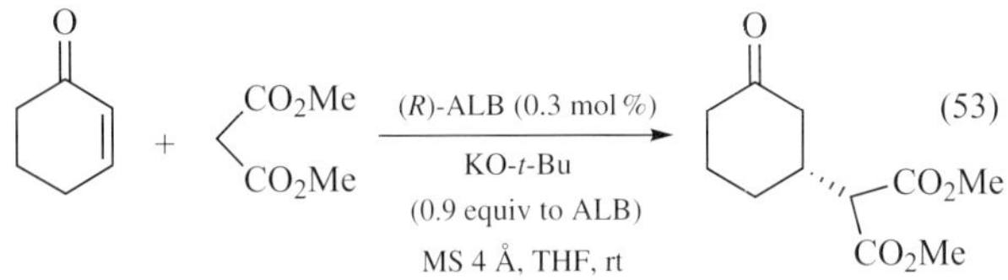

(53)

94% yield
99% ee

(*R*)-ALB =

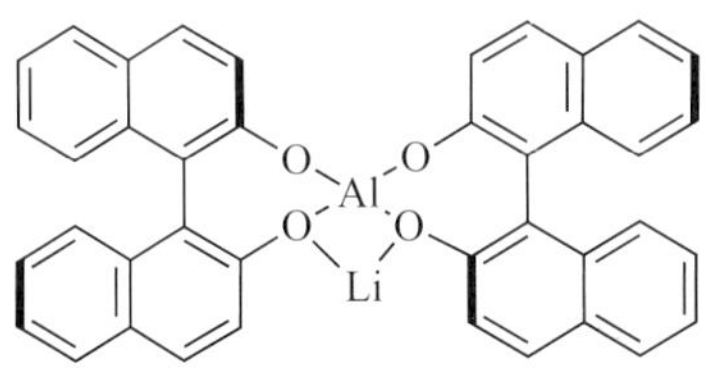

A calcium-BINOL catalyst has been also developed for asymmetric Michael addition reactions of enones and enals in good yields but with moderate enantioselectivities up to 87% ee (eq 54).[114]

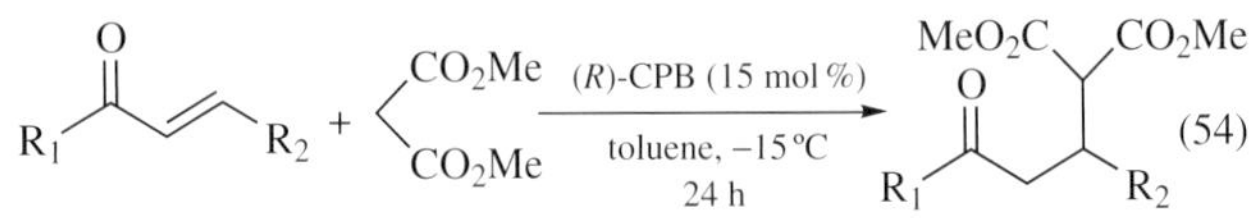

(54)

R_1, R_2 = aryl, alkyl

0–88% ee

(*R*)-CPB =

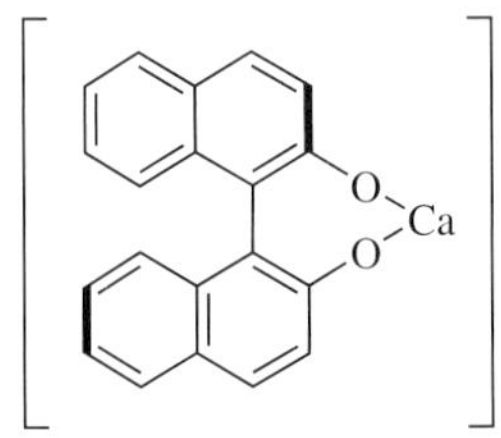

2 KCl

Enantioselective Michael additions of α-nitro esters with α,β-unsaturated ketones were carried out in the presence of a catalytic amount of (*R*)-ALB with enantiomeric excesses varying from 5 to 80% ee depending on the nature of the substrate (eq 55).[115,116]

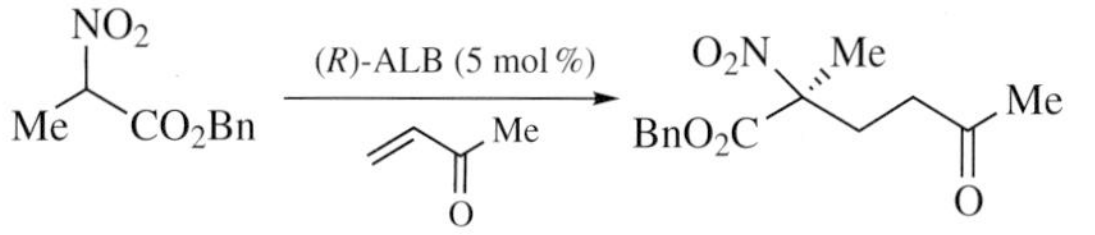

(55)

81% yield
80% ee

Synthesis of α-Hydroxy- and Aminophosphonates. $LaLi_3$tris(binaphthoxide) catalyst (LLB) prepared from $LaCl_3 \cdot 7H_2O$, (*R*)-BINOL dilithium salt, and NaO-*t*-Bu, is effective for the hydrophosphonylation of various aldehydes in up to 95% ee (eq 56).[117–123]

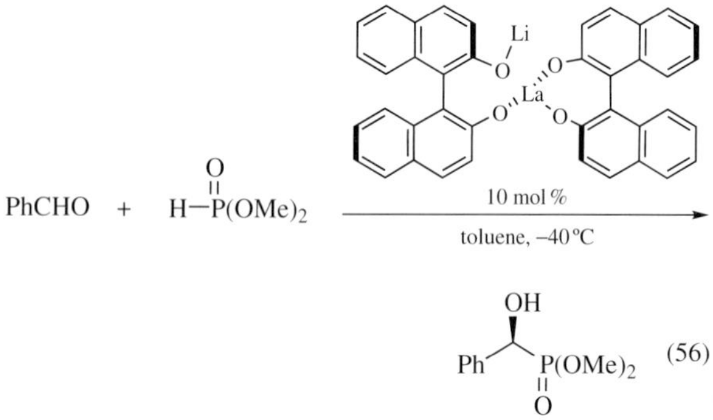

(56)

90% yield
85% ee

***meso*-Epoxide Ring Opening Reaction.** Chiral β-amino alcohols were prepared in enantiomeric excesses up to 80% by ring opening of *meso*-epoxides catalyzed by $Yb(OTf)_3$ and (*R*)-BINOL (eq 57).[124,125]

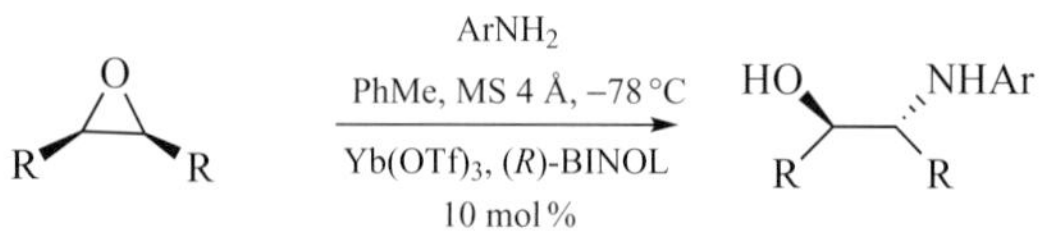

(57)

R = Me, Ph, $-(CH_2)_3-$
Ar = Ph, *p*-BrC_6H_4, *o*-EtC_6H_4

64–98% yield
up to 80% ee

Using 10–20 mol % of gallium heterobimetallic multifunctional complexes such as (*R*)-GaLB catalyst, a variety of epoxides were smoothly cleaved with phenolic oxygen nucleophiles in 67–93% ee (eq 58).[126,127]

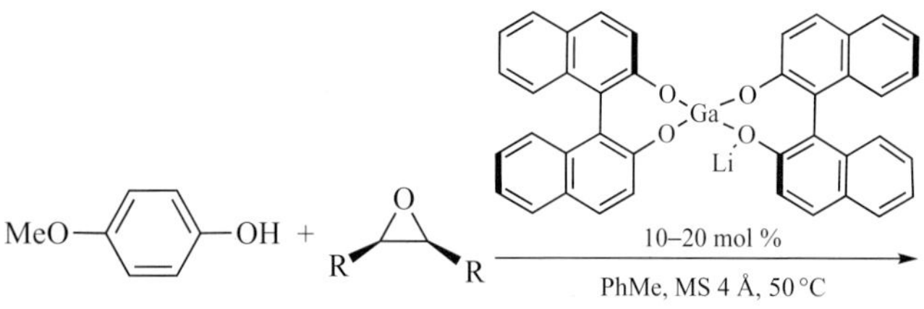

R = $-(CH_2)_3-$, $-(CH_2)_4-$, $-(CH_2)_5-$

(58)

31–75% yield
up to 93% ee

Second Update

Xavier Mollat du Jourdin
BASF SE, Ludwigshafen, Am Rhein, Germany

Reduction of Prochiral Imines. As a result of the development of novel catalysts for hydrogenation, this transformation was revisited in the past few years. Catalytic asymmetric hydrogenation of imines was originally performed with biphosphine ligands, but, contrary to 30 years of experience, it was discovered that the more economical monodentate ligands could successfully promote this reaction, but with moderate enantioselectivity.[128,129] On the basis of this preliminary work, Feringa and coworkers developed a highly enantioselective catalytic hydrogenation of substituted quinolines[130] (eq 59) and acyclic imines (eq 60),[131] using iridium-based catalysis and (*S*)-BINOL derivative (*S*)-PipPhos as a chiral ligand to afford the corresponding amines with good to excellent enantiomeric excess (71–99% ee, 1.5–26 h for full conversion).

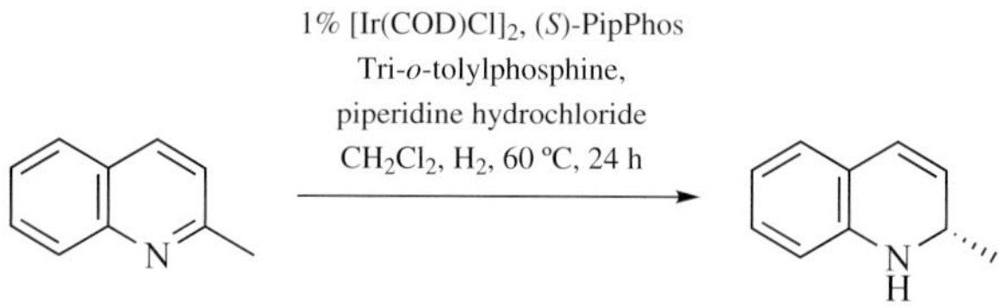

(59)

100% conversion, 89% ee (*S*)

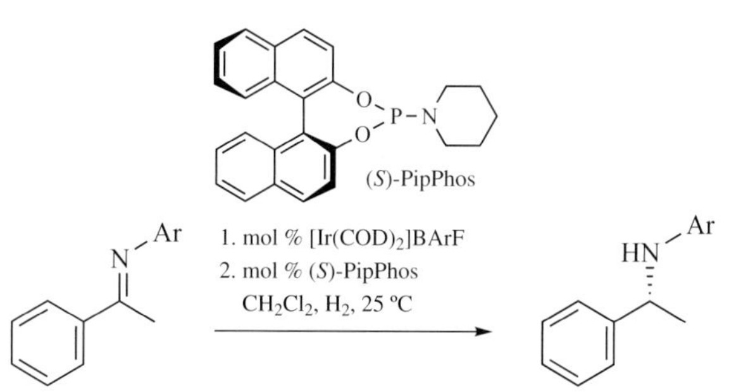

(60)

Ar = Ph, 4-MeO-Ph,
2-MeO-Ph, 3,5-$(CH_3)_2$-Ph
3,4,5-$(CH_3)_3$-Ph,
3,5-$(CH_3)_2$-4-OMe-Ph

1.5–26 h for full conversion
71–99% ee (*R*)

Reduction of Prochiral Ketones. Similar to imines, ketones, α-keto esters, and β-keto esters could undergo a similar transition metal-based hydrogenative process. Most known catalysts used for this purpose need a bidentate chiral ligand and a directing/coordinating group (i.e., carbonyl moiety) to allow good enantioselectivity. Thus, ketones bearing no neighboring esters are usually reduced with lower enantioselectivity similarily to α-keto esters and β-keto esters.[132] More recently, the usage of cheaper and more flexible monodentate ligands was further developed. For instance, β-keto esters could be reduced with a Ru catalyst and a phosphine derivative of (*S*)-BINOL, whose synthesis was previously reported[133] and which could also be utilized for the hydrogenation of itaconic esters, enamides, and α-amino acid precursors.[134] In the aforementioned report, β-hydroxy esters could be obtained in up to 99% yield with excellent ee (95%, eq 61).[135] In a similar fashion, α-keto esters could be hydrogenated with Ru catalyst to deliver α-hydroxy esters with medium enantioselectivity (65–78% ee, eq 62).[136] Finally, prochiral ketones could be hydrogenated with a Ru catalyst in the presence of a (*S*)-BINOL derivative to provide the corresponding secondary alcohols with good to excellent conversion (68–100% after 4–24 h) and good to excellent enantioselectivity (54–99% ee, eq 63).[137] However, a chiral diamine ligand was necessary to boost the enantioselectivity.

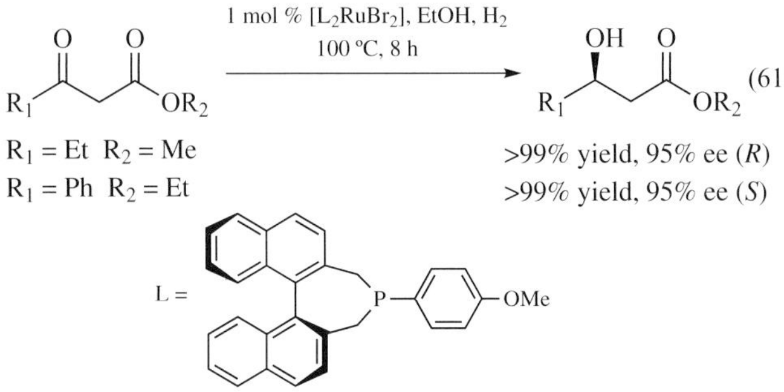

(61)

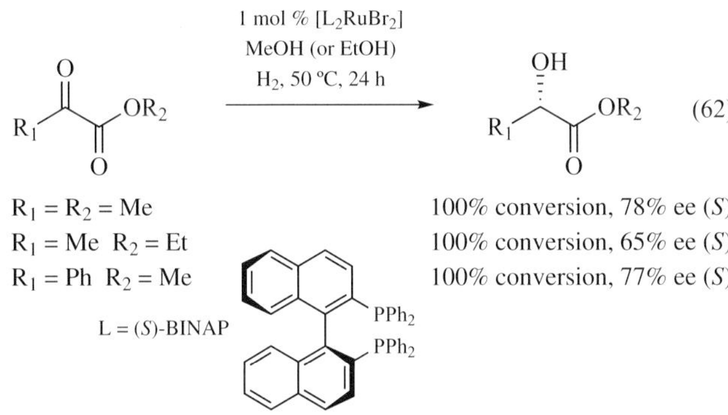

(62)

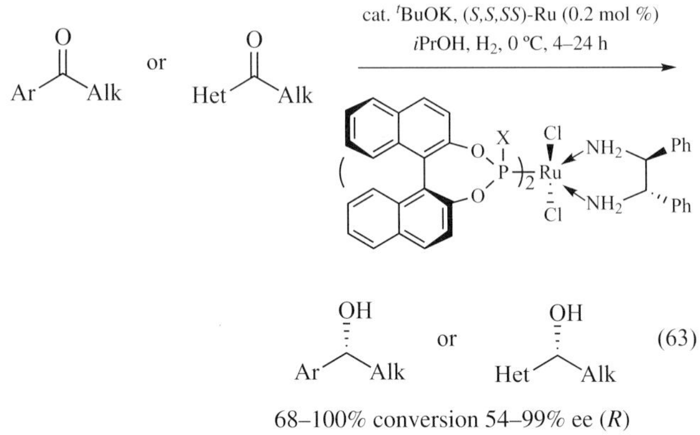

(63)

Addition to Pyridine *N*-Oxides. Almqvist and coworkers reported in 2009 the first reliable and scalable synthesis of substituted piperidines via Grignard addition. Although high yielding and highly regio- and diastereoselective, only a racemic mixture of products was obtained.[138] The first asymmetric addition of Grignard reagents to pyridine *N*-oxides was reported by the same group in 2013 and required noncatalytic amount of (*R*)-BINOL. Pyridine *N*-oxides could be used as carbonyl surrogates to allow asymmetric addition at the C2 position with (*R*)-BINOL ligand and thus to provide enantiomerically enriched C2-functionalized *N*-hydroxypiperidines in various yield depending on the Grignard reagent (30–94% yield) with moderate enantioselectivity (33–80% ee, eq 64).[139]

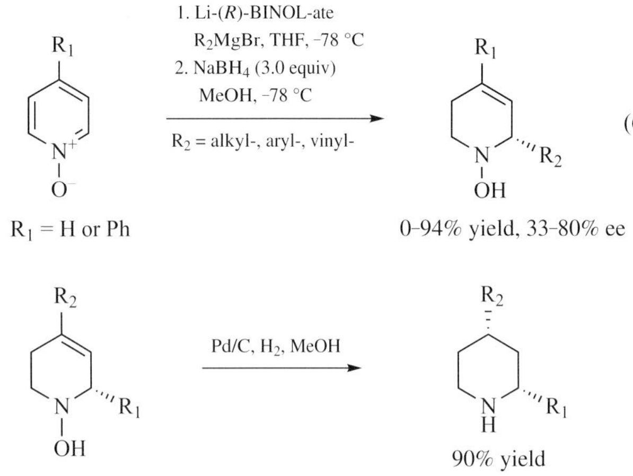

(64)

Aldol Reaction. BINOL ligands have been extensively used to promote aldol or vinylogous aldol reactions (i.e., Mukaiyama aldol condensation) and thus provide β-hydroxy carbonyls.[30,31] Such reactions were already described earlier in this report. More recently, the synthesis of β,γ-unsaturated esters via dynamic kinetic asymmetric transformation (DYKAT) was reported by Shibasaki. β,γ- conjugated esters were used as substrates and underwent direct aldol/retro-aldol reactions in the presence of $Ba(Oi\text{-}Pr)_2$ and BINOL to provide β-hydroxy esters in moderate to good yield (53–85% yield) with good to excellent α- vs. γ-regioselectivity (15:1->20:1) and enantioselectivity (87–99% ee, eq 65).[140]

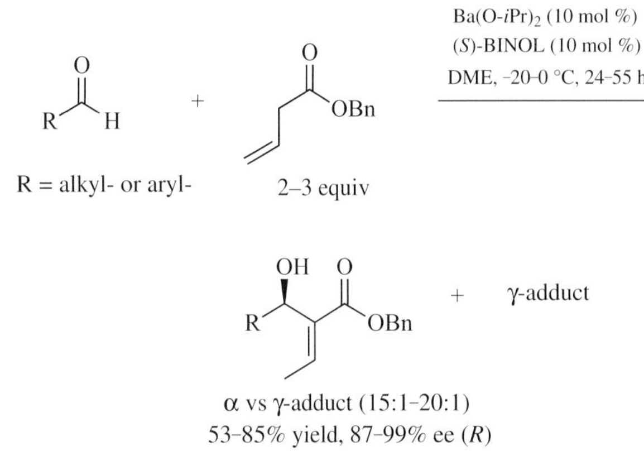

(65)

Michael Addition. The usage of BINOL ligands for the addition of soft carbon nucleophiles to α,β-enones via Michael addition has been described in this article. The scope of this reaction was further extended to various nucleophiles such as amines that can be added to butenoyl-*N*-imides with Pd(II) catalyst and (*R*)-BINAP in low to good yield (22–89% yield) and with medium to excellent enantioselectivity (9–99% ee, eq 66).[141] In a similar fashion, oximes could be added to α,β-unsaturated *N*-acyloxazolidinones to afford, in the presence of (*R*)-BINOL·SmI_2 complex, the corresponding amines in moderate to good yield (83–88% yield) and with relatively modest enantioselectivity (76–84% ee, eq 67).[142] Other nucleophiles such as thiols could also be added to enones in low to good yield (35–80% yield) with good to excellent enantioselectivity (92–99% ee) in the presence of catalytic (*S*)-BINOL derivative (eq 68).[143]

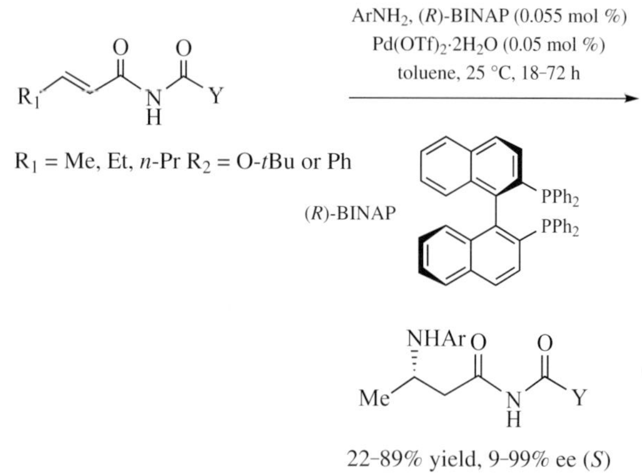

(66)

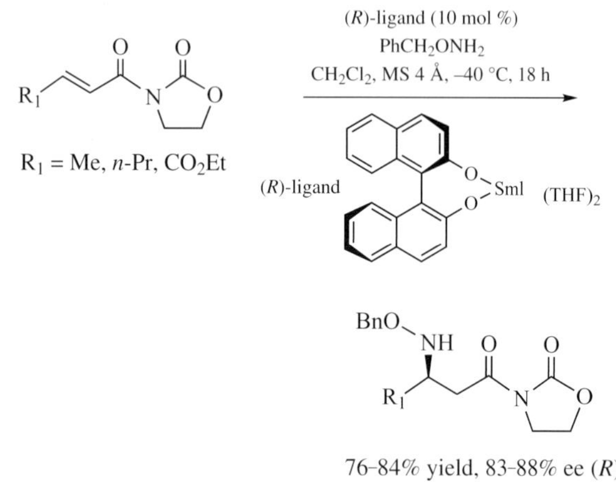

(67)

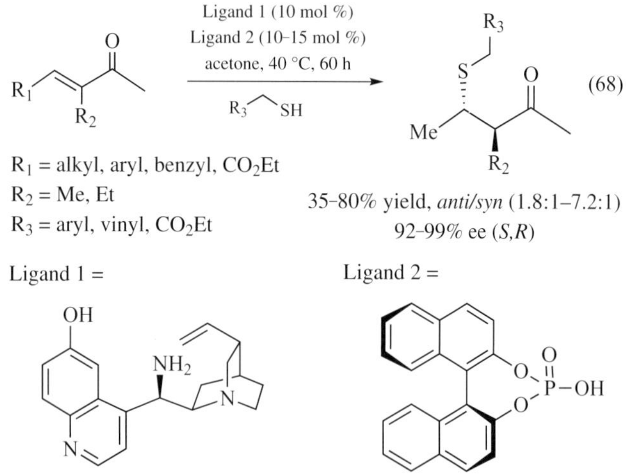

(68)

The Rh-catalyzed asymmetric conjugate addition of alkenyl- and arylboronic acids to α,β-unsaturated carbonyls, also referred to as Hayashi–Miyaura reaction[144,145] used in the total synthesis of biologically active substances.[146] A Pd-catalyzed asymmetric version of the Hiyashi–Miyaura reaction was also developed, but only using cyclic α,β-unsaturated ketones that are generally more reactive than their acyclic counterparts.[147] More recently, Chong and coworkers developed a novel protocol using BINOL derivatives as catalysts to trigger conjugate addition of alkenyl-, alkynyl-, and aryl boronates to linear α,β-unsaturated ketones.[148–150] In practice, enones undergo conjugate addition with alkenyl boronates (81–98% yield, 94–99% ee), alkynyl boronates (78–97% yield, 82–96% ee), and aryl boronates (65–90% yield, 80–98% ee) with BINOL ligand as chiral catalysts (eqs 69–71). Conjugate addition of boronates could also be performed with high regioselectivity with α,β–γ,δ-dienones to afford adducts in moderate to good yield (46–85% yield) with high regioselectivity and good to excellent enantioselectivity (80–96% ee, eq 72).[151]

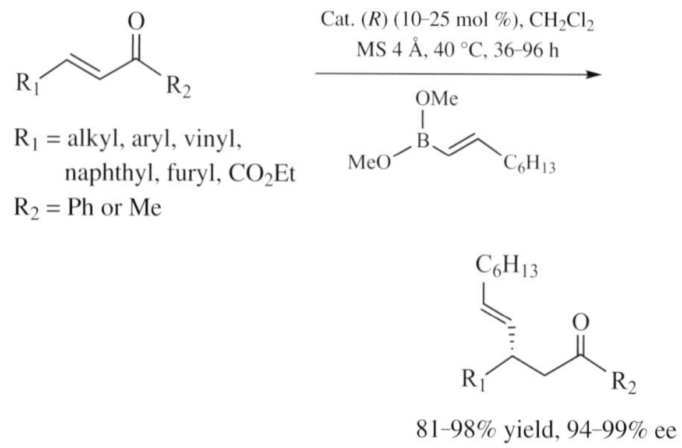

(69)

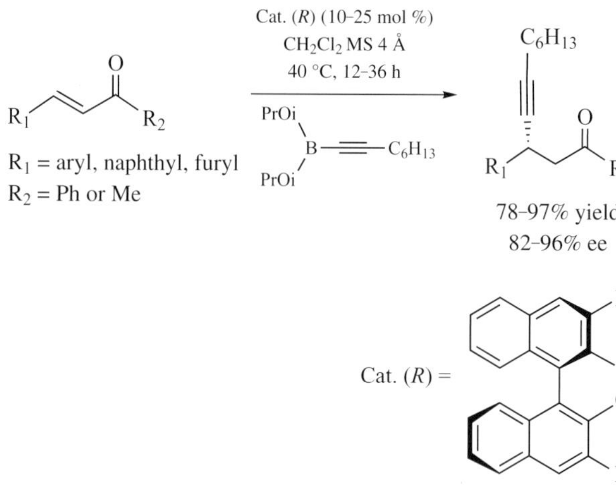

(70)

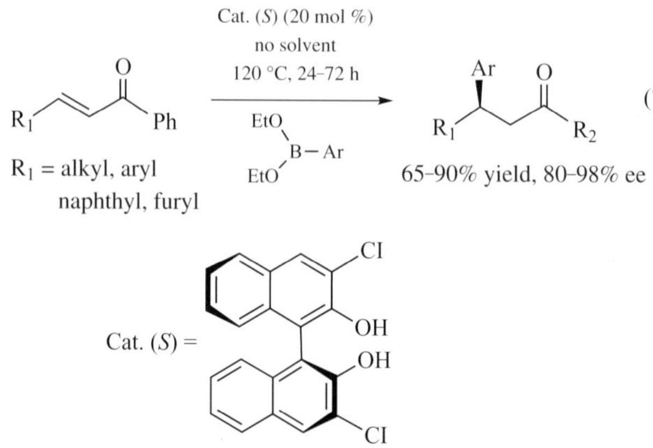

(71)

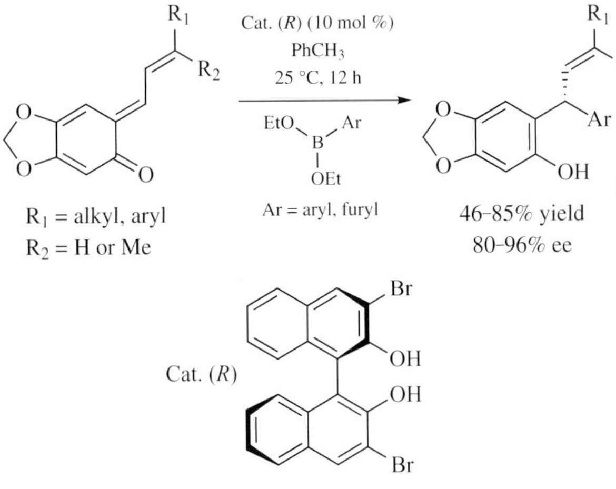

(72)

Allylic Substitution. Roggen and Carreira reported the Ir-based catalytic asymmetric allylic substitution of racemic allyl alcohols with thiols to provide a single enantiomer of the allylic sulfide via direct enantioconvergent transformation (DET). Racemic allyl alcohols underwent allylic substitution with various thiols to provide allyl sulfides in 79–92% yield and with 80–96% ee.[152] A BINOL derivative is used as chiral ligand (eq 73).

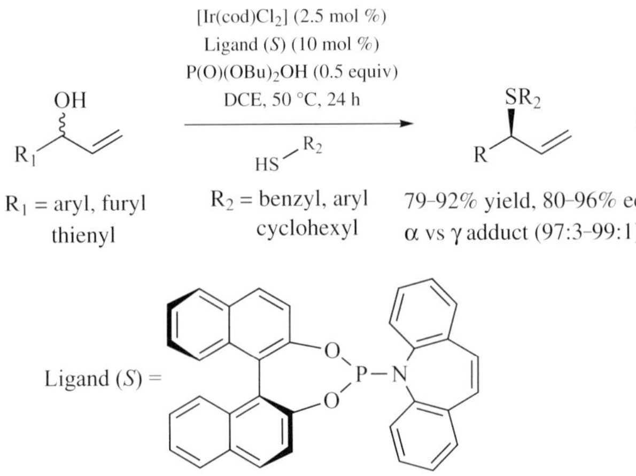

(73)

In a related report from Roggen and Carreira, a catalytic asymmetric kinetic resolution was performed with racemic allyl alcohol in the presence of an Ir catalyst and a chiral ligand. In the presence of a racemic mixture of allylic alcohols and various aliphatic alcohols, a large variety of enantiopure allylic ethers were obtained in moderate to excellent yield (36–99% yield) and with good to excellent enantioselectivity (81–99% ee, eq 74).[153]

Developments in transition metal-catalyzed [3+2] cycloadditions (especially the trimethylenemethane cycloaddition or "TMM" reaction) based on catalytic asymmetric allylation were recently published by Trost et al. The TMM reaction is an important reaction that allows the chemo-, regio-, and diastereoselective construction of highly functionalized five-membered rings.[154] The first catalytic asymmetric version of the TMM reaction was reported by Hayashi and coworkers in 1989 (4–78% ee)[155] and then further developed by Trost et al. in 2006.[156] Following this preliminary work, the latter further extended the scope of the Pd-catalyzed asymmetric TMM reaction using a phosphoramidite BINOL derivative to afford five-membered rings in low to excellent yield (0–100% yield) and with moderate to excellent enantioselectivity (66–98% ee, eq 75).[157]

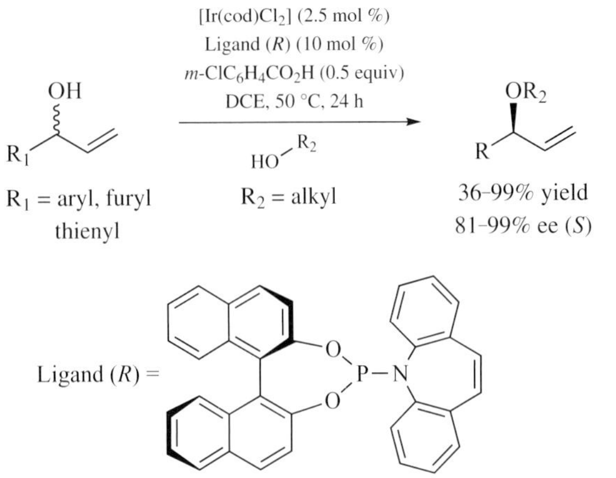

(74)

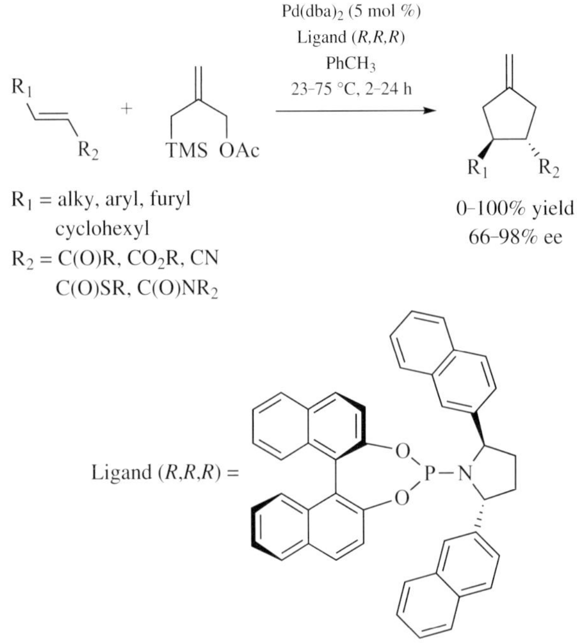

(75)

Opening of Aziridines. Enantiomerically pure *N*-sulfonyl, *N*-acyloxy, *N*-alkyl, and *N*-phosphoryl aziridines are key synthetic intermediates that can be synthesized via catalytic asymmetric aziridination.[158–161] However, this technique is currently not suited for *N*-acyl aziridines that are generally obtained via chiral synthons. Catalytic asymmetric kinetic resolution can however be utilized to obtain enantioenriched *N*-acyl aziridines. Treatment of a racemic mixture of aziridines with a catalytic amount of triphenylborate in the presence of a catalytic amount of (*R*)-BINOL provides an enantioenriched aziridine. In parallel, the other enantiomer of the aziridine is obtained in its open form that can be further treated with BCl_3 to provide the corresponding β-chloro acylamine via a S_N2 mechanism. Depending on the substrate, aziridines could be obtained in low to good yield (43–87% yield) and with low to excellent enantioselectivity (5.6–99.4% ee) (eq 76).

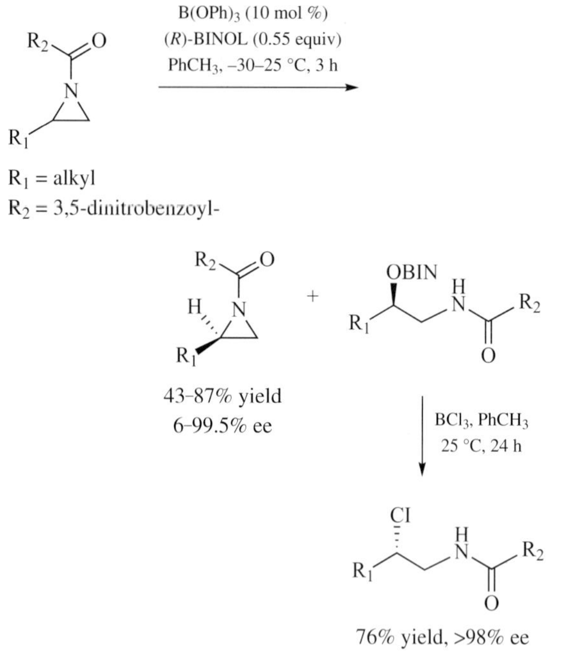

(76)

Epoxidation of Ketones. The use of BINOL ligands for epoxidation reactions of α,β-unsaturated ketones, amides, and esters was recently reviewed.[37] The epoxidation of alkenes with such ligands was also attempted, but only low to medium enantioselectivity could be obtained (0–22% ee). Recent reports described BINOL derivatives for such transformations, providing improved enantioselectivity (12–95% ee).[162,163]

In addition to the aforementioned epoxidation reactions, other techniques were developed to allow the synthesis of epoxides. Among them, the Corey–Chaykovsky epoxidation of ketones allows the synthesis of 2,2-functionalized epoxides, which are valuable building blocks in organic synthesis.[164] The production of these motifs via the Corey–Chaykovsky reaction under asymmetric conditions could be very valuable. The first attempt of asymmetric Corey–Chaykovsky was reported by Zhang and Wu in 1997, but provided only low enantioselectivities (21–57% ee).[165] More recently, Shibasaki and coworkers reported an optimized version of this reaction based on the use of La-Li_3-tris(binaphthoxide), also called LBB catalyst. A large collection of ketones were treated to provide the corresponding epoxides in high yield (88–100% yield) and with excellent enantioselectivity (91–97% ee, eq 77).[166]

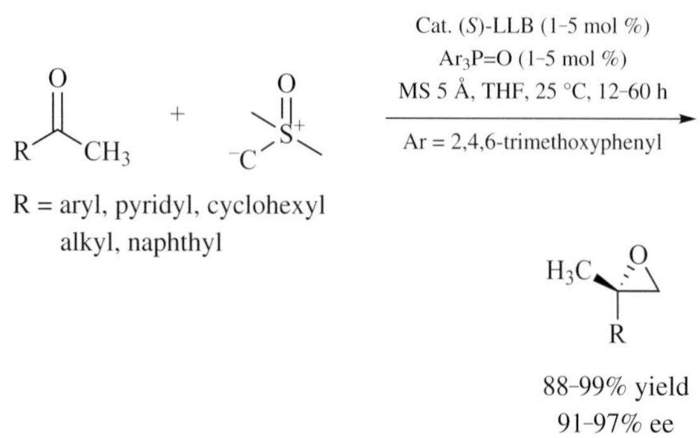

(77)

Epoxidation via Darzens Reactions. The Darzens reaction is an elegant alternative to the epoxidation reaction.[167] The asymmetric version of the Darzens reaction was known, but only moderate ee obtained.[168–171] The first catalytic highly enantioselective Darzens reaction of diazoacetamides with aldehydes with chiral catalyst (*R*)-BINOL/Ti(O*i*Pr)$_4$ (10 mol %) to provided *cis*-epoxides with excellent enantioselectivity (88–99% ee) and high yield (52–95% yield, eq 78).[172]

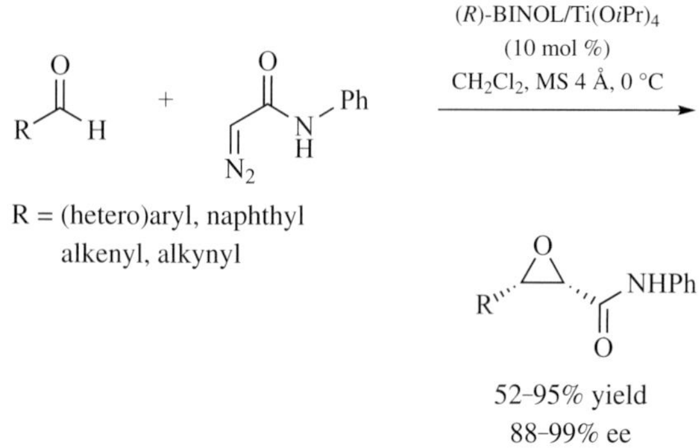

(78)

ype="main"One-pot Epoxidation–Epoxide Opening The usage of BINOL chiral ligands for epoxidation reactions was previously described by Shibasaki and coworkers and was discussed in the first update.[54] An extension of this reaction was recently described by Shibasaki and coworkers who describe a one-pot asymmetric epoxidation–epoxide opening reaction. The domino sequence uses a Samarium based (*S*)-BINOL complex as catalyst to provide α-hydroxy amides in moderate to high yield (45–99% yield), excellent enantioselectivity (>98% ee) and complete regioselectivity during the epoxide-opening step (exclusive opening at the β-position) with various nucleophiles ($PhSSiMe_3$, Me_3SiN_3, and CN^-, eq 79).[173]

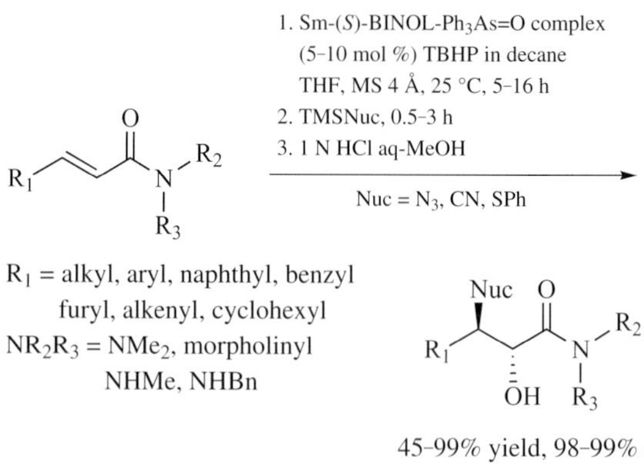

(79)

Polyene Cyclization. As already described, Carreira and coworkers developed methodologies allowing Ir-catalyzed asymmetric substitution with allyl alcohols as substrate and a (*R*)-phosphoramidite BINOL derivative to induce the chirality. This methodology was extended to polyene cyclization using the Ir catalyst as Lewis acid activator. In practice, a series of allyl alcohols provided tricyclic products in good to excellent yield (69–93% yield) and with excellent enantioselectivity (99–99.5% ee, eq 80).[174]

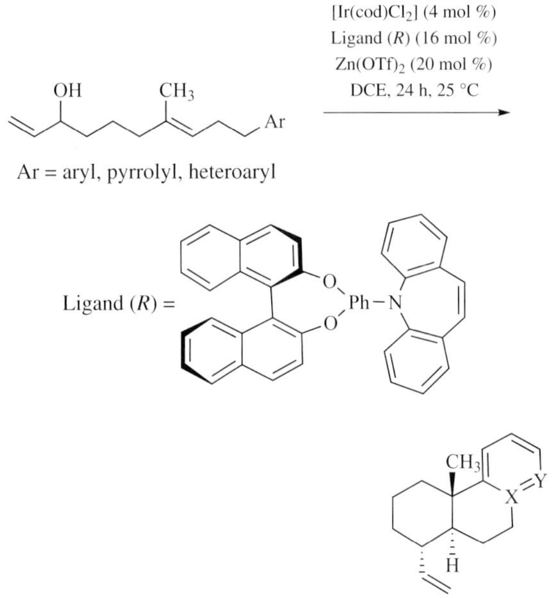

(80)

X,Y = C or N
69–93% yield, >99% ee

Hydrogenation of Olefins. The Rh-catalyzed asymmetric hydrogenation of olefins is one of the most powerful reactions in the toolbox of the organic chemist. It was believed for more than 30 years that bidentate phosphorous ligands synthesized from the chiral BINOL ligand (BINAP, etc.) were the most efficient ligands for high enantioselectivity. Contrary to this belief, appropriate monodentate phosphorous ligands can also provide excellent enantioselectivities. Until recently, MonoPhos[175] was considered the most efficient monodentate BINOL derivative for asymmetric Rh-catalyzed hydrogenation.[176] However, PipPhos and MorfPhos showed improved performance in the asymmetric Rh-catalyzed hydrogenation of *N*-acyl dehydroamino esters, itaconic acid derivatives, and *N*-acyl enamides, as shown below (eqs 81–85).[7]

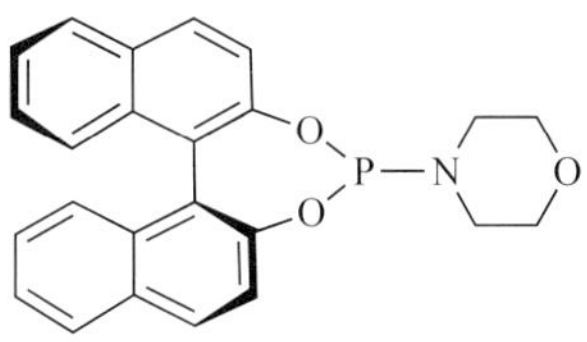

(*S*)-MorfPhos

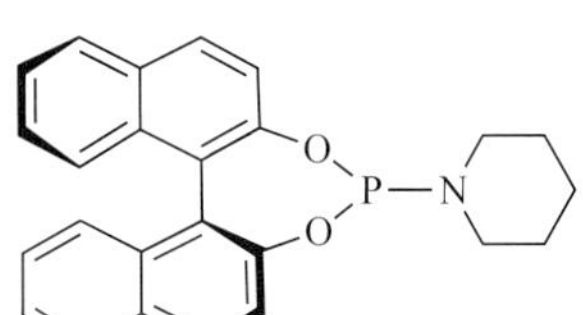

(*S*)-PipPhos

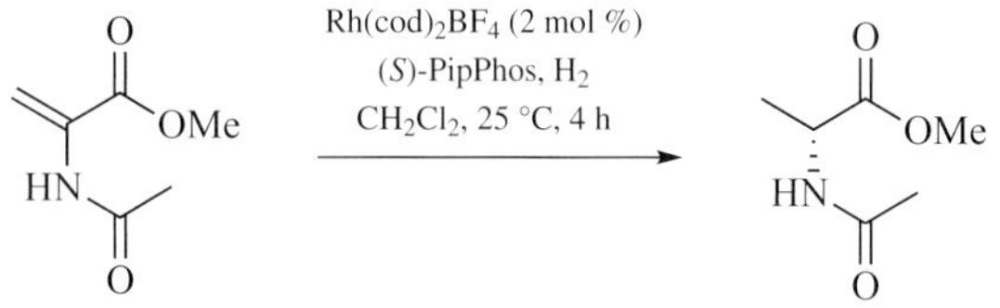

(81)

100% completion, 99% ee

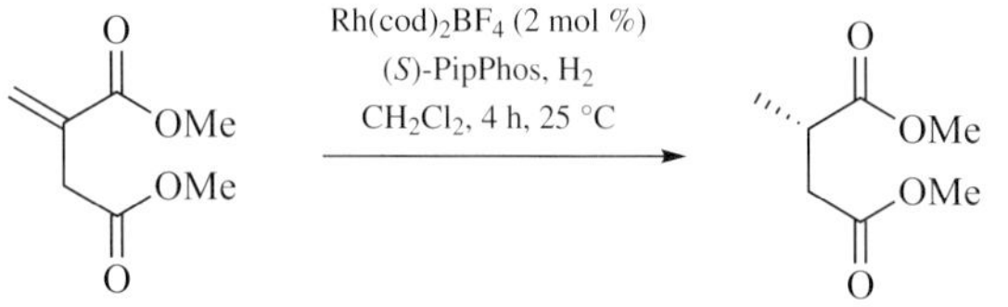

(82)

100% completion, >99% ee

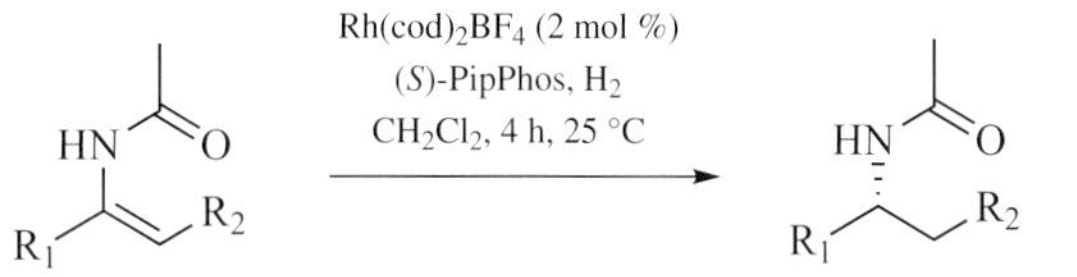

(83)

R_1 = aryl
R_2 = H or alkyl

100% completion
96–99% ee

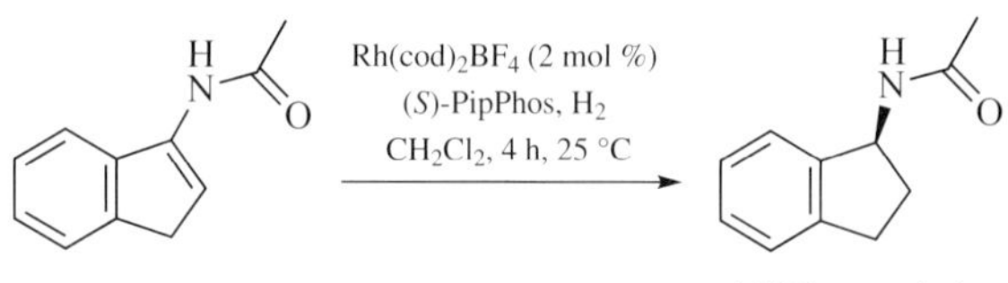

(84)

100% completion
89% ee

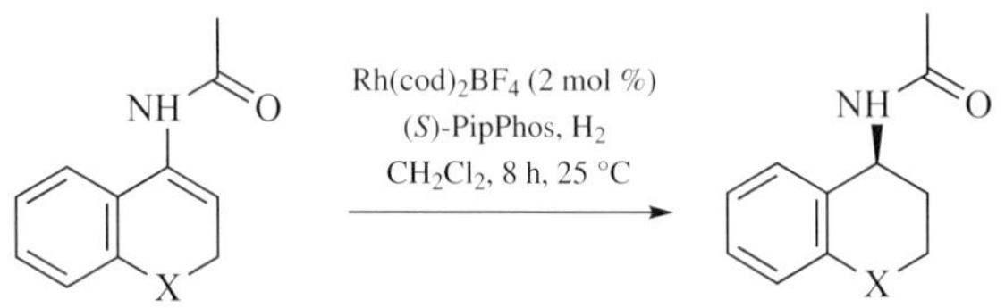

(85)

X = O, CH_2

94–100% completion
99% ee

Asymmetric Hydroboration. The regioselective Rh-based catalytic asymmetric hydroboration was originally limited to styrene derivatives, but was extended to cyclopropene substrates by Gevorgyan and coworkers in 2003 using Noyori's BINAP reagent as chiral ligand and carbonyl moieties as directing group for regioselectivity of the transformation.[177] In fact, boronates could be obtained with excellent regioselectivity (>99:1 *cis/trans*), good enantioselectivity (92–97% ee) in excellent yield (94–99% yield) and could be further utilized for Suzuki cross-coupling (eq 86).

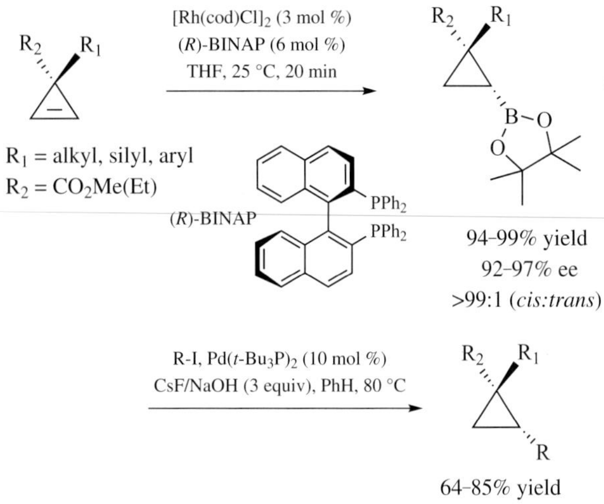

(86)

This reaction was further extended to acyclic β,γ-unsaturated amides. Similar to Gevorgyan and coworkers' protocol, the amide moiety was utilized as directing group for the hydroboration. However, in this instance, a simple phosphoramidite ligand easily obtainable from BINOL[178] provided the best enantioselectivity. Alcohols were obtained after hydroboration and oxidation in good yield (76–80% yield) with good to excellent enantioselectivity (93–99% ee, eq 87).[179]

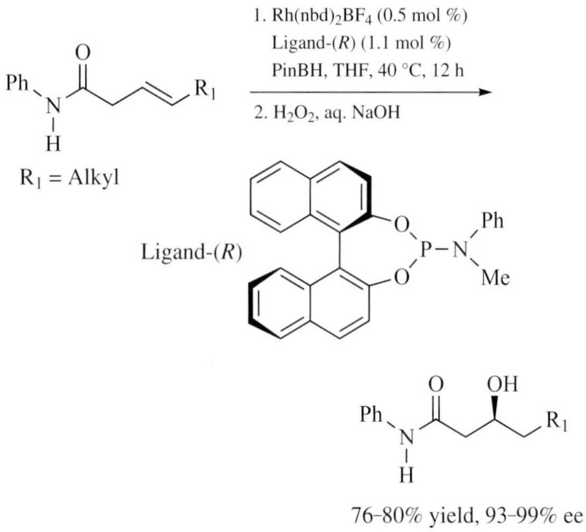

(87)

Related Reagents. (R)-1,1′-Bi-2,2′-naphthotitanium Dichloride; (R)-1,1′-Bi-2,2′-naphthotitanium Diisopropoxide; (R,R)-[Ethylene-1,2-Bis(η5-4,5,6,7-tetrahydro-1-indenyl)]titanium (R)-1,1′-Bi-2,2′-naphtholate; (−)-[Ethylene-1,2-Bis(η5-4,5,6,7-tetrahydro-1-indenyl)]zirconium (R)-1,1′-Bi-2,2′-naphtholate; Lithium Aluminum Hydride–2,2′-Dihydroxy-1,1′-binaphthyl; (*R*)-(+)-(3,5-Dioxa-4-phosphacyclohepta [2,1-α:3,4-α'] dinaphthalen-4-yl) piperidine;(*R*)-(+)-(3,5-dioxa-4-phosphacyclohepta[2,1-α:3,4-α'] dinaphthalen-4-yl) morpholine; (*R*)-(+)-2,2(-bis(diphenylphosphino)-1,1′-binaphthalene; (*R*)-(+)-3,3(-dibromo-1,1′-bi-2-naphthol; (*R*)-(+)-3,3′-diido-1,1′-bi-2-naphthol; (*R*)-(+)-3,3′-dichloro-1,1′-bi-2-naphthol; (*R*)-(−)-1,1′-binaphthyl-2,2′-diyl hydrogen phosphate; (*R,R,R*)-1-(3,5-dioxa-4-phosphacyclohepta[2,1-α:3,4-α']dinaphthalen-4-yl)-2,5-dinaphthalenylpyrrolidine; (*R*)-(+)-phenyl(3,5-dioxa-4-phospha-cyclohepta[2,1-α;3,4-α']dinaphthalen-4-yl)methylamine.

Third Update

Jean Michel Brunel
Aix-Marseille Université, Marseille, France

Enantioselective Petasis Reaction. The catalytic enantioselective three-component Petasis reaction among salicylaldehydes, amines, and organoboronic acids has been developed with readily available chiral BINOL as catalyst. Various salicylaldehydes, secondary amines, and organoboronic acids were found to be suitable substrates for this reaction and a broad range of desired alkylaminophenol products were obtained in good yields (up to 89%) and moderate to good enantioselectivities (34–86% ee, eq 88).[180]

Ring Opening of *meso*-Aziridines. A highly enantioselective ring-opening reaction of *meso*-aziridines with anilines to furnish a broad range of *trans*-1,2-diamines in good to excellent yields and up to 99% ee has been achieved by using a chiral catalyst easily prepared in situ from the commercially available components Ti(O*i*-Pr)$_4$ or Ti(O*t*-Bu)$_4$ and (*R*)-BINOL in a 1 : 2 ratio. Investigations concerning the structure of the metal point to an oligomeric complex with dual activation of both components which are further supported through the observation of a positive nonlinear effect (eq 89).[181,182]

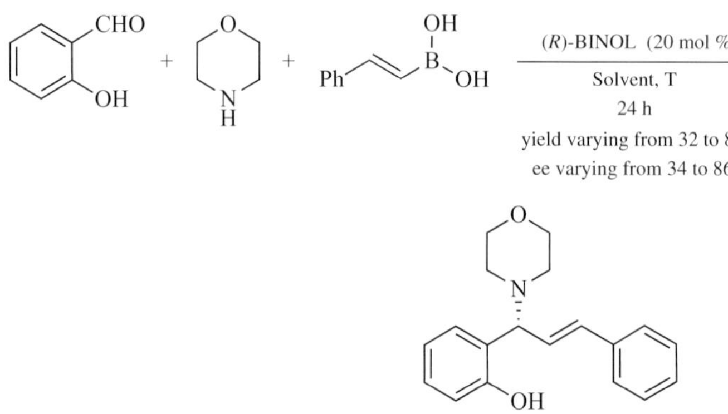

(88)

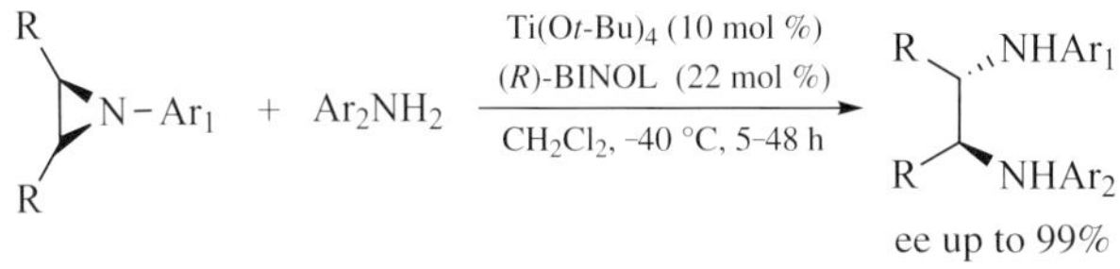

(89)

Recently, Cockrell et al. reported a unique method for kinetic resolution of *N*-acylaziridines with up to 99% ee by stereoselective ring opening on treatment with a stoichiometric amount of (*R*)-BINOL as chiral modifier and a catalytic amount of $B(OPh)_3$. However, (*R*)-BINOL is easily recovered in 85% yield in a large-scale reaction without loss of enantiopurity. The utility of the resulting chiral aziridines was demonstrated by further ring-opening and rearrangement reactions of (*R*)-1-(3,5-dinitrobenzyl)-2-hexyl-aziridine with the formation of a series of functionalized diamines and oxazolines (eq 90).[183]

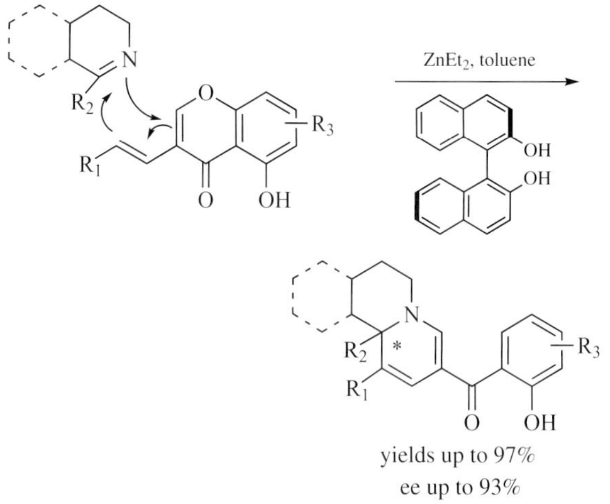

(92)

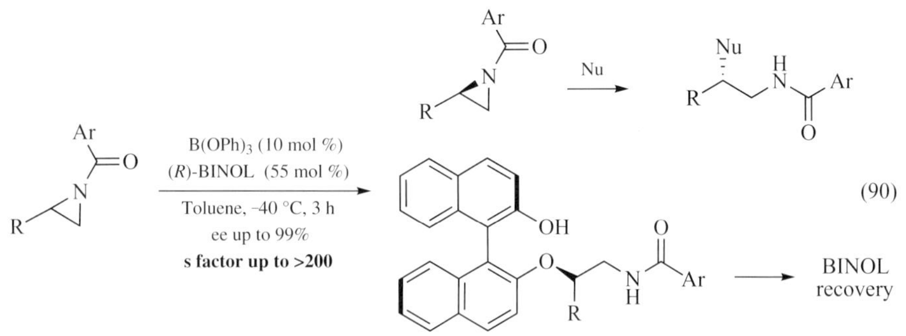

(90)

Ring Opening of *meso*-Epoxides. Recently, a simple and optimized protocol was developed for the asymmetric ring opening of a *meso*-epoxide with various amines using a chiral scandium complex affording a wide variety of enantiomerically pure β-amino alcohols in high yields with excellent ee (eq 91).[184]

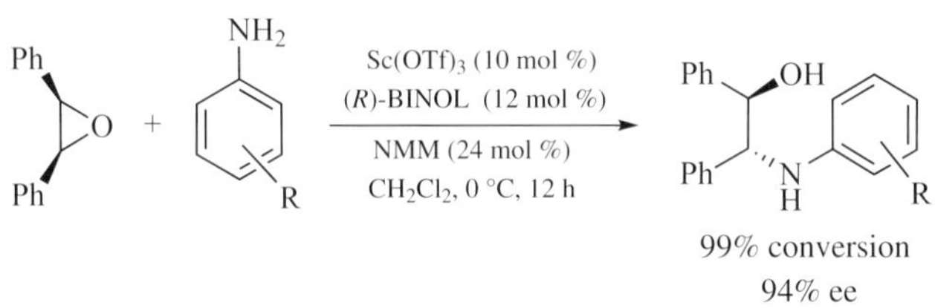

(91)

Enantioselective Inverse-Electron-Demand Imino Diels–Alder Reaction. The first enantioselective imino Diels–Alder reaction involving electron-poor dienes (R_1 = CN, CO $_2$ Me; R_2 = H, Cl, Me; R_3 = H, Cl) and cyclic imines catalyzed by Zn/BINOL complexes has been developed and provided a facile entry to a natural-product-inspired collection of ring-fused quinolizines (eq 92).[185]

Difunctionalization of Stable Tertiary Enamides. Various stable tertiary enamides may react in the presence of a catalytic amount of a chiral BINOL–Ti(O*i*Pr)$_4$ complex with salicylaldehydes to afford diverse *cis,trans*-configured 4-chromanols containing three continuous stereogenic centers in good yields with excellent diastereoselectivity and enantioselectivity. The reaction involved the addition of an enamide to aldehyde followed by an intramolecular interception of the resulting iminium by the hydroxy group (eq 93).[186]

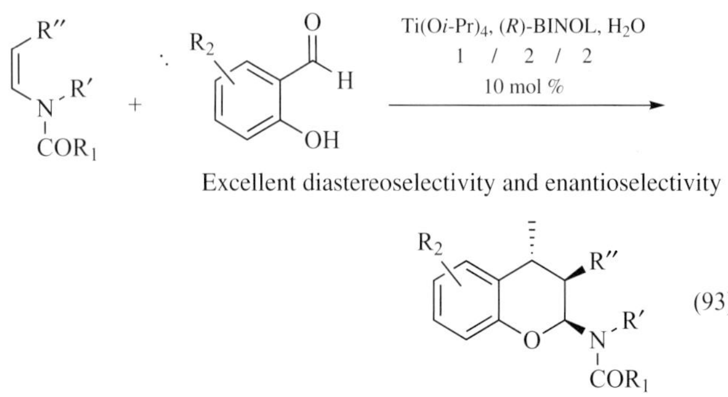

(93)

Enantioselective Pictet–Spengler Reaction. An enantioselective Pictet–Spengler-type reaction has been performed in 2013 by using a catalytic amount of $B(OMe)_3$/(*R*)-BINOL (1:1, 10 mol °) complex and a range of 2-(1*H*-pyrrol-1-yl)aniline derivatives reacting with aldehydes to give structurally diverse 4,5-dihydropyrrolo[1,2-a]quinoxaline compounds in good to excellent yield and enantiomeric excess up to 95% (eq 94).[187]

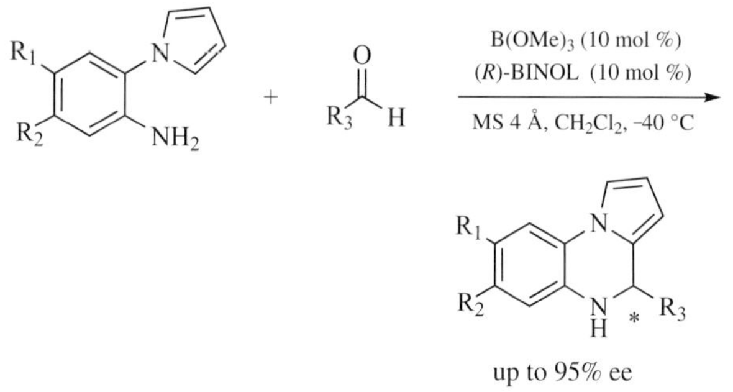

(94)

Enantioselective Alkynylation. A Zn/BINOL-catalyzed procedure for enantioselective alkynylation of a variety of ketoimines, including α-fluoroalkyl α-imino esters, α-aryl α-imino esters, and trifluoromethyl aryl ketimines, has been envisioned and provide their corresponding quaternary propargyl amines in excellent yields with high ee values (up to 99% ee). The notable features of this methodology are its broad substrate scope, high reaction efficiency (up to 99%) under mild experimental reaction conditions with a low catalyst loading (5 mol ° of BINOL), and enantioselectivity (up to 99% ee) (eq 95).[188]

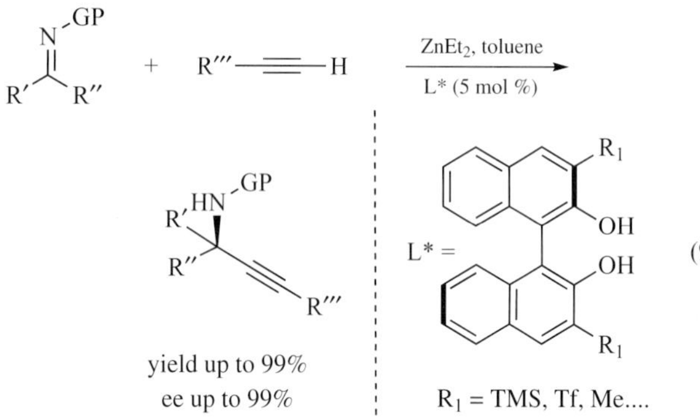

(95)

On the other hand, an asymmetric linear alkyl alkyne addition to aromatic aldehydes catalyzed by the Ti-(*R*)-BINOL system has been reported with high enantioselectivities and yields leading to propargylic alcohols as potent intermediates for the synthesis of various natural products (eq 96).[189]

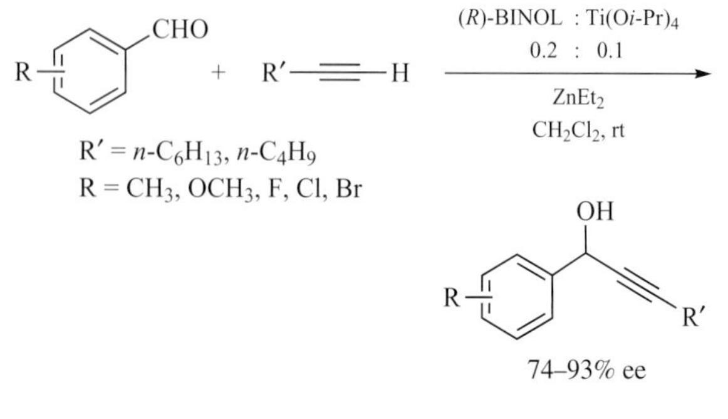

(96)

In 2005, Shibasaki et al. developed a new catalytic asymmetric alkynylation of aldehydes promoted by In(III)/BINOL complex and Cy_2NMe. Dual activation of both substrates due to the bifunctional character of In(III) would make possible a broad range of substrate generality with high enantioselectivity (eq 97).[190]

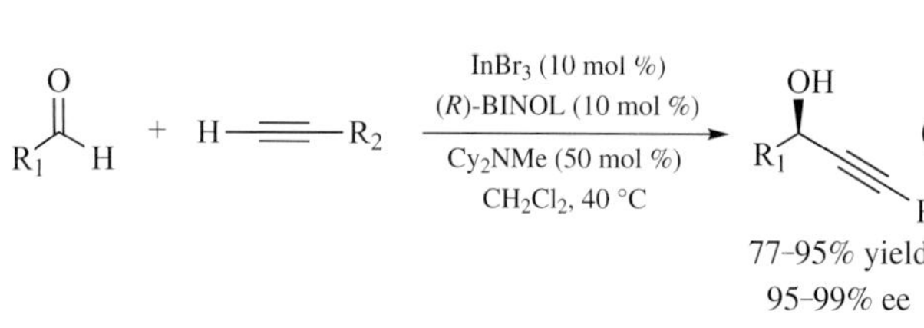

(97)

Epoxidation. Combination of (*R*)-BINOL **1** and Et_2Zn affords in situ, a catalyst for homogeneous epoxidation of (*E*)-α,β-enones to the corresponding *trans*-epoxy ketones using cumylhydroperoxide (CHP) as oxidant. Enantiomeric excess of up to 96% can be achieved conveniently at room temperature (eq 98).[191]

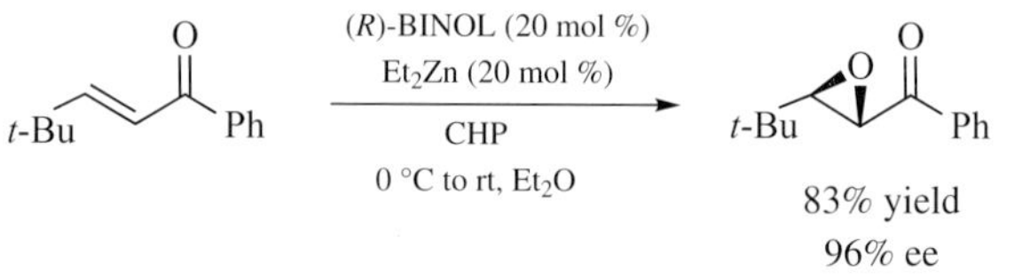

(98)

A mechanistic rationale has been postulated suggesting an electrophilic activation of the substrate by the chiral BINOL–Zn catalyst and a subsequent nucleophilic attack of the oxidant.

Enantioselective Friedel–Crafts Reaction. The first efficient enantioselective Friedel–Crafts hydroxyalkylation of pyrroles possessing one electron-withdrawing group with alkyl glyoxylates catalyzed by readily available chiral BINOL–Ti(IV) complexes (1–5 mol %) has been reported. The reaction regioselectively could lead to the desired pyrrole–hydroxyacetic acid derivatives with good yields (70–96%) and enantiomeric excesses up to 96% (eq 99).[192]

Enantioselective Direct Mannich-Type Reaction. A highly enantioselective direct Mannich-type reaction of aldimines with dialkyl malonates was developed with the use of a Mg(II)–BINOLate salt as a cooperative acid–base catalyst activating both aldimines and malonates. Under these conditions, optically active β-aminoesters and α-halo-β-aminoesters have been synthesized in high yields and with high enantioselectivities (eq 100).[193]

Aldol Reaction. Although the aldol reaction has been widely studied, some recent results need to be mentioned. Thus, the first highly enantio- and diastereoselective vinylogous aldol reaction between propionyl acetate-derived Brassard's diene and aldehydes was accomplished by titanium–lithium combined Lewis acid, affording δ-hydroxy-γ-methyl-β-methoxy acrylates permitting to achieve the convenient and concise construction of the polypropionate moiety in melithiazole C (eq 101).[194,195]

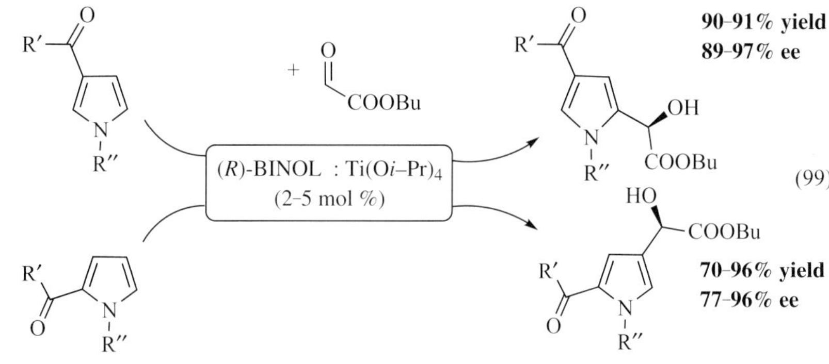

(99)

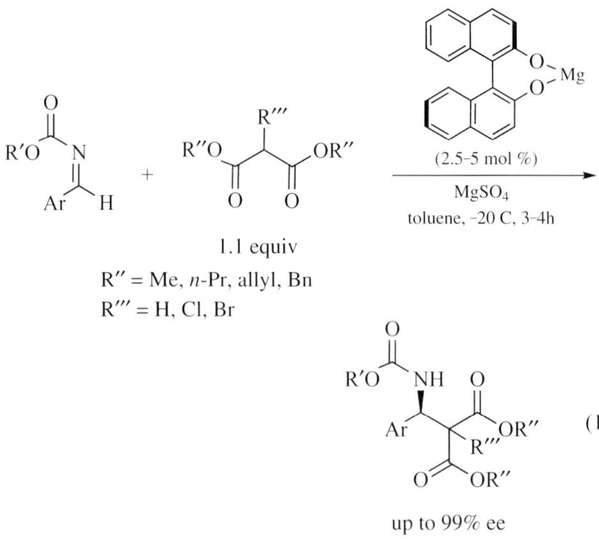

(100)

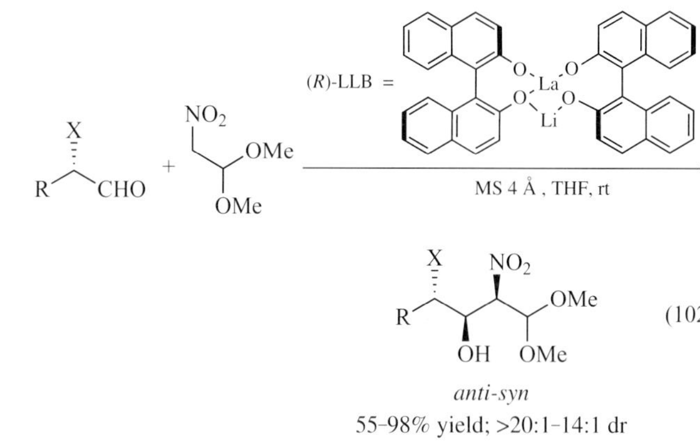

(102)

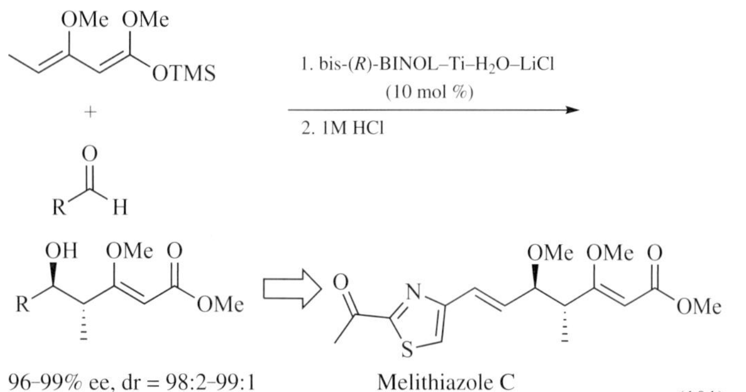

(101)

In another context, a stereodivergent construction of three contiguous stereocenters in a catalytic doubly diastereoselective nitroaldol reactions of α-chiral aldehydes (R = Me, $PhCH_2$, $EtO_2CCH_2CH_2$), with nitroacetaldehyde dimethyl acetal using a La–Li–BINOL catalyst afforded *anti-syn*-nitroaldol products with a >20:1 to 14:1 selectivity (eq 102).[196–198]

Recently, Gnadadesikan et al. have achieved the catalytic aldol–Tishchenko reaction by coupling an irreversible Tishchenko reaction to a reversible aldol reaction affording the products in high enantio- and diastereoselectivity using 10 mol % of the chiral catalyst. A variety of ketones and aldehydes, including propyl and butyl ketones, were coupled efficiently, yielding the corresponding aldol–Tishchenko products in up to 96% yield and 95% ee (eq 103).[199]

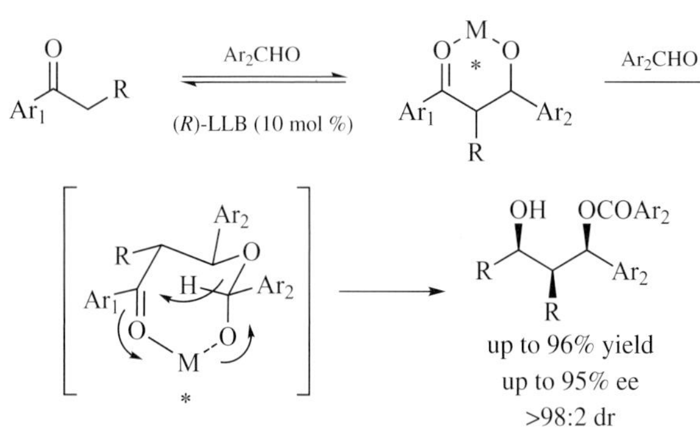

(103)

Hetero-Diels–Alder Reaction. Recently, an efficient catalytic, enantioselective hetero-Diels–Alder reaction of Danishefsky's diene with aldehydes using the magnesium binaphthoxide system has been developed, affording a variety of 2-substituted 2,3-dihydro-4*H*-pyran-4-ones in high yields and with excellent ee values. On the basis of the structure of the isolated reaction intermediate, the stereochemistry of the products, and the information attained from a study of the nonlinear effect, a plausible asymmetric induction model has been proposed (eq 104).[200]

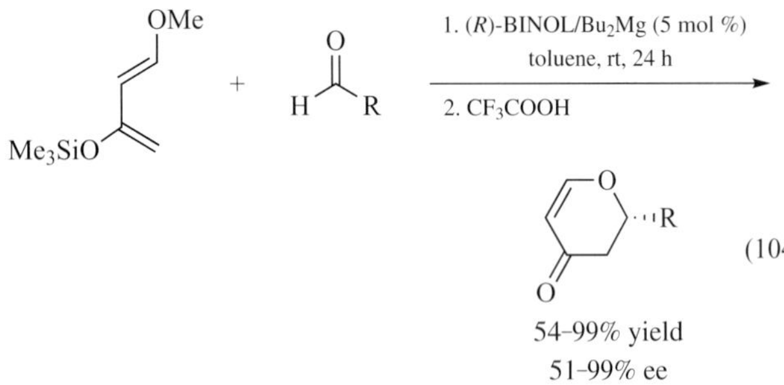

(104)

In this context, an efficient catalytic asymmetric hetero-Diels–Alder reaction of Brassard's diene with aliphatic aldehydes has also been reported. The catalyst, which was generated from (*R*)-BINOL, Ti(O*i*-Pr)$_4$, and 4-picolyl chloride hydrochloride, promoted the reaction smoothly to afford the corresponding α,β-unsaturated δ-lactone derivatives in moderate to good yields (46–79%) with high enantioselectivities (up to 88% ee). As an illustration of this methodology, natural products such as (*R*)-kavain (70% ee) and (*S*)-dihydrokavain (84° ee) were successfully prepared (eq 105).[201]

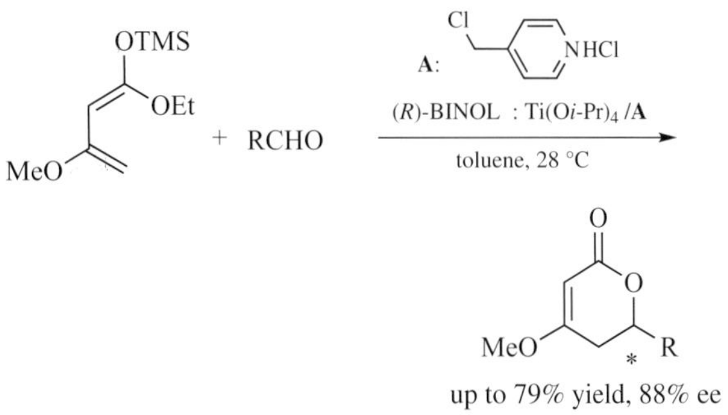

(105)

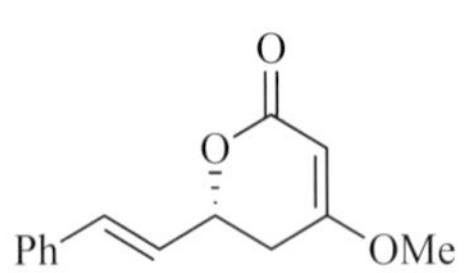

(*R*)-kavain

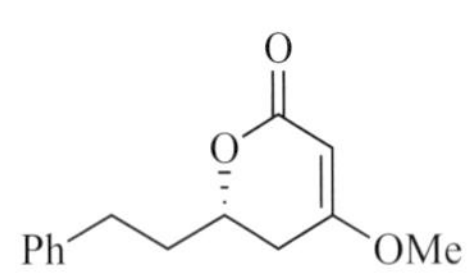

(*S*)-dihydrokavain

Another approach has been investigated for the enantioselective synthesis of 2,5-disubstituted dihydropyrone derivatives involving a [(*R*)-BINOL]$_2$-Ti(O*i*-Pr)$_4$ complex. Thus, 99% yield and 99% ee were encountered for the hetero-Diels–Alder reaction between *trans*-1-methoxy-2-methyl-3-trimethylsiloxy-1,3-butadiene and aldehydes in the presence of 5 mol% of catalyst. The potential and generality of this catalyst was evaluated by a variety of aldehydes including aromatic, heteroaromatic, α,β-unsaturated and aliphatic aldehydes (eq 106).[202]

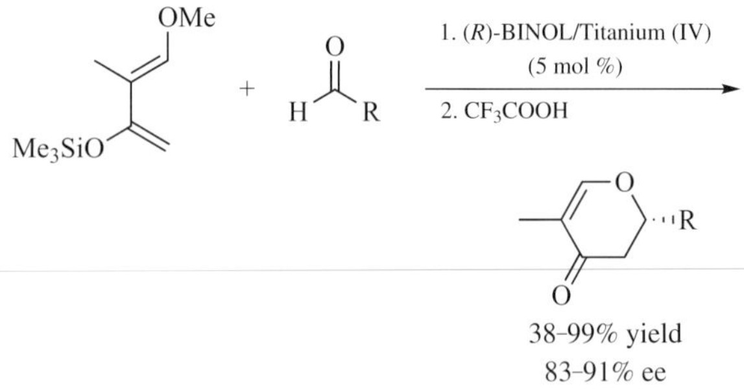

(106)

1,3-Dipolar Cycloaddition. In 2008, a regio- and stereoselective 1,3-dipolar cycloaddition of nitrile oxides to internal 2-pentenols, α,β-unsaturated esters, and amides catalyzed by (*R*)-BINOL–lanthanide complexes was reported affording the corresponding 3-aryl-2-isoxazolines with >89% ee (eq 107).[203]

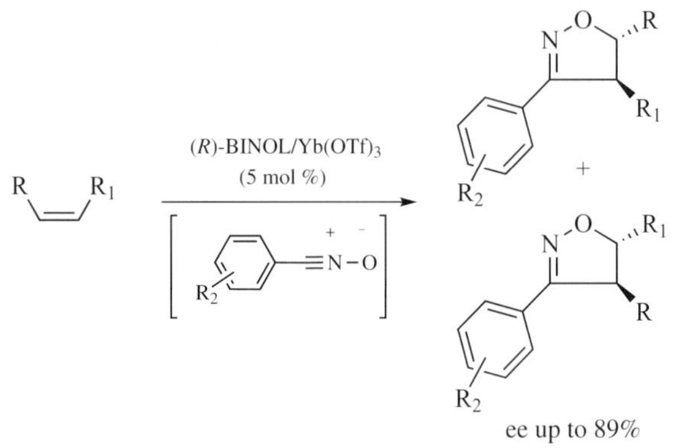

(107)

1. (a) Rosini, C.; Franzini, L.; Raffaelli, A.; Salvadori, P., *Synthesis* **1992**, 503. (b) Miyano, S.; Hashimoto, H., *J. Synth. Org. Chem. Jpn.* **1986**, *44*, 713.
2. (a) Toda, F.; Tanaka, K.; Iwata, S., *J. Org. Chem.* **1989**, *54*, 3007. (b) Pirkle, W. H.; Schreiner, J. L., *J. Org. Chem.* **1981**, *46*, 4988. (c) McKillop, A.; Turrell, A. G.; Young, D. W.; Taylor, E. C., *J. Am. Chem. Soc.* **1980**, *102*, 6504. (d) Carrick, W. L.; Karapinka, G. L.; Kwiatkowski, G. T., *J. Org. Chem.* **1969**, *34*, 2388. (e) Dewar, M. J. S.; Nakaya, T., *J. Am. Chem. Soc.* **1968**, *90*, 7134. (f) Pummerer, R.; Prell, E.; Rieche, A., *Ber. Dtsch. Chem. Ges./Chem. Ber.* **1926**, *59*, 2159.
3. (a) Jacques, J.; Fouquey, C., *Org. Synth.* **1988**, *67*, 1. (b) Jacques, J.; Fouquey, C.; Viterbo, R., *Tetrahedron Lett.* **1971**, 4617. (c) For the use of phenethylamine, see: Gong, B.; Chen, W.; Hu, B., *J. Org. Chem.* **1991**, *56*, 423.
4. (a) Brussee, J.; Groenendijk, J. L. G.; Koppele, J. M.; Jansen, A. C. A., *Tetrahedron* **1985**, *41*, 3313. (b) Smrcina, M.; Polakova, J.; Vyskocil, S.; Kocovsky, P., *J. Org. Chem.* **1993**, *58*, 4534.
5. Yamamoto, K.; Fukushima, H.; Nakazaki, M., *Chem. Commun./J. Chem. Soc., Chem. Commun.* **1984**, 1490.
6. Alper, H.; Hamel, N., *J. Am. Chem. Soc.* **1990**, *112*, 2803.
7. (a) Cram, D. J.; Cram, J. M., *Acc. Chem. Res.* **1978**, *11*, 8. (b) Helgeson, R. C.; Timko, J. M.; Moreau, P.; Peacock, S. C.; Mayer, J. M.; Cram, D. J., *J. Am. Chem. Soc.* **1974**, *96*, 6762.
8. Cram, D. J.; Sogah, G. D. Y., *Chem. Commun./J. Chem. Soc., Chem. Commun.* **1981**, 625.
9. Cram, D. J.; Sogah, G. D. Y., *J. Am. Chem. Soc.* **1985**, *107*, 8301.
10. Sepulchre, M.; Spassky, N., *Makromol. Chem. Rapid Commun.* **1981**, *2*, 261.

11. (a) Miyano, S.; Tobita, M.; Hashimoto, H., *Bull. Chem. Soc. Jpn.* **1981**, *54*, 3522. (b) Miyano, S.; Fukushima, H.; Handa, S.; Ito, H.; Hashimoto, H., *Bull. Chem. Soc. Jpn.* **1988**, *61*, 3249.
12. (a) Noyori, R., *Chem. Soc. Rev.* **1989**, *18*, 187. (b) Noyori, R.; Tomino, I.; Yamada, M.; Nishizawa, M., *J. Am. Chem. Soc.* **1984**, *106*, 6717. (c) Noyori, R.; Tomino, I.; Tanimoto, Y.; Nishizawa, M., *J. Am. Chem. Soc.* **1984**, *106*, 6709. (d) Noyori, R.; Tomino, I.; Tanimoto, Y., *J. Am. Chem. Soc.* **1979**, *101*, 3129.
13. (a) Olivero, A. G.; Weidmann, B.; Seebach, D., *Helv. Chim. Acta* **1981**, *64*, 2485. (b) Seebach, D.; Beck, A. K.; Roggo, S.; Wonnacott, A., *Chem. Ber.* **1985**, *118*, 3673. (c) Wang, J. T.; Fan, X.; Feng, X.; Qian, Y. M., *Synthesis* **1989**, 291.
14. Reetz, M. T.; Kyung, S.-H.; Bolm, C.; Zierke, T., *Chem. Ind. (London)* **1986**, 824.
15. Seebach, D.; Beck, A. K.; Imwinkelried, R.; Roggo, S.; Wonnacott, A., *Helv. Chim. Acta* **1987**, *70*, 954.
16. (a) Mikami, K.; Terada, M.; Narisawa, S.; Nakai, T., *Org. Synth.* **1992**, *71*, 14. (b) Mikami, K.; Terada, M.; Nakai, T., *J. Am. Chem. Soc.* **1990**, *112*, 3949.
17. Dijkgraff, C.; Rousseau, J. P. G., *Spectrochim. Acta* **1968**, *2*, 1213.
18. (a) Mikami, K.; Terada, M.; Motoyama, Y.; Nakai, T., *Tetrahedron: Asymmetry* **1991**, *2*, 643. (b) Mikami, K.; Motoyama, Y.; Terada, M., *J. Am. Chem. Soc.*, **1994**, *116*, 2812. (c) For the use of a modified BINOL-derived titanium complex, see: Maruoka, K.; Murase, N.; Yamamoto, H., *J. Org. Chem.* **1993**, *58*, 2938.
19. Kelly, T. R.; Whiting, A.; Chandrakumar, N. S., *J. Am. Chem. Soc.* **1986**, *108*, 3510.
20. Chapuis, C.; Jurczak, J., *Helv. Chim. Acta* **1987**, *70*, 436.
21. Kaufmann, D.; Boese, R., *Angew. Chem., Int. Ed. Engl.* **1990**, *29*, 545.
22. (a) Hattori, K.; Yamamoto, H., *Tetrahedron* **1993**, *49*, 1749. (b) For the use in imine-aldol reactions, see: Hattori, K.; Miyata, M.; Yamamoto, H., *J. Am. Chem. Soc.* **1993**, *115*, 1151.
23. (a) Maruoka, K.; Itoh, T.; Shirasaka, T.; Yamamoto, H., *J. Am. Chem. Soc.* **1988**, *110*, 310. (b) For the D.–A. reaction, see: Maruoka, K.; Concepcion, A. B.; Yamamoto, H., *Bull. Chem. Soc. Jpn.* **1992**, *65*, 3501.
24. Terada, M.; Mikami, K.; Nakai, T., *Tetrahedron Lett.* **1991**, *32*, 935.
25. (a) Glyoxylate-ene reaction with vinylic sulfides and selenides: Terada, M.; Matsukawa, S.; Mikami, K., *Chem. Commun./J. Chem. Soc., Chem. Commun.* **1993**, 327. (b) For the ene reaction of chloral, see: Maruoka, K.; Hoshino, Y.; Shirasaka, T.; Yamamoto, H., *Tetrahedron Lett.* **1988**, *29*, 3967.
26. (a) Mikami, K.; Terada, M., *Tetrahedron* **1992**, *48*, 5671. (b) Terada, M.; Mikami, K.; Nakai, T., *Chem. Commun./J. Chem. Soc., Chem. Commun.* **1990**, 1623.
27. Mikami, K.; Narisawa, S.; Shimizu, M.; Terada, M., *J. Am. Chem. Soc.* **1992**, *114*, 6566.
28. Sakane, S.; Maruoka, K.; Yamamoto, H., *Tetrahedron* **1986**, *42*, 2203.
29. Sakane, S.; Fujiwara, J.; Maruoka, K.; Yamamoto, H., *J. Am. Chem. Soc.* **1983**, *105*, 6154.
30. Mikami, K.; Matsukawa, S., *J. Am. Chem. Soc.* **1993**, *115*, 7039.
31. Mikami, K.; Matsukawa, S., *J. Am. Chem. Soc.* **1994**, *116*, 4077.
32. (a) Sasai, H.; Suzuki, T.; Itoh, N.; Shibasaki, M., *Tetrahedron Lett.* **1993**, *34*, 851. (b) Sasai, H.; Suzuki, T.; Arai, S.; Arai, T.; Shibasaki, M., *J. Am. Chem. Soc.* **1992**, *114*, 4418.
33. Aoki, S.; Mikami, K.; Terada, M.; Nakai, T., *Tetrahedron* **1993**, *49*, 1783.
34. (a) Costa, A. L.; Piazza, M. G.; Tagliavini, E.; Trombini, C.; Umani-Ronchi, A., *J. Am. Chem. Soc.* **1993**, *115*, 7001. (b) Keck, G. E.; Tarbet, K. H.; Geraci, L. S., *J. Am. Chem. Soc.* **1993**, *115*, 8467. (c) Keck, G. E.; Krishnamurthy, D.; Grier, M. C., *J. Org. Chem.* **1993**, *58*, 6543.
35. Maruoka, K.; Banno, H.; Yamamoto, H., *J. Am. Chem. Soc.* **1990**, *112*, 7791.
36. (a) Fuji, K.; Node, M.; Tanaka, F., *Tetrahedron Lett.* **1990**, *31*, 6553. (b) Fuji, K.; Node, M.; Tanaka, F.; Hosoi, S., *Tetrahedron Lett.* **1989**, *30*, 2825.
37. Brunel, J. M., *Chem. Rev.* **2005**, *105*, 857.
38. Brunel, J. M.; Buono, G., *J. Org. Chem.* **1993**, *58*, 7313.
39. Cai, J. X.; Zhou, Z. H.; Li, K. Y.; Yeung, C. H.; Tang, C. C., *Chin. J. Chem.* **2002**, *13*, 617.
40. Cai, J. X.; Zhou, Z. H.; Li, K. Y.; Yeung, C. H.; Tang, C. C., *Phosphorus, Sulfur and Silicon and Rlted Elements* **2002**, *177*, 189.
41. Periasamy, M.; Venkatraman, L.; Sivakumar, S.; Sampathkumar, N.; Ramanathan, C. R., *J. Org. Chem.* **1999**, *64*, 7643.
42. Shan, Z.; Wang, G.; Duan, B.; Zhao, D., *Tetrahedron: Asymmetry* **1996**, *7*, 2847.
43. Shan, Z.; Cheng, F.; Huang, S.; Zhao, D.; Jing, Z., *Tetrahedron: Asymmetry* **1997**, *8*, 1175.
44. Luo, Z.; Liu, Q.; Gong, L.; Cui, X.; Mi, A.; Jiang, Y., *Angew. Chem. Int. Ed. Engl.* **2002**, *41*, 4532.
45. Luo, Z.; Liu, Q.; Gong, L.; Cui, X.; Mi, A.; Jiang, Y., *Chem. Commun.* **2002**, 914.
46. Chu, C. Y.; Hwang, D. R.; Wang, S. K.; Uang, B. J., *Chem. Commun.* **2001**, 980.
47. Chu, C. Y.; Uang, B. J., *Tetrahedron: Asymmetry* **2003**, *14*, 53.
48. Barhate, N. B.; Chen, C. T., *Org. Lett.* **2002**, *4*, 2529.
49. Hon, S. W.; Li, C. H.; Kuo, J. H.; Barhate, N. B.; Liu, Y. H.; Wang, Y.; Chen, C. T., *Org. Lett.* **2001**, *3*, 869.
50. Fu, I. P.; Uang, B. J., *Tetrahedron: Asymmetry* **2001**, *12*, 45.
51. Lin, Y.-M.; Fu, I. P.; Uang, B.-J., *Tetrahedron: Asymmetry* **2001**, *12*, 3217.
52. Imma, H.; Mori, M.; Nakai, T., *Synlett* **1996**, 1229.
53. Nishikori, H.; Yoshihara, R.; Hosomi, A., *Synlett* **2003**, 561.
54. Bougauchi, M.; Watanabe, S.; Arai, T.; Sasai, H.; Shibasaki, M., *J. Am. Chem. Soc.* **1997**, *119*, 2329.
55. Kakei, H.; Nemoto, T.; Ohshima, T.; Shibasaki, M., *Angew. Chem. Int. Ed. Engl.* **2004**, *43*, 317.
56. Chen, R.; Qian, C. ; de Vries, J. G., *Tetrahedron* **2001**, *57*, 9837.
57. Chen, R. F.; Qian, C. T.; de Vries, J. G.; Sun, P. P.; Wang, L. M., *Chin. J. Chem.* **2001**, *19*, 1225.
58. Daikai, K.; Kamaura, M.; Junji, I., *Tetrahedron Lett.* **1998**, *39*, 7321.
59. Daikai, K.; Kamaura, M.; Inanaga, J., *Kidorui* **1999**, *34*, 196.
60. Watanabe, S.; Kobayashi, Y.; Arai, T.; Sasai, H.; Bougauchi, M.; Shibasaki, M., *Tetrahedron Lett.* **1998**, *39*, 7353.
61. Nemoto, T.; Ohshima, T.; Yamaguchi, K.; Shibasaki, M., *J. Am. Chem. Soc.* **2001**, *123*, 2725.
62. Komatsu, N.; Hashizume, M.; Sugita, T.; Uemura, S., *J. Org. Chem.* **1993**, *58*, 4529.
63. Komatsu, N.; Nishibayashi, Y.; Sugita, T.; Uemura, S., *Tetrahedron Lett.* **1992**, *33*, 5391.
64. Capozzi, M. A. M.; Cardellicchio, C.; Naso, F.; Rosito, V., *J. Org. Chem.* **2002**, *67*, 7289.
65. Capozzi, M. A. M.; Cardellicchio, C.; Fracchiolla, G.; Naso, F.; Tortorella, P., *J. Am. Chem. Soc.* **1999**, *121*, 4708.
66. Massa, A.; Lattanzi, A.; Siniscalchi, F. R.; Scettri, A., *Tetrahedron: Asymmetry* **2001**, *12*, 2775.
67. Massa, A.; Siniscalchi, F. R.; Bugatti, V.; Lattanzi, A.; Scettri, A., *Tetrahedron: Asymmetry* **2002**, *13*, 1277.
68. Lattanzi, A.; Scettri, A., *Current Org. Chem.* **2004**, *8*, 607.
69. Mikami, K.; Matsukawa, S., *J. Am. Chem. Soc.* **1994**, *116*, 4077.
70. Matsukawa, S.; Mikami, K., *Tetrahedron: Asymmetry* **1995**, *6*, 2571.
71. Zimmer, R.; Peritz, A.; Czerwonka, R.; Schefzig, L.; Reissig, H.-U., *Eur. J. Org. Chem.* **2002**, 3419.
72. Bluet, G.; Campagne, J.-M., *Tetrahedron Lett.* **1999**, *40*, 5507.

73. Sato, M.; Sunami, S.; Sugita, Y.; Kaneko, C., *Heterocycles* **1995**, *41*, 1435.
74. De Rosa, M.; Acocella, M. R.; Soriente, A.; Scettri, A., *Tetrahedron: Asymmetry* **2001**, *12*, 1529.
75. De Rosa, M.; Acocella, M. R.; Villano, R.; Soriente, A.; Scettri, A., *Tetrahedron: Asymmetry* **2003**, *14*, 2499.
76. De Rosa, M.; Acocella, M. R.; Villano, R.; Soriente, A.; Scettri, A., *Tetrahedron Lett.* **2003**, *44*, 6087.
77. De Rosa, M.; Dell'Aglio, R.; Soriente, A.; Scettri, A., *Tetrahedron: Asymmetry* **1999**, *10*, 3659.
78. De Rosa, M.; Soriente, A.; Scettri, A., *Tetrahedron: Asymmetry* **2000**, *11*, 3187.
79. Soriente, A.; De Rosa, M.; Stanzione, M.; Villano, R.; Scettri, A., *Tetrahedron: Asymmetry* **2001**, *12*, 959.
80. Soriente, A.; De Rosa, M.; Dovinola, P.; Sodano, G.; Scettri, A., *Tetrahedron: Asymmetry* **1998**, *9*, 2197.
81. Soriente, A.; De Rosa, M.; Villano, R.; Scettri, A., *Tetrahedron: Asymmetry* **2000**, *11*, 2255.
82. Villano, R.; De Rosa, M.; Salerno, C.; Soriente, A.; Scettri, A., *Tetrahedron: Asymmetry* **2002**, *13*, 1949.
83. Martin, S. F.; Lopez, O. D., *Tetrahedron Lett.* **1999**, *40*, 8949.
84. Yu, C. M.; Yoon, S. K.; Baek, K.; Lee, J. Y., *Angew. Chem. Int. Ed. Engl.* **1998**, *37*, 2392.
85. Yu, C. M.; Lee, S. L.; Jeon, M., *J. Chem. Soc., Perkin Trans. 1* **1999**, 3557.
86. Yu, C. M.; Jeon, M.; Lee, J. Y.; Jeon, J., *Eur. J. Org. Chem.* **2001**, 1143.
87. Yu, C. M.; Kim, J. M.; Shin, M. S.; Cho, D., *Tetrahedron Lett.* **2003**, *44*, 5487.
88. Yu, C. M.; Kim, J. M.; Shin, M. S.; Yoon, M. O., *Chem. Commun.* **2003**, 1744.
89. Yu, C. M.; Le, J. Y.; So, B.; Hong, J., *Angew. Chem. Int. Ed. Engl.* **2002**, *41*, 161.
90. Bedeschi, P.; Casolari, S.; Costa, A. L.; Tagliavini, E.; Umani-Ronchi, A., *Tetrahedron Lett.* **1995**, *36*, 7897.
91. Hanawa, H.; Kii, S.; Asao, N.; Maruoka, K., *Tetrahedron Lett.* **2000**, *41*, 5543.
92. Kurosu, M.; Lorca, M., *Tetrahedron Lett.* **2002**, *43*, 1765.
93. Casolari, S.; Cozzi, P. G.; Orioli, P., *Chem. Commun.* **1997**, 2123.
94. Lu, G.; Li, X.; Chan, W. L.; Chan, A. S. C., *Chem. Commun.* **2002**, 172.
95. Lu, G.; Li, X.; Chen, G.; Chan, W. L.; Chan, A. S. C., *Tetrahedron: Asymmetry* **2003**, *14*, 449.
96. Moore, D.; Pu, L., *Org. Lett.* **2002**, *4*, 1855.
97. Li, X.; Lu, G.; Kwok, W. H.; Chan, A. S. C., *J. Am. Chem. Soc.* **2002**, *124*, 12636.
98. Gao, G.; Moore, D.; Xie, R.-G.; Pu, L., *Org. Lett.* **2002**, *4*, 4143.
99. Gao, G.; Xie, R. G.; Pu, L., *Proc. Nat. Acad. Sci.* **2004**, *101*, 5417.
100. Zhou, Y.; Wang, R.; Xu, Z.; Yan, W.; Liu, L.; Kang, Y.; Han, Z., *Org. Lett.* **2004**, *6*, 4147.
101. Hattori, K.; Yamamoto, H., *J. Org. Chem.* **1992**, *57*, 3264.
102. Cros, J. P.; Perez-Fuertes, Y.; Thatcher, M. J.; Arimori, S.; Bull, S. D.; James, T. D., *Tetrahedron: Asymmetry* **2003**, *14*, 1965.
103. Hattori, K.; Yamamoto, H., *Tetrahedron* **1993**, *49*, 1749.
104. Hattori, K.; Yamamoto, H., *Synlett* **1993**, 129.
105. Ishitani, H.; Kobayashi, S., *Tetrahedron Lett.* **1996**, *37*, 7357.
106. Mori, M.; Nakai, T., *Tetrahedron Lett.* **1997**, *38*, 6233.
107. Zhang, F. Y.; Yip, C. W.; Cao, R.; Chan, A. S. C., *Tetrahedron: Asymmetry* **1997**, *8*, 585.
108. Holmes, I. P.; Kagan, H. B., *Tetrahedron Lett.* **2000**, *41*, 7453.
109. Nakamura, S.; Sato, N.; Sugimoto, M.; Toru, T., *Tetrahedron: Asymmetry* **2004**, *15*, 1513.
110. Nakamura, S.; Sato, N.; Sugimoto, M.; Toru, T., *Tetrahedron: Asymmetry* **2004**, *15*, 1513.
111. Shimizu, S.; Ohori, K.; Arai, T.; Sasai, H.; Shibasaki, M., *J. Org. Chem.* **1998**, *63*, 7547.
112. Sasai, H.; Arai, T.; Satow, Y.; Houk, K. N.; Shibasaki, M., *J. Am. Chem. Soc.* **1995**, *117*, 6194.
113. Xu, Y.; Ohori, K.; Ohshima, T.; Shibasaki, M., *Tetrahedron* **2002**, *58*, 2585.
114. Kumaraswamy, G.; Sastry, M. N. V.; Jena, N., *Tetrahedron Lett.* **2001**, *42*, 8515.
115. Keller, E.; Veldman, N.; Spek, A. L.; Feringa, B. L., *Tetrahedron: Asymmetry* **1997**, *8*, 3403.
116. Sasai, H.; Kim, W.-S.; Suzuki, T.; Shibasaki, M.; Mitsuda, M.; Hasegawa, J.; Ohashi, T., *Tetrahedron Lett.* **1994**, *35*, 6123.
117. Yokomatsu, T.; Yamagishi, T.; Shibuya, S., *J. Chem. Soc., Perkin Trans. 1* **1997**, 1527.
118. Yokomatsu, T.; Yamagishi, T.; Shibuya, S., *Tetrahedron: Asymmetry* **1997**, *4*, 1783.
119. Rath, N. P.; Spilling, C. D., *Tetrahedron Lett.* **1994**, *35*, 227.
120. Sasai, H.; Arai, S.; Tahara, Y.; Shibasaki, M., *J. Org. Chem.* **1995**, *60*, 6656.
121. Yamagishi, T.; Suemune, K.; Yokomatsu, T.; Shibuya, S., *Tetrahedron* **2002**, *58*, 2577.
122. Yamagishi, T.; Yokomatsu, T.; Suemune, K.; Shibuya, S., *Tetrahedron* **1999**, *55*, 12125.
123. Arai, T.; Bougauchi, M.; Sasai, H.; Shibasaki, M., *J. Org. Chem.* **1996**, *61*, 2926.
124. Hou, X. L.; Wu, J.; Dai, L. X.; Xia, L. J.; Tang, M. H., *Tetrahedron: Asymmetry* **1998**, *9*, 1747.
125. Fu, X. L.; Wu, S. H., *Synth. Commun.* **1997**, *27*, 1677.
126. Matsunaga, S.; Ohshima, T.; Shibasaki, M., *Tetrahedron Lett.* **2000**, *41*, 8473.
127. Matsunaga, S.; Das, J.; Roels, J.; Vogl, E. M.; Yamamoto, N.; Iida, T.; Yamaguchi, K.; Shibasaki, M., *J. Am. Chem. Soc.* **2000**, *122*, 2252.
128. Faller, J. W.; Milheiro, S. C.; Parr, J., *J. Organomet. Chem.* **2006**, *691*, 4945.
129. Murai, T.; Inaji, S.; Morishita, K.; Shibahara, F.; Tokunaga, M.; Obora, Y.; Tsuji, Y., *Chem. Lett.* **2006**, *35*, 1424.
130. Mrsic, N.; Lefort, L.; Boogers, J. A. F.; Minnaard, A. J.; Feringa, B. L.; De Vries, J. G., *Adv. Synth. Catal.* **2008**, *350*, 1081.
131. Mrsic, N.; Minnaard, A. J.; Feringa, B. L.; De Vries, J. G., *J. Am. Chem. Soc.* **2009**, *131*, 8358.
132. (a) Jiang, Q.; Jiang, Y.; Xiao, D.; Cao, P.; Zhang, X., *Angew. Chem., Int. Ed.* **1998**, *37*, 1100. (b) Fehring, V.; Selke, R., *Angew. Chem., Int. Ed.* **1998**, *37*, 1827.
133. Junge, K.; Hagemann, B.; Enthaler, S.; Oehme, G.; Michalik, M.; Monsees, A.; Riermeier, T.; Dingerdissen, U.; Beller, B., *Angew. Chem., Int. Ed.* **2004**, *43*, 5066.
134. Bernsmann, H.; van den Berg, M.; oen, R.; Minnaard, A. J.; Mehler, G.; Reetz, M. T.; De Vries, J. G.; Feringa, B. L., *J. Org. Chem.* **2005**, *3*, 943.
135. Hagemann, B.; Junge, K.; Enthaler, S.; Michalik, M., Riermeier, T.; Monsees, A.; Beller, M., *Adv. Synth. Catal.* **2005**, *15*, 1978.
136. Jahjah, M.; Alame, M.; Pellet-Rostaing, S.; Lemaire, M., *Tetrahedron : Asymmetry* **2007**, *19*, 2305.
137. Xu, Y.; Clarkson, G. C.; Docherty, G.; North, C. L.; Woodward, G.; Wills, M., *J. Org. Chem.* **2005**, *20*, 8079.
138. Andersson, H.; Gustafsson, M.; Boström, D.; Olsson, R.; Almqvist, F., *Angew.Chem., Int. Ed.* **2009**, *48*, 3288.
139. Hussain, M.; Sainte-Luce Banchelin, T.; Andersson, H.; Olsson, R.; Almqvist, F., *Org. Lett.* **2013**, *15*, 54.
140. Yamaguchi, A.; Matsunaga, S.; Shibasaki, M., *J. Am. Chem. Soc.* **2009**, *131*, 10842.
141. Phua, P. H.; White, A. J. P.; De Vries, J. G.; Hii, K. K., *Adv. Synth. Catal.* **2010**, *348*, 587.

142. Didier, D.; Meddour, A.; Bezzenine-Laffolee, S.; Collin, J., *Eur. J. Org. Chem.* **2011**, *14*, 2678.
143. Tian, X.; Cassani, C.; Liu, Y.; Moran, A.; Urakawa, A.; Galzerano, P.; Arceo, E.; Melchiorre, P., *J. Am. Chem. Soc.* **2011**, *133*, 17934.
144. Takaya, Y.; Ogasawara, M.; Hayashi, T.; Sakai, M.; Miyaura, N., *J. Am. Chem. Soc.* **1998**, *120*, 5579.
145. (a) Hayashi, T.; Yamasaki, K., *Chem. Rev.* **2003**, *103*, 2829. (b) Edwards, H. J.; Hargrave, J. D.; Penrose, S. D.; Frost, C. G., *Chem. Soc. Rev.* **2010**, *39*, 2093.
146. Hickmann, V.; Kondoh, A.; Gabor, B.; Alcarazo, M.; Fürstner, A., *J. Am. Chem. Soc.* **2011**, *133*, 13471.
147. (a) Lin, S.; Lu, X., *Org. Lett.* **2010**, *12*, 2536. (b) Kikushima, K.; Holder, J. C.; Gatti, M.; Stoltz, B. M., *J. Am. Chem. Soc.* **2011**, *133*, 6902.
148. Wu, T. R.; Chong, J. M., *J. Am. Chem. Soc.* **2007**, *129*, 4908.
149. Wu, T. R.; Chong, J. M., *J. Am. Chem. Soc.* **2005**, *127*, 3244.
150. Turner, H. M.; Patel, J.; Niljianskul, N.; Chong, J. M., *Org. Lett.* **2011**, *13*, 5796.
151. Luan, Y.; Schaus, S. E., *J. Am. Chem. Soc.* **2012**, *134*, 19965.
152. Roggen, M.; Carreira, E. M., *Angew. Chem., Int. Ed.* **2012**, *51*, 8652.
153. Roggen, M.; Carreira, E. M., *Angew.Chem., Int. Ed.* **2011**, *24*, 5568.
154. (a) Trost, B. M., *Angew. Chem., Int. Ed.* **1986**, *25*, 1. (b) Trost, B. M., *Pure Appl. Chem.* **1988**, *60*, 1615. (c) Lautens, M.; Klute, W.; Tam, W., *Chem. Rev.* **1996**, *96*, 49. (d) Yamago, S.; Nakamura, E., *Org. React.* **2002**, *61*, 1.
155. Yamamoto, A.; Ito, Y.; Hayashi, T., *Tetrahedron Lett.* **1989**, *30*, 375.
156. Trost, B. M.; Stambuli, J. P.; Silverman, S. M.; Schwörer, U., *J. Am.Chem. Soc.* **2006**, *128*, 13328.
157. Trost, B. M.; Silverman, S. M.; Stambuli, J. P., *J. Am. Chem. Soc.* **2011**, *133*, 19483.
158. (a) Evans, D. A.; Faul, M. M.; Bilodeau, M. T.; Anderson, B. A.; Barnes, D. M., *J. Am. Chem. Soc.* **1993**, *115*, 5328. (b) Li, Z.; Conser, K. R.; Jacobsen, E. N., *J. Am. Chem. Soc.* **1993**, *115*, 5326. (c) Omura, K.; Murakami, M.; Uchida, T.; Irie, R. ; Katsuki, T., *Chem. Lett.* **2003**, *32*, 354. (d) Subbarayan, V.; Ruppel, J. V.; Zhu, S.; Perman, J. A.; Zhang, X. P., *Chem. Commun.* **2009**, 4266.
159. (a) Vesely, J.; Ibrahem, I.; Zhao, L.; Rios, R.; Cordova, A., *Angew. Chem., Int. Ed.* **2007**, *46*, 778. (b) Hashimoto, T.; Uchiyama, N.; Maruoka, K., *J. Am. Chem. Soc.* **2008**, *130*, 14380. (c) Minakata, S.; Murakami, Y.; Tsuruoka, R.; Kitanaka, S.; Komatsu, M., *Chem. Commun.* **2008**, 6363. (d) Huang, L.; Wulff, W. D., *J. Am. Chem. Soc.* **2011**, *133*, 8892.
160. Antilla, J. C.; Wulff, W. D., *J. Am. Chem. Soc.* **1999**, *121*, 5099.
161. Jones, J. E.; Ruppel, J. V.; Gao, G. Y.; Moore, T. M.; Zhang, X. P., *J. Org. Chem.* **2008**, *73*, 7260.
162. Bulman Page, P.C.; Buckley, B. B. R.; Farah, M. M.; Blacker, A. J., *Eur. J. Org. Chem.* **2009**, 3413.
163. Bulman Page, P.C.; Farah, M. M.; Buckley, B. R.; Blacker, A. J., *J. Org. Chem.* **2007**, *72*, 4424.
164. Corey, E. J.; Chaykovsky, M., *J. Am. Chem. Soc.* **1965**, *87*, 1353.
165. Zhang, Y.; Wu, W., *Tetrahedron: Asymmetry*. **1997**, *8*, 2723.
166. Sone, T.; Yamaguchi, A.; Matsunaga, S.; Shibasaki, M., *J. Am. Chem. Soc.* **2008**, *130*, 10078.
167. Rosen, T.*Comprehensive Organic Synthesis*; Pergamon: Oxford, 1991, Vol. 2, p. 409.
168. (a) Arai, S.; Shioiri, T., *Tetrahedron Lett.* **1998**, *39*, 2145. (b) Arai, S.; Shirai, Y.; Ishida, T.; Shioiri, T., *Tetrahedron Lett.* **1998**, *39*, 8299.
169. (a) Bakó, P.; Szöllõsy, A.; Bombicz, P.; Tõke, L., *Synlett* **1997**, 291. (b) Bakó, P.; Vizvárdi, K.; Bajor, Z.; Vizvárdi, K.; Tõke, L., *Chem. Commun.* **1998**, 1193.
170. Achard, T. J. R.; Belokon, Y. N.; Ilyin, M.; Moskalenko, M.; North, M.; Pizzato, F., *Tetrahedron Lett.* **2007**, *48*, 2965.
171. (a) Imashiro, R.; Yamanaka, T.; Seki, M., *Tetrahedron: Asymmetry* **1999**, *10*, 2845. (b) Aggarwal, V. K.; Hynd, G.; Picoul, W.; Vasse, J. L., *J. Am. Chem. Soc.* **2002**, *124*, 9964. (c) Aggarwal, V. K.; Charmant, J. P. H.; Fuentes, D.; Harvey, J. N.; Hynd, G.; Ohara, D.; Picoul, W.; Smith, C.; Vasse, J. L.; Winn, C. L., *J. Am. Chem. Soc.* **2006**, *128*, 2105.
172. Liu, W. J.; Lv, B. D.; Gong, L. Z., *Angew. Chem., Int. Ed.* **2009**, *48*, 6503.
173. Tosaki, S. Y.; Tsuji, R.; Ohshima, T.; Shibasaki, M., *J. Am. Chem. Soc.* **2005**, *7*, 2147.
174. Schafroth, M. A.; Sarlah, D.; Krautwald, S.; Carreira, E. M., *J. Am. Chem. Soc.* **2012**, *134*, 20276.
175. Hulst, R.; De Vries, N. K.; Feringa, B. L., *Tetrahedron: Asymmetry* **1994**, *5*, 699.
176. Van den Berg, M.; Minnaard, A. J.; Schudde, E. P.; Van Esch, J.; DeVries, A. H. M.; De Vries, J. G.; Feringa, B. L., *J. Am. Chem. Soc.* **2000**, *122*, 11539.
177. Rubina, M.; Rubin, M.; Gevorgyan, V., *J. Am. Chem. Soc.* **2003**, *125*, 7198.
178. Bernsmann, H.; van den Berg, M.; Hoen, R.; Minnaard, A. J.; Mehler, G.; Reetz, M. T.; De Vries, J. G.; Feringa, B. L., *J. Org. Chem.* **2005**, *70*, 943.
179. Smith, S. M.; Thacker, N. C.; Takacs, J. M., *J. Am. Chem. Soc.* **2008**, *130*, 3734.
180. Han, W. Y.; Zuo, J.; Zhang, X. M.; Yuan, W. C., *Tetrahedron* **2013**, *69*, 537.
181. Peruncheralathan, S.; Aurich, S.; Teller, H.; Schneider, C., *Org. Biomol. Chem.* **2013**, *11*, 2787.
182. Peruncheralathan, S.; Aurich, S.; Teller, H.; Schneider, C., *Angew. Chem., Int. Ed.* **2009**, *48*, 4849.
183. Cockrell, J.; Wilhelmsen, C.; Rubin, H.; Martin, A.; Morgan, J. B., *Angew. Chem., Int. Ed.* **2012**, *51*, 9842.
184. More, G. V.; Bhanage, B. M., *Eur. J. Org. Chem.* **2013**, *30*, 6900.
185. Eschenbrenner-Lux, V.; Kuechler, P.; Ziegler, S.; Kumar, K.; Waldmann, H., *Angew. Chem., Int. Ed.* **2014**, *53*, 2134.
186. He, L.; Zhao, L.; Wang, D. X.; Wang, M. X., *Org. Lett.* **2014**, *16*, 5972.
187. Li, Y.; Su, Y. H.; Dong, D. J.; Wu, Z.; Tian, S. K., *RSC Adv.* **2013**, *3*, 18275.
188. Huang, G.; Yin, Z.; Zhang, X., *Chem. Eur. J.* **2013**, *19*, 11992.
189. Du, X.; Wang, Q.; He, X.; Peng, R. G.; Zhang, X.; Yu, X. Q., *Tetrahedron: Asymmetry* **2011**, *22*, 1142.
190. Takita, R.; Yakura, K.; Ohshima, T.; Shibasaki, M., *J. Am. Chem. Soc.* **2005**, *127*, 13760.
191. Minatti, A.; Doetz, K. H., *Eur. J. Org.* **2005**, 268.
192. Majer, J.; Kwiatkowski, P.; Jurczak, J., *Org. Lett.* **2011**, *13*, 5944.
193. Hatano, M.; Horibe, T.; Ishihara, K., *Org. Lett.* **2010**, *12*, 3502.
194. Blay, G.; Ceballos, E.; Monleon, A.; Pedro, J. R., *Tetrahedron* **2012**, *68*, 2128.
195. Wang, G.; Zhao, J.; Zhou, Y.; Wang, B.; Qu, J., *J. Org. Chem.* **2010**, *75*, 5326.
196. Hara, K.; Tosaki, S.; Gnanadesikan, V.; Morimoto, H.; Harada, S.; Sugita, M.; Yamagiwa, N.; Matsunaga, S.; Shibasaki, M., *Tetrahedron* **2009**, *65*, 5030.
197. Sohtome, Y.; Kato, Y.; Handa, S.; Aoyama, N.; Nagawa, K.; Matsunaga, S.; Shibasaki, M., *Org. Lett.* **2008**, *10*, 2231.
198. Villano, R.; Acocella, M. R.; De Rosa, M.; Soriente, A.; Scettri, A., *Tetrahedron: Asymmetry* **2004**, *15*, 2421.
199. Gnanadesikan, V.; Horiuchi, Y.; Ohshima, T.; Shibasaki, M., *J. Am. Chem. Soc.* **2004**, *126*, 7782.
200. Du, H.; Zhang, X.; Wang, Z.; Bao, H.; You, T.; Ding, K., *Eur. J. Org. Chem.* **2008**, 2248.
201. Lin, L.; Chen, Z.; Yang, X.; Liu, X.; Feng, X., *Org. Lett.* **2008**, *10*, 1311.
202. Gao, B.; Fu, Z.; Yu, Z.; Yu, L.; Huang, Y.; Feng, X., *Tetrahedron* **2005**, *61*, 5822.
203. Golebiewski, W. M.; Gucma, M., *J. Heterocycl. Chem.* **2008**, *45*, 1687.

1,1′-Binaphthyl-2,2′-disulfonic Acid

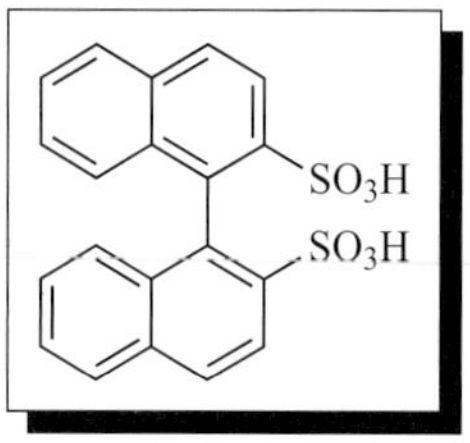

[122361-59-5]
[852127-05-0]
[1092934-18-3] $C_{20}H_{14}O_6S_2$ (MW 414.4516)
InChI = 1S/C20H14O6S2/c21-27(22,23)17-11-9-13-5-1-3-7-15(13)19(17)20-16-8-4-2-6-14(16)10-12-18(20)28(24,25)26/h1-12H,(H,21,22,23)(H,24,25,26)
InChIKey = ZSWODTYSMOCDKE-UHFFFAOYSA-N

(organocatalyst and ligand for metal complexes)

Alternate Names: BINSA; 1,1′-binaphthalene-2,2′-disulfonic acid.

Physical Data: mp 156–160 °C; $[\alpha]_D^{23} = +61.4°$ (*c* 2.2, MeOH).

Solubility: sol THF, DMF, CH_3CN, CH_2Cl_2, $CHCl_3$, MeOH, EtOH, H_2O; less soluble in Et_2O, EtOAc, toluene.

Handling, Storage, and Precautions: moisture sensitive. Keep tightly closed under an inert gas atmosphere; store in a cool dark place.

Preparative Methods: the potassium salt of racemic 1,1′-binaphthyl-2,2′-disulfonic acid (BINSA) was obtained by reacting potassium 1-iodonaphthalene-2-sulfonate with copper powder in an Ullmann coupling reaction.[1] Similarly, the sodium salt of racemic BINSA was obtained from diphenyl 1,1′-binaphthalene-2,2′-disulfonate, which was prepared from phenyl 1-iodonaphthalene-2-sulfonate with copper powder in an Ullmann coupling reaction, by treatment with sodium/1-butanol under reflux conditions.[2] Ishihara and coworkers reported the practical asymmetric synthesis of optically pure (*R*)-BINSA from commercially available (*R*)-BINOL ((*R*)-1,1′-bi(2-naphthol)) in four steps (eq 1).[3,4] *O*-Thiocarbamoylation of (*R*)-BINOL, followed by thermolysis in Newman–Kwart rearrangement to the *S*-thiocarbamoyl compound by a microwave technique at milder temperature (200 °C) relative to traditional thermal conditions (270–400 °C),[5–7] reduction to the dithiol by $LiAlH_4$, oxidation by 10 atm of O_2/KOH in HMPA, and protonation by ion exchange provided (*R*)-BINSA on a >10 g scale without loss of optical purity.

Synthesis of Chiral 1,1′-Binaphthyl-2,2′-disulfonimide. List and coworkers synthesized (*R*)-3,3′-[3,5-$(CF_3)_2C_6H_3$]$_2$-BINSA and the corresponding chiral disulfonimide from (*R*)-3,3′-[3,5-$(CF_3)_2C_6H_3$]$_2$-BINOL in five steps (eq 2).[8] By taking advantage of bulky electron-withdrawing substituents at the 3,3′-positions, an intramolecular nucleophilic attack at the 2,2′-*ipso*-carbons in Newman–Kwart rearrangement was facilitated, and the *S*-thiocarbamoyl product was obtained in 95% yield.[9,10] Subsequent oxidation was conducted by treatment with 30 equiv of H_2O_2/HCO_2H at room temperature for 1 h, and (*R*)-3,3′-[3,5-$(CF_3)_2C_6H_3$]$_2$-BINSA was obtained in 81% yield. Moreover, after the reaction with NH_3/MeOH, the desired (*R*)-3,3′-[3,5-$(CF_3)_2C_6H_3$]$_2$-BINSA-derived disulfonimide was obtained in 70% yield.

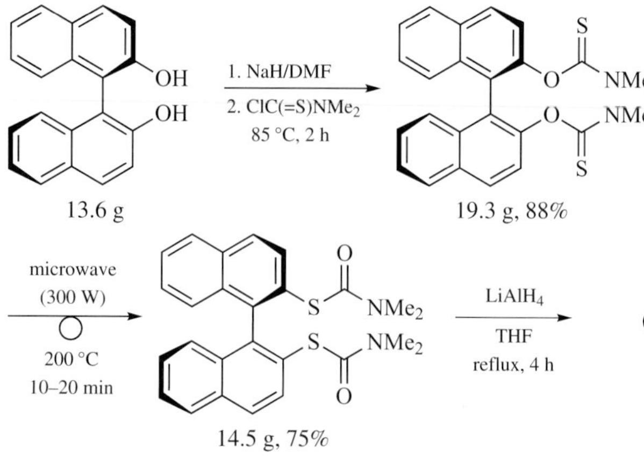

(1)

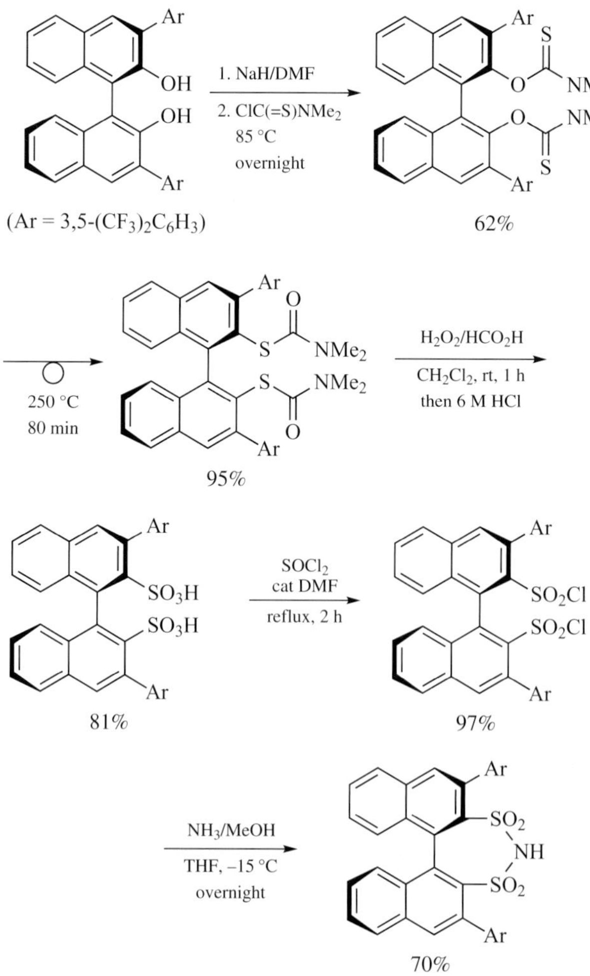

(2)

Giernoth and coworkers reported the alternative oxychlorination of *S*-thiocarbamoyl compound to (*R*)-1,1′-binaphthyl-2,2′-disulfonyl dichloride using *N*-chlorosuccinimide (NCS) (eq 3).[11] Subsequent treatment with gaseous ammonia in benzene at room temperature provided the desired (*R*)-BINSA-derived disulfonimide in 91% yield.

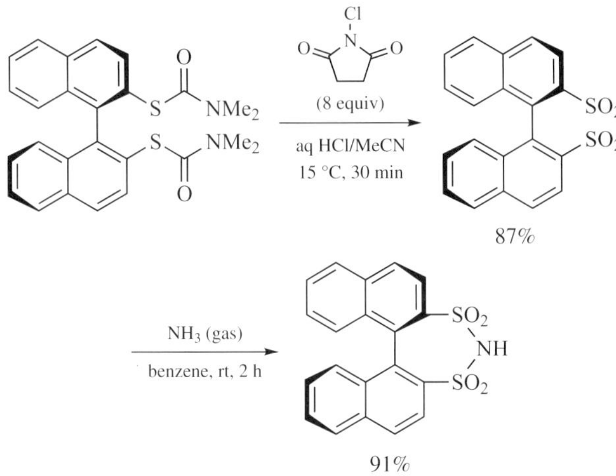

(3)

Synthesis of Chiral 3,3′-Disubstituted BINSAs. A convenient synthesis of chiral 3,3′-disubstituted BINSAs was developed by Ishihara and coworkers.[4] The key step involved a directed *ortho*-lithiation of (*R*)-BINSA methyl ester, which was prepared from (*R*)-BINSA and trimethyl orthoacetate in dichloromethane in 88% yield (eq 4), followed by reaction with *n*-BuLi and a subsequent reaction with an electrophile. The reaction of (*R*)-3,3′-dilithiated BINSA methyl ester with electrophiles such as Br_2 and I_2 gave the corresponding (*R*)-3,3′-dihalo-BINSA derivatives in high yields (eqs 5 and 6).[4]

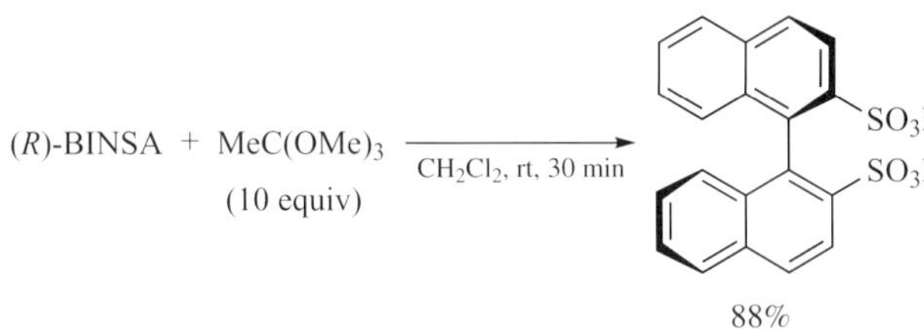

(4)

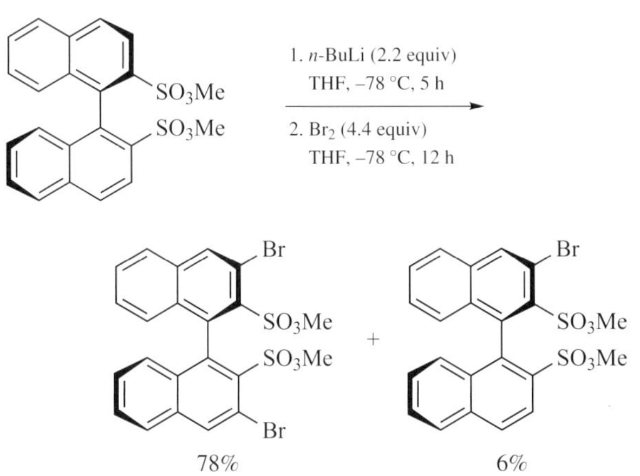

(5)

The reaction of (*R*)-BINSA methyl ester with *n*-BuLi (2.2 equiv) and then trimethylsilyl chloride (TMSCl) (4.4 equiv) in THF at –78 °C gave the desired product in 21% yield along with monosubstituted compound in 20% yield (eq 7).[4] The yield was further improved when trimethylsilyl triflate (TMSOTf) was used in place of TMSCl, and the corresponding hydrolytically cleaved 3,3′-disubstituted compound was obtained from the aqueous phase in 62% yield (eq 8).[4]

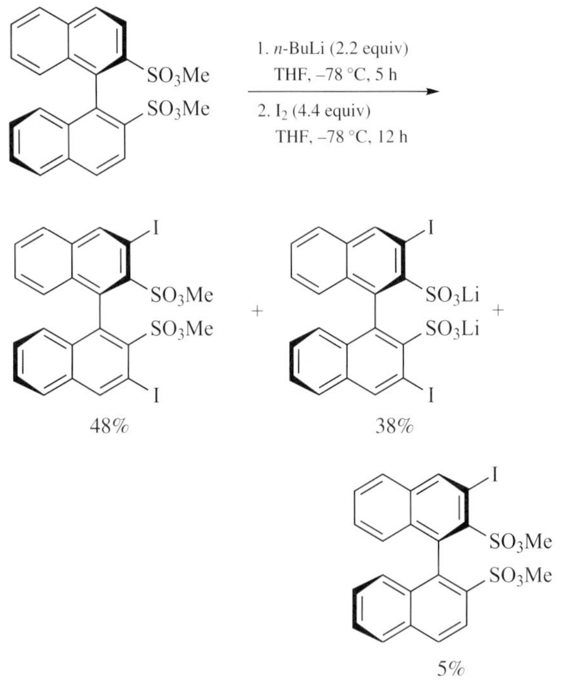

(6)

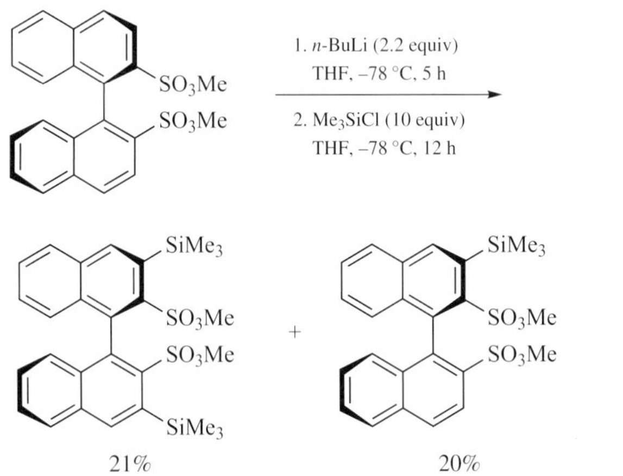

(7)

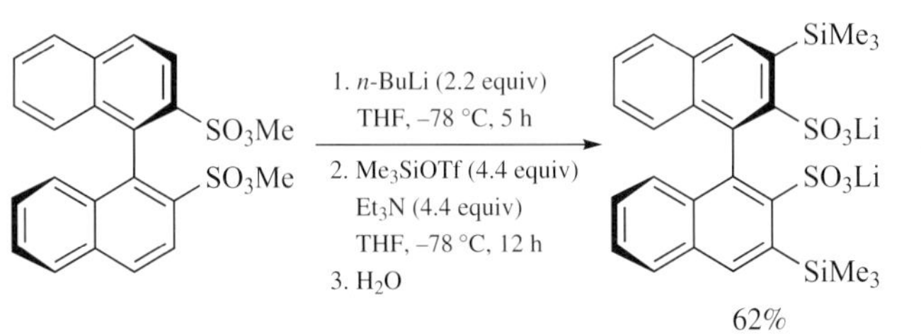

(8)

The 3,3′-borylation of (*R*)-BINSA methyl ester with isopropyl pinacol borate proceeded smoothly in THF at –78 °C for 2 h, and the desired compound was isolated in 71% yield (eq 9).[4] Moreover, the compound obtained was readily transformed to the corresponding boronic acid in quantitative yield by treatment with $HSiEt_3$ in the presence of 20 mol % of $B(C_6F_5)_3$ in dichloromethane at room temperature and was accompanied by hydrolysis of the SO_3Me moieties.

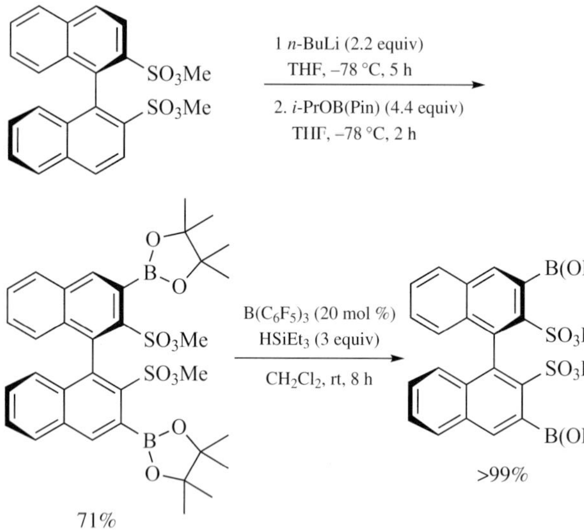

(9)

A 3,3′-diiodo compound and a 3,3′-diboryl compound were transformed to the 3,3′-diphenyl compound by Suzuki–Miyaura cross-coupling. The coupling reaction in the presence of $PdCl_2$(dppf) (20 mol %), dppf (1,1′-bis(diphenylphosphino)ferrocene) (10 mol %), and K_2CO_3 (2 equiv) in DMF at 100 °C gave the desired 3,3′-diphenyl compound in respective yields of 7 and 13% (eqs 10 and 11).[4]

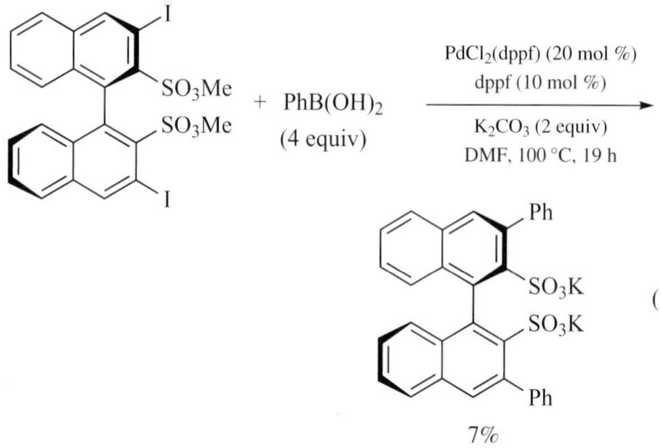

(10)

The Newman–Kwart rearrangement of a bulky but electron-donating 3,3′-Ph_2-substituted *O*-thiocarbamoyl compound under microwave conditions (200 °C for 25 min or 250 °C for 45 min) provided a *S*-thiocarbamoyl compound in yields of only 10–25% (eq 12).[4] A significant amount of a thiophene by-product and a black, "burned"-looking material were obtained, although the starting material was completely converted. Moreover, subsequent reduction and oxidation steps also provided unfavorable results. Therefore, in sharp contrast to unsubstituted BINSA, the presence of substitutions at the 3,3′-positions appears to prevent the chemoselective Newman–Kwart rearrangement and subsequent reactions.

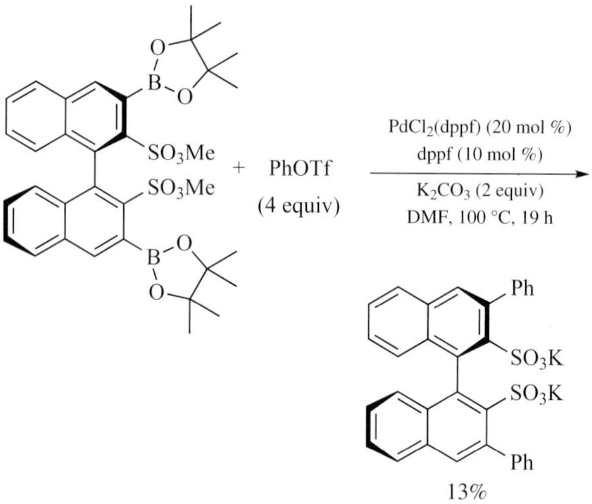

(11)

Asymmetric Organocatalysis by Chiral BINSA. List and coworkers reported the catalytic asymmetric acylcyanation of aldimine with some chiral Brønsted acid catalysts.[12] In contrast to chiral thiourea catalysts, BINSA promoted the reaction, but the racemic product was obtained in 68% yield (eq 13).

Pan and List also reported Brønsted acid-catalyzed three-component Hosomi–Sakurai reactions among aldehydes, silylethers, and allylsilanes.[13] BINSA provided the desired homoallyl benzyl ether in 80% yield, but no measurable enantioselectivity was observed (eq 14).

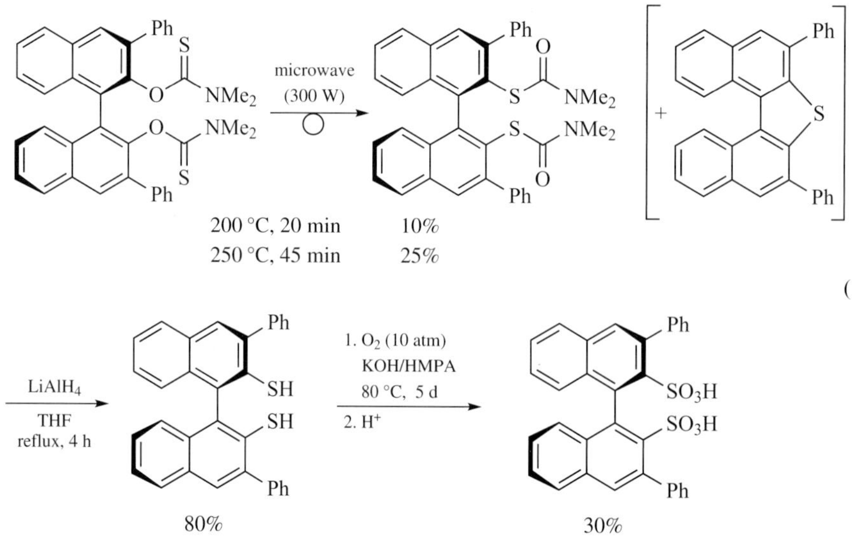

(12)

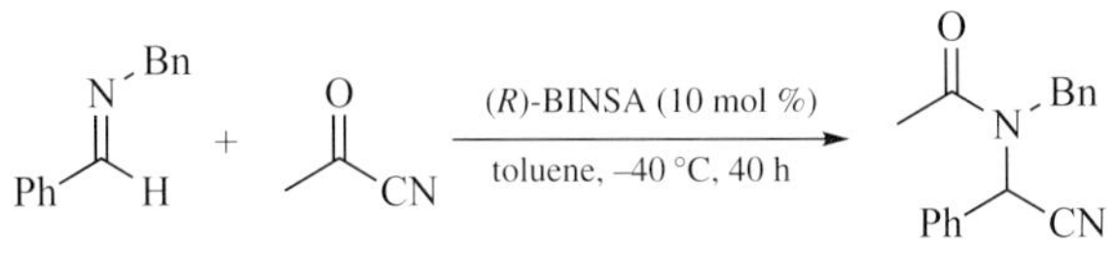

(13)

68%, 0% ee

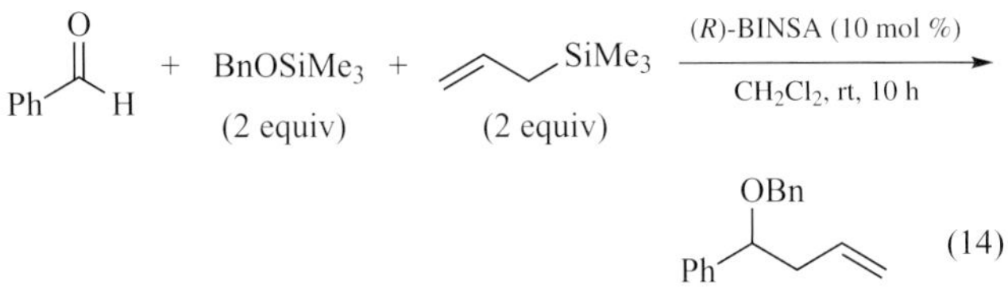

(14)

80%, <5% ee

Asymmetric Organocatalysis by Chiral BINSA–Pyridinium Salts. In general, acid–base combined salts have several advantages over single-molecule catalysts, with regard to flexibility in the design of their dynamic complexes. Ishihara and coworkers reported that BINSA was a highly effective chiral Brønsted acid that could be combined with 2,6-biarylpyridine as an achiral Brønsted base, and the acidity as well as the bulkiness could be tuned in situ (eq 15).[3] The combination of the achiral bulky 2,6-biarylpyridine with the structurally simple BINSA circumvented the tedious synthesis of bulky substituents at the 3,3′-position, as is normally required for the analogous binaphthyl phosphoric acid catalysts. In the presence of 1 mol % of (*R*)-BINSA and 2 mol % of 2,6-diphenylpyridine, highly enantioselective direct Mannich-type reactions of a variety of 1,3-diketones with arylaldimines proceeded smoothly with high enantioselectivities (eq 16).

A suitable chiral ammonium salt was easily tailor-made for a ketoester equivalent such as 3-acetoacetyl-2-oxazolidinone (eq 17). Although the chiral ammonium salt (*R*)-BINSA–2,6-diphenylpyridine was not so effective in this reaction, high enantioselectivities (93% and 90% ee) were observed when 2,6-dimesitylpyridine was used in place of 2,6-diphenylpyridine. In this way, these salts made it possible to avoid preparing single-molecule catalysts in advance and offered a quick solution to this type of optimization problem. The product obtained was easily transformed to the β-amino carbonyl compound via deprotection of the oxazolidinone moiety without loss of enantioselectivity.

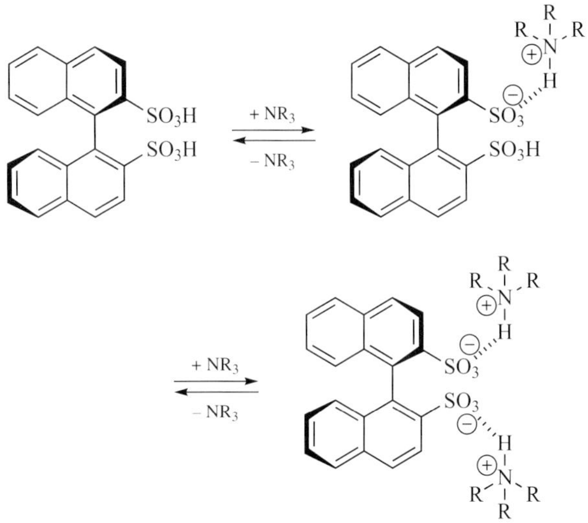

(15)

Asymmetric Catalysis by Chiral BINSA–La(III) Complexes. BINSA was thought to be a potentially promising chiral ligand for metal-mediated enantioselective catalysis. Ishihara and coworkers reported the catalytic enantioselective Strecker reaction of aldimines with trimethylsilyl cyanide (TMSCN) catalyzed by novel chiral (*R*)-BINSA–lanthanum(III) complexes (eq 18).[14] The key to promoting the reactions was a semistoichiometric amount (50 mol %) of AcOH or *i*-$PrCO_2H$, which takes advantage of the in situ generation of HCN. The corresponding cyanation products from aromatic and heteroaromatic aldimines bearing electron-withdrawing and electron-donating groups were obtained in high yields and with high enantioselectivities. ESI-MS analysis of the catalyst suggested the predominant generation of monomeric La(III) complexes, $[Ar^*(SO_3)_2La(MeCN)_n]^+$ ($Ar^*(SO_3H)_2$ = BINSA, n = 1–4) (eq 19).

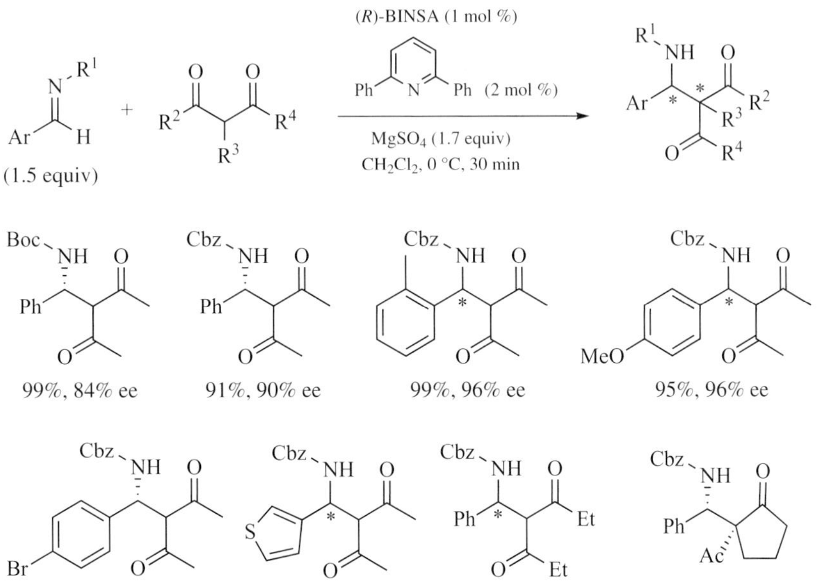

(16)

99%, 84% ee; 91%, 90% ee; 99%, 96% ee; 95%, 96% ee

92%, 98% ee; 98%, 98% ee; 95%, 95% ee; 98% (dr = 83/17), 91% ee

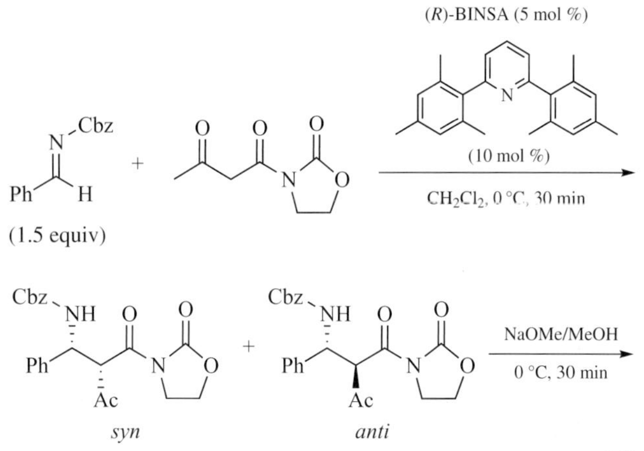

81% (*syn*/*anti* = 60/40)
93% ee (*syn*) and 90% ee (*anti*)

62% (*syn*/*anti* = 58/42)

(17)

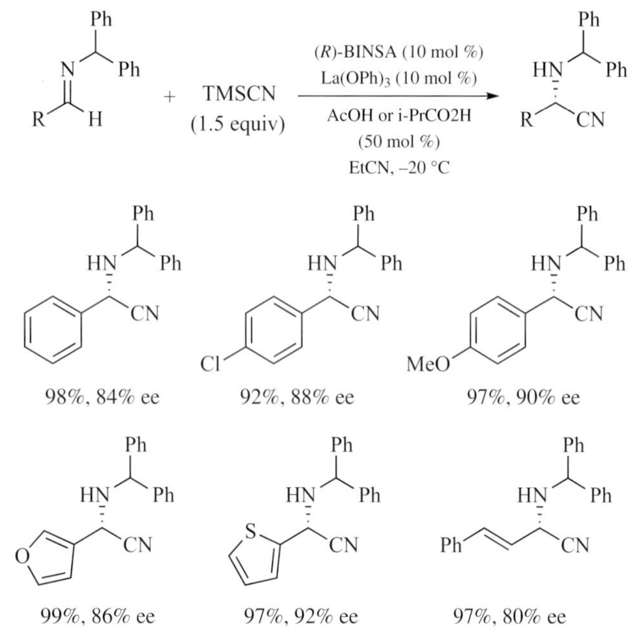

(18)

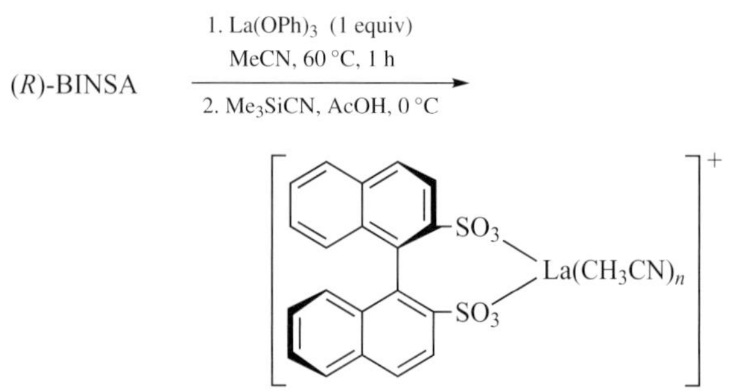

(19)

m/*z* (calc): 591.94 (*n* = 1), 632.97 (*n* = 2), 673.99 (*n* = 3), 715.02 (*n* = 4)
m/*z* (found): 591.94 (*n* = 1), 632.97 (*n* = 2), 673.99 (*n* = 3), 715.01 (*n* = 4)

Asymmetric Catalysis by Chiral BINSA–Au(I) Complexes. Toste and coworkers reported the enantioselective intramolecular hydroalkoxylation of allenes with hydroxylamines catalyzed by chiral gold(I) chloride/chiral silver phosphate.[15] During screening of the chiral gold(I) catalysts, they found that the combined use of [dppm(AuCl)$_2$] (dppm = bis(diphenylphosphino)methane) and silver(I) salt of (*R*)-3,3′-(3,5-di(trifluoromethyl)phenyl)$_2$-BINSA could also promote the reaction (eq 20). The desired isoxazolidine was obtained in quantitative yield with 65% ee, while other chiral gold(I) precursors such as [(*R*)-BINAP(AuCl)$_2$] and [(*S*)-BINAP(AuCl)$_2$] were less effective.

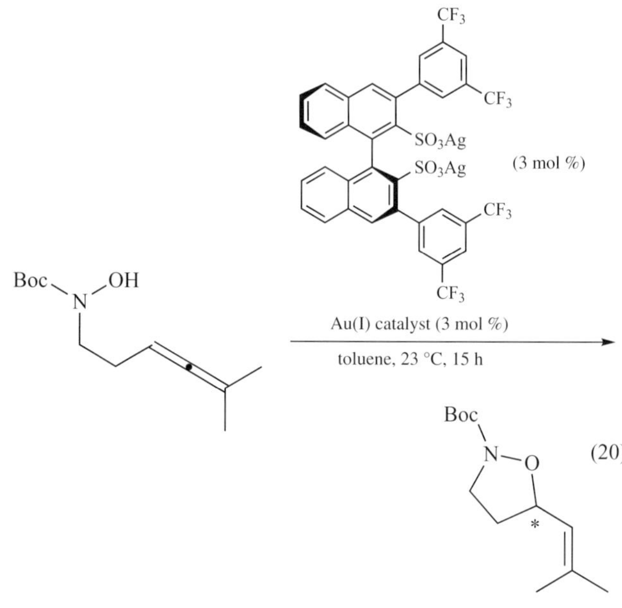

(20)

Au(I) catalyst	Yield and enantioselectivity
[dppm(AuCl)$_2$]	98%, 65% ee
[(*R*)-BINAP(AuCl)$_2$]	98%, 8% ee
[(*S*)-BINAP(AuCl)$_2$]	98%, 42% ee

Asymmetric Catalysis by Chiral BINSA-derived Disulfonimide. List and coworkers reported that the (*R*)-BINSA-derived disulfonimide was a chiral counteranion motif for the catalytic asymmetric Mukaiyama aldol reaction.[8] The comparison of the disulfonimide catalyst and the disulfonic acid catalyst in the Mukaiyama aldol reaction between naphthaldehyde and ketene silyl acetal revealed that the former catalyst gave the desired product in >99% yield with 80% ee, while the latter catalyst did not promote the reaction (eq 21). The actual catalyst was assumed to be *N*-silyl imide, which was generated upon the initial reaction of the disulfonimide catalyst with the ketene silyl acetal (eq 22). Aromatic, aliphatic, and α,β-unsaturated aldehydes could be used with TMS- and TBS-protected ketene acetal, and the corresponding products were obtained in high yields with high enantioselectivities (eq 22).

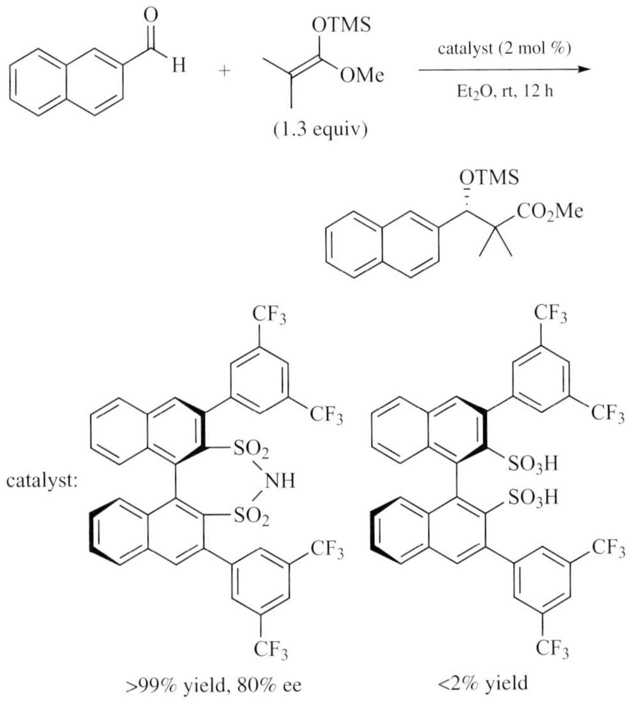

(21)

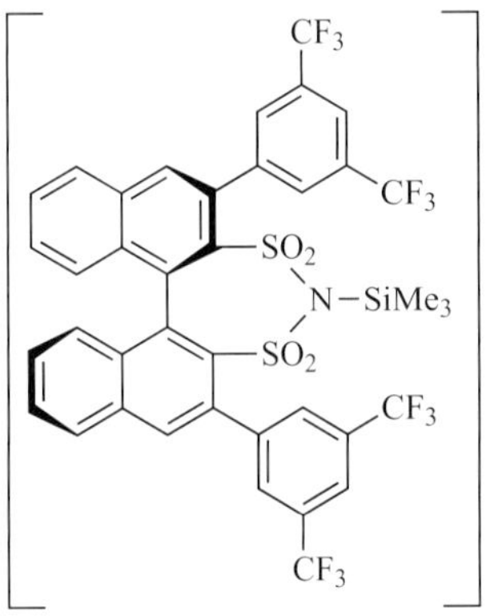

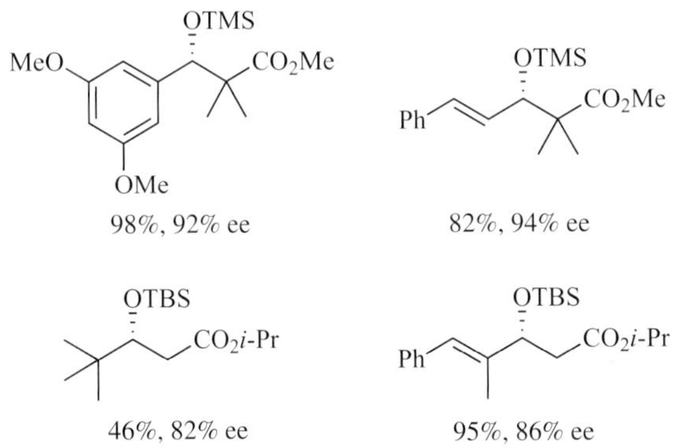

(22)

1. Barber, H. J.; Smiles, S., *J. Chem. Soc.* **1928**, 1141.
2. Armarego, W. L. F.; Turner, E. E., *J. Chem. Soc.* **1957**, 13.
3. Hatano, M.; Maki, T.; Moriyama, K.; Arinobe, M.; Ishihara, K., *J. Am. Chem. Soc.* **2008**, *130*, 16858.
4. Hatano, M.; Sugiura, Y.; Ishihara, K., *Tetrahedron: Asymmetry*, in press (doi:10.1016/j.tetasy.2010.03.014).
5. Fabbri, D.; Delogu, G.; De Lucchi, O., *J. Org. Chem.* **1993**, *58*, 1748.
6. Bandarage, U. K.; Simpson, J.; Smith, R. A. J.; Weavers, R. T., *Tetrahedron* **1994**, *50*, 3463.
7. Cossu, S.; De Lucchi, O.; Fabbri, D.; Valle, G.; Painter, G. F.; Smith, R. A. J., *Tetrahedron* **1997**, *53*, 6073.
8. García-García, P.; Lay, F.; García-García, P.; Rabalakos, C.; List, B., *Angew. Chem., Int. Ed.* **2009**, *48*, 4363.
9. Lloyd-Jones, G. C.; Moseley, J. D.; Renny, J. S., *Synthesis* **2008**, 661.
10. Mondragón, A.; Monsalvo, I.; Regla, I.; Castillo, I., *Tetrahedron Lett.* **2010**, *51*, 767.
11. Treskow, M.; Neudörfl, J.; Giernoth, R., *Eur. J. Org. Chem.* **2009**, 3693.
12. Kampen, D.; Ladépêche, A.; Claßen, G.; List, B., *Adv. Synth. Catal.* **2008**, *350*, 962.
13. Pan, S. C.; List, B., *Chem. Asian J.* **2008**, *3*, 430.
14. Hatano, M.; Hattori, Y.; Furuya, Y.; Ishihara, K., *Org. Lett.* **2009**, *11*, 2321.
15. LaLonde, R. L.; Wang, Z. J.; Mba, M.; Lackner, A. D.; Toste, F. D., *Angew. Chem., Int. Ed.* **2010**, *49*, 598.

Manabu Hatano & Kazuaki Ishihara
Nagoya University, Nagoya, Japan

1,1′-Binaphthyl-2,2′-diyl Hydrogen Phosphate

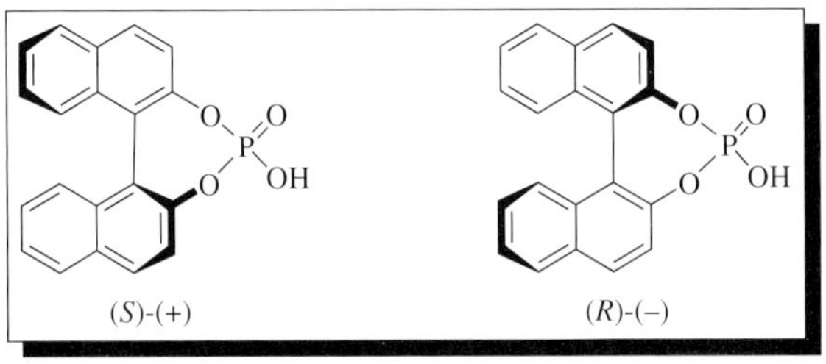

(*S*)-(+)
[35193-64-7] $C_{20}H_{13}O_4P$ (MW 348.30)
InChI = 1/C20H13O4P/c21-25(22)23-17-11-9-13-5-1-3-7-15(13)19(17)20-16-8-4-2-6-14(16)10-12-18(20)24-25/h1-12H,(H,21,22)/f/h21H
InChIKey = JEHUZVBIUCAMRZ-PKSOQXRJCE
(*R*)-(−)
[39648-67-4]
InChI = 1/C20H13O4P/c21-25(22)23-17-11-9-13-5-1-3-7-15(13)19(17)20-16-8-4-2-6-14(16)10-12-18(20)24-25/h1-12H,(H,21,22)/f/h21H
InChIKey = JEHUZVBIUCAMRZ-PKSOQXRJCE
(±)
[50574-52-2]

InChI = 1/C20H13O4P/c21-25(22)23-17-11-9-13-5-1-3-7-15(13)19(17)20-16-8-4-2-6-14(16)10-12-18(20)24-25/h1-12H,(H,21,22)/f/h21H
InChIKey = JEHUZVBIUCAMRZ-PKSOQXRJCE

(reagent for optical resolution of a variety of organic bases[2] and helicenes;[3] NMR shift reagent for determining the enantiomeric purity of secondary and tertiary amines;[4] chiral ligand for homogeneous asymmetric catalysts[5,6])

Alternate Name: BNPPA.
Physical Data: (*S*)-(+)-BNPPA: $[\alpha]_D^{20}$ +605° (*c* 1.35, MeOH). (*R*)-(−)-BNPPA: $[\alpha]_D^{20}$ −605° (*c* 1.35, MeOH). Decomposes without melting at ca. 300 °C.
Solubility: racemate: sol ethanol (10 g/100 mL), sol methanol (3 g/100 mL), slightly sol water and other organic solvents. Enantiomer: sol ethanol (6 g/100 mL), sol methanol (2 g/100 mL), slightly sol water and other organic solvents.
Form Supplied in: white solid; racemic and optically active BNPPA are commercially available.
Preparative Methods: racemic BNPPA is obtained by condensation of (±)-binaphthol and Phosphorus Oxychloride followed by hydrolysis (eq 1).[1] Racemic BNPPA can be resolved by recrystallization of its cinchonine salt.[1]

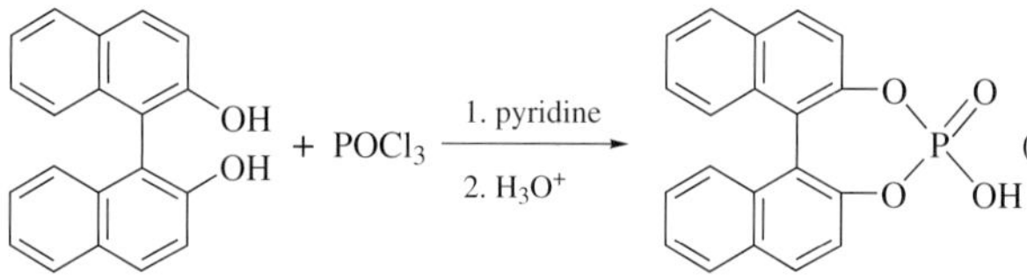

(1)

Original Commentary

Ryoji Noyori & Kazuhiko Sato
Nagoya University, Nagoya, Japan

Chiral Ligand for Asymmetric Catalysts. (*S*)-(+)- and (*R*)-(−)-BNPPA are efficient chiral ligands for the Pd-catalyzed hydrocarboxylation of alkenes.[5] Naproxen can be obtained regioselectively in 91% ee (eq 1).

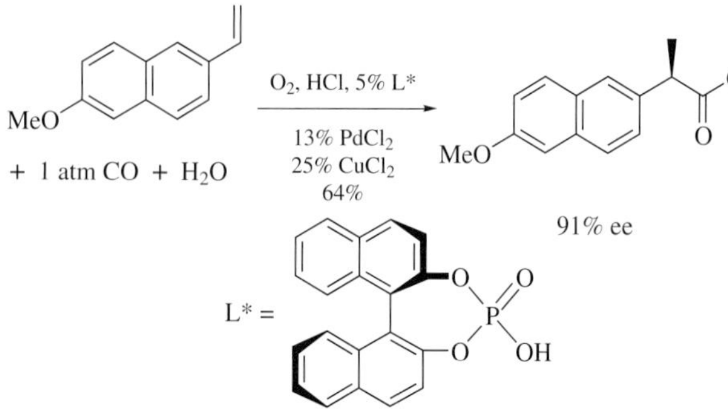

91% ee (2)

Dinuclear Rh complexes modified with optically active BNPPA catalyze asymmetric carbene reactions with moderate enantioselectivity (eqs 3 and 4).[6]

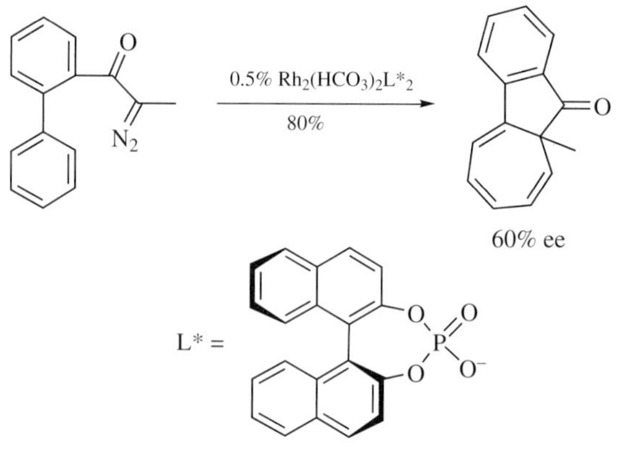

(3)

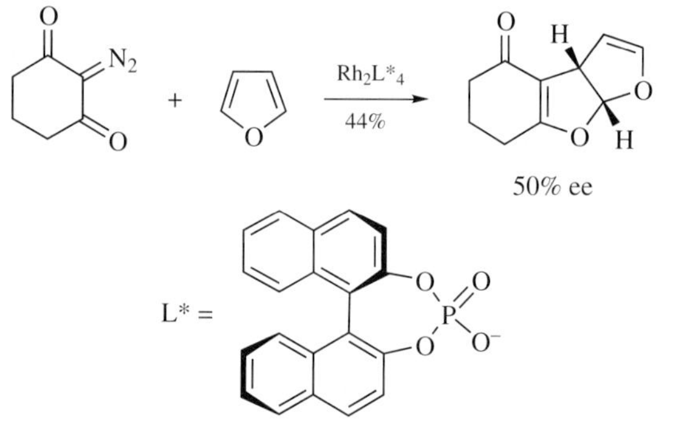

(4)

First Update

Qingwei Yao
Northern Illinois University, DeKalb, IL, USA

Chiral Resolution of Organic Bases. A large scale of resolution of the cardioprotective drug aryl amino thiazolidinone **1** was achieved via diastereomeric salt formation of racemic **1** with (*R*)-(−)-BNPPA (eq 5).[7] Repeated crystallization from toluene gave the pure (−)-**1**·(*R*)-(−)-BNPPA diastereomer which upon treatment with 10% K_2CO_3 solution gave enantiomerically pure (−)-**1** (also known as CP-060*S*) (>99% ee). Interestingly, (*R*)-(−)-BNPPA proved to be the most effective resolving agent for **1**, since attempted resolution using other chiral acids such as -camphor-10-sulphonic acid and mandelic acid all failed.

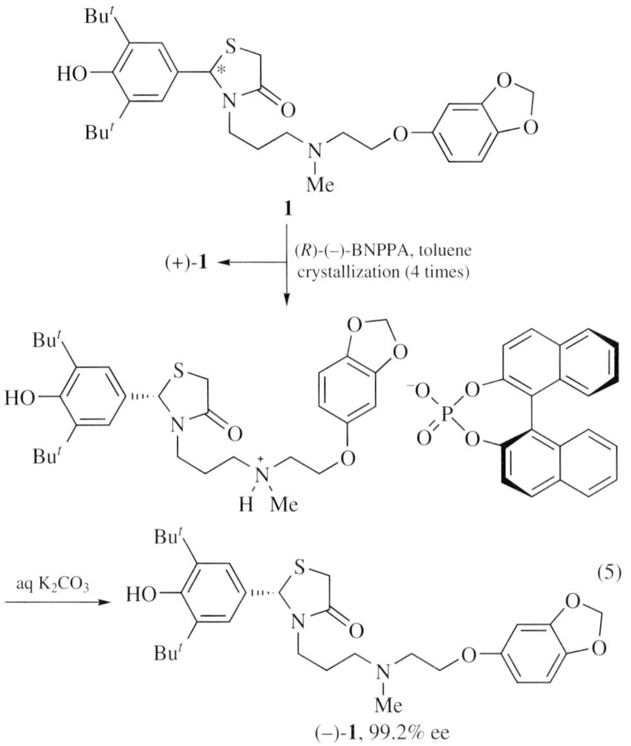

(±)-*threo*-Methylphenidate hydrochloride (**2**) is a mild nervous system stimulant marketed for the treatment of children with attention deficit hyperactivity disorder (ADHD). (2*R*,2*R*)-(+)-**2** has been reported to be 52–38 times more active than its enantiomer. A highly reproducible and scalable resolution of racemic **2** using (*R*)-(−)-BNPPA as the resolving agent has been developed.[8] As described in eq 6, the use of 0.5 molar equiv of (*R*)-(−)-BNPPA was reported to give more efficient resolution than 1 equiv of (*R*)-(−)-BNPPA. The overall yield of recrystallized (2*R*,2*R*)-(+)-**2** was 31.4%. The resolving agent (*R*)-(−)-BNPPA can be recovered in 83% yield and was recyclable.

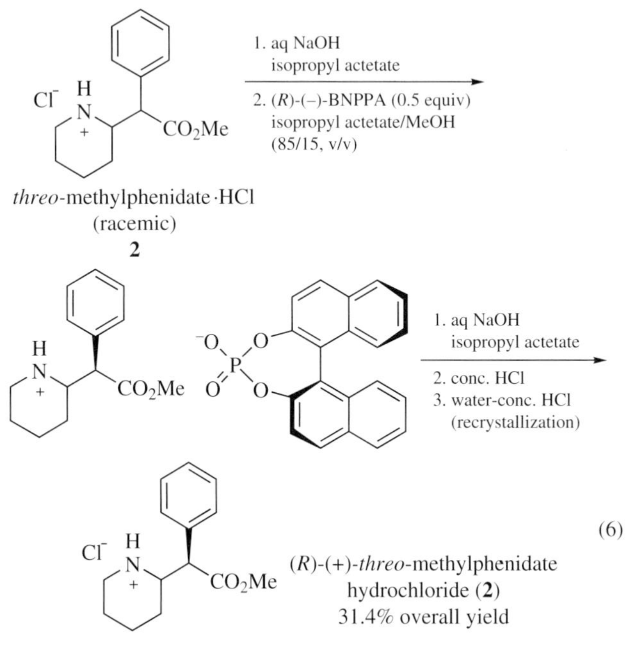

The tricyclic benzodiazepine **3** was resolved to yield (*S*)-(+)-**3** and (*R*)-(−)-**3** through recrystallization with (*R*)-(−)-BNPPA and (*S*)-(+)-BNPPA, respectively, from methanol (eq 7).[9] The absolute stereochemistry was established by X-ray crystallographic structure determination of the (*S*)-(+)-**3**·(*R*)-(−)-BNPPA adduct.

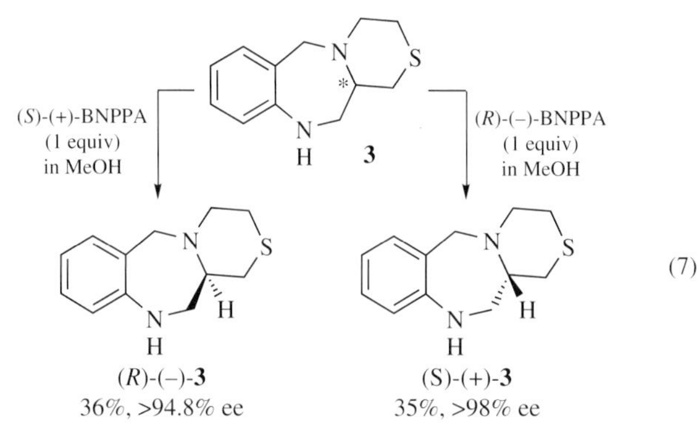

Resolution of quinolizidine (±)-**4** using (*R*)-(−)-BNPPA provided enantiomerically pure (−)-**4** in 44% overall yield from the racemate (eq 8).[10] The latter was used as a key intermediate for the synthesis of quinolizidine alkaloid (−)-217A.

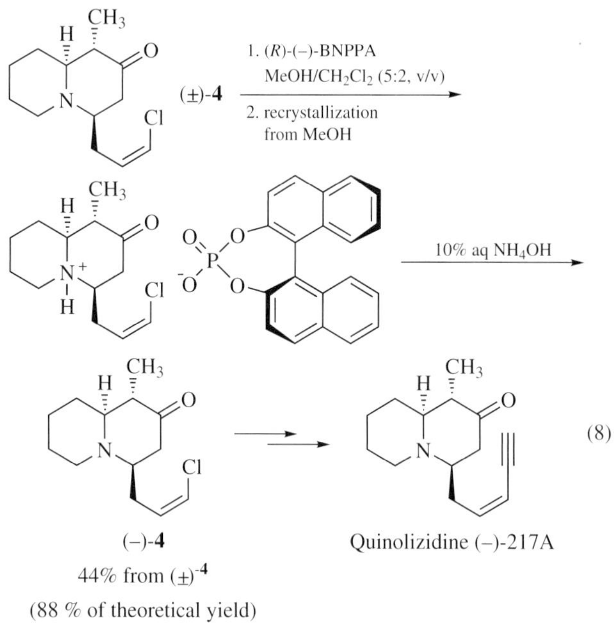

Other chiral organic bases that have been resolved using (*R*)-(−)-BNPPA or (*S*)-(+)-BNPPA include the *cis*-aziridine **5**[11] and the quaternary germane **6**[12] as shown in eq 9 and eq 10, respectively.

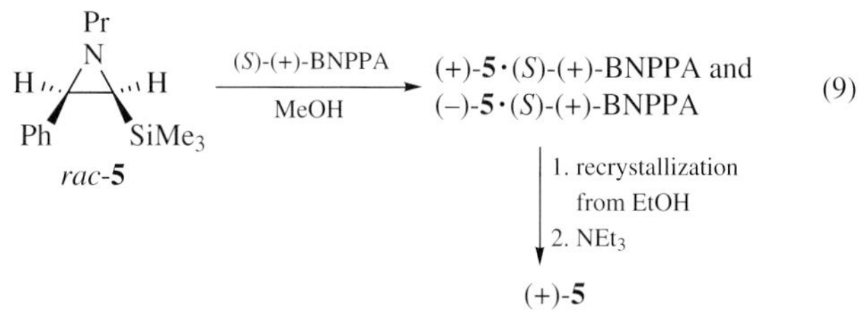

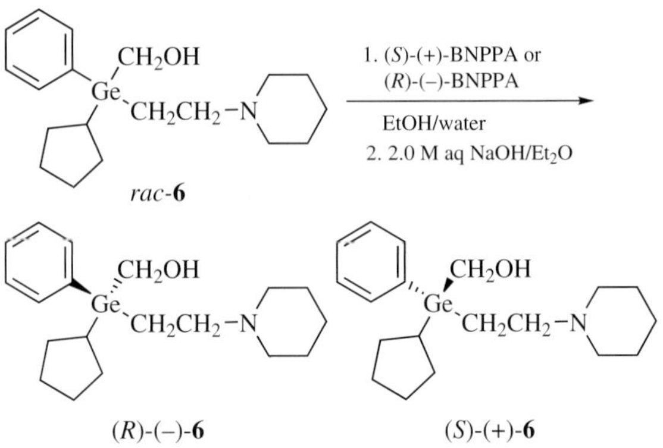

(*R*)-(−)-**6** from (*S*)-(+)-BNPPA (*S*)-(+)-**6** from (*R*)-(−)-BNPPA (10)

Ligand for the Synthesis and Application of Chiral Lewis Acids. A series of trivalent rare earth metal salts derived from (*R*)-BNPPA were synthesized as isolable chiral Lewis acids.[13–17] For example, the chiral Lewis acid Yb[(*R*)-BNP]$_3$ was readily synthesized by treating the sodium or potassium salt of (*R*)-BNPPA with $YbCl_3$ (eq 11). Hetero Diels–Alder reaction of benzaldehyde with Danishefsky's diene in the presence of Yb[(*R*)-BNP]$_3$ as the Lewis acid catalyst and 2,6-lutidine as an additive delivered the dihydropyranone **7** after acidic workup in excellent yield and high ee (eq 12).[14,17] The same Lewis acid also catalyzes the hetero Diels–Alder reaction of α-keto esters such as methyl phenyglyoxylate with Danishefsky's diene to yield dihydropyranone **8** with excellent yield and ee (eq 13).[17]

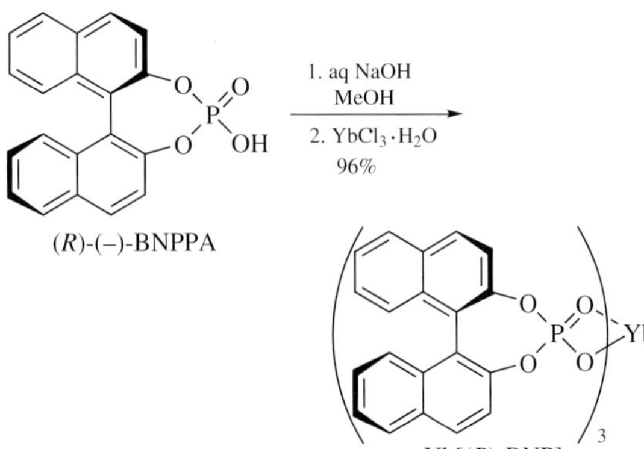

Yb[(*R*)-BNP]$_3$ (11)

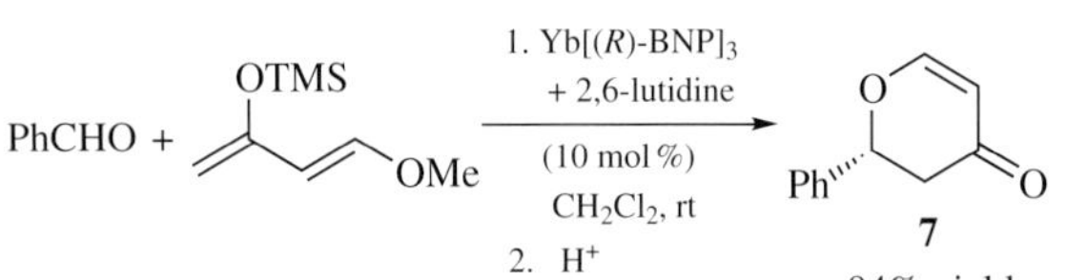

7 94% yield 89% ee (12)

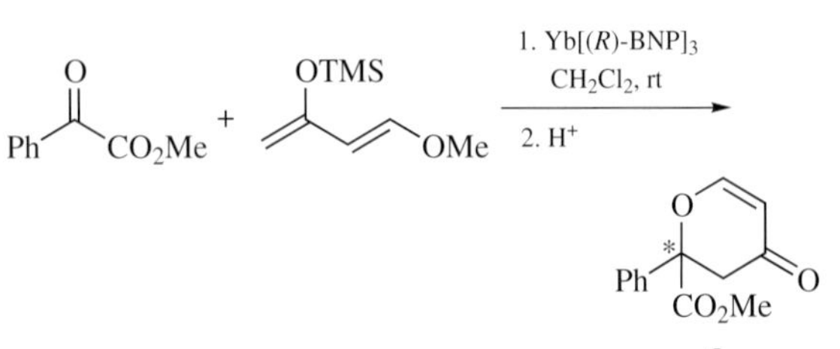

8 99% yield 98% ee (13)

A chiral cerium(III) Lewis acid **9** was prepared in a similar way from the cheaply available ceric ammonium nitrate (eq 14).[16] This chiral cerium complex was found to be a highly efficient catalyst for hetero Diels–Alder reactions (eq 15).[17] Unlike the case of the ytterbium reagent, inclusion of an additive such as 2,6-lutidine did not improve the enantioselectivity of the Diels–Alder reactions catalyzed by the cerium complex.

(*R*)-(−)-BNPPA —(1. aq NaOH, MeOH; 2. 1/4Ce$(NO_3)_6(NH_4)_2$; 59%)→ Ce[(*R*)-BNP]$_3$ $(NaNO_3)_{0.5}$ (14)
9

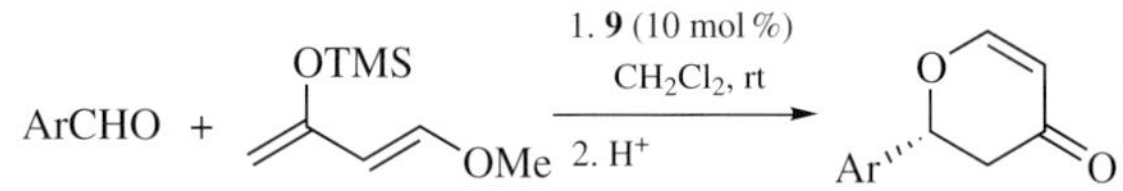

(Ar = 4-Et-C_6H_4) **10** 95% yield 93% ee (15)

A stable scandium(III) complex Sc[(*R*)-BNP]$_3$ (**11**) was prepared from (*R*)-BNPPA (eq 16).[18,19] The conjugate addition of *O*-alkylhydroxylamines to α,β-unsaturated ketones can be efficiently catalyzed by **11**. Excellent enantioselectivity was obtained in the Micheal addition of *O*-diphenylmethylhydroxylamine with various α,β-unsaturated ketones as shown by the example in eq 17.[19]

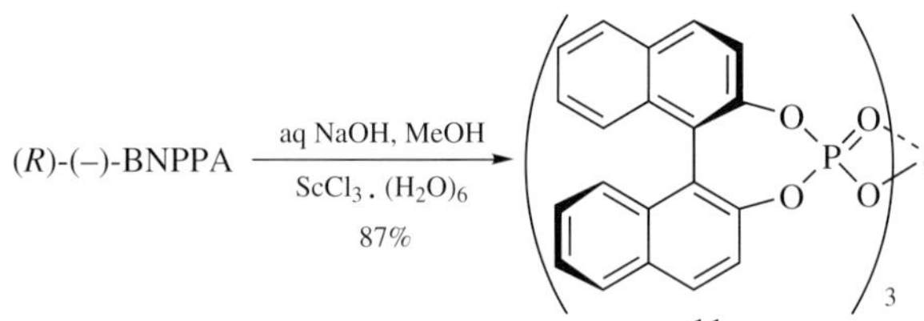

11 Sc[(*R*)-BNP]$_3$ (16)

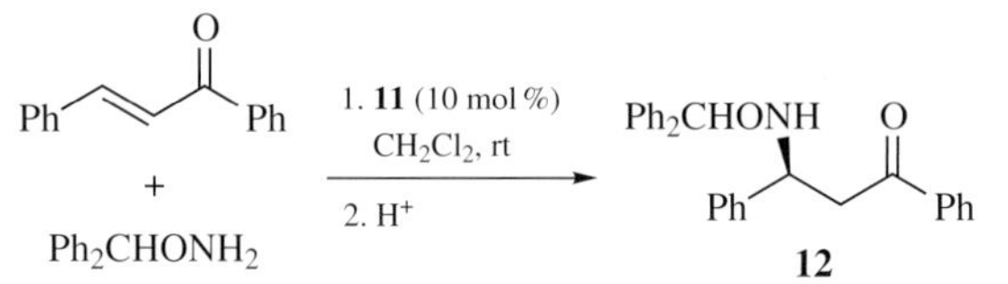

12 94% yield, 99% ee (17)

Chiral Brønsted Acid as an Organo- or Hydrogen-bonding Catalyst. Oxidation of sulfide **13** with 3 equiv of 30% H_2O_2 in the presence of 1 equiv of (*R*)-BNPPA in CH_2Cl_2 for 24 h at room temperature gave the salt of (*R*)-**14**·(*R*)-BNPPA in 51% conversion and 31% de (eq 18).[20]

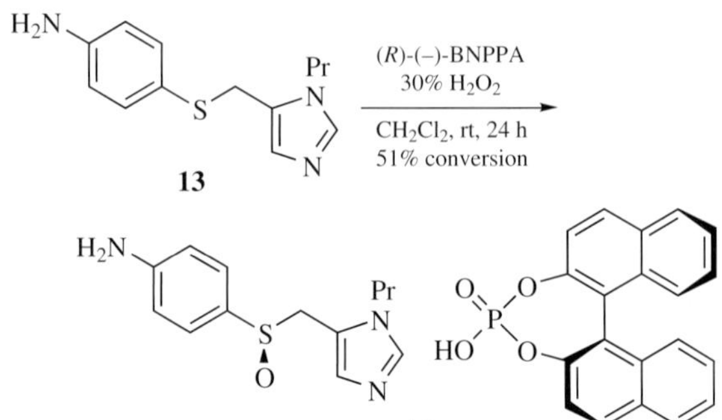

14 31% de (18)

Mannich-type reaction between an *N*-Boc-protected imine such as **15** and acetylacetone to form a β-aminoketone such as **16** can be catalyzed by (*R*)-(−)-BNPPA albeit in low ee (eq 19).[21] Structural modification on the binaphthyl skeleton by introduction of aryl groups on its 3,3′-positions (cf. catalyst **17**) was found to bring about very significant improvement on the enantioselectivity of this reaction.

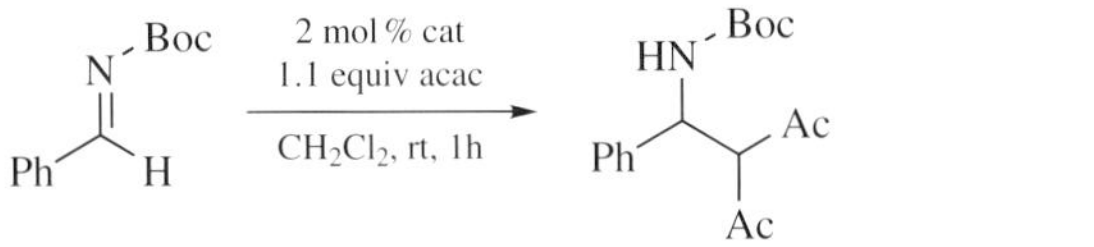

(19)

15 **16**

with (*R*)-(−)-BNPPA:
92% yield, 12% ee
with cat **17**:
99% yield, 95% ee

cat.17 = 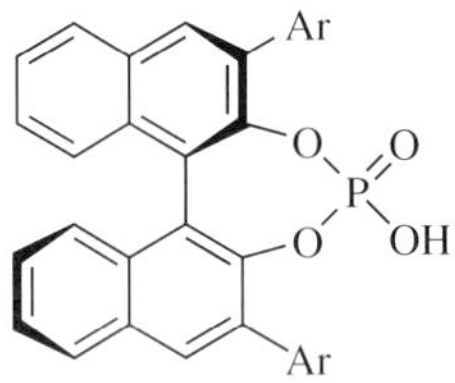

Related Reagents. 3,3′-Bis(triphenylsilyl)-1,1′-binaphthyl-2,2′-diyl Hydrogen Phosphate;[22] 3,3′-Bis(aryl)-1,1′-binaphthyl-2,2′-diyl Hydrogen Phosphate;[21, 23–25] 6,6′-bis(aryl)-1,1′-binaphthyl-2,2′-diyl hydrogen phosphate.[26]

1. Jacques, J.; Fouquey, C., *Org. Synth.* **1989**, *67*, 1.
2. (a) Jacques, J.; Fouquey, C.; Viterbo, R. *Tetrahedron Lett.* **1971**, 4617. (b) Arnold, W.; Daly, J. J.; Imhof, R.; Kyburz, E. *Tetrahedron Lett.* **1983**, *24*, 343.
3. Mikes, F.; Boshart, G., *J. Chem. Soc., Chem. Commun.* **1978**, 173.
4. Shapiro, M. J.; Archinal, A. E.; Jarema, M. A., *J. Org. Chem.* **1989**, *54*, 5826.
5. Alper, H.; Hamel, N., *J. Am. Chem. Soc.* **1990**, *112*, 2803.
6. (a) McCarthy, N.; McKervey, M. A.; Ye, T.; McCann, M.; Murphy, E.; Doyle, M. P., *Tetrahedron Lett.* **1992**, *33*, 5983. (b) Pirrung, M. C.; Zhang, J. *Tetrahedron Lett.* **1992**, *33*, 5987.
7. Kato, T.; Ozaki, T.; Ohi, N., *Tetrahedron: Asymmetry* **1999**, *10*, 3963.
8. Prashad, M.; Hu, B.; Repič, O.; Blacklock, T. J.; Giannousis, P., *Org. Proc. Res. Devel.* **2000**, *4*, 55.
9. Matthews, J. M.; Dyatkin, A. B.; Evangelisto, M. E.; Gauthier, D. A.; Hecker, L. R.; Hoekstra, W. J.; Liu, F.; Poulter, B. L.; Sorgib, K. L.; Maryanoff, B. E., *Tetrahedron: Asymmetry* **2004**, *15*, 1259.
10. Maloney, K. M.; Danheiser, R. L., *Org. Lett.* **2005**, *7*, 3115.
11. Bassindale, A. B.; Kyle, P. A.; Soobramanien, M.-C.; Taylor, P. G., *J. Chem. Soc., Perkin Trans. 1* **2000**, 439.
12. Tacke, R.; Kornek, T.; Heinrich, T.; Burschka, C.; Penka, M.; Pülm, M.; Keim, C.; Mutschler, E.; Lambrecht, G., *J. Organomet. Chem.* **2001**, *640*, 140.
13. Inanaga, J.; Sugimoto, Y.; Hanamoto, T., *New J. Chem.* **1995**, *19*, 707.
14. Hanamoto, T.; Furuno, H.; Sugimoto, Y.; Inanaga, J., *Synlett* **1997**, 79.
15. Furuno, H.; Hanamoto, T.; Sugimoto, Y.; Inanaga, J., *Org. Lett.* **2000**, *2*, 49.
16. Hayano, T.; Sakaguchi, T.; Furuno, H.; Ohba, M.; Okawa, H.; Inanaga, J., *Chem. Lett.* **2003**, *32*, 608.
17. Furuno, H.; Hayano, T.; Kambara, T.; Sugimoto, Y.; Hanamoto, T.; Tanaka, Y.; Jin, Y. Z.; Kagawa, T.; Inanaga, J., *Tetrahedron* **2003**, *59*, 10509.
18. Sugihara, H.; Daikai, K.; Jin, X. L.; Furuno, H.; Inanaga, J., *Tetrahedron Lett.* **2002**, *43*, 2735.
19. Jin, X. L.; Sugihara, H.; Daikai, K.; Tateishi, H.; Jin, Y. Z.; Furuno, H.; Inanaga, J., *Tetrahedron* **2002**, *58*, 8321.
20. Ikemoto, T.; Nishiguchi, A.; Ito, T.; Tawada, H., *Tetrahedron* **2005**, *61*, 5043.
21. Uraguchi, D.; Terada, M., *J. Am. Chem. Soc.* **2004**, *126*, 5356.
22. Storer, R. I.; Carrera, D. E.; Ni, Y.; MacMillan, D. W. C., *J. Am. Chem. Soc.* **2006**, *128*, 84.
23. Akiyama, T.; Itoh, J.; Yokota, K.; Fuchibe, K., *Angew. Chem., Int. Ed.* **2004**, *43*, 1566.
24. Rueping, M.; Sugiono, E.; Azap, C.; Theissmann, T.; Bolte, M., *Org. Lett.* **2005**, *7*, 3781.
25. Uraguchi, D.; Sorimachi, K.; Terada, M., *J. Am. Chem. Soc.* **2005**, *127*, 9360.
26. Furuno, H.; Kambara, T.; Tanaka, Y.; Hanamoto, T.; Kagawab, T.; Inanaga, J., *Tetrahedron Lett.* **2003**, *44*, 6129.

N,N′-Bis[3,5-bis(trifluoromethyl)-phenyl]thiourea

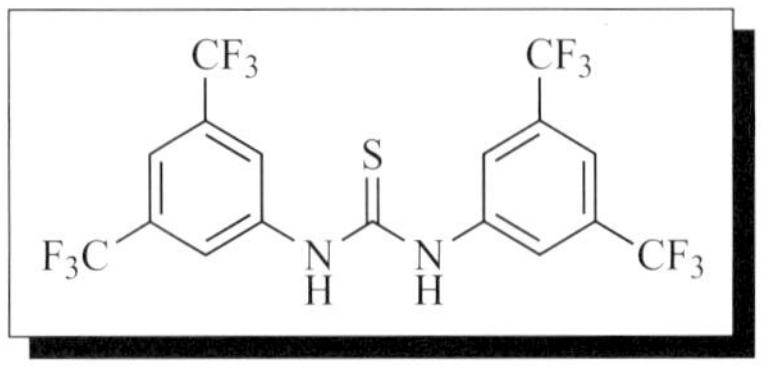

[1060-92-0] $C_{17}H_8F_{12}N_2S$ (MW 500.30)

InChI = 1S/C17H8F12N2S/c18-14(19,20)7-1-8(15(21,22)23)4-11(3-7)30-13(32)31-12-5-9(16(24,25)26)2-10(6-12)17(27,28)29/h1-6H,(H2,30,31,32)

InChIKey = RWXWQJYJWJNJNW-UHFFFAOYSA-N

(used as hydrogen-bonding organocatalyst in a wide variety of reactions)

Alternate Name: Schreiner's thiourea.

Physical Data: mp 172–173 °C,[1] pK_a (DMSO) 8.4/8.5.[2,3]

Solubility: soluble in methanol, ethanol, Et_2O, CH_2Cl_2, EtOAc, MeCN, DMSO, and toluene; insoluble in H_2O and hexane.

Form Supplied in: commercially available as a white solid.

Purification: recrystallization from $CHCl_3$ then either hexane/Et_2O[1] or hexane/EtOAc.[4]

Handling, Storage, and Precautions: bench stable; keep container tightly closed and store in a cool and dark place, away from incompatible materials such as oxidizing agents.

Acetal Formation. The reaction between aldehydes or ketones (including aliphatic, aromatic, and acid-sensitive substrates) and alcohols to afford acetals in high yields can be catalyzed by the use of Schreiner's thiourea.[1] Schreiner's thiourea can also catalyze the addition of alcohols to enol ethers affording acetals under mild acid-free conditions. This was utilized for the reaction between dihydropyran (DHP) and various alcohols (including primary, secondary, tertiary and hindered alcohols, phenols, and alcohols bearing acid-sensitive functionalities) to afford the corresponding tetrahydropyran (THP)-protected alcohols in high yields, and in many cases under solvent-free conditions (eq 1).[5] The addition of alcohols to functionalized galactals to afford 2-deoxygalactosides can also be catalyzed by Schreiner's thiourea in refluxing CH_2Cl_2 (eq 2).[6] The 2-deoxygalactosides were obtained in high yield and exclusively as the α-anomers; the mild conditions employed were shown to be tolerant of a wide range of common protecting groups in carbohydrate chemistry.

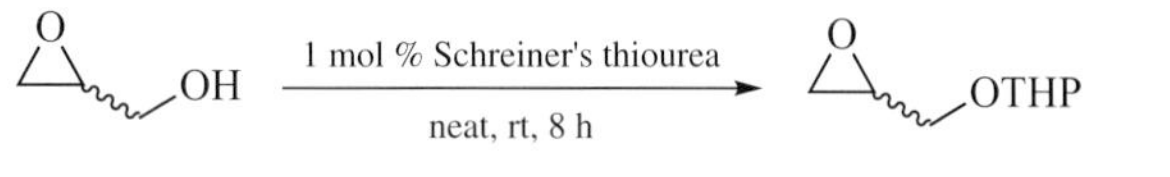

93% (1)

Aldol Reactions. Schreiner's thiourea and proline can form a host–guest complex that in turn can catalyze the enantioselective aldol reaction between ketones and aldehydes in high yield and enantioselectivity, with respectable *syn*:*anti* ratios.[7] This reaction can also be catalyzed by the amino acid threonine (with the hydroxyl group protected as the *tert*-butyldimethylsilyl ether)[8] instead of proline. Schreiner's thiourea helps to improve both the yield and timescale of the reaction, and in some cases the enantioselectivity. In addition, Schreiner's thiourea has been shown to increase the yield, dr, and ee of the cobalt-catalyzed aldol reaction between α-isocyanoacetate and aldehydes (eq 3).[9]

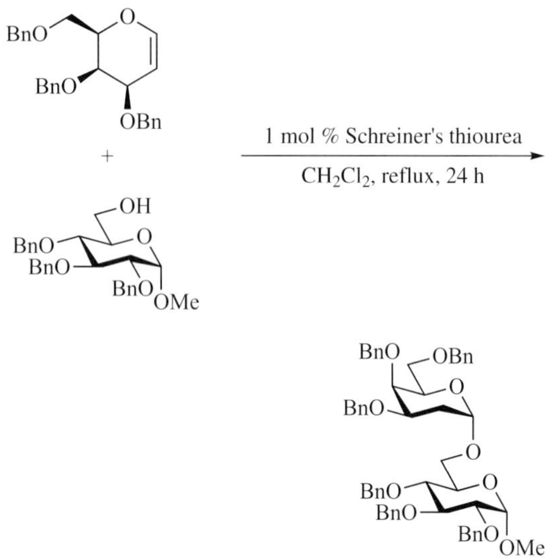

98% (2)

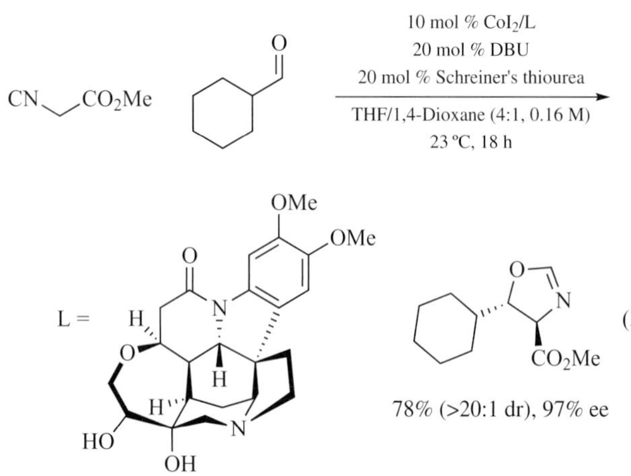

78% (>20:1 dr), 97% ee (3)

Cycloadditions. Diels–Alder reactions have been shown to be accelerated by the use of Schreiner's thiourea as a catalyst with moderate to good *endo*:*exo* selectivity.[10] The use of Schreiner's thiourea in the hetero-Diels–Alder reaction between activated ketones and dienes to afford racemic precursors to δ-lactone derivatives has also been reported.[11] When used in combination with a proline-based cocatalyst, Schreiner's thiourea can be used to catalyze the formal [2 + 2] cycloaddition between enals and nitroalkenes (eq 4).[12] The proline derivative activates the enal while Schreiner's thiourea increases the electrophilicity of the nitroalkene through hydrogen bonding, which helps to increase the overall yield of the reactions. The intramolecular [5 + 2] cycloaddition of oxidopyrylium ions with two-carbon dipolarophiles can be catalyzed by Schreiner's thiourea as a cocatalyst with a chiral thiourea to afford complex, chiral 8-oxabicyclo[3.2.1]octanes (eq 5).[13] Schreiner's thiourea helps to increase both the yield and enantioselectivity of this [5+2] reaction.

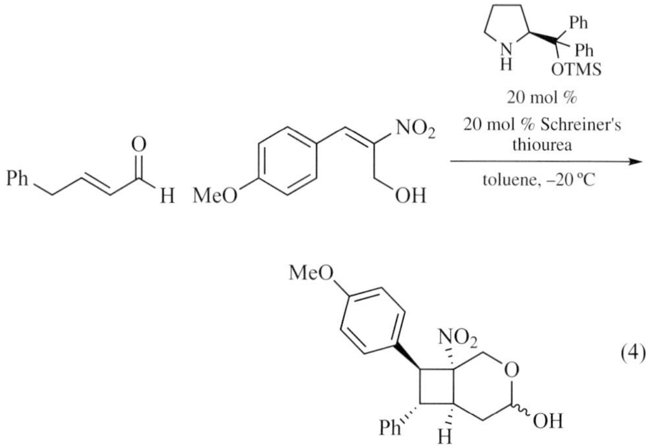

72%, 95% ee (4)

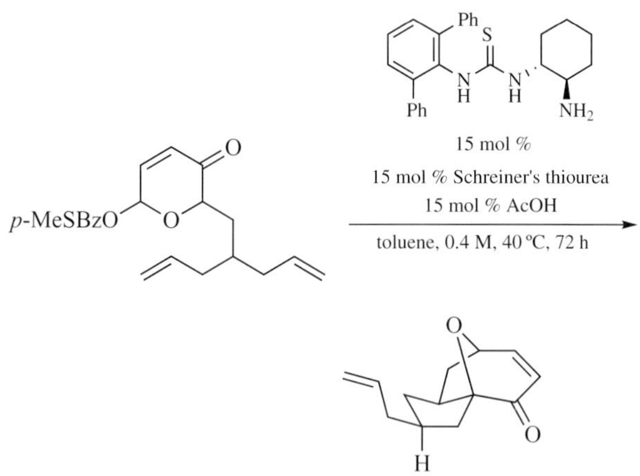

77%, 90% ee (5)

Epoxide Ring Opening and Rearrangements. The alcoholysis of styrene oxides can be achieved through cooperative catalysis of Schreiner's thiourea and a Bronsted acid with high regioselectivity affording β-alkoxy alcohols (eq 6).[14] The aminolysis of epoxides can also be catalyzed by Schreiner's thiourea under solvent-free conditions[15] as well as in water (a rate amplification was observed)[16]; however, the regioselectivity for ring opening varies depending on both the epoxide and amine. In addition, Schreiner's thiourea has been shown to react in situ with $SiCl_4$ in the presence of a base to form a silicon-thiourea Lewis acid that can be used to catalyze the rearrangement of epoxides forming quaternary aldehydes (eq 7).[17]

Multicomponent Reactions. The use of Schreiner's thiourea to catalyze the acyl-Strecker reaction between a broad range of aldehydes and amines has been reported using acetyl cyanide to afford racemic α-amino nitriles that can be used as α-amino acid precursors (eq 8).[18] The asymmetric acyl-Strecker reaction can be achieved by employing a chiral thiourea.[19] The three component reaction between pyruvates, aldehydes, and anilines can also be catalyzed by Schreiner's thiourea affording racemic 3-amino-1,5-dihydro-2*H*-pyrrol-2-ones (eq 9).[2]

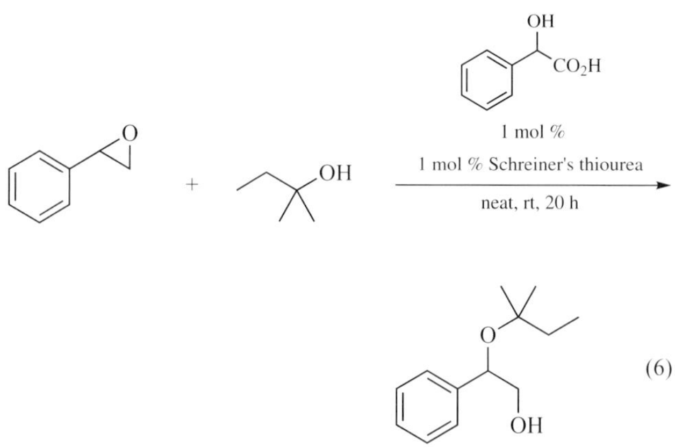

73% (6)

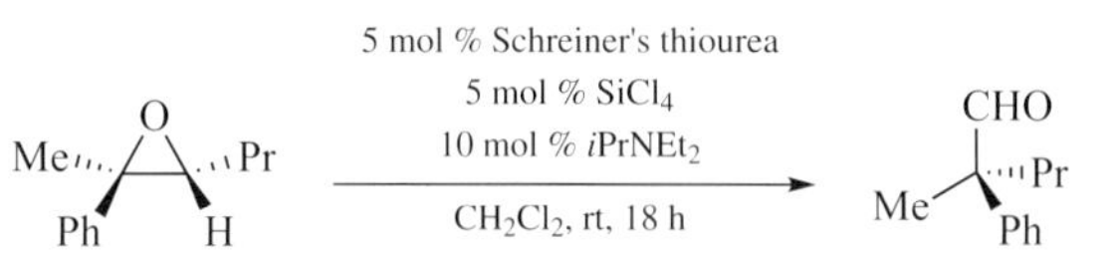

95% ee 86%, 93% ee (7)

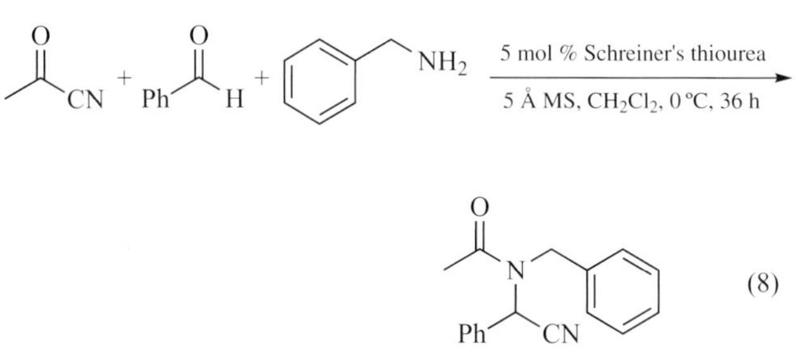

80% (8)

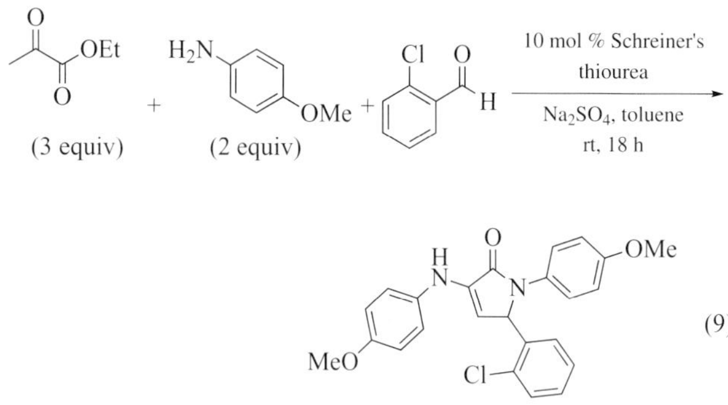

79% (9)

Nitroalkenes and Nitrones. The activation of nitroalkenes by catalytic amounts of Schreiner's thiourea toward the conjugate addition of both hydrazones[20] and nitroalkanes (moderate to good dr) (eq 10)[21] as well as Friedel–Crafts alkylation[22] of aromatic and heteroaromatic systems can be achieved. Enantioselective reactions between nitroalkenes and aldehydes have been shown with Schreiner's thiourea when L-proline is employed as a cocatalyst, obtaining moderate to good enantioselectivity with a range of aliphatic aldehydes.[23] It has also been demonstrated that the nucleophilic addition of both TMSCN and ketene silyl acetals to nitrones can be catalyzed by the use Schreiner's thiourea, in good yields.[24]

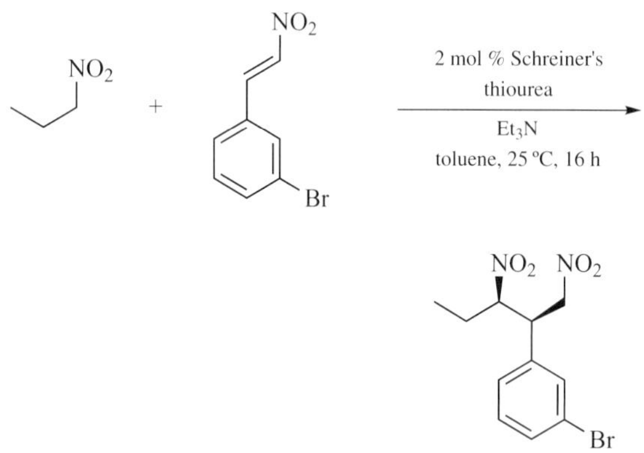

75%, 9:1 dr (10)

Oxidations. The chemoselective metal-free oxidation of sulfides to racemic sulfoxides can be catalyzed by Schreiner's thiourea using *tert*-butyl hydroperoxide as the oxidant.[25] An asymmetric organocatalytic protocol for this transformation has been developed in which a chiral Brønsted acid is used as the catalyst with H_2O_2 as the oxidant.[26] The achiral Baeyer–Villiger oxidation of cyclobutanones can also be achieved under mild, acid- and metal-free conditions using Schreiner's thiourea and employing H_2O_2 as the oxidant.[27] Oxidation of alcohols with NBS (*N*-bromosuccinimide) can also be achieved through the use of Schreiner's thiourea (eq 11).[28] In contrast to the more standard mode of activation through hydrogen bonding, in this case, the polarizable sulfur center of the thiourea acts as a Lewis base and catalyzes the transfer of bromine to the alcohol. The selective oxidation of secondary alcohols in the presence of primary alcohols can also be achieved (eq 11).[28]

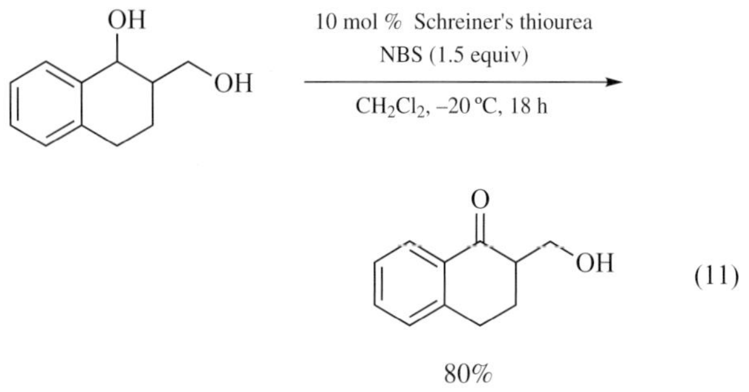

Reductions. The reduction of aldimines (and reductive amination of aldehydes) can be catalyzed by Schreiner's thiourea with the Hantzsch ester as the source of hydrogen to afford secondary anilines (eq 12).[29] Both electron-rich and electron-poor aromatic aldimines, as well as aliphatic aldimines, can be reduced in high yields with a catalyst loading of 1 mol %.[29] Nitroalkenes can also be reduced to their corresponding nitroalkanes under similar mild conditions, with the Hantzsch ester and 10 mol % of Schreiner's thiourea in refluxing CH_2Cl_2. These conditions were found to be applicable to both aromatic and aliphatic conjugated nitroalkenes; it is noteworthy that enones are not reduced.[30]

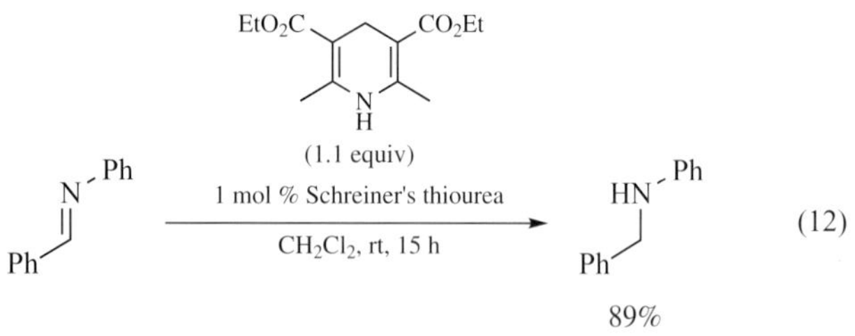

Transesterification. The transesterification of methyl esters with alcohols can be catalyzed by Schreiner's thiourea in combination with 4-pyrrolidinopyridine (PPY) under azeotropic conditions for methanol removal (eq 13).[31] PPY reacts with Schreiner's thiourea displacing 3,5-bis(trifluoromethyl)aniline forming the catalytically active zwitterionic compound.[31]

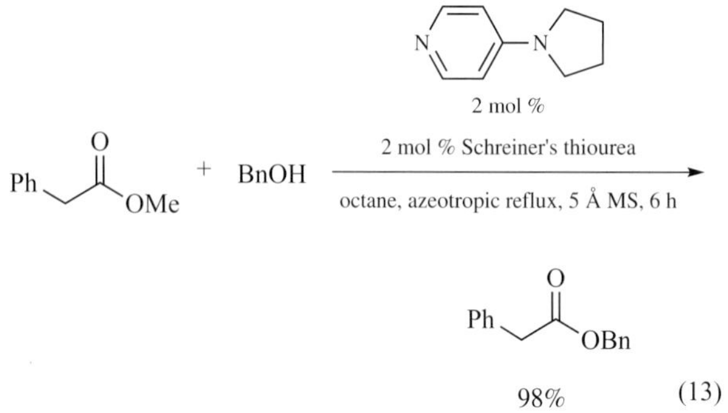

1. Kotke, M.; Schreiner, P. R., *Tetrahedron* **2006**, *62*, 434.
2. Li, X.; Deng, H.; Luo, S.; Cheng, J. P., *Eur. J. Org. Chem.* **2008**, 4350.
3. Jakab, G.; Tancon, C.; Zhang, Z. G.; Lippert, K. M.; Schreiner, P. R., *Org. Lett.* **2012**, *14*, 1724.
4. Kirsten, M.; Rehbein, J.; Hiersemann, M.; Strassner, T., *J. Org. Chem.* **2007**, *72*, 4001.
5. Kotke, M.; Schreiner, P. R., *Synthesis* **2007**, 779.
6. Balmond, E. I.; Coe, D. M.; Galan, M. C.; McGarrigle, E. M., *Angew. Chem., Int. Ed.* **2012**, *51*, 9152.
7. Reis, O.; Eymur, S.; Reis, B.; Demir, A. S., *Chem. Commun.* **2009**, 1088.
8. Ma, G. N.; Bartoszewicz, A.; Ibrahem, I.; Córdova, A., *Adv. Synth. Catal.* **2011**, *353*, 3114.
9. Kim, H. Y.; Oh, K., *Org. Lett.* **2011**, *13*, 1306.
10. (a) Wittkopp, A.; Schreiner, P. R., *Chem. Eur. J.* **2003**, *9*, 407. (b) Schreiner, P. R.; Wittkopp, A., *Org. Lett.* **2002**, *4*, 217.
11. Mori, K.; Maddaluno, J.; Nakano, K.; Ichikawa, Y.; Kotsuki, H., *Synlett* **2009**, 2346.
12. Talavera, G.; Reyes, E.; Vicario, J. L.; Carrillo, L., *Angew. Chem., Int. Ed.* **2012**, *51*, 4104.
13. Burns, N. Z.; Witten, M. R.; Jacobsen, E. N., *J. Am. Chem. Soc.* **2011**, *133*, 14578.
14. Weil, T.; Kotke, M.; Kleiner, C. M.; Schreiner, P. R., *Org. Lett.* **2008**, *10*, 1513.
15. Chimni, S. S.; Bala, N.; Dixit, V. A.; Bharatam, P. V., *Tetrahedron* **2010**, *66*, 3042.
16. Kleiner, C. M.; Schreiner, P. R., *Chem. Commun.* **2006**, 4315.
17. Hrdina, R.; Muller, C. E.; Wende, R. C.; Lippert, K. M.; Benassi, M.; Spengler, B.; Schreiner, P. R., *J. Am. Chem. Soc.* **2011**, *133*, 7624.
18. (a) Pan, S. C.; List, B., *Synlett* **2007**, 318. (b) Pan, S. C.; Zhou, J.; List, B., *Synlett* **2006**, 3275.
19. (a) Pan, S. C.; Zhou, J.; List, B., *Angew. Chem., Int. Ed.* **2007**, *46*, 612. (b) Pan, S. C.; List, B., *Org. Lett.* **2007**, *9*, 1149.
20. Pettersen, D.; Herrera, R. P.; Bernardi, L.; Fini, F.; Sgarzani, V.; Fernandez, R.; Lassaletta, J. M.; Ricci, A., *Synlett* **2006**, 239.
21. Rabalakos, C.; Wulff, W. D., *Synlett* **2008**, 2826.
22. Dessole, G.; Herrera, R. P.; Ricci, A., *Synlett* **2004**, 2374.
23. Demir, A. S.; Eymur, S., *Tetrahedron: Asymmetry* **2010**, *21*, 112.
24. Okino, T.; Hoashi, Y.; Takemoto, Y., *Tetrahedron Lett.* **2003**, *44*, 2817.
25. Russo, A.; Lattanzi, A., *Adv. Synth. Catal.* **2009**, *351*, 521.
26. Liao, S. H.; Coric, I.; Wang, Q. G.; List, B., *J. Am. Chem. Soc.* **2012**, *134*, 10765.
27. Sasakura, N.; Nakano, K.; Ichikawa, Y.; Kotsuki, H., *RSC Adv* **2012**, *2*, 6135.
28. Tripathi, C. B.; Mukherjee, S., *J. Org. Chem.* **2012**, *77*, 1592.
29. Zhang, Z. G.; Schreiner, P. R., *Synlett* **2007**, 1455.
30. Zhang, Z. G.; Schreiner, P. R., *Synthesis* **2007**, 2559.
31. Ishihara, K.; Niwa, M.; Kosugi, Y., *Org. Lett.* **2008**, *10*, 2187.

Edward I. Balmond & M. Carmen Galan
University of Bristol, Bristol, UK

Eoghan M. McGarrigle
University College Dublin, Belfield, Ireland

1-[3,5-Bis(*tert*-butyl)phenyl]octahydro-3,6,6-trimethylcyclopenta[*b*]phosphole

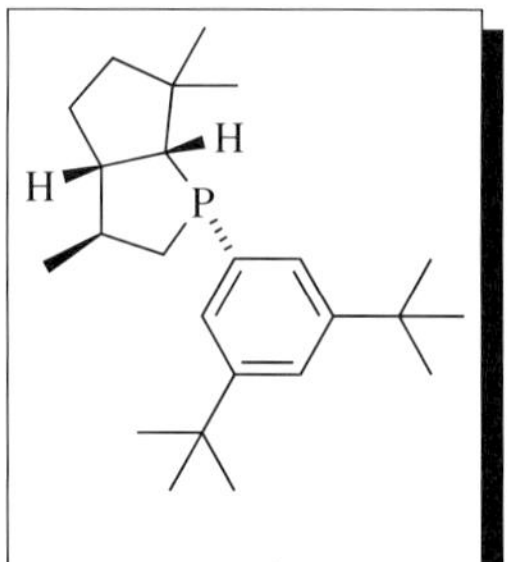

(+)-**1**

[MW232923-66-9] $C_{24}H_{39}P$ (MW 358.54)

InChI = 1/C24H39P/c1-16-15-25(21-20(16)10-11-24(21,8)9)19-13-17(22(2,3)4)12-18(14-19)23(5,6)7/h12-14,16,20-21H,10-11,15H2,1-9H3/t16-,20+,21-/m1/s1

InChIKey = LTXIQCIVTVZGOA-TYCQWZJGBQ

(reagent used as an asymmetric nucleophilic catalyst in kinetic resolution[1] of alcohols)

Alternate Name: (+)-1-[3,5-bis(*t*-butyl)phenyl]-octahydro-3,6,6-trimethylcyclopenta[b]phosphole, (+)-(1*R*,2*R*,4*S*,5*S*)-4,8,8- trimethyl-2-(3′,5′-di-*tert*-butylphenyl)-2-phosphabicyclo[3.3.0]-octane (^{t}Bu$_2$Ph-PBO).

Physical Data: [2](+)-^{t}Bu$_2$Ph-PBO/BH$_3$ adduct **6**: colorless crystals; mp 114–115 °C (MeOH/H$_2$O); $\alpha_D = +15.1$ (c, 3.2, EtOAc); ^{1}H NMR (300 MHz, CDCl$_3$) δ 7.65 (dd, $J = 1.7$, 11.5 Hz, 2H), 7.51 (q, $J = 1.7$ Hz, 1H), 2.62 (d, $J = 10.2$ Hz, 1H), 2.63–2.41 (m, 2H), 2.33–2.16 (m, 1H), 2.07–1.82 (m, 2H), 1.55–1.38 (m, 3H), 1.55–0.37 (br m, 3H), 1.33 (s, 18H), 1.21 (d, $J = 6.3$ Hz, 3H), 0.98 (s, 3H), 0.42 (s, 3H); ^{31}P NMR (121.4 MHz, {H}, CDCl$_3$) δ 33.3–30.5 (br m); IR (KBr, cm^{-1}) 2361 (B-H); analysis calculated: C, 77.39; H, 11.39, found: C, 77.27; H, 11.30. (+)-^{t}Bu$_2$Ph-PBO (**1**): obtained and used crude following heating the above borane adduct **6** in pyrrolidine at 50–55 °C for 90 min and evaporation of solvent under a stream of N$_2$ to leave a colorless viscous oil (N.B. can be crystallized for analysis, see *Purification* below). $\alpha_D = +6.3$ (c, 3.2, THF); ^{1}H NMR (300 MHz, C$_6$D$_6$) δ 7.78 (dd, $J = 1.8$, 7.7 Hz, 2H), 7.45 (dt, $J = 0.4$, 1.8 Hz, 1H), 2.68 (dd, $J = 9.9$, 26.0 Hz, 1H), 2.52–2.34 (m, 2H), 2.25–2.11 (m, 1H), 1.95–1.81 (m, 1H), 1.75–1.56 (m, 1H), 1.46–1.22 (m, 3H), 1.30 (s, 18H), 1.07 (d, $J = 6.3$ Hz, 3H), 1.04 (s, 3H), 0.79 (s, 3H); ^{31}P NMR (121.4 MHz, {H}, C$_6$D$_6$) δ -1.6; HRMS calculated for $C_{24}H_{39}P$: 358.2790, found 358.2786.

Analysis of Reagent Purity: reagent purity is assessed for borane adduct **6**. Enantiomeric excess is determined by conversion of cyclic sulfate **5** to the related Ph-PBO borane adduct and ee determined by CSP-HPLC; ee for ^{t}Bu$_2$Ph-PBO is assumed to be the same.[2,3] Isamagilov's mathematical treatment[4] of the results of kinetic resolution (KR) reactions supports this assumption.[2]

Preparative Method: [2] prepared as borane adduct **6** via an eight-step stereoselective synthesis from methyl (*S*)-lactate in an overall yield of 4.0% (eq 1). Borane adduct **6** is warmed with pyrrolidine to give the active catalyst (+)-**1** (>99.7% ee). The crude product, after removal of the solvent under a stream of N$_2$, is then generally used directly.

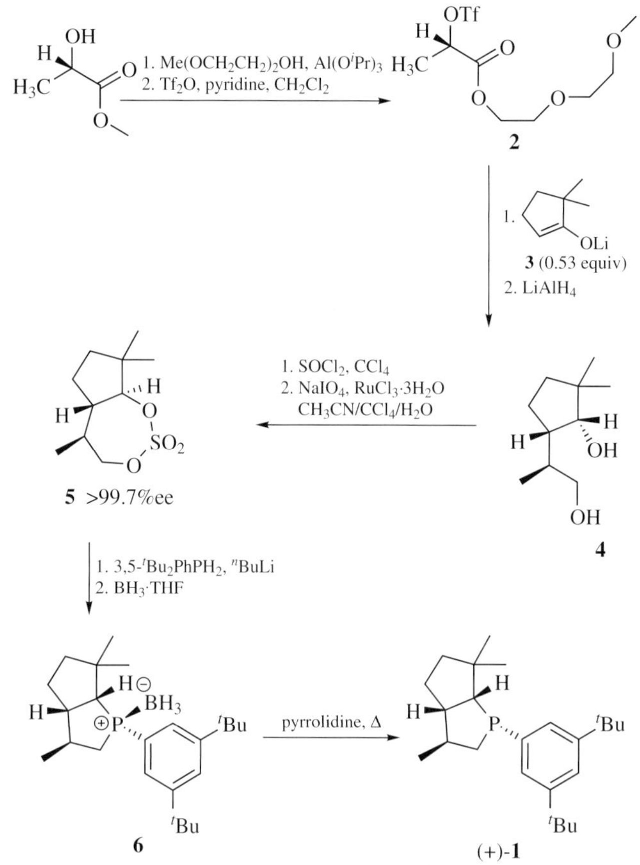

4.0% from methyl (*S*)-lactate

(1)

Purity: the borane adduct can be crystallized from methanol/water. Recrystallization of the active catalyst (+)-**1** is also possible from acetonitrile at −20 °C.[3]

Handling, Storage, and Precautions: no special precautions are taken to exclude oxygen from the KR reactions, although this does result in a 15–30% loss of catalyst due to phosphine oxide formation.[3]

Introduction. 1-[3,5-Bis(*t*-butyl)phenyl]-octahydro-3,6,6-trimethylcyclopenta[b]phosphole (*t*-Bu$_2$Ph-PBO) is one of the members of a family of chiral phosphine catalysts based on the phosphabicyclo[3.3.0]octane (PBO) core,[3] which have been found to be highly effective for the catalytic KR of various secondary alcohols.

KR of Aryl-alkyl Carbinols. The discovery of the PBO family of catalysts resulted from a search for more reactive phosphines following the discovery that KR of aryl-alkyl carbinols could be catalyzed by chiral phosphines[5] in which chirality results from a configurationally stable phosphorus center.

Table 1 shows representative examples for KR of aryl-alkyl carbinols (eq 2), the selectivity (*s*) is defined as the ratio of the rates of reaction of the fast reacting enantiomer to the slow reacting enantiomer. An *s* value of >7 is considered synthetically useful since it allows recovery of 20% unreacted starting material in

Table 1 KR of aryl-alkyl carbinols

Entry	Ar	R	*s*	Temp	Time (h)	% cat.	% conv	ee alcohol (%)	ee ester (%)
1	Ph	Me	22	rt	1	4	42.4	61.9	84.0
2	Ph	*n*-Bu	18	rt	2	3.7	39.3	53.6	83.0
3	Ph	*n*-Bu	57	−40°C	8	3.9	51.3	93.3	88.6
4	Ph	*i*-Bu	31	−40°C	7	3.5	41.9	63.6	88.2
5	*o*-Tolyl	Me	145	−40°C	4	3.5	50.1	95.3	94.9
6	α-Naphthyl	Me	99[a]	−40°C	7	3.9	29.8	41.2	97.0
7	Mesityl	Me	369[a,b]	−40°C	16	12.1	44.4	78.8	98.7

[a] Toluene is used as a solvent due to solubility limitations.

[b] Reaction performed with twice recrystallized catalyst (cf. 99.7% ee for all other entries).

>99% ee.[1]

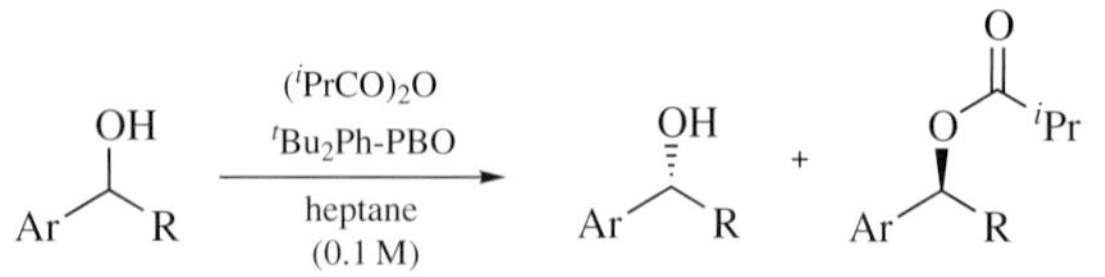

(2)

t-Bu$_2$Ph-PBO shows useful levels of selectivity for reaction at room temperature (entry 1), but an unusual temperature/reactivity profile, which results in a decreased rate of acylation by only a factor of 10 when the temperature is lowered to −40 °C (entries 2 and 3), means that very impressive levels of enantioselectivity may be obtained without a significant detrimental effect on reactivity. The best selectivities are obtained when there is a large difference between the steric bulk of the two substituents (entries 3–7). The selectivities are in general better when the reaction is performed in heptane solution, so toluene is used only when necessary due to alcohol solubility.

KR of Allylic Alcohols. *t*-Bu$_2$Ph-PBO has also been tested as a catalyst in the KR of various allylic alcohols (see eq 3, Table 2, and structures below).[6] Alcohols with significant conformational constraints (e.g., **7–9**) are poor substrates ($s < 2$). Those with more flexibility showed better results.

Table 2 KR of allylic alcohols

Alcohol	*s*	Time (h)	% conv	ee alcohol (%)	ee ester (%)
10	10	12	55.1	77.7	63.3
11	21	27	45.1	67.3	81.0
12	4	41	47.9	41.7	45.4
13	12	19	48.1	66.4	71.7
14	11	19	38.1	45.7	74.3
15	52[a]	14	50.4	89.8	88.2
16	34[a]	7	53.0	90.0	81.4
17	55	72	52.6	96.1	86.7
18	61	128	34.0	48.9	94.8
19	82[a]	25	40.3	64.2	95.3
20	25[a]	46	67.2	99.9	48.8
21	52	46	37.7	56.4	93.5

[a] Reaction performed in heptane solution.

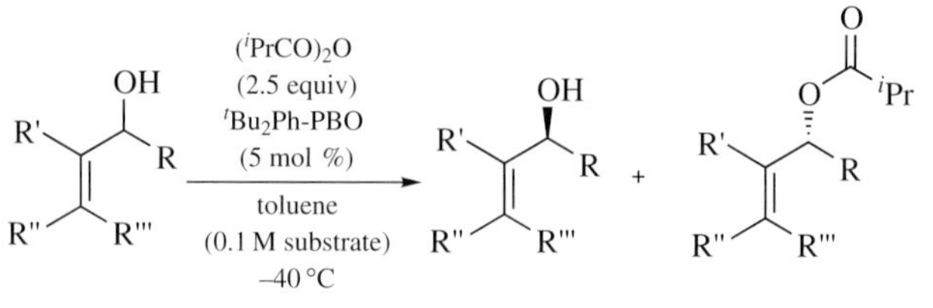

(3)

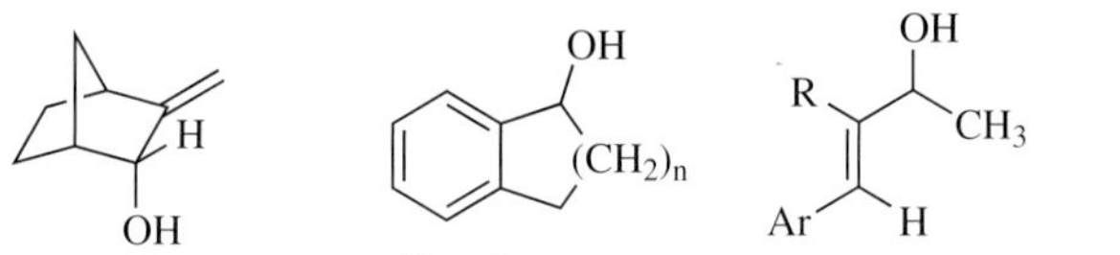

7 **8**, *n*=1 **9**, *n*=2 **10**, Ar = *p*-Cl-Ph, R = H **11**, Ar = Ph, R = Me

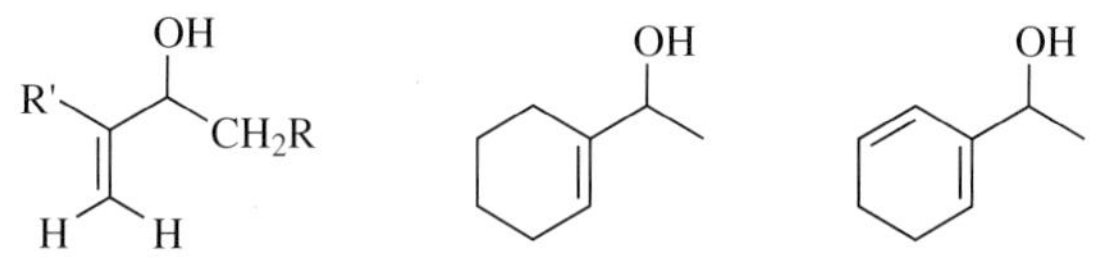

12, R = Bn, R' = H
13, R = Bn, R' = Me
14, R = H, R' = Ph
15 **16**

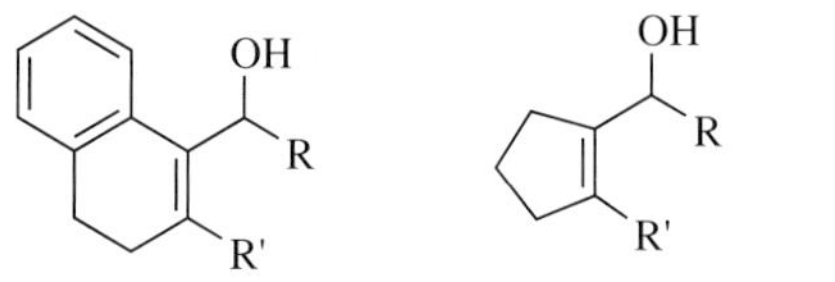

17, R = Me, R' = H
18, R = Et, R' =H
19, R = R' = Me
20, R = R' = Me
21, R = Et, R' = H

Parallel Kinetic Resolution Using a Three-phase System. One of the drawbacks of standard kinetic resolution is that selectivity often decreases with conversion, since there is an ever decreasing concentration of the most reactive enantiomer. Vedejs et al. have circumvented this limitation to some extent by performing catalytic parallel kinetic resolutions (PKRs)[7] using *t*-Bu$_2$Ph-PBO in combination with a solid supported mixed anhydride coupled with a commercial cross linked lipase (ChiroCLEC-PC).[8]

t-Bu$_2$Ph-PBO selectively activates the polymer supported mixed anhydride (eq 4) to give a presumed polymer supported acylphosphonium salt that reacts with the (*S*)-enantiomer of the aryl-alkyl carbinol (Ar = 1-naphthyl). Mesitoate ester formation was not detected. In the same pot, the enzyme selectively activates vinyl pivaloate (eq 5) which then reacts with the

(*R*)-enantiomer of the aryl-alkyl carbinol (Ar = 1-naphthyl). The two pseudo-enantiomeric products are easily separated since one is bound to the polymer support.

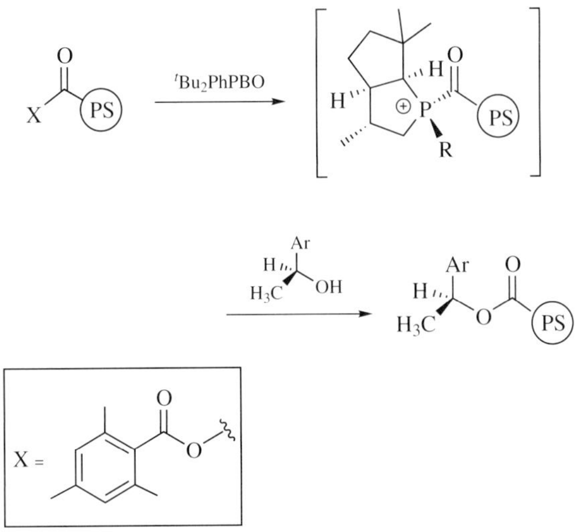

(4)

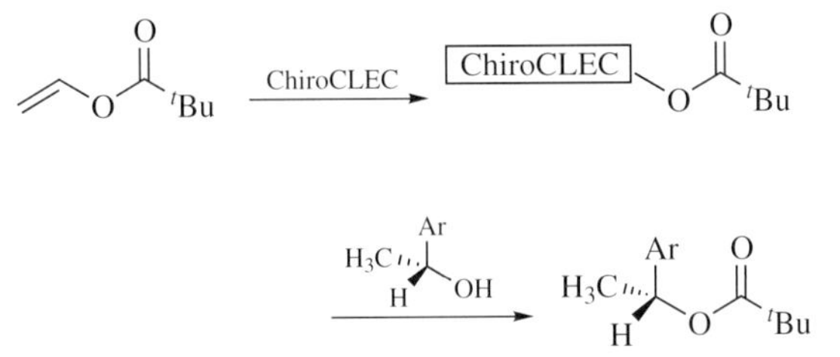

(5)

Related Reagents. Related non-enzymatic enantioselective acylation catalysts have been reported by Fu,[9] Oriyama,[10] Fuji,[11] Spivey,[12] Miller,[13] and Campbell.[14] For a review of asymmetric catalysis of acyl transfer, see Spivey et al.[15]

1. Kagan, H. B.; Fiaud, J. C., *Top. Stereochem.* **1988**, *18*, 249.
2. Vedejs, E.; Daugulis, O., *J. Am. Chem. Soc.* **2003**, *125*, 4166.
3. Vedejs, E.; Daugulis, O., *J. Am. Chem. Soc.* **1999**, *121*, 5813.
4. Isamagilov, R. F., *J. Org. Chem.* **1998**, *63*, 3772.
5. Vedejs, E.; Daugulis, O.; Diver, S. T., *J. Org. Chem.* **1996**, *61*, 430.
6. Vedejs, E.; Mackay, J. A., *Org. Lett.* **2001**, *3*, 535.
7. Vedejs, E.; Chen, X., *J. Am. Chem. Soc.* **1997**, *119*, 2584.
8. Vedejs, E.; Rozners, E., *J. Am. Chem. Soc.* **2001**, *123*, 2428.
9. Ruble, J. C.; Fu, G. C., *J. Org. Chem.* **1996**, *61*, 7230.
10. Oriyama, T.; Imai, K.; Hosoya, T.; Sano, T., *Tetrahedron Lett.* **1998**, *39*, 397.
11. Kawabata, T.; Nagato, M.; Takusu, K.; Fuji, K., *J. Am. Chem. Soc.* **1997**, *119*, 3169.
12. Spivey, A. C.; Zhu, F.; Mitchell, M. B.; Davey, S. G.; Jarvest, R. L., *J. Org. Chem.* **2003**, *68*, 7379.
13. Miller, S. J.; Copeland, G. T.; Papaioannou, N.; Horstmann, T. E.; Ruel, E. M., *J. Am. Chem. Soc.* **1998**, *120*, 1629.
14. Priem, G.; Pelotier, B.; Macdonald, S. J. F.; Anson, M. S.; Campbell, I. B., *J. Org. Chem.* **2003**, *68*, 3844.
15. Spivey, A. C.; Maddaford, A.; Redgrave, A. J., *Org. Prep. Proc. Int.* **2000**, *32*, 331.

Alan C. Spivey & Stephen G. Davey
Imperial College, London, UK

1,3-bis(2,6-diisopropylphenyl)-4,5-dihydroimidazol-2-ylidene

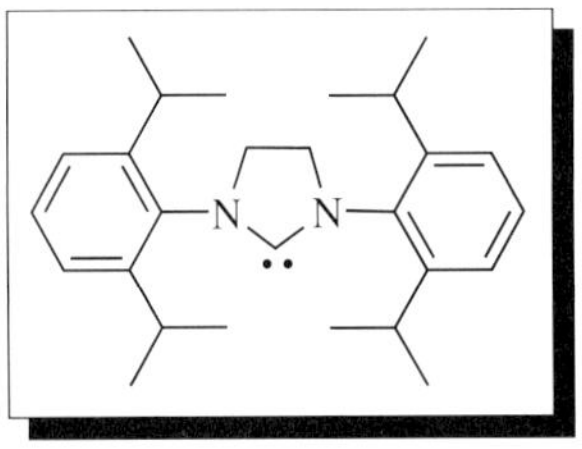

[258278-28-3] $C_{27}H_{38}N_2$ (MW 390.60)

InChI = 1S/C27H38N2/c1-18(2)22-11-9-12-23(19(3)4)26(22)28-15-16-29(17-28)27-24(20(5)6)13-10-14-25(27)21(7)8/h9-14,18-21H,15-16H2,1-8H3

InChIKey = XZDYFCGKKKSOEY-UHFFFAOYSA-N

(SIPr shows catalytic activity in the cyclotrimerization of Isocyanates.[65] Primary use is as a ligand for a variety of transition metal-catalyzed reactions such as olefin metathesis, hydroamination, hydrogenation of olefins, Ni-catalyzed cycloaddition, and Pd-, Ni-, and Fe- catalyzed cross-coupling reactions. Another potential application is in reactions employing phosphine mimics)

Physical Data: colorless solid, mp 167–168 °C.

Solubility: soluble in polar and nonpolar aromatic and aliphatic solvents, such as ether. Stable in air-free dichloromethane for days at room temperature.

Form Supplied in: commercially available as the imidazolium tetrafluoroborate salt.

Analysis of Reagent Purity: NMR.

Preparative Method: a simple three-step, high yield preparation of imidazolium chlorides using glyoxal with primary amines as starting materials has been described.[1] An alternative procedure gives a tetrafluoroborate salt.[2] Free imidazol-2-ylidene can be produced by stirring the imidazolium chloride with KH in THF;[1] with tetrafluoroborate salts good conversions can be achieved with $KOCMe_3$.[3,4]

Handling, Storage, and Precautions: stable indefinitely as the imidazolium salt. Pure imidazolidene is extremely sensitive to air and moisture and should be stored under a nitrogen atmosphere. It is incompatible with chloroform.

Physical Data: colorless solid; mp 170–175 °C.

Form Supplied in: commercially available as imidazolidin-2-ylidene as well as in form of the imidazolium tetrafluoroborate salt ($C_{27}H_{39}BF_4N_2$, MW 478.42, *282109-83-5*) or imidazolium chloride salt ($C_{27}H_{39}ClN_2$, MW427.06, *258278-25-0*).

Preparative Method: a simple three-step, high yield preparation of imidazolium chlorides using glyoxal with primary amines as starting materials has been described.[1] An alternative procedure gives a tetrafluoroborate salt.[2] Free imidazol-2-ylidene can be produced by stirring the imidazolium chloride with KH in THF[1]; with tetrafluoroborate salts good conversions can be achieved with $KOCMe_3$.[3,4]

Handling, Storage, and Precaution: stable indefinitely as the imidazolium salt. Pure imidazolidene is extremely sensitive to air and moisture and should be stored under a nitrogen atmosphere at (20 °C. It is incompatible with chloroform. Skin contact should be avoided in all forms of the reagent (GHS05, GHS07).

Original Commentary

Michael A. Zhuravel & SonBinh T. Nguyen
Northwestern University, Evanston, IL, USA

Olefin Metathesis. Imidazol-2-ylidenes are most commonly used as ligands for olefin metathesis catalysts. Imidazol-2-ylidenes readily replace either one or both of the phosphines in Grubbs catalyst, $(PCy_3)_2RuCl_2(=CHPh)$, and the resulting complexes, so called super-Grubbs catalysts,[5] exhibit improved activities and longer lifetimes.[3,6–7] To date, however, 1,3-bis(2,6-diisopropylphenyl)-4,5-dihydroimidazol-2-ylidene (SIPr) has not been used extensively in olefin metathesis compared to the other members of this family of ligands, [8,9]

C–Coupling Reactions. SIPr has seen greater utility in Pd- and Ni-catalyzed cross-coupling than in olefin metathesis.[10] The $Pd(OAc)_2$/SIPr·HCl system can be used to efficiently cross-couple aryl, vinyl, and alkyl boronates (or boronic acids) with aryldiazonium tetrafluoroborate substrates under base-free conditions (eq 1).[11] These reactions can be carried out at either 0 °C or rt for 2–4 h and give products in 80–99% isolated yields. The diazonium substrate can be formed in situ prior to the cross-coupling reaction.

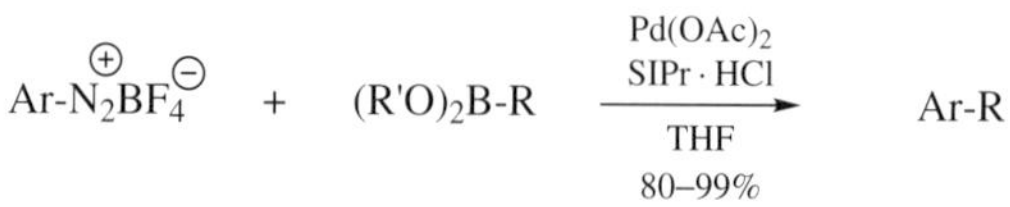
(1)

R = Ar, alkyl, vinyl
R' = H, catechol

$Pd(OAc)_2$/SIPr·HCl is also an effective catalyst in the Heck coupling of olefins and aryl diazonium tetrafluoroborate substrates under conditions similar to those described above.[12] $Pd(OAc)_2$/SIPr·HCl catalyzes the carbonylative coupling of aryl diazonium tetrafluoroborate salts and aryl boronic acids to form aryl ketones.[13] Optimal conditions involve stirring the reagents and catalysts under CO (1 atm) in 1,4-dioxane at 100 °C for 5 h to produce the aryl ketone products in 76–90% yield.

The Heck coupling of bromotoluene with *tert*-butyl acrylate gives better yields with $Pd(OAc)_2$/SIPr·HI than with a $Pd(dba)_2$/SIPr·HCl system.[14] In the Suzuki coupling of aryl chlorides with boronic acids, SIPr is inferior to IPr [IPr = 1,3-bis(2,6-diisopropylphenyl)imidazol-2-ylidene], giving only 70% yield.[15] $Pd(OAc)_2$/SIPr·HCl fails as a catalyst in the coupling of aryl chlorides with aryl tin compounds.[16] However, it is effective in the coupling of aryl bromides with alkynylsilanes (eq 2), albeit not as efficiently as the analogous $Pd(OAc)_2$/IPr·HCl system.[17] This coupling proceeds smoothly without the use of CuI which is often required to facilitate Pd/phoshine-catalyzed Sonogashira reactions.[18]

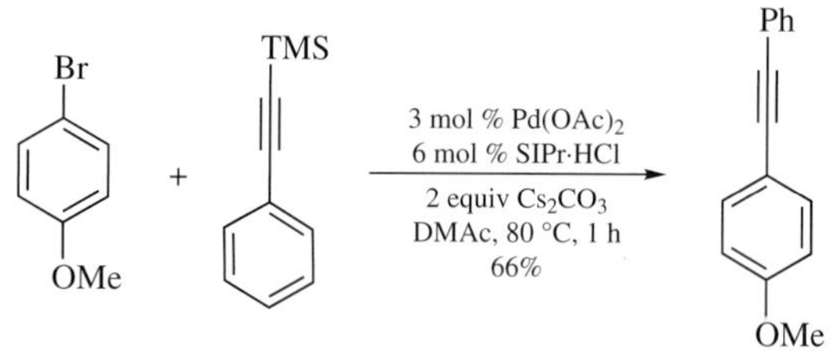
(2)

An attempt to dimerize 1-heptyne using $Pd(OAc)_2$/SIPr·HCl as a catalyst resulted in the formation of the mixture of three products in 14% combined yield.[19] This is quite interesting considering that the $Pd(OAc)_2$/IMes·HCl [IMes=1,3-bis(2,6,4-trimethylphenyl)-2-ylidene] catalyst system is quite efficient for this reaction, giving 87% yield of the major product.

C-Heteroatom Coupling. A general and simple Ni-catalyzed coupling of aryl chlorides and amines was made possible by using Ni(0)/SIPr·HCl as the catalyst (eq 3).[20,21] The transformations catalyzed by this system include reaction of aryl chlorides with morpholines and other secondary amines, coupling of anilines with aryl chlorides, and arylation of primary amines.

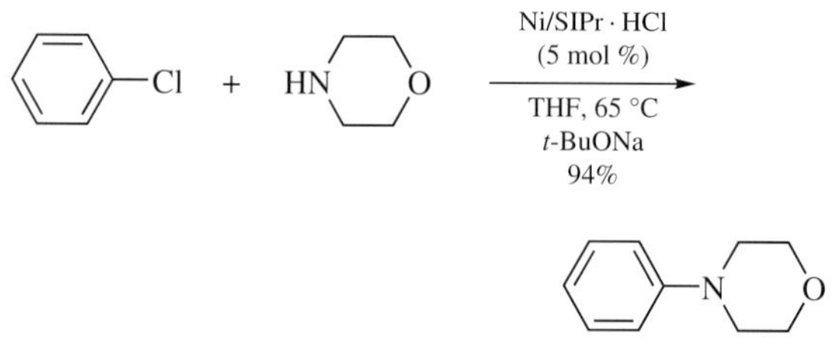
(3)

The amination of aryl chlorides under mild conditions is possible using $Pd(dba)_2$/SIPr·HBF_4 as the catalyst.[22] Aryl and heteroaryl chlorides react with secondary amines and anilines within hours at room temperature. The yields reported for these in situ generated catalysts are slightly lower than those for reactions conducted using the preformed catalysts $Pd(SIPr)_2$ and Pd(SIPr)(P(*o*-tolyl)$_3$) at 100 °C for 1 h.[23] The $Pd_2(dba)_3$/SIPr system is efficient for the *N*-arylation of diverse indoles with aryl bromides.[24] In the palladium-catalyzed conversion of aryl halides to anilines using lithium bis(trimethylsilyl)amide, SIPr is inferior to P(*tert*-Bu)$_3$, giving only a modest yield at 90 °C.[25] A catalyst comprising $Pd(OAc)_2$ and SIPr·HCl provides fast rates for a palladium-catalyzed synthesis of oxindoles by amide α-arylation (eq 4).[26] The reactivity of SIPr·HCl-based catalyst is similar to that of $Pd(OAc)_2/PCy_3$ and considerably better than that of the $Pd(OAc)_2$/P(*tert*-Bu)$_3$ system.

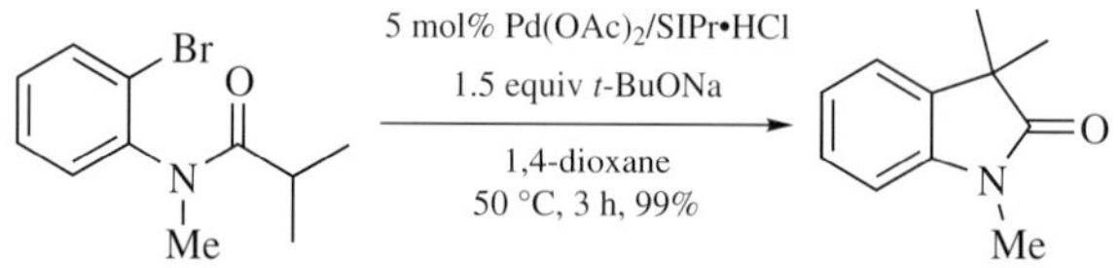
(4)

Other Uses. (SIPr)Pd(allyl)(Cl) is an extremely efficient catalyst for arylation of ketones using aryl chlorides and aryl triflates.[27] Excellent yields are achieved in 1–1.5 h at 50–70 °C, and a wide range of substrates are compatible with the reaction conditions. Dehalogenation of chlorobenzene with sodium isopropoxide in the presence of Ni(0)/SIPr·HCl catalyst gave 58% yield of benzene.[28] IMes·HCl was shown to be a better ligand for this reaction under similar conditions, but since the yield is very sensitive to the nature of both Ni(0) and alkoxide, SIPr·HCl remains a suitable ligand. In another example of the dehalogenation of chloroarenes, the reaction of chlorotoluene with potassium methoxide and $Pd(dba)_2$/SIPr·HCl catalyst gave toluene in 56% yield.[29] SIMes [SIMes = 1,3-bis(2,6,4-trimethylphenyl)-4,5-dihydroimidazol-2-ylidene] worked the best for this transformation; however, as in the case of Ni-based catalysts, the reaction is sensitive to the choice of reagents. SIPr·HCl, SIPr·HBF_4, and SIPr all catalyze the alkylation of a variety of meso epoxides in the presence of $AlEt_3$ under mild conditions. Imidazolinium salts are better catalysts than their free-carbene derivatives, but can lead to dimerization side reactions under extended reaction times.[30] Aldehydes can be successfully arylated with boronic acids using Rh(acac)$(coe)_2$/SIPr·HCl as a catalyst (eq 5), however, both IPr·HCl and IMes·HCl give higher yields.[31]

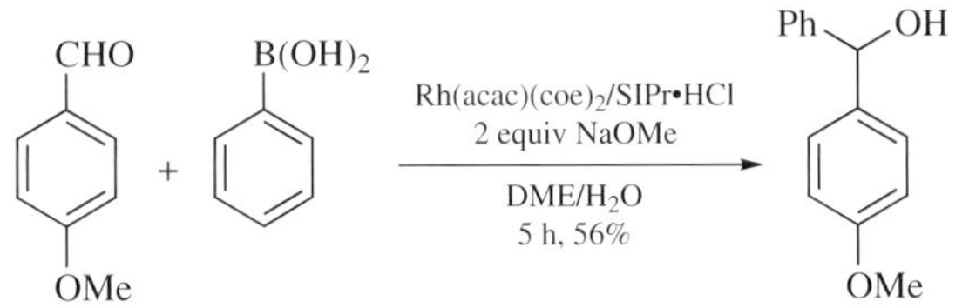

(5)

Related Reagents. 1,3-Bis(2,6-diisopropylphenyl)imidazol-2-ylidene (IPr); 1,3-bis(2,6,4-trimethylphenyl)-4, 5-dihydroimidazol-2-ylidene (SIMes); 1,3-bis(2,6,4-trimethylphenyl)-2-ylidene (IMes).

First Update

Stephanie Felten & Janis Louie
University of Utah, Salt Lake City, UT, USA

Olefin Metathesis. [(SIPr)$RuCl_2(PCy_3)$Ind] shows a high activity in the RCM (ring-closing metathesis) of unhindered and slightly hindered dienes and enynes at room temperature and short reaction times.[32] The enyne substrate (1-(allyloxy)prop-2-yne-1,1-diyl)dibenzene can be converted in 25 min at room temperature with 1 mol % catalytic loading to the corresponding product 2,2-diphenyl-3-vinyl-2,5-dihydrofuran in 96% yield (eq 6). No reaction is observed with sterically demanding substrates like tetrasubstituted dienes. In general, the reaction is tolerant of a wide range of functionalized diene and enyne substrates. SIPr as a ligand is superior to the IPr, IMes, SIMes, and PCy_3 analog catalysts.[32]

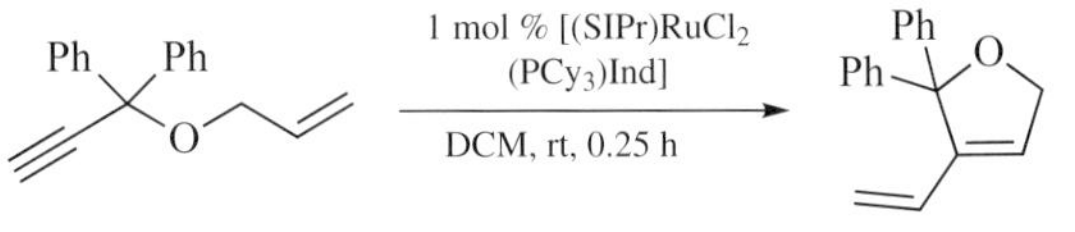

(6)

96%

[(SIPr)$RuCl_2$(Py)Ind] and [(SIPr)$RuCl_2$(3-BrPy)Ind] are reported catalysts for the RCM of enynes and the CM (cross metathesis) of olefins under mild reaction conditions. Both catalysts showed a high catalytic activity for less hindered substrates to afford the corresponding products in good yields. Nitrogen heterocycles can be converted in a RCM to afford good to excellent yields with catalytic loadings varying from 100 to 500 ppm. Olefins and methyl acrylates underwent cross metathesis by 250–500 ppm catalytic loading.[33]

[$RuCl_2$(Ind)(SIPr){$P(OR)_3$}] (R = Et, ^{i}Pr) is a reported catalytic system that shows high activity for the enyne and cross metathesis. Good yields are obtained for a variety of substrates under short reaction times and low catalyst loadings (eq 7).[34]

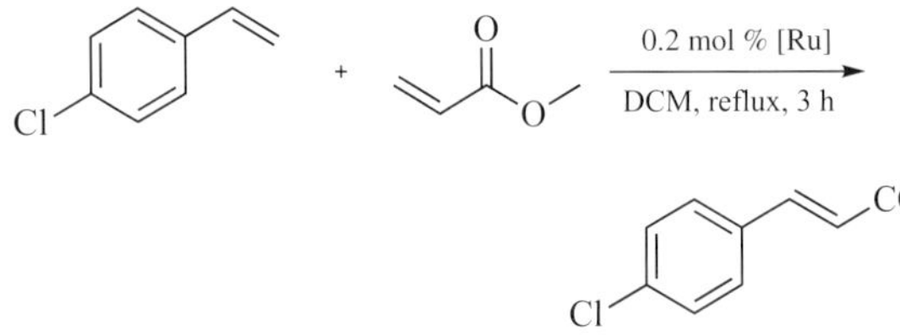

(7)

[Ru]= **a** [$RuCl_2$(Ind)(SIPr){$P(O^iPr)_3$}] **a** 69%
b [$RuCl_2$(Ind)(SIPr){$P(OEt)_3$}] **b** 58%

Highly active catalysts for the olefin cross metathesis are the ruthenium-based complexes [(SIPr)$(PCy_3)(Cl)_2$Ru·CHPh] and [(SIPr)$(Cl)_2$Ru·CH(o-^{i}PrOC_6H_4)].[35,36] The latter is effective for the cross metathesis of trisubstituted olefins in excellent yields (>98%). An exception is the reaction of 1,1-disubstituted allylic olefins in which the IMes-analog was found to be superior.[36,37]

C–C Coupling Reactions. Different catalytic systems for the C–C-bond formation by metal-catalyzed cross-coupling reactions have been recently discovered. In 2002, (SIPr)Pd(allyl)Cl was reported to be an efficient catalyst for the α-arylation of ketones.[38] High catalytic activity was presented for the coupling of aryl halides and triflates with a wide range of ketones (eq 8). Features of the catalytic system are low catalyst loadings (1 mol %), short reaction times, moderate temperatures (50–70 °C), and good to excellent conversions. SIPr as a ligand is superior in comparison to other NHC ligands (IAd, I^{t}Bu, IPr, and IMes).[38,39]

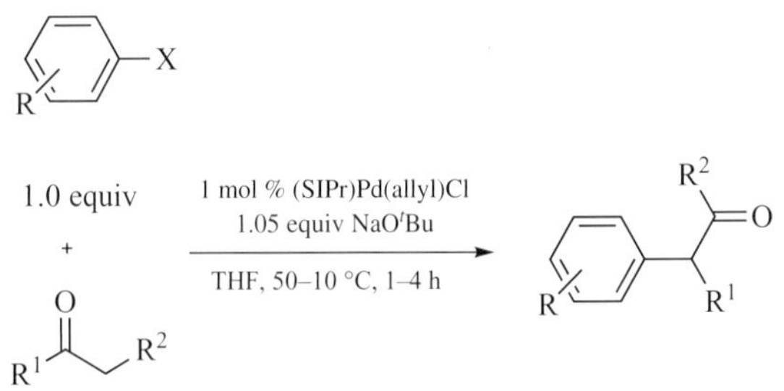

(8)

1.0 equiv

R^1, R^2 = -alkyl, -aryl.
R= -H, -COPh, -Me, -OMe, $-CF_3$.
X= -Cl, -Br, -OTf.

Another effective catalyst for the α-arylation of aryl or alkyl ketones is the commercially available (SIPr)Pd(Py)Cl_2 (Py = Pyridine) also known as the PEPPSI™SIPr catalyst. High catalytic activity was shown for a wide range of activated, unactivated, and sterically hindered aryl chlorides. The alkyl ketone 3-pentanone selectively reacted with chlorobenzene in dry THF at room temperature to 72% of monoarylated 2-phenyl-3-pentanone product. At 70 °C the system favors the α,α'-diarylated product in 87% yield.[40] The cross-coupling reaction between aryl and heteroaryl halides with aryl Grignard substrates selectively afforded unsymmetrical biaryl products in excellent yields.[41] The cross-coupling reaction is facilitated by the fluorinated iron, cobalt, and nickel complexes of N-heterocyclic carbene (NHC) (NHC = IPr, SIPr). Hydrous or anhydrous FeF_3/SIPr·HCl catalyzes the cross-coupling of electrophilic and nucleophilic aryl chlorides with arylmagnesium substrates to afford biaryl products in high yield. The coupling between 1-chloro-4-methoxy-benzene and methylmagnesium bromide in THF at 80 °C demonstrates the reaction of electron-rich aryl chlorides with alkyl or methyl Grignard reagents (eq 9). Other NHC chloride salts (NHC = IPr, SIMes) afforded the desired product in decreased yield. Under similar conditions, $CoF_2{\cdot}4H_2O$ and NiF_2 catalyzed heteroaromatic cross-coupling as well as cross-coupling reactions of sterically hindered substrates and aryl bromides. However, IPr·HCl was superior to SIPr·HCl for both reactions.[41,42]

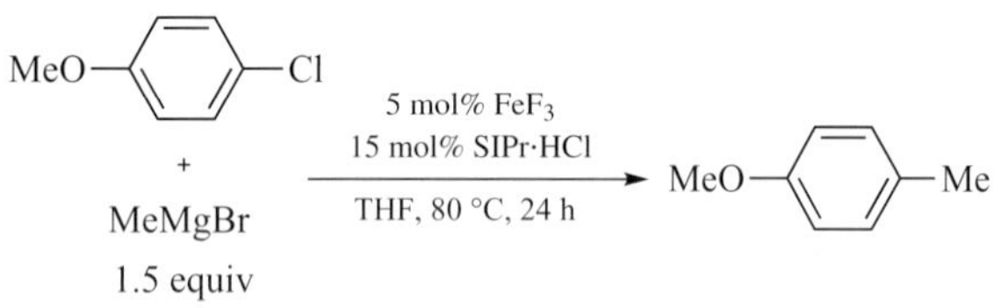

(9)

$Ni(COD)_2$/SIPr·HCl is an active catalyst for the cross-coupling reaction of alkyl and aryl sulfides with aryl Grignard reagents to afford alkenyl–aryl products in good yields.[43] A wide range of alkyl and aryl sulfides are tolerated affording products in high yields and with good *E*/*Z* selectivity. However, reactions of trisubstituted olefins displayed poor selectivity (86%, *E*:*Z* = 46:54). Commercially available $[Pd(\mu\text{-Cl})Cl(SIPr)]_2$ has high catalytic activity in the Mizoroki–Heck cross-coupling reaction of aryl and heterocyclic bromides yielding products in moderate to excellent yields (42–99%).[44] The reaction is highly efficient with a catalytic loading of 100–200 ppm Pd/SIPr and $KHCO_3$ in DMF at 120–140 °C for 20 h. The use of the SIPr ligand afforded the highest conversions and turnover numbers as compared to other $[Pd(\mu\text{-Cl})Cl(NHC)]_2$ catalysts containing IPr, IMes, or SIMes as a ligand.[44] $PdCl_2(SIPr)(AsPh_3)$ and $PdCl_2(SIPr)(SbPh_3)$ are reported to catalyze the Hiyama cross-coupling to form symmetrical and unsymmetrical biaryls. The reaction must be run in dry toluene to couple aryl or heterocyclic chlorides with organosilanes to yield biphenyl products (eq 10). In general the Pd/IPr catalyst was superior to the Pd/SIPr system.[45]

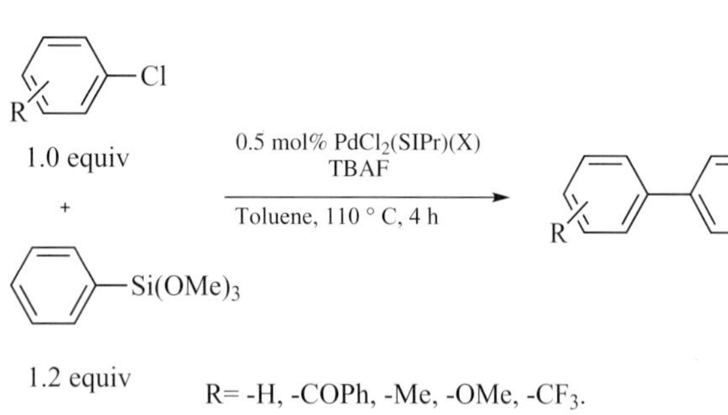

(10)

R = -H, -COPh, -Me, -OMe, $-CF_3$.

X = $-AsPh_3$, $-SbPh_3$.

A wide range of enol ethers underwent highly selective Ni/SIPr-catalyzed (5 mol %) cross-coupling reaction with $LiCH_2SiMe_3$ at 60 °C in toluene. Aromatic and aliphatic heterocycles were effective substrates. Furthermore, the same conditions were used to catalyze silylenol ethers and enol ether ring-opening reactions. Substituted and unsubstituted furan and pyran ring systems afforded primary and secondary alcohols in good yields.[46] (SIPr)Pd(acac)Cl displayed high catalytic activity in the α-arylation of ketones. The reaction required only 0.1 mol % catalyst loading in toluene at 60 °C to afford 97% yield with short reaction times (eq 11).[47]

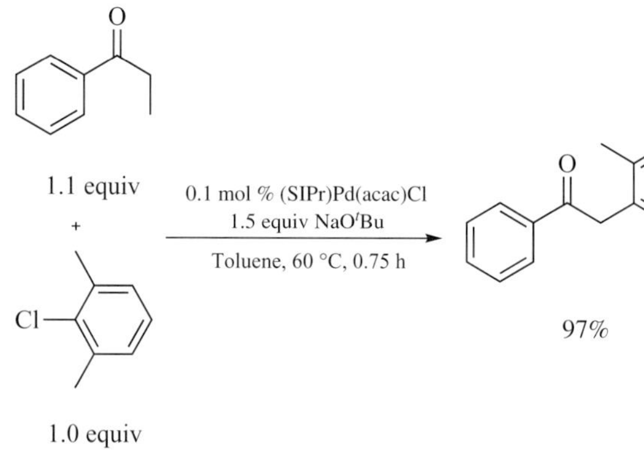

(11)

$Pd(OAc)_2$ with 1–3 mol % SIPr ligand and 0.6 equiv *p*-benzoquinone (pbq) in methanol is an effective catalyst system for the homocoupling of arylboronic acids to give symmetrical biaryl products. Several *para*- and *meta*-substituted arylboronic acids were tolerated (eq 12). Poor yields were obtained in the homocoupling of 1-phenylvinylboronic acid and benzofuran derivatives under the same conditions. IPr ligated to $Pd(OAc)_2$ was superior in this type of reaction.[48]

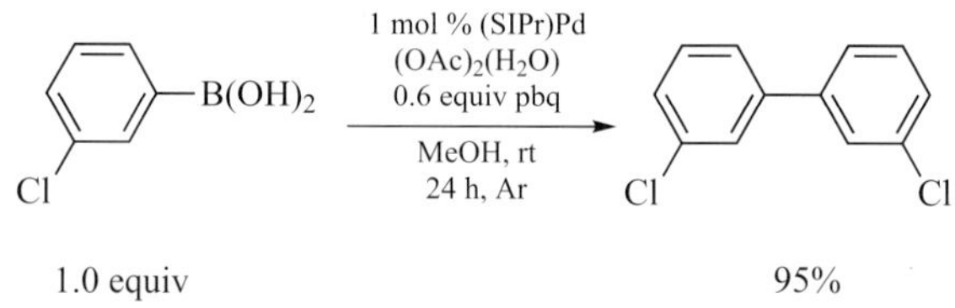

(12)

C–Heteroatom Coupling Reactions. 2 mol % Ni/SIPr catalyzes the intramolecular *N*-arylation of amino aryl chlorides in dioxane at 100 °C to afford indolines and tetrahydroquinolines (eq 13).[49] Slightly higher yields were obtained for the cyclization to five- and six-membered ring products than seven-membered ring systems. The same catalytic system was used for the intramolecular coupling reaction of benzazepines, benzoxazines, and benzoxazepines and the intramolecular arylation of cyclic amines. However, Ni/bpy (bpy = 2,2′-bipyridine) was a superior catalyst than Ni/SIPr.[49]

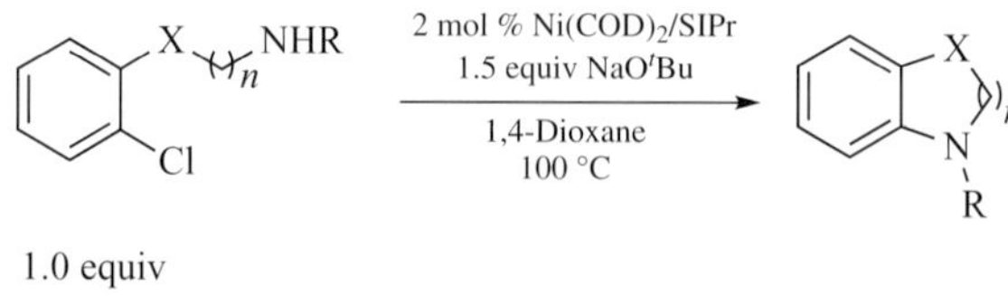

(13)

n = 1–3.
X = $-CH_2$, -O.

(SIPr)Pd(cin)Cl (cin = cinnamyl) is an efficient catalyst for the *N*-aryl amination of aryl, naphthyl, and anthryl halides (chloride, bromide) with anilines and amines to afford the corresponding

C–N coupling product.[50] (SIPr)Pd(cin)Cl is also an effective catalyst for C–H-functionalization of benzothiophene, thiophene, and imidazopyridine with a wide range of aryl bromides at low catalytic loadings (eq 14).[51]

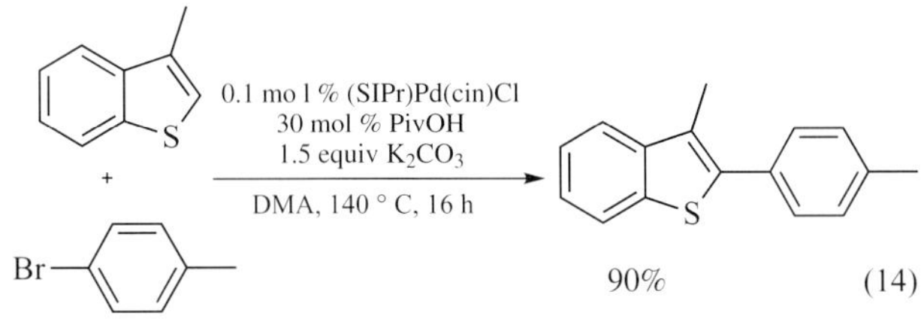
(14)

Another active catalytic system for the Buchwald–Hartwig amination containing SIPr as a ligand is (SIPr)Pd(acac)Cl.[47] The reaction of 2,6-diisopropylphenylaniline with 1-chloro-2-methylbenzene is catalyzed by 1 mol% catalyst loading at 50 °C (eq 15). Although the IPr analog is also an effective catalyst, reaction times are shorter with (SIPr)Pd(acac)Cl. In contrast, reactions with (SIPr)PdCl$_2$(TEA) required longer reaction times and gave products in slightly lower yield.[52]

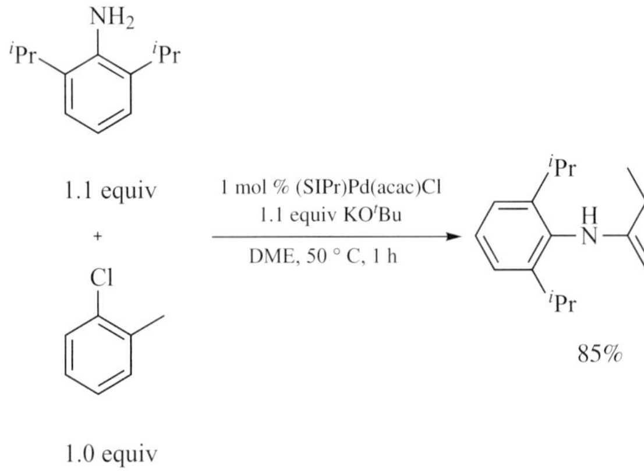
(15)

The Buchwald–Hartwig coupling of different aryl chlorides with aromatic and aliphatic amines afforded products in good yields. The reaction is performed with 1 mol % (SIPr)Pd(sal)Cl (sal = salicylaldimine) under aerobic reaction conditions (eq 16).[53]

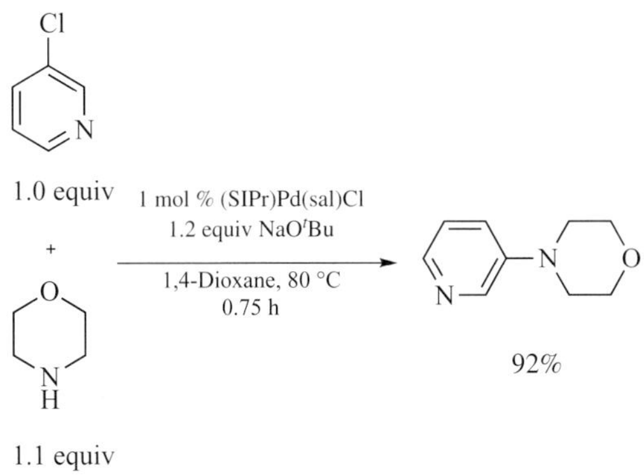
(16)

Another catalyst for the Buchwald–Hartwig amination is PdCl$_2$(SIPr)(X) [X = AsPh$_3$, SbPh$_3$]. The reaction tolerated electronic and steric effects at the aryl chloride substrates. The Pd/IPr containing catalyst gave higher product yields than the saturated analog.[45]

Cycloaddition Reactions. Nickel-catalyzed cycloaddition reactions afford a wide range of carbocycles and heterocycles. The Ni(COD)$_2$/SIPr system is particular effective for a variety of cycloaddition reactions.[54] The Ni(COD)$_2$/SIPr system require an 6 h equilibrium reaction between Ni(COD)$_2$ and SIPr at room temperature in the respective solvent prior to use as a catalyst.[53,54] In 2004, the inter and intra Ni-catalyzed cycloaddition of alkyne and isocyanate substrates to 2-pyridones was reported (eq 17). The cycloaddition was performed with 0.1 M substrate concentration and a catalyst loading of 3–5 mol % Ni(COD)$_2$/SIPr to afford 2-pyridones in high yields. Sterically hindered substrates were tolerated. SIPr as a ligand was superior to phosphine or other NHC ligands (ICy, IMes, IAd, and I^{t}Bu).[54,55]

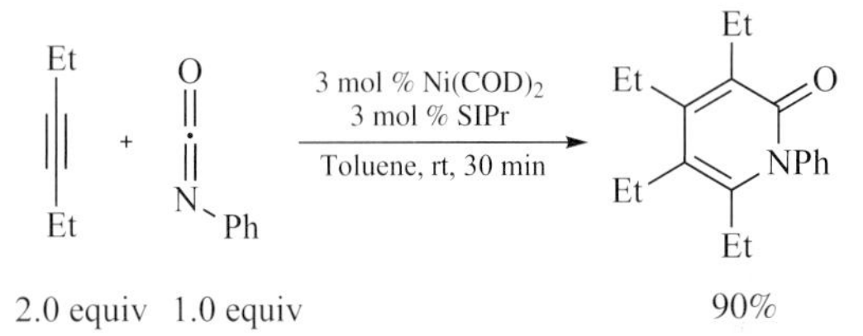
(17)

Replacing isocyanates with nitriles in Ni-catalyzed cycloaddition affords pyridines in high yields under the same mild conditions (56–97%). Several diynes reacted with alkyl, aryl, and heteroaryl nitriles. In addition, intermolecular cycloaddition of 3-hexyne with benzonitrile was effective (eq 18).[54,56]

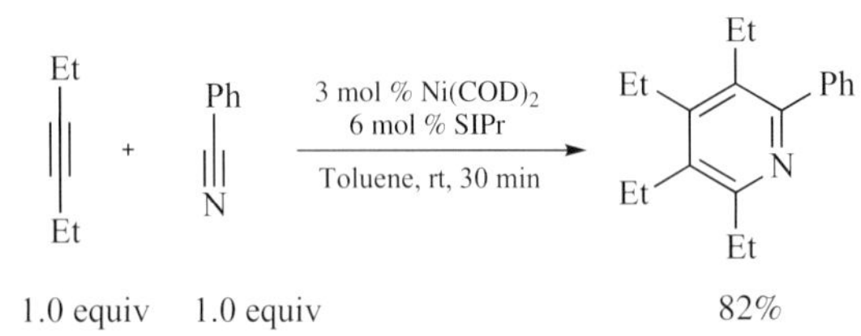
(18)

Furthermore, Ni(COD)$_2$/SIPr catalyzed the cycloaddition of internal and terminal diynes with cyanamides. Optimized conditions were 3–5 mol % Ni(COD)$_2$ with 6–10 mol % NHC ligand (NHC = SIPr/IMes) at room temperature. Ni/IMes is an active catalyst system for internal diynes but lowered yields were obtained.[54,56,57] The intermolecular cycloaddition of alkyls, terminal alkynes, and cyanamides can be catalyzed by the Ni/SIPr system. Reactions were moderately selective although yields ranged from good to low. The reaction between pent-1-yne and morpholine-4-carbonitrile afforded the major product 4-(2,4-dipropylphenyl)morpholine, and 4-(3,5-dipropylphenyl)morpholine as the minor product (eq 19).[54,58] Alkyl chlorides, alkynyl esters, and aryl acetylenes cannot be converted to the cycloaddition product.

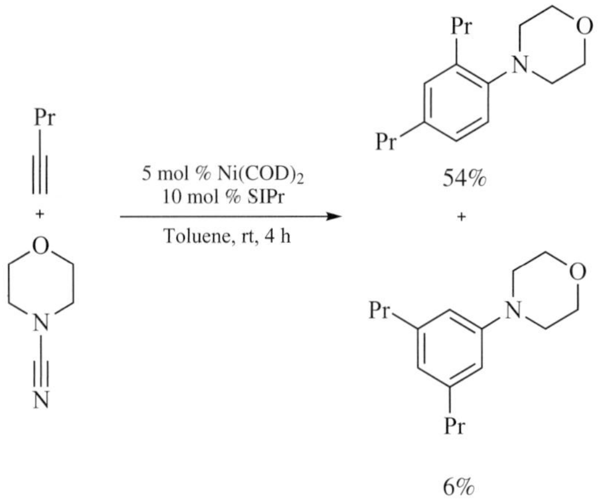
(19)

The cycloadditions involving carbonyl functional groups (aldehydes and ketones) can be efficiently catalyzed by $Ni(COD)_2$/SIPr. Activated and unactivated aldehydes undergo cycloaddition with diynes to afford dienones under milder reaction conditions in comparison to $Ni(COD)_2/PR_3$. Electron-donating aldehydes afforded better yields than electron-withdrawing-containing aldehyde.[54,59] Enynes also underwent smooth and selective cycloaddition with activated and unactivated aldehydes under similar conditions to the corresponding enone products (eq 20).[54,60]

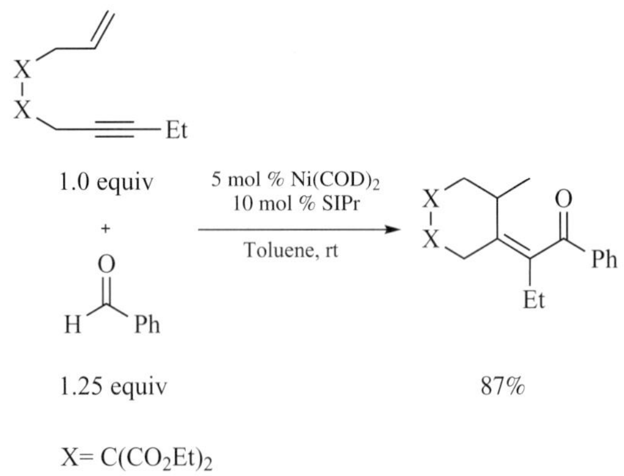
(20)

The cycloaddition of enynes and ketones selectively afforded pyrans in good to excellent yields (eq 21).[54] The Ni/SIPr system displayed increased conversion compared to the Ni/IPr system.

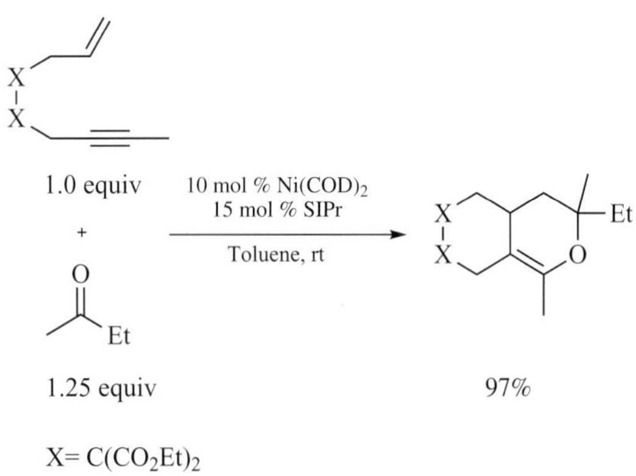
(21)

The challenging cycloaddition of diynes and tropones was efficiently catalyzed by the Ni/SIPr system to afford tricyclic products. Unsymmetrical aryl-substituted diynes reacted regioselectively to afford products where the carbonyl resides next to the aryl substituent (eq 22).[54,61]

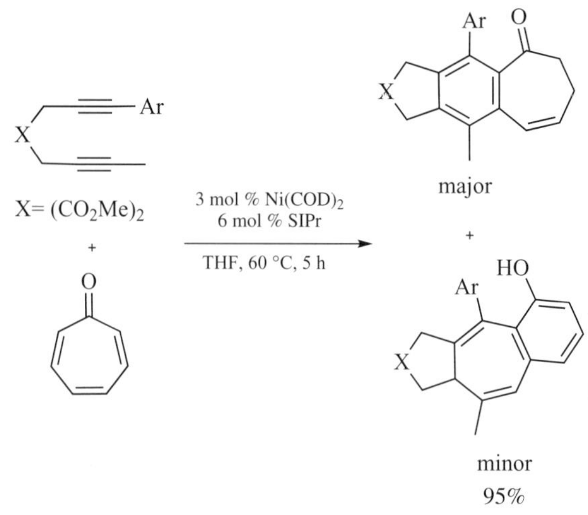
(22)

$Ni(COD)_2$/SIPr also catalyzes the reaction of acrylates with isocyanate to yield fumarate products (1,3,5-trisubstituted hydantoin). The reaction tolerated sterically hindered and electronically varied aryl isocyanates. However, no desired products were observed in reactions with cyclohexyl isocyanate and *tert*-butyl isocyanate. Different primary and secondary alkyl acrylates were amenable substrates with the exception of *tert*-butyl acrylate that gave poor conversion (35%).[62] [(SIPr)CuCl] effectively catalyzes the [3 + 2] cycloaddition of benzylic or alkyl azides with alkynes to afford triazole products. However, the reaction showed no activity under ambient conditions.[63]

Rearrangement Reactions. Cyclopropylenynes can be converted to five- and seven-membered heterocycles under mild reaction conditions for a wide range of substrates. The Ni/SIPr system demonstrates excellent performance in reaction time, scope, and yield in comparison to other Ni/L catalysts (L = phosphines, NHC ligands). Cyclopropylenynes with small substituents (e.g., -Me) afforded only the five-membered heterocycle in comparison to larger substituents (e.g., ^{i}Pr, ^{t}Bu) that afforded both heterocycles. The use of Ni/^{t}Bu allowed selective formation of the cyclopentane heterocycle regardless of substrate, whereas the use of Ni/SIPr afforded the mixture of heterocycles (eq 23).[64]

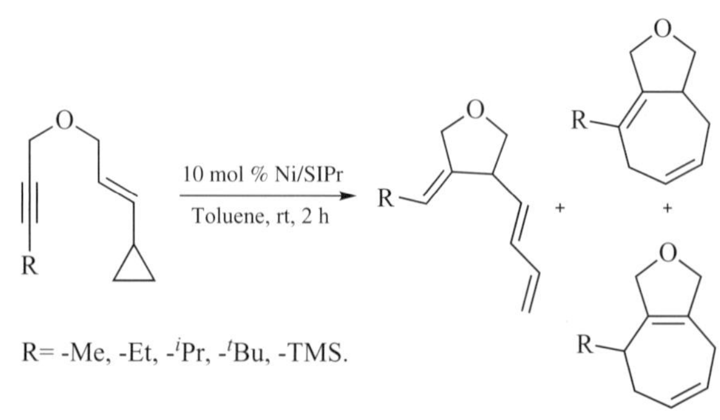
(23)

SIPr-catalyzed Reactions. The catalytic activity of SIPr on its own was demonstrated by the cyclotrimerization of alkyl, aryl, and allyl isocyanates to afford isocyanurates in high yields (85–99%).[65] Other NHC catalysts displayed lower reactivity or selectivity in the reaction. Specifically cyclodimerization side products were observed. Only ICy and imidazolium carboxylate were effective catalysts for this reaction. Depending on the isocyanate substrate, 0.001–1 mol % SIPr was necessary to catalyze the reaction (eq 24). The isolated yields were not affected by steric or electronic changes of the substrate.[65]

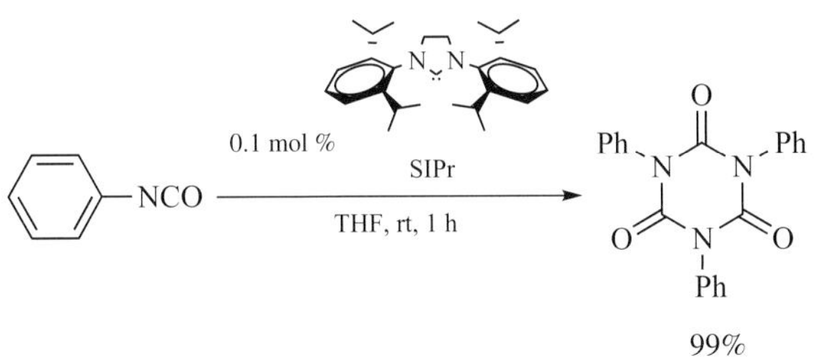

99% (24)

SIPr displayed a lower selectivity in organocatalyzed reactions compared to other N-heterocyclic (NHC) carbenes (e.g., IMes, ICy). Transesterifications and acylations,[66,67] epoxide ring-opening alkylations,[68] and the aza-Morita–Baylis–Hillman reaction[69] are reportedly catalyzed by the SIPr-carbene in moderate yields; however, other N-heterocyclic carbenes were superior in these reactions.

Other Uses. Anaerobic oxidation of secondary aryl- and alkyl-substituted alcohols is catalyzed by the commercially available (SIPr)Pd(cin)Cl catalyst to afford ketones in good to excellent yields (70–99%).[70] The reaction is performed in toluene at 40 °C with low catalytic loadings (0.25–1 mol %) and short reaction times (0.25–5 h). Pd/SIPr is superior to the Pd/IPr in this oxidation reaction. Metal-catalyzed hydrogenation shows a high activity with [Pd(SIPr)(PCy_3)] catalyst for a variety of functionalized and nonfunctionalized substrates. Hydrogenation of 9-ethynyl-9*H*-fluoren-9-ol with 0.25 mol % catalytic loading and 1 bar H_2 in methanol at room temperature afforded 9-vinyl-9*H*-fluoren-9-ol (eq 25).[71]

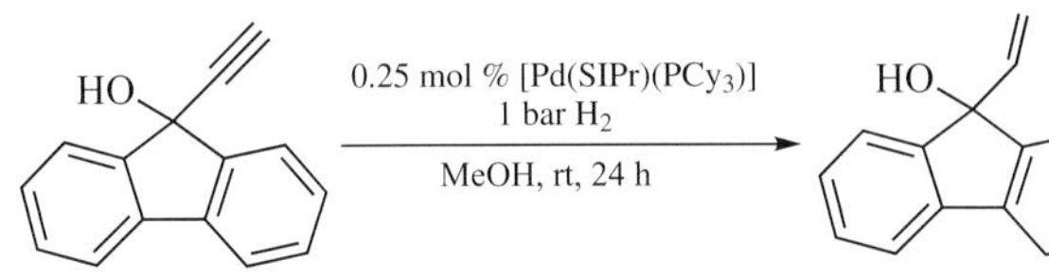

95%

(25)

1. Arduengo, A. J., III; Krafczyk, R.; Schmutzler, R.; Craig, H. A.; Goerlich, J. R.; Marshall, W. J.; Unverzagt, M., *Tetrahedron* **1999**, *55*, 14523–14534.
2. Saba, S.; Brescia, A. M.; Kaloustian, M. K., *Tetrahedron Lett.* **1991**, *32*, 5031–5034.
3. Scholl, M.; Ding, S.; Lee, C. W.; Grubbs, R. H., *Org. Lett.* **1999**, *1*, 953–956.
4. Arduengo, A. J., III; Dias, H. V. R.; Harlow, R. L.; Kline, M., *J. Am. Chem. Soc.* **1992**, *114*, 5530–5534.
5. Stinson, S., *Chem. & Eng. News* **2000**, *78*, 6–7.
6. Weskamp, T.; Schattenmann, W. C.; Spiegler, M.; Herrmann, W. A., *Angew. Chem. Int. Ed. Engl.* **1998**, *37*, 2490–2493.
7. Huang, J.; Stevens, E. D.; Nolan, S. P.; Petersen, J. L., *J. Am. Chem. Soc.* **1999**, *121*, 2674–2678.
8. Fürstner, A.; Ackermann, L.; Gabor, B.; Goddard, R.; Lehmann, C. W. ; Mynott, R.; Stelzer, F.; Thiel, O. R., *Chem. Eur. J.* **2001**, *7*, 3236–3253.
9. Bell, A.; Grubbs, R. H.; Morgan, J. P.; Moore, J. L., *PCT Int. Appl.*; California Institute of Technology and Cymetech, LLC: WO, **2002**, pp 100.
10. Hillier, A. C.; Grasa, G. A.; Viciu, M. S.; Lee, H. M.; Yang, C.; Nolan, S. P., *J. Organomet. Chem.* **2002**, *653*, 69–82.
11. Andrus, M. B.; Song, C., *Org. Lett.* **2001**, *3*, 3761–3764.
12. Andrus, M. B.; Song, C.; Zhang, J., *Org. Lett.* **2002**, *4*, 2079–2082.
13. Andrus, M. B.; Ma, Y.; Zang, Y.; Song, C., *Tetrahedron Lett.* **2002**, *43*, 9137–9140.
14. Yang, C.; Nolan, S. P., *Synlett* **2001**, 1539–1542.
15. Fürstner, A.; Leitner, A., *Synlett* **2001**, 290–292.
16. Grasa, G. A.; Nolan, S. P., *Org. Lett.* **2001**, *3*, 119–122.
17. Yang, C.; Nolan, S. P., *Organometallics* **2002**, *21*, 1020–1022.
18. Sonogashira, K., In *Metal-Catalyzed Cross-Coupling Reactions*; Diederich, F.; Stang, P. J., Eds.; Wiley-VCH: Weinheim, **1998**, pp 203–229.
19. Yang, C.; Nolan, S. P., *J. Org. Chem.* **2002**, *67*, 591–593.
20. Desmarets, C.; Schneider, R.; Fort, Y., *J. Org. Chem.* **2002**, *67*, 3029–3036.
21. Gradel, B.; Brenner, E.; Schneider, R.; Fort, Y., *Tetrahedron Lett.* **2001**, *42*, 5689–5692.
22. Stauffer, S. R.; Lee, S.; Stambuli, J. P.; Hauck, S. I.; Hartwig, J. F. , *Org. Lett.* **2000**, *2*, 1423–1426.
23. Titcomb, L. R.; Caddick, S.; Cloke, F. G. N.; Wilson, D. J.; McKerrecher, D., *Chem. Commun.* **2001**, 1388–1389.
24. Grasa, G. A.; Viciu, M. S.; Huang, J.; Nolan, S. P., *J. Org. Chem.* **2001**, *66*, 7729–7737.
25. Lee, S.; Jorgensen, M.; Hartwig, J. F., *Org. Lett.* **2001**, *3*, 2729–2732.
26. Lee, S.; Hartwig, J. F., *J. Org. Chem.* **2001**, *66*, 3402–3415.
27. Viciu, M. S.; Germaneau, R. F.; Nolan, S. P., *Org. Lett.* **2002**, *4*, 4053–4056.
28. Desmarets, C.; Kuhl, S.; Schneider, R.; Fort, Y., *Organometallics* **2002**, *21*, 1554–1559.
29. Viciu, M. S.; Grasa, G. A.; Nolan, S. P., *Organometallics* **2001**, *20*, 3607–3612.
30. Zhou, H.; Campbell, E. J.; Nguyen, S. T., *Org. Lett.* **2001**, *3*, 2229–2231.
31. Fürstner, A.; Krause, H., *Adv. Synth. Catal.* **2001**, *343*, 343–350.
32. Clavier, H.; Urbina-Blanco, C. A.; Nolan, S. P., *Organometallics* **2009**, *28*, 2848.
33. Urbina-Blanco, C. A.; Leitgeb, A.; Slugovc, C.; Bantreil, X.; Clavier, H.; Slawin, A. M. Z.; Nolan, S. P., *Chem. Eur. J.* **2011**, *17*, 5045.
34. Urbina-Blanco, C. A.; Bantreil, X.; Wappel, J.; Schmid, T. E.; Slawin, A. M. Z.; Slugvoc, C.; Cazin, C. S. J., *Organometallics* **2013**, *32*, 6240.
35. Dinger, M. B.; Mol, J. C., *Adv. Synth. Catal.* **2002**, *344*, 671.
36. Courchay, F. C.; Sworen, J. C.; Wagener, K. B., *Macromolecules* **2003**, *36*, 8231.
37. Stewart, I. C.; Douglas, C. J.; Grubbs, R. H., *Org. Lett.* **2008**, *10*, 441.
38. Viciu, M. S.; Germaneau, R. F.; Nolan, S. P., *Org. Lett.* **2002**, *4*, 4053.
39. Viciu, M. S.; Navarro, O.; Germaneau, R. F.; Kelly III, R. A.; Sommer, W.; Marion, N.; Stevens, E. D.; Cavallo, L.; Nolan, S. P., *Organometallics* **2004**, *23*, 1629.
40. Cao, C.; Wang, L.; Cai, Z.; Zhang, L.; Guo, J.; Pang, G.; Shi, Y., *Eur. J. Org. Chem.* **2011**, 1570.
41. Hatakeyama, T.; Hashimoto, S.; Ishizuka, K.; Nakamura, M., *J. Am. Chem. Soc.* **2009**, *131*, 11949.

42. Agata, R.; Iwamoto, T.; Nakagawa, N.; Isozaki, K.; Hatakeyama, T.; Takaya, H.; Nakamura, M., *Synlett* **2015**, *47*, 1733.
43. Ishizuka, K.; Seike, H.; Hatakeyama, T.; Nakamura, M., *J. Am. Chem. Soc.* **2010**, *132*, 13117.
44. Tessin, U. I.; Bantreil, X.; Songis, O.; Cazin, C. S. J., *Eur. J. Inorg. Chem.* **2013**, 2007.
45. Yang, J.; Li, P.; Zhang, Y.; Wang, L., *Dalton Trans.* **2014**, *43*, 14114.
46. Guo, L.; Leiendecker, M.; Hsiao, C.-C.; Baumann, C.; Rueping, M., *Chem. Commun.* **2015**, *51*, 1937.
47. Winkelmann, O. H.; Riekstins, A.; Nolan, S. P.; Navarro, O., *Organometallics* **2009**, *28*, 5809.
48. Yamamoto, Y., *Synlett* **2007**, *12*, 1913.
49. Omar-Amrani, R.; Thomas, A.; Brenner, E.; Schneider, R.; Fort, Y., *Org. Lett.* **2003**, *5*, 2311.
50. Marion, N.; Navarro, O.; Mei, J.; Stevens, E. D.; Scott, N. M.; Nolan, S. P., *J. Am. Chem. Soc.* **2006**, *128*, 4101.
51. Martin, A. R.; Chartoire, A.; Slawin, A. M. Z.; Nolan, S. P., *Beilstein J. Org. Chem.* **2012**, *8*, 1637.
52. Chen, M.-T.; Vicic, D. A.; Turner, M. L.; Navarro, O., *Organometallics* **2011**, *30*, 5052.
53. Jin, Z.; Qiu, L.-L.; Li, Y.-Q.; Song, H.-B.; Fang, J.-X., *Organometallics* **2010**, *29*, 6578.
54. Thakur, A.; Louie, J., *Acc. Chem. Res.* **2015**, *48*, 2354.
55. Duong, H. A.; Cross, M. J.; Louie, J., *J. Am. Chem. Soc.* **2004**, *126*, 11438.
56. McCormick, M. M.; Duong, H. A.; Zuo, G.; Louie, J., *J. Am. Chem. Soc.* **2005**, *127*, 5030.
57. Stolley, R. M.; Maczka, M. T.; Louie, J., *Eur. J. Org. Chem.* **2011**, *2011*, 3815.
58. Zhong, Y.; Spahn, N. A.; Stolley, R. M.; Nguyen, M. H.; Louie, J., *Synlett* **2015**, *26*, 307.
59. Tekavec, T. N.; Louie, J., *Org. Lett.* **2005**, *7*, 4037.
60. Tekavec, T. N.; Louie, J., *J. Org. Chem.* **2008**, *73*, 2641.
61. Kumar, P.; Thakur, A.; Hong, X.; Houk, K. N.; Louie, J., *J. Am. Chem. Soc.* **2014**, *136*, 17844.
62. Miura, T.; Mikano, Y.; Murakami, M., *Org. Lett.* **2011**, *13*, 3560.
63. Díez-González, S.; Stevens, E. D.; Nolan, S. P., *Chem. Commun.* **2008**, 4747.
64. Zuo, G.; Louie, J., *J. Am. Chem. Soc.* **2005**, *127*, 5798.
65. Duong, H. A.; Cross, M. J.; Louie, J., *Org. Lett.* **2004**, *6*, 4679.
66. Grasa, G. A.; Kissling, R. M.; Nolan, S. P., *Org. Lett.* **2002**, *4*, 3583.
67. Singh, R.; Kissling, R. M.; Letellier, M.-A.; Nolan, S. P., *J. Org. Chem.* **2004**, *69*, 209.
68. Zhou, H.; Campbell, E. J.; Nguyen S. T., *Org. Lett.* **2001**, *3*, 2229.
69. He, L.; Jian, T.-Y.; Ye, S., *J. Org. Chem.* **2007**, *72*, 7466.
70. Landers, B.; Berini, C.; Wang, C.; Navarro, O., *J. Org. Chem.* **2011**, *76*, 1390.
71. Jurčík, V.; Nolan, S. P.; Cazin, C. S. J., *Chem. Eur. J.* **2009**, *15*, 2509.

1,3-Bis(2,6-diisopropylphenyl)imidazolidinium Chloride

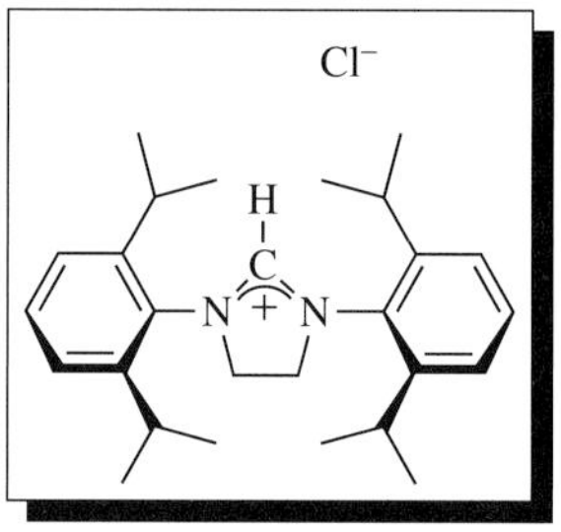

[258278-25-0] $C_{27}ClH_{39}N_2$ (MW 427.06)

InChI = 1S/C27H39N2.ClH/c1-18(2)22-11-9-12-23(19(3)4)26(22)28-15-16-29(17-28)27-24(20(5)6)13-10-14-25(27)21(7)8;/h9-14,17-21H,15-16H2,1-8H3;1H/q+1;/p-1

InChIKey = LWPXTYZKAWSRIP-UHFFFAOYSA-M

(used as a precursor to the free carbene, 1,3-bis(2,6-diisopropylphenyl)-2-imidazolidinylidene, and as an in situ formed catalyst)

Physical Data: mp 289–293°C.

Solubility: highly soluble: CH_2Cl_2; moderately soluble: $CHCl_3$, H_2O; slightly soluble: THF.

Form Supplied in: microcrystalline white or light cream powder.

Purification: recrystallization from acetonitrile and ether, dried in a well-ventilated oven at 120°C.

Handling, Storage, and Precautions: compound is air stable over prolonged periods of time but is hygroscopic. When handling large quantities use dust mask.

Synthesis. Following Wanzlick's initial work on 1,3-bisaryl-limidazolidinium rings,[1–3] Arduengo first published a general method to synthesize 1,3-bis(2,6-diisopropylphenyl)imidazolidinium chloride (SIPr·HCl) (Scheme 1).[4] In a condensation reaction involving glyoxal with 2 equiv of 2,6-diisopropylphenylamine, the diimine is obtained as a yellow solid in moderate yield. Further treatment with sodium borohydride and the addition of a few drops of hydrochloric acid precipitates the diamine dihydrochloride product. Ring closure is completed with triethyl orthoformate as both the C_1-building block and a sacrificial base. Purification can be achieved by recrystallization from acetonitrile and ether.

Delaude and coworkers have disclosed an improved method for ring closure using microwave-assisted organic synthesis (MAOS).[5,6] Using conductive heating, the final step in Scheme 1 requires 45 h.[4] When using MAOS, the reaction mixture is heated at 145°C for 5 min and the yield is improved from 59% to 72%.[5] Unfortunately, the reaction involving 1,3-bis(2,6-diisopropylphenyl)imidazolidinium chloride was performed only on a 500 mg scale. Synthesis of 1,3-bis(2,4,6-triphenyl)imidazolidinium chloride (SIMes·HCl) was, however, achieved in 80% yield at greater than 10 g scale using MAOS.

Expanding on a method first envisioned by Bertrand,[7,8] Grubbs has published a one-pot synthesis in which SIPr·HCl is made directly from substituted anilines, triethyl orthoformate, and dichlo-romethane (Scheme 2).[9] The formamidine formed in situ

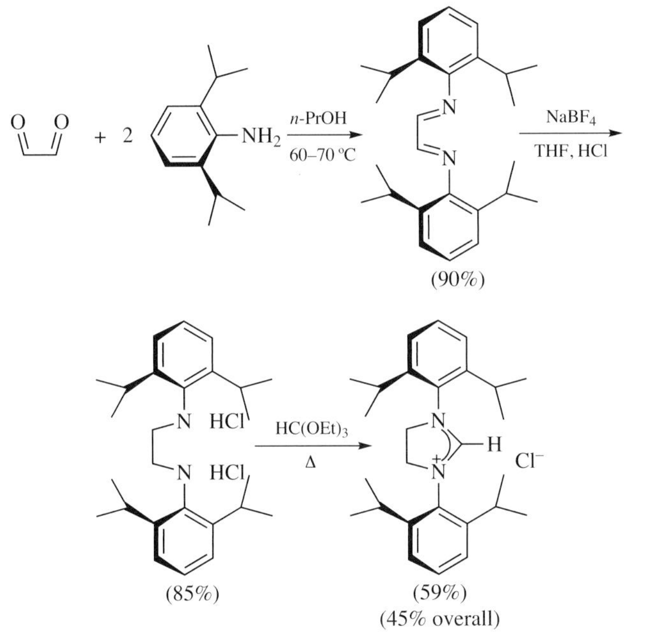

Scheme 1 The Arduengo synthesis of SIPr·HCl.

acts both as a base and as an intermediate to the final imidazolidinium salt. The reaction time from Arduengo's preparation is decreased from 45 to 36 h and results in a similar overall yield of 45% (90% if a 50% yield is viewed as a maximum due to half of the substrate being used as the in situ base).

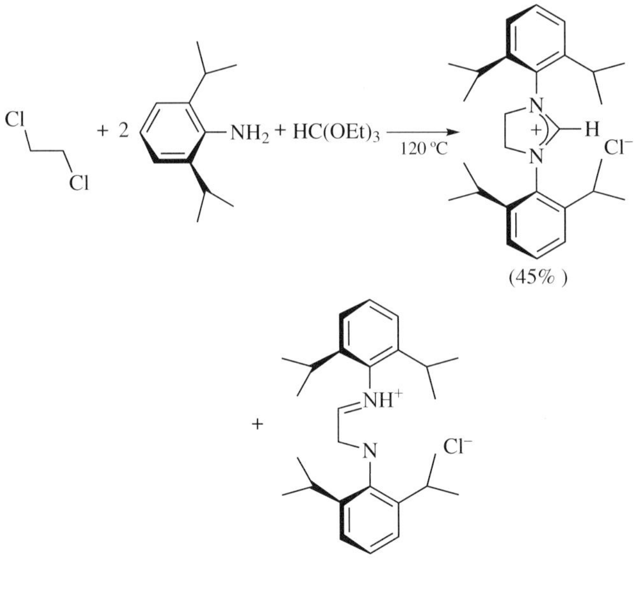

Scheme 2 Grubbs synthesis of SIPr·HCl.

Purification was achieved by simple filtration and subsequent washing. Again, no large-scale preparation for SIPr·HCl was reported, but the similar SIMes·HCl salt was synthesized on a 20 g scale with no significant loss in yield.

Precursor to Crystalline Free Carbene. Under an inert atmosphere, the addition of sodium hydride or potassium *tert*-butoxide to THF solutions of SIPr·HCl produces the free carbene 1,3-bis(2,6-diisopropylphenyl)-2-imidazolidinylidene (SIPr) (eq 1). Unfortunately reaction times are long due to the low solubility of SIPr·HCl in THF.[4] Following deprotonation, the carbenes are utilized in a myriad of reactions and as ligands on organometallic complexes.[10,11]

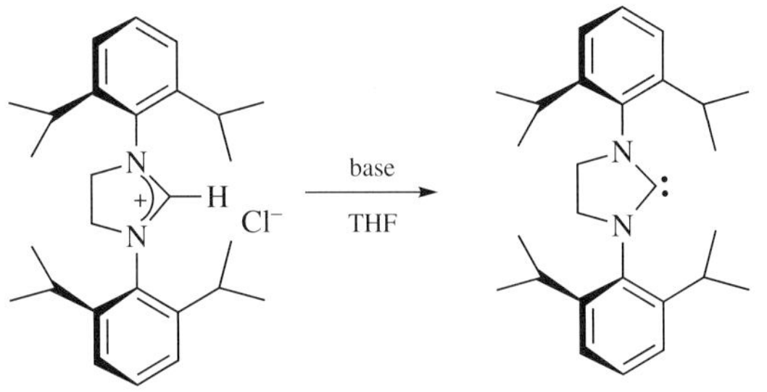

(1)

Reactivity. SIPr·HCl is predominately used as a precursor to form free carbene. This report aims at illustrating the uses of the salt and not of the free carbene. As such we will only discuss syntheses involving the use of SIPr·HCl in which the catalysts are formed in situ. Generally, deprotonation of SIPr·HCl is performed by adding either a strong base (such as KO^{t}Bu or NaH) or an organometallic complex (such as $Pd(OAc)_2$). The following sections will highlight the strengths and weaknesses of the reactions in which 1,3-bis(2,6-diisopropylphenyl)imidazolidinium chloride has played a role in organic synthesis.

Organocatalysis. NHC salts have been examined as catalyst in which no supporting organometallic complex is required in a variety of reactions.[12] To date, however, SIPr·HCl has been reported to catalyze only a very limited number of reactions efficiently. Gravel and coworkers have reported the use of SIPr·HCl as a catalyst to synthesize lactones via ring expansion of oxacycloalkanes-2-carboxaldehydes as shown in eq 2.[13] Surprisingly, even under non-inert atmosphere conditions, the reaction produced lactones consisting of five-, six-, and seven-membered rings.

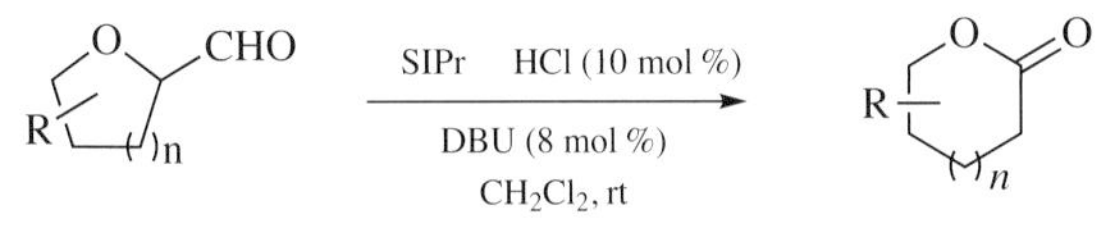

(2)

Gravel postulates that the reaction proceeds via the mechanism shown in Scheme 3. The active catalyst is formed by addition of a base to the NHC salt. Nucleophilic addition of the free NHC to the oxacycloalkane-2-carboxaldehydes primes species **I** for rearrangement to the ring-opened homoenolate (**II**). The final lactone is formed after intramolecular alkoxide attack, which subsequently regenerates the catalytically active free carbene.

The homoenolate intermediate **II** has been observed to play an important role in several organocatalytic reactions involving NHCs.[12b] Gravel's article is, to the authors' knowledge, the only report on the subject of NHC organocatalysis in which SIPr·HCl resulted in the most favorable outcome. Of the symmetrical NHCs, IMes·HCl seems to most often be the catalyst of choice.[14–19]

Amination. Coupling reactions between amines and aryl chlorides to form new C–N bonds have been successfully accomplished using an in situ generated NHC-Ni catalyst.[20–22] Aryl, primary, and secondary amines used as substrates gave rise to good yields (eq 3). The active catalyst is believed to be a Ni(0)

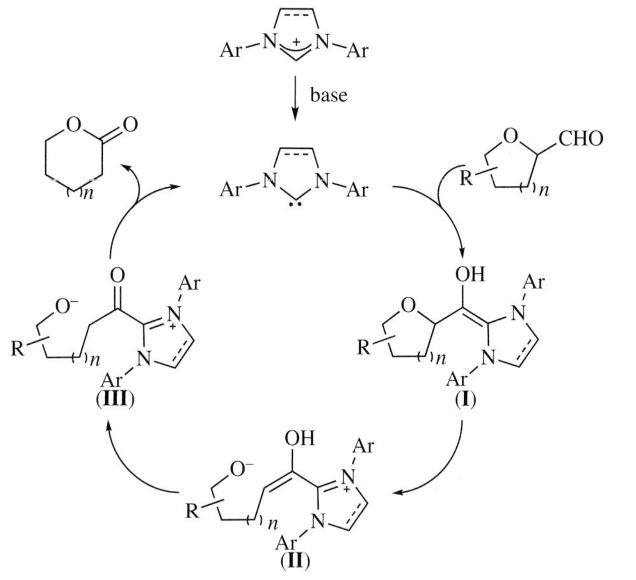

Scheme 3 Proposed mechanism of lactone formation via homoenolate intermediate (**II**).

species.[21] Good yields were obtained for both electron-rich and electron-poor aryl chlorides. In addition, heterocyclic halides also proved to undergo successful coupling reactions.

$$\mathrm{Ar{-}Cl} + \mathrm{HNR_2} \xrightarrow[\mathrm{NaO^{\mathit{t}}Bu,\ THF}]{\mathrm{SIPr\cdot HCl,\ Ni(acac)_2}} \mathrm{Ar{-}NR_2} \qquad (3)$$

Couplings between indoles and aryl bromides have been achieved using $Pd(OAc)_2$ with SIPr·HCl in good yields.[23] Of the carbene salts screened, the more electron-donating SIPr·HCl, salt proved to promote completion over those of IPr·HCl, SIMes·HCl and ICy·HCl. Unfortunately, the scope of the amination was limited to aryl bromides, as aryl chlorides proved unreactive.

Coupling Reactions. Andrus et al. have performed the Mizoroki–Heck reaction at room temperature using a catalyst formed in situ from $Pd(OAc)_2$ and SIPr·HCl with olefins and aryldiazonium tetrafluoroborate substrates (eq 4).[24] Catalyst loading was maintained at 2 mol % and reaction times generally ranged from 2 to 4 h. In addition, no added base was needed to regenerate the active catalyst. Andrus further utilized this catalyst in a four-step synthesis of resveratrol, a natural antioxidant and radical scavenger.[25] In this synthesis, one step involves a Heck-based reaction in which an aryl acid chloride is coupled to a styrene derivative. It was, however, reported that the nature of the added base was critical to the reaction and that noncoordinating amines such as *N*-ethylmorpholine and *N,N*-dimethylbenzylamine gave the desired product while added phosphine caused retardation of the reaction.

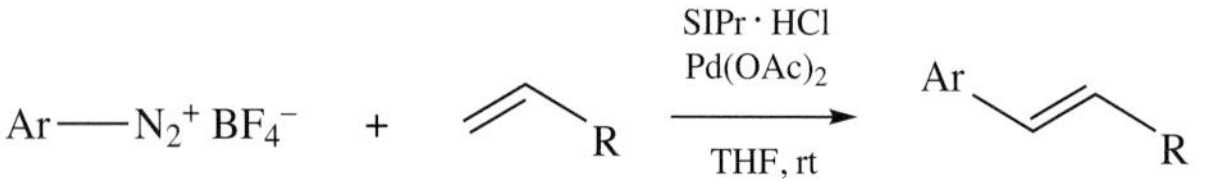

R = Ar, CO_2Me, CN (4)

In situ formed NHC/palladium systems have also been reported to facilitate the Songashira reactions of aryl halides with alkynylsilanes as shown in eq .[26] Once again, SIPr·HCl was found to lead to poor performance in catalysis. While the yield of reaction 5 was found to be 60% for SIPr·HCl, IMes·HCl produced yields of 87%. Similarly, this samc catalytic systcm was also used to dimerize terminal alkynes.[27] Analogous shortcomings of SIPr·HCl were found.

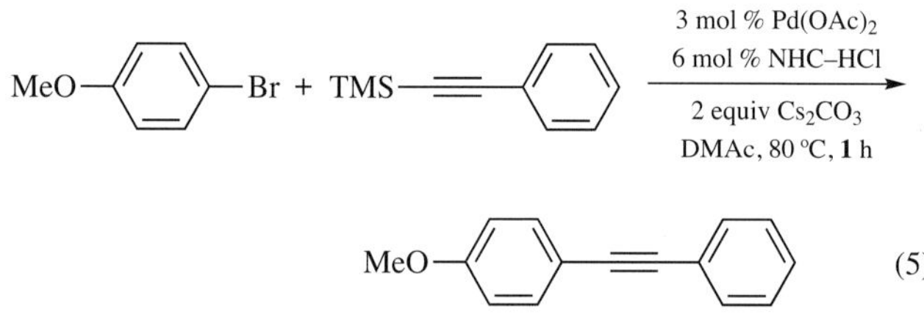

(5)

In 2002, both Noels[28] and Dixneuf[29] reported on the use of an in situ formed NHC-Ru catalyst used for the ring-opening metathesis polymerization (ROMP) of cyclooctene. Both reports employed $[RuCl_2(p\text{-cymene})]_2$ as a ruthenium source and a variety of bases to deprotonate the imidizol(in)ium salt to form the active catalyst (eq 6). Both studies included SIPr·HCl as part of the three-part catalyst and both found this imidazolidinium salt to not be the optimal choice. Dixneuf studied only saturated NHC salts and found IMes·HCl to be the NHC of choice. Noels disclosed finding reactions involving both saturated and unsaturated NHCs and found 1,3-biaryl saturated carbenes with substitution at the *ortho* position on the aryl rings to give the optimal transformations.

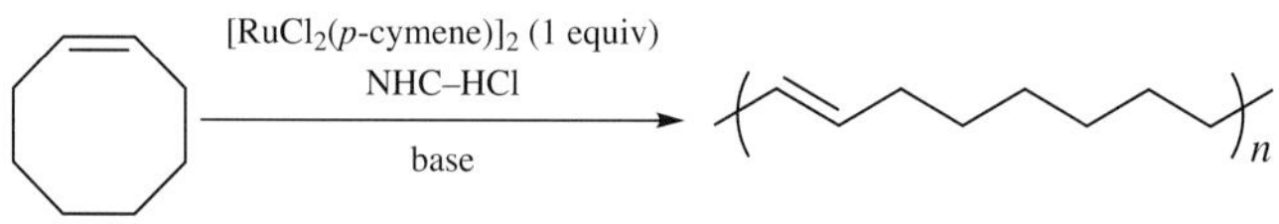

(6)

Hydrogenation. Elsevier and coworkers have used $Pt(\eta^2\text{-ma})(\eta^4\text{-nbd})$ (ma = maleic anhydride; nbd = norbornadiene) and imidizol(in)ium salts to form an in situ Pt(0) catalyst for the hydrogenation of 1-phenyl-1-propyne (eq 7).[30] SIPr·HCl as a cocatalyst did show higher activity than SIMes·HCl; however, this was at the expense of reaction selectivity. Unsaturated NHCs, and specifically 1,3-bis(2,6-diethylphenyl)imidazolidium chloride, produced the highest activity and selectivity, affording 95% of *Z*-alkene.

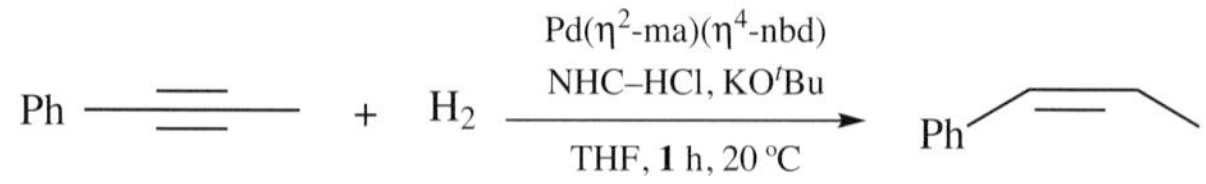

(7)

Closing Remarks. While SIPr·HCl has been proven to be an efficient catalyst precursor in a limited array of reactions, in general, this compound both as an organocatalyst and as part of an in situ formed organometallic catalyst typically produces non-optimum reaction conditions. Generation of the free carbene and subsequent substitution onto metal complexes has, however, led to some very promising systems. Several complexes of the form $(SIPr)ML_n$ have been shown to be extremely active in catalysis.[31–34]

Related Reagents. 1,3-Bis(2,6-diisopropylphenyl)-2-imidazolidinylidene; 1,3-bis(2,6-diisopropylphenyl)imidazolium chloride; 1,3-bis(2,4,6-trimethylphenyl)imidazolium chloride.

---

1. Wanzlick, H.W., *Angew. Chem., Int. Ed.* **1962**, *1*, 75.
2. Wanzlick, H.-W.; Schikora, E., *Angew. Chem.* **1960**, *72*, 494.
3. Wanzlick, H.-W.; Schikora, E., *Chem. Ber.* **1961**, *94*, 2389.
4. Arduengo, A.J., III; Krafczyk, R.; Schmutzler, R.; Craig, H.A.; Goerlich, J.R.; Marshall, W.J.; Unverzagt, M., *Tetrahedron* **1999**, *55*, 14523.
5. Aidouni, A.; Bendahou, S.; Demonceau, A.; Delaude, L., *J. Comb. Chem.* **2008**, *10*, 886.
6. Aidouni, A.; Demonceau, A.; Delaude, L., *Synlett* **2006**, 493.
7. Jazzar, R.; Bourg, J.-B.; Dewhurst, R.D.; Donnadieu, B.; Bertrand, G., *J. Org. Chem.* **2007**, *72*, 3492.
8. Jazzar, R.; Liang, H.; Donnadieu, B.; Bertrand, G., *J. Organomet. Chem.* **2006**, *691*, 3201.
9. Kuhn, K.M.; Grubbs, R.H., *Org. Lett.* **2008**, *10*, 2075.
10. *N-Heterocyclic Carbenes in Synthesis*, 1st ed.; Nolan, S.P., Ed.; Wiley-VCH Verlag GmbH & Co. KGaA: Weinheim, 2006.
11. *N-Heterocyclic Carbenes in Transition Metal Catalysis*, 1st ed.; Glorius, F., Ed.; Springer, 2007.
12. For recent reviews on NHCs as organocatalysts see: (a) Enders, D.; Niemeier, O.; Henseler, A., *Chem. Rev.* **2007**, *107*, 5606. (b) Marion, N.; Díez-González, S.; Nolan, S.P., *Angew. Chem., Int. Ed.* **2007**, *46*, 2988.
13. Wang, L.; Thai, K.; Gravel, M., *Org. Lett.* **2009**, *11*, 891.
14. Nair, V.; Vellalath, S.; Poonoth, M.; Suresh, E., *J. Am. Chem. Soc.* **2006**, *128*, 8736.
15. Burstein, C.; Glorius, F., *Angew. Chem., Int. Ed.* **2004**, *43*, 6205.
16. Sohn, S.S.; Rosen, E.L.; Bode, J.W., *J. Am. Chem. Soc.* **2004**, *126*, 14370.
17. He, M.; Bode, J.W., *Org. Lett.* **2005**, *7*, 3131.
18. Zeitler, K.; *Org. Lett.* **2006**, *8*, 637.
19. Nair, V.; Vellalath, S.; Poonoth, M.; Mohan, R.; Suresh, E., *Org. Lett.* **2006**, *8*, 507.
20. Desmarets, C.; Kuhl, S.; Schneider, R.; Fort, Y., *Organometallics* **2002**, *21*, 1554.
21. Desmarets, C.; Schneider, R.; Fort, Y., *J. Org. Chem.* **2002**, *67*, 3029.
22. Gradel, B.; Brenner, E.; Schneider, R.; Fort, Y., *Tetrahedron Lett.* **2001**, *42*, 5689.
23. Grasa, G.A.; Viciu, M.S.; Huang, J.; Nolan, S.P., *J. Org. Chem.* **2001**, *66*, 7729.
24. Andrus, M.B.; Song, C.; Zhang, J., *Org. Lett.* **2002**, *4*, 2079.
25. Andrus, M.B.; Liu, J.; Meredith, E.L.; Nartey, E., *Tetrahedron Lett.* **2003**, *44*, 4819.
26. Yang, C.; Nolan, S.P., *Organometallics* **2002**, *21*, 1020.
27. Yang, C.; Nolan, S.P.; *J. Org. Chem.* **2002**, *67*, 591.
28. Delaude, L.; Szypa, M.; Demonceau, A.; Noels, A.F., *Adv. Synth. Catal.* **2002**, *344*, 749.
29. Castarlenas, R.; Alaoui-Abdallaoui, I.; Semeril, D.; Mernari, B.; Guesmi, S.; Dixneuf, P.H., *New J. Chem.* **2003**, *27*, 6.
30. Sprengers, J.W.; Wassenaar, J.; Clement, N.D.; Cavell, K.J.; Elsevier, C.J., *Angew. Chem., Int. Ed.* **2005**, *44*, 2026.
31. Winkelmann, O.H.; Riekstins, A.; Nolan, S.P.; Navarro, O., *Organometallics* **2009**, *28*, 5809.
32. Viciu, M.S.; Germaneau, R.F.; Nolan, S.P., *Org. Lett.* **2002**, *4*, 4053.
33. Viciu, M.S.; Navarro, O.; Germaneau, R.F.; Kelly, R.A.; Sommer, W.; Marion, N.; Stevens, E.D.; Cavallo, L.; Nolan, S.P., *Organometallics* **2004**, *23*, 1629.
34. Clavier, H.; Urbina-Blanco, C.A.; Nolan, S.P.; *Organometallics* **2009**, *28*, 2848.

George C. Fortman & Steven P. Nolan
University of St. Andrews, St. Andrews, UK

1,3-Bis(2,6-diisopropylphenyl)imidazol-2-ylidene

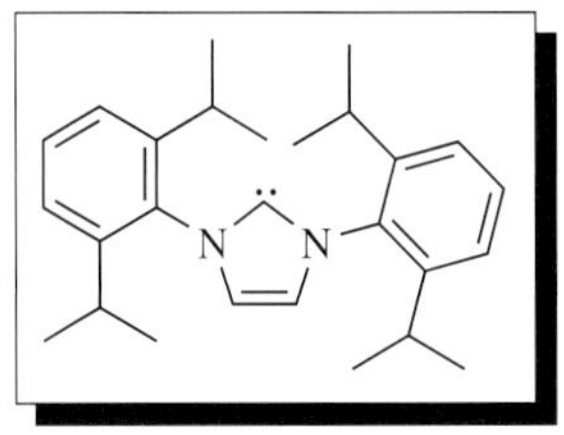

[244187-81-3] $C_{27}H_{36}N_2$ (MW 388.59)
InChI = 1S/C27H36N2/c1-18(2)22-11-9-12-23(19(3)4)26(22)28-15-16-29(17-28)27-24(20(5)6)13-10-14-25(27)21(7)8/h9-16,18-21H,1-8H3
InChIKey = VYCIHDBIKGRENI-UHFFFAOYSA-N

(ligand for Pd complexes for amination reaction of aryl halides under mild conditions; organocatalyst; ligand for Pd complexes for cross-coupling, Au and Cu complexes for a number of transformations; ligand for the stabilization of reactive *p*-block species such as B_2 and P_2)

Alternate Name: IPr.

Solubility: soluble in THF, hexane, toluene, ether, and most organic solvents.

Analysis of Reagent Purity: elemental microanalysis, NMR spectroscopy.

Preparative Method: can be prepared under inert atmosphere[3] utilizing dry, oxygen-free solvents.[4] IPr·HCl[2] and KO*t*Bu were added to a NaH suspension in THF in the glove box (eq 1). The mixture was stirred for 1 h at room temperature, for 2 h at 60 °C, followed by overnight stirring at room temperature. THF was removed in vacuo, the product was extracted with hot hexanes, filtered through a pad of Celite, and the solvent removed in vacuo resulting in a white solid.[2]

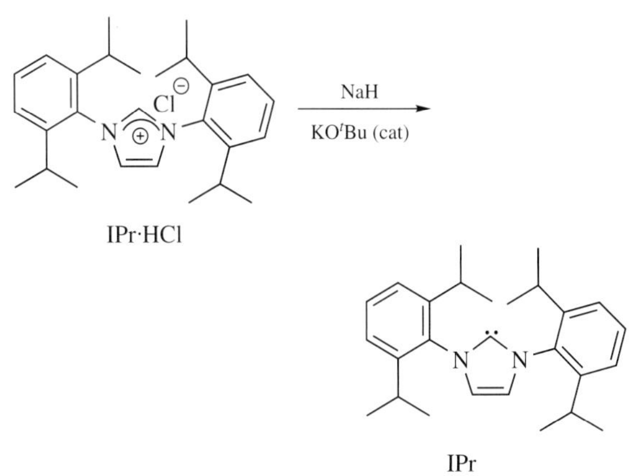

(1)

Purity: recrystallization from cold pentane. Wash with cold pentane.

Handling, Storage, and Precautions: air- and moisture-sensitive in both solid state and solution. Handle and store under inert atmosphere. Thermally stable. No toxicity or hazard data available.

Disposal: organic waste.

Original Commentary

Roy A. Kelly, Gabriela A. Grasa & Steven P. Nolan
University of New Orleans, LA, USA

Catalytic Transformations. The title compound (IPr) represents a very efficient supporting ligand for a dimeric palladium catalyst $[Pd(IPr)Cl_2]_2$ in the amination reaction of various aryl halides and primary and secondary alkyl and aryl amines (eq 2).[5] The ligand (generated in situ) is also useful in catalytic cross coupling reactions involving aryl halides with aryl Grignards (the Kumada–Tamao–Corriu reaction),[2] aryl boronic acids (Suzuki–Miyaura),[6] and organotin reagents (Stille)[7](eq 3). The steric and electronic properties of this ligand not only allow the activation of the less active aryl chlorides under mild conditions, but also allow the synthesis of an air and moisture stable catalyst, which permits the aryl amination reaction to be performed in air with reagent grade solvents.

$$\text{Ar—X} + \underset{(1.2\text{ equiv})}{\text{NRR'H}} \xrightarrow[\text{1.5 equiv KO}^t\text{Am, DME, 80 °C, Air}]{0.5\text{ mol \% }[\text{Pd(IPr)Cl}_2]_2} \underset{85\text{–}100\%}{\text{Ar—NRR'}} \quad (2)$$

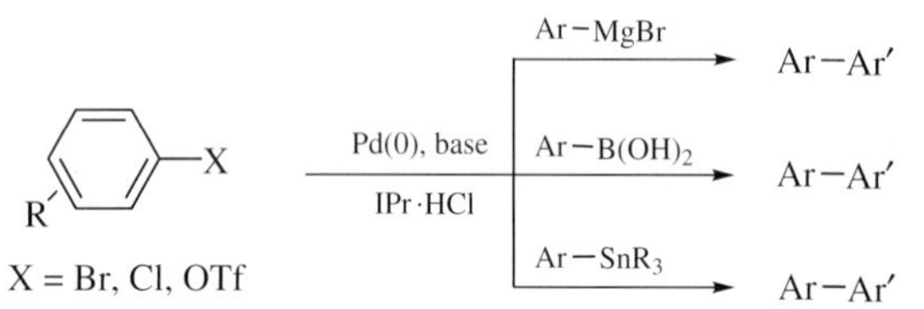
(3)

Related Reagents. ICy [1,3-bis(cyclohexyl)imidazol-2-ylidene]; ITol [1,3-bis(4-methylphenyl)imidazol-2-ylidene]; IAd [1,3-bis(adamantyl)imidazol-2-ylidene];[8] IMes [1,3-bis(2,4,6-trimethylphenyl)imidazol-2-ylidene].[7,9]

First Update

David J. Nelson
University of Strathclyde, Glasgow, UK
Steven P. Nolan
King Saud University, Riyadh , Saudi Arabia

The use of NHCs across many fields of chemistry has significantly increased during the past 10 years, especially in three key areas: organocatalysis,[10] as ligands for transition metal catalysts,[11] and as stabilizing ligands to allow the isolation of reactive species such as those containing multiple bonds between heavier elements.[12]

Organocatalysis. IPr is not routinely used in organocatalysis, due to its bulky *N*-substituents and achirality. It can perform a number of organocatalytic transformations, some of which are discussed here.

IPr can catalyze transesterification reactions, although not as well as the less bulky IMes or ICy congeners (IMes = 1,3-bis(2,4,6-trimethylphenyl)imidazol-2-ylidene; ICy = 1,3-dicyclohexylimidazol-2-ylidene).[13]

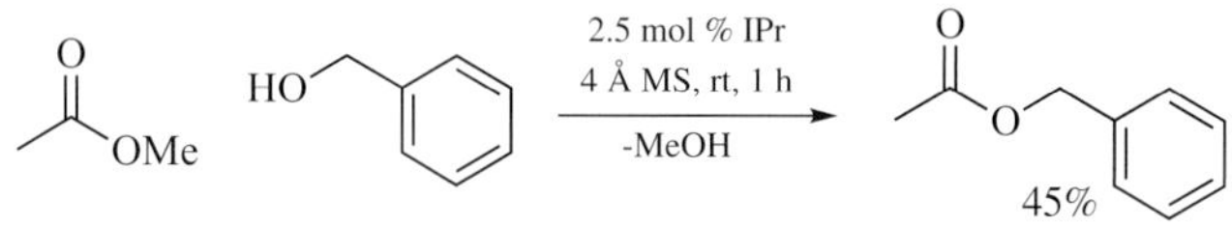

A number of literature reports using IPr as an organocatalyst involve the activation of enal substrates. For example, this NHC has been reported to effect the annulation of nitrosoarenes with enals, leading to interesting products.[14] Yields ranged from modest to good. The NHC was generated in situ from the NHC salt and a base.

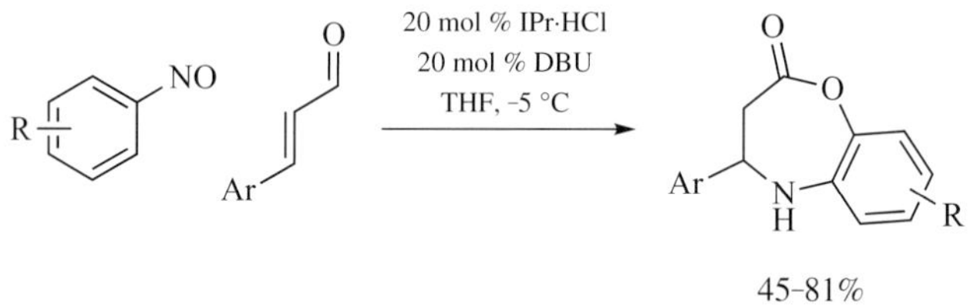

The synthesis of dihydropyrans from divinylesters, catalyzed by IMes, has been reported;[15] IPr was used to catalyze an intermolecular variant that deployed a silyl enol ether and an α,β-unsaturated acyl fluoride. One example of a similar product was prepared by the reaction of a-bromocinnamaldehyde with acetylacetone, catalyzed by IPr.[16]

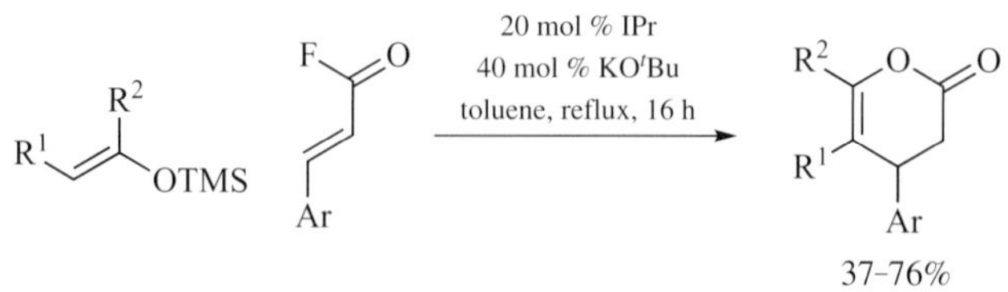

IPr can be used to effect the oxidative coupling of alcohols and enals (with TEMPO the terminal oxidant).[17] The tandem esterification/oxidation of alkanals was also achieved, although yields and selectivity (for ester versus α, β-unsaturated ester) were variable.

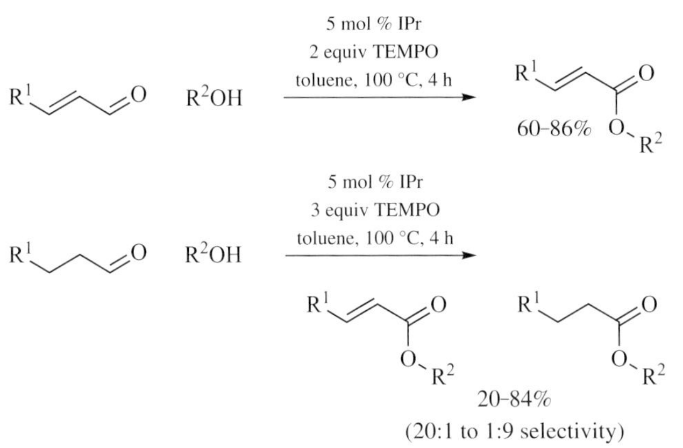

IPr has also been used to effect the addition of $C_6F_5SiMe_3$ to aldehydes by activating the silane nucleophile, yielding the corresponding secondary alcohols.[18] The reaction forms the TMS-protected alcohol, with acidic hydrolysis yielding the alcohol itself during the workup.

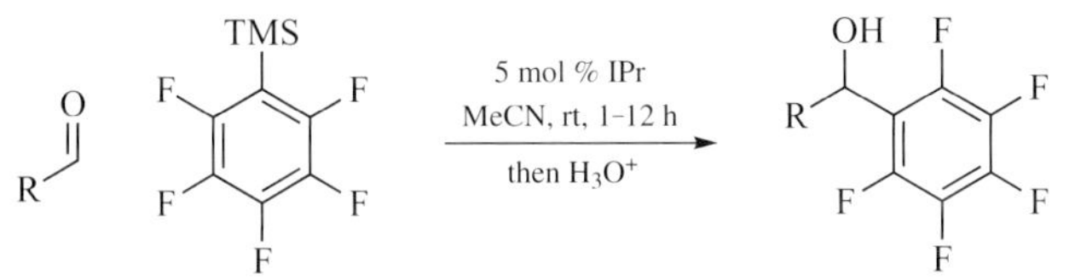

Transition Metal Catalysis. IPr is a ubiquitous NHC in transition metal catalysis, primarily due to its expedient, cheap, and scalable synthetic route (three steps from a readily available aniline). A full discussion of the use of this carbene is beyond the scope of this article, but selected examples are noted, with the aim of outlining the use of IPr in state-of-the-art processes.

Palladium–NHC complexes, either generated in situ or prepared and isolated beforehand, have found widespread application in synthetic chemistry.[19] The considerable electron-donating ability of NHCs allows catalysts bearing these ligands to activate aryl chloride substrates, which are more challenging than aryl bromides yet more readily available. Organ's PEPPSI-IPr complex has been used to conduct the Negishi cross-coupling of alkyl and aryl chlorides,[20] the Buchwald-Hartwig amination,[21] and the Suzuki-Miyaura[22] and Kumada–Tamao–Corriu[23] coupling of aryl chlorides. [Pd(η^3-cinnamyl)Cl(IPr)] has been used to effect the Suzuki—Miyaura and Buchwald—Hartwig cross-coupling of aryl chlorides at room temperature.[24]

A second area where IPr is widely used is gold catalysis,[25] where complexes such as [AuCl(IPr)],[26] [Au(IPr)(NCMe)]-$[BF_4]$,[27] [Au(NTf_2)(IPr)],[28] and [{Au(IPr)}$_2$(μ-OH)][29] are valuable (pre)catalysts. Reactions catalyzed by these precatalysts include the additions of various nucleophiles to alkenes, allenes, and alkynes.[30]

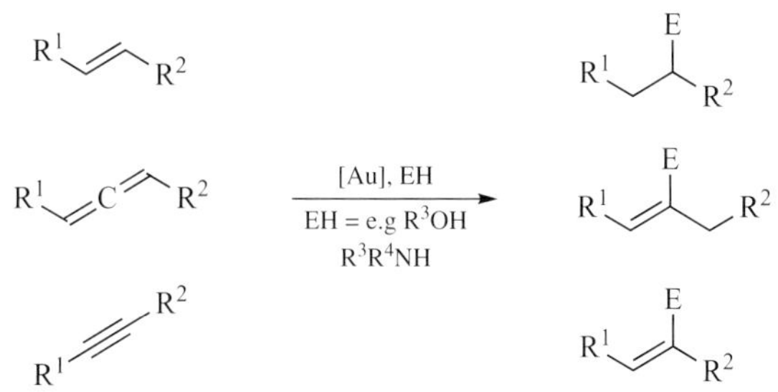

Stabilization of Reactive Species. IPr is frequently used to stabilize reactive main group compounds, particularly coordinatively unsaturated main group species featuring multiple bonds. For example, IPr has been used to stabilize boron-boron triple bonds in B_2,[31] silicon–silicon double bonds in Si_2,[32] and phosphorus-phosphorus bonds in P_2.[33] This is due to the ability of IPr to form strong bonds and the steric shielding that it affords fragments to which it is bound. While these stabilized species currently have limited use in synthetic organic chemistry, there is the potential to use these in the future as sources of interesting fragments that are otherwise very difficult to construct or incorporate into organic molecules.

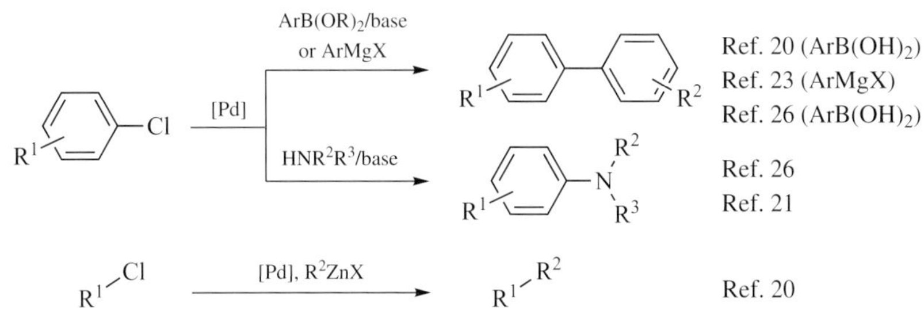

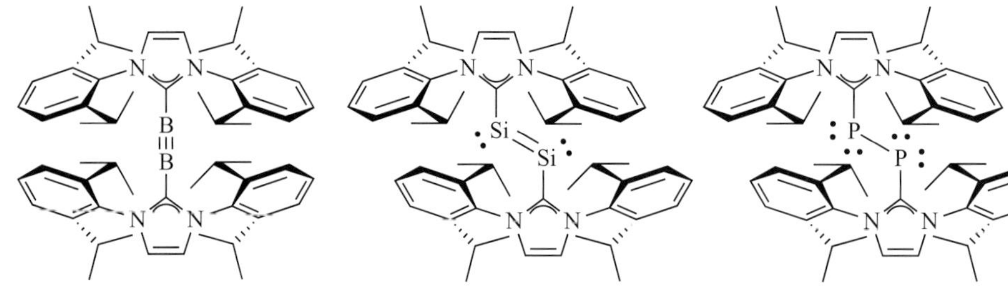

1. Jafarpour, L.; Stevens, E. D.; Nolan, S. P., *J. Organomet. Chem.* **2000**, *606*, 49–54.
2. Huang, J.; Nolan, S. P., *J. Am. Chem. Soc.* **1999**, *121*, 9889–9890.
3. Shriver, D. F.; Drezdzon, M. A., *The Manipulation of Air-Sensitive Compounds*, 2nd ed.; Wiley: New York, **1986**.
4. Perrin, D. D.; Armarego, W. L. F.; Perrin, D. R., *Purification of Laboratory Chemicals*, 3rd ed.; Pergamon Press: Oxford, **1988**.
5. Viciu, M. S.; Kissling, R. M.; Stevens, E. D.; Nolan, S. P., *Org. Lett.* **2002**, *4*, 2229–2231.
6. (a) Zhang, C.; Huang, J.; Trudell, M. L.; Nolan, S. P., *J. Org. Chem.* **1999**, *64*, 3804–3805. (b) Grasa, G. A.; Viciu, M. S.; Huang, J.; Zhang, C.; Trudell, M. L.; Nolan, S. P., *Organometallics* **2002**, *21*, 2866–2873.
7. Grasa, G. A.; Nolan, S. P., *Org. Lett.* **2001**, *3*, 119–122.
8. Arduengo, A. J., III; Harlow, R. L.; Kline, M. J., *J. Am. Chem. Soc.* **1991**, *113*, 361–363.
9. Arduengo, A. J., III; Gamper, S. F.; Calabrese, J. C.; Davidson, F., *J. Am. Chem. Soc.* **1994**, *116*, 4391–4393.
10. Flanigan, D. M.; Romanov-Michailidis, F.; White, N. A.; Rovis, T., *Chem. Rev.* **2015**.
11. Dáz-González, S.; Marion, N.; Nolan, S. P., *Chem. Rev.* **2009**, *109*, 3612.
12. Hopkinson, M. N.; Richter, C.; Schedler, M.; Glorius, F., *Nature* **2014**, *510*, 485.
13. Grasa, G. A.; Kissling, R. M.; Nolan, S. P., *Org. Lett.* **2002**, *4*, 3583.
14. Yang, L.; Tan, B.; Wang, F.; Zhong, G., *J. Org. Chem.* **2009**, *74*, 1744.
15. Ryan, S. J.; Candish, L.; Lupton, D. W., *J. Am. Chem. Soc.* **2009**, *131*, 14176.
16. Sun, F.-G.; Sun, L.-H.; Ye, S., *Adv. Synth. Catal.* **2011**, *353*, 3134.
17. Ji, M.; Wang, X.; Lim, Y. N.; Kang, Y.-W.; Jang, H.-Y., *Eur. J. Org. Chem.* **2013**, *2013*, 7881.
18. Du, G.-F.; Xing, F.; Gu, C.-Z.; Dai, B.; He, L., *RSC Adv.* **2015**, *5*, 35513.
19. Fortman, G. C.; Nolan, S. P., *Chem. Soc. Rev.* **2011**, *40*, 5151.
20. Organ, M. G.; Ajvola, S.; Dubovyk, I.; Hadei, N.; Kantchev, E. A. B.; O'Brien, C. J.; Valente, C., *Chem. Eur. J.* **2006**, *12*, 4749.
21. Organ, M. G.; Abdel-Hadi, M.; Ajvola, S.; Dubovyk, I.; Hadei, N.; Kantchev, E. A. B.; O'Brien, C. J.; Sayah, M.; Valente, C., *Chem. Eur. J.* **2008**, *14*, 2443.
22. O'Brien, C. J.; Kantchev, E. A. B.; Valente, C.; Hadei, N.; Chass, G. A.; Lough, A.; Hopkinson, A. C.; Organ, M. G., *Chem. Eur. J.* **2006**, *12*, 4743.
23. Organ, M. G.; Abdel-Hadi, M.; Ajvola, S.; Hadei, N.; Nasielski, J.; O'Brien, C. J.; Valente, C., *Chem. Eur. J.* **2007**, *13*, 150.
24. Marion, N.; Navarro, O.; Mei, J.; Stevens, E. D.; Scott, N. M.; Nolan, S. P., *J. Am. Chem. Soc.* **2006**, *128*, 4101.
25. Nolan, S. P., *Acc. Chem. Res.* **2011**, *44*, 91.
26. Fructos, M. R.; Belderrain, T. R.; de Frémont, P.; Scott, N. M.; Nolan, S. P.; Díaz-Requejo, M. M.; Pérez, P. J., *Angew. Chem., Int. Ed.* **2005**, *44*, 5284.
27. de Frémont, P.; Marion, N.; Nolan, S. P., *J. Organomet. Chem.* **2009**, *694*, 551.
28. Ricard, L.; Gagosz, F., *Organometallics* **2007**, *26*, 4704.
29. Ramón, R. S.; Gaillard, S.; Poater, A.; Cavallo, L.; Slawin, A. M. Z.; Nolan, S. P., *Chem. Eur. J.* **2011**, *17*, 1238.
30. Fürstner, A.; Davies, P. W., *Angew. Chem., Int. Ed.* **2007**, *46*, 3410.
31. Braunschweig, H.; Dewhurst, R. D.; Hammond, K.; Mies, J.; Radacki, K.; Vargas, A., *Science* **2012**, *336*, 1420.
32. Wang, Y.; Xie, Y.; Wei, P.; King, R. B.; Schaefer, H. F.; von R. Schleyer, P.; Robinson, G. H., *Science* **2008**, *321*, 1069.
33. Wang, Y.; Xie, Y.; Wei, P.; King, R. B.; Schaefer, I. I. I. H. F.; Schleyer, P. v. R.; Robinson, G. H., *J. Am. Chem. Soc.* **2008**, *130*, 14970.

(−)-(4R,5R)-4,5-Bis[hydroxy(diphenyl)-methyl]-2,2-dimethyl-1,3-dioxolane (TADDOL)

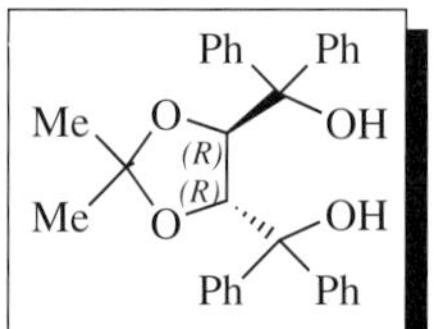

1

[93379-48-7] $C_{31}H_{30}O_4$ MW 466.57

InChI = 1S/C31H30O4/c1-29(2)34-27(30(32,23-15-7-3-8-16-23)24-17-9-4-10-18-24)28(35-29)31(33,25-19-11-5-12-20-25)26-21-13-6-14-22-26/h3-22,27-28,32-33H,1-2H3/t27-,28-/m1/s1

InChIKey = OWVIRVJQDVCGQX-VSGBNLITSA-N

(chiral shift reagent; chiral host in inclusion compounds; chiral reagent; chiral ligand for asymmetric catalysis)

Alternate Name: TADDOL.

Physical Data: mp: 190–192 °C[1a]; 195–196.5 °C[1b]; 192–193 °C;[1c,e] 193.5–195 °C;[1d] $[\alpha]_d^{RT} = -68.5$ ° (c = 1, CH_3Cl);[1a] $[\alpha]_d = -60.6$ ° (c = 1, CH_3Cl);[1b] $[\alpha]_d^{20} = -67$ ° (c = 1, CH_3Cl);[1c] $[\alpha]_d^{20} = -65.1$° (c = 1, CH_3Cl);[1d] $[\alpha]_d^{20} = -64.6$° (c = 1, CH_3Cl)[1e]

Solubility: soluble in toluene, cyclohexane, dichloromethane and tetrahydrofuran.

Form Supplied in: colorless powder.

Preparative Methods: commercially available, TADDOL can also be obtained from dimethyl or diethyl tartrate (eq 1).[1]

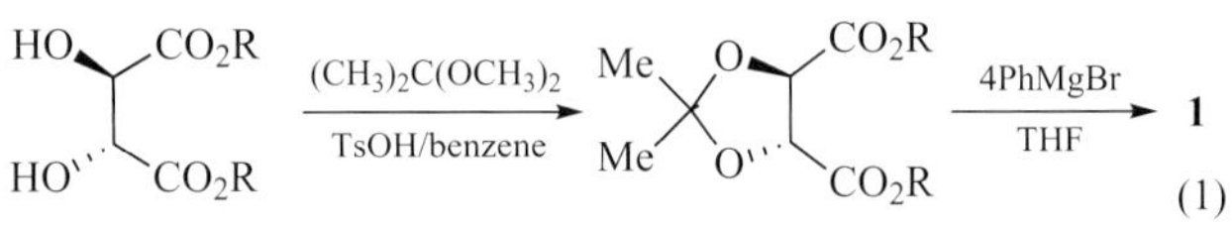

R = Et, Me

(1)

Original Commentary

Antonio Simón Fuentes & Carlos del Pozo Losada
Universidad de Valencia, Burjassot, Spain

Derivatives and Substitution Products. The OH groups in TADDOL are subject to the usual chemical reactions: ether formation, esterification, silylation, and treatment with $ClPR_2$, Cl_2PR, Cl_3P, or $SOCl_2$. They can be initially transformed into mono-and dichlorides (**2**) and (**3**), respectively (eq 2). Reactions of these intermediates with nucleophiles such as alcohols, phenols, ammonia and amines, anilines, azide, phosphates, phosphinites, thiols, thiocyanate, and thiourea lead to a variety of products. The resulting compounds can then be used as chiral ligands with metallic centers, as auxiliaries, and as reagents for EPC syntheses.[2]

Chiral Shift Reagent in NMR Spectroscopy. TADDOL is a chiral hydrogen-bond donor that contains four phenyl groups in the immediate vicinity of the donor OH groups. As a consequence, hydrogen-bond acceptors that dock at these OH groups enter either the shielding or deshielding regions of nearby aromatic rings. This is of special interest as diastereomeric hydrogen-bonding complexes can be distinguished with the aid of NMR spectroscopy. In contrast to spectra resulting from the use of lanthanide shift reagents, the signals observed in this case are not subject to line-broadening effects, a fact that makes TADDOL extremely useful for determining the enantiomeric purity of secondary alcohol derivatives,[3] acyclic and cyclic amines,[4] cyanohydrins derived from aliphatic and aromatic aldehydes and ketones,[4a,5] and esters of amino acids.[4a]

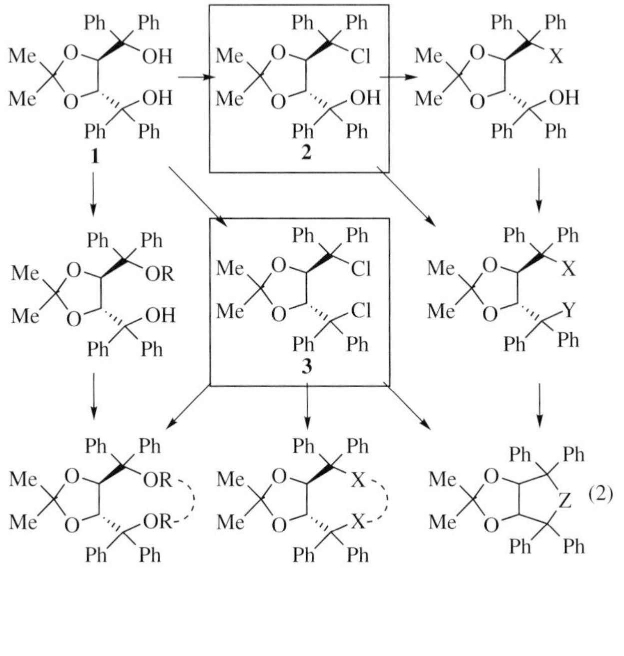

(2)

TADDOL organophosphorus derivatizing agents (**4**) and (**5**) (Figure 1) have also been used for determining the enantiomeric excess of chiral alcohols and carboxylic acids.[6] It is worth noting that TADDOL phosphorus derivatives (**4**) from carboxylic acids (**6**–**9**) show $\Delta\delta = 0.44$–2.31 ppm in their ^{31}P NMR.

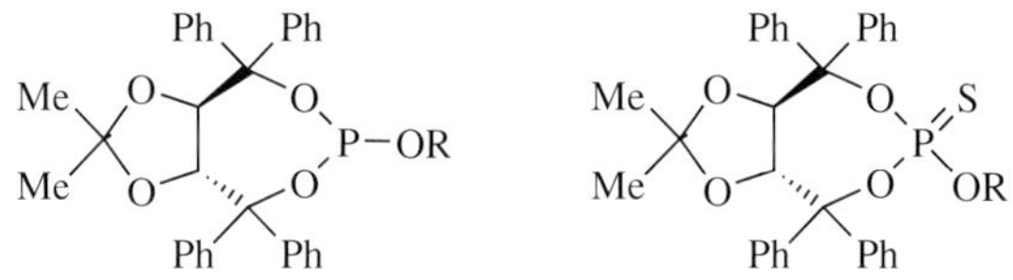

4 **5**

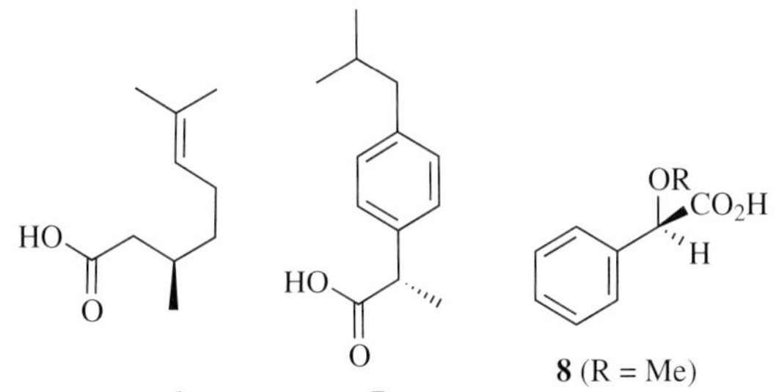

6 **7** **8** (R = Me) **9** (R = Ac)

Chiral Host in Inclusion Compounds: Resolution of Racemates. TADDOL crystallizes in the presence of hydrogen-bond acceptors, with which it forms inclusion compounds. The differential crystallization of TADDOL inclusion compounds has proved to be a valuable tool for the resolution of racemates, especially in situations in which the incorporated partners cannot be classified as bases in the usual sense. In addition to enantiomers, diastereomers and even compounds with a different constitution altogether lend themselves to separation by means of TADDOL inclusion compounds.[1c,7] In fact, this method has led to the successful resolution (ep > 98%) of a variety of nitrogen- (piperidine,[4b] *N*-nitrosopiperidine,[8] imide,[9] and *N*-imidazole derivatives[10]), oxygen- (alcohols,[11] epoxides,[11b] ketones,[1b,12] and anhydrides[9]), sulfur- (sulfoxides[13]), and phosphorus compounds.[14]

A recent example of the successful application of this strategy is the optical resolution of the *anti*-head-to-head racemic coumarin dimer (**10**) via molecular complexation with TADDOL, which afforded enantiomerically pure enantiomers (eq 3).[15]

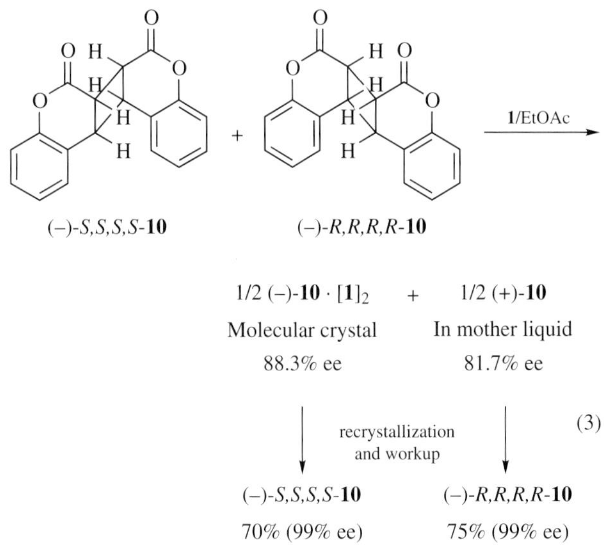

(3)

An optical resolution of medium-size lactones through inclusion crystallization with TADDOL has also been achieved, albeit with poorer selectivity (up to 59% ee).[16]

In addition, inclusion compounds made with TADDOL have also facilitated the distillatory separation of racemates such as *rac* 3-phenyl-2-propanamine (amphetamine)[17] and *rac* 1-phenylethanol.[11b]

Even more recently, a novel enantioseparation process combining the highly enantioselective nature of inclusion complexation with the subsequent separation of enantiomers from the chiral host through solvent decomplexation and organic solvent nanofiltration has also been developed. This technology replaces distillation as a method for recovering enantiomers from enantioenriched solid complexes and expands the application of enantioselective inclusion complexation to the resolution of nonvolatile racemates. The process, in which TADDOL serves as the chiral host, has been successfully applied to the resolution of *rac* 1-phenylethanol.[18]

Selective Transformations in TADDOL Inclusion Compounds. Another important process, namely that of stereoselective transformations, has also been observed in TADDOL host–guest compounds. It is worth noting that within TADDOL inclusion compounds, the migration tendency of various aryl groups in the photorearrangement of 4,4-diaryl-2-cyclohexenones (**11**) into 5,6-diarylbicyclo[3.1.0]hexan-2-ones (**12**) occurs in reverse order to that observed in solution (eq 4).[12c]

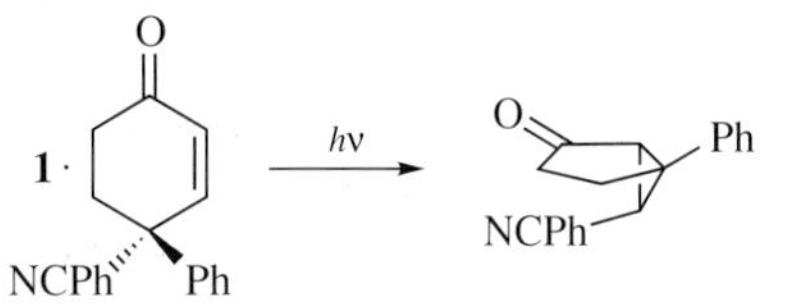

(*R*)-**11** **12a**

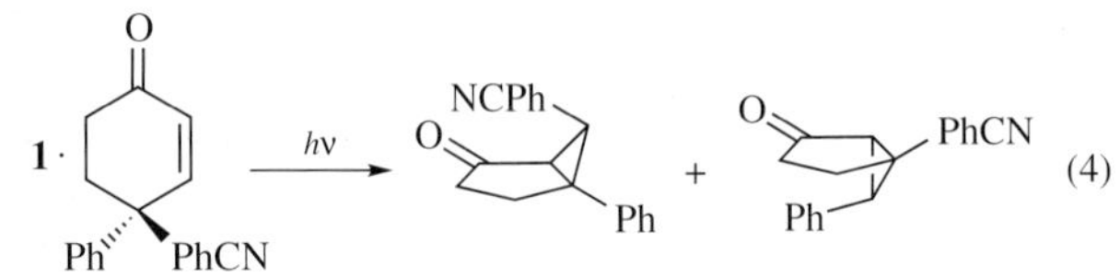

(*S*)-**11** **12b** **12c**

Selective reduction of the 7-keto group of (*R*)-6-methylbicyclo[4.4.0]dec-1-en-3,7-dione **13** and (*R*)-6-methylbicyclo[4.3.0]non-1-en-3,7-dione **14** occurs with > 99% diastereo- and regioselectivity when the inclusion compound formed between TADDOL and the dione is triturated with $NaBH_4$ in an agate mortar and the mixture is subsequently allowed to stand at room temperature for 3 days with exclusion of moisture (eq 5).[19]

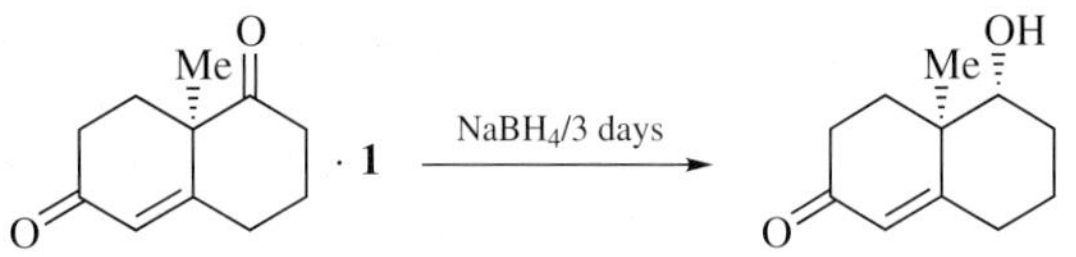

13 54%

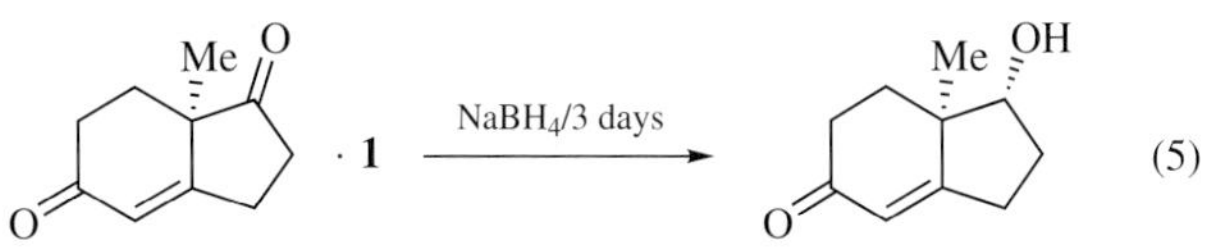

14 55%

Enantioselective photoreactions in TADDOL inclusion compounds have also been investigated. Indeed, the classic photochemical valence isomerization that converts pyridone into azabicyclo[2.2.0]hexene leads to an enantiomerically pure product when the inclusion compound is made with TADDOL.[20] A similar photoisomerization reaction leads to *N*-methylhexahydro-4-carbazolones **15** from *N*-methyl-*N*-phenyl-3-amino-2-cyclohexen-1-one derivatives **16** (eq 6)[21] and *β*-lactam derivatives **17** from 2-(*N*-acyl-*N*-alkylamino)cyclohex-2-enones **18** in inclusion crystals with TADDOL (eq 7).[22]

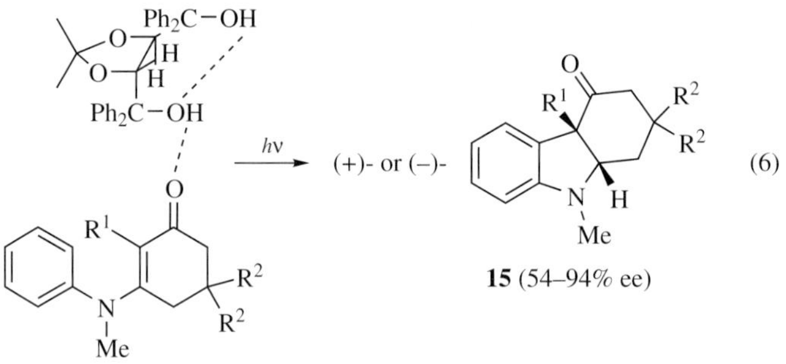

16 **15** (54–94% ee)

R^1, R^2 = H, Me

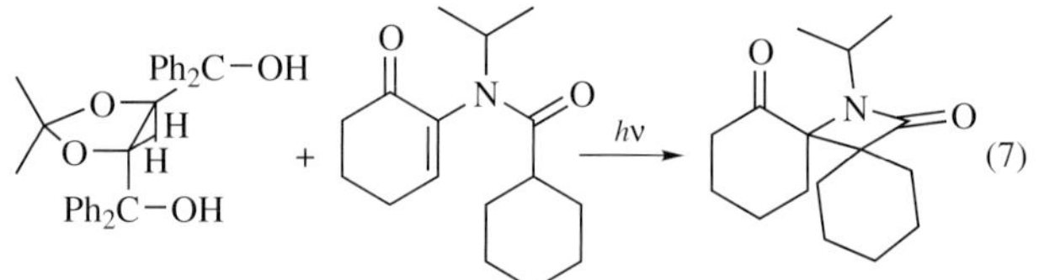

(1:1 complex) (−)-**17** (>99% ee)

Photochemical intermolecular [2+2] cycloaddition of TADDOL-coumarin inclusion compound **19** leads to the *anti*-head-to-head dimerization of coumarin in the solid-state, in which a single-crystal to single-crystal conversion has been accomplished (eq 8).[23] This reaction also proceeds efficiently and with high enantioselectivity in a homogeneous solution in the presence of TADDOL.[24]

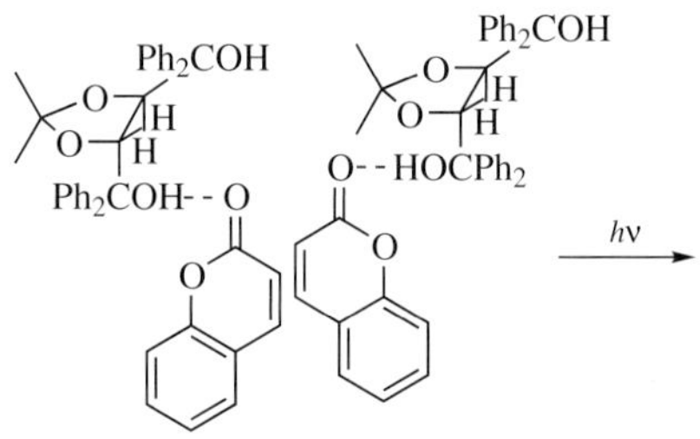

19

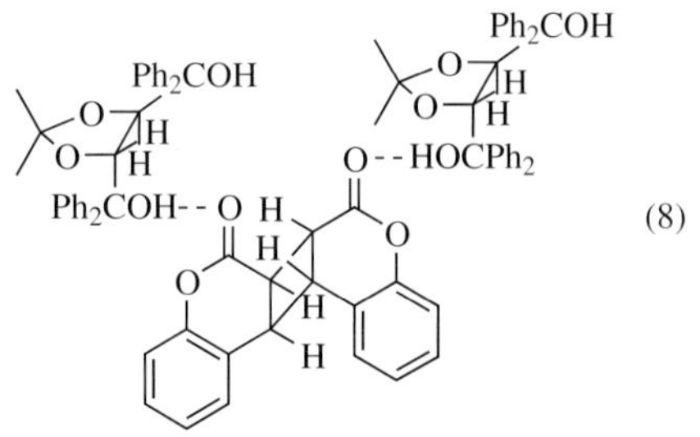

Other solid-state reactions with TADDOL have proven less successful. For example, a Wittig–Horner reaction of a 1:1

inclusion compound of 4-methylcyclohexanone and TADDOL with (carbethoxymethylene)triphenylphosphorane affords optically active 4-methyl-1-(carbethoxymethylene)cyclohexanone in 50.8% yield and with only 42.8% ee,[25] while borane reduction of unsymmetrical ketones occurs with even poorer yields and enantioselectivities.[26]

Chiral Reagent. TADDOL has been employed as a chiral acid in the enantioselective protonation of the achiral enolate **20** (eq 9).[27]

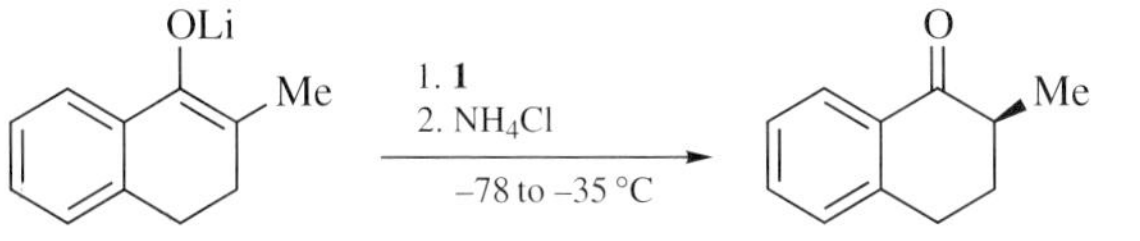

(9)

20 75% ee

In addition, TADDOL derivatives of main-group metals have been used as additives in enantioselective C–C bond formations. For example, in the synthesis of α-branched amino acids **21**, sodium TADDOLate (at a ratio of 1:1 or 1:10 to the substrates) functions as a phase-transfer catalyst (eq 10)[28] whereas in additions of primary alkyl Grignard reagents to aromatic, heteroaromatic, and α,β-unsaturated ketones **22**, magnesium TADDOLate behaves as a chiral Lewis acid (eq 11).[29] In these additions, tertiary alcohols **23** are formed through nucleophilic attack from the *Re* face of the keto carbonyl group.

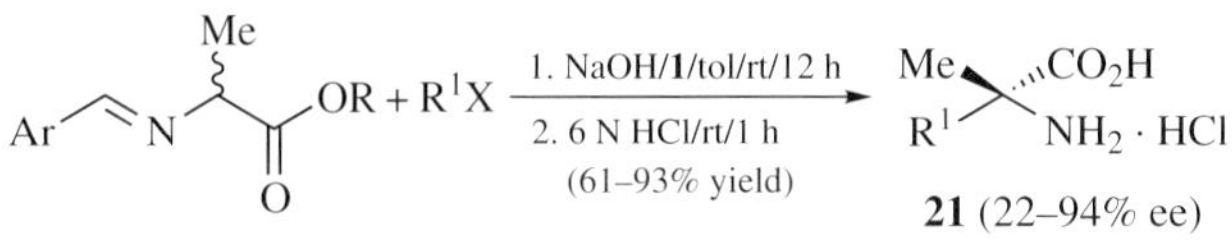

21 (22–94% ee)

R = Me, ^{i}Pr, ^{t}Bu (10)
Ar = Ph, 4-ClC$_6$H$_4$, 4-FC$_6$H$_4$
R^1 = Bn, allyl, α-naphthCH$_2$
X = Br, Cl

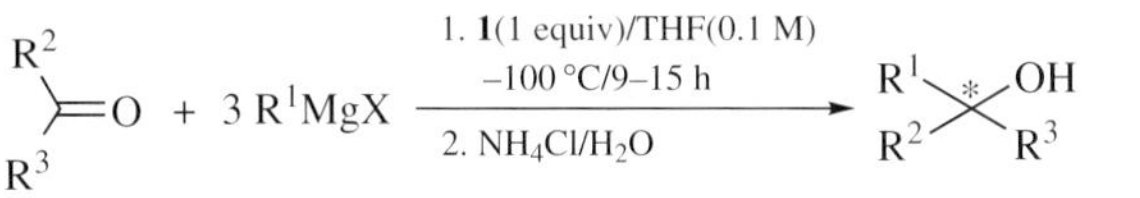

(11)

22 **23** (up to 98% de)

Chiral Chelating Ligand. Enantioenriched (*S*)-phosphoglutaminic acid derivatives **24** are obtained with up to 72% ee in the TADDOL-catalyzed, enantioselective Michael reaction of the achiral phosphoglycine synthon **25** when solid alkali metal *tert*-butoxides are used as bases (eq 12). Results indicate that the sodium TADDOLate is not the deprotonating agent, but that it takes part in the process as a chelating agent.[30]

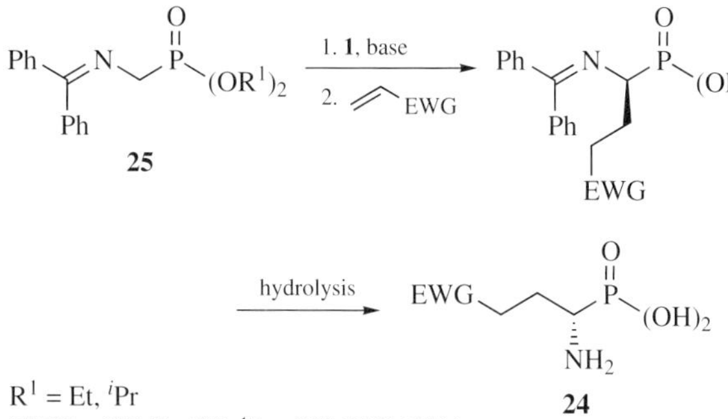

(12)

R^1 = Et, ^{i}Pr
EWG = CO$_2$Et, CO$_2{}^t$Bu, CN, P(O)(OEt)$_2$

In the stereoselective synthesis of amino acid-derived β-lactams **26**(eq 13), TADDOL seems to enhance memory of chirality instead of acting as a real catalyst. One proposed mechanism to rationalize the function of TADDOL in these reactions involves the formation of complexes between the starting amino acid derivative (or its derived enolate intermediate) and the chiral additive through H-bonds and π–π interactions.[31]

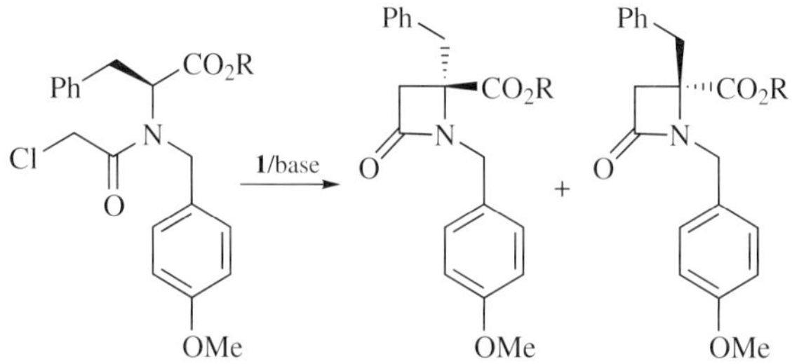

(13)

R = ^{t}Bu **26a** (R = ^{t}Bu) **26c** (R = ^{t}Bu)
R = Me **26b** (R = Me) **26d** (R = Me)

ratio: up to 99:1
ee: up to 82%

Nevertheless, the principal field of application for TADDOL as a chelating ligand is enantioselective syntheses mediated by titanium TADDOLates,[32] although enantioselective transformations also occur when complexes of TADDOL with other metallic centers are used. For example, moderate to good enantioselectivities are achieved in the Zr(OBu^t)$_4$-promoted asymmetric Meerwein–Ponndorf–Verley cyanation of aldehydes **27** when TADDOL is used as a chiral ligand (eq 14).[33]

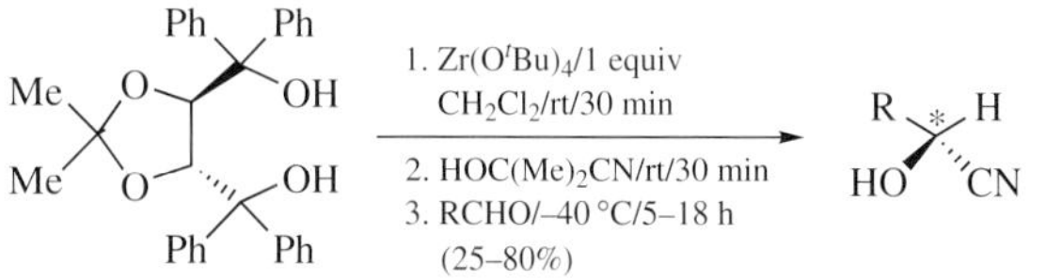

(14)

1 **27** (29–85% ee)

R = Ph(CH$_2$)$_2$, CH$_3$(CH$_2$)$_8$, *c*-Hex, ^{t}Bu, Bn, Ph thiophenyl, *trans*-PhCH=CH

Zr-TADDOLate also catalyzes the aldol-Tishchenko reactions of ketone aldols **28** with a range of aldehydes **29**, giving rise to differentiated 1,3-*anti*-diol monoesters **30** in good yields, with complete diastereocontrol and moderate enantioselectivities (eq 15).[34]

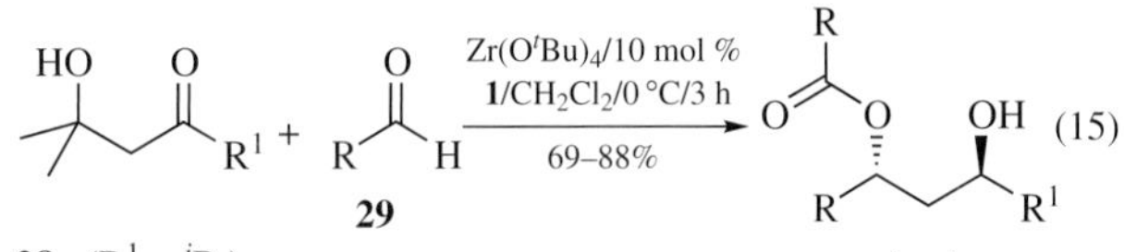

28a (R^1 = ^{i}Pr)
28b (R^1 = ^{t}Bu)
R = n-C_6H_{13}, ^{i}Pr, ^{t}Bu, 2-ethylpropyl

30a (R^1 = ^{i}Pr)
30b (R^1 = ^{t}Bu)
(30–57% ee)

In contrast, only low enantioselectivity (46% ee) is achieved in the allylation of benzaldehyde with the (dialkoxyallyl)chromium(III) complex derived from TADDOL.[35] The addition of the butyl group to benzaldehyde or cyclohexanecarboxaldehyde through a TADDOL-butylcerium reagent **31** results in ee's of 66 and 70%, respectively.[36] The selectivity of the reaction can be increased even further—up to 92% ee—by increasing the number of equivalents of TADDOL per cerium, with the optimum ratio being 3:1 (eq 16).[37]

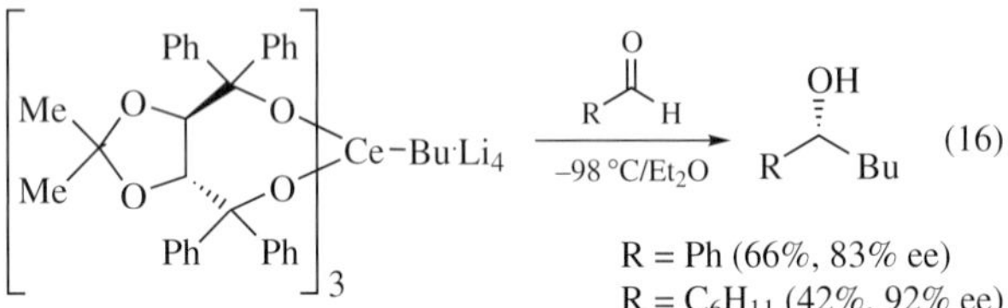

31

R = Ph (66%, 83% ee)
R = C_6H_{11} (42%, 92% ee)

With TADDOL as a chiral proton source, samarium-mediated reduction of propargylic phosphate to its allenic ester derivative occurs with 41% ee; the asymmetric environment is provided by chelation of the chiral alcohol and samarium(III).[38]

Phosphorus derivatives of TADDOL, including phosphite **32**, phosphite derivatives (**33a–e**), phosphoramidites **34**, and bis(diphenylphosphanyl) **35** (Figure 2),have also been employed as ligands with the metals Zn, Rh, Pd, Ir, and Cu in enantioselective reactions. For example, TADDOL-phosphite **32** reacts with aromatic nitroalkenes **36** in the presence of Et_2Zn and TMEDA to afford β-nitrophosphonates **37** in very high diastereomeric excess; the cleavage of the chiral auxiliary leads to optically active β-nitrophosphonic acids **38** in good yields and with high enantiomeric excesses (eq 17).[39]

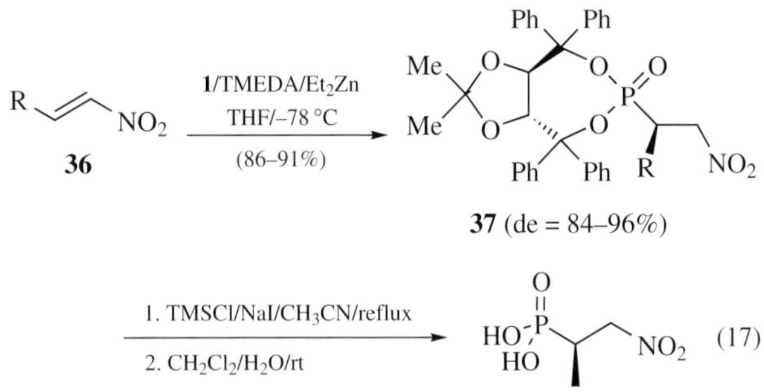

37 (de = 84–96%)

38 (ee = 81–95%)

Rhodium complexes bearing ligands (**33a**) (R^1,R^2 = H, Ph, i-Pr) have been utilized for hydrosilylation of dialkyl and aryl alkyl ketones **39** with enantiomer ratios as high as 97:3. In this case, the configuration of the secondary alcohol products **40** is mainly determined by the oxazoline moiety (eq 18).[40] The chiral phosphite P,N-ligand derived from (+)-N-methylephedrine (**33b**) has also proven adequate for this reduction reaction, providing ee's of up to 94%.[41]

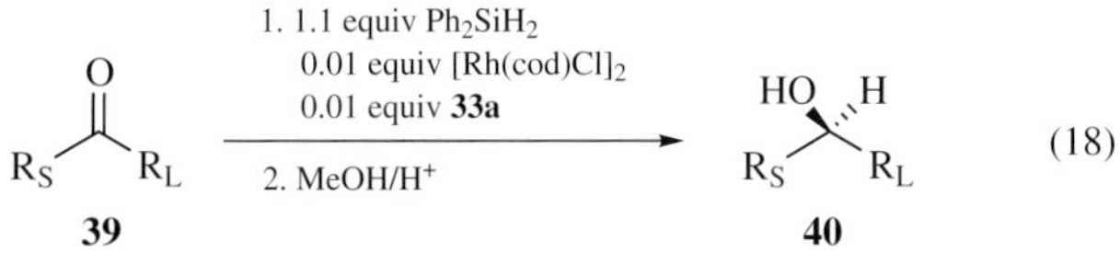

33a (R^1 = ^{i}Pr, R^2 = H)

(Hydrogen transfer with ligand from the Re face, CIP priority O>R_L >R_S).

High levels of regio- (up to 96:4) and enantioselectivity (up to 91% ee) are achieved in the Rh-catalyzed hydroboration of styrene **41** when chiral phosphite P,P-ligands (**33c,d**) derived from diphenylphosphanes are used (eq 19).[42]

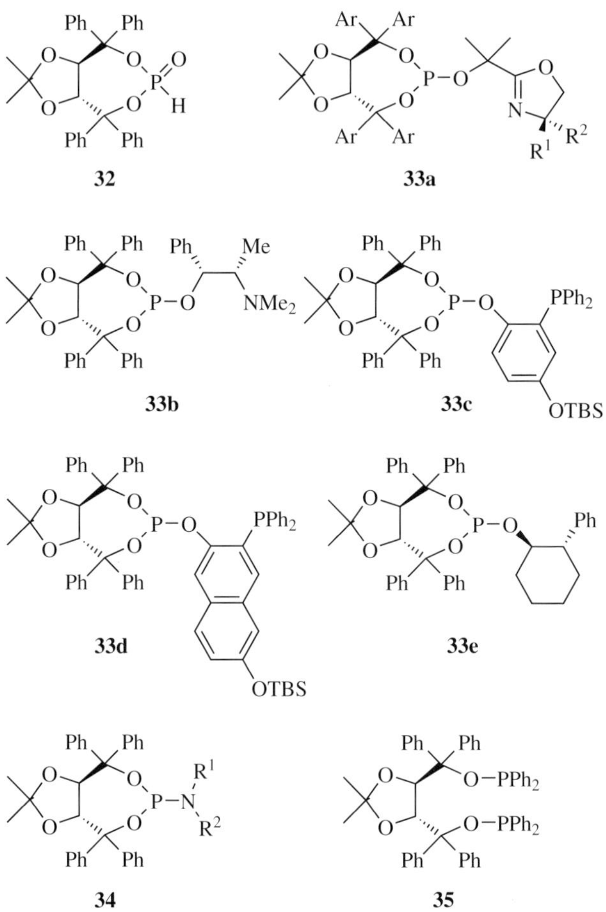

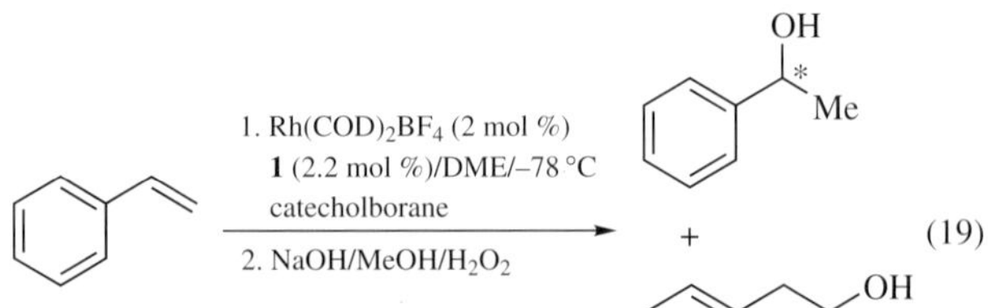

41

When either phosphite (**33e**) or phosphoramidite **34**(R^1 =Ph, R^2 = Bn) derivatives are used as chiral ligands, the Rh-catalyzed hydroboration of substituted styrenes **42** bearing either electron-donating or electron-withdrawing substituents also occurs with high levels of enantioselectivity (eq 20).[43]

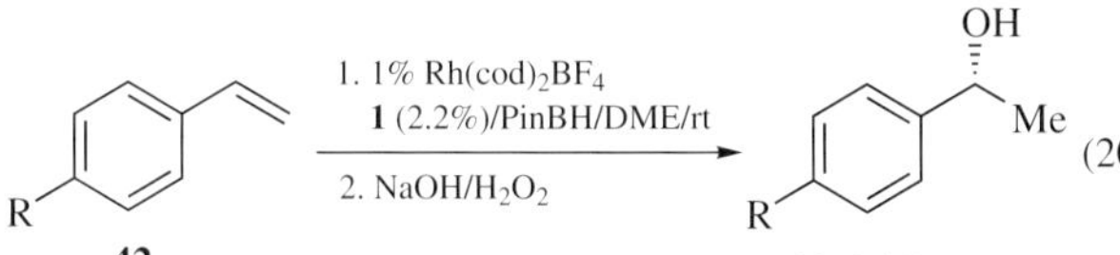

42 90–96% ee (20)

R = H, F, Cl, CH_3, CF_3, OMe
PinBH = pinacolborane

TADDOL-derived phosphoramidites **34** (R^1,R^2 = Me, c-C_4H_8, c-C_5H_{10}) have been found to be the most efficient ligands in the enantioselective Rh-catalyzed [2+2+2] cycloadditions of alkenyl isocyanates **43** and terminal alkynes **44** (eq 21). This reaction has been applied to the total synthesis of (+)-lasubine II.[44]

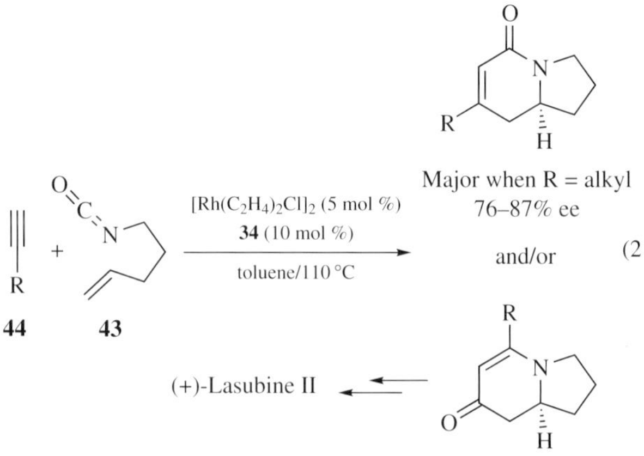

44 **43** (21)

The complex TADDOL forms with **35** and $PdCl_2$ has been used for Pd-catalyzed enantioselective allyl substitutions of 1,3-diphenylallyl acetate and carbonate **45** with various nucleophiles to afford products with enantiomer ratios of up to 88:12 (eq 22).[45]

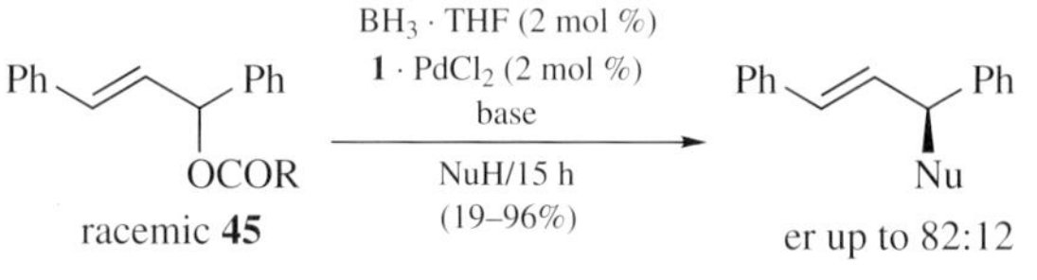

racemic **45** er up to 82:12 (22)

R = Me, OMe
Nu = $CH(CO_2Me)_2$, $CH(COMe)_2$, $CH(CO_2{}^tBu)_2$, $CH(SO_2Ph)_2$ $C(NO_2)Me_2$, $PhCH_2NH$, p-Tol-SO_2

TADDOL-phosphoramidite **34** (R^1,R^2 = i-Pr) induces moderate enantioselectivity (73% ee) in the Pd-catalyzed allylic alkylation of 1,3-diphenylallyl acetate with dimethyl malonate.[46] Higher enantioselectivities are achieved in the Pd-catalyzed allylic alkylation of (*E*)-3-phenyl-2-propen-1-yl and (*E*)-3-(1′-naphthyl)-2-propen-1-yl acetates **46** and **4746** when (**33a**) (R^1 = H, R^2 = t-Bu) is used as a chiral ligand (eq 23).[47]

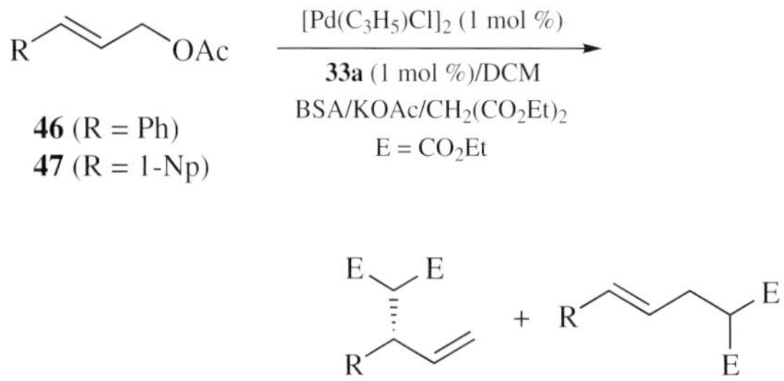

46 (R = Ph)
47 (R = 1-Np)

(23)

48 (br/l = 26:74; 87% ee)
49 (br/l = 66:34; 94% ee)

TADDOL-phosphite-oxazoline (**33a**) has also proven to be an adequate ligand for enantioselective iridium-catalyzed hydrogenation of several types of olefins (62–90% ee).[47]

The Cu-catalyzed enantioselective 1,4-addition reactions of Et_2Zn to cyclic enones in the presence of TADDOL-phosphoramidite **34**(R^1,R^2 = Me) result in ee's of up to 71% for cyclohexenone and 62% for cyclopentenone.[48] α-Unsaturated imines **48** derived from α-amino acids also undergo a similar addition to afford α-dehydroamino acids **49** with a stereogenic center in the γ-position (eq 24).[49]

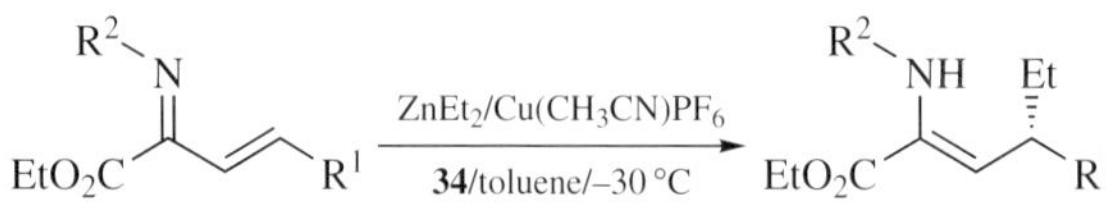

48 **49** (24)

R^1	R^2	Yield (%)	1,4:1,2:double	er
p-Tol	p-NO_2-C_6H_4	89	92:<1:8	94:6
p-MeO-C_6H_4	p-NO_2-C_6H_4	87	98:1:1	89:11
p-Tol	p-Tol	85	99:<1:1	88:12
p-NO_2-C_6H_4	p-Tol	83	96:3:1	91:9
p-NO_2-C_6H_4	p-Tol	85	95:3:2	88:12

With complexes of Cu and **34** (R^1,R^2 = Me or i-Pr), the ring opening of cyclohexene N-p-toluenesulfonylimine with methylmagnesium halides[50] and the alkylation of ethynyl cyclohexene epoxide with Et_2Zn[51] both occur with poorer enantioselectivities (29% and 36% ee, respectively).

Related Reagents. (+)-TADDOL; 4,5-bis[hydroxy(diphenyl) methyl]-2-cyclopentyl-1,3-dioxolane; 4,5-bis[hydroxy (diphenyl) methyl]-2-cyclohexyl-1,3-dioxolane; 4,5-bis [hydroxy(diphenyl) methyl]-1-methyl-2-cyclopentyl-1,3-dioxolane; 4,5-bis{hydroxy[bis(1-naphthyl)] methyl}-2,2-dimethyl-1,3-dioxolane; 4,5-bis{hydroxy[bis(2-naphthyl)]methyl}-2,2-dimethyl-1,3-dioxolane; 4,5-bis[hydroxy(diphenyl)methyl]-2-phenyl-1,3-dioxolane; 4,5-bis[hydroxy(diphenyl)methyl]-2,2-diphenyl-1,3-dioxolane; 4,5-bis[hydroxy (diphenyl)methyl]-2-(*tert*-butyl)-1,3-dioxolane; 4,5-bis{hydroxy[bis(3,5-dimethylphenyl)] methyl}-2,2-dimethyl-1,3-dioxolane; 4,5-bis[hydroxy(diphenyl)methyl]-2,2-diethyl-1,3-dioxolane; 4,5-bis{hydroxy[bis(3,5-dimethylphenyl)]methyl}-2,2-diethyl-1,3-dioxolane; 4,5-bis{hydroxy[bis(4-hydroxyphenyl)]methyl}-2,2-dimethyl-1,3-dioxolane; 4,5-bis{hydroxy[bis(4-

diaminomethylphenyl)]methyl}-1,3-dioxolane; 4,5-bis[hydroxy(diphenyl)methyl]-2-(4-hydroxymethylphenyl)-1,3-dioxolane; 4,5-bis[hydroxy(diphenyl)methyl]-2-(4-ethenylphenyl)-1,3-dioxolane.

First Update

Harit U. Vora
Merck Research Laboratories, West Point, PA

Chiral Chelating Reagents (Figure 3). TADDOL derivatives have found extensive use as chiral chelating ligands for metal-catalyzed reactions. In particular, focus has largely been on phosphite and phosphoramidite ligands in conjunction with transition metals. TADDOL-derived phosphoramidite ligand (**34**) (R^1 = Ph, R^2 = 2-pyridyl) has been found to be efficient in catalyzing the enantioselective Pd-catalyzed Nazarov-type cyclization to provide pentasubstituted cyclopentanones (**56**). Additionally, it was observed that activation of the substrate (**55**) with aryl substituents is not required to obtain products in high yields and with good enantioselectivities (eq 25).[52]

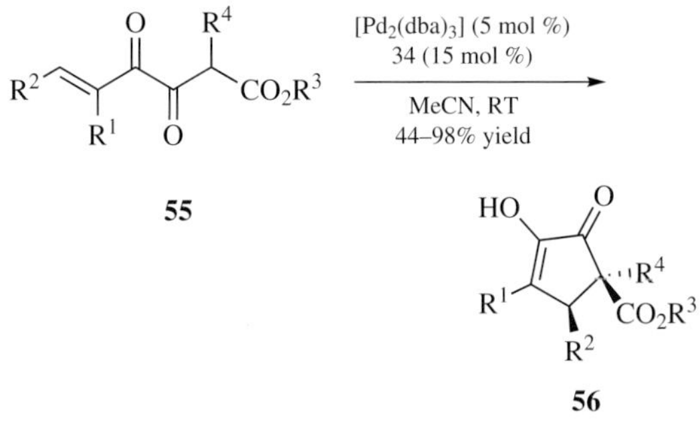

(25)

50–96% ee

R^1, R^2 = alkyl and aryl
R^3 = Et
R^4 = Me, Et

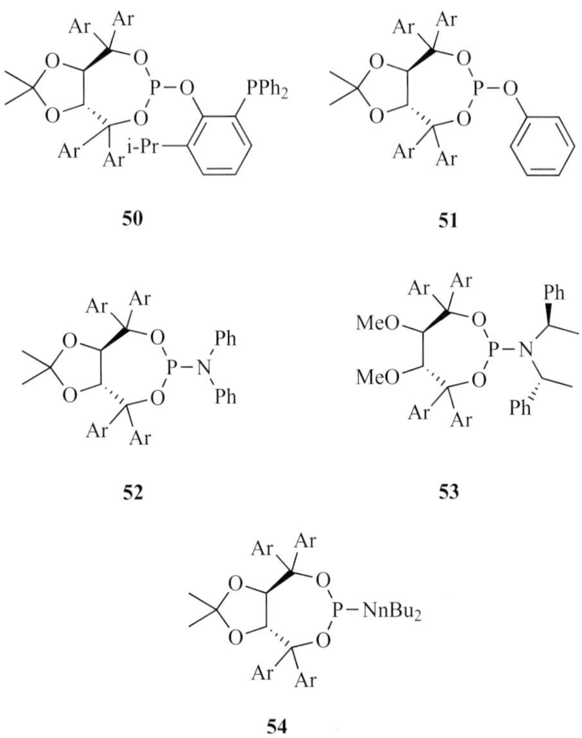

Nickel and a TADDOL–phosphoramidite ligand (**52**) (Ar = C_6H_5) have been reported to catalyze the addition of an alkyne C–H bond (**57**) across 1,3-dienes (**58**). The subsequent hydroalkylation products (**59**) can be further functionalized to the respective ketones (**60**) via a two-step procedure without epimerization of the chiral center (eq 26).[53]

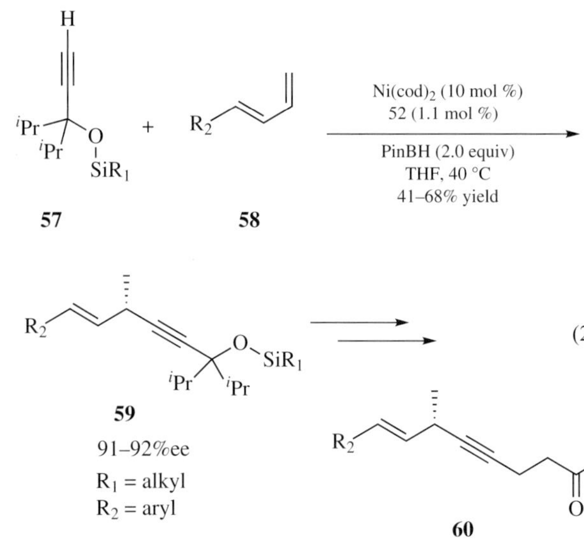

(26)

91–92%ee
R_1 = alkyl
R_2 = aryl

The use of gold TADDOL–phosphoramidite (**53**)) (Ar = C_6H_5) complex has been utilized in a [2+2] cycloaddition reaction toward the synthesis of [2,3,0] fused carbocycles (**62**), which are obtained with high enantioselectivities (87–96% ee). The intramolecular reaction between the alkene and allene is facilitated by the Thorpe–Ingold effect that is required to obtain reactivity. The use of TADDOL ligands provides a more robust method for imparting enantioselectivity with gold catalysis (eq 27).[54]

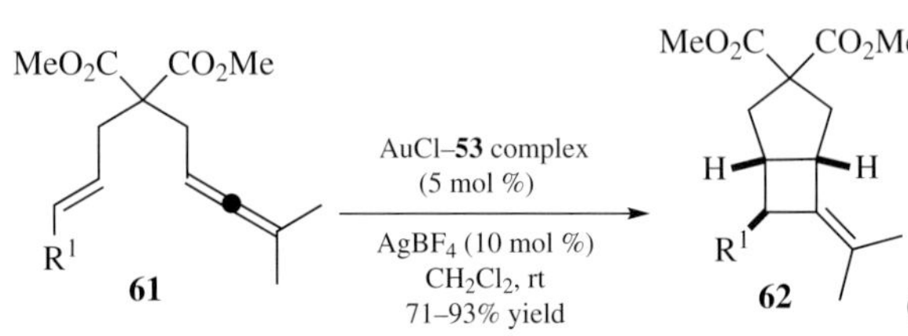

(27)

87–96% ee
R_1 = Aryl

The use of palladium TADDOL–phosphoramidite complex (**54**) (Ar = C_6H_5) has been shown to catalyze a direct arylation of vinyl triflates (**63**) to furnish trisubstituted carbocycles (**64**). The reaction is tolerant of substitution at the *ortho-* and *para-*positions of the arene without any loss in yield or enantioselectivity. Halogens on the arene are tolerated and electron-poor aromatic substrates possess higher reactivity and furnish high diastereoselectivities (eq 28).[55]

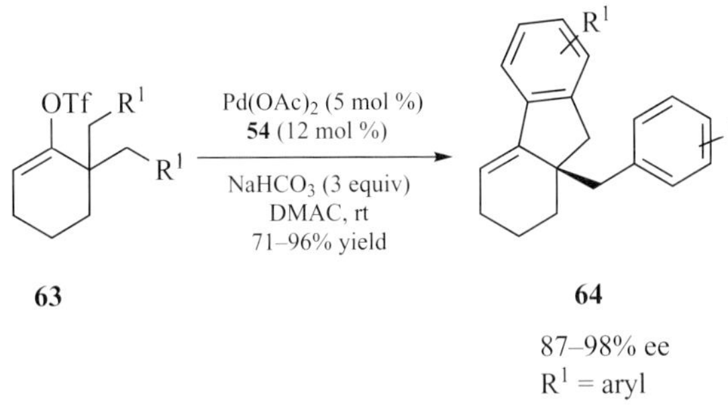

(28)

An enantioselective Nickel-catalyzed hydrocyanation of vinyl arenes (**65**) occurs where a TADDOL-derived phosphite ligand (**50**) (Ar = 3,5-diMe, R^1 = ^{i}Pr) was shown effective to provide the respective benzylic nitriles (**66**) in 15–100% conversion and 67–92% ee (eq 29). [56]

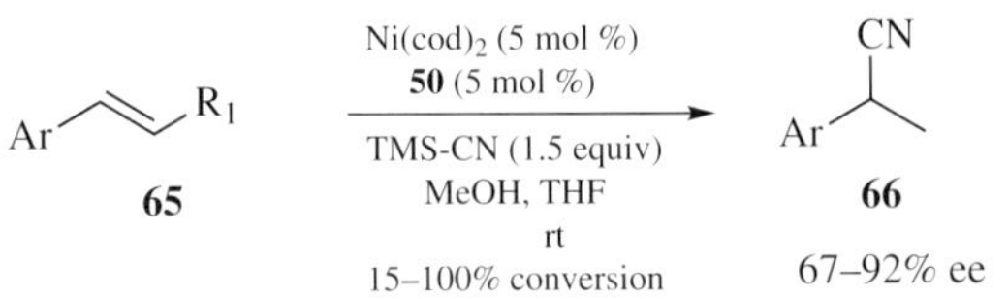

(29)

Trisubstituted alkenes have also been reported to undergo an asymmetric hydroboration reaction of β,γ-unsaturated amides (**67** and**69**) catalyzed by a rhodium–TADDOL-derived phosphite ligand (**51**) (Ar = C_6H_5). The reaction is stereospecific as the (*E*)- and (*Z*)-alkenes give rise to the respective *anti-* and *syn*-products (**68** and **70**), affording excellent enantioselectivities (91–95%) of the β-hydroxy amides. The high levels of enantioselectivities are attributed to coordination of the carbonyl group of the amide and the rhodium–TADDOL complex (eq 30). [57]

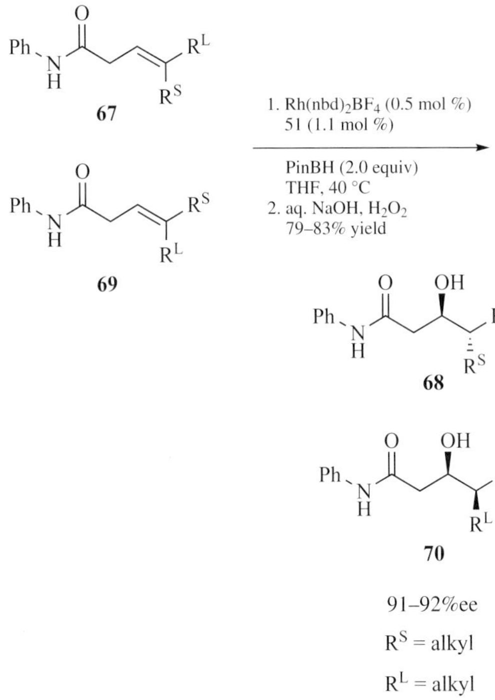

(30)

Chiral Reagent. TADDOL-derived iminophosphoranes (**73**) (Ar = p-^{t}BuC_6H_4) were recently shown to be effective in substoichiometric quantities to effect chlorination of oxindoles (**71**) under mild conditions to provide the desired chlorinated oxindoles (**72**) in 93–99% yield and with 90–90% ee (eq 31).[58]

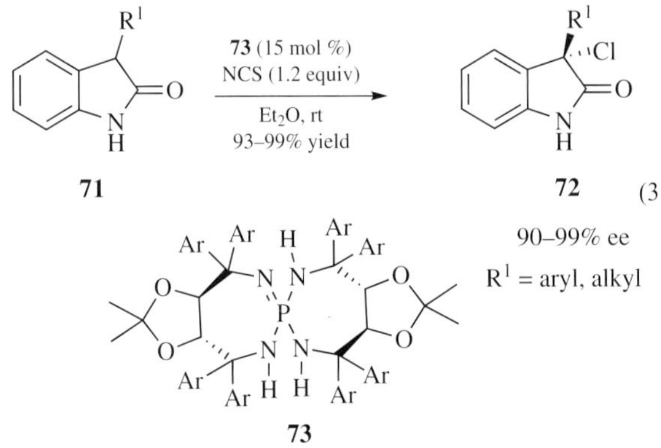

(31)

A similar approach was applied toward the hydroxylation of β-dicarbonyl derivatives. The chiral guanidine (**76**) (Ar = p-PhC_6H_4), a derivative of TADDOL, is efficient to catalyze an asymmetric hydroxylation via a chiral Brønsted activation of the 1,3-dicarbonyl compound (**74**) to provide the hydroxylation products (**75**) in 90–99% yield and with 55–94% ee (eq 32). [59]

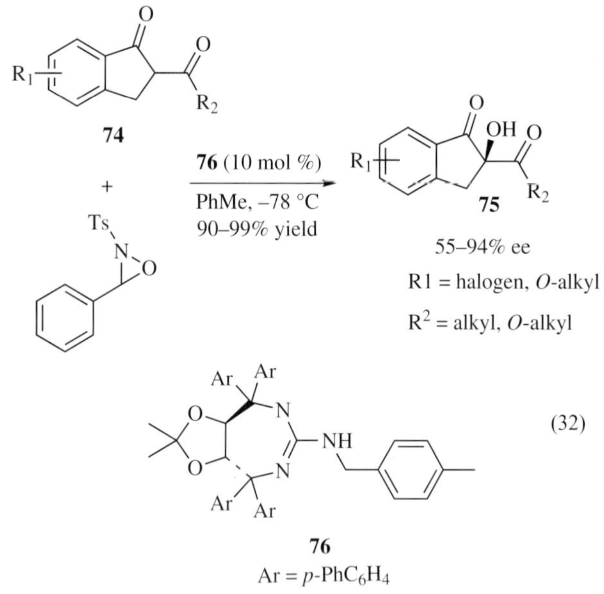

1. (a) Seebach, D.; Beck, A. K.; Imwinkelried, R; Roggo, S.; Wonnacott, A., *Helv. Chim. Acta* **1987**, *70*, 954. (b) Toda, F.; Tanaka, K., *Tetrahedron Lett.* **1988**, *29*, 551. (c) Weber, E.; Dörpinghaus, N.; Goldberg, I., *J. Chem Soc., Chem. Commun.* **1988**, 1566. (d) Beck, A. K.; Bastani, B.; Plattner, D. A.; Petter, W.; Seebach, D.; Braunschweiger, H.; Gysi, P.; La Vecchia, L., *Chimia* **1991**, *45*, 238. (e) Weber, E.; Dörpinghaus, N.; Wimmer, C.; Stein, Z.; Krupitsky, H.; Goldberg, I., *J. Org. Chem.* **1992**, *57*, 6825.
2. (a) Seebach, D.; Beck, A. K.; Heckel, A., *Angew. Chem., Int. Ed. Engl.* **2001**, *40*, 92 and references cited therein. (b) Gerbino, D. C.; Mandolesi, S. D.; Koll, L. C.; Podestá, J.C., *Synthesis* **2005**, 2491.
3. von dem Bussche-Hünnefeld, C.; Beck, A. K.; Lengweiler, U.; Seebach, D., *Helv. Chim. Acta* **1992**, *75*, 438.
4. (a) Tanaka, K.; Ootani, M.; Toda, F., *Tetrahedron: Asymmetry* **1992**, *3*, 709. (b) Toda, F.; Tanaka, K.; Ootani, M.; Hayashi, A.; Miyahara, I.; Hirotsu, K., *J. Chem. Soc., Chem. Commun.* **1993**, 1413.
5. Schrader, T., *Angew. Chem. Int. Ed. Engl.* **1995**, *34*, 917.
6. Alexakis, A.; Chauvin, A.-S., *Tetrahedron: Asymmetry* **2000**, *11*, 4245.
7. Goldberg, I.; Stein, Z.; Weber, E.; Dörpinghaus, N.; Franken, S., *J. Chem. Soc., Perkin Trans. 2* **1990**, 953.
8. Olszewska, T.; Milewska, M. J.; Gdaniec, M.; Polonski, T., *J. Chem. Soc., Chem. Commun.* **1999**, 1385.
9. Zaderenko, P.; López, P.; Ballesteros, P.; Takumi, H.; Toda, F., *Tetrahedron: Asymmetry* **1995**, *6*, 381.
10. (a) Toda, F.; Matsuda, S.; Tanaka, K., *Tetrahedron: Asymmetry* **1991**, *2*, 983. (b) Toda, F.; Tohi, Y., *J. Chem. Soc., Chem. Commun.* **1993**, 1238.
11. (a) Toda, F.; Tanaka, K.; Marks, D.; Goldberg, I., *J. Org. Chem.* **1991**, *56*, 7332. (b) Takemoto, Y.; Kuraoka, S.; Hamaue, N.; Aoe, K.; Hiramatsu, H.; Iwata, C., *Tetrahedron* **1996**, *52*, 14177. (c) Zimmerman, H. E.; Alabugin, I. V.; Smolenskaya, V.N., *Tetrahedron* **2000**, *56*, 6821.
12. Toda, F.; Miyamoto, H.; Ohta, H., *J. Chem. Soc., Perkin Trans. 1* **1994**, 1601.
13. (a) Toda, F.; Tanaka, K.; Okada, T., *J. Chem. Soc., Chem. Commun.* **1995**, 639. (b) Zhu, J.; Qin, Y.; He, Z.; F, F.-M.; Zhou, Z.-Y.; Deng, J.-G.; Jiang, Y.-Z.; Chau, T.-Y., *Tetrahedron: Asymmetry* **1997**, *8*, 2505.
14. Novák, T.; Schindler, J.; Ujj, V.; Czugler, M.; Fogassy, E.; Keglevih, G., *Tetrahedron: Asymmetry* **2006**, *17*, 2599.
15. Zhao, D.; Sun, J.; Ding, K., *Org. Lett.* **2003**, *5*, 1349.
16. Tanaka, K.; Kuchiki, D.; Caira, M. R., *Tetrahedron: Asymmetry* **2006**, *17*, 1678.
17. Ács, M.; Mravik, A.; Fogassy, E.; Böcskei, Z., *Chirality* **1994**, *6*, 314.
18. Ghazali, N. F.; Ferreira, F. C.; White, A. J. P.; Livington, A. G., *Tetrahedron: Asymmetry* **2006**, *17*, 1846.
19. Toda, F.; Kisyoshige, K.; Yagi, M., *Angew. Chem., Int. Ed. Engl.*. **1989**, *28*, 320.
20. Toda, F.; Tanaka, K., *Tetrahedron Lett.* **1988**, *29*, 4299.
21. Toda, F.; Miyamoto, H.; Tamashima, T.; Kondo, M.; Ohashi, Y., *J. Org. Chem.* **1999**, *64*, 2690.
22. Toda, F.; Miyamoto, H.; Inoue, M.; Yasaka, S.; Matijasic, I., *J. Org. Chem.* **2000**, *65*, 2728.
23. Tanaka, K.; Toda, F.; Mochizuki, E.; Yasui, N.; Kai, Y.; Miyahara, I.; Hirotsu, K., *Angew. Chem., Int. Ed. Engl.* **1999**, *38*, 3523.
24. Tanaka, K.; Fujiwara, T., *Org. Lett.* **2005**, *7*, 1501.
25. Toda, F.; Akai, H., *J. Org. Chem.* **1990**, *55*, 3446.
26. Toda, F.; Mori, K., *J. Chem. Soc., Chem. Commun.* **1989**, 1245.
27. Cuenca, A.; Medio-Simón, M.; Asensio Aguilar, G.; Weibel, D.; Beck, A. K.; Seebach, D., *Helv. Chim. Acta* **2000**, *83*, 3153.
28. (a) Belokon, Y. N.; Kochetkov, K. A.; Churkina, T. D.; Ikonnikov, N. S.; Chesnokov, A. A.; Larionov, O. V.; Parmar, V. S.; Kumar, R.; Kagan, H.B., *Tetrahedron: Asymmetry* **1998**, *9*, 851. (b) Belokon, Y. N.; Kochetkov, K. A.; Churkina, T. D.; Ikonnikov, N. S.; Chesnokov, A. A.; Larionov, O. V.; Kagan, H. V., *Russ. Chem. Bull.* **1999**, *48*, 917. (c) Belokon, Y. N.; Kochetkov, K. A.; Churkina, T. D.; Ikonnikov, N. S.; Chesnokov, A. A.; Larionov, O. V.; Singh, I.; Parmar, V. S.; Vyskocil, S.; Kagan, H. B., *J. Org. Chem.* **2000**, *65*, 7041.
29. (a) Weber, B.; Seebach, D., *Angew. Chem., Int. Ed. Engl.* **1992**, *31*, 84. (b) Weber, B.; Seebach, D., *Tetrahedron* **1994**, *50*, 6117.
30. Jászay, Z. M.; Németh, G.; Pham, T. S.; Petneházy, I.; Grön, A.; Töke, L., *Tetrahedron:Asymmetry* **2005**, *16*, 3837.
31. Bonache, M. A.; López, P.; Martín-Martinez, M.; Garcáa-López, M. T.; Cativiela, C.; González-Muñiz, R., *Tetrahedron* **2006**, *62*, 130.
32. See: 2,2-Dimethyl-$\alpha,\alpha,\alpha',\alpha'$-tetraphenyl-1,3-dioxolane-4,5-dimethanol atotitanium diisopropoxide in this encyclopedia.
33. (a) Ooi, T.; Takaya, K.; Miura, T.; Ichikawa, H.; Marouka, K., *Synlett* **2000**, 1133. (b) Ooi, T.; Miura, T.; Takaya, K.; Ichikawa, H.; Marouka, K., *Tetrahedron* **2001**, *57*, 867.
34. Schneider, C.; Hansch, M., *Synlett* **2003**, 837.
35. Sugimoto, K.; Aoyagi, S.; Kibayashi, C., *J. Org. Chem.* **1997**, *62*, 2322.
36. Greeves, N.; Pease, J. E.; Bowden, M. C.; Brown, S. M., *Tetrahedron Lett.* **1996**, *37*, 2675.
37. Greeves, N.; Pease, J. E., *Tetrahedron Lett.* **1996**, *37*, 5821.
38. Mikami, K.; Yoshida, A., *Tetrahedron* **2001**, *57*, 889.
39. Enders, D.; Tedeschi, L.; Bats, J. W., *Angew. Chem., Int. Ed.* **2000**, *39*, 4605.
40. Heldmann, D. K.; Seebach, D., *Helv. Chim. Acta* **1999**, *82*, 1096.
41. Yao, S.; Meng, J.-C.; Siuzdak, G.; Finn, M. G., *J. Org. Chem.* **2003**, *68*, 2540.
42. Blume, F.; Zemolka, S.; Fey, T.; Kranich, R.; Schmalz, H.-G., *Adv. Synth. Catal.* **2002**, *344*, 868.
43. Moteki, S. A.; Wu, D.; Chandra, K. L.; Reddy, D. S.; Takacs, J. M., *Org. Lett.* **2006**, *8*, 3097.
44. Yu, R. T.; Rovis, T., *J. Am. Chem. Soc.* **2006**, *128*, 12370.
45. Seebach, D.; Devaquet, E.; Ernst, A.; Hayakawa, M.; Kühnle, F. N. M.; Schweizer, W. B.; Weber, B., *Helv. Chim. Acta* **1995**, *78*, 1636.
46. Boele, M. D. K.; Kamer, P. C. J.; Lutz, M.; Spek, A. L.; de Vries, J. G.; van Leeuwen, P. W. N. M.; van Strijdonck, G. P. F., *Chem. Eur. J.* **2004**, *10*, 6232.
47. (a) Hilgraf, R.; Pfaltz, A., *Synlett* **1999**, 1814. (b) Hilgraf, R.; Pfaltz, A., *Adv. Synth. Catal.* **2005**, *347*, 61.
48. Keller, E.; Maurer, J.; Naasz, R.; Schader, T.; Meetsma, A.; Feringa, B. L., *Tetrahedron: Asymmetry* **1998**, *9*, 2409.
49. Palacios, F.; Vicario, J., *Org. Lett.* **2006**, *8*, 5405.
50. Müller, P.; Nury, P., *Org. Lett.* **1999**, *1*, 439.

51. Bertozzi, F.; Crotti, P.; Macchia, F.; Pineschi, M.; Arnold, A.; Feringa, B.L., *Tetrahedron Lett.* **1999**, *40*, 4893.
52. Kitamura, K.; Shimada, N.; Stewart, C.; Atesin, A. C.; Atesin, T. A.; Tius, M. A., *Angew. Chem., Int. Ed.* **2015**, *54*, 6288.
53. Shirakura, M.; Suginome, M., *Angew. Chem., Int. Ed.* **2010**, *49*, 3827.
54. Teller, H.; Flügge, S.; Goddard, R.; Fürstner, A., *Angew. Chem., Int. Ed.* **2010**, *49*, 1949.
55. Albicker, M. R.; Cramer, N., *Angew. Chem., Int. Ed.* **2009**, *48*, 9139.
56. Falk, A.; Göderz, A.-L.; Schmalz, H.-G., *Angew. Chem., Int. Ed.* **2013**, *52*, 1576.
57. Smith, S. M.; Takacs, J. M., *J. Am. Chem. Soc.* **2010**, *132*, 1740.
58. Gao, X.; Han, J.; Wang, L., *Org. Lett.* **2015**, *17*, 4596.
59. Zou, L.; Wang, B.; Mu, H.; Zhang, H.; Song, Y.; Qu, J., *Org. Lett.* **2013**, *15*, 3106.

(2*S*,5*S*)-2-(Bis(4-methoxyphenyl) ((trimethylsilyl)oxy)methyl)-5-(4-methoxy-3,5-dimethylphenyl)pyrrolidine

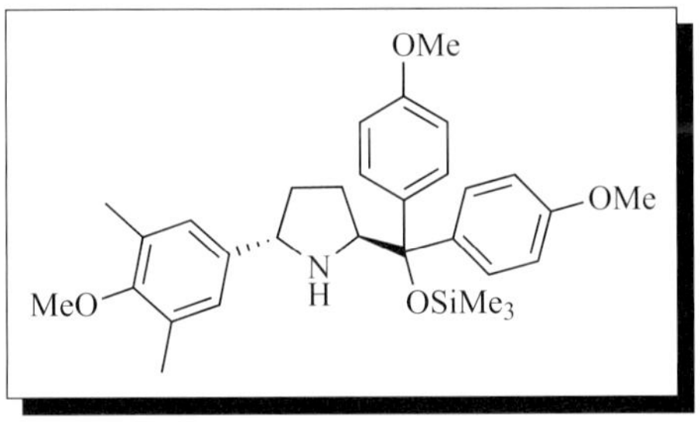

$C_{31}H_{41}NO_4Si$ (MW 519.74)

InChI = 1S/C31H41NO4Si/c1-21-19-23(20-22(2)30(21)35-5)28-17-18-29(32-28)31(36-37(6,7)8,24-9-13-26(33-3)14-10-24)25-11-15-27(34-4)16-12-25/h9-16,19-20,28-29,32H,17-18H2,1-8H3/t28-,29-/m0/s1

InChIKey = QKGUAHHIDNAIHI-VMPREFPWSA-N

(iminium catalyst used for Mukaiyama–Michael addition of enolsilanes to acrolein)

Physical Data: thick oil or glass.

Solubility: sol alcohol, ether, and most organic solvents, insoluble in H_2O.

Purification: purified by flash chromatography (80:20 hexanes:EtOAc eluent).

Handling, Storage, and Precautions: hydrolytic loss of silyl group may occur under highly basic or acidic conditions. Stored at 2–8 °C or at −20 °C for prolonged periods of time.

Iminium Catalysis. This catalyst, in a combination with 4-nitrobenzoic acid, can provide excellent enantioselectivities in Mukaiyama–Michael reactions between acrolein and thioester-derived enolsilanes such as ((1-(*tert*-butylthio)prop-1-en-1-yl)oxy)trimethylsilane, providing α-substituted chiral thioesters bearing a pendant aldehyde group (eq 1).[1] In contrast with usual iminium catalysts,[2] this catalyst controls the face selectivity on the nucleophile (enolsilane), not on the iminium electrophile. The aldehyde group can be protected in situ as an acetal with, for example, 2,2-dimethyl-1,3-propanediol.

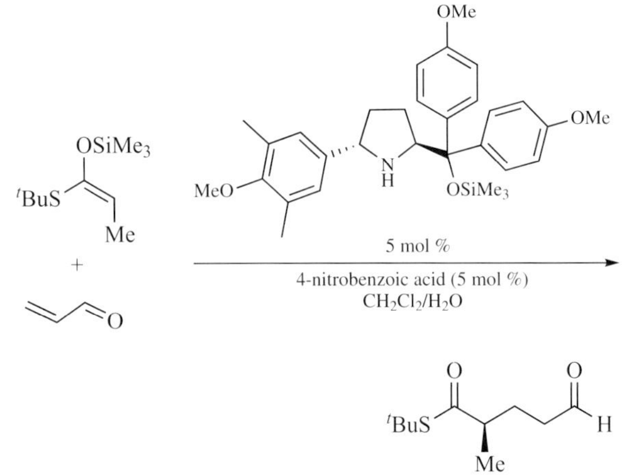

(1)

76% yield, 95:5 er

The products can be useful in natural product synthesis, as exemplified by the synthesis of the C_4–C_{13} segment of bistramide A[3] (eq 2).[1]

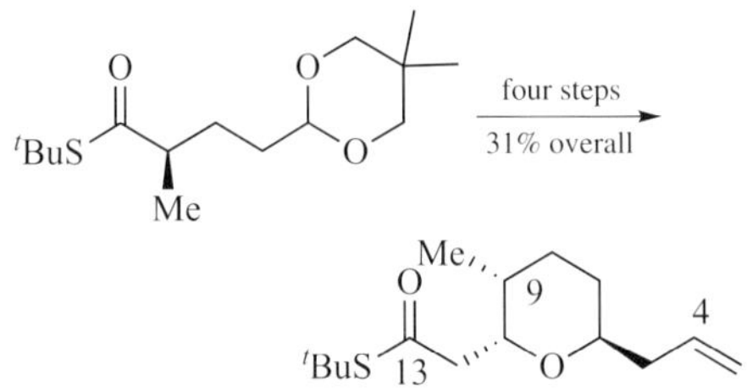

(2)

C_4–C_{13} segment of (−)-bistramide A

The modular structure of the catalyst allows steric and electronic optimization of this catalyst family, suggesting that it may find use in other transformations as well where the classical iminium catalysts fail to elicit useful selectivity.

---

1. Claraz, A.; Sahoo, G.; Berta, D.; Madarász, Á.; Pápai, I.; Pihko, P. M., *Angew. Chem., Int. Ed.* **2016**, *55*, 669.
2. Claraz, A.; Siitonen, J. H.; Pihko, P. M. In *Lewis Base Catalysis in Organic Synthesis*; Vedejs, E.; Denmark, S., Eds.; Wiley-VCH Verlag GmbH: Weinheim, Germany, 2016, Vol. 2, Chapter 16.
3. Statsuk, A. V.; Liu, D.; Kozmin, S. A., *J. Am. Chem. Soc.* **2004**, *126*, 9546.

Petri M. Pihko
University of Jyväskylä, Jyvaskyla, Finland

(3*R*,5*R*,8*R*)- 3,8-Bis(1-methylethyl)-2,2,7,7-tetrakis(4-methylphenyl)-1,4,6,9-tetraaza-5λ5-phosphaspiro[4,4]non-5-ene, Hydrochloride salt (1:1)

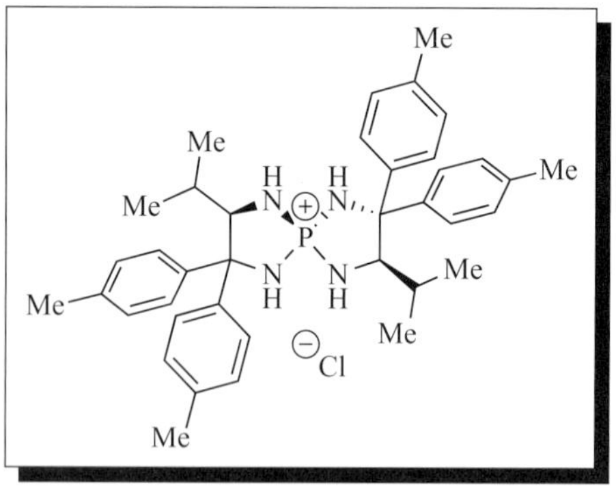

[1258886-30-4]
[1259007-52-7] $C_{38}H_{48}ClN_4P$ (MW 627.2413)

InChI = 1S/C38H48N4P.ClH/c1-25(2)35-37(31-17-9-27(5)10-18-31,32-19-11-28(6)12-20-32)41-43(39-35)40-36(26(3)4)38(42-43,33-21-13-29(7)14-22-33)34-23-15-30(8)16-24-34;/h9-26,35-36,39-42H,1-8H3;1H/q+1;/p-1/t35-,36-,43?;/m1./s1

InChIKey = FOSSSRXVDQXUGK-XQBLMWPJSA-M

(reagent used as a precatalyst for an asymmetric Henry reaction)

Physical Data: mp 198–202 °C.
Solubility: soluble in MeOH, EtOH, CH_2Cl_2, $CHCl_3$, acetone, ethyl acetate, and DMF. Insoluble in hexane.
Form Supplied in: white solid.
Analysis of Reagent Purity: NMR analysis (^{1}H, ^{31}P).
Preparative Method:[1] *N*-Boc-D-valine methyl ester is reacted with excess 4-MeC_6H_4MgBr in THF at 0 °C. Subsequent treatment of the crude product with 1 M HCl in methanol at 50 °C furnished hydrochloride salt of (*R*)-2-amino-3-methyl-1,1-di-*p*-tolylbutan-1-ol. This salt is treated with sodium azide in trifluoroacetic acid at 0 °C, furnishing the 1,2-amino azide that is converted to (*R*)-3-methyl-1,1-di-*p*-tolylbutane-1,2-diamine by reduction of the azide moiety with zinc powder and ammonium formate in methanol. To a solution of the 1,2-diamine and triethylamine in refluxing toluene was added a solution of PCl_5 in toluene, giving the *P*-spiro tetraaminophosphonium chloride as a mixture of (3*R*,5*R*,8*R*)- and (3*R*,5*S*,8*R*)-isomers. The major (3*R*,5*R*,8*R*)-isomer is isolated by silica gel column chromatography with $CHCl_3$/MeOH as eluent followed by recrystallization from EtOH/hexane or acetone/hexane solvent system.
Handling, Storage, and Precautions: Stable at ambient conditions under air as solid. May decompose on the treatment with an aqueous solution of a strong base.

Henry Reaction. The *P*-spiro chiral tetraaminophosphonium chloride has been used as a precursor of a chiral organic base catalyst for a highly *anti*- and enantioselective Henry reaction of dodec-2-ynal with nitroethane (eq 1).[2] Treatment of the salt with potassium *tert*-butoxide in the presence of nitroethane in THF/DMF (95:5) at −78 °C cleanly generated the corresponding triaminoiminophosphorane. Subsequent slow addition of dodec-2-ynal to the catalyst solution initiates the Henry reaction to give the product in 94% yield with virtually complete stereoselectivity (*anti*/*syn* = >20:1, 98% ee for *anti*-isomer). The stereochemically homogeneous Henry adduct is quantitatively converted into a potential reverse transcriptase inhibitor, (+)-xestoaminol C,[3] via simple hydrogenation of the nitro and triple bond moieties with palladium on charcoal. Consequently, this protocol is attractive as an alternative method of *anti*-selective asymmetric Henry reaction with simple aliphatic aldehydes, which still remains an important problem to be solved (eq 1).[1a,4]

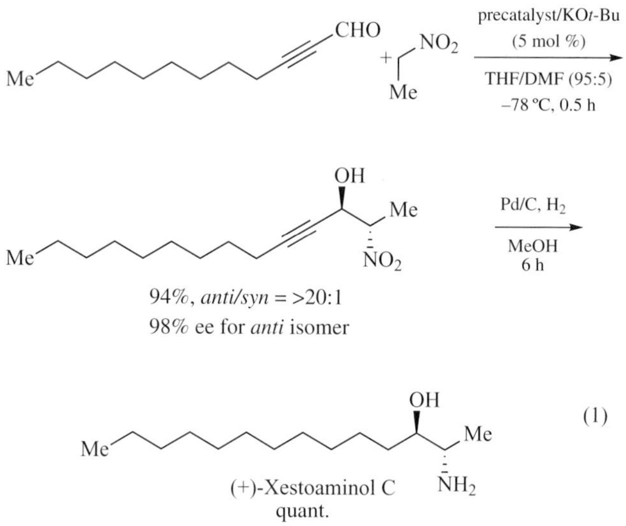

(1)

The enantiomer of the salt and related compounds, which are synthesized from *N*-Boc-L-valine methyl ester, have been utilized as precatalysts for Henry reactions of aldehydes.[1a,2] and hydrophosphonylations of aldehydes[5a] and ynones.[5b]

---

1. (a) Uraguchi, D.; Sakaki, S.; Ooi, T., *J. Am. Chem. Soc.* **2007**, *129*, 12392. (b) Uraguchi, D.; Sakaki, S.; Ueki, Y.; Ito, T.; Ooi, T., *Heterocycles* **2008**, *76*, 1081.
2. Uraguchi, D.; Nakamura, S.; Ooi, T., *Angew. Chem., Int. Ed.* **2010**, *49*, 7562.
3. (a) Jiménez, C.; Crews, P., *J. Nat. Prod.* **1990**, *53*, 978. (b) Garrido, L.; Zubía, E.; Ortega, M. J.; Naranjo, S.; Salvá, J., *Tetrahedron* **2001**, *57*, 4579.
4. (a) Nitabaru, T.; Kumagai, N.; Shibasaki, M., *Tetrahedron Lett.* **2008**, *49*, 272. (b) Handa, S.; Nagawa, K.; Sohtome, Y.; Matsunaga, S.; Shibasaki, M., *Angew. Chem., Int. Ed.* **2008**, *47*, 3230. (c) Blay, G.; Domingo, L. R.; Hernández-Olmos, V.; Pedro, J. R., *Chem. Eur. J.* **2008**, *14*, 4725. (d) Nitabaru, T.; Nojiri, A.; Kobayashi, M.; Kumagai, N.; Shibasaki, M., *J. Am. Chem. Soc.* **2009**, *131*, 13860. (e) Lang, K.; Park, J.; Hong, S., *Angew. Chem., Int. Ed.* **2012**, *51*, 1620.
5. (a) Uraguchi, D.; Ito, T.; Ooi, T., *J. Am. Chem. Soc.* **2009**, *131*, 3836. (b) Uraguchi, D.; Ito, T.; Nakamura, S.; Ooi, T., *Chem. Sci.* **2010**, *1*, 488.

Daisuke Uraguchi & Takashi Ooi
Nagoya University, Nagoya, Japan

1,2-Bis(phenylsulfonyl)ethylene[1]

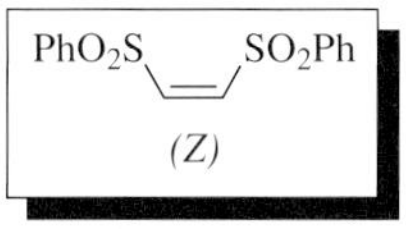

(Z)
[963-15-5] $C_{14}H_{12}O_4S_2$ (MW 308.37)
InChI = 1S/C14H12O4S2/c15-19(16,13-7-3-1-4-8-13)11-12-20(17,18)14-9-5-2-6-10-14/h1-12H/b12-11-
InChIKey = YGBXMKGCEHIWMO-QXMHVHEDSA-N
(E) *[963-16-6]*
InChI = 1S/C14H12O4S2/c15-19(16,13-7-3-1-4-8-13)11-12-20(17,18)14-9-5-2-6-10-14/h1-12H/b12-11+
InChIKey = YGBXMKGCEHIWMO-VAWYXSNFSA-N

(powerful dienophiles and synthetic equivalents of acetylene and ethylene in Diels–Alder cycloadditions;[2] react as Michael acceptors with substitution of a phenylsulfonyl group[3])

Alternate Name: BPSE.
Physical Data: (Z) mp 100–101 °C; (E) mp 219.5 °C.
Solubility: the (Z)-isomer is more soluble than the (E)-isomer in most solvents. In Diels–Alder cycloadditions conducted in dichloromethane, the (Z)-isomer affords quite insoluble adducts; conversely, the (E)-compound leads to more soluble products.
Form Supplied in: colorless solids; commercially available.
Analysis of Reagent Purity: purity with respect to one another can be checked by ^{1}H NMR. The vinyl protons of the (Z)-isomer appear at 6.82 ppm and those of the (E)-isomer at 7.37 ppm in $CDCl_3$.
Preparative Methods: (Z)-1,2-bis(phenylsulfonyl)ethylene is prepared by oxidation of the corresponding thioether,[2] which is prepared from (Z)-1,2-dichloroethylene[4] or, for large quantities, from 1,1-dichloroethylene.[4b] A mixture of the (Z)- and (E)-1,2-dichloroethylenes can also be used, but the (E)-isomer is unreactive under standard conditions.[4b] Under more drastic reaction conditions (DMF, 70–90 °C), (E)-1,2-dichloroethylene also affords the substitution product [57% (E), 14% (Z)].[4c] The preparation of the (E)-isomer is accomplished by sunlight–iodine catalyzed isomerization of the (Z)-isomer.[2]
Handling, Storage, and Precautions: the (Z)-isomer should be stored protected from direct light because it isomerizes to the (E)-form.

Original Commentary

Ottorino De Lucchi
Università di Venezia, Venezia, Italy
Lucia Pasquato
CNR-CMRO, Padova, Padova, Italy

Cycloadditions. 1,2-Bis(phenylsulfonyl)ethylenes [(Z)- and (E)-BPSE] are most commonly used as synthetic equivalents of acetylene in cycloaddition reactions (eq 1).[2,5] A number of dienes have been prepared via this route which are otherwise not readily accessible.[6]

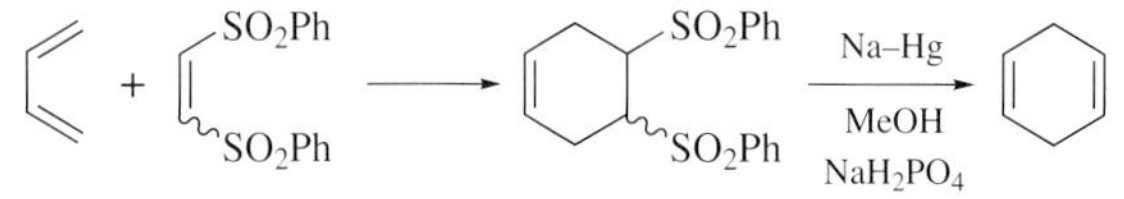

(1)

The preparation of both the *syn*- and the *anti*-sesquinorbornatrienes was possible because of the different stereochemistry by which the (Z)- and (E)-isomers add to 8,9-dehydroisodicyclopentadiene (eq 2).[7] Further studies on these molecules have been reported, including the preparation of deutero derivatives via the deuterated (Z)-BPSE.[8] Use of (E)-BPSE in the preparation of a natural product is described in eq 3.[9]

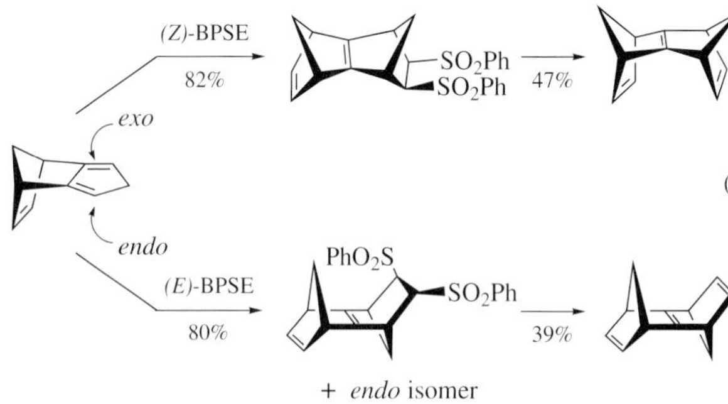

(2)

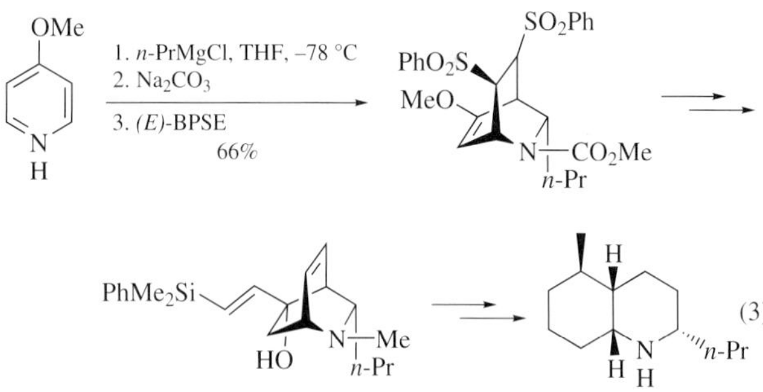

(3)

The choice of the solvent for cycloadditions is often crucial. While both reagents are not reactive towards furan in refluxing benzene,[10] they react smoothly at rt in quantitative yields in dichloromethane.[2,11]

Generally, the (E)-isomer is more reactive in cycloadditions than the (Z)-isomer. For example, the reaction of (E)-BPSE with cyclopentadiene in dioxane is 100 times faster than that of the (Z)-isomer.[12] A few dienes (1,3-cycloheptadiene and cyclooctatetraene are examples) react only with the former. With cyclopentadiene in dioxane, (E)-BPSE is twice as reactive as Maleic Anhydride,[12] but cases have been reported where (E)-BPSE failed to react while maleic anhydride did react. Styrene, homobarrelene, and other polycyclic dienes are examples.[13] High pressure (10–12 kbar) can be used to increase reactivity.[14] Examples are the preparations of derivatives of *syn*-sesquinorbornatriene[14a] and of the $C_{4}v$ [4]peristylane tetraketone (eq 4).[14c]

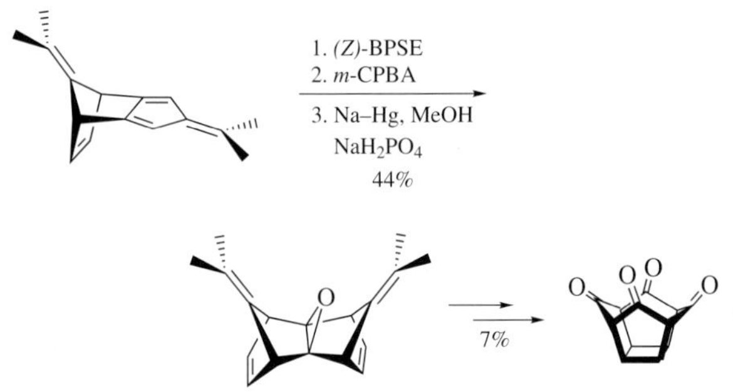

(4)

The desulfonylation of cycloadducts can be performed with Sodium Amalgam (1.5–6%) in methanol buffered with suspended sodium dihydrogen phosphate.[2] Because the reaction is heterogeneous, sonication has been proved beneficial.[13a] Samarium(II) Iodide in HMPA has also been used.[15] Other reducing agents and reaction conditions, such as Magnesium-Methanol,[16] afford mixtures of saturated and unsaturated hydrocarbons. Lithium or Sodium metal in liquid ammonia effectively gives solely the saturated adduct, making the title reagent equivalent to ethylene in cycloaddition reactions.[17]

When treated with base, the Diels–Alder adducts give vinyl sulfones. In one example, the vinyl sulfone was further reacted with nucleophiles in a synthesis of ibogamine (eq 5).[18]

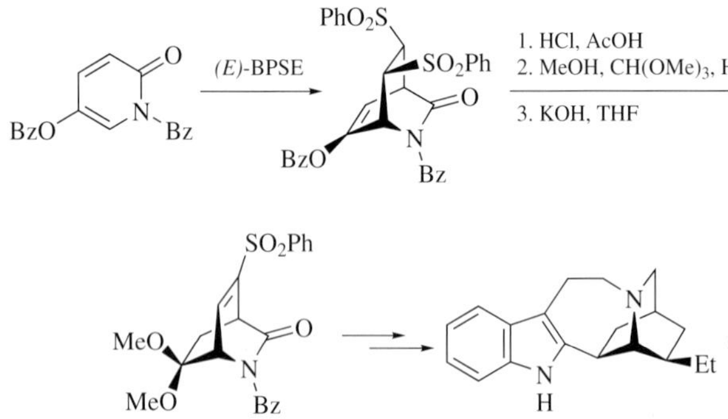

(5)

Ibogamine
(overall 14%)

Substitution of one vinyl hydrogen with chlorine gives reagents that can be considered equivalents of the unstable bis(arylsulfonyl) alkynes (eq 6).[19]

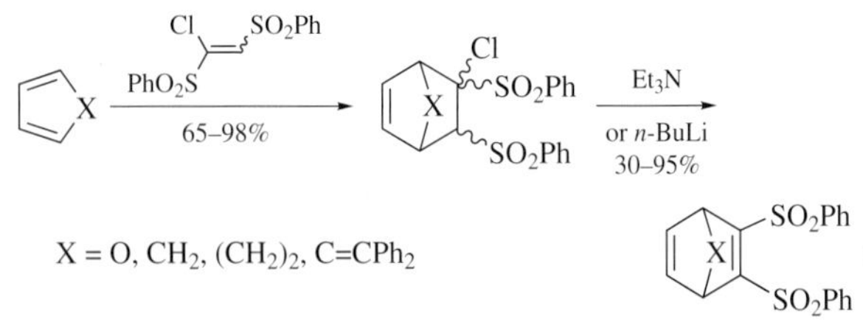

(6)

The parent and substituted bis(arylsulfonyl)alkenes, such as the one prepared in eq 6, are good Michael acceptors[20] and react with heteroatomic nucleophiles and Grignard reagents.[3,21] The products correspond to the formal substitution of one of the phenylsulfonyl groups. In eq 7, the product of the Michael addition cyclizes to a dihydrofuran upon deprotonation with Lithium Diisopropylamide.[21a]

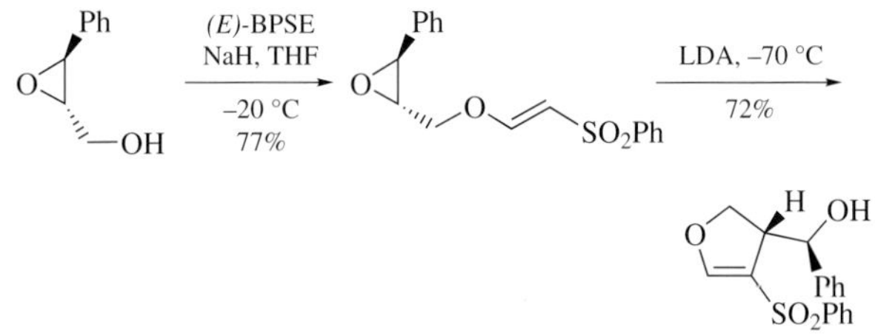

(7)

Both isomeric reagents are also reactive dipolarophiles.[22] For example, they react with azomethine ylides,[22a,b] nitrones,[22c] and diazoalkanes (eq 8).[22d]

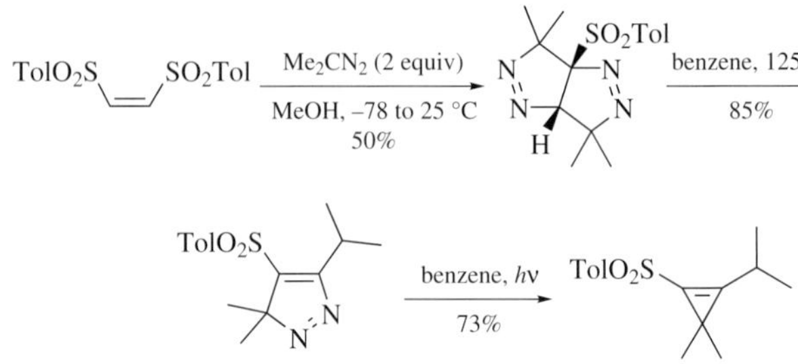

(8)

C_2-Symmetrical chiral versions of these reagents (**1**) and (**2**), which maintain the chemical properties associated with the achiral reagents, are available.[23]

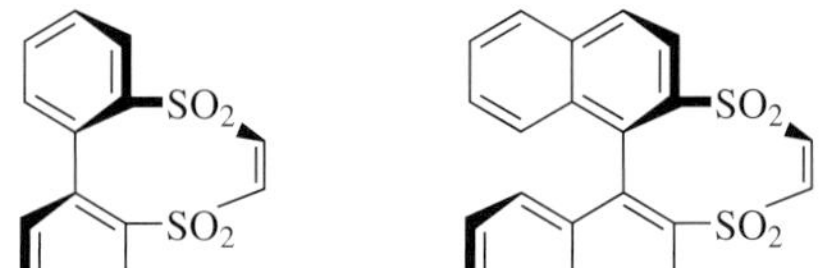

(**1**) (**2**)

These reagents are stable towards racemization and are readily available from the respective dibenzo- or dinaphtho-2,2′-dithiols. In cycloadditions with nonsymmetrical dienes they afford a single diastereoisomer in most cases.

First Update

Patrick Rollin & Arnaud Tatibouët
Université d'Orléans, Orléans, France

Cycloadditions. The use of 1,2-bis(phenylsulfonyl)ethylenes [(*Z*)- and (*E*)-BPSE] as synthetic equivalents of acetylene in cycloaddition reactions was further developed, with applications to either simple[24a] or more complex[24b–e] dienic templates providing sophisticated polycyclic systems. Cycloaddition of BPSE—or closely related reagents—to heterocyclic dienes was also performed, using furanic[25] or pyrrolic systems[26] as well as vinyl pyridones,[27] vinyl indoles,[28] and indolo-pyrones.[29] Finally, some dipole cycloadditions were explored by Bhaskar Reddy et al. using diazomethane,[30] nitrile oxides, and nitrilimines.[31]

Michael Additions. The strong electrophilic character of BPSE has been exploited in Michael-type additions, generally followed by an elimination step (eq 9).

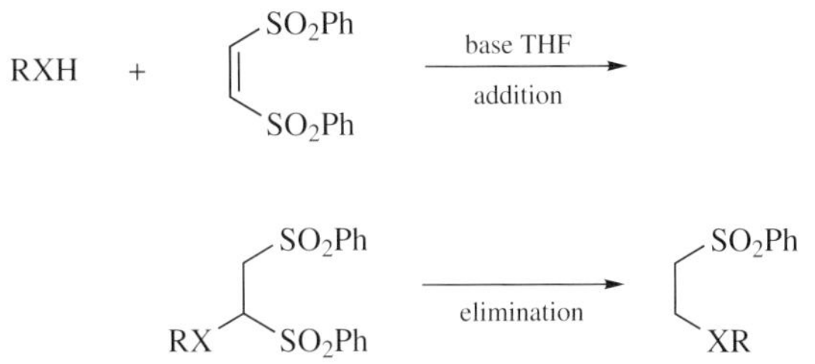

(9)

With a view to constructing new acceptors for radical cyclizations, Evans et al. have developed a controlled process to β-alkoxyvinyl sulfones—also called "vinylogous sulfonates"—involving addition to BPSE of an alkoxide anion generated by Lithium or Potassium Hexamethyldisilyl Azide in **THF** at low temperature and subsequent expulsion of a phenylsulfonate anion (eq 10).[32] Application of such a radical cyclization was considered in the course of a synthetic route to the antitumor agent mucocin[33] and in a synthesis of garsubellin A, a potent inducer of choline acetyltransferase.[34]

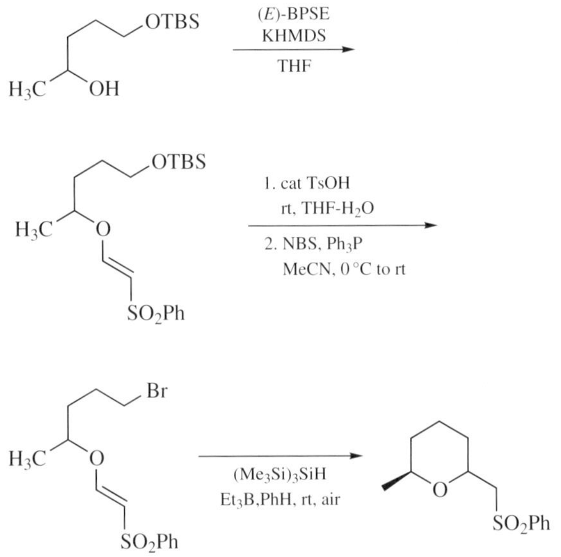

(10)

The above vinylogous sulfonates also function as leaving groups in palladium-catalyzed allylic alkylations and rearrangements (eq 11).[35]

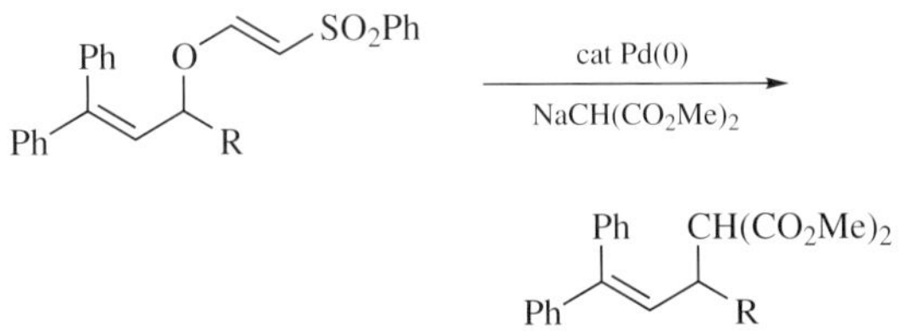

(11)

Selective Tributyltin Hydride reduction of BPSE was demonstrated by De Lucchi as an alternative preparation of Phenyl Vinylsulfone.[36] Further reduction of the previously obtained β-alkoxyvinyl sulfones using single-electron transfer reagents—Sodium Amalgam in particular—smoothly afforded the corresponding vinyl ethers: this process was applied by Rollin and co-workers to both simple alcohols[37] and many diverse carbohydrate derivatives.[38] Sugar-based vinyl ethers were investigated by the same group as templates for stereoselective dipolar cycloadditions (eq 12).[39]

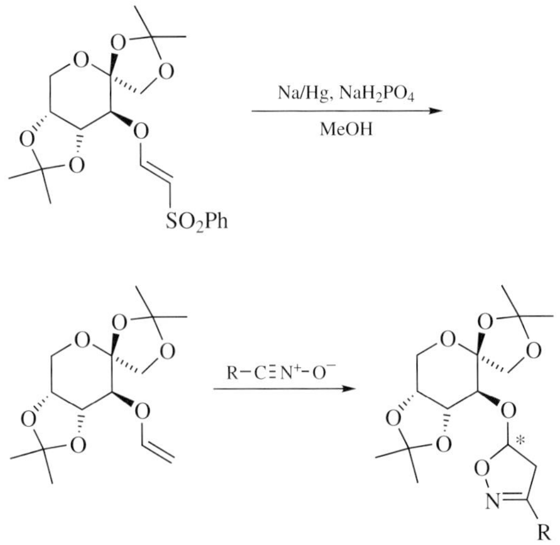

(12)

Some nitrogen nucleophiles were also shown to afford *N*-phenylsulfonylvinyl) derivatives, precursors to the *N*-vinylated counterparts: such a reaction was first observed serendipitously on a γ-lactam by Knapp[40] and later applied, with triethylamine catalysis, to simple or sugar-derived 1,3-oxazolidine-2-thiones (eq 13).[41]

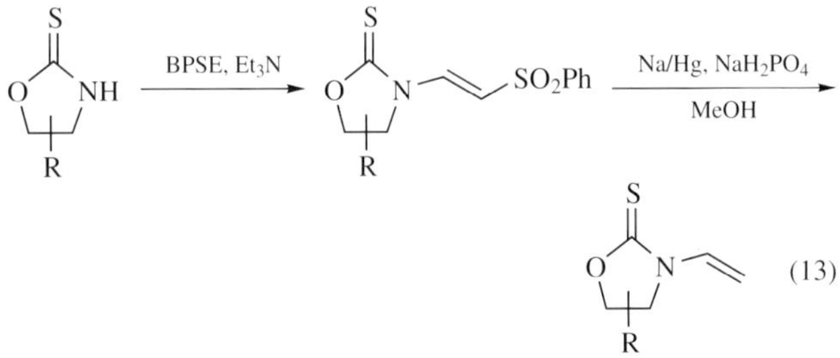

(13)

The case of thiols was mentioned[25b] then thoroughly examined: under triethylamine catalysis, it was shown that more complex reactions occurred (eq 14).[42]

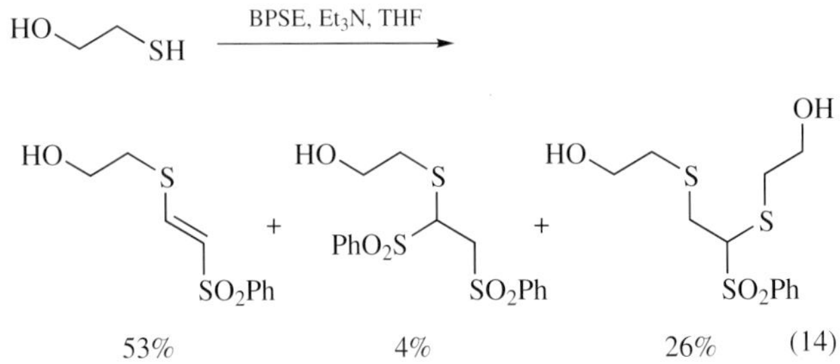

(14)

Addition of a sulfone-stabilized carbanion to BPSE constituted the initial step in the synthesis of highly substituted naphthalenes (eq 15).[43]

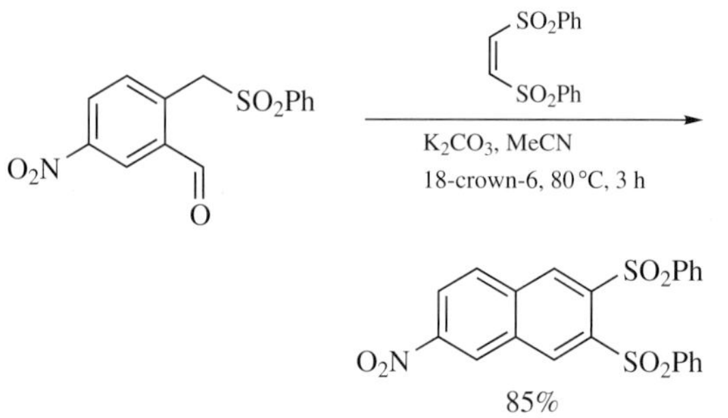

(15)

Radical monoaddition resulting in a vinylsulfonyl group transfer to alkyl radicals generated from alkyl allyl sulfones was demonstrated by Kim (eq 16).[44]

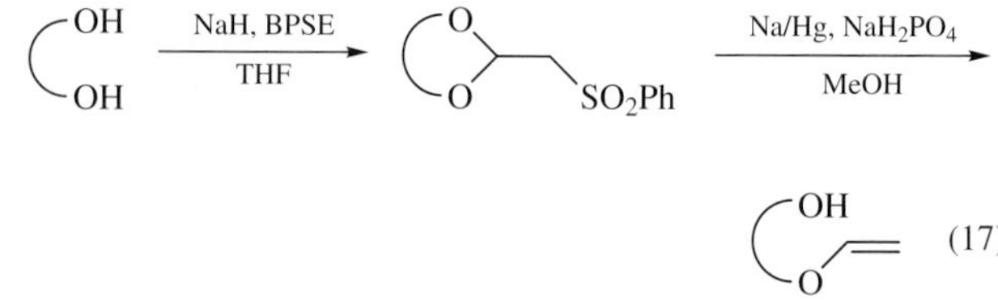

(17)

PSE acetals—which are prepared in basic media—show very unusual properties in that they are strongly resistant to acidic conditions but can in contrast undergo ring-opening reactions under basic or reductive conditions (eq 18). A general selective synthesis of ω-hydroxyethenyl ethers was devised through sodium amalgam reductive opening of these novel protecting groups (eq 17).[47] PSE acetals can be applied as a protecting group particularly in carbohydrate chemistry: for example, regioselective de-*O*-benzylation using Triisobutylaluminum (TIBAL) was successfully performed on monosaccharidic glycosides (eq 18).[48]

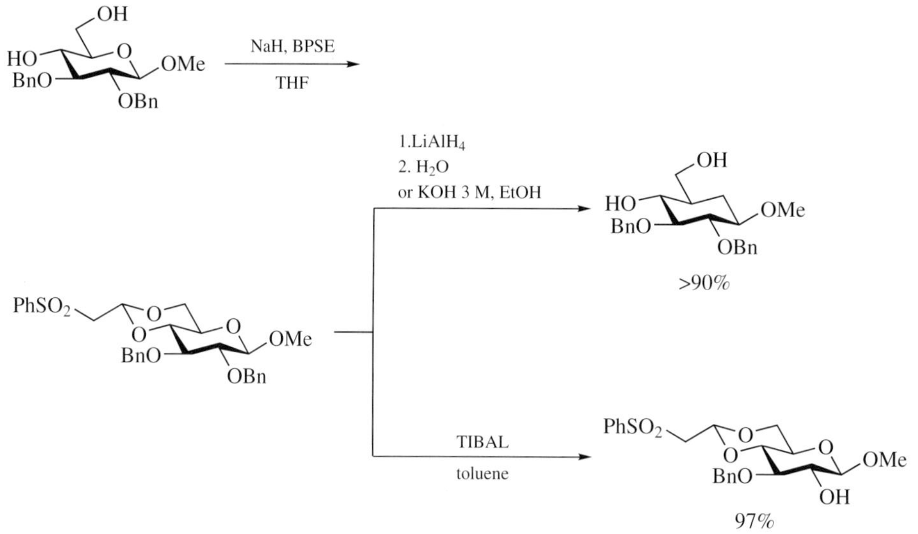

(18)

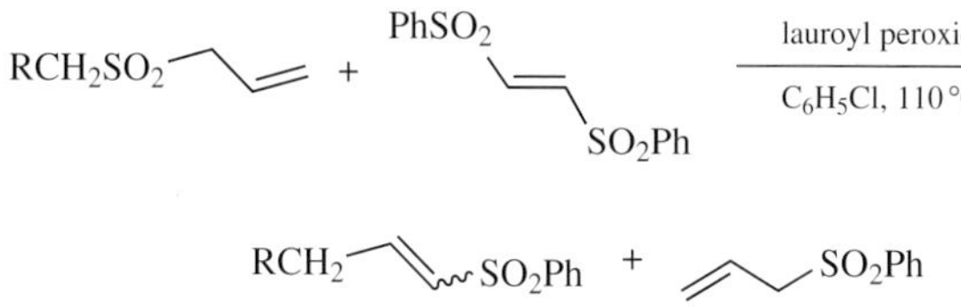

(16)

Michael Double Additions. Following a double Michael addition process on BPSE, diols were readily converted into a novel family of cyclic acetals: 2-phenylsulfonylethylidene acetals (PSE acetals) (eq 17). Pioneering results from De Lucchi's group[45] were extended to the carbohydrate field, in which both 1,3-dioxolanes and 1,3-dioxanes could be prepared in good yield and in most cases with high regio- and stereoselectivity.[46] Syntheses of thio-analog (PSE thioacetals) were similarly investigated from simple substrates, then adapted to the case of carbohydrate–derived 1,3-oxathianes.[42]

Second Update

Fedor Romanov-Michailidis
Colorado State University, Fort Collins, CO, USA

Cycloadditions

(*E*)- and (*Z*)-BPSE are commonly used as dipolarophiles in [2+3] cycloaddition reactions. A number of 1,3-dipoles were employed, azomethine ylides being among the most popular, leading to formation of five-membered nitrogen heterocycles.

An ingenious method for making oligopyrroles was disclosed by Carretero and coworkers.[49,50] The 1,3-dipolar cycloaddition of both (*E*)- and (*Z*)-BPSE with azomethine ylides derived from α-imino esters, followed by DBU-promoted elimination of the sulfonyl groups, afforded 2,5-disubstituted pyrroles in good to high yields (27–97%) (eq 19). An iterative protocol for making α, α'-linked oligopyrroles resulted when converting the ester moiety of the pyrrole unit into a new α-imino ester functionality and subjecting it to a subsequent 1,3-dipolar cycloaddition.

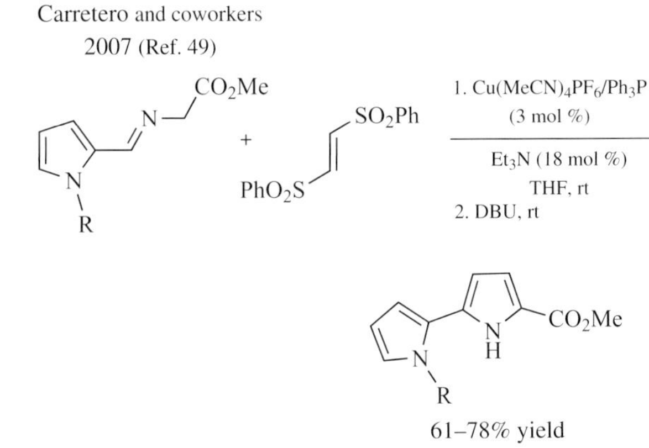

(19)

The same research group described a practical approach to catalytic, asymmetric synthesis of 3-pyrrolidines by means of an enantioselective Cu-mediated 1,3-dipolar cycloaddition of azomethine ylides to (*E*)-BPSE, followed by reductive desulfonylation (eq 20).[51] The products are obtained in high yields (65–93%) and with excellent enantioselectivities (26–98% ee). The synthetic utility of this protocol was demonstrated by performing enantioselective synthesis of biologically active trihydroxylated pyrrolidines.

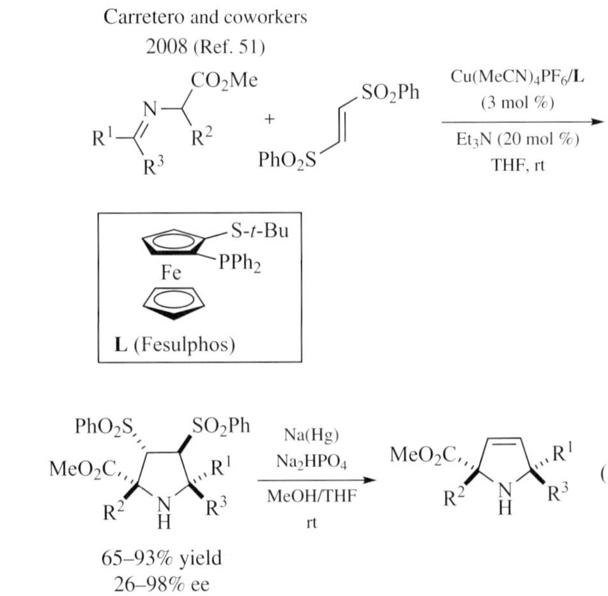

(20)

An extension of the methodology by Carretero and coworkers dealt with a highly enantioselective procedure for the synthesis of (-heteroarylpyrrolidines.[52] As shown in eq 21, a Cu/Walphos-catalyzed [2+3] cycloaddition of heteroarylsilylimines with (*Z*)-BPSE furnished heterocyclic products in high yields (72%) and with exquisite enantioselectivities (>99% ee). The sulfonyl groups were easily removed under reductive conditions.

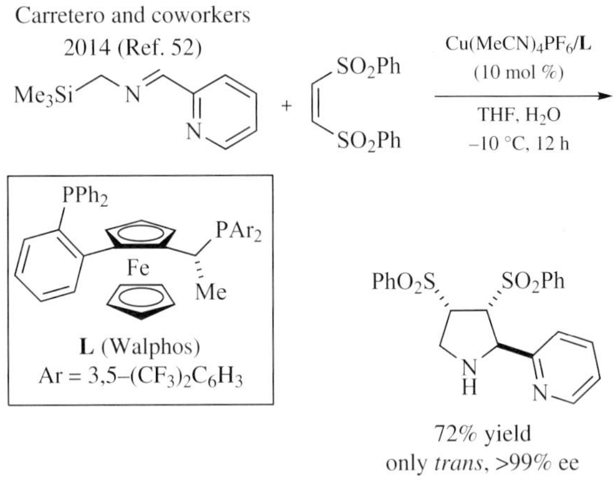

(21)

Along similar lines, Guo and coworkers developed a phosphine-catalyzed [2+3] cycloaddition of azomethine imines with (*E*)- and (*Z*)-BPSE (eq 22), providing dinitrogen-fused heterocyclic compounds in high yields (69–93%).[53] Moreover, the two phenylsulfonyl groups on the heterocyclic products could be conveniently removed or transformed to other functional groups.

An interesting Diels–Alder [4+2] cycloaddition between (*E*)-BPSE and in situ generated isobenzofuran and *N*-Boc-isoindole was reported by Rincon and Plumet (eq 23).[54] The reaction leads to the formation of useful bicyclic structures in moderate to good yields (41–51%).

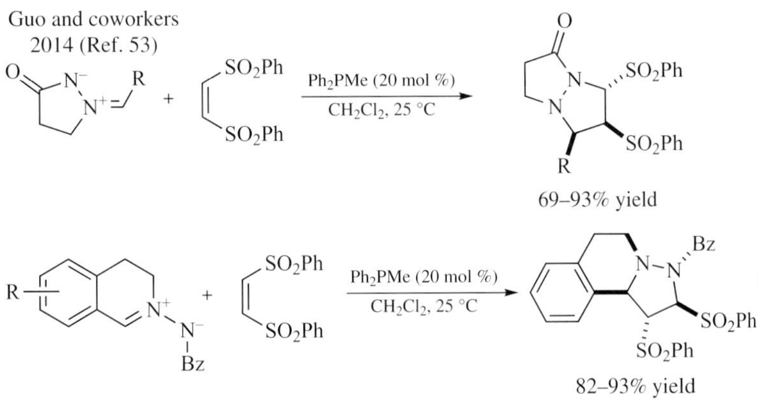

(22)

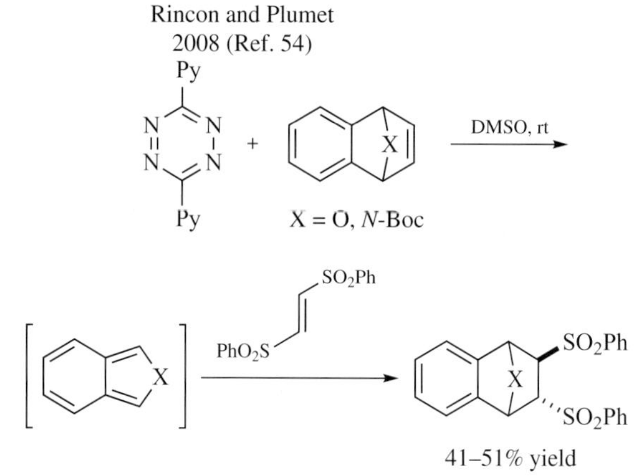

(23)

Sulfonyl Migrations

Brønsted base-catalyzed reactions that lead to intramolecular [1,2]-migration of the sulfonyl group in (*E*)-BPSE have been recently discovered.

An unprecedented rearrangement of (*Z*)-BPSE was reported by Quintard and Alexakis to take place in the course of organocatalytic additions to these Michael acceptors.[55,56] Enamines generated in situ from aldehydes and ketones add smoothly to (*Z*)-BPSE with concomitant [1,2]-migration of the sulfonyl group (eq 24). The reactions occur in high yields (49–89%) and with high enantioselectivities (64–87% ee) when the chiral aminal–pyrrolidine catalyst **L** is employed. (*E*)-BPSE afforded mixtures of rearranged and nonrearranged Michael adducts. After reductive desulfonylation, the method constitutes a practical, indirect α-alkylation of carbonyl compounds.

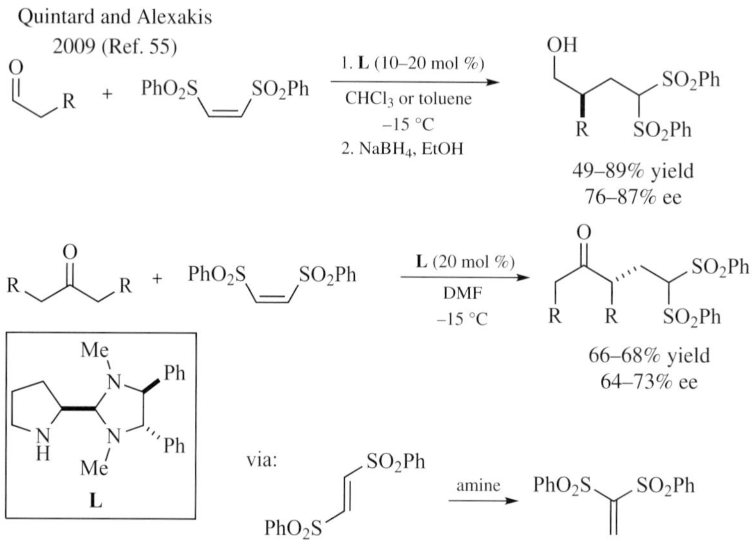

(24)

In the field of NHC organocatalysis, Scheidt and coworkers discovered that triazolium-derived carbenes promote rearrangement of 1,1-bis(arylsulfonyl)ethylenes to (*E*)-BPSE, which were subsequently reacted with nitrones through a stereoselective [2+3] cycloaddition (eq 25).[57] Heavily substituted isoxazolines were obtained in high yields (61–94%). This reaction is unusual in that the substrates are vinylsulfones and not the standard NHC reaction partners such as α,β-unsaturated carbonyl compounds, ketones, or aldehydes.

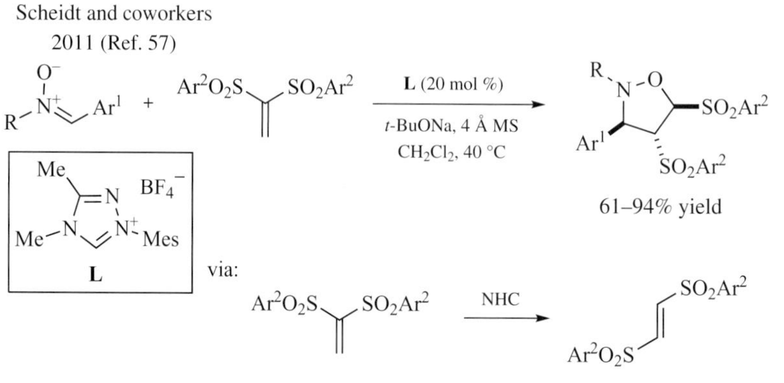

(25)

Free-radical Additions

A series of multicomponent free-radical reactions involving (*E*)-BPSE were recently reported by Landais and other researchers. These reactions represent valuable synthetic tools as they lead to functionalization of unreactive sp^3 C–H bonds and to formation of two new carbon–carbon bonds in a single step.

Landais and coworkers reported a three-component carboalkynylation/alkenylation process between (*E*)-BPSE, olefins, and bromides or xanthates (eq 26).[58] Free-radical addition across the olefin double bond gives rise to the formation of two new carbon–carbon bonds in high yields (53–96%).

A similar three-component reaction involving enamides, xanthates, and (*E*)-BPSE was reported by the same research group.[59] The reaction proceeds through the addition of free-radical species derived from a xanthate and a vinylsulfone across the olefinic bond of enamides and ene-carbamates, giving rise to cyclic or acyclic nitrogen-containing products with concomitant formation of two new carbon–carbon bonds (eq 27). Good yields are observed (10–88%) and the products obtained represent valuable intermediates for organic synthesis.

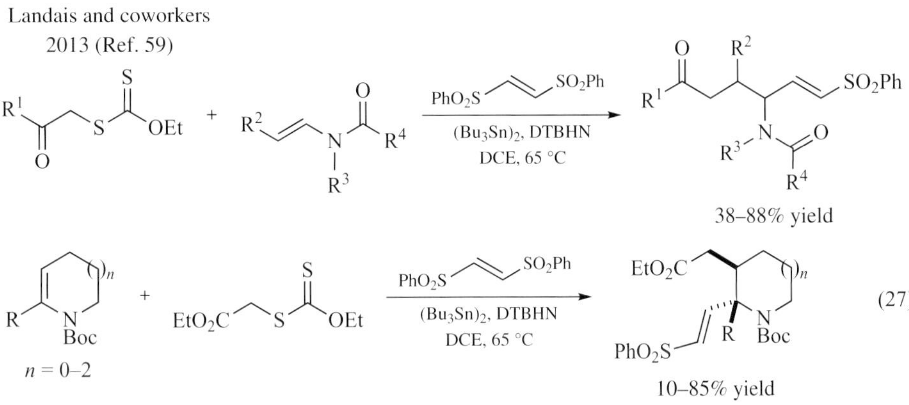

(27)

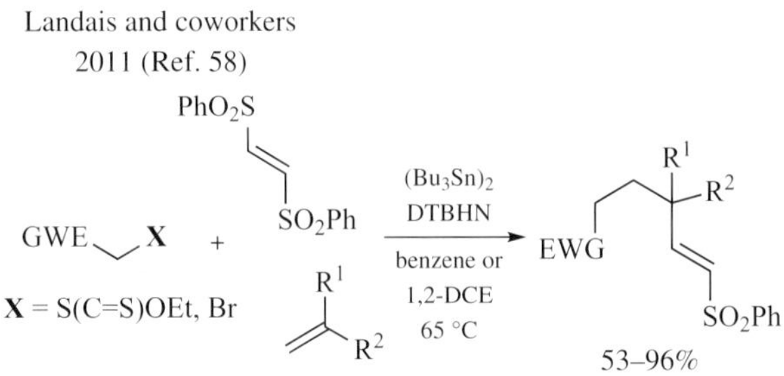

(26)

Inoue and coworkers developed a mild and metal-free alkenylation of sp^3 C–H bonds using benzophenone and (*E*)-BPSE under photoirradiation conditions (eq 28).[60] The reaction occurs with predictable chemoselectivity toward the most electron-rich sp^3 C–H bond (X = nitrogen > oxygen > aliphatic carbon) and furnishes products in good yields (27–99%). Furthermore, the products were converted in a single step to prenyl and pyrrole derivatives, which are substructures of numerous bioactive molecules.

In a related reaction, Renaud and coworkers demonstrated that hydroboration of alkenes with commercially available catecholborane, followed by treatment with (*E*)-BPSE in the presence of a radical initiator (di-*tert*-butylhyponitrite or oxygen), occurs readily and represents a one-pot procedure for direct vinylation of alkylboronates (eq 29).[61]

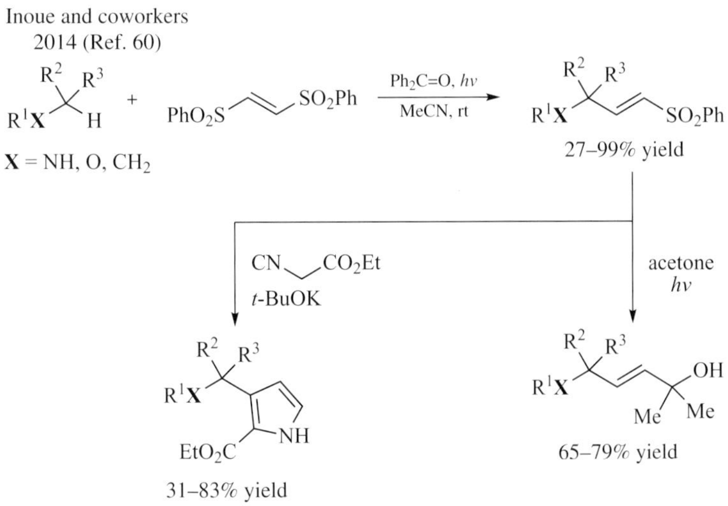

(28)

Renaud and coworkers
2006 (Ref. 61)

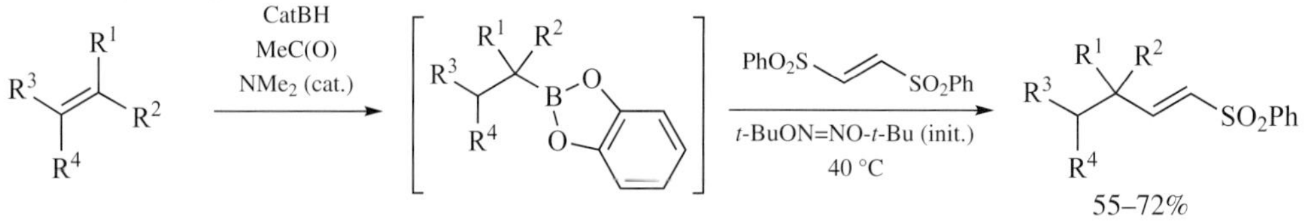

(29)

Michael Additions

The strong electrophilic character of BPSE makes of it a potent Michael acceptor. In most of these reactions, one of the two sulfonyl groups is lost as a leaving group.

Vilarrasa and coworkers described new types of protecting groups based on the vinylsulfonyl motif for imides, azinones, nucleosides, sultams, and lactams.[62] These groups (called *E*-besvinyl or *E*-Bsv and *Z*-besvinyl or *Z*-Bsv) are installed in high yields (90–96%) by reaction of the corresponding nitrogen nucleophiles with (*E*)- and (*Z*)-BPSE, respectively (eq 30). The Bsv groups are stable in acidic media and nonnucleophilic bases, but can be removed efficiently by treatment with good nucleophiles such as sodium dodecanethiolate or pyrrolidine.

Vilarrasa and coworkers
2014 (Ref. 62)

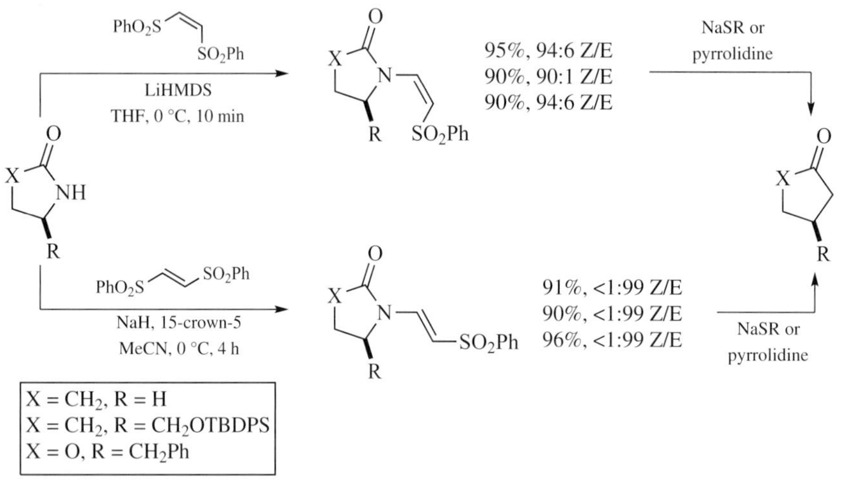

(30)

A new class of pyrazolidinediones, isoxazolidinediones, pyrimidinetriones, and thioxopyrimidinediones were synthesized by Mahesh and coworkers by the reaction of (*E*)-BPSE with various nucleophiles (eq 31).[63]

Mahesh et al, 2011

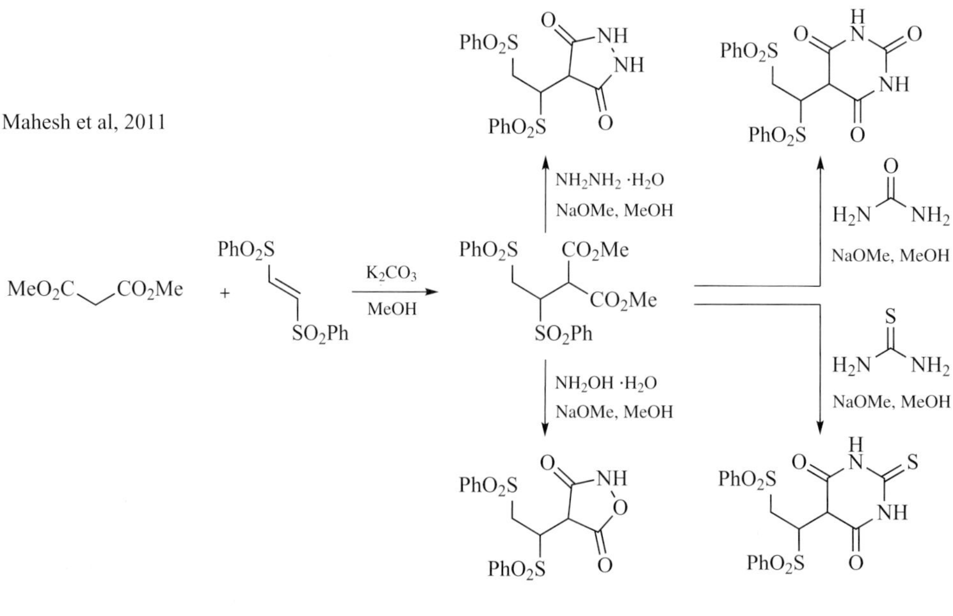

(31)

1. (a) De Lucchi, O.; Pasquato, L., *Tetrahedron* **1988**, *44*, 6755. (b) Cossu, S.; De Lucchi, O.; Fabbri, D.; Licini, G.; Pasquato, L., *Org. Prep. Proced. Int.* **1991**, *23*, 571. (c) Tanaka, K.; Kaji, A. In *The Chemistry of Sulphones and Sulphoxides*, Patai, S.; Rappoport, Z.; Stirling, C. J. M., Eds.; Wiley: Chichester, 1988; Chapter 15, pp 791 (d) Simpkins, N. S., *Sulphones in Organic Synthesis*; Pergamon: Oxford, 1993.
2. De Lucchi, O.; Lucchini, V.; Pasquato, L.; Modena, G., *J. Org. Chem.* **1984**, *49*, 596.
3. Meek, J. S.; Fowler, J. S., *J. Org. Chem.* **1968**, *33*, 985.
4. (a) Cusa, N. W.; McCombie, H., *J. Chem. Soc.* **1937**, 767. (b) Truce, W. E.; Boudakian, M. M.; Heine, R. F.; McManimie, R. J., *J. Am. Chem. Soc.* **1956**, *78*, 2743. (c) Tanimoto, S.; Taniyasu, R.; Takahashi, T.; Miyake, T.; Okano, M., *Bull. Chem. Soc. Jpn.* **1976**, *49*, 1931. (d) Tiecco, M.; Testaferri, L.; Tingoli, M.; Chianelli, D.; Montanucci, M., *J. Org. Chem.* **1983**, *48*, 4795.
5. De Lucchi, O.; Modena, G., *Tetrahedron* **1984**, *40*, 2585.
6. (a) Mirsadeghi, S.; Rickborn, B., *J. Org. Chem.* **1985**, *50*, 4340. (b) Kazimirchik, I. V.; Lukin, K. A.; Taranyuk, I. L.; Bebikh, G. F.; Zefirov, N. S., *Zh. Org. Khim.* **1985**, *21*, 313. (c) Paquette, L. A.; Racherla, U. S., *J. Org. Chem.* **1987**, *52*, 3250. (d) Kobayashi, T.; Miki, S.; Yoshida, Z.; Asako, Y.; Kajimoto, C., *J. Am. Chem. Soc.* **1988**, *110*, 5622. (e) Krebs, J.; Guggisberg, D.; Stämpfli, U.; Neuenschwander, M., *Helv. Chim. Acta* **1986**, *69*, 835. (f) Azzena, U.; Cossu, S.; De Lucchi, O.; Melloni, G., *Synth. Commun.* **1988**, *18*, 351. (g) Patney, H. K., *J. Org. Chem.* **1988**, *53*, 6106. (h) Paquette, L. A.; Gugelchuk, M.; McLaughlin, M. L., *J. Org. Chem.* **1987**, *52*, 4732.
7. (a) De Lucchi, O.; Licini, G.; Pasquato, L., *Chem. Commun./J. Chem. Soc., Chem. Commun.* **1985**, 418. (b) Paquette, L. A.; Künzer, H.; Green, K. E., *J. Am. Chem. Soc.* **1985**, *107*, 4788. (c) Paquette, L. A.; Künzer, H.; Green, K. E.; De Lucchi, O.; Licini, G.; Pasquato, L.; Valle, G., *J. Am. Chem. Soc.* **1986**, *108*, 3453.
8. (a) Künzer, H.; Cottrell, C. E.; Paquette, L. A., *J. Am. Chem. Soc.* **1986**, *108*, 8089. (b) Künzer, H.; Litterst, E.; Gleiter, R.; Paquette, L. A., *J. Org. Chem.* **1987**, *52*, 4740. (c) Paquette, L. A.; Künzer, H.; Kesselmayer, M. A., *J. Am. Chem. Soc.* **1988**, *110*, 6521.
9. Polniaszek, R. P.; Dillard, L. W., *J. Org. Chem.* **1992**, *57*, 4103.
10. Snyder, H. R.; Hallada, D. P., *J. Am. Chem. Soc.* **1952**, *74*, 5595.
11. Cossu, S.; De Lucchi, O.; Dilillo, F., *Gazz. Chim. Ital.* **1989**, *119*, 519.
12. (a) Sauer, J.; Wiest, H.; Mielert, A., *Ber. Dtsch. Chem. Ges./Chem. Ber.* **1964**, *97*, 3183. (b) Sauer, J.; Lang, D.; Wiest, H., *Ber. Dtsch. Chem. Ges./Chem. Ber.* **1984**, *97*, 3208.
13. (a) De Lucchi, O.; Pasquato, L. unpublished. (b) Fessner, W.-D.; Sedelmeier, G.; Spurr, P. R.; Rihs, G.; Prinzbach, H., *J. Am. Chem. Soc.* **1987**, *109*, 4626. (c) Fessner, W.-D.; Grund, C.; Prinzbach, H., *Tetrahedron Lett.* **1989**, *30*, 3133.
14. (a) Paquette, L. A.; Waykole, L.; Shen, C.-C.; Racherla, U. S.; Gleiter, R.; Litterst, E., *Tetrahedron Lett.* **1988**, *29*, 4213. (b) Paquette, L. A.; Shen, C.-C., *J. Am. Chem. Soc.* **1990**, *112*, 1159. (c) Paquette, L. A.; Shen, C.-C.; Engel, P., *J. Org. Chem.* **1989**, *54*, 3329.
15. Künzer, H.; Stahnke, M.; Sauer, G.; Wiechert, R., *Tetrahedron Lett.* **1991**, *32*, 1949.
16. Brown, A. C.; Carpino, L. A., *J. Org. Chem.* **1985**, *50*, 1749.
17. De Lucchi, O.; Pasquato, L., *Gazz. Chim. Ital.* **1984**, *114*, 349.
18. (a) Herdeis, C.; Hartke, C., *Heterocycles* **1989**, *29*, 287. (b) Herdeis, C.; Hartke-Karger, C., *Justus Liebigs Ann. Chem./Liebigs Ann. Chem.* **1991**, 99.
19. Pasquato, L.; De Lucchi, O.; Krotz, L., *Tetrahedron Lett.* **1991**, *32*, 2177.
20. (a) Azzena, U.; Cossu, S.; De Lucchi, O.; Melloni, G., *Tetrahedron Lett.* **1989**, *30*, 1845. (b) Cossu, S.; De Lucchi, O., *Gazz. Chim. Ital.* **1990**, *120*, 569. (c) Azzena, U.; Cossu, S.; De Lucchi, O.; Melloni, G., *Gazz. Chim. Ital.* **1989**, *119*, 357.
21. (a) McCombie, S. W.; Shankar, B. B.; Ganguly, A. K., *Tetrahedron Lett.* **1985**, *26*, 6301. (b) McCombie, S. W.; Shankar, B. B.; Ganguly, A. K., *Tetrahedron Lett.* **1987**, *28*, 4123. (c) Padwa, A.; Bullock, W. H.; Dyszlewski, A. D.; McCombie, S. W.; Shankar, B. B.; Ganguly, A. K., *J. Org. Chem.* **1991**, *56*, 3556. (d) Flitsch, W.; Lubisch, W., *Ber. Dtsch. Chem. Ges./Chem. Ber.* **1982**, *115*, 1547.
22. (a) Padwa, A.; Fryxell, G. E.; Gasdaska, J. R.; Venkatramanan, M. K.; Wong, G. S. K., *J. Org. Chem.* **1989**, *54*, 644. (b) Pandey, G.; Lakshmaiah, G.; Kumaraswamy, G., *Chem. Commun./J. Chem. Soc., Chem. Commun.* **1992**, 1313. (c) Burdisso, M.; Gandolfi, R.; Grünanger, P.; Rastelli, A., *J. Org. Chem.* **1990**, *55*, 3427. (d) Padwa, A.; Wannamaker, M. W., *Tetrahedron* **1991**, *47*, 6139.
23. (a) De Lucchi, O., *Phosphorus Sulfur/Phosphorus Sulfur Silicon* **1993**, *74*, 195. (b) Cossu, S.; Delogu, G.; De Lucchi, O.; Fabbri, D.; Licini, G., *Angew. Chem., Int. Ed. Engl.* **1989**, *28*, 766. (c) De Lucchi, O.; Fabbri, D.; Cossu, S.; Valle, G., *J. Org. Chem.* **1991**, *56*, 1888 and unpublished results.
24. (a) Cossu, S.; Battaggia, S.; De Lucchi, O., *J. Org. Chem.* **1997**, *62*, 4162. (b) Griesbeck, A. G.; Deufel, T.; Peters, K.; Peters, E.-M.; von Schnering, H. G., *Angew. Chem. Int. Ed. Engl.* **1993**, *32*, 97. (c) Shen, C. C.; Paquette, L. A., *Tetrahedron* **1994**, *50*, 4949. (d) Lin, C.-T.; Wang, N.-J.; Yeh, Y.-L.; Chou, T.-C., *Tetrahedron* **1995**, *51*, 2907. (e) Hirai, Y.; Ito, K.; Nagaoka, H., *Heterocycles* **1998**, *48*, 235.
25. (a) Arjona, O.; Iradier, F.; Mañas, R. M.; Plumet, J.; Grabuleda, X.; Jaime, C., *Tetrahedron* **1998**, *54*, 9095. (b) Arjona, O.; Iradier, F.; Medel, R.; Plumet, J., *J. Org. Chem.* **1999**, *64*, 6090.
26. Ando, K.; Kankake, M.; Suzuki, T.; Takayama, H., *Synlett* **1994**, 741.
27. Kuethe, J. T.; Brooks, C. A.; Comins, D. L., *Org. Lett.* **2003**, *5*, 321.
28. Pindur, U.; Lutz, G.; Fischer, G.; Schollmeyer, D.; Masser, W.; Schroeder, L., *Tetrahedron* **1993** *49*, 2863.
29. Pindur, U.; Erfanian-Abdoust, H., *Heterocycles* **1990**, *31*, 1751.
30. Bashkar Reddy, D.; Rajagopala Sarma, M.; Padmaja, A.; Padmavathi, V., *Phosphorous Sulfur Silicon* **2000**, *164*, 23.
31. Padmavathi, V.; Sumathi, R. P.; Venugopal Reddy, K.; Somasekar Reddy, A.; Bashkar Reddy, D., *Synth. Commun.* **2000**, *30*, 4007.
32. (a) Evans, P. A.; Manangan, T., *Tetrahedron Lett.* **1997**, *38*, 8165. (b) Evans, P. A.; Manangan, T., *J. Org. Chem.* **2000**, *65*, 4523.
33. Evans, P. A.; Roseman, J. D., *Tetrahedron Lett.* **1997**, *38*, 5249.
34. Nicolaou, K. C.; Pfefferkorn, J. A.; Kim, S.; Wei, H. X., *J. Am. Chem. Soc.* **1999**, *121*, 4724.
35. (a) Evans, P. A.; Brandt, T. A., *Org. Lett.* **1999**, *1*, 1563. (b) Evans, P. A.; Brandt, T. A.; Robinson, J. E., *Tetrahedron Lett.* **1999**, *40*, 3105.
36. Cossu, S.; De Lucchi, O.; Durr, R.; Fabris, F., *Synth. Commun.* **1996**, *26*, 211.
37. Cabianca, E.; Chéry, F.; Rollin, P.; Cossu, S.; De Lucchi, O., *Synlett* **2001**, 1962.
38. Chéry, F.; Desroses, M.; Tatibouët, A.; De Lucchi, O.; Rollin, P., *Tetrahedron* **2003**, *59*, 4563.
39. Desroses, M.; Chéry, F.; Tatibouët, A.; De Lucchi, O.; Rollin, P., *Tetrahedron: Asymmetry* **2002**, *13*, 2535.
40. Knapp, S.; Levorse, A. T., *J. Org. Chem.* **1988**, *53*, 4006.
41. Girniene, J.; Tardy, S.; Tatibouët, A.; Sackus, A.; Rollin, P., *Tetrahedron Lett.* **2004**, *45*, 6443.
42. Cabianca, E.; Tatibouët, A.; Chéry, F.; Pillard, C.; De Lucchi, O.; Rollin, P., *Tetrahedron Lett.* **2003**, *44*, 5723.
43. Tyrala, A.; Makosza, M., *Synthesis* **1994**, 264.
44. Kim, S.; Lim, C. J., *Angew. Chem. Int. Ed. Engl.* **2002**, *41*, 3265.
45. Cossu, S.; De Lucchi, O.; Fabris, F.; Ballini, R.; Bosica, G., *Synthesis* **1996**, 1481.
46. Chéry, F.; Rollin, P.; De Lucchi, O.; Cossu, S., *Synthesis* **2001**, 286.
47. Cabianca, E.; Chéry, F.; Rollin, P.; Tatibouët, A.; De Lucchi, O., *Tetrahedron Lett.* **2002**, *43*, 585.
48. Chevalier-Du Roizel, B.; Cabianca, E.; Rollin, P.; Sinaÿ, P., *Tetrahedron* **2002**, *58*, 9579.
49. Lopez-Perez, A.; Robles-Machin, R.; Adrio, J.; Carretero, J. C., *Angew. Chem., Int. Ed.* **2007**, *46*, 9261.
50. Robles-Machin, R.; Lopez-Perez, A.; Gonzalez-Esguevillas, M.; Adrio, J.; Carretero, J. C., *Chem. Eur. J.* **2010**, *16*, 9864.

51. Lopez-Perez, A.; Adrio, J.; Carretero, J. C., *J. Am. Chem. Soc.* **2008**, *130*, 10084.
52. Pascual-Escudero, A.; Gonzalez-Esguevillas, M.; Padilla, S.; Adrio, J.; Carretero, J. C., *Org. Lett.* **2014**, *16*, 2228.
53. Li, Z.; Yu, H.; Liu, H.; Zhang, L.; Jiang, H.; Wang, B.; Guo, H., *Chem. Eur. J.* **2014**, *20*, 1731.
54. Rincon, R.; Plumet, J., *Synlett* **2008**, 911.
55. Quintard, A.; Alexakis, A., *Chem. Eur. J.* **2009**, *15*, 11109.
56. Quintard, A.; Alexakis, A., *Org. Biomol. Chem.* **2011**, *9*, 1407.
57. Atienza, R. L.; Roth, H. S.; Scheidt, K. A., *Chem. Sci.* **2011**, *2*, 1772.
58. Liautard, V.; Robert, F.; Landais, Y., *Org. Lett.* **2011**, *13*, 2658.
59. Poittevin, C.; Liautard, V.; Beniazza, R.; Robert, F.; Landais, Y., *Org. Lett.* **2013**, *15*, 2814.
60. Amaoka, Y.; Nagatomo, M.; Watanabe, M.; Tao, K.; Kamijo, S.; Inoue, M., *Chem. Sci.* **2014**, *5*, 4339.
61. Schaffner, A.-P.; Darmency, V.; Renaud, P., *Angew. Chem., Int. Ed.* **2006**, *45*, 5847.
62. Petit, E.; Bosch, L.; Font, J.; Mola, L.; Costa, A. M.; Vilarrasa, J., *J. Org. Chem.* **2014**, *79*, 8826.
63. Liautard, V.; Robert, F.; Landais, Y., *Org. Lett.* **2011**, *13*, 2658.

3,3′-Bis(trifluoromethanesulfonyl)-2,2′-dihydroxy-1,1′-binaphthalene[1]

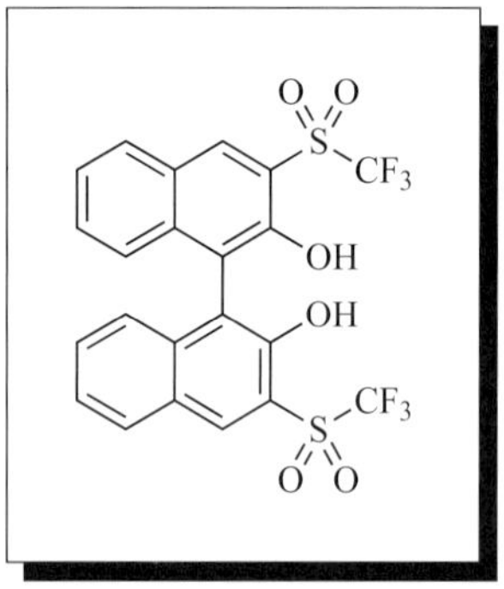

(±)

[551929-90-9] $C_{22}H_{12}F_6O_6S_2$ (MW 550.45)

InChI = 1S/C22H12F6O6S2/c23-21(24,25)35(31,32)15-9-11-5-1-3-7-13(11)17(19(15)29)18-14-8-4-2-6-12(14)10-16(20(18)30)36(33,34)22(26,27)28/h1-10,29-30H

InChIKey = NPGLJCZCRLOUSS-UHFFFAOYSA-N

(*R*)

[936234-98-9] $C_{22}H_{12}F_6O_6S_2$ (MW 550.45)

InChI = 1S/C22H12F6O6S2/c23-21(24,25)35(31,32)15-9-11-5-1-3-7-13(11)17(19(15)29)18-14-8-4-2-6-12(14)10-16(20(18)30)36(33,34)22(26,27)28/h1-10,29-30H

InChIKey = NPGLJCZCRLOUSS-UHFFFAOYSA-N

(*S*)

[1029916-98-0] $C_{22}H_{12}F_6O_6S_2$ (MW 550.45)

InChI = 1S/C22H12F6O6S2/c23-21(24,25)35(31,32)15-9-11-5-1-3-7-13(11)17(19(15)29)18-14-8-4-2-6-12(14)10-16(20(18)30)36(33,34)22(26,27)28/h1-10,29-30H

InChIKey = NPGLJCZCRLOUSS-UHFFFAOYSA-N

(chiral ligand for use in In-mediated imine allylation,[2] catalyst in asymmetric Petasis reactions,[3] and a precursor to chiral BINAPHOS ligands for use in Pd-catalyzed hydrophosphorylation[4])

Alternate Name: BINOL-3,3′-bistriflone.

Physical Data: mp 194–196 °C.

Solubility: sol in methylene chloride, chloroform, THF, diethyl ether, acetone, ethyl acetate, acetonitrile, and methanol. Sparingly sol in toluene. Insol in hexanes and water.

Form Supplied in: white solid.

Analysis of Reagent Purity: ^{1}H NMR ($CDCl_3$, 300 MHz) δ 8.69 (2H, s), 8.04 (2H, m), 7.52 (4H, m), 7.15 (2H, m); ^{13}C NMR ($CDCl_3$, 75 MHz) δ 138.5, 136.7, 132.2, 131.9, 130.4, 127.5, 126.0, 125.8, 124.7, 120.6, 116.5 (CF_3 q, J = 309 Hz); ^{19}F NMR ($CDCl_3$, 282 MHz) δ −78.6 (3F, s); $[\alpha]_D^{20}$ (*R* enantiomer) = +80 (*c* 0.4, ethyl acetate, 24 °C); IR υ 3427, 2923, 1620, 1584, 1504, 1456, 1350, 1299, 1199, 1110, 1080, 1040, 983, 874, 822, 779, 697; *m/z* (EI) = 550 (M^+, 100%), 532 (8%), 418 (12%), 399 (10%), 284 (24%), 267 (15%), 255 (20%), 239 (16%), 226 (22%), 207 (20%); HRMS (EI) calculated for $C_{22}H_{12}F_6O_6S_2$: 549.9979, found 549.9986.

Preparative Method:[1] in the published procedure,[2] *n*-BuLi (0.156 mL, 1.6 M in hexanes, 0.25 mmol) is added to a stirred solution of diisopropylamine (DIPA, 0.037 mL, 0.26 mmol) in anhydrous THF (1 mL) under nitrogen atmosphere at −78 °C. The LDA solution is warmed to room temperature and stirred for 1 h, before cooling back down to −78 °C. (*R*)-1,1′-Binaphthalene-2,2′-bis(trifluoromethanesulfonate) (137 mg, 0.249 mmol in dry THF, 1 mL) is added and the reaction mixture is warmed up to room temperature. *n*-BuLi (0.156 mL, 1.6 M in hexanes, 0.25 mmol) is added to another stirred solution of DIPA (0.037 mL, 0.26 mmol) in anhydrous THF (1 mL) under nitrogen atmosphere at −78 °C. The second LDA solution is warmed to room temperature and stirred for 1 h, and then added to the reaction mixture dropwise at −78 °C. When addition is complete, the reaction mixture is allowed to warm to room temperature and stirred for an additional 1 h. The reaction is quenched with 10% aqueous HCl solution and extracted three times with methylene chloride. The organic extracts are combined and washed, first with 10% aqueous HCl and then with brine, before drying over anhydrous magnesium sulfate. Filtration, solvent removal in vacuo, and subsequent purification (as described below) gives the (*R*)-enantiomer of the title compound (71 mg, 51%).

Purification: column chromatography on silica gel, eluting with methylene chloride (R_f = 0.64), followed by recrystallization from hot toluene.

Handling, Storage, and Precautions: indefinitely stable at room temperature in air. No known safety hazards – take standard precautions while handling (wear suitable personal protection, for example, gloves, safety spectacles, and lab coat).

Notes on Preparation and Purification. If using commercial *n*-BuLi, it is *essential* that it has been recently prepared, has minimal halide content[5] (an issue that can substantially differ between suppliers), and is titrated prior to use. The DIPA must be freshly distilled prior to preparation of the LDA for the synthesis of the title compound. This minimizes the generation of complicated product mixtures arising via aryne intermediates. It is also important to monitor the reaction mixture by TLC 1 h after the addition of the second equivalent of LDA, as the rearrangement does not always go to completion by this stage and may require a further 2–3 h reaction time before quenching. If the reaction mixture is reasonably clean, then the purification method described above should suffice to obtain the title compound. However, purification can be somewhat complicated if there is any unreacted starting material and/or the mono-rearrangement product in the crude reaction mixture. In these cases, column chromatography, first with hexanes/ethyl acetate (30:1) (to elute the unreacted starting material and mono-rearrangement product) and then with hexanes/ethyl acetate (5:1), allows isolation of the desired bis-rearrangement product. This can also be achieved by elution with mixtures of hexanes and methylene chloride (3:2 → 1:1). If there are impurities left after column chromatography, then the desired bis-rearrangement product can be isolated by triturating the crude product with hexanes (this may be assisted by sonication). The starting material and mono-rearrangement product dissolve and the title compound precipitates as a white solid.

Related Compounds. The title compound is part of a range of compounds, which have been prepared by anionic thia-Fries rearrangement of aryl triflates.[1,6] Aryl triflates rearrange to form *ortho*-trifluoromethanesulfonyl phenols most readily when there are electron-withdrawing substituents on the aryl ring. Electron-donating substituents cause elimination, giving rise to electrophilic benzynes instead. These can be trapped by either nucleophiles – for example, DIPA (formed by protonation of LDA under reaction conditions)[1] – or dienes, to give Diels–Alder adducts.[6] Apart from simple phenyl, naphthyl, and binaphthyl triflones, triflones of tricarbonylchromium aryl complexes and ferrocenyl complexes have also been prepared.[7,8] In these cases, only thia-Fries rearrangement products are obtained, irrespective of the electronic properties of other groups attached to the aryl ring. Higher fluoroalkylated analogs of the title compound (such as BINOL-3,3′-bisnonaflone) have been prepared,[2] which could facilitate applications of this class of compounds in fluorous solvents. Alkyl sulfone analogs of the title compound can be prepared by simple substitution of the triflone CF_3 group with alkyl Grignard reagents (eq 1).[2]

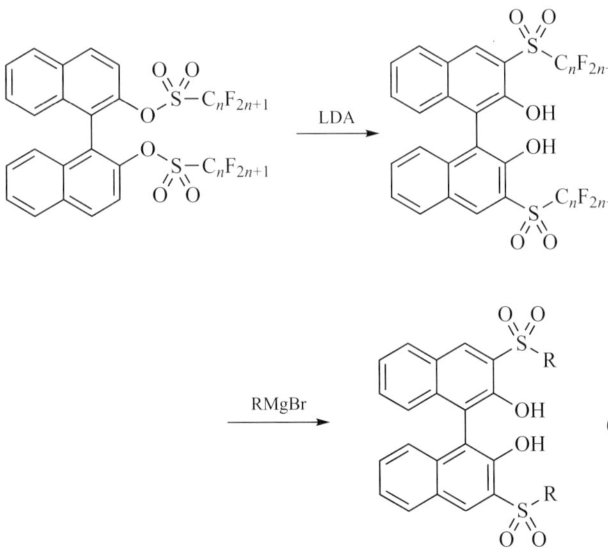

(1)

In-mediated Imine Allylation. Substantial improvements to enantioselectivities were achieved (compared with other BINOL catalysts) when the title compound was used in indium-mediated imine allylation.[2] Under optimized conditions, reactions of hydrazones with allyl bromide (3 equiv) in the presence of indium (2 equiv) and the title compound (10 mol %) gave allylic amine products in yields of up to 98% and enantiomeric excesses of up to 99% (eq 2). Even in lower loadings (> 3 mol %), the title compound gave an ee of 98% for one set of substrates. It was found that the C_2 symmetry of the ligand was vitally important, and that product ee enormously diminishes when this symmetry is broken. For instance, with one set of substrates, the mono-rearrangement analog gave an ee of only 9% (versus 97% for the title compound). The title compound was also found to be useful for catalyzing the asymmetric allylation of aliphatic imines, affording ee's of up to 74%. Obtaining good enantioselectivities with these substrates is often problematic due to their higher reactivity in the uncatalyzed background reaction.

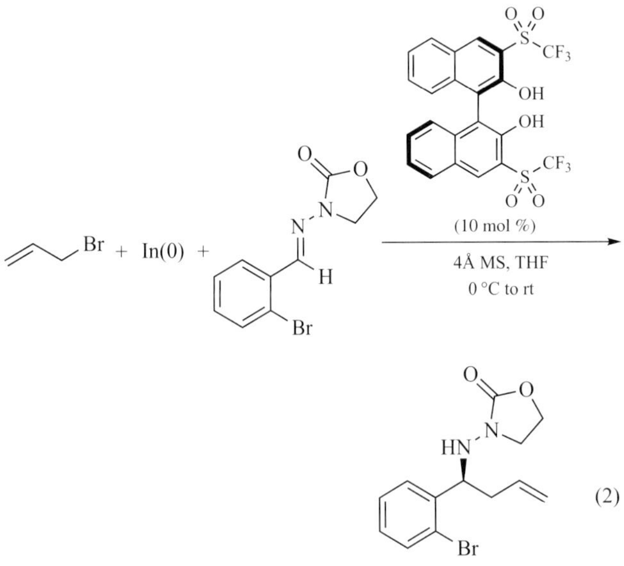

(2)

96%, 99% ee

Asymmetric Petasis Reaction. The utility of the title compound and its mono-rearrangement analog as catalysts in asymmetric Petasis reactions was demonstrated (eq 3).[3] Both compounds gave good product yields, but the title compound gave poor enantioselectivity (10% ee), and the mono-triflone analog gave moderate enantioselectivity (44% ee) in an initial substrate screen.

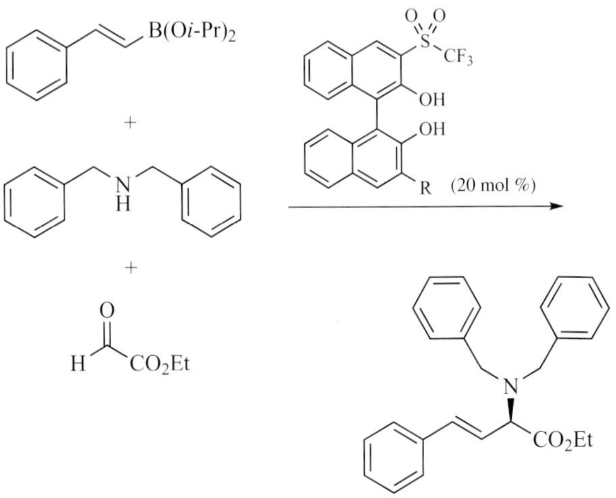

(3)

R = SO_2CF_3, 70%, 10% ee
R = H, 67%, 44% ee

BINAPHOS[9] Analogs for Hydrophosphorylation.[4] Analogs of BINAPHOS[9] have been prepared from BINOL-2,2′-bistriflate in a sequence that includes a thia-Fries rearrangement (eq 4).[4] Further functionalization of the ligands can be achieved by substituting the CF_3 of the triflone group with Grignard reagents. Some of the sulfonated BINAPHOS analogs reported gave higher activity and branched (*iso-*) product ee in Pd-catalyzed hydrophosphorylation than the parent BINAPHOS ligand (up to 74%) (eq 5).

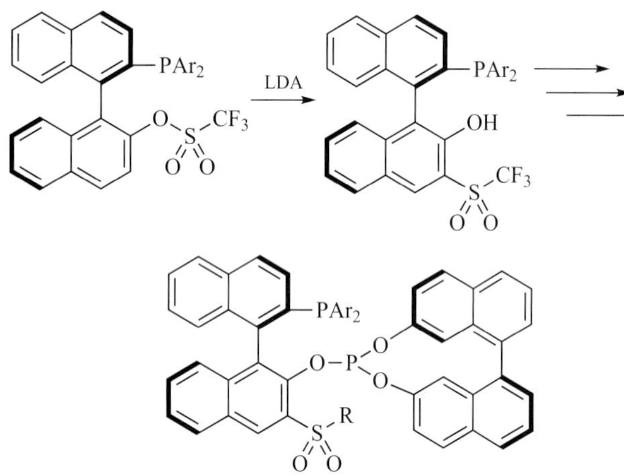

(4)

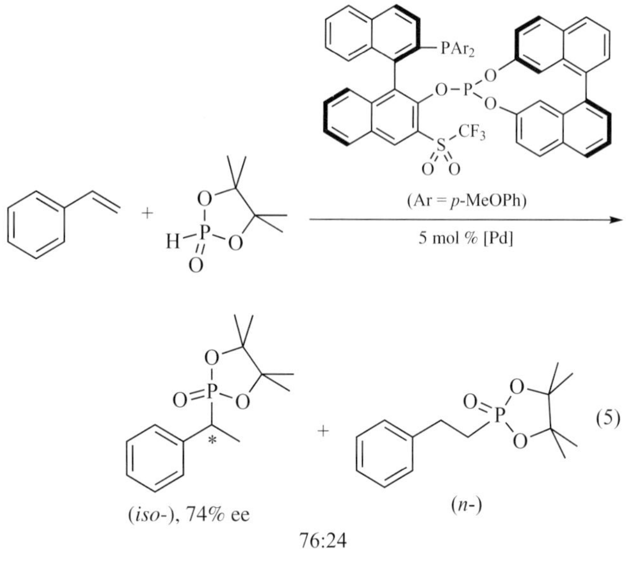

(5)

Phosphoramidite Derivatives.[10] The title compound (and related sulfones) can easily be converted into phosphoramidite ligands by reaction with HMPT.[10] These compounds showed substantial, solvent internal pressure-dependent, through-space ^{31}P–^{19}F NMR couplings between the phosphoramidite and the triflone, which facilitated spectral assignment of the diastereomers (eq 6). These ligands have been effectively utilized in Rh-catalyzed asymmetric hydrogenation. Dehydroamino acid esters were reduced with selectivities of up to 99% ee (eq 7).[11]

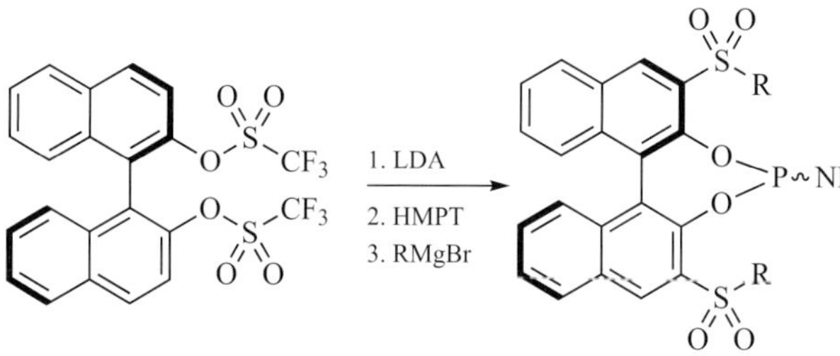

(6)

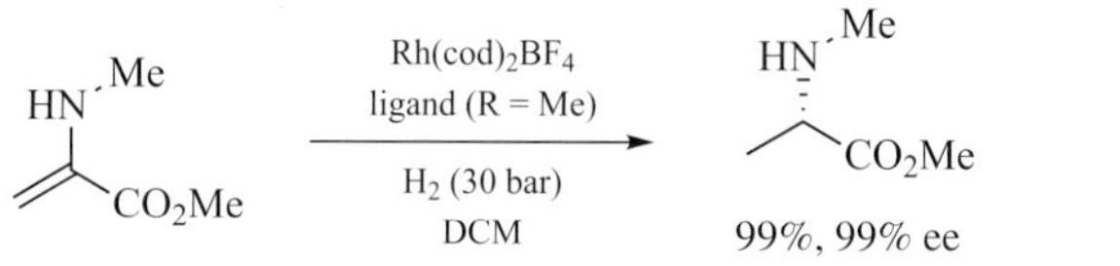

(7)

1. Charmant, J. P. H.; Dyke, A. M.; Lloyd-Jones, G. C., *Chem. Commun.* **2003**, 380.
2. Kargbo, R.; Takahashi, Y.; Bhor, S.; Cook, G. R.; Lloyd-Jones, G. C.; Shepperson, I. R., *J. Am. Chem. Soc.* **2007**, *129*, 3846.
3. Lou, S.; Schaus, S. E., *J. Am. Chem. Soc.* **2008**, *130*, 6922.
4. Barta, K.; Franciò, G.; Leitner, W.; Lloyd-Jones, G. C.; Shepperson, I. R., *Adv. Synth. Catal.* **2008**, *350*, 2013.
5. Gupta, L.; Hoepker, A. C.; Singh, K. J.; Collum, D. B., *J. Org. Chem.* **2009**, *74*, 2231.
6. Dyke, A. M.; Gill, D.; Harvey, J. N.; Hester, A. J.; Lloyd-Jones, G. C.; Muñoz, M. P.; Shepperson, I. R., *Angew. Chem., Int. Ed.* **2008**, *47*, 5067.
7. Zhao, Z.; Messinger, J.; Schönn, U.; Wartchow, R.; Butenschön, H., *Chem. Commun.* **2006**, 3007.
8. Werner, G.; Lehmann, C. W.; Butenschön, H., *Adv. Synth. Catal.* **2010**, *352*, 1345.
9. Nozaki, K.; Sakai, N.; Nanno, T.; Higashijima, T.; Mano, S.; Horiuchi, T.; Takaya, H., *J. Am. Chem. Soc.* **1997**, *119*, 4413.
10. Kruck, M.; Muñoz, M. P.; Bishop, H. L.; Frost, C. G.; Chapman, C. J.; Kociok-Köhn, G.; Butts, C. P.; Lloyd-Jones, G. C., *Chem. Eur. J.* **2008**, *14*, 7808.
11. Franciò, G.; Leitner, W.; Lloyd-Jones, G. C.; Muñoz, M. P.; Shepperson, I. R. Unpublished results.

Guy C. Lloyd-Jones & Gareth J. J. Owen-Smith
University of Bristol, Bristol, UK

1-[Bis(trifluoromethanesulfonyl)methyl]-2,3,4,5,6-pentafluorobenzene[1]

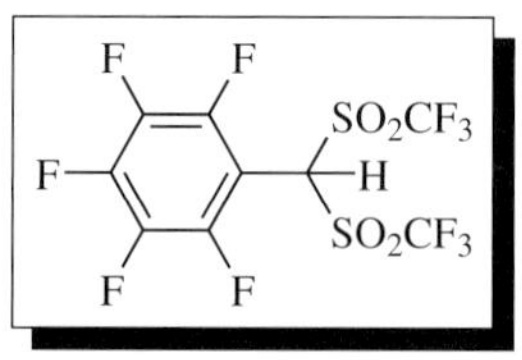

[405074-81-9] $C_9HF_{11}O_4S_2$ (MW 446.21)
InChI = 1S/C9HF11O4S2/c10-2-1(3(11)5(13)6(14)4(2)12)7(25(21,22)8(15,16)17)26(23,24)9(18,19)20/h7H
InChIKey = RLLDXJXYMKTGPV-UHFFFAOYSA-N

(an acid catalyst for organic transformations; an acid activator for chiral Lewis acid catalysts; a precursor for metal catalysts; a sterically bulky counter anion; a nucleophilically weak counter anion)

Physical Data: mp 86–87 °C; IR (KBr) 1522, 1501, 1347, 1321, 1198, 1127, 1024, 988, 613 cm^{-1}; ^{1}H NMR (300 MHz, $CDCl_3$) δ 6.21(br s, 1H); ^{13}C NMR (125 MHz, $CDCl_3$) δ 70.4, 98.0, 119.2 (q, 2C, J_{CF} = 330 Hz), 137.8 (d, 1C, J_{CF} = 258 Hz), 138.6 (d, 1C, J_{CF} = 257 Hz), 144.7 (d, 1C, J_{CF} = 264 Hz), 145.4 (d, 1C, J_{CF} = 262 Hz), 147.2 (d, 1C, J_{CF} = 262 Hz); ^{19}F NMR (282 MHz, $CDCl_3$) δ −157.9 (dt, 1F, J = 6.2, 21.5 Hz), −156.8 (dt, 1F, J = 6.2, 21.5 Hz), −142.6 (tt, 1F, J = 5.9, 21.5 Hz), −140.3 (br, 1F), −127.7 (ddd, 1F, J = 5.9, 15.2, 21.5 Hz), −75.2 (s, 6F).

Solubility: soluble in most organic solvents such as alcohols, acetonitrile, and dichloromethane.

Form Supplied in: white solid.

Preparative Method:[1] To a solution of pentafluorobenzyl bromide (0.5 mmol) in dry Et_2O (3 mL) was added *t*-BuLi (0.34 mL, 0.55 mmol, 1.6 M solution in hexane) dropwise at −78 °C, and the resulting mixture was stirred for 20 min. Triflic anhydride (92 μL, 0.55 mmol) was then added, and the resulting mixture was allowed to warm to room temperature over a period of 1 h. After the reaction mixture was cooled to −78 °C, *tert*-BuLi (0.34 mL, 0.55 mmol, 1.6 M solution in hexane) was added dropwise, and the resulting mixture was stirred for 20 min. Triflic anhydride (0.55 mmol) was then added, and the resulting mixture was allowed to reach room temperature over a period of 1 h before the reaction was quenched with water. The resultant mixture was neutralized and washed with hexane. The aqueous phase was acidified with 4 M aqueous HCl and extracted with ether twice. The organic layers were dried over magnesium sulfate, filtered, and concentrated under reduced pressure. The residue was subjected to bulb-to-bulb distillation.

Purification: recrystallization from hexane-AcOEt.

Handling, Storage, and Precautions: this compound is a stable solid in air, but is highly absorbent. Use of gloves and protective clothing is recommended.

Original Commentary

Aiko Hasegawa
Shionogi & Co., Ltd., Hyogo, Japan

Kazuaki Ishihara
Nagoya University, Nagoya, Japan

Properties. 1-[Bis(trifluoromethanesulfonyl)methyl]-2,3,4,5,6-pentafluorobenzene (**1**) is one of the strongest Brønsted acids. Its pK_a value in CD_3CO_2D is 1.5 as measured by the ^{1}H NMR method.[1,2] It is stronger than sulfuric acid, which has a pK_a value of 7.0 as measured by the same method.

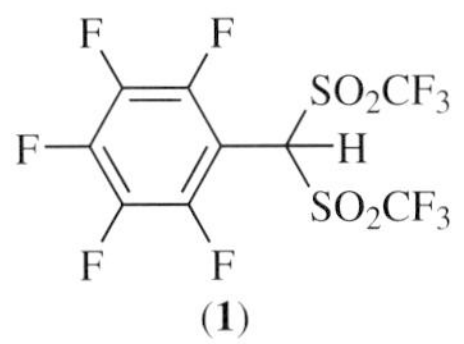

(**1**) pK_a (CD_3CO_2D) 1.5

H_2SO_4 pK_a (CD_3CO_2D) 7.5

Benzoylation of (−)-Menthol. (−)-Menthol is benzoylated with benzoic anhydride using a catalytic quantity of **1** (eq 1).[3]

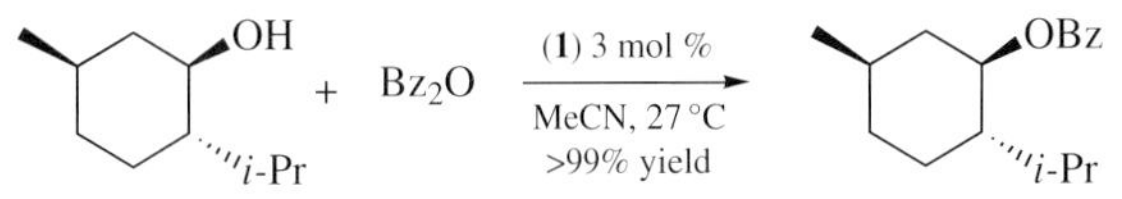

(1)

Synthesis of (±)-α-Tocopherol (Vitamin E) in scCO₂. (±)-α-Tocopherol (vitamin E) was synthesized by the condensation reaction of trimethylhydroquinone with isophytol in supercritical CO_2 ($scCO_2$) in the presence of (**1**) (eq 2).[4]

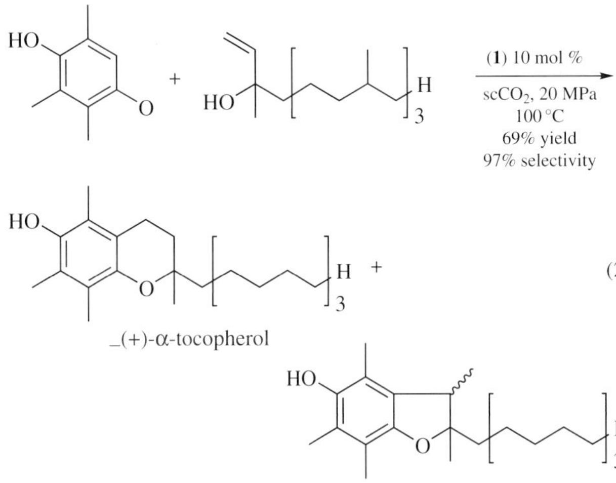

(2)

Chemoselective Diels–Alder Reaction with α,β-unsaturated Aldehydes and Ketones. Brønsted acid (**1**) is an effective catalyst for the chemoselective Diels–Alder reaction of cyclopentadiene with acrolein and ethyl vinyl ketone (eq 3).[5] The reaction catalyzed by (**1**) proceeds smoothly and shows high chemoselectivity for ethyl vinyl ketone, although the catalytic activities of methanesulfonic acid and *p*-toluenesulfonic acid are poor. However, some Lewis acids such as $SnCl_4$ and $B(C_6F_5)_3$ show high chemoselectivity for acrolein.

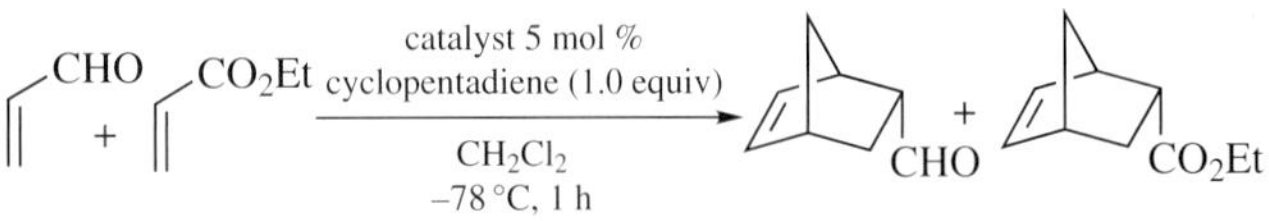

$SnCl_4$	73% yield	97 : 3	
p-TsOH·H_2O	45% yield	29 : 71	(3)
$C_6F_5CHTf_2$	86% yield	8 : 92	
$C_6F_5CHTf_2$ (−96 °C)	84% yield	1 : 99	

Design of Brønsted Acids. The nucleophilic *para*-substitution reaction occurs with some nucleophiles.[1] This makes it possible to design Brønsted acids such as polystyrene resin-bound Brønsted acid (**2**), fluorous Brønsted acid (**3**), and dibasic acid (**4**) (eqs 4–6).

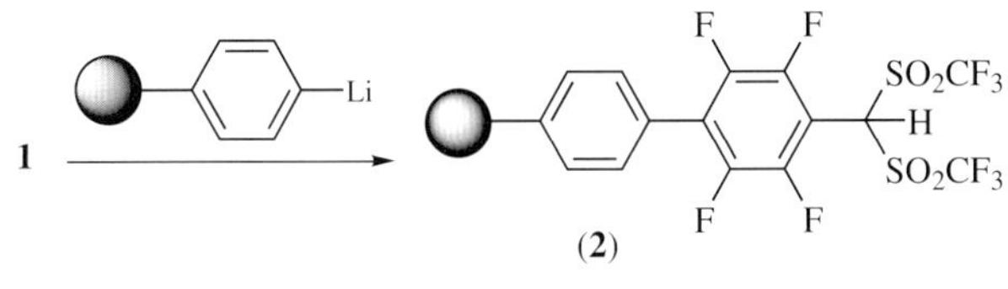

(4)

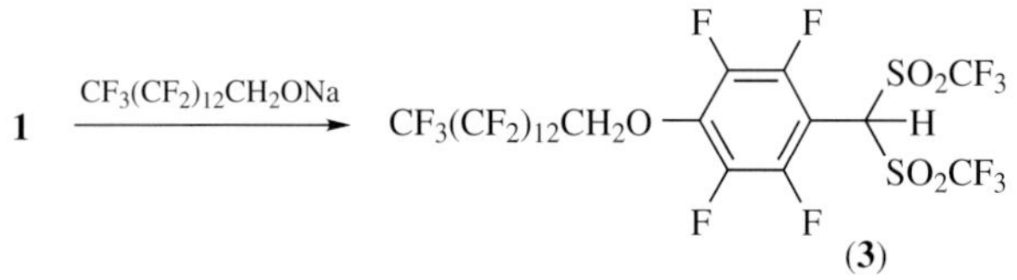

(5)

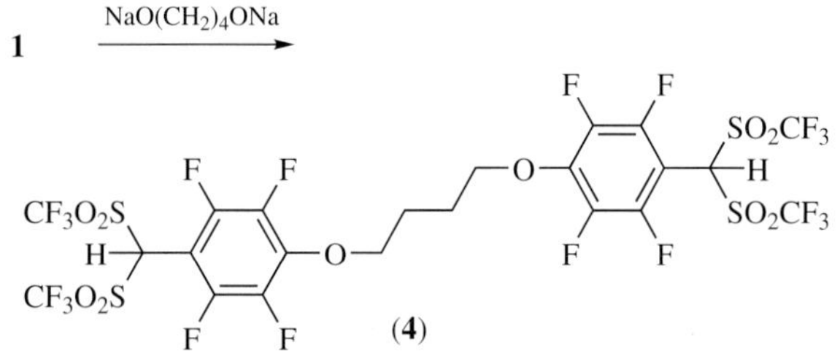

(6)

Highly Acidic Heterogeneous Catalyst. Both soluble Brønsted acid (**1**) and polystyrene resin-bound Brønsted acid (**2**) are widely used acid catalysts in many common acid-catalyzed reactions, including esterification (eq 7), Friedel-Crafts acylation (eq 8), acetalization (eq 9), Mukaiyama aldol reaction (eq 10), and Sakurai–Hosomi allylation (eq 11).[3] The polymer catalyst (**2**) can be quantitatively recovered by filtration and reused.

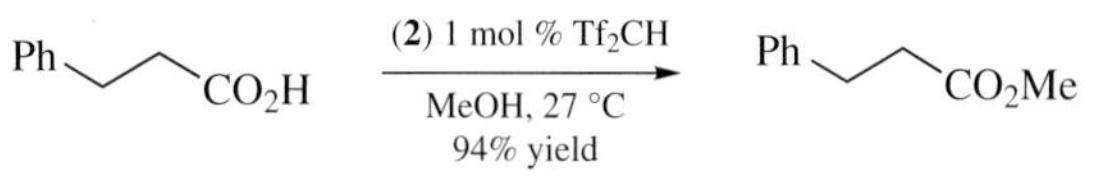

(7)

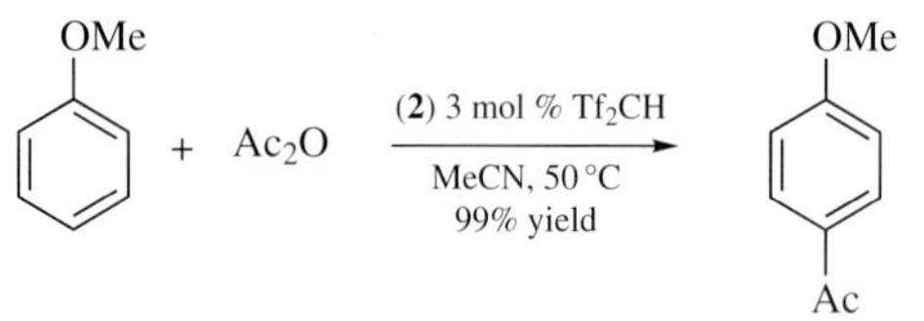

(8)

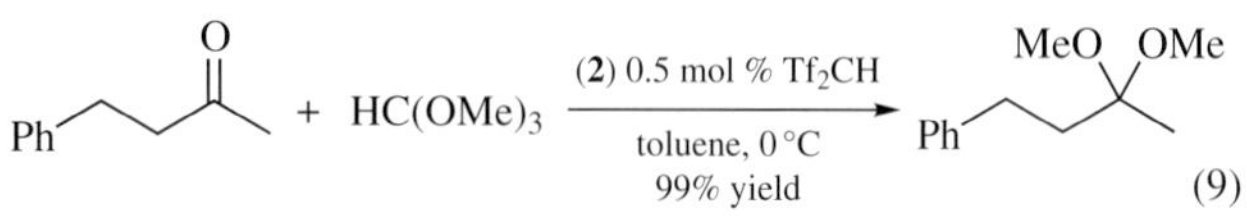

(9)

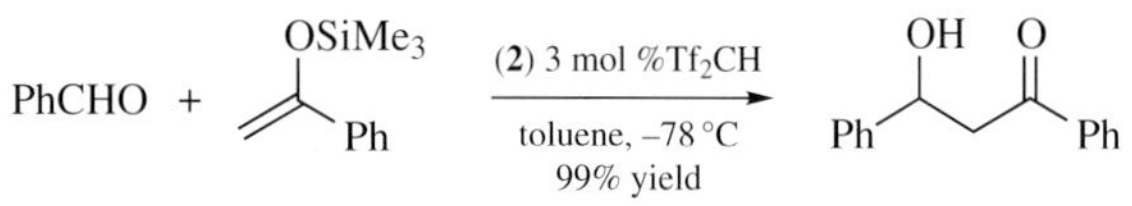

(10)

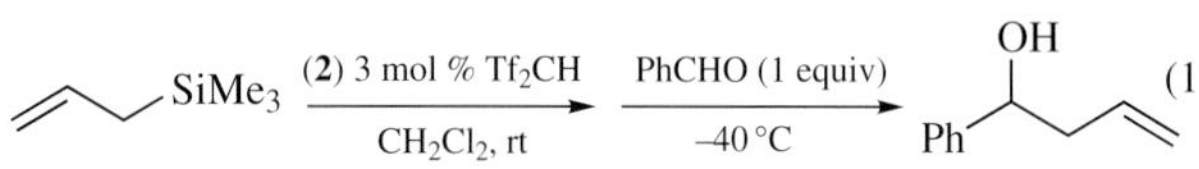

(11)

Furthermore, various acid-promoted reactions proceed to give the desired products in high yields by passing a solution of reactants through a reaction column packed with (**2**) and Celite® (eqs 12–14).[6] The polymer catalyst can be isolated from a mixture of (**2**) and Celite® based on the difference in their specific gravity: while (**2**) floats on water, Celite® sinks.

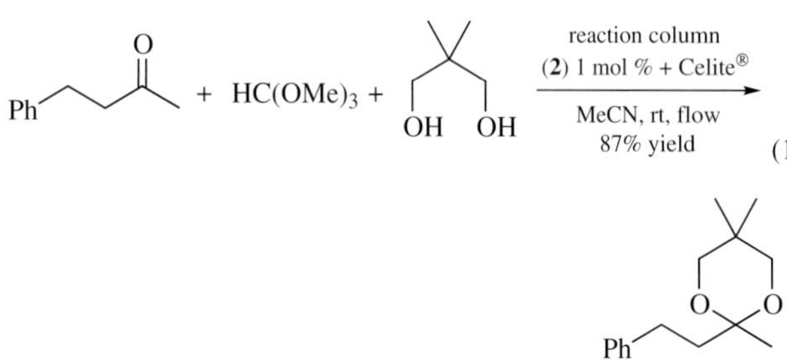

(12)

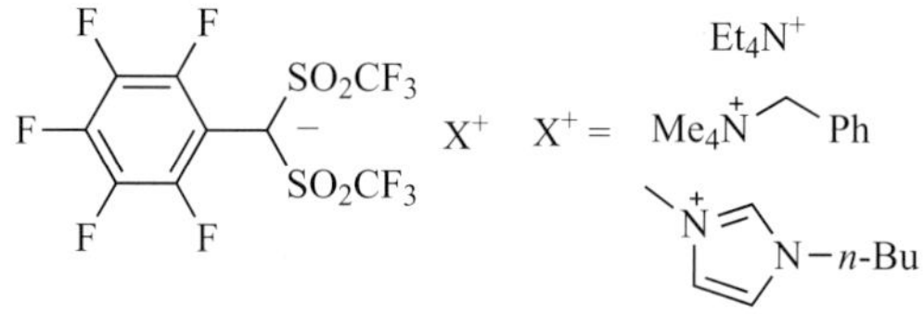

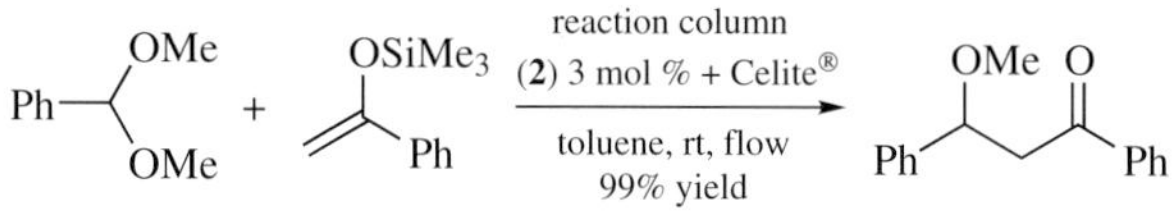

(13)

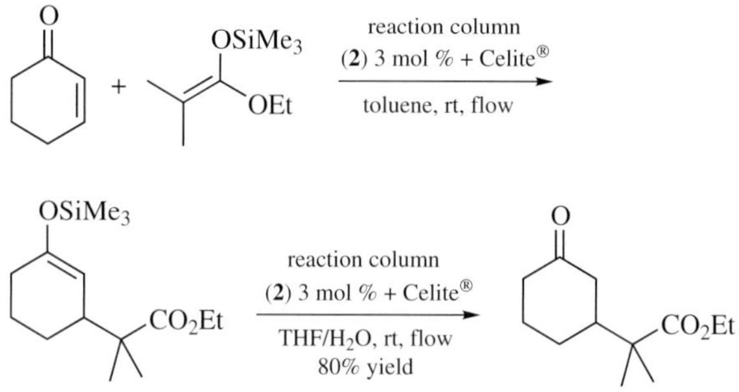

(14)

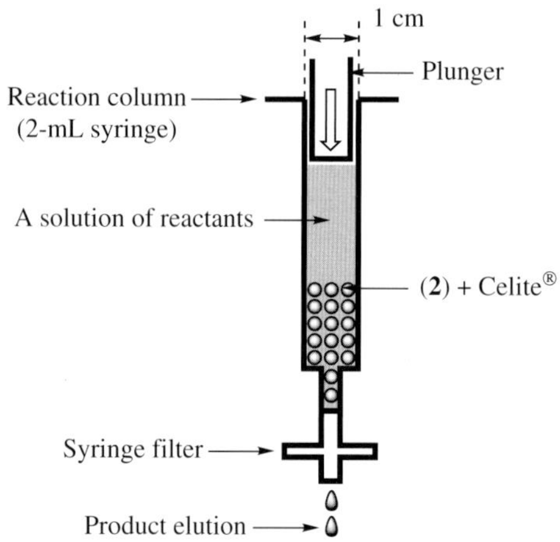

Fluorous Acid Catalyst. Fluorous catalysts are generally recycled by dissolving catalysts in a fluorous solvent that is not miscible with an organic solvent.[7] Recently, however, it has been shown that some fluorous catalysts do not need a fluorous solvent and can be recovered based on the use of temperature-dependent solubility in an organic solvent.[8,9] While (**1**) is soluble in most organic and fluorous solvents at any temperature, (**3**) appended to (**1**) by $OCH_2(CF_2)_nCF_3$ shows temperature-dependent solubility in some organic solvents.[10,11] Fluorous Brønsted acid (**3**) catalyzes acetalization (eq 15) and benzoylation (eq 16), and is soluble in an organic solvent under reaction conditions and can be recovered by precipitation at room temperature.

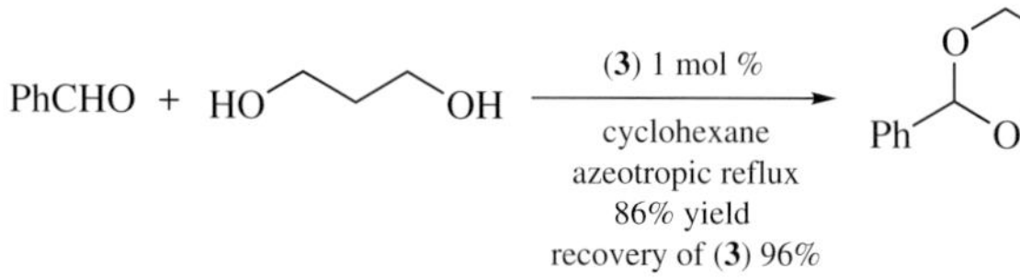

(15)

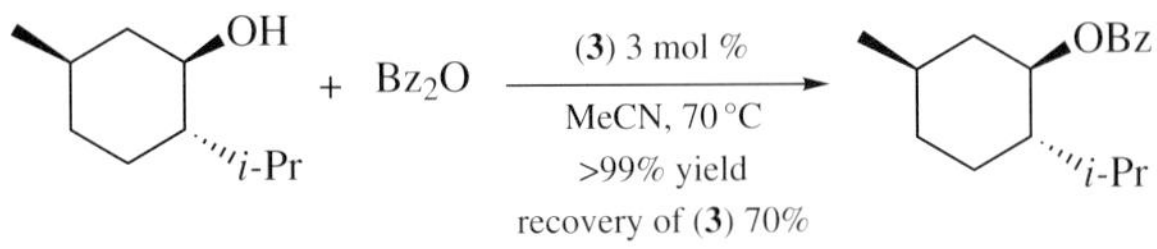
(16)

Moreover, (**3**) has high catalytic activity even under heterogeneous conditions (eq 17).

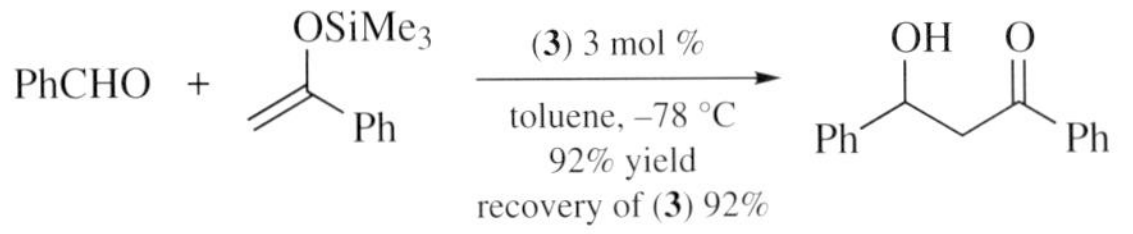
(17)

Application of Conjugate Base. Brønsted acid (**1**) is the starting material for the preparation of the super Lewis acid catalyst (**5**) (eq 18).[12]

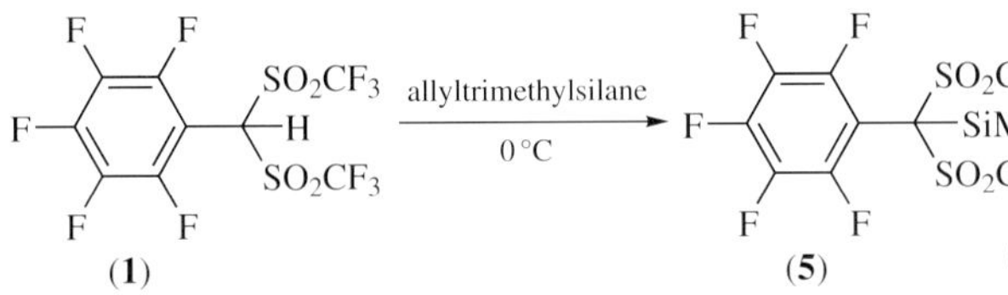
(18)

This catalyst is highly effective for the regioselective condensation of trimethylhydroquinone with isophytol to afford (±)-α-tocopherol (eq 19).[12]

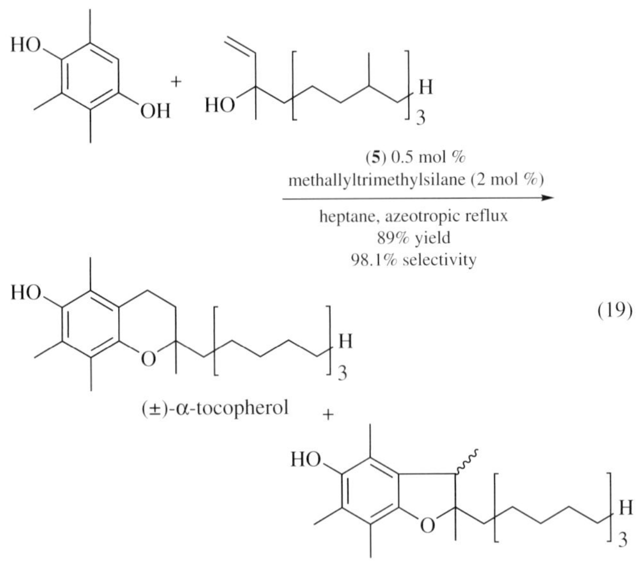
(19)

Brønsted acid (**1**) can be used for materials chemistry. The conjugate base of (**1**) is used as a counter anion of ionic liquids bearing quaternary ammonium, imidazolium, and pyridinium cations.[12] These ionic liquids are not only useful reaction solvents but also electrolytic solutions because of their low melting points, viscosity, and water solubility.

Diiminium compounds can be synthesized from the corresponding amine, silver nitrate, and (**1**). These are near-infrared ray-absorbing compounds, which are free from antimony and arsenic, and show excellent stability, particularly with regard to heat resistance, light stability, and moist heat resistance.[13]

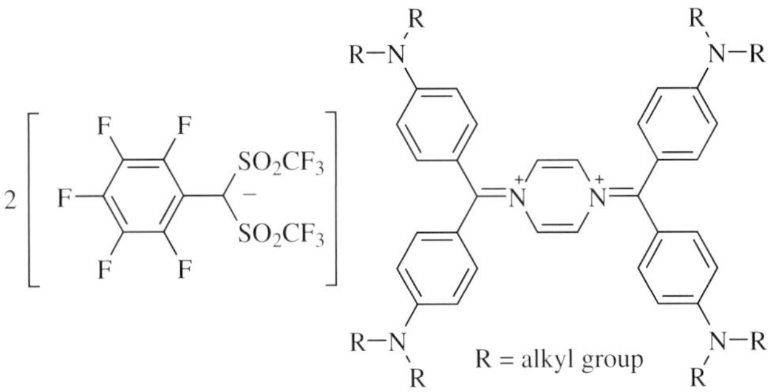

A polymer electrolyte is made from (**1**) supported on a polymer such as polystyrene. The film shows high proton condensation in the absence of water, thereby eliminating the need for a humidifier for a fuel cell gas.[14]

The conjugate base of (**1**) is also used as a weak nucleophile. The reactivity of hydride-bridged borane complex (**6**) with weak nucleophiles is attributed to the presence of a small amount of borenium ion (**9**). Some weak nucleophiles, $^-$OTf and $^-NTf_2$, which are conjugate bases of strong acids, can be used to demonstrate the presence of borenium ion. These two nucleophiles cleave the B–H–B bond. However, hydride-bridged borane complex is only partly converted to the corresponding tetravalent adduct (**7**) when conjugate base (**1**) is used (eq 20). In this case, unreacted borane complex (**8**) competes with weakly nucleophilic anion $^-C(C_6F_5)Tf_2$ for coordination into the unoccupied orbital of borenium ion (**9**).[15]

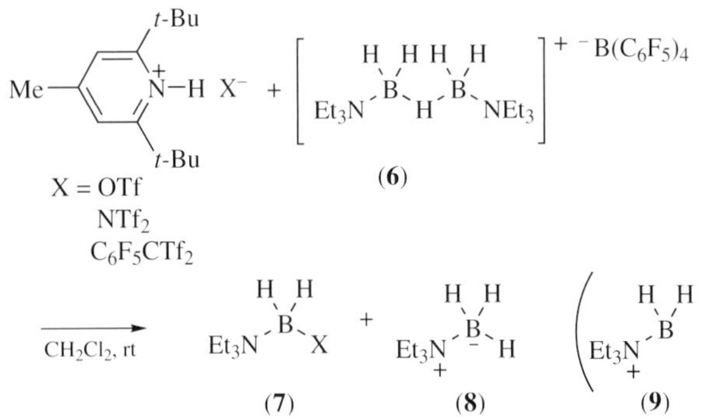
(20)

Brønsted Acid-activated Chiral Oxazaborolidine. The regio- and enantioselective Diels–Alder reaction of 1- and 2-substituted cyclopentadienes catalyzed by chiral oxazaborolidine is activated by (**1**).[16] Brønsted acid (**1**) is the most effective among various Brønsted acid activators ($C_6F_5CHTf_2$ > Tf_2NH > TfOH). This trend indicates that the larger steric dimensions of the counter anion confer reactivity, presumably due to its lower capacity to coordinate to boron, rather than pK_a of the Brønsted acid (eq 21).[3,17]

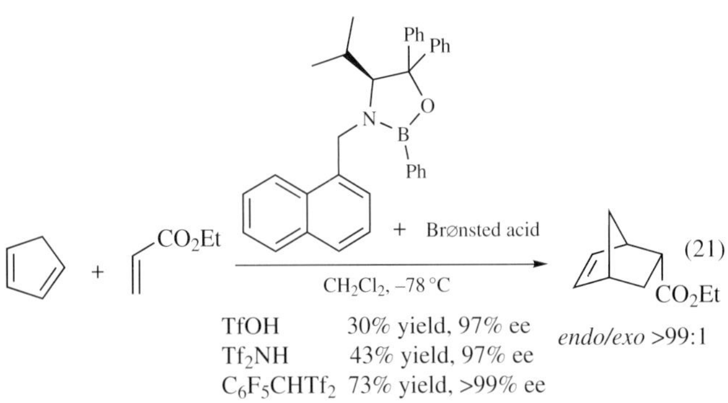
(21)

Related Reagents. **Trifluoromethanesulfonic acid**; trifluoromethanesulfonimide; bis(trifluoromethanesulfonyl)methane; tris(trifluoromethanesulfonyl)methane; [bis(trifluoromethane sulfonyl)methyl]benzene.

First Update

Aiko Hasegawa
Shionogi & Co., Ltd., Osaka, Japan

Kazuaki Ishihara
Nagoya University, Nagoya, Japan

Properties. The pK_a value of 1-[bis(trifluoromethane sulfonyl)methyl]-2,3,4,5,6-pentafluorobenzene (**1**) in DCE (1,2-dichloroethane) has been determined as −9.0 as measured by the UV-vis spectrophotometric titration method[18a] and estimated by computational calculation.[18b] Sulfuric acid has a pK_a value of −2.5 in DCE as measured by the above method.

Synthesis of Polymer. Brønsted acid **1** acts as an organocatalyst to control the ring-opening polymerizations of ε-caprolactone (ε-CL) and δ-valerolactone (δ-VL) with hydroxy compounds as the initiator, affording the well-defined poly(ε-caprolactone)s and poly(δ-valerolactone)s with narrow polydispersities (eq 22).[19]

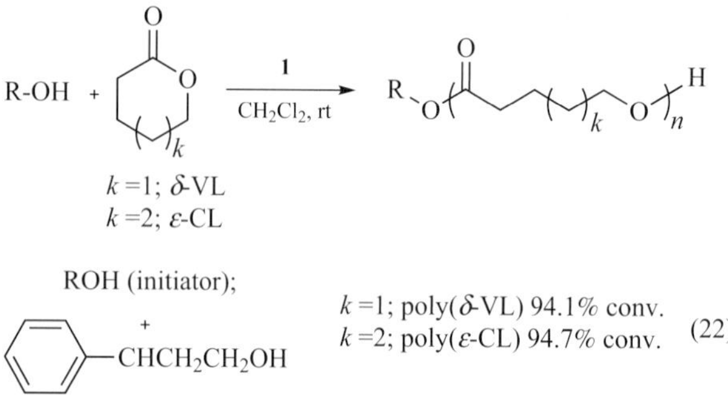

(22)

Brønsted acid **1** and its derivatives can be used as polymerization initiator to provide organic thin film that is effective for organic EL device.[20] Brønsted acid **1** also activates group-transfer polymerization starting agents that initiate polymerization of β-substituent-containing α,β-unsaturated carboxylate ester.[21]

Synthesis of β-Substituted Pyrrole Derivatives. Brønsted acid **1** catalyzes the reaction of pyrrole with carbonyl groups regioselectively (eq 23).[22]

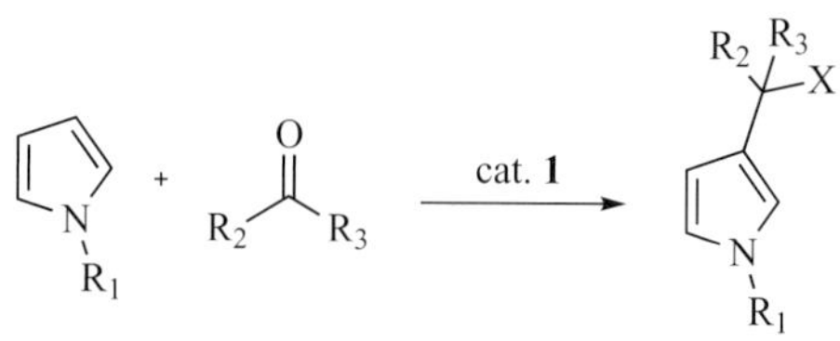

(23)

Application of Conjugate Base. Brønsted acid **1** is also an effective precatalyst for 1,4-addition of silyl dienolates to α,β-unsaturated aldehydes (eq 24).[23] In this reaction system, the in situ generation of silicon Lewis acid from 2-silyloxyfuran and Brønsted acid **1** initially occurs and the resultant silyl methide strongly activates the carbonyl substrates.

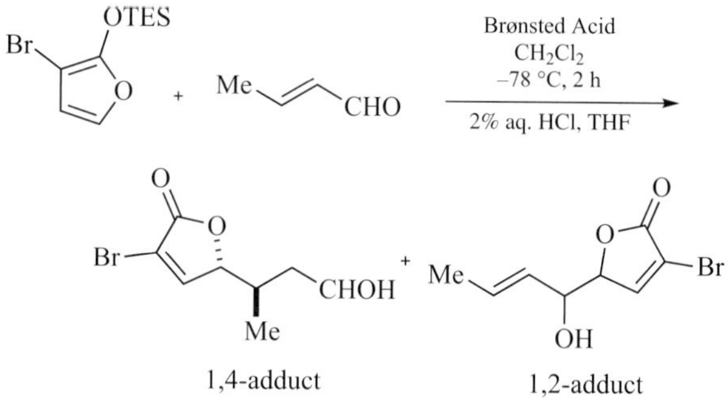

	Yield (%)	1,4-/1,2-	*anti/syn* (1,4-adduct)
$Tf_2CHCH_2CHTf_2$ (0.5 mol %)	96	20:1	11:1
TfOH (2.0 mol %)	91	4.9:1	9.5:1
1 (0.5 mol %)	92	19:1	12:1

(24)

The in situ generation of silicon Lewis acid from Brønsted acid **1** also promotes transfer polymerization of *N,N*-dimethylacrylamide (eq 25).[24] The equilibrium constant for the monomer activation (K) and the rate constant for the propagation reaction (k_p) are determined from the integrated rate equation derived from the presumed polymerization mechanism. Both of the constants for the polymerization using Brønsted acid **1** are higher than those of Tf_2NH.

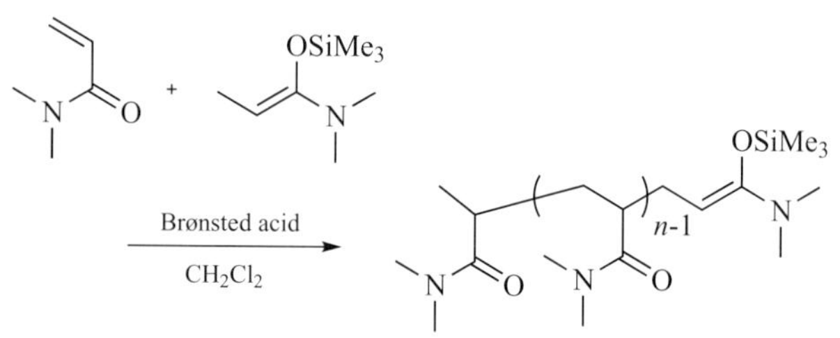

Temp (°C)	Tf_2NH K (L mol^{-1})	Tf_2NH k_p (L mol^{-1}s^{-1})	1 K (L mol^{-1})	1 k_p (L mol^{-1}s^{-1})
−40	17.3	272	29.4	387
−30	13.4	457	25.6	581
−20	10.6	770	22.4	931
−10	7.6	1230	19.6	1260

(25)

The in situ generation of silicon Lewis acid from **1** has moderate catalytic activity in the olefination reaction of lactones with ketene silyl acetals (eq 26).[25]

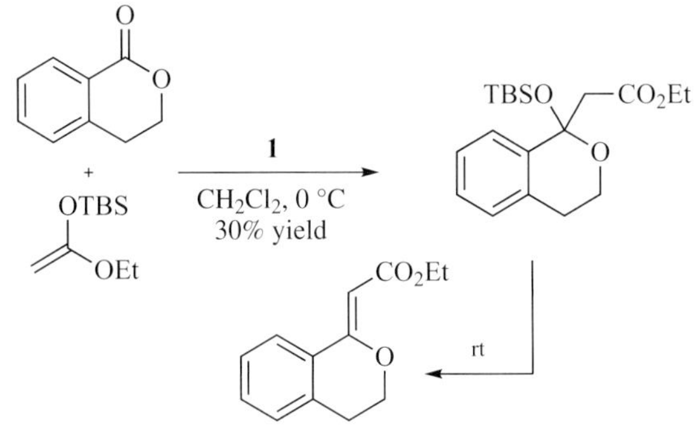

(26)

The cross-aldol reaction of α-haloacetaldehyde-derived silyl enol ethers catalyzed by Brønsted acid **1**, affording β-siloxy-α-haloaldehydes with excellent *anti*-stereoselectivities, and the sequential double-aldol reaction of such enol ethers catalyzed by $C_6F_5CTf_2AlMe_2$, furnishing polyketide-like fragments in one-pot reaction (eq 27).[26a,b]

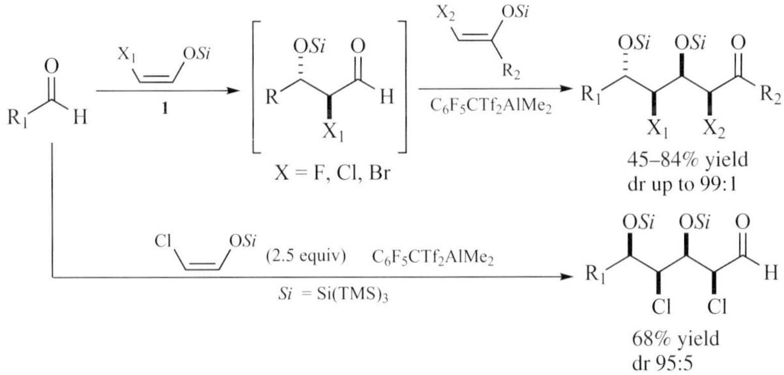

(27)

The *E*/*Z* geometry of tris(trimethylsilyl)silyl enol ethers derived from propionaldehyde controls diasteroselectivity in the reactions of aldehydes, thus allowing the synthesis of four different dipropionate stereotetrads (eq 28).[20c]

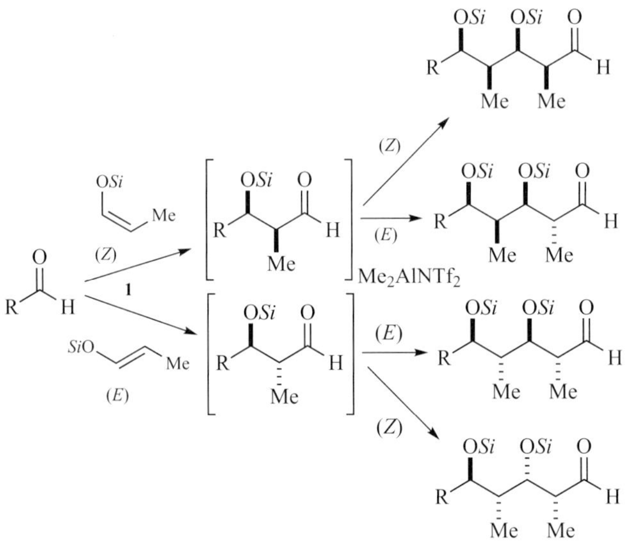

(28)

Conjugate base of **1** is also a useful anionic part of not only an acid generator for the resist composition in the field of discrete semiconductor devices and liquid crystal display,[27] but also a salt contained in an actinic ray-sensitive or radiation-sensitive resin composition forming film which is used in electronic device manufacturing method and an electronic device.[28]

Brønsted Acid-activated Chiral Oxazaborolidine. The cationic oxazaborolidine catalyzes Diels–Alder reaction of acetylenic ketones with various cyclic and acyclic dienes with excellent levels of asymmetric induction (eq 29).[29]

Related Reagents. Trifluoromethanesulfonic Acid; trifluoromethanesulfonimide; bis(trifluoromethanesulfonyl)methane; tris(trifluoromethanesulfonyl)methane; [bis(triflloromethane sulfonyl)methy]benzene.

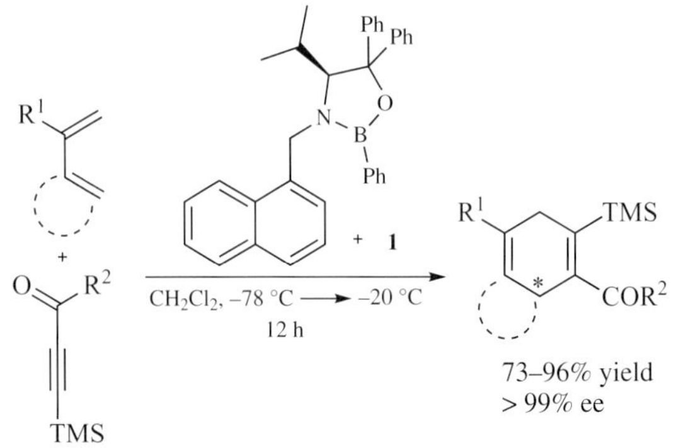

(29)

1. Hasegawa, A.; Ishikawa, T.; Ishihara, K.; Yamamoto, H., *Bull. Chem. Soc. Jpn.* **2005**, *78*, 1401.
2. Rode, B. M.; Engelbrecht, A.; Schantl, J., *Z. Phys. Chem.* **1973**, *253*, 17.
3. Ishihara, K.; Hasegawa, A.; Yamamoto, H., *Angew. Chem., Int. Ed.* **2001**, *40*, 4077.
4. Kokubo, Y.; Hasegawa, A.; Kuwata, S.; Ishihara, K.; Yamamoto, H.; Ikariya, T., *Adv. Synth. Catal.* **2005**, *347*, 220.
5. Nakashima, D.; Yamamoto, H., *Org. Lett.* **2005**, *7*, 1251.

6. Ishihara, K.; Hasegawa, A.; Yamamoto, H., *Synlett* **2002**, 1296.
7. (a) Horváth, I. T., *Acc. Chem. Res.* **1998**, *31*, 641. (b) Cavazzini, M.; Montanari, F.; Pozzi, G.; Quici, S., *J. Fluorine Chem.* **1999**, *94*, 183. (c) Bhattacharyya, P.; Croxtall, B.; Fawcett, J.; Gudmunsen, D.; Hope, E. G.; Kemmitt, R. D. W.; Paige, D. R.; Srauart, A. M.; Wood, D. R. W., *J. Fluorine Chem.* **2000**, *101*, 247.
8. Ishihara, K.; Kondo, S.; Yamamoto, H., *Synlett* **2001**, 1371.
9. Wende, M.; Meier, R.; Gladysz, J. A., *J. Am. Chem. Soc.* **2001**, *123*, 11490.
10. Ishihara, K.; Hasegawa, A.; Yamamoto, H., *Synlett* **2002**, 1299.
11. (a) Ishihara, K.; Yamamoto, H. In *Handbook of Fluorous Chemistry*; Gladysz, J. A.; Curran., D. P., Horváth, I. T., Eds.; Wiley: Weinheim, **2004**; p. 386. (b) Ishihara, K.; Hasegawa, A.; Yamamoto, H., *Angew. Chem., Int. Ed.* **2003**, *42*, 5731.
12. Suzuki, N.; Toda, N.; Kato, K.; Onosawa, T.; Tahara, K., *Jpn. Pat.* 2004307446 (**2004**).
13. Kurata, T., WO 2006120888 (**2006**).
14. Tsuzuki, T.; Nakajima, K.; Yamamoto, T.; Ishihara, K., *Jpn. Pat.* 2003272665 (**2003**).
15. Vreis, T. S. D.; Vedejs, E., *Organometallics* **2007**, *26*, 3079.
16. (a) Li, P.; Payette, J. N.; Yamamoto, H., *J. Am. Chem. Soc.* **2007**, *129*, 9534. (b) Payette, J. N.; Yamamoto, H., *J. Am. Chem. Soc.* **2007**, *129*, 9536.
17. Mathieu, B.; Ghosez, L., *Tetrahedron* **2002**, *58*, 8219.
18. (a) Kutt, A.; Rodima, T.; Saame, J.; Raamat, E.; Maemets, V.; Kaljurand, I.; Koppel, I. A.; Garlyauskayte, R. Y.; Yagupolskii, Y. L.; Yagupolskii, L. M.; Bernhardt, E.; Willner, H.; Leito, I., *J. Org. Chem.*, **2011**, *76*, 391. (b) Trummal, A.; Rummel, A.; Lippmaa, E.; Koppel, I.; Koppel, I. A., *J. Phys. Chem. A* **2011**, *115*, 6641.
19. (a) Makiguchi, K.; Satoh, T.; Kakuchi, T., *J. Polym. Sci. Part A: Polym. Chem.* **2011**, *49*. 3769. (b) Makiguchi, K.; Kaga, H.; Satoh, T.; Kakuchi, T., *Polym. Prepr.* **2012**, *53*, 195.
20. (a) Ishizuka, K.; Funyu, S.; Asano, N., Jpn. Pat. 2012072310 (**2012**). (b) Ishizuka, K.; Funano, S.; Asano, N., Jpn. Pat. 2015015178 (**2015**).
21. Takenaka, Y.; Abe, H., Jpn. Pat. 2014231601 (**2014**).
22. Tsuchimoto, A.; Nomiyama, S., Jpn. Pat. 2013184920 (**2013**).
23. Yanai, H.; Takahashi, A.; Taguchi, T., *Chem. Commun.* **2010**, *46*, 8728.
24. (a) Fuchise, K.; Chen, Y.; Tanaka, K.; Satoh, T.; Kakuchi, T., *Macromol. Chem. Phys.* **2012**, *213*, 1604. (b) Takada, K.; Fuchise, K.; Chen, Y.; Satoh, T.; Kakuchi, T., *J. Polym. Sci. Part A: Polym. Chem.* **2012**, *50*. 3560. (c) Takeda, K.; Fuchise, K.; Satoh, T.; Kakuchi, T., *Polym. Prepr.* **2012**, *53*, 197.
25. Yanai, H.; Taguchi, T., *Chem. Commun.* **2012**, *48*, 8967.
26. (a) Saadi, J.; Akakura, M.; Yamamoto, H., *J. Am. Chem. Soc.* **2011**, *133*, 14248. (b) Saadi, J.; Yamamoto, H., *Chem. Eur. J.* **2013**, *19*, 3842. (c) Brady, P. B.; Yamamoto, H., *Angew. Chem., Int. Ed.* **2012**, *51*, 1942.
27. (a) Kawaue, A.; Utsumi, Y.; Seshimo, T.; Nakamura, T.; Moritake, N.; Shimizu, H.; Matsuzawa, K.; Hada, H. US 20100015552. (b) Kawaue, A.; Utsumi, Y.; Abe, S. US20100081088. (c) Kawakami, K.; Utsumi, Y.; Matsuzawa, K.; Seshita, T. JP 2010077103. (d) Utsumi, Y.; Kawakami, A.; Hirayama, H.; Haneda, H.; Seshita, T.; Matsuzawa, K. JP 2010097081. (e) Utumi, Y.; Haneda, H.; Shimizu, H.; Nakamura, T.; Kawakami, K.; Seshita, T.; Matsuzawa, K.; Michigai, K. JP 2010120923. (f) Kawaue, A.; Utsumi, Y.; Hada, H.; Seshimo, T.; Matsuzawa, K.; Ishiduka, K.; Endo, K. US 20100136478. (g) Haneda, H.; Utsumi, Y.; Seshita, T.; Kawakami, K. JP 2010138158. (h) Kawakami, K.; Utsumi, Y.; Hidesaka, S.; Maruyama, N. JP 2010139662. (i) Utsumi, Y.; Irie, M. JP 2010215600. (j) Seshimo, T.; Utsumi, Y.; Komuro, Y.; Mimura, T.; Takaki, D. US 20100297560. (k) Utsumi, Y.; Irie, M. JP 2011043783. (l) Haneda, H.; Utsumi, Y. JP 2011081044. (m) Kawakami, K.; Seshita, T.; Utsumi, Y. JP 2011081146.
28. (a) Yamaguchi, S. JP 2013257468. (b) Yamaguchi, S. WO 2013187520.
29. Payette, J. N.; Yamamoto, H., *Angew. Chem., Int. Ed.* **2009**, *48*, 8060.

3-[[3,5-Bis(trifluoromethyl)phenyl]-amino]-4-[[(1*R*,2*R*)-2-(dimethylamino)-cyclohexyl]amino]-3-cyclobutene-1,2-dione

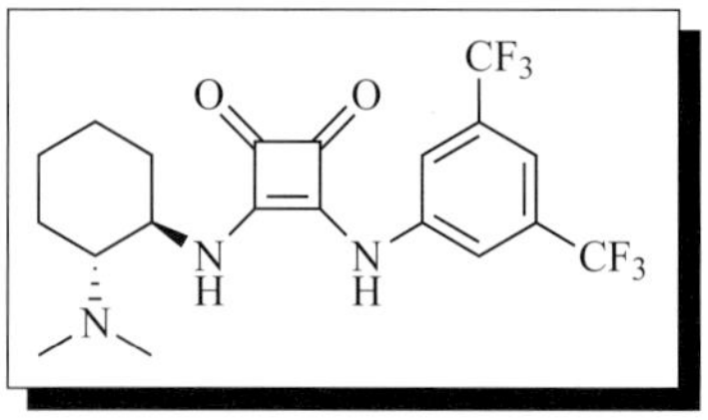

[1211565-07-9] $C_{20}H_{21}F_6N_3O_2$ (MW 449.39)

InChI = 1S/C20H21F6N3O2/c1-29(2)14-6-4-3-5-13(14)28-16-15(17(30)18(16)31)27-12-8-10(19(21,22)23)7-11(9-12)20(24,25)26/h7-9,13-14,27-28H,3-6H2,1-2H3/t13-,14-/m1/s1

InChIKey = QQXCDABMANNFAA-ZIAGYGMSSA-N

(reagent used as bifunctional organocatalyst in asymmetric catalysis)

Physical Data: mp 196–198 °C (decomp.); $[\alpha]_D^{28}$ −50.76 (*c* 1.00, DMSO).

Solubility: dimethyl sulfoxide.

Form Supplied in: white solid.

Purification: recrystallized from *i*-PrOH and H_2O.

Analysis of Reagent Purity: ^{1}H NMR (500 MHz, DMSO-d_6) δ 10.36 (br s, 1*H*), 8.07 (s, 2*H*), 7.85 (br s, 1*H*), 7.65 (s, 1*H*), 3.78–3.93 (m, 1*H*), 2.36–2.46 (m, 1*H*), 2.19 (s, 6*H*), 2.04–2.14 (m, 1*H*), 1.80–1.90 (m, 1*H*), 1.71–1.79 (m, 1*H*), 1.63–1.70 (m, 1*H*), 1.13–1.41 (m, 4*H*); ^{13}C NMR (125 MHz, DMSO-d_6) δ 184.4, 179.9, 169.4, 162.2, 141.2, 131.7, 131.4, 131.1, 130.9, 126.4, 124.2, 122.0, 119.9, 117.9, 114.5, 66.1, 54.7, 39.9, 34.4, 24.3, 21.2; IR (film) 3207, 3149, 2944, 2866, 2788, 1790, 1671, 1654, 1579, 1531, 1488, 1473, 1454, 1379, 1276, 1174, 1121 cm^{-1}; HRMS (ESI) $[M+H]^+$ calcd. for $C_{20}H_{21}F_6N_3O_2$: 450.1611, found 450.1605.

Handling, Storage, and Precautions: can be stored in air at room temperature.

Introduction. 3-[[3,5-Bis(trifluoromethyl)phenyl]amino]-4-[[(1*R*,2*R*)-2-(dimethylamino)cyclohexyl]amino]-3-cyclobutene-1,2-dione is considered to be an important chiral bifunctional organocatalyst and is used in the hydrogen bonding catalysis for a series of asymmetric transformations. While the more extensive substrate–squaramide hydrogen bonding possibilities compared to those with urea/thiourea analogues may lead to a broader substrate scope, the rigidity, offered by the partially aromatic cyclobutenedione ring of the squaramide, in specific cases may result in improved molecular recognition of substrates, thus providing superior rate enhancements and stereoselective controls in these catalytic transformations.[1] An advantage of chiral squaramides as organocatalysts is the simple and modular preparation, which offers a facile route to a wide range of bifunctional acid–base catalysts.

Preparation. 3-[[3,5-Bis(trifluoromethyl)phenyl]amino]-4-[[(1*R*,2*R*)-2-(dimethylamino)cyclohexyl]amino]-3-cyclobutene-1,2-dione was prepared for the first time by Rawal and coworkers[2–4] in 2010 from the easily available or commercial starting materials in three steps (eq 1). A mixture of 3,4-dimethoxy-3-cyclobutene-1,2-dione and 3,5-bis(trifluoromethyl)aniline in MeOH was stirred at room temperature for 2 days (a yellow precipitate was generated). The reaction mixture was filtered and washed with MeOH and then dried *in vacuo* to give the desired squaramide **3**. Subsequently, **3** was added to a solution of (1*R*,2*R*)-1,2-diaminocyclohexane in CH_2Cl_2, and the resulting mixture was stirred at room temperature for 2 h and a white precipitate was generated. The crude amine **4** was obtained through filtration. Finally, to a solution of the crude amine in MeOH was added 37% aqueous HCHO and HCO_2H at room temperature. The reaction mixture was stirred at 65 °C for 12 h and then basified by 6 mol L^{-1} aqueous NaOH to pH 12 and extracted with CH_2Cl_2. The combined organic layers were dried over $MgSO_4$ and then concentrated. The residue was purified by recrystallization from *i*-PrOH and H_2O to give squaramide organocatalyst **5** as a white solid.

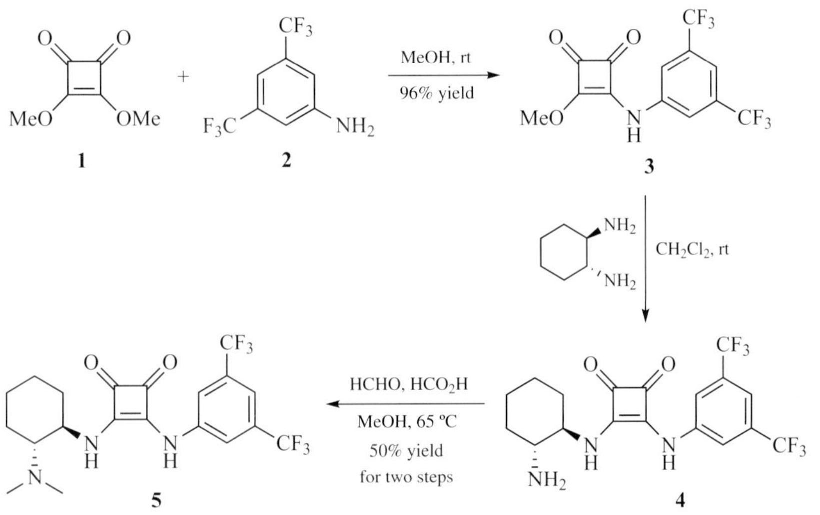

(1)

Michael Addition Reactions. This squaramide catalyst exhibited excellent catalytic activity in the asymmetric Michael addition reaction of diphenyl phosphinate to *trans*-*β*-nitrostyrene affording the corresponding adduct in high conversion and with good enantioselectivity (eq 2).[2] This method leads to chiral *β*-nitro phosphonates, which are precursors to biologically active *β*-amino phosphonic acids.

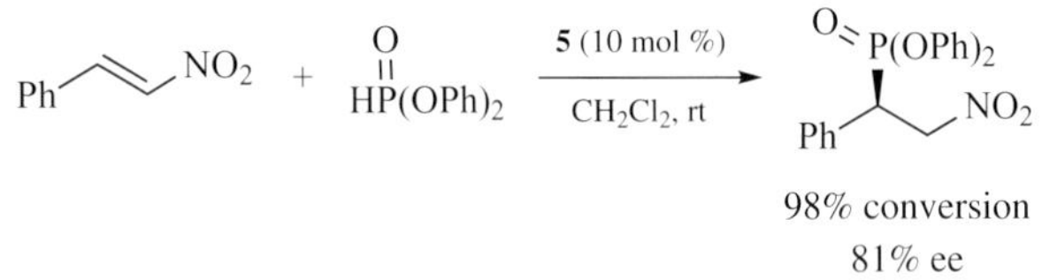

(2)

Friedel–Crafts Reactions. The Friedel–Crafts reaction represents the main thoroughfare by which C–C bonds are introduced into the indole framework. Chiral squaramide **5** displayed excellent asymmetric induction for the asymmetric Friedel–Crafts reaction of indoles with *N*-tosyl imines, affording 3-indolyl methanamine products with good to excellent enantioselectivities with as low as 2.5 mol % catalyst loading (eq 3).[3] The scope of the reaction was evaluated by using different imines and indoles. Broad tolerance of imines was demonstrated, while only unsubstituted or electron-rich indoles were applied as nucleophiles.

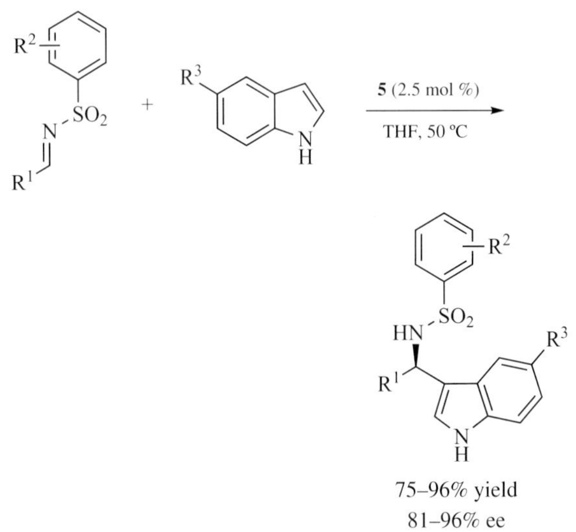

(3)

***α*-Amination Reactions.** The asymmetric *α*-hydrazination of carbonyl compounds with azodicarboxylates is a powerful tool with which to install protected amino groups with control of enantioselectivity. When *α*-monosubstituted-1,3-dicarbonyl compounds are used as nucleophiles, quaternary *α*-amino acid derivatives are obtained as adducts, which are of special interest as building blocks in organic synthesis. Squaramide organocatalyst **5** possessing the (*R*,*R*)-1,2-diaminocyclohexane unit successfully promoted the addition reaction of *β*-ketoesters to diethyl azodicarboxylate in high yield and with high enantioselectivity with low catalyst loading under mild conditions (eq 4).[4]

Desymmetrization Reactions. The enantioselective desymmetrization of cyclic *meso*-anhydrides offers access to optically active compounds, which represent important chiral intermediates in the synthesis of many natural products and biologically active substances. Squaramide organocatalyst **5** is a highly suitable catalyst for the enantioselective methanolysis of glutaric anhydride.

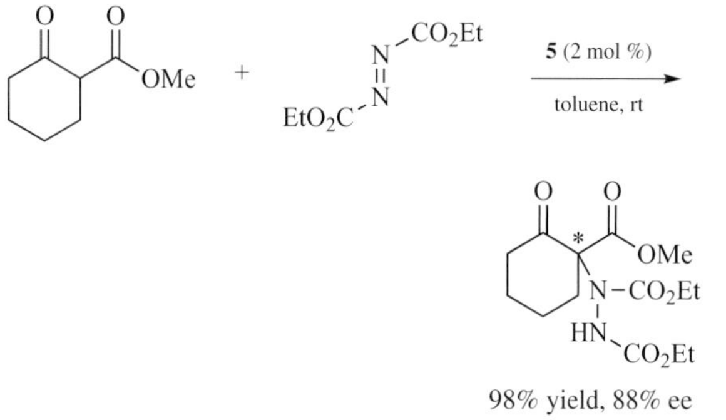

(4)

The reaction showed a very strong dilution effect, and decreasing the concentration of the anhydride from 0.2 to 0.0125 mol L^{-1} led to a significant increase in enantioselectivity (eq 5).[5]

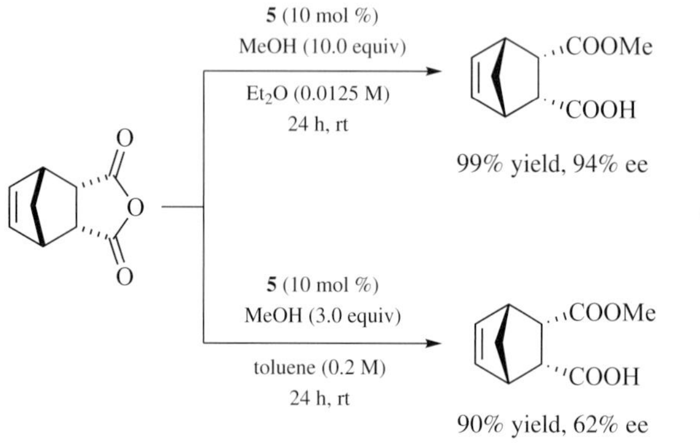

(5)

Neber Reactions. 2*H*-Azirines are the smallest heterocycles that contain a C=N double bond in a three-membered ring. Several naturally occurring 2*H*-azirines such as azirinomycin and dysidazirine have been isolated as promising antibiotics. The enantioselective Neber reaction of *O*-Ts-oxime *tert*-butyl ester catalyzed by bifunctional squaramide **5** gave the 2*H*-azirine carboxylic esters in moderate yield and with low enantioselectivity (eq 6).[6]

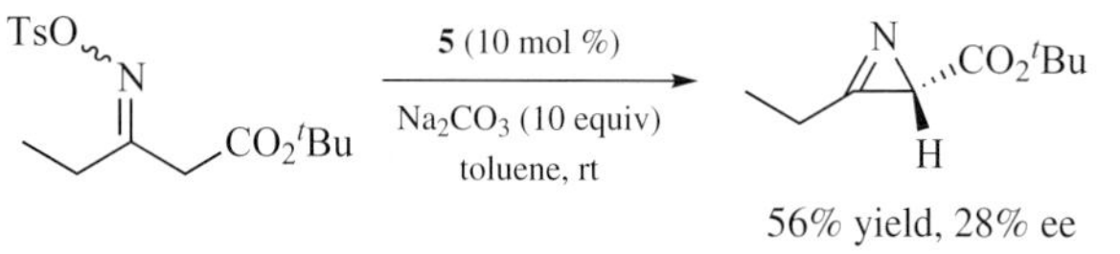

(6)

Sulfa-Michael Addition Reactions. Enantioselective protonation of prochiral enolates has become one of the most intensively studied reactions in the field of asymmetric catalysis over the last several years. This reaction provides an efficient means of access to a wide range of optically active carbonyl compounds that possess a tertiary stereogenic carbon at the α-position. The tandem Michael–protonation reaction of benzyl thiol to α-methyl *N*-benzoyl acrylamide catalyzed by squaramide **5** proceeded smoothly to give the corresponding adduct in good yield and with moderate enantioselectivity (eq 7).[7]

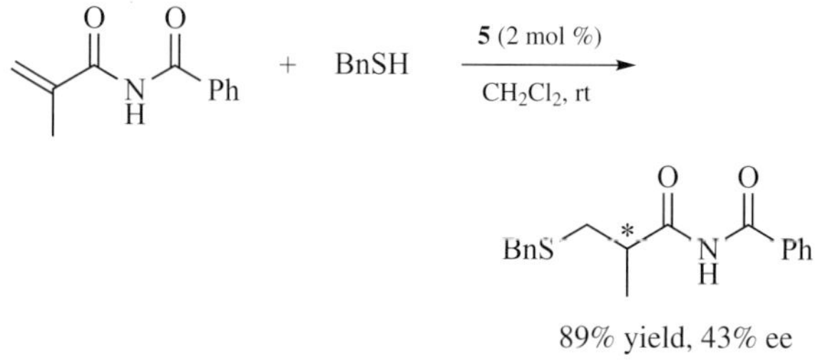

(7)

Sulfenylation Reactions. 3,3-Disubstitued oxindoles are important precursors for the synthesis of hexahydropyrrolo[2,3-*b*]indoles, which exist in a series of biologically active natural products. Oxindoles with a heteroatom at the C-3 position have received considerable attention owing to their potential applications in medicinal chemistry. Many oxindoles with a thio group at the carbon stereocenter have been found to have anticancer, antifungal, or antitubercular activities. The first asymmetric sulfenylation of *N*-Boc-protected oxindoles by using *N*-(sulfanyl)phthalimides as the sulfenylating agents was efficiently promoted by the low catalyst loading of squaramide **5**, furnishing the products bearing a tetrasubstituted stereocenter in high yields and with good to excellent enantioselectivities (eq 8).[8]

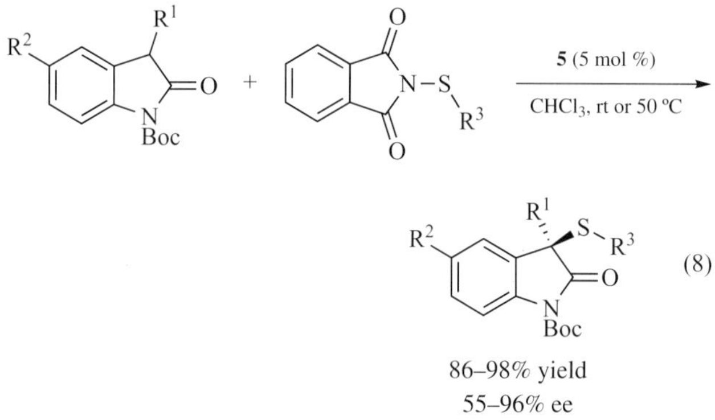

(8)

Other Reactions. Chiral *N,S*-acetals are common structural motifs present in numerous biologically active natural products and pharmaceuticals. The trifluoromethylated *N,S*-acetal motif may introduce some unprecedented benefits and is expected to find valuable applications in medicinal chemistry. The catalytic asymmetric addition of benzyl mercaptan to trifluoromethylated aldimine catalyzed by chiral squaramide **5** proceeded efficiently affording the expected adducts in high yields and with moderate enantioselectivity (eq 9).[9] This present method serves as a powerful approach to a new family of trifluoromethylated *N,S*-acetals.

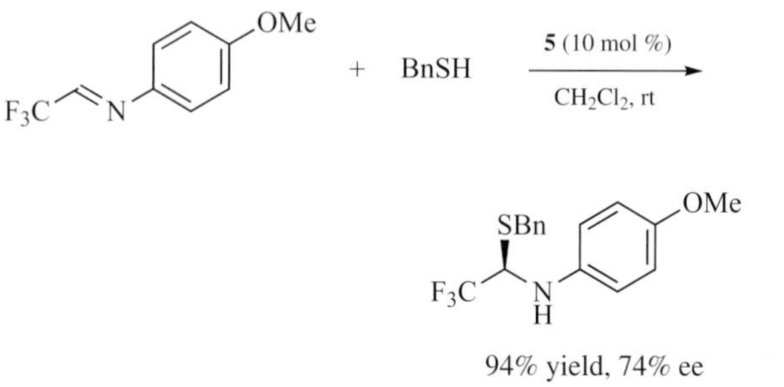

(9)

1. Alemán, J.; Parra, A.; Jiang, H.; Jørgensen, K. A., *Chem. Eur. J.* **2011**, *17*, 6890.
2. (a) Zhu, Y.; Malerich, J. P.; Rawal, V. H., *Angew. Chem.* **2010**, *122*, 157. (b) Zhu, Y.; Malerich, J. P.; Rawal, V. H., *Angew. Chem., Int. Ed.* **2010**, *49*, 153.
3. Qian, Y.; Ma, G.; Lv, A.; Zhu, H.-L.; Zhao, J.; Rawal, V. H., *Chem. Commun.* **2010**, *46*, 3004.
4. Konishi, H.; Lam, T. Y.; Malerich, J. P.; Rawal, V. H., *Org. Lett.* **2010**, *12*, 2028.
5. Schmitt, E.; Schiffers, I.; Bolm, C., *Tetrahedron* **2010**, *66*, 6349.
6. Sakamoto, S.; Inokuma, T.; Takemoto, Y., *Org. Lett.* **2011**, *13*, 6374.
7. Dai, L.; Yang, H.; Niu, J.; Chen, F.-E., *Synlett* **2012**, *23*, 314.
8. Wang, C.; Yang, X.; Loh, C. C. J.; Raabe, G.; Enders, D., *Chem. Eur. J.* **2012**, *18*, 11531.
9. Fang, X.; Li, Q.-H.; Tao, H.-Y.; Wang, C.-J., *Adv. Synth. Catal.* **2013**, *355*, 327.

Xin Fang & Chun-Jiang Wang
College of Chemistry and Molecular Sciences, Wuhan University, Wuhan, China

N-[3,5-Bis(trifluoromethyl)phenyl]-*N′*-(9*R*)-cinchonan-9-yl-thiourea[1]

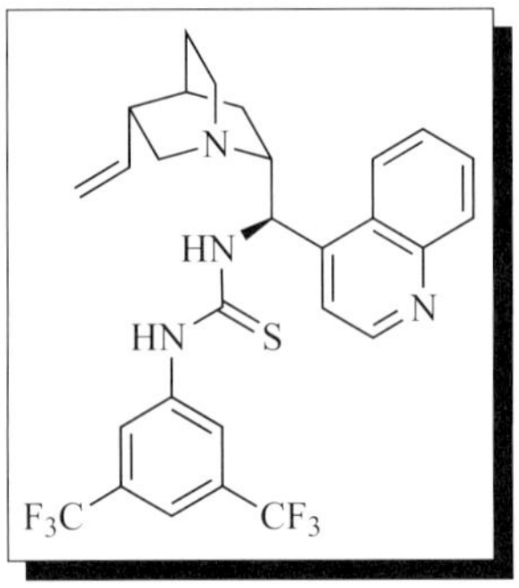

[851477-19-5] $C_{28}H_{26}F_6N_4S$ (MW 564.59)

InChI = 1S/C28H26F6N4S/c1-2-16-15-38-10-8-17(16)11-24(38)25(22-7-9-35-23-6-4-3-5-21(22)23)37-26(39)36-20-13-18(27(29,30)31)12-19(14-20)28(32,33)34/h2-7,9,12-14,16-17,24-25H,1,8,10-11,15H2,(H2,36,37,39)/t16-,17-,24+,25+/m0/s1

InChIKey = IIQBCCXHYIEQTC-WAJMBDEPSA-N

(cinchonine-derived bifunctional organocatalyst for stereoinduction in a range of reactions)

Physical Data: mp 186–189 °C;[2] $[\alpha]_D^{25}$ + 168.2 °C (c = 1.03, $CHCl_3$); IR; ^1H NMR; ^{13}C NMR.[3]

Solubility: DCE, DCM, $CHCl_3$ THF, Et_2O, TBME,[3] MeOH, toluene,[4] MeCN,[5] poor in *n*-hexane[3] at room temperature.

Form Supplied in: solid.

Analysis of Reagent Purity: NMR.

Preparative Methods: made in two steps. *Cinchonine* was converted into 9-amino(9-deoxy)epicinchonine (AECN)[6] via Mitsunobu reaction and then added to commercially available 3,5-bis(trifluoromethyl) phenyl isothiocyanate in THF to form thiourea (**1**) in 94% yield.[3]

Purification: column chromatography (elution gradient: EtOAc:MeOH:NEt_3 = 100:2:3 to 100:10:3)

Handling, Storage, and Precautions: bench stable, structurally similar commercial quinidine is toxic.

Like other cinchona thioureas, this one can be used as an organocatalyst in a variety of reactions from Michael to Mannich additions and Friedel–Crafts alkylations. One example shows its use in an enantioselective decarboxylation of α-aminomalonate.[7] Generally, enantioinduction can be significantly enhanced compared to cinchonine-catalyzed reactions; the aromatic residue of the thiourea, the inverted stereochemical configuration at C9 and the extended sites of hydrogen bonding are all important in this regard.

Michael Additions. Thiourea (**1**) can be used as a stereoinducing agent in Michael addition reactions. In detail, *phenylthiol* was added to *cinnamic acid*-derived imides, but only a low enantiomeric excess of 7% was observed.[4]

Many other examples involve the use of nitro olefins as electrophiles due to their ability to bind the catalyst through H-bonding interactions in the course of the reaction. Based on theoretical calculations, several models of hydrogen bonding of the thiourea moiety were discussed.[8]

Also 1,3-dicarbonyl pronucleophiles like *dimethylmalonate* (**2**) could be added to nitro olefins (**3**) with high enantioselectivity and in good yields. This provides a valuable γ-amino acid precursor (**4**) (eq 1), even in the presence of air and moisture.[3] By means of a solvent screen, high catalytic activity of **1** was revealed in ethereal and chlorinated solvents; however, DCM turns out to have the best solubilizing properties for both the starting materials and the products. Aryl-substituted substrates react faster than alkyl-substituted compounds in this type of reaction. A variation of this reaction was used to synthesize γ-butyrolactone autoregulators[9] and (*R*)-Rolipram.[10]

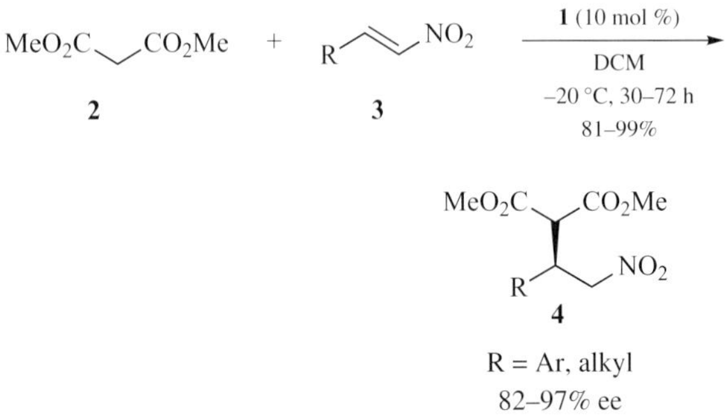

(1)

The procedure was expanded to create a one-pot Michael/nitro-Mannich/lactamization cascade to yield polycyclic piperidinone (**8**) with excellent enantioselectivity (eqs 2 and 3).[11]

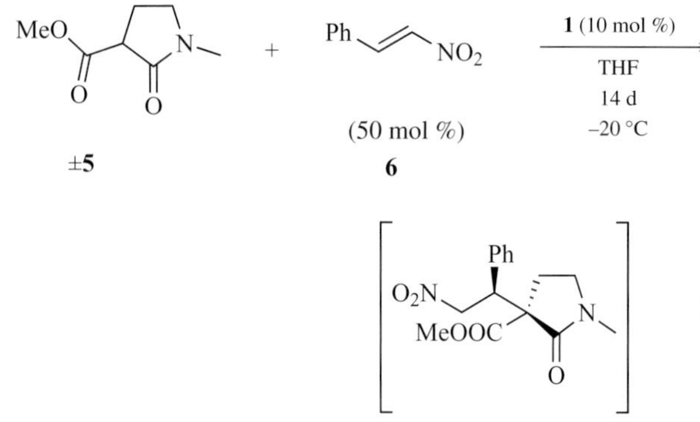

±5 6 7 (2)

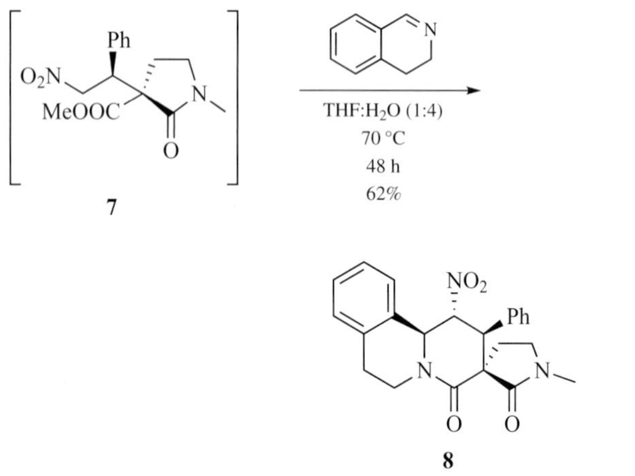

7 8 (3)

90% ee

A one-pot three-component reaction catalyzed by **1** using an aromatic aldehyde (**9**) (eq 4) and nitro olefin (**12**) yielded the Michael adduct (**13**) with high enantioselectivity (eq 5) through in situ formation of imine (**11**).[12]

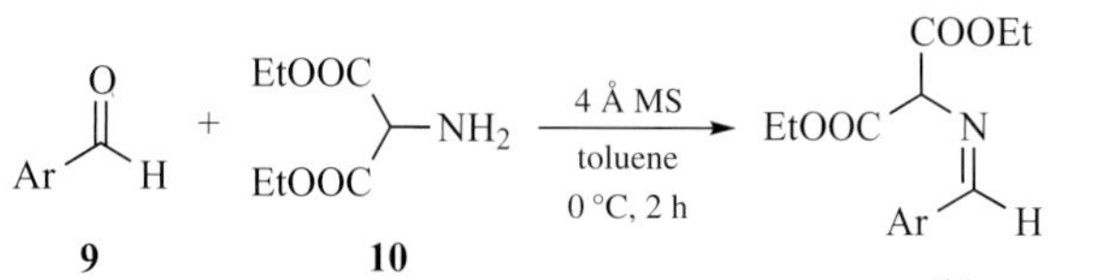

9 10 11 (4)

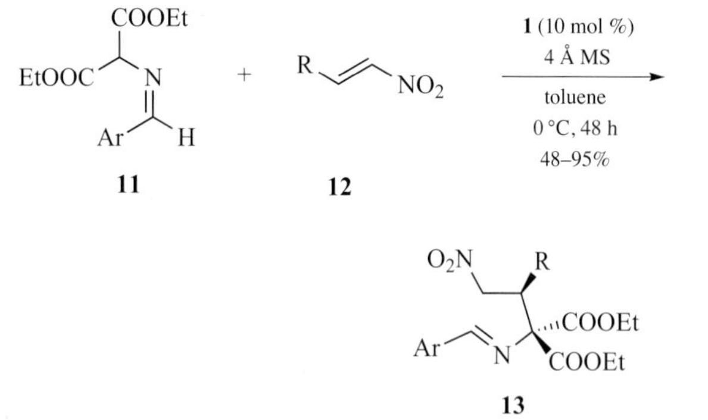

11 12 13 (5)

R = Ar, Alkyl

94–98% ee

Excellent enantio- and diastereoselectivities were achieved by the addition of α,β-unsaturated γ-butyrolactams (**14**) to chalcones (**15**) catalyzed by (**1**) (eq 6) at elevated temperatures.[5] When the aromatic residue situated next to the carbonyl was replaced with an alkyl residue, no product was formed.

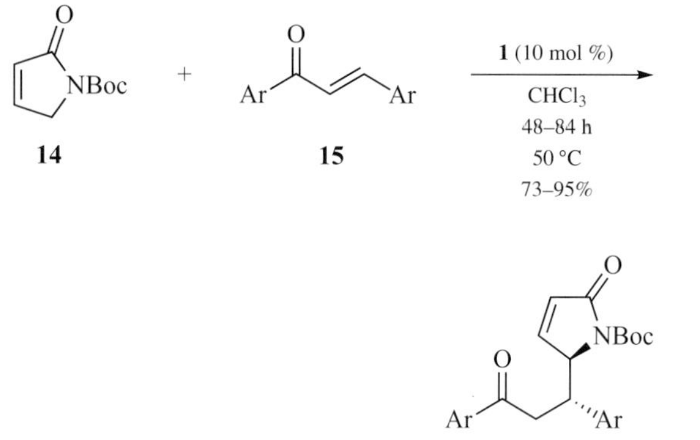

14 15 16 (6)

R = Ar, Alkyl

dr: 10:1–40:1; 94–99% ee

The Michael addition of C-succinimidyl ester (**17**) to nitrostyrene derivative (**18**) catalyzed by thiourea (**1**) (eq 7) resulted in excellent enantioselectivities for both diastereomers **19** and **20**, which could easily be separated by flash column chromatography.[2]

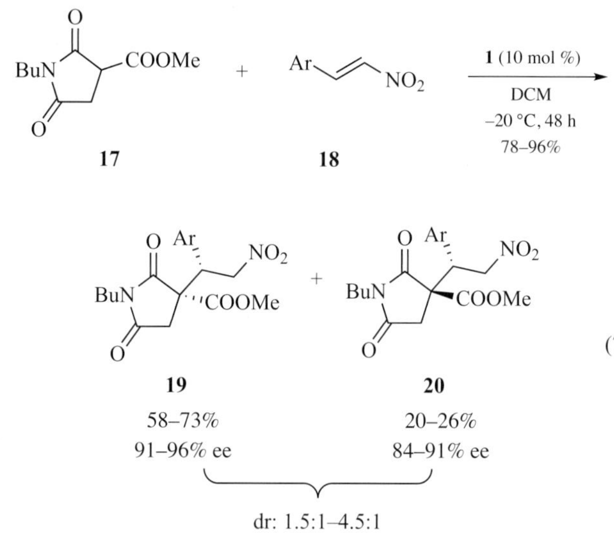

17 18 19 20 (7)

19: 58–73%, 91–96% ee; 20: 20–26%, 84–91% ee

dr: 1.5:1–4.5:1

Derivatives of acrylic acid[13] or indole-derived vinylogous imines[14] could also be used as Michael acceptors in the enantioselective additions catalyzed by **1**.

Catalyst **1** was screened in several other conjugate additions with pronucleophiles such as β-ketosulfones[15] or β-cyanoketones,[16] but did not prove to be optimal in these transformations.

Thiourea (**1**) was also screened against several other thiourea catalysts containing a tertiary amino group for the Michael addition of α-phenyl cyanoacetates to phenyl vinyl sulfone. This gave a high yield (96%), but only moderate enantioselectivity (43% ee) in the formation of the quaternary center.[17]

Two adjacent stereocenters were formed by Michael addition of racemic mandelic acid derivatives (**21**) to aromatic nitro olefins (**22**) under complete diastereo- and good enantiocontrol (eq 8).[18] A comparable approach afforded α-amino acid precursors with further control of two adjacent stereocenters.[19]

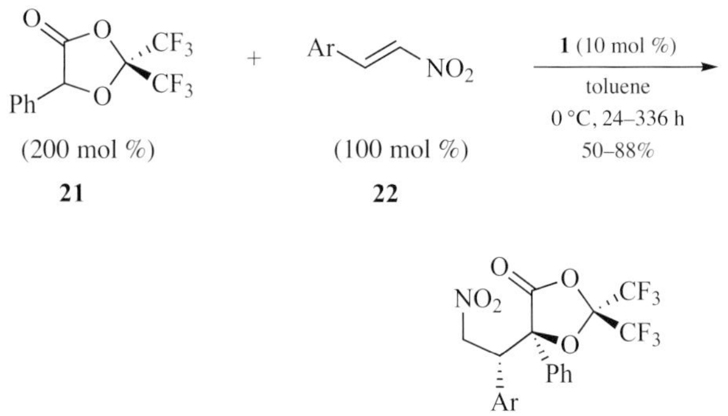

(200 mol %) **21** (100 mol %) **22**

23 93–98% de 60–89% ee (8)

Cyclohexadienones were desymmetrized in an aza-Michael reaction catalyzed by **1**. High yields and enantioselectivities were achieved (eq 9). A slightly modified protocol was used for the synthesis of (−)-Mesembrine.[20]

NMR studies provided insight into the complicated conformational dynamics of thiourea (**1**).[21] Thiourea (**1**) exists as a dimeric species in concentrated solutions bound together by hydrogen bonding and other interactions. Other similar thiourea catalysts were also screened.

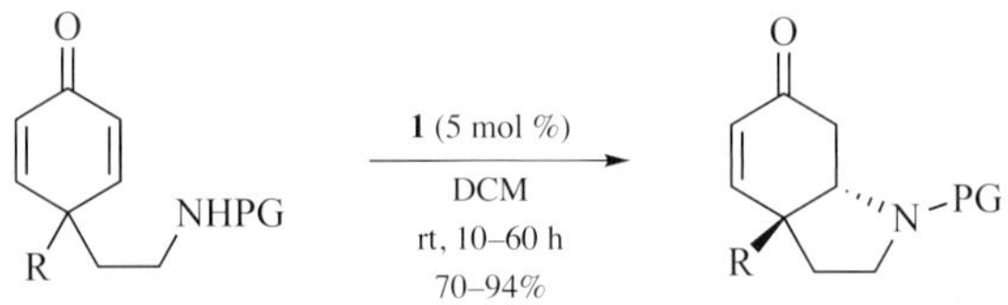

24 **25** (9)

R = OR, OH, NHAc
PG = Ts, *p*-Ns, Ms
60–99% ee

Mannich Reactions. Thiourea (**1**) could be used as an enantioselective catalyst in Mannich reactions. Several 1,3-dicarbonyl compounds (**26**) were used as pronucleophiles in the addition to *N*-protected aldimines (**27**).[22] Catalyst **1** afforded excellent yields and enantioselectivities, even where a quaternary stereocenter was formed (eq 10).

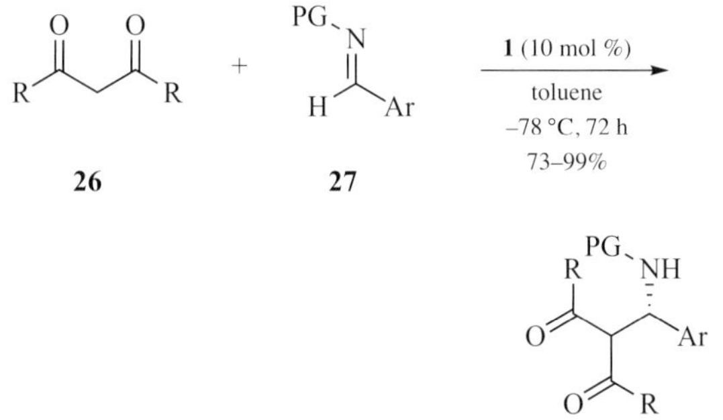

26 **27**

28 (10)

R = Me, OMe
PG = Cbz, Boc
64–97% ee

Thiourea (**1**) was compared to Takemoto's asymmetric thiourea in a domino Mannich/aza-Michael reaction.[23]

Friedel–Crafts Alkylation. Compound **1** was effective as an enantioselective catalyst in Friedel–Crafts alkylations of hydroxy arenes with electrophilic *trans*-β-nitrostyrene derivatives (**30**).[24] Both naphthol derivatives (**29**) and sesamol could act as nucleophiles to form **31** (eq 11).[25]

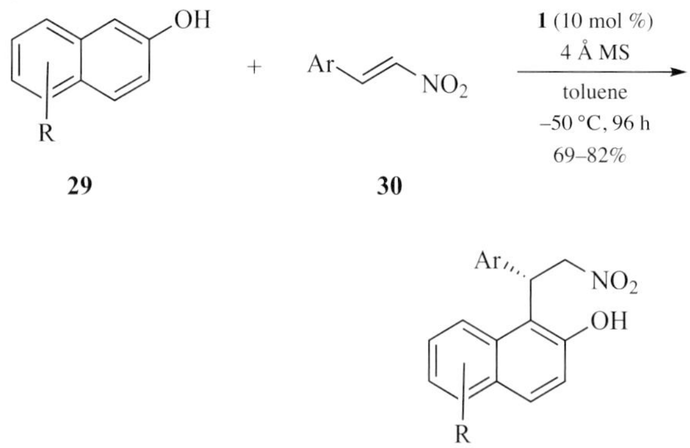

29 **30**

31 (11)

R = H, 7-MeO, 6-Br
85–95% ee

Related Reagents. *N*-[3,5-Bis(trifluoromethyl)phenyl]-*N*′-[(9*R*)-6′-methoxycinchonan-9-yl]-thiourea; *N*-[3,5-Bis(trifluoromethyl)phenyl]-*N*′-(8α,9*S*)-cinchonan-9-yl-thiourea; *N*-[3,5-bis(trifluoromethyl)phenyl]-*N*′-[(8α,9*S*)-6′-methoxycinchonan-9-yl]-thiourea.

---

1. Kotke, M.; Schreiner, P. R. In *Hydrogen Bonding in Organic Synthesis*; Pihko, P. M., Ed.; Wiley-VCH: Weinheim, **2009**.
2. Jakubec, P.; Cockfield, D. M.; Hynes, P. S.; Cleator, E.; Dixon, D. J., *Tetrahedron: Asymmetry* **2011**, *22*, 1147.
3. Ye, J.; Dixon, D. J.; Hynes, P. S., *Chem. Commun.* **2005**, 4481.
4. Li, B.-J.; Jiang, L.; Liu, M.; Chen, Y.-C.; Ding, L.-S.; Wu, Y., *Synlett* **2005**, 603.
5. Zhang, Y.; Shao, Y.-L.; Xu, H.-S.; Wang, W., *J. Org. Chem.* **2011**, *76*, 1472.
6. Brunner, H.; Bügler, J.; Nuber, B., *Tetrahedron: Asymmetry* **1995**, *6*, 1699.
7. Amere, M.; Lasne, M.-C.; Rouden, J., *Org. Lett.* **2007**, *9*, 2621.
8. (a) Hamza, A.; Schubert, G.; Soós, T.; Pápai, I., *J. Am. Chem. Soc.* **2006**, *128*, 13151. (b) Yalalov, D. A.; Tsogoeva, S. B.; Schmatz, S., *Adv. Synth. Catal.* **2006**, *348*, 826.
9. Elsner, P.; Jiang, H.; Nielsen, J. B.; Pasi, F.; Jorgensen, K. A., *Chem. Commun.* **2008**, 5827.
10. Hynes, P. S.; Stupple, P. A.; Dixon, D. J., *Org. Lett.* **2008**, *10*, 1389.
11. Jakubec, P.; Helliwell, M.; Dixon, D. J., *Org. Lett.* **2008**, *10*, 4267.
12. Liu, Y.-K.; Liu, H.; Du, W.; Yue, L.; Chen, Y.-C., *Chem. Eur. J.* **2008**, *14*, 9873.
13. Rigby, C. L.; Dixon, D. J., *Chem. Commun.* **2008**, 3798.
14. Jing, L.; Wei, J.; Zhou, L.; Huang, Z.; Li, Z.; Wu, D.; Xiang, H.; Zhou, X., *Chem. Eur. J.* **2010**, *16*, 10955.
15. Garcia, M. O.; Tangen, P.; Rohlmann, R.; Froehlich, R.; Aleman, J., *Chem. Eur. J.* **2011**, *17*, 984.
16. Wang, H.-F.; Li, P.; Cui, H.-F.; Wang, X.-W.; Zhang, J.-K.; Liu, W.; Zhao, G., *Tetrahedron* **2011**, *67*, 1774.
17. Liu, T.-Y.; Long, J.; Li, B.-J.; Jiang, L.; Li, R.; Wu, Y.; Ding, L.-S.; Chen, Y.-C., *Org. Biomol. Chem.* **2006**, *4*, 2097.

18. Hynes, P. S.; Stranges, D.; Stupple, P. A.; Guarna, A.; Dixon, D. J., *Org. Lett.* **2007**, *9*, 2107.
19. Aleman, J.; Milelli, A.; Cabrera, S.; Reyes, E.; Joergensen, K. A., *Chem. Eur. J.* **2008**, *14*, 10958.
20. Gu, Q.; You, S.-L., *Chem. Sci.* **2011**, *2*, 1519.
21. Kiraly, P.; Soos, T.; Varga, S.; Vakulya, B.; Tarkanyi, G., *Magn. Reson. Chem.* **2010**, *48*, 13.
22. Tillman, A. L.; Ye, J.; Dixon, D. J., *Chem. Commun.* **2006**, 1191.
23. Enders, D.; Goeddertz, D. P.; Beceno, C.; Raabe, G., *Adv. Synth. Catal.* **2010**, *352*, 2863.
24. Liu, T.-Y.; Cui, H.-L.; Chai, Q.; Long, J.; Li, B.-J.; Wu, Y.; Ding, L.-S.; Chen, Y.-C., *Chem. Commun.* **2007**, 2228.
25. Zhang, H.; Liao, Y.-H.; Yuan, W.-C.; Zhang, X.-M., *Eur. J. Org. Chem.* **2010**, 3215.

Darren J. Dixon & Linus Stegbauer
University of Oxford, Oxford, UK

N-[3,5-Bis(trifluoromethyl)phenyl]-*N*′-[(1*R*,2*R*)-2-(dimethylamino)cyclohexyl]-thiourea

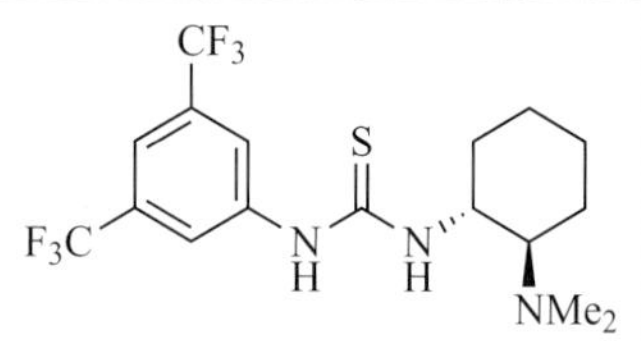

(1*R*,2*R*-**1**)

[620960-26-1] $C_{17}H_{21}F_6N_3S$ (MW 413.42)
InChI = 1S/C17H21F6N3S/c1-26(2)14-6-4-3-5-13(14)25-15(27)24-12-8-10(16(18,19)20)7-11(9-12)17(21,22)23/h7-9,13-14H,3-6H2,1-2H3,(H2,24,25,27)/t13-,14-/m1/s1
InChIKey = NQRCAVZHOLYEBJ-ZIAGYGMSSA-N
[851477-20-8] $C_{17}H_{21}F_6N_3S$ (MW 413.42)
InChI = 1S/C17H21F6N3S/c1-26(2)14-6-4-3-5-13(14)25-15(27)24-12-8-10(16(18,19)20)7-11(9-12)17(21,22)23/h7-9,13-14H,3-6H2,1-2H3,(H2,24,25,27)/t13-,14-/m1/s1
InChIKey = NQRCAVZHOLYEBJ-ZIAGYGMSSA-N

(asymmetric catalyst for conjugate additions, 1,2-additions, and cycloadditions; catalyst for condensation polymerization)

Physical Data: mp 111–113 °C; $[\alpha]^{D}_{16}$ −32.7 (*c* 0.99, $CHCl_3$).[1]
Solubility: soluble in toluene, ether, ethyl acetate, and most organic solvents, but insoluble in water.
Form Supplied in: colorless powder; commercially available in either enantiomeric form.
Purification: silica gel chromatography.
Analysis of Reagent Purity: ^{1}H NMR and ^{13}C NMR ($CDCl_3$).[1]
Handling, Storage, and Precautions: store in a tightly sealed container. Keep in a cool, dry, well-ventilated place.

Enantioselective Conjugate Addition to Electron-deficient C=C Bonds. This chiral reagent possesses a Brønsted acidic thiourea function and a basic amine function placed so as not to deactivate each other. Thus, it can synergistically activate electrophiles and nucleophiles and catalyze bond-forming reactions with asymmetric induction. Nitroolefins (eqs 1–4),[1,2] α,β-unsaturated ketoesters (eq 5),[3] α,β-unsaturated imides (eq 6),[4] maleimides,[5] vinylsulfones,[6] and simple ketones/aldehydes (eq 7)[7] can be used as Michael acceptors. Malonates, β-ketoesters, malononitriles, oxindoles, and several carbonyl compounds having an activated proton can serve as nucleophilic partners. Thiols, thioacetic acid, phosphites, and phenols also react as nucleophiles to form a carbon–heteroatom bond. In general, the stereogenic center is formed at the β-carbon of the Michael acceptor with high enantioselectivity. In some cases, chirality is induced at the α-carbon of the Michael acceptor and/or the nucleophilic carbon of the nucleophile. There are some reports on the total synthesis of naturally occurring substances, such as esermethole[2a] and epibatidine,[2f] and synthetic drugs, such as baclofen[2d] and pregabalin,[2j] using this asymmetric catalyst.

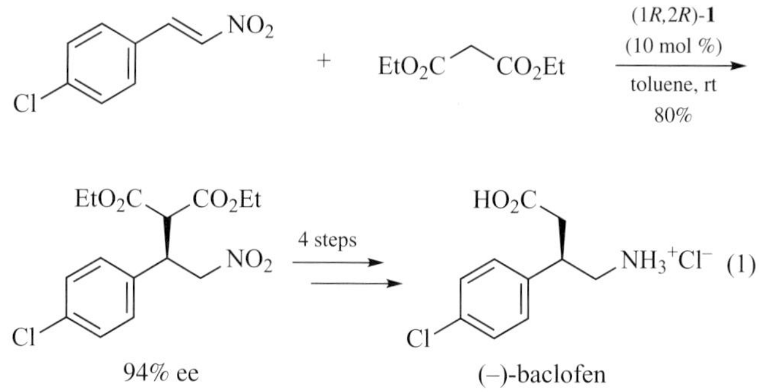

(1)

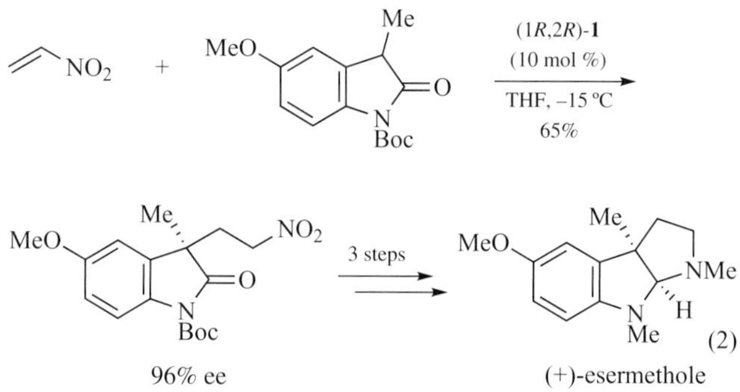

(2)

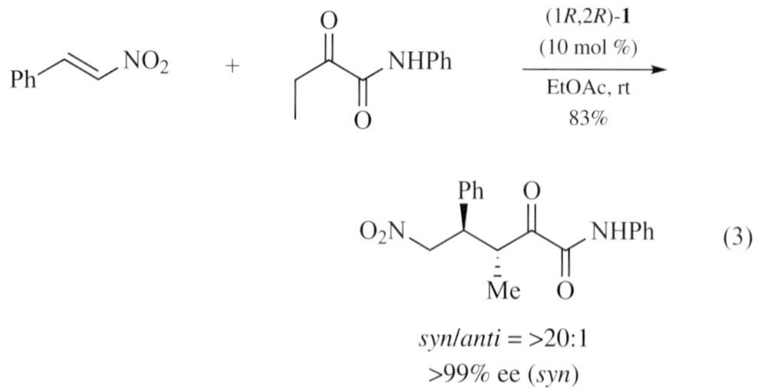

(3)

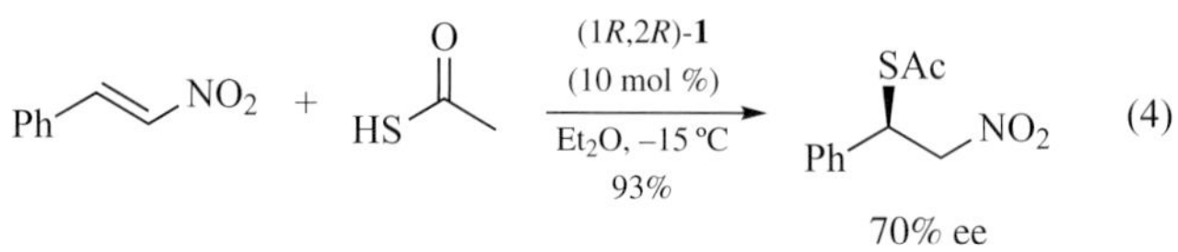

(4)

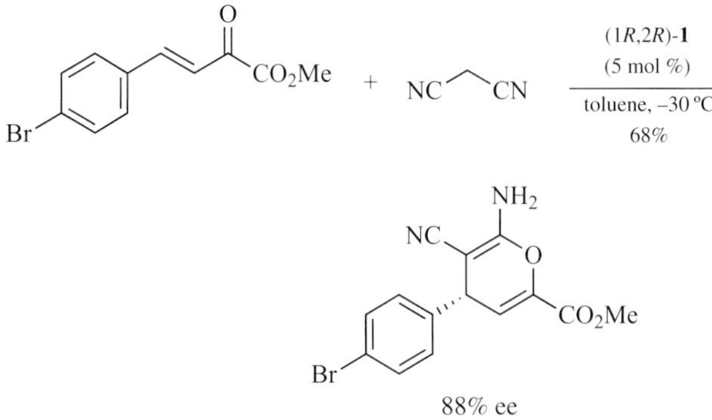

(5)

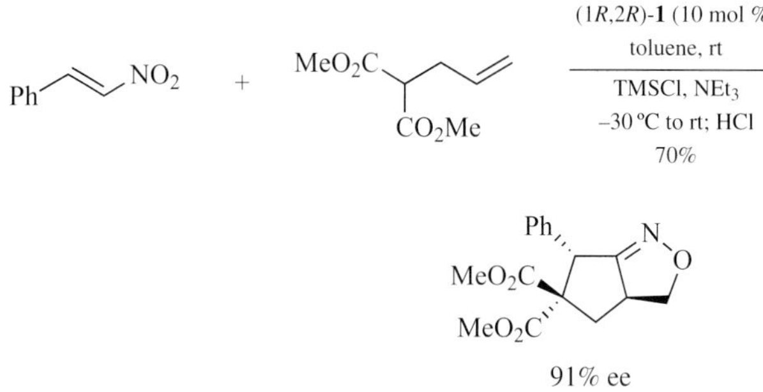

(9)

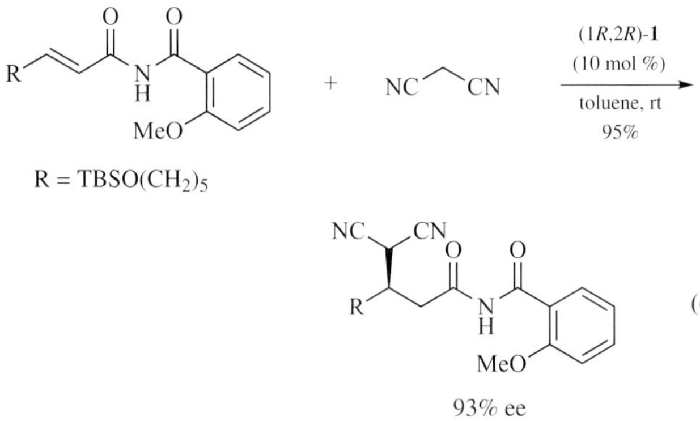

(6)

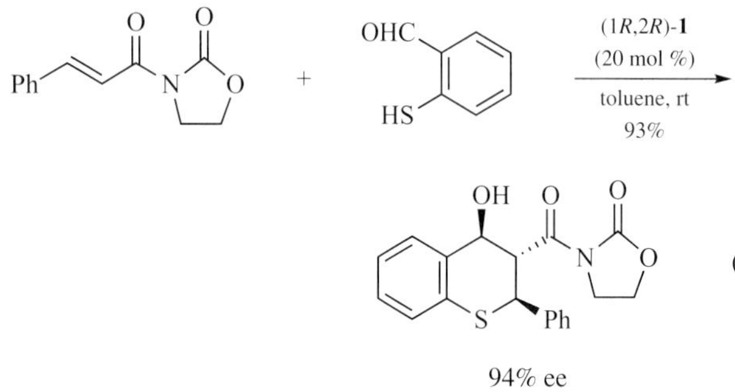

(10)

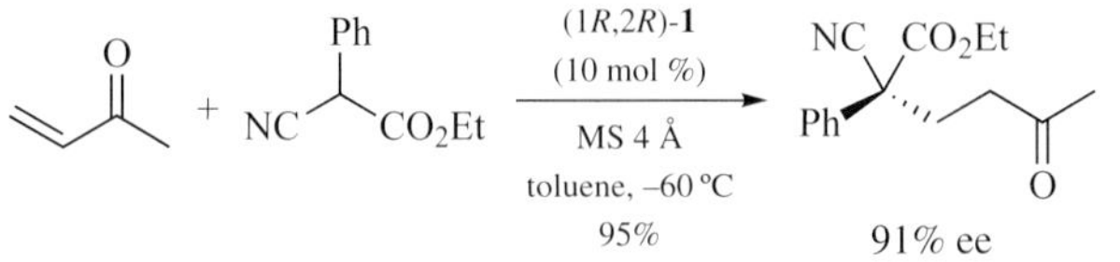

(7)

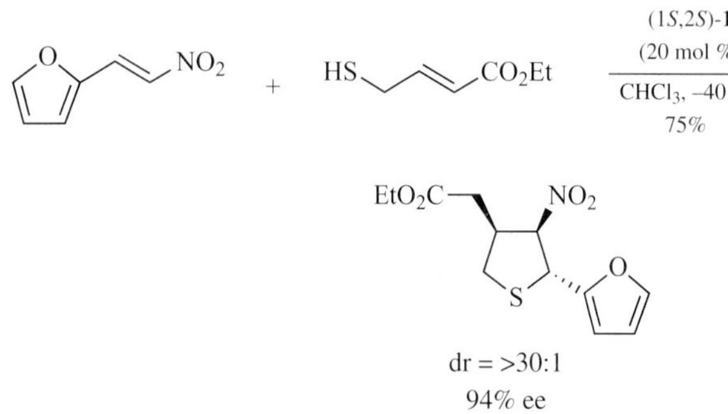

(11)

When the Michael adduct contains both nucleophilic and electrophilic moieties in the molecule, further intramolecular bond-forming reactions can successively occur to give a cyclic product. There are several reports of one-pot reactions (eqs 8 and 9)[8] and domino reactions (eqs 10 and 11)[9] beginning with the asymmetric 1,4-addition catalyzed by this reagent. The one-pot and domino reactions afford highly substituted cyclic compounds bearing more than two stereogenic centers with high enantio- and diastereoselectivity.

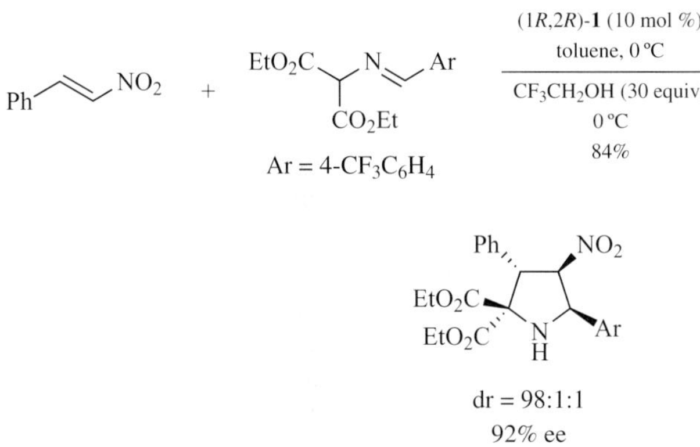

(8)

Enantioselective 1,2-Addition to Imines. There are several reports of the use of this reagent to catalyze 1,2-additions of a variety of nucleophiles to imines such as nitroalkanes, β-ketoesters, and cyanide anion (eqs 12–14).[10] The reaction of nitroalkanes with imines is called the aza-Henry reaction. The catalyst allows the preparation of amine derivatives bearing one or more stereogenic centers with high enantioselectivity.

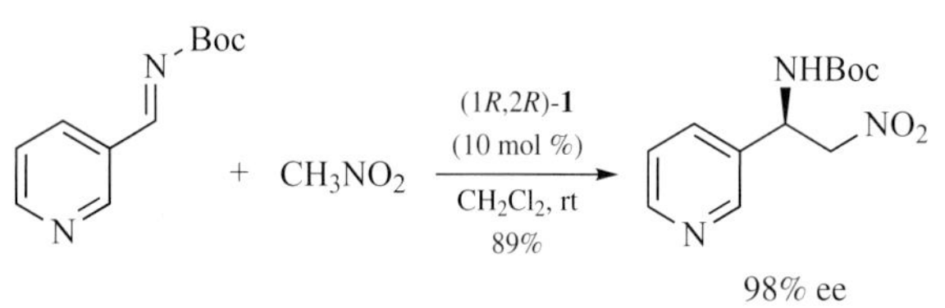

(12)

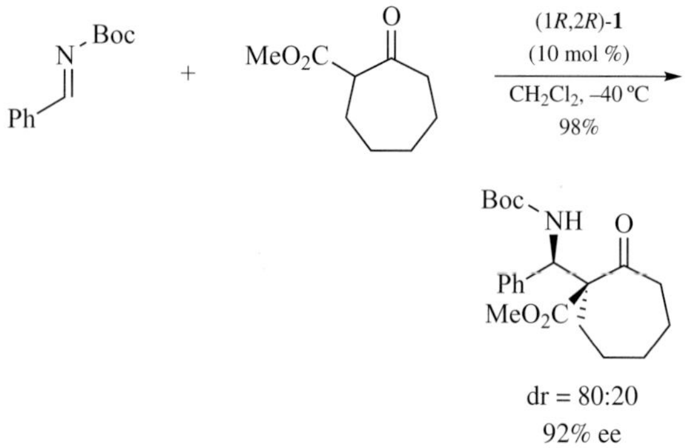

(13)

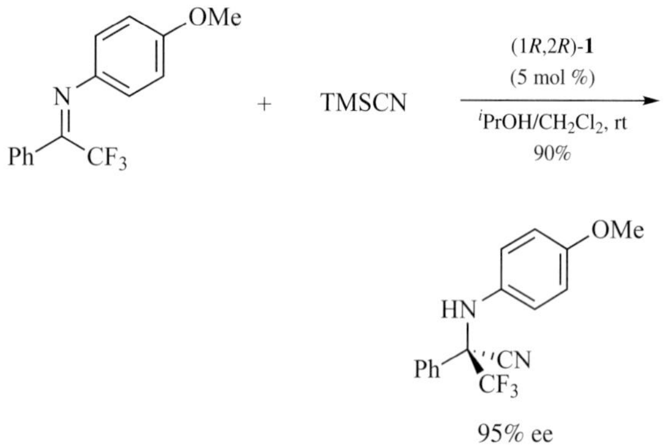

(14)

Domino reactions are applicable when the Mannich adducts have further reactive moieties (eq 15).[11] Although the possibility of the catalytic multicomponent reactions has been reported, the enantioselectivity needs to be improved (eq 16).[12]

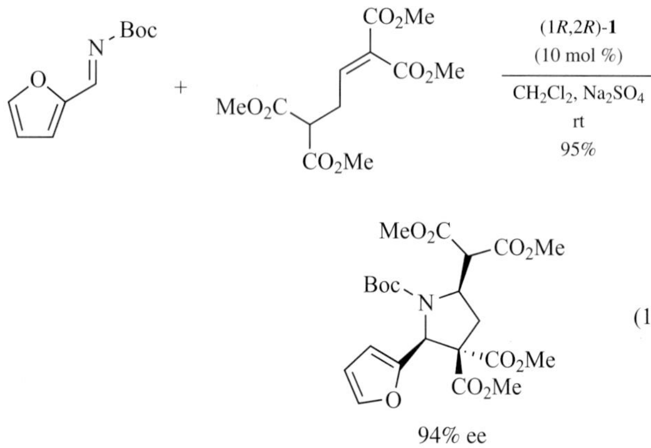

(15)

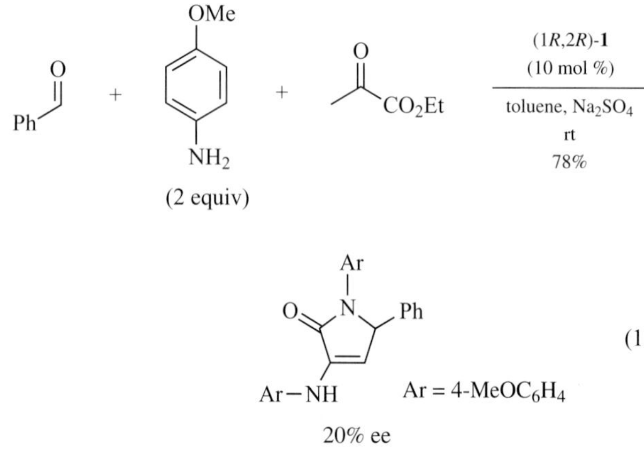

(16)

Enantioselective α-Hydrazination of β-Ketoesters. The reagent shows catalytic activity for the enantioselective α-hydrazination of β-ketoesters with azodicarboxylates (eq 17).[13] However, it was found that slow decomposition of the catalyst occurred under the reaction conditions. The N–N bond of the hydrazine adducts can be cleaved by treatment with SmI_2 in the presence of HMPA.

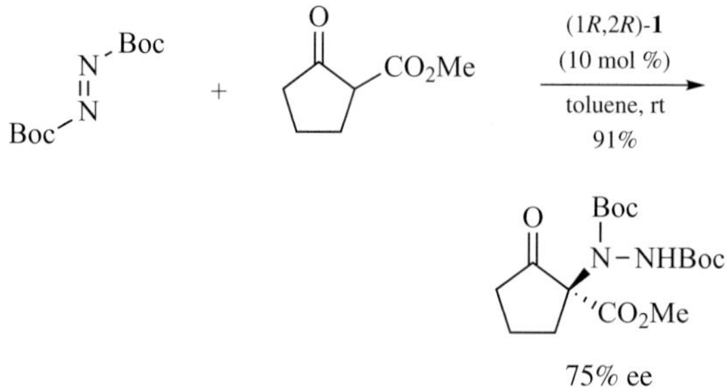

(17)

Living Polymerization of Lactide. The catalyst displays excellent catalytic activity for the ring-opening polymerization and copolymerization of lactide (eq 18).[14] The catalytic polymerization shows characteristics of a living polymerization giving polylactides with narrow polydiversity. However, the organocatalyst exhibits lower reactivity than some transition metal catalysts.

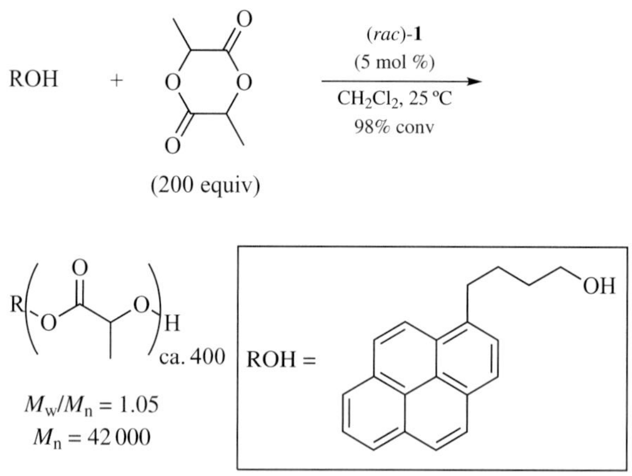

(18)

Formation of Chiral Esters. The reagent catalyzes the dynamic kinetic resolution of azlactones with alcohols giving chiral amino acid derivatives with high enantioselectivity (eq 19).[15] The catalyst is also applicable to the desymmetrization of cyclic *meso*-anhydrides providing enantiomerically enriched hemiesters (eq 20).[16]

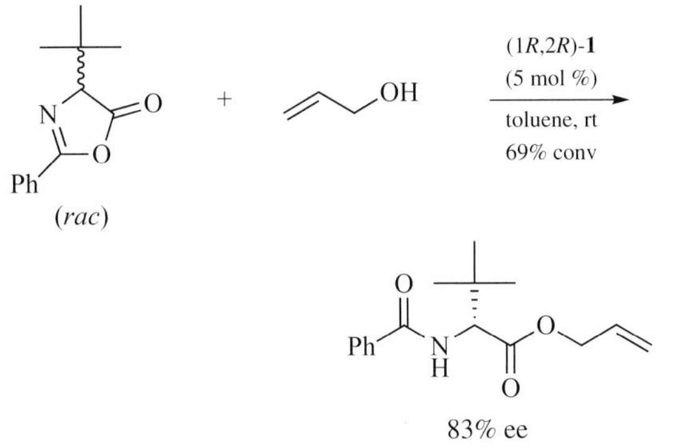

(19)

83% ee

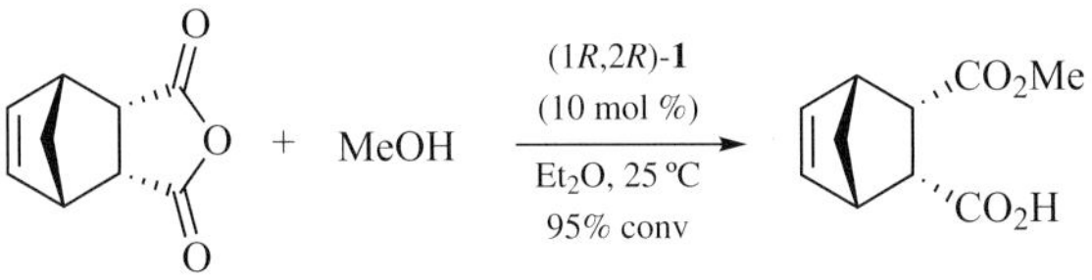

(20)

92% ee

1. Okino, T.; Hoashi, Y.; Takemoto, Y., *J. Am. Chem. Soc.* **2003**, *125*, 12672.
2. (a) Bui, T.; Syed, S.; Barbas, C. F., III, *J. Am. Chem. Soc.* **2009**, *131*, 8758. (b) Baslé, O.; Raimondi, W.; Sanchez Duque, M. M.; Bonne, D.; Constantieux, T.; Rodriguez, J., *Org. Lett.* **2010**, *12*, 5246. (c) Li, H.; Wang, J.; Zu, L.; Wang, W., *Tetrahedron Lett.* **2006**, *47*, 2585. (d) Okino, T.; Hoashi, Y.; Furukawa, T.; Xu, X.; Takemoto, Y., *J. Am. Chem. Soc.* **2005**, *127*, 119. (e) Zhoh, W.-M.; Liu, H.; Du, D.-M., *Org. Lett.* **2008**, *10*, 2817. (f) Hoashi, Y.; Yabuta, T.; Takemoto, Y., *Tetrahedron Lett.* **2004**, *45*, 9185. (g) Fan, R.; Ye, Y.; Li, W.; Wang, L., *Adv. Synth. Catal.* **2008**, *350*, 2488. (h) Liao, Y.-H.; Chen, W.-B.; Wu, W.-J.; Du, X.-L.; Cun, L.-F.; Zhang, X.-M.; Yuan, W.-C., *Adv. Synth. Catal.* **2010**, *352*, 827. (i) Alcaine, A.; Marquéz-López, E.; Merino, P.; Tejero, T.; Herrera, R. P., *Org. Biomol. Chem.* **2011**, *9*, 2777. (j) Liu, J.-M.; Wang, X.; Ge, Z.-M.; Sun, Q.; Cheng, T.-M.; Li, R.-T., *Tetrahedron* **2011**, *67*, 636.
3. (a) Zhao, S.-L.; Zheng, C.-W.; Zhao, G., *Tetrahedron: Asymmetry* **2009**, *20*, 1046. (b) Wang, X.-S.; Zheng, C.-W.; Zhao, S.-L.; Chai, Z.; Zhao, G.; Yang, G.-S., *Tetrahedron: Asymmetry* **2008**, *19*, 2699.
4. (a) Inokuma, T.; Hoashi, Y.; Takemoto, Y., *J. Am. Chem. Soc.* **2006**, *128*, 9413. (b) Hoashi, Y.; Okino, T.; Takemoto, Y., *Angew. Chem., Int. Ed.* **2005**, *44*, 4032. (c) Li, B.-J.; Jiang, L.; Liu, M.; Chen, Y.-C.; Ding, L.-S.; Wu, Y., *Synlett* **2005**, 603.
5. Li, X.; Hu, S.; Xi, Z.; Zhang, L.; Luo, S.; Cheng, J.-P., *J. Org. Chem.* **2010**, *75*, 8697.
6. Bravo, N.; Alba, A.-N. R.; Valero, G.; Companyó, X.; Moyano, A.; Rios, R., *New J. Chem.* **2010**, *34*, 1816.
7. (a) Liu, T.-Y.; Li, R.; Chai, O.; Long, J.; Li, B.-J.; Wu, Y.; Ding, L.-S.; Chen, Y.-C., *Chem. Eur. J.* **2007**, *23*, 319. (b) Li, X.; Xi, Z.; Luo, S.; Cheng, J.-P., *Adv. Synth. Catal.* **2010**, *352*, 1097. (c) Zhang, Y.; Yu, C.; Ji, Y.; Wang, W., *Chem. Asian J.* **2010**, *5*, 1303. (d) Zea, A.; Alba, A.-N. R.; Bravo, N.; Moyano, A.; Rios, R., *Tetrahedron* **2011**, *67*, 2513.
8. (a) Xie, J.; Yoshida, K.; Takasu, K.; Takemoto, Y., *Tetrahedron Lett.* **2008**, *49*, 6910. (b) Raimondi, W.; Lettieri, G.; Dulcère, J.-P.; Bonne, D.; Rodriguez, J., *Chem. Commun.* **2010**, *46*, 7247.
9. (a) Zu, L.; Wang, J.; Li, H.; Xie, H.; Jiang, W.; Wang, W., *J. Am. Chem. Soc.* **2007**, *129*, 1036. (b) Xie, J.-W.; Fan, L.-P.; Su, H.; Li, X.-S.; Xu, D.-C., *Org. Biomol. Chem.* **2010**, *8*, 2117. (c) Yu, C.; Zhang, Y.; Song, A.; Ji, Y.; Wang, W., *Chem. Eur. J.* **2011**, *17*, 770. (d) Zu, L.; Xie, H.; Li, H.; Wang, J.; Jiang, W.; Wang, W., *Adv. Synth. Catal.* **2007**, *349*, 1882. (e) Inokuma, T.; Sakamoto, S.; Takemoto, Y., *Synlett* **2009**, 1627.
10. (a) Okino, T.; Nakamura, S.; Furukawa, T.; Takemoto, Y., *Org. Lett.* **2004**, *6*, 625. (b) Yamaoka, Y.; Miyabe, H.; Yasui, Y.; Takemoto, Y., *Synthesis* **2007**, 2571. (c) Enders, D.; Gottfried, K.; Raabe, G., *Adv. Synth. Catal.* **2010**, *352*, 3147. (d) Wang, X.; Chen, Y.-F.; Niu, L.-F.; Xu, P.-F., *Org. Lett.* **2009**, *11*, 3310.
11. Enders, D.; Göddertz, D. P.; Beceño, C.; Raabe, G., *Adv. Synth. Catal.* **2010**, *352*, 2863.
12. Li, X.; Deng, H.; Luo, S.; Cheng, J.-P., *Eur. J. Org. Chem.* **2008**, 4350.
13. Xu, X.; Yabuta, T.; Yuan, P.; Takemoto, Y., *Synlett* **2006**, 137.
14. (a) Dove, A. P.; Pratt, R. C.; Lohmeijer, B. G. G.; Waymouth, R. M.; Hedrick, J. L., *J. Am. Chem. Soc.* **2006**, *127*, 13798. (b) Lu, J.; Shoichet, M. S., *Macromolecules* **2010**, *43*, 4943.
15. Berkessel, A.; Cleemann, F.; Mukherjee, S.; Müller, T. N.; Lex, J., *Angew. Chem., Int. Ed.* **2005**, *44*, 807.
16. Schmitt, E.; Schiffers, I.; Bolm, C., *Tetrahedron* **2010**, *66*, 6349.

Kiyosei Takasu & Yoshiji Takemoto
Kyoto University, Kyoto, Japan

1-[3,5-Bis(trifluoromethyl)phenyl]-3-{(2*S*)-3,3-dimethyl-1-[(triphenylphosphoranylidene)amino]butan-2-yl}thiourea

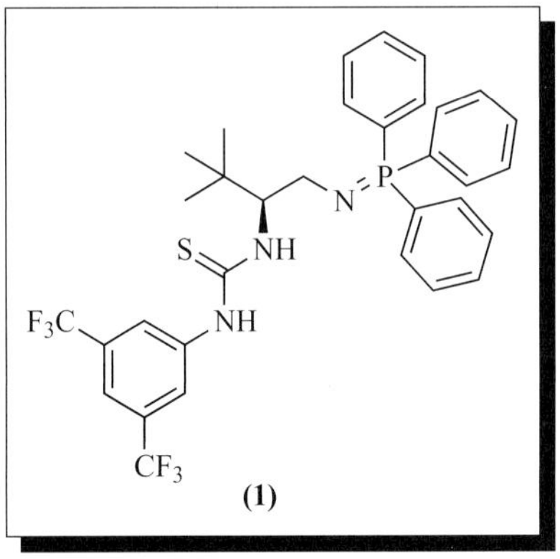

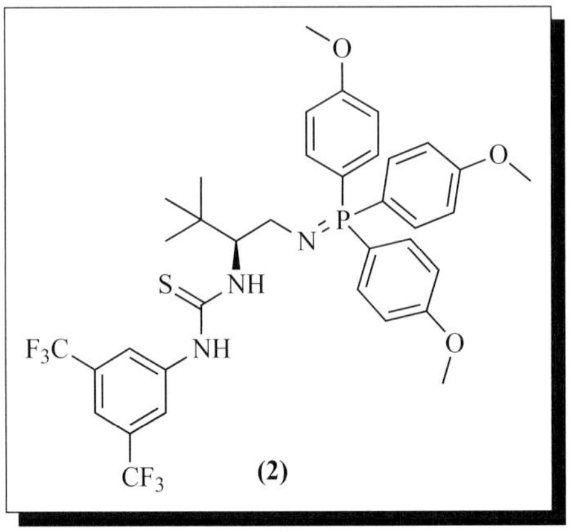

[1476029-73-8] $C_{33}H_{32}F_6N_3P$ (MW: 647.20)

InChI = 1S/C33H32F6N3PS/c1-31(2,3)29(42-30(44)41-25-20-23(32(34,35)36)19-24(21-25)33(37,38)39)22-40-43(26-13-7-4-8-14-26,27-15-9-5-10-16-27)28-17-11-6-12-18-28/h4-21,29H,22H2,1-3H3,(H2,41,42,44)/t29-/m1/s1

InChIKey = CVQMQPMZVJBALJ-GDLZYMKVSA-N

(bifunctional iminophosphorane (BIMP) organocatalysts for stereoinduction in a range of reactions)
1-[3,5-Bis(trifluoromethyl)phenyl]-3-{(2*S*)-3,3-dimethyl-1-[(triphenylphosphoranylidene)amino]butan-2-yl}thiourea (**1**)

Physical Data: mp 96–99 °C; $[\alpha]_D^{24}$ −64.4 (c = 1.40, $CHCl_3$); IR; ^{1}H NMR; ^{13}C NMR; ^{31}P NMR.[1]
Solubility: Et_2O, $MeNO_2$.[1]
Form Supplied in: solid or in solution.
Analysis of Reagent Purity: NMR.
Preparative Method:[1] (a) To the azide [*476029-54-5*] (300 mg, 0.726 mmol) in Et_2O (1.8 mL) under argon atmosphere was added triphenylphosphine (190 mg, 0.726 mmol) at room temperature. Stirring was maintained at room temperature for 26 h and the reaction mixture was then concentrated *in vacuo* to afford a colorless foam. Pentane (4 mL) was added under N_2 and the mixture stirred vigorously for 2 h. The resultant suspension was filtered and the precipitate dried *in vacuo* to obtain compound **1** (344 mg, 73%) as a colorless solid. (b) To the corresponding organoazide (0.020 mmol, 1.0 equiv) and phosphine (0.020 mmol, 1.0 equiv) under an Ar atmosphere was added Et_2O (0.1 mL), and the reaction mixture was stirred at room temperature for 24 h (eq 1). The iminophosphorane product was confirmed by LRMS and TLC, and the volatiles were removed by a N_2 stream to yield the crude product, which was used as a catalyst without further purification.

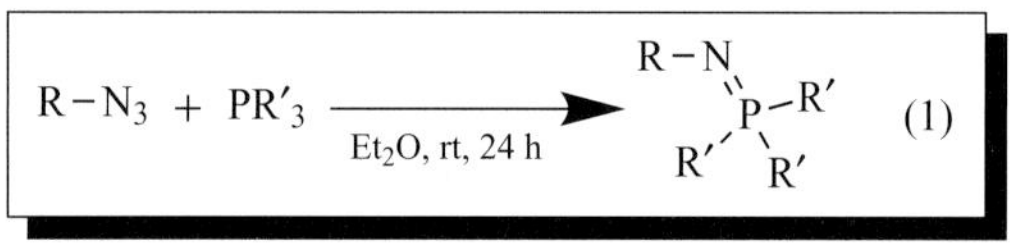

(1)

1-[3,5-Bis(trifluoromethyl)phenyl]-3-[(2*S*)-3,3-dimethyl-1-{[tris(4-methoxyphenyl)phosphoranylidene]amino}butan-2-yl]thiourea (**2**)
[*1476029-72-7*] $C_{36}H_{38}F_6N_3OP$ (MW: 737.23)
InChI = InChI=1S/C36H38F6N3O3PS/c1-34(2,3)32(45-33(50)44-25-20-23(35(37,38)39)19-24(21-25)36(40,41)42)22-43-49(29-13-7-26(46-4)8-14-29,30-15-9-27(47-5)10-16-30)31-17-11-28(48-6)12-18-31/h7-21,32H,22H2,1-6H3,(H2,44,45,50)/t32-/m1/s1
InChIKey = CIHOEOIYPQGYIV-JGCGQSQUSA-N
Physical Data: mp 148–150 °C; $[\alpha]_D$ −54.4° (c = 1.00, $CHCl_3$); IR; ^{1}H NMR; ^{13}C NMR; ^{31}P NMR; X-ray diffraction (CCDC 966013).[1]
Solubility: Et_2O,[1] $MeNO_2$,[2] DMF,[2] EtOAc,[2] THF,[2] CH_2Cl_2,[2] $CHCl_3$,[2] TBME,[2] toluene,[2] benzene,[2] cyclohexane,[2] hexane (poorly soluble).[2]
Form Supplied in: solid or in solution.
Analysis of Reagent Purity: NMR.
Preparative Method:[1] To azide [*476029-54-5*] (500 mg, 1.21 mmol) in Et_2O (3.0 mL) under an argon atmosphere was added tris(4-methoxyphenyl)phosphine (426 mg, 1.21 mmol) at rt. Stirring was maintained at rt for 24 h and the reaction mixture was concentrated under a stream of N_2 until 1 mL of solvent remained. Pentane (1 mL) was added and the resultant thick precipitate was filtered. The precipitate was washed with pentane/Et_2O 1:1 (1 mL) and dried *in vacuo* to obtain compound **2** (786 mg, 88%) as a colorless solid.
Handling, Storage, and Precautions: stable for >2 years at −20 °C, no other stability tests have been performed; no toxicity data have been reported for this class of compounds.

The bifunctional iminophosphorane catalysts are a novel class of organocatalysts that can be used in a variety of reactions. Their salient features include a hydrogen-bond donor group such as the 3,5-bis(trifluoromethyl)phenyl thiourea and a triaryliminophosphorane moiety as the Brønsted base linked by a chiral linker derived from, for example, an amino acid (e.g., *tert*-leucine). The modular design of the catalyst family allows for rapid library generation with modifications to the chiral linker, the H-bond donor group and the Brønsted base and both enantiomers of a catalyst are readily available. The basicity of two triaryliminophosphoranes have been measured and found to be comparable to other organosuperbases such as amidines and guanidines (eq 2).[1]

The BIMP catalysts have also been immobilized using polystyrene-supported (PS) triphenyl phosphine and have demonstrated similar levels of performance to their homogeneous analogs.[3]

Additions to Imines. Catalysts **1** and **2** can be used in the ketimine nitro-Mannich reaction using *N*-DPP diphenylphosphinoylketimines and nitromethane - a reaction that necessitates the use of organosuperbases to proceed.[4] After optimization of various catalyst structures, Dixon and co-workers performed the first general organocatalytic ketimine nitro-Mannich methodology with catalyst **1** which provided the best compromise between rate and enantioselectivity. A range of aromatic, heteroaromatic, and aliphatic ketimines were well tolerated in the reaction using 10 mol % **1** at either −15 or 0 °C with nitromethane as the solvent.[1] Comparable results were also obtained using PS-**1** in toluene (eq 3).[3] The recyclability of catalyst PS-**1** was also demonstrated in the ketimine nitro-Mannich reaction over 10 cycles with maintained conversion and enantioselectivity.[3]

trialkyl-amine	triarylimino-phosphorane	guanidine	amidine	triarylimino-phosphorane	P1 phos-phazene

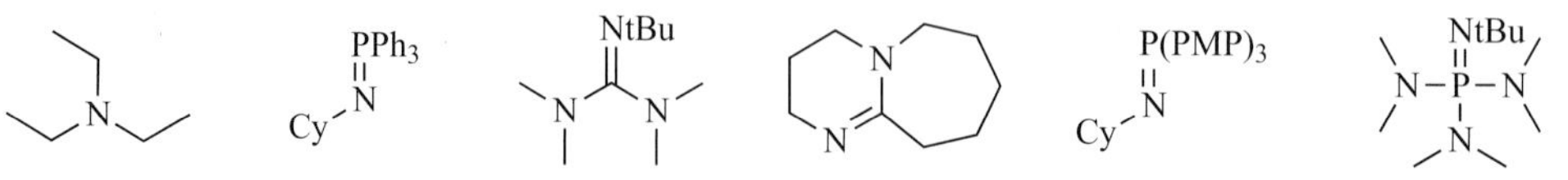

(2)

TEA	**1a**	BTMG	DBU	**1b**	P_1-^{t}Bu
pK_{BH+} 18.82	22.7	23.56	24.34	25.0	26.98

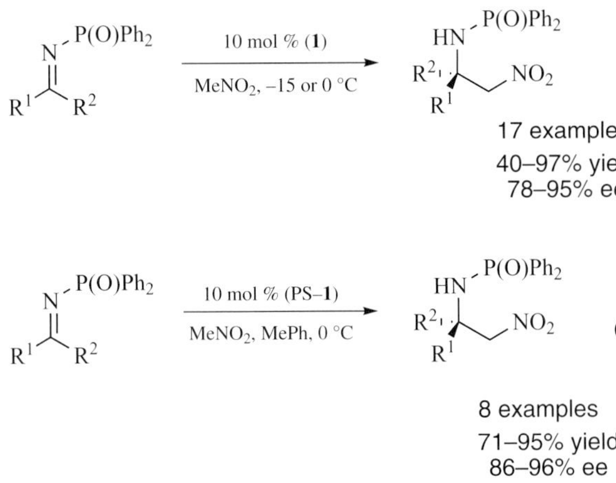

(3)

The BIMP catalysts have also been used on preparative scale in a demonstration of their high reactivity. The ketimine nitro-Mannich reaction was performed on a 10 g scale using 1 mol % of the more strongly Brønsted basic catalyst (**2**) and the reaction reached full conversion after 1 day at rt. A single recrystallization afforded the product in 98% ee and 70% isolated yield and the β-nitroamine product was derivatized to the α-amino acid and 1,2-diamine (eq 4)[1].

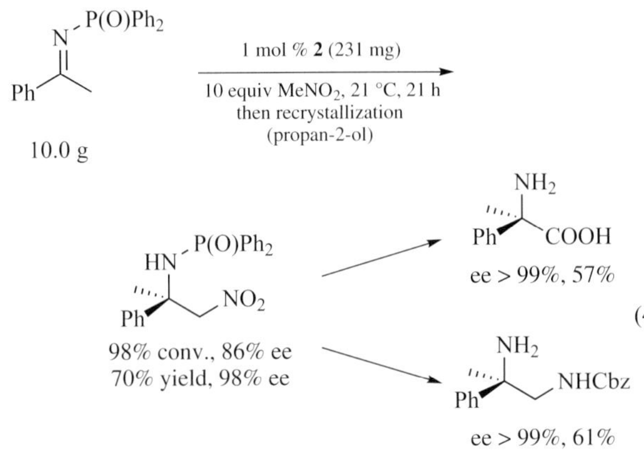

(4)

Catalyst **2** has also been used to promote the first organocatalytic addition of diethyl phosphite to *N*-DPP protected ketimines with excellent yields and moderate to good levels of enantiocontrol (eq 5).[5]

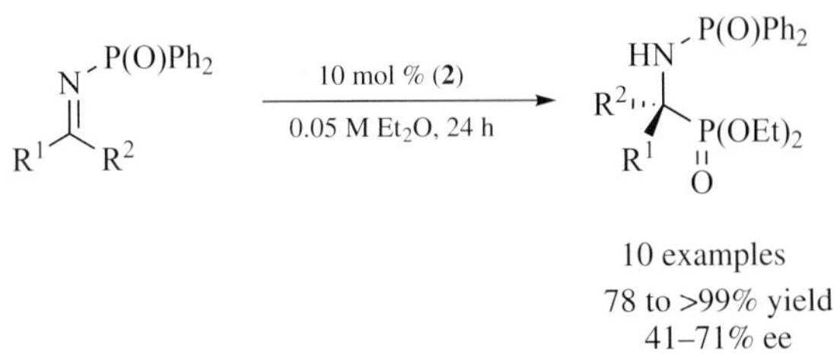

(5)

Conjugate Addition Reactions to Nitro-Olefins. The BIMP catalysts are also efficacious in conjugate addition reactions to reactive electrophiles such as nitro-olefins. Dixon and co-workers reported that PS-**1** promoted the Michael addition between nitrostyrene and substituted malonate esters[3] that are a class of synthetically useful high pK_a pronucleophiles (the pK_a value of dimethyl methylmalonate is 18.0, whereas the pK_a of dimethylmalonate is 15.9).[6] In a similar vein, ketoamides (pK_a value of *N*,*N*-dimethyl-3-oxobutanamide is 18.2)[7] underwent a Michael addition reaction with nitrostyrene using 10 mol % PS-**1** at −20 °C (eq 6).

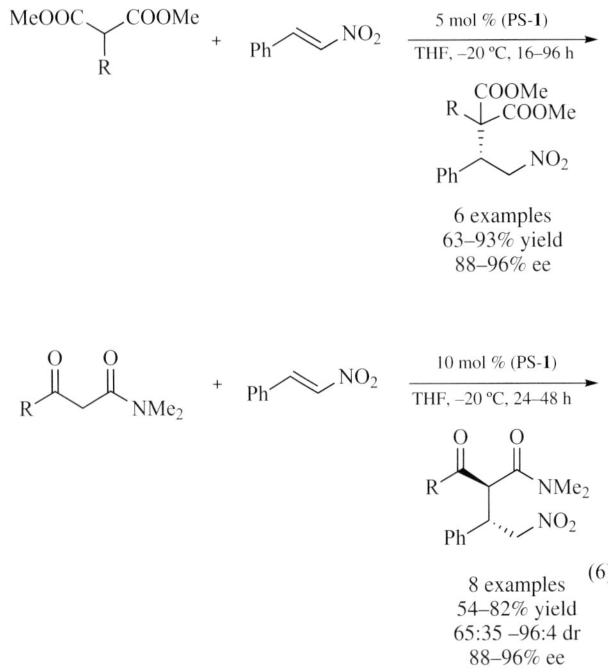

(6)

Catalysts **1** and **2** (and related catalysts) are also highly effective in the Michael addition of amido esters to nitrostyrene. Dixon and co-workers had previously reported the reaction with a cinchona-derived tertiary amine bifunctional organocatalyst [*851477-19-5*] required a 2 week reaction time.[8] Under otherwise identical reactions conditions, catalyst **2** afforded the addition product in just 15 min with comparable levels of enantio- and diastereocontrol (eq 7).[9]

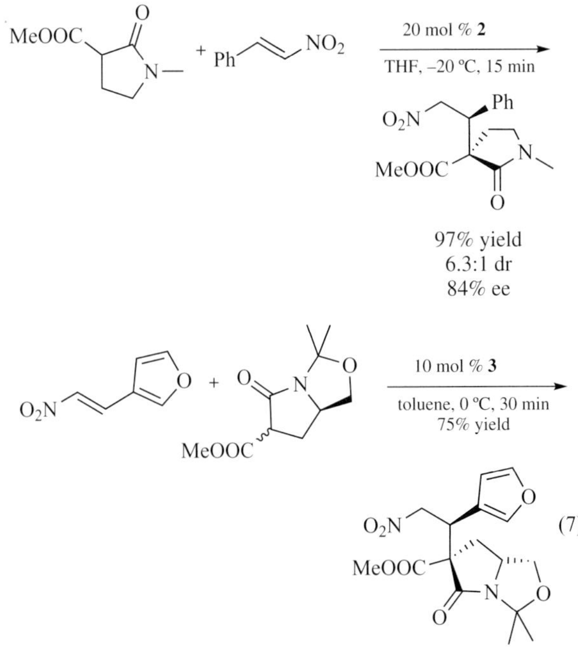

(7)

The amido ester nitro-olefin Michael addition reaction has been successfully applied to the total synthesis of nakadomarin A utilizing the cinchona-derived urea catalyst [*945985-98-8*] albeit with a 48 h reaction time at 30 °C.[10] The reaction with the catalyst **3**, in the enantiomeric series of **2**, containing a benzhydryl group at the stereogenic center afforded the addition product in just 30 min at 0 °C (eq 8).[9]

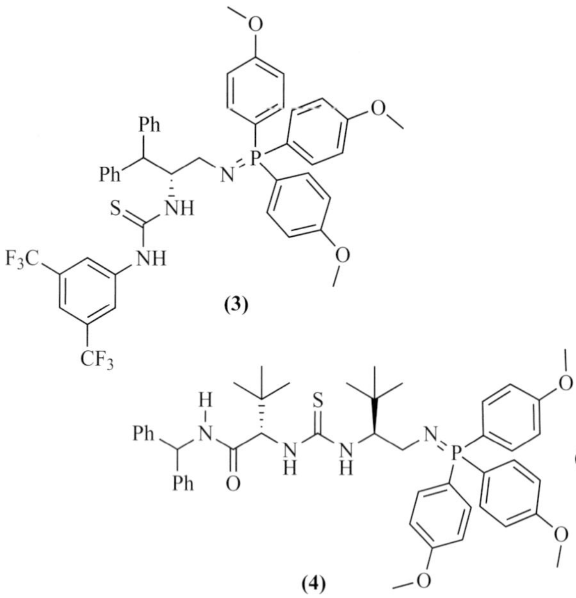

(8)

Conjugate Addition Reactions to α,β-Unsaturated Esters. The BIMP family of catalysts has also found application in the asymmetric union of other challenging pronucleophile/electrophile combinations such as the sulfa-Michael addition (SMA) of alkyl thiols to unactivated α-substituted α,β-unsaturated esters.[2] Using 10 mol % catalyst (**2**), the addition of 1-propanethiol to methyl methacrylate afforded the β-mercaptoester in quantitative yield and 72% ee. Optimization of the H-bond donor group of the catalyst to encompass an amide/thiourea group (catalyst **4**) improved the enantioselectivity in the formation of the β-mercaptoester to 94% ee. The reaction was performed on 24 examples with variations to the thiol, the ester group, and the α-position of the a,β-unsaturated ester in yields up >99% and up to 96% ee (eq 9).

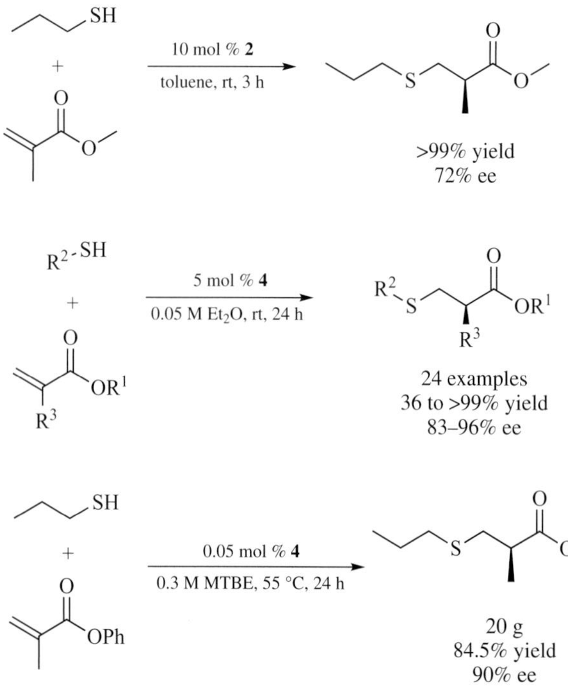

(9)

To demonstrate the potential industrial applicability of the BIMP catalysts, the SMA addition reaction of 1-propanethiol to phenylmethacrylate was performed on a 100 mmol scale to afford 20 g of the β-mercaptoester product with just 0.05 mol % catalyst.

1. Núñez, M. G.; Farley, A. J. M.; Dixon, D. J., *J. Am. Chem. Soc.* **2013**, *135*, 16348.
2. Farley, A. J. M.; Sandford, C.; Dixon, D. J., *J. Am. Chem. Soc.* **2015**, *137*, 15992.
3. Goldys, A. M.; Núñez, M. G.; Dixon, D. J., *Org. Lett.* **2014**, *16*, 6294.
4. Pahadi, N. K.; Ube, H.; Terada, M., *Tetrahedron Lett.* **2007**, *48*, 8700.
5. Robertson, G. P.; Farley, A. J. M.; Dixon, D. J., *Synlett* **2016**, *27*, 21.
6. Arnett, E. M.; Maroldo, S. G.; Schilling, S. L.; Harrelson, J. A., *J. Am Chem. Soc.* **1984**, *106*, 6759.
7. Bordwell, F. G.; Fried, H. E., *J. Org. Chem.* **1981**, *46*, 4327.
8. Jakubec, P.; Helliwell, M.; Dixon, D. J., *Org. Lett.* **2008**, *10*, 4267.
9. Farley, A. J. M., DPhil Thesis, University of Oxford, **2015**.
10. Kyle, A. F.; Jakubec, P.; Cockfield, D. M.; Cleator, E.; Skidmore, J.; Dixon, D. J., *Chem. Commun.* **2011**, *47*, 10037.

Alistair J. M. Farley
University of Oxford, Oxford, UK

Darren J. Dixon
University of Oxford, Oxford, UK

2-[Bis[3,5-bis(trifluoromethyl)phenyl] [(trimethylsilyl)oxy]methyl]pyrrolidine

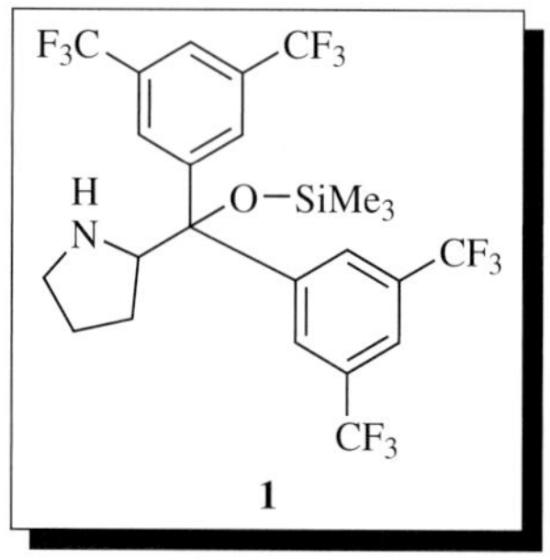

1

[R-908303-26-4]
[S-848821-61-4] $C_{24}H_{23}F_{12}NOSi$ MW 597.51

InChI = 1/C24H23F12NOSi/c1-39(2,3)38-20(19-5-4-6-37-19, 13-7-15(21(25,26)27)11-16(8-13)22(28,29)30)14-9-17 (23(31,32)33)12-18(10-14)24(34,35)36/h7-12,19,37H, 4-6H2,1-3H3/t19-/m1/s1

InChIKey = MOHRGTBNEJKFMB-LJQANCHMBS

InChI = 1S/C24H23F12NOSi/c1-39(2,3)38-20(19-5-4-6-37-19, 13-7-15(21(25,26)27)11-16(8-13)22(28,29)30)14-9-17 (23(31,32)33)12-18(10-14)24(34,35)36/h7-12,19,37H, 4-6H2,1-3H3/t19-/m0/s1

InChIKey = MOHRGTBNEJKFMB-IBGZPJMESA-N

(reagent used as organocatalyst in asymmetric catalysis)

Physical Data: mp 57–58 °C. $[\alpha]_D = -6.8$ (c = 1.2, CH_2Cl_2) for *S*-isomer.[1]

Solubility: sol in a wide range of organic solvents.
Form Supplied in: colorless solid.
Analysis of Reagent Purity: ^{1}H NMR, ^{13}C NMR.
Preparative Method:[1] the title compound is prepared from the corresponding alcohol by silylation. TMSOTf (10 mmol, 1.3 equiv) is added at 0 °C to a solution of the alcohol (7.6 mmol) and triethylamine (10 mmol, 1.3 equiv) in methylene chloride (50 mL). The reaction is then allowed to reach ambient temperature and is stirred for 1 h until full conversion of the starting material is achieved. The reaction is quenched with water. The product is extracted with methylene chloride and dried over sodium sulfate. The concentrated crude product can be purified by flash chromatography to give the title compound.
Purification: flash chromatography (gradient pentane/Et_2O 1/1).
Handling, Storage, and Precautions: to help prevent decomposition, store under inert atmosphere and refrigerate.

Original Commentary

Chongjie Zhu & Ming Xian
Washington State University, Pullman, WA, USA

α-Functionalization of Aldehydes. 2-[Bis[3,5-bis(trifluoromethyl)phenyl][(trimethylsilyl)oxy]methyl]pyrrolidine **1** can serve as an organocatalyst capable of inducing high enantioselectivity by controlling the geometry of the enamine intermediate and through efficient facial shielding. Unlike the case of the more generally used proline catalyst hydrogen bonding to the incoming electrophile is not necessary. **1** has been applied in various stereoselective α-functionalizations of aldehydes.[1–3] It is able to catalyze α-amination, fluorination, bromination, and sulfenylation of aldehydes, with excellent ee from achiral precursors in various solvents. For example, butyraldehyde reacts with 1-benzylsulfanyl-1,2,4-triazole in the presence of catalyst **1** at room temperature to give the corresponding α-sulfenylation product in 96% ee and 85% yield (eq 1).[3] The newly functionalized aldehyde racemized on silica gel, therefore the product was isolated after in situ reduction with $NaBH_4$ to the corresponding optically active alcohol.

Mechanistic study suggests that **1** forms a reactive enamine intermediate with the aldehyde substrates, in which the conjugation of the nitrogen lone-pair electrons with the double bond restricts the number of possible conformations. The bulky substituent of **1** efficiently shields the *Re*-face of the intermediate. As a consequence, the electrophilic attack must occur from the *Si*-face providing the excellent enantioselectivity.

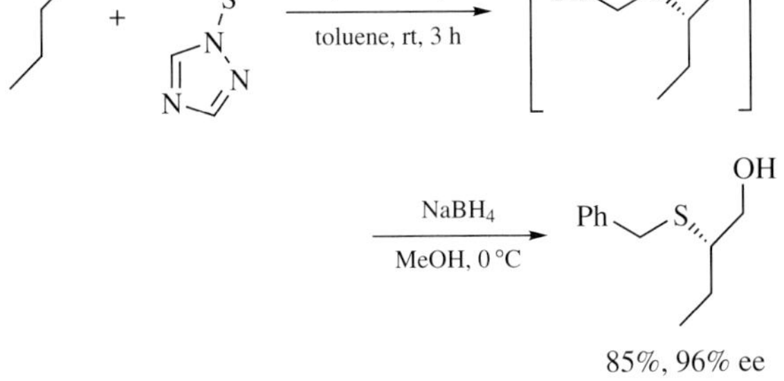

(1)

Asymmetric Multicomponent Domino Reactions. α,β-Unsaturated aldehydes undergo asymmetric Michael addition with various nucleophiles in the presence of catalyst **1**. In several one-pot domino reactions, **1** catalyzes Michael additions of α,β-unsaturated aldehydes with multifunctionalized molecules to form a first stereogenic center at the β-position with excellent ee.[4–6] This can be followed by a second intermolecularMichael addition, intramolecularaldol reaction or ester formation to afford highly functionalized cyclic structures (eq 2).[7,8]

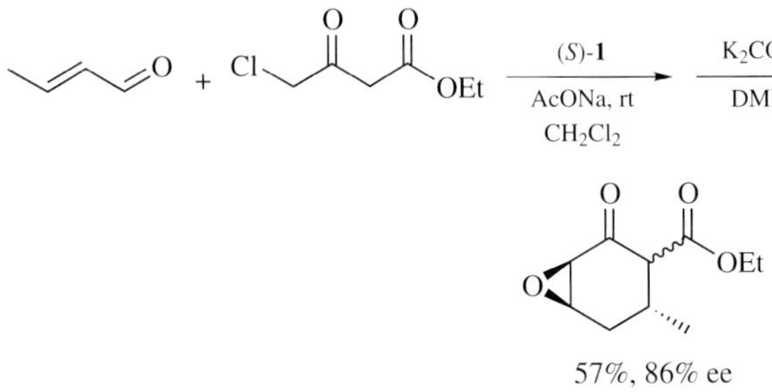

(2)

Other applications include multicomponent-conjugated thiol addition–amination reactions, β-hydroxylations using aromatic oximes as the hydroxylating agent, and the addition of nitrogen heterocycles.[9–11] For example, on reaction with a thiol, diethyl azodicarboxylate, and an α,β-unsaturated aldehyde, the nucleophilic thiol first reacts with the iminium ion intermediate derived from the α,β-unsaturated aldehyde. This is followed by addition of the electrophilic azodicarboxylate to the resulting enamine intermediate, giving nearly enantiopure products when **1** is used as the catalyst (eq 3). It is worthwhile to note both (***S***)-**1** and (***R***)-**1** work well toward different optically active products.

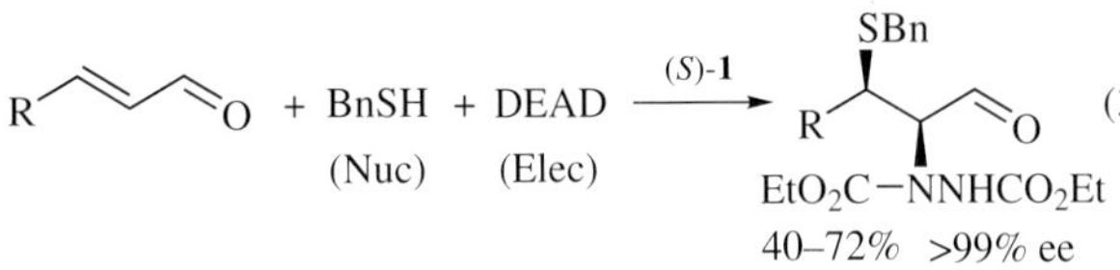

(3)

γ-Amination of α,β-Unsaturated Aldehydes. Reagent **1** can invert the common reactivity pattern of α,β-unsaturated aldehydes.[12] The enamine intermediate arising from condensation of **1** with α,β-unsaturated aldehydes transforms an electron-poor alkene into an electron-rich diene, thereby enabling an electrophilic addition to the γ-position. In the γ-amination reaction of 2-pentenal and DEAD, catalyst **1** is currently the only one in its category to give excellent ee and moderate yield (eq 4). Polymerization of the starting material and product are believed to be the reasons for the moderate yield. The presence of 10 mol % benzoic acid enhanced the rate and yield of the reaction markedly, but the use of a stronger acid or of a stoichiometric amount of benzoic acid led to extensive polymerization. The reaction is general for α,β-unsaturated aldehydes that are β-monosubstituted with linear aliphatic, olefinic, or aromatic substituents. The γ-amination is considered to be the result of a [4+2]-cycloaddition reaction between DEAD and the chiral dienamine formed in situ with the catalyst.

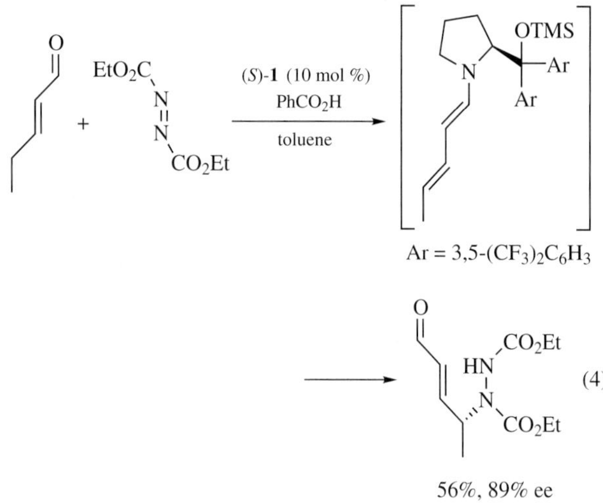

(4)

Epoxidation of α,β-Unsaturated Aldehydes. Reagent **1** has been used in the direct catalytic asymmetric epoxidation of α,β-unsaturated aldehydes to α,β-epoxy aldehydes.[13,14] The reaction proceeds with high ee and high to moderate yield using hydrogen peroxide, urea-hydrogen peroxide(UHP), *tert*-butyl hydroperoxide, or cumene hydroperoxide as the oxidant (eq 5). A variety of solvents, such as methylene chloride, toluene, methanol, ethanol, and THF, are compatible with the reaction. Various aromatic, alkyl, heteroatom functionalities, protected alcohols and ester mono-substituents in the β-position of α,β-unsaturated aldehydes are tolerated. Furthermore, regioselectiveepoxidation of α,β-unsaturated aldehydes in the presence of another olefinic group is also possible (eq 6).

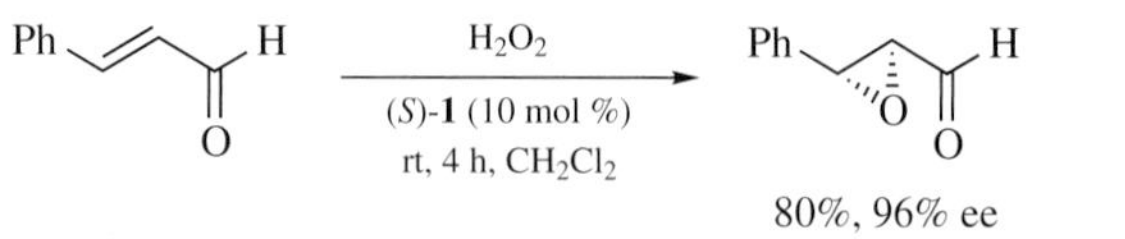

(5)

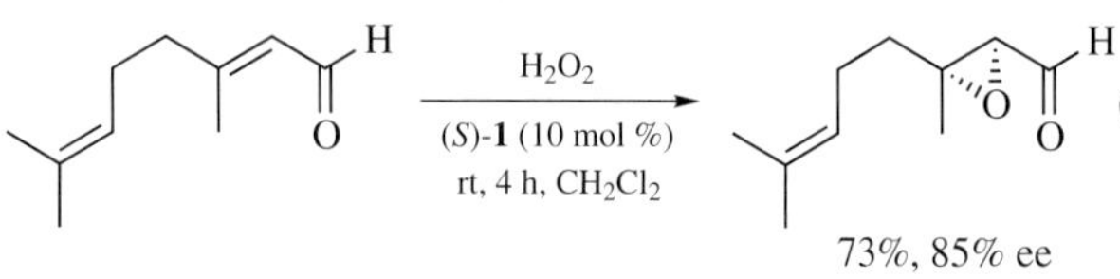

(6)

Related Reagents. **(S)-Proline; (S)-2-(5-1H- Tetrazolyl).**

First Update

Katherine L. Thomas
Ohio Wesleyan University, Delaware, OH, USA

2-[Bis(trifluoromethyl)phenyl][trimethylsilyl)oxy]methylpyrrolidine (**1**) as an efficient, highly enantioselective organocatalyst has contributed greatly toward the progress in the development of asymmetric reactions, especially in the areas of α-functionalization reactions and conjugate addition reactions giving high to excellent enantioselectivity.[15]

α-Functionalization of Aldehydes. A wide array of asymmetric α-functionalized aldehydes has been reported in the recent literature using catalyst **1**. Hayashi et al. reported the asymmetric Mannich reaction[16] between aliphatic and aromatic *N*-tosylamidosulfones and aldehydes catalyzed by **1** to give the alkylated product in low to good yields and with excellent enantioselectivities. From this intermediate, functionalized chiral aziridines can be easily obtained in good yields as well.[17] Yamamoto[18] reported enantioselective fluorination with *N*-fluorosuccinimide (NFSI) of α-chloro aldehydes to afford aromatic and aliphatic substituted α,α-chlorofluoro aldehydes in high enantioselectivity using catalyst **1** (eq 7). In situ reduction of the aldehydes with $NaBH_4$ afforded β,β-chlorofluoro alcohols in high enantioselectivity and good yield. Similarly, Brenner-Moyer reported[19] an organocascade approach to *gem*-chlorofluoro compounds from unfunctionalized aldehydes in a cascade reaction in good yields and excellent enantioselectivities.

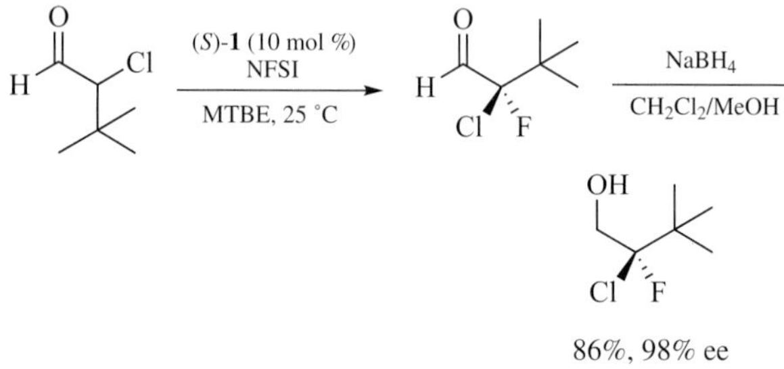

(7)

In 2009, Jørgensen and coworkers reported[20] a simple, direct one-pot organocatalytic approach to give optically active propargylic fluorides. The products were obtained from linear saturated and unsaturated aldehydes of different length in good yields with excellent enantioselectivities via catalyst **1** using an in situ generated Ohira–Bestmann reagent (eq 8). Further elaboration of this work has extended to the synthesis of chiral γ-seleno-$\alpha\beta$-unsaturated esters.[21]

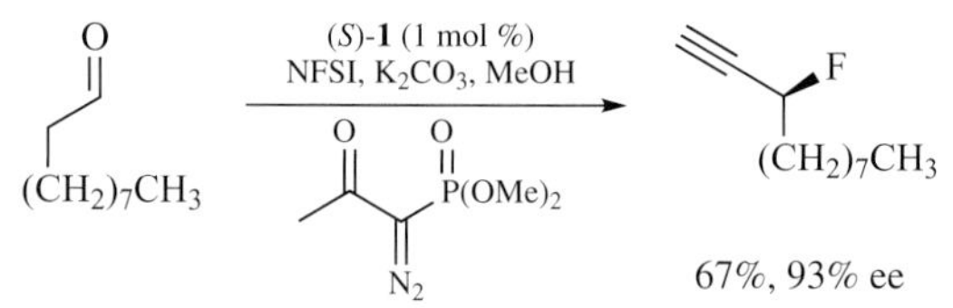

(8)

Metal-catalyzed enantioselective propargylic alkylation can be obtained from a wide range of propargylic alcohols with aldehydes[22] and α,β-unsaturated aldehydes[23] in the presence of a thiolate-bridged diruthenium complex and catalyst **1** (eq 9). The combination of organocatalyst **1** and metal Lewis acids has efficiently catalyzed enantioselective intermolecular α-alkylations of aldehydes with previously inactive alcohols.[24] Additional metal/organocatalyzed α-alkylation reactions have been reported[25] with good yields and high enantioselectivity as well. An "acid-free" method has been developed[26] in combination with **1** for use with acid-sensitive substrates, yet requires high temperatures (eq 10). Christmann et al. reported γ-alkylation of α-unsubstituted enals promoted by the catalyst **1** (eq 11).[27] The reaction led to the formation of separable mixtures of γ- and α-alkylation products with γ-alkylation products prevailing in the case of linear and β-substituted enals (γ-selectivity up to >99:1).

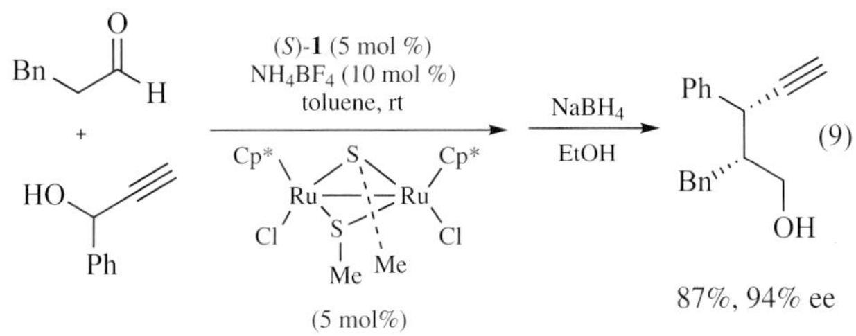

(9)

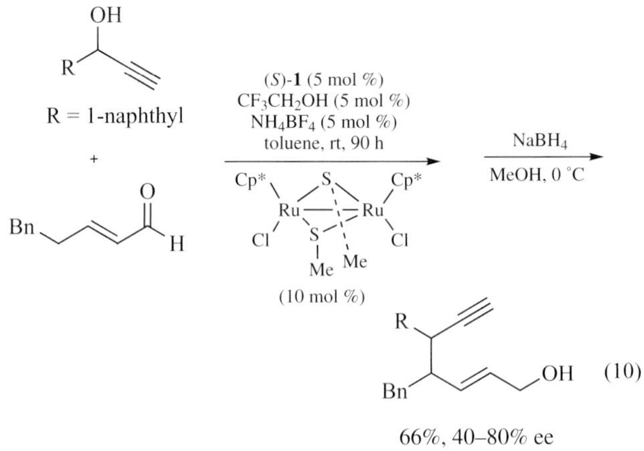

(10)

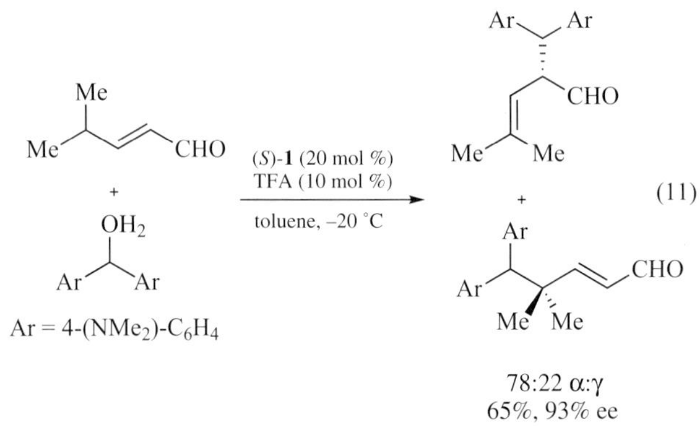

(11)

In 2011, Nishibayashi and coworkers[28] demonstrated Cu(I)/**1** catalyzed enantioselective propargylic alkylation of aldehydes with propargylic esters in good yields as a mixture of two diastereoisomers with high enantioselectivity with a much narrower substrate scope (eq 12).

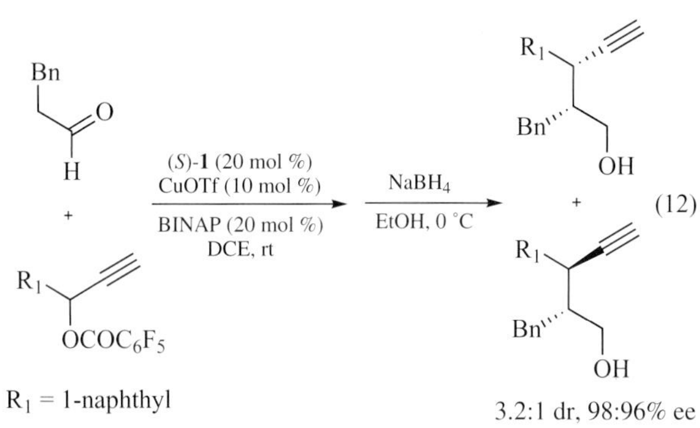

(12)

Cozzi and coworkers recently described[29] the activation of various isoquinolines by Boc_2O or CbzCl in an organocatalytic alkylation promoted by catalyst **1** in moderate yield and with high enantiomeric excess (eq 13). They found that bulky linear aliphatic aldehydes reduced the product yields with moderate enantioselectivities and a substituent at the 6-position on the isoquinoline caused a drop in both enantioselectivity and diastereoselectivity.

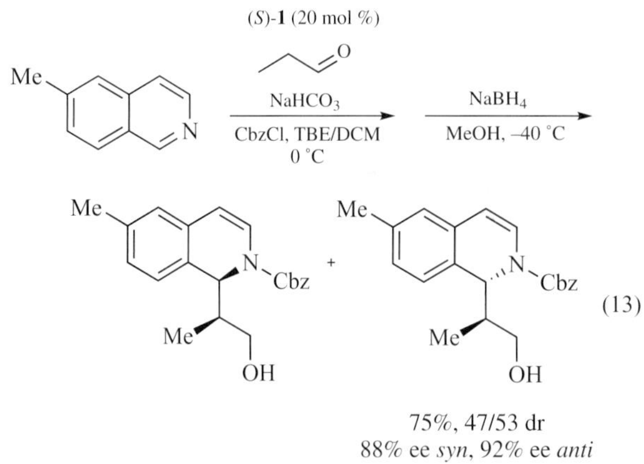

(13)

Asymmetric Multicomponent Domino Reactions. The enantioselective organocatalytic domino and one-pot strategies have had tremendous growth in recent years. These useful procedures allow efficient synthesis of complex molecules under mild reaction conditions. Benzopyrans,[30] pyranones,[31] and pyranonaphthoquinones[32] can be obtained from a tandem Michael addition/acetalization reaction in good yields and with excellent enantioselectivities. A novel domino Friedel–Crafts/Michael/aromatization reaction, reported by Xiao et al.[33] provided a range of chiral highly functionalized tetrahydrocarbazoles upon catalysis with **1** in good to high yields and with high enantio and diastereoselectivities (eq 14).

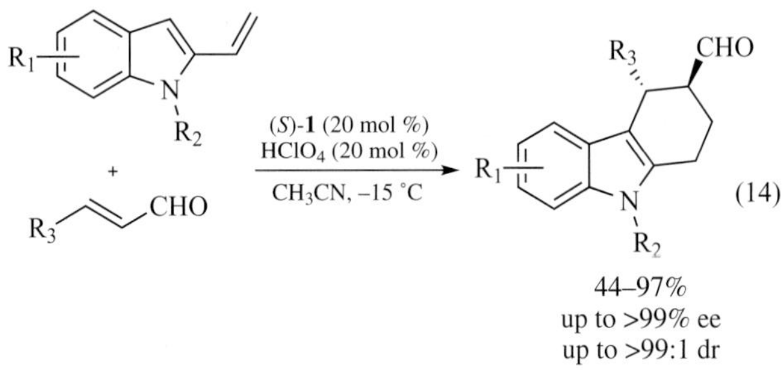

An interesting domino reaction sequence involving nitro-Michael to α,β-unsaturated aldehydes followed by an intramolecular Horner–Wadsworth–Emmons reaction[34] gave high levels of stereocontrol of cyclohexene derivatives in the presence of copromoters (DABCO and $LiClO_4$) with aromatic and aliphatic enals (eq 15).

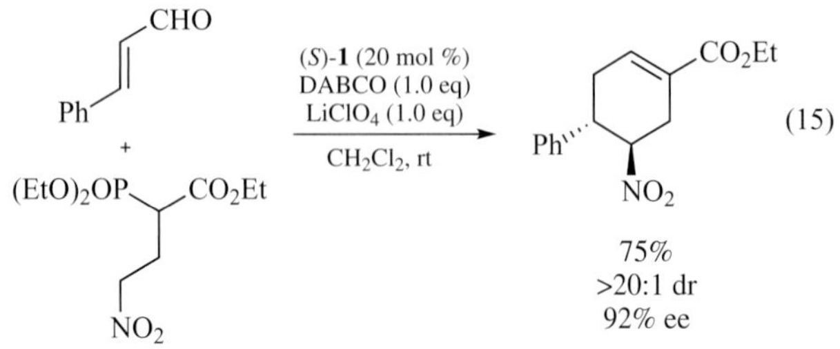

Chiral 3,4-dihydrocoumarins were synthesized by Jørgensen et al.[35] in an initial NHC-catalyzed isomerization followed by a Michael reaction of aryloxyacetaldehydes in good yields and moderate to high enantioselectivities of up to 96% ee with both electron-donating and electron-withdrawing substituents (eq 16). Another organocatalyzed annulation was reported to produce optically active cyclohepta[*b*]indoles[36] with high enantioselectivities and in good yields through a Michael/Friedel–Crafts cascade reaction.

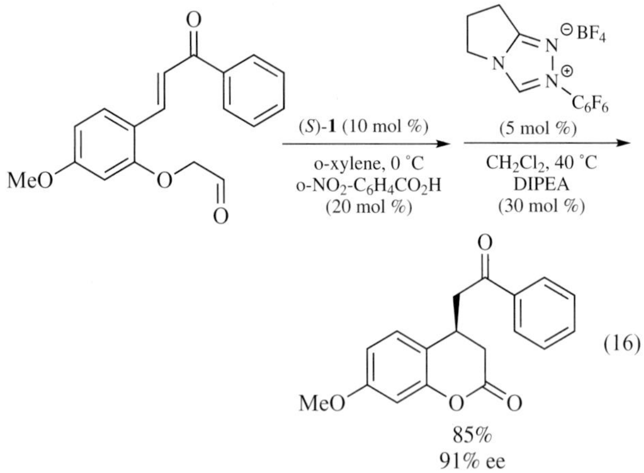

In 2010, Jørgensen and coworkers[37] presented the reaction of α,β-unsaturated aldehyde with alkyne-tethered carbon-based nucleophiles, such as propargylmalonotrile and cyanoacetates, in a highly stereoselective iminium ion/enamine/Lewis acid-catalyzed sequence furnishing cyclopentene carbaldehydes in good yields and excellent enantio- and diastereoselectivities (eq 17). Aromatic aldehydes and nonconjugated unsaturated substituents provided the products in good yields and excellent enantioselectivities.

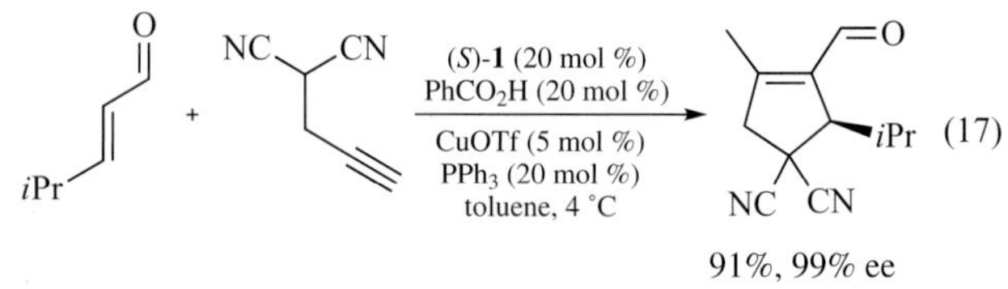

An intramolecular aza-Michael reaction was reported by Fustero and coworkers in 2008; their optimized approach was carried out at −30 °C with **1** that gave the isoindoline with good yields and excellent enantioselectivities (eq 18).[38] When Boc-protecting groups were used, the process took place with a similar yield, but lower enantioselectivity. Under these reaction conditions, enantiomerically enriched indolines, isoindolines, tetrahydroquinolines, and tetrahydroisoquinolines can be obtained in good yield and with high enantioselectivity.

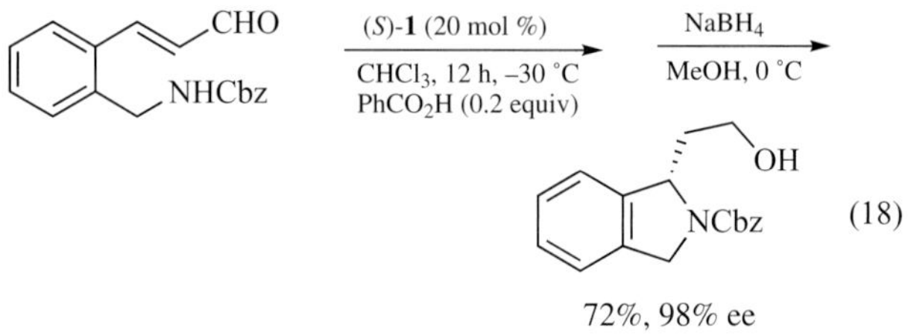

Jørgensen et al. reported[39] an organocatalytic Michael addition of α,β-unsaturated aldehydes and racemic oxazolones with **1** to produce chiral quaternary amino acids from the diastereo- and enantioselective nucleophilic addition in good enantioselective control and with high ee (eq 19). The azlactone product can be hydrolyzed to provide the corresponding protected α,α-disubstituted α-amino acids. Particularly high diastereoselectivities were obtained when a bulky benzhydryl substituent was present at C-2 of the oxazolone.

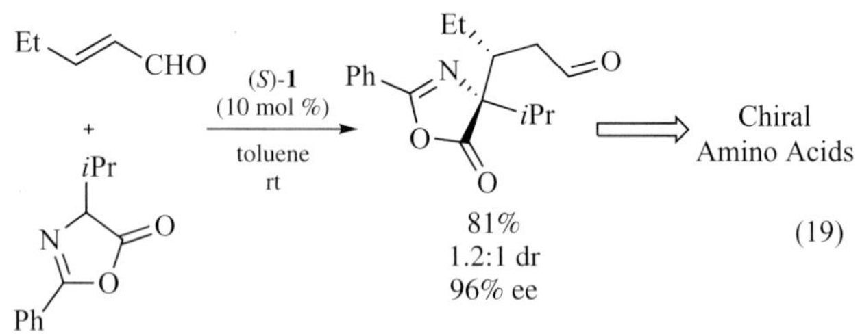

In 2009, Rovis and Lathrop[40] developed a multicatalytic asymmetric domino reaction with **1**, α,β-unsaturated aldehydes and β-dicarbonyl compounds to produce a chiral α-hydroxycyclopentanones with three contiguous stereocenters in high yields and enantioselectivities with moderate diastereoselectivities (eq 20). A wide range of substituents were tolerated, with branched aliphatic aldehydes giving lower yields. When the transformation is performed in stepwise manner, the intermediate aldehydes undergo retro-Michael reaction in the presence of **1** and are prone to epimerization during purification by column chromatography.

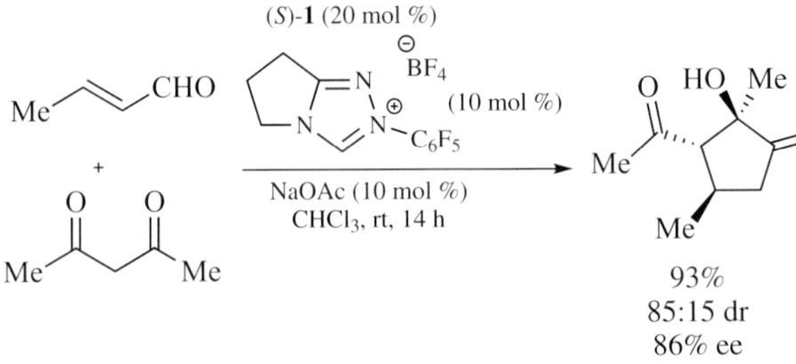

(20)

Epoxidation of α,β-Unsaturated Aldehydes. Recently, asymmetric one-pot organocatalysis strategies have been developed to create an easy stereoselective assembly of molecular and stereochemical complexity. Asymmetric epoxidation of enals with **1** has been applied to the synthesis of a number of building blocks, natural products,[41] and bioactive molecules[42]. Jørgensen and coworkers have extensively published various one-pot synthetic methods incorporating their epoxidation process and other transformations, providing access to a wide range of synthetically useful molecules, which include β-hydroxy and β-amino esters,[43] *trans*-disubstituted-2,3-dihydrobenzofurans,[44] substituted furans,[45] propargylic epoxides,[46] isoxazoline-*N*-oxides,[47] *trans*-dihydroxylated α,β-unsaturated aldehydes,[48] imidazo[1,2-α]pyridines,[49] 1,3-azoles,[50] and substituted thiophenes.[51]

Cross-dehydrogenative Coupling Reactions. The atom economical cross-dehydrogenative coupling of two C–H bonds is an increasingly attractive route to C–C bond formation in transition mental and amine catalysis. Chi and coworkers[52] demonstrated an enantioselective oxidative cross-dehydrogenative coupling of tertiary amines with aldehydes using **1** with $CuBr_2$ to produce optically active β-amino alcohols that contain tetrahydroquinoline with both electron-withdrawing and electron-donating R_1 substituents (eq 21).

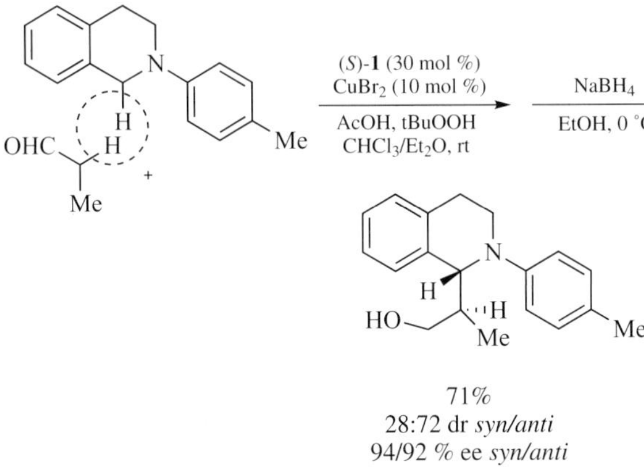

(21)

Aziridination of α,β-Unsaturated Aldehydes. There have been recent reports of several organocatalytic procedures for the enantioselective aziridination of alkenes. Among them, aziridination of α,β-unsaturated aldehydes and enones with *N*-tosyloxycarbamates has been widely investigated. In 2013, Jørgensen and coworkers reported[53] a novel tandem vinylogous iminium-ion–dienamine cascade reaction in moderate yields and with high enantiomeric excess. Organocatalyst **1** and TsONHBoc, as "nitrene equivalent" allowed for the regioselective formation of *N*-Boc-protected aziridines in high yields and with excellent enantioselectivity (eq 22).

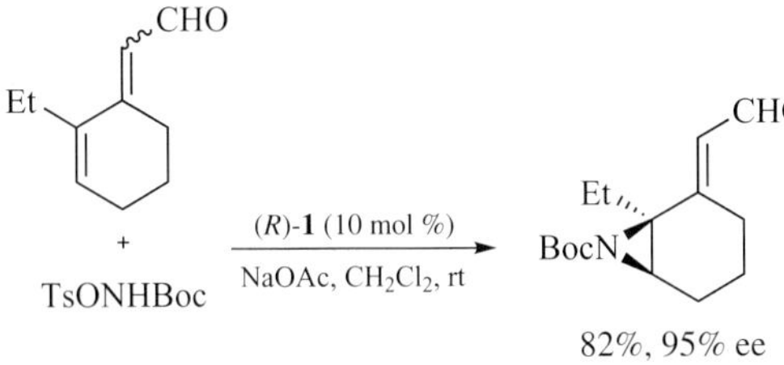

(22)

Greck, de Figueiredo and coworkers[54] showed that α-branched enals were aziridinated in 69–86% yield and up to 90% ee with **1** and TsNHOTs (eq 23). Different alkyl groups at the α-position of the aldehyde (R = Bn, Me, Pent, and *i*Pr) were well tolerated and functionalized aldehydes can also efficiently participate in these reactions.

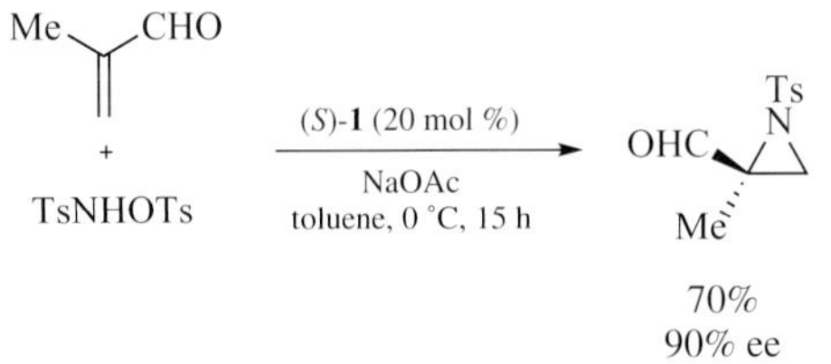

(23)

Friedel–Crafts Reaction. The asymmetric Friedel–Crafts reaction of 4,7-dihydroindoles with enals occurred smoothly under the promotion of the combined catalyst of **1** and TEA[55] (eq 24) leading to the production of 2-alkylated 4,7-dihydroindoles in high yields and with excellent enantioselectivity. The substituent on the aromatic rings of α,β-unsaturated aldehydes had limited influence on the stereochemical outcome as neutral, electron-donating, and electron-withdrawing groups were well tolerated.

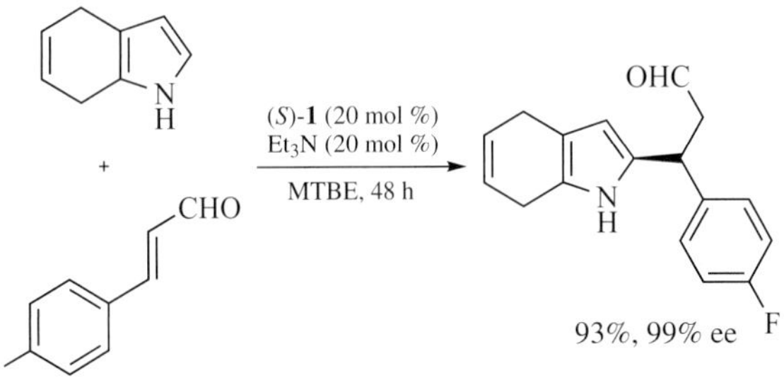

(24)

A highly enantioselective intramolecular Friedel–Crafts-type alkylation of indolyl α,β-unsaturated aldehydes has been reported[56] by using **1** to give tetrahydrocarbazoles in good yields and high enantioselectivities (eq 25). Additionally, incorporation of an electron-withdrawing or electron-donating group, such as bromo, chloro, fluoro, or methyl at the C5 or C6 position of the indole ring was well tolerated without obvious influence on the enantioselectivity.

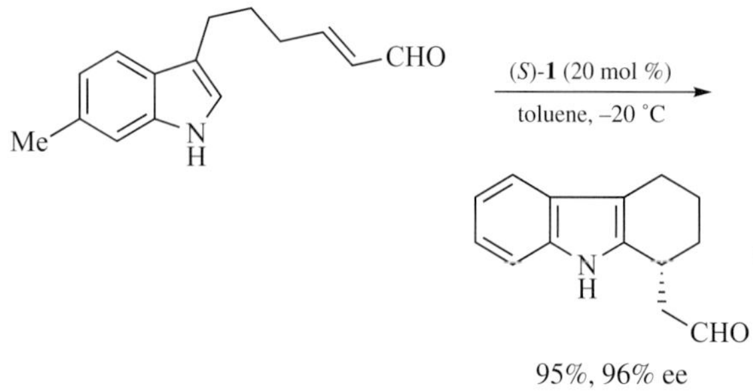

(25)

1. Marigo, M.; Wabnitz, T. C.; Fielenbach, D.; Jørgensen, K. A., *Angew. Chem., Int. Ed.* **2005**, *44*, 794.
2. Marigo, M.; Fielenbach, D.; Braunton, A.; Kjøersgaard, A.; Jørgensen, K. A., *Angew. Chem., Int. Ed.* **2005**, *44*, 3703.
3. Franzén, J.; Marigo, M.; Fielenbach, D.; Wabnitz, T. C.; Kjøersgaard, A.; Jørgensen, K. A., *J. Am. Chem. Soc.* **2005**, *127*, 18296.
4. Carlone, A.; Marigo, M.; North, C.; Landa, A.; Jørgensen, K. A., *Chem. Commun.* **2006**, 4928.
5. Marigo, M.; Bertelsen, S.; Landa, A.; Jørgensen, K. A., *J. Am. Chem. Soc.* **2006**, *128*, 5475.
6. Brandau, S.; Maerten, E.; Jørgensen, K. A., *J. Am. Chem. Soc.* **2006**, *128*, 14986.
7. Carlone, A.; Cabrera, S.; Marigo, M.; Jørgensen, K. A., *Angew. Chem., Int. Ed.* **2007**, *46*, 1101.
8. Brandau, S.; Landa, A.; Franzén, J.; Marigo, M.; Jørgensen, K. A., *Angew. Chem., Int. Ed.* **2006**, *45*, 4305.
9. Marigo, M.; Schulte, T.; Franzén, J.; Jørgensen, K. A., *J. Am. Chem. Soc.* **2005**, *127*, 15710.
10. Bertelsen, S.; Dinér, P.; Johansen, R. L.; Jørgensen, K. A., *J. Am. Chem. Soc.* **2007**, *129*, 1536.
11. Dinér, P.; Nielsen, M.; Marigo, M.; Jørgensen, K. A., *Angew. Chem., Int. Ed.* **2007**, *46*, 1983.
12. Bertelsen, S.; Marigo, M.; Brandes, S.; Dinér, P.; Jørgensen, K. A., *J. Am. Chem. Soc.* **2006**, *128*, 12973.
13. Marigo, M.; Franzén, J.; Poulsen, T. B.; Zhuang, W.; Jørgensen, K. A., *J. Am. Chem. Soc.* **2005**, *127*, 6964.
14. Zhuang, W.; Marigo, M.; Jørgensen, K. A., *Org. Biomol. Chem.* **2005**, *3*, 3883.
15. Donslund, B. S.; Johansen, T. K.; Poulsen, P. H.; Halskov, K. S.; Jørgensen, K. A., *Angew. Chem., Int. Ed.* **2015**, *54*, 13860.
16. Urushima, T.; Ishikawa, H.; Hayashi, Y., *Chem. Eur. J.* **2011**, *17*, 8273.
17. Hayashi, Y.; Urushima, T.; Sakamoto, D.; Torii, K.; Ishikawa, H., *Chem. Eur. J.* **2011**, *17*, 11715.
18. Yamamoto, H.; Shibatomi, K., *Angew. Chem., Int. Ed.* **2008**, *47*, 5796.
19. Brenner-Moyer, S.E.; Hayes, M.D.; Rodrígues-Alvarado, M., *Tetrahedron Lett.* **2015**, *56*, 4718.
20. Jiang, H.; Falcicchio, A.; Jensen, K.L.; Paixaõ, M.W.; Bertelsen, S.; Jørgensen, K.A., *J. Am. Chem. Soc.* **2009**, *131*, 7153.
21. Genna, D.T.; Hencken, C.P.; Siegler, M.A.; Posner, G.H., *Org. Lett.* **2010**, *12*, 4694.
22. Ikeda, M.; Miyake, Y.; Nishibayashi, Y., *Angew. Chem., Int. Ed.* **2010**, *49*, 7289.
23. Ikeda, M.; Miyake, Y.; Nishibayashi, Y., *Organometallics* **2012**, *31*, 3810.
24. Xiao, J., *Org. Lett.* **2012**, *14*, 1716.
25. Carreira, E.M.; Zipfel, H.F.; Krautwald, S.; Sandmeier, T., *Angew. Chem., Int. Ed.* **2015**, *54*, 14363.
26. Xiao, J.; Zhao, K.; Loh, T.P., *Chem. Commun.* **2012**, *48*, 3548.
27. Stiller, J.; Marqués-López, E.; Herrera, R.P.; Frohlich, R.; Strohmann, C.; Christmann, M., *Org. Lett.* **2011**, *13*, 70.
28. Yoshida, A.; Ikeda, M.; Hattori, G.; Miyake, Y.; Nishibayashi, Y., *Org. Lett.* **2011**, *13*, 592.
29. Mengozzi, L.; Gualandi, A.; Cozzi, P.G., *Chem. Sci.* **2014**, *5*, 3915.
30. Rueping, M.; Sugiono, E.; Merino, E., *Chem. Eur. J.* **2008**, *14*, 6329.
31. Rueping, M.; Merino, E.; Sugiono, E., *Adv. Synth. Catal.* **2008**, *350*, 2127.
32. Rueping, M.; Sugiono, E.; Merino, E., *Angew. Chem., Int. Ed.* **2008**, *47*, 3046.
33. Cao, Y.J.; Chen, H-G.; Lu, L-Q.; Zhang, J-J.; Cheng, Y.; Chen, J-R.; Xiao, W-J., *Adv. Synth. Catal.* **2011**, *353*, 617.
34. Weng, J.; Li, J-M.; Li, F-Q.; Xie, Z-S.; Lu, G., *Adv. Synth. Catal.* **2012**, *354*, 1961.
35. Jacobsen, C.B.; Albrecht, Ł.; Udmark, J.; Jørgensen, K.A., *Org. Lett.* **2012**, *14*, 5526.
36. Dange, N.S.; Hong, B-C.; Lee, C-C.; Lee, G-H., *Org. Lett.* **2013**, *15*, 3914.
37. Jensen, K.L.; Franke, P.T.; Arróniz, C.; Kobblegaard, S.; Jørgensen, K.A., *Chem. Eur. J.* **2010**, *16*, 1750.
38. Fustero, S.; Moscardó, J.; Jiménez, D.; Pérez-Carrión, M.D.; Sánchez-Roselló, M.; del Pozo, C., *Chem. Eur. J.* **2008**, *14*, 9868.
39. Cabrera, S.; Reyes, E.; Alemán, J.; Milelli, A.; Kobbelgaard, S.; Jørgensen, K.A., *J. Am. Chem. Soc.* **2008**, *130*, 12031.
40. Lathrop, S.P.; Rovis, T., *J. Am. Chem. Soc.* **2009**, *131*, 13628.
41. Mukaiyama, T.; Ogata, K.; Sato, I.; Hayashi, Y., *Chem. Eur. J.* **2014**, *20*, 13583.
42. Armstrong, A.; Emmerson, D.P.G.; Milner, H.J.; Sheppard, R.J., *J. Org. Chem.* **2014**, *79*, 3895.
43. Jiang, H.; Gschwend, B.; Albrecht, Ł.; Jørgensen, K.A., *Org. Lett.* **2010**, *12*, 5052.
44. Albrecht, Ł.; Ransborg, L.K.; Lauridsen, V.; Overgaard, M.; Zweifel, T.; Jørgensen, K.A., *Angew. Chem., Int. Ed.* **2011**, *50*, 12496.
45. Albrecht, Ł.; Ransborg, L.K.; Gschwend, B.; Jørgensen, K.A., *J. Am. Chem. Soc.* **2010**, *132*, 17886.
46. Jiang, H.; Holub, N.; Paixaõ, M.W.; Tiberi, C.; Falcicchio, A.; Jørgensen, K.A., *Chem. Eur. J.* **2009**, *15*, 9638.
47. Jiang, H.; Elsner, P.; Jensen, K.L.; Falcicchio, A.; Marcos, V.; Jørgensen, K.A., *Angew. Chem., Int. Ed.* **2009**, *48*, 6844.
48. Jiang, H.; Gschwend, B.; Albrecht, Ł.; Dickmeiss, G.; Hansen, S.G.; Jørgensen, K.A., *J. Am. Chem. Soc.* **2010**, *132*, 9188.
49. Albrecht, Ł.; Albrecht, A.; Ransborg, L.K.; Jørgensen, K.A., *Chem. Sci.* **2011**, *2*, 1273.
50. Albrecht, Ł.; Ransborg, L.K.; Albrecht, A.; Lykke, L.; Jørgensen, K.A., *Chem. Eur. J.* **2011**, *17*, 13240.
51. Ransborg, L.K.; Albrecht, Ł.; Weise, C.F.; Bak, J.R.; Jørgensen, K.A., *Organic Lett.* **2012**, *14*, 724.
52. Zhang, J.; Tiwari, B.; Xing, C.; Chen, X.; Chi, Y.R., *Angew. Chem., Int. Ed.* **2012**, *51*, 3649.
53. Halskov, K.S.; Naicker, T.; Jensen, M.E.; Jørgensen, K.A., *Chem. Commun.* **2013**, *49*, 6382.
54. Desmarchelier, A.; de Sant'Ana, D.P.; Terrasson, V.; Campagne, J.M.; Moreau, X.; Greck, C.; de Figueiredo, R.M., *Eur. J. Org. Chem.* **2011**; 4046.
55. Hong, L.; Liu, C.; Sun, W.; Wang, L.; Wong, K.; Wang, R., *Organic Lett.* **2009**, *11*, 2177.
56. Zhu, X-Y.; An, X-L.; Li, C-F.; Zhang, F-G.; Hua, Q-L.; Chen, J-R.; Xiao, W-J., *ChemCatChem* **2011**, *3*, 679.

(1*R*)-3,3′-Bis(2,4,6-trimethylphenyl)-[1,1′-binaphthalene]-2,2′-dicarboxylic acid

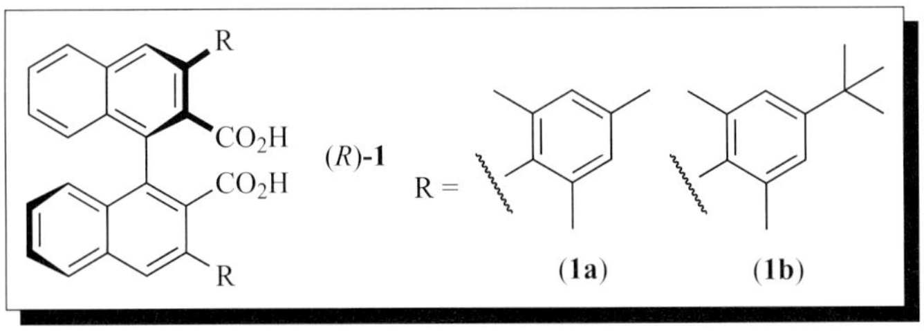

[(1*R*)-3,3′-Bis(2,4,6-trimethylphenyl)-[1,1′-binaphthalene]-2,2′-dicarboxylic acid, (*R*)-**1a**]
[950889-65-3] $C_{40}H_{34}O_4$ (MW 578.70)
InChI = 1S/C40H34O4/c1-21-15-23(3)33(24(4)16-21)31-19-27-11-7-9-13-29(27)35(37(31)39(41)42)36-30-14-10-8-12-28(30)20-32(38(36)40(43)44)34-25(5)17-22(2)18-26(34)6/h7-20H,1-6H3,(H,41,42)(H,43,44)
InChIKey = ZJELRGBNUDBQQZ-UHFFFAOYSA-N
[(*R*)-3,3′-Bis(4-*tert*-butyl-2,6-dimethylphenyl)-1,1′-binaphthyl-2,2′-dicarboxylic acid, (*R*)-**1b**]
[950889-66-4] $C_{46}H_{46}O_4$ (MW 662.86)
InChI = 1S/C46H46O4/c1-25-19-31(45(5,6)7)20-26(2)37(25)35-23-29-15-11-13-17-33(29)39(41(35)43(47)48)40-34-18-14-12-16-30(34)24-36(42(40)44(49)50)38-27(3)21-32(22-28(38)4)46(8,9)10/h11-24H,1-10H3,(H,47,48)(H,49,50)
InChIKey = NEPNLMCRZKWRSD-UHFFFAOYSA-N

(Chiral Brønsted acid catalyst used in various C–C bond formation reactions)

Preparation.[1] (*1R*)-3,3′-Bis(2,4,6-trimethylphenyl)-[1,1′-binaphthalene]-2,2′-dicarboxylic acid (*R*)-**1a** as well as (*R*)-3,3′-bis(4-*tert*-butyl-2,6-dimethylphenyl)-1,1′-binaphthyl-2,2′-dicarboxylic acid (*R*)-**1b** can be synthesized starting from commercially available (*R*)-1,1′-bis-2-naphthol following the procedure summarized below. The compounds are stable on the benchtop, and can be stored for a long term without special precautions (eqs 1 and 2).

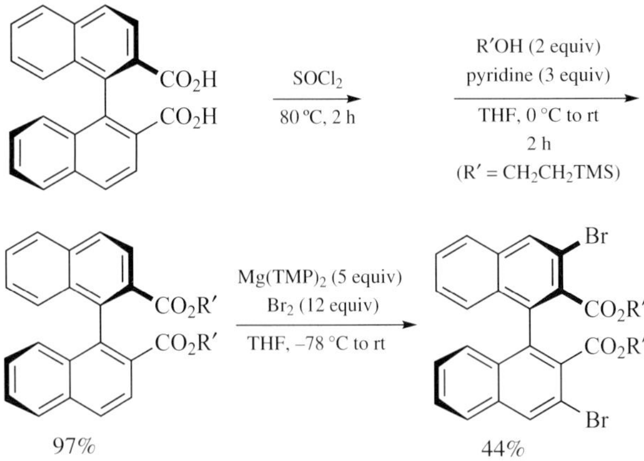

(1)

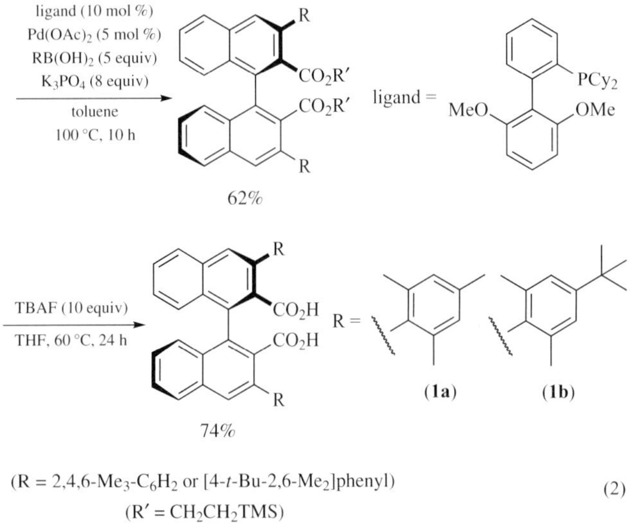

(2)

***trans*-Selective Asymmetric Aziridination of Diazoacetamides and *N*-Boc Imines.**[2] The reaction should be conducted in the presence of molecular sieves 4 Å to prevent imine hydrolysis. Both electron-donating and electron-withdrawing *N*-Boc imines led to moderate yields and high enantioselectivities. Compounds, such as **2h**, bearing *para*-substituted electron-withdrawing groups, as well as those containing electron-donating methoxy groups (**2e**) led to poor conversion (Table 1).

Table 1 ***Trans***-selective asymmetric aziridination of diazoacetamides and *N*-Boc imines

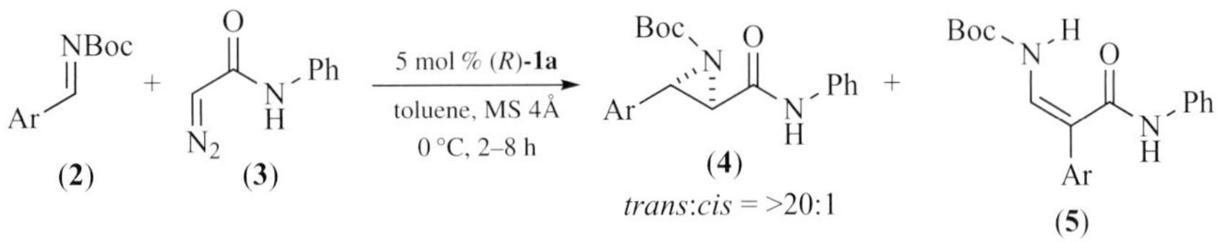

Ar	4:5	Yield (%)	ee (%)	Ar	4:5	Yield (%)	ee (%)
Ph (**2a**)	90:10	61	97	*m*-MeOC$_6$H$_4$ (**2f**)	72:28	55	96
p-tolyl (**2b**)	73:27	51	99	*p*-PivOC$_6$H$_4$ (**2g**)	77:23	49	96
m-tolyl (**2c**)	81:19	52	98	*p*-FC$_6$H$_4$ (**2h**)	72:28	31	89
2-Np (**2d**)	82:18	66	99	*m*-ClC$_6$H$_4$ (**2i**)	77:23	50	91
p-MeOC$_6$H$_4$ (**2e**)	56:44	<20	ND				

Conversely, electron-donating and electron withdrawing groups on *N*-aryl substituents of diazoacetamide were well tolerated with a range of *N*-Boc imines (Table 2).

Table 2 Scope of *trans*-selective asymmetric aziridination of diazoacetamides and *N*-Boc imines

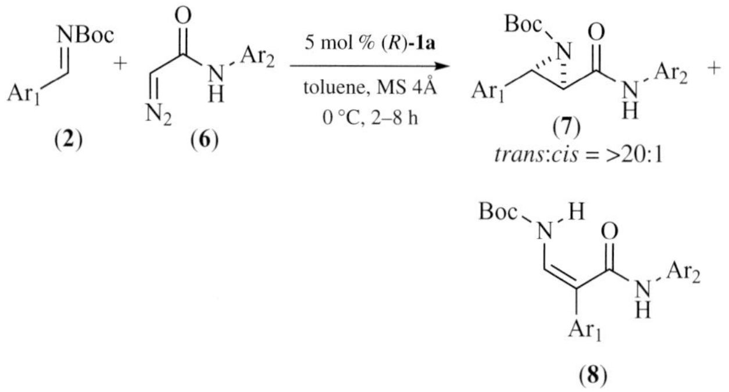

Ar_1	Ar_2	**7:8**	Yield (%)	ee (%)
Ph (**2a**)	*p*-$MeOC_6H_4$ (**6a**)	87:13	61	97
2-Np (**2d**)	*p*-$MeOC_6H_4$ (**6a**)	88:12	71	99
p-$PivOC_6H_4$ (**2g**)	*p*-$MeOC_6H_4$ (**6a**)	74:26	57	97
Ph (**2a**)	*p*-ClC_6H_4 (**6b**)	85:15	60	97
2-Np (**2d**)	*p*-ClC_6H_4 (**6b**)	81:19	70	99

Asymmetric Mannich Reactions of Arylaldehyde *N*-Boc Imines and *tert*-Butyl Diazoacetate.[1] The catalyst provides excellent enantioselectivities over the wide range of substrates, regardless of the substituent pattern of the substrate aryl group. A slight decrease of ee was observed with imine **2n** bearing a heteroaromatic group (Table 3).

Asymmetric Mannich Reactions of Arylaldehyde *N*-Boc Imines and Dimethyl Diazomethylphosphonate.[1] The reaction conditions described above can be applied to the synthesis of various *α*-diazo-*β*-aminophosphonates through an asymmetric Mannich reaction between *N*-Boc imines and dimethyl diazomethylphosphonate. Although the reaction turnover is slightly lower, enantioselectivities remain high over the wide range of arylaldehyde *N*-Boc imines (Table 4).

Asymmetric Additions of Formaldehyde *N,N*-Dialkylhydrazone to *N*-Boc Imines.[3] The catalyst (*R*)-**1b** tolerates the various substituent patterns and electronic properties of the aryl moiety, providing the desired products with excellent yields and enantioselectivities. The heteroarylaldehyde *N*-Boc imine **2n** can be applied to this reaction, though a slight decrease in enantioselectivity was observed (Table 5).

Asymmetric Additions of Arylaldehyde *N,N*-Tetramethylenehydrazones to Benzaldehyde *N*-Boc Imines.[3] Both electron-donating and electron-withdrawing aromatic groups on hydrazones reacted equally well, resulting in excellent enantioselectivities (Table 6).

Asymmetric Addition of Arylaldehyde *N,N*-Tetramethylenehydrazones to *N*-Boc Imines.[3] The electronic nature of the aryl moieties of *N*-Boc imines affects the outcome of the reaction. For example, the electron-rich *p*-methoxybenzaldehyde *N*-Boc imine suffers from lower yield and ee values (eq 3).

Electronically unbiased *m*-tolyl- and *o*-naphthyl-substituted *N*-Boc imines (**2c** and **2d**) provided the products in moderate yields with excellent selectivities (eq 4).

Table 3 Asymmetric mannich reaction of arylaldehyde *N*-Boc imines and ***tert***-butyl diazoacetate

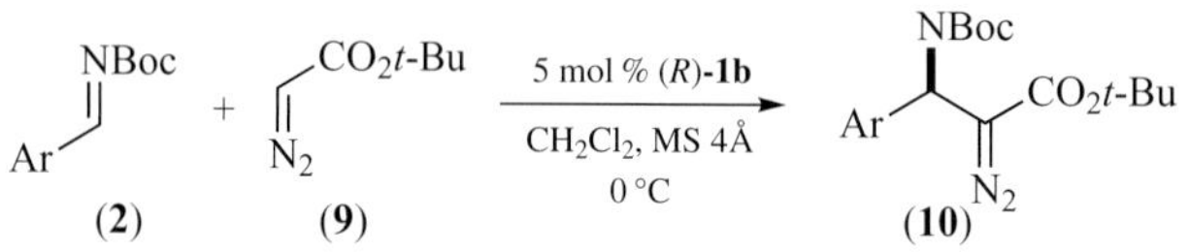

Ar	Time (h)	Yield (%)	ee (%)	Ar	Time (h)	Yield (%)	ee (%)
Ph (**2a**)	24	80	95	*p*-FC_6H_4 (**2h**)	20	83	95
o-tolyl (**2b**)	72	53	90	*p*-ClC_6H_4 (**2k**)	26	89	96
m-tolyl (**2c**)	20	74	92	*p*-$NO_2C_6H_4$ (**2l**)	72	38	95
p-tolyl (**2j**)	18	79	95	*p*-$MeOC_6H_4$ (**2m**)	20	72	95
2-Np (**2d**)	17	77	94	2-furyl (**2n**)	5	84	85

Table 4 Asymmetric mannich reaction of arylaldehyde *N*-Boc imines and dimethyl diazomethylphosphonate

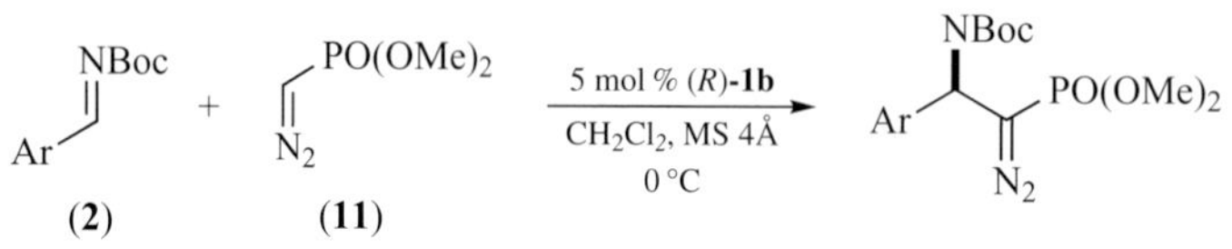

Ar	Time (h)	Yield (%)	ee (%)	Ar	Time (h)	Yield (%)	ee (%)
Ph (**2a**)	85	68	96	*p*-ClC_6H_4 (**2k**)	68	81	96
p-tolyl (**2j**)	68	68	96	*p*-$MeOC_6H_4$ (**2m**)	50	40	95
p-FC_6H_4 (**2h**)	52	85	96	2-furyl (**2n**)	46	89	92

The *N*-Boc imine bearing the electron-withdrawing *p*-chlorophenyl substituent significantly improved the yield of the transformation (eq 5).

Table 5 Asymmetric addition of formaldehyde *N*,*N*-dialkylhydrazone to *N*-Boc imines

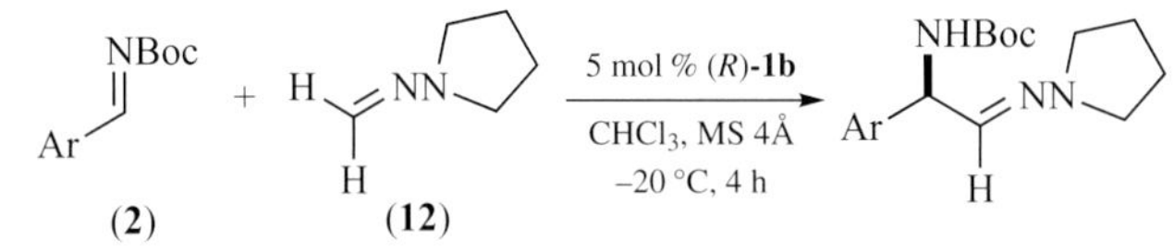

Ar	Yield (%)	ee (%)	Ar	Yield (%)	ee (%)
Ph (**2a**)	89	96	*p*-ClC_6H_4 (**2k**)	76	97
m-tolyl (**2c**)	80	93	*o*-BrC_6H_4 (**2o**)	86	97
2-Np (**2d**)	83	93	2-furyl (**2n**)	82	89
p-$MeOC_6H_4$ (**2e**)	70	97			

Table 6 Asymmetric addition of arylaldehyde *N*,*N*-tetramethylenehydrazones to benzaldehyde *N*-Boc imine

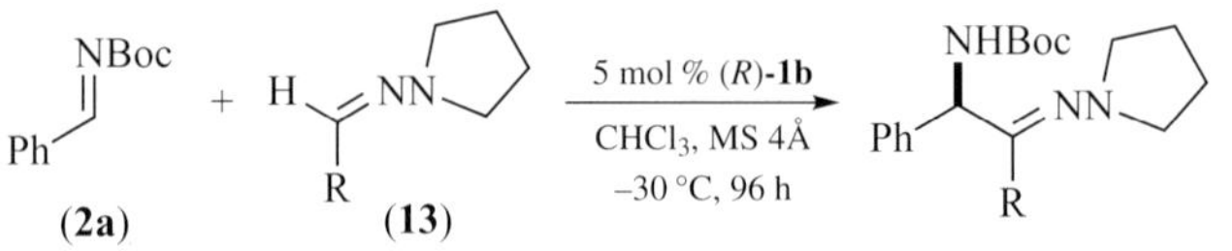

R	Yield (%)	ee (%)	R	Yield (%)	ee (%)
Ph (**13a**)	66	95	*p*-$MeOC_6H_4$ (**13d**)	55	92
p-tolyl (**13b**)	51	92	*p*-ClC_6H_4 (**13e**)	60	91
2-Np (**13c**)	56	92			

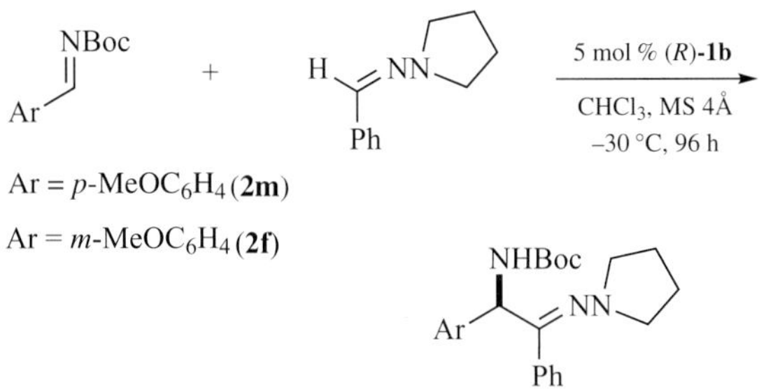

(3)

Ar = *p*-$MeOC_6H_4$ (**2m**)

Ar = *m*-$MeOC_6H_4$ (**2f**)

Ar = *p*-$MeOC_6H_4$: 35%, 84% ee

Ar = *m*-$MeOC_6H_4$: 54%, 91% ee

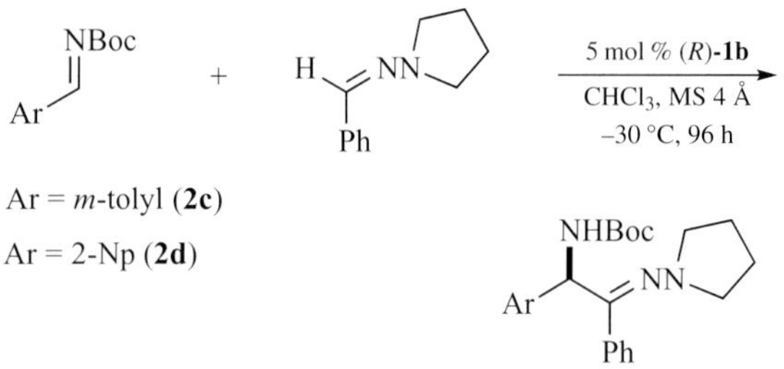

(4)

Ar = *m*-tolyl (**2c**)

Ar = 2-Np (**2d**)

Ar = *m*-tolyl: 55%, 91% ee

Ar = 2-Np: 51%, 92% ee

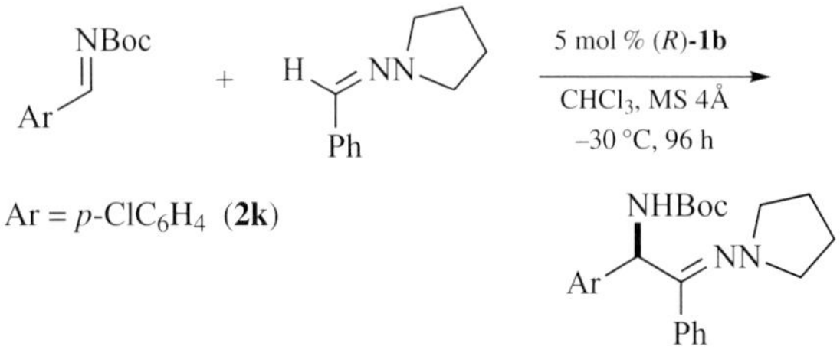

(5)

Ar = *p*-ClC_6H_4 (**2k**)

Ar = *p*-ClC_6H_4: 77%, 90% ee

Reactions combining *p*-chlorobenzaldehyde *N*-Boc imine and other hydrazones can be performed successfully regardless of the electronic nature of the substituents in hydrazones (eq 6).

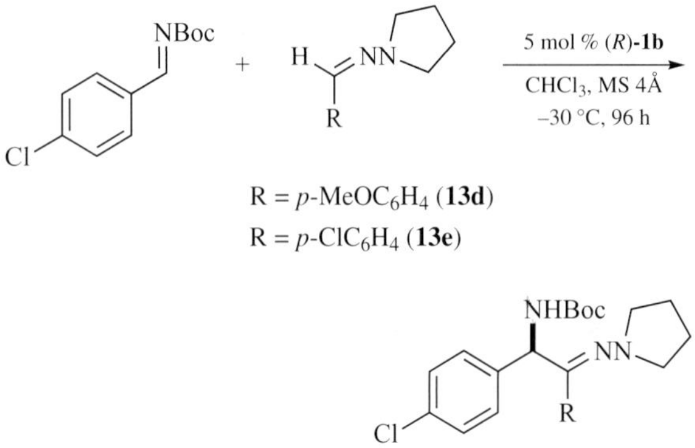

(6)

R = *p*-$MeOC_6H_4$ (**13d**)

R = *p*-ClC_6H_4 (**13e**)

R = *p*-$MeOC_6H_4$: 67%, 91% ee

R = *p*-ClC_6H_4: 73%, 89% ee

1. Hashimoto, T.; Maruoka, K., *J. Am. Chem. Soc.* **2007**, *129*, 10054.
2. Hashimoto, T.; Uchiyama, N.; Maruoka, K., *J. Am. Chem. Soc.* **2008**, *130*, 14380.
3. Hashimoto, T.; Hirose, M.; Maruoka, K., *J. Am. Chem. Soc.* **2008**, *130*, 7556.

Shin A. Moteki, Seiji Shirakawa & Keiji Maruoka
Kyoto University, Kyoto, Japan

Bromamine-T

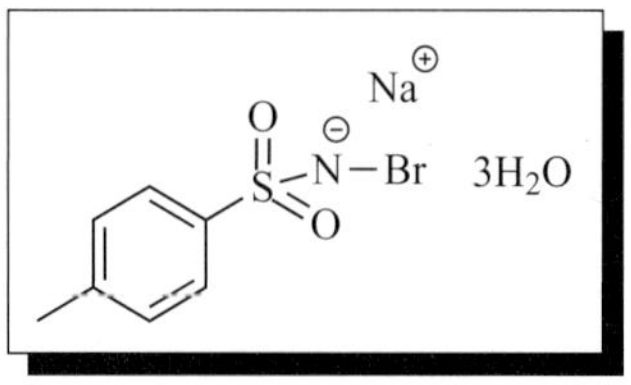

[41085-71-6] $C_7H_7BrNO_2S{\cdot}Na\,(3H_2O)$ (MW 324.96)
InChI = 1S/C7H7BrNO2S.Na.3H2O/c1-6-2-4-7(5-3-6)12(10,11)9-8;;;;/h2-5H,1H3;;3*1H2/q-1;+1;;;
InChIKey = COPFQPVBHZRRAE-UHFFFAOYSA-N

(reagent used as an oxidizing agent, brominating agent, and analytical reagent)

Alternate Names: N-bromo-*N*-sodio-*p*-toluenesulfonamide; BAT.
Physical Data: mp 150–152 °C.
Preparation: prepared by the partial debromination of dibromamine-T (p-$CH_3C_6H_4SO_2NBr_2$ or DBT), which is obtained from chloramine-T (CAT) and bromine. It can also be prepared by the bromination of tosylamine, with the later being treated with sodium hydroxide.
Solubility: insoluble in ether, benzene, and chloroform; soluble in cold H_2O; soluble in hot H_2O; less soluble in EtOH.
Analysis of Reagent Purity: by iodometric titration.
Handling, Storage, and Precautions: irritant; keep containers tightly sealed and protect from light; although the trihydrate is stable.

Bromamine-B [16917-09-2] is the sodium salt of *N*-bromobenzenesulfonamide ($C_6H_5SO_2NBrNa{\cdot}1.5H_2O$ or BAB). It can be prepared from chloramine-B (CAB) and bromine. The properties of bromamine-B have found to be similar to that of bromamine-T.

Introduction. The sodium salts of *N*-bromo-*p*-toluenesulfonamide (p-$CH_3C_6H_4SO_2NBrNa{\cdot}3H_2O$ or bromamine-T; BAT) and *N*-bromobenzenesulfonamide($C_6H_5SO_2NBrNa{\cdot}1.5H_2O$ or bromamine-B; BAB) are an addition to the class of organic haloamines and were employed as an oxidimetric titrants in aqueous media.[1,2] The chemistry of these haloamines has revealed profound interest, as they are oxidants, analytical regents, and brominating agents employed for a number of synthetic purposes. These bromamines are the precursor of bromonium cations, hypobromites, and *N*-anions, which are capable of acting both as electrophiles and as nucleophiles.[3] These compounds resemble hypobromites in their oxidative behavior.[4,5]

Investigations of Morris et al.,[5] Bishop and Jennings,[4] Pryde and Soper,[6] and Hardy and Johnston[7] have shown the existence of similar equilibria of *N*-metallo-*N*-arylhalosulfonamides (*N*-haloamines) in aqueous media. Depending on the pH, these bromamines exhibit the following equilibria in aqueous solution (eqs 1–7):

$$\mathrm{RNBrNa} \rightleftharpoons \mathrm{RNBr^- + Na^+} \quad (1)$$

$$\mathrm{RNBr^- + H^+} \rightleftharpoons \mathrm{RNHBr} \quad (2)$$

$$\mathrm{2RNHBr} \rightleftharpoons \mathrm{RNH_2 + RNBr_2} \quad (3)$$

$$\mathrm{RNBr_2 + H_2O} \rightleftharpoons \mathrm{RNHBr + HOBr} \quad (4)$$

$$\mathrm{RNHBr + H_2O} \rightleftharpoons \mathrm{RNH_2 + HOBr} \quad (5)$$

$$\mathrm{HOBr + H^+} \rightleftharpoons \mathrm{H_2OBr^+} \quad (6)$$

$$\mathrm{HOBr} \rightleftharpoons \mathrm{H^+ + OBr^-} \quad (7)$$

(R=$CH_3C_6H_4$ for Bromamine-T; and C_6H_5 for Bromamine-B)

The above species furnished in an aqueous medium act as the reactive species involved in the oxidation and the halogenation reactions.

Bromamine-T and bromamine-B have been served as the reagents, oxidants, brominating agents as well as analytical reagents for many organic processes. Generally these bromamines are more reactive than the chlorine analogs.

Aziridination of Olefins. Aziridines are valuable intermediates for the synthesis of useful nitrogen-containing compounds.[8,9] Generally, bromamine-T has served as the nitrene source for the aziridination of olefins.

Branco and coworkers[10] have performed palladium(II)-promoted aziridination of the olefins with bromamine-T as the nitrogen transfer reagent. A variety of *N*-tosyl-2-substituted aziridines have been synthesized under mild reaction conditions (eq 8).

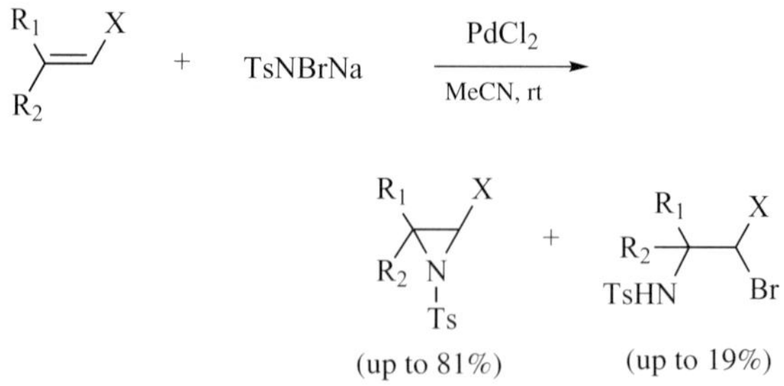

(8)

(R_1 = H, $(CH_2)_4$, $CO(CH_3)_2$; R_2 = H, Me; X = COOMe, $CONMe_2$, CN, COMe, SOPh, Ph, CH_2OH)

A promising application of iron catalysts for the aziridination of olefins with bromamine-T was reported by Zhang and coworkers.[11] Varieties of aziridines have been synthesized applying different iron metalloporphyrin catalysts from both internal and terminal alkenes (eq 9).

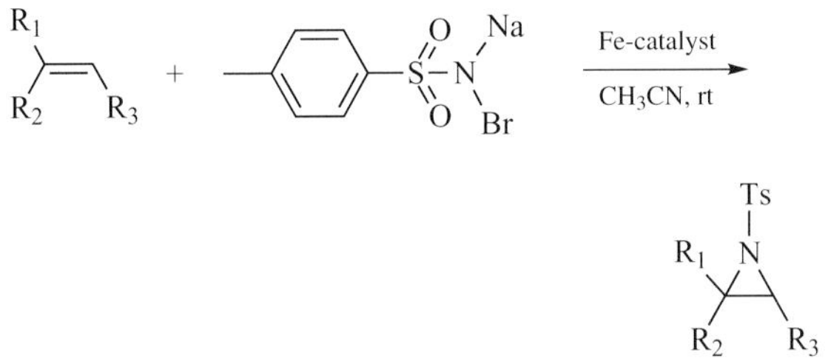

(9)

(up to 73%)

Fe-catalyst = Fe(TPP)Cl, Fe(TMOPP)Cl, Fe(TPFPP)Cl, Fe(TDClPP)Cl.

(R_1 = H, Me, Ph; R_2 = Ph, XPh (X = halide), MePh, NO_2Ph, $ClCH_2$Ph, *n*-heptyl; R_3 = H, Ph, COOEt, COO(*t*-Bu))

Interestingly, aromatic, aliphatic, cyclic, and acyclic alkenes have all produced the corresponding aziridines using bromamine-T. All these reactions can be performed under mild conditions with alkenes as limiting reagents and proceed with moderate to low stereospecificity for 1,2-disubstituted alkenes.

Further, Chanda and coworkers[12] have performed the aziridination of olefins with bromamine-T using polymer-supported Mn catalysts. Notably, this heterogeneous Mn catalyst was recycled and reused (eq 10).

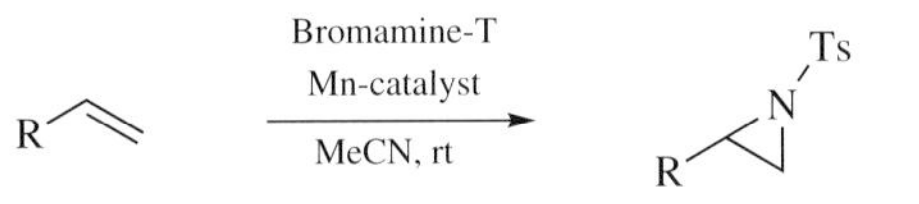

(10)

(up to 73%)

Mn-catalyst

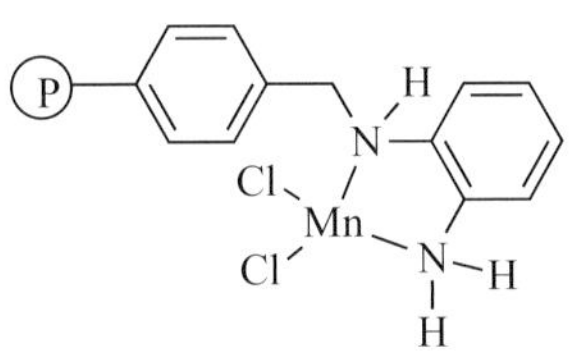

(R = Ph, *n*-octyl, cyclooctyl, cyclobenzyl)

C–H Amination Reactions. Amines are central intermediates and key precursors for the synthesis of variety of chemicals such as dyes, pharmaceuticals, and agrochemicals. These nitrogen-containing molecules are abundant in nature and their importance in biology and medicine has been widely documented. Among modern amination methods,[13] metal-catalyzed nitrene insertion provides an attractive strategy for direct transformation of ubiquitous C–H bonds to value-added amine functionalities through the use of appropriate nitrene sources.[14]

Cobalt-supported porphyrins are capable of catalyzing intermolecular nitrene insertion of sp^3 C–H bonds with bromamine-T as the nitrene source, forming the desired tosyl-protected amines with NaBr as the only by-product (eq 11).[15]

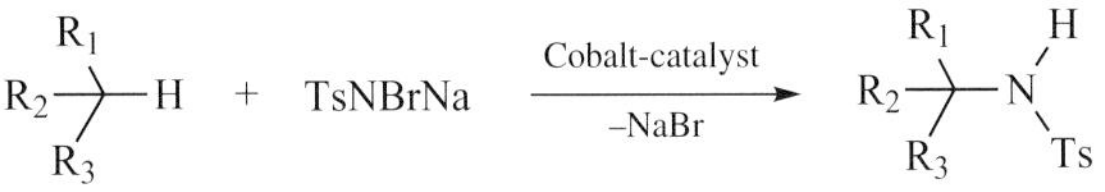

(11)

(14–74%)

(This is the general scheme, in the original article, Rs have not defined)

Cobalt-catalyst = Co(TPP)Cl, Co(TMP)Cl, Co(TPFPP)Cl, Co(TDClPP)Cl.

Using indane as a model substrate, the authors[15] have systematically evaluated catalytic C–H nitrene insertion with bromamine-T under different reaction conditions by cobalt–porphyrin complexes possessing varied electronic and steric environments, thereby producing amination products up to 73% yields (eq 12).

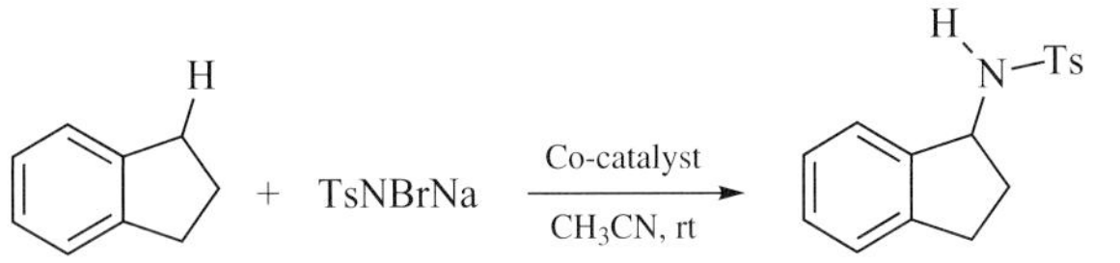

(12)

(up to 73%)

Co-catalysts = Co(TPP)Cl, Co(TMP)Cl, Co(TPFPP)Cl, Co(TDClPP)Cl.

A convenient and practical procedure for the aminobromination of electron-deficient olefins using bromamine-T as nitrogen and bromine source promoted by (diacetoxyiodo)benzene has been developed.[16] This metal-free protocol is highly efficient and affords the vicinal bromamines with excellent stereoselectivities (eq 13).

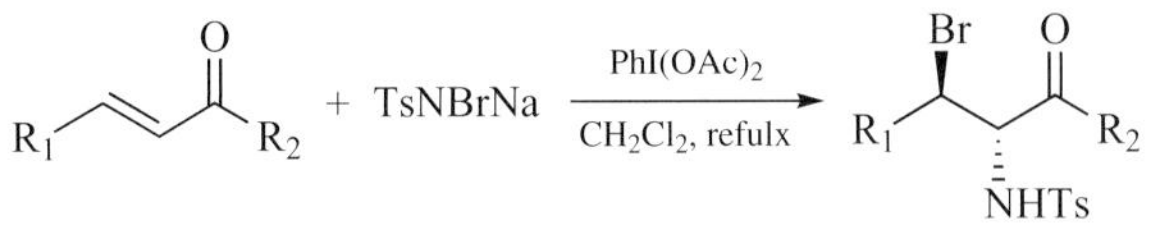

(13)

(up to 74%) (±)

(R_1 = Ph, 4-Ph, 3-ClPh, 3,4-Cl_2Ph, 4-NO_2Ph; R_2 = Ph, 4-MeOPh, 4-ClPh, Me, OMe, NEt_3)

Irrespective of the substituent pattern and the electronic property of the aryl moiety, the products are obtained in good yields with very promising diastereoselectivities when chalcones were employed as reactants. Furthermore, other electron-deficient double bond systems were also examined. Enones with methyl substituents on the carbonyl side afforded the bromamines in 57% yield with high diastereoselectivity. Cinnamates and cinnamides were also tolerated in the present system; both substrate classes exclusively gave the *anti*-products in moderate yields.

A copper-catalyzed aminobromination–elimination provides easy access to α,β-unsaturated vicinal haloamides[17] from readily available α,β-unsaturated ketones and esters in good to excellent yields using *N*-alkyl, *N*-bromo-*p*-toluenesulfonamide. The isolated intermediate reveals that the reaction proceeds through the aminobromination process.

To improve the yield, metal catalysts, such as $Ni(OAc)_2$, $CuCl_2$, $CuBr_2$, CuCl, CuI, $Mn(OAc)_2$, and $MnSO_4$ were tested. $CuCl_2$ was found to be most efficient giving up to 81% yield at 35 °C in CH_2Cl_2 (eq 14).

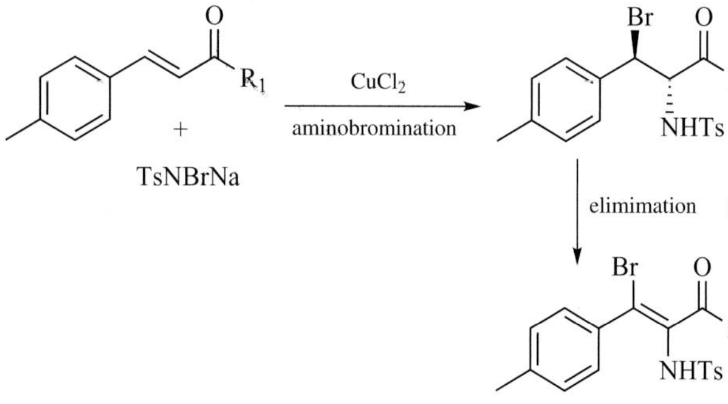

(14)

(up to 81%)

The carbon dioxide (CO_2)-induced amidobromination of olefins with bromamine-T was used in reactions with a wide range of olefins, both aromatic and aliphatic, as well as electron-rich and electron-deficient olefins, leading to the regioselective formation of amido-brominated compounds.[18] The *p*-substituted styrene derivatives were subjected to the amidobromination, in an attempt to determine the active species for the reaction (eq 15).

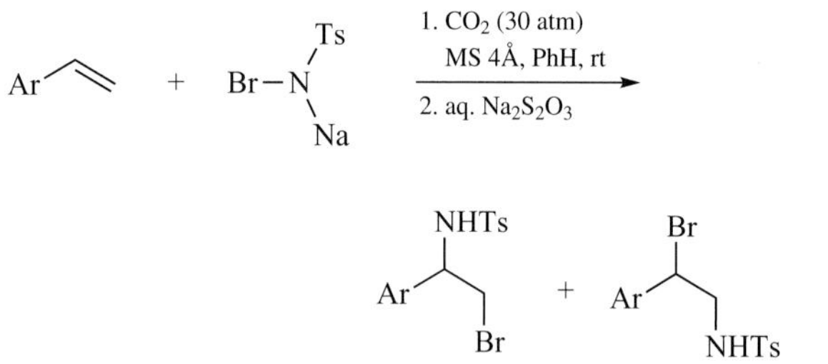

(15)

(up to 90%)

Ar = $NO_2C_6H_4$, ClC_6H_4, $MeOC_6H_4$, MeC_6H_4.

Further investigations have been explored on the diastereoselective amidobromination of an olefin bearing a chiral auxiliary (Evans' oxazolidinone) (eq 16).[18] A 2-acryloyloxazolidinone without a stereocenter was first employed in the reaction. A phenyl-substituted chiral olefin also reacted, affording the α-bromo-β-amino product with excellent diastereoselectivity. The diastereomers were readily separated and the major isomer was treated with 4-di(methylamino)pyridine (DMAP) in MeOH leading to the formation of the enantiopure (*S*)-aziridine **B**. This chiral aziridine should be a good precursor of amino acid derivatives via ring opening. From the absolute configuration of aziridine **B**, the stereocenter α to the carbonyl group in major diastereomer of **A** would be predicted to have an *R* configuration.

_N_-Oxidation Processes. *N*-Oxides have been extensively used as oxidants and in the synthesis of various organic compounds and as bioreductive drugs. A variety of tertiary amines were efficiently and selectively oxidized to the corresponding *N*-oxides by bromamine-T using ruthenium(III) trichloride as catalyst in alkaline (pH 8.4) acetonitrile–water (1:1) at 80 °C (eq 17).[19]

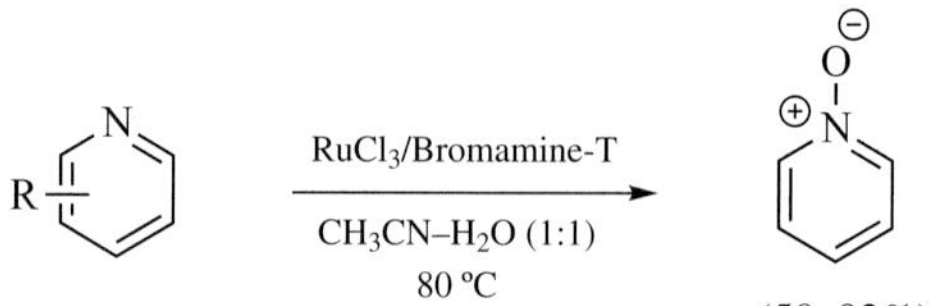

(17)

(50–92%)

(R = NH_2, CH_3, CN, $CONH_2$)

N-oxidation of pyrazines by bromamine-B in perchloric acid has been studied from the view point of kinetic and mechanistic implications (eq 18).[20]

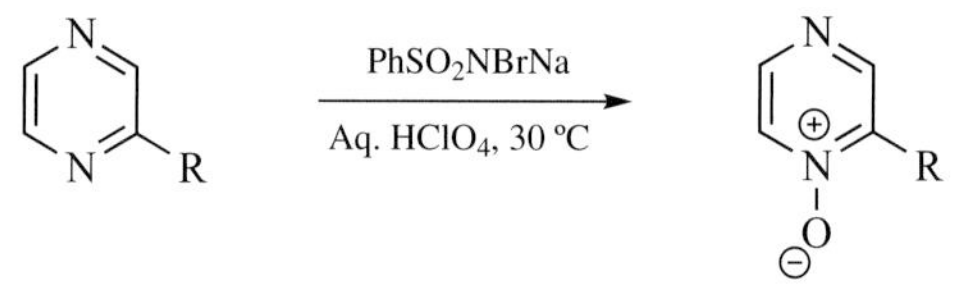

(18)

(R = Me, OMe, C_2H_5, NH_2)

The reactions show identical kinetics being first-order each in [bromamine-B] and [pyrazine] and a fractional-order dependence on [H^+]. The ionic strength of the medium and addition of benzenesulfonamide or halide ions showed no significant effect on the reaction rate. The dielectric effect is positive. The solvent isotope effect was studied using D_2O.

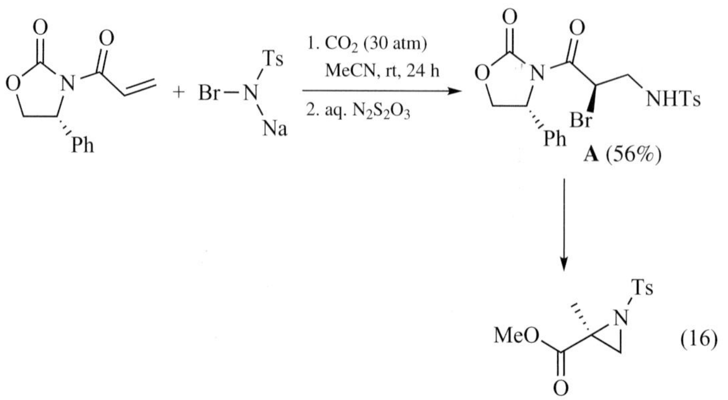

(16)

(*S*), **B** (64%)

1. Nair, C. G. R.; Kumari, R. L.; Indrasenan, P., *Talanta* **1978**, *25*, 525.
2. Ahmed, M. S.; Mahadevappa, D. S., *Talanta* **1980**, *27*, 669.
3. Campbell, M. M.; Johnson, G., *Chem. Rev.* **1978**, *78*, 65.
4. Bishop, E.; Jennings, V. J., *Talanta* **1958**, *1*, 197.
5. Morris, J. C.; Salazar, J. A.; Wineman, M. A., *J. Am. Chem. Soc.* **1948**, *70*, 2036.
6. Pryde, B. G.; Soper, F. G., *J. Chem. Soc.* **1931**, 1514.
7. Hardy, F. F.; Johnston, J. P., *J. Chem. Soc., Perkin Trans. 2* **1973**, 742.
8. Tanner, D., *Angew. Chem., Int. Ed. Engl.* **1994**, *33*, 599.
9. Ibuka, T., *Chem. Soc. Rev.* **1998**, *27*, 145.
10. Antunes, A. M. M.; Marto, S. J. L.; Branco, P. S.; Prabhakar, S.; Lobo, A. M., *Chem. Commun.* **2001**, 405.
11. Vyas, R.; Gao, G.-Y.; Harden, J. D.; Zhang, X. P., *Org. Lett.* **2004**, *6*, 1907.
12. Vyas, R.; Chanda, B. B.; Belhekar, A. A.; Patel, D. R.; Ram, R. N.; Bedekar, A. V., *J. Mol. Catal A: Chem.* **2000**, *160*, 237.
13. (a) Modern Amination Methods; Racci, A., Ed.; Wiley-VCH: Weinheim, **2000**; (b) Salvatore, R. N.; Yoon, C. H.; Jung, K. W., *Tetrahedron* **2001**, *57*, 7785. (c) Johannsen, M.; Jorgensen, K. A., *Chem. Rev.* **1998**, *98*, 1689. (d) Muller, T. E.; Beller, M., *Chem. Rev.* **1998**, *98*, 675.
14. (a) Muller, P.; Fruit, V., *Chem. Rev.* **2003**, *103*, 2905. (b) Dauban, P.; Dodd, R. H., *Synlett* **2003**, 1571. (c) Koser, G. F., *Top. Curr. Chem.* **2003**, *224*, 137. (d) Halfen, J. A., *Curr. Org. Chem.* **2005**, *9*, 657.
15. Harden, J. D.; Ruppel, J. V.; Gao, G.-Y.; Zhang, X. P., *Chem. Commun.* **2007**, 4644.
16. Xia, J.-J.; Xue-Liang Wu, X.-L.; Wang, G.-W., *ARKIVOC* **2008**, *xvi*, 22.
17. Sun, H.; Zhang, G.; Zhi, S.; Han, J.; Lib, G.; Pan, Y., *Org. Biomol. Chem.*, **2010**, *8*, 4236.
18. Hayakawa, J.; Kuzuhara, M.; Minakata, S., *Org. Biomol. Chem.* **2010**, *8*, 1424.
19. Sharma, V. B.; Jain, S. L.; Sain, B., *Tetrahedron Lett.* **2004**, *45*, 4281.
20. Puttaswamy; Shubha, J. P., *Bull. Korean Chem. Soc.* **2009**, *30*, 1399.

Rajenahally V. Jagadeesh
Leibniz Institute for Catalysis (LIKAT), Rostock, Germany

Puttaswamy
Bangalore University, Bangalore, India

Bromonitromethane

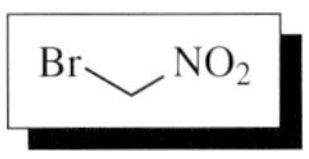

[563-70-2] CH_2BrNO_2 (MW 139.94)
InChI = 1S/CH2BrNO2/c2-1-3(4)5/h1H2
InChIKey = DNPRVXJGNANVCZ-UHFFFAOYSA-N

(reagent used as both nucleophile and electrophile, and for the synthesis of the pharmaceutically important cyclic compounds)

Physical Data: mp −28°C; bp 146–148°C/750 mmHg; *d* 2.007 g cm^{-3}; Fp 113 °C.
Solubility: sol toluene, ether, and most organic solvents.
Form Supplied in: yellow liquid; widely available.
Purification: purified by distilling (40–52°C/16 mmHg).
Handling, Storage, and Precautions: keep container tightly closed. Keep container in a cool, well-ventilated area. Separate from acids, bases, reducing reagents and combustibles. Empty containers pose a fire risk, evaporate the residue under a fume hood. Do not ingest. Avoid contact with skin and eyes.

Original Commentary

Tsubasa Inokuma & Yoshiji Takemoto
Kyoto University, Kyoto, Japan

1,2-Additions. Due to the acidity of the α-proton ($p_{Ka} = 12.5 \pm 0.1$ in DMSO)[1] of bromonitromethane, this reagent serves as a nucleophile in the presence of proper additives. Aldehydes and azetidine-2,3-diones can be used as electrophilic partners (eqs 1–5).[2] When the substrate has chiral centers, high diastereoselectivity is obtained in some cases. In the presence of a copper catalyst and chiral ligands, good enantioselectivity can be achieved in this reaction. Although the diastereoselectivity is low, the bromo group can be easily removed under radical conditions.

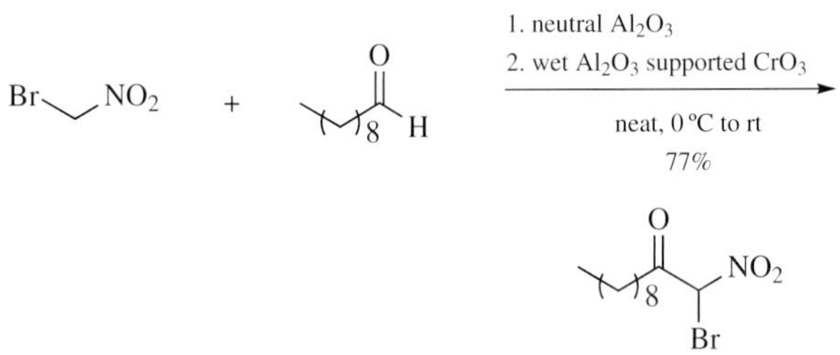

(1)

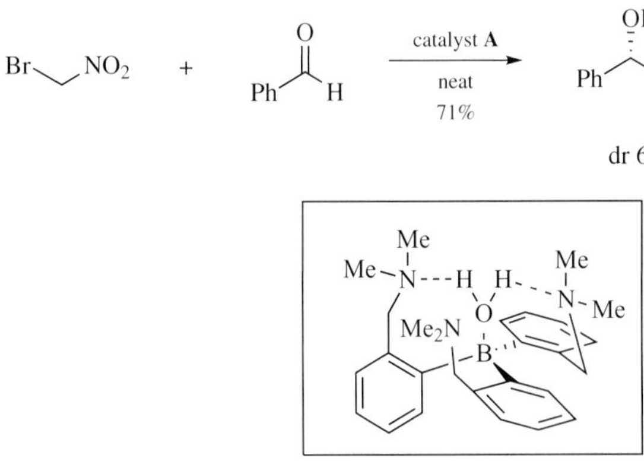

(2)

catalyst **A**

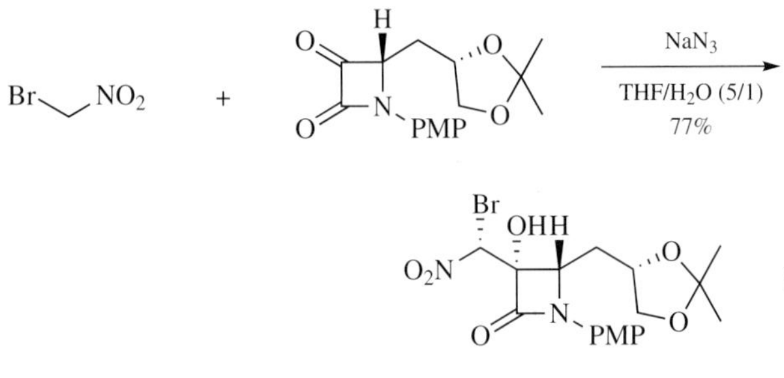

(3)

dr 85:15

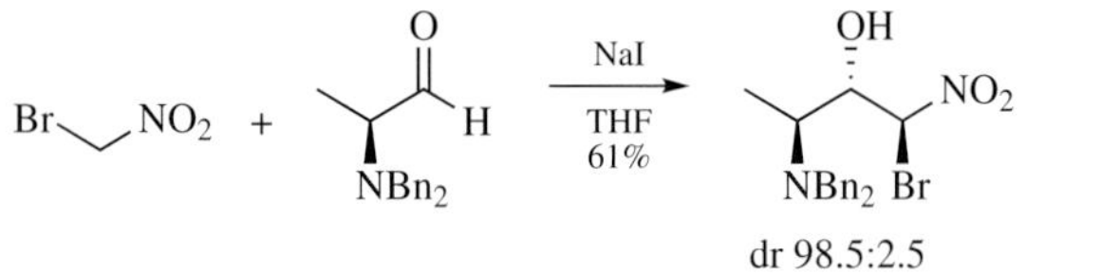

(4)

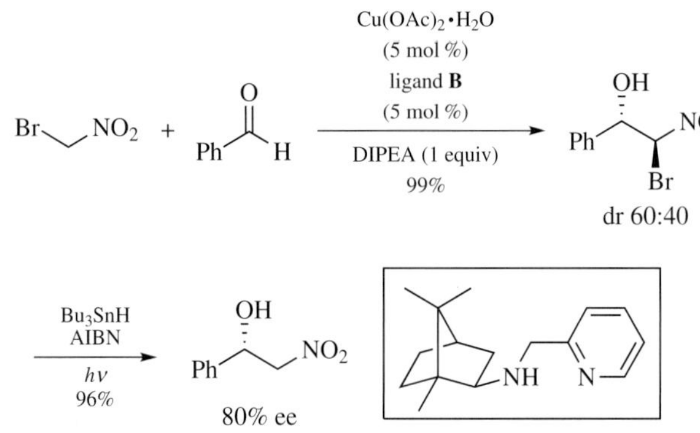

(5)

ligand **B**

In addition, the resulting bromoalkane derivatives have an electrophilic nature, so they are suitable substrates for the synthesis of some fascinating heterocyclic compounds. Until now, aziridines, thiophenes, furopyrimidinones, and 4,5,6,7-tetrahydrothieno [3,2,c]pyridines have been synthesized by this procedure (eqs 6–9).[3]

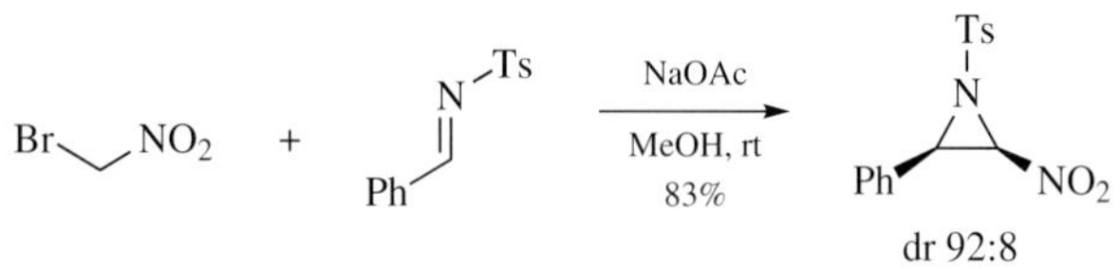

(6)

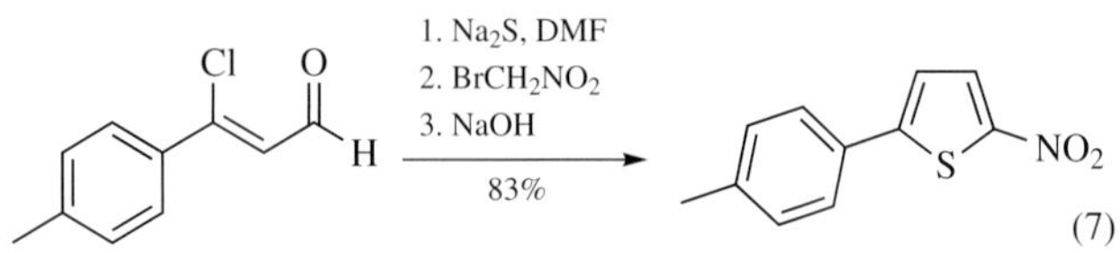

(7)

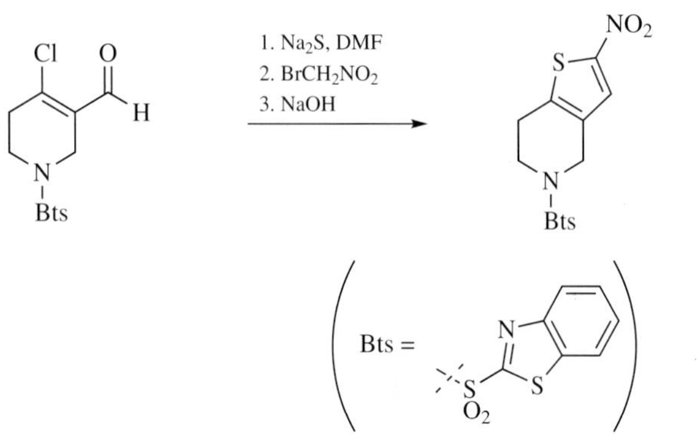

(8)

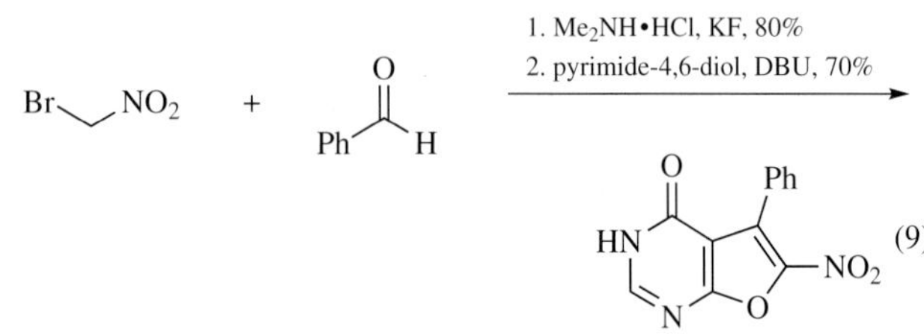

(9)

1,4-Additions. There are several reports of 1,4-additions of bromonitromethane to a variety of electrophiles such as enones, enals, and α,β-unsaturated α-cyanoimides (eqs 10–15).[4] Different from 1,2-addition, several asymmetric catalytic 1,4-additions have already been reported. Chiral amines, pyrrolidines, and aminothioureas can be used as asymmetric catalysts in these reactions. The products are useful precursors for the synthesis of nitrocyclopropanes under basic conditions. However, there are still problems with low diastereoselectivities when acyclic compounds are used as substrates.

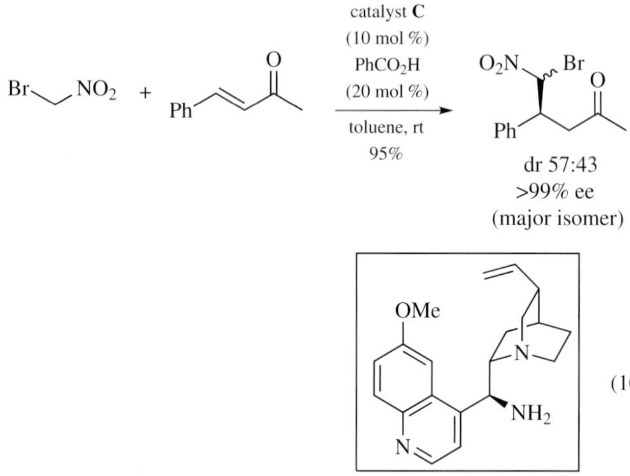

(10)

catalyst **C**

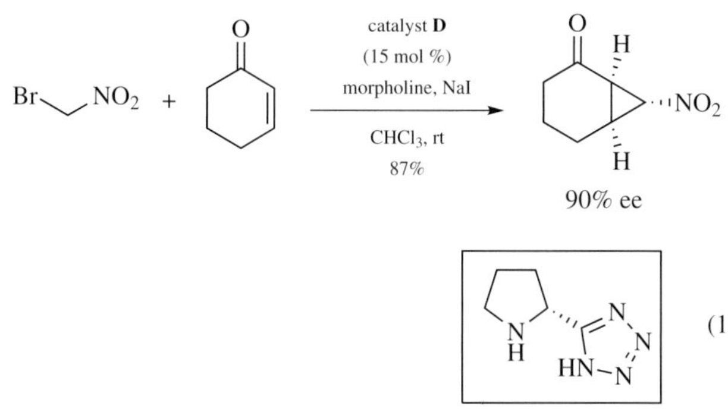

(11)

catalyst **D**

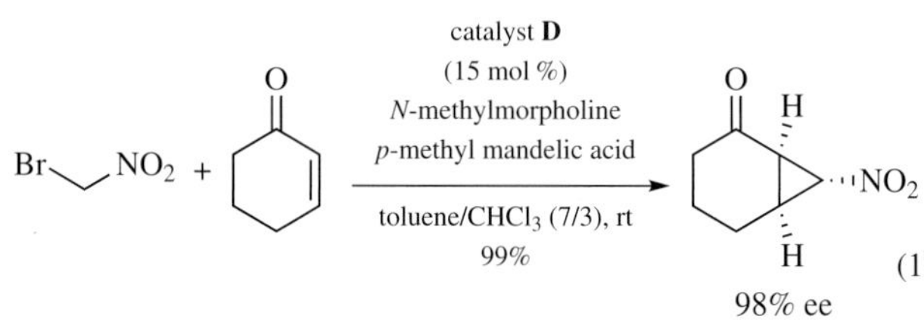

(12)

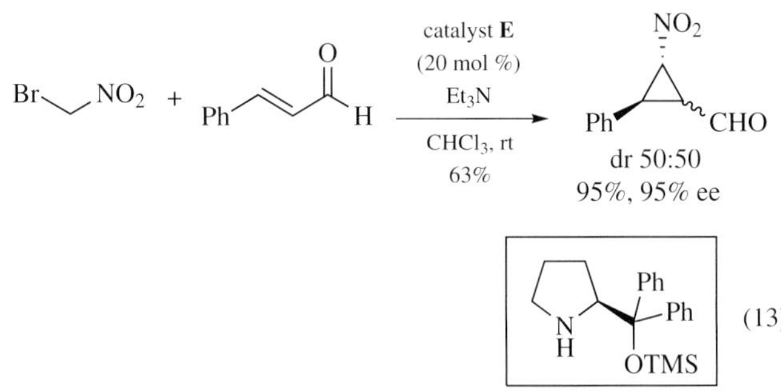

catalyst **E**

(13)

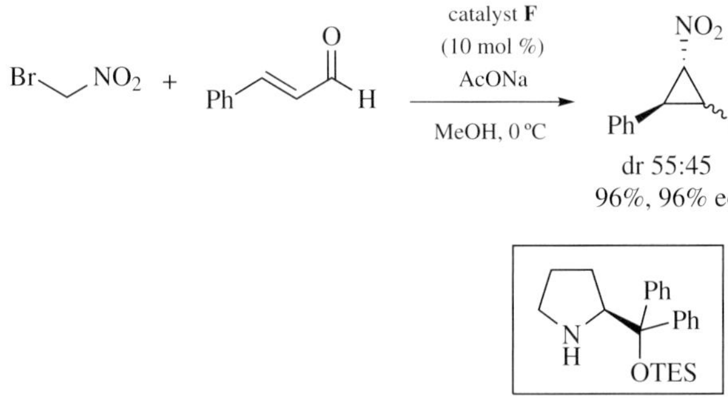

catalyst **F**

(14)

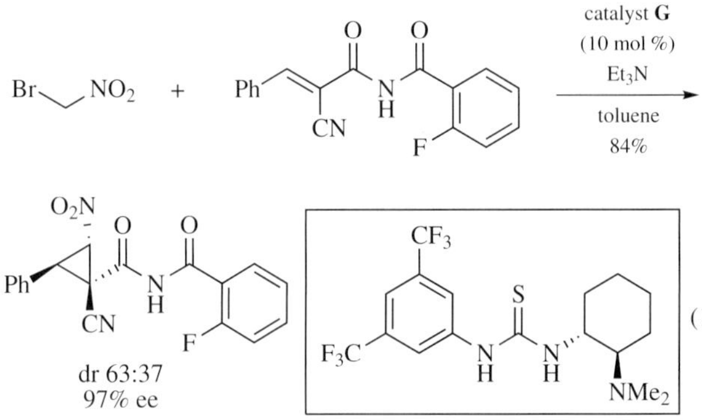

dr 63:37
97% ee

catalyst **G**

(15)

Electrophilic substitutions. Several nucleophiles such as thiolates and alkyllithium compounds react with bromonitromethane to give the corresponding nitroalkane derivatives (eqs 16–18).[5] The products can subsequently be used as nucleophiles for further transformations.

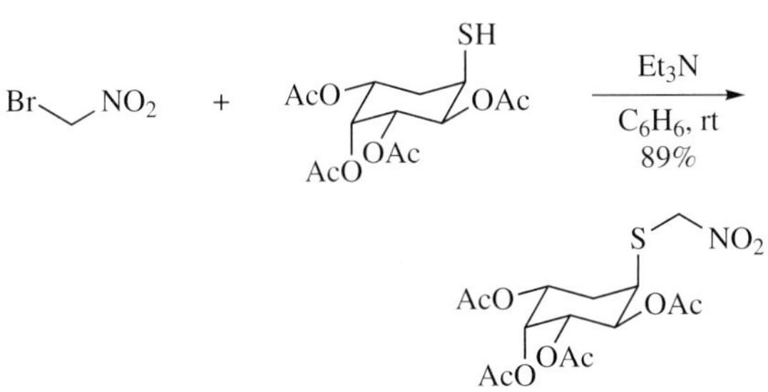

(16)

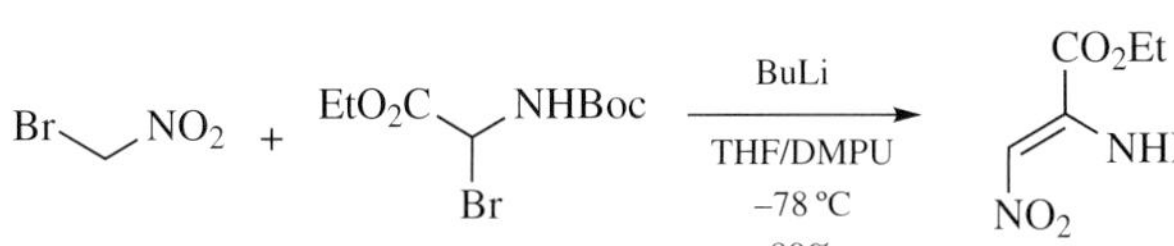

(17)

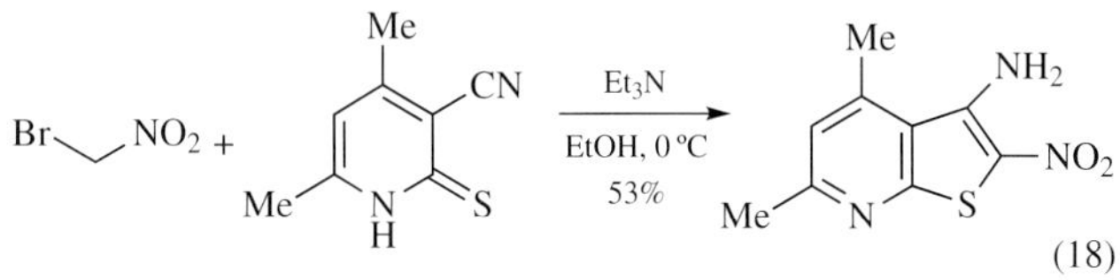

(18)

1,3-Dipolar cycloadditions. When bromonitromethane is treated with Lewis acids, the corresponding 1,3-dipole is generated and it can be used as substrate for 1,3-dipolar cycloadditions. However, examples of these reactions are quite limited until now.

$BF_3{\cdot}OEt_2$ can be used as a Lewis acid for generating the 1,3-dipole (eq 19)[6]. The reaction sequence consists of a [3+2] cycloaddition with dihydrofuran, fragmentation of the heterobicyclic compound, intramolecular cyclization, ring opening reaction, and a Nef reaction of the resultant nitronate compound.

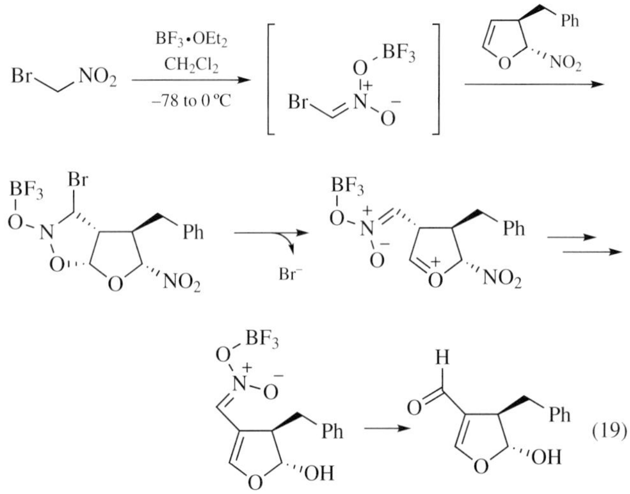

72%

(19)

First Update

Sergey V. Tsukanov & Jeffrey N. Johnston
Vanderbilt University, Nashville, TN, USA

Barbier-type Additions to Aldehydes and Imines. Reductive procedures using indium or samarium metals have been reported as alternatives to the usual generation of nitronate by deprotonation under basic conditions (eqs 20–24).[7] In these cases, bromonitromethane is a functional equivalent to nitromethane, and amino alcohols and nitroamines are prepared under mild conditions and with a broad scope of substrates. The ability to prepare products with aliphatic (enolizable) aldehydes and imines is notable. These conditions were also successfully applied toward more complex carbohydrates to furnish a variety of nitro and amino sugars with moderate diastereoselectivity and in good yield (eq 24).

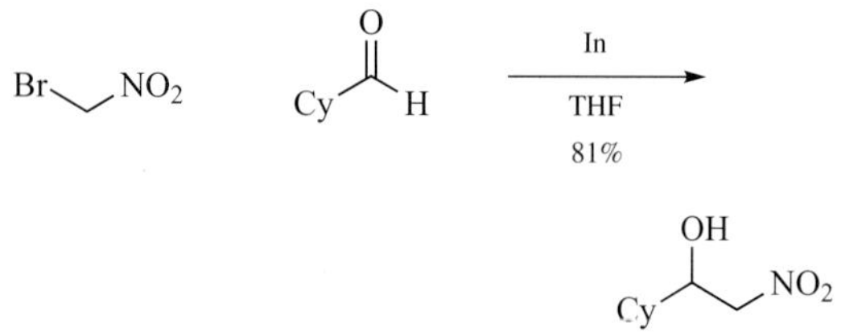
(20)

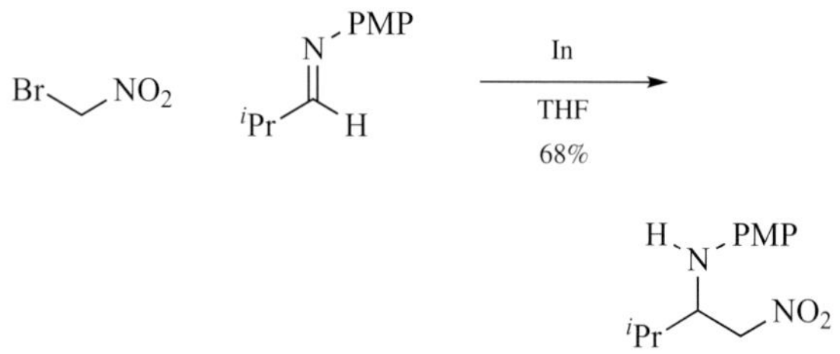
(21)

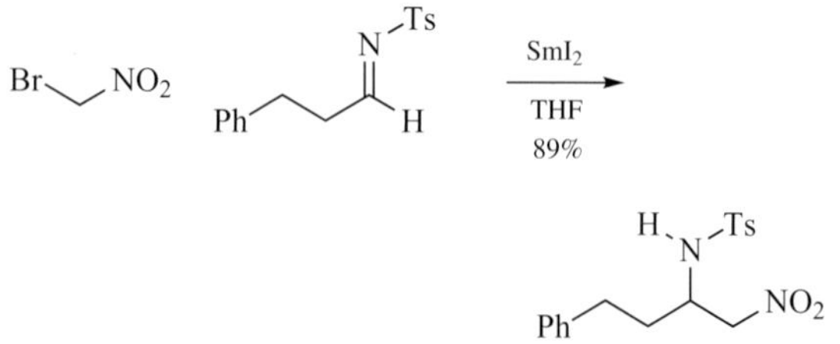
(22)

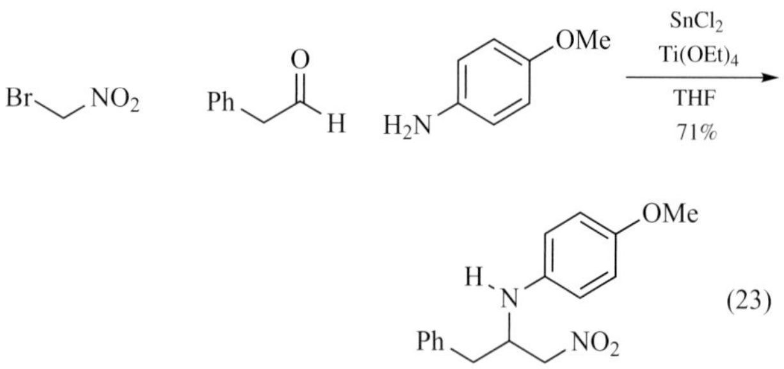
(23)

Chiral Bromonitroalkanes as Precursors to Natural and Nonnatural Amino Acids. The discovery of umpolung amide synthesis (UmAS)[8] has further improved the utility of bromonitromethane as a reagent and has enhanced its synthetic impact. These developments have largely highlighted the advantage of amide formation using a unique mechanism wherein the carbon and nitrogen polarities are reversed (umpolung) at the carbon–nitrogen bond-forming stage.[9] Several methods to prepare enantioenriched chiral α-bromonitroalkane donors have been reported (eqs 25–28),[10,11] and UmAS allows these donors to function as surrogates for chiral α-hydroxy carboxylic acids (eq 26) and aryl glycine α-amino acids (eq 25). Bromonitromethane behaves like nitromethane in BAM-catalyzed additions to *N*-Boc imines,[12] but the bromine-substituted carbon is typically produced as a mixture of epimers. This feature is of little consequence since the bromine-substituted carbon becomes the sp^2-hybridized carbon of the amide product resulting from UmAS. The scope of substrates is broad and most aromatic aldimines provide the amide precursors in high yields and with excellent enantioselectivity. A diverse group of *N*-protecting groups can now be accessed by the addition of bromonitromethane to trimethylsilyl imines (eq 27).[10] Natural amino acids can participate in UmAS coupling reactions with α-bromonitroalkanes to provide peptidic products (protecting groups are required for certain side chain functionalities),[8] and amide bond formation substoichiometric in NIS is effective using oxygen as a terminal oxidant.[13] Finally, it was reported that the addition products of bromonitromethane can be converted to their corresponding diamines by a cobalt-mediated reduction providing a high-yielding and an efficient alternative to a direct nitromethane addition in challenging cases (eq 28).[14]

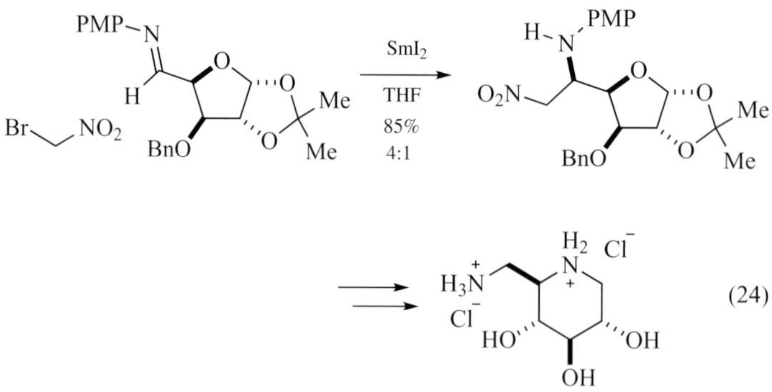
(24)

6-amino-1,6-dideoxynojirimycin

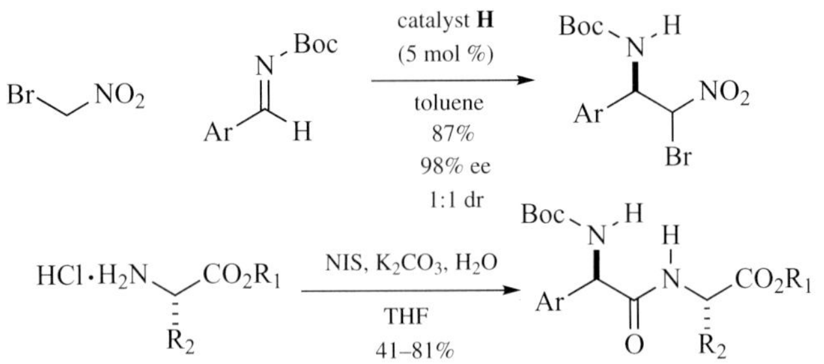

Ar = $^{p}ClC_6H_4$; R^1 = Me, Et, or ^{t}Bu
Amino acids: Gly, Ala, Val, Leu, Ile, Ser, Thr
Phe, Tyr($O^{t}Bu$), Asp(CO_2Me), Asn, Glu(CO_2Et)
Gln, Lys(NHCbz), Pro, Cys(SBn), Met, His(NTr)
Arg(N-NO_2), Tyr(NBoc)

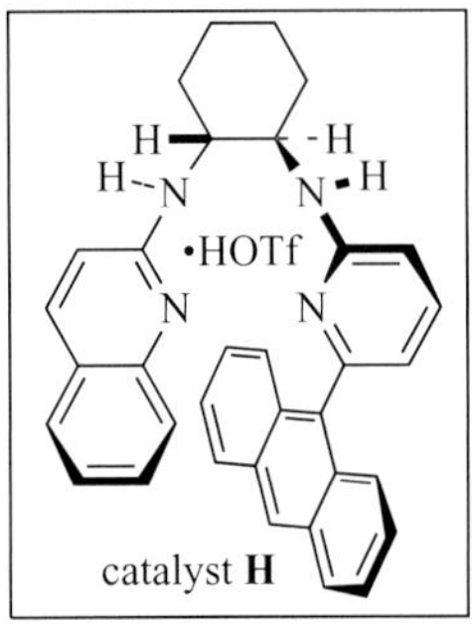
catalyst **H**

(25)

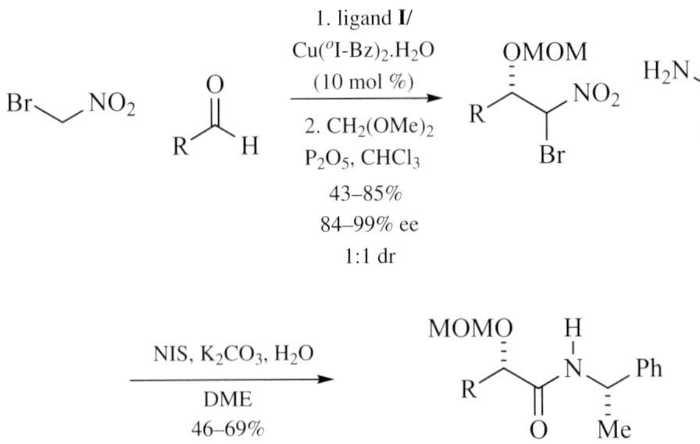

(26)

R = C_6H_5, $^{p}MeC_6H_4$, $^{p}PhC_6H_4$, $^{p}MeSC_6H_4$
$^{p}MeOC_6H_4$, $^{p}FC_6H_4$, $^{p}CF_3C_6H_4$ $^{m}BrC_6H_4$
$^{m}MeOC_6H_4$, $^{o}MeOC_6H_4$ $^{o}BrC_6H_4$ $^{o}MeC_6H_4$
$^{o}MOMOC_6H_4$, *N*-Ts-^{2}indole, *N*-Ts-^{2}pyrrole
C_6H_{11}, $PhCH_2CH_2$ ^{3}thiophene

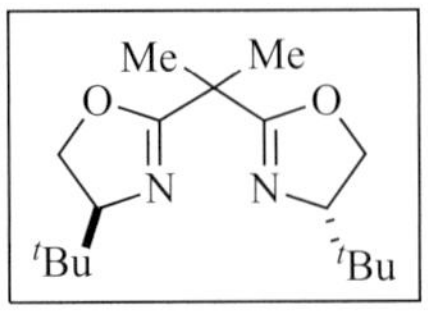
ligand **I**

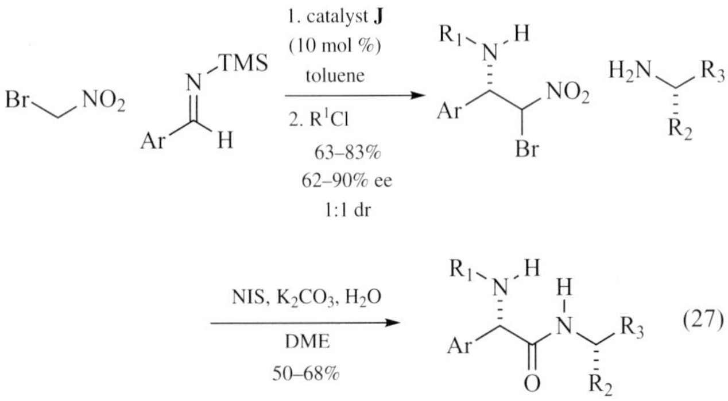

(27)

R^1 = Ac, Bz, Fmoc, Piv, N_3Ac, Cbz, Alloc
Ar = C_6H_5, $^{p}MeOC_6H_4$, $^{p}CF_3C_6H_4$, $^{p}IC_6H_4$
$^{m}ClC_6H_4$, $^{p}BrC_6H_4$, $^{p}FC_6H_4$, $^{o,p}Cl_2C_6H_3$, $^{p}MeC_6H_4$
R_2 = Me, R_3 = Ph; R_2 = ^{t}Bu, R_3 = $CO_2{}^{t}Bu$
R_2 = Me, R_3 = CO_2Me

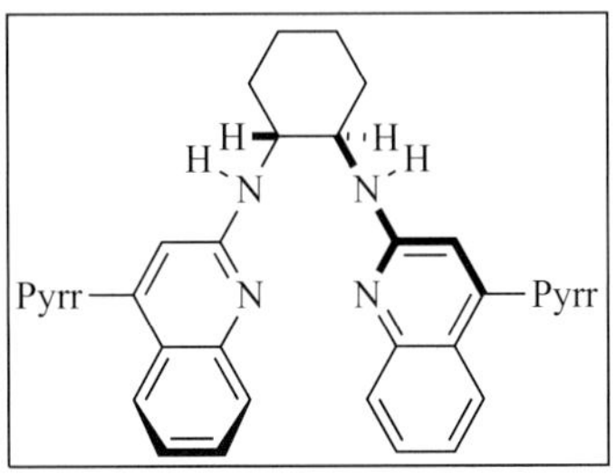
catalyst **J**

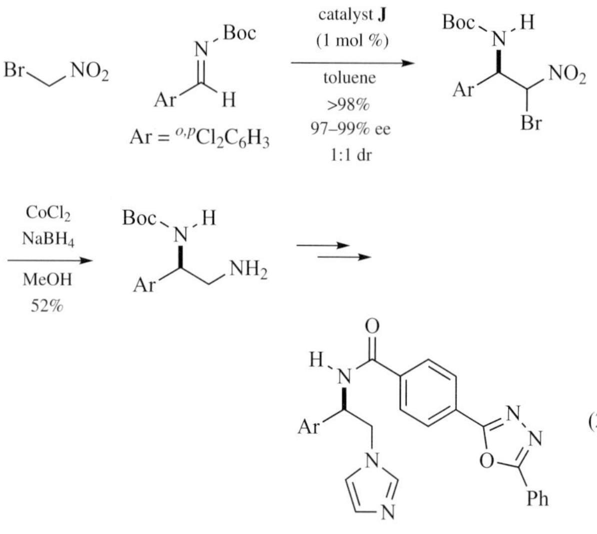

(28)

(+) -VNI

1. Bordwell, F. G.; Bartmess, J. E.; *J. Org. Chem.* **1978**, *43*, 3101.
2. (a) Ballini, R.; Bosica, G.; Parrini, M.; *Tetrahedron Lett.* **1998**, *39*, 7963. (b) Yoshimoto, J.; Sandoval, C. A.; Saito, S.; *Chem. Lett.* **2008**, *37*, 1294. (c) Alcaide, B.; Almendros, P.; Luna, A.; Torres, M. R.; *Org. Biomol. Chem.* **2008**, *6*, 1635. (d) Concellon, J. M.; Rodriguez-Solla, H.; Concellon, C.; Garcia-Granda, S.; Diaz, M. R.; *Org. Lett.* **2006**, *8*, 5979. (e) Blay, G.; Hernandez-Olmos, V.; Pedro, J. R.; *Chem. Commun.* **2008**, 4840.
3. (a) Yadav, L. D.; Kapoor, G. R.; *Tetrahedron Lett.* **2009**, *50*, 5420. (b) Rodriguez-Dominguez, J. C.; Thomae, D.; Seck, P.; Kirsch, G.; *Synlett* **2008**, 286. (c) Dimauro, E. F.; Newcomb, J.; Nunes, J. J.; Bemis, J. E.; Boucher, C.; Buchanan, J. L.; Buckner, W. H.; Cheng, A.; Faust, T.; Hsieh, F.; Huang, X.; Lee, J. H.; Marshall, T. L.; Martin, M. W.; Mcgowan, D. C.; Schneider, S.; Turci, S. M.; White, R. D.; Zhu, X.; *Bioorg. Med. Chem. Lett.* **2007**, *17*, 2305. (d) Grunewald, G. L.; Seim, M. R.; Bhat, S. R.; Wilson, M. E.; Criscione, K. R.; *Bioorg. Med. Chem.* **2008**, *16*, 542.
4. (a) Dong, L.; Lu, R.; Du, Q.; Zhang, J.; Liu, S.; Xuan, Y.; Yan, M.; *Tetrahedron* **2009**, *65*, 4124. (b) Hansen, H. M.; Longbottom, D. A.; Ley, S. V.; *Chem. Commun.* **2006**, 4838. (c) Waschowolski, V.; Hansen, H. M.; Longbottom, D. A.; Ley, S. V.; *Synthesis* **2008**, *8*, 1269. (d) Lv, J.; Zhang, J.; Lin, Z.; Wang, Y.; *Chem. Eur. J.* **2009**, *15*, 972. (e) Zhang, J.; Hu, Z.; Zhao, S.; Yan, M.; *Tetrahedron* **2009**, *65*, 802. (f) Vesely, J.; Zhao, G.; Bartoszewicz, A.; Cordova, A.; *Tetrahedron Lett.* **2008**, *49*, 4209. (g) Zhang, J.; Hu, Z.; Dong, L.; Xuan, Y.; Lou, C.; Yan, M.; *Tetrahedron: Asymmetry* **2009**, *20*, 355. (h) Inokuma, T.; Sakamoto, S.; Takemoto, Y.; *Synlett* **2009**, *10*, 1627.
5. (a) Witczak, Z. J.; Boryczewski, D.; *Bioorg. Med. Chem. Lett.* **1998**, *8*, 3265. (b) Caputo, F.; Clerici, F.; Gelmi, M. L.; Nava, D.; Pellegrino, S.; *Tetrahedron* **2006**, *62*, 1288. (c) Kislyi, V. P.; Nesterov, V. N.; Semenov, V. V.; *Russ. Chem. Bull.* **1999**, *48*, 1139.
6. Hwu, J. R.; Sambaiah, T.; Chakraborty, S. K.; *Tetrahedron Lett.* **2003**, *44*, 3167.
7. (a) Soengas, R. G.; Estevez, A. M., *Eur. J. Org. Chem.* **2010**, 5190. (b) Soengas, R. G.; Silva, S.; Estevez, A. M.; Estevez, J. C.; Estevez, R. J.; Rodriguez-Solla, H., *Eur. J. Org. Chem.* **2012**, 4339. (c) Rodriguez-Solla, H.; Concellon, C.; Alvaredo, N.; Soengas, R. G., *Tetrahedron* **2012**, *68*, 4339. (d) Soengas, R. G.; Silva, A. M. S., *Synlett* **2012**, *23*, 873. (e) Rodriguez-Solla, H.; Alvaredo, N.; Soengas, R. G., *Synlett* **2012**, *23*, 2083. (f) Soengas, R. G.; Silva, A. M. S., *Eur. J. Org. Chem.* **2013**, 5022. (g) Soengas, R. G.; Silva, A. M. S., *Synlett* **2013**, *24*, 1949. (h) Soengas, R. G.; Silva, A. M. S., *Tetrahedron Lett.* **2013**, *54*, 2156. (i) Soengas, R. G.; Silva, A. M. S., *Tetrahedron* **2013**, *69*, 3425.
8. Shen, B.; Makley, D. M.; Johnston, J. N., *Nature* **2010**, *465*, 1027.
9. Shackleford, J. P.; Shen, B.; Johnston, J. N., *Proc. Natl. Acad. Sci. USA* **2012**, *109*, 44.
10. Makley, D. M.; Johnston, J. N., *Org. Lett.* **2014**, *16*, 3146.
11. Leighty, M. W.; Shen, B.; Johnston, J. N., *J. Am. Chem. Soc.* **2012**, *134*, 15233.
12. (a) Nugent, B. M.; Yoder, R. A.; Johnston, J. N., *J. Am. Chem. Soc.* **2004**, *126*, 3418. (b) Davis, T. A.; Wilt, J. C.; Johnston, J. N., *J. Am. Chem. Soc.* **2010**, *132*, 2880.
13. Schwieter, K. E.; Shen, B.; Shackleford, J. P.; Leighty, M. W.; Johnston, J. N., *Org. Lett.* **2014**, *16*, 4714.
14. Dobish, M. C.; Villalta, F.; Waterman, M. R.; Lepesheva, G. I.; Johnston, J. N., *Org. Lett.* **2012**, *14*, 6322.

tert-Butyl 2-(*tert*-butyldimethylsilyl)-2-oxoacetate[1]

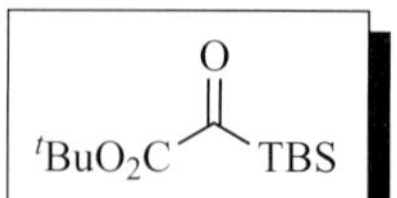

(1)

[852447-17-7] $C_{12}H_{24}O_3Si$ (MW 244.40)

InChI = 1S/C12H24O3Si/c1-11(2,3)15-9(13)10(14)16(7,8)12(4,5)6/h1-8H3
InChIKey = GGYHTNUVBJRLDM-UHFFFAOYSA-N

(reagent used in various multicomponent coupling reactions typically for the synthesis of substituted glycolic acid derivatives)

Physical Data: bright yellow oil.
Solubility: soluble in most organic solvents.
Form Supplied in: bright yellow liquid. The most common impurity is silanol remaining from the final oxidation step.
Analysis of Reagent Purity: R_f: 0.49 (25% Et_2O/Petroleum Ether); IR (thin film) cm^{-1} 2931, 2860, 1716, 1657, 1464, 1369, 1252, 1159, 993, 841, 785; ^1H NMR (300 MHz, $CDCl_3$) δ: 0.27 (s, 6*H*), 0.96 (s, 9*H*), 1.55 (s, 9*H*); ^{13}C NMR (75 MHz, $CDCl_3$) δ: −6.6, 17.1, 26.6, 28.1, 83.6, 163.0, 232.9; Anal. Calcd. for $C_{12}H_{24}O_3Si$: *C*, 58.97; *H*, 9.90. Found: *C*, 58.81; *H*: 9.98.
Preparative Methods: the reagent is prepared[2] by a three-step sequence starting from commercially available *t*-butyl acetoacetate. Treating *t*-butyl acetoacetate with *p*-acetamidobenzenesulfonyl azide (1.0 equiv), tetrabutylammonium bromide (0.02 equiv), and sodium hydroxide (2.8 equiv) yields the requisite diazoacetate as yellow oil. A silylation is performed on the unpurified diazoacetate with the desired silyl triflate (1.2 equiv) in the presence of Hünig's base (1.2 equiv) to provide the silyl diazoacetates as yellow oils for a series of silyl triflates (TMS, TES, TBS, TBDPS). It is typically most convenient to proceed without purification after the *C*-silylation, since impurities are more easily rejected after the final step. Oxo transfer to the silyl diazoacetate is mediated by dimethyldioxirane formed in situ[3] from Oxone® and acetone to provide the silyl glyoxylate as a yellow oil. The product is purified by flash chromatography in yields typically around 50% for the three-step sequence. This

method allows for preparative quantities of silyl glyoxylates with varied silyl functionalities to be prepared. Other methods exist,[4] but this preparation of silyl glyoxylates appears to be the most widely applicable at this time. Silyl glyoxylates with a benzyl ester (**3**)[5] or an allylic ester[6] functionality can be prepared by a modified procedure.

Purification: the final step in the protocol is an Oxone oxidation of the diazoacetate after which the product is purified by flash chromatography eluting with 5% EtOAc/hexanes. Since the product is yellow, the progress of elution can be visually monitored. No purification of the intermediates from the first two steps of the synthesis is required.

Handling, Storage, and Precautions: silyl glyoxylates can be stored for several months without significant decomposition if stored in a freezer away from light. The reagents decompose upon prolonged exposure to light and should be wrapped in aluminum foil. These reagents typically are stored at 0 °C to prevent thermal decomposition. The reagent is somewhat volatile and some may be lost by prolonged exposure to high vacuum (particularly the TMS silyl glyoxylates). Silyl glyoxylate toxicity has not been evaluated, but the reagents should be used with common precautions to limit exposure.

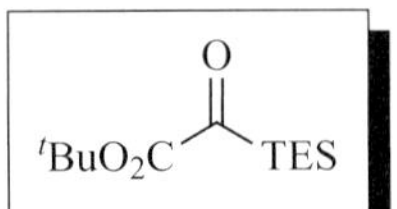

(**2**)

[852447-25-7] $C_{12}H_{24}O_3Si$ (MW 244.40)

InChI = 1S/C12H24O3Si/c1-7-16(8-2,9-3)11(14)10(13)15-12(4,5)6/h7-9H2,1-6H3
InChIKey = UEYABJBDWBGSNH-UHFFFAOYSA-N

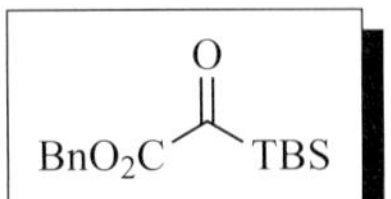

(**3**)

[443988-56-5] $C_{15}H_{22}O_3Si$ (MW 278.42)

InChI = 1S/C15H22O3Si/c1-15(2,3)19(4,5)14(17)13(16)18-11-12-9-7-6-8-10-12/h6-10H,11H2,1-5H3
InChIKey = XVYKODFPDWKLLK-UHFFFAOYSA-N

Introduction to Silyl Glyoxylate Chemistry and Reactivity Patterns. Silyl glyoxylates are a specific subclass of acyl silanes that possess an attached ester; they are vicinal dicarbonyls. The effect of this functionality is to significantly increase the electrophilicity of the acyl silane and the rate at which tetrahedral addition intermediates undergo 1,2-Brook rearrangement. The unique reactivity of silyl glyoxylates has led to their use as conjunctive dipolarophiles in multicomponent couplings. Although silyl glyoxylate typically functions as a dipolar glycolic acid synthon for the union of nucleophilic and electrophilic partners, the scope of the reagent's reactivity has been expanded to include glyoxylate anion[7] and α-keto ester homoenolate synthons.[8,9] An asset of silyl glyoxylate chemistry is the ability to "mix and match" a variety of complementary components into the reaction manifold enabling access to a diverse collection of complex molecular motifs.

Reacting silyl glyoxylates with nonstabilized nucleophiles results in addition to the acyl silane portion of the reagent, which triggers a facile [1,2]-Brook rearrangement to provide a transient secondary nucleophile that can react with a secondary electrophile (eq 1). Due to the high electrophilicity of the silyl glyoxylate, secondary electrophiles can often be present in the reaction mixture without significant side reactions occurring. Depending on the identity of the nucleophile and associated metal countercation, silyl glyoxylates can undergo [1,2]-Brook rearrangement very rapidly, even at low temperatures. For this reason, reactions with silyl glyoxylates are typically performed below 0 °C and often at −78 °C. Performing couplings at higher temperatures can result in uncontrolled oligomerization of the reagent.[8] It is noteworthy that the controlled stereoselective dimerization of silyl glyoxylates is possible and was employed in the synthesis of zaragozic acid C.[10]

Silyl Glyoxylate Reactivity

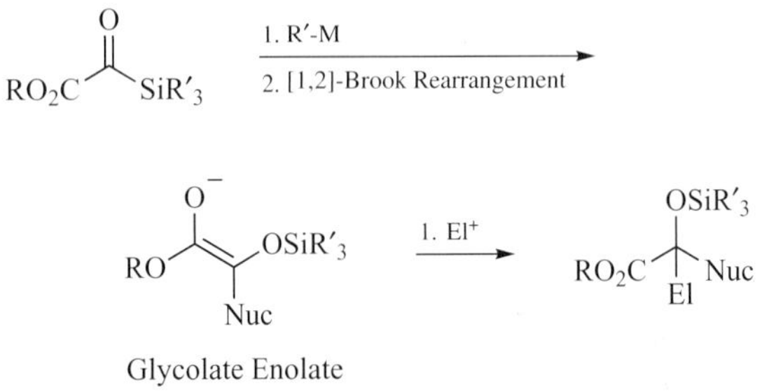

(1)

Silyl glyoxylates can typically be varied at the ester and the silyl portion of the reagent without significantly affecting the yield for various couplings. TES- and TBS-substituted silyl glyoxylates can often be used interchangeably. TMS-substituted silyl glyoxylates tend to perform poorest, since the [1,2]-Brook rearrangement is much faster for smaller silyl groups, which leads to uncontrolled oligomerization. The facile deprotection and/or migration of the TMS silyl ether can also lead to product mixtures.

α-Coupling Aldol and Aldol-type Cascades. Silyl glyoxylate **1** reacts with various aldehydes and alkynylzinc reagents to provide an alkynylation/[1,2]-Brook rearrangement/glycolate aldol sequence to afford orthogonally protected diol products with good diastereoselectivity (eq 2).[11] Originally, the TMS-substituted silyl glyoxylate was utilized; however, this reagent resulted in mixtures of secondary and tertiary silyl ethers. Opting for the bulkier TBS group inhibited the undesired silyl migration and improved the diastereoselectivity. A single example employing vinyl Grignard instead of an alkynylzinc reagent was also performed that provided the analogous product with high diastereoselectivity.

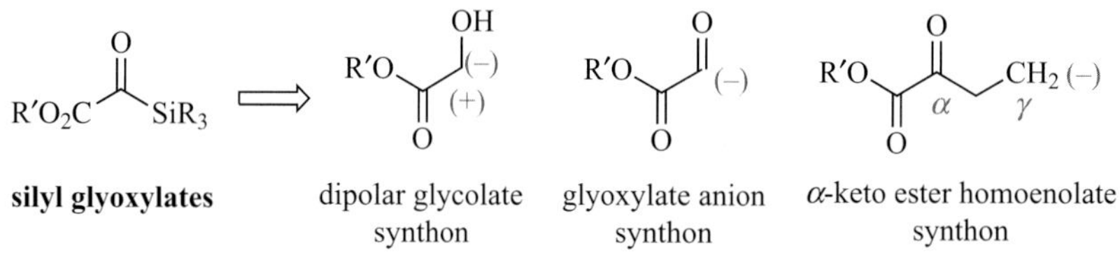

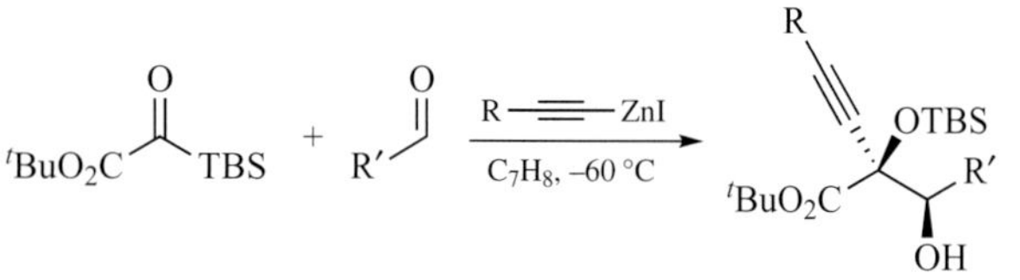

(2)

Another aldol-based three component coupling cascade (Reformatsky reaction/[1,2]-Brook rearrangement/aldol/lactonization) was realized when silyl glyoxylate **3** is reacted with Reformatsky reagents and ketones. Pentasubstituted γ-butyrolactones, which arise from lactonization of the zinc-aldolate, were obtained with high levels of diastereoselectivity (eq 3).[5] Metalloenolates other than Reformatsky reagents were ineffective, highlighting the importance of the enolate counterion. Various alkyl, aryl, and heteroaryl ketones were amenable to the reaction conditions to provide yields ranging from 41% to 73% and diastereomeric ratios often >20:1.

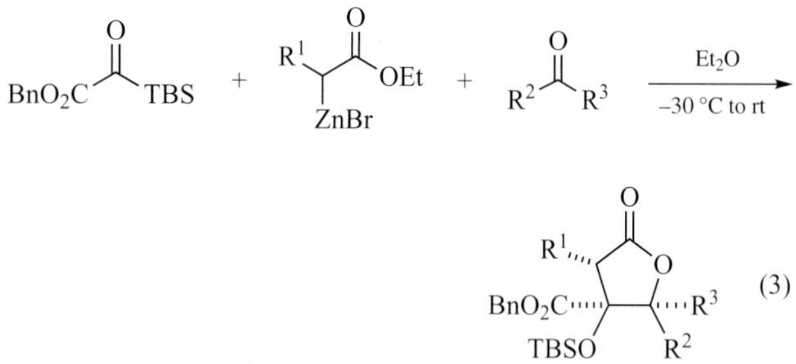

(3)

An adaptation of this aldol-type silyl glyoxylate cascade involves replacing the ketone secondary electrophile with a stereodefined β-lactone; a terminating quaternary Claisen-type bond construction ensues (eq 4). In most cases, the reaction proceeds with high levels of 1,4-stereoinduction.[12] The Reformatsky/Claisen methodology was also exploited in the formal synthesis of leustroducsin B.[13]

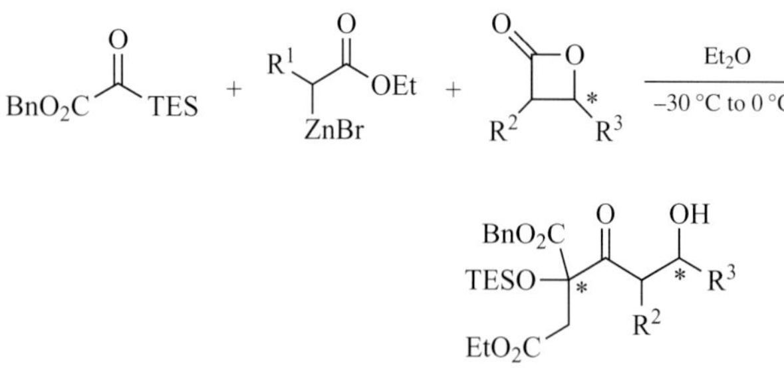

(4)

Reactions of silyl glyoxylate **1** and **3** with metallated sulfonyl imidates and enantiopure sulfinyl aldimines provided good yields of cyclic *N*-sulfonylamidines with complete stereocontrol (eq 5).[14] The reaction was amenable to scale-up and only a minor decrease in yield occurred when the reaction was run on a gram scale.

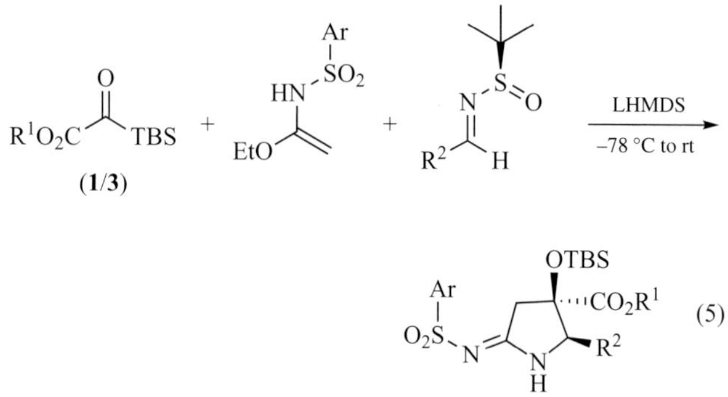

(5)

Silylglyoxylates and magnesium alkoxides react to provide α,β-dihydroxyesters via a hydride addition/[1,2]-Brook rearrangement/aldol sequence (eq 6).[15] The nucleophilic initiator becomes the secondary electrophile via a Meerwein–Ponndorf–Verley/Oppenauer redox process. The magnesium alkoxide serves as the hydride source that triggers the [1,2]-Brook rearrangement to form the glycolate enolate; the latter attacks the newly formed carbonyl. Yields ranged from 63% to 97% with diastereomeric ratios up to 10:1.

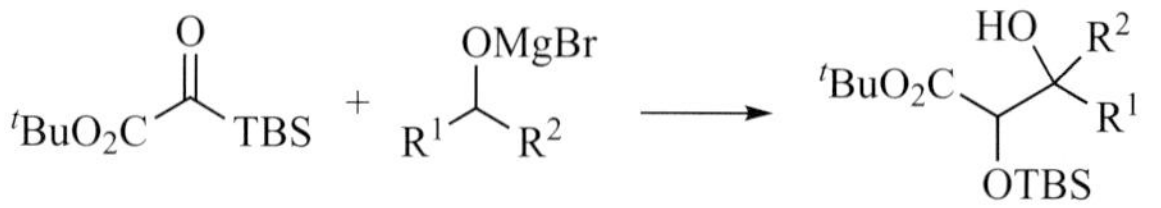

(6)

A catalytic variant of this aldol reaction utilizing strain release silacycles and silyl glyoxylates was developed (eq 7).[16] The reductive aldol was catalyzed by various lanthanide triisopropoxides. This is one of the few catalytic silyl glyoxylate cascades reported to date.

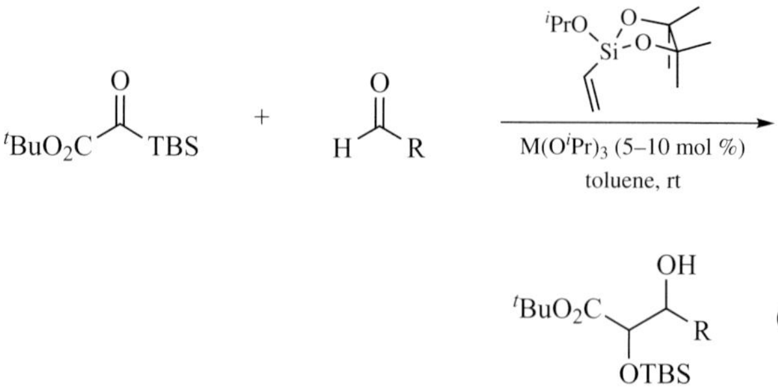

(7)

α-Coupling Michael Cascades.. Silyl glyoxylates react with enolates and enones in a Michael-terminated cascade reaction to provide products with two contiguous stereocenters (eq 8).[17] When lithium is used as the enolate countercation, the regioselectivity for this coupling is excellent (20:1) in favor of the Michael adduct over the 1,2-addition adduct. Replacing the enolate countercation with zinc provides complete regioselectivity for the 1,2-addition cascade.

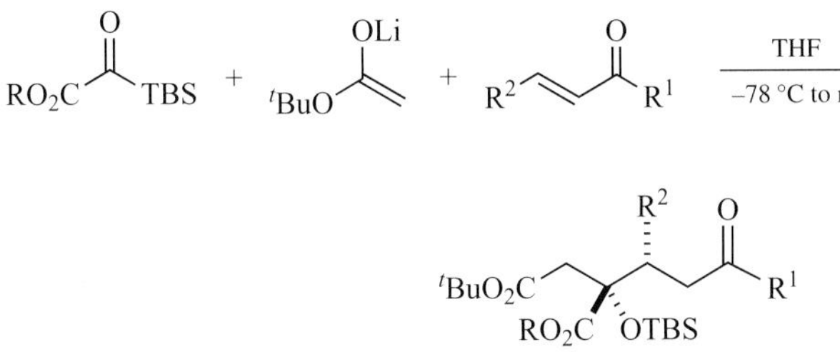

(8)

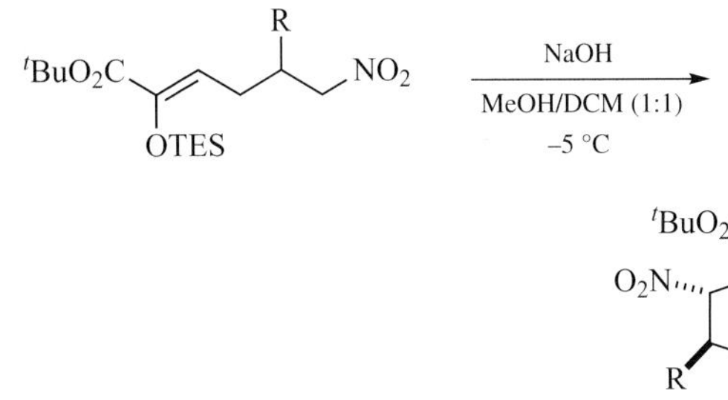

(10)

The Michael-terminated cascade is amenable to aryl and heteroaryl substitution on the enone electrophile. Yields ranged from 40% to 80% with diastereomeric ratios typically ~4:1. The diastereoselectivity is believed to hinge on the glycolate enolate geometry. A (*Z*)-glycolate enolate is proposed to provide the major diastereomer, a hypothesis that is in agreement with the observations of other silyl glyoxylate couplings.[6,8] This Michael sequence was employed in studies directed toward the synthesis of the natural product trachyspic acid.[18]

Vinylogous Michael Coupling Cascades. When π-nucleophiles (vinyl Grignard or alkynylzinc reagents) are added to silyl glyoxylates the [1,2]-Brook rearrangement provides a silyloxydienolate intermediate. This dienolate intermediate can function as an α-keto ester homoenolate synthetic equivalent by engaging an electrophile in vinylogous fashion. The identity of the secondary electrophile dictates whether an α-coupling or vinylogous (γ-coupling) pathway will occur. If the secondary electrophile is a carbonyl the α-coupling pathway predominates (vide supra); however, when electrophilic alkenes such as α,α-dicyanoolefins and nitroalkenes are used, a γ-trapping or vinylogous[8] mechanism predominates. In this manner, the secondary electrophile can function as a regioselective switch that determines which type of reactivity (α vs. γ) will be observed.

When treated with vinyl Grignard as the nucleophilic initiator and various nitroalkenes as the secondary electrophile, silyl glyoxylates undergo a vinylation/[1,2]-Brook rearrangement/vinylogous Michael cascade to provide (*Z*)-enolsilanes with complete regioselectivity (eq 9).[8] As with the α-couplings, an excess of silyl glyoxylate is required to compensate for the oligomerization of silyl glyoxylate. The enolsilane products can be converted to nitrocyclopentanols via a silyl deprotection/Henry cyclization with complete diastereoselectivity in many cases (eq 10). This methodology enables the efficient construction of densely functionalized cyclopentanols in two steps from silyl glyoxylate.

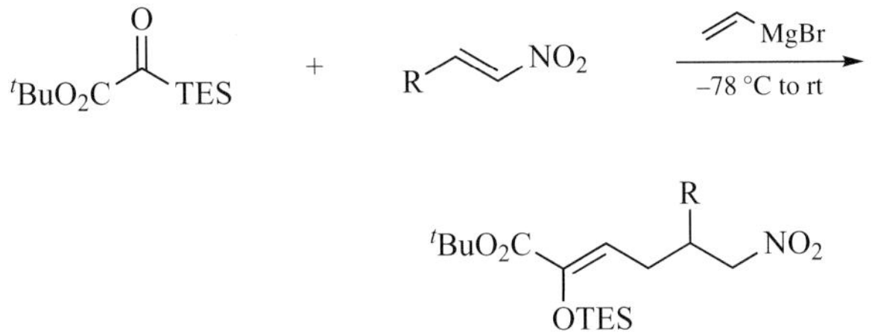

(9)

A related methodology utilizes alkynylmagnesium bromide reagents instead of vinyl magnesium bromide reagents to achieve a tandem alkynylation/[1,2]-Brook rearrangement/vinylogous Michael reaction to form silyloxyallenes with complete regio- and diastereoselectivity (eq 11).[9] Density functional theory (DFT) calculations and control experiments suggested the regio- and diastereoselectivity arose from bidentate coordination of the nitro group to the MgBr cation of the (*Z*)-glycolate enolate. A Lewis acid-assisted Henry cyclization was also developed to access fully substituted cyclopentenols (eq 12).

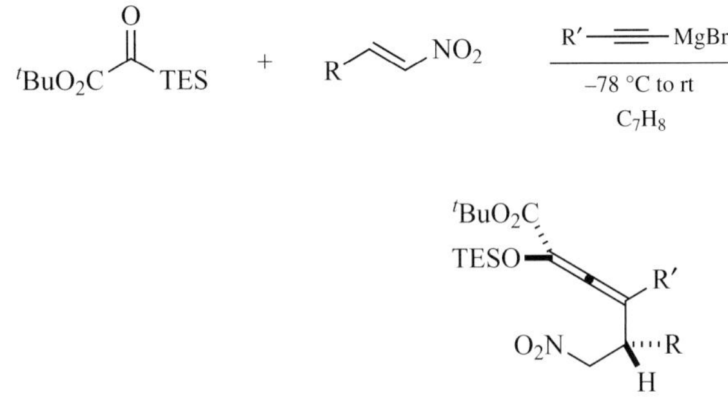

(11)

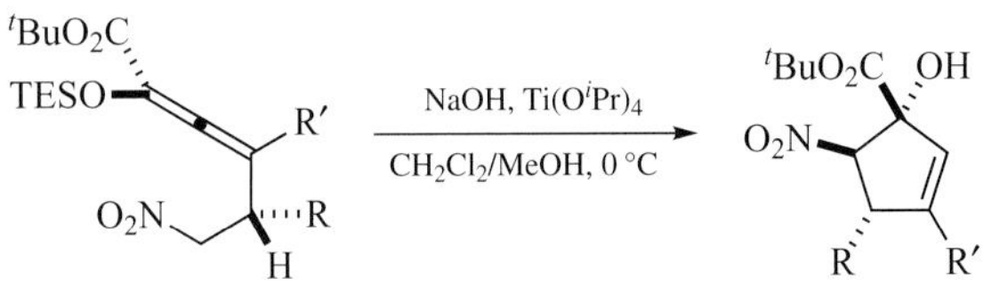

(12)

Miscellaneous Reactions. The reaction of silyl glyoxylate **1** with aldehydes in the presence of a cyanide catalyst produces β-silyloxy-α-keto ester via a benzoin-type reaction.[7] In this manner, the silyl glyoxylate functions as a glyoxylate anion synthetic equivalent. The coupling provides excellent yields ranging from 80% to 96% for aryl-, heteroaryl-, and alkyl-substituted aldehydes. The products can be converted to amino esters via a reductive amination with moderate levels of diastereoselectivity in favor of the *syn*-diastereomer.

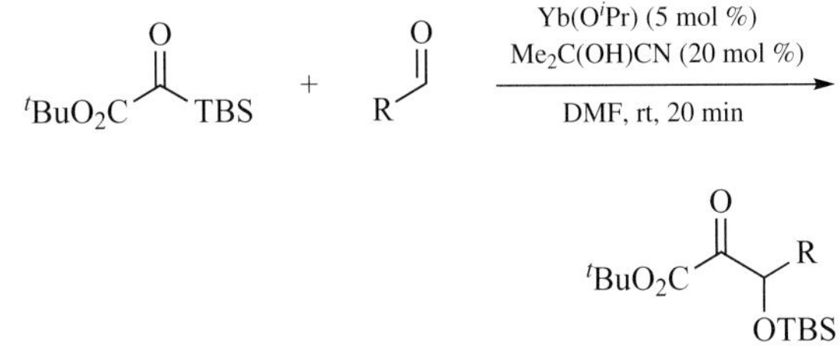

(13)

Silyl glyoxylates that contain an allylic pendant ester can undergo a [1,2]-Brook rearrangement/Ireland-Claisen rearrangement leading to γ,δ-unsaturated α-silyloxy acid derivatives.[6] The sequence provides the [3,3]-rearrangement products in moderate yields with variable diastereoselectivities. This cascade can also be used to form contiguous quaternary stereocenters with moderate diastereoselectivity. The predominance of the (*Z*)-glycolate enolate intermediate was inferred from the stereochemical outcome of the rearrangement.

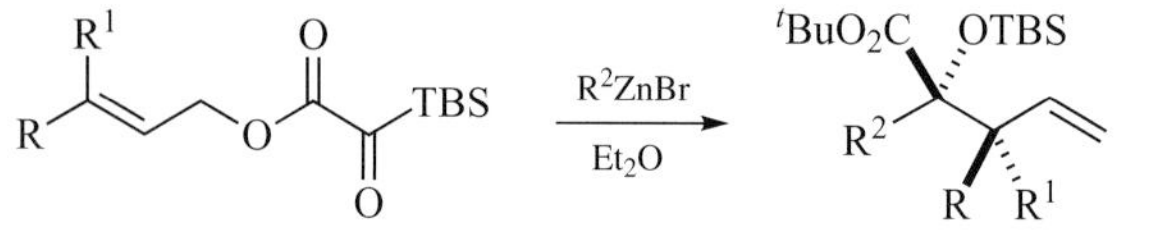

(14)

1. Boyce, G. R.; Greszler, S. N.; Johnson, J. S.; Linghu, X.; Malinowski, J. T.; Nicewicz, D. A.; Satterfield, A. D.; Schmitt, D. C.; Steward, K. M., *J. Org. Chem.* **2012**, *77*, 4503.
2. Nicewicz, D. A.; Brétéché, G.; Johnson, J. S., *Org. Synth.* **2008**, *85*, 278.
3. Yao, W.; Wang, J., *Org. Lett.* **2003**, *5*, 1527.
4. For alternative means for preparing silyl glyoxylates, see (a) Sekiguchi, A.; Kabe, Y.; Ando, W., *J. Org. Chem.* **1982**, *47*, 2900. (b) Bolm, C.; Kasyan, A.; Heider, P.; Saladin, S.; Drauz, K.; Guenther, K.; Wagner, C., *Org. Lett.* **2002**, *4*, 2265.
5. Greszler, S. N.; Johnson, J. S., *Angew. Chem., Int. Ed.* **2009**, *48*, 3689.
6. Schmitt, D. C.; Johnson, J. S., *Org. Lett.* **2010**, *12*, 944.
7. Steward, K. M.; Johnson, J. S., *Org. Lett.* **2010**, *12*, 2864.
8. Boyce, G. R.; Johnson, J. S., *Angew. Chem., Int. Ed.* **2010**, *49*, 8930.
9. Boyce, G. R.; Liu, S.; Johnson, J. S., *Org. Lett.* **2012**, *14*, 652.
10. Nicewicz, D. A.; Satterfield, A. D.; Schmitt, D. S.; Johnson, J. S., *J. Am. Chem. Soc.* **2008**, *130*, 17281.
11. Nicewicz, D. A.; Johnson, J. S., *J. Am. Chem. Soc.* **2005**, *127*, 6170.
12. Greszler, S. N.; Malinowski, J. T.; Johnson, J. S., *J. Am. Chem. Soc.* **2010**, *132*, 17393.
13. Greszler, S. N.; Malinowski, J. T.; Johnson, J. S., *Org. Lett.* **2011**, *13*, 3206.
14. Yao, M.; Lu, C.-D., *Org. Lett.* **2011**, *13*, 2782.
15. Linghu, X.; Satterfield, A. D.; Johnson, J. S., *J. Am. Chem. Soc.* **2006**, *128*, 9302.
16. Greszler, S. N.; Johnson, J. S., *Org. Lett.* **2009**, *11*, 827.
17. Schmitt, D. S.; Malow, E. J.; Johnson, J. S., *J. Org. Chem.* **2012**, *77*, 3246.
18. Schmitt, D. S.; Lam, L.; Johnson, J. S., *Org. Lett.* **2011**, *13*, 5136.

Gregory R. Boyce & Jeffrey S. Johnson
University of North Carolina at Chapel Hill, Chapel Hill, NC, USA

(*S*)-5-((*tert*-Butyldimethylsilyloxy)-diphenylmethyl)-2-phenyl-6,7-dihydro-5*H*-pyrrolo[2,1-*c*][1,2,4]triazol-2-ium Tetrafluoroborate 1a (and Related Reagents)

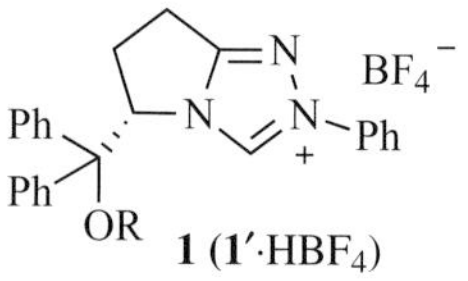

1 (**1′**·HBF_4)

1a, R = TBS
1b, R = TMS

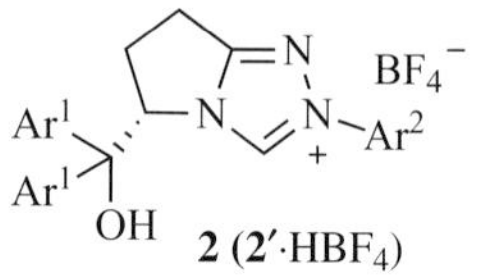

2 (**2′**·HBF_4)

2a, Ar^1 = Ar^2 = Ph
2b, Ar^1 = 3,5-$(CF_3)_2C_6H_3$, Ar^2 = Bn
2c, Ar^1 = 3,5-$(CF_3)_2C_6H_3$, Ar^2 = 2-*i*PrC_6H_4

1a

[1005210-14-9] $C_{30}H_{36}N_3OSi^+BF_4^-$ (MW 569.525)

InChI = 1S/C30H36N3OSi.BF4/c1-29(2,3)35(4,5)34-30(24-15-9-6-10-16-24,25-17-11-7-12-18-25)27-21-22-28-31-33(23-32(27)28)26-19-13-8-14-20-26;2-1(3,4)5/h6-20,23,27H,21-22H2,1-5H3;/q+1;-1/t27-;/m0./s1

InChIKey = WACIZCGUCKGRFQ-YCBFMBTMSA-N

1b

[1005209-71-1] $C_{27}H_{30}N_3OSi^+BF_4^-$ (MW 527.445)

InChI = 1S/C27H30N3OSi.BF4/c1-32(2,3)31-27(22-13-7-4-8-14-22,23-15-9-5-10-16-23)25-19-20-26-28-30(21-29(25)26)24-17-11-6-12-18-24;2-1(3,4)5/h4-18,21,25H,19-20H2,1-3H3;/q+1;-1/t25-;/m0./s1

InChIKey = KEOIBAIUDFKSPT-UQIIZPHYSA-N

2a

[1092807-31-2] $C_{24}H_{22}N_3O^+BF_4^-$ (MW 455.263)

InChI = 1S/C24H22N3O.BF4/c28-24(19-10-4-1-5-11-19,20-12-6-2-7-13-20)22-16-17-23-25-27(18-26(22)23)21-14-8-3-9-15-21;2-1(3,4)5/h1-15,18,22,28H,16-17H2;/q+1;-1/t22-;/m0./s1

InChIKey = JPOKVMIJBNOZQV-FTBISJDPSA-N

2b

[1472647-01-0] $C_{29}H_{20}F_{12}N_3O^+BF_4^-$ (MW 741.283)

InChI = 1S/C29H20F12N3O.BF4/c30-26(31,32)19-8-17(9-20(12-19)27(33,34)35)25(45,18-10-21(28(36,37)38)13-22(11-18)29(39,40)41)23-6-7-24-42-43(15-44(23)24)14-16-4-2-1-3-5-16;2-1(3,4)5/h1-5,8-13,15,23,45H,6-7,14H2;/q+1;-1/t23-;/m0./s1

InChIKey = ACWRWVHJWQQRJO-BQAIUKQQSA-N

2c

[1248518-63-9] $C_{31}H_{24}F_{12}N_3O^+ BF_4^-$ (MW 769.337)

InChI = 1S/C31H24F12N3O.BF4/c1-16(2)23-5-3-4-6-24(23)46-15-45-25(7-8-26(45)44-46)27(47,17-9-19(28(32,33)34)13-20(10-17)29(35,36)37)18-11-21(30(38,39)40)14-22(12-18)31(41,42)43;2-1(3,4)5/h3-6,9-16,25,47H,7-8H2,1-2H3;/q+1;-1/t25-;/m0./s1

InChIKey = NXHCMWDDLCCBAC-UQIIZPHYSA-N

(the *N*-heterocyclic carbenes (NHCs) **1′–2′**; generated from the triazolium salts **1–2** in the presence of base, are efficient organocatalysts for a variety of reactions of aldehydes, ketenes, and carboxylic acid derivatives)

Physical Data: (**1a**) a pale yellow solid; mp 210–211 °C; $[\alpha]_D^{25}$ = −112.4 (*c* 0.5, CH_3CN); (**1b**) a pale yellow solid; mp 196–198 °C; $[\alpha]_D^{25}$ = −113.8 (*c* 0.5, CH_3CN); (**2a**) a white solid; mp 197–180 °C; = (58.6 (*c* 0.5, CH_3CN); (**2b**) a white solid; mp 113–115 °C; $[\alpha]_D^{20}$ = −70.9 (*c* 1.0, CH_2Cl_2); (**2c**) a pale yellow solid; mp 120–122 °C; $[\alpha]_D^{25}$ = −112.5 (*c* 1.0, $CHCl_3$).

Solubility: soluble in CH_2Cl_2, CH_3Cl, and THF.

Preparative Methods: the addition of Grignard reagent to the methyl ester **3** of L-pyroglutamic acid gave the diarylmethyl alcohols **4**,[1] and subsequent protection of the hydroxy group with TBSOTf or TMSOTf gave the corresponding ethers **5**.[2] Methylation with Meerwein's reagent, reaction with arylhydrazine, and cyclization with triethyl orthoformate gave the triazolium salts **1**.[3] Desilylation of the triazolium salts under acidic conditions gave the triazolium salts **2a**,[4] **2b**,[5] and **2c**[6] with a free hydroxyl group (Scheme 1).

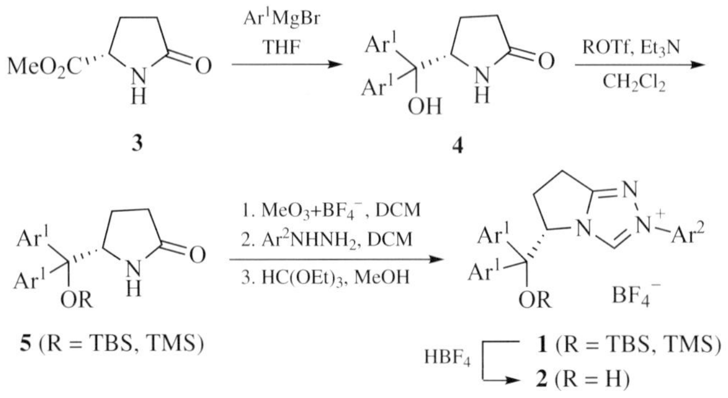

Scheme 4

Purification: recrystallization from hot MeOH.

Handling, Storage, and Precautions: the compound can be handled in air and is generally stable for several months in air.

[2 + 2] Cycloaddition Reaction of Ketenes. The chiral *N*-heterocyclic carbene **1a**′, generated from triazolium salt **1a** in the presence of Cs_2CO_3, is an efficient organocatalyst for the formal [2 + 2] cycloaddition reaction of ketenes with imines,[3] 2-oxoaldehydes,[7] and *N*-sulfinylanilines,[8] giving the corresponding *cis*-*β*-lactams, *β*-lactones, and 1,2-thiazetidin-3-one-1-oxides, respectively, in good yields with excellent enantioselectivities (eqs 1–3).

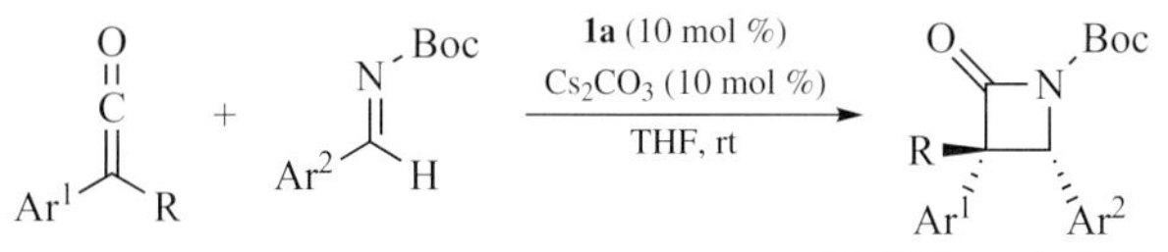 (1)

53–78%, up to 99% ee

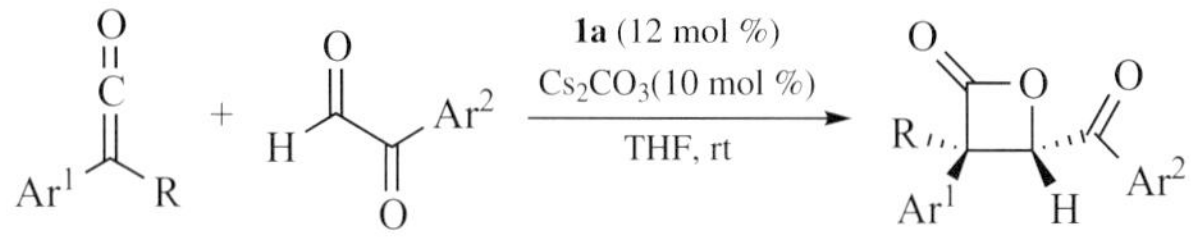 (2)

73–99%, up to 99% ee

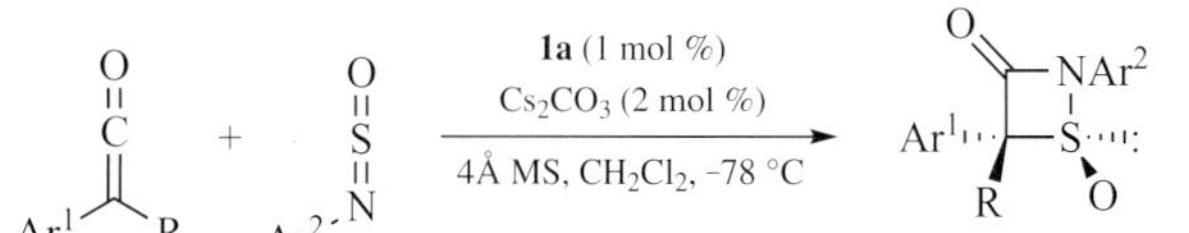 (3)

up to 95% yield
up to 99% ee

The bifunctional NHCs **2**′, which may interact with the substrates via a H-bonding, showed better activity and/or enantioselective than the one with silyl ether. For example, the bifunctional triazolium salt **2b** was demonstrated as working precatalyst for the reaction of ketenes with isatin-derived ketimines, affording the corresponding spirocyclic oxindolo-*β*-lactams in high yields with excellent diastereo- and enantioselectivities (eq 4).[9]

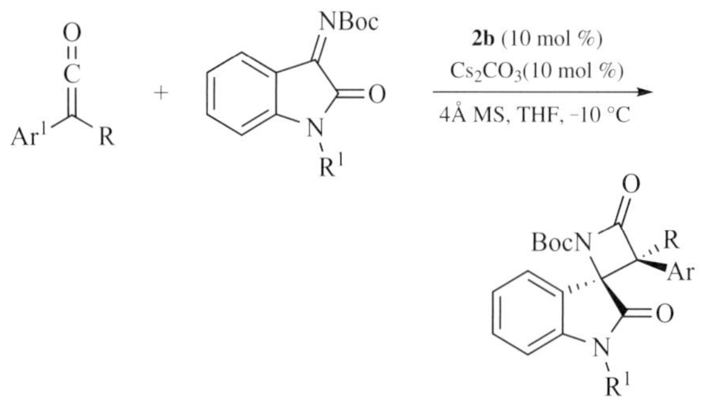 (4)

70–92%
up to >20:1 dr, 99% ee

[4 + 2] Cycloaddition Reaction of Ketenes. The [4 + 2] cycloaddition reaction of ketenes is a straightforward approach to six-membered heterocycles. The triazolium salt **1a** has been demonstrated as an efficient organocatalyst for the formal [4 + 2] cycloaddition reaction of disubstituted ketenes with enones,[10] *N*-benzoyldiazenes,[11] and *o*-quinone methides,[12] providing the corresponding *δ*-lactones, 1,3,4-oxadiazin-6-ones, and dihydrocoumarin derivatives, respectively, in good yield and with good diastereoselectivity and high enantioselectivity (eqs 5–7).

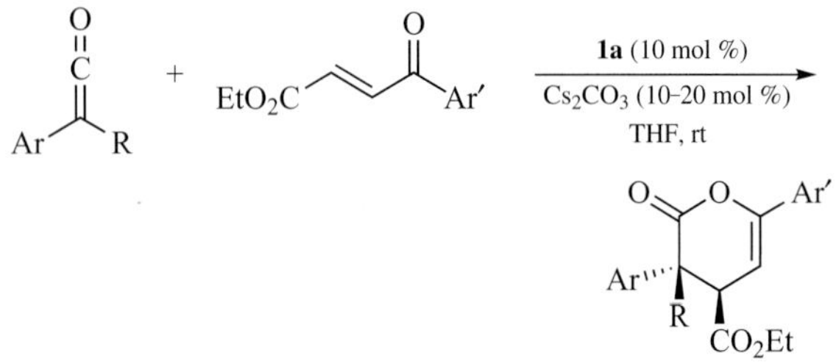

(5)

up to 93% yield
up to 92% ee

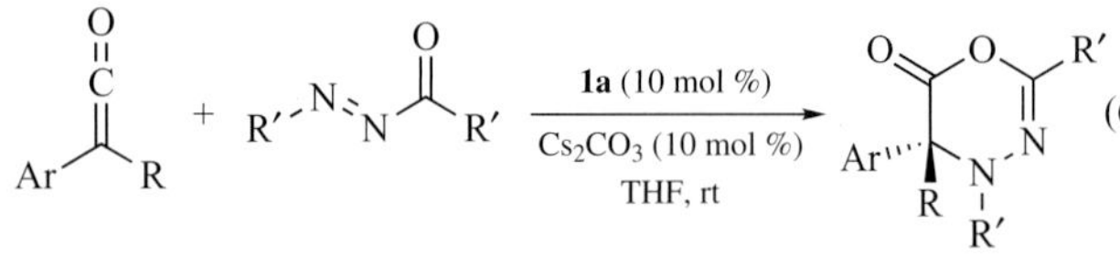

(6)

up to 85% yield
up to 96% ee

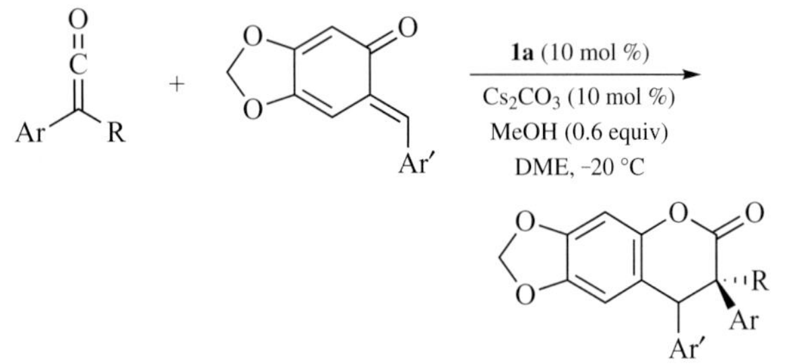

(7)

up to 96% yield
up to 99% ee

[3+2] Cyclization Reaction of Ketenes. In the presence of 10 mol % of bifunctional chiral triazolium salt **2c**, ketenes reacted smoothly with oxaziridine to give the oxazolidin-4-ones in good yield and with good diastereoselectivity and high enantioselectivity (eq 8).[6] The oxazolidin-4-one scaffold could be further transformed to α-hydroxy acid and 1,2-diol in good yields and without erosion of enantiopurity.

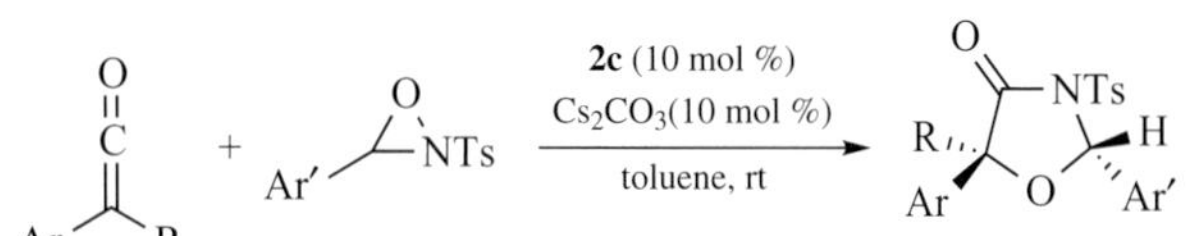

(8)

up to 78% yield
up to 95% ee

[3 + 4] Cycloaddition of Enals. Bifunctional NHC **2c** is also efficient catalyst for the [3 + 4] cycloaddition of enals with *o*-quinone methides[13] and aurones,[14] affording the corresponding benzo-ε-lactones and benzofuran-ε-lactones, respectively, in high yields and with excellent diastereoselectivities and enantioselectivities (eqs 9 and 10).

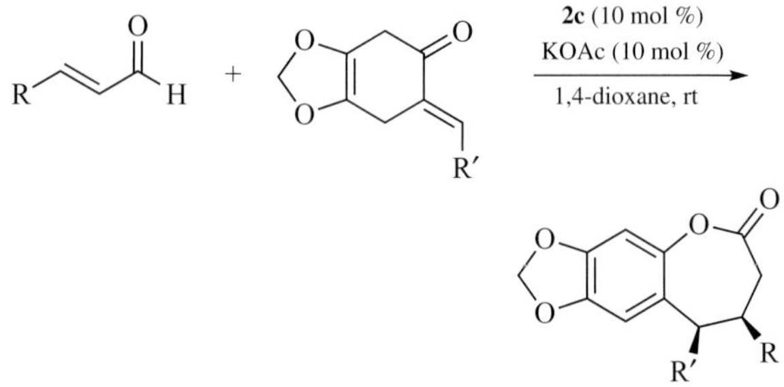

(9)

up to 96% yield
up to 98% ee

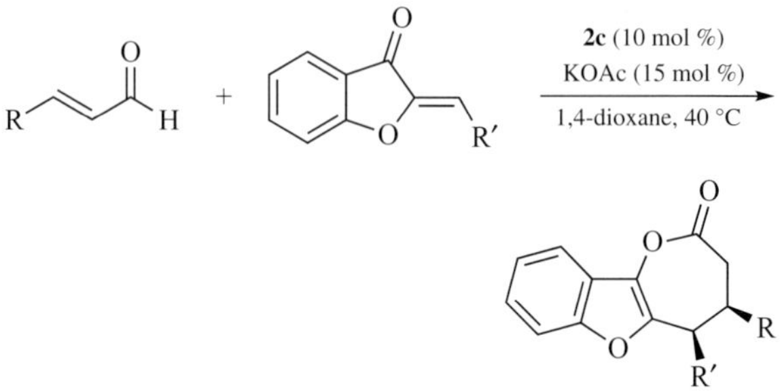

(10)

up to 99% yield
up to 99% ee

Aza-Benzoin Reaction of Enals. The bifunctional NHC **2b′** is an efficient catalyst for the enantioselective aza-benzoin reaction of enals with trifluoromethyl ketone-derived ketimines.[15] Enals with electron-withdrawing and electron-donating substituents all gave the aza-benzoin products in good yields and with high enantioselectivities (eq 11).

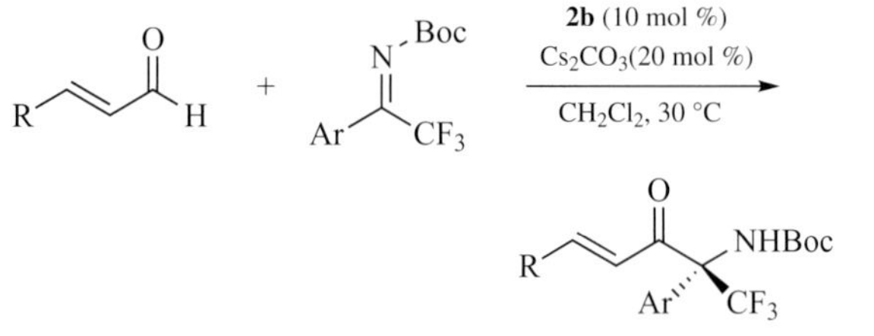

(11)

up to 89% yield
up to 95% ee

[3 + 3]/[3 + 2] Cycloaddition of Bromoenals. The bifunctional NHC **2b′** catalyzed the [3 + 3] cyclization reaction of α-bromoenals and 1,3-dicarbonyl compounds to afford dihydropyranones in good yield with good enantioselectivities (eq 12).[5] The bifunctional NHC **2a′** was efficient catalyst for the [3 + 2] cycloaddition of β-bromoenals with isatins to give the spirooxindole-butenolides (eq 13).[16]

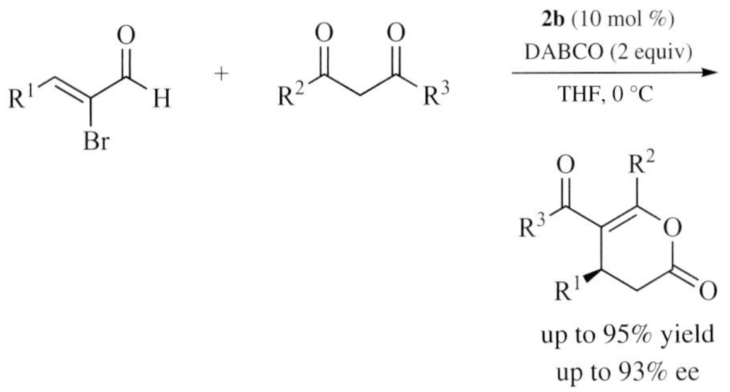

(12)

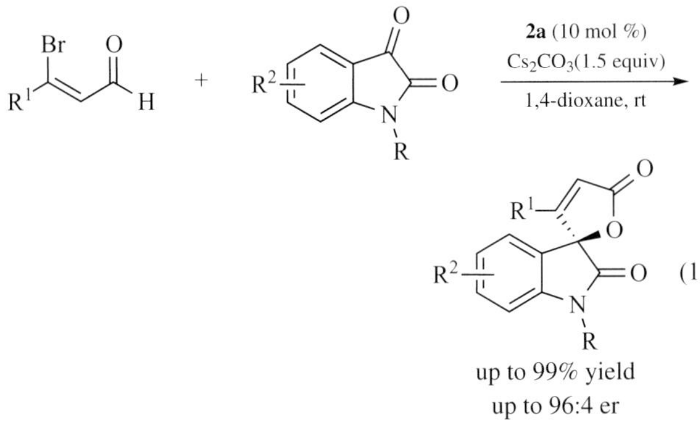

(13)

1. (a) Acevedo, C. M.; Kogut, E. F.; Lipton, M. A., *Tetrahedron* **2001**, *57*, 6353. (b) Ostendorf, M.; Dijkink, J.; Rutjes, F. P. J. T.; Hiemstra, H., *Eur. J. Org. Chem.* **2000**, *1*, 115.
2. Marigo, M.; Wabnitz, T. C.; Fielenbach, D.; Jorgensen, K. A., *Angew. Chem., Int. Ed.* **2005**, *44*, 794.
3. Zhang, Y. R.; He, L.; Wu, X.; Shao, P. L.; Ye, S., *Org. Lett.* **2008**, *10*, 277.
4. He, L.; Zhang, Y. R.; Huang, X. L.; Ye, S., *Synthesis* **2008**, 2825.
5. Sun, F. G.; Sun, L. H.; Ye, S., *Adv. Synth. Catal.* **2011**, *353*, 3134.
6. Shao, P. L.; Chen, X. Y.; Ye, S., *Angew. Chem., Int. Ed.* **2010**, *49*, 8412.
7. He, L.; Lv, H.; Zhang, Y. R.; Ye, S., *J. Org. Chem.* **2008**, *73*, 8101.
8. Jian, T. Y.; He, L.; Tang, C.; Ye, S., *Angew. Chem., Int. Ed.* **2011**, *50*, 9104.
9. Zhang, H. M.; Gao, Z. H.; Ye, S., *Org. Lett.* **2014**, *16*, 3079.
10. Zhang, Y. R.; Lv, H.; Zhou, D.; Ye, S., *Chem. Eur. J.* **2008**, *14*, 8473.
11. Huang, X. L.; He, L.; Shao, P. L.; Ye, S., *Angew. Chem., Int. Ed.* **2009**, *48*, 192.
12. Lv, H.; You, L.; Ye, S., *Adv. Synth. Catal.* **2009**, *351*, 2822.
13. Lv, H.; Jia, W. Q.; Sun, L. H.; Ye, S., *Angew. Chem., Int. Ed.* **2013**, *52*, 8607.
14. Liang, Z. Q.; Gao, Z. H.; Jia, W. Q.; Ye, S., *Chem. Eur. J.* **2014**, *20*, 1868.
15. Sun, L. H.; Liang, Z. Q.; Jia, W. Q.; Ye, S., *Angew. Chem., Int. Ed.* **2013**, *52*, 5803.
16. Zheng, C.; Yao, W.; Zhang, Y.; Ma, C., *Org. Lett.* **2014**, *16*, 5028.

Song Ye
Chinese Academy of Sciences, Beijing, China

C

Chloramine-T[1]

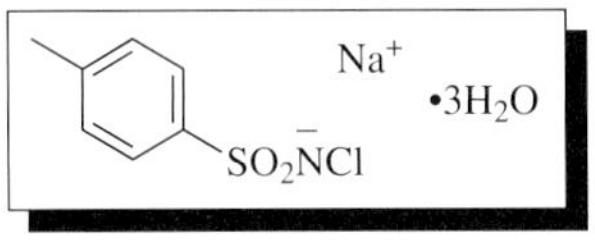

[127-65-1] $C_7H_{13}ClNNaO_5S$ (MW 281.69)
InChI = 1S/C7H7ClNO2S.Na.3H2O/c1-6-2-4-7(5-3-6)12(10,11)9-8;;;;/h2-5H,1H3;;3*1H2/q-1;+1;;;
InChIKey = NZYOAGBNMCVQIV-UHFFFAOYSA-N

(source of chloronium cations and/or nitrogen anions; oxidizing agent, chlorinating agent, analytical reagent; in conjunction with OsO_4 can hydroxyaminate alkenes)

Alternative Names: *N*-chloro-*N*-sodio-*p*-toluenesulfonamide; CAT.
Physical Data: mp 167–170 °C (dec).
Solubility: insoluble in ether, benzene, and chloroform; soluble in cold H_2O; soluble in hot H_2O; less soluble in EtOH.
Form Supplied in: white or pale yellow solid; widely available. While other *N*-halo-*N*-sodiosulfonamides and -carbamates have been reported in the literature and used in organic synthesis, only chloramine-T and its demethyl analog, chloramine-B, are commercially available.
Drying: in some reactions, anhydrous chloramine-T [chloramine-T (anhyd)] is required; the drying of the trihydrate at 80 °C under vacuum has been described,[2] although precautions should be taken (see below).
Preparation: prepared from *p*-toluenesulfonamide and sodium hypochlorite, with the latter being produced in situ from sodium hydroxide and Cl_2.
Analysis of Reagent Purity: by iodometric titration.[3]
Handling, Storage, and Precautions: irritant; keep containers tightly sealed and protect from light; although the trihydrate is stable, there have been reports of chloramine-T (anhyd) decomposing with explosive force;[2,4] to minimize potential risks it is advised that chloramine-T (anhyd) be prepared in minimal amounts and all manipulations be performed behind safety shields (full-face and free-standing); use in a fume hood.

Original Commentary

R. Richard Goehring
Scios Nova, Baltimore, MD, USA

Reaction with Chalcogens and their Derivatives. Chloramine-T reacts with a variety of sulfur-containing compounds to form sulfur–nitrogen bonds.[1] A partial listing includes thiols, sulfides, sulfoxides, disulfides, sulfimides, xanthates, thioketones, sulfenyl chlorides, and sulfinyl chlorides. The reaction of chloramine-T with sulfides forms the basis for a deprotection of thioacetal groups. Thus 1,3-oxathiolanes,[5] 1,3-dithiolanes,[6] and 1,3-dithianes[7] are all cleaved by chloramine-T to regenerate the carbonyl compounds. However, with chloramine-T (anhyd), 1,3-dithianes yield ketene thioacetals (eq 1).[8]

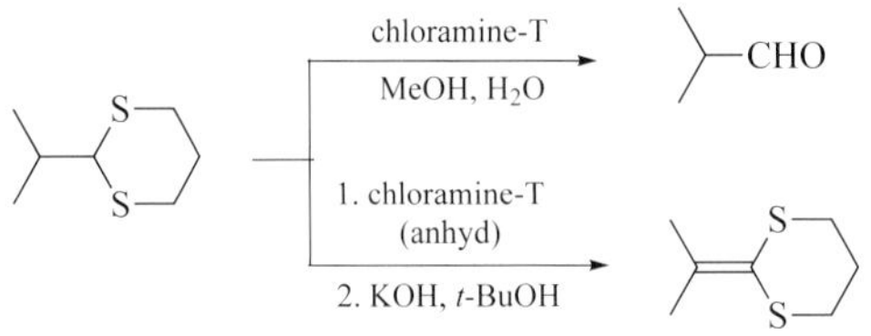
(1)

Selenides react with chloramine-T (anhyd) to give selenimides (eq 2).[9] In the case of primary alkyl phenyl selenides, the selenimide further undergoes β-elimination to give the terminal alkene (eq 3).[10] Tellurides behave similarly.[11] Allyl phenyl selenides can be converted to protected allylic amines via [2,3]-sigmatropic rearrangement of the selenimides (eq 4).[12]

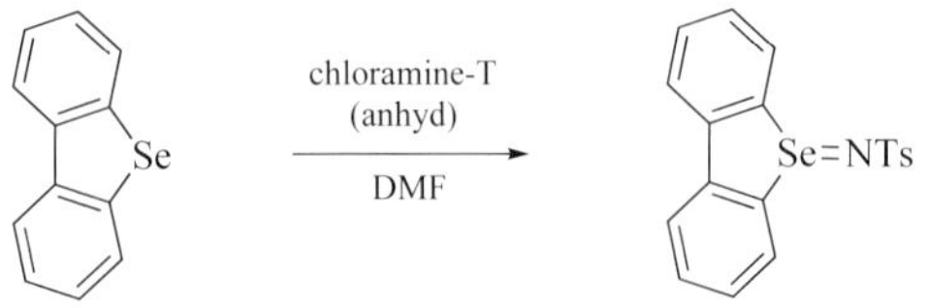
(2)

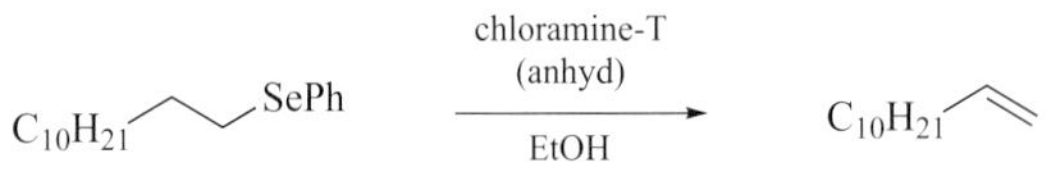
(3)

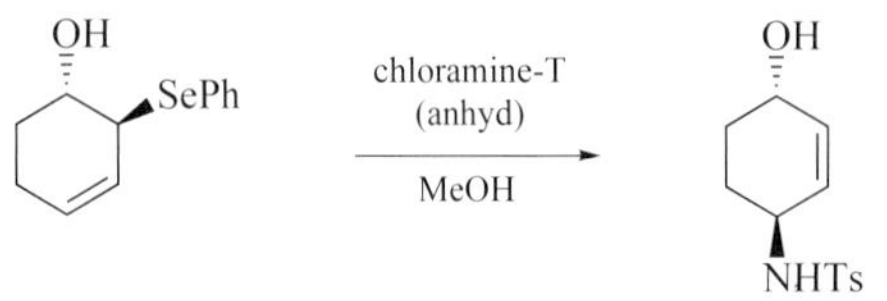
(4)

Diphenyl disulfide and diphenyl diselenide react with chloramine-T (anhyd) to give reagents that convert alkenes to *trans*-vicinal phenylthio(seleno)-*N*-tosylamines (eq 5).[13] Extension of this to the synthesis of α-phenylthio(seleno) ketones from silyl enol ethers has been reported.[14]

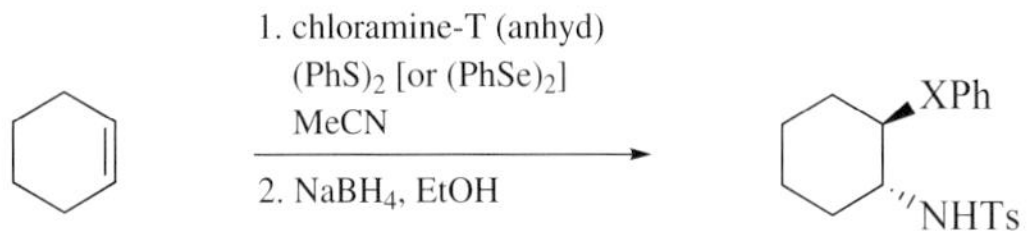
(5)

X = S, Se

Chloramine-T (anhyd) converts selenium and tellurium metal to the diimide species (**1**) and (**2**), respectively. The selenium analog (**1**) has been used for the allylic amination of alkenes and alkynes (eq 6),[2,15] while its tellurium counterpart (**2**) has found utility in the synthesis of *N*-tosylimines from aldehydes.[16] Compound (**1**) has also been used in the 1,2-diamination of 1,3-dienes, but the yields were poor.[17]

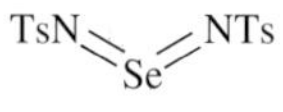

(**1**)

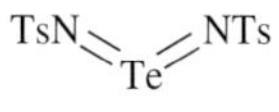

(**2**)

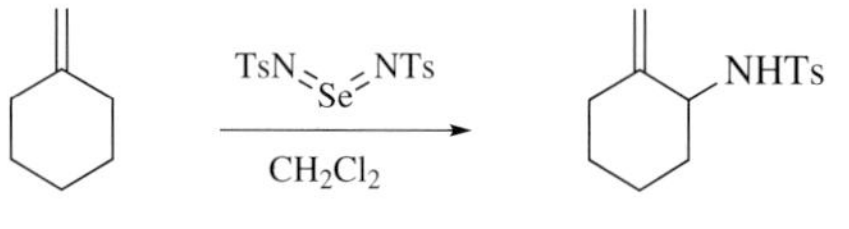

(6)

Other Reactions with Alkenes. Chloramine-T, either alone or in conjunction with other agents, reacts with a variety of carbon–carbon double bonds to generate useful products. Under acidic conditions in aqueous acetone, alkenes give chlorohydrins,[18] while in an acetic acid medium the acetylated products are produced.[19] An intramolecular version allows for the synthesis of chlorolactones (eq 7).[20] Electron-rich alkenes, such as enamines, react to give α-aminated aldehydes (eq 8).[21] Vinyl ethers give the α-(*N*-tosylamino)aldehydes.

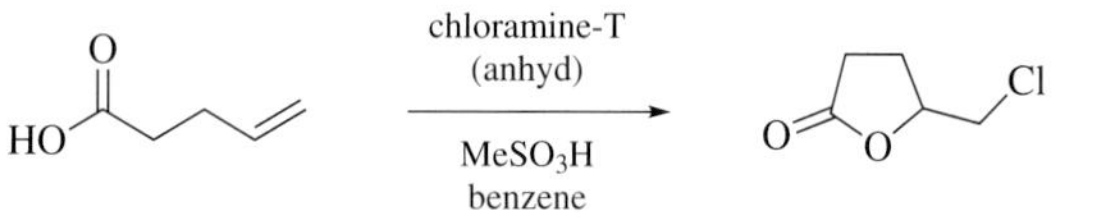

(7)

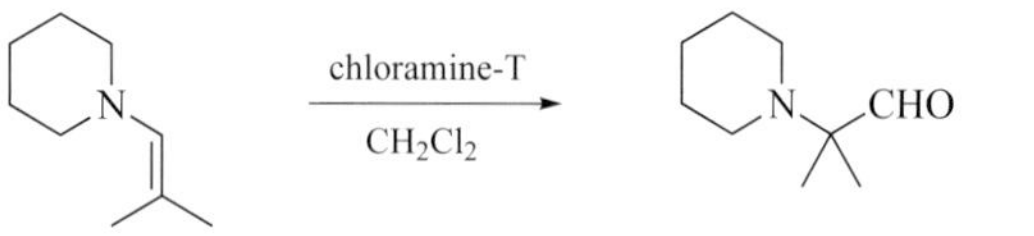

(8)

The reaction of alkenes with chloramine-T in the presence of catalytic amounts of ***Osmium Tetroxide*** provides a convenient and general method for the vicinal oxyamination of alkenes (eq 9).[22] Phase transfer conditions have also been employed.[23]

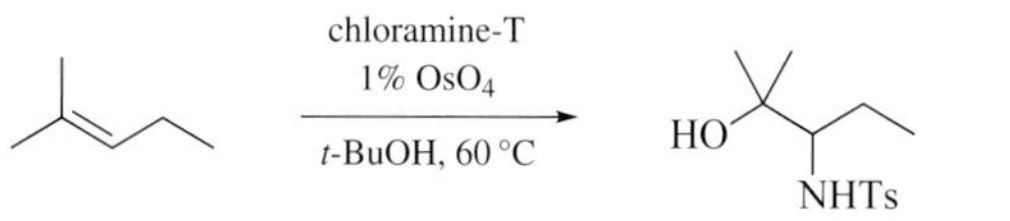

(9)

Other Oxidations. Chloramine-T reacts with NaBr and NaI to provide a convenient method of generating BrCl and ICl, respectively, in situ (see ***Bromine Chloride***).[24,25] Chloramine-T has also been used to oxidize aldoximes to nitrile oxides as part of a synthesis of isoxazolines (eq 10).[26] A similar oxidation of aldehyde hydrazones to nitrilimines has also been reported.[27]

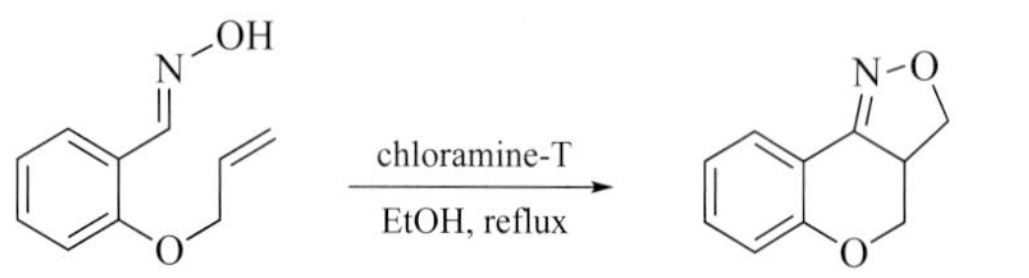

(10)

First Update

Harry Heaney
Loughborough University, Loughborough, UK

The chemistry of chloramine-T continues to be widely exploited and a significant number of papers involve the use of the reagent, especially in the conversion of alkenes into aziridines or β-aminoalkanols. The use of anhydrous chloramine-T is sometimes advocated: it is worth reemphasizing the potential explosion hazard of this material.[2,4]

Reactions with Chalcogens and their Derivatives. The oxidation of sterically crowded thiones and selones using chloramine-T have been reported.[28] Reactions of some unsymmetrical diaryl sulfides with chloramine-T in aqueous potassium hydrogen carbonate lead directly to the related racemic sulfoxides in 70–80% yields.[29] High diastereoselectivity has been reported in reactions of diaryl sulfides that have an oxazolinyl moiety in an *ortho* position, as illustrated in the formation of the related *N*-sulfonylsulfilimine (eq 11).[30] There has been some debate concerning the best procedure for the preparation of *N*-sulfonylsulfilimines, as well as the role of copper(II) salts and additional ligands such as acetonitrile.[31] Almost quantitative yields of *N*-sulfonylsulfilimines have been reported using a variety of sulfides by carrying out the reactions in acetonitrile in the presence of a slight excess of chloramine-T.[32] Reactions of allylic phenyl selenides and tellurides lead to toluene sulfonamides following [2,3]-sigmatropic rearrangements of the intermediates and hydrolysis,[33] and chirality transfer has been recorded in reactions of some chiral ferrocenyl selenides.[34] Thiochroman-4-ones[35] and benzo[b]thiophene[36] have been shown to react with chloramine-T to give, eventually, ring expanded products as a result a sequence of reactions that involve ring opening followed by ring closure.

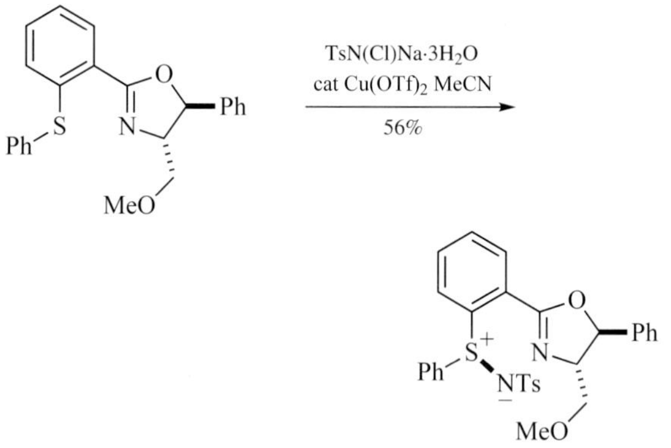

(11)

The results of reactions of 2-vinylthiacycloalkanes with chloramine-T are particularly interesting.[37] Treatment of 2-vinylthiacyclopentane and the six-membered ring analog with chloramine-T in methanol at room temperature gave the corresponding *N*-sulfonylsulfilimines in 60–70% yields; those products undergo [2,3]-sigmatropic rearrangement reactions in xylene at 140 °C to give ring-expanded derivatives, illustrated in eq 12. In the case of the reaction of chloramine-T with the seven-membered ring derivative, the *N*-sulfonylsulfilimine was not isolated and the product of [2,3]-sigmatropic rearrangement (eq 13) was isolated in 66% yield.

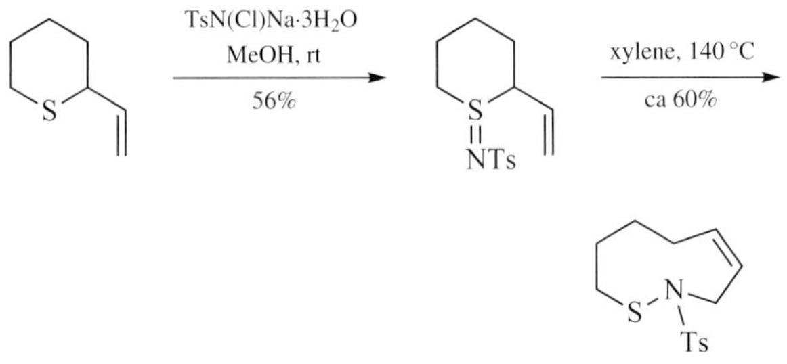

(12)

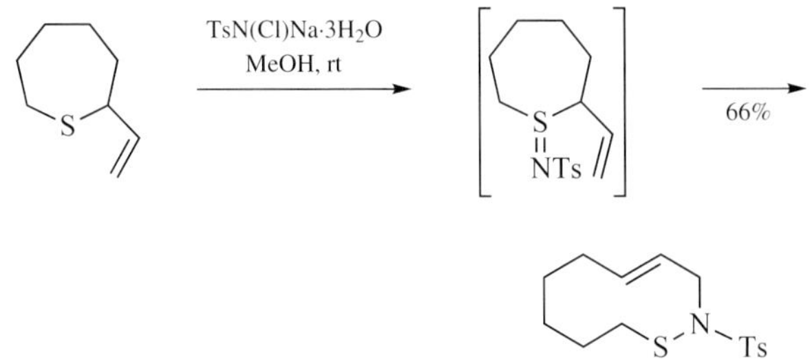

(13)

Amino-hydroxylation Reactions. Aminohydroxylation reactions of alkenes are mentioned in the earlier entry. Using the original protocol,[22] that is in the achiral mode, where prochiral alkenes yield racemic products, poor regioselectivity and the formation of significant amounts of diols as by-products were additional problems that needed to be addressed. Significant developments have centered on the control of regiochemistry and stereochemistry; they have particularly involved the use of cinchona alkaloid based ligands. The hydroxysulfonamide derivatives are frequently crystalline and so recrystallization can lead to high enantiomeric excesses in the osmium-catalyzed asymmetric process.[38] An early example that involved the synthesis of the enantiomers of the taxol side chain from methyl cinnamate, is shown in eq 14.[39]

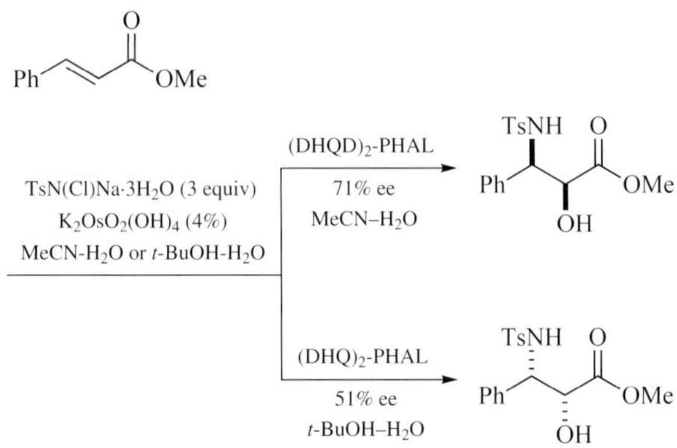

(14)

Improvements in the yields or yields and enantiomeric excesses have been reported using chloramine-M,[40] ethyl sodium *N*-chlorocarbamate[41] and sodium *N*-chloro-*t*-butylsulfonamide,[42] reagents with a smaller substituents on nitrogen. The process is improved still further by using *N*-halocarbamate salts.[43] A study of the aminohydroxyaltion reactions using chloramine-T and carboxamides as substrates showed that high reactivity was exhibited whether a ligand was present or not. Those latter reactions gave very high yields of hydroxysulfonamides. However, although carboxamides are excellent substrates in the related asymmetric dihydroxylation reactions, the products of aminohydroxylation were, surprisingly, always racemic.[44] The best substrate was found to be the *N*-methoxy-*N*-methylcinnamide; it gave the best regioselectivity and the product was formed in the shortest reaction time. The osmium catalyzed aminohydroxylation of α,β-unsaturated carboxylates using chloramine-T proceeds efficiently with a much lower loading of the catalyst and only a stoichiometric amount of the chlorosulfonamide, even in the presence of a chiral ligand: a consequence is that these reactions produce racemates.[45] Osmium tetroxide has been used as the catalyst in aminohydroxylation reactions of terpenes using chloramine-T,[46] and also in the enantioselective synthesis of α-amino ketones in reactions with silyl enol ethers using, for example dihydroquinidine-*p*-chlorobenzoate [$(DHQD)_2$-CLB] as the chiral ligand as exemplified in eq 15.[47] Reaction of a number of alkenes with chloramine-T in the presence of an atmosphere of carbon dioxide at high pressure has been shown to result in β-chloroamination,[48a] and enamides have been shown to undergo diamination.[48b]

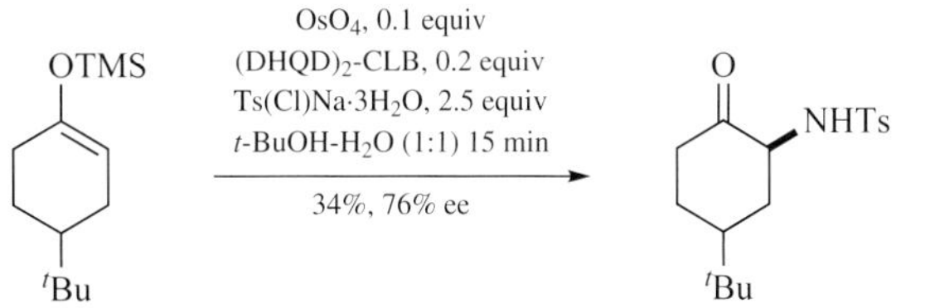

(15)

The harsh conditions that are required for the removal of the *p*-toluenesulfonyl moiety, for example the use of sodium in liquid ammonia, remains, in many cases, as a major problem when using chloramine-T. Removal of the 2-nitrobenzenesulfonyl[49] and 2-methoxybenzenesulfonyl[50] residues can be achieved under significantly milder reaction conditions. However, the corresponding *N*-chloroamide salts do not appear to be as useful as chloramine-T in catalytic aminohydroxylation processes.

Aziridine Forming Reactions. Although the chemistry of aziridines has been well established for many years,[51] the use of chloramine-T as a nitrene or a nitrenoid precursor in cycloaddition reactions with alkenes has only been reported since the earlier entry. The use of [*N*-(*p*-toluenesulfonyl)imino]phenyliodinane (PhI=NTs) as a nitrene precursor has been reported many times.[52] There are a number of disadvantages to the use of the latter reagent, namely, commercial unavailability due to a short shelf life, high cost as compared to the relatively inexpensive chloramine-T, and the production of the relatively high molecular weight iodobenzene as a by-product of the aziridination process.

Early studies established the possibility of using chloramine-T as a source of the nitrogen atom in the formation of aziridines from alkenes both, for example, using copper(I) chloride, eq 16,[53a] or bromine eq 17[53b] or iodine as the catalysts.[53c] Alternative sources of bromine have been used as catalysts together with chloramine-T; they include *N*-bromoamides,[54a,b] bromine, in the form of pyridinium hydrobromide perbromide[53c] and *N*-methylpyrrolidine-2-one hydrobromide perbromide,[54d] as well as aqueous hydrogen peroxide and hydrobromic acid.[54e] The use of iodine as a catalyst has been reported using a phase transfer[55a] or a silica–water reaction medium,[55b] and the use of iodine together with chloramine-T has been extended to the diamination of glycals in reactions that proceed via aziridines.[55c] A number of environmental friendly heteropoly acids have been reported as catalysts together with chloramine-T; of these 12-phosphomolybdic acid was found to be the most efficient one.[56]

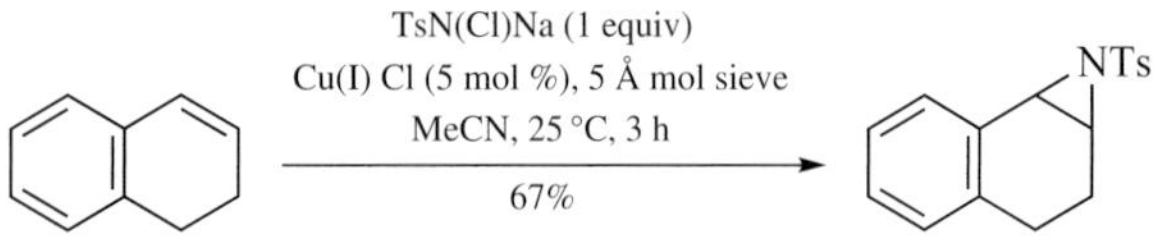

(16)

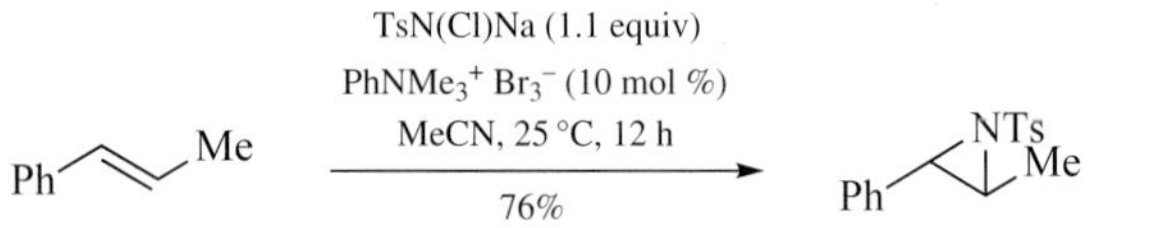

(17)

In addition to the use of anhydrous chloramine-T and copper(I) chloride mentioned above,[53a] a large number of papers have also been published on the synthesis of aziridines from alkenes in which various metal catalysts have been used together with chloramine-T. A variety of copper catalysts have been reported as alternatives to copper(I) chloride,[52] they include the interesting complex that is formed by the interaction of copper(I) triflate with the imine derived from *n*-pentylamine and 2-formylpyridine.[57] Silver nitrate has also been used as a catalyst but the reactions mostly showed poor diastereoselectivity.[58] Reactions with *trans*-β-methylstyrene were shown to proceed with variable diastereoselectivity dependent on the ratios of the coreactants (eq 18) and a reaction with 1,6-heptadiene resulted in the formation of the bicyclic pyrrolidine derivative (eq 19), which was obtained in 42% yield. This is an indicative of a biradical process that involves a triplet nitrene.[58b] A comparison of chloramine-T with bromamine-T in aziridination and other nitrene insertion reactions indicates that bromamine-T is an excellent reagent. Various other metal derivatives have been used as catalysts together with chloramine-T in aziridination reactions.[59] Reactions of chloramine-T that involve reactions with bromine and iodine have been mentioned earlier: *N*,*N*-dichloro-*p*-toluenesulfonamide has also been used in palladium catalyzed aziridination reactions.[60] Bromamine-T has been shown to be superior to chloramine-T as the source of the tosylnitrene, particularly in reactions with electron deficient alkenes,[61] and the use of bromamine-T in palladium catalysed aziridination reactions have also been reported.[62]

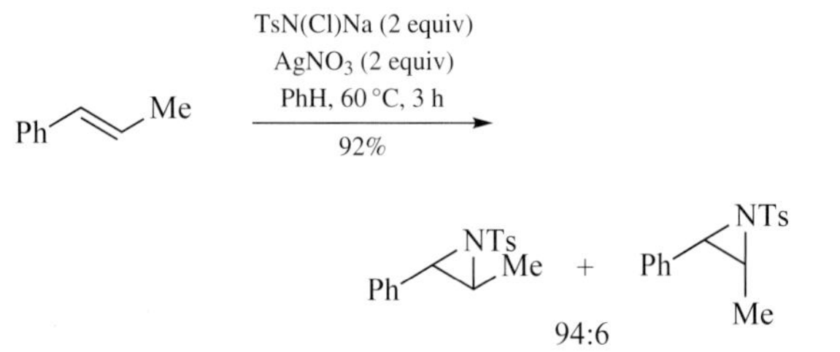

(18)

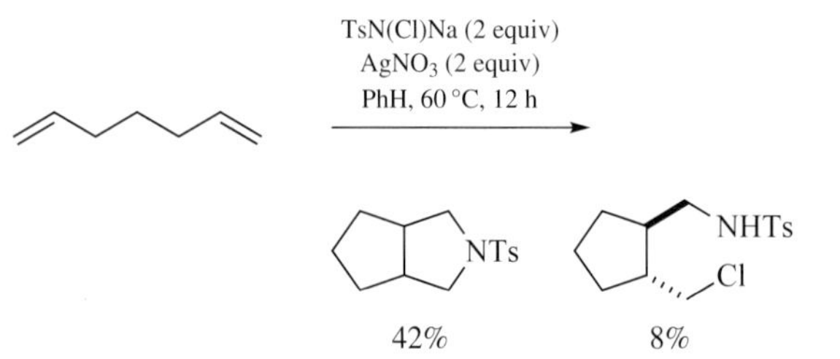

(19)

Other Reactions that Involve the Formation of Heterocycles. Epoxide forming reactions have been reported that involve the use of chloramine-T,[63] for example, a wide range of alkenes can be epoxidized using chloramine-T or chloramine-M together with benzaldehyde and benzyltriethylammonium chloride.[63b] Electrophilic cyclization reactions of *N*-alkenylamides have been carried out using chloramine-T and iodine in acetonitrile. It was concluded that the reactions involve the intermediacy of iodonium species and are exemplified in eq 20.[64a] The protocol can also be extended to give thiazoline or oxazoline derivatives as shown in eq 21,[64a] and similar reactions of γ-iodoalkenes give pyrrolidine derivatives.[64b] The synthesis of pyrazolines that involve the oxidation of aryl hydrazones with chloramine-T followed by dipolar cycloaddition reactions with alkenes have been reported;[65] they include a one-pot procedure using a polyethylene glycol supported acrylate, an aldehyde, an aryl hydrazine, and chloramine-T in methanol.[65c]

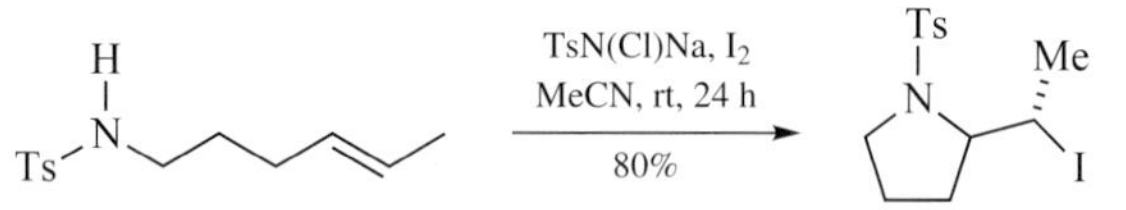

(20)

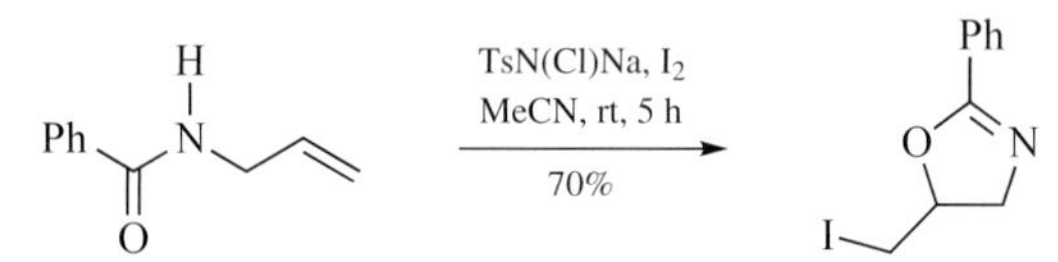

(21)

Nitrile oxide cycloaddition reactions were mentioned in the earlier entry. Additional examples of syntheses of ioxazolines have been reported,[65a,66] for example using polymer supported acrylates.[66a] and of sugar derivatives without hydroxyl group protection similar to the example shown in eq 10.[66b] Nitrile oxides are involved in reactions of araldoximes with 4 equiv. of chloramine-T that eventually result in the formation of benzamide derivatives.[67]

Other Oxidation Reactions. Chloramine-T has been used in a number of amidation reactions, for example α- to oxygen in cyclic ethers and at benzylic positions,[68a,b] including phenylacetone and propionaldehyde derivatives as shown in eq 22.[68c] In reactions of a number of alkylbenzene derivatives, the system chloramine-T, oxygen and an iron-porphyrin catalyst led to the formation of aryl ketones.[69]

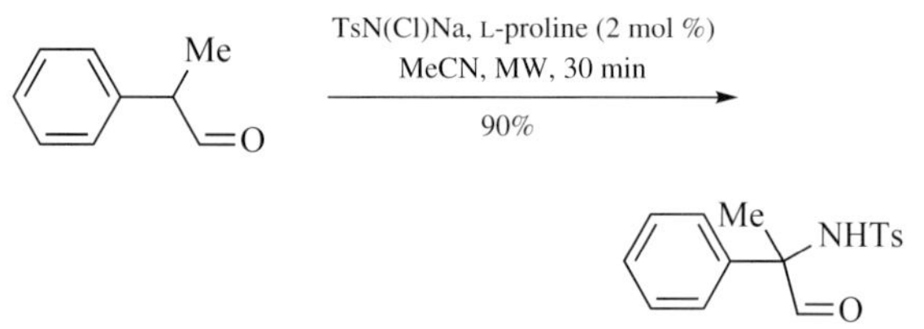

(22)

The conversion of organotrifluoroborates into the related bromides and iodides has been reported using chloramine-T and sodium bromide or iodide,[70] and the introduction of [^{125}I] has been achieved by iododestannylation reactions using Na [^{125}I] together with chloramine-T.[71] The interesting semipinacol rearrangement shown in eq 23 involves the use of chloramine-T and zinc halides.[72] The oxidation of ketohydrazones that contain an α-methylene group using chloramine-T provides a route to α-chloroazo- derivatives that are converted into azoalkenes by base treatment,[73] and the three-component reaction of isonitriles with arylamines and chloramines-T provides a route to *N*-tosylguanidines.[74]

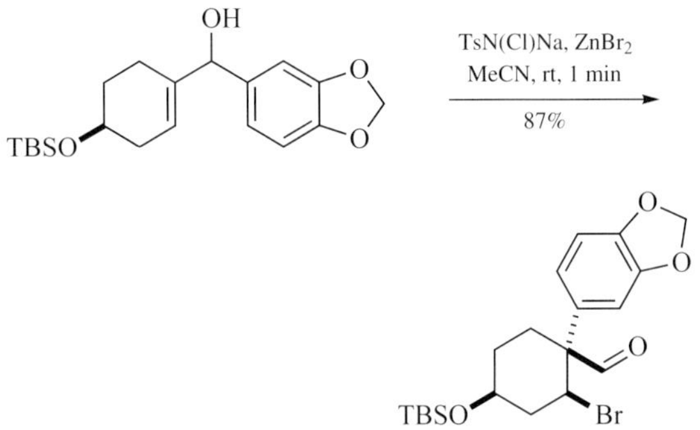

(23)

Second Update

Rajenahally V. Jagadeesh
Leibniz-Institut für Katalyse e. V. an der Universität Rostock, Rostock, Germany

Puttaswamy
Bangalore University, Central College Campus, Bangalore, India

Chloramine-T is a prominent member of sodium *N*-haloarene sulfonamidates (organic haloamines). It is a by-product of saccharin manufacture and was first prepared by Chattaway.[75] The chemistry of chloramine-T has evinced keen interest, as it is an oxidant, analytical regent, and halogenating agent employed for a myriad of synthetic purposes. The benzene analog of this haloamine is chloramine-B (CAB; sodium *N*-chloro-benzenesulfonamide). Chloramine-T is the precursor of halonium cations, hypohalite species, and *N*-anions, which are capable of acting both as bases and as nucleophiles.[76] These compounds resemble hypohalites in their oxidative behavior and, though less familiar, are more stable.[77,78]

Investigations of Morris et al.,[78] Bishop and Jennings,[77] Pryde and Soper,[79] and Hardy and Johnston[80] have shown the existence of similar equilibria of *N*-metallo-*N*-arylhalosulfonamides (*N*-haloamines) in aqueous media. Depending on the pH, chloramine-T exhibits the following equilibria in aqueous solutions (eqs 24–30):

$$\mathrm{RNXNa} \rightleftharpoons \mathrm{RNCl^-} + \mathrm{Na^+} \quad (24)$$

$$\mathrm{RNCl^-} + \mathrm{H^+} \rightleftharpoons \mathrm{RNHCl} \quad (25)$$

$$\mathrm{2RNHCl} \rightleftharpoons \mathrm{RNH_2} + \mathrm{RNCl_2} \quad (26)$$

$$\mathrm{RNCl_2} + \mathrm{H_2O} \rightleftharpoons \mathrm{RNHCl} + \mathrm{HOCl} \quad (27)$$

$$\mathrm{RNHCl} + \mathrm{H_2O} \rightleftharpoons \mathrm{RNH_2} + \mathrm{HOCl} \quad (28)$$

$$\mathrm{HOCl} + \mathrm{H^+} \rightleftharpoons \mathrm{H_2OCl^+} \quad (29)$$

$$\mathrm{HOCl} \rightleftharpoons \mathrm{H^+} + \mathrm{OCl^-} \quad (30)$$

(R = -$CH_3C_6H_4$ for chloramine-T; and -C_6H_5 for chloramine-B)

Depending on the kinetics and reaction conditions, the above species act as the reactive species involved in oxidation and halogenation reactions.

Catalytic Oxidation of Dyes. Dyes are useful materials and have been used extensively for dyeing cotton, textile, and fabrics. While dyes are useful synthetic and analytical reagents, they form hazardous effluents that affect both aerobic and aquatic living systems. Therefore, suitable methods to degrade these dyes are of considerable importance.

Puttaswamy et al.[81–83] have extensively studied the oxidation of azo dyes with transition metal catalysts using chloramine-T and chloramine-B as oxidants. The oxidation of *p*-hydroxyazobenzene[81] and *p*-aminoazobenzene[82] have been studied using a ruthenium[III] catalyst in acidic and alkaline medium. Notably, the azo group of the dye molecules have been selectively oxidized to yield corresponding nitroso compounds. The oxidation of *p*-hydroxyazobenzene was performed using four *N*-haloamines such as bromamines and chloramines. Clearly, the reactivity indicates that the bromamines are highly reactive compared to chloramines. The related reactivity of these *N*-haloamines follows the order: bromamine-B (BAB) > bromamine-T (BAT) > chloramine-B (CAB) > chloramine-T (CAT). The kinetics and mechanism of oxidation of azo dye with all these oxidants follow similar kinetics and proceed with a common mechanism.

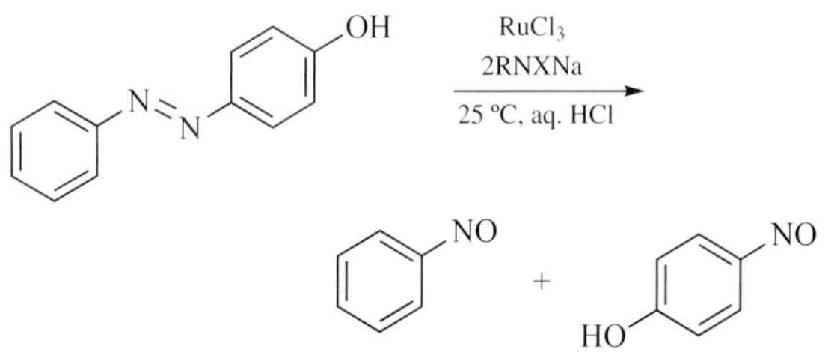

(31)

R = *p*-$CH_3C_6H_4SO_2$ for CAT and BAT and $C_6H_5SO_2$ for CAB and BAB; X = Cl and Br

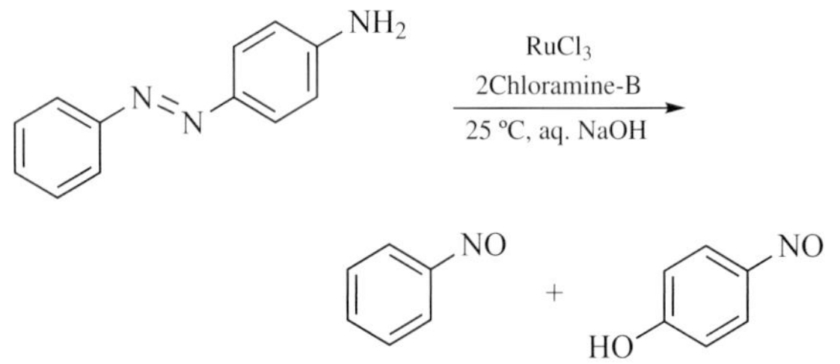

(32)

Osmium-catalyzed oxidative conversion of indigo carmine to isatins and anthranilic acids has been studied using chloramine-B.[84] This is an important transformation because both isatins and anthranilic acids are precursors for the synthesis of dyes, agro chemicals, and pharmaceutical ingredients. The indigo carmine reacts with chloramine-B and produced isatin first. The so formed isatin again reacts with chloramine-T to produce anthranilic acids as the final products (eq 33).

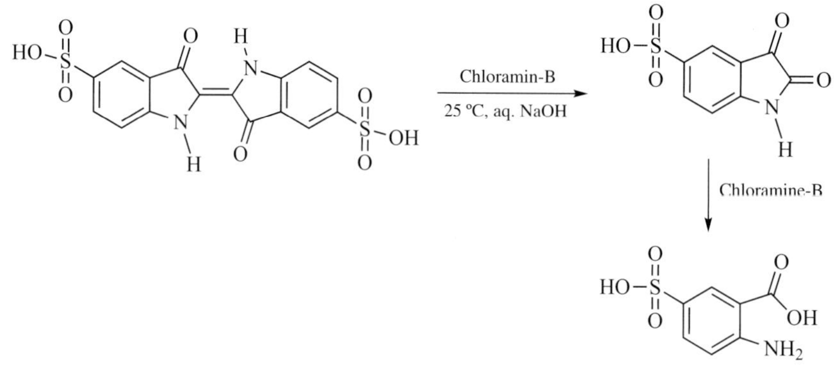

(33)

Oxidation of Heterocyclic Compounds. Notable achievements have been made by Puttaswamy et al. and Jagadeesh et al.[85–88] on the selective oxidation of heterocyclic compounds. Ruthenium-catalyzed oxidation of benzimidazoles and imidazole[85] and osmium-catalyzed oxidation of benzimidazoles[86] have been studied with chloramine-T in acidic and alkaline media. The preferential oxidation occurs in the heterocyclic ring without affecting benzene ring. Oxidation of all the studied heterocyclic compounds yields the corresponding *o*-phenylene diamnes, while imidazole produced amino acid such as glycine (eqs 34–36). Here the yields were not quantified, but the transformation is very important.

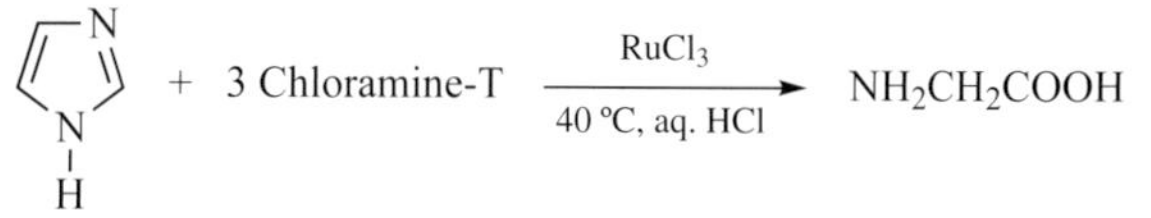

(34)

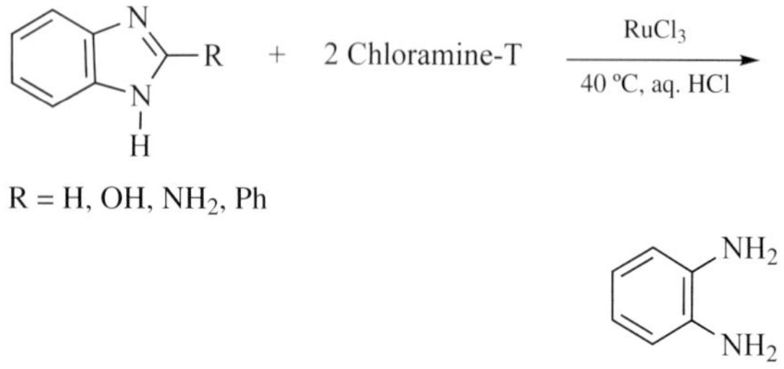

(35)

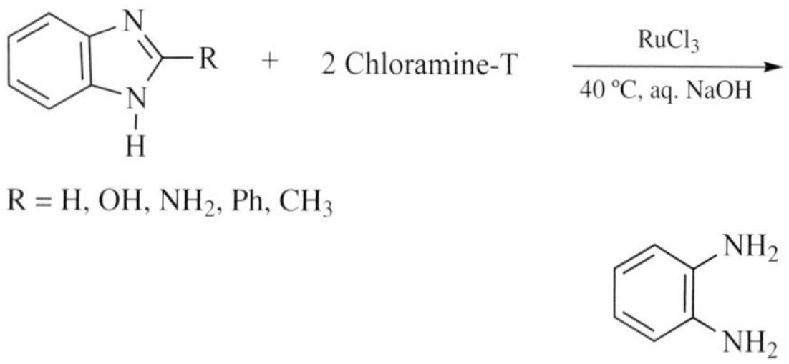

(36)

Oxidative conversion of isatines to anthranilic acids has been performed using chloramine-B at 35 °C in alkaline medium (eq 37).[87] The detailed kinetics and mechanism for oxidation of isatins reveal reactivity in the order: 5-methylisatin > isatin > 5-bromoisatin > 5-nitroisatin.

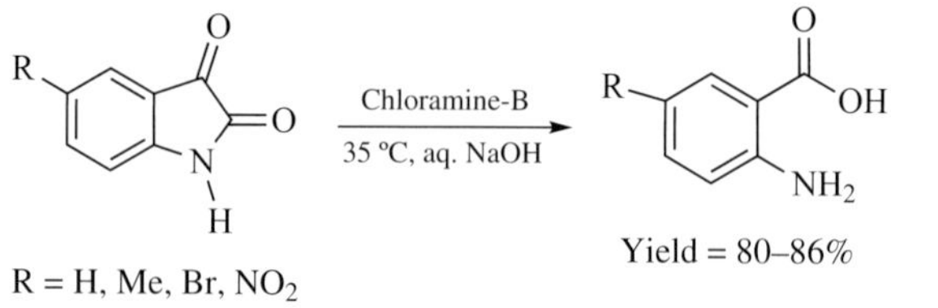

(37)

An efficient ruthenium-Schiff base-catalyzed protocol has been developed[88] for the synthesis of orthanilic acids from benzothiazoles in good to excellent yields using *N*-haloamines (CAT, CAB, BAT, and BAB) (eq 38). The detailed catalytic, mechanistic, and kinetic investigations have been made for these reactions.

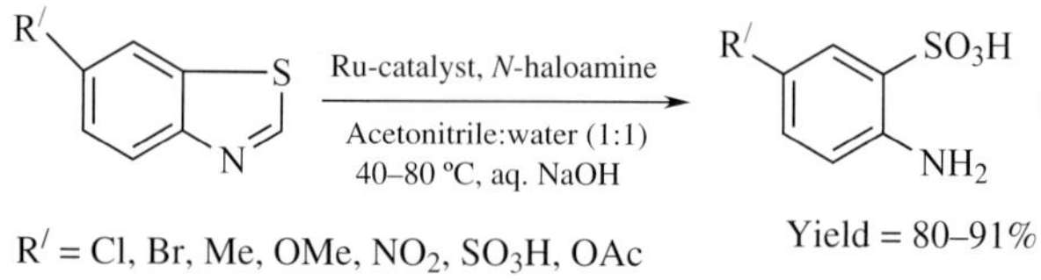

(38)

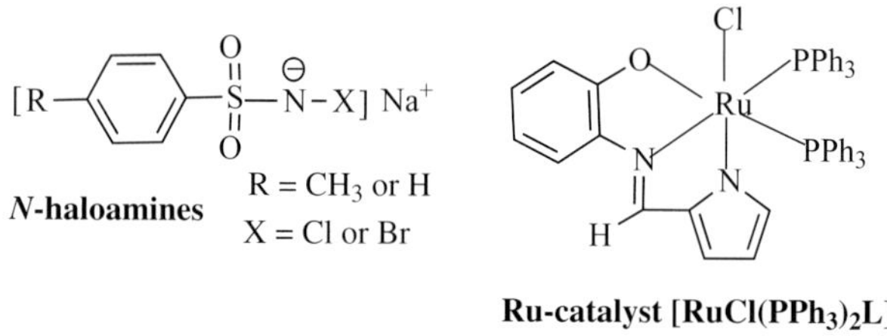

Ru-catalyst [RuCl(PPh$_3$)$_2$L]

Oxidation of Sulfur Compounds. The oxidation of five alkane thiols,[89] such as 1-ethanethiol, 1-propanethiol, 1-butanethiol, 1-pentanethiol, and 1-hexanethiol, to symmetrical disulfides by sodium *N*-chloro-*p*-toluenesulfonamide or chloramine-T (CAT) in the presence of NaOH has been investigated at 15 °C (eq 39). Here, the chloramine-T chemoselectively oxidized thiols to disulfides without further oxidation.

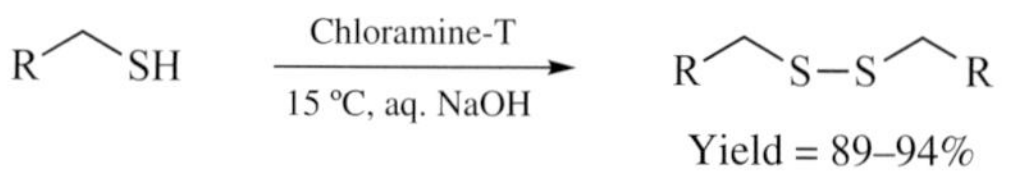

(39)

The osmium-catalyzed and osmium-uncatalyzed oxidations of biotin using chloramine-T were performed in alkaline medium.[90] The controlled oxidation of biotin to biotin sulfoxide provides a very important synthetic route in biochemical reactions. Optimum conditions have been developed for the oxidation of biotin to biotin sulfoxide. The kinetics of Os(VIII)-catalyzed and Os(VIII)-uncatalyzed oxidations of biotin by CAT have been studied in NaOH medium at 30 °C under identical experimental conditions (eq 40).

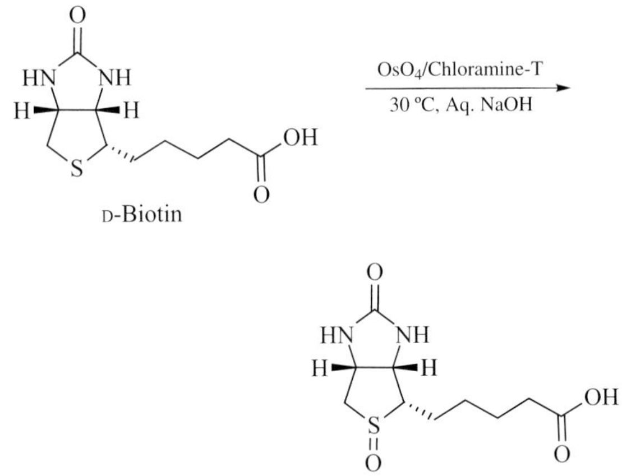

D-Biotin

D-biotin sulfoxide
Yield = 78%

(40)

The platinum group metals (Ru(III), Os(VIII), Rh(II), Ir(III) Pd(II), Pt(IV)) catalyzed oxidation of thiazine dyes, such as methylene blue, has been performed using chloramine-T in alkaline medium (eq 41).[91] The sulfide group of the methylene dye has been preferentially and selectively oxidized to sulfone without further oxidation of any other functionality.

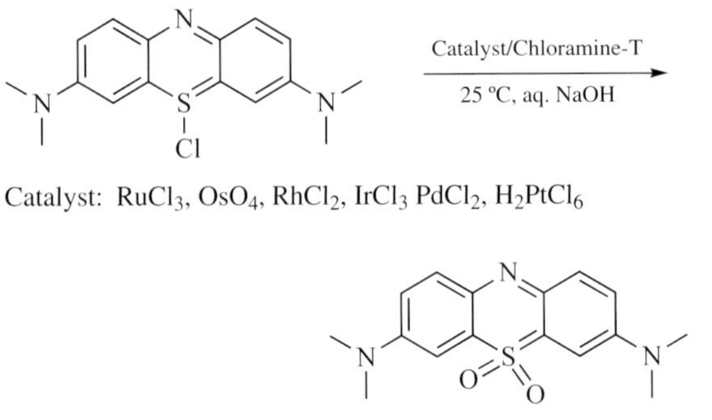

Catalyst: $RuCl_3$, OsO_4, $RhCl_2$, $IrCl_3$ $PdCl_2$, H_2PtCl_6

Yield = 75–82%

(41)

1. (a) Campbell, M. M.; Johnson, G., *Chem. Rev.* **1978**, *78*, 65. (b) Bremner, D. H. In *Synthetic Reagents*; Pizey, J. S., Ed.; Wiley: New York, **1985**; Vol. *6*, p 9.
2. Sharpless, K. B.; Hori, T.; Truesdale, L. K.; Dietrich, C. O., *J. Am. Chem. Soc.* **1976**, *98*, 269.
3. *Reagent Chemicals: American Chemical Society Specifications*, 8th ed.; American Chemical Society: Washington, **1993**; p 242.
4. Klundt, I. L., *Chem. Eng. News* **1977**, *55*(49), 56.
5. Emerson, D. W.; Wynberg, H., *Tetrahedron Lett.* **1971**, 3445.
6. Huurdeman, W. F. J.; Wynberg, H.; Emerson, D. W., *Tetrahedron Lett.* **1971**, 3449.
7. Huurdeman, W.; Wynberg, H.; Emerson, D. W., *Synth. Commun.* **1972**, *2*, 7.
8. Yoshida, H.; Ogata, T.; Inokawa, S., *Synthesis* **1976**, 552.
9. Hellwinckel, D.; Fahrbach, G., *Justus Liebigs Ann. Chem.* **1968**, *715*, 68.
10. Sharpless, K. B.; Gordon, K. M.; Lauer, R. F.; Patrick, D. W.; Singer, S. P.; Young, M. W., *Chem. Scr.* **1975**, *8A*, 9 (*Chem. Abstr.* **1976**, *85*, 4632k).
11. Otsubo, T.; Ogura, F.; Yamaguchi, H.; Higuchi, H.; Sakata, Y.; Misumi, S., *Chem. Lett.* **1981**, 447.
12. (a) Fankhauser, J. E.; Peevey, R. M.; Hopkins, P. B., *Tetrahedron Lett.* **1984**, *25*, 15. (b) For the use of *N*-chloro-*N*-sodiocarbamates to furnish the synthetically more useful Boc- or Cbz-protected primary allylic amines, see: Shea, R. G.; Fitzner, J. N.; Fankhauser, J. E.; Hopkins, P. B., *J. Org. Chem.* **1984**, *49*, 3647.
13. Barton, D. H. R.; Britten-Kelly, M. R.; Ferreira, D., *J. Chem. Soc., Perkin Trans. 1* **1978**, 1090.
14. Magnus, P.; Rigoller, P., *Tetrahedron Lett.* **1992**, *33*, 6111.
15. The corresponding sulfur analog appears to be superior to (**1**) as an aminating agent. However, chloramine-T is not used in its synthesis; see: Sharpless, K. B.; Hori, T., *J. Org. Chem.* **1976**, *41*, 176.
16. Trost, B. M.; Marrs, C., *J. Org. Chem.* **1991**, *56*, 6468.
17. Sharpless, K. B.; Singer, S. P., *J. Org. Chem.* **1976**, *41*, 2504.
18. Damin, B.; Garapon, J.; Sillion, B., *Synthesis* **1981**, 362.
19. Damin, B.; Garapon, J.; Sillion, B., *Tetrahedron Lett.* **1980**, 1709.
20. Damin, B.; Forestiere, A.; Garapon, J.; Sillion, B., *J. Org. Chem.* **1981**, *46*, 3552.
21. Dyong, I.; Lam-Chi, Q., *Angew. Chem., Int. Ed. Engl.* **1979**, *18*, 933.
22. (a) Sharpless, K. B.; Chong, A. O.; Oshima, K., *J. Org. Chem.* **1976**, *41*, 177. (b) Herranz, E.; Sharpless, K. B., *Org. Synth., Coll. Vol.* **1990**, *7*, 375.
23. Herranz, E.; Sharpless, K. B., *J. Org. Chem.* **1978**, *43*, 2544.
24. Kabalka, G. W.; Sastry, K. A. R.; Hsu, H. C.; Hylarides, M. D., *J. Org. Chem.* **1981**, *46*, 3113.
25. (a) Kabalka, G. W.; Gooch, E. E., *J. Org. Chem.* **1981**, *46*, 2582. (b) Kometani, T.; Watt, D. S.; Ji, T., *Tetrahedron Lett.* **1985**, *26*, 2043.
26. Hassner, A.; Lokanatha Rai, K. M., *Synthesis* **1989**, 57.
27. Lokanatha Rai, K. M.; Hassner, A., *Synth. Commun.* **1989**, *19*, 2799.
28. Shimada, K.; Kodaki, K.; Nanae, T.; Aoyagi, S.; Takikawa, Y.; Kabuto, C., *Tetrahedron Lett.* **2000**, *41*, 6833.
29. Szabó, D.; Szendeffy, S.; Kapovits, I.; Kucsman, I.; Argay, G.; Kálmaán, A.; Párkânyi, L., *Tetrahedron: Asymmetry.* **1997**, *8*, 2403.
30. (a) Takada, H.; Ohe, K.; Uemura, S., *Angew Chem., Int. Ed.* **1999**, *38*, 1288. (b) Takada, H.; Oda, M.; Oyamada, A.; Ohe, K.; Uemura, S., *Chirality* **2000**, *12*, 299.
31. (a) Takada, H.; Nishibayashi, Y.; Ohe, K.; Uemura, S., *J. Chem. Soc., Chem. Commun.* **1996**, 931. (b) Takada, H.; Nishibayashi, Y.; Ohe, K.; Uemura, S.; Baird, C. P.; Sparey, T. J.; Taylor, P. C., *J. Org. Chem.* **1997**, *62*, 6512. (c) Aujia, P. S.; Baird, C. P.; Taylor, P. C.; Mauger, H.; Valée, Y., *Tetrahedron Lett.* **1997**, *38*, 7453.
32. (a) Marzinzik, A. L.; Sharpless, K. B., *J. Org. Chem.* **2001**, *66*, 594.
33. (a) Nishibayashi, Y.; Srivistava, S. K.; Obe, K.; Uemura, S., *Tetrahedron Lett.* **1995**, *36*, 6725. (b) Uzarewicz, A.; Scianowski, J.; Bakowska-Janiszewska, J., *Polish J. Chem.* **1999**, *73*, 1791; (c) **2000**, *74*, 1079. (d) Bakowska-Janiszewska, J.; Scianowski, J.; Uzarewicz, A., *Polish J. Chem.* **2001**, *75*, 649.
34. Nishibayashi, Y.; Chiba, T.; Ohe, K.; Uemura, S., *J. Chem. Soc., Chem. Commun.* **1995**, 1243.
35. Takebe, Y.; Bayomi, S. M. M.; Mukai, C.; Ikeda, M.; Murasi, M.; Kise, M., *J. Chem. Soc., Perkin Trans. 1* **1981**, 1037.
36. Archibald, S. J.; Boa, A. N.; Pesa, N., *Chem. Commun.* **2003**, 1736.
37. Sahida, H.; Tsuchiya, T., *Heterocycles* **1984**, *22*, 1303.
38. (a) Reiser, O., *Angew. Chem., Int. Ed. Engl.* **1996**, *35*, 1308. (b) O'Brien, P., *Angew. Chem. Int. Ed.* **1999**, *38*, 326. (c) Kolb, H. C.; Sharpless, K. B. In *Transition Metals for Fine Chemicals and Organic Synthesis*; Beller, M.; Bolm, C.; Eds. Wiley-VCH: Weinheim, **1998**, Vol. 2. p 243. (d) Schlingloff, G.; Sharpless, K. B. In *Asymmetric Oxidation Reactions: A Practical Approach*; Katsuki, T., Ed.; Oxford University Press: Oxford, **2001**.

39. Li, G.; Chang, H.-T.; Sharpless, K. B., *Angew. Chem., Int. Ed. Engl.* **1996**, *35*, 451.
40. Li, G.; Angert, H. H.; Sharpless, K. B., *Angew. Chem., Int. Ed. Engl.* **1996**, *35*, 2810.
41. Angelaud, R.; Landais, Y.; Schenk, K., *Tetrahedron Lett.* **1997**, *38*, 1407.
42. Goncharov, A. V.; Liu, H.; Sharpless, K. B., *Org. Lett.* **1999**, *1*, 783.
43. Li, G.; Angert, H. H.; Sharpless, K. B., *Angew. Chem., Int. Ed. Engl.* **1996**, *35*, 2813.
44. Rubin, A. E.; Sharpless, K. B., *Angew. Chem., Int. Ed. Engl.* **1997**, *36*, 2637.
45. Folkin, V. V.; Sharpless, K. B.; *Angew. Chem., Int. Ed.* **2001**, *40*, 3455.
46. Pinheiro, S.; Pedraza, S. F.; Farias, F. M. C.; Goncalves, A. S.; Costa, P. R. R., *Tetrahedron: Asymmetry*. **2000**, *11*, 3845.
47. Phukan, P.; Sudalai, A., *Tetrahedron: Asymmetry* **1998**, *9*, 1001.
48. (a) Minakata, S.; Yoneda, Y.; Oderaotoshi, Y.; Komatsu, M., *Org. Lett.* **2006**, *8*, 967. (b) Zöllinger, M.; Mayer, P.; Lindel, T., *J. Org. Chem.* **2006**, *71*, 9431.
49. Fukuyama, T.; Jow, C.-K.; Cheung, M., *Tetrahedron Lett.* **1995**, *36*, 6373.
50. Milburn, R. R.; Snieckus, V., *Angew. Chem., Int. Ed.* **2004**, *43*, 892.
51. (a) Rai, K. M. L.; Hassner, A. In *Comprehensive Heterocyclic Chemistry II*, Padwa, A., Ed.; Pergamon Press: Oxford, **1996**. Vol. **1A**, p 61. (b) Cheng, X.-X.; Zhang, Y.-Q.; Lan, H.; Chen, Q.-H., *Chinese J. Org. Chem.* **2006**, *26*, 1173.
52. see for example, Mairena, M. A.; Díaz-Requejo, M. M.; Belderraín, T. R.; Nicasio, M. C.; Trofimenko, S.; Pérez, P. J., *Organometallics* **2004**, *23*, 253, and references cited therein.
53. (a) Ando, T.; Minakata, S.; Ryu, I.; Komatsu, M., *Tetrahedron Lett.* **1998**, *39*, 309. (b) Jeong, J. U.; Tao, B.; Sagasser, I.; Henniges, H.; Sharpless, K. B., *J. Am. Chem. Soc.* **1998**, *120*, 6844. (c) Ando, T.; Kano, D.; Minakata, S.; Ryu, I.; Komatsu, M., *Tetrahedron* **1998**, *54*, 13485.
54. (a) Thakur, V. V.; Sudulai, A., *Tetrahedron Lett.* **2003**, 44, 989. (b) Jain, S. L.; Sain, B., *Chem. Commun.* **2006**, 943. (c) Ali, S. I.; Nikalje, M. D.; Sudalai, A., *Org. Lett.* **1999**, *1*, 705. (d) Jain, S. L.; Joseph, J. K.; Sain, B., *J. Mol. Catal. A: Chem.* **2006**, *256*, 16. (e) Jain, S. L.; Sharma, V. B.; Sain, B., *Tetrahedron Lett.* **2004**, *45*, 8731.
55. (a) Kano, D.; Minakata, S.; Komatsu, M., *J. Chem. Soc., Perkin Trans. 1* **2001**, 3186. (b) Minakata, S.; Kano, D.; Oderaotoshi, Y.; Komatsu M., *Angew. Chem. Int. Ed.* **2004**, *43*, 79. (c) M. Kumar, V.; Ramesh, N. G., *Chem. Commun.* **2006**, 4952.
56. Kumar, G. D. K.; Baskaran, S., *Chem. Commun.* **2006**, 1026.
57. Albone, D. P.; Aujla, P. S.; Taylor, P. C.; Challenger, S.; Derrick, A. M., *J. Org. Chem.* **1998**, *63*, 9569.
58. (a) Kumar, K. A.; Rai, K. M. L.; Umesha, K. B., *Tetrahedron* **2001**, *57*, 6993. (b) Minakata, S.; Kano, D.; Fukuoka, R.; Oderaotoshi, Y.; Komatsu, M., *Heterocycles* **2003**, *60*, 289.
59. (a) Vyas, R.; Gao, G.-Y.; Harden, J. D.; Zhang, X. P., *Org. Lett.* **2004**, *6*, **1907**. (b) Simkhovich, L.; Gross, Z., *Tetrahedron Lett.* **2001**, *42*, 8089. (c) Gao, G.-Y., Harden, J. D.; Zhang, X. P., *Org. Lett.* **2005**, *7*, 3191. (d) Masuyama, Y.; Ohtsuka, M.; Harima, M.; Kurusu, Y., *Heterocycles* **2006**, *67*, 503.
60. Han, J.; Li, Y.; Zhi, S.; Pan, Y.; Timmons, C.; Li, G., *Tetrahedron Lett.* **2006**, *47*, 7225.
61. Chanda, B. M.; Vyas, R.; Bedekar, A. V., *J. Org. Chem.* **2001**, *66*, 30.
62. Antunes, A. M. M.; Marto, S. J. L.; Branco, P. S.; Prabhakar, S.; Lobo, A. M., *Chem. Commun.* **2001**, 405.
63. (a) von Riesen, C.; Jones, P. G.; Hoffmann, H. M. R., *Chem. Eur. J.* **1996**, *2*, 673. (b) Ynag. D.; Zhang, C.; Wang, X.-C., *J. Am. Chem. Soc.* **2000**, *122*, 4039.
64. (a) Morino, Y.; Hidaka, I.; Oderaotoshi, Y.; Komatsu, M.; Minakata, S., *Tetrahedron* **2006**, *62*, 12247. (b) Minakata, S.; Kano, D.; Oderaotoshi, Y.; Komatsu, M., *Org. Lett.* **2002**, *4*, 2097.
65. (a) Padmavathi, V.; Reddy, K. V.; Balaiah, A.; Reddy, T. V. R.; Reddy, D. B., *Heteroatom Chem.* **2002**, *13*, 677. (b) Deng, L.; Hu, Y., *J. Heterocyclic Chem.* **2006**, *43*, 1705. (c) Wang, Y.-G.; Zhang, J.; Lin, X.-F.; Ding, H.-F., *Synlett* **2003**, 1467.
66. (a) Tao, L.; Zhang, P. F.; Gu, Y. X., *Chinese Chem. Lett.* **2005**, *16*, 440. (b) Shing, T. K. M.; Wong, W. S.; Cheng, H. M.; Kwok, W. S.; So, K. H., *Org. Lett.* **2007**, *9*, 753.
67. Padmavathi, V.; Reddy, K. V.; Padmaja, A.; Venugopalan, P., *J. Org. Chem.* **2003**, *68*, 1567.
68. (a) Albone, D. P.; Challenger, S.; Derrick, A. M.; Fillery, S. M.; Irwin, J. L.; Parsons, C. M.; Takeda, H.; Taylor, P. C.; Wilson, D. J., *Org. Biomol. Chem.* **2005**, *3*, 107. (b) Fructos, M. R.; Trofimenko, S.; Díaz-Requejo, M. M.; Pérez, P. J., *J. Am. Chem. Soc.* **2006**, *128*, 11784. (c) Baumann, T.;. Bächie, M.; Btäse, S., *Org. Lett.* **2006**, *8*, 3797.
69. Li, S.-J.; Wang, Y.-G., *Tetrahedron Lett.* **2005**, *46*, 8013.
70. (a) Kabalka, G. W.; Mereddy, A. R., *Organometallics* **2004**, *23*, 4519. (b) Kabalka, G. W.; Mereddy, A. R., *Tetrahedron Lett.* **2004**, *45*, 1417.
71. Ali, H.; Rousseau, J.; Ghaffari, M. A.; van Lier, J. E., *J. Med. Chem.* **1991**, *34*, 854.
72. Wang, B. M.; Song, Z. L.; Fan, C. A.; Tu, Y. Q.; Chen, W. M., *Synlett* **2003**, 1497.
73. Gaonkar, S. L.; Rai, K. M. L., *Tetrahedron Lett.* **2005**, *46*, 5969.
74. Bossio, R.; Marcaccini, S.; Pepino, R., *Tetrahedron Lett.* **1995**, *36*, 2325.
75. Chattaway, F. D., *J. Chem. Soc., Perkin Trans. 1* **1905**, 145.
76. Campbell, M. M; Johnson, G., *Chem. Rev.* **1978**, *78*, 65.
77. Bishop, E; Jennings, V. J, *Talanta* **1958**, *1*, 197.
78. Morris, J. C; Salazar, J. A; Wineman, M. A, *J. Am. Chem. Soc.* **1948**, *70*, 2036.
79. Pryde, B. G.; Soper, F. G., *J. Chem. Soc.* **1931**, 1514.
80. Hardy, F. F.; Johnston, J. P., *J. Chem. Soc., Perkin Trans. 2* **1973**, 742.
81. Puttaswamy; Jagadeesh, R. V.; *Appl. Catal. A: Gen.* **2005**, *292*, 259.
82. Puttaswamy; Shubha, J. P.; Jagadeesh, R. V., *Transition Metal Chem.* **2007**, *32*, 991.
83. Puttaswamy; Vinod, K. N.; Ninge Gowda, K. N., *Dyes and Pigments* **2008**, *78*, 131.
84. Cholkar, K.; Kouassi, G. K.; Ananda, S.; Veeraiah, M. K.; Gowda, N. M. M., *Synth. React. Inorg., Met-Org. Nano-Met. Chem.* **2011**, *41*, 1126.
85. Puttaswamy; Jagadeesh, R. V.; Vaz, N.; Radhakrishna, A.; *J. Mol. Catal. A: Chem.* **2005**, *229*, 211.
86. Ramalingaiah; Jagadeesh, R. V.; Puttaswamy, *Catal. Commun.* **2008**, *9*, 1442.
87. Puttaswamy; Jagadeesh, R. V., *Cent. Eur. J. Chem.* **2005**, *3*, 482.
88. Jagadeesh, R. V.; Karthikeyan, P.; Nithya, N.; Sandhya, Y. S; Reddy, S. S.; Reddy, P. P. K.; Kumar, M. V.; Charan, K. T. P.; Nerender, R.; Bhagat, P. R., *J. Mol. Catal. A: Chem.* **2010**, *328*, 99.
89. Puttaswamy; Jagadeesh, R. V., *Ind. Eng. Chem. Res.* **2006**, *45*, 1563.
90. Ramalingaiah; Jagadeesh, R. V.; Puttaswamy, *J. M. Catal. A: Chem.* **2007**, *265*, 70.
91. Vinod, K. N.; Puttaswamy; Gowda, K. N. N., *Ind. Eng. Chem. Res.* **2010**, *49*, 3137.

Cinchonanium, 1-(9-Anthracenyl-methyl)-9-(propen-1-yloxy)-, Bromide (1:1), (8α,9R)

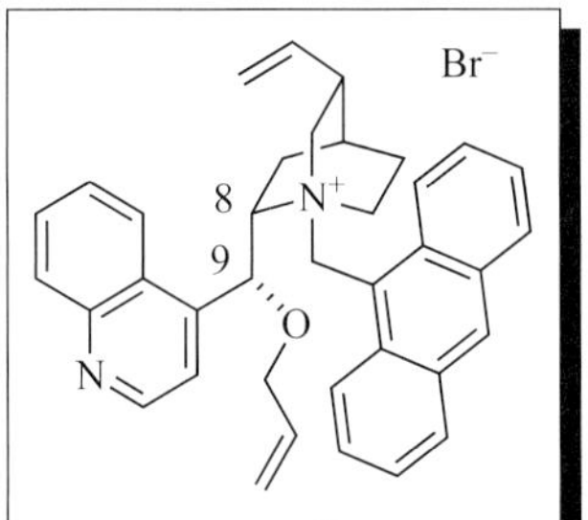

(1)

Cinchonanium, 1-(9-anthracenylmethyl)-9-(2-propen-1-yloxy)-, bromide (1:1), (8*α*,9*R*) (**1**)

[200132-54-3] $C_{37}H_{37}BrN_2O$ (MW 605.61)

InChI = 1S/C37H37N2O.BrH/c1-3-21-40-37(33-17-19-38-35-16-10-9-15-32(33)35)36-23-27-18-20-39(36,24-26(27)4-2)25-34-30-13-7-5-11-28(30)22-29-12-6-8-14-31(29)34;/h3-17,19,22,26-27,36-37H,1-2,18,20-21,23-25H2;1H/q+1;/p-1/t26-,27-,36-,37+,39+;/m0./s1

InChIKey = QOWNPAUSLGATNL-MNSTUQKISA-M

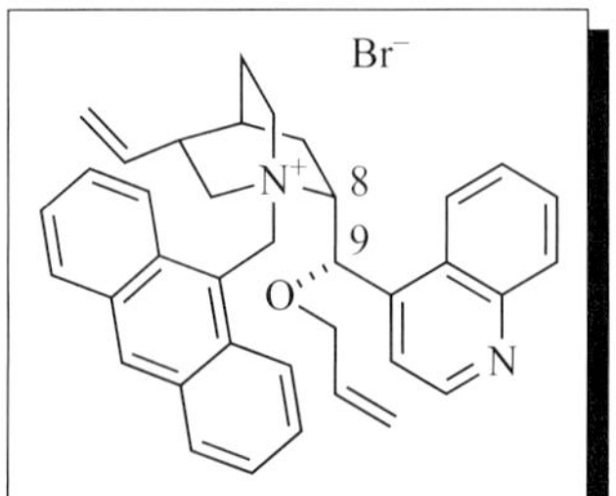

(2)

Cinchonanium, 1-(9-anthracenylmethyl)-9-(2-propen-1-yloxy)-, bromide (1:1), (9*S*) (**2**)

[219601-70-4] $C_{37}H_{37}BrN_2O$ (MW 605.61)

InChI = 1S/C37H37N2O.BrH/c1-3-21-40-37(33-17-19-38-35-16-10-9-15-32(33)35)36-23-27-18-20-39(36,24-26(27)4-2)25-34-30-13-7-5-11-28(30)22-29-12-6-8-14-31(29)34;/h3-17,19,22,26-27,36-37H,1-2,18,20-21,23-25H2;1H/q+1;/p-1/t26-,27-,36+,37-,39+;/m0./s1

InChIKey = QOWNPAUSLGATNL-APXOJCFISA-M

(reagents used as phase-transfer catalysts for a wide range of asymmetric transformations)

Physical Data: (**1**) mp 194–197 °C, $[\alpha]_D^{23}$ −320 (*c* 0.45, $CHCl_3$); (**2**) mp 140–142 °C, $[\alpha]_D^{23}$ +286 (*c* 1.6, $CHCl_3$).

Solubility: sparingly soluble in H_2O and most organic solvents.

Form Supplied in: yellow crystalline solid.

Purification: suspended in methylene chloride, washed with water, dried with sodium sulfate, filtered, rinsed with methylene chloride, and dried under vacuum.

Handling, Storage, and Precautions: store the light-sensitive yellow crystalline material under nitrogen in a darkened glass bottle. Avoid skin, eye, and respiratory contact. Not compatible with strong acid, base, oxidizing, or reducing agents. Toxicity, LD_{50} values have not been determined. Quaternary ammonium salts have a range of biological activities including antibiotic, cytotoxic, anaphylactic shock induction, and neurotransmitter disruption.

Original Commentary

Merritt B. Andrus
Brigham Young University, Provo, UT, USA

Preparation. The cinchonidinium PTC catalyst **1** is readily prepared from inexpensive cinchonidine in two steps (eq 1).[1] Treatment with 9-chloromethylanthracene gives the intermediate quaternary ammonium chloride with a free hydroxyl group remaining at the C9 position. This intermediate can be converted to the corresponding cinchonanium, 1-(9-anthracenylmethyl)-9-(hydroxyl)-, bromide (1:1), (8*α*,9*R*), a catalyst [199588-80-2] developed independently by Lygo, which has also found use as a phase-transfer catalyst.[2] The 9-allyloxy group is installed using allyl bromide with sodium hydroxide followed by ion exchange with sodium bromide. Drying and filtration gives *O*-allyl-*N*-(9-anthracenylmethyl)cinchonidinium bromide, **1** (CD) (eq 1). The corresponding pseudoenantiomeric cinchoninium (CN) catalyst **2**, with the C8 and C9 stereocenters inverted, is also produced using similar *N*- and *O*-alkylation steps. CD and CN cinchona catalyst variations are not enantiomeric in that the vinyl bearing carbons in **1** and **2** have the same absolute stereochemistry.

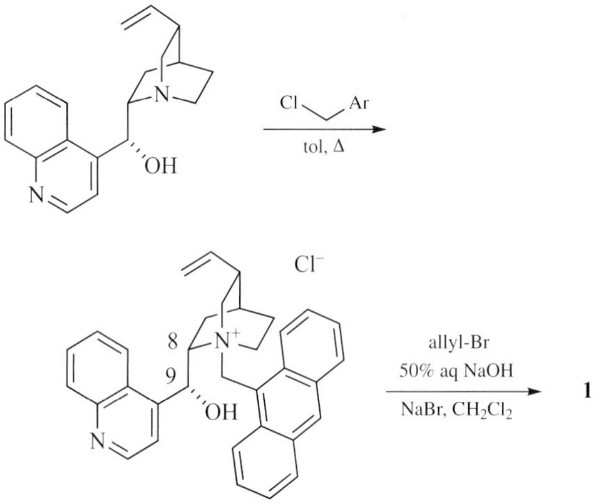

(1)

Alkylation. Cinchona-based compounds **1** and **2** have been widely used as asymmetric phase-transfer catalysts (PTC) for a variety of substrates and applications with alkylation being the most common. Pioneering efforts, using the less sterically hindered *N*-benzyl-*O*-allyl-cinchonadinium bromide catalyst (10 mol %), allowed for *N*-(diphenylmethylene) glycine *t*-butyl ester alkylation in modest selectivity (66% ee).[3] An early PTC study at Merck reported a specific indanone methylation leading to a drug candidate in high selectivity using this early catalyst (94% ee).[4] These examples employed established liquid:liquid PTC conditions using

50% aqueous sodium hydroxide and organic solvents (CH_2Cl_2 or toluene) at room temperature. *N*-Anthracenylmethyl-*O*-allyl-CD **1** was developed by Corey to give highly selective PTC glycine Schiff base alkylation (92–99% ee) using novel solid:liquid conditions with insoluble $CsOH{\cdot}H_2O$ (5 equiv) employed as base in methylene chloride at −78 °C (eq 2).[5] A wide range of primary alkyl, allyl, and benzyl halide electrophiles were successful giving the *S*-glycine alkylation products with **1**. Shortly thereafter, Lygo independently reported the use of the *N*-anthracenylmethyl-9-hydroxy-CD catalyst, with free hydroxyl in place of allyloxy group at C9, for glycine alkylation under liquid:liquid conditions with more modest selectivity (80–90% ee).[6] The low-temperature conditions of Corey using CD catalyst **1** to access *S*-products have become a standard approach to asymmetric amino acid synthesis with many examples and applications now reported.[7]

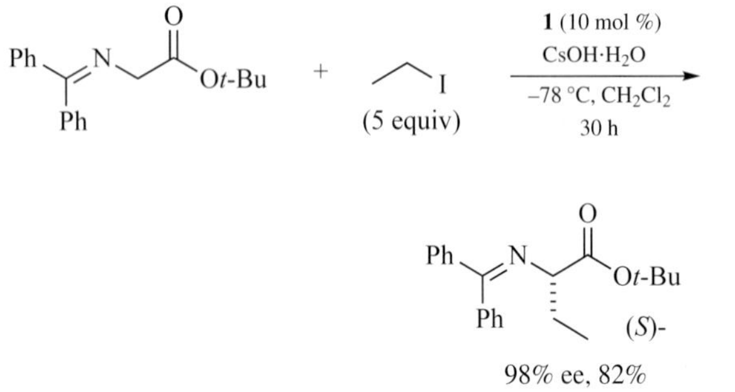

(2)

The complementary *R*-enantiomeric alkylation products can be obtained using the pseudoenantiomeric cinchonine (CN) catalyst **2**. The corresponding 9-hydroxy-*N*-anthracenylmethyl-CN gave the *R*-glycine product using liquid:liquid conditions (86% ee).[8] Other examples using **2** have now been reported that reliably provide the *R*-product in comparable high selectivity using the low-temperature solid:liquid conditions (eq 3).[9] Large-scale

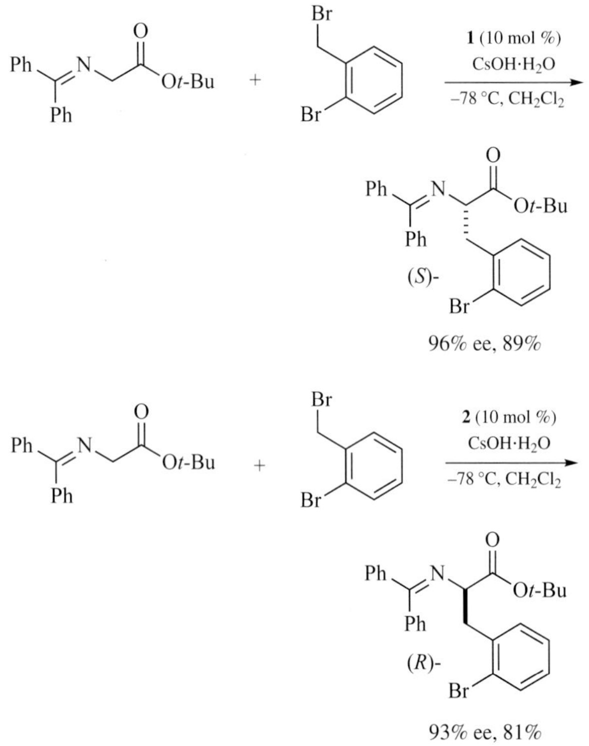

(3)

PTC alkylations have also been performed with **2** using secondary benzhydryl bromide electrophiles (1.5 equiv).[10] High selectivities (99% ee) were obtained in modest yields (56%) when the catalyst and base were added last. Hoffman elimination was determined to be a pathway for catalyst **2** degradation with extended time with base at 0 °C. Ring-opened enol ether products were identified for catalyst **2** degradation in this case.

Conjugate Addition. The enolate of benzophenone imine glycine *t*-butyl ester also adds to methacrylate under low-temperature PTC conditions with catalyst **1** (10 mol %, 5 h) giving conjugate addition products in high selectivity (95% ee) (eq 4).[11] Enones were also shown to be suitable for this transformation. Chalcone substrates also participate in PTC transformations using catalysts of this type for nitromethane anion conjugate addition (eq 5).[12] In this case, the analogous 10,11-dihydro-*O*-benzyl-CD catalyst, with ethyl in place of vinyl, gave higher yield and selectivity.

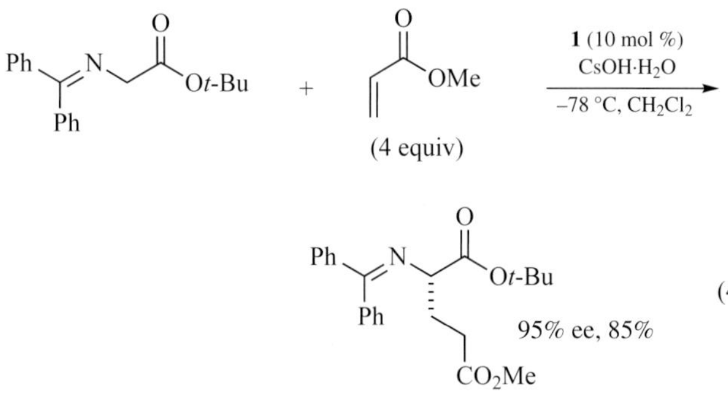

(4)

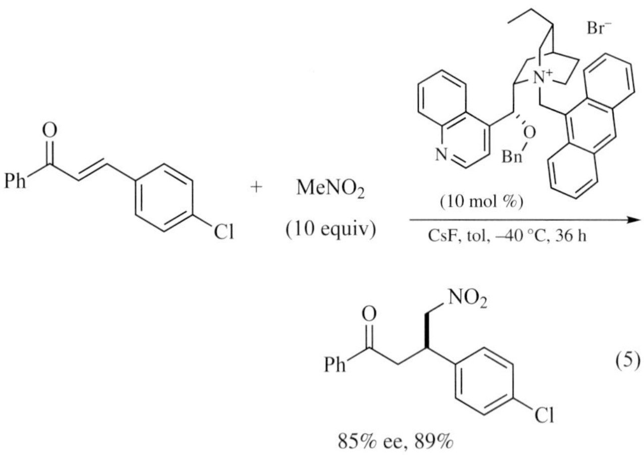

(5)

Tandem conjugate addition–elimination was reported under low-temperature PTC with catalyst **1** for racemic allylic acetates and benzophenone glycine (eq 6).[13] High yields and selectivities were obtained with a range of acrylate electrophiles of this type.

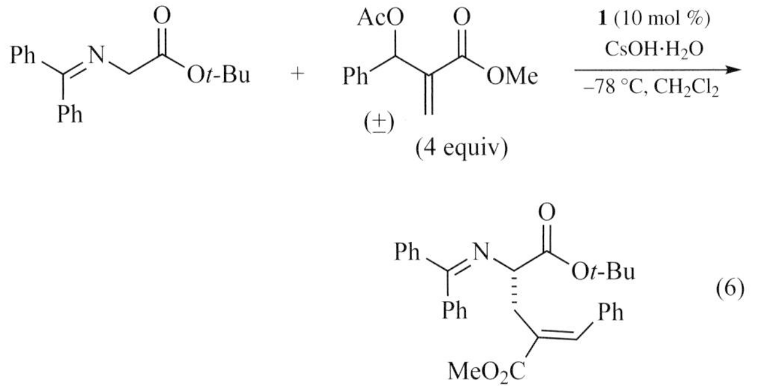

92% ee, 92% (6)

Jørgensen reported various PTC conjugate additions using catalysts **1** and **2** together with the closely related hindered ester 9-*O*-adamantoyl cinchonine (CN) catalysts (3 mol %) with chloroenones, dienones, and activated allene electrophiles (1.3 equiv). *β*-Ketoester and glycine enolates give various products in high selectivity (eq 7).[14] In these cases, the larger 9-*O*-adamantoyl group led to higher selectivities using liquid:liquid conditions with aqueous potassium carbonate at −20 °C compared to **1** and **2**.

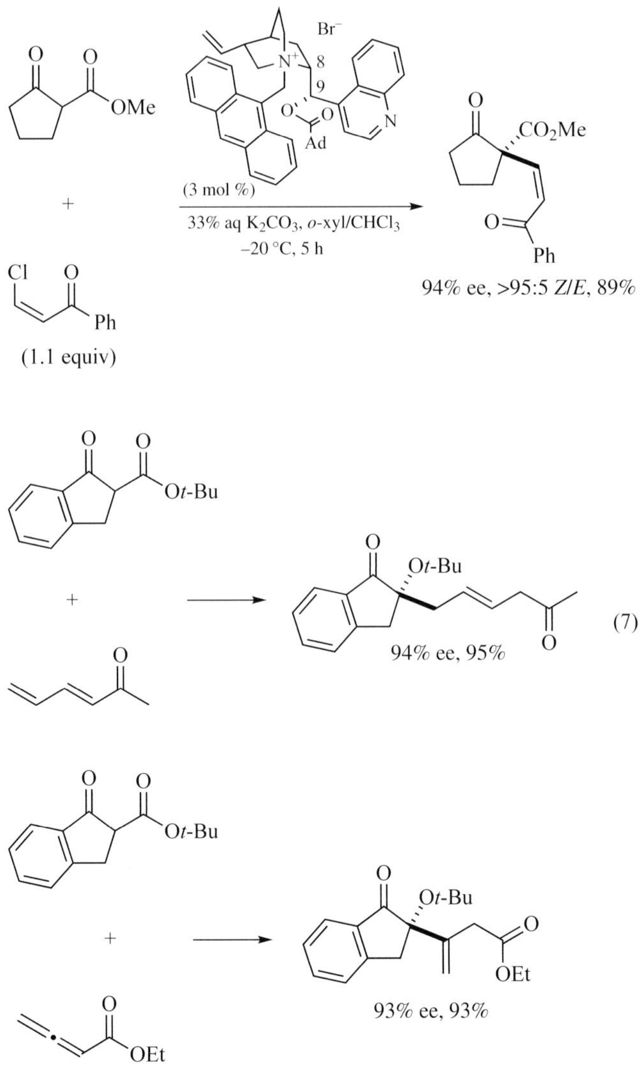

(7)

11,10-Dihydrocinchonium (CN) catalyst **2** was found to give good to high selectivities (80–90% ee) for acrylate and acrylonitrile electrophiles using a nonstabilized glycolate enolate substrate under aqueous potassium hydroxide PTC conditions (eq 8).[15] Very good selectivities and useful yields were obtained with a range of unsaturated electrophiles. The diphenylmethyl protecting group of the product is easily removed and the aryl ketone is converted to the corresponding ester under Baeyer–Villiger conditions.

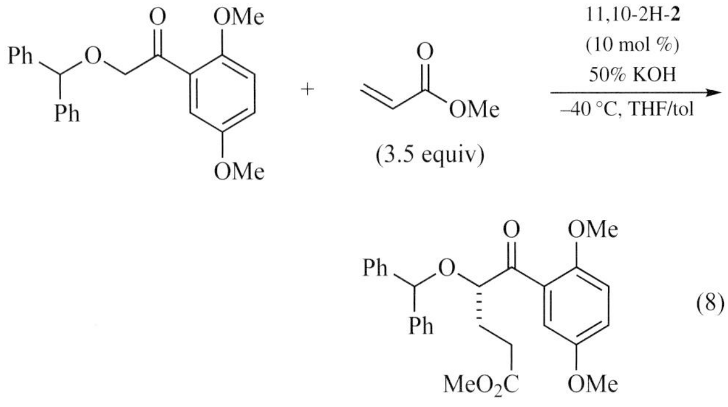

86% ee, 75% (8)

Aldol Reactions. PTC aldol reactions have been reported using the closely related 9-*O*-benzyloxy-CD hydrogen difluoride catalyst (10 mol %) with the silyl ketene acetal of benzophenone glycine (eq 9).[16] Reaction with isobutyraldehyde at −78 °C for 5 h, followed by citric acid treatment, gave the *syn*-amino alcohol product in high selectivity (95% ee) with modest diastereocontrol (6:1) and yield (70%). Castle reported a similar aldol process using **1** as catalyst with free benzophenone imine glycine and a strong phosphazene base to give a mixture of diastereomers in good selectivity (89% ee *syn*).[17] Nitroaldol reactions have also proven successful using **1** (10 mol %) with nitromethane and (*S*)-amino aldehydes with KF as base in methylene chloride (eq 10).[18] The amino alcohol product shown was obtained with very good diastereoselectivity, 9:1 *anti/syn*, in 88% yield.

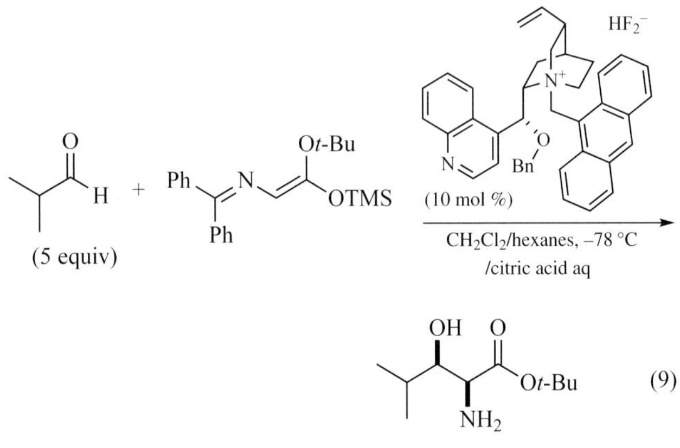

6:1 *syn/anti*, 95% ee, 70% (9)

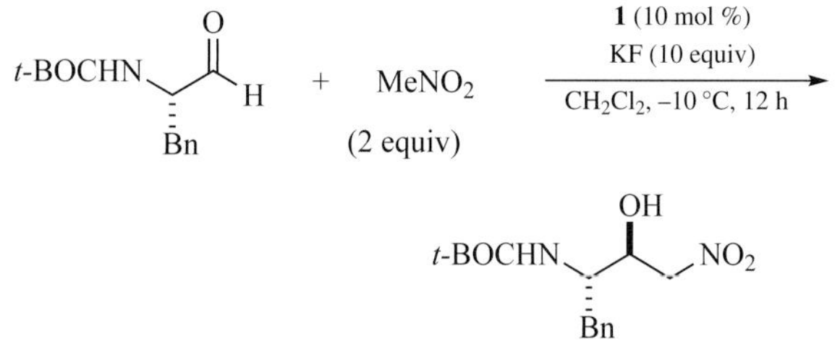

(10)

9:1 *anti/syn*, 88%

Aldol reactions were also performed with a silylenol ether glycolate substrate using hydrogen difluoride **2** in THF to give the *syn* diastereomer in near-complete selectivity with modest enantiocontrol (78% ee) (eq 11).[19] PTC aldol reactions have also been reported for diazoacetate esters and aldehydes using aqueous hydroxide bases with catalyst **1**. Products were obtained in modest yield and selectivity (50–70% ee).[20]

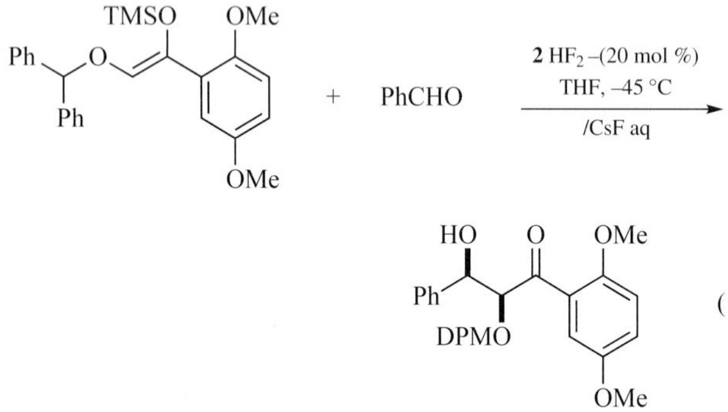

(11)

syn only, 78% ee, 52%

Epoxidation. PTC conditions using cinchonium catalysts have been developed for the epoxidation of activated alkenes. Shioiri reported the use of the early *N*-benzyl-9-*O*-hydroxy cinchonium catalyst (5 mol %) with 30% hydrogen peroxide and lithium hydroxide as base for chalcone (97% yield, 84% ee).[21] Use of *N*-anthracenylmethyl-9-*O*-benzyl cinchonadinium **1** with aqueous sodium hypochlorite and toluene at 25 °C produces the epoxide product with improved selectivity (86% ee) (eq 12).[22] Corey reported a similar system using **1** (10 mol %) and potassium hypochlorite in heptane at −40 °C giving chalcone epoxides with excellent selectivity (91–99% ee).[23] The 9-*O*-benzyl-**1** catalyst has also been reported using sodium hypochlorite for chiral dioxolane enone substrates providing highly diastereoselective epoxidation.[24]

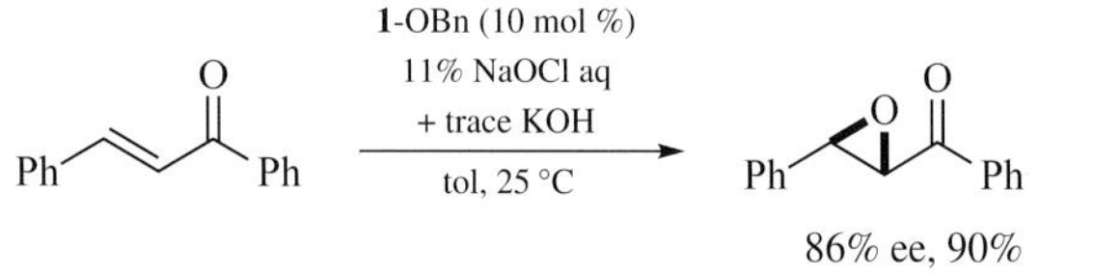

(12)

86% ee, 90%

Hydroxylation. Use of PTC conditions for asymmetric hydroxylation has been reported using **1** in the presence of oxygen and triethyl phosphite or cumyl hydroperoxide as stoichiometric oxidants.[25] When the racemic *N*-*p*-methoxybenzyl (PMB) allyloxindole substrate was treated with the chloride salt of **1** and phosphite under 50% aqueous base/toluene conditions, the alcohol was generated with very good selectivity (83% ee) (eq 13).[25a] Interestingly, when the catalyst was changed to the bromide salt **1**, the selectivity dropped to 20% ee in this case.

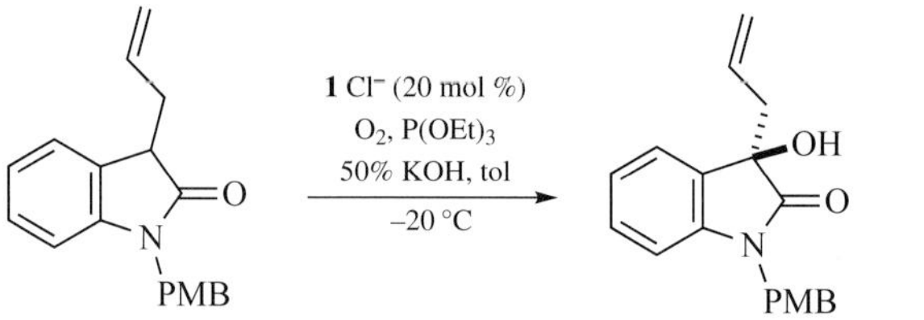

(13)

83% ee, 86%

Miscellaneous Reactions. Other promising transformations have been performed under PTC conditions with the cinchonanium catalysts **1** and **2**, yet to date these have involved only limited study. Asymmetric fluorination was reported with *N*-fluorobenzenesulfonimide and *β*-ketoester substrates at room temperature with potassium carbonate in toluene (eq 14).[26] The fluorinated product was generated in 90% yield and only modest 66% ee (absolute stereochemistry not determined) using catalyst **2** (10 mol %).

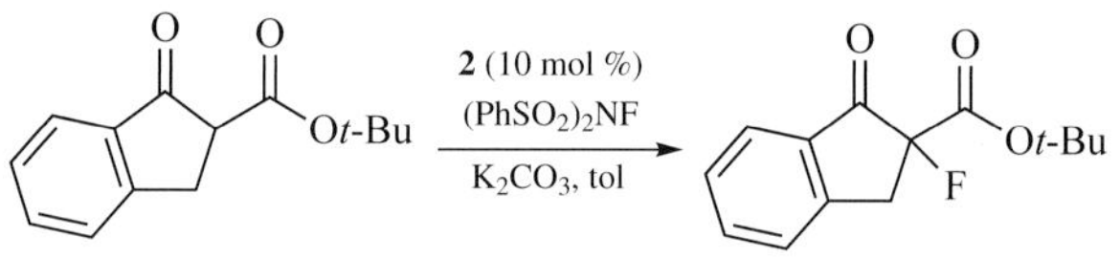

(14)

66% ee, 90%

An intramolecular malonate–benzylimine-type addition, involving a novel cation-directed electrocyclization, has been disclosed using PTC with *N*-anthracenyl catalyst **1**-9-OH-CD (10 mol %) in aqueous potassium carbonate/toluene.[27] The cyclized diester indoline product was obtained in very good selectivity (85% ee) in 96% yield (eq 15). The less hindered *N*-benzyl catalyst version of **1** proved to be more selective in this case at 97% ee.

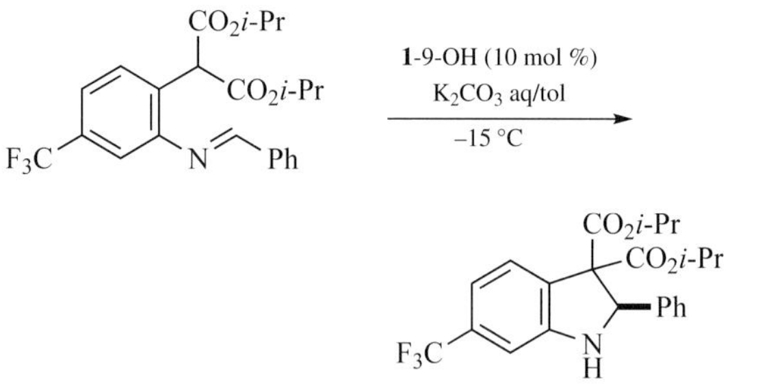

(15)

85% ee, 96%

An ingenious multistep alkylative dearomatization transformation was developed by Porco and coworkers for the total synthesis of a polyprenylated acylphloroglucinol natural product, hyperibone K, using asymmetric PTC.[28] When the precursor clusiaphenone B was treated with the **α**-acetoxy enal shown with **1** (25 mol %) and cesium hydroxide hydrate at low temperature (15 h), the dearomatized adamantane product was obtained with 75% ee in 68% yield (eq 16). A conjugate addition–elimination–intramolecular addition sequence followed by an intramolecular aldol reaction appears to be operative to give the product. An *N,N*-dimeric version of CD catalyst **1**, made from 1,3-bis(bromomethyl)benzene alkylation, was ultimately shown to be more selective at 90% ee for this remarkable multistep process.

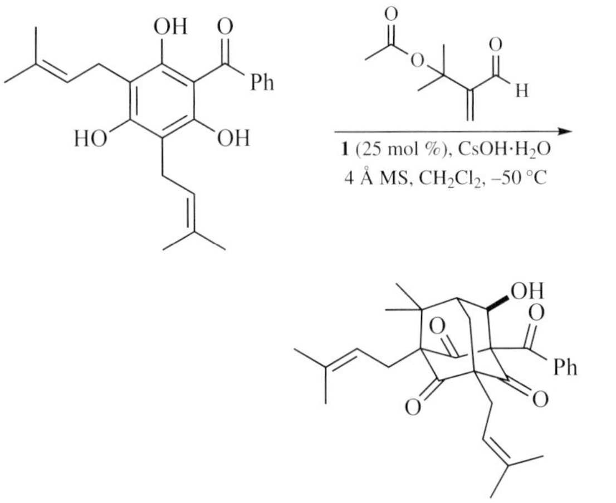

(16)

Sulfenate salt alkylation was disclosed as a route to chiral sulfoxides using PTC conditions with catalyst **1**.[29] The sulfone undergoes β-elimination and the resultant sulfenate anion is methylated to give the product sulfoxide in modest selectivity, 59% ee, and 88% yield (eq 17). An interesting aza-Diels–Alder reaction has been shown to operate under PTC conditions with a dramatic rate enhancement (eq 18).[30] Danishefsky's diene reacted with *N*-phenyl benzylimine at room temperature in acetonitrile with catalyst **1** (10 mol %). The pyridone product was obtained in 97% yield after reacting for only 8 h. In contrast, without catalyst **1**, the yield was only 10% after 72 h at room temperature. Use of tetra-*n*-butylammonium iodide as catalyst gave product in 92% yield after 24 h. Unfortunately, products were racemic in all cases reported using this system.

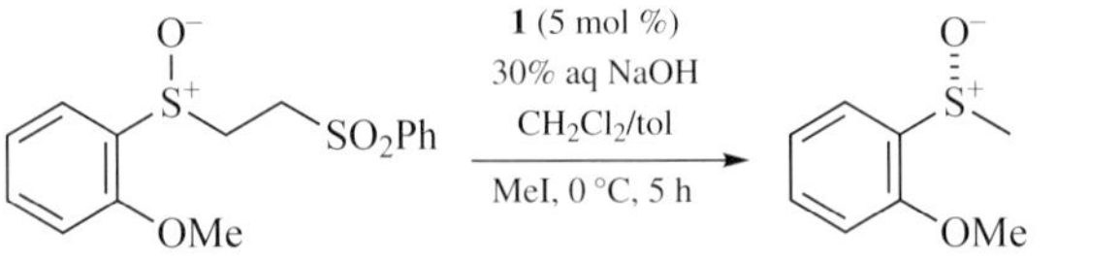

(17)

58% ee, 96%

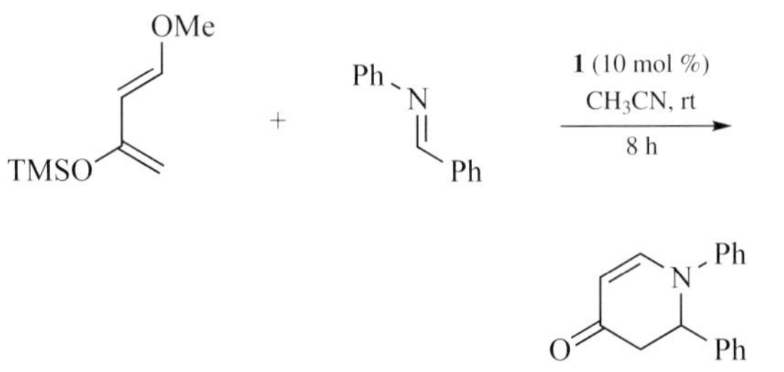

(18)

(±) 97%

Related Reagents. Cinchonanium, 1-(9-phenylmethyl)-9-(propen-1-yloxy)-, bromide(1:1), (8**α**,9*R*); cinchonanium, 1-(9-anthracenylmethyl)-9-(hydroxy)-, bromide(1:1), (8**α**,9*R*); cinchonanium, 1-(9-methyl-(2,3,4-trifluorophenyl)-9-(propen-1-yloxy)-, bromide(1:1), (8**α**,9*R*); tetra-*n*-butylammonium iodide.

First Update

Merritt B. Andrus
Brigham Young University, Provo, UT, USA

Alkylation. Cinchona-based compounds **1** and **2** continue to be widely used as asymmetric phase-transfer catalysts (PTC) for a variety of substrates, and applications with alkylation-based transformations remaining the most common. Low temperature solid–liquid PTC conditions with catalyst **1** were employed for the synthesis of nitro-substituted glycine products leading to quinolinone antibacterial agents (eq 19).[31] A range of *ortho*-nitro benzyl bromides reacted at −30 °C for 18 h with the O'Donnell benzophenone Schiff base glycine with isolated yields from 50 to 92% and 75 to 100% ee. Higher selectivities were obtained at −78 °C; however, greatly extended reactions times were required (72 h) for low to modest yields.

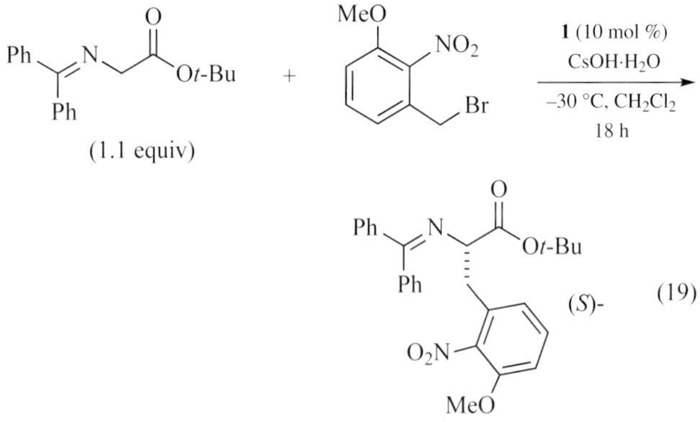

(19)

92%, 96% ee

Asymmetric PTC was employed to access novel pyrazolopyridinones using nitro methyl-bromo-pyrazoles with *tert*-butyl ester glycine in reported high yields and % ee (eq 20).[32] Again −30 °C was used under solid–liquid conditions with insoluble cesium hydroxide hydrate as base in methylene chloride for 18 h. The final pyrazolopyridionones, accessed in five additional steps from the PTC product, were evaluated as kynurenine aminotransferase inhibitors to treat neurological disorders.

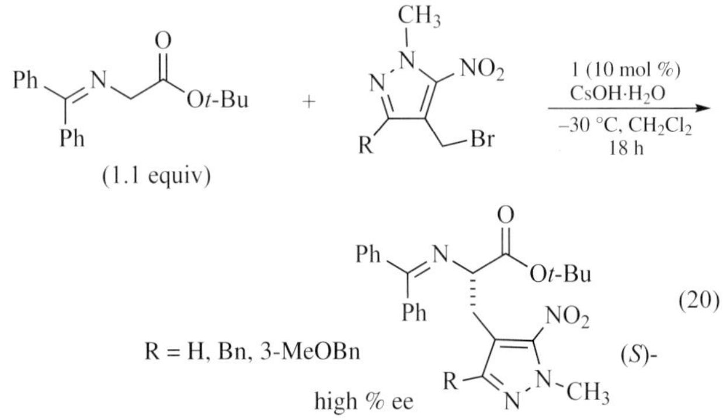

(20)

Ball mill based solvent free conditions were developed for convenient room temperature PTC alkylation using **1** (10 mol %) and solid KOH. Grinding with stainless steel balls operating at 20 Hz converted the glycine imine with allyl bromide to product in 1 h (eq 21).[33] High yields and moderate enantioselectivities (75% ee) were obtained following standard extraction and purification with a series of allyl and benzyl bromide electrophiles. The amount of added base and milling frequency had significant effects on the efficiency of this process.

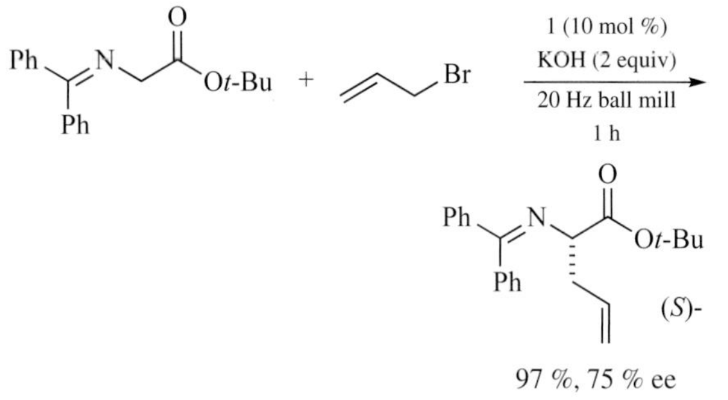

(21)

PTC alkylation of Schiff base glycine was used to access ^{18}F-L-dopa, an important radio-labeled agent to evaluate presynaptic dopaminergic function via PET (positron emission tomography).[34] The key step in an improved automated production employed alkylation of ^{18}F-benzyl iodide under liquid–liquid conditions with the C8-benzyl variation of catalyst **1** shown (eq 22). The PTC labeled L-dopa product was formed after only 5 min at room temperature with high selectivity (97% ee). Catalyst **1** was only slightly inferior at 93% ee. Other cinchona-derived and expensive synthetic biarylammonium catalyst were far less effective. The imine and *tert*-butyl ester were then removed using aqueous HI (15 min) to give the desired ^{18}F-L-dopa radiolabel.

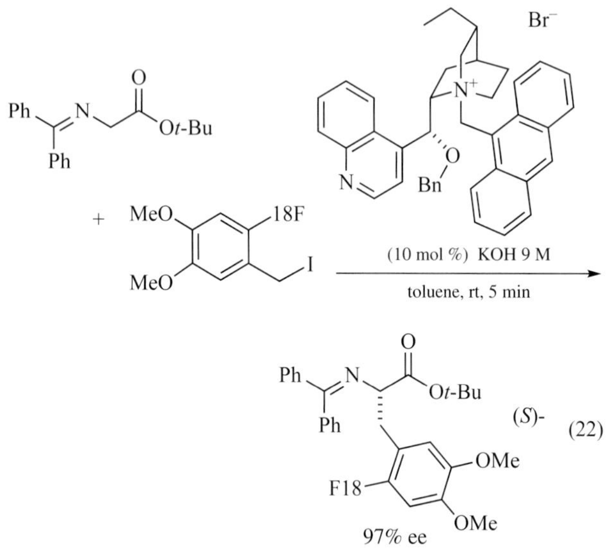

(22)

A photoaffinity trifluoromethyldiazirine-based phenylalanine probe was produced using PTC alkylation (eq 23).[35] Low temperature conditions with BTPP superbase (*N*-*tert*-butyl-tris(pyrrolidino)phosphinimine) where shown to be optimal for the diazirine benzyl bromide electrophile giving product with high selectivity and poor yield. TFA (trifluoroacetic acid) was then used to generate the phenylalanine label target.

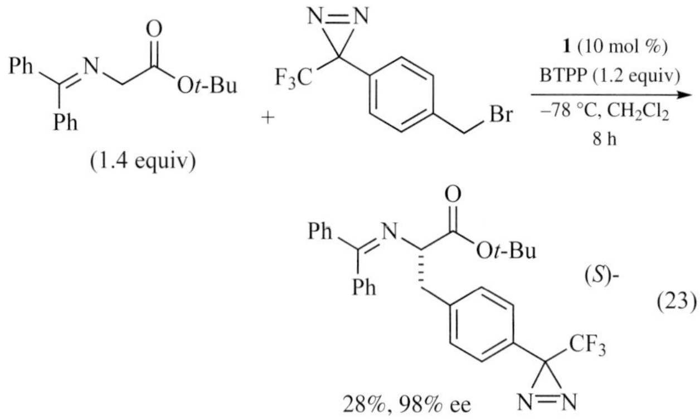

(23)

A new cumyl (dimethylbenzyl) ester glycine Schiff base PTC substrate was developed to allow for more convenient access to the free carboxylic acid following alkylation (eq 24).[36] The cumyl glycine ester was employed in slight excess (1.4 equiv) with a 3-bromomethylpyridine under solid–liquid conditions with catalyst **1** giving product in good yield and high selectivity (94% ee). Other electrophiles reacted in similar manner with good to high yields (56–91%) and very good selectivities (80–90% ee) under these conditions. The cumyl ester product was then treated with palladium hydroxide and hydrogen allowing for *N*-Fmoc carbamate formation without removal of the acid-sensitive TBS ether.

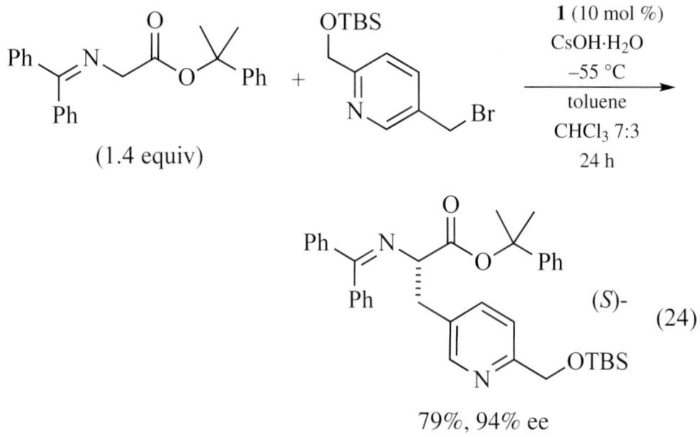

(24)

o-Trifluoromethylphenyl *N*-sulfonylaziridine was used to alkylate hydrindanone *tert*-butyl ester using the C9-adamantoyl chloride variation of cinchonine catalyst **2** (eq 25). Aqueous potassium phosphate in toluene and chloroform at room temperature generated the ring opened *R*-sulfonamide product in good yield and high selectivity (97% ee).[37] Other *β*-keto esters gave products with this sulfonylaziridine electrophile with very good yields (78–88%) and selectivity (82–97% ee). Use of catalyst **2** for the example shown was less selective at 51% ee. A racemic phenyl-substituted substrate allowed for a novel kinetic resolution PTC alkylation with the sulfonylaziridine using cesium carbonate in xylenes at –20 °C (eq 26).[37] The resolved product was obtained in good yield (75%), excellent diastereo- (>95% de), and enantioselectivity (92% ee). Isomerization appears to be faster than alkylation under these conditions. Use of a cyclic sulfamidate as an alternative electrophile also allowed for two carbon ligation along with direct incorporation of an *N*-BOC carbamate group (eq 27).[37] The product was again obtained with good yield and high selectivity using the hydrindanone ester. These products allow for access to a number of alkaloid natural products.

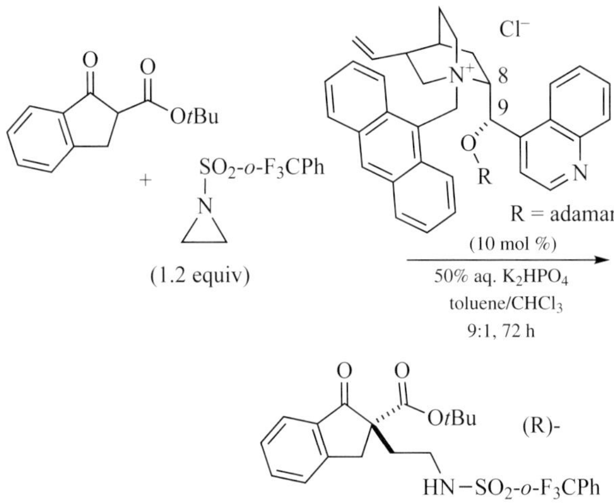

(25)

80%, 97% ee

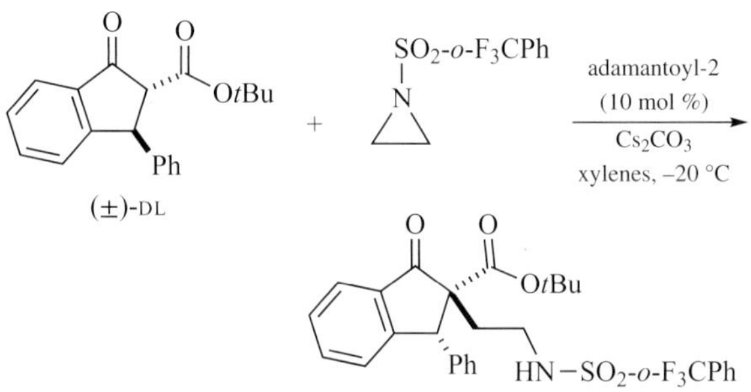

(26)

75%, >95% de, 92% ee

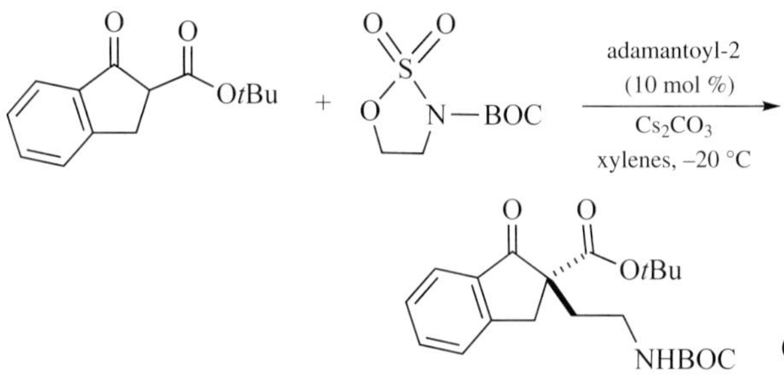

(27)

75%, 93% ee

Tryptophans were produced under asymmetric PTC conditions using quaternary ammonium ions formed in situ from gramines (3-*N,N*-dimethylaminoindoles) reacted with 4-trifluoromethoxybenzyl bromide as electrophiles (eq 28).[38] The benzyl bromide was allowed to react with gramine for 30 min for ammonium salt formation followed by addition of Schiff base glycine, catalyst **1** (20 mol %), and aqueous base (KOH). Protected tryptophan products were produced in good yield and high selectivity. Interestingly, PTC appears to favor enolate•**1** ion pairing and alkylation in this case, even in the presence of the gramine ammonium ion, which could promote racemic alkylation.

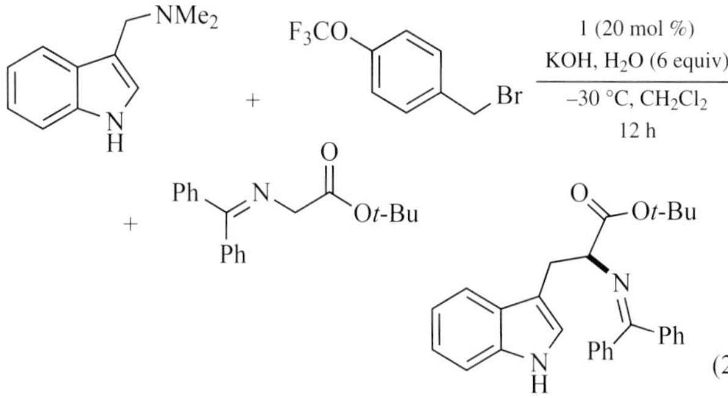

(28)

80%, 92% ee

Alkynylation of activated fluoronitromethylsulfone was performed under PTC condition with **1** using the hypervalent iodine agent 1-[(trimethylsilyl)ethynyl]-1,2-benziodoxol-3(1*H*)-one (TMS-EBX) without added base (eq 29).[39] The substituted propyne product was produced in high yield and modest selectivity (61% ee). In addition to the fluoro group, the appended nitro and sulfone activating groups proved critical to the success of the reaction.

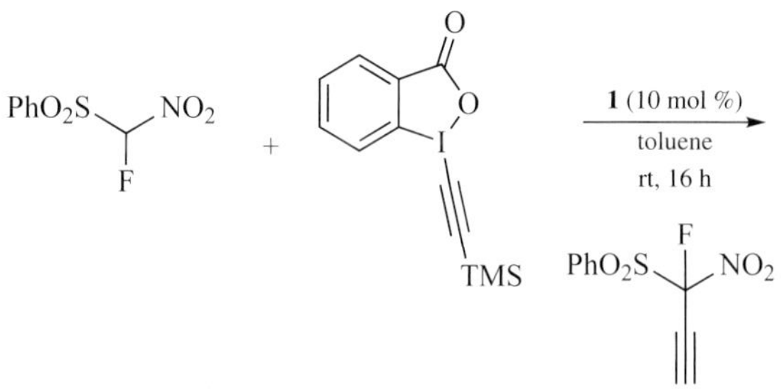

(29)

91%, 61% ee

Protonation. The C9-hydroxyl chloride variation of **1** was shown to be effective for an asymmetric PTC hydrolysis/protonation process with vinyl ester substrates (eq 30).[40] Six- and seven-ring vinyl esters were most effective using aqueous base and chloroethanol at –40 °C in mixed chloroform–mesitylene solvent giving $\boldsymbol{\alpha}$-alkylated ketones in high yield and selectivity (90% ee). Subsequent efforts using silylenol ethers for asymmetric protonation under similar conditions were far less effective.[41]

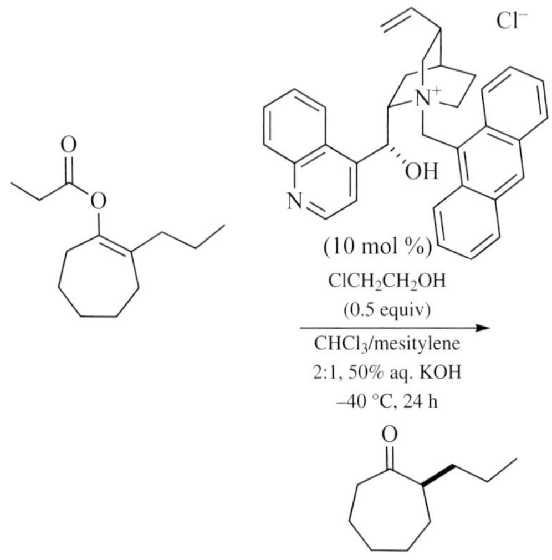

(30)

91%, 90% ee

Aziridination. A combined auxiliary catalyst-based approach was developed for asymmetric acrylamide PTC aziridination (eq 31).[42] A (−)-menthone-derived acylpyrazole reacted with *N*-chloro-*N*-sodio-*tert*-butylcarbamate in a matched manner with **1** at −40 °C to generate aziridine product in good yield and very high diastereomeric ratio (dr). Achiral acrylamides and other chiral auxiliaries in combination with PTC catalysts proved to be less selective. Treatment of the product with methanol and DMAP produced the corresponding aziridine methyl ester (78%) without racemization.

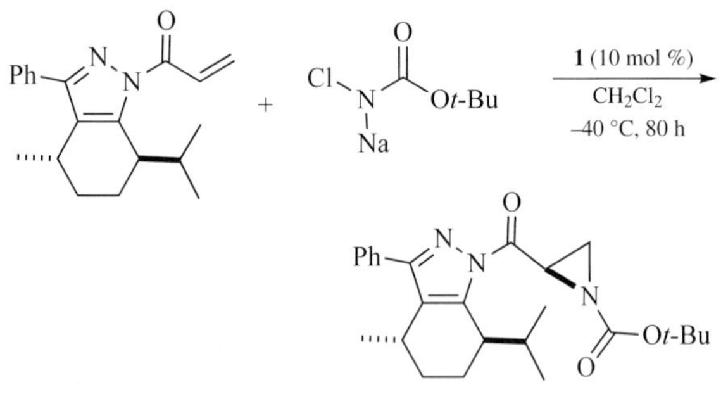

(31)

72%, >99:1 dr

Conjugate Addition. Asymmetric conjugate additions of 3-ester *N-n*-butyl isoindolinones to methylvinyl ketone were performed using the C9-hydroxy chloride variation of **1** (76% ee) (eq 32). Use of **1** as catalyst was far less effective (11% ee).[43] Other unsaturated electrophiles were explored with similar outcomes giving various 3,3-disubstituted isoindolinone products (60–97%, 0–76% ee).

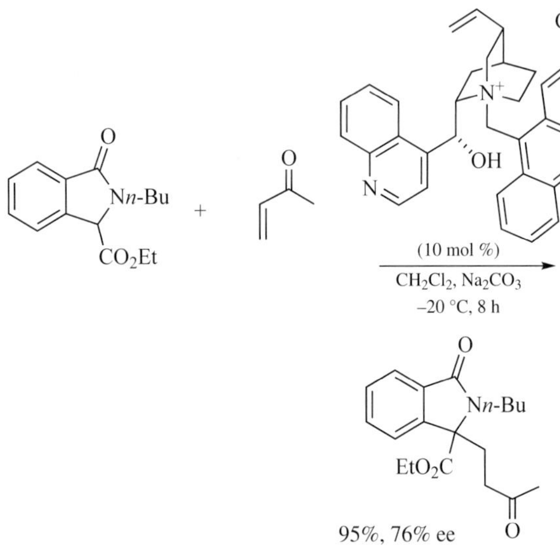

(32)

95%, 76% ee

Hydroxylation. The C9-adamantoyl version of cinchonine catalyst **2** was employed in asymmetric PTC hydroxylations of cyclic β-keto esters using cumyl peroxide (eq 33).[44] Liquid–liquid conditions with aqueous potassium phosphate in toluene provided hydroxylated products in very good yield (83%) and selectivity (90% ee). Other substrate variations, including acyclic β-keto adamantly esters gave similar results (35–95%, 63–90% ee).

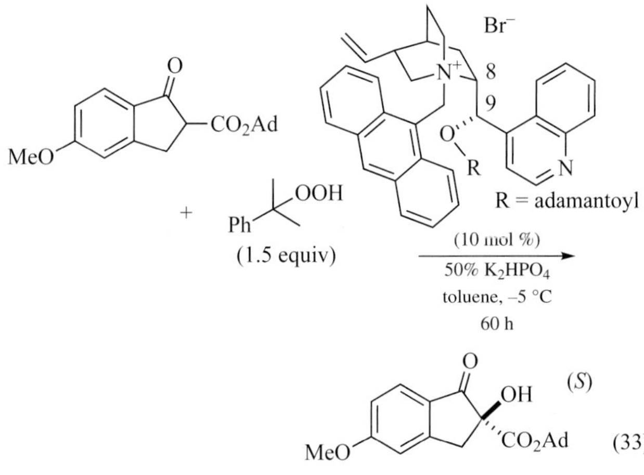

(33)

83%, 90% ee

Cycloaddition. Catalyst **1** was employed for intramolecular nitrile oxide cycloaddition to form 16-membered intermediates leading to macrosphelide natural products with reversed, modest diastereoselectivity (eq 34).[45] These compounds which possess considerable therapeutic potential were produced from oxime acrylate substrates and **1** (1.2 equiv) generating isoxazole products in good yield (72%) and modest selectivity (1.2:1.0 de) as shown. In the absence of PTC under comparable sodium hypochlorite conditions, the selectivity was reversed to 1.0:2.0 (96%). Known Lewis-acid titanium, magnesium, and zinc catalysts also gave product favoring the reversed diastereomer. Interestingly, only the PTC catalyst **1** favored the isomer indicated.

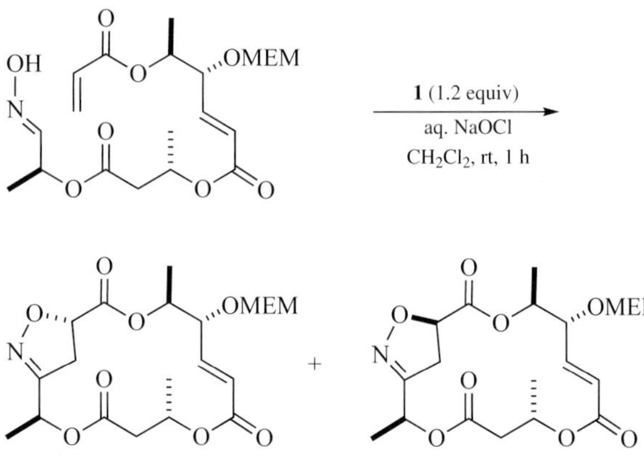

(34)

72%, 1.2:1

1. Corey, E. J.; Noe, M. C., *Org. Synth. Vol.* **2009**, *11*, 404.
2. Lygo, B.; Crosby, J.; Lowdon, T. R.; Wainwright, P. G., *Tetrahedron* **2001**, *57*, 2391.
3. O'Donnell, M. J.; Bennett, W. D.; Wu, S., *J. Am. Chem. Soc.* **1989**, *111*, 2353.
4. Hughes, D. L.; Dolling, U. H.; Ryan, K. M.; Schoenewaldt, E. F.; Grabowski, E. J. J., *J. Org. Chem.* **1987**, *52*, 4745.
5. Corey, E. J.; Xu, F.; Noe, M. C., *J. Am. Chem. Soc.* **1997**, *119*, 12414.
6. Lygo, B.; Wainwright, P. G., *Tetrahedron Lett.* **1997**, *38*, 8595.
7. (a) Boeckman, R. K., Jr.; Clark, T. J.; Shook, B. C., *Org. Lett.* **2002**, *4*, 2109. (b) Lépine, R.; Zhu, J., *Org. Lett.* **2005**, *7*, 2981. (c) Waldmann, H.; He, Y. P.; Tan, H.; Arve, L.; Arndt, H. D., *Chem. Commun.* **2008**, 5562.

(d) Wu, Y. C.; Liron, M.; Zhu, J., *J. Am. Chem. Soc.* **2008**, *130*, 7148. (e) Benohoud, M.; Leman, L.; Cardoso, S. H.; Retailleau, P.; Dauban, P.; Thierry, J.; Dodd, R. H., *J. Org. Chem.* **2009**, *74*, 5331.

8. Lygo, B.; Crosby, J.; Lowdon, T. R.; Peterson, J. A.; Wainwright, P. G., *Tetrahedron* **2001**, *57*, 2403.
9. (a) Viwanathan, R.; Prabhakaran, E. N.; Plotkin, M. A.; Johnston, J. N., *J. Am. Chem. Soc.* **2003**, *125*, 163. (b) Armstron, A.; Scutt, J. N., *Org. Lett.* **2003**, *5*, 2331. (c) Lemaire, C.; Gillet, S.; Guillouet, S.; Plenevaux, A.; Aerts, J.; Luxen, A., *Eur. J. Org. Chem.* **2004**, 2899.
10. Patterson, D. E.; Xie, S.; Jones, L. A.; Osterhout, M. H.; Henry, C. G.; Roper, T. D., *Org. Proc. Res. Devel.* **2007**, *11*, 624.
11. Corey, E. J.; Noe, M. C.; Xu, F., *Tetrahedron Lett.* **1998**, *39*, 5347.
12. (a) Corey, E. J.; Zhang, F. Y., *Org. Lett.* **2000**, *2*, 4257. (b) Zhang, F. Y.; Corey, E. J., *Org. Lett.* **2004**, *6*, 3397.
13. Ramachandran, V. P.; Madhi, S.; Bland-Berry, L.; Reddy, M. V. R., O'Donnell, M. J., *J. Am. Chem. Soc.* **2005**, *127*, 13450.
14. (a) Poulsen, T. B.; Bernardi, L.; Bell, M.; Jørgensen, K. A., *Angew. Chem., Int. Ed.* **2006**, *45*, 6551. (b) Bernardi, L.; Lopez-Cantarero, J.; Niess, B.; Jørgensen, K. A. *J. Am. Chem. Soc.* **2007**, *129*, 5772. (c) Elsner, P.; Bernardi, L.; Salla, G. D.; Overgaard, J.; Jørgensen, K. A., *J. Am. Chem. Soc.* **2008**, *130*, 4897.
15. Andrus, M. B.; Ye, Z., *Tetrahedron Lett.* **2008**, *49*, 534.
16. Horikawa, M.; Busch-Petersen, J.;Corey, E. J., *Tetrahedron Lett.* **1999**, *40*, 3843.
17. Mettath, S.; Srikanth, G. S. C., Dangerfield, B. S.; Castle, S. L., *J. Org. Chem.* **2004**, *69*, 6489.
18. Corey, E. J.; Zhang, F. Y., *Angew. Chem., Int. Ed.* **1999**, *38*, 1931.
19. Andrus, M. B.; Liu, J.; Ye, Z.; Cannon, J. F., *Org. Lett.* **2005**, *7*, 3861.
20. Hasegawa, K.; Arai, S.; Nishida, A., *Tetrahedron* **2006**, *62*, 1390.
21. Arai, S.; Tsuge, H.; Shioiri, T., *Tetrahedron Lett.* **1998**, *39*, 7563.
22. Lygo, B.; Wainwright, P. G., *Tetrahedron* **1999**, *55*, 6289.
23. Corey, E. J.; Zhang, F. Y., *Org. Lett.* **1999**, *1*, 1287.
24. Lygo, B.; Gardiner, S. D.; McLeod, M. C.; To, D. C. M., *Org. Biomol. Chem.* **2007**, *5*, 2283.
25. (a) Sano, D.; Nagata, K.; Itoh, T., *Org. Lett.* **2008**, *10*, 1593. (b) Lian, M.; Li, Z.; Meng, Q.; Gao, Z., *Eur. J. Org. Chem.* **2010**, 6525.
26. Kim, D. Y.; Park, E. J., *Org. Lett.* **2002**, *4*, 545.
27. Maciver, E. E.; Thompson, S.; Smith, M. D., *Angew. Chem., Int. Ed.* **2009**, *48*, 9979.
28. Qi, J.; Beeler, A. B.; Zhang, Q.; Porco, J. A., Jr., *J. Am. Chem. Soc.* **2010**, *132*, 13642.
29. Gelat, F.; Jayashankaran, J.; Lohier, J. F.; Gaumont, A. C.; Perrio, S., *Org. Lett.* **2011**, *13*, 3170.
30. Park, Y.; Park, E.; Jung, H.; Lee, Y. J.; Jew, S. s.; Park, H. g., *Tetrahedron* **2011**, *67*, 1166.
31. McAllister, L. A.; Bechle, B. M.; Dounay, A. B.; Evrard E.; Gan, X.; Ghosh, S.; Kim, J. -Y.; Parikh, V. D.; Tuttle, J. B.; Verhoest, P. R., *J. Org. Chem.* **2011**, *76*, 3484.
32. Dounay, A. B.; Anderson, M; Bechle, B. M.; Evrard, E.; Gan, X.; Kim, J. -Y.; McAllister, L. A.; Pandit, J.; Rong, S.; Salafia M. A.; Tuttle, J. B.; Zawadzke, L. E.; Verhoest, P. R., *Bioorg. Med. Chem. Lett.* **2013**, *23*, 1961.
33. Nun, P.; Pérez, V.; Calmès, M.; Martinez, J.; Lamaty, F., *Chem. Eur. J.* **2012**, *18*, 3773.
34. Libert, L. C.; Franci, X.; Plenevaux, A. R.; Ooi, T.; Maruoka, K.; Luxen, A. J.; Lemaire, C. F., *J. Nuc. Med.* **2013**, *54*, 1154.
35. Wang, L.; Hisano, W.; Murai, Y.; Sakurai, M.; Muto, Y.; Ikemoto, H.; Okamoto, M.; Murotani, T.; Isoda, R.; Kim, D.; Sakihama, Y.; Sitepu, I. R.; Hashidoko, Y.; Hatanaka, Y.; Hashimoto, M., *Molecules* **2013**, *18*, 8393.
36. Respondek, T.; Cueny, E.; Kodanko, J. J., *Org. Lett.* **2012**, *14*, 150.
37. Moss, T. A.; Barber, D. M.; Kyle, A. F.; Dixon, D. J., *Chem. Eur. J.* **2013**, *19*, 3071.
38. Todd, R.; Huisman, M.; Uddin N.; Oehm, S.; Hossain, M. M., *Synlett* **2012**, *23*, 2687.
39. Kamlar, M.; Putaj, P.; Veselý, J., *Tetrahedron Lett.* **2013**, *54*, 2097.
40. Yamamoto, E.; Nagai, A.; Hamasaki, A.; Tokunaga, M., *Chem. Eur. J.* **2011**, *17*, 7178.
41. Claraz, A.; Landelle, G.; Oudeyer, S.; Levacher, V., *Eur. J. Org. Chem.* **2013**, 7693.
42. Murakami, Y.; Takeda, Y.; Minakata, S., *J. Org. Chem.* **2011**, *76*, 6277.
43. Scorzelli, F.; Di Mola, A.; Palombi, L.; Massa, A., *Molecules* **2015**, *20*, 8484.
44. Yao, H.; Lian, M.; Li, Z.; Wang, Y.; Meng, Q., *J. Org. Chem.* **2012**, *77*, 9601.
45. Paek, S.-M.; Suh, Y.-G., *Molecules* **2011**, *16*, 4850.

1,2-Cyclohexanediamine, N^1,N^2-Bis[4-(1-pyrrolidinyl)-2-quinolinyl]-, (1*R*,2*R*)-

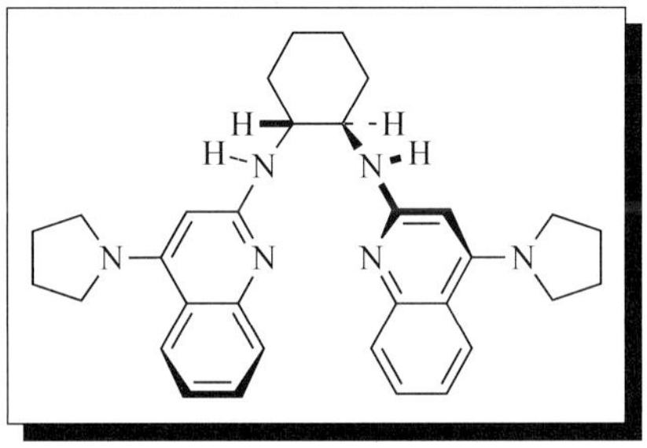

[1214287-91-8] $C_{32}H_{38}N_6$ MW 506.68

InChI = 1S/C32H38N6/c1-3-13-25-23(11-1)29(37-17-7-8-18-37)21-31(33-25)35-27-15-5-6-16-28(27)36-32-22-30(38-19-9-10-20-38)24-12-2-4-14-26(24)34-32/h1-4,11-14,21-22,27-28H,5-10,15-20H2,(H,33,35)(H,34,36)/t27-,28-/m1/s1

InChIKey = LAROGWWYQDAUPL-VSGBNLITSA-N

(reagent used as a bifunctional organocatalyst for asymmetric catalysis, typically in combination with a Brønsted acid)

Alternate Names: (*R*,*R*)-PBAM, PBAM.
Physical Data: (mp 241–246 °C, $[\alpha]_D^{25}$ +398 (*c* 1.2, $CHCl_3$))
Solubility: dichloromethane, toluene.
Form Supplied in: light brown solid.
Analysis of Reagent Purity: IR, ^{1}H NMR, ^{13}C NMR, mp, optical rotation.
Preparative Methods: 1,2-cyclohexanediamine, N^1,N^2-bis[4-(1-pyrrolidinyl)-2-quniolinyl]-, (1*R*,2*R*)- (PBAM) was prepared first in 2010 by Johnston and coworkers in two steps (eqs 1 and 2) from commercially available starting materials.[1] A flame-dried flask was charged with $Pd(dba)_2$ (217 mg, 378 μmol), *rac*-BINAP (235 mg, 378 μmol), sodium *tert*-butoxide (2.180 g, 22.68 mmol), (*R*,*R*)-diaminocyclohexane (865 mg, 7.57 mmol), and 2,4-dichloroquinoline (3.000 g, 15.22 mmol), and the reaction vessel was placed under an argon atmosphere. Toluene (90 mL) was added, and the resulting red-brown solution was heated at 85 °C. The reaction was monitored by TLC; after 3 h, nearly complete conversion was observed. The reaction was cooled to 25 °C, diluted with EtOAc, and washed with saturated aqueous NH_4Cl. The organic layer was dried (Na_2SO_4)

and concentrated. Column chromatography (SiO_2, 10% ethyl acetate in hexanes) of the residue provided **1** (1.8 g, 54%) as a yellow powder. A 10–20 mL microwave vial was charged with **1** (600 mg, 1.37 mmol), pyrrolidine (2.24 mL, 27.3 mmol), and trifluoromethylbenzene (9 mL). This suspension was heated at 200 °C and stirred in the microwave for 3.5 h. The reaction was then concentrated and purified by column chromatography (5–20% methanol in dichloromethane) to provide a light brown solid. This material was dissolved in dichloromethane and then washed with 3 M aqueous NaOH. The combined organic layers were dried over sodium sulfate and concentrated. The material was then triturated with hexanes to afford a light brown powder (635 mg, 92%). A more detailed procedure for the synthesis of a 3–4 g batch of (*R*,*R*)-PBAM has been published.[2] The opposite enantiomer of catalyst can be obtained through the same sequence starting with (*S*,*S*)-diaminocyclohexane.

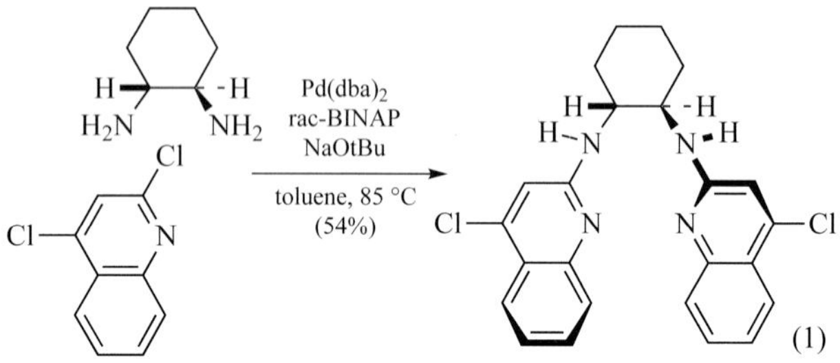

(1)

1

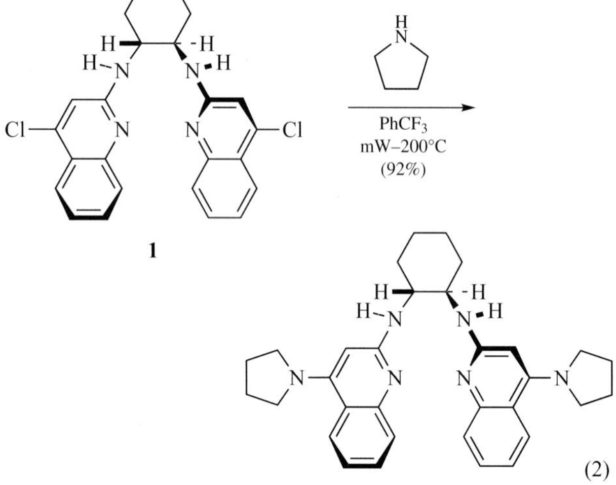

(2)

(*R*,*R*)-PBAM

Purification: recrystallization from 90:10 acetone:toluene, followed by extended warming under vacuum to remove residual acetone

Handling, Storage, and Precautions: indefinitely stable in air at room temperature

---

Introduction. The chiral bisamidine (BAM) catalyst family was introduced in 2004. The acid salts of these catalysts feature a unique polar ionic hydrogen bond. These organocatalysts promote asymmetric reactions through hydrogen bonding, as both donors and acceptors of hydrogen donors. In this initial study, the triflic acid salt of a bisquinoline catalyst (H,QuinBAM·HOTf, eq 3) promoted the asymmetric aza-Henry (nitro-Mannich) reaction of nitromethane and nitroethane to aryl *N*-Boc imines with good enantioselectivity, diastereoselectivity, and yield.[3] An advantage of these bisamidine catalysts is that they are amenable to steric and electronic modification. The addition of electron-withdrawing or electron-donating groups to the quinoline ring can respectively decrease or increase the pK_a of the protonated BAM catalyst.[4] The installation of a pyrrolidine at the 4-position of the quinoline led to the introduction of 2-cyclohexanediamine, N^1,N^2-bis[4-(1-pyrrolidinyl)-2-quinolinyl]-, (1*R*,2*R*)- ((*R*,*R*)-PBAM) in 2010. PBAM is a more Brønsted basic catalyst that has expanded the utility and scope of this catalyst.

For the sake of a complete discussion of the capabilities of this catalyst, this article will address the catalysis of the unprotonated (free base) and various protonated forms (PBAM·HX) of PBAM. For simplicity, (*R*,*R*)-PBAM will be referred to as PBAM. Any use of the other enantiomer of PBAM will be explicitly stated.

Increased Reactivity in the Aza-Henry Reaction. The original report of the PBAM catalyst demonstrated its increased reactivity in the aza-Henry reaction.[1] Organocatalysts, including chiral thioureas, cinchona alkaloids, phosphoric acids, and sulfoxides, have been used to achieve the diastereo- and enantioselective aza-Henry reaction.[5] The vast majority of these reactions featured the use of 5–10 equiv of nitroalkane and reaction times ≥2 days. The previous generation of BAM catalyst (H,QuinBAM·HOTf) required the nitroalkane nucleophile to be used as solvent (56 equiv) and relatively long reaction times to obtain good conversion and yields of **3** (eq 3). With 1.5 equiv of nitroethane in toluene, full conversion was achieved in less than 24 h with PBAM·HOTf, whereas H,QuinBAM·HOTf provided only a trace of product by GC (Table 1). Furthermore, good diastereo- and enantioselectivity (7:1 dr, 90% ee) were observed with PBAM·HOTf.

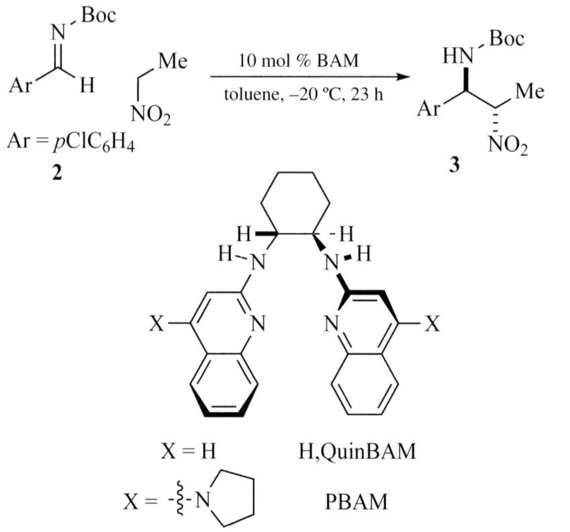

(3)

BAM	3 h conversion (%)	23 h conversion (%)	dr *anti:syn*	ee(%)	yield(%)
H,QuinBAM	n/a	1	n/a	n/a	n/a
H,QuinBAM·HOTf	n/a	2	n/a	n/a	n/a
PBAM	21	100	1:1	71	73
PBAM·HOTf	35	100	7:1	90	78

Aza-Henry Addition of Simple, Unactivated Nitroalkanes. The increased reactivity of PBAM allowed a wider scope of nitroalkanes to be utilized in the asymmetric aza-Henry reaction under more desirable reaction conditions. The triflic acid salt of PBAM with a 2:3 ligand:triflic acid ratio was found to be optimal for the highest levels of diastereoselection. Nitroethane addition products **4** with a variety of aryl and heteroaryl *N*-Boc aldimines were formed with good diastereoselectivity, enantioselectivity, and yield (eq 4). The use of near stoichiometric amounts of nitroalkane (1.5 equiv) allows the practical use of a wider range of nitroalkanes. PBAM·HOTf promotes the addition of four different primary nitroalkanes (eq 5) to give products **5** in 7–23:1 dr, 88–91% ee, and 61–85% yield.

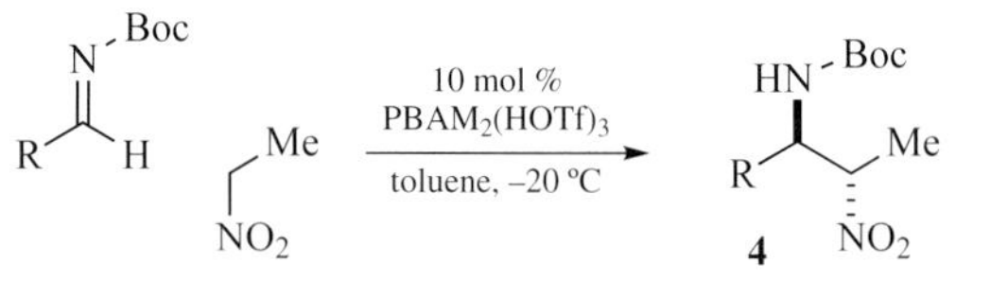

(4)

R = aryl, heteroaryl — 8–35:1 dr, 82–95% ee, 63–100%

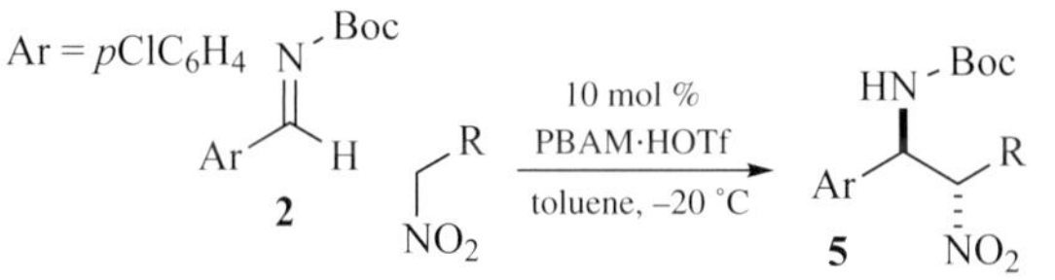

(5)

R = Et, ^{n}Pr, ^{i}Pr, C_6H_{11} — 7-23:1 dr, 88–91% ee, 61–85%

Addition of Secondary Nitroalkanes. The aza-Henry addition of secondary nitroalkanes leads to tertiary nitroalkane products that have the potential for conversion to tertiary amines or denitration. Unactivated secondary nitroalkanes are much less reactive than primary nitroalkanes as nucleophiles. Due to this fact, only one other example (that of Jacbosen and Yoon[6]) of an addition of an unactivated secondary nitroalkane exists in the literature. No reactivity was observed when attempting the addition of 2-nitropropane to aryl *N*-Boc imines with H,QuinBAM or H,QuinBAM·HOTf. The reactivity of PBAM·HOTf allows this reaction to occur at room temperature to give promising enantioselectivity and yield of **6** (eq 6).[1] PBAM·HOTf also promoted the addition of *N*-protected 3-nitroazetidine **8** to *N*-Boc imine **7** (eq 7).[7] Appreciable enantioselectivity of addition product **9** was observed at room temperature (78% ee), but the high reactivity of PBAM·HOTf permitted a lowering of the reaction temperature to increase enantioselection. At −20 °C, the product **9** was formed with 86% ee and in 87% yield. The triflic acid salt of a 7-methoxy- derivative of PBAM provided even higher yield and enantioselectivity (92% ee, 93% yield). Azetidine **9** can be converted into the potent GlyT1 inhibitor **10** after denitration and two deprotection/acylation sequences (eq 8).

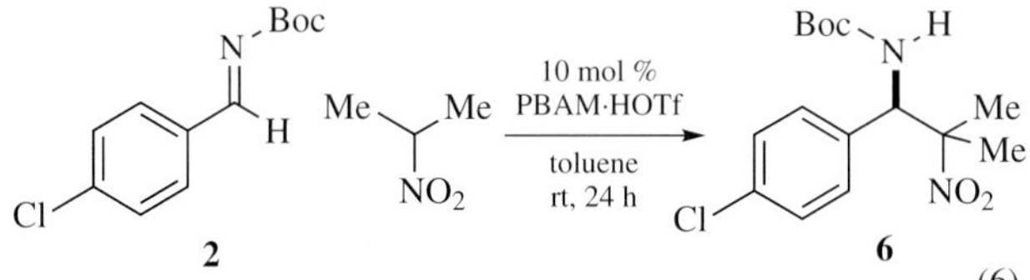

(6)

71% ee, 67% yield

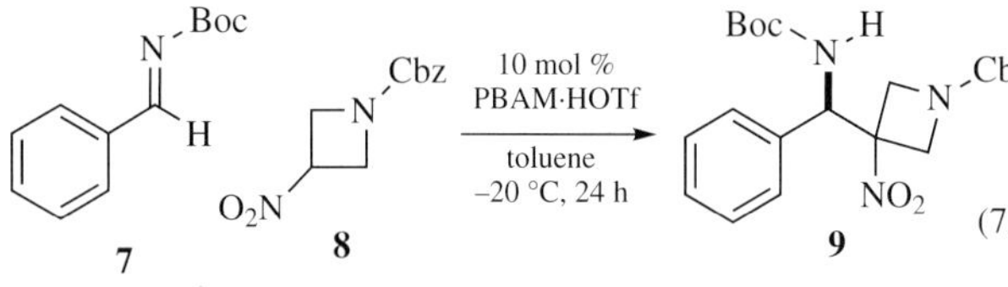

(7)

86% ee, 87% yield

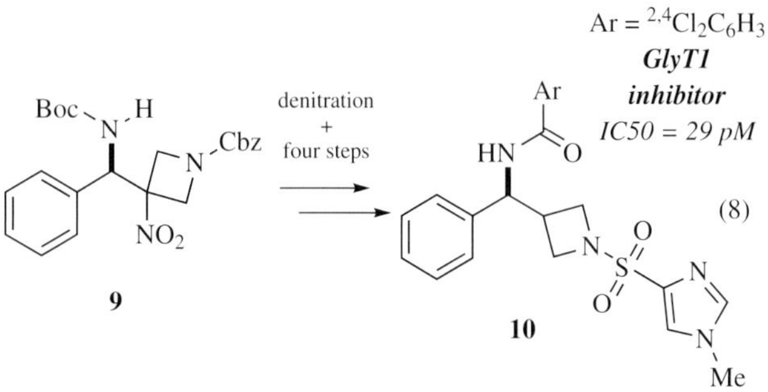

(8)

Aza-Henry Addition of Bromonitromethane. The increased reactivity of PBAM resulted in the aza-Henry addition products that became competitive nucleophiles themselves. $PBAM_4(HOTf)_5$ promoted this reaction of nitromethane to imine **11** to provide a 2.7:1 mixture of mono- (**12**) and di-addition (**13**) with 90% ee of the desired product (eq 9).[8] This double addition could be somewhat overridden by the use of excess nitromethane, but a decrease in enantioselectivity (80% ee) accompanied this when using 10 equiv of nitromethane. A solution to this problem was to use bromonitromethane as a nucleophile. Just 1 mol% PBAM promoted this bromonitromethane addition to give **14** (eq 10) in exceptional yield (99%) and with excellent enantioselectivity (98% ee for both diastereomers) on considerable scale (19.2 g product). No double addition product was formed and the bromine could be reduced concomitantly with the nitro group (with $CoCl_2$ and $NaBH_4$) to provide the desired 1,2-diamine **15** (eq 11). In three further steps, this amine was converted to the potent inhibitor of CYP51 and potential therapeutic for Chagas disease, (+)-VNI.[9]

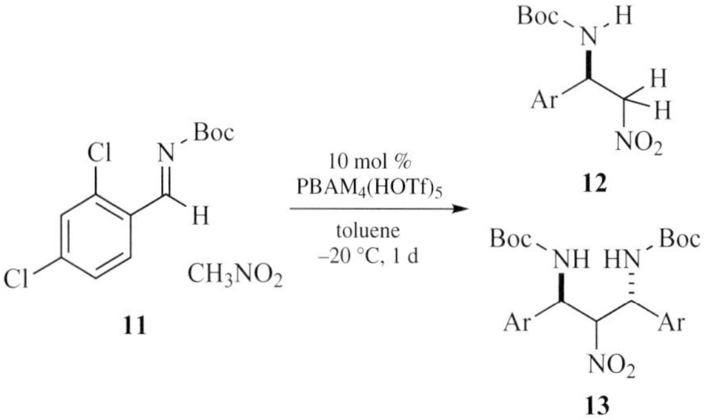

(9)

CH_3NO_2 (2 equiv) — 90% ee, 2.7:1 (**12**:**13**)

CH_3NO_2 (10 equiv) — 80% ee, 6.5:1 (**12**:**13**)

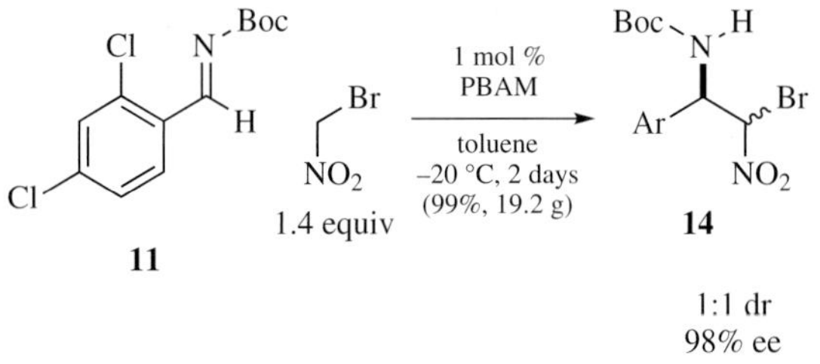

(10)

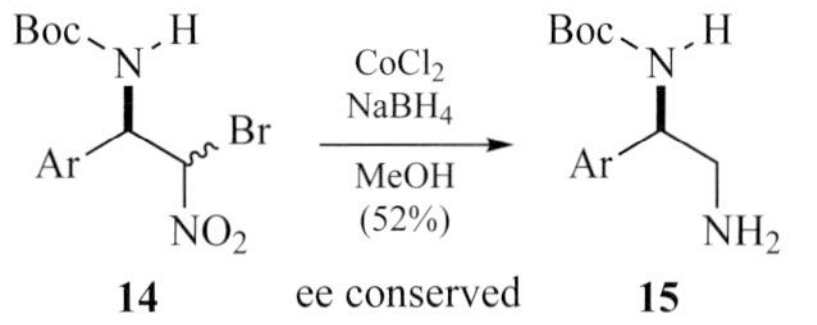

(11)

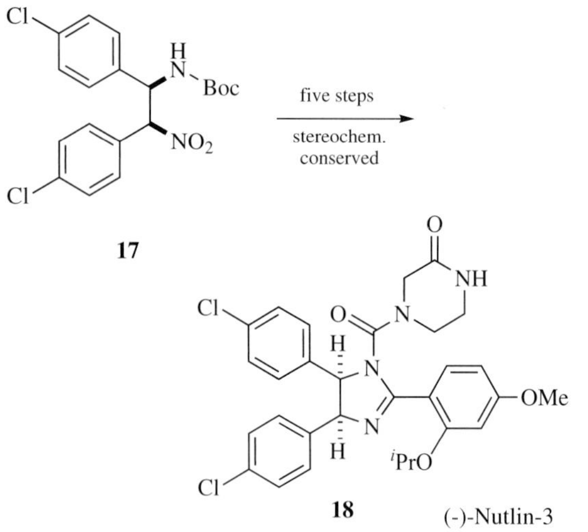

(13)

(-)-Nutlin-3
p53-MDM2 inhibitor

Aza-Henry Addition of Aryl Nitromethanes. The addition of aryl nitroalkanes to aryl *N*-Boc imines gives access to unsymmetrical masked *cis*-stilbene diamines.[10] Asymmetric examples of this addition are limited to a few examples with moderate stereoselectivity,[11–13] with the exception of a very recent example using chiral, ammonium betaines by Ooi and coworkers.[14]

Aryl nitromethanes are considerably more acidic than simple aliphatic nitroalkanes like nitroethane. These nucleophiles are therefore more reactive, such that 5 mol % PBAM promotes this reaction to completion in less than 2 h at −20 °C. Also, unlike the addition of aliphatic nitroalkanes, PBAM·HOTf or other acid salts of PBAM do not outperform the free base. The increased reactivity of this system permitted the temperature to be lowered to −78 °C to achieve higher levels of enantioinduction. PBAM promotes the addition of phenylnitromethane **16** to *N*-Boc imine **2** to give **17** (eq 12) with good dr (9:1), ee (86%), and yield (97%). A bis(MeO) derivative of PBAM was found to be more enantioselective and was applied to an array of imines and aryl nitroalkanes to show a general scope for this reaction. This masked stilbene diamine is then orthogonally functionalized and cyclized in five steps to provide the p53-MDM2 inhibitor (−)-Nutlin-3[15] with full conservation of the stereochemistry (eq 13). More recently, this synthesis was accomplished on 17 g scale featuring another derivative of PBAM (8-MeOPBAM) that provides 23.1 g of **17** (200:1 dr, 91% ee, 90% yield) at −20 °C and with 0.5 mol % catalyst loading.[16]

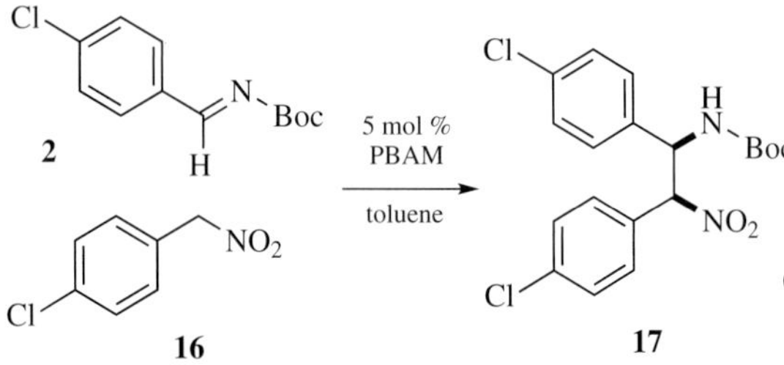

(12)

−20 °C, 2 h 6:1 dr, 83% ee, 99% yield
−78 °C, 23 h 9:1 dr, 86% ee, 97% yield

Other methods to approach these *cis*-stilbene diamines include a route developed by Hoffmann-La Roche featuring a diaza-Cope rearrangement[16–18] and an enantioselective desymmetrization of *meso*-diamines developed by Siedel and Kanta De.[19] A distinct advantage of this PBAM-catalyzed aryl nitromethane addition over these two methods is the ability to access these *cis*-stilbene diamines containing different aryl groups. This advantage has been exploited and applied toward Nutlin derivatives bearing different aryl groups on the stilbene backbone.[20,21] (*S,S*)-PBAM was used to promote the addition of **16** to **19** giving **20** in 2:1 dr, 52% ee, and 40% yield. Ultimately, a new class of unsymmetrical mono(amidine) catalysts was developed and catalyst **21** was found to be optimal for this particular reaction (6:1 dr, 94% ee, 62% yield) (eq 14).

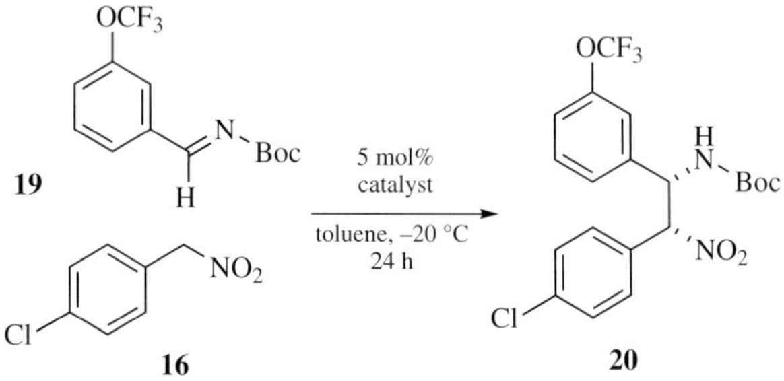

(*S,S*)-PBAM 2:1 dr, 52% ee, 40% yield
21 6:1 dr, 94% ee, 62% yield

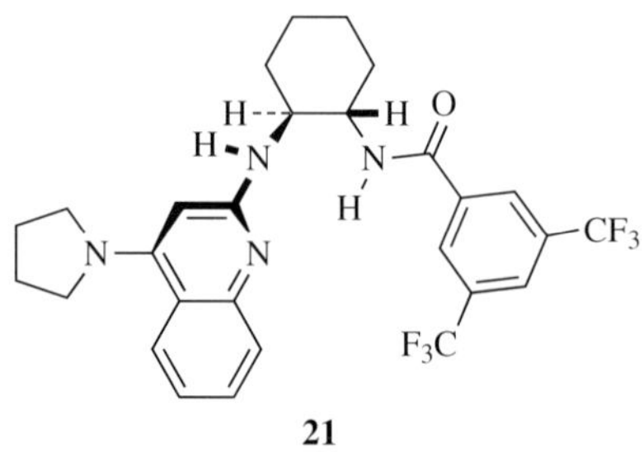

(14)

Synthesis of Enantioenriched Arylglycine Equivalents. Johnston and coworkers have developed umpolung amide synthesis (UmAS) that uses *α*-halo nitroalkanes as carboxylic acid surrogates for the synthesis of amides.[22] The addition of bromonitromethane to aryl *N*-(trimethylsilyl)imines such as **22** and subsequent acylation provide access to aryl glycine donors for UmAS. A post-reaction acylation serves to prevent racemization from the *retro*-aza-Henry reaction and provide the desired *N*-protected products. A survey of catalysts was initially conducted for this enantioselective bromonitromethane addition (eq 15).[23] Chiral thioureas **24** and **25** both promoted this reaction to form **23**, but gave little enantioselection.

Slightly higher ee values (33, 32% ee) were observed with H,QuinBAM. PBAM was found to be a superior catalyst for this transformation delivering **23** with 80/81% ee and in 56% yield. The 1:1 and 2:3 PBAM:HOTf salts gave substantially lower enantioselection and yield than PBAM.

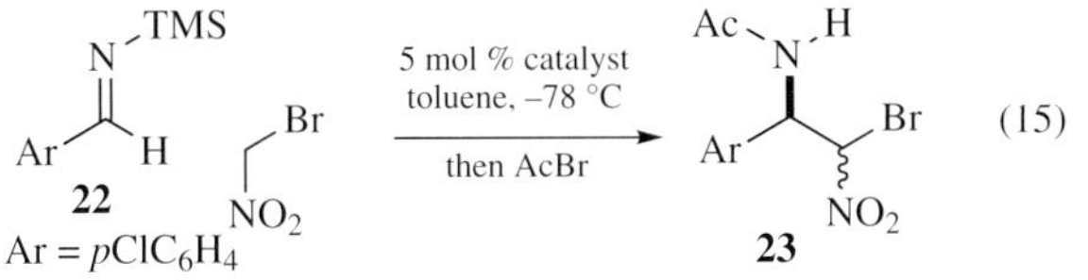

catalyst	ee (%)	yield (%)
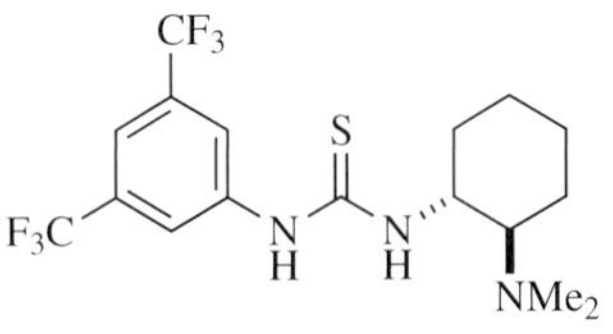 **24**	−16, −17	42
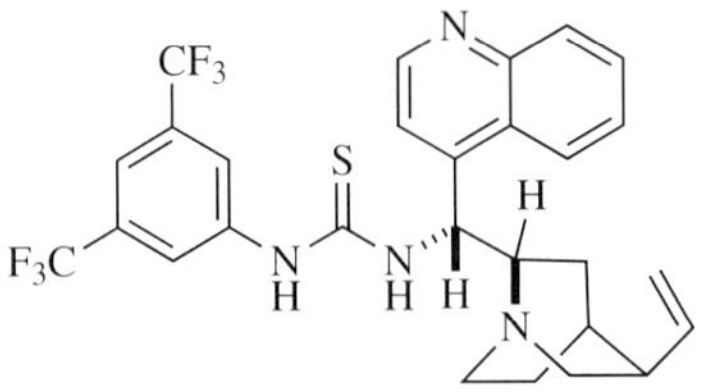 **25**	8, 3	34
H,QuinBAM	33, 32	18
PBAM	80, 81	56
PBAM·HOTf	60, 61	19
$PBAM_2(HOTf)_3$	38, 40	10

This post-addition quench can be performed with a variety of acid chlorides allowing the synthesis of a variety of aryl glycine donors with different nitrogen protecting groups. The newly formed *α*-bromonitroalkanes **27** from a variety of aryl *N*-(trimethylsilyl)imines **26** are highly enantioenriched using 10 mol% PBAM (most examples with 82–93% ee) (eq 16). No control is exerted on the other stereocenter, but this stereochemical information is removed during the UmAS step. Different amines can then be used under UmAS conditions to form new amide products (eq 17).

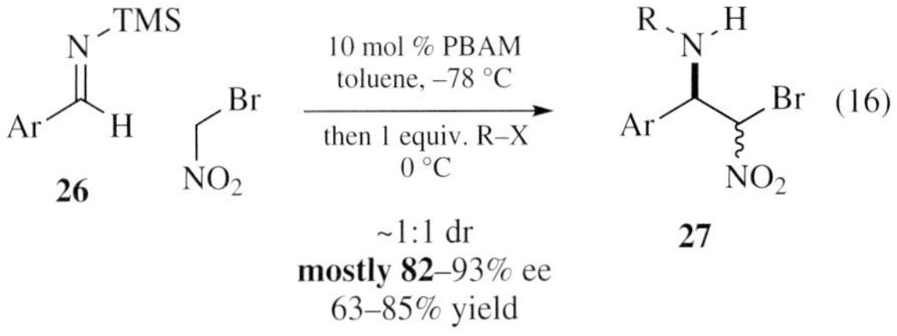

R = Ac, Bz, Piv, Cbz, Alloc, Fmoc, N_3Ac

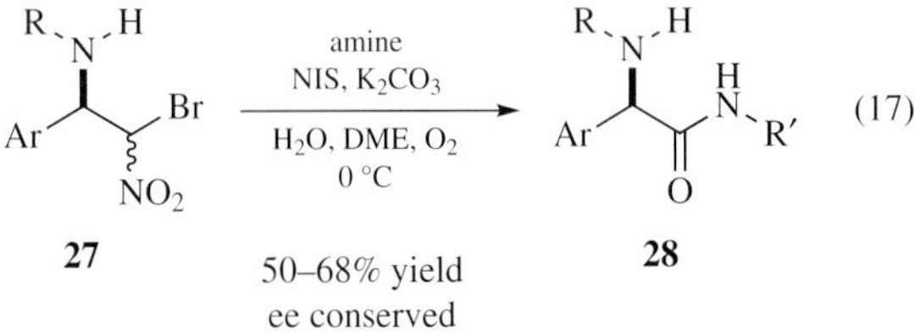

Addition of Nitroalkanes to Indolenines. PBAM exhibited unprecedented reactivity by promoting a nitroalkane addition to *in situ* formed indolenines.[24] Sulfones **29** eliminate in the presence of PBAM, K_2CO_3, and nitroalkane to form indolenine electrophiles **30**, which then react with the nitroalkane to provide alkylated enantioenriched indoles (**31**) (eq 18). Interestingly, the use of water as a cosolvent had a positive effect on enantioselectivity. These products were synthesized in good yield and with low dr but high ee in most cases (75–90% ee). The nitro group of these products is easily removed under hydrogenative conditions with full conservation of enantiopurity.

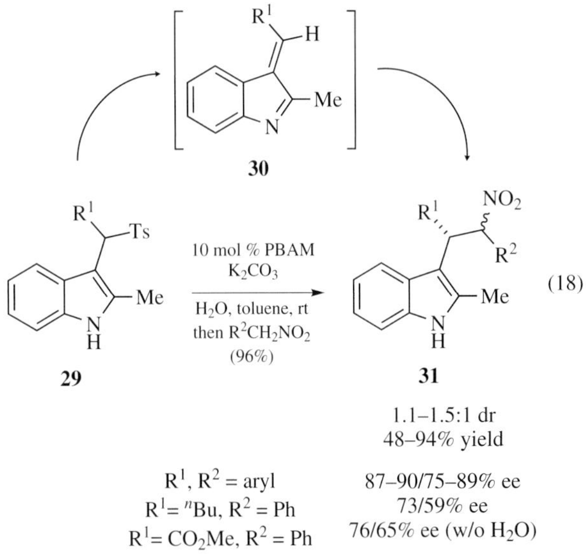

Iodolactonizations. Various acid salts of PBAM have proven effective at the iodolactonization of δ-unsaturated carboxylic acids.[25] With PBAM, a pronounced counterion effect was seen with this reaction. Ten counterions were examined for the cyclization of **32** to form **33** with ee as low as 20% and as high as 84% with PBAM·$HN(SO_2CF_3)_2$ (90% yield) (eq 19). A caveat of this reaction is that the catalyst itself is prone to iodination at the 3- and 3′-positions, but this had no discernible effect on the reaction besides the requirement to use 1.2 equiv of NIS to account for I^+ consumption. Ultimately, substitution of the cyclohexane backbone of PBAM for a stilbene backbone provided a catalyst that was most enantioselective for this reaction (95–97% ee for most substrates). This particular transformation has also been achieved organocatalytically in a highly enantioselective fashion (>90% ee) with a chiral urea and a chiral squaramide catalyst.[26–28] In comparison with these other organocatalysts, the bisamidine-catalyzed reactions have an advantage of a warmer reaction temperature (−20 °C).

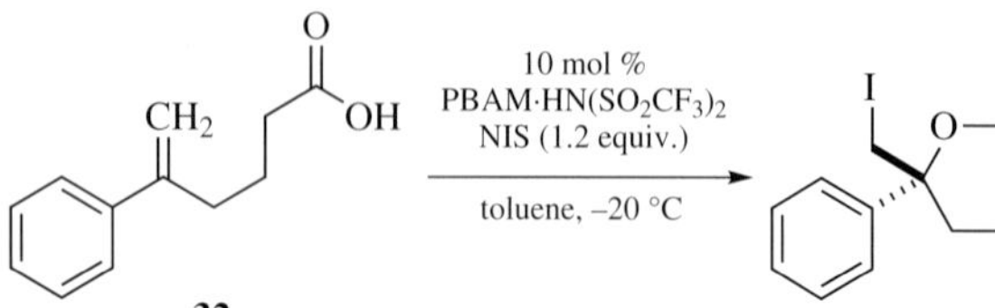

32 33 (19)

84% ee, 90% yield

1. Davis, T. A.; Wilt, J. C.; Johnston, J. N., *J. Am. Chem. Soc.* **2010**, *132*, 2880.
2. Davis, T. A.; Dobish, M. C.; Schwieter, K. E.; Chun, A. C.; Johnston, J. N., *Org. Synth.* **2012**, *89*, 380.
3. Nugent, B. M.; Yoder, R. A.; Johnston, J. N., *J. Am. Chem. Soc.* **2004**, *126*, 3418.
4. Hess, A. S.; Yoder, R. A.; Johnston, J. N., *Synlett* **2006**, 147.
5. Noble, A.; Anderson, J. C., *Chem. Rev.*, **2013**, *113*, 2887.
6. Yoon, T. P.; Jacobsen, E. N., *Angew. Chem., Int. Ed.* **2005**, *44*, 466.
7. Davis, T. A.; Danneman, M. W.; Johnston, J. N., *Chem. Commun.* **2012**, *48*, 5578.
8. Dobish, M. C.; Villalta, F.; Waterman, M. R.; Lepesheva, G. I.; Johnston, J. N., *Org. Lett.* **2012**, *14*, 6322.
9. Villalta, F.; Dobish, M. C.; Nde, P. N.; Kleshchenko, Y. Y.; Hargrove, T. Y.; Johnson, C.A.; Waterman, M. R.; Johnston, J. N.; Lepesheva, G. I., *J. Infect. Dis.* **2013**, *208*, 504.
10. Davis, T. A.; Johnston, J. N., *Chem. Sci.* **2011**, *2*, 1076.
11. Rueping, M.; Antonchick, A. P., *Org. Lett.* **2008**, *10*, 1731.
12. Gothelf, K. V.; Jorgensen, K. A., *Angew. Chem., Int. Ed.*, **2001**, *40*, 2992.
13. Walvoord, R. R.; Kozlowski, M. C., *Tetrahedron Lett.* **2015**, doi: 10.1016/j.tetlet.2014.12.105.
14. Uraguchi, D.; Oyaizu, K.; Noguchi, H.; Ooi, T., *Chem. Asian J.* **2015**, *10*, 334.
15. Vassilev, L. T.; Vu, B. T.; Graves, B.; Carvajal, D.; Podlaski, F.; Filipovic, Z.; Kong, N.; Kammlott, U.; Lukacs, C.; Klein, C.; Fotouhi, N.; Liu, E. A., *Science* **2004**, *303*, 844.
16. Davis, T. A.; Vilgelm, A. E.; Richmond, A.; Johnston, J. N., *J. Org. Chem.* **2013**, *78*, 10605.
17. Shu, L.; Wang, P.; Liu, W.; Gu, C., *Org. Process Res. Dev.* **2012**, *16*, 1866.
18. See also the supporting information of Ref. 10.
19. De, C. K.; Seidel, D., *J. Am. Chem. Soc.* **2011**, *133*, 14538.
20. Vara, B. A.; Mayasundari, A.; Tellis, J. C.; Danneman, M. W.; Arredondo, V.; Davis, T.A.; Min, J.; Finch, K.; Guy, R. K.; Johnston, J. N., *J. Org. Chem.* **2014**, *79*, 6913.
21. Qin, L.; Yang, F.; Zhou, C.; Chen, Y.; Zhang, H.; Su, Z., *J. Am. Chem. Soc.* **2014**, *136*, 18023.
22. Shen, B.; Makley, D. M.; Johnston, J. N., *Nature* **2010**, *465*, 1027.
23. Makley, D. M.; Johnston, J. N., *Org. Lett.* **2014**, *16*, 3146.
24. Dobish, M. C.; Johnston, J. N., *Org. Lett.* **2010**, *12*, 5744.
25. Dobish, M. C.; Johnston, J. N., *J. Am. Chem. Soc.* **2012**, *134*, 6068.
26. Denmark, S. E.; Kuester, W. E., Burk, M. T., *Angew. Chem., Int. Ed.* **2012**, *51*, 10938.
27. Cheng, Y. A.; Yu, W. Z.; Yueng, Y.-Y., *Org. Biomol. Chem.*, **2014**, *12*, 2333.
28. Hansen, T. V.; Nolsøe, J. M. J., *Eur. J. Org. Chem.* **2014**, *15*, 3051.

Tyler A. Davis
Merck Research Laboratories, Rahway, NJ, USA

D

1,4-Diazabicyclo[2.2.2]octane[1]

[280-57-9] $C_6H_{12}N_2$ (MW 112.20)
InChI = 1/C6H12N2/c1-2-8-5-3-7(1)4-6-8/h1-6H2
InChIKey = IMNIMPAHZVJRPE-UHFFFAOYAP

(forms crystalline organometallic complexes;[2] selective base catalyst for Baylis–Hillman reaction;[3] can activate sodamide;[4] cleaves esters selectively;[5] transalkylating agent;[6] dehydrohalogenating agent;[7] pyrolysis catalyst;[8] nucleophilic acylating agent;[9] additive for trialkyl borohydrides;[10] reagent in miscellaneous reactions[11–13])

Alternate Names: DABCO; triethylenediamine; TED.
Physical Data: mp 161 °C (subl. sample); bp 174 °C; fp 62 °C; *d* 1.14 g cm^{-3}.
Solubility: (g/100g of the solvent at 25 °C): water 45, ethanol 77, benzene 51, acetone 13, ethyl methyl ketone 26.
Form Supplied in: colorless crystalline compound; commercially available.
Analysis of Reagent Purity: HCl titration.
Purification: purified by either vacuum sublimation or azeotropic distillation with benzene followed by recrystallization from petroleum ether or methanol–ethyl ether (1:1).
Handling, Storage, and Precautions: is extremely hygroscopic, corrosive, and readily sublimes at room temperature. This reagent should be stored at 0–5 °C under nitrogen. Moderately toxic by ingestion or inhalation; skin and eye irritant. Use in a fume hood.

Original Commentary

Uppuluri V. Mallavadhani
Regional Research Laboratory, Bhubaneswer, India

Crystalline Organometallic Complexes. DABCO forms stable and crystalline complexes with organometallic compounds of Li, Zn, and Mg in hydrocarbon solvents.[2] The organolithium complexes are particularly useful in promoting inert metalations of aromatic hydrocarbons.[14] An interesting application of the DABCO–BuLi complex is the quantitative lithiation of thioanisole to phenylthiomethyllithium and its subsequent use in homologation of primary alkyl bromides and iodides (eq 1).[15,16]

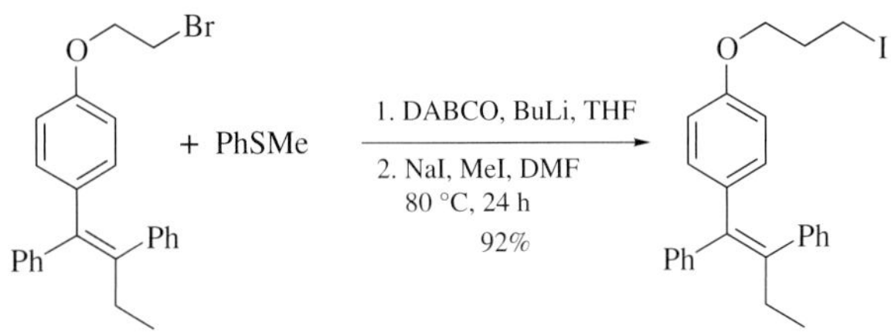

(1)

Catalyst for Baylis–Hillman Reaction. DABCO is a very effective catalyst for the Baylis–Hillman reaction[17] of α,β-unsaturated vinylic systems with electrophiles (e.g. aldehydes) to produce α-substituted acrylates (eq 2).[3,18,19]

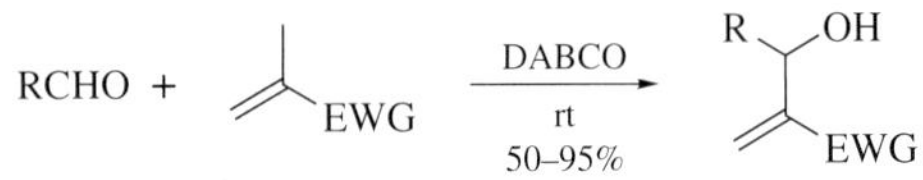

(2)

R = alkyl, aryl
EWG = electron withdrawing group: CO_2R, COR, CN, SO_2R

These typically slow reactions can be accelerated by increasing the molar proportions of DABCO or by varying experimental conditions, or even by use of microwave energy.[19–21] The asymmetric Baylis–Hillman reaction with a variety of chiral auxiliaries is also effectively catalyzed by DABCO (eqs 3–5).[22–24] The use of DABCO with chiral (arylaldehydeimine)tricarbonylchromium complexes significantly enhances the yields and diastereoselectivity (eq 6).[25]

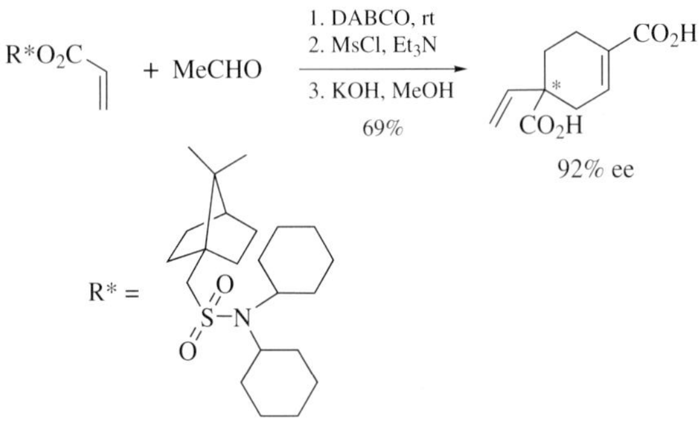

(3)

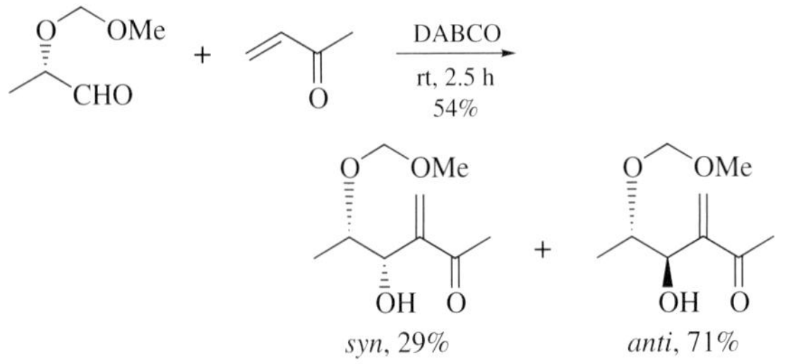

(4)

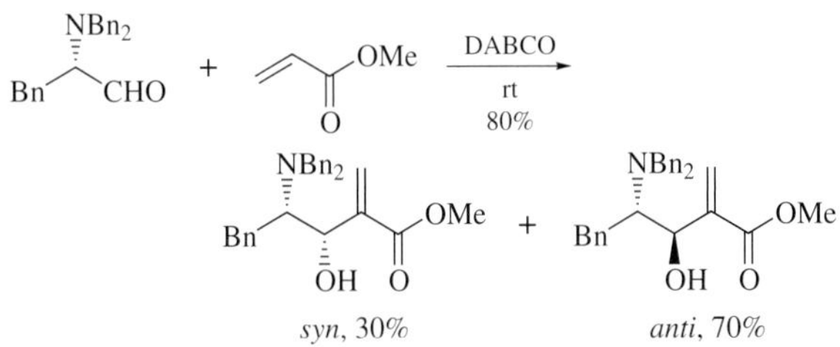

(5)

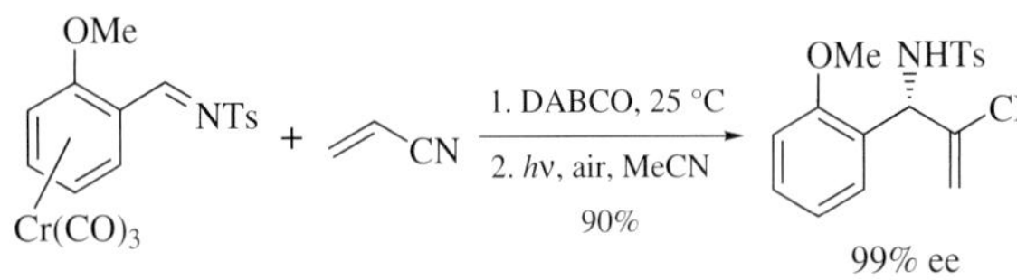

(6)

Activation of Sodamide. DABCO remarkably enhances the reactivity of sodamide in the Haller–Bauer cleavage of nonenolizable ketones to the corresponding carboxamides (eq 7).[4,26]

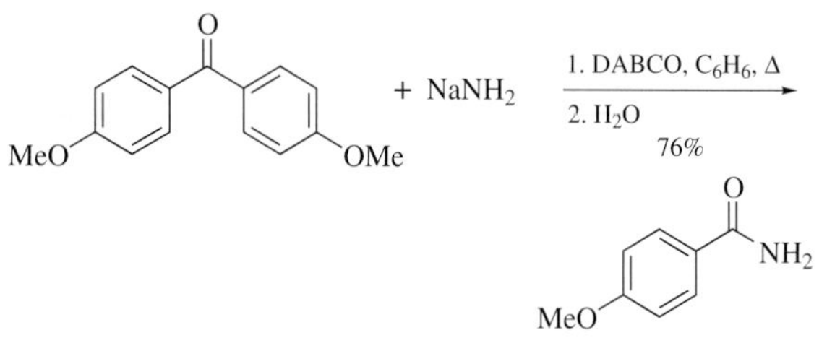

(7)

Selective Cleavage of Esters. DABCO is a selective reagent for the single-pot dealkoxycarbonylation of β-keto esters, geminal diesters, and α-sulfonyl(alkyl)malonic esters.[5,27,28] Good selectivity was noticed with β-keto esters (eq 8) and geminal diesters (eq 9), which gave ketones and monoesters, respectively. The deethoxycarbonylation of α-sulfonyl(alkyl)malonic esters in presence of DABCO is a competitive reaction with desulfonylation and is dependent on the size of alkyl groups (eqs 10–11).[28] As the alkyl group changes from methyl to benzyl to phenyl, more desulfonation than deethoxycarbonylation is observed. This method is very effective for the cleavage of esters over conventional methods, which are essentially multi-step.

(8)

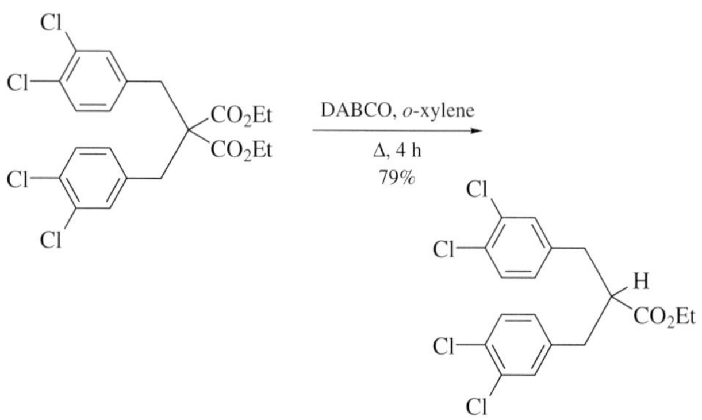

(9)

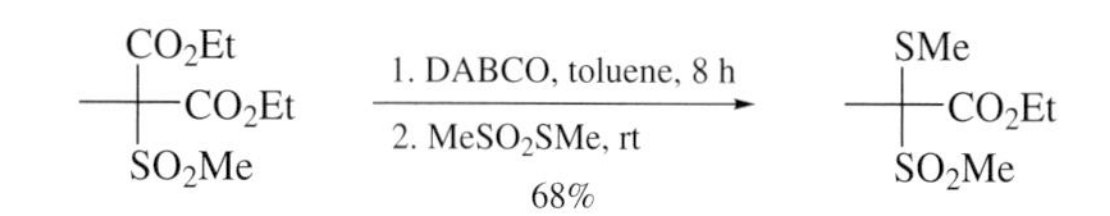

(10)

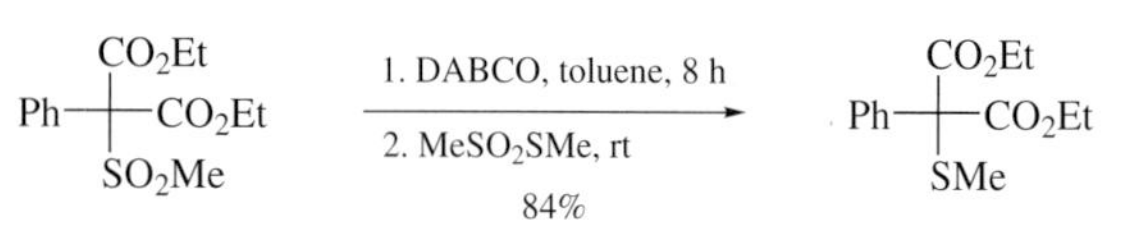

(11)

Transalkylating Agent. The exposed lone pair of electrons and interactions between the nitrogen atoms of DABCO, which are separated by only 2.5 Å, are the key factors in the transalkylation reactions with quaternary ammonium salts (eq 12).[6]

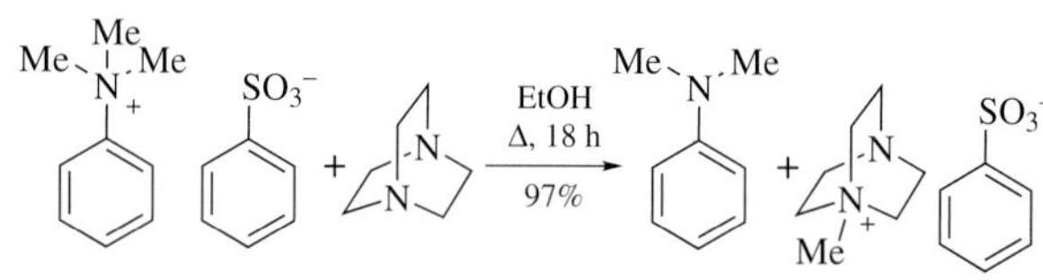

(12)

Dehydrohalogenating Agent. The Ramberg–Bäcklund rearrangement of α,α-dichlorobisbenzyl sulfones to the corresponding 2,3-diphenylthiirene 1,1-dioxides involves dehydrohalogenation and can be achieved very effectively by stirring with DABCO in DMSO at rt (eq 13).[7]

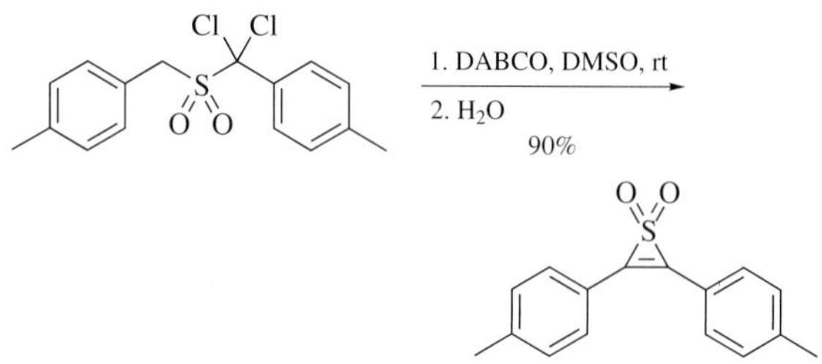

(13)

DABCO is also effective in the dehydrobromination of substituted dibromocyclohexanones to the corresponding cyclohexadienes (eq 14).[29]

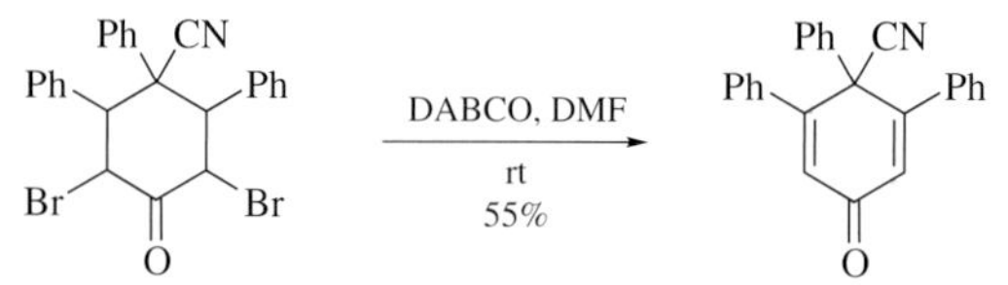

(14)

As a Pyrolysis Catalyst. DABCO is very effective among the tertiary amines studied for the pyrolysis of vinyl azides.[8] The products are 1-azirines and are obtained in 44–95% yields (eq 15). Use of aprotic solvents with DABCO decreases the yields.

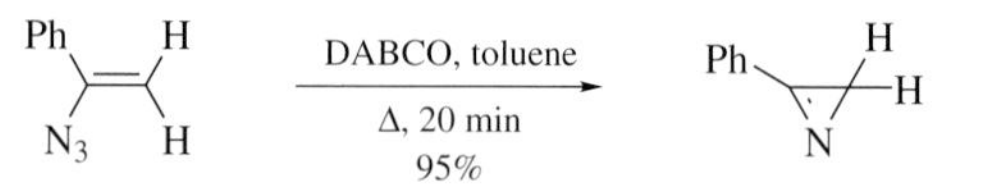

(15)

As a Nucleophilic Acylating Agent. Its action as a nucleophilic acylation catalyst is significant in the formation of a large number of cyanohydrin esters from α-keto nitriles and aldehydes (eqs 16–17).[9] The homogeneous, mild, and safe experimental conditions make this method very attractive over the two-phase PTC methods with use of excess cyanide ion.[30] Commercially useful 6′-*O*-acylsucroses were synthesized by the DABCO-catalyzed regioselective acylation of sucrose with 3-acyl-5-methyl-1,3,4-thiadiazole-2(3*H*)-thiones (eq 18).[31]

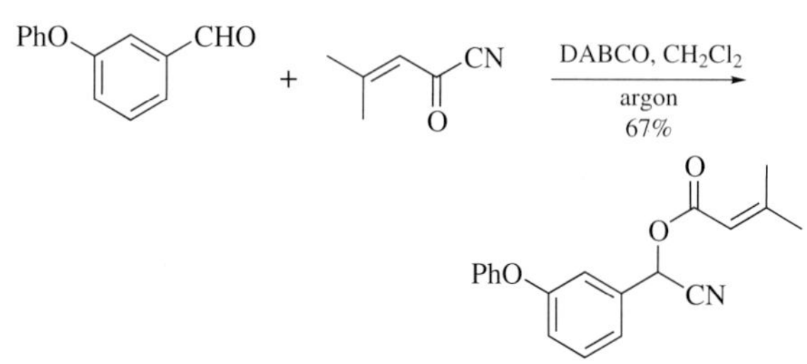

(16)

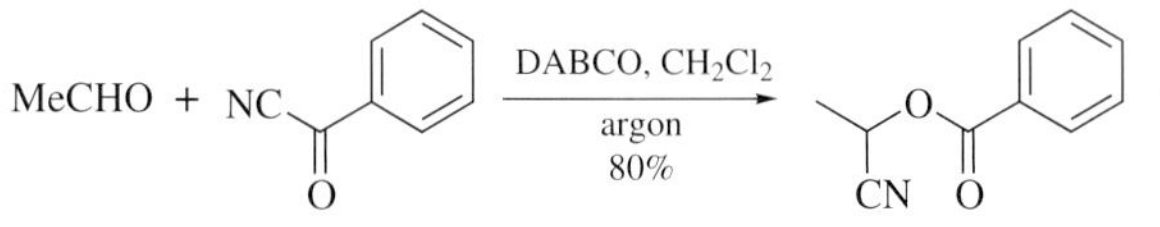
(17)

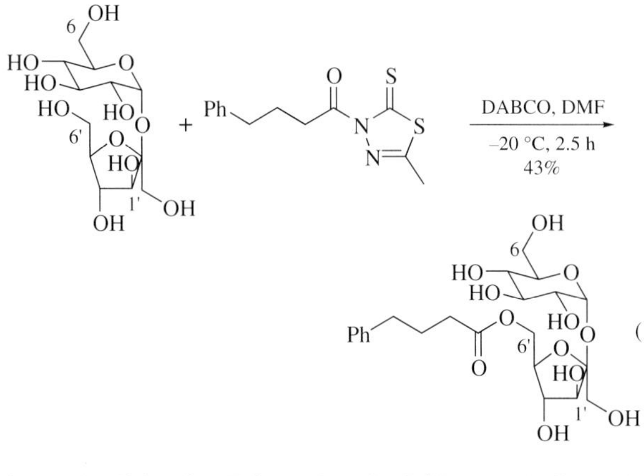
(18)

As an Additive for Trialkylborohydrides. Alkali metal trialkylborohydrides (Li and Na) with bulky alkyl groups are useful chiral reducing agents for carbonyl compounds. These hydrides can be prepared in high yield by the reaction of trialkylboranes with ***Lithium Aluminum Hydride*** or with $NaEt_2AlH_2$ in presence of DABCO (eqs 19–20). The role of DABCO is to precipitate AlH_3 and Et_2AlH from the reaction medium.[10,32]

$$LiAlH_4 + R_3B \xrightarrow[\substack{0\text{–}25\,^\circ C\\ 90\%}]{\text{DABCO, ether}} LiR_3BH + DABCO{\bullet}AlH_3 \quad (19)$$

$$NaEt_2AlH_2 + R_3B \xrightarrow[\substack{0\text{–}25\,^\circ C\\ 96\%}]{\text{DABCO, ether}} NaR_3BH + DABCO{\bullet}Et_2AlH \quad (20)$$

As a Catalyst in Miscellaneous Reactions. DABCO effects singlet oxygen quenching very efficiently without undergoing any chemical change.[11] It is a selective basic catalyst for the reaction of phenyl isocyanate with polyhydric alcohols to form polyurethanes.[12] DABCO is particularly effective as a base catalyst in the Michael addition reactions of α,β-unsaturated ketones with β-keto thiolesters and *S,S′*-diethyl dithiomalonate (eqs 21–22).[13] The reaction is so specific to the base catalyst that the use of strong bases did not result in any product formation. Recently its use as a base in the phospha-Witting reaction to form C–P double bonds was also reported (eq 23).[33]

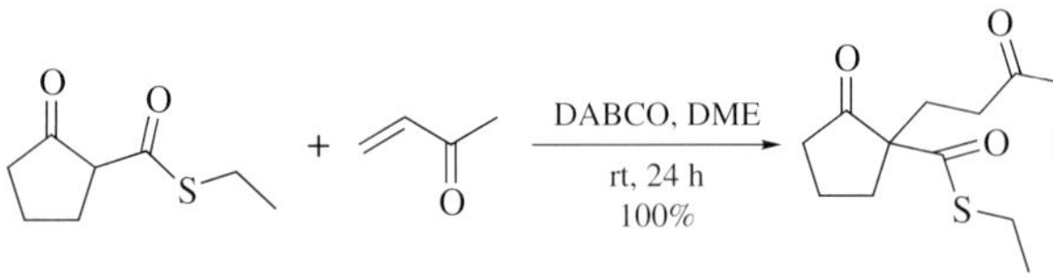
(21)

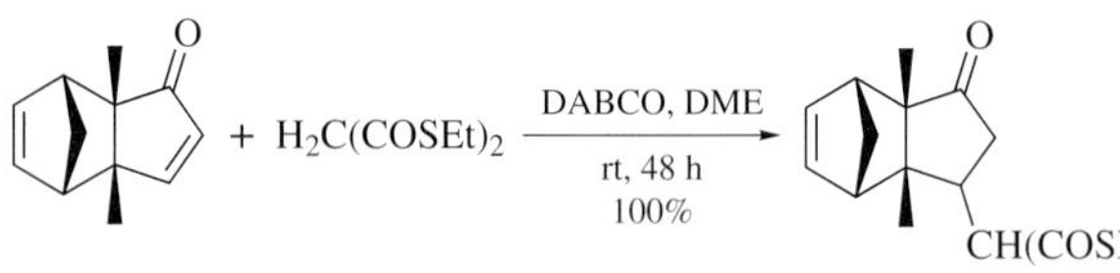
(22)

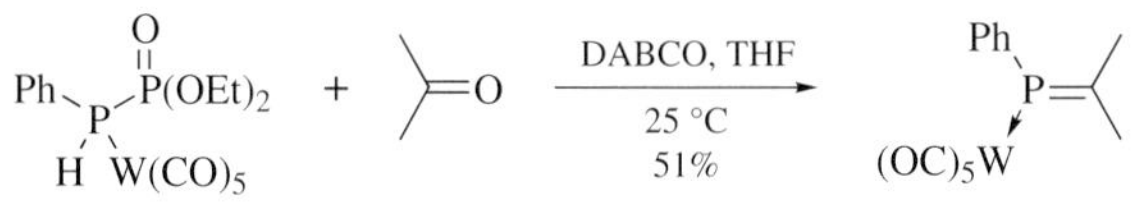
(23)

As a electron-donor reagent it forms a variety of useful adducts and complexes. The DABCO–$2Br_2$ complex (mp 155–60 °C, stable to air, light, and water) is a selective reagent for the oxidation of benzylic and secondary hydroxyls,[34] and for sulfides.[35] The DABCO–$2H_2O_2$ complex is an equivalent to anhydrous ***Hydrogen Peroxide***.[36] This on treatment with ***Chlorotrimethylsilane*** gives ***Bis(trimethylsilyl) Peroxide***, a useful reagent for the electrophilic hydroxylation of aromatic and heteroaromatic systems[37] and also for the facile conversion of sulfones to carbonyl compounds.[38]

First Update

Nicolas Fleury-Bregeot
University of Pennsylvania, Philadelphia, PA, USA

Although the main applications have been listed above, DABCO continues to be used extensively by chemists and broadens its field of application.

Organocatalyst. One of the most important areas in which DABCO is represented is organocatalysis, a considerably expanding domain. It is widely used as a strong Lewis base organocatalyst owing to its ability to add to activated unsaturated bonds, triggering various chemical transformations.[39]

Baylis–Hillman. DABCO is utilized in the Baylis–Hillman reaction as depicted above (see eq 22) [40,41] However, this reaction is known to be very slow and therefore was not widely used for years despite its very powerful synthetic potential. Recent studies focused on the mechanism and kinetics of this transformation and helped provide insight to overcome some limitations of this reaction. For instance, the use of alcoholic solvents enables accelerated proton transfer, favoring formation of intermediate D and thus considerably improving the reaction rate (eq 24).

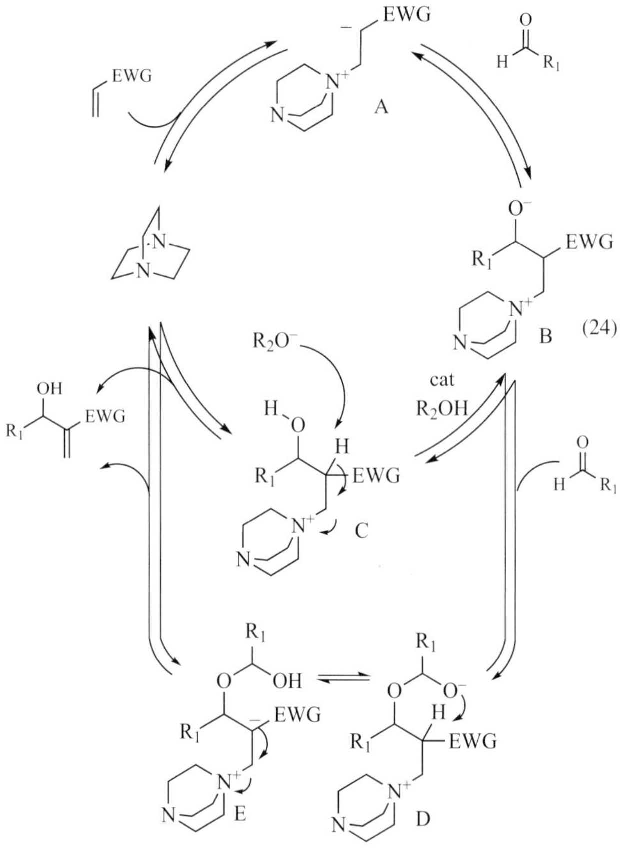
(24)

The efficiency of Lewis base catalysts is linked to their basicity (pK_a value in DMSO)[42] as well as other factors: Mayr et al. introduced a scale of nucleophilicity and electrophilicity,[43] quantifying that DABCO is both a better nucleophile and a better nucleofuge than DMAP.[44] Concerning the asymmetric versions of the Baylis–Hillman reaction, recent investigations focus on the use and design of chiral organocatalysts.[45,46] DABCO can be used in some enantioselective versions when used in combination with chiral co catalysts. For example, Shi and Liu recently reported an efficient Baylis–Hillman reaction of aryl aldehydes with 2-cyclohexen-1-one using a chiral bis(thio)urea and DABCO (eq 25).[47]

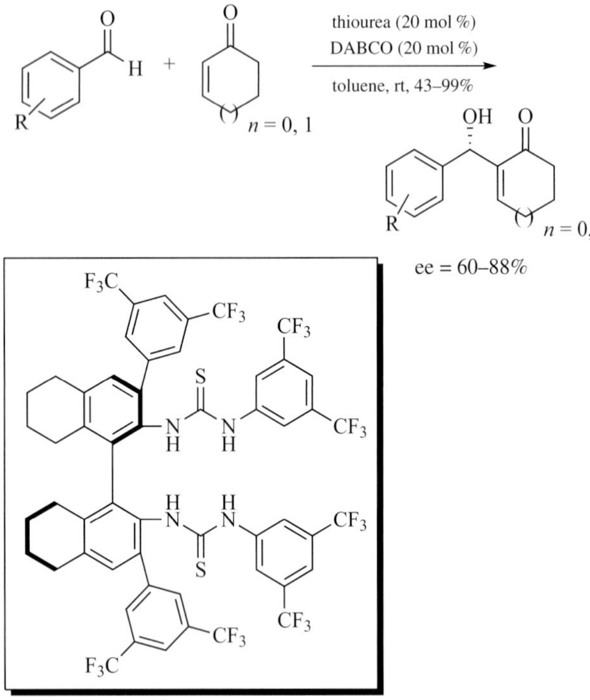

(25)

DABCO can also be used in the aza-Baylis–Hillman reaction, where the aldehyde is replaced by an imine, leading to highly functionalized amines (eq 26).[48,49]

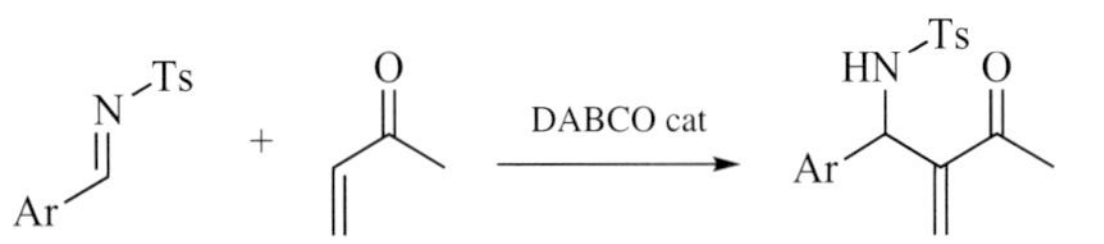

(26)

Other Reactions. During the course of a study toward the use of tosylated imines and 2,3-butadienoates in the aza-Baylis–Hillman reaction using DABCO and DMAP, Shi and coworkers observed the formation of unusual products. Indeed, instead of the regular Baylis–Hillman adducts, they observed the formation of azetidine rings when using DABCO (and the formation of dihydropyridines when using DMAP).[50,51] Interestingly, other catalysts such as phosphines do not lead to the formation of the same products (eq 27).

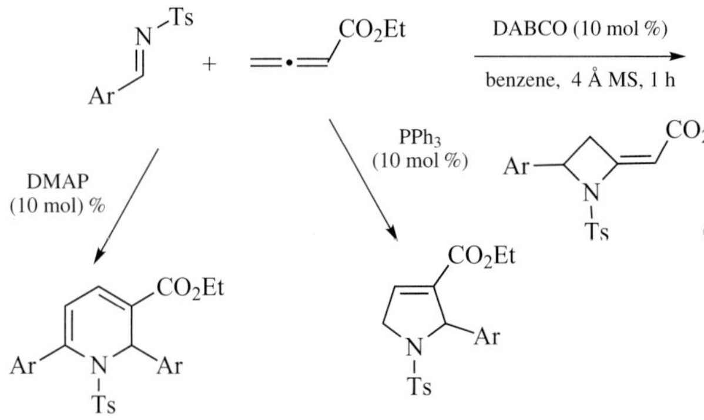

(27)

The use of DABCO was extended to efficiently catalyze the synthesis of highly functionalized chromenes starting from salicyl *N*-tosylimines and activated allenes or alkynes (eq 28).[52–54]

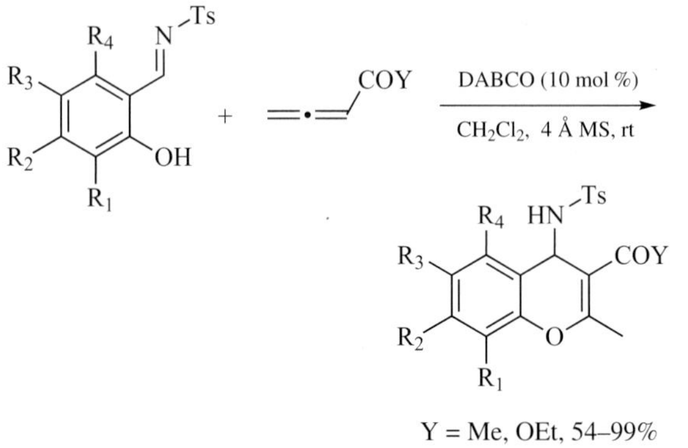

(28)

The synthesis of other heterocycles such as dihydrofurans[55] can also be catalyzed by DABCO. The role of DABCO is to activate the alkyl halide, generating an ammonium halide in situ that then upon treatment with another base can react with benzylidene-oxobutanoates (eq 29).

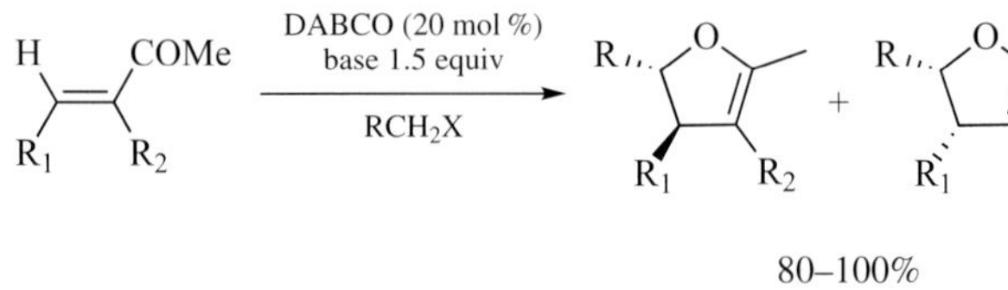

(29)

In 2005, Wu et al. reported that DABCO can be used as a catalyst for the opening of aziridines with amines or thiols. The reaction proceeds smoothly for aromatic nucleophiles, with low catalyst loading (eq 30).[56]

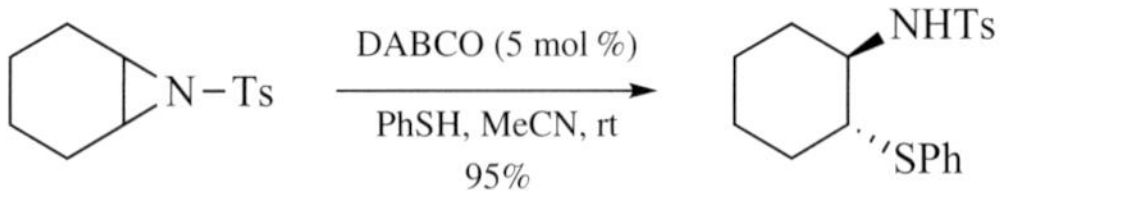

(30)

DABCO has also been used as a catalyst for the N-methylation of indoles, showing a dramatic superiority to other amines examined in this transformation. The authors hypothesized a dual nucleophilic catalysis by DABCO, inducing alkylation and acylation processes (eq 31).[57]

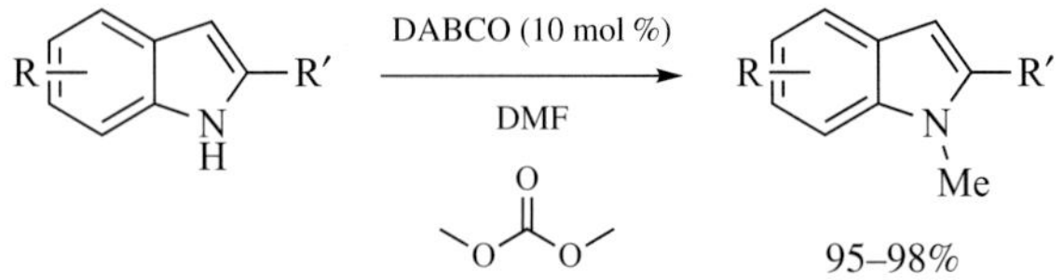

(31)

Ligand in Cross-coupling Reactions. DABCO can be used as an inexpensive and efficient ligand in several transition metal-catalyzed reactions. Li and coworkers reported its use in the Suzuki–Miyaura cross-coupling reaction catalyzed by palladium[58–62] or copper.[63,64] They demonstrated very efficient catalytic systems (recyclable when using PEG-400 as solvent) using palladium acetate with TONs up to 960 000, offering an interesting alternative to the use of more traditional phosphine ligands (eq 32).[60,62]

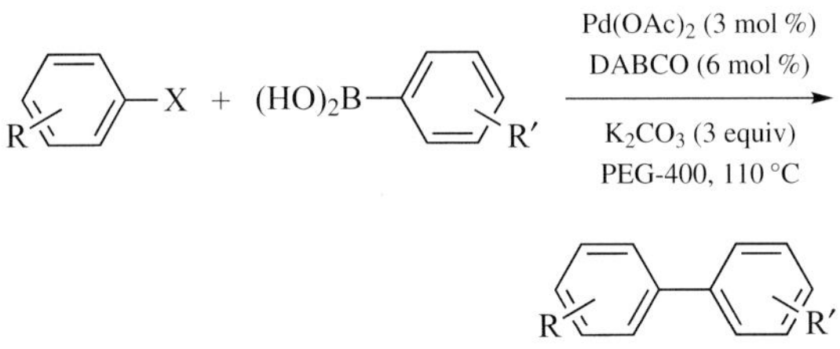

(32)

TONs up to 960 000

The catalytic system has been extended to other cross-coupling reactions including the Sonogashira reaction.[64–66] In this case, DABCO can play the role of a base in a copper- and ligand-free version of the reaction (eq 33).[66]

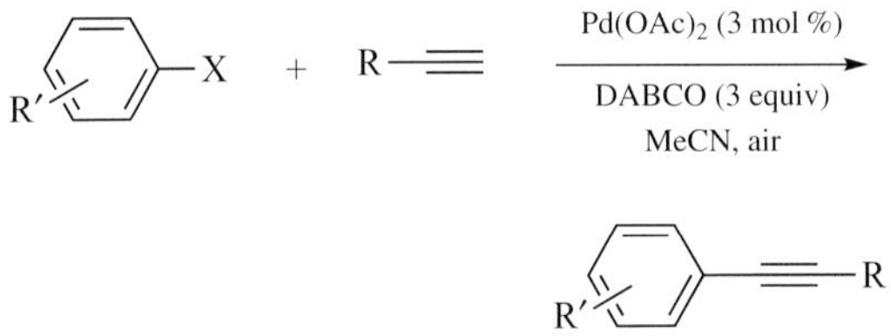

(33)

TONs up to 9200

DABCO is also a good ligand for other coupling reactions such as the Heck,[67,68] the Stille,[69] and the Hiyama cross-coupling reactions (eq 34).[70]

Ligand in Supramolecular Chemistry. DABCO has been widely used as a ligand in supramolecular chemistry because of its ability to bind metals. DABCO is known to coordinate Zn porphyrins, inducing the formation of supramolecular structures by self-assembly. Depending on the concentration of metalloporphyrin, 1:1 complexes or 2:1 sandwich complexes can be formed (eq 35).[71,72]

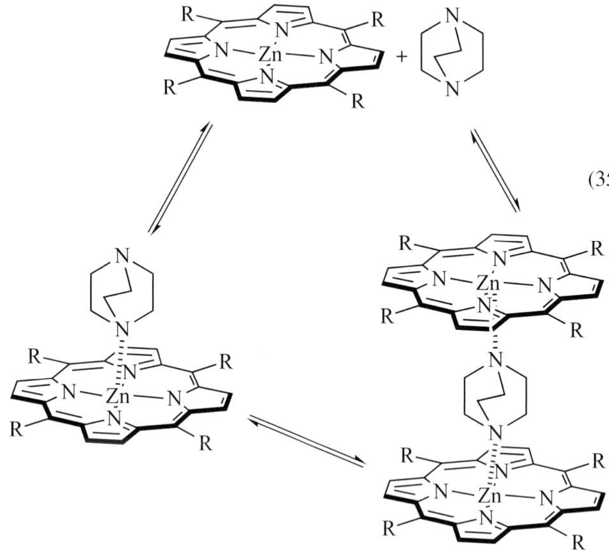

(35)

This behavior has also been extended to polyporphyrins and other porphyrin-based molecules such as calixarenes, increasing the complexity of the self-assembled structures.[73–75] In 2004, porphyrin–fullerene assemblies were synthesized, and DABCO was used to transform them from rotaxanes to a new kind of catenanes.[76]

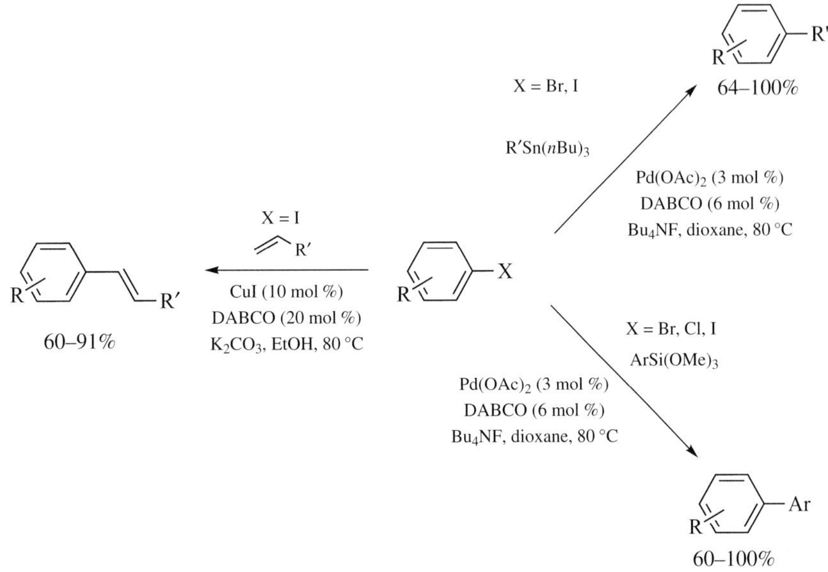

(34)

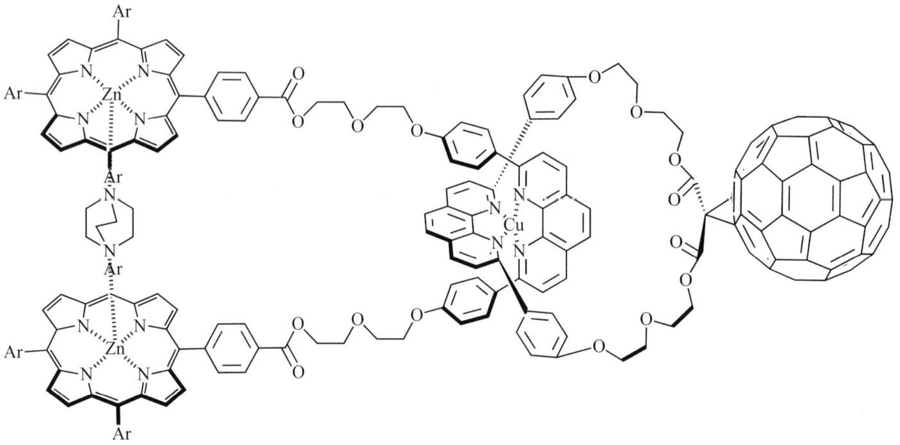

Stabilizing Agent to Create Stable Organometallic Aluminium Complex. The complex formed by the addition of DABCO and Me_3Al forms a sensitive but 'bench handable' complex: $((Me_3Al)_2 \cdot DABCO)$.[77] The latter, known as DABAL·Me_3, can be used as a methyl anion equivalent. (see the e-EROS article dedicated to DABAL·Me3).

Selective Dechloroacetylation Reagent. DABCO has been shown to be an efficient, highly selective dechloroacetylating agent used in carbohydrate synthesis. Indeed, the use of an excess of DABCO allows the selective removal of the chloroacetyl group, leaving other protecting groups such as benzyl groups and allyl groups intact (eq 36).[78]

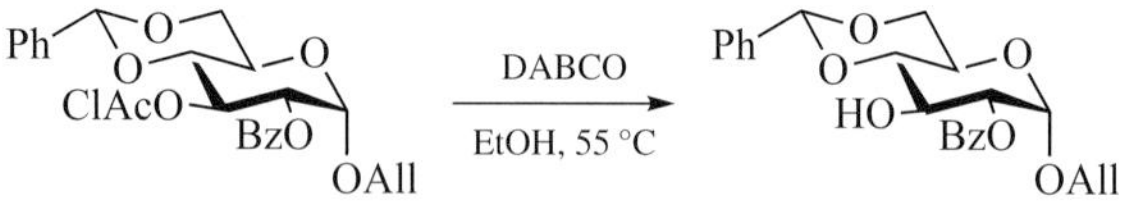

(36)

1. (a) Mosby, W. L. *Heterocyclic Systems with Bridgehead Nitrogen Atoms*; Interscience: New York, 1961; Part 2, Chapter XVII, pp 1372–75. (b) Yakhontov, L. N.; Mrachkovskaya, L. B., *Chem. Heterocycl. Compd. (Engl. Transl.)* **1976**, *12*, 607. (c) *Beilstein*; 23, 487.
2. Screttas, G. C.; Eastham, J. F., *J. Am. Chem. Soc.* **1965**, *87*, 3276.
3. Drewes, S. E.; Roos, G. H. P., *Tetrahedron* **1988**, *44*, 4653.
4. Kaiser, E. M.; Warner, C. D., *Synthesis* **1975**, 395.
5. Huang, S-B.; Parish, E. J.; Miles, D. H., *J. Org. Chem.* **1974**, *39*, 2647.
6. Ho, T.-L., *Synthesis* **1972**, 702.
7. Philips, J. C.; Swisher, J. V.; Haidukewych, D.; Morales, O., *J. Chem. Soc., Chem. Commun.* **1971**, 22.
8. Komatsu, M.; Ichijima, S.; Ohshiro, Y.; Agawa, T., *J. Org. Chem.* **1973**, *38*, 4341.
9. Hoffmann, H. M. R.; Ismail, Z. M.; Hollweg, R.; Zein, A. R., *Bull. Chem. Soc. Jpn.* **1990**, *63*, 1807.
10. Hubbard, J. L.; Fuller, J. C.; Jackson, T. C.; Singaram, B., *Tetrahedron* **1993**, *49*, 8311.
11. (a) Ovannes, C.; Wilson, T., *J. Am. Chem. Soc.* **1968**, *90*, 6527. (b) Clennan, E. L.; Noe, L. J.; Wen, T.; Szneler, E., *J. Org. Chem.* **1989**, *54*, 3581.
12. (a) Hartshorn, S. R.; Thind, S. S. In *Comprehensive Heterocyclic Chemistry*; Katritzky, A. R.; Rees, C. W., Eds; Pergamon: Oxford, 1984; Vol. 1, p 405. (b) Nakamura, H.; Takata, T.; Endo, T., *Macromolecules* **1990**, *23*, 3032.
13. (a) Liu, H. J.; Oppong, I. V., *Can. J. Chem.* **1982**, *60*, 94. (b) Liu, H.-J.; Ho, L.-K.; Lai, H. K., *Can. J. Chem.* **1981**, *59*, 1685.
14. Mallan, J. M.; Bebb, R. L., *Chem. Rev.* **1969**, *69*, 693.
15. Corey, E. J.; Jautelat, M., *Tetrahedron Lett.* **1968**, 5787.
16. Robertson, D. W.; Katzenllenbogen, J. A., *J. Med. Chem.* **1982**, *25*, 167.
17. Baylis, A. B.; Hillman, M. E. D. Ger. Patent 2 155 113, 1972 (*Chem. Abstr.* **1972**, *77*, 34 174q).
18. Fort, Y.; Berthe, M. C.; Caubere, P., *Tetrahedron* **1992**, *48*, 6371.
19. Van Rozendaal, E. L. M.; Voss, B. M. W.; Scheermen, H. W., *Tetrahedron* **1993**, *49*, 6931.
20. Ameer, F.; Drewes, S. E.; Freese, S.; Kaye, P. T., *Synth. Commun.* **1988**, *18*, 495.
21. Kundu, M. K.; Mukherjee, S. B.; Balu, N.; Padmakumar, R.; Bhat, S. V., *Synlett* **1994**, 444.
22. Basavaiah, D.; Pandiaraju, S.; Sarma, P. K. S., *Tetrahedron Lett.* **1994**, *35*, 4227.
23. Drewes, S. E.; Manickum, T.; Roos, G. H. P., *Synth. Commun.* **1988**, *18*, 1065.
24. Manickum, T.; Roos, G., *Synth. Commun.* **1991**, *21*, 2269.
25. (a) Kundig, E. P.; Xu, L.-H.; Romanens, P.; Bernardinelli, G., *Tetrahedron Lett.* **1993**, *34*, 7049. (b) Kundig, E. P.; Xu, L.-H.; Schnell, B., *Synlett* **1994**, 413.
26. Gilday, J. P.; Paquette, L. A., *Org. Prep. Proced. Int.* **1990**, *22*, 169.
27. Miles, D. H.; Huang, B.-S., *J. Org. Chem.* **1976**, *41*, 208.
28. Waldislaw, B.; Marzorati, L.; Donnici, C. L., *J. Chem. Soc., Perkin Trans. 1* **1993**, 3167.
29. Charaf, A.; Otto, H. H., *Arch. Pharm. (Weinheim, Ger.)* **1983**, *316*, 887.
30. Chenevert, R.; Plante, R.; Voyer, N., *Synth. Commun.* **1983**, *13*, 403.
31. Chauvin, C.; Plusquellec, D., *Tetrahedron Lett.* **1991**, *32*, 3495.
32. Singaram, B.; Pai, G. G., *Heterocycles* **1982**, *18*, 387.
33. (a) Marinetti, A.; Mathey, F., *Angew. Chem., Int. Ed. Engl.* **1988**, *27*, 1382. (b) Marinetti, A.; Richard, L.; Mathey, F., *Organometallics* **1990**, *9*, 788.
34. Blair, K. L.; Baldwin, J.; Smith, Jr. W. C., *J. Org. Chem.* **1977**, *42*, 1816.
35. Oae, S.; Ohinishi, Y.; Kozuka, S.; Tagaki, W., *Bull. Chem. Soc. Jpn.* **1966**, *39*, 364.
36. Cookson, P. G.; Davies, A. G.; Fazal, N., *J. Organomet. Chem.* **1975**, *99*, C31.
37. Taddei, M.; Ricci, A., *Synthesis* **1986**, 633.
38. Hwu, J. R., *J. Org. Chem.* **1983**, *48*, 4433.
39. Denmark, S. E.; Beutner, G. L., *Angew. Chem., Int. Ed.* **2008**, *47*, 1560.
40. Basavaiah, D.; Rao, P. D.; Hyma, R. S., *Tetrahedron* **1996**, *52*, 8001.
41. Basavaiah, D.; Rao, A. J.; Satyanarayana, T., *Chem. Rev.* **2003**, *103*, 811.

42. Aggarwal, V. K.; Emme, I.; Fulford, S. Y., *J. Org. Chem.* **2003**, *68*, 692.
43. Mayr, H.; Kempf, B.; Ofial, A. R., *Acc. Chem. Res.* **2003**, *36*, 66.
44. Baidya, M.; Kobayashi, S.; Brotzel, F.; Schmidhammer, U.; Riedle, E.; Mayr, H., *Angew. Chem. Int. Ed.* **2007**, *46*, 6176.
45. Langer, P., *Angew. Chem., Int. Ed.* **2000**, *39*, 3049.
46. Masson, G.; Housseman, C.; Zhu, J. P., *Angew. Chem., Int. Ed.* **2007**, *46*, 4614.
47. Shi, M.; Liu, X. G., *Org. Lett.* **2008**, *10*, 1043.
48. Shi, M.; Xu, Y. M., *Angew. Chem., Int. Ed.* **2002**, *41*, 4507.
49. Raheem, I. T.; Jacobsen, E. N., *Adv. Synth. Catal.* **2005**, *347*, 1701.
50. Zhao, G.-L.; Huang, J.-W.; Shi, M., *Org. Lett.* **2003**, *5*, 4737.
51. Zhao, G.-L.; Shi, M., *J. Org. Chem.* **2005**, *70*, 9975.
52. Shi, Y. L.; Shi, M., *Org. Lett.* **2005**, *7*, 3057.
53. Shi, Y. L.; Shi, M., *Chem. Eur. J.* **2006**, *12*, 3374.
54. Guo, Y. W.; Shi, Y. L.; Li, H. B.; Shi, M., *Tetrahedron* **2006**, *62*, 5875.
55. Yang, Z.; Fan, M.; Mu, R.; Liu, W.; Liang, Y., *Tetrahedron* **2005**, *61*, 9140.
56. Wu, J.; Sun, X. Y.; Li, Y. Z., *Eur. J. Org. Chem.* **2005**, 4271.
57. Shieh, W.-C.; Dell, S.; Bach, A.; Repič, O.; Blacklock, T. J., *J. Org. Chem.* **2003**, *68*, 1954.
58. Li, J. H.; Zhu, Q. M.; Xie, Y. X., *Tetrahedron* **2006**, *62*, 10888.
59. Li, J.-H.; Liu, W.-J., *Org. Lett.* **2004**, *6*, 2809.
60. Li, J.-H.; Liu, W.-J.; Xie, Y.-X., *J. Org. Chem.* **2005**, *70*, 5409.
61. Li, J. H.; Hu, X. C.; Liang, Y.; Xie, Y. X., *Tetrahedron* **2006**, *62*, 31.
62. Li, J. H.; Hu, X. C.; Me, Y. X., *Tetrahedron Lett.* **2006**, *47*, 9239.
63. Li, J. H.; Wang, D. P., *Eur. J. Org. Chem.* **2006**, 2063.
64. Li, J.-H.; Li, J.-L.; Wang, D.-P.; Pi, S.-F.; Xie, Y.-X.; Zhang, M.-B.; Hu, X.-C., *J. Org. Chem.* **2007**, *72*, 2053.
65. Li, J. H.; Zhang, X. D.; Xie, Y. X., *Synthesis* **2005**, 804.
66. Li, J.-H.; Liang, Y.; Xie, Y.-X., *J. Org. Chem.* **2005**, *70*, 4393.
67. Li, J. H.; Wang, D. P.; Xie, Y. X., *Synthesis* **2005**, 2193.
68. Li, J. H.; Wang, D. P.; Xie, Y., *Tetrahedron Lett.* **2005**, *46*, 4941.
69. Li, J.-H.; Liang, Y.; Wang, D.-P.; Liu, W.-J.; Xie, Y.-X.; Yin, D.-L., *J. Org. Chem.* **2005**, *70*, 2832.
70. Li, J. H.; Deng, C. L.; Liu, W. J.; Xie, Y. X., *Synthesis* **2005**, 3039.
71. Hunter, C. A.; Meah, M. N.; Sanders, J. K. M., *J. Am. Chem. Soc.* **1990**, *112*, 5773.
72. Anderson, H. L.; Hunter, C. A.; Meah, M. N.; Sanders, J. K. M., *J. Am. Chem. Soc.* **1990**, *112*, 5780.
73. Baldini, L.; Ballester, P.; Casnati, A.; Gomila, R. M.; Hunter, C. A.; Sansone, F.; Ungaro, R., *J. Am. Chem. Soc.* **2003**, *125*, 14181.
74. Ballester, P.; Costa, A.; Castilla, A. M.; Deya, P. M.; Frontera, A.; Gomila, R. M.; Hunter, C. A., *Chem. Eur. J.* **2005**, *11*, 2196.
75. Ballester, P.; Oliva, A. I.; Costa, A.; Deya, P. M.; Frontera, A.; Gomila, R. M.; Hunter, C. A., *J. Am. Chem. Soc.* **2006**, *128*, 5560.
76. Schuster, D. I.; Li, K.; Guldi, D. M.; Ramey, J., *Org. Lett.* **2004**, *6*, 1919.
77. Bradford, A. M.; Bradley, D. C.; Hursthouse, M. B.; Motevalli, M., *Organometallics* **1992**, *11*, 111.
78. Lefeber, D. J.; Kamerling, J. P.; Vliegenthart, J. F. G., *Org. Lett.* **2000**, *2*, 701.

1,3-Di-*tert*-butylimidazol-2-ylidene

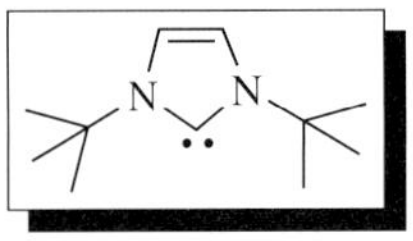

[157197-53-0] $C_{11}H_{20}N_2$ (MW 180.29)

InChI = 1/C11H20N2/c1-10(2,3)12-7-8-13(9-12)11(4,5)6/h7-8H, 1-6H3

InChIKey = FENRCIKTFREPGS-UHFFFAOYAG

(catalyst for organic transformations, ligand for transition metal complexes, base or substrate for organometallic chemistry)

Alternate Name: I^{t}Bu.

Physical Data: white to yellow solid.

Solubility: soluble in common organic solvents such as THF, DMF, methylene chloride, and acetonitrile.

Handling, Storage, and Precautions: air- and moisture-sensitive in both solid state and solution. Handle and store under inert atmosphere. Refrigeration is recommended for storage. MSDS is available online from Strem Chemicals.

Form Supplied in: commercially available as a solid.

Analysis of Reagent Purity: ^{1}H NMR and elemental analysis (C, H, N).

Preparative Methods: it can be prepared[1] under an inert atmosphere utilizing anhydrous, oxygen-free solvents. A mixture of I^{t}Bu-HCl and THF was treated with KO^{t}Bu at room temperature. The mixture was stirred for 0.5 h at room temperature and filtered through a pad of Celite, and solvents were removed under vacuum giving the crude product that could be further purified via recrystallization from cold hexane or sublimation.[2]

General Characteristics. I^{t}Bu has a relatively high basicity. The measured p*Ka* value of I^{t}Bu was 20 in THF.[3] It was found to be more basic in DMSO with a p*Ka* value of 22.7, which is in good agreement with a predicted p*Ka* value.[4]

Stability of I^{t}Bu toward oxygen and hydrolysis was studied.[5,6] I^{t}Bu was found to be quite stable to oxygen, but it can be oxidized in the presence of CuI or light. Hydrolysis of I^{t}Bu by water is very slow, therefore, I^{t}Bu can be conveniently handled in the laboratory.

As Catalyst for Organic Transformations. Owing to its strong nucleophilicity and basicity, I^{t}Bu can act as a stand-alone catalyst for a number of fundamental organic transformations. It has been used as an organocatalyst to catalyze transesterification of both carboxylic and phosphorus esters (eq 1).[7,8] A theoretical study indicated that I^{t}Bu serves as a base to facilitate the proton transfer.[9]

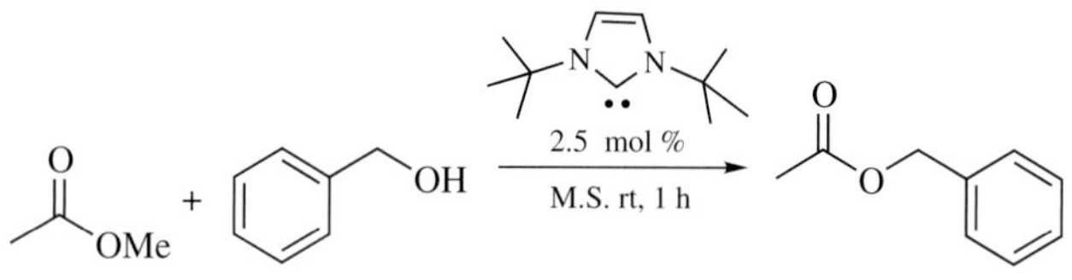

100% GC yield

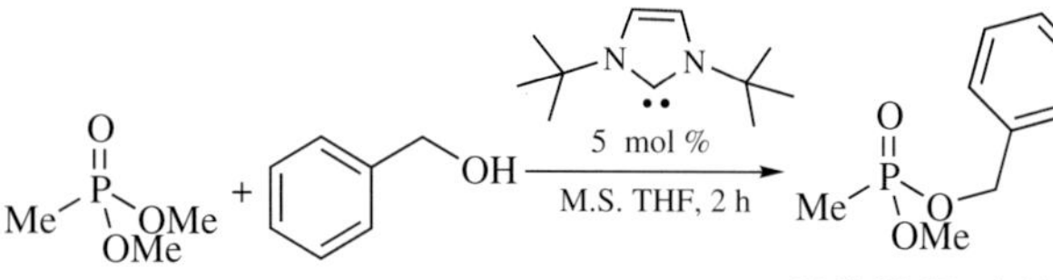

(1)

32% NMR yield
monobenzyl ester only

Recently imidazol-2-ylidenes, including I^{t}Bu were shown to be capable of catalyzing silicon-based transformations (eq 2). For example, I^{t}Bu was used to activate $TMSCF_3$ and TMSCN for trifluoromethylation[10] and cyanosilylation[11] of carbonyl compounds at low catalyst loadings (0.01–0.5 mol %). Furthermore, I^{t}Bu was reported to be an effective catalyst for Mukaiyama aldol reactions.[12]

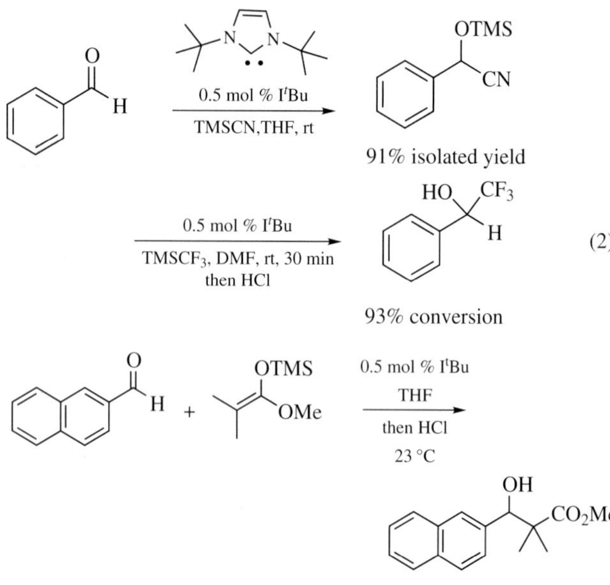

(2)

78% isolated yield

I^{t}Bu was also employed as a catalyst for ring opening polymerization of cyclic siloxanes, such as cyclotetrasiloxane D_4 (eq 3).[2]

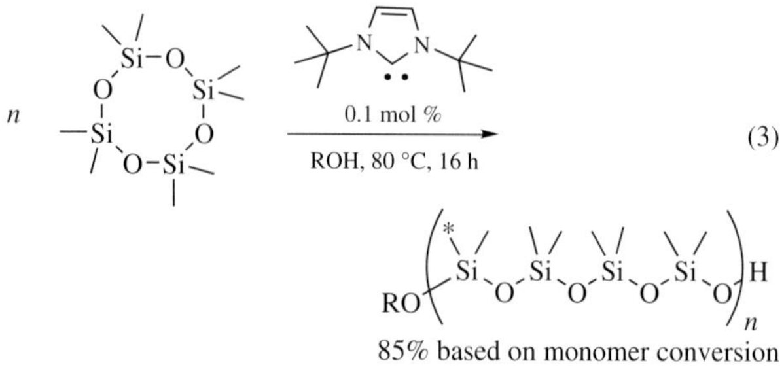

(3)

85% based on monomer conversion

Complexation with Ni. The zerovalent nickel carbene complex [Ni(I^{t}Bu)$_2$] **1** was first synthesized in 10% yield by co-condensation of nickel vapor with excess I^{t}Bu (eq 4).[13]

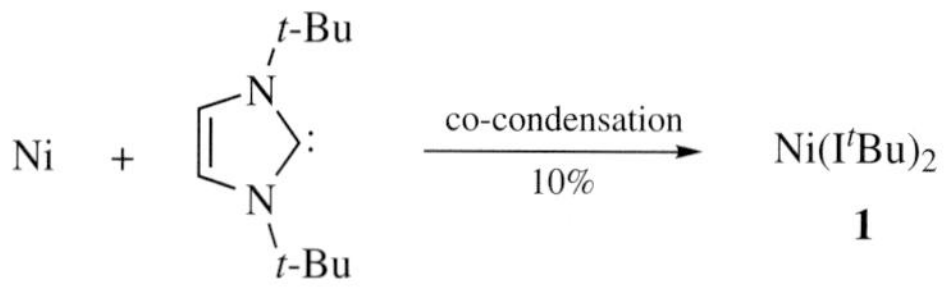

(4)

Attempts to develop a solution-phase synthesis of **2** by treatment of Ni(cod)$_2$ with excess I^{t}Bu in THF at 25 °C gave the binuclear complex **3** in 38% yield (eq 5).[14] The intermediate monocarbene complex **2**, which was also isolated, undergoes dissociation of a cod ligand, C–H insertion into a *t*-butyl group followed by loss of isobutene en route to complex **3**.

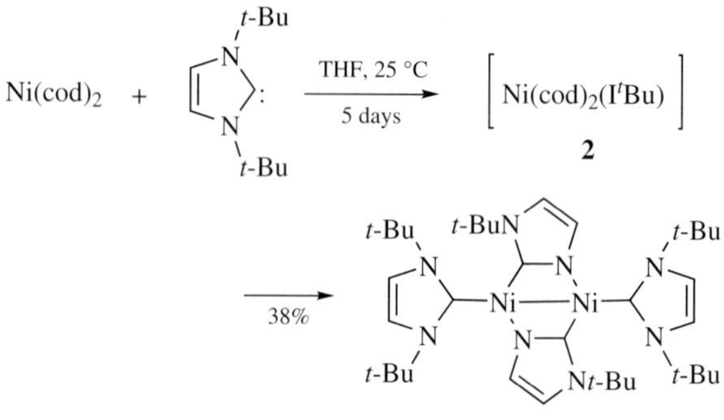

(5)

3

The treatment of Ni(CO)$_4$ with I^{t}Bu led to the isolation of the stable three-coordinate nickel carbonyl complex Ni(CO)$_2$(I^{t}Bu) **4** (eq 6).[15] A Ni-I^{t}Bu bond dissociation energy value of 39 ± 3 kcal/mol was determined from variable temperature FT-IR data.

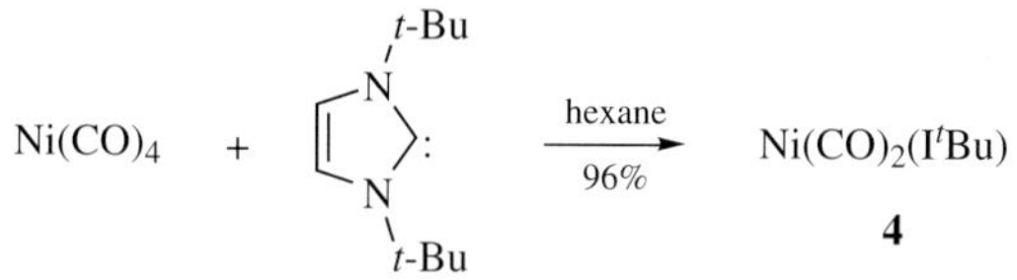

(6)

Studies of the electronic absorption and emission spectra of the related Ni(0) complex Ni(CO)$_2$(I^{t}Bu)$_2$, prepared from Ni(CO)$_4$ and I^{t}Bu in a 1:2 stoichiometry, showed a metal to ligand charge transfer absorption at 365 nm and a room temperature phosphorescence at 510 nm.[16]

The treatment of Ni(dme)Cl$_2$ and Ni(cod)$_2$ with I^{t}Bu (1:1:2 stoichiometry) gave the unexpected bis-μ-ethoxonickel(II) dimer **5** in 21% yield (eq 7).[17] The EtOH required for the formation of **5** was found to be present as a contaminant in the Ni(dme)Cl$_2$.

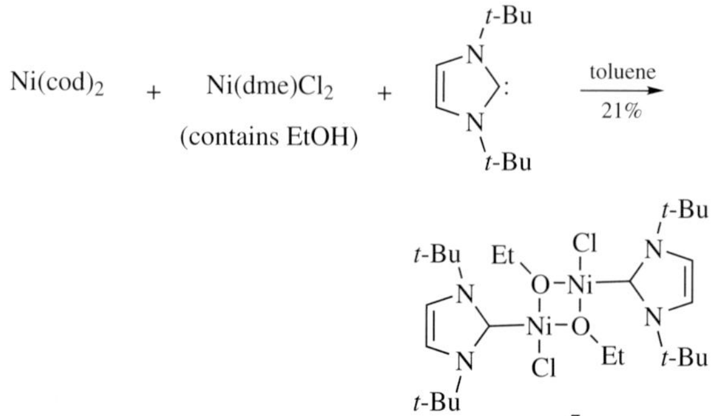

(7)

5

The π-allylchloro(I^{t}Bu)nickel(II) complex **6** was prepared from Ni(cod)$_2$ in a two-step, one-pot procedure in high yield (eq 8).[18]

Complex **6** was found to be resistant to reaction with O_2. This lack of reactivity was ascribed to blocking of the axial sites above and below the square plane by I^{t}Bu and/or the lack of conformational freedom of the complex. The related π-allylbromo(I^{t}Bu)nickel(II) complex was prepared from $[Ni(\eta^3\text{-}C_3H_5)Br]_2$ and 2 equiv of I^{t}Bu in 89% yield.[19]

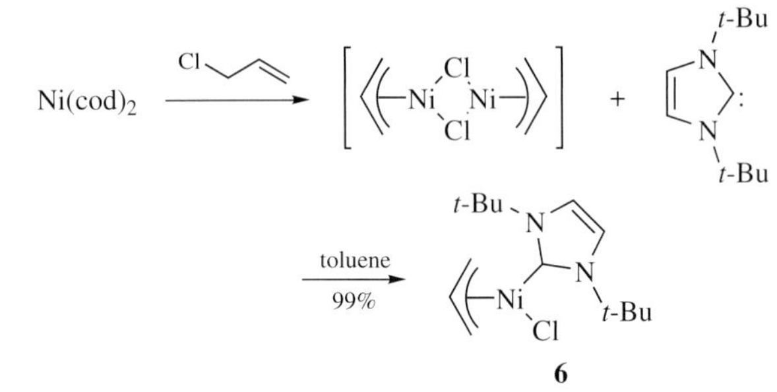

(8)

The dimeric nickelalactone (I^{t}Bu) complex **7** was synthesized from the monomeric bis(pyridyl) nickelalactone and I^{t}Bu and characterized by NMR, IR, and X-ray diffraction (eq 9).[20]

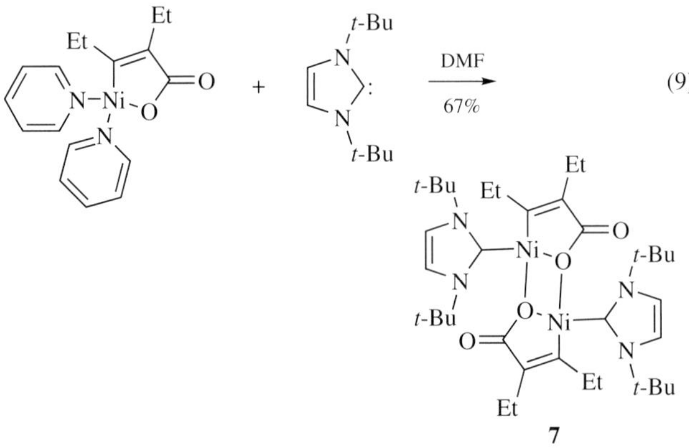

(9)

The reductive coupling of a 1,3-diene and an aldehyde using an in situ generated Ni(0)–I^{t}Bu complex has been reported (eq 10).[21] In contrast to other NHCs, the reaction using I^{t}Bu gave primarily the product **8** with an *E*-olefin conjugated to the aromatic ring, while other NHCs gave product with an internal *Z*-olefin.

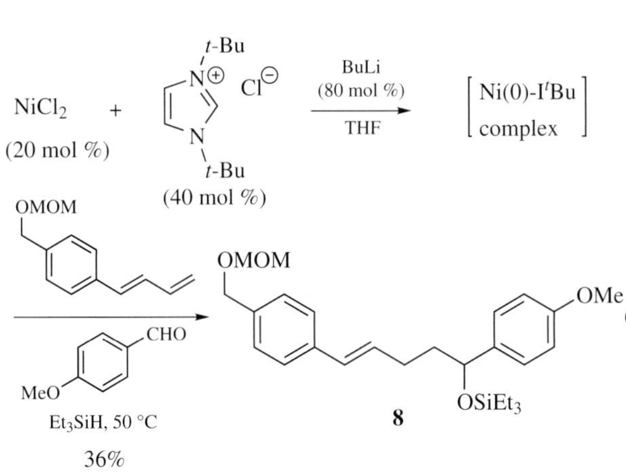

(10)

The intramolecular cycloisomerization of alkynes and vinylcyclopropanes using catalytic $Ni(cod)_2$ and I^{t}Bu (1:1 stoichiometry) has been reported (eq 11).[22] The reaction was performed using a variety of NHC ligands, but was unsuccessful using tertiary phosphine ligands.

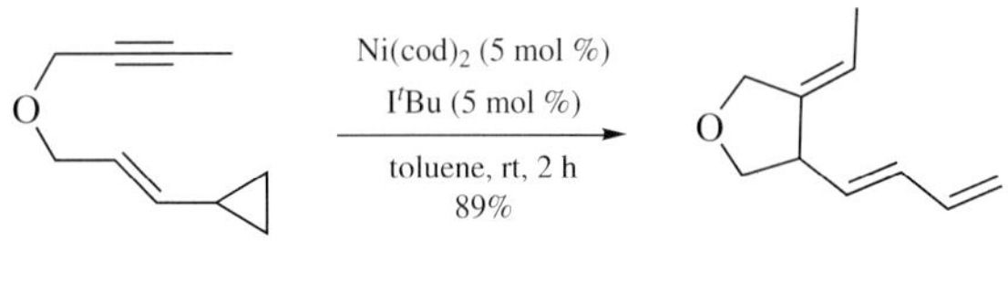

(11)

Complexation with Pd. NHCs have emerged as important alternative ligands to phosphines for palladium complexes and catalysts. NHCs such as I^{t}Bu act as a 2 electron neutral ligand through a strong σ bond with minimal π bonding.[23,24] The steric factor of I^{t}Bu for complexation with palladium can be conceptualized through a buried volume of 32.25%.[25]

One of the first synthesis of the pale yellow crystalline $[Pd(0)(I^tBu)_2]$ was by a metal vapor technique through the co-condensation of palladium vapor with I^{t}Bu in 32% yield (eq 12).[26] The crystal structure of **9** exhibited a linear carbene-metal-carbene vector (C-Pd-C angle 180.0°) with a torsion angle of 90° between the planes of the ligands.

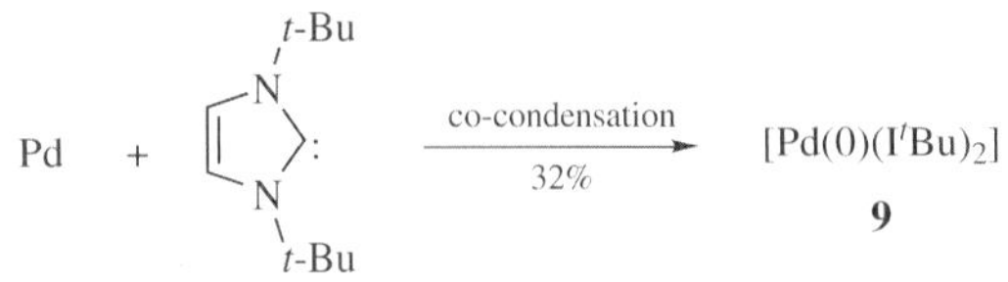

(12)

Alternatively, complex **9** was prepared by a more efficient ligand exchange procedure with $[(o\text{-tol})_3P)_2Pd(0)]$ in 61% yield (eq 13).[27] Also utilizing ligand substitution

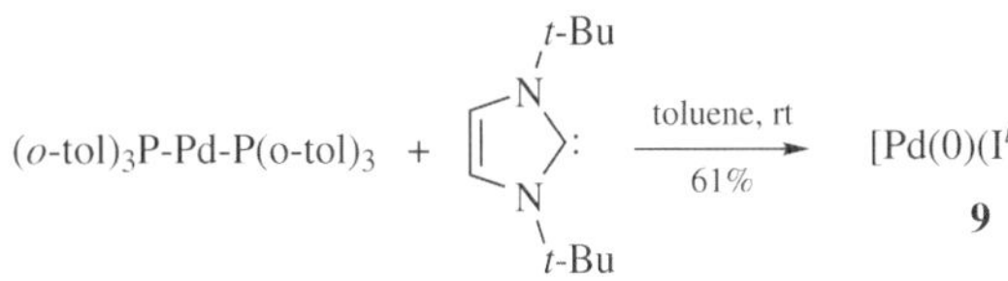

(13)

the respective mono(carbene) complex **10** was prepared in reasonable yields by simply controlling the reaction stoichiometry (eq 14).[28] The exchange is reversible and with

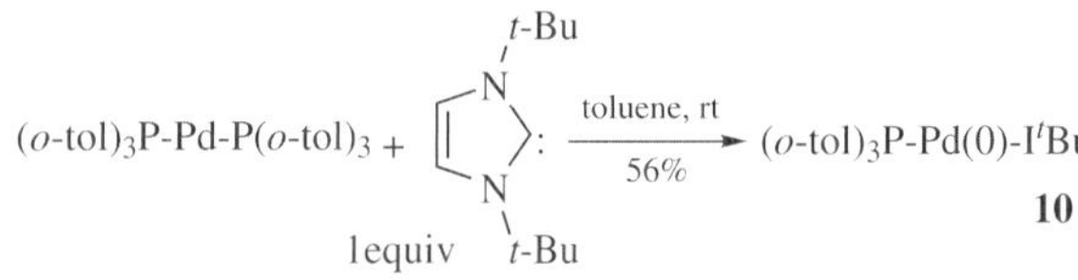

(14)

specific phosphines the mono(carbene) adduct can be prepared in good yield by treating the bis(carbene) complex **9** with an appropriate phosphine ligand (eq 15).[28]

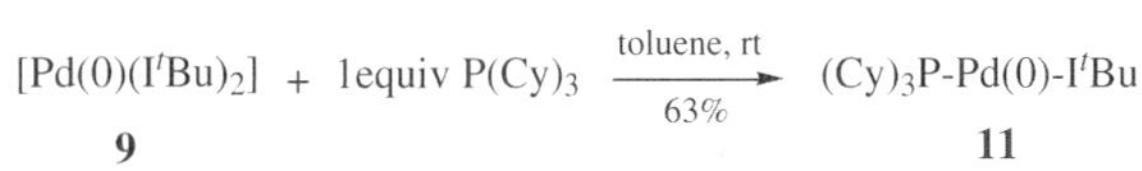

(15)

On the contrary to the above synthesis of $[Pd(0)(I^tBu)_2]$ that requires the use of a palladium(0) precursor, complex **9** was prepared by treating the more stable $[(\eta^2C_3H_5)PdCl]_2$ with the carbene and sodium malonate in good yield (eq 16).[29] Without the malonate nucleophile, the reaction afforded the π-allyl-Pd(II)-NHC complexes **12a** and **12b** in 90% and 75% yields, respectively (eq 17).[25,29–31] The iodo complex **12b** was also prepared in good yield from the chloro-adduct **12a** by treatment with an excess of sodium iodide.[31] Alternatively, the free carbene I^{t}Bu can be generated in situ and employed in the synthesis of the Pd(II) complex **12a**. The crystal structure of the Pd(II) complex **12a** exhibited a pseudo-square-planar geometry. Additionally, the NHC trans effect was demonstrated by a significantly longer C_{allyl}–Pd bond trans to the carbene in comparison to the C_{allyl}–Pd bond trans to the chloride. Treatment of the allyl–I^{t}Bu–Pd(II) complex **12a** with $TlPF_6$ or TlOTf furnished the cationic dinuclear complexes **13**(eq 18).[31] Complex (**13a**) also exhibited a pseudo-square-planar geometry about each palladium atom with the NHC trans effect elongating the C_{allyl}–Pd bond trans to the carbene ligand.

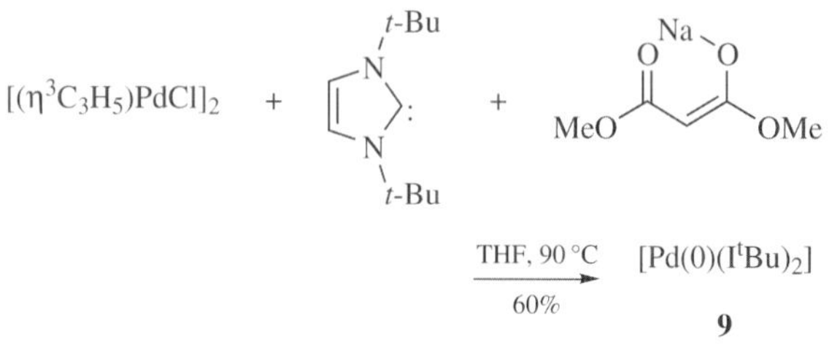

(16)

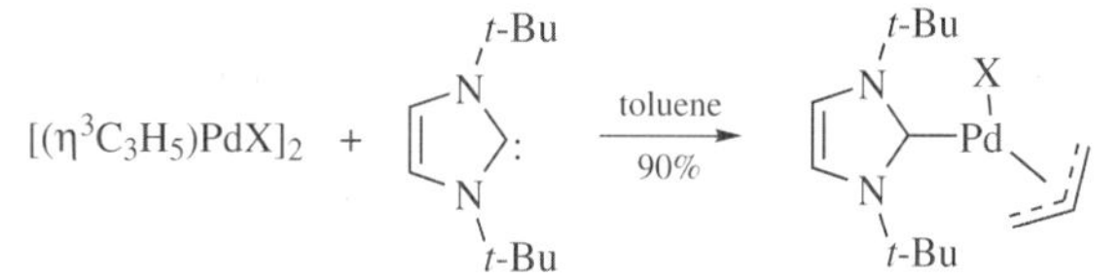

(17)

12a X = Cl 70–90%
12b X = I 75%

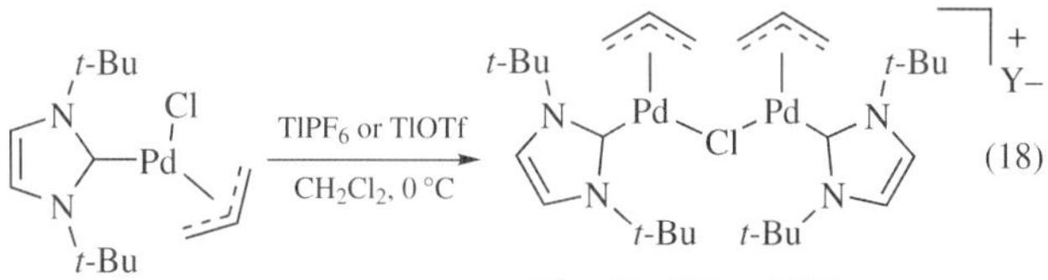

(18)

12a

13a Y = PF_6 76%
13b Y = OTf 85%

Herrmann and co-workers introduced a new class of palladacycles by the synthesis of the NHC complex **14**.[32] This complex was easily prepared in high yield by treatment of Herrmann's palladacycle with I^{t}Bu (eq 19).

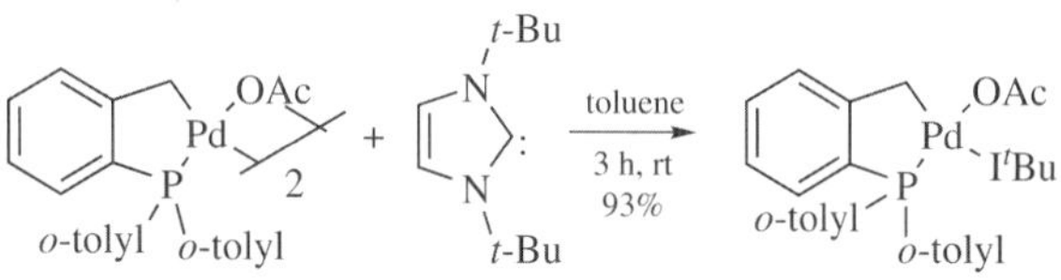

(19)

14

During mechanistic studies on palladium catalyzed aminations, Caddick and co-workers prepared the oxidative addition adduct **15** of complex **9** quantitatively by NMR (eq 20).[33] Crystallographic studies of complex **15** revealed a square-planar geometry with the carbene ligands orientated trans to each other. On the contrary to the structure of complex **9**, the carbene ligands are nearly coplanar to accommodate the additional two ligands.

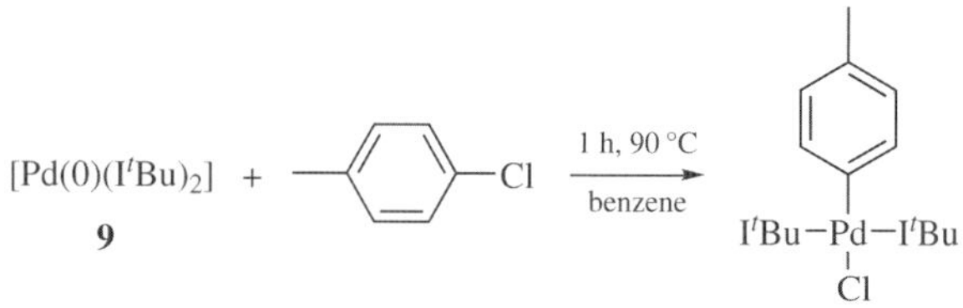

(20)

15

NHCs have been found to be an important alternative ligands to phosphines in numerous palladium catalyzed cross couplings involving aryl chloride and other deactivated substrates. The improved properties of these catalysts have been attributed to the strong σ-donor nature of NHC over phosphine ligands and robust NHC-palladium bond.[34] One of the first examples of a catalyzed transformation with [Pd(0)(I^{t}Bu)$_2$] **9** was in a Suzuki–Miyaura coupling (eq 21).[27] Although a moderate yield was obtained, a high initial turnover frequency of 552 (mole product x mole $Pd^{-1} \times h^{-1}$) was observed. Beller and co-workers also reported the efficient Heck and Suzuki–Miyaura reactions of aryl chlorides with the I^{t}Bu Pd(0) complex **16**; however, use of other carbene ligands were shown to afford more efficient catalysts in these transformations (eq 22).[35]

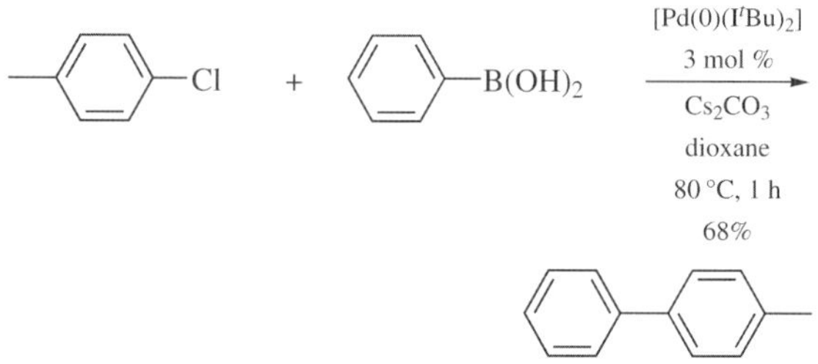

(21)

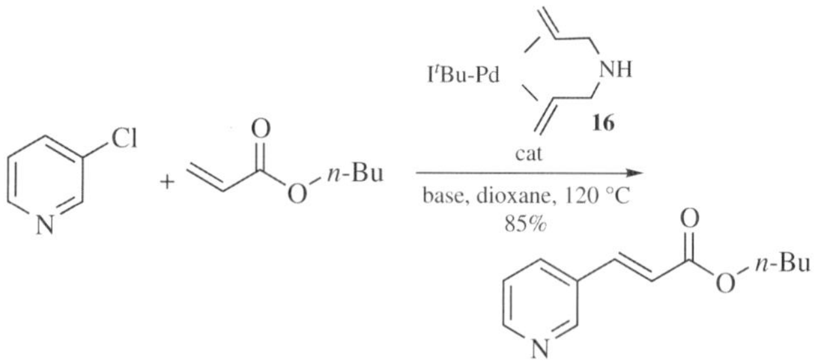

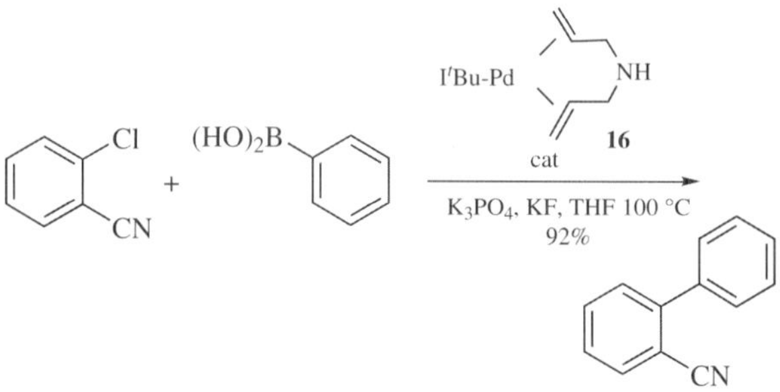

(22)

I^{t}Bu was demonstrated as an effective ligand for the palladium catalyzed Buchwald–Hartwig aminations of aryl chlorides. Utilizing 2 mol % of complex **9**, the amination of *p*-tolylchloride proceeded in high yield (eq 23).[29] The Pd(II) precatalyst **12a** is also effective for this specific amination; however, other carbene ligands were shown to afford more reactive catalysts.[25] Kinetic studies by Lewis and co-workers demonstrated that the oxidative addition proceeds through a dissociative mechanism, wherein the dissociation of complex **9** is rate determining under typical amination conditions.[34] Computational studies by Green and co-workers further support the oxidative addition occurring through the mono-carbene Pd(0) intermediate.[36]

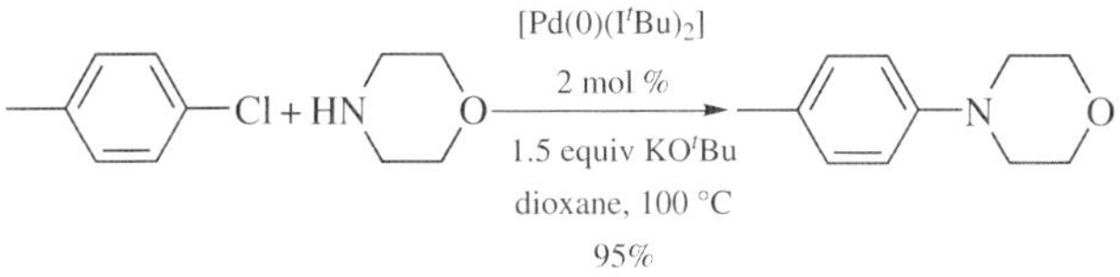
(23)

Faller and co-workers briefly examined the effect of the I^{t}Bu ligand in a palladium catalyzed allylic alkylation.[37] The alkylation with scalemic carbonate proceeded with only a modest memory effect and a poor branched to linear selectivity (eq 24).

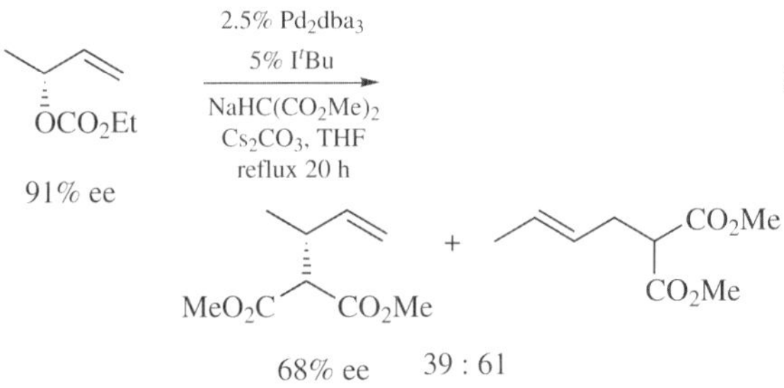
(24)

The palladium(0) carbene complex **9** was shown to be an effective substrate for the preparation of Pd(0) silylene complexes. Zeller and co-workers demonstrated by NMR the complete ligand substitution of I^{t}Bu with silylenes (eq 25).[38] DFT calculations indicated that steric interactions are important factors favoring the facile ligand exchange.

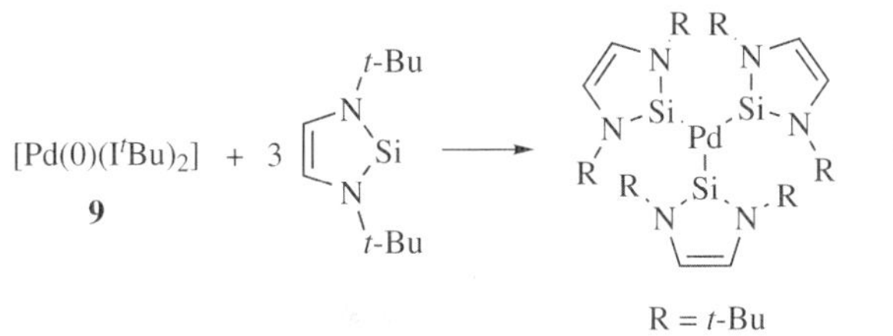
(25)

Complexation with Other Metals. Carbene I^{t}Bu has found broad utility as a ligand for numerous transition metals. For example, the mono-carbene gold (I) complex **17** was prepared by Singh and co-workers in good yield through ligand substitution with Au(CO)Cl (eq 26).[39] Reaction of this complex with ethynyl magnesium chloride provided the corresponding ethynyl Au(I) complex **18**. The crystal structures of both Au(I) complexes revealed monomeric carbene adducts with no Au(I)–Au(I) interactions.

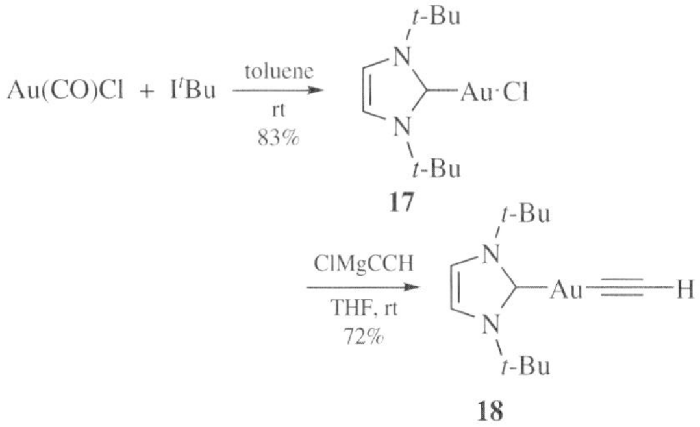
(26)

Herrmann and co-workers prepared both mono- and bis-carbene Rh(I) complexes.[40] Treatment of the carbene with $[(COD)RhCl]_2$ provided the mono-carbene Rh adduct **19** in low yield (eq 27). Utilizing in situ generated $[(COD)RhOEt]_2$, the bis-carbene complex **20** was prepared in good yield.

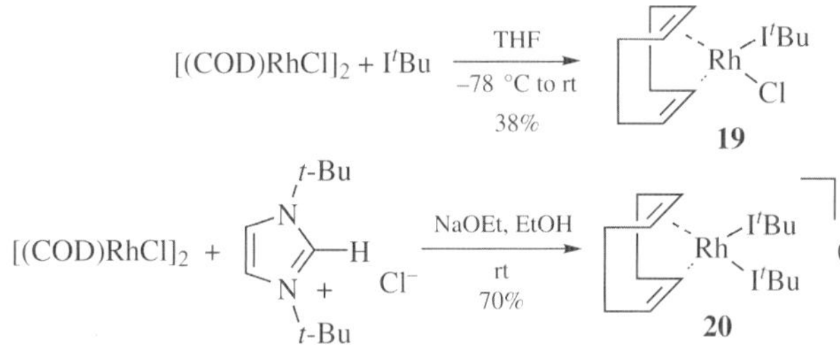
(27)

By examining the reactivity of I^{t}Bu with rhodium complexes, Nolan and co-workers prepared stable, unsaturated, and high valent NHC rhodium complexes.[41] Depending on the conditions employed, the precursors and the doubly C–H activated adduct **23** were isolated (eq 28). In the mono-cyclometalated adduct **22**, X-ray analysis showed the carbenes are positioned trans to each other with an agostic interaction with one ligand. Crystallographic studies on the doubly cyclometalated complex **23** depicted a distorted square pyramidal structure with the carbenes orientated trans to each other at a 167.43° angle. Treatment of the pentane solution of complex **21** with benzene and excess NHC also enabled the formation of the doubly cyclometalated complex **23** to demonstrate that complex **21** is a precursor to complex **23**. The cationic 14-electron Rh(III) complex **24** was also prepared by reacting complex **23** with $AgPF_6$. This four-coordinate doubly cyclometalated complex adopts a highly distorted cis-divacant octahedral structure. The vacant sites in complexes **23** and **24** are accessible to carbon monoxide, and after treatment with CO, the corresponding saturated complexes **25** and **26** were prepared in high yields.

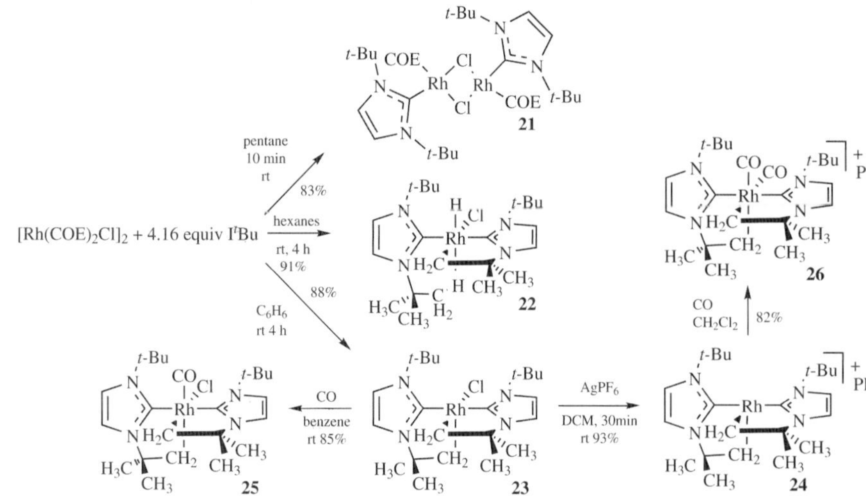

(28)

Ruthenium alkylidene complexes have become the reagents of choice for a variety of olefin metathesis reactions.[42] The incorporation of an NHC ligand into these complexes in place of a phosphine has led to catalysts with excellent stability, reactivity and substrate scope.[43] The treatment of $RuCl_2(=CHPh)(Ph_3P)_2$ with 1.2 equiv of I^tBu allowed formation of the mixed NHC/phosphine complex **27** in 70% yield (eq 29).[44]

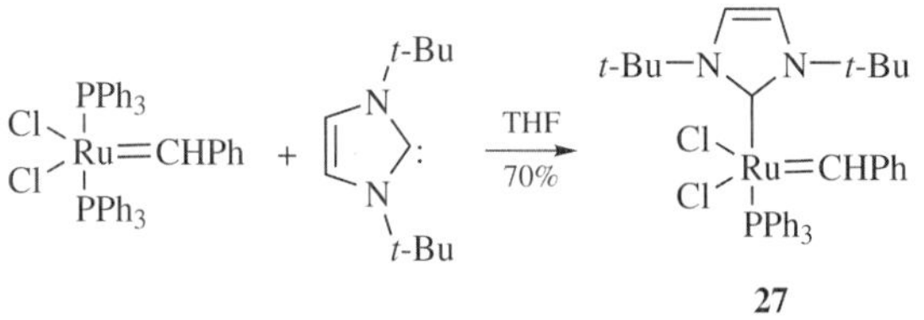

(29)

A ruthenium hydride complex containing an I^tBu ligand has been prepared.[45] Treatment of $RuH(Cl)(PPh_3)_3$ with 2 equiv of I^tBu in refluxing THF and subsequent exposure to hydrogen gave the isomeric complexes **28** and **29** in 50% combined yield and a 7:3 ratio (eq 30). A hydrogen from one of the *tert*-butyl methyl groups forms an agostic bond with Ru.

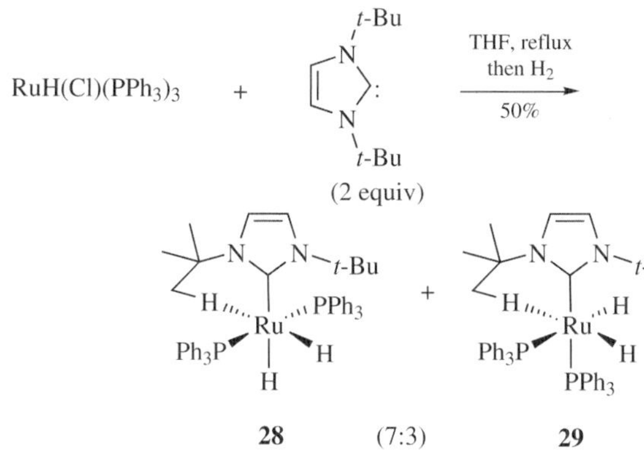

(30)

In analogy to Nolan and co-workers' previously synthesized NHC Rh complexes, they demonstrated related chemistry with Ir.[41b] Both the mono- and doubly cyclometalated adducts **30** and **31** were prepared and isolated in high yield (eq 31). The cationic 14 electron Ir complex **32**, and the saturated adducts **33** and **34** were also prepared in an analogous manner to Rh. Complex **32** exhibited a similar highly distorted divacant octahedral geometry as observed with the Rh analog **24**.

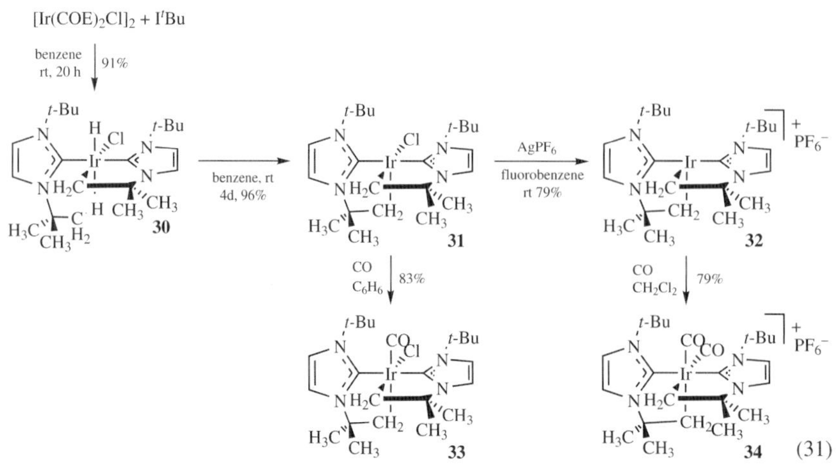

(31)

By analogy to the Rh complexes (eq 27), Herrmann and co-workers also prepared both the mono- and bis- carbene Ir(I) analogs.[40] Accordingly, utilizing either $[(COD)IrCl]_2$ or $[(COD)IrOEt]_2$, the respective complexes **35** and **36** were prepared in low to moderate yields (eq 32).

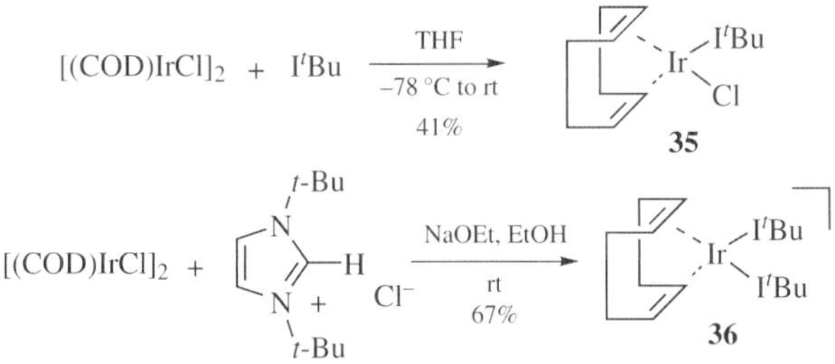

(32)

Arnold and co-workers prepared the analogous NHC platinum(0) complex to palladium with use of their metal vapor technique.[26] The co-condensation of platinum vapor with I^tBu provided the dark yellow crystalline $[Pt(0)(I^tBu)_2]$ **37** in 25% yield (eq 33). Similar to the palladium(0) complex, the crystal structure of complex **37** exhibited a linear carbene-metal-carbene vector (C-Pt-C angle 180.0°) with the planes of the ligands staggered by 90°.

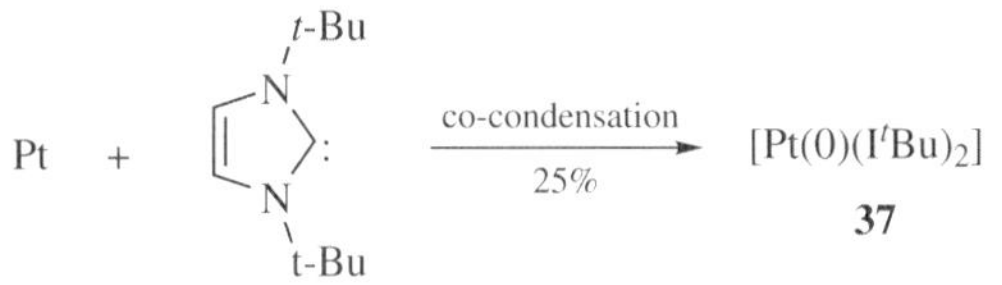

(33)

The dimeric I^tBu zinc complex **38** (eq 34) was briefly reported by Jensen and co-workers for the stereoselective polymerization of D,L-lactide.[46]

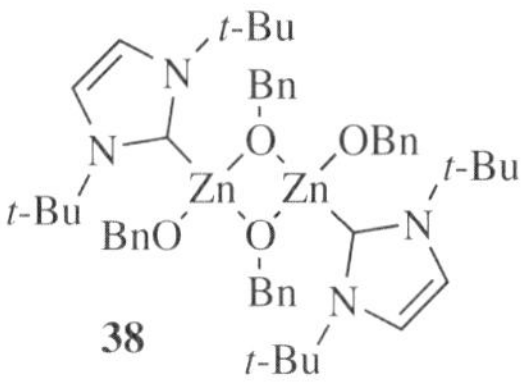

(34)

Reactions with Azides. Tamm and co-workers utilized I^tBu as a substrate for the preparation of imidazolin-2-iminato ligands for both rhenium and titanium. The silylated 2-aminoimidazoline ligand **39** was prepared in high yield by treatment of I^tBu with trimethylsilylazide (eq 35).[47] Reaction of the ligand precursor with Re_2O_7[47] and $CpTiCl_3$[48] provided the respective imidazolin-2-iminato complexes **40** and **41**.

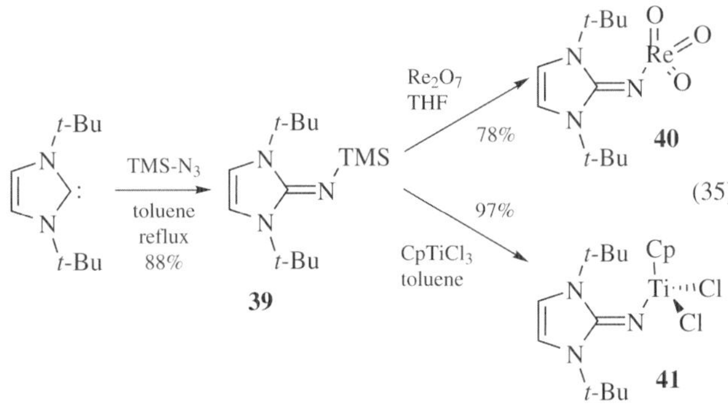

(35)

Bielawaski et al. further demonstrated that alkyl, aryl, acyl, and tosyl azides could all react in a similar manner with imidazol-2-ylidenes such as I^tBu to give a range of substituted triazenes.[49]

As Base for Organometallic Chemistry. Due to the strongly basic nature of I^tBu, the carbene has found utility in the preparation of aluminum and gallium complexes. Both the aluminum amide **43** and hydroxide complexes **44** were prepared in good yield with use of 2 equiv of I^tBu (eq 36).[50] The analogous gallium mono-hydroxide complex **46** was also prepared in good yield with use of 1 equiv of I^tBu (eq 37).[51,52]

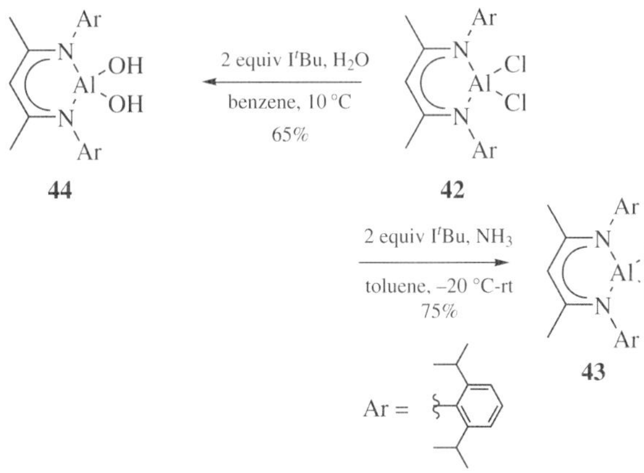

(36)

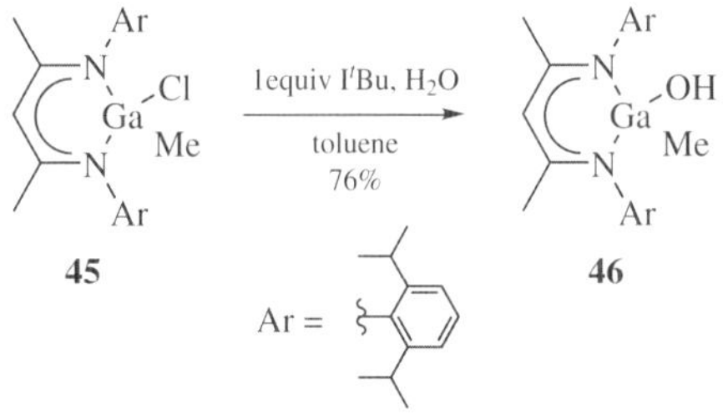

(37)

Related Reagents. ICy [1,3-bis(cyclohexyl)imidazol-2-ylidene]; ITol [1,3-bis(4-methylphenyl)imidazol-2-ylidene]; IAd [1,3-bis(adamantyl)imidazol-2-ylidene]; IMes [1,3-bis(2,4,6-trimethylphenyl)imidazol-2-ylidene]; IPr [1,3-bis(2,6-diisopropylphenyl)-imidazol-2-ylidene].

1. Arduengo, A. J., III; Bock, H.; Chen, H.; Denk, M.; Dixon, D. A.; Green, J. C.; Herrmann, W. A.; Jones, N. L.; Wagner, M.; West, R., *J. Am. Chem. Soc.* **1994**, *116*, 6641.
2. Rodriguez, M.; Marrot, S.; Kato, T.; Sterin, S.; Fleury, E.; Baceiredo, A., *J. Organomet. Chem.* **2007**, *692*, 705.
3. Kim, Y.-J.; Streitwieser, A., *J. Am. Chem. Soc.* **2002**, *124*, 5757.
4. Magill, A. M.; Cavell, K. J.; Yates, B. F., *J. Am. Chem. Soc.* **2004**, *126*, 871.
5. Ishiguro, K.; Hirabayashi, K.; Nojima, T.; Sawaki, Y., *Chem. Lett.* **2002**, 796.
6. Denk, M. K.; Rodezno, J. M.; Gupta, S.; Lough, A. J., *J. Organomet. Chem.* **2001**, *617*, 242.
7. Grasa, G. A.; Gueveli, T.; Singh, R.; Nolan, S. P., *J. Org. Chem.* **2003**, *68*, 2812.
8. Singh, R.; Nolan, S. P., *Chem. Commun.* **2005**, 5456.
9. Lai, C.-L.; Lee, H. M.; Hu, C.-H., *Tetrahedron Lett.* **2005**, *46*, 6265.

10. Song, J. J.; Tan, Z.; Reeves, J. T.; Gallou, F.; Yee, N. K.; Senanayake, C. H., *Org. Lett.* **2005**, *7*, 2193.
11. Song, J. J.; Gallou, F.; Reeves, J. T.; Tan, Z.; Yee, N. K.; Senanayake, C. H., *J. Org. Chem.* **2006**, *71*, 1273.
12. Song, J. J.; Tan, Z.; Reeves, J. T.; Yee, N. K.; Senanayake, C. H., *Org. Lett.* **2007**, *9*, 1013.
13. Arnold, P. L.; Cloke, F. G. N.; Geldbach, T.; Hitchcock, P. B., *Organometallics* **1999**, *18*, 3228.
14. Caddick, S.; Cloke, F. G. N.; Hitchcock, P. B.; Lewis, A. K. K., *Angew. Chem. Int. Ed.* **2004**, *43*, 5824.
15. (a) Dorta, R.; Stevens, E. D.; Hoff, C. D.; Nolan, S. P. J., *Am. Chem. Soc.* **2003**, *125*, 10490. (b) Dorta, R.; Stevens, E. D.; Scott, N. M.; Costabile, C.; Cavallo, L.; Hoff, C. D.; Nolan, S. P., *J. Am. Chem. Soc.* **2005**, *127*, 2485.
16. Kunkely, H.; Vogler, A., *J. Organomet. Chem.* **2003**, *684*, 113.
17. Dible, B. R.; Sigman, M. S.; Arif, A. M., *Inorg. Chem.* **2005**, *44*, 3774.
18. Dible, B. R.; Sigman, M. S., *Inorg. Chem.* **2006**, *45*, 8430.
19. Silva, L. C.; Gomes, P. T.; Veiros, L. F.; Pascu, S. I.; Duarte, M. T.; Namorado, S.; Ascenso, J. R.; Dias, A. R., *Organometallics* **2006**, *25*, 4391.
20. Langer, J.; Walther, D.; Görls, H., *J. Organomet. Chem.* **2006**, *691*, 4874.
21. Sato, Y.; Sawaki, R.; Mori, M., *Organometallics* **2001**, *20*, 5510.
22. Zuo, G.; Louie, J., *J. Am. Chem. Soc.* **2005**, *127*, 5798.
23. Crabtree, R.H.; *The Organometallic Chemistry of the Transition Metals*; Whiley & Sons: New York, **2001**; Ch 11.
24. Green, J. C.; Scurr, R. G.; Arnold, P. L.; Cloke, F. G. N., *Chem. Commun.* **1997**, 1963.
25. Viciu, M. S.; Navarro, O.; Germaneau, R. F.; Kelly, R. A., III; Sommer, W.; Marion, N.; Stevens, E. D.; Cavallo, L.; Nolan, S. P., *Organometallics* **2004**, *23*, 1629.
26. Arnold, P. L.; Cloke, F. G. N.; Geldbach, T.; Hitchcock, P. B., *Organometallics* **1999**, *18*, 3228.
27. Böhm, V. P. W.; Gstöttmayr, C. W. K.; Weskamp, T.; Herrmann, W. A., *J. Organomet. Chem.* **2000**, *595*, 186.
28. Titcomb, L. R.; Caddick, S.; Cloke, F. G. N.; Wilson, D. J.; McKerrecher, D., *Chem. Commun.* **2001**, 1388.
29. Caddick, S.; Cloke, F. G. N.; Clentsmith, G. K. B.; Hitchcock, P. B.; McKerrecher, D.; Ticomb, L. R.; Williams, M. R. V., *J. Organomet. Chem.* **2001**, *617*, 635.
30. Viciu, M. S.; Germaneau, R. F.; Navarro-Fernandez, O.; Stevens, E. D.; Nolan, S. P., *Organometallics* **2002**, *21*, 5470.
31. Ding, Y.; Goddard, R.; Pörschke, K.-R., *Organometallics* **2005**, *24*, 439.
32. (a) Frey, G. D.; Schütz, J.; Herdtweck, E.; Herrmann, W. A., *Organometallics* **2005**, *24*, 4416. (b) Frey, G. D.; Schütz, J.; Herrmann, W. A., *J. Organomet. Chem.* **2006**, *691*, 2403.
33. Caddick, S.; Cloke, F. G. N.; Hitchcock, P. B.; Leonard, J.; Lewis, A. K.; de K. McKerrecher, D.; Titcomb, L. R., *Organometallics* **2002**, *21*, 4318.
34. Lewis, A. K.; de K. Caddick, S.; Cloke, F. G. N.; Billingham, N. C.; Hitchcock, P. B.; Leonard, J., *J. Am. Chem. Soc.* **2003**, *125*, 10066 and references therein.
35. Beller, M.; Andreo, M. G.; Zapf, A.; Karch, R.; Kleinwaechter, I.; Briel, O.; Preparation of carbene diene complexes of nickel, palladium, and platinum as catalysts for Heck or coupling reactions. Ger. Offen. (**2002**).
36. Green, J. C.; Herbert, B. J.; Lonsdale, R., *J. Organomet. Chem.* **2005**, *690*, 6054.
37. Faller, J. W.; Sarantopoulos, N., *Organometallics* **2004**, *23*, 2179.
38. Zeller, A.; Bielert, F.; Haerter, P.; Herrmann, W. A.; Strassner, T., *J. Organomet. Chem.* **2005**, *690*, 3292.
39. Singh, S.; Kumar, S. S.; Jancik, V.; Roesky, H. W.; Schmidt, H.-G.; Noltemeyer, M., *Eur. J. Inorg. Chem.* **2005**, 3057.
40. Frey, G. D.; Rentzsch, C. F.; Preysing, D. V.; Scherg, T.; Mühlhofer, M.; Herdtweck, E.; Herrmann, W. A., *J. Organomet. Chem.* **2006**, *691*, 5725.
41. (a) Dorta, R.; Stevens, E. D.; Nolan, S. P., *J. Am. Chem. Soc.* **2004**, *126*, 5054. (b) Scott, N. M.; Dorta, R.; Stevens, E. D.; Correa, A.; Cavallo, L.; Nolan, S. P., *J. Am. Chem. Soc.* **2005**, *127*, 3516.
42. Schwab, P.; Grubbs, R. H.; Ziller, J. W., *J. Am. Chem. Soc.* **1996**, *118*, 100.
43. (a) Scholl, M.; Trnka, T. M.; Morgan, J. P.; Grubbs, R. H., *Tetrahedron Lett.* **1999**, *40*, 2247. (b) Scholl, M.; Ding, S.; Lee, C. W.; Grubbs, R. H., *Org. Lett.* **1999**, *1*, 953.
44. (a) Hermann, W. A.; Schattenmann, W. C.; Weskamp, T.; DE Pat. 19,815,275.2 (April 6, **1998**). (b) Weskamp, T.; Kohl, F. J.; Herrmann, W. A., *J. Organomet. Chem.* **1999**, *582*, 362. (c) Hermann, W. A.; Schattenmann, W. C.; Weskamp, T.; U.S. Pat. 2005/0107626A1, (May 19, **2005**)
45. Abdur-Rashid, K.; Fedorkiw, T.; Lough, A. J.; Morris, R. H., *Organometallics* **2004**, *23*, 86.
46. Jensen, T. R.; Breyfogle, L. E.; Hillmyer, M. A.; Tolman, W. B., *Chem. Commun.* **2004**, 2504.
47. Tamm, M.; Beer, S.; Herdtweck, E., *Z. Naturforsch., B.: Chem. Sci.* **2004**, *59*, 1497.
48. Tamm, M.; Randoll, S.; Herdtweck, E.; Kleigrewe, N.; Kehr, G.; Erker, G.; Rieger, B., *Dalton Trans.* **2006**, 459.
49. Khramov, D. M.; Bielawski, C. W., *Chem. Commun.* **2005**, 4958.
50. Jancik, V.; Pineda, L. W.; Pinkas, J.; Roesky, H. W.; Neculai, D.; Neculai, A. M.; Herbst-Irmer, R., *Angew. Chem., Int. Ed. Engl.* **2004**, *43*, 2142.
51. Singh, S.; Jancik, V.; Roesky, H. W.; Herbst-Irmer, R., *Inorg. Chem.* **2006**, *45*, 949.
52. Jancik, V.; Pineda, L. W.; Stueckl, A. C.; Roesky, H. W.; Herbst-Irmer, R., *Organometallics* **2005**, *24*, 1511.

Jinhua J. Song, Jonathan T. Reeves, Daniel R. Fandrick, Zhulin Tan, Nathan K. Yee & Chris H. Senanayake
Boehringer Ingelheim Pharmaceuticals, Inc., Ridgefield, CT, USA

Diethyl 1,4-Dihydro-2,6-dimethyl-3,5-pyridinedicarboxylate[1]

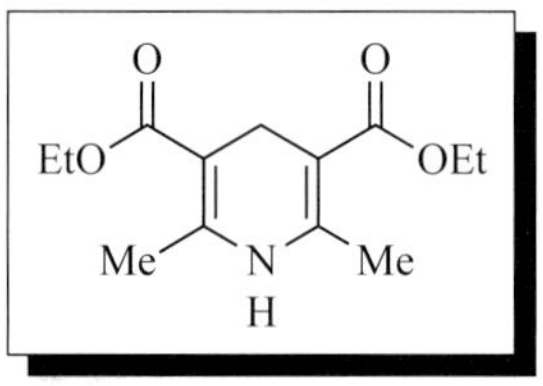

[1149-23-1] $C_{13}H_{19}NO_4$ (MW 253.29)
InChI = 1S/C13H19NO4/c1-5-17-12(15)10-7-11(13(16)18-6-2)9(4)14-8(10)3/h14H,5-7H2,1-4H3
InChIKey = LJXTYJXBORAIHX-UHFFFAOYSA-N

(reagent used as a source of hydride)

Alternate Names: HEH; Hantzsch ester hydride; Hantzsch ester; Hantzsch dihydropyridine; 1,4-dihydropyridine.
Physical Data: mp 175–180 °C, bright yellow solid.[2] Large variations in the melting point have been reported depending on the method of synthesis and purification.
Solubility: sol in most organic solvents.
Form Supplied in: solid (95% purity).

Analysis of Reagent Purity: ^{1}H NMR (1:1 DMSO/$CDCl_3$, 400 MHz): δ 7.99 (s, 1H), 4.04 (q, J = 6.8 Hz, 4H), 3.09 (s, 2H), 2.09 (s, 6H), 1.20 (t, J = 7.2 Hz, 6H). ^{13}C NMR ($CHCl_3$, 100 MHz): δ 168.3, 145.1, 99.7, 59.9, 25.0, 19.3, 14.7. FTIR (cm^{-1}) (neat): 3346, 3108, 2953, 2924, 2854, 1697, 1646, 1506.

Preparative Method: the reagent is most often prepared by a multicomponent reaction involving formaldehyde, ethyl acetoacetate, and ammonium acetate or ammonia.[2–6]

Purification: recrystallization from EtOH or flash chromatography with petroleum/ethyl acetate (7:3).[4,5]

Handling, Storage, and Precautions: wear suitable gloves. Store compound in a dry container and in a cool place. Avoid contact with eyes and skin. Do not breathe dust.

Preparation Methods. The Hantzsch dihydropyridine synthesis is the most common approach used to prepare HEH. It is a multicomponent reaction between formaldehyde or paraformaldehyde, 2 equiv of a β-ketoester, such as ethyl acetoacetate, and a nitrogen donor, such as ammonia or ammonium acetate (eq 1).[2–5] A recent improvement is to run the reaction under neat conditions (80–88% on a 1 mol scale).[6] The reaction can be applied to the preparation of various substituted 1,4-dihydropyridines (eqs 2–4).[7]

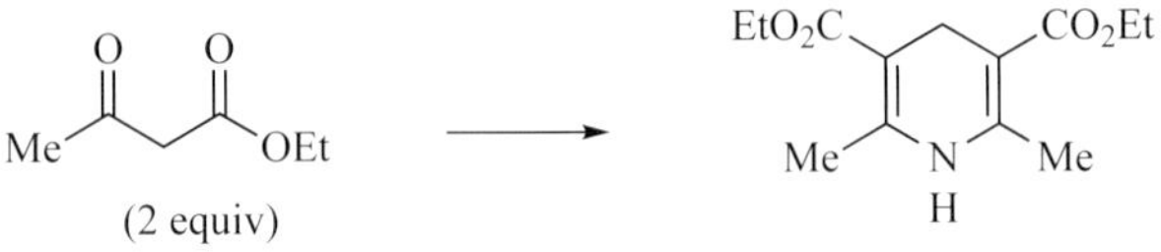

(1)

HCHO (1.04 equiv), Et_2NH (cat), H_2O, 0 °C, 45 h, then NH_3 (excess), EtOH, 0 °C, 50 h — 84–89%

HCHO (1 equiv), NH_4OAc (1.5 equiv), 80 °C, 45 min — 87%

HCHO (1 equiv), NH_4OAc (1–2 equiv), EtOH or H_2O, Bu_4NBr (cat), mwave, 0.75–3 min — 92–96%

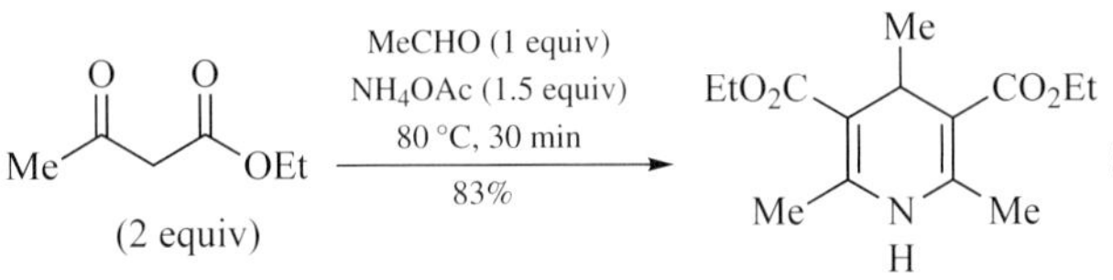

(2)

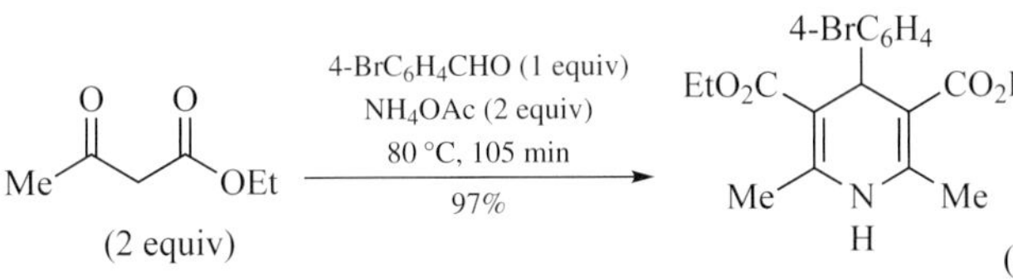

(3)

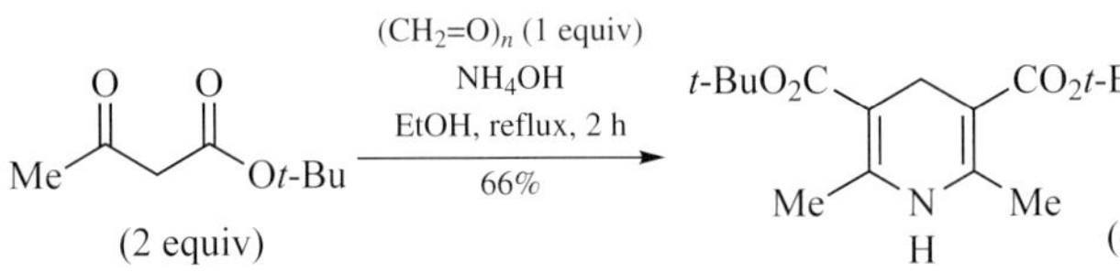

(4)

HEH can also be synthesized by the reduction of the corresponding tetrasubstituted pyridine **1** with $NaBH_3CN$ (eq 5).[8]

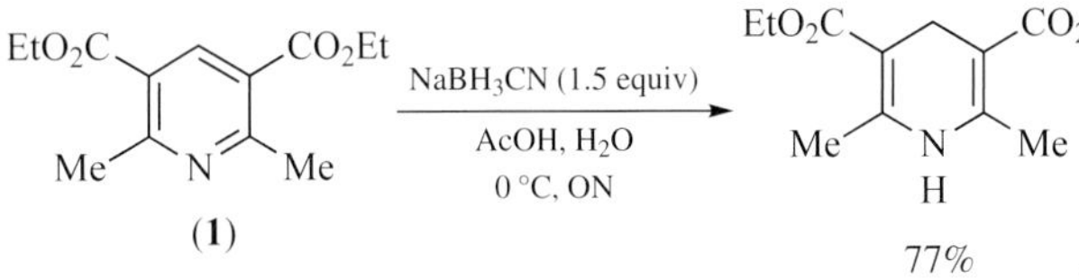

(5)

An alternative method is the acid-catalyzed reaction between tetrahydro-1,3-oxazine (**2**) and 2 equiv of an enamine (**3**), forming a bisenaminylmethane derivative. This then undergoes cycloelimination to give the corresponding 1,4-dihydropyridine (eq 6).[9,10]

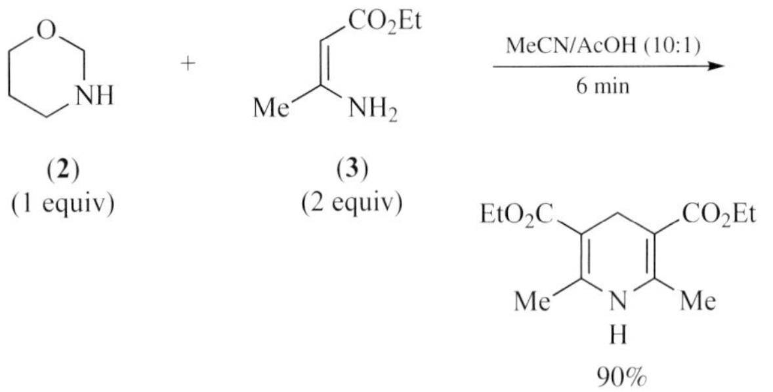

(6)

It is also possible to synthesize substituted Hantzsch ester analogs in enantioenriched forms by a multicomponent process catalyzed by a chiral substituted prolinol derivative.[11]

Biomimetic Characteristics of Hantzsch Ester Hydride. HEH is a biomimetic analog of NADH (reduced nicotinamide adenine dinucleotide).[12–14] Thus, HEH is a mild, stoichiometric, and nonmetallic reducing agent suited for transfer hydrogenation reactions and is frequently used in enantioselective organocatalytic reductions of many unsaturated functionalities.[1,15–17]

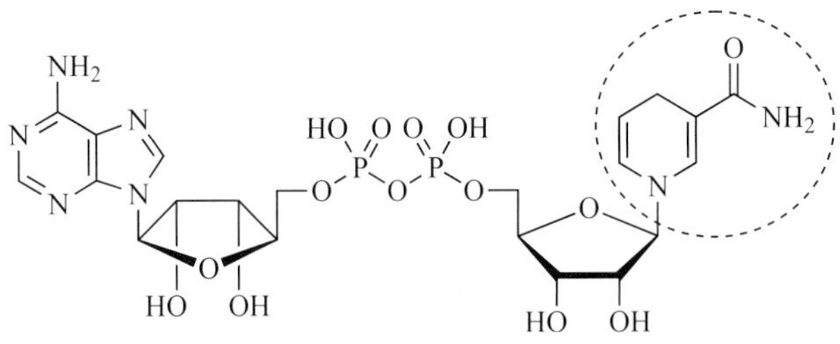

NADH

Reduction of Imines. It was found that Hantzsch ester hydride can be used for the enantioselective transfer hydrogenation of imines in the presence of catalytic quantities of chiral phosphoric acid **4** (eq 7).[18–22]

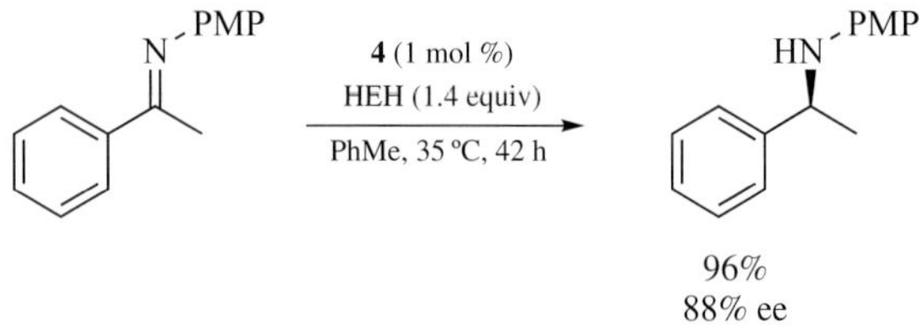

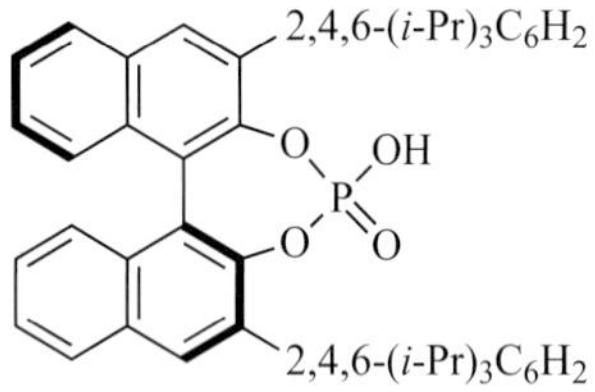

(7)

(**4**)

The organocatalytic reduction of acyclic α-imino esters with HEH to provide α-amino esters in high yields and with excellent enantioselectivities using catalyst **5** has been reported (eq 8).[23]

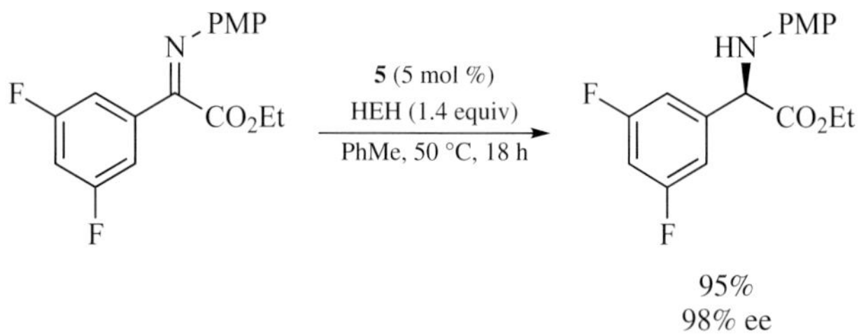

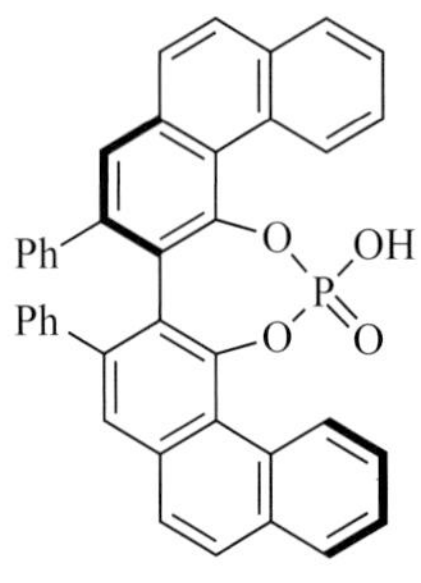

(8)

(**5**)

Reductive Amination Reactions. Direct reductive amination of aldehydes and ketones was described for the efficient synthesis of structurally diverse amines. This acid- and metal-free process requires only catalytic amounts of thiourea as hydrogen-bond donor and utilizes the Hantzsch ester for transfer hydrogenation (eqs 9–11).[24–26]

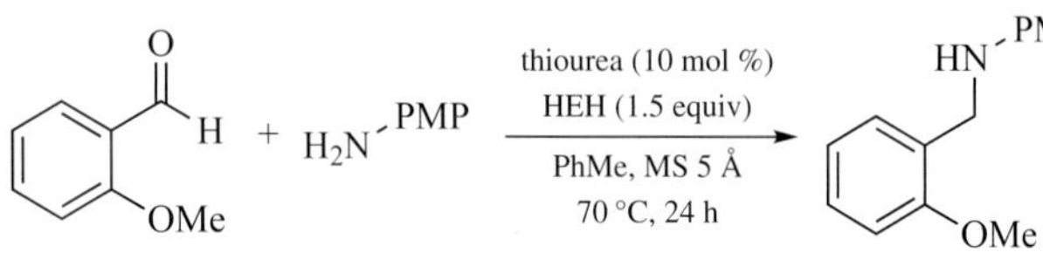

(9)

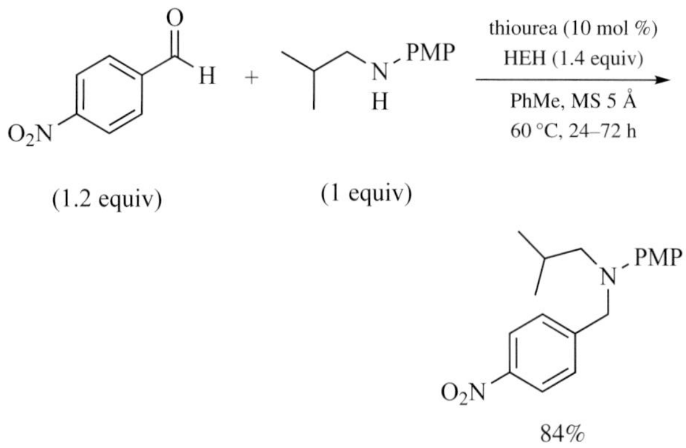

(10)

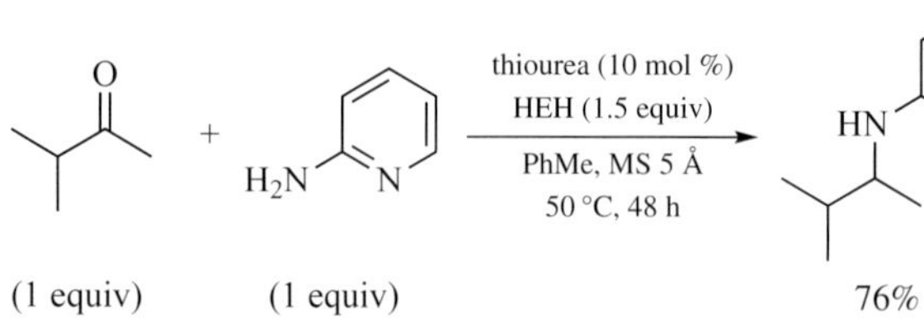

(11)

The reductive amination of aldehydes and ketones by a Hantzsch dihydropyridine can also be performed in the presence of a catalytic amount of Lewis acid, such as the scandium triflate ($Sc(OTf)_3$) (eq 12).[27]

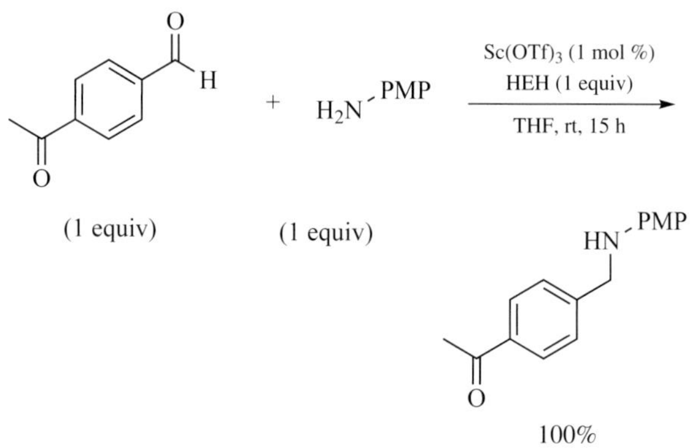

(12)

Chiral, nonracemic α-branched amines can be prepared by the enantioselective organocatalytic reductive amination of ketones catalyzed by an *ortho*-triphenylsilyl-substituted phosphoric acid **6** and using stoichiometric amounts of the Hantzsch ester (eq 13).[28]

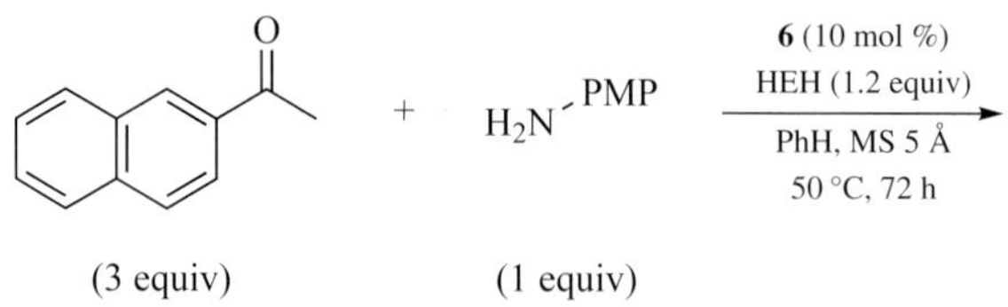

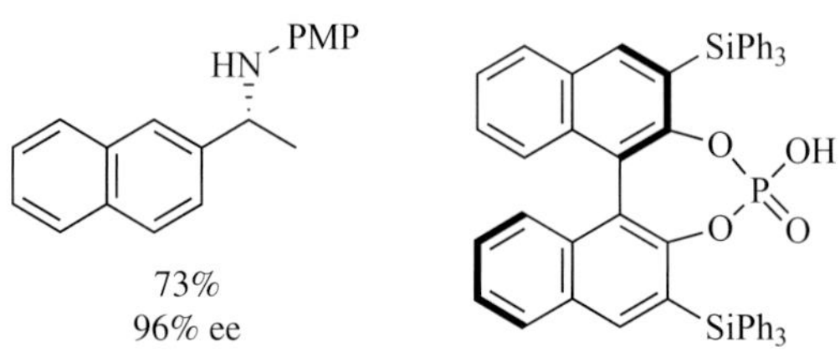

(13)

(**6**)

A catalytic asymmetric reductive amination of racemic α-branched ketones using a dynamic kinetic resolution in the presence of phosphoric acid **7** provides high yields of the corresponding *cis*-2-substituted cyclohexylamine (eq 14).[29]

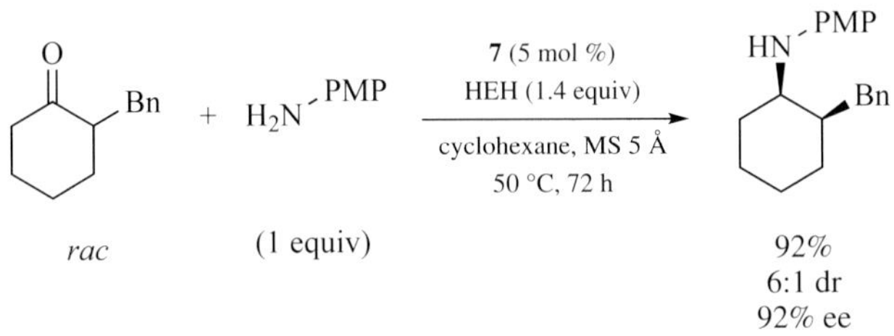

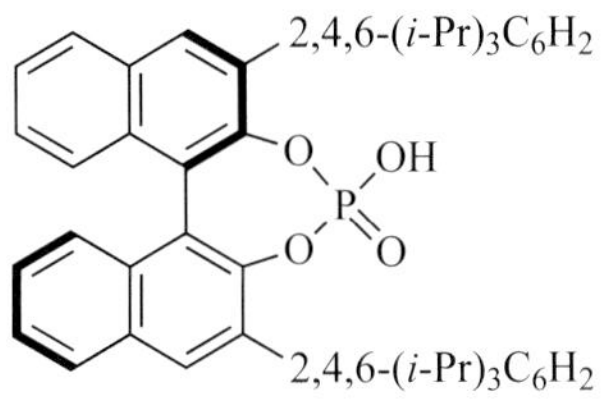

(7)

(14)

Reduction of Amides. Two highly chemoselective metal-free reduction processes for the reduction of carboxamides have been described. Amides are initially treated with triflic anhydride (Tf_2O) to generate a highly electrophilic imidoyl triflate, which is subsequently reduced to the corresponding amine via reduction of the iminium intermediate. Tertiary amides are reduced to the corresponding amine using 2.5 equiv of HEH (eq 15), and secondary amides via the sequential reduction with triethylsilane and HEH (eq 16).[2,30]

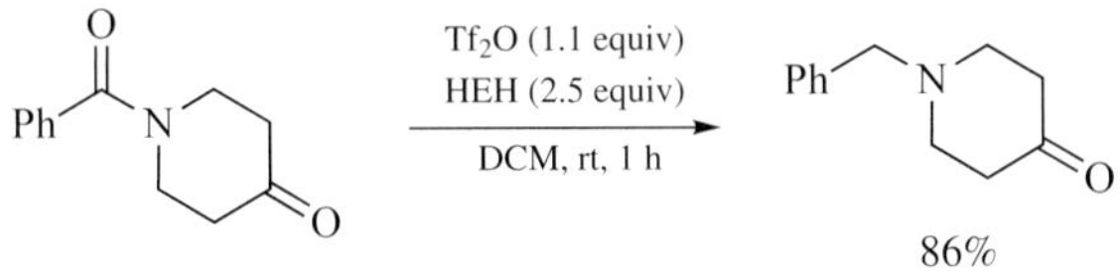

(15)

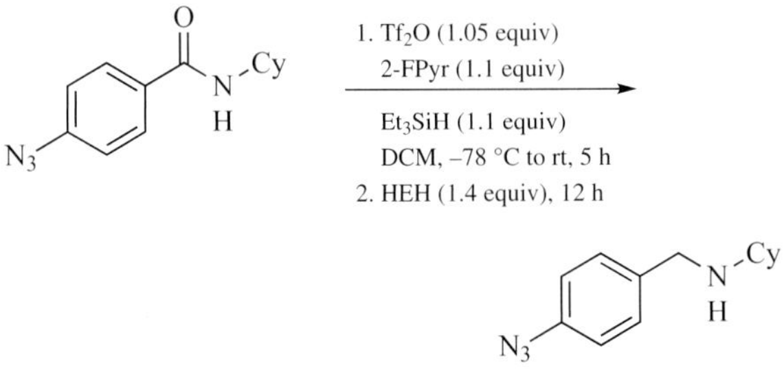

(16)

Transfer Hydrogenation of Quinolines and Pyridines. A mild Brønsted acid-catalyzed transfer hydrogenation of quinolines has been developed. This process provides direct access to a variety of 2-substituted tetrahydroquinolines in good yields and with excellent enantioselectivities (eq 17).[31–33]

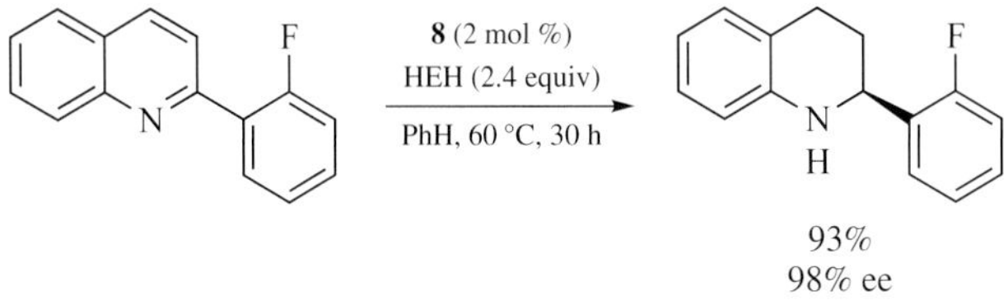

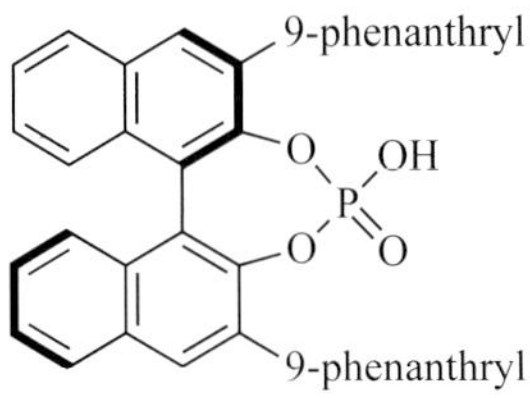

(8)

(17)

The enantioselective reduction of trisubstituted pyridine derivatives catalyzed by phosphoric acid **9** to afford hexahydroquinolinones and tetrahydropyridines has also been described (eq 18).[34]

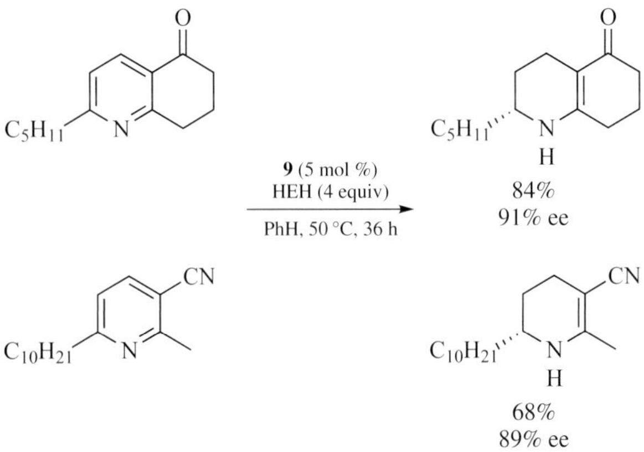

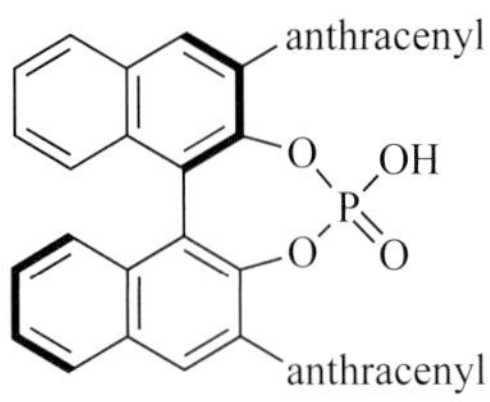

(9)

(18)

Transfer Hydrogenation of Quinoxalines Derivatives. The metal-free transfer hydrogenation of quinoxalines, benzoxazines, and benzothiazines in the presence of HEH and catalytic amounts of a Brønsted acid has also been reported (eqs 19–21).[35,36]

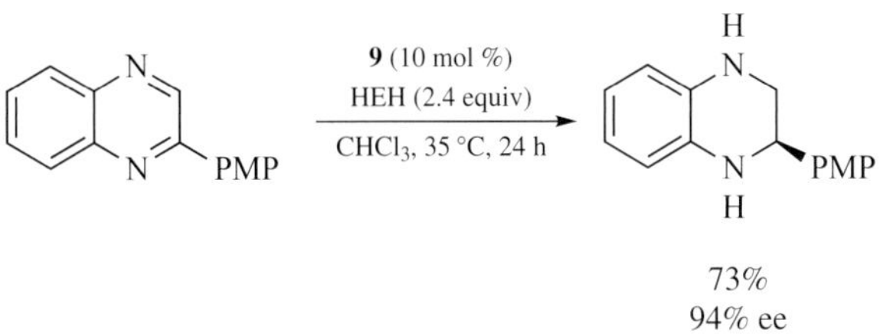

(19)

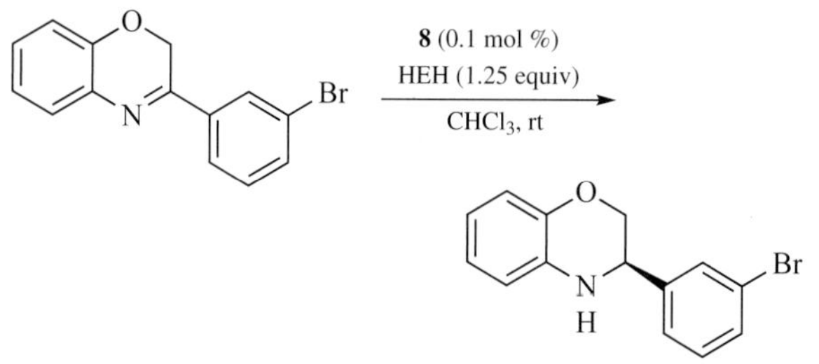

(20)

93%
98% ee

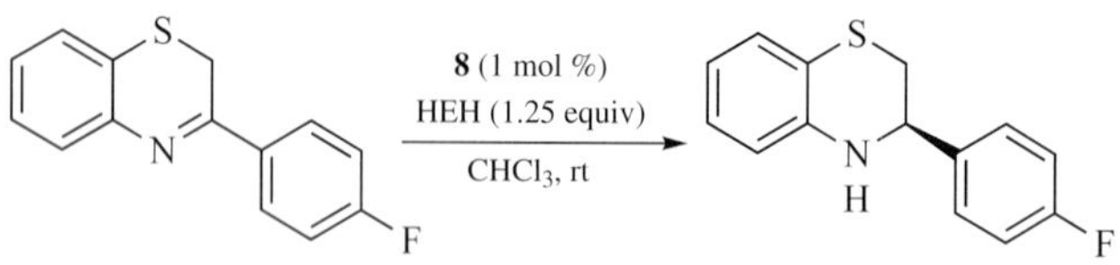

70%
>99% ee

(21)

Hydrogenation of α,β-Unsaturated Aldehydes and Ketones. The Hantzsch ester is also used to reduce α,β-unsaturated compounds. The chiral imidazolidinone catalyst **10** has been employed to provide an organocatalytic strategy for the enantioselective reduction of α,β-unsaturated aldehydes to generate chiral β,β′-disubstituted aldehydes (eq 22). In addition, fast *E-Z* isomerization prior to selective hydride delivery to the *E*-olefin isomer allowed the use of an E/Z mixture of enals.[37]

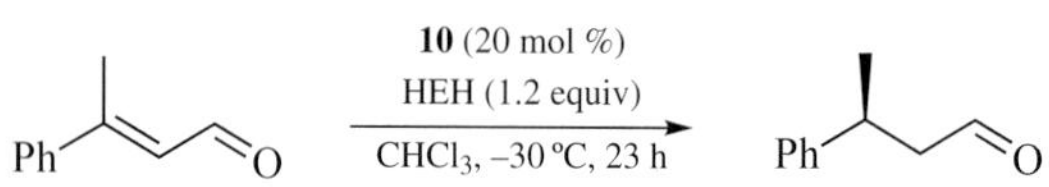

(22)

91%
93% ee

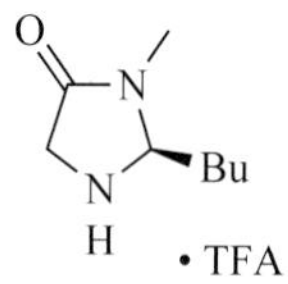

(**10**)

Alternative methodologies for the transfer hydrogenation of α,β-unsaturated aldehydes and cyclic ketones in the presence of different organocatalysts have also been reported (eqs 23 and 24).[38–41]

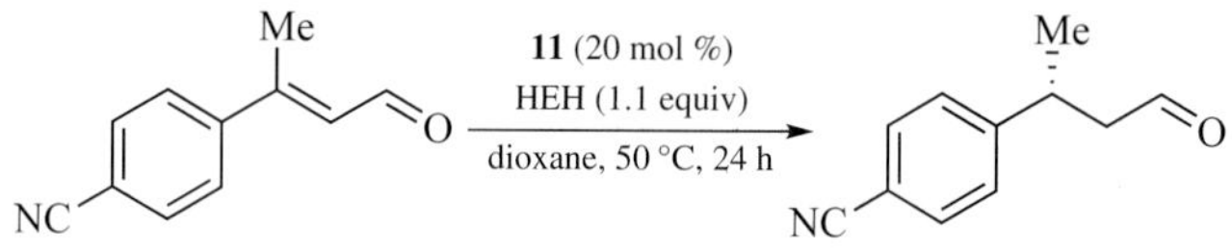

84%
98% ee

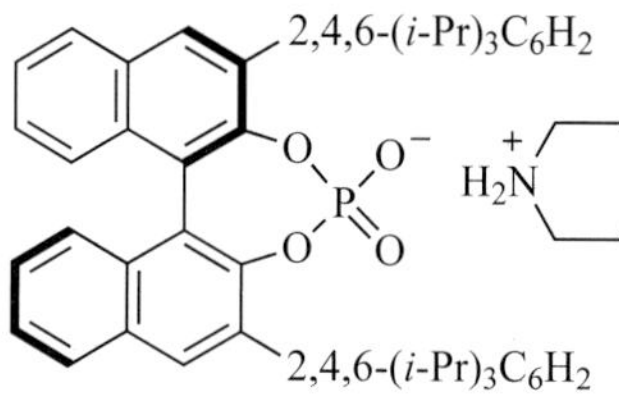

(23)

(**11**)

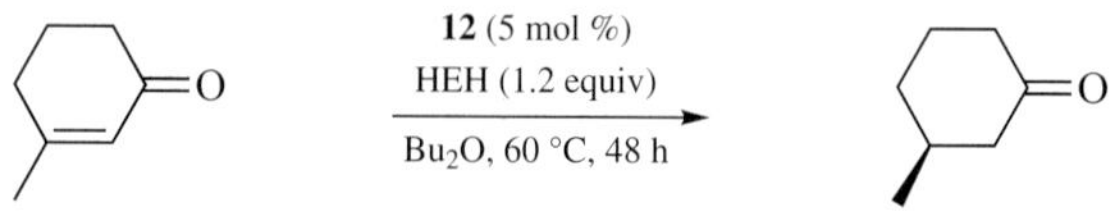

99%
94% ee

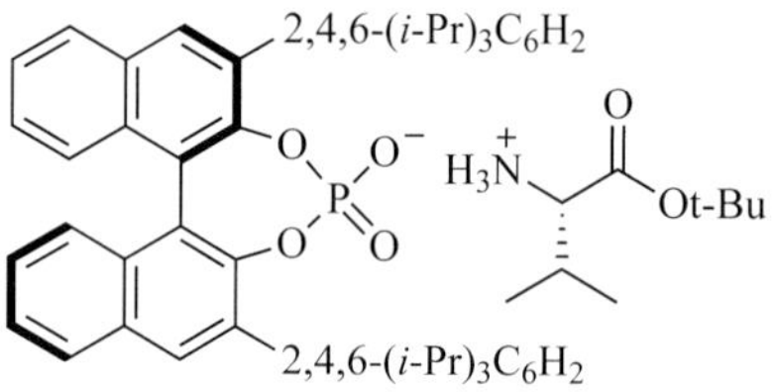

(24)

(**12**)

Transfer Hydrogenation of Nitroolefins. Chiral thioureas, such as **13**, can be employed in the conversion of nitroolefins and β-nitroacrylates to their saturated analogs by an organocatalytic asymmetric transfer hydrogenation in the presence of the Hantzsch ester derivative **14** (eq 25).[42–44]

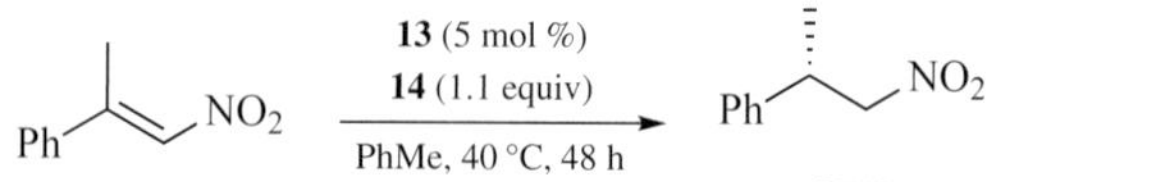

97%
94% ee

(25)

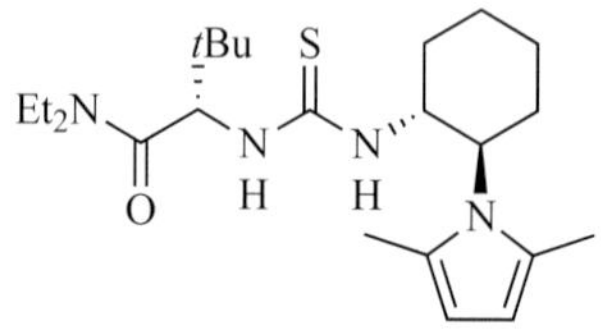

(**13**)

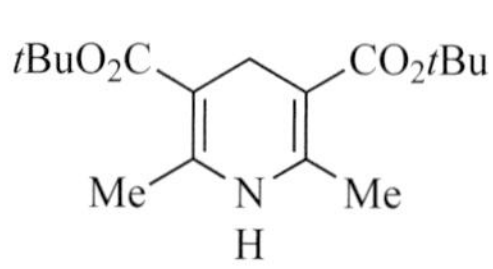

(**14**)

Cascade Reactions. Organocatalytic asymmetric cascade reactions are also of high interest due to their economical potential and efficiency.[45–47] A highly enantioselective triple organocatalytic cascade reaction for the synthesis of *cis*-3-substituted cyclohexylamines derivatives from 2,6-diones has been reported in the presence of HEH (eq 26).[48]

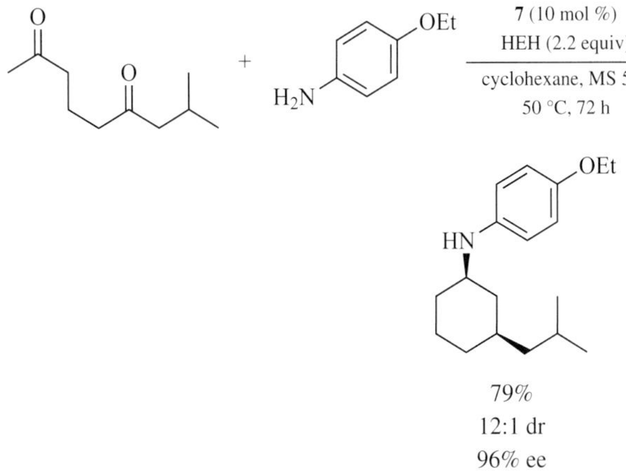

(26)

A parallel organocatalytic cascade reaction employing the imidazolidinone salt **15** as an efficient catalyst was also developed. The reaction proceeds via a catalytic conjugate reduction of the iminium intermediate with HEH followed by an in situ enamine catalytic asymmetric Michael cyclization (eq 27).[49,50]

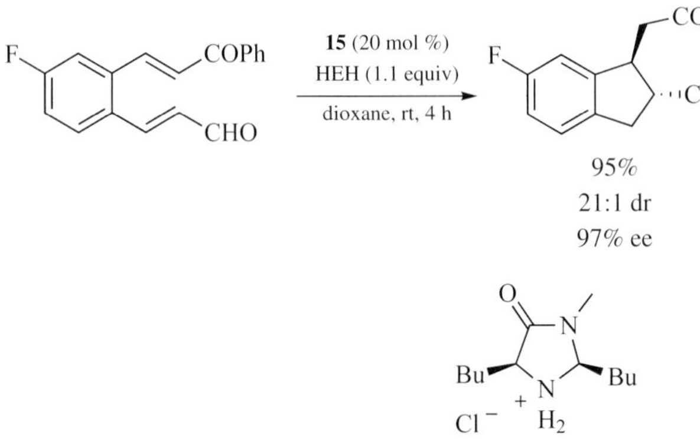

(27)

(15)

It was also found that compounds bearing two *gem*-electron-withdrawing groups, such as Meldrum's acid and other derivatives, are efficient precursors in a metal-free and multicatalytic cascade process for the asymmetric synthesis of chiral building blocks starting from aldehydes (eq 28).[51–55]

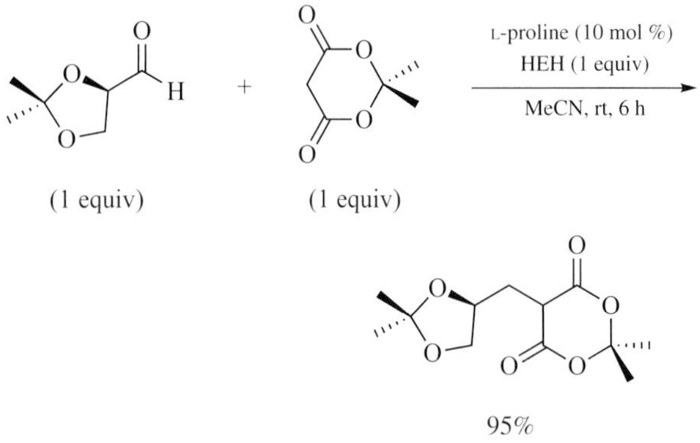

(28)

Reduction of Isoxazolones and Pyrazolones. The selective reduction of the exocyclic double bond of isoxazolones and pyrazolones by the Hantzsch 1,4-dihydropyridine can also be performed in excellent yields (eq 29).[56]

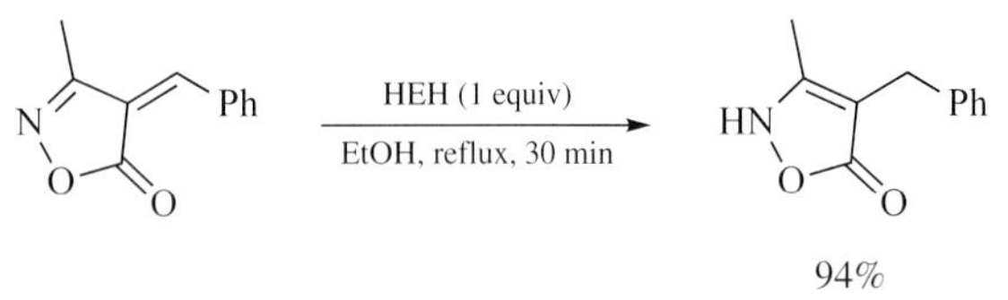

(29)

Alkene Hydrogenation. The hydrogenation of unactivated olefins can be performed using the Hantzsch ester hydride when the reaction is catalyzed by palladium on carbon (eq 30).[57] The scope of the reaction also includes α,β-unsaturated ketones.

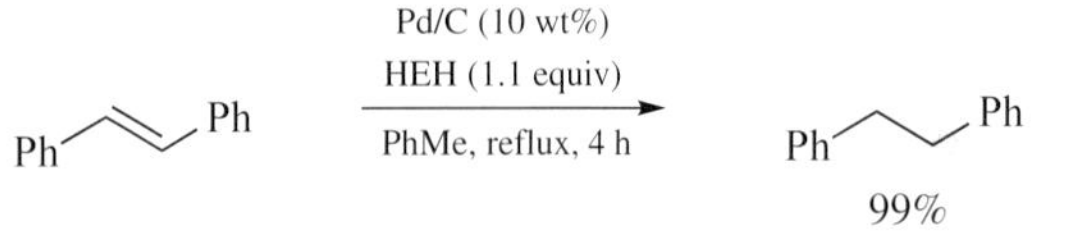

(30)

Transfer Hydrogenation of α-Ketoesters. The copper-catalyzed transfer hydrogenation of α-ketoesters with the Hantzsch ester gives access to α-hydroxy esters with excellent enantioselectivities when the bis(oxazoline) derivative **16** is used as chiral ligand (eq 31).[58]

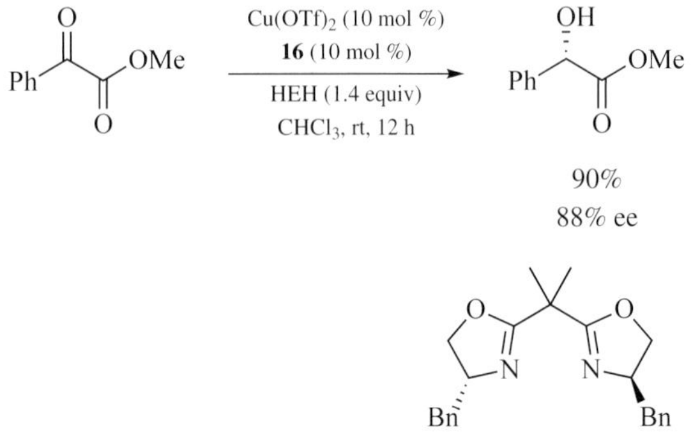

(31)

(16)

Reduction of Nitrobenzenes. HEH is also used in the oxazine synthesis via the palladium-catalyzed reduction of the nitro functionality forming a free amine, which proceeds via a ring opening of the epoxide moiety in a 6-*exo* fashion (eq 32).[59]

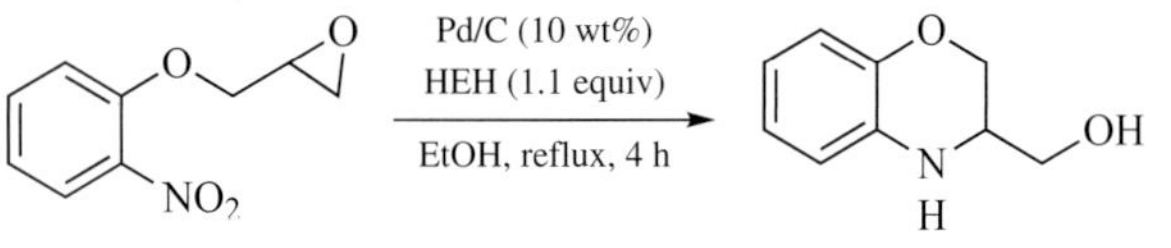

rac 84% (32)

Dehalogenation Reactions. In a preliminary study on electron-transfer photoredox catalysis for the development of a reductive debromination reaction, it was found that treating compound **17** with $Ru(bpy)_3Cl_2$,$i$$Pr_2NEt$, and Hantzsch ester hydride under irradiation afforded the corresponding dehalogenation product in high yields (eq 33).[60]

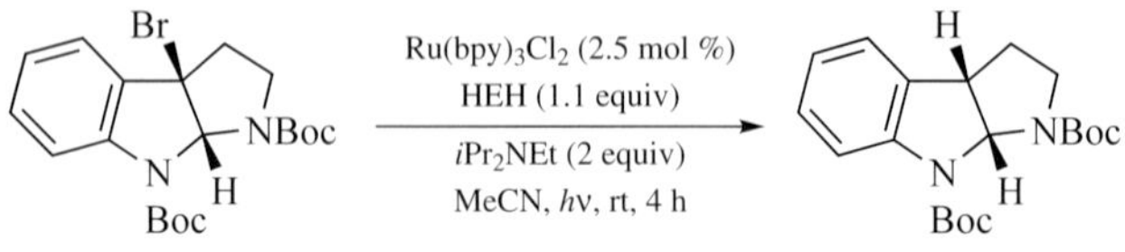

(**17**) (**18**) 95% (33)

1. Wang, Z.; Jiang, Z., *Asian J. Chem.* **2010**, *22*, 41411.
2. Singer, A.; McElvain, S. M., *Org. Synth.* **1943**, *2*, 214.
3. Hantzsch, A., *Justus Liebigs Ann. Chem.* **1882**, *215*, 1.
4. Salehi, H.; Guo, Q. X., *Synth. Commun.* **2004**, *34*, 4349.
5. Anniyappan, M.; Muralidharan, D.; Perumal, P. T., *Synth. Commun.* **2002**, 659.
6. Zolfigol, M. A.; Safaiee, M., *Synlett* **2004**, 827.
7. Roomi, M. W., *J. Med. Chem.* **1975**, *18*, 457.
8. Booker, E.; Eisner, U., *J. Chem. Soc., Perkin Trans 1* **1974**, 929.
9. Singh, H.; Singh, K., *Tetrahedron* **1989**, *45*, 3967.
10. Singh, H.; Sarin, R.; Singh, K., *Heterocycles* **1986**, *24*, 3039.
11. Franke, P. T.; Johansen, R. L.; Bertelsen, S.; Jørgensen, K. A., *Chem. Asian J.* **2008**, *3*, 216.
12. Jones, J. B., *Tetrahedron* **1986**, *42*, 3351.
13. Eisner, U.; Kuthan, J., *Chem. Rev.* **1972**, *72*, 1.
14. Ouellet, S. G.; Walji, A. M.; MacMillan, D. W. C., *Acc. Chem. Res.* **2007**, *40*, 1327.
15. You, S.-L., *Chem. Asian J.* **2007**, *2*, 820.
16. Rueping, M.; Sugiono, E.; Schoepke, F. R., *Synlett* **2010**, 852.
17. Stout, D. M.; Meyers, A. I., *Chem. Rev.* **1982** *82*, 223.
18. Hoffmann, S.; Seayad, A. M.; List, B., *Angew. Chem., Int. Ed.* **2005**, *44*, 7424.
19. Rueping, M.; Sugiono, E.; Azap, C.; Theissman, T.; Bolte, M., *Org. Lett.* **2005**, *7*, 3781.
20. Li, G.; Antilla, J. C., *Org. Lett.* **2009**, *11*, 1075.
21. Zhang, Z.; Schreiner, P. R., *Synlett* **2007**, 1455.
22. Han, Z.-Y.; Xiao, H.; Gong, L.-Z., *Bioorg. Med. Chem. Lett.* **2009**, *19*, 3729.
23. Li, G.; Liang, Y.; Antilla, J. C., *J. Am. Chem. Soc.* **2007**, *129*, 5830.
24. Menche, D.; Arikan, F., *Synlett* **2006**, 841.
25. Menche, D.; Bohm, S.; Li, J.; Rudolph, S.; Zander, W., *Tetrahedron Lett.* **2007**, *48*, 365.
26. Menche, D.; Hassfeld, J.; Li, J.; Menche, G.; Ritter, A.; Rudolph S., *Org. Lett.* **2006**, *8*, 741.
27. Itoh, T.; Nagata, K.; Kurihara, A.; Miyazaki, M.; Ohsawa, A., *Tetrahedron Lett.* **2002**, *43*, 3105.
28. Storer, I. R.; Carrera, D. E.; Ni, Y.; MacMillan, D. W. C., *J. Am. Chem. Soc.* **2006**, *128*, 84.
29. Wakchaure, V. N.; Zhou, J.; Hoffmann, S.; List, B., *Angew. Chem., Int. Ed.* **2010**, *49*, 4612.
30. (a) Pelletier, G.; Bechara, W. S.; Charette, A. B., *J. Am. Chem. Soc.* **2010**, *132*, 12817. (b) Barbe, G.; Charette, A. B., *J. Am. Chem. Soc.* **2008**, *130*, 18.
31. Rueping, M.; Antonchick, A. P.; Theissmann, T., *Angew. Chem., Int. Ed.* **2006**, *45*, 3683.
32. Bruckmann, A.; Pena, M. A.; Bolm, C., *Synlett* **2008**, 900.
33. Rueping, M.; Theissmann, T.; Antonchick, A. P., *Synlett* **2006**, 1071.
34. Rueping, M.; Antonchick, A. P., *Angew. Chem., Int. Ed.* **2007**, *46*, 4562.
35. Rueping, M.; Tato, F.; Schoepke, F. R., *Chem. Eur. J.* **2010**, *16*, 2688.
36. Rueping, M.; Antonchick, A. P.; Theissmann, T., *Angew. Chem., Int. Ed.* **2006**, *45*, 6751.
37. Ouellet, S. G.; Tuttle, J. B.; MacMillan, D. W. C., *J. Am. Chem. Soc.* **2005**, *7*, 32.
38. Mayer, S.; List, B., *Angew. Chem., Int. Ed.* **2006**, *45*, 4193.
39. Martin, N. J. A.; List, B., *J. Am. Chem. Soc.* **2006**, *128*, 13368.
40. Yang, J. W.; Fonseca, M. T. H.; Vignola, N. List, B., *Angew. Chem., Int. Ed.* **2005**, *44*, 108.
41. Yang, J. W.; Fonseca, M. T. H.; List, B., *Angew. Chem., Int. Ed.* **2004**, *43*, 6660.
42. Martin, N. J. A.; Ozores, L.; List, B., *J. Am. Chem. Soc.* **2007**, *129*, 8976.
43. Martin, N. J. A.; Chen, X.; List, B., *J. Am. Chem. Soc.* **2008**, *130*, 13862.
44. Zhang, Z.; Schreiner, P. R., *Synthesis* **2007**, 2559.
45. Grondal, C.; Jeanty, M.; Enders, D., *Nat. Chem.* **2010**, *2*, 167.
46. List, B., *Chem. Commun.* **2006**, 819.
47. Simmons, B.; Walji, A. M.; MacMillan, D. W. C., *Angew. Chem., Int. Ed.* **2009**, *48*, 4349.
48. Zhou, J.; List, B., *J. Am. Chem. Soc.* **2007**, *129*, 7498.
49. Yang, J. W.; Fonseca, M. T. H.; List, B., *J. Am. Chem. Soc.* **2005**, *127*, 15036.
50. Zhao, G.-L.; Cordova, A., *Tetrahedron Lett.* **2006**, *47*, 7417.
51. Ramachary, D. B.; Reddy, Y. V., *J. Org. Chem.* **2010**, *75*, 74.
52. Ramachary, D. B.; Kishor, M., *J. Org. Chem.* **2007**, *72*, 5056.
53. Ramachary, D. B.; Kishor, M.; Ramakumar, K., *Tetrahedron Lett* **2006**, *47*, 651.
54. Liu, Z.; Liu, Q.; Zhang, W.; Mu, R.; Yang, L.; Liu, Z.-L.; Yu, W., *Synthesis* **2006**, 771.
55. Garden, S. J.; Guimaraes, C. R. W.; Correa, M. B.; Oliveira, C. A. F.; Pinto, A. C.; Alencastro, R. B., *J. Org. Chem.* **2003**, *68*, 8815.
56. Liu, Z.; Han, B.; Liu, Q.; Zhang, W.; Yang, L.; Liu, Z.-L.; Yu, W., *Synlett* **2005**, 1579.
57. Liu, Q.; Li, J.; Shen, X.-X.; Xing, R.-G.; Yang, J.; Liu, Z.; Zhou, B., *Tetrahedron Lett.* **2009**, *50*, 1026.
58. Yang, J. W.; List, B., *Org. Lett.* **2006, ***8*, 5653.
59. Meng, Q.-Y.; Liu, Q.; Li, J.; Xing, R.-G.; Shen, X.-X.; Zhou, B., *Synlett* **2009**, 3283.
60. Narayanam, J. M. R.; Tucker, J. W.; Stephenson, C. R. J., *J. Am. Chem. Soc.* **2009**, *131*, 8756.

William S. Bechara & André B. Charette
Université de Montréal, Montréal, Québec, Canada

Diethyl Oxomalonate

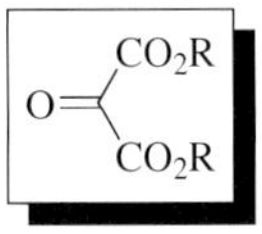

(R = Et)

[609-09-6] $C_7H_{10}O_5$ (MW 174.15)

InChI = 1/C7H10O5/c1-3-11-6(9)5(8)7(10)12-4-2/h3-4H2,1-2H3

InChIKey = DBKKFIIYQGGHJO-UHFFFAOYAK

(R = Me)

[3298-40-6] $C_5H_6O_5$ (MW 146.10)

InChI = 1/C5H6O5/c1-9-4(7)3(6)5(8)10-2/h1-2H3

InChIKey = VVOCYQDSDXLBEA-UHFFFAOYAG

(electron-deficient heterodienophile; Diels–Alder reaction;[1–4] heterodipolarophile; [3 + 2] cycloaddition;[5] 1,3-diyl trapping agent;[6] 1,4-dipolar cycloaddition;[7] [2 + 2] cycloaddition;[8] enophile; ene reaction;[9,10] electrophilic addition;[11–17] Wittig reactions;[18–22] condensation with amines;[23] Pictet–Spengler reactions[24])

Physical Data: bp 95–98 °C/12 mmHg; *d* 1.144 g cm^{-3}.

Solubility: sol ethanol and diethyl ether.

Form Supplied in: colorless liquid; widely available.

Analysis of Reagent Purity: n_D^{20} 1.425.

Preparative Method: diethyl oxomalonate can be prepared by bromination of Diethyl Malonate to the dibromomalonic ester, exchange of one bromine atom for acetate and α-elimination of acetyl bromide.[25] Alternatively, diethyl malonate is condensed with benzaldehyde to afford the Knoevenagel product which is cleaved by ozonolysis.[26]

Purification: fractional vacuum distillation.

Handling, Storage, and Precautions: diethyl oxomalonate is stable for at least 6 months if stored in the refrigerator.

Original Commentary

Lutz F. Tietze & Christoph Schneider

Georg-August-Universität zu Göttingen, Germany

Diethyl oxomalonate (**1**) undergoes Diels–Alder reactions as an electron-poor heterodienophile.[1a] Thus reactions with electron-rich butadienes furnish the *gem*-diesters (**2**)[1] (eq 1), which can be converted to the monoacid (**3**) by decarboxylation[1d] or to the unsaturated δ-valerolactones (**4**) either by a Curtius degradation sequence[1b,c] or Cerium(IV) Ammonium Nitrate (CAN) oxidation (Scheme 1).[7]

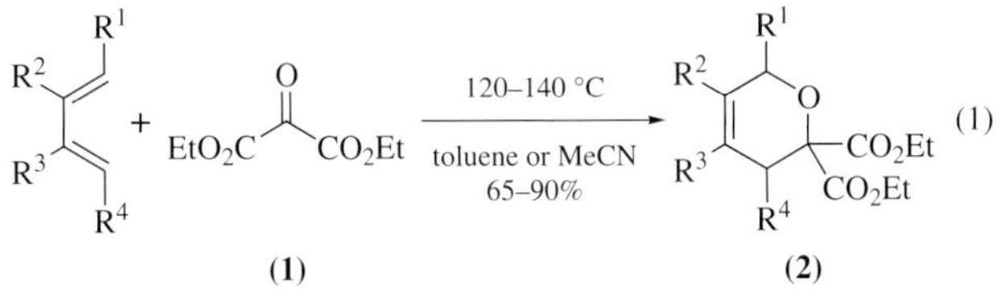

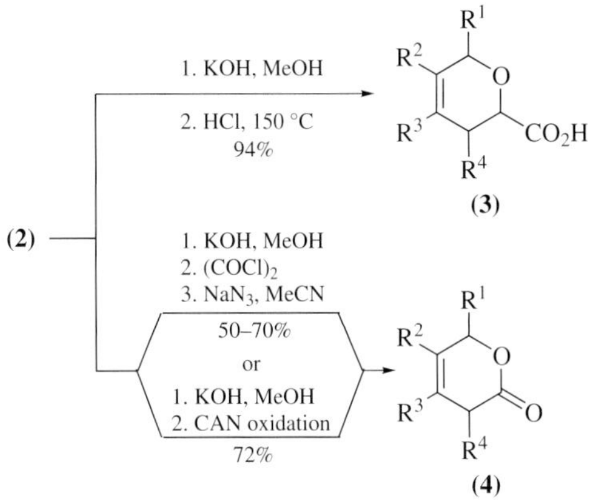

Scheme 1

The reaction is run either at elevated temperature (120–140 °C, MeCN or toluene),[1b–d] under high pressure conditions (10 kbar)[1e,f] or under Lewis acid catalysis.[1a] Chiral Lewis acids have also been applied, but gave only very modest enantioselectivities (<15% ee).[2]

Azadienes (**5**) undergo hetero Diels–Alder reactions with diethyl oxomalonate (**1**) in THF at 80 °C (eq 2).[3] 2,3-Dihydropyran-4-ones (**6**) are formed in the reaction with Danishefsky's diene (**7**) under Lewis acid catalysis at ambient temperature (eq 3).[4]

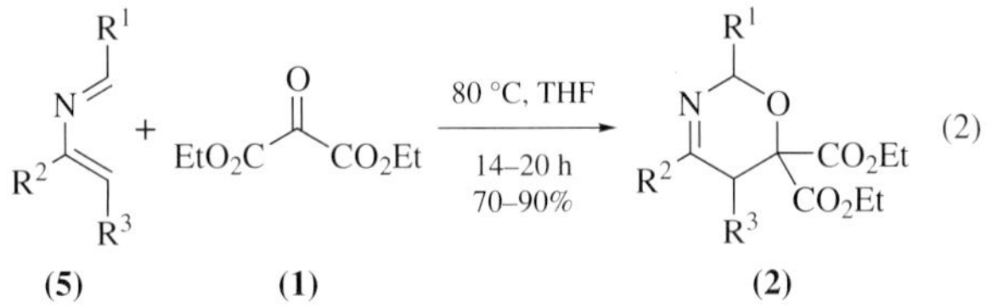

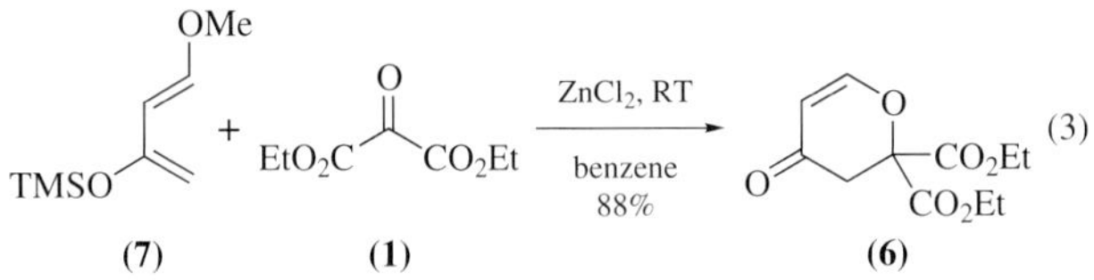

Reagent (**1**) has been used in [3 + 2] dipolar cycloadditions as a heterodipolarophile.[5] *N*-(Silylmethyl)-substituted ketene *N,S*-acetals (**8**) are converted to the 1,3-dipolar reagent (**9**) with Cesium Fluoride in MeCN, which undergoes the cycloaddition with (**1**) to afford the alkylideneoxazolidines (**10**) in modest yield, but with complete regioselectivity (eq 4).

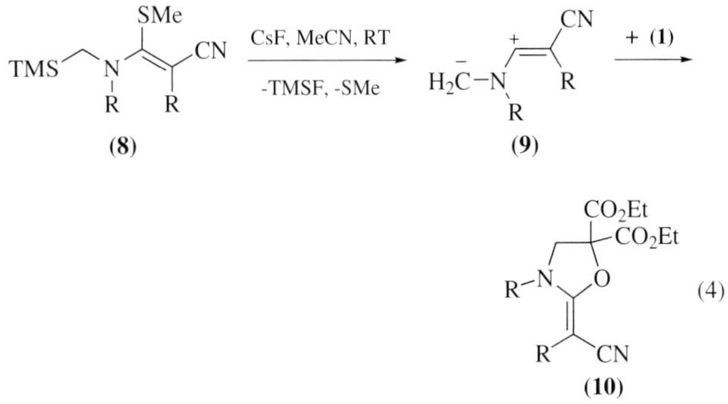

Diethyl oxomalonate has been shown to be reactive towards 1,3-diyls.[6] Heating the diazene (**11**) produces the 2-alkylidene-1,3-diyl (**12**), which is trapped in situ by (**1**) to form the fused cycloadduct (**13**) as a sole isomer in good yield (eq 5). Other diyls, however, give mixtures of isomers.

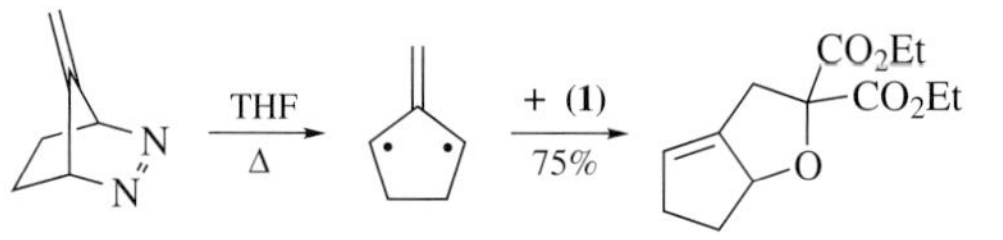

(5)

(**11**) (**12**) (**13**)

An example of a 1,4-dipolar cycloaddition of (**1**) with 6-oxo-3,6-dihydro-1-pyrimidinium-4-olate (**14**) has been described which leads to diazabicyclo[2.2.2]octanediones (**15**) (eq 6).[7]

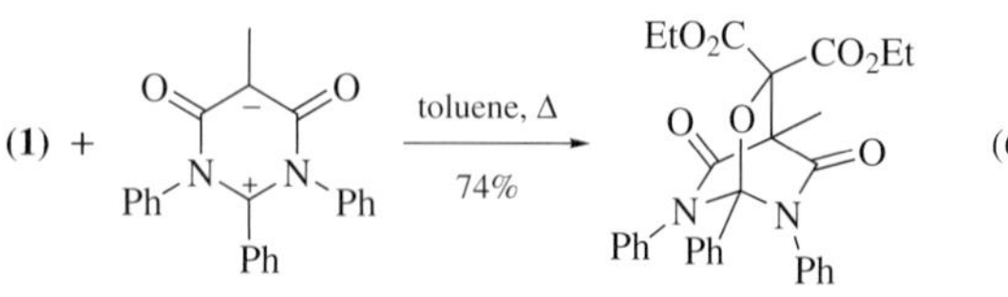

(6)

(**14**) (**15**)

[2 + 2] Cycloadditions of (**1**) and electron-rich alkenes like 1,1-dimethoxy-2-propene (**16**) have also been investigated.[8] Oxetanes (**17**) are formed in good yield with complete regioselectivity in a thermal reaction (eq 7). Under photochemical conditions, addition does not take place.

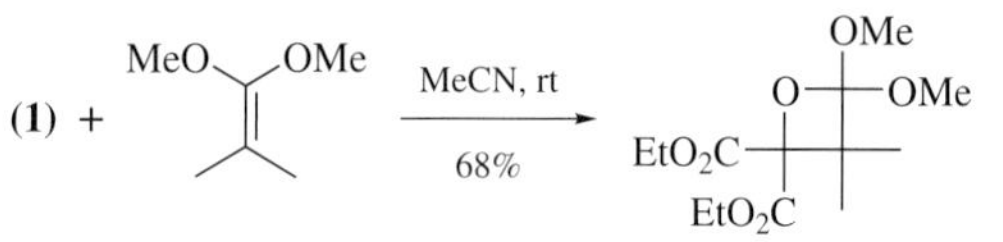

(7)

(**16**) (**17**)

Diethyl oxomalonate reacts with alkenes that have an allylic hydrogen atom in an ene reaction (eq 8).[9] The reaction is run either thermally or under Lewis acid catalysis. Occasionally the ene reaction is followed by a cyclization to form a tetrahydrofuran diester (**18**), which can be converted to the butyrolactone (**19**) by CAN oxidation (eq 8).[10]

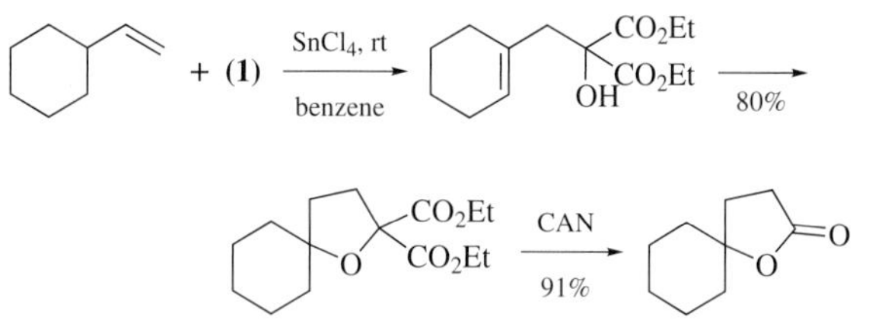

(8)

(**18**) (**19**)

Due to the adjacent ester groups, (**1**) is a highly reactive carbonyl compound which reacts with a variety of nucleophiles. Thus benzoxazinones (**20**) are formed in the reaction with aminophenols (eq 9).[11] Similarly, 6-ethoxycarbonyl-7-oxopyrazinothiadiazines (**21**) are accessible by the reaction of 3,4,5-triamino-2*H*-1,2,6-thiadiazine 1,1-dioxide (**22**) with diethyl oxomalonate (eq 10).[12]

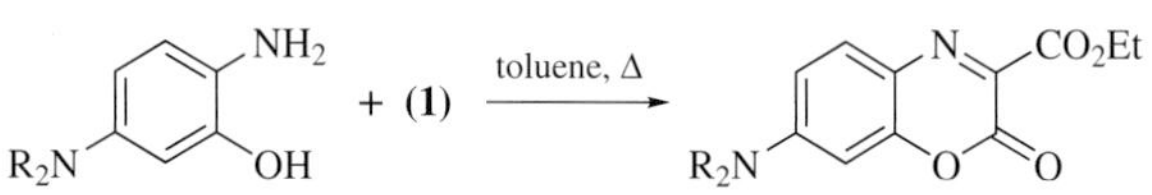

(9)

(**20**)

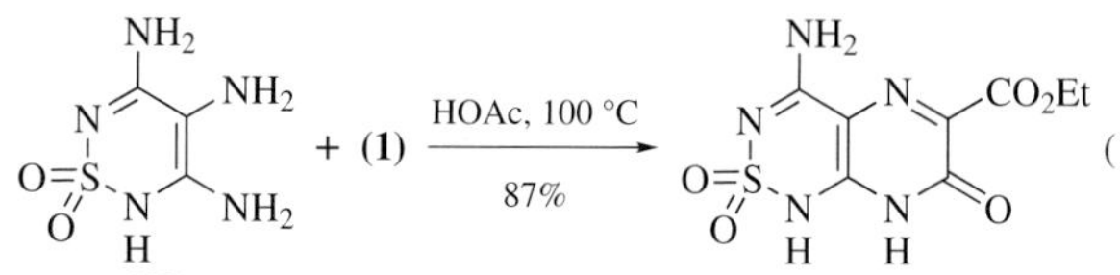

(10)

(**22**) (**21**)

Enamines (**23**) react with diethyl oxomalonate to furnish aldol products which can be dehydrated and cyclized to the butenolides (**24**) by acid catalysis (eq 11).[13] *N*-Methylindole (**25**) smoothly adds to (**1**) to afford 3-indolylhydroxymalonate (**26**) (eq 12).[14] In the presence of Ethylaluminum Dichloride a second addition takes place and the 3,3′-bisindolyl derivative (**27**) is formed (eq 12).

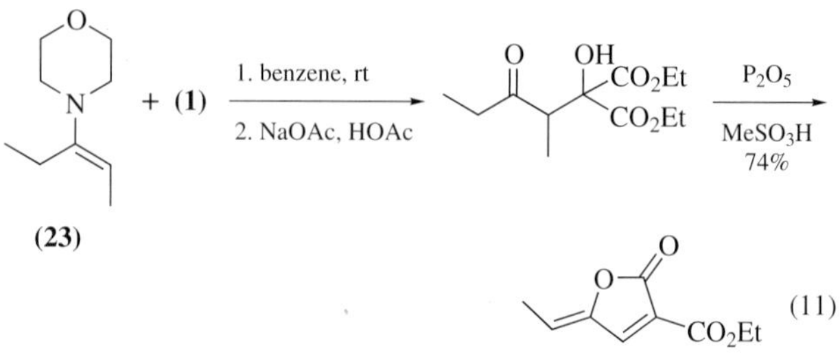

(**23**) (**24**) (11)

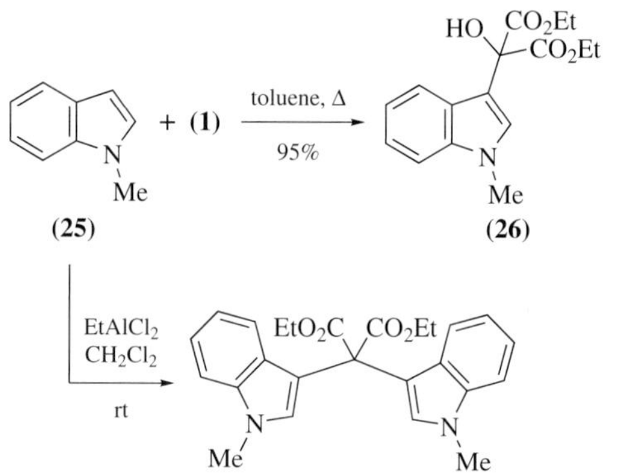

(**25**) (**26**) (**27**) (12)

The high-pressure addition of 2,5-dialkylfurans (**28**) to (**1**) very selectively affords the addition product (**29**) (eq 13).[15] No double addition is observed. The reaction has been shown to proceed through a [1,5]-suprafacial shift of the diethyl oxomalonate moiety.

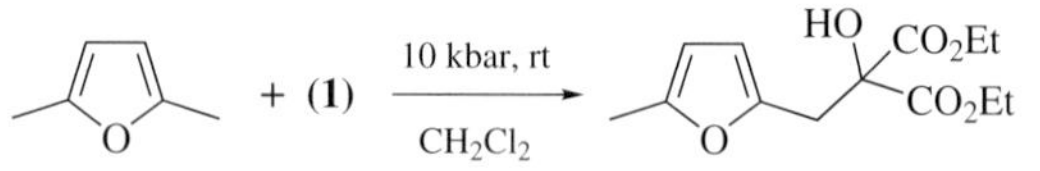

(13)

(**28**) (**29**)

Diethyl trimethylsilyl phosphite (**30**) adds to diethyl oxomalonate in a redox process to yield α-hydroxyphosphonic ester (**31**) (eq 14).[16] Only very electron-poor ketones display such a reactivity towards trimethylsilyloxyphosphorus(III) derivatives.

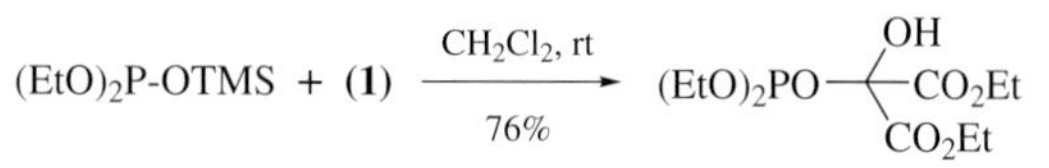

(14)

(**30**) (**31**)

Even aromatic compounds are sufficiently reactive to add to (**1**). The reaction of *p*-xylene (**32**) with (**1**) is promoted by Tin(IV) Chloride at ambient temperature (eq 15).[17]

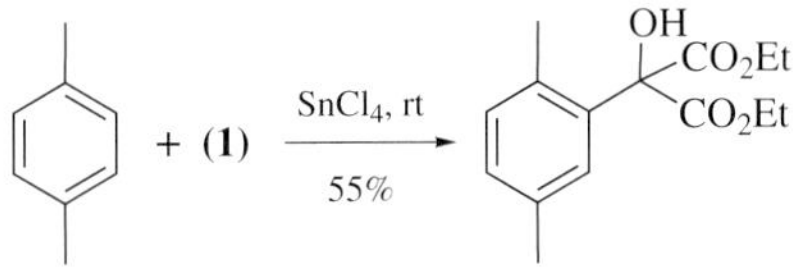
(15)

(32)

Diethyl oxomalonate has frequently been used in Wittig reactions. Thus the phosphorus ylide (**33**) of a cephalosporin derivative is smoothly converted to the unsaturated diester (**34**) (eq 16).[18] The ylide (**35**) of a dithioacetal furnishes the ketene dithioacetal (**36**) (eq 17).[19] An imino–Wittig-type reaction has been used for the construction of the side chain of the mycosporins.[20] The iminophosphorane (**37**), obtained by a Staudinger reduction of the corresponding azide, is condensed with (**1**); the resulting imine (**38**) is reduced in situ with Sodium Cyanoborohydride (eq 18).

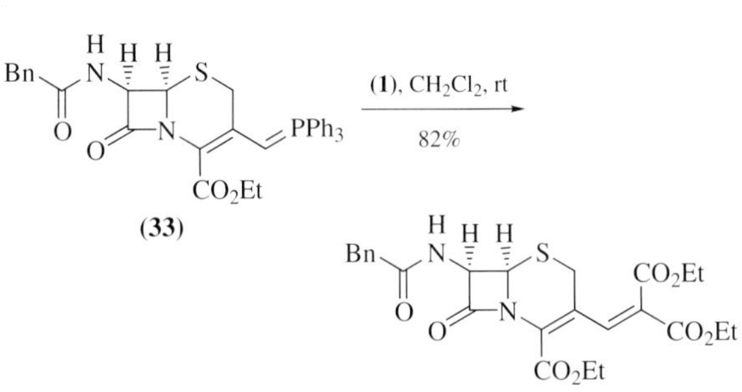
(16)

(33) **(34)**

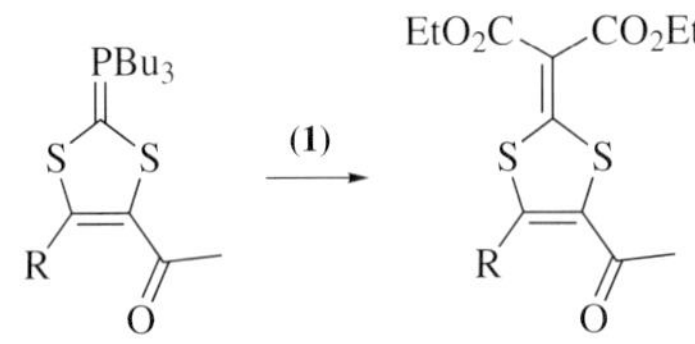
(17)

(35) **(36)**

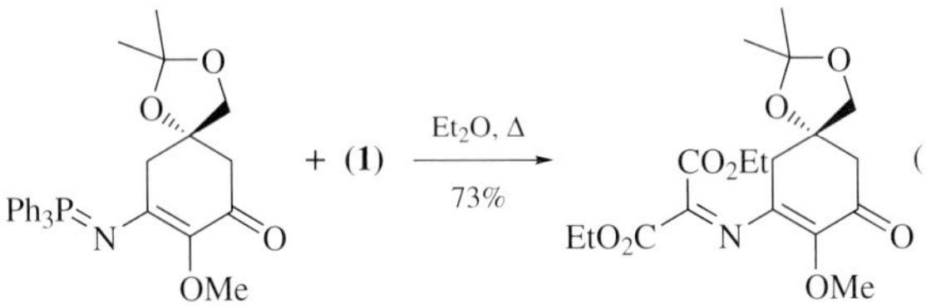
(18)

(37) **(38)**

Reagent (**1**) has also been applied to a phospha–Wittig reaction with (**39**) to yield the phospha-alkene complex (**40**) (eq 19).[21] These complexes have been shown to take part in [2+2] processes with enol ethers, enamines, and ethoxyacetylene to give phosphetanes.[22]

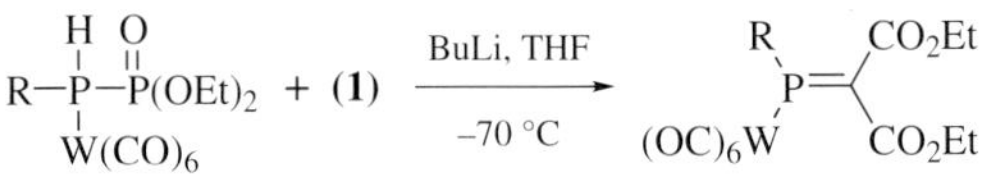
(19)

(39) **(40)**

Amines can easily be condensed with diethyl oxomalonate by heating at reflux with water removal (Dean–Stark trap). The imines (**41**) undergo electrophilic cyclization with an internal electron-rich double bond under Lewis acid or Trimethylsilyl Trifluoromethanesulfonate catalysis to produce various nitrogen heterocycles, including piperidines, octahydroisoquinolines, azepines, and pyrrolidines (eq 20).[23] Depending on the reaction conditions *cis*-lactones (**42**) are formed in a second step very selectively (eq 20).

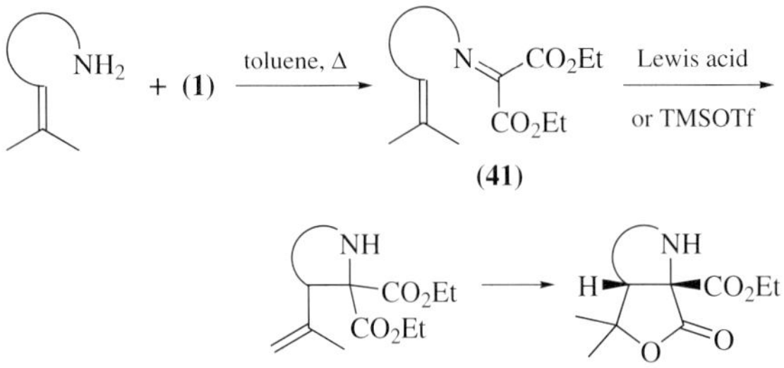
(20)

(41) **(42)**

Reagent (**1**) has also been used in Pictet–Spengler reactions of tryptamine-2-carboxylic acids (**43**) (eq 21).[24] The reaction is run in refluxing benzene/dioxane/trifluoroacetic acid solution with water removal (Dean–Stark trap). Rearrangement of the initially produced iminium ion gives rise to the 1,2,3,4-tetrahydro-β-carbolin (**44**).

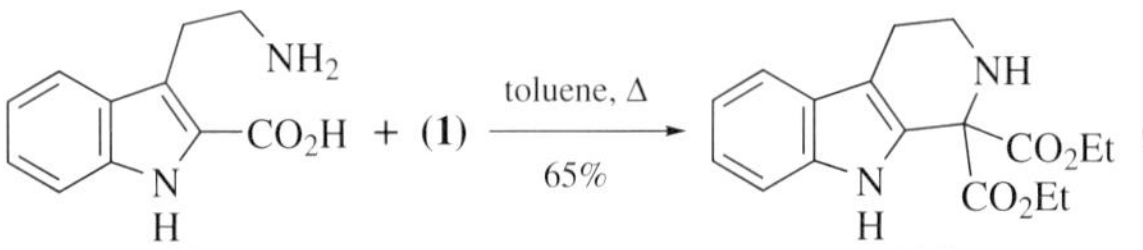
(21)

(43) **(44)**

In addition to the transformations of diethyl oxomalonate described above, several achiral and chiral dialkyl oxomalonates, such as dimethyl oxomalonate,[27–40] have been used in related reactions. The esters are best prepared by ozonolysis of the Knoevenagel products of benzaldehyde and the corresponding dialkyl malonates.[23,26]

First Update

Daniel J. Coughlin
Johnson Matthey Pharma Services, Devens, MA, USA

Recent Examples of Cycloaddition Reactions of (1). Several recent examples of heterocycloaddition reactions of diethyl oxomalonate (**1**) have recently appeared.[41–46] Diethyl oxomalonate is a highly activated electron deficient-dieneophile shown to react with (trialkylsilyl)vinyl ketenes (TAS-vinyl ketenes) to give lactone products by [4+2] cycloadditions.[41] Thus, photolysis of diazo ketone (**45**) (eq 22) gives rise to the stable but reactive TAS-vinyl ketene (**46**), which reacted with (**1**) in CH_3CN at reflux to give the silyl lactone (**47**) in good yield. Lactone (**47**) can be protodesilylated quantitatively to give desilylylated lactone (**48**). Unactivated carbonyl compounds were not reactive with the TAS-vinyl ketenes under these conditions.

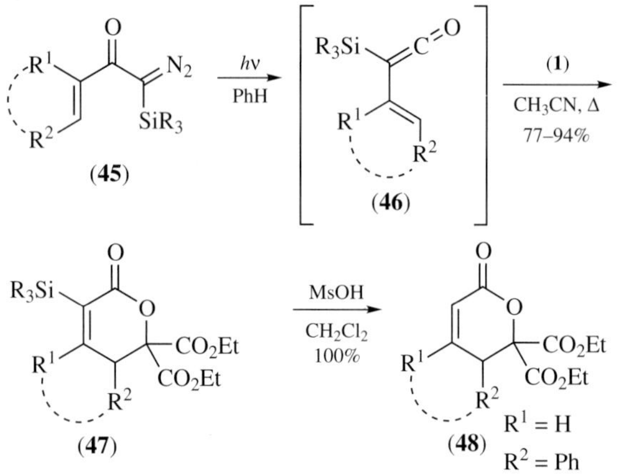

(22)

A study of consecutive cycloisomerization–[4 + 2] cycloaddition reactions was reported using various reactive dienophiles.[42] In the case of bromo ene-yne (**49**), cycloisomerization using *t*-BuLi at low temperature gave rise to the *exo*-cyclopentyl diene (**50**), which was isolated in good yield (eq 23). Treating (**1**) with diene (**50**) in toluene at reflux with added hydroquinone afforded highly substituted bicyclic ether (**51**) in 86% yield.

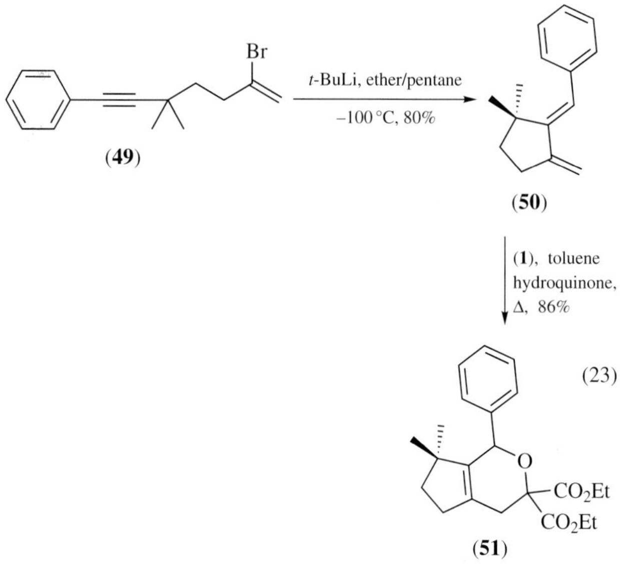

(23)

The use of (**1**) in catalytic asymmetric hetero-Diels–Alder cycloadditions, catalyzed by chiral metal complexes, has been reported recently.[43] A variety of chiral methylene bis-oxazolidine and biaryl ligands were used with Al, Cu, and Zn salts to form the chiral catalysts (eq 24). Optimization of the reaction conditions with regard to metal, ligand, catalyst loading, temperature, and reaction time was examined with cyclohexadiene as the model diene. Thus, cyclohexadiene reacts with (**1**) to give the chiral bicyclic ether (**52**) in moderate to good yield and high enantioselectivity, as a function of conditions. The best yields and enantioselectivities were obtained using $Zn(OTf)_2$ or $Cu(OTf)_2$ with chiral ligand (**53**). The chiral bicyclic ether (**52**) can be easily transformed to the interesting chiral building block (**54**).

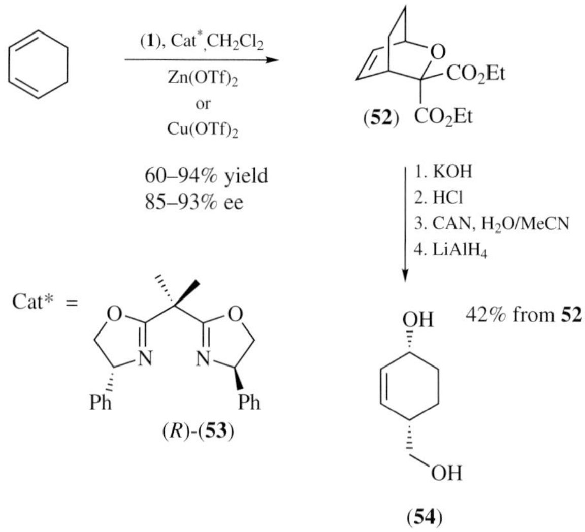

(24)

The formation of pyran rings has been demonstrated from the reaction of zirconacyclopentadienes (**55**) with (**1**) in the presence of $BiCl_3$ (eq 25).[44] This represents a novel approach to α-pyrans (**56**) and gives a variety of substituted and ring-fused products in high yield. $BiCl_3$ is necessary for this transformation. The starting zirconacyclopentadienes (**55**) are readily obtained from the intramolecular cross-coupling of two alkynes with zirconocene.

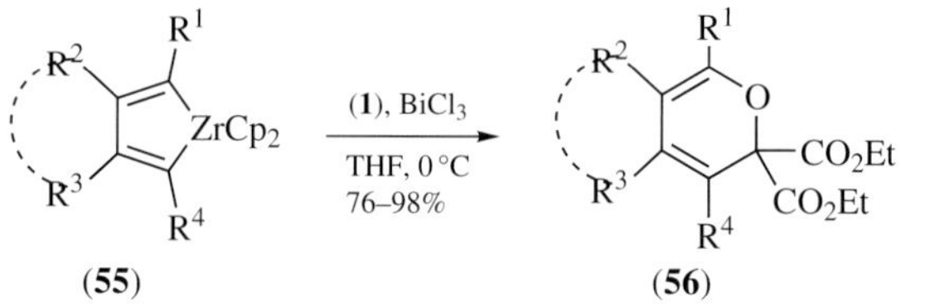

(25)

The activated carbonyl group of (**1**) has been shown to enter into [3 + 2] cycloadditions with various phenyl-substituted methylene cyclopropanes (**57**) in the presence of a catalytic amount of Lewis acid (eq 26).[45a,b] The regioselectivity of the addition is excellent and differs from that observed with the Pd-catalyzed [3 + 2] cycloaddition of similar substrates. Ytterbium triflate (5 mol %) in 1,2-dichloroethane was found to give the best yields of the product tetrahydrofuran derivatives (**58**). When R^1 and R^2 are differentially substituted, good *E/Z* stereoselectivity is observed in some but not all cases, depending on the electronic nature of the substituents.

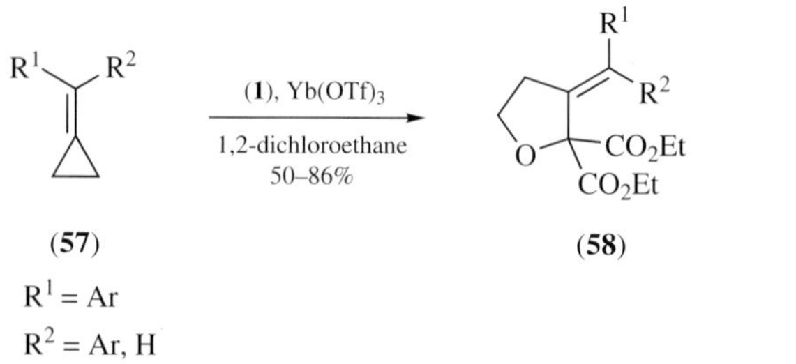

(26)

Cycloadditions of oxomalonate (**1**) have been extended to a novel ruthenium(II)-catalyzed [2 + 2 + 2] cycloaddition with 1,6-diynes to give dieneals and dieneones.[46] This ruthenium(II)-catalyzed reaction is one of the first examples of this type of cyclotrimerization utilizing C sp^2-heteroatom double bonds and a catalytic metal. Thus, diethyl oxomalonate (**1**) reacted with a series of substituted 1,6-diynes (**59**) in the presence of 5–10 mol %

$Cp^*Ru(cod)Cl$ in 1,2-dichloroethane at 90 °C to give moderate to good yields of dieneones (R = alkyl, Ph) and dienals (R = H) (**61**) (eq 27). The formation of (**61**) presumably goes through an intermediate bicyclic pyran (**60**), which undergoes electrocylic ring opening to give the carbonyl products. When R = CH_3, ring closure of (**61**) to the bicyclic cyclopentenones (**62**) can be performed via an intramolecular Michael reaction using strong base.

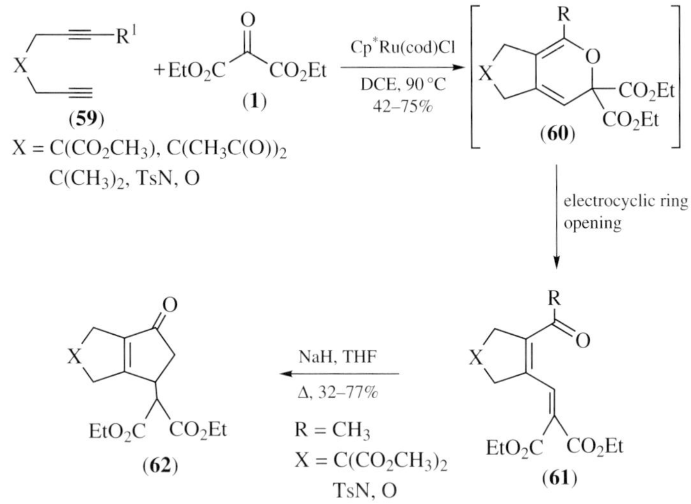

(27)

Recent Examples of Heterocycloadditions of Diethyl Oxomalonate-Derived Imines.[47,48] A report on lanthanide salt Lewis acid-catalyzed electrophilic cyclization of diethyl oxomalonate (**1**) derived imines has been published.[47] Reaction of (**1**) with alkenyl amine (**63**) under dehydrating conditions gives rise to the reactive imine (**64**), which was not isolated (eq 28). Subsequent treatment of (**64**) with 10 mol % ytterbium triflate as a Lewis acid catalyst promotes the bicyclization via addition across the double bond to *trans*-fused lactone (**65**). This is complimentary to previous studies where *cis* ring fusion was observed.[24]

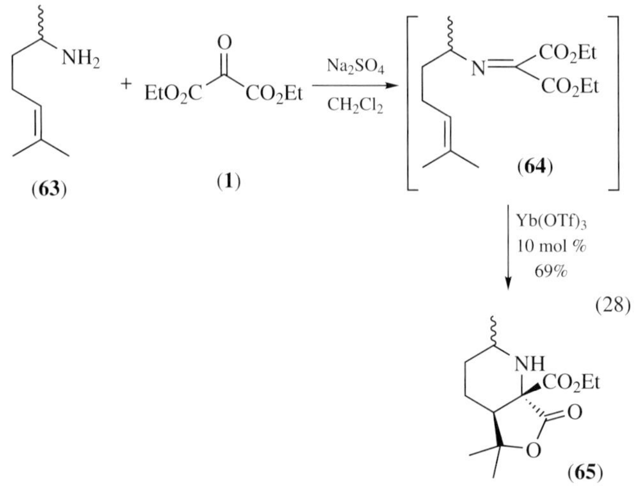

(28)

Another recent report utilizes diethyl oxomalonate (**1**) derived imines in [4 + 2] cycloadditions with enol ethers in the presence of lithium perchlorate to give highly substituted pyridine and tetrahydropyridine derivatives.[48] These electron-poor azadienes such as (**67**) (eq 29), have not been utilized to any great extent for hetero-Diels–Alder reactions with enol ethers. Diethyl oxomalonate-derived azadienes such as (**67**) are generated from the Aza–Wittig reaction of phosphoranes such as (**66**) with (**1**).[49] Reaction of (**67**) with the methyl enol ether (**68**) and lithium perchlorate gives the substituted tetrahydropyridine derivative (**69**) in good yield and excellent regio- and stereoselectivity. The *endo* stereoselectivity of this cycloaddition was determined by NOE difference experiments on (**69**). Alternatively, reaction of azadiene (**67**) with phenyl-substituted silyl enol ether (**70**) gives tetrahydropyridine (**71**) (eq 30). Subsequently, when (**71**) is exposed to tetra-*n*-butyl ammonium fluoride desilylation conditions, aromatization to the 1,4,5-trisubstituted pyridine (**72**) occurs in near quantitative yield, from loss of CO_2, EtOH, and H_2O.

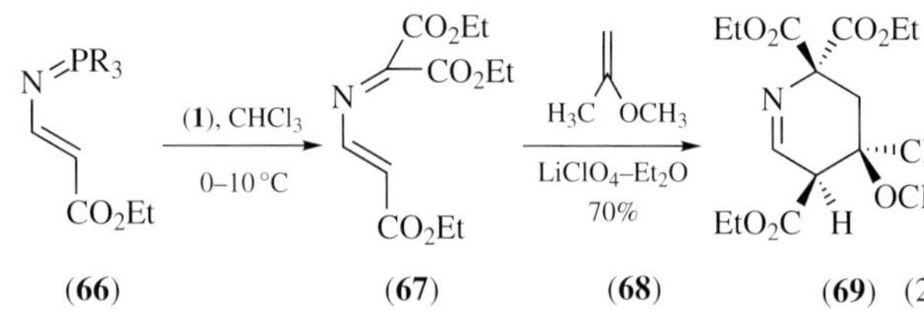

(29)

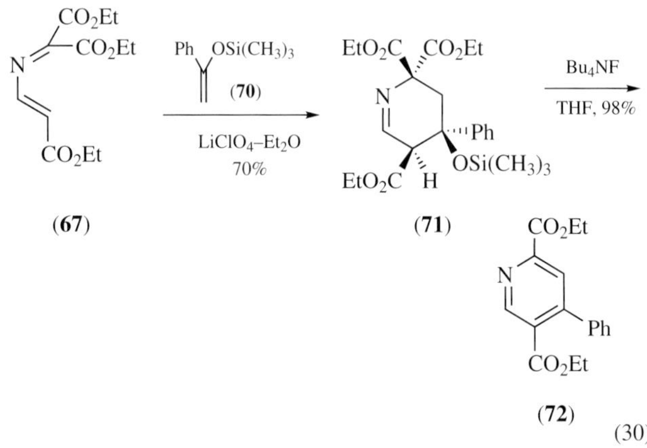

(30)

Wittig Adducts of Diethyl Oxomalonate (1). A Wittig adduct of diethyl oxomalonate (**1**) has been utilized in a novel annulation procedure to produce highly substituted cyclopentenones.[50] Reaction of phosphoranes (**73**) with diethyl oxomalonate (**1**) gives the reactive acylmethylene malonates (**74**) (eq 31). Conjugated phosphorane (**75**) reacts with (**74**) at low temperature to give phosphorane intermediate (**76**) that is not isolated, but undergoes intramolecular Wittig reaction upon warming to room temperature to give the highly substituted cyclopentadiene (**77**) in excellent yields. Treatment of (**78**) with aqueous HCl hydrolyzes the enol ether to give the functionalized cyclopentenones (**79**) in excellent yields, displaying a *trans* arrangement of the ring substituents. These cyclopentenones (**79**) are useful building blocks for a variety of natural products such as prostaglandins or jasmones.

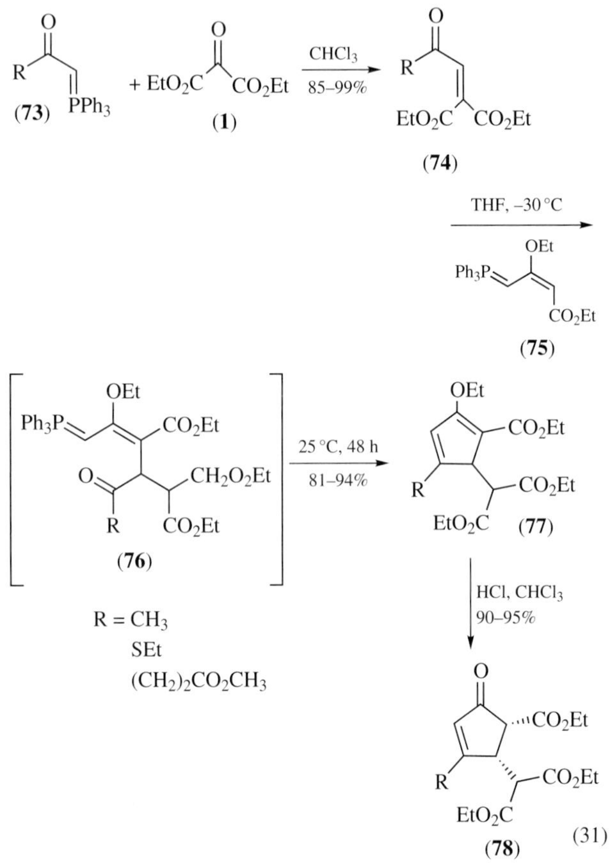

(31)

The electrophilic oxomalonate Wittig adduct (**74**) has also been observed to undergo an unusual [2 + 1] cycloaddition with electron-rich olefins (**79**) in the presence of a Lewis acid to give highly substituted cyclopropanes (**80**) (eq 32).[51] Stereochemical control of this cycloaddition is excellent. Compound (**80**) has been transformed into a cyclopropane aldehyde that is a useful intermediate for synthesis of pyrethroid insecticides.

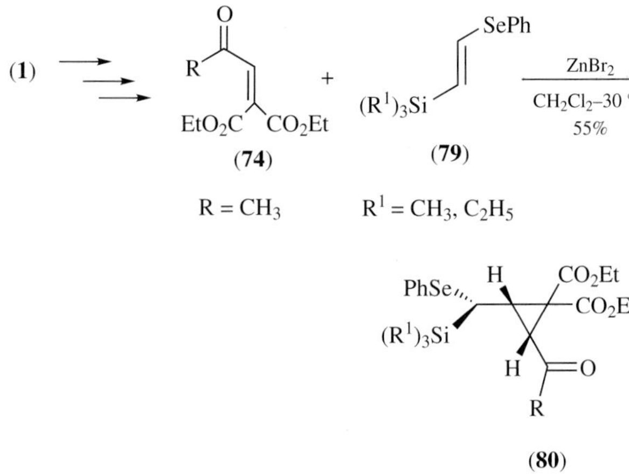

(32)

The ethenetricarboxylate derivative (**80**), another Wittig adduct of diethyl oxomalonate (**1**), reacts with a variety of substituted olefins (**81**) in the presence of Lewis acids to give highly functionalized γ-lactones (**82**) (eq 33).[52] Presumably, the reaction proceeds through the free acid intermediate (**83**) formed by removal of the *t*-Bu ester under the Lewis acidic conditions.

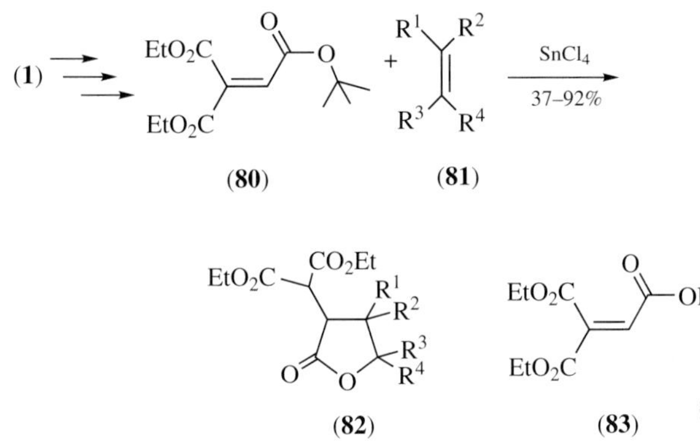

(33)

In another study, oxomalonate adduct acid (**83**) has been isolated from acidolysis of (**80**) and subsequently reacted with a variety of symmetrically and unsymmetrically substituted carbodiimides to give a series of 1,3,5-trisubstituted hydantoins (**84**) in good yields (eq 34).[53] This reaction involves a regiospecific domino condensation/aza-Michael /O→N acyl migration, with the electron-poor olefin of the adduct from (**83**) serving as a Michael acceptor to close the heterocyclic ring.

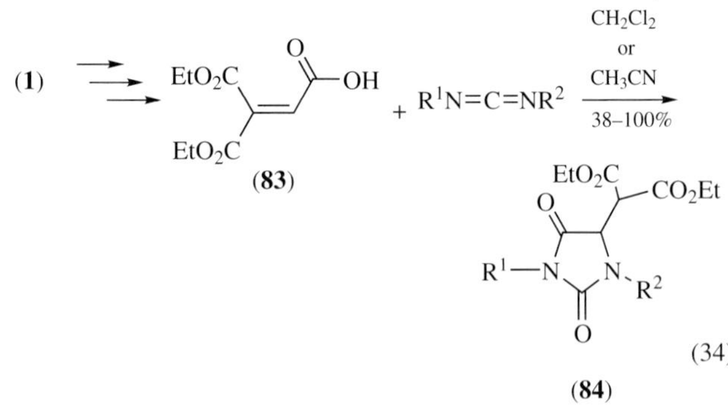

(34)

Nucleophilic Additions to Diethyl Oxomalonate (1). A one-step synthesis of a pyran-fused tetrahydropyranone (**86**) containing geminal carboxylic esters was achieved via addition of metalated tetrahydopyran (**85**) to (**1**) (eq 35).[54]

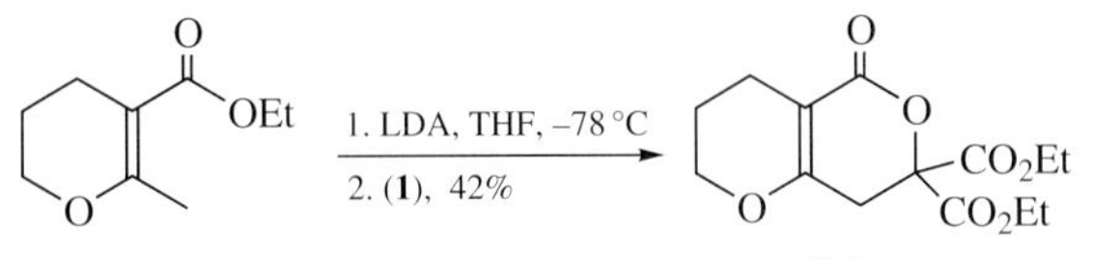

(85) (86) (35)

Enamines have been shown to react as nucleophiles with (**1**),[13] and recent reports have expanded the scope of nucleophilic additions to (**1**) to include aldol reactions.[55–57] In one example, the lithium enolate of cyclohexanone added to (**1**) to give the expected aldol product (**87**) in good yield (eq 36).[55] Acetylation and acetic acid elimination from (**87**) followed by isomerization gives (**88**), which is a key synthon in the synthesis of the natural product squalene synthase inhibitors CP-225,917 and CP-263,114.

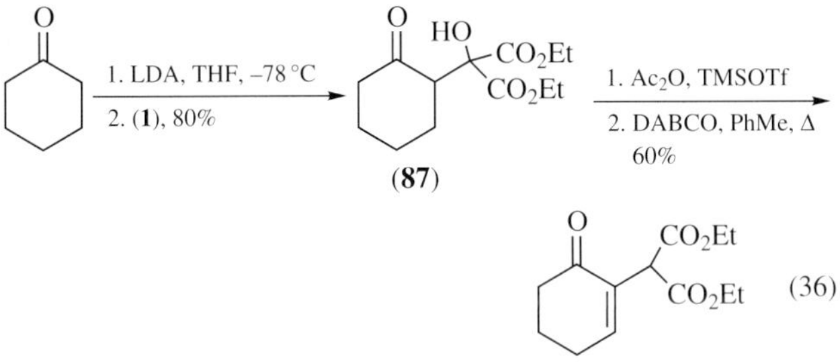

(87) (88) (36)

A *trans* diastereoselective aldol reaction of (**1**) with alcohol-protected cyclohexeneones (**89**) to give the products (**90**) in good yield and excellent diastereoselectivity was reported (eq 37).[56] This aldol product was transformed by a series of reactions to tricyclic intermediate (**91**), which is a key intermediate in the synthesis of the antibiotic natural products luminamicin and lustromycin.

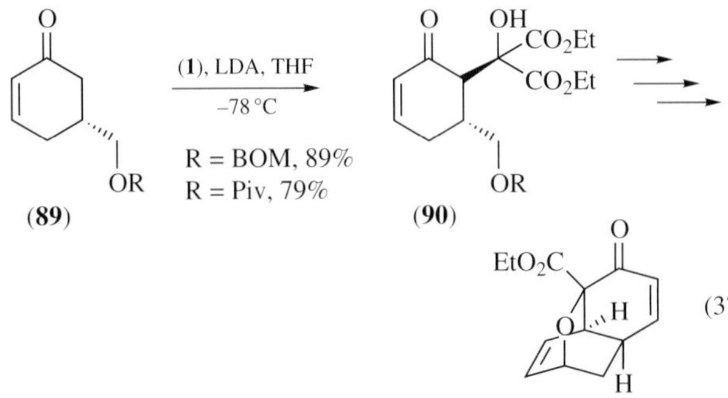

(89) (90) (91) (37)

Chiral aldol additions to (**1**) have been reported using organocatalysis and metal–chiral ligand-based systems.[57,58] Direct catalytic enantioselective additions of a series of aldehydes (**92**) to (**1**) using 50% L-proline as organocatalyst have been shown to give the chiral aldol products (**93**) in excellent yield and enantioselectivity (eq 38).[57] The configuration of the product enantiomer was (*S*) as shown. The excellent enantioselectivity results from the highly ordered orientation of the resultant enamine intermediate and (**1**) in the transition state.

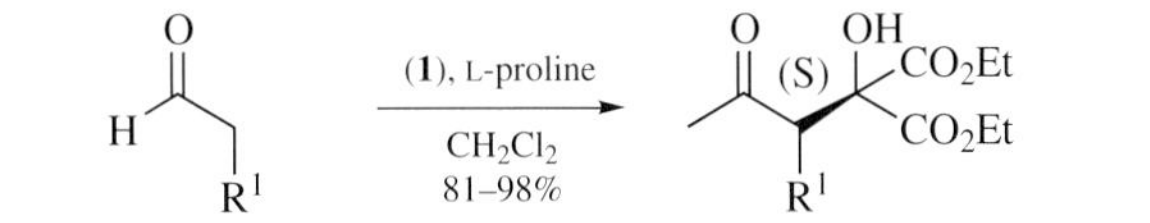

(92) (93) 84–90% ee (38)

R^1 = Me, Et, *i*-Pr, allyl, *n*-hexyl

C_2-symmetric copper (II) bisoxazoline Lewis acid complexes have been shown to catalyze the additions of silyl enol ethers to (**1**), with good yields and enantioselectivities (eq 39).[58] The chiral ligand (**53**) with 10 mol% $Cu(OTf)_2$ was shown to give optimal yields and high enantioselectivities when silyl enol ether (**94**) reacted with (**1**) at low temperature to give the (*R*) enantiomer of aldol product (**95**).

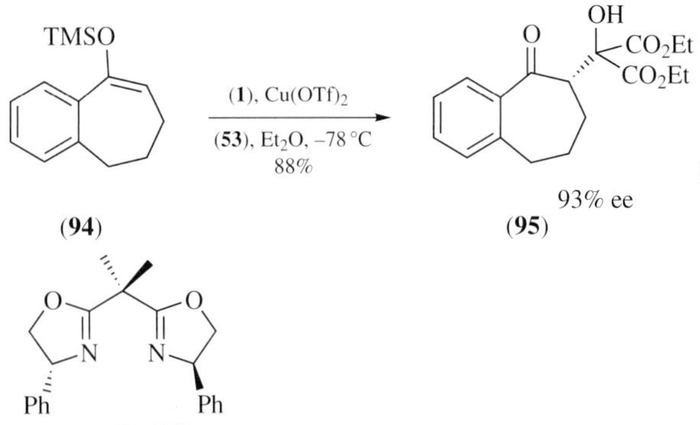

(94) (95) 93% ee (39)

(*R*)-(**53**)

Under Lewis acidic conditions, cyclopropyl ketone (**96**) adds to (**1**) (eq 40) to give the 3-phenyl-1,6-dioxaspiro[4.4]non-3-ene-2-one (**98**)[59] with the *trans* configuration. The reaction presumably proceeds via the rearrangement of the cyclopropyl ketone to the linear terminal hydroxy ketone (**97**), which enolizes and adds to (**1**) to give the observed product in good yield via spirocyclization.

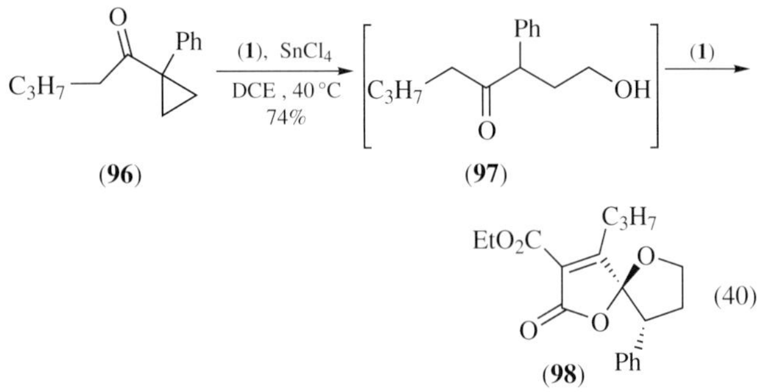

(96) (97) (98) (40)

There are few reports of organometallic additions to (**1**). Presumably, the high reactivity of the oxo and ester functions contained in (**1**) can give rise to multiple additions and complicated product mixtures with reactive organometallic reagents. There are two recent reports of lithium acetylides giving addition products with (**1**), albeit in moderate yields.[60,61] The lithium acetylide derived from propargyl chloride reacts with (**1**) to give the tertiary alcohol (**99**) (eq 41) in moderate yield. This alcohol was converted to the propargyl tungsten species (**100**), which after a series of five steps, including an intramolecular alkoxycarbonylation, gave the mold metabolite isoavenociolide.[61] In a related report, the lithium dianion of propiolic acid reacted with (**1**), to give the tertiary alcohol addition product in 45% yield.[61]

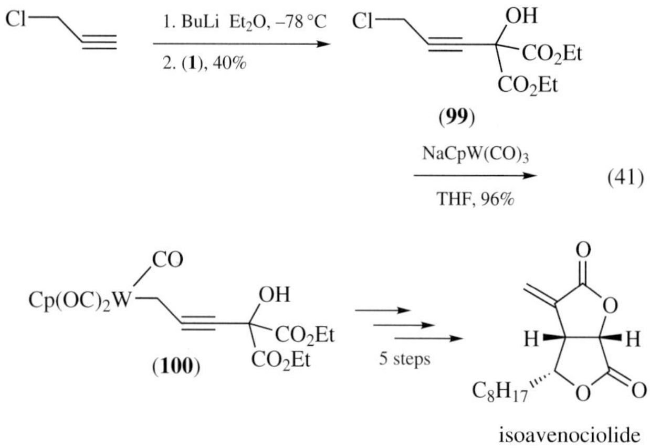

Ene Reactions of (1). Ene reactions of (**1**) are well precedented.[8,9] A recent report examining the Lewis acid-promoted ene reaction of (**1**) with allylbenzene and 3-phenylbut-1-ene by $SnCl_4$ shows the Lewis acid-promoted reaction mechanism to be ionic in nature in contrast to the thermal concerted reaction done without Lewis acid.[62] Strong Lewis acids such as $SnCl_4$ are used in stoichiometric amounts and often promote side reactions. A ruthenium(II) (salen) complex has been shown to be a milder Lewis acid promoter of ene reactions of (**1**) and can be used catalytically.[63] Another report describes a noncatalyzed aza-ene reaction of heterocyclic ketene aminals (**101**) with (**1**) to give γ-lactam fused pyrimidine and imidazole derivatives (**102**) in moderate to good yield (eq 42).[64]

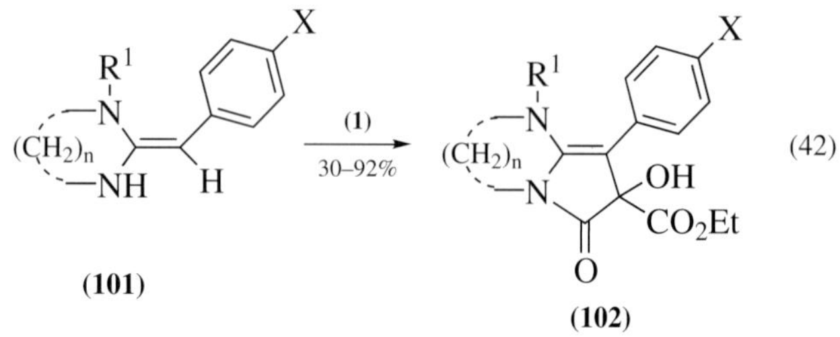

Miscellaneous Reactions of (1) and Its Derivatives. The electrophilic carbonyl group of oxomalonate (**1**) participates in the Passerini reaction with isonitrile (**103**) and LiBr in toluene to give the oxazole product (**104**) in moderate yield (eq 43).[65]

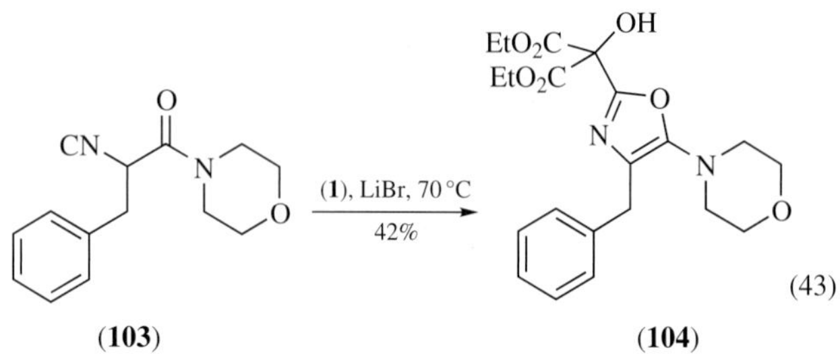

The *N*-Boc-oxaziridine (**105**) derived from (**1**) and *N*-Boc phosphinimine, followed by oxidation, is an efficient heteroatom transfer agent.[66,67] (**105**) serves to aminate sulfides to give good to excellent yields of *N*-Boc sulfimides (**106**) (eq 44).[66] Similarly, reaction of (**105**) with amines gives the corresponding *N*-Boc hydrazines (**107**) (eq 44).[67]

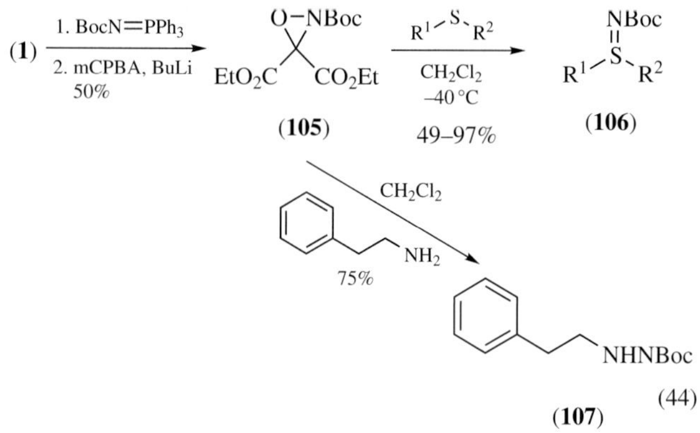

Another electrophilic amination reagent derived from (**1**) is the *p*-anisidine-derived iminomalonate (**108**).[68] This iminomalonate reacts with organo-aluminum, magnesium, and zinc reagents at nitrogen (eq 45) to give the amino malonates (**109**), which can be oxidatively cleaved to the secondary amines (**110**), in moderate to good yields.

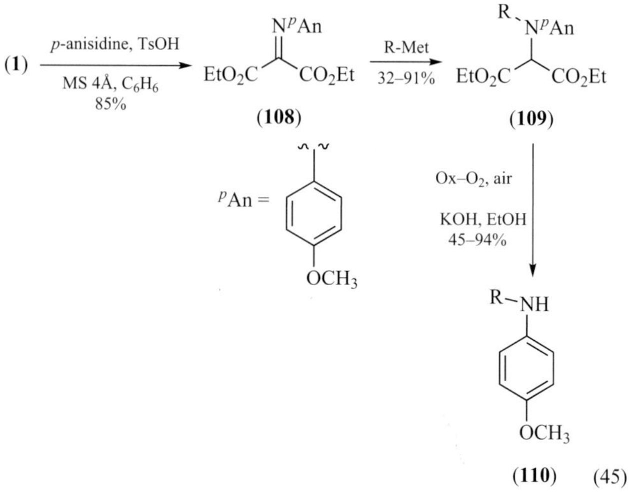

An unusual reduction of (**1**) to the corresponding hydroxyl malonate in 85% yield has been reported to occur in THF in the presence of trimethyl phosphine.[69]

1. (a) Boger, D. L.; Weinreb, S., *Hetero Diels–Alder Methodology in Organic Synthesis*; Academic: New York, 1987. (b) Bonjouklian, R.; Ruden, R. A., *J. Org. Chem.* **1977**, *42*, 4095. (c) Bonjouklian, R.; Ruden, R. A., *J. Am. Chem. Soc.* **1975**, *97*, 6892. (d) Knapp, S.; Levorse, A. T.; Potenza, J. A., *J. Org. Chem.* **1988**, *53*, 4773. (e) Achmatowicz, O.; Bialecka-Florjanczyk, E., *Tetrahedron* **1990**, *46*, 5317. (f) Daniewski, W. M.; Kubak, E.; Jurczak, J., *J. Org. Chem.* **1985**, *50*, 3963.
2. Quimpere, M.; Jankowski, K., *J. Chem. Soc., Chem. Commun.* **1987**, 676.
3. Barluenga, J.; Gonzalez, F. J.; Fustero, S., *Tetrahedron Lett.* **1989**, *30*, 2685.

4. Page, P. C. B.; Williams, P. H.; Collington, E. W.; Finch, H., *J. Chem. Soc., Chem. Commun.* **1987**, 756.
5. Hosomi, A.; Miyashiro, Y.; Yoshida, R.; Tominaga, Y.; Yanagi, T.; Hojo, M., *J. Org. Chem.* **1990**, *55*, 5308.
6. Little, R. D.; Bode, H.; Stone, K. J.; Wallquist, O.; Dannecker, R., *J. Org. Chem.* **1985**, *50*, 2400.
7. Gotthardt, H.; Schenk, K.-H., *Angew. Chem.* **1985**, *97*, 604.
8. Mattay, J.; Buchkremer, K., *Heterocycles* **1988**, *27*, 2153.
9. (a) Salomon, M. F.; Pardo, S. N.; Salomon, R. G., *J. Am. Chem. Soc.* **1984**, *106*, 3797. (b) Pardo, S. N.; Salomon, R. G., *J. Org. Chem.* **1981**, *46*, 2598. (c) Salomon, M. F.; Pardo, S. N.; Salomon, R. G., *J. Am. Chem. Soc.* **1980**, *102*, 2473. (d) Tidwell, J. H.; Senn, D. R.; Buchwald, S. L., *J. Am. Chem. Soc.* **1991**, *113*, 4685. (e) Concerning the mechanism of the ene reaction: Song, Z.; Beak, P., *J. Am. Chem. Soc.* **1990**, *112*, 8126.
10. Salomon, R. G.; Roy, S.; Salomon, S. F., *Tetrahedron Lett.* **1988**, *29*, 769.
11. Trebaul, C.; Roncali, J.; Garnier, F.; Guglielmetti, F., *Bull. Chem. Soc. Jpn.* **1987**, *60*, 2657.
12. Goya, P.; Paez, J. A., *Liebigs Ann. Chem.* **1988**, 121.
13. Schultz, A. G.; Yee, Y. K., *J. Org. Chem.* **1976**, *41*, 561.
14. Pindur, U.; Kim, M. H., *Tetrahedron* **1989**, *45*, 6427.
15. Jurczak, J.; Belniak, S.; Chmielewski, M.; Kozluk, T.; Pikul, S.; Jenner, G., *J. Org. Chem.* **1989**, *54*, 4469.
16. Afarinkia, K.; Rees, C. W.; Cadogan, J. I. G., *Tetrahedron* **1990**, *46*, 7175.
17. Riebsomer, J. L.; Irvine, J., *Org. Synth.* **1955**, *3*, 326.
18. Fell, S. C. M.; Pearson, M. J.; Burton, G.; Bateson, J. H., *J. Chem. Soc., Perkin Trans. 1* **1991**, 1361.
19. Taylor, E. C.; Dötzer, R., *J. Org. Chem.* **1991**, *56*, 1816.
20. White, J. D.; Cammack, J. H.; Sakuma, K., *J. Am. Chem. Soc.* **1989**, *111*, 8970.
21. Marinetti, A.; Mathey, F., *Angew. Cheml.* **1988**, *100*, 1435.
22. Marinetti, A.; Mathey, F., *J. Chem. Soc., Chem. Commun.* **1990**, 153.
23. (a) Tietze, L. F.; Bratz, M.; Pretor, M., *Chem. Ber.* **1989**, *122*, 1955; (b) Tietze, L. F.; Bratz, M., *Liebigs Ann. Chem.* **1989**, 559.
24. (a) Narayanan, K.; Cook, J. M., *Tetrahedron Lett.* **1990**, *31*, 3397; (b) Narayanan, K.; Schindler, L.; Cook, J. M., *J. Org. Chem.* **1991**, *56*, 359.
25. Pardo, S. N.; Salomon, R. G., *J. Org. Chem.* **1981**, *46*, 2598.
26. Tietze, L. F.; Bratz, M., *Org. Synth.* **1992**, *71*, 214.
27. Neunhoeffer, H.; Reichel, D.; Cullmann, B.; Rehn, I., *Liebigs Ann. Chem.* **1990**, 631.
28. Achmatowicz, O.; Cieplak, P.; Rozwadowski, J., *Pol. J. Chem.* **1989**, *63*, 531.
29. Tietze, L. F.; Bratz, M., *Chem. Ber.* **1989**, *122*, 997.
30. Pawlak, J. L.; Padykula, R. E.; Kronis, J. D.; Aleksejczyk, R. A.; Berchtold, G. A., *J. Am. Chem. Soc.* **1989**, *111*, 3374.
31. Cameron, D. W.; Read, R. W.; Stavrakis, J., *Aust. J. Chem.* **1987**, *40*, 1831.
32. Gajewski, J. J.; Jurayi, J.; Kimbrough, D. R.; Gande, M. E.; Ganem, B.; Carpenter, B. K., *J. Am. Chem. Soc.* **1987**, *109*, 1170.
33. Gotoh, T.; Padias, A. B.; Hall, H. K., *J. Am. Chem. Soc.* **1986**, *108*, 4920.
34. Jenner, G.; Papadopoulos, M., *J. Org. Chem.* **1986**, *51*, 585.
35. Jenner, G.; Papadopoulos, M.; Jurczak, J.; Kozluk, T., *Tetrahedron Lett.* **1984**, *25*, 5747.
36. Jurczak, J.; Kozluk, T.; Pikul, S.; Salanski, P., *J. Chem. Soc., Chem. Commun.* **1983**, 1447.
37. Hoare, J. H.; Policastro, P. P.; Berchtold, G. A., *J. Am. Chem. Soc.* **1983**, *105*, 6264.
38. Papadopoulos, M.; Jost, R.; Jenner, G., *J. Chem. Soc., Chem. Commun.* **1983**, 221.
39. Abdallah, H.; Gree, R.; Carrie, R., *Can. J. Chem.* **1983**, *61*, 217.
40. McGowan, D. A.; Berchtold, G. A., *J. Am. Chem. Soc.* **1982**, *104*, 7036.
41. Bennett, D. M.; Okamoto, I.; Danheiser, R. L., *Org. Lett.* **1999**, *1*, 641.
42. Bailey, W. F.; Wachter-Jurcsak, N. M.; Pineau, M. R.; Ovaska, T.V.; Warren, R. R.; Lewis, C. E., *J. Am. Chem. Soc.* **1996**, *61*, 8216.
43. Yao, S.; Roberson, M.; Reichel, F.; Hazell, R. G.; Jorgensen, K. A., *J. Am. Chem. Soc.* **1999**, *64*, 6677.
44. Takahashi, T.; Li, Y.; Ito, T.; Xu, X.; Nakajima, K.; Liu, Y., *J. Am. Chem. Soc.* **2002**, *124*, 1144.
45. (a) Shi, M.; Xu, B., *Tetrahedron Lett.* **2003**, *44*, 3839. (b) Liu, L.-P.; Shi, M., *Tetrahedron* **2007**, *63*, 4535.
46. Yamamoto, Y.; Takagishi, H.; Itoh, K., *J. Am. Chem. Soc.* **2002**, *124*, 6844.
47. Jia, X.; Xie, W.; Zhang, W.; Janczuk, A.; Lou, S.; Zhang, B.; Cheng, J. P.; Ksebati, M. B.; Wang, P. G., *Tetrahedron Lett.* **2002**, *43*, 2339.
48. Palacios, F.; Harran, E.; Rubiales, G.; Ezpeleta, J. M., *J. Org. Chem.* **2002**, *67*, 2131.
49. Palacios, F.; de Heredia, I. P.; Rubiales, G., *J.Org. Chem.* **1995**, *60*, 2384.
50. Himeda, Y.; Tanaka, Y.; Udea, I.; Hatanaka, M., *J. Chem. Soc. Perkin, Trans. 1* **1998**, 1389.
51. Yamazaki, S.; Kumagai, H.; Takada, T.; Yamabe, S.; Yamamoto, K., *J. Org. Chem.* **1997**, *62*, 2968.
52. Yamazaki, S.; Ohmitsu, K.; Ohi, K.; Otsubo, T.; Moriyama, K., *Org. Lett.* **2005**, *7*, 759.
53. Volonterio, A.; Ramirez de Arellano, C.; Zanda, M., *J. Org. Chem.* **2004**, *69*, 2161.
54. Armstrong, A.; Goldberg, F. W.; Sandham, D. A., *Tetrahedron Lett.* **2001**, *42*, 4585.
55. Armstrong, A.; Critchley, T.; Gourdel-Martin, M.-E.; Kelsey, R. D.; Mortlock, A. A., *J. Chem. Soc., Perkin Trans. 1* **2002**, 1344.
56. Sunazuka, T.; Handa, M.; Hirose, T.; Matsumaru, T.; Togashi, Y.; Nakamura, K.; Iwai, Y.; Omura, S., *Tetrahedron Lett.* **2007**, *48*, 5297
57. Bogevig, A.; Kumaragurubaran, N.; Jorgensen, K. A., *Chem. Commun.* **2002**, 620.
58. Reichel, F.; Fang, X.; Yao, S.; Ricci, M.; Jorgensen, K. A., *Chem. Commun.* **1999**, 1505.
59. Shi, M.; Yang, Y.-H., *Org. Lett.* **2006**, *8*, 1709.
60. Chen, M.-J.; Narkunan, K.; Liu, R.-S., *J. Org. Chem.* **1999**, *64*, 8311.
61. Nagao, Y.; Sano, S.; Morimoto, K.; Kakegawa, H.; Takatani, T.; Shiro, M.; Katunuma, N., *Tetrahedron Lett.* **2000**, *41*, 2419.
62. Achmatowicz, O.; Florjanczyk, E.-B., *Tetrahedron* **1996**, *52*, 8827.
63. Ellis, W. W.; Odenkirk, W.; Bosnich, B., *Chem. Commun.* **1998**, 1311.
64. Zhang, J.-H.; Wang, M.-X.; Huang, Z.-T., *J. Chem. Soc. Perkin, Trans. 1* **1999**, 321.
65. Cuny, G.; Gamez-Montano, R.; Zhu, J., *Tetrahedron* **2004**, *60*, 4879.
66. Armstrong, A.; Cooke, R. S., *Chem. Commun.* **2002**, 904.
67. Armstrong, A.; Jones, L. H.; Knight, J. D.; Kelsey, R. D., *Org. Lett.* **2005**, *7*, 713.
68. Niwa, Y.; Takayama, K.; Shimizu, M., *Tetrahedron Lett.* **2001**, *42*, 5473.
69. Zhang, W.; Shi, M., *Chem. Commun.* **2006**, 1218.

1,3-Dihydro-1,3-bis(tricyclo[3.3.1.13,7] dec-1-yl)-2*H*-imidazol-2-ylidene

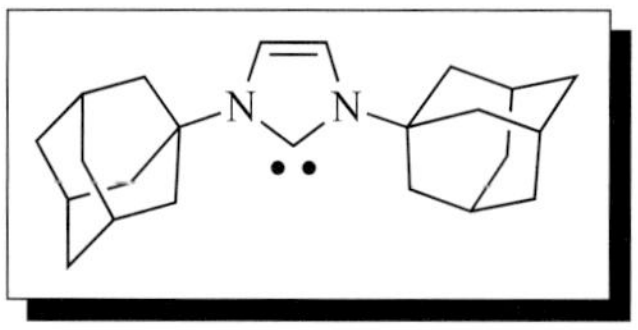

[131042-77-8] $C_{23}H_{32}N_2$ (MW 336.51)

InChI = 1S/C23H32N2/c1-2-25(23-12-19-6-20(13-23)8-21(7-19)14-23)15-24(1)22-9-16-3-17(10-22)5-18(4-16)11-22/h1-2,16-21H,3-14H2

InChIKey = HPESDZAFNBMVII-UHFFFAOYSA-N

(*N*-heterocyclic carbene, carbene ligand, organocatalysis)

Alternate Name: IAd.

Physical Data: mp: 240–241 °C; pK_a(H_2O): 25.4.

Solubility: soluble in THF.

Form Supplied in: white crystalline solid.

Analysis of Reagent Purity: ^{1}H NMR (C_6D_6) δ 1.58 ppm (s, 2*H*), 2.01 (s, 6 *H*), 2.29 (s, 12 *H*), 6.91 (s, 2 *H*); ^{13}C NMR (C_6D_6) δ 211.43, 113.88, 55.99, 44.8, 37.7, 30.3.

Preparative Method: the free carbene is prepared by deprotonation of the corresponding imidazolium salt.[1] The salt can be prepared in a single pot from commercially available starting materials (eq 1).[2]

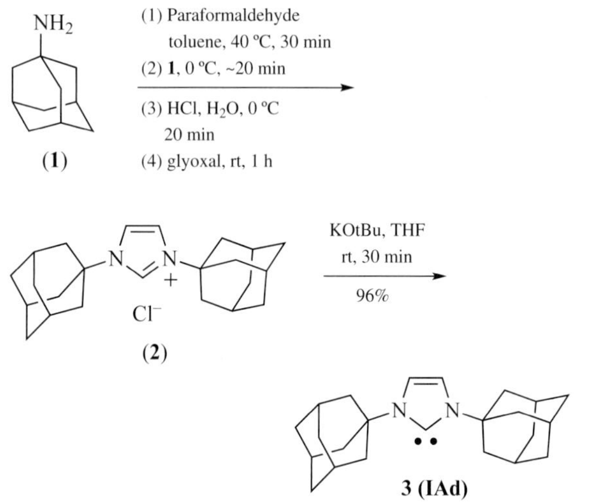

(1)

Purification: recrystallization from dry toluene.

Handling, Storage, and Precautions: Air and moisture sensitive, store under an inert atmosphere.

General Reactivity. The reactivity of IAd is centered on the carbene at position 2. Due to the nitrogen atoms in the adjacent positions, the electronic configuration is singlet. The carbene is nucleophilic, which allows for interesting reactivity with a variety of electrophiles. IAd has played a role in catalysis both as an organocatalyst and as a ligand for a variety of transition metals.

IAd can act as a base in the presence of methyl acetate to yield the corresponding Claisen condensation product anion **4** with the imidazolium as the cation (eq 2).[3]

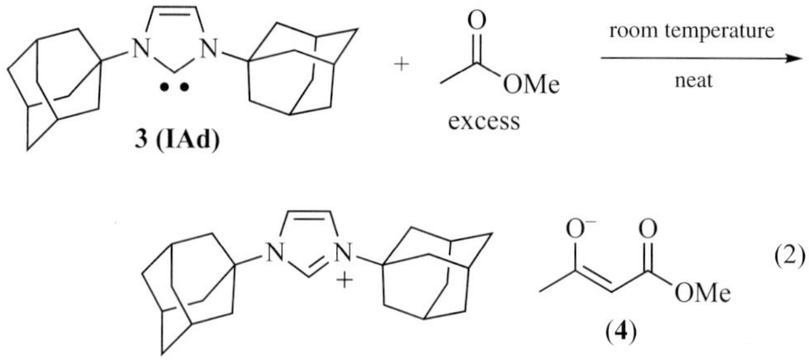

(2)

Organocatalysis. IAd has been used as an organocatalyst for several transformations. *N*-Heterocyclic carbenes have been used to catalyze transesterification. IAd is effective in transesterification; however, the scope was explored using a different carbene.[4]

IAd can be employed as a catalyst for the Mukaiyama aldol reaction (eq 3).[5] 2-Naphthylenecarboxaldehyde **5** reacts with silyl enol ether **6** in the presence of a catalytic amount of **3** to generate aldol adduct **7** in 77% yield. It was proposed that the carbene is acting as a Lewis base in order to activate the silyl group and facilitate nucleophilic addition of the enol to the aldehyde.

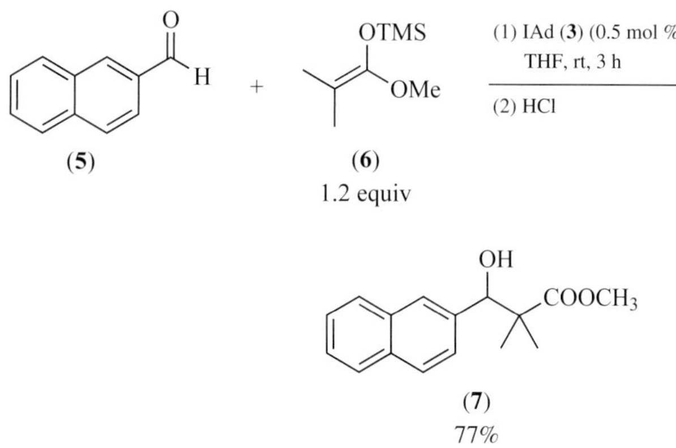

(3)

IAd can also be employed as a catalyst for silyl enol ether formation (eq 4). When ketone **8** is treated with enol ether **6** in the presence of a catalytic amount of **3**, the silyl group is transferred to form enol ether **9** in 92% yield.

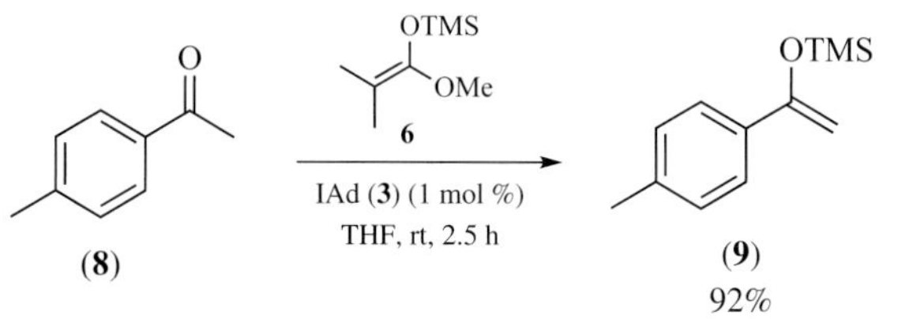

(4)

The new reactivity allows for generation of an enol ether followed by a Mukaiyama aldol reaction in tandem (eq 5).[6] Acetophenone **10** is transformed in situ by enol ether **6** and catalyst **3** to enol ether **11**. Addition of aldehyde **8** followed by protodesilylation generates aldol product **12** in 75% yield.

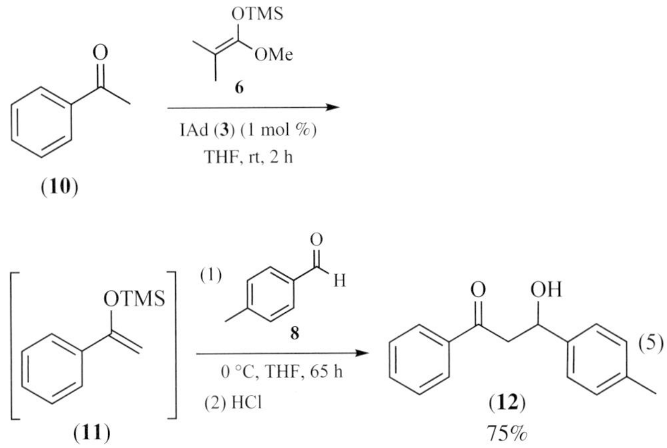

(5)

IAd is a catalyst in the trifluoromethylation of aldehydes on a protected β-cyclodextrin with $TMSCF_3$ (eq 6).[7] Treatment of benzylated β-cyclodextrin **13** with $TMSCF_3$ and 30 mol % of IAd (**3**) leads to formation of the double trifluoromethylated product **14** in 27% yield.

As a Ligand for Transition Metals. IAd has been installed as a ligand on a variety of transition metals, including nickel,[8,9] copper,[10] gold,[11,12] and palladium.[13] For applications in catalysis, it is important to be able to compare the steric properties of ligands when bound to a metal center. Due to the planarity of *N*-heterocyclic carbenes, Tolman cone angles are not an apt comparison. A new computation known as buried volume has been derived.[14] The buried volumes of IAd and many other NHC and phosphine ligands while they are bound to metal centers have been compared (eq 7).[15]

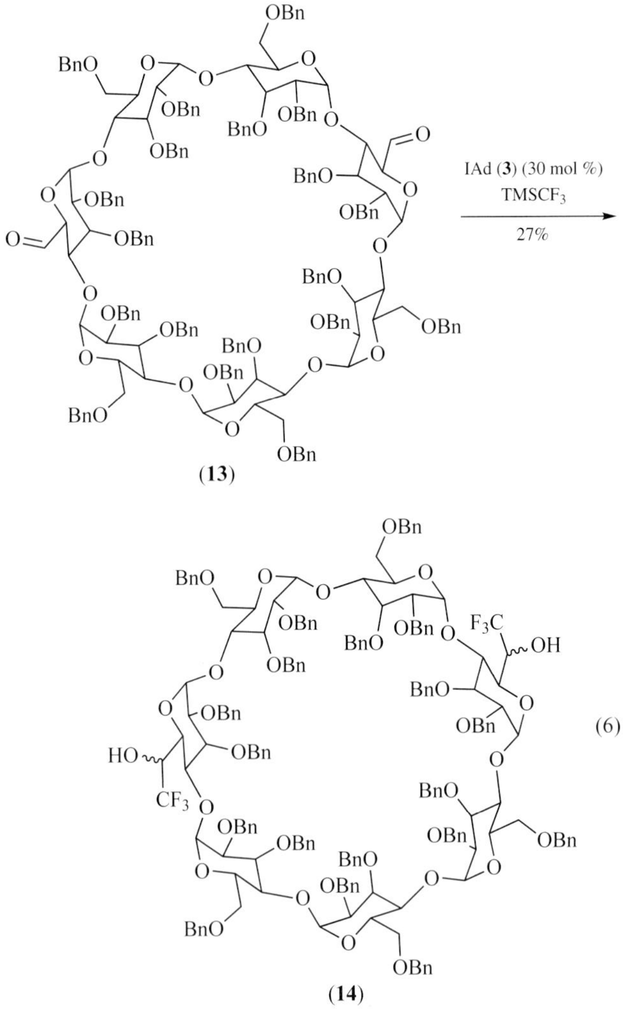

(6)

Buried volume for an assortment of ligands bound to AuCl (2 Å)

Ligand	Buried Volume (%)	Ligand	Buried Volume (%)
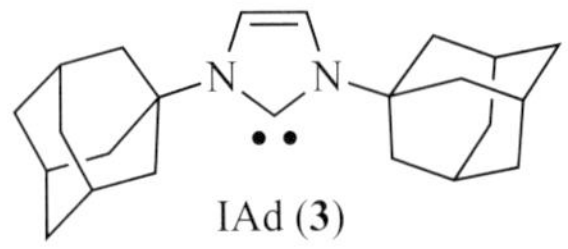 IAd (**3**)	39.8	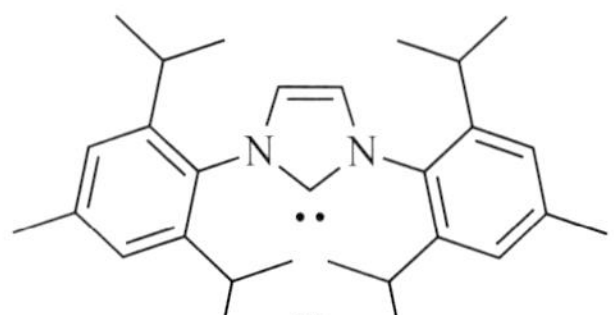 IPr	44.5
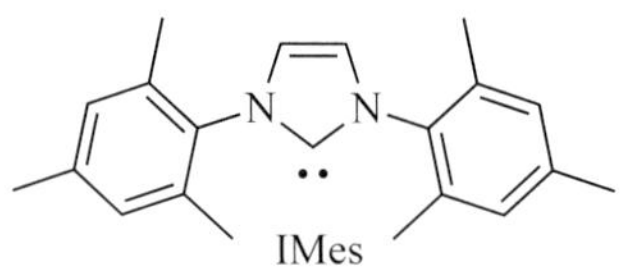 IMes	36.5	$PtBu_3$	43.9
		PPh_3	34.8
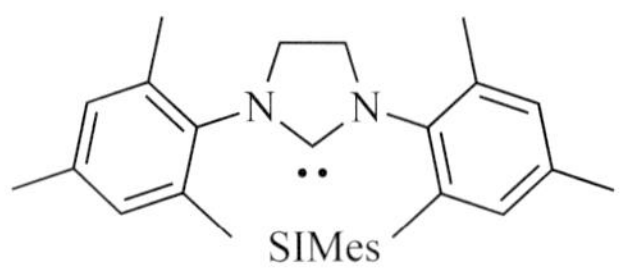 SIMes	36.9	PMe_3	27.3

(7)

A nickel(0) IAd complex allows the regioselective hydrocarbamoylation of terminal olefins (eq 8).[16] The nickel complex $Ni(cod)_2$ forms a nickel IAd complex in situ in the presence of **3**. The complex then catalyzes the reaction of formamide **15** and terminal olefin **16** in the presence of triethylaluminum to form amide **17** in 69% yield. The mechanism is postulated to begin with oxidative addition of the carbamate C–H bond, followed by hydronickelation of the olefin and reductive elimination.

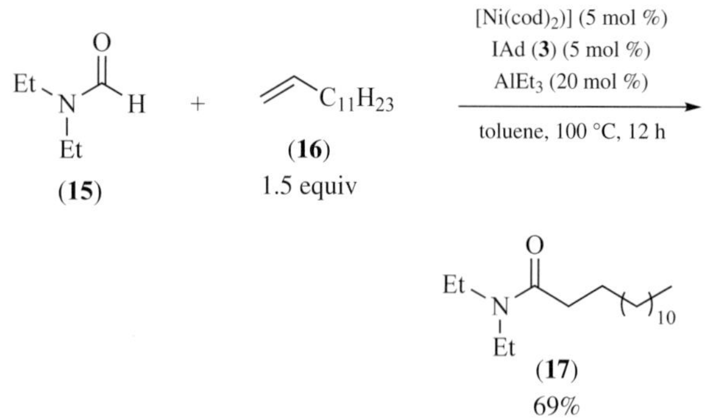

(8)

Cu(IAd)I **20** catalyzes the Huisgen [3+2] cycloaddition of azides and terminal alkynes (eq 9).[10] Benzyl azide **18** and phenyl acetylene **19** provide near quantitative yield of triazole **21** in under 10 min when treated with solvent-free **20** at room temperature.

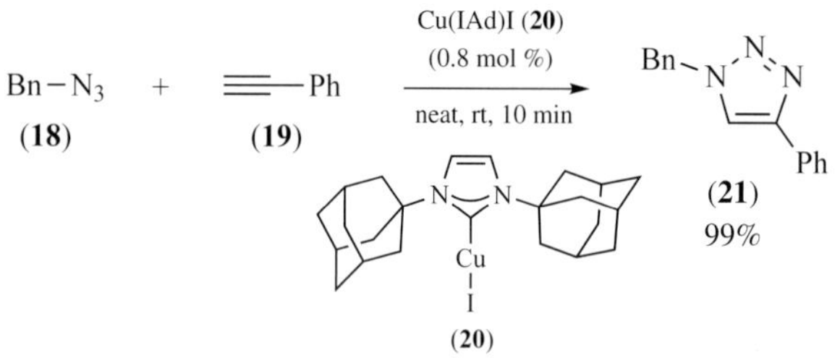

(9)

An IAd gold complex **23** has been employed for the synthesis of 2,3-disubstituted indoles (eq 10) and 3-indolones (eq 11).[12] Complex **23** catalyzes the cyclization of **22** and nucleophilic addition of ethanol to the intermediate species leads to formation of indole **24** in 99% yield.

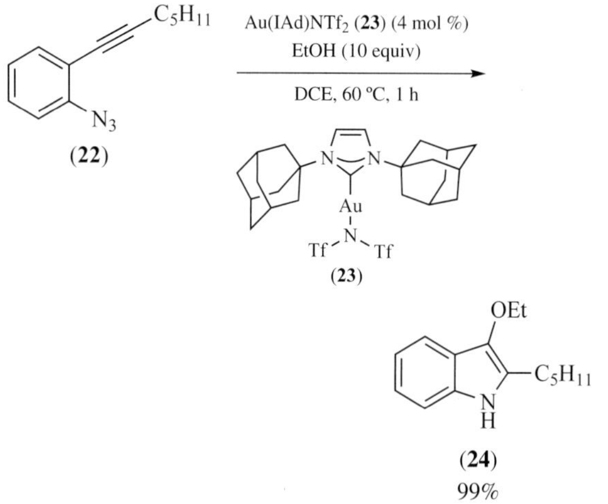

(10)

When the indole-forming reaction was conducted using the allylic alcohol **25**, the product undergoes Claisen rearrangement to form a 2,2-disubstituted 3-indolone. Following cyclization of **22**, nucleophilic addition of **25** leads to the formation of an intermediate 3-alkoxy indole that following rearrangement gives 92% yield of 3-indolone **26** in a 3:1 ratio of diastereomers.

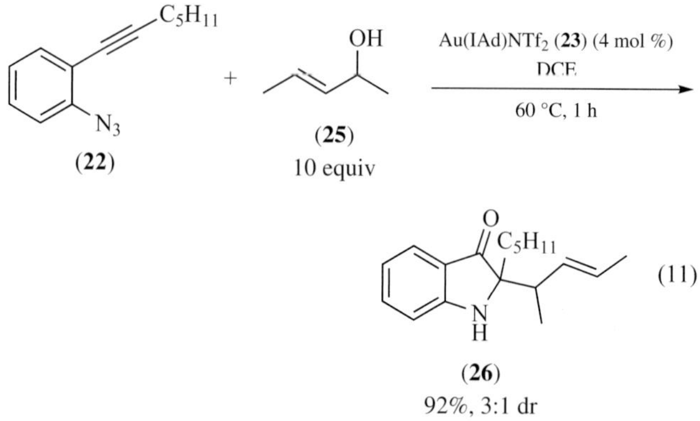

(11)

Mixing IAd (**3**) and allylpalladium chloride dimer **28** in dioxane forms a complex in situ that has been employed to synthesize cyclobutanones from tethered aryl chlorides bearing a pendant aldehyde (eq 12).[17] The catalyst cyclizes aldehyde **27** in the presence of allyl ether and cesium carbonate to give cyclobutanone **29** in 74% yield.

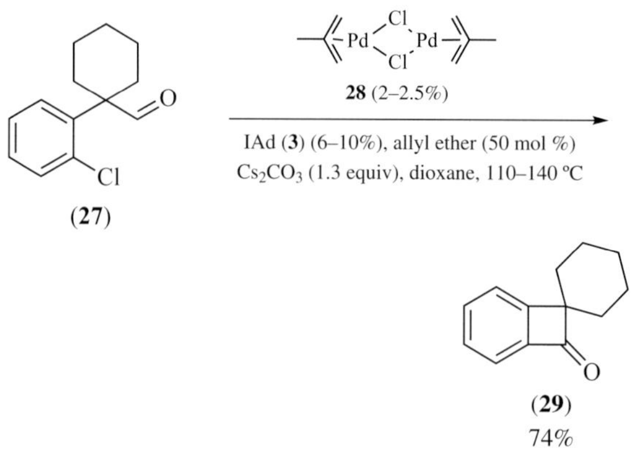

(12)

Bis IAd palladium **32** facilitates the oxidative addition of aryl chlorides for Suzuki coupling (eq 13).[18] Aryl chloride **30** couples with a slight excess of boronic acid **31** in the presence of 2 equiv of cesium fluoride and catalyst **32** to afford biphenyl **33** in 97% yield.

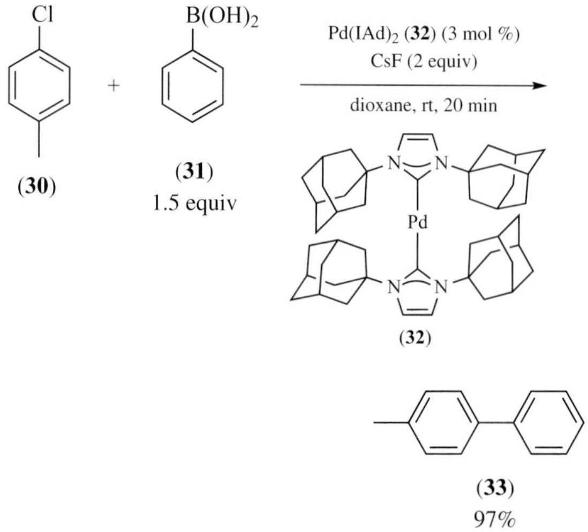

(13)

1. Arduengo, III, A. J.; Harlow, R. L.; Kline, M., *J. Am. Chem. Soc.* **1991**, *113*, 361.
2. Arduengo, III, A. J., U.S. PAT. 5,077,414 (1991).
3. Grasa, G. A.; Singh, R.; Scott, N. M.; Stevens, E. D.; Nolan, S. P., *Chem. Commun.* **2004**, 2890.
4. Grasa, G. A.; Guveli, T.; Singh, R.; Nolan, S. P., *J. Org. Chem.* **2003**, *68*, 2812.
5. Song, J. J.; Tan, Z.; Reeves, J. T.; Yee, N. K.; Senanayake, C. H., *Org. Lett.* **2007**, *9*, 1013.
6. Song, J. J.; Tan, Z.; Reeves, J. T.; Fandrick, D. R.; Yee, N. K.; Senanayake, C. H., *Org. Lett.* **2008**, *10*, 877.
7. Bjerre, J.; Fenger, T. H.; Marinescu, L. G.; Bols, M., *Eur. J. Org. Chem.* **2007**, 704.
8. Dorta, R.; Stevens, E. D.; Scott, N. M.; Costabile, C.; Cavallo, L.; Hoff, C. D.; Nolan, S. P., *J. Am. Chem. Soc.* **2006**, *127*, 2485.
9. Dorta, R.; Stevens, E. D.; Hoff, C. D.; Nolan, S. P., *J. Am. Chem. Soc.* **2003**, *125*, 10490.
10. Diez-Gonzalez, S.; Escudero-Adan, E. C.; Benet-Buchholz, J.; Stevens, E. D.; Slawin, A. M.; Nolan, S. P., *Dalton Trans.* **2010**, *39*, 7595.
11. de Fremont, P.; Scott, N. M.; Stevens, E. D.; Nolan, S. P., *Organometallics* **2005**, *24*, 2411.
12. Wetzel, A.; Gagosz, F., *Angew. Chem., Int. Ed.* **2011**, *50*, 7354.
13. Viciu, M. S.; Navarro, O.; Germaneau, R. F.; Kelly, III, R. A.; Sommer, W.; Marion, N.; Stevens, E. D.; Cavallo, L.; Nolan, S. P., *Organometallics* **2004**, *23*, 1629.
14. Poater, A.; Cosenza, B.; Correa, A.; Giudice, S.; Ragone, F.; Scarano, V.; Cavallo, L., *Eur. J. Inorg. Chem.* **2009**, 1759.
15. Clavier, H.; Nolan, S. P., *Chem. Commun.* **2010**, *46*, 841.
16. Miyazaki, Y.; Yamada, T.; Nakao, Y.; Hiyama, T., *Chem. Lett.* **2012**, *41*, 298.
17. Flores-Gaspar, A.; Gutierrez-Bonet, A.; Martin, R., *Org. Lett.* **2012**, *14*, 5234.
18. Gstottmayr, C. W. K.; Bohm, V. P. W.; Herdtweck, E.; Grosche, M.; Herrmann, W. A., *Angew. Chem., Int. Ed.* **2002**, *41*, 1363.

Michael Holtz-Mulholland & Shawn K. Collins
Université de Montréal, QC, Canada

1,3-Dihydro-4,5-dimethyl-1,3-bis(1-methylethyl)-2*H*-imidazol-2-ylidene

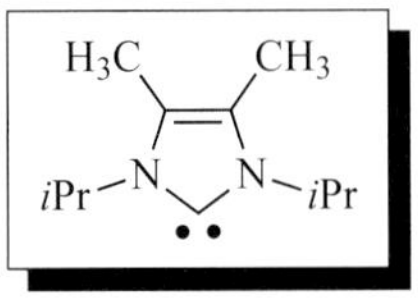

[151515-23-0] $C_{11}H_{20}N_2$ (MW 180.29)
InChI = 1S/C11H21IN2/c1-8(2)12-14-7-13(9(3)4)10(5)11(14)6/h8-9,12H,1-6H3
InChIKey = NOOSYWWJOSYXOF-UHFFFAOYSA-N

Physical Data: pale yellow solid, mp 190 °C (dec.).
Solubility: soluble in toluene, benzene, THF, CH_2Cl_2, and $scCO_2$.
Form Supplied in: not commercially available.
Analysis of Reagent Purity: not reported.
Preparative Methods: the title compound (IPrMe (**1**)) is most commonly prepared using procedures reported by Kuhn that involve the reduction of 1,3-diisopropyl-4,5-dimethylimidazole-2(3*H*)-thione using potassium in boiling THF.[1] The thione itself is prepared by the condensation of diisopropyl thiourea and 3-hydroxy-2-butanone. In addition, deprotonation of the imidazolium salt using modifications to procedures originally reported by Arduengo have also been used.[2,3]
Handling, Storage, and Precautions: indefinitely stable under an inert atmosphere at ambient temperature.

Introduction. The first isolated NHCs, the imidazolylidenes,[4] are highly versatile Lewis bases acting as either ligands[5] or organocatalysts[6] in a range of reactions. While in many cases *N*-aryl-substituted NHCs, such as IMes (**2**) have proven useful, smaller,[7] and more electron-rich[8] *N*-alkyl imidazolylidenes such as IPrMe (**1**) can lead to orthogonal or enhanced reactivity. Herein, we examine applications of IPrMe (**1**) as polymerization catalysts, organocatalysts, and ligands in metal-mediated organic synthesis.

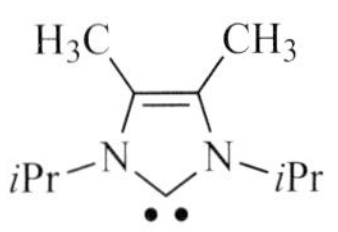

1

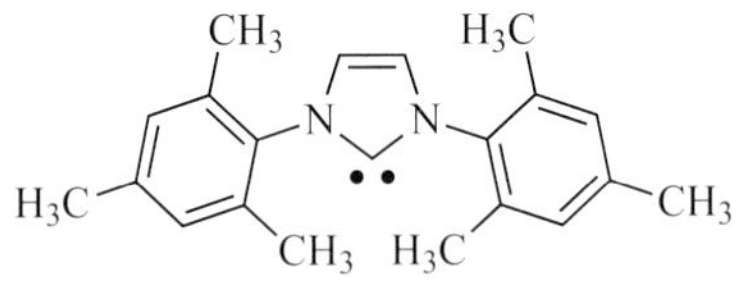

2

Polymerization. Atom transfer radical polymerization (ATRP) utilizes redox-active transition metal catalysts to allow controlled radical polymerization. In 2000, Louie and Grubbs described the efficient ATRP of styrene (eq 1), catalyzed by iron–NHC complexes $[FeX_2(\mathbf{1})_2]$ (X = Cl, Br) and initiated by 1-phenylethyl bromide.[9] The significant Lewis basicity of **1** enabled high polymerization rates accompanied by low polydispersities (PDI ~1.1). Although related ATRP's have been demonstrated with a range of transition metals, use of $[FeX_2(\mathbf{1})_2]$ (X = Cl, Br) introduces advantages relating to cost and toxicity.

Group transfer polymerization (GTP) of acrylates and methacrylates using silyl ketene acetal initiators (i.e., **3**) allows mild and controlled polymerization in the presence of nucleophilic or Lewis acidic catalysts. While Lewis acid catalysts require relatively high catalyst loadings, and anionic nucleophilic catalysts can induce undesired side reactions, NHCs represent a class of neutral nucleophiles that address such disadvantages. In 2008, studies by Hedrick and coworkers revealed that IPrMe (**1**) is an efficient catalyst for the GTP of methyl methacrylate, providing polymers with narrow dispersity and molecular weights up to 21 kg mol^{-1} (eq 2).[10] In addition, the GTP of *tert*-butyl acrylate at room temperature using IPrMe (**1**) was reported, again achieving low polydispersities (PDI = 1.18) and predictable molecular weights. Gnanou and coworkers concurrently reported similar observations using *N*-alkyl imidazolylidene NHCs such as **1**.[3,11]

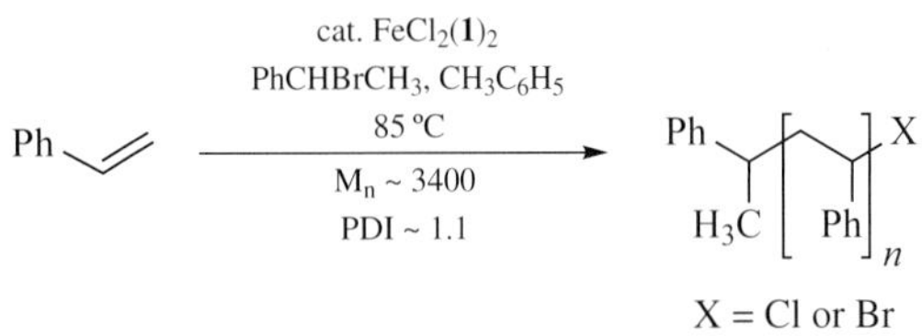

(1)

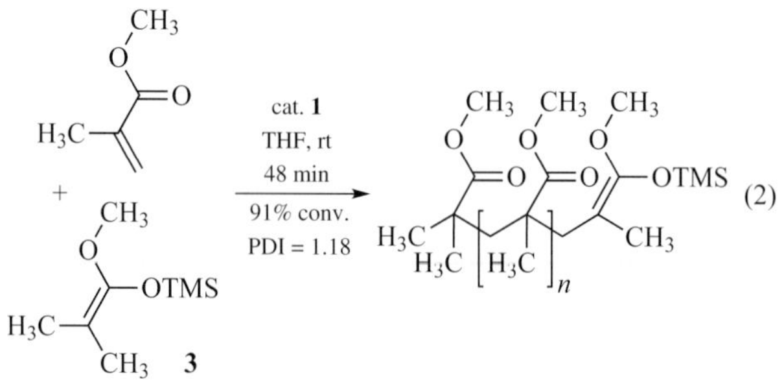

Hedrick and coworkers have also exploited IPrMe (**1**) in the ring-opening polymerization (ROP) of cyclic siloxanes initiated by 4-pyrenebutan-1-ol (PyBuOH) (eq 3). IPrMe (**1**) was extremely active in this reaction, with complete polymerization achieved in 1 min (eq 3).[12] While IPrMe (**1**) produced narrowly dispersed polycarbosiloxanes (PDI = 1.19) of predictable molecular weight and high end-group fidelity, narrower polydispersity was achieved using guanidine **4** (PDI = 1.04). Use of **1** was also found to result in polymer scrambling by transetherification upon monomer consumption.

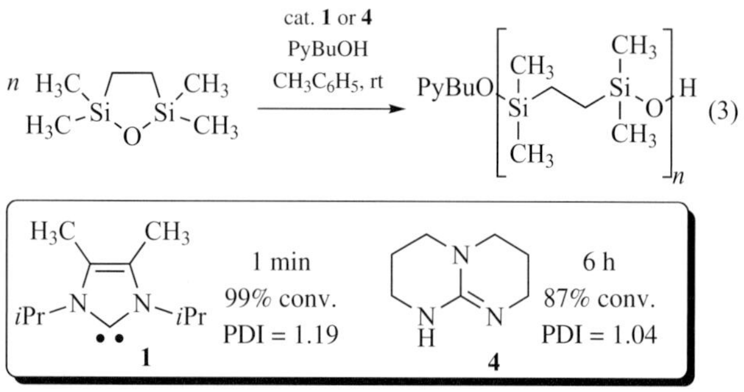

IPrMe (**1**) also mediates an extremely rapid BnOH-initiated ROP of cyclic carbonate **5**, however at the expense of polydispersity (eq 4).[13] A higher level of control for this polymerization was attained with the *N*-aryl NHC (IPr, **6**) or amidine **7**.

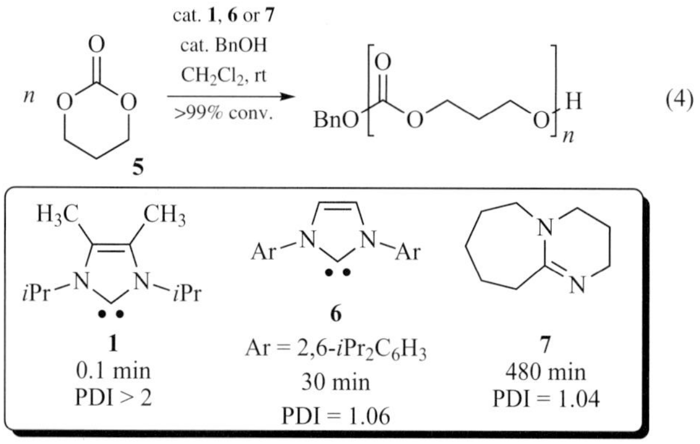

When the BnOH-initiated ROP of lactone **8** was investigated, IPr (**6**) was found to be inactive, while the more electron-rich IPrMe (**1**) allowed controlled polymerization up to approximately 70% monomer conversion (eq 5).[14] Increased polydispersities were observed beyond this stage, suggesting that competing NHC-catalyzed transesterification reactions were occurring. The use of multifunctional alcohol initiators allowed branched polymers with narrow polydispersity to be accessed. When lactone **8** was reacted with NHC **1** in the absence of an initiator, high molecular weight (41–114 kg mol^{-1}) cyclic polylactones were produced (eq 6).[15]

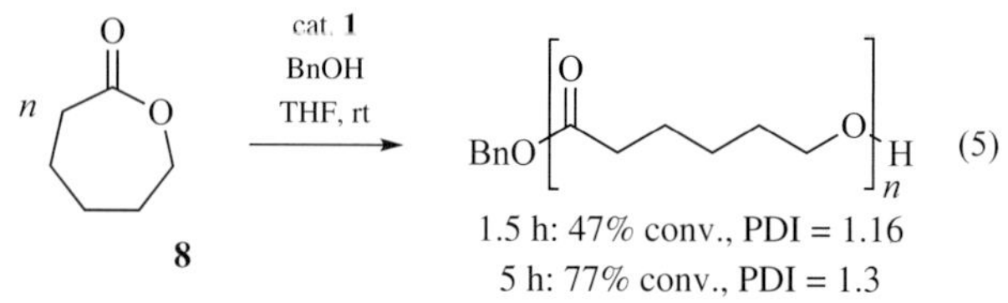

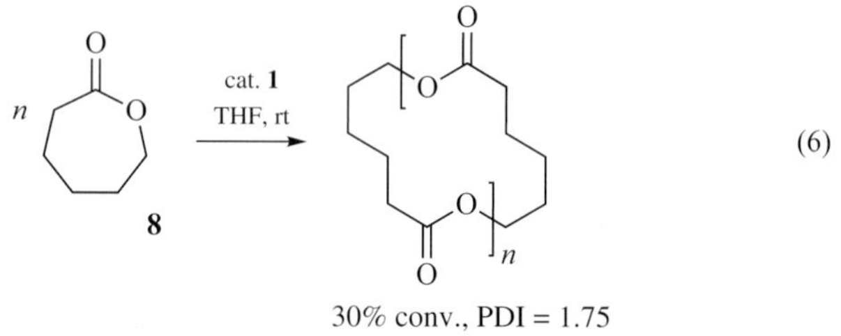

Organocatalysis with CO_2 and Ester Oxidation State Substrates. In 2009, Ikariya and coworkers reported the application of *N*-alkyl NHC carboxylates, (i.e., **9** and **10**) as efficient organocatalysts in CO_2 fixation reactions.[16] These reactions involved nucleophilic addition of the zwitterionic NHC carboxylate into alkyne and subsequent intramolecular cyclization to generate cyclic carbonates while liberating the NHC (eq 7). Finally the NHC upon reaction with CO_2 regenerates the NHC carboxylate. While the I*t*Bu catalyst **10** was most effective for this transformation, IPrMe carboxylate **9** gave comparable yields. Ikariya also reported CO_2 fixation using epoxides, a transformation previously described by Lu with *N*-aryl NHCs (i.e., IPr, (**6**)).[17]

In 2010, Candish and Lupton reported the use of IPrMe (**1**) for the rearrangement of α,β-unsaturated enol ester **11** to dihydropyranone **12** (eq 8), a key step in the total synthesis of (−)-7-deoxyloganin.[18] A variety of less electron rich, and bulkier, *N*-aryl NHC catalysts (for example IMes (**2**) and IPr (**6**)) failed to effect the rearrangement, while IPrMe (**1**) afforded **12** in good yield and moderate diastereoselectivity. The total synthesis was completed in four additional steps.

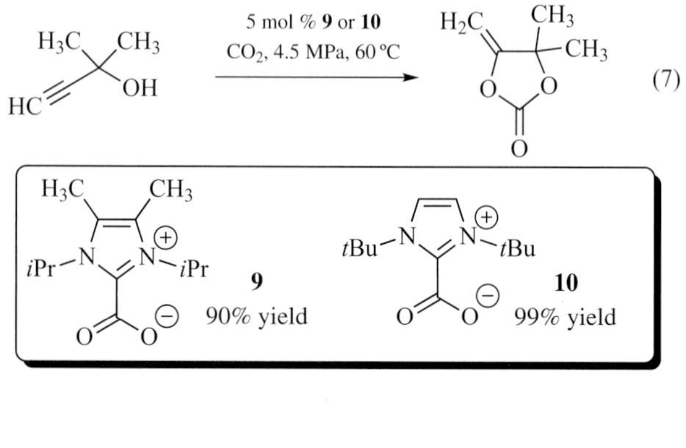

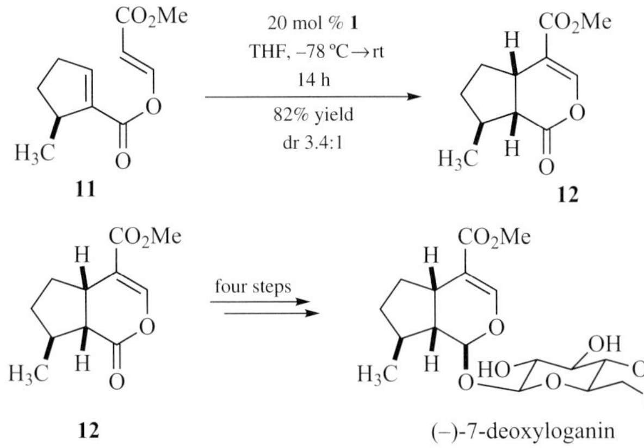

(8)

An all carbon [4+2] cycloaddition between TMS-protected dienol ether **13** and α,β-unsaturated acyl fluorides **14** has been reported using NHC catalysis (eq 9).[19] While other imidazolylidene NHC catalysts were effective (e.g., IMes (**2**)), IPrMe (**1**) afforded **15** in maximal yield. Kinetic isotope analysis and computational studies led to the proposal of a stepwise vinylogous Michael–aldol–lactonization–decarboxylation sequence between dienolate **16** and α,β-unsaturated acyl azolium **17** to provide diene **15**. This reaction was later modified to allow the β-lactone intermediate **18** to be trapped under reductive conditions (eq 10) or with a range of alkyl lithium reagents (eq 11).[20] Pivotal to these reactions was the use of temperatures below −20 °C, above which decarboxylation occurs. All of these reactions (eqs 9–11)

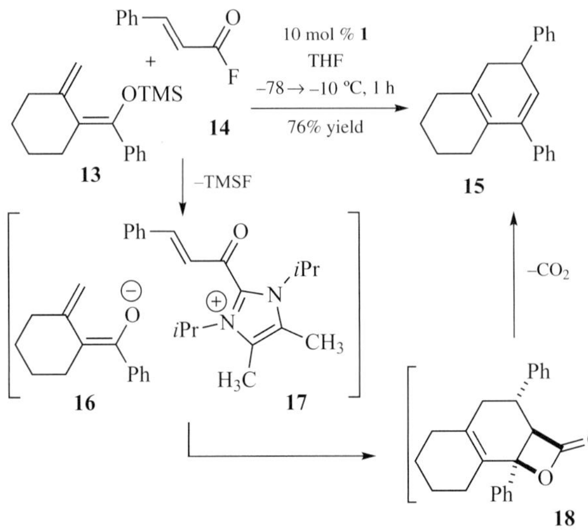

(9)

were found to proceed with complete diastereoselectivity (dr > 20:1), providing products with up to six contiguous stereocenters as a single diastereoisomer.

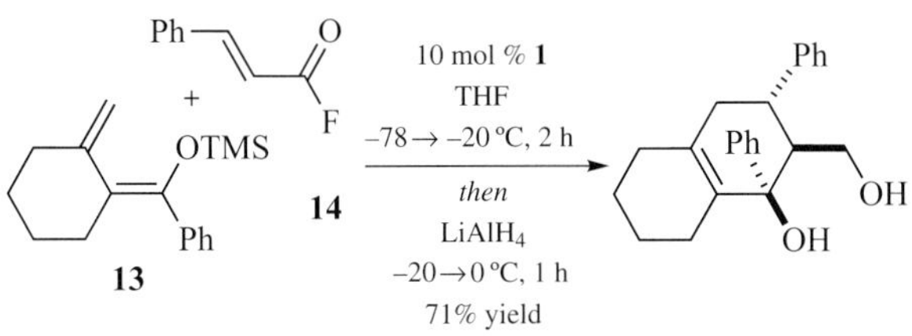

(10)

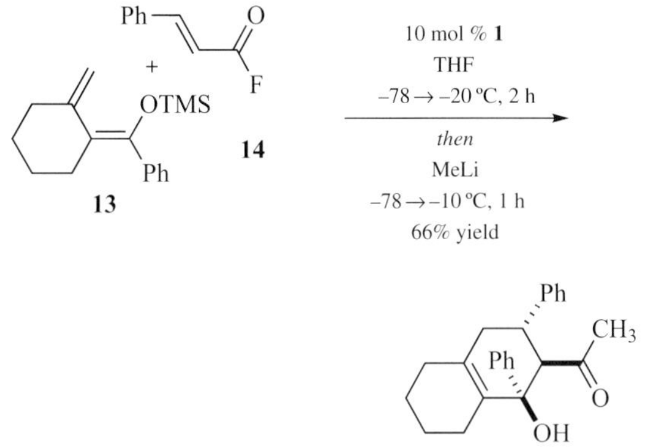

(11)

Formation of Si–C and Si–N Bonds. The coupling of silanes to amines has been examined by Cui and coworkers using NHC ytterbium complexes **19** and **20**.[21] The performance of these catalysts was heavily affected by the NHC ligand, with the more sterically encumbered **19** optimal in most cases, although when bulky substrates were used the complex derived from NHC **1** (namely **20**) was ideal (eqs 12 and 13).

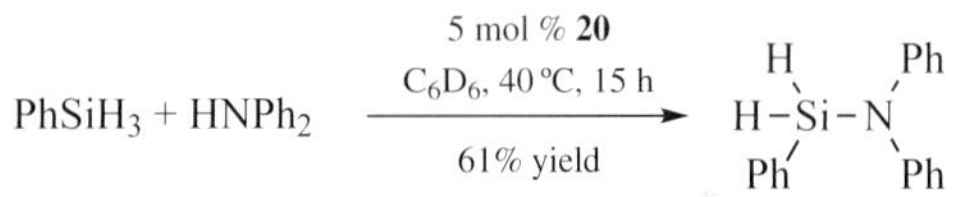

(12)

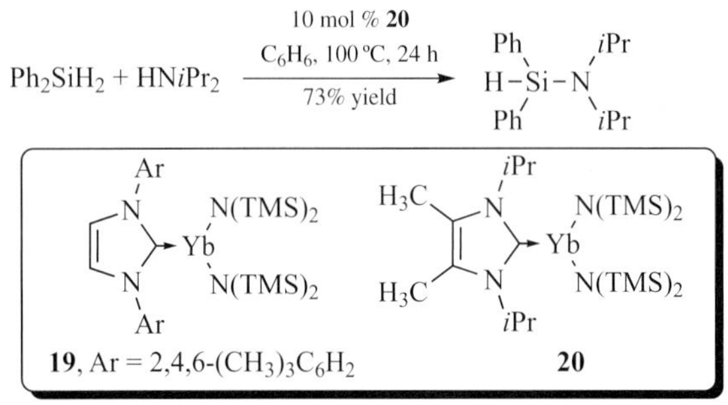

(13)

In 2012, Cui et al. reported the bis-silylation of alkynes, using NHC-stabilized silylene **21** as a silylating agent.[22] For terminal alkynes with an electron-withdrawing group, such as 4-ethynylbenzonitrile, bis-silylation proceeds in a stereo- and regiospecific manner to provide bis-silylated alkenes (i.e., **22**) as a single isomer (eq 14). Subsequent alcoholysis enables preparation of the bis-silylated alkene **23** bearing two distinct silyl groups (eq 15).

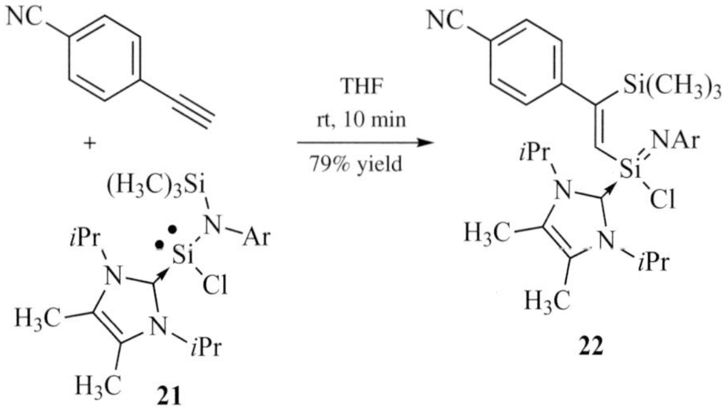

(14)

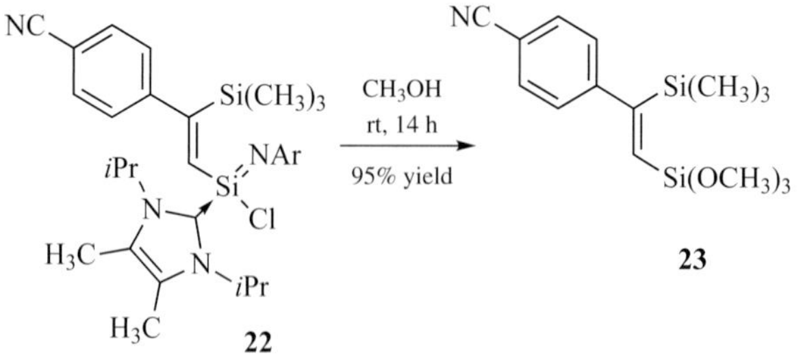

(15)

Nolan and coworkers have communicated the hydrosilylation of terminal alkenes catalyzed by [Rh(cod)(**1**)(OH)] (**24**) (eq 16).[23] NHC **1** proved to be the optimal ligand for the transformation, with catalyst **24** displaying the highest selectivity for the target alkylsilane **25**, over the vinylsilane by-product **26**, compared to other *N*-alkyl and *N*-aryl NHC rhodium complexes.

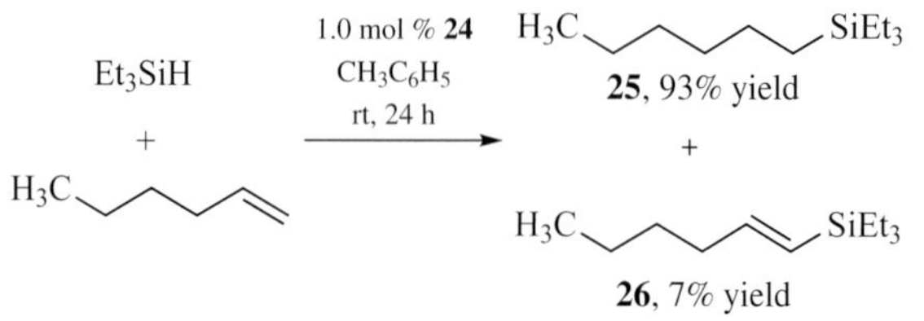

(16)

Miscellaneous C–C Bond Formation. Whittlesey and coworkers demonstrated oxidation–Wittig–reduction cascades using ruthenium NHC complex **27**, a sequence that proceeds by reversible C–H activation of the *o*-methyl groups of the IMes NHC ligand.[24] In a comprehensive study examining the role of the NHC ligand, it was found that complexation of NHC **1** to $Ru(PPh_3)_3(CO)H_2$ leads to a C–H activated complex **28** that is the most active species for the tandem oxidation–Wittig–reduction (eq 17).[25] The transformation occurs by oxidation of the alcohol to the aldehyde, with conversion of **28** to complex **29**. Following Wittig reaction the resultant olefin is reduced by **29**, reforming catalyst **28** and providing the product.

In 2010, Nolan and coworkers reported a series of ruthenium complexes bearing a small and large NHC (i.e., **30**) and determined their catalytic activity in ring closing methathesis reactions.[26] While in most cases the use of IMeMe (**31**) as the small ligand was optimal, RCM with tosyl amine **32** was found to be best suited to IPrMe (**1**) ruthenium complex **30** (eq 18).

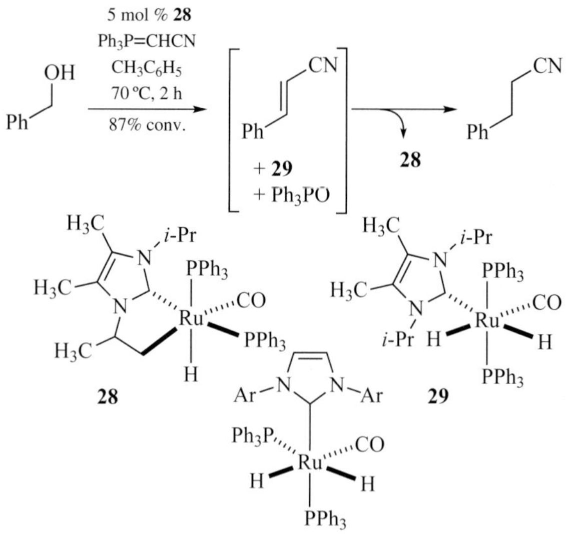

(17)

27, Ar = 2,4,6-$(CH_3)_3C_6H_2$

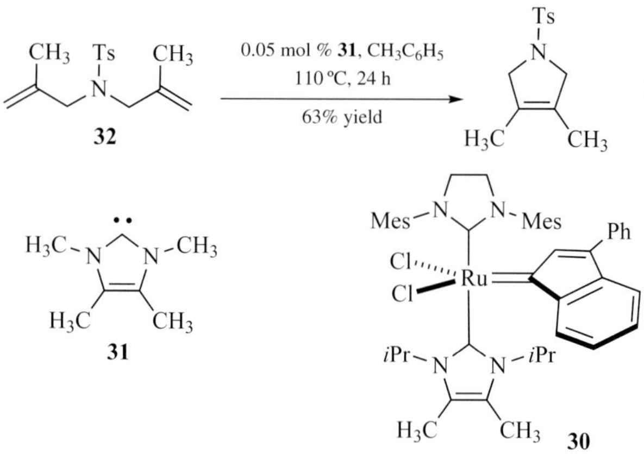

(18)

The trimerization of dimethyl barbituric acid to provide **33** was reported using electrochemical conditions by Dryhurst.[27] In 2012, Kuhn and Maier reported that IPrMe (**1**) provided the trimer **33** under mild conditions and in excellent yield from dimethylbarbituric acid dichloride (eq 19).[28] Although mechanistic studies have not been reported, the authors suggest product formation via the trimerization of dimethylbarbituric acid carbene **34**, or the monochloro anionic intermediate **35**.

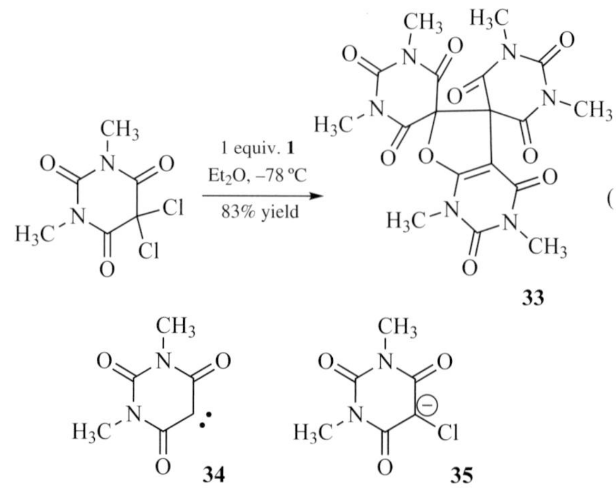

(19)

Related Reagents. 1,3-Dimesityl-1*H*-imidazolium salts; 1,3-dimesityl-imidazolylidene; 1,3-dimesityl-1*H*-imidazolinium salts; 1,3-bis(2,6-diisopropylphenyl)-1*H*-imidazolium salts; 1,3-dimesityl-4,5-dihydroimidazol-2-ylidene; 1,3-bis(2,6-diisopropylphenyl) imidazol-2-ylidene.

1. Kuhn, N.; Kratz, T., *Synthesis* **1993**, 561.
2. Arduengo, III, A. J.; Dias, H. V. R.; Harlow, R. L.; Kline, M., *J. Am. Chem. Soc.* **1992**, *114*, 5530.
3. Raynaud, J.; Liu, N.; Gnanou, Y.; Taton, D., *Macromolecules* **2010**, *43*, 8853.
4. Arduengo, III, A. J.; Harlow, R. L.; Kline, M., *J. Am. Chem. Soc.* **1991**, *113*, 361.
5. Díez-González, S.; Marion, N.; Nolan, S. P., *Chem. Rev.* **2009**, *109*, 3612.
6. Enders, D.; Niemeier, O.; Henseler, A., *Chem. Rev.* **2007**, *107*, 5606.
7. For a discussion of *N*-alkyl imidazolylidene sterics see: Dorta, R.; Stevens, E. D.; Scott, N. M.; Costabile, C.; Cavallo, L.; Hoff, C. D.; Nolan, S. P., *J. Am. Chem. Soc.* **2005**, *127*, 2485.
8. For a useful review discussing NHC electronics see: Dröge, T.; Glorius, F., *Angew. Chem., Int. Ed.* **2010**, *49*, 6940.
9. Louie, J.; Grubbs, R. H., *Chem. Commun.* **2000**, 1479.
10. Scholten, M. D.; Hedrick, J. L.; Waymouth, R. M., *Macromolecules* **2008**, *41*, 7399.
11. Raynaud, J.; Ciolino, A.; Baceiredo, A.; Destarac, M.; Bonnette, F.; Kato, T.; Gnanou, Y.; Taton, D., *Angew. Chem., Int. Ed.* **2008**, *47*, 5390.
12. Lohmeijer, B. G. G.; Dubois, G.; Leibfarth, F.; Pratt, R. C.; Nederberg, F.; Nelson, A.; Waymouth, R. M.; Wade, C.; Hedrick, J. L., *Org. Lett.* **2006**, *8*, 4683.
13. Nederberg, F.; Lohmeijer, B. G. G.; Leibfarth, F.; Pratt, R. C.; Choi, J.; Dove, A. P.; Waymouth, R. M.; Hedrick, J. L., *Biomacromolecules* **2007**, *8*, 153.
14. Kamber, N. E.; Jeong, W.; Gonzalez, S.; Hedrick, J. L.; Waymouth, R. M., *Macromolecules* **2009**, *42*, 1634.
15. Shin, E. J.; Jeong, W.; Brown, H. A.; Koo, B. J.; Hedrick, J. L.; Waymouth, R. M., *Macromolecules* **2011**, *44*, 2773.
16. Kayaki, Y.; Yamamoto, M.; Ikariya, T., *Angew. Chem., Int. Ed.* **2009**, *48*, 4194.
17. Zhou, H.; Zhang, W.-Z.; Liu, C.-H.; Qu, J.-P.; Lu, X.-B., *J. Org. Chem.* **2008**, *73*, 8039.
18. Candish, L.; Lupton, D. W., *Org. Lett.* **2010**, *12*, 4836.
19. Ryan, S. J.; Candish, L.; Lupton, D. W., *J. Am. Chem. Soc.* **2011**, *133*, 4694.
20. Ryan, S. J.; Stasch, A.; Paddon-Row, M. N.; Lupton, D. W., *J. Org. Chem.* **2012**, *77*, 1113.
21. Xie, W.; Hu, H.; Cui, C., *Angew. Chem., Int. Ed.* **2012**, *51*, 11141.
22. Cui, H.; Ma, B.; Cui, C., *Organometallics* **2012**, *31*, 7339.
23. Truscott, B. J.; Slawin, A. M. Z.; Nolan, S. P., *Dalton Trans.* **2013**, *42*, 270.
24. Edwards, M. G.; Jazzar, R. F. R.; Paine, B. M.; Shermer, D. J.; Whittlesey, M. K.; Williams, J. M. J.; Edney, D. D., *Chem. Commun.* **2004**, 90.
25. Burling, S.; Paine, B. M.; Nama, D.; Brown, V. S.; Mahon, M. F.; Prior, T. J.; Pregosin, P. S.; Whittlesey, M. K.; Williams, J. M. J., *J. Am. Chem. Soc.* **2007**, *129*, 1987.
26. Bantreil, X.; Randall, R. A. M.; Slawin, A. M. Z.; Nolan, S. P., *Organometallics* **2010**, *29*, 3007.
27. Kato, S.; Poling, M.; van der Helm, D.; Dryhurst, G., *J. Am. Chem. Soc.* **1974**, *96*, 5255.
28. Mallah, E.; Sweidan, K.; Engelmann, J.; Steinmann, M.; Kuhn, N.; Maier, M. E., *Tetrahedron* **2012**, *68*, 1005.

Alison Levens, Marcin Kowalczyk & David W. Lupton
Monash University, Victoria, Australia

4,5-Dihydro-4-phenyl-3*H*-dinaphtho[2,1-c:1′,2′-*e*]phosphepin

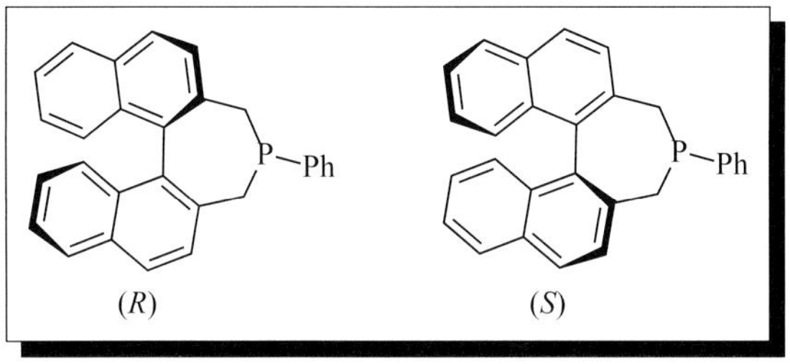

(*R*)
[174390-35-3] $C_{28}H_{21}P$ (MW 388.44)
InChI = 1S/C28H21P/c1-2-10-24(11-3-1)29-18-22-16-14-20-8-4-6-12-25(20)27(22)28-23(19-29)17-15-21-9-5-7-13-26(21)28/h1-17H,18-19H2
InChIKey = NHIVGNHRQMUGNK-UHFFFAOYSA-N
(*S*)
[155898-25-2] $C_{28}H_{21}P$ (MW 388.44)
InChI = 1S/C28H21P/c1-2-10-24(11-3-1)29-18-22-16-14-20-8-4-6-12-25(20)27(22)28-23(19-29)17-15-21-9-5-7-13-26(21)28/h1-17H,18-19H2
InChIKey = NHIVGNHRQMUGNK-UHFFFAOYSA-N
[155825-57-3] $C_{28}H_{21}P$ (MW 388.44)
InChI = 1S/C28H21P/c1-2-10-24(11-3-1)29-18-22-16-14-20-8-4-6-12-25(20)27(22)28-23(19-29)17-15-21-9-5-7-13-26(21)28/h1-17H,18-19H2
InChIKey = NHIVGNHRQMUGNK-UHFFFAOYSA-N

axially chiral monophosphanes-phosphepine derivatives, enantioselective catalysis, asymmetric organocatalysis

Alternative Name: Ph-BINEPINE.

Physical Data: mp 189–190 °C (CH_2Cl_2/hexane); $[\alpha]_D^{25}$ −90.2 (*c* 0.8; $CHCl_3$) for (*S*) enantiomer.[1]

Solubility: sol THF, dichloromethane, chloroform, ethyl acetate, acetone; modestly sol diethyl ether, toluene; slightly sol hexane; insol water.

Form Supplied in: white powder.

Analysis of Reagent Purity: TLC analysis (E. Merck Kiesel-gel 60 PF_{254}, hexane- ethyl acetate 2:1) R_f0.15 (Ph-BINEPINE oxide) and 0.9 (Ph-BINEPINE). ^{1}H NMR (CDCl3) δ 2.71–3.06 (m, 4H), 6.90 (d, J = 8.4 Hz, Ar, 1H), 7.20–7.33 (m, Ar, 9H), 7.38–7.47 (m, Ar, 2H), 7.67 (dd, J = 0.9, 8.4 Hz, Ar, 1H), 7.71 (d, J = 8.4 Hz, Ar, 1H), 7.88 (d, J = 8.4 Hz, Ar, 1H), 7.94 (dd, J = 1.5, 8.4 Hz, Ar, 2H). ^{13}C NMR ($CDCl_3$) (aliphatic carbons only) δ 30.31 (d, $^1J_{C-P}$ = 16.5 Hz), 32.20 (d, $^1J_{C-P}$ = 22.0 Hz); ^{31}P-NMR ($CDCl_3$) δ 6.97 (s).

Preparative Methods: a solution of *n*-BuLi (12.8 mmol, 8 mL, 1.6 M in *n*-hexane) was concentrated under vacuum and the residual oil dissolved in diethyl ether (10 mL). After cooling to 0 °C, a solution of (*S*)-2,2′-dimethyl-1,1′-binaphthyl (5.3 mmol, 1.5 g) in diethyl ether (5 mL) was added via a dropping funnel over 20 min, giving a red solution. Afterward, TMEDA (13 mmol, 1.96 mL, distilled over CaH_2) was added slowly and the resulting solution was kept for 24 h at room temperature and deep red crystals were formed. The supernatant solution was decanted via cannula and the crystals were washed three times with dry *n*–hexane (3 mL each time, removed by cannula).

n-Hexane was then added (10 mL) and the suspension was cooled to −78 °C while a solution of phenyldichlorophosphine (5.3 mmol, 0.72 mL) in *n*-hexane (5 mL) was added dropwise. The white-yellowish suspension was warmed up slowly to room temperature and was stirred overnight. The reaction was quenched with water, and the organic layer was separated and dried over $MgSO_4$. Ph-BINEPINE was purified by flash chromatography (hexane–dichloromethane 5:1) to provide the (*S*)-enantiomer as a white powder (0.7–0.9 g; 35–45%).[2]

Purification: the product may contain some phosphine oxide as the main impurity, as can be determined by TLC or by NMR (31P-NMR (CDC13) δ 54–55). The oxide can be removed by flash chromatography.

Handling, Storage, and Precautions: Ph-BINEPINE is substantially stable to air in the solid state. Slow oxidation to the oxide can occur upon standing in solution. Use of deareated solvents for solutions and tightly closed containers flushed with nitrogen or argon is recommended for prolonged storage.

Preparation of the Rh Complexes. Cationic Rh(I) complexes of general formula $[Rh(nbd)(Ph\text{-}BINEPINE)_2]^+X^-$ where X^- is a weakly coordinating anion (tetrafluoroborate, hexafluorophosphate, BaRF, etc.) can be prepared by reaction of the corresponding cationic diolefin rhodium species $[Rh(nbd)_2]^+X^-$ with 2.1 equiv of ligand in dichloromethane.[3] These complexes are active catalysts in the asymmetric hydrogenation and in the hydrogen transfer reduction of a range of unsaturated substrates.[4]

Hydrogenation Reactions. α,β-Unsaturated acid derivatives are hydrogenated in the presence of preformed, or in situ prepared, Ph-BINEPINE/Rh catalysts at room temperature under atmospheric pressure to give the corresponding saturated products in quantitative yields and high stereoselectivities.

The enantioselective hydrogenation of α-acylaminoacrylic acids or esters is better performed in toluene, affording the corresponding α-acylamino acid derivatives in greater than 90% ee (eq 1).[5]

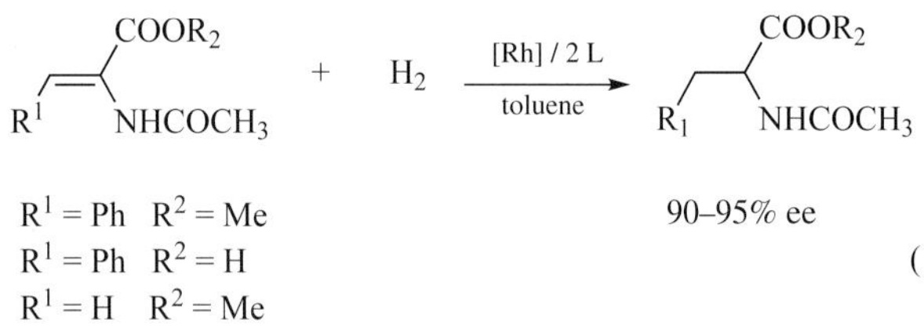

(1)

The enantioselective reduction of itaconic acid, or its dimethylester equivalent, represents a convenient entry to enantioenriched 2-methylsuccinic acid derivatives. The best results for the hydrogenation are obtained in dichloromethane (eq 2).[6]

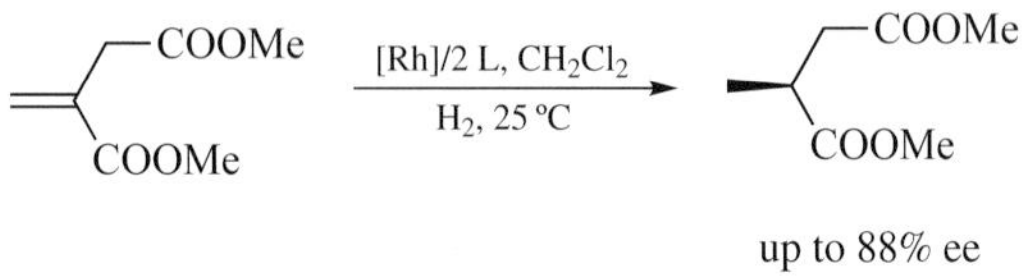

(2)

The asymmetric reduction of the free itaconic acid is best performed by transfer hydrogenation using formic acid as the hydrogen donor in the presence of triethylamine. The reaction proceeds under very mild conditions, providing methyl succinic acid with excellent ee (97%) (eq 3).[3]

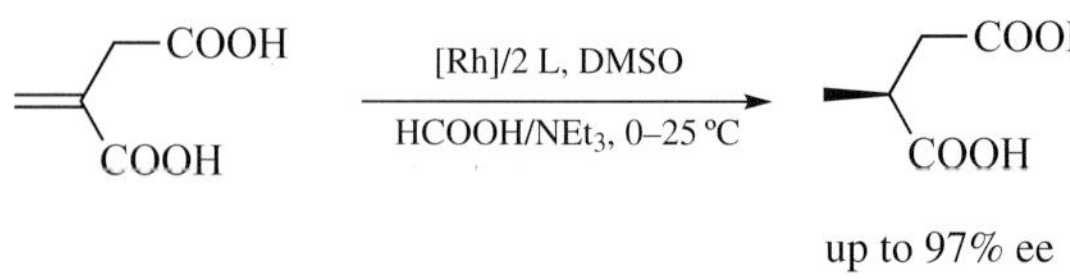

(3)

The asymmetric hydrogenation of β-acylaminoacrylic esters by Rh/Ph-BINEPINE catalysts provides protected β-amino acid derivatives in high enantioselectivities. The reaction is carried out in ethanol or *iso*-propanol under a modest positive pressure. The influence of the C–C double bond geometry on the outcome of the reaction has been assessed on the (*E*)- and (*Z*)-isomers of the substrates. In general, the *Z*-isomer provides higher ee's than the *E* isomer, and is about three to five times more reactive (eq 4).[2]

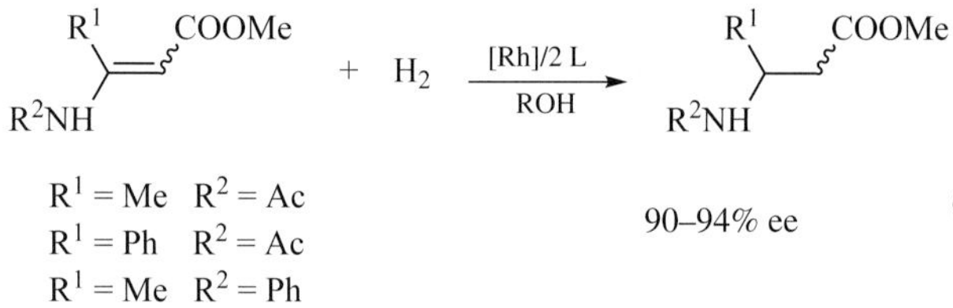

(4)

The Rh-catalyzed asymmetric hydrogenation of acyclic and cyclic *N*-acetyl enamides gives *N*-protected amines in quantitative yield. The enantioselectivity largely depends on the nature of the substituent within the enamide substrate. Under the optimal conditions, up to 95% ee and catalyst turnovers up to 2000 h^{-1} (TOF) can be achieved (eq 5).[7]

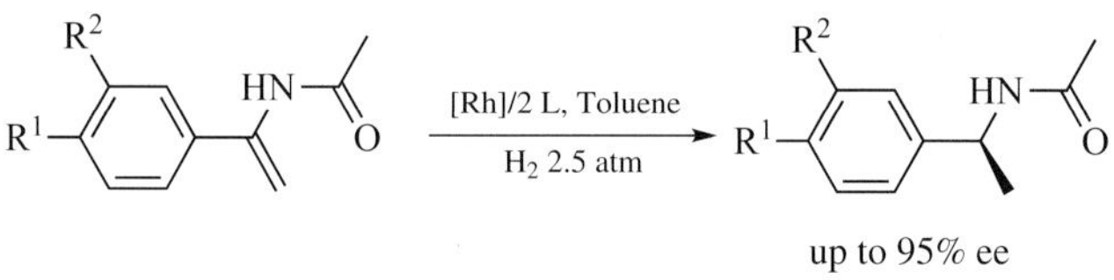

(5)

The hydrogenation of enol carbamates provides an alternative approach to chiral phenyl carbinols. In this reaction, Ph-BINEPINE/Rh catalysts enable high stereoselectivities in a wide range of temperatures (10–90 °C).[8] Notably, the catalyst gives similar ee's over the entire range of temperatures (eq 6).

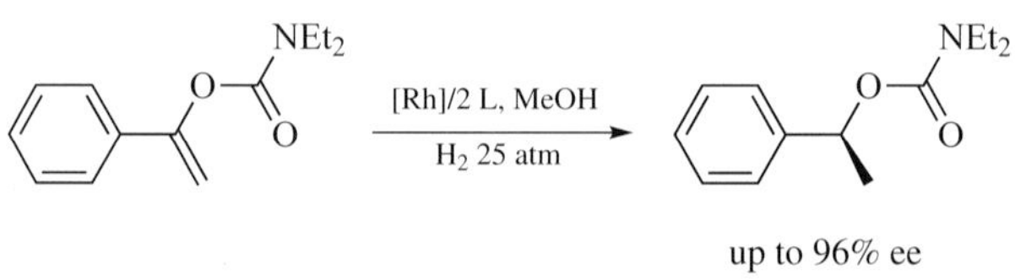

(6)

The Ru complexes with BINEPINE ligands are active catalysts for the asymmetric hydrogenation of β-ketoesters. The reaction requires moderately high temperatures and hydrogen pressures (100–120 °C and 60 atm in MeOH). The ee′s obtained with Ph-BINEPINE (84%) improve substantially (up to 95%) when a *p*-fluorine or a *p*-methoxy group is installed on the *P*-phenyl substituent (eq 7).[9]

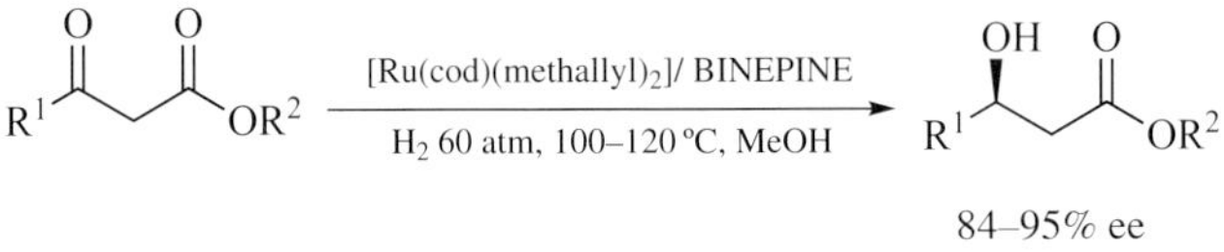

(7)

Hydroformylation, Hydroboration, and Hydrosilylation Reactions. The hydroformylation of styrene was the first catalytic reaction tested in the presence of Ph-BINEPINE. The branched aldehyde was obtained with a high selectivity (95%), but the enantiomeric excess was found to be only 20%. Hydroboration[10] and hydrosilylation[11] reactions have then been studied in the presence of Ph-BINEPINE ligand (eq 8). Although low enantiomeric excesses were obtained in the case of the Rh-catalyzed hydroboration of styrene leading to alcohol, a major improvement has been described in the case of the hydrosilylation of the acetophenone. The alcohol is isolated in 90% yield and 84% ee.

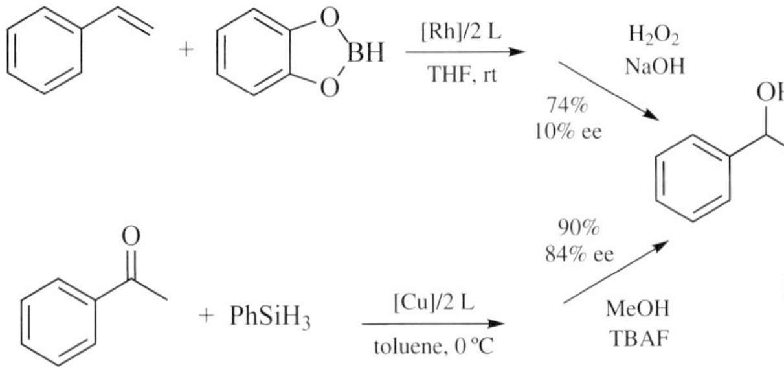

(8)

1,2-Addition of Trimethylaluminum Derivatives to α,β-Unsaturated Aldehydes by Ni Catalysts. The reaction of (*E*)-cinamaldehyde with $AlMe_3$ reagents under the catalytic action of Ph-BINEPINE/$Ni(acac)_2$ complexes provides the relevant secondary alcohol in up to 80% ee (eq 9).[12]

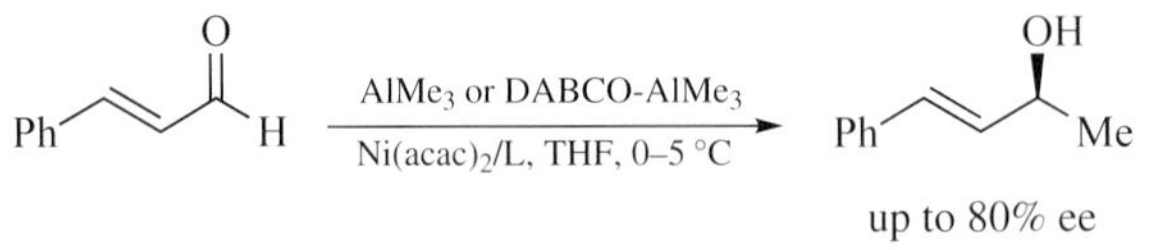

(9)

Asymmetric Allylic Alkylation Coupling by Pd Catalysts. In the asymmetric allylic alkylation of 1,3-diphenylallyl esters by dimethylmalonate anion, stereoselectivities up to 86% have been scored with in situ prepared Ph-BINEPINE/Pd catalysts (eq 10). Improved ee′s (up to 92%) can be obtained when a *p*-methoxy group is installed on the *P*-phenyl substituent.[13]

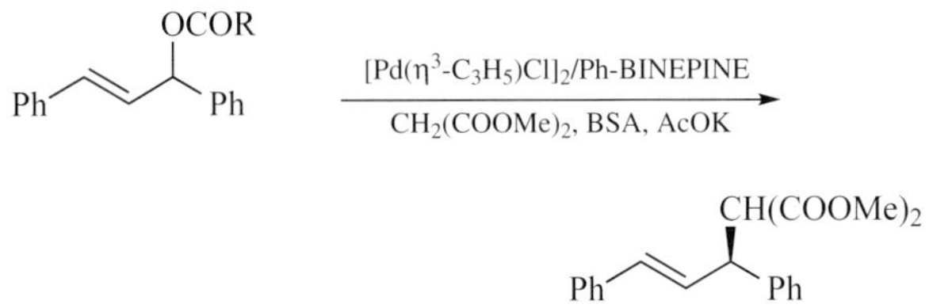

(10)

Allylation of benzaldehyde with cinnamyl acetate by umpolung of π-allyl palladium complexes was performed in the presence of 5 equiv of diethylzinc at −30 °C, giving the addition product (1*R*,2*S*)-*anti* in 70% yield and 70% ee (eq 11).

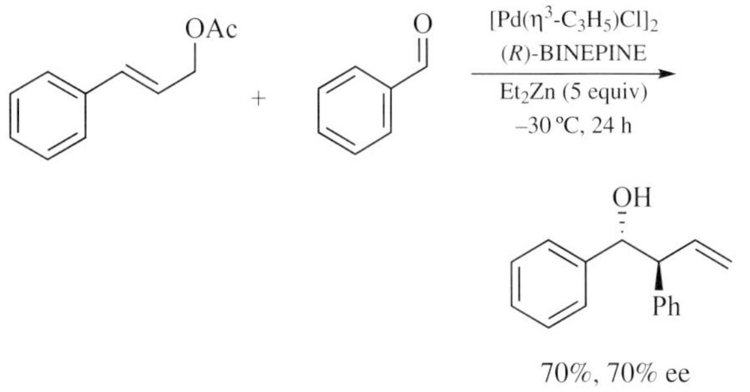

(11)

The Pd/Ph-BINEPINE system was also tested in the asymmetric Suzuki coupling, but led to low enantiomeric excesses.[10]

Cycloisomerization of 1,6-Enynes. The cycloisomerization of 1,6-enynes has been studied in the presence of Ph-BINEPINE ligand. The first enantioselective hydroxycyclization of 1,6-enynes in the presence of water or alcohol has been performed using an in situ catalytic system prepared from $PtCl_2$, 3 equiv of phosphorus ligand Ph-BINEPINE, and 2.5 equiv of $AgSbF_6$ (eq 12).[14] Carba- or azacycles are obtained in good yields and complete diastereoselectivity in up to 85% ee's. According to the same methodology, the Pt-catalyzed asymmetric tandem addition/cyclization reaction was extended to the addition of electron-rich aromatic and heteroaromatic rings (eq 13).[15] Good to excellent ee values (70–96%) were obtained for cyclic derivatives depending on the carbon-tethered enyne and the nucleophile.

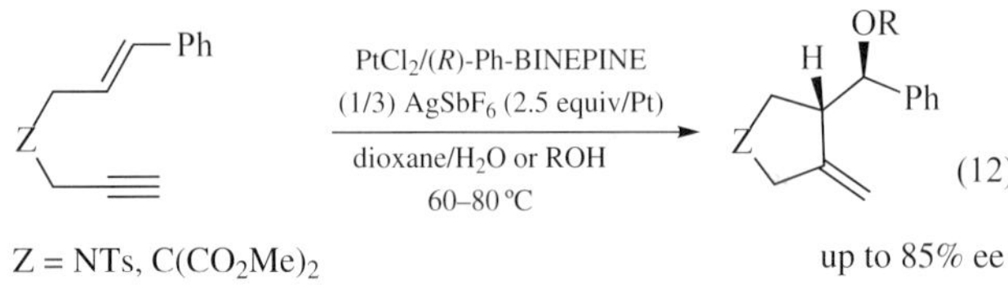

(12)

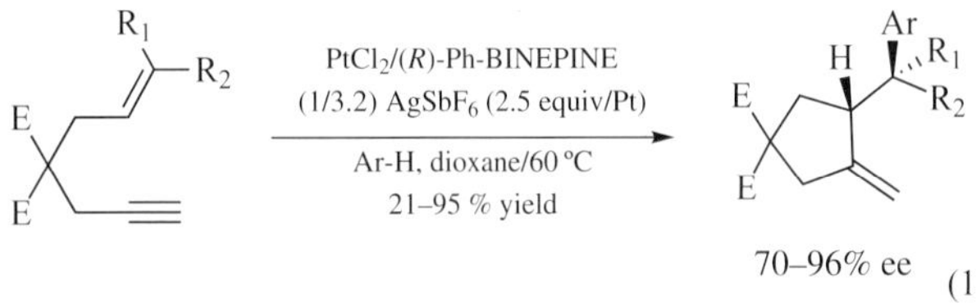

(13)

E = CO_2Me, CO_2i-Pr, CO_2t-Bu, CO_2Bn, SO_2Ph
R_1, R_2 = H, Ph, Me, 3,4-$(OCH_2O)C_6H_3$

Given these results on cycloisomerization of 1,6-enynes in the presence of an external nucleophile, the cycloisomerization of nitrogen-tethered enynes in the presence of Pt(II) complexes (eq 14) has been investigated in the absence of any external nucleophiles.[16] The higher efficiency of Ph-BINEPINE was once again demonstrated as enantiomeric excesses up to 97% were obtained.

Organocatalysis. The first reported example on the use of Ph-BINEPINE ligand as organocatalyst is the enantioselective monoacylation with benzoic anhydride: the monoprotected alcohol was obtained in 22% enantiomeric excess.[17] The BINEPINE ligands have then emerged as excellent catalysts for the [4 + 2] annulation of imines with allenes (eq 15).[18] The enantiomeric excess of 21% and the *cis:trans* ratio obtained in the presence of

Ph-BINEPINE ligand were highly improved by switching to the *P-t*.butyl substituted BINEPINE.

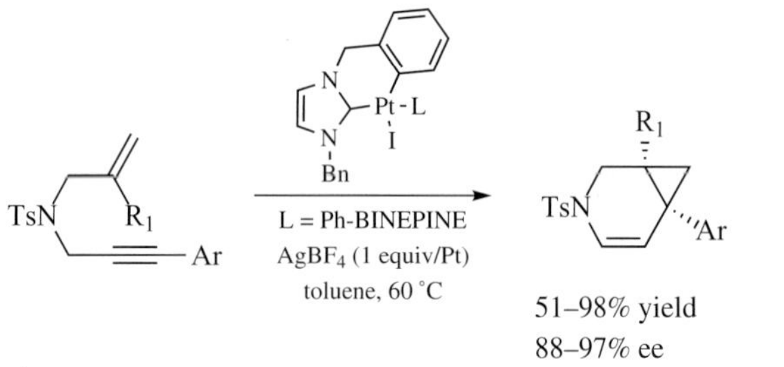

(14)

R^1 = H, Me
Ar = Ph, 3,5-$(Me)_2C_6H_3$, 4-MeO-C_6H_4, 4-NO_2-C_6H_4

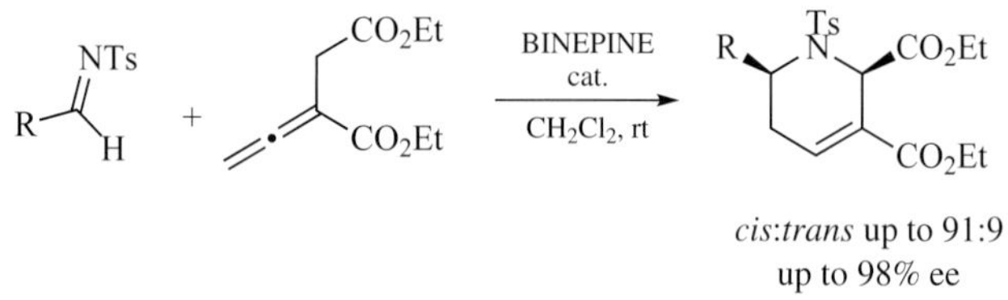

(15)

The synthesis of oxygen heterocycles according to phosphine-catalyzed cyclizations of hydroxy-2-alkynoates was also described[19] in the presence of Ph-BINEPINE ligand and led to moderate enantioselectivities as compared with chiral spiro phosphines.

1. Gladiali, S.; Dore, A.; Fabbri, D.; De Lucchi, O.; Manassero, M., *Tetrahedron: Asymmetry* **1994**, *5*, 511.
2. Enthaler, S.; Erre, G.; Junge, K.; Holz, J.; Börner, A.; Alberico, E.; Nieddu, I.; Gladiali, S.; Beller, M., *Org. Proc. Res. Dev.* **2007**, *11*, 568.
3. Alberico, E.; Nieddu, I.; Taras, R.; Gladiali, S., *Helv. Chim. Acta* **2006**, *89*, 1716.
4. Gladiali, S.; Alberico, E. In *Trivalent Phosphorus Compounds in Asymmetric Catalysis: Synthesis and Applications*; Börner, A., Ed.; Wiley-VCH: Weinheim, 2008, Vol. 2, p 177.
5. (a) Junge, K.; Oehme, G.; Monsees, A.; Riermeier, T.; Dingerdissen, U.; Beller, M., *Tetrahedron Lett.* **2002**, *43*, 4977. (b) Chi, X.; Zhang, X., *Tetrahedron Lett.* **2002**, *43*, 4849. (c) Junge, K.; Oehme, G.; Monsees, A.; Riermeier, T.; Dingerdissen, U.; Beller, M., *J. Organomet. Chem.* **2003**, *675*, 91.
6. Junge, K.; Hagemann, B.; Enthaler, S.; Spannenberg, A.; Michlaik, M.; Oehme, G.; Monsees, A.; Riermeier, T.; Beller, M., *Tetrahedron: Asymmetry* **2004**, *15*, 2621.
7. Enthaler, S.; Hagemann, B.; Junge, K.; Erre, G.; Beller, M., *Eur. J. Org. Chem.* **2006**, 2912.
8. Enthaler, S.; Erre, G.; Junge, K.; Michalik, D.; Spannenberg, A.; Marras, F.; Gladiali, S.; Beller, M., *Tetrahedron: Asymmetry* **2007**, *18*, 1288.
9. (a) Junge, K.; Hagemann, B.; Enthaler, S.; Oehme, G.; Michalik, M.; Monsees, A.; Riermeier, T.; Dingerdissen, U.; Beller, M., *Angew. Chem., Int. Ed.* **2004**, *43*, 5066. (b) Hagemann, B.; Junge, K.; Enthaler, S.; Michalik, M.; Riermeier, T.; Monsees, A.; Beller, M., *Adv. Synth. Catal.* **2005**, *347*, 1978.
10. Kasak, P.; Mereiter, K.; Widhalm, M., *Tetrahedron: Asymmetry* **2005**, *16*,' 3416.
11. Junge, K.; Wendt, B.; Addis, D.; Zhou, S.; Das, S.; Beller, M., *Chem. Eur. J.* **2010**, *16*, 68.
12. Biswas, K.; Chapron, A.; Cooper, T.; Fraser, P. K.; Novak, A.; Prieto, O.; Woodward, S., *Pure Appl. Chem.* **2006**, *78*, 511.
13. Alberico, E.; Gladiali, S.; Taras, R.; Junge, K.; Beller, M., *Tetrahedron: Asymmetry* **2010**, *21*, 1406.
14. (a) Charruault, L.; Michelet, V.; Taras, R.; Gladiali, S.; Genêt, J.-P., *Chem. Commun.* **2004**, 850. (b) Michelet, V.; Charruault, L.; Gladiali, S.; Genêt, J.-P., *Pure Appl. Chem.* **2006**, *78*, 397.
15. Toullec, P. Y.; Chao, C.-M.; Chen, Q.; Gladiali, S.; Genêt, J.-P.; Michelet, V., *Adv. Synth. Catal.* **2008**, *250*, 2401.
16. Brissy, D.; Skander, M.; Jullien, H.; Retailleau, P.; Marinetti, A., *Org. Lett.* **2009**, *11*, 2137.
17. Vedejs, E.; Daugulis, O.; Diver, S. T., *J. Org. Chem.* **1996**, *61*, 430.
18. Wurz, R. P.; Fu, G. C., *J. Am. Chem. Soc.* **2005**, *127*, 12234.
19. Chung, Y. K.; Fu, G. C., *Angew. Chem., Int. Ed.* **2009**, *48*, 2225.

Serafino Gladiali
Università di Sassari, Sassari, Italy

Patrick Yves Toullec, Jean-Pierre Genêt, &
Véronique Michelet
Ecole Nationale Supérieure de Chimie, Paris, France

2,3-Dihydro-2-phenylimidazo[2,1-*b*]benzothiazole

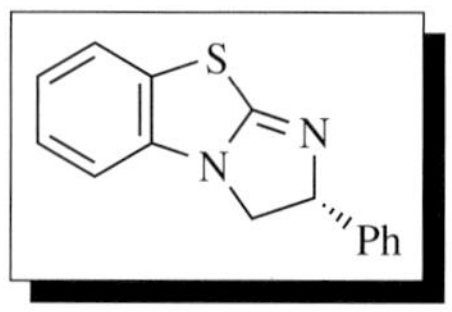

[885051-07-0] $C_{15}H_{12}N_2S$ (MW 252.33)
InChI = 1S/C15H12N2S/c1-2-6-11(7-3-1)12-10-17-13-8-4-5-9-14(13)18-15(17)16-12/h1-9,12H,10H2/t12-/m0/s1
InChIKey = YGCWPCVAVSIFLO-LBPRGKRZSA-N
[950194-37-3] (2*S*-enantiomer)
InChI = 1S/C15H12N2S/c1-2-6-11(7-3-1)12-10-17-13-8-4-5-9-14(13)18-15(17)16-12/h1-9,12H,10H2/t12-/m1/s1
InChIKey = YGCWPCVAVSIFLO-GFCCVEGCSA-N

chiral Lewis base catalyst useful for acylative kinetic resolution (KR)[1] of secondary alcohols,[2–3,15–24] lactams and thiolactams,[4,25–26] acylative desymmetrization of a *meso*-diol,[5] KR of α-substituted alkanoic acids,[27–29] and *N*-aroyl-β-lactams[33]; dynamic kinetic resolution (DKR)[30] of 2-aryl-propionic acids[31] and azlactones[32] via enantioselective alcoholysis; enantioselective hetero-Diels–Alder,[34] [3+2] cycloadditions with azomethine ylides[35,36] and Diels–Alder[37] reactions, asymmetric synthesis of 1,5-benzothiazepines,[38] [2,3]-rearrangement of allylic ammonium ylides,[39] and silylative KR of alcohols.[40,41]

Alternate Names: Benzotetramisole (BTM).
Physical Data: white solid; mp 94.5–95.0 °C (from Et_2O-hexanes); $[\alpha]_D$ = + 256.7° (*c* 1.00, MeOH).
Solubility: soluble in common solvents, including $CHCl_3$, CH_2Cl_2, Et_2O, THF, MeCN, PhMe, DMF, *tert*-amyl alcohol, and MeOH. Slightly soluble in hexanes. Insoluble in H_2O.
Preparative Method: prepared by heating 2-chlorobenzothiazole with (*R*)-(+)-phenylglycinol in the presence of diisopropylethy-

lamine at 130 °C, followed by treatment of the resulting intermediate with methanesulfonyl chloride and triethylamine at 0 °C in CH_2Cl_2 and subsequent reflux.[2]

Purification: flash chromatography (5% *i*-PrOH, 1% NEt_3/hexanes) and recrystallization (Et_2O/hexanes).

Handling, Storage, and Precautions: may be stored at room temperature. Should be protected from excessive humidity. Toxicity data are unavailable. The racemic compound has been shown to possess pharmacological activity.[6]

Original Commentary

Ximin Li, Hui Jiang, & Vladimir B. Birman
Washington University, St. Louis, MO, USA

KR of Alcohols. KR of alcohols has traditionally relied on enzyme-catalyzed enantioselectivetransesterification and hydrolysis.[7–9] BTM and the closely related (2*R*)-7-chloro-1,2-dihydro-2-phenyl-imidazo[1,2-*a*]quinoline (Cl-PIQ)[10] are among the most easily obtainable and effective nonenzymatic enantioselective acylation catalysts[11] reported to date and thus represent a useful alternative to enzymes.

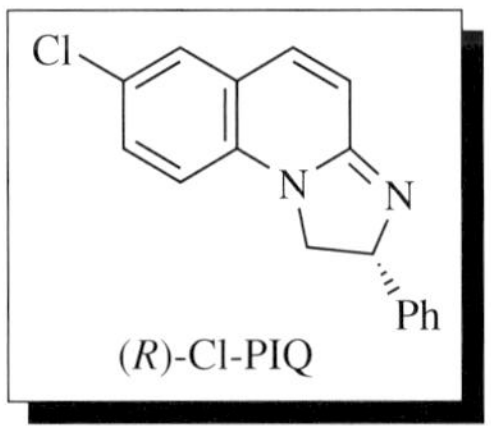

Both catalysts are especially suitable for the KR of benzylic alcohols (eq 1). Cl-PIQ is generally more catalytically active and achieves good to excellent enantioselectivity (selectivity factors *sec* in the 30–100 range), while BTM is much more enantioselective (*sec* in the 100–350 range).[12] Both propionic and isobutyric anhydrides have proved to be useful as achiral acyl donors in BTM-catalyzed acylations, the latter usually leading to higher enantioselectivities, but somewhat lower reaction rates. Cl-PIQ has been used primarily in combination with propionic anhydride. Chloroform is the solvent of choice in most cases. BTM-catalyzed reactions are sensitive to moisture (rapid catalyst deactivation) and, therefore, should be carried out in the presence of anhydrous Na_2SO_4.

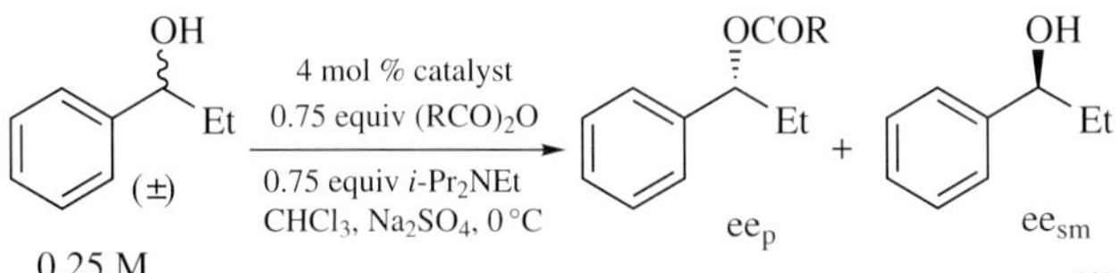

(1)

Catalyst	Anhydride	Time (h)	ee_p(%)	ee_{sm}(%)	*C*(%)	*S*
(*R*)-Cl-PIQ	$(EtCO)_2O$	3.5	88	90	51	46
(*R*)-BTM	$(EtCO)_2O$	33	95	85	47	104
(*R*)-BTM	$(i\text{-}PrCO)_2O$	36	97	82	46	144

A combination of free aliphatic carboxylic acids and benzoic anhydride derivatives may be employed to generate a variety of enantioenriched esters of benzylic alcohols via BTM-catalyzed KR (eq 2).[13]

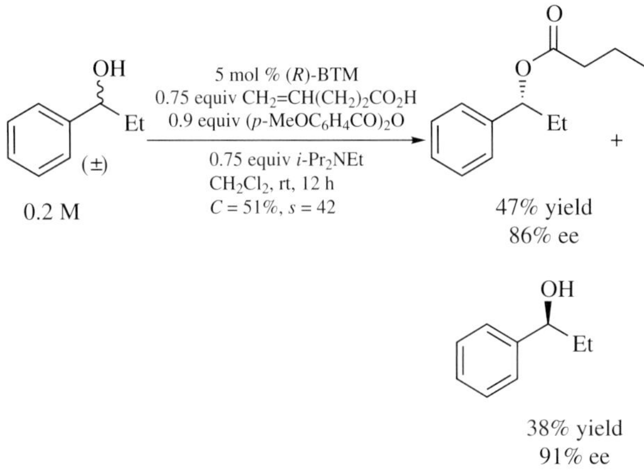

(2)

Cl-PIQ affords useful levels of enantioselectivity in KR of allylic alcohols (*sec* in the 17–56 range) (eq 3).[10] BTM is generally less effective in this process.[2]

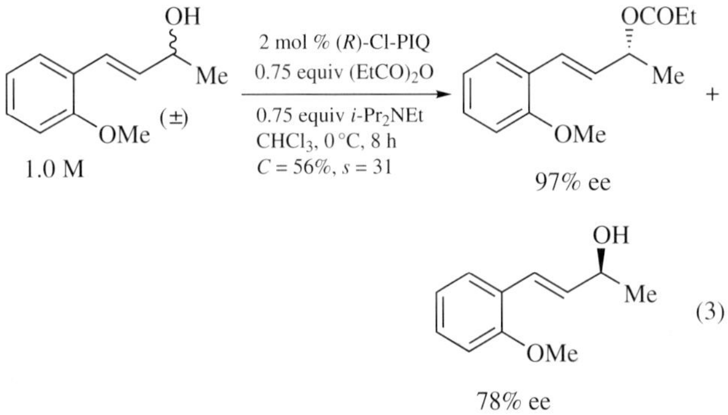

(3)

Propargylic alcohols can be resolved using BTM with selectivity factors up to 32, which is the highest ever achieved with nonenzymatic catalysts for this class of substrates (eq 4). Cl-PIQ affords lower enantioselectivities and reaction rates in this process.[3]

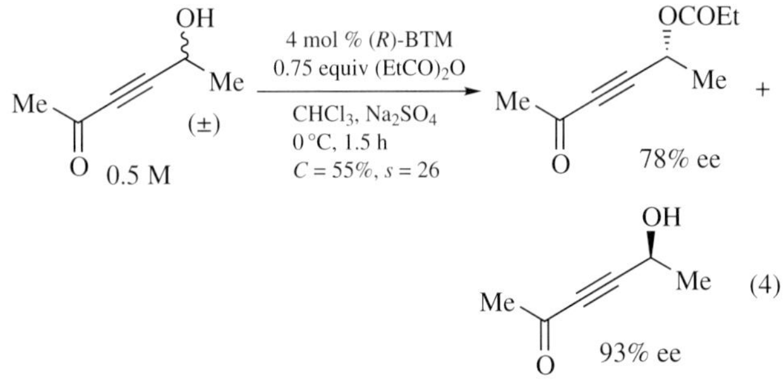

(4)

KR of Oxazolidinones. Unlike alcohols and amines, racemic oxazolidinones and other secondary lactams have never been resolved via enzymatic acylation. BTM has proved to be especially effective in KR of 4-aryl-substituted 2-oxazolidinones, affording selectivity factors in the 50–500 range in chloroform at room temperature (eq 5). Cl-PIQ is also suitable, but gives lower enantioselectivities (*sec* in the 16–110 range in *tert*-amyl alcohol at 0 °C).[4]

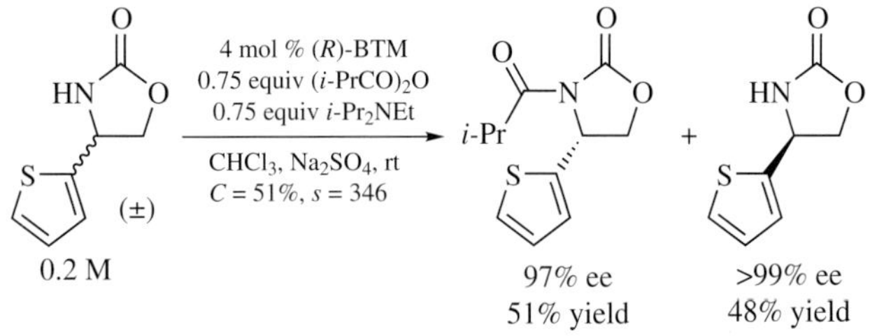

(5)

Desymmetrization of meso-diols. BTM-catalyzed desymmetrization of *meso*-diol lobelanidine has been employed as the key step in an asymmetric synthesis of alkaloid lobeline (eq 6).[5]

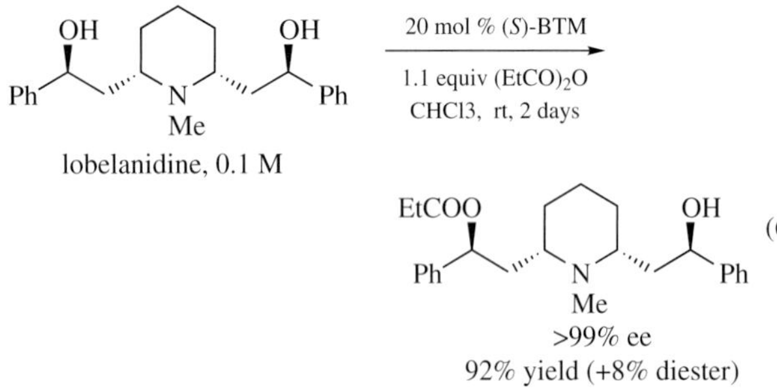

(6)

First Update

Vladimir B. Birman
Washington University, St. Louis, MO, USA

Introduction. Since the publication of the original commentary in 2008, the synthetic utility of BTM has expanded considerably, as evidenced by the many new entries below. Both enantiomers of BTM are now available commercially. A scalable, chromatography-free preparation of BTM has been developed.[14] Cl-PIQ, although less widely used, has also found several new applications (see below).

Acylative Kinetic Resolution of Alcohols. Birman and coworkers provided additional examples of the KR of benzylic and allylic alcohols via BTM- and Cl-PIQ-catalyzed enantioselective acylation.[15] Several other groups utilized BTM in the KR of related classes of secondary alcohols. Chen and coworkers resolved benzylic alcohols with fluorinated side chains (eqs 7 and 8).[16,17] Shiina et al. developed an alternative acylation method—diphenylacetylation via mixed anhydrides—that produces higher selectivity factors than using propionic and isobutyric anhydrides. They successfully applied this methodology to the KR of α-hydroxyalkanoates (eq 9),[18] heteroaryl alcohols (eq 10),[19] a propargylic alcohol (eq 11),[20] α-hydroxyphosphonates (eq 12),[21] and 2-hydroxy-γ-butyrolactones (eq 13[AQ1]).[22] A computational study published by Birman and coworkers explained the origin of enantioselectivity in the KR of benzylic alcohols in terms of cation–π interactions.[23] Similar effects appear to operate in the chiral recognition of other classes of substrates shown below.

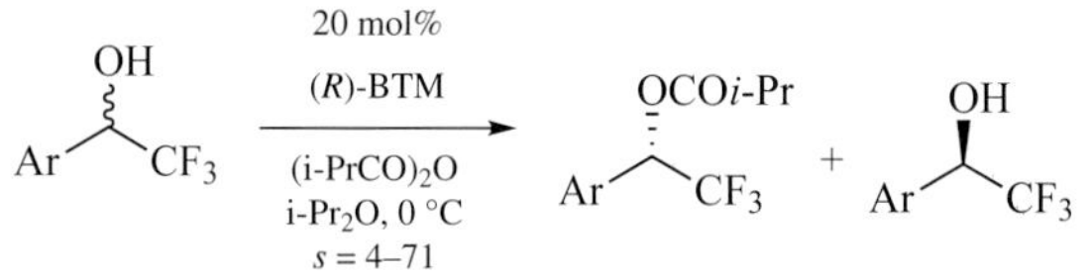

(7)

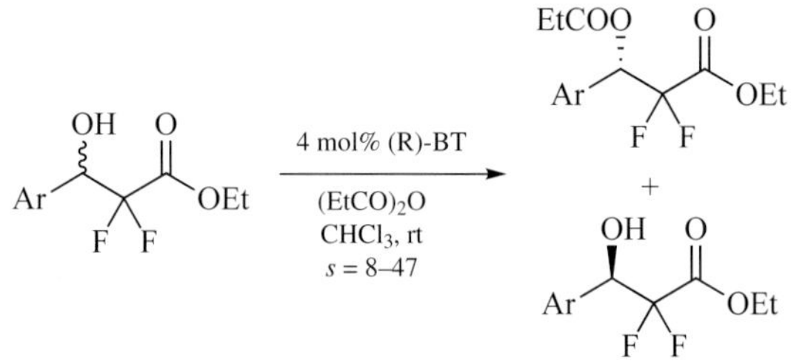

(8)

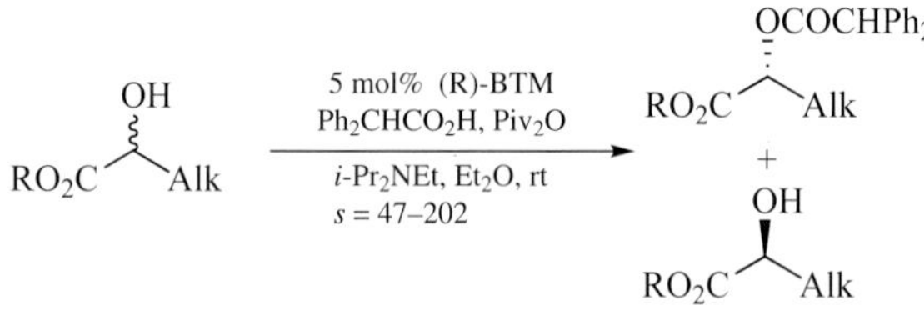

(9)

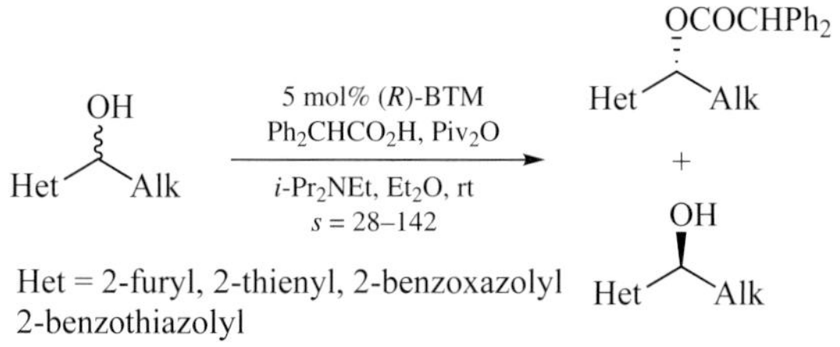

Het = 2-furyl, 2-thienyl, 2-benzoxazolyl 2-benzothiazolyl

(10)

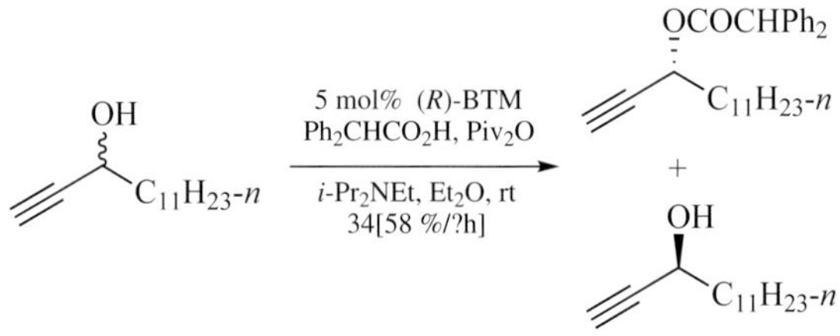

(11)

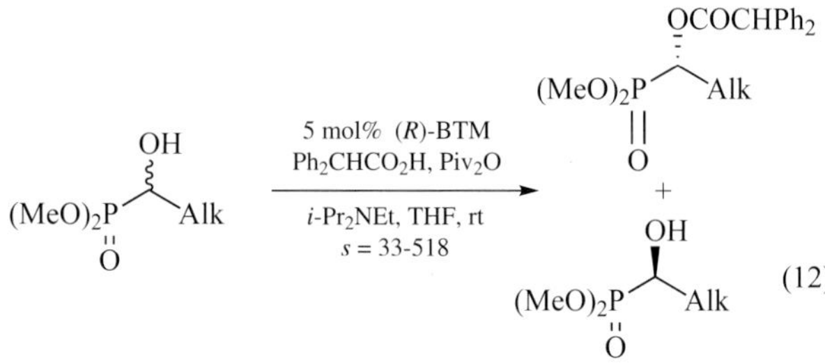

(12)

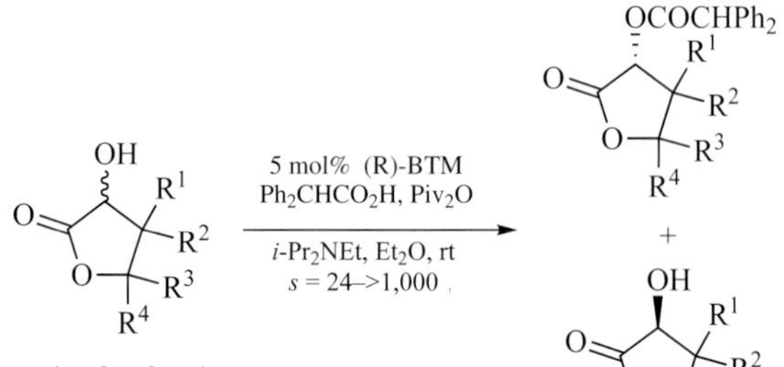

R^1, R^2, R^3, R^4 = H, alkyl, cycloalkyl

(13)

Dynamic Kinetic Diastereomeric Acylation of Lactols. Tang and coworkers transformed homochiral, epimeric mixtures of lactols into diastereomerically enriched lactol esters via acylation in the presence of BTM (eq 14).[24]

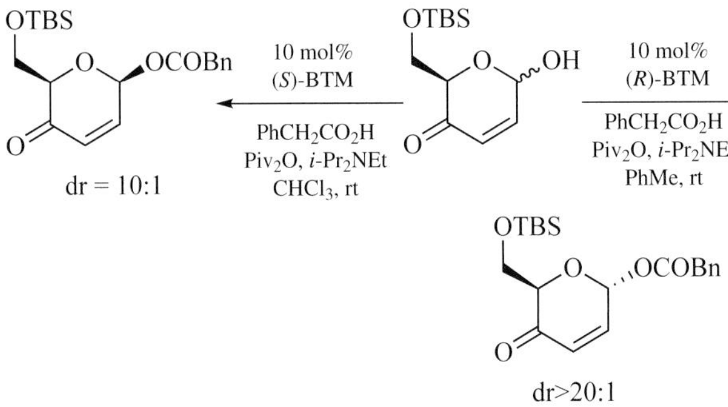

(14)

Kinetic Resolution of Lactams and Thiolactams Besides 4-aryl-oxazolidinones covered in the original commentary, a number of structurally related lactams and thiolactams have been resolved by Birman and coworkers via enantioselective *N*-acylation catalyzed by BTM, often with very high selectivity factors (eqs 15–17). Cl-PIQ is also competent, but less enantioselective in this process (eq 18). Interestingly, however, Cl-PIQ demonstrated its efficacy in the KR of several types of substrates that could not be acylated in the presence of BTM.[25] A later study demonstrated that *N*-acyl-thiazolidine-2-thiones and *N*-acyl-oxazolidine-2-thiones can be resolved via the opposite process: catalytic *N*-deacylation (eq 19).[26]

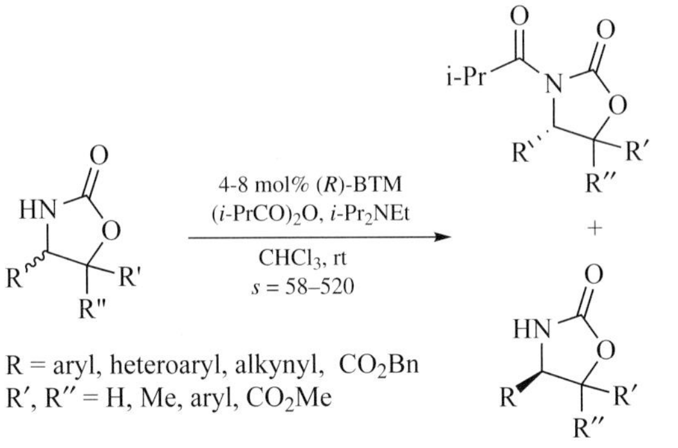

(15)

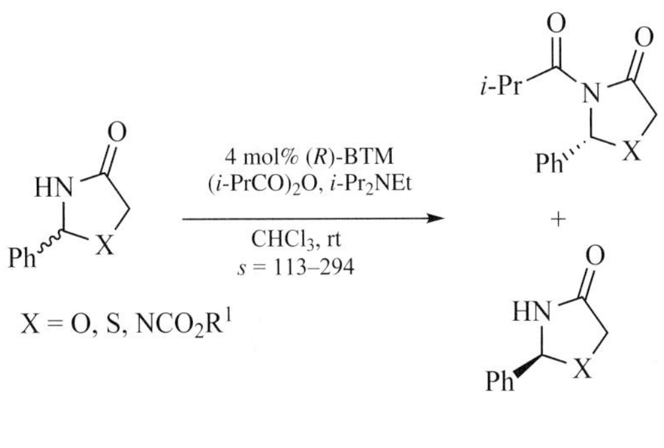

(16)

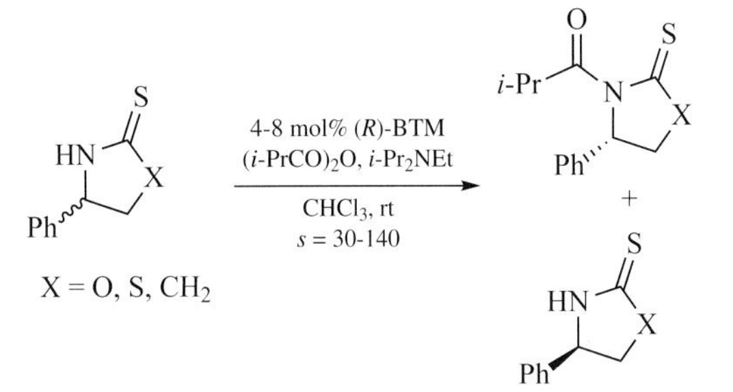

(17)

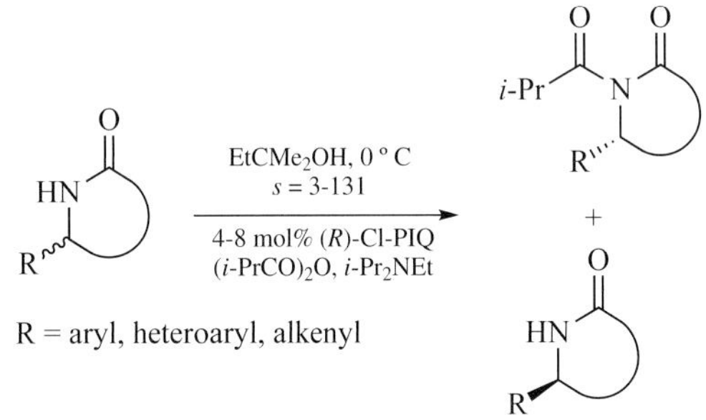

(18)

Substrates successfully resolved by Cl-PIQ, but not BTM:

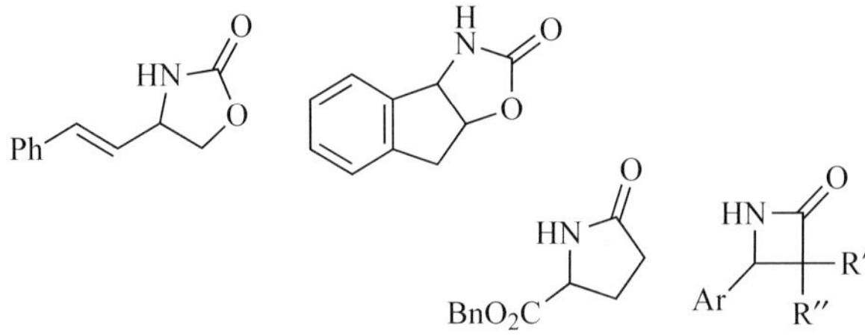

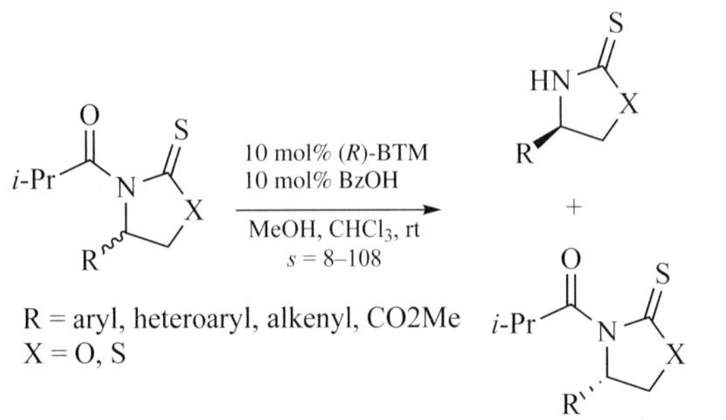

(19)

Kinetic Resolution of 2-Aryl-Alkanoic Acids. Shiina and coworkers developed a protocol for the KR of 2-aryl-propionic acids (eq 20),[27] 2-aryloxy-propionic acids (eq 21),[28] and 2-aryl-2-fluoropropionic acids (eq 22)[29] via enantioselective esterification with di-(1-naphthyl)-methanol. Furthermore, dynamic kinetic resolution (DKR)[30] of 2-aryl-propionic acids was achieved by Shiina et al. under similar conditions using DMF as a solvent (eq 23).[31]

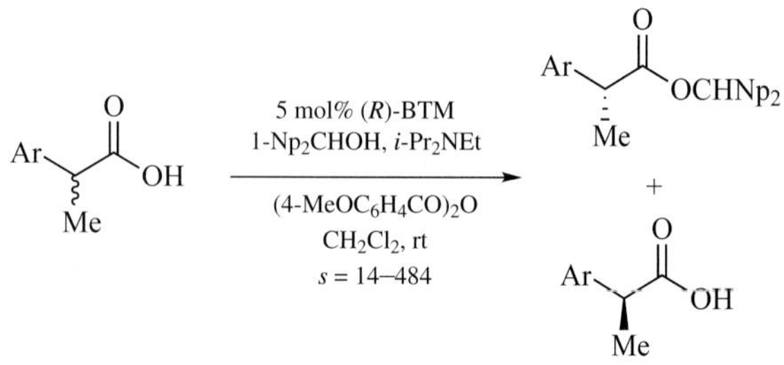

(20)

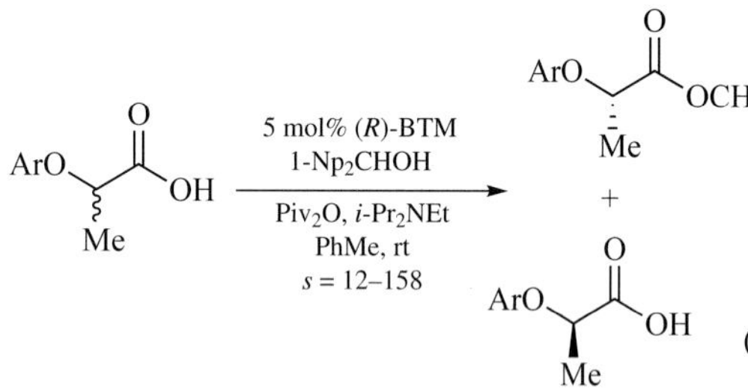

(21)

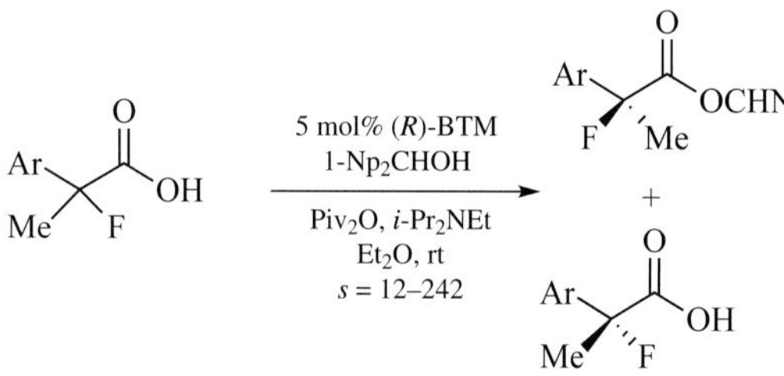

(22)

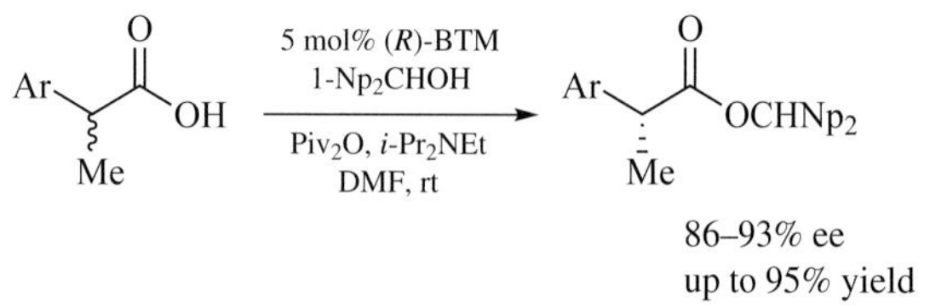

(23)

Dynamic Kinetic Resolution of Azlactones. Birman and coworkers achieved DKR of azlactones via enantioselective alcoholysis with di-(1-naphthyl)-methanol in the presence of (*S*)-BTM and benzoic acid as a cocatalyst (eq 24).[32]

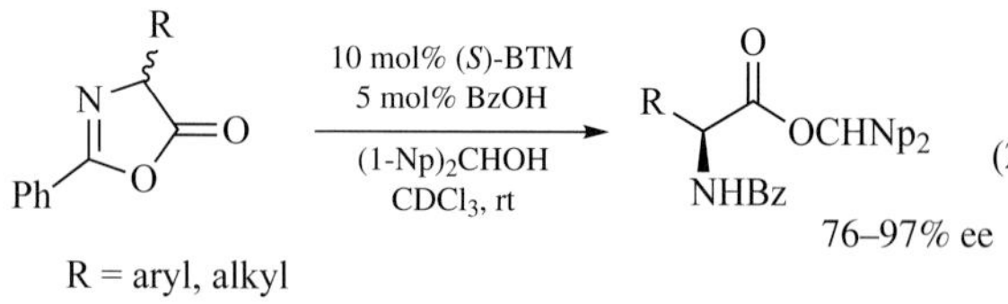

(24)

Kinetic Resolution of *N*-Aroyl-*β*-lactams. Bumbu and Birman developed KR of *N*-*p*-chlorobenzoyl-*β*-lactams via highly enantioselective methanolysis (eq 25).[33]

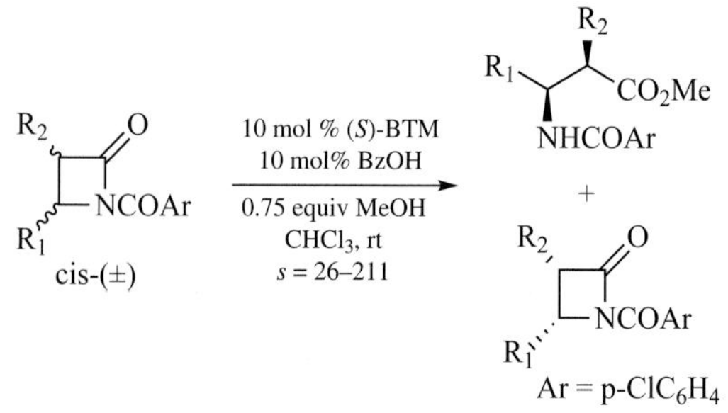

(25)

Asymmetric Synthesis of Dihydropyridones. Smith and coworkers achieved a formal hetero-Diels–Alder reaction between *α*,*β*-unsaturated *N*-sulfonylimines and carboxylic acids activated in situ as mixed anhydrides (eq 26).[34]

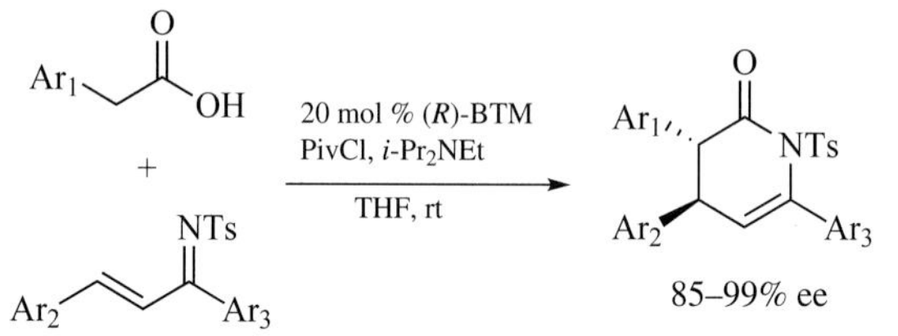

(26)

Asymmetric [3+2] Cycloadditions. Studer and coworkers achieved cycloaddition of azomethine ylides with carboxylic acids activated in situ as mixed anhydrides (eq 27).[35] Chi and coworkers developed a similar process using 4-chlorocyclobutenones as vinylketene equivalents (eq 28).[36]

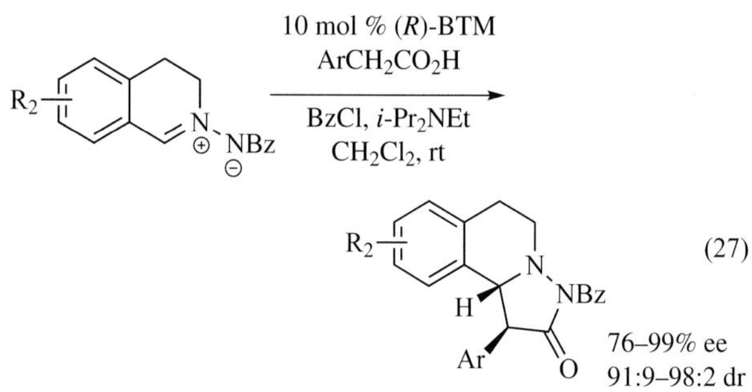

(27)

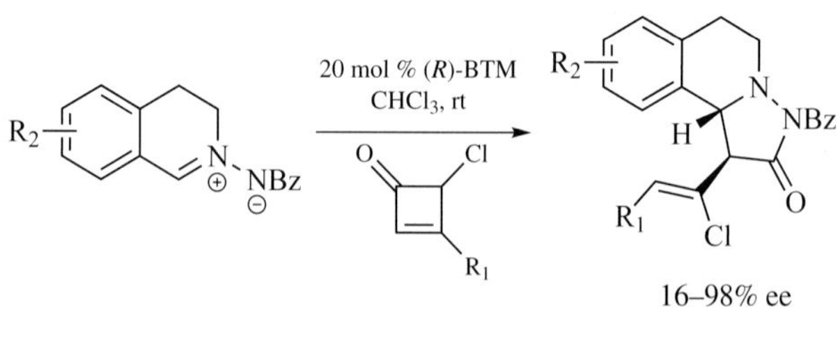

(28)

Asymmetric Diels–Alder Reaction. Romo and coworkers developed a highly enantioselective intermolecular Diels–Alder lactonization cascade using BTM to activate *α*,*β*-unsaturated acyl chlorides (eq 29).[37]

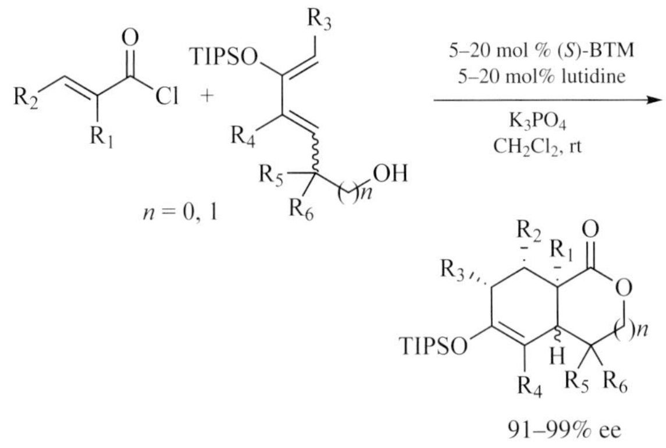

(29)

Asymmetric Synthesis of Benzothiazepines. Matsubara and coworkers disclosed a BTM-catalyzed cascade proceeding via enantioselective thiolate conjugate addition and subsequent intramolecular *N*-acylation giving rise to highly enantioenriched 1,5-benzothiazepines (eq 30).[38]

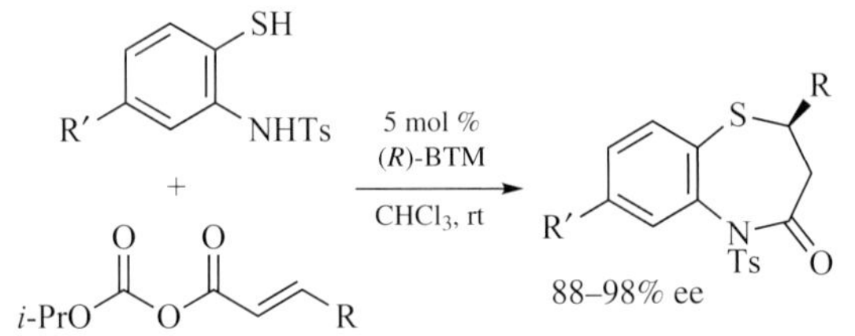

(30)

Rearrangement of Ammonium Ylides. Smith and coworkers have devised a highly enantio- and diastereoselective [2,3]-rearrangement of quaternary allylic ammonium salts induced by activation of an adjacent nitrophenyl ester by BTM. The resulting activated ester intermediates are converted into esters or amides (eq 31).[39]

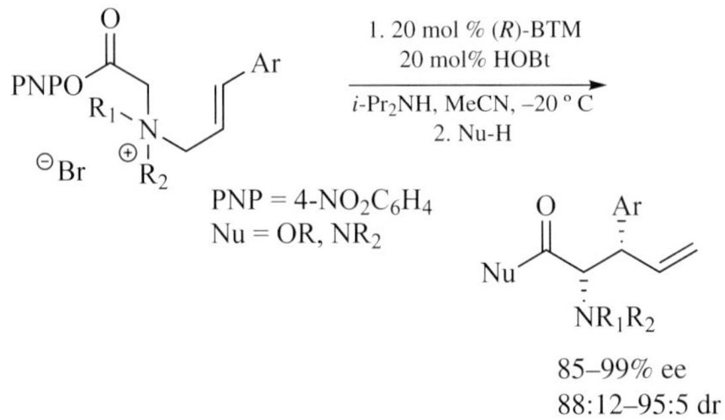

(31)

Silylative Kinetic Resolution of Alcohols. Wiskur and coworkers achieved KR of cyclic alcohols via BTM-catalyzed enantioselective silylation (eqs 32 and 33).[40,41]

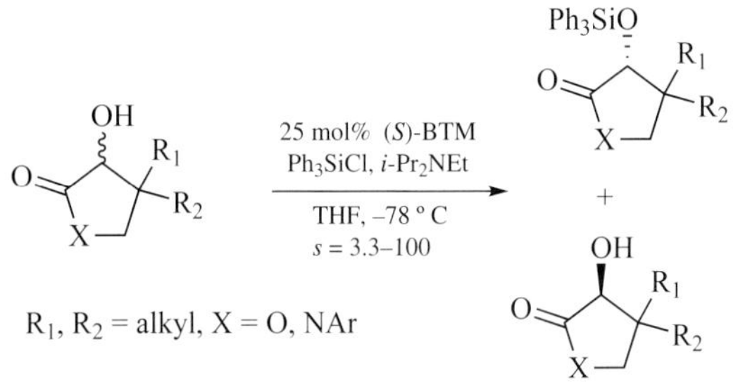

(32)

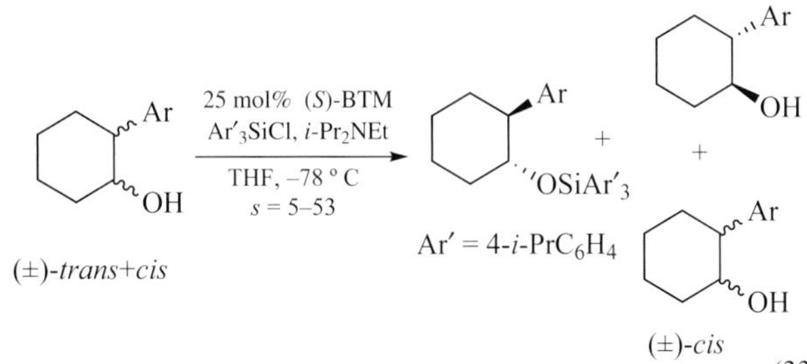

(33)

Related Reagents. 2,3,5,6-tetrahydro-6-phenyl-imidazo[2,1-*b*]thiazole (tetramisole, (*S*)-enantiomer: [14769-73-4], (*R*)-enantiomer: [14769-74-5])[2]; 7-chloro-1,2-dihydro-2-phenyl-imidazo[1,2-*a*]quinoline (Cl-PIQ, (*R*)-enantiomer: [863319-04-4])[10]; 3,4-dihydro-2-phenyl-2*H*-pyrimido[2,1-*b*]benzothiazole (HBTM)[43]; 3,4-dihydro-3-methyl-2-phenyl-2*H*-pyrimido[2,1-*b*]benzothiazole (HBTM-2, (2*S*,3*R*)-enantiomer: [1239194-96-7])[43]; 3,4-dihydro-3-(1-methylethyl)-2-phenyl-2*H*-pyrimido[2,1-*b*]benzothiazole (HBTM-2.1, (2*S*,3*R*)-enantiomer: [1203507-02-1], (2*R*,3*S*)-enantiomer: [1699751-03-5]).[44] 2-(*N*-benzyl, *N*-methylaminomethyl)-1-methylpyrrolidine ((*R*)-enantiomer: [540474-12-2], (*S*)-enantiomer: [208833-00-5])[15]; 4-dimethylaminopyrindinyl(pentaphenylcyclopentadienyl)iron (C_5Ph_5-DMAP, (*R*)-enantiomer: [187682-64-0], (*S*)-enantiomer: [187596-69-6]).[16]

---

1. Kagan, H. B.; Fiaud, J. C., *Top. Stereochem.* **1988**, *18*, 249.
2. Birman, V. B.; Li, X., *Org. Lett.* **2006**, *8*, 1351.
3. Birman, V. B.; Guo, L., *Org. Lett.* **2006**, *8*, 4859.
4. Birman, V. B.; Jiang, H.; Li, X.; Guo, L.; Uffman, E. W., *J. Am. Chem. Soc.* **2006**, *128*, 6536.
5. Birman, V. B.; Jiang, H.; Li, X., *Org. Lett.* **2007**, *9*, 3237.
6. Amarouch, H.; Loiseau, P. R.; Bonnafous, M.; Caujolle, R.; Payard, M.; Loiseau, P. M.; Bories, C.; Gayral, P., *Farmaco Ed. Sci.* **1988**, *43*, 421.
7. Birman, V. B.; Jiang, H., *Org. Lett.* **2005**, *7*, 3445.
8. Sih, C. J.; Wu, S.-H., *Top. Stereochem.* **1989**, *19*, 63.
9. Klibanov, A. M., *Acc. Chem. Res.* **1990**, *23*, 114.
10. Schoffers, E.; Golebiowski, A.; Johnson, C. R., *Tetrahedron*, **1996**, *52*, 3769.
11. Vedejs, E.; Jure, M., *Angew. Chem., Int. Ed.* **2005**, *44*, 3974.
12. Selectivity factor is defined as $s = k$ (fast-reacting enantiomer)/k (slow-reacting enantiomer). Both conversion C and selectivity factor s are calculated from ee values of the product (ee_p) and the recovered starting material (ee_{sm}) according to Kagan's equations (Ref. 1): $C = ee_{sm}/(ee_p + ee_{sm})$; $s = \ln[(1 - C)(1 - ee_A)]/\ln[(1 - C)(1 + ee_A)]$.
13. Shiina, I.; Nakata, K., *Tetrahedron Lett.* **2007**, *48*, 8314.

14. Daniels, D. S. B.; Smith, S. R.; Lebl, T.; Shapland, P.; Smith, A. D., *Synthesis* **2015**, *47*, 34.
15. Li, X.; Jiang, H.; Uffman, E. W.; Guo, L.; Zhang, Y.; Yang, X.; Birman, V. B., *J. Org. Chem.* **2012**, *77*, 1722.
16. Xu, Q.; Zhou, H.; Geng, X.; Chen, P., *Tetrahedron* **2009**, *65*, 2232.
17. Zhou, H.; Xu, Q.; Chen, P., *Tetrahedron* **2008**, *64*, 6494.
18. Shiina, I.; Nakata, K.; Ono, K.; Sugimoto, M.; Sekiguchi, A., *Chem. Eur. J.* **2010**, *16*, 167.
19. Shiina, I.; Ono, K.; Nakata, K., *Chem. Lett.* **2011**, *40*, 147.
20. Shiina, I.; Umezaki, Y.; Kuroda, N.; Iizumi, T.; Nagai, S.; Katoh, T., *J. Org. Chem.* **2012**, *77*, 4885.
21. Shiina, I.; Ono, K.; Nakahara, T., *Chem. Commun.* **2013**, *49*, 10700.
22. Nakata, K.; Gotoh, K.; Ono, K.; Futami, K.; Shiina, I., *Org. Lett.* **2013**, *15*, 1170.
23. Li, X.; Liu, P.; Houk, K. N.; Birman, V. B., *J. Am. Chem. Soc.* **2008**, *130*, 13836.
24. Wang, H.-Y.; Yang, K.; Yin, D.; Liu, C.; Glazier, D. A.; Tang, W., *Org. Lett.* **2015**, *17*, 5272.
25. Yang, X.; Bumbu, V. D.; Liu, P.; Li, X.; Jiang, H.; Uffman, E. W.; Guo, L.; Zhang, W.; Jiang, X.; Houk, K. N.; Birman, V. B., *J. Am. Chem. Soc.* **2012**, *134*, 17605.
26. Bumbu, V. D.; Yang, X.; Birman, V. B. *Org. Lett.* **2013**, *15*, 2790.
27. Shiina, I.; Nakata, K.; Ono, K.; Onda, Y.-S.; Itagaki, M., *J. Am. Chem. Soc.* **2010**, *132*, 11629.
28. Tengeiji, A.; Nakata, K.; Ono, K.; Shiina, I., *Heterocycles* **2012**, *86*, 1227.
29. Tengeiji, A.; Shiina, I., *Molecules* **2012**, *17*, 7356.
30. Pellissier, H., *Adv. Synth. Catal.* **2011**, *353*, 659.
31. Shiina, I.; Ono, K.; Nakata, K., *Catal. Sci. Technol.* **2012**, *2*, 2200.
32. Yang, X.; Lu, G.; Birman, V. B., *Org. Lett.* **2010**, *12*, 892.
33. Bumbu, V. D.; Birman, V. B., *J. Am. Chem. Soc.* **2011**, *133*, 13902.
34. Simal, C.; Lebl, T.; Slawin, A. M. Z.; Smith, A. D., *Angew. Chem., Int. Ed.* **2012**, *51*, 3653.
35. Hesping, L.; Biswas, A.; Daniliuc, C. G.; Mueck-Lichtenfeld, C.; Studer, A., *Chem. Sci.* **2015**, *6*, 1252.
36. Li, B.-S.; Wang, Y.; Jin, Z.; Chi, Y. R., *Chem. Sci.* **2015**, *6*, 6008.
37. Abbasov, M. E.; Hudson, B. M.; Tantillo, D. J.; Romo, D., *J. Am. Chem. Soc.* **2014**, *136*, 4492.
38. Fukata, Y.; Asano, K.; Matsubara, S., *J. Am. Chem. Soc.* **2015**, *137*, 5320.
39. West, T. H.; Daniels, D. S. B.; Slawin, A. M. Z.; Smith, A. D., *J. Am. Chem. Soc.* **2014**, *136*, 4476.
40. Clark, R. W.; Deaton, T. M.; Zhang, Y.; Moore, M. I.; Wiskur, S. L., *Org. Lett.* **2013**, *15*, 6132.
41. Wang, L.; Akhani, R. K.; Wiskur, S. L., *Org. Lett.* **2015**, *17*, 2408.
42. Birman, V. B.; Li, X., *Org. Lett.* **2008**, *10*, 1115.
43. Zhang, Y.; Birman, V. B., *Adv. Synth. Catal.* **2009**, *351*, 2525.
44. Belmessieri, D.; Joannesse, C.; Woods, P. A.; MacGregor, C.; Jones, C.; Campbell, C. D.; Johnston, C. P.; Duguet, N.; Concellon, C.; Bragg, R. A.; Smith, A. D., *Org. Biomol. Chem.* **2011**, *9*, 559.
45. Sano, T.; Imai, K.; Ohashi, K.; Oriyama, T., *Chem. Lett.* **1999**, 265.
46. Fu, G. C., *Acc. Chem. Res.* **2004**, *37*, 542.

6,7-Dihydro-2-phenyl-5-(phenylmethyl)-5*H*-pyrrolo[2,1-*c*]-1,2,4-triazolium Chloride

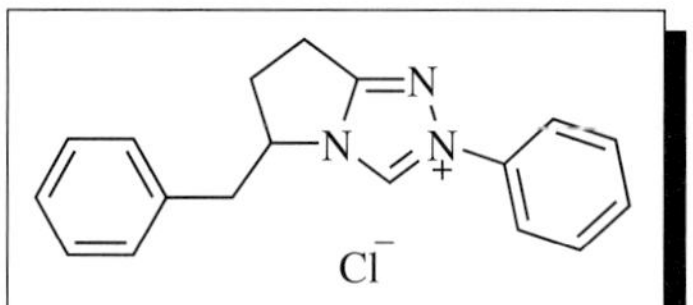

1a

[640737-16-2] $C_{18}H_{18}ClN_3$ (MW 311.8086)

InChI = 1S/C18H18N3.ClH/c1-3-7-15(8-4-1)13-17-11-12-18-19-21(14-20(17)18)16-9-5-2-6-10-16;/h1-10,14,17H,11-13H2;1H/q+1;/p-1

InChIKey = JOMVXOWRDSZYJW-UHFFFAOYSA-M

(catalyst used in a variety of transformations, mostly involving umpolung reactivity)

Physical Data: mp 198.4–199.6 (BF_4 salt).
Solubility: common organic solvents—low solubility in hexanes, benzene, and toluene.
Form Supplied in: anhydrous solid.
Analysis of Reagent Purity: NMR, MS.
Purification: precipitation from chlorobenzene followed by recrystallization from MeOH (BF_4 catalyst).
Handling, Storage, and Precaution: støred at ambient temperature with no special precautions.

Original Commentary

Tomislav Rovis
Colorado State University, Fort Collins, CO, USA

Introduction. *N*-Arylpyrrolotriazolium salts such as **1a** are effective precursors to triazolinylidene carbenes, cousins of the more extensively studied imidazolinylidene carbenes. Their application in organic synthesis has revolved primarily around their role as catalysts for a variety of organic transformations and as ligands. Both of these roles exploit the propensity of triazolium salts to form ylides or carbenes upon treatment with base. (eq 1)

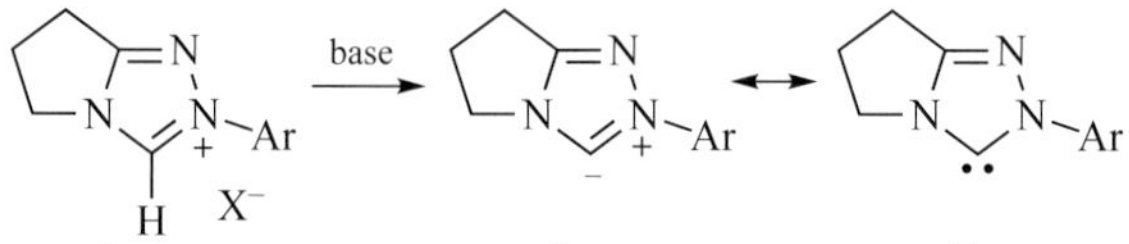

2 **3a** **3b** (1)

Both chiral and achiral versions of these salts have been used as reagents and catalysts for a number of organic transformations. Indeed, the application of these catalysts in the asymmetric Stetter reaction[1] and redox chemistry[2] has been recently reviewed. The title compound **1a** is typically formed as the chloride or tetrafluoroborate salt. Modification of this core structure is trivial due to the modularity of its assembly,[3] and literature suggests that substitution around this scaffold may have a profound impact on associated reactivity (vide infra). Extensive work has been reported with related structures **1**–**6**. The impetus for the development of these catalysts has been their promotion of umpolung chemistry, notably the benzoin and Stetter reactions. These have also led to investigations of other transformations.

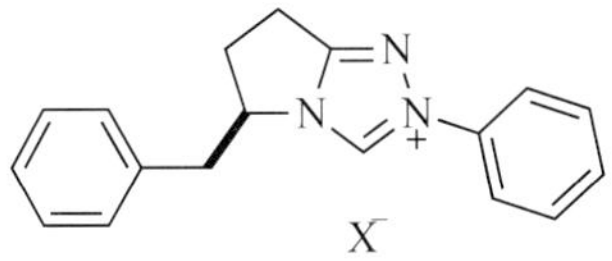

1a X = Cl
1b X = BF_4

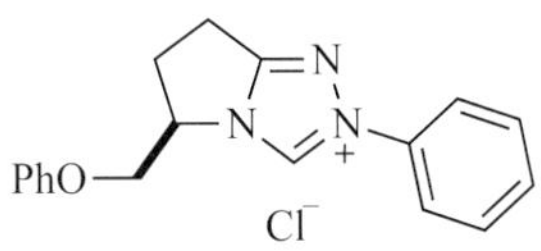

4

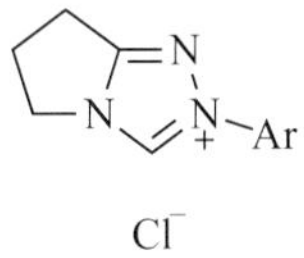

2a Ar = Ph
2b Ar = Mes

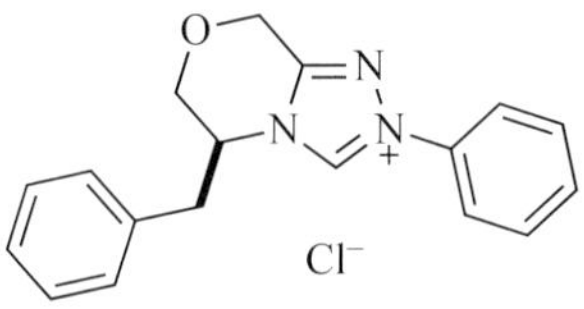

5

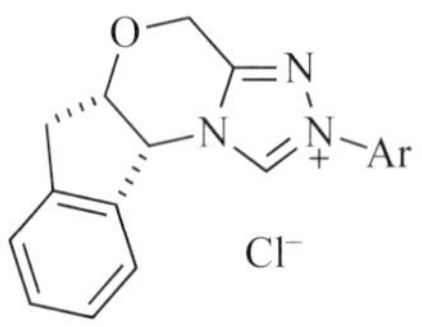

6a Ar = Ph
6b Ar = 4-MeOPh
6c Ar = C_6F_5
6d Ar = Mes

Benzoin Reaction. The benzoin reaction, historically demonstrated to be subject to asymmetric catalysis using chiral thiazolium salts,[4] has been shown also to be catalyzed by triazolium salts.[5] It was demonstrated in 1997 that the use of chiral nonracemic bicyclic triazolium salts affords much improved selectivities for the benzoin reaction. Catalysts **4** and **5** provide ee's up to 83% for the dimerization of aryl aldehydes, although the chemical yields remained problematic (eq 2).[6]

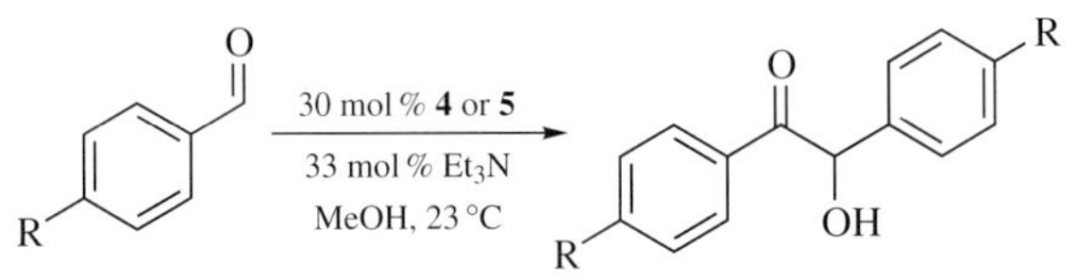

(2)

R = H, Cl, Me

Catalyst	Yield (%)	ee (%)
4	27–47	40–61
5	11–45	76–83

To date, the most useful catalyst reported for the asymmetric benzoin reaction is triazolium salt **7**. In the presence of KO*t*-Bu, this catalyst mediates the benzoin reaction of a variety of aryl aldehydes affording the corresponding benzoins in good yields and generally high enantioselectivity (eq 3).

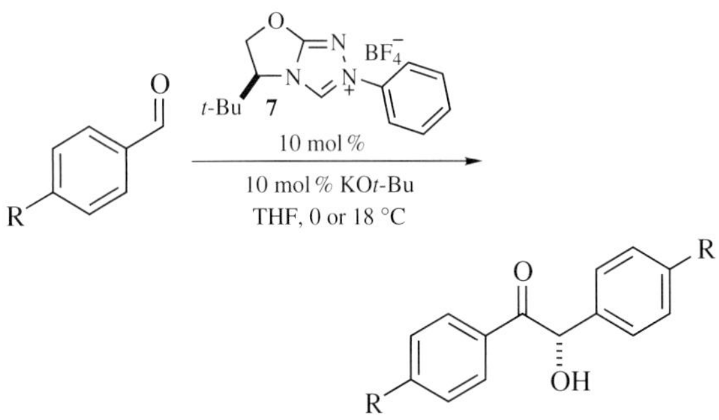

(3)

R = H, F, Cl, Br, Me, MeO
(also 2-Fu, 2-Nap)
36–85% yield
80–91% ee

Intramolecular crossed-benzoin reactions using triazolium-based catalysts have been reported by two groups. Enders showed that bicyclic triazolium salt (**8**) is an efficient precatalyst for this transformation, affording α-hydroxyketones in good yields and excellent enantioselectivities (eq 4).[7]

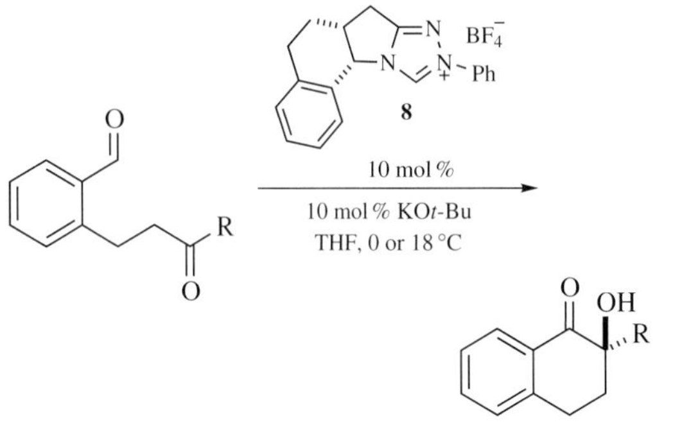

(4)

R = Me, Et, *n*-Bu, *i*-Bu, Bn
43–93% yield
93–98% ee

Suzuki has shown that triazolium catalyst (**6a**) affects a similar transformation with broad substrate scope. Of note is the tolerance of the reaction to a variety of tethers and steric requirements, especially adjacent to the ketone (eq 5).[8]

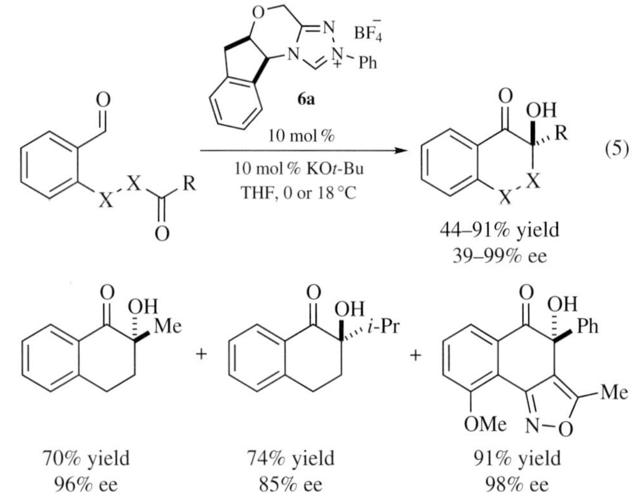

(5)

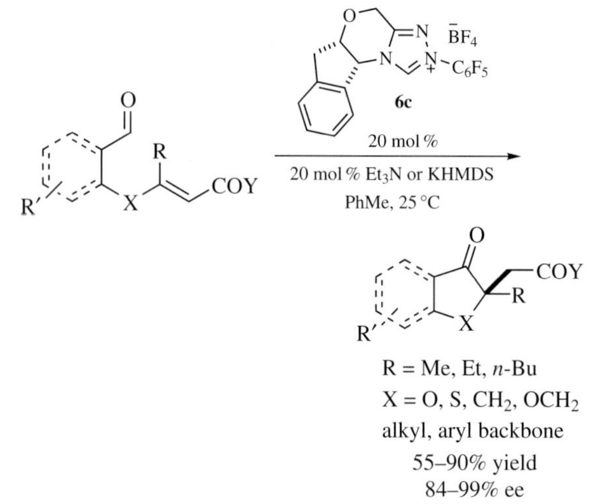

(7)

Stetter Reaction. The Stetter reaction may be viewed as a cross-benzoin reaction involving a Michael acceptor as the electrophile. While thiazolium salts in the presence of base have been historically used as catalysts for this transformation,[9] it has been illustrated that triazolium salts behave analogously.[10] The first highly enantioselective Stetter reaction was reported in 2002. A variety of aldehydes were found to undergo an intramolecular Stetter reaction with tethered Michael acceptors affording five- and six-membered rings in high enantioselectivity and excellent yield (eq 6).[11] A variety of Michael acceptors are tolerated, including enones, enoates, and enenitriles, with the former two providing highest enantioselectivities.[12]

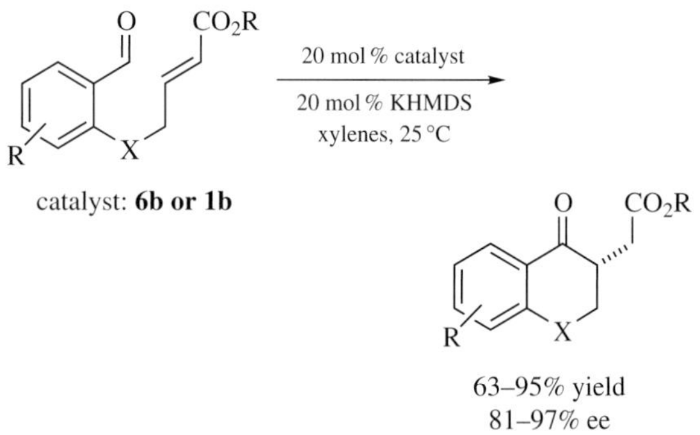

(6)

This family of catalysts has also been shown to cyclize aldehydes onto β,β-disubstituted Michael acceptors, forming tertiary ethers and sulfides as well as all-carbon quaternary stereocenters in excellent yield and selectivity (eq 7).[13,14]

Diastereoselective cyclizations may be achieved through the use of α,α,β-trisubstituted Michael acceptors such as **10**. Enolate protonation must proceed faster than C–C bond rotation as the relative stereochemistry corresponds to *syn*-addition of the aldehyde carbonyl and the proton to the Michael acceptor (eq 8).[15] Intriguingly, optimal reaction conditions for diastereoselective cyclizations involve the use of the free carbene (**9**) as catalyst.

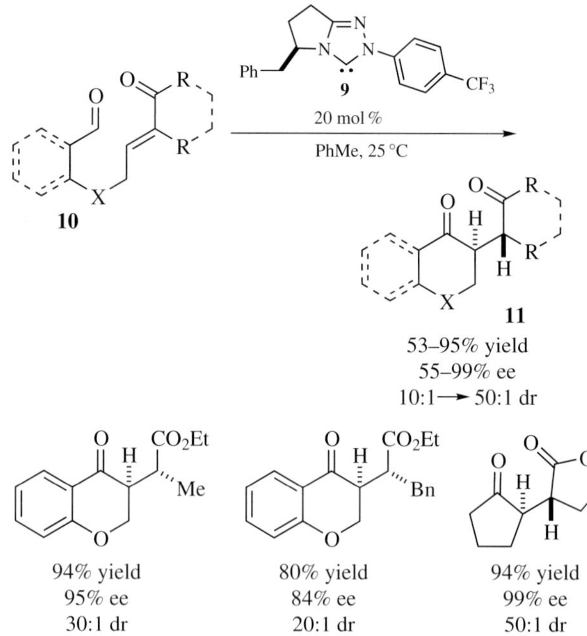

(8)

The reaction is tolerant of backbone substitution and the selectivity is dominated by the catalyst and not by the preexisting stereocenters, except when these are alpha to the aldehyde.[16] Aldehydes tethered to dienone acceptors allow for an enantioselective and diastereoselective intramolecular Stetter reaction to provide benzofuranone scaffolds in high yield and enantioselectivity (eq 9).[17]

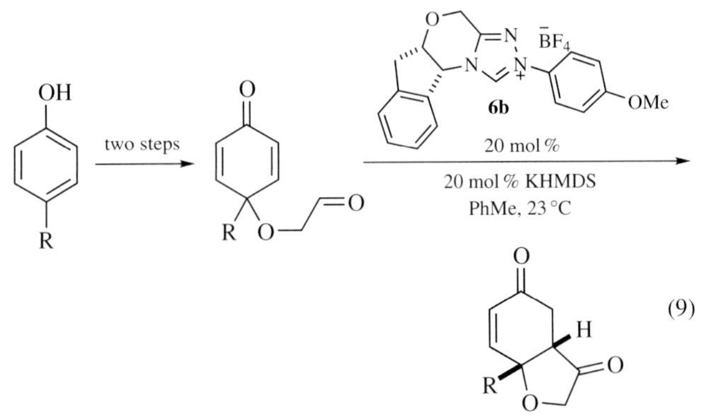

R = Me, Et, *i*-Pr, *t*-Bu, Ph, CH_2X
78–94% yield
82–94% ee
>19:1 dr

(9)

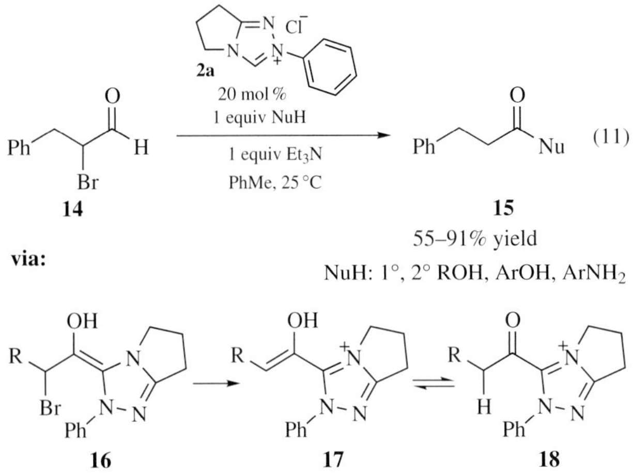

55–91% yield
NuH: 1°, 2° ROH, ArOH, $ArNH_2$

(11)

Acyl Transfer. Triazolium salts in the presence of base have been demonstrated to be effective acyl transfer catalysts. Treatment of acyloxyoxazoles with catalyst **2a** results in an effective transfer to form the acyl azlactone **13** (eq 10). Using a crossover experiment, the authors have demonstrated that the reaction proceeds via ion-pair dissociation. Triazolium salts were found to be superior to imidazolium salts as precatalysts.[18] Impressive catalyst loading levels may be achieved using these precursors, but the present technique does not yet compete with the best asymmetric catalysts for this process.[19,20]

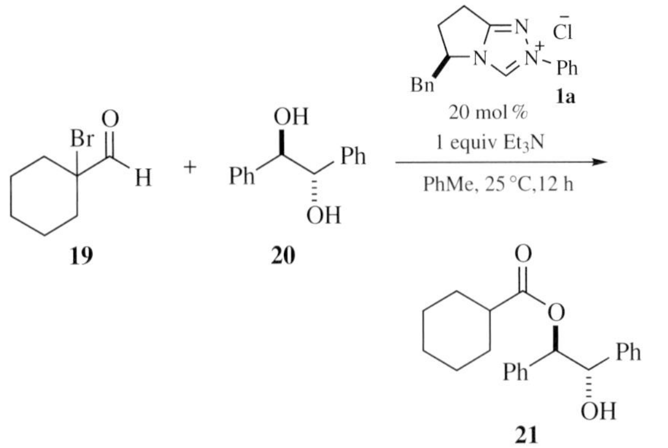

75% yield
83% ee

(12)

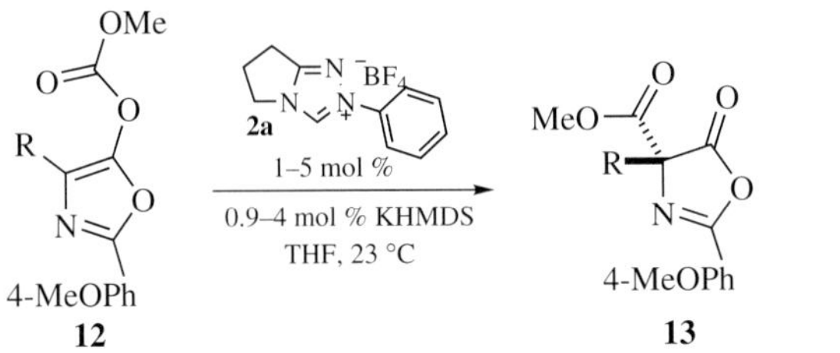

67–84% yield
R = Me, Bn, *i*-Bu, *i*-Pr

(10)

Triazolium-derived carbenes have been shown to effect the transformation of enals into alkanoates by a similar redox process at slightly elevated temperatures (eq 13).[23] Benzimidazolium salts also appear to be competent precatalysts for this chemistry.[24] Ynals provide enoates when subjected to similar reaction conditions, catalyzed by triazolium salts as well as imidazolium precursors, the latter affording higher selectivities for the *E*-alkene.[25]

Redox Catalysis. A variety of other reaction manifolds may be accessed by intercepting the Breslow intermediate (**16** in eq 11). The use of α-halo aldehydes as substrates in the presence of alcohols results in an internal redox reaction to form α-reduced esters in excellent yields (eq 11). The use of chiral triazolium salts allows a desymmetrization of a *meso*-diol to proceed in good ee (eq 12).[21] Thiazolium salts are also competent catalysts for related redox reactions, and α,β-epoxy- and aziridinoaldehydes have been shown to be appropriate substrates.[22]

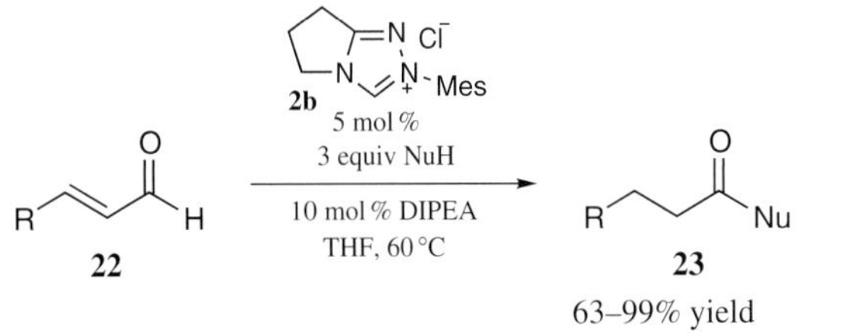

63–99% yield
NuH: 1°, 2° ROH, ArOH

(13)

This chemistry has been extended to the redox reaction of cyclopropyl carboxaldehydes bearing electron-withdrawing groups. The apparently facile C–C bond cleavage under very mild conditions is particularly noteworthy. The process generates desymmetrized glutarates and analogs and maintains the integrity of the C-3 stereocenter (eq 14).[26]

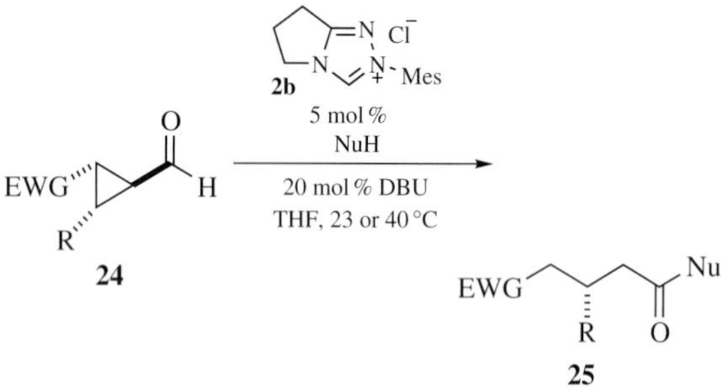

(14)

81–99% yield
NuH: ROH, RSH

The redox process catalyzed by these carbenes proceeds through an azolium enol/enolate which tautomerizes to the acyl azolium (**17** to **18** in eq 11). One can envision that the use of a chiral catalyst may allow for an enantioselective bond-formation at the alpha position. Two strategies have been advanced illustrating this concept. The use of α,α-dichloroaldehydes allows for an asymmetric protonation/alcohol trapping of the prochiral enol/enolate providing α-chloroesters in good yield and high enantioselectivities (eq 15). Catalyst **6c** provides optimal results.

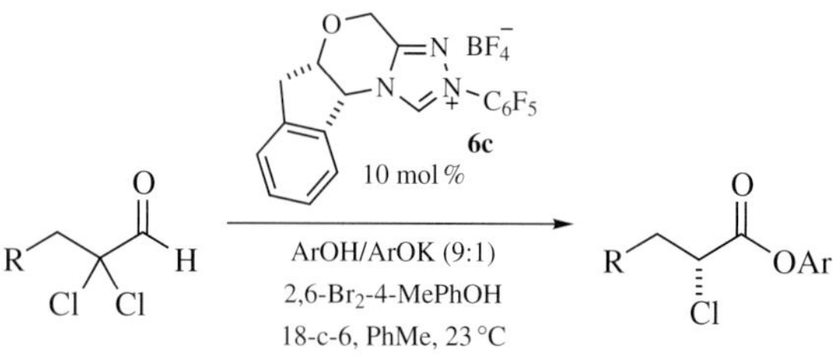

(15)

65–78% yield
84–92% ee
R = alkyl, aryl
Ar = various phenols

A complementary strategy uses a heterodiene to trap the enol/enolate intermediate via a cycloaddition manifold. The selectivities exhibited by catalyst **6d** are spectacular, as is the level of chemoselectivity observed (eq 16).[27]

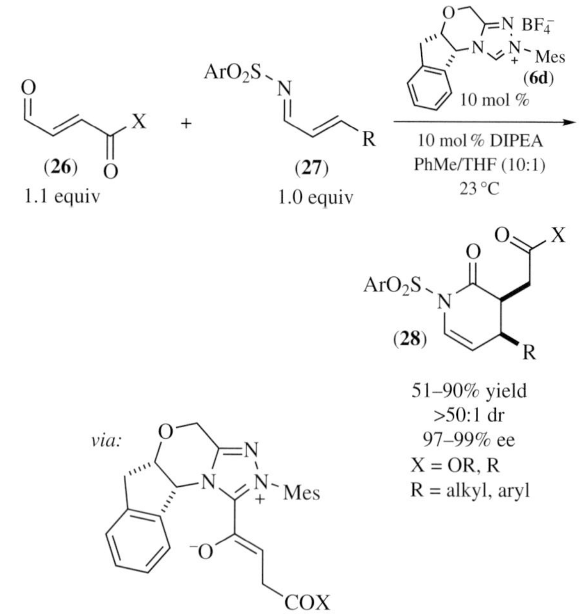

(16)

51–90% yield
>50:1 dr
97–99% ee
X = OR, R
R = alkyl, aryl

Metal Complexes. Achiral triazolium salts have been appended to Au and Ag and the resultant complexes characterized crystallographically (eq 17). The metal complexes appear to behave analogously to their imidazolinylidene counterparts. Application of these complexes as catalysts has not been reported.

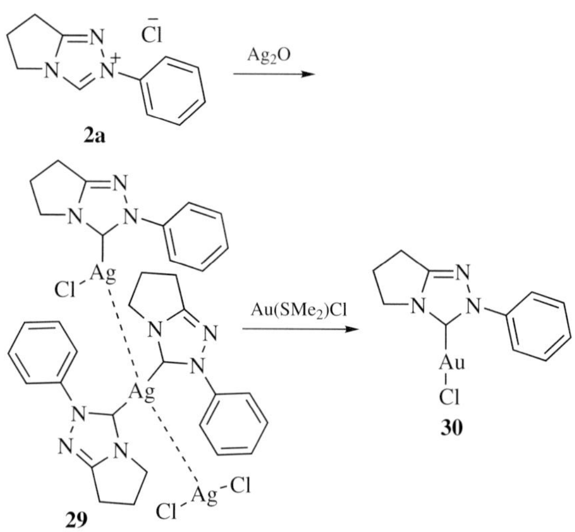

(17)

determined by
X-ray analysis

Related Reagents. (4S,5S)-4-(2,2-Dimethyl-4-phenyl-1,3-dioxan-5-yl)-1-phenyl-4H-1,2,4-triazol-1-ium Perchlorate.

First Update

Alberto Munoz & Tomislav Rovis
Colorado State University, Fort Collins, CO, USA

Stetter Reaction. The asymmetric intramolecular Stetter reaction has already been described herein (vide supra), although in 2008 a more complete study was reported by Rovis and coworkers.[28] The reaction was found to be functional with both electron-rich and electron-poor aromatic and aliphatic tethered aldehydes as catalyzed by several triazolium-based precatalysts in the presence of base. Various Michael acceptors have been found to perform well, including amides, esters, thioesters, ketones, aldehydes, and nitriles. It was found that triazolium precatalysts that bear electron-deficient aryl groups tend to give higher yields and selectivities for this particular transformation.

Also in 2008, Rovis and coworkers disclosed the catalytic asymmetric intermolecular Stetter reaction of glyoxamides with alkylidenemalonates.[29] It was found that catalyst **1b** gave only trace amounts of product. Thus, the aryl substituent off the N_1 nitrogen was replaced with the highly electron-deficient pentafluorophenyl. This change led to increased reactivity and enantioselectivity; the same hypothesis applied to catalyst **6e** did not yield the same results (eq 18). This reaction is tolerant of a number of aliphatic β-substituted alkylidenemalonates, delivering product with high levels of asymmetric induction while also managing to increase the reaction scope of the intermolecular Stetter reaction.

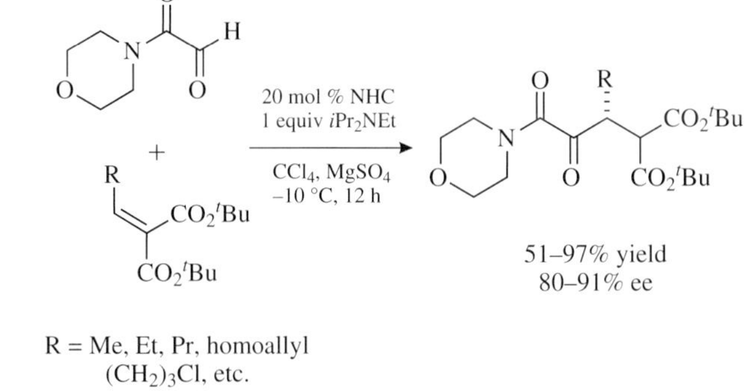

1c
Up to 100% yield, 91% ee

6e
12% yield, 65% ee

(18)

A complementary methodology was concurrently disclosed by the same authors but instead of alkylidenemalonates, alkylidene ketoamides were employed (eq 19).[30] The diasteresoselectivity here is attributed to insulation of the newly formed β-ketoamide stereocenter due to developing $A_{1,3}$ strain in the enolization event.[31]

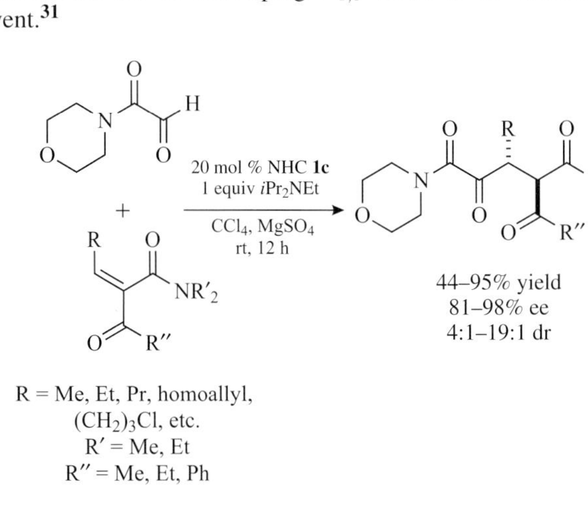

(19)

A study on the mechanism of the intramolecular Stetter reaction was reported by Rovis and coworkers in 2011.[32] The reaction was found to be first order in aldehyde and carbene. A primary kinetic isotope effect was found for the aldehydic proton, suggesting that proton transfer from the tetrahedral intermediate is the first irreversible and turnover limiting step in the catalytic cycle for the intramolecular Stetter reaction. Catalyst **1b** was used for most of these experiments.

As previously stated, the intermolecular Stetter reaction suffers from the relative narrow scope of Michael acceptors that participate in the reaction. In 2009, Rovis and coworkers discovered that backbone fluorination (catalyst **7**) on the NHC scaffold improves both reactivity and selectivity relative to catalyst **1a**.[33] This catalyst allowed for the reaction of heteroaryl aldehydes and nitro olefins (eq 20). Unfortunately, this reactivity was once again limited to aromatic substituents off the aldehyde that features as an α-heteroatom. A solution to this limitation was found almost 2 years later, when it was found that the addition of catechol allowed the reaction to proceed with α,β-unsaturated aldehydes.[34] The role of the catechol additive here was posited to promote the proton-transfer event forming the Breslow intermediate (eq 21).

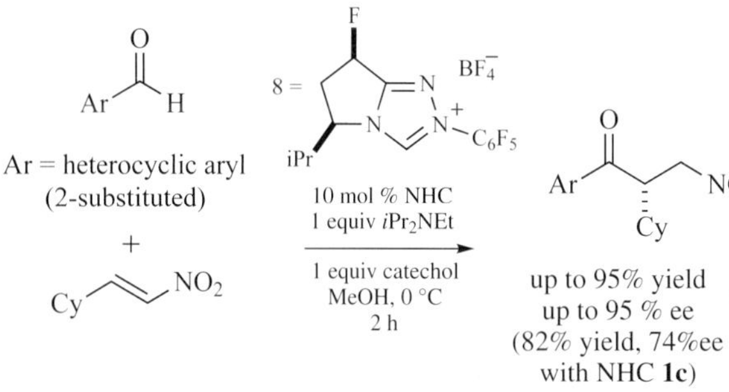

(20)

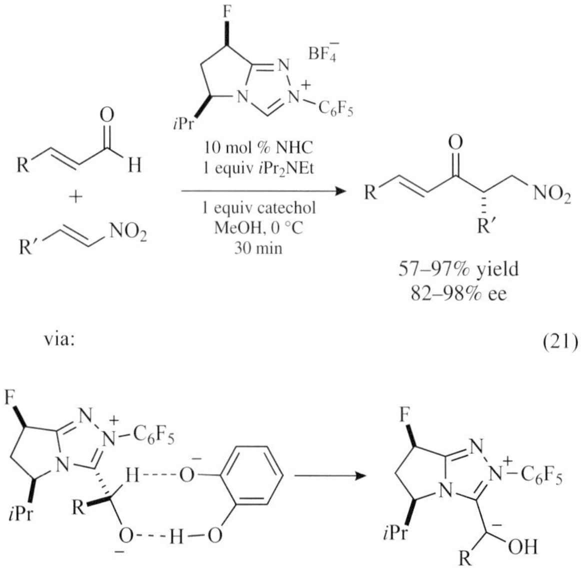

(21)

The scope of Michael acceptors in the intermolecular Stetter reaction was once again further expanded with work by the Gravel group in 2011. They found that β,γ-unsaturated-α-ketoesters served well with various aldehydes in the synthesis of 1,2,5-tricarbonyl compounds (eq 22).[35] The reaction is high yielding and highly enantioselective, a feat previously unknown with β-aryl substituted acceptors with this reaction manifold.

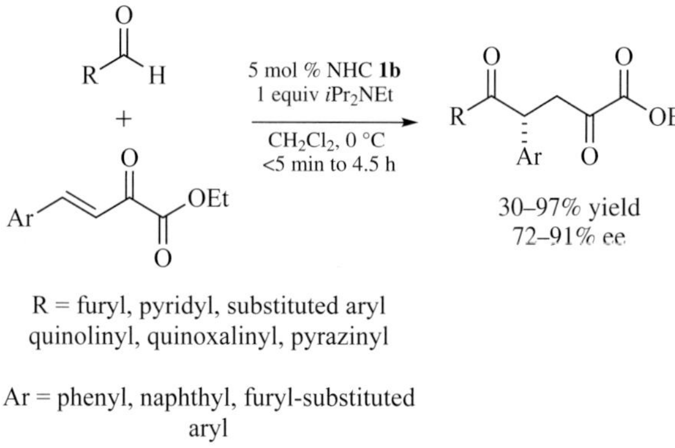

(22)

The scope was further expanded via thework of Glorius and coworkers. Although previous studies were shown to work with olefin that had two activating groups[35] and then one,[36,37] simple acrylates were shown to work with eq 23.[38] The main feature of this reaction is that a number of chiral and achiral catalysts work with this reaction, but the key feature is the highly electron-rich *N*-aryl substituent 2,6-dimethoxyphenyl aryl group (eq 23).

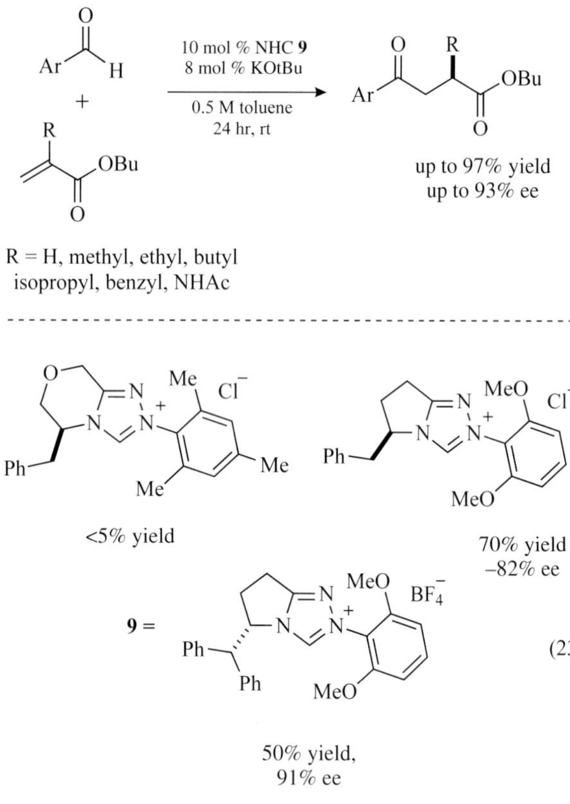

(23)

In 2013, Smith and coworkers reported a series of mechanistic studies on the NHC-catalyzed Stetter and benzoin reactions with the aim of deciphering the role of the *N*-aryl substituent.[39] The authors report spectroscopic observations of the aldehyde-NHC intermediates in both reaction profiles. They found that the equilibrium constants of the reactions are affected by both the electronic and steric effects of the *N*-aryl substituent. Electron-withdrawing substituents result in significantly larger K values, substitution at the 4-position leads to minimal perturbation of K, and both the benzoin and Stetter reactions work better with 2,6-electron-withdrawing substituents (the result of benefiting from both high K values and rapid k_{ex} values).

In 2013, Rovis and Lathrop reported efforts toward the synthesiz of (−)-cephalimysin A.[40] The key step here was a novel tandem photoisomerization/intramolecular Stetter reaction to set the desired spirofuranone lactam core. This reaction gives high yields and enantioselectivities (eq 24). Supplementary to this is a late-stage oxidation strategy that allows for the synthesis of three of the four diastereomers of cephalimysin A.

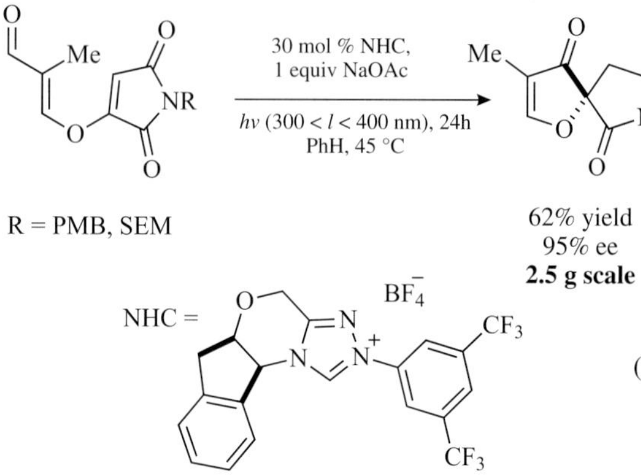

(24)

Homoenolate Reactivity. In 2012, Bode and coworkers reported the development of a chiral NHC-catalyzed Ireland–Coates Claisen rearrangement.[41] The reaction proceeds with a rather broad substrate scope and good enantioselectivities (eq 25). Either enals or ynals work for this methodology, but ynals proved to be the more reliable substrate.

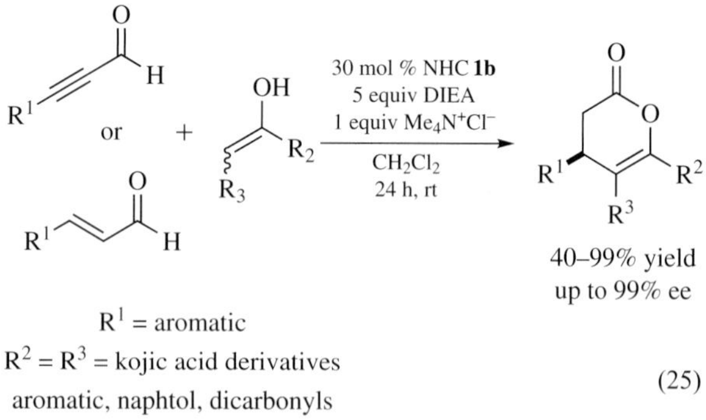

(25)

A diastereoselective synthesis of β-lactam-fused spirocyclic oxindoles from enals and unsaturated imines was described by Chi and coworkers in 2012 (eq 26).[42] The reaction proceeds through an enal-derived NHC-bound homoenolate intermediate bearing the two carbon centers that are involved in this reaction. The reaction was rendered enantioselective with the use of catalyst **1b**.

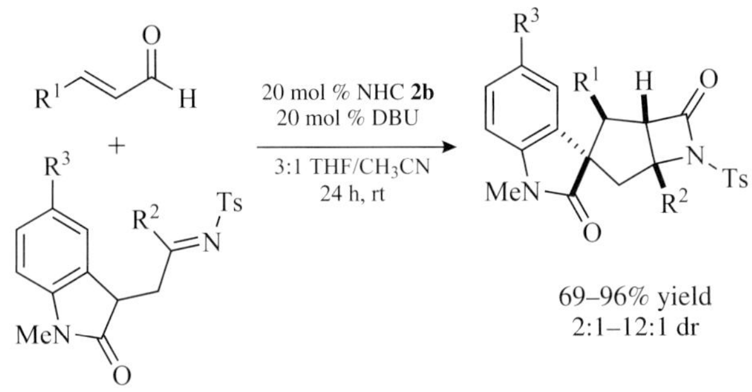

R^1 = aromatic, aliphatic
R^2 = aromatic
R^3 = H, methyl, chloro, bromo

(26)

The first successful and intentional generation of *cis*-homoenolate equivalents from *cis*-enals via NHC catalysis was accomplished in 2013 by Chi and coworkers.[43] The catalytically generated *cis*-homoenolate intermediates undergo opposite diastereoselectivities and reactivities relative to their trans counterparts (eq 27). The catalyst used for the majority of these reactions is **1b**, and both yields and ee values were high (> 70% and >90%, respectively).

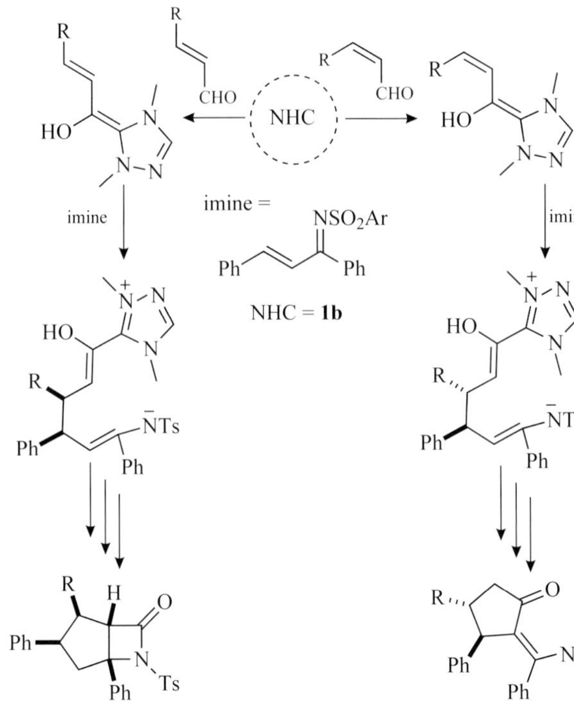

(27)

Enolate Reactivity. In 2012, Chi and coworkers disclosed a clever way to access the enolate intermediate using activated stable esters rather than the typical aldehyde, enal, or ketene substrate.[44] Chi used a 4-nitroester to form NHC-bound enolate intermediates, which underwent enantioselective formal [4 + 2] cyclizations with α,β-unsaturated imines (eq 28). This reaction manifold represents the first NHC-catalyzed activation of stable carboxylate esters to form enolate intermediates. Catalyst **1b** was used for these purposes.

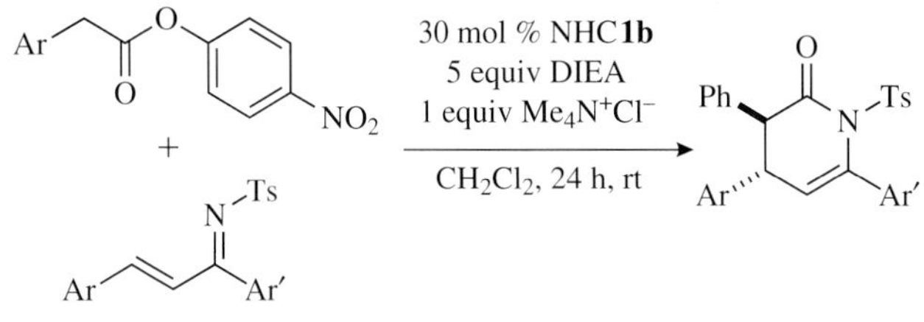

Ar = Ar′ = phenyl, substituted phenyl heteroaromatic

(28)

Subsequent to that seminal publication, the reactivity was further extended toward the activation of simple alkylacetic esters (eq 29).[45] The reaction is highly enantioselective and represents a new method to access enolate intermediates from normally unactivated aliphatic substrates. The scope of participating α,β-unsaturated imines was also expanded with this work.

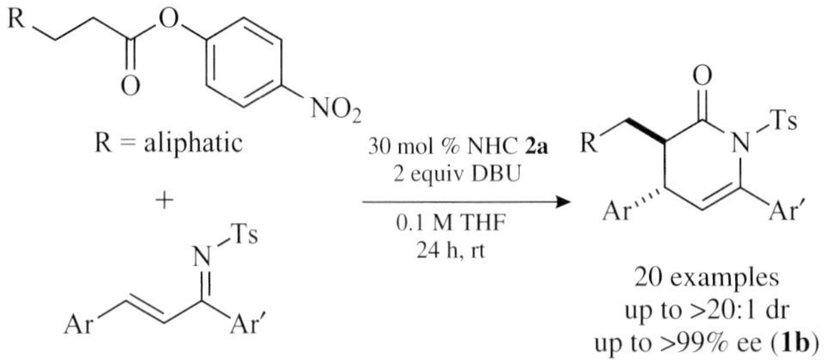

(29)

In 2015, Scheidt and Lee introduced an stereoselective annulation between in situ generated *ortho*-quinone methides and acyl imidazoles for the synthesis of dihydrocoumarins (eq 30).[46] This reaction uses chiral catalyst **1d**, is tolerant of both electron-rich and electron-deficient acyl imidazoles, and proceeds with good enantioselectivities.

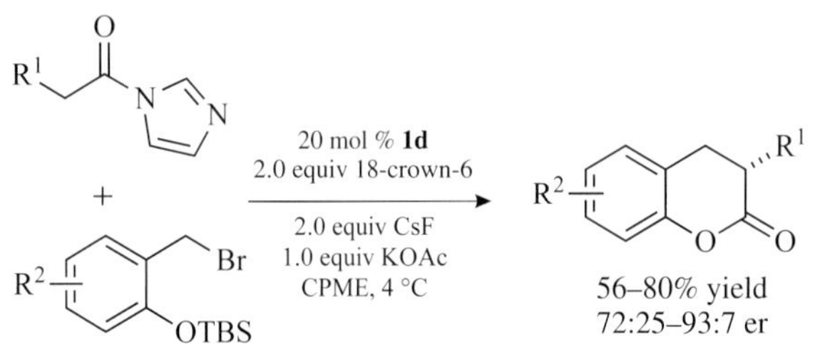

R^1 = aliphatic, aromatic
R^2 = Me, F, Cl

(30)

Desymmetrizations. In 2007, Scheidt and coworkers discovered the oxidation of allylic, propargylic, and benzylic alcohols to esters in a one-pot tandem oxidation using *N*-heterocyclic carbenes and manganese(IV) oxide (eq 31).[47] This process represents one of the first organocatalytic oxidations of alcohols to esters, being preceded by reports by Corey and Togo, respectively.[48,49]

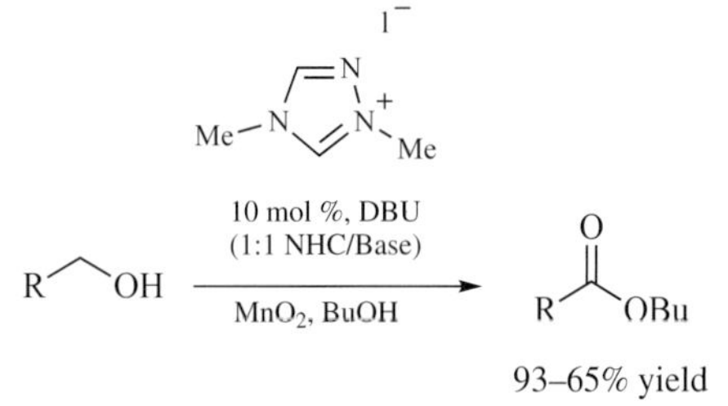

(31)

R = allyl, propargyl
benzylic, hydrocinnamaldehyde

This process differentiates itself from other NHC redox modes of catalysis in that the substrate undergoes oxidation without reduction at the α,β-unsaturated moiety, obviating the need for substrates that feature α-reducible functionalities in order to oxidize at the carbonyl center. A closer look at the proposed mechanism reveals an opportunity for the generation of a chiral-activated ester. As such, the desymmetrization of *cis*-1,2-cyclohexane diol was achieved with the use of several chiral catalysts, such as catalyst **1a** (eq 32). The yield and enantiomeric excess varied with the alteration of conditions and chiral NHC used, ranging from 38 to 78% yield and 20 to 80% ee.

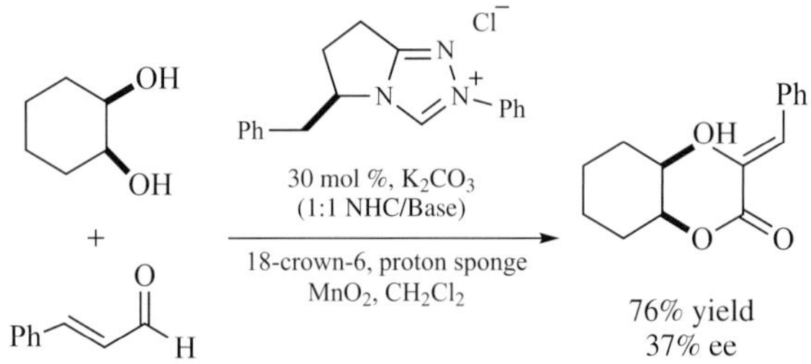

(32)

Tremendous advances have been made recently in the field of kinetic resolution, particularly with regard to dynamic kinetic resolution (DKR).[50,51] Methodology that uses DKR to set three stereocenters is limited in the field of NHC catalysis, but has been explored on a substrate that reacts in an intramolecular fashion by Scheidt and coworkers.[52] In 2015, Johnson and coworkers published a DKR that sets three stereocenters in an *intermolecular* fashion.[53] The catalyst induces the formation of the homoenolate equivalent from the enal that subsequently undergoes addition to the α-keto esters, the latter bearing an epimerizable stereocenter that is labile under the reaction conditions allowing for the DKR. This reaction proceeds in high enantio- and diastereoselectivities, as well as high yields (eq 33). Ye's catalyst **2b** performed the best was used for this reaction, where catalyst **1b** gave moderate to low diastereoselectivities.

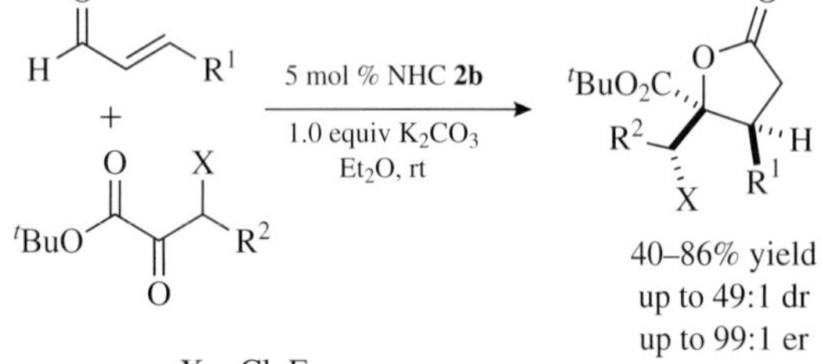

(33)

X = Cl, F
R^1 = H, aryl, furyl, thiophenyl, ester
R^2 = allyl, propargyl, aryl, aliphatic

Formal Cycloadditions. In 2008, Ye and coworkers disclosed the synthesis of *N*-Boc β-lactams via a formal [2 + 2] NHC-catalyzed cycloaddition with ketenes.[54] With this particular reaction, catalyst **1b** showed great reactivity with a 93% yield, but delivered low diastereoselectivity. This prompted the synthesis of TMS- and TBS-protected chiral catalysts **7a** and **7b** from L-pyroglutamic acid. With these catalysts, a diastereoselectivities were improved and a relatively wide substrate scope was accomplished (eq 34).

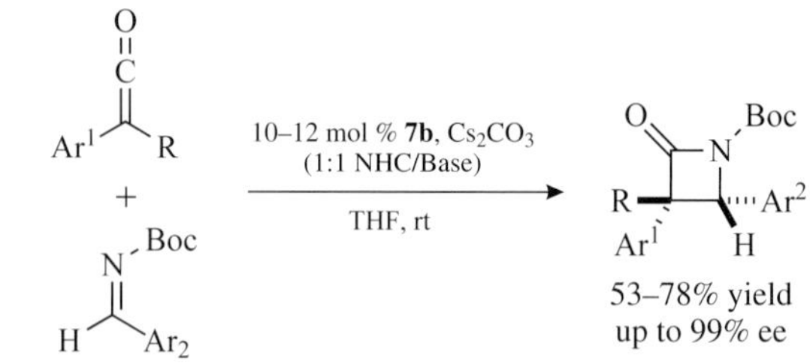

Ar^1 = 2-ClC_6H_4, Ph, $(CH_2)_6$
Ar^2 = aryl, naphtyl
R = Et, ^{i}Pr

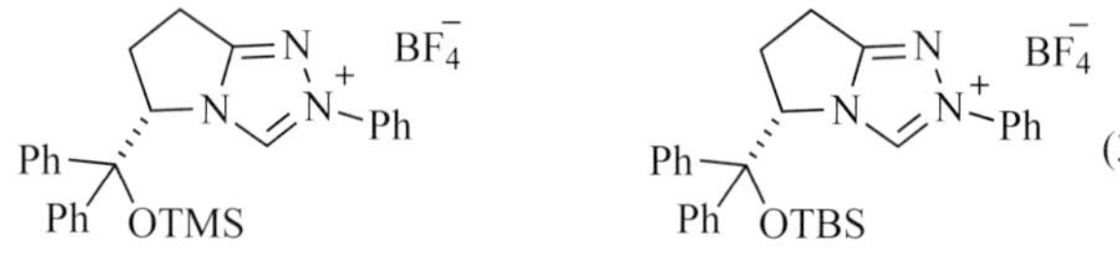

7a **7b**

(34)

Shortly after this, Song Ye disclosed a strategy complementary to the aforementioned one, synthesizing β-lactones bearing α-quaternary-β-stereocenters.[55] Unfortunately, simply aryl aldehydes did not work in the presence of either achiral or chiral NHCs, but a variety of 2-oxoaldehydes worked quite well with several ketenes (eq 35).

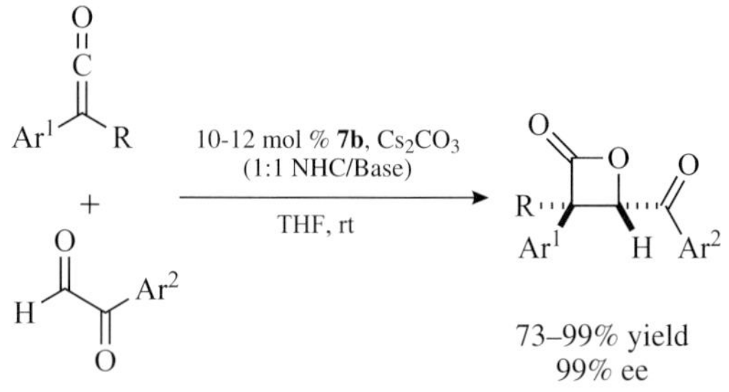

(35)

Ar^1 = 2-ClC_6H_4, Ph, $(CH_2)_6$
Ar^2 = aryl, naphtyl
R = Et, ^{i}Pr

In 2009, Bode and coworkers disclosed that α'-hydroxyenones serve as scope-expanding surrogates for α,β-unsaturated aldehydes in NHC-catalyzed reactions,[56] formed by an aldol condensation between aryl aldehydes and 3-hydroxy-3-methylbutanone. The authors claim that these substrates are almost always stable crystalline solids and are more easily stored than their α,β-unsaturated aldehyde counterparts. The Breslow intermediate/homoenolate equivalent is accessed by these substrates via a retro-benzoin reaction that furnishes the catalytically active intermediate and 1 equiv of acetone (eq 36).

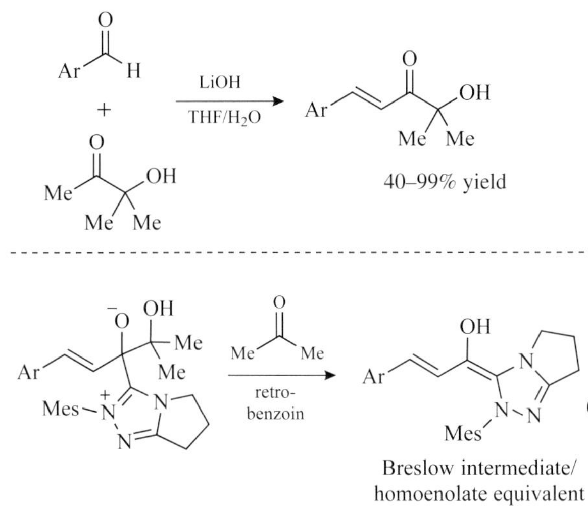

(36)

The formation of cyclopentenes, a reaction originally discovered by Nair[57] and rendered enantioselective by Bode,[58] was found to proceed quite well with catalyst **2b**, but other catalysts were studied as well (eq 37). It was found that the reaction was sensitive to sterics around the NHC backbone. Due to this, the use of α'-hydroxyenones is limited to achiral catalyst **2b**. Catalyst **1b** delivered product in 9% yield with 90% ee and 3:1 dr.

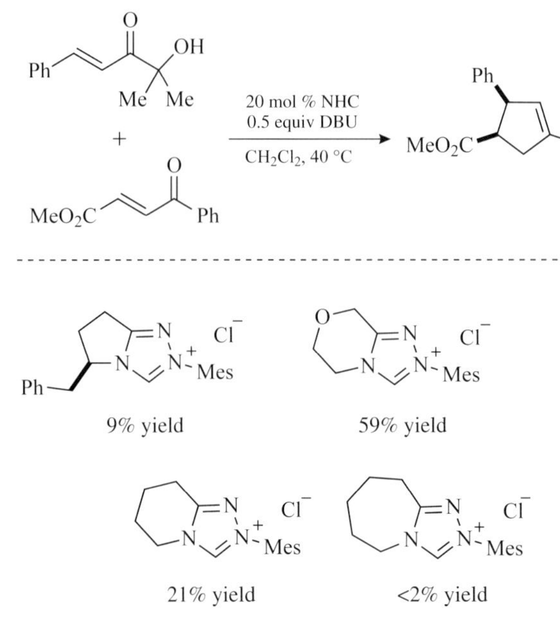

(37)

In 2013, Lupton and coworkers developed a series of *N-tert*-butyl triazolium salts to serve as precatalysts for the enantioselective [3 + 2] annulation of α,β-unsaturated acyl azoliums (eq 38).[59] The triazolium precatalysts here are relatively new in the sense that the N1-substituent is an aliphatic sigma donor as opposed to the more classical aryl substituent. These precatalysts serve well for this reaction, and ^{13}C-NMR analysis suggests that the electronic nature of these catalysts plays a significant role in this.

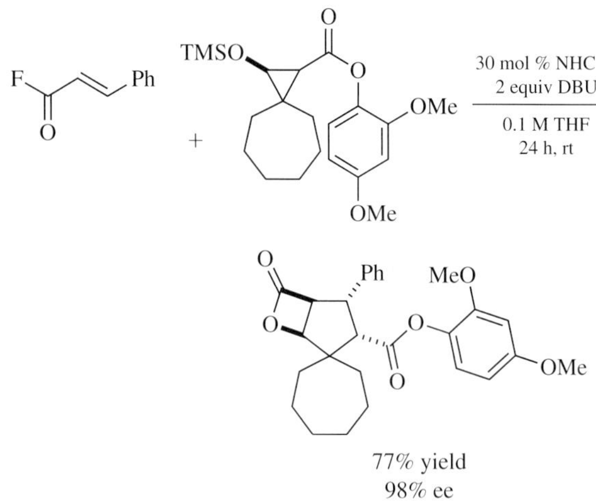

(38)

In 2013, Chi and coworkers[60] reported an oxidative enolization of heteroaryl carbaldehydes, forming a heterocyclic *ortho*-quinodimethane as an intermediate (eq 39). This then undergoes a formal [4 + 2] cyclization with activated ketones to form the quaternary carbon center containing polycyclic lactone.

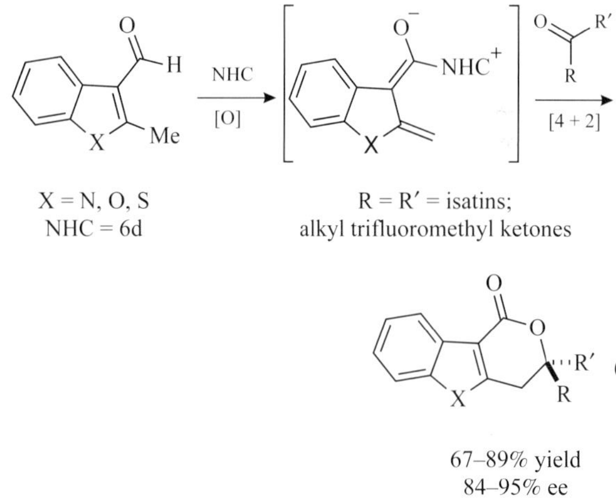

(39)

The first NHC-mediated synthesis of dihydroquinolones was achieved via a formal [4 + 2] cyclization between carboxylic acids and 2-substituted Boc-protected anilines (eq 40).[61] Both substrates are activated within the reaction medium, where the acid is converted to an acyl imidazole that then reacts with the triazolilydene catalyst. An aza-*o*-quinonemethide is also formed in situ, where the 2-substituent on the aryl features a benzylic leaving group. At this point the substrates react to furnish the enantioenriched dihydroquinolone.

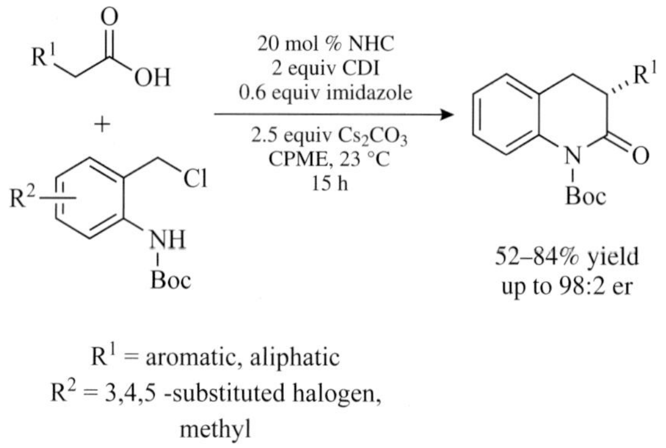

(40)

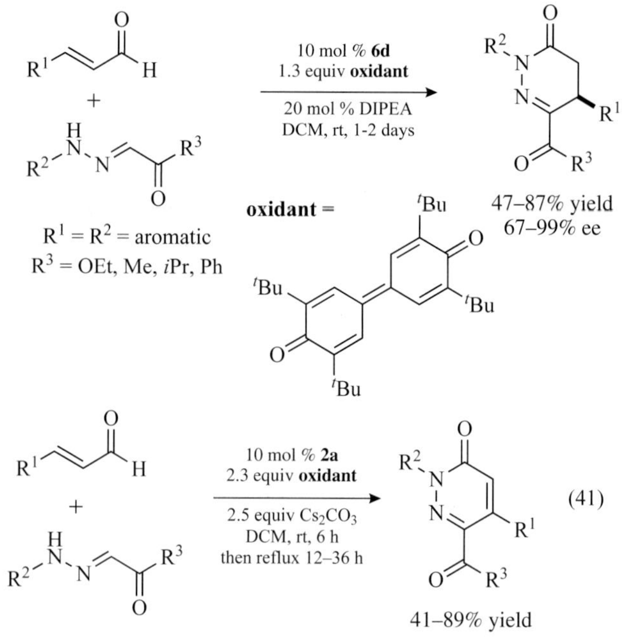

(41)

A high yielding and stereoselective method for the synthesis of chiral 4,5-dihydropyridazin-3-one derivatives has been described by Cheng and coworkers.[62] This reaction proceeds via an NHC-catalyzed oxidative annulation between α,β-unsaturated aldehydes and hydrazones (eq 41). The synthesis of pyridazin-3-one derivatives is also possible from the same reactants by varying the catalytic and reaction conditions.

Annulations. The NHC-catalyzed divergent annulation of enals with heterocyclic enones was developed in 2014 by Zhao and coworkers.[63] The diverging products are either ϵ-lactones or spiroheterocycles, the chemoselectivity being dictated by the backbone of the chiral catalyst (eq 42). The reaction proceeds with high diastereo- and enantioselectivities.

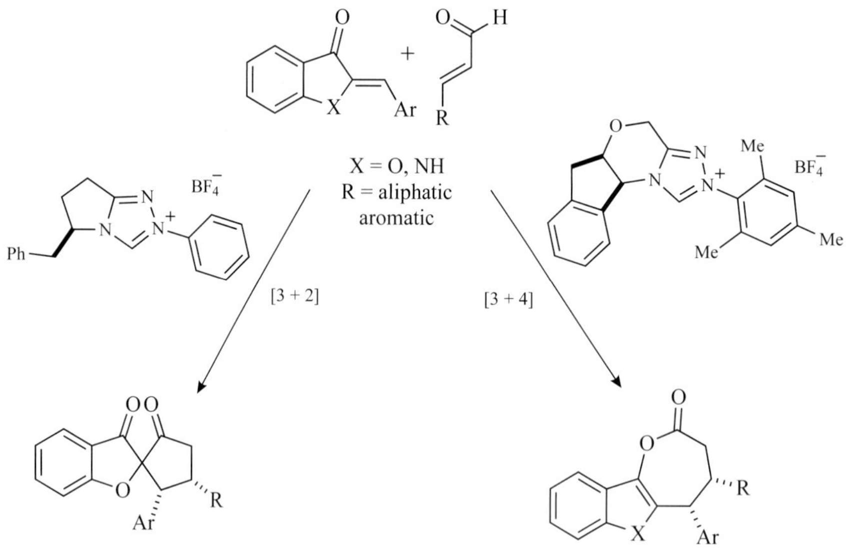

(42)

A merger of metal catalysis and organocatalysis was published by Chi and coworkers in 2014, involving the copper-catalyzed activation of alkynes to synthesize ketenimine intermediates, which can then undergo an NHC-catalyzed transformation for the stereoselective formation of spirocyclic products (eq 43).[64] In this case, CsF works well for this one-pot, multistep reaction.

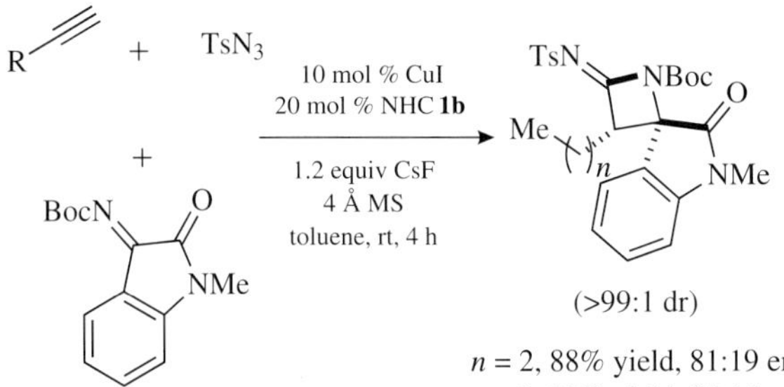

(>99:1 dr)

(43)

$n = 2$, 88% yield, 81:19 er
$n = 3$, 83% yield, 85:15 er
$n = 5$, 74% yield, 79:21 er

NHC-catalyzed cascades are powerful tools to access complex structures. Another such example is the synthesis of pyrrolo[3,2-*c*]quinolones from tosyl-protected *o*-amino aromatic aldimines and 2-bromoenals in a NHC-catalyzed reaction disclosed by Hui and coworkers.[65] The reaction manages to set three contiguous stereocenters in good yields and excellent stereoselectivities (eq 44).

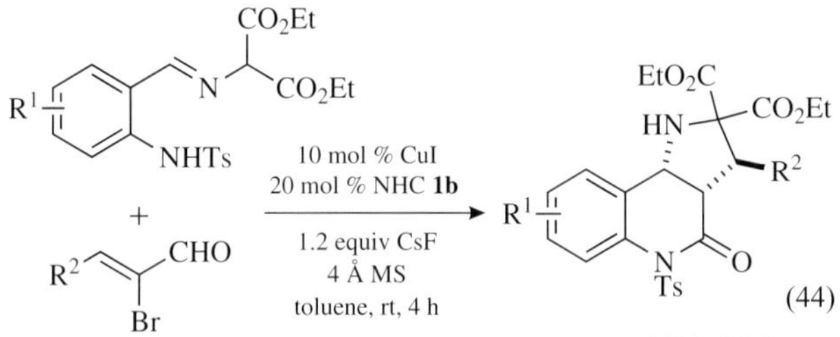

(44)

up to 94% yield
up to 99% ee
>25:1 dr

An NHC-catalyzed Mannich/lactamization domino reaction of α-chloroaldehydes and *N*-(benzothiazolyl)imines to synthesize benzothiazolo–pyrimidinones was disclosed in 2015 by the Enders group. This group uses catalyst **1d** to accomplish this in good yields and excellent stereoselectivities (eq 45),[66] a feat relatively unknown prior to the application of NHC catalysis toward the synthesis of this structural motif.

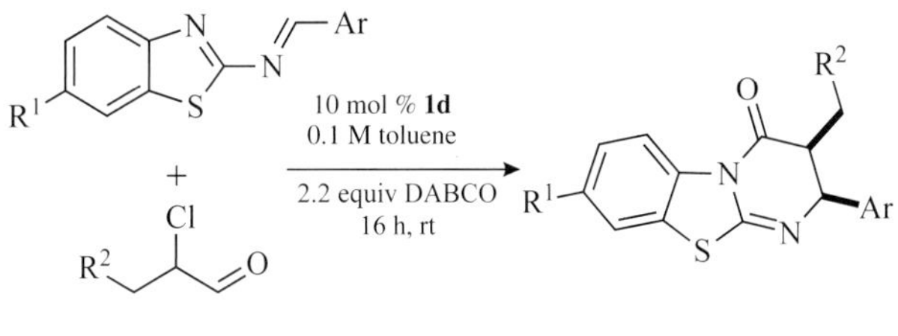

R^1 = H, methyl, methoxy, chloro, fluoro
R^2 = phenyl, *n*-propyl, substituted aryl

(45)

1d =

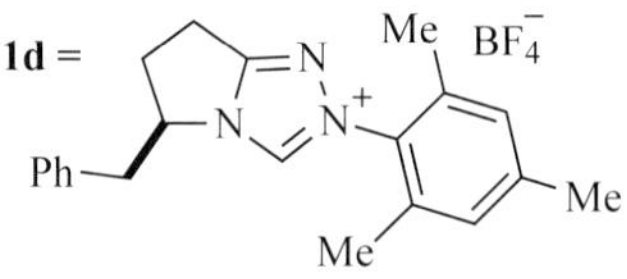

The divergent synthesis of indeno[2,1-*c*]pyran-1-one and indenocyclopentan-1-one derivatives with the reaction of 2-aroylvinylcinnamaldehydes and α,β-unsaturated imines was achieved by Cheng and coworkers in 2015 (eq 46).[67] This reaction is highly diastereoselective and furnishes either derivative when switching between NHCs **6b** and **6d**.

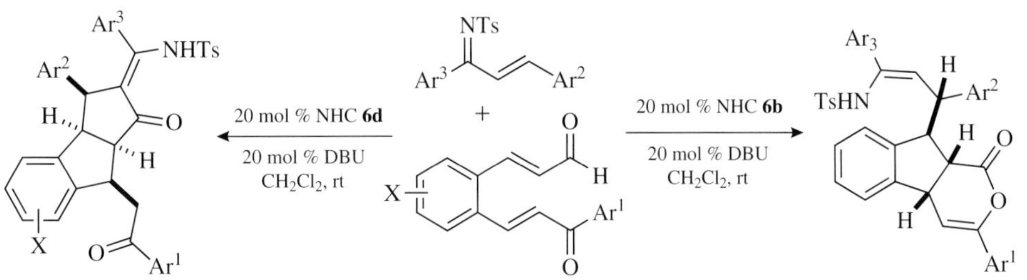

(46)

42–68% yield
>99% ee

37–58% yield
90–99% ee

An approach to the synthesis of functionalized indolo[2,3-α]quinolizidines via an annulative pathway from β-enamino esters and enals was described in 2015 by Du and Lu and coworkers (eq 47).[68] The importance of this reaction lends itself to the relative ubiquity of the synthesized core structure in natural products.

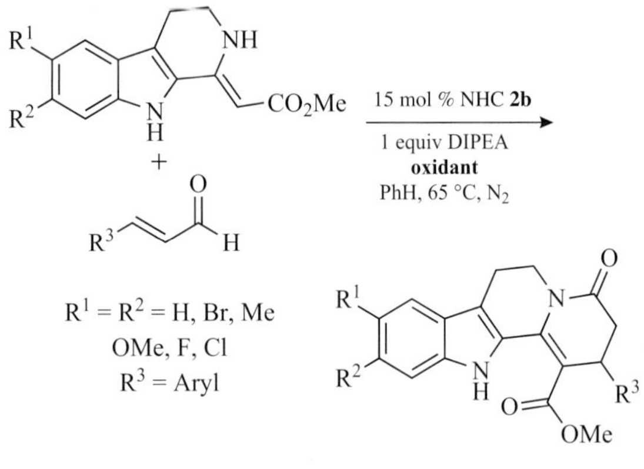

(47)

up to 81% yield

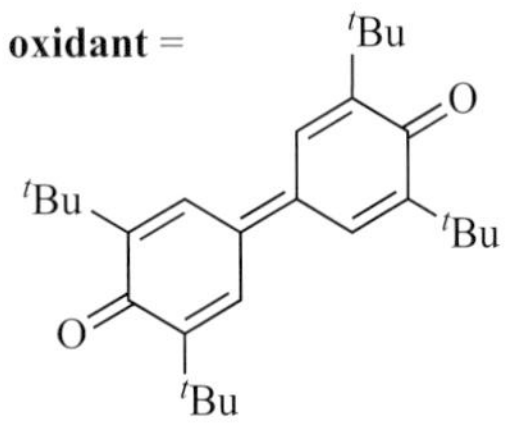

A formal [3 + 3] cyclization with 2-bromoenals and benzothiazolyl acetates furnishing dihydro-1*H*-benzothiazolopyridinones has been described by the Enders group in 2015.[69] The reaction proceeds in good yields and moderate enantioselectivities (eq 48), enriched by recrystallization. Catalyst **1b** can be used to give the opposite antipode.

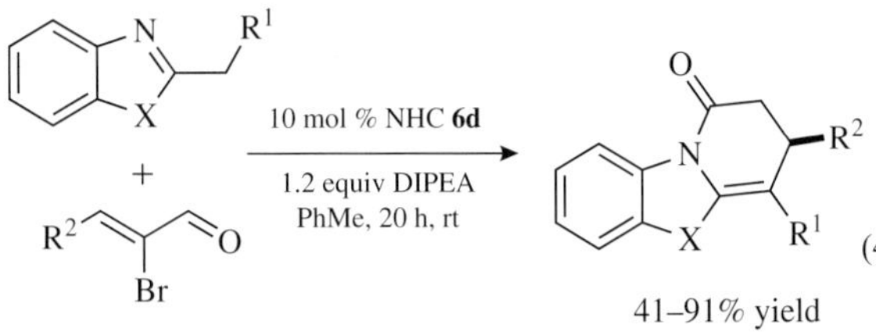

(48)

41–91% yield
32–73% ee

$R^1 = CO_2Et, CO_2Me$, CN
R^2 = aryl, furyl

Metal Complexes. Marinetti and coworkers published a new synthetic approach that allows access to axially chiral square-planar platinum(II) complexes that are based on carbene–diphosphine pairs.[70] The synthetic route (eq 49) is highlighted by its relative simplicity, where the previously disclosed[71] NHC–Pt(0) complex is first oxidized with iodine, upon which the labile diolefin is replaced with 1,2-bis(diphenylphosphino)ethane (dppe), giving the desired complex in 82% yield.

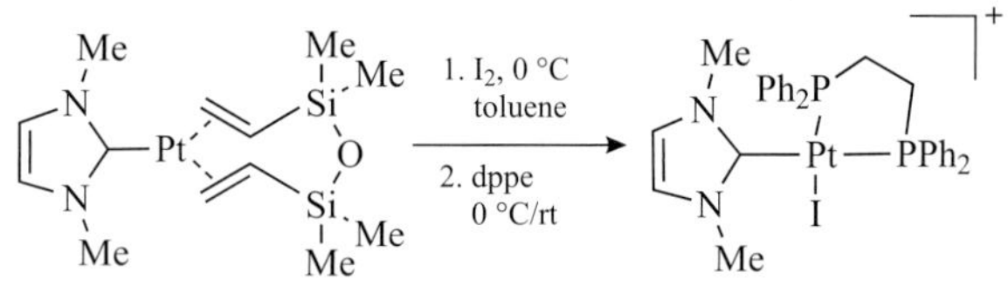

(49)

82% yield

Using this route, an axially chiral NHC–Pt complex using chiral NHC **1b** was synthesized in 68% yield. The purpose of this was to investigate the possibility of using these NHC–Pt complexes in enantioselective transformations. For this purpose, the cycloisomerization of a 1,6-enyne containing sulfonamide was explored (eq 50). The use of achiral metal-based complexes such as $PtCl_2$ has been explored,[72] but no enantioselective variants were known at the time. It was found that the NHC–Pt complex required activation by $AgBF_4$, but served well in promoting enantioselectivities up to 56% ee using a chiraphos variant of the complex. The aforementioned chiral NHC–Pt complex was also tried with this reaction, but did not promote enantioselectivities nearly as well as other complexes.

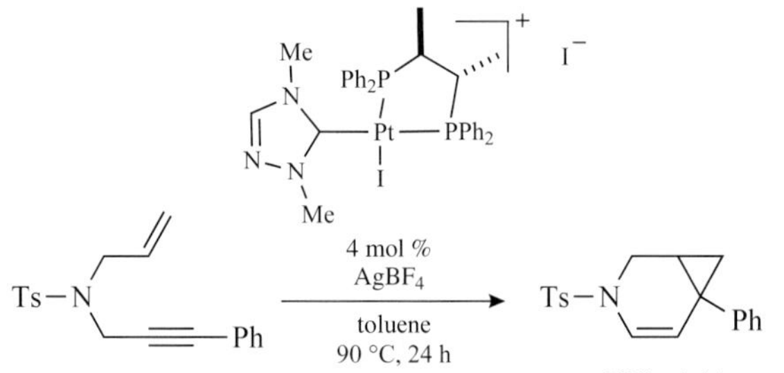

(50)

90% yield
56% ee
(<15% ee with NHC **1b**)

Miscellaneous. An NHC-catalyzed aza-Morita–Baylis–Hillman reaction between cyclic enones and imines was described (eq 51).[73] It was found that Ye's catalysts served better than a number of previously established imidazolium and triazolium NHC chiral precatalysts. Of particular interest was the impact of a free hydroxyl group on this reaction, leading to enhaned enantioselectivities. Ye proposed a hydrogen-bond interaction to create a stronger interaction between the NHC and the imine as the cause for the improved selectivities.

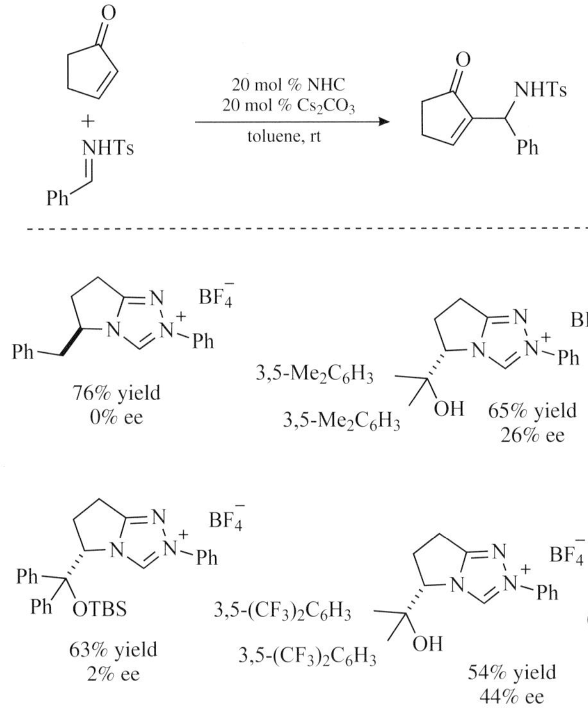

(51)

Smith and O'Donoghue disclosed the pK_a's of the C2-proton on various thiazolium, imidazolium, and triazolium precatalysts.[74] A thorough battery of experiments was performed in order to elucidate these numbers, including deuterium-exchange reactions, substituent effects on kinetic acidities toward the deuteroxide ion, the effect of counterion on kinetic acidities, and the potential for additional pathways for deuterium exchange. It was ultimately found that triazolium precatalysts have protons that are more acidic by about 5 pK_a units relative to their imidazolium and dihydroimidazolium counterparts (eq 52). It was also proven that the *N*-aryl substituent has the ability to impact the pK_a of the C2-proton.

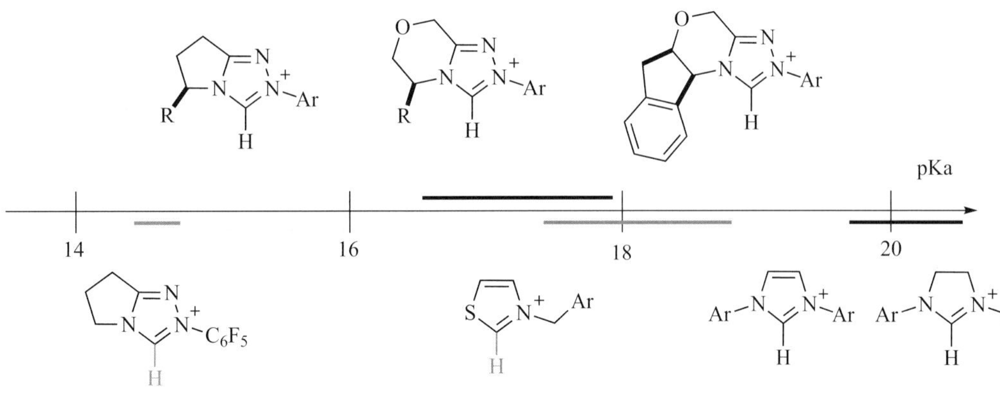

(52)

The ability to modify triazolium salts as NHC precatalysts is an important one, as these modifications and alterations to and around the catalyst may impact the reactivity and selectivity. Normally, modifications to the NHC scaffold come before the installation of the azolium core, as the C2 proton is relatively acidic and the position is prone to hydrolysis, thus rendering the catalyst useless for carbene-type catalysis. Modes of manipulation after installation of the azolium moiety thus become necessary in order to save steps. Waser and coworkers installed ureas and thioureas from azide moieties in the presence of the azolium core.[75] Bode and coworkers later used a similar strategy utilizing azides for the synthesis of pyrroles without disturbing the azolium center,[76] but these strategies were lacking in the sense that they required prior installation of the azide moiety. A methodology that circumvented this apparent limitation was described in 2014 by Rovis and coworkers. The method involves reduction of the azolium core down to its azoline form, followed by a given transformation around the periphery of the catalyst, and subsequent oxidation of the azoline back to its azolium state using a trityl salt (eq 53).[77] This strategy not only facilitated cross-coupling type strategies, but also allowed for a quick and reliable way to exchange counterions. In one such example, catalyst **1a** had its BF_4 counterion switched to the slightly more lipophilic PF_6 counterion in high yields.

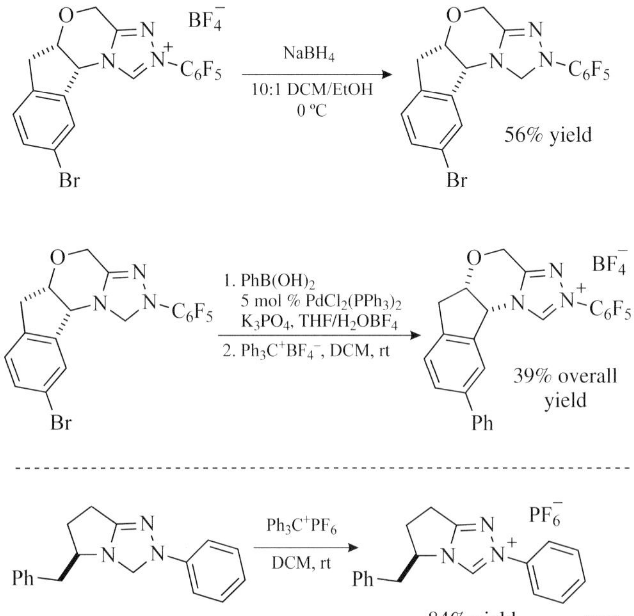

(53)

1. Christmann, M., *Angew. Chem. Int. Ed.* **2005**, *44*, 2632.
2. Zeitler, K., *Angew. Chem. Int. Ed.* **2005**, *44*, 7506.
3. Kerr, M. S.; Read de Alaniz, J.; Rovis, T., *J. Org. Chem.* **2005**, *70*, 5725.
4. Hassner, A. In *Comprehensive Organic Synthesis*; Trost, B. M.; Fleming, I., Eds.; **1991**; Vol. 1, 541.
5. Enders, D.; Breuer, K.; Teles, J. H., *Helv. Chim. Acta* **1996**, *79*, 1217.

6. Knight, R. L.; Leeper, F. J., *J. Chem. Soc., Perkin Trans.* **1998**, *1*, 1891.
7. Enders, D.; Niemeier, O.; Balensiefer, T., *Angew. Chem. Int. Ed.* **2006**, *45*, 1463.
8. Takikawa, H.; Hachisu, Y.; Bode, J. W.; Suzuki, K., *Angew. Chem. Int. Ed.* **2006**, *45*, 3492.
9. Stetter, H.; Kuhlmann, H., *Org. React.* **1991**, *40*, 407.
10. Enders, D.; Breuer, K.; Runsink, J.; Teles, J. H., *Helv. Chim. Acta* **1996**, *79*, 1899.
11. Kerr, M. S.; Read de Alaniz, J.; Rovis, T., *J. Am. Chem. Soc.* **2002**, *124*, 10298.
12. Kerr, M. S.; Rovis, T., *Synlett* **2003**, 1934.
13. Kerr, M. S.; Rovis, T., *J. Am. Chem. Soc.* **2004**, *126*, 10298.
14. Moore, J. L.; Kerr, M. S.; Rovis, T., *Tetrahedron* **2006**, *62*, In Press.
15. Read de Alaniz, J.; Rovis, T., *J. Am. Chem. Soc.* **2005**, *127*, 6284.
16. Reynolds, N. T.; Rovis, T., *Tetrahedron* **2005**, *61*, 6368.
17. Liu, Q.; Rovis, T., *J. Am. Chem. Soc.* **2006**, *128*, 2552.
18. Thomson, J. E.; Rix, K.; Smith, A. D., *Org. Lett.* **2006**, *8*, 3785.
19. Ruble, J. C.; Fu, G. C., *J. Am. Chem. Soc.* **1998**, *120*, 11532.
20. Shaw, S. A.; Aleman, P.; Christy, J.; Kampf, J. W.; Va, P.; Vedejs, E., *J. Am. Chem. Soc.* **2006**, *128*, 925.
21. Reynolds, N. T.; Read de Alaniz, J.; Rovis, T., *J. Am. Chem. Soc.* **2004**, *126*, 9518.
22. Chow, K. Y. K.; Bode, J. W., *J. Am. Chem. Soc.* **2004**, *126*, 8126.
23. Sohn, S. S.; Bode, J. W., *Org. Lett.* **2005**, *7*, 3873.
24. Chan, A.; Scheidt, K. A., *Org. Lett.* **2005**, *7*, 905.
25. Zeitler, K., *Org. Lett.* **2006**, *8*, 637.
26. Sohn, S. S.; Bode, J. W., *Angew. Chem. Int. Ed.* **2006**, *45*, 6021.
27. He, M.; Struble, J. R.; Bode, J. W., *J. Am. Chem. Soc.* **2006**, *128*, 8418.
28. Read de Alaniz, J.; Kerr, M. S.; Moore, J. L.; Rovis, T., *J. Org. Chem.* **2008**, *73*, 2033.
29. Liu, Q.; Perreault, S.; Rovis, T., *J. Am. Chem. Soc.* **2008**, *130*, 14066.
30. Liu, Q.; Rovis, T., *Org. Lett.* **2009**, *11*, 2856.
31. Evans, D. A.; Ennis, M. D.; Le, T.; Mandel, N.; Mandel, G., *J. Am. Chem. Soc.* **1984**, *106*, 1154.
32. Moore, J. L.; Silvestri, A. P.; Read de Alaniz, J.; DiRocco, D. A.; Rovis, T., *Org. Lett.* **2011**, *13*, 1742.
33. DiRocco, D. A.; Oberg, K. M.; Dalton, D. M.; Rovis, T., *J. Am. Chem. Soc.* **2009**, *131*, 10872.
34. DiRocco, D. A.; Rovis, T., *J. Am. Chem. Soc.* **2011**, *133*, 10402.
35. Sanchez-Larios, E.; Thai, K.; Bilodeau, F.; Gravel, M., *Org. Lett.* **2011**, *13*, 4942.
36. (a) Liu, Q.; Perreault, S.; Rovis, T., *J. Am. Chem. Soc.* **2008**, *130*, 14066. (b) Jousseaume, T.; Wurz, N. E.; Glorius, F., *Angew. Chem.* **2011**, *123*, 1446. (c) Fang, X.; Chen, X.; Lv, H.; Chi, Y. R., *Angew. Chem.* **2011**, *123*, 11986;

    Angew. Chem., Int. Ed. **2011**, *50*, 11782.
37. (a) Enders, D.; Han, J.; Henseler, A., *Chem. Commun.* **2008**, 3989. (b) DiRocco, D. A.; Oberg, K. M.; Dalton, D. M.; Rovis, T., *J. Am. Chem. Soc.* **2009**, *131*, 10872.
38. Wurz, N. E.; Daniliuc, C. G.; Glorius, F., *Chem. Eur. J.* **2012**, *18*, 16297.
39. Collett, C. J.; Massey, R. S.; Maguire, O. R.; Batsanov, A. S.; O'Donoghue, A. C.; Smith, A. D., *Chem. Sci.* **2013**, *4*, 1514.
40. Lathrop, S. P.; Rovis, T., *Chem. Sci.* **2013**, *4*, 1668.
41. Mahatthananchai, J.; Kaeobamrung, J.; Bode, J. W., *ACS Catal.* **2012**, *2*, 494.
42. Jiang, K.; Tiwari, B.; Chi, Y. R., *Org. Lett.* **2012**, *14*, 2382.
43. Chen, X.; Fang, X.; Chi, Y. R., *Chem. Sci.* **2013**, *4*, 2613.
44. Hao, L.; Du, Y.; Lv, H.; Chen, X.; Jiang, H.; Shao, Y.; Chi, Y. R., *Org. Lett.* **2012**, *14*, 2154.
45. Hao, L.; Chen, S.; Xu, J.; Tiwari, B.; Fu, Z.; Li, T.; Lim, J.; Chi, Y. R., *Org. Lett.* **2013**, *15*, 4956.
46. Lee, A.; Scheidt, K. A., *Chem. Commun.* **2015**, *51*, 3407.
47. Maki, B. E.; Chan, A.; Phillips, E. M.; Scheidt, K. A., *Org. Lett.* **2007**, *9*, 371.
48. Corey, E. J.; Gilman, N. W.; Ganem, B. E., *J. Am. Chem. Soc.* **1968**, *90*, 5618.
49. Mori, N.; Togo, H., *Tetrahedron* **2005**, *61*, 5915.
50. Pellissier, H., *Adv. Synth. Catal.* **2011**, *353*, 659.
51. For one stereocenter: (a) Dinh, P. M.; Howarth, J. A.; Hudnott, A. R.; Williams, J. M. J.; Harris, W., *Tetrahedron Lett.* **1996**, *37*, 7623. (b) Larsson, A. L. E.; Persson, B. A.; Backvall, J. E., *Angew. Chem., Int. Ed.* **1997**, *36*, 1211.

    For two stereocenters: (c) Noyori, R.; Ikeda, T.; Ohkuma, T.; Wildhalm, M.; Kitamura, M.; Takaya, H.; Akutagawa, S.; Sayo, N.; Saito, T., *J. Am. Chem. Soc.* **1989**, *111*, 9134. (d) Fernandez, R.; Ros, A.; Magriz, A.; Dietrich, H.; Lassaletta, J. M., *Tetrahedron* **2007**, *63*, 6755.

    For three stereocenters: (e) Steward, K. M.; Gentry, E. C.; Johnson, J. S., *J. Am. Chem. Soc.* **2012**, *134*, 7329.
52. (a) Cohen D. T.; Eichmann, C. C.; Phillips, E. M.; Zafersky, E. R.; Scheidt, K. A., *Angew. Chem., Int. Ed.* **2012**, *51*, 7309. (b) Johnston, R. C.; Cohen, D. T.; Eichman, C. C.; Scheidt, K. A.; Cheong, P. H.-Y., *Chem. Sci.* **2014**, *5*, 1974.
53. Goodman, C. G.; Walker, M. M.; Johnson, J. S., *J. Am. Chem. Soc.* **2015**, *137*, 122.
54. Zhang, Y. R.; He, L.; Wu, X.; Shao, P. L.; Ye, S., *Org. Lett.* **2008**, *10*, 277.
55. He, L.; Lv, H.; Zhang, Y. R.; Ye, S., *J. Org. Chem.* **2008**, *73*, 8101.
56. Chiang, P. C.; Rommel, M.; Bode, J. W., *J. Am. Chem. Soc.* **2009**, *131*, 8714.
57. Nair, V.; Vellalath, S.; Poonoth, M.; Suresh, E., *J. Am. Chem. Soc.* **2006**, *128*, 8736.
58. Chiang, P. C.; Kaeobamrung, J.; Bode, J. W., *J. Am. Chem. Soc.* **2007**, *129*, 3520.
59. Candish, L.; Forsyth, C. M.; Lupton, D. W., *Angew. Chem., Int. Ed.* **2013**, *52*, 9149.
60. Chen, X.; Yang, S.; Song, B. A.; Chi, Y. R., *Angew. Chem., Int. Ed.* **2013**, *52*, 11134.
61. Lee, A.; Younai, A.; Price, C. K.; Izquierdo, J.; Mishra, R. K.; Scheidt, K. A., *J. Am. Chem. Soc.* **2014**, *136*, 10589.
62. Mao, J. H.; Wang, Z. T.; Wang, Z. Y.; Cheng, Y., *J. Org. Chem.*, **2015**, *80*, 6250.
63. Wang, M.; Rong, Z. Q.; Zhao, Y., *Chem. Commun.* **2014**, *50*, 15309.
64. Namitharan, K.; Zhu, T.; Cheng, J.; Zheng, P.; Li, X.; Yang, S.; Song, B. A.; Chi, Y. R., *Nat. Commun.* **2014**, *5*, 3982.
65. Yang, Y. J.; Zhang, H. R.; Zhu, S. Y.; Zhu, P.; Hui, X. P., *Org. Lett.* **2014**, *16*, 5048.
66. Ni, Q.; Song, X.; Xiong, J.; Raabe, G.; Enders, D., *Chem. Commun.* **2015**, *51*, 1263.
67. Wang, Z. T.; Zhao, Y.; Wang, Z. Y.; Cheng, Y., *J. Org. Chem.* **2015**, *80*, 1727.
68. Hu, S.; Wang, B.; Zhang, Y.; Tang, W.; Fang, M.; Lu, T.; Du, D., *Org. Biomol. Chem.* **2015**, *13*, 4661.
69. Ni, Q.; Xiong, J.; Song, X.; Raabe, G.; Enders, D., *Synlett* **2015**, *26*, 1465.
70. Brissy, D.; Skander, M.; Retailleau, P.; Marinetti, A., *Organometallics* **2007**, *26*, 5782.
71. Buisine, O.; Berthon-Gelloz, G.; Briere, J. G.; Sterin, S.; Mignani, G.; Branlard, P.; Tinant, B.; Declercq, J. P.; Marko, I. E., *Chem. Commun.* **2005**, 3856.

72. (a) Cho, E. J.; Kim, M.; Lee, D., *Org. Lett.* **2006**, *8*, 5413. (b) Hercouet, A.; Berree, F.; Lin, C. H.; Toupet, L.; Carboni, B., *Org. Lett.* **2007**, *9*, 1717.
73. He, L.; Zhang, Y. R.; Huang, X. L.; Ye, S., *Synthesis* **2008**, *17*, 2825.
74. Massey, R. S.; Collett, C. J.; Lindsay, A. G.; Smith, A. D.; O'Donoghue, A. C., *J. Am. Chem. Soc.* **2012**, *134*, 20412.
75. Brand, J. P.; Osuna Siles, J. I.; Waser, J., *Synlett* **2010**, *6*, 881.
76. Zheng, P.; Gondo, C. A.; Bode, J. W., *Chem. Asian J.* **2011**, *6*, 614.
77. Ozboya, K.; Rovis, T., *Synlett* **2014**, *25*, 2665.

(*R*)-2,2′-Dihydroxy-3,3′diphenyl-1,1′-binaphthyl

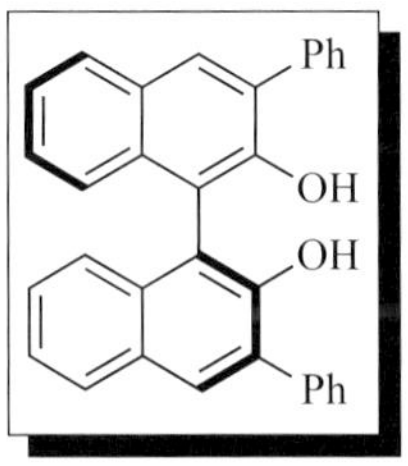

[102490-05-1] $C_32H_22O_2$ (MW 438.5159)
InChI = 1S/C32H22O2/c33-31-27(21-11-3-1-4-12-21)19-23-15-7-9-17-25(23)29(31)30-26-18-10-8-16-24(26)20-28(32(30)34)22-13-5-2-6-14-22/h1-20,33-34H
InChIKey = UXSXQZYUHXUTTI-UHFFFAOYSA-N

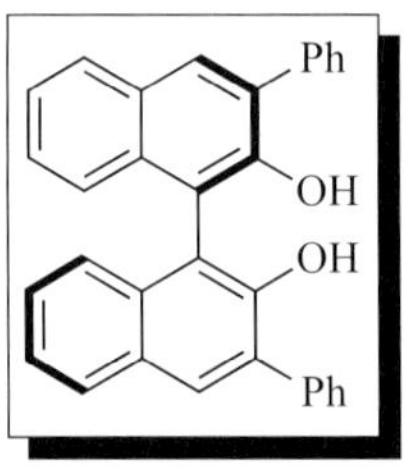

[684-93-4] $C_32H_22O_2$ (MW 438.5159)
InChI = 1S/C32H22O2/c33-31-27(21-11-3-1-4-12-21)19-23-15-7-9-17-25(23)29(31)30-26-18-10-8-16-24(26)20-28(32(30)34)22-13-5-2-6-14-22/h1-20,33-34H
InChIKey = UXSXQZYUHXUTTI-UHFFFAOYSA-N

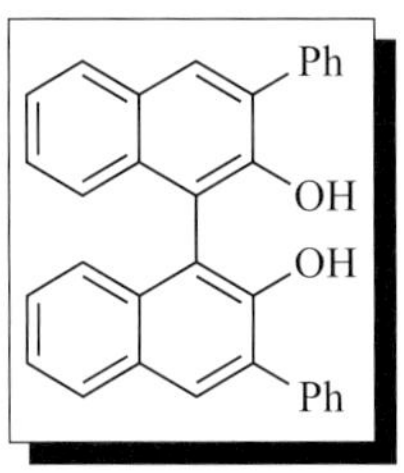

[75640-70-9] $C_32H_22O_2$ (MW 438.5159)
InChI = 1S/C32H22O2/c33-31-27(21-11-3-1-4-12-21)19-23-15-7-9-17-25(23)29(31)30-26-18-10-8-16-24(26)20-28(32(30)34)22-13-5-2-6-14-22/h1-20,33-34H
InChIKey = UXSXQZYUHXUTTI-UHFFFAOYSA-N

(chiral ligand and auxiliary)

Alternate Name: 3,3′-Ph_2-BINOL.
Physical Data: mp 197–199 °C; $[\alpha]_D^{21}$ +135 (c = 1, THF, *R* form).
Solubility: sol toluene, CH_2Cl_2.
Form Supplied in: white solid; widely available.
Preparative Methods: (*S*)-BINOL was methylated with iodomethane to give dimethoxybinaphthyl (*S*)-2 in 98% yield. Compound (*S*)-**2** was treated with *n*-BuLi in the presence of *N*,*N*,*N*′,*N*′-tetramethylenediamine (TMEDA) at room temperature to generate the dilithiated compound. This compound reacted with bromine at −78 °C to afford the dibromide (*S*)-**3** in 71% yield. Dichlorobis(triphenylphosphine) nickel(II)-catalyzed coupling of (*S*)-**3** with phenylmagnesium bromide provided the dimethylether derived from **4**. The resulting product was demethylated to give phenyl-substituted binaphthol in 69% yield (eq 1).[1]

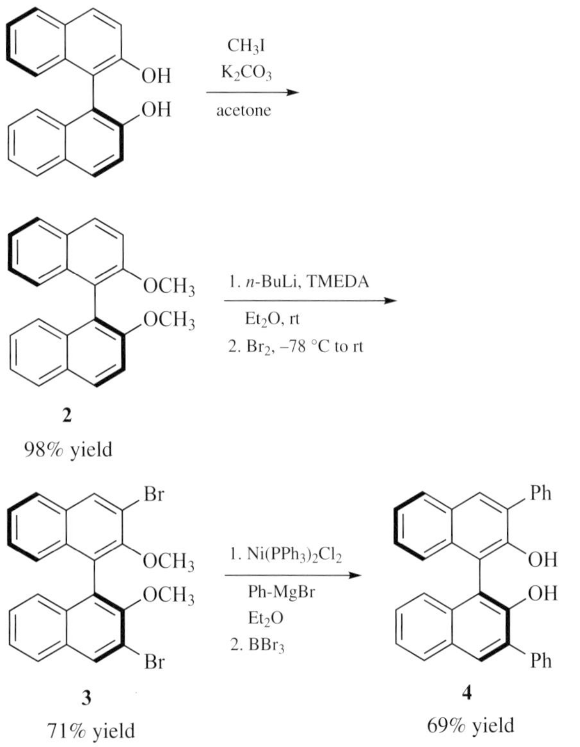

(1)

Iron-based salan complexes were determined to be efficient catalysts for the asymmetric aerobic oxidative coupling of phenyl-substituted 2-naphthol. This reaction can be carried out in air at 60 °C with high levels of enantioselectivity (eq 2).[2]

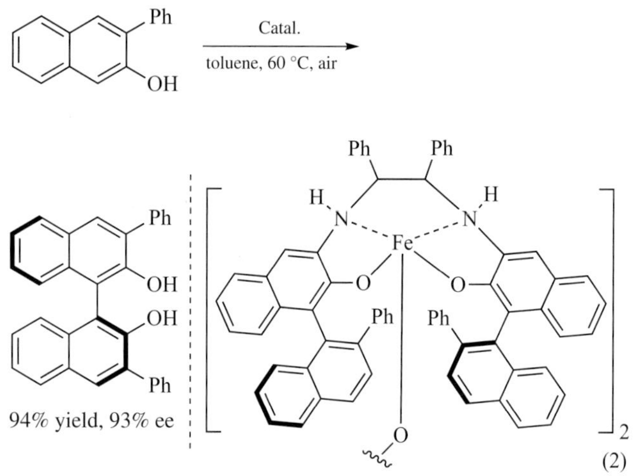

(2)

Handling, Storage, and Precautions: keep tightly closed, store in a cool dark place.

Asymmetric Diels–Alder Reaction. (*S*)-3,3′-Ph_2-BINOL was used for promoting an asymmetric Diels–Alder reaction between juglone and a diene to produce the adduct with excellent enantiomeric excesses (eq 3).[1]

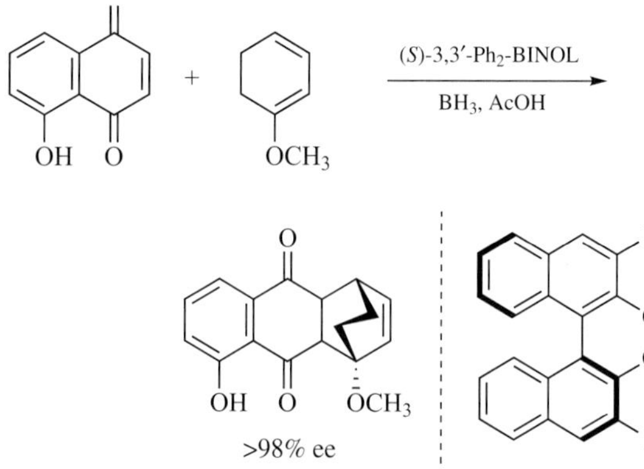

(3)

Asymmetric Vinylogous Michael Addition. An (*R*)-3,3′-Ph_2-BINOL-Mg-catalyzed asymmetric vinylogous Michael addition of α,β-unsaturated γ-butyrolactam was developed. High diastereoselectivities (up to 20:1) and excellent enantioselectivities were disclosed (eq 4).[3]

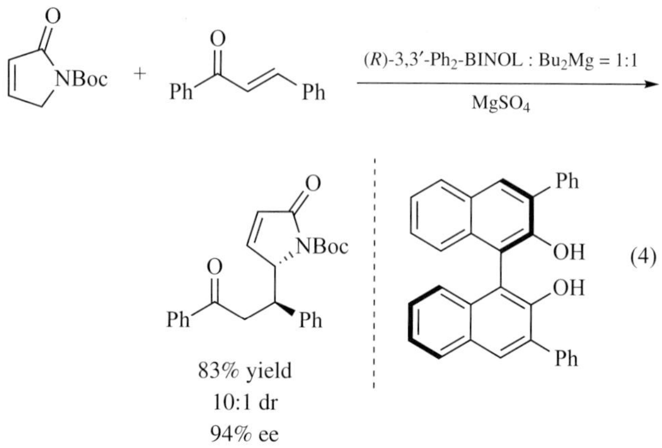

(4)

Ion-Paired Chiral Ligands for Asymmetric Palladium Catalysis. An achiral cationic ammonium–phosphine hybrid ligand paired with a chiral binaphtholate anion was developed. This ion-paired chiral ligand catalyzes a highly enantioselective allylic alkylation of α-nitrocarboxylates (eq 5).[4]

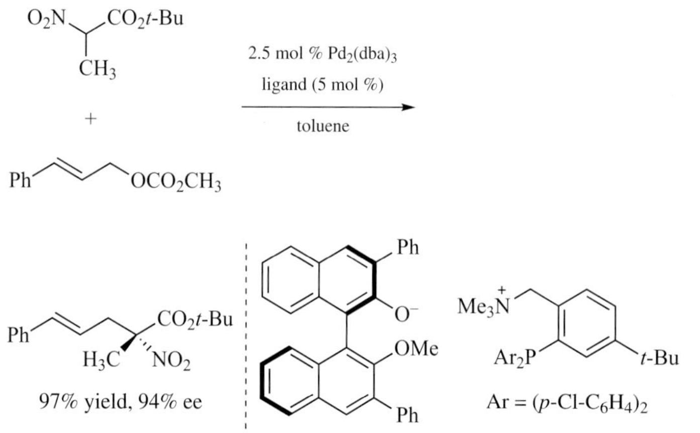

(5)

Enantioselective Aldol–Tishchenko Reaction.[5] Chiral lithium diphenylbinaphtholate is an effective catalyst for the enantioselective aldol–Tishchenko reaction. 1,3-Diol derivatives are produced with three contiguous chiral centers and high stereoselectivities (eq 6).

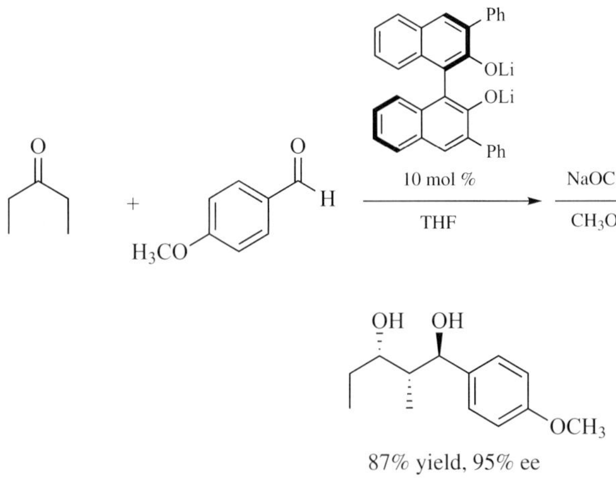

(6)

Asymmetric Allylboration of Acyl Imines. An enantioselective asymmetric allylboration of acyl imines using diisopropyl allylboronate and (*S*)-3,3′-Ph_2-BINOL was reported.[6] The corresponding products are obtained in good yields (75–94%) with high enantiomeric ratios (95:5–99.5:0.5) (eq 7).

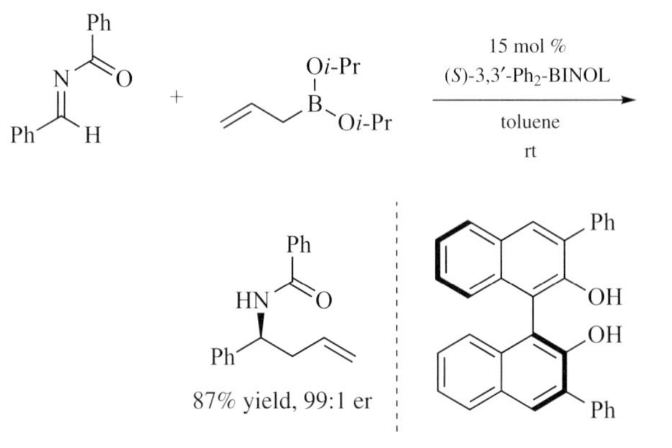

(7)

Asymmetric Conjugate Addition of Alkynylboronates to Enones. Alkynylboronates derived from the title compound add efficiently to enones in excellent yields and with high enantioselectivities (eq 8).[7]

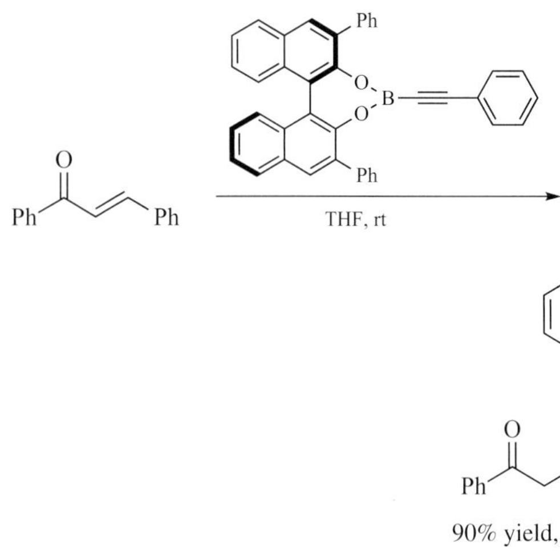

(8)

Asymmetric Conjugate Alkenylation of Enones Catalyzed by Chiral Diols. 3,3′-Ph_2-BINOL was also used in the addition of dimethyl alkenylboronates to α,β-unsaturated ketones. This reaction provides conjugate alkenylation products in good yields and with generally high (99:1 er) stereoselectivities (eq 9).[8]

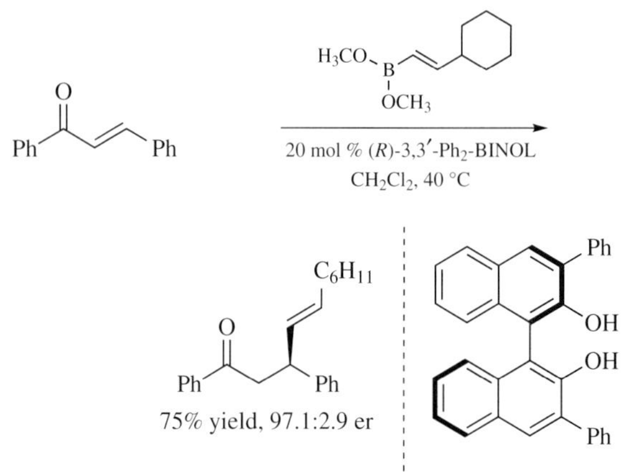

(9)

1. Wipf, P.; Jung, J. K., *J. Org. Chem.* **2000**, *65*, 6319.
2. Egami, H.; Katsuki T., *J. Am. Chem. Soc.* **2009**, *131*, 6082.
3. Lin, L.; Zhang, J. L.; Ma, X. J.; Fu, X.; Wang, R., *Org. Lett.* **2011**, *13*, 6410.
4. Ohmatsu, K.; Ito, M.; Kunieda, T.; Ooi, T., *Nat. Chem.* **2012**, *4*, 473.
5. Ichibakase, T.; Nakajima, M., *Org. Lett.* **2011**, *13*, 1579.
6. Lou, S.; Moquist, P. N.; Schaus, S. E., *J. Am. Chem. Soc.* **2007**, *129*, 15398.
7. Chong, J. M.; Shen, L. X.; Taylor, N. J., *J. Am. Chem. Soc.* **2000**, *122*, 1822.
8. Wu, T. R.; Chong, J. M., *J. Am. Chem. Soc.* **2007**, *129*, 4908.

Yi Luan & Scott E. Schaus
Boston University, Boston, MA, USA

1,3-Dimesityl-4,5-dihydroimidazol-2-ylidene

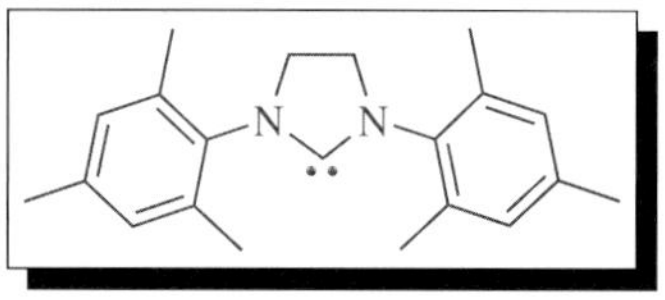

[173035-11-5], $C_{21}H_{26}N^2$ (MW 306.44)

InChI = 1/C21H26N2/c1-14-9-16(3)20(17(4)10-14)22-7-8-23(13-22)21-18(5)11-15(2)12-19(21)6/h9-12H,7-8H2,1-6H3

InChIKey = LSMWOQFDLBIYPM-UHFFFAOYAT

Alternate Name: SIMes.

Physical Data: mp = 107–109 °C (hexane).[1]

Solubility: soluble in THF, benzene, and toluene; insoluble in pentane and hexane; reacts with chlorinated solvents, alcohols, acetone, and acetonitrile.

Form Supplied in: white crystalline solid, crystal structure available.[1]

Analysis of Reagent Purity: ^{1}H and ^{13}C NMR spectroscopies.

Preparative Methods: 1,3-Dimesityl-4,5-dihydroimidazol-2-ylidene can be synthesized following several procedures; the most widely used being the deprotonation of the corresponding imidazolinium salt by a strong base (typically, sodium hydride, potassium hydride, potassium *tert*-butoxide, sodium *tert*-butoxide, DBU, etc). Formation of free N-heterocyclic carbenes can also be carried out by elimination of methanol or chloroform from the corresponding adducts at high temperature under vacuum.

Purification: recrystallization from THF/hexane or sublimation.

Handling, Storage, and Precautions: store under inert atmosphere (nitrogen or argon). For practical reasons, handling in a dry box is advised. Decomposes on exposure to water and is susceptible to oxidative decomposition.

Reactions with Halogenated Reagents, Alcohols, and Water. SIMes is highly sensitive to chlorinated solvents such as chloroform and carbon tetrachloride.[2] Its reactivity with dichloromethane is due to the acidic impurities this solvent usually contains. SIMes reacts rapidly with carbon tetrachloride at room temperature to reduce it and produce subsequently an olefin (eq 1). On the contrary, SIMes reacts with chloroform and results in the formation of an adduct that can be used to regenerate the free carbene by heating under vacuum (eq 1). Of note, similar reactions occur with pentafluorobenzene[3] and alcohols.[4]

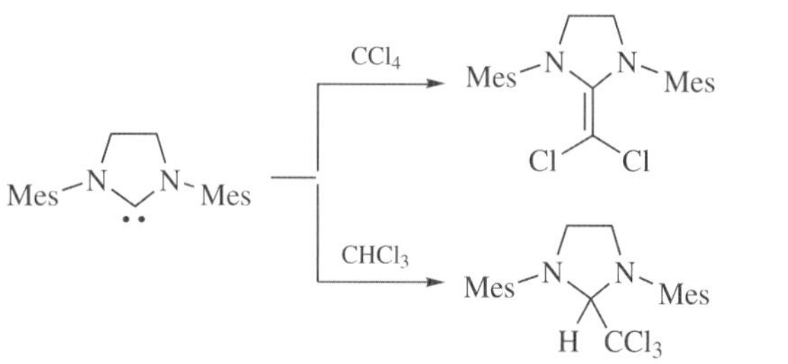

(1)

The reaction of SIMes with alkyl halides proceeds rapidly to alkylate the carbene center, and after hydrolysis, produces a stable 2-alkylimidazolinium salt (eq 2).[2] On the contrary, SIMes is highly sensitive to water; its hydrolysis into formamide proceeds through a polar, stepwise mechanism or as insertion of the carbene into H_2O.[5]

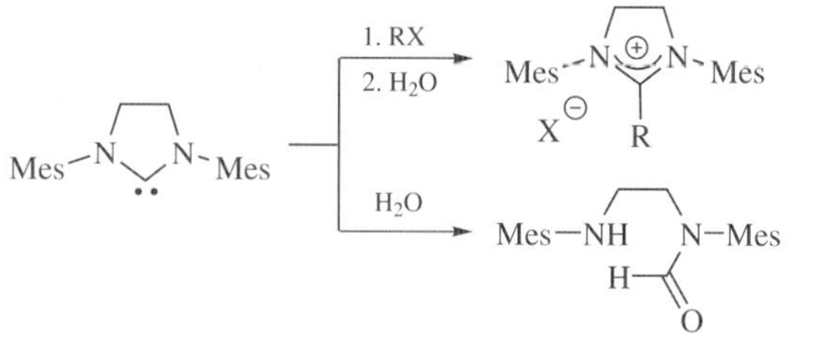

(2)

Organocatalysis. Nucleophilic carbenes are well recognized to promote efficiently organocatalysis and numerous applications have been developed based on their properties. These transformations include the benzoin condensation, Stetter, and transesterification reactions.[6] For those reactions, triazolydidenes, thiazolydidenes, and even imidazolylidene have been found more efficient than SIMes.[7] Nevertheless, SIMes was found particularly competent for the ring-opening polymerization of cyclic esters.[4]

Ruthenium Complexes. Commonly, SIMes- and more generally N-heterocyclic carbene (NHC)-containing late transition metal complexes can be synthesized following two different procedures: by ligand substitution or by simple cleavage of a dimeric metal complex.[8] The first route was used to yield the SIMes-containing Grubbs catalyst (eq 3)[9] that exhibits high activity in many metathesis transformations ranging from ring-closing metathesis to ring-opening metathesis polymerization, cross metathesis, and enyne metathesis.[10] Noteworthy, SIMes is the NHC which is most widely used for NHC-containing ruthenium-based metathesis catalysts.[11]

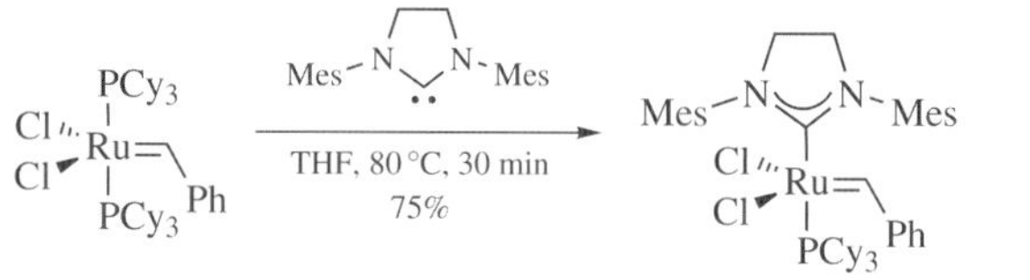

(3)

Few others, SIMes-containing ruthenium complexes have been synthesized;[12] the one reported by Fogg and Nolan has been used in catalysis.[13] The substitution of labile phosphine from $[RuHCl(CO)(PPh_3)_3]$ complex by in situ generated SIMes led cleanly to the product without activation of the mesityl group, in contrast with previous complexes (eq 4).[12] This complex performed efficiently the hydrogenation of cyclodecene as well as the isomerization of allylbenzene.

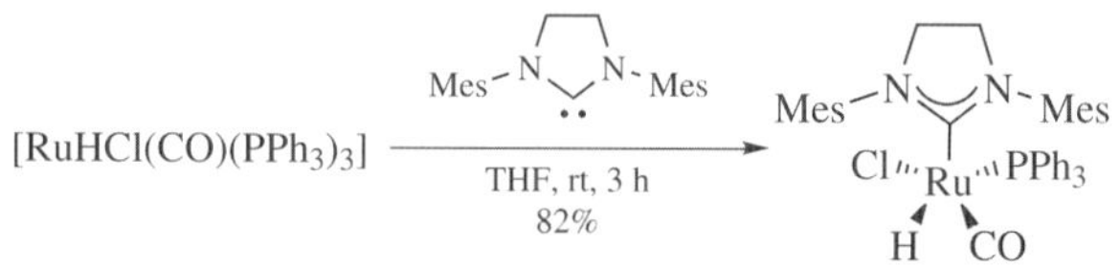

(4)

Palladium Complexes. Palladium complexes are well known to promote cross-coupling reactions and also telomerization and dehalogenation reactions. A catalytic system composed of imidazolium salt, palladium(II) dibenzylideneacetone, and potassium methoxide has been evaluated for the dehalogenation of aryl halide. SIMes, generated in situ, was found to be the best NHC ligands for this catalytic system, performing dehalogenation of both aryl chloride and bromide at 100 °C in 1 h.[14]

Copper Complexes. The proposal that NHC ligands, via strong σ-donation, could provide an accelerating effect in the copper-catalyzed conjugate addition reactions to enones, led Woodward and co-workers to use SIMes instead of phosphorus-containing donors in this process.[15] This work employed SIMes, generated in situ, for the conjugate addition of diethylzinc to various enones. Using 4.5 mol % of $Cu(OTf)^2$ and 5 mol % of SIMes, complete conversion was reached in 30 min at −10 °C whereas only 10% conversion was obtained under the same conditions without SIMes.

On the other hand, NHC-containing copper complexes have been found extremely efficient to perform the hydrosilylation of hindered and functionalized ketones.[16] The synthesis of well-defined copper(I) chloride complex bearing an SIMes ligand was achieved in good yield by the reaction of the corresponding imidazolium salt, CuCl, and sodium *tert*-butoxide (eq 5). SIMes is the best performing ligand for the hydrosilylation of heteroaromatic ketones. Complete conversions were obtained in a few hours at 80 °C using only 3 mol % of copper.

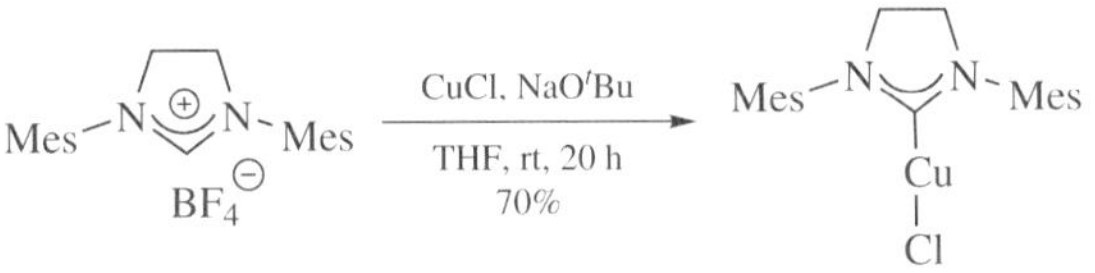

(5)

Iridium Complexes. Starting from the bispyridine complex $[Ir(COD)(py)_2]PF_6$, the substitution of one pyridine ligand by SIMes led to the monopyridine adduct $[Ir(COD)(SIMes)(py)]PF_6$ (eq 6).[17] This catalyst was investigated for the hydrogenation of simple, aliphatic olefins. The hydrogenation activity was found lower than other systems; nevertheless this complex is thermally significantly more stable. Less sterically demanding NHCs accomplish this transformation more efficiently.

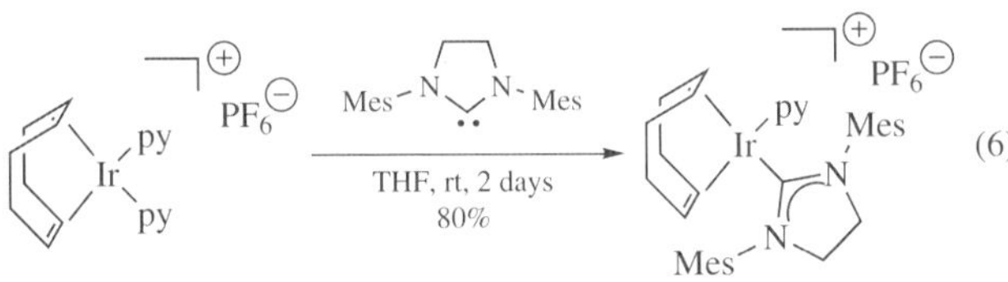

(6)

Nickel Complexes. In order to quantify the steric and electronic effects of NHC ligands by spectroscopic studies, Nolan reported the synthesis and the characterization of Ni(0)–NHC complexes.[18] This study, involving the substitution of carbon monoxide from $Ni(CO)_4$ by NHC ligands (eq 7), showed that SIMes exhibits a moderate steric hindrance and its electronic properties are comparable to its unsaturated analog.

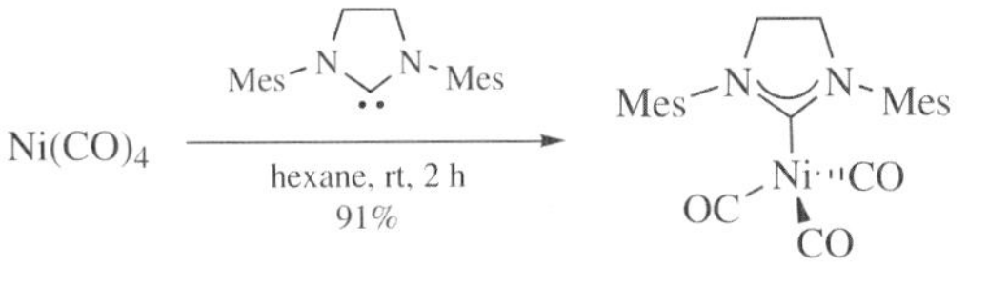

(7)

Nickel-based complexes bearing NHCs can mediate organic transformations such as the cycloaddition of CO_2 and diynes, cycloaddition of alkynes and isocyanates, or rearrangements of cyclopropylen-ynes,[19] these catalytic applications have led to the development of well-defined NHC-containing Ni-based catalysts.[20] The straightforward reaction of SIMes·HCl with nickelocene allowed the clean formation of CpNi(SIMes)Cl (eq 8).[20] The activity of this complex has been investigated in the dehalogenation of aryl halide and the aryl amination reactions. Whereas the dehalogenation of *p*-bromotoluene gave moderate results; low to good conversions depending of the substrates were obtained for C–N bond between aryl bromide or bromide and morpholine.

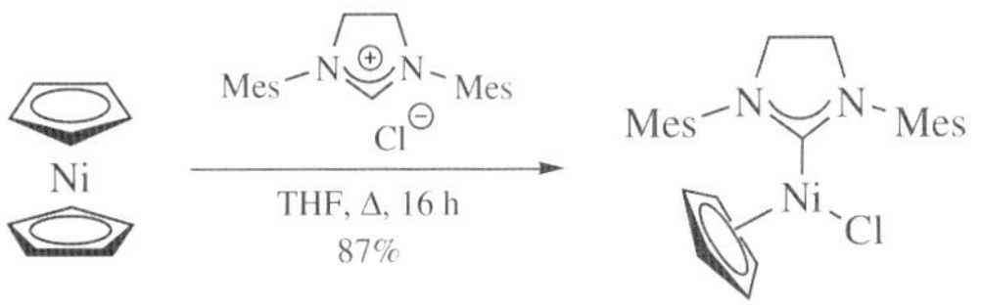

(8)

This Ni-based complex has also been used to perform the selective transfer of a single arylthio group in the catalytic hydrothiolation of alkynes.[21] The addition of thiophenol to 1-heptyne was carried out with a high regioselectivity and good isolated yield.

Related Reagents. 1,3-Dimesitylimidazol-2-ylidene; 1,3-Bis-(2,6-Diisopropylphenyl)imidazol-2-ylidene; 1,3-dicyclohexylimidazol-2-ylidene; 1,3-Di-*tert*-butylimidazol-2-ylidene; 1,3-Bis(1-adamantyl) imidazol-2-ylidene.

1. Arduengo, A. J. III.; Goerlich, J. R.; Marshall, W. J., *J. Am. Chem. Soc.* **1995**, *117*, 11027.
2. Arduengo, A. J. III; Davidson, F.; Dias, H. V. R.; Goerlich, J. R.; Khasnis, D.; Marshall, W. J.; Prakasha, T. R., *J. Am. Chem. Soc.* **1995**, *119*, 12742.
3. Nyce, G. W.; Csihony, S.; Waymouth, R. M.; Hedrick, J. L., *Chem. Eur. J.* **2004**, *10*, 4073.
4. Csihony, S.; Culkin, D. A.; Sentman, A. C.; Dove, A. P.; Waymouth, R. M.; Hedrick, J. L., *J. Am. Chem. Soc.* **2005**, *127*, 9079.
5. Denk, M. K.; Rodezno, J. M.; Gupta, S.; Lough, A. J., *J. Organomet. Chem.* **2001**, *617*, 242.
6. (a) Enders, D.; Gielen, H., *J. Organomet. Chem.* **2001**, *617*, 70. (b) Enders, D.; Balensiefer, *Acc. Chem. Res.* **2004**, *37*, 534. (c) Grasa, G. A.; Singh, R.; Nolan, S. P., *Synthesis* **2004**, 971. (d) Marion, M.; Díez-González, S.; Nolan, S. P., *Angew. Chem. Int. Ed.* **2007** *46*.
7. (a) Teles, J. H.; Melder, J. P.; Ebel, K.; Schnieder, R.; Gehrer, E.; Harder, W.; Enders, D.; Breuer, K.; Raabe, G., *Helv. Chim. Acta* **1996**, *79*, 61. (b) Grasa, G. A.; Güveli, T.; Singh, R.; Nolan, S. P., *J. Org. Chem.* **2003**, *68*, 2812. (c) Pesh, J.; Harms, K.; Bach, T., *Eur. J. Org. Chem.* **2004**, 2025.
8. (a) Nolan, S. P.; *N-Heterocyclic Carbenes in Synthesis*; Wiley-VCH: Weinheim, 2006. (b) Glorius, F.; *N-Heterocyclic Carbenes in Transition Metal Catalysis*, Springer-Verlag: Berlin Heidelberg, 2007.
9. Scholl, M.; Ding, S.; Lee, C. W.; Grubbs, R. H., *Org. Lett.* **1999**, *1*, 953.
10. Grubbs, R. H.; *Handbook of Metathesis*; Wiley-VCH: Weinheim, 2003.
11. (a) Garber, S. B.; Kingsbury, J. S.; Gray, B. L.; Hoveyda, A. H., *J. Am. Chem. Soc.* **2000**, *122*, 8168. (b) Gessler, S.; Randl, S.; Blechert, S., *Tetrahedron Lett.* **2000**, *41*, 9973. (c) Wakamatsu, H.; Blechert, S., *Angew. Chem. Int. Ed.* **2002**, *41*, 794. (d) Grela, K.; Harutyunyan, S.; Michrowska, A., *Angew. Chem. Int. Ed.* **2002**, *41*, 4038.

12. (a) Hillier, A. C.; Sommer, W. J.; Yong, B. S.; Petersen, J. L.; Cavallo, L.; Nolan, S. P., *Organometallics* **2003**, *22*, 4322. (b) Abdur-Rashid, K.; Fedorkiw, T.; Lough, A. J.; Morris, R. H., *Organometallics* **2004**, *23*, 86.

13. Dharmasena, U. L.; Foucault, H. M.; dos Santos, E. N.; Fogg, D. E.; Nolan, S. P., *Organometallics* **2005**, *24*, 1056.

14. Vicui, M. S.; Grasa, G. A.; Nolan, S. P., *Organometallics* **2001**, *20*, 3607.

15. Fraser, P. K.; Woodward, S., *Tetrahedron Lett.* **2001**, *42*, 2747.

16. (a) Kaur, H.; Zinn, F. K.; Stevens, E. D.; Nolan, S. P., *Organometallics* **2004**, *23*, 1157. (b) Díez-González, S.; Kaur, H.; Zinn, F. K.; Stevens, E. D.; Nolan, S. P., *J. Org. Chem.* **2005**, *70*, 4787.

17. Lee, H. M.; Jiang, T.; Stevens, E. D.; Nolan, S. P., *Organometallics* **2001**, *20*, 1255.

18. Dorta, R.; Stevens, E. D.; Scott, N. M.; Costabille, C.; Cavallo, L.; Hoff, C. D.; Nolan, S. P., *J. Am. Chem. Soc.* **2005**, *127*, 2485.

19. (a) Louie, J.; Gibby, J. E.; Farnworth, M. V.; Tekavec, T. N., *J. Am. Chem. Soc.* **2002**, *124*, 15188. (b) Duong, H. A.; Cross, M. J.; Louie, J., *J. Am. Chem. Soc.* **2004**, *126*, 11438. (c) Zuo, G.; Louie, J., *J. Am. Chem. Soc.* **2005**, *127*, 5798.

20. Kelly, III, R. A.; Scott, N. M.; Díez-González, S.; Stevens, E. D.; Nolan, S. P., *Organometallics* **2005**, *24*, 3442.

21. Malyshev, D. A.; Scott, N. M.; Marion, M.; Stevens, E. D.; Ananikov, V. P.; Beletskaya, I. P.; Nolan, S. P., *Organometallics* **2006**, *25*, 4462.

Hervé Clavier & Steven P. Nolan
Institute of Chemical Research of Catalonia (ICIQ), Tarragona, Spain

1,3-Dimesityl-1*H*-imidazolium Salts and 1,3-Dimesityl-imidazolylidene

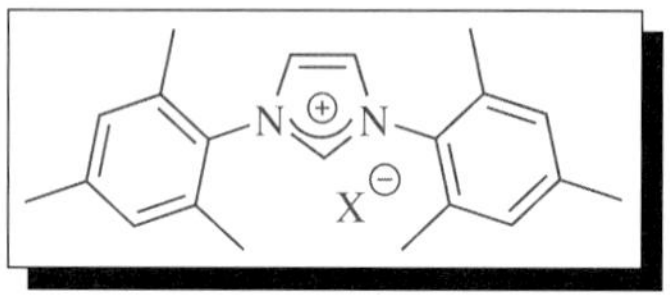

X= Cl
[141556-45-8] IMes·HCl (MW 340.89)
InChI = 1/C21H25N2.ClH/c1-14-9-16(3)20(17(4)10-14)22-7-8-23(13-22)21-18(5)11-15(2)12-19(21)6;/h7-13H,1-6H3;1H/q+1;/p-1;/p-1/fC21H25N2.Cl/h;1h/qm;-1
InChIKey = OTOSIXGMLYKKOW-LAEMHCMKCC
X = I
[632366-35-9] IMes·HI (MW 482.34)
InChI = 1/C21H25N2.HI/c1-14-9-16(3)20(17(4)10-14)22-7-8-23(13-22)21-18(5)11-15(2)12-19(21)6;/h7-13H,1-6H3;1H/q+1;/p-1
InChIKey = TZBPEYKJQGUOKE-WKDGGWEFCE
X = BF_4
[286014-53-7] IMes·HBF_4 (MW 392.24)
InChI = 1/C21H25N2.BF4/c1-14-9-16(3)20(17(4)10-14)22-7-8-23(13-22)21-18(5)11-15(2)12-19(21)6;2-1(3,4)5/h7-13H,1-6H3;/q+1;-1
InChIKey = VMNIOVNFFKZSAL-UHFFFAOYAJ
X = PF_6,
[160256-32-6] IMes·HPF_6 (MW 450.40)
InChI = 1/C21H25N2.F6P/c1-14-9-16(3)20(17(4)10-14)22-7-8-23(13-22)21-18(5)11-15(2)12-19(21)6;1-7(2,3,4,5)6/h7-13H,1-6H3;/q+1;-1
InChIKey = WSPQKEWKAZQLBF-UHFFFAOYAQ

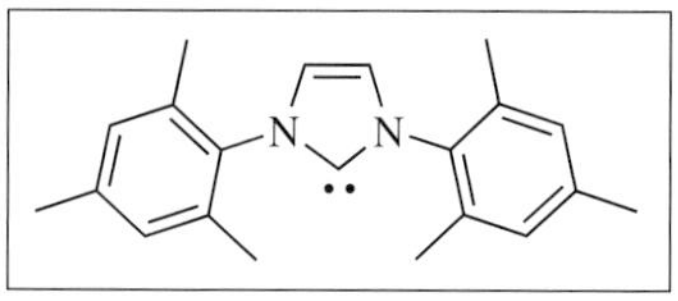

[160256-31-5] $C_{21}H_{24}N_2$ (MW 304.43)
InChI = 1/C21H24N2/c1-14-9-16(3)20(17(4)10-14)22-7-8-23(13-22)21-18(5)11-15(2)12-19(21)6/h7-12H,1-6H3
InChIKey = JCYWCSGERIELPG-UHFFFAOYSAJ

Physical Data: IMes·HCl: mp = 350–352 °C;[1] IMes: mp = 150–155 °C (hexane).[2]

Solubility: IMes·HCl: soluble in water, acetone and dichloromethane, insoluble in ethyl acetate, diethyl ether and hexane; IMes·HBF_4 and IMes·HPF_6: soluble in most organic solvents, insoluble in water; IMes, soluble in THF, benzene, and toluene, insoluble in pentane and hexane; reacts with chlorinated solvents, alcohols, acetone, and acetonitrile.

Form Supplied in: white crystalline solids, crystal structure of IMes available.[2]

Preparative Method: IMes·HBF_4 and IMes·HPF_6 are easily obtained from IMes·HCl by anionic metathesis using tetrafluoroboric acid and hexafluorophosphoric acid, respectively. Free IMes can be synthesized following several procedures; the most widely used being the deprotonation of the corresponding imidazolinium salt by a strong base (typically, sodium hydride, potassium hydride, potassium *tert*-butoxide, sodium *tert*-butoxide, DBU, etc). Formation of free *N*-heterocyclic carbenes can also be carried out by elimination of methanol or chloroform from the corresponding adducts at high temperature under vacuum.

Purification: imidazolium salts: recrystallization from dichloromethane/hexane or filtration over a plug of silica; IMes: recrystallization from THF/hexane or sublimation.

Handling, Storage, and Precautions: the use of imidazolium salts does not require special conditions. However, IMes should be stored preferably under inert atmosphere (nitrogen or argon). For practical reasons, IMes handling in a drybox is suggested. IMes decomposes on exposure to water and is susceptible to oxidative decomposition.

Reactions of IMes with Solvents. The choice of the solvent to perform reactions dealing with *N*-heterocyclic carbenes (NHC) is crucial since they react with numerous common organic solvents. IMes reacts with carbon tetrachloride to lead to the chlorination of the 4- and 5-positions of the imidazole ring (eq 1).[1] This unusual NHC is exceptionally stable and has been used to synthesize transition metal-based complexes active in catalysis.[3] Addition of alcohol to IMes yields cleanly its corresponding adduct (eq 1). The reverse reaction is possible by heating the adduct under vacuum.[4]

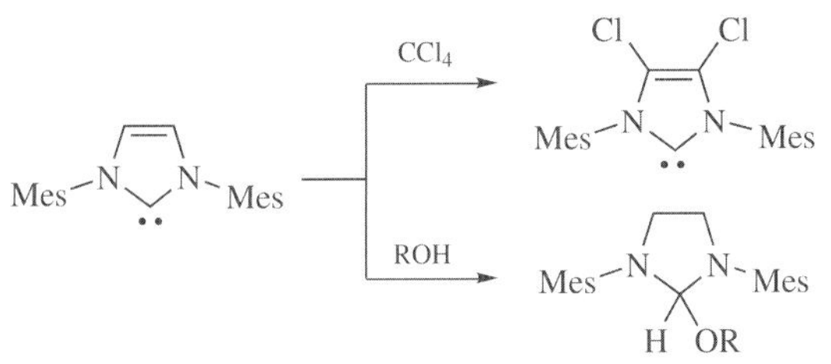

(1)

Of note is the fact that IMes is sensitive to water. Its hydrolysis into formamide proceeds through a polar, stepwise mechanism or as insertion of the carbene into H_2O (eq 2).[5] Nevertheless, IMes was found less sensitive than its unsaturated counterpart, and its hydrolysis occurs slowly.

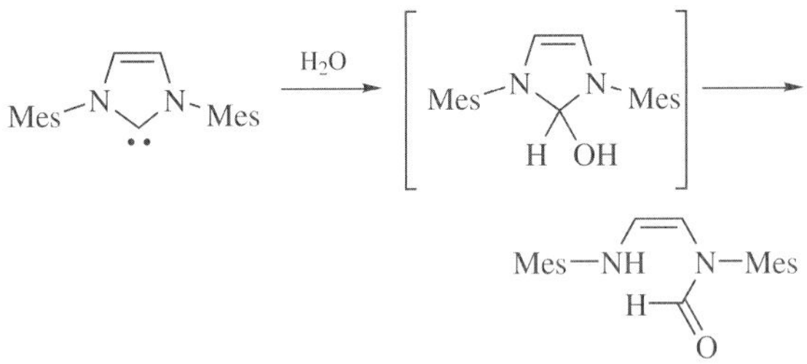

(2)

Organocatalysis. Generally known as excellent ligands for metal-based catalysis, there is also an increasing interest in the role of nucleophilic carbenes as organocatalysts.[6] Even if triazolydidenes and thiazolydidenes are the most frequently used NHCs, notably because they possess asymmetric versions. IMes was found to be very effective for a few reactions. Acylation of benzyl alcohol with vinyl acetate (eq 3)[7] and amidation of unactivated esters with amino alcohols (eq 4)[8] could be performed in excellent yield using a catalytic amount of IMes.

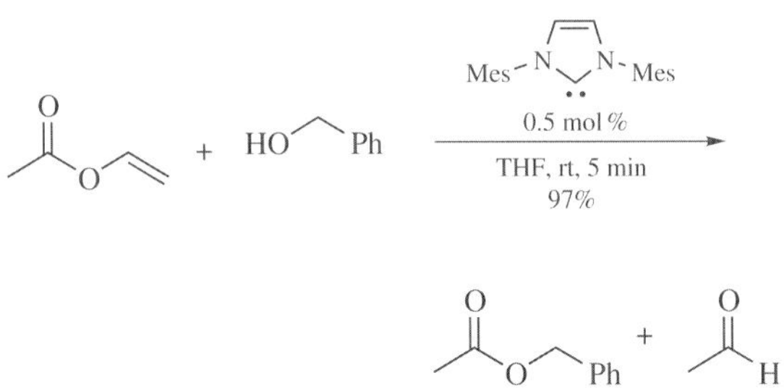

(3)

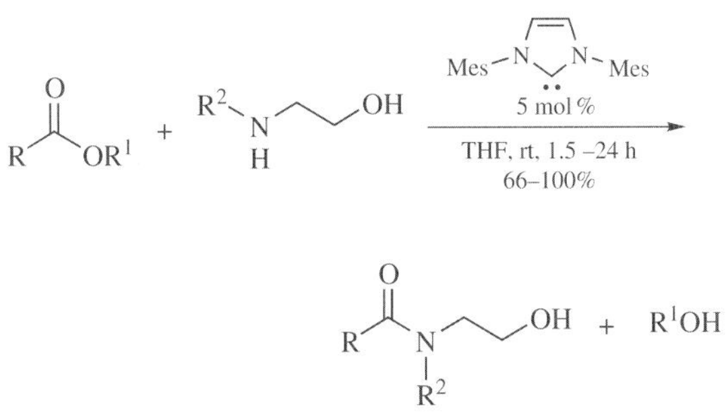

(4)

Moreover, IMes catalyzes efficiently the formation of γ–butyrolactones from conjugated enals and aromatic aldehydes (eq 5).[9] These studies highlighted the necessity for very specific steric bulk of the NHC. Thiazolium precursors, with only one side of the ring substituted, provided only trace amounts of benzoin condensation while imidazolium salts, properly sterically encumbered, resulted in selective formation of the desired lactones. The IMes, flanked with mesityl groups, was found to be an excellent catalyst, unlike 1,3-dicyclohexylimidazol-2-ylidene (ICy) (unreactive) or 1,3-bis(2,6-diisopropylphenyl)imidazol-2-ylidene (IPr) (nonselective), and produced *cis*-lactones in high yields and selectivity.

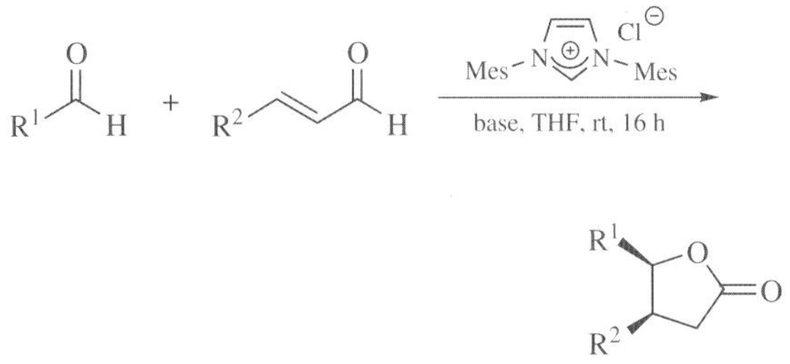

(5)

Ruthenium Complexes. Groundbreaking applications of NHC ligands to organometallic catalysis have led to the development of precatalyst complexes possessing higher activity, improved stability, or/and different reactivity than their phosphine analogues.[10] One of the foremost examples is the synthesis of NHC-containing ruthenium-based complexes (eq 6).[11] This second-generation metathesis catalyst is more stable and permits reactions requiring higher temperature such as the ring-closing metathesis of tetrasubstituted dienes.

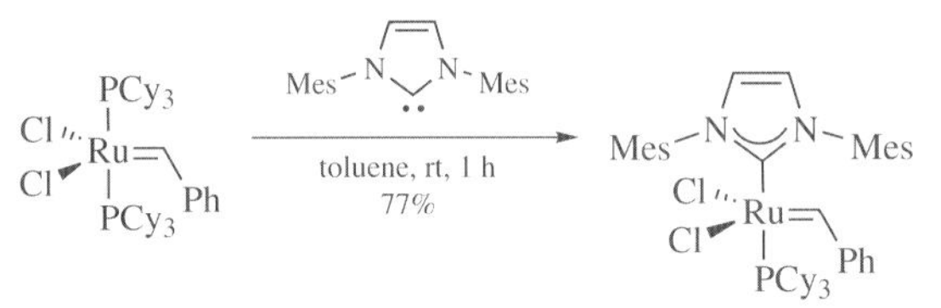

(6)

Several other ruthenium-based complexes have been developed and found very active. As an example, the $Ru(IMes)(PPh_3)(CO)H_2$ complex reported by Whittlesey, Macgregor et al. was found capable of either C–C or C–H bond activation as a function of the reaction conditions (eq 7).[12] Interestingly, direct insertion of ruthenium into an unstrained C–C bond of the IMes is thermodynamically more favorable than insertion into a C–H bond.

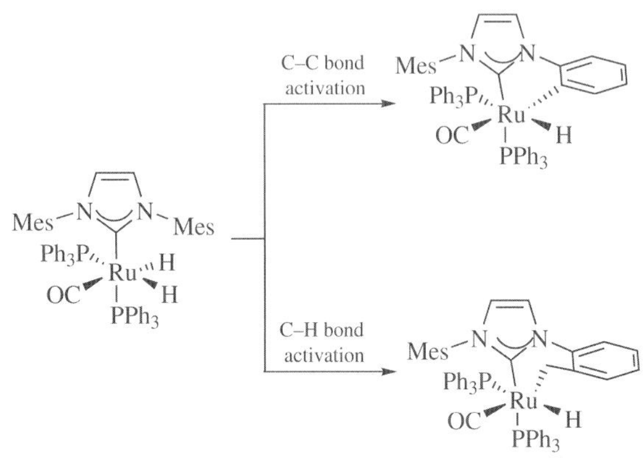

(7)

Palladium Complexes. Pd(NHC) species have been widely studied because they are known to promote palladium cross coupling reactions widely used in academic and industrial laboratories.[10] Following pioneering work by Herrmann[13] and

Nolan[14] who use, respectively, a bis-NHC complex and a system involving a Pd(II) source, imidazolium salt, and base to perform Suzuki-Miyaura and Kumada reactions, highly active and well-defined Pd(NHC) complexes have been synthesized. Beller reported the synthesis of monocarbenepalladium(0) complexes (eq 8) that allow efficient Heck and Suzuki reactions.[15] Monomeric Pd(NHC) complexes have been developed by Nolan et al. and tested in arylamination and α-arylation of ketones.[16] The catalytic activity of this air stable palladacycle was found to be good, nevertheless the 1,3-bis(2,6-diisopropylphenyl)imidazol-2-ylidene-containing analogue yielded better results.

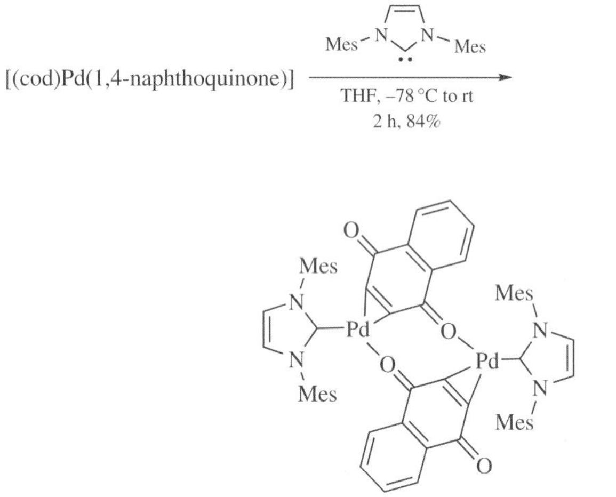

(8)

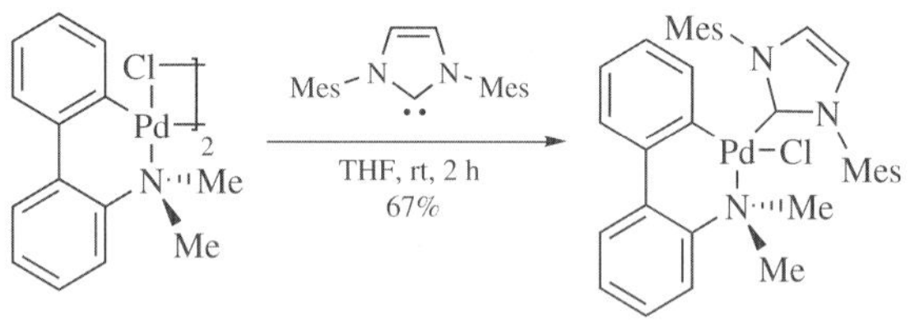

(9)

A new generation of well-defined, air-, and moisture-stable palladium-based complexes have been more recently synthesized and used in several catalytic reactions.[17] The synthetic approach of these precatalysts uses a simple palladium source $[(allyl)PdCl]_2$ and involves the simple cleavage of a dimeric metal complex (eq 10). This precatalyst was found extremely efficient in Suzuki-Miyaura cross-coupling reactions. The 1,3-bis(2,6-diisopropylphenyl)imidazol-2-ylidene-containing analog has been more frequently studied and used.

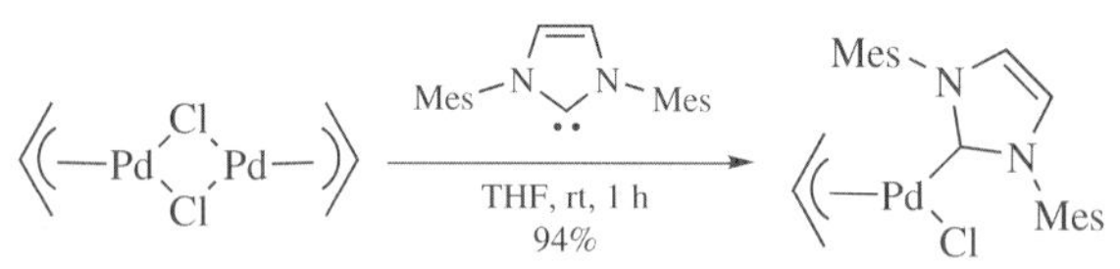

(10)

To illustrate the crucial role played by the reaction conditions used for complex generation, Lebel et al. reported the "unusual" coordination of the imidazole moiety.[18] Whereas in the presence of base, a bis-NHC palladium complex is formed with NHCs bound at the C2 position, the absence of base in the reaction mixture leads to one of the NHCs binding at the C5 position (eq 11). This unusual bonding was found crucial since only the "unusually" bound complex exhibited activity in palladium-catalyzed Heck reactions.

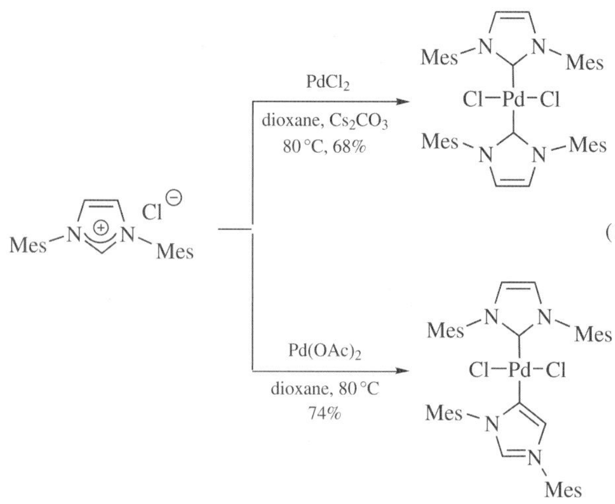

(11)

Silver and Gold Complexes. Formation of NHC-containing complexes could be achieved using the well-known transmetalation reaction from silver to other late-transition metals.[19] As an example, the straightforwardly synthesized IMes-silver(I) complex from IMes·HCl and silver oxide has subsequently been employed to form, in good yield, the corresponding Au(I) chloride compound (eq 12).[20]

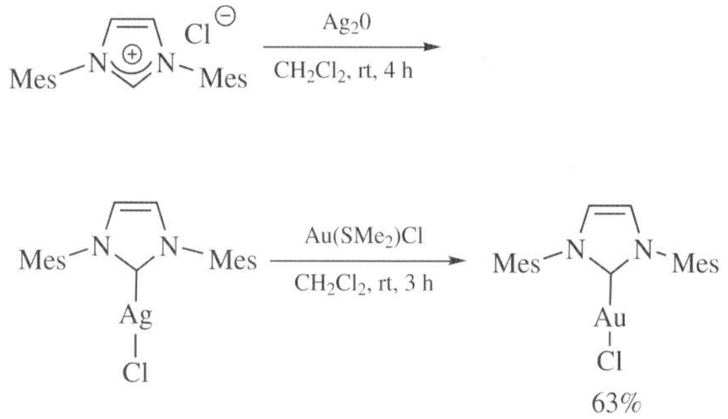

(12)

This gold-based complex has been used for tandem [3,3] rearrangement-intramolecular hydroarylation leading to the formation of substituted indenes and in the cycloisomerization of 1,5-enyes bearing a propargylic acetate leading to novel structural motifs.[21]

Related Reagents. 1,3-Dimesityl-1*H*-imidazolinium salts; 1,3-Bis(2,6-diisopropylphenyl)-1*H*-imidazolium salts, 1,3-Dimesityl-4,5-dihydroimidazol-2-ylidene; 1,3-Bis(2,6-diisopropylphenyl)Imidazol-2-ylidene.

1. Arduengo, A. J., III; Ktrafczyk, R.; Schmutzler, R.; Craig, H. A.; Goerlich, J. R., *Tetrahedron* **1999**, *55*, 14523.
2. (a) Arduengo, A.J., III; Davidson, F.; Dias, H. V. R., Goerlich, J. R.; Khasnis, D.; Marshall, W. J.; Prakasha., *J. Am. Chem. Soc.* **1992**, *114*, 5530. (b) Arduengo, A. J., III; Davidson, F.; Dias, H. V. R., Goerlich, J. R.; Khasnis, D.; Marshall, W. J.; Prakasha. *J. Am. Chem. Soc.* **1997**, *119*, 12742.
3. (a) Fürstner, A.; Ackermann, L.; Gabor, B.; Goddard, R.; Lehmann, C. W.; Mynott, R.; Stelzer, F.; Thiel, O. R.;. *Chem. Eur. J.* **2001**, *7*, 3236. (b) Harkal, S.; Jackstell, R.; Nierlich, F.; Ortmann, D.; Beller, M., *Org. Lett.* **2005**, *7*, 541.

4. Scholl, M.; Ding, S.; Lee, C. W.; Grubbs, R. H., *Org. Lett.* **1999**, *1*, 953.
5. Denk, M. K.; Rodezno, J. M.; Gupta, S.; Lough, A. J., *J. Organomet. Chem.* **2001**, *617*, 242.
6. (a) Enders, D.; Gielen, H., *J. Organomet. Chem.* **2001**, *617*, 70. (b) Enders, D.; Balensiefer, *Acc. Chem. Res.* **2004**, *37*, 534. (c) Grasa, G. A.; Singh, R.; Nolan, S. P., *Synthesis* **2004**, 971. (d) Marion, M.; Díez-González, S.; Nolan, S. P., *Angew. Chem. Int. Ed.* **2007**, *46*, doi:10.1016/j.ccr.2006.10.004.
7. Grasa, G. A.; Güveli, T.; Singh, R.; Nolan, S. P, *J. Org. Chem.* **2003**, *68*, 2812.
8. Sohn, S. S.; Rosen, E. L.; Bode, J. W., *J. Am. Chem. Soc.* **2004**, *126*, 14370.
9. (a) Burstein, C.; Glorius, F., *Angew. Chem. Int. Ed.* **2004**, *43*, 6205. (b) Movassaghi, M.; Schmidt, M. A., *Org. Lett.* **2005**, *7*, 2453.
10. (a) Nolan, S. P., *N-Heterocyclic Carbenes in Synthesis*, Wiley-VCH: Weinheim, **2006**. (b) Glorius, F. *N-Heterocyclic Carbenes in Transition Metal Catalysis*, Springer-Verlag: Berlin Heidelberg, **2007**.
11. Huang, J.; Stevens, E. D.; Nolan, S. P.; Petersen, J. L., *J. Am. Chem. Soc.* **1999**, *121*, 2674.
12. Jazzar, R. F. R.; Macgregor, S. A.; Mahon, M. F.; Richards, S. P.; Whittlesey, M. K., *J. Am. Chem. Soc.* **2002**, *124*, 4944.
13. Herrmann, W. A.; Reisinger, C.-P.; Spielgler, M., *J. Organomet. Chem.* **1998**, *557*, 93.
14. (a) Huang, J.; Nolan, S. P., *J. Am. Chem. Soc.* **1999**, *121*, 9889. (b) Zhang, C.; Huang, J.; Trudell, M. L.; Nolan, S. P., *J. Org. Chem.* **1999**, *64*, 3804. (c) Yang, C.; Nolan, S. P., *Organometallics* **2002**, *21*, 1020. (d) Grasa, G. A.; Viciu, M. S.; Huang, J.; Zhang, C.; Trudell, M. L.; Nolan, S. P., *Organometallics* **2002**, *21*, 2866.
15. (a) Selvakumar, K.; Zapf, A.; Spannenberg, A.; Beller, M., *Chem. Eur. J.* **2002**, *8*, 3901. (b) Selvakumar, K.; Zapf, A.; Beller, M., *Org. Lett.* **2002**, *4*, 3031.
16. (a) Viciu, M. S.; Kelly, R.A., III.; Stevens, E. D.; Naud, F.; Studer, M. S.; Nolan, S. P., *Org. Lett.* **2003**, *5*, 1479. (b) Navarro, O.; Kelly, R.A., III.; Nolan, S. P., *J. Am. Chem. Soc.* **2003**, *125*, 16194.
17. (a) Viciu, M. S.; Gremaneau, R. F.; Navarro-Fernandez, O.; Stevens, E. D.; Nolan, S. P., *Organometallics* **2002**, *21*, 5470. (b) Navarro, O.; Kaur, H.; Mahjoor, P.; Nolan, S. P., *J. Org. Chem.* **2004**, *64*, 3804.
18. Lebel, H.; Janes, M. K.; Charette, A. B.; Nolan, S. P., *J. Am. Chem. Soc.* **2004**, *126*, 5046.
19. Wang, H. M. J.; Lin, I. J. B., *Organometallics* **1998**, *17*, 972.
20. de Frémont Scott, N. M.; Stevens, E. D.; Nolan, S. P., *Organometallics* **2005**, *24*, 2411.
21. (a) Marion, N.; Díez-González, S.; de Frémont, P.; Noble, A. R.; Nolan, S. P., *Angew. Chem. Int. Ed.* **2006**, *45*, 3647. (b) Marion, N.; de Frémont, P.; Lemiére, G.; Stevens, E. D.; Fensterbank, M. M.; Nolan, S. P., *Chem. Commun.* **2006**, 2048.

Hervé Clavier & Steven P. Nolan
Institute of Chemical Research of Catalonia (ICIQ), Tarragona, Spain

4-Dimethylaminopyridine[1]

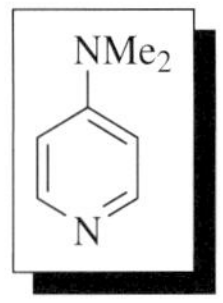

[1122-58-3] $C_7H_{10}N_2$ (MW 122.19)
InChI = 1S/C7H10N2/c1-9(2)7-3-5-8-6-4-7/h3-6H,1-2H3
InChIKey = VHYFNPMBLIVWCW-UHFFFAOYSA-N

(catalyst for acylation of alcohols or amines,[1–11] especially for acylations of tertiary or hindered alcohols or phenols[12] and for macrolactonizations;[13–15] catalyst for direct esterification of carboxylic acids and alcohols in the presence of dicyclohexylcarbodiimide (Steglich–Hassner esterification);[5] catalyst for silylation or tritylation of alcohols,[9,10] and for the Dakin–West reaction[20])

Alternate Name: DMAP.
Physical Data: colorless solid; mp 108–110 °C; pK_a 9.7.
Solubility: sol MeOH, $CHCl_3$, CH_2Cl_2, acetone, THF, pyridine, HOAc, EtOAc; partly sol cold hexane or water.
Form Supplied in: colorless solid; commercially available.
Preparative Method: prepared by heating 4-pyridone with HMPA at 220 °C, or from a number of 4-substituted (Cl, OPh, SO_3H, $OSiMe_3$) pyridines by heating with DMA.[2] Prepared commercially from the 4-pyridylpyridinium salt (obtained from pyridine and $SOCl_2$) by heating with DMF at 155 °C.[1,2]
Purification: can be recrystallized from EtOAc.
Handling, Storage, and Precautions: skin irritant; corrosive, toxic solid.

Original Commentary

Alfred Hassner
Bar-Ilan University, Ramat Gan, Israel

Acylation of Alcohols. Several 4-aminopyridines speed up esterification of hindered alcohols with acid anhydrides by as much as 10 000 fold; of these, DMAP is the most commonly used but 4-pyrrolidinopyridine (PPY) and 4-tetramethylguanidinopyridine are somewhat more effective.[11] DMAP is usually employed in 0.05–0.2 mol equiv amounts.

DMAP catalyzes the acetylation of hindered 11β- or 12α-hydroxy steroids. The alkynic tertiary alcohol acetal in eq 1 is acetylated at rt within 20 min in the presence of excess DMAP.[3]

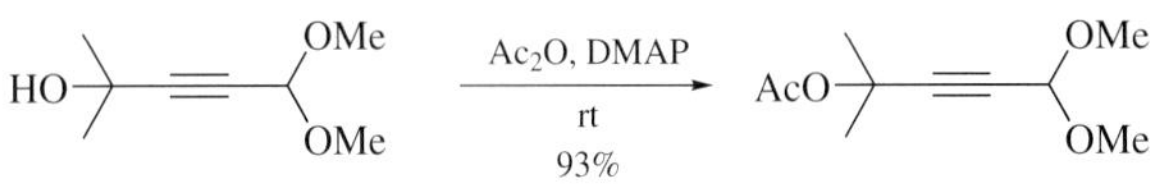

(1)

Esterifications mediated by 2-Chloro-1-methylpyridinium Iodide also benefit from the presence of DMAP.[22]

DMAP acts as an efficient acyl transfer agent, so that alcohols resistant to acetylation by Acetic Anhydride–Pyridine usually react well in the presence of DMAP.[4a] Sterically hindered phenols can be converted into salicylaldehydes via a benzofurandione prepared by DMAP catalysis (eq 2).[4b]

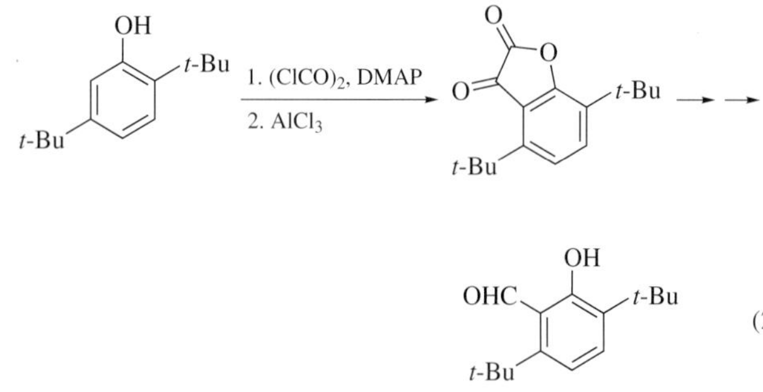
(2)

Direct Esterification of Alcohols and Carboxylic Acids. Instead of using acid anhydrides for the esterification of alcohols, it is possible to carry out the reaction in one pot at rt by employing a carboxylic acid, an alcohol, 1,3-Dicyclohexylcarbodiimide, and DMAP.[5,6] In this manner, *N*-protected amino acids and even hindered carboxylic acids can be directly esterified at rt using DCC and DMAP or 4-pyrrolidinopyridine (eq 3).[5]

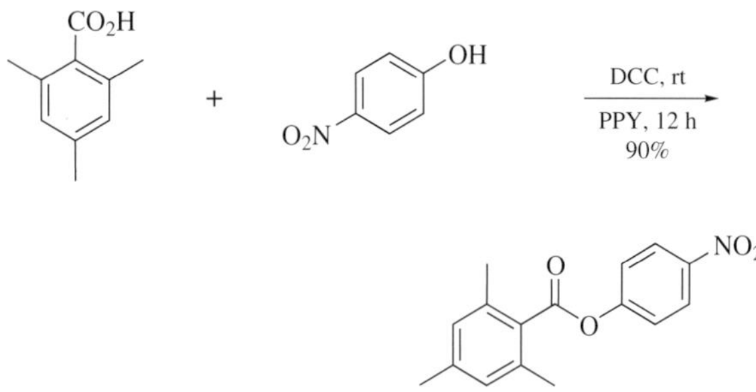
(3)

DCC–DMAP has been used in the synthesis of depsipeptides.[6b] Macrocyclic lactones have been prepared by cyclization of hydroxy carboxylic acids with DCC–DMAP. The presence of salts of DMAP,[13a] such as its trifluoroacetate, is beneficial in such cyclizations, as shown for the synthesis of a (9*S*)-dihydroerythronolide.[13b] Other macrolactonizations have been achieved using 2,4,6-Trichlorobenzoyl Chloride[1] and DMAP in Triethylamine at rt[14] or Di-2-pyridyl Carbonate (6 equiv) with 2 equiv of DMAP at 73 °C.[15]

Acylation of Amines. Acylation of amines is also faster in the presence of DMAP,[7] as is acylation of indoles,[8a] phosphorylation of amines or hydrazines,[2,8] and conversion of carboxylic acids into anilides by means of Phenyl Isocyanate.[1] β-Lactam formation from β-amino acids has been carried out with DCC–DMAP, but epimerization occurs.[8b]

Silylation, Tritylation, and Sulfinylation of Alcohols. Tritylation, including selective tritylation of a primary alcohol in the presence of a secondary one,[9] silylation of tertiary alcohols, selective silylation to *t*-butyldimethylsilyl ethers,[6] and sulfonylation or sulfinylation[10] of alcohols proceed more readily in the presence of DMAP. Silylation of β-hydroxy ketones with Chlorodiisopropylsilane in the presence of DMAP followed by treatment with a Lewis acid gives diols (eq 4).[16]

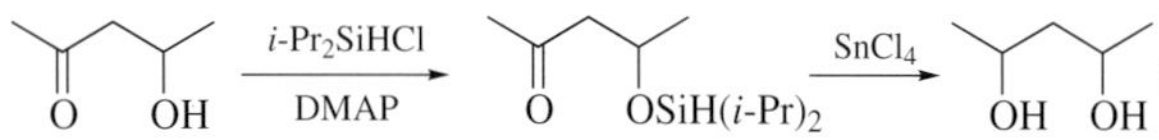
(4)

Miscellaneous Reactions. Alcohols, including tertiary ones, can be converted to their acetoacetates by reaction with Diketene in the presence of DMAP at rt.[17] Decarboxylation of β-keto esters has been carried out at pH 5–7 using 1 equiv of DMAP in refluxing wet toluene (eq 5).[18]

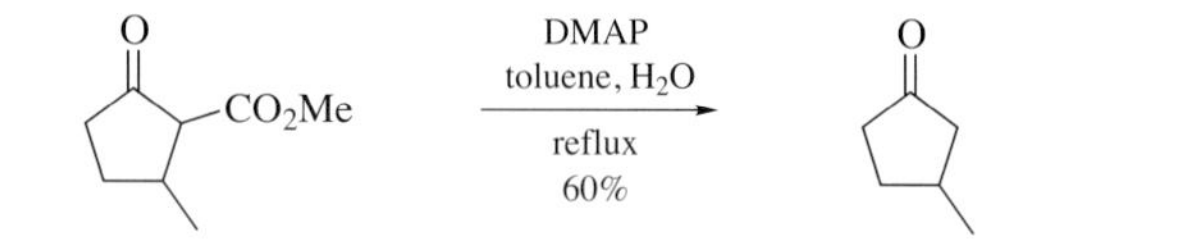
(5)

Elimination of water from a *t*-alcohol in a β-hydroxy aldehyde was carried out using an excess of Methanesulfonyl Chloride–DMAP–H_2O at 25 °C.[23]

Glycosidic or allylic alcohols (even when *s*-) can be converted in a 80–95% yield to alkyl chlorides by means of p-Toluenesulfonyl Chloride–DMAP. Simply primary alcohols react slower and secondary ones are converted to tosylates.[24]

Aldehydes and some ketones can be converted to enol acetates by heating in the presence of TEA, Ac_2O, and DMAP.[2] DMAP catalyzes condensation of malonic acid monoesters with unsaturated aldehydes at 60 °C to afford dienoic esters (eq 6).[19]

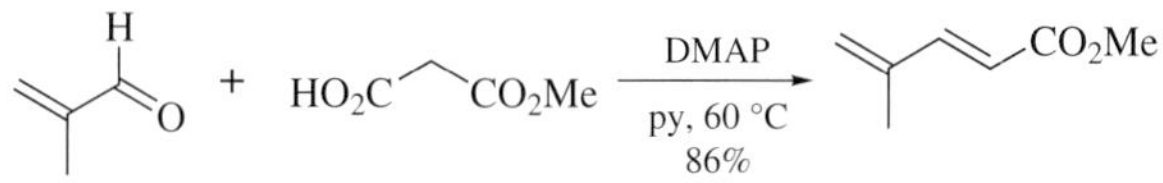
(6)

The conversion of α-amino acids into α-amino ketones by means of acid anhydrides (Dakin–West reaction)[20] also proceeds faster in the presence of DMAP (eq 7).

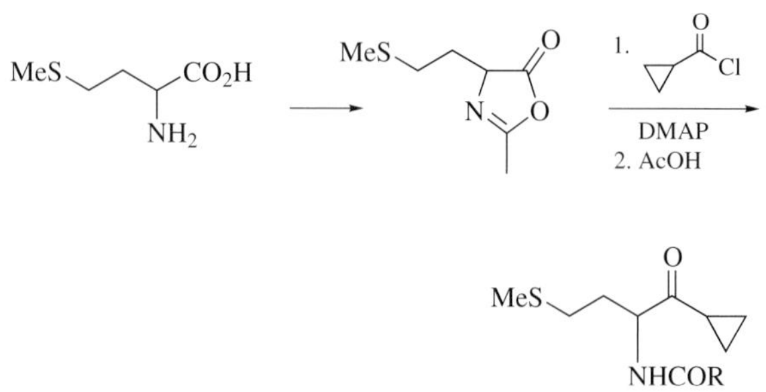
(7)

Ketoximes can be converted to nitrimines which react with Ac_2O–DMAP to provide alkynes (eq 8).[21]

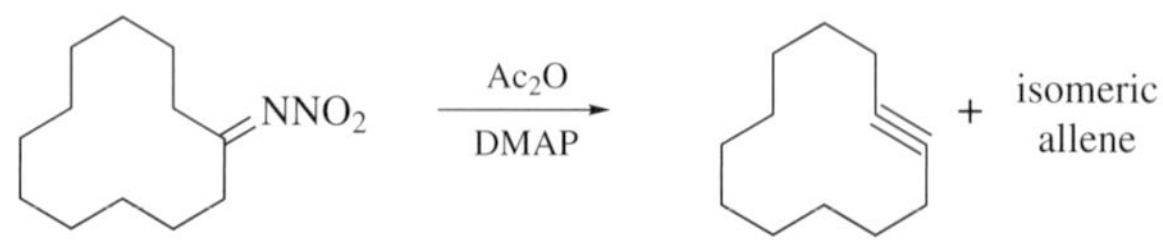
(8)

For the catalysis by DMAP of the *t*-butoxylcarbonylation of alcohols, amides, carbamates, NH-pyrroles, etc., see Di-t-butyl Dicarbonate.

First Update

Alison Hart and Julie A. Pigza
University of Southern Mississippi, Hattiesburg, MS, USA

Since the original article, DMAP has had further investigation into its thermal and magnetic properties,[25] as well as a thermal analysis and calorimetric study.[26] An in-depth review of the factors affecting the basicity and nucleophilicity of common nitrogen-containing catalysts, including DMAP and its derivatives, has also been presented.[27]

Esterification Reactions. While esterification of carboxylic acids and alcohols via a coupling reaction utilizing DCC and DMAP has already been covered, some new aspects will be discussed here. A DCC/DMAP esterification has been shown between heteroaromatic carboxylic acids and phenols to form functionalized (*E*)-2-arylvinyl bromides in excellent yields.[28] Other coupling agents such as diisopropylcarbodiimide (DIC) have been used with DMAP to form tyrosine esters bound to a solid support (Marshall resin).[29] The resulting amino esters were then subjected to a Pictet–Spengler reaction to form the tetrahydroisoquinoline ring. Another reagent, 1-ethyl-3-(3-dimethylaminopropyl)carbodiimide (EDC), can also be used in tangent with DMAP to perform a coupling reaction between a caffeic acid derivative and an alcohol as a precursor to the synthesis of zanamivir analogs, which display anti-inflammatory activity and are potent influenza virus neuraminidase inhibitors.[30] The combination of EDC and DMAP has also been shown to be effective for the synthesis of amino esters.[31] The benefit of EDC is that the byproduct urea is water soluble and can be removed in the aqueous workup, resulting in an easier purification by chromatography.

Other methods of esterification that do not utilize a carbodiimide coupling method have also been explored. For example, the stereoselective esterification of 20-(*S*)-camptothecin, a hindered 3° alcohol, with amino acid derivatives was accomplished with the use of scandium triflate ($Sc(OTf)_3$) in high yields while retaining the integrity of the amino acid stereocenter (eq 9).[32]

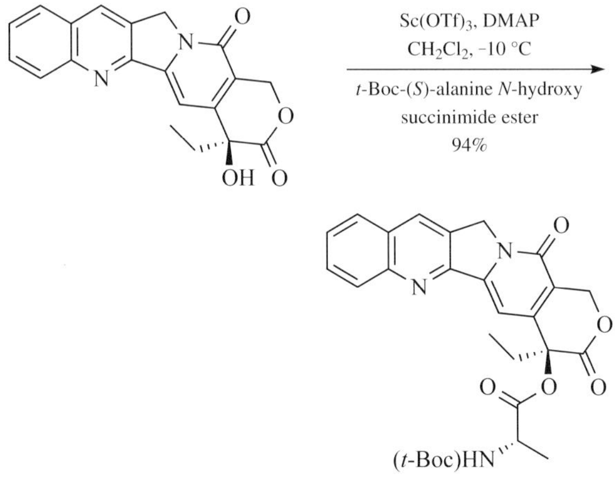

(9)

Another acylating system utilizing Et_3N, DMAP, and acetic anhydride was used in the formation of mixed acetals from an *in situ*-generated hemiacetal in good yields (eq 10).[33] Similar acetate-protected acetals have also been formed from the reductive acetylation of esters, using DIBAL to reduce the ester and form the hemiacetal which is then treated with acetic anhydride in the presence of DMAP, Et_3N, and pyridine.[34]

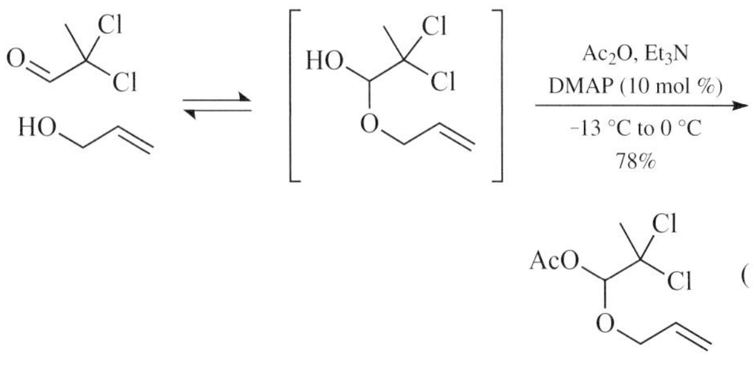

(10)

Amidation Reactions. Formation of an amide bond is also possible with the same coupling reagents as used for esterification reactions. Jeffamines ED®, diamines containing a polyether chain, were coupled with two molecules of chelidamic acid, a heteroaromatic acid, using a DCC/DMAP catalyst system to form oligo(ether-amide)s.[35] These were used as a water-soluble chelating agent with the ability to complex with many metal ions, including Fe^{3+}. DCC and DMAP were also used in the formation of phenylethynyl-terminated bis(amide) derivatives in an effort to form heat-resistant polymers.[36] Additionally, DCC/DMAP was used in the formation of a thiazolidinethione amide as part of the synthesis of iron(III)-specific fluorescent probes (eq 11).[37]

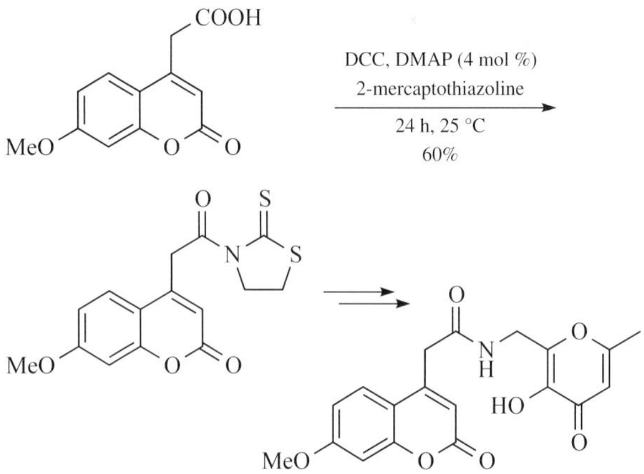

(11)

Trifluoroacetylation of primary anilines with ethyl trifluoroacetate can be accomplished in the presence of DMAP to afford good overall yields (49–98%).[38] In addition, a selectivity was observed for trifluoroacetylation of primary anilines in the presence of secondary anilines and even in the presence of hindered secondary amines.

The reaction between acid chlorides and a urea nitrogen has also been demonstrated. Hydrocinnamoyl chloride and an imidazolidinone were reacted in the presence of DMAP, leading to the formation of an imidazolidinone analog, which has attracted attention for its applicability as an HIV protease inhibitor and a

5-HT_3 receptor antagonist (eq 12).[39] In addition, this compound can also serve as a useful chiral auxiliary.

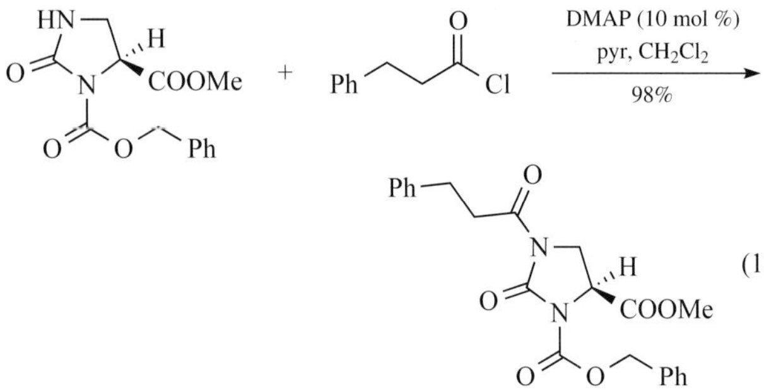

(12)

Acylation Reactions. Acylation reactions can also be accomplished using coupling reagents such as DCC/DMAP. For example, *C*-acylation of 1,3-diones with unactivated carboxylic acids can be accomplished with an excess of DMAP using a variety of carboxylic acids and cyclic 1,3-diones.[40] In a different example involving tetramic acids, either *C*- or *O*-acylation can be observed by changing the amount of DMAP. Substoichiometric amounts of DMAP lead to *O*-acylation of the ketone oxygen, while a slight excess of DMAP leads to the *C*-acylation product (eq 13).[41]

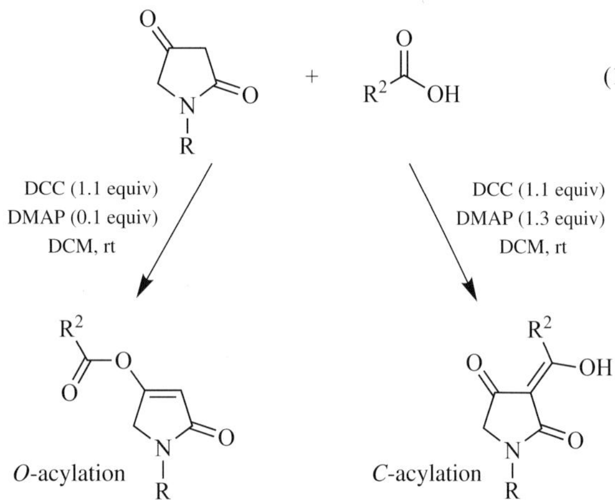

(13)

DMAP Catalysis to Form Key Substructures. DMAP has been used as a catalyst in the Morita–Baylis–Hillman (MBH) reaction. Conjugated nitroalkenes can be reacted with an amine base to form a zwitterionic species that can add to many types of ketones and aldehydes.[42] Remarkably, only DMAP and imidazole were able to effectively catalyze this reaction, likely due to their small size and ability to stabilize the intermediate. In another example, a unique catalyst system involving 1:1:1 DMAP:TMEDA:MgI_2 (each in 10 mol %) allowed the MBH reaction to proceed in fairly high yields between aldehydes and α,β-unsaturated ketones, esters, or a thioester.[43]

The catalyzed generation of a chiral enolate equivalent, formed from an enal and a triazolium-derived chiral *N*-heterocyclic carbene, can undergo a hetero-Diels–Alder reaction with enones with excellent enantioselectivity and diastereoselectivity (eq 14, left side).[44] Alternatively, the authors have previously shown that stronger bases, such as DBU, instead result in a homoenolate addition to the unsaturated ester, resulting in a cyclopentene ring (eq 14, right side).[45]

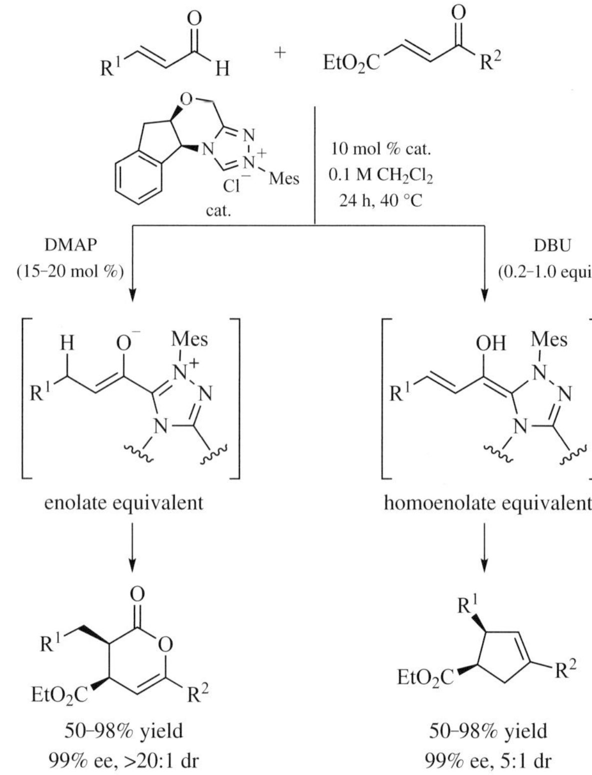

(14)

DMAP is also key in the formation of hindered, enantioenriched Michael adducts between α-branched aldehydes and nitroalkanes in conjunction with an L-threonine derivative and sulfamide (eq 15).[46] A transition state is proposed involving extensive hydrogen bonding between the chiral enamine, formed from threonine and the aldehyde, the nitroalkene, sulfamide, and protonated DMAP in which the sulfur atom acquires a distorted tetrahedral geometry. This allows one of the oxygen atoms of sulfamide to always be in a position to take part in an electrostatic interaction with the protonated nitrogen atom of DMAP, resulting in a cyclic coordination of the four components and providing 96–98% ee in all of the examples.

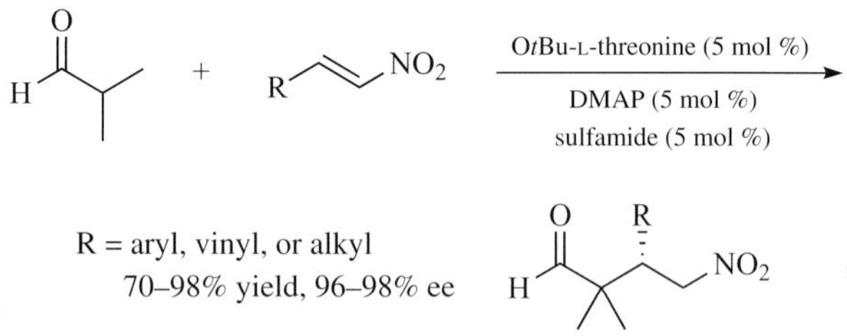

(15)

DMAP acts as a catalyst in the sulfinylation of diacetone glucose (DAG) with *tert*-butanesulfinyl chloride in the presence of DIPEA or Et_3N, resulting in the formation of the (*R*)-sulfinate esters in high yields (~90%, still bound to glucose).[47] Interestingly, the selectivity can be reversed to (*S*) by using Et_3N alone. The enantiopure DAG-sulfinate esters can then serve as important precursors to chiral sulfoxides, sulfinamides, or sulfinylimines via a displacement reaction with Grignards, amines, or LiHMDS, respectively.

Polysubstituted homoallylic alcohols with a trifluoromethyl group are highly sought after for their use in pharmaceutical and agrochemical applications. They can be synthesized from the allylation of trifluoropyruvate with alkenes activated by two electron-withdrawing groups in the presence of bases such as *N*-methylmorpholine, quinine, diethylamine, and DMAP, of which DMAP has the highest yields, using only 0.2 equiv (52–88%).[48]

In a final example, α-chloro-β-hydroxyalkanones are reacted with acetic anhydride and pyridine in the presence of catalytic DMAP to afford selectively (*Z*)-α-chloroenones in good yield via an elimination reaction.[49]

Cycloadditions Involving DMAP Catalysis. Bicyclic lactones containing a spirooxindole can be formed through a DMAP-catalyzed [4 + 2] Diels–Alder cycloaddition of pyrones and indole derivatives, resulting in both *endo* and *exo* products in approximately 2:1, 1:1, or 1:2 ratios in good yields. DMAP is proposed to abstract the proton on the pyrone, causing the initial cyclization to occur (eq 16).[50]

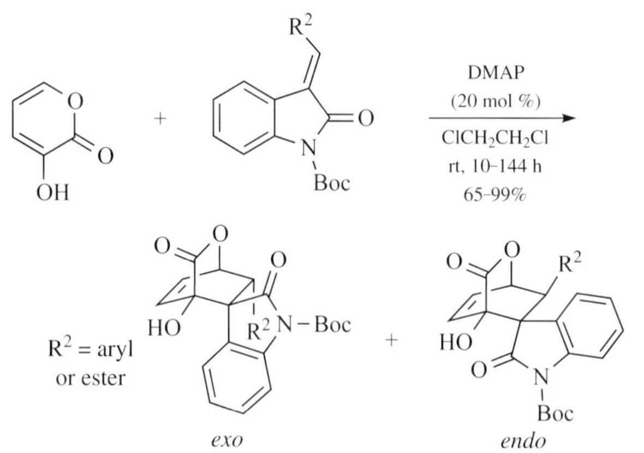

(16)

DMAP can be used to facilitate the formal [2 + 2] annulation reaction of alkynyl ketones with *N*-tosylimines for complete stereoselectivity of highly functionalized *trans*-substituted azetines.[51] DMAP initially adds to the ynone via a Michael-type addition, generating an anion at the 4-position via proton transfer and resonance, which then adds to the imine and can undergo a cyclization via the nitrogen. Interestingly, if Bu_3P is used as the catalyst instead, pyrrolidine products result (five-membered ring) from a formal [3 + 2] annulation due to an anion formation at the 2-position of the ynone. Other [3 + 2] and [4 + 2] cycloadditions incorporating C_{60} fullerenes and Morita–Baylis–Hillman adducts have also been catalyzed by DMAP. When reacted in the presence of Ac_2O, either cyclopentene- or cyclohexene-fused C_{60} fullerenes can be formed just by changing the temperature (room temperature vs. 120 °C) and conditions (DMAP/Bu_3P vs. DMAP alone).[52]

Multicomponent Coupling Reactions Catalyzed by DMAP. Multicomponent coupling reactions can also be facilitated with the use of a catalytic amount of DMAP. A one-pot, three-component domino coupling of β-ketodithioesters, aldehydes, and nitriles yielded 4*H*-thiopyrans in high yields via a cascading Knoevenagel condensation/Michael addition/cyclization sequence, forming three new bonds with high atom economy.[53] A similar strategy utilized β-diketones instead of the β-ketodithioesters resulting in pyran-annulated heterocycles.[54] Moderate to good yields are observed in these reactions as well.

DMAP-Initiated Protecting Group Strategies. DMAP has also been utilized as a key component in several protecting group strategies. The formation of cyclic acetals from 1,2-diols and alkyl propynoates in the presence of DMAP has been demonstrated.[55] These acetals are stable to acid-catalyzed hydrolysis, unlike other acetals, and methanolysis. These acetals can be deprotected by heating in neat pyrrolidine.

Demethylation of acid-sensitive aryl methyl ethers, such as 2,4-dimethoxyquinolines, can be accomplished using a trimethylsilyl iodide/DMAP reagent system, where DMAP is used as a Lewis base to prevent protonation of the alkene from the HI by-product.[56] This reaction was utilized to form the quinoline-containing alkaloid natural product, atanine, from the *Rutaceae* family.

DMAP can also be used in the removal of the thiazolidinethione moiety from *N*-acylthiazolidinethiones to the corresponding benzyl esters.[57] DMAP was found to be superior to other bases including imidazole, 2,6-lutidine, and DBU (eq 17).

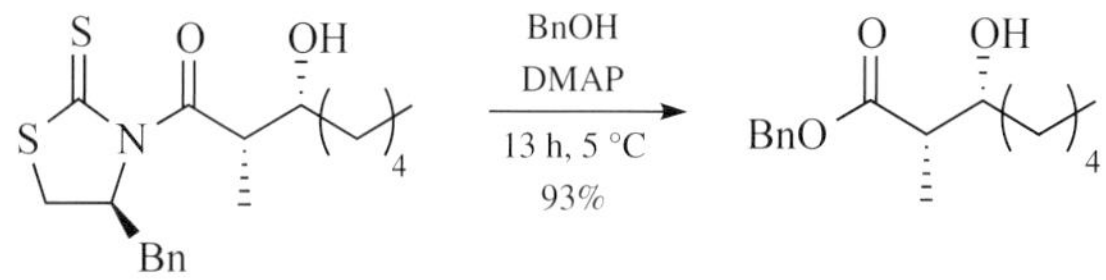

(17)

Miscellaneous Non-catalytic DMAP Reactions. Iodoacetylenes, which are useful intermediates in organic synthesis and in medicinal and pharmaceutical research, can be prepared in high yields through the reaction of terminal acetylenic ketones or aryl-substituted acetylenes with molecular iodine in the presence of DMAP as a mild base.[58]

Nucleophile-promoted aldol-lactonization of ketoacids using tosyl chloride, DIPEA, and K_2CO_3 in the presence of DMAP as a Lewis base creates constrained bicyclic β-lactones with high diastereoselectivity (eq 18).[59] The β-lactone substructure is found in a numerous bioactive natural products and pharmaceutical agents.

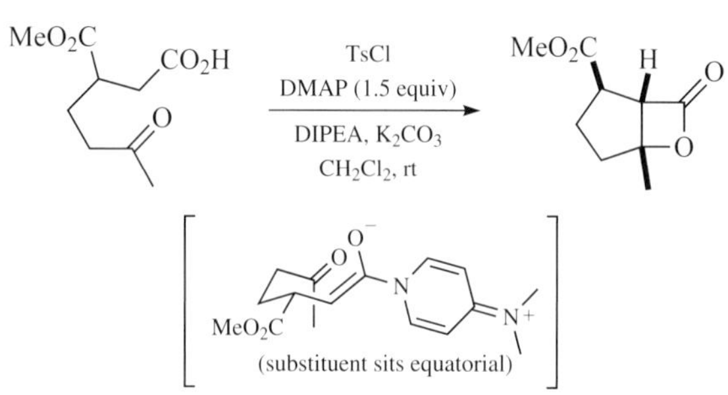

(18)

The allylic alcohol products from Morita–Baylis–Hillman reactions were shown to participate in a DMAP-mediated Tsuji–Trost-type reaction with β-diketones or β-ketoesters, forming the *C*-allylation product without requiring the use of palladium. Previously, it was shown that allylic alcohols combined with β-ketoesters and DMAP afforded the transesterification products, in which the allylic alcohol displaced the ester substituent.[60] The difference between these diverging reaction pathways is likely due to the electron-withdrawing group on the allylic alcohol in the MBH adducts vs. just alkyl substituents in the latter case.

The reaction of pressurized CO_2 with substituted epoxides to yield cyclic carbonates has been demonstrated in the presence of cobalt(II) tetraphenylporphyrin chloride (CoTTP(Cl)) and DMAP as a cocatalyst (eq 19).[61] The same author has shown that this reaction can also be carried out in the presence of chromium(III) salen complexes.[62] DMAP gave superior activity to other Lewis bases tested and resulted in high turnover numbers and short reaction times for terminal monosubstituted epoxides containing alkyl, vinyl, aryl, and even α-halo substituents. Stereochemistry on 1,2-disubstituted epoxides was retained in the product and one example of a 1,1-disubstituted epoxide was also shown.

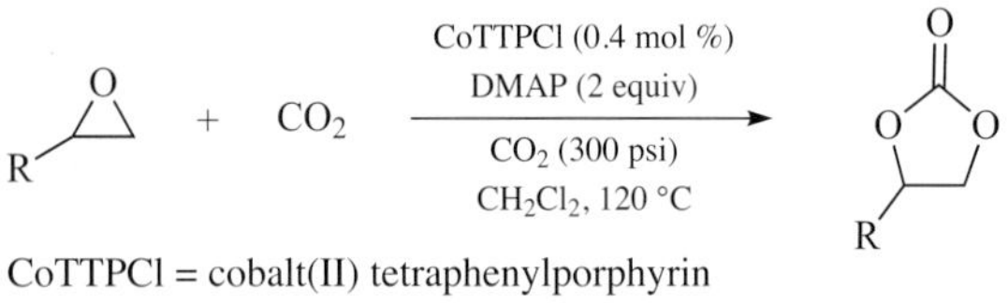

(19)

CoTTPCl = cobalt(II) tetraphenylporphyrin

R = alkyl, aryl, α-halo
conversions >92% by NMR

DMAP has also been applied to the regioselective ring opening of epoxides and oxetanes with $POCl_3$ or PCl_3, providing α-chloroalcohols.[63] DMAP attacks the phosphorus atom and releases a chloride ion, which can then undergo a nucleophilic attack to open the ring.

Thionation or selenation of amides can be accomplished through the use of elemental sulfur or selenium, respectively, along with a hydrochlorosilane and an amine base, where DMAP gave the highest yields.[64]

Nitrilium ions are highly reactive intermediates that serve as imine synthons and are used in a multitude of synthetic named reactions; however, their lack of stability hampers their use. DMAP has been shown to stabilize nitrilium ions in the presence of trimethylsilyl triflate to create the base-stabilized triflate salts, which are stable in air with only 3% decomposition over a period of 1 month (eq 20).[65]

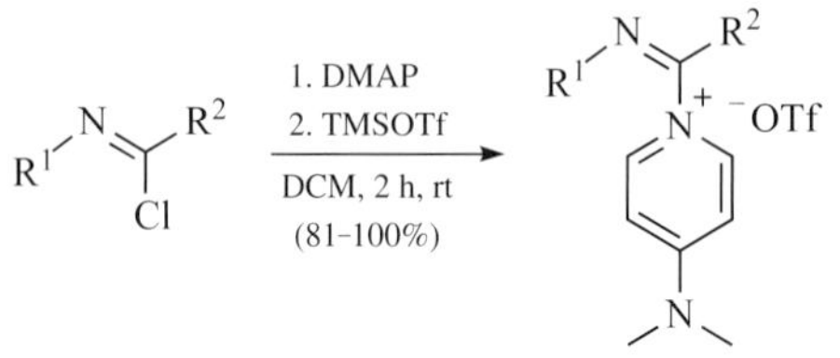

R^1	R^2
Mes	*t*Bu
Ph	*t*Bu
*i*Pr	Ph
Cy	Ph

(20)

DMAP and alkyl bromides are used to make homologs of 1-alkyl-(*N*,*N*-dimethylamino)pyridinium bromide salts by undergoing a simple nucleophilic substitution.[66] These compounds were tested for antifungal and antibacterial capacities and show promising activity against bacterial strains such as *Mycobacterium*, *E. coli*, *Staphylococcus*, and *Salmonella*.

DMAP has been incorporated into dispirooxindole-fused heterocycles via a domino 1,4-dipolar addition and Diels–Alder reaction of DMAP, acetylenedicarboxylates, and 3-phenacylideneoxindoles.[67] The reactive intermediate is an *in situ* generated Huisgen 1,4-dipole.

DMAP has also proved useful in releasing an imine-bound borane complex formed after an alkene hydroboration reaction, thereby providing higher yields than the usual workup (56% vs. 35%).[68]

1. Hoefle, G.; Steglich, W.; Vorbrueggen, H., *Angew. Chem., Int. Ed. Engl.* **1978**, *17*, 569.
2. Scriven, E. F. V., *Chem. Soc. Rev.* **1983**, *12*, 129.
3. (a) Steglich, W.; Hoefle, G., *Angew. Chem., Int. Ed. Engl.* **1969**, *8*, 981. (b) Hoefle, G.; Steglich, W., *Synthesis* **1972**, 619.
4. (a) Salomon, R. G., Salomon, M. F.; Zagorski, M. G.; Reuter, J. M.; Coughlin, D. J., *J. Am. Chem. Soc.* **1982**, *104*, 1008. (b) Zwanenburg, D. J.; Reynen, W. A. P., *Synthesis* **1976**, 624.
5. (a) Neises, B.; Steglich, W., *Angew. Chem., Int. Ed. Engl.* **1978**, *17*, 522. (b) Hassner, A.; Alexanian, V., *Tetrahedron Lett.* **1978**, 4475.
6. (a) Ziegler, F. E.; Berger, G. D., *Synth. Commun.* **1979**, 539. (b) Gilon, C.; Klausner, Y.; Hassner, A., *Tetrahedron Lett.* **1979**, 3811.
7. Litvinenko, L. M., Kirichenko, A. C.; *Dokl. Akad. Nauk SSSR* **1967**, *176*, 97. (b) Kirichenko, A. C.; Litvinenko, L. M.; Dotsenko, I. N.; Kotenko, N. G.; Nikkel'sen, E.; Berestetskaya, V. D., *Dokl. Akad. Nauk SSSR* **1969**, *244*, 1125 (*Chem. Abstr.* **1979**, *90*, 157 601).
8. (a) Nickisch, K.; Klose, W.; Bohlmann, F., *Chem. Ber.* **1980**, *113*, 2036. (b) Kametami, T.; Nagahara, T.; Suzuki, Y.; Yokohama, S.; Huang, S.-P.; Ihara, M., *Tetrahedron* **1981**, *37*, 715.
9. (a) Chaudhary, S. K.; Hernandez, O., *Tetrahedron Lett.* **1979**, *95*, 99. (b) Hernandez, O.; Chaudhary, S. K.; Cox, R. H.; Porter, J., *Tetrahedron Lett.* **1981**, *22*, 1491.
10. Guibe-Jampel, E.; Wakselman, M.; Raulais, D., *J. Chem. Soc., Chem. Commun.* **1980**, 993.
11. Hassner, A.; Krepski, L. R.; Alexanian, V., *Tetrahedron* **1978**, *34*, 2069.
12. Vorbrueggen, H. (Schering AG) Ger. Offen. 2 517 774, **1976** (*Chem. Abstr.* **1977**, *86*, 55 293).
13. (a) Boden, E. P.; Keck, G. E., *J. Org. Chem.* **1985**, *50*, 2394. (b) Stork, G.; Rychnovsky, S. D., *J. Am. Chem. Soc.* **1987**, *109*, 1565.
14. Hikota, M.; Tone, H.; Horita, K.; Yonemitsu, O., *J. Org. Chem.* **1990**, *55*, 7.
15. (a) Kim, S.; Lee, J. I.; Ko, Y. K., *Tetrahedron Lett.* **1984**, *25*, 4943. (b) Denis, J.-N.; Greene, A. E.; Guenard, D.; Gueritte-Voegelein, F.; Mangatal, L.; Potier, P., *J. Am. Chem. Soc.* **1988**, *110*, 5917.
16. Anwar, S.; Davis, A. P., *Tetrahedron* **1988**, *44*, 3761.
17. Nudelman, A.; Kelner, R.; Broida, N.; Gottlieb, H. E., *Synthesis* **1989**, 387.
18. Taber, D. F.; Amedio, J. C., Jr.; Gulino, F., *J. Org. Chem.* **1989**, *54*, 3474.
19. Rodriguez, J.; Waegell, B., *Synthesis* **1988**, 534.
20. (a) Buchanan, G. L., *Chem. Soc. Rev.* **1988**, *17*, 91. (b) McMurry, J., *J. Org. Chem.* **1985**, *50*, 1112. (c) Hoefle, G., Steglich, W., *Chem. Ber.* **1971**, *104*, 1408.
21. Buechi, G.; Wuest, H., *J. Org. Chem.* **1979**, *44*, 4116.
22. Nicolaou, K. C.; Bunnage, M. E.; Koide, K., *J. Am. Chem. Soc.* **1994**, *116*, 8402.
23. Furukawa, J.; Morisaki, N.; Kobayashi, H.; Iwasaki, S.; Nozoe, S.; Okuda, S., *Chem. Pharm. Bull.* **1985**, *33*, 440.
24. Hwang, C. K.; Li, W. S.; Nicolaou, K. C., *Tetrahedron Lett.* **1984**, *25*, 2295.
25. Balachandran, V.; Rajeswari, S.; Lalitha, S., *Spectrochim. Acta A* **2014**, *124*, 277.
26. Shi, Q.; Tan, Z.-C.; Di, Y.-Y.; Tong, B.; Li, Y.-S.; Wang, S.-X., *J. Chem. Eng. Data* **2007**, *52*, 941.
27. De Rycke, N.; Couty, F.; David, O. R. P., *Chem. Eur. J.* **2011**, *17*, 12852.
28. Jiang, Y.; Kuang, C., *J. Chem. Sci.* **2009**, *121*, 1035.
29. Kane, T. R.; Ly, C. Q.; Kelly, D. E.; Dener, J. M., *J. Comb. Chem.* **2004**, *6*, 564.
30. Liu, K.-C.; Fang, J.-M.; Jan, J.-T.; Cheng, T.-J. R.; Wang, S.-Y.; Yang, S.-T.; Cheng, Y.-S. E.; Wong, C.-H., *J. Med. Chem.* **2012**, *55*, 8493.
31. Dhaon, M. K.; Olsen, R. K.; Ramasamy, K., *J. Org. Chem.* **1982**, *47*, 1962.
32. (a) Greenwald, R. B.; Pendri, A.; Zhao, H., *Tetrahedron: Asymmetry* **1998**, *9*, 915. (b) Liu, Y.-Q.; Tian, X.; Yang, L.; Zhan, Z.-C., *Eur. J. Med. Chem.* **2008**, *43*, 2610.
33. Roncaglia, F.; Parsons, A. F.; Bellesia, F.; Ghelfi, F., *Synth. Commun.* **2011**, *41*, 1175.

34. (a) Kopecky, D. J.; Rychnovsky, S. D., *Org. Synth.* **2003**, *80*, 177. (b) Dahanukar, V. H.; Rychnovsky, S. D., *J. Org. Chem.* **1996**, *61*, 8317. (c) Jaber, J. J.; Mitsui, K.; Rychnovsky, S. D., *J. Org. Chem.* **2001**, *66*, 4679.
35. Ignatova, M.; Sepulchre, M.; Manolova, N., *Eur. Polym. J.* **2001**, *38*, 33.
36. Wang, D. H.; Baek, J.-B.; Nishino, S. F.; Spain, J. C.; Tan, L.-S., *Polymer* **2006**, *47*, 1197.
37. Luo, W.; Ma, Y. M.; Quinn, P. J.; Hider, R. C.; Liu, Z. D., *J. Pharm. Pharmacol.* **2004**, *56*, 529.
38. Prashad, M.; Hu, B.; Har, D.; Repic, O.; Blacklock, T. J., *Tetrahedron Lett.* **2000**, *41*, 9957.
39. Doyle, M. P.; Colyer, J. T., *Tetrahedron: Asymmetry* **2003**, *14*, 3601.
40. Goncalves, S.; Nicolas, M.; Wagner, A.; Baati, R., *Tetrahedron Lett.* **2010**, *51*, 2348.
41. Jeong, Y.-C.; Moloney, M. G., *J. Org. Chem.* **2011**, *76*, 1342.
42. Deb, I.; Shanbhag, P.; Mobin, S. M.; Namboothiri, I. N. N., *Eur. J. Org. Chem.* **2009**, 4091.
43. Bugarin, A.; Connell, B. T., *J. Org. Chem.* **2009**, *74*, 4638.
44. Kaeobamrung, J.; Kozlowski, M. C.; Bode, J. W., *Proc. Natl. Acad. Sci. USA* **2010**, *107*, 20661.
45. Kaeobamrung, J.; Bode, J. W., *Org. Lett.* **2009**, *11*, 677.
46. Nugent, T. C.; Shoaib, M.; Shoaib, A., *Org. Biomol. Chem.* **2011**, *9*, 52.
47. Chelouan, A.; Recio, R.; Alcudia, A.; Khiar, N.; Fernández, I., *Eur. J. Org. Chem.* **2014**, *2014*, 6935.
48. Zhang, F.; Wang, X.-J.; Cai, C.-X.; Liu, J.-T., *Tetrahedron* **2009**, *65*, 83.
49. Concellon, J. M.; Huerta, M., *Tetrahedron* **2002**, *58*, 7775.
50. Zhou, M.-Q.; Zhao, J.-Q.; You, Y.; Xu, X.-Y.; Zhang, X.-M.; Yuan, W.-C., *Tetrahedron* **2015**, doi: 10.1016/j.tet.2015.04.032.
51. Meng, L.-G.; Cai, P.; Guo, Q.; Xue, S., *J. Org. Chem.* **2008**, *73*, 8491.
52. Yang, H.-T.; Ren, W.-L.; Miao, C.-B.; Dong, C.-P.; Yang, Y.; Xi, H.-T.; Meng, Q.; Jiang, Y.; Sun, X.-Q., *J. Org. Chem.* **2013**, *78*, 1163.
53. Khan, A. T.; Lal, M.; Ali, S.; Khan, M. M., *Tetrahedron Lett.* **2011**, *52*, 5327.
54. Verma, R. K.; Verma, G. K.; Shukla, G.; Nagaraju, A.; Singh, M. S., *ACS Comb. Sci.* **2012**, *14*, 224.
55. Ariza, X.; Costa, A. M.; Faja, M.; Pineda, O.; Vilarrasa, J., *Org. Lett.* **2000**, *2*, 2809.
56. Jones, K.; Roset, X.; Rossiter, S.; Whitfield, P., *Org. Biomol. Chem.* **2003**, *1*, 4380.
57. Wu, Y.; Sun, Y.-P.; Yang, Y.-Q.; Hu, Q.; Zhang, Q., *J. Org. Chem.* **2004**, *69*, 6141.
58. Meng, L.-G.; Cai, P.-J.; Guo, Q.-X.; Xue, S., *Synth. Commun.* **2008**, *38*, 225.
59. Liu, G.; Shirley, M. E.; Romo, D., *J. Org. Chem.* **2012**, *77*, 2496.
60. Gilbert, J. C.; Kelly, T. A., *J. Org. Chem.* **1988**, *53*, 449.
61. (a) Paddock, R. L.; Hiyama, Y.; McKay, J. M.; Nguyen, S. T., *Tetrahedron Lett.* **2004**, *45*, 2023. (b) Darensbourg, D. J.; Mackiewicz, R. M., *J. Am. Chem. Soc.* **2005**, *127*, 14026.
62. Paddock, R. L.; Nguyen, S. T., *J. Am. Chem. Soc.* **2001**, *123*, 11498.
63. Sartillo-Piscil, F.; Quintero, L.; Villegas, C.; Santacruz-Juarez, E.; Anaya de Parrodi, C., *Tetrahedron Lett.* **2002**, *43*, 15.
64. Shibahara, F.; Sugiura, R.; Murai, T., *Org. Lett.* **2009**, *11*, 3064.
65. van Dijk, T.; Bakker, M. S.; Holtrop, F.; Nieger, M.; Slootweg, J. C.; Lammertsma, K., *Org. Lett.* **2015**, *17*, 1461.
66. Sundararaman, M.; Kumar, R. R.; Venkatesan, P.; Ilangovan, A., *J. Med. Microbiol.* **2013**, *62*, 241.
67. Sun, J.; Sun, Y.; Gong, H.; Xie, Y.-J.; Yan, C.-G., *Org. Lett.* **2012**, *14*, 5172.
68. (a) Pigza, J. A.; Han, J.-S.; Chandra, A.; Mutnick, D; Pink, M.; Johnston, J. N., *J. Org. Chem.* **2013**, *78*, 822. (b) Chandra, A.; Pigza, J. A.; Han, J.-S.; Mutnick, D.; Johnston, J. N., *J. Am. Chem. Soc.* **2009**, *131*, **3470**.

2,2-Dimethyl-1,3-dioxan-5-one[1]

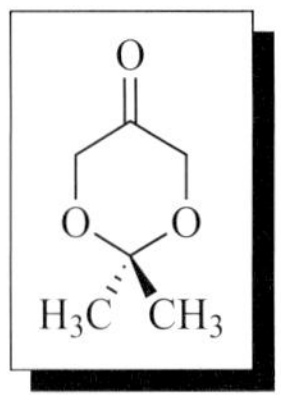

[74181-34-3] $C_6H_{10}O_3$ (MW 130.14)
InChI = 1S/C6H10O3/c1-6(2)8-3-5(7)4-9-6/h3-4H2,1-2H3
InChIKey = ASFQDNDZFGFMMP-UHFFFAOYSA-N

(reagent utilized as a C_3 building block equivalent of dihydroxyacetone (DHA) in syntheses, often used with the SAMP/RAMP methodology,[2] in asymmetric aldol reactions,[3] Mannich reactions,[4] Michael additions,[5] or as starting material in the synthesis of different carbohydrates[6])

Physical Data: bp 50–54 °C/11 Torr; *d* 1.09 g cm^{-3}.
Solubility: soluble in H_2O and most organic solvents.
Form Supplied in: colorless to pale yellow liquid.
Analysis of Reagent Purity: NMR.
Preparative Methods: synthesized from 2-amino-2-hydroxymethyl-1,3-propanediol hydrochloride in two steps.[7]
Purification: fractional distillation at 11 Torr over a Vigreux column.
Handling, Storage, and Precaution: 2,2-dimethyl-1,3-dioxan-5-one is hygroscopic; store at 4 °C under argon atmosphere to prevent polymerization and decomposition. It is irritating to eyes, skin, and the respiratory system; use personal protective equipment and work in a well-ventilated hood. Incompatible with oxidizing agents and acids.

Synthesis. The first synthesis of 2,2-dimethyl-1,3-dioxan-5-one was published in 1989 by Hoppe et al. based on previously unpublished results by Woodward and Vorbrüggen.[7] Commercially available 2-amino-2-hydroxymethyl-1,3-propanediol hydrochloride is first transformed into the corresponding acetal and then oxidized by sodium periodate to 2,2-dimethyl-1,3-dioxan-5-imine that affords 2,2-dimethyl-1,3-dioxan-5-one after in situ hydrolization with an overall yield of 68% (eq 1). Slightly

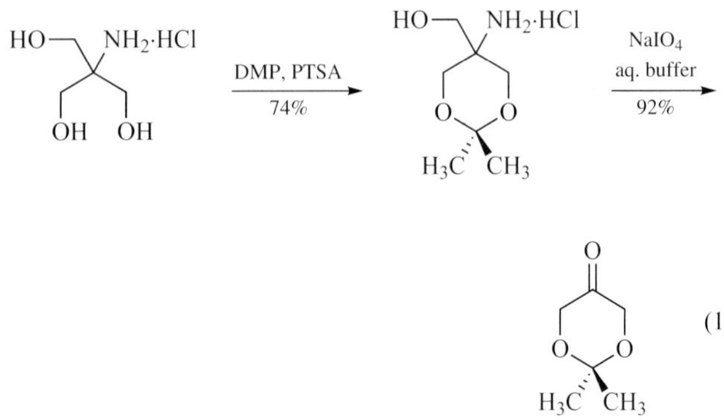

(1)

modified version,[2,8,9] as well as other routes towards 2,2-dimethyl-1,3-dioxan-5-one were published later.[10] It is used as a synthetic equivalent to dihydroxyacetone phosphate (DHAP) an important building block in nature.[11] Other cyclic (e.g., 2-phenyl-1,3-dioxane-5-one used by Ulven and Carlsen[12,13]) and acyclic derivatives (e.g., unprotected dihydroxyacetone) are also known but 2,2-dimethyl-1,3-dioxan-5-one is by far the most widely used.

SAMP/RAMP – Hydrazone Methodology. Soon after the publication of the synthesis in 1989 other groups started using 2,2-dimethyl-1,3-dioxan-5-one as a starting material. One of the first utilizations, the synthesis of the corresponding SAMP or RAMP hydrazones was first published by Enders in 1989 (eq 2a).[14] Later another synthesis using milder conditions was published by K. C. Nicolaou et al. (eq 2b).[15]

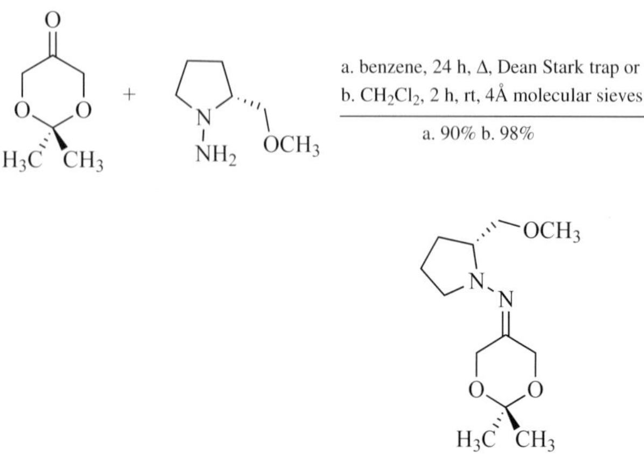

(2)

The α-alkylation of those hydrazones gave the substituted dioxanones in good yields and very good enantioselectivities within three steps. First the hydrazone was deprotonated with *tert*-butyl lithium at $-78\,^{\circ}$C, which affords the lithium enolate. After cooling to $-100\,^{\circ}$C a wide variety of electrophiles could be added leading to the corresponding α-substituted 2,2-dimethyl-1,3-dioxan-5-ones after cleavage of the hydrazones (eq 3).[14] This could be accomplished by ozonolysis, aqueous oxalic acid, or a wide variety of other methods in good yields.[16,17] The substituted products could generally be isolated in good yields with excellent enantiomeric excesses. In a later paper it was shown that the synthesis of α,α'-disubstituted ketones was also a possibility. In this case before the hydrazone was cleaved, the product was again deprotonated and a second electrophile was added to the reaction mixture. After cleavage of the hydrazone the products could be isolated with excellent enantiomeric excesses and diastereoselectivities (eq 4).[18] This methodology was used in the total synthesis of a variety of natural products[19–22] as well as carbohydrates.[6]

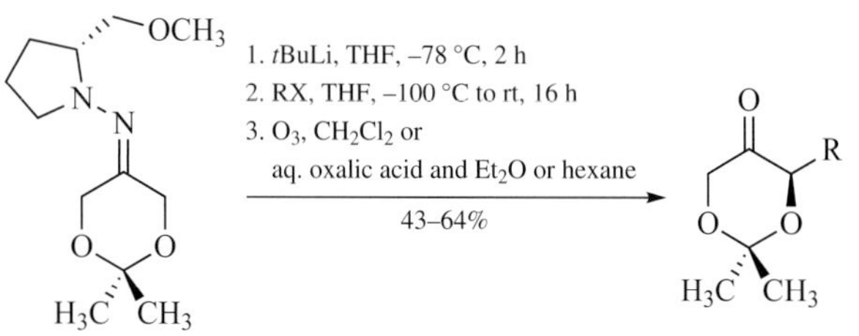

(3)

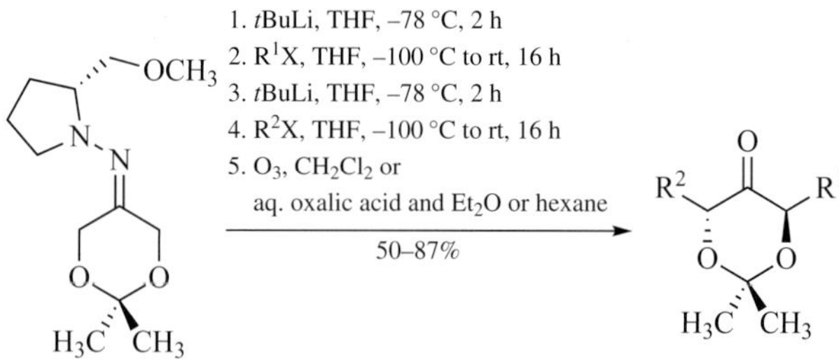

(4)

Other hydrazones can be synthesized as well that can lead to a different stereoselectivity. For example the dimethylhydrazone affords the racemic α,α'-*syn*-disubstituted products after bisalkylation. Without the influence from the chiral group the second electrophile prefers the equatorial over the axial position (eq 5).[23]

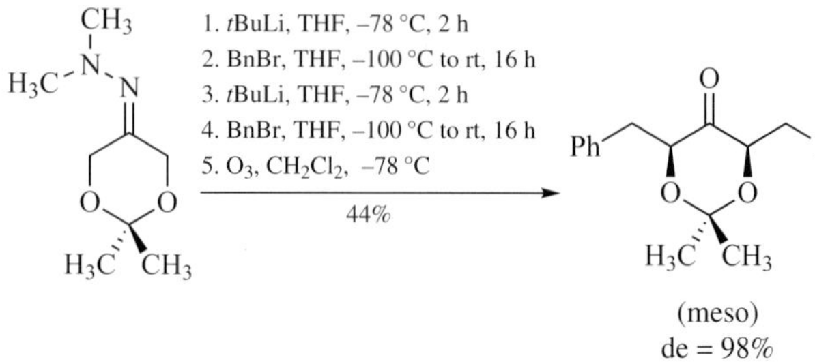

(5)

1,3-Dioxin Methodology. The 1,3-dioxin methodology is another method using 2,2-dimethyl-1,3-dioxan-5-one as a starting material. Developed by Funk and Yost, it was for example used in the synthesis of functionalized taxol-A-rings. In the first step (acyloxy)dioxins are synthesized by *O*-acylation of the ketone. This substrates underwent a retro-cycloaddition at high temperatures, which led to 2-(acyloxy)acroleins (eq 6). To avoid polymerization the 2-(acyloxy)acroleins had to be stored at low temperatures but were stable enough to be a very useful starting material for Diels–Alder reactions.[24]

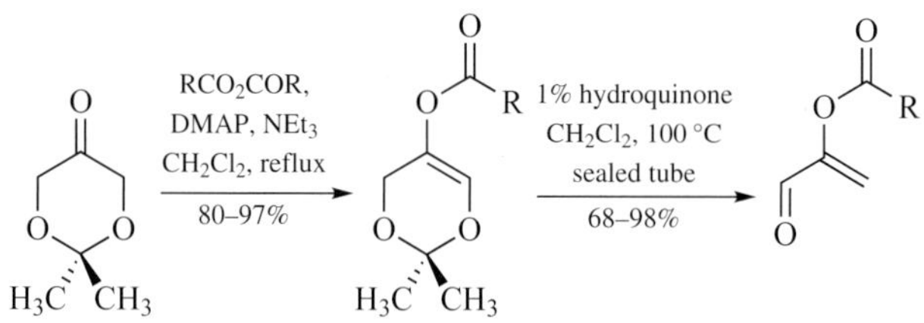

(6)

As a good dienophile they could be reacted with a variety of different dienes. After starting with simple dienes such as cyclopentadiene (eq 7) later more complex substrates were used to synthesize highly functionalized synthons for the synthesis of the desired ring system (eq 8). The Diels–Alder reaction gave yields ranging from 68 to higher than 90% under optimized conditions using lanthanide catalysis.

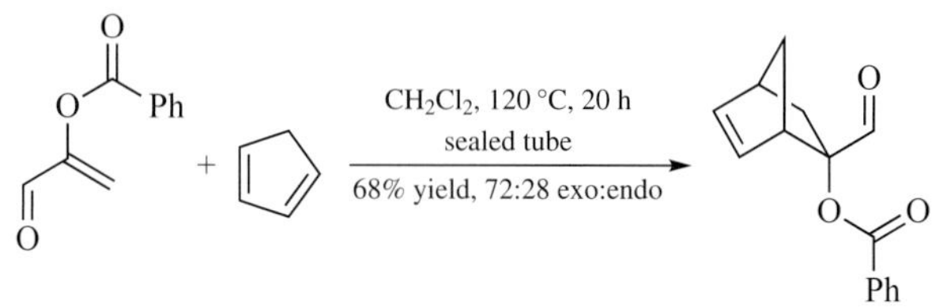

(7)

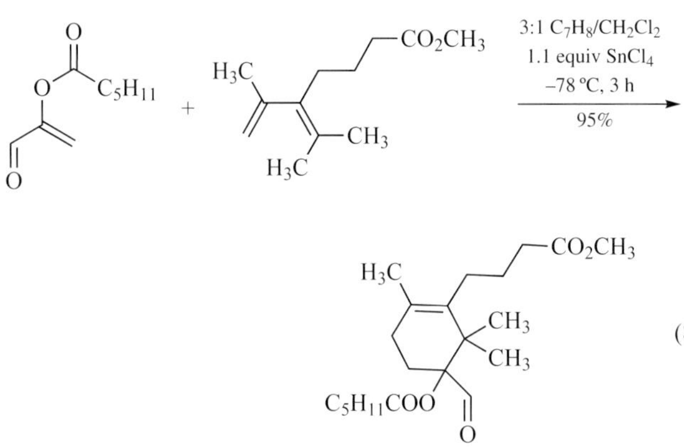

(8)

4-Alkyl-5-trialkylsiloxy-4*H*-1,3-dioxins, which could be synthesized from 2,2-dimethyl-1,3-dioxan-5-one, not only gave good results for the Diels–Alder reactions but could additionally be used as starting materials in Sasaki-type [4+3] cycloadditions (eq 9).[25]

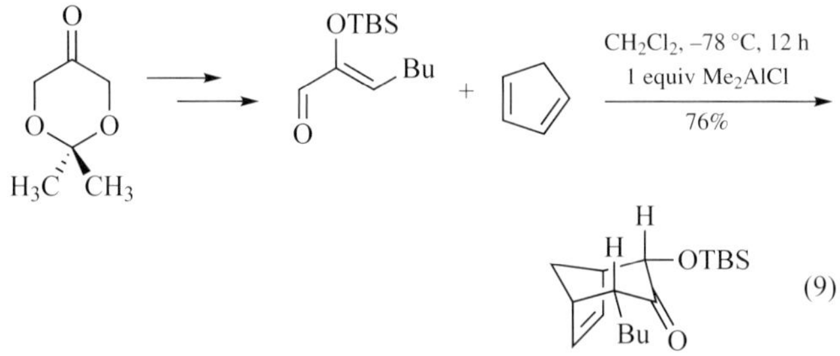

(9)

Aldol Reactions. The first aldol reactions of 2,2-dimethyl-1,3-dioxan-5-one were carried out with preformed enolates generated by bases such as LDA.[8,10] The deprotonation with LDA followed by aqueous work up often resulted in the formation of the self-aldol product as major product (eq 10).

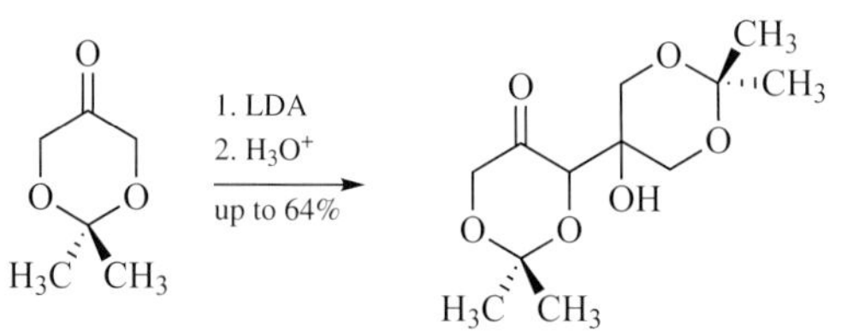

(10)

However the aldol reaction of the lithium enolate with aldehydes proceeds very well (eq 11).

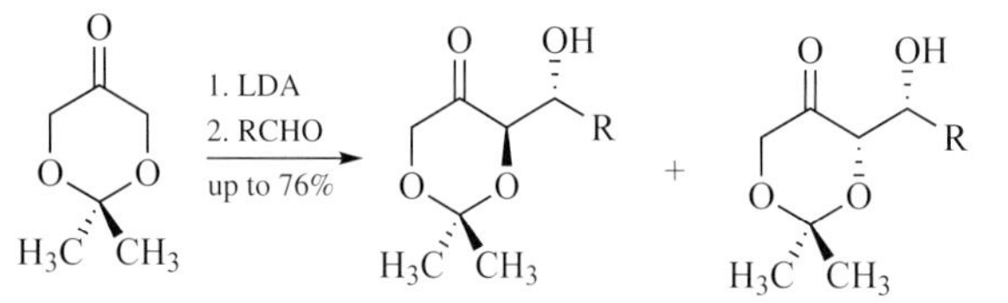

(11)

In 2005 the asymmetric aldol reaction with 2,2-dimethyl-1,3-dioxan-5-one catalyzed by proline was developed for the synthesis of carbohydrates by Enders (eq 12),[26] who showed that for this biomimetic organocatalytic $C_3 + C_n$ strategy a broad range of aldehydes was appropriate to access diastereo- and enantioselectively orthogonal doubly protected sugars and amino sugars in one step.[3]

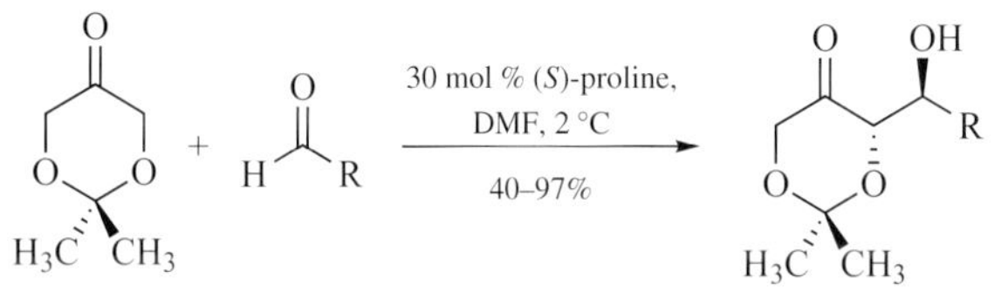

(12)

88–96% *de*, 90–98% *ee*

These ketal protected carbohydrates were usually deprotected with the acidic ion-exchange resin DOWEX® (W50X2-200) or with TBAF.[3,27] The optimization of these reactions was also studied by other groups. Córdova and coworkes for example showed that the addition of water not only accelerates the synthesis but also improves the enantioselectivity.[28] Majewski et al. addressed the use of additives, but due to the sensitivity of dioxanones towards acids only weak acids can be used. Whereas the addition of amines or amine salts does not lead to any benefits in the aldol reaction,[29] the addition of Brønsted acids does; pyridinium-*p*-toluenesulfate (PPTS) and lithium chloride are especially useful additives.[30]

Several natural products such as jaspin B[31] or α-D-galactosyl ceramide (KRN7000)[32], carba-, and azasugars could be obtained via this reaction protocol.[27,33–38]

Furthermore stereoselective syntheses with two consecutive aldol reactions were carried out.[39,40] For example the group of Majewski developed a α,α'-bisaldol approach to higher carbohydrates (eq 13). They synthesized 6-*C*-phenyl-D-glycero-D-allo-hexopyranose and D-erythro-D-allo-octapyranose. Whereas the first aldol reaction was catalyzed by proline the second one involved the lithium enolate of the protected aldol product of the first step.[41]

Also syntheses with two dioxanones[40] and aldol reactions with other catalysts than proline are known.[42–44]

Sello and coworkers reported in 2009 on the first biomimetic synthesis of the newly discovered class of bacterial signalling molecules from *Streptomyces coelicolor* starting from 2,2-dimethyl-1,3-dioxan-5-one as DHA equivalent. They described a scandium triflate-catalyzed domino reaction (aldol reaction, cyclization, and dehydrative aromatization) of β-ketoesters and the

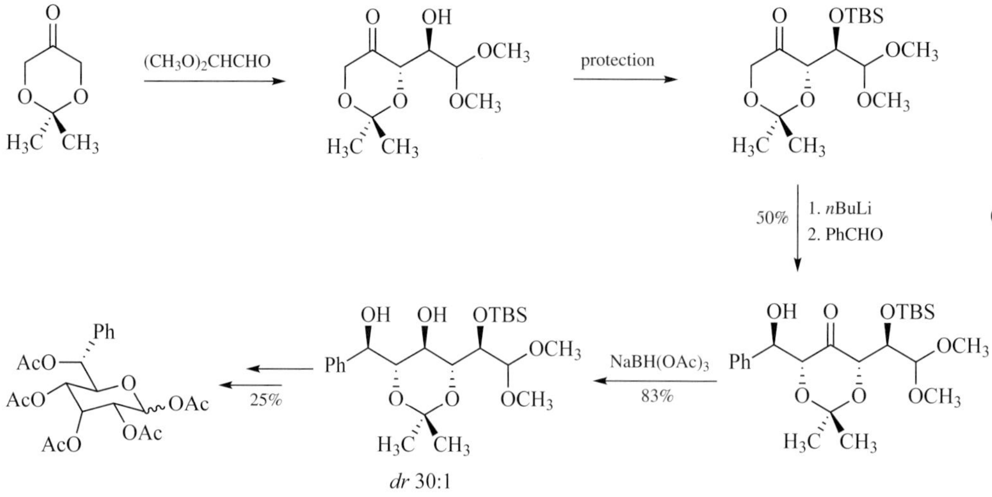

(13)

dioxanone resulting in 2,3,4-substituted furans in good yields (eq 14).[45]

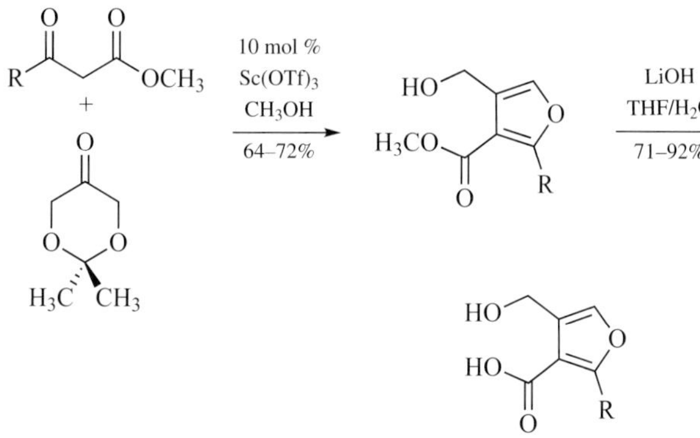

(14)

In 2008 Cheng and coworkers developed simple primary-tertiary diamine-Brønsted acid conjugates and showed that both *syn*- and *anti*-aldol reactions of DHAs were catalyzed obtaining high diastereo- and enantioselectivities. For example, they carried out the aldol reaction of 2,2-dimethyl-1,3-dioxan-5-one with *p*-nitrobenzaldehyde.[46] Later several groups reported on this reaction with different catalysts.[44,47–51] For example, Fernándes-Mayorales and coworkers published the same reaction catalyzed by a heterogenized proline on a mesoporous support. A wide range of solvents, even hydrophobic ones like toluene, could be used and in this way they obtained precursors of azasugars.[42]

Michael Reactions. In 2006 Córdova and coworkers reported a new entry to polyhydroxylated pyrrolidines via an amino acid–catalyzed Michael reaction of 2,2-dimethyl-1,3-dioxan-5-one and nitrostyrenes. Yet the Michael product was isolated in only low enantioselectivities of 11%.[52] Also in 2006 they reported that simple dipeptides containing a catalytic primary amine residue catalyze the Michael addition of ketones and nitroalkenes yielding the corresponding γ-nitro ketones with high stereoselectivities.[53] At the same time Enders published the proline-based catalyzed 1,4-addition of 2,2-dimethyl-1,3-dioxan-5-one to nitroolefines resulting in polyfunctionalized nitro ketones (eq 15).[5]

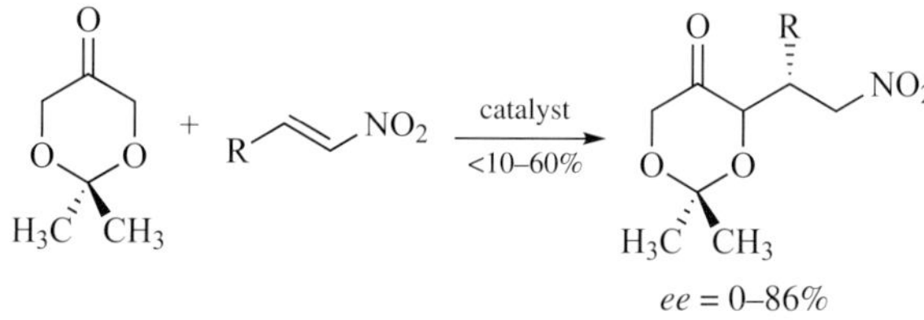

(15)

Mannich Reactions. The first Mannich or Mannich-type reactions of 2,2-dimethyl-1,3-dioxan-5-one with various presynthesized imines were contemporaneously carried out organocatalytically in 2005 by our group[4] (eq 16) and Westermann and Neuhaus.[54] We developed a (*S*)-proline catalyzed three-component Mannich reaction to yield amino sugars and their derivatives. This led to a biomimetic $C_3 + C_n$ entry to amino sugars and derivatives employing the direct asymmetric organocatalytic Mannich reaction of 2,2-dimethyl-1,3-dioxan-5-one and *N*-PMP or *N*-Boc protected imines.[55]

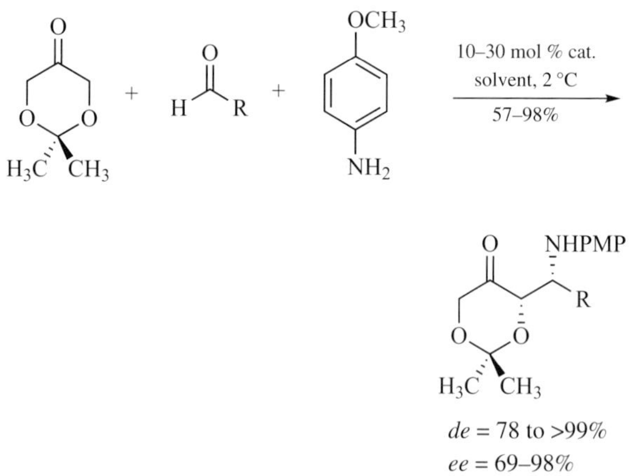

(16)

Westermann and Neuhaus also described the use of microwave irradiation to accelerate the reactions. In 2007 Kappe and coworkers accomplished detailed investigations on microwave-assisted asymmetric organocatalysis.[56]

In 2007 the group of Córdova published the synthesis of carbohydrates[29,36] and amino acid derivatives[28] via Mannich reactions, too. In 2008 Barbas et al. reported on the first primary amine-containing amino acid-catalyzed anti-Mannich reactions of dioxanone with a variety of imines.[57] In the same year Hayashi et al. developed an organocatalyst-mediated, asymmetric Mannich reaction in water catalyzed by a siloxy–tetrazole hybrid catalyst. Hereby dioxanone was successfully used as the methylene component affording enantioselectively a polyoxyamine derivative as a crucial synthetic intermediate of amino sugars.[58]

Córdova and coworkers presented an organocatalytic enantioselective one-pot three-component α-aminomethylation of ketones (eq 17). By this method virtually enantiomerically pure Mannich bases could be synthesized.[59]

Other Reactions. 2,2-Dimethyl-1,3-dioxan-5-one has also been used as a substrate in a variety of different reactions that were not mentioned so far and are not as widely used as the previously described methodologies. For example, a Wittig olefination with 2,2-dimethyl-1,3-dioxan-5-one is described by Ogura and Watanabe as a starting point for the synthesis of the lysidicin framework (eq 18).[60] It is also possible to perform a Wittig reaction on previously functionalized dioxanones, for example, by using the SAMP/RAMP hydrazone methodology.[61]

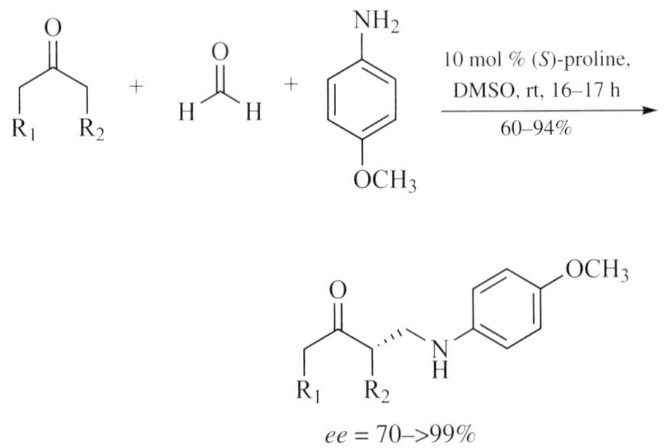

(17)

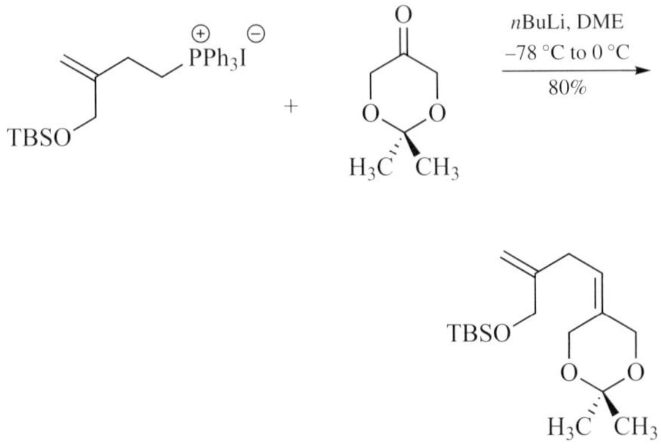

(18)

Another interesting reaction, a novel [3+3] annulation, using 2,2-dimethyl-1,3-dioxan-5-one was developed by Alonso and coworkers in 2006.[62] They synthesized functionalized cyclohexanes bearing five stereogenic centers by using a one-pot cascade annulation of β-(hetero)aryl-α-nitro-α,β-enals starting from 2,2-dimethyl-1,3-dioxan-5-one. The reaction takes place via a pyrrolidine based enamine activation and after addition of the β-(hetero)aryl-α-nitro-α,β-enals the racemic products were obtained in good yields (eq 19). Some of the substrates created by this reaction, were then used in the total synthesis of different natural products by the same group.[63,64]

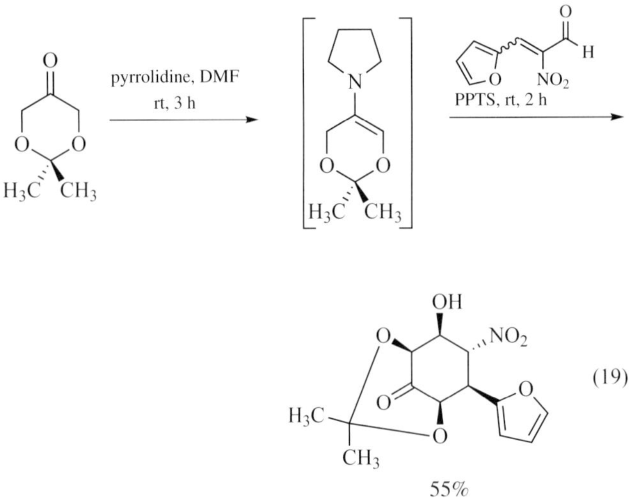

(19)

In 2001 Funk and coworkers reported the total syntheses of several tricyclic alkaloids in a three-component assembly of 2,2-dimethyl-1,3-dioxan-5-one, a primary amine, and an acylating agent.[65–69] By this methodology they synthesized erythroidine[70] in 2006 and (±)-cortistatin[71] in 2011. Recently they published the synthesis of (±)-isophellibiline, a tetracyclic natural product (eq 20).[72]

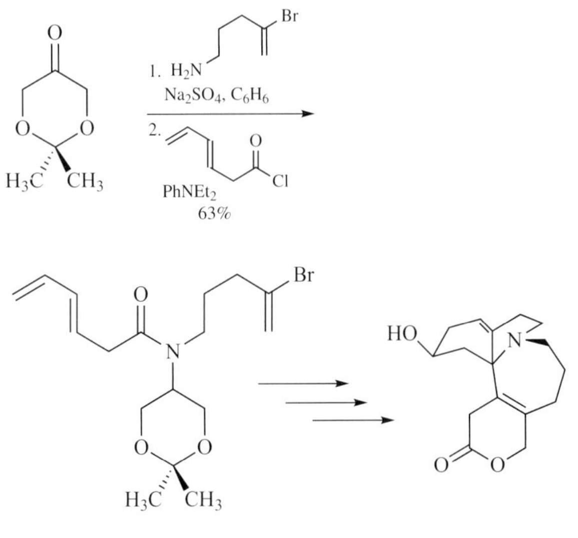

(20)

In 2008 Abaee et al. described the synthesis of 4,6-bis(arylmethylidene)dioxan-5-ones employing magnesium bromide diethyl etherate and diethylamine starting from 2,2-imethyl-1,3-dioxan-5-one (eq 21).[73]

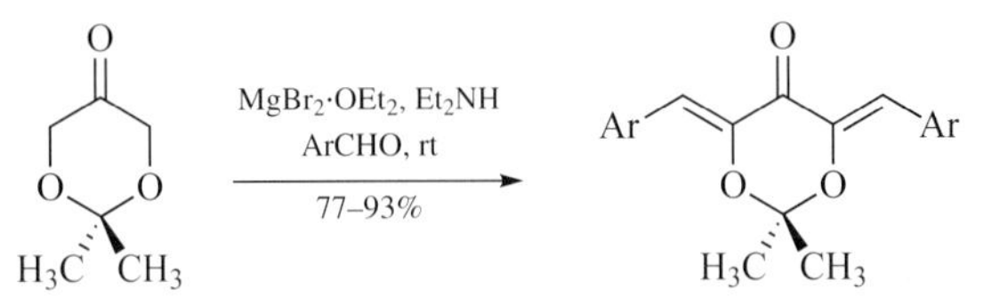

(21)

1. Enders, D.; Voith, M.; Lenzen, A., *Angew. Chem., Int. Ed.* **2005**, *44*, 1304.
2. Enders, D.; Voith, M.; Ince, S. J., *Synthesis* **2002**, 1775.
3. Grondal, C.; Enders, D., *Tetrahedron* **2006**, *62*, 329.

4. Enders, D.; Grondal, C.; Vrettou, M.; Raabe, G., *Angew. Chem., Int. Ed.* **2005**, *44*, 4079.
5. Enders, D.; Chow, S., *Eur. J. Org. Chem.* **2006**, 4578.
6. Enders, D.; Jegelka, U., *Tetrahedron Lett.* **1993**, *34*, 2453.
7. Hoppe, D.; Schmincke, H.; Kleemann, H.-W., *Tetrahedron* **1989**, *45*, 687.
8. Forbes, D. C.; Ene, D. G.; Doyle, M. P., *Synthesis* **1998**, 879.
9. Carman, R. M.; Handley, P. N., *Aust. J. Chem.* **2001**, *54*, 769.
10. Majewski, M.; Gleave, D. M.; Nowak, P., *Can. J. Chem.* **1995**, *73*, 1616.
11. Enders, D.; Narine, A. A., *J. Org. Chem.* **2008**, *73*, 7857.
12. Ulven, T.; Sorbye, K. A.; Carlsen, P. H. J., *Acta Chem. Scand.* **1997**, *51*, 1041.
13. Ulven, T.; Carlsen, P. H. J., *Synth. Commun.* **2000**, *30*, 2275.
14. Enders, D.; Bockstiegel, B., *Synthesis* **1989**, 493.
15. Nicolaou, K. C.; Brenzovich, W. E.; Bulger, P. G.; Francis, T. M., *Org. Biomol. Chem.* **2006**, *4*, 2119.
16. Enders, D.; Wortmann, L.; Peters, R., *Acc. Chem. Res.* **2000**, *33*, 157.
17. Enders, D.; Hundertmark, T.; Lazny, R., *Synlett* **1998**, 721.
18. Enders, D.; Gatzweiler, W.; Jegelka, U., *Synthesis* **1991**, 1137.
19. Enders, D.; Barbion, J., *Chem. Eur. J.* **2008**, *14*, 2842.
20. Enders, D.; Lenzen, A.; Müller, M., *Synthesis* **2004**, 1486.
21. Nicolaou, K. C.; Bulger, P. G.; Brenzovich, W. E., *Org. Biomol. Chem.* **2006**, *4*, 2158.
22. Majewski, M.; Nowak, P., *Tetrahedron: Asymmetry* **1998**, *9*, 2611.
23. Enders, D.; Jegelka, U.; Dücker, B., *Angew. Chem. Int., Ed. Engl.* **1993**, *32*, 423.
24. Funk, R. L.; Yost, K. J., *J. Org. Chem.* **1996**, *61*, 2598.
25. Aungst, R. A.; Funk, R. L., *Org. Lett.* **2001**, *3*, 3553.
26. Enders, D.; Grondal, C., *Angew. Chem., Int. Ed.* **2005**, *44*, 1210.
27. Suri, J. T.; Ramachary, D. B.; Barbas, C. F., *Org. Lett.* **2005**, *7*, 1383.
28. Ibrahem, I.; Córdova, A., *Tetrahedron Lett.* **2005**, *46*, 3363.
29. Pihko, P. M.; Laurikainen, K. M.; Usano, A.; Nyberg, A. I.; Kaavi, J. A., *Tetrahedron* **2006**, *62*, 317.
30. Majewski, M.; Niewczas, I.; Palyam, N., *Synlett* **2006**, 2387.
31. Enders, D.; Terteryan, V.; Paleček, J., *Synthesis* **2008**, 2278.
32. Enders, D.; Terteryan, V.; Paleček, J., *Synthesis* **2010**, 2979.
33. Grondal, C.; Enders, D., *Synlett* **2006**, 3507.
34. Enders, D.; Palecek, J.; Grondal, C., *Chem. Commun.* **2006**, 655.
35. Suri, J. T.; Mitsumori, S.; Albertshofer, K.; Tanaka, F.; Barbas, C. F., *J. Org. Chem.* **2006**, *71*, 3822.
36. Calderón, F.; Doyagüez, E. G.; Fernández-Mayoralas, A., *J. Org. Chem.* **2006**, *71*, 6258.
37. Palyam, N.; Majewski, M., *J. Org. Chem.* **2009**, *74*, 4390.
38. Hanessian, S.; Mi, X., *Synlett* **2010**, 761.
39. Das, B.; Balasubramanyam, P.; Veeranjaneyulu, B.; Chinna Reddy, G., *Helv. Chim. Acta* **2011**, *94*, 881.
40. Palyam, N.; Niewczas, I.; Majewski, M., *Tetrahedron Lett.* **2007**, *48*, 9195.
41. Niewczas, I.; Majewski, M., *Eur. J. Org. Chem.* **2009**, 33.
42. Doyagüez, E. G.; Calderón, F.; Sánchez, F.; Fernández-Mayoralas, A., *J. Org. Chem.* **2007**, *72*, 9353.
43. Aratake, S.; Itoh, T.; Okano, T.; Nagae, N.; Sumiya, T.; Shoji, M.; Hayashi, Y., *Chem. Eur. J.* **2007**, *13*, 10246.
44. Lombardo, M.; Pasi, F.; Easwar, S.; Trombini, C., *Adv. Synth. Catal.* **2007**, *349*, 2061.
45. Davis, J. B.; Bailey, J. D.; Sello, J. K., *Org. Lett.* **2009**, *11*, 2984.
46. Luo, S.; Xu, H.; Zhang, L.; Li, J.; Cheng, J.-P., *Org. Lett.* **2008**, *10*, 653.
47. Pearson, A. J.; Panda, S., *Tetrahedron* **2011**, *67*, 3969.
48. Yang, H.; Mahapatra, S.; Cheong, P. H.-Y.; Carter, R. G., *J. Org. Chem.* **2010**, *75*, 7279.
49. Yang, H.; Carter, R. G., *Org. Lett.* **2008**, *10*, 4649.
50. Córdova, A., Zou, W.; Dziedzic, P.; Ibrahem, I.; Reyes, E.; Xu, Y., *Chem. Eur. J.* **2006**, *12*, 5383.
51. Cordova, A.; Zou, W.; Ibrahem, I.; Reyes, E.; Engqvist, M.; Liao, W.-W., *Chem. Commun.* **2005**, 3586.
52. Ibrahem, I.; Zou, W.; Xu, Y.; Córdova, A., *Adv. Synth. Catal.* **2006**, *348*, 211.
53. Xu, Y.; Zou, W.; Sundén, H.; Ibrahem, I.; Córdova, A., *Adv. Synth. Catal.* **2006**, *348*, 418.
54. Westermann, B.; Neuhaus, C., *Angew. Chem., Int. Ed.* **2005**, *44*, 4077.
55. Enders, D.; Grondal, C.; Vrettou, M., *Synthesis* **2006**, 3597.
56. Hosseini, M.; Stiasni, N.; Barbieri, V.; Kappe, C. O., *J. Org. Chem.* **2007**, *72*, 1417.
57. Zhang, H.; Ramasastry, S. S. V.; Tanaka, F.; Barbas, C. F., *Adv. Synth. Catal.* **2008**, *350*, 791.
58. Hayashi, Y.; Urushima, T.; Aratake, S.; Okano, T.; Obi, K., *Org. Lett.* **2007**, *10*, 21.
59. Ibrahem, I.; Zou, W.; Casas, J.; Sundén, H.; Córdova, A., *Tetrahedron* **2006**, *62*, 357.
60. Ogura, Y.; Watanabe, H., *Tetrahedron Lett.* **2010**, *51*, 3294.
61. Enders, D.; Haas, M., *Synlett* **2003**, 2182.
62. Ortiz, J. C.; Ozores, L.; Cagide-Fagin, F.; Alonso, R., *Chem. Commun.* **2006**, 4239.
63. Martínez-Bescos, P.; Cagide-Fagín, F.; Roa, L. F.; Ortiz-Lara, J. C.; Kierus, K.; Ozores-Viturro, L.; Fernández-González, M.; Alonso, R., *J. Org. Chem.* **2008**, *73*, 3745.
64. Cagide-Fagín, F.; Alonso, R., *Eur. J. Org. Chem.* **2010**, 6741.
65. Maeng, J.-H.; Funk, R. L., *Org. Lett.* **2001**, *3*, 1125.
66. Fuchs, J. R.; Funk, R. L., *Org. Lett.* **2001**, *3*, 3349.
67. Greshock, T. J.; Funk, R. L., *Org. Lett.* **2001**, *3*, 3511.
68. Fuchs, J. R.; Funk, R. L., *Org. Lett.* **2001**, *3*, 3923.
69. Maeng, J.-H.; Funk, R. L., *Org. Lett.* **2002**, *4*, 331.
70. He, Y.; Funk, R. L., *Org. Lett.* **2006**, *8*, 3689.
71. Nilson, M. G.; Funk, R. L., *J. Am. Chem. Soc.* **2011**, *133*, 12451.
72. Funk, R. L.; Belmar, J., *Tetrahedron Lett.* **2012**, *53*, 176.
73. Abaee, M. S.; Mojtahedi, M. M.; Hamidi, V.; Mesbah, A. W.; Massa, W., *Synthesis* **2008**, 2122.

Dieter Enders, Stefanie Breukers & Linus F. Reichenbach
RWTH Aachen University, Aachen, Germany

Dimethyldioxirane[1]

[74087-85-7] $C_3H_6O_2$ (MW 74.09)
InChI = 1/C3H6O2/c1-3(2)4-5-3/h1-2H3
InChIKey = FFHWGQQFANVOHV-UHFFFAOYAF

(selective, reactive oxidizing agent capable of epoxidation of alkenes and arenes,[11] oxyfunctionalization of alkanes,[19] and oxidation of alcohols,[23] ethers,[21] amines, imines,[32] and sulfides[35])

Alternate Name: DDO.
Physical Data: known only in the form of a dilute solution.
Solubility: soluble in acetone and CH_2Cl_2; soluble in most other organic solvents, but reacts slowly with many of them.

Form Supplied in: dilute solutions of the reagent in acetone are prepared from Oxone and acetone, as described below.

Analysis of Reagent Purity: the concentrations of the reagent can be determined by classical iodometric titration or by reaction with an excess of an organosulfide and determination of the amount of sulfoxide formed by NMR or gas chromatography.

Preparative Methods: the discovery of a convenient method for the preparation of dimethyldioxirane has stimulated important advances in oxidation technology.[1] The observation[2] that ketones enhance the decomposition of the monoperoxysulfate anion prompted mechanistic studies that implicated dioxiranes as intermediates.[3] Ultimately, these investigations led to the isolation of dilute solutions of several dioxiranes.[4] DDO is by far the most convenient of the dioxiranes to prepare and use (eq 1). Several experimental set-ups for the preparation of DDO have been described,[4–6] but reproducible generation of high concentration solutions of DDO (ca 0.1M) is aided by a well-formulated protocol.[6] The procedure involves the portionwise addition of solid Oxone (Potassium Monoperoxysulfate) to a vigorously stirred solution of $NaHCO_3$ in a mixture of reagent grade Acetone and distilled water at 5–10 °C. The appearance of a yellow color signals the formation of DDO, at which point the cooling bath is removed and the DDO–acetone solution is distilled into a cooled (−78 °C) receiving flask under reduced pressure (80–100 Torr). After preliminary drying over reagent grade anhydrous $MgSO_4$ in the cold, solutions of DDO are stored over molecular sieves in the freezer of a refrigerator at −10 to −20 °C. In instances where the concentration of DDO is crucial, analysis is typically based on reaction with an excess of an organosulfide monitored by NMR.[4,7,8]

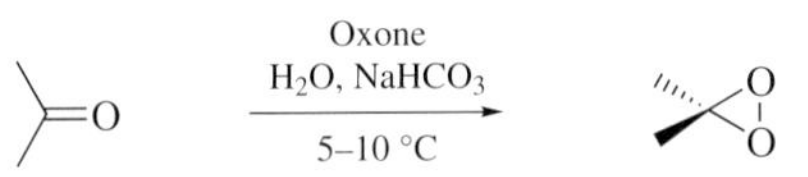

(1)

Concentrated solutions of DDO in chlorinated solvents may be obtained by a simple extraction technique.

A fresh solution (50 mL) of isolated DDO (0.06–0.08 M in acetone), prepared as reported, is diluted with an equal volume of cold water (0–5 °C) and extracted in a chilled separatory funnel with four 10 mL portions of cold CH_2Cl_2, $CHCl_3$, or CCl_4 to yield a total volume of ca. 35 mL of extract (pale yellow DDO solution). In order to concentrate this DDO solution, the combined extracts in chlorinated solvent are washed three times in a separatory funnel at 0–5 °C with an equal volume of cold 0.01 M phosphate buffer (pH 7). The resulting solution is 0.19–0.36 M in DDO. Its concentration can be estimated by iodometry; the recovery of dioxirane from the initial acetone solution is 35–45% in most cases. ^{1}H NMR spectroscopy analysis reveals that initial solvent acetone is not completely eliminated.

Handling, Storage, and Precaution: solutions of the reagent can be kept in the freezer of a refrigerator (−10 to −20 °C) for as long as a week. The concentration of the reagent decreases relatively slowly, provided solutions are kept from light and traces of heavy metals. These dilute solutions are not known to decompose violently, but the usual precautions for handling peroxides should be applied, including the use of a shield. All reactions should be performed in a fume hood to avoid exposure to the volatile oxidant.

Original Commentary

Jack K. Crandall
Indiana University, Bloomington, IN, USA

Introduction. Reactions with DDO are typically performed by adding the cold reagent solution to a cold solution of a reactant in acetone or some other solvent. CH_2Cl_2 is a convenient solvent which facilitates reaction in a number of cases. After the reactant has been consumed, as monitored by TLC, etc., the solvent and excess reagent are simply removed to provide a nearly pure product. An excess of DDO is often used to facilitate conversion, provided further oxidation is not a problem. Where the product is especially sensitive to acid, the reaction can be run in the presence of solid Potassium Carbonate as an acid scavenger and drying agent. When it is important to minimize water content, the use of powdered molecular sieves in the reaction mixture is recommended. Reactions can be run from ambient temperatures down to −78 °C.

Dimethyldioxirane is a powerful oxidant, but shows substantial selectivity in its reactions. It has been particularly valuable for the preparation of highly reactive products, since DDO can be employed under neutral, nonnucleophilic conditions which facilitate the isolation of such species. Whereas DDO performs the general conversions of more classic reagents like m-Chloroperbenzoic Acid, it generates only an innocuous molecule of acetone as a byproduct. This is to be contrasted with peracids whose acidic side-products can induce rearrangements and nucleophilic attack on products. Although several other dioxiranes have been prepared, these usually offer no advantage over DDO. An important exception is Methyl(trifluoromethyl)dioxirane, whose greater reactivity is advantageous in situations where DDO reacts sluggishly, as in the oxyfunctionalization of alkanes.

The need to prepare DDO solutions beforehand, the low yield of the reagent based on Potassium Monoperoxysulfate (Oxone) (ca. 5%),[6] and the inconvenience of making DDO for large-scale reactions are drawbacks that can be avoided when the product has good stability. In these instances, an in situ method for DDO oxidations is recommended.

Oxidation of Alkenes and Other Unsaturated Hydrocarbons. The epoxidation of double bonds has been the major area for the application of DDO methodology and a wide range of alkenes are effectively converted to epoxides by solutions of DDO.[4,7] Epoxidation is stereospecific with retention of alkene stereochemistry, as shown by the reactions of geometrical isomers; for example, (*Z*)-1-phenylpropene gives the *cis*-epoxide cleanly (eq 2), whereas the (*E*) isomer yields the corresponding *trans*-epoxide. Rate studies indicate that this reagent is electrophilic in nature and that alkyl substitution on the double bond enhances reactivity.[7] Interestingly, *cis*-disubstituted alkenes react 7–9 times faster than the *trans* isomers, an observation that has been interpreted in terms of a 'spiro' transition state.[9]

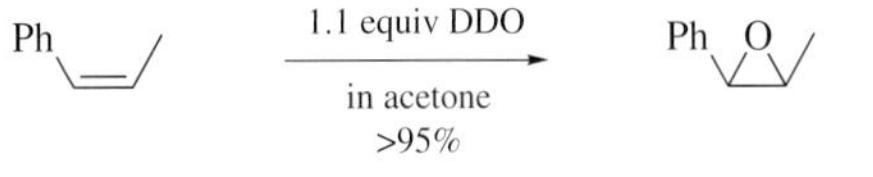

(2)

From a preparative viewpoint, the use of DDO solutions, while efficient and easy to perform, are generally not needed for simple alkenes that give stable epoxides. Rather, in situ methodology is suggested. However, the extraordinary value of isolated DDO has been amply demonstrated for the generation of unstable epoxides that would not survive most epoxidation conditions.[1] A good example of this sort of application is the epoxidation of precocenes, as exemplified in eq 3.[10] A number of impressive epoxidations have been reported for oxygen substituted alkenes, including enol ethers, silyl enol ethers, enol carboxylates, etc.[1] Examples include a number of 1,2-anhydro derivatives of monosaccharides.[11] Steric features often result in significant stereoselection in the epoxidation, as illustrated in eq 4.[11] Conversions of alkenes with two alkoxy substituents have also been achieved (eq 5), even when the epoxides are not stable at rt.[12]

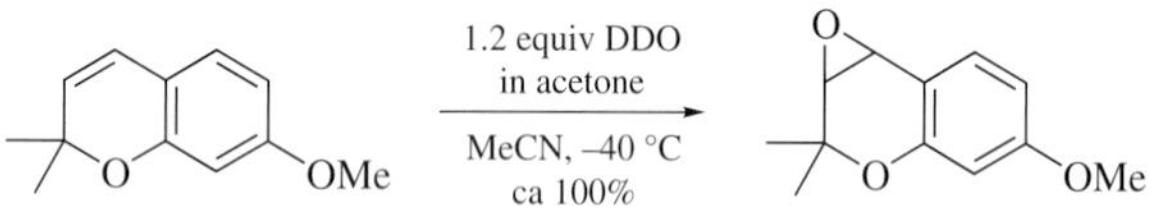

(3)

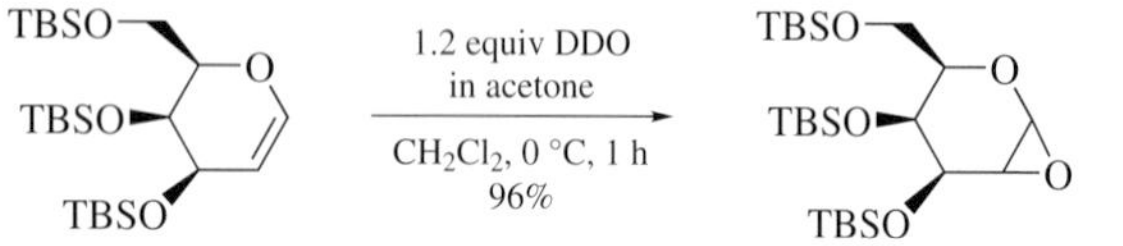

(4)

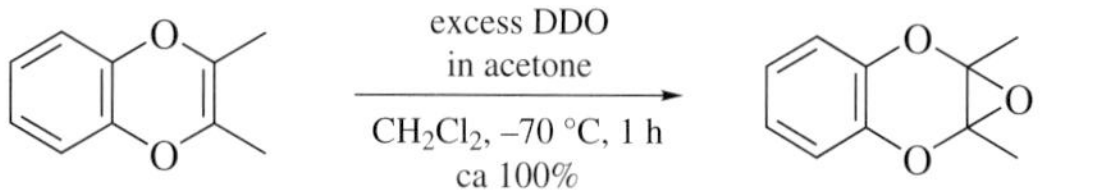

(5)

Although reactions are much slower with conjugated carbonyl compounds, DDO is still effective for the epoxidation of these electron-deficient double bonds (eq 6).[13] Alkoxy-substitution on such conjugated alkenes can also be tolerated (eq 7).[14]

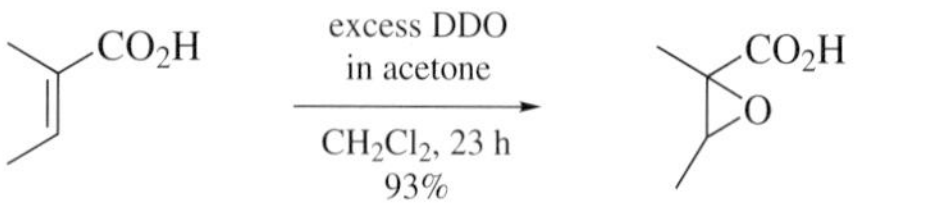

(6)

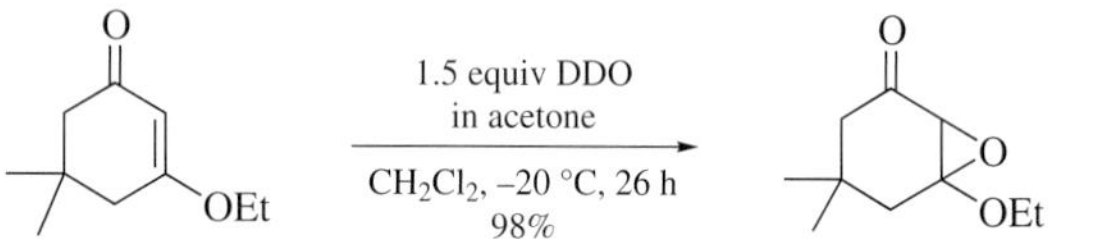

(7)

Allenes react with DDO by sequential epoxidation of the two double bonds to give the previously inaccessible, highly reactive allene diepoxides.[15] In the case of the *t*-butyl-substituted allene shown in eq 8, a single diastereomer of the diepoxide is generated, owing to steric control of the *t*-butyl group on reagent attack.

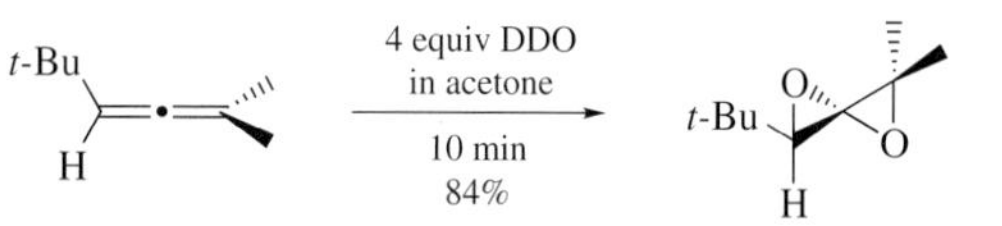

(8)

Certain polycyclic aromatic hydrocarbons can be converted to their epoxides, as typified by the reaction of phenanthrene with DDO (eq 9).[4] Aromatic heterocycles like furans and benzofurans also give epoxides, although these products are quite susceptible to rearrangement, even at subambient temperatures (eq 10).[16] The oxidation of heavily substituted phenols by DDO leads to quinones, as shown in eq 11, which illustrates the formation of an orthoquinone.[17] The corresponding hydroquinones are intermediates in these reactions, but undergo ready oxidation to the quinones.

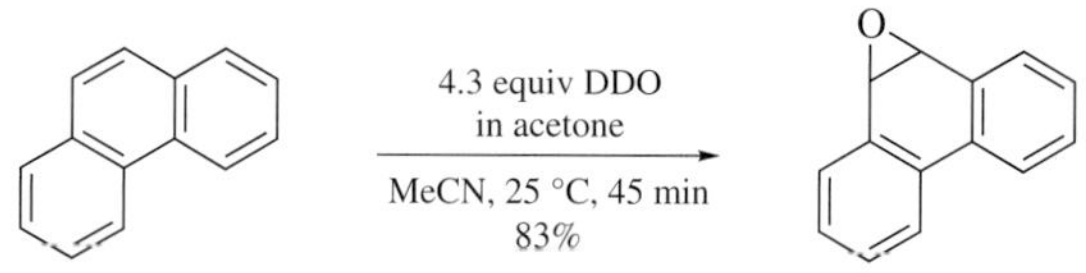

(9)

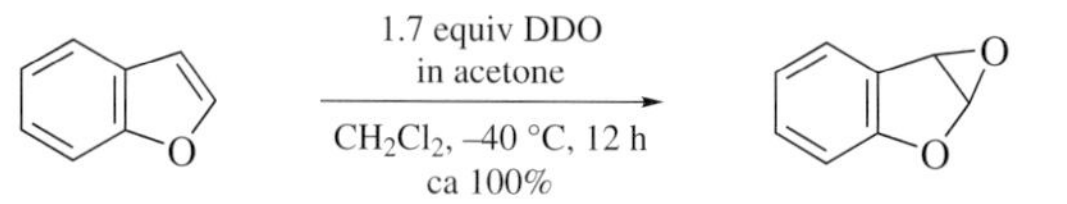

(10)

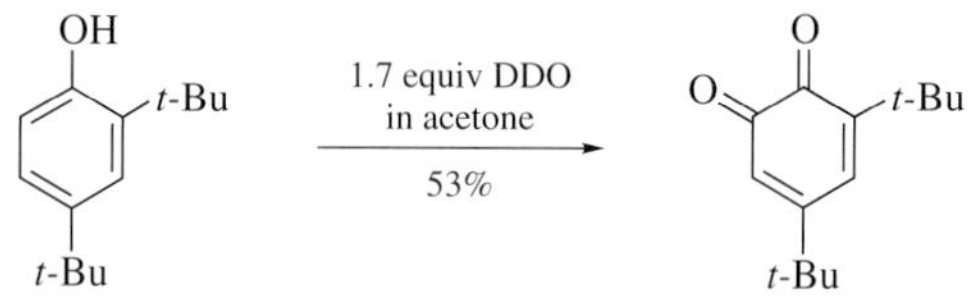

(11)

Finally, preformed lithium enolates are converted to α-hydroxy ketones by addition to a cold solution of DDO (eq 12).[18]

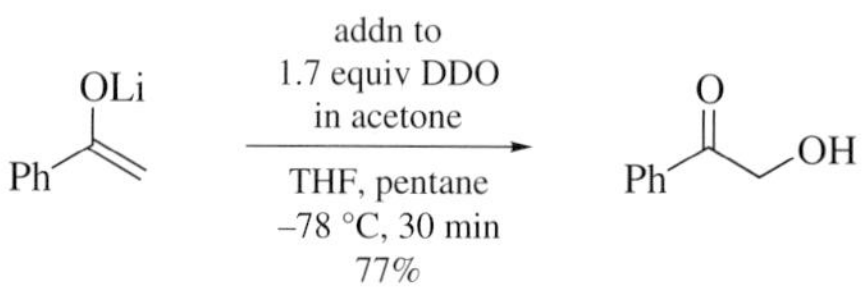

(12)

Oxidation of Saturated Hydrocarbons, Ethers, and Alcohols. Surely the most striking reaction of dioxiranes is their ability to functionalize unactivated C–H bonds by the insertion of an oxygen atom into this σ-bond. This has opened up an important new area of oxidation chemistry.[1] While DDO has been used in a number of useful transformations outlined below, the more reactive Methyl(trifluoromethyl)dioxirane is often a better reagent for this type of conversion, despite its greater cost and difficulty of preparation.

The discrimination of DDO for tertiary > secondary > primary C–H bonds of alkanes is more pronounced than that of the *t*-butoxide radical.[19] Good yields of tertiary alcohols can be secured in favorable cases, as in the DDO oxidation of adamantane to 1-adamantanol, which occurs with only minor reaction at C-2 (eq 13). Of major significance is the observation that these reactions are stereospecific with high retention of configuration, as illustrated by the oxidation of *cis*-dimethylcyclohexane shown in (eq 14); the *trans* isomer gives exclusively the diastereomeric alcohol. This and other data have been interpreted in terms of an 'oxenoid' mechanism for the insertion into the C–H bond. Several interesting applications in the steroid field involve significant site selectivity as well.[20] The slower reactions of DDO with hydrocarbons without tertiary hydrogens are less useful and lead to ketones owing to a rapid further oxidation of the initially formed secondary alcohol. For example, cyclododecane is converted to cyclododecanone.

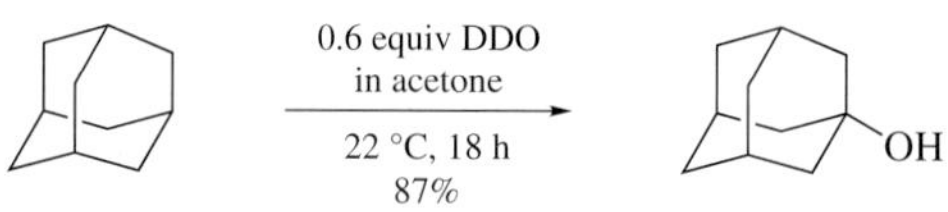

(13)

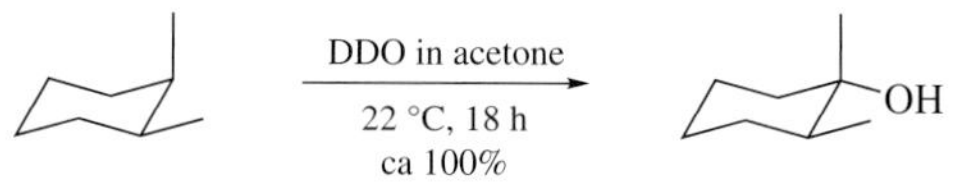

(14)

Ethers and acetals are slowly converted by DDO to carbonyl compounds. This serves as a nontraditional method for deprotection of these derivatives, an example of which is shown in (eq 15).[21,22] Hemiacetals are presumed intermediates in these transformations.

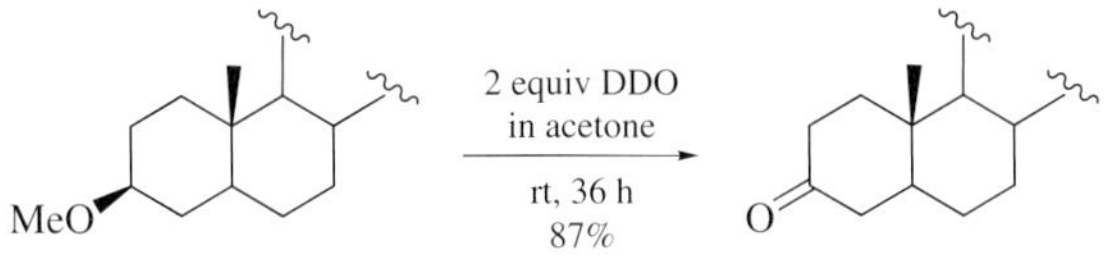
(15)

While DDO has been little used for the oxidation of simple alcohols, it has found application in useful conversions of vicinal diols. The oxidation of tertiary–secondary diols to α-hydroxy ketones occurs without the usual problem of oxidative cleavage between the two functions (eq 16).[23] DDO has also been used to convert appropriate optically active diols selectively into α-hydroxy ketones of high optical purity; for example, see (eq 17).[24]

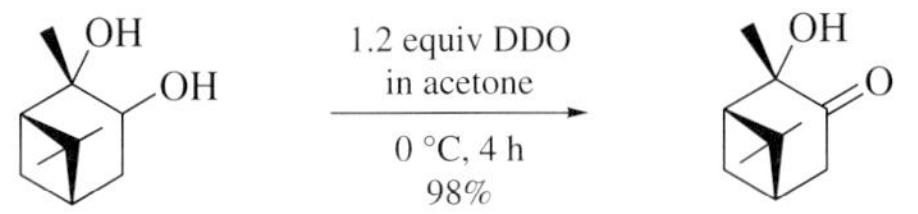
(16)

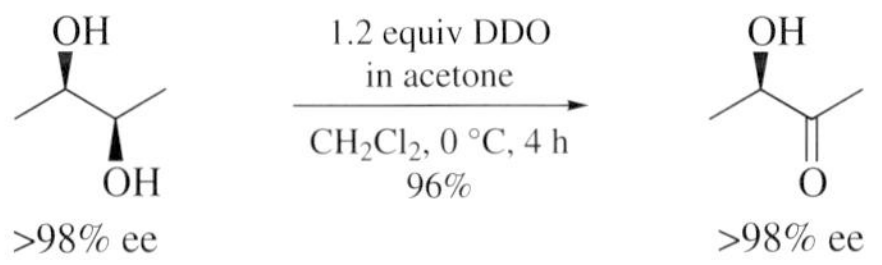
(17)

Finally, the Si–H bond of silanes suffers analogous oxidation to silanols upon reaction with DDO. This reaction takes place with retention of configuration and is, as expected, more facile than C–H oxidations.[25]

Oxidation of Nitrogen Functional Groups. Selective oxidations of nitrogen compounds are often difficult to achieve, but DDO methodology has been shown to be very useful in a number of instances. For example, one of the first applications of this reagent was in the conversion of primary amines to the corresponding nitro compounds (eq 18).[26] This process probably proceeds by successive oxidation steps via hydroxylamine and nitroso intermediates. Complications arise with unhindered primary aliphatic amines, owing to dimerization of the intermediate nitrosoalkanes and their tautomerization to oximes.[27] In oxidations of amino sugar and amino acid derivatives, it is possible to isolate the initially formed hydroxylamines (eq 19).[28]

(18)

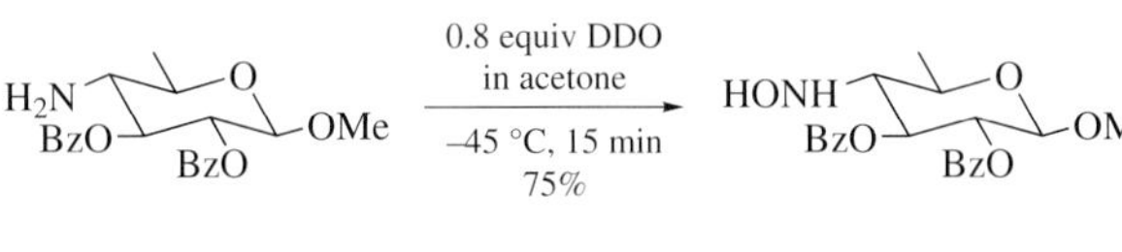
(19)

The oxidation of secondary amines to hydroxylamines is readily achieved with 1 equiv of DDO (eq 20).[29] The use of 2 equiv of DDO results in further oxidation, the nature of which depends on the structure of the amine. Thus cyclic secondary amines which do not possess α-hydrogens are converted to nitroxides,[30] as illustrated in (eq 21). Secondary benzylamines give nitrones (eq 22).[31]

$$(PhCH_2)_2NH \xrightarrow[\text{0 °C, 15 min; 98\%}]{\text{1 equiv DDO, in acetone}} (PhCH_2)_2NOH \quad (20)$$

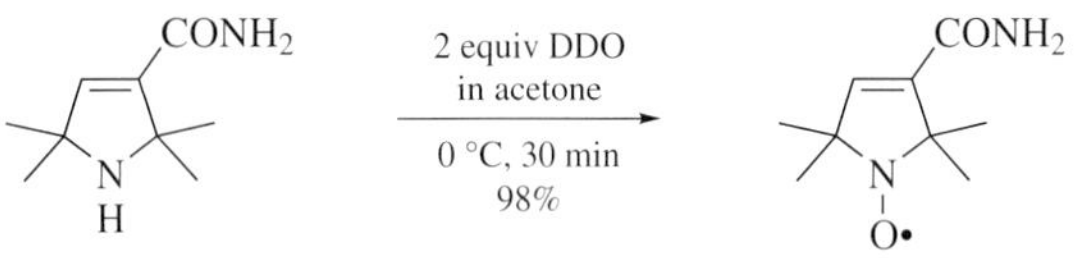
(21)

$$PhCH_2NH\text{-}t\text{-}Bu \xrightarrow[\text{0 °C, 10 min; 96\%}]{\text{2 equiv DDO, in acetone}} PhCH{=}N(O)\text{-}t\text{-}Bu \quad (22)$$

A related transformation is the oxidation of imines to nitrones by DDO (eq 23).[32] It is interesting that the isomeric oxaziridines are not produced here, given that peracids favor these heterocycles.

$$C_6Me_5CH{=}NMe \xrightarrow[\text{CH}_2\text{Cl}_2\text{, 0 °C, 2 h; 71\%}]{\text{1.1 equiv DDO, in acetone}} C_6Me_5CH{=}N(O)Me \quad (23)$$

Reaction of α-diazo ketones with DDO leads to α-keto aldehyde hydrates (eq 24).[33] Oximes are converted to the free ketones by DDO.[34]

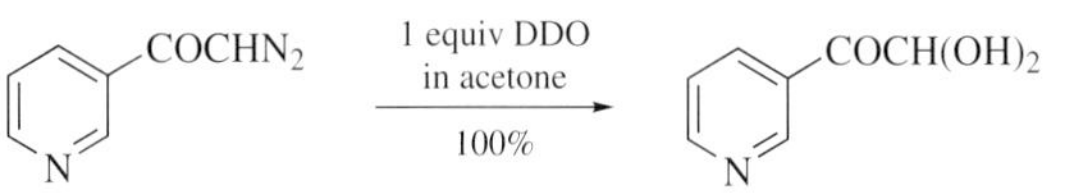
(24)

Oxidation of Sulfur Functional Groups. Dimethyldioxirane rapidly oxidizes sulfides to sulfoxides and converts sulfoxides to sulfones (eq 25).[4,35] The partial oxidation of sulfides to sulfoxides can be controlled by limiting the quantity of DDO. Since Oxone is one of the many reagents that can perform these reactions, the extra effort involved in preparing DDO solutions is often not warranted. An exception involves the transformation of thiophenes to the corresponding sulfones (eq 26).[36] A similar procedure gives α-oxo sulfones by DDO oxidation of thiol esters (eq 27).[37]

$$PhSMe \xrightarrow{\text{DDO}} PhSOMe \xrightarrow{\text{DDO}} PhSO_2Me \quad (25)$$

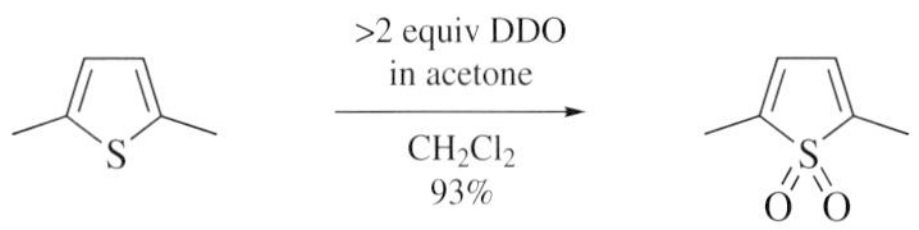
(26)

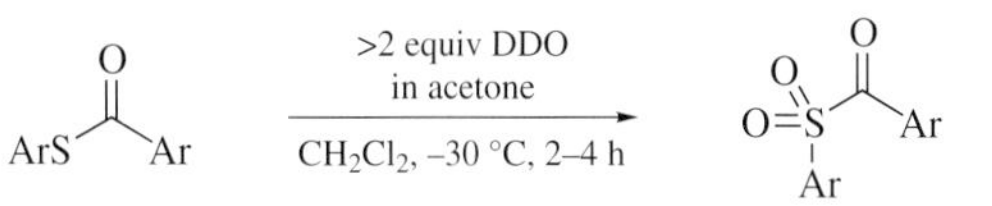
(27)

Alkanethiols are selectively oxidized to alkanesulfinic acids by DDO (eq 28).[38] Air oxidation of an intermediate species appears to be important in this transformation.

$$Me(CH_2)_4SH \xrightarrow{\text{DDO, in acetone}} \xrightarrow{O_2} Me(CH_2)_4SO_2H \quad (28)$$

First Update

Ruggero Curci, Lucia D'Accolti & Caterina Fusco
Universitá di Bari, Bari, Italy

Direct Functionalization of C–H Bonds by Dimethyldioxirane.[1,39] The efficient oxyfunctionalization of simple, "unactivated" C–H bonds of alkanes under extremely mild conditions undoubtedly is one of the major highlights of dioxirane chemistry.[1,39]

Although less effective than methyl(trifluoromethyl)dioxirane (TFDO), oxyfunctionalization of unactivated methine C–Hs with dimethyldioxirane (DDO) is feasible for various substituted steroids related to the 5β-cholane and 5α-cholestane series to give novel mono- and dihydroxylated steroids.[40] The reactivity and site selectivity of oxyfunctionalization is affected conspicuously by the structural and steric environments of the target methine carbon atoms. This nonenzymatic procedure may be advantageously applied to selective and short-course syntheses of bioactive steroids. Thus, the major reaction product of methyl 3α-acetoxy-5α-cholan-24-oate with DDO in 36 h was the corresponding 5β-hydroxylated compound in 48% yield, while concurrent double oxyfunctionalization at the C-5 and C-17 produced the corresponding dihydroxylated derivative in comparable yield (36%) (eq 29).[40]

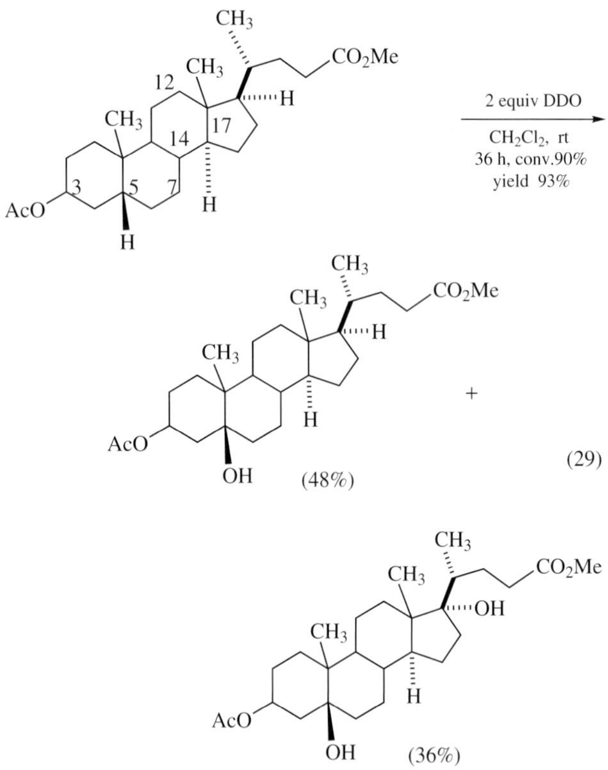

(29)

Both the 5β- and 17α-hydroxylation proceeded stereoselectively and the configuration of the resulting C–OH was the same as that of the original methine C–H bond. In order to accelerate the reaction of *O*-insertion (otherwise rather sluggish), instead of the usual DDO solution (up to 0.11 M) in acetone, the aformentioned concentrated DDO solutions (0.33–0.35 M) in $CHCl_3$ were employed.[40] These are obtained by following a procedure of extraction (outlined above) of the dioxirane into the chlorinated solvent,[41] similar to that devised to obtain ketone-free TFDO solutions.[42]

In another study where DDO was employed for *tert*-C–H hydroxylation of several di- and triacetates of (5β)-bile acid methyl esters, it was reported that derivatives bearing a 7-acetoxy group give 17α- or 14α-hydroxylated products in addition to the 5β-hydroxylated ones.[43]

For bile acid esters, the DDO oxidation of *sec*-CHOH groups of hydroxy cholate methyl esters to the corresponding carbonyls (via *gem*-diols) occurs readily; the positional order of reactivity C3 $\simeq$; C7 > C6 > C12 was established.[44]

In line with the effective oxyfunctionalizations recorded for steroids, the DDO hydroxylation of cephalostatin derivatives provide another interesting example. In fact, the DDO oxidation hecogenin acetate β-hydroxyketone in (eq 30) results in *O*-insertion into C16–H to yield the corresponding hemiketal with amazing site-selectivity.[45a] This key transformation paves the road for synthesis of the cephalostatin North 1 hemisphere.[45]

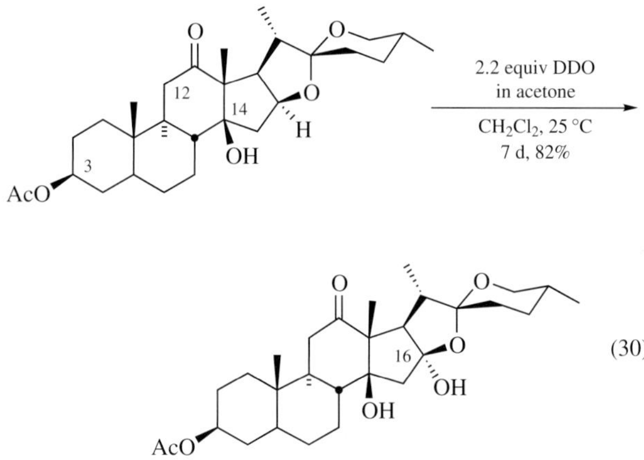

(30)

Remarkable site-selectivity is often observed in the hydroxylation of protected *N*-Boc derivatives of α-amino acid esters bearing an alkyl side chain (Boc-Gly-OMe, Boc-Ala-OMe, Boc-Val-OMe, Boc-Ile-OMe, Boc-Phe-OMe); this results in different products depending on the structure.[46] Although these reactions are rather sluggish, requiring long reaction times for sizable substrate conversion, they offers a novel entry to side-chain modified α-amino acids and peptides that avoids multi-step synthetic approaches. The reactivity and site-selectivity depends on the steric environments and electron density of the target C–H bonds in the side chain; high regioselectivity for the *O*-insertion into the γ-C–H bond of leucine (Leu) residues with respect to the weaker α,C–H bonds is observed. Thus, Boc-Leu-OMe was found to yield the corresponding 4,4-dimethyl-4-butanolide derivative; the latter is formed by selective *O*-insertion into the tertiary γ-C–H bond of Leu followed by cyclization (eq 31).

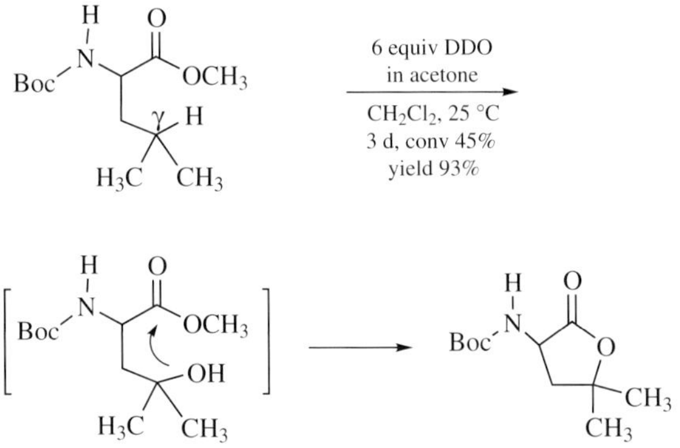

(31)

A position selectivity in the oxidation of peptides containing more than one Leu residue was also reported.[46] However, it should be noted that the same transformation in eq 31 gave a markedly lower yield in γ-butanolide using a lower DDO excess.[47] The reaction occurs more rapidly (6 h, 91% conv) using the powerful TFDO (6 equiv); it gives the *N*-hydroxy derivative of the butanolide in 21% yield, along with the uncyclized *N*-hydroxy derivative of the starting Boc-Leu-OMe as the major product (57%).[47]

In general, benzhydrylic C–H bonds (only slightly more reactive than *tert*-C–Hs) are distinctly more reactive than benzylic C–H bonds toward dioxirane *O*-insertion.[39,48,49] However, special situations arise in the selective hydroxylation of complex polycyclic indan hydrocarbons, i.e., the centropolyindans.[50] For instance, the rigidity of the polycyclic framework and steric factors in the angular centrotriindan 1,1′-(*o*-phenylene)-2,2′-spirobiindan moderate the otherwise distinct selectivity for benzhydrylic vs. benzylic C–H *O*-insertion by DDO; so then, spirobiindanone is formed along with the tertiary 4bα-alcohol (eq 32).[50]

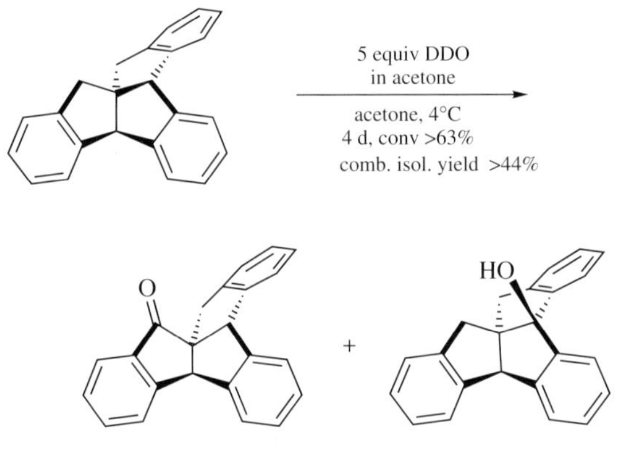

(32)

(28%) (16%)

Triptindane, another propellane-type centropolyindan, was found to react with excess DDO yielding triptindan-9-one as the major product (37% yield) at the conditions given in (eq 33).[50]

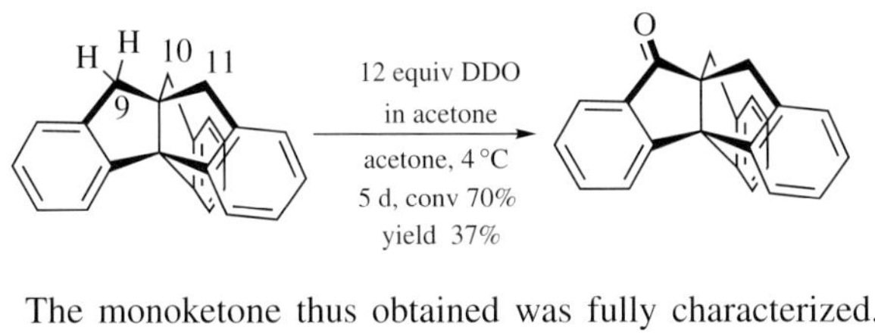

(33)

The monoketone thus obtained was fully characterized. The formation of sizeable amounts of more highly oxidized products (triptindan-9,10-dione and triptindan-9,10,11-trione) was detected by mass spectrometry.

Turning to polycyclic saturated hydrocarbons, the shown (eq 13) selective oxyfunctionalization of adamantane (*requiring 6 equiv of the powerful TFDO for exhaustive bridgehead hydroxylation*)[11] serves well to illustrate the high tertiary vs. secondary selectivities that are customarily observed. The ratio of tertiary to secondary carbons and the different reactivity towards oxidation for each type of C–H bond in Binor S renders it an attractive probe for the study of regioselectivity of oxyfunctionalization. In fact, this saturated heptacyclic hydrocarbon (the head-to-head dimer of norbornadiene) consists of two nortricyclane units, each containing one three-membered ring and three five-membered rings; it presents two symmetric methylene groups and 12 tertiary carbons ordered in four different geometries.

Treatment of Binor S in CH_2Cl_2 with aqueous (pH 7) monoperoxosulfate (caroate)/acetone (DDO in situ) afforded Binor S 1-ol in 98% yield.[51] Further oxidation of this material with isolated DDO gives the symmetrical 1,9-diol as the major product (eq 34).[51]

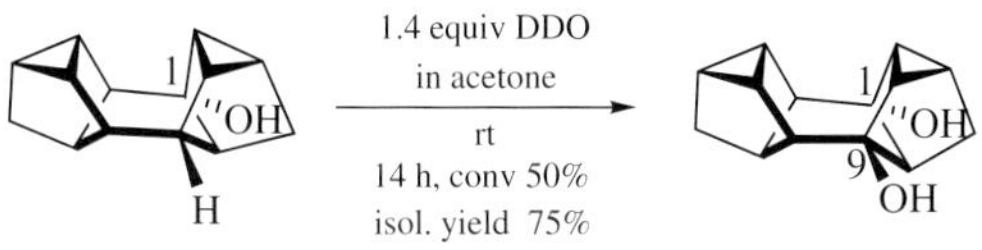

(34)

The examples above further demonstrate that dioxirane reactions are characterized by high selectivity and that *tert*-C–H bonds are considerably more reactive towards dioxirane *O*-insertion than their *sec*-C–H complements.[39] However, the cyclopropane moiety, if suitably oriented,[52,53] can have a marked influence in activating proximal α-C–H bonds towards dioxirane oxyfunctionalization.[52]

This is illustrated by the application of dioxiranes to polycyclic alkanes possessing a sufficiently rigid framework, such as the 2,4-didehydroadamantane case. For this substrate the reaction with DDO proceeds with 82% conversion during 12 h, yielding the expected 2,4-didehydroadamantan-7-ol, but also the precious 2,4-didehydroadamantan-10-one in comparable yield (eq 35).[54] The ketone derives from competitive dioxirane attack at the methylene positions α to the cyclopropyl moiety; the latter is encompassed in the rigid 2,4-didehydroadamantane framework to lay constrained into a "bisected"[52] orientation relative to the neighboring methylene C–H bonds.

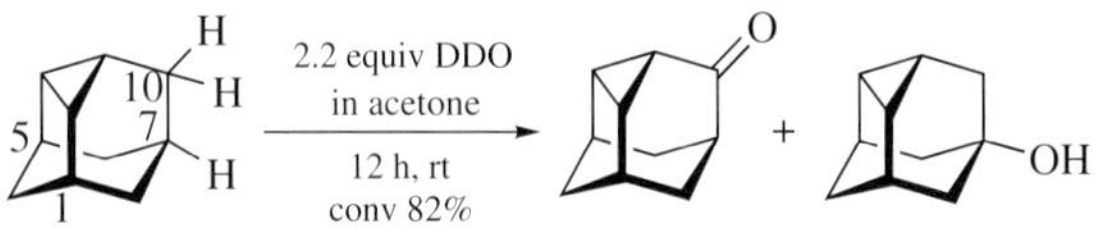

(35)

(29%) (21%)

Hydroxylation at the bridgehead C1 and C5 does not take place because the bridgehead *tert*-C–H bonds are deactivated by the proximal cyclopropyl moiety lying perpendicularly. It is worth of note that, at variance with what observed with other oxidants (e.g., dry ozone), no rearranged products are observed in the reaction of DDO with this target compound.[54]

It was mentioned that oxidation by DDO allows the clean conversion of secondary alcohols into carbonyls under mild conditions. In this transformation the dioxiranes rank high with respect to transition-metal oxidants because of their efficiency, superior versatility, and ease of operations. Based on kinetic data and the application of reaction probes,[55] the oxidation proceeds via a substantially concerted *O*-insertion by the dioxirane into the C–H bond "alpha" to the OH functionality generating a *gem*-diol $C(OH)_2$, hence the carbonyl. As shown by the example in eq 36, remarkable chemoselectivity is achieved in the oxidation of epoxy alcohols in that the corresponding epoxy ketones are formed in high yield, while the epoxy functionality remains untouched.[56] The epoxy ketone in eq 36 is a key intermediate in the convergent synthesis of active 1α,25-dihydroxyvitamin D_3 analogs.[57]

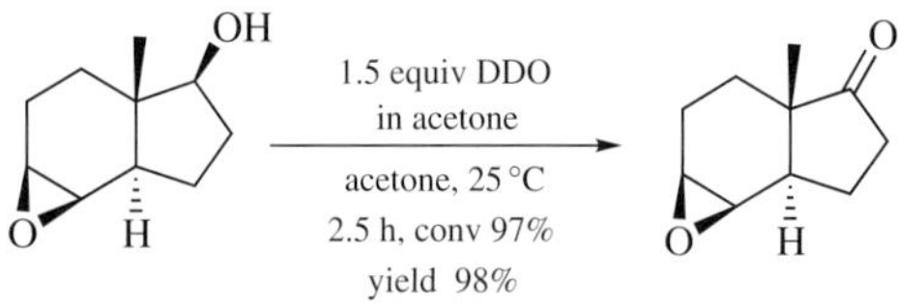

(36)

Both open chain and cyclic epoxy alcohols can be neatly transformed into the corresponding epoxy ketones with high conversions and yields using just 1.1–1.5 equiv of DDO oxidant. Also, the conversion of optically active epoxy alcohols into epoxy ketones occurs selectively leaving the configuration at the chiral center(s) at the oxirane ring unaffected.[56]

Related Reagents. Potassium Monoperoxosulfate (Oxone); Potassium Monoperoxosulfate (Oxone)/Acetone (DDO in situ); Methyl(trifluoromethyl)Dioxirane.

---

1. (a) Adam, W.; Hadjiarapoglou, L. P.; Curci, R.; Mello, R. In *Organic Peroxides*; Ando, W., Ed.; Wiley: New York,1992; Chapter 4, p 195. (b) Murray, R. W., *Chem. Rev.* **1989**, *89*, 1187. (c) Curci, R. In *Advances in Oxygenated Processes*; Baumstark, A., Ed; JAI: Greenwich, CT, 1990; Vol. 2, Chapter 1, p 1. (d) Adam, W.; Edwards, J. O.; Curci, R., *Acc. Chem. Res.* **1989**, *22*, 205. (e) Adam, W.; Hadjiarapoglou, L., *Top. Curr. Chem.* **1993**, *164*, 45.
2. Montgomery, R. E., *J. Am. Chem. Soc.* **1974**, *96*, 7820.
3. Edwards, J. O.; Pater, R. H.; Curci, P. R.; Di Furia, F., *Photochem. Photobiol.* **1979**, *30*, 63.
4. Murray, R. W.; Jeyaraman, R., *J. Org. Chem.* **1985**, *50*, 2847.
5. Eaton, P. E.; Wicks, G. E., *J. Org. Chem.* **1988**, *53*, 5353.
6. Adam, W.; Bialas, J.; Hadjiarapoglou, L., *Chem. Ber.* **1991**, *124*, 2377.
7. Baumstark, A. L.; Vasquez, P. C., *J. Org. Chem.* **1988**, *53*, 3437.
8. Murray, R. W.; Shiang, D. L., *J. Chem. Soc., Perkin Trans. 2* **1990**, *2*, 349.
9. Baumstark, A. L.; McCloskey, C. J., *Tetrahedron Lett.* **1987**, *28*, 3311.
10. Bujons, J.; Camps, F.; Messeguer, A., *Tetrahedron Lett.* **1990**, *31*, 5235.
11. Halcomb, R. L.; Danishefsky, S. J., *J. Am. Chem. Soc.* **1989**, *111*, 6661.
12. Adam, W.; Hadjiarapoglou, L.; Wang, X., *Tetrahedron Lett.* **1991**, *32*, 1295.
13. Adam, W.; Hadjiarapoglou, L.; Nestler, B., *Tetrahedron Lett.* **1990**, *31*, 331.
14. Adam, W.; Hadjiarapoglou, L., *Chem. Ber.* **1990**, *123*, 2077.
15. (a) Crandall, J. K.; Batal, D. J.; Sebesta, D. P.; Lin, F., *J. Org. Chem.* **1991**, *56*, 1153. (b) Crandall, J. K.; Batal, D. J.; Lin, F.; Reix, T.; Nadol, G. S.; Ng, R. A., *Tetrahedron* **1992**, *48*, 1427.
16. (a) Adger, B. M.; Barrett, C.; Brennan, J.; McGuigan, P.; McKervey, M. A.; Tarbit, B., *J. Chem. Soc., Chem. Commun.* **1993**, 1220. (b) Adger, B. M.; Barrett, C.; Brennan, J.; McKervey, M. A.; Murray, R. W., *J. Chem. Soc., Chem. Commun.* **1991**, 1553. (c) Adam, W.; Bialas, J.; Hadjiarapoglou, L.; Sauter, M., *Chem. Ber.* **1992**, *125*, 231.
17. (a) Crandall, J. K.; Zucco, M.; Kirsch, R. S.; Coppert, D. M., *Tetrahedron Lett.* **1991**, *32*, 5441. (b) Altamura, A.; Fusco, C.; D'Accolti, L.; Mello, R.; Prencipe, T.; Curci, R., *Tetrahedron Lett.* **1991**, *32*, 5445. (c) Adam, W.; Schönberger, A., *Tetrahedron Lett.* **1992**, *33*, 53.
18. Guertin, K. R.; Chan, T. H., *Tetrahedron Lett.* **1991**, *32*, 715.
19. Murray, R. W.; Jeyaraman, R.; Mohan, L., *J. Am. Chem. Soc.* **1986**, *108*, 2470.
20. (a) Bovicelli, P.; Lupattelli, P.; Mincione, E.; Prencipe, T.; Curci, R., *J. Org. Chem.* **1992**, *57*, 2182. (b) Bovicelli, P.; Lupattelli, P.; Mincione, E.; Prencipe, T.; Curci, R., *J. Org. Chem.* **1992**, *57*, 5052.
21. Curci, R.; D'Accolti, L.; Fiorentino, M.; Fusco, C.; Adam, W.; González-Nuñez, M. E.; Mello, R., *Tetrahedron Lett.* **1992**, *33*, 4225.
22. van Heerden, F. R.; Dixon, J. T.; Holzapfel, C. W., *Tetrahedron Lett.* **1992**, *33*, 7399.
23. Curci, R.; D'Accolti, L.; Detomaso, A.; Fusco, C.; Takeuchi, K.; Ohga, Y.; Eaton, P. E.; Yip, Y. C., *Tetrahedron Lett.* **1993**, *34*, 4559.
24. D'Accolti, L.; Detomaso, A.; Fusco, C.; Rosa, A.; Curci, R., *J. Org. Chem.* **1993**, *58*, 3600.
25. Adam, W.; Mello, R.; Curci, R., *Angew. Chem., Int. Ed. Engl.* **1990**, *102*, 890.
26. Murray, R. W.; Rajadhyaksha, S. N.; Mohan, L., *J. Org. Chem.* **1989**, *54*, 5783.
27. Crandall, J. K.; Reix, T., *J. Org. Chem.* **1992**, *57*, 6759.
28. Wittman, M. D.; Halcomb, R. L.; Danishefsky, S. J., *J. Org. Chem.* **1990**, *55*, 1981.
29. Murray, R. W.; Singh, M., *Synth. Commun.* **1989**, *19*, 3509.
30. Murray, R. W.; Singh, M., *Tetrahedron Lett.* **1988**, *29*, 4677.
31. Murray, R. W.; Singh, M., *J. Org. Chem.* **1990**, *55*, 2954.
32. Boyd, D. R.; Coulter, P. B.; McGuckin, M. R.; Sharma, N. D., *J. Chem. Soc., Perkin Trans. 1* **1990**, 301.
33. (a) Ihmels, H.; Maggini, M.; Prato, M.; Scorrano, G., *Tetrahedron Lett.* **1991**, *32*, 6215. (b) Darkins, P.; McCarthy, N.; McKervey, M. A.; Ye, T., *J. Chem. Soc., Chem. Commun.* **1993**, 1222.
34. Olah, G. A.; Liao, Q.; Lee, C.-S.; Prakash, G. K. S., *Synlett* **1993**, 427.
35. Murray, R. W.; Jeyaraman, R.; Pillay, M. K., *J. Org. Chem.* **1987**, *52*, 746.
36. Miyahara, Y.; Inazu, T., *Tetrahedron Lett.* **1990**, *31*, 5955.
37. Adam, W.; Hadjiarapoglou, L., *Tetrahedron Lett.* **1992**, *33*, 469.
38. Gu, D.; Harpp, D. N., *Tetrahedron Lett.* **1993**, *34*, 67.
39. Curci, R.; Dinoi, A.; Rubino, M. F., *Pure Appl. Chem.* **1995**, *67*, 811. See references therein.
40. Iida, T.; Yamaguchi, T.; Nakamori, R.; Hikosaka, M.; Mano, N.; Goto, J.; Nambara, T., *J. Chem. Soc., Perkin Trans. 1* **2001**, 2229.
41. Gilbert, M.; Ferrer, M.; Sánchez-Baeza, F.; Messeguer, M., *Tetrahedron* **1997**, *53*, 8643.
42. Adam, W.; Curci, R.; Gonzalés-Nuñez, M. E.; Mello, R., *J. Am. Chem. Soc.* **1991**, *113*, 7654.
43. Cerré, C.; Hofmann, A. F.; Schteingart, C. D., *Tetrahedron* **1997**, *53*, 435.
44. Buxton, P. C.; Marples, B. A.; Toon, R. C.; Waddington, V. L., *Tetrahedron Lett.* **1999**, *40*, 4729.
45. (a) Lee, J. S.; Fuchs, P. L., *Org. Lett.* **2003**, *5*, 2247. (b) La Cour, T. G.; Guo, C.; Boyd, M. R.; Fuchs, P. L., *Org. Lett.* **2000**, *2*, 33.
46. Saladino, R.; Mezzetti, M.; Mincione, E.; Torrini, I.; Paglialunga-Paradisi, M.; Mastropietro, G., *J. Org. Chem.* **1999**, *64*, 8468.
47. Detomaso, A.; Curci, R., *Tetrahedron Lett.* **2001**, *42*, 755.
48. Kuck, D.; Schuster, A., *Z. Naturforsch.* **1991**, *46*, 1223.
49. Fusco, C.; Fiorentino, M.; Dinoi, A.; Curci, R.; Krause, R. A.; Kuck, D., *J. Org. Chem.* **1996**, *61*, 8681.
50. Kuck, D.; Schuster, A.; Fusco, C.; Fiorentino, M.; Curci, R., *J. Am. Chem. Soc.* **1994**, *116*, 2375. See also references therein.
51. Pramod, K.; Eaton, P. E.; Gilardi, R.; Flippen-Anderson, J. L., *J. Org. Chem.* **1990**, *55*, 6105.

52. D'Accolti, L.; Dinoi, A.; Fusco, C.; Russo, A.; Curci, R., *J. Org. Chem.* **2003**, *68*, 7806.
53. Rhodes, Y. E.; DiFate, V. G., *J. Am. Chem. Soc.* **1972**, *94*, 7582.
54. Murray, R. K.; Teager, D. S., *J. Org. Chem.* **1993**, *58*, 5548.
55. Mello, R.; Cassidei, L.; Fiorentino, M.; Fusco, C.; Hümmer, W.; Jäger, V.; Curci, R., *J. Am. Chem. Soc.* **1991**, *113*, 2205.
56. D'Accolti, L.; Fusco, C.; Annese, C.; Rella, M. R.; Turteltaub, J. S.; Williard, P. G.; Curci, R., *J. Org. Chem.* **2004**, *69*, 8510.
57. Nancy E. Lee, N. E.; Reddy, G. S.; Brown, A. J.; Williard, P. G., *Biochemistry* **1997**, *36*, 9429.

(11*bR*)-4-(1,1-Dimethylethyl)-4,5-dihydro-3*H*-dinaphtho[2,1-*c*:1′,2′-*e*]phosphepin

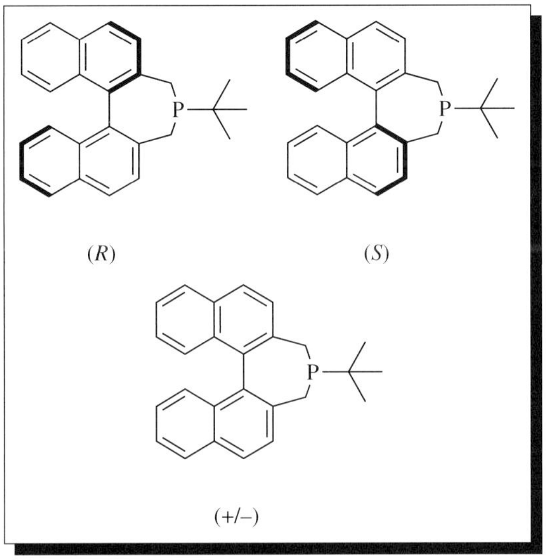

(*R*)
[680220-67-1] $C_{26}H_{25}P$ (MW 368.45)
InChI = 1S/C26H25P/c1-26(2,3)27-16-20-14-12-18-8-4-6-10-22(18)24(20)25-21(17-27)15-13-19-9-5-7-11-23(19)25/h4-15H,16-17H2,1-3H3
InChIKey = ZSJGLMPHZYLIKA-UHFFFAOYSA-N
(*S*)
[475992-24-6]
InChI = 1S/C26H25P/c1-26(2,3)27-16-20-14-12-18-8-4-6-10-22(18)24(20)25-21(17-27)15-13-19-9-5-7-11-23(19)25/h4-15H,16-17H2,1-3H3
InChIKey = ZSJGLMPHZYLIKA-UHFFFAOYSA-N
(+/–)
[515811-32-2]
InChI = 1S/C26H25P/c1-26(2,3)27-16-20-14-12-18-8-4-6-10-22(18)24(20)25-21(17-27)15-13-19-9-5-7-11-23(19)25/h4-15H,16-17H2,1-3H3
InChIKey = ZSJGLMPHZYLIKA-UHFFFAOYSA-N

(chiral and optically active phosphine reagent suitable for application as ligand for homogeneous transition metal catalysis and as chiral organocatalyst for enantioselective transformations)

Physical Data: (*S*)-1: $[\alpha]_D^{24} = +304$ (*c* 0.5, toluene); other physical data unknown.

Solubility: soluble in common organic solvents such as CH_2Cl_2, THF, $PhCH_3$, and EtOAc.

Form Supplied in: unknown.

Preparative Methods: (*S*)-**1** can be prepared from (*S*)-2,2′-dimethyl-1,1′-binaphthyl by double lithiation with *n*-BuLi/TMEDA followed by quenching with *tert*-butyl dichlorophosphine in 60% yield (eq 1).[1] (*S*)-**1** can also be synthesized by the reaction of *tert*-butyl Grignard reagent with the crystalline material (*S*)-4-chloro-4,5-dihydro-3*H*-4-phosphacyclohepta-[2,1-*a*;3,4-*a*′] binaphthalene; the latter can be accessed from the reaction of the dilithium salt with trichlorophosphine (eq 2).[2]

Purification: checked by TLC and purified by column chromatography or recrystallization from toluene.

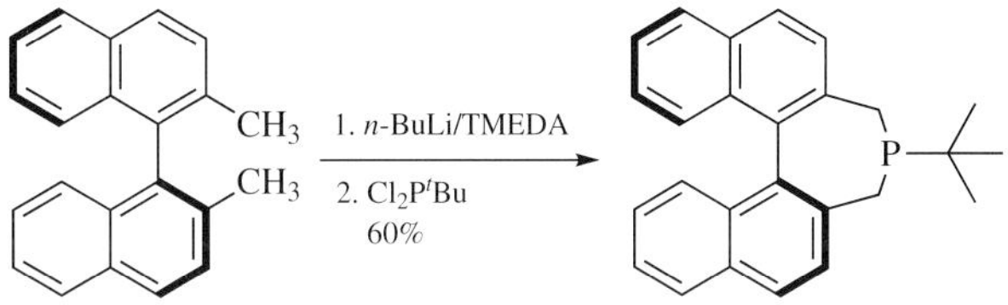

(*S*)-(**1**)

(1)

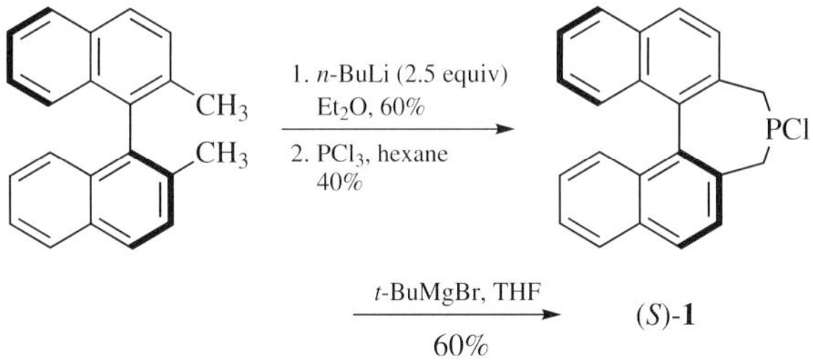

(2)

Handling, Storage, and Precautions: electron-rich phosphine reagents are generally air sensitive and may undergo slow air oxidation; stored in inert atmospheres; phosphine reagents could be irritant and potentially toxic and care should be taken when handling to avoid skin contact and vapor inhalation.

Catalytic Enantioselective Hydrogenation of C=C Double Bonds. (*S*)-**1** has been used as a chiral ligand in the rhodium-catalyzed asymmetric hydrogenation of C=C bond of substituted dehydroamino acid derivatives and enamides. For the hydrogenation of methyl (*Z*)-α-acetamidocinnamate catalyzed by $[Rh(cod)_2]BF_4$/(*S*)-**1**, the reaction proceeded with poor enantioselectivity (20% ee) (eq 3).[1a,b] The aryl-substituted phosphine analog (*S*)-**2** proved to be a more selective ligand. Similarly, poor enantioselectivity was observed for *N*-(1-phenylvinyl)acetamide when (*S*)-**1** was used alone as the ligand (eq 4).[3] Interestingly, when the ligand system was changed to a heterocombination of (*S*)-**1** and phosphite ligand (*S*)-**3**, a remarkable improvement in enantioselectivity was observed.

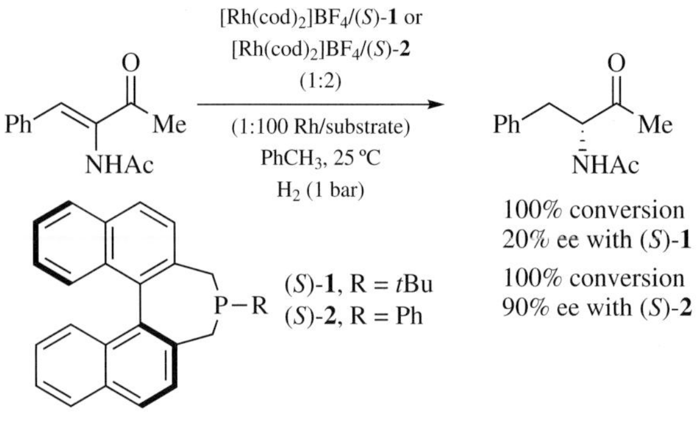

(3)

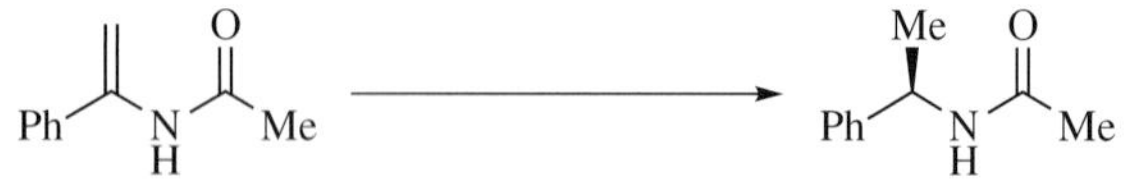

(4)

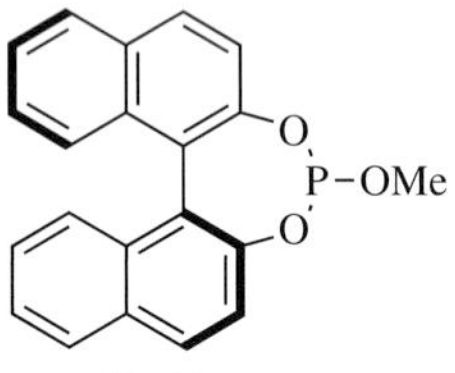

(*S*)-(**3**)

L	ee (% conversion)
(*S*)-**1**	24.4% (94)
(*S*)-**3**	76% (100)
(*S*)-**1** + (*S*)-**3** (1:1)	89% (100)

High enantioselectivity was achieved for the hydrogenation of β-dehydroamino acid derivatives when (*S*)-**1** was used as the ligand (eq 5).[1c] Both the ester group and the group at the β-position were found to have strong influence on the enantioselectivity.

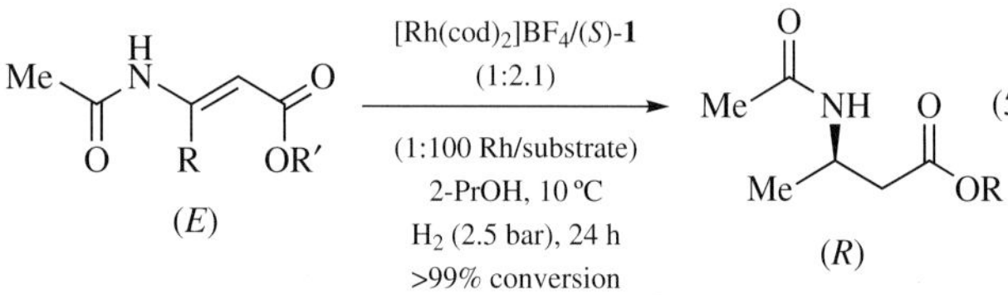

(5)

R = Me, R′ = Me: 87% ee
R = Me, R′ = Et: 80% ee
R = Me, R′ = *i*Pr, 76% ee
R = Et, R′ = Me: 84% ee
R = *i*Pr, R′ = Me: 90% ee

Organocatalysis. Both (*R*)- and (*S*)-**1** have been successfully used as chiral organocatalysts in several synthetically useful transformations. Thus, (*R*)-**1** catalyzes the formation of highly functionalized piperidine derivatives by a [4 + 2]-annulation of imines with allenes (eqs 6 and 7).[4] Enantioselectivities of up to 97% were achieved. The corresponding ethyl and isopropyl analogs of (*R*)-**1** were found to be much less effective for this reaction.

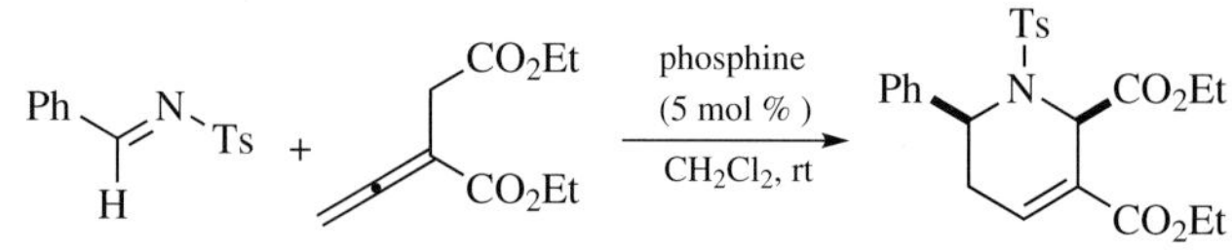

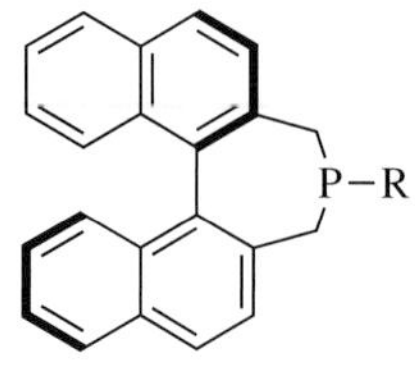

phosphine

R	ee[a]	*cis*:*trans*	isolated yield (%)
Et	0	70:30	46
*i*Pr	51	69:31	49
t-Bu	98	91:9	93

[a]The ee value is for the *cis* diastereomer.

(6)

The synthetic utility of this transformation was demonstrated by the highly diastereo- and enantioselective synthesis of a piperidine derivative containing an indole moiety (eq 8).

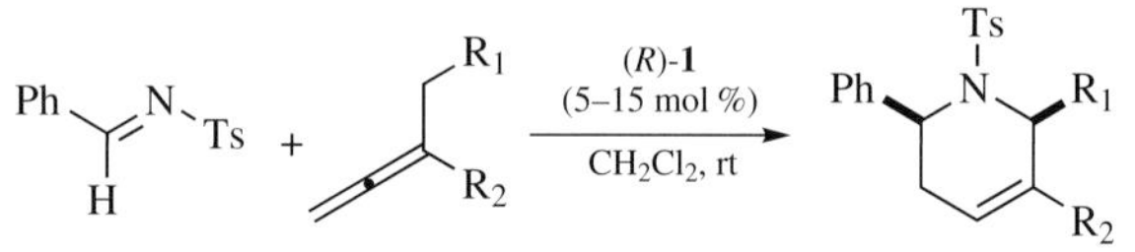

(7)

R_1	R_2	ee	*cis*:*trans*	isolated yield (%)
Ph	CO_2Et	87	99:1	78
4-$(CF_3)(C_6H_4)$	CO_2Et	88	99:1	81
H	CO_2Et	68	–	72
H	COPh	76	–	97

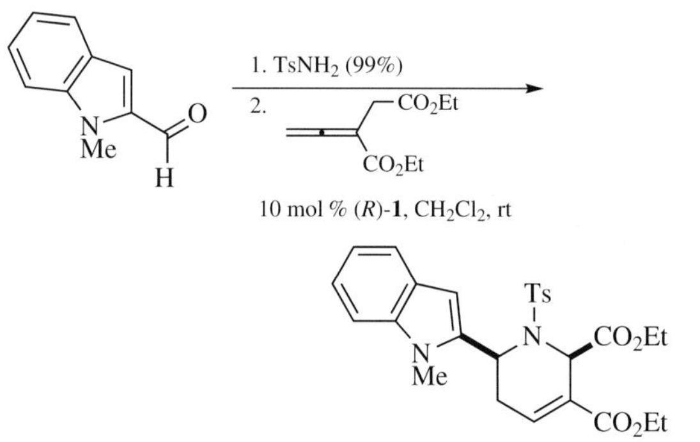

(8)

99% yield
93:7 dr, 97% ee

The reaction of *N*-tosylimines with allenic esters in the presence of catalytic amounts of chiral phosphines can also proceed in a [3 + 2]-fashion to form chiral pyrroline derivatives. Among the large number of chiral phosphines that were screened, (*S*)-**1** was found be the most selective (eq 9).[5] (*S*)-**1** was also used in a highly enantioselective annulation of *N*-diphenylphosphinoyl imines with 2-butynoates (eq 10).[6]

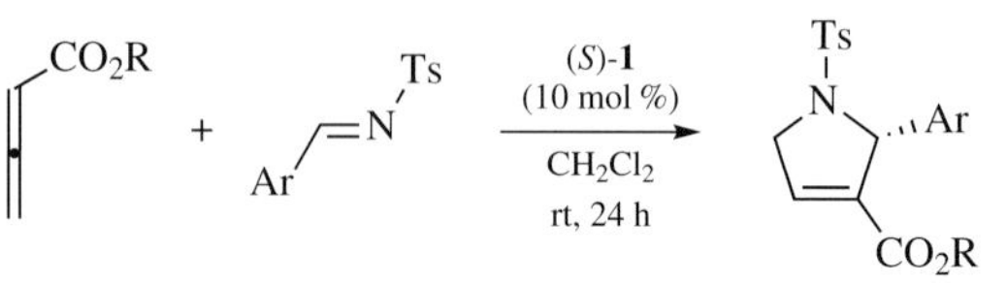

(9)

for R = Et, Ar = 1-Naph:
70% yield, 54% ee
for R = Cy, Ar = *p*-MeO-Ph:
60% yield, 80% ee

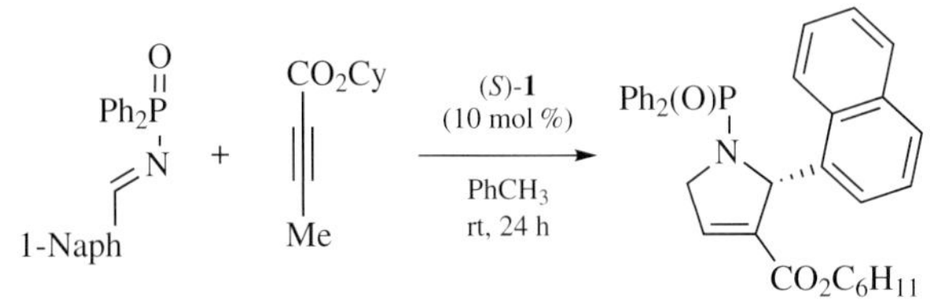

(10)

74% yield
85% ee

(*R*)-**1** was employed as a chiral organocatalyst for the [3 + 2]-cycloaddition of allenes with enones to form functionalized cyclopentenes.[7] For the cycloaddition of ethyl 2,3-butadienoate and chalcone in the presence of (*R*)-**1** (10 mol %), the reaction gave the cyclopentene products in good yield, ee, and regioselectivity (eq 11). The use of other mono- or bisphosphines such as (*R*)-BINAP, (*R,R*)-Me-BPE, and (*R,R*)-Et-DuPhos as the chiral ligand gave much inferior results. Dienones are also suitable substrates and undergo a single [3 + 2]-cycloaddition in the presence of (*R*)- **1** (eqs 12 and 13).

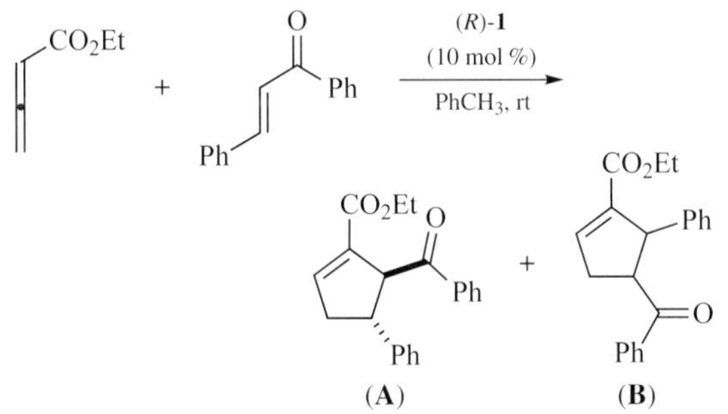

(11)

64% isolated yield
(**A**:**B** = 13:1)
88% ee (**A**)

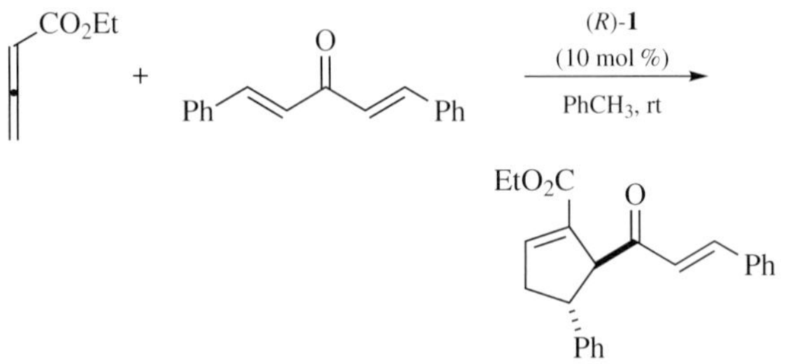

(12)

75% yield, 89% ee

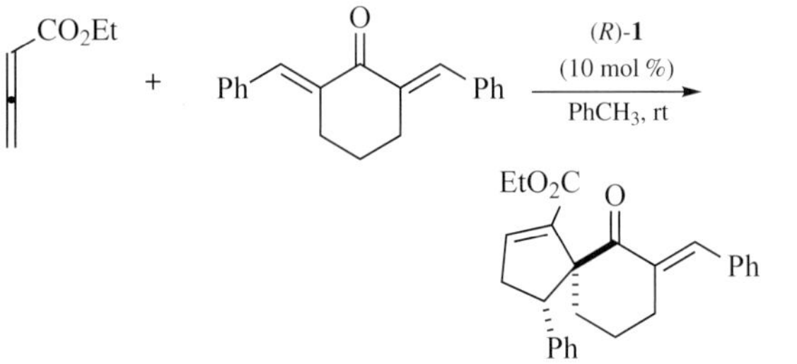

(13)

57% yield, 93% ee

This transformation was extended to the reaction between diethyl allenylphosphonate and diethyl fumarate or other enones using (*S*)-**1** as the catalyst for the synthesis of chiral cyclic phosphonates (eq 14).[8]

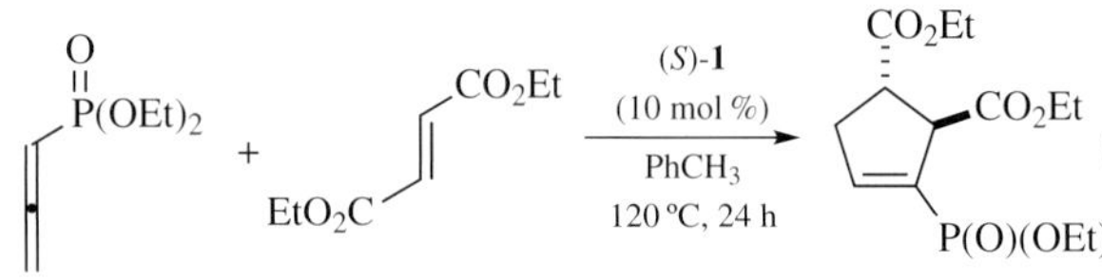

(14)

55% yield
91% ee

In a similar vein, (*S*)-**1** was used as catalyst in the diastereo- and enantioselective synthesis of a highly functionalized diquinane by an intramolecular Michael addition (eq 15).[9]

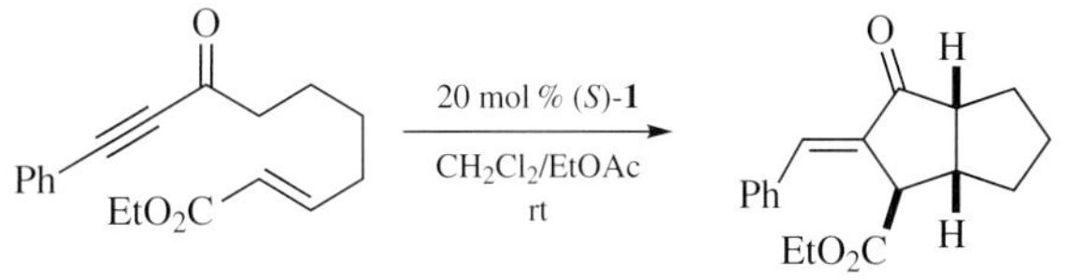

(15)

76% yield, 60% ee

Phosphine-catalyzed cyclization of hydroxy-2-alkynoate delivers cyclic ethers[10] and can be rendered asymmetric when chiral phosphines are used as catalyst. Thus, ethyl 7-hydroxyhept-2-ynoate underwent enantioselective cyclization to give the oxygen heterocycle in good yield and enantioselectivity with several monophosphines as the catalyst, including (*S*)-**1** and its phenyl analog (*S*)-**2** (eq 16).[11]

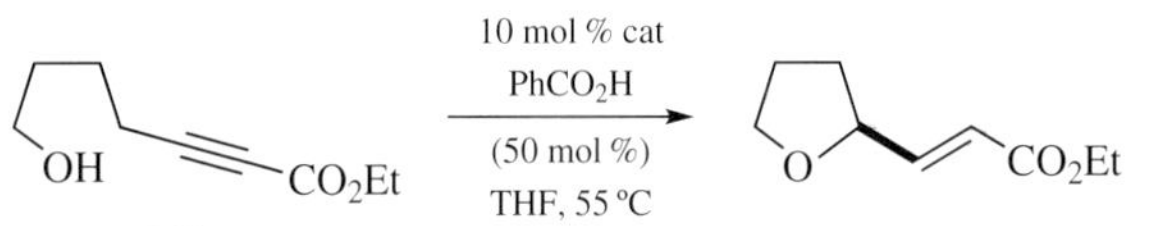

(16)

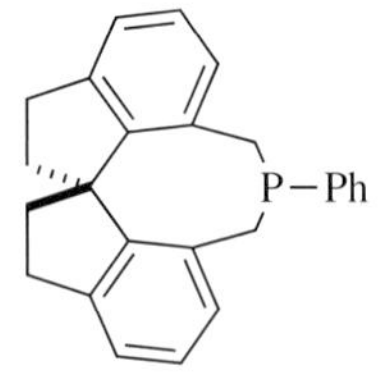

(*S*)-(**4**)

cat:	(*S*)-**1**	(*S*)-**2**	(*S*)-**4**
ee:	66	45	–87
Yield (%):	72	65	80

Related Reagents. (11b*R*)- and (11b*S*)-4-phenyl-4,5-dihydro-3*H*-dinaphtho[2,1-*c*:1′,2′-*e*]phosphepin.

1. (a) Junge, K.; Oehme, G.; Monsees, A.; Riermeier, T.; Dingerdissen, U.; Beller, M., *Tetrahedron Lett.* **2002**, *43*, 4977. (b) Junge, K.; Hagemann, B.; Enthaler, S.; Spannenberg, A.; Michalik, M.; Oehme, G.; Monsees, A.; Riermeier, T.; Beller, M., *Tetrahedron: Asymmetry* **2004**, *15*, 2621. (c) Enthaler, S.; Erre, G.; Junge, K.; Holz, J.; Börner, A.; Alberico, E.; Nieddu, I.; Gladiali, S.; Beller, M., *Org. Process Res. Dev.* **2007**, *11*, 568.
2. Chi, Y.; Zhang, X., *Tetrahedron Lett.* **2002**, *43*, 4849.
3. Reetz, M. T.; Mehler, G.; Meiswinkel, A., *Tetrahedron: Asymmetry* **2004**, *15*, 2165.
4. Wurz, R. P.; Fu, G. C., *J. Am. Chem. Soc.* **2005**, *127*, 12234.

5. Fleury-Brégeot, N.; Jean, L.; Retailleau, P.; Marinetti, A., *Tetrahedron* **2007**, *63*, 11920.
6. Pinto, N.; Fleury-Brégeot, N.; Marinetti, A., *Eur. J. Org. Chem.* **2009**, 146.
7. Wilson, J. E.; Fu, G. C., *Angew. Chem., Int. Ed.* **2006**, *45*, 1426.
8. Panossian, A.; Fleury-Brégeot, N.; Marinetti, A., *Eur. J. Org. Chem.* **2008**, 3826.
9. Wilson, J. E.; Sun, J.; Fu, G. C., *Angew. Chem., Int. Ed.* **2010**, *49*, 161.
10. Trost, B. M.; Li, C.-J., *J. Am. Chem. Soc.* **1994**, *116*, 10819.
11. Chung, Y. K.; Fu, G. C., *Angew. Chem., Int. Ed.* **2009**, *48*, 2225.

Qingwei Yao
Northern Illinois University, DeKalb, IL, USA

[4S-(4α,5α)]-1-(1,3-Dimethyl-2-oxido-4,5-diphenyl-1,3,2-diazaphospholidine-2-yl)piperidine

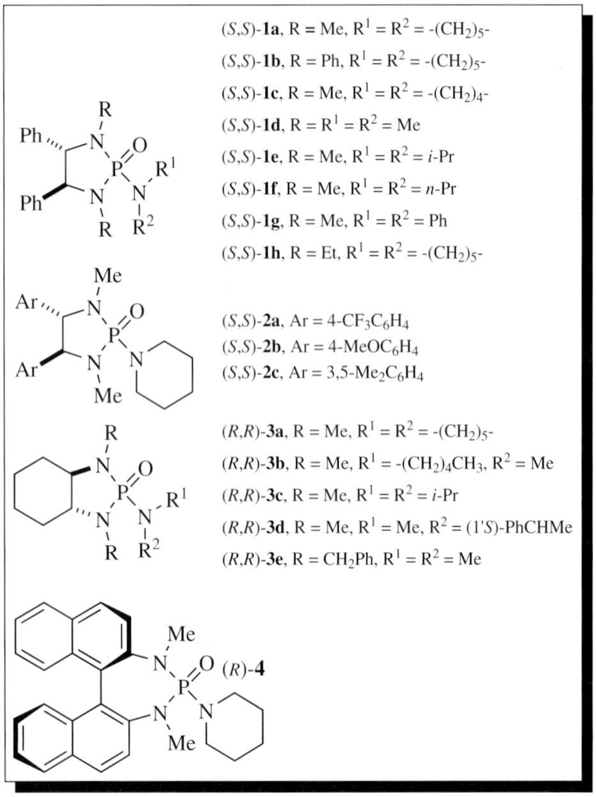

[180475-25-6] $C_{21}H_{28}N_3OP$ (MW 369.45)

InChI = 1/C21H28N3OP/c1-22-20(18-12-6-3-7-13-18)21(19-14-8-4-9-15-19)23(2)26(22,25)24-16-10-5-11-17-24/h3-4,6-9,12-15,20-21H,5,10-11,16-17H2,1-2H3/t20-,21-/m0/s1

InChIKey = HZVXHACBRXLPRG-SFTDATJTBN

(catalyst for asymmetric aldol additions of trichlorosilyl enolates to aldehydes,[1] catalyst for asymmetric allylations and crotylations of allylic trichlorosilanes to aldehydes,[2] catalyst for asymmetric ring opening of epoxides using silicon tetrachloride [3])

Physical Data: mp 110 °C; $[\alpha]_D^{22}$ +18.2 (c=1.3, $CHCl_3$).

Solubility: soluble in most organic solvents except hydrocarbons.

Form Supplied in: white needles.

Preparative Methods: The enantiopure phosphoramide (*S*,*S*)-**1a** is prepared from treatment of (*S*,*S*)-*N*,*N*′-dimethyl-1,2-diphenyl-1,2-ethanediamine (**5**) with piperidinylphosphoric dichloride (**6**) and triethylamine (eq).[4] Synthesis of related phosphoramides is easily accomplished by substituting the desired diamine backbone in the former procedure. Recrystallization of phosphoramide (*S*,*S*)-**1a** from hexane followed by drying over P_2O_5 allows for isolation of analytically pure material. Enantiopure 1,2-diphenyl-1,2-ethanediamine (**5**) can be prepared from benzil and cyclohexanone according to the method of Pikul & Corey.[5] Bis-formylation of the diamine with acetic formic anhydride followed by $LiAlH_4$ reduction affords *N*,*N*′-dimethyldiamine **5**. Alternatively, enantiopure **5** can be prepared directly from benzaldehyde and methylamine according to the method of Alexakis et al.[6] Piperidinylphosphoric dichloride (**6**) is prepared from piperidine, triethylamine and phosphorous oxychloride according to the modified procedure of Peyronel et al.[7]

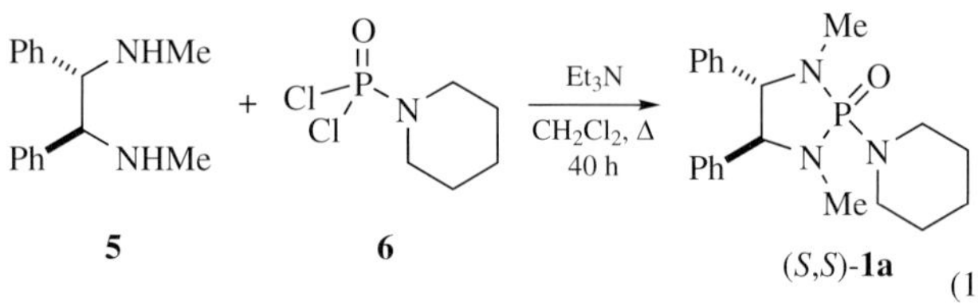

(1)

Handling, Storage, and Precautions: The phosphoramide is a stable, hygroscopic compound which is best stored in a dessicator or dry-box. Avoid prolonged exposure to air or moisture.

Introduction. [4*S*-(4*α*,5*β*)]-1-(1,3-Dimethyl-2-oxido-4,5-diph-enyl-1,3,2-diazaphospholidine-2-yl)piperidine (**1a**) and its derivatives are an important class of phosphoramides which have seen much success as Lewis-basic catalysts for aldol additions,[8] allylations,[9] crotylations, and epoxide openings.[10] The three nitrogen subunits of the phosphoramide provide the opportunity for a large number of structurally diverse analogs, allowing a wide spectrum of properties and shapes to be customized.[11]

Ester Enolate Aldol Additions to Aldehydes. Among the first examples of aldol additions employing chiral Lewis bases as catalysts were the additions of trichlorosilyl ketene acetals to aldehydes.[12] Silyl ketene acetal **7** could be generated by metathesis of methyl tributylstannylacetate with $SiCl_4$. Treatment of **7** with benzaldehyde and 10 mol % of a phosphoramide in CH_2Cl_2 at −78 °C afforded aldol products in good to high yields with moderate enantioselectivities for all phosphoramides employed. Reaction of **7** with pivalaldehyde provided aldol products in similar yields and with slightly improved enantioselectivities. The increase in stereoselection is presumably attributed to a less competent background reaction inherent with additions to more sterically encumbered aldehydes (eq , Table 1).

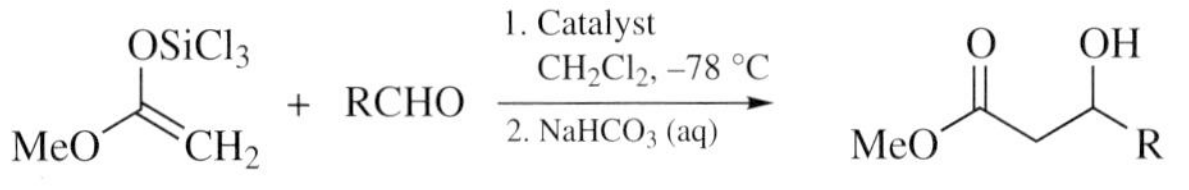

(2)

7 **8** (70–90%)

Table 1 Phosphoramide-catalyzed addition of ketene acetal **6** to aldehydes

Entry	Catalyst	Aldehyde	er (configuration)
1	(*S*,*S*)-**1**	PhCHO	2.0:1 (*S*)
2	(*R*,*R*)-**3**	PhCHO	1.5:1 (*R*)
3	(*R*)-**4**	PhCHO	1.6:1 (*S*)
4	(*S*,*S*)-**1**	*t*-BuCHO	2.2:1 (*S*)
5	(*R*,*R*)-**3**	*t*-BuCHO	1.4:1 (*R*)
6	(*R*)-**4**	*t*-BuCHO	2.9:1 (*S*)

er, enantiomeric ratio.

Ketone Enolate Aldol Additions to Aldehydes. Addition of methyl ketone trichlorosilyl enolate **9** to benzaldehyde in the presence of catalytic amounts of phosphoramide (*S*,*S*)-**1** affords aldol products in excellent yields.[13] The level of enantioselectivity was found to be dependent upon the amount of (*S*,*S*)-**1a** used, with higher loadings providing better selectivities. A typical procedure involves equilibration of a solution of trichlorosilyl enolate and (*S*,*S*)-**1a** in CH_2Cl_2 at −78 °C followed by addition of aldehyde. Reactions are quenched upon completion by quickly pouring the cold reaction mixture into a rapidly stirring aqueous $NaHCO_3$ solution placed in an ice bath. Alternatively, completed reactions may be quenched using a cold 1:1 mixture of a saturated aqueous KF solution and an aqueous 1 M KH_2PO_4 solution. In either case, rapid stirring of the cold quench mixture prevents acid-catalyzed β-elimination of the aldol products to give undesired α,β-unsaturated ketones (eq 3, Table 2).

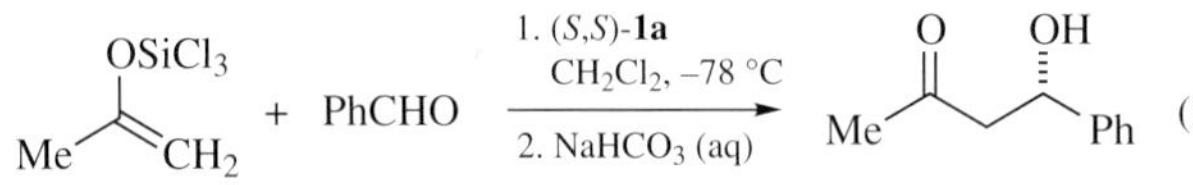

(3)

9 (−)-**10**

Table 2 Effect of catalyst loading on aldol additions with enolate **9**

Entry	Loading (mol %)	Conc. (M)	er	Yield (%)
1	10	0.1	11.3:1	90
2	20	0.1	12.3:1	89
3	10[a]	0.5	1:11.7	87
4	5	0.5	11.5:1	88
5	3	0.5	8.60:1	86
6	2	0.5	7.77:1	88
7	1	0.5	5.10:1	92

er, enantiomeric ratio.

[a] Performed with 10 mol % (*R*,*R*)-**1a.**

The generality of methyl ketone enolate additions to benzaldehyde was demonstrated by varying the spectator portion of the nucleophile.[13] In all cases, high yields were achieved using as little as 5 mol % (*S*,*S*)-**1a**. The enantioselectivity of the process, however, was sensitive to the enolate structure with larger groups, such as *t*-butyl and phenyl (eq 4, Table 3, entries 5 and 6), providing lower enantioselectivities. The success of a functionalized enolate (eq 4, Table 3, entry 7) demonstrated the tolerance for oxygenated substituents in the phosphoramide-catalyzed aldol addition, whereby no deleterious effect was observed on either the yield or enantioselectivity as compared with unfunctionalized enolates.

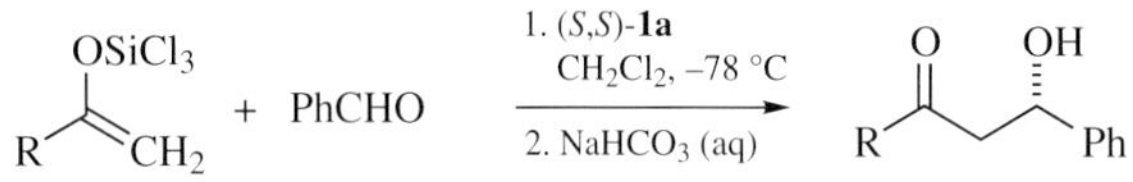

(4)

11 (−)-**12**

Table 3 Aldol additions of methyl ketone trichlorosilyl enolates to benzaldehyde catalyzed by (*S*,*S*)-**1a**

Entry	R	er	Yield (%)
1	Me	14.6:1	98
2	*n*-Bu	12.0:1	98
3	*i*-Bu	10.1:1	95
4	*i*-Pr	9.75:1	97
5	*t*-Bu	3.17:1	95
6	Ph	2.92:1	93
7	$TBSOCH_2$	13.5:1	94

er, enantiomeric ratio.

The phosphoramide-catalyzed aldol addition was less sensitive to variations on the electrophile. Using 2-hexanone trichlorosilyl enolate **13** as an example aldol additions to various aldehydes affords aldol products in good to high yields and with excellent enantioselectivities, using as little as 5 mol % catalyst.[13] Additions to branched aliphatic aldehydes, such as cyclohexanone and pivalaldehyde (eq 5, Table 4, entries 5 and 6), were sluggish when using 5 mol % catalyst. However, increasing the catalyst loading to 10 mol % provides good yields of the aldol products after 6 h with excellent enantioselectivities.

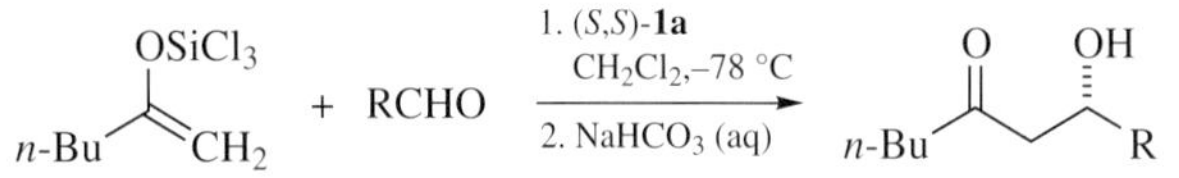

(5)

13 (−)-**14**

Table 4 Aldol additions of enolate **13** to various aldehydes catalyzed by (*S*,*S*)-**1a**

Entry	Aldehyde	Loading (mol %)	Time (h)	er	Yield (%)
1	(*E*)-PhCH=CHCHO	5	2	14.6:1	94
2	(*E*)-PhCH=C(Me)CHO	5	2	21.7:1	95
3	1-NaphthylCHO	5	2	13.1:1	92
4	4-PhC_6H_4CHO	5	2	12.7:1	95
5	$C_6H_{11}CHO$	10	6	17.5:1	79
6	*t*-BuCHO	10	6	24.0:1	81

er, enantiomeric ratio.

Recently, the synthetic utility of phosphoramide-catalyzed aldol additions of trichlorosilyl enolates to aldehydes has been enhanced by the ability to generate the delicate trichlorosilyl enolates in situ.[13] Treatment of the corresponding TMS enol ether with $Hg(OAc)_2$ followed by addition of a solution phosphoramide and aldehyde afford aldol products in high yields over two steps without affecting stereoselectivities. This is exemplified by the addition of chiral α-oxygenated trichlorosilyl enolates to benzaldehyde. Using the (*S*)-lactate-derived methyl ketone enol ether **15**, addition to benzaldehyde afforded aldol products in high diastereoselectivity for the matched-sense using (*R*,*R*)-**1a**.[13,14] Although the reaction was compatible with various oxygen functionalities, the stereoselectivity was highly dependent upon the nature of the protecting group, with the OTBS group superior to both the benzyloxy or pivaloyl protecting groups (eq 6, Table 5).

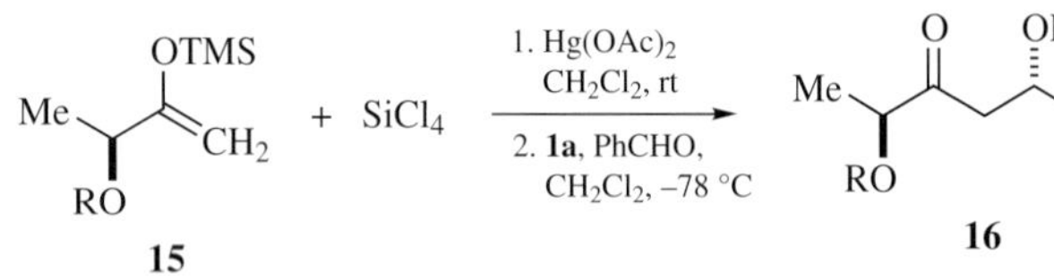
(6)

Table 5 Addition of TMS enol ether **15** catalyzed by 5 mol % **1a**

Entry	R	Catalyst	*syn*/*anti*	Yield (%)
1	TBS	(*S*,*S*)-**1a**	1.5:1	85
2	Piv	(*S*,*S*)-**1a**	3.4:1	78
3	Bn	(*S*,*S*)-**1a**	1:1.1	78
4	TBS	(*R*,*R*)-**1a**	73:1	85
5	Piv	(*R*,*R*)-**1a**	20:1	78
6	Bn	(*R*,*R*)-**1a**	11:1	77

To further exemplify the dependence of the stereochemical course of chiral enolate additions catalyzed by phosphoramides, a detailed survey of additions to aldehydes using chiral β-hydroxyenolate **17** and 10 mol % of **1a** was performed.[15] Unlike additions of **15**, diastereoselectivities of the resultant aldol products using **17** could be switched depending upon the configuration of the phosphoramide employed (eq 7, Table 6).

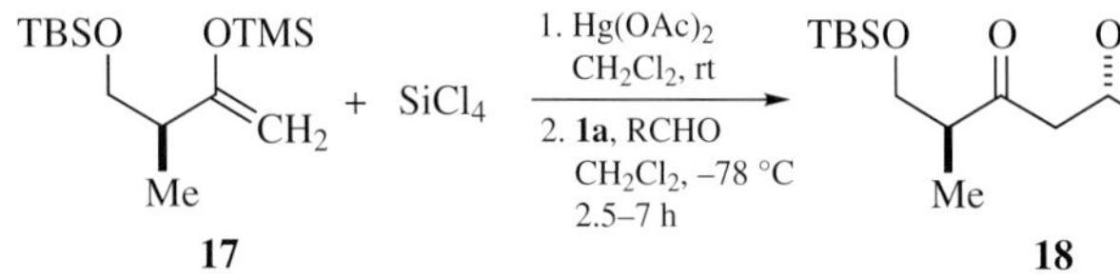
(7)

Table 6 Addition of TMS enol ether **17** to aldehydes catalyzed by **1a**

Entry	Catalyst	Aldehyde	*syn*/*anti*	Yield (%)
1	(*R*,*R*)-**1a**	PhCHO	19.0:1	80
2	(*S*,*S*)-**1a**	PhCHO	1:7.33	75
3	(*R*,*R*)-**1a**	1-NaphthylCHO	15.7:1	85
4	(*S*,*S*)-**1a**	1-NaphthylCHO	1:8.09	83
5	(*R*,*R*)-**1a**	(*E*)-PhCH=CHCHO	8.00:1	81
6	(*S*,*S*)-**1a**	(*E*)-PhCH=CHCHO	1:4.26	82
7	(*R*,*R*)-**1a**	*t*-BuCHO	27.9:1	73
8	(*S*,*S*)-**1a**	*t*-BuCHO	1:6.51	78

The phosphoramide-catalyzed aldol addition of substituted enolates to aldehydes provides the opportunity to generate disubstituted aldol products with high stereoselectivity. The (*S*,*S*)-**1a**-catalyzed addition of cyclohexanone trichlorosilyl enolate **19** to aldehydes has been well studied and demonstrates good substrate generality.[16] The aldol reactions of this *E*-enolate are rapid when 10 mol % of (*S*,*S*)-**1a** is used. Reactions are typically complete within 2 h affording *anti*-aldol products in high yields with excellent diastereo- and enantioselectivities. Although the typical procedure using methyl ketone enolates can be followed, higher diastereoselectivities may be achieved by slow addition of a solution of aldehyde to the mixture of enolate and catalyst in CH_2Cl_2 at −78 °C over several minutes.[17] Remarkably, a dramatic switch in diastereoselectivity is observed when phosphoramide **1b** (10 mol %) is employed as a catalyst (eq 8, Table 7, entry 2). Although the level of enantioselectivity was modest, the change in diastereoselectivity demonstrates that small structural changes in the phosphoramide can have profound effects on the stereochemical course of the reaction. This is further exemplified through a systematic substitution of the internal (phospholidino) nitrogens (eq 8, Table 7, entries 1 and 8), the external nitrogen (eq 8, Table 7, entries 1 and 3–7),[18,19] as well as changes to the electronic demand of the aryl substituent (eq 8, Table 7, entries 1 and 9–11).[20]

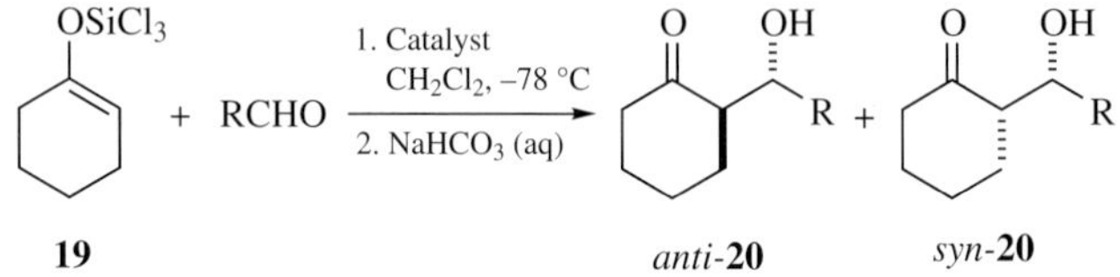
(8)

Table 7 Aldol addition of enolate **19** catalyzed by 10 mol % (*S*,*S*)-**1** and (*S*,*S*)-**2**

Entry	Catalyst	Aldehyde	*anti*/*syn*	er (major)	Yield (%)
1	**1a**	C_6H_5CHO	61:1	27.6:1	95
2	**1b**	C_6H_5CHO	1:97	3.08:1	94
3	**1c**	C_6H_5CHO	8.1:1	7.33:1	89
4	**1d**	C_6H_5CHO	19:1	14.4:1	91
5	**1e**	C_6H_5CHO	1:32	1.00:1	90
6	**1f**	C_6H_5CHO	1:16	1.11:1	89
7	**1g**	C_6H_5CHO	1:1.9	2.39:1	86
8	**1h**	C_6H_5CHO	2.0:1	17.2:1	92
9	**2a**	C_6H_5CHO	2.5:1	12.0:1	80
10	**2b**	C_6H_5CHO	53:1	24.0:1	80
11	**2c**	C_6H_5CHO	10:1	9.50:1	98
12	**1a**	1-NaphthylCHO	>99:1	65.7:1	94
13	**1a**	(*E*)-PhCH=CHCHO	>99:1	15.7:1	94
14	**1a**	(*E*)-PhCH=C(Me)CHO	>99:1	24.0:1	98
15	**1a**	PhC≡CCHO	5.3:1	10.1:1	90

er, enantiomeric ratio.

Table 8 Aldol addition of enolate **21** catalyzed by 15 mol % (*S*,*S*)-**1a**

Entry	Aldehyde	*syn*/*anti*	er (*syn*)	Yield (%)
1	C_6H_5CHO	18:1	39.0:1	95
2	1-NaphthylCHO	3.0:1	11.5:1	96
3	(*E*)-PhCH=CHCHO	9.4:1	24.0:1	97
4	PhC≡CCHO	1:3.5	3.76:1	92

er, enantiomeric ratio.

In contrast to the *E*-enolates derived from cyclic ketones, addition of propiophenone trichlorosilyl enolate (*Z*)-**21** to aldehydes requires longer reaction times and higher loadings of catalyst.[16] Although the yields of the aldol products remain high, both the diastereo- and enantioselectivities are attenuated as compared to their *E* counterparts (eq 9, Table 8).

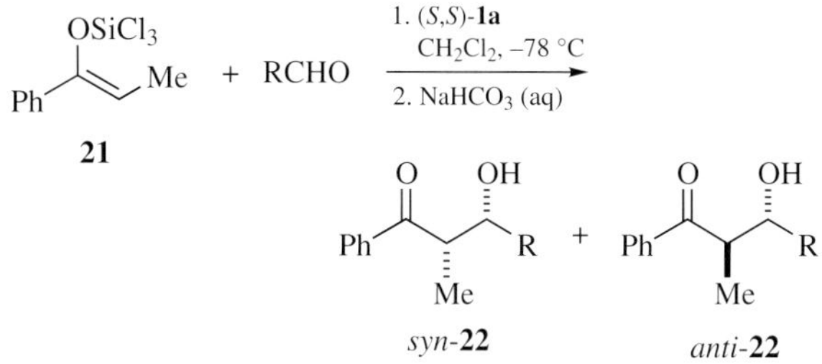

(9)

Allylic Additions to Aldehydes. Although phosphoramide **1** is capable of promoting the addition of allyltrichlorosilane to aldehydes, the enantioselectivities were modest at best compared with the cyclohexyldiamine analogs **3**.[21] Preliminary results showed the allylations to be high yielding but giving only moderate enantioselectivities when using one equivalent of phosphoramide as a promoter. As observed in the stereoselectivealdol additions with phosphoramides, subtle changes to the nitrogen substituents also dramatically affect the stereochemical course of allylations using allylic trichlorosilanes (eq 10, Table 9, entries 7–9).[22] Under optimized conditions, as few as 0.25 equivalents of phosphoramide may be used with little erosion in yield or stereoselectivity (eq 10, Table 9, entries 3–5).

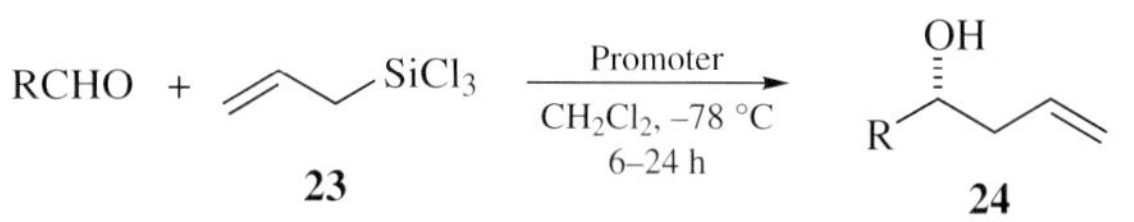

(10)

Table 9 Asymmetric allylation of aldehydes

Entry	Promoter	Equiv.	Conc. (M)	Aldehyde	er	Yield (%)
1	(*S*,*S*)-**1a**	1.0	1.0	C_6H_5CHO	2.0:1	NA
2	(*R*,*R*)-**3a**	1.0	1.0	C_6H_5CHO	4.0:1	81
3	(*R*,*R*)-**3a**	0.5	0.5	C_6H_5CHO	3.7:1	78
4	(*R*,*R*)-**3a**	0.25	0.5	C_6H_5CHO	3.9:1	74
5	(*R*,*R*)-**3a**	0.1	0.5	C_6H_5CHO	3.3:1	40
6	(*R*,*R*)-**3b**	1.0	1.0	C_6H_5CHO	3.1:1	73
7	(*R*,*R*)-**3c**	1.0	1.0	C_6H_5CHO	1.4:1	NA
8	(*R*,*R*)-**3d**	1.0	1.0	C_6H_5CHO	1.5:1	NA
9	(*R*,*R*)-**3e**	1.0	1.0	C_6H_5CHO	1.9:1	NA
10	(*R*,*R*)-**3a**	1.0	1.0	$2\text{-}MeC_6H_4CHO$	4.7:1	81
11	(*R*,*R*)-**3a**	1.0	1.0	$4\text{-}NO_2C_6H_4CHO$	1.5:1	76
12	(*R*,*R*)-**3a**	1.0	1.0	$4\text{-}MeOC_6H_4CHO$	3.0:1	80
13	(*R*,*R*)-**3a**	1.0	1.0	$4\text{-}NMe_2C_6H_4CHO$	2.0:1	69
14	(*R*,*R*)-**3a**	1.0	1.0	PhC≡CCHO	2.2:1	67

er, enantiomeric ratio; NA, not available.

The phosphoramide-catalyzed allylic additions to aldehydes have also been extended to include stereoselective crotylations. When using geometrically defined 2-butenyltrichlorosilanes, additions to benzaldehyde afford homoallylic alcohols in good yields and excellent diastereoselectivity (eq 11).[21a] Enantioselectivities, however, remain comparable to the results for allylic additions. Interestingly, the crotylation results suggest that addition proceeds through a cyclic transition structure.[23]

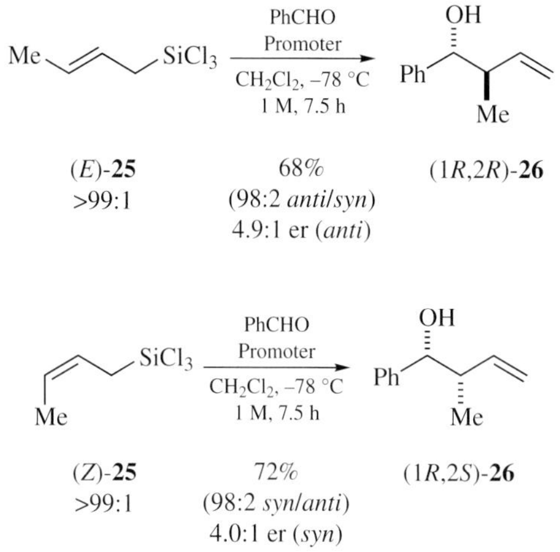

(11)

Epoxide Openings Using $SiCl_4$. The binaphthyldiamine-derived phosphoramide (*R*)-**4** is an effective catalyst for the opening of various epoxides with $SiCl_4$ to afford chlorohydrins in high yields.[24] The rate and stereoselectivity for the ring opening of *meso*-epoxides are extremely dependent upon the structure of the epoxide. In the examples shown for epoxycycloalkanes, only cyclohexene oxide [10 mol % (*R*)-**4**] affords a chlorohydrin in high yield and good enantioselectivity (eq 12, Table 10, entry 2). In contrast, cyclopentene oxide and cyclooctene oxide give essentially racemic chlorohydrin (eq 12, Table 10, entries 1 and 3) with cyclooctene oxide requiring more than 5 days for complete reaction. In comparison to the cyclic epoxides, the acyclic substrates afford chlorohydrins with increased enantioselectivities (eq 12, Table 10, entries 4 and 5). In a typical procedure, a solution of epoxide in CH_2Cl_2 is added to a mixture of $SiCl_4$ and phosphoramide at −78 °C in CH_2Cl_2. To avoid opening of epoxides by adventitious HCl,[25] it is essential to maintain $SiCl_4$ acid free.[26]

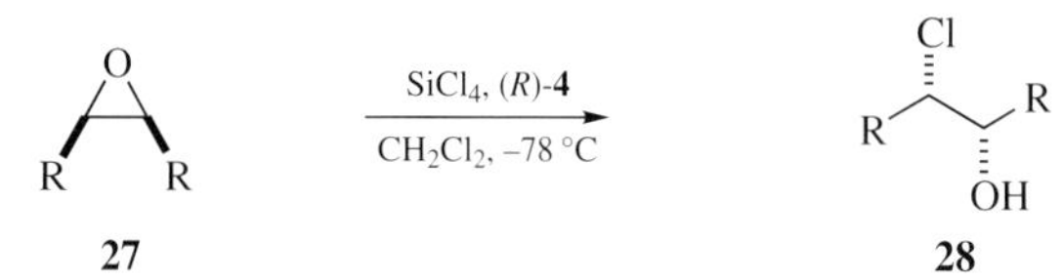

(12)

Table 10 Stereoselectivering opening of *meso*-epoxides catalyzed by (*R*)-**4**

Entry	Epoxide (R)	Time (h)	er	Yield (%)
1	$\text{-}(CH_2)_3\text{-}$	0.3	1.16:1	87
2	$\text{-}(CH_2)_4\text{-}$	0.3	3.13:1	90
3	$\text{-}(CH_2)_6\text{-}$	132	1.04:1	95
4	Ph	3	14.4:1	94
5	$BnOCH_2$	4	5.94:1	95

er, enantiomeric ratio.

1. (a) Paterson, I.; Cowden C. J.; Wallace, D. J. In *Modern Carbonyl Chemistry*; Otera, J., Ed.: Wiley-VCH: Weinheim, **2000**; Ch. 9. (b) Cowden, C. J.; Paterson, I., *Org. React.* **1997**, *51*, 1. (c) Heathcock, C. H.; Kim, B. M.; Williams, S. F.; Masamune, S.; Paterson, I.; Gennari, C. In *Comprehensive Organic Synthesis*; Trost, B. M.; Fleming, I., Eds.; Vol. 2, Pergamon Press: Oxford, **1991**.

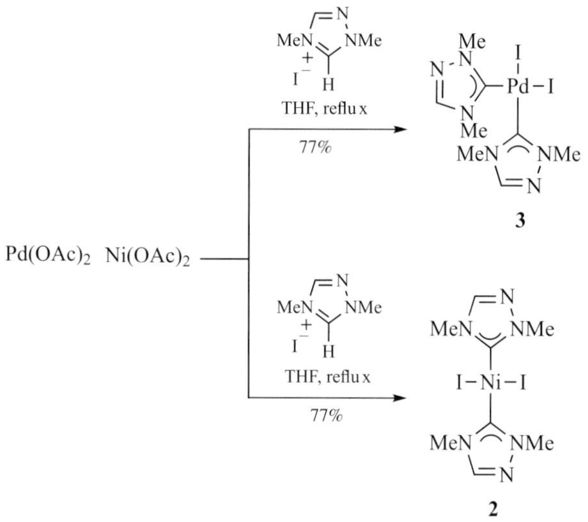

(2)

The azolium salt **1** has also been incorporated into complexes with iridium; however, in each case, the *N*-heterocyclic carbene of **1** was first generated and captured as a silver complex **4**. The silver complex **4** was then employed as a carbene transfer agent to form the iridium complex **6** (eq 3).[5]

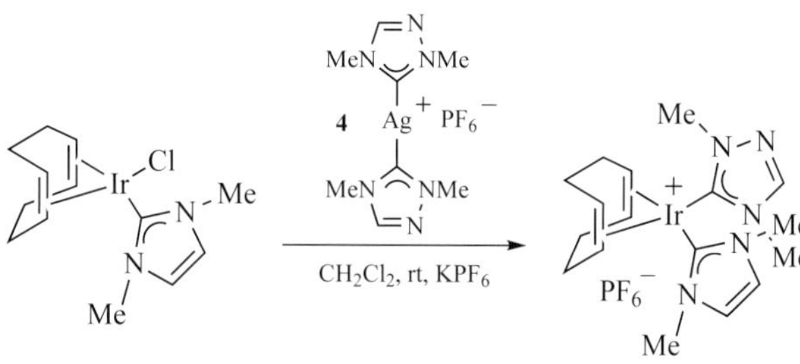

5 **6** (3)

Despite the fact that many complexes of the azolium salt **1** with transition metals have been formed, in-depth investigations of their catalytic behavior have yet to be reported. In a recent study, Ni intermediates bearing the azolium salt **1** as a ligand have been shown both experimentally and in silico as intermediates in an atom efficient coupling of azolium salts and ethylene.[6]

Application in Organocatalysis. The azolium salt **1** can be used as an organocatalyst to promote tandem oxidation processes, particularly those that allow the transformation of alcohols or aldehydes to esters. In these reactions, the azolium salt is deprotonated to generate a carbene intermediate that acts as a Lewis base catalyst to promote mild oxidations. The addition of acyl silanes to chalcone catalyzed by thiazolium salts was investigated (eq 4).[7] However, azolium salt **1** was identified as a poor catalyst for this transformation relative to a thiazolium analog.

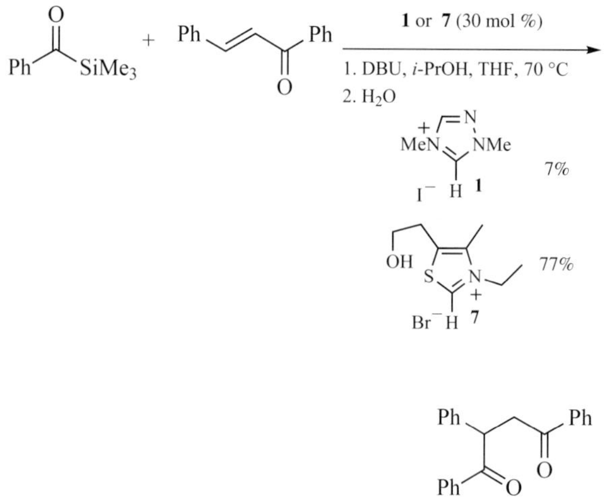

(4)

In addition, the azolium salt **1** was found to be an efficient catalyst for hydroacylation of activated ketones (eq 5).[8] Interestingly, the hydroacylation was highly dependent on the nature of the solvent. When the reaction was performed in CH_2Cl_2 as solvent, the expected ester product was isolated. However, in MeOH, only a reduction of the activated ketone was observed. In these processes, the carbene generated from azolium salt **1** is believed to be responsible for two key bond forming steps in the catalytic cycle.

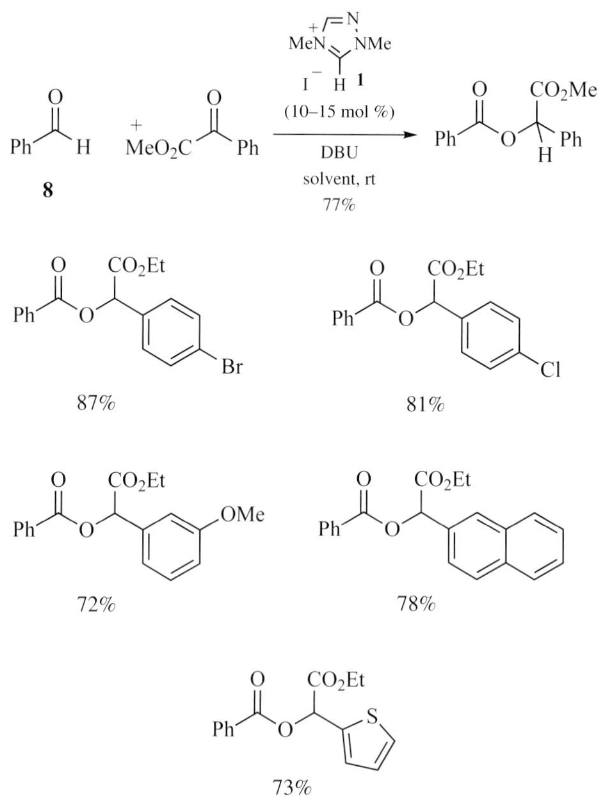

(5)

The azolium salt **1** was employed as an efficient catalyst in a number of other oxidation processes. The tandem oxidation of activated allylic alcohols to various esters can be promoted by **1** (eq 6).[9]

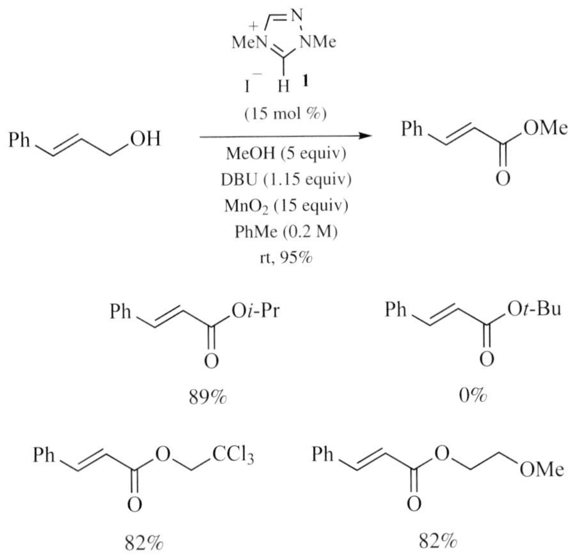

(6)

In addition, it was later reported that **1** can also promote the same oxidation of unactivated aldehydes to esters (eq 7).[10] During the oxidation, manganese(IV) oxide acts as a stoichiometric oxidant. The oxidation is again notable for its wide substrate scope: both electron-poor and electron-rich substrates are easily oxidized and potentially sensitive stereocenters are not racemized in the process.

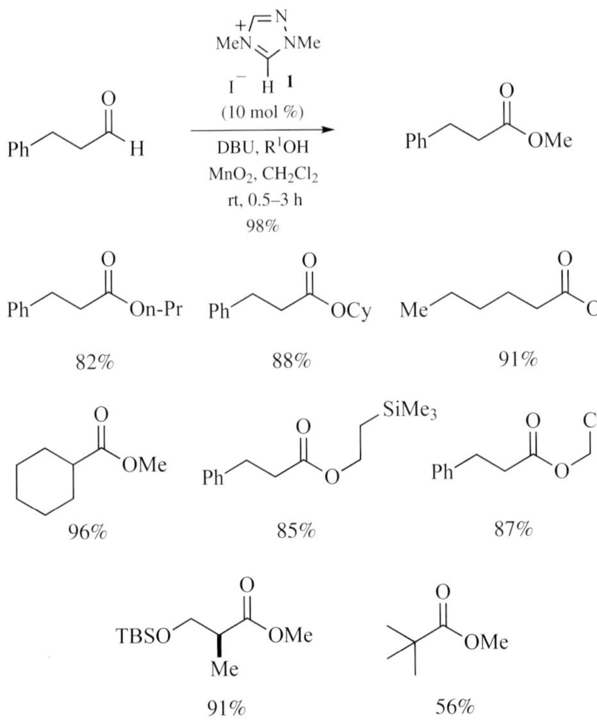

(7)

As shown above, the oxidation of α,β-unsaturated aldehydes to esters with azolium salt **1** proceeds in good yields. However, a unique solvent dependence on the reaction that can be used to rationally design new reactions with **1** whereby activated aldehydes could be converted into reduced esters was highlighted. While the results were encouraging, a detailed account and substrate scope are yet to be reported (eq 8).[11]

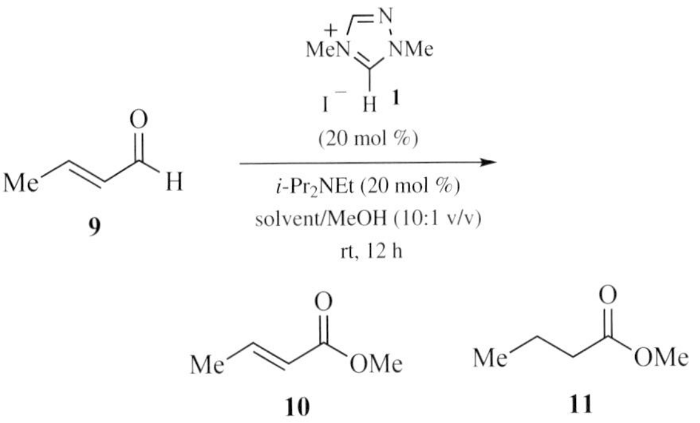

(8)

Solvent	10:11	Yield (%)
MeOH	1:7.4	33
THF	1:1.4	18
CH_2Cl_2	1:1.9	29
PhMe	2.1:1	20

Similar oxidations were studied, and it was found that the azolium salt **1** can be used as a reagent to promote chemoselective acylations of alcohols in the presence of amines (eqs 9 and 10).[12] In these reactions, it was proposed that the chemoselectivity is derived from the carbene generated from **1**, which selectively activates alcohol over the amine. This mechanistic proposal was supported by quantum chemical calculations. Only a brief substrate scope was reported.

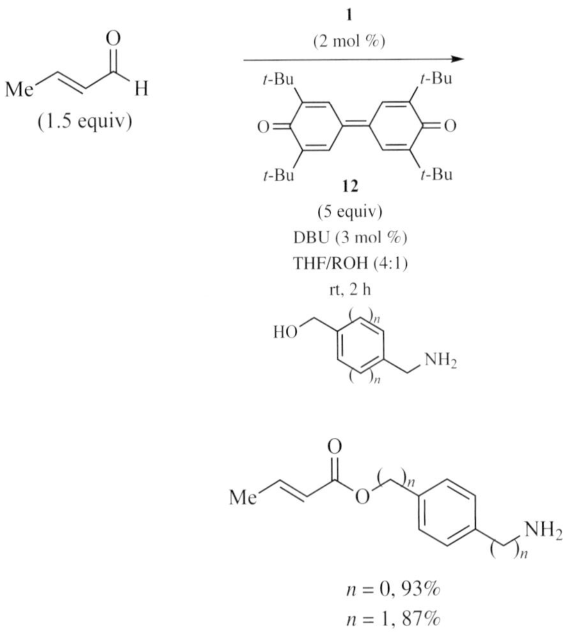

(9)

The corresponding carbene generated from azolium **1** could be employed to promote oxidations of aldehydes into aminated products such as amides (eq 11) and acyl azides (eq 12).[13] Azolium **1** could be used along with **12** as an organic stoichiometric oxidant to promote oxidation of aldehydes with $(CF_3)_2CHOH$ (HFIP) to their corresponding HFIP esters. Following the formation of these activated esters, the addition of an amine led to the formation of the amide in an overlap two step one-pot process. Acyl azides could be formed from aldehydes using the same organic oxidant **12** and TMS azide.

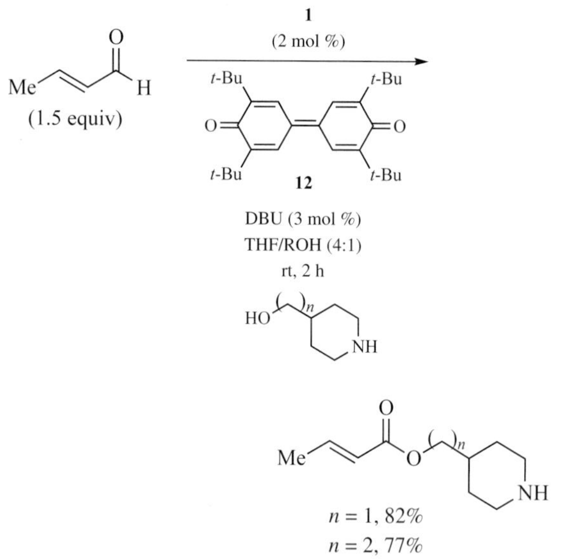

(10)

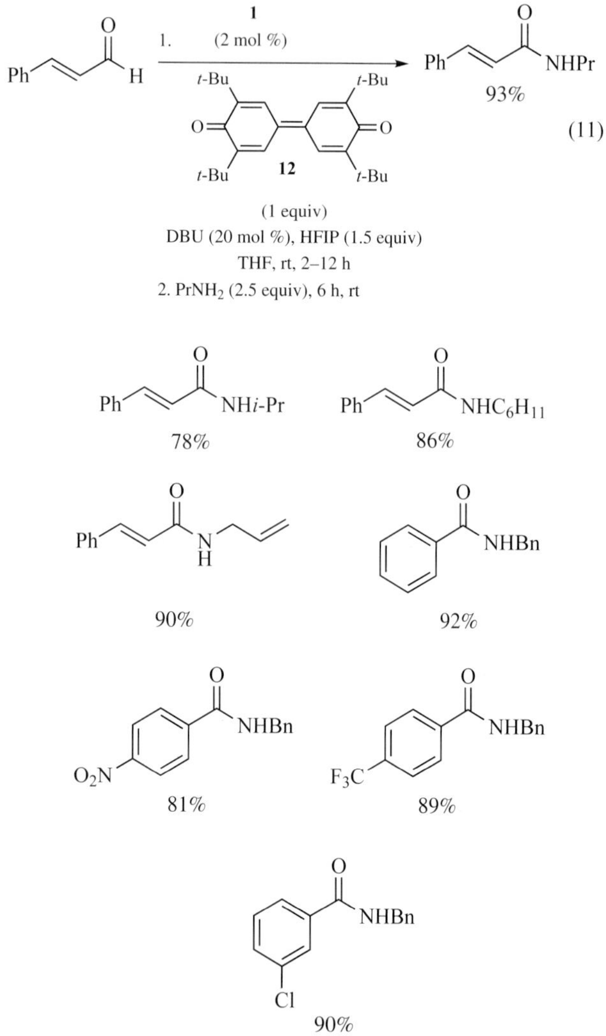

(11)

converted to the corresponding carboxylic acid or ester through acidic hydrolysis or treatment with an alcohol in acidic media.

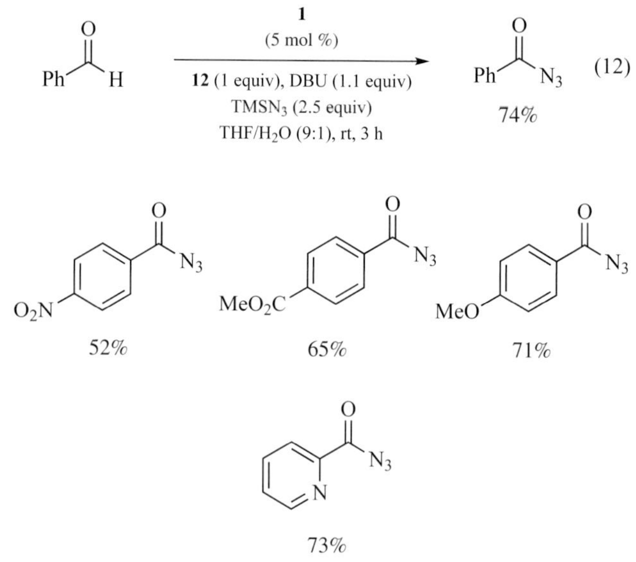

(12)

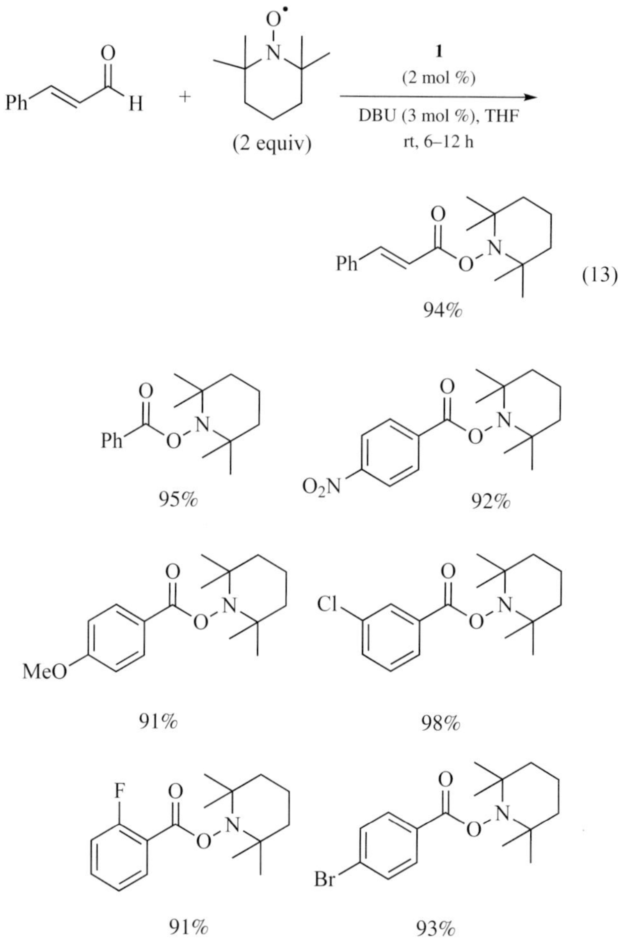

(13)

A biomimetic oxidation of aldehydes can be carried out with TEMPO to afford the corresponding TEMPO esters (eq 13).[14] It was proposed that the carbene generated from azolium salt **1** activates aldehydes to react as electron transfer reagents. Due to the volatility of TEMPO, TEMPO-OH can be used as an alternative oxidant and the resulting TEMPO esters can be subsequently converted to the corresponding carboxylic acid or ester through acidic hydrolysis or treatment with an alcohol in acidic media.

It has been shown that azolium salt **1** can act as an oxidative catalyst to combine enals and 1,3-dicarbonyl compounds into dihydropyranones (eq 14).[15] The quinone oxidant **12** was once again used and most of the substrates are commercially available and only a low loading of a catalyst was required. A diastereoselective example using a chiral β-keto amide provided a dihydropyranone with high diastereoselectivity.

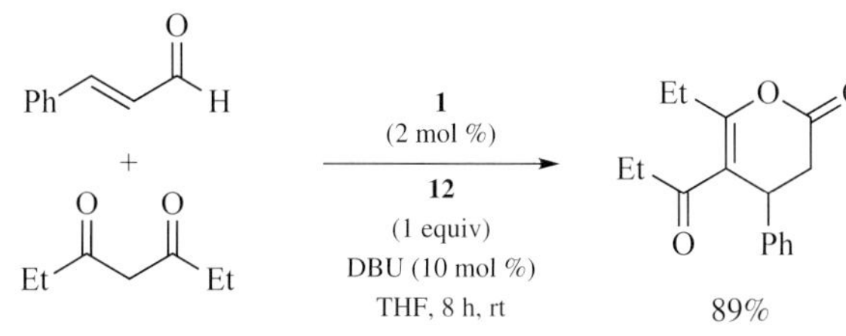

(14)

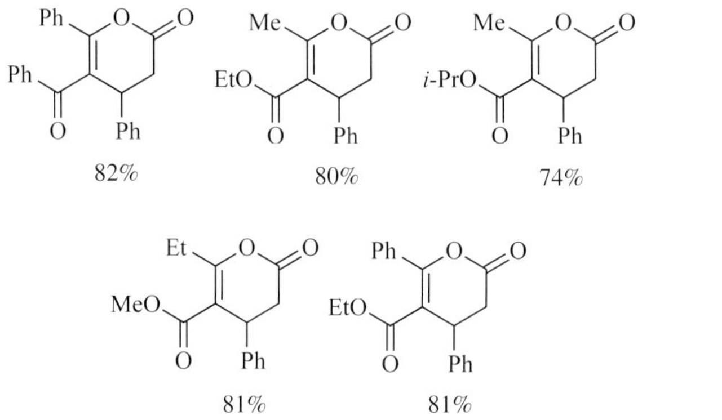

A detailed account of oxidations using **1** and the details into developing enantioselective versions have been reported.[16] The oxidation processes catalyzed by **1** has found application in the total synthesis of cyanolide A.[17]

1. (a) Miyashita, A.; Suzuki, Y.; Nagasaki, I.; Ishiguro, C.; Iwamoto, K.; Higashino, T., *Chem. Pharm. Bull.* **1997**, *45*, 1254. Other procedures have also been developed: (b) Belletire, J. L.; Bills, R. A.; Shackelford, S. A., *Synth. Commun.* **2008**, *38*, 738.
2. Ueng, S.-H.; Fensterbank, L.; Lacote, E.; Malacria, M.; Curran, D. P., *Org. Lett.* **2010**, *12*, 3002.
3. (a) Wang, R.; Guo, Y.; Zeng, Z.; Twamley, B.; Shreeve, J. M., *Chem. Eur. J.* **2009**, *15*, 2625. (b) Wang, R.; Gao, H.; Ye, C.; Twamley, B.; Shreeve, J. M., *Inorg. Chem.* **2007**, *46*, 932. (c) Jin, C.-M.; Ye, C.; Piekarski, C.; Twamley, B.; Shreeve, J. M., *Eur. J. Inorg. Chem.* **2005**, *18*, 3760. (d) Xue, H.; Twamley, B.; Shreeve, J. M., *J. Mater. Chem.* **2005**, *15*, 3459. (e) Ye, Ch.; Xiao, J.-C.; Twamley, B.; Shreeve, J. M., *Chem. Commun.* **2005**, *21*, 2750.
4. (a) Herrmann, W. A.; Gerstberger, G.; Spiegler, M., *Organometallics* **1997**, *16*, 2209. (b) Herrmann, W. A.; Fischer, J.; Oefele, K.; Artus, G. R. J., *J. Organomet. Chem.* **1997**, *530*, 259.
5. (a) Rentzsch, C. F.; Tosh, E.; Herrmann, W. A.; Kuehn, F. E., *Green Chem.* **2009**, *11*, 1610. (b) Herrmann, W. A.; Schuetz, J.; Frey, G. D.; Herdtweck, E., *Organometallics* **2006**, *25*, 2437. (c) Koecher, C.; Herrmann, W. A., *J. Organomet. Chem.* **1997**, *532*, 261. (d) Appelhans, L. N.; Incarvito, C. D.; Crabtree, R. H., *J. Organomet. Chem.* **2008**, *693*, 2761.
6. Normand, A. T.; Hawkes, K. J.; Clement, N. D.; Cavell, K. J.; Yates, B. F., *Organometallics* **2007**, *26*, 5352.
7. Mattson, A. E.; Bharadwaj, A. R.; Zuhl, A. M.; Scheidt, K. A., *J. Org. Chem.* **2006**, *71*, 5715.
8. Chan, A.; Scheidt, K. A., *J. Am. Chem. Soc.* **2006**, *128*, 4558.
9. Maki, B. E.; Chan, A.; Phillips, E. M.; Scheidt, K. A., *Org. Lett.* **2007**, *9*, 371.
10. Maki, B. E.; Scheidt, K. A., *Org. Lett.* **2008**, *10*, 4331.
11. Maki, B. E.; Patterson, E. V.; Cramer, C. J.; Scheidt, K. A., *Org. Lett.* **2009**, *11*, 3942.
12. De Sarkar, S.; Grimme, S.; Studer, A., *J. Am. Chem. Soc.* **2010**, *132*, 1190.
13. De Sarkar, S.; Studer, A., *Org. Lett.* **2010**, *12*, 1992.
14. Guin, J.; De Sarkar, S.; Grimme, S.; Studer, A., *Angew. Chem., Int. Ed.* **2008**, *47*, 8727.
15. De Sarkar, S.; Studer, A., *Angew. Chem., Int. Ed.* **2010**, *49*, 9266.
16. (a) Maki, B. E.; Chan, A.; Phillips, E. M.; Scheidt, K. A., *Tetrahedron* **2009**, *65*, 3102. (b) Suzuki, Y.; Muramatsu, K.; Yamauchi, K.; Morie, Y.; Sato, M., *Tetrahedron* **2006**, *62*, 302. (c) Wei, S.; Liu, B.; Zhao, D.; Wang, Z.; Wu, J.; Lan, J.; You, J., *Org. Biomol. Chem.* **2009**, *7*, 4241.
17. Kim, H.; Hong, J., *Org. Lett.* **2010**, *12*, 2880.

Shawn K. Collins
Université de Montréal, Montréal, Québec, Canada

(11*bR*)-Dinaphtho[2,1-*d*:1′,2′-*f*][1,3,2]dioxaphosphepin, 2,6-Bis[3,5-bis (trifluoromethyl)phenyl]-4-hydroxy-, 4-Oxide

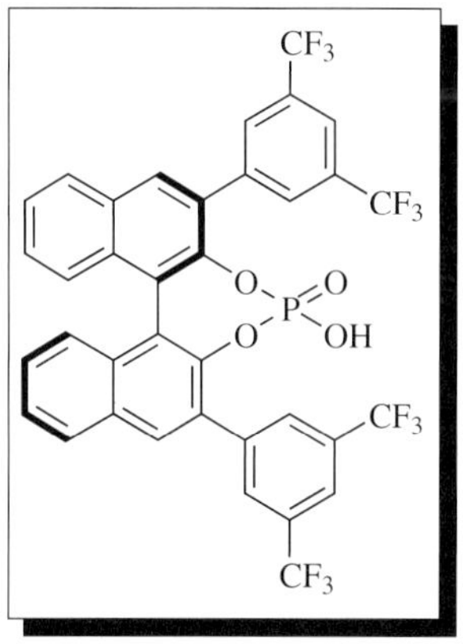

[791616-62-1] $C_{36}H_{17}F_{12}O_4P$ (MW 772.47)
InChI = 1S/C36H17F12O4P/c37-33(38,39)21-9-19(10-22(15-21)34(40,41)42)27-13-17-5-1-3-7-25(17)29-30-26-8-4-2-6-18(26)14-28(32(30)52-53(49,50)51-31(27)29)20-11-23(35(43,44)45)16-24(12-20)36(46,47)48/h1-16H,(H,49,50)
InChIKey = DQORDVSQWPKAQJ-UHFFFAOYSA-N

(a versatile chiral Brønsted acid catalyst has been widely used in enantioselective addition reactions of nucleophiles to imines and carbonyl compounds).[1]

Alternate Name: none.

Physical Data: mp 163.5–180.0 °C (dec), $[\alpha]_D^{26} = -197.5$ (*c* 0.97, $CHCl_3$).

Solubility: soluble in common organic solvents such as chloroform, toluene, THF, ethyl acetate.

Preparative Methods: the phosphoric acid, (11b*R*)-dinaphtho[2,1-*d*:1′,2′-*f*][1,3,2]dioxaphosphepin, 2,6-bis[3,5-bis-(trifluoromethyl)phenyl]-4-hydroxy-, 4-oxide, **1** has been accessed by a reaction of (*R*)-3,3′-di(2-naphthyl)-1,1′-binaphthyl-2,2′-diol, which was prepared via a procedure developed by either Jørgensen or Wipf,[2] with phosphoryl trichloride, followed by hydrolysis (eq 1).[3]

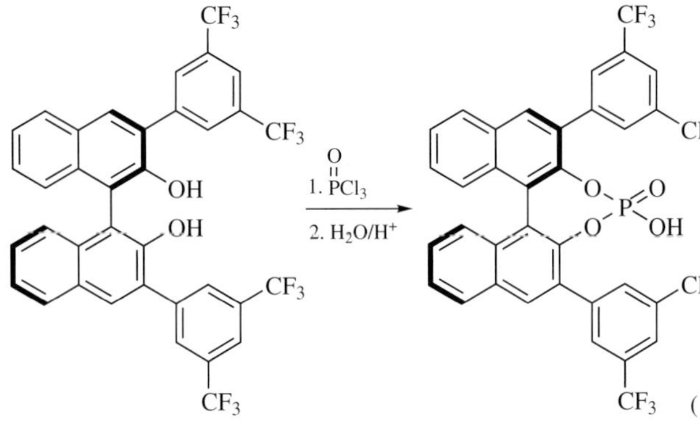

(1)

Form Supplied in: white solid.

Handling, Storage, and Precautions: the compound is stable and does not require special storage and handling.

Asymmetric Hydrophosphonylation of Imines. α-Amino phosphonic acid derivatives have attracted much attention because they are broadly applicable to the synthesis of biologically active compounds. Asymmetric hydrophosphonylation of imines with phosphites represents a shortcut to produce enantioenriched α-amino phosphonates. Akiyama reported that the phosphoric acid catalyst **1** exhibited much higher ability in controlling the stereochemistry than its structural analogs in the phosphonylation of PMP-protected imines with diisopropyl phosphite, providing good enantioselectivities (52–77% ee) for *ortho*-substituted benzaldimines and even higher enantioselectivities (81–90% ee) for imines derived from cinnamaldehyde derivatives (eqs 2 and 3).[3]

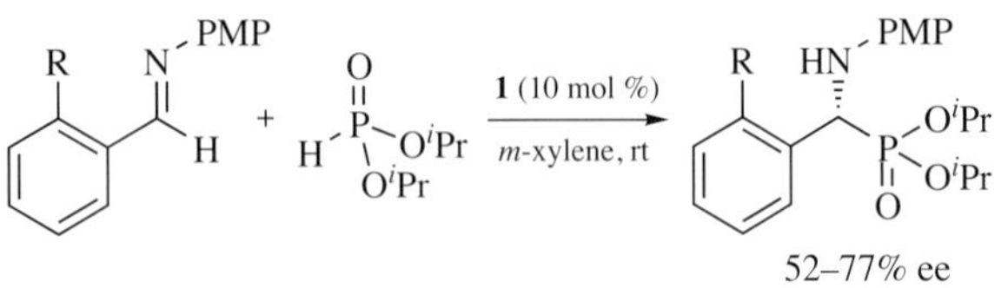

(2)

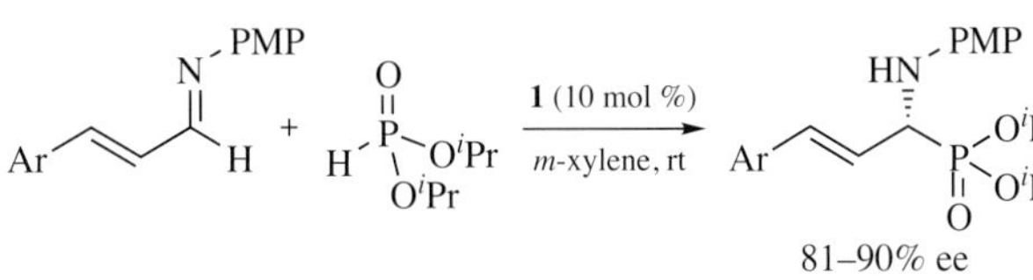

(3)

Asymmetric Transfer Hydrogenation Reaction. Three research groups independently presented the asymmetric transfer hydrogenation of ketimines with Hantzsch ester promoted by chiral phosphoric acids derived from BINOL derivatives.[4] Rueping found that **1** is the best catalyst for the reduction of an imine derived from 1-(2-naphthalenyl)ethanone and benzene is the solvent of choice in terms of conversion and stereoselectivity (eq 4).[4a]

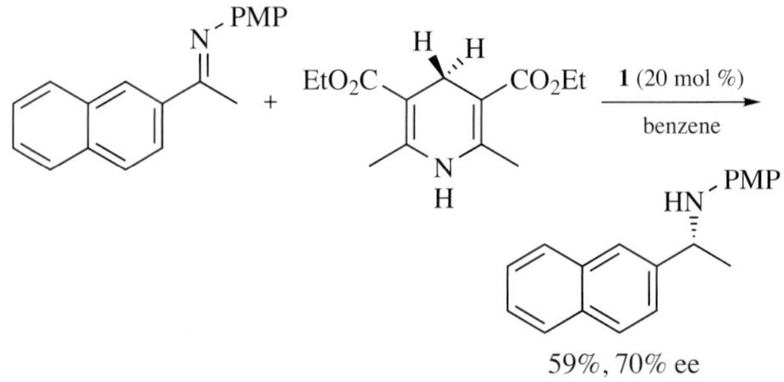

(4)

Under the optimal conditions, **1** delivered fairly good stereochemical outcomes for a variety of ketimines derived from various acetophenone derivatives (68–84% ee, eq 5). Notably, the use of phosphoric acids analogous to **1** gave even higher enantioselectivity as demonstrated by MacMillan and List.[4b,c]

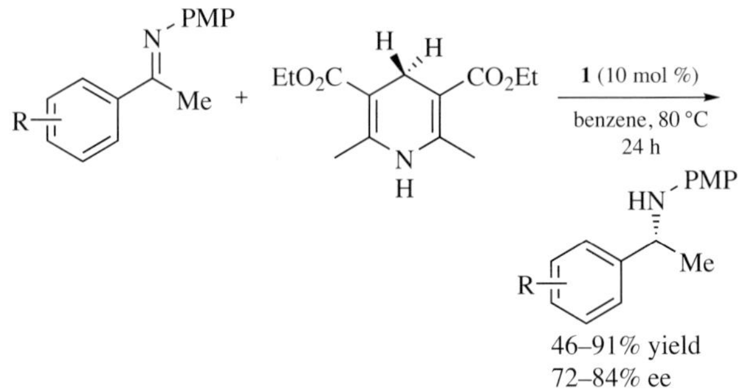

(5)

Expansion of the protocol to reduction of nitrogen-containing heterocycles has also been attempted.[5–7] The conversion and enantioselectivity highly depend on the substrate structure. Fairly good enantioselectivity of 72% ee was obtained in the reduction of 2-phenylquinoline.[5] However, catalyst **1** provided much lower enantioselectivities for benzoxazine,[6] 2-methyl-6-pentylnicotinonitrile,[7] and 3-phenylquinoline[8] under identical conditions (eqs 6-9).

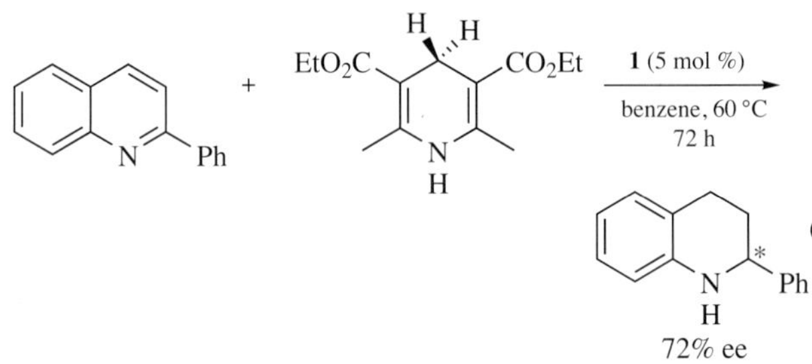

(6)

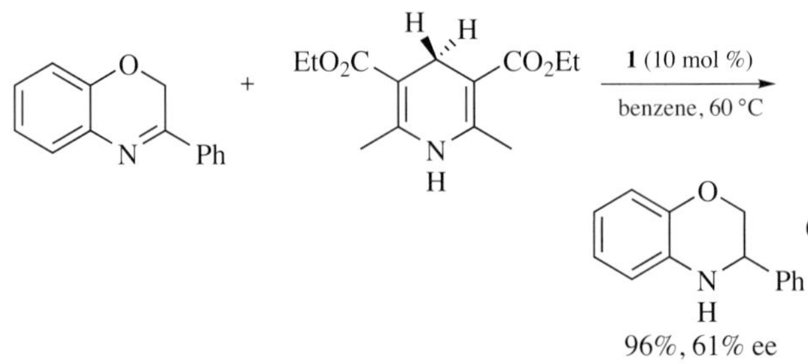

(7)

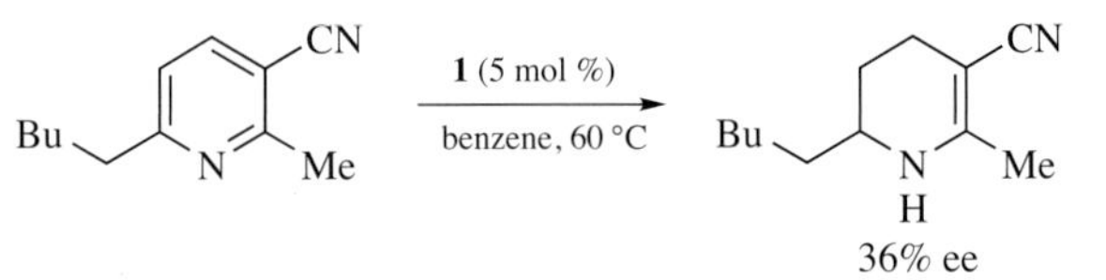

(8)

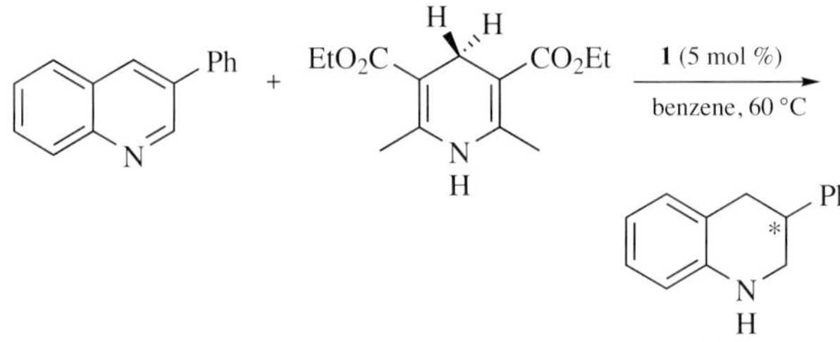

(9)

10% ee

The Brønsted acid **1** exhibited lower stereoselectivity in the reduction of imino esters than for ketimines as seen with the transfer hydrogenation of α-phenylimino ester (eq 10).[9]

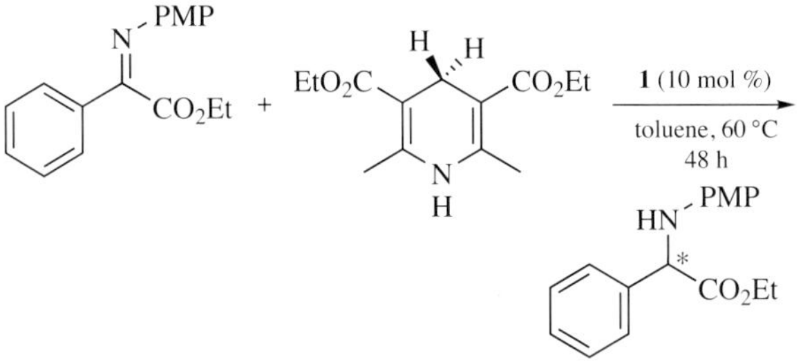

(10)

39% conversion, 32% ee

Asymmetric Pictet–Spengler Reaction. The asymmetric Pictet–Spengler reaction represents an important organic transformation for the synthesis of tetrahydro-β-carbolines and tetrahydroisoquinolines. Brønsted acids are efficient catalysts of this reaction. List reported that **1** furnished the Pictet–Spengler reaction of a tryptamine with propionaldehyde in excellent yield (95%), but with low enantioselectivity (18% ee) (eq 11).[10]

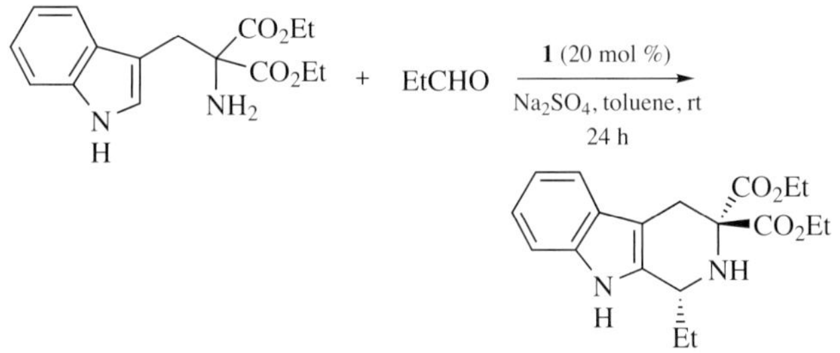

95%, 18% ee (11)

Interestingly, the Brønsted acid **1** could provide superior stereoselectivity in the Pictet–Spengler reaction involving sulfenyliminium ions derived from tryptamine.[11] Aromatic and aliphatic aldehydes delivered clean Pictet–Spengler reaction with *N*-(2-(1*H*-indol-3-yl)ethyl)-*S*-tritylthiohydroxylamine, affording tetrahydro-β-carbolines in high yields with enantioselectivities ranging from 30 to 88% ee after deprotection (eq 12).

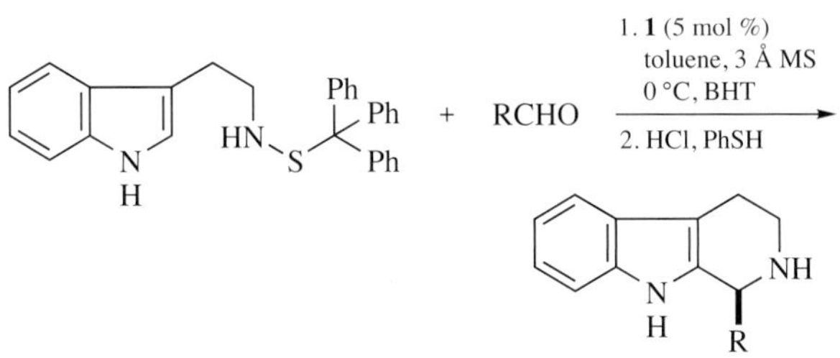

(12)

77–88% yield
30–88% ee

Asymmetric Friedel–Crafts Reaction. Reaction of indoles with imines generates 3-indolyl methanamines that serve as key components in many biologically active natural products. You reported the use of phosphoric acids in the Friedel–Crafts reaction of indoles with imines. **1** provided high enantioselectivity for the reaction of indole with N-TS protected benzaldimine,[12] but gave a lower ee for reaction with *N*-(1-phenylvinyl)acetamide using dichloromethane as a solvent (eqs 13 and 14).[13]

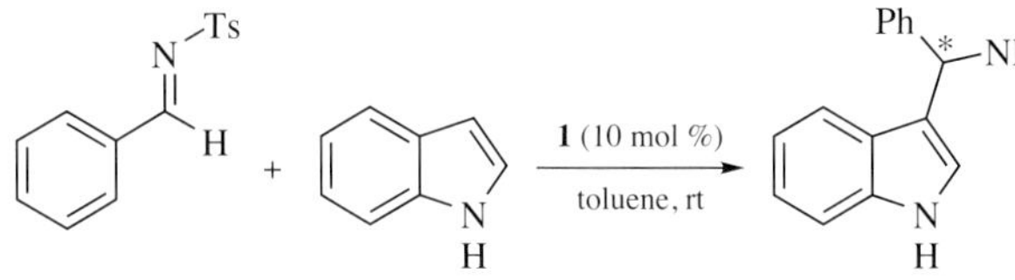

(13)

80%, 83% ee

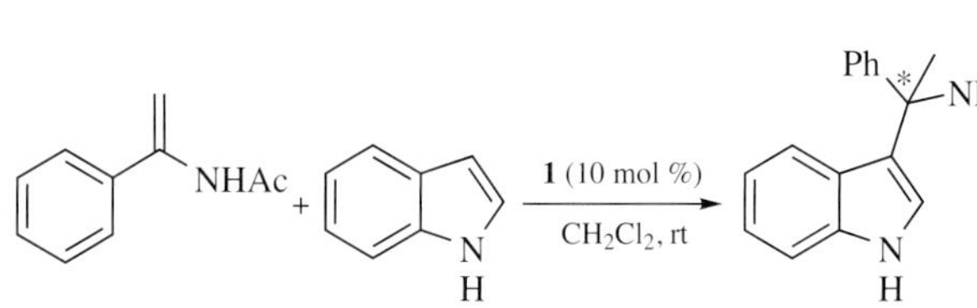

(14)

72%, 38% ee

Asymmetric Three–component Cyclization Reaction. The phosphoric acid-catalyzed Biginelli reaction is an important multicomponent cyclization for the direct assembly of aldehydes, urea or thiourea, and 1,3-dicarbonyls into multifunctionalized 3,4-dihydropyrimidin-2-(1*H*)-ones (DHPMs), which show a wide scope of pharmacological properties and make up a large family of medicinally relevant compounds.[14] The catalyst **1** delivers a moderate enantioselectivity but low conversion (eq 15).[15]

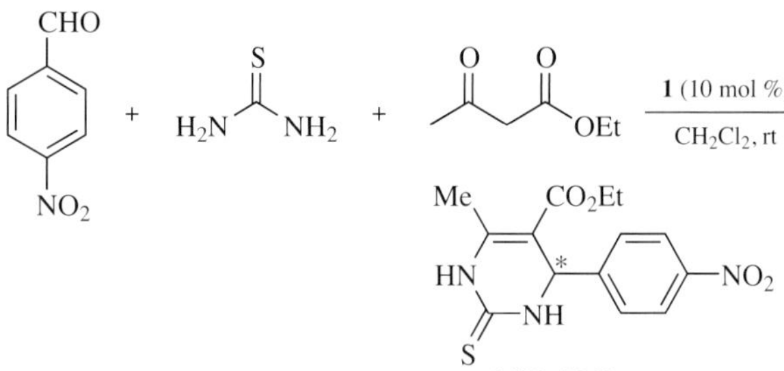

(15)

25%, 52% ee

Dihydropyridines (DHPs), particularly the 4-aryl-substituted 1,4-dihydropyridines, have been recognized as calcium channel modulators for the treatment of cardiovascular diseases.[16] Compounds containing the 1,4-dihydropyridine moiety showed a wide scope of other pharmaceutical activities.[17] The asymmetric catalytic three-component cyclization reaction of cinnamaldehydes, an aromatic primary amine, and 1,3-dicarbonyls proceeded in the presence of catalytic phosphoric acid to allow the formation of 4-aryl-substituted 1,4-dihydropyridines with enantioselectivity highly sensitive to the structure of the catalyst. Phosphoric acid **1** afford a high yield but low enantiomeric excess (eq 16).[18]

The 1,3-dipolar cycloaddition of azomethine ylides to electron-deficient olefins yields chiral pyrolidines, having widespread applications. The phosphoric acid **1** enables *p*-nitrobenzaldehye, diethyl 2-aminomalonate, and dimethyl maleate to afford a pyrolidine derivative in high yield but low enantioselectivity (eq 17).[19]

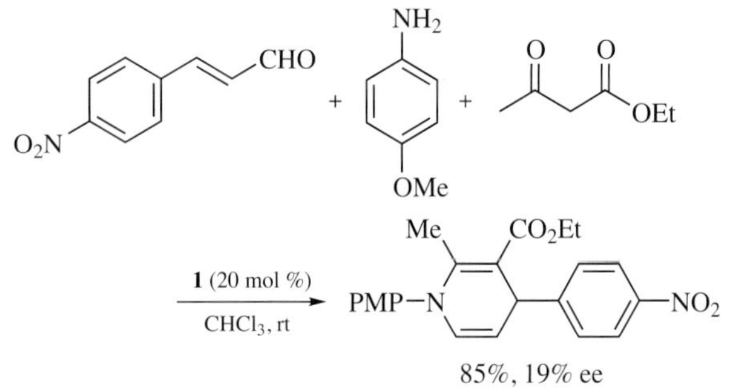

(16)

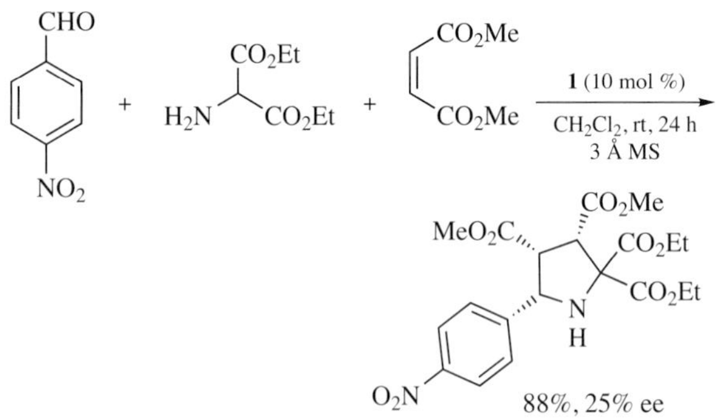

(17)

Asymmetric *anti*-Mannich Reaction. A one-pot Mannich reaction of cyclohexanone, 4-methoxyaniline, and *p*-nitrobenzaldehyde with 5 mol % of **1** generated the β-amino ketone in good yield favoring the *anti*-diastereomer, but neither diastereo- nor enantioselectivity reached synthetically useful levels (eq 18).[20]

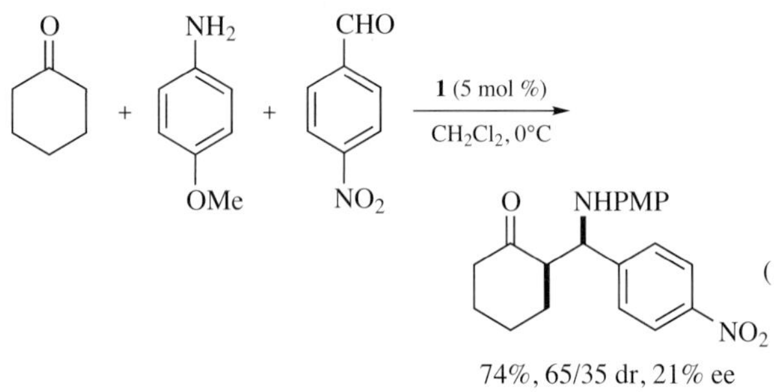

(18)

Asymmetric Baeyer–Villiger Reaction. Phosphoric acid **1** catalyzed the asymmetric Baeyer–Villiger reaction of 3-phenylcyclobutanone using aqueous hydrogen peroxide, affording 3-phenyl-γ-butyrolactone in nearly quantitative yield, but with only 12% ee (eq 19).[21]

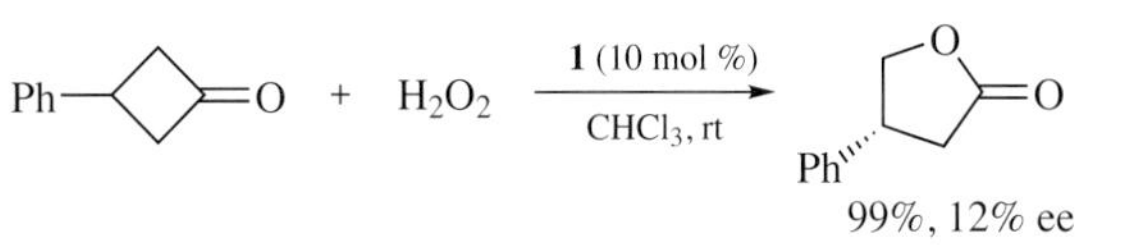

(19)

Asymmetric Cyanation of Imines. Strecker reaction of the benzyl-protected imine of *p*-anisaldehyde with HCN catalyzed by **1** yielded an amino nitrile with 38% ee (eq 20).[22]

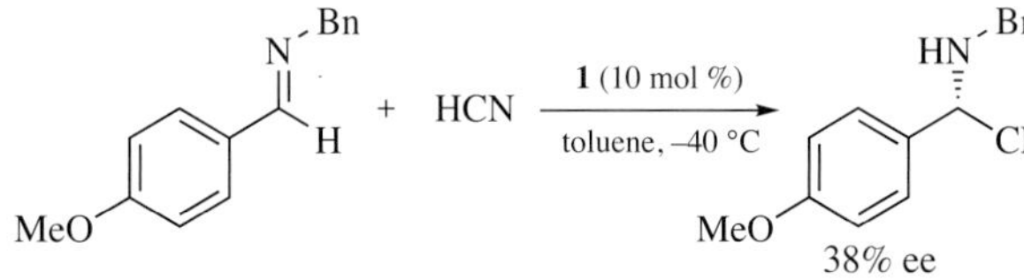

(20)

Asymmetric Alkynylation of α-Imino Esters. A combined catalyst system of phosphoric acid **1** and silver acetate was applied to the asymmetric addition of phenylacetylene to ethyl 2-(4-methoxyphenylimino)acetate, but provided a low enantioselectivity (eq 21). The phosphoric acid was assumed to function as a counterion after a reaction with sliver acetate, and thus providing an interesting example of counterion-controlled enantioselectivity.[23]

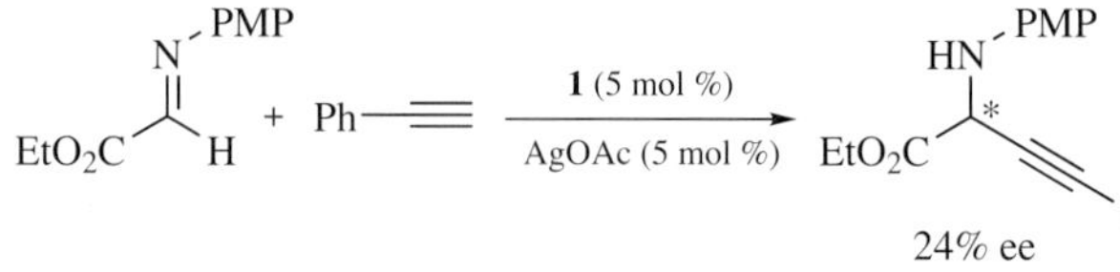

(21)

1. For reviews on the chiral phosphoric acid catalysis, see (a) Akiyama, T., *Chem. Rev.* **2007**, *107*, 5744. (b) Doyle, A. G.; Jacobsen, E. N., *Chem. Rev.* **2007**, *107*, 5713 (c) Akiyama, T.; Itoh, J.; Fuchibe, K., *Adv. Synth. Catal.* **2006**, *348*, 999.
2. (a) Simonsen K. B.; Gothelf, K. V.; Jørgensen, K. A., *J. Org. Chem.* **1998**; *63*, 7536. (b) Wipf, P.; Jung, J. K., *J. Org. Chem.* **2000**, *65*, 6319.
3. Akiyama, T.; Morita, H.; Itoh, J.; Fuchibe, K., *Org. Lett.* **2005**, *7*, 2583.
4. (a) Rueping, M.; Sugiono, E.; Azap, C.; Theissmann, T.; Bolte, M., *Org. Lett.* **2005**, *7*, 3781. (b) Hoffmann, S.; Seayad A. M.; List, B., *Angew. Chem., Int. Ed.* **2005**, *44*, 7424. (c) Storer, R. I.; Carrera, D. E.; Ni, Y.; MacMillan, D. W. C., *J. Am Chem. Soc.* **2006**, *128*, 84.
5. Rueping, M.; Antonchick, A. P.; Theissmann, T., *Angew. Chem., Int. Ed.* **2006**, *45*, 3683.
6. Rueping, M.; Antonchick, A. P.; Theissmann, T., *Angew. Chem., Int. Ed.* **2007**, *46*, 6751.
7. Rueping, M.; Antonchick, A. P., *Angew. Chem., Int. Ed.* **2007**, *46*, 4562.
8. Rueping, M.; Theissmann, T.; Raja, S.; Bats, J. W., *Adv. Synth. Catal.* **2008**, *350*, 1001.
9. Kang, Q.; Zhao, Z. A.; You, S. L., *Adv. Synth. Catal.* **2007**, *349*, 1657.
10. Seayad, J.; Seayad, A. M.; List, B., *J. Am. Chem. Soc.* **2006**, *128*, 1086.
11. Wanner, M.; van der Haas, R. N. S.; de Cuba, K. R.; van Maarseveen, J. H.; Hiemstra, H., *Angew. Chem., Int. Ed.* **2007**, *46*, 7485.
12. Kang, Q.; Zhao, Z. A.; You, S. L., *J. Am. Chem. Soc.* **2007**, *129*, 1484.
13. Jia, Y. X.; Zhong, J.; Zhu, S. F.; Zhang, C. M.; Zhou, Q. L., *Angew. Chem., Int. Ed.* **2007**, *46*, 5565.
14. Kappe, C. O., *Eur. J. Med. Chem.* **2000**, *35*, 1043.
15. Chen, X. H.; Xu, X. Y.; Liu, H.; Cun, L. F.; Gong, L. Z., *J. Am. Chem. Soc.* **2006**, *128*, 14802.
16. (a) Janis, R. A.; Silver, P. J.; Triggle, D. J., *Adv. Drug Res.* **1987**, *16*, 309. (b) Bossert, F.; Vater, W., *Med. Res. Rev.* **1989**, *9*, 291.
17. Kappe, C. O., *Eur. J. Med. Chem.* **2000**, *35*, 1043.
18. Jiang, J.; Yu, J.; Sun, X. X.; Rao, Q. Q.; Gong, L. Z., *Angew. Chem., Int. Ed.* **2008**, *47*, 2458.
19. Chen, X. H.; Zhang, W. Q.; Gong, L. Z., *J. Am. Chem. Soc.* **2008**, *130*, 5652.
20. Guo, Q.-X.; Liu, H.; Guo, C.; Luo, S.-W.; Gu, Y.; Gong, L.-Z., *J. Am. Chem. Soc.* **2007**, *129*, 3790.
21. Xu, S.; Wang, Z.; Zhang, X.; Ding, K., *Angew. Chem., Int. Ed.* **2008**, *47*, 2880.
22. Rueping, M.; Sugiono, E.; Azap, C., *Angew. Chem., Int. Ed.* **2006**, *45*, 2617.
23. Rueping, M.; Antonchick, A. P.; Brinkmann, C., *Angew. Chem., Int. Ed.* **2007**, *46*, 6903.

Liu-Zhu Gong
University of Science and Technology of China, Hefei, China

(11b*R*)-Dinaphtho[2,1-*d*:1′,2′-*f*][1,3,2] dioxaphosphepin, 4-Hydroxy-2,6-di-2-naphthalenyl-, 4-Oxide

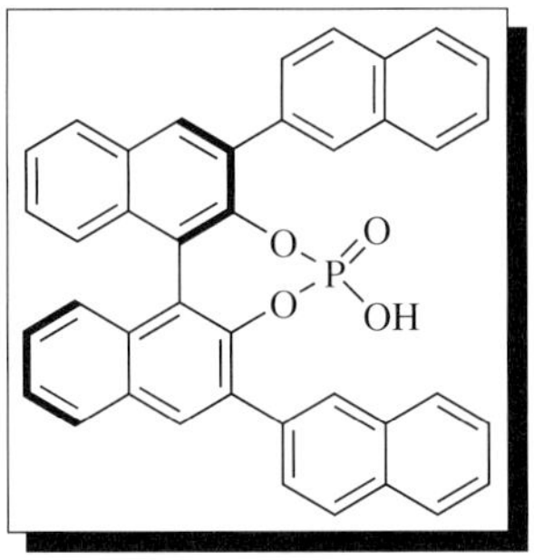

[791616-56-3] $C_{40}H_{25}O_4P$ (MW 600.60)
InChI = 1S/C40H25O4P/c41-45(42)43-39-35(31-19-17-25-9-1-3-11-27(25)21-31)23-29-13-5-7-15-33(29)37(39)38-34-16-8-6-14-30(34)24-36(40(38)44-45)32-20-18-26-10-2-4-12-28(26)22-32/h1-24H,(H,41,42)
InChIKey = IHLZSTKPUGDSET-UHFFFAOYSA-N

(a versatile chiral Brønsted acid catalyst has been widely used in enantioselective addition reactions of nucleophiles to imines and carbonyl compounds)[1]

Alternate Name: none.
Physical Data: mp 250–260 °C (dec), $[\alpha]_D^{23} = -322.8°$ (*c* 1.02, $CHCl_3$).
Solubility: it is soluble in common organic solvents such as chloroform, toluene, THF, ethyl acetate.
Preparative Method: the phosphoric acid, (11b*R*)-dinaphtho[2,1-*d*:1′,2′-*f*][1,3,2]dioxaphosphepin, 4-hydroxy-2,6-di-2-naphthalenyl-, 4-oxide, **1** has been accessed by a reaction of (*R*)-3,3′-bis(3,5-bis(trifluoromethyl)phenyl)-1,1′-binaphthyl-2,2′-diol, which was prepared following a procedure developed by either Jørgensen or Wipf,[2] with phosphoryl trichloride, followed by hydrolysis under acidic conditions (eq 1).[3]

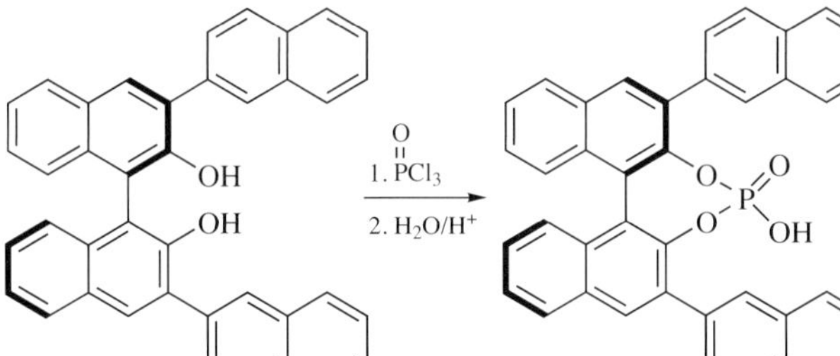

(1)

Form Supplied in: white solid.
Handling, Storage, and Precautions: the phosphoric acid **1** is a stable compound that does not require special precaution for storage and handling.

Asymmetric Transfer Hydrogenation. Asymmetric transfer hydrogenation of prochiral imines with Hantzsch ester catalyzed by the chiral phosphoric acid was independently described by three groups.[3] **1** does not deliver efficient catalytic activity for the transfer hydrogenation reaction of ketimines, only affording moderate conversions and enantioselectivities as exemplified by the reduction of ketimines derived from 1-(2-naphthalenyl)ethanone[3a] and acetophenone (eqs 2 and 3).[3b,c] Moreover, the transfer hydrogenation of α-phenylimino ester with Hantzsch ester under the catalysis of **1** led to the formation of phenylglycine ethyl ester with 56% conversion and 40% ee (eq 4).[4]

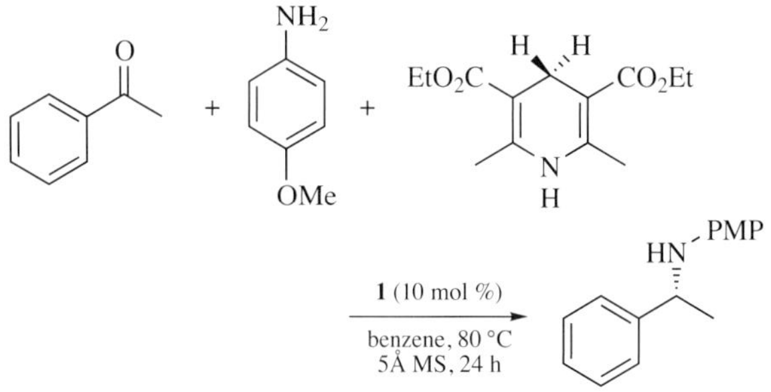

(2)

41% Conversion, 45% ee

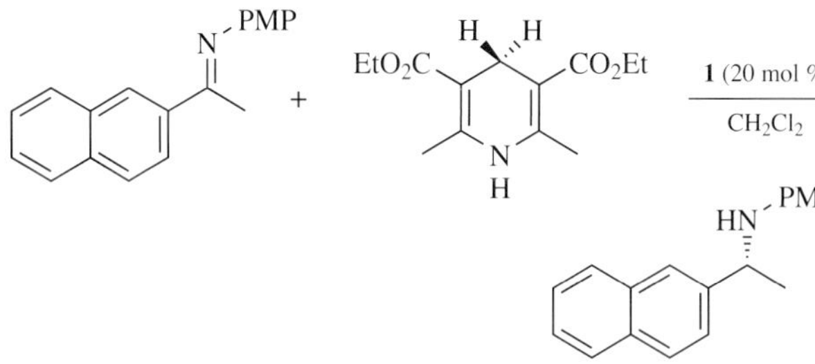

(3)

54%, 40% ee

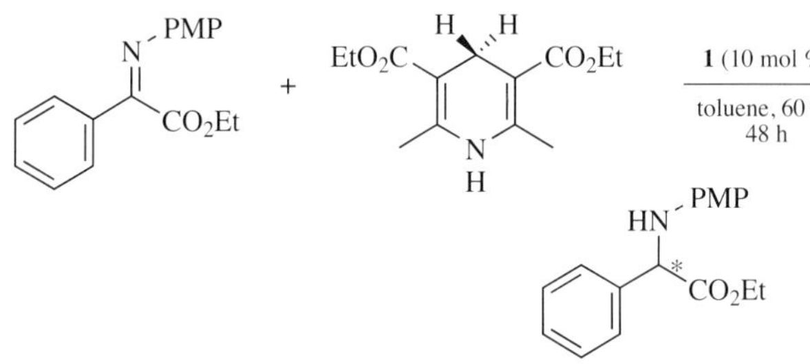

(4)

56% conversion, 40% ee

Optically active α-branched amines could be obtained by the reductive amination of aldehydes via asymmetric dynamic kinetic resolution. Once again, phosphoric acid **1** showed low catalytic activity for the reductive amination of 2-phenylpropanal with Hantzsch ester with quite low stereochemical outcome (eq 5).[5]

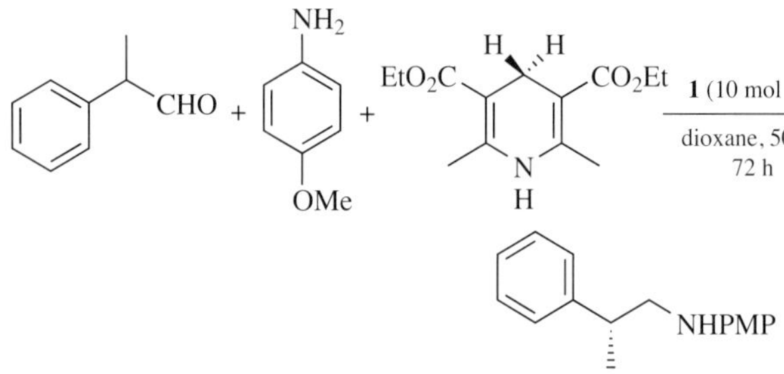

(5)

48% yield, 14% ee

Enantioselective reductions of 2-phenylquinoline,[6] benzoxazine,[7] and 2-methyl-6-pentylnicotinonitrile[8] with Hantzsch ester underwent smoothly using benzene at 60 °C in the presence of catalytic amount of **1**, but gave corresponding chiral N-containing heterocycles with moderate enantiomeric excesses (eqs 6-8).

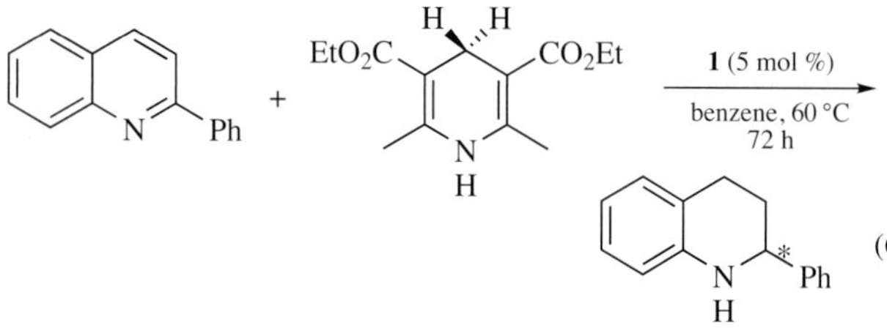

(6)

26% ee

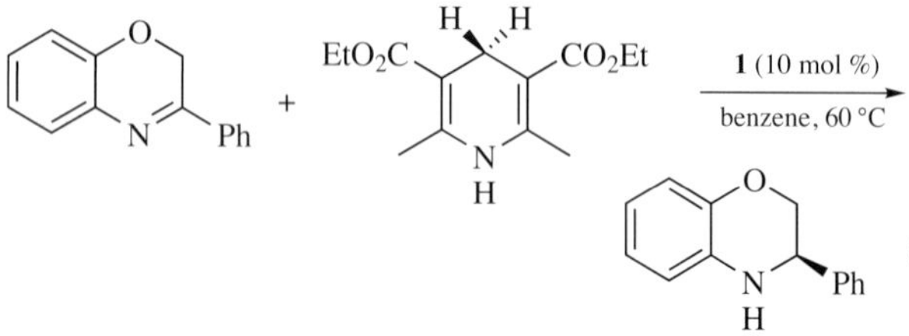

(7)

93%, 38% ee

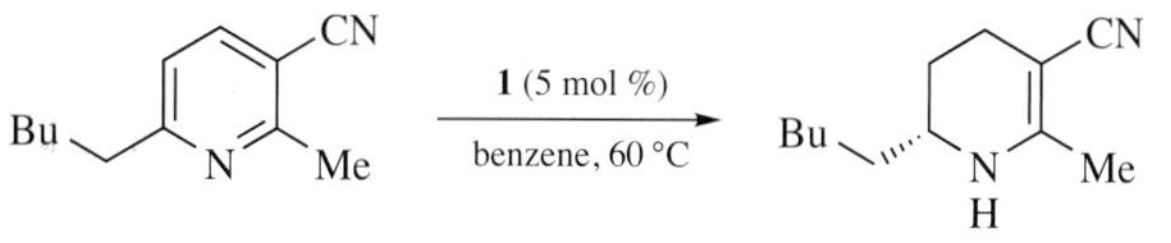

(8)

32% ee

Asymmetric Cyanation of Imines. Asymmetric catalytic cyanation of imines provides a straightforward method for the synthesis of α-amino acids. The Strecker reaction of a benzyl-protected imine derived from *p*-anisaldehyde with HCN catalyzed by **1** yielded an amino nitrile with 32% ee.[9] Expansion of the protocol to PMP-protected ketimine derived from 4′-methylacetophenone resulted in 30% ee.[10] In addition, the phosphoric acid **1** only afforded 14% ee for an acylcyanation of benzyl-protected benzaldimine (eqs 9-11).[11]

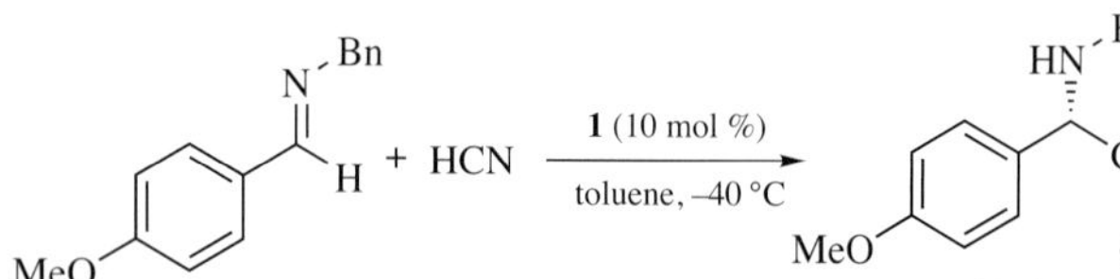

(9)

32% ee

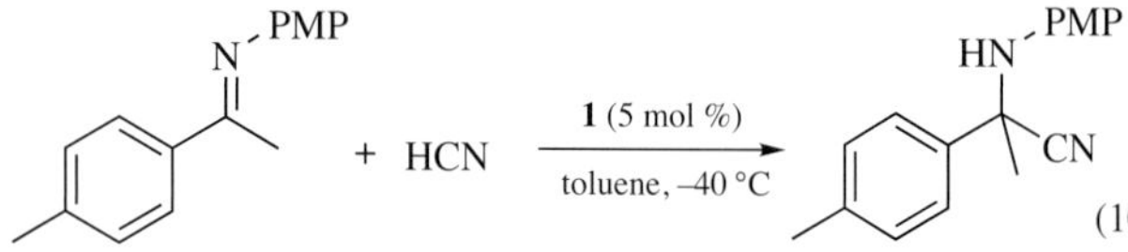

(10)

30% ee

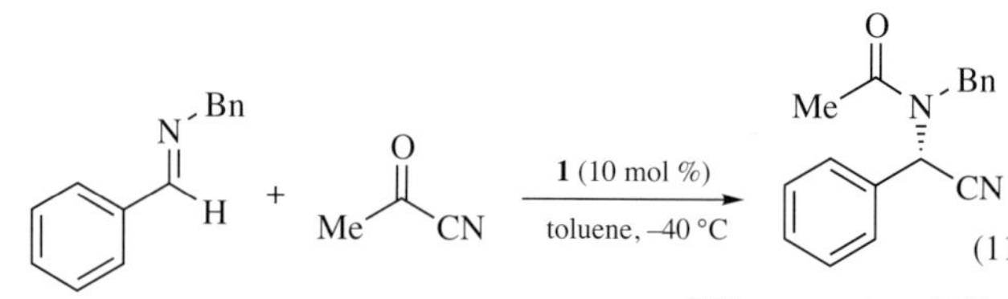

(11)

95% conversion, 14% ee

Asymmetric Ene-type Reactions. Phosphoric acid **1** catalyzes an asymmetric imino ene reaction of methyleneaminopyrrolidine with Boc-protected benzaldimine to afford an α-aminohydrazone in 72% yield with 25% ee (eq 12).[12]

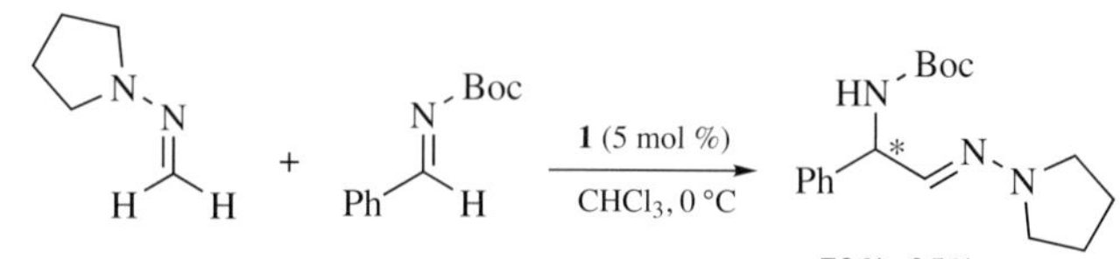

72%, 25% ee

(12)

The aza-ene reaction of ethyl glyoxylate with methyl 1-phenylvinylcarbamate proceeded cleanly with 5 mol % **1** to generate a β-hydroxy imine in 80% yield with 91% ee (eq 13).[13]

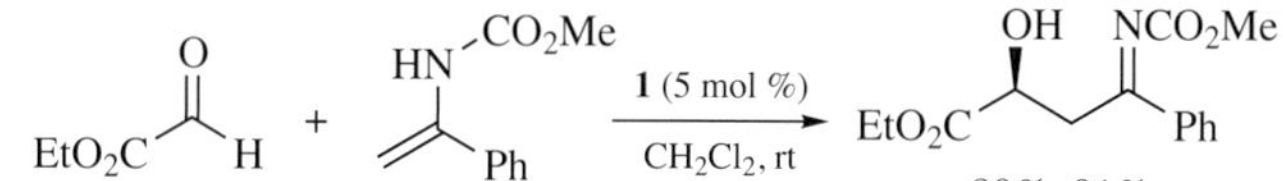

80%, 91% ee

(13)

Asymmetric Cyclization. Isoquinuclidines have been accessed by Brønsted acid-catalyzed cyclization between cyclohexenone and aryl aldimines. Phosphoric acid **1** showed superior ability in stereocontrol in comparison to its structural analogs. Using 20 mol % acetic acid as cocatalyst, isoquinuclidines are obtained in good yields and high stereoselectivities (eq 14).[14]

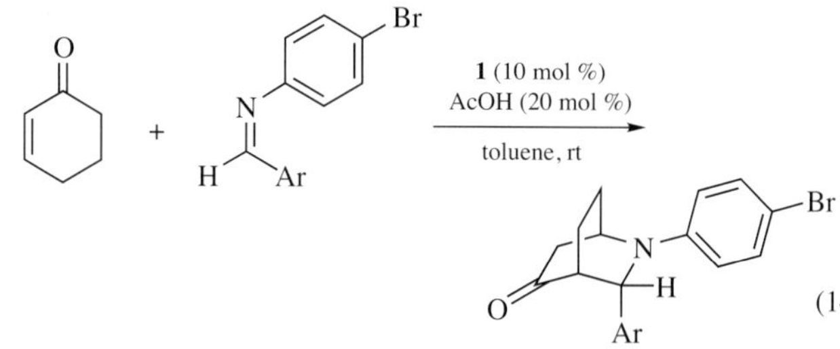

(14)

54–84% yield
1/3–1/8 dr
82–88% ee

Nazarov cyclization of (*E*)-1-(3,4-dihydro-6-pyranyl)-2-methyl-3-phenylpropen-1-one accelerated by 10 mol % of **1** gave cyclopentenones with 1:1 diastereomeric ratio. The *cis*-cyclopentenone was isolated in 54% ee (eq 15).[15]

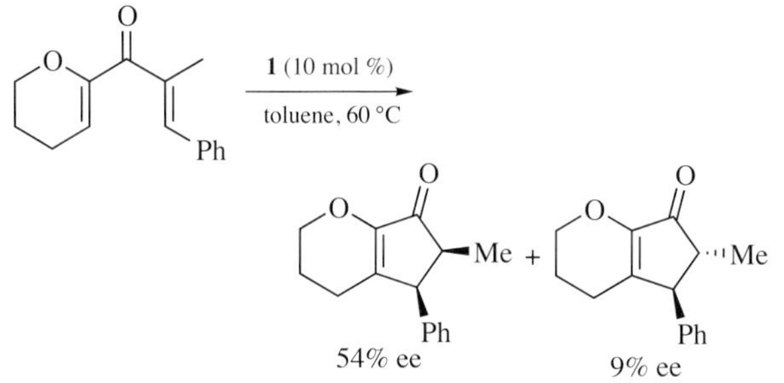

(15)

Asymmetric Baeyer–Villiger Reaction. Phosphoric acid **1** enables the asymmetric Baeyer–Villiger reaction of 3-phenylcyclobutanone using aqueous hydrogen peroxide as an oxidant to produce 3-phenyl-γ-butyrolactone in 72% yield with 37% ee (eq 16).[16]

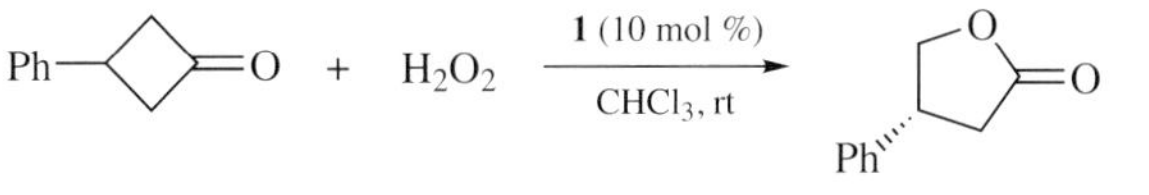

(16)

72%, 37% ee

Asymmetric Pictet–Spengler Reaction. List et al. first evaluated chiral phosphoric catalysis for asymmetric Pictet–Spengler synthesis of tetrahydro-β-carbolines and tetrahydroisoquinolines. While the reaction of a tryptamine with propionaldehyde proceeded readily using 20 mol % of **1** to afford the Pictet–Spengler product in 75% yield, the ee was only 30% (eq 16).[17]

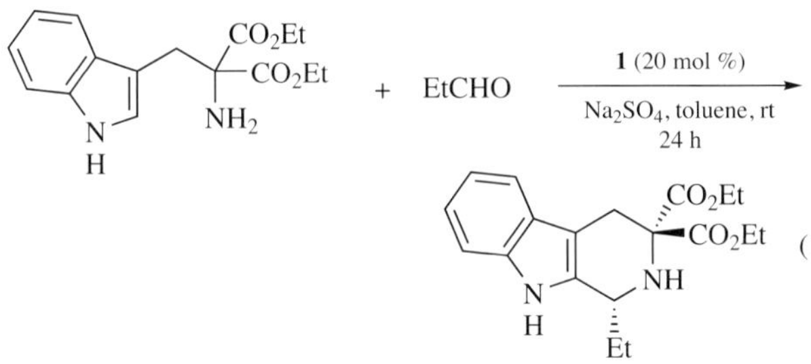

(17)

75%, 30% ee

1. For reviews on the chiral phosphoric acid catalysis, see (a) Akiyama, T., *Chem. Rev.* **2007**, *107*, 5744. (b) Doyle, A. G.; Jacobsen, E. N., *Chem. Rev.* **2007**, *107*, 5713. (c) Akiyama, T.; Itoh, J.; Fuchibe, K., *Adv. Synth. Catal.* **2006**, *348*, 999.
2. (a) Simonsen, K. B.; Gothelf, K. V.; Jørgensen, K. A., *J. Org. Chem.* **1998**, *63*, 7536. (b) Wipf, P.; Jung, J. K., *J. Org. Chem.* **2000**, *65*, 6319.
3. (a) Rueping, M.; Sugiono, E.; Azap, C.; Theissmann, T.; Bolte, M., *Org. Lett.* **2005**, *7*, 3781. (b) Hoffmann, S.; Seayad, A. M.; List, B., *Angew. Chem., Int. Ed.* **2005**, *44*, 7424. (c) Storer, R. I.; Carrera, D. E.; NiY; MacMillan, D. W. C., *J. Am Chem. Soc.* **2006**, *128*, 84.
4. (a) Kang, Q.; Zhao, Z. A.; You, S. L., *Adv. Synth. Catal.* **2007**, *349*, 1657. (b) Li, G.; Liang, Y.; Antilla, J. C., *J. Am Chem. Soc.* **2007**, *129*, 5830.
5. Hoffmann, S.; Nicoletti, M.; List, B., *J. Am Chem. Soc.* **2006**, *128*, 13074.
6. Rueping, M.; Antonchick, A. P.; Theissmann, T., *Angew. Chem., Int. Ed.* **2006**, *45*, 3683.
7. Rueping, M.; Antonchick, A. P.; Theissmann, T., *Angew. Chem., Int. Ed.* **2006**, *45*, 6751.
8. Rueping, M.; Antonchick, A. P., *Angew. Chem., Int. Ed.* **2007**, *46*, 4562.
9. Rueping, M.; Sugiono, E.; Azap, C., *Angew. Chem., Int. Ed.* **2006**, *45*, 2617.
10. Rueping, M.; Sugiono, E.; Moreth, S. A., *Adv. Synth. Catal.* **2007**, *349*, 759.
11. Pan, S. C.; List, B., *Chem. Asian J.* **2008**, *3*, 430.
12. Rueping, M., et al., *Org. Lett.* **2007**, *9*, 1065.
13. Terada, M.; Soga, K.; Momiyama, N., *Angew. Chem., Int. Ed.* **2008**, *47* 4122.
14. (a) Rueping, M.; Azap, C., *Angew. Chem., Int. Ed.* **2006**, *45*, 7832. (b) Liu, H.; Cun, L. F.; Mi, A. Q.; Jiang, Y. Z. Gong, L. Z. *Org. Lett.* **2006**, *8*, 6023.
15. Rueping, M.; Ieawsuwan, W.; Antonchick, A. P.; Nachtsheim, B. J., *Angew. Chem., Int. Ed.* **2007**, *46*, 2097.
16. Xu, S.; Wang, Z.; Zhang, X.; Zhang, X.; Ding, K., *Angew. Chem., Int. Ed.* **2008**, *47*, 2880.
17. Seayad, J.; Seayad, A. M.; List, B., *J. Am Chem. Soc.* **2006**, *128*, 1086.

Liu-Zhu Gong
University of Science and Technology of China, Hefei, China

Dinaphtho[2,1-*d*:1′,2′-*f*][1,3,2]dioxaphosphepin, 4-Hydroxy-2,6-diphenyl-, 4-Oxide, (11b*R*)- (Family of Reagents)

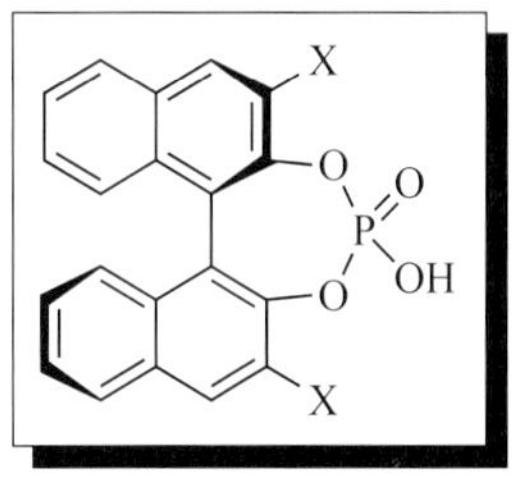

(**1**: X = H)
[39648-67-4] $C_{20}H_{13}O_4P$ (MW 348.29)
InChI = 1S/C20H13O4P/c21-25(22)23-17-11-9-13-5-1-3-7-15(13)19(17)20-16-8-4-2-6-14(16)10-12-18(20)24-25/h1-12H,(H,21,22)
InChIKey = JEHUZVBIUCAMRZ-UHFFFAOYSA-N

(**2**: X = 4-$NO_2C_6H_4$)
[695162-89-1] $C_{32}H_{19}N_2O_8P$ (MW 590.48)
InChI = 1S/C32H19N2O8P/c35-33(36)23-13-9-19(10-14-23)27-17-21-5-1-3-7-25(21)29-30-26-8-4-2-6-22(26)18-28(20-11-15-24(16-12-20)34(37)38)32(30)42-43(39,40)41-31(27)29/h1-18H,(H,39,40)
InChIKey = ZAVIDRFWCFHRJK-UHFFFAOYSA-N

(**3**: X = 4-(2-naphthyl)C_6H_4)
[699006-55-8] $C_{52}H_{33}O_4P$ (MW 752.79)
InChI = 1S/C52H33O4P/c53-57(54)55-51-47(37-23-17-35(18-24-37)41-27-21-33-9-1-3-11-39(33)29-41)31-43-13-5-7-15-45(43)49(51)50-46-16-8-6-14-44(46)32-48(52(50)56-57)38-25-19-36(20-26-38)42-28-22-34-10-2-4-12-40(34)30-42/h1-32H,(H,53,54)
InChIKey = FRXFYAVXIXQHHE-UHFFFAOYSA-N

(**4**: X = 4-ClC_6H_4)
[922711-71-5] $C_{32}H_{19}C_{12}O_4P$ (MW 569.37)
InChI = 1S/C32H19Cl2O4P/c33-23-13-9-19(10-14-23)27-17-21-5-1-3-7-25(21)29-30-26-8-4-2-6-22(26)18-28(20-11-15-24(34)16-12-20)32(30)38-39(35,36)37-31(27)29/h1-18H,(H,35,36)
InChIKey = WHRXFMNUFLRKPO-UHFFFAOYSA-N

(**5**: X = C_6H_3-3,5-$(CF_3)_2$)
[791616-62-1] $C_{36}H_{17}F_{12}O_4P$ (MW 772.47)
InChI = 1S/C36H17F12O4P/c37-33(38,39)21-9-19(10-22(15-21)34(40,41)42)27-13-17-5-1-3-7-25(17)29-30-26-8-4-2-6-18(26)14-28(32(30)52-53(49,50)51-31(27)29)20-11-23(35(43,44)45)16-24(12-20)36(46,47)48/h1-16H,(H,49,50)
InChIKey = DQORDVSQWPKAQJ-UHFFFAOYSA-N

(**6**: X = 3,5-dimesityphenyl)
[780780-94-1] $C_{68}H_{61}O_4P$ (MW 973.18)
InChI = 1S/C68H61O4P/c1-37-21-41(5)61(42(6)22-37)53-29-51(30-54(33-53)62-43(7)23-38(2)24-44(62)8)59-35-49-17-13-15-19-57(49)65-66-58-20-16-14-18-50(58)36-60(68(66)72-73(69,70)71-67(59)65)52-31-55(63-45(9)25-39(3)26-46(63)10)34-56(32-52)64-47(11)27-40(4)28-48(64)12/h13-36H,1-12H3,(H,69,70)
InChIKey = LEXGAMZWYFXSHC-UHFFFAOYSA-N

(**7**: X = 1-naphthyl)
[864943-23-7] $C_{40}H_{25}O_4P$ (MW 600.60)
InChI = 1S/C40H25O4P/c41-45(42)43-39-35(33-21-9-15-25-11-1-5-17-29(25)33)23-27-13-3-7-19-31(27)37(39)38-32-20-8-4-14-28(32)24-36(40(38)44-45)34-22-10-16-26-12-2-6-18-30(26)34/h1-24H,(H,41,42)
InChIKey = SLYOKZNBLPQLJA-UHFFFAOYSA-N

(**8**: X = $SiPh_3$)
[791616-55-2] $C_{56}H_{41}O_4PSi_2$ (MW 865.07)
InChI = 1S/C56H41O4PSi2/c57-61(58)59-55-51(62(43-25-7-1-8-26-43,44-27-9-2-10-28-44)45-29-11-3-12-30-45)39-41-23-19-21-37-49(41)53(55)54-50-38-22-20-24-42(50)40-52(56(54)60-61)63(46-31-13-4-14-32-46,47-33-15-5-16-34-47)48-35-17-6-18-36-48/h1-40H,(H,57,58)
InChIKey = BDQOCXQVIFQJRK-UHFFFAOYSA-N

(**9**: X = 3,5-diphenylphenyl)
[361342-55-4] $C_{56}H_{37}O_4P$ (MW 804.24)
InChI = 1S/C56H37O4P/c57-61(58)59-55-51(47-31-43(37-17-5-1-6-18-37)29-44(32-47)38-19-7-2-8-20-38)35-41-25-13-15-27-49(41)53(55)54-50-28-16-14-26-42(50)36-52(56(54)60-61)48-33-45(39-21-9-3-10-22-39)30-46(34-48)40-23-11-4-12-24-40/h1-36H,(H,57,58)
InChIKey = QFROSIXQYJFKDW-UHFFFAOYSA-N

(**10**: X = C_6H_2(*i*-$Pr)_3$)
[791616-63-2] $C_{50}H_{57}O_4P$ (MW 590.48)
InChI = 1S/C50H57O4P/c1-27(2)35-23-39(29(5)6)45(40(24-35)30(7)8)43-21-33-17-13-15-19-37(33)47-48-38-20-16-14-18-34(38)22-44(50(48)54-55(51,52)53-49(43)47)46-41(31(9)10)25-36(28(3)4)26-42(46)32(11)12/h13-32H,1-12H3,(H,51,52)
InChIKey = AGQAQYPGJDBIQR-UHFFFAOYSA-N

(**11**: X = 9-anthryl)
[361342-51-0] $C_{48}H_{29}O_4P$ (MW 700.72)
InChI = 1S/C48H29O4P/c49-53(50)51-47-41(43-35-19-7-1-13-29(35)25-30-14-2-8-20-36(30)43)27-33-17-5-11-23-39(33)45(47)46-40-24-12-6-18-34(40)28-42(48(46)52-53)44-37-21-9-3-15-31(37)26-32-16-4-10-22-38(32)44/h1-28H,(H,49,50)
InChIKey = QYJHYRSRVCOHMY-UHFFFAOYSA-N

(**12**: X = 9-phenanthryl)
[864943-22-6] $C_{48}H_{29}O_4P$ (MW 700.72)
InChI = 1S/C48H29O4P/c49-53(50)51-47-43(41-25-29-13-1-5-17-33(29)37-21-9-11-23-39(37)41)27-31-15-3-7-19-35(31)45(47)46-36-20-8-4-16-32(36)28-44(48(46)52-53)42-26-30-14-2-6-18-34(30)38-22-10-12-24-40(38)42/h1-28H,(H,49,50)
InChIKey = BVICVEIKBNVROI-UHFFFAOYSA-N

(**13**: X = 4-biphenyl)
[699006-54-7] $C_{44}H_{29}O_4P$ (MW 652.67)
InChI = 1S/C44H29O4P/c45-49(46)47-43-39(33-23-19-31(20-24-33)29-11-3-1-4-12-29)27-35-15-7-9-17-37(35)41(43)42-38-18-10-8-16-36(38)28-40(44(42)48-49)34-25-21-32(22-26-34)30-13-5-2-6-14-30/h1-28H,(H,45,46)
InChIKey = SKOZHEMVYVCNLO-UHFFFAOYSA-N

(acid catalysts for enantioselective addition and cycloaddition reaction toward imines, and hydrogenation reactions)

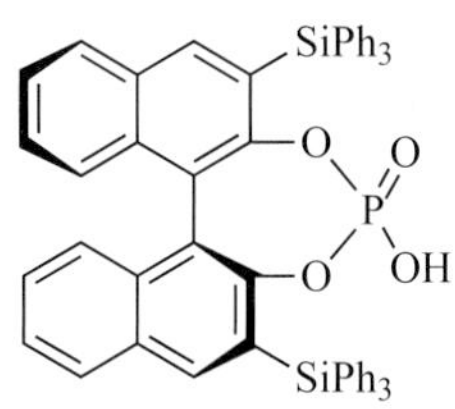

Dinaphtho[2,1-*d*:1′,2′-*f*][1,3,2]dioxaphosphepin, 4-hydroxy-2,6-bis(triphenylsilyl)-, 4-oxide, (11b*R*)-. [791616-55-2] and Dinaphtho[2,1-*d*:1′,2′-*f*][1,3,2]dioxaphosphepin, 4-hydroxy-2,6-bis(triphenylsilyl)-, 4-oxide, (11b*S*)-. [929097-92-7]
InChI = 1S/C56H41O4PSi2/c57-61(58)59-55-51(62(43-25-7-1-8-26-43,44-27-9-2-10-28-44)45-29-11-3-12-30-45)39-41-23-19-21-37-49(41)53(55)54-50-38-22-20-24-42(50)40-52(56(54)60-61)63(46-31-13-4-14-32-46,47-33-15-5-16-34-47)48-35-17-6-18-36-48/h1-40H,(H,57,58)
InChIKey = BDQOCXQVIFQJRK-UHFFFAOYSA-N

InChI = 1S/C56H41O4PSi2/c57-61(58)59-55-51(62(43-25-7-1-8-26-43,44-27-9-2-10-28-44)45-29-11-3-12-30-45)39-41-23-19-21-37-49(41)53(55)54-50-38-22-20-24-42(50)40-52(56(54)60-61)63(46-31-13-4-14-32-46,47-33-15-5-16-34-47)48-35-17-6-18-36-48/h1-40H,(H,57,58)
InChIKey = BDQOCXQVIFQJRK-UHFFFAOYSA-N

(reagent used in organocatalytic reductive amination, chiral hydrogen-bonding catalyst, and transfer hydrogenation)

Characterization Data: ^{1}H and ^{13}C NMR, IR (film), specific rotation.[31]
Solubility: benzene, dichloromethane, and toluene.
Preparative Method: preparation of (*R*)-2,6-bis-(triphenylsilyl)-4-oxo-3,5-dioxa-4λ5-phospha-cyclohepta [2,1*a*;3,4-*a*′] dinaphthalen-4-ol requiring three steps from commercially available MOM-protected (*R*)-BINOL [173831-50-0].[32]

Original Commentary

Takahiko Akiyama
Gakushuin University, Tokyo, Japan

Introduction.

Chiral Resolution. Parent phosphoric acid **1** is readily prepared starting from 1,1′-bi-2-naphthol[1] and has been employed as a chiral resolving agent for amines.[2] The phosphoric acid **1** has been employed for the preparation of both enantiomers of 1,1′-bi-2-naphthol.[3]

Asymmetric Mannich and Mannich-type Reactions. The efficiency of phosphoric acid was demonstrated as a chiral catalyst for the Mannich-type reaction for the first time.[4] Introduction of the 3,3′-groups had a beneficial effect on the reactivity and enantioselectivity. Thus, Mannich-type reaction of ketene silyl acetals with aldimines bearing *N*-*o*-hydroxyphenyl substituents was catalyzed by phosphoric acid **2** to give the corresponding β-amino α-alkyl substituted amino acid esters with high *syn* selectivity and excellent enantioselectivities (eq 1).[5]

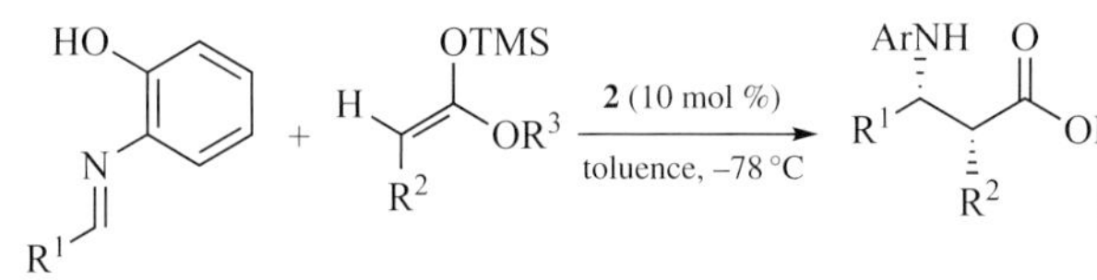

(1)

Direct Mannich reaction of acetyl acetone with *N*-Boc aldimines was catalyzed by **3** to give the corresponding adducts with excellent enantioselectivities.[6] The three-component direct Mannich reaction of cyclohexanone, aldehyde, and aniline also proceeded with excellent enantioselectivity by means of **4**.[7]

Asymmetric Addition to Imines. Hydrophosphonylation of aldimines with dialkyl phosphite proceeded by means of **5** to give α-amino phosphonates with good enantioselectivity.[8]

The phosphoric acid turned out to be a catalyst of choice for the Friedel–Crafts alkylation reaction of *N*-Boc aldimines. For example, methoxyfuran reacts with aldimines by means of 2 mol % of **6** gave the corresponding adducts with excellent enantioselectivity.[9] Indole (cat. **7–9**)[10] and pyrrole (cat. **8**)[11] also proved to be suitable substrates for phosphoric acid catalyzed Friedel–Crafts alkylation reactions.

The Pictet–Spengler reaction starting from tryptamine provides a useful method for the preparation of tetrahydro-β-carbolines. Although use of disubstituted tryptamine derivatives was required under the catalysis with **10** (eq 2),[12] use of *N*-sulfenyl derivatives as substrates in combination with **5** obviated the requirement.[13]

Direct alkylation of aldimines with α-diazo esters by means of **11** gave β-amino-α-diazoesters with excellent enantioselectivities.[13]

Strecker reaction of aldimines with HCN gave α-cyano nitriles by use of **12** as a catalyst.[14]

Addition of enecarbamates to *N*-acylimines by means of 0.5 mol % of **12** gave β-amino ketones with excellent enantioselectivities.[15] Reduction of the in situ generated β-amino imine gave 1,3-diamines with excellent diastereo- and enantioselectivities. The phosphoric acid **10** activated the enecarbamate to generate an iminium salt, which was trapped by indole to furnish 1-indolyl-1-alkylamine derivatives.[16]

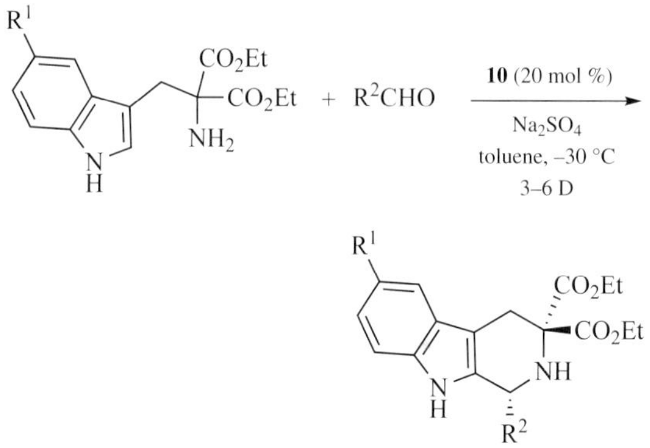

(2)

Asymmetric Diels–Alder Reactions. Aza Diels–Alder reactions of Danishefsky's diene with aldimine by means of **10** gave piperidinone derivatives with high enantioselectivity (eq 3).[17] The

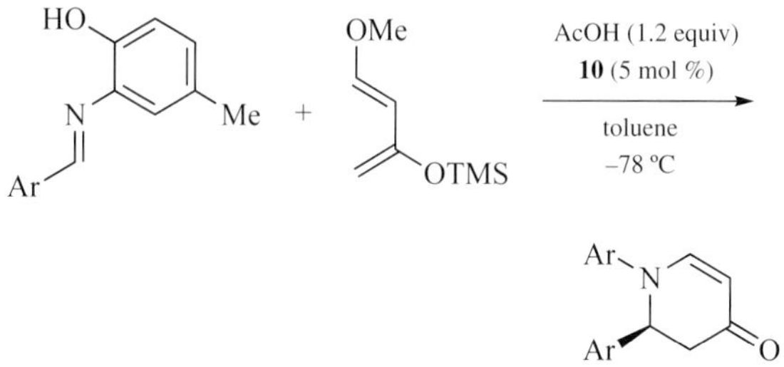

(3)

Brassard diene also participated as a diene to give δ-lactams under the influence of **11** with excellent enantioselectivity.[18] Inverse electron-demand aza Diels–Alder reactions of *N*-aryl aldimine with alkyl vinyl ethers gave tetrahydroquinoline derivatives by use of **11** as a catalyst with excellent diastereoselectivity.[19]

The aza Diels–Alder reaction of aldimine with cyclohexenone by **13** gave isoquinuclidines with good enantioselectivities.[20] Cyclohexenone underwent enolization to generate dienes in situ under acid catalysis.

Asymmetric Hydrogenations. Recently, organocatalytic transfer hydrogenation of α,β-unsaturated carbonyl compounds and imines has been discovered by the use of Hantzsch ester as a hydrogen source.[21] In addition to iminium catalysis, phosphoric acid also proved to be an efficient catalyst for the reduction reaction. Phosphoric acid catalyzed the transfer hydrogenation of imines to give amines with high to excellent enantioselectivities; both **5** and **10** exhibited high to excellent enantioselectivities.[22,23]

The three-component transfer hydrogenation reaction starting from ketone, aniline, and Hantzsch ester proceeded smoothly by **8** in the presence of 5 ÅMS to give secondary amines with excellent enantioselectivity (eq 4).[24]

The transfer hydrogenation reaction was successfully applied to the reduction of quinolines,[25] benzoxazines,[26] benzothiazine, and benzoxazinones. Reduction of benzoxazines, in particular, is highly stereoselective, reducing the catalyst load of **12** to as low as 0.01 mol % without considerable loss of reactivity and selectivity. They applied the method to the enantioselective synthesis of biologically active tetrahydroquinoline alkaloids such as (+)-galipinine.

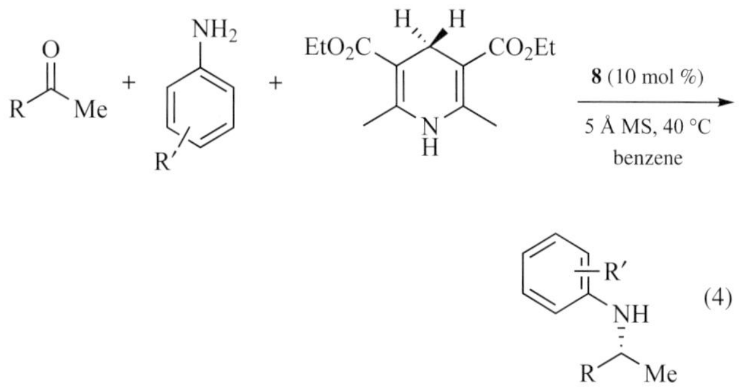

(4)

Asymmetric Cascade Reactions. Catalytic reductive amination of α-branched aldehydes by means of **10** gave β-substituted amines with high chemical yields and with good to excellent enantioselectivities by dynamic kinetic resolution.[27] Transfer hydrogenation of α-imino esters under the catalysis of **11** was also reported.[28] Elegant cascade intramolecular aldol-condensation/Michael/reductive-amination process by means of **10** gave substituted cyclohexylamines (eq 5).[29] The cascade aza-ene-type/cyclization process was promoted by **12** to give piperidines with excellent diastereo- and enantioselectivities.[30]

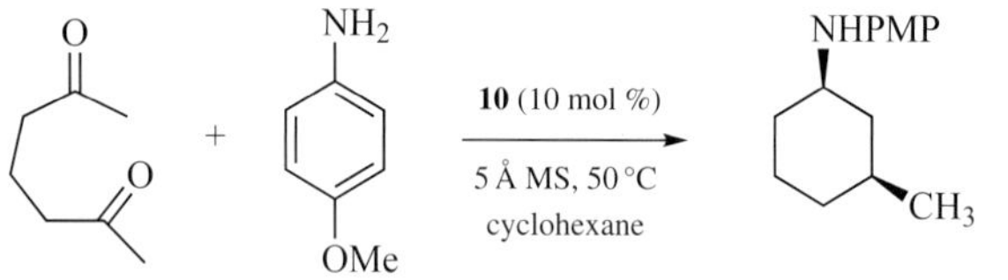

(5)

First Update

David J. Meyers
Johns Hopkins University School of Medicine, Baltimore, MD, USA

Chiral Phosphoric Acid-catalyzed Transfer Hydrogenation. It had previously been reported that chiral phosphoric acids catalyze the reduction of α-imino esters to their respective α-amino esters via transfer hydrogenation using Hantzsch esters as the hydrogen donor. You and coworkers[33] further expanded on this finding by the reduction of β,γ-alkynyl α-imino esters to *trans*-alkenyl α-amino esters with up to 96% ee (eq 6). The reaction utilizes low catalyst loadings (1 mol %) under mild conditions (25 °C) and tolerates a variety of common solvents (toluene, CH_2Cl_2, THF, and *t*-BuOMe). Unfortunately, the isolated yields were dependent on the steric size of the ester group contained in the α-imino ester. It is interesting to note that substitution on the phosphoric acid could result in products with reversed optical rotations (i.e., **8** R_2 = $SiPh_3$ (85% ee) vs. **11** R_2 = 9-anthryl (92% ee).

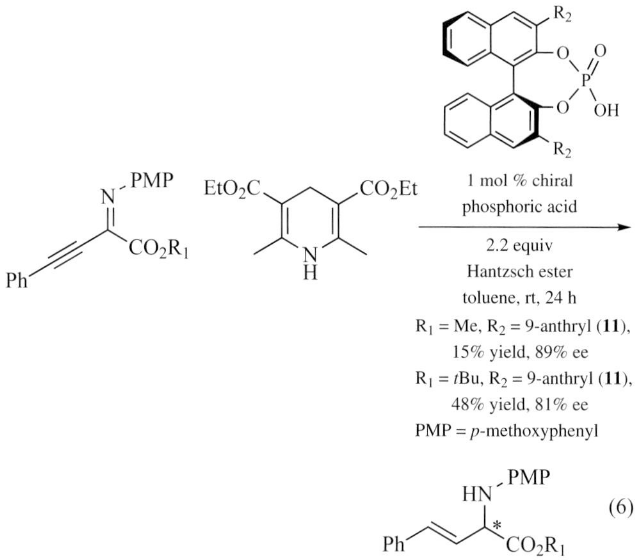

(6)

Rueping previously reported the reduction of quinolones via Brønsted acid-catalyzed transfer hydrogenation to the corresponding 2- or 4-substituted tetrahydroquinolines. It is thought that the reduction of these substrates proceeds through the addition of hydride in a 1,4-addition to the chiral ion pair. This is followed by protonation and a chiral 1,2-hydride addition to generate the tetrahydroquinoline product. Reductions of 3-substituted quinolines follow a similar mechanism, but the enantioselectivity is induced in the protonation step, rather than either of the hydride additions (eq 7).[34] Reaction conditions were optimized for various chiral BINOL phosphate catalysts, solvents, and Hantzsch esters.

Although there are several reports of using chiral BINOL-derived phosphoric acids for the enantioselective reduction of *N*-protected imines, Wang and coworkers provided the first example of the reduction of unprotected *ortho*-hydroxyaryl N–H ketimines using the conditions shown in eq 8.[35] The authors found that aromatic rings bearing either electron-donating or electron-withdrawing groups did not significantly influence enantioselectivity. There was a trend that the enantioselectivity decreased as the steric size of the alkyl group bearing the N–H ketimine was increased, and the yield decreased as the 3-position was substituted.

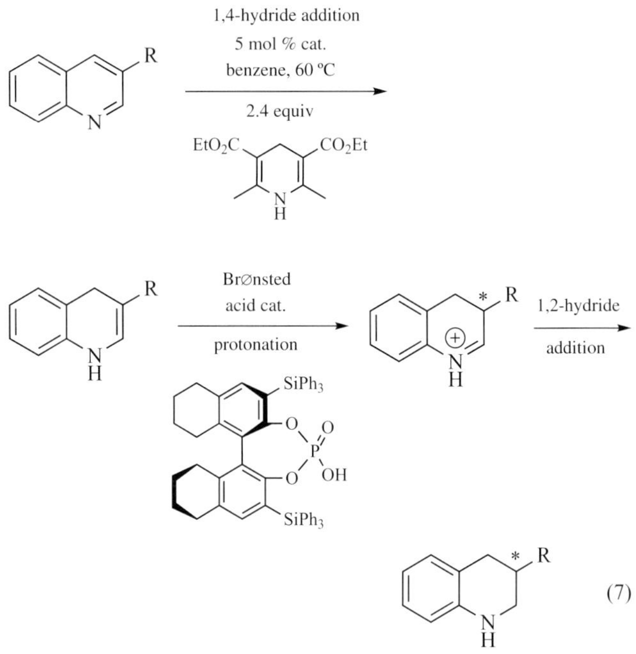

(7)

R = Ph, 76% yield, 84% ee
R = 1-naphthyl, 84%, 77% ee
R = 3,5-Me-C_6H_3, 59%, 80% ee

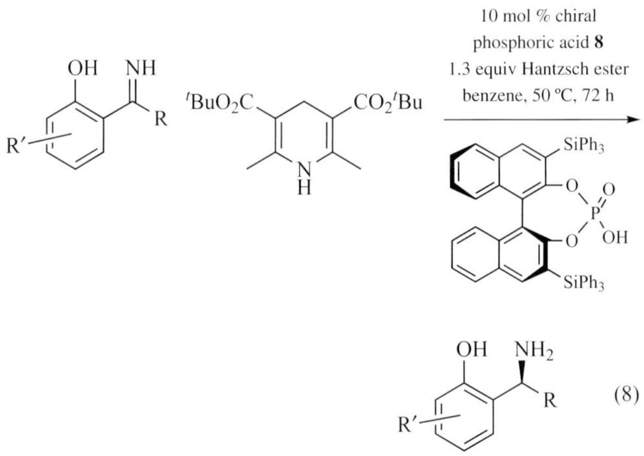

(8)

R = Me, R′ = H, 94%, 92% ee
R = *i*Pr, R′ = H, 98%, 81% ee
R = Me, R′ = 3-Me, 56%, 97% ee
R = Me, R′ = 4-OMe, 85%, 94% ee
R = Me, R′ = 4-NO_2, 88%, 96% ee

A racemic mixture of 2*H*-chromen-2-ols is exposed to the chiral phosphoric acid **10** (Y = 2,4,6-(iPr)$_3$C$_6$H$_2$−) to generate the 1-benzopyrylium ion in situ, which undergoes exclusive 1,4-reduction via hydride delivery from the Hantzsch ester (eq 9).[36] The resulting 4*H*-chromenes are synthesized in good to excellent yields with the highest enantioselectivities obtained when a bulky alkyl ether was at the 6-position. The electronic and steric properties of Ar_1 and Ar_2 had little influence over the chemical yield of the reduction, but enantioselectivity suffered slightly when the Ar_2 group was substituted in the *ortho*-position.

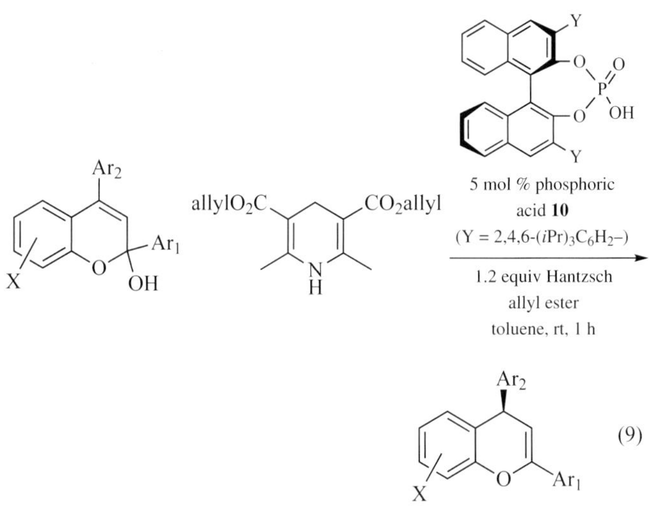

(9)

$Ar_1 = Ar_2$ = Ph
X = H, 90%, 43% ee
X = 6-MeO, 89%, 72% ee
X = 7-MeO, 99%, 19% ee
X = 6-*t*BuO, 91%, 93% ee

Chiral aryl phosphonic acids were also found to catalyze the intramolecular reductive coupling of aryl ketones and dimethyl hydrazones to provide *cis*-amino alcohols, referred to as the aza-pinacol cyclization.[37] Using a Hantzsch ester as the stoichiometric reductant, blue light excites the iridium photocatalyst and proton-coupled electron transfer leads to C–C bond formation followed by H-atom transfer to provide the cyclized product (eq 10). The reductive coupling is tolerant of electron-rich and electron-poor substrates as well as *ortho*-, *meta*-, and *para*-substitution on the aryl ring. The unprotected amino alcohol can be furnished in high yield by reductive cleavage of the N-N bond with Raney nickel.

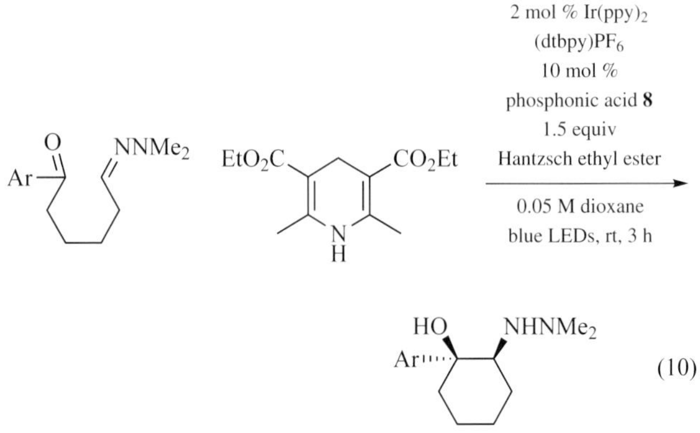

(10)

Ar (%yield,% ee); Ph (83,93)
p-OMe (86,94); *p*-CF_3 (58,90)
o-Me (63,94); *m*-Me (78,90); *p*-Me (81,95)

Although the Pictet–Spengler reaction is a convenient synthesis of tetrahydro-β-carbolines, there are a limited number of catalytic enantioselective examples of this reaction. An alternative route reported by You and coworkers[38] is via the hydroxy-lactam, which then generates an *N*-acyliminium ion formed in situ by a catalytic amount of the chiral Bronsted acid (eq 11). Hydride addition transferred from a Hantzsch ester provides the tetrahydro-β-carbolines with good enantioselectivity and in good yield.

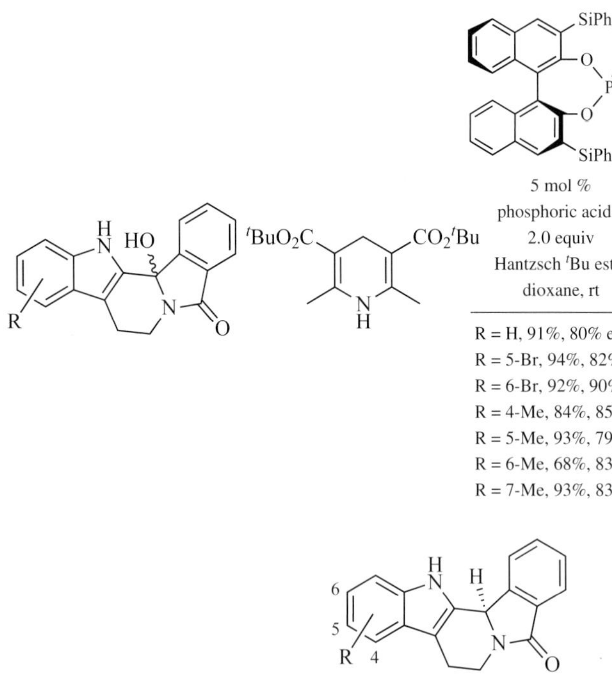

(11)

Dixon and coworkers reported that a combination of a chiral phosphoric acid catalyst and a gold(I) Lewis acid catalyst facilitated the 5-*exo*-dig hydroamination/*N*-sulfonyliminium cyclization cascade with yields up to 85% and up to 96% ee (eq 12).[39]

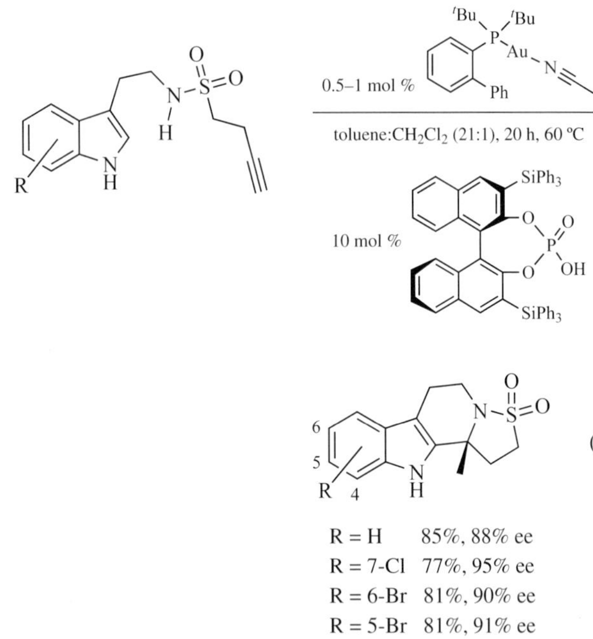

(12)

The authors also demonstrated that the application could be extended to the 6-*exo*-dig cyclization of δ-lactams to provide the corresponding products in similar yields and with similar enantioselectivities (eq 13). Dixon was also able to produce polyheterocyclic products via a Michael addition/iminium ion cyclization cascade of enones and ureas catalyzed by the same chiral phosphoric acid catalyst (eq 14).[40]

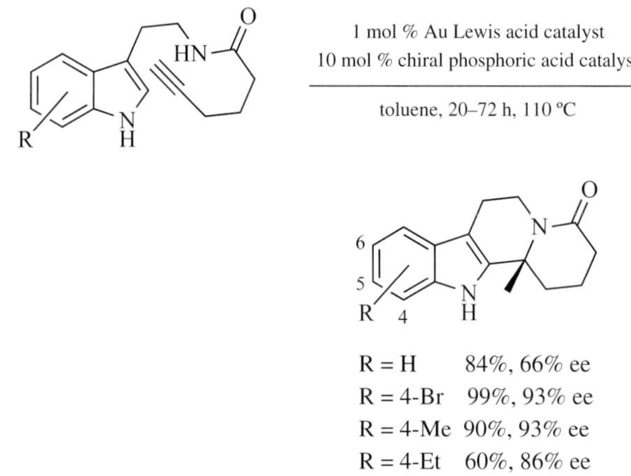

(13)

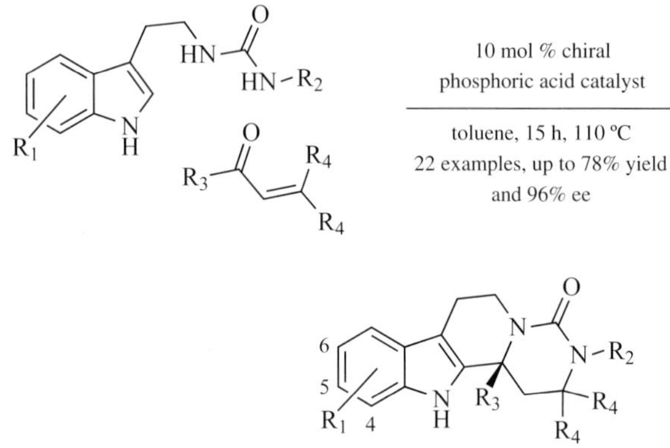

(14)

Sano et al. explored the ring opening of symmetric cyclic carbonates with catalytic amounts of acid ($TsOH{\bullet}H_2O$ and $(PhO)_2P(O)OH$).[41] BINOL-based phosphoric acids were screened with benzyl alcohol and were found to provide good yields (71–99%) and enantioselectivities (55–81% ee) (eq 15).

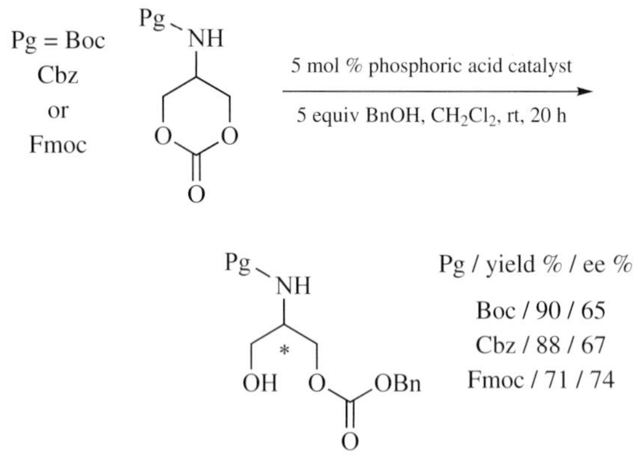

(15)

1. Jacques, J.; Fouquey, C., *Org. Synth.* **1989**, *67*, 1.
2. For an example, see: Wilen, S. H.; Qi, J. Z., *J. Org. Chem.* **1991**, *56*, 487.
3. Cai, D.; Payack, J. F.; Bender, D. R.; Hughes, D. L.; Verhoeven, T. R.; Reider, P. J., *Org. Synth.* **1999**, *76*, 1.
4. For reviews on chiral phosphoric acid, see: (a) Akiyama, T.; Itoh, J.; Fuchibe, K., *Adv. Synth. Catal.* **2006**, *348*, 999. (b) Connon, S. J., *Angew. Chem. Int. Ed.* **2006**, *45*, 3909. (c) Akiyama, T., *Chem. Rev.* **2007**, *107*, 5744. (d) Yu, X.; Wang, W., *Chem. Asian J.* **2008**, *3*, 516.
5. (a) Akiyama, T.; Itoh, J.; Yokota, K.; Fuchibe, K., *Angew. Chem. Int. Ed.* **2004**, *43*, 1566. (b) Yamanaka, M.; Itoh, J.; Fuchibe, K.; Akiyama, T., *J. Am. Chem. Soc.* **2007**, *129*, 6756.
6. Uraguchi, D.; Terada, M., *J. Am. Chem. Soc.* **2004**, *126*, 5356.

7. Guo, Q.-X.; Liu, H.; Guo, C.; Luo, S.-W.; Gu, Y.; Gong, L.-Z., *J. Am. Chem. Soc.* **2007**, *129*, 3790.
8. Akiyama, T.; Morita, H.; Itoh, J.; Fuchibe, K., *Org. Lett.* **2005**, *7*, 2583.
9. Uraguchi, D.; Sorimachi, K.; Terada, M., *J. Am. Chem. Soc.* **2004**, *126*, 11804.
10. (a) Kang, Q.; Zhao, Z.-A.; You, S.-L., *J. Am. Chem. Soc.* **2007**, *129*, 1484. (b) Rowland, G. B.; Rowland, E. B.; Liang, Y.; Perman, J. A.; Antilla, J. C., *Org. Lett.* **2007**, *9*, 2609. (c) Terada, M.; Yokoyama, S.; Sorimachi, K.; Uraguchi, D., *Adv. Synth. Catal.* **2007**, *349*, 1863.
11. Li, G.; Rowland, G. B.; Rowland, E. B.; Antilla, J. C., *Org. Lett.* **2007**, *9*, 4065.
12. Seayad, J.; Seayad, A. M.; List, B., *J. Am. Chem. Soc.* **2006**, *128*, 1086
13. Wanner, M. J.; van der Haas, R. N. S.; de Cuba, K. R.; van Maarseveen, J. H.; Hiemstra, H., *Angew. Chem. Int. Ed.* **2007**, *46*, 7485.
14. Rueping, M.; Sugiono, E.; Azap, C., *Angew. Chem., Int. Ed.* **2006**, *45*, 2617.
15. Terada, M.; Machioka, K.; Sorimachi, K., *Angew. Chem., Int. Ed.* **2006**, *45*, 2254.
16. Terada, M.; Sorimachi, K., *J. Am. Chem. Soc.* **2007**, *129*, 292.
17. Akiyama, T.; Tamura, Y.; Itoh, J.; Morita, H.; Fuchibe, K., *Synlett* **2006**, 141.
18. Itoh, J.; Fuchibe, K.; Akiyama, T., *Angew. Chem., Int. Ed.* **2006**, *45*, 4796.
19. Akiyama, T.; Morita, H.; Fuchibe, K., *J. Am. Chem. Soc.* **2006**, *128*, 13070.
20. Rueping, M.; Azap, C., *Angew. Chem., Int. Ed.* **2006**, *45*, 7832.
21. Ouellet, S. G.; Walji, A. M.; Macmillan, D. W. C., *Acc. Chem. Res.* **2007**, *40*, 1327.
22. Rueping, M.; Sugiono, E.; Azap, C.; Theissmann, T.; Bolte, M., *Org. Lett.* **2005**, *7*, 3781.
23. Hoffmann, S.; Seayad, A. M.; List, B., *Angew. Chem., Int. Ed.* **2005**, *44*, 7424.
24. Storer, R. I.; Carrera, D. E.; Ni, Y.; MacMillan, D. W. C., *J. Am. Chem. Soc.* **2006**, *128*, 84.
25. Rueping, M.; Antonchick, A. P.; Theissmann, T., *Angew. Chem., Int. Ed.* **2006**, *45*, 3683.
26. Rueping, M.; Antonchick, A. P.; Theissmann, T., *Angew. Chem., Int. Ed.* **2006**, *45*, 6751.
27. Hoffmann, S.; Nicoletti, M.; List, B., *J. Am. Chem. Soc.*, **2006**, *128*, 13074.
28. Kang, Q.; Zhao, Z.-A.; You, S.-L., *Adv. Synth. Catal.* **2007**, *349*, 1657
29. Zhou, J.; List, B., *J. Am. Chem. Soc.* **2007**, *129*, 7498.
30. Terada, M.; Machioka, K.; Sorimachi, K., *J. Am. Chem. Soc.* **2007**, *129*, 10336.
31. Itoh, J.; Fuchibe, K.; Akiyama, T., *Angew. Chem., Int. Ed.* **2008**, *47*, 4016.
32. Storer, R.I.; Carrera, D.E.; Ni, Y.; MacMillan, D.W.C., *J. Am. Chem. Soc.* **2006**, *128*, 84.
33. Kang, Q.; Zhao, Z.-A.; You, S.-L., *Org. Lett.* **2008**, *10*, 2031.
34. Rueping, M.; Theissmann, T.; Raja, S.; Bats, J. W., *Adv. Synth. Catal.* **2008**, *350*, 1001.
35. Nguyen, T.B.; Bousserouel, H.; Wang, Q.; Guerette, F., *Org. Lett.* **2010**, *12*, 4705.
36. Terada, M.; Yamanaka, T.; Toda, Y., *Chem. Eur. J.* **2013**, *19*, 13658.
37. Rono, L.J.; Yayla, H.G.; Wang, D.Y.; Armstrong, M.F.; Knowles, R.R., *J. Am. Chem. Soc.* **2013**, *135*, 17735.
38. Yin, Q.; Wang, S.-G.; You, S.-L., *Org. Lett.* **2013**, *15*, 2688.
39. Gregory, A.W.; Jakubec, P.; Turner, P.; Dixon, D.J., *Org. Lett.* **2013**, *15*, 4330.
40. Aillaud, I.; Barber, D.M.; Thompson, A.L.; Dixon, D.J., *Org. Lett.* **2013**, *15*, 2946.
41. Sano, S.; Tsumura, T.; Horibe, M.; Nakao, M., *Synlett* **2013**, *24*, 2302.

5,6-Dioxaspiro[2.4]heptane-4,7-dione

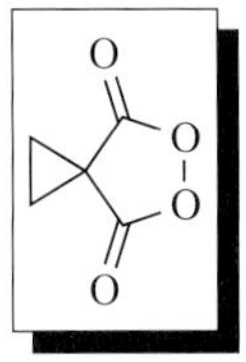

[34867-88-4] $C_5H_4O_4$ (MW 128.08)
InChI = 1S/C5H4O4/c6-3-5(1-2-5)4(7)9-8-3/h1-2H2
InChIKey = DCWJVLCAHFTROS-UHFFFAOYSA-N

(reagent used as an electrophilic oxidant of carbon π-systems)

Physical Data: mp 90 °C.
Solubility: soluble in organic solvents.
Preparative Methods: prepared in one step from cyclopropane-1,1-dicarboxylic acid and isolated as a white crystalline solid.
Handling, Storage, and Precautions: the peroxide melts at 89.9 °C and decomposes with a maximum at 181 °C (1593 J g^{-1}). The material was found to have a limiting impact energy of 5–10 J. Risk of explosion by shock, friction, or heating.

Synthesis. 5,6-Dioxaspiro[2.4]heptane-4,7-dione (**1**) is prepared in one step by treatment of cyclopropane-1,1-dicarboxylic acid[1] with hydrogen peroxide–urea complex[2] in the presence of methanesulfonic acid followed by purification through aqueous workup (eq 1).[3] Synthesis of **1** from cyclopropane-1,1-diethyl carboxylate has also been described,[4] although potential for the preparation of neutral impurities may detract from this method.

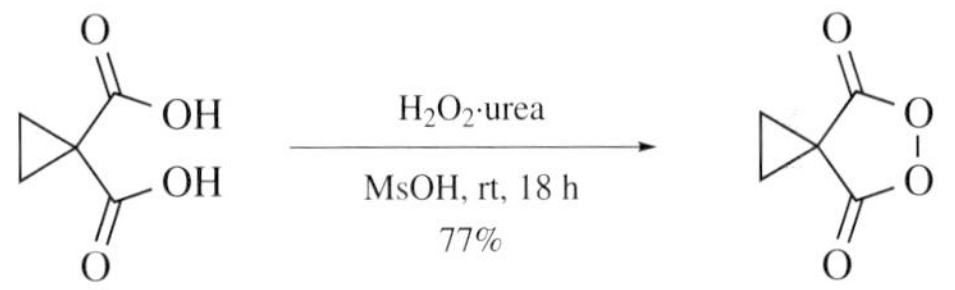

(1)

Malonoyl peroxide (**1**) is a bench-stable white solid that can be used as an electrophilic oxidant of electron-rich π-systems. X-ray crystallography (CCCD 783242) shows the five-membered ring to be planar with an O–O bond length of 1.476 Å. A O=C–C–C=O bond angle of 107.56° reveals a strain on the ring that provides a driving force for reactivity (Scheme 1). Reaction of **1** with alkenes provides a metal-free method for *syn*-dihydroxylation.[5] Reaction with electron-rich aromatic species leads to the corresponding phenol.[6]

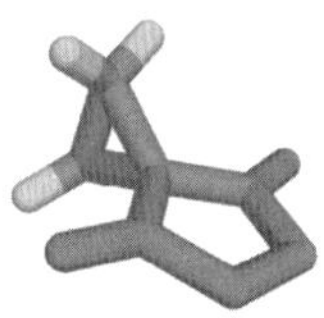

Alkene Dihydroxylation. Malonoyl peroxide (**1**) is an effective reagent for the *syn*-dihydroxylation of alkenes (eq 2).[3,7] Reaction of the alkene with a slight excess of **1** (1.2 equiv) leads to an intermediate ester that is hydrolyzed to deliver the diol product. A majority of reactions proceed at room temperature in the presence of water and air, greatly adding to the practicality of the procedure. Cyclopropyl-1,1-dicarboxylic acid can also be recovered and recycled after hydrolysis (87%) through aqueous workup. For 1,2-disubstituted alkenes, *syn*-selectivity >10:1 was observed in most examples reported.

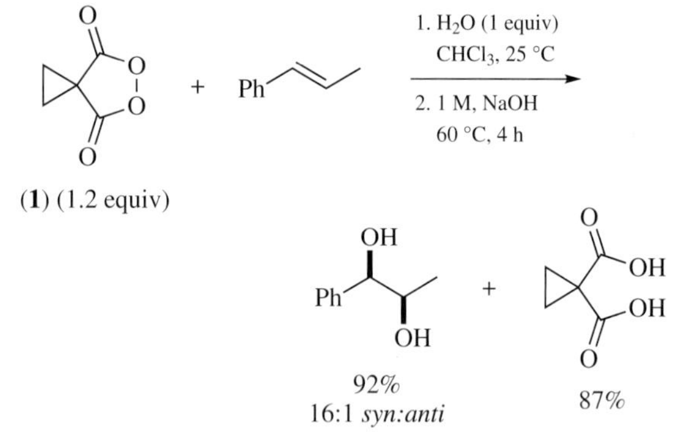

(2)

The mechanistic pathway for the transformation has been thoroughly examined (eq 3).[8] Reaction of stilbene **2** with peroxide

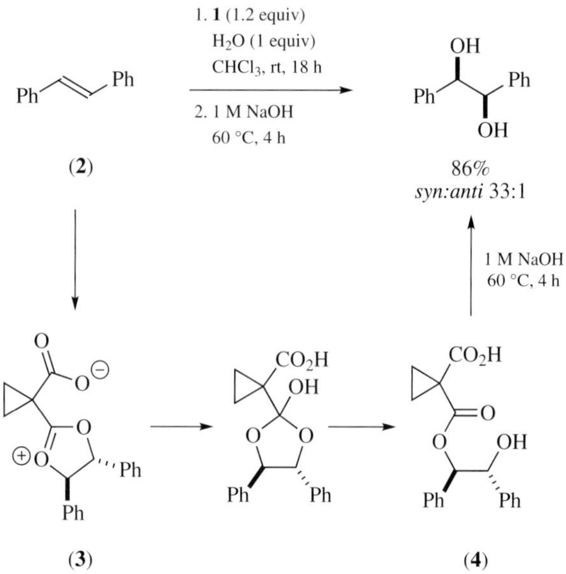

(3)

1 leads to a Woodward-type 1,3-dioxolan-2-yl cation **3** that on reaction with water gives ester **4**. Hydrolysis of the crude reaction mixture under basic conditions gives the *syn*-diol in excellent yield (86%) and very high levels of stereoselectivity (33:1 *syn:anti*). An ionic mechanism was proposed supported by isotopic labeling studies, DFT calculations, use of a Newcomb clock substrate, and Hammett analysis.

Enhancement of both the reaction rate and the stereoselectivity can be achieved through the addition of H-bond donors such as fluorinated alcohols.[9]

Arene Oxidation. Malonoyl peroxide (**1**) efficiently oxidizes electron-rich arenes providing access to the corresponding phenol.[6] Reaction of **1** (1.3 equiv) with an arene at room temperature in hexafluoroisopropanol (HFIP) leads to an ester (40–98%) that on treatment with methylamine (or through aqueous hydrolysis) gives the phenol product (eq 4). The reaction of substituted arenes follows an electrophilic aromatic substitution profile. Isotopic labeling studies and Hammett analysis were consistent with the oxygenation reaction proceeding by an ionic process.

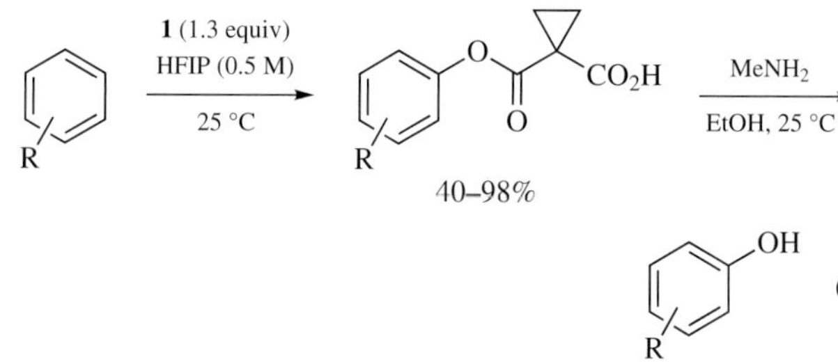

(4)

Of particular note are the mild reaction conditions under which the reaction proceeds. The ability to conduct the transformation in the presence of moisture and air also adds to the practicality of the procedure and high functional group tolerance suggests that the method should be applicable to late-stage functionalization. A drawback is that substrates containing electron-withdrawing groups are unreactive under the standard reaction conditions (25 °C, 1.3 equiv **1**, HFIP (0.5 M)).

Further Reactions. In addition to its use as an electrophilic oxidizing agent, transformations of **1** have also been explored. Photolysis of **1** at 77 K leads to the α-lactone product **5** that on prolonged irradiation gives the allene **6** (eq 5).[10] Flash vacuum pyrolysis at 450 °C also leads to the allene **6** (50% yield).[11]

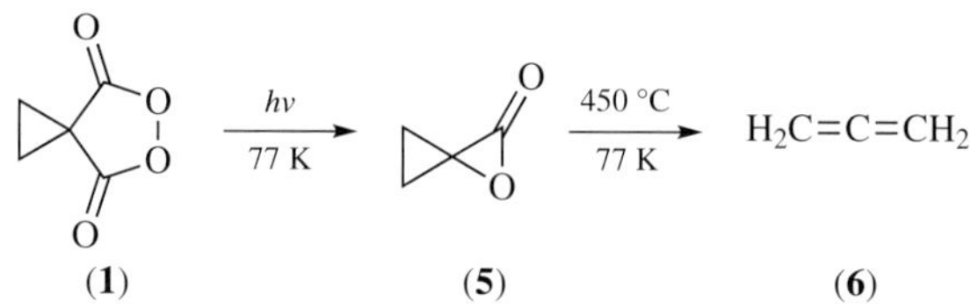

(5)

1. Singh, R. K.; Danishefsky, S., *Org. Synth.* **1981**, *60*, 66.
2. Heaney, H., *Aldrichim. Acta* **1993**, *26*, 35.
3. Griffith, J. C.; Jones, K. M.; Picon, S.; Rawling, M. J.; Kariuki, B. M.; Campbell, M.; Tomkinson, N. C. O., *J. Am. Chem. Soc.* **2010**, *132*, 14409.
4. Terent'ev, A.; Vil, V. A.; Mulina, O. M.; Pivnitsky, K. K.; Nikishin, G. I., *Mendeleev Commun.* **2014**, *24*, 345.
5. Rawling, M. J.; Tomkinson, N. C. O., *Org. Biomol. Chem.* **2013**, *11*, 1434.
6. Dragan, A.; Kubczyk, T. M.; Rowley, J. H.; Sproules, S.; Tomkinson, N. C. O., *Org. Lett.* **2015**, *17*, 2618.
7. Jones, K. M.; Tomkinson, N. C. O., *J. Org. Chem.* **2012**, *77*, 921.
8. Rawling, M. J.; Rowley, J. H.; Campbell, M.; Kennedy, A. R.; Parkinson, J. A.; Tomkinson, N. C. O., *Chem. Sci.* **2014**, *5*, 1777.
9. Picon, S.; Rawling, M.; Campbell, M.; Tomkinson, N. C. O., *Org. Lett.* **2012**, *14*, 6250.
10. Chapman, O. L.; Wojtkowski, P. W.; Adam, W.; Rodriquez, O.; Rucktaschel, R., *J. Am. Chem. Soc.* **1972**, *94*, 1365.
11. Adam, W.; Cadiz, C.; Mazenod, F., *Tetrahedron Lett.* **1981**, *22*, 1203.

Nicholas C. O. Tomkinson
University of Strathclyde, Glasgow, UK

2,2′-Diphenyl-[3,3′-biphenanthrene]-4,4′-diol (VAPOL) and 3,3′-Diphenyl-[2,2′-binaphthrene]-1,1′-diol (VANOL)

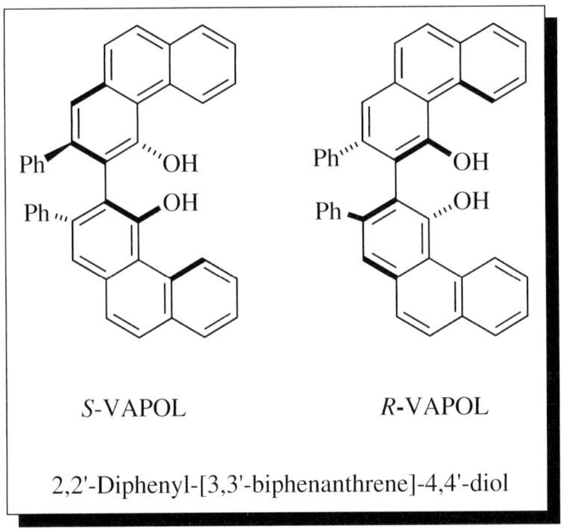

2,2'-Diphenyl-[3,3'-biphenanthrene]-4,4'-diol

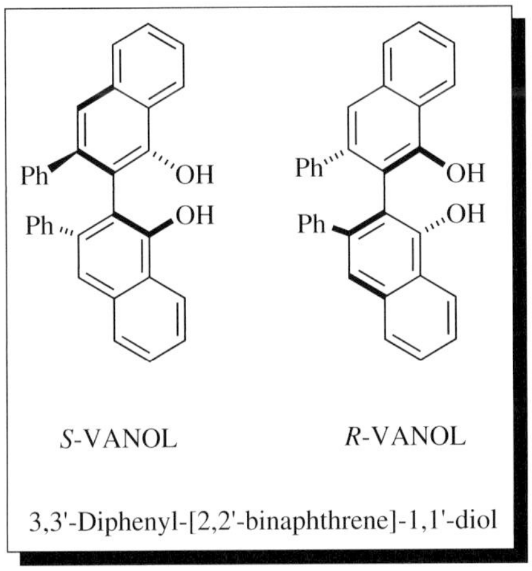

3,3'-Diphenyl-[2,2'-binaphthrene]-1,1'-diol

[147702-16-7] $C_{40}H_{26}O_2$ (MW 538.19)

InChI = 1S/C40H26O2/c41-39-35-29(21-19-27-15-7-9-17-31(27)35)23-33(25-11-3-1-4-12-25)37(39)38-34(26-13-5-2-6-14-26)24-30-22-20-28-16-8-10-18-32(28)36(30)40(38)42/h1-24,41-42H

InChIKey = UFYXKDMLGBKHIC-UHFFFAOYSA-N

[147702-14-5] $C_{32}H_{22}O_2$ (MW 438.52)

InChI = 1S/C32H22O2/c33-31-25-17-9-7-15-23(25)19-27(21-11-3-1-4-12-21)29(31)30-28(22-13-5-2-6-14-22)20-24-16-8-10-18-26(24)32(30)34/h1-20,33-34H

InChIKey = NDTDVKKGYBULHF-UHFFFAOYSA-N

(reagent used as a chiral catalyst for asymmetric Diels–Alder, aziridination reaction and for imino-aldol reaction)

Alternate Name: VAPOL (VAulted PhenanthrOLs), VANOL (VAulted NapthOLs).

Physical Data: VAPOL: mp >250 °C; VANOL: mp 231–233 °C.

Solubility: soluble in toluene, hexane, benzene, ethyl acetate, dichloromethane, and chloroform.

Form Supplied in: R and S isomers as white solid.

Handling, Storage, and Precautions: solid is air stable while its organic solutions turn yellow upon standing after 1 day. To help prevent oxidation and decomposition, store in a tight-lidded container, preferably under nitrogen at <5 °C.

Original Commentary

Aniruddha P. Patwardhan & William D. Wulff
Michigan State University, East Lansing, MI, USA

Synthesis and Resolution of 2,2′-Diphenyl-[3,3′-biphenanthrene]-4,4′-diol (VAPOL).[1c] 2,2′-Diphenyl-[3,3′-biphenanthrene]-4,4′-diol is prepared in four steps (eq 1) starting from 1-bromonaphthlene (**1**) and chromium hexacarbonyl via the corresponding Fischer carbene complex (**2**).

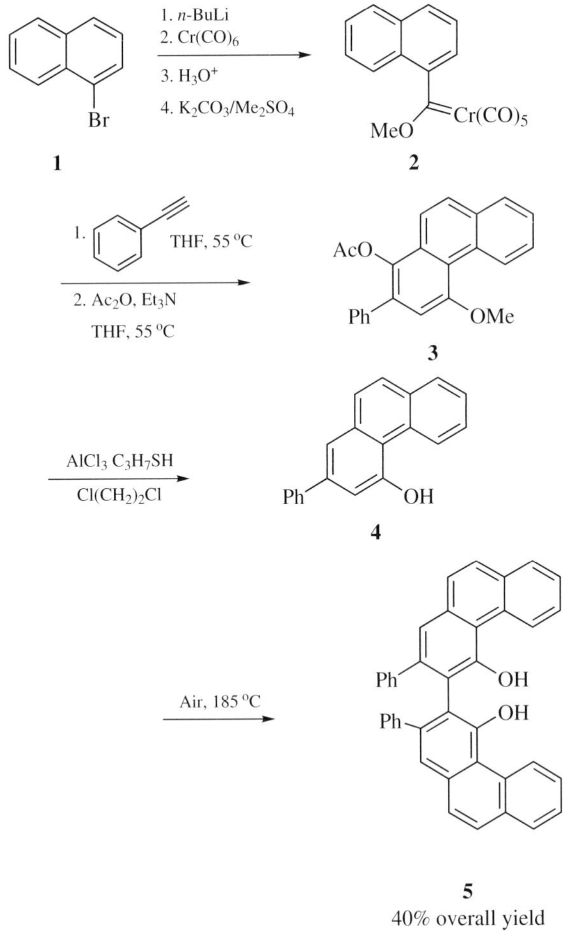

(1)

Complex **2** is reacted with phenyl acetylene to yield the benzannulated product **3** after acetylation. Demethylation and reductive cleavage of the acetoxy group is carried out in a single step by employing aluminum trichloride and propane thiol to give the phenanthrene monomer unit **4**. Racemic VAPOL **5** is obtained in 40% overall yield from **1** by oxidative coupling of **4** in air at 185 °C. The racemic VAPOL is then converted to its phosphoric acid derivative **6** with phosphorus oxychloride. The *R* and *S* enantiomers are then resolved with (−)-cinchonidine (eq 2). VANOL is prepared in a similar fashion, but is resolved with (−)-brucine (eq 3).

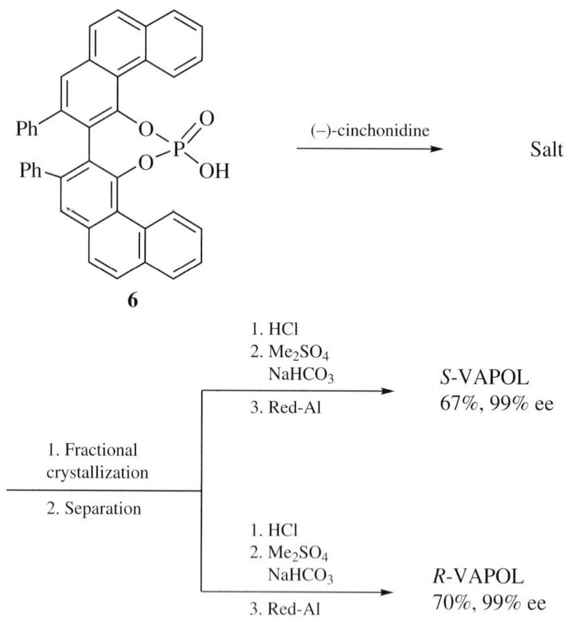

6

(2)

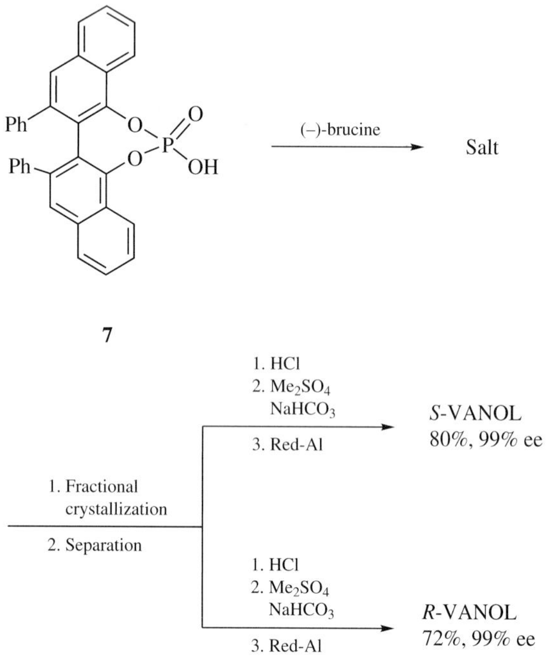

7

(3)

Catalytic Asymmetric Diels–Alder Reaction. One of the most significant reactions in terms of synthetic utility is the Diels–Alder reaction. Since the first example of a chiral Lewis acid-catalyzed enantioselective Diels–Alder reaction,[2] great progress has been made in the development of catalytic enantioselective versions of the homo[3] and the heteroatom Diels–Alder reactions.[4] The asymmetric induction in chiral cycloadducts from reaction via catalytic methods is strongly influenced by the type of chiral ligands employed. Bis(oxazoline)[3b] and 1,1′-binaphth-2-ol (BINOL) have been two of the most popular chiral ligands to be employed for these reactions. In 1993, VAPOL and VANOL were introduced by Wulff and co-workers[1] as new ligands for Lewis acid catalyzed reactions (eq 4). They have reported both aluminum[1a] and boron[1b] complexes of these ligands as chiral Lewis acids for Diels–Alder reactions. The catalyst is generated in situ by reacting VAPOL or VANOL with diethyl aluminum chloride in anhydrous dichloromethane at 25 °C for 30 min. It was observed that 10 mol % of the VAPOL-Al catalyst gave optimal results with 98:2 *exo*/*endo* ratio and with high enantioselectivity (97.8% ee) (Table 1, entry 3).[1a]

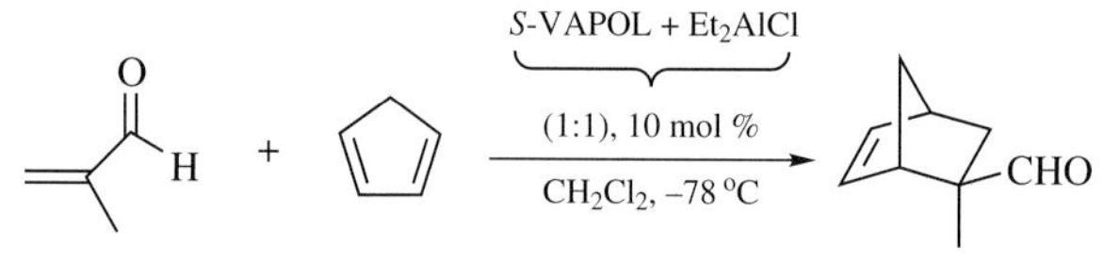

(4)

8 9 10

The same enantioselectivity was maintained in the reaction even when the catalyst loading was decreased from 10 to 0.5 mol % (Table 1, entry 4). It should be noted, however, that higher induction values are obtained upon slow addition of the dienophile (acrolein) to the reaction pot. While the VANOL-derived catalyst under the same conditions gave comparable *exo*/*endo* ratios, it gave poorer enantioselectivity than the VAPOL ligand. When BINOL (Table 1, entries 7 and 8) was used as a ligand, poor yields and poor enantioselectivities (41% ee) were observed.[1a]

Autoinduction Phenomenon. An autoinduction is observed[5] with the VAPOL-Al catalyst in the Diels–Alder reactions for both aldehydes and ester dienophiles, but is more pronounced for those of esters (eq 5). The induction at earlier stages of the reaction are much lower than the induction at the end of the reaction. The basic principle of the proposed mechanism for this asymmetric induction is that more than one carbonyl compound coordinates to the aluminum catalyst. The mechanism (eq 6) shows that the product can be incorporated into the catalyst and can account for the autoinduction if this catalyst gives a higher induction. This suggested that other carbonyl compounds could be added to increase the asymmetric induction for the reaction. Indeed, a remarkable increase in induction is observed when both simple carbonyl and 1,3-dicarbonyl compounds are used as additives in the reaction mixture. (The reaction was performed at 0 °C, while the reactions in absence of the 1,3-diacarbonyl compounds were performed at −80 °C.) This suggests that both five- and six-coordinated aluminum species may be involved in this auto-induction process. Greater than 99% ee was observed for cycloadduct **12** if 0.5 equiv of di-*tert*-butyl-2,2-dimethylmalonate is added at the beginning of the reaction.

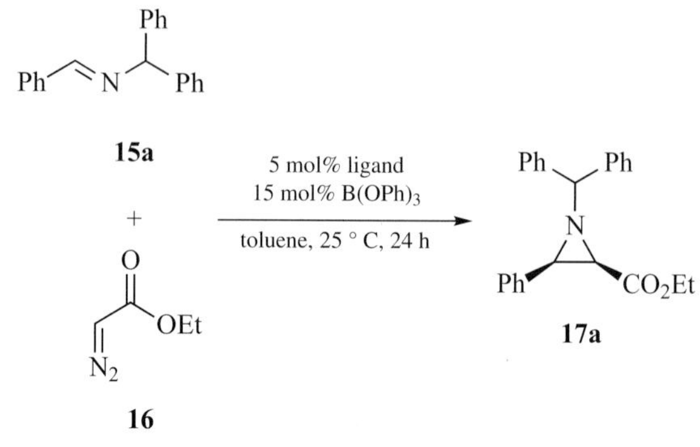

(5)

Table 1 Asymmetric catalytic Diels-Alder reaction catalyzed by VAPOL-Al catalyst

Entry	Ligands	Conc. of **8**	mmol **8**	mol % cat.[a]	Reaction + time (h)	*exo/endo*[b]	ee[c] *exo*	(−)**10**/(+) **10**	Yield (%)[e] *endo+exo*
1	(+)-*S*-VAPOL	1.0	2.0	10	16	98.2	91.4	22.3:1	100
2	(+)-*S*-VAPOL	0.2	2.0	10	21	98.2	94.7	23.9:1	99
3	(+)-*S*-VAPOL	0.05	2.1	10	24	98:2	97.8	88.0:1	100
4	(+)-*S*-VAPOL	1.0[d]	40.9	0.5	4	97:3	97.4	84.9:1	100
5	(+)-*R*-VAPOL	0.05	2.0	10	23	92:8	17	1:1.4	85
6	(+)-*R*-VAPOL	1.0	39.9	0.5	4	96:4	29	1:1.8	100
7	(+)-*R*-BINOL	1.0[d]	42.6	0.5	4	95:5	41	1:2.4	5
8	(−)-*S*-BINOL	0.05	2.0	10	24	95:5	13	1.3:1	45

[a] Catalyst prepared from ligands and 0.85–1.0 equiv of Et_2AlCl in CH_2Cl_2 at 25°C for 30 min.

[b] The *exo/endo* ratio and the total yields were determined by ^{1}H NMR on the crude reaction mixture.

[c] The % ee was determined by capillary GC after conversion of **1** in the crude reaction mixture of acetals with (2*R*, 4*R*)-(−)-2,4-pentanediol.

[d] Ten percent of dienophile was first added and the rest added slowly by syringe pump over 3 h.

[e] Isolated yield by column chromatography.

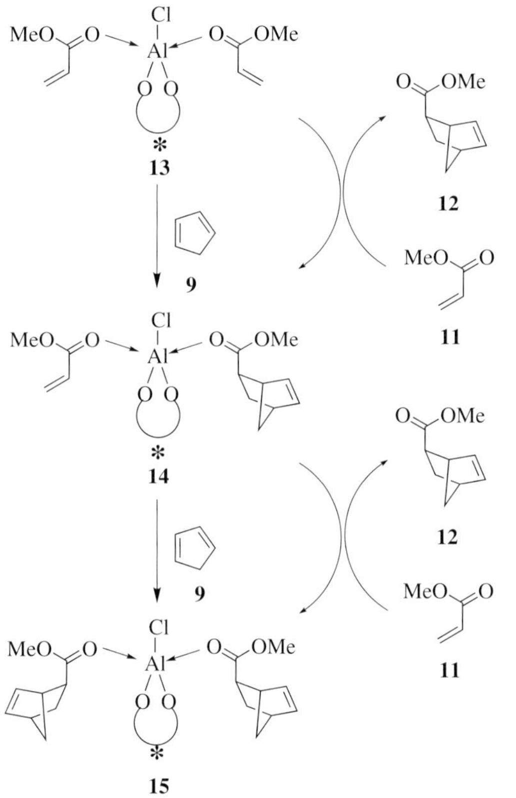

(6)

Catalytic Asymmetric Aziridination Reaction. Aziridines are very important synthetic targets due to their utility in providing access to unnatural optically pure amino acids[6] and amines for preparation of biologically active molecules in the pharmaceutical industry. This strained three-membered ring functionality is structurally analogous to epoxides and cyclopropanes.[7] In recent years mastery over enantioselective catalytic synthesis of epoxides[8a–e] and cyclopropanes[8f–h] has beens made possible by the development of several effective catalysts. Success in the asymmetric catalyst preparation of chiral aziridines lags far behind that of epoxides despite the recent activity.[9]

Catalysts derived from boron Lewis acids and both the VAPOL and VANOL ligands have been recently reported to provide the first general asymmetric catalytic synthesis of aziridines.[10] This methodology involves the addition of ethyl diazoacetate to imine catalyzed by a VAPOL–boron or VANOL–boron catalyst. Initially $BH_3·THF$ complex[10a] was used as the boron source for the preparation of the catalyst. This VAPOL–BH_3 catalyst exhibit high asymmetric inductions (>50:1 de; 94–97% ee) for imines prepared from aryl aldehydes as well as from aliphatic aldehydes. It should be mentioned that only benzhydryl imines formed from the corresponding aldehydes gave the best results. (Benzyl and trityl amine were previously used to prepare imines, but gave poor asymmetric inductions.) Later, it was found that the success of the reaction was dependent on the batch of $BH_3·THF$ complex employed. Older bottles of the boron reagent gave higher induction values. Upon careful analysis it was found that the impurity present was $B(OBu)_3$. The use of tributylborate as the boron source gave identical induction values observed for older $BH_3·THF$ bottles. Eventually, after screening several borate esters, a catalyst prepared from triphenyl borate [$B(OPh)_3$] and VAPOL was found to give exceptionally good asymmetric inductions (90–98% ee) for aryl as well as for aliphatic substrates. (Typically the catalyst is prepared by treating three equiv of the boron reagent with VAPOL/VANOL in orgainic solvents (for example, CH_2Cl_2, $PhCH_3$, C_6H_5) followed by heating at 55 °C. The solvent is then removed under vacuum and the solid mass heated at 55 °C for an additional 30 min. The catalyst is dissolved in solvent, the imine, and then the diazo compound are added.) A catalyst prepared from the VANOL ligand was also effective and, in fact, a remarkably small variation was found between the two catalysts for all of the substrates (Table 2, eq 7). When the BINOL ligand was compared with VANOL and VAPOL, the BINOL-derived catalyst gave poor induction (20% ee), low *cis*/*trans* ratio and an increased proportion of the noncyclized side products **18** and **19**. This newly developed methodology has been applied to the enantioselective synthesis of L-DOPA[10b] and chloroamphenicol[10c] and for the latter with the shortest route ever reported (eqs 8 and 9).

Table 2 Asymmetric catalytic aziridination reaction using VAPOL-B(OPh)$_3$ and VANOL-B(OPh$_3$) catalysts

Entry	Imine	R	Ligand	Time (h)	Yield (%) *cis* **17**	% ee *cis* **17**	*cis* **17**/*trans* **17**	Yield (%) **18**	Yield (%) **19**
1	15a	Ph	*S*-VAPOL	48	77	95	>50:1	4.4	1.7
2			*S*-VAPOL	0.5	85	96	>50:1	3.3	1.0
3	15b	*p*-BrC$_6$H$_4$	*S*-VAPOL	5	91	98	>50:1	3.4	2.2
4			*S*-VANOL	1	85	98	>50:1	8	3
5	15c	2-MeC$_6$H$_4$	*S*-VAPOL	14	69	94	40:1	11.1	7.1
6			*S*-VANOL	16	65	91	>50:1	6	4
7	15d	3,4-(OAc)$_2$C$_6$H$_4$	*R*-VAPOL	20	85	96	>50:1	<1	<1
8			*S*-VAPOL	5	83	97	>50:1	<1	<1
9	15e	2-Furyl	*R*-VANOL	16	55	93	>50:1	<1	<1
10	15f	1-Naphthyl	*S*-VAPOL	12	87	92	>50:1	3.8	0.3
11	15g	*n*-Pr	*S*-VAPOL	20	54	91	>50:1	8.3	9.2
12			*S*-VANOL	5	60	90	>50:1	6.7	9.2
13	15h	-Bu	*S*-VAPOL	12	78	91	40:1	<1	<1
14			*S*-VANOL	5	77	97	>50:1	<1	<1
15	15i	*c*-C$_6$H$_{11}$	*S*-VAPOL	8	74	94	38:1	<1	<1

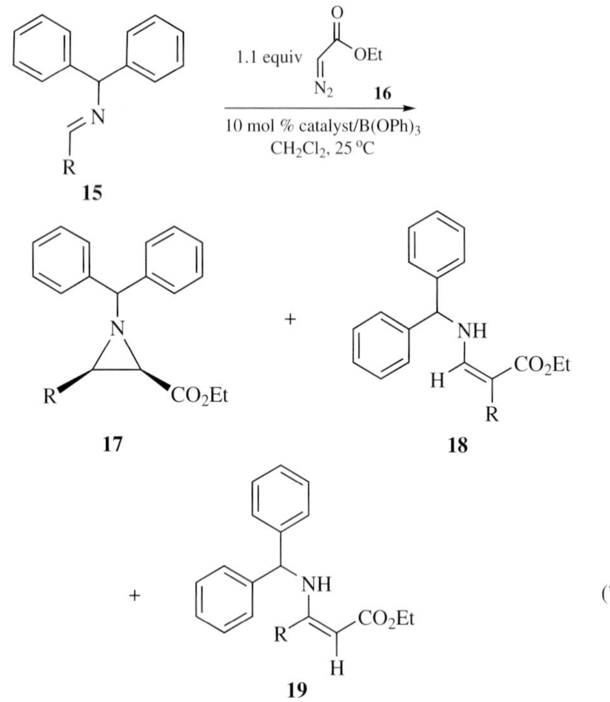

(7)

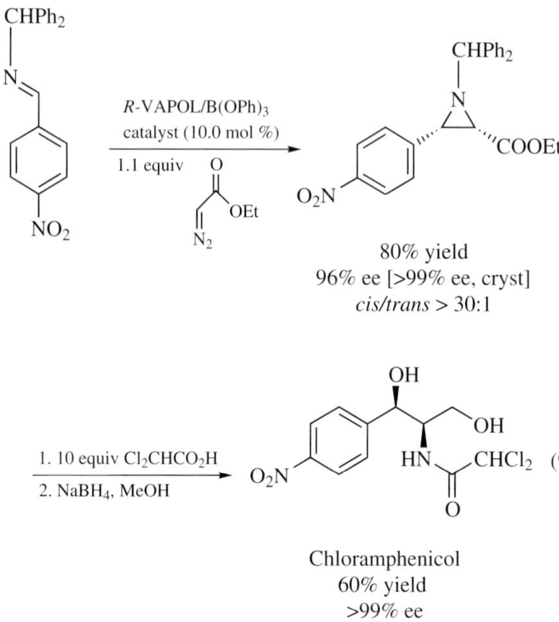

(9)

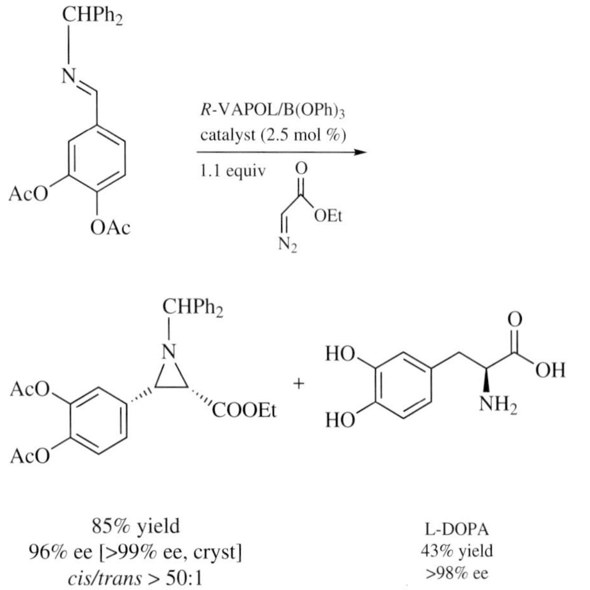

(8)

Catalytic Asymmetric Imino-aldol Reaction. The synthetic utility of this reaction lies in its ability to directly access β-amino acids. Kobayashi et al.[11] recently developed the first catalytic asymmetric imino-aldol reaction using 3,3′-dibromoBINOL[11a] and 6,6′-dibromoBINOL[11b] as chiral ligands and zirconium (IV) *tert*-butoxide as the Lewis acid.[11] Their method involves the catalytic reaction of imines derived from *o*-aminophenols and ketene acetals to generate the imino-aldol product. The use of VAPOL as a chiral ligand was investigated and was compared with 6,6′-dibromoBINOL and BINOL based catalysts on the imino-aldol reaction mediated by zirconium (eq 10).[12] The catalyst is prepared by reaction of the ligand with 0.5 equiv of zirconium tetralkoxide in the presence of 0.6 equiv of *N*-methyl imidazole at room temperature for 1 h.

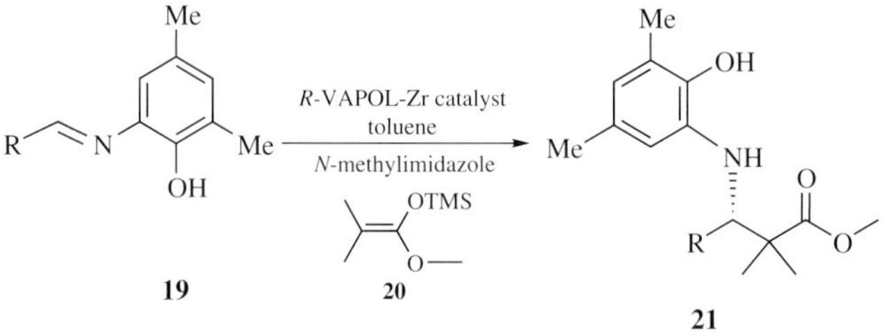

(10)

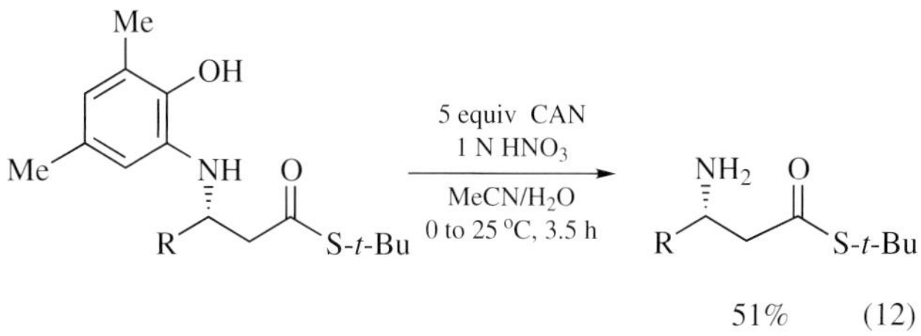

(12)

It was found that VAPOL and 6,6-diboromoBINOL catalysts are superior to BINOL at −45 °C.[15] When the reaction was performed at room temperature, the asymmetric induction for VAPOL remained unchanged, while that for dibromoBINOL and BINOL decreased. Additionally, a series of imines **19** were prepared where the position of the methyl group on the arene ring was changed. It was observed that, there was a profound change in asymmetric inductions (>98% ee), when the *ortho*-position was methylated.[12] The VAPOL–zirconium catalyst is remarkably temperature independent giving an induction of >98% ee at 25 °C and at 100 °C. Moreover, the inductions could be maintained when 2 mol % of the catalyst was employed at 100 °C (Table 3, entries 6–10) and this is true for a variety of imines. Chiral β-amino esters can be then obtained from the corresponding imino-aldol products by treatment with five equiv of cerric ammonium nitrate in 1 M nitric acid (eqs 11 and 12).

Table 3 Temperature dependence and catalyst loading on the induction of **21**

Entry	R	mol % catalyst	Temperature (°C)	Yield (%) **21**	% ee **21**
1	Ph	20	25	100[a]	98.0
2		20	65	93[b]	98.4
3		20	85	94[b]	98.5
4		20	100	91[b]	98.5
5		5	100	90[b]	98.4
6		2	100	95[c]	98.5
7	4-Cl-C_6H_4	2	100	90[d]	95.4
8	4-MeO-C_6H_4	2	100	85[d]	99.8
9	3,4-$(MeO)_2$-C_6H_4	2	100	85[d]	96.4
10	1-napthyl	2	100	83[d]	93

[a] Reaction time 15 h.

[b] *S*-VAPOL used.

[c] Reaction time 5 h.

[d] Reaction time 24 h.

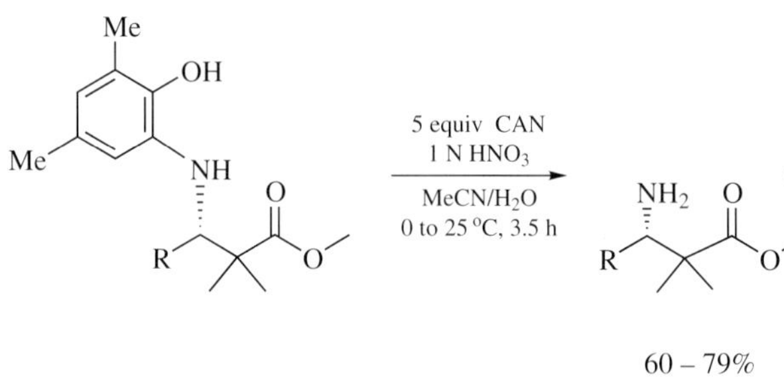

(11)

First Update

William D. Wulff
Michigan State University, East Lansing, MI, USA

Synthesis of VANOL and VAPOL. Different approaches to the synthesis of the VANOL monomer 3-phenyl-1-naphthol **23** were evaluated by Wulff and coworkers.[13,14] The most efficient approach involves a $AlCl_3$-assisted tautomerization and a 1,2-migration of the phenyl group in 4-phenyl-1-naphthol (not shown), which is generated in situ from the reaction of 4-chloro-1-naphthol **27** with $AlCl_3$ and benzene (eq 13). The VANOL monomer **23** can be obtained with high overall yield in two steps from 1-naphthol **26** using very inexpensive reagents. Another scalable synthetic approach[14] to **23** is via a cycloaddition/electrocyclization cascade (CAEC) that begins with a [2 + 2] cycloaddition of phenylacetylene with an aryl-substituted ketene generated from acid chloride **22** (eq 14). The VAPOL monomer **4** can also be prepared by CAEC with the commercially available 2-naphthaleneacetic acid **25** in high yields (eq 15). The oxidative phenol coupling reactions that are used to generate racemic VANOL and VAPOL derivatives were also improved by employing mineral oil as solvent. The reactions go to completion in shorter periods of time and the crude product is easier to purify compared to that for the neat reaction. This approach has been successfully extended to the synthesis of a number of substituted VANOL and VAPOL derivatives.[15,16]

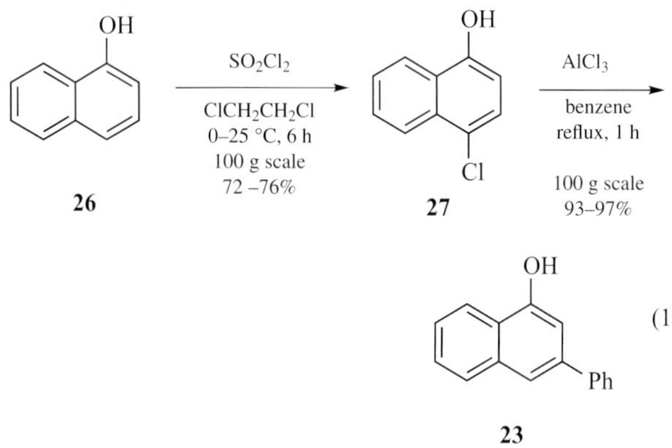

(13)

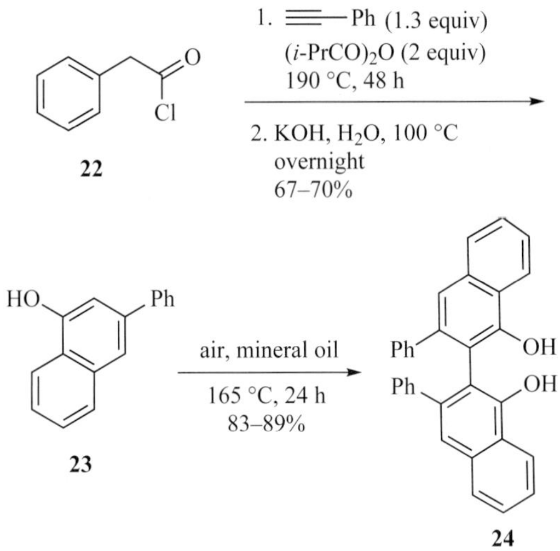

(14)

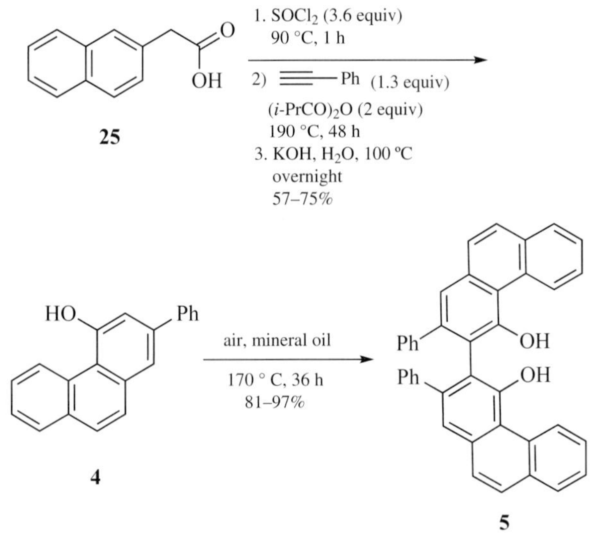

(15)

Stability of VANOL and VAPOL. The stability of these ligands was investigated[14] by their performance in asymmetric aziridination of imines **15a** with ethyl diazoacetate **16** as a function of storage time and conditions. As shown in Table 4, neither the yield nor the asymmetric induction of aziridine **17a** was affected by storage at room temperature under air after 2.5 years.

Deracemization of VANOL and VAPOL. A copper(II)-mediated deracemization of the VANOL and VAPOL ligands was developed[17] and further improved[18] by Wulff and coworkers. As shown in eqs 16 and 17, 1.4 equiv of CuCl and 1.4 equiv of (−)-sparteine were first sonicated in air in methanol for 30 min to give Cu(II)–sparteine complex. After being purged with argon, the green Cu–diamine solution was transferred to a suspension

Table 4 Evaluation of stabilities of **VANOL** and **VAPOL**

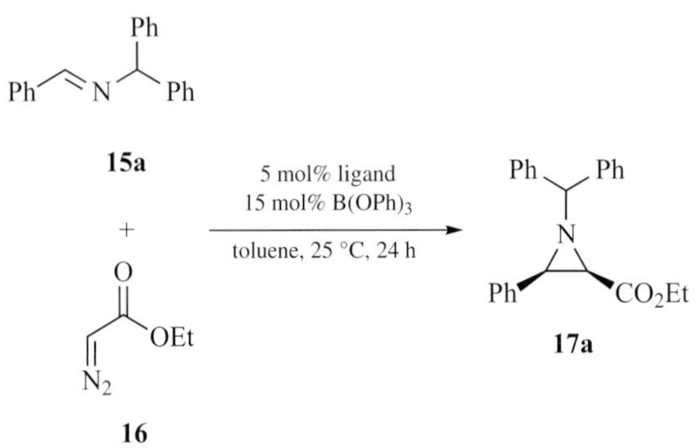

Storage conditions		Ligand	Timespan = 0 year **17a** % Yield[c]	% ee	2.5 year % Yield[c]	% ee
A[a]	3 °C, under N_2	VANOL	90	93	87	95
		VAPOL	84	90	86	85
B[b]	25 °C, under argon	VANOL	83	90	84	93
		VAPOL	83	90	87	92
C[b]	25 °C, under air	VANOL	83	94	91	94
		VAPOL	93	90	91	95

[a] Ligands are stored in a refrigerator in a brown bottle and sealed with parafilm.

[b] Ligands are stored on a bench in a brown bottle and sealed with parafilm.

[c] Isolated yield.

of the racemic ligand and 1.4 equiv of Na_2CO_3 in deoxygenated CH_2Cl_2, and the mixture was allowed to stir at room temperature for 2–29 h. Either VANOL or VAPOL can be deracemized into the corresponding (*S*)-ligands with high mass conversion and >99% ee. The (*R*)-ligands can also be accessed in the same manner using (+)-sparteine[15]or O'Brian's diamine.[18] Optically active substituted VANOL and VAPOL are also accessible with this deracemization procedure.[15,16]

$$(\pm)\text{-VANOL} \xrightarrow[\text{CH}_2\text{Cl}_2\text{/MeOH, argon, 25 °C, 2 h, gram scale}]{\text{Cu(II)/(}-\text{)-sparteine (1.4 equiv), Na}_2\text{CO}_3\text{ (1.4 equiv)}} (S)\text{-VANOL, 84\% yield, >99\% ee} \quad (16)$$

$$(\pm)\text{-VAPOL} \xrightarrow[\text{CH}_2\text{Cl}_2\text{/MeOH, argon, 25 °C, 29 h, gram scale}]{\text{Cu(II)/(}-\text{)-sparteine (1.4 equiv), Na}_2\text{CO}_3\text{ (1.4 equiv)}} (S)\text{-VAPOL, 92\% yield, >99\% ee} \quad (17)$$

Catalytic Asymmetric *cis*-Aziridination Reactions.. A major application of the VANOL/VAPOL ligands is the asymmetric aziridination reaction of imines with ethyl diazoacetate (EDA) developed by the Wulff lab.[10] One of the modified methods[19] to generate the active catalytic BOROX catalyst is shown in eq 18. VANOL or VAPOL is first heated with 4 equiv commercial $B(OPh)_3$ in toluene at 80 °C for 1 h. The solvent is then removed by exposure to high vacuum at 80 oC for an additional 30 min. Induced by the imine substrate, the assembly of the BOROX catalyst rapidly occurs and catalyzes the aziridination reactions. The catalyst was crystallized, which revealed the structure of the chemzyme–substrate complex upon X-ray diffraction analysis and a ^{11}B NMR study.[20] The effect of the protecting group on the imine was studied.[19] As shown in Table 5, for imines derived from six aryl aldehydes and three aliphatic aldehydes, the tetra-*tert*-butyldianisylmethyl (BUDAM) and tetramethyldianisylmethyl (MEDAM) imines are both superior substrates when compared with the benzhydryl imines. In addition, the methoxyl substituents on BUDAM and MEDAM enable the clean deprotection (72–97%) of the aziridine product by treatment with 5 equiv of triflic acid in anisole.

Table 5 Averaged ligand and ***N***-substituent effects over nine substrates[a]

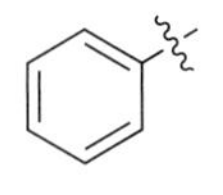

N-substituent	Ar	Ligand	Average % yield of **26**	Average % ee of **26**
benzhydryl	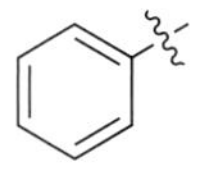	VAPOL	70	88
		VANOL	77	88
BUDAM	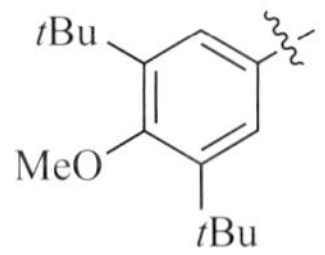	VAPOL	88	95
		VANOL	90	94
MEDAM	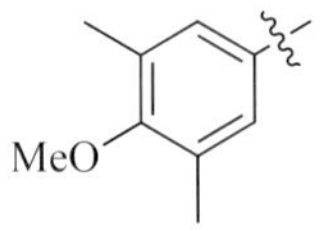	VAPOL	92	97
		VANOL	91	96

[a] For details of imine substrates and the reaction conditions, see Ref. 19.

$$\text{VANOL or VAPOL} \xrightarrow[\text{toluene, 80 °C, 1 h}]{\text{4 equiv B(OPh)}_3\text{, 1 equiv H}_2\text{O}} \xrightarrow[\text{80 °C, 1 h}]{\text{0.1 mm Hg}} \xrightarrow{\text{imine}} \quad (18)$$

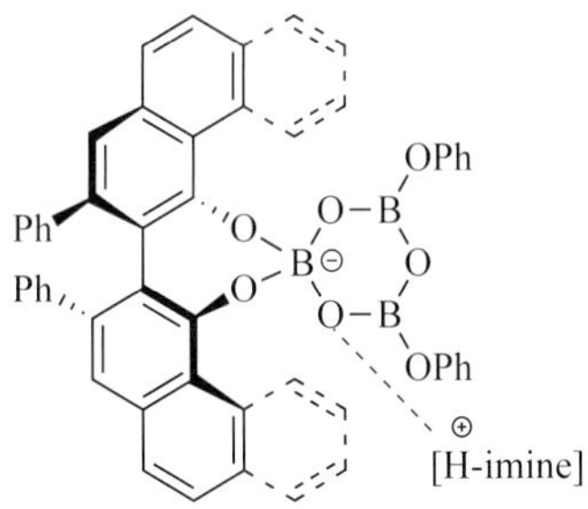

BOROX catalyst

A multicomponent protocol developed later by the Wulff lab simplifies the process in avoiding preparation of preformed imines and allows for the aziridination reaction with aliphatic aldehyde substrates that failed to give stable preformed imines.[21] As shown in eq 19, the VAPOL ligand was mixed with 3 equiv of $B(OPh)_3$ and 10 equiv of amine **27** and stirred in toluene at room temperature for 1 h to ensure complete formation of the BOROX–amine catalyst. This was followed by the addition of 4 Å MS and aldehyde substrate followed directly by addition of EDA **16**. The resulting mixture was stirred at −10 °C for 24 h to afford the aziridine **26a** in 80% yield and 98% ee.

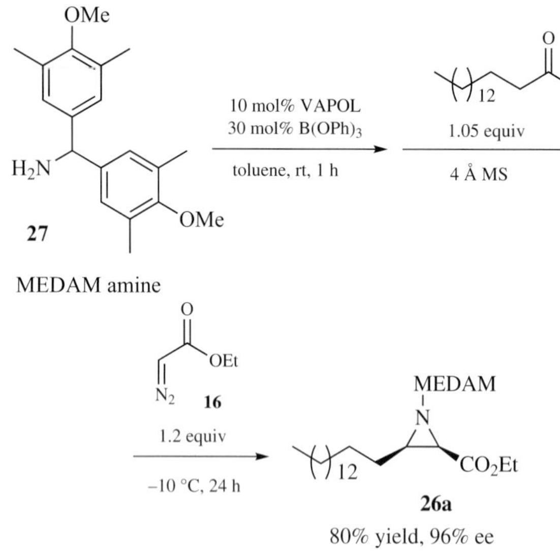

(19)

Catalytic Asymmetric *trans*-Aziridination Reactions. An important finding is that the asymmetric aziridination could be extended to the synthesis of *trans*-aziridines.[22] Treatment of imine **15** with diazoacetamide **28** in the presence of (*S*)-VANOL BOROX catalyst in toluene at −20 °C for 24 h affords *trans*-aziridine **29** in good yield and with excellent enantioselectivity (eq 20). In the following communication,[23] DFT calculations were performed to elucidate the reaction mechanism and the sense of diastereoselection for the *cis*- and *trans*-aziridination reactions. A H-bonding interaction between the amidic hydrogen of diazoacetamide and an oxygen atom of the BOROX catalyst was identified as the key interaction responsible for *trans* selectivity. The calculated TS energy difference was in good agreement with the experimental diastereoselectivity.

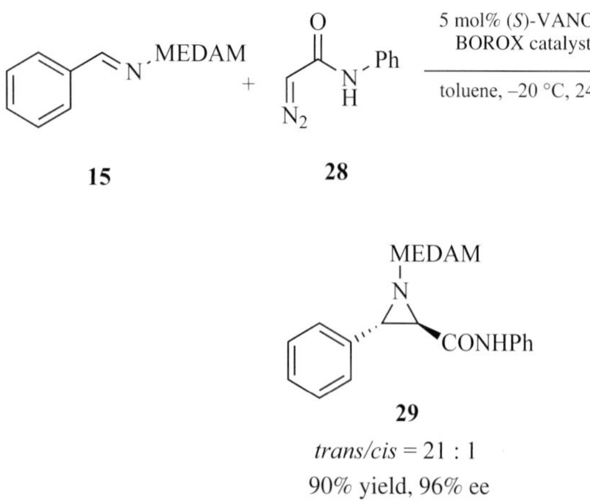

(20)

Asymmetric Catalytic Heteroatom Diels–Alder Reactions. Wulff and coworkers showed that a catalytic amount (5–10 mol %) of VAPOL with a stoichiometric amount of B(OPh)$_3$ could promote asymmetric reactions between imines and Danishefsky's diene (eq 21).[24] A control experiment showed that the reaction gave higher yield when the amount of B(OPh)$_3$ was increased from 15 to 150 mol %. Surprisingly, the asymmetric induction did not decrease until 500 mol % was added. These results together with binding constants determined from NMR studies suggest that the VAPOL BOROX catalyst can catalyze the reaction 100 times faster than the nonchiral B(OPh)$_3$. They also found that BINOL under the same conditions only gave 26% yield with 23% ee.

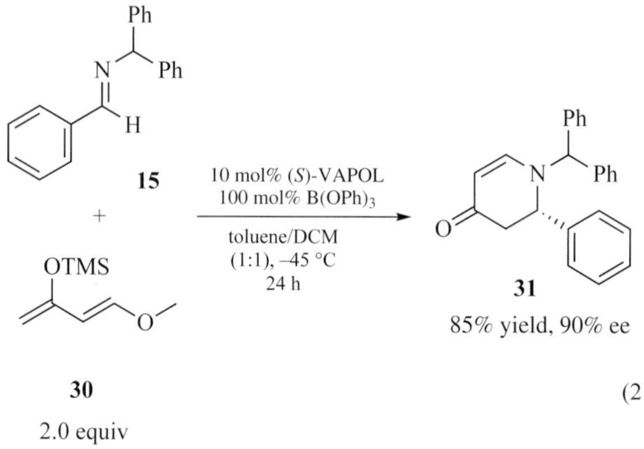

(21)

aza-Cope Rearrangements of Imines. Ren and Wulff discovered[25] an enantioselective aza-Cope rearrangement of imines catalyzed by a VANOL BOROX catalyst with benzoic acid as an additive (eq 22). Without benzoic acid, the reaction gave the rearranged product in 77% yield with only 25% ee, which indicates a synergistic interaction of these two Brønsted acids. VAPOL, BINOL, and 3,3′-diphenylBINOL proved not to be effective in this reaction. After optimization, they extended this methodology to a direct aminoallylation of aldehydes as shown in eq 23. number of aromatic and aliphatic aldehydes could be converted to homoallylic amines with good to excellent enantioselectivities (72-96%) on gram scale.

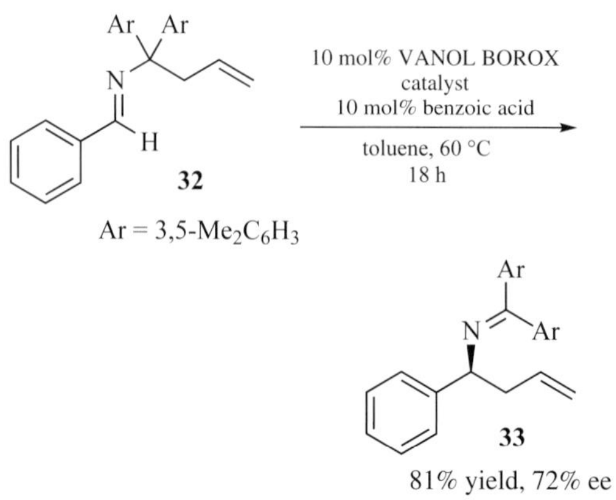

(22)

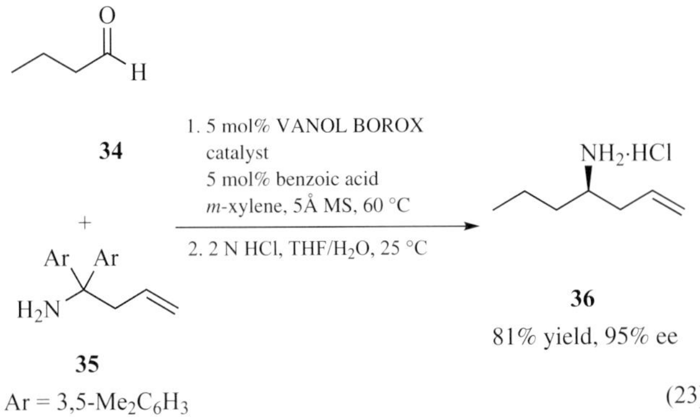

(23)

Three-Component Asymmetric Catalytic Ugi Reactions. After screening a large set of different BOROX catalysts by varying the ligand and phenol components, Wulff and Zhao reported the first effective chiral catalyst for the three-component Ugi reaction (eq 24).[26] The precatalyst was prepared by heating a mixture of a substituted VAPOL ligand (20 mol %), 2,6-dimethyl-4-methoxyphenol (40 mol %), H_2O (60 mol %), and $BH_3(SMe_2$(60 mol %) in toluene at 100 °C for 1 h. After removal of volatiles under high vacuum, the BOROX catalyst **38** was generated in situ by addition of amine **36** at room temperature. Aldehyde **35** and isonitrile **37** were then introduced to the mixture at room temperature for 24 h to afford the α-amino amide **39** in 85% yield with an er of 93:7. The enantioselectivity could be improved to 95:5 by running the reaction at 0 °C, although the rate was sluggish (75% in 48 h). This process seems fairly general for aryl aldehydes with both electron-withdrawing and electron-donating groups, giving the corresponding α-amino amides with an er > 90:10 in most cases. The enantiomeric purity of the product could be enhanced by recrystallization.

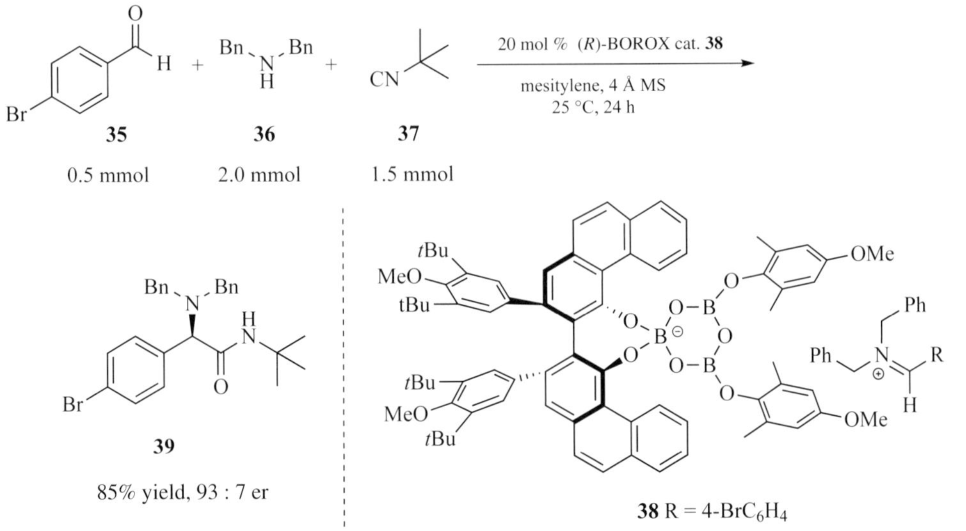

(24)

Asymmetric Petasis Reactions and Propargylation of Ketones. The laboratory of Schaus reported[27] asymmetric Petasis reactions between alkenyl boronates, secondary amines, and glyoxylates using biphenol catalysts to afford chiral α-amino esters. After optimization, the reaction of (E)-diisopropyl styrylboronate **40**, dibenzylamine **36**, and ethyl glyoxylate **41** in the presence of 15 mol % of (S)-VAPOL and 3 Å molecular sieves in toluene at −15 °C provided α-amino ester **42** in 81% yield with an er of 95.5:4.5 (eq 25). It was found that BINOL, substituted BINOLs, and VANOL were inferior chiral catalysts for this process. Similar results were obtained from alkenyl boronates with other substituents. Other secondary amines gave asymmetric inductions in the range of er 89:11 to 98.5:1.5 depending on the nature of the amine.

Further work by Schaus showed[28] that VAPOL could promote the enantioselective formation of homopropargyl alcohols from ketones and allenylboronates. As shown in eq 26, the neat reaction of allenyldioxaborinane **43** with acetophenone **44** using 15 mol % VAPOL as the catalyst under microwave irradiation for 1 h afforded the alcohol **45** in 80% yield and 92% ee. However, 3,3′-Br_2-BINOL gave slightly higher yield (85%) and enantioselectivity (94% ee) for this reaction, and thus it was chosen for use in the assessment of the substrate scope.

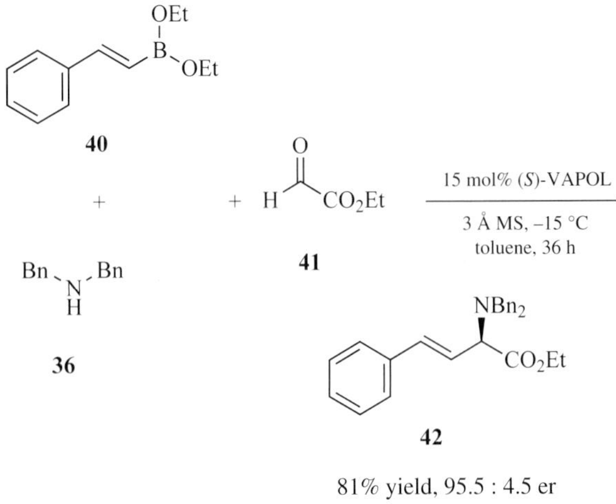

(25)

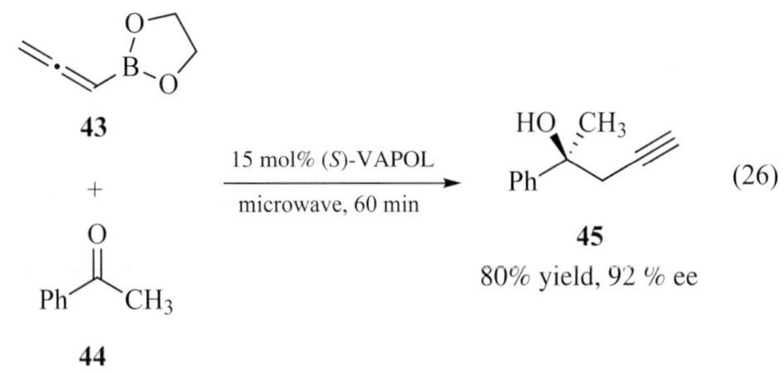

Asymmetric Diels–Alder Reactions. Lei and Wulff have reported[29] asymmetric Diels–Alder reactions promoted by substituted VANOL boron complexes as key steps in the total syntheses of kuwanons I and J. In the model reaction shown in eq 27, a VANOL boron complex was generated by mixing the ligand, BH_3($SMe_2$and acetic acid in THF for 25 min. Dienophile **46** was then introduced with 5 Å MS, and then after 1.5 h the addition of diene **47**. *Endo*-**48** was obtained in 80% yield and 97% ee after 20 h. TADDOL, BINOL, and VAPOL gave unsatisfactory enantioselectivity for this reaction. After further screening, they found that (*S*)-8,8(-dimethyl-VANOL was the optimal ligand and gave good enantioselectivity with the fully functionalized substrates to give both *endo*- and *exo*-products (86% and 84% ee), which were subjected to deprotection to furnish the natural products kuwanons I and J, respectively.

Determination of Enantiomeric Excess in Amine Derivatives. Another application of VANOL/VAPOL boron complexes is the determination of the enantiomeric excess of amines. As shown in eq 28, Anzenbacher found[30] that chiral amine

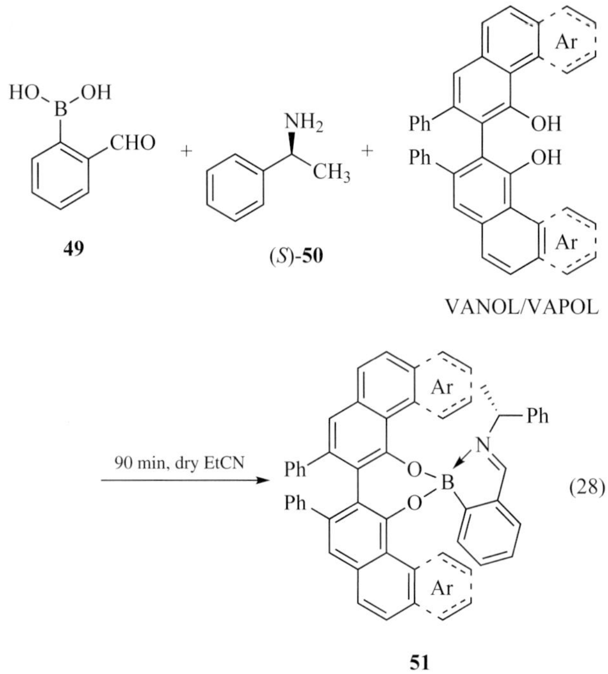

50 could form molecular self-assembled complexes **51** with 1 equiv VANOL/VAPOL and 1 equiv 2-formylphenylboronic acid (FPBA) **49**. Structures of enantiomeric VAPOL iminoboronate esters **51** (*S, S* and *R, R*) were determined by single-crystal X-ray diffraction analysis. Fluorescence titrations showed a red-shift of the fluorescence arising from the formation of an oxazolidine

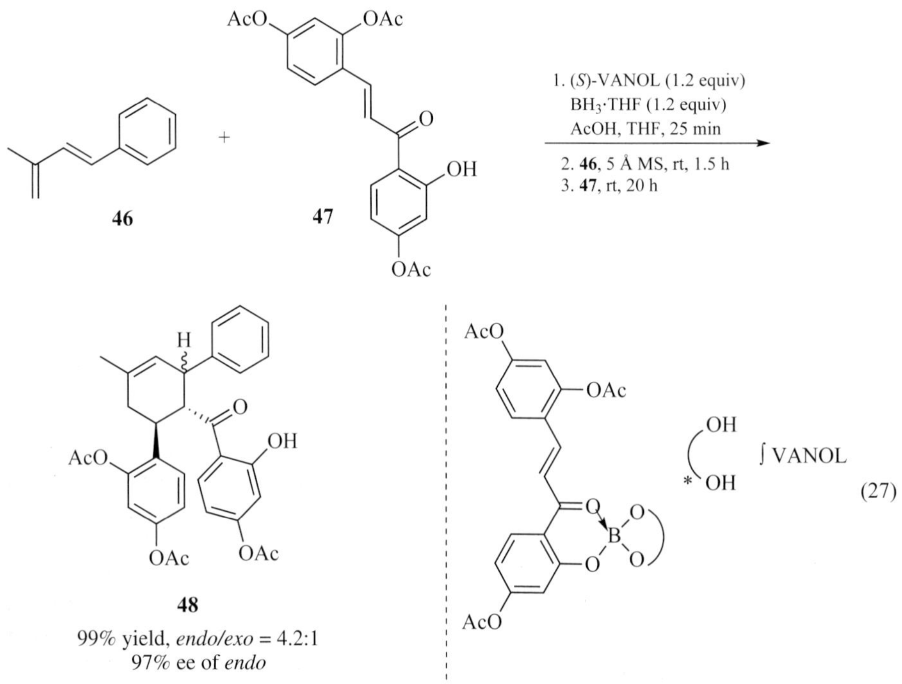

boronate ester. Interestingly, the change of the fluorescence for diastereomers generated from (*S*)-**50** and (*R*)-**50** was different. Determination of the %ee values of various amine derivatives was successfully achieved by monitoring the changes in the fluorescence signal.

Aluminum-catalyzed Asymmetric Baeyer–Villiger Reactions. Bolm et al. found[31] that 20 mol % of a 1:1 mixture of Me_2AlCl and VANOL could catalyze the Baeyer–Villiger reactions of 3-substituted cyclobutanones with cumene hydroperoxide (CHP) in toluene. As shown in eq 29, lactone **53** was obtained in 92% yield and 84% ee. With other substrates, the enantioselectivities were only moderate (34–69% ee).

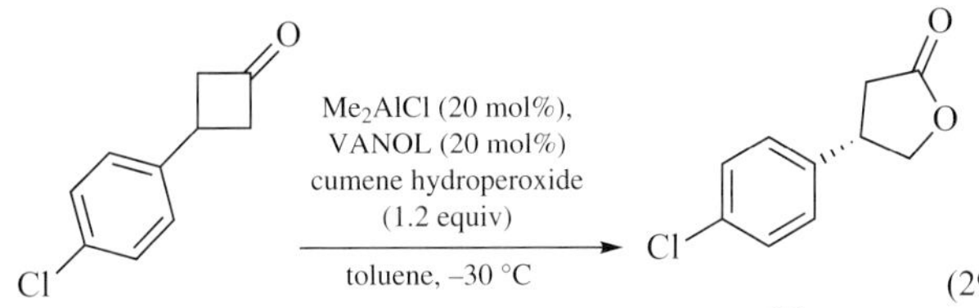

(29)

52 **53**

92% yield, 84% ee

Enantioselective Zinc-mediated Michael Addition of Alkynes to Enones. Blay et al. showed[32] that a catalytic amount of VANOL could be applied to the zinc-mediated asymmetric conjugate alkynylations as a chiral inducer. As shown in eq 30, treatment of enone **54** with phenylacetylene **55** in the presence of 20 mol % (*R*)-VANOL and 2 equiv diethylzinc afforded the corresponding alkynylation product in good yield and 87% ee. Results from reactions between other enones bearing different aryl substituents and terminal alkynes with different aromatic or heteroaromatic groups were equally effective.

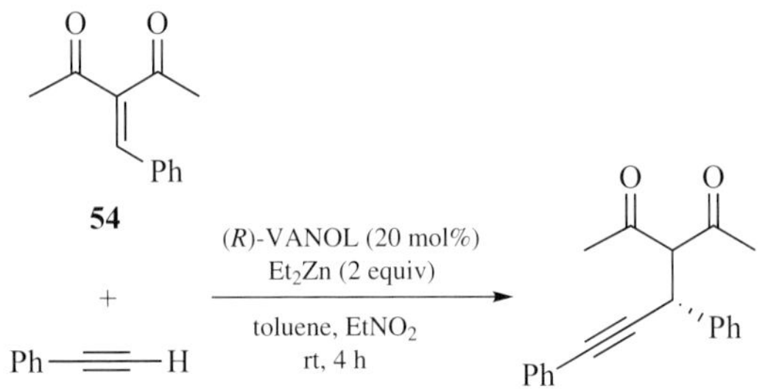

(30)

55 **56**

72% yield, 87% ee

Enantioselective Intramolecular Hydroarylation of Alkenes. Ellman and Bergman have reported[33] that $[RhCl(coe)_2]_2$ together with VANOL- and VAPOL-based isopropylphosphoramidite ligands promoted the asymmetric intramolecular *ortho*-alkylation of aromatic imines containing pendant alkenyl groups. As shown in eq 31, alkene **57** cyclized in the presence of 10 mol % of $[RhCl(coe)_2]_2$ and 20 mol % of the VANOL-based ligand **58** or the VAPOL-based ligand **59** to give (*R*)-**60** in high yield and good enantioselectivity. It was found that the optimal ligand for this substrate was an (*S*)-octahydrobinaphthol-based ligand, which gave 82% yield and 91% ee.

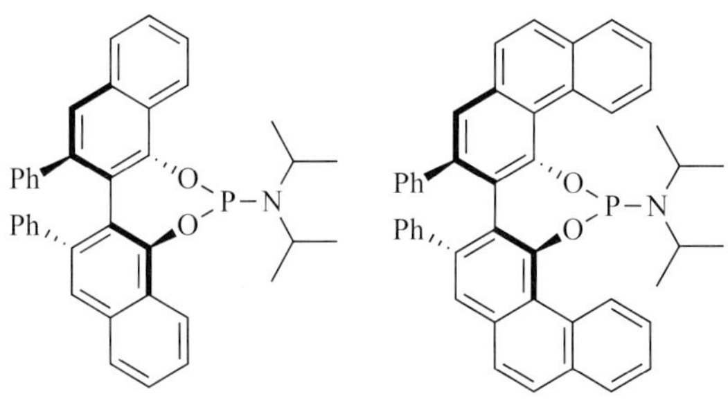

58 **59**

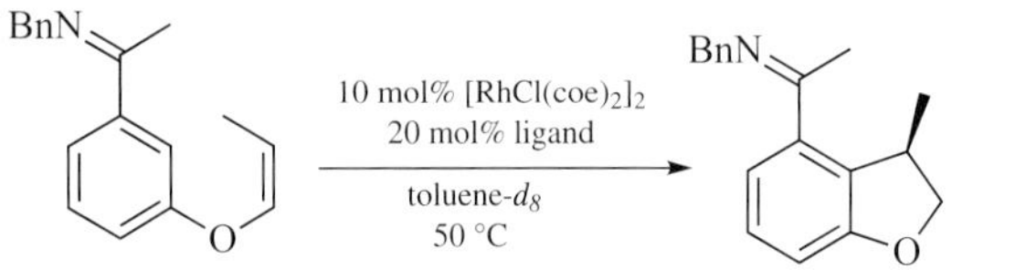

(31)

57 **60**

ligand = **58**, 21 h, 81% yield, 86% ee
ligand = **59**, 43 h, 76% yield, 88% ee

Catalytic Asymmetric α-Iminol Rearrangements. Wulff and Zhang found that an unexpected VANOL zirconium complex could catalyze the asymmetric rearrangement of α-hydroxy imines shown in eq 32.[34] Anticipating the formation of a 2:1 VANOL to zirconium complex, the catalyst **62** was prepared by stirring a mixture of VANOL, Zr-$(O^iPr)_4(HO^iPr)$, and *N*-methylimidazole in a 2:1:1 ratio at room temperature for 30 min. The single crystals obtained from the catalyst solution revealed that the zirconium exists as a six-coordinate homoleptic complex with three VANOL ligands and is charge balanced with two protonated *N*-methyl imidazoles. A number of aromatic or aliphatic groups could be induced to undergo a 1,2-shift to produce chiral α-amino ketones in excellent yields (74–100%) and good to excellent enantioselectivities (54 to >99%). It was shown that the reaction of aldehyde **64** and aniline **65** gave the same result as that with a preformed imine.

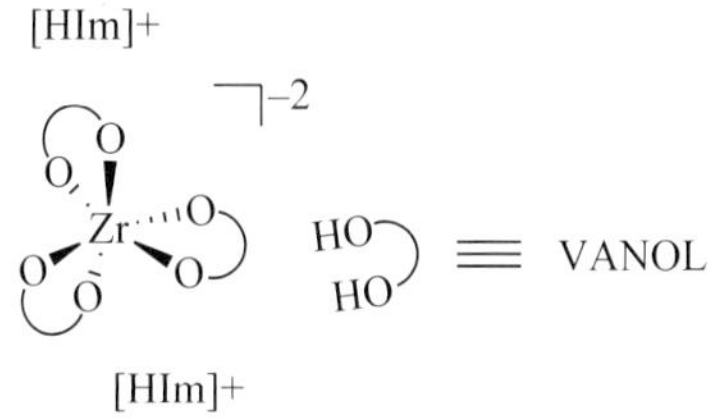

62

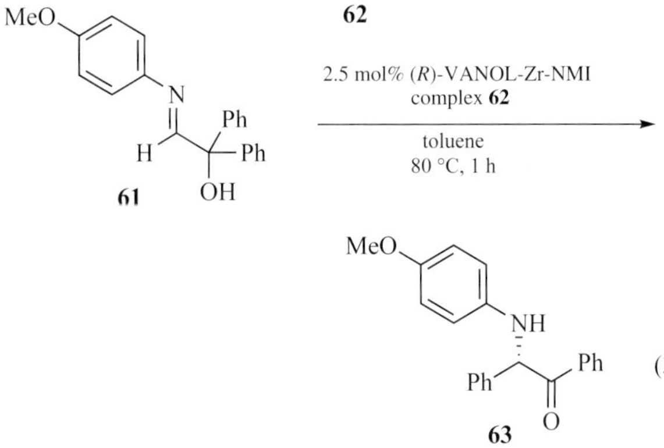

(32)

61 **63**

98% yield, 98% ee

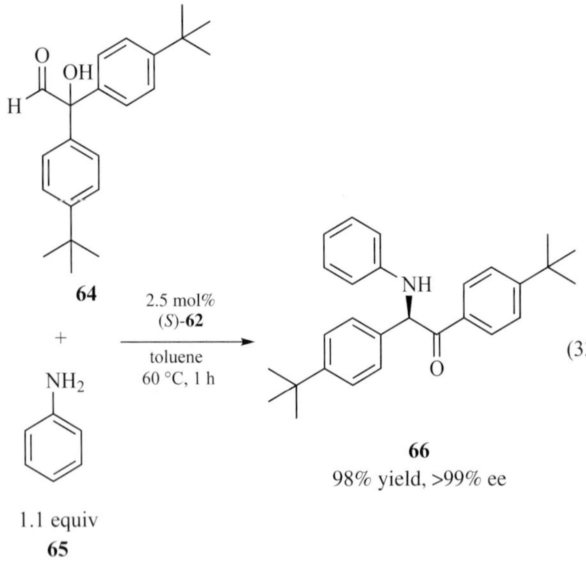

(33)

1. (a) Bao, J.; Wulff, W. D., *J. Am. Chem. Soc.* **1993**, *115*, 3814. (b) Bao, J.; Wulff, W., *Tet. Lett.* **1995**, *36*, 3321. (c) Bao, J.; Wulff, W. D.; Dominy, J. B.; Fumo, M, J.; Grant, E. B.; Rob, A. C.; Whitcomb, M. C.; Yeung, S.-M.; Ostrander, R. L.; Rheingold, A. L., *J. Am. Chem. Soc.* **1996**, *118*, 3392.
2. (a) Guseinov, M. M.; Akhmedov, I. M.; Mamadov, E. G., *Azerb. Khim. Zh.* **1976**, *1*, 46. (b) Hashimoto, S.-L.; Komeshima, N.; Koga, K., *J. Chem. Soc., Chem. Commum.* **1979**, *10*, 437. (c) Takemuraa, H.; Komeshima, N.; Takahashi, I.; Hashimoto, S.-I.; Ikota, N.; Tomioka, K.; Koga, K., *Tet. Lett.* **1987**, *28*, 5687.
3. (a) Bear, B. R.; Sparks, S. M.; Shea, K. J., *Angew. Chem., Int. Ed.* **2001**, *40*, 820. (b) Carmona, D.; Lamata, M. P.; Oro, L. A., *Coord. Chem. Rev.* **2000**, *200–202*, 717. (c) *Lewis acids and Selectivity in Organic Synthesis*; Santelli, M.; Pons, J.-M., Eds.; CRC Press: New York, **1996**, p 300. (d) Kagan, H. B.; Riant, O., *Chem. Rev.* **1992**, *92*, 1007. (e) Diaz, L. C., *J. Braz. Chem. Soc.* **1997**, *8*, 289.
4. (a) Kohsuke, A.; Irie, R.; Katsuki, T., *Tetrahedron* **2001**, *57*, 845. (b) Motoyama, Y.; Koga, Y.; Nishiyama, H., *Tetrahedron* **2001**, *57*, 853. (c) Mihara, J.; Aikawa, K.; Uchida, T.; Irie, R.; Katsuki, T., *Heterocycles* **2001**, *54*, 395. (d) Jorgenson, K. A., *Angew. Chem., Int. Ed.* **2000**, *39*, 3558. (e) Evans, D. A.; Rovis, T.; Johnson, J. S., *Pure Appl. Chem.* **1999**, *71*, 1407.
5. Heller, D. P.; Goldberg, D. R.; Wulff, W. D., *J. Am. Chem. Soc.* **1997**, *119*, 10551.
6. (a) Atkinson, R. S., *Tetrahedron* **1999**, *55*, 1519. (b) Sweeney, J.; Osborn, H. M. I., *Tet. Asymm.* **1997**, *8*, 1689. (c) Tanner, D., *Angew. Chem., Int. Ed. Engl.* **1994**, *33*, 599.
7. *Small Ring Compounds in Organic Synthesis VI: Topics in Current Chemistry*; De Meijere, A., Ed.; Springer Verlag: Berlin, **2000**, p 210.
8. (a) Peukert, S.; Jacobsen, E., *Org. Lett.* **1999**, *1*, 1245. (b) Adolfsson, H.; Converso, A.; Sharpless, K. B., *Tet. Lett.* **1999**, *40*, 3991. (c) Berrisford, D. J.; Bolm, C.; Sharpless, K. B., *Angew. Chem., Int. Ed. Engl.* **1995**, *34*, 1059. (d) Burns, C. J.; Martin, C. A.; Sharpless, K. B., *J. Org. Chem.* **1989**, *54*, 2826. (e) Carlier, P. R.; Sharpless, K. B., *J. Org. Chem.* **1989**, *54*, 4016. (f) Taylor, R. E.; Engelhardt, F. C.; Yuan, H., *Org. Lett.* **1999**, *1*, 1257. (g) Nakamura, A.; Konishi, A.; Tatsuno, Y.; Otsuka, S., *J. Am. Chem. Soc.* **1978**, *100*, 3443. (h) Nakamura, A.; Konishi, A.; Tsujitani, R.; Kudo, M.-A.; Otsuka, S., *J. Am. Chem. Soc.* **1978**, *100*, 3449.
9. (a) Doyle, M. P.; Hu, W.; Timmons, D. J., *Org. Lett.* **2001**, *3*, 933. (b) Llewellyn, D. B.; Adamson, D.; Arndtsen, B. A., *Org. Lett.* **2000**, *2*, 4165. (c) Kim, D. Y.; Suh, K. H.; Choi, J. S.; Mang, J. Y.; Chang, S. K., *Syn. Commun.* **2000**, *30*, 87. (d) Zhu, W.; Li, Y.; Chen, Z.; Li, D.; Yang, G., *Syn. Commun.* **2000**, *30*, 1075. (e) Adam, W.; Roschmann, K. J.; Saha-Moller, C. R., *Eur. J. Org. Chem.* **2000**, *3*, 557–561. (f) Nishimura, M.; Minakata, S.; Thongchant, S.; Ruy, I.; Komatsu, M., *Tet. Lett.* **2000**, *41*, 7089. (g) Davis, F. A.; McCoull, W., Titus, D. D., *Org. Lett.* **1999**, *1*, 1053. (h) Davis, F. A.; McCoull, W., *Tet. Lett.* **1999**, *40*, 249. (i) McLaren, A. B.; Sweeny, J. B., *Org. Lett.* **1999**, *1*, 1339. (j) Ho, C.-M.; Lau, T.-C.; Kwong, H.-L.; Wong, W.-T., *J. Chem. Soc., Dalton Trans.* **1999**, 2411. (k) Simonato, J.-P.; Pecaut, J.; Scheidt, W. R.; Marchon, J.-C., *Chem. Commun.* **1999**, *11*, 989. (l) Li, A.-H.; Zhou, Y.-G.; Dai, L.-X.; Hou, X.-L.; Xia, L.-J.; Lin, L., *J. Org. Chem.* **1998**, *63*, 4338. (m) Cardillo, G.; Gentilucci, L.; Bastardas, I. R.; Tolomelli, A., *Tetrahedron* **1998**, *54*, 8217.
10. (a) Antilla, J.; Wulff W. D., *Angew. Chem., Int. Ed.* **2000**, *39*, 4518. (b) Antilla, J.; Wulff W. D., *J. Am. Chem. Soc.* **1999**, *121*, 5099. (c) Loncaric, C.; Wulff W. D., *Org. Lett.* **2001**, *3*, 3675–3678.
11. (a) Kobayashi, S.; Ishitani, H.; Yamashita, Y.; Ueno, M.; Shimizu, H., *Tetrahedron* **2001**, *57*, 861. (b) Ishitani, H.; Ueno, M.; Kobayashi, S., *J. Am. Chem. Soc.* **2000**, *122*, 8180. (c) Kobayashi, S.; Ishitani, H., *Chem. Rev.* **1999**, *99*, 1069. (d) Ishitani, H.; Kitazawa, S.; Kobayashi, S., *Tet. Lett.* **1999**, *40*, 2161. (e) Kobayashi, S.; Kusakabe, K.-I.; Komiyama, S.; Ishitani, H., *J. Org. Chem.* **1999**, *64*, 4220.
12. Xue, S.; Yu, S.; Deng, Y.; Wulff, W. D., *Angew. Chem., Int. Ed. Engl.* **2001**, *40*, 2271–2274.
13. Ding, Z.; Xue, S.; Wulff, W. D., *Chem. Asian J.* **2011**, *6*, 2130.
14. Ding, Z.; Osminski, W. E. G.; Ren, H.; Wulff, W. D., *Org. Process Res. Dev.* **2011**, *15*, 1089.
15. Guan, Y.; Ding, Z.; Wulff, W. D., *Chem. Eur. J.* **2013**, *19*, 15565.
16. Gupta, A. K.; Zhang, X.; Staples, R. J.; Wulff, W. D., *Catal. Sci. Technol.* **2014**, *4*, 4406.
17. Zhang, Y.; Yeung, S.-M.; Wu, H.; Heller, D. P.; Wu, C.; Wulff, W. D., *Org. Lett.* **2003**, *5*, 1813.
18. Hu, G.; Holmes, D.; Gendhar, B. F.; Wulff, W. D., *J. Am. Chem. Soc.* **2009**, *131*, 14355.
19. Mukherjee, M.; Gupta, A. K.; Lu, Z.; Zhang, Y.; Wulff, W. D., *J. Org. Chem.* **2010**, *75*, 5643.
20. Hu, G.; Gupta, A. K.; Huang, R. H.; Mukherjee, M.; Wulff, W. D., *J. Am. Chem. Soc.* **2010**, *132*, 14669.
21. Gupta, A. K.; Mukherjee, M.; Wulff, W. D., *Org. Lett.* **2011**, *13*, 5866.
22. Desai, A. A.; Wulff, W. D., *J. Am. Chem. Soc.* **2010**, *132*, 13100.
23. Vetticatt, M. J.; Desai, A. A.; Wulff, W. D., *J. Am. Chem. Soc.* **2010**, *132*, 13104.
24. Newman, C. A.; Antilla, J. C.; Chen, P.; Predeus, A. V.; Fielding, L.; Wulff, W. D., *J. Am. Chem. Soc.* **2007**, *129*, 7216.
25. Ren, H.; Wulff, W. D., *J. Am. Chem. Soc.* **2011**, *133*, 5656.
26. Zhao, W.; Huang, L.; Guan, Y.; Wulff, W. D., *Angew. Chem., Int. Ed.* **2014**, *53*, 3436.
27. Lou, S.; Schaus, S. E., *J. Am. Chem. Soc.* **2008**, *130*, 6922.
28. Barnett, D. S.; Schaus, S. E., *Org. Lett.* **2011**, *13*, 4020.
29. Han, J.; Li, X.; Guan, Y.; Zhao, W.; Wulff, W. D.; Lei, X., *Angew. Chem., Int. Ed.* **2014**, *53*, 9257.
30. Shcherbakova, E. G.; Minami, T.; Brega, V.; James, T. D.; Anzenbacher, P., *Angew. Chem., Int. Ed.* **2015**, *54*, 7130.
31. Bolm, C.; Frison, J.-C.; Zhang, Y.; Wulff, W. D., *Synlett* **2004**, *2004*, 1619.
32. Blay, G.; Cardona, L.; Pedro, J. R.; Sanz-Marco, A., *Chem. Eur. J.* **2012**, *18*, 12966.
33. Harada, H.; Thalji, R. K.; Bergman, R. G.; Ellman, J. A., *J. Org. Chem.* **2008**, *73*, 6772.
34. Zhang, X.; Staples, R. J.; Rheingold, A. L.; Wulff, W. D., *J. Am. Chem. Soc.* **2014**, *136*, 13971.

α,α-Diphenyl-2-pyrrolidinemethanol[1]

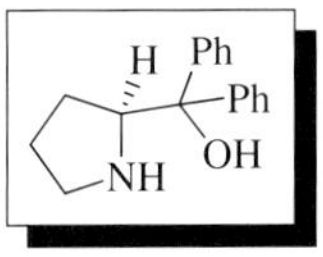

(*S*)
[112068-01-6] $C_{17}H_{19}NO$ (MW 253.37)
InChI = 1S/C17H19NO/c19-17(16-12-7-13-18-16,14-8-3-1-4-9-14)15-10-5-2-6-11-15/h1-6,8-11,16,18-19H,7,12-13H2/t16-/m0/s1
InChIKey = OGCGXUGBDJGFFY-INIZCTEOSA-N
(*R*)
[22348-32-9]
InChI = 1S/C17H19NO/c19-17(16-12-7-13-18-16,14-8-3-1-4-9-14)15-10-5-2-6-11-15/h1-6,8-11,16,18-19H,7,12-13H2/t16-/m1/s1
InChIKey = OGCGXUGBDJGFFY-MRXNPFEDSA-N
((*S*) ·HCl)
[16226-54-3] $C_{17}H_{20}ClNO$ (MW 289.83)
InChI = 1S/C17H19NO.ClH/c19-17(16-12-7-13-18-16,14-8-3-1-4-9-14)15-10-5-2-6-11-15;/h1-6,8-11,16,18-19H,7,12-13H2;1H/t16-;/m0./s1
InChIKey = XYSYRXOJSOXZPT-NTISSMGPSA-N

(precursor to several chiral oxazaborolidine catalysts[1] used for the enantioselective reduction of prochiral ketones[2,3])

Alternate Name: diphenylprolinol.
Physical Data: mp 79–79.5 °C (hexane); 80–82 °C (EtOH). $[\alpha]_{589}$ −54.3° (*c* 0.261, MeOH); −68.1° (*c* 3.17, $CHCl_3$) for the (*S*)-enantiomer.
Solubility: very sol THF, CH_2Cl_2, MeOH, toluene.
Preparative Methods: addition of ***Phenylmagnesium Bromide*** to (*S*)-proline-*N*-carboxyanhydride (73% overall from proline).[4] Addition of *N*-benzyl-(*S*)-proline ethyl ester to phenylmagnesium chloride followed by catalytic hydrogenolysis (49% overall from proline).[5] Addition of *N*-(benzyloxycarbonyl)-(*S*)-proline methyl ester to phenylmagnesium chloride (50% overall yield from proline).[3a,c] Addition of phenylmagnesium bromide to *N*-(ethyloxycarbonyl)-(*S*)-proline methyl ester followed by alkaline hydrolysis (65% overall yield from proline).[6] Addition of (*S*)-proline ethyl ester hydrochloride to phenylmagnesium chloride (20–26% overall yield from proline, ca. 80% ee).[7] Enantioselective deprotonation of *N*-Boc-pyrrolidine with ***sec-Butyllithium/(−)-Sparteine*** followed by reaction with benzophenone to give (*R*)-diphenylprolinol (63% yield from pyrrolidine).[8] Addition of phenylmagnesium chloride to methyl pyroglutamate followed by reduction with borane to give racemic diphenylprolinol (51% yield) which can be resolved as its *O*-acetylmandelate salt to give the (*R*)- and (*S*)-enantiomers (30% yield from the racemate).[3b] Addition of lithiated *N*-nitrosopyrrolidine to benzophenone to give racemic diphenylprolinol (58–60% yield based on benzophenone).[9]
Purification: recrystallization from hexane, ethanol, or methanol/water.
Handling, Storage, and Precautions: no special information available. In general, however, it is advisable that all reactions with this reagent be conducted in a well ventilated fume hood. Care should be exercised to avoid contact of this reagent and the derived oxazaborolidine catalyst with the eyes and skin.

Original Commentary

David J. Mathre and Ichiro Shinkai
Merck Research Laboratories, Rahway, NJ, USA

Enantioselective Ketone Reduction. Following Itsuno's lead for enantioselective reductions using diphenylvalinol,[10] Kraatz was the first to describe the use of a 1:2 mixture of (*S*)-diphenylprolinol (**1**) and ***Borane–Tetrahydrofuran*** for the stoichiometric enantioselective reduction of ketone (**2**) to obtain the plant growth regulator triapenthenol (**3**) (eq 1).[2] Although not characterized at the time, the species responsible for the enantioselectivity observed was presumed to be an oxazaborolidine–borane complex.[10b]

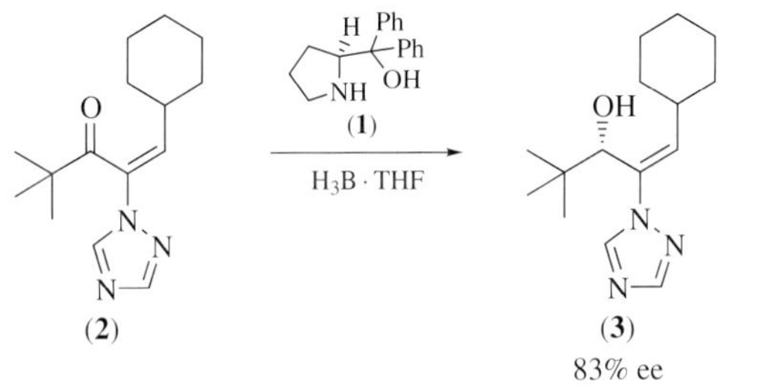
(1)

Diphenylprolinol (**1**) and borane–THF react in a multistep process to give the unsubstituted oxazaborolidine–borane complex (**7**) (eq 2). Formation of amine–borane complex (**4**) is exothermic, and this intermediate can be isolated as a stable crystalline solid.[4b] Subsequent conversion to oxazaborolidine (**6**) requires heating the THF solution under pressure (1.7 bar) at 70–75 °C for 48–72 h. Corey isolated and characterized free oxazaborolidine (**6**) as a solid (mp 107–124 °C), which was reported to be a mixture of monomer and dimer (NMR).[3a] Finally, addition of borane–THF to a solution of oxazaborolidine (**6**) affords oxazaborolidine–borane complex (**7**) which was not isolated and was identified based on ^{11}B NMR evidence.

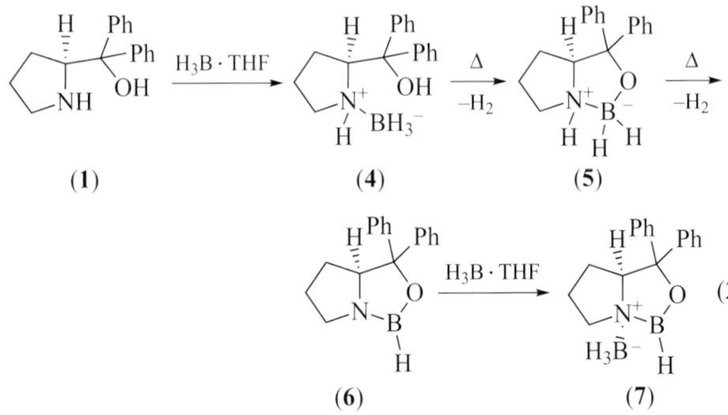
(2)

Corey demonstrated that oxazaborolidine (**6**) can be used catalytically (2.5–100 mol %) with excess borane (60–200 mol %) for the enantioselective reduction of prochiral ketones (eq 3 Table 1).[3a,c]

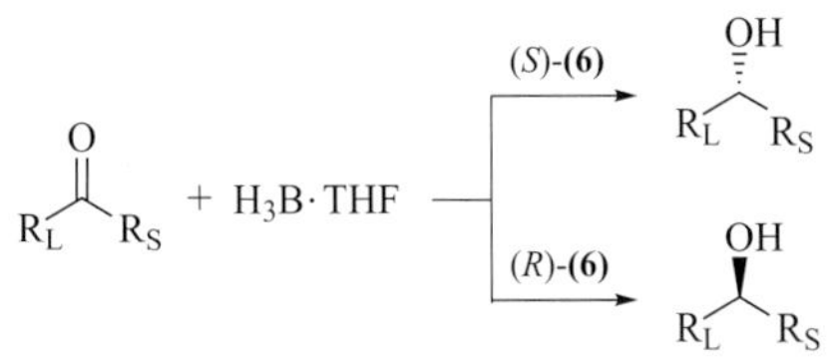

(3)

Table 1 Catalytic Enantioselective Reduction of Ketones using Oxazaborolidine (*S*)-(**6**)

Ketone	$H_3B{\cdot}THF$ (mol %)	Catalyst (*S*)-(**6**) (mol %)	Enantiomeric purity[a]	Absolute configuration
PhCOMe	100	10	97%	(*R*)
PhCOEt	60	5	90%	(*R*)
t-BuCOMe	60	10	92%	(*R*)
1-Tetralone	60	5	89%	(*R*)
$PhCOCH_2Cl$	60	5	97%	(*S*)

[a]Enantiomeric purities determined by capillary GC analysis of the (−)-menthyloxycarbonyl derivative, and have been corrected by adding 1% to the experimentally observed values to correspond to values for optically pure catalyst.[3a]

The following catalytic cycle was proposed to explain the behavior of the oxazaborolidine catalyst (eq 4).[3a,11] Oxazaborolidine (**6**) reacts with borane to give oxazaborolidine–borane complex (**7**) Coordination between the Lewis acidic ring boron and the carbonyl oxygen activates the ketone toward reduction. Intramolecular hydride transfer from the BH_3 coordinated to the ring nitrogen then occurs via a six-membered ring chair transition state.[12] Following hydride transfer, the alkoxy-BH_2 dissociates, and oxazaborolidine (**6**) is free to begin the cycle again. For a more detailed discussion, see the entry for ***Tetrahydro-1-methyl-3,3-diphenyl-1H,3H-pyrrolo[1,2-c][1,3,2]oxazaborole***.

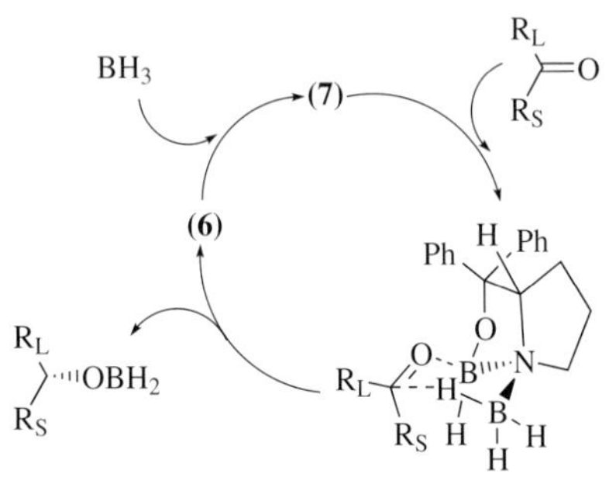

(4)

Problems with the preparation and stability of oxazaborolidine (**6**) led to the development of a series of *B*-substituted oxazaborolidines derived from diphenylprolinol. The *B*-methyl substituted oxazaborolidine (**9a**) was first prepared (eq 5) by reaction of diphenylprolinol (**1**) with methylboronic acid under dehydrating conditions (toluene at 23 °C in the presence of 4 Å molecular sieves or toluene at reflux using a Dean–Stark trap) followed by vacuum distillation (0.1 mmHg, 170 °C)[3a,c] Based on NMR evidence, the product (mp 74–87 °C) was reported to be a mixture of monomer and dimer.[3a] The corresponding *B*-butyloxazaborolidine (**9c**), prepared in a similar manner from *n*-butylboronic acid, was also reported to be a mixture of monomer and dimer.[13] Subsequent investigations demonstrated that the reported 'dimers' were in fact the intermediate (**8**) and the more stable disproportionation product (**10**) (eq 6).[4] Furthermore, the presence of (**8**) or (**10**) was demonstrated to be deleterious to the enantioselectivity of the catalyst.[14]

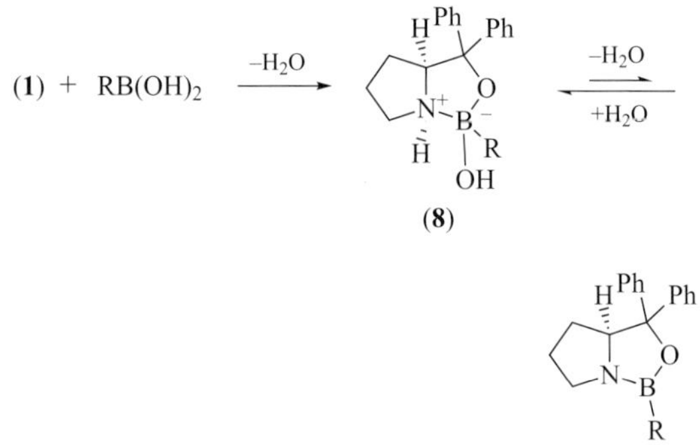

(5)

(**9a**) R = Me
(**9b**) R = Et
(**9c**) R = Bu
(**9d**) R = Ph

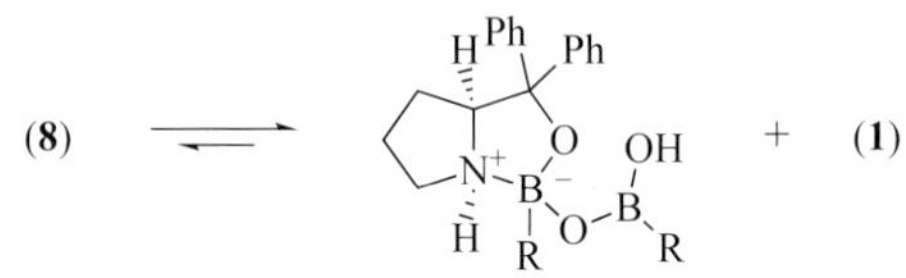

(6)

(**10**)

A small-scale procedure, based on the reaction of bis(trifluoroethyl) alkylboronates with diphenylprolinol (eq 7) was reported for the preparation of the *B*-ethyl- (**9b**) and *B*-butyl- (**9c**) oxazaborolidines.[15] The bis(trifluoroethyl) alkylboronates were prepared from the disproportionation of tris(trifluoroethyl) borate and the corresponding trialkylborane. Since trimethylborane is not commercially available, this procedure is not applicable for the preparation of *B*-methyloxazaborolidine (**9a**).

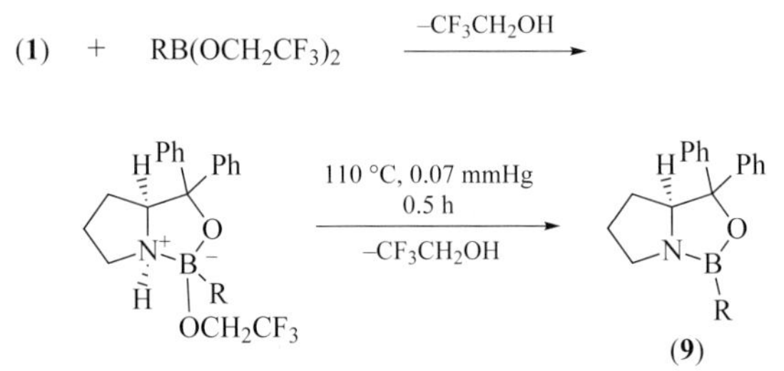

(7)

(**9**)

A practical large-scale process for the synthesis of *B*-methyloxazaborolidine (**9a**) was developed using commercially available trimethylboroxine (eq 8), which affords the product as an analytically pure, colorless crystalline solid (mp 79–81 °C).[4,16] The *B*-ethyl- (**9b**), *B*-butyl- (**9c**), and *B*-phenyloxazaborolidines (**9d**) were also prepared from the corresponding triethyl-, tributyl-, or triphenylboroxine. The free oxazaborolidines, thus prepared, are stable if rigorously protected from moisture.

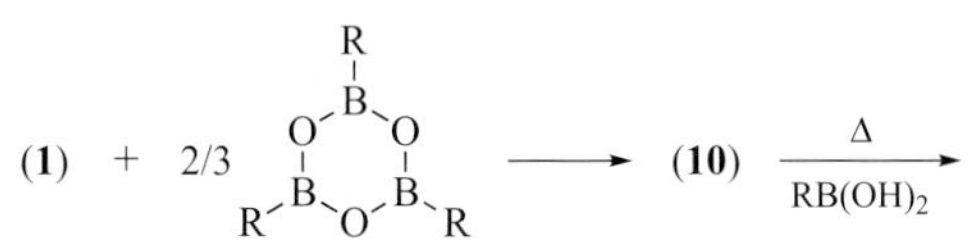

(8)

A significantly more stable form of the catalyst is the crystalline oxazaborolidine–borane complex (**11**).[4b,16,17] This borane complex is readily prepared from oxazaborolidine (**9a**) and ***Borane–Dimethyl Sulfide*** complex (BMS) (eq 9).

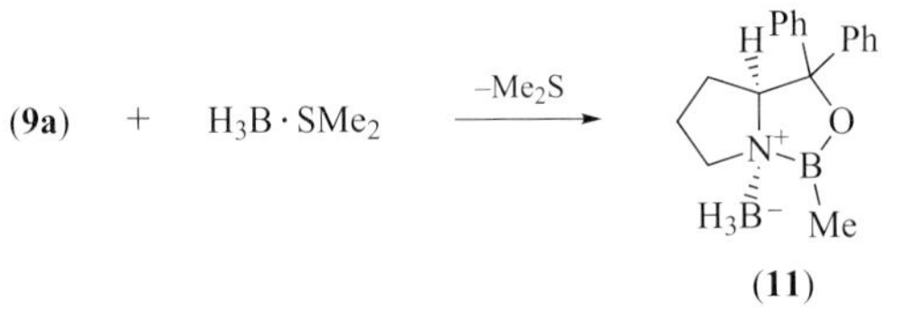

(11) (9)

The enantioselectivities reported for the reduction of acetophenone and α-tetralone using the different catalysts (5–10 mol %) and borane–THF or BMS are summarized in Table 2. The best results are obtained by slowly adding the substrate (neat or as a solution in dichloromethane) to a solution of borane complex (**11**) (5 mol %) and BMS (0.6–1.0 mol equiv) in dichloromethane at $-20\,°C$.[16]

Table 2 Enantioselective Reduction of Acetophenone and α-Tetralone

Catalyst	Prep.	mol %	Acetophenone	α-Tetralone
(**9a**) (R = Me)	eq 5	10	96.5% ee	83.3% ee
(**9a**) (R = Me)	eq 8	10	98% ee	94% ee
(**9b**) (R = Et)	eq 7	10	96% ee	–
(**9c**) (R = Bu)	eq 7	10	96% ee	–
(**9d**) (R = Ph)	eq 8	10	72% ee	94% ee
(**11**)	eqs 8 and 9	5	97.6% ee	99.2% ee
(**11**)	eqs 8 and 9	100	>99.8% ee	99.2% ee

In addition to these simple examples, oxazaborolidines derived from diphenylprolinol have been used as enantioselective catalysts for the preparation of prostaglandins,[3a] PAF antagonists,[3a] a key intermediate of ginkgolide B,[18] a key intermediate of forskolin,[19] (*R*)- and (*S*)-fluoxetine,[20] (*R*)- and (*S*)-isopreterenol,[21] vitamin D analogs,[22] the carbonic anhydrase inhibitor MK-0417,[14] the dopamine D1 agonist A-77636,[23] taxol,[24] the LTD4 antagonists L-695,499 and L-699,392,[25] the β-adenergic agonist CL 316,243,[26] and MK-0499.[27] Recently, Bringmann employed oxazaborolidines (**9a**) and (**9c**) to catalyze the atropo-enantioselective ring opening of achiral biaryl lactones (eq 10).[28]

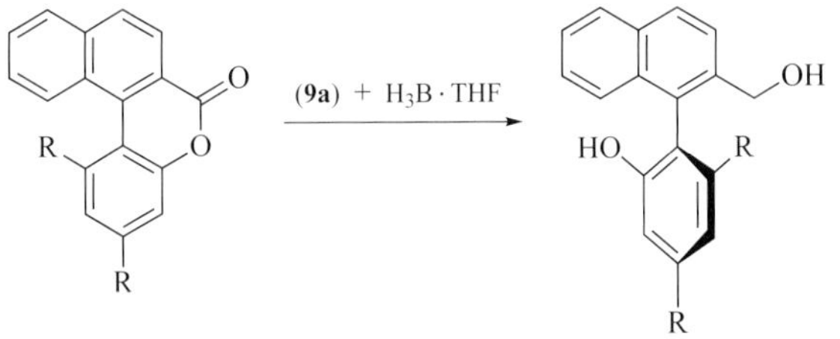

(10)

R = OMe, 94% ee
R = Me, 97% ee

First Update

Aiguo Song and Wei Wang
University of New Mexico, Albuquerque, NM, USA

General Considerations. In the past few years, the application of α, α-diphenylprolinol-derived CBS catalysts in Lewis acid catalysis has been growing steadily. It has been successfully employed in catalytic asymmetric reduction, cycloaddition, Michael, and epoxidation reactions. Moreover, α,α-diphenylprolinol has gone beyond the territory and serves as an effective organocatalyst in several asymmetric transformations.

Reduction of Diones. Chiral oxazaborolidines, derived from α,α-diphenylprolinol, known as CBS catalysts[29] have been used for the stereoselective reduction of α,β-ynones,[30] allenyl ketones,[31] glycosyl α-ketoesters,[32] and trifluoroacetophenones[33] as well as simple ketones to chiral secondary alcohols since the pioneering work of Corey, Bakshi, and Shibata. As a result of the relative low cost, ease of use, and high selectivity, its popularity has also grown dramatically in the application of total synthesis of natural products.[34–39]

In addition to simple ketones, diones can also be reduced regio- and enantioselectively by the CBS catalytic system. For example, Joshi and colleagues reported the stereoselective reduction of benzils (**12**), providing a convenient preparation of the corresponding enantiomerically pure 1,2-diarylethanediols (**13**) (eq 11).[40]

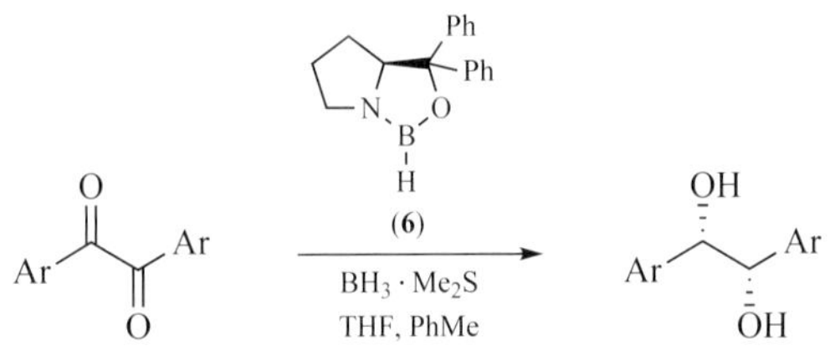

(12) (13) (11)

70–85%, >99% ee
Ar = phenyl, 4-anisyl, 4-tolyl, 3-tolyl, 2-furyl, 2-thienyl

Moreover, highly enantioselective reduction of five, six, and seven-membered prochiral 1,3-cycloalkanediones (**14**) possessing a methyl group and a protected hydroxymethyl group at their C2 position with CBS has been reported by Nakada and coworkers (eq 12).[41]

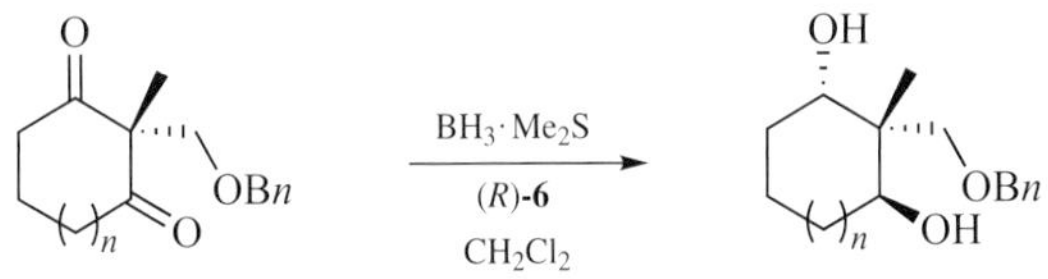

(14) (15) (12)

$n = 0–2$,
84–91%, >99% ee

Corey and coworkers reported the (*S*)-**9c**-catalyzed enantioselective reduction of single carbonyl group of 1,3-cycloalkanediones (**16**) to five, six, seven, and eight-membered keto-alcohols (**18**) (eq 13).[42] Excellent enantioselectivities were observed with five, six, and eight-membered substrates, and yields of chiral β-hydroxyketone (**18**) were generally good (ca. 70%) in the presence of essential $PhNEt_2$.

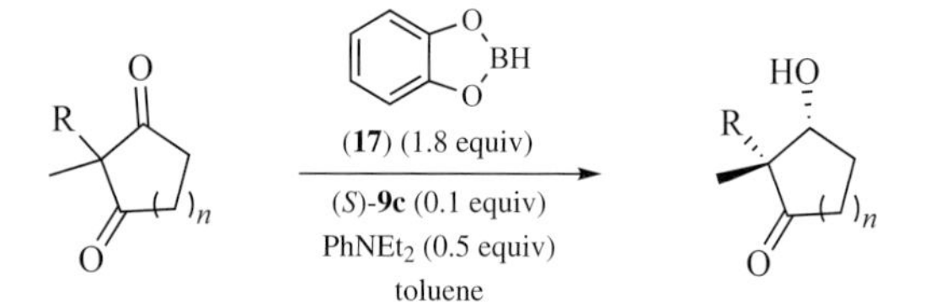

(16) (18)

n = 1, 2, 4, 92–94% ee
n = 3, 76% ee

(13)

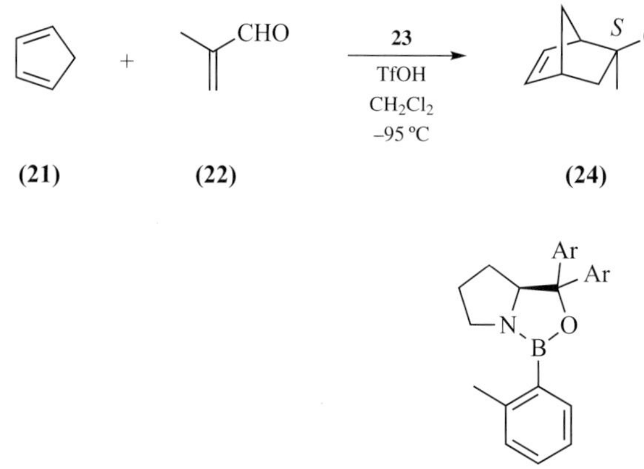

(21) (22) (24)

(23)

(15)

Ar = Ph, 91%, *exo*:*endo* = 89:11, 90% ee
Ar = 3,5-Me_2-Ph, 99%, *exo*:*endo* = 89:11, 96% ee

The preparation of 1,4- and 1,5-diols (**20**) was also achieved by highly enantioselective reduction of 1,4- and 1,5-diones (**19**) (eq 14). Essentially slow addition of the dione to a solution of *in situ* prepared boron-methoxy oxazaborolidine derived from α,α-diphenylprolinol and borane dimethyl sulfide complex afforded the desired diol in excellent yield and with high stereoselectivity.[43]

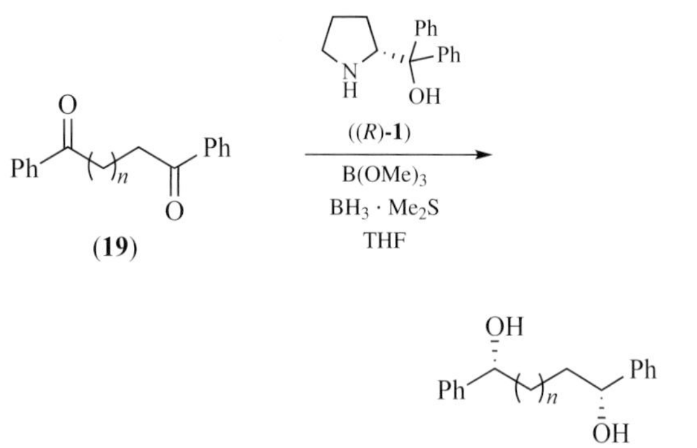

(19) (20)

(14)

n = 1, 96%, 95% de, 99% ee
n = 2, 91%, 96% de, 98% ee

It is noteworthy that the CBS-mediated reduction reaction has also been applied to the kinetic resolution of racemic lactones.[44]

Cycloaddition Reactions. On the basis of CBS catalytic systems, Corey pioneered a new type of catalytic enantioselective Diels–Alder (DA) reaction and a series of DA variants have been developed. In the initial study, they employed oxazaborolidine as precatalyst and triflic acid as activator to generate a potent, real cationic Lewis acid catalyst for the DA reaction between cyclopentadiene (**21**) and 2-methylacrolein (**22**) (eq 15).[45] Notably, the system was extended to 2-bromoacrolein and a variety of 1,3-dienes in excellent yields and with high enantioselectivities. Moreover, subsequent studies demonstrated that an even broader range of the dienophile components, including α,β-unsaturated esters, lactones, and ketones could be applied.[46] It was found that the catalyst complex decomposed at an appreciable rate at or above 0°C. However, the change of the counterion from triflate $CF_3SO_3^-$ to triflimide $(CF_3SO_2)_2N^-$ had a remarkably beneficial effect on catalyst stability, resulting in no loss of catalytic activity at a higher temperature.[47,48]

In their continuing effort, Snyder and Corey also successfully applied this catalytic system for the first intramolecular enantioselective Diels–Alder reaction in the efficient construction of the scaffold of bioactive dolabellane-type marine natural products, α-araneosene, dolabellatrienone, and palominol.[49] The same group also reported a novel enantioselective [3+2] cycloaddition of 1,4-benzoquinone (**25**) with vinyl ether (**26**), catalyzed by chiral oxazaborolidinium ion (**23**) (eq 16).[50]

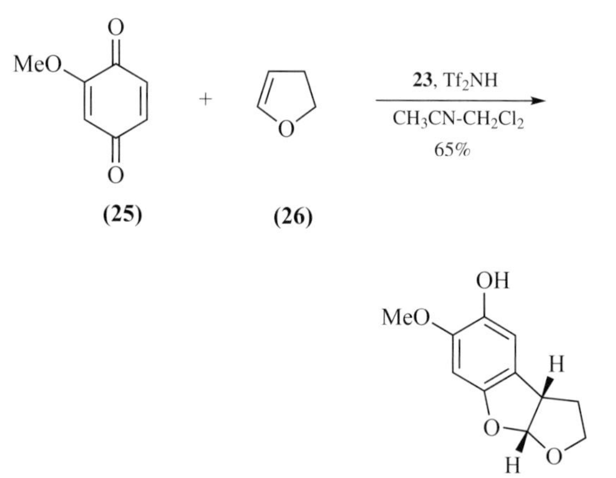

(25) (26) (27)

(16)

92% ee
99% ee, recrystallized

In addition to the complexation of α,α-diphenylprolinol with Lewis acids to form effective catalysts, recently, the ligand itself has been found as a chiral organocatalyst for reactions of enolizable aldehydes by using its "N" basicity to form active enamine species. The resulting chiral nucleophilic species subsequently reacted with electrophiles for new bond formation. As shown, Juhl and Jøgensen developed an organocatalytic asymmetric hetero-Diels–Alder reaction of aldehyde (**28**) with enone (**29**) for the first time (eq 17).[51]

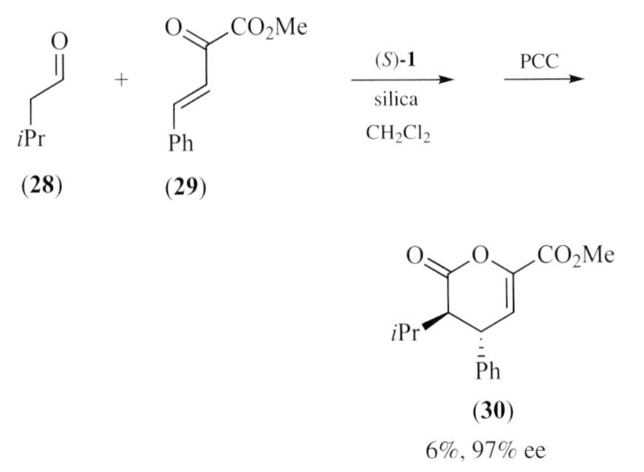

(17)

Furthermore, the chiral secondary amine could also react with an enal to form an iminium species by decreasing the energy of the lowest unoccupied molecular orbital (LUMO) of the dienophile. The activated dienophile could undergo [3 + 2] cycloaddition reactions. Vicario and coworkers demonstrated the possibility between enals (**31**) and azomethine ylides (**32**) with complete control of regioselectivity and high diastereo- and enantioselectivities in good yields (eq 18).[52,53]

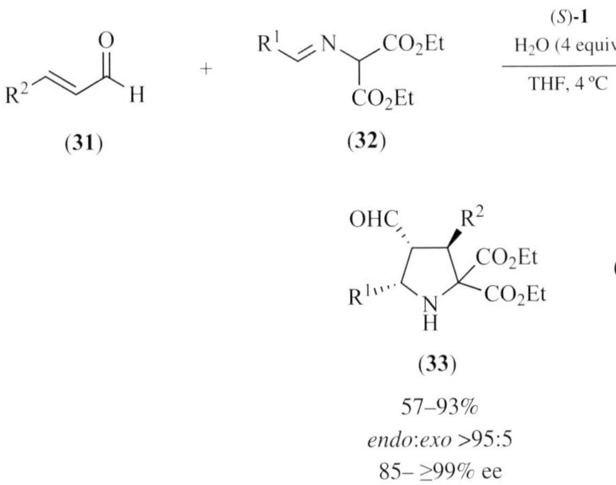

(18)

Michael Addition Reactions. The chiral cationic oxazaborolidinium catalysts, which have been extensively applied for enantioselective Diels–Alder reactions across a broad range of substrates, were also found to be effective for an enantioselective Michael addition reaction.[54] The study was carried out with 2-cyclohexenone (**36**) and 1-methoxy-2-methyl-1-(trimethylsilyloxy) (**35**) (eq 19). It was found that the reaction proceeded with high efficiency and enantioselectively when conducted in toluene solution at −20°C in the presence of 0.25 equiv of triphenylphosphine oxide as a trap for electrophilic Me_3Si species to give the desired Michael adduct. With the catalyst *S*-**28**, Michael adduct *S*-**37** was obtained in 91% yield and with about 20:1 enantioselection (90% ee). Similarly, the catalyst ent-*R*-**23** provided the Michael adduct *R*-**34** with the same efficiency and enantioselectivity.

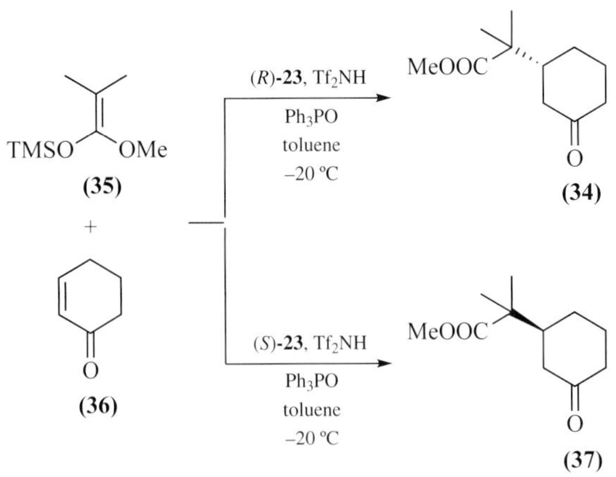

(19)

When this new enantioselective process was extended to ketene acetal (**35**) and 2-cycloheptenone (**38**), lower enantioselectivity and yield were obtained (eq 20). However, when 1 equiv of 2,6-diisopropylphenol (DIPP) was added as an additional scavenger of electrophilic Me_3Si species, the reaction proceeded well to give the Michael adduct (**39**) in 99% yield and with 99% ee.

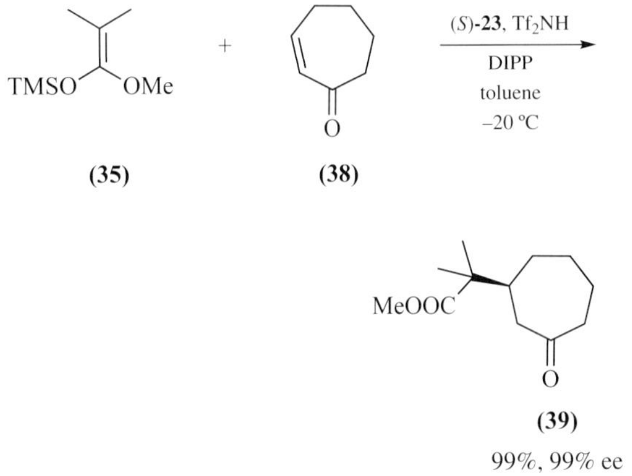

(20)

The direct enantioselective Michael addition of simple aldehydes such as (**40**) to vinyl ketones (**41**) could also be realized using (*S*) α,α-diphenylprolinol **1** through enamine catalysis (eq 21).[55]

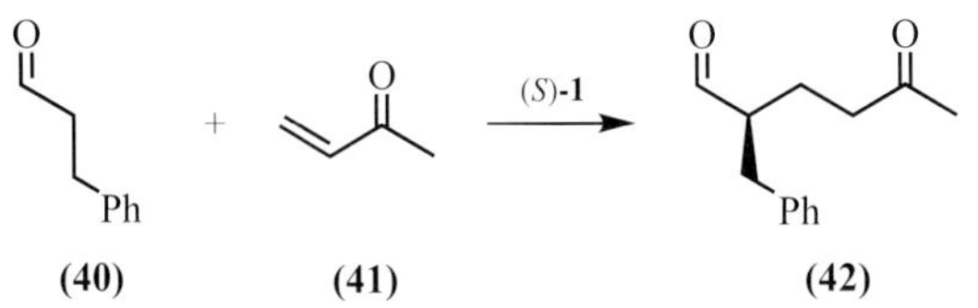

(21)

(*S*)-α,α-Diphenylprolinol (**1**) was also used to catalyze the Michael reaction of aldehydes and nitrostyrenes.[56] Although excellent enantioselectivity (up to 95% ee) was observed, the process was slow and the yield was unsatisfactory (29%) even after 24 h.

In addition, chiral α,α-diarylprolinol salts were found to be effective iminium organocatalysts for the enantioselective direct vinylogous Michael addition of α,α-dicyanoolefins (**43**) to α,β-unsaturated aldehydes (**44**) (eq 22).[57] The catalyst (*S*)-**1** exhibited excellent catalytic activity in combination with PNBA. The scope of this conjugate addition reaction proved to be general with respect to both the α,α-dicyanoolefins and β-substitution on the electrophiles with high level of ee values.

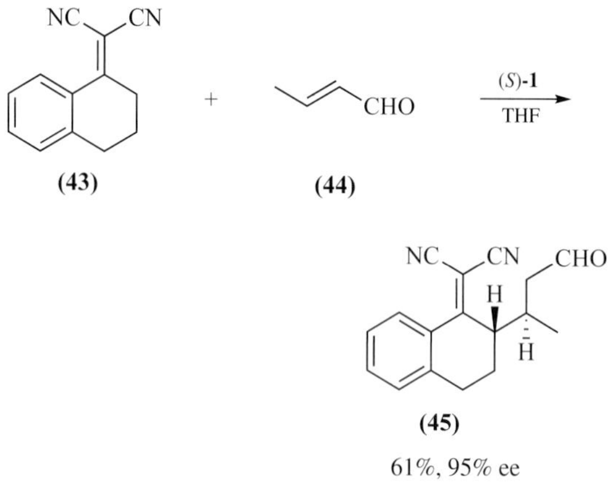

(22)

Shi and coworkers reported an interesting dual activation approach, where Lewis base (LB) activation of nitroalkenes (**46**) and carbonyl activation via an iminium ionoccur, for the enantioselective nitroalkene/enal cross-coupling mediated by (*S*)-α,α-diphenylprolinol **1** (eq 23).[58] It is noted that mechanistic studies revealed that the secondary amine served as the LB catalyst in a 1,4-addition fashion to the nitroalkenes to form nucleophilic species for subsequent addition to the carbonyl group in enals (**47**), a type of Morita–Baylis–Hillman pathway. The use of cat-**1** at lower temperatures gave an improved enantioselectivity; however, further lowering of the reaction temperature caused a significant decrease in the reaction rate. When $P(OMe)_3$ was used as cocatalyst, both the reaction rate and the enantioselectivity were enhanced with the addition of AcOH for facilitating the iminium formation, leading to the products in excellent yields and with high enantioselectivities.

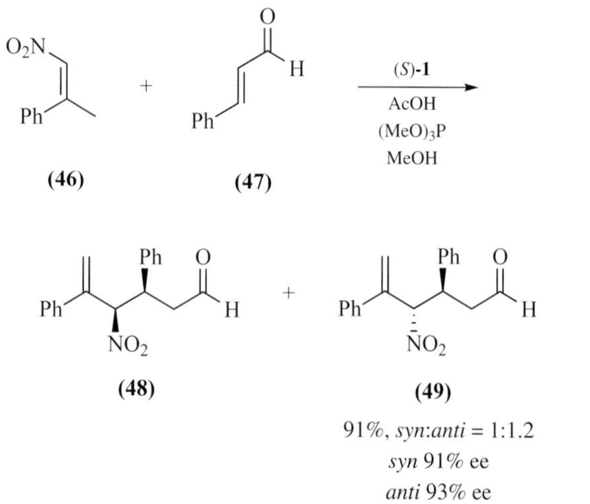

(23)

Epoxidation Reactions. It was found that α,α-diphenylprolinol (**1**) could also be exploited as bifunctional catalyst for the enantioselective epoxidation of α,β-enones (**50**) (eq 24).[59,60] The reaction was operationally simple and did not require inert atmospheres or temperature manipulation, and the epoxides were readily isolated by flash chromatography without further workup with good enantioselectivity. With modified prolinol as the organocatalyst, improved enantioselectivities (up to 96% ee) were obtained.[61]

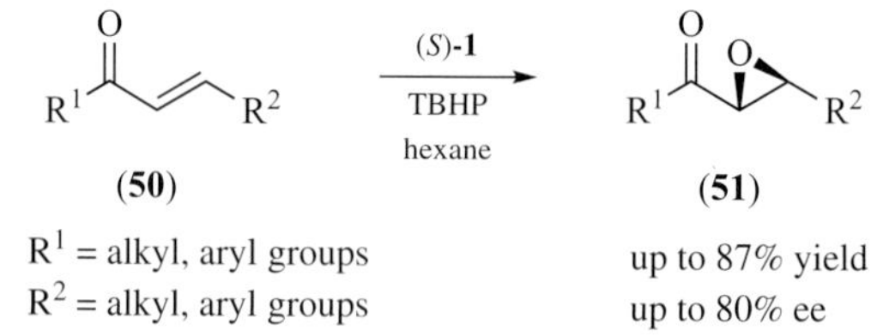

(24)

In situ asymmetric epoxidation of styrenes by molybdenum complexes with (*S*)-α,α-diphenylprolinol (**1**) as ligand was reported with moderate to good ee values up to 84%.[62]

Copolymerization Reactions. Nozaki reported the first example of asymmetric copolymerization of *meso* epoxide with CO_2, initiated by a chiral Zn-α,α-diphenylprolinol catalyst (eq 25).[63,64] (*S*)-α,α-Diphenylprolinol (**1**) or its C_2-symmetric bis-β-amino alcohols were also found to catalyze the enantioselective addition of zinc reagents to aldehydes with high enantioselectivities.[65,66]

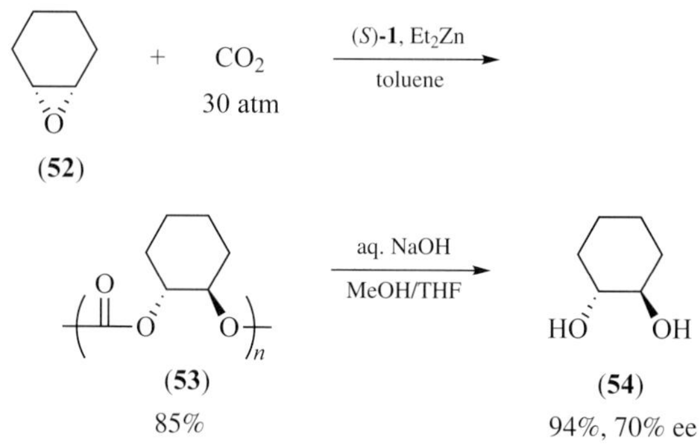

(25)

Related Reagents. 2-Amino-3-methyl-1,1-diphenyl-1-butanol; Ephedrine-borane; Norephedrine–Borane; Tetrahydro-1-methyl-3,3-diphenyl-1*H*,3*H*-pyrrolo[1,2-*c*][1,3,2]oxazaborole.

1. (a) Wallbaum, S.; Martens, J., *Tetrahedron: Asymmetry* **1992**, *3*, 1475. (b) Singh, V. K., *Synthesis* **1992**, 607. (c) Deloux, L.; Srebnik, M., *Chem. Rev.* **1993**, *93*, 763.
2. Kraatz, U. Ger. Patent 3 609 152, 1986 (*Chem. Abstr.* **1978**, *108*, 56 111c).
3. (a) Corey, E. J.; Bakshi, R. K.; Shibata, S., *J. Am. Chem. Soc.* **1987**, *109*, 5551. (b) Corey, E. J.; Bakshi, R. K.; Shibata, S.; Chen, C. P.; Singh, V. K., *J. Am. Chem. Soc.* **1987**, *109*, 7925. (c) Corey, E. J.; Shibata, S.; Bakshi, R. K., *J. Org. Chem.* **1988**, *53*, 2861.
4. (a) Mathre, D. J.; Jones, T. K.; Xavier, L. C.; Blacklock, T. J.; Reamer, R. A.; Mohan, J. J.; Jones, E. T. T.; Hoogsteen, K.; Baum, M. W.; Grabowski, E. J. J., *J. Org. Chem.* **1991**, *56*, 751. (b) Blacklock, T. J.. Jones, T. K.; Mathre, D. J.; Xavier, L. C. U.S. Patent 5 039 802, 1991. (c) Blacklock, T. J.; Jones, T. K.; Mathre, D. J.; Xavier, L. C. U.S. Patent 5 264 585, 1993.

5. Enders, D.; Kipphardt, H.; Gerdes, P.; Brena-Valle, L. J.; Bhushan, V., *Bull. Soc. Chim. Belg.* **1988**, *97*, 691.
6. Kanth, J. V. B.; Periasamy, M., *Tetrahedron* **1993**, *49*, 5127.
7. (a) Kapfhammer, J.; Matthes, A. *Hoppe-Seylers, Z., Physiol. Chem.* **1933**, *223*, 43. (b) Roussel-Uclaf Fr. Patent 3638M (*Chem. Abstr.* **1969**, *70*, 106 375m).
8. Kerrick, S. T.; Beak, P., *J. Am. Chem. Soc.* **1991**, *113*, 9708.
9. (a) Seebach, D.; Enders, D.; Renger, B., *Chem. Ber.* **1977**, *110*, 1852. (b) Enders, D.; Pieter, R.; Renger, B. Seebach, D., *Org. Synth., Coll. Vol.* **1988**, *6*, 542.
10. (a) Itsuno, S.; Ito, K.; Hirao, A.; Nakahama, S., *Chem. Commun.* **1983**, 469. (b) Itsuno, S.; Hirao, A.; Nakahama, S.; Yamazaki, N., *J. Chem. Soc., Perkin Trans. 1* **1983**, 1673. (c) Itsuno, S.; Ito, K.; Hirao, A.; Nakahama, S., *J. Org. Chem.* **1984**, *49*, 555. (d) Itsuno, S.; Nakano, M.; Miyazaki, K.; Masuda, H.; Ito, K.; Hirao, A.; Nakahama, S., *J. Chem. Soc., Perkin Trans. 1* **1985**, 2039. (e) Itsuno, S.; Nakano, M.; Ito, K.; Hirao, A.; Owa, M.; Kanda, N.; Nakahama, S., *J. Chem. Soc., Perkin Trans. 1* **1985**, 2615.
11. Evans, D. A., *Science* **1988**, *240*, 420.
12. Jones, D. K.; Liotta, D. C.; Shinkai, I.; Mathre, D. J., *J. Org. Chem.* **1993**, *58*, 799.
13. Corey, E. J.; Bakshi, R. K., *Tetrahedron Lett.* **1990**, *31*, 611.
14. (a) Jones, T. K.; Mohan, J. J.; Xavier, L. C.; Blacklock, T. J.; Mathre, D. J.; Sohar, P.; Jones, E. T. T.; Reamer, R. A.; Roberts, F. E.; Grabowski, E. J. J., *J. Org. Chem.* **1991**, *56*, 763. (b) Shinkai, I., *J. Heterocycl. Chem.* **1992**, *29*, 627.
15. Corey, E. J.; Link, J. O., *Tetrahedron Lett.* **1992**, *33*, 4141.
16. (a) Mathre, D. J.; Thompson, A. S.; Douglas, A. W.; Hoogsteen, K.; Carroll, J. D.; Corley, E. G.; Grabowski, E. J. J., *J. Org. Chem.* **1993**, *58*, 2880. (b) Blacklock, T. J.; Jones, T. K.; Mathre, D. J.; Xavier, L. C. U. S. Patent 5 189 177, **1993**. (c) Carroll, J. D.; Mathre, D. J.; Corley, E. G.; Thompson, A. S. U. S. Patent 5 264 574, **1993**.
17. Corey, E. J.; Azimioara, M.; Sarshar, S., *Tetrahedron Lett.* **1992**, *33*, 3429.
18. Corey, E. J.; Gavai, A. V., *Tetrahedron Lett.* **1988**, *29*, 3201.
19. Corey, E. J.; Jardine, P. D. S.; Mohri, T., *Tetrahedron Lett.* **1988**, *29*, 6409.
20. Corey, E. J.; Reichard, G. A., *Tetrahedron Lett.* **1989**, *30*, 5207.
21. Corey, E. J.; Link, J. O., *Tetrahedron Lett.* **1990**, *31*, 601.
22. (a) Kabat, M.; Kiegiel, J.; Cohen, N.; Toth, K.; Wovkulich, P. M.; Uskokovic, M. R., *Tetrahedron Lett.* **1991**, *32*, 2343. (b) Lee, A. S.; Norman, A. W.; Okamura, W. H., *J. Org. Chem.* **1992**, *57*, 3846.
23. DeNinno, M. P.; Perner, R. J.; Morton, H. E.; DiDomenico, S. Jr., *J. Org. Chem.* **1992**, *57*, 7115.
24. Nicolaou, K. C.; Hwang, C.-K.; Sorensen, E. J.; Clairborne, C. F., *Chem. Commun.* **1992**, 1117.
25. (a) Labelle, M.; Prasit, P.; Belley, M.; Blouin, M.; Champion, E.; Charette, L.; DeLuca, J. G.; Dufresne, C.; Frenette, R.; Gauthier, J. Y.; Grimm, E.; Grossman, S. J.; Guay, D.; Herold, E. G.; Jones, T. R.; Lau, Y.; Leblanc, Y.; Leger, S.; Lord, A.; McAuliffe, M.; McFarlane, C.; Masson, P.; Metters, K. M.; Ouimet, N.; Patrick, D. H.; Perrier, H.; Piechuta, H.; Roy, P.; Williams, H.; Wang, Z.; Xiang, Y. B.; Zamboni, R. J.; Ford-Hutchinson, A. W.; Young, R. N., *Bioorg. Med. Chem. Lett.* **1992**, *2*, 1141. (b) King, A. O.; Corley, E. G.; Anderson, R. K.; Larsen, R. D.; Verhoeven, T. R.; Reider, P. J.; Xiang, Y. B.; Belley, M.; Leblanc, Y.; Labelle, M.; Prasit, P.; Zamboni, R. J., *J. Org. Chem.* **1993**, *58*, 3731.
26. Bloom, J. D.; Dutia, M. D.; Johnson, B. D.; Wissner, A.; Burns, M. G.; Largis, E. E.; Dolan, J. A.; Claus, T. H., *J. Med. Chem.* **1992**, *35*, 3081.
27. Cai, D.; Tschaen, D.; Shi, Y.-J.; Verhoeven, T. R.; Reamer, R. A.; Douglas, A. W., *Tetrahedron Lett.* **1993**, *34*, 3243.
28. Bringmann, G.; Hartung, T., *Angew. Chem., Int. Ed.* **1992**, *31*, 761.
29. Corey, E. J.; Helal, C. J., *Angew. Chem., Int. Ed.* **1998**, *37*, 1986.
30. Helal, C. J.; Magriotis, P. A.; Corey, E. J., *J. Am. Chem. Soc.* **1996**, *118*, 10938.
31. Yu, C.-M.; Kim, C.; Kweon, J.-H., *Chem. Commun.* **2004**, 2494.
32. Harb, W.; Ruiz-Lopez, M. F.; Coutrot, F.; Grison, C.; Coutrot, P., *J. Am. Chem. Soc.* **2004**, *126*, 6996.
33. Korenaga, T.; Nomura, K.; Onoue, K.; Sakai, T., *Chem. Commun.* **2010**, 8624.
34. Corey, E. J.; Guzman-Perez, A.; Lazerwith, S. E., *J. Am. Chem. Soc.* **1997**, *119*, 11769.
35. Nicolaou, K. C.; Safina, B. S.; Funke, C.; Zak, M.; Zecri, F. J., *Angew. Chem., Int. Ed.* **2002**, *41*, 1937.
36. Trost, B. M.; Gunzner, J. L.; Dirat, O.; Rhee, Y. H., *J. Am. Chem. Soc.* **2002**, *124*, 10396.
37. Williams, D. R.; Heidebrecht, R. W., Jr., *J. Am. Chem. Soc.* **2003**, *125*, 1843.
38. Herzon, S. B.; Myers, A. G., *J. Am. Chem. Soc.* **2005**, *127*, 5342.
39. Evans, D. A.; Kvaernø, L.; Dunn, T. B.; Beauchemin, A.; Raymer, B.; Mulder, J. A.; Olhava, E. J.; Juhl, M.; Kagechika, K.; Favor, D. A., *J. Am. Chem. Soc.* **2008**, *130*, 16295.
40. Prasad, K. R. K.; Joshi, N. N., *J. Org. Chem.* **1996**, *61*, 3888.
41. Watanabe, H.; Iwamoto, M.; Nakada, M., *J. Org. Chem.* **2005**, *70*, 4652.
42. Yeung, Y.-Y.; Chein, R.-J.; Corey, E. J., *J. Am. Chem. Soc.* **2007**, *129*, 10346.
43. Aldous, D. J.; Duttona, W. M.; Steel, P. G., *Tetrahedron: Asymmetry* **2000**, *11*, 2455.
44. Bringmann, G.; Pabst, T.; Henschel, P.; Kraus, J.; Peters, K.; Peters, E.-M.; Rycroft, D. S.; Connolly, J. D., *J. Am. Chem. Soc.* **2000**, *122*, 9127.
45. Corey, E. J.; Shibata, T.; Lee, T. W., *J. Am. Chem. Soc.* **2002**, *124*, 3808.
46. Ryu, D. H.; Lee, T. W.; Corey, E. J., *J. Am. Chem. Soc.* **2002**, *124*, 9992.
47. Ryu, D. H.; Core, E. J., *J. Am. Chem. Soc.* **2003**, *125*, 6388.
48. Zhou, G.; Hu, Q.-Y.; Corey, E. J., *Org. Lett.* **2003**, *5*, 3979.
49. Snyder, S. A.; Corey, E. J., *J. Am. Chem. Soc.* **2006**, *128*, 740.
50. Zhou, G.; Corey, E. J., *J. Am. Chem. Soc.* **2005**, *127*, 11958.
51. Juhl, K.; Jøgensen, K. A., *Angew. Chem., Int. Ed.* **2003**, *42*, 1498.
52. Vicario, J. L.; Reboredo, S.; Bada, D.; Carrillo, L., *Angew. Chem., Int. Ed.* **2007**, *46*, 5168.
53. Iza, A.; Carrillo, L.; Vicario, J. L.; Badia, D.; Reyes, E.; Martinez, J. I., *Org. Biomol. Chem.* **2010**, *8*, 2238.
54. Liu, D.; Hong, S.; Corey, E. J., *J. Am. Chem. Soc.* **2006**, *128*, 8160.
55. Melchiorre, P.; Jørgensen, K. A., *J. Org. Chem.* **2003**, *68*, 4151.
56. Hayashi, Y.; Gotoh, H.; Hayashi, T.; Shoji, M., *Angew. Chem., Int. Ed.* **2005**, *44*, 4212.
57. Xie, J-W.; Yue, L.; Xue, D.; Ma, X.-L.; Chen, Y.-C.; Wu, Y.; Zhu, J.; Deng, J.-G., *Chem. Commun.* **2006**, 1563.
58. Zhong, C.; Chen, Y.; Petersen, J. L.; Akhmedov, N. G.; Shi, X., *Angew. Chem., Int. Ed.* **2009**, *48*, 1279.
59. Lattanzi, A., *Org. Lett.* **2005**, *7*, 2579.
60. Lattanzi, A.; Russo, A., *Tetrahedron* **2006**, *62*, 12264.
61. Li, Y.; Liu, X.; Yang, Y.; Zhao, G., *J. Org. Chem.* **2007**, *72*, 288.
62. Wang, Y.; Wu, Z.; Li, Z.; Zhou, X.-G., *Tetrahedron Lett.* **2009**, *50*, 2509.
63. Nozaki, K.; Nakano, K.; Hiyama, T., *J. Am. Chem. Soc.* **1999**, *121*, 11008.
64. Nakano, K.; Nozaki, K.; Hiyama, T., *J. Am. Chem. Soc.* **2003**, *125*, 5501.
65. Long, J.; Ding, K., *Angew. Chem., Int. Ed.* **2001**, *40*, 544.
66. He, A.; Falck, J. R., *Angew. Chem., Int. Ed.* **2008**, *47*, 6586.

Dispiro[2*H*-pyran-2,4′-[4*H*-5,6,8b]-triazaacenaphthylene-7′(5′*H*),2″-[2*H*]-pyran], 1′,2′,2′a,3,3′,3″,4,4″,5,5″,6,6″,8′, 8′a-tetradecahydro-1′,2′-dimethoxy-6, 6″-dimethyl-, monohydrochloride, (1′*S*,2*R*,2′*S*,2″*R*,2′a*S*,6*S*,6″S,8′a*S*)-

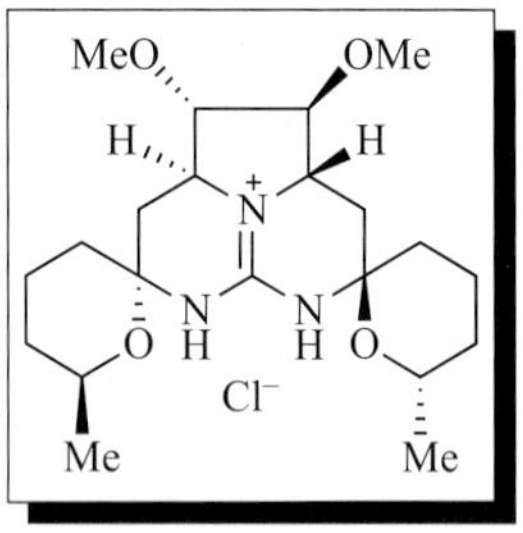

[393109-20-1] $C_{21}H_{35}N_3O_4{\cdot}ClH$ (MW 429.99)

InChI = S/C21H35N3O4.ClH/c1-13-7-5-9-20(27-13)11-15-17(25-3)18(26-4)16-12-21(10-6-8-14(2)28-21)23-19(22-20)24(15)16;/h13-18H,5-12H2,1-4H3,(H,22,23);1H/t13-,14-,15+,16+,17+,18+,20-,21-;/m1./s1

InChIKey = NNRYVRZHVDWGAK-WAKVXNIGSA-N

Dispiro[2*H*-pyran-2,4′-[4*H*-5,6,8b]triazaacenaphthylene-7′(5′*H*),2″-[2*H*]pyran], 1′,2′,2′a,3,3′,3″,4,4″,5,5″,6,6″,8′,8′a-tetradecahydro-1′,2′-dimethoxy-6,6″-dimethyl-, monohydrochloride, (1′*S*,2*R*,2′*S*,2″*R*,2′a*S*,6*R*,6″*R*,8′a*S*)-

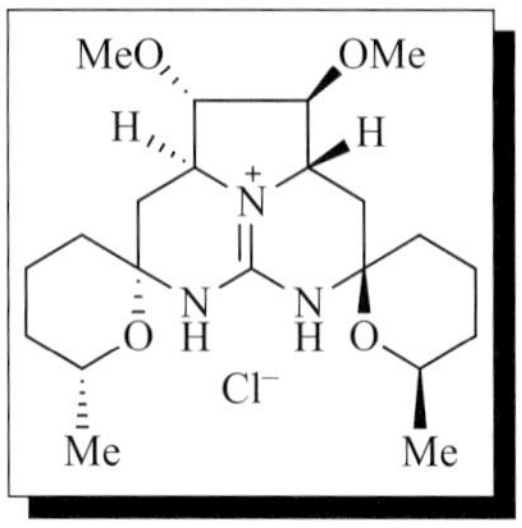

[393109-05-2] $C_{21}H_{35}N_3O_4{\cdot}ClH$ (MW 429.99)

InChI = 1S/C21H35N3O4.ClH/c1-13-7-5-9-20(27-13)11-15-17(25-3)18(26-4)16-12-21(10-6-8-14(2)28-21)23-19(22-20)24(15)16;/h13-18H,5-12H2,1-4H3,(H,22,23);1H/t13-,14-,15-,16-,17-,18-,20+,21+;/m0./s1

InChIKey = NNRYVRZHVDWGAK-YUKWCZSMSA-N

(C_2-symmetric chiral pentacyclic guanidine-based organocatalysts)

Physical Data: for [393109-20-1]: white crystal, mp 189–190 °C (decomposition). $[\alpha]_D^{24} = +92.5$ (*c* 0.6, $CHCl_3$); for [393109-05-2]: white crystal, mp 245–246 °C (decomposition). $[\alpha]_D^{24} = +148.7$ (*c* 1.2, $CHCl_3$).

Solubility: most organic solvents for both 393109-20-1 and 393109-05-2.

Form Supplied in: no commercial sources available.

Analysis of Reagent Purity: determined by ^{1}H NMR.

Preparative Method: Prepared from the optically active nitrone via a six-step sequence.[1]

Handling, Storage, and Precautions: no information available.

As Catalysts to Accelerate the Hetero–Michael Reaction. X-ray crystallographic analysis has revealed that pentacyclic guanidine **393109-05-2** has a closed-type cavity around the guanidine function group, while the analog **393109-20-1** has an open-type cavity. Both pentacyclic guanidines catalytically accelerated the hetero–Michael reaction of pyrrolidine with γ-crotonolactone in 2.3 and 8.3-fold increase in rate, respectively, over the uncatalyzed reaction, however, no asymmetric inductions were observed (eq 1).[1]

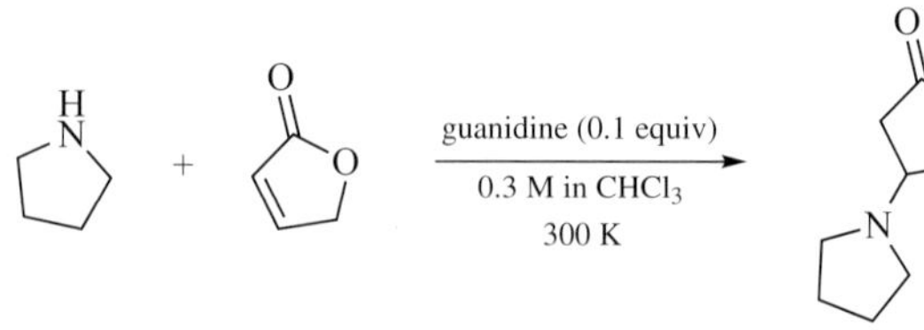

(1)

As Phase-Transfer Catalysts for the Asymmetric Alkylation of *tert*-Butyl Glycinate Schiff Base. Both pentacyclic guanidines **393109-05-2** and **393109-20-1** have a C_2-symmetric chiral reaction cavity around the guanidine group, a site for substrate recognition and activation. The closed-type cavity inside PTC **393109-05-2** induces high asymmetry with 81% ee, while open-type catalyst **393109-20-1** only provides 12% ee (eq 2). The high enantioselectivity was obtained in the alkylation of the *tert*-butyl glycinate Schiff base with various alkyl halides (methyl iodide, octyl iodide, methallyl bromide, crotyl bromide, propargyl bromide, 4-nitrobenzyl bromide, 2-bromomethylnaphthalene) in the presence of PTC **393109-05-2** with enantiomeric excesses of 76–90% (eq 3).[2]

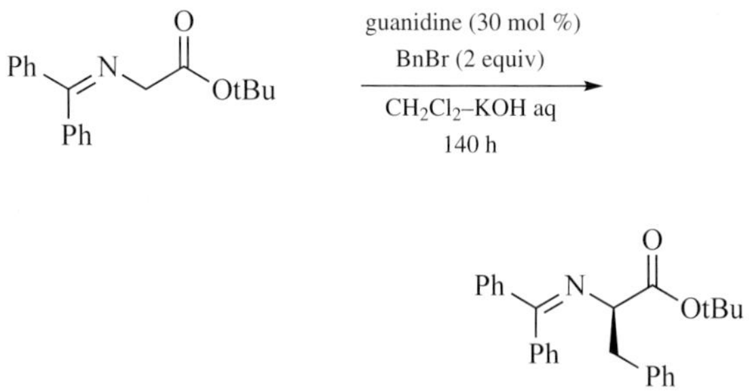

(2)

393109-05-2: RT, 90% yield, 81% ee
393109-20-1: 0 °C, 65% yield, 12% ee

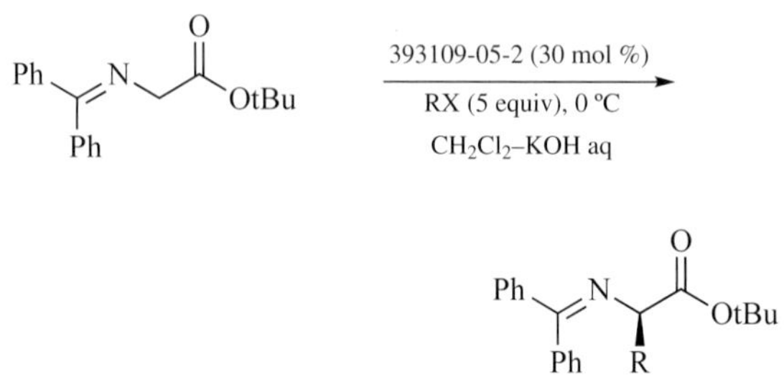

(3)

61–85% yield
76–90% ee

As Catalysts for Nucleophilic Asymmetric Epoxidation. The pentacyclic guanidine **393109-05-2** with a characteristic closed-type cavity catalyzes the asymmetric nucleophilic epoxidation of chalcone and its derivatives to form epoxides with 39–60% ee (eqs 4 and 5).[3]

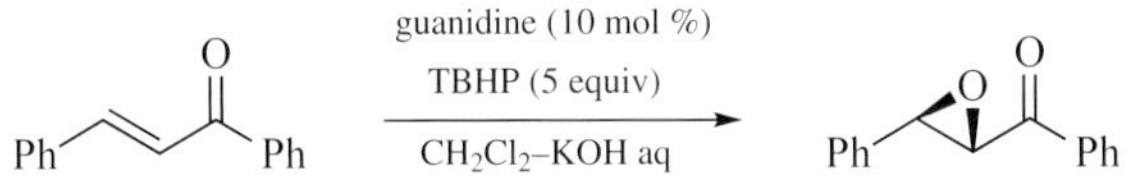

(4)

393109-05-2: 0 °C, 110 h, 35% yield, 39% ee
393109-20-1: 0 °C, 130 h, 95% yield, 10% ee

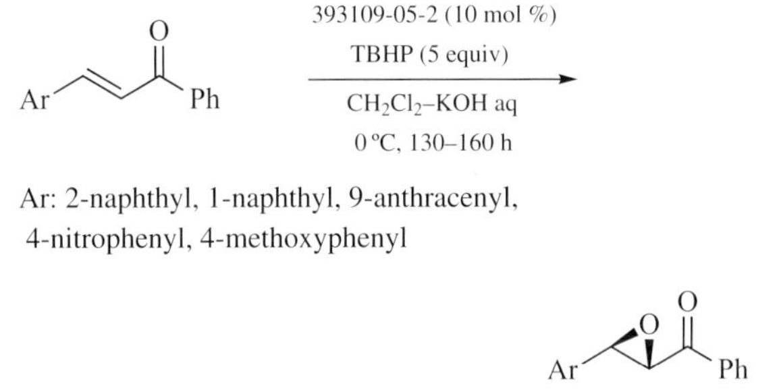

(5)

22–99% yield, 35–60% ee

1. (a) Nagasawa, K.; Georgieva, A.; Takahashi, H.; Nakata, T., *Tetrahedron* **2001**, *57*, 8959. (b) Nagasawa, K., *Yakugaku Kenkyu no Shinpo* **2004**, *20*, 63.
2. Kita, T.; Georgieva, A.; Hashimoto, Y.; Nakata, T.; Nagasawa, K., *Angew. Chem., Int. Ed.* **2002**, *41*, 2832.
3. Kita, T.; Shi, B.; Hashimoto, Y.; Nagasawa, K., *Heterocycles* **2007**, *73*, 241.

Yuzhong Chen
DuPont Crop Protection, Newark, DE, USA

N-[(4-Dodecylphenyl)sulfonyl]-2-pyrrolidinecarboxamide (Hua Cat)

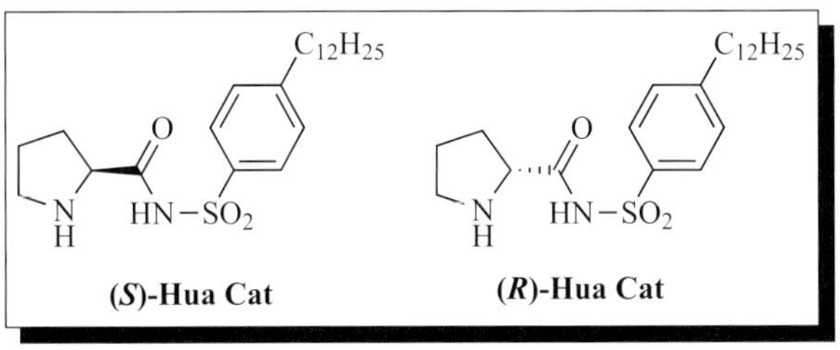

[1068139-38-7]
[1134192-85-0] $C_{23}H_{38}N_2O_3S$ (MW 422.62)

N-[(4-Dodecylphenyl)sulfonyl]-2-pyrrolidinecarboxamide (Hua Cat)

InChI = 1S/C23H38N2O3S/c1-2-3-4-5-6-7-8-9-10-11-13-20-15-17-21(18-16-20)29(27,28)25-23(26)22-14-12-19-24-22/h15-18,22,24H,2-14,19H2,1H3,(H,25,26)/t22-/m0/s1

InChIKey = FAGYODAJNLXMRF-QFIPXVFZSA-N

InChI = 1S/C23H38N2O3S/c1-2-3-4-5-6-7-8-9-10-11-13-20-15-17-21(18-16-20)29(27,28)25-23(26)22-14-12-19-24-22/h15-18,22,24H,2-14,19H2,1H3,(H,25,26)/t22-/m1/s1

InChIKey = FAGYODAJNLXMRF-JOCHJYFZSA-N

(reagent used as an organocatalyst for asymmetric carbon–carbon bond-forming reactions)

Physical Data: **(*S*) isomer:** mp: 203–208 °C; $[\alpha]_D^{20}$ −110.2° (c = 0.5 in $CHCl_3$).
(*R*) isomer: mp: 201–206 °C; $[\alpha]_D^{20}$ +102.5° (c = 0.5 in $CHCl_3$).

Solubility: 300 mg mL^{-1} in CH_2Cl_2, soluble in most organic solvents.

Form Supplied in: white solid; commercially available from Sigma-Aldrich.

Purification: silica gel chromatography.

Handling, Storage, and Precautions: stored in glass or plastic bottles at room temperature.

Preparation. Hua Cat can be conveniently prepared in three steps from commercially available *p*-dodecylphenylsulfonyl chloride (eq 1).[1] The sulfonyl chloride can also be prepared from the corresponding sodium sulfonate salt by treatment with oxalyl chloride (3 equiv) or thionyl chloride at room temperature in quantitative yield.[2] Treatment of the sulfonyl chloride with ammonium hydroxide followed by coupling with Cbz-proline and Pd/C-catalyzed hydrogenation yields Hua Cat. The ratio of isomers of the dodecyl moiety (e.g., 1-dodecyl and 2-dodecyl) can vary depending on the source of the starting material; however, no meaningful variation in performance of the catalyst has been reported in organocatalyzed processes.

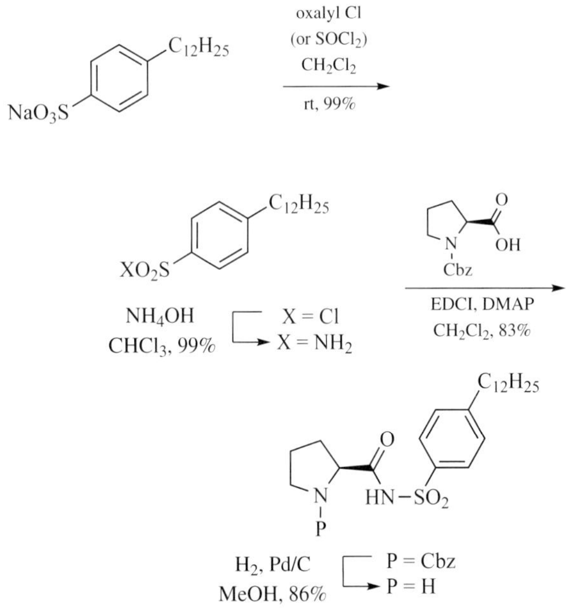

(1)

Aldol Reactions. Hua Cat has been shown to be a highly effective catalyst for the asymmetric aldol reactions.[1a,3] In most cases, these reactions are highly diastereoselective and provide products with excellent levels of enantiomeric excess. A range of reaction conditions can be employed including the use of a variety of organic solvents (e.g., DCE, CH_2Cl_2, and 2-Me-THF), aqueous conditions, or the absence of any solvent. The addition of 1 equiv of water is critical to the reaction rate and selectivity. A typical catalyst loading of 20 mol % is used in most cases; however, catalyst loadings as low as 2 mol % have been demonstrated using solvent-free conditions on a scale suitable to produce 1 mol of the aldol product (eq 2). 4-Substituted cyclohexanones and heterocyclic ketones have also proven effective in this transformation (eqs 3–5). Electron-rich aldehydes typically proceeded in reduced chemical yields. Acyclic ketones such as acetone have also been utilized in this process (eq 6).

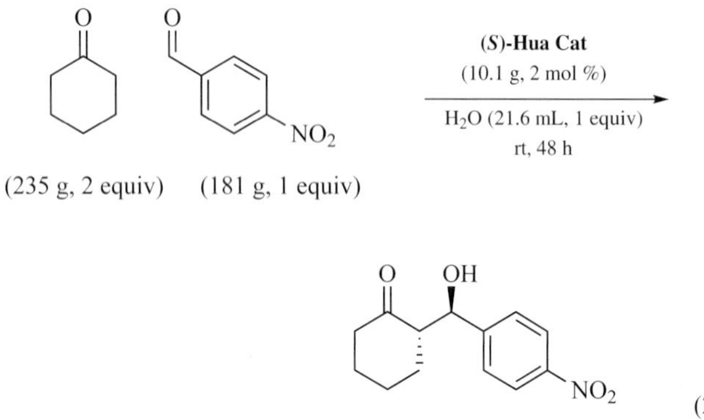

263 gm (1.06 mol) prepared
88% yield, 97% ee, 98:1 dr (2)

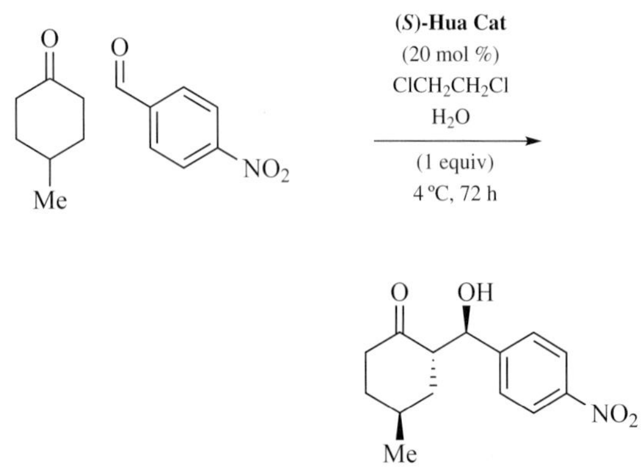

93% yield
49:1 dr, 97% ee (3)

Mannich Reactions. Hua Cat is also effective at mediating Mannich reactions to generate the *syn*-β-amino ketones with high stereoselectivity.[4] Cyclohexanone reacts with the PMP-imine of ethyl glycolate to produce the Mannich product in high yield and with excellent stereoselectivity (eq 7). The reaction proceeds equally well in DCE or 2-Me-THF. This process was performed on 49 g of cyclohexanone, 20.7 g of the imine and 15 mol% of Hua Cat in 2-Me-THF with comparable yields and selectivity with 92% recovery of Hua Cat. Acyclic ketones (e.g., acetone) can also be employed in this process (eq 8). The use of acetone as the solvent led to diminished enantioselectivity as compared to DCE. Propanal was effectively reacted with the imine of 3-substituted pyridine to produce the β-amino aldehyde in reasonable yield and stereoselectivity (eq 9). It should be noted that proline, in a side-by-side comparison of this transformation, gave diminished diastereoselectivity (57% yield, 1.2 : 1 dr, 99% ee).

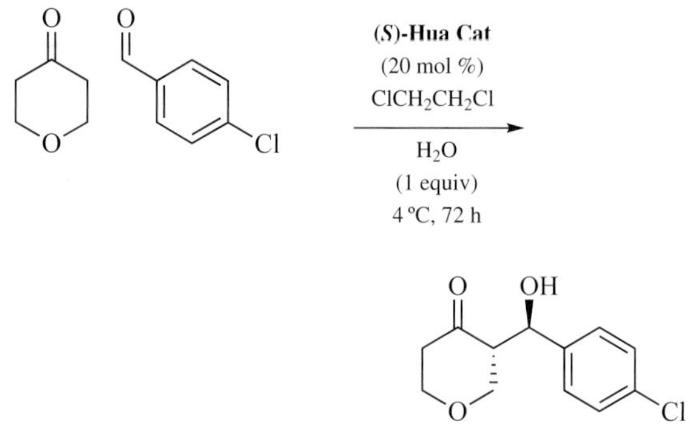

60% yield
>99:1 dr, 97% ee (4)

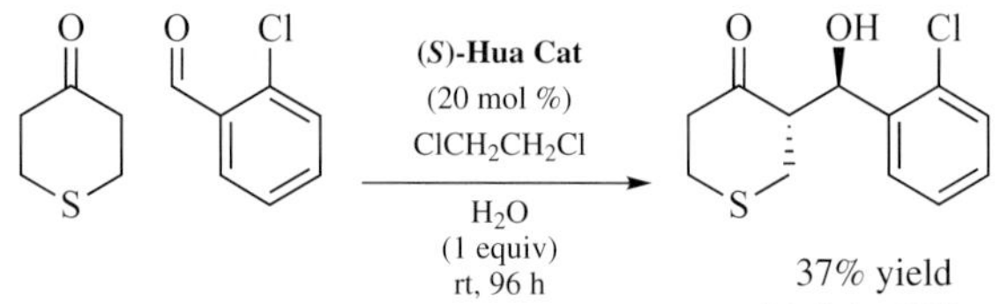

37% yield
>99:1 dr, 99% ee (5)

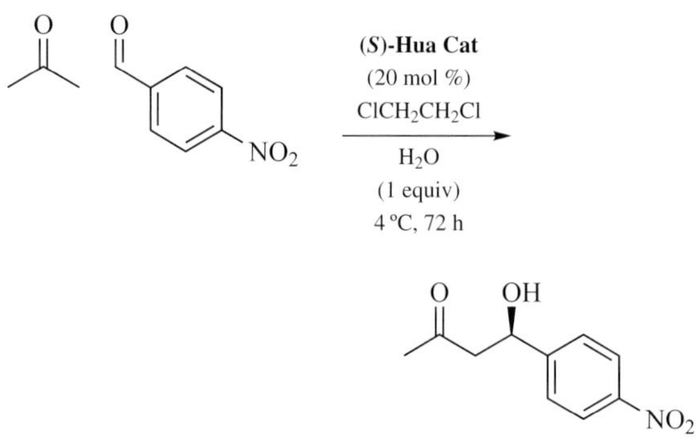

64% yield
83% ee (6)

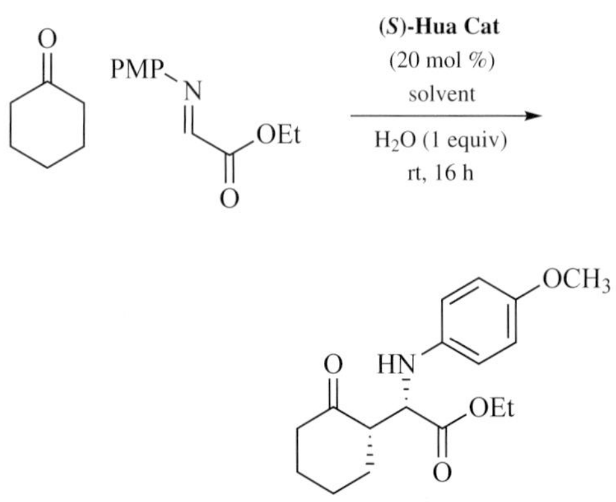

DCE: 93% yield, >20:1 dr, 96% ee[b]
2-Me-THF: 85% yield, >20:1 dr, 98% ee (7)

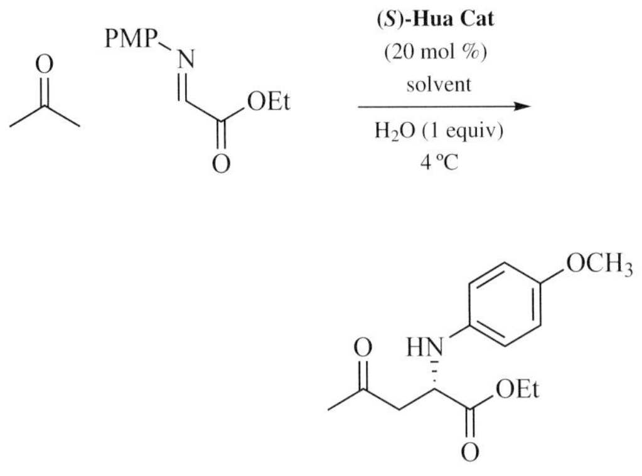

DCE (4 d): 62% yield, 92% ee[b]
Acetone (3 d): 64% yield, 75% ee (8)

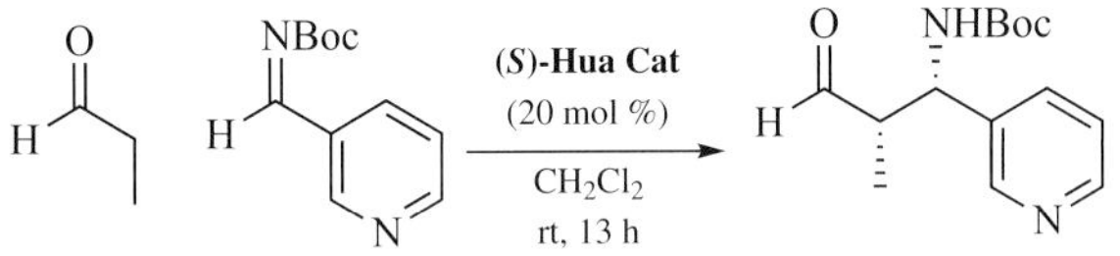

(9)

58% yield
15:1 dr, 99% ee

Formal [4+2] Aza-Diels–Alder Reactions. Construction of [2.2.2]-azabicycles is feasible utilizing neat conditions to provide the *exo*-cycloadduct in reasonable yields and with excellent stereoselectivity.[5] This *exo*-stereoadduct is the opposite of that produced using chiral Brønsted acid catalysis—thereby providing complementary and orthogonal reactivity. The use of electron-rich anilines for imine formation was key to optimum reactivity. Both 3,4-dimethoxyaniline-derived (eq 10) and 4-methoxyaniline-derived (eq 11) reactants were used in this process.

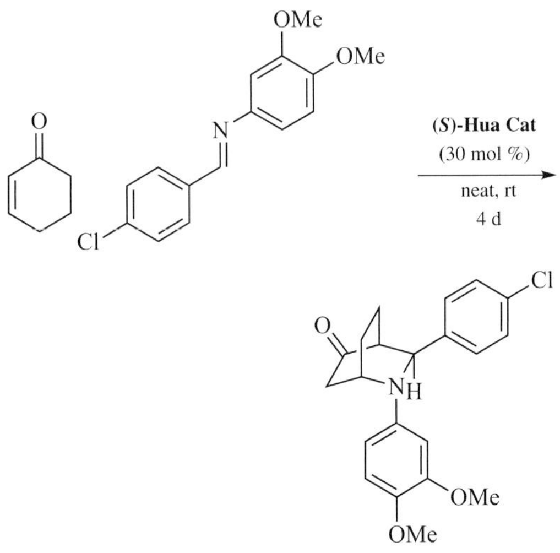

(10)

63% yield, 99% ee
>99:1 *exo:endo*

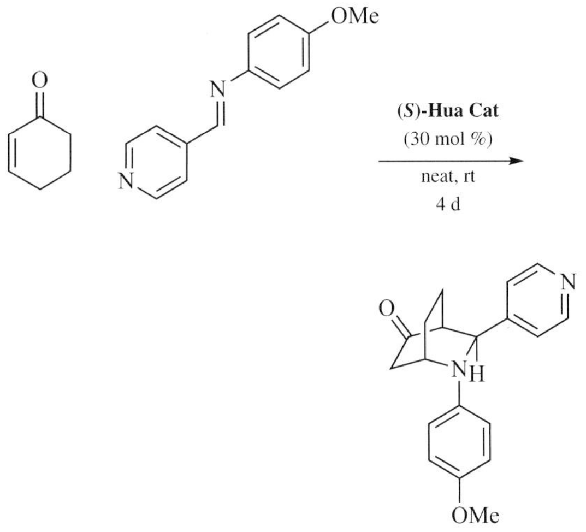

(11)

52% yield, 99% ee
>99:1 *exo:endo*

Synthesis of Bicyclo[2.2.2]octan-2-ones. Hua Cat can also be used to access [2.2.2] all-carbon bicycles through a tandem Michael/Mannich process. Preformed imines can be employed in this transformation (eqs 12 and 13) under neat conditions with cyclohexenone to provide access to bicycles containing up to four contiguous stereogenic centers (eq 13).[5] Alternatively, the aldehyde can be simply mixed with benzylamine for a short period of time prior to addition of the enone and Hua Cat. In either protocol, the formed imine is believed to be in slow equilibrium with its enamine tautomer. Benzylic aldehydes proved to be particularly effective substrates in this transformation (eq 14).[6] The imine derived from condensation between the benzylic aldehyde and the benzylamine is more readily able to tautomerize to the corresponding enamine due to the increased acidity of the neighboring α-protons.

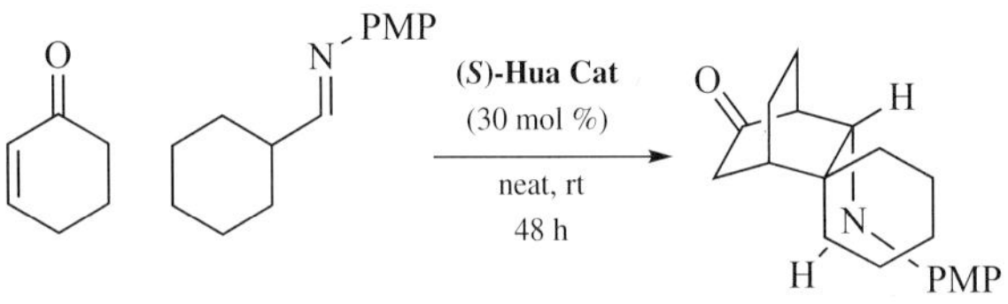

(12)

55% yield, 88% ee
>20:1 *endo:exo*

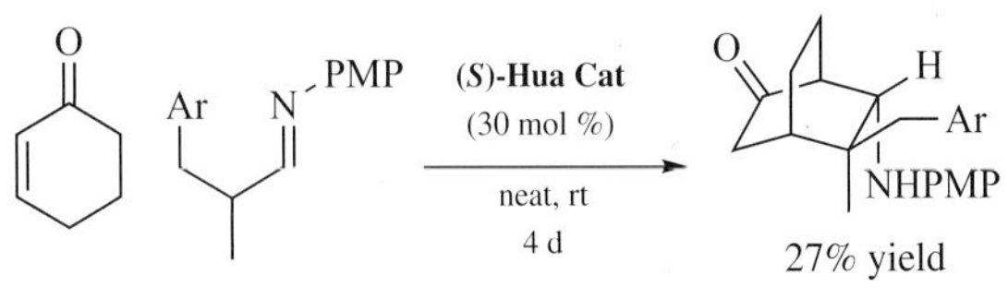

(13)

Ar = 4-*i*Pr-C_6H_4

27% yield
>20:1 dr
91% ee

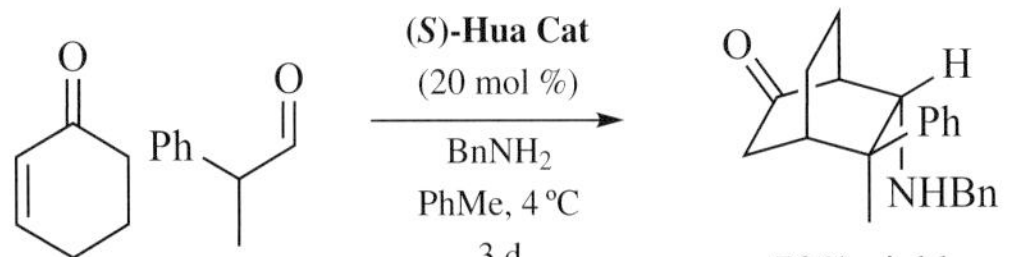

(14)

72% yield
>20:1 dr
87% ee

Intramolecular Michael Additions. Organocatalyzed cyclizations of keto sulfones can also be facilitated by Hua Cat. This process provides access to both five- and six-membered rings in generally good yield and with good stereoselectivity (eqs 15 and 16).[1b] It is important to note that the mixed solvent system (DCE/EtOH 99:1) and the presence of 1 equiv of piperidine proved critical to this transformation. Hua Cat was recently used in a diastereoselective, intramolecular Michael addition during the synthesis of 10-hydroxylated lycopodines (eq 17).[7] In this transformation, the catalyst was particularly important to increase the reaction rate and conversion for the transformation.

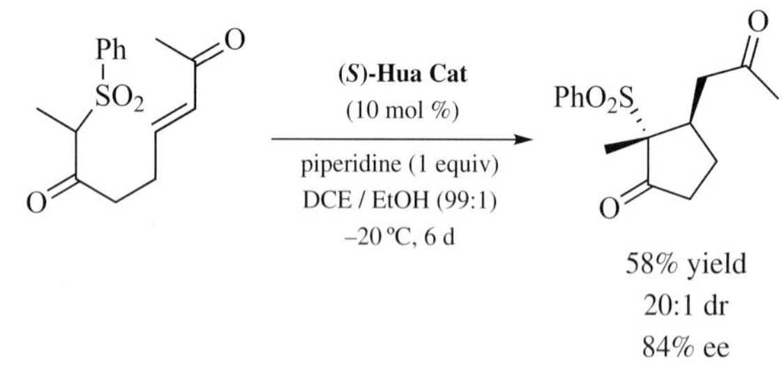

(15)

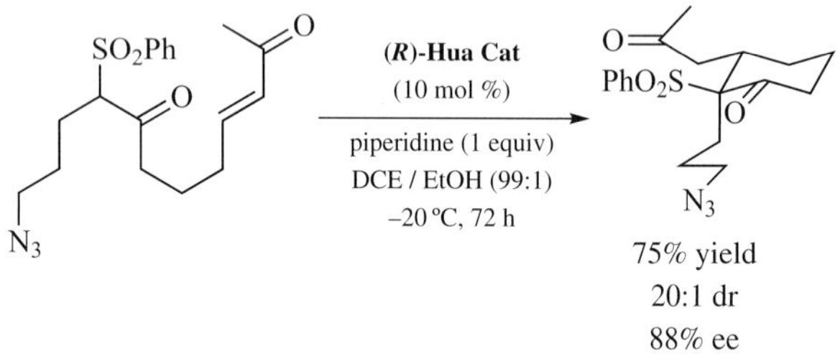

(16)

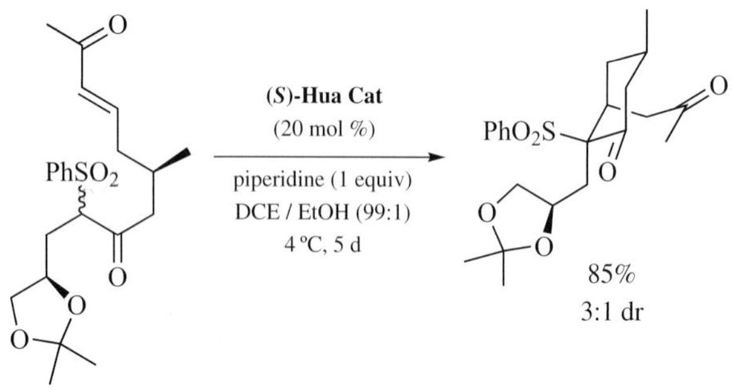

(17)

Related Reagents. Dodecyl 4-[[[(2S)-2-pyrrolidinylcarbonyl]amino]sulfonyl]benzoate (Hua Cat-II); proline.

1. (a) Yang, H.; Carter, R. G., *Org. Lett.* **2008**, *10*, 4649. (b) Yang, H.; Carter, R. G., *J. Org. Chem.* **2010**, *75*, 4929.
2. Yang, H.; Carter, R. G., *Synlett* **2010** 2827.
3. Yang, H.; Mahapatra, S.; Cheong, P. Y.-H.; Carter, R. G., *J. Org. Chem.* **2010**, *75*, 7279.
4. Yang, H.; Carter, R. G., *J. Org. Chem.* **2009**, *74*, 2246.
5. Yang, H.; Carter, R. G., *J. Org. Chem.* **2009**, *74*, 5151.
6. Yang, H.; Carter, R. G., *Tetrahedron* **2010**, *66*, 4854.
7. Saha, M.; Carter, R. G., *Org. Lett.* **2013**, *15*, 736.

Rich G. Carter & Rajinikanth Lingampally
Oregon State University, Corvallis, OR, USA

Hua Yang
Central South University, Changsha, Hunan, China

Dodecyl 4-[[[(2S)-2-pyrrolidinylcarbonyl]- amino]sulfonyl]benzoate (Hua Cat II)

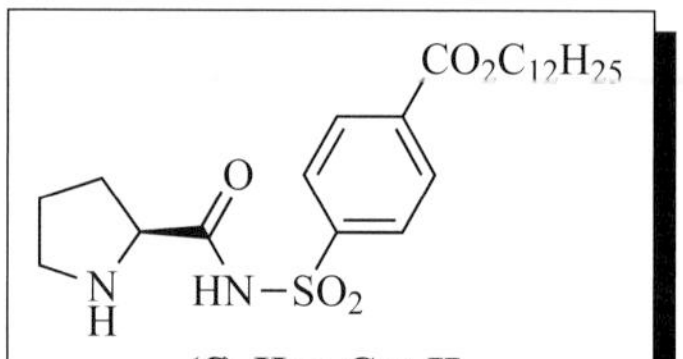

(S)-Hua Cat-II

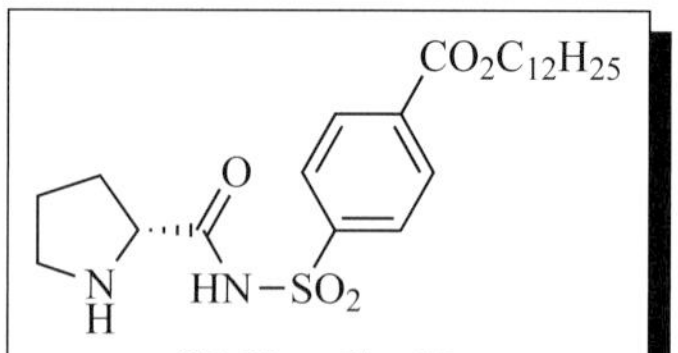

(R)-Hua Cat-II

Dodecyl 4-[[[(2S)-2-pyrrolidinylcarbonyl]amino]sulfonyl]-benzoate (Hua Cat II)
[134193-36-4]
[1235835-45-6] $C_{24}H_{38}N_2O_5S$ (MW 466.63)
InChI = 1S/C24H38N2O5S/c1-2-3-4-5-6-7-8-9-10-11-19-31-24(28)20-14-16-21(17-15-20)32(29,30)26-23(27)22-13-12-18-25-22/h14-17,22,25H,2-13,18-19H2,1H3,(H,26,27)/t22-/m0/s1
InChIKey = VKMQKPSUHQIIOA-QFIPXVFZSA-N
InChI = 1S/C24H38N2O5S/c1-2-3-4-5-6-7-8-9-10-11-19-31-24(28)20-14-16-21(17-15-20)32(29,30)26-23(27)22-13-12-18-25-22/h14-17,22,25H,2-13,18-19H2,1H3,(H,26,27)/t22-/m1/s1
InChIKey = VKMQKPSUHQIIOA-JOCHJYFZSA-N

(reagent used as an organocatalyst for asymmetric carbon–carbon bond-forming reactions)

Physical Data: **(S) isomer:** mp: 166–168 °C. $[\alpha]_D^{20}$ −88.1° (c = 0.7 in $CHCl_3$).
(R) isomer: mp: 166–168 °C. $[\alpha]_D^{20}$ +85° (c = 0.7 in $CHCl_3$).
Solubility: 20 mg mL^{-1} in CH_2Cl_2, soluble 15 mg mL^{-1} in THF, DMF, and DCE.
Form Supplied in: white solid.
Purification: silica gel chromatography.
Handling, Storage, and Precautions: stored in glass or plastic bottle at room temperature.

Preparation. Hua Cat-II can be prepared in three steps from commercially available 4-sulfamoylbenzoic acid (eq 1).[1] Selective esterification with 1-dodecanol and EDCI followed by amide formation using Cbz-proline provided the *N*-protected prolinamide derivative. Hydrogenation over Pd/C produced Hua Cat-II.

Yamada–Otani Reaction. This sulfonamide catalyst was specifically developed for the enantioselective Yamada–Otani cyclization to access stereochemically dense cyclohexenones in high levels of stereoselectivity.[1] These reactions were typically performed using benzylic aldehydes and α,β-unsaturated methyl ketones. The parent reaction of 2-phenylpropanal and 3-penten-2-one proceeds in good yield and with excellent stereoselectivity (eq 2). A wide variety of groups were employed on the β-position of the enone such as the phthalamide moiety (eq 3), a benzyl ether (eq 4), and aryl rings (eq 5). Variation of the benzylic aldehyde was typically limited to p-substitution (eq 6). An example employing an ester moiety has been reported in good yield; however, no significant level of diastereoselectivity was observed (eq 7). The mechanism of this reaction has been studied computationally and is considered to follow an unusual duumvirate stereocontrol model in which the initial Michael addition controls the enantioselectivity, while the subsequent Mannich step controls diastereoselectivity.[2]

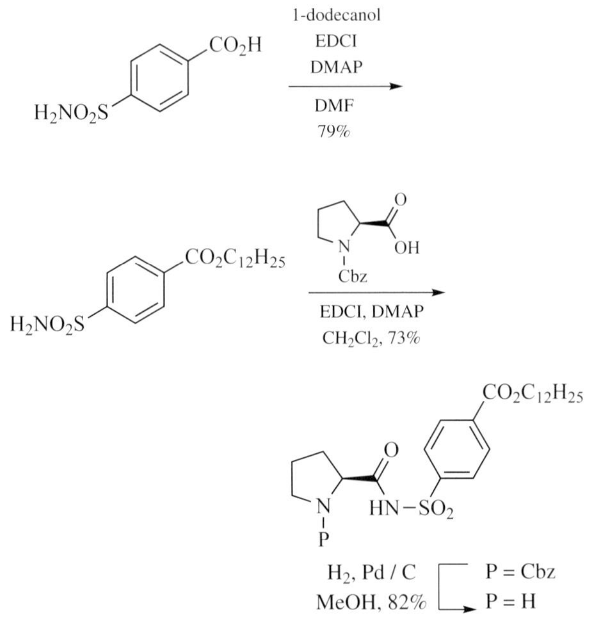

(1)

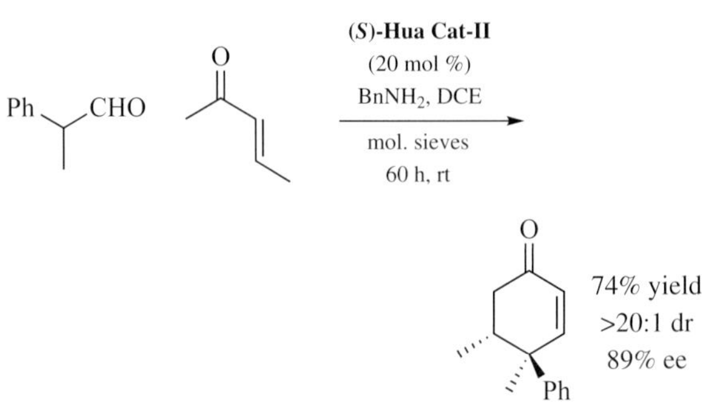

(2)

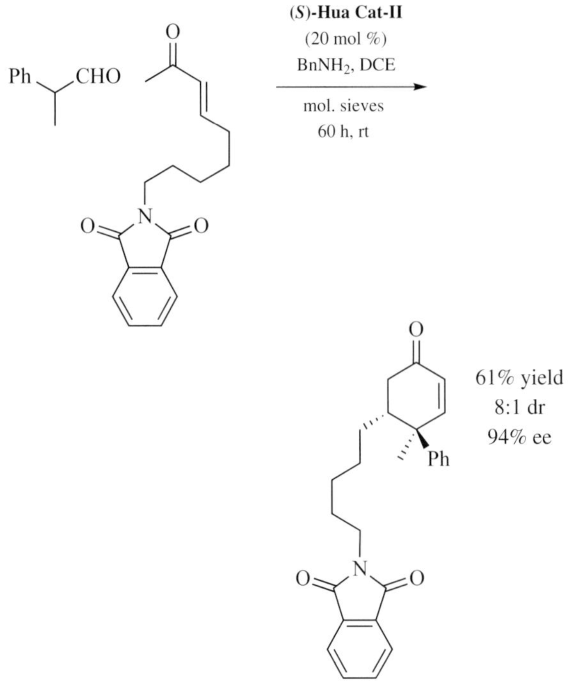

(3)

Aldol Reactions. Hua Cat-II has also been employed as an organocatalyst for enantioselective aldol reactions (eq 8).[3] Interestingly, the reaction rate of this aldol reaction is noticeably slower than that of the parent Hua Cat catalyst (60 h with Hua Cat-II vs. 16 h with Hua Cat). Additionally, the aldol reaction was slightly less diastereoselective (26:1 dr with Hua Cat-II vs. 36:1 dr with Hua Cat).

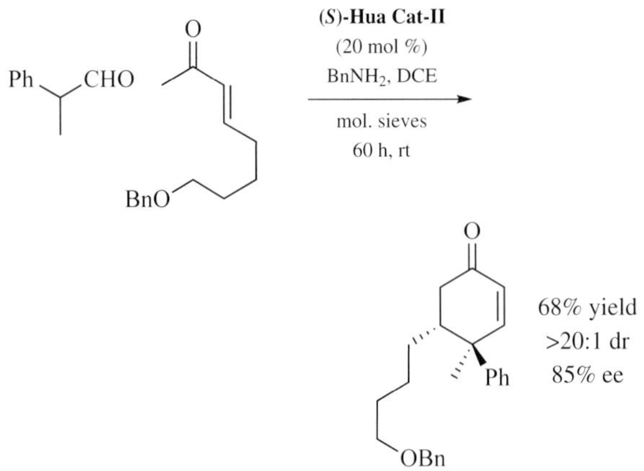

(4)

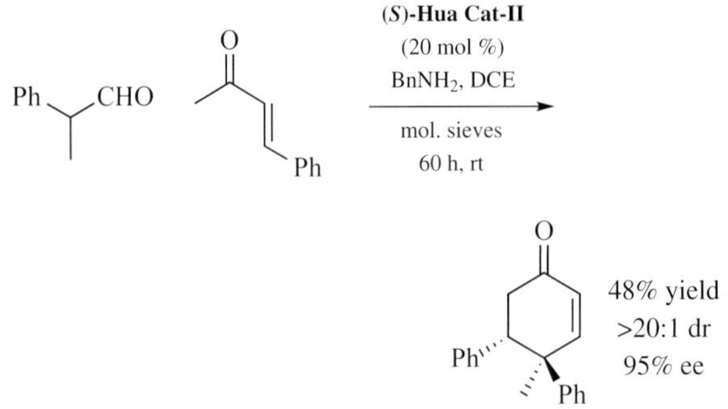

(5)

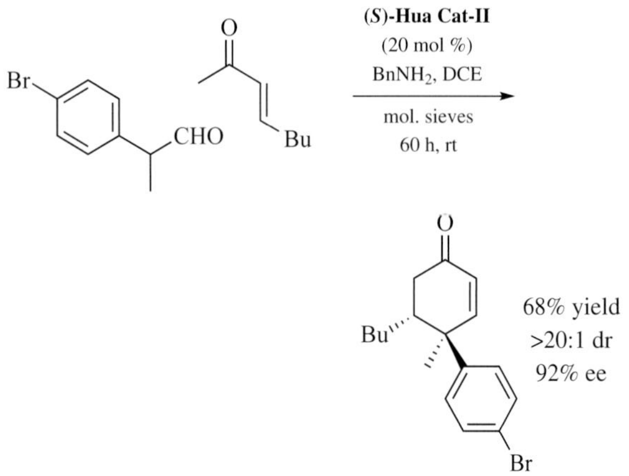

(6)

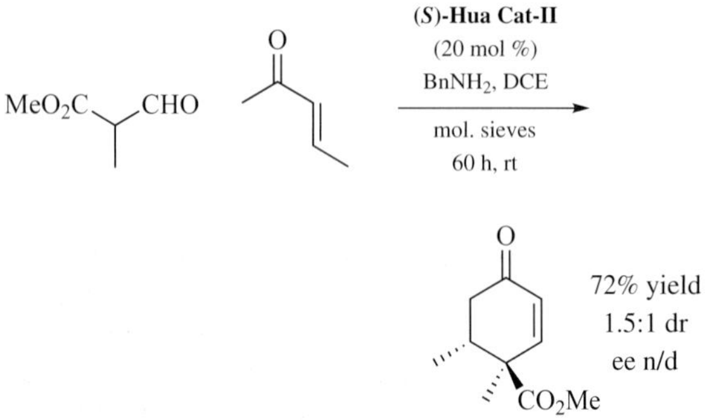

(7)

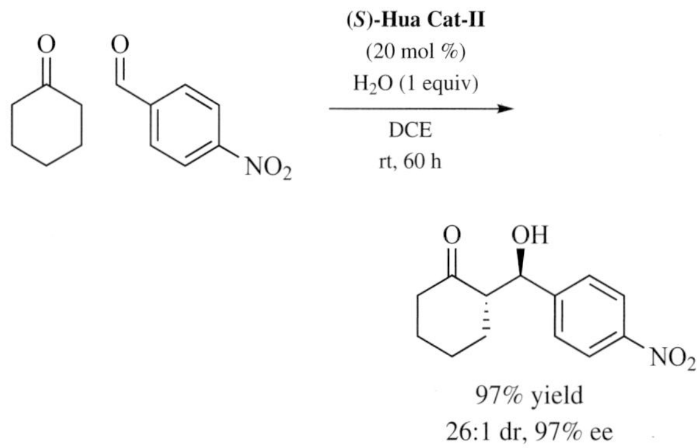

(8)

Related Reagents. *N*-[(4-Dodecylphenyl)sulfonyl]-2-pyrrolidinecarboxamide (Hua Cat-I); proline.

1. (a) Yang, H.; Carter, R. G., *Org. Lett.* **2010**, *12*, 3108. (b) Yang, H.; Banerjee, S.; Carter, R. G., *Org. Biomol. Chem.* **2012**, *10*, 4851.
2. Pierce, M.; Johnston, R. C.; Mahapatra, S.; Yang, H.; Carter, R. G.; Cheong, P. H.-Y., *J. Am. Chem. Soc.* **2012**, *134*, 13624.
3. Yang, H.; Mahapatra, S.; Cheong, P. Y.-H.; Carter, R. G., *J. Org. Chem.* **2010**, *75*, 7279.

Rich G. Carter & Rajinikanth Lingampally
Oregon State University, Corvallis, OR, USA

Hua Yang
Central South University, Changsha, Hunan, China

H

(11b*R*)-4-Hydroxy-2,6-bis[2,4,6-tris-(1-methylethyl)phenyl]-dinaphtho[2,1-*d*: 1′,2′-*F*][1,3,2]dioxaphosphepin-4-oxide, [(*R*)-TRIP] & (11b*S*)-4-Hydroxy-2,6-bis[2,4,6-tris(1-methylethyl)phenyl]-dinaphtho[2,1-*d*: 1′,2′-*f*][1,3,2]dioxaphosphepin-4-oxide, [(*S*)-TRIP]

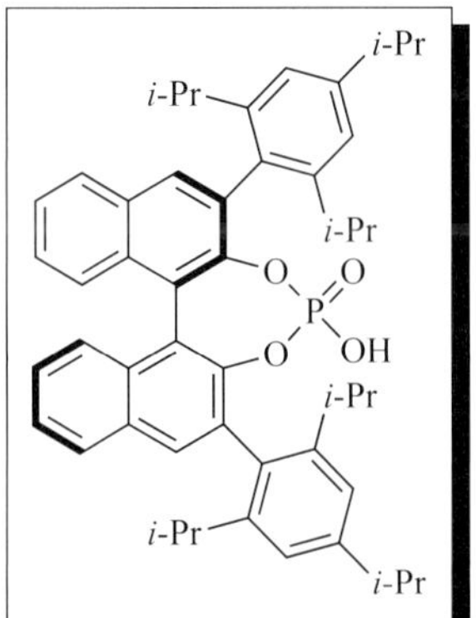

(*R*)-TRIP (**1**)

(11b*R*)-4-Hydroxy-2,6-bis[2,4,6-tris(1-methylethylphenyl]-dinaphtho[2,1-*d*: 1′,2′-*f*][1,3,2]dioxaphosphepin-4-oxide
[791616-63-2] $C_{50}H_{57}O_4P$ (MW 752.96)
InChI = 1S/C50H57O4P/c1-27(2)35-23-39(29(5)6)45(40(24-35)30(7)8)43-21-33-17-13-15-19-37(33)47-48-38-20-16-14-18-34(38)22-44(50(48)54-55(51,52)53-49(43)47)46-41(31(9)10)25-36(28(3)4)26-42(46)32(11)12/h13-32H,1-12H3,(H,51,52)
InChIKey = AGQAQYPGJDBIQR-UHFFFAOYSA-N

Solubility: soluble in dichloromethane, dioxane, toluene, and methanol.
Form Supplied in: white crystalline solid.
Preparative Methods: most preparative methods used for chiral phosphoric acid catalysts were followed in analogy to those developed by Akiyama[1] and Uraguchi[2] that follow the original substituted BINOL synthesis of Wipf[3] from (*R*)-BINOL (**3**), as described in eq 1.[4]
Purification: following phosphate formation it is very important that the hydrolysis product is carefully stirred with hydrochloric acid solution to make sure that the product is not contaminated with salts or heavy metal impurities from earlier steps in the synthesis. The dichloromethane extract is then dried over anhydrous magnesium sulfate, concentrated, and then recrystallized from acetonitrile.
Handling, Storage, and Precautions: while no major hazards with the handling of *(R)*- or *(S)*-TRIP are known, these reagents should be handled with the customary precautions used for other laboratory chemicals.

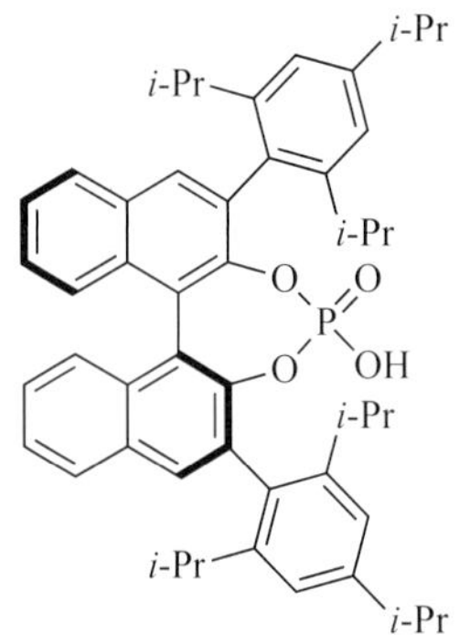

(*S*)-TRIP (**2**)

(11b*S*)-4-Hydroxy-2,6-bis[2,4,6-tris(1-methylethyl)phenyl]-Dinaphtho[2,1-*d*: 1′,2′-*f*][1,3,2]dioxaphosphepin-4-oxide
[874948-63-7] (MW 752.96)
InChI = 1S/C50H57O4P/c1-27(2)35-23-39(29(5)6)45(40(24-35)30(7)8)43-21-33-17-13-15-19-37(33)47-48-38-20-16-14-18-34(38)22-44(50(48)54-55(51,52)53-49(43)47)46-41(31(9)10)25-36(28(3)4)26-42(46)32(11)12/h13-32H,1-12H3,(H,51,52)
InChIKey = AGQAQYPGJDBIQR-UHFFFAOYSA-N

Solubility: soluble in dichloromethane, dioxane, toluene, and methanol.
Form Supplied in: white crystalline solid.
Preparative Methods: preparative methods used *(S)*-TRIP are in direct analogy to those described above for *(R)*-TRIP. These methods were developed by Akiyama[1] and Uraguchi[2] that follow the original substituted BINOL synthesis of Wipf[3] from *(S)*-BINOL (**3**), as described in eq 2.[4]

Enantioselective Reductive Amination with Hantzsch Esters and Related Work. The first application of TRIP catalysts was in the enantioselective reduction of ketone imines using Hantzsch esters as hydrogen-transfer agents. As shown in eq 3, a number of *p*-anisidine imines were reduced in the presence of *(S)*-TRIP and Hantzsch ester **10** in toluene solution at 35 °C in yields of 84–91% and enantiomeric ratios (er) of 90:10 to 96.5:3.5.[5]

Preliminary experiments included in this study demonstrated that the tris(isopropyl)phenyl group in the flanking *ortho*-positions of the BINOL phosphate of (*S*)-TRIP gave the highest enantioselectivities of all the catalysts examined.

Follow-up experiments from the laboratory of Akiyama have shown that very similar results to those in eq 3 could be obtained using benzothiazoline (**11**) as a hydrogen source.[6a] This hydrogen source afforded the α-iminoester in high yield and with excellent enantioselectivity (eq 4).[6b]

Further exploitation of the above methodology demonstrated that α-branched aldehydes could be reductively aminated with concurrent dynamic kinetic resolution as shown in eq 5.[7]

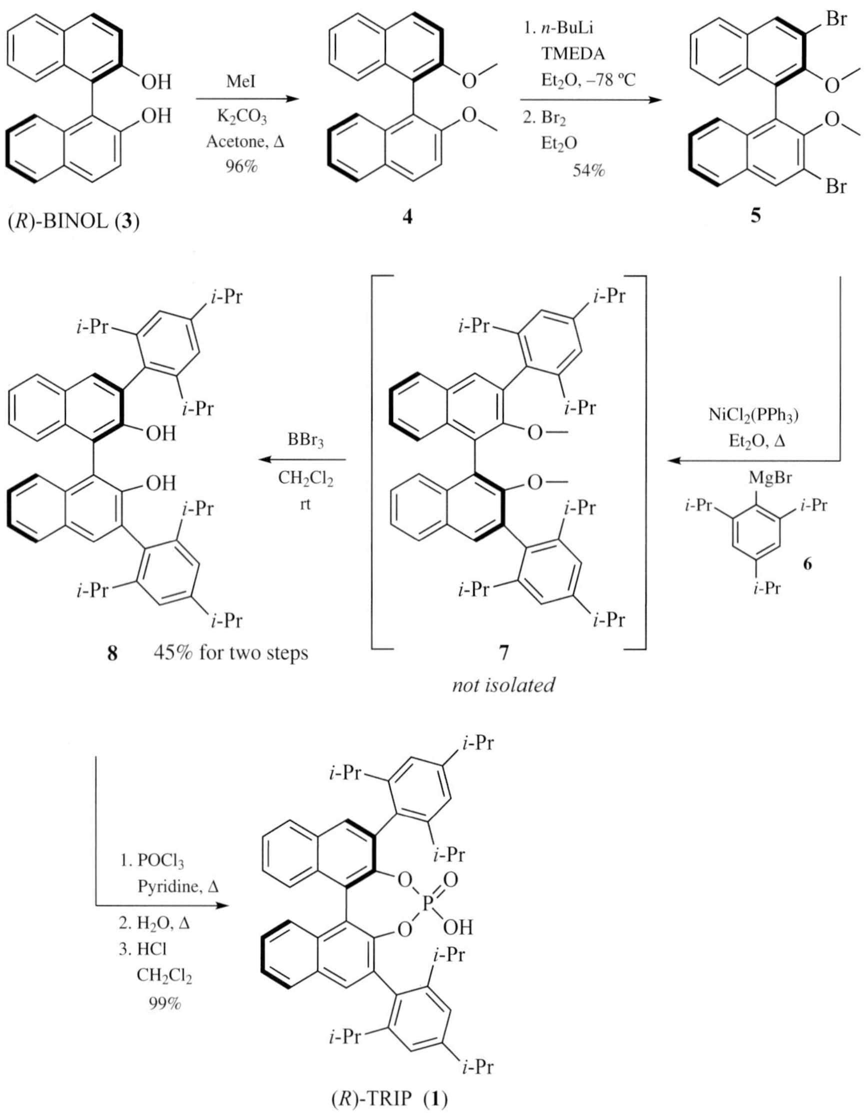

(1)

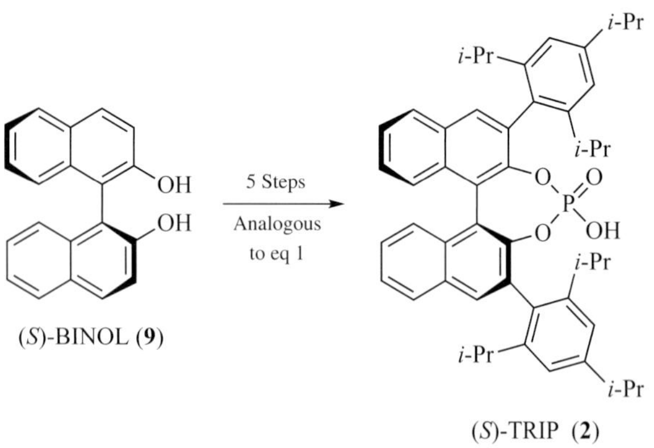

(2)

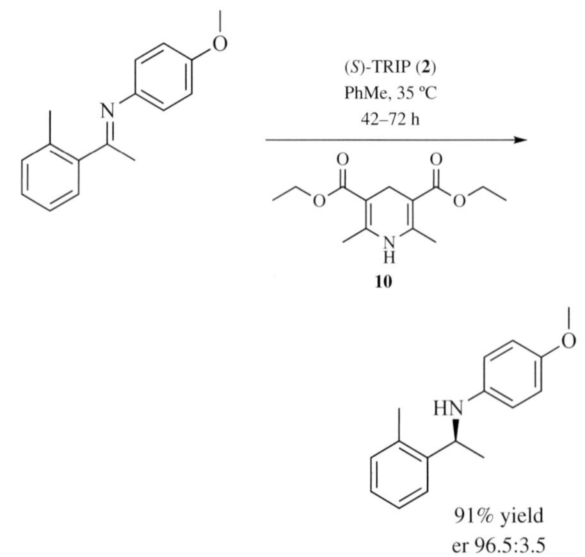

(3)

This method proved high yielding and provided amines in high enantiomeric ratios, except in cases of completely aliphatic α-branched aldehydes. Preliminary experiments described in this paper again demonstrated that the superiority of the tris(isopropyl)phenyl group of the (*R*)-TRIP catalyst in providing the highest enantiomeric ratios.

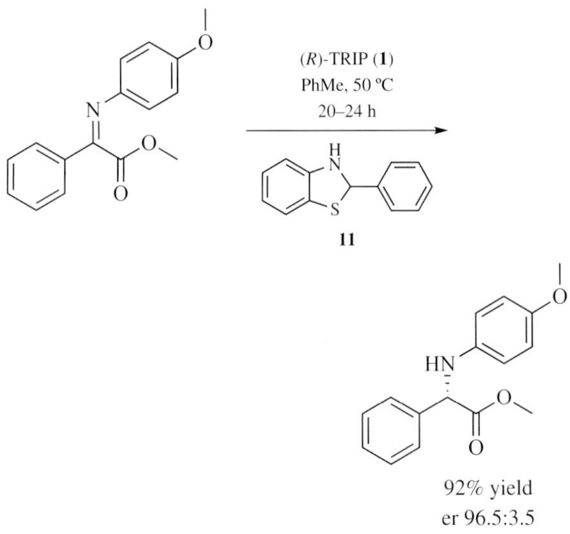

(4)

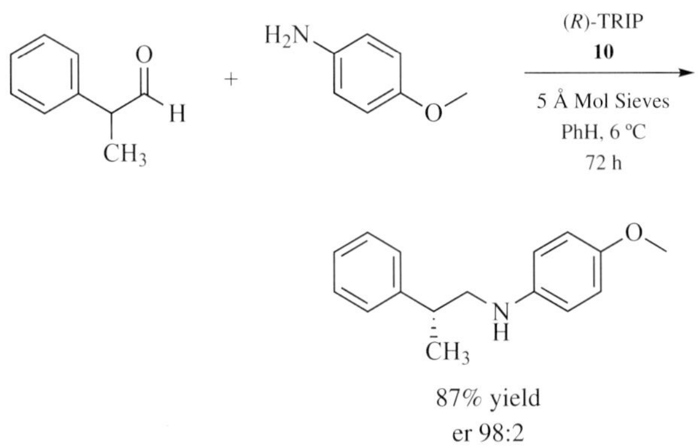

(5)

These reductive amination/kinetic resolution results were further extended to reaction cascades resulting in the enantioselective syntheses of cyclohexylamine derivatives. As shown in eq 6, 1,5-diketones could be cyclized to give *cis*-3-alkylcyclohexylamine derivatives in good yield and with high diastereo- and enantioselectivities. This process also can be used to prepare enantioselectively the corresponding tetrahydropyran (or even tetrahydrothiopyran) derivatives as shown in eq 7.[8] Further work by List has served to characterize intermediates involved in this reaction cascade to elucidate mechanistic details.[9]

List has also found racemic α-branched cyclohexanone derivatives can undergo reductive alkylation with concurrent dynamic kinetic resolution.[10] As shown in eq 8, treatment of the racemic 2-allylcyclohexanone with *p*-anisidine in the presence of (*R*)-TRIP and Hantzsch ester **10** afforded the reductive amination product as largely a single *cis*-diastereomer with very high enantioselectivity. Remarkably this process even works extremely well for 2-chlorocyclohexanone, as shown in eq 9. While this process seems fairly general for 2-substituted cyclohexanones, diastereoselectivity and enantioselectivity fall off considerably for 2-subsituted *cyclopentanones*.

With the simple expedient of increasing the amounts of Hantzsch ester **10** in the reaction mixture, α,β-unsaturated cyclohexanones can be reductively aminated with concomitant double bond reduction. As shown in eq 10, 2-phenylcyclohex-2-enone is reduced to the saturated cyclohexanone in situ, followed by the reductive amination. While the yield is lower than for reaction on saturated cyclohexanones and uses higher amounts of the TRIP catalyst, the diastereoselectivity and enantioselectivity are comparable to other examples.

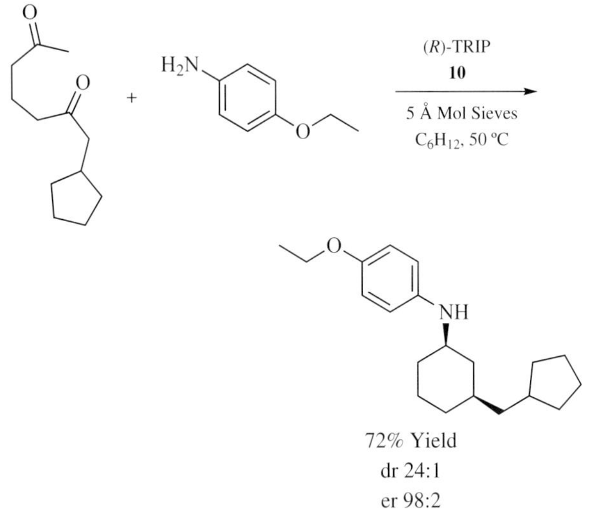

(6)

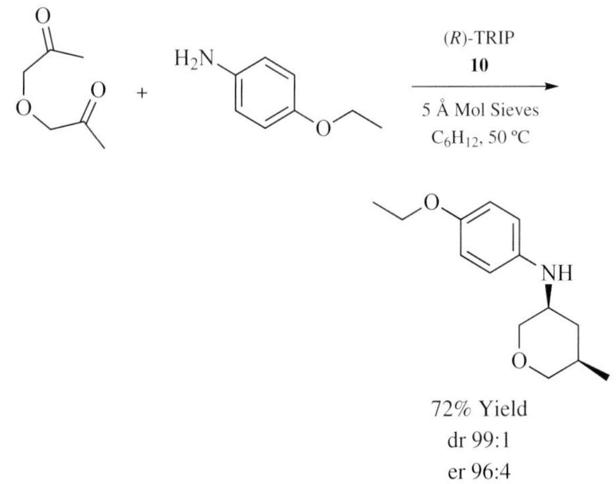

(7)

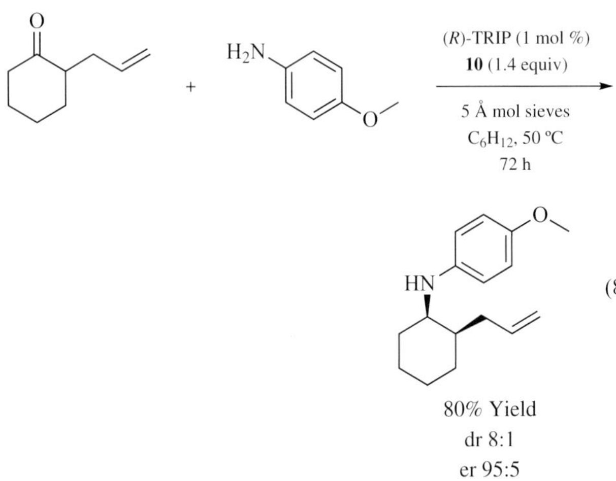

(8)

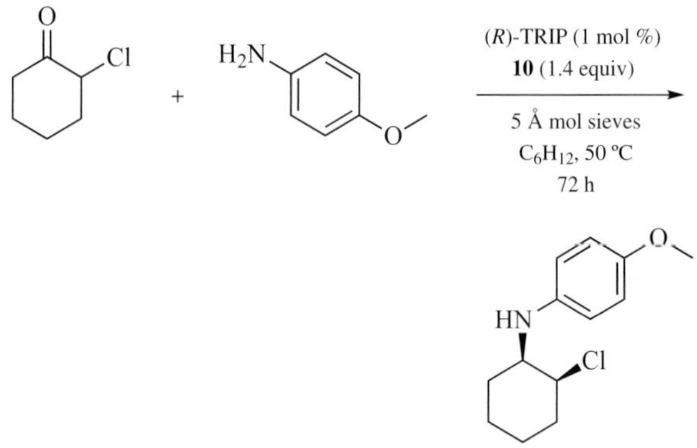

96% Yield
dr 10:1
er 98:2

(9)

Another variant on the enantioselective reductive amination followed by a cyclization was provided by Enders et al.[11] As shown in eq 11, treatment of the keto ester shown with *p*-anisidine in the presence of (*R*)-TRIP and the less usual reducing agent **12** affords an intermediate amine that undergoes intramolecular conjugate addition when treated with potassium *tert*-butoxide. The whole cascade of reactions proceeds in high yield with almost complete diasteoselectivity and excellent enantioselectivity.

(*R*)- and (*S*)-TRIP catalysts have also been found to be essential to the success of iridium-catalyzed hydrogenations of acyclic imines.[12] Iridium(III) catalyst (*S*,*S*)-**13**, shown in eq 12, provided no enantioselectivity in reductive amination experiments. Interestingly, protonation of (*S*,*S*)-**13** with (*R*)-TRIP affords catalyst **14**, which gave moderate enantioselectivity in reductive amination of acyclic ketones. Further optimization showed that addition of 5 mol % of (*R*)-TRIP along with 4Å molecular sieves to promote ketone imine formation allowed for outstanding enantioselectivity and yields in the reductive aminations shown in eq 13. However, reductive aminations of aliphatic ketones did not require additional (*R*)-TRIP for optimum results, as shown in eq 14.

Finally, the equivalent of an enantioselective reduction of a ketone imine can be accomplished by appending List's Hantzsch ester-mediated reduction to a gold(I)-mediated hydroamination of an acetylene in the same flask.[13] As shown in eq 15, treatment of a mixture of *m*-anisidine and an acetylene with Hantzsch ester **10**, 5–10 mol % of (*S*)-TRIP, and 1–2 mol % of gold catalyst **15** in benzene, in the presence of 5 Å molecular sieves, produces the corresponding (*S*)- amine derivative in good yield and with excellent enantioselectivity.

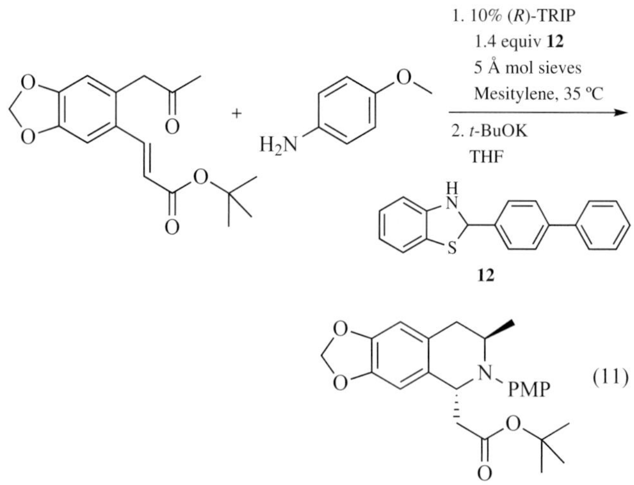

(11)

93% Yield
dr >95:5
er 98:2

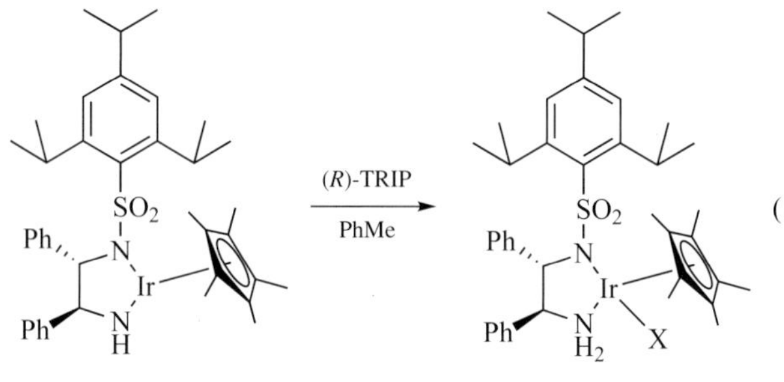

(12)

(*S*,*S*)-**13** **14** X = (*R*)-TRIP⁻

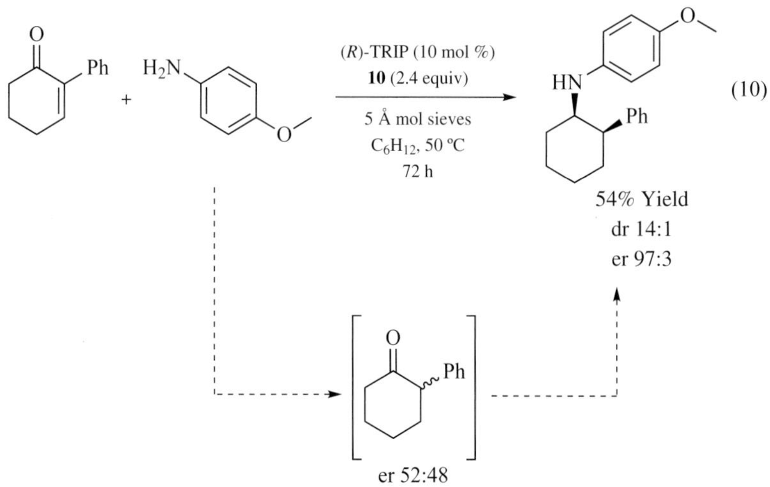

(10)

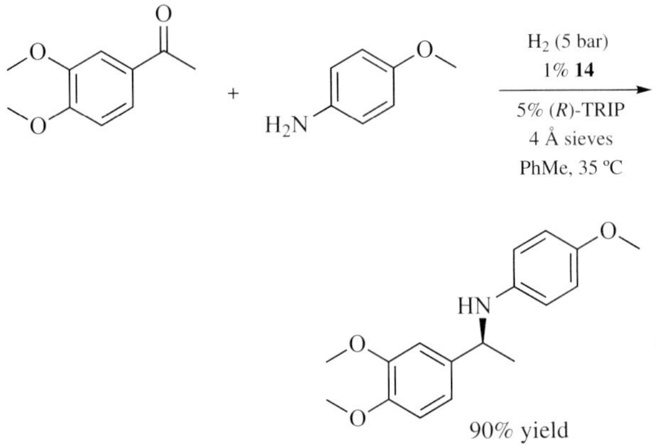

(13)

90% yield
er 96.5:3.5

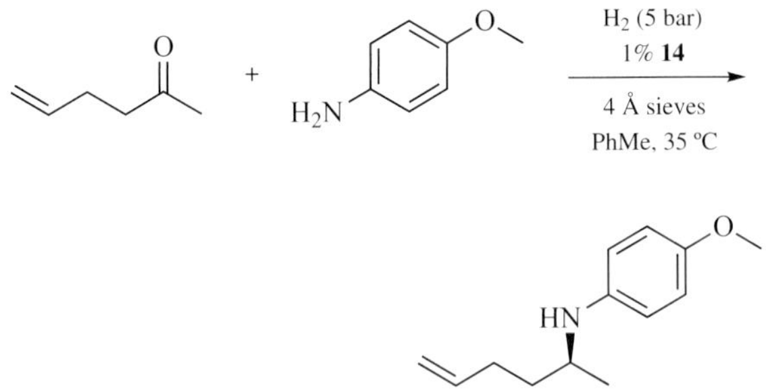

(14)

80% yield
er 96:4

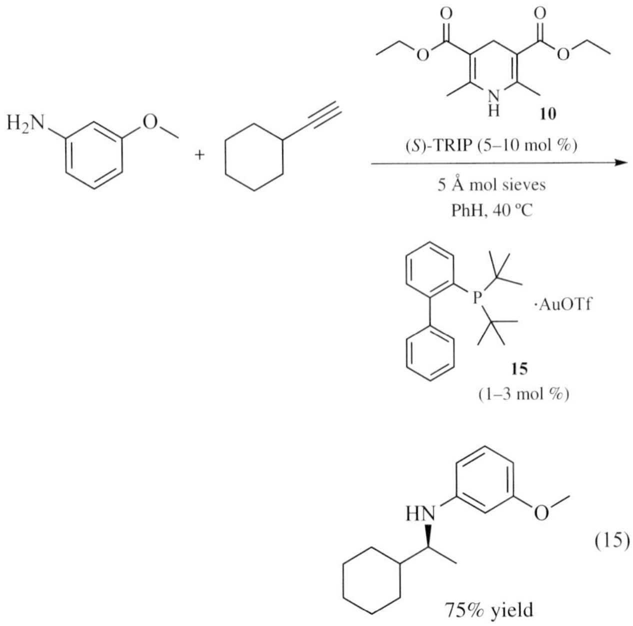

(15)

75% yield
er 97.5:2.5

Enantioselective Friedel–Crafts, Pictet–Spengler, and Related Reactions. A second major application of the TRIP catalysts was the enantioselective Friedel–Crafts alkylation of activated arenes with enecarbamates. As shown in eq 16, Terada[14] found that substituted indole derivatives react with enecarbamates in the presence of (*R*)-TRIP in acetonitrile to give the Boc-protected amines in good yield and with excellent enantioselectivity. Even greater enantioselectivity could be obtained by running the reaction at −20 °C instead of 0 °C, but prolonged reaction times were needed for completion of the reaction. Interestingly, electron-donating or electron-withdrawing substituents on the benzene ring of the indole had little effect on either the yields or the enantioselectivity of this process.

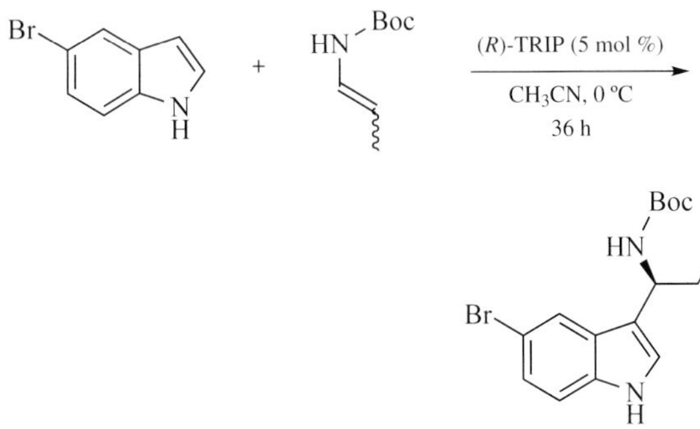

(16)

91% Yield
er 96.5:3.5

A variation on this process developed by Zhou[15] allows for the selective preparation of quaternary stereocenters. As shown in eq 17, starting with an indole derivative and an aryl enamide, in the presence of (*S*)-TRIP and molecular sieves in toluene, the corresponding diaryl methyl amine is produced in the (*S*) configuration in good yield and with excellent stereocontrol. As with the previous process, results were generally not affected by substituents on the indole ring system. Moreover, either electron-donating or electron-withdrawing substituents in the 3- or 4-position of the aryl ring of the enamide had little effect on the enantioselectivity or yield of the reaction. However, substituents in the 2-position of the aryl ring slow the reaction precipitously and give products of significantly lower enantioselectivity.

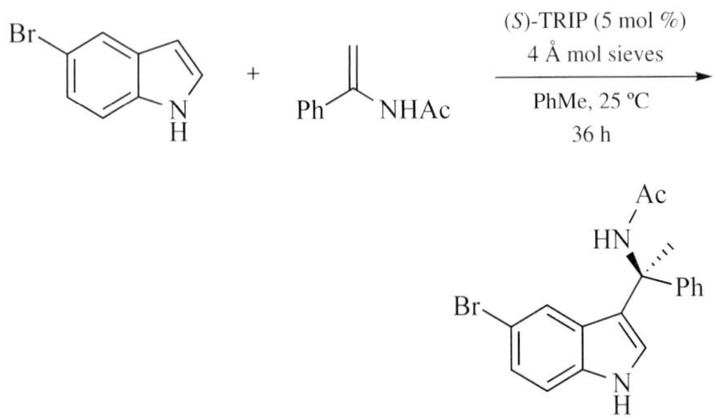

(17)

91% Yield
er 96.5:3.5

Terada[16] has extended this work to a one-pot enecarbamate two-carbon homologation followed by an enantoselective Friedel–Crafts alkylation, shown in eq 18.

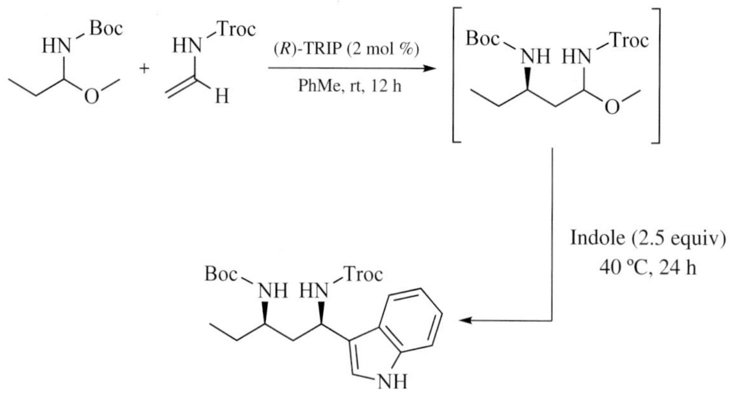

(18)

70% Yield
76% *syn*-
er >99:1

Even more surprisingly, TRIP catalysts can promote the enantioselective Friedel–Crafts reactions of highly electron-deficient substrates containing the trifluoromethyl group. Ma and coworkers[17] demonstrated that the treatment of indole derivatives with trifluoroacetaldehyde hemiacetal and trimethoxyaniline in the presence of (*S*)-TRIP afforded the trifluoromethyl amine derivatives with the (*S*)configuration, as determined by single-crystal X-ray diffraction analysis (eq 19). Here, as with other TRIP-mediated Friedel–Crafts reactions, electron-donating or electron-withdrawing substituents on the benzene ring of the indole had little effect on the yields or enantioselectivites of this process. However, a 2-position substituent on the indole, as in 2-methylindole, had a deleterious effect on the reaction rate and especially the enantioselectivity, causing the yields to drop to 80% and the enantiomeric ratio to fall to er 90:10.

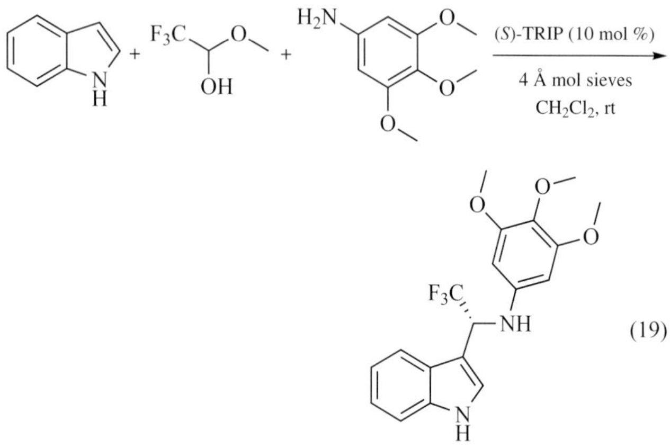

(19)

99% Yield
er 97:3

Further work by Ma[18] showed that TRIP catalysts could promote the enantioselective formation of tertiary alcohols from trifluoromethyl ketones. As shown in eq 20, treatment of indole with trifluoroacetophenone in the presence of (*S*)-TRIP afforded the tertiary alcohol in high yield and with excellent enantioselectivity. Indoles and acetophenone derivatives with either electron-donating or electron-withdrawing groups tended to give similar results as the parent unsubstituted system.

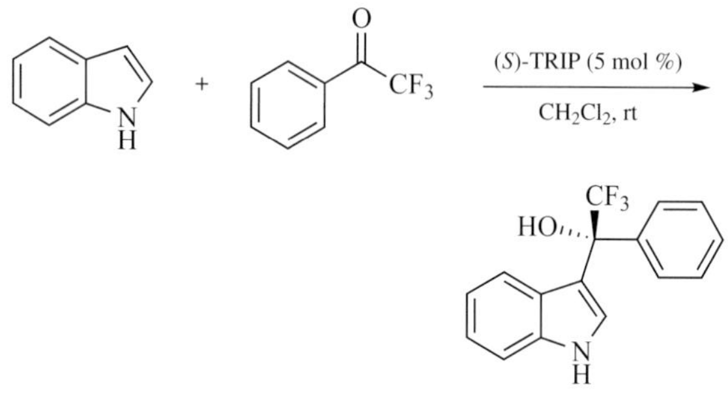

(20)

99% Yield
er 96:4

Following up on these results, Ma[19] then showed that the TRIP catalysis could be applied to the Friedel–Crafts reactions of ethyl trifluoracetoacetate and ethyl trifluoropyruvate derivatives. As shown in eq 21, treatment of indole derivatives with ethyl trifluoroacetoacetate in the presence of (*S*)-TRIP afforded the tertiary alcohol in good yield and with the modest enantioselectivity of er 82:18. Other indole derivatives, depending upon substituents, gave enantioselectivities in the range of er 77.5:22.5 (for 6-chloroindole) to er 89:11 (for 7-ethylindole).

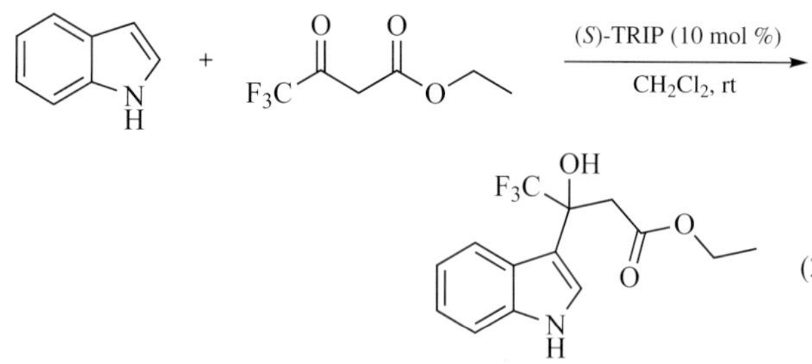

(21)

99% Yield
er 82:18

When ethyl trifluoropyruvate was substituted for ethyl trifluoroacetoacetate in this process even poorer enantioselectivities were obtained. As shown in eq 22, treatment of indole with ethyl trifluoropyruvate in the presence of 5 mol % (*S*)-TRIP afforded the tertiary alcohol in good yield and with enantioselectivity of er 67:33. Results for other indole derivatives ranged from er 63:37 (for 5-fluoroindole) to er 81:19 (for 7-ethylindole). Efforts to improve the enantioselectivity by lowering the temperature of the reaction were unsuccessful. Efforts to determine the absolute sense of the enantioselectivity in either the reaction with ethyl trifluoroacetoacetate or with ethyl trifluoropyruvate by X-ray crystallographic studies were also unsuccessful.

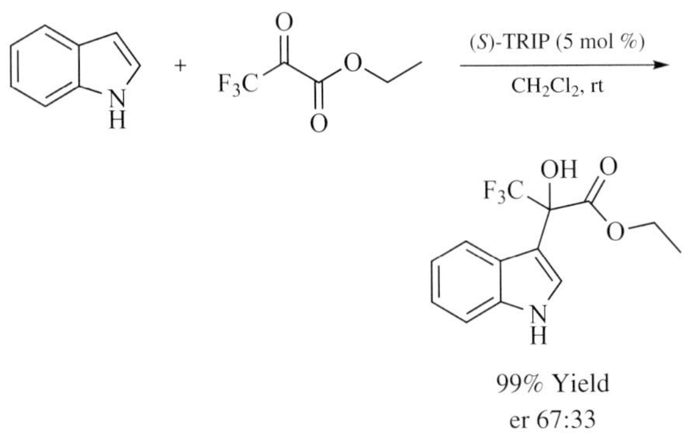

(22)

Bolm and coworkers[20] have reported more successful results in Friedel–Crafts reactions of *N*-Boc-protected trifluoropyruvate imine derivatives. As shown in eq 23, treatment of indole with the pyruvate imine in the presence of (*R*)-TRIP at −78 °C afforded very high yield of the amino acid derivative and with excellent enantioselectivity. Results for other substituted indoles were similar, regardless of the presence of electron-withdrawing or donating substituents.

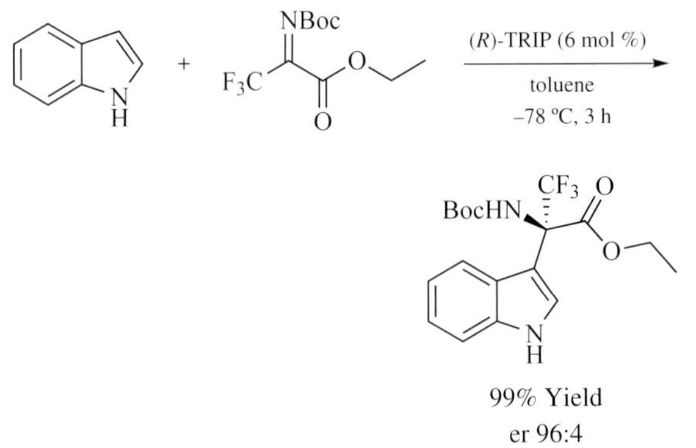

(23)

Finally, Goodman[21] has published DFT studies designed to more carefully elucidate the mechanism and sense of enantioselectivity of the Friedel–Crafts reactions of both *N*-acyl and *N*-tosyl imines catalyzed by BINOL-phosphoric acid derivatives, such as TRIP.

In addition to the intermolecular Friedel–Crafts reaction, (*R*)- and (*S*)-TRIP can be used to catalyze the intramolecular Pictet–Spengler reaction.[22] The case shown in eq 24 is typical of the results obtained: treatment of the tryptamine derivative shown with cyclohexanecarboxaldehyde in the presence of (*S*)-TRIP and anhydrous sodium sulfate affords the cyclization product in good yield and with high enantioselectivity. The enantioselectivity can be improved to er 97:3 by running the reaction at −45 °C, although the yield drops to 64% due to the rate of reaction becoming sluggish at that temperature. The reaction works well with straight chain and with branched aldehydes, as well as with aromatic aldehydes. The primary limitation to this process is the need for the *gem*-carboethoxy groups on the indole side chain: without these, the Pictet–Spengler cyclization does not occur, with aldol condensation of the aldehyde becoming the main product of the reaction.

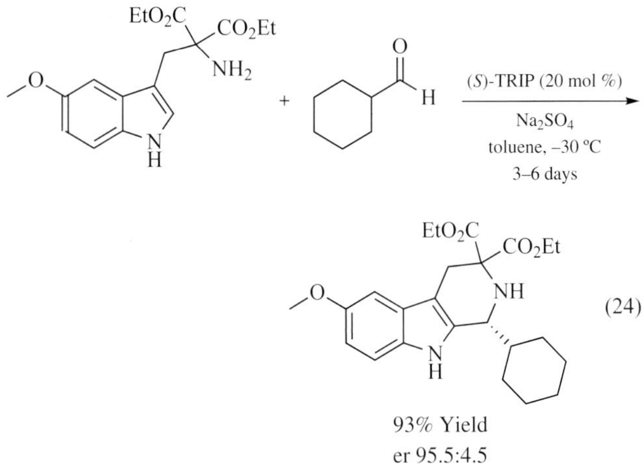

(24)

Ricci[23] has reported that (*S*)-TRIP promotes the enantioselective Povarov reaction of 2- and 3-vinylindoles. As shown in eq 25, the aldimine undergoes cyclization with 2-vinylindole in the presence of (*S*)-TRIP to give the tetrahydroisoquinoline in high yield and with excellent enantiocontrol. Similar results were obtained with imines with other substitution patterns, but these reaction conditions did not translate to cyclizations with the more polymerization-prone 3-vinyl indole.

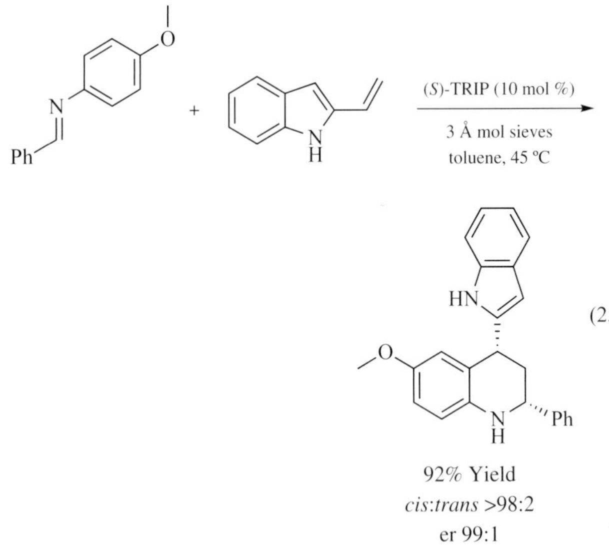

(25)

After some experimentation, it was found that replacement of toluene with the coordinating solvent tetrahydrofuran and slowly adding the indole over 13 h gave results comparable to those with 2-vinyl indole (eq 26).

Enantioselective Hetero-Diels–Alder Reactions. After screening a number of BINOL-derived catalysts, Akiyama showed that (*R*)-TRIP could promote the inverse electron demand Diels–Alder reaction between imines and Danishefsky's diene.[24] After optimization, it was found that 1.2 equiv of acetic acid provided the hetero-Diels–Alder product in good yield and with moderate enantioselectivity (eq 27). Only slight improvements in the enantioselectivity were possible by replacing the phenyl group on the imine with 4-bromophenyl (er 92:8) or 1-naphthyl (er 95.5:4.5).

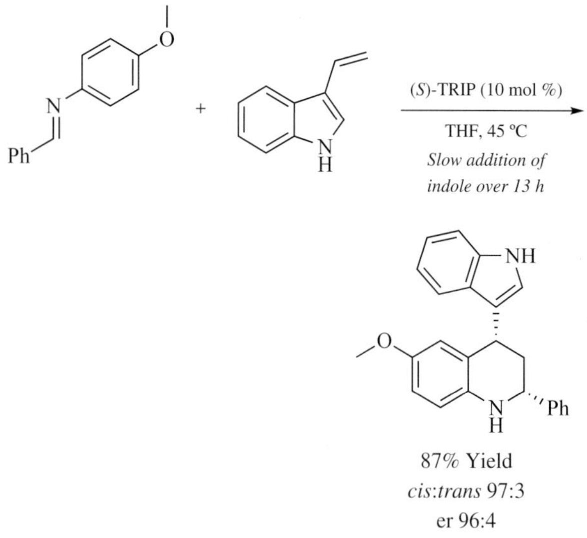

(26)

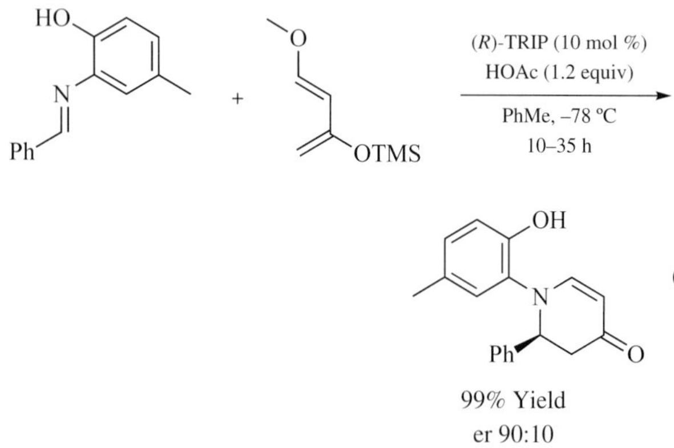

(27)

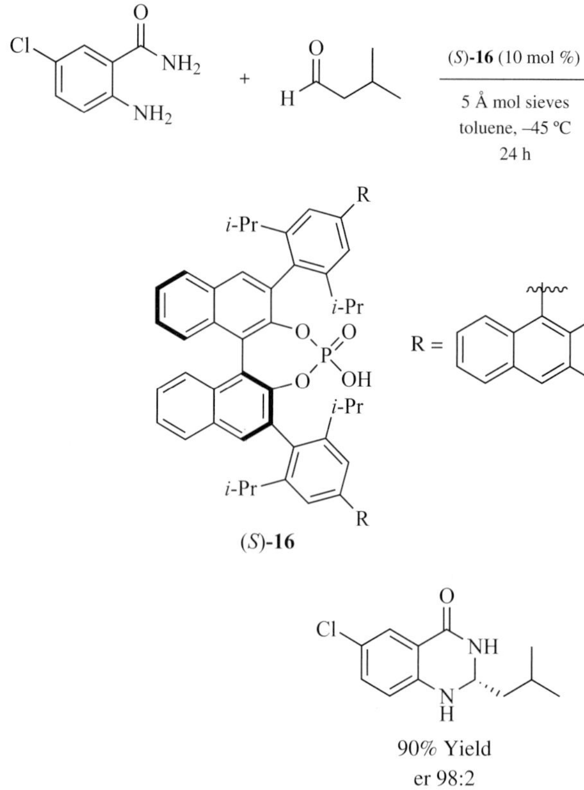

(28)

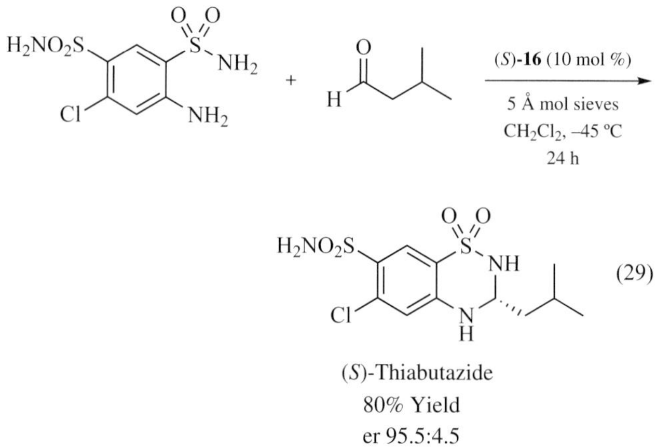

(29)

Unfortunately, although a number of examples of normal[25] or inverse-electron demand[26] Diels–Alder reactions promoted by BINOL-type catalysts have appeared, the TRIP catalysts have seldom been the optimum reagents for these transformations.

Asymmetric Aminal and Acetal Formation. TRIP catalysts, albeit in sometimes modified form, have been successfully applied to the enantioselective formation of aminals, hemiaminals, and in some cases, acetals. List[27] reported that by using the slightly modified TRIP catalyst (*S*)-**16**, amides such as 5-chloroanthranilamide (eq 28) form aminals with aldehydes as shown, in good yield and with high enantioselectivity. In this case, the sense of the enantioselectivity was confirmed by single-crystal X-ray diffraction studies. These reaction conditions were sufficiently general that List was able to prepare a number of known benzothiadiazine diuretics, such as (*S*)-thiabutazide (eq 29), with moderate to good enantioselectivity. It must be noted that these compounds had previously been obtained in configurationally pure form only by chiral HPLC separation.[28]

In a last example of enantioselective aminal formation, Gong and coworkers[29] have used (*R*)-TRIP to catalyze the highly diastereoselective and enantioselective formation of imidazolidinones via a 1,3-dipolar cycloaddition as shown in eq 30.

In this case, aliphatic and alicyclic aldehydes failed to react with useful reaction rates, and aldehydes with less strongly electron-withdrawing groups tended to give lower diastereoselectivities. Interestingly, despite this, the enantioselectivity for the major diastereomer obtained was still in the ratio of 90:10. Single crystal X-ray studies served to determine the sense of the enantiocontrol.

Using another modified version of (*S*)-TRIP, (*S*)-**17**, (eq 31), List[30] was able to prepare N,O-acetals. In general, the yields and the enantioselectivities were not affected by the nature of the substituents on the benzene ring of the amides examined. In the case at hand in eq 32, the absolute configuration of the products obtained was determined by single crystal X-ray diffraction analysis, and the remainder of the examples assigned by analogy.

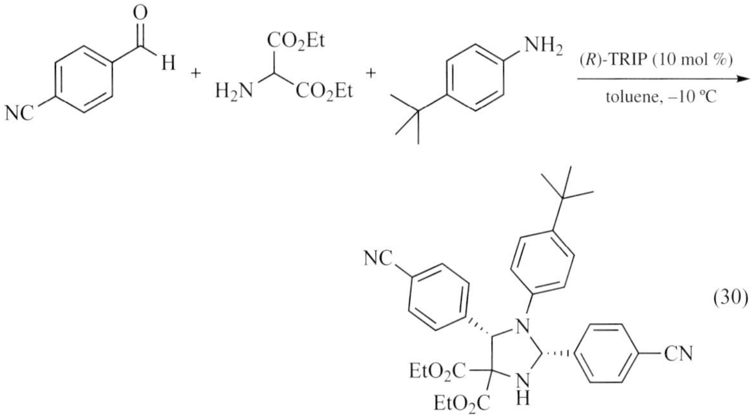

(30)

90% Yield
dr 90:10
er 96:4

List has also demonstrated that the TRIP catalysts can promote the enantioselective acetal exchange reaction of diethyl acetals containing a tertiary alcohol.[31a,b] As shown in eq 33, the diethyl acetal cyclizes to the tetrahydrofuran acetal in high yield and with excellent enantioselectivity. The dibromide shown in eq 33 was used in a single-crystal X-ray diffraction study to determine the sense of the enantioselectivity in this process. These results could be extended to the formation of tetrahydropyran acetals, as shown in eq 34, albeit with lower enantioselectivity.

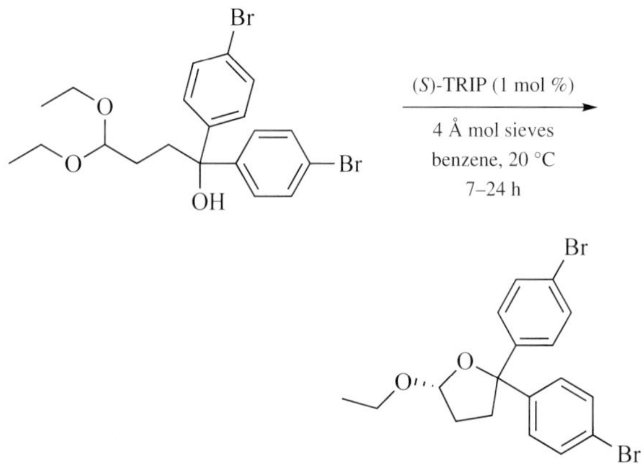

(33)

99% Yield
er 97:3

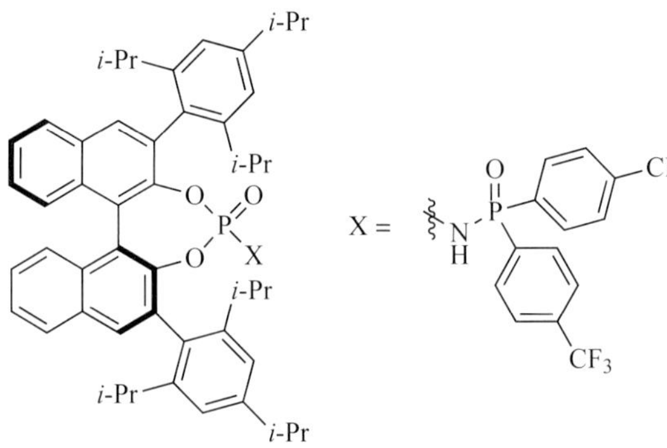

(S)-**17**

(31)

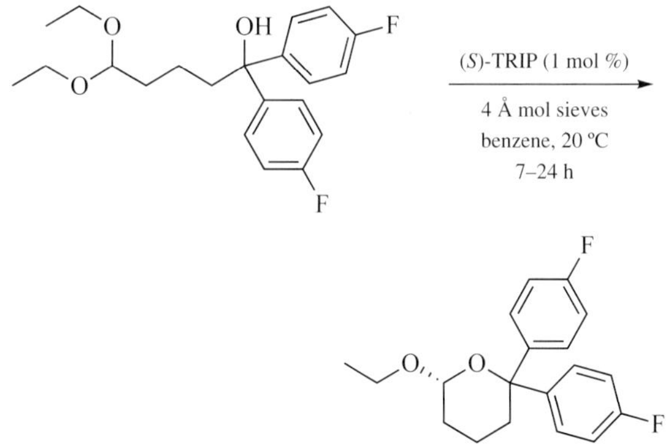

(34)

96% Yield
er 82:18

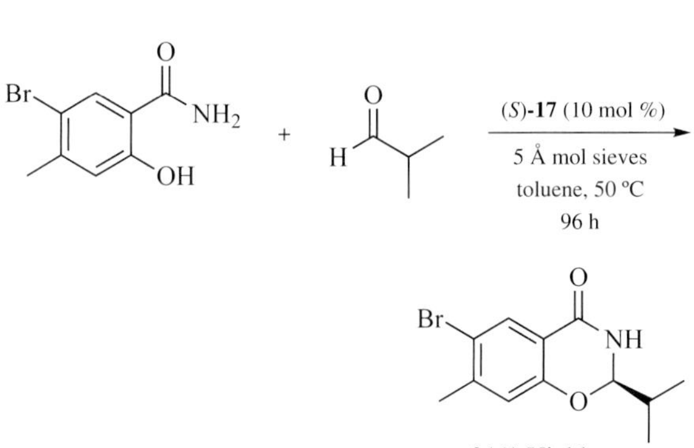

(32)

91% Yield
er 96:4

In making attempts to extend this methodology to the kinetic resolution of secondary homoaldols, List found that the TRIP catalysts gave mediocre results, but results could be improved by the use of the spirocyclic TRIP analog, (*S*)-STRIP (**18**), shown in eq 35.[31c] This modified catalyst then afforded results such as shown in eq 36 where one enantiomer of the homoaldol forms the tetrahydrofuran acetal with high diastereoselectivity and the remaining open chain homoaldol is also obtained in highly enantiomerically enriched form.

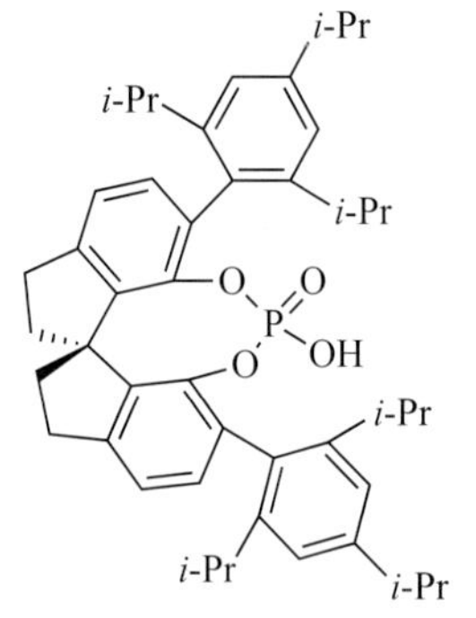

(S)-STRIP (**18**) (35)

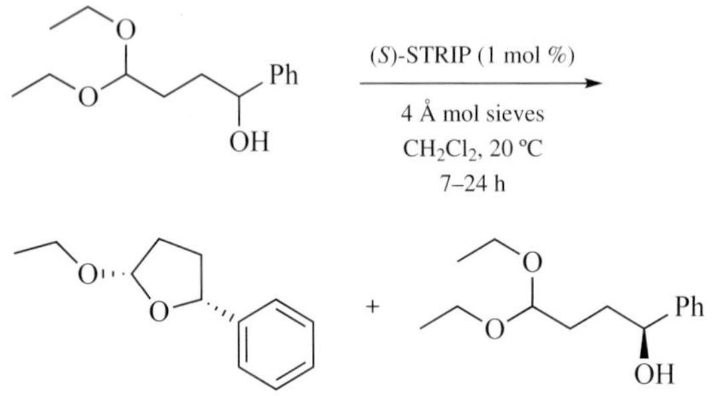

55% Yield er 97:3 dr 13:1 — er 98.5:1.5 (36)

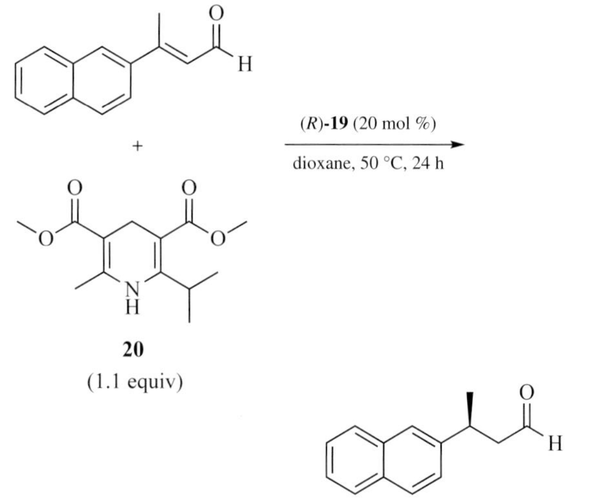

20 (1.1 equiv)

72% Yield er >99:1 (38)

Even aliphatic systems, such as (*E*)-citral (eq 39) can be reduced under slightly milder conditions with good enantioselectivity.

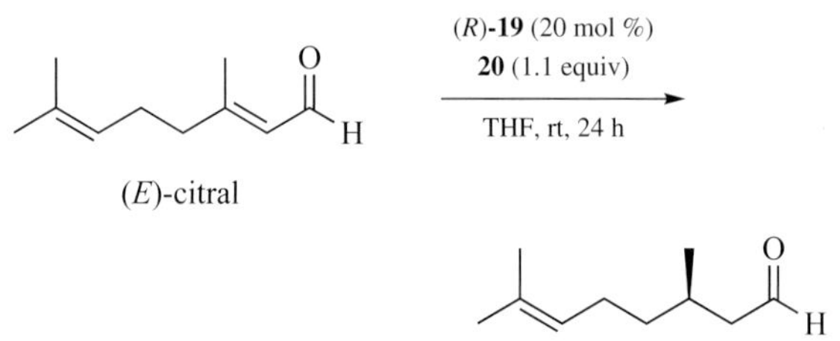

(*E*)-citral

71% Yield er 95:5 (39)

Asymmetric Transfer Hydrogenations. After demonstrating that salts of chiral amines could be used to effect the metal-free reduction of α,β-unsaturated aldehydes,[32] List examined the ability of salts of TRIP with achiral amines to bring about the same transformation.[33] After examining a number of achiral amine salts with (*R*)-TRIP, it was found that the morpholine provided optimum results. The catalyst thus prepared, (*R*)-**19** (eq 37) in combination with the Hantzsch ester **20**, was found to reduce the α,β-unsaturated aldehyde shown (eq 38) in good yield and with excellent enantioselectivity, Other substituted aromatic rings at the β-position gave similar yields and with high enantioselectivities.

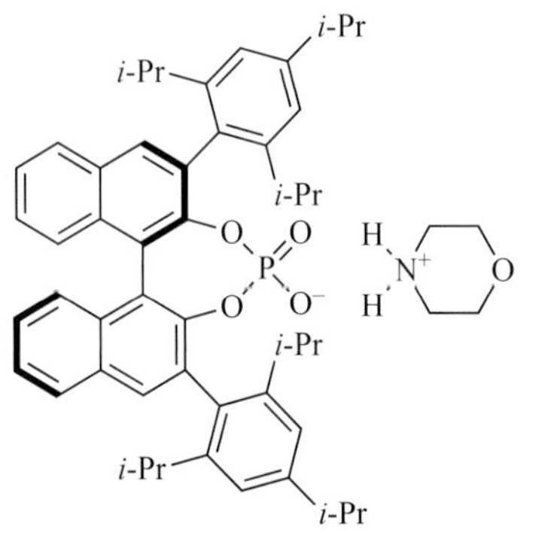

(*R*)-**19** (37)

In the course of efforts to apply this methodology to cyclic α,β-unsaturated ketones, List found that the morpholine of salt (*R*)-**19** had to be replaced with a primary amine for maximum efficiency.[34] After screening a number of primary amines, it was found that the salt made from *(R)*-TRIP and *tert*-butyl (*S*)-valine (*(R)*-**21**), shown in eq 40, afforded optimum results. Thus, as shown in eq 41, treatment of the 3-substituted cyclohex-2-enone with Hantzsch ester **10** in the presence of 5 mol % (*R*)-**21** afforded the saturated ketone in essentially quantitative yield with excellent enantioselectivity. These results were general not only for 3-alkylcylclohex-2-enones, but also for the analogous cyclopent-2-enones and cyclohept-2-enones. Acyclic enones, such as in eq 42, tended to give lower enantioselectivities than cyclic enones.

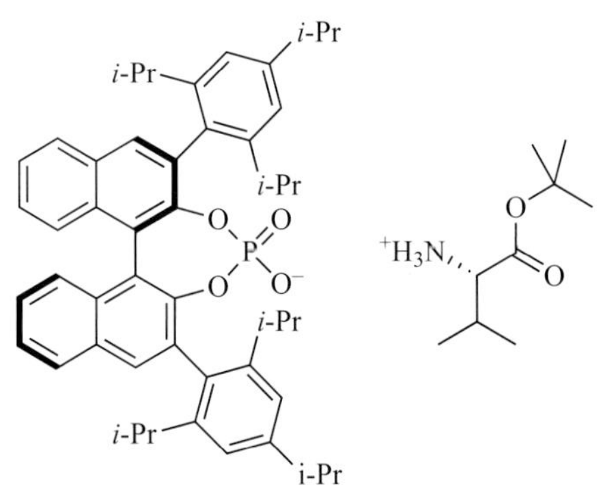

(*R*)-**21** (40)

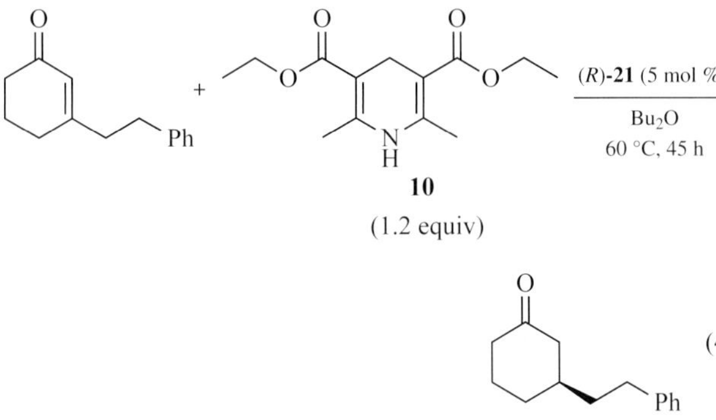

99% Yield
er 98:2

(41)

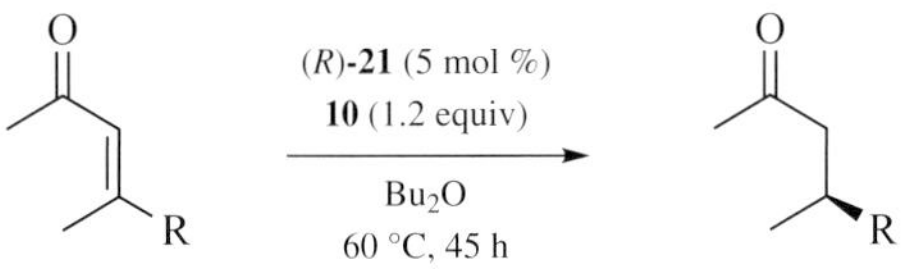

(42)

R = CO_2Et: 99% yield; er 92:8

R = Ph: 81% yield; er 85:15

Epoxidations and Baeyer–Villiger Oxidations. List then found that the secondary amine salts of TRIP catalyze the epoxidation of α,β-unsaturated aldehydes and ketones with *tert*-butylhydroperoxide.[35] After screening a series of secondary amine salts of (*R*)-TRIP, it was found that the trifluoromethyl-containing bis-benzylic amine salt (*R*)-**22**, (eq 43) was optimum for this purpose. As shown in eq 44, treatment of the cinnamaldehyde derivative with *tert*-butyl hydroperoxide in the presence of catalyst (*R*)-**22** in warm dioxane afforded the epoxide as essentially one diastereomer with good enantioselectivity. Other substituted cinnamaldehyde derivatives were epoxidized with similar results. A slight modification of these conditions allowed successful application to β,β-disubstituted enals, as shown in eq 45. Attempts by List to extend these results to cyclic enones, such as cyclohex-2-enone, required a chiral amine cocatalyst, and the results suggest the nature of the chiral amine used is far more important than the coincidental presence of TRIP.[36]

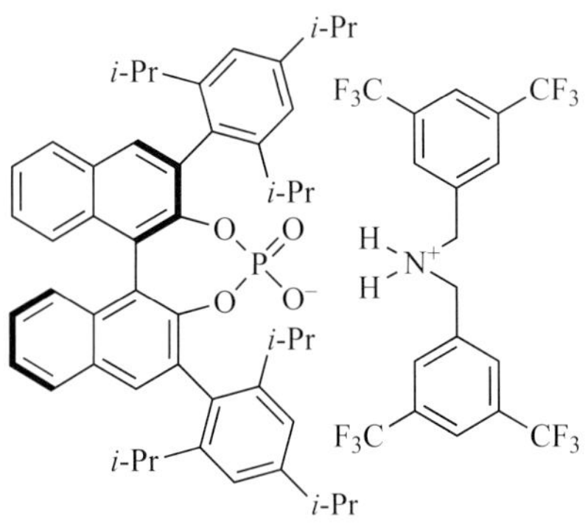

(*R*)-**22**

(43)

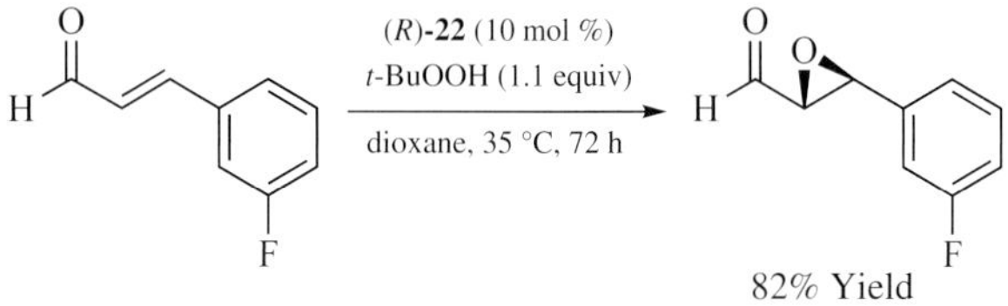

82% Yield
dr >99:1
er 92:8

(44)

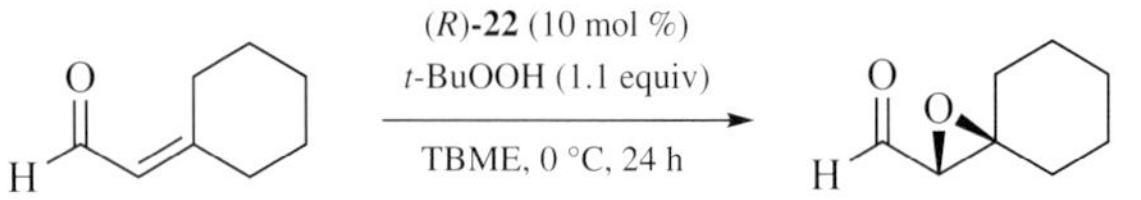

75% Yield
er 95:5

(45)

In a general exploration of the use of BINOL-type catalysts for promoting enantioselective Baeyer–Villiger oxidations, Ding[37] found that (*R*)-TRIP was the ideal catalyst to promote the hydrogen peroxide-mediated Baeyer–Villiger reaction of the tricyclic ketone shown in eq 46. Unfortunately, with other substrates, other BINOL phosphates gave better results than (*R*)-TRIP, suggesting that the optimum catalyst would have to be discovered by experimental evaluation of a library of potential catalysts.

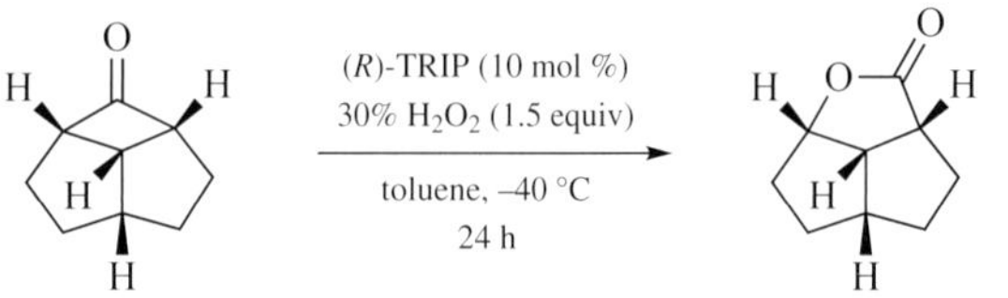

99% Yield
er 97.5:2.5

(46)

Enantioselective Condensations, Annulations, and Cycloadditions. Antilla and coworkers found that (*R*)-TRIP was the choice of catalyst for the enantioselective condensation of dihydropyran with *N*-acyl imines as shown in eq 47.[38] This process seems general for *N*-acyl imines of a number of electron-rich and deficient benzaldehyde derivatives. The product of the condensation can subsequently be transformed to the spirocyclic product by oxidation with MCPBA, with retention of the original stereochemistry. Single-crystal X-ray diffraction studies on the oxidation product in eq 47 provided evidence for the sense of the enantioselectivity of this process.

Akiyama examined a number of BINOL phosphate catalysts in order to develop an enantioselective variant on the Robinson annulation.[39] As shown in eq 48, the Michael addition of the keto ester to ethyl vinyl ether was best promoted by BINOL phosphate (*R*)-**23** providing the adduct in good yield and with moderate enantioselectivity. After workup and purification by chromatography, the Michael adduct is then cyclized by treatment with (*R*)-TRIP in refluxing toluene to afford the cyclohexenone derivative in highly enriched enantiomeric purity. The yields were moderate, the balance of the product mixture being the uncyclized minor enantiomer from the Michael addition step.

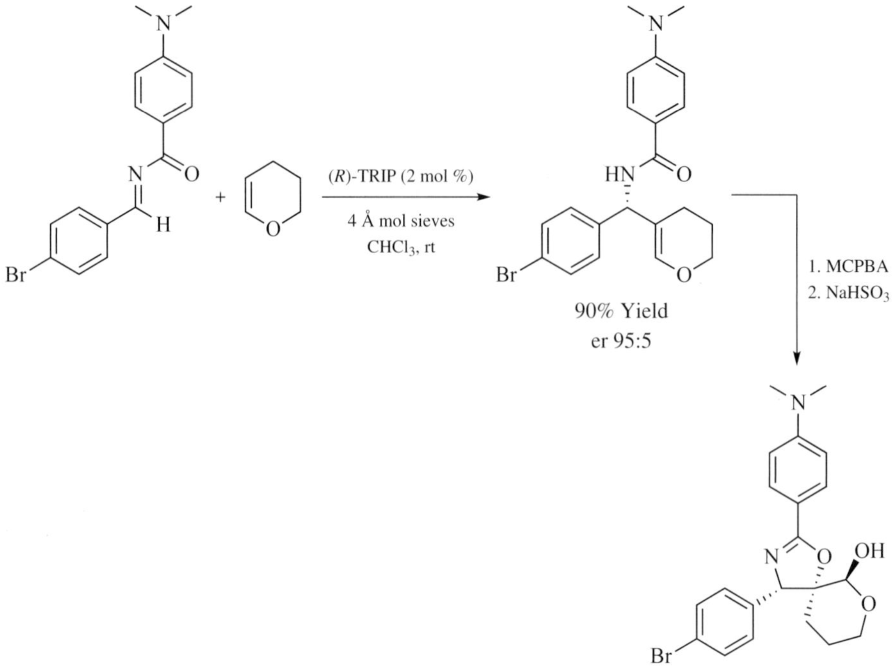

90% Yield er 95:5

91% Yield er 96:4 (47)

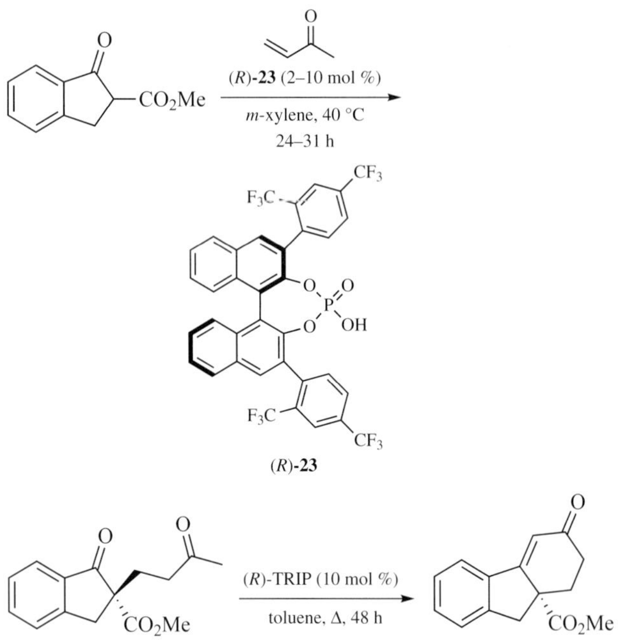

(R)-**23**

99% Yield er 89:11

64% Yield er 98:2 (48)

Akiyama then proceeded to extend these results to the desymmetrization of *meso*-substrates that had previously been used in Robinson annulations with (*S*)-proline.[40] As shown in eq 49, treatment of the triketone with (*R*)-TRIP in toluene, or gently refluxing hexane, afforded excellent yields of the cyclohexenone derivative in high enantiomeric purity in a single step. In contrast, cyclization of this substrate with (*S*)-proline, followed by dehydration with *p*-toluenesulfonic acid in benzene, afforded the opposite enantiomer in 57% yield with only an enantiomeric ratio of 60:40. Other substrates with variation at the position of the methyl group afforded virtually identical enantioselectivities as those for the case of eq 49. These conditions were also applied to desymmetrization of the prototypical substrate of the Hajos–Parrish–Eder–Sauer–Weichert reaction,[41] affording the cyclohexenone derivative in good yield with a slightly lower enantioselectivity (eq 50).

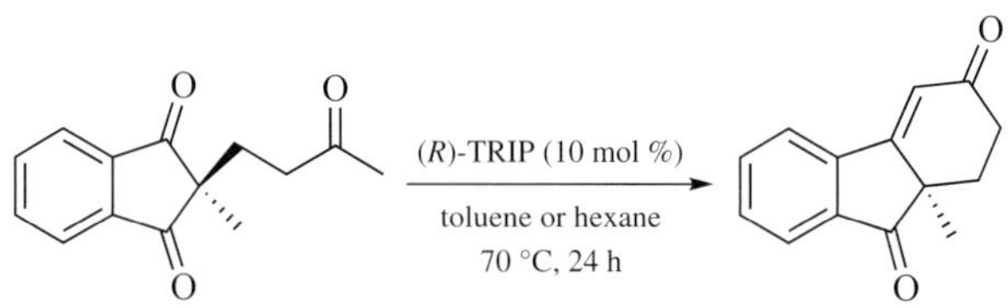

(49)

95% Yield er 95:5

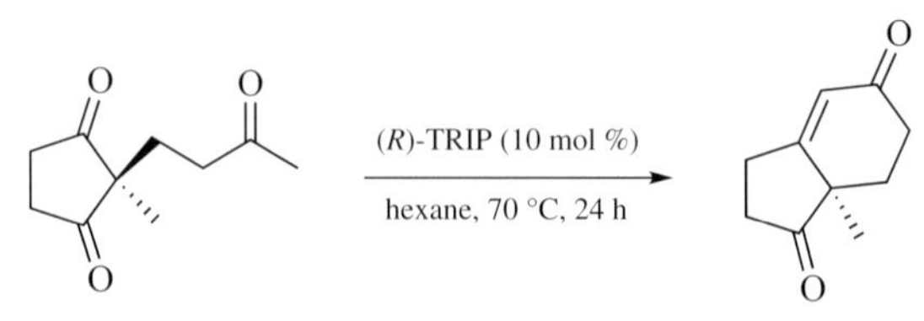

(50)

82% Yield er 85:15

Enantioselective Transition Metal-mediated Allylation Reactions. Krische[42] reported one of the earlier examples of Bronsted acid catalysis by (*R*)-TRIP in a rhodium-mediated addition of enyne systems to heterocyclic carboxaldehydes. As shown in eq 51, under an atmosphere of hydrogen, the rhodium catalyst promotes the addition of the enyne in the presence of catalyst (*R*)-**24** and (*R*)-TRIP. In general, yields and enantioselectivities are similar for other heterocyclic derivatives, but the reaction fails for 3- and 4-pyridinecarboxaldehydes.

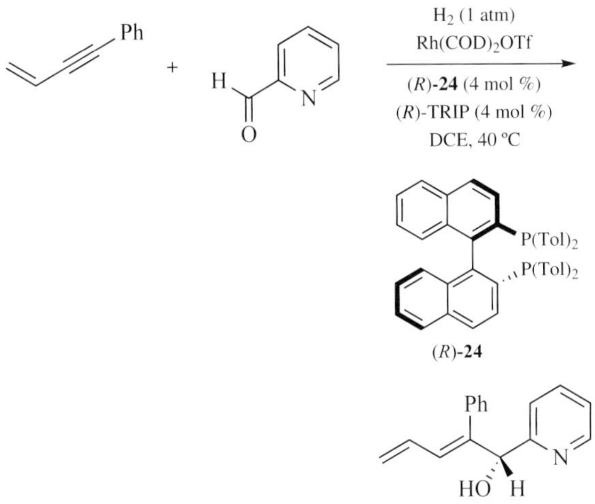

(51)

56% Yield
er 91:1

Toste made use of (*R*)-TRIP in the gold(I)-mediated hydroamination and hydroalkoxylation of allenes.[43] In both cases, the silver salt of (*R*)-TRIP was used with an achiral gold(I) precatalyst to generate the gold(I)-(*R*)-TRIP active catalyst in situ by precipitation of silver chloride. As shown in eq 52, cyclization of the sulfonamide onto the allene system proceeds under exceedingly mild conditions in good yield with virtually complete enantioselectivity. Yields and enantioselectivity are very similar for compounds having other substitution patterns on the allene system or other alkyl groups on the hydrocarbon chain between the sulfonamide and the allene.

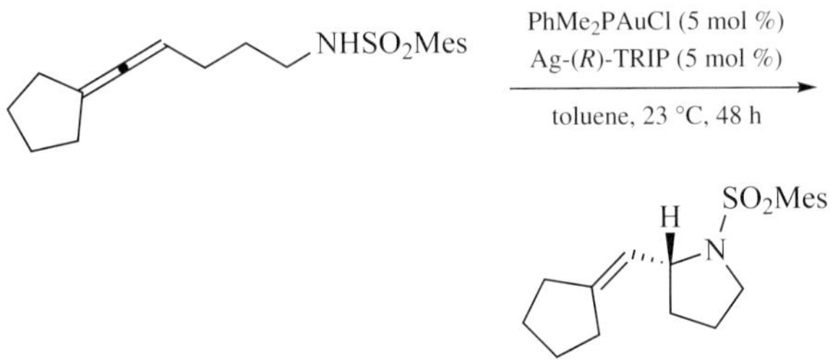

(52)

88% Yield
er 99:1

Using a slightly different Au(I) source under otherwise similar conditions, the allenic alcohol (eq 53) cyclizes to the tetrahydrofuran derivative with high enantioselectivity. As with the cyclization of eq 52, differing substitution patterns on either the allene or on the hydrocarbon chain caused no loss of enantioselectivity. In one case examined, lengthening the carbon chain by one carbon allowed for cyclization to a tetrahydropyran derivative with no loss of enantioselectivity in comparison to the tetrahydrofuran examples.

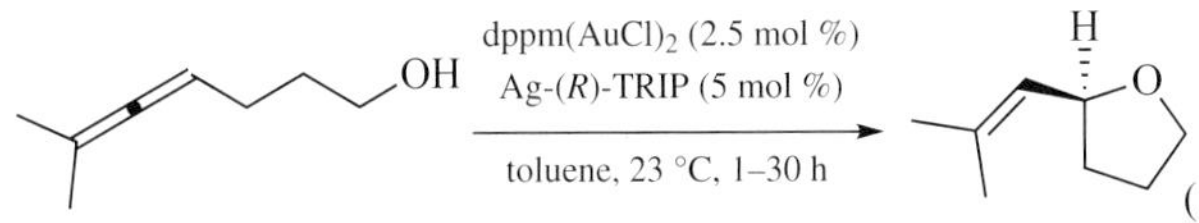

(53)

91% Yield
er 97.5:2.5

List has also found that (*R*)-TRIP can mediate the enantioselective formation of quaternary stereocenters by Trost–Tsuji allyllation.[44] As shown in eq 54, 2-arylpropionaldehydes are smoothly allylated using allyl-benzhydryl-amine in the presence of (*R*)-TRIP and a Pd(0) catalyst. In general, yields and enatioselectivity are high, although the stereocontrol falls off considerably for 2-alkylpropionaldehydes.

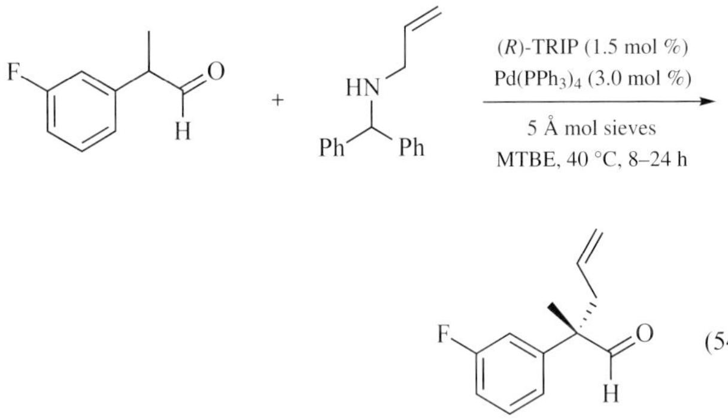

(54)

85% Yield
er 98:2

Antilla has also shown that (*R*)-TRIP will catalyze the enatioselective addition of allyl boronates to benzaldehyde derivatives.[45] As shown in eq 55, treatment of a number of substituted benzaldehyde derivatives react with the allyl boronate affording the corresponding alcohols in good yields and with excellent enantioselectivity. These results were general for benzaldehydes with either electron-withdrawing or electron-donating substituents, as well as cinnamaldehyde derivatives. However, aliphatic aldehydes often gave the opposite sense of asymmetric induction, with much diminished stereocontrol.

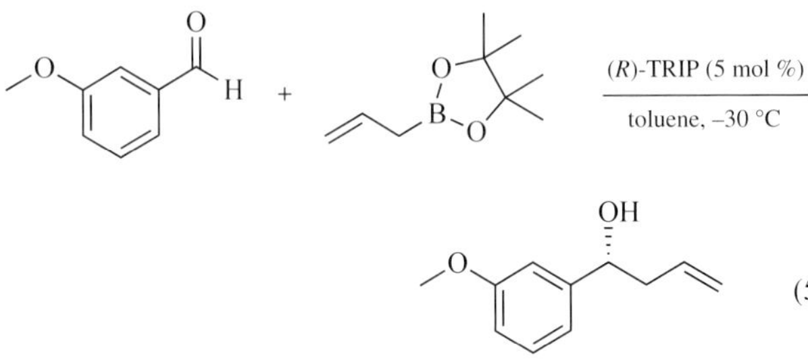

(55)

96% Yield
er 98.5:1.5

These results were extended to the (*E*)- and (*Z*)-crotyl boronates, as shown in eqs 56 and 57. The (*E*)-crotyl boronate (eq 56) reacted at 0 °C to give the desired *anti*-isomer with both extremely high diasteroselectivity and enantioselectivity. The (*Z*)-crotyl boronate reacted under the standard reaction conditions (eq 55) at −30 °C to give the *syn*-isomer with nearly the same level of stereocontrol (eq 57).

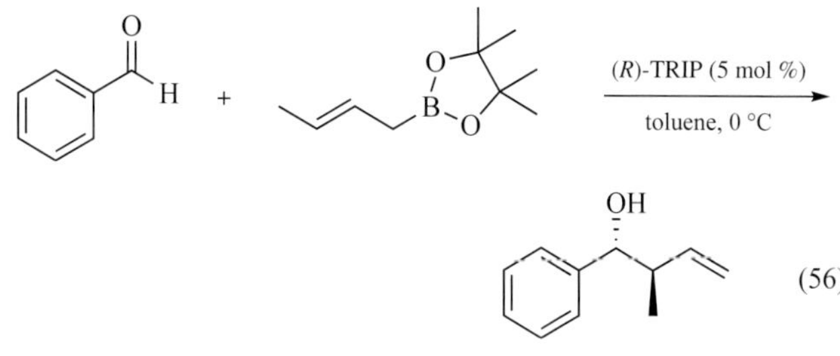

(56)

96% Yield
syn:anti 2:98
er 99.5:0.5

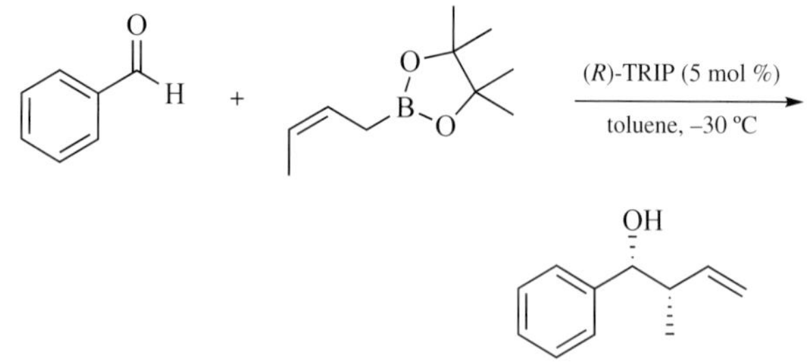

(57)

95% Yield
syn:anti 98:2
er 97:3

Two other silver ion/TRIP-mediated processes have appeared, which do not fit neatly into the previous general categories. The first of these, due to Toste,[46] uses (*S*)-TRIP to enantioselectively open a *meso*-aziridinum ion to produce chiral amino alcohols. Treatment of the racemic amino chloride with (*S*)-TRIP in the presence of silver carbonate and an alcohol (eq 58) produces the protected amino ethers in generally good yields and with high enantioselectivities. In general, results were similar for both primary and secondary alcohols and for amine moieties incorporated into larger rings, such as a piperidine or piperazine.

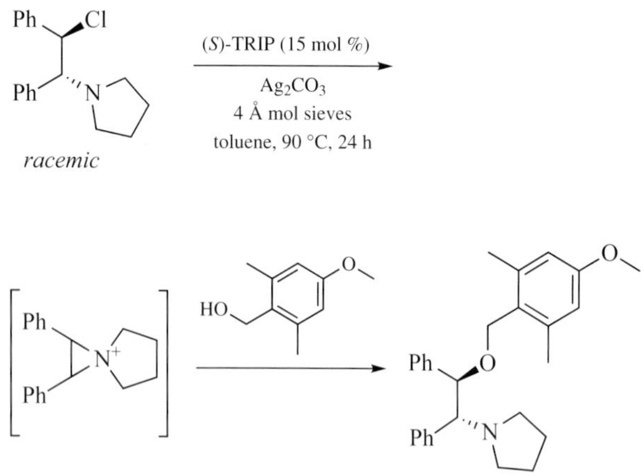

(58)

81% Yield
er 96:4

A silver-free analog of this process was possible in the case of the hydroxy sulfides. As shown in eq 59, the trichloroacetimidate derivative of the hydroxy sulfide is treated with (*S*)-TRIP and an alcohol to afford the corresponding alkoxy sulfide in high yield and with good enantioselectivity.

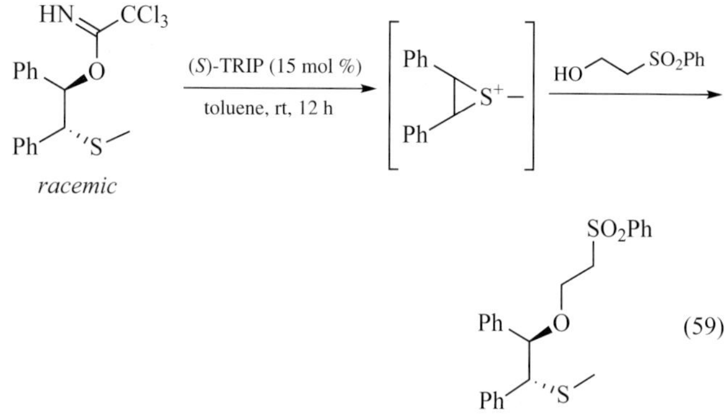

(59)

90% Yield
er 94.5:5.5

In a final example, Tu and coworkers[47] have reported that the TRIP catalysts can promote the semipinacol rearrangement of cyclobutanol derivatives (eq 60). In this process, use of the silver salt of (*R*)-TRIP in many cases gave better enantioselectivities than (*R*)-TRIP by itself. Alkyl substitution on the dihydropyran ring tended to slightly lower the enantioselectivity of the rearrangement.

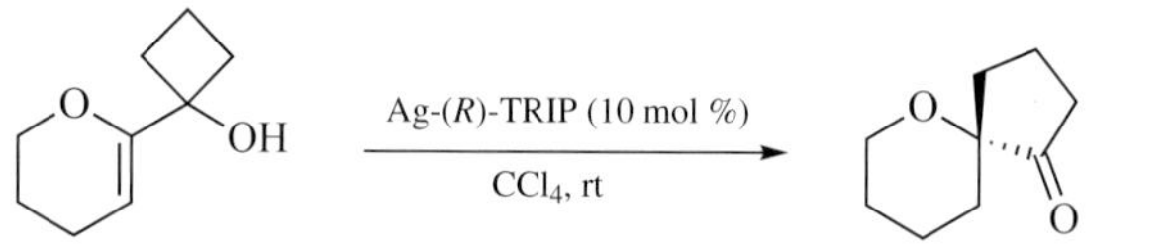

(60)

90% Yield
er 99:1

Earliest results with the TRIP catalysts have been reviewed.[48]

---

1. (a) Akiyama, T.; Itoh, J.; Yokota, K.; Fuchibe, K., *Angew. Chem., Int. Ed.* **2004**, *43*, 1566. (b) Akiyama, T.; Morita, H.; Itoh, J.; Fuchibe, K., *Org. Lett.* **2005**, *7*, 2583.
2. (a) Uraguchi, D.; Terada, M., *J. Am. Chem. Soc.* **2004**, *126*, 5356. (b) Uriguchi, D.; Sorimachi, K.; Terada, M., *J. Am. Chem. Soc.* **2004**, *126*, 11804. (c) Uriguchi, D.; Sorimachi, K.; Terada, M., *J. Am. Chem. Soc.* **2005**, *127*, 9360.
3. Wipf, P.; Jung, J.-K., *J. Org. Chem.* **2000**, *65*, 6319.
4. Klussmann, M.; Ratjen, L.; Hoffmann, S.; Wakchaure, V.; Goddard, R.; List, B., *Synlett* **2010**, 2189.
5. Hoffmann, S.; Seayad, A. M.; List, B., *Angew. Chem., Int. Ed.* **2005**, *44*, 7424.
6. (a) Zhu, C.; Akiyama, T., *Org. Lett.* **2009**, *11*, 4180. (b) Zhu, C.; Akiyama, T., *Adv. Synth. Catal.* **2010**, *352*, 1846.
7. Hoffmann, S.; Nicoletti, M.; List, B., *J. Am. Chem. Soc.* **2006**, *128*, 13075.
8. Zhou, J.; List, B., *J. Am. Chem. Soc.* **2007**, *129*, 7498.
9. Schrader, W.; Handayani, P. P.; Zhou, J.; List, B., *Angew. Chem., Int. Ed.* **2009**, *48*, 1463.
10. Wakchaure, V. N.; Zhou, J.; Hoffmann, S.; List, B., *Angew. Chem., Int. Ed.* **2010**, *49*, 4612.
11. Enders, D.; Liebich, J. X.; Raabe, G., *Chem. Eur. J.* **2010**, *16*, 9763.
12. (a) Li, C.; Wang, C.; Villa- Marcos, B.; Xiao, J., *J. Am. Chem. Soc.* **2008**, *130*, 14450. (b) Li, C.; Villa-Marcos, B.; Xiao, J., *J. Am. Chem. Soc.* **2009**, *131*, 6967.
13. Liu, X.-Y.; Che, C.-M., *Org. Lett.* **2009**, *11*, 4204.

14. Terada, M.; Sorimachi, K., *J. Am. Chem. Soc.* **2007**, *129*, 292.
15. Jia, Y.-X.; Zhong, J.; Zhu, S.-F.; Zhang, C.-M.; Zhou, Q.-L., *Angew. Chem., Int. Ed.* **2007**, *46*, 5565.
16. Terada, M.; Machioka, K.; Sorimachi, K., *Angew. Chem., Int. Ed.* **2009**, *48*, 2553.
17. Zhang, G.-W.; Wang, L.; Nie, J.; Ma, J.-A., *Adv. Syn. Catal.* **2008**, *350*, 1457.
18. Nie, J.; Zhang, G.-W.; Wang, L.; Fu, A.; Ma, J.-A., *Chem. Commun.* **2009**, 2356.
19. Nie, J.; Zhang, G.-W.; Wang, L.; Zheng, D.-H.; Zheng, Y.; Ma, J.-A., *Eur. J. Org. Chem.* **2009**, 3145.
20. Husmann, R.; Sugiono, E.; Mersmann, S.; Raabe, G.; Rueping, M.; Bolm, C., *Org. Lett.* **2011**, *13*, 1044.
21. Simón, L.; Goodman, J. M., *J. Org. Chem.* **2010**, *75*, 589.
22. Seayad, J.; Seayad, A. M.; List, B., *J. Am. Chem. Soc.* **2006**, *128*, 1086.
23. Bergonzini, G.; Gramigna, L.; Mazzanti, A.; Fochi, M.; Bernardi, L.; Ricci, A., *Chem. Commun.* **2012**, *46*, 327.
24. Akiyama, T.; Tamura, Y.; Itoh, J.; Morita, H.; Fuchibe, K., *Synlett* **2006**, 141.
25. (a) Itoh, J.; Fuchibe, K.; Akiyama, T., *Angew. Chem., Int. Ed.* **2006**, *45*, 4796. (b) Rueping, M.; Azap, C., *Angew. Chem., Int. Ed.* **2006**, *45*, 7832. (c) Liu, H.; Cun, L.-F.; Mi, A.-Q.; Jiang, Y.-Z.; Gong, L.-Z., *Org. Lett.* **2006**, *8*, 6023. (d) Momiyama, N.; Tabuse, H.; Terada, M., *J. Am. Chem. Soc.* **2009**, *131*, 12882.
26. Akiyama, T.; Morita, H.; Fuchibe, K., *J. Am. Chem. Soc.* **2006**, *128*, 13070.
27. Cheng, X.; Vellalath, S.; Goddard, R.; List, B., *J. Am. Chem. Soc.* **2008**, *130*, 15786.
28. (a) Uzunov, D. P.; Zivkovich, I.; Pirkle, W. H.; Costa, E.; Guidotti, A., *J. Pharm. Sci.* **1995**, *84*, 937. (b) Hyun, M. H.; Pirkle, W. H., *J. Chromatogr. A* **2000**, *876*, 221.
29. Liu, W.-J.; Chen, X.-H.; Gong, L.-Z., *Org. Lett.* **2008**, *10*, 5357.
30. Vellalath, S.; Čorić, I.; List, B., *Angew. Chem., Int. Ed.* **2010**, *49*, 9749.
31. (a) Čorić, I.; Vellalath, S.; List, B., *J. Am. Chem. Soc.* **2010**, *132*, 8536. (b) Čorić, I.; Vellalath, S.; List, B., *J. Am. Chem. Soc.* **2010**, *132*, 12155. (c) Čorić, I.; Müller, S.; List, B., *J. Am. Chem. Soc.* **2012**, *132*, 17370.
32. Yang, J. W.; HechavarriaFronseca, M. T.; Vignola, N.; List, B., *Angew. Chem., Int. Ed.* **2005**, *44*, 108.
33. Mayer, S.; List, B., *Angew. Chem., Int. Ed.* **2006**, *45*, 4193.
34. Martin, N. J. A.; List, B., *J. Am. Chem. Soc.* **2006**, *128*, 13368.
35. Wang, X.; List, B., *Angew. Chem., Int. Ed.* **2008**, *47*, 1119.
36. Wang, X.; Reisinger, C. M.; List, B., *J. Am. Chem. Soc.* **2008**, *130*, 6070.
37. Xu, S.; Wang, Z.; Zhang, X.; Ding, K., *Eur. J. Chem.* **2011**, 110.
38. Li, G.; Kaplan, M. J.; Wojtas, L.; Antilla, J. C., *Org. Lett.* **2010**, *12*, 1960.
39. Akiyama, T.; Katoh, T.; Mori, K., *Angew. Chem., Int. Ed.* **2009**, *48*, 4226.
40. Mori, K.; Katoh, T.; Suzuki, T.; Noji, T.; Yamanaka, M.; Akiyama, T., *Angew. Chem., Int. Ed.* **2009**, *48*, 9652.
41. Hajos, Z. G.; Parrish, D. R., *Org. Synth.* **1990**, *7*, 363.
42. Komanduri, V.; Krische, M. J., *J. Am. Chem. Soc.* **2006**, *128*, 16448.
43. Hamilton, G. L.; Kang, E. J.; Mba, M.; Toste, F. D., *Science* **2007**, *317*, 496.
44. Mukherjee, S.; List, B., *J. Am. Chem. Soc.* **2007**, *129*, 11336.
45. Jain, P.; Antilla, J. C., *J. Am. Chem. Soc.* **2010**, *132*, 11884.
46. Hamilton, G. L.; Kanai, T.; Toste, F. D., *J. Am. Chem. Soc.* **2008**, *130*, 14984.
47. Zhang, Q.-W.; Fan, C.-A.; Zhang, H.-J.; Tu, Y.-Q.; Zhao, Y.-M.; Gu, P.; Chen, Z.-M., *Angew. Chem., Int. Ed.* **2009**, *48*, 8572.
48. Adair, G.; Mukherjee, S.; List, B., *Aldrichchimica Acta* **2008**, *41*, 31.

Douglas K. Hutchinson
Bedoukian Research, Danbury, CT, USA

N-[(1*S*)-1-(Hydroxydiphenylmethyl)-3-methylbutyl]-2-pyrrolidinecarboxamide

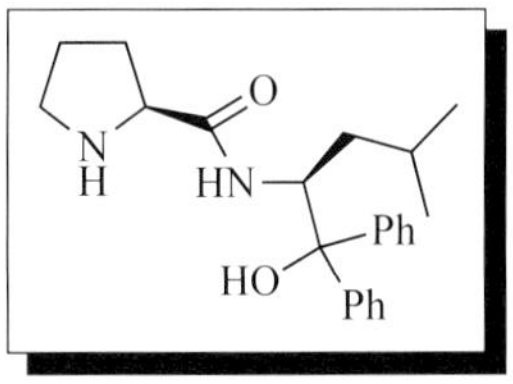

[1088110-92-2] $C_{23}H_{30}N_2O_2$ (MW 366.49)

InChI = 1S/C23H30N2O2/c1-17(2)16-21(25-22(26)20-14-9-15-24-20)23(27,18-10-5-3-6-11-18)19-12-7-4-8-13-19/h3-8,10-13,17,20-21,24,27H,9,14-16H2,1-2H3,(H,25,26)/t20-,21-/m0/s1

InChIKey = PWBULUUPVDXEQS-SFTDATJTSA-N

(organocatalyst used in the enantioselective direct aldol reaction in both organic[1] and aqueous media[2])

Alternate Name: (*S*)-*N*-((*S*)-1-hydroxy-4-methyl-1,1-diphenylpentan-2-yl)pyrrolidine-2-carboxamide.

Physical Data: mp 184–187 °C; $[\alpha]_D^{25}$ −45.9 (*c* 1.2, $CHCl_3$).

Solubility: soluble in organic solvents such as CH_2Cl_2, $CHCl_3$, and acetone.

Form Supplied in: white amorphous solid.

Analysis of Reagent Purity: NMR and elemental analysis.[1]

Preparative Methods: this proline-derived ligand is prepared in two steps starting from *N*-(*tert*-benzyloxycarbonyl)-L-proline (**1**) and (−)-*α*,*α*-diphenyl-leucinol (**2**) following Singh's procedure (eq 1).[1]

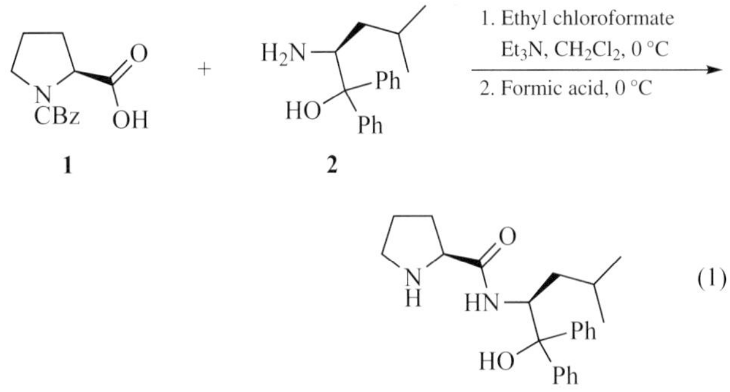

(1)

Purification: during synthesis, the *N*-(*tert*-benzyloxycarbonyl)-L-prolinamide intermediate is purified by recrystallization from ethyl acetate, while the final product is obtained pure after basic workup.

Handling, Storage, and Precautions: no special handling or storage precautions are necessary. There is no known toxicity, however the compound may be harmful by inhalation, ingestion, or skin absorption and may cause eye and skin irritation.

Enantioselective Direct Aldol Reaction. The homochiral bifunctional prolinamide[3] catalyst (**3**) was shown to promote the direct aldol reaction[4] between acetone and a variety of aldehydes with a high level of enantioselectivity (eq 2).[1] Interestingly, the enamine-type activation[5] involved during the process is thought to take advantage of a double hydrogen bonding[6] that guides the approach of the ketone.

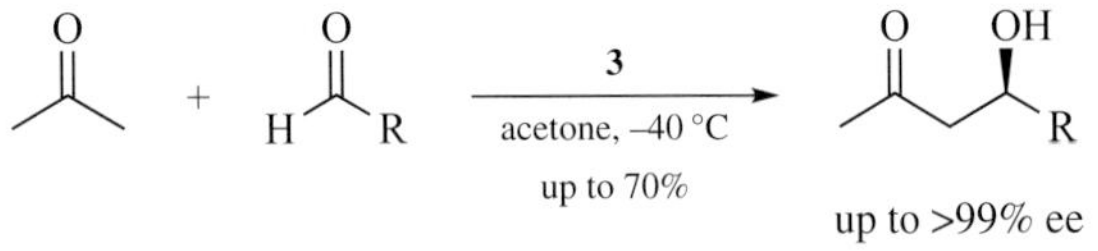
(2)

Higher reactivities in conjunction with excellent enantio- (up to >99% ee) and diastereoselectivities (*syn/anti* up to 4/96) are observed when the reaction is performed in an aqueous medium, thus allowing the method to be generalized to a variety of ketones (eq 3).[2,7]

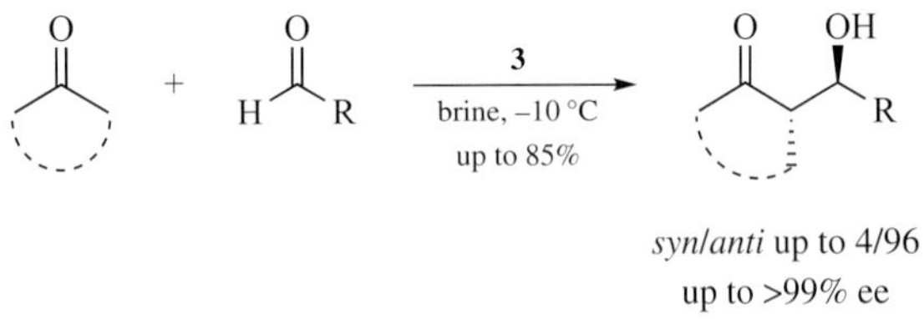
(3)

The enantioselective direct aldol combines with a lactonization reaction to afford a straightforward access to chiral 3-substituted phthalides with up to 99% ee (eq 4).[8] It is worth noting that the use of an acidic additive, such as benzoic acid, significantly enhances the efficiency of the reaction without altering the enantioselectivity.

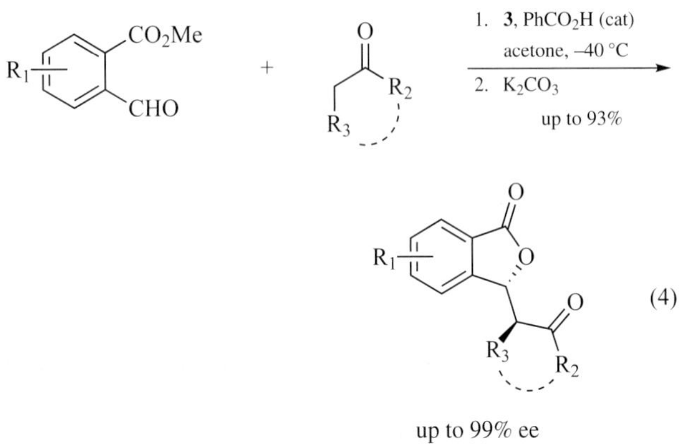
(4)

Chiral prolinamide (**3**) has also been used to prepare 1,3-diols via a multicatalytic process that features a direct aldol reaction followed by a biocatalytic reduction.[9] This highly enantioselective one-pot transformation involves the use of **3** to generate a β-hydroxyketone that is subsequently reduced to the corresponding 1,3-diol by addition of an enzyme, alcohol dehydrogenase (ADH), and its cofactor, nicotinamide adenine dinucleotide (NADH) (eq 5).

Enantioselective Friedel–Craft Reaction of Indoles with Isatins. The proline-derived organocatalyst (**3**) has recently been used in a Friedel–Craft reaction between indole and isatin to afford 3-indolyl-3-hydroxy-2-oxindole; however, the enantioselectivity remained rather modest (22% ee) compared to that obtained with other organocatalysts under otherwise identical conditions (eq 6).[10]

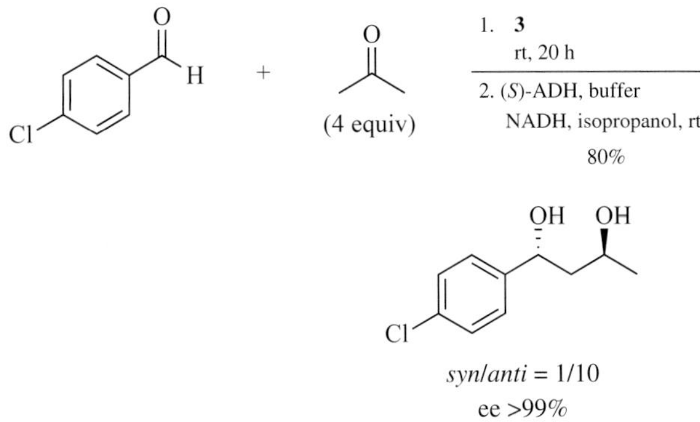
(5)

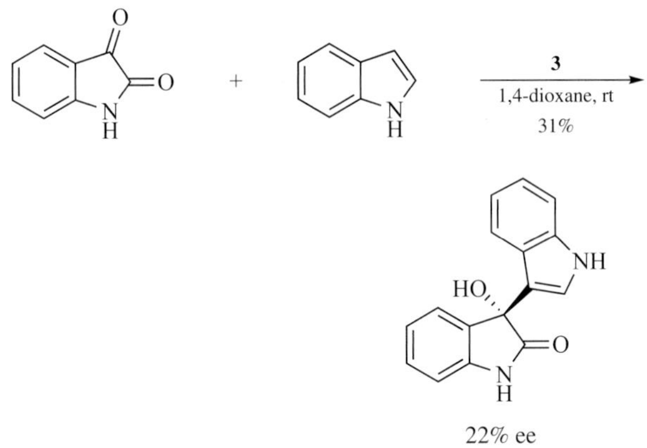
(6)

Related Reagent. (*S*)-*N*-((*S*)-1-Hydroxy-2-phenyl-1,1-diphenylethan-2-yl)pyrrolidine-2-carboxamide.

1. Raj, M.; Maya, V.; Ginotra, S. K.; Singh, V. K., *Org. Lett.* **2006**, *8*, 4097.
2. Maya, V.; Raj, M.; Singh, V. K., *Org. Lett.* **2007**, *9*, 2593.
3. For a review, see Liu, X.; Lin, L.; Feng, X., *Chem. Commun.* **2009**, 6145.
4. For a review, see (a) Guillena, G.; Najera, C.; Ramon, D. J., *Tetrahedron: Asymmetry* **2007**, *18*, 2249. (b) Trost, B. M.; Brindle, C. S., *Chem. Soc. Rev.* **2010**, *39*, 1600.
5. For a review, see Mukherjee, S.; Yang, J. W.; Hoffmann, S.; List, B., *Chem. Rev.* **2007**, *107*, 5471.
6. Chen, X.-H.; Yu, J.; Gong, L.-Z., *Chem. Commun.* **2010**, *46*, 6437.
7. Vishnumaya, R. M.; Singh, V. K., *J. Org. Chem.* **2009**, *74*, 4289.
8. Zhang, H.; Zhang, S.; Liu, L.; Luo, G.; Duan, W.; Wang, W., *J. Org. Chem.* **2010**, *75*, 368.
9. Baer, K.; Krauber, M.; Burda, E.; Hummel, W.; Berkessel, A.; Gröger, H., *Angew. Chem., Int. Ed.* **2009**, *48*, 9355.
10. Deng, J.; Zhang, S.; Ding, P.; Jiang, H.; Wang, W.; Li, J., *Adv. Synth. Catal.* **2010**, *352*, 833.

Stellios Arseniyadis
ESPCI ParisTech, Paris, France

Cyril Bressy
Université Paul Cezanne, Marseille, France

trans-4-Hydroxy-L-proline

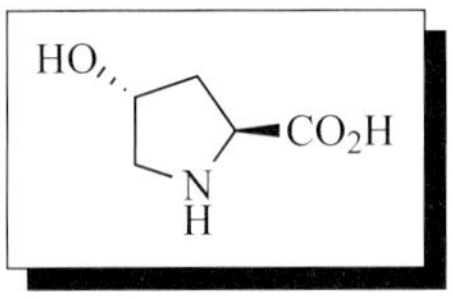

(L-*trans*-4-hydroxy-proline)

[51-35-4] $C_5H_9NO_3$ (MW 131.13)

InChI = 1S/C5H9NO3/c7-3-1-4(5(8)9)6-2-3/h3-4,6-7H,1-2H2,(H,8,9)/t3-,4+/m1/s1

InChIKey = PMMYEEVYMWASQN-DMTCNVIQSA-N

Alternate Names: (2*S*,4*R*)-4-hydroxypyrrolidine-2-carboxylic acid.

Physical Data: mp 274°C.

$[\alpha]_D^T$ −75.5 (c=5, H_2O).

Solubility: soluble in H_2O, 357.8 g L^{-1} (20 °C).

Form Supplied in: white crystalline powder, widely available.

Handling, Storage, and Precautions: wash thoroughly after handling. 4-hydroxy-L-proline is nonhazardous, nontoxic, and nonflammable. Store in a cool and dry place, a well-closed container. Avoid contact with skin and eyes.

Significance. Hydroxyproline is a common nonproteinogenic amino acid, which plays a key role in collagen stability and has many synthetic applications.

Used as a starting material for the preparation of important synthetic intermediates and nucleopeptides.

Used as chiral ligand in combination with different metal halides particularly copper for metal-catalyzed synthetic transformations.

As Substrate.

cis-4-Hydroxy-D-proline. 4-Hydroxy-L-proline is one of the most well-known substrate for various naturally occurring compounds as well as biologically active synthetic intermediates.[1] Treatment of *N*-acyl-*trans*-4-hydroxy-L-prolines with acetic anhydride leads to the formation of *cis*-4-hydroxy-D-proline by acidic cleavage of (1*R*,4*R*)-*N*-acyl-2-oxa-5-aza-bicyclo[2.2.1]-heptan-3-ones intermediate (eq 1).[2] Similarly *cis*- and *trans*-4-azido-L- and D-prolines,[3] protected *cis*- and *trans*-3-azido-L-prolines[4] have been synthesized from the corresponding hydroxyproline precursors by using diphenylphosphoryl azide under Mitsunobu conditions.

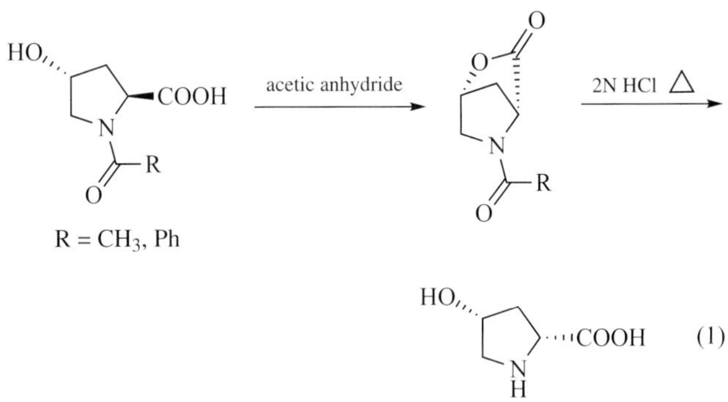

(1)

Natural Product Synthesis. A series of antiviral pyrroline nucleosides,[5] CNS stimulator Kainic acid (eq 2),[6] and Acromelic acid[7] were stereoselectively synthesized from 4-hydroxy-L-proline. Side chains of the antibiotic doripenem hydrate (S-4661)[8] and HCVNS3/4a protease inhibitor vaniprevir (MK-7009)[9] have been synthesized from 4-hydroxy L-proline.

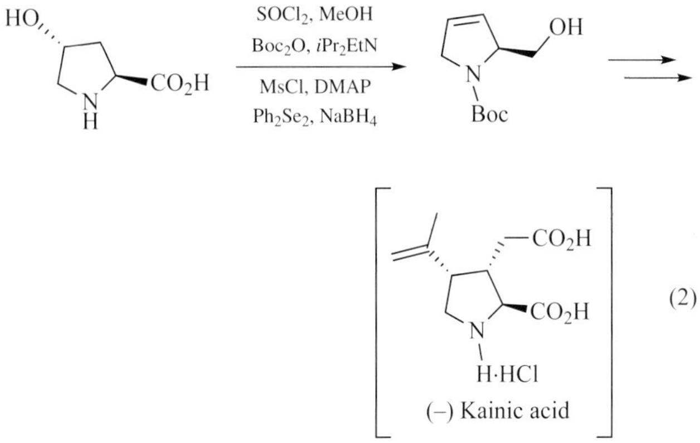

(2)

Condensation Reaction. *Trans*-4-hydroxy-L-proline is an important precursor for the formation of C ring in synthetic analogs of Pyrrolo[2,1-c][1,4]benzodiazepines (PBDs).[10,11] *Trans*-4-hydroxy-L-proline undergoes cyclocondensation reaction with isatoic anhydrides resulting in the formation of PBD dilactam analogs (eq 3).[12]

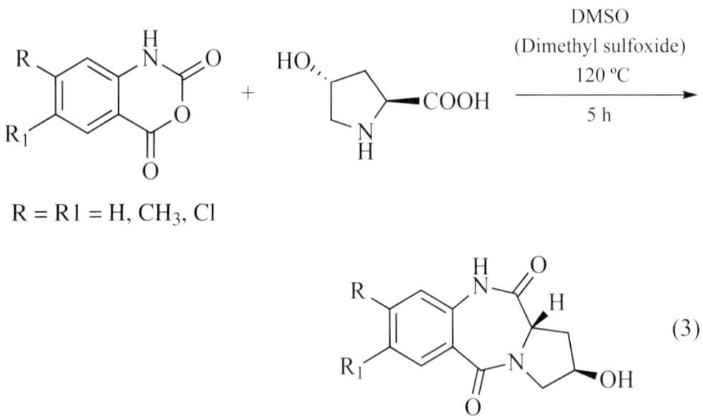

(3)

Dimerization Reaction. *Trans*-4-hydroxy-L-proline is the starting material for the synthesis of C2/C2′-linked PBD dimers (eq 4)[13]; and a sequence-selective PBD-dimer SJG-136 has been synthesized that possesses promising anticancer activity.[14]

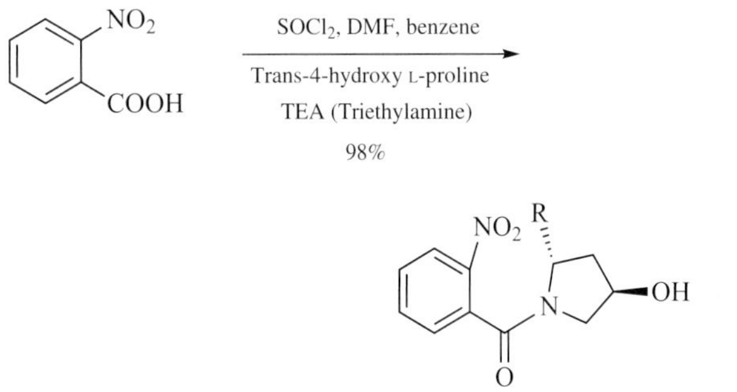

(4)

R = CHO, CH(SEt)$_2$

R = COOH, COOCH$_3$

Trans-4-hydroxy-L-proline undergoes regioselective cyclization in microwave conditions to afford cyclodipeptides (eq 5).[15]

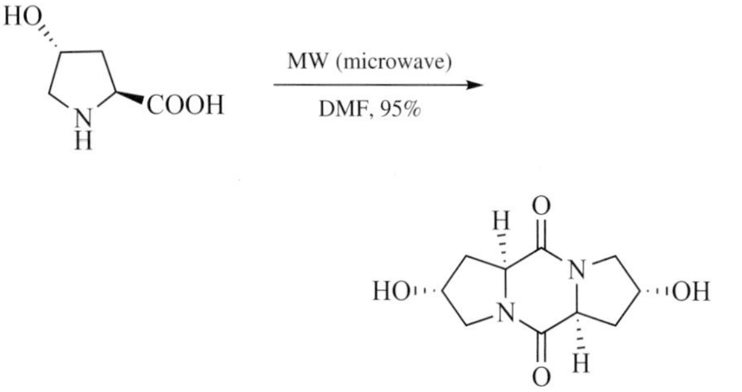

(5)

Dimer of two novel *O*-glycosides linked *trans*-4-hydroxyproline has been reported recently (eq 6).[16]

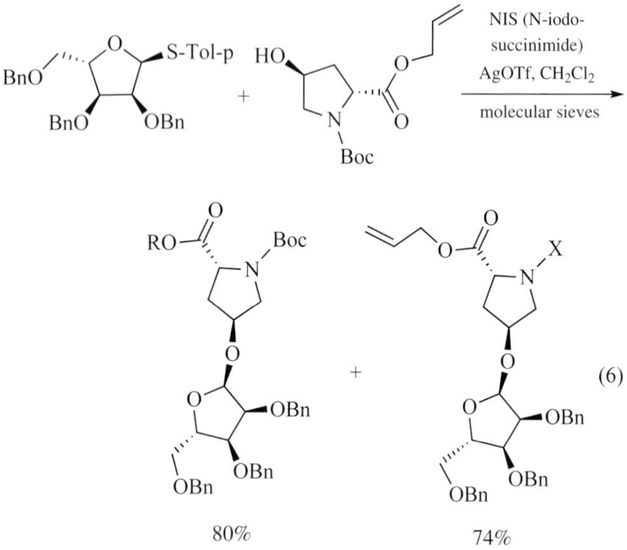

(6)

N-Alkyl and N-Aryl Pyrrole Synthesis. Azomethine ylide formation takes place by the reaction of aldehydes and *trans*-4-hydroxy-L-proline via decarboxylation followed by a redox isomerization reaction for the formation of *N*-alkyl pyrroles under neutral conditions (eq 7).[17]

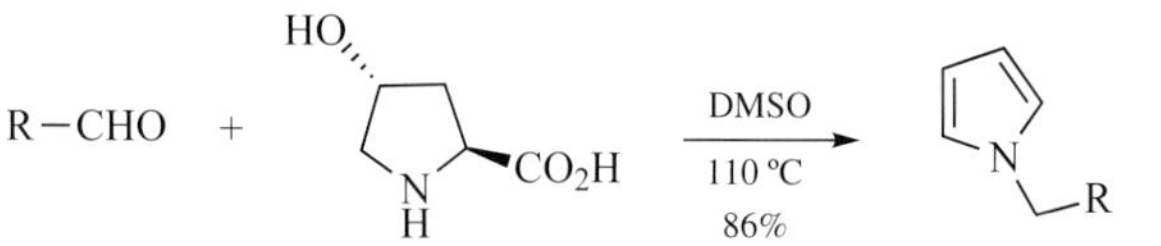

(7)

Similarly *trans*-4-hydroxy-L-proline acts as an effective nucleophilic coupling partner in Cu-catalyzed C–N cross-coupling reaction with aryl halides to produce *N*-aryl pyrroles (eq 8).[18]

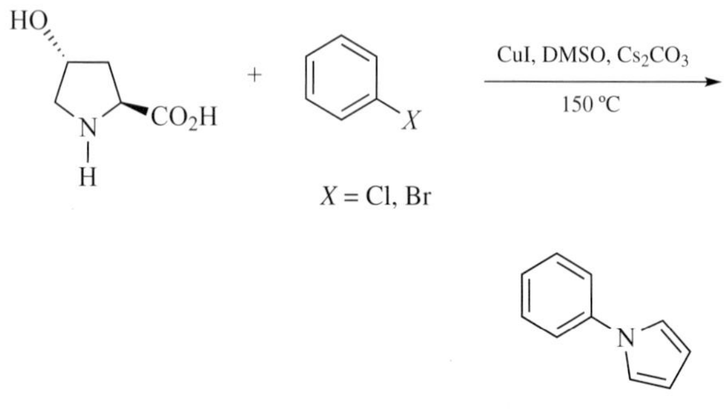

(8)

Trans-4-hydroxy-L-proline is converted to 4-flouro pyrrolidine derivatives by reacting with fluolead reagent ($ArSF_3$)(eq 9).[19] These fluoropyrrolidines behave as DPP-IV inhibitors.[20]

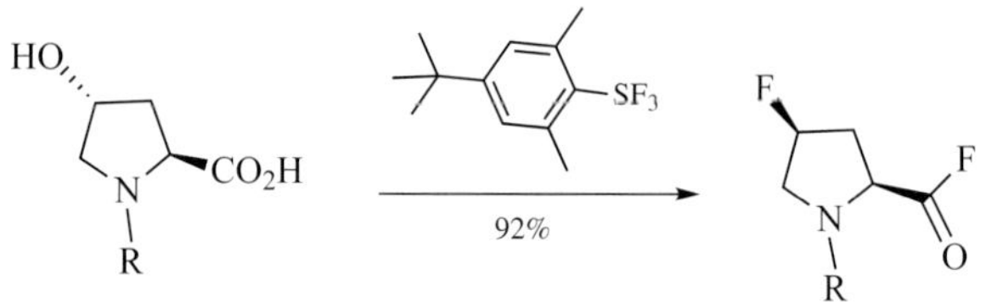

(9)

R = Boc (t-butoxycarbonyl)
Fmoc (a-fluoronylmethoxycarbonyl)
cbz (benyloxycarbonyl), $COCH_2Cl$

Polymer-supported Synthesis (Diversity to library design). Polymer-supported synthesis of bicyclic *γ-lactams* based on 2,5-diazabicylco[2.2.1]heptan-3-one skeleton with various *N*-substituents were successfully synthesized on solid phase using *trans*-4-hydroxy-L-proline as starting material. *N*-Benzyloxycarbonyl-4-hydroxy-*trans*-L-proline was treated with *O*-*tert*-butyl *N,N*-diisopropylisourea to furnish the corresponding *tert*-butyl ester followed by hydrogenolysis over Pd/C to deprotect Cbz group. The obtained secondary amine was immobilized onto REM resin via Michael addition to synthesize a library of bicyclic *γ-lactams* (eq 10).[21]

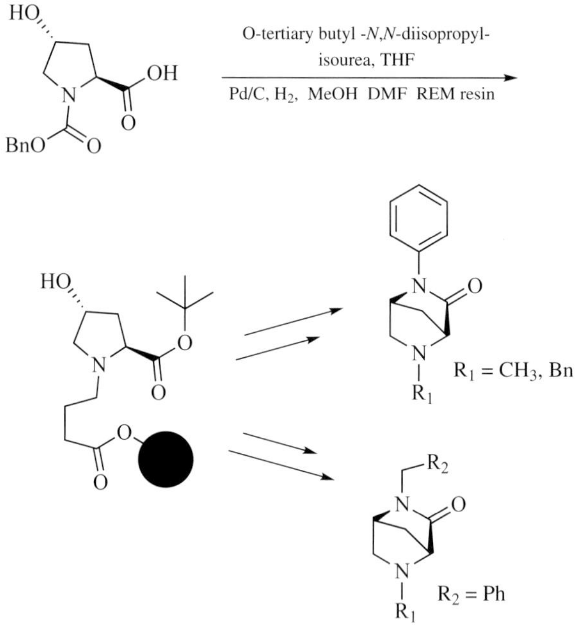

(10)

Highly substituted hydroxyproline-based 2,5-diketopiperazines have also been synthesized from 4-hydroxy proline on solid support.[22] A library of PBD derivatives has been constructed by using Merrifield resin-coupled *trans*-4-hydroxy-L-proline (eq 11).[23]

Preparation of Useful Polymers. Polyamide of *trans*-4-hydroxy-L-proline can be used as porogen filler in nondegradable acrylic bone cement. These polymers are prepared from commercially available *trans*-4-hydroxy-L-proline by using esterification and polycondensation.[24] Poly(*trans*-4-hydroxy-L-proline ester) (PHP ester), a self-destroying, biodegradable, and polycationic polyester, is prepared from naturally occurring *trans*-4-hydroxy-L-proline (eq 12).[25]

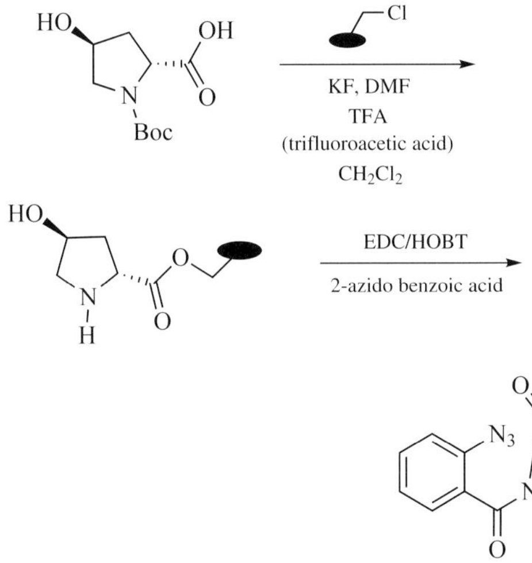

(11)

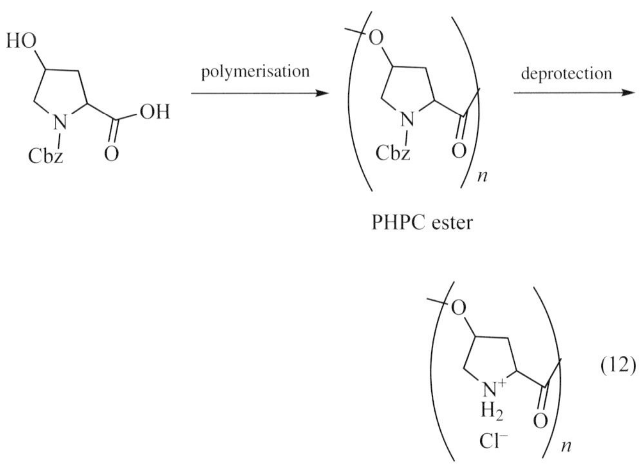

PHPC ester

(12)

PHP ester

Recently polystyrene-based copolymers with tunable incorporations of 4-hydroxy L-proline have been synthesized. These copolymers are used as organocatalysts with high turnover number and with excellent diastereo- and enantioselectivity (eq 13).[26]

As Catalyst. *Trans*-4-hydroxy-L-proline catalyzes asymmetric aldol, Michael addition reactions resulting in the formation of β-hydroxyl ketones, Knoevenagel reaction of dimethyl malonate and 3-methyl butyraldehyde catalyzed by hydroxy proline provide excellent yields of the products.[27] Immobilized *trans*-4-hydroxy-L-proline on polystyrene shows a remarkable improvement in enantioselectivity of asymmetric aldol reactions.[28]

CuI/*trans*-4-hydroxy-L-proline-catalyzed Ullmann-type biaryl etherification of 2-iodo-trifluoroacetanilide with ethyl 2-methylacetoacetate showed significant improvement in the yield and enantioselectivity (eq 14).[29]

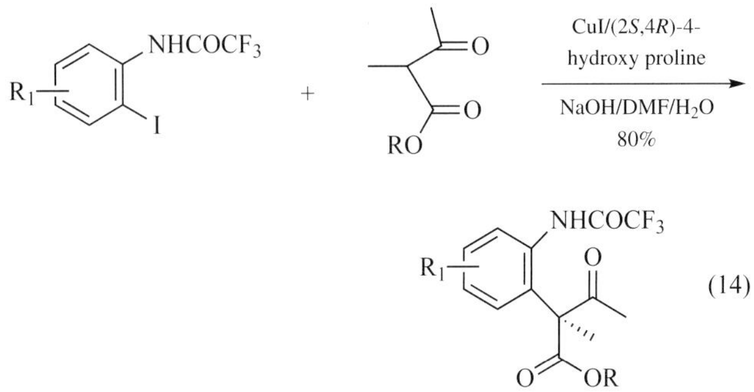

(14)

N-Substituted 1,3-dihydrobenzimidazol-2-ones were obtained from methyl *o*-haloarylcarbamates and *trans*-4-hydroxy-L-proline/CuI-catalyzed coupling with amines (eq 15).[30]

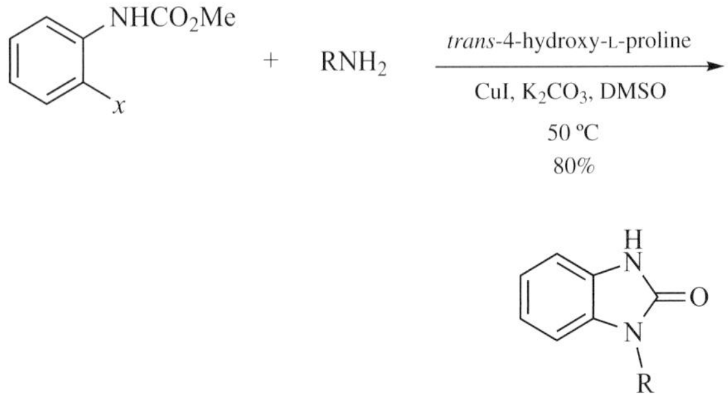

(15)

CuI/4-hydroxy-L-proline-catalyzed coupling of aryl bromides with *N*-Boc hydrazine takes place in DMSO at 80 °C to give *N*-aryl hydrazides. It is an example of amino acid-promoted Ullmann-type reaction, in which 4-hydroxy-L-proline was used as ligand, leading to the coupling of aryl bromides with *N*-Boc hydrazine to afford *N*-arylation products in good yields (eq 16).[31]

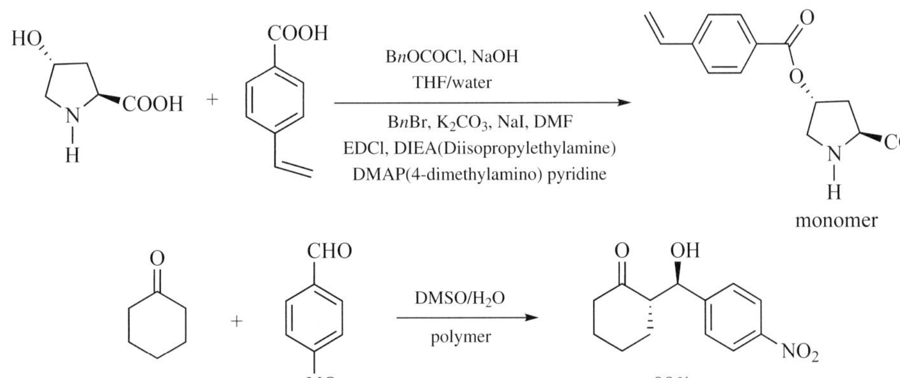

monomer

98% ee

(13)

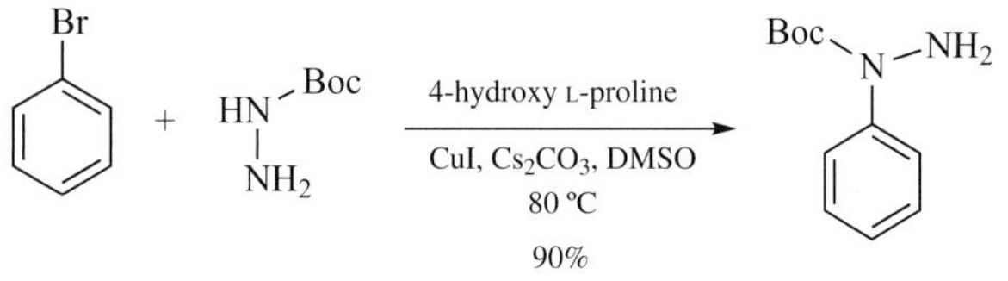

(16)

CuI/4-hydroxy-L-proline-catalyzed reactions of *N*-substituted *o*-bromobenzamides with formamide affords 3-substituted quinazolinones at 80°C (eq 17).[32]

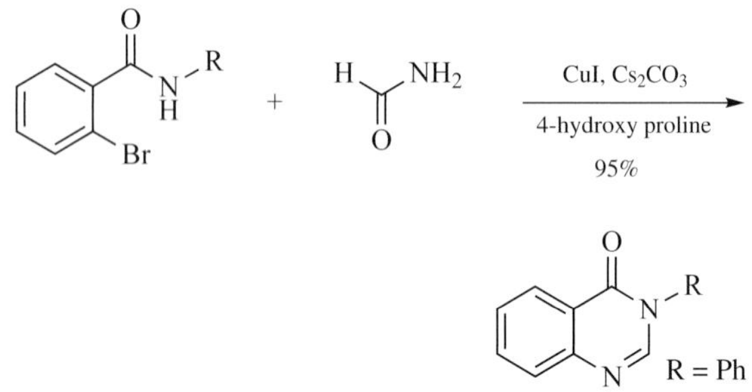

(17)

1. Murphy, A. C.; Mitova, M. I.; Blunt, J. W.; Munro, M. H. G., *J. Nat. Prod.* **2008**, *71*, 806.
2. Croce, P. D.; Rosa, C. L., *Tetrahedron: Asymmetry* **2002**, *13*, 197.
3. Silverman, R. B., *Acc. Chem. Res.* **2009**, *42*, 439.
4. Gomez-Vidal, J. A.; Silverman, R. B., *Org. Lett.* **2001**, *3*, 2481.
5. Gregson, S. J.; Howard, P. W.; Hartley, J. A.; Brooks, N. A.; Adams, L. J., *J. Med. Chem.* **2001**, *44*, 5.
6. Poisson, J. F.; Orellana, A.; Greene, A. E., *J. Org. Chem.* **2005**, *70*, 10860.
7. Mauger, A. B., *J. Nat. Prod.* **1996**, *59*, 1205.
8. (a) Shin, K. J.; Koo, D. K.; Yoo, K. H.; Kim, D. C.; Kim, D. J.; Park, S. W., *Bioorg. Med. Chem. Lett.* **2000**, *10*, 1421. (b) Niphakis, M. J.; Turunen, B. J.; Georg, G. L., *J. Org. Chem.* **2010**, *75*, 6793.
9. Kong, J.; Chen, C.; Padros, J. B.; Cao, Y.; Dunn, R. F.; Dolman, S. J.; Janey, J.; Li, H.; Zacuto, M. J., *J. Org. Chem.* **2012**, *77*, 3820.
10. Kamal, A.; Bharathi, E. V.; Ramaiah, M. J.; Reddy, J. S.; Dastagiri, D.; Viswanath, A.; Sultana, F.; Pushpavalli, S. N. C. V. L.; Bhadra, M. P.; Subrata, A. J. S.; Zingde, S., *Bioorg. Med. Chem. Lett.* **2010**, *20*, 3310.
11. Kamal, A.; Reddy, P. S. M. M. D.; Reddy, R.; Laxman, E.; Murthy, Y. L. N., *Bioorg. Med. Chem. Lett.* **2004**, *14*, 5699.
12. Clark, R. L.; Carter, K. C.; Mullen, A. B.; Coxon, G. D.; Owusu-Dapaah, G.; McFarlane, E.; Duong Thi, M. D.; Grant, M. H.; Tettey, J. N.; Mackay, S. P., *Bioorg. Med. Chem. Lett.* **2007**, *17*, 624.
13. Reddy, B. S. P.; Damayanthi, Y.; Reddy, B. S. N.; Lown, J. W., *Anticancer Drug Des.* **2000**, *15*, 225.
14. Gregson, S. J.; Howard, P. W.; Hartley, J. A.; Brooks, N. A.; Adams, L. J.; Jenkins, T. C.; Kelland, L. R.; Thurston, D. E., *J. Med. Chem.* **2001**, *44*, 737.
15. Nonappa, N.; Ahonen, K.; Lahtinen, M.; Kolehmainen, E., *Green Chem.* **2011**, *13*, 1203.
16. Xie, N.; Taylor, C. M., *Org. Lett.* **2010**, *12*, 4968.
17. Kumar, A. V.; Rao, K. R., *Tetrahedron Letters* **2011**, *52*, 3237.
18. Reddy, V. P.; Kumar, A. V.; Rao, K. R., *Tetrahedron Letters* **2011**, *52*, 777.
19. Haffner, C. D.; McDougald, D. L.; Reister, S. M.; Thompson, B. D.; Conlee, C.; Fang, J.; Bass, J.; Lenhard, J. M.; Croom, D.; Secosky-Chang, M. B.; Tomaszek, T.; McConn, D.; Wells-Knecht, K.; Johnson, P. R., *Bioorg. Med. Chem. Lett.* **2005**, *15*, 5257.
20. Singh, R. P.; Umemoto, T., *J. Org. Chem.* **2011**, *76*, 3113.
21. Zhu, T.; Yan, Z.; Chucholowski, A.; Li, R., *J. Comb. Chem.* **2005**, *7*, 520.
22. Bianco, A.; Sonksen, C. P.; Roepstorff, P.; Briand, J. P., *J. Org. Chem.* **2000**, *65*, 2179.
23. Kamal, A.; Reddy, K. L.; Devaiah, V.; Shankaraiah, N.; Reddy, G. S. K.; Raghavan, S., *J. Comb. Chem.* **2007**, *9*, 29.
24. Puska, M.; Yliurpo, A.; Vallittu, P., *J. biomater. appl.* **2005**, *19*, 287.
25. Beom Lim, Y.; Hun Choi, Y.; sang Park, J., *J. Am. Chem. Soc.* **1999**, *121*, 5633.
26. Evans, A. C.; Lu, A.; Ondeck, C.; Longbottom, D. A.; O'Reilly, R. K., *Macromolecules* **2010**, *43*, 6374.
27. Al-Momani, A., *Arkivoc* **2012**, *6*, 101.
28. Kehat, T.; Portnoy, M., *Chem. Commun.* **2007**, *27*, 2823.
29. Xie, X.; Chen, Y.; Dawei Ma., *J. Am. Chem. Soc.* **2006**, *128*, 16050.
30. Zou, B.; Yuan, Q.; Ma, D., *Org. Lett.* **2007**, *9*, 4291.
31. Jiang, L.; Lu, X.; Zhang, H.; Jiang, Y.; Ma, D., *J. Org. Chem.* **2009**, *74*, 4542.
32. Xu, L.; Jiang, Y.; Ma, D., *Org. Lett.* **2012**, *14*, 1150.

Ahmed Kamal & Farheen Sultana
CSIR—Indian Institute of Chemical Technology
Hyderabad, India

(4a*R*,9a*S*)-4-Hydroxy-4,4a,9,9a-tetrahydroindeno[2,1-*b*][1,4]oxazin-3(2*H*)-one

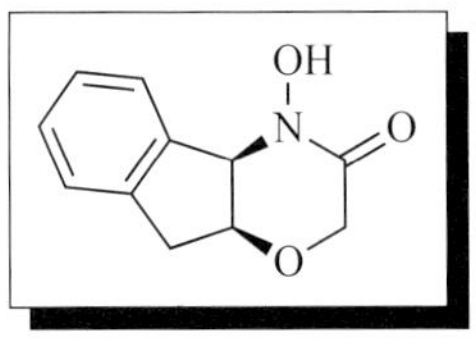

[1346681-70-6]
[1346681-72-8] $C_{11}H_{11}NO_3$ (MW 205.21)
InChI = 1S/C11H11NO3/c13-10-6-15-9-5-7-3-1-2-4-8(7)11(9)12(10)14/h1-4,9,11,14H,5-6H2/t9-,11+/m0/s1
InChIKey = RTMHYIADHDEGTI-GXSJLCMTSA-N

(enantioselective amidation catalyst for kinetic resolution of *N*-heterocycles[1])

Physical Data: colorless needle (recrystallized from MeOH); compound decomposes at temperatures above 210 °C; $[\alpha]_D^{23}$ −42.7 (*c* 0.5, EtOAc).

Solubility: soluble in THF and DMSO; partially soluble in CH_2Cl_2, EtOAc, and MeOH.

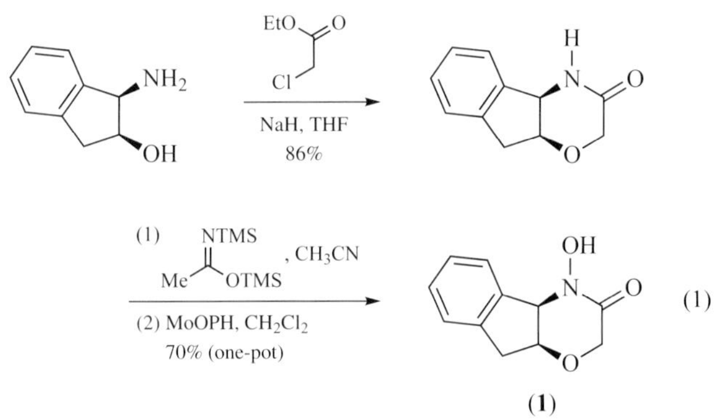

Preparative Methods: the chiral hydroxamic acid can be prepared from (1*R*,2*S*)-1-amino-2,3-dihydro-1*H*-inden-2-ol (eq 1).[1] This lactam (4,4a,9,9a-tetrahydro-1-oxa-4-aza-fluoren-3-one) is activated by *N,O*-bis(trimethylsilyl)acetamide, followed by oxidation with oxodiperoxymolybdenum(pyridine)-(hexamethylphosphorictriamide) (MoO_5·Py·HMPA, MoOPH)[2] to afford the desired hydroxamic acid (**1**).

Purification: column chromatography (silica gel, 100% EtOAc) and recrystallization (MeOH).

Handling, Storage, and Precautions: air and moisture stable; can be stored at room temperature in a desiccator with drying agent. Toxicity data is not available.

Kinetic Resolution of *N*-Heterocycles. To obtain optically pure *N*-heterocycles, classical resolution by formation of diastereomers or separation by chromatography on chiral supports are still the state-of-the-art approaches in the industry, but these methods are usually tedious for rapid screening and not general.[3] The chiral hydroxamic acid (**1**) can serve as an effective acyl-transfer agent from acyl azoliums, which are catalytically generated from α′-hydroxyenones and *N*-heterocyclic carbenes, to perform catalytic kinetic resolution. This method can be easily applied to various classes of cyclic secondary amines, such as piperidines, piperazines, morpholines, tetrahydroisoquinolines, and azepanes, providing selectivity factors (*S*) between 8 and 74 (eq 2).[1,4]

The catalytic kinetic resolution of *N*-heterocycles with a bromo derivative of chiral hydroxamic acid (**2**) with *i*PrOAc or THF as solvent showed improved selectivities and a broader substrate scope.[5] In addition, this bromo chiral hydroxamic acid can be immobilized on a solid support to obtain a robust, reusable stoichiometric reagent for kinetic resolution of chiral *N*-heterocycles.[6]

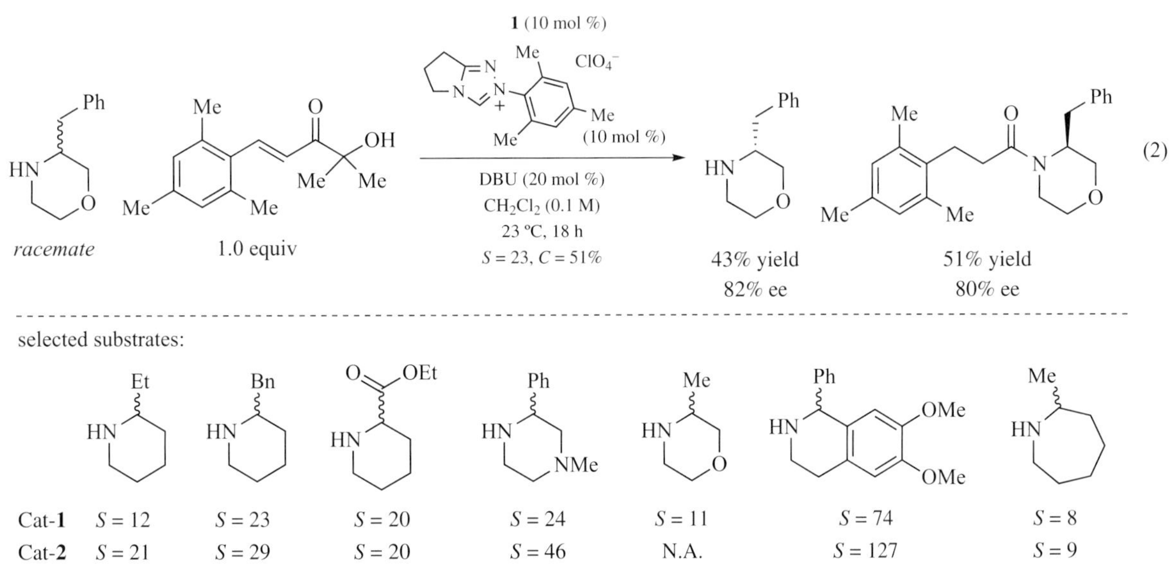

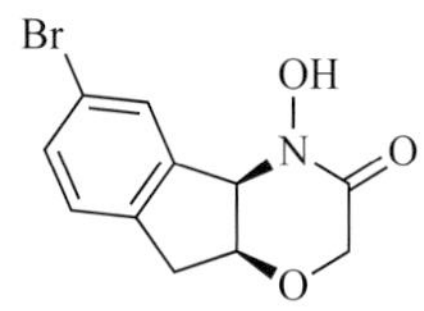

2

Bromohydroxamic acid

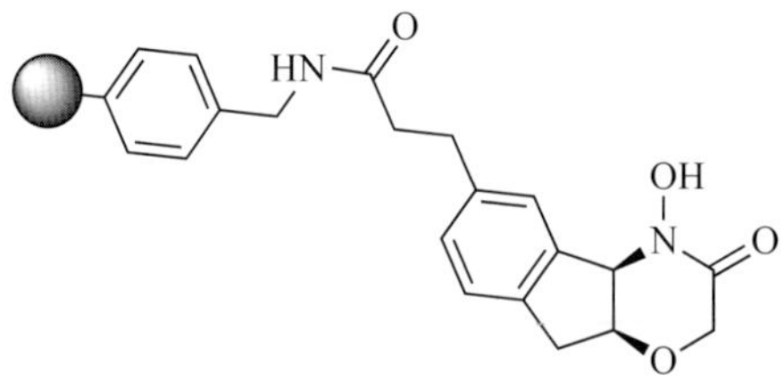

Polymer-supported hydroxamic acid

1. Binanzer, M.; Hsieh, S.-Y.; Bode, J. W., *J. Am. Chem. Soc.* **2011**, *133*, 19698.
2. Vedejs, E.; Larsen, S., *Org. Synth.* **1985**, *64*, 127.
3. Baxter, C. A.; Cleator, E.; Brands, K. M. J.; Edwards, J. S.; Reamer, R. A.; Sheen, F. J.; Stewart, G. W.; Strotman, N. A.; Wallace, D. J., *Org. Process Res. Dev.* **2011**, *15*, 367.
4. (a) Selectivity factor, *S* = (rate of fast-reaching enantiomer)/(rate of slow-reaching enantiomer). (b) Kagan, H. B.; Fiaud, J. C., *Top. Stereochem.* **1988**, *18*, 249. (c) Vedejs, E.; Jure, M.,*Angew. Chem., Int. Ed.* **2005**, *44*, 3974.
5. Hsieh, S.-Y.; Binanzer, M.; Kreituss, I.; Bode, J. W., *Chem. Commun.* **2012**, *48*, 8892.
6. Kreituss, I.; Murakami, Y.; Binanzer, M.; Bode, J. W., *Angew. Chem., Int. Ed.* **2012**, *51*, 10660.

Sheng-Ying Hsieh & Jeffrey W. Bode
Eidgenössische Technische Hochschule, Zürich, Switzerland

1*H*-Imidazolium, 1,3-Bis[2,6-bis-(1-methylethyl)phenyl], tetrafluoroborate(1-)

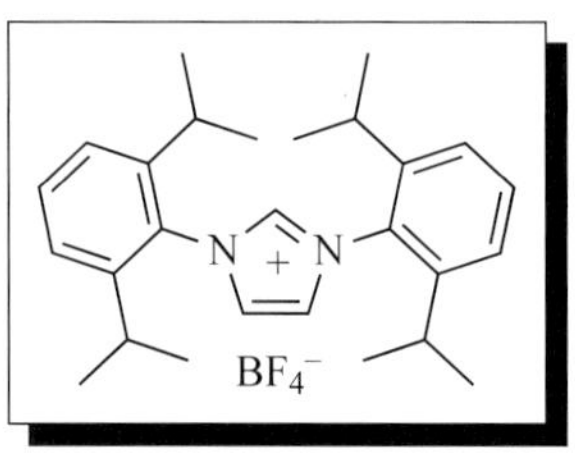

[286014-25-3] $C_{27}H_{37}N_2BF_4$ (MW 476.40)

InChI = 1S/C27H37N2.BF4/c1-18(2)22-11-9-12-23(19(3)4)26(22)28-15-16-29(17-28)27-24(20(5)6)13-10-14-25(27)21(7)8;2-1(3,4)5/h9-21H,1-8H3;/q+1;-1

InChIKey = NTJNPOPDNJTGKC-UHFFFAOYSA-N

(used as a precursor to the free carbene, 1,3-bis(2,6-diisopropyl-phenyl)imidazol-2-ylidene, which is typically used as ligand for the preparation of various transition metal catalysts)

Alternate Name: IPr-HBF_4.

Solubility: highly soluble: CH_2Cl_2, slightly soluble: THF.

Form Supplied in: microcrystalline white solid, not widely available.

Analysis of Reagent Purity: NMR.

Preparative Methods: IPr-HBF_4 can be prepared through a straightforward anion metathesis starting with the corresponding chloride salt (eq 1).[1] This can be accomplished by dissolving IPr-HCl in a minimal amount of water. Next, HBF_4 (40 wt% in water) is added resulting in a white slurry. The desired product is extracted into dichloromethane, dried over $MgSO_4$, and filtered. Removal of the solvent under vacuum and drying yields a white solid.

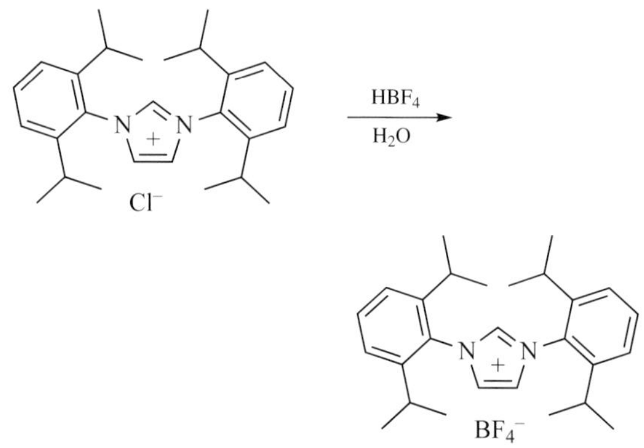

(1)

Purification: recrystallize from dichloromethane by the addition of diethyl ether.

Handling, Storage, and Precautions: compound is air stable and can be stored under air for years.

Free Carbene Preparation. IPr-HBF_4 is primarily used as an alternative precursor to IPr-HCl[2] for the preparation of the corresponding free carbene (IPr).[3] The main advantage of using the HBF_4 versus the HCl salt is due to the ease of handling of IPr-HBF_4, and a more efficient deprotonation resulting in higher yields of IPr.[1] The deprotonation is typically accomplished by treating a THF solution of IPr-HBF_4 with potassium *tert*-butoxide or NaH under an inert atmosphere at ambient temperature (eq 2). After stirring for more than 4 h, the free carbene is isolated by polish filtering and removing the solvent.[1] IPr has also been isolated by first removing the THF followed by dissolving the residual solid in toluene, addition of hexanes to precipitate excess potassium *tert*-butoxide, polish filtration, and removal of the solvent.[4]

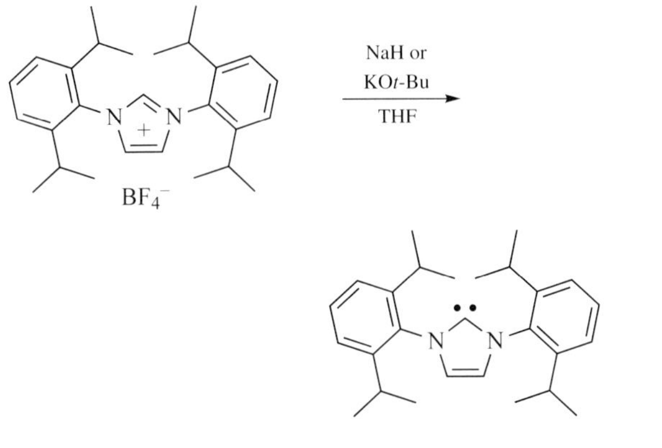

(2)

Although the primary use of IPr-HBF_4 is as an alternative starting material for the synthesis of the corresponding free carbene, this article focuses on the uses of the salt instead of the free carbene. There are a variety of examples of IPr-BF_4 being applied to form metal complexes directly without the isolation of the air-sensitive IPr carbene. In addition, IPr-BF_4 has been used to form catalysts in situ with no isolation of a discrete organometallic species. This is typically accomplished by treatment of IPr-BF_4 with a base, such as NaH or NaO*t*-Bu, to form the carbene followed by direct exposure to a transition metal. The following sections offer select examples of application of IPr-BF_4 for in situ catalyst formation.

Ni-Catalyzed Cross-Coupling with Grignard Reagents. One of the earliest reports of IPr-BF_4 for in situ formation of an active catalyst was reported by Herrmann and coworkers for the selective activation of C–F bonds (eq 3).[5] The in situ preparation of a Ni catalyst performed slightly better than the corresponding $Ni(IPr)_2$ catalyst for the cross-coupling of both electron-rich and electron-deficient aryl fluorides with aryl Grignard reagents.

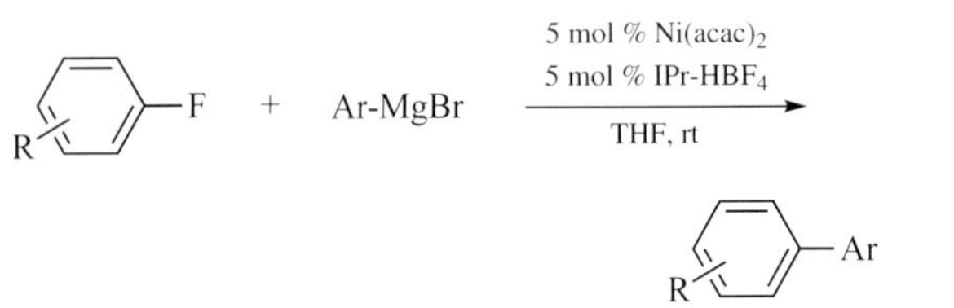

(3)

More recently, IPr-BF_4 has also been used in the in situ formation of a nickel catalyst for the cross-coupling of neopentylarylsulfonates with a variety of different Grignard reagents (eq 4).[6] This was accomplished by charging 1.5 equiv of the Grignard reagent into a solution of the neopentylarylsulfonates, $Ni(acac)_2$, and IPr-BF_4 in THF at ambient temperature. The cross-coupling was successful using a variety of aryl and heteroaryl neopentylarylsulfonates in combination with alkyl and benzyl Grignard reagents.

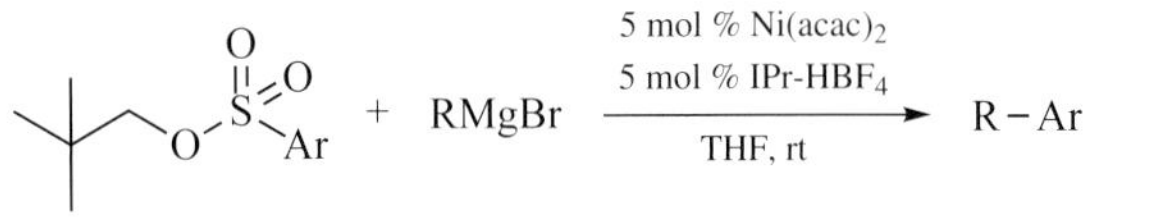

(4)

Palladium-Catalyzed Alcohol Oxidation. IPr has been used as a ligand in a variety of different catalyst systems for the aerobic oxidation and oxidative kinetic resolution of alcohols.[7,8] While IPr-HBF_4 is typically used as a reagent in the synthesis of the catalyst precursor, the active Pd catalyst can also be formed by mixing $Pd(OAc)_2$, IPr-HBF_4, and KO*t*-Bu in THF followed by addition of the substrate in toluene (eq 5).[9] While the oxidation was slightly slower compared to the preformed catalyst, it eliminated the two-step catalyst synthesis.

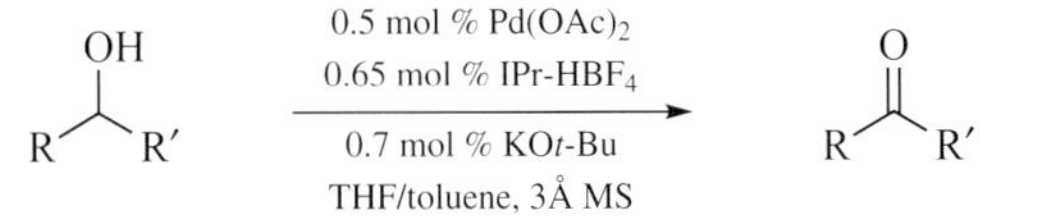

(5)

IPr-CuCl-Catalyzed Hydrosilylation of Ketones. The use of IPr-BF_4 for in situ preparation and application of the IPr-CuCl complex for hydrosilylations was recently disclosed by Nolan and coworkers.[10] The IPr-BF_4 salt resulted in a more efficient catalytic hydrosilylation than the similar IPr-Cl salt. This system was applied to the reduction of a variety of cyclic, aliphatic, and aryl ketones (eq 6). It was recently shown that this system could be applied to the reduction of encumbered ketones by increasing the temperature of the reaction.[11] Other *N*-heterocyclic carbene ligands such as ICy and SIMes proved even more efficient than the IPr-derived catalyst.

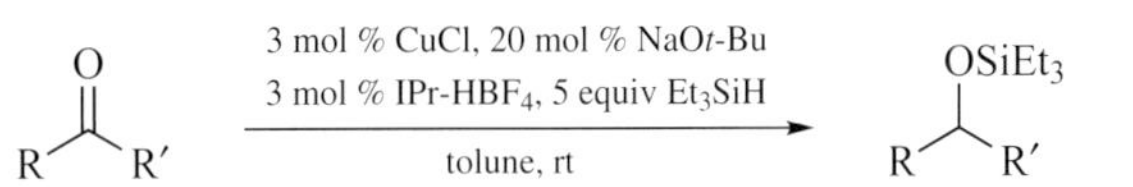

(6)

Palladium-Catalyzed Cyclization of *o*-Bromobenzyl Ketones. Faragó and Kotschy reported that an in situ prepared Pd-IPr catalyzed the ring closure of *o*-bromobenzyl ketones.[12] IPr outperformed a variety of other *N*-heterocyclic carbene ligands. These reactions were conducted by mixing all reagents together in *o*-xylenes under an inert atmosphere at 110 °C (eq 7).

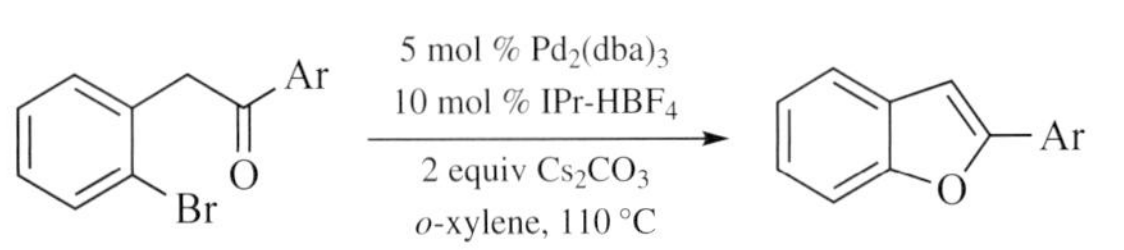

(7)

1. Bantreil, X.; Nolan, S. P., *Nat. Protocols* **2011**, *6*, 69.
2. Fortment, G. C.; Nolan, S. P., *e-EROS*, 2010.
3. Kelly, R. A.; Grasa, G. A.; Nolan, S. P., *e-EROS*, 2010.
4. Dible, B. R.; Sigman, M. S., *J. Am. Chem. Soc.* **2003**, *125*, 872.
5. Böhm, V. P. W.; Gstöttmayr, C. W. K.; Weskamp, T.; Herrmann, W. A., *Angew. Chem., Int. Ed.* **2001**, *40*, 3387.

6. Kim, C.-B.; Jo, H.; Ahn, B.-K.; Kim, C. K.; Park, K., *J. Org. Chem.* **2009**, *74*, 9566.
7. Schultz, M. J.; Hamilton, S. S.; Jensen, D. R.; Sigman, M. S., *J. Org. Chem.* **2005**, *70*, 3343.
8. Jensen, D. R.; Sigman, M. S., *Org. Lett.* **2003**, *5*, 63.
9. Jensen, E. R.; Schultz, M. J.; Mueller, J. A.; Sigman, M. S., *Angew. Chem., Int. Ed.* **2003**, *42*, 3810.
10. Kaur, H.; Zinn, F. K.; Stevens, E. D.; Nolan, S. P., *Organometallics* **2004**, *23*, 1157.
11. Díez-González, S.; Kaur, H.; Zinn, F. K.; Stevens, E. D.; Nolan, S. P., *J. Org. Chem.* **2005**, *70*, 4784.
12. Faragó, J.; Kotschy, A., *Synthesis* **2009**, 85.

Mitchell J. Schultz
Archer Daniels Midland Company, Decatur, IL, USA

I

Iron(2+), tris(2,2′-bipyridine-*N,N*′)-, Dibromide, (OC-6-11)

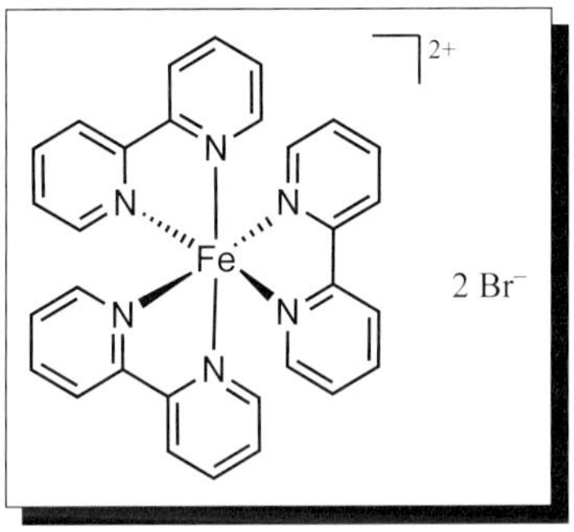

[15388-40-6; 14751-83-8; 15025-74-8; 14586-54-0; 70811-29-9; 28966-86-1] $C_{30}H_{24}Br_2FeN_6$ (MW 684.22)
InChI = 1S/3C10H8N2.2BrH.Fe/c3*1-3-7-11-9(5-1)10-6-2-4-8-12-10;/h3*1-8H;2*1H;/q;+2/p-2
InChIKey = OFTQCNRTJBDRBZ-UHFFFAOYSA-L

(reagent used in electron-transfer reactions, spin crossover, magnetic properties, and for photophysical studies)

Alternate Name: tris(2,2′-bipyridyl)-Fe(II) complex; dipyridyl ferrous dibromide; $Fe(bpy)_3Br_2$; tris(2,2′-bipyridine)iron dibromide; tris(2,2′-bipyridine)iron(2+) dibromide.
Solubility: sol H_2O, poorly soluble in DMSO, DMF, and CH_3CN, insoluble in DCM, toluene, and hexanes.
Form Supplied in: red crystal obtained from chemical supply or prepared by chemical reaction or electrochemical reaction.
Purification: recrystallization from CH_3CN.
Handling, Storage, and Precautions: $Fe(bpy)_3Br_2$ can be stored without any particular precautions in closed bottles or vials. The compound is however quite sensitive to oxidation to $Fe(bpy)_3^{3+}$ in solution (DMF, DMSO, or water). Toxicity (injection) mice LD_{50}: 3.6 mg kg^{-1}. Incompatible with strong acids, strong bases, and oxidizing and reducing agents.

Introduction. The use of iron complexes as catalysts in organic synthesis is becoming an important and attractive research topic[1] for a wide number of reasons. Iron is the second most abundant metal in the earth's crust, and it is accessible and cheap. Recently, many transition and rare metals have suffered an increase of price, due to the large demand and to geopolitical problems. The ability of iron to sustain interesting catalytic processes in biological systems is well known, and it is related to the low toxicity of the metal. The capability of iron to adopt various oxidation states can give iron chameleonic possibilities: in function of the ligands and oxidation state, iron can be used either for oxidative or nucleophilic processes or as a Lewis acid. Most importantly, iron has also the ability to transfer one electron to substrates opening the wide possibility to radical reactions assisted by light. Lastly, the application of the couple Fe(I)/Fe(III) in cross-coupling reaction is gaining more fundamental and applicative studies. For the future, iron will be surely used as a primary metal for organic synthesis.

Iron(II) salts form stable complexes with polypyridyl ligands. The most used polypyridyl ligands are 2,2′-bipyridine *[366-18-7]* (bpy) (1), 1,10-phenanthroline *[66-71-7]* (phen), and terpyridine *[1148-79-4]* (terpy). Polypyridyl ligands are good σ-donor and π-acceptor, a property that is translated in a high-field complex. Iron(2+), tris(2,2′-bipyridine-*N,N*′)-, dibromide $[Fe(bpy)_3Br_2]$ is a low-spin complex.[2] Perichon has reported[3] that the addition of 2,2′-bipyridine to a solution containing $FeBr_2$ leads to an equilibrium mixture of $[Fe(bpy)_3]^{2+}$ and free iron(II) species. If 3 equiv of bpy are added, $[Fe(bpy)_3]^{2+}$ is the only detectable species. Synthesis of dissymmetrical iron (bpy) complexes was also described.[4] The complex has an intense red color, with absorption maximum at 522 nm and molar extinction coefficient of 8650 (M cm)$^{-1}$.[5] Solution of the $[Fe(bpy)_3]^{2+}$complexes is sensitive to oxygen and can be easily oxidized to the corresponding iron(III) complexes, that are blue and nearly colorless if compared to the iron(II) complexes. The $[Fe(bpy)_3]^{3+}$ complexes are unstable with respect to reduction and to ligand dissociation, due to the hardness of the Fe(III). For most counterions, $[Fe(bpy)_3]^{2+}$ has a good solubility in water, but it is moderately soluble in organic solvent. It has good solubility in methanol, DMF, and DMSO; it is hardly soluble in hexane, toluene, or related solvents. The dissociation of the ligands is observed in strong acid or basic solutions. Crystal structure of $[Fe(bpy)_3]Br_2$[6] indicates the D_3 symmetry of the FeN_6 unit with the five-membered chelate ring almost coplanar with the organic ligand. $[Fe(bpy)_3]^{2+}$ can be resolved into the corresponding Δ and Λ stereoisomers, but the stereoisomers are configurationally labile cations,[7] although diastereoisomeric salts can show stability in nonpolar solvents.[8] Electrochemistry of the $[Fe(bpy)3]^{2+}$[9] reported that the complex is more readily oxidized than its ruthenium analog (E° $[Fe(bpy)_3]^{3+}/[Fe(bpy)_3]^{2+}$ 1.05 V; E_{red} for $[Fe(bpy)_3]^{3+}/[Fe(bpy)_3]^{2+}$ in DMF is 0.80 V versus SCE).[10] $[Fe(bpy)_3]^{2+}$ is capable to bind DNA via electrostatic interaction.[11]

Electrochemistry. The incorporation of $[Fe(bpy)_3]^{2+}$ in simple constructed amperometric devices permitted the construction of sensor for NO_2^-.[12] This practical use of the complex is based on the construction of an electrode based on Nafion®, a perfluorosulfonated ion-exchange polymer. The capacity of Nafion® to incorporate the positively charged $[Fe(bpy)_3]^{2+}$ could be considered also for other application in catalysis. Similarly, the heteropolyanion, $[P_2W_{18}O_{62}]_6$, could be used for the same purpose.[13] Electroactive solutes as ferrocene or decamethylferrocene can also be combined with $[Fe(bpy)_3]^{2+}$ and used for the quantitative determination of electroactive compounds, showing linear relationship between the catalytic reduction current and the concentration of reactant in the solution.[14] The use of a simple electrochemical procedure to produce robust thin films containing the $[Fe(bpy)_3]^{2+}$ has been recently reported using bispyridyl ligands.[15] Apart from the keen strategy, this work gives some indication on the possibility to produce polymeric material containing active $[Fe(bpy)_3]^{2+}$ for application in heterogeneous reactions.

Theoretical Investigations. $[Fe(bpy)_3]^{2+}$ complexes have been investigated because they are good candidates for high-density magnetic storage materials. For this purpose, it is of primary importance to establish if the complex exhibits spin-state transition that can cause the formation of a high-spin (HS) complex from a low-spin (LS) one. The stability of the complex and the temperature that causes this transition also influences the application of the complex in photocatalytic reactions, as it alters the gap between a possible photoinduced transition. The spin crossover was investigated by a number of techniques, namely Mössbauer spectroscopy,[16] nuclear inelastic scattering (NIS),[17] X-ray diffraction (XRD),[18] magnetization measurements,[19] Raman,[20] X-ray absorption (XAS),[21] and emission (XES)[22] spectroscopies and neutron scattering.[23] Also theoretical investigations by DFT, TDDFT, and CASPT2 calculations were carried out.[24] The theoretical investigation was also carried out to investigate the influence of supramolecular interactions for complexes embedded in zeolites.[25] Density functional theory (DFT) calculations were also useful to understand the fundamental properties of $[Fe(bpy)_3]^{2+}$: they show that the higher energy HOMO orbitals of different iron(II) diimine complexes are metal centered, while the lower energy LUMO orbitals are ligand centered. The metal-to-ligand charge-transfer (MLCT) band occurs at higher energy compared to d–d transitions, causing the quite short lifetime of the excited states.[26] This quite peculiar difference respect to other $M(bpy)_3$ complexes has a pronounced influence on the photochemistry of $[Fe(bpy)_3]^{2+}$. Also, the redox potential of $[Fe(bpy)_3]^{2+}$ can be calculated by density functional theory.[27] The importance of theoretical investigations in these complexes stems in their possible application in dye-sensitized solar cells (DSSCs).[28] In these dispositive based on TiO_2, Ru(II)-polypyridines are often utilized as chromophores thanks to their chemical stability, long-lived metal-to-ligand charge-transfer (MLCT) excited states, tunable redox potentials, and near perfect quantum efficiency of interfacial electron transfer (IET) into TiO_2. However, recently, an iron complex has been developed with effective photoluminescent properties.[29] These recent breakthroughs in the use of Fe(II) complexes will probably renew interest in application of iron salts for solar cells.

Physical Chemistry of $[Fe(bpy)_3]^{2+}$. The spin transition from a low-spin (LS, $S=0$) to a high-spin (HS, $S=2$) state of $[Fe(bpy)_3]^{2+}$ can be triggered by temperature, pressure, or light irradiation toward the formation of HS metal-centered (MC) ^5T quintet state. This finding has stimulated an intense research activity aimed at determining the timescale, pathways, and mechanisms of the light-induced formation of the quintet state using various ultrafast optical and X-ray spectroscopy techniques.[30] The importance in these studies stems in the deep understanding of the photoexcited states and the possibility to stabilize them through a tailored design of the ligands for Fe(II) for photocatalytic-induced electron transfer.

Material Science Applications. Recently, systems consisting of metal bipyridyl [metal = Fe, Ru, Os, etc.] complexes immobilized on carbon nanotubes (CNTs) have been reported.[31] Nanotube cannot only function as support, as junction for electrode scan be constructed with them, taking advantage of their conductive properties. π–π Stacking interactions were also used to link metal complexes bearing a pyrene functionality on a CNT surface through noncovalent interaction.[32] The derivatization of an intrinsic iron impurity in a carbon nanotube (CNT-*Fe, *Fe-intrinsic, and redox-active iron impurity) with bpy as ligand was recently realized.[33] This multi walled MWCNT-$[Fe(bpy)_3]^{2+}$ hybrid modified electrode was further used as an electrochemical detector for selective analysis of DNA's purine bases, adenine and guanine, without interference from pyrimidine bases, cytosine and thymine. $[Fe(bpy)_3]^{2+}$ has been incorporated in other materials like silica.[34]

Zeolites and Solid-Supported $[Fe(bpy)_3]^{2+}$. As discussed in modified electrodes, $[Fe(bpy)_3]^{2+}$ can be incorporated in zeolites[35] due to the possibilities offered by zeolites as a regular, characterizable medium for potential nonaqueous, aqueous, or gas-phase photochemistry. As different alkaline earth metal cations can be present, these can play a relevant influence on the physicochemical properties and catalytic activity of $[Fe(bpy)_3]^{2+}$.[36] In particular the intensity of the MLCT band was influenced, enhancing the possibility to employ the $[Fe(bpy)_3]^{2+}$ in photocatalytic applications. Moreover, the stability of the material can offer some interesting application as selective oxidant.[37]

Photocatalysis and Photocatalytic Reactions. The most promising and interesting use of $[Fe(bpy)_3]^{2+}$ complexes is in photocatalytic applications. Iron(II) polypyridyls complexes are not considered as good photosensitizers, due to very short lifetime of the MLCT states (picoseconds to nanoseconds). This is due to the high energy of the MLCT states relative to the *eg* orbitals. However, Ferrere has reported a relevant study in which Fe(II) polypyridyl complexes were found able to inject electrons in TiO_2.[38] The use of $[Fe(bpy)_3]^{2+}$ could in principle replace the commonly employed photosensitizers, based on rare and expensive ruthenium and iridium complexes. The employment of $[Fe(bpy)_3]^{2+}$ as alternative, earth-abundant, first-row transition metal for photoinduced synthetic organic transformations has been recently applied to photocatalytic oxidations[39] in which the $[Fe(bpy)_3]^{2+}$ species, encapsulated in solid support, are supposedly able to generate $O_2^{\bullet-}$. In certain studies, the presence of the species $O_2^{\bullet-}$ was characterized by EPR.[38]

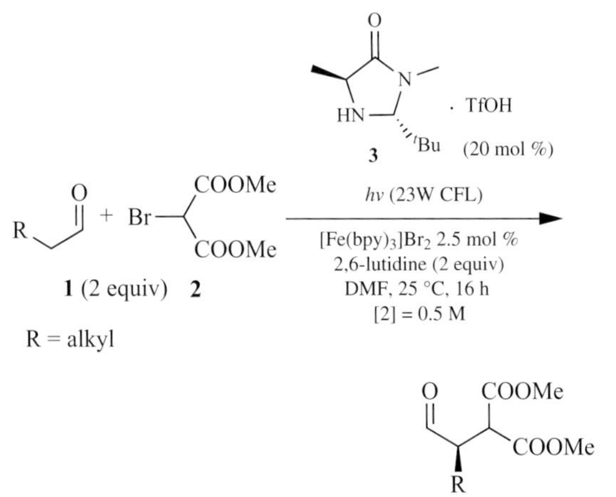

(1)

4, 60–83%, 89–97% ee

It is also important to mention that the complexation of OH groups present on the solid support with $[Fe(bpy)_3]^{2+}$, proved by FTIR, XPS, and UV-vis diffuse reflectance spectroscopy measurements, seemed to enhance the lifetime of the excited state.[39d] $[Fe(bpy)_3]Br_2$ was used to promote a stereoselective photocatalytic reaction[40] in an organocatalytic alkylation of aldehydes[41] in the presence of a MacMillan catalyst[42] (eq 1). The employment of other bromo derivatives was possible, and application towards the synthesis of a natural product was also reported (eq 2).

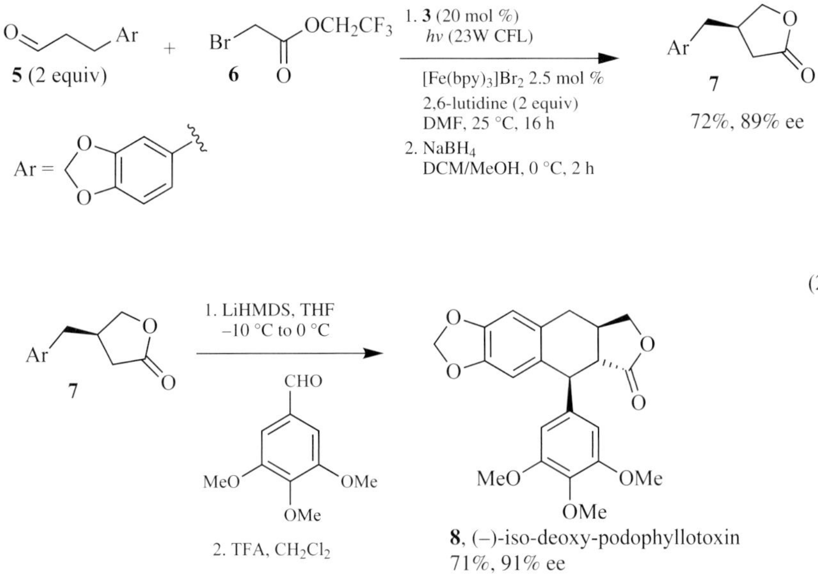

(2)

A deep photophysical investigation carried out with various techniques was able to suggest a radical chain reaction for the stereoselective transformation. Formation of carbon radicals was proven by a radical trap and was induced by $[Fe(bpy)_3]Br_2$ and visible light. The perspective to use different complexes and to enhance the lifetime of excited state of related iron complexes can open new possible application in this field.[43]

Other Reactivity and Miscellaneous Reactions. Electrochemically generated iron(I) complexes in presence of bpy ligand were reported to mediate the synthesis of β-hydroxyesters by Perichon.[4] The ability of the $[Fe(bpy)_3]^{2+}$ to promote the catalytic oxidation of organic species by dioxygen was reported,[44] while the oxidation of the complexes with other species was also investigated.[45] Finally, the generation of radical species for ATRP polymerization[46] was initiated by the combination of CBr_4/ $[Fe(bpy)_3]^{3+}$ initiation system in the presence of 2,2′-azobisisobutyronitrile (AIBN). In the mechanistic picture proposed for the reactions, the authors have supposed that a reductant is able to form the reactive $[Fe(bpy)_3]^{2+}$ that is then forming stabilized radical species.[47]

1. For a recent review, see Bauer, I.; Knölker, H. J., *Chem. Rev.*, **2015**, *115*, 3170.
2. For the preparation of the $Fe(bpy)_3Br_2$ complex, see (a) Jaeger, F.; van Dijk, J. A., *Z. Anorg. Allgem. Chem.* **1936**, *227*, 273. (b) Palmer, R. A.; Piper, T. S., *Inorg. Chem.*, **1965**, *5*, 865.
3. Buriez, O.; Durandetti, M.; Périchon J., *J. Electroanal. Chem.*, **2005**, *578*, 63.
4. Collomb, M. N.; Deronzier, D.; Gorgy, K.; Leprêtre, J.-C., *New J. Chem.*, **2000**, *24*, 455.
5. (a) Braterman, P. S.; Song, J.-S.; Peacock, R. D., *Inorg. Chem.* **1992**, *31*, 555. (b) Kober, E. M.; Meyer, T. J., *Inorg. Chem.*, **1982**, *21*, 3967.
6. Heilmann, J.; Lerner, H.-W.; Bolte, M., *Acta Cryst. E* **2012**, 2006; E62: m1477. For the structure of tris (4,4′-dimethoxy-2,2′-bipyridine) iron(II) see: Shklover, V.; Nesper, R.; Zakeeruddin, S. M.; Fraser, D. M.; Grätzel, M., *Inorg. Chim. Acta* **1996**, *247*, 237.
7. Lacour, J.; Hebbe-Viton, V., *Chem. Soc. Rev.*, **2003**, *32*, 373.
8. Jodry, J. J.; Frantz, R.; Lacour, J., *Inorg. Chem.*, **2004**, *43*, 3329.
9. (a) Abruna H. D., *J. Electroanal. Chem. Interfacial Electrochem.* **1984**, *175*, 321. (b) Sutin, N.; Creutz, C., *Pure Appl. Chem.*, **1980**, *52*, 2717.
10. Braterman, P. S.; Song, J. I.; Peacock, R. D., *Inorg. Chem.* **1992**, *31*, 555.
11. (a) Carter, M. T.; Rodriguez, M.; Bard, A. J., *J. Am. Chem. Soc.* **1989**, *111*, 8901. (b) Baik, M-H.; Silverman, J.S.; Yang, I.V.; Ropp, P.A.; Szalai, V.A.; Yang, W.; Thorp, H.H., *J. Phys. Chem. B* **2001**, *105*, 6437.
12. Azad, U.P.; Ganesan, V.; *Chem. Commun.* **2010**, *46*, 6156.
13. Fay, N.; Dempsey, E.; McCormac, T., *J. Electroanal. Chem.* **2005**, *574*, 359.
14. Winkler, K.; Plonska, M.E.; Basa, A.; Lach, M.; Balch, A.L., *Electroanalysis* **2003**, *15*, 55.
15. Lombard, J.; Leprêtre, J.C.; Jouvenot, D.; Deronzier, A.; Collomb, M.N., *New J. Chem.* **2008**, *32*, 1117.
16. (a) Müller, E.W.; Ensling, J.; Spiering, H.; Gütlich, P., *Inorg. Chem.* **1983**, *22*, 2074; (b) Elhaïk, J.; Evans, D.J.; Kilner, C.A.; Halcrow, M.A., *J. Chem. Soc., Dalton Trans.* **2005**, 1963; (c) Bhattacharjee, A.; Kusz, J.; Ksenofontov, V.; Sugiyarto, K.H.; Goodwin, H.A.; Gütlich, P., *Chem. Phys. Lett.* **2006**, *431*, 72; (d) England, J.; Scarborough, C.C.; Weyhermüller, T.; Sproules, S.; Wieghardt, K., *Eur. J. Inorg. Chem.* **2012**, 4605.
17. (a) Ronayne, K.L.; Paulsen, H.; Höfer, A.; Dennis, A.C.; Wolny, J.A.; Chumakov, A.I.; Schünemann, V.; Winkler, H.; Spiering, H.; Bousseksou, A.; Gütlich, P.; Trautwein, A.X.; McGarvey, J., *J. Phys. Chem. Chem. Phys.* **2006**, *8*, 4685; (b) Boöttger, L.H.; Chumakov, A.I.; Grunert, C.M.; Gütlich, P.; Kusz, J.; Paulsen, H.; Ponkratz, U.; Rusanov, V.; Trautwein, A.X.; Wolny, X., *Chem. Phys. Lett.* **2006**, *429*, 189; (c) Wolny, J.A.; Diller, R.; Schünemann, V., *Eur. J. Inorg. Chem.* **2012**, 2635.
18. (a) Kusz, J.; Spiering, H,; Gütlich, P., *Appl. Crystallogr.* **2001**, *34*, 229; (b) Elhaïk, J.; Kilner, C.A.; Halcrow, M.A., *J. Chem.Soc., Dalton Trans.* **2006**, 823.
19. (a) Stassen, A.F.; Grunert, M.; Dova, E.; Müller, M.; Weinberger, P.; Wiesinger, G.; Schenk, H.; Linert, W.; Haasnoot, J.G.; Reedijk, J., *Eur. J. Inorg. Chem.* **2003**, 2273.
20. Moussa, N.O.; Molnár, G.; Ducros, X.; Zwick, A.; Tayagaki, T.; Tanaka, K.; Bousseksou, A., *Chem. Phys. Lett.* **2005**, *402*, 503.
21. Briois, V.; Moulin, C.; Sainctavit, P.; Brouder, C.; Flank, A-M., *J. Am. Chem. Soc.* **1995**, *117*, 1019.
22. Vankó, G.; Neisius, T.; Molnár, G.; Renz, F.; Kárpáti, S.; Shukla, A.; de Groot, F.M.F., *J. Phys. Chem. B* **2006**, *110*, 11647.
23. Goujon, A.; Gillon, B.; Gukasov, A.; Jeftić, J.; Nau, Q.; Codjovi, E.; Varret, F., *Phys. Rev.* B **2003**, *67*, 220401.

24. (a) Swart, M.J., *Chem. Theory Comput.* **2008**, *4*, 2057; (b) Pápai, M.; Vankó, G.; de Graaf, C.; Rozgonyi, T., *J. Chem. Theory Comput.* **2013**, *9*, 509; (c) Bowman, D.N.; Jakubikova, E., *Inorg. Chem.* **2012**, *51*, 6011; (d) de Graaf, C.; Sousa, C., *Chem. Eur. J.* **2010**, *16*, 4550.
25. Vargas, A.; Hauser, A.; Daku, L.M.L, *J. Chem. Theory Comput.* **2009**, *5*, 97.
26. (a) Kirgan, R.A.; Rillema, D.P., *J. Phys Chem. A* **2007**, *111*, 13157.
27. (a) Kim, H.; Park, J.; Lee, Y.S., *J. Comput. Chem.* **2013**, *34*, 2233; (b) Kim, H.; Park, J.; Lee, Y.S., *J. Comput. Chem.* **2015**, *36*, 33.
28. Jakubikova, E.; Bowman, D.N., *Acc. Chem. Res.* **2015**, *48*, 1441.
29. Harlang, T.C.B.; Liu, Y.; Gordivska, O.; Fredin, L.A.; Ponseca, C.S.Jr.; Huang, P.; Chábera, P.; Kjaer, K.S.; Mateos, H.; Uhlig, J.; Lomoth, R.; Wallenberg, R.; Styring, S.; Persson, P.; Sundström, V.; Wärnmark, K., *Nat. Chem.* **2015**, *7*, 883.
30. (a) Cannizzo, A.; Milne, C.J.; Consani, C.; Gawelda, W.; Bressler, C.H.; van Mourik, F.; Chergui, M., *Coord. Chem. Rev.* **2010**; *254*, 2677; (b) Consani, C.; Prmont-Schwarz, M.; El Nahhas, A.; Bressler, C.; van Mourik, F.; Cannizzo, A.; Chergui, M., *Angew. Chem., Int. Ed.* **2009**, *48*, 7184; (c) Zamponi, F.; Freyer, B.; Juvé, V.; Stingl, J.; Woerner, M.; Chergui, M., Elsaesser, T., *EPJ. Web Conf.* **2013**; DOI: 10.1051/epjconf/2014105046; (c) Bressler, Ch.; Milne, C.; Pham, V-T.; ElNahhas, A.; van der Veen, R.M.; Gawelda, W.; Johnson, S.; Beaud, P.; Grolimund, D.; Kaiser, M.; Borca, C.N.; Ingold, G.; Abela, R.; Chergui, M., *Science* **2009**, *323*, 489; (d) Auböck, G.; Chergui, M., *at. Chem.* **2015**: DOI: 10.1038/NCHEM.2305; (e) Sousa, C.; de Graaf, C.; Rudavskyi, A.; Broer, R.; Tatchen, J.; Etinski, M., *Marian CM.Chem. Eur. J.* **2013**, *19*, 17541.
31. (a) Khairoutdinov, R.F.; Doubova, L.V.; Haddon, R.C.; Saraf, L.J., *Phys. Chem. B* **2004**, *108*, 19976; (b) Chen, S; Yuan, R; Chai, Y; Min, L; Li, W; Xu, Y., *Electrochim. Acta* **2009**, *54*, 7242; (c) Goff, A.L.; Cosnier, S., *J. Mater. Chem.* **2011**, *21*, 3910.
32. McQueen, E.W.; Goldsmith, J.I., *J. Am. Chem. Soc.* **2009**, *131*, 17554.
33. Mayuri, P.; Kumar, A.S., *Langmuir* **2015**, *31*, 5945.
34. Nakanishi, C.; Ikeda, S.; Isobe, T.; Senna, M., *Mater. Res. Bull.* **2002**, *37*, 647.
35. (a) Quayle, W.H.; Peeters, G.; De Roy, G.L.; Vansant, E.F.; Lunsford, J.H., *J. Am. Chem. Soc.* **1982**, *21*, 2226; (b) Ege, D.; Ghosh, P.K.; White, J.R.; Equey, J-F.; Bard, A.J., *J. Am. Chem. Soc.* **1985**; *107*, 5664.
36. (a) Martis, M.; Mori, K.; Yamashita, H.J., *Chem. Soc., Dalton Trans* **2014**, *43*, 1132; (b) Fukuzumi, S.; Yoshida, Y.; Urano, T.; Suenobu, T.; Imahori, H.J., *Am. Chem. Soc.* **2001**, *23*, 11331; (c) Sewell, G.; Forster, R.J.; Keyes, T.E., *J. Phys. Chem. A* **2008**; *112*, 880; (d) Bessel, C.A.; Rolison, D.R., *J. Phys. Chem. B* **1997**, *101*, 1148;
37. Yamaguchia, S.; Fukura, T.; Takiguchi, T.; Fujita, T.; Nishibori, M.; Teraoka, Y.; Yahiro, H., *Catal. Today* **2015**, *242*, 261.
38. (a) Ferrere, S., *Chem. Mater.* **2000**, *12*, 1083; (b) Ferrere, S.; Gregg, B.J., *Am. Chem. Soc.* **1998**, *120*, 843.
39. (a) Ma, W.; Li, J.; Tao, X.; He, J.; Xu, Y.; Yu, J.C.; Zhao, J., *Angew. Chem., Int. Ed.* **2003**, *42*, 1029; (b) Li, J.; Maa, W.; Huanga, Y.; Tao, X.; Zhao, J.; Xu, Y., *Appl. Catal. B* **2004**, *48*, 17; (b) Shiraishi, Y.; Manabe, K.; Hirai, T., *Appl. Catal. B* **2010**, *93*, 292; (c) Han, Z.; Guo, J.; Li, W., *Chem. Eng. J.* **2013**, *228*, 36; (d) Bossmann, S.H.; Shahin, N.; Le Thanh, H.; Bonfill, A.; Wêrner, M.; Braun, A.M., *ChemPhysChem* **2002**, *3*, 401.
40. For photocatalytic organocatalysis, see Prier, C.K.; Rankic, D.A.; MacMillan, D.W.C., *Chem. Rev.* **2013**, *113*, 5322.
41. Gualandi, A.; Marchini, M.; Mengozzi, L.; Natali, M.; Lucarini, M., Ceroni, P.; Cozzi, P.G., *ACS Catal.* **2015**, *5*, 5927.
42. (a) Nicewicz, D.A.; MacMillan, D.W.C., *Science* **2008**, *322*, 77; (b) Nagib, D.A.; Scott, M.E.; MacMillan, D.W.C., *J. Am. Chem. Soc.* **2009**, *131*, 10875; (c) Shih, H-W.; Vander Wal, M.N.; Grange, R.L.; MacMillan, D.W.C., *J. Am. Chem. Soc.* **2010**, *132*, 13600.
43. (a) Liu, Y.; Kjáer, K.S.; Fredin, L.A.; Chábera, P.; Harlang, T.; Canton, S.E.; Lidin, S.; Zhang, J.; Lomoth, R.; Bergquist, K-E.; Persson, P.; Wärnmark, K.; Sundstrçm, V., *Chem. Eur. J.* **2015**, *21*, 3628; (b) Duchanois, T.; Etienne, T.; Cebrián, C.; Liu, L.; Monari, A.; Beley, M.; Assfeld, X.; Haacke, S.; Gros, P.C., *Eur. J. Inorg. Chem.* **2015**, *2015*, 2469.
44. Sobkowiak, A.; Naróg, D.; Sawyer, D.T., *J. Mol. Catal. A* **2000**, *156*, 247.
45. Sánchez, F.; Moya, M.L.; Jiménez, R.; Gómez-Herrera, C., Carmona, M.C.; López-Cornejo, P., *J. Chem. Soc., Faraday Trans.* **1997**, *93*, 2181.
46. Wang, J.S.; Matyjaszewski, K., *J. Am. Chem. Soc.* **1995**, *117*, 5614.
47. Saikia, P.J.; Baruah, S.D., *Polym. Bull.* **2013**, *70*, 3291.

Pier Giorgio Cozzi, Andrea Gualandi, Luca Mengozzi & Elisabetta Manoni
Universitá degli Studi di Bologna, Bologna, Italy

β-Isocupreidine (β-ICD)

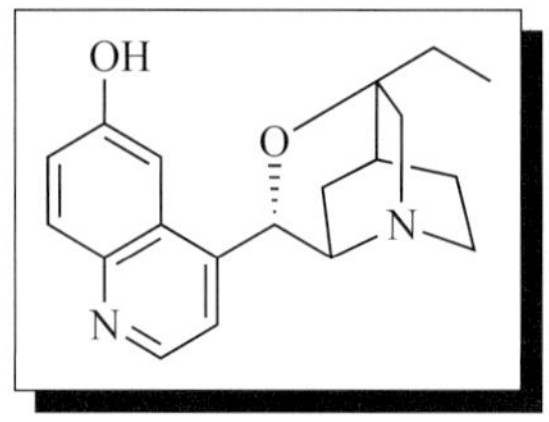

[253430-48-7] $C_{19}H_{22}N_2O_2$ (MW 310.40)

InChI = 1S/C19H22N2O2/c1-2-19-11-21-8-6-12(19)9-17(21)18(23-19)14-5-7-20-16-4-3-13(22)10-15(14)16/h3-5,7,10,12,17-18,22H,2,6,8-9,11H2,1H3/t12-,17+,18-,19+/m0/s1

InChIKey = ULTQUFCPWDEOND-JHQYFNNDSA-N

(reagent used as a chiral Lewis basic catalyst for asymmetric reactions)

Physical Data: mp 259 °C; $[\alpha]_D^{23}$ −13 (*c* 1, EtOH).

Solubility: soluble in methanol, CH_2Cl_2, $CHCl_3$, DMF, DMSO, and THF.

Form Supplied in: colorless needles.

Preparation: [1,2] quinidine (10.0 g, 30.8 mmol) is heated together with KBr (36.7 g, 308 mmol) in 85% H_3PO_4 (150 mL) at 100 °C for 10 days (eq 1). After being cooled to room temperature, the mixture is added dropwise to an ice-cooled 25% KOH. The pH is adjusted to about 8 with 25% NH_4OH, and the mixture is extracted with $CHCl_3$. The extract is washed with brine, dried over K_2CO_3, and chromatographed (SiO_2, $CHCl_3$/MeOH = 9/1-4/1) to give β-isocupreidine (5.85 g, 61%) as a pale yellow amorphous solid. β-ICD thus obtained was dissolved in NH_3-MeOH and filtered to remove insoluble precipitates. Concentration of the filtrate gave a crystalline solid, which was recrystallized from MeOH-H_2O to afford colorless needles of β-ICD·MeOH·H_2O. The dried β-ICD is obtained by azeotropic removal of the water and methanol by repeating evaporation of its THF solution.

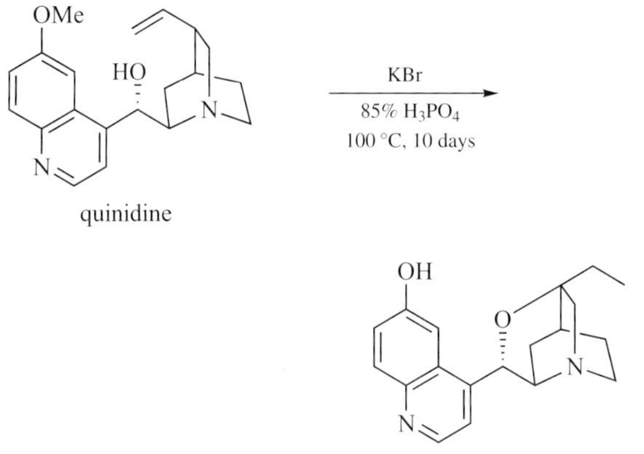

(1)

β-ICD

Handling, Storage, and Precautions: sealed under inert gas and then stored in a cool dark place avoiding moisture.

Asymmetric Morita–Baylis–Hillman Reactions. The nucleophilic amine- or phosphine-catalyzed C–C bond formation of aldehydes with activated alkenes such as acrylates is widely referred to as the Morita–Baylis–Hillman reaction. β-ICD serves as an efficient catalyst for the asymmetric Morita–Baylis–Hillman reaction.[2] β-ICD catalyzes the reactions of various aliphatic and aromatic aldehydes with hexafluoroisopropyl acrylate (HFIPA) to give the (α-methylene-β-hydroxy) esters of *R*-configuration with excellent ee (eq 2). In the case of aliphatic aldehydes, the yields become moderate because of the production of the *S*-enriched dioxanones with moderate to low ee. On the other hand, in the case of aromatic aldehydes apart from reactive *p*-nitrobenzaldehyde, the dioxanones are not formed. This β-ICD-HFIPA method is the first example of a highly enantioselective catalytic Morita–Baylis–Hillman reaction. The enantioselectivity can be rationalized by the formation of a betaine intermediate, which is stabilized by intramolecular hydrogen bonding between the oxy anion and the phenolic hydroxy group. The β-ICD-HFIPA method has been utilized in the syntheses of natural products, mycestericin E,[3] epopromycin B,[4] fostriecin,[5] and phoslactomycin B.[5] The Morita–Baylis–Hillman reaction employing β-ICD can also be applied to the asymmetric synthesis of substituted 3-hydroxy-2-oxindoles (eq 3).[6]

Asymmetric Aza–Morita–Baylis–Hillman Reactions. Aza–Morita–Baylis–Hillman reactions of *N*-(diphenylphosphinoyl)imines with HFIPA are also catalyzed by β-ICD to give the corresponding adducts enantioselectively (eq 4). These reactions proceed with *S*-selectivity in contrast to reactions of aldehydes that afford *R*-selectivity.[7] Interestingly, reactions of *N*-tosylimines with acrolein or methyl acrylate proceed with *S*-selectivity, while the reactions with methyl vinyl ketone (MVK) or ethyl vinyl ketone (EVK) show *R*-selectivity (eq 5).[8] Reactions of the *N*-tosylimines derived from salicylaldehydes with MVK or EVK give *S*-enriched products in high yields and with excellent ee (eq 6).[9] The high enantioselectivity observed can be explained by assuming the intermediate stabilized by hydrogen bonding network between the amidate ion, quinolinol, and *o*-phenol moieties.

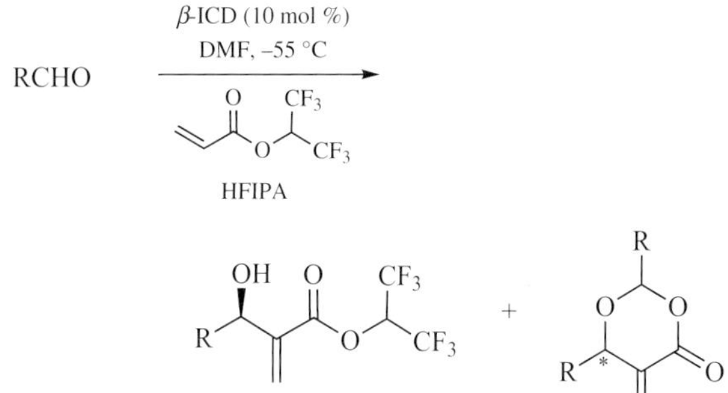

(2)

R	Yield (%), Configuration (% ee)	
	Ester	Dioxanone
$PMBO(CH_2)_2$	58, *R* (99)	14, *S* (38)
i-Bu	58, *R* (99)	18, *S* (78)
c-Hex	36, *R* (99)	22, *S* (65)
p-$(NO_2)C_6H_4$	58, *R* (95)	17, *R* (49)
Ph	75, *R* (97)	0
(*E*)-PhCH=CH	64, *R* (94)	0
2-naphthyl	82, *R* (97)	0

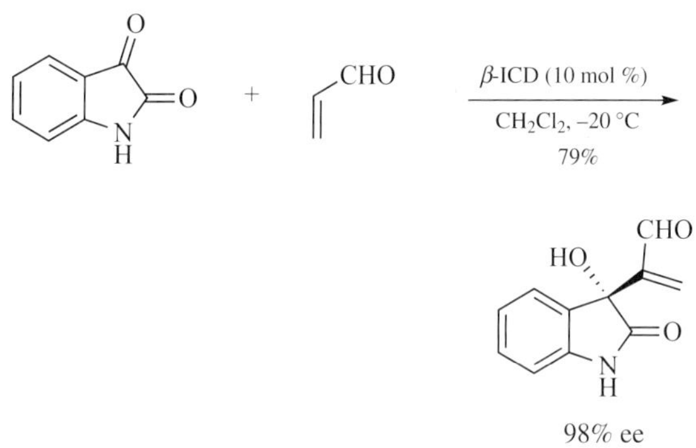

(3)

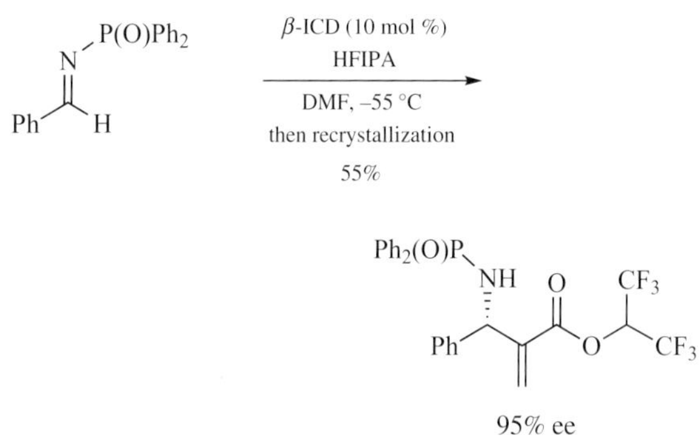

(4)

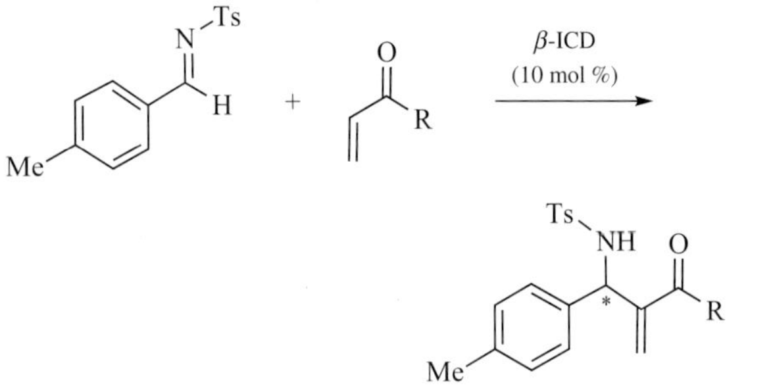

(5)

R	Conditions	Yield (%)	% ee
Me	MeCN/DMF (1:1), –30 °C	80	96 (*R*)
Et	MeCN/DMF (1:1), –30 °C	49	87 (*R*)
H	THF, –25 °C	55	83 (*S*)
OMe	CH_2Cl_2, 0 °C	67	80 (*S*)

Asymmetric Aminations of α-Substituted α-Cyanoacetates and β-Dicarbonyl Compounds. β-ICD catalyzes the reactions of carbon nucleophiles such as α-substituted α-cyanoacetates, β-dicarbonyl compounds, and nitroacetates with di-*tert*-butyl azodicarboxylate with a remarkable catalyst turnover to produce the amination products having a quaternary stereocenter in high yields and with excellent ee (eqs 7–9).[10,11]

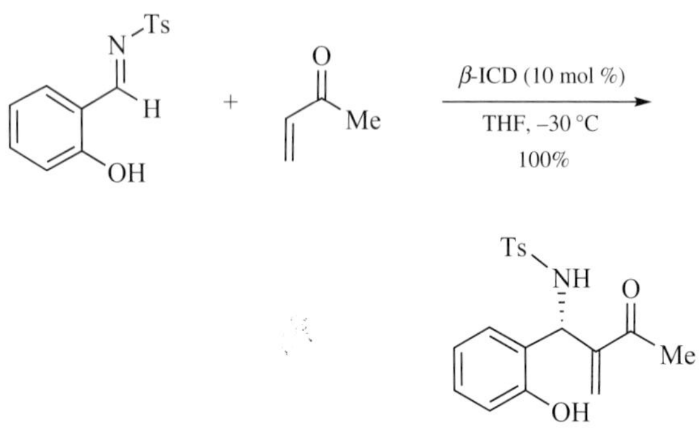

91% ee

(6)

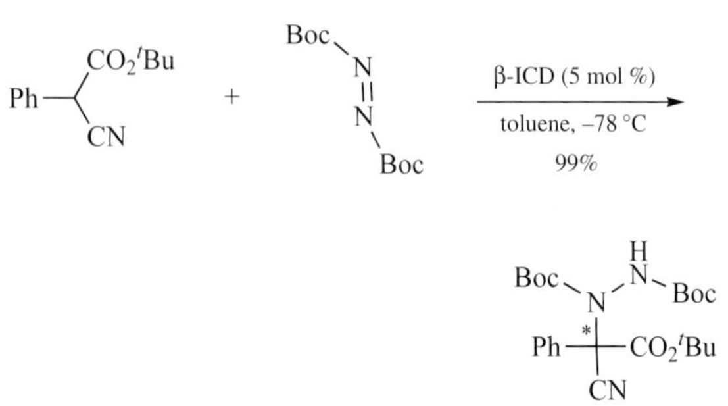

>98% ee

(7)

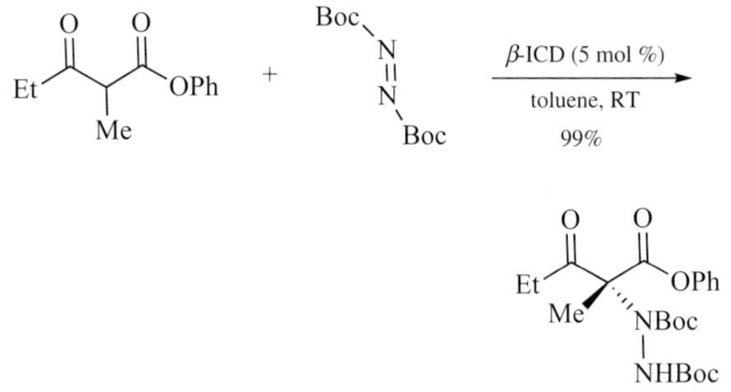

90% ee

(8)

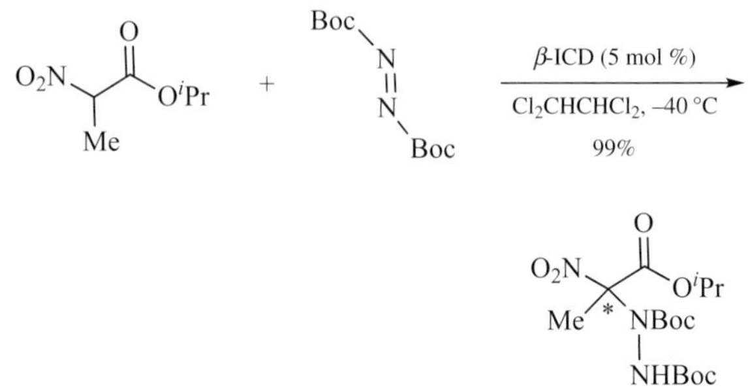

90% ee

(9)

Asymmetric Intermolecular Rauhut-Currier Reactions. Vinylogous Morita–Baylis–Hillman reactions are referred to as the Rauhut–Currier reaction. β-ICD effectively catalyzes the reactions of *N*-benzylmaleimide and electron-deficient allenes to give functionalized allenic products in high yields and with excellent ee (eq 10).[12]

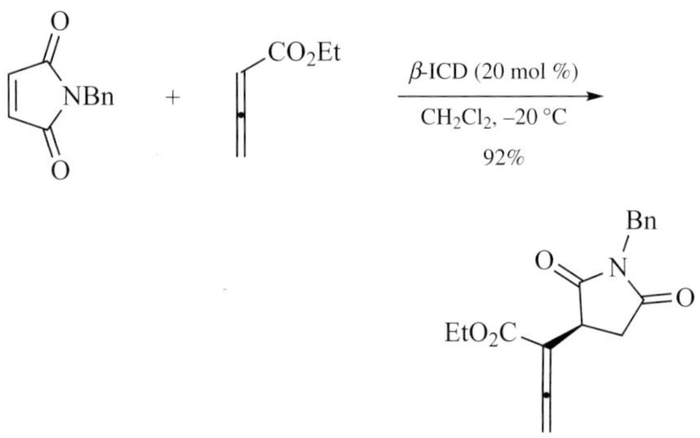

96% ee

(10)

Asymmetric Intramolecular Cyclocondensation Reactions. β-ICD promotes the intramolecular cyclocondensation reaction of 6-oxohexanoic acid with high enantioselectivity but in moderate yield (eq 11).[13] In this particular reaction, *o*-acetyl quinidine shows the enantioselectivity opposite to that observed for β-ICD.

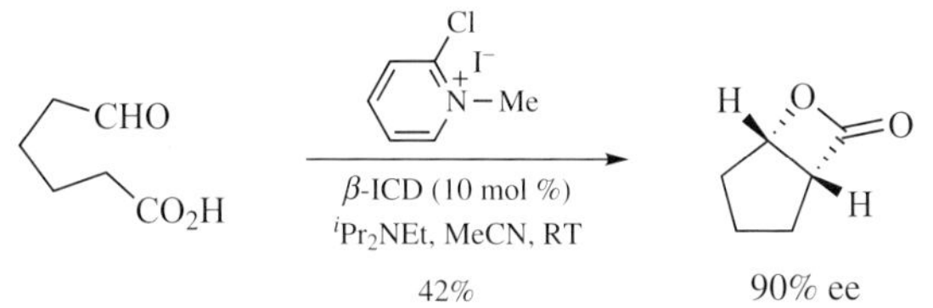

(11)

Enantioselective Allylic Nucleophilic Substitution Reactions. Racemic Boc-protected Morita–Baylis–Hillman adducts undergo enantioselective allylic substitution reaction with various nucleophiles such as naphthols and cyanoacetates under β-ICD-catalyzed conditions (eqs 12 and 13).[14,15]

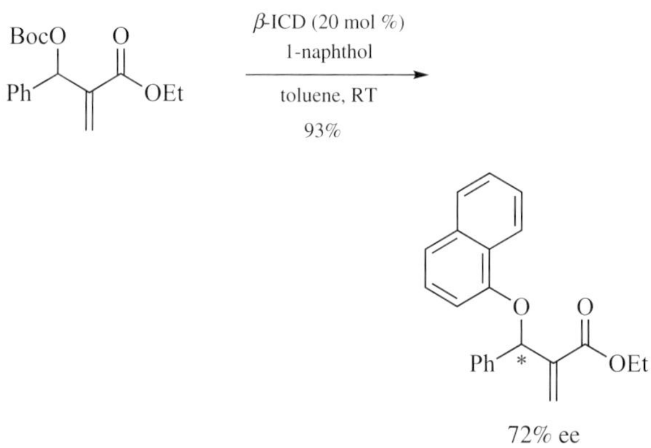

(12)

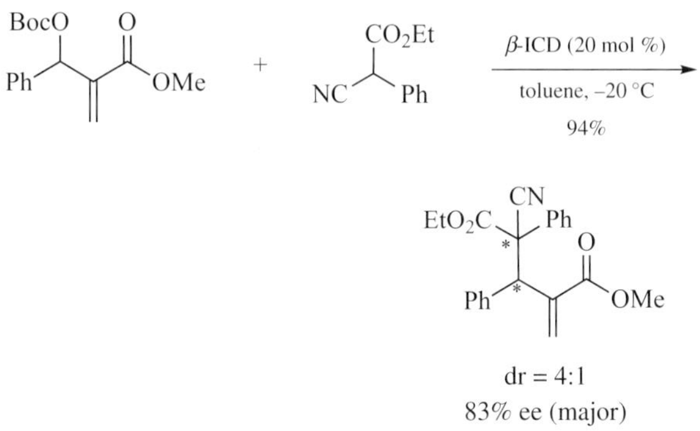

(13)

β-Isocupreidine-derived Catalysts. The derivatization of β-ICD often confers new catalytic functions on the molecule. For example, the β-ICD-derived catalyst possessing a phenyl group, prepared in two steps from β-ICD, catalyzes a [4+2] cycloaddition reaction between a heterodiene and an acetylene carboxylate with high enantioselectivity (eq 14).[16] The β-ICD-derived catalyst possessing a (*N*-Boc-glycinyl)amino group catalyzes aza–Baylis–Hillman reactions of aromatic sulfonylimines with 2-naphthyl acrylate to give the corresponding adducts in good yields and with high enantioselectivities (eq 15).[17] In this reaction, the addition of 2-naphthol as a cocatalyst markedly improves the yield and the enantioselectivity. The β-ICD-derived catalyst possessing a (9-anthracenyl)carbonylamino group also catalyzes aza–Morita–Baylis–Hillman reactions between *N*-sulfonylimines and alkyl vinyl ketones to give *R*-enriched adducts (eq 16).[18] In this reaction, the addition of 2-naphthol inverts the enantioselectivity to give *S*-enriched adducts of high ee.

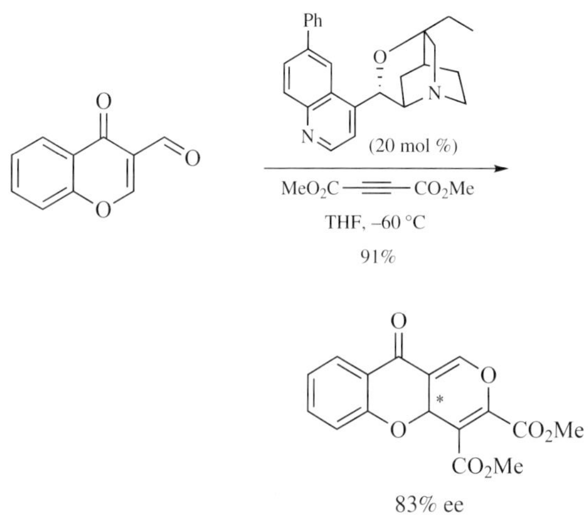

(14)

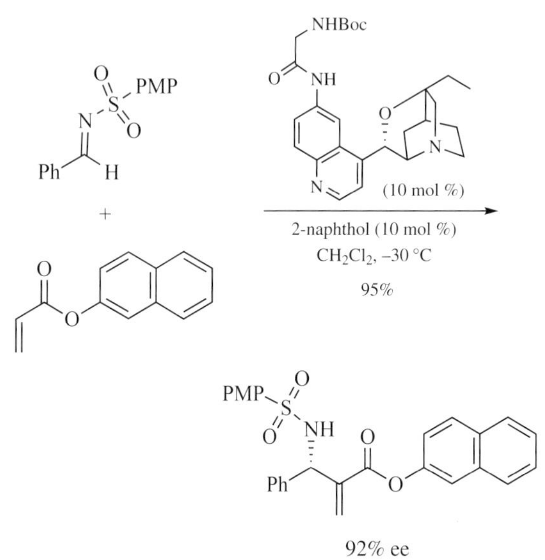

(15)

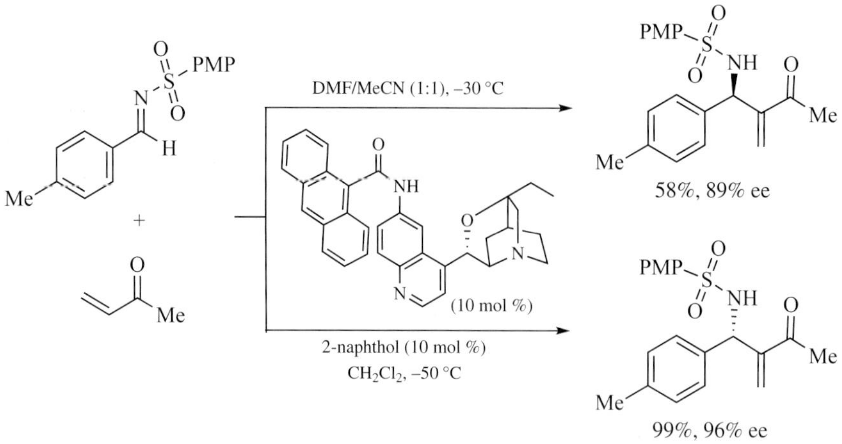

(16)

Related Reagents. hexafluoroisopropyl acrylate (HFIPA); acrolein; methyl vinyl ketone; quinidine; quinine.

1. Braje, W.; Frackenpohl, J., Langer, P.; Hoffmann, H. M. R., *Tetrahedron* **1998**, *54*, 3495.
2. (a) Iwabuchi, Y.; Nakatani, M.; Yokoyama, N.; Hatakeyama, S., *J. Am. Chem. Soc.* **1999**, *121*, 10219. (b) Nakano, A.; Kawahara, S.; Akamatsu, S.; Morokoma, K.; Nakatani, M.; Iwabuchi, Y.; Takahashi, K.; Ishihara, J.; Hatakeyama, S., *Tetrahedron* **2006**, *62*, 381. (c) Hatakeyama, S., *J. Synth. Org. Chem. Jpn.* **2006**, *11*, 1132.
3. Iwabuchi, Y.; Furukawa, M.; Esumi, T.; Hatakeyama, S., *Chem. Commun.* **2001**, 2030.
4. Iwabuchi, Y.; Sugihara, T.; Esumi, T., Hatakeyama, S., *Tetrahedron Lett.* **2001**, *42*, 7867.
5. Sarkar, S. M.; Wanzala, E. N.; Shibahara, S.; Takahashi, K.; Ishihara, J.; Hatakeyama, S., *Chem. Commun.* **2009**, 5907.
6. (a) Liu, Y. L.; Wang, B. L.; Cao, J. J.; Chen, L.; Zhang, Y. X.; Wang, C.; Zhou, J., *J. Am. Chem. Soc.* **2010**, *132*, 15176. (b) Guan, X. Y.; Wei, Y.; Shi, M., *Chem. Eur. J.* **2010**, *16*, 13617.
7. Kawahara, S.; Nakano, A.; Esumi, T.; Iwabuchi, Y.; Hatakeyama, S., *Org. Lett.* **2003**, *5*, 3103.
8. (a) Shi, M.; Xu, Y. M., *Angew. Chem., Int. Ed.* **2002**, *41*, 4507. (b) Shi, M.; Xu, Y. M.; Shi, Y. L., *Chem. Eur. J.* **2005**, *11*, 1794.
9. Shi, M.; Qi, M. J.; Liu, X. G., *Chem. Commun.* **2008**, 6025.
10. Saaby, S.; Bella, M.; Jϕrgensen, K. A., *J. Am. Chem. Soc.* **2004**, *126*, 8120.
11. Ji, C. B.; Liu, Y. L.; Zhao, X. L.; Guo, Y. L.; Wang, H. Y.; Zhou, J., *Org. Biomol. Chem.* **2012**, 1158.
12. Zhao, Q. Y.; Pei, C. K.; Guan, X. Y.; Shi, M., *Adv. Synth. Catal.* **2011**, *353*, 1973.
13. Cortez, G. S.; Oh, S. H.; Romo, D., *Synthesis* **2001**, 1731.
14. Du, Y.; Han, X.; Lu, X., *Tetrahedron Lett.* **2004**, *45*, 4967.
15. van Steenis, D. J. V. C.; Marcelli, T.; Lutz, M.; Spek, A. L.; van Maarseveen, J. H.; Hiemstra, H., *Adv. Synth. Catal.* **2007**, *349*, 281.
16. Waldmann, H.; Khedkar, V.; Dückert, H.; Schürmann, M.; Oppel, I. M.; Kumar, K., *Angew. Chem., Int. Ed.* **2008**, *47*, 6869.
17. Abermil, N.; Masson, G.; Zhu, J., *J. Am. Chem. Soc.* **2008**, *130*, 12596.
18. (a) Abermil, N.; Masson, G.; Zhu, J., *Org. Lett.* **2009**, *11*, 4648. (b) Abermil, N.; Masson, G.; Zhu, J., *Adv. Synth. Calal* **2010**, *352*, 656.

Keisuke Takahashi & Susumi Hatakeyama
Nagasaki University, Nagasaki, Japan

M

1-Mesityl-3,4-dimethyl-4*H*-1,2,4-triazol-1-ium Tetrafluoroborate

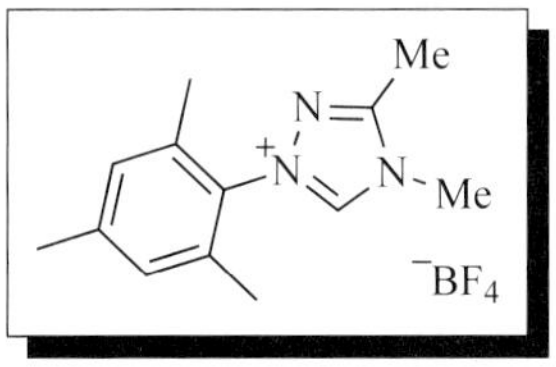

[1012335-23-7] $C_{13}H_{18}BF_4N_3$ (MW 303.11)
InChI = 1S/C13H18N3.BF4/c1-9-6-10(2)13(11(3)7-9)16-8-15(5)12(4)14-16;2-1(3,4)5/h6-8H,1-5H3;/q+1;-1
InChIKey = FTMHTTDMZWLISN-UHFFFAOYSA-N

(organocatalysis, *N*-heterocyclic carbene, homoenolate)

Physical Data: mp 130–132 °C.
Form Supplied in: light yellow solid.
Solubility: soluble in dichloromethane, chloroform, and acetonitrile.
Analysis of Reagent Purity: IR (film) 3107, 2972, 1597, 1456, 1062 cm^{-1}; ^1H NMR (500 MHz, $CDCl_3$) δ 9.51 (s, 1H), 6.97 (s, 2H), 4.03 (s, 3H), 2.67 (s, 3H), 2.64 (s, 3H), 2.02 (s, 6H), ^{13}C NMR (125 MHz, $CDCl_3$) δ 154.5, 144.9, 141.9, 135.3, 131.2, 129.8, 33.7, 21.4, 17.3, 10.3, LRMS (electrospray): mass calculated for $C_{13}H_{18}BF_4N_3$ $[M-BF_4]^+$: 216. Found: 216.
Preparative Methods: the synthesis of the triazolium salt **1** from *N*-methyl acetamide[1] is adapted from previously reported work for the synthesis of achiral triazolium salts (eq 1).[2]

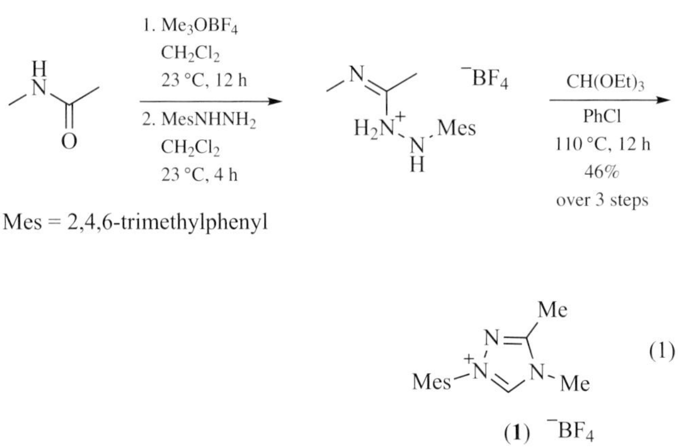

Mes = 2,4,6-trimethylphenyl

(1)

Purification: the triazolium salt is purified by column chromatography with 50–100% gradient of ethyl acetate/hexane.

Catalytic Electrophilic Amination of Homoenolates. Upon deprotonation of triazolium salt **1** by DBU, the generated *N*-heterocyclic carbene reacts with α,β-unsaturated aldehydes to form a homoenolate, which, in the presence of a diazene, affords a pyrazolidinone *via* a formal intermolecular [3 + 2]-cycloaddition. When compared to other NHC precursors, it is found that triazolium salt **1** is the best catalyst for the formation of pyrazolidinones (eq 2).[1]

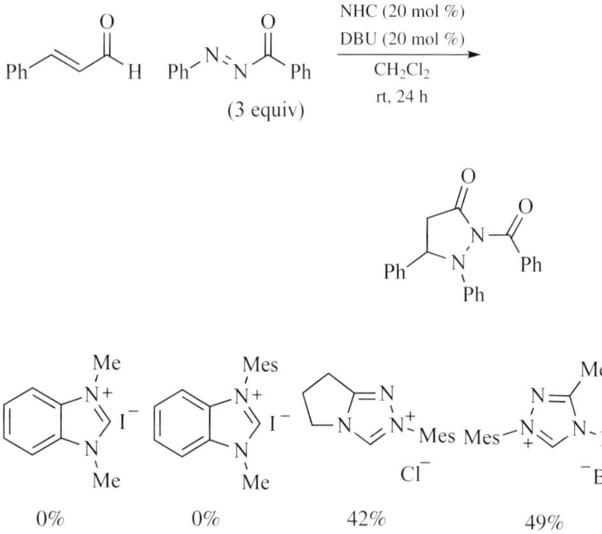

(2)

Upon optimization of the reaction conditions, one can see that the reaction tolerates electron-poor and electron-rich β-aryl aldehydes, as well as β-alkyl-substituted aldehydes. In the case of the diazene, the benzoyl moiety can bare both electron-poor and electron-rich substituents. However, substituting the aryl moiety with electron-poor substrates results in moderate to poor yields with regards to the generation of the desired pyrazolidinone (eq 3).

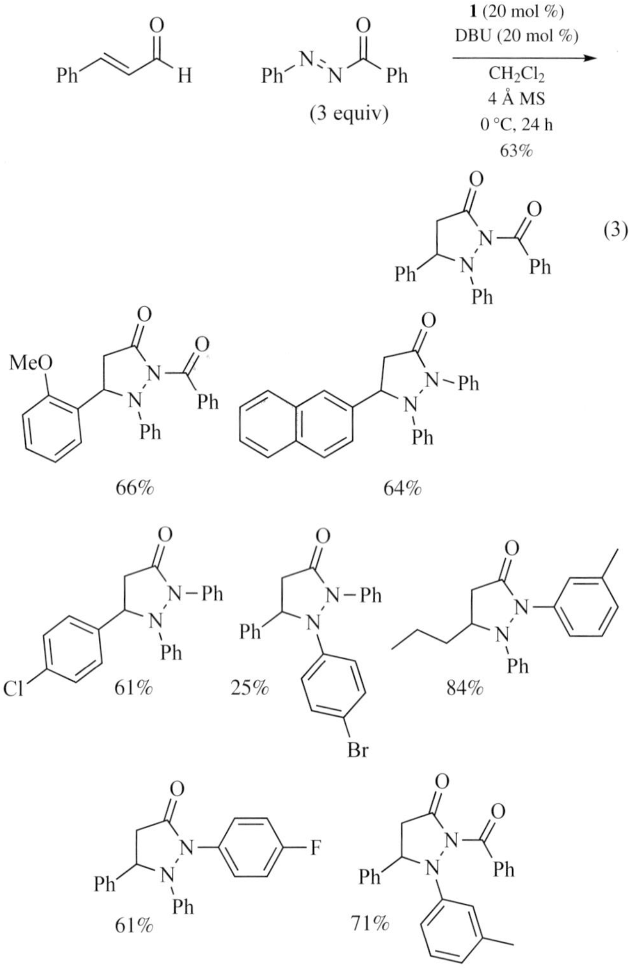

(3)

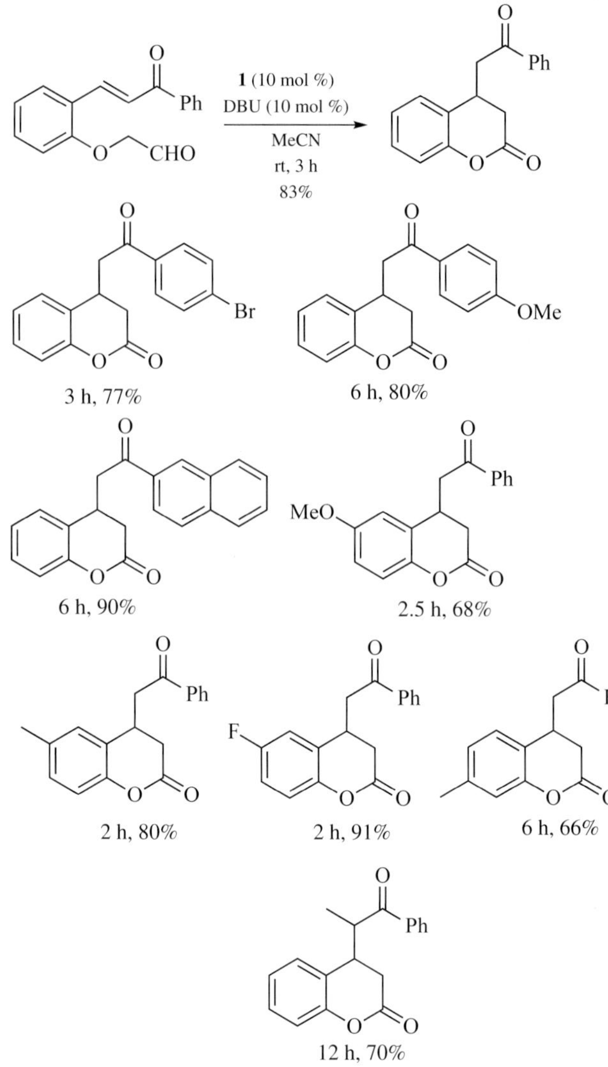

(4)

Domino Process for the Synthesis of Substituted Coumarins. Triazolium salt **1** is also successful in catalyzing the synthesis of substituted coumarins.[3] In the presence of DBU, the *N*-heterocyclic carbene is formed and reacts with aryloxyacetaldehydes to yield 3,4-dihydrocoumarins via a domino elimination of the aryloxy moiety/Michael addition onto the α,β-unsaturated ketone moiety/acylation process (eq 4).

The reaction tolerates both electron-poor and electron-rich aryl and ketone moieties. However, substitution at the α-position of the aldehyde is not suitable for this domino process.

1. Chan, A.; Scheidt, K., *J. Am. Chem. Soc.* **2008**, *130*, 2740.
2. Kerr, M.; Alaniz, J.; Rovis, T., *J. Org. Chem.* **2005**, *70*, 5725.
3. Phillip, E.; Wadamoto, M.; Roth, H.; Ott, A.; Scheidt, K., *Org. Lett.* **2009**, *11*, 105.

Tatiana Le Gall-Diop & Shawn K. Collins
Université de Montréal, Montréal, Québec, Canada

9-Mesityl-10-methylacridinium Perchlorate[1]

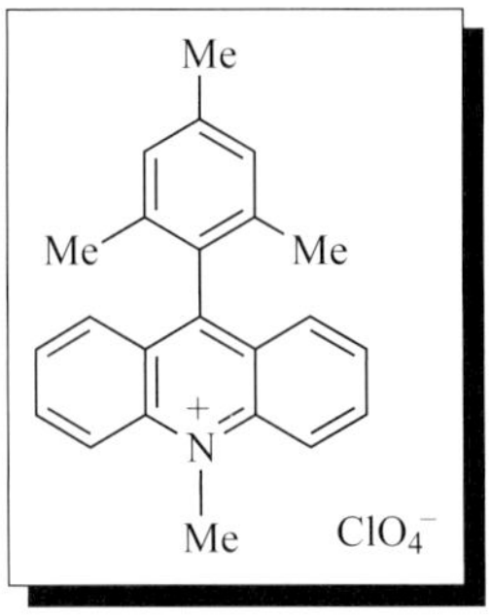

(ClO_4^- Salt)
[674783-97-2] $C_{23}H_{22}ClNO_4$ (MW 411.88)
InChI = 1S/C23H22N.ClHO4/c1-15-13-16(2)22(17(3)14-15)23-18-9-5-7-11-20(18)24(4)21-12-8-6-10-19(21)23;2-1(3,4)5/h5-14H,1-4H3;(H,2,3,4,5)/q+1;/p-1
InChIKey = LFMBERYWDLWXNO-UHFFFAOYSA-M

(BF_4^- Salt)
9-Mesityl-10-methylacridinium tetrafluoroborate
[$C_{23}H_{22}NBF_4$] (MW 399.24)

(photocatalyst, particularly useful for oxygenation and bromination of aromatic compounds)

Alternate Names: 10-Methyl-9-(2,4,6-trimethylphenyl)acridinium perchlorate; [Mes-Acr]$^+$.
Physical Data: solid at 25 °C.; m.p. (BF_4 Salt): 267–274 °C (dec.); λ_{max} = 360, 430 nm; $E^{red}_{1/2}$ = −0.49 V vs. SCE (MeCN); $^*E^{red}_{1/2}$ = +2.18 V vs. SCE (MeCN; singlet); $^*E^{red}_{1/2}$ = +1.45 V vs. SCE (MeCN; triplet).[12]
Solubility: X = ClO_4^-: insol (ClO_4^- Salt): hexane, ether, benzene; sol acetonitrile, dichloromethane, chloroform; slightly sol H_2O; X = BF_4^-: soluble in acetone, MeCN, DMF, DMSO, nitromethane, MeOH, EtOH, chlorinated hydrocarbon solvents such as 1,2-dichloroethane, CH_2Cl_2, and $CHCl_3$; sparingly soluble in toluene and tetrahydrofuran; insoluble in nonpolar aprotic solvents, Et_2O and H_2O.
Form Supplied in: yellow powder. The perchlorate salt is available from both TCI and Sigma-Aldrich. The tetrafluoroborate salt is available from Sigma-Aldrich.
Analysis of Reagent Purity: UV-Vis (λ_{max} 360, 423 [log ε 4.20, 3.75 acetonitrile]). ^{1}H NMR.[2]
Purification: recrystallization from ethanol; recrystallization from MeOH/Et_2O.[2]
Preparative Method: for X = ClO_4^- : the acridinium salt is made from *N*-methylacridone and mesityl magnesium bromide, followed by the addition of perchloric acid.[2] X = BF_4^- : same procedure as ClO_4^- salt preparation, but with $HBF_4{\cdot}Et_2O$ as the acid.[13]
Handling, Storage, and Precautions: potential skin irritant; this reagent should be handled in a fume hood. It may decompose violently or explode by mixing with combustible/flammable, heat, shock, or friction; keep away from strong reducing agents. General perchlorate safety should be exercised when handling [Mes-Acr]$^+ClO_4^-$, though an incident has not been reported.

Original Commentary

Kei Ohkubo & Shunichi Fukuzumi
Osaka University, Osaka, Japan

Oxygenation Reactions. Photoexcitation of 9-mesityl-10-methylacridinium perchlorate (Acr^+–Mes ClO_4^-) results in the formation of the electron transfer state ($Acr^\bullet$–$Mes^{\bullet+}$), which has an extremely long lifetime (e.g., 2 h at 203 K) as well as both high oxidizing and reducing ability.[1] In such a case, Acr^+–Mes acts as an efficient electron-transfer photocatalyst for highly selective oxygenation of various substrates with O_2 via selective radical coupling of the electron donor radical cation and $O_2^{\bullet-}$ under visible light irradiation.[2] For example, photooxygenation reactions of anthracene and tetraphenylethylene with Acr^+–Mes and O_2 occur efficiently to form the O_2 adducts, 9,10-epidioxyanthracene (eq 1)[2] and 1,2-dioxetane (eq 2),[3] respectively, via radical coupling of the radical cation of substrate with $O_2^{\bullet-}$. When tetramethylethylene is used as a substrate, the allylic hydroperoxide is formed in 75% yield (eq 3).[4]

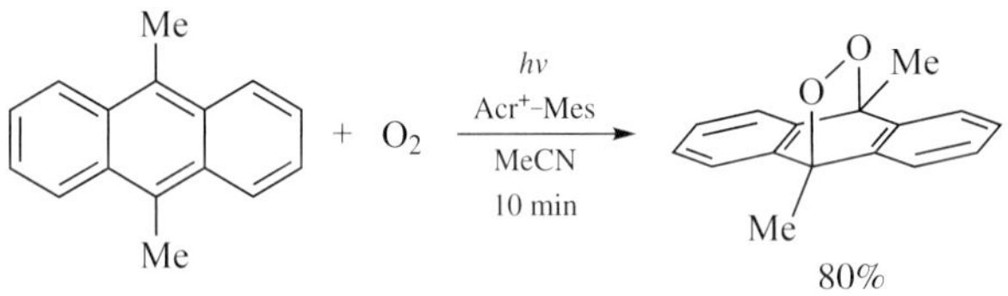

(1)

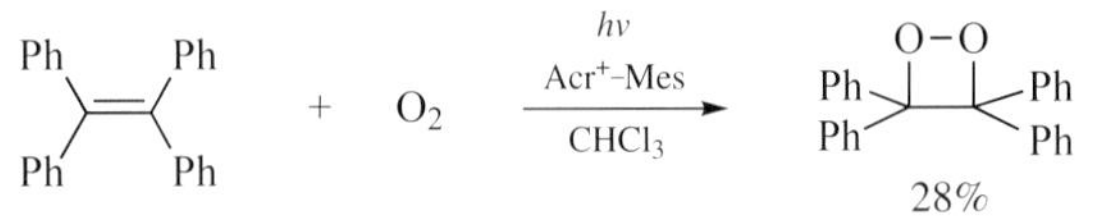

(2)

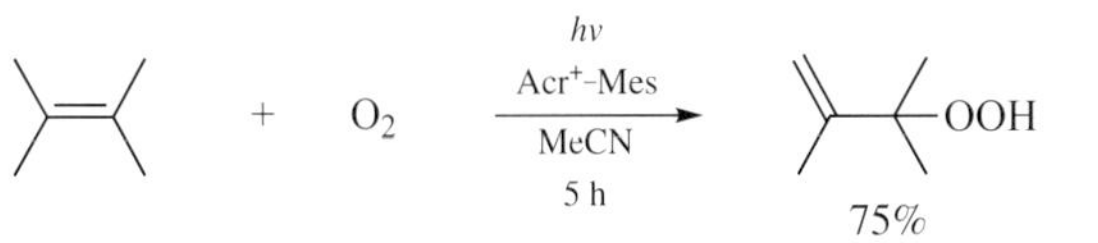

(3)

9-Mesityl-2,7,10-trimethylacridinium ion (Me_2Acr^+–Mes) is effective for selective photooxygenation of *p*-xylene to the corresponding benzaldehyde and hydrogen peroxide (H_2O_2) (eq 4).[5]

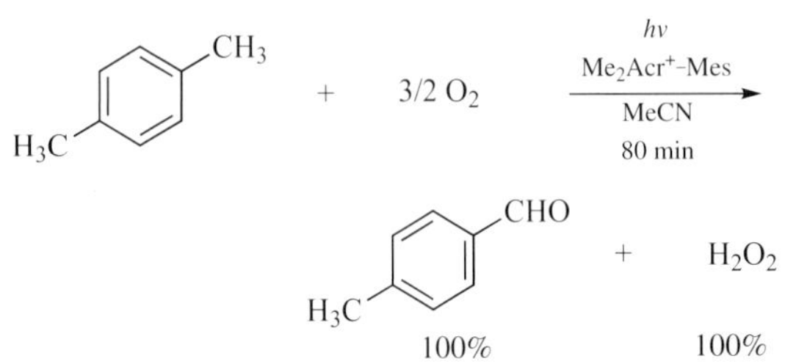

(4)

The photocatalytic oxygenation of triphenylphosphine also occurs quantitatively to triphenylphosphine oxide (eq 5).[6]

$$Ph_3P + 1/2\ O_2 \xrightarrow[\text{MeCN, 30 min}]{h\nu,\ Acr^+\text{–Mes}} Ph_3P{=}O\ (98\%) \quad (5)$$

The photocatalytic reactions are initiated by intramolecular electron transfer from the mesitylene moiety to the acridinium ion moiety to form the electron transfer state, $Acr^{\bullet}$–$Mes^{\bullet+}$. The substrates that can be oxygenated with oxygen using Acr^+–Mes Mes have been limited to those that have lower one-electron oxidation potentials than the one-electron oxidation potential of the mesityl moiety of Acr^+–Mes. The combination of Acr^+–Mes and hydrogen chloride (HCl) has made it possible to oxygenate toluene to benzaldehyde (eq 6)[7] and cyclohexane to cyclohexanone and cyclohexanol (eq 7).[8]

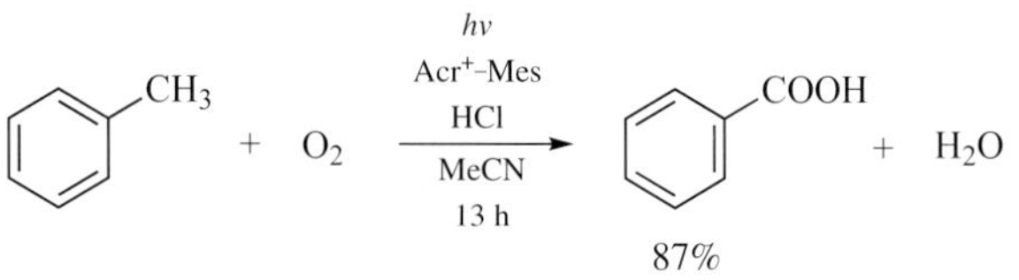
(6)

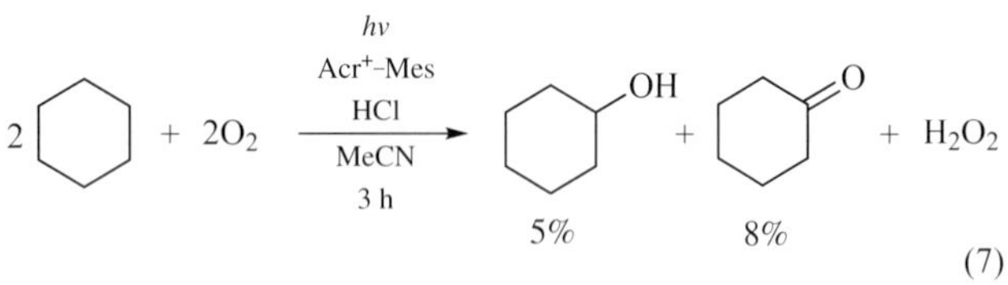
(7)

After photoirradiation of Acr^+–Mes with O_2 in the absence of substrate for 5 days, the mesitylene moiety is gradually oxidized to form 10-methyl-9-(4-formyl-2,6-dimethyl)acridinium ion (eq 8).[9]

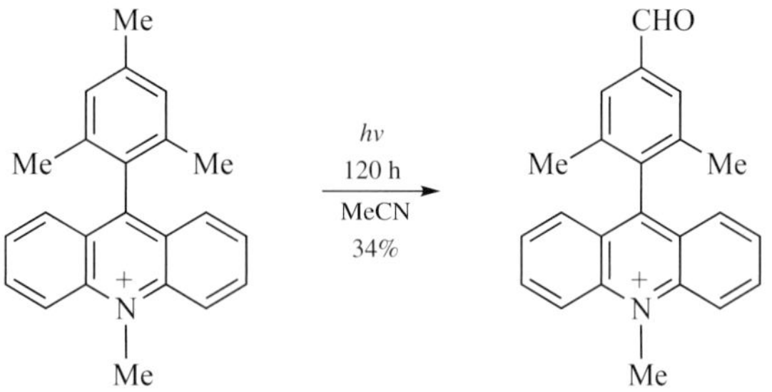
(8)

Halogenation Reactions. Radical cations of alkylbenzenes, which are generated by the electron-transfer oxidation with $Acr^{\bullet}$–$Mes^{\bullet+}$, act as efficient electrophiles and readily react with nucleophiles to form various adducts. Electrophilic addition of Br^- to the radical cations enables selective bromination following electron transfer. Visible light irradiation of Acr^+–Mes in O_2-saturated acetonitrile solution containing 1,2,4-trimethoxybenzene and aqueous HBr with a Xe lamp for 20 min resulted in the formation of a brominated product (eq 9).[10] When HBr is replaced by HCl, the reaction provides the corresponding chlorinated product.[7]

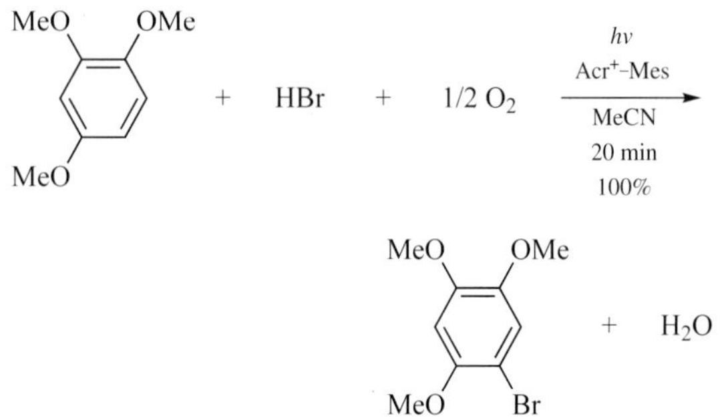
(9)

C–C Bond Formation. The Acr^+–Mes-catalyzed radical coupling reactions between radical cations and $O_2^{\bullet-}$ have been extended to the development of carbon–carbon bond formation reactions such as dimerization of anthracene and oligomerization of fullerene. When photochemical reaction of 9,10-dimethylanthracene with Acr^+–Mes is carried out without O_2 in chloroform, photocatalytic carbon–carbon bond formation occurs efficiently to give dimethyllepidopterene (5,12-dimethyl-5,6,11,12-tetrahydro-4*b*,12[1′,2′],6,10*b*[1″,2″]-dibenzeno-chrysene) (eq 10).[11]

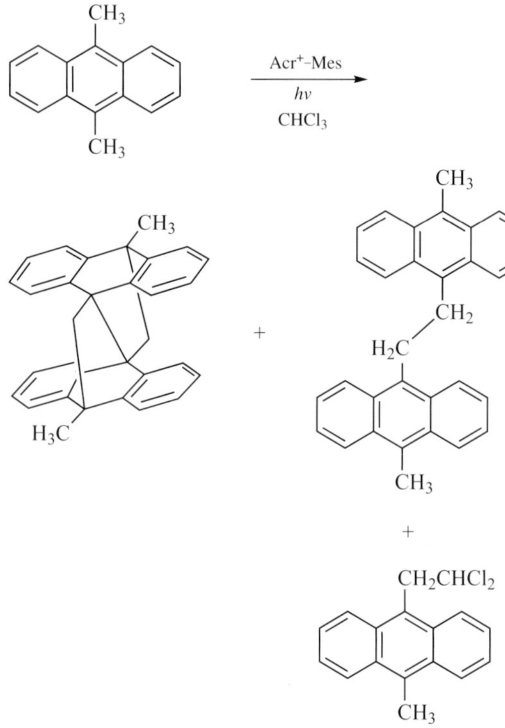
(10)

Photocatalytic oligomerization of fullerene in toluene/acetonitrile solution also occurs efficiently via electron-transfer reactions with photogenerated electron transfer state of Acr^+–Mes, followed by the radical coupling reaction between fullerene radical cation and radical anion (eq 11).[12] The MALDI-TOF-MS analysis of the crude mixture shows the formation of fullerene oligomers, C_{120}, C_{180}, and C_{240}.

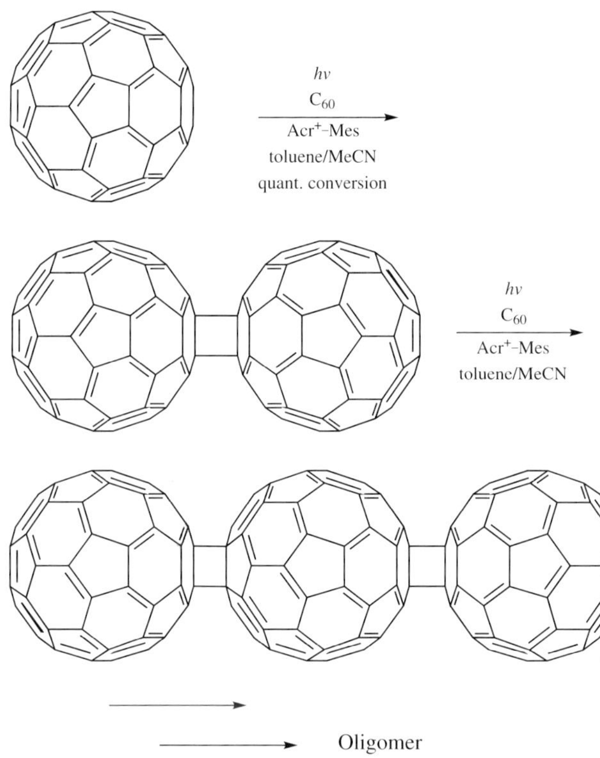

(11)

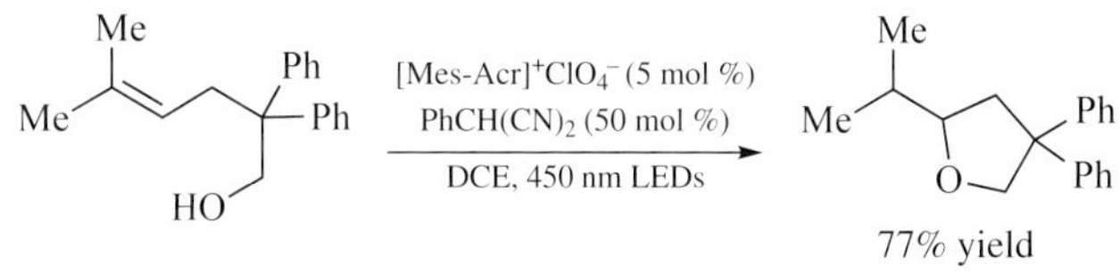

(12)

Anti-Markovnikov Alkene Hydroacetoxylation. A second-generation catalyst system comprised of $[\text{Mes-Acr}]^+\text{BF}_4^-$ and either thiophenol (PhSH; 20 mol %) or sodium benzenesulfinate ($PhSO_2Na$; 25 mol %) promotes the intermolecular reaction of carboxylic acids with alkenes (eq 13).[15] The reactions all give complete anti-Markovnikov regioselectivity and in the case of β-alkyl styrenes, addition of the carboxylic acid occurs at the β-position. Enamides also undergo addition of the carboxylic acid at the β-position and both aryl and aliphatic carboxylic acids can be employed as substrates.

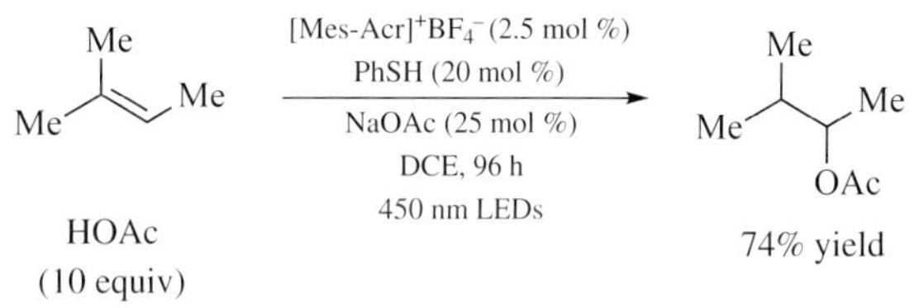

(13)

First Update

David A. Nicewicz
University of North Carolina at Chapel Hill, Chapel Hill, NC, USA

$[\text{Mes-Acr}]^+$ is a powerful oxidant in the singlet or triplet excited states *via* a reductive quenching mechanism. The primary use of this catalyst involves reactions commencing with an oxidation event. This catalyst can be employed in either net two-electron oxidation chemistry or redox-neutral chemistry, as the reduced form is a reasonably persistent radical species and can act as a reductant to return back to the ground state. The oxidative reactivity of $*[\text{Mes-Acr}]^+$ is complimentary to the commonly used photoredox catalyst, $[\text{Ru(bpy)}_3]^{2+}$ ($*E^{\text{red}}_{1/2}$ = +0.84 V vs. SCE; $E^{\text{red}}_{1/2}$ = +1.29 V vs. SCE for $[\text{Ru(bpy)}_3]^{3+}$), which is not as powerful of an oxidant as photoexcited $[\text{Mes-Acr}]^+$ and should be employed in cases where the oxidation potential of a substrate exceeds +1.29 V vs. SCE.

Anti-Markovnikov Alkene Hydroalkoxylation. A dual catalytic system of $[\text{Mes-Acr}^+]\text{ClO}_4^-$ and phenylmalononitrile ($PhCH(CN)_2$) provides a method for intramolecular anti-Markovnikov hydroalkoxylation of oxidizable alkenes (eq 12).[14] Single electron oxidation of alkenes using $[\text{Mes-Acr}]^+$ generates reactive alkene cation radicals, to which the alcohol adds predictably to the less-substituted position on trisubstituted alkenes and at the homobenzylic position of β-substituted styrenes to form five- to seven-membered cyclic ethers. The role of the $PhCH(CN)_2$ is presumed to act as a redox-active hydrogen-atom donor to complete the catalytic cycle.

Anti-Markovnikov Alkene Hydroamination. Both intramolecular[16] and intermolecular[17] hydroamination of alkenes can be accomplished employing the $[\text{Mes-Acr}]^+\text{BF}_4^-$ photooxidation catalyst in combination with either thiophenol (20 mol %) or phenyldisulfide (10 mol %) as the cocatalyst (eq 14). Chlorinated hydrocarbon solvents such as 1,2-dichloroethane (DCE) are best suited for these transformations, which afford reliable anti-Markovnikov selectivity. For the intramolecular reactions, *tert*-butoxycarbonyl (Boc) or *p*-toluenesulfonyl (Ts) protecting groups are necessary, while for the intermolecular examples, a trifluoromethanesulfonyl (Tf) group is optimal for high reactivity. In addition to triflylamide as the nucleophile in the intermolecular reactions, nitrogen-containing heteroaromatics such as indazole, pyrazole, imidazole, and 1,2,3-triazole are reactive in this context.

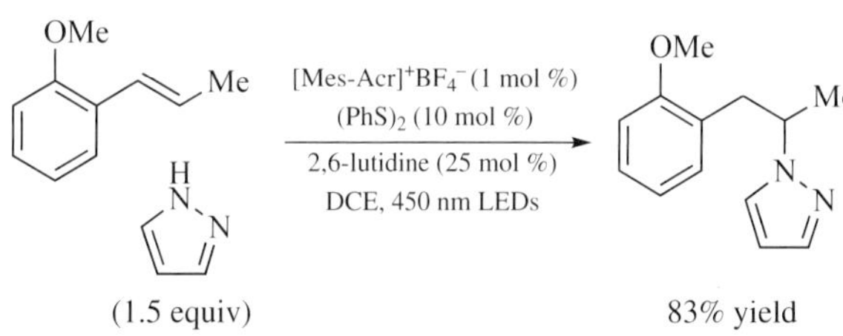

(14)

Anti-Markovnikov Alkene Hydrohalogenation. Using the $[\text{Mes-Acr}^+]\text{BF}_4^-$ catalyst in conjunction with 20 mol % of 4-methoxythiophenol, styrenes can undergo anti-Markovnikov hydrochlorination using lutidinium chloride as the stoichiometric source of HCl (eq 15).[18] Both unsubstituted, α- and β-methylated styrenes bearing various functional groups are competent substrates for the reported photolytic conditions. A related derivative of $[\text{Mes-Acr}]^+\text{BF}_4^-$, 9-mesityl-2,7-dimethyl-

10-phenylacridinium tetrafluoroborate is employed for the anti-Markovnikov addition of HF, $HOP(O)(OPh)_2$, and $HOSO_2CH_3$, using $Et_3N{\cdot}3HF$, $HOP(O)(OPh)_2$, and lutidinium mesylate, respectively.

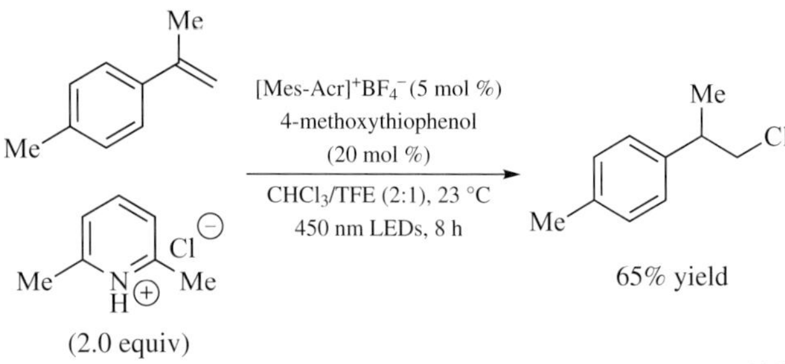

(15)

Alkene Hydrotrifluoromethylation. Trifluoromethyl radicals can be accessed by single-electron oxidation of sodium trifluoromethanesulfinate ($NaSO_2CF_3$– Langlois' reagent $-E_{p/2}$ = +1.05 V vs. SCE) employing the $[Mes\text{-}Acr]^+BF_4^-$ catalyst. Irradiation of catalyst system comprised of 5 mol % of $[Mes\text{-}Acr]^+BF_4^-$ and 20 mol % of a thiophenol allows for hydrotrifluoromethylation of alkenes using 1.5–3.0 equiv of $NaSO_2CF_3$ (eq 16).[19] The trifluoromethyl group is selectively added to the less-substituted position of the alkene. A 9:1 mixture of $CHCl_3$:trifluoroethanol is the key solvent mixture for this transformation as it is believed that trifluoroethanol may play a role as a hydrogen atom shuttle in the reaction. Suitable substrates for this transformation include aliphatic and styrenyl olefins. The scope of this reaction is consistent with oxidation of $NaSO_2CF_3$ as the operative mechanism as alkenes with oxidation potentials out of range of the $^*[Mes\text{-}Acr]^+$ ($E_{p/2}$ > +2.0 V vs. SCE), such as terminal aliphatic olefins, can be employed.

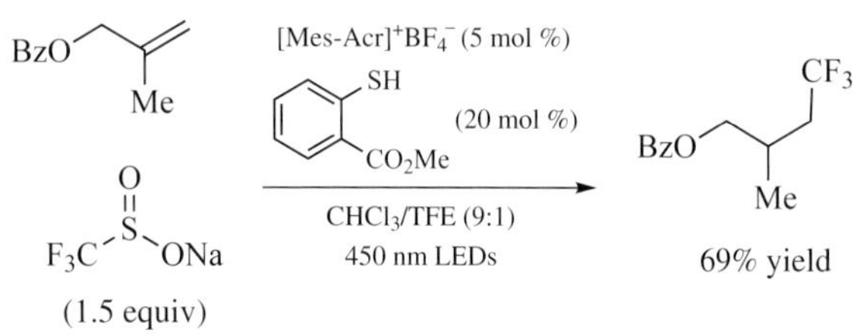

(16)

Cycloadditions. A non-concerted cycloaddition of allyl alcohols and oxidizable alkenes ($E_{p/2}$ < +2.0 V vs. SCE) can be accomplished using $[Mes\text{-}Acr^+]ClO_4^-$ as the photoredox catalyst and 1.0 equiv of the redox-active hydrogen atom donor, $PhCH(CN)_2$ (eq 9).[20] A number of highly substituted tetrahydrofurans bearing functional groups including free alcohols, protected amines and halides can be used in this reaction. The mechanism proceeds *via* single electron oxidation of the alkene by $^*[Mes\text{-}Acr^+]$, followed by addition of the allyl alcohol. The resultant radical undergoes a 5-*exo* cyclization followed by hydrogen-atom abstraction from $PhCH(CN)_2$. Propargyl alcohols can also be employed to furnish tetrahydrofuran adducts bearing *exo*-methylene groups.

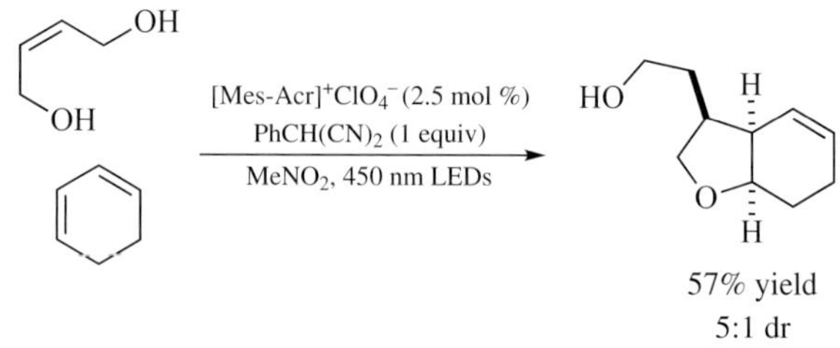

(17)

The $[Mes\text{-}Acr]^+ClO_4^-$ photooxidant can also be used to catalyze cycloaddition reactions between azirines and alkynes to furnish highly substituted pyrrolidines (eq 18).[21] The mechanism likely proceeds *via* single electron oxidation of the azirine by $^*[Mes\text{-}Acr]^+$ and *C–C* bond fissure to a 1,3-imine cation radical. Cycloaddition with the alkyne followed by reduction affords the corresponding pyrrole adducts. Electron-deficient alkynes, such as dimethylacetylene dicarboxylate, are reaction partners with azirines bearing combinations of aryl and alkyl groups under irradiation with 450–460 nm LEDs.

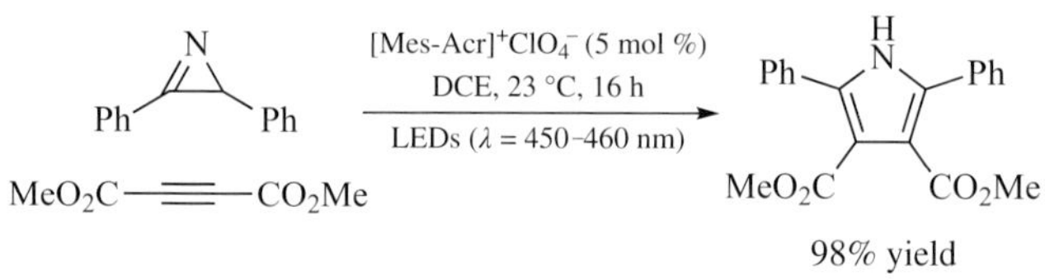

(18)

1. (a) Fukuzumi, S.; Kotani, H.; Ohkubo, K.; Ogo, S.; Tkachenko, N. V.; Lemmetyinen, H., *J. Am. Chem. Soc.* **2004**, *126*, 1600. (b) Fukuzumi, S.; Kotani, H.; Ohkubo, K., *Phys. Chem. Chem. Phys.* **2008**, *10*, 5159.
2. Kotani, H.; Ohkubo, K.; Fukuzumi, S., *J. Am. Chem. Soc.* **2004**, *126*, 15999.
3. (a) Ohkubo, K.; Nanjo, T.; Fukuzumi, S., *Org. Lett.* **2005**, *7*, 4265; (b) Ohkubo, K.; Nanjo, T.; Fukuzumi, S., *Catal. Today* **2006**, *117*, 356.
4. Griesbeck, A. G.; Cho, M., *Org. Lett.* **2007**, *9*, 611.
5. Ohkubo, K.; Mizushima, K.; Iwata, R.; Souma, K.; Suzuki, N.; Fukuzumi, S., *Chem. Commun.* **2010**, *46*, 601.
6. Ohkubo, K.; Nanjo, T.; Fukuzumi, S., *Bull. Chem. Soc. Jpn.* **2006**, *79*, 1489
7. Ohkubo, K.; Mizushima, K.; Fukuzumi, S., *Res. Chem. Intermed.* **2012**, *32*, in press.
8. Ohkubo, K.; Fujimoto, A.; Fukuzumi, S., *Chem. Commun.* **2011**, *47*, 8515.
9. Benniston, A. C.; Elliott, K. J.; Harrington, R. W.; Clegg, W., *Eur. J. Org. Chem.* **2009**, 253.
10. Ohkubo, K.; Mizushima, K.; Iwata, R.; Fukuzumi, S., *Chem. Sci.* **2011**, *2*, 715.
11. Ohkubo, K.; Iwata, R.; Miyazaki, S.; Kojima, T.; Fukuzumi, S., *Org. Lett.* **2006**, *8*, 6079.
12. Ohkubo, K.; Iwata, R.; Yanagimoto, T.; Fukuzumi, S., *Chem. Commun.* **2007**, 3139
13. Benniston, A. C.; Harriman, A.; Li, P.; Rostron, J. P.; van Ramesdonk, H. J.; Groeneveld, M. M.; Zhang, H.; Verhoeven, J. W., *J. Am. Chem. Soc.* **2005**, *127*, 16054.
14. Hamilton, D. S.; Nicewicz, D. A., *J. Am. Chem. Soc.* **2012**, *134*, 18577.

15. Perkowski, A. J.; Nicewicz, D. A., *J. Am. Chem. Soc.* **2013**, *135*, 10334.
16. Nguyen, T. M.; Nicewicz, D. A., *J. Am. Chem. Soc.* **2013**, *135*, 9588.
17. Nguyen, T. M.; Manohar, N.; Nicewicz, D. A., *Angew. Chem., Int. Ed.* **2014**, *53*, 6198.
18. Wilger, D. J.; Grandjean, J.-M. M.; Lammert, T. R.; Nicewicz, D. A., *Nat. Chem.* **2014**, *6*, 720.
19. Wilger, D. J.; Gesmundo, N. J.; Nicewicz, D. A., *Chem. Sci.* **2013**, *4*, 3160.
20. Grandjean, J.-M. M.; Nicewicz, D. A., *Angew. Chem., Int. Ed.* **2013**, *52*, 3967.
21. Xuan, J.; Xia, X.-D.; Zeng, T.-T.; Feng, Z.-J.; Chen, J.-R.; Lu, L.-Q.; Xiao, W.-J., *Angew. Chem., Int. Ed.* **2014**, *53*, 5653.

6′-Methoxy-(8*S*,9*R*)-cinchonan-9-yl *N-tert*-butoxycarbonyl-2-pyrrolidine-carboxylate

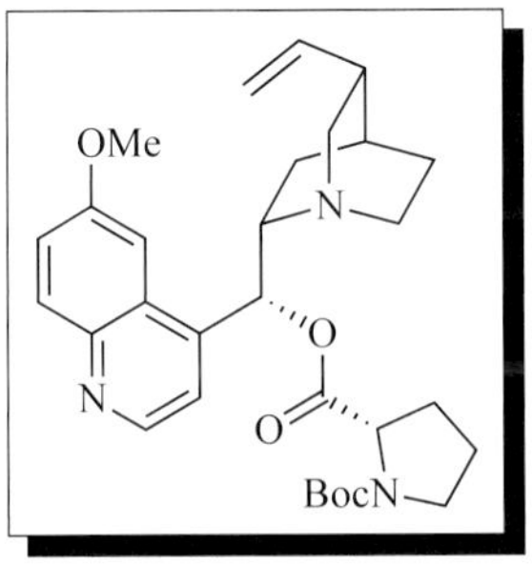

6′-Methoxy-(8*S*,9*R*)-cinchonan-9-yl *N-tert*-butoxycarbonyl-2-pyrrolidinecarboxylate

[916748-40-8] $C_{30}H_{39}N_3O_5$ (MW 521.65)

InChI = 1S/C30H39N3O5/c1-6-19-18-32-15-12-20(19)16-26(32)27(22-11-13-31-24-10-9-21(36-5)17-23(22)24)37-28(34)25-8-7-14-33(25)29(35)38-30(2,3)4/h6,9-11,13,17,19-20,25-27H,1,7-8,12,14-16,18H2,2-5H3/t19-,20-,25-,26-,27+/m0/s1

InChIKey = KXDUYNVLODLRCV-QJORRVHFSA-N

(asymmetric organocatalyst, used for asymmetric α-bromination of acyl chlorides)[1,2]

Alternative Names: (*R*)-ProQ; *N*-Boc-L-Proline Quinine.
Physical Data: mp = 116 °C; $[\alpha]_D^{25} = +17$ (c = 0.01, $CHCl_3$).[3]
Form Supplied in: white crystalline solid.
Analysis of Reagent Purity: IR, NMR, Elemental analysis.[3]
Preparative Method: prepared according to the procedure of Lectka and coworkers.[3]
Purification: purified through a short plug of silica with 99.5% EtOAc/0.5% triethylamine to yield a white solid (foam) and recrystallized from boiling Et_2O/hexanes.

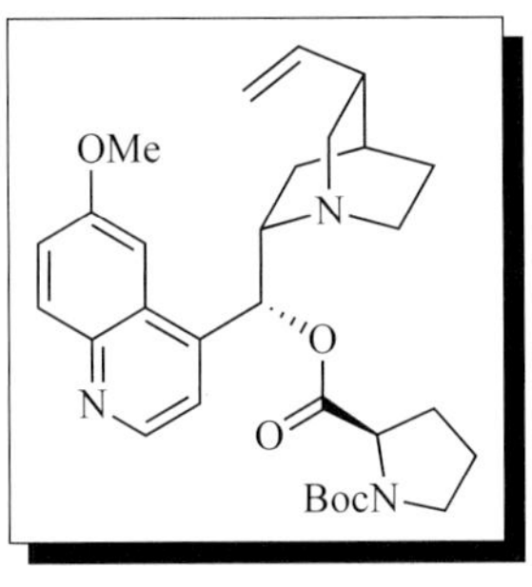

6′-Methoxy-(8*S*, 9*S*)-cinchonan-9-yl *N-tert*-butoxycarbonyl-(2*S*)-2-pyrrolidinecarboxylate

[916748-39-5] $C_{30}H_{39}N_3O_5$ (MW 521.65)

InChI = 1S/C30H39N3O5/c1-6-19-18-32-15-12-20(19)16-26(32)27(22-11-13-31-24-10-9-21(36-5)17-23(22)24)37-28(34)25-8-7-14-33(25)29(35)38-30(2,3)4/h6,9-11,13,17,19-20,25-27H,1,7-8,12,14-16,18H2,2-5H3/t19-,20-,25+,26-,27+/m0/s1

InChIKey = KXDUYNVLODLRCV-RJGJBJKLSA-N

(asymmetric organocatalyst, used for asymmetric α-bromination of acyl chlorides)[1,2]

Alternative Names: (*S*)-ProQ; *N*-Boc-D-Proline Quinine.
Physical Data: mp = 134 °C; $[\alpha]_D^{25} = -78$ (c = 0.01,$CHCl_3$).[3]
Form Supplied in: white crystalline solid.
Analysis of Reagent Purity: IR, NMR, Elemental analysis.[3]
Preparative Methods: prepared according to the procedure of Lectka and coworkers.[3]
Purification: purified through a short plug of silica with 99.5% EtOAc/0.5% triethylamine to yield a white solid (foam) and recrystallized from boiling Et_2O/hexanes.

General Procedure for the Synthesis of 6′-Methoxy-(8*S*,9*R*)-cinchonan-9-yl-*N-tert*-butoxycarbonyl-(2*R*)-2-pyrrolidinecarboxylate(3).[3] Quinine (**1**) (1 equiv), *N*-Boc-L-Proline (**2**) (1 equiv), and 4-(dimethylamino)pyridine (0.1 equiv) are dissolved in dichloromethane (0.5 M solution of (**1**)) and the solution is cooled to 0 °C. *N,N′*-Dicyclohexylcarbodiimide (1.1 equiv) in dichloromethane (1.7 M) is added and the reaction warmed to room temperature overnight. The reaction mixture is then filtered and the resulting filtrate is concentrated before the crude product is purified by column chromatography on a plug of silica eluting with 99.5% EtOAc/0.5% triethylamine. (*R*)-ProQ (**3**) is recrystallized from boiling Et_2O/hexanes to afford pure (**3**) in high yield (94%) (eq 1).[3]

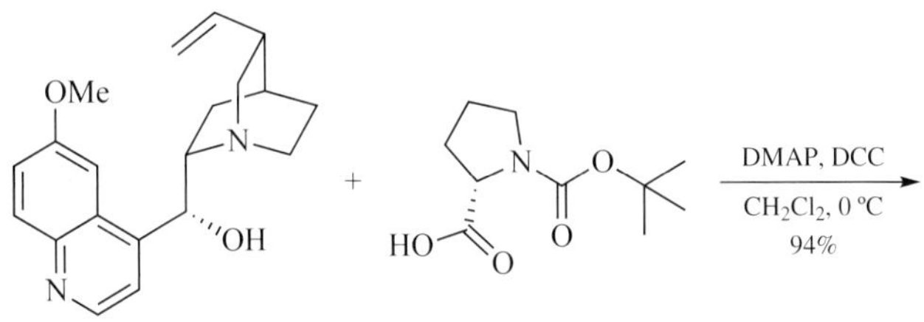

(**1**) (**2**)

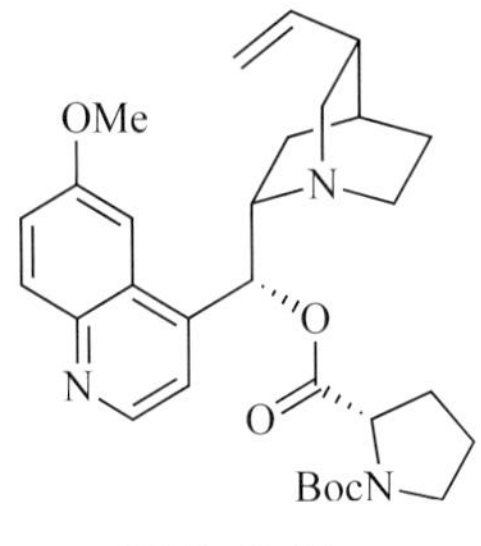

(*R*)-ProQ (**3**) (1)

Catalytic Asymmetric α-Bromination of Acid Chlorides. Cinchona alkaloid catalytic systems have seen considerable success over the last fifty years as the catalysts of choice for many asymmetric reactions of ketenes.[1,2] For example, cinchona alkaloid systems (see quinine, TMS-quinine, and benzoylquinine) have been used to catalyze the enantioselective synthesis of chiral esters, β-lactones, β-lactams, ketene homodimer β-lactones, ketene heterodimer β-lactones, as well as a host of other important processes.[4–14]

α-Halogenated carbonyl compounds are a very important class of molecules due to their potential as intermediates for the construction of complex molecules. Of this class, α-brominated carbonyl compounds are arguably the most useful as they typically undergo nucleophilic substitutions through a purely S_N2 mechanistic pathway. Traditionally, asymmetric halogenation of carbonyl compounds has been carried out using chiral halogenating agents, metal complexes with chiral ligands, or chiral auxiliaries.[15]

In 2000, Lectka and coworkers introduced a cinchona alkaloid system (benzoylquinine, BQ) for the asymmetric synthesis of β-lactams from in situ generated ketenes and α-iminoesters.[6,7] This system provides access to a range of *cis*-β-lactams with high enantioselectivity (> 95%).[7] The same system was also applied to the asymmetric α-bromination of acid chlorides.[12] While the benzoylquinine catalyst does catalyze the α-bromination of acid chlorides in moderate to good yields and with excellent enantioselectivity on a small scale (0.065 mmol), enantioselectivity and yields decrease significantly upon scale-up of the reaction. To overcome these problems Lectka and coworkers investigated other alkaloid derivatives.[3,13] Lectka and coworkers observed an improvement in enantioselection when a (*R*)-Boc-prolyl group was conjugated to quinine instead of the benzoyl group. The choice of the (*R*)-Boc-prolyl group was based on a molecular mechanics computational study (Macromodel program), which showed an increase in the energy gap between the *re-* and *si*-face exposed conformations of the ammonium enolate intermediate when the (*R*)-Boc-prolyl group was utilized (instead of the benzoyl group).[3] Bromination at the α-carbon with high enantiomeric excess (83–99% ee for 12 examples) is accomplished through two related methods depending on whether the *R* group of acyl chloride (**4**) is an aryl or alkyl group (eq 2).[3,13]

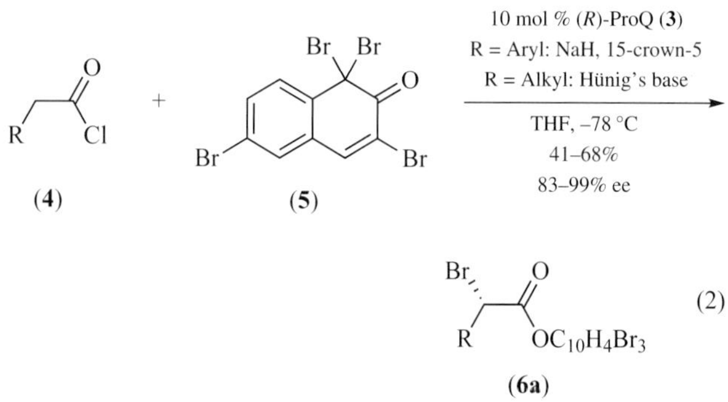

(2)

The general procedure for aryl-substituted acetyl chlorides involves stirring 15-crown-5 (0.1 equiv) with sodium hydride (1.2 equiv) and (*R*)-ProQ(**3**) (0.1 equiv) in THF (0.015 M solution of (**3**)) at –78 °C. Acyl chloride (1.4 equiv) dissolved in THF (0.61 M) is added dropwise to the stirred suspension, followed by a 5 min dropwise addition of brominating agent (**5**) (1.0 equiv) dissolved in THF (0.15 M). The solution is stirred for 5 h at −78 °C and then quenched with methanolic HCl. The reaction is extracted with ether, the combined organics dried and purified by column chromatography with hexanes to afford (**6a**) as a white solid (44–59%).[3,13]

The general procedure for aliphatic acyl chlorides differs from that of aryl-substituted acyl chlorides only in the base used. Hünig's base (1.0 equiv) and (*R*)-ProQ(**3**) (0.1 equiv) in THF (0.015 M solution of (**3**)) are stirred at −78 °C. Acyl chloride (1.2 equiv) dissolved in THF (0.52 M) is added dropwise followed by a 5 min dropwise addition of brominating agent (**5**) (1.0 equiv) dissolved in THF (0.15 M). The solution is stirred for 5 h at −78 °C and quenched with methanolic HCl. The reaction is extracted with ether, the combined organics dried and purified by column chromatography with hexanes to afford (**6a**) as a white solid (41–68%).[3,13]

Reaction Mechanism of the Asymmetric α-Bromination of Acid Chlorides. The mechanism of the asymmetric α-bromination reaction is proposed to involve the following steps.[3,13] In the presence of an appropriate shuttle base system (NaH/15-crown-5 or Hünig's base along with (*R*)-ProQ (**3**)), dehydrohalogenation of acyl chloride (**4**) takes place to generate ketene (**7**) in situ.[7] (*R*)-ProQ (**3**), acting as a nucleophilic catalyst, then adds to the less sterically hindered side of (**7**), to generate a chiral zwitterionic ammonium enolate intermediate (**8**) in stereoselective fashion. Ammonium enolate (**8**) then abstracts Br^+ from the brominating agent (**5**) to form the enantioenriched ion-paired intermediate (**9**). Finally, the desired α-brominated product (**6**) is formed through transacylation of the leaving group (LG = $^-OC_{10}H_4Br_3$) with the acyl ammonium of (**9**), along with simultaneous regeneration of free (*R*)-ProQ (**3**) (eq 3).

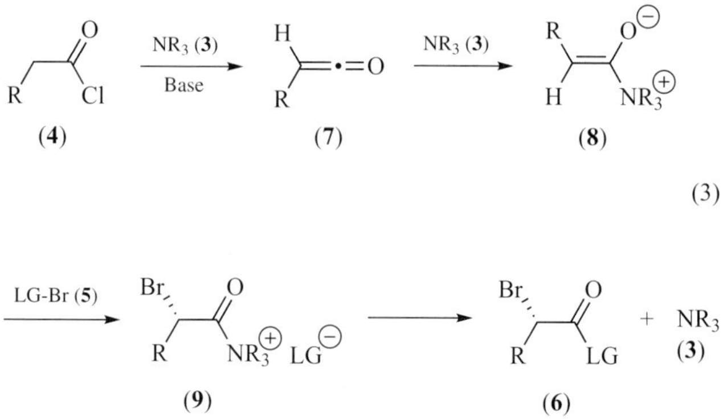

(3)

A limitation of the procedure is demonstrated by the decrease in yield and enantioselectivity when electron withdrawing groups are placed on the aromatic R group of the acyl chloride (**4**). The resulting increased acidity of the α-proton in (**6**) is presumed to be responsible for the lower yield (31% in the case of *p*-nitrophenylacetyl chloride), as well as being implicated in racemization of the product.[13]

Early experiments conducted by Lectka and coworkers showed that selection of the brominating agent is also an important factor in maximizing enantioselectivity. Under the conditions of the asymmetric α-bromination reaction, a stoichiometric amount of base (NaH or Hünig's base) is employed, and so racemization of intermediates is possible. This was encountered as a problem when reactions were scaled-up to the 1-mmol level.[3,13] As the acyl ammonium intermediate (**9a**) is expected to be the most racemization-prone intermediate, limiting the lifetime of (**9a**) was regarded as the key to minimizing racemization (eq 4). With this in mind, it was considered beneficial to have the Br^+ leave from a position *ortho* (rather than *para*) to the carbonyl of the brominating agent (**5**) in order to shorten the lifetime of (**9a**) (eq 4). This was reflected in the experimental results whereby use of an *ortho*-brominating agent (**5**) led to much higher enantioselectivity (99% ee) than when the analogous *para*-brominating agent (**10**) (78% ee) was used in the asymmetric bromination of phenylacetyl chloride.[3]

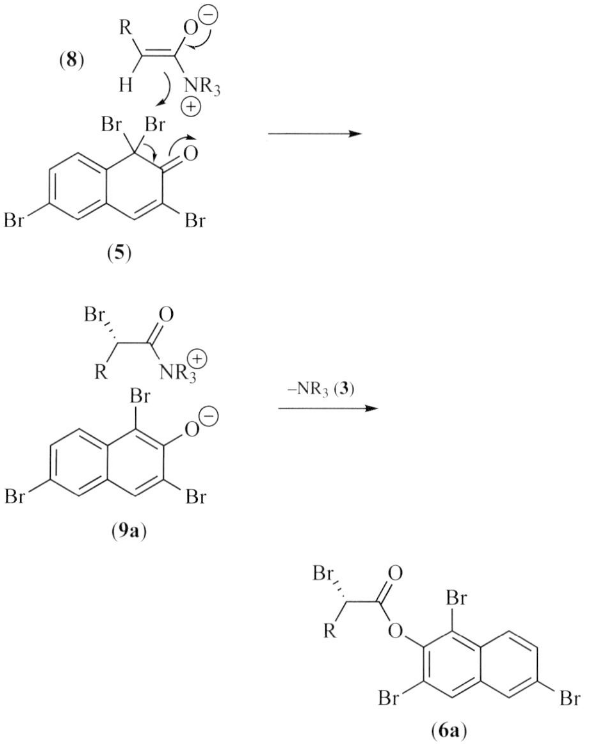

(4)

An analysis of the reaction of the *para*-brominating agent (**10**) with ammonium enolate **8** shows how the acyl ammonium intermediate (**9b**) is more susceptible to racemization (eq 5). In order for the phenolate leaving group to displace the (*R*)-ProQ (**3**) from (**9b**), it must first rotate 180°to give (**9b’**) (unlike when (**5**) is used as brominating agent). It is this requirement for rotation to occur that causes the lifetime of the racemization-prone acyl ammonium intermediate (**9b**) to be extended, and hence more racemization is observed when the *para*-brominating agent (**10**) is used instead of the *ortho*-brominating agent (**5**) (eq 5).

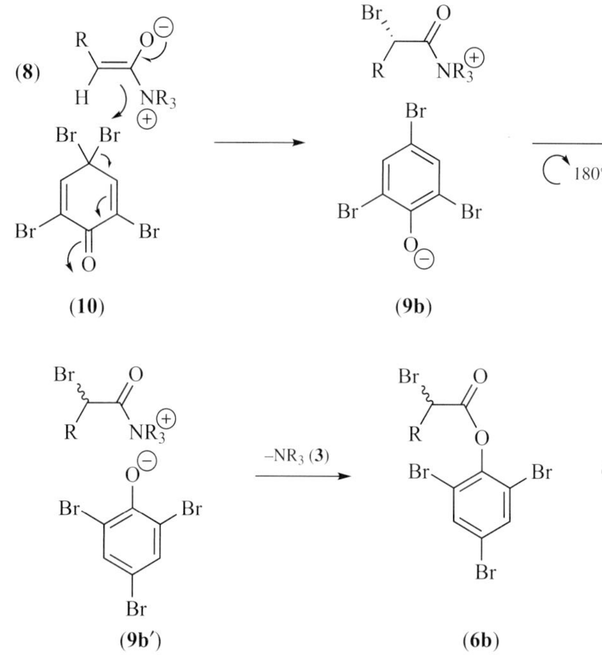

(5)

Overall, (*R*)-ProQ (**3**) has been a rarely used reagent thus far but it is likely to see increasing use as a catalyst for the development of new reactions of ketenes, especially in situations where the more commonly used cinchona alkaloid catalysts (quinine, TMS-quinine, or benzoylquinine) fail or give unsatisfactory enantioselectivity. The (*R*)-ProQ(**3**)-catalyzed asymmetric α-bromination method reported by Lectka and co-workers compares very favorably with other asymmetric methods for the α-bromination of carbonyl compounds, in particular giving consistently higher levels of enantioselectivity than any other catalytic system.[16,17]

---

1. Paull, D. H.; Weatherwax, A.; Lectka, T., *Tetrahedron* **2009**, *65*, 6771.
2. Chen, S.; Salo, E. C.; Kerrigan, N. J. In *Science of Synthesis Reference Library, Asymmetric Organocatalysis, Vol. 1, Lewis Base and Acid Catalysts*; List, B., Ed.; Thieme: Stuttgart, **2012**; p 455.
3. Dogo-Isonagie, C.; Bekele, T.; France, S.; Wolfer, J.; Weatherwax, A.; Taggi, A. E.; Lectka, T., *J. Org. Chem.* **2006**, *71*, 8946.
4. Pracejus, H., *Justus Liebigs Ann. Chem.* **1960**, *634*, 9.
5. Wynberg, H.; Staring, E. G. J., *J. Am. Chem. Soc.* **1982**, *104*, 166.
6. Taggi, A. E.; Hafez, A. M.; Wack, H.; Young, B.; Drury, III, W. J.; Lectka, T., *J. Am. Chem. Soc.* **2000**, *122*, 7831.
7. Taggi, A. E.; Hafez, A. M.; Wack, H.; Young, B.; Ferraris, D.; Lectka, T., *J. Am. Chem. Soc.* **2002**, *124*, 6626.
8. Calter, M. A.; Orr, R. K.; Song, W., *Org. Lett.* **2003**, *5*, 4745.
9. Ibrahim, A. A.; Nalla, D.; Van Raaphorst, M.; Kerrigan, N. J., *J. Am. Chem. Soc.* **2012**, *134*, 2942.
10. Wack, H.; Taggi, A. E.; Hafez, A. M.; Drury, III, W. J.; Lectka, T., *J. Am. Chem. Soc.* **2001**, *123*, 1531.
11. Paull, D. H.; Scerba, M. T.; Alden-Danforth, E.; Widger, L. R.; Lectka, T., *J. Am. Chem. Soc.* **2008**, *130*, 17260.
12. Hafez, A. M.; Taggi, A. E.; Wack, H.; Esterbrook, J.; Lectka, T., *Org. Lett.* **2001**, *3*, 2049.
13. Dogo-Isonagie, C.; Bekele, T.; France, S.; Wolfer, J.; Weatherwax, A.; Taggi, A. E.; Paull, D. H.; Dudding, T.; Lectka, T., *Eur. J. Org. Chem.* **2007**, 1091.
14. Bekele, T.; Shah, M. H.; Wolfer, J.; Abraham, C. J.; Weatherwax, A.; Lectka, T., *J. Am. Chem. Soc.* **2006**, *128*, 1810.

15. Shibata, N.; Kohno, J.; Takai, K.; Ishimaru, T.; Nakamura, S.; Toru, T.; Kanemasa, S., *Angew. Chem., Int. Ed.* **2005**, *44*, 4204.
16. Bertelsen, S.; Halland, N.; Bachmann, S.; Marigo, M.; Braunton, A.; Jørgensen, K. A., *Chem. Commun.* **2005**, 4821.
17. Marigo, M.; Kumaragurubaran, N.; Jørgensen, K. A., *Chem. Eur. J.* **2004**, *10*, 2133.

Nicholas J. Peraino & Nessan J. Kerrigan
Oakland University, Rochester, MI, USA

1-Methyl-2-azaadamantane *N*-Oxyl

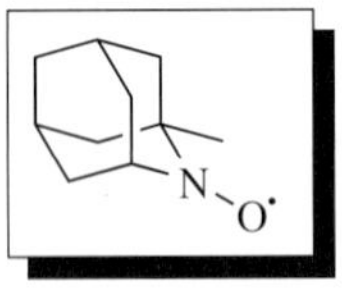

[872598-44-2] $C_{10}H_{16}NO$ (MW 166.24)
InChI = 1/C10H16NO/c1-10-5-7-2-8(6-10)4-9(3-7)11(10)12/h7-9H,2-6H2,1H3/t7-,8+,9-,10+
InChIKey = IBNXYCCLPCGKDM-YNFQOJQRBA

(stable nitroxyl radical, highly efficient organocatalyst for the oxidation of both primary and secondary alcohols to aldehydes and ketones, respectively)

Solubility: soluble in common organic solvents such as CH_2Cl_2, $CHCl_3$ and EtOAc.
Form Supplied in: red solid.
Preparative Methods: the title reagent can be prepared in two steps from 1-iodomethyl-2-azaadamantane, which itself is accessible through a five-step synthesis starting from 1,3-admantanediol (eq 1):[1]

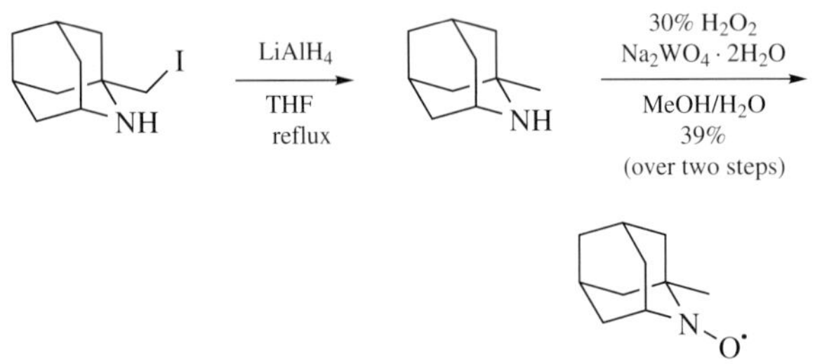

(1)

To a solution of 1-iodomethyl-2-azaadamantane (3.35 g, 12 mmol) in THF (60 mL) at 0 °C [$LiAlH_4$ (550 mg, 14.4 mmol)] was added portionwise. After stirring at 70 °C for 30 min, the reaction was cooled to 0 °C, diluted with Et_2O and quenched with 30% aq. NH_3. The resulting mixture was filtered through *Celite®* and the filtrate concentrated under reduced pressure. The residue was diluted with $CHCl_3$. Saturated Na_2CO_3 was then added and extracted with $CHCl_3$. The organic layer was dried over Na_2CO_3 and concentrated under reduced pressure to give the crude product as pale yellow oil. A mixture of the crude amine and $Na_2WO_4{\cdot}2H_2O$ (404 mg, 1.23 mmol) in H_2O–MeOH (20 mL, 1:4 v/v) was stirred at room temperature for 30 min. The reaction mixture was treated with 30% H_2O_2 (5.6 mL, 51.2 mmol) at 0 °C and stirred for 30 min. After stirring at room temperature for 3 h, MeOH was removed by rotary evaporation. The residue was poured into saturated $NaHCO_3$ and extracted with Et_2O. The organic layer was washed with brine, dried over $MgSO_4$, and concentrated under reduced pressure. The crude material was purified by flash column chromatography on silica (1:4 AcOEt:hexanes) to give the title compound (777 mg, 39%) as a red solid.
Purification: checked by TLC and purified by column chromatography on silica as needed.
Handling, Storage, and Precautions: care should be taken in handling this free radical reagent due to potential toxicity; should be stored in dark.

Oxidation of Alcohols to Aldehydes and Ketones. 1-Me-AZADO was shown to be a superior catalyst to 2,2,6,6-tetramethylpiperidin-1-oxyl (TEMPO) for the oxidation of primary alcohols to aldehydes. As shown in eqs and, 1-Me-AZADO is significantly more reactive than TEMPO when either sodium hypochlorite in combination with KBr (Anelli's conditions[2]) or $PhI(OAc)_2$ (Margarita's conditions[3]) was used as the stoichiometric oxidant. The loading of 1-Me-AZADO could be reduced to 0.004 mol % in the oxidation of 3-phenyl propanol when the reaction was run for 30 min at 0 °C, after which the aldehyde was isolated in 88% yield.

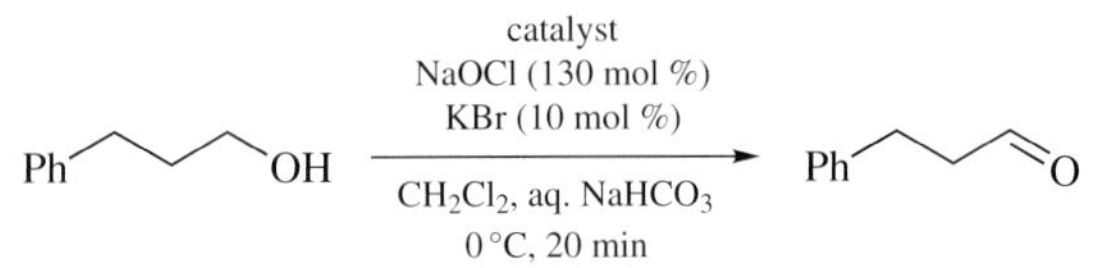

(2)

catalyst (0.01 mol %):	yield:
1-Me-AZADO	91%
TEMPO	23%

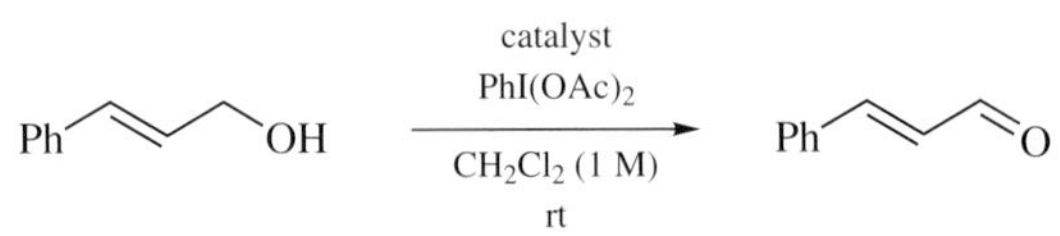

catalyst (1 mol %):	yield reaction time:
1-Me-AZADO	93%/42 min
TEMPO	42%/360 min

(3)

Unlike TEMPO, which is known to catalyze the selective oxidation of primary alcohols in the presence of secondary alcohols,[4] 1-Me-AZADO is capable of catalyzing the oxidation of a variety of secondary alcohols to ketones. Thus, in the presence of 1 mol % of 1-Me-AZADO, a variety of structurally hindered secondary alcohols were oxidized with NaOCl to the corresponding ketones in excellent yields (eq 4).

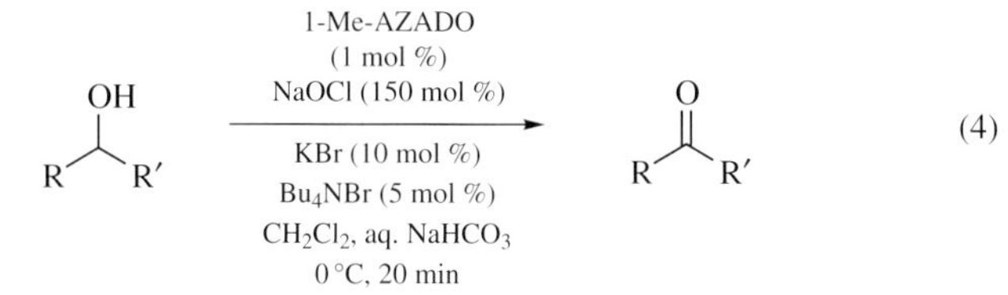

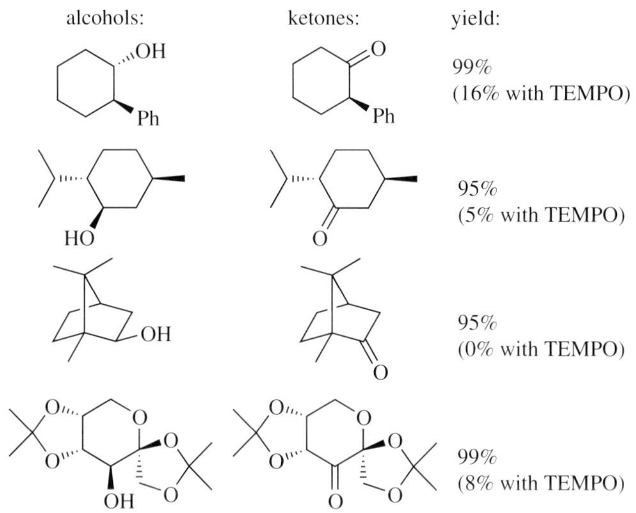

1-Me-AZADO was recently used in the catalytic oxidation of a sensitive alcohol during an enantio- and diastereocontrolled synthesis of (+)-juvabione (eq 5).[5] The reaction gave a higher yield (94%) than oxidation employing Dess–Martin periodinane.[6]

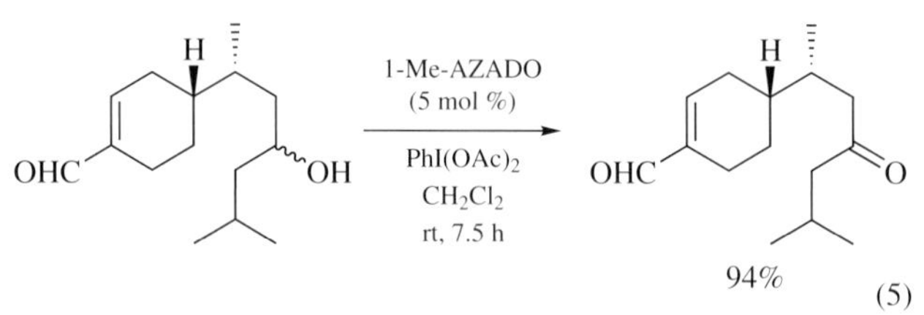

Related Reagents. 2,2,6,6-Tetramethylpiperidin-1-oxyl (TEMPO); *N*-hydroxyphthalimide (NHPI).

---

1. Shibuya, M.; Tomizawa, M.; Suzuki, I.; Iwabuchi, Y., *J. Am. Chem. Soc.* **2006**, *128*, 8412.
2. (a) Anelli, P. L.; Banfi, C.; Montanari, F.; Quici, S., *J. Org. Chem.* **1987**, *52*, 2559. (b) Anelli, P. L.; Banfi, S.; Montanari, F.; Quici, S., *J. Org. Chem.* **1989**, *54*, 2970.
3. De Mico, A.; Margarita, R.; Parlanti, L.; Vescovi, A.; Piancatelli, G., *J. Org. Chem.* **1997**, *62*, 6974.
4. (a) Siedlecka, R.; Skarzewski, J.; Michowski, J., *Tetrahedron Lett.* **1990**, *31*, 2177. (b) de Nooy, A. E. J.; Besemer, A. C.; van Bekkum, H., *Tetrahedron* **1995**, *51*, 8023.
5. Itagaki, N.; Iwabuchi, Y., *Chem. Commun.* **2007**, 1175.
6. Dess, D. B.; Martin, J. C., *J. Org. Chem.* **1983**, *48*, 4155.

Qingwei Yao
Northern Illinois University, DeKalb, IL, USA

(*S*)-1-Methyl-2-[(dihydroisoindol-2-yl)methyl]pyrrolidine

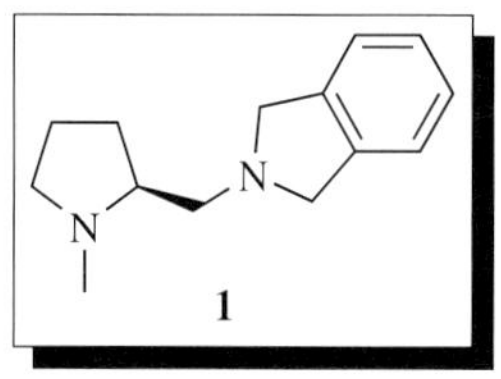

[*159497-37-7*] $C_{14}H_{20}N_2$ (MW 216.32)
InChI = 1/C14H20N2/c1-15-8-4-7-14(15)11-16-9-12-5-2-3-6-13(12)10-16/h2-3,5-6,14H,4,7-11H2,1H3/t14-/m0/s1
InChIKey = MKFAAQPMZNXLET-AWEZNQCLBM

(asymmetric catalyst for the Mukaiyama aldol reaction,[1,2] kinetic resolution of secondary alcohols,[3] acylation of *meso*-diols,[4] and palladium-catalyzed rearrangement of allylic imidates to allylic amides[5])

Alternate Name: (*S*)-2-(isoindolinylmethyl)-*N*-methylpyrrolidine.
Physical Data: bp 98–99 °C/ 0.08 mm Hg, $[\alpha]_D^{28}$ −76.5 (*c* 1.3, EtOH).
Solubility: soluble in CH_2Cl_2, CH_3CN.
Form Supplied in: not commercially available; colorless oil.
Analysis of Reagent Purity: by NMR and specific rotation determination.
Preparative Method: can be prepared by DCC coupling of *N*-Boc-(*S*)-proline and isoindoline followed by $LiAlH_4$ reduction.[5]
Purification: can be purified by distillation.

Asymmetric Aldol Reaction.[1,2] A number of proline-derived chiral diamines have been prepared and utilized in stereoselective syntheses since they can effectively create chiral environments by forming structurally rigid chelate complexes. Optically active *syn*-2,3-dihydroxy thioesters are prepared from (*Z*)-1-(ethylthio)-1-(trimethylsiloxy)-2-(*tert*-butyldimethylsiloxy)-ethene and various aldehydes in the presence of $Sn(OTf)_2$, chiral diamine **1**, and $Bu_2Sn\ (OAc)_2$ (eq 1). This reaction is general with respect to the aldehyde component and both diastereo and enantioselectivities are excellent. The absolute configuration of the product diol is 2*R*,3*S*. The use of (*S*)-1-methyl-2-[(indolinyl)methyl]pyrrolidine as a chiral ligand gives the *syn* aldol adducts with the reversed absolute configuration (2*S*,3*R*).

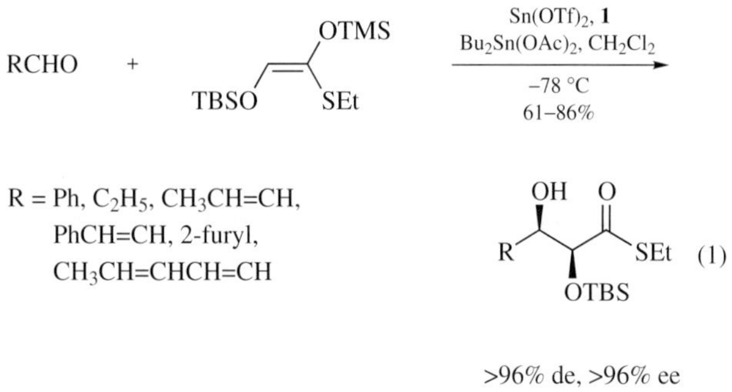

Kinetic Resolution of Racemic Secondary Alcohols. Racemic cyclic[3] and acyclic[3] secondary alcohols and β-halohydrins[6] are kinetically resolved in good chemical yields with modest-to-excellent enantioselectivity (eqs 2 and 3).

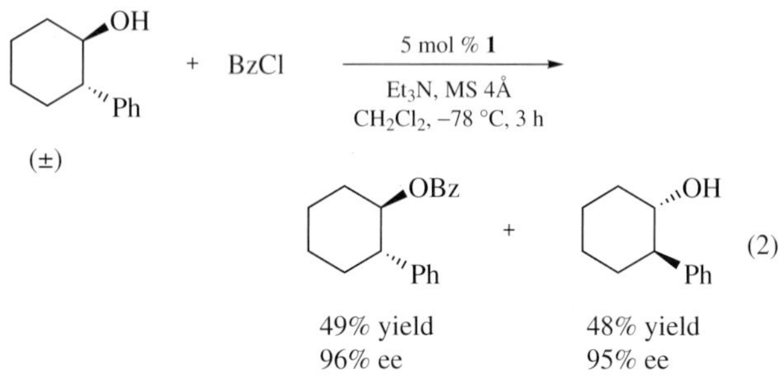

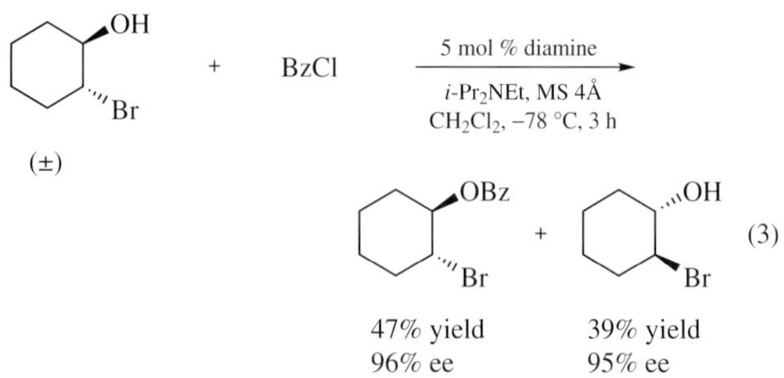

Asymmetric Acylation of *meso*-Diols. Both cyclic[4] and acyclic[4] *meso*-1,2-diols are desymmetrized by acylation in the presence of a stoichiometric amount of this ligand with modest-to-excellent enantioselectivity (eq 4). In a special case, *cis*-5,5-dimethyl-2-cyclopentene-1, 4-diol[7] was monobenzoylated in the presence of a catalytic amount of this ligand in good yield and with perfect enantioselection (87%, > 99.5% ee).

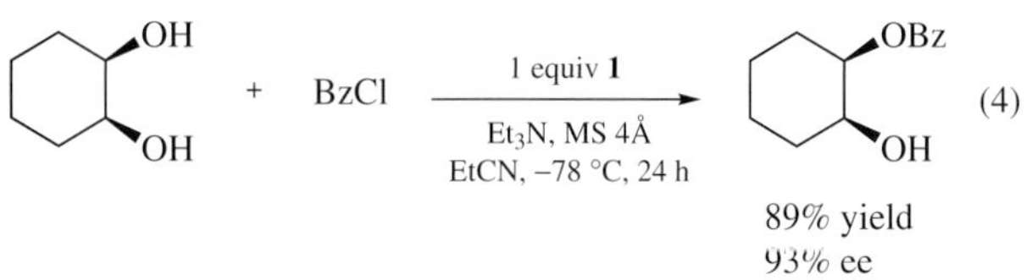

Rearrangement of Allylic Imidates.[5] Overman and co-workers reported an interesting application of this chiral diamine to induce enantioselectivity in Pd(II)-catalyzed rearrangement of allylic imidates to allylic amides (eq 5).

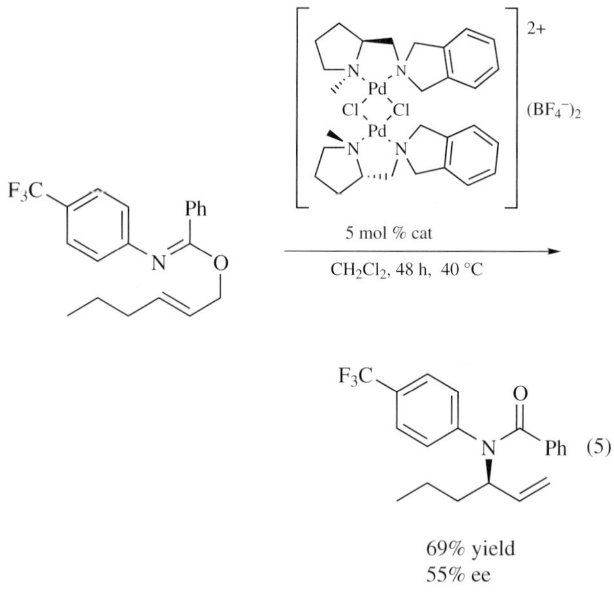

Related Reagents. (*S*)-1-Methyl-2-[(indolinyl)methyl] pyrrolidine; (*S*)-1-methyl-2- [(1-benz[*cd*]indolinyl) methyl] pyrrolidine; (*S*)-1-methyl-2-[(1-naphthylamino)methyl]pyrroli dine; (*S*)-1-methyl-2-[(1-piperidin-1-yl)methyl]pyrrolidine; (*S*)-1-ethyl-2- [(1-piperidin-1-yl)methyl]pyrrolidine; (*S*)-1-propyl-2-[(1-piperidin-1-yl)methyl]pyrrolidine.

1. Kobayashi, S.; Horibe. M., *J. Am. Chem. Soc.* **1994**, *116*, 9805.
2. Kobayashi, S.; Horibe. M., *Chem. Eur. J.* **1997**, *3*, 1472.
3. Sano, T.; Imai, K.; Ohashi, K.; Oriyama., T., *Chem. Lett.* **1999**, 265.
4. Oriyama, T.; Imai, K.; Hosoya, T.; Sano, T., *Tetrahedron Lett.* **1998**, *39*, 397.
5. Calter, M.; Hollis, T. K.; Overman, L. E.; Ziller, J.; Zipp, G. G., *J. Org. Chem.* **1997**, *62*, 1449.
6. Sano, T.; Miyata, H.; Oriyama, T., *Enantiomer* **2000**, *5*, 119.
7. Oriyama, T.; Hosoya, T.; Sano, T., *Heterocycles* **2000**, *52*, 1065.

Erik J. Sorensen, Hirofumi Seike & Jason M. Rohde
The Scripps Research Institute, La Jolla, CA, USA

2-Methyltetrahydrofuran[1–3]

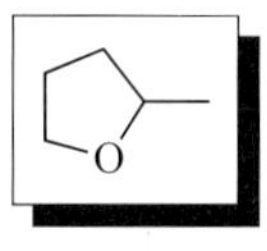

[96-47-9] $C_5H_{10}O$ (MW 86.13)
InChI = 1/C5H10O/c1-5-3-2-4-6-5/h5H,2-4H2,1H3
InChIKey = JWUJQDFVADABEY-UHFFFAOYSA-N

(eco-friendly aprotic ethereal-type solvent with advantages over THF, used in a wide array of reactions ranging from organic synthesis to biocatalysis)

Physical Data: mp −136 °C; bp 80.2 °C; *d* 0.854 g mL^{-1}.

Solubility: practically immiscible with water (14 g/100 g at 20 °C); as a consequence, addition of organic solvents for aqueous workup with extractive procedures can be avoided. Miscible with organic solvents.

Form Supplied in: colorless liquid; widely available.

Preparative Method: 2-MeTHF can be produced from renewable sources via substrates such as furfural or levulinic acid.

Purification: since it forms an azeotrope with water (10.6%) and considering the fact that the solubility of water in 2-MeTHF is 14 g /100 g, dry 2-MeTHF can be obtained via a simple batch distillation.[3]

Handling, Storage, and Precautions: the formation of peroxides which often plague the use of ethereal solvents can be conveniently alleviated by adding stabilizers such as BHT (2,6-di-*tert*-butyl-4-methylphenol, 150–400 ppm). Recent toxicology studies pointed out the absence of genotoxicity and mutagenicity associated with exposure to 2-MeTHF.[4]

General Considerations. The presence of a methyl group on the THF-skeleton accounts for a series of interesting chemical properties that notably differentiate 2-MeTHF from THF: (a) the so-called α-cleavage[5] in the presence of organolithium reagents (C-2 lithiation followed by a reverse [3+2] cycloaddition) is considerably reduced[6] thus, 2-MeTHF can be considered an optimal medium in which to carry out such processes at higher temperatures;[7] (b) the coordinating ability of 2-MeTHF is decreased compared to THF thus, reactions involving species stabilized through a solvent effect (e.g., lithium carbenoids) can be affected by this solvent;[8] (c) stability toward acids is about nine times increased compared to THF; (d) based on the dielectric constant values 2-MeTHF has polarity between that of THF (more polar) and diethyl ether (less polar).

Significantly, bromide-based Grignard reagents tend to be much more soluble in 2-MeTHF than in THF (up to four times) so that highly concentrated homogeneous solutions can be prepared and safely stored.[3]

2-MeTHF in Organocatalysis. In recent years, 2-MeTHF has been often employed in asymmetric processes. Although there is no correlation between the presence of the methyl group on the C-2 position of the THF-skeleton (perhaps racemic) and the enantioinduction, the beneficial effects displayed by this green solvent compared to other both polar and apolar ones are frequently observed.

Yang et al. disclosed the successful substitution of 1,2-dichloroethane (DCE) by 2-Me-THF in organocatalytic aldol reactions. In this case, the proline-mimetic organocatalyst (**1**) was assessed (eq 1).[9]

2-Me-THF is also useful for the asymmetric vinylogous Mannich-type reactions of aldimines (**2**) with nonactivated natural α-angelica lactone (**3**) (eq 2). This process is catalyzed by Sc(OTf)$_3$ complexed with (**5**) in the presence of molecular sieves.[10]

Johnson described the first enantioselective metallophosphite (**6**)-catalyzed addition of acylsilanes (**7**) to nitrone electrophiles (**8**), pointing out the remarkable asymmetric induction achieved in 2-MeTHF compared to other solvents (eq 3).[11]

2-MeTHF has also been employed as the solvent in stereoselective additions of a wide range of Grignard reagents to cyclic α,β-unsaturated compounds (ketones and lactones), catalyzed by Cu/TADDOL-derived phosphine–phosphite complexes; again Schmalz et al. noticed the best enantioselectivities were observed when working in this solvent (eqs 4–6).[12–14]

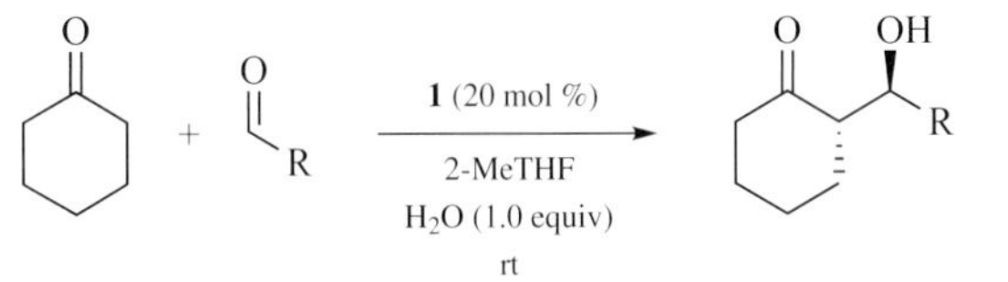

(1)

R = 2,4-dinitrophenyl (48 h), 81% yield, 15:1 dr, 96% ee

R = 4-pyridyl (13 h), 71% yield, 48:1 dr, 99% ee

cat =

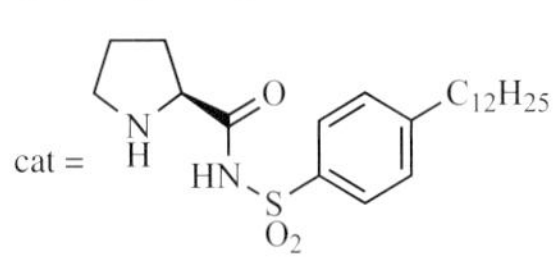

(**1**)

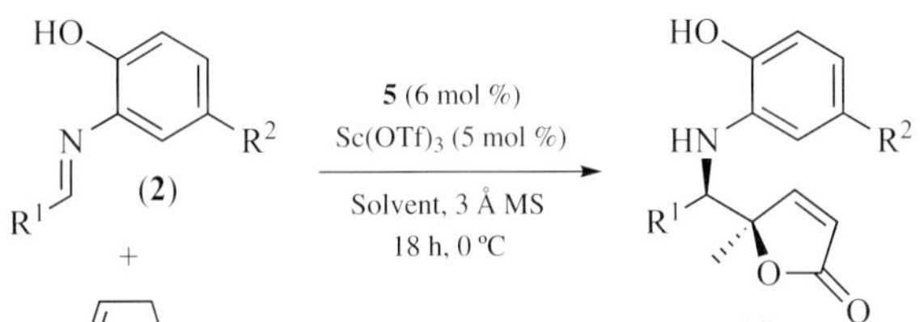

(2)

with THF: 68% yield, 82% ee

with 2-MeTHF: 81% yield, 95% ee

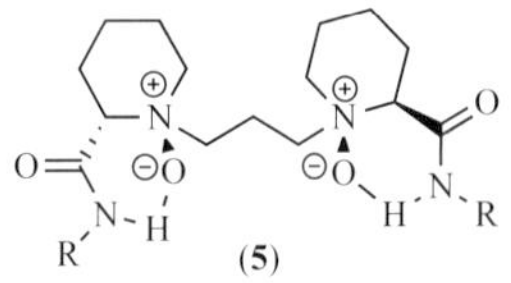

(**5**)

R = 2,6-diisopropylphenyl

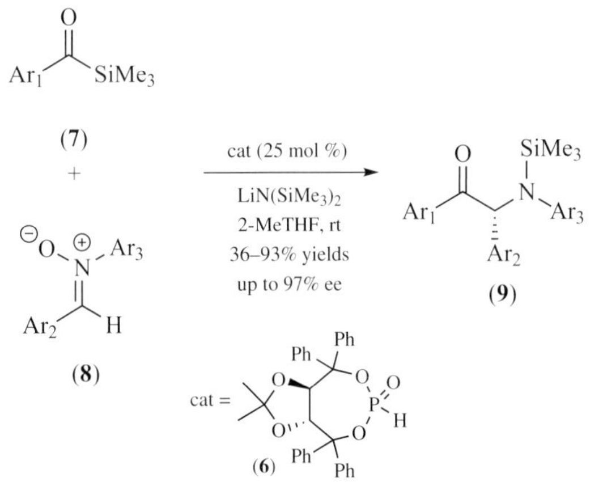

(3)

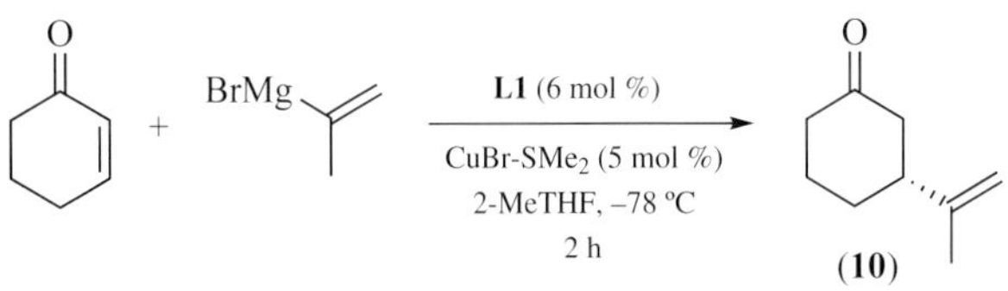

(4)

(**10**)

49%, 92% ee

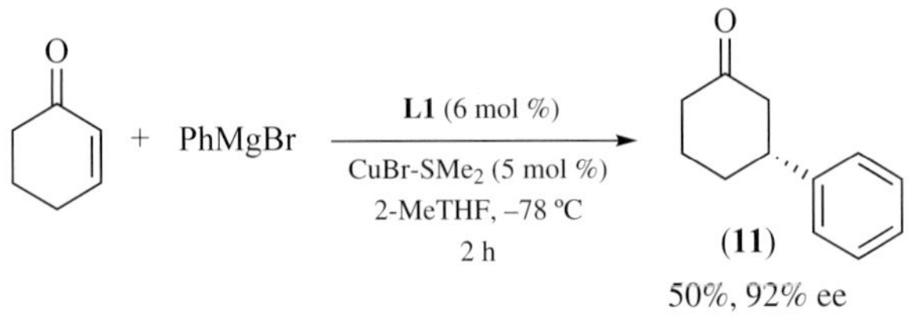

(5)

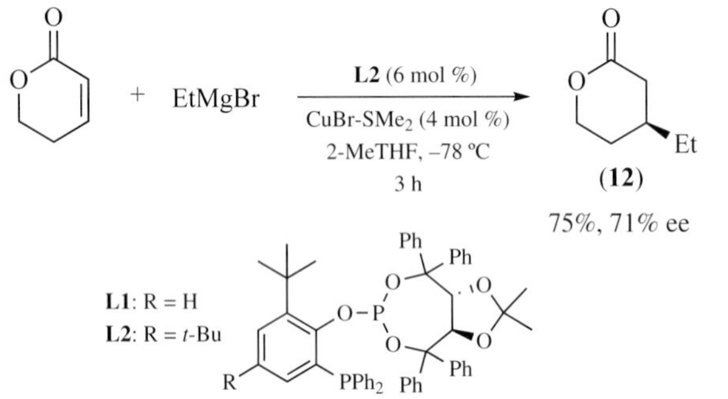

(6)

Zhou and Yang used 2-MeTHF for the Pd-catalyzed Mizoroki–Heck reaction of benzylic electrophiles (**13**) employing phosphoramidites as the chiral ligands (eq 7).[15]

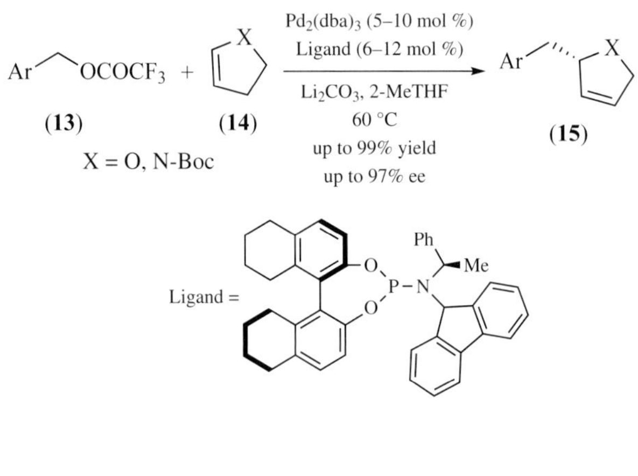

(7)

Carreira and coworkers selected 2-MeTHF as the solvent in which to realize the direct enantioselective Ir-catalyzed substitution of racemic secondary allylic alcohols (**16**) to give optically active primary amines (**17**) with sulfamic acid in the presence of a phosphoramidite ligand (eq 8).[16]

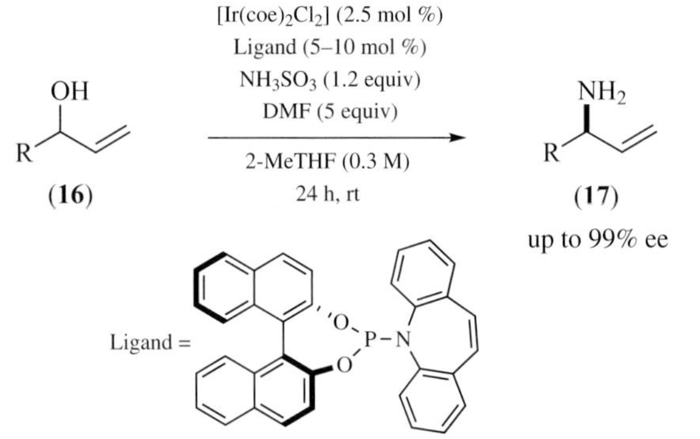

(8)

Use of 2-MeTHF in Organometallic Chemistry. Due to the high stability of concentrated solutions of Grignard reagents in 2-MeTHF, Milton and Clarke developed a protocol for the Pd-catalyzed cross-coupling of haloarenes (**18**) with ArMgBr (5 M in 2-MeTHF) at high temperatures (50–75 °C) (eq 9).[17] These reactions did not require any added solvents and noticeably, thle obligatory Pd precatalysts have been also prepared in 2-MeTHF under MW conditions.

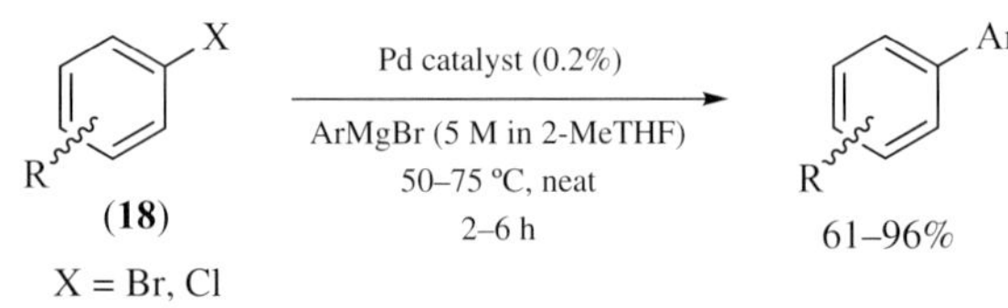

(9)

As a consequence of its higher stability compared to THF or diethyl ether versus strong basic conditions such as those caused by the presence of organolithium reagents, 2-MeTHF has been efficiently employed for the selective 1,2-addition of these reagents to various α,β-unsaturated systems (**19**) (ketones, aldehydes, and imines) at 0 °C.[18] Notably, the concomitant presence of LiBr as a mild Lewis acid and a disaggregating agent allows the efficiency of the process to be maximized (eq 10).

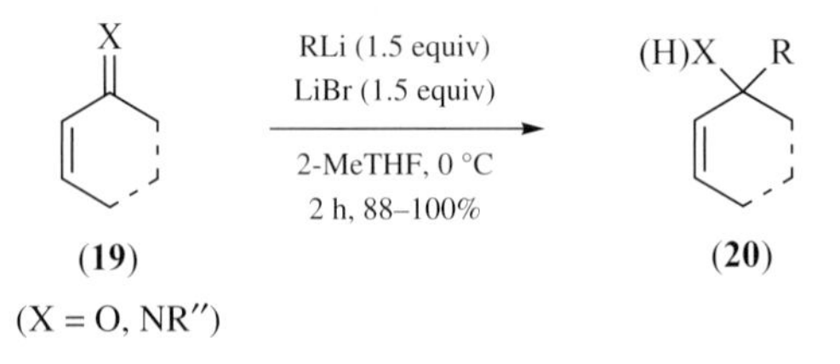

(10)

Pace et al. employed 2-MeTHF as the optimal solvent to successfully perform the chemoselective protection of anilines (**21**) as *N-tert*-butyldimethylsilyl derivatives via their deprotonation with MeLi followed by the trapping with TBDMSCl at 0 °C (eq 11).[19]

Moreover, 2-MeTHF proved to be the ideal solvent for the obtainment of γ-hydroxy-α, β-acetylenic esters by the sequential one-pot elimination-addition of 2-bromoacrylates (**23**) to aldehydes promoted by LTMP at (40 °C (eq 12).[20]

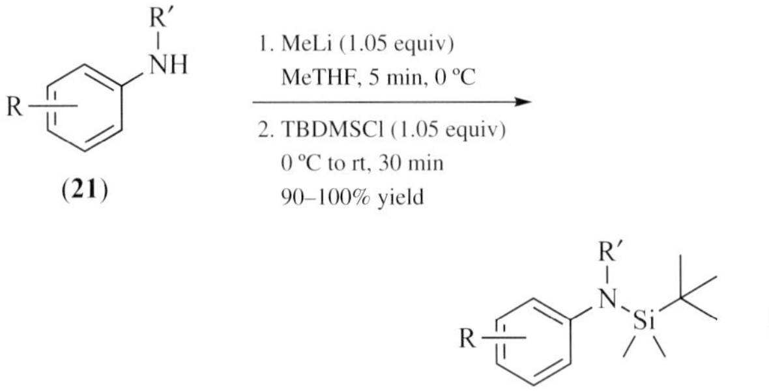

(21) (22) (11)

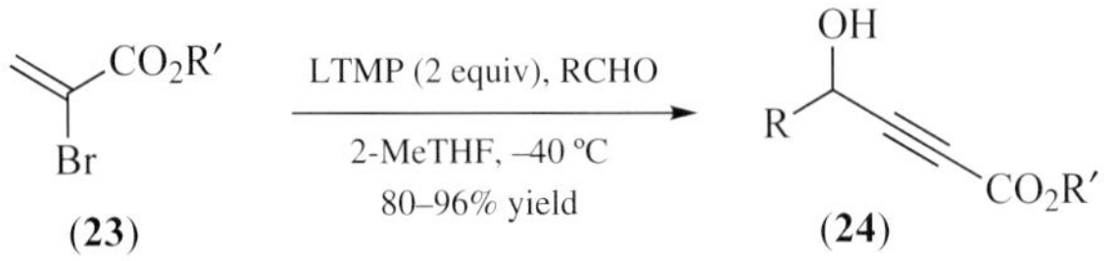

(23) (24) (12)

Interestingly, Grellepois found in 2-MeTHF the best conditions to accomplish the asymmetric synthesis of β-sulfinylamino-β-trifluoromethylesters (**25**) via a highly stereoselective imino-Reformatsky reaction with bench-stable trifluoromethyl *N-tert*-butanesulfinylketoimines surrogates (**26**) (eq 13).[21]

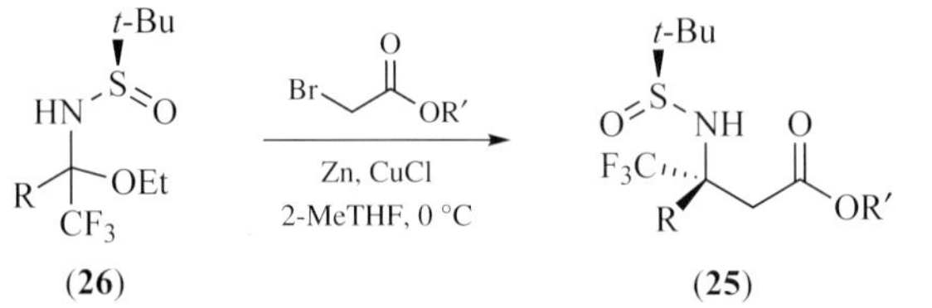

(26) (25) (13)

dr up to > 99 : 1
yield up to 87%

The weak coordinating properties of 2-MeTHF allow the preparation of a simplified iron-containing analog of hydroxyeicosatetraenoic acid (HETE (**28**), eq 14).[22] The key step was the introduction of the hydrophilic chain in a stereoselective fashion by coupling an aldehyde (**27**) with γ-metalated ethyl butyrate. 2-MeTHF enabled the preparation of a Zn/Ti species (a Yoshida-type reagent) that was smoothly added to the aldehyde, affording the secondary alcohol (**28**) with the desired stereochemistry (Ψ-*exo* diastereomer) in 50% yield, as well as a 10% of the corresponding Ψ-*endo* isomer.

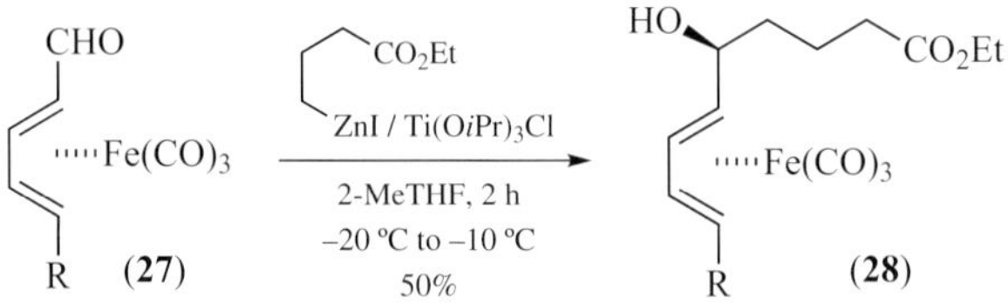

(27) (28) (14)

2-MeTHF in C-Heteroatom Bond Formation. Because of the practically immiscibility of 2-MeTHF with water, it has been employed in Schotten-Baumann-type processes to functionalize (alkylation, acylation) heteroatoms such as nitrogen or oxygen.[23] Very recently, Pace et al. reported a high-yielding synthesis of Weinreb amides (**30**) through the reaction of acyl halides (**29**) and *N,O*-dimethylhydroxylamine hydrochloride (DMHA) in the biphasic medium 2-MeTHF-basic water (eq 15).[24] The procedure is particularly simple since no purification is required to obtain analytically pure compounds.

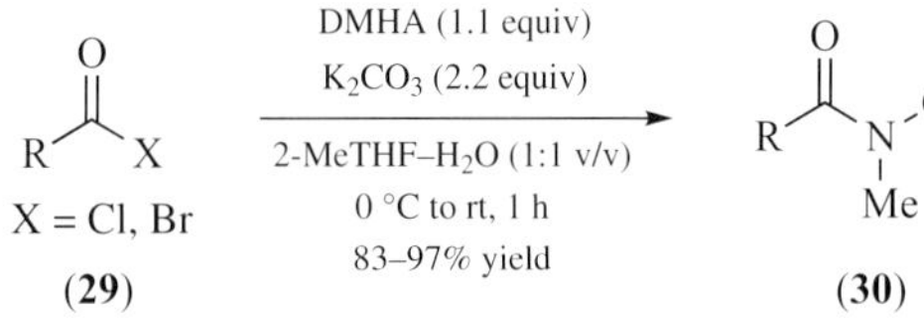

(29) (30) (15)

Analogous acylation-type reactions under stoichiometric conditions can be carried out on (enantiopure) alcohols and amines (**31**) by adding CaO as a heterogeneous acid scavenger thus, avoiding the use of classical bases (eq 16).[25] Notably, also the DBU-catalyzed amidation of CDI-activated carboxylic acids has been reported in 2-MeTHF.[26] The higher boiling point of 2-MeTHF allows the alkylation of phthalimide derivatives in the presence of supported KF.[27]

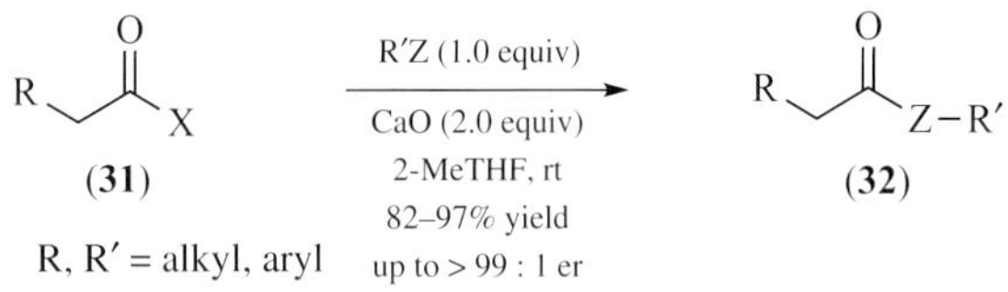

(31) (32) (16)

R, R′ = alkyl, aryl
X = halogen
Z = -OH, -NH_2, -NH

Use of 2-MeTHF in Biocatalysis. The use of biocatalysts for developing organic reactions is one of the greenest strategies in organic synthesis.[28] In this sense, combining biotransformations with the use of biomass-derived solvents such as 2-MeTHF can even increase the overall sustainability of the process.[29]

The ability of lipases to efficiently work in organic solvents is well known.[30] Normally, solvents with logP values >2 are considered as the best ones, provided that they do not induce distortion of the active conformation of the enzyme.[31] Nevertheless, the use of polar solvents is needed in the biotransformation of polar substrates, and in this sense 2-MeTHF is a good alternative.[2]

Although Nakamura et al. in 1994 assessed 2-MeTHF in a biocatalytic screening,[32] the first reported process of this type was the regioselective acylation of the nucleosides (**33**) 1-β-arabinofuranosyl uracil, 9-β-arabinofuranosyl adenosine, 2′-*O*-(2-methoxyethyl)-5-methyluridine, adenosine, and uridine catalyzed by lipase B from *Candida antarctica* (CAL-B) described by Simeó et al.[33] (eq 17). Reactions proceeded in 6 h to render either 3′- or 5′-esters (**34**), depending on the substrate, avoiding any protection-deprotection steps, and reached 95% conversion in 6 h.

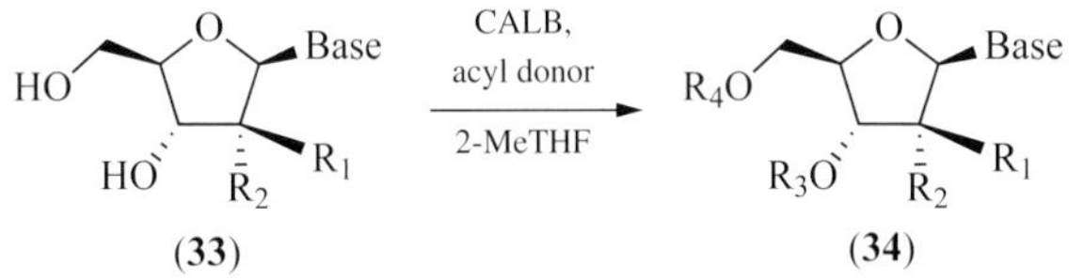

(33) (34) (17)

R_1, R_2 = H, OH, $O(CH_2)_2$-OCH_3
R_3, R_4 = H, COR

After this seminal paper, the use of 2-MeTHF for nucleoside acylation has been extended; thus, Gao et al. have reported the regioselective undecylenoylation of purine nucleosides (**35**) and different chloro-substituted derivatives by using *Penicillium expansum* lipase and 2-MeTHF as a biomass-derived reaction medium.[34] In some of these examples, high regioselectivities (94 to >99%) and yields (up to 95%) were achieved in the 5′-position in short reaction times (typically, 3 h) (eq 18).

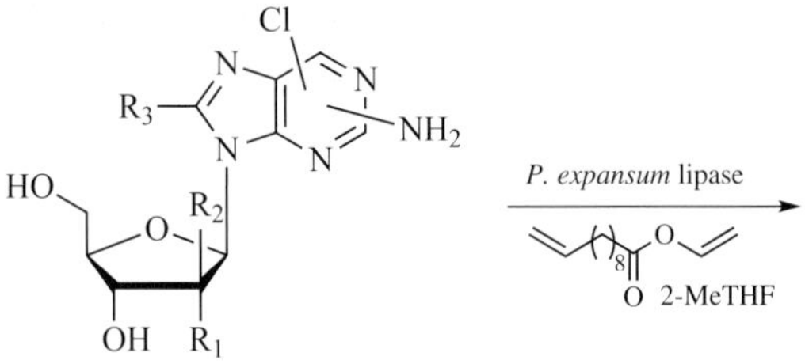

(**35**)

R_1, R_2 = H, OH
R_3 = H, Cl

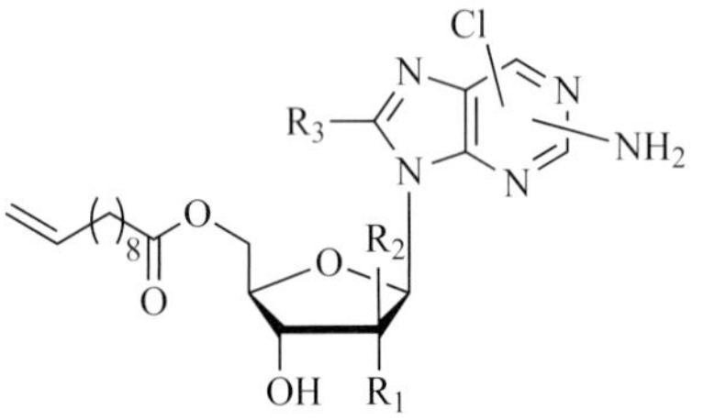

(18)

Very recently, Gao et al. have also reported the enzymatic regioselective acylation of pyrimidine nucleosides (**36**) under the same conditions,[35] affording 5′-esters (**37**) with conversions of 96–99% and 5′-regioselectivities of 96 to >99% (eq 19). However, in enzymatic acylation of 2′-deoxynucleosides, 5′-regioselectivities showed a clear dependence on the 5-substituents present in the base moiety, although the substrate conversions reached >98% within 1–3 h. In the cases of 2′,3′-dideoxynucleoside analogs, the reaction rate decreased markedly due to the lack of 3′-hydroxyl group.

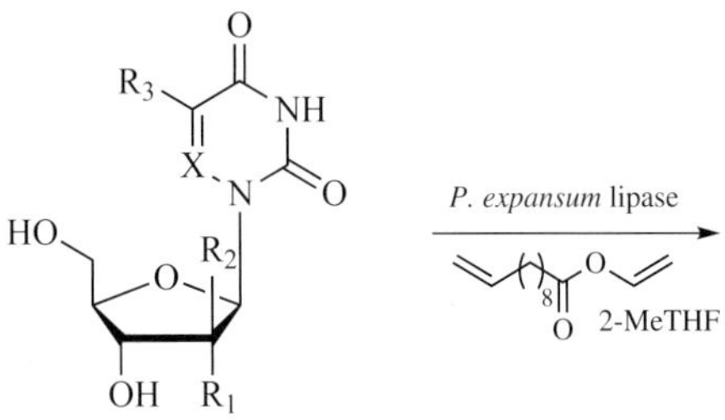

(**36**)

R_1, R_2 = H, OH, OMe
R_1, R_2 = H, OH
R_3 = H, F, Cl, Br, I, Me
X = N, CH

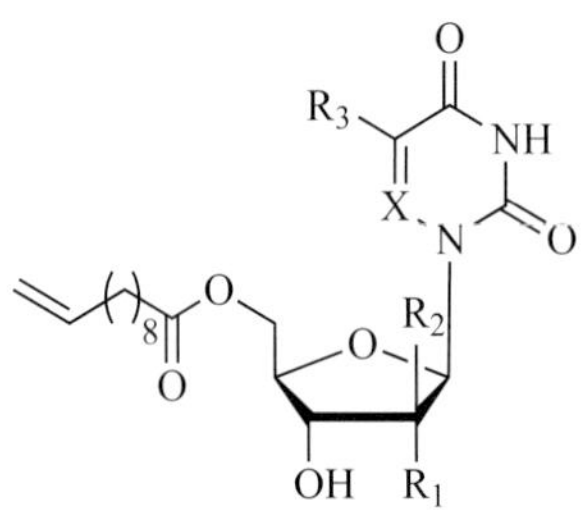

(19)

(**37**)

Despite the moderate dissolution ability of 2-MeTHF for nucleosides, the initial enzymatic reaction rate was much higher in this eco-friendly solvent than in other commonly used organic solvents. This observation has been rationalized by a kinetic study, as resulting from a lower apparent activation energy of the enzymatic reaction and a higher catalytic efficiency of the enzyme in 2-MeTHF.

In a similar way, the highly efficient and regioselective acylation of cordycepin (3′-deoxydenosine) (**38**) with vinyl acetate catalyzed by an immobilized *Candida antarctica* lipase B (Novozym 435) in 2-MeTHF has been recently described on 25 g scale,[36] rendering cordycepin 5′-acetate (**39**) as the sole detectable product from each acylation reaction (96% isolated yield). Under these conditions, the immobilized enzyme could be recycled so that 63% of its original activity was maintained after being reused for seven batches.

Concerning lipases and 2-MeTHF, Alcántara's group has also disclosed the enantioselective resolution of racemic benzoins (1,2-diaryl-2-hydroxyethanones (**40**)) in 2-MeTHF (eq 20). Herein, the setup of efficient dynamic kinetic resolutions (DKR) by combining *Pseudomonas stutzeri* lipase (PSL) and Shvo's catalyst (ruthenium-based catalyst) was highly successful. In 2-MeTHF, excellent enantioselectivities (ee_p > 99%) with 85% conversion were reached, and also reaction rates were faster than those reported in THF.[37]

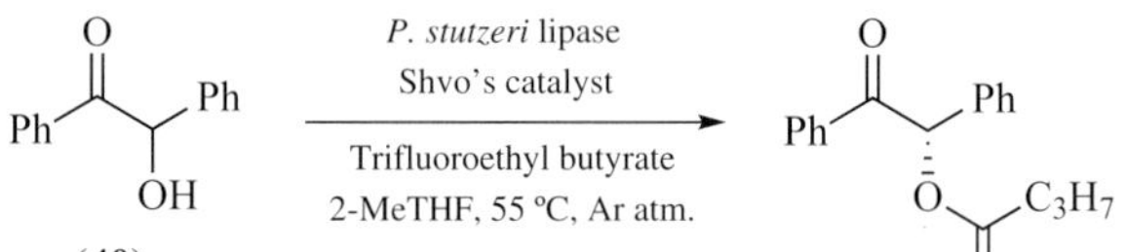

(**40**) (**41**)

48 h
85% conv.
> 99% ee_p

(20)

Furthermore, using the same lipase in 2-MeTHF, the aminolysis reaction between methyl 3-phenylpropanoate (**42**) and cyclohexylamine (**43**) (eq 21) proceeds successfully, reaching high conversions (>80%), yet in still relatively long reaction times (44 h).[38]

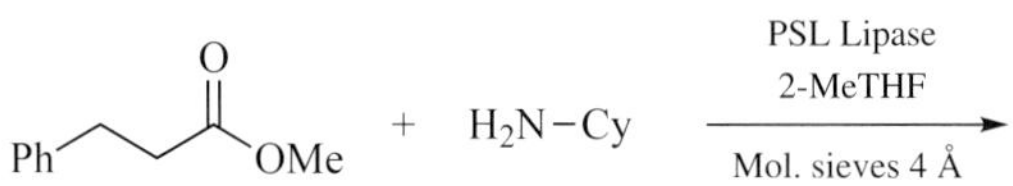

(**42**) (**43**)

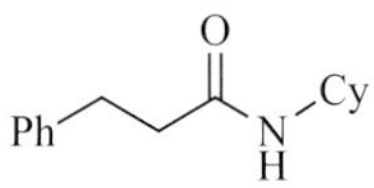

(21)

(**44**)

44 h
conv. > 80%

2-MeTHF has also been tested as solvent for transaminase-catalyzed synthesis of the oral antihyperglycemic sitagliptin (**45**),[39] and as cosolvent in the preparation of montelukast (**46**)-a drug used for treatment of asthma and allergies-catalyzed by oxidoreductases.[40] Finally, also acting as cosolvent, 2-MeTHF has proven to be very efficient as a greener alternative vesus DMSO or MTBE for benzaldehyde lyase (BAL, a thiamine diphosphate (ThDP)-dependent lyase) catalyzed preparation of enantiopure acyloins (**47**) and benzoins (**40**),[41] where it also showed excellent properties as extractive solvent during workup (eq 22).

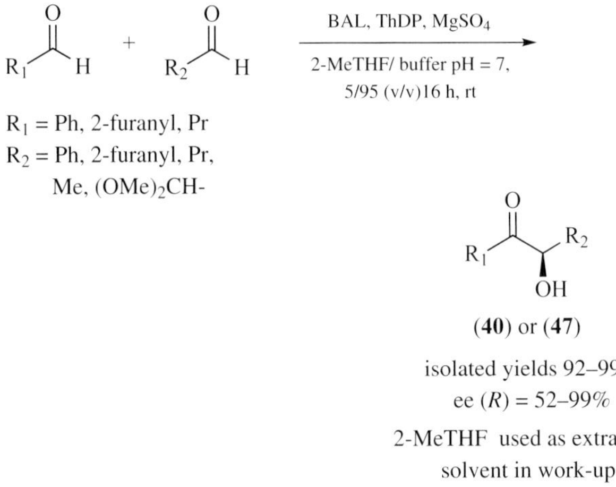

(22)

Similarly, adding 2-MeTHF (5% vol.) to an aqueous buffer has been shown to increase phenylacetylcarbinol (PAC) (**48**) production mediated by another ThDP-dependent carboligase, pyruvate decarboxylase from *Acetobacter pasteurianus* (*Ap*PDC).[42]

Finally, 2-MeTHF has been also used as cosolvent for domino enzymatic reactions, namely the combination of several enzymes in one-pot procedure: thus, examples of lipase–lyase[43] (eq 23) and oxidase-lyase-oxidoreductase[44] (eq 24) combinations have been reported.

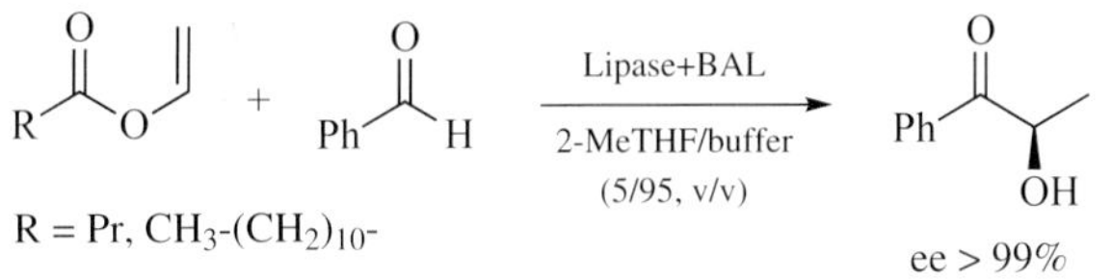

(23)

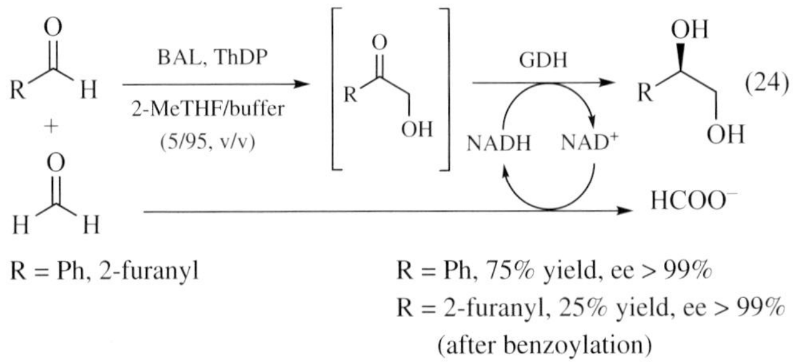

(24)

Related Reagents. THF; diethyl ether; 1,4-dioxane; *tert*-butylmethyl ether; cyclopentyl methyl ether.

1. Pace, V., *Aust. J. Chem.* **2012**, *65*, 301.
2. Pace, V.; Hoyos, P.; Castoldi, L.; Dominguez de María, P.; Alcántara, A. R., *ChemSusChem* **2012**, *5*, 1369.
3. Aycock, D. F., *Org. Process Res. Dev.* **2007**, *11*, 156.
4. Antonucci, V.; Coleman, J.; Ferry, J. B.; Johnson, N.; Mathe, M.; Scott, J. P.; Xu, J., *Org. Process Res. Dev.* **2011**, *15*, 939.
5. Maercker, A., *Angew. Chem., Int. Ed.* **1987**, *26*, 972.
6. Bates, R. B.; Kroposki, L. M.; Potter, D. E., *J. Org. Chem.* **1972**, *37*, 560.
7. Slavov, N.; Cvengroš, J.; Neudörfl, J.-M.; Schmalz, H.-G., *Angew. Chem., Int. Ed.* **2010**, *49*, 7588.
8. Pace, V.; Holzer, W.; Verniest, G.; Alcántara, A. R.; De Kimpe, N., *Adv. Synth. Catal.* **2013**, *355*, 919.
9. Yang, H.; Mahapatra, S.; Cheong, P. H.-Y.; Carter, R. G., *J. Org. Chem.* **2010**, *75*, 7279.
10. Zhou, L.; Lin, L.; Ji, J.; Xie, M.; Liu, X.; Feng, X., *Org. Lett.* **2011**, *13*, 3056.
11. Garrett, M. R.; Tarr, J. C.; Johnson, J. S., *J. Am. Chem. Soc.* **2007**, *129*, 12944.
12. Naeemi, Q.; Robert, T.; Kranz, D. P.; Velder, J.; Schmalz, H.-G., *Tetrahedron: Asymmetry* **2011**, *22*, 887.
13. Robert, T.; Velder, J.; Schmalz, H. G., *Angew. Chem., Int. Ed.* **2008**, *47*, 7718.
14. Drusan, M.; Lölsberg, W.; Škvorcová, A.; Schmalz, H.-G.; Šebesta, R., *Eur. J. Org. Chem.* **2012**, 6285.
15. Yang, Z.; Zhou, J., *J. Am. Chem. Soc.* **2012**, *134*, 11833.
16. Lafrance, M.; Roggen, M.; Carreira, E. M., *Angew. Chem., Int. Ed.* **2012**, *51*, 3470.
17. Milton, E. J.; Clarke, M. L., *Green Chem.* **2010**, *12*, 381.
18. Pace, V.; Castoldi, L.; Hoyos, P.; Sinisterra, J. V.; Pregnolato, M.; Sánchez-Montero, J. M., *Tetrahedron* **2011**, *67*, 2670.
19. Pace, V.; Alcántara, A. R.; Holzer, W., *Green Chem.* **2011**, *13*, 1986.
20. Pace, V.; Castoldi, L.; Alcántara, A. R.; Holzer, W., *Green Chem.* **2012**, *14*, 1859.
21. Grellepois, F., *J. Org. Chem.* **2013**, *78*, 1127.
22. Kausch-Busies, N.; Kater, B.; Neudörfl, J. M.; Prokop, A.; Schmalz, H.-G., *Eur. J. Org. Chem.* **2011**, 1133.
23. Brown Ripin, D. H.; Vetelino, M., *Synlett* **2003**, 2353.
24. Pace, V.; Castoldi, L.; Alcántara, A. R.; Holzer, W., *RSC Adv.* **2013**, *3*, 10158.
25. Pace, V.; Hoyos, P.; Alcántara, A. R.; Holzer, W., *ChemSusChem* **2013**, *6*, 905.
26. Larrivée-Aboussafy, C.; Jones, B. P.; Price, K. E.; Hardink, M. A.; McLaughlin, R. W.; Lillie, B. M.; Hawkins, J. M.; Vaidyanathan, R., *Org. Lett.* **2010**, *12*, 324.
27. Pace, V.; Hoyos, P.; Fernández, M.; Sinisterra, J. V.; Alcántara, A. R., *Green Chem.* **2010**, *12*, 1380.
28. Clouthier, C. M.; Pelletier, J. N., *Chem. Soc. Rev.* **2012**, *41*, 1585.
29. Hernáiz, M.; Alcántara, A. R.; García, J. I.; Sinisterra, J. V., *Chem. Eur. J.* **2010**, *16*, 9422.
30. Bornscheuer, U. T.; Flickinger, M. C.,Lipases, Synthesis of Chiral Compounds, Aqueous and Organic Solvents. In *Encyclopedia of Industrial Biotechnology*; Wiley: New York, 2009.
31. Adlercreutz, P.,Fundamentals of Biocatalysis in Neat Organic Solvents. In *Organic Synthesis with Enzymes in Non-Aqueous Media*; Carrea, G., Riva, S., Eds.; WILEY-VCH Verlag GmbH & Co. KGaA: Weinheim, **2008**; p 3.
32. Nakamura, K.; Kinoshita, M.; Ohno, A., *Tetrahedron* **1994**, *50*, 4681.
33. Simeó, Y.; Sinisterra, J. V.; Alcántara, A. R., *Green Chem.* **2009**, *11*, 855.
34. Gao, W. L.; Liu, H.; Li, N.; Zong, M. H., *Bioresour. Technol.* **2012**, *118*, 82.
35. Gao, W. L.; Li, N.; Zong, M. H., *J. Biotechnol.* **2013**, *164*, 91.
36. Chen, Z.-G.; Zhang, D.-N.; Cao, L.; Han, Y.-B., *Bioresour. Technol.* **2013**, *133*.
37. Hoyos, P.; Quezada, M. A.; Sinisterra, J. V.; Alcántara, A. R., *J. Mol. Catal. B: Enzym.* **2011**, *72*, 20.

38. van Pelt, S.; Teeuwen, R. L. M.; Janssen, M. H. A.; Sheldon, R. A.; Dunn, P. J.; Howard, R. M.; Kumar, R.; Martinez, I.; Wong, J. W., *Green Chem.* **2011**, *13*, 1791.
39. Truppo, M. D.; Strotman, H.; Hughes, G., *ChemCatChem* **2012**, *4*, 1071.
40. Liang, J.; Lalonde, J.; Borup, B.; Mitchell, V.; Mundorff, E.; Trinh, N.; Kochrekar, D. A.; Cherat, R. N.; Pai, G. G., *Org. Process Res. Dev.* **2010**, *14*, 193.
41. Shanmuganathan, S.; Natalia, D.; van den Wittenboer, A.; Kohlmann, C.; Greiner, L.; de Maria, P. D., *Green Chem.* **2010**, *12*, 2240.
42. Gerhards, T.; Mackfeld, U.; Bocola, M.; von Lieres, E.; Wiechert, W.; Pohl, M.; Rother, D., *Adv. Synth. Catal.* **2012**, *354*, 2805.
43. Perez-Sanchez, M.; de Maria, P. D., *Chemcatchem* **2012**, *4*, 617.
44. Shanmuganathan, S.; Natalia, D.; Greiner, L.; de Maria, P. D., *Green Chem.* **2012**, *14*, 94.

Vittorio Pace & Wolfgang Holzer
University of Vienna, Vienna, Austria

Pilar Hoyos, María José Hernáiz & Andrés R. Alcántara
Complutense University, Madrid, Spain

Methyl(trifluoromethyl)dioxirane[1]

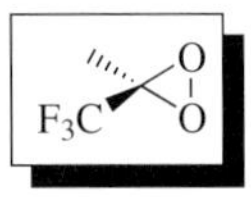

[115464-59-0] $C_3H_3F_3O_2$ (MW 128.06)
InChI = 1S/C3H3F3O2/c1-2(7-8-2)3(4,5)6/h1H3
InChIKey = NEGUMGNBGIFXAL-UHFFFAOYSA-N

(selective, reactive oxidizing agent capable of epoxidation of unreactive alkenes[5] and arenes,[7] oxyfunctionalization of alkanes,[3] oxidation of alcohols[16] and ethers[19])

Alternate Name: TFDO.
Physical Data: known only as a dilute solution.
Solubility: sol in acetone and CH_2Cl_2; sol in most other organic solvents, but reacts slowly with many of these.
Form Supplied in: dilute solutions of the reagent in 1,1,1-trifluoro-2-propanone are prepared from Oxone (***Potassium Monoperoxysulfate***) and this ketone as described below.
Drying: initial drying of reagent solutions is accomplished with reagent grade anhyd $MgSO_4$ in the cold. After filtration, solutions are typically stored over molecular sieves.
Analysis of Reagent Purity: concentrations of the reagent can be determined by classical iodometric titration or by reaction with an excess of an organosulfide and determination of the amount of sulfoxide formed by NMR or gas chromatography.
Preparative Method: TFDO solutions are prepared by mixing 1,1,1-trifluoro-2-propanone (TFP) and aqueous buffered Oxone in the cold and collecting the volatile TFDO–TFP mixture by transfer into a cold trap in a stream of inert gas under reduced pressure (eq 1).[3]

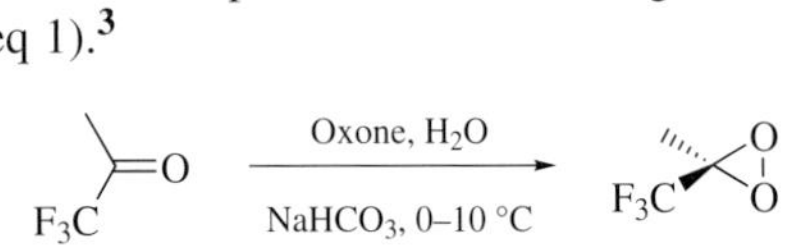

(1)

Handling, Storage, and Precautions: solutions of the reagent can be kept in the freezer of a refrigerator at −20 °C for as long as a week. The concentration of the reagent decreases relatively slowly (ca. 6–8% in 48 h), provided solutions are properly stored. It is particularly important that solutions be kept from light and traces of heavy metals, since the reagent is particularly susceptible to their influence. These dilute solutions are not known to decompose explosively, but the usual precautions for handling peroxides should be applied, including the use of a shield. All reactions should be performed in a hood to avoid exposure to the powerful, volatile oxidant.

Original Commentary

Jack K. Crandall
Indiana University, Bloomington, IN, USA

Introduction. Methyl(trifluoromethyl)dioxirane is a fluorinated derivative of the extremely useful oxidant ***Dimethyldioxirane*** (DDO), over which it has several advantages as a reagent. However, it is more expensive and difficult to prepare, since it is obtained from the volatile (bp 22 °C) ketone TFP.[2] Nonetheless, the much greater reactivity of TFDO can be of major advantage for less reactive substrates. Furthermore, the concentration of TFDO (ca. 0.8 M) normally obtained is several times greater than that of DDO. Solutions of TFDO in halocarbon solvents free of starting ketone can be obtained by extraction of the TFP into water, owing to the water solubility of the ketone hydrate.[4] Oxidations with TFDO are performed by adding the reagent to the reactant, often in CH_2Cl_2 solution, and then simply removing the volatile solvents when reaction is complete (usually minutes).

Alkenes and Arenes. Like DDO, TFDO performs rapid epoxidations of alkenes with retention of alkene stereochemistry.[1,2] Since DDO is adequately reactive towards most double bonds, there is ordinarily no reason to employ TFDO in reactions of this type. One important exception is the epoxidation of trifluoromethyl-substituted alkenes, which are resistant to classic epoxidizing reagents and which require a large excess of DDO during two weeks to complete epoxidation.[5] TFDO rapidly converts such alkenes to epoxides (eq 2). Another example of the beneficial effect of TFDO is in the preparation of the sensitive epoxides of enol ethers, where the short reaction times and facile product isolation are crucial.[6]

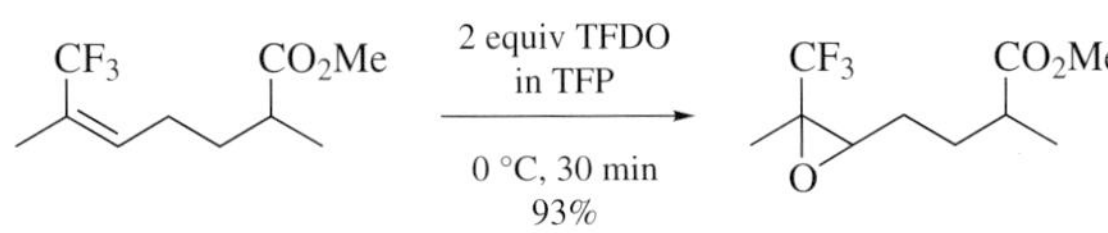

(2)

Polycyclic aromatic hydrocarbons like phenanthrene are converted to epoxides by TFDO.[7] This reactive reagent even permits the epoxidation of naphthalene, which undergoes fast sequential diepoxidation as shown in eq 3. Interestingly, the second epoxidation shows high *anti* stereoselectivity. Catechol is oxidatively cleaved by TFDO into (*Z*,*Z*)-muconic acid in good yield (eq 4).[8] This reagent also oxidizes 2,6-di-*t*-butylphenol to the corresponding quinone and a hydroxyquinone derivative (eq 5).[8]

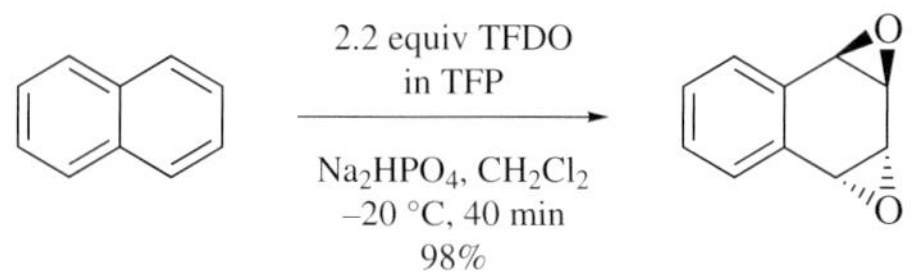

(3)

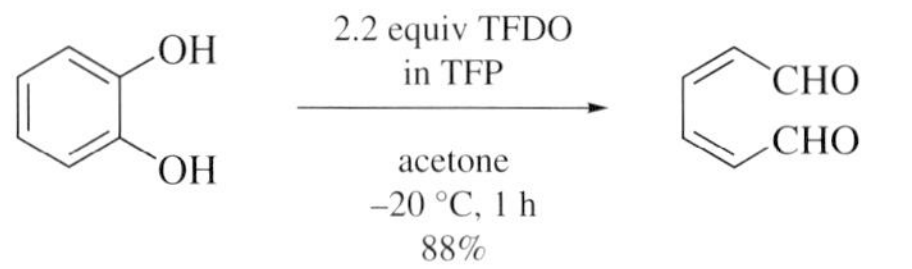

(4)

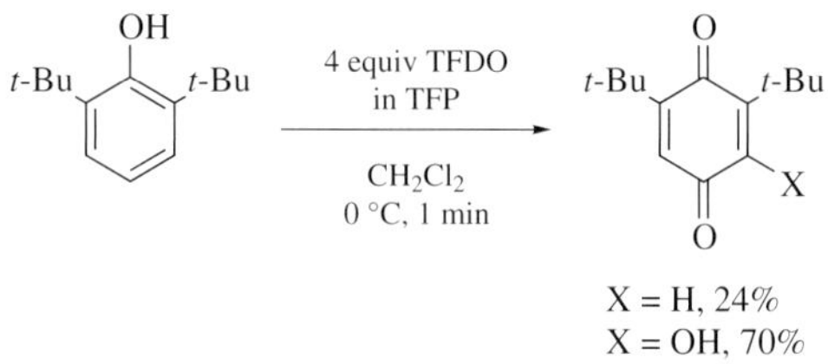

(5)

Alkynes are oxidized by TFDO to give a variety of different products, depending on the nature of the alkynic substituents.[9] Reaction is thought to proceed via an oxirene intermediate. This highly unstable, anti-aromatic heterocycle quickly rearranges to the isomeric α-ketocarbene, which then leads to the observed products by typical carbenoid transformations. For example, cyclodecyne gives bicyclic ketones by transannular insertion processes common to this medium-ring system (eq 6).

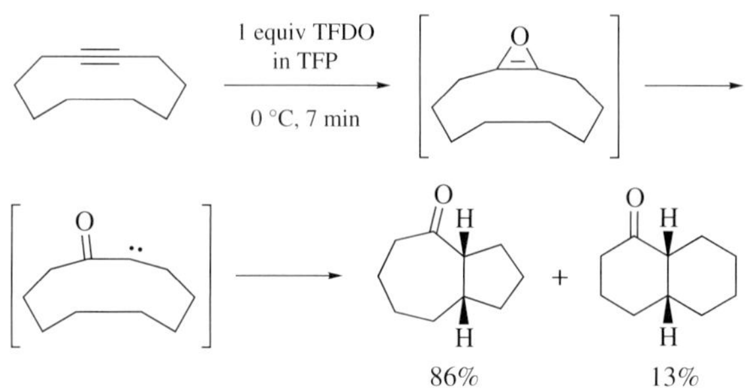

(6)

Alkanes. The most impressive applications of TFDO to date have involved the hydroxylation of unactivated C–H bonds.[1,3] While dimethyldioxirane performs similar reactions, the much greater reactivity of TFDO permits higher conversions of alkanes with less oxidant and in minutes rather than hours. Despite its more reactive nature, TFDO displays selectivity for attack at tertiary > secondary > primary C–H bonds almost as advantageous as that for DDO.[3] Examples include the exclusive formation of the tertiary alcohol from 2,3-dimethylbutane (eq 7) and the stereospecific oxidation of *cis*-1,2-dimethylcyclohexane (eq 8). Cyclohexane is oxidized to cyclohexanone by TFDO in a slower process that involves a fast second oxidation of the initially formed cyclohexanol. Heptane generates a 41:41:18 mixture of 2-, 3-, and 4-heptanone.

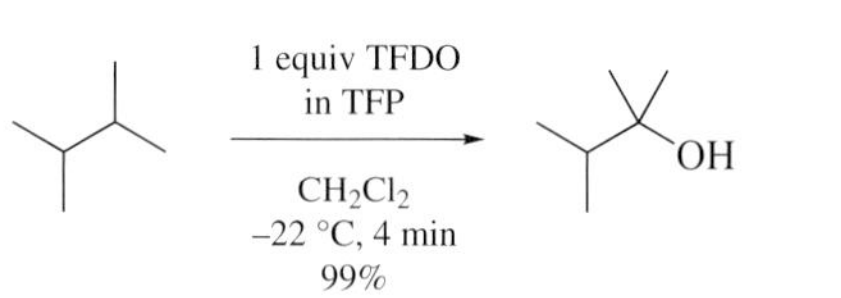

(7)

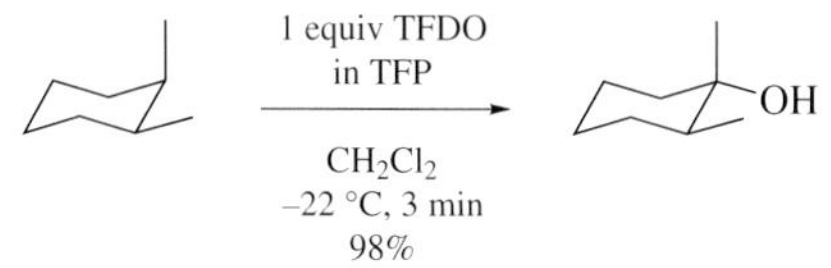

(8)

The stereospecificity of these C–H oxidations is illustrated in eq 9 by the benzylic oxidation of optically active (*S*)-(−)-2-phenylbutane, which gives only the tertiary alcohol with total retention of configuration.[10] Interestingly, ketone-free TFDO in CH_2Cl_2 is three times more reactive than TFDO–TFP.

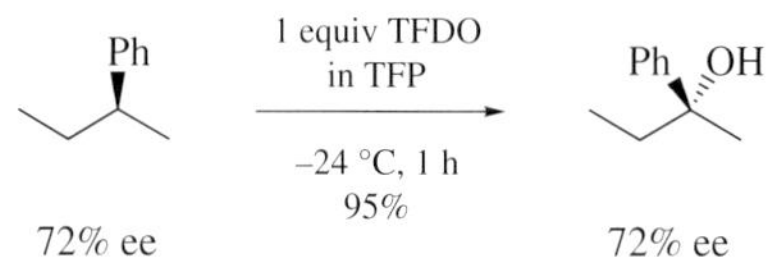

(9)

Adamantane is an informative substrate for oxidation by TFDO; it not only shows very high bridgehead selectivity, but can be converted to mono-, di-, tri-, and even tetrahydroxylated adamantane (eq 10) in good yields by varying the ratio of oxidant to hydrocarbon.[11]

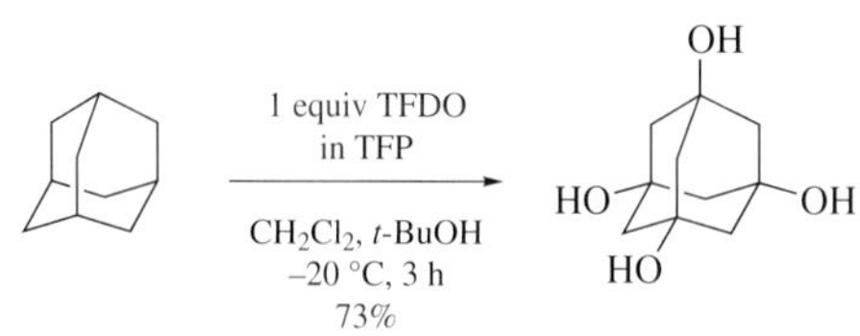

(10)

Several impressive oxyfunctionalization reactions on steroid substrates illustrate the enormous potential of this process. Not only is there a preference for tertiary C–H oxidation, but there is also significant site selectivity, presumably governed by steric features. For example, the cholestane derivative in eq 11 gives rapid, selective C-25 side-chain oxidation without appreciable reaction at other tertiary carbons.[12] Coprostane[13] and estrone[14] derivatives undergo C-5 and C-10 hydroxylation, respectively.

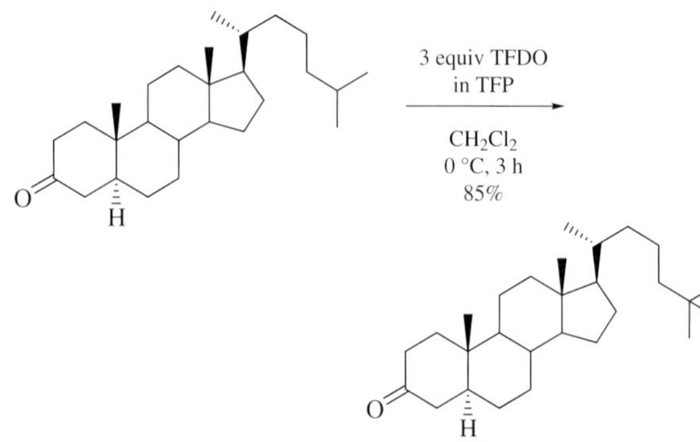

(11)

The stability of TFDO to strong acid adds a further dimension to its chemistry, as illustrated by the reactions of amines in the form of their fluoroborate salts with ketone-free solutions of TFDO.[15] The amino group is deactivated under these conditions and remote C–H hydroxylation takes place (eq 12).

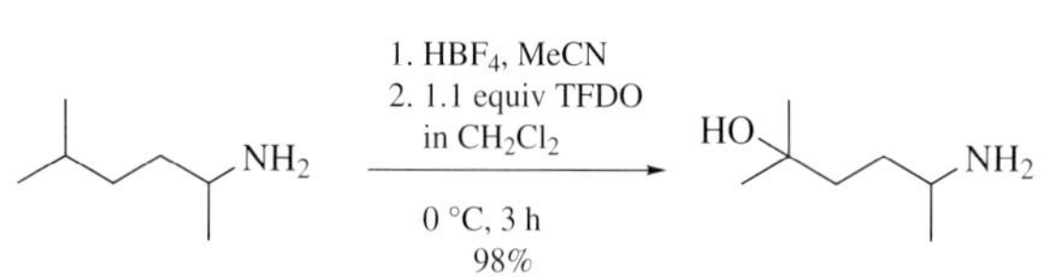

(12)

Alcohols. Secondary alcohols are smoothly oxidized to the corresponding ketones by TFDO, whereas primary alcohols are converted to acids in a slower process.[16] In view of the many methods for alcohol oxidations, this reagent will be advantageous for such conversions only in special situations. However, TFDO does perform sensitive oxidations under favorable experimental conditions that may be useful with problem cases. Cyclobutanol is oxidized by TFDO without the ring cleavage that is often problematic (eq 13). Other reactive functions can be accommodated as shown for the epoxy alcohol in eq 14.

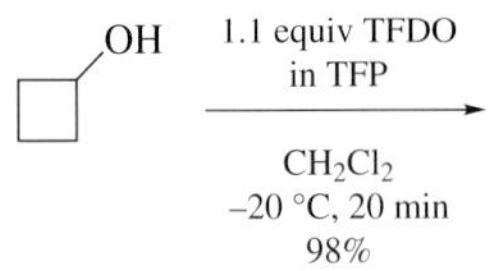

(13)

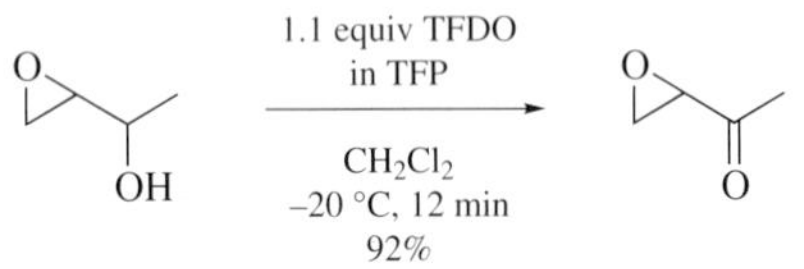

(14)

Vicinal diols are oxidized to α-hydroxy ketones by TFDO without cleavage between the two functional groups. Thus, tertiary–secondary *vic*-diols are usefully oxidized by TFDO (eq 15).[17] Optically active secondary–secondary diols have been converted to α-hydroxy ketones without racemization.[18]

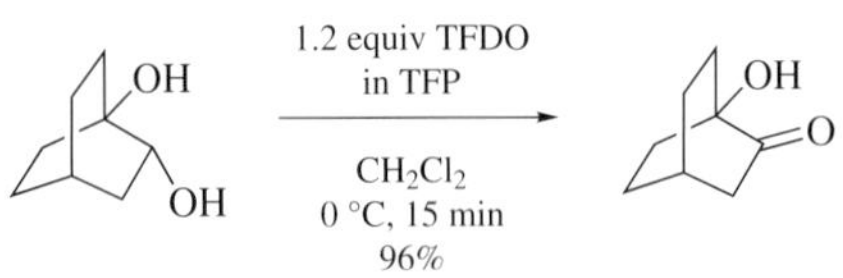

(15)

Ethers and Acetals. TFDO selectively oxidizes adjacent to an oxygen atom in compounds of these types.[19] The initially generated hemiacetals decompose to carbonyl compounds and alcohols, which may be subject to further oxidation. Ethylene glycol acetals are attacked at the dioxolane ring, leading to an interesting deprotection of the carbonyl unit under nonacid conditions (eq 16). Cyclic ethers like tetrahydropyran yield lactones by subsequent oxidation of a stable cyclic hemiacetal intermediate (eq 17). Even methyl *t*-butyl ether is oxidatively liberated to *t*-butanol upon reaction with TFDO.

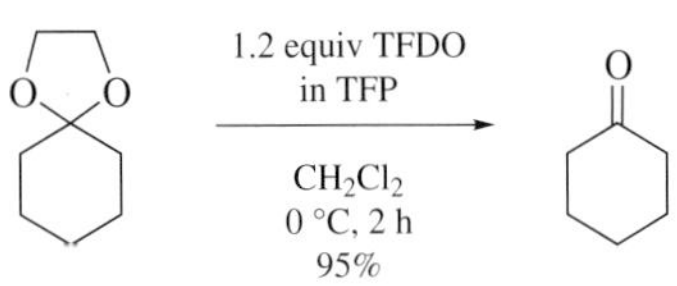

(16)

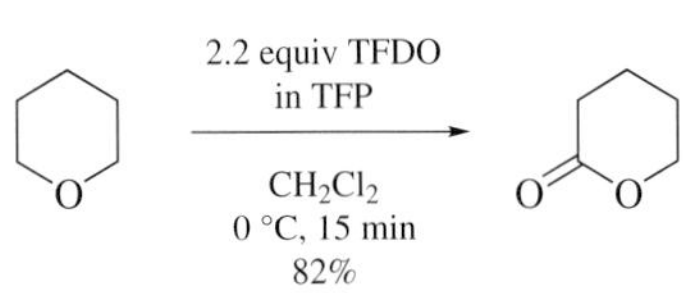

(17)

First Update

Ruggero Curci, Lucia D'Accolti & Caterina Fusco
Università di Bari, Bari, Italy

Direct Functionalization of C–H Bonds by Methyl-(trifluoro-methyl)dioxirane. With most existing methods of direct oxyfunctionalization of saturated hydrocarbons, the major problem is not mainly low reactivity of the alkane molecules, but rather the difficulty of achieving selective oxidations. Indeed, selective oxidations of nonactivated alkane C–H bonds are limited to a few biological processes, which are hard to imitate.

Thus, it is remarkable that methyl(trifluoromethyl)dioxirane (TFDO) has proven to be an outstanding reagent for the selective oxidation of a variety of alkanes, including polycyclic saturated hydrocarbons, even at subambient temperatures.[1,20] Reaction kinetics and the careful application of radical-clock probes[20,21] have shown that the facile *O*-insertion into unactivated C–H bonds is likely to occur by a substantially concerted oxenoid-type mechanism. Despite the high reactivity, TFDO oxidations of alkanes can be remarkably selective.[1,20] In the alkane functionalizations by TFDO, the energetic and mechanistic characteristics differ significantly from radical oxidations more common with traditional reagents.[3] Careful theoretical studies are in accord with the finding that tertiary C–H bonds are considerably more prone to *O*-insertion by dioxiranes than their secondary and primary counterparts.[22]

A FMO model for dioxirane tertiary C–H *O*-insertion[20,22] is suited for explaining the stringent steric and stereoelectronic alignment factors which result in the observed astounding selectivity. An example is reaction of the tetraepoxide of 3β-acetyl vitamin D_2 with TFDO to afford its 25-hydroxy derivative in good isolated yield (eq 18).[23a] The tetraepoxide starting material (having the *R* configuration at all of the seven newly generated stereocenters) was obtained in high yield as a single diastereoisomer upon reaction of 3β-acetyl vitamin D_2 with TFDO at −40 °C.[23a]

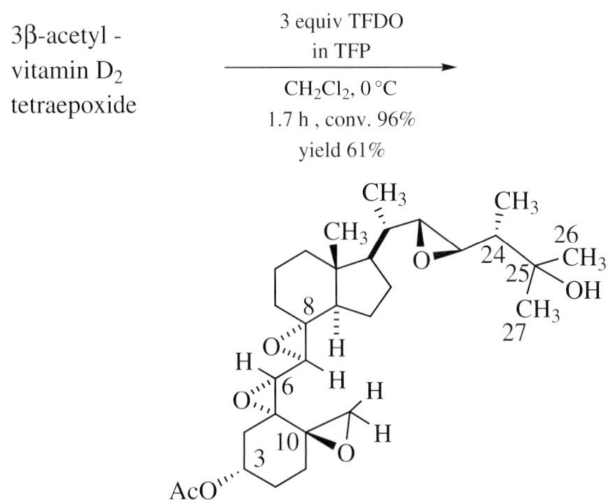

(18)

This surprisingly high site-selectivity for dioxirane *O*-insertion at the tertiary C25–H parallels that already recorded for the analogous triepoxide of vitamin D_3.[23b] Analogous to the case reported for some cholestane derivatives,[12–14] the site-selectivity is likely to derive from a distinct preference displayed by dioxiranes in attacking tertiary C–H centers bearing *geminal* methyl groups; in fact, these offer less steric opposition to optimal stereoalignment as compared to the other tertiary C–H positions available in these target molecules.

Along the same lines, treatment of the 2,3,22,23-tetraacetyl derivative of brassinolide — a plant hormone with a steroidal skeleton — with TFDO using the conditions given in eq 19 led to the nearly exclusive formation of its C25–OH analog.[24]

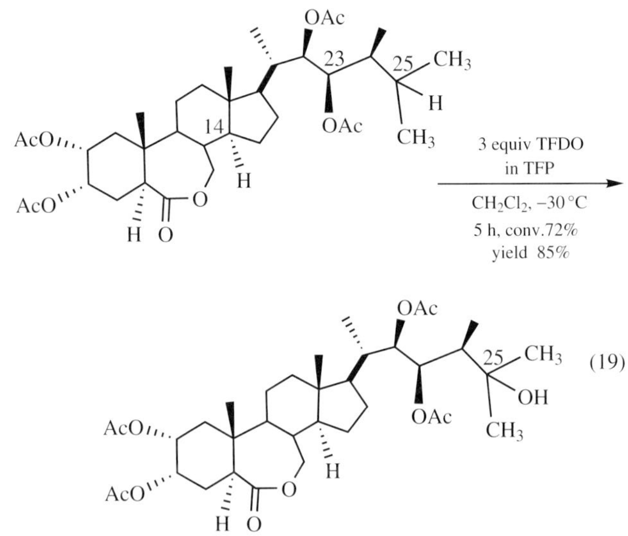

(19)

Treatment of the same starting material with a larger excess of TFDO (6 equiv) at room temperature and prolonged reaction time (24 h) gave the 14,25-dihydroxy tetraacetyl brassinolide as the major product (78% yield).[24]

The examples above illustrates the facile hydroxylation of unactivated methine carbons of several 5β- and 5α-steroids achieved by TFDO. This reagent uses simple procedures and mild conditions to selectively yield the corresponding mono- and/or dioxygenated derivatives in reasonable isolated yields. The method enables site-selective oxyfunctionalization in side chain and ring D in bile acids and sterols from natural sources, is a key transformation into bioactive products such as 25-hydroxyvitamin D derivatives, brassinolides, ecdysonic compounds, and cardiotonic steroids.[24] TFDO is the reagent of choice because of its reactivity and high selectivity. Concerns about the cost of the reagent are chiefly linked to that of its parent ketone, i.e., 1,1,1-trifluoropropanone (TFP) (industrial price: ca. $ 150/100 mL). However, this concern fades considering that the expensive TFP is volatile (bp 22 °C) and it is easily recovered from spent solutions by careful low-temperature fractional distillation over granular P_2O_5.[25]

The regioselective functionalization of complex compounds is an important goal in organic synthesis; in particular, the bridgehead functionalization of polycyclic compounds can provide an access to derivatives bearing quaternary carbon centers or strained bridgehead double bonds. Controlled functionalization of centropolyindans[26] at their benzylic and/or benzhydrylic bridgehead positions is of particular interest since it allows the synthesis of complex three-dimensionally fused polyquinane carbon skeletons. Representative centropolyindanes undergoes selective oxygen atom insertion into their bridgehead C–H bonds by dioxiranes.[27] Varying the excess of the TFDO reagent, allows conversion of tetrabenzo[5.5.5.5]fenestrane (fenestrindane) into either the monoalcohol or the all-bridgehead tetraalcohol (eq 20).[27a]

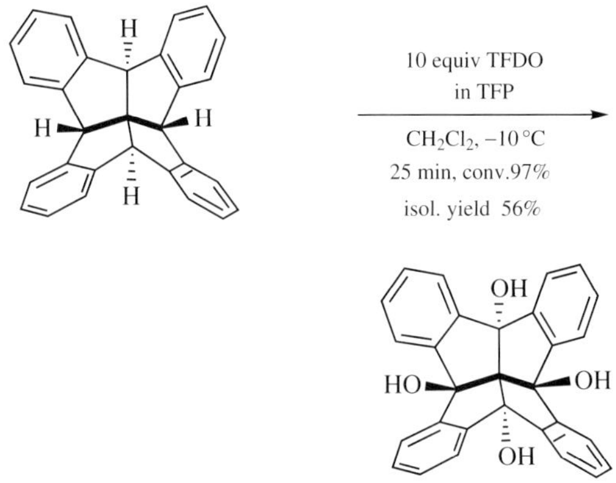

(20)

The versatile use of TFDO is demonstrated by the stepwise hydroxylation of the centropolyindane with regio- and stereocontrol. Treatment of fenestrindan with nearly stoichiometric TFDO at the conditions given in eq 21 rapidly afforded the monoalcohol as the only product. Under similar conditions, the monoalcohol gave a single dialcohol regioisomer bearing both hydroxyl groups on the same indane unit, i.e., 4b,8b-dihydroxyfenestrindane (eq 21).[27b]

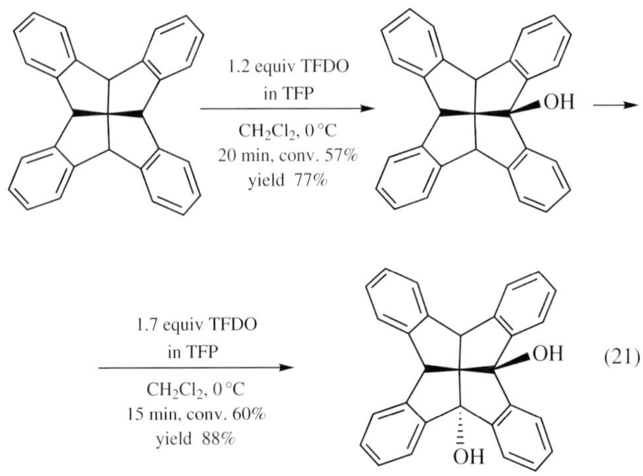

(21)

The second hydroxylation occurs preferentially at the proximal 8b position with *anti* stereochemistry, perhaps because the orientation of the initial-OH moiety disfavors dioxirane *O*-insertion at the benzhydrylic C–H from the same concave *syn* face by electrostatic dipole-dipole repulsion.[27b]

The efficiency of *O*-atom transfer by TFDO is demonstrated by its ability to oxidize a variety of saturated hydrocarbons into alcohols. Examples are provided by the transformation of *trans*- and *cis*-1,2-dimethylcyclohexanes into the corresponding 1,2-dimethylcyclohexan-1-ols (eq 8) stereospecifically, by the conversion of *trans*- and *cis*-decalin into the corresponding 1-decalol with no epimerization at the reaction center,[3] and by the oxidation of optically active 2-phenylbutane to 2-phenyl-2-butanol in over 90% yield and with *complete retention* of configuration (eq 9),[10] all serve well to illustrate the efficiency and selectivities attainable with TFDO. With stoichiometric or a modest excess of TFDO, these oxidations entail reaction times of minutes rather than hours, and avoid excess oxidant, as is required when using the milder DDO (dimethyldioxirane).

All the reactions above are characterized by high tertiary vs. secondary selectivities. The selective bridgehead hydroxylation of adamantane by *excess* (6 equiv) TFDO affords adamantan-

1,3,5,7-tetraol along with the 1,3,5-triol, in 73% and 24% yield, respectively (eq 10). In these oxidations, kinetic data have shown that TFDO is more reactive than DDO by a factor of ca. 10^3, with no loss of selectivity.[11]

Data reported on oxidation of hydrocarbons bearing cyclopropyl moieties suggest that alkane C–H bonds positioned "alpha" to a cyclopropane ring are "activated" towards dioxirane *O*-insertion.[28] Product distributions indicate that usually cyclopropyl activation of α-C–H bonds dominates when no tertiary C–H is present (eq 22).

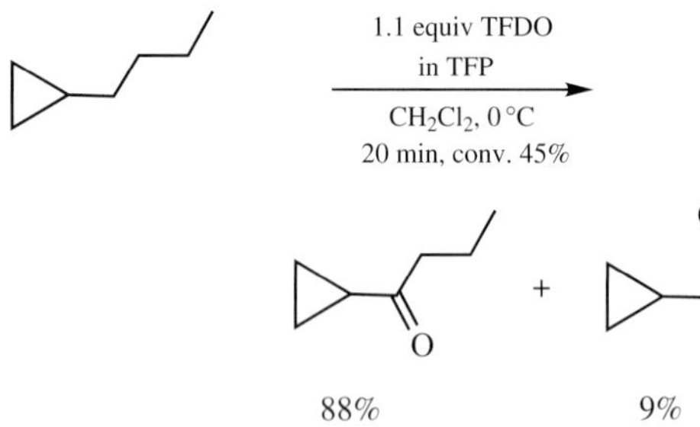

(22)

In the absence of serious steric constrains, the cyclopropane moiety is free to adopt orientations with angles between the extremes of 0 and 90 degrees with respect to the p-orbital component of the proximal C–H bonds.[29] However, oxidation of (3-methyl-butyl)-cyclopropane (eq 23) clearly demonstrates that cyclopropyl activation competes poorly with oxidation at the tertiary C–H.[28]

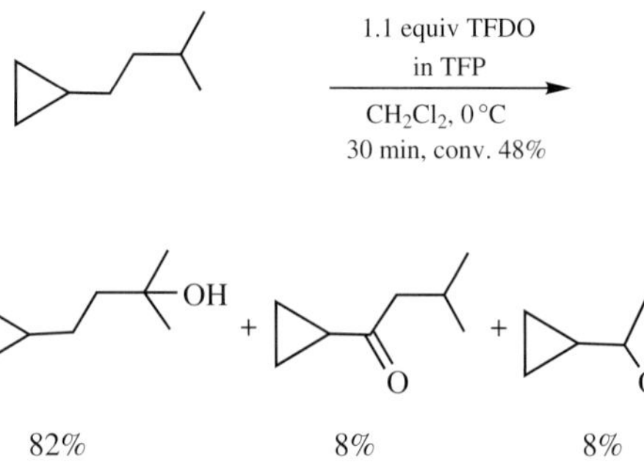

(23)

Thus, in the absence of steric constraints which prevent the cyclopropane moiety from adopting the specific favorable orientations, the normal order of dioxirane reactivity towards alkane C–H bonds may be established as: tertiary C–H $\cong$ benzhydrylic C–H > α-cyclopropyl C–H > benzylic C–H > secondary-C–H > primary C–H.

Cyclopropyl activation can again be invoked to rationalize the preferential oxyfunctionalization at the α-CH_2 of spirooctane (eq 24).[28]

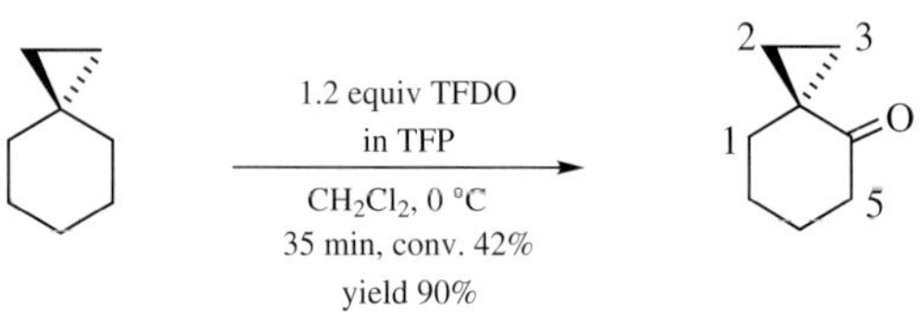

(24)

The cyclopropyl moiety — if suitably oriented — can exert a marked activating effect on dioxirane oxidation at alpha C–H bonds as seen in the hydroxylation of 2,4-didehydroadamantane (eq 25).[28]

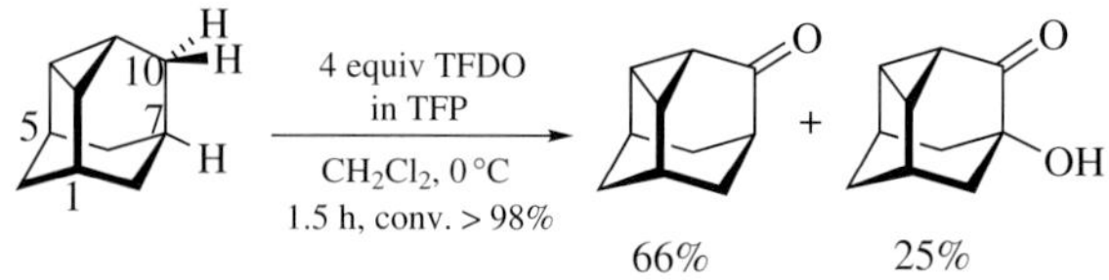

(25)

For this substrate, Murray et al. have reported that reaction with 2.2 equiv of dimethyldioxirane (DDO) proceeds at room temperature with 82% conversion, yielding 2,4-didehydroadamantan-10-one and 2,4-didehydroadamantan-7-ol in 29% and 21% yield, respectively, during 12 h.[30] Comparable selectivity is obtained with the more powerful TFDO, although the oxidation is considerably faster, giving 80% conversion in 1.5 h at 0 °C. Alternatively, treatment with TFDO excess (4 equiv) using the conditions in eq 25 results in practically complete conversion into ketone and the valuable 7-hydroxy-2,4-didehydroadamantan-10-one. Control experiments show that the latter is generated by the oxidation of the initially formed 7-hydroxy-2,4-didehydroadamantane.[28]

Similarly, in the TFDO oxyfunctionalization of Binor S, the valuable ketone resulting from the oxidation of methylene CH_2 alpha to the cyclopropyl moiety is obtained, along with the product expected from hydroxylation at tertiary bridgehead C–H bonds (eq 26).[31]

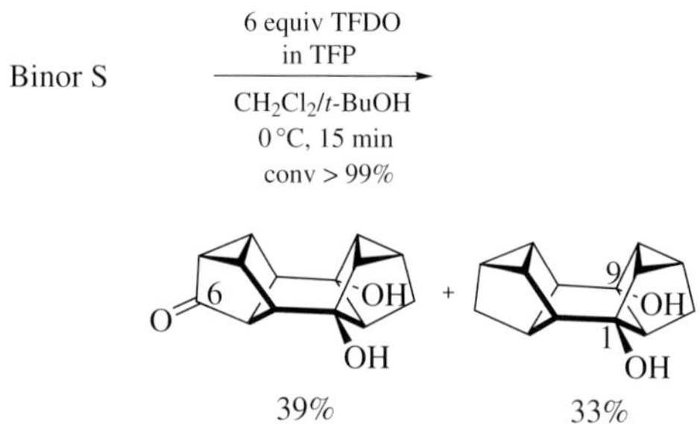

(26)

Reaction monitoring, and control experiments using the isolated diol, shows that oxyfunctionalization at the C-6 methylene only takes place in the facing nortricyclane subunit after hydroxylation at the bridgehead C(1)–H and C(9)-H.[31]

In both cases above, the effective competition with the preferential dioxirane *O*-insertion at tertiary bridgehead C–H's stems from the favorable "bisected" orientation of the methylene C–H bonds relative to the neighboring cyclopropyl ring.

The "deactivating" arrangement of the cyclopropyl moiety in spiro adamantanylcyclopropane makes this compound an informative probe; in this substrate TFDO hydroxylation of occurs at C-5 exclusively, yielding the bridgehead alcohol shown in eq 27.[28]

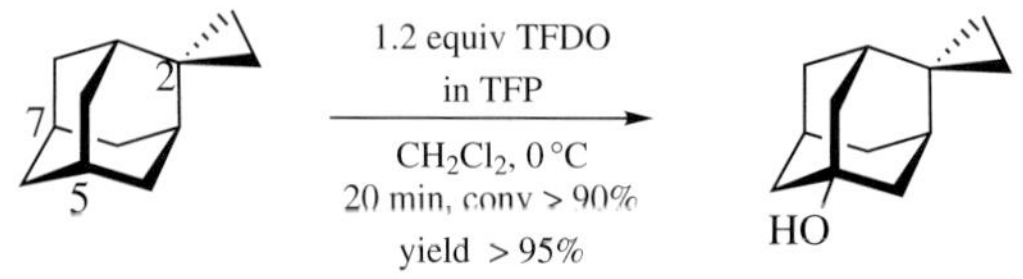

(27)

In this case, the proximal tertiary C–H's at C-3 and C-1 becomes deactivated since their p-orbital component is constrained in the unfavorable "eclipsed" arrangement with respect to the cyclopropane ring.[28]

A similar regioselectivity is observed for methyleneadamentane oxide bearing a spiro fused oxiranyl ring (eq 28).[32]

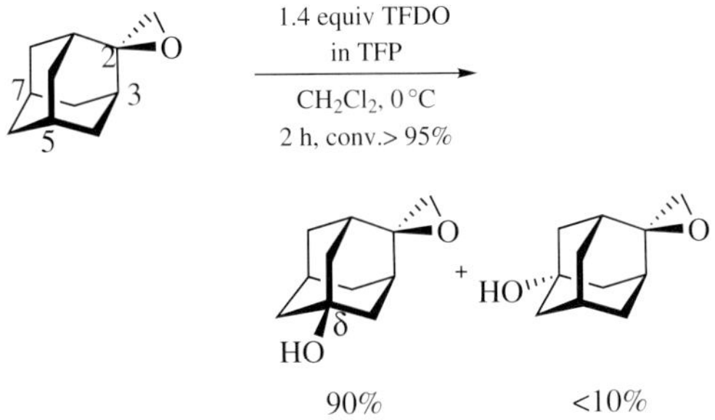

(28)

Here it is apparent that the oxiranyl ring has a marked deactivating effect on the proximal C–H bonds, leading to exclusive hydroxylation at bridgehead C5–H and C7–H with surprisingly high *Z/E* diastereoselectivity. This is parallel to the stereodiscrimination observed for the TFDO oxidation of adamantanes carrying electron-withdrawing substituents at C-2.[33]

In the transformation presented in eq 28, the epoxide ring remains untouched; thus, the reaction is also earmarked by valuable chemoselectivity. This feature of TFDO oxidation is confirmed by the finding that a variety of epoxy alcohols can be cleanly converted into the corresponding epoxy ketones in high yields. Epoxy ketones could also be obtained by the *direct* oxidation of bicyclic epoxides using TFDO. The oxyfunctionalization usually occurs at the remote γ and/or δ C–H bonds with respect to the epoxide ring. An example is given by the transformation in eq 29.[34]

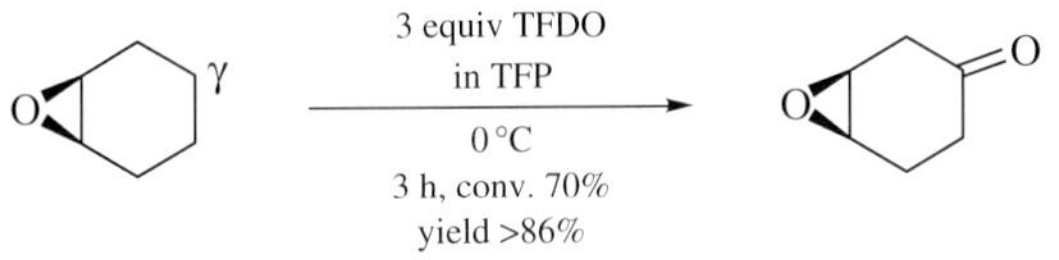

(29)

Other interesting examples of remote oxyfunctionalization using TFDO are provided by the oxidation of several open-chain, cyclic, and bicyclic esters.[35] One clear-cut case is represented by the transformation of *exo*-2-[(*p*-chlorobenzoyl)oxy]norbornane into its 6-norbornanone with high regioselectivity (eq 30).

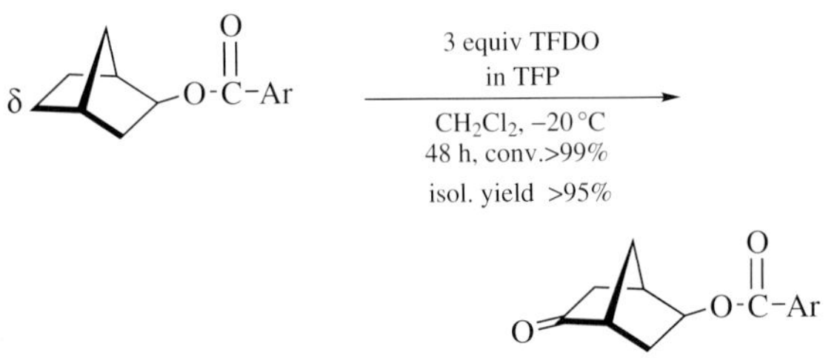

(30)

The collection of the examples reported herein convincingly suggest that TFDO should be considered the reagent of choice for the direct oxidation of C–H bonds because of its efficiency, remarkable selectivity, and simplicity of operations.

Related Reagents. Potassium monoperoxosulfate (Oxone); potassium monoperoxosulfate (Oxone)/1,1,1-trifluoropropanone (TFDO in situ); Dimethyldioxirane (DDO).

Second Update

Caterina Fusco, Lucia D'Accolti & Cosimo Annese
Institute of Chemistry of OrganoMetallic Compounds, University of Bari, Via Orabona, Bari, Italy

Further Progress on Reactivity of Methyl(trifluoromethyl)-dioxirane. The development of new strategies for the effective functionalization of organic molecules continues to play a central role in the sciences and innovation that spans medicines, agrochemicals, and high-performance materials.

The efficient oxidation of simple and/or structurally complex molecules under mild conditions is one of the most important methods for the introduction and manipulation of functional groups. These transformations are undoubtedly the best highlight of dioxirane chemistry.[1,36]

The applications in synthesis of the currently popular dimethyldioxirane (DDO, 1a)[1a] and of methyl(trifluoromethyl)dioxirane (TFDO, 1b)[3] *in situ* and/or isolated form have proliferated the access to key products that are useful in organic synthesis.[1–5] Their unique ability ranges from the transfer of an oxygen atom to unactivated C–H bonds of saturated hydrocarbons, to epoxidations, to the oxidation of heteroatoms containing a lone pair of electrons such as amines and sulfides. Noticeably, all these reactions occur under very mild conditions, that is, ambient or subambient temperature and pH close to neutrality.

This update provides an overview of the most recent strategies of natural products, non-natural products, and targets that use the selective oxyfunctionalization with TFDO.

Selectivity in the Oxyfunctionalization of Natural Products. A major highlight of dioxirane chemistry consists precisely in their ability to yield site-selective hydroxylation of target substrates without the need for a catalyst and for elaborate procedures. The great potential of these reagents can be found in recent literature by a number of impressive oxyfunctionalizations on steroid substrates.

A facile synthetic strategy was recently reported by Lee and Fuchs to obtain cephalostatin analogs.[37] The key steps for this synthetic approach include TFDO-mediated site-selective aliphatic C–H oxidation, where in Cr(VI)-catalyzed E-ring opening and a cyclization cascade are used.

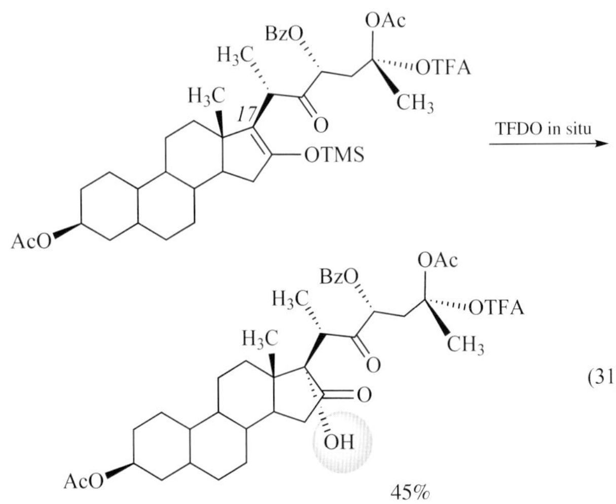

(31)

The C17-OH group was introduced in 45% yield over two steps under condition reported in eq 31.

A remarkable regioselective synthetic procedure for the total synthesis of natural product pericosine E has been described by Usami and coworker.[38] Treatment of the key intermediate diene in pericosine synthesis with DDO (1a) gave an improved 4:3 ratio of (−)-2a and (+)-2b. Furthermore, oxidation of intermediate diene with TFDO *in situ* provided a 16:1 mixture of 2a and 2b in 61% combined yield (eq 32).

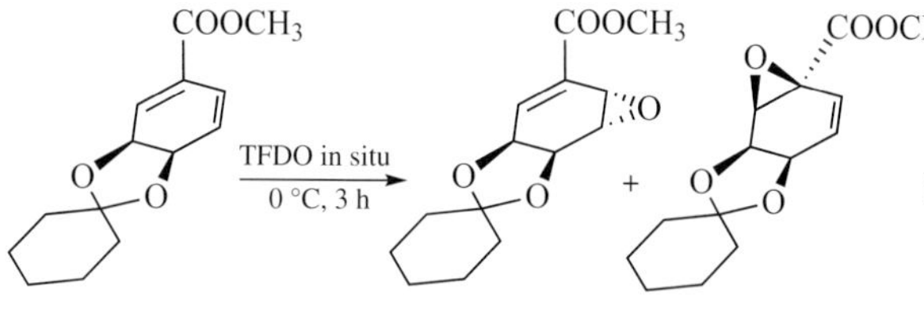

(-) 2a 65% (+) 2b (32)

The authors suggested two explanations for better selectivity with TFDO against DDO. The bulkiness of the trifluoromethyl group in TFDO might inhibit oxidation of the double bond bearing a methoxycarbonyl group in the key diene, or the increased electrophilicity of TFDO from inductive effect of the trifluoromethyl group might accelerate the reaction to double bond that does not bear an electron-withdrawing methoxycarbonyl group in the key diene.[38]

Many other remarkable examples of tertiary C–H bond hydroxylation with TFDO in steroid substrates can be found in the literature, including the biomimetic oxidation of derivatives of vitamin D as well 3β-acetyl vitamin D2 (eq 18).[23a]

En route to the total synthesis of vitamin D analogs, Fuchs and coworker reported a highly convergent synthesis of Vitamin D3 analogs that involve the coupling of an A-ring synthon with a CD-ring/side-chain fragment.[39] In this context, 25-hydroxy-Grundmann's ketone is the key intermediate for the preparation of the CD-ring/side-chain part of 1α,25(OH)2D3 derivatives.[39]

Treatment of Grundman's ketone with TFDO generated *in situ* from Caroat® (6 equiv) and 1,1,1-trifluoroacetone (TFP) (20 equiv) allowed the hydroxylation of tertiary C–H bond in 6 h at 10 °C under the reaction conditions given in eq 33.

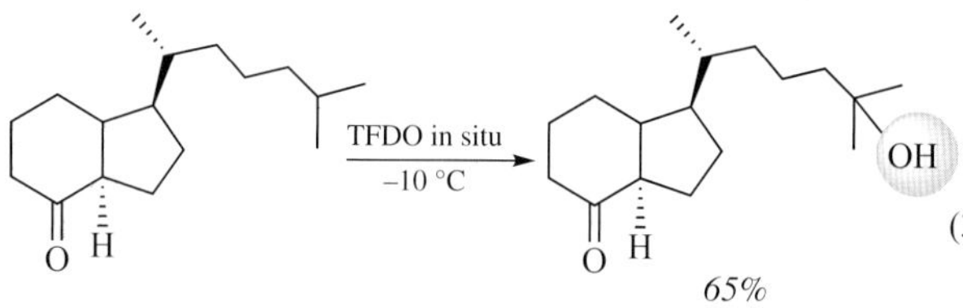

65% (33)

The feat of the direct C-25 side-chain hydroxylation of vitamin D derivatives is remarkable in view of the biological significance and uses of 25-hydroxy-vitamin D3 and D2 and their metabolites.

The site-selective functionalization of steroid substrates is impressive due to the many other potential sites of oxidation. More recently, Lei and coworkers have completed an effective bioinspired synthetic approach of a trideuterated derivative of (25*S*)-Δ^7-dafachronic acid (5,24,25-D3) (3), a useful tool in drug discovery and biotechnology research.[40]

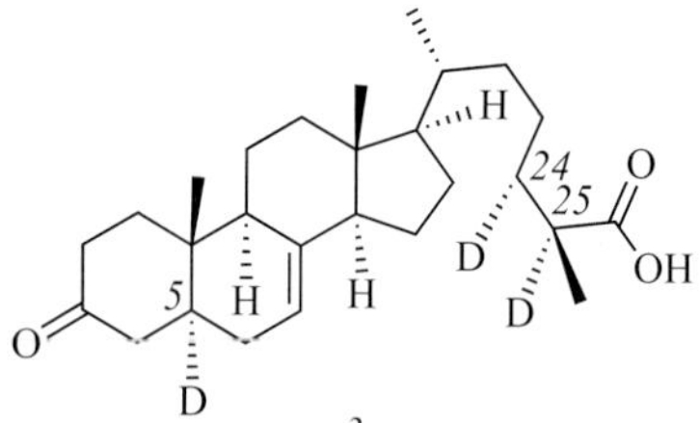

3

Their approach begins from cholesterol and involves acetyl protection, allylic oxidation, and followed by hydrogenation. Of particular interest is the TFDO-mediated site-selective oxidation of the *unactivated* tertiary C–H bond at C25 position (eq 34). The oxidation product was obtained in 52% yield (70%) based on recovered starting material showing high site-selectivity as confirmed by Curci.[12]

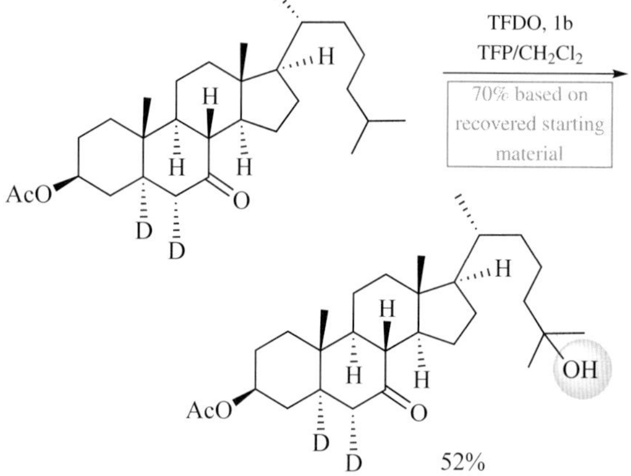

(34)

In 2009, Chen and Baran developed a useful strategy to access steroids, diterpenes, and sesquiterpenes.[41] The approach includes a laboratory oxidase phase, where a series of site-selective oxidations furnishes highly oxidized terpene targets. Their key reaction is the site-specific intermolecular oxidation of carbamate using TFDO (1.0 equiv), which led to the formation of intermediate in 82% isolated yield on a gram scale (eq 35).[41]

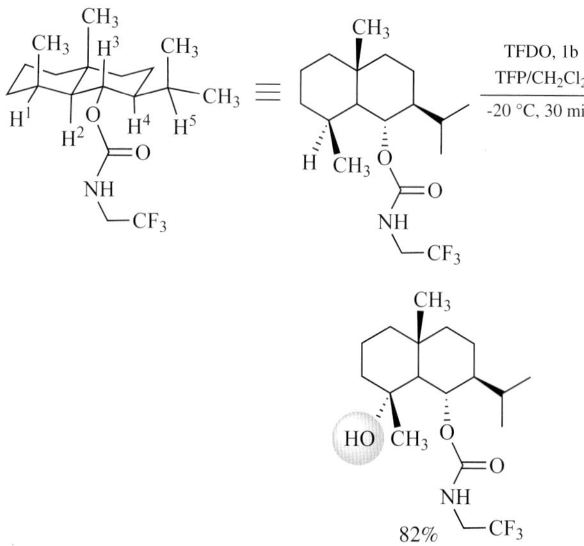

(35)

Of the five *unactivated* tertiary C–H bonds in the substrate, only the equatorial C–H[1] is oxidized to an alcohol by the dioxirane reagent (eq 35). These direct oxidations represent an outstanding example of site-selective C–H functionalization reaction, which gave access to the four eudesmane terpenes in good yields and with excellent selectivities.

Further examples in intermolecular C–H oxidation were described in 2014, when Baran and coworkers used the direct C–H oxidation in the field of drug discovery as a tool to improve not only the potency of a pharmaceutical lead but also its physicochemical properties.[42]

They applied the TFDO to a derivative of the pentacyclic triterpene betulin, which afforded selectively the C16 ketone product in 30% yield as a major product (eq 36).

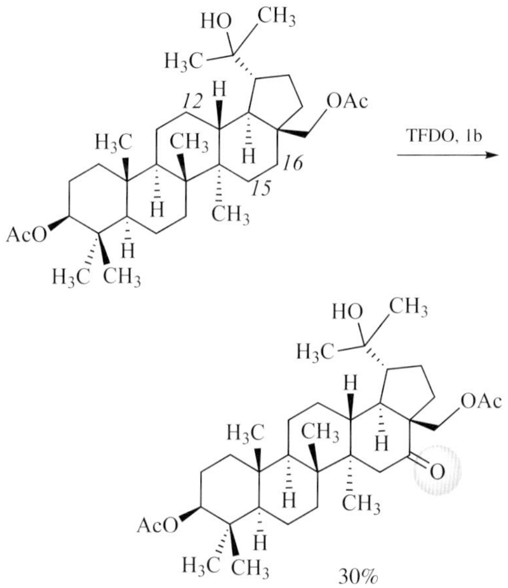

(36)

No minor products were isolated in quantities sufficient for characterization, but it is of note that none of these products contained a ketone motif. This reaction constitutes a first example of selective methylene oxidation in the context of a complex triterpene substrate.[42]

The origin of selectivity observed may result from a cooperation of reactivity factors that involve the C–H bond nucleophilicity, steric hindrance to reagent approach, and strain release in transition-state formation.[43,44]

Supramolecular host–guest chemistry provides a promising means to achieve selectivity in a wide range of organic reactions.[45] The selectivity enhancement by directing the target site to the reaction centre has been widely investigated in the C–H bond oxidation by cyclodextrin-based catalysts.[45]

In 2012, Wong et al. developed efficient methods for C–H bond oxidation by dioxiranes generated *in situ* in aqueous medium.[45] The reaction was conducted with 3,7-dimethyloctyl esters, TFP, and β-cyclodextrin (β-CD) in the presence of Caroat® and $NaHCO_3$ in H_2O at room temperature for 8 h under the reaction conditions given in eq 37.

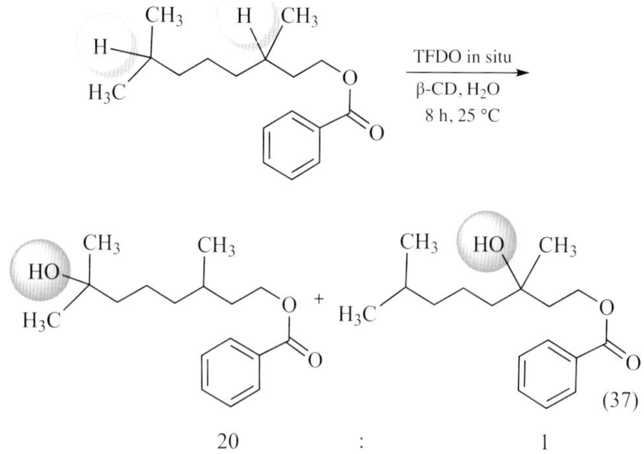

(37)

The authors demonstrated that the β-CD could provide enhanced site-selectivity for C–H bond oxidation (20:1 vs. 7:1), probably by obstructing the approach of dioxirane to the internal C–H bond of 3a through inclusion complex formation in H_2O. As a result, the unbound C–H bonds far away from CD would be preferentially oxidized.[45]

The TFDO oxidation of Boc-protected di- and tripeptide esters bearing alkyl side chains showed that hydroxylation of the terminal N–H can be achieved efficiently with no appreciable loss of configuration at the chiral centers.[46]

The oxidation procedure merely involved the addition of TFDO to an acetone solution of the tripeptide (eq 38).

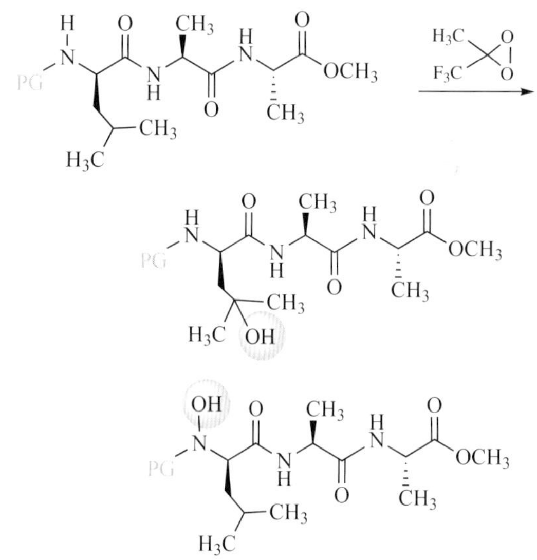

(38)

Our research group established that the C–H hydroxylation with TFDO of peptides protected with amide groups (PG = Ac-, Tfa-, Piv-) undergo exclusive hydroxylation at the side chain, while their analogs bearing a carbamate group (PG = Cbz-, Moc-, Boc-, TcBoc-) give competitive and/or concurrent hydroxylation at the terminal N–H moiety.[46]

As a result a powerful approach to novel drug design and to the development of peptide-based nanomaterials his presented.

In 2011, our research group achieved a synthesis of representative monohydroxy derivatives of valinomycin (VLM, 4), a cyclic dodecadepsipeptide able to coordinate alkali metal ions, allowing the permeability of K^+ through biological membranes (e.g., mitochondria, erythrocytes, chloroplasts, phospholipid micelles, etc).[47]

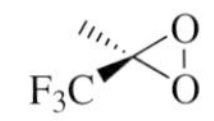

The oxidation reaction of VLM with TFDO produces selectively representative monohydroxy analogs, achieved under mild conditions by direct hydroxylation at the side chains of the macrocyclic substrate (eq 39).

The direct oxidation of methane to liquid oxygenated compounds, such as methanol is an attractive strategy. The direct oxidation of methane to methanol under mild conditions would have enormous economic and environmental impact. Methanol, for instance, is a desirable product because it is an important chemical intermediate, used for the synthesis of other oxygenated compounds such as formaldehyde and acetic acid.

One of the remaining challenges in dioxirane chemistry was the direct functionalization of methane to methanol. Our research group reports the application of TFDO to the direct, remarkably low-temperature oxidation of methane to methanol with high reactivity and selectivity.[51]

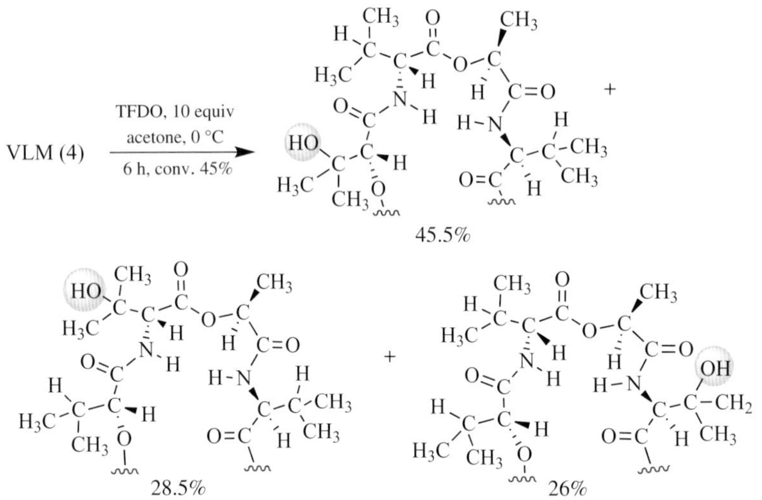

(39)

Selectivity in the Oxyfunctionalization of Non-natural Products. Another application that is different from those described the above, takes advantage from selective C–H oxidation by TFDO of organic substrates covalently bonded to solid surface, such as silica gel, both in homogeneous and in heterogeneous reactions. As an example, Mello et al. described a facile method for the oxygenation of organic substrates covalently bonded to solid surface, as silica gel, using TFDO as probe.[48]

Our group has long investigated the reactivity of strained polycyclic hydrocarbons having a peculiar architecture, that is, 2,4-didehydroadamantane, BinorS, along with bicyclics that function as fast radical-clock probes.[31] In 2009, our research group demonstrated the first direct *O*-insertion in to the C–H bond of cubane with the potent TFDO. Treatment of cubane in acetone or CH_2Cl_2 at subambient temperature with TFDO performed a mono- and subsequent bishydroxylation to give mono- and bishydroxycubane, respectively (eq 40).[49]

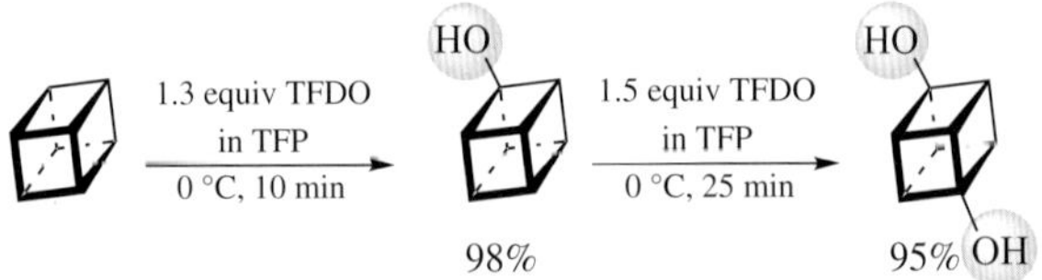

(40)

The reactions proceed through a concerted oxenoid mechanism, in light of ease of formation of cubyl cations.[50]

The dioxirane generated in solution can accept attack by further caroate ions, yielding sulfate ion and molecular oxygen. However, in a competitive process, it can then also be attacked by a variety of electron-rich substrates, viz., "S", yielding the oxidation product "SO" (Scheme 1). In both reactions, the parent ketone is regenerated, so it returns to the catalytic cycle.

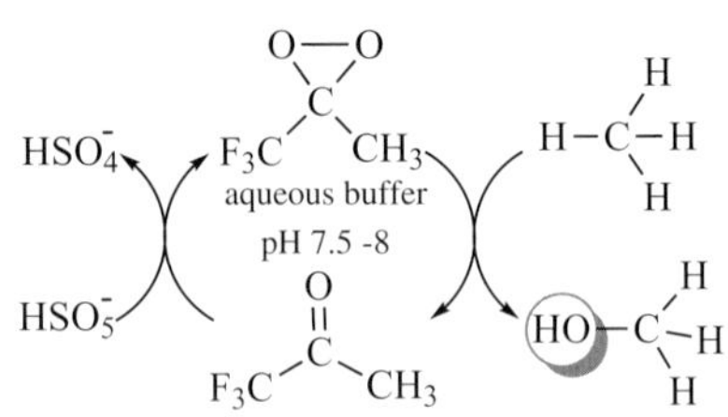

Scheme 1

Results under the reaction conditions given in eq 41 provide a significant breakthrough for this key transformation (eq 41).[51]

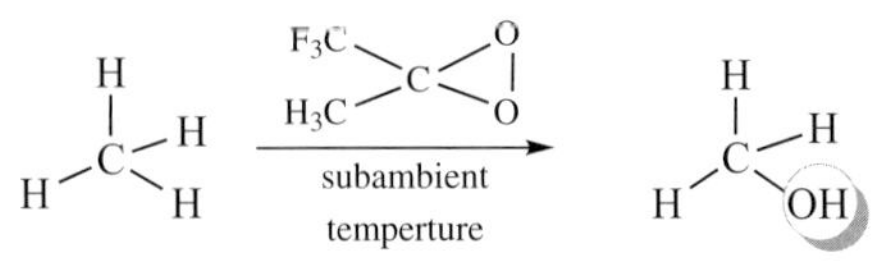

(41)

The data show that, using the powerful TFDO in acetone under given conditions, the CH_3OH yield is > 95% with 15% conversion after a 24 h reaction time.[51]

Selectivity in the Oxyfunctionalization of CNTs. Carbon nanotubes (CNTs) are new carbon-based nanomaterials that possess a quite unique range of mechanical, electrical, and magnetic properties, making them particularly interesting for numerous bionanotechnology applications.

Chemical modification with different functional groups directly to the CNTs' sidewalls is a well-established approach for facilitating solubility and has broad applicability in prominent fields such as biomedicine and catalysis.[52]

Interesting examples of selective surface modification of CNTs for the production of oxidized carbon nanotubes in the form of epoxy-decorated nanosystems (*ox*-CNT) have been reported.

Indeed, the efficient epoxidation of simple and/or structurally complex alkenes under mild and neutral conditions is the highlight of the chemistry of dioxirane,[20] and can be successfully applied as an easily approach to the sidewall decoration of single-walled carbon nanotubes (SWCNTs) with epoxides and their subsequent derivatization (ring opening) upon treatment with reactive nucleophiles.

Billups and coworkers demonstrated that the epoxidation of SWCNTs can be carried out by the reaction with either *in situ* generated TFDO or with *m*-CPBA (eq 42).[53]

Our research group describes the efficient chemical oxidation of SWCNTs using isolated TFDO. Unlike the *in situ* prepared TFDO,[1] our approach allows a simple tuning of the oxidation degree throughout a systematic treatment of the SWCNTs within successive oxidation cycles.[54]

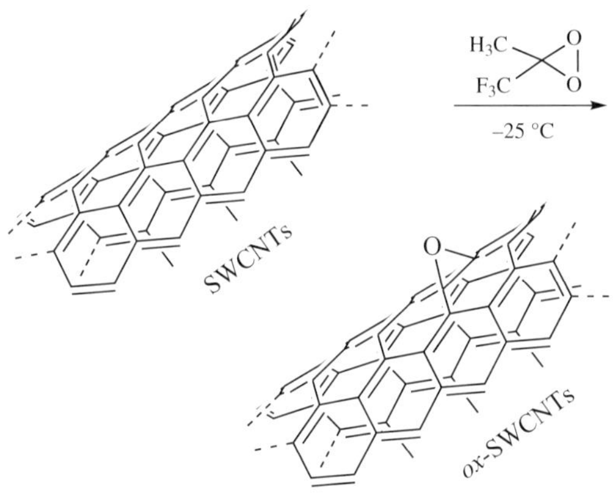

(42)

Although the key epoxidation step has already been conducted by Billups en route to the synthesis of SWCNTs oxide, it was valuable to find that on similar structures the oxidation reaction precedes under mild conditions and in excellent yield in the presence of **1b** in isolated form.

As part of a continuing effort to devise oxidations of CNT, our research group was able to apply TFDO to the direct introduction of epoxide groups on the multi-walled carbon nanotubes (MWCNTs) surface (eq 43).[55]

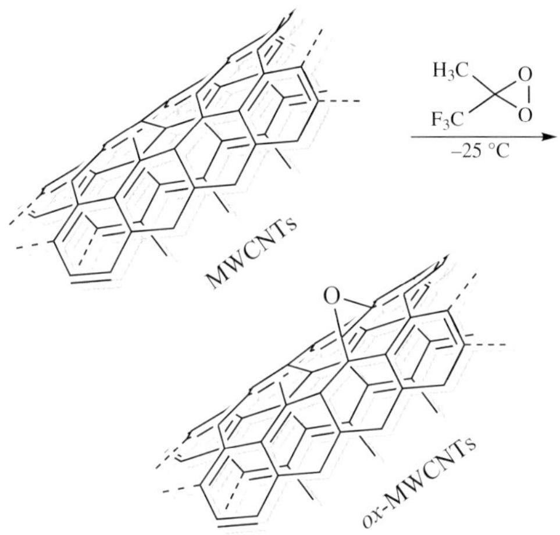

(43)

Concerns about the cost of the parent reagent, that is, 1,1,1-trifluoropropanone (TFP), subside when one considers that the expensive TFP (bp 22 °C) is easily recovered from spent reaction mixtures by fractional distillation.[36]

---

1. (a) Adam, W.; Hadjiarapoglou, L. P.; Curci, R.; Mello, R. In *Organic Peroxides*, Ando W., Ed.; Wiley: New York, **1992**; Chapter 4, pp 195. (b) Murray, R. W., *Chem. Rev.* **1989**, *89*, 1187: (c) Curci, R. In *Advances in Oxygenated Processes*; Baumstark, A., Ed; JAI Press: Greenwich, CT, **1990**; Vol. *2*, Chapter 1, pp 1, (d) Adam, W.; Edwards, J. O.; Curci, R., *Acc. Chem. Res.* **1989**, *22*, 205. (e) Adam, W.; Hadjiarapoglou, L., *Top. Curr. Chem.* **1993**, *164*, 45.
2. Mello, R.; Fiorentino, M.; Sciacovelli, O.; Curci, R., *J. Org. Chem.* **1988**, *53*, 3890.
3. Mello, R.; Fiorentino, M.; Fusco, C.; Curci, R., *J. Am. Chem. Soc.* **1989**, *111*, 6749.
4. Adam, W.; Curci, R.; González-Núñez, M. E.; Mello, R., *J. Am. Chem. Soc.* **1991**, *113*, 7654.
5. Lluch, A.-M.; Sanchez-Baeza, F.; Messeguer, A.; Fusco, C.; Curci, R., *Tetrahedron* **1993**, *49*, 6299.
6. Troisi, L.; Cassidei, L.; Lopez, L.; Mello, R.; Curci, R., *Tetrahedron Lett.* **1989**, *30*, 257.
7. Mello, R.; Ciminale, F.; Fiorentino, M.; Fusco, C.; Prencipe, T.; Curci, R., *Tetrahedron Lett.* **1990**, *31*, 6097.
8. Altamura, A.; Fusco, C.; D'Accolti, L.; Mello, R.; Prencipe, T.; Curci, R., *Tetrahedron Lett.* **1991**, *32*, 5445.
9. Curci, R.; Fiorentino, M.; Fusco, C.; Mello, R.; Ballistreri, F. P.; Failla, S.; Tomaselli, G. A., *Tetrahedron Lett.* **1992**, *33*, 7929.
10. Adam, W.; Asensio, G.; Curci, R.; González-Núñez, M. E.; Mello, R., *J. Org. Chem.* **1992**, *57*, 953.
11. Mello, R.; Cassidei, L.; Fiorentino, M.; Fusco, C.; Curci, R., *Tetrahedron Lett.* **1990**, *31*, 3067.
12. Bovicelli, P.; Lupattelli, P.; Mincione, E.; Prencipe, T.; Curci, R., *J. Org. Chem.* **1992**, *57*, 5052.
13. Bovicelli, P.; Gambacorta, A.; Lupattelli, P.; Mincione, E., *Tetrahedron Lett.* **1992**, *33*, 7411.
14. Bovicelli, P.; Lupattelli, P.; Mincione, E.; Prencipe, T.; Curci, R., *J. Org. Chem.* **1992**, *57*, 2182.
15. Asensio, G.; González-Nuñez, M. E.; Bernardini, C. B.; Mello, R.; Adam, W., *J. Am. Chem. Soc.* **1993**, *115*, 7250.
16. Mello, R.; Cassidei, L.; Fiorentino, M.; Fusco, C.; Hümmer, W.; Jäger, V.; Curci, R., *J. Am. Chem. Soc.* **1991**, *113*, 2205.
17. Curci, R.; D'Accolti, L.; Detomaso, A.; Fusco, C.; Takeuchi, K.; Ohga, Y.; Eaton, P.; Yip, C. Y., *Tetrahedron Lett.* **1993**, *34*, 4559.

18. D'Accolti, L.; Detomaso, A.; Fusco, C.; Rosa, A.; Curci, R., *J. Org. Chem.* **1993**, *58*, 3600.
19. Curci, R.; D'Accolti, L.; Fiorentino, M.; Fusco, C.; Adam, W.; González-Núñez, M. E.; Mello, R., *Tetrahedron Lett.* **1992**, *33*, 4225.
20. Curci, R.; Dinoi, A.; Rubino, M. F., *Pure Appl. Chem.* **1995**, *67*, 811. See references therein.
21. (a) Curci, R.; D'Accolti, L.; Fusco, C., *Tetrahedron Lett.* **2001**, *42*, 7087. (b) Newcomb, M.; Choi, S.-Y.; Simakov, P. A., *Tetrahedron Lett.* **1998**, *39*, 8187. (c) Vanni, R.; Garden, S. J.; Banks, J. T.; Ingold, K. U., *Tetrahedron Lett.* **1995**, *36*, 7999.
22. (a) Glukhovtsev, M. N.; Canepa, C.; Bach, R. D., *J. Am. Chem. Soc.* **1998**, *120*, 10528. (b) Du, X. H.; Houk, K. N., *J. Org. Chem.* **1998**, *63*, 6480. (c) Shustov, G. V.; Rauk, A., *J. Org. Chem.* **1998**, *63*, 5413. (d) Bach, R. D.; Andres, J. L.; Owensby, A. L.; Schlegel, H. B.; McDouall, J. J. W., *J. Am. Chem. Soc.* **1992**, *114*, 7207. See also references therein.
23. (a) Curci, R.; Detomaso, A.; Lattanzio, M. E.; Carpenter, G. B., *J. Am. Chem. Soc.* **1996**, *118*, 11089. (b) Curci, R.; Detomaso, A.; Prencipe, T.; Carpenter, G. B., *J. Am. Chem. Soc.* **1994**, *116*, 8112.
24. Seto, H.; Fujioka, S.; Koshino, H.; Yoshida, S.; Tsubuki, M.; Honda, T., *Tetrahedron* **1999**, *55*, 8341. See also references therein.
25. D'Accolti, L.; Fusco, C.; Rella, M. R.; Curci, R., *Synth. Commun.* **2003**, *33*, 3009.
26. Kuck, D. In *Quasicrystals, Networks, and Molecules of Fivefold Symmetry*; Hargittai, I., Ed.; VCH: New York, **1990**; Chapter 19.
27. (a) Kuck, D.; Schuster, A.; Fusco, C.; Fiorentino, M.; Curci, R., *J. Am. Chem. Soc.* **1994**, *116*, 2375. (b) Fusco, C.; Fiorentino, M.; Dinoi, A.; Curci, R.; Krause, R. A.; Kuck, D., *J. Org. Chem.* **1996**, *61*, 8681.
28. D'Accolti, L.; Dinoi, A.; Fusco, C.; Russo, A.; Curci, R., *J. Org. Chem.* **2003**, *68*, 7806.
29. Rhodes, Y. E.; DiFate, V. G., *J. Am. Chem. Soc.* **1972**, *94*, 7582.
30. Murray, R. K.; Teager, D. S., *J. Org. Chem.* **1993**, *58*, 5548.
31. D'Accolti, L.; Fusco, C.; Lucchini, V.; Carpenter, G. B.; Curci, R., *J. Org. Chem.* **2001**, *66*, 9063.
32. D'Accolti, L.; Kang, P.; Khan, S.; Curci, R.; Foote, C. S., *Tetrahedron Lett.* **2002**, *43*, 4649.
33. Gonzalès-Nuñez, M. E.; Royo, J.; Castellano, G.; Andreu, C.; Boix, C.; Mello, R.; Asensio, G., *Org. Lett.* **2000**, *2*, 831.
34. D'Accolti, L.; Fusco, C.; Annese, C.; Rella, M. R.; Turteltaub, J. S.; Williard, P. G.; Curci, R., *J. Org. Chem.* **2004**, *69*, 8510.
35. Asensio, G.; Castellano, G.; Mello, R.; Gonzalès-Nuñez, M. E., *J. Org. Chem.* **1996**, *61*, 5564.
36. Curci, R.; D'Accolti, L.; Fusco, C., *Acc. Chem. Res.* **2006**, *39*, 1. See also references therein.
37. Lee, S.; Jamieson, D.; Fuchs, P. L., *Org. Lett.* **2009**, *11*, 5.
38. Mizuki, K.; Iwahashi, K.; Murata, N.; Ikeda, M.; Nakai, Y.; Yoneyama, H.; Harusawa, S.; Usami, Y., *Org. Lett.* **2014**, *16*, 3760.
39. Sikervar, V.; Fleet, J. C.; Fuchs, P. L., *J. Org. Chem.* **2012**, *77*, 5132.
40. Liu, W.; Li, X.; Chen, J.; Li, T.; Dong, M.; Lei, X., *Chem. Eur. J.* **2015**, *21*, 5345.
41. Chen, K.; Baran, P. S., *Nature* **2009**, *459*, 824.
42. Michaudel, Q.; Journot, G.; Regueiro-Ren, A.; Goswami, A.; Guo, Z.; Tully, T. P.; Zou, L.; Ramabhadran, R. O.; Houk, K. N.; Baran, P. S., *Angew. Chem., Int. Ed.* **2014**, *53*, 12091.
43. Newhouse, T.; Baran, P. S., *Angew. Chem., Int. Ed.* **2011**, *50*, 3362.
44. Chen, K.; Eschenmoser, A.; Baran, P. S., *Angew. Chem., Int. Ed.* **2009**, *48*, 9705.
45. Fung, Y.-S.; Yan, S.-C.; Wong, M.-K., *Org. Biomol. Chem.* **2012**, *10*, 3122.
46. Annese, C.; D'Accolti, L.; De Zotti, M.; Fusco, C.; Toniolo, C.; Williard, P. G.; Curci, R., *J. Org. Chem.* **2010**, *75*, 4812
47. Annese, C.; Fanizza, I.; Calvano, C. D.; D'Accolti, L.; Fusco, C.; Curci, R, Williard, P. G., *Org. Lett.* **2011**, *13*, 5096.
48. Mello, R.; Martinez-Ferrer, J.; Alcalde-Aragonés, A.; Varea, T.; Acerete, R.; González-Núñez, M. E.; Asensio, G., *J. Org. Chem.* **2011**, *76*, 10129.
49. Annese, C.; D'Accolti, L.; Fusco, C.; Gandolfi, R.; Eaton, P. E.; Curci, R., *Org. Lett.* **2009**, *11*, 3574.
50. Biegasiewicz, K. F.; Griffiths, J. R.; Savage, G. P.; Tsanaktsidis, J.; Priefer, R., *Chem. Rev.* **2015**, *115*, 6719.
51. Annese, C.; D'Accolti, L.; Fusco, C.; Curci, R., *Org. Lett.* **2011**, *13*, 2142.
52. Tasis, D.; Tagmatarchis, N.; Bianco, A.; Prato, M., *Chem. Rev.* **2006**, *106*, 1105.
53. Ogrin, D.; Chattopadhyay, J.; Sadana, A. K.; Billups, W. E.; Barron, A. R., *J. Am. Chem. Soc.* **2006**, *128*, 11322.
54. Annese, C.; D'Accolti, L.; Giambastiani, G.; Mangone, A.; Milella, A.; Tuci, G.; Fusco, C., *Eur. J. Org. Chem.* **2014**, 1666.
55. Annese, C.; D'Accolti, L.; Armuzza, V.; Da Ros, T.; Fusco, C., *Eur. J. Org. Chem.* **2015**, 3063.

Methyl Vinyl Ketone[1]

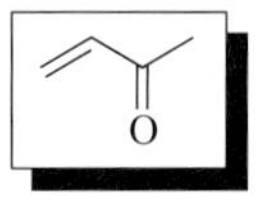

[78-94-4] C_4H_6O (MW 70.10)
InChI = 1/C4H6O/c1-3-4(2)5/h3H,1H2,2H3
InChIKey = FUSUHKVFWTUUBE-UHFFFAOYAV
[59120-04-6]
InChI = 1/C4H6O/c1-3-4(2)5/h3H,1H2,2H3
InChIKey = FUSUHKVFWTUUBE-UHFFFAOYAV

(reagent for attachment of 3-oxobutyl side chains, i.e. Michael additions;[2] reagent for the annulation of cyclohexenones[3])

Alternate Name: MVK; 3-buten-2-one.

Physical Data: fp -6°C; bp 81.4°C, 36.5-36.8°C/145 mmHg, 33-34°C/130 mmHg, 32-34°C/60 mmHg; *d* (20°C) 0.8636 g cm^{-3}, (25°C) 0.8407 g cm^{-3}; n_D^{20} 1.4086; liquid with pungent odor.

Solubility: sol water, methanol, ethanol, ether, acetone, glacial acetic acid; slightly sol hydrocarbons; forms binary azeotrope with water, bp 75°C (12% water).

Form Supplied in: clear liquid stabilized with 0.1% acetic acid and 0.05% or 1% hydroquinone.

Handling, Storage, and Precautions: should be kept cold. Polymerizes upon standing in the pure form. Readily absorbed through skin; lachrymator; highly toxic and flammable. For best results, dry over K_2CO_3 and freshly distill at reduced pressure. Use in a fume hood.

Original Commentary

John A. Colapret & Paul T. Buonora
Lamar University, Beaumont, TX, USA

Introduction As a synthetic reagent, methyl vinyl ketone is commonly viewed as a 3-oxobutyl synthon which undergoes predominantly 1,4 conjugate addition (Michael addition) resulting in the attachment of a 3-ketoalkyl side chain. The resulting products

Table 1 Methyl vinyl ketone syntetic equivalents (3′-oxobutyl synthons)

Alkylating agents and Michael acceptors	Conditions to regenerate ketone	Nucleophilic agents (umpolung type)	Conditions to regenetrate ketone
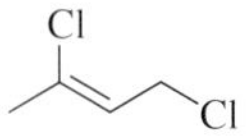 **(1)**	H_2SO_4 (conc.) or $Hg(OAc)_2$, AcOH	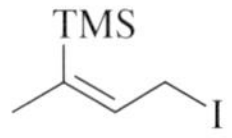 **(2)**	1 N HCl, rt
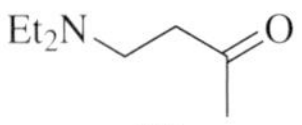 **(3)**	*m*-CPBA		
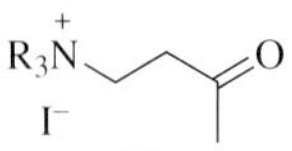 **(4)**	None	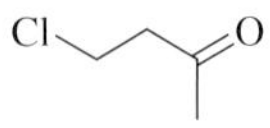 **(5)**	1 N HCl, rt
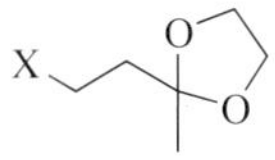 **(6)** X = Br, I, OTs	None		
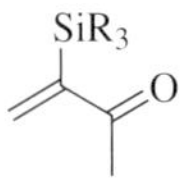 **(7)** R = Me, Et	None		
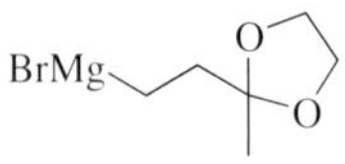 **(8)**	1 N HCl, rt	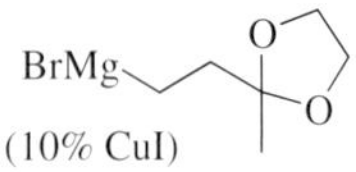 (10% CuI) **(9)**	1 N HCl, rt
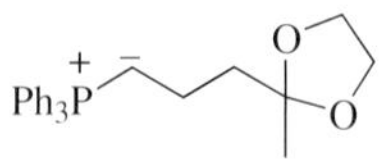 **(10)**	NaOMe, MeOH	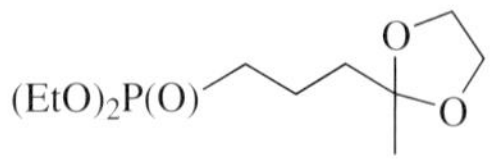 **(11)**	1 N HCl, rt

are often poised for a subsequent cyclization step, thus resulting in a cyclohexenone derivative; this tandem is the well-known Robinson annulation.[4] In most cases, MVK is the electrophilic partner, but there are a few important nucleophilic (umpolung) derivatives of MVK which also are useful for accomplishing ring construction (see below). MVK itself is particularly susceptible to anionic polymerization, which limited its early use under basic conditions; however, it should be noted that MVK will polymerize under certain acidic and radical conditions as well.[5] The many efforts to improve the Robinson ring synthesis have led to monumental synthetic advances with respect to both nucleophilic reactants and MVK synthetic equivalents (3-oxobutyl synthons).

Aside from inducing the anionic polymerization of MVK, enolates as nucleophilic partners in alkylations are limited by *O*- vs. *C*-alkylation, mono- vs. polyalkylation, rapid proton transfer, and by the ability to control the regioselectivity of their generation. These problems have found solutions through the development of trimethylsilyl enol ethers,[6] in the utilization of enamine chemistry,[7] and by employing metalloenamines (imine anions).[8] With each of these developments in the nucleophilic partner, improvements in MVK alkylation have resulted in a greatly expanded use of the reagent. Nevertheless, a number of MVK synthetic equivalents have been successfully developed and are briefly reviewed in Table 1. Note that in this context, 'MVK equivalent'; is defined as any reagent which contains all four carbon atoms; also the 3′-oxo functionality is in place or is one transformation away.

Methyl Vinyl Ketone versus its Synthetic Equivalents (as Alkylating Agents). Primarily to circumvent the polymerization problem, numerous MVK synthetic equivalents have been developed, and these reagents should also be explored as possible sources of the 3-oxobutyl synthon. However, each reagent is handicapped by the fact that there are additional synthetic steps involved in its preparation, and the newly alkylated adduct often must be transformed into the 3-oxobutyl side chain. It should be noted that the manipulation of the side chain to arrive at the keto side chain may require harsh conditions not compatible with the other functional groups present in the molecule.

One of the first reagents developed as an MVK equivalent was 1,3-Dichloro-2-butene (**1**) (the Wichterle reagent).[9] Relative to MVK, it is generally regarded as a superior alkylating agent, but the stringent hydrolytic conditions for conversion of the vinyl chloride to its corresponding ketone limits its application. An excellent alternative reagent of the same type (i.e. allylic halides) is the iodovinylsilane (**2**).[10] The use of this reagent leads to high yields of alkylated adducts, and the subsequent conversion of the alkylated vinylsilanes to their corresponding ketones is achieved by oxidative rearrangement with m-Chloroperbenzoic Acid. These vinylsilanes can be used with enolates and enamines alike.

Robinson himself recognized that MVK polymerization was compromising the efficiency of the annulation protocol and developed the first in a series of compounds which serve as precursors to MVK. The synthetic equivalents (**3**),[11](**4**),[12] and (**5**)[13] are characterized by β-elimination to generate MVK in situ, but often the results are still not satisfactory (albeit improved) over the use of MVK itself. The principal problems involve poor yields and polyalkylated adducts and, if the synthetic sequence is extended further, unusual cyclization products.

The strategy of carbonyl protection in MVK in order to retard base-catalyzed polymerization fostered interest in a group of β-halo acetals, mixed acetals, and thioacetals of the type (**6**). After alkylation with such reagents, the 3-oxo functionality can be easily regenerated under mildly acidic conditions. The haloacetals (**6**; X = Br, I)[14,15] are prepared in two steps by the conjugate addition of the halo acid to MVK followed by acetalization of the β-halo ketone. The corresponding tosylate (**6**; X = OTs)[16] has also been prepared by a similar, albeit longer, route. At first glance, these reagents would seem ideal for appending the 3-oxobutyl side chain, by alkylation followed by mild acidic deprotection of the carbonyl. Unfortunately, in most cases these reagents are poor alkylating agents. A notable exception is the general annulation procedure (eq 1) which employs imine anions (metalloenamines) as the nucleophilic partner and the bromo acetal (**6**; X = Br) as the MVK equivalent. The sequence commences with the Wadsworth–Emmons alkenation of ketones to afford 2-azadienes, which can be isolated at this point but are usually treated with n-Butyllithium to generate the highly reactive metalloenamine. The imine anions thus generated can be trapped with the bromoacetal (**6**; X = Br). Hydrolytic workup followed by aldol cyclization–dehydration completes the preparation of 4,4-disubstituted cyclohexenones in yields of 55–65% from the ketones.[17]

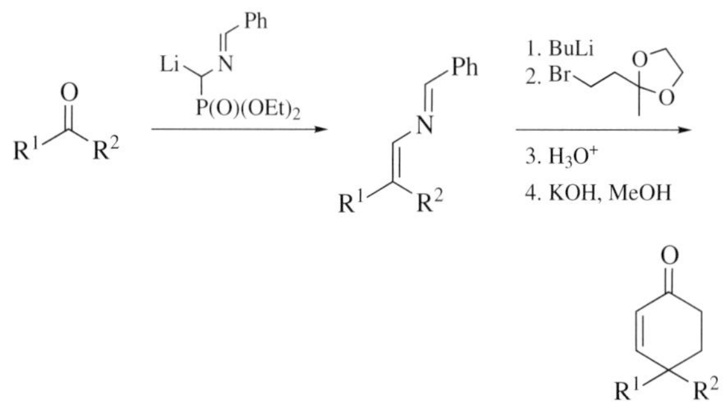

(1)

Another pair of MVK equivalents are the α-trialkylsilyl derivatives of MVK itself (**7**; R = Me, Et).[18,19] The trialkylsilyl group stabilizes the incipient negative charge after Michael addition and provides some steric bulk to impede polymerization. The silyl group can then be removed under basic conditions, usually those employed for a cyclization step. These MVK equivalents are especially important in that they allow effective Michael additions to vinyl ketones under aprotic conditions. The major disadvantage of these reagents is the multistep linear synthesis required for their preparation.

Methyl Vinyl Ketone versus its Synthetic Equivalents (Umpolung Reagents). In those instances in which the 3-oxobutyl side chain is to be connected to pre-existing carbonyl (or other electrophilic) centers, several umpolung-type MVK equivalents have been developed. Although somewhat capricious in its stability, the Grignard reagent (**8**)[20], derived from the bromo acetal (**6**; X = Br) has been successfully employed in a number of synthetic applications. Particularly useful is the fact that the acetal moiety may be unmasked in the usual hydrolytic workup or carried on as an acetal for removal at a later stage in the synthesis.[21] Special mention should be made of the Grignard reagent's use in the presence of catalytic copper salts (**9**).[22] This technique allows for the conjugate addition of the acetal moiety to α,β-unsaturated ketones as well as the direct attachment of the side chain to less complex systems. The related homologous Wittig reagent (**10**)[23] was developed in the late 1960s and the Horner–Emmons variant (**11**) appeared shortly thereafter.[24] Both compounds have been employed as reacting partners with lactones, ultimately leading to cyclohexanone products via an intramolecular Wittig alkenation process.

1,4-Conjugate Additions (Michael Addition Reactions, Achiral). There are numerous examples of the use of MVK in conjugate 1,4-addition reactions, i.e. the Michael reaction. Although most of these applications are framed in the context of an annulation procedure (e.g. the Robinson annulation), many of these reactions are worthy of discussion in their own right. The prototypical Michael addition of an enolate with MVK is, in fact, the first step of the Robinson annulation (eq 2).

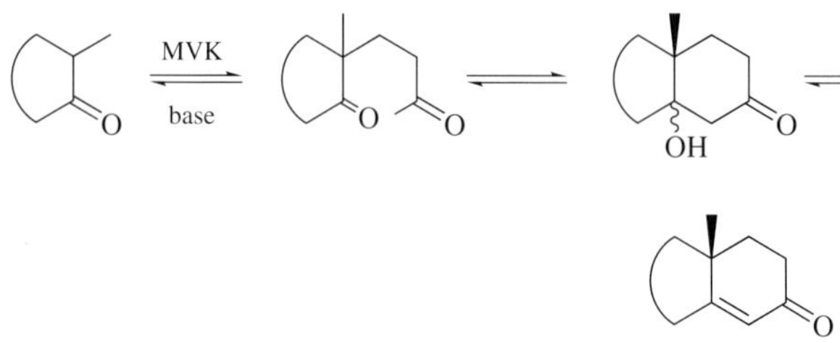

(2)

Under basic conditions (e.g. alkylation reactions employing enolates), MVK suffers a high degree of polymerization, leading to the development of the aforementioned MVK equivalents. However, there are derivatives of ketones which are particularly suitable for Michael addition reactions, most notably the silyl enol ethers.[25] The development of silyl enol ethers addressed the most serious problems encountered in the alkylations of enolates. Firstly, the reactions are conducted in aprotic media under anhydrous conditions, thus avoiding proton transfer side reactions. Secondly, by judicious choice of the base used to generate the intermediate enolate, the regiochemistry of the alkylation can be controlled. However, the strongly basic conditions required for the regeneration of the enolates themselves preclude successful additions to MVK, resulting in the formation of polymer. Only recently have trimethylsilyl enol ethers been efficiently added to MVK with the assistance of Lewis acid catalysis (eqs 3 and 4).[26,27] The acidic conditions not only serve to activate MVK towards Michael addition, but assist in unmasking the enolate as well. These procedures are usually conducted under relatively mild aprotic conditions which also retard MVK polymerization.

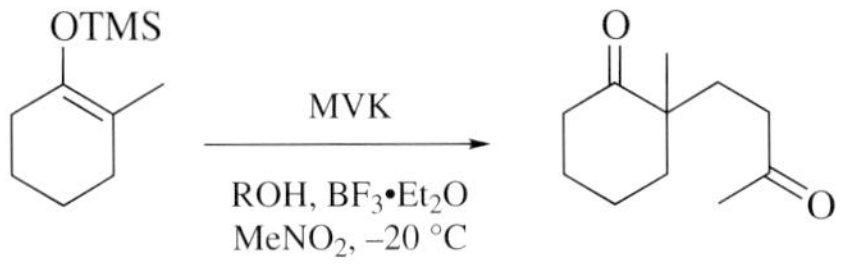

(3)

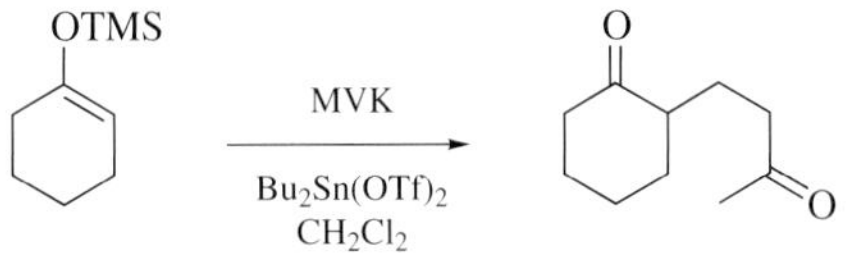

(4)

β-Keto esters, α-nitro ketones, and α-keto sulfones are but a few of the stabilized enolate derivatives which have been used successfully in efficient Michael additions to MVK. One of the most troublesome examples involving β-keto esters was the condensation of cyclohexanone methyl carboxylate with MVK under equilibrating conditions,[28] but this reaction has been developed into a viable synthetic procedure with the use of high pressure techniques under otherwise mild conditions (eq 5).[29]

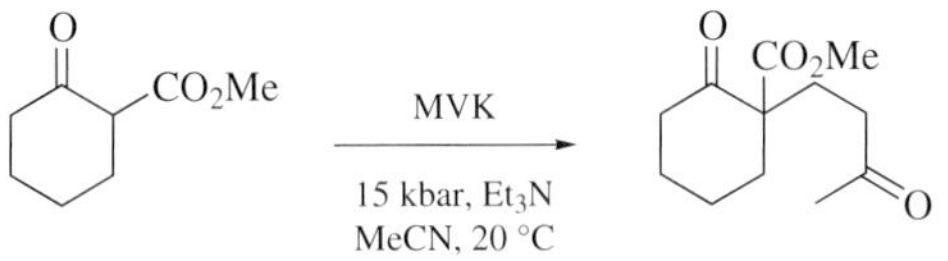

(5)

Another stabilized enolate family includes the α-nitro ketones; although the nitro group activates the formation of the enolate, it is usually necessary to remove it in a subsequent procedure (eq 6).[30] Nevertheless, the overall yields of the products are good.

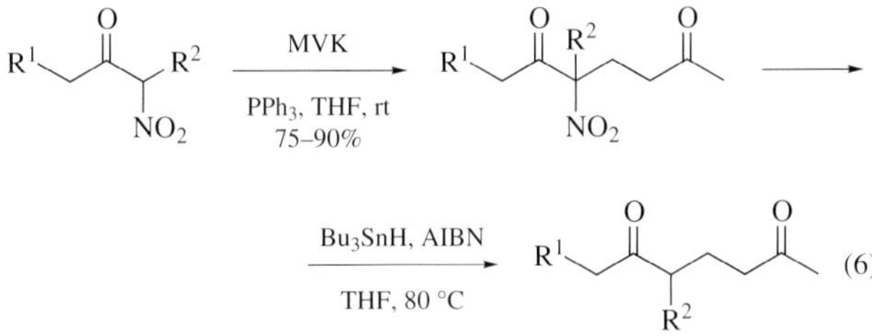

(6)

Simple nitro alkanes have recently been used in an exceptionally mild procedure for the attachment of the 3-oxobutyl side chain using MVK (eq 7). This protocol highlights the use of Alumina at room temperature for the conditions of the Michael addition, and the yields are good to moderate.[31]

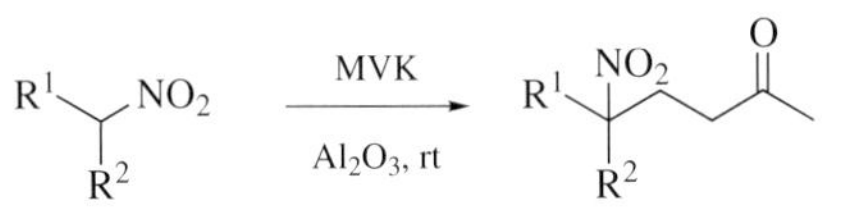

(7)

The Michael addition of α-cyanoamines to MVK (eq 8) has also been used for the preparation of 1,4-dione systems,[32] which are historically difficult to prepare.[33] The reaction occurs under mild conditions; however, the use of Hexamethylphosphoric Triamide as the solvent limits the scale and the widespread use of this procedure. In this sequence, acidic hydrolysis of the alkylated α-cyanoamine leads to the dione system.

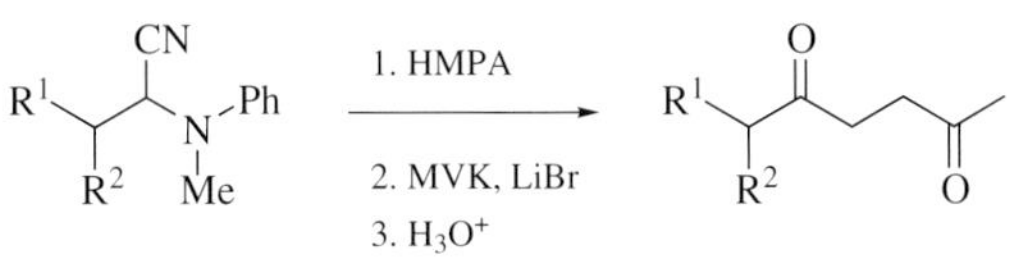

(8)

β-Methylthio allylic systems have also served as the nucleophilic partner in reactions with MVK (eq 9).[34] The procedure employs the 'hyperbasic media'; of the Lithium Diisopropylamide/HMPA complex[35] and generates a quaternary carbon center in the product.

(9)

Recently, dithioenamines have emerged as useful synthetic intermediates in general alkylation methods.[36] In their Michael additions with MVK in the presence of Lewis acid catalysts, an intermediate iminium dithiane is generated which is somewhat sluggishly hydrolyzed; however, the overall yields of alkylated adducts are impressive (eq 10).[37]

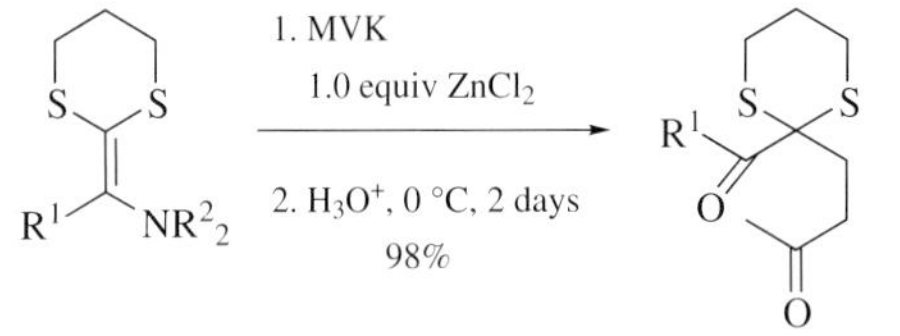

(10)

1,4-Conjugate Additions (Michael Addition Reactions, Chiral) In recent years, perhaps the most significant development in synthetic organic chemistry has been the explosion in the field of asymmetric induction. As the reactive intermediates of the enolate, enamine, and imine anions were developed into their chiral counterparts, not only were standard alkylation reactions studied but also Michael additions to α,β-unsaturated systems. It is not surprising that the prototypical model studies involved MVK as the Michael acceptor. While most of these methods are used in asymmetric ring synthesis (see below), some examples terminate at the alkylation step with MVK. All of these methods feature the chiral auxiliary in the nucleophilic reagent. One of the earliest examples of asymmetric induction using MVK was developed by Yamada and utilizes chiral pyrrolidino enamines (eq 11).[38] Although the yields of the Michael additions are moderate and the asymmetric induction itself is not particularly impressive by today's standards, this methodology laid the groundwork for many improvements which were to follow. A notable advancement in chiral imine alkylations with MVK is found in the method of d'Angelo, in which the conjugate addition is catalyzed by Titanium(IV) Chloride (eq 12).[39] A particularly attractive feature of this method is that the chiral auxiliary, **(S)-α-Methylbenzylamine**, is readily available, and the chemical yields are good with high asymmetric induction.

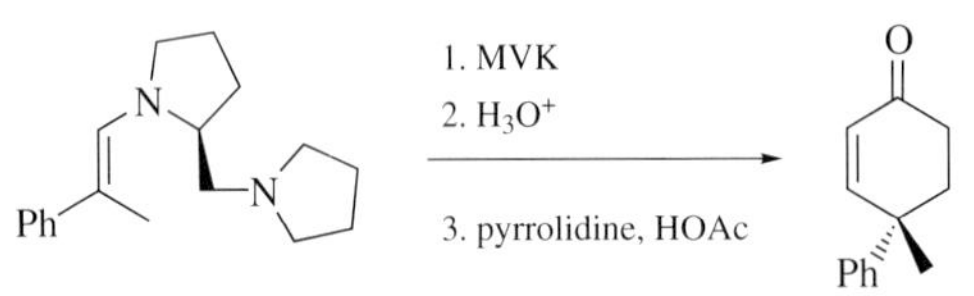

(11)

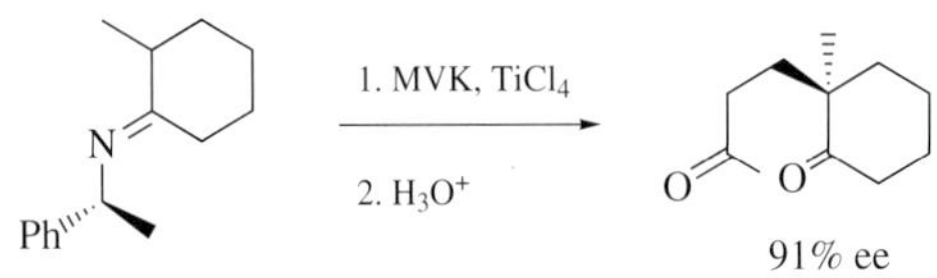

(12)

Another chiral enamine system which has been used with MVK takes advantage of the high nucleophilicity of imine anions (i.e. metalated enamines). In this method the intermediate enolate from the Michael addition is trapped as its TMS enol ether, which is subsequently hydrolyzed in the workup (eq 13).[40] The yields are fair and the asymmetric induction is effective.

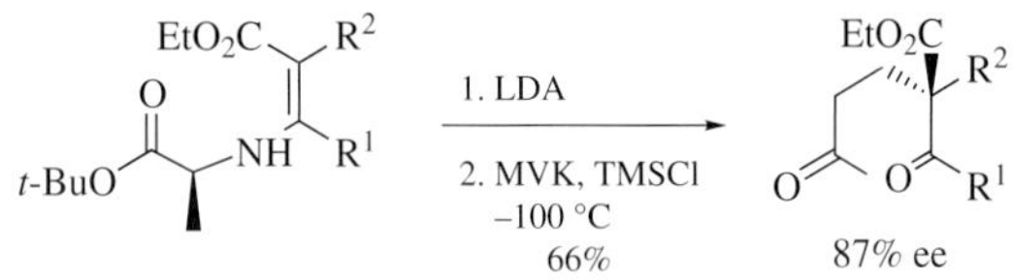

(13)

A chiral reagent within the family of Horner–Emmons phosphonates has been used with MVK, but the results were disappointing in that the major product was the dialkylated adduct (eq 14). In addition, preparations of these reagents are often multistep and tedious.[41]

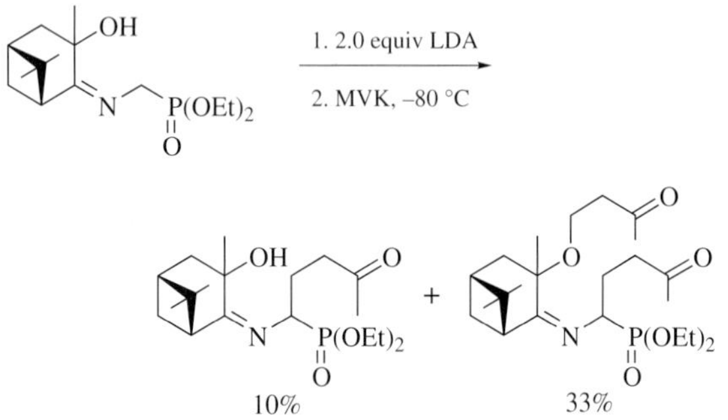

(14)

In a recent synthesis of (+)-*O*-methyljoubertiamine, Taber utilized very mild conditions for the addition of a chiral enolate to MVK in which the chiral auxiliary is the effective naphthyl camphor system (eq 15).[42]

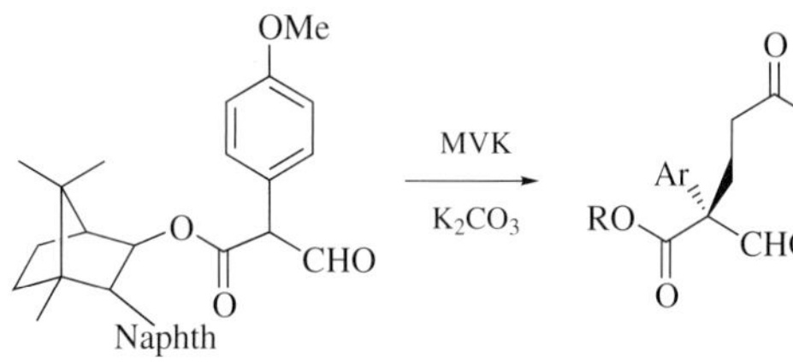

(15)

There are several polymer-based methods which induce some asymmetry to certain substrates. The degree of asymmetric induction is not particularly high (66% ee) and as yet does not appear to be very general. One such example is the cobalt-based diamine polymer which provides modest yields of the Michael adduct (eq 16).[43]

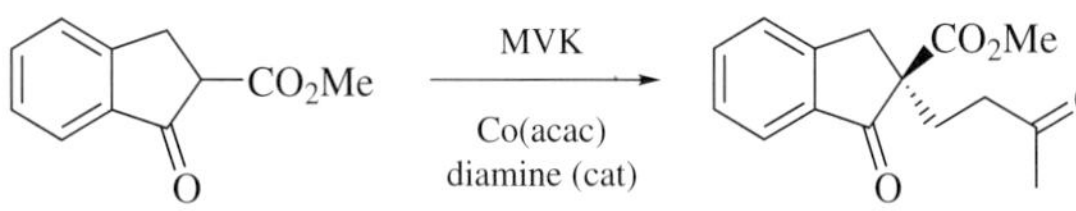

(16)

In terms of high yields and efficiency of asymmetric induction, the Evans procedure using chiral acyl oxazolidones stands apart as a method of choice. In the MVK example the chiral reagent adds, with the assistance of titanium tetrachloride activation at 0 °C, with an impressive 88% chemical yield and 99% ee (eq 17). The oxazolidone can be recycled and the method is general for a variety of α,β-unsaturated systems.[44]

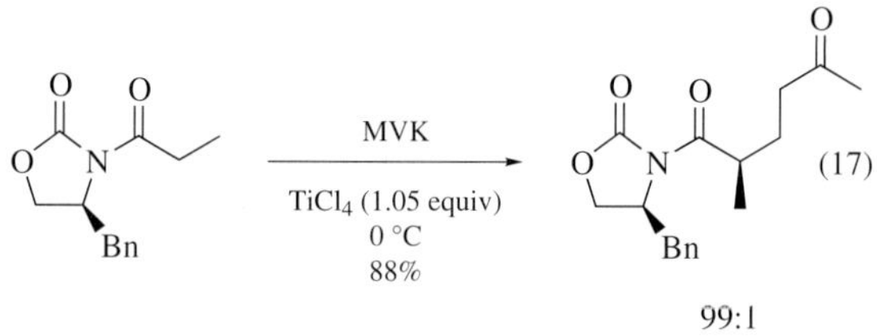

(17)

Alkylation/Cyclization Tandems: The Robinson Annulation The most prevalent use of MVK in modern synthesis is the annulation of a six-membered ring onto a pre-existing ketone: the Robinson annulation (see above). In its original format,[45] the sequence involves the generation of an enolate under basic conditions followed by Michael addition to MVK, thus affording a 1,5-dione. The dione can be subjected to basic conditions which induce aldol cyclization–dehydration, resulting in a newly formed cyclohexenone appended to the starting ketone (eq 2). Since its inception, the sequence has been handicapped by a variety of problems: (1) proton transfer; (2) regiochemistry of Michael addition to MVK; (3) *O*- vs. *C*-alkylation; (4) polyalkylation; (5) poor yields in the conjugate addition step; and (6) base-promoted polymerization of MVK itself. In spite of these drawbacks, the procedure is so synthetically important that development and refinement of Robinson's protocol continues to this very day. Most of the problems which are unique to MVK are often overcome by the use of a synthetic equivalent (see above). This section will deal with the annulation procedures which involve MVK itself and can be grouped into linear vs. spiroannulation sequences. In addition, some of these methods involve asymmetric induction, resulting in chiral cyclohexenones. When chirality is involved, it is usually induced in the alkylation (Michael addition) step, but there are a few examples in which the aldol cyclization of the dione or keto aldehyde features the chiral auxiliary.

The Linear Robinson Annulation This was the earliest class of Robinson annulation procedure and represents the majority of examples. When simple ketones are used in this sequence with MVK, the results are often unsatisfactory. However, when acidic dicarbonyl compounds are employed, the results are quite viable and practical. For example, 1,3-diones may be treated with a catalytic amount of base to generate a stable enolate which is subsequently trapped with MVK. The alkylated adduct is then cyclized–dehydrated under standard Knoevenagel conditions (eq 18). The acidic conditions of the cyclization help retard reversible β-elimination of MVK.[46]

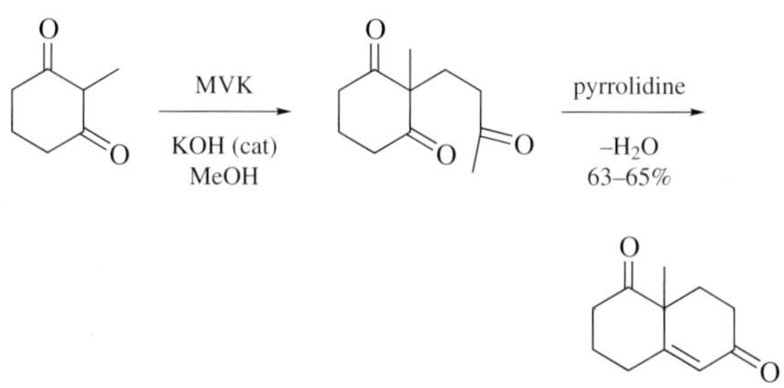

(18)

In a classic case of stereochemical control, chiral auxiliaries were not involved, but rather a simple reversal of order in alkylations of enolates resulting in the preparation of isomeric decalenones (eq 19).[47] Although this is a dialkylation procedure leading to a specific product, the overall strategy of stereochemical control in the Robinson annulation has been extensively studied and discussed.[48]

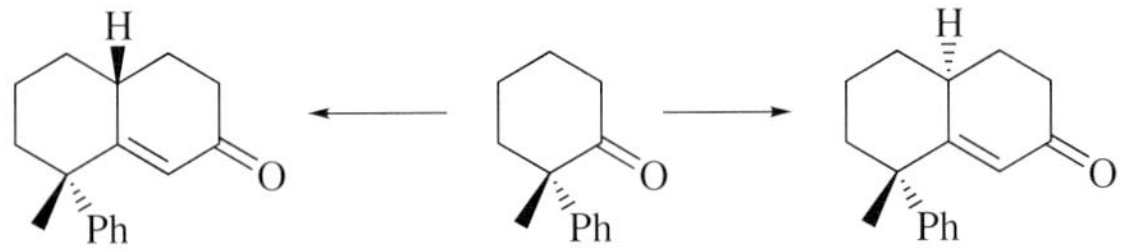

(19)

With the advent of kinetic, sterically hindered bases (such as LDA), which could be used under aprotic conditions, the regiochemistry of enolates could be controlled. In the following sequence, a slight excess of the ketone allows the Michael addition to occur at the most hindered site via the thermodynamic enolate, thus resulting in a quaternary carbon center with stereochemical control. Aldol cyclization–dehydration then affords the cyclohexenone in 60% yield (eq 20).[49]

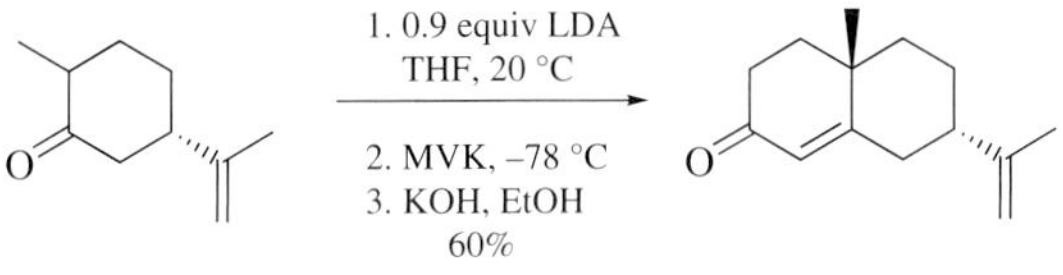

(20)

A highly successful approach to Robinson annulation involves the use of trimethylsilyl enol ethers with MVK under Lewis acid conditions, which thwarts MVK polymerization. The yields of the Michael addition itself are greatly improved (see above), and the resulting diones can be cyclized–dehydrated under standard aldol (basic) conditions. This particular example affords the decalenones in 89% overall yield from MVK (eq 21).[27]

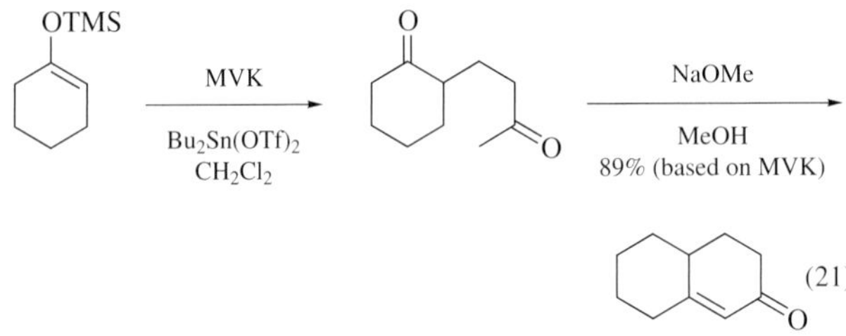

(21)

The Chiral Robinson Annulation Perhaps the most important extensions involving the Robinson annulation protocol have concerned the production of chiral cyclohexenones. In general, the asymmetric induction takes place at the Michael addition step with a chiral reactive intermediate and MVK, although there are a few examples in which the aldol cyclization–dehydration step affords the stereochemical control. However, great care must be exercised in the cyclization step in order to avoid epimerization at the newly created chiral center. One of the earliest methods for chiral Robinson annulation is a general procedure in which a chiral Pyrrolidine enamine is condensed with MVK; the resulting keto imines are then hydrolyzed to give the chiral δ-keto aldehydes, which in turn are cyclized under (acidic) Knoevenagel conditions, finally affording chiral 4,4-disubstituted cyclohexenones (eq 22).

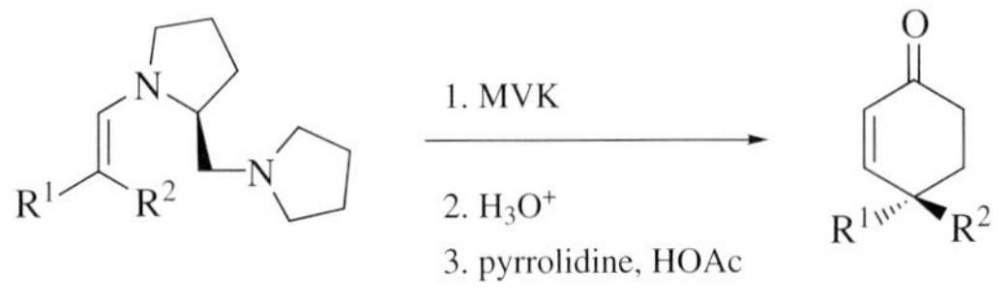

(22)

This methodology was highlighted in the chiral total synthesis of the *Sceletium* alkaloid (+)-mesembrine.[50] More recently, an even milder, shorter sequence was used in the key step for the synthesis of the *Sceletium* alkaloid (+)-*O*-methyljoubertiamine. In this case, the chiral auxiliary was the naphthylcamphor system (see above), which governed the condensation of the enolate with MVK using mild Potassium Carbonate as the base (eq 23).[42]

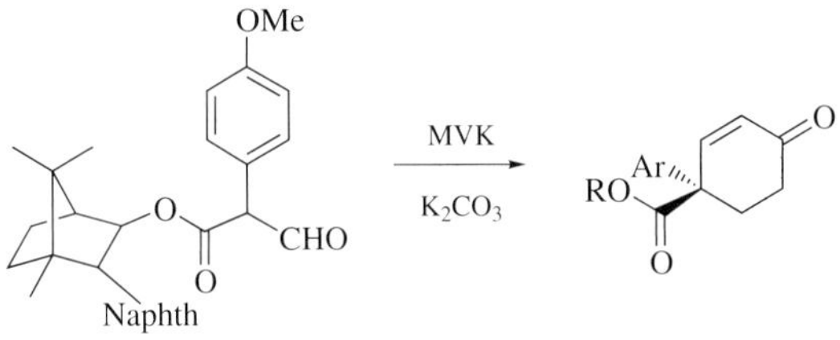

(23)

The use of chiral imines has recently received attention as reactive intermediates for Michael additions to MVK. Their advantage is that they are derived from α-methylbenzylamine, which is affordable and can be recycled. The Michael addition itself is activated with titanium tetrachloride and gives good yields (61%) of the Michael adduct with high ee (91%). Hydrolysis concomitant with cyclization provides the decalenone (eq 24).[51]

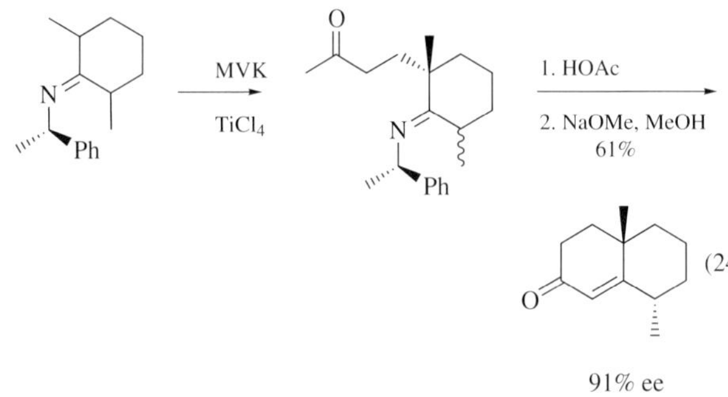

(24)

A very useful procedure for inducing chirality in the cyclization step was reported some years ago but has recently been optimized for scale-up. 2-Methyl-1,3-cyclopentanedione serves as the reactive enolate system for the Michael addition with MVK, providing the trione in quantitative yield. The chiral cyclization is a modified Knoevenagel reaction using (S)-Proline as the asymmetric catalyst. This sequence provides the enedione in 93% optical purity (eq 25).[52]

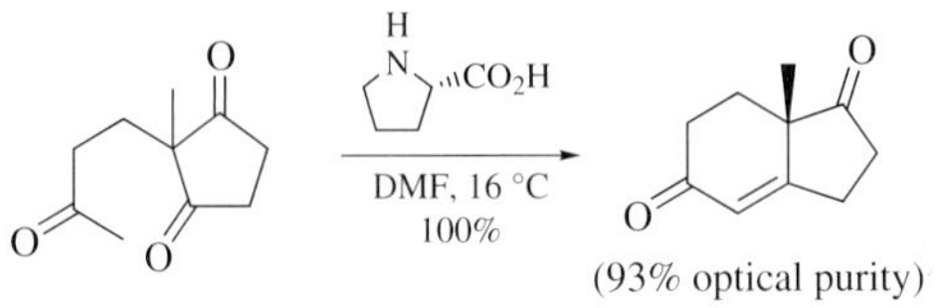

(25)

The Spirocyclic Robinson Annulation Procedures which attach two carbocyclic rings to one another by a single carbon atom (spiroannulation)[53] are not commonplace. The major synthetic problem to overcome in such an operation is the construction of the common, quaternary carbon center. Although there are a number of ingenious ways to construct quaternary carbon centers,[54] those involving spiroannulations with MVK are generally limited to the reactions of trisubstituted enamines with MVK as the conjugate addition step of the sequence (eq 26).[55] The starting carbonyl compounds are usually aldehydes, but there is also a highly efficient homologation procedure which commences from ketones.[56] In either case, the intermediate at the alkylation stage is a δ-keto imine which is subjected to mild acid hydrolysis, thus generating a δ-keto aldehyde, which is then cyclized–dehydrated with base. Enamines from Piperidine and Morpholine have been used successfully in this procedure. There is flexibility in the carbonyl starting materials as well: five-, six-, seven-, and eight-membered carbonyl systems have all been satisfactory substrates in this methodology. The spiroannulation procedure of MVK with enamines is summarized below (Table 2).

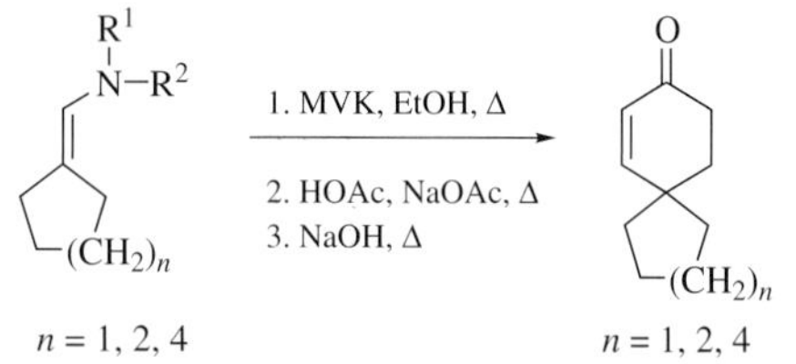

(26)

Table 2 Spiroannulation procedures involving enamines and methyl vinyl ketone

Ketone enamine	Yield of spirocycle (%)	Ref.	
Cyclopentane-carbaldehyde	Piperidino	59	55c
Isopropylidene-cyclopentane-carbaldehyde	Piperidino	45–65	55a
Cyclohexane-carbaldehyde	Piperidino	45–65	55b,c
Cyclohexanone	Morpholino	41	56
2-Methylcyclo-hexanone	Morpholino	36	56
4-*t*-Butylcyclo-hexanone	Morpholino	41	56
Isophorone	Morpholino	30	56
2-Norbornanone	Morpholino	41	56
Cycloheptane-carbaldehyde	Piperidino	45–65	55b,c
Cyclooctane-carbaldehyde	Piperidino	45–65	55b,c

First Update

Kohei Inomata
Tohoku Pharmaceutical University, Sendai, Japan

Morita–Baylis–Hillman Reaction Methyl vinyl ketone is a good reagent for the Morita–Baylis–Hillman reaction,[57,58] which is one of the most important reactions to build a new C–C bond on α-vinyl carbon in methyl vinyl ketone and to afford α-subtituted methyl vinyl ketones.[59,65] The reaction is first initiated by a tertiary amine such as 1,4-diazabicyclo[2.2.2]octane (DABCO) to generate a zwitterionic enolate, which can be easily trapped by electrophiles such as aldehydes, ketones, α,β-unsaturated compounds, alkyl halides, and imines (eq 27). Phosphines and calcogens instead of tertiary amines can also be used as a catalyst for the reaction.[66] For an example, the Morita–Baylis–Hillman reaction of methyl vinyl ketone and benzaldehyde in the presence of DABCO affords the phenylhydroxymethyl derivative in 51% yield (eq 28).[60]

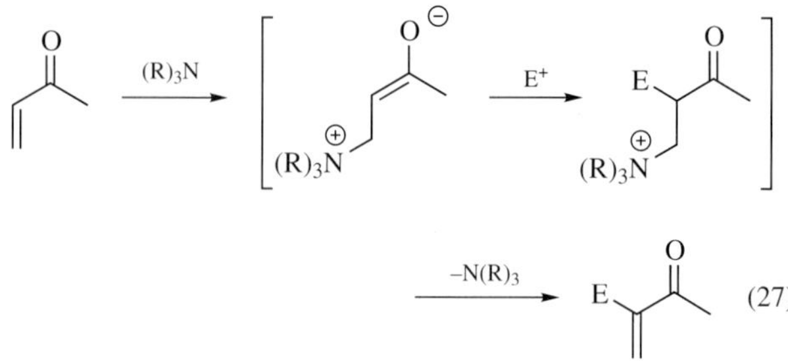

(27)

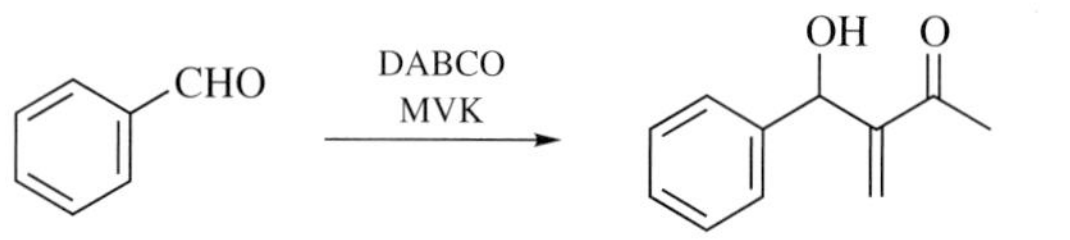

(28)

Asymmetric Morita–Baylis–Hillman reactions using chiral amines and chiral phosphines as the catalyst are also available.[67–69] The most successful recent one involves an aza-type reaction mediated by a chiral phosphine catalyst bearing a binaphthyl moiety (eq 29).[70,71] Although the Morita–Baylis–Hillman reaction usually requires long reaction timeis. This reaction is complete in only 5 min when using proazaphosphatrane as a catalyst.[72]

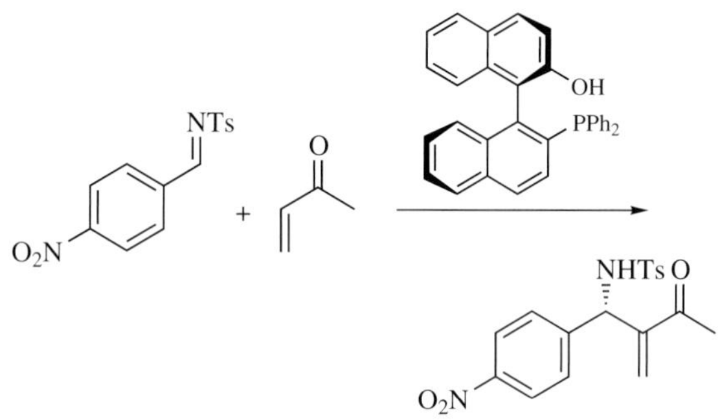

(29)

(96% yield, 95% ee)

Catalytic Asymmetric 1,4-Addition Methy vinyl ketone is one of the most important reagents in the area of catalytic asymmetric 1,4-addition reaction.[73–77] Enantioselective conjugate addition of 2-cyanopropionate to methyl vinyl ketone can be achieved with rhodium complexes containin binaphthalene ligands (eq 30).[78]

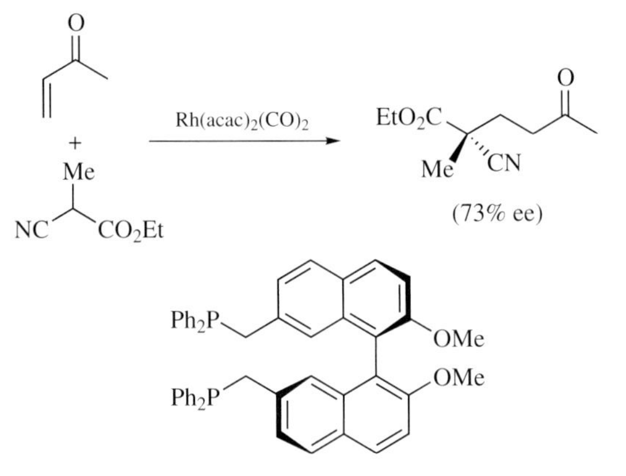

(30)

Recently, highly enantioselective catalytic 1,4-addition of 2-hydroxy-2′-methoxypropiophenone to methyl vinyl ketone mediated by diethylzinc and linked BINOL complex has been achieved to construct a chiral quaternary carbon (eq 31).[79]

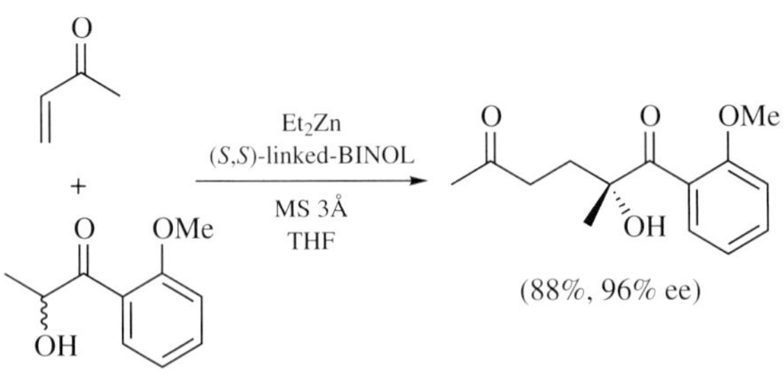

(31)

Catalytic asymmetric Michael reactions mediated by chiral alkaloids such as quinine, quinidine, and cinchonine derivatives have been widely developed.[80–82] Michael addition of indenone derivatives with methyl vinyl ketone in the presence of a cinchonine derivative afford the corresponding Michael adducts in 95% yield with 80% ee (eq 32).[83]

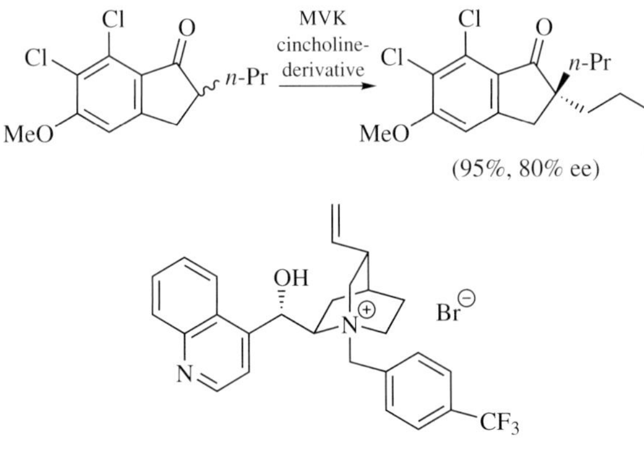

(32)

Recently, the catalytic asymmetric Michael addition of 2-carboxycyclopentanone to methyl vinyl ketone using a chiral phase transfer catalyst bearing spiro binaphthyl units has succeeded to deliver the adduct in 92% yield with 90% ee (eq 33).[83]

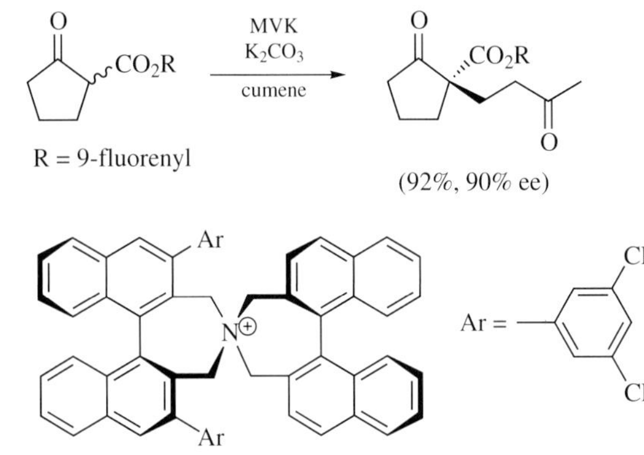

(33)

Organocatalysis mediated cyclopropanation of methyl vinyl ketone has also been reported.[84] Thus, α-chloroacetophenone and methyl vinyl ketone in the presence of catalytic amounts of DABCO and Na_2CO_3 in acetonitrile affords diastereoselectively a *trans*-cyclopropanation product in 82% yield (eq 34).

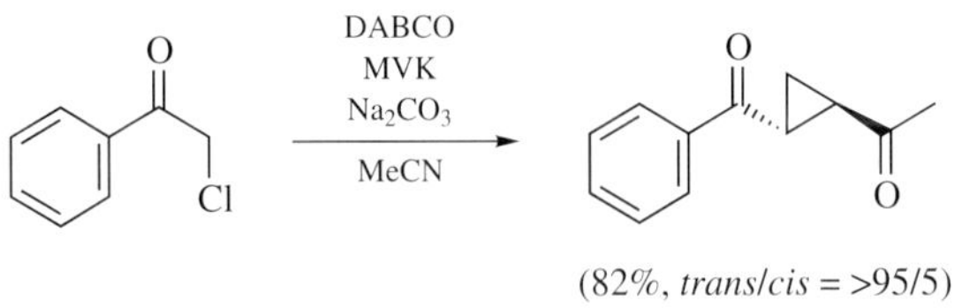

(34)

Metathesis Methyl vinyl ketone can be reactive with Grubbs catalyst as trapping reagent of ruthenium carbene intermediate. For an example, ene-yne ring closing metathesis in the presence of methyl vinyl ketone has been developed to construct a medium ring bearing a methyl butenone moiety (eq 35).[85]

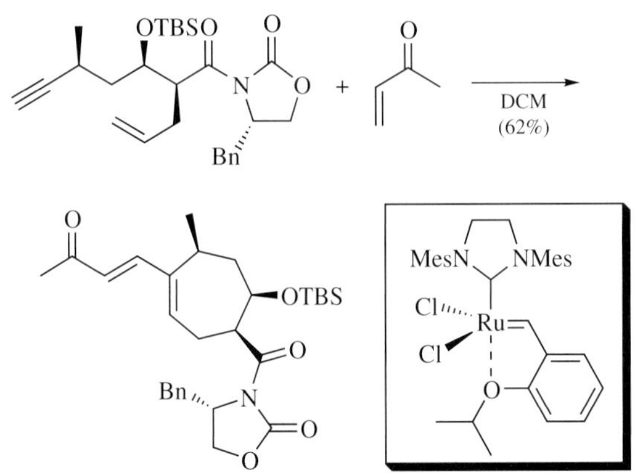

(35)

Heck Reaction Methyl vinyl ketone can be a reagent in an alkenylation reaction such as the Heck reaction.[86] The catalytic asymmetric domino Wacker–Heck reaction has been used as a key reaction in the total synthesis of vitamine E (eq 36).[87] In this reaction, methyl vinyl ketone plays the role of an acceptor of a palladated intermediate to attach the butenone.

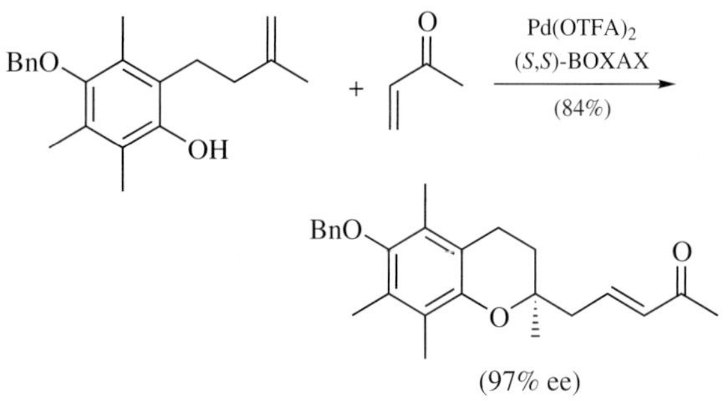

(36)

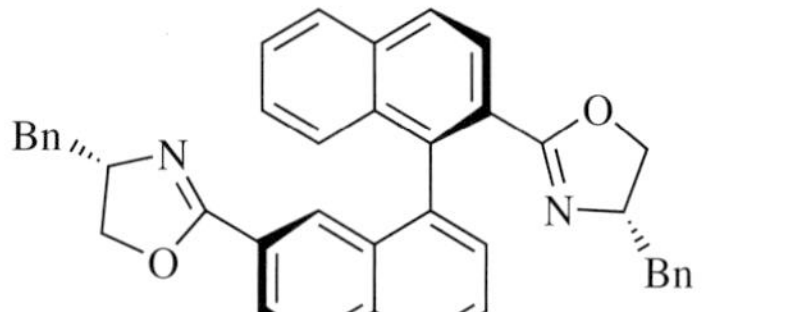

BOXAX: 2,2′-bis(oxazoyl)-1,1′-binaphthyl

The Heck-type reaction between a borate and MVK is also reported to construct a highly conjugated enone moiety (eq 37).[88]

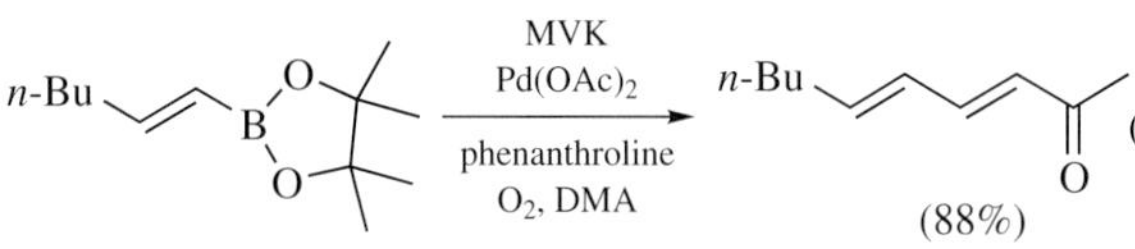

(37)

Cycloaddition Reaction as a Precursor of Siloxy Diene Methyl vinyl ketone is one of the most important precursors of 2-siloxy-1,3-butadiene in cycloaddition reactions. Typically, *O*-silylation reaction of methyl vinyl ketone in the presence of a base such as triethylamine or lithium diisopropylamide, and sililated reagent such as TMSCl, TBSOTf, and TIPSOTf can be used to generate the diene bearing a corresponding siloxy substituent.[89–91] The generated diene is reactive toward dienophiles including methyl vinyl ketone itself to construct a cyclohexanone derivative in Diels–Alder reaction (or double Michael reaction, eq 38).[92]

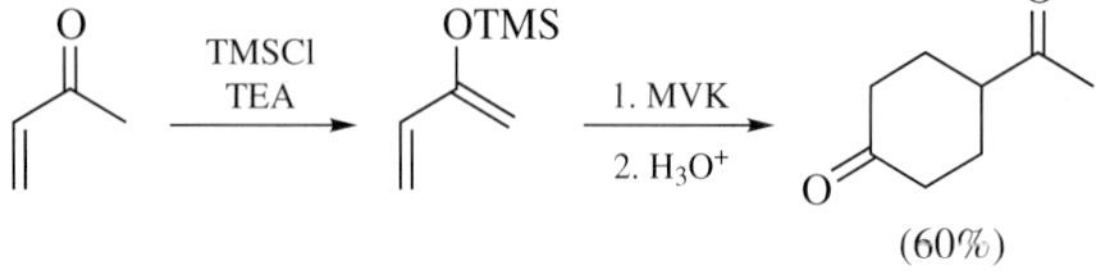

(38)

Catalytic regio- and enantioselective Diels–Alder reaction between a siloxydiene, generated from methyl vinyl ketone, and an unsymmetrical quinone has also been reported (eq 39).[93–95]

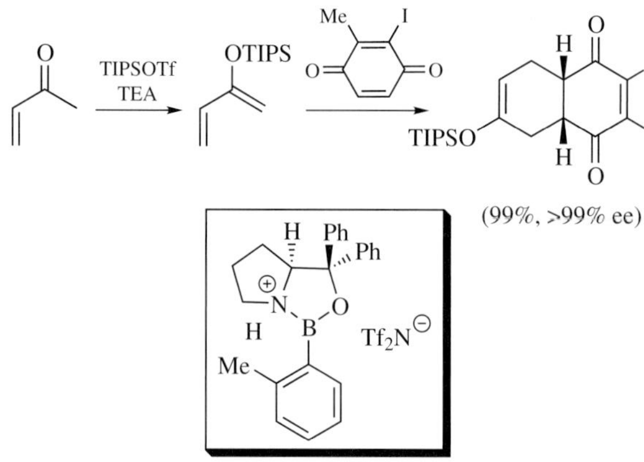

(39)

2-*tert*-Butyldimethylsilyloxy-1,3-butadiene[96] is also available from methyl vinyl ketone. Obtained diene readily reacts with methyl acrylate in the presence of a Lewis acid to afford a cyclobutane derivative via tandem [4+2]/[2+2] reaction (eq 40).[97]

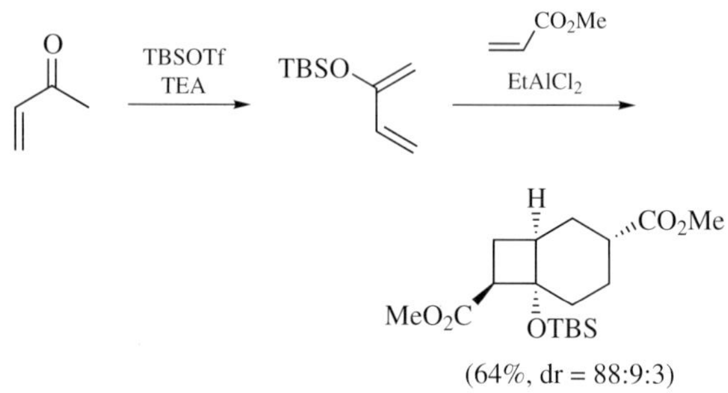

(40)

1. *Fieser & Fieser* **1967**, *1*, 697; **1969**, *2*, 283; **1975**, *5*, 464; **1977**, *6*, 407; **1979**, *7*, 247; **1982**, *10*, 272; **1986**, *12*, 329.
2. Review: Bergmann, E. D.; Ginsburg, D.; Pappo, R., *Org. React.* **1959**, *10*, 179.
3. Review: Jung, M. E., *Tetrahedron* **1976**, *32*, 3.
4. Review: Gawley, R. E., *Synthesis* **1976**, 777.
5. Nicholson, J. W. *The Chemistry of Polymers*; Royal Society of Chemistry: Cambridge, **1991**.
6. House, H. O.; Czuba, L. J.; Gall, M.; Olmstead, H. D., *J. Org. Chem.* **1969**, *34*, 2324.
7. (a) Stork, G.; Brizzolara, A.; Landesman, H.; Szmuszkovicz, J.; Terrell, R., *J. Am. Chem. Soc.* **1963**, *85*, 207. (b) Heathcock, C. H.; Ellis, J. E.; McMurry, J. E.; Coppolino, A., *Tetrahedron Lett.* **1971**, 4995.
8. Martin, S. F.; Phillips, G. W.; Puckette, T. A.; Colapret, J. A., *J. Am. Chem. Soc.* **1980**, *102*, 5866 and references therein.
9. (a) Wichterle, O.; Procházka, J.; Hofman, J., *Collect. Czech. Chem. Commun.* **1948**, *13*, 300. (b) Prelog, V.; Barman, P.; Zimmermann, M., *Helv. Chim. Acta* **1949**, *32*, 1284. (c) Julia, M., *Bull. Soc. Chim. Belg.* **1954**, *21*, 780. (d) Marshall, J. A.; Schaeffer, D. J., *J. Org. Chem.* **1965**, *30*, 3642. (e) House, H. O. *Modern Synthetic Reactions*, 2nd ed.; Benjamin: Menlo Park, CA, **1972**.
10. (a) Stork, G.; Jung, M. E., *J. Am. Chem. Soc.* **1974**, *96*, 3682. (b) Stork, G.; Jung, M. E.; Colvin, E.; Noel, Y., *J. Am. Chem. Soc.* **1974**, *96*, 3684.
11. Balasubramanian, K.; John, J. P.; Swaminathan, S., *Synthesis* **1974**, 51.
12. (a) Cornforth, J. W.; Robinson, R., *J. Chem. Soc* **1949**, 1855. (b) McQuillin, F. J.; Robinson, R., *J. Chem. Soc* **1938**, 1097.
13. (a) Taylor, D. A. H., *J. Chem. Soc* **1961**, 3319. (b) Pinder, A. R.; Williams, R. A., *J. Chem. Soc* **1963**, 2773. (c) Halsall, T. G.; Theobald, D. W.; Walshaw, K. B., *J. Chem. Soc* **1964**, 1029. (d) Theobald, D. W., *Tetrahedron* **1966**, *22*, 2869.

14. (a) Stork, G., *Pure Appl. Chem.* **1964**, *9*, 131. (b) Stork, G.; Borch, R., *J. Am. Chem. Soc.* **1964**, *86*, 935. (c) Brown, E.; Dahl, R., *Bull. Soc. Claim. Fr., Part 2* **1972**, 4292. (d) Sato, T.; Kawara, T.; Sakata, K.; Fujisawa, T., *Bull. Chem. Soc. Jpn.* **1981**, *54*, 505. (e) Stowell, J. C.; Keith, D. R.; King, B. T., *Org. Synth., Coll. Vol.* **1990**, *7*, 59.

15. (a) Stowell, J. C.; King, B. T.; Hauck, H. F., Jr., *J. Org. Chem.* **1983**, *48*, 5381. (b) Murai, A.; Ono, M.; Masamune, T., *J. Chem. Soc.* **1977**, 573. (c) Crombie, L.; Tuchinda, P.; Powell, M. J., *J. Chem. Soc., Perkin Trans. 1* **1982**, 1477. (d) Trost, B. M.; Kunz, R. A., *J. Org. Chem.* **1974**, *39*, 2475. (e) Trost, B. M.; Conway, W. P.; Strege, P. E.; Dietsche, T. J., *J. Am. Chem. Soc.* **1974**, *96*, 7165. (f) Trost, B. M.; Bridges, A. J., *J. Org. Chem.* **1975**, *40*, 2014. (g) Solas, D.; Wolinsky, J., *J. Org. Chem.* **1983**, *48*, 670. (h) Trost, B. M.; Kunz, R. A., *J. Am. Chem. Soc.* **1975**, *97*, 7152. (i) Kametani, T.; Suzuki, Y.; Furuyama, H.; Honda, T., *J. Org. Chem.* **1983**, *48*, 31.

16. (a) Yardley, J. P.; Rees, R. W.; Smith, H., *J. Med. Chem.* **1967**, *10*, 1088. (b) Danishefsky, S.; Cavanaugh, R., *J. Am. Chem. Soc.* **1968**, *90*, 520.

17. Martin, S. F.; Phillips, G. W.; Puckette, T. A.; Colapret, J. A., *J. Am. Chem. Soc.* **1980**, *102*, 5866.

18. (a) Stork, G.; Singh, J., *J. Am. Chem. Soc.* **1974**, *96*, 6181. (b) Boeckman, R. K., Jr., *J. Am. Chem. Soc.* **1973**, *95*, 6867. (c) Boeckman, R. K., Jr., *J. Am. Chem. Soc.* **1974**, *96*, 6179. (d) Boeckman, R. K., Jr.; Blum, D. M.; Ganem, B.; Halvey, N., *Org. Synth., Coll. Vol.* **1988**, *6*, 1033.

19. Stork, G.; Ganem, B., *J. Am. Chem. Soc.* **1973**, *95*, 6152.

20. (a) Ponaras, A. A., *Tetrahedron Lett.* **1976**, 3105. (b) Büchi, G.; Wüest, H., *J. Org. Chem.* **1969**, *34*, 1122.

21. Martin, S. F.; Puckette, T. A.; Colapret, J. A., *J. Org. Chem.* **1979**, *44*, 3391.

22. (a) Snider, B. B.; Cartaya-Marin, C. P., *J. Org. Chem.* **1984**, *49*, 153. (b) Fujisawa, T.; Sato, T.; Kawara, T.; Noda, A., *Tetrahedron Lett.* **1982**, *23*, 3193. (c) Gras, J.-L., *J. Org. Chem.* **1981**, *46*, 3738. (d) Paquette, L. A.; Galemmo, R. A., Jr.; Caille, J.-C.; Valpey, R. S., *J. Org. Chem.* **1986**, *51*, 686.

23. Henrick, C. A.; Böhme, E.; Edwards, J. A.; Fried, J. H., *J. Am. Chem. Soc.* **1968**, *90*, 5926.

24. Sturtz, G., *Bull. Soc. Claim. Fr., Part 2* **1964**, 2340.

25. Review: Kuwajima, I.; Nakamura, E., *Acc. Chem. Res.* **1985**, *18*, 181.

26. Duhamel, P.; Hennequin, L.; Poirier, N.; Poirier, J.-M., *Tetrahedron Lett.* **1985**, *26*, 6201.

27. (a) Sato, T.; Wakahara, Y.; Otera, J.; Nozaki, H., *Tetrahedron* **1991**, *47*, 9773. (b) Sato, T.; Wakahara, Y.; Otera, J.; Nozaki, H., *J. Am. Chem. Soc.* **1992**, *113*, 4028.

28. (a) Metzger, J. D.; Baker, M. W.; Morris, R. J., *J. Org. Chem.* **1972**, *37*, 789. (b) Marshall, J. A.; Warne, T. M., Jr., *J. Org. Chem.* **1971**, *36*, 178.

29. Dauben, W. G.; Bunce, R. A., *J. Org. Chem.* **1983**, *48*, 4642.

30. Ono, N.; Miyake, H.; Kaji, A., *J. Chem. Soc.* **1983**, 875.

31. (a) Rosini, G.; Marotta, E.; Ballini, R.; Petrini, M., *Synthesis* **1986**, 237. (b) Ballini, R.; Petrini, M.; Rosini, G., *Synthesis* **1987**, 711. (c) Ballini, R.; Petrini, M.; Marcantoni, E.; Rosini, G., *Synthesis* **1988**, 231. (d) Bergbreiter, D. E.; LaLonde, J. J., *J. Org. Chem.* **1987**, *52*, 1601.

32. (a) Ahlbrecht, H.; Kompter, H.-M., *Synthesis* **1983**, 645. (b) Ivanov, I. C.; Sulay, P. B.; Dantchev, D. K., *Justus Liebigs Ann. Chem./Liebigs Ann. Chem.* **1983**, 753.

33. For a review see: Miyakoshi, T., *Org. Prep. Proced. Int.* **1989**, *21*, 661.

34. Kende, A. S.; Constantinides, D.; Lee, S. J.; Liebeskind, L., *Tetrahedron Lett.* **1975**, 405.

35. Review: Mordini, A. In *Advances in Carbanion Chemistry*; Snieckus, V., Ed.; J A I: Greenwich, CT, **1992**; Vol. *1*.

36. Page, P. C. B.; Harkin, S. A.; Marchington, A. P.; van Niel, M. B., *Tetrahedron* **1989**, *45*, 3819.

37. Page, P. C. B.; Harkin, S. A.; Marchington, A. P., *Synth. Commun.* **1989**, *19*, 1655.

38. Sone, T.; Terashima, S.; Yamada, S., *Chem. Pharm. Bull.* **1976**, *24*, 1273.

39. (a) Pfau, M.; Revial, G.; Guingant, A.; d'Angelo, J., *J. Am. Chem. Soc.* **1985**, *107*, 273. (b) Hickmott, P. W.; Rae, B., *Tetrahedron Lett.* **1985**, *26*, 2577. (c) Fourtinon, M.; De Jeso, B.; Pommier, J.-C., *J. Organomet. Chem.* **1985**, *289*, 239.

40. Tamioka, K.; Seo, W.; Ando, K.; Koga, K., *Tetrahedron Lett.* **1987**, *28*, 6637.

41. El Achqar, A.; Boumzebra, M.; Roumestant, M.-L.; Viallefont, P., *Tetrahedron* **1988**, *44*, 5319.

42. Taber, D. F.; Mack, J. F.; Reingold, A. L.; Geib, S. J., *J. Org. Chem.* **1989**, *54*, 3831.

43. (a) Cram, D. J.; Sogah, G. D. Y., *J. Chem. Soc.* **1981**, 625. (b) Brunner, H.; Hammer, B., *Angew. Chem., Int. Ed. Engl.* **1984**, *23*, 312. (c) Li, T.-T.; Wu, Y.-L., *Tetrahedron Lett.* **1988**, *29*, 4039. (d) Polymer based: Hodge, P.; Khoshdel, E.; Waterhouse, J., *J. Chem. Soc., Perkin Trans. 1* **1983**, 2205.

44. Evans, D. A.; Bilodeau, M. T.; Somers, T. C.; Clardy, J.; Cherry, D.; Kato, Y., *J. Org. Chem.* **1991**, *56*, 5750.

45. (a) Rapson, W. S.; Robinson, R., *J. Chem. Soc* **1935**, 1285. (b) Du Feu, E. C.; McQuillin, F. J.; Robinson, R., *J. Chem. Soc* **1937**, 53.

46. (a) Ramachandran, S.; Newman, M. S., *Org. Synth.* **1961**, *41*, 38. (b) Ramachandran, S.; Newman, M. S., *Org. Synth., Coll. Vol.* **1973**, *5*, 486. (c) Mekler, A. B.; Ramachandran, S.; Swaminathan, S.; Newman, M. S., *Org. Synth.* **1961**, *41*, 56. (d) Meckler, A. B.; Ramachandran, S.; Swaminathan, S.; Newman, M. S., *Org. Synth., Coll. Vol.* **1973**, *5*, 743. (e) Newman, M. S.; Mekler, A. B., *J. Am. Chem. Soc.* **1960**, *82*, 4039.

47. Ireland, R. E.; Kierstead, R. C., *J. Org. Chem.* **1966**, *31*, 2543.

48. (a) See ref. **9**(e). (b) Conia, J.-M., *Rec. Chem. Prog.* **1963**, *24*, 43. (c) House, H. O., *Rec. Chem. Prog.* **1967**, *28*, 99.

49. (a) Huffman, J. W.; Rowe, C. D.; Matthews, F. J., *J. Org. Chem.* **1982**, *47*, 1438. (b) Ziegler, F. E.; Wang, K. J., *J. Org. Chem.* **1983**, *48*, 3349. (c) Pariza, R. J.; Fuchs, P. L., *J. Org. Chem.* **1983**, *48*, 2306. (d) Chen, E. Y., *Synth. Commun.* **1983**, *13*, 927. (e) Dauben, W. G.; Bunce, R. A., *J. Org. Chem.* **1983**, *48*, 4642.

50. Yamada, S.; Otani, G., *Tetrahedron Lett.* **1971**, 1133.

51. (a) Pfau, M.; Revial, G.; Guingant, A.; d'Angelo, J., *J. Am. Chem. Soc.* **1985**, *107*, 273. (b) Volpe, T.; Revial, G.; Pfau, M.; d'Angelo, J., *Tetrahedron Lett.* **1987**, *28*, 2367. (c) Revial, G., *Tetrahedron Lett.* **1989**, *30*, 4121.

52. Hajos, Z. G.; Parrish, D. R., *Org. Synth., Coll. Vol.* **1990**, *7*, 363.

53. Review: Krapcho, A. P., *Synthesis* **1974**, 383.

54. Review: Martin, S. F., *Tetrahedron* **1979**, *36*, 419.

55. (a) Hutchins, R. O.; Natale, N. R.; Taffer, I. M.; Zipkin, R., *Synth. Commun.* **1984**, *14*, 445. (b) Kane, V. V., *Synth. Commun.* **1976**, *6*, 237. (c) Kane, V. V.; Jones, M., Jr., *Org. Synth.* **1983**, *61*, 129. (d) Kane, V. V.; Jones, M., Jr., *Org. Synth., Coll. Vol.* **1990**, *7*, 473.

56. (a) Martin, S. F., *J. Org. Chem.* **1976**, *41*, 3337. (b) Martin, S. F.; Gompper, R., *J. Org. Chem.* **1974**, *39*, 2814.

57. For a review of Morita–Baylis–Hillmann reaction: Basavaiah, D.; Rao, P. D.; Hyma, R. S., *Tetrahedron* **1996**, *52*, 8001.

58. For a review of Morita-Baylis-Hillmann reaction: Basavaiah, D.; Rao, A. J.; Satyanarayana, T., *Chem. Rev.* **2003**, *103*, 811.

59. Amri, H.; Villieras, J., *Tetrahedron Lett.* **1986**, *27*, 4307.

60. Basavaiah, D.; Gowriswari, V. V. L., *Tetrahedron Lett.* **1986**, *27*, 2031.

61. Basavaiah, D.; Bharathi, T. K.; Gowriswari, V. V. L., *Synth. Commun.* **1987**, *17*, 1893.

62. Hill, J. S.; Issacs, N. S., *Tetrahedron Lett.* **1986**, *27*, 5007.

63. Basavaiah, D.; Gowriswari, V. V. L.; Bharathi, T. K., *Tetrahedron Lett.* **1987**, *28*, 4591.

64. Drewes, S. E.; Emsile, N. D.; Karodia, N., *Synth. Commun.* **1990**, *20*, 1915.

65. Basavaiah, D.; Shrada, D. S.; Kumaragurubaran, N.; Reddy, R. M., *J. Org. Chem.* **2002**, *67*, 7135.

66. Iwama, T.; Kinoshita, H.; Kataoka, T., *Tetrahedron Lett.* **1999**, *40*, 3741.

67. Marko, I. E.; Giles, P. R.; Hindley, N. J., *Tetrahedron* **1997**, *53*, 1015.
68. Alcaide, B.; Almendros, P.; Aragoncillo, C., *Chem. Commun.* **1999**, 1913.
69. Alcaide, B.; Almendros, P.; Aragoncillo, C., *J. Org. Chem.* **2001**, *66*, 1612.
70. Shi, M.; Chen, L.-H.; Li, C.-Q., *J. Am. Chem. Soc.* **2005**, *127*, 3790.
71. Mastui, K.; Takizawa, S.; Sasai, H., *J. Am. Chem. Soc.* **2005**, *127*, 3680.
72. You, J.; Xu, J.; Verkade, J. G., *Angew. Chem. Int. Ed.* **2003**, *42*, 5054.
73. For a review of enantioselective conjugate addition: Sibi, M. P.; Manyem, S., *Tetrahedron* **2000**, *56*, 8033.
74. For a review of enantioselective conjugate addition: Krause, N.; Hoffmann-Röder, A., *Synthesis* **2001**, 171.
75. Sawamura, M.; Hamashima, H.; Ito, Y., *Tetrahedron* **1994**, *50*, 4439.
76. Sawamura, M.; Hmashima, H.; Shinoto, H.; Ito, Y., *Tetrahedron Lett.* **1995**, *36*, 6479.
77. Sawamura, M.; Hamashima, H.; Ito, Y., *J. Am. Chem. Soc.* **1992**, *114*, 8295.
78. Inagaki, K.; Nozaki, K.; Takaya, H., *Synlett* **1997**, 119.
79. Harada, S.; Kumagai, T.; Kinoshita, S.; Matsunaga, S.; Shibasaki, M., *J. Am. Chem. Soc.* **2003**, *125*, 2582.
80. Wynberg, H.; Helder, R., *Tetrahedron Lett.* **1975**, *46*, 4057.
81. Hermann, K.; Wynberg, H., *J. Org. Chem.* **1979**, *44*, 2238.
82. Conn, R. S. E.; Lovell, A. V.; Karady, L. M.; Weinstock, L. M., *J. Org. Chem.* **1986**, *51*, 4710.
83. Ooi, T.; Miki, T.; Taniguchi, M.; Shiraishi, M.; Takeuchi, M.; Maruoka, K., *Angew. Chem. Int. Ed.* **2003**, *42*, 3796.
84. Parageorgiou, C. D.; Ley, S. V., *Angew. Chem. Int. Ed.* **2003**, *42*, 828.
85. Kummer, D. A.; Brenneman, J. B.; Martin, S. F., *Tetrahedron* **2006**, *62*, 11437.
86. For a review of Heck reaction: Beletskaya, I. P.; Cheprakov, A. V., *Chem. Rev.* **2000**, *100*, 3009.
87. Tietze, L. F.; Stecker, F.; Zinngrebe, J.; Sommer, K. M., *Chem. Eur. J.* **2006**, *12*, 8770.
88. Yoo, K. S.; Yoon, C. H.; Jung, K. W., *J. Am. Chem. Soc.* **2006**, *128*, 16384.
89. Pennanen, S. I., *Synth. Commun.* **1985**, *15*, 865.
90. Cazeau, P.; Duboudin, F.; Moulines, F.; Babot, O.; Dunogues, J., *Tetrahedron* **1987**, *43*, 2089.
91. Beifuss, U.; Ledderhose, S., *Synlett* **1995**, 938.
92. Jung, M. E.; McCombs, C. A.; Takeda, Y.; Pan, Y.-G., *J. Am. Chem. Soc.* **1981**, *103*, 6677.
93. Liu, D.; Canales, E.; Corey, E. J., *J. Am. Chem. Soc.* **2007**, *129*, 1498.
94. Ryu, D. H.; Zhou, G.; Corey, E. J., *J. Am. Chem. Soc.* **2004**, *126*, 4800.
95. Ryu, D. H.; Corey, E. J., *J. Am. Chem. Soc.* **2003**, *125*, 6388.
96. Tochtermann, W.; Pahl, A.; Peters, E.-M.; Peters, K.; Schnering, H. G., *Chem. Ber.* **1988**, *121*, 493.
97. Inanaga, K.; Takasu, K.; Ihara, M., *J. Am. Chem. Soc.* **2004**, *126*, 1352.

N

D-2-Naphthylalanine, *N,N'*-[[(2*S*, 5*S*)-1-(4-pyridinyl)-2,5-pyrrolidinediyl]dicarbonyl]bis-, 3,3'-Dioctyl Ester[1] and L-2-Naphthylalanine, *N,N'*-[[(2*S*,5*S*)-1-(4-pyridinyl)-2,5-pyrrolidinediyl]dicarbonyl]bis-, 3,3'-Dioctyl Ester

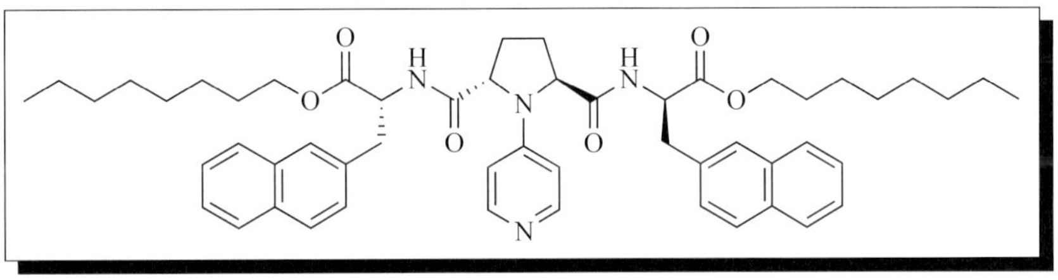

D-Reagent
[1384476-09-8] $C_{53}H_{66}N_4O_6$ (MW 855.13)
L-Reagent
[958229-32-8] $C_{53}H_{66}N_4O_6$ (MW 855.13)
InChI = 1S/C53H66N4O6/c1-3-5-7-9-11-17-33-62-52(60)46(37-39-23-25-41-19-13-15-21-43(41)35-39)55-50(58)48-27-28-49(57(48)45-29-31-54-32-30-45)51(59)56-47(53(61)63-34-18-12-10-8-6-4-2)38-40-24-26-42-20-14-16-22-44(42)36-40/h13-16,19-26,29-32,35-36,46-49H,3-12,17-18,27-28,33-34,37-38H2,1-2H3,(H,55,58)(H,56,59)/t46-,47-,48+,49+/m1/s1
InChIKey = KJXLSDLFHSGZTD-QTIZPXICSA-N

(reagent used for chemo- and regioselective acylation of amino alcohols)

Physical Data: colorless solid crystals with mp 136–137 °C and $[\alpha]_D^{20} = -95$ (0.83, $CHCl_3$).
Solubility: soluble in common organic solvents.
Form Supplied in: not commercially available.
Preparative Methods: prepared in six steps from commercially available L-pyroglutamic acid (eq 1).[1,2]

Chemoselective Acylation of 2-Aminopentane-1,5-diols. Reversal of chemoselectivity is observed in acylation of 2-aminopentane-1,5-diol derivatives catalyzed by **1**.[1] DMAP-catalyzed acylation of **3** takes place at the C(1)–OH predominantly with concomitant formation of a significant amount of the diacylate (eq 2). On the other hand, 5-OAc is obtained in the acetylation of **3** catalyzed by **1** highly selectively in 95% yield without concomitant formation of the diacylate. The distance between the OH and the NHNs (Ns = 2-nitrobenzenesulfonyl) groups seems to be important for the regioselectivity of the acylation. Acylation of the corresponding 1,4-diol **4** catalyzed by **1** provides a mixture of regioisomers (eq 3, 1-OAc:4-OAc = 1:1) under the identical conditions. This observation may indicate that catalyst **1** promotes regioselective acylation based on the recognition of the distance between the functionalities in the substrates. Catalyst **1** also enables selective acylation of a secondary hydroxy group in the presence of an otherwise more reactive primary hydroxy group (eq 4). DMAP-catalyzed acetylation of **5** takes place predominantly at the primary hydroxy group, indicating the intrinsic high reactivity of the primary hydroxy group. On the other hand, the secondary hydroxy group is selectively acetylated in the presence of catalyst **1** by overcoming its lower intrinsic reactivity. A surprising example is observed in the acylation of **5** (R = *i*-Pr). The secondary hydroxy group located in the sterically congested environment undergoes selective acylation by catalyst **1** in the presence of a primary hydroxy group.

Geometry-selective Acylation of Tri- and Tetrasubstituted α,α'-Alkenediols. Catalyst **2** can discriminate the geometry of tri- and tetrasubstituted α,α'-alkenediols by their acylation.[3,4] While DMAP-catalyzed acetylation of trisubstituted α,α'-alkenediol **6** gives almost 1:1 mixture of *E*-OAc and *Z*-OAc in 56% yield, *E*-OAc forms selectively in the acetylation of **6** promoted by **2** (eq 5, 90% yield, *E*-OAc/*Z*-OAc = 95/5). Catalyst **2** is also effective in differentiating the geometry of tetrasubstituted α,α'-alkenediols **7**. Acylation of **7** takes place to give *E*-OAc selectively and in good yields in the presence of **2** (eq 6, 72–96% yield, *E*-OAc/*Z*-OAc = 98/2). The N*H*Ns group seems to play an important role for the *E*-OH-selective acylation because the acetylation proceeds with much less selectivity in the corresponding N*Me*Ns derivative **8** to give a regioisomeric mixture (*E*-OAc/*Z*-OAc = 64/36) (eq 7). These results indicate that the catalyst may recognize the geometry of the substrate on the basis of catalyst–substrate interaction, in which the N–H group of the substrate serves as a key hydrogen bond donor. A transition state model is proposed based on the theoretical calculations.[5]

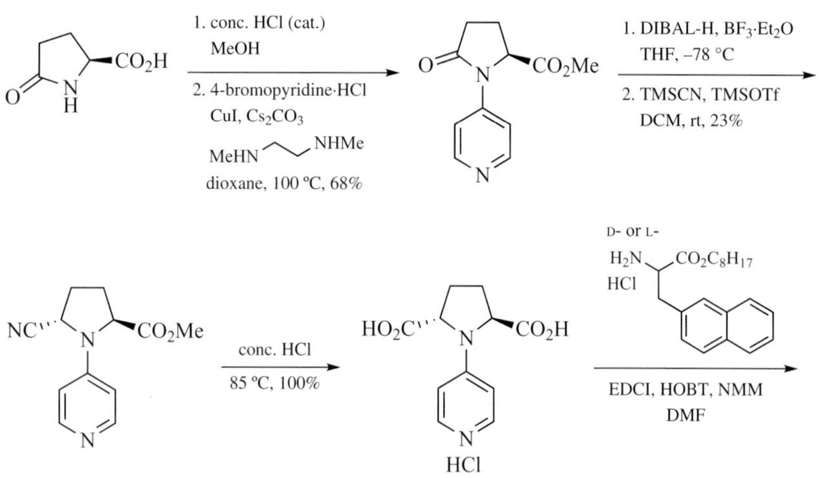

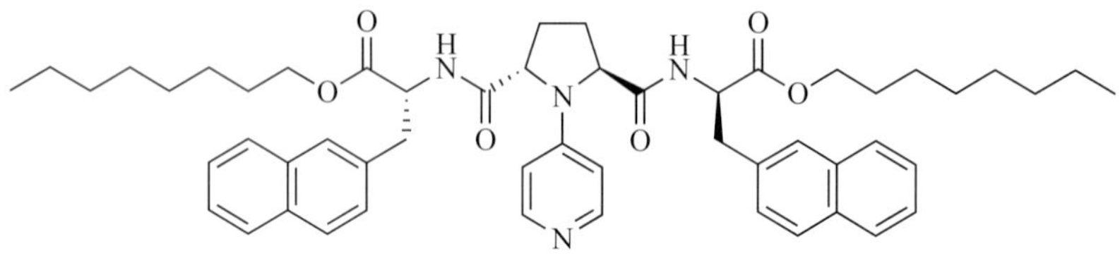

(1) from (*R*)-isomer

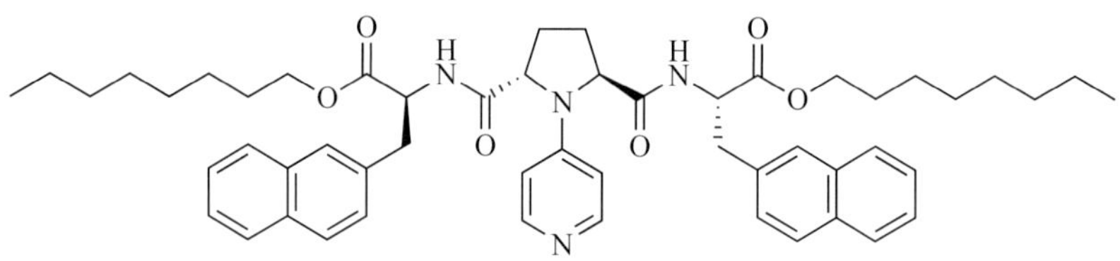

(2) from (*S*)-isomer (1)

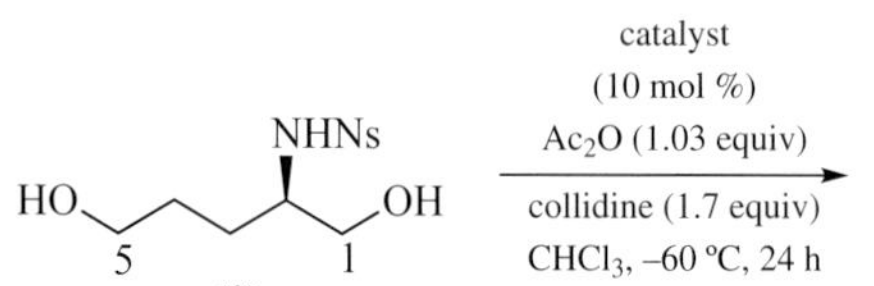

(3)

5-OAc | 1-OAc (2)

	5-OAc	1-OAc
with DMAP	6%	32% + 1,5-diacylate (30%)
with **1**	95%	<1%

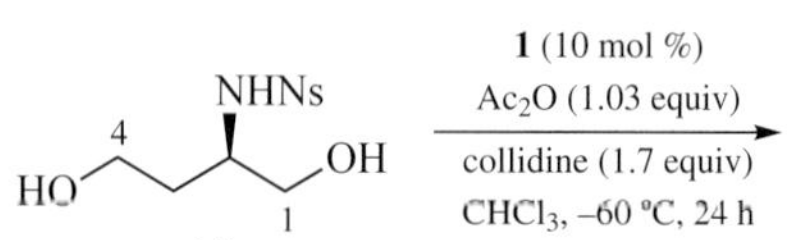

(4)

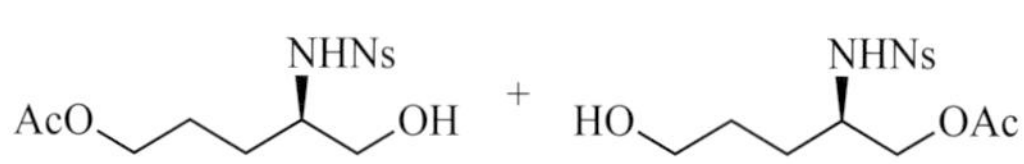

5-OAc	1-OAc
37%	37%

(3)

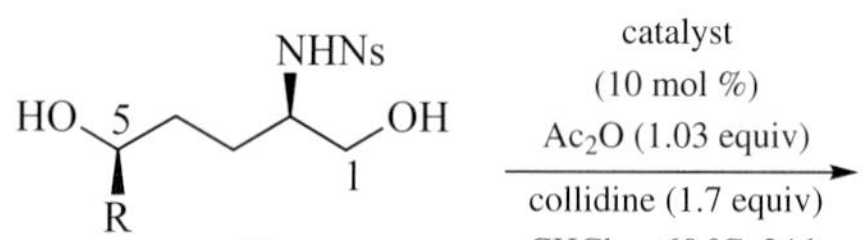

(5)

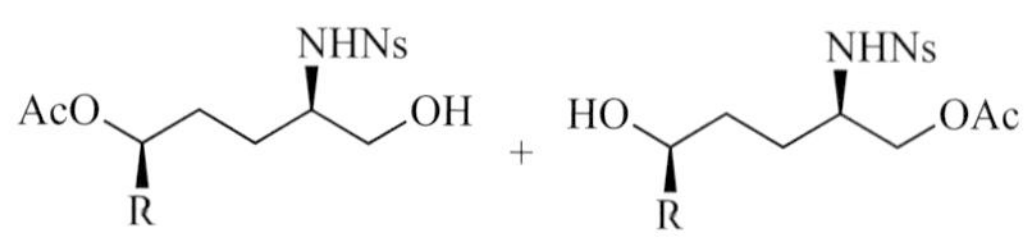

sec-OAc + *pri*-OAc (4)

	R	*sec*-OAc	*pri*-OAc
with DMAP	R = Me	<1%	90%
with **1**	R = Me	77%	1%
	R = Et	76%	1%
	R = *i*-Pr	79%	2%
	R = *n*-Bu	74%	1%

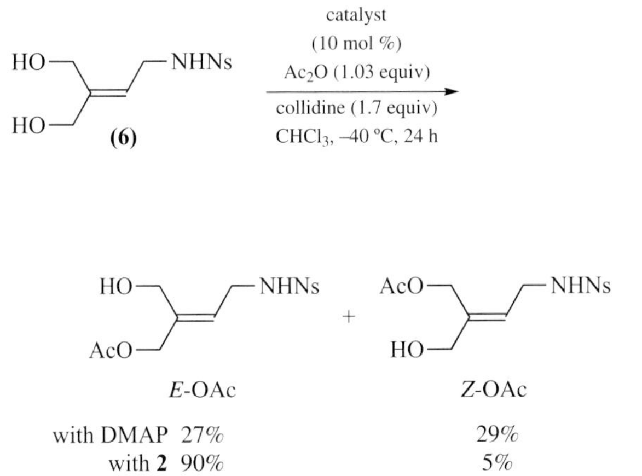

(5)

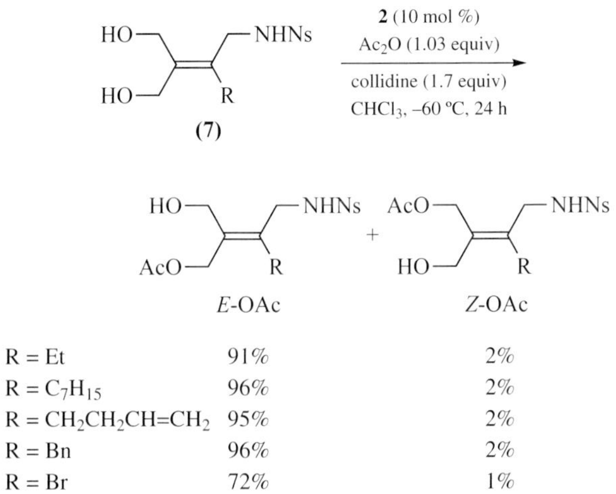

(6)

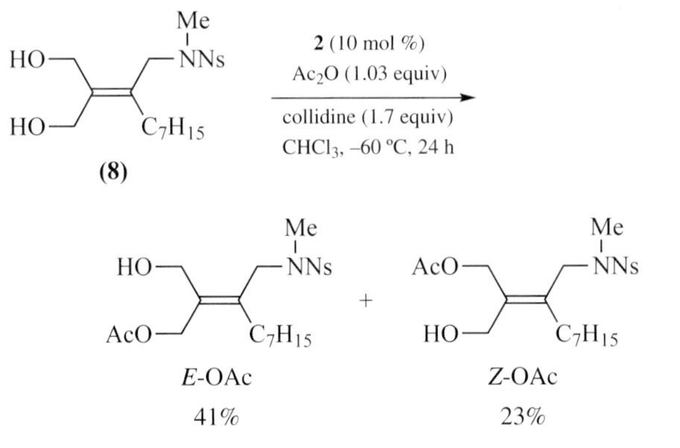

(7)

1. Yoshida, K.; Shigeta, T.; Furuta, T.; Kawabata, T., *Chem. Commun.* **2012**, *48*, 6981.
2. Kawabata, T.; Muramatsu, W.; Nishio, T.; Shibata, T.; Schedel, H., *J. Am. Chem. Soc.* **2007**, *129*, 12890.
3. Yoshida, K.; Mishiro, K.; Ueda, Y.; Shigeta, T.; Furuta, T.; Kawabata, T., *Adv. Synth. Catal.* **2012**, *354*, 3291.
4. Furuta, T.; Kawabata, T. Science of Synthesis. *Asymmetric Organocatalysis 1: Lewis Base and Acid Catalysts*; List, B., Ed.; Georg Thieme Verlag KG: Stuttgart, 2012.
5. Yamanaka, M.; Yoshida, U.; Sato, M.; Shigeta, T.; Yoshida, K.; Furuta, T.; Kawabata, T., *J. Org. Chem.* **2015**, *80*, 3075.

Yoshihiro Ueda & Takeo Kawabata
Kyoto University, Uji, Japan

Nitroethane

$EtNO_2$

[79-24-3] $C_2H_5NO_2$ (MW 75.08)
InChI = 1S/C2H5NO2/c1-2-3(4)5/h2H2,1H3
InChIKey = MCSAJNNLRCFZED-UHFFFAOYSA-N

(building block in synthesis; polar solvent)

Physical Data: mp $-90\,°C$; bp $115\,°C$; d 1.04 g cm^{-3}; pK_a 8.5.
Solubility: completely misc most organic solvents; slightly sol water; sol alkaline solution.
Form Supplied in: colorless liquid, widely available; contains 2-nitropropane as the chief contaminant.
Purification: by drying over $MgSO_4$ and distillation at reduced pressure; a small acidic forerun is discarded.
Handling, Storage, and Precautions: stable compound. Flammable; toxic.

Original Commentary

Kurt B. G. Torssell & Kurt V. Gothelf
Aarhus University, Aarhus, Denmark

Introduction. The reactions of nitroethane discussed here can, in principle, be applied to most primary nitroalkanes. Nitroethane reacts with metal alkoxides and hydroxides to form metal nitronates which are often applied as in situ generated reagents. They are unstable and it is advisable not to isolate them.

The Nitroaldol Reaction (Henry Reaction).[1,2]

Reactions with Aldehydes. In the presence of base, nitroethane reacts with aliphatic aldehydes in an aldol-type reaction with formation of 2-nitro alcohols (nitroaldols). Due to its reversibility, the reaction is normally carried out in the presence of only catalytic quantities of base. The reaction of nitroethane with an aldehyde in the presence of one equivalent of base gives the salt of the *aci*-nitro tautomer of the product, which must be carefully acidified to avoid the Nef reaction. Alkali metal hydroxides, alkoxides, or carbonates and tertiary amines have been applied as catalysts.[2a] Elimination of water, the aldol reaction, and the Cannizzaro reaction are competing side reactions.[2a,b] 2-Nitro alcohols are unstable compounds and care has to be exercised during workup (decomposition during distillation). Higher yields are often obtained when the reaction is catalyzed in heterogeneous systems with catalysts such as powdered NaOH,[3] Al_2O_3,[4] and Al_2O_3-supported KF.[5] The 2-nitro alcohols are readily converted into a variety of functionalities (eq 1).[1b,4–10,13–17]

Aromatic aldehydes react with nitroethane under the same conditions.[11] Elimination of water to give β-nitrostyrenes takes place on acidification.[2a] Synthesis of β-nitrostyrenes can be accomplished in one step by heating aromatic aldehydes in acetic acid with NH_4OAc as catalyst.[12] In the presence of NaOAc in refluxing acetic acid, aldehydes are converted into nitriles (eq 2).[12b]

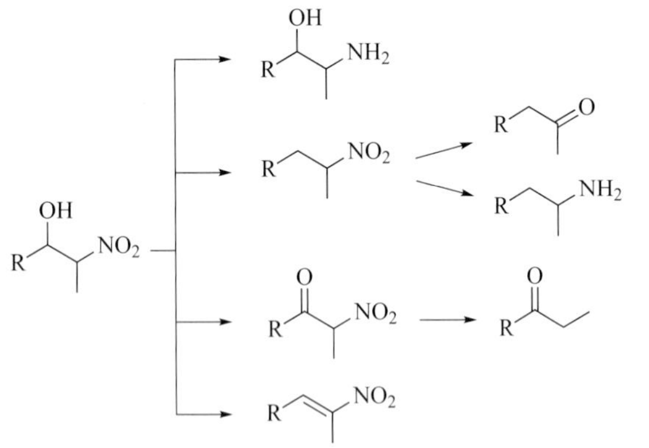

(1)

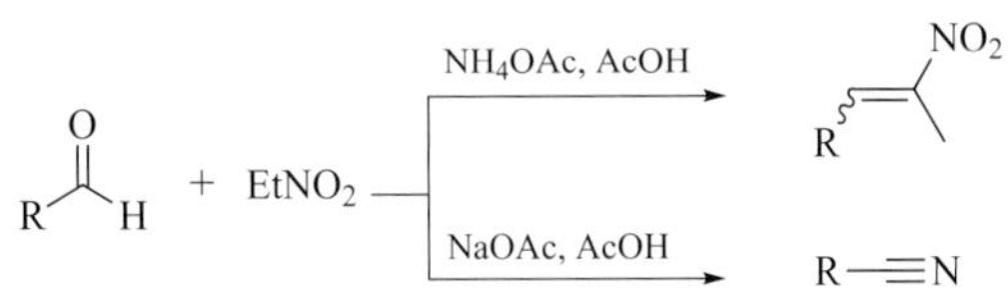

(2)

Dehydration of 2-nitro alcohols to nitroalkenes has been accomplished with Methanesulfonyl Chloride,[13] phthalic anhydride,[14] 1,3-Dicyclohexylcarbodiimide,[15] pivaloyl chloride,[16] Trifluoroacetic Anhydride,[17] and Acetic Anhydride.[17] Heating 2-nitro alcohols in dichloromethane in the presence of basic aluminium oxide is another mild method for the synthesis of nitroalkenes.[18] Nitroalkenes can be converted into a variety of functionalities (eq 3),[2,19a,21–26] and have found utility as heterodienes in hetero Diels–Alder reactions and as reactants in Lewis acid promoted tandem [4+2]/[3+2] cycloadditions (eq 4).[20]

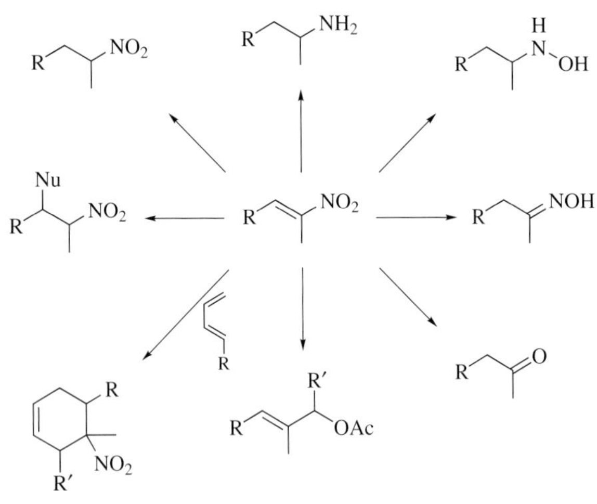

(3)

Nitroethane can be silylated to form stable trimethylsilyl or *t*-butyldimethylsilyl nitronates.[27] In the presence of a catalytic amount of Tetrabutylammonium Fluoride at −78 °C in THF, the silyl nitronates react with aldehydes to give 2-nitro alcohol *O*-silyl ethers in high yields.[6b,28] The TBDMS nitronate reacts in the presence of freshly dried Bu_4NF to give the *erythro*-isomer of the product in >95% de (eq 5).

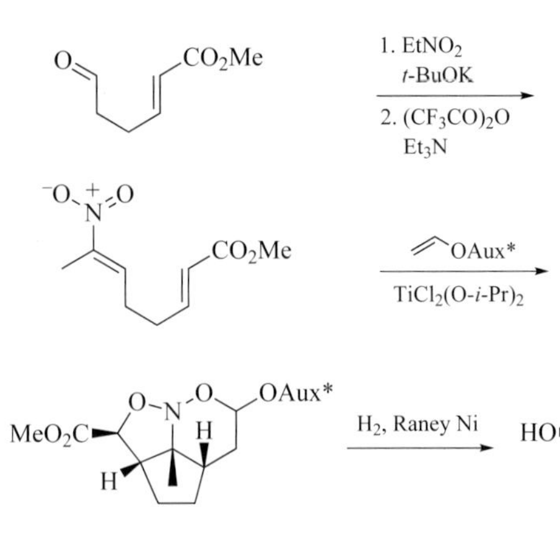

(4)

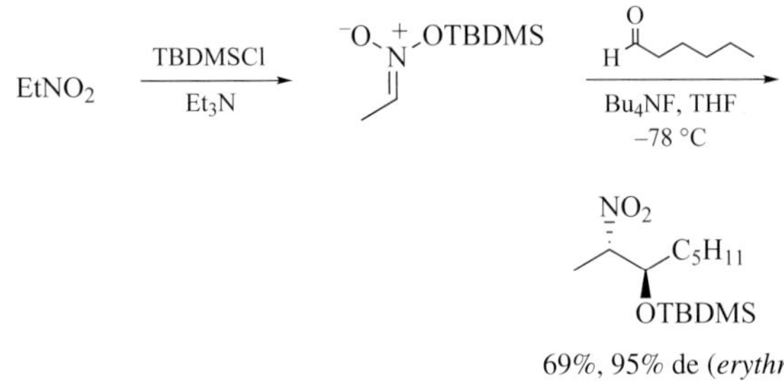

(5)

The dilithium salt of nitroethane is formed by treatment of nitroethane with 2 mol equiv of Butyllithium in THF/HMPA at −90 °C.[28,29] It reacts with aldehydes at −70 to −60 °C to form the doubly deprotonated nitro alcohols which must be carefully acidified at −90 °C with acetic acid. The dilithium salt is a much harder carbon nucleophile than the sodium nitronate and the reaction with carbonyl compounds is irreversible; this often leads to higher yields of 2-nitro alcohols than the conventional Henry reaction. The reaction proceeds diastereoselectively to give the *threo* and *erythro* isomers in a ratio of characteristically 3–5:1 (eq 6).[28]

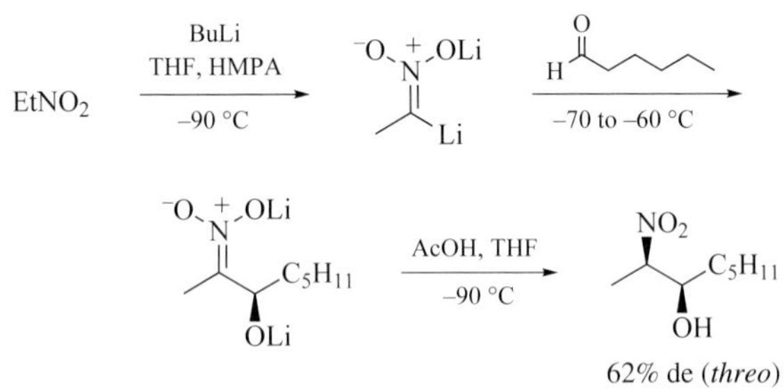

(6)

Diastereomeric mixtures of nitro alcohols derived from nitroethane can be transformed to the *threo*-isomer, predominantly, by double deprotonation and protonation or to the *erythro*-isomer by double deprotonation, silylation with tert-Butyldimethylchlorosilane, protonation, and desilylation (eq 7).[28]

Nitroethane can react with two mol equiv of formaldehyde.[2a] With acetaldehyde and higher analogs, the reaction with a second molecule is difficult.[2b,d]

Dialdehydes derived from periodate cleavage of a sugar react with nitroethane in an one-pot cyclization reaction to give predominantly one stereoisomer (eq 8).[30]

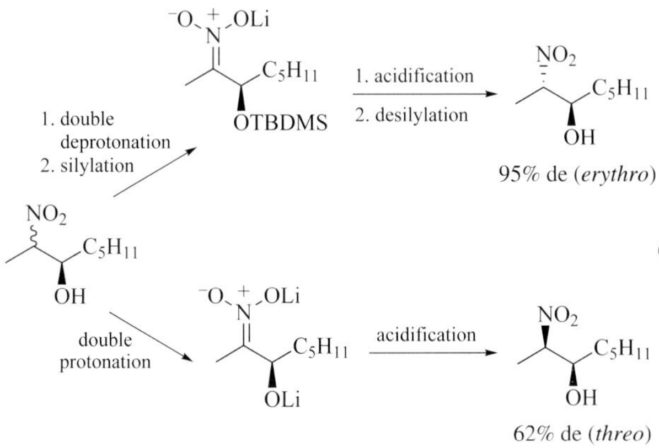

(7)

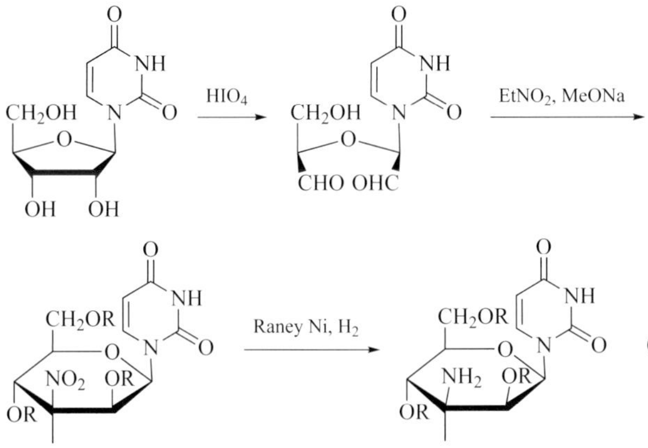

(8)

meso-1,5-Dialdehydes react with nitroethane in methanol in the presence of a catalytic amount of NaOH to give a crude mixture of 2,6-dihydroxynitrocyclohexanes from which the major *trans,trans*-isomer precipitates (eq 9).[31] The diacetate of 2,6-dihydroxy-1-methyl-1-nitrocyclohexane formed by this reaction can be saponified enantioselectively with pig liver esterase (PLE) to give the monoacetate of >95% ee (eq 9).[31]

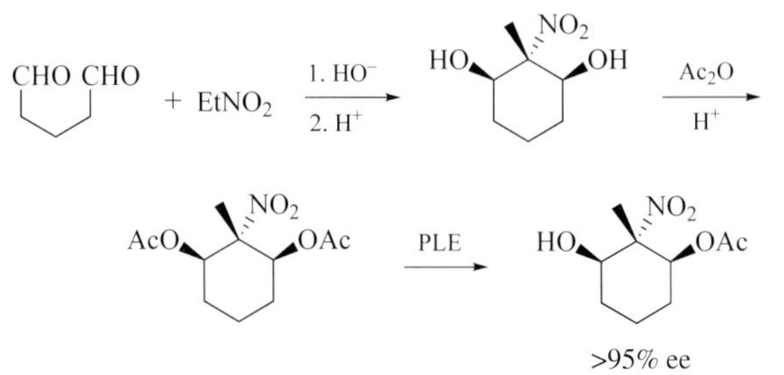

(9)

In the presence of primary or secondary amines, aldehydes react with nitroethane in a Mannich-type reaction.[2d]

Reactions with Ketones. The reaction is complex and depends on the ratio of reactants, base, temperature, and time (eq 10).[2a]

With alkali metal hydroxides or alkoxides, quaternary ammonium hydroxides, or primary or tertiary amines, low yields (10–30%) of nitro alcohols are normally obtained.[2a,b] Nitroethane reacts with methylcyclohexanones under pressure with TBAF as catalyst to give nitro alcohols in 41–87% yields.[32] With *N,N*-dimethylethylenediamine as base, it is possible to selectively synthesize allylic nitro compounds in moderate to high yields from both acyclic and alicyclic ketones (eq 11).[33] The doubly deprotonated forms of primary nitroalkanes give (with nitromethane as an exception) higher yields of 2-nitro alcohols in the reaction with ketones.[34]

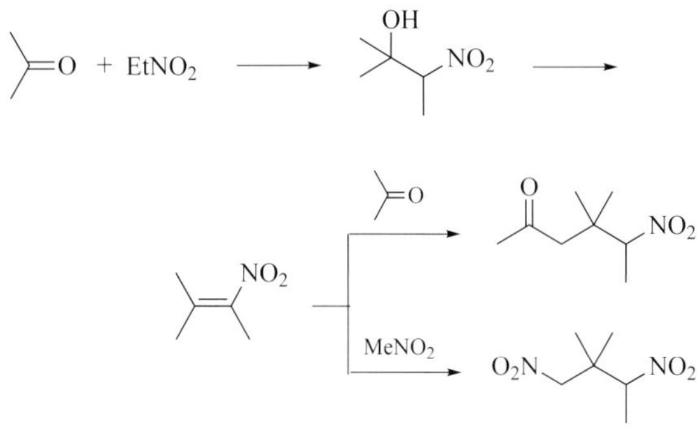

(10)

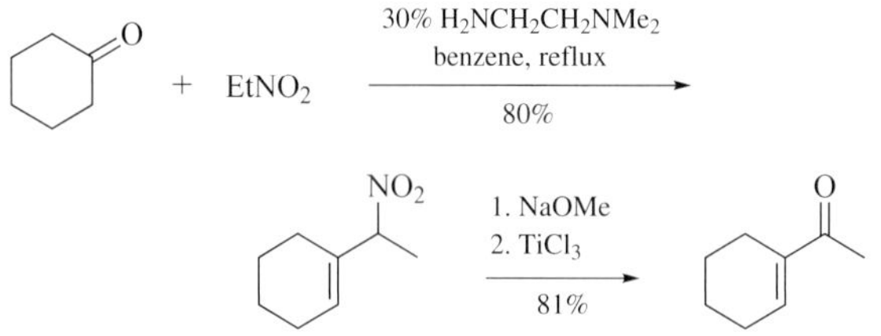

(11)

Michael Reactions. Conjugate addition of nitroethane to activated double bonds is an important C–C bond forming reaction (eq 12).

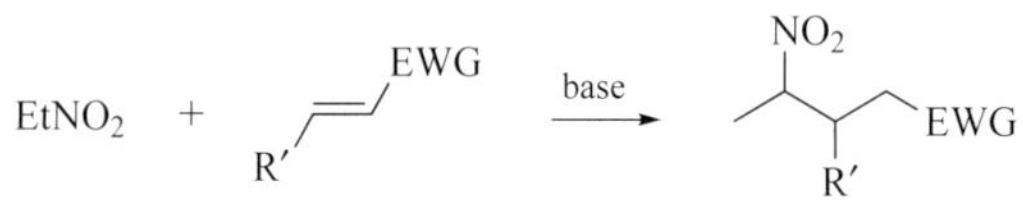

(12)

EWG = CHO, COR, CO_2R, CN, SOR, SO_2R, NO_2

Michael additions with nitroethane are catalyzed in homogeneous solution with catalysts such as alkali metal hydroxides in alcohol,[35] organic nitrogen bases,[36] fluoride ion,[37] and sodium hydride.[8] The reaction is also catalyzed in heterogeneous systems with alumina[38] and Al_2O_3-supported KF or CsF.[39] When combined with the Nef reaction,[1a,9] synthetically useful 1,4-dicarbonyl compounds can be synthesized. Nitroethane reacts with 2 mol equiv of Michael acceptors to give a variety of coupling products (eq 13).[37]

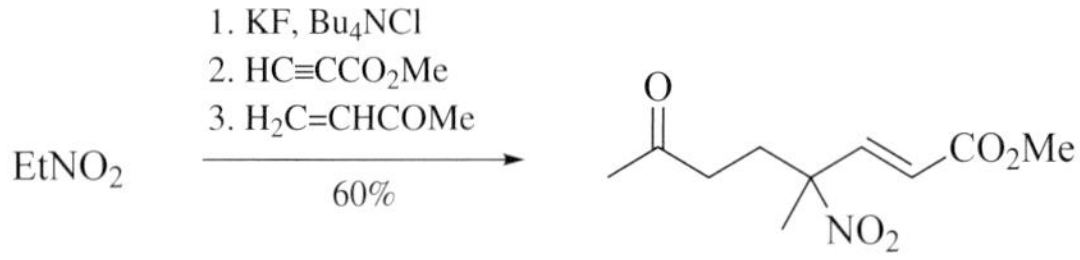

(13)

Acylation of Nitroethane. With few exceptions, acylation of sodium ethanenitronate with acyl halides or anhydrides occurs on oxygen. The unstable products rearrange into hydroxamic acid derivatives.[40] *C*-Acylation can be accomplished with acylimidazoles[41] and acyl cyanide (eq 14).[42] Carboxylation can be accomplished by reaction with methoxymagnesium methyl carbonate (Stiles reagent).[43]

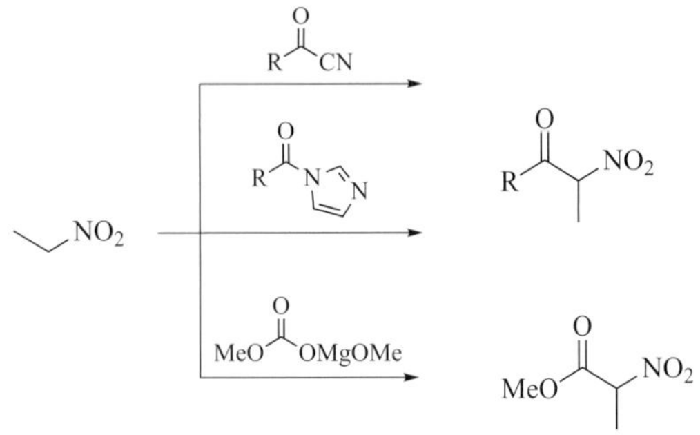

(14)

The dilithium salts of nitroethane and higher primary nitroalkanes react with carboxylic esters, anhydrides, and dialkyl carbonates to form β-oxo nitronates or β-carboxy nitronates, which must be carefully acidified at −90 °C to avoid Nef reaction with acetic acid.[34]

Alkylation of Nitroethane. The ambident ethanenitronate ion reacts with alkyl iodides to give a mixture of *C*-alkylated and, primarily, *O*-alkylated compounds. The alkyl nitronates of nitroethane are unstable and decompose to carbonyl compounds and acetaldoxime.[44] The reaction has found some utility in the oxidation of primary alkyl iodides to aldehydes.[45] The methyl and ethyl nitronate of nitroethane are stable below −78 °C and can be synthesized from the sodium nitronate and Trimethyloxonium Tetrafluoroborate and Triethyloxonium Tetrafluoroborate, respectively, at −60 °C in excellent yields.[44] The sodium salt of nitroethane can selectively be *C*-benzylated with 1-benzyl-2,4,6-triphenylpyridinium tetrafluoroborate in moderate yields.[46] The pyridinium cations are readily available from 2,4,6-Triphenylpyrylium Tetrafluoroborate and the corresponding benzylamines. The dilithium salts of primary nitroalkanes are *C*-alkylated with primary iodo- and bromoalkanes.[47] A homoallylic nitro compound is formed when nitroethane, after treatment with lithium methoxide in methanol, reacts with cinnamyl acetate in the presence of 10 mol % Tetrakis(triphenylphosphine)-palladium(0) as catalyst (eq 15).[48]

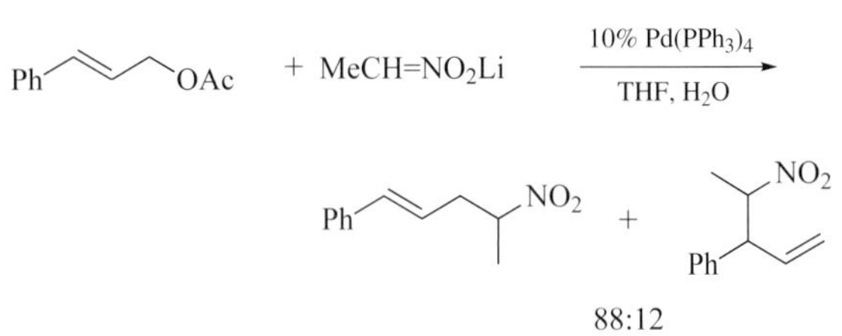

(15)

1,3-Dipolar Cycloadditions of Nitroethane.[49] The in situ generated trimethylsilyl ethanenitronate reacts with activated terminal alkenes in a 1,3-dipolar cycloaddition to form 5-substituted 3-methyl-*N*-trimethylsilyloxyisoxazolidines which, upon acid treatment or heating, give isoxazolines.[27a,50] Asymmetric isoxazolines have been synthesized in this manner. The cycloaddition of *N*-acryloyl-(2*R*)-bornane-10,2-sultam (acrylate of 10,2-Camphorsultam) with the silyl nitronate, followed by acid-catalyzed elimination of trimethylsilyl alcohol, gives the corresponding isoxazoline in high yields and 78% ee of the C-5 (*R*)-isomer (eq 16).[51]

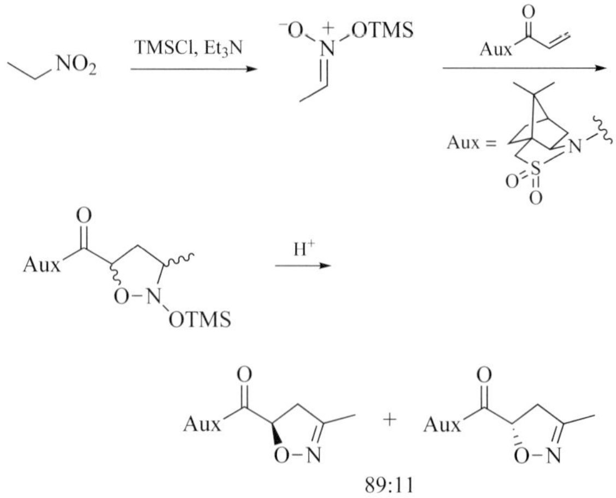

(16)

Acetonitrile N-Oxide is formed when nitroethane is treated with Phenyl Isocyanate in the presence of Triethylamine in the Mukayiama reaction.[52] The nitrile oxide is a potent dipole and reacts with alkenes and alkynes to give isoxazolines and isoxazoles, respectively. With terminal alkenes or alkynes, the reaction is regioselective, giving 5-substituted 3-methylisoxazolines or -isoxazoles. 1,2-Disubstituted *cis*- and *trans*-alkenes react in a stereospecific manner giving *cis*- and *trans*-4,5-substituted isoxazolines, respectively.[49] Acetonitrile oxide reacts stereoselectively with optically active acrylic esters and amides to give the isoxazolines illustrated in eq 17.[53–56] With Curran's bis-lactam chiral auxiliaries, products with about 98% ee are obtained (eq 17).[53]

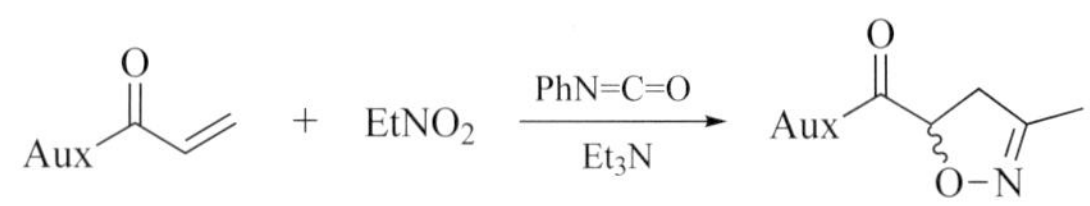

(17)

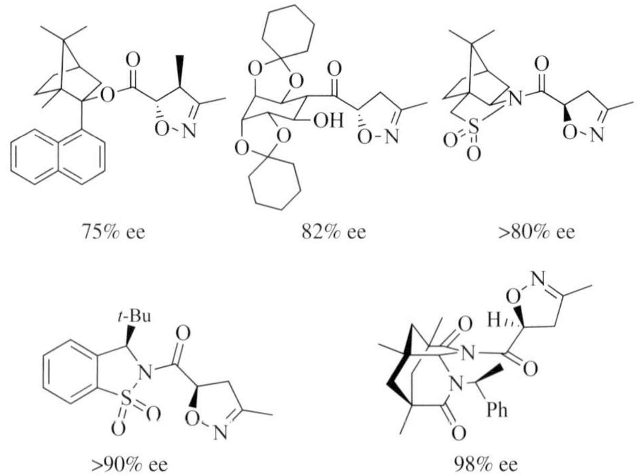

3-Methylisoxazolines can easily be reduced to β-hydroxy ketones (an alternative route to the products from aldol reactions with acetone)[57] and 3-amino alcohols (eq 18).[58]

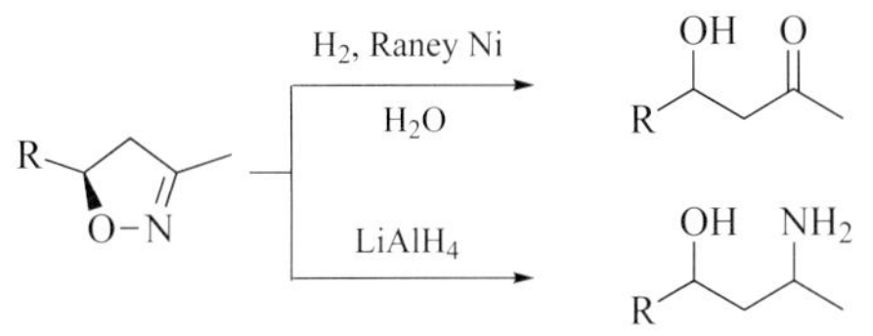

(18)

First Update

Marie-Noelle Roy
Université de Montréal, Montreal, Québec, Canada

The Nitroaldol reaction (Henry Reaction).

Reactions with Aldehydes. The reaction of nitroethane with aldehydes or ketones under basic conditions leads to the formation of secondary or tertiary 2-nitroalcohols. Diverse systems involving different categories of bases have been used for the completion of the reaction. Surrogate systems, such as phosphorus ylides[59] or basic phosphines,[60] have also been used to catalyze the reaction.

In the past two decades, efforts on the Henry reaction involving nitroethane have been focused on the asymmetric variants of the reaction furnishing systems that can be divided into two categories: metal-based chiral catalysts (Cu, Co, Zn, Pd, rare-earth metal)[61–63] and organocatalysts (guanidine, thiourea, cinchona alkaloid, phosphonium salt).[64] The control of the diastereoselectivity remains a challenge, though efficient systems have been developed to access either high selective *syn-* or *anti-*nitroalkanols with excellent enantioselectivities. The *syn-* diastereoisomer is obtained by establishing a transition state that chelates the nucleophile and the electrophile. For example, the use of a chiral bis(imidazoline), in the presence of $Cu(OTf)_2$ and *N*-methylmorpholine, gives access to a wide range of aliphatic and aromatic *syn*-nitroalcohols (eq 19).[65] Other *syn*-selective Cu-catalyzed procedures using nitroethane involve chiral ligands such as quinine,[66] alkyldiamine,[67] or sulfonyldiamine.[68]

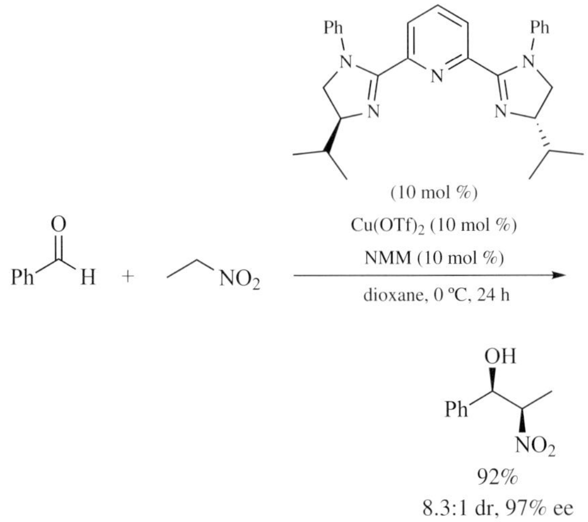

(19)

The organocatalysis approach, using a bifunctional guanidine/thiourea-based catalyst, also leads to highly *syn*-selective nitroalcohols with aliphatic aldehydes (eq 20).[69] Potassium iodide is used as additive to inhibit the reversibility of the reaction, enabling high asymmetric induction. It is suggested that the bifunctionality present on the organocatalyst simultaneously activates both the nucleophile (by the guanidinium) and the electrophile (by the thiourea).

The *anti*-diastereoisomer may be obtained by using a bimetallic dual activation or a H-bond/metal bifunctional activation. This second approach is well illustrated by the development of a Co-[bisurea-salen] catalyst.[70] In the presence of nitroethane, a wide range of aromatic aldehydes yield highly enriched *anti*-nitroalcohols. Superior results are obtained with *ortho*-substituted aryl aldehydes (eq 21). Conversely, aliphatic aldehydes lead to lower diastereoselectivities.

A Nd/Na bimetallic system may be applied to control the selectivity of the addition of nitroethane to aldehydes (eq 22).[71] In the presence of a leucine-derived ligand, the heterogeneous conditions allow the preparation of a variety of nitroalkanols. Aromatic aldehydes furnish higher diastereomeric ratios than their aliphatic analogs. The nitroaldol reaction of benzyl-protected hydroxy phenyl aldehyde has been used for the synthesis of a β_3-adrenoreceptor agonist.

A Pd/La bimetallic system has also been used to achieve the *anti*-stereoselective synthesis of nitroalkanols derived from nitroethane.[72]

Anti-β-nitro propargylic alcohols are readily prepared from ynals and nitroethane via an organocatalyzed Henry reaction using chiral phosphonium salt **1** and potassium *tert*-butoxide.[73] The synthetic potential of the method has been demonstrated by short syntheses of natural compounds, as illustrated with (−)-2-*epi*-codonopsinine (eq 23).

Reaction with Ketones. The nitroaldol reaction involving ketones is more limited. Due to its poor reactivity, only a few reports using activated ketones, such as keto esters or keto amides, have been reported. In the presence of a bifunctional guanidine-thiourea catalyst, the following keto esters react with nitroethane to form a nitroalkanols bearing a quaternary center (eq 24).[74]

The reaction catalyzed by phosphonium salt **1** has also been reported.[75] Isatin undergoes a Henry reaction with nitroethane in the presence of a catalytic amount of cupreine (eq 25).[76] The racemic reaction has been reported by 1,4-diazabicyclo[2.2.2]-octane (DABCO) catalysis.[77]

Reaction with Imines (Aza-Henry Reaction). The application of imines in the base-catalyzed addition of nitroalkanes leads to two vicinal centers bearing nitro and amino functional groups. Further reduction of the nitro results in the formation of a vicinal diamine. As its parent reaction, efforts in the past two decades have been mainly focused on the asymmetric variation of the reaction, leading to two contiguous chiral centers. Metal-based catalysts involving Cu, Al, Zn, rare-earth metal, or organocatalysts, such as thioureas, phosphoric acids, ammonium, and phosphonium salts, have been used as chiral inductor, leading to high diastereo- and enantioselectivities.[78] One recent example is a system catalyzed by a cinchona-derived bifunctional Brønsted base/H-bond donor catalyst. Nitroethane reacts with an in situ generated imine (from

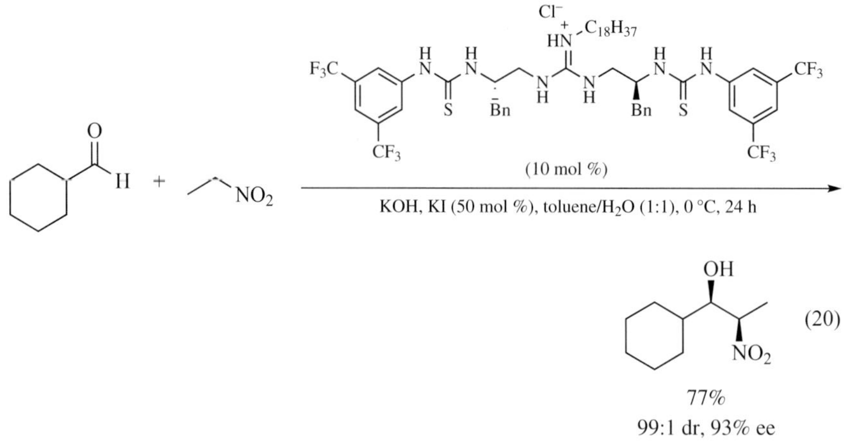
(20)

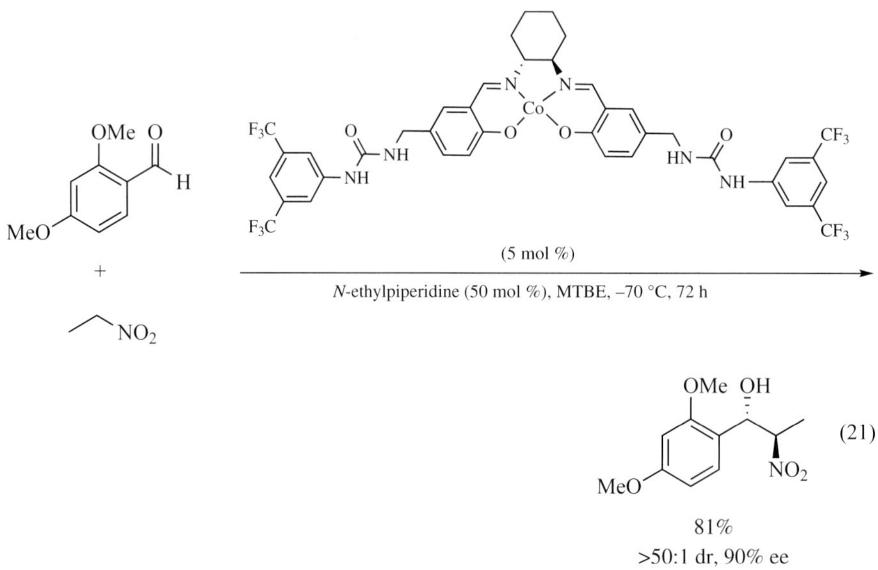
(21)

α-sulfonylamine) in the presence of only 5 mol % of the cinchona-derived asymmetric phase transfer catalyst (eq 26).[79]

A *trans*-selective addition of nitroethane to Boc-protected imines using a Cu/Sm-catalyzed system has also been reported.[80] Finally, racemic addition of nitroethane to ketimide is possible using catalytic amount of either tetramethylguanidine or a phosphazene base.[81]

Michael Additions. The asymmetric base-catalyzed addition of nitroethane has been reported with a wide range of Michael acceptors, including, but not limited to, indolylidenes,[82] β-nitroalkenes (eq 27),[83,84] alkylidenemalonates,[85,86] vinyl sulfones,[87] α,β-unsaturated thioamides (eq 28).[88]

The latter example, leading to γ-nitrothioamides, is interesting, since the α,β -unsaturated thioamide moiety has the same oxidation state as its conjugated carboxylic acid analog, which is not reactive in the reaction due to its poor electrophilicity. The thioamide can be further transformed into a thioester and then hydrolyzed to give the corresponding carboxylic acid. It can also be converted into an amide by treatment with trifluoroacetic anhydride (TFAA) or reduced into an amine. Furthermore, the Michael addition of nitroethane on α,β-unsaturated thioamides is chemoselective; competitive reactions in the presence of conjugated amide, ester, and ketone give only the desired γ-nitrothioamide.

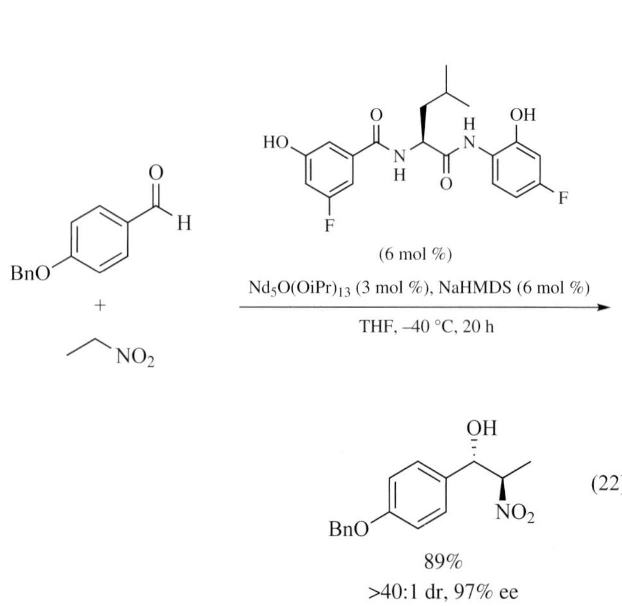
(22)

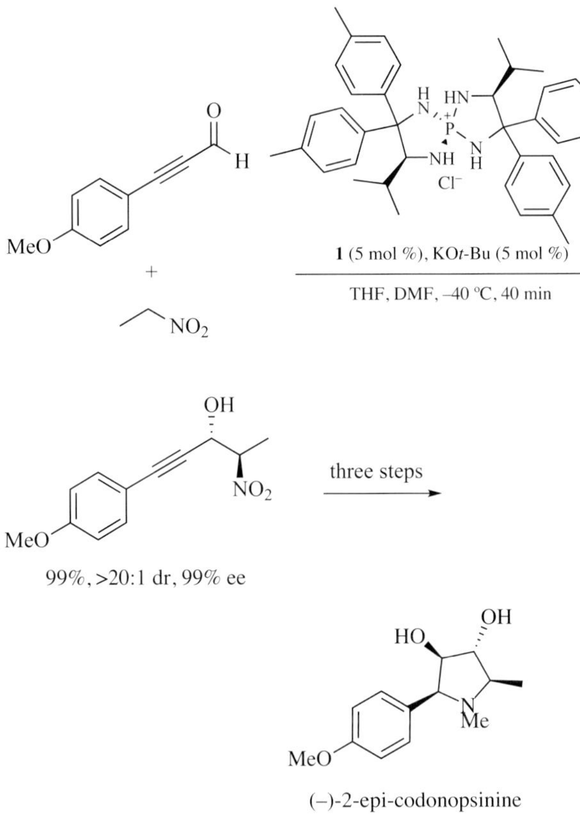

(23)

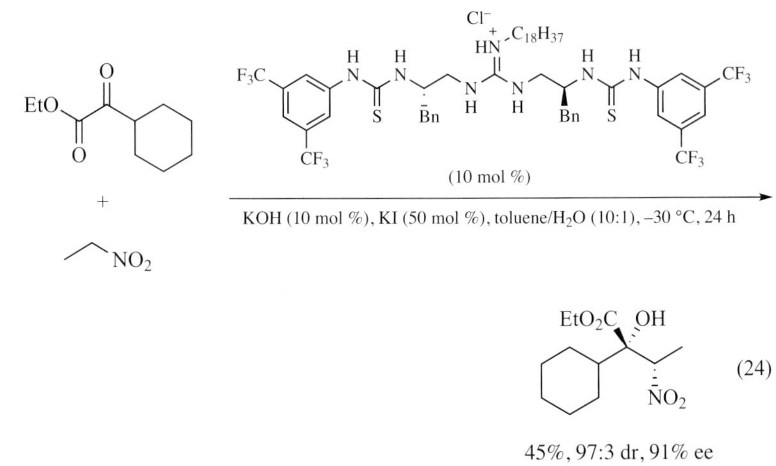

(24)

C–C Bond Coupling Involving Nitroethane. The oxidative alkylation of tetrahydroquinoline with nitroethane has been reported with different metal complexes. One of the most efficient examples involves an aerobic Au-catalyzed system in which a variety of *N*-aryl substituted tetrahydroquinolines are alkylated in high yields (eq 29).[89] Light-sensitive dyes, such as Eosin Y[90] or Rose Bengal,[91] have also been used to catalyze the same transformation.

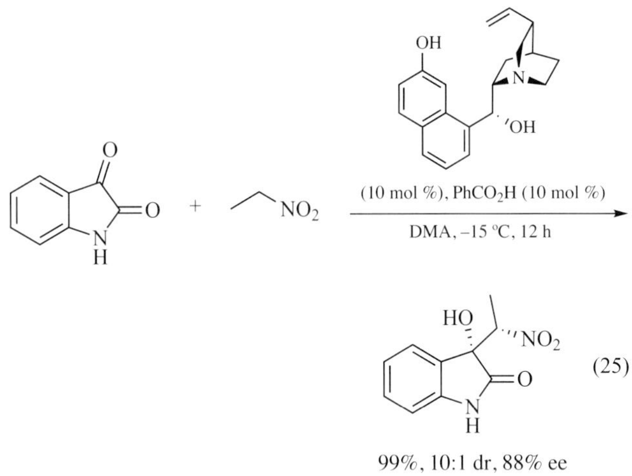

(25)

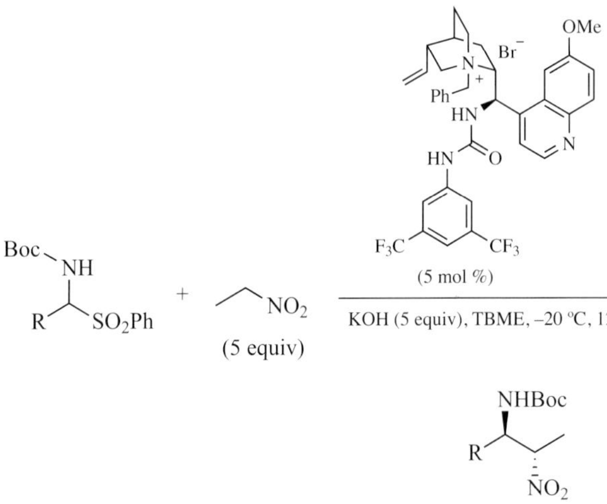

(26)

R	Yield (%)	dr	ee (%)
Ph	83	24:1	94
C_6H_{11}	75	9:1	95
i-Bu	100	13:1	90

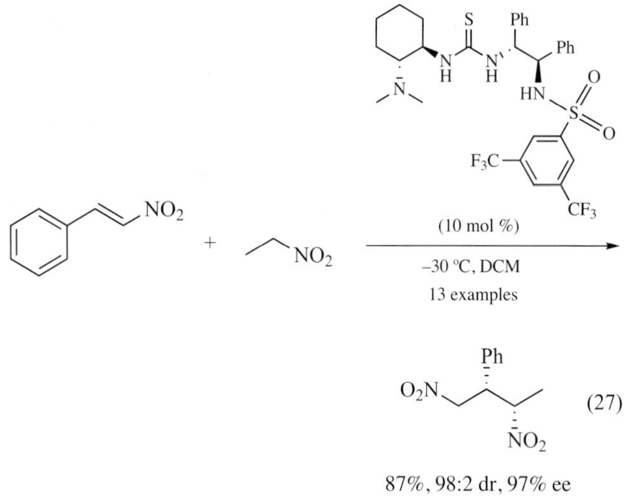

(27)

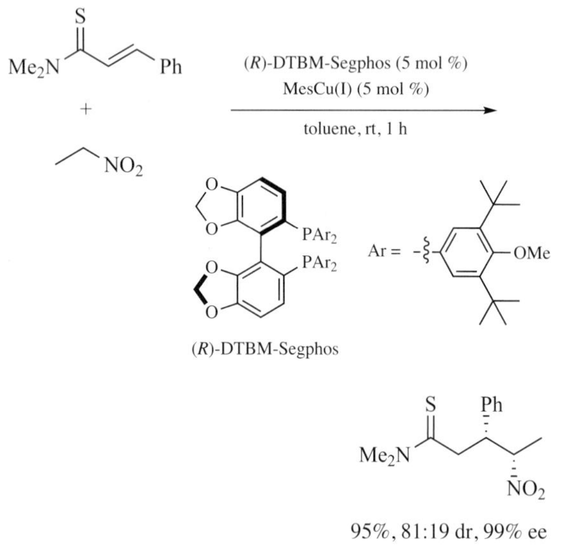

(28)

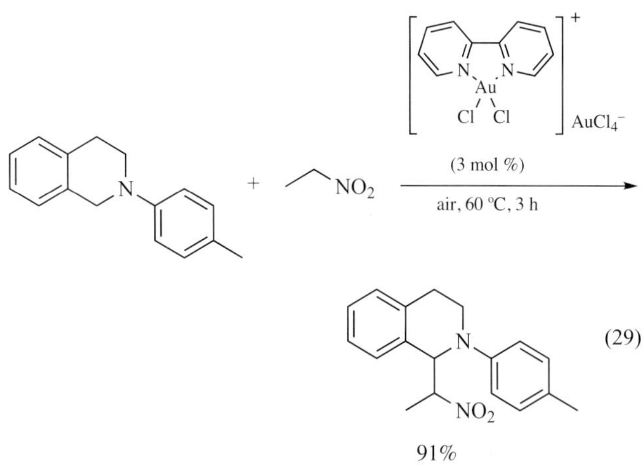

(29)

The formation of E-nitrostyrene has been achieved by the coupling reaction of *ortho*-substituted benzoic acid with nitroethane via a decarboxylation and dehydrogenation process (eq 30).[92] The reaction is possible with either electron-rich or electron-deficient arenes.

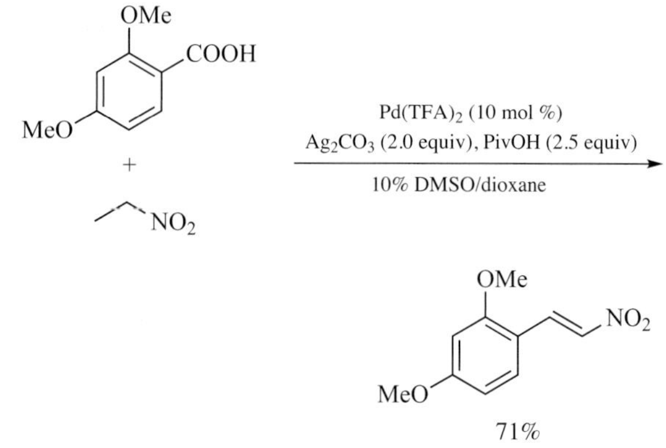

(30)

Other C–C bond couplings involving nitroethane have been reported such as an Ir-catalyzed enantioselective and regioselective allylic substitution[93] and a Pd-catalyzed monoarylation of nitroalkene.[94]

1,3-Dipolar Cycloadditions of Nitroethane. The cycloaddition of nitroethane with an alkene[95] or an alkyne[96] continues to be applied in the synthesis of isoxazoline or isoxazole derivatives. The reaction is induced either by a Cu/NMP catalytic system[95] or by the use of Mukaiyama conditions.[52]

Miscellaneous Reactions. The use of nitroethane in polyphosphoric acid allows the acetamidation of arenes (eq 31).[97] The authors suggest that nitroethane forms a diphosphorylated species that undergoes acylation of arenes via a Vilsmeier-type reaction followed by a Beckmann rearrangement.

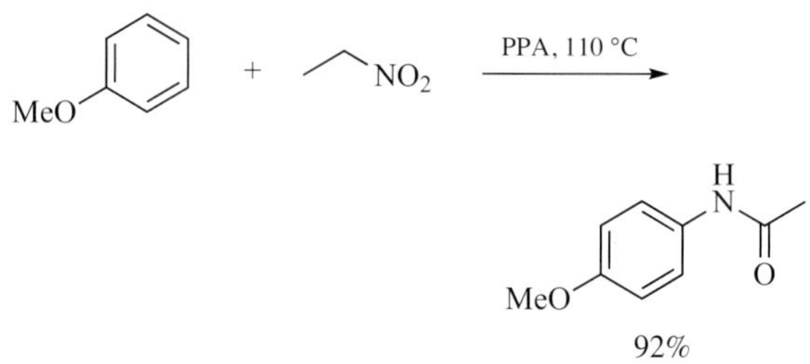

(31)

Nitroalkenes are commonly synthesized via a two-step sequence involving a Henry condensation between a carbonyl compound and a nitroalkane, followed by dehydration of the resulting β-nitroalkanol. The transformation can be done in a one-pot process, with complete stereocontrol, by simply changing the solvent and the temperature of the reaction (eq 32).[98]

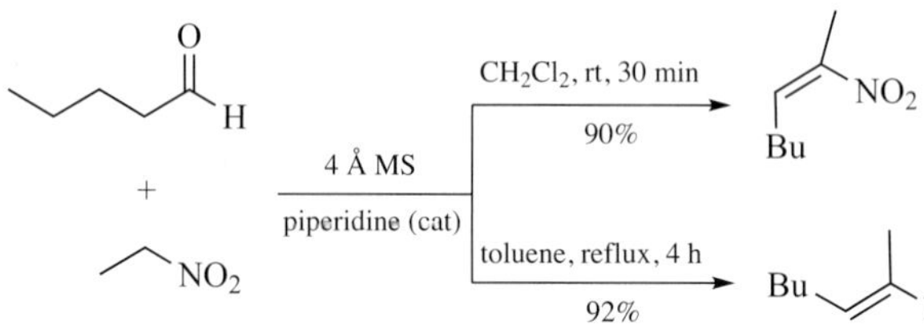

(32)

Finally, nitroethane has been used as a Michael acceptor in a Fe-catalyzed multicomponent reaction leading to the formation of functionalized pyrroles.[99]

Related Reagents. *t*-butyldimethylsilyl ethylnitronate; lithium α-lithiomethanenitronate; *O,O*-dilithio-1-nitropropene; nitromethane; 1-nitropropane; 1-nitro-1-propene.

---

1. (a) Seebach, D.; Colvin, E. W.; Lehr, F.; Weller, H., *Chimia* **1979**, *33*, 1. (b) Rosini, G.; Ballini, R., *Synthesis* **1988**, 833.
2. (a) v. Schickh, O.; Apel, G.; Padeken, H. G.; Schwarz, H. H.; Segnitz, A., *Methoden Org. Chem. (Houben-Weyl)* **1971**, *10/1*. (b) Rosini, G., *Comprehensive Organic Synthesis* **1991**, *2*, 321. (c) Jones, G., *Org. React.* **1967**, *15*, 204. (d) Baer, H. H.; Urbas, L. In *The Chemistry of the Nitro and Nitroso Groups*; Feuer, H., Ed.; Wiley: New York, 1970; Part 2, p 75. (e) Hass, H. B.; Riley, E. F., *C. R. Hebd. Seances Acad. Sci.* **1943**, *32*, 373.
3. Bachman, G. B.; Maleski, R. J., *J. Org. Chem.* **1972**, *37*, 2810.
4. (a) Rosini, G.; Ballini, R.; Petrini, M.; Sorrenti, P., *Synthesis* **1985**, 515. (b) Rosini, G.; Ballini, R.; Sorrenti, P., *Synthesis* **1983**, 1014.
5. Melot, J.-M.; Texier-Boullet, F.; Foucaud, A., *Tetrahedron Lett.* **1986**, *27*, 493.
6. (a) Dauben, H. J.; Ringold, H. J.; Wade, R. H.; Pearson, D. L.; Anderson, A. G., *J. Am. Chem. Soc.* **1951**, *73*, 2359. (b) Colvin, E. W.; Seebach, D., *J. Chem. Soc., Chem. Commun.* **1978**, 689. (c) Colvin, E. W.; Beck, A. K.; Seebach, D., *Helv. Chim. Acta* **1981**, *64*, 2264.
7. (a) Noland, W. E., *C. R. Hebd. Seances Acad. Sci.* **1955**, *55*, 137. (b) Ballini, R.; Petrini, M., *Tetrahedron Lett.* **1989**, *30*, 5329. (c) Clark, J. H.; Cork, D. G., *J. Chem. Soc., Perkin Trans. 1* **1983**, 2253.
8. Petrini, M.; Ballini, R.; Rosini, G., *Synthesis* **1987**, 713.
9. Rosini, G.; Ballini, R.; Sorrenti, P.; Petrini, M., *Synthesis* **1984**, 607.
10. Ono, N.; Kaji, A., *Synthesis* **1986**, 693.
11. Schales, O.; Graefe, H. A., *J. Am. Chem. Soc.* **1952**, *74*, 4486.
12. (a) Raiford, L. C.; Fox, D. E., *J. Org. Chem.* **1944**, *9*, 170. (b) Karmarkar, S. N.; Kelkar, S. L.; Wadia, M. S., *Synthesis* **1985**, 510.
13. Melton, J.; McMurry, J. E., *J. Org. Chem.* **1975**, *40*, 2138.
14. Buckley, G. D.; Scaife, C. W., *J. Chem. Soc.* **1947**, 1471.
15. Knochel, P.; Seebach, D., *Synthesis* **1982**, 1017.
16. Knochel, P.; Seebach, D., *Tetrahedron Lett.* **1982**, *23*, 3897.
17. Denmark, S. E.; Moon, Y.-C.; Cramer, C. J.; Dappen, M. S.; Senanayake, C. B. W., *Tetrahedron* **1990**, *46*, 7373.
18. Ballini, R.; Castagnani, R.; Petrini, M., *J. Org. Chem.* **1992**, *57*, 2160.
19. (a) Barrett, A. G.; Graboski, G. G. M., *C. R. Hebd. Seances Acad. Sci.* **1986**, *86*, 751. (b) Posner, G. H.; Crouch, R. D., *Tetrahedron* **1990**, *46*, 7509.
20. (a) Denmark, S. E.; Senanayake, C. B. W.; Ho, G.-D., *Tetrahedron* **1990**, *46*, 4857. (b) Denmark, S. E.; Senanayake, C. B. W., *J. Org. Chem.* **1993**, *58*, 1853.
21. (a) Kabalka, G. W.; Guindi, L. H. M. Varma, R. S., *Tetrahedron* **1990**, *46*, 7443. (b) Rylander, P. *Catalytic Hydrogenation in Organic Synthesis*; Academic: New York, 1979. (c) Erne, M.; Ramirez, F., *Helv. Chim. Acta* **1950**, *33*, 912.
22. Aizpurua, J. M.; Oiarbide, M.; Palomo, C., *Tetrahedron Lett.* **1987**, *28*, 5365.
23. Torii, S.; Tanaka, H.; Katoh, T., *Chem. Lett.* **1983**, 607.
24. Varma, R. S.; Varma, M.; Kabalka, G. W., *Tetrahedron Lett.* **1985**, *26*, 3777.
25. Ono, N.; Kamimura, A.; Kaji, A., *Tetrahedron Lett.* **1984**, *25*, 5319.
26. Ono, N.; Miyake, H.; Kaji, A., *J. Chem. Soc., Chem. Commun.* **1982**, 33.
27. (a) Torssell, K. B. G.; Zeuthen, O., *Acta Chem. Scand.* **1978**, *B32*, 118. (b) Colvin, E. W.; Beck, A. K.; Bastani, B.; Seebach, D.; Kai, Y.; Dunitz; J. D., *Helv. Chim. Acta* **1980**, *63*, 697.
28. Seebach, D.; Beck, A. K.; Mukhopadhyay, T.; Thomas, E., *Helv. Chim. Acta* **1982**, *65*, 1101.
29. Seebach, D.; Lehr, F., *Angew. Chem.* **1976**, *88*, 540.
30. (a) Lichtenthaler, F. W.; Zinke, H., *Angew. Chem., Int. Ed. Engl.* **1966**, *5*, 737. (b) Baer, H. H.; Rao, G. V., *Justus Liebigs Ann. Chem.* **1965**, *686*, 210.
31. Eberle, M.; Egli, M.; Seebach, D., *Helv. Chim. Acta* **1988**, *71*, 1.
32. Matsumoto, K., *Angew. Chem.* **1984**, *96*, 599.
33. Tamura, R.; Sato, M.; Oda, D., *J. Org. Chem.* **1986**, *51*, 4368.
34. Lehr, F.; Gonnermann, J.; Seebach, D., *Helv. Chim. Acta* **1979**, *62*, 2258.
35. Asaoka, M.; Mukuta, T.; Takei, H., *Tetrahedron Lett.* **1981**, *22*, 735.
36. (a) Bäckvall, J.-E.; Ericsson, A. M.; Plobeck, N. A.; Juntunen, S. K., *Tetrahedron Lett.* **1992**, *33*, 131. (b) Ono, N.; Kamimura, A.; Miyake, H.; Hamamoto, I.; Kaji, A., *J. Org. Chem.* **1985**, *50*, 3692. (c) Ono, N.; Kamimura, A.; Kaji, A., *Synthesis* **1984**, 226.
37. (a) Anderson, D. A.; Hwu, J. R., *J. Org. Chem.* **1990**, *55*, 511. (b) Anderson, D. A.; Hwu, J. R., *J. Chem. Soc., Perkin Trans. 1* **1989**, 1694. (c) Clark, J. H.; Miller, J. M.; So, K. H., *J. Chem. Soc., Perkin Trans. 1* **1978**, 941. (d) Colonna, S.; Hiemstra, H.; Wynberg, H., *J. Chem. Soc., Chem. Commun.* **1978**, 238. (e) Belsky, I., *J. Chem. Soc., Chem. Commun.* **1977**, 237.
38. Rosini, G.; Ballini, R.; Petrini, M.; Marotta. E., *Angew. Chem.* **1986**, *98*, 935.
39. (a) Bergbreiter, D. E.; Lalonde, J. J., *J. Org. Chem.* **1987**, *52*, 1601. (b) Clark, J. H.; Cork, D. G.; Robertson, M. S., *Chem. Lett.* **1983**, 1145.
40. McKillop, A.; Kobylecki, R. J., *Tetrahedron* **1974**, *30*, 1365.
41. Baker, D. C.; Putt, S. R., *Synthesis* **1978**, 478; Crumbie, R. L.; Nimitz, J. S.; Mosher, H. S., *J. Org. Chem.* **1982**, *47*, 4040.
42. Backman, G. B.; Hokama, T., *J. Am. Chem. Soc.* **1959**, *81*, 4882.
43. Stiles, M.; Finkbeiner, H. L., *J. Am. Chem. Soc.* **1959**, *81*, 505.
44. (a) Kerber, R. C.; Urry, G. W.; Kornblum, N., *J. Am. Chem. Soc.* **1965**, *87*, 4520. (b) Kornblum, N.; Brown, R. A., *J. Am. Chem. Soc.* **1964**, *86*, 2681. (c) Kornblum, N.; Brown, R. A., *J. Am. Chem. Soc.* **1963**, *85*, 1359.
45. Liebermann, S. V., *J. Am. Chem. Soc.* **1955**, *77*, 1114.
46. Katritzky, A. R.; De Ville, G.; Patel, R. C., *J. Chem. Soc., Chem. Commun.* **1979**, 602.
47. Seebach, D.; Henning, R.; Lehr, F.; Gonnermann, J., *Tetrahedron Lett.* **1977**, 1161.
48. Wade, P. A.; Morrow, S. D.; Hardinger, S. A., *J. Org. Chem.* **1982**, *47*, 365.
49. (a) Torssell, K. B. G. *Nitrile Oxides, Nitrones and Nitronates in Organic Synthesis*; VCH: New York, **1988**. (b) Caramella, P.; Grünhanger, P. *1,3-Dipolar Cycloaddition Chemistry*; Padwa, A., Ed.; Wiley: New York, **1984**.
50. Andersen, S. H.; Das, N. B.; Jørgensen, R. D.; Kjeldsen, G.; Knudsen, J. S.; Sharma, S. C.; Torssell, K. B. G., *Acta Chem. Scand.* **1982**, *B36*, 1.
51. (a) Kim, B. H.; Lee, J. Y.; Kim, K.; Whang, D., *Tetrahedron: Asymmetry* **1991**, *2*, 27. (b) Kim, B. H.; Lee, J. Y., *Tetrahedron: Asymmetry* **1991**, *2*, 1359.
52. Mukaiyama, T.; Hoshino, T *J. Am. Chem. Soc.* **1960**, *82*, 5339.
53. (a) Stack, J. A.; Heffner, T. A.; Geib, S. J.; Curran, D. P., *Tetrahedron* **1993**, *49*, 995. (b) Curran, D. P.; Jeong, K.-S.; Heffner, T. A.; Rebek, Jr., J., *J. Am. Chem. Soc.* **1989**, *111*, 9238.
54. Olsson, T.; Stern, K., *J. Org. Chem.* **1988**, *53*, 2468.
55. Akiyama, T.; Okada, K.; Ozaki, S., *Tetrahedron Lett.* **1992**, *33*, 5763.
56. (a) Oppolzer, W.; Kingma, A. J.; Pillai, S. K., *Tetrahedron Lett.* **1991**, *32*, 4893. (b) Kim, K. S.; Kim, B. H.; Park, W. M.; Cho, S. J.; Mhin, B. J., *J. Am. Chem. Soc.* **1993**, *115*, 7472.
57. Curran, D. P.; Zhang, J., *J. Chem. Soc., Perkin Trans. 1* **1991**, 2613.
58. Jäger, V.; Buss, V.; Schwab, W., *Liebigs Ann. Chem.* **1980**, 122.
59. Wang, X.; Fang, F.; Zhao, C.; Tian, S.-K., *Tetrahedron Lett.* **2008**, *49*, 6442.
60. Hirata, N.; Hayashi, M., *Synth. Commun.* **2007**, *37*, 1653.
61. Boruwa, J.; Gogoi, N.; Saikia, P. P.; Barua, N. C., *Tetrahedron: Asymmetry* **2007**, *17*, 3315.

62. Palomo, C.; Oiarbide, M.; Laso, A., *Eur. J. Org. Chem.* **2007**, 2561.
63. Blay, G.; Hernandez-Olmos, V.; Pedro, J. R., *Synlett* **2011**, 1195.
64. Alvarez-Casao, Y.; Marques-Lopez, E.; Herrera, R. P., *Symmetry* **2011**, *3*, 220.
65. Cheng, L.; Dong, J.; You, J.; Gao, G.; Lan, J., *Chem. Eur. J.* **2010**, *16*, 6761.
66. Zielinska-Blajet, M.; Skarzewski, J., *Tetrahedron: Asymmetry* **2011**, *22*, 351.
67. Zhou, Y.; Dong, J.; Zhang, F.; Gong, Y., *J. Org. Chem.* **2011**, *76*, 588.
68. Arai, T.; Takashita, R.; Endo, Y.; Watanabe, M.; Yanagisawa, A., *J. Org. Chem.* **2008**, *73*, 4903.
69. Sohtome, Y.; Takemura, N.; Takada, K.; Takagi, R.; Iguchi, T.; Nagasawa, K., *Chem. Asian J.* **2007**, *2*, 1150.
70. Lang, K.; Park, J.; Hong, S., *Angew. Chem., Int. Ed.* **2012**, *51*, 1620.
71. Nitabaru, T.; Nojiri, A.; Kobayashi, M.; Kumagai, N.; Shibasaki, M., *J. Am. Chem. Soc.* **2009**, *131*, 13860.
72. Handa, S.; Nagawa, K.; Sohtome, Y.; Matsunaga, S.; Shibasaki, M., *Angew. Chem., Int. Ed.* **2008**, *47*, 3230.
73. Uraguchi, D.; Nakamura, S.; Ooi, T., *Angew. Chem., Int. Ed.* **2010**, *49*, 7562.
74. Takada, K.; Takemura, N.; Cho, K.; Sohtome, Y.; Nagasawa, K., *Tetrahedron Lett.* **2008**, *49*, 1623.
75. Uraguchi, D.; Ito, T.; Nakamura, S.; Sakaki, S.; Ooi, T., *Chem. Lett.* **2009**, *38*, 1052.
76. Liu, L.; Zhang, S.-L.; Xue, F.; Lou, G.-S.; Zhang, H.-Y.; Ma, S.-C.; Duan, W.-H.; Wang, W., *Chem. Eur. J.* **2011**, *17*, 7791.
77. Meshram, H. M.; Ramesh, P.; Sanjeeva, K. A.; Swetha, A., *Tetrahedron Lett.* **2011**, *52*, 5862.
78. Marques-Lopez, E.; Merino, P.; Tejero, T.; Herrera, R. P., *Eur. J. Org. Chem.* **2009**, 2401.
79. Johnson, K. M.; Rattley, M. S.; Sladojevich, F.; Barber, D. M.; Nuñez, M. G.; Goldys, A. M.; Dixon, D. J., *Org. Lett.* **2012**, *14*, 2492.
80. Handa, S.; Gnanadesikan, V.; Matsunaga, S.; Shibasaki, M., *J. Am. Chem. Soc.* **2010**, *132*, 4925.
81. Pahadi, N. K.; Ube, H.; Terada, M., *Tetrahedron Lett.* **2007**, *48*, 8700.
82. Liu, L.; Wu, D.; Zheng, S.; Li, T.; Li, X.; Wang, S.; Li, J.; Li, H.; Wang, W., *Org. Lett.* **2012**, *14*, 134.
83. Dong, X.-Q.; Teng, H.-L.; Wang, C.-J., *Org. Lett.* **2009**, *11*, 1265.
84. Yang, X.; Zhou, X.; Lin, L.; Chang, L.; Liu, X.; Feng, X., *Angew. Chem., Int. Ed.* **2008**, *47*, 7079.
85. Ooi, T.; Fujioka, S.; Maruoka, K., *J. Am. Chem. Soc.* **2004**, *126*, 11790.
86. Chiarucci, M.; Lombardo, M.; Trombini, C.; Quintavalla, A., *Adv. Synth. Catal.* **2012**, *354*, 364.
87. Zhu, Q.; Lu, Y., *Org. Lett.* **2009**, *11*, 1721.
88. Ogawa, T.; Mouri, S.; Yazaki, R.; Kumagai, N.; Shibasaki, M., *Org. Lett.* **2012**, *14*, 110.
89. Xie, J.; Li, H.; Zhou, J.; Cheng, Y.; Zhu, C., *Angew. Chem., Int. Ed.* **2012**, *51*, 1252.
90. Hari, D. P.; Koenig, B., *Org. Lett.* **2011**, *13*, 3852.
91. Pan, Y.; Kee, C. W.; Chen, L.; Tan, C.-H., *Green Chem.* **2011**, *13*, 2682.
92. Zhang, M.; Zhou, J.; Kan, J.; Wang, M.; Su, W.; Hong, M., *Chem. Commun.* **2010**, *46*, 5455.
93. Dahnz, A.; Helmchen, G., *Synlett* **2006**, 697.
94. Vogl, E. M.; Buchwald, S. L., *J. Org. Chem.* **2002**, *67*, 106.
95. Cecchi, L.; De, S. F.; Machetti, F., *Chem. Eur. J.* **2008**, *14*, 7903.
96. Khalaf, A.; Gree, D.; Abdallah, H.; Jaber, N.; Hachem, A.; Gree, R., *Tetrahedron* **2011**, *67*, 3881.
97. Aksenov, A. V.; Aksenov, N. A.; Nadein, O. N.; Aksenova, I. V., *Synlett* **2010**, 2628.
98. Fioravanti, S.; Pellacani, L.; Tardella, P. A.; Vergari, M. C., *Org. Lett.* **2008**, *10*, 1449.
99. Maiti, S.; Biswas, S.; Jana, U., *J. Org. Chem.* **2010**, *75*, 1674.

4-Nitropyridine *N*-oxide

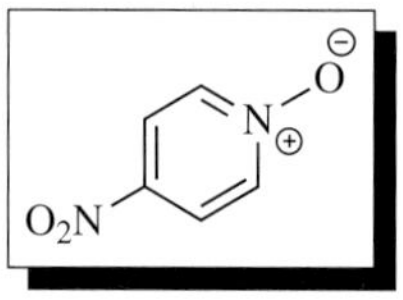

[1124-33-0] $C_5H_4N_2O_3$ (MW 140.10)
InChI = 1S/C5H4N2O3/c8-6-3-1-5(2-4-6)7(9)10/h1-4H
InChIKey = RXKNNAKAVAHBNK-UHFFFAOYSA-N

(reagent used as oxidant, base, oxygen transfer reagent, building block for pyridine synthesis)

Physical Data: mp 160.5 °C.
Solubility: sol DMSO, acetone, CH_2Cl_2, THF, Et_2O, toluene, MeOH, H_2O.
Form Supplied in: yellow solid, widely available.
Purification: recrystallization from acetone.
Handling, Storage, and Precautions: avoid contact with skin, eyes, and clothing. Avoid formation of dust and aerosols. Keep container tightly closed in a well-ventilated and dry place.

Reduction of 4-Nitropyridine *N*-Oxide. A variety of methods have been developed for the reduction of 4-nitropyridine *N*-oxide. Methods for the deoxygenation of the *N*-oxide moiety, as well as methods for the global reduction of both the *N*-oxide and the nitro group have been reported. 4-Nitropyridine *N*-oxide may be deoxygenated with triethylamine-sulfur dioxide complex,[1] diphosphorus tetraiodide,[2] nickel-aluminum alloy,[3] photochemical oxygen transfer to dimethylaniline,[4] oxygen transfer to triphenylphosphine via an oxorhenium(V) catalyst,[5] sodium borohydride/lithium chloride (eq 1),[6] and Raney nickel.[7] Global reduction to 4-aminopyridine may be accomplished with Pd-catalyzed transfer hydrogenation[8,9] or with zinc dust in conjunction with ammonium formate.[10]

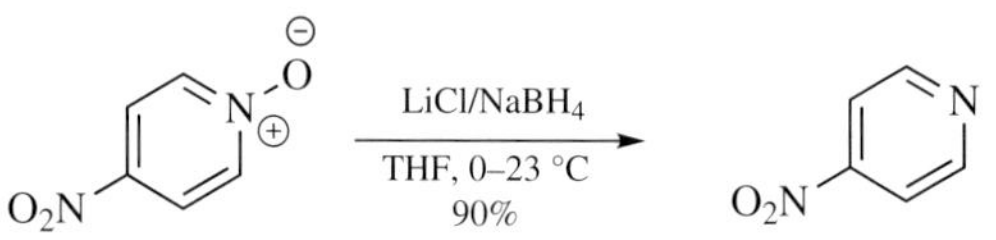

(1)

In Pyridine Synthesis. 4-Nitropyridine *N*-oxide has been used as a synthetic precursor in a number of pyridine syntheses. 4-Nitropyridine *N*-oxide may be chlorinated at the 2-position with $POCl_3/PCl_5$ in a chlorination/deoxygenation reaction to generate 2-chloro-4-nitropyridine, which is capable of undergoing further manipulation to form a variety of substituted pyridines.[11] It has been shown that 2-arylsulfonyl-4-nitropyridine *N*-oxide is formed by treatment of *N*-oxide with chloromethyl phenyl sulfone in DMSO with KOH. 2-Arylsulfonyl-4-nitropyridine *N*-

oxide may then be further manipulated to form bicyclic pyridine derivatives.[12] Amides react with 4-nitropyridine *N*-oxide in the presence of trifluoromethane sulfonic anhydride and 2-fluoropyridine to effect a deoxygenation of the *N*-oxide with a concomitant C–N bond formation at the 2-position to generate functionalized pyridines (eq 2).[13] 2-Cyano-4-chloropyridine is generated when 4-nitropyridine *N*-oxide is treated with trimethylsilyl cyanide and ethyl chloroformate.[14] The nitro group of 4-nitropyridine *N*-oxide is capable of undergoing S_NAr chemistry with benzyl alcohols to furnish 4-benzyloxypyridine *N*-oxide derivatives, which may be reduced to form the functionalized pyridine.[15] Finally, 4-nitropyridine *N*-oxide is capable of being coupled with 1-chloro-4-isocyanobenzene in the presence of trimethylsilyl trifluoromethane sulfonate in a deoxygenation/C-N bond forming reaction to form *N*-(4-chlorophenyl)-4-nitropyridin-2-amine.[16]

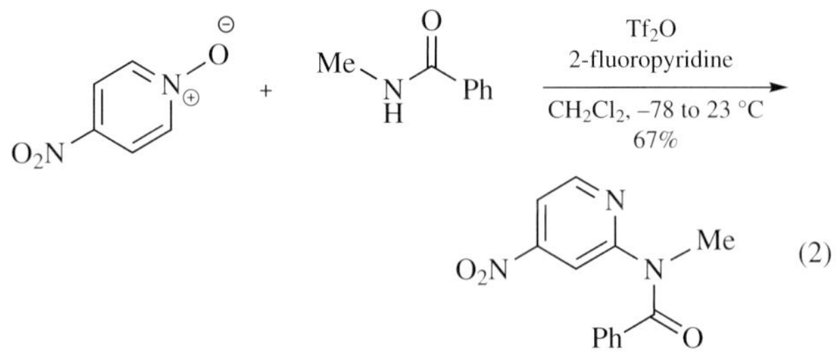

(2)

Oxidation in *N*-Heterocyclic Carbene Catalysis. Within the context of *N*-heterocyclic carbene (NHC) catalysis, 4-nitropyridine *N*-oxide has been used as an oxidant in the formation of β-hydroxy ester from enals and in the annulation of enals to form cyclopentanones. 4-Nitropyridine *N*-oxide behaves as a source of oxygen for an atom transfer wherein an oxygen from the nitro group is transferred to the β-carbon of the Breslow intermediate. A variety of aryl and aliphatic enals react with 4-nitropyridine *N*-oxide in the presence of a chiral NHC catalyst in methanol to generate β-hydroxy esters in high yields and with good enantioselectivities (eq 3).[17]

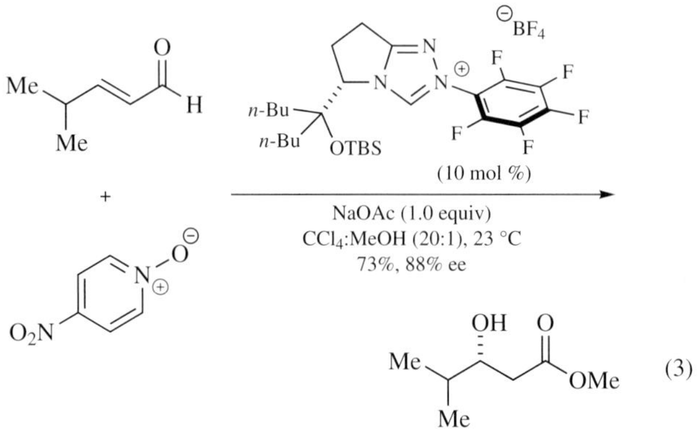

(3)

4-Nitropyridine *N*-oxide has also been used as an oxidant to facilitate the annulation of enals to furnish cyclopentanones in high yields and with good enantioselectivities. In this chemistry, 4-nitropyridine *N*-oxide is believed to act as a single-electron oxidant to generate the radical cation of the Breslow intermediate. This radical cation is then attacked by a native Breslow intermediate to furnish the carbon-carbon bond between the β-carbons of the enals. Aryl enals participate in this chemistry to generate symmetrical cyclopentanones via a dimerization. A cross-annulation has also been developed wherein the more reactive enal is used in excess to bias the reaction toward the cross-product (eq 4).[18]

Terminal Oxidant in Metal-catalyzed Reactions. In the osmium(VIII)-catalyzed oxidative cyclization of *N*-Ts amino alcohols, 4-nitropyridine *N*-oxide was shown to be the optimal terminal oxidant to reoxidize osmium(IV) to osmium(VI) (eq 5).[19]

Oxidation of α, β-(Epoxyalkyl)trimethylsilane via Oxygen Transfer. 4-Nitropyridine *N*-oxide is capable of oxidizing α, β-(epoxyalkyl)trimethyl silanes to the corresponding glyceraldehyde derivative. This reaction proceeds via nucleophilic attack of the *N*-oxide into the α-carbon of the epoxide. The resultant alkoxide is silylated and the N–O bond of the *N*-oxide is cleaved with concomitant loss of trimethylsilyl cation. A variety of pyridine *N*-oxides participate in this reaction. In this reaction it was found that the nucleophilicity of the *N*-oxide is not the decisive factor; instead, the ability of the pyridine to act as a leaving group appears to be more important. For this reason, 4-nitropyridine *N*-oxide was found to be the optimal oxidant (eq 6).[20]

4-Nitropyridine *N*-Oxide as a Weak Base. In the silver-catalyzed [4+2] cyclization of 2-alkylbenzaldehydes and alkenes, 4-nitropyridine *N*-oxide was found to be an optimal noncoordinating weak base to effect the enolization of the 2-alkylbenzaldehyde component to generate the reactive *ortho*-quinone methide intermediate. A variety of bases were screened in this reaction and it was found that bases capable of coordinating to the silver catalyst lead to catalyst poisoning. 4-Nitropyridine *N*-oxide proved basic enough to generate the *ortho*-quinone methide, but weakly coordinating to avoid poisoning of the silver catalyst. A variety of tetrahydronaphthalenones are generated in high yields and with good diastereoselectivities using this protocol (eq 7).[21]

In C–H Activation. 4-Nitropyridine *N*-oxide has been used as a substrate in transition-metal-catalyzed C–H activation with the *N*-oxide moiety acting as a directing group. In 2009, Fagnou and coworkers showed the Pd-catalyzed directed arylation of 4-nitropyridine *N*-oxide at the 2-position with 4-methylbromobenzene to furnish 4-nitro-2-(*p*-tolyl)pyridine 1-oxide (eq 8).[22] A similar reaction has been reported wherein 4-nitropyridine *N*-oxide undergoes C-H activation with Pd and is then coupled to 2-bromo-6-methoxypyridine.[23] Finally, 4-nitropyridine *N*-oxide has been used in a Pd-catalyzed dual C–H activation oxidative coupling with thiazole.[24]

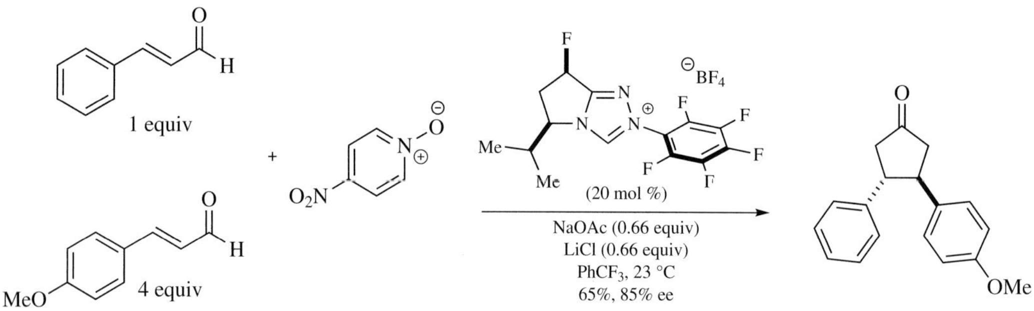

(4)

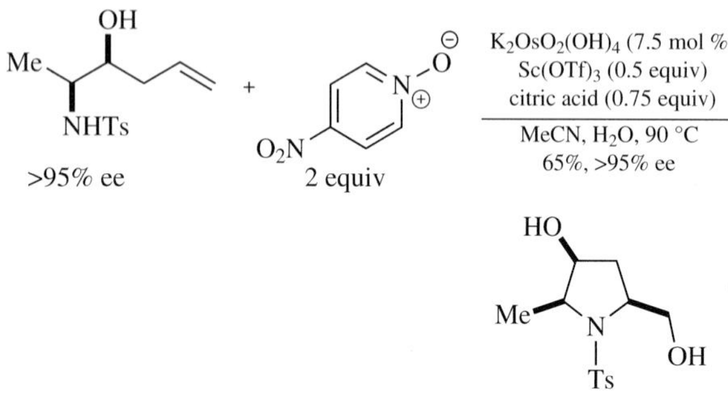

(5)

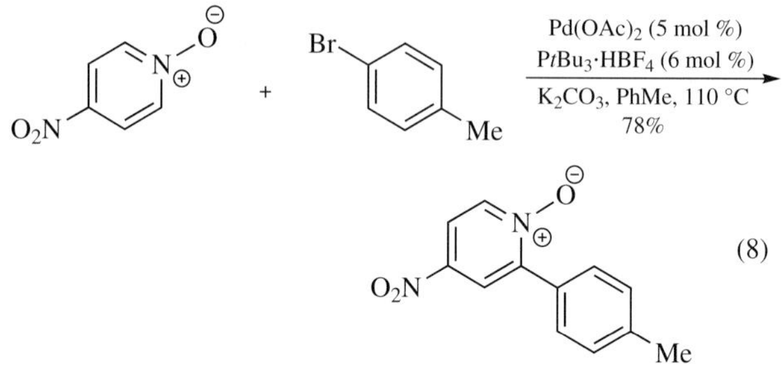

(8)

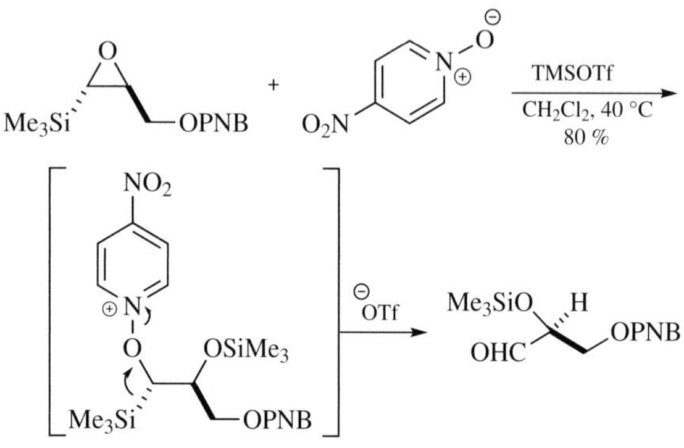

(6)

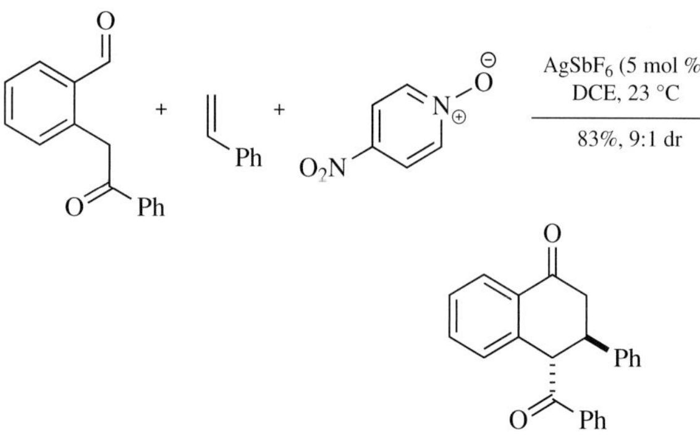

(7)

Related Reagents. Pyridine *N*-oxide; 6-methoxypyridine *N*-oxide; 4-methoxypyridine *N*-oxide; 3,5-dichloropyridine *N*-oxide; *N*-methylmorpholine *N*-oxide; 2,6-pyridine dicarboxylic acid, 1-oxide.

1. Olah, G. A.; Arvanaghi, M.; Vankar, Y. D., *Synthesis* **1980**, *8*, 660.
2. Suzuki, H.; Sato, N.; Osuka, A., *Chem. Lett.* **1980**, 459.
3. Lunn, G.; Sansone, E. B., *J. Org. Chem.* **1986**, *51*, 513.
4. Sako, M.; Shimada, K.; Hirota, K.; Maki, Y., *J. Am. Chem. Soc.* **1986**, *108*, 6039.
5. Wang, Y.; Espenson, J. H., *Org. Lett.* **2000**, *2*, 3525.
6. Ram, S. R.; Chary, K. P.; Iyengar, D. S., *Synth. Commun.* **2000**, *30*, 3511.
7. Balicki, R.; Maciejewski, G., *Synth. Commun.* **2002**, *32*, 1681.
8. Fuentes, J. A.; Clarke, M. L., *Synlett* **2008**, *17*, 2579.
9. Quinn, J. F.; Bryant, C. E.; Golden, K. C.; Gregg, B. T., *Tetrahedron Lett.* **2010**, *51*, 786.
10. Balicki, R.; Cybulski, M.; Maciejewski, G., *Synth. Commun.* **2003**, *33*, 4137.
11. Walters, M. A.; Shay, J. J., *Tetrahedron Lett.* **1995**, *36*, 7575.
12. Bernotas, R. C.; Antane, S. A.; Lenicek, S. E.; Haydar, S. N.; Robichaud, A. J.; Harrison, B. L.; Zhang, G. M.; Smith, D.; Coupet, J.; Schechter, L. E., *Bioorg. Med. Chem. Lett.* **2009**, *19*, 6935.

13. Medley, J. W.; Movassaghi, M., *J. Org. Chem.* **2009**, *74*, 1341.
14. Veerareddy, A.; Surendrareddy, G.; Duby, P. K., *J. Heterocycl. Chem.* **2011**, *48*, 961.
15. Ichihara, Y.; Fujimura, R.; Tsuneki, H.; Wada, T.; Okamoto, K.; Gouda, H.; Hirono, S.; Sugimoto, K.; Matsuya, Y.; Sasaoka, T.; Toyooka, N., *Eur. J. Med. Chem.* **2013**, *62*, 649.
16. Vamos, M.; Cosford, N. D. P., *J. Org. Chem.* **2014**, *79*, 2274.
17. White, N. A.; Rovis, T., *J. Am. Chem. Soc.* **2014**, *136*, 14674.
18. White, N. A.; Rovis, T., *J. Am. Chem. Soc.* **2015**, *137*, 10112–10115.
19. Donohoe, T. J.; Lindsay-Scott, P. J.; Parker, J. S.; Callens, C. K. A., *Org. Lett.* **2010**, *12*, 1060.
20. Raubo, P.; Wicha, J., *Tetrahedron Lett.* **1994**, *35*, 3387.
21. Zhu, S.; Liang, R.; Jian, H.; Wu, W., *Angew. Chem., Int. Ed.* **2012**, *51*, 10861.
22. Campeau, L.-C.; Stuart, D. R.; Leclerc, J.-P.; Bertand-Laperle, M.; Villemure, E.; Sun, H.-Y.; Lasserre, N. G.; Guimond, N.; Lecavallier, M.; Fagnou, K., *J. Am. Chem. Soc.* **2009**, *131*, 3291.
23. Duric, S.; Sypaseuth, F. D.; Hoof, D.; Svensson, E.; Tzschucke, C. C., *Chem. Eur. J.* **2013**, *19*, 17456.
24. Fu, X.-P.; Xuan, Q.-Q.; Liu, L.; Wang, D.; Chen, Y.-J.; Li, C.-J., *Tetrahedron* **2013**, *69*, 4436.

Nicholas A. White
Colorado State University, Fort Collins, CO, USA

P

(2*S*)-*N*-(2,3,4,5,6-Pentafluorophenyl)-2-pyrrolidinecarboxamide

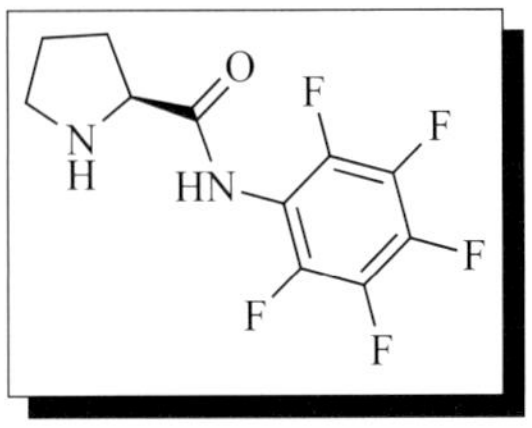

[(2*S*)-*N*-(2,3,4,5,6-Pentafluorophenyl)-2-pyrrolidinecarboxamide]

[1021188-17-9] $C_{11}H_9F_5N_2O$ (MW 280.19)

InChI = 1S/C11H9F5N2O/c12-5-6(13)8(15)10(9(16)7(5)14)18-11(19)4-2-1-3-17-4/h4,17H,1-3H2,(H,18,19)/t4-/m0/s1

InChIKey = QACVCVKUWCBPRE-BYPYZUCNSA-N

Physical Data: mp 82–84 °C; $[\alpha]^{27}_D = -36.46$ ($c = 0.5$, EtOH).

Solubility: soluble in H_2O, alcohol, THF, HMPA, and most organic solvents.

Form Supplied in: crystalline.

Preparative Method: prepared in three steps from commercially available Boc-protected L-proline (**1**). The Boc derivative **3** was synthesized from **1** via the formation of a mixed anhydride with ethyl chloroformate, followed by treatment with pentafluoroaniline (**2**). Deprotection (eq 1) then gives pentafluorophenyl prolinamide (**4**).[1]

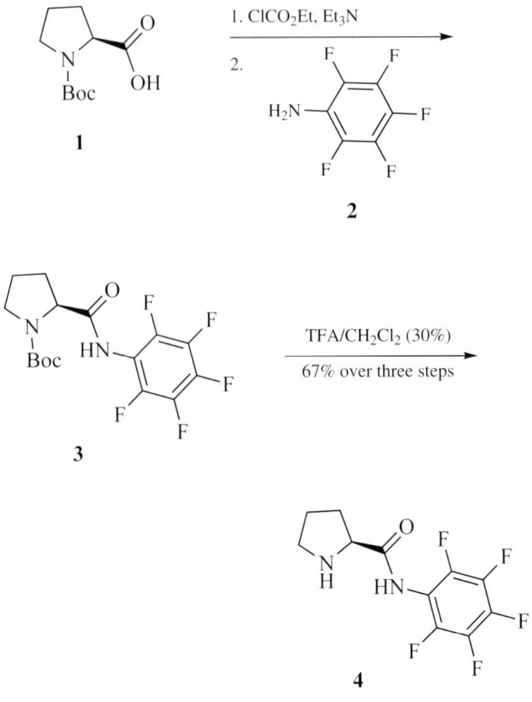

(1)

Handling, Storage, and Precautions: storage in a cool, dry, and ventilated place is recommended.

Introduction. (2*S*)-*N*-(2,3,4,5,6-Pentafluorophenyl)-2-pyrrolidine carboxamide is a bifunctional organocatalyst analog of L-proline. The presence of a perfluorophenyl moiety increases the acidity of the amide NH, which increases its hydrogen bond donating capabilities and so leads to more compact transition states and better enantioselectivity.

Asymmetric Aldol Reactions. Although, the first observation[2] of **4** catalyzing an asymmetric aldol reaction between *p*-nitrobenzaldehyde and acetone employed HMPA as a solvent (eq 2), it was found later that the catalyst performs efficiently in most of the common organic solvents, both nonpolar and polar, including brine.

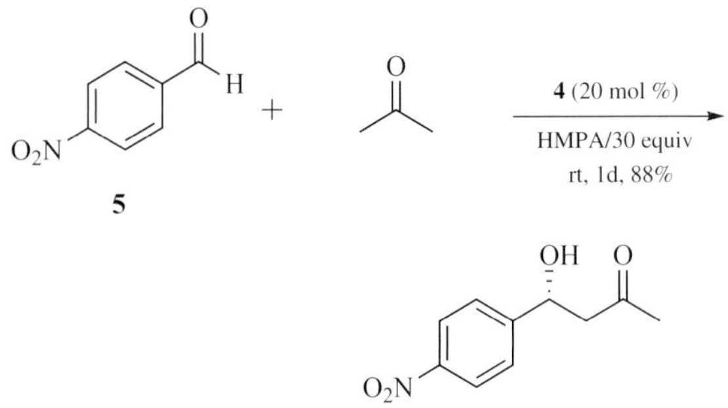

(2)

The catalyst facilitates the reaction between cyclohexanone and *p*-nitrobenzaldehyde in diverse conditions yielding the aldol adduct with excellent de's and ee's in the presence of TFA as an additive (eq 3). This reagent is one of the few known organocatalysts that catalyze asymmetric aldol reactions at room temperature with very high enantiopurity in both nonpolar solvents such as cyclohexane and polar solvents such as brine (eq 3).

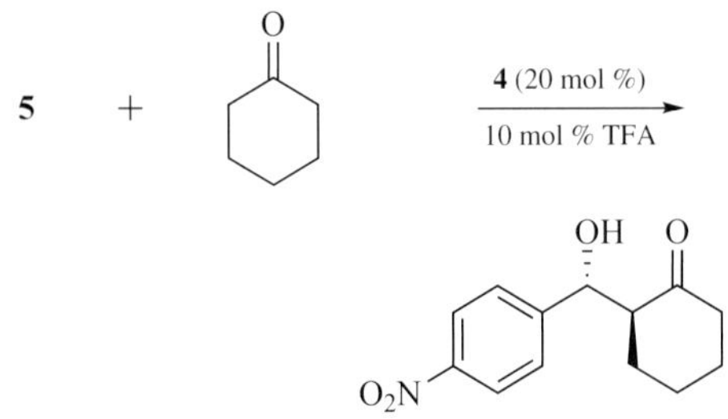

(3)

Solvent	Temp (°C)	Time	Yield (%)	dr	ee (%)
Cyclohexane	3	36	88	98:2	95
Dichloromethane	3	36	86	98:2	95
Acetonitrile	22	22	88	95:5	93
Toluene	3	20	76	98:2	93
Water	3	20	79	98:2	96
Brine	3	20	75	99:1	97
Methanol	3	40	72	99:1	95
Ethyl formate	3	40	80	98:2	94

Apart from a variety of substituted benzaldehydes, the reagent also performs well with heteroaromatic aldehydes as electrophilic partners in aldol reactions with cyclohexanone (eq 4).[1] Pyridine-3-carboxaldehyde and pyridine-4-carboxaldehyde react quite rapidly with cyclohexanone in the presence of 20 mol % of **4** and 10 mol % of TFA in DMF leading to the formation of 70–90% isolated yields of the aldol products within 3–4 h with enantioselectivities >85%. So far, *N*-perfluorophenyl prolinamide **4** gave the best results in comparison with the other prolinamides in the enantioselective catalysis of the reaction between cyclohexanone and aromatic aldehydes. The X-ray crystal structure of **4** revealed the near orthogonal orientation of the pentafluorophenyl ring with the amide bond, which is in contrast to the conformations of other similar *N*-aryl prolinamides. The enhanced reactivity of **4** has been attributed to this structural difference as well as to the increased acidity of the -NH group. It is presumed that the near orthogonal orientation of the pentafluorophenyl ring allows better orientation of the electrophilic aldehyde resulting in higher enantioselectivity.

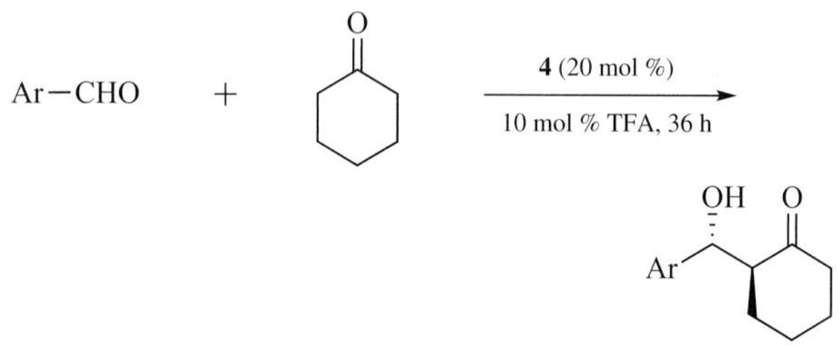

Ar =	Yield (%)	dr	ee (%)
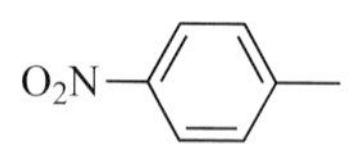	88	98:2	98
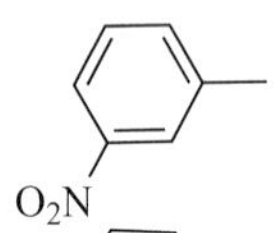	88	98:2	98
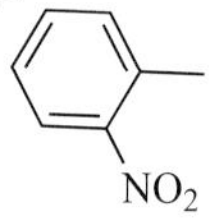	86	99:1	98
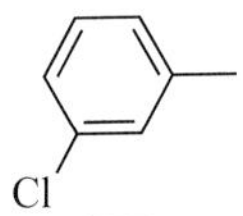	90	98:2	96
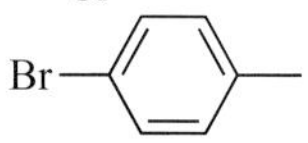	90	98:2	96
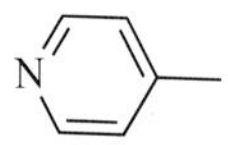	86	98:2	97
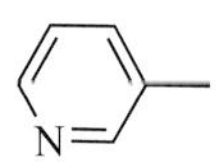	70	98:2	90

(4)

Asymmetric Biginelli Reaction. The reagent catalyzes the multicomponent Biginelli reaction (eq 5)[3] to yield the product in 45% in the presence of 10 mol % of TFA in dichloromethane albeit with very low enantioselectivity (9%).

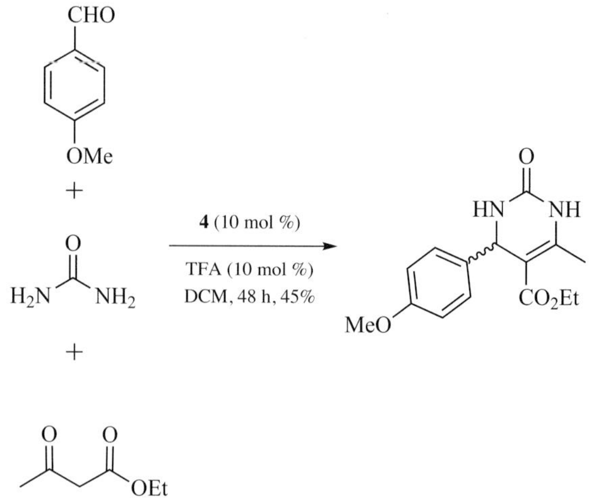

(5)

1. Moorthy, J. N.; Saha, S., *Eur. J. Org. Chem.* **2009**, 739.
2. Sato, K.; Kuriyama, M.; Shimazawa, R.; Morimoto, T.; Kakiuchi, K.; Shirai, R., *Tetrahedron Lett.* **2008**, *49*, 2402.
3. Saha, S.; Moorthy, J. N., *J. Org. Chem.* **2011**, *76*, 396.

Pavan K. Kancharla
Wayne State University, Michigan, USA

1,5-Pentanediamine, *N,N′*-Bis[(11b*R*)-3,5-dihydro-3,5-dimethyl-4-oxido-4*H*-dinaphtho[2,1-*d*:1′,2′-*f*]-[1,3,2]diazaphosphepin-4-yl]-*N,N′*-dimethyl

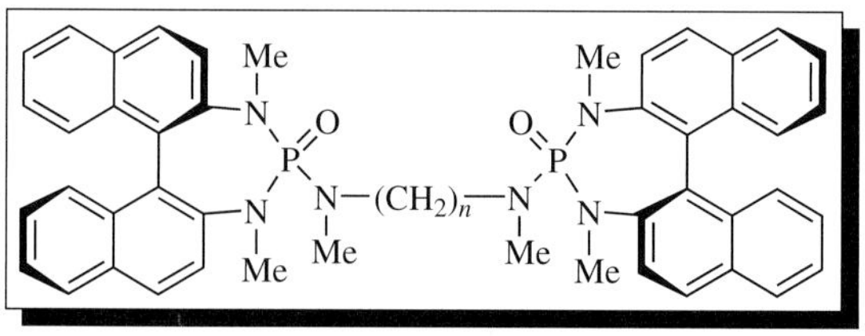

[352310-87-3] $C_{51}H_{52}N_6O_2P_2$ (MW 842.95)

InChI = 1/C51H52N6O2P2/c1-52(60(58)54(3)44-30-26-36-18-8-12-22-40(36)48(44)49-41-23-13-9-19-37(41)27-31-45(49)55(60)4)34-16-7-17-35-53(2)61(59)56(5)46-32-28-38-20-10-14-24-42(38)50(46)51-43-25-15-11-21-39(43)29-33-47(51)57(61)6/h8-15,18-33H,7,16-17,34-35H2,

InChIKey = HTIRTTKKZTYOST-UHFFFAOYAH

(catalyst for the asymmetric allylations of allylic tributylstannanes to aldehydes,[1] catalyst for the asymmetric aldol additions of aldehyde-derived trichlorosilyl enolates to aldehydes,[2,3] catalyst for the asymmetric aldol additions of aldehyde-,[4] ketone-,[5,6] and ester-derived[7] trialkylsilyl enal derivatives to aldehydes, catalyst for the vinylogous aldol addition of ketone-,[8] ester-,[7] and amide-derived[9] trialkylsilyl dienolates to aldehydes, catalyst for the asymmetric additions of isocyanides to aldehydes[10])

Physical Data: mp 216–218 °C (benzene); $[\alpha]_{24}^{D}$ −390° (*c*=1.07, MeOH).

Solubility: soluble in most organic solvents except hydrocarbons and diethyl ether.

Form Supplied in: fine white crystals.

Analysis of Reagent Purity: EA, IR, MS, NMR.

Preparative Method: bisphosphoramide (*R,R*)-**1** is prepared in a one-pot, three-step procedure[1] beginning with (*R*)-*N,N'*-dimethyl-1,1'-binaphthyl-2,2'diamine **2**[11] (eq 1). Treatment of the dilithium dianion of **2** with PCl_3, followed by the coupling with the appropriate *N,N'*-dimethylalkanediamine **3** in the presence of Et_3N, and finally oxidation with *m*-CPBA affords bisphosphoramides (*R,R*)-**1**. (*R,R*)-**1c** is commercially available from Obiter Research, LLC. Contact waboulanger@obiterresearch.com

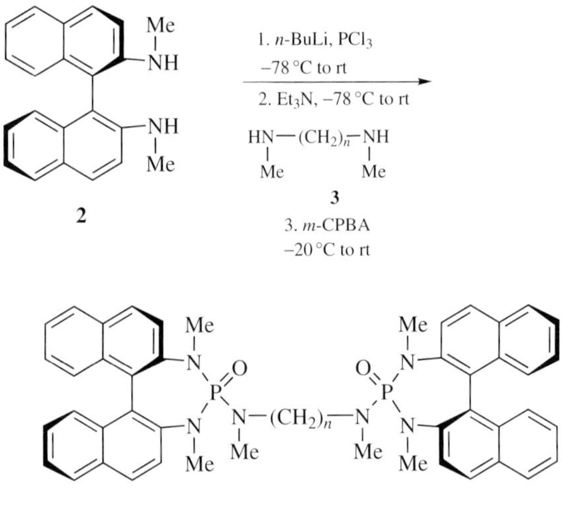

(*R,R*)-1 (1)

Handling, Storage, and Precautions: the phosphoramide is a stable, hygroscopic compound which is best stored in a dessicator or dry-box. Avoid prolonged exposure to air or moisture.

Introduction. *N,N'*-Bis[(11*bR*)-3,5-dihydro-3,5-dimethyl-4-oxido-4*H*-dinaphtho[2,1-*d*:1',2'-*f*][1,3,2]diazaphosphepin-4-yl]-*N,N'*-dimethyl-1,5-pentanediamine and its derivatives are catalysts for two types of Lewis-base-catalyzed addition reactions that differ by the nucleophilic addition partner. The first of these methods employs trichlorosilyl enolates.[12] Coordination of bisphosphoramide **1** to the silicon atom of the trichlorosilyl enolate results in ionization of a chloride ion and generation of a cationic species. This cationic silicon species binds to the carbonyl oxygen of an aldehyde, generating a ternary complex that undergoes reaction through a closed, six-membered transition structure.

The second method involves the activation of the weak Lewis acid $SiCl_4$ for Mukaiyama-type aldol additions of trialkylsilyl enol derivatines.[7] In the presence of chiral bisphosphoramide **1** (0.01–0.15 equiv), $SiCl_4$ is activated to form a strongly Lewis acidic, chiral trichlorosilyl cation through coordination and ionization of chloride. This species can then bind and activate the aldehyde, allowing for addition of the trialkylsilyl enolate through an open transition structure.

To obtain the highest yields and selectivities when employing the bisphosphoramide **1**/$SiCl_4$ catalyst system, it is essential that the $SiCl_4$ be free of HCl. This is accomplished by heating the $SiCl_4$ at reflux for 24 h and then distilling it prior to use. The need to purify the $SiCl_4$ can be obviated if substoichiometric amounts of acid-scavenging amines such as *i*-Pr_2NEt are included in the reaction mixture.

Allylic Additions to Aldehydes. The ability of the bisphosphoramide (*R,R*)-**1**/$SiCl_4$ catalyst system to promote the addition of latent nucleophiles to carbonyl compounds was first demonstrated in the addition of tri-*n*-butylallylstannane (**4**) to aldehydes.[1] Chiral bisphosphoramides with differing tether lengths were initially evaluated to determine the optimal arrangement of the two phosphoramide units via testing by the addition of allylic stannane **4** to benzaldehyde **5** (eq 2, Table 1). Although all four bisphosphoramides are effective catalysts for the allylation, catalyst (*R,R*)-**1c**, bearing a five-methylene linker, afforded the highest enantioselectivity.

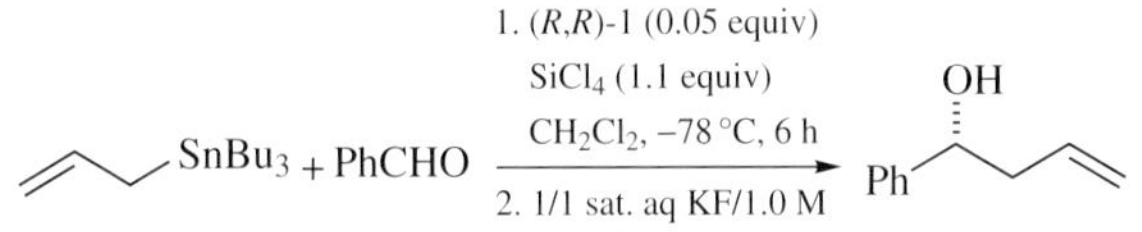

4 5 6 (2)

Table 1 Allylation of benzaldehyde **5** promoted by phosphoramides

Entry	Catalyst	*n*	Yield (%)	er
1	**1a**	3	81	92.0:8.0
2	**1b**	4	86	90.5:0.5
3	**1c**	5	89	96.5:3.5
4	**1d**	6	84	84.5:15.5

The catalytic action of 0.05 equiv of bisphosphoramide (*R,R*)-**1c** and 1.1 equiv of $SiCl_4$ at −78 °C promotes the addition of a variety of aromatic and conjugated aldehydes to allylic stannane **4** in high yields and moderate to good enantioselectivity (eq 3, Table 2, entries 1–5). However, aliphatic aldehydes did not react under these optimized conditions (entry 6).

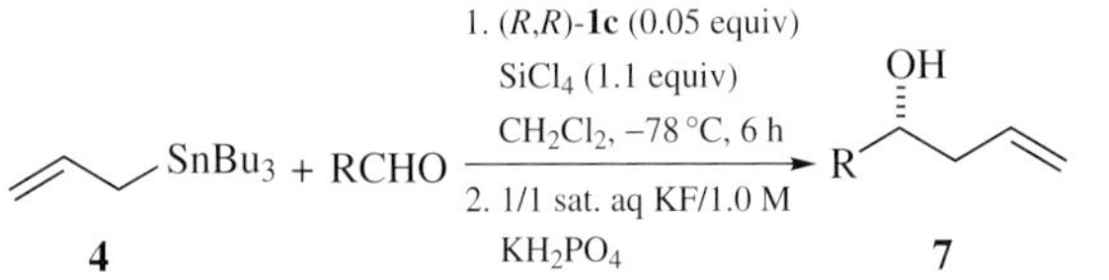

4 7 (3)

Aldehyde Enolate Aldol Additions to Aldehydes. Among the first examples of aldol additions employing bisphosphoramides **1** as catalysts was the addition of aldehyde-derived trichlorosilyl enolates to aldehydes.[2] In the presence of 0.05 equiv of (*R,R*)-**1c**, the reaction provides *β*-hydroxy aldehydes as protected dimethyl acetals in high yields after quenching with dry

Table 2 Allylations of aldehydes with **4**

Entry	R	Yield (%)	er
1	Phenyl	91	97.0:3.0
2	2-Furyl	65	81.0:19.0
3	4-$NO_2C_6H_4$	90	91.5:8.5
4	1-Naphthyl	94	97.0:3.0
5	(*E*)-PhCH=CH	91	82.5:17.5
6	$PhCH_2CH_2$	nr	nd

methanol (eq 4, Table 3). The diastereoselectivity of the reaction is uniformly excellent and correlates well with the geometrical composition of the trichlorosilyl enolates derived from heptanal **8** (entries 1 and 2) or propanal **9** (entries 3–10). This feature allows access to all possible stereoisomers of the aldolate architecture by changing enolate geometry and catalyst configuration. On the contrary, the enantioselectivity is highly variable and dependent on both enolate and aldehyde structure.

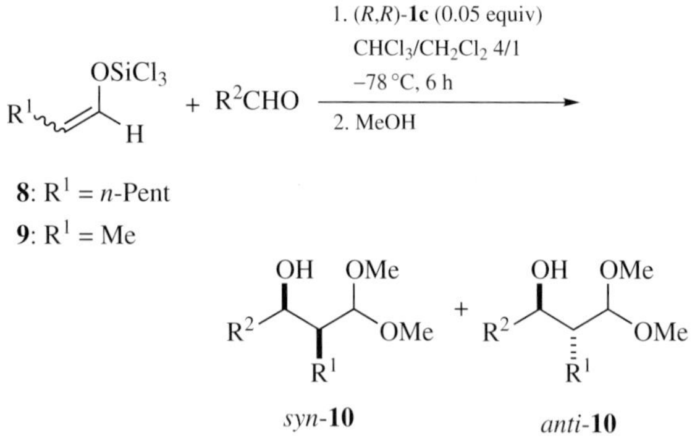

(4)

8: R^1 = *n*-Pent
9: R^1 = Me
syn-**10** *anti*-**10**

Table 3 Aldol additions of aldehyde trichlorosilyl enolates to various aldehydes

Entry	Enolate	R^2	Yield (%)	*syn*:*anti*	er (major)
1	(*Z*)-**8**	Phenyl	92	99:1	95.0:5.0
2	(*E*)-**8**	Phenyl	91	3:97	91.0:9.0
3	(*Z*)-**9**	Phenyl	95	98:2	90.5:9.5
4	(*E*)-**9**	Phenyl	97	1:99	79.5:20.5
5	(*Z*)-**9**	(*E*)-PhCH=CH	86	99:1	71.0:29.0
6	(*E*)-**9**	(*E*)-PhCH=CH	88	1:99	63.0:37.0
7	(*Z*)-**9**	(*E*)-PhCH=C(Me)	91	97:3	84.0:16.0
8	(*E*)-**9**	(*E*)-PhCH=C(Me)	89	1:99	95.0:5.0
9	(*Z*)-**9**	$PhCH_2CH_2$	47	95:5	54.0:46.0
10	(*E*)-**9**	$PhCH_2CH_2$	79	1:99	83.0:17.0

Aldehyde-derived trichlorosilyl enolates that are disubstituted at the α-carbon are also competent nucleophiles.[3] The addition of isobutyraldehyde-derived trichlorosilyl enolate **11** to various aldehydes in the presence of 0.10 equiv of bisphosphoramide (*R,R*)-**1c** affords β-hydroxy aldehydes as protected dimethyl acetals in high yields and moderate to good enantioselectivity (eq 5, Table 4). This reaction works well with a variety of aromatic, olefinic, and aliphatic aldehydes.

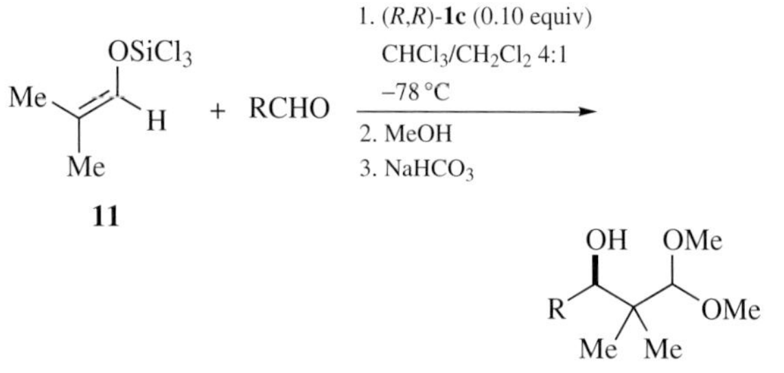

(5)

11 **12**

Table 4 Aldol additions of the isobutyraldehyde-derived trichlorosilyl enolate **11** to various aldehydes

Entry	R	Time (h)	Yield (%)	er
1	Phenyl	8	86	70.0:30.0
2	3,4,5-$(MeO)_3C_6H_2$	26	80	87.5:12.5
3	4-$CF_3C_6H_4$	8	86	91.0:9.0
4	(*E*)-PhCH=CH	12	90	67.5:32.5
5	PhC≡C	12	85	81.5:18.5
6	*n*-Butyl	30	80	91.0:9.0

Trialkylsilyl enol ethers derived from aldehydes can also be used when the catalyst system of $SiCl_4$ is activated by bisphosphoramide (*R,R*)-**1**.[4] In the presence of 0.15 equiv of bisphosphoramide (*R,R*)-**1c** and 2.0 equiv of $SiCl_4$, the addition of acetaldehyde-derived TMS enol ether **13** to aromatic and conjugated aldehydes affords aldol products in moderate to good yields and high enantioselectivities (eq 6, Table 5, entries 1–5). Aliphatic aldehydes were found to be unreactive under these reaction conditions (entry 6).

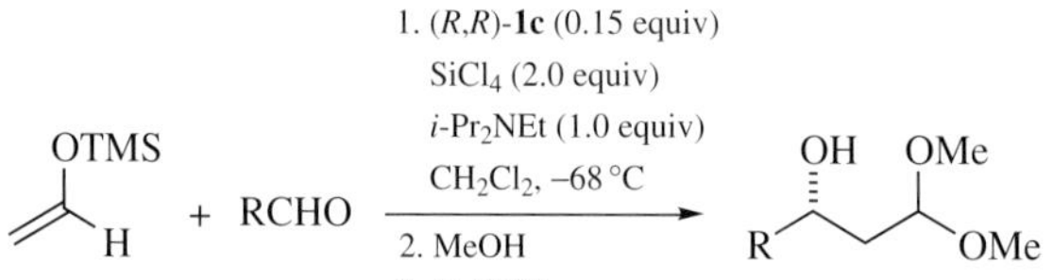

(6)

13 **14**

Table 5 Aldol additions of acetaldehyde-derived **TMS** enol ether **13** to various aldehydes

Entry	R	Yield (%)	er
1	Phenyl	80	97.1:2.9
2	4-$MeOC_6H_4$	30	91.2:8.8
3	4-$CF_3C_6H_4$	84	98.1:1.9
4	1-Naphthyl	78	92.2:7.8
5	(*E*)-PhCH=CH	60	98.2:1.8
6	*n*-Butyl	nr	nd

Ketone Enolate and Dienolate Aldol Additions to Aldehydes. The $SiCl_4$/bisphosphoramide (*R,R*)-**1** catalyst system is also capable of promoting the addition of methyl ketone-derived trialkylsilyl enol ethers to aldehydes.[5] Good generality in the spectator portion of the nucleophile in the additions to benzaldehyde **5** has been demonstrated (eq 7, Table 6). With 0.05 equiv of bisphosphoramide (*R,R*)-**1c** and 1.5 equiv of $SiCl_4$, high yields and enantioselectivities can be obtained in short reaction times (3 h) with all enolates surveyed with the exception of the pinacolone-derived TMS enol ether (entry 5).

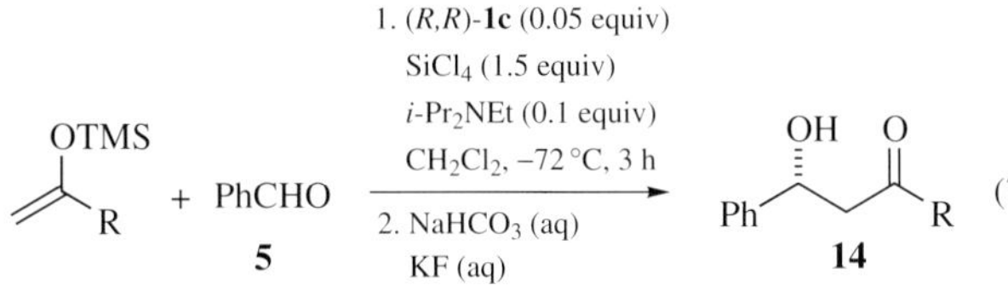

(7)

Table 6 Aldol additions of methyl ketone derived **TMS** enol ethers to benzaldehyde **5**

Entry	R	Yield (%)	er
1	Methyl	97	98.0:2.0
2	*n*-Butyl	99	99.5:0.5
3	Isobutyl	98	99.0:1.0
4	Isopropyl	95	99.5:0.5
5	*tert*-Butyl	nr	nd
6	Phenyl	98	99.5:0.5

The high yields and selectivities obtained in the reaction of methyl ketones are not limited to benzaldehyde (**5**). In the presence of 0.05 equiv of (*R,R*)-**1c** and 1.5 equiv of $SiCl_4$, the aldol addition of the TMS enolate of 2-hexanone (**15**) to a variety of aromatic and conjugated aldehydes affords the aldol product in high yields and excellent enantioselectivities (eq 8, Table 7, entries 1–5). However, aliphatic aldehydes were found to be unreactive with enolate **15** (entry 6).

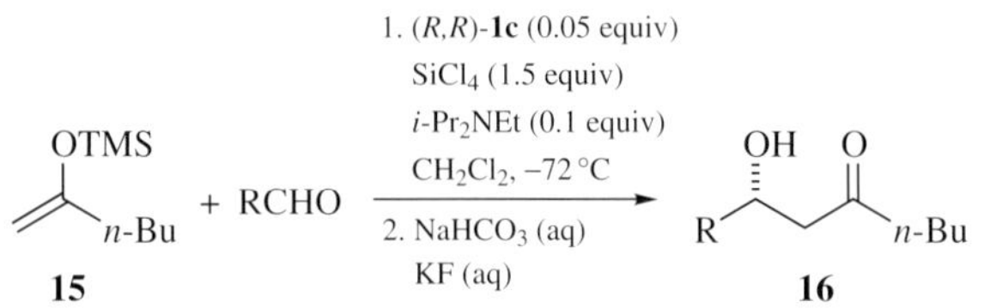

(8)

Table 7 Aldol additions of methyl ketone **TMS** enolate **15** to various aldehydes

Entry	R	Time (h)	Yield (%)	er
1	4-$CH_3OC_6H_4$	4	97.5	99.5:0.5
2	4-$CF_3C_6H_4$	4	96	99.5:0.5
3	2-Furyl	6	88	95.0:5.0
4	1-Naphthyl	10	95	96.0:4.0
5	(*E*)-PhCH=CH	4	97.5	99.5:0.5
6	$PhCH_2CH_2$	24	nr	nd

The $SiCl_4$/bisphosphoramide **1** catalyzed additions to aldehydes have also been extended to include chiral TMS enol ethers in double diastereodifferentiating aldol reactions.[6] The addition of TMS enol ether **17** to aromatic and conjugated aldehydes was efficiently catalyzed by bisphosphoramide **1c** and afforded products in high yields with modest to excellent diastereoselectivities (eq 9, Table 8, entries 1–8). Aliphatic aldehydes did not afford any product (entry 9). In most cases, (*R,R*)-**1c** yielded the *syn*-aldol product selectively, whereas the *anti*-isomer was obtained in the matched case using (*S,S*)-**1c** as the catalyst. These results show that the influence from the resident stereogenic center in the α-position is small and the catalyst-controlled diastereoselection dominates, allowing for the selective preparation of both diastereomers in the addition to variety of conjugated aldehydes.

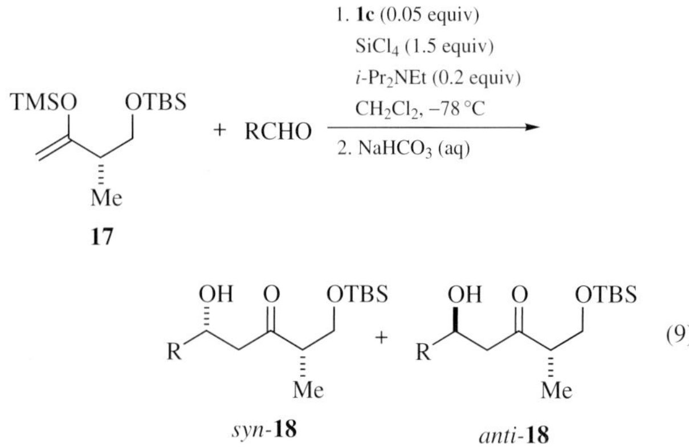

(9)

Table 8 Aldol additions of chiral methyl ketone **TMS** enolate **17** to various aldehydes

Entry	R	Catalyst	Time (h)	Yield (%)	*syn*:*anti*
1	Phenyl	(*R,R*)-**1c**	4	91	24:1
2	Phenyl	(*S,S*)-**1c**	4	94	1:49
3	1-Naphthyl	(*R,R*)-**1c**	24	85	19:1
4	1-Naphthyl	(*S,S*)-**1c**	24	83	1:43
5	(*E*)-PhCH=CH	(*R,R*)-**1c**	5	85	32:1
6	(*E*)-PhCH=CH	(*S,S*)-**1c**	5	83	1:34
7	PhC≡C	(*R,R*)-**1c**	6	74	1:1.5
8	PhC≡C	(*S,S*)-**1c**	6	72	1:4
9	$PhCH_2CH_2$	(*S,S*)-**1c**	24	nr	nd

The stereoinduction of the β-stereogenic center is also negligible in the aldol addition of TMS enol ether **19** to aromatic and conjugated aldehydes catalyzed by 0.05 equiv of bisphosphoramide **1c** and 1.5 equiv of $SiCl_4$ (eq 10, Table 9, entries 1–8)[6]. The diastereoselectivity is primarily controlled by the catalyst configuration and typically (*R,R*)-**1c** provided the 1,5-*anti*-aldol product, whereas the 1,5-*syn*-aldol products were obtained using (*S,S*)-**1c** as catalyst.

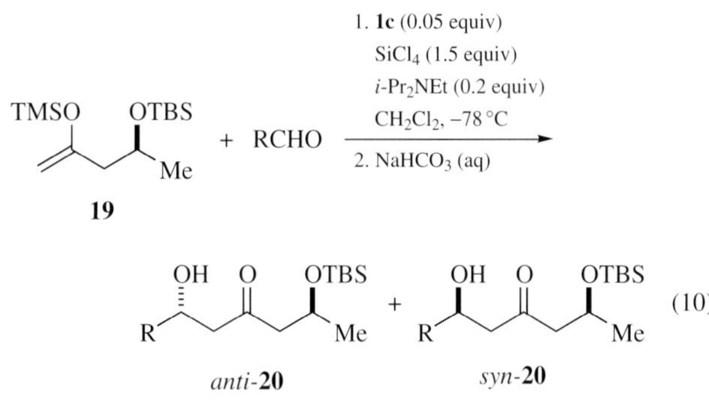

(10)

Table 9 Aldol additions of methyl ketone **TMS** enolate **19** to various aldehydes

Entry	R	Catalyst	Time (h)	Yield (%)	*syn*:*anti*
1	Phenyl	(*R,R*)-**1c**	5	92	1:36
2	Phenyl	(*S,S*)-**1c**	5	94	37:1
3	1-Naphthyl	(*R,R*)-**1c**	12	74	1:21
4	1-Naphthyl	(*S,S*)-**1c**	12	79	61:1
5	(*E*)-PhCH=CH	(*R,R*)-**1c**	7	88	1:30
6	(*E*)-PhCH=CH	(*S,S*)-**1c**	7	90	>99:1
7	PhC≡C	(*R,R*)-**1c**	24	74	1.3:1
8	PhC≡C	(*S,S*)-**1c**	24	72	1.2:1
9	$PhCH_2CH_2$	(*S,S*)-**1c**	24	nr	nd

Although the normal aldol addition provides access to 1,3-difunctional relationships, the vinylogous extension of this reaction allows 1,5-difunctional subunits to be constructed when γ-enolizable α,β-unsaturated carbonyl substrates are employed as "extended dienolates.CC The combination of bisphosphoramide **1c** and $SiCl_4$ has been demonstrated to afford exclusive γ-site selectivity in the vinylogous aldol addition of simple, acyclic α,β-unsaturated ketone-derived dienol ethers to a variety of aldehydes.[8] In the presence of 0.05 equiv of (*R,R*)-**1c** and 1.5 equiv of $SiCl_4$ at −50 °C, the addition of dienolates **21** and **22** to conjugated aldehydes affords δ-hydroxy-α,β-unsaturated ketones in high yield and enantioselectivity with exclusive *E*-double bond configuration (eq 11, Table 10, entries 1–5). Even using elevated temperatures did not allow dienol ether **22** to undergo addition to aliphatic aldehydes (entry 6).

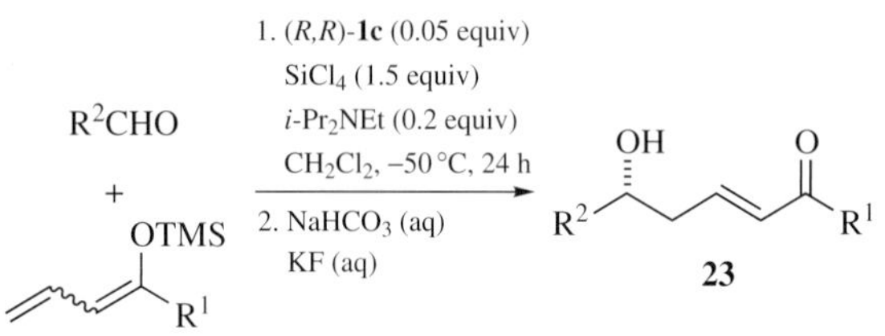

(11)

21: R^1 = Ph
22: R^1 = *t*-Bu

Table 10 Aldol additions of acyclic ketone **TMS** dienolates to various aldehydes

Entry	R^1	R^2	Yield (%)	γ:α	er
1	Phenyl	Phenyl	80	>99:1	99.0:1.0
2	*tert*-Butyl	Phenyl	94	>99:1	99.5:0.5
3	*tert*-Butyl	2-Furyl	90	>99:1	96.0:4.0
4	*tert*-Butyl	1-Naphthyl	85	>99:1	99.0:1.0
5	*tert*-Butyl	(*E*)-PhCH=CH	82	>99:1	>99.5:0.5
6	*tert*-Butyl	$PhCH_2CH_2$	nr	nd	nd

γ-Substituted, ketone-derived TMS dienol ethers are also susceptible to catalysis by the $SiCl_4$/bisphosphoramide **1** system.[8] The addition of the "extended dienolateCC of cyclohexenone **24** to various aromatic and conjugated aldehydes catalyzed by 0.05 equiv of bisphosphoramide **1c** and 1.5 equiv of $SiCl_4$ at −72 °C affords vinylogous aldol addition product in high yields with exclusive γ-site selectivity and moderate to good diastereo- and enantioselectivities (eq 12, Table 11, entries 1–4). However, as in the case of the acyclic-ketone-derived dienol ethers, **24** did not react with aliphatic aldehydes (entry 5).

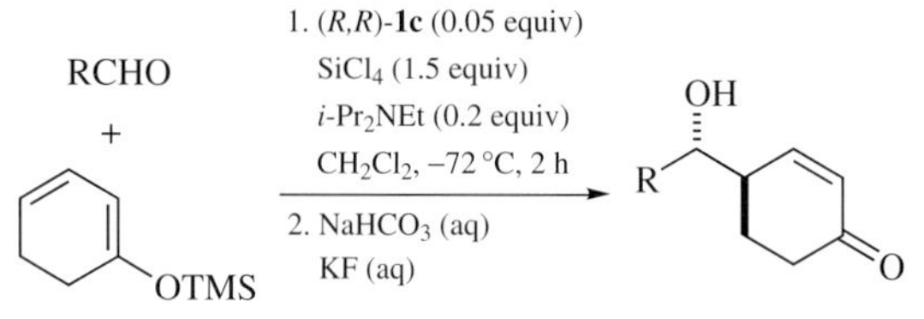

(12)

24 **25**

Table 11 Aldol additions of ketone **TMS** dienolate **24** to various aldehydes

Entry	R	Yield (%)	γ:α	*anti*:*syn*	er (*anti*)
1	Phenyl	90	>99:1	97.5:2.5	97.5:2.5
2	2-Furyl	94	>99:1	95.5:4.5	81.5:18.5
3	1-Naphthyl	80	>99:1	89.0:11.0	95.0:5.0
4	(*E*)-PhCH=CH	74	>99:1	98.0:2.0	84.5:15.5
5	$PhCH_2CH_2$	nr	nd	nd	nd

Ester Enolate and Dienolate Aldol Additions to Aldehydes. Addition of the acetate-derived TBS ketene acetal **26** to a wide variety of aldehydes in the presence of (*R,R*)-**1c** (0.05 equiv) and $SiCl_4$ (1.5 equiv) at −78 °C affords aldol products in high yields and enantioselectivities in short reaction times (15 min to 6 h) (eq 13, Table 12).[7] Notably, aliphatic aldehydes are competent electrophiles with these highly reactive nucleophiles affording adducts in high yields and enantioselectivities (entries 7 and 8).

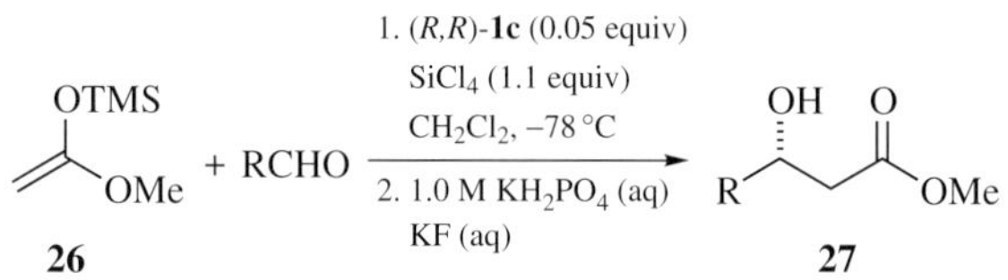

(13)

26 **27**

Table 12 Aldol additions of **TBS** ketene acetal **26** to various aldehydes

Entry	R	Time (min)	Yield (%)	er
1	Phenyl	15	97	96.3:3.7
2	4-$MeOC_6H_4$	15	97	98.7:1.3
3	4-$CF_3C_6H_4$	15	97	95.7:4.3
4	2-Furyl	15	94	93.4:6.6
5	1-Naphthyl	15	98	90.2:9.8
6	(*E*)-PhCH=CH	15	95	96.9:3.1
7	Cyclohexyl	360	86	94.2:5.8
8	$PhCH_2CH_2$	360	72	90.7:9.3

The high yields and selectivities obtained in the addition of acetate-derived ketene acetal (**26**) are maintained when propanoate-derived ketene acetals are employed.[7] Initially, a survey of ketene acetal structure in the addition to benzaldehyde (**5**)

revealed that the alkoxy substituent had little effect on the diastereoselectivity but did have a major influence of enantioselectivity such that er increased with the steric bulk of the alkoxy substituent in the order of Me < Et < *i*-Pr < Ph << *t*-Bu (eq 14, Table 13, entries 1–5). Remarkably, nearly identical levels of diastereo- and enantioselectivities are maintained regardless of the geometry of the silyl ketene acetal (entries 5 and 6).

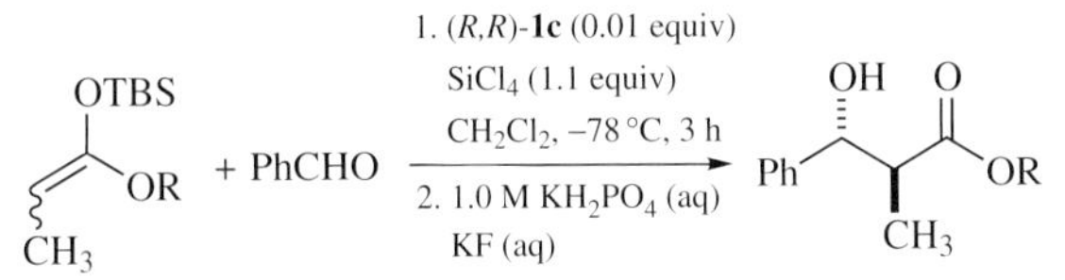

5 **28** (14)

Table 13 Aldol additions of various propanoate-derived **TBS** ketene acetals to benzaldehyde

Entry	R	*E*:*Z*	Yield (%)	*anti*:*syn*	er
1	Methyl	82:18	98	99:1	85.4:14.6
2	Ethyl	95:5	78	98:2	88.8:11.2
3	Isopropyl	71:29	80	98:2	91.1:8.9
4	Phenyl	94:6	98	94:6	93.9:6.1
5	*tert*-Butyl	95:5	93	99:1	>99.9:0.1
6	*tert*-Butyl	12:88	73	99:1	99.5:0.5

When the *tert*-butyl propanoate-derived ketene acetal **29** is employed, catalyst loadings of 0.01 equiv of (*R,R*)-**1c** are sufficient to provide aldol adducts in high yields, diastereo-, and enantioselectivities in the addition to a variety of aromatic and conjugated aldehydes (eq 15, Table 14).[7]

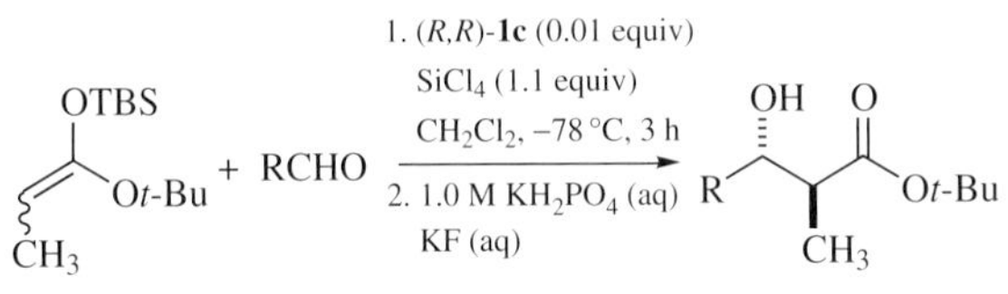

29 **30** (15)

Table 14 Aldol additions of **TBS** ketene acetal **29** to aromatic and conjugated aldehydes

Entry	R	Yield (%)	*anti*:*syn*	er
1	Phenyl	93	99:1	>99.9:0.1
2	4-$MeOC_6H_4$	88	>99:1	99.2:0.8
3	4-$CF_3C_6H_4$	93	>99:1	96.0:4.0
4	2-Furyl	96	98:2	96.9:3.1
5	1-Naphthyl	98	>99:1	97.2:2.8
6	(*E*)-PhCH=CH	98	>99:1	>99.0:0.1

The reaction of TBS ketene acetal **29** with aliphatic aldehydes did not afford any product, even after extended reaction times (24 h). However, the less sterically encumbered silyl ketene acetal **31** was found to react with aliphatic aldehydes in the presence of 0.05 equiv of bisphosphoramide (*R,R*)-**1c** and 1.1 equiv of $SiCl_4$ to provide aldol product in moderate to good yield after 24 h (eq 16, Table 15).[7] Furthermore, commonly used silyl-, ester-, and ether-protecting groups were found to be compatible with this catalyst system, affording products in similar yields, diastereo- and enantioselectivities compared to simple, unfunctionalized aliphatic aldehydes (entries 3–5).

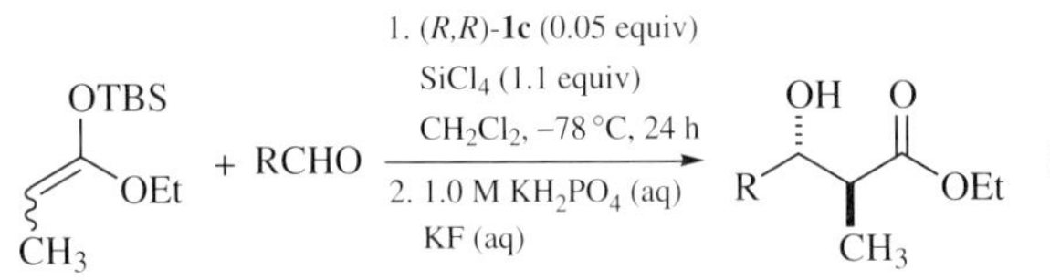

31 **32** (16)

Table 15 Aldol additions of propanoate-derived **TBS** ketene acetal **31** to aliphatic aldehydes

Entry	R	Yield (%)	*anti*:*syn*	er
1	$PhCH_2CH_2$	71	91:9	94.9:5.1
2	$(CH_3)_2CHCH_2$	65	89:11	93.0:7.0
3	$TBSO(CH_2)_5$	51	96:4	96.7:3.3
4	$BnO(CH_2)_5$	70	95:5	96.7:3.3
5	$BzO(CH_2)_5$	60	95:5	95.4:4.5

The $SiCl_4$/bisphosphoramide **1c** catalyst system also effectively promotes the vinylogous aldol addition of unsubstituted ethyl crotonate derived as well as α-, β-, and γ-substituted vinyl ketene acetals to a variety of aldehydes with excellent γ-site selectivity and *E*-double bond geometry for all substitution patterns on the dienolate (eq 17, Table 16).[7] Furthermore, this catalyst system maintains the high level of *anti*-diastereoselectivity (>99:1) observed in the addition of propanoate-derived silyl ketene acetals. Catalyst loadings of 0.01 equiv of (*R,R*)-**1c** are sufficient to provide vinylogous aldol adduct in high yields and selectivity in the addition to aromatic and conjugated aldehydes (entries 1, 2, 4, 5, 7, 8, 10, and 11). Aliphatic aldehydes require higher catalyst loadings (0.05 equiv) and longer reaction times (24 h) in the addition of the crotonate and β-substituted vinyl ketene acetals (entries 3 and 9), whereas vinyl ketene acetals with methyl groups in the α- and γ-positions were found to be unreactive (entries 6 and 12).

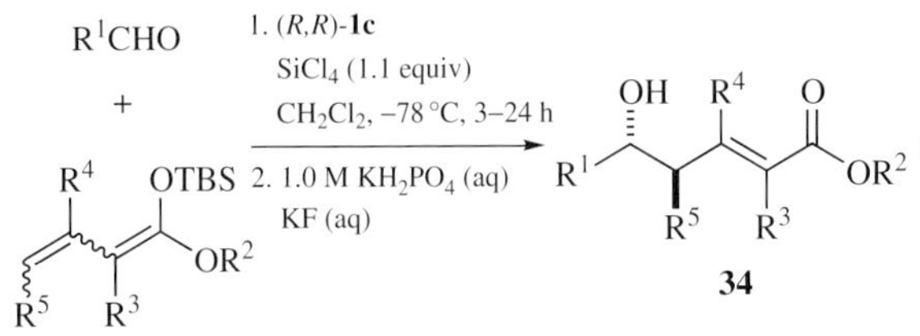

33 (17)

Exclusive γ site-selectivity and excellent enantioselectivity have been achieved in the vinylogous aldol addition of the ethyl crotonate-derived TBS vinyl ketene acetal **35** to

Table 16 Vinylogous aldol additions of ester-derived **TBS** vinyl ketene acetals with various aldehydes

Entry	R^1	R^2	R^3	R^4	R^5	Yield (%)	γ:α	*anti*:*syn*	er (major)
1	Phenyl	Et	H	H	H	89	>99:1	—	98.8:1.2
2	PhCH=CH	Et	H	H	H	84	>99:1	—	98.3:1.7
3	$PhCH_2CH_2$	Et	H	H	H	68	>99:1	—	95.3:4.7
4	Phenyl	Me	Me	H	H	93	>99:1	—	99.5:0.5
5	PhCH=CH	Me	Me	H	H	88	>99:1	—	99.6:0.4
6	$PhCH_2CH_2$	Me	Me	H	H	nr	nd	—	nd
7	Phenyl	Et	H	Me	H	91	>99:1	—	96.1:3.9
8	PhCH=CH	Et	H	Me	H	97	>99:1	—	94.0:6.0
9	$PhCH_2CH_2$	Et	H	Me	H	73	>99:1	—	97.5:2.5
10	Phenyl	*t*-Bu	H	H	Me	92	99:1	>99:1	94.5:5.5
11	PhCH=CH	*t*-Bu	H	H	Me	71	99:1	>99:1	91.2:8.8
12	$PhCH_2CH_2$	*t*-Bu	H	H	Me	nr	nd	nd	nd

3-benzyldimethylsilyl-2-propenal **36** using the combination of 0.05 equiv of bisphosphoramide (*R,R*)-**1c** and 1.1 equiv of $SiCl_4$ (eq 18).[13] Notably, this reaction could be carried out on 50 mmol scale without any loss in yield or enantioselectivity, and was exploited for construction of a key intermediate in the total synthesis of the polyene macrolide RK-397.

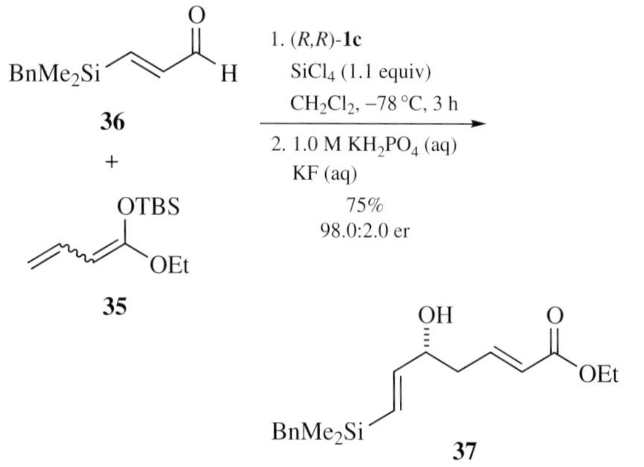

(18)

The catalytic action of bisphosphoramide (*R,R*)-**1c** and $SiCl_4$ also promotes the addition of the dioxanone-derived silyl ketene acetal **38** to a variety of aldehydes in high yields (eq 19, Table 17).[7] Catalyst loadings of 0.01 equiv of (*R,R*)-**1c** are sufficient to afford aldol adduct in high yields and moderate enantioselectivity in the addition to aromatic and conjugated aldehydes (entries 1 and 2), whereas aliphatic aldehydes required higher catalyst loadings (0.05 equiv) and longer reaction times (24 h) but still afford high yield and enantioselectivity (entry 3).

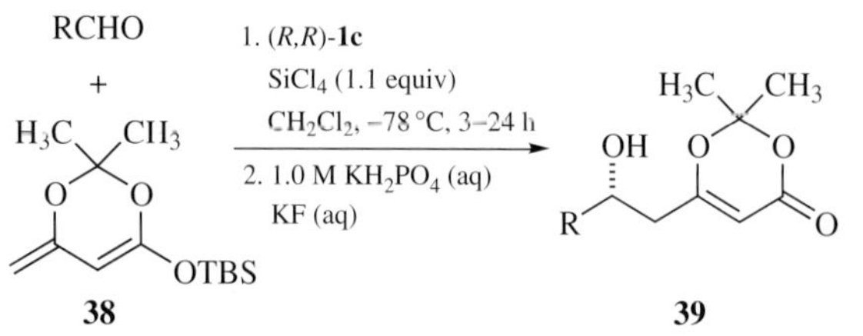

(19)

The addition of dioxanone-derived silyl ketene acetal **38** to conjugated aldehyde **40** was found to proceed in moderate yield (67%)

Table 17 Vinylogous aldol additions of dioxanone-derived silyl ketene acetal **38** to various aldehydes

Entry	R	Yield (%)	γ:α	er
1	Phenyl	92	>99:1	87.2:12.8
2	(*E*)-PhCH=CH	88	>99:1	88.9:11.1
3	$PhCH_2CH_2$	83	>99:1	94.6:5.4

and excellent enantioselectivity (96.5:3.5 er) in the presence of only 0.01 equiv of bisphosphoramide (*S,S*)-**1c** and 1.1 equiv of $SiCl_4$ (eq 20).[14] Notably, this reaction proceeds with higher enantioselectivity than with cinnamaldehyde (eq 19, Table 17, entry 2), demonstrating compatibility with more complex aldehydes. Whereas the experimental protocol used in Table 17 involved the addition of **38** neat in one portion to a solution of the mixture of aldehyde, $SiCl_4$, and (*R,R*)-**1c** in CH_2Cl_2, in the reaction with **40** a 1.3 M solution of **38** in CH_2Cl_2 was added via syringe pump over 6 h resulting in improved enantioselectivity.

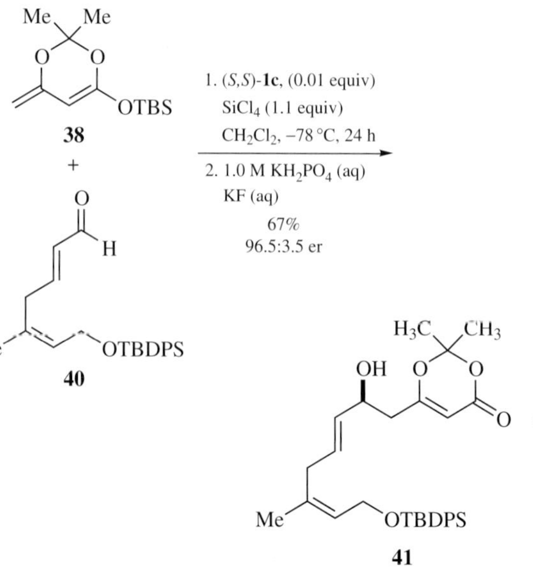

(20)

Vinyl ketene *N, O*- Acetals Additions to Aldehydes. The excellent site γ-selectivity obtained in the vinylogous aldol addition of simple ketone- and ester-derived silyl dienolates promoted by the $SiCl_4$/phosphoramide **1** catalyst system is maintained in the addition of the morpholine amide-derived silyl dienolate **42** to aldehydes (eq 21, Table 18).[9] In the presence of 0.05 equiv of (*R,R*)-**1c** and 1.1 equiv of $SiCl_4$, the unsubstituted crotonate-derived TBS dienolate as well as α- and β-substituted TBS dienolates reacted with a variety of aldehydes to afford δ-hydroxy-α,β-unsaturated amides with excellent γ-site selectivity and enantioselectivities. Notably, all of the substitution patterns on the dienyl unit surveyed allowed for reaction with aliphatic aldehydes affording product in good to high yields and excellent enantioselectivity (entries 3–5, 8, and 11). To achieve maximal selectivity, slow addition of a 0.2 M solution of dienolate in CH_2Cl_2 to a mixture of aldehyde, 0.05 equiv (*R,R*)-**1c**, and 1.1 equiv of $SiCl_4$ in CH_2Cl_2 over several minutes is required.

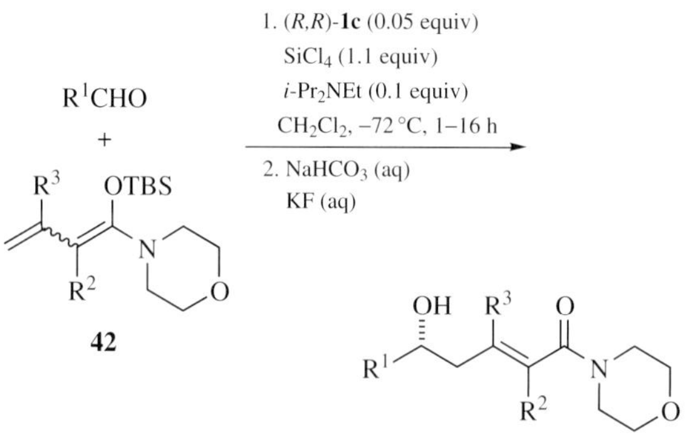

42 **43** (21)

Table 18 Vinylogous aldol additions of morpholine amide-derived **TBS** vinyl ketene acetals to various aldehydes

Entry	R^1	R^2	R^3	Yield (%)	γ:α	er (major)
1	Phenyl	H	H	95	>99:1	97.2:2.8
2	(*E*)-PhCH=CH	H	H	94	>99:1	98.2:1.8
3	$PhCH_2CH_2$	H	H	80	>99:1	99.0:1.0
4	$(CH_3)_2CHCH_2$	H	H	84	>99:1	99.7:0.3
5	Cyclohexyl	H	H	63	>99:1	99.4:0.6
6	Phenyl	Me	H	98	>99:1	98.5:1.5
7	(*E*)-PhCH=CH	Me	H	91	>99:1	99.1:0.8
8	$PhCH_2CH_2$	Me	H	66	>99:1	97.0:3.0
9	Phenyl	H	Me	89	>99:1	98.4:1.6
10	(*E*)-PhCH=CH	H	Me	90	>99.1	98.6:1.4
11	$PhCH_2CH_2$	H	Me	68	>99:1	97.1:2.9

Additions of Isocyanides to Aldehydes. The $SiCl_4$/bisphosphoramide **1** catalyst system effectively promotes the addition of isocyanides to aldehydes affording α-hydroxy amides in high yields (eq 22, Table 19).[10] Notably, products arising from multiple additions of the isocyanide to the primary adduct are not isolated. The level of enantioselectivity was found to be highly dependent on the rate of addition of the *t*-BuNC **44** to the reaction mixture. Under the optimized conditions established for this reaction, the addition of *t*-BuNC **44** as a 1.0 M CH_2Cl_2 solution over 4 h to a mixture of an aldehyde, $SiCl_4$, 0.05 equiv of bisphosphoramide **1**, and 0.01 equiv *i*-Pr_2NEt in CH_2Cl_2 (1.0 M) afforded excellent enantioselectivities with conjugated and aromatic aldehydes (entries 1–5). With aliphatic aldehydes, the presence of *i*-Pr_2NEt is deleterious to the yields of the product; however, moderate to good yields and enantioselectivities are obtained in its absence (entries 6 and 7).

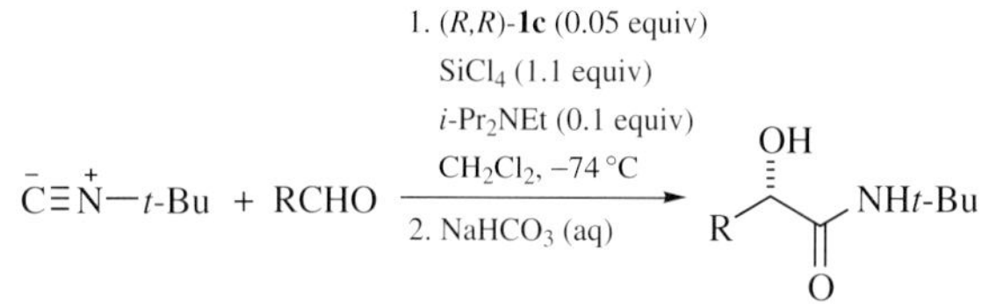

44 **45** (22)

Table 19 Addition of **44** to various aldehydes

Entry	R	Yield (%)	er
1	4-$MeOC_6H_4$	89	98.3:1.7
2	4-$CF_3C_6H_4$	89	96.5:3.5
3	2-furyl	83	95.9:4.1
4	1-Naphthyl	92	92.2:7.8
5	(*E*)-PhCH=CH	81	97.8:2.2
6	Cyclohexyl	53	87.1:12.9
7	$PhCH_2CH_2$	92	81.9:18.1

Other isocyanide structures have also been examined in the addition to aldehydes (eq 23, Table 20). High yields are obtained in the addition to hydrocinnamaldehyde (entry 1) and benzaldehyde (entries 2–4); however, the enantioselectivities are highly dependent of the structure of the isocyanide.

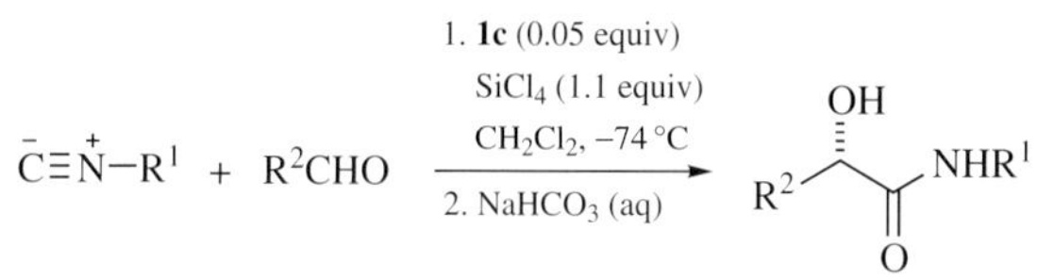

46 **47** (23)

Table 20 Addition of isocyanides to various aldehydes

Entry	R^1	R^2	Yield (%)	er
1	1,1,3,3-Tetramethyl	$PhCH_2CH_2$	87	70.0:30.0
2	Phenyl	Phenyl	82	73.2:26.8
3	Ethyl acetate	Phenyl	83	83.3:16.7
4	4-Tosylmethyl	Phenyl	80	88.5:11.5

The immediate products of these addition reactions catalyzed by the $SiCl_4$/bisphosphoramide **1** catalyst system are likely (trichlorosilyloxy)imidoyl chlorides (**48**), which are versatile intermediates and can be converted to a variety of useful products. The conversion of in situ generated **48** to a carboxylic ester **49** is possible when quenching the reaction with MeOH at low temperature followed by workup with sat. aq $NaHCO_3$ solution (eq 24, Table 21.[15] This quench/workup procedure affords high yields with all three classes of aldehydes without any detectable racemization.

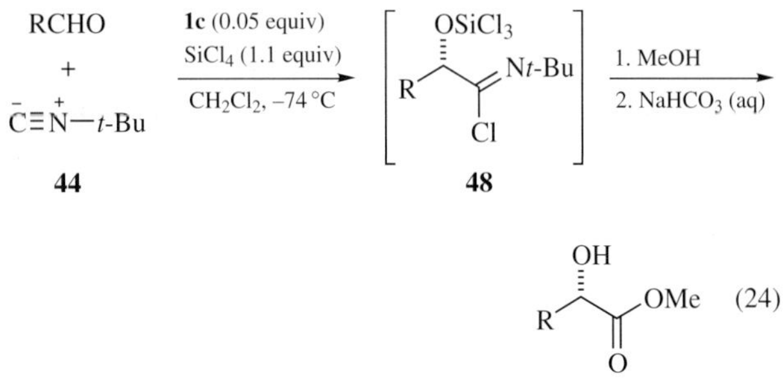

44 **48** **49** (24)

Table 21 Conversion of imidoyl chlorides **48** to methyl esters

Entry	R	Yield (%)	er
1	Phenyl	97	>99.0:1.0
2	(*E*)-PhCH=CH	71	97.9:2.1
3	$PhCH_2CH_2$	88	81.9:18.2

1. Denmark, S. E.; Wynn, T., *J. Am. Chem. Soc.* **2001**, *123*, 6199.
2. Denmark, S. E.; Ghosh, S. K., *Angew. Chem. Int. Ed.* **2001**, *40*, 4759.
3. Denmark, S. E.; Bui, T., *Proc. Nat. Acad. Sci. U.S.A.* **2004**, *101*, 5439.
4. Denmark, S. E.; Bui, T., *J. Org. Chem.* **2005**, *70*, 10190.
5. Denmark, S. E.; Heemstra, J. R., Jr., *Org. Lett.* **2003**, *5*, 2303.
6. Denmark, S. E.; Fujimori, S.; Pham, S. M., *J. Org. Chem.* **2005**, *70*, 10823.
7. Denmark, S. E.; Beutner, G. L.; Wynn, T.; Eastgate, M. D., *J. Am. Chem. Soc.* **2005**, *127*, 3774.
8. Denmark, S. E.; Heemstra, J. R., Jr, *Synlett* **2004**, 2411.
9. Denmark, S. E.; Heemstra, J. R., Jr, *J. Am. Chem. Soc.* **2006**, *128*, 1038.
10. Denmark, S. E.; Fan, Y., *J. Org. Chem.* **2005**, *70*, 9667.
11. Miyano, S.; Nawa, M.; Mori, A.; Hashimoto, H., *Bull. Chem. Soc. Jpn.* **1984**, *57*, 2171.
12. Denmark, S. E.; Stavenger, R. A., *Acc. Chem. Res.* **2000**, *33*, 432.
13. Denmark, S. E.; Fujimori, S., *J. Am. Chem. Soc.* **2005**, *127*, 8971.
14. Aubele, D. L.; Wan, S.; Floreancig, P. E., *Angew. Chem., Int. Ed.* **2005**, *44*, 3485.

Scott E. Denmark & John R. Heemstra Jr.
University of Illinois, Urbana, IL, USA

2-Phenyl-6,10b-dihydro-4*H*,5a*H*-5-oxa-3,10c-diaza-2-azonia-cyclopenta[*c*] fluorene tetrafluoroborate; 2-Pentafluorophenyl-6,10b-dihydro-4*H*, 5a*H*-5-oxa-3,10c-diaza-2-azoniacyclopenta[*c*]fluorene tetrafluoroborate; 6,7-Dihydro-2-phenyl-5-(phenylmethyl)-5*H*-pyrrolo[2,1-*c*]-1,2,4-triazolium chloride

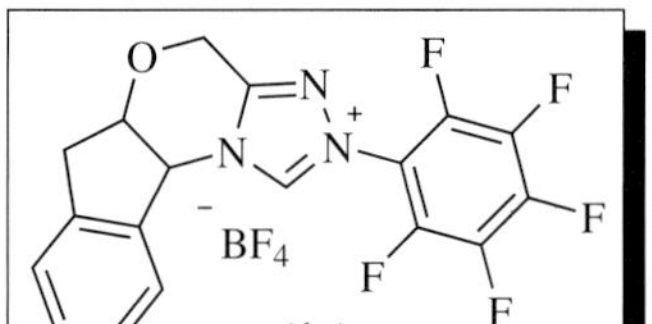

(**1a**)

2-Pentafluorophenyl-6,10b-dihydro-4*H*,5a*H*-5-oxa-3,10c-diaza-2-azoniacyclopenta[*c*]fluorene tetrafluoroborate

[740816-14-2] $C_{18}H_{11}BF_9N_3O$ (MW 467.0961)

InChI = 1/C18H11F5N3O.BF4/c19-12-13(20)15(22)18(16(23)14(12)21)26-7-25-11(24-26)6-27-10-5-8-3-1-2-4-9(8)-17(10)25;2-1(3,4)5/h1-4,7,10,17H,5-6H2;/q+1;-1

InChIKey = CPCMDOOTVHDRTM-UHFFFAOYSA-N

(catalyst used in a variety of transformations, mostly involving umpolung reactivity)

Physical Data: mp 222–226 °C.
Solubility: common organic solvents – low solubility in hexanes, benzene, and toluene.
Form Supplied in: anhydrous solid.
Analysis of Reagent Purity: NMR, MS.
Purification: trituration 95:5 ethyl acetate/methanol.
Handling, Storage, and Precautions: stored at ambient temperature with no special precautions.

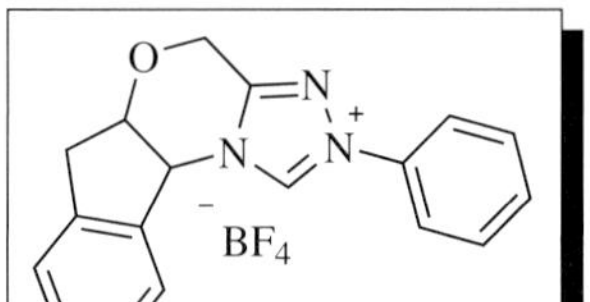

(**1b**)

2-phenyl-6,10b-dihydro-4*H*,5a*H*-5-oxa-3,10c-diaza-2-azonia-cyclopenta[*c*]fluorene tetrafluoroborate

[463326-74-1] $C_{18}H_{16}BF_4N_3O$ (MW 377.1438)

InChI = 1/C18H16N3O.BF4/c1-2-7-14(8-3-1)21-12-20-17(19-21)11-22-16-10-13-6-4-5-9-15(13)18(16)20;2-1(3,4)5/h1-9,12,16,18H,10-11H2;/q+1;-1

InChIKey = WHAPHEJUSIRKAF-UHFFFAOYAN

(catalyst used in a variety of transformations, mostly involving umpolung reactivity)

Physical Data: mp 203–206 °C.
Solubility: common organic solvents – low solubility in hexanes, benzene, and toluene.
Form Supplied in: anhydrous solid.
Analysis of Reagent Purity: NMR, MS.
Purification: trituration 95:5 ethyl acetate/methanol.
Handling, Storage, and Precautions: stored at ambient temperature with no special precautions.

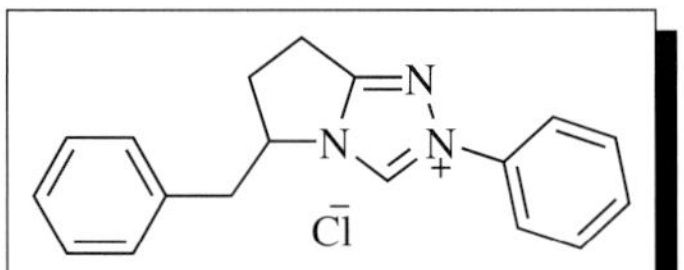

(**2a**)

6,7-Dihydro-2-phenyl-5-(phenylmethyl)-5*H*-pyrrolo[2,1-*c*]-1,2,4-triazolium chloride

[640737-16-2] $C_{18}H_{18}ClN_3$ (MW 311.8086)
InChI = 1/C18H18N3.ClH/c1-3-7-15(8-4-1)13-17-11-12-18-19-21(14-20(17)18)16-9-5-2-6-10-16;/h1-10,14,17H,11-13H2;1H/q+1;/p-1/fC18H18N3.Cl/h;1h/qm;-1
InChIKey = JOMVXOWRDSZYJW-RUJLSZRICT

(catalyst used in a variety of transformations, mostly involving umpolung reactivity)

Physical Data: mp 198.4–199.6 °C (BF_4 salt).
Solubility: common organic solvents – low solubility in hexanes, benzene, and toluene.
Form Supplied in: anhydrous solid.
Analysis of Reagent Purity: NMR, MS.
Purification: precipitation from chlorobenzene followed by recrystallization from MeOH (BF_4 catalyst).
Handling, Storage, and Precautions: stored at ambient temperature with no special precautions.

Original Commentary

Tomislav Rovis, Stephen P. Lathrop & Harit U. Vora
Colorado State University, Fort Collins, CO, USA

Introduction. *N*-Aryltriazolium salts such as **1a**, **1b**, and **2a** are effective precursors to triazolinylidene carbenes, cousins of the more extensively studied imidazolinylidene carbenes. Their application in organic synthesis has revolved primarily around their role as catalysts for a variety of organic transformations and as ligands. Both of these roles exploit the propensity of triazolium salts to form ylides or carbenes upon treatment with base (eq 1).

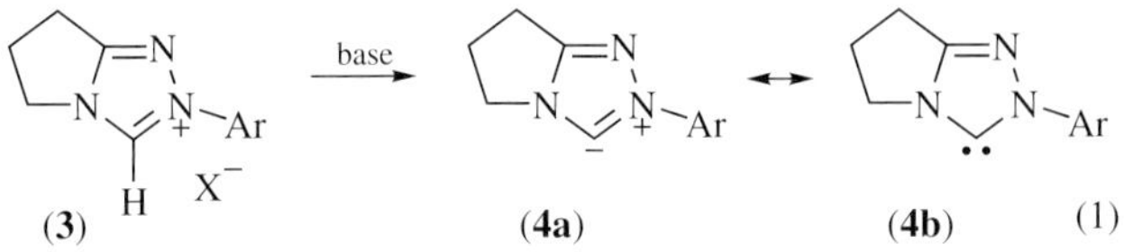

(3) (4a) (4b) (1)

Both chiral and achiral versions of these salts have been used as reagents and catalysts for a number of organic transformations. Indeed, the application of these catalysts in the asymmetric Stetter reaction[1,2] and redox chemistry[3] has been recently reviewed. The title compounds **1a**–**2a** are typically formed as the chloride or tetrafluoroborate salts. Modification of this core structure is trivial due to the modularity of its assembly,[4] and literature suggests that substitution around this scaffold may have a profound impact on associated reactivity (see below). Extensive work has been reported with related structures **1**–**6**. The impetus for the development of these catalysts has been their promotion of umpolung chemistry, notably the benzoin and Stetter reactions. These have also led to investigations of other transformations.

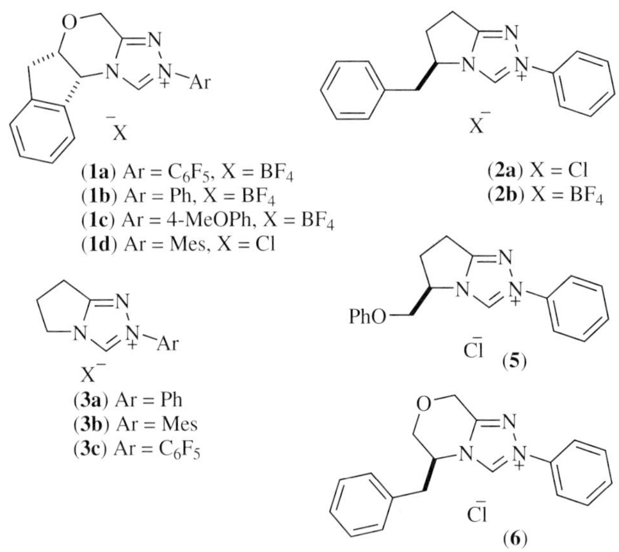

(**1a**) Ar = C_6F_5, X = BF_4
(**1b**) Ar = Ph, X = BF_4
(**1c**) Ar = 4-MeOPh, X = BF_4
(**1d**) Ar = Mes, X = Cl
(**2a**) X = Cl
(**2b**) X = BF_4
(**3a**) Ar = Ph
(**3b**) Ar = Mes
(**3c**) Ar = C_6F_5
(**5**)
(**6**)

Benzoin Reaction. The benzoin reaction, historically demonstrated to be subject to asymmetric catalysis using chiral thiazolium salts,[5] has also been shown to be catalyzed by triazolium salts.[6] It was demonstrated in 1997 that the use of chiral nonracemic bicyclic triazolium salts affords much improved selectivities for the benzoin reaction. Catalysts **5** and **6** provide ee's up to 83% for the dimerization of aryl aldehydes, although chemical yields remained problematic (eq 2).[7]

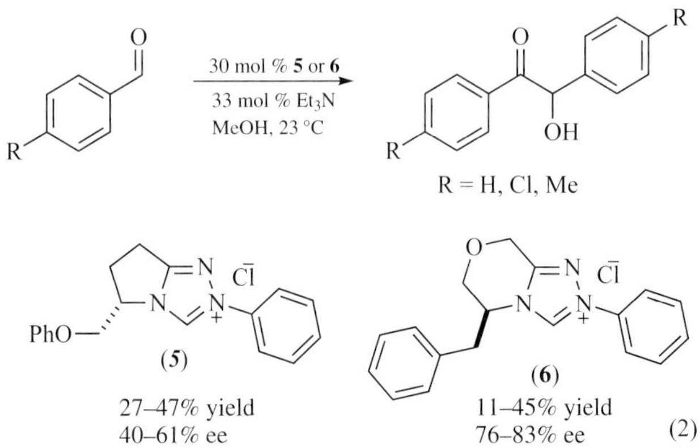

(**5**) (**6**) (2)

To date, the most useful catalyst reported for the asymmetric benzoin reaction is triazolium salt **7**. In the presence of KO*t*-Bu, this catalyst mediates the benzoin reaction of a variety of aryl aldehydes affording the corresponding benzoin adducts in good yields and generally high enantioselectivity (eq 3).

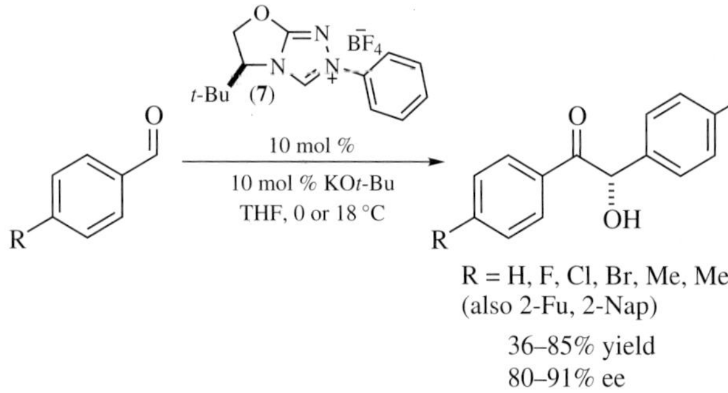

(3)

Intramolecular crossed-benzoin reactions using triazolium-based catalysts have been reported by two groups. Enders showed that bicyclic triazolium salt **8** is an efficient precatalyst for this transformation, affording α-hydroxyketones in good yields and excellent enantioselectivities (eq 4).[8]

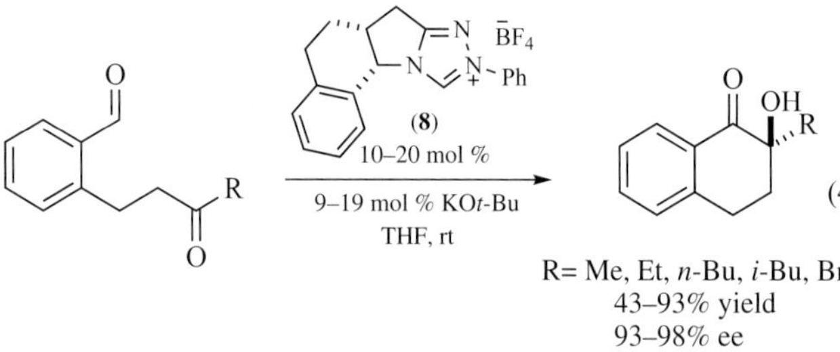

(4)

Suzuki has shown that triazolium catalyst **9** effects a similar transformation with broad substrate scope. Of note is the tolerance of the reaction to a variety of tethers and steric requirements, especially adjacent to the ketone (eq 5).[9]

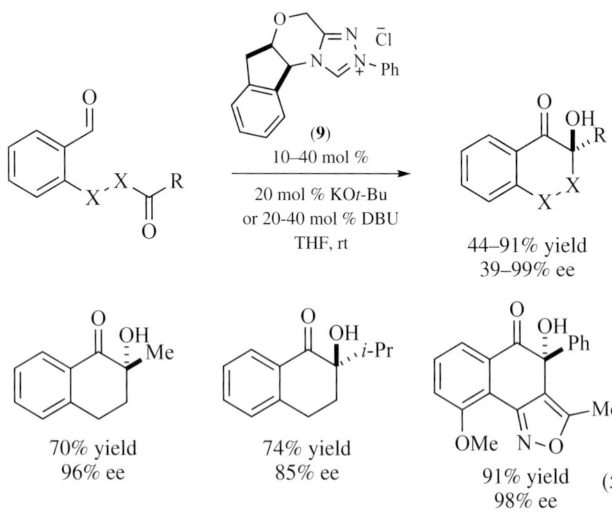

(5)

Suzuki has also shown that triazolium catalyst **10** can be utilized for the benzoin cyclization of enolizable keto-aldehydes. This catalyst proved fruitful for a number of substrates and was used as the key step in the total synthesis of (+)-sappanone B (eq 6).[10]

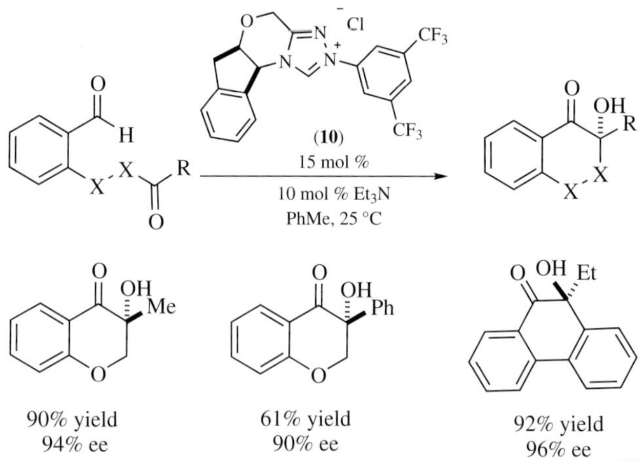

(6)

Benzoin-oxy-Cope reactivity can be coupled when utilizing triazolium salt in conjunction with α-reducible aldehydes. Using 10 mol % of triazolium salt **1d**, a variety of aryl *cis*-substituted cyclopentene products are obtained. Products are believed to arise through a crossed-benzoin reaction followed by an oxy-Cope rearrangement via intermediate **11**. Finally, intramolecular aldol followed by acyl addition and decarboxylation gives the observed cyclopentenes. The authors also observed that methyl esters gave higher enantioselectivities than the corresponding ethyl esters (eq 7).[11]

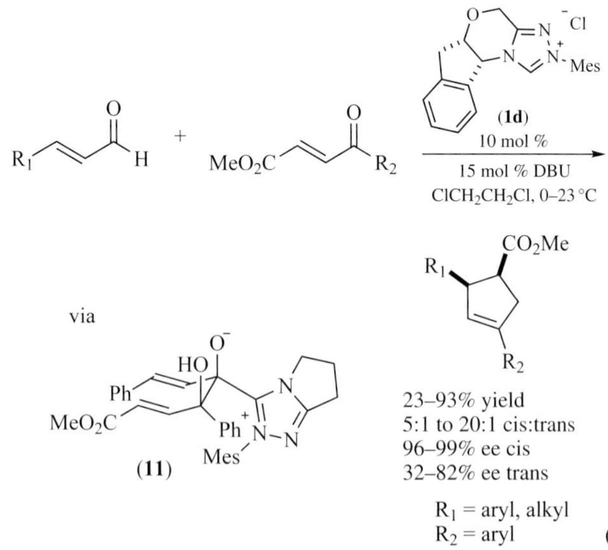

(7)

Stetter Reaction. The Stetter reaction may be viewed as a crossed-benzoin reaction involving a Michael acceptor as the electrophile. While thiazolium salts in the presence of base have been historically used as catalysts for this transformation,[12] it has been illustrated that triazolium salts behave analogously.[13] The first highly enantioselective Stetter reaction was reported in 2002.

A variety of aldehydes were found to undergo an intramolecular Stetter reaction with tethered Michael acceptors affording five- and six-membered rings in high enantioselectivity and excellent yield (eq 8).[14] A variety of Michael acceptors are tolerated, including enones, enoates, and enenitriles, with the former two providing highest enantioselectivities.[15,16]

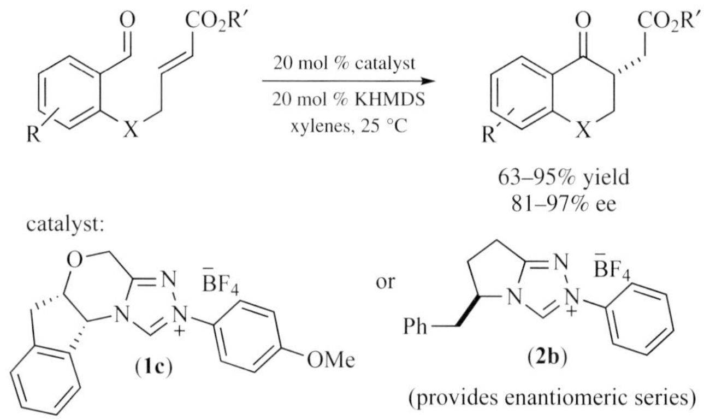

(8)

This family of catalysts has also been shown to cyclize aldehydes onto β,β-disubstituted Michael acceptors, forming tertiary cyclic ethers and sulfides as well as all-carbon quaternary stereocenters in excellent yield and selectivity (eq 9).[17,18]

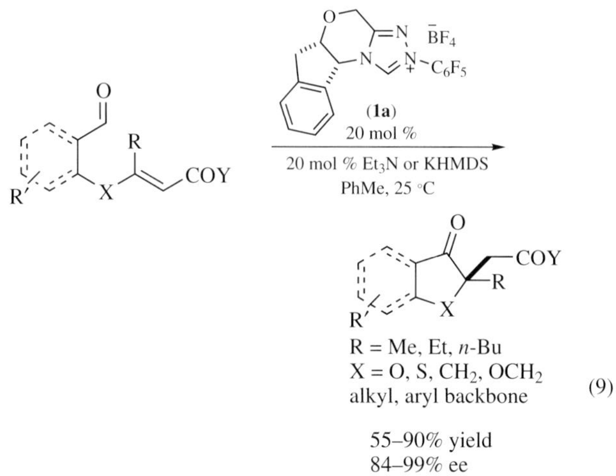

(9)

Diastereoselective cyclizations may be achieved through the use of α,α,β-trisubstituted Michael acceptors such as **12**. Enolate protonation must proceed faster than C–C bond rotation as the relative stereochemistry corresponds to *syn* addition of the aldehyde carbonyl and the proton to the Michael acceptor (eq 10).[19] Intriguingly, optimal reaction conditions for diastereoselective cyclizations involve the use of the free carbene **13** as catalyst.

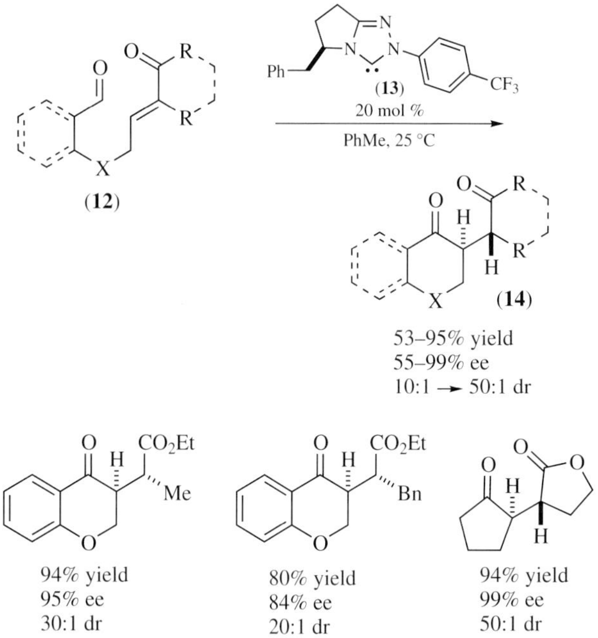

(10)

The reaction is tolerant of backbone substitution and the selectivity is dominated by the catalyst and not by the preexisting stereocenters, except when these are alpha to the aldehyde.[20] Aldehydes tethered to dienone acceptors allow for an enantioselective and diastereoselective intramolecular Stetter reaction to provide benzofuranone scaffolds in high yield and enantioselectivities (eq 11).[21,22]

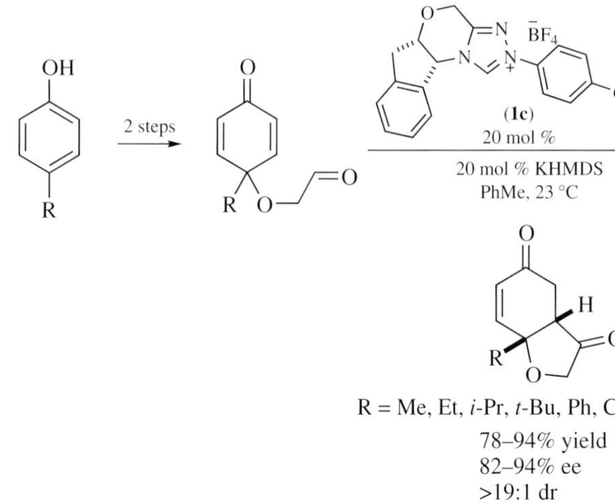

(11)

Vinylphosphine oxides and vinylphosphonates have also been shown to be competent Michael acceptors for the intramolecular Stetter reaction. Triazolium catalyst **1a** provides the desired organophosphorus compounds in yields up to 99% and ee values up to 96% (eq 12). This catalyst has proved useful for both aromatic and aliphatic tethered vinylphosphine oxides and vinylphophonates.[23]

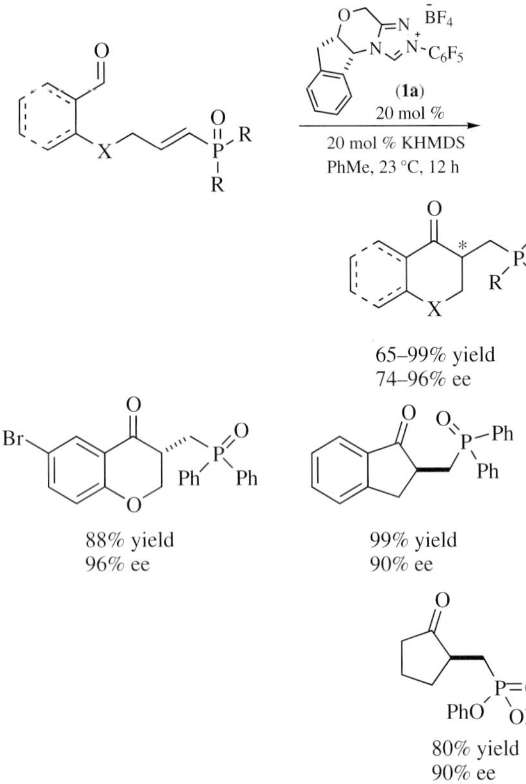

(12)

Triazolium catalyst **1a** has also been used to assemble the spirocyclic core of natural product FD-838.[24] Treatment of aldehyde-tethered maleimide **13** with catalyst **1a** in the presence of KHMDS gives the desired spirocycle in good yield and excellent enantioselectivity. This spirocyclic core could be further functionalized to advanced intermediate**15** (eq 13).

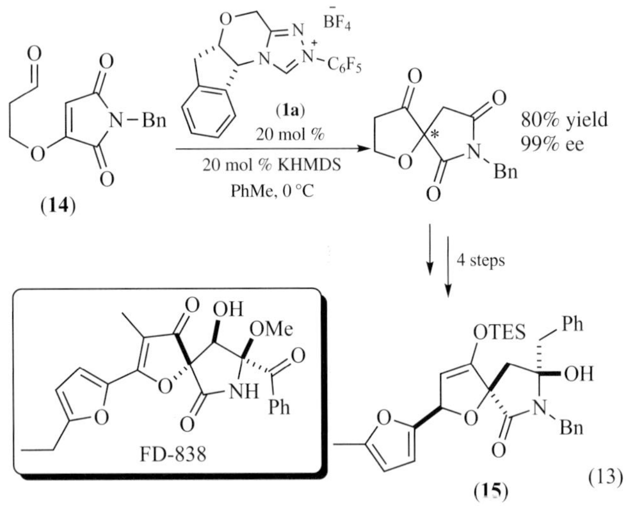

(13)

Acyl Transfer. Triazolium salts in the presence of base have been demonstrated to be effective acyl transfer catalysts. Treatment of acyloxyoxazoles with catalyst **3a** results in an effective transfer to form the acyl azlactone **17** (eq 14). Using a crossover experiment, the authors have demonstrated that the reaction proceeds via ion-pair dissociation. Triazolium salts were found to be superior to imidazolium salts as precatalysts.[25] Impressive catalyst loading levels may be achieved using these precursors, but the present technique does not yet compete with the best asymmetric catalysts for this process.[26,27]

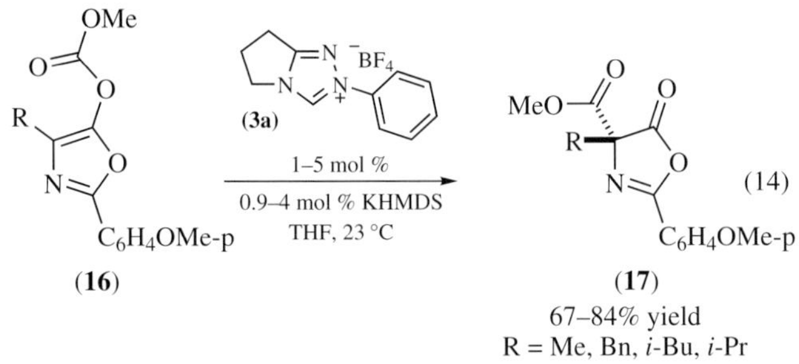

(14)

Redox Catalysis. A variety of other reaction manifolds may be accessed by intercepting the Breslow intermediate (**20** in eq 15). The use of α-halo aldehydes as substrates in the presence of alcohols results in an internal redox reaction to form α-reduced esters in excellent yields (eq 15). The use of chiral triazolium salts allows a desymmetrization of a *meso*-diol to proceed in good ee (eq 16).[28] Thiazolium salts are also competent catalysts for related redox reactions, and α,β-epoxy- and aziridinoaldehydes have been shown to be appropriate substrates.[29]

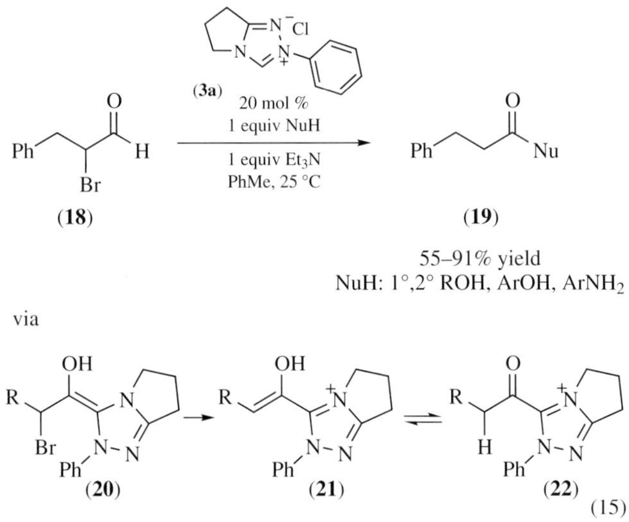

(15)

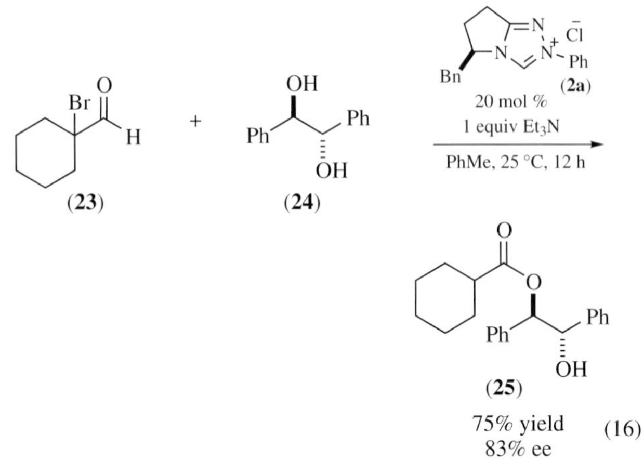

(16)

Triazolium-derived carbenes have been shown to effect the transformation of enals into alkanoates by a similar redox process at slightly elevated temperatures (eq 17).[30] Benzimidazolium salts also appear to be competent precatalysts for this chemistry.[31] Ynals provide enoates when subjected to similar reaction conditions, catalyzed by triazolium salts as well as imidazolium precursors, the latter affording higher selectivities for the *E*-alkene.[32]

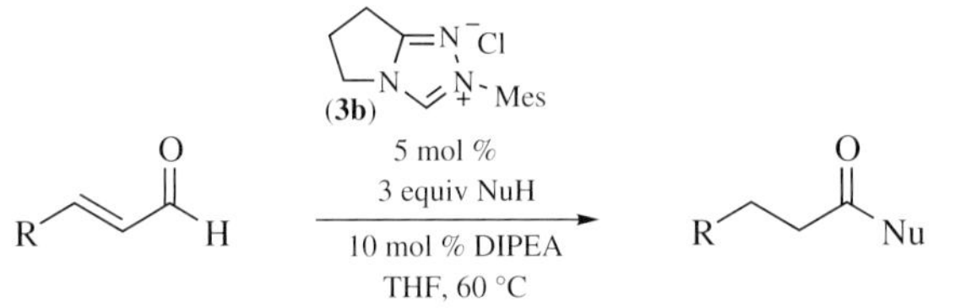

(26) (27)

63–99% yield
NuH: 1°, 2° ROH, ArOH

(17)

This chemistry has been extended to the redox reaction of cyclopropyl carboxaldehydes bearing electron-withdrawing groups. The apparently facile C–C bond cleavage under very mild conditions is particularly noteworthy. The process generates desymmetrized glutarates and analogs and maintains the integrity of the C-3 stereocenter (eq 18).[33]

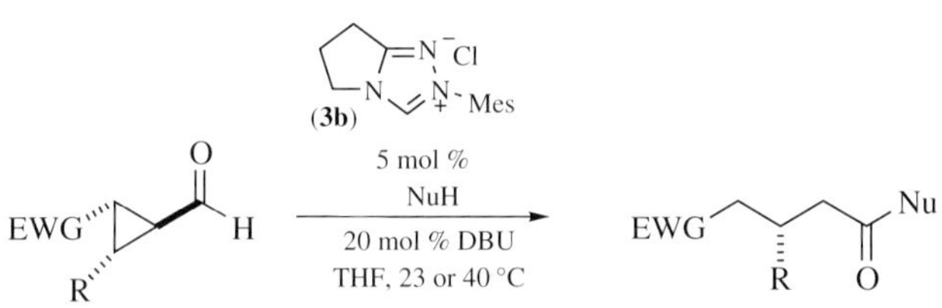

(28) (29)

81–99% yield
NuH: ROH, RSH

(18)

The redox process catalyzed by these carbenes proceeds through an azolium enol/enolate, which tautomerizes to the acyl azolium (**21** to **22** in eq 15). One can envision that the use of a chiral catalyst may allow for an enantioselective bond formation at the alpha position. A few strategies have been advanced illustrating this concept. The use of α,α-dichloroaldehydes allows for an asymmetric protonation/alcohol trapping of the prochiral enol/enolate providing α-chloroesters in good yield and high enantioselectivities (eq 19). Catalyst **1a** provides optimal results.

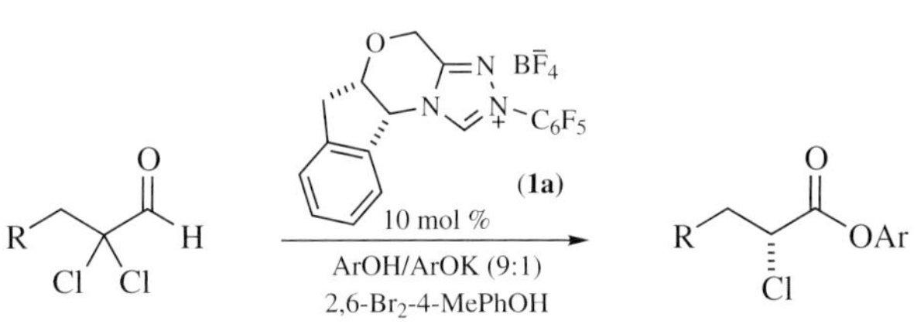

65–78% yield
84–92% ee

R = alkyl, aryl
Ar = various phenols

(19)

Treatment of α,α-dichloroaldehydes with catalyst **3c** has also been shown to undergo amidation reactions in the presence of cocatalyst and amine. The cocatalyst is believed to intercept the acyl azolium (**22** in eq 15) to generate an activated ester (**30** in eq 20). In the presence of an amine nucleophile, this activated ester undergoes an in situ amidation to generate α-chloroamides. The aldehyde component may be varied from haloaldehydes to epoxy and aziridino aldehydes as well as enals. The use of triazolium catalyst **1a** generates the desired α-choroamide in good yield and good ee (eq 21).[34]

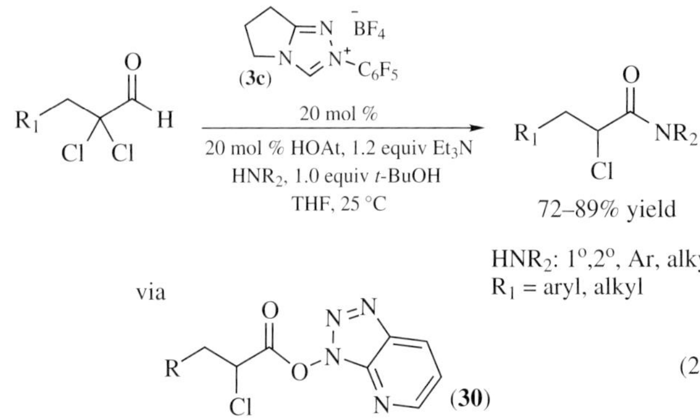

72–89% yield

HNR_2: 1°,2°, Ar, alkyl
R_1 = aryl, alkyl

(20)

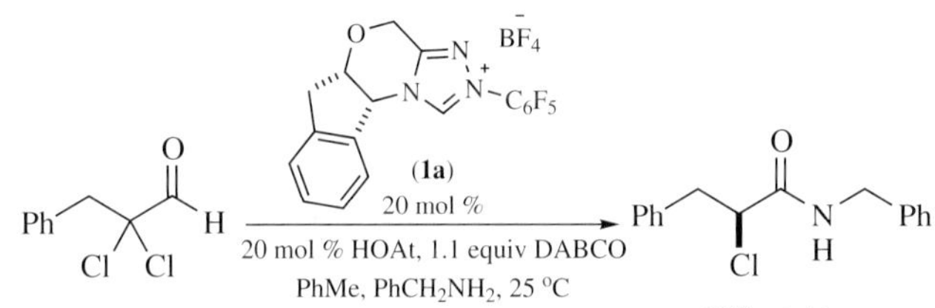

62% yield
80% ee

(21)

It has been shown that triazolium catalyst **3c** effects and directs amidation of aldehydes with nitroso compounds to give the corresponding *N*-arylhydroxamic acids in good yields (eq 22). Triazolium catalyst **1a** can be utilized to generate the enantioenriched *N*-arylhydroxamic acid via kinetic resolution of α-methyl phenylacetaldehyde (eq 23).[35]

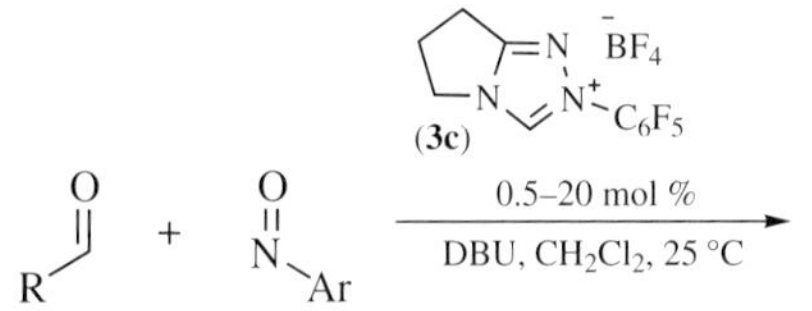

(22)

55–99% yield
R = aryl, alkyl, alkenyl

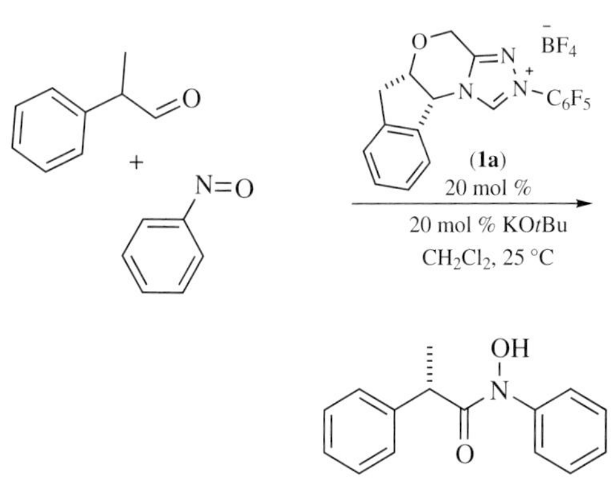

48% yield
68% ee

(23)

A complementary strategy uses a heterodiene to trap the enol/enolate intermediate via a cycloaddition manifold. The selectivities exhibited by catalyst **1d** are spectacular, as is the level of chemoselectivity observed (eq 24).[36]

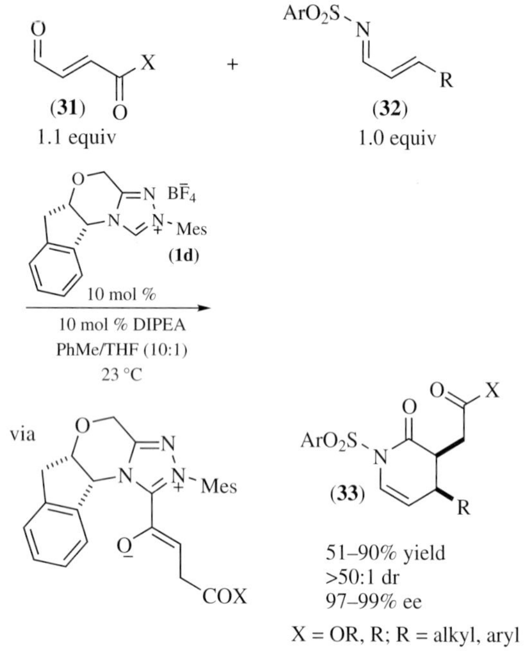

(24)

Bicyclo-β-lactams can also be accessed when enals are used as the redox partner followed by chalcone-derived imines in the presence of azolium salt **1d**. Using 10 mol % of azolium salt and 30 mol % of DMAP, the authors are able to obtain a variety of substituted bicyclo-β-lactams with excellent selectivity (eq 25). These products arise from a similar pathway as that shown in eq 7. A crossed aza-benzoin followed by an oxy-Cope reaction generates an intermediate similar to **11**. Tautomerization followed by an intramolecular Mannich reaction and β-lactam formation affords the observed products.[37]

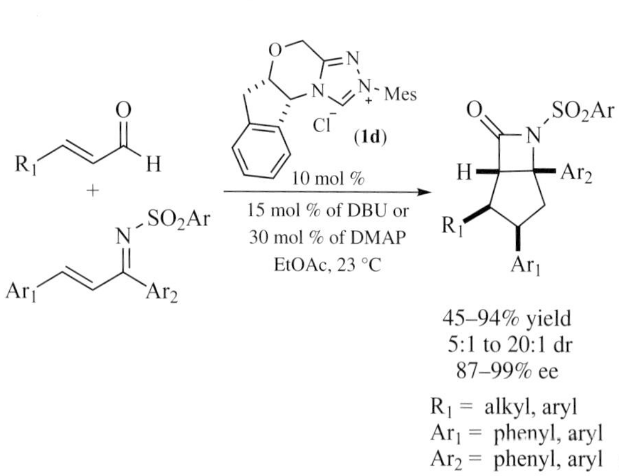

(25)

This cycloaddition strategy was also extended toward the synthesis of 1-oxodiene Diels–Alder reactions. α-Chloro aldehydes are used as the dienophile precursor with a variety of enones to yield the cycloaddition products in excellent yields (eq 26). The reaction is facilitated with 0.5 mol % catalyst loading and high diastereo- and enantioselectivities are observed with a variety of substrates.[38]

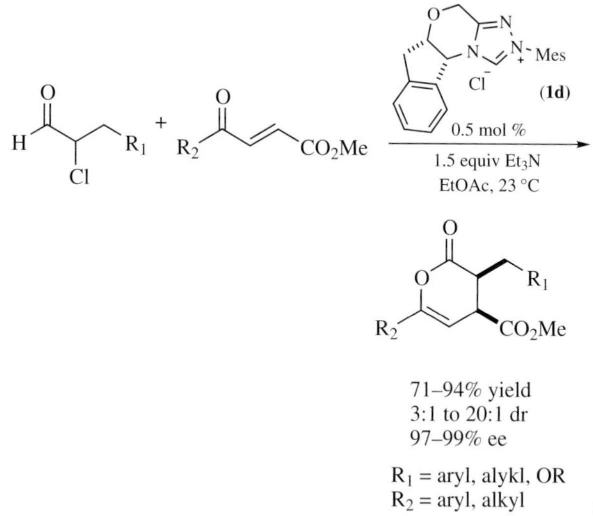

(26)

This reactivity was also applied to β,γ-unsaturated α-ketoesters as the Diels–Alder partner. Optimal catalyst loading was found to be 2 mol % to yield the desired products in good yield and selectivities. The authors propose that high *cis*-diasteroselectivities are a result of stereoselective formation of the (*Z*)-enolate in the redox reaction with a high preference for an *endo* cycloaddition (eq 27).

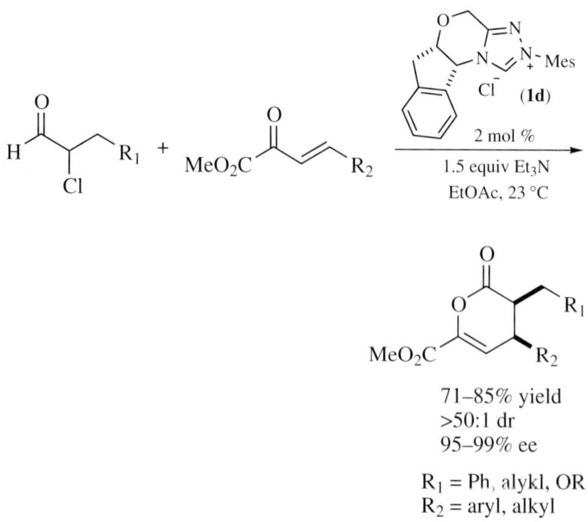

(27)

Due to the capricious nature of α-chloroaldehydes, the authors describe a process improvement in which bisulfite derivatives can be utilized in the reaction as the dienophile precursor. Using biphasic conditions under which the bisulfite salt could be liberated and

1 mol % of azolium salt, the reaction proceeds to the desired product with no observable epimerization (eq 28).[39]

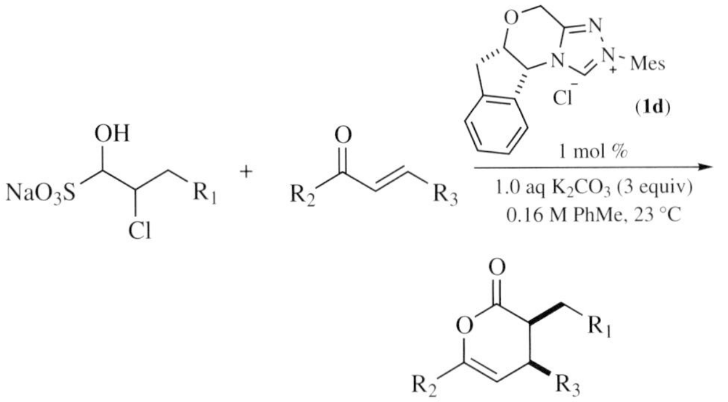

55–98% yield
>50:1 dr
91–99% ee

R_1 = Ph, alkyl
R_2 = aryl, alkyl, CO_2Et
R_3 = aryl, alkyl, CO_2Et (28)

Triazolium catalyst **1b** promotes a formal [2 + 2] cycloaddition of ketenes with *N*-tosylimines to generate *β*-lactams in excellent yields and moderate enantioselectivities (eq 29). High levels of enantioselectivity can be achieved after crystallization. The reaction is believed to proceed via addition of the carbene into the ketene to generate the azolium enolate. Enolate addition to the *N*-tosyl imine followed by cyclization delivers the *β*-lactam.[40]

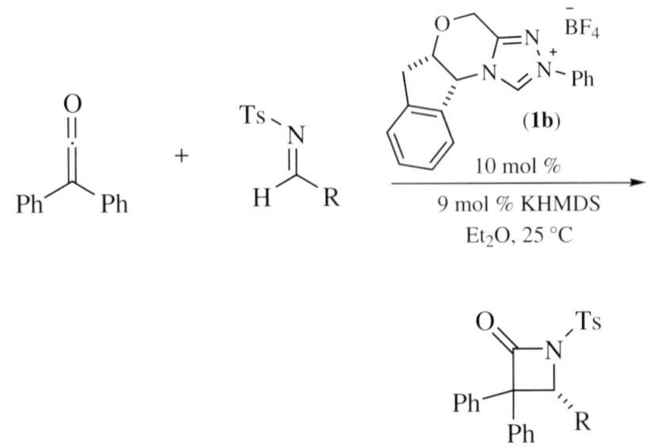

92–96% yield
55–75% ee
>92% ee after crystallization

R = Ph, 2-napthyl, 2-furyl, 4-BrC_6H_4 (29)

Metal Complexes. Achiral triazolium salts have been appended to Au and Ag and the resultant complexes characterized crystallographically (eq 30). The metal complexes appear to behave analogously to their imidazolinylidene counterparts. Application of these complexes as catalysts has not been reported.

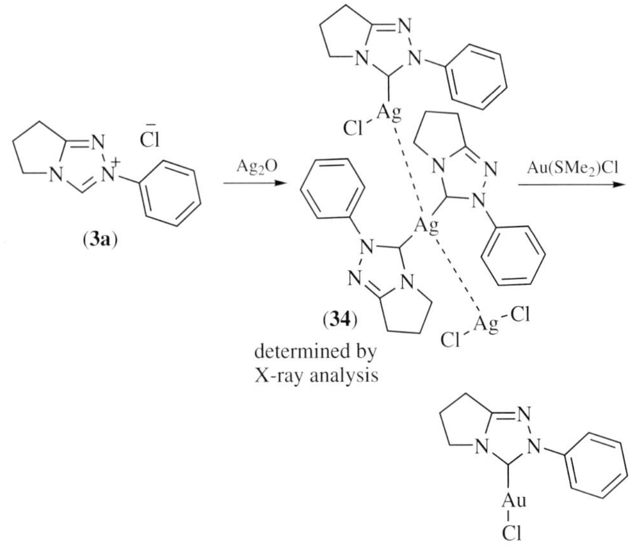

(30)

First Update

Tomislav Rovis & Dian-Feng Chen
Colorado State University, Fort Collins, CO, USA

Benzoin Reaction. A novel synthesis of (±)-glycinol featuring intramolecular benzoin reaction was reported.[41] The use of triazolium salt **1a** provides high yield, but with no enantioselectivity (eq 31).

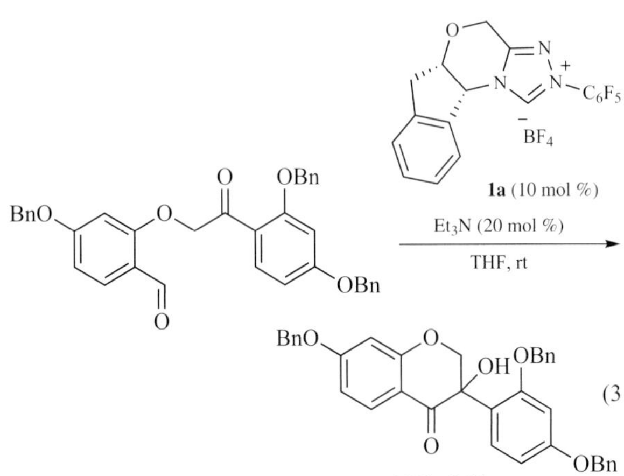

90% yield
racemic (31)

Triazolium salts can be part of multiple catalyst systems[42] for the rapid construction of molecular complexity through single-pot operations. The combination of prolinol ether catalyst **36** and carbene was first reported in 2009.[43] Intramolecular crossed benzoin reaction provides α-hydroxycyclopentanones containing three contiguous stereocenters via chiral enolate **37** generated from the initial enantioselective Michael addition step (eq 32). Control experiments revealed that two-step procedure eroded ee due to a prolinol-mediated retro-Michael process that could be limited by carbene catalyst via shuttling **38** to stable final product (eq 33).

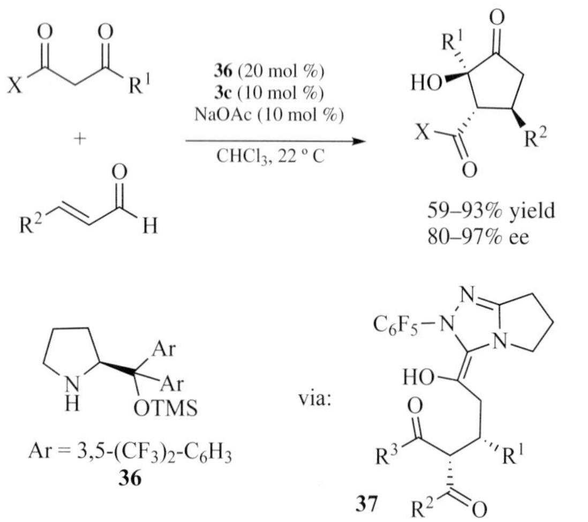

(32)

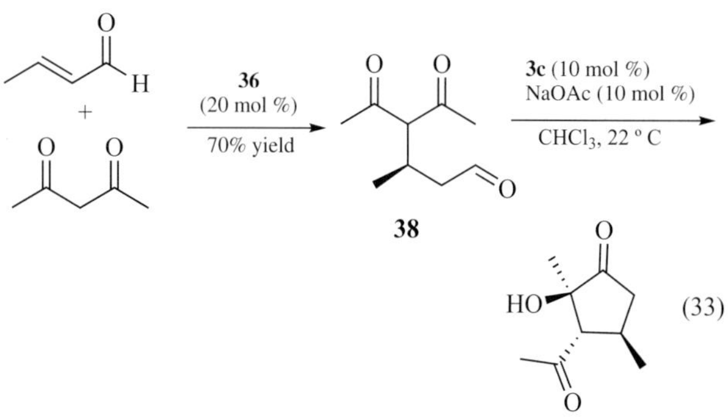

(33)

65% yield
5:1 dr, 58% ee

Prolinol ether/carbene multiple catalyst system was then extended to enamine activation/cross-benzoin reaction sequence (eq 34).[44] Prolinol **39** is responsible for high ee through the first Michael addition step. Catalyst match/mismatch relationship (eq 34) and prolinol's role of interconverting two diastereomers of **40** are essential for high dr (eq 35).

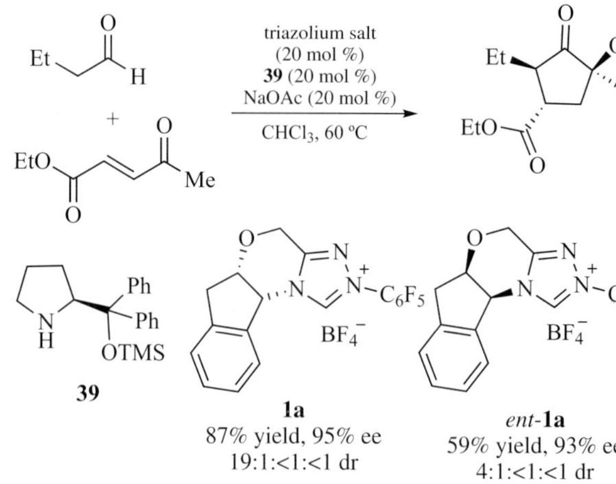

(34)

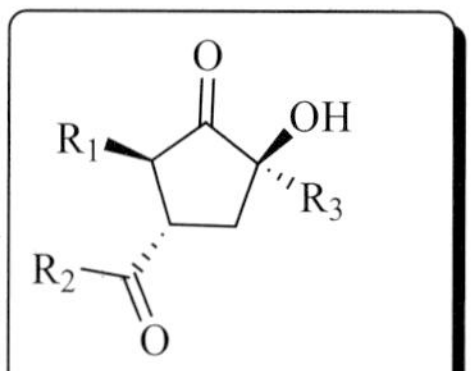

35–99% yield
51–98% ee
up to 96:2:2:1 dr

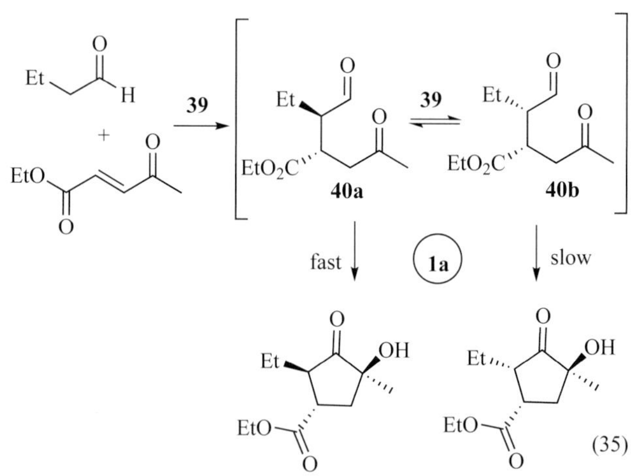

(35)

Stetter Reaction. Acetaldehyde has been found to be a donor for intermolecular Stetter reaction with *trans*-chalcones.[45]

10 mol % of triazolium catalyst **1a** provides up to 85% yield and moderate enantioselectivities (eq 36).

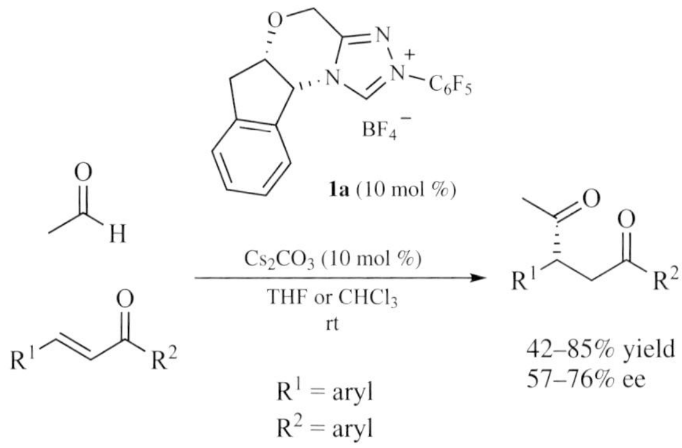

(36)

Crossed-benzoin reaction involving Michael acceptor with larger conjugated system might result in a vinylogous Stetter reaction. Treatment of **27** with triazolium salt **1a** selectively provides chromone derivatives in excellent yields and up to 96% ee (eq 37).[46]

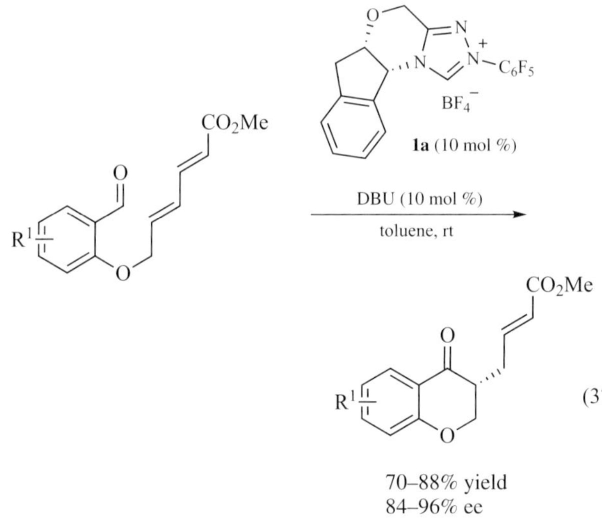

(37)

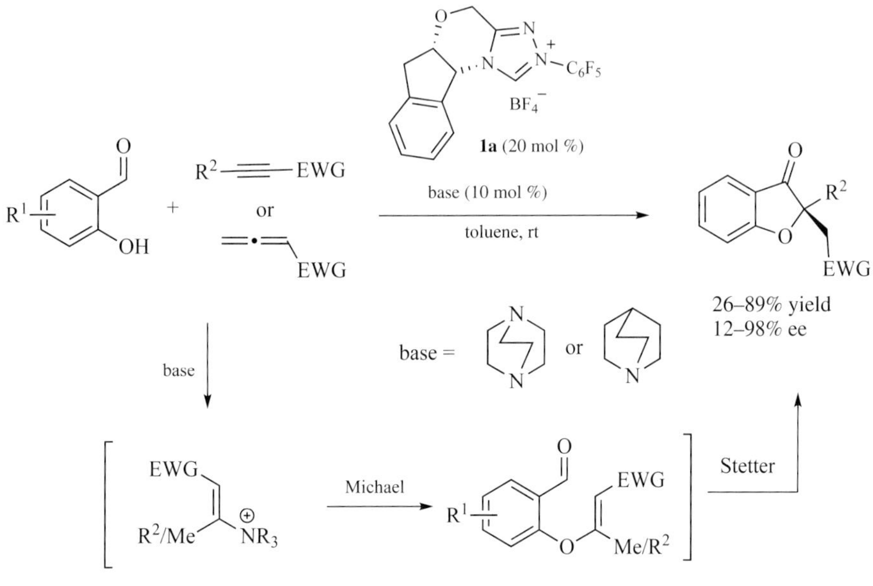

(38)

One-pot base-catalyzed Michael/carbene-catalyzed enantioselective Stetter reaction was also reported in 2010.[47] Both electron-deficient alkynes and allenoates can be reaction partners of salicylaldehydes to provide a unique access to chiral benzofuranones (eq 38). Strong H-bond donors, such as the starting material salicylaldehyde and by-product catechol, made this one-pot reaction superior to two-step procedure with respect to enantioselectivity.

Acyl Transfer. In the presence of external oxidants, the Breslow intermediate can be intercepted to generate the acyl azolium (**41**–**42** in eq 39),[48] which allows alcohol/amine trapping process.

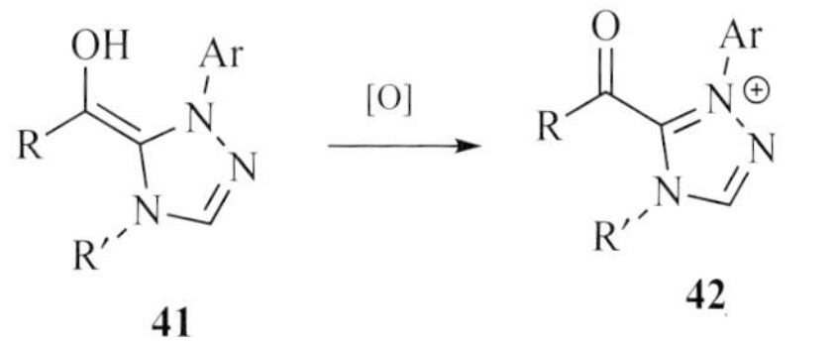

(39)

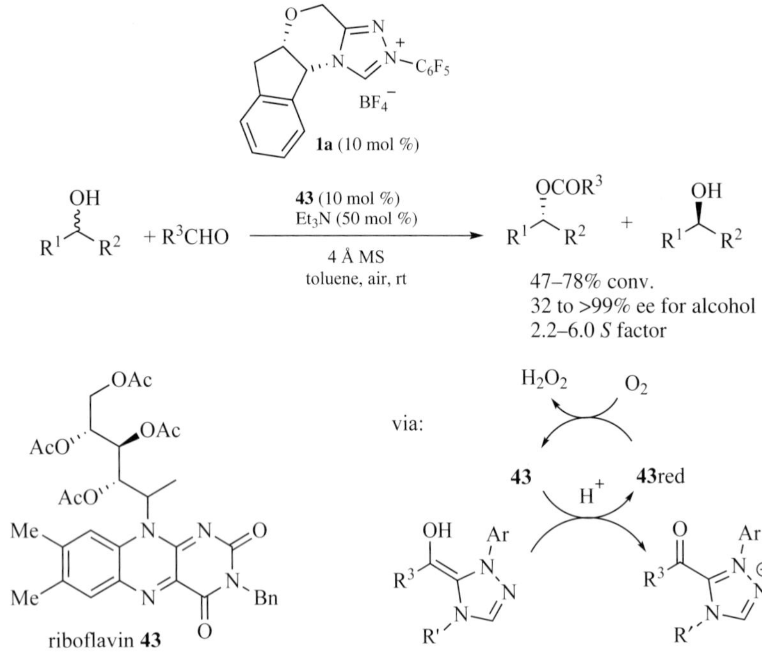

(40)

Iida and Yashima reported a chiral triazolium **1a**/riboflavin **43** binary catalyst system for a kinetic resolution of secondary alcohols, providing moderate *S* factors and moderate to excellent ee values for recovered alcohols (eq 40).[49] A catalytic amount of riboflavin was the in situ oxidant with molecular oxygen as the terminal oxidant.

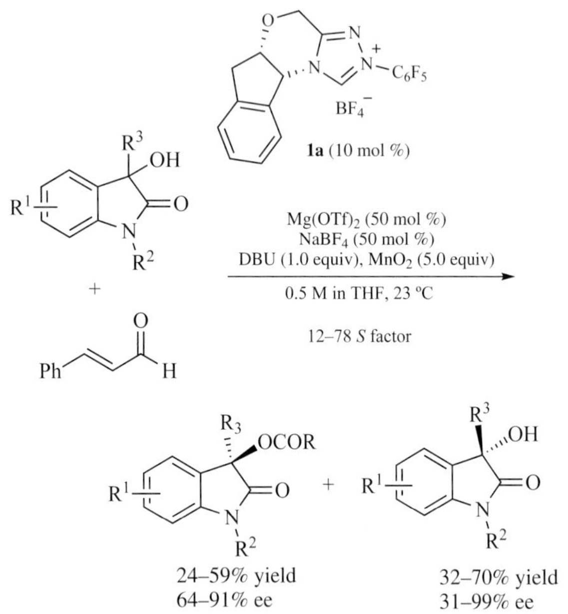

(41)

Kinetic resolution of tertiary 3-hydroxy oxindole alcohols was reported in 2013 by Zhao and coworkers.[50] Using 10 mol % of triazolium salt **1a**, 50 mol % of $Mg(OTf)_2$ as cocatalyst, and excess MnO_2 as oxidant, both the ester products and recovered alcohols are obtained with moderate to excellent enantioselectivities (eq 41).

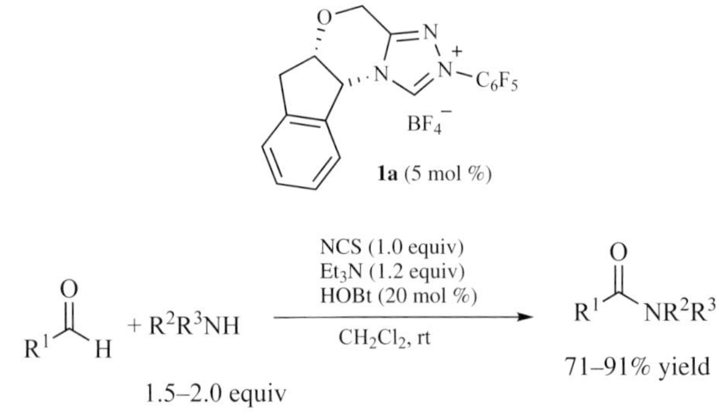

(42)

Amine trapping of acyl azolium generally requires an additional cocatalyst.[34] In addition to aromatic and unsaturated aldehydes, unmodified alkyl aldehydes can also be easily converted into amides via triazolium salt **1a**/HOBt catalyst system (eq 42).[51] A carbene-dependent nucleophile selectivity switch was observed in this reaction (eq 43).

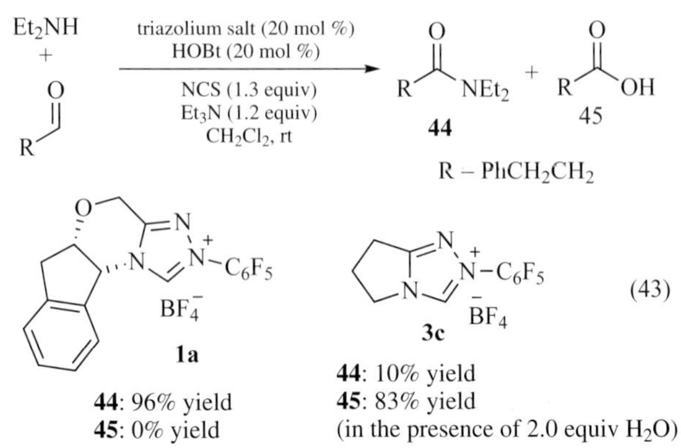

(43)

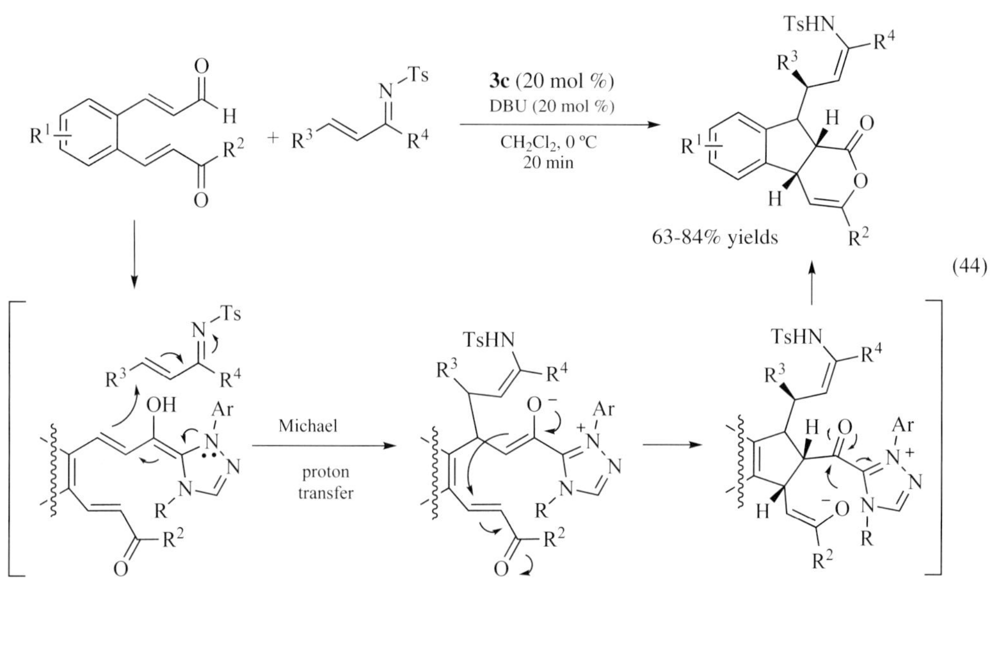

(44)

Redox Catalysis. An access to functionalized indane derivatives via trapping of enol/enolate intermediate was reported in 2014.[52] Triazolium salts **3c** can accelerate the whole process and provide the indane products in good yields (eq 44). Chiral triazolium salt **1b** is effective for an enantioselective process (eq 45).

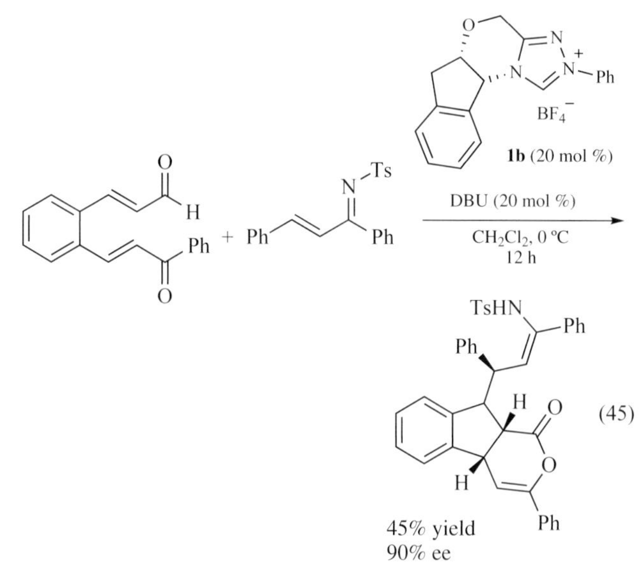

(45)

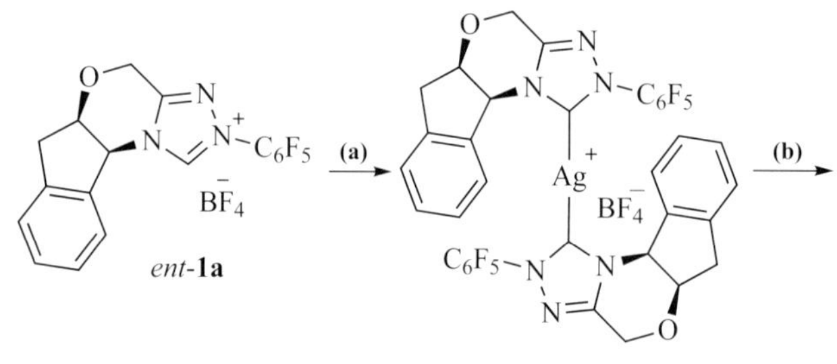

(a) Ag_2O, CH_2Cl_2, rt, 81% yield;
(b) $Cl_2(PCy_3)_2Ru{=}CHPh$, THF, rt, 52% yield

(46)

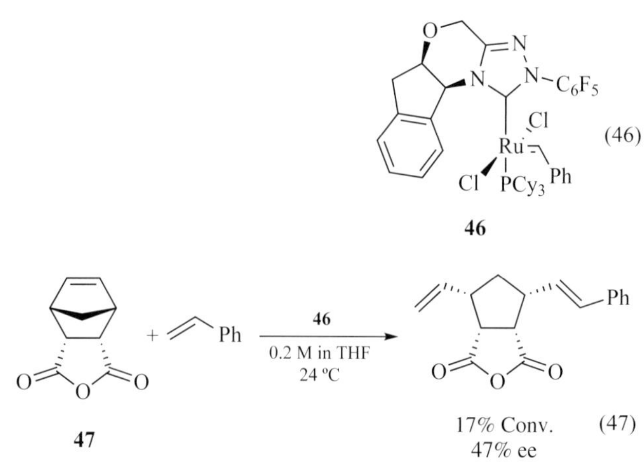

(47)

Metal Complex. Chiral 1,2,4-triazolium salts were recently applied as chiral ligand precursor for transition-metal catalysis.[53] Starting from *ent*-**1a**, Grela and co-workers report a two-step synthesis of Ru(II) complex **46**, which is unfortunately not very stable for storage (eq 46).[54] In situ generated **46** is available to be evaluated for an asymmetric ring-opening cross metathesis of **47** with styrene but only low conversion and moderate enantioselectivity are observed (eq 47).

1. Christmann, M., *Angew. Chem., Int. Ed.* **2005**, *44*, 2632.
2. Enders, D.; Niemeier, O.; Henseler, A., *Chem. Rev.* **2007**, *107*, 5606.
3. Zeitler, K., *Angew. Chem., Int. Ed.* **2005**, *44*, 7506.
4. Kerr, M. S.; Read de Alaniz, J.; Rovis, T., *J. Org. Chem.* **2005**, *70*, 5725.
5. Hassner, A., In *Comprehensive Organic Synthesis*; Trost, B. M.; Fleming, I., Eds.; Pergamon: Oxford, **1991**, vol. 1, 541.

6. Enders, D.; Breuer, K.; Teles, J. H., *Helv. Chim. Acta* **1996**, *79*, 1217.
7. Knight, R. L.; Leeper, F. J., *J. Chem. Soc., Perkin Trans. 1* **1998**, 1891.
8. Enders, D.; Niemeier, O.; Balensiefer, T., *Angew. Chem., Int. Ed.* **2006**, *45*, 1463.
9. Takikawa, H.; Hachisu, Y.; Bode, J. W.; Suzuki, K., *Angew. Chem., Int. Ed.* **2006**, *45*, 3492.
10. Takikawa, H.; Suzuki, K., *Org. Lett.* **2007**, *9*, 2713.
11. Chiang, P.; Kaeobamrung, J.; Bode, J., *J. Am. Chem. Soc.* **2007**, *129*, 3520.
12. Stetter, H.; Kuhlmann, H., *Org. React.* **1991**, *40*, 407.
13. Enders, D.; Breuer, K.; Runsink, J; Teles, J. H., *Helv. Chim. Acta* **1996**, *79*, 1899.
14. Kerr, M. S.; Read de Alaniz, J.; Rovis, T., *J. Am. Chem. Soc.* **2002**, *124*, 10298.
15. Kerr, M. S.; Rovis, T., *Synlett* **2003**, 1934.
16. Read de Alaniz, J.; Kerr, M. S.; Moore, J. L.; Rovis, T., *J. Org. Chem.* **2008**, *73*, 2033.
17. Kerr, M. S.; Rovis, T., *J. Am. Chem. Soc.* **2004**, *126*, 10298.
18. Moore, J. L.; Kerr, M. S.; Rovis, T., *Tetrahedron* **2006**, *62*, 11477.
19. Read de Alaniz, J.; Rovis, T., *J. Am. Chem. Soc.* **2005**, *127*, 6284.
20. Reynolds, N. T.; Rovis, T., *Tetrahedron* **2005**, *61*, 6368.
21. Liu, Q.; Rovis, T., *J. Am. Chem. Soc.* **2006**, *128*, 2552.
22. Liu, Q.; Rovis, T., *Org. Proc. Res. Dev.* **2007**, *11*, 598.
23. Cullen, S. C.; Rovis, T., *Org. Lett.* **2008**, *10*, 3141.
24. Orellana, A.; Rovis, T., *Chem. Commun.* **2008**, 730.
25. Thomson, J. E.; Rix, K.; Smith, A. D., *Org. Lett.* **2006**, *8*, 3785.
26. Ruble, J. C.; Fu, G. C., *J. Am. Chem. Soc.* **1998**, *120*, 11532.
27. Shaw, S. A.; Aleman, P.; Christy, J.; Kampf, J. W.; Va, P.; Vedejs, E., *J. Am. Chem. Soc.* **2006**, *128*, 925.
28. Reynolds, N. T.; Read de Alaniz, J.; Rovis, T., *J. Am. Chem. Soc.* **2004**, *126*, 9518.
29. Chow, K. Y. K.; Bode, J. W., *J. Am. Chem. Soc.* **2004**, *126*, 8126.
30. Sohn, S. S.; Bode, J. W., *Org. Lett.* **2005**, *7*, 3873.
31. Chan, A.; Scheidt, K. A., *Org. Lett.* **2005**, *7*, 905.
32. Zeitler, K., *Org. Lett.* **2006**, *8*, 637.
33. Sohn, S. S.; Bode, J. W., *Angew. Chem., Int. Ed.* **2006**, *45*, 6021.
34. Vora, H. U.; Rovis, T., *J. Am. Chem. Soc.* **2007**, *129*, 13796.
35. Wong, F. T.; Patra, P. K.; Seayad, J.; Zhang, Y.; Ying, J. Y., *Org. Lett.* **2008**, *10*, 2333.
36. He, M.; Struble, J. R.; Bode, J. W., *J. Am. Chem. Soc.* **2006**, *128*, 8418.
37. He, M.; Bode, J. W., *J. Am. Chem. Soc.* **2008**, *130*, 418.
38. He, M.; Uc, G. J.; Bode, J. W., *J. Am. Chem. Soc.* **2006**, *128*, 15088.
39. He, M.; Beahm, B. J.; Bode, J. W., *Org. Lett.* **2008**, *10*, 3817.
40. Duguet, N.; Campell, C. D.; Slawin, A. M. Z.; Smith, A. D., *Org. Biomol. Chem.* **2008**, *6*, 1108.
41. Malik, N.; Erhardt, P., *Tetrahedron Lett.* **2013**, *54*, 4121.
42. (a) Lee, J. M.; Na, Y.; Han, H.; Chang, S., *Chem. Soc. Rev.* **2004**, *33*, 302. (b) Wasilke, J.-C.; Obrey, S. J.; Baker, R. T.; Bazan, G. C., *Chem. Rev.* **2005**, *105*, 1001.
43. Lathrop, S. P.; Rovis, T., *J. Am. Chem. Soc.* **2009**, *131*, 13628.
44. Ozboya, K. E.; Rovis, T., *Chem. Sci.* **2011**, *2*, 1835.
45. Kim, S. M.; Jin, M. Y.; Kim, M. J.; Cui, Y.; Kim, Y. S.; Zhang, L.; Song, C. E.; Ryu, D. H.; Yang, J. W., *Org. Biomol. Chem.* **2011**, *9*, 2069.
46. Law, K. R.; McErlean, C. S. P., *Chem. Eur. J.* **2013**, *19*, 15852.
47. Filloux, C. M.; Lathrop, S. P.; Rovis, T., *Proc. Natl. Acad. Sci.* **2010**, *107*, 20666.
48. De Sarkar, S.; Biswas, A.; Samanta, R. C.; Studer, A., *Chem. Eur. J.* **2013**, *19*, 4664.
49. Iwahana, S.; Iida, H.; Yashima, E., *Chem. Eur. J.* **2011**, *17*, 8009.
50. Lu, S.; Poh, S. B.; Siau, W.-Y.; Zhao, Y., *Angew. Chem., Int. Ed.* **2013**, *52*, 1731.
51. Kuwano, S.; Harada, S.; Oriez, R.; Yamada, K.-i., *Chem. Commun.* **2012**, *48*, 145.
52. Tang, M.-S.; Zhao, Y.; Cheng, Y., *Synthesis* **2014**, *46*, 87.
53. (a) Strand, R. B.; Helgerud, T.; Solvang, T.; Dolva, A.; Sperger, C. A.; Fiksdahl, A., *Tetrahedron: Asymmetry* **2012**, *23*, 1350.
(b) Zhao, L.; Ma, Y.; Duan, W.; He, F.; Chen, J.; Song, C., *Org. Lett.* **2012**, *14*, 5780.
54. Gawin, R.; Pieczykolan, M.; Malińska, M.; Woźniak, K.; Grela, K., *Synlett* **2013**, *24*, 1250.

(8α,9*R*)-(8″α,9″*R*)-1,1″-[1,3-Phenylene-bis(methylene)]bis[9-hydroxy-cinchonanium] Dibromide and (9*S*)-(9″*S*)-1,1″-[1,3-Phenylene-bis(methylene)]bis[9-hydroxy-cinchonanium] Dibromide

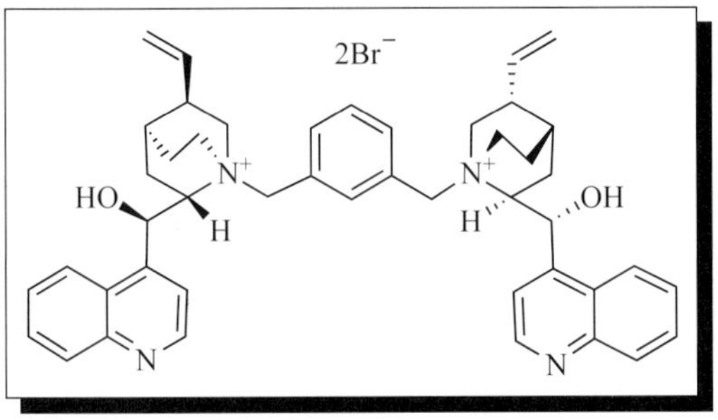

(8α,9*R*)-(8″α,9″*R*)-1,1″-[1,3-Phenylene-bis(methylene)]bis[9-hydroxycinchonanium] dibromide

[473410-07-0] $C_{46}H_{52}Br_2N_4O_2$ (MW 852.74)

InChI = 1S/C46H52N4O2.2BrH/c1-3-33-29-49(22-18-35(33)25-43(49)45(51)39-16-20-47-41-14-7-5-12-37(39)41)27-31-10-9-11-32(24-31)28-50-23-19-36(34(4-2)30-50)26-44(50)46(52)40-17-21-48-42-15-8-6-13-38(40)42;;/h3-17,20-21,24,33-36,43-46,51-52H,1-2,18-19,22-23,25-30H2;2*1H/q+2;;/p-2/t33-,34-,35-,36-,43-,44-,45+,46+,49+,50+;;/m0../s1

InChIKey = MMRDCPWCZCXEPN-WBXZERQYSA-L

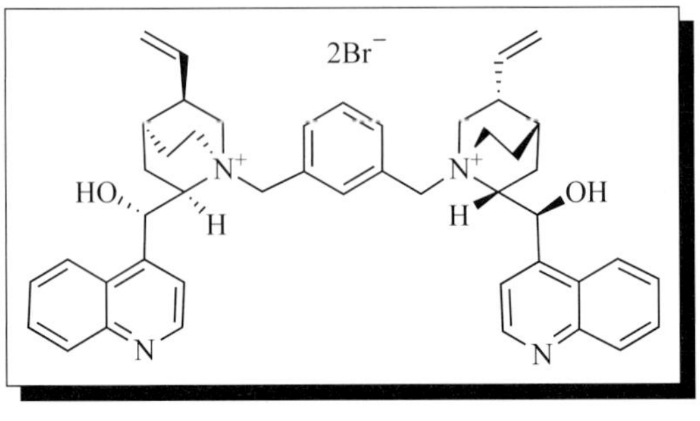

(9S)-(9″S)-1,1″-[1,3-Phenylene-bis(methylene)]bis[9-hydroxycinchonanium] dibromide

[479070-35-4] $C_{46}H_{52}Br_2N_4O_2$ (MW 852.74)

InChI = 1S/C46H52N4O2.2BrH/c1-3-33-29-49(22-18-35(33)25-43(49)45(51)39-16-20-47-41-14-7-5-12-37(39)41)27-31-10-9-11-32(24-31)28-50-23-19-36(34(4-2)30-50)26-44(50)46(52)40-17-21-48-42-15-8-6-13-38(40)42;;/h3-17,20-21,24,33-36,43-46,51-52H,1-2,18-19,22-23,25-30H2;2*1H/q+2;;/p-2/t33-,34-,35-,36-,43+,44+,45-,46-,49+,50+;;/m0../s1

InChIKey = MMRDCPWCZCXEPN-XRBLQDLUSA-L

(ammonium cinchona alkaloids, asymmetric phase transfer catalysis)[1]

Alternative Name: α,α′-biscinchoninium-*m*-xylene dibromide.

Physical Data: α,α′-biscinchonidinium-*m*-xylene dibromide: mp 205 °C (dec.); $[\alpha]_D^{25} = -171°$ ($c = 0.276$ in methanol).
α,α′-biscinchoninium-*m*-xylene dibromide: mp 222–225 °C (dec.), $[\alpha]_D^{22} = +165°$ ($c = 0.53$ in methanol).

Solubility: sol methanol, ethanol, DMF, chloroform, and other common solvents.

Form Supplied in: both dibromides are light pink solids.

Preparative Methods: dimeric ammonium salts are readily prepared by the reaction of (−)-cinchonidine or (+)-cinchonine with 1,3-bis(bromomethyl)benzene at room temperature or at occasionally elevated temperatures in ethanol–DMF–chloroform in 96% and 73% yields, respectively.[2]

Cinchona alkaloids α,α′-biscinchonidinium-*m*-xylene dibromide [$(CD)_2Xy$] and α,α′-biscinchoninium-*m*-xylene dibromide [$(CN)_2Xy$] are related to optically active quaternary ammonium salts, which can be used in various kinds of asymmetric transformations as chiral organocatalysts.

The preparation of optically active α-amino acid derivatives by use of asymmetric PTC such as cinchoninium salts is one of the most successful asymmetric PTC applications to date.

The catalytic efficiency of the ammonium salts $(CD)_2Xy$ and $(CN)_2Xy$ was examined by the benzylation of *N*-(diphenylmethylene)glycine *tert*-butyl ester under phase transfer conditions (eq 1).[3,4]

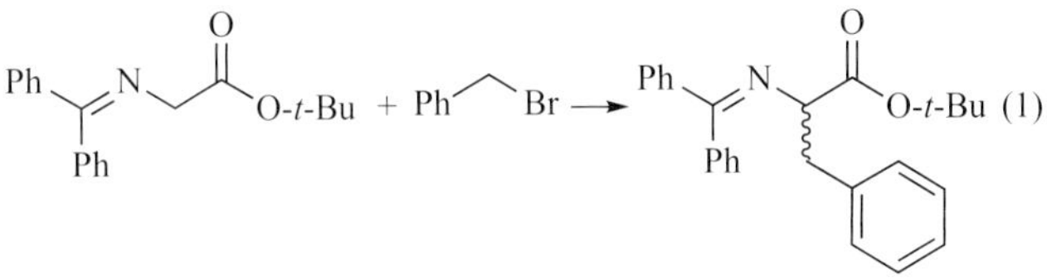

A: 89–90%, 76–84% ee (S)

B: 88%, 74% ee (R)

C: 75% 60% ee (S)

Metod A: $(CD)_2Xy$, KOH (or NaOH), PhMe/$CHCl_3$ (or CH_2Cl_2), −10 –0 °C, 4–10 h

Metod B: $(CN)_2Xy$, KOH, PhMe/$CHCl_3$, 0 °C, 10 h

Metod C: ionic polymer based on $(CD)_2Xy$, KOH, PhMe, 20 °C, 12 h

The authors[3] found that the 1,3-xylene-linked $(CD)_2Xy$ catalyst showed the highest enantioselectivity among other xylene-linked dimeric PTC (eq 1, method A). The order of enantioselectivity of the three PTC was as follows: 1,3-xylene-linked > 1,4-xylene-linked >> 1,2-xylene-linked catalysts.

The cinchonine-derived catalyst $(CN)_2Xy$ that bears a pseudoenantiomeric relationship with the cinchonidine-derived $(CD)_2Xy$ gave the opposite enantioselectivity, while the enantiomeric excess was somewhat lower (eq 1, method B).[3]

The study[5] reported incorporation of $(CD)_2Xy$ into ionic network by counterion exchange with 2,6-naphthalenedisulfonate, resulting in a polymer with an average molecular weight ~7400 Da and anionic bonds in the main chain.

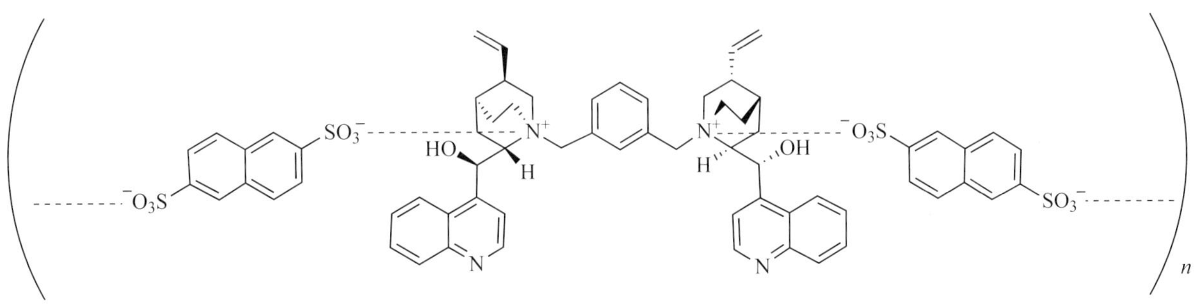

The chiral ionic polymer served as an organocatalyst in the asymmetric benzylation of *N*-(diphenylmethylidene)glycine *tert*-butyl ester (eq 1, method C). In a biphasic system of toluene and an aqueous alkaline solution, the reaction occurred smoothly to afford the corresponding chiral phenylalanine derivative. After the asymmetric benzylation, the polymeric catalyst was easily separated from the reaction mixture. The recovered polymer could be reused.

Reference 6 describes the nucleophilic enantioselective trifluoromethylation of ketones using the Ruppert–Prakash reagent (Me_3SiCF_3) and the combination of $(CN)_2Xy$ with tetramethylammonium fluoride (TMAF) (eq 2).

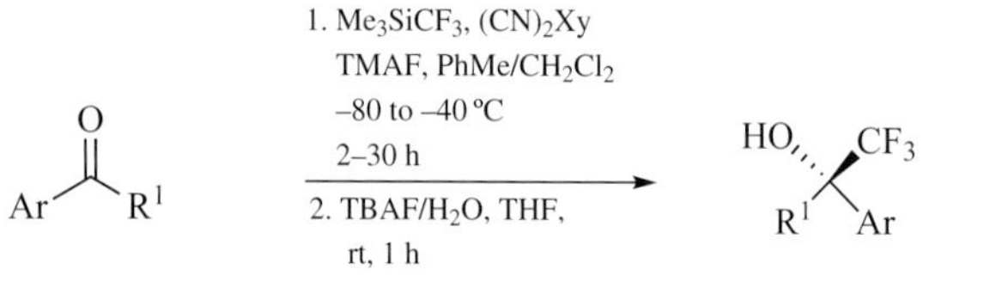

(2)

34–96% 10–94% ee

R^1 = H, Me, Et, Pr

Ar = Ph, 3-XC_6H_4 (X = Cl, Br, NO_2), 4-XC_6H_4 (X = Me, F, Cl, Br, NO_2, MeO) Np, 2-Np, 6-MeONp, 6-Me-2Np, $PhCH_2CH_2$, PhCH=CH, 1-indanone 1-tetralone, 6-MeO-1-tetralone, 1-benzosuberon

This approach enables nonfluorinated prochiral substrates to be directly transformed to chiral trifluoromethylated aryl alcohols bearing tetrasubstituted carbon centers. The authors found that the $(CN)_2Xy$/TMAF combination is an extremely general catalyst for the asymmetric trifluoromethylation of a broad range of alkyl aryl ketones. The desired products were obtained in high yields with up to 94% ee (in case of 1-tetralone). The selective 1,2-addition occurred in the trifluoromethylation of styryl methyl ketone to furnish chiral *trans-R*-trifluoromethyl allylic alcohol in 85% yield with 70% ee. The $(CN)_2Xy$/TMAF catalyst system was not effective for the enantioselective trifluoromethylation of both the 2-naphthalenealdehyde and the aliphatic ketone 4-phenylbutan-2-one (only 41% and 10% ee, correspondingly). Kawai et al.[7] have developed the enantioselective trifluoromethylation of alkynyl ketones with Me_3SiCF_3 by employing $(CN)_2Xy$ as catalyst to produce trifluoromethylated propargyl alcohols (eq 3).

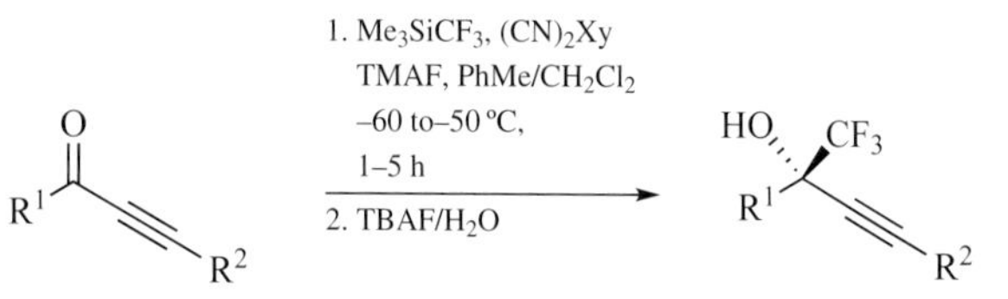

(3)

65–99% 40–96% ee

R^1 = Ph, *p*-Tol, *m*-Tol, 3-$MeOC_6H_4$, 4-XC_6H_4 (X = Me, *t*-Bu, F, Cl, MeO, 4-Ph) 3,4-$Me_2C_6H_3$, 2-Np, PhCH=CH

R^2 = *t*-Bu, Ph, $C(Me)_2OBn$, $C(Me)_2OCHPh_2$ $C(Me)_2OCH_2(4\text{-}BrC_6H_4)$, Si-*i*-$Pr_3$

The combination of $(CN)_2Xy$/TMAF was found to be very general for chiral transformation with enantioselectivity up to 96% ee (eq 3, R^1 = *p*-Tol, R^2 = *t*-Bu). In all cases, the initial adduct was a trimethylsilyl ether that was deprotected by the treatment with *n*-Bu_4NF to provide the corresponding alcohol. All the reactions proceeded in essentially quantitative yields as confirmed by the TLC analysis, but the isolated yields of final products were often lower due to loss of the product during deprotection of the trimethylsilyl group. The reaction is applicable not only for *tert*-butyl-substituted alkynones but also for other sterically demanding acetylenic ketones. Namely, aryl alkynyl ketones with hydroxypropyl ethers, $R^2 = C(Me)_2OCHAr_2$, were smoothly converted to the corresponding trifluoromethylated propargyl alcohols in yields 66–99% with enantioselectivities up to 91% ee.

Importantly, (triisopropylsilyl)ethynyl aryl ketones (R^2 = *i*-Pr_3Si) were also found to provide trifluoromethylated adducts in high yields and with 86–88% ee. The triisopropylsilyl group at the ethynyl position and the *O*-trimethylsilyl group were simultaneously removed by *n*-Bu_4NF treatment to generate nonsilylated propargyl alcohols in yields 53–64%.

The same authors[8] attempted the reaction of alkynone (eq 3, R^1 = 2-nitro-5-chlorophenyl, R^2 = cyclopropyl) in the presence of $(CD)_2Xy$/TMAF at −60 °C for 48 h. However, this attempt gave a disappointing result: after treatment by *n*-Bu_4NF, trifluoromethylated arylcyclopropyl butyn-2-ol was obtained in 72% yield with 4% ee (*S*).

In another study,[9] the PTC catalyst $(CD)_2Xy$ was applied to asymmetric epoxidation. Initially, the authors performed epoxidation of *trans*-chalcone by using 5 mol % of the $(CD)_2Xy$ along with H_2O_2 and KOH in diisopropyl ether (eq 4, condition (a)).

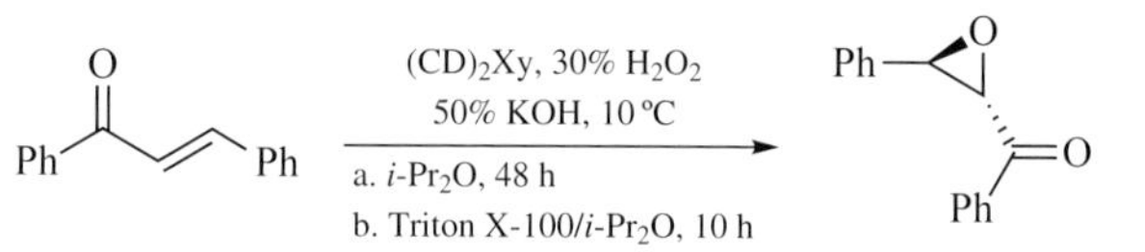

(4)

(a) 85%, 48% ee
(b) 89%, 82% ee

The use of just 5 mol % of surfactant Triton X-100 dramatically increased not only the rate of the reaction (five times) but also the enantioselectivity (eq 4, condition (b)). Presumably, when the nucleophile and electrophile were not in the same phase, the PTC reaction was considerably slower than the reaction with both components in the organic phase.

1. Kacprzak, K.; Gawronski, J., *Synthesis* **2001**, *7*, 961.
2. Qi, J.; Beeler, A. B.; Zhang, Q.; Porco, Jr., J. A., *J. Am. Chem. Soc.* **2010**, *132*, 13642.
3. Lee, J.-H.; Yoo, M.-S.; Jung, J.-H.; Jew, S.-S.; Park, H.-G.; Jeong, B.-S., *Tetrahedron* **2007**, *63*, 7906.
4. Siva, A.; Murugan, E., *J. Mol. Catal. A* **2005**, *241*, 111.
5. Itsuno, S.; Paul, D. K.; Salam, M. A.; Haraguchi, N., *J. Am. Chem. Soc.* **2010**, *132*, 2864.
6. Mizuta, S.; Shibata, N.; Akiti, S.; Fujimoto, H.; Nakamura, S.; Toru, T., *Org. Lett.* **2007**, *9*, 3707.
7. Kawai, H.; Tachi, K.; Tokunaga, E.; Shiro, M.; Shibata, N., *Org. Lett.* **2010**, *12*, 5104.
8. Kawai, H.; Kitayama, T.; Tokunaga, E.; Shibata, N., *Eur. J. Org. Chem.* **2011**, 5959.
9. Jew, S.-S.; Lee, J.-H.; Jeong, B.-S.; Yoo, M.-S.; Kim, M.-J.; Lee, Y.-J.; Lee, J.; Choi, S.-H.; Lee, K.; Lah, M. S.; Park H.-G., *Angew. Chem., Int. Ed.* **2005**, *44*, 1383.

Andrey Yu. Platonov
Saint Petersburg, Russia

2-(Phenylmethoxy)acetaldehyde

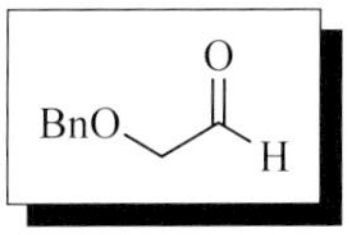

[60656-87-3] $C_9H_{10}O_2$ (MW 150.18)
InChI = 1S/C9H10O2/c10-6-7-11-8-9-4-2-1-3-5-9/h1-6H,7-8H2
InChIKey = NFNOAHXEQXMCGT-UHFFFAOYSA-N

(versatile aldehyde in chemical synthesis)

Alternate Name: α-benzyloxyacetaldehyde; 2-benzyloxyacetaldehyde; benzyloxyacetaldehyde; benzyloxyethanal; *O*-benzyl-α-hydroxyacetaldehyde.
Physical Data: bp 118–120 °C/13 mm Hg.[1]
Solubility: soluble in most commonly used organic solvents.
Form Supplied in: clear oil, commercially available with 0.5% hydroquinone as stabilizer.
Analysis of Reagent Purity: ^{1}H NMR.
Preparative Method: most commonly synthesized by ozonolysis of (*Z*)-1-((4-(benzyloxy)but-2-enyloxy)methyl)benzene[2] or by oxidative cleavage of *meso*-1,4-bis(benzyloxy)butane-2,3-diol.[3]
Purification: the oil can be purified by column chromatography on silica gel and/or distillation.
Handling, Storage, and Precautions: store in a freezer, under an inert atmosphere; after some weeks of storage, purification by distillation is recommended. Use in a fume hood.

Introduction. 2-(Phenylmethoxy)acetaldehyde is often applied in the total syntheses of natural products. It is a protected equivalent of glycolaldehyde. In contrast to glycolaldehyde, which exists as a mixture of its hydrate and dimer in water, the title compound is monomeric and does not form a stable hydrate. Synthetic equivalents of this protected aldehyde include its TBS-protected analog[4] and the parent *p*-methoxybenzyl derivative.[5,6] In biological systems, the phosphorylated derivative is also important.[7,8] With its ease of synthesis, monomeric nature, and commercial availability, 2-(phenylmethoxy)acetaldehyde has proven to be an effective reagent and, most recently, an organocatalyst. As the reagent is prochiral, emphasis will be placed on reactions in which the formation of chiral products is stereocontrolled. However, its application in reactions that do not generate stereocenters will be first discussed.

Acetal and Oxonium Ion Formation. Acetal formation is typically not facile due to the inductive effect of the *benzyloxy* functional group that destabilizes the transient oxocarbenium ion. The corresponding cyclic acetals and thioacetals are rare, often formed under forcing conditions.[9,10] Furthermore, examples of acylic derivatives are also uncommon. Such a derivative has been prepared utilizing cerium(III) chloride heptahydrate and trimethyl orthoformate (eq 1).[11] 2-(Phenylmethoxy)acetaldehyde has also been used in the presence of alcohols to generate oxocarbenium ions in situ, which can then undergo further reaction processes (eq 2).[12]

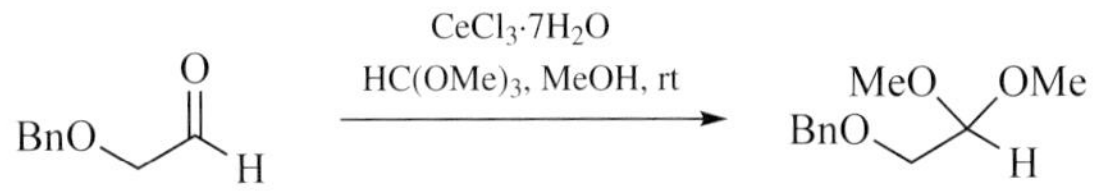

(1)

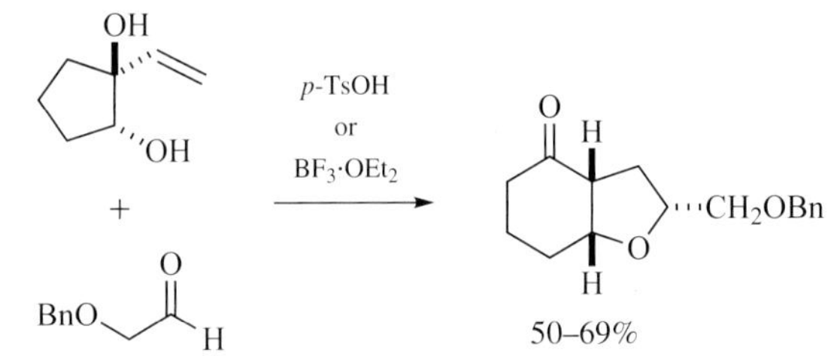

(2)

Enolate Derivatives. 2-(Phenylmethoxy)acetaldehyde has been applied in the synthesis of reactive dienophiles. For example, acetylation proceeds with predominately *Z*-selectivity in moderate yield (eq 3). These dienophiles have been employed in a variety of inverse electron-demand Diels–Alder reactions, furnishing substituted carbohydrates and 1,2,3,4-tetrahydropyridines.[13–16] The parent silyl enol ethers have also been reported.[17,18]

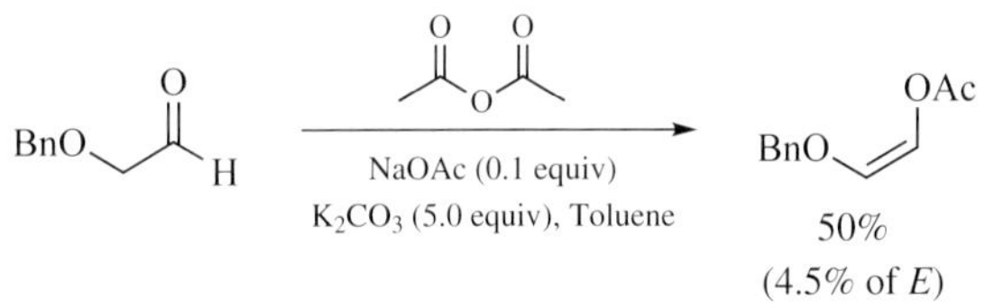

(3)

Olefinations. The title compound has been used as a reagent for various olefination processes. Several examples include the Horner—Wadsworth—Emmons (eq 4),[19] the Wittig (eq 5),[20] and the Julia-Kocienski (eq 6) olefination reactions.[21] These transformations proceed with moderate to excellent *E*/*Z* selectivity.

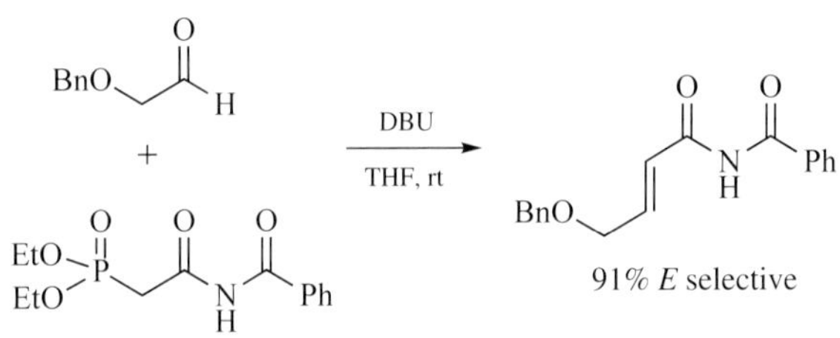

(4)

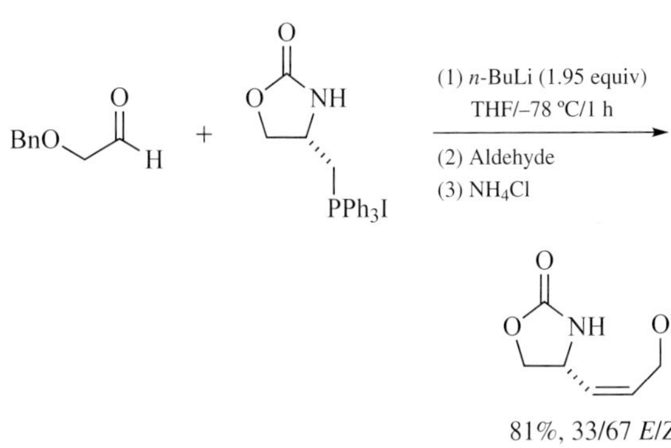

(5)

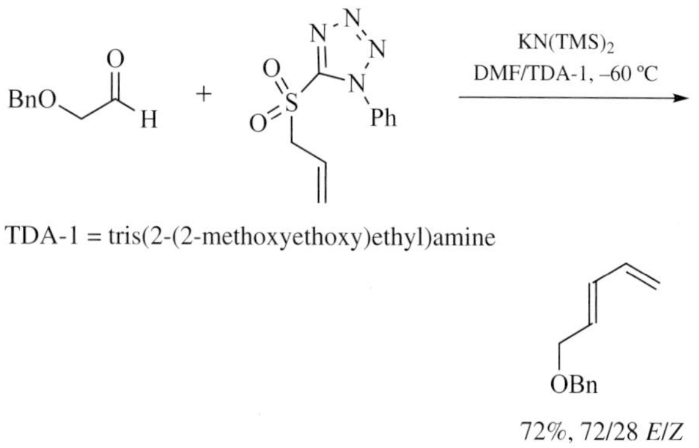

(6)

2-(Phenylmethoxy)acetaldehyde has also been used to access (*E*)-1,3-dienes via Lewis acid-promoted addition of 1,3-bis(silyl)propenes to aldehydes, providing the corresponding (*E*)-1,3-dienes (eq 7)—a sequence that relies on a Peterson olefination. This methodology is highly selective for the formation of terminal *E* dienes.[22]

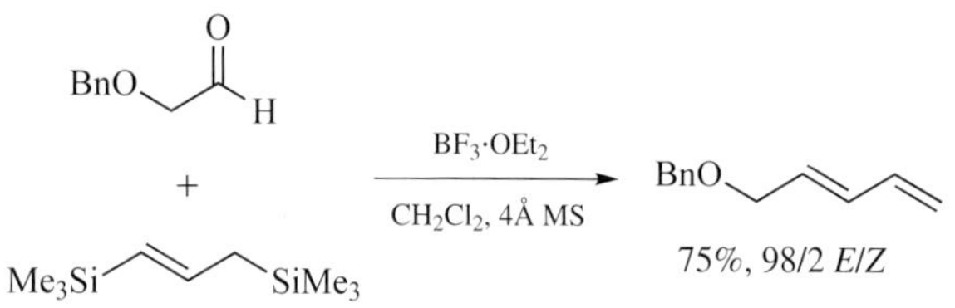

(7)

Homologation. The title compound may be homologated using diazo-based reagents. For example, the valuable ketoester derivative can be generated under simple reaction conditions, which in this case allowed the "in-flow" synthesis of a substituted pyrimidine derivative (eq 8).[23]

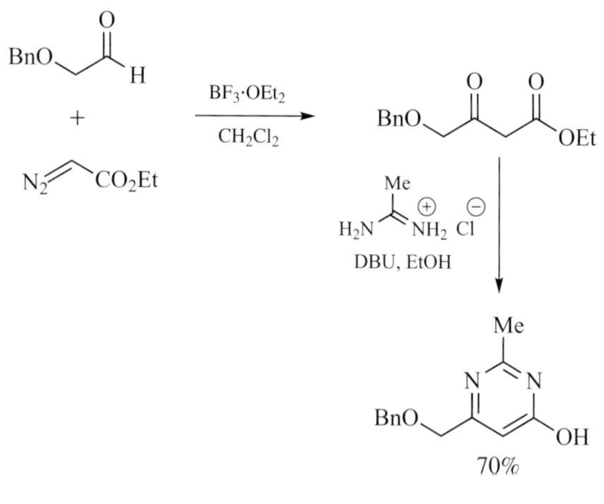

(8)

Cross-coupling of Aldehydes with Acylimines. The synthesis of a Boc-protected ketoamine has been accomplished in good yield using 2-(phenylmethoxy)acetaldehyde (75%, eq 9). This reaction operates via thiazolium carbene catalysis to generate an acyl anion equivalent providing substrates that have been used in heterocyclic synthesis and the assembly of more complex targets.[24,25]

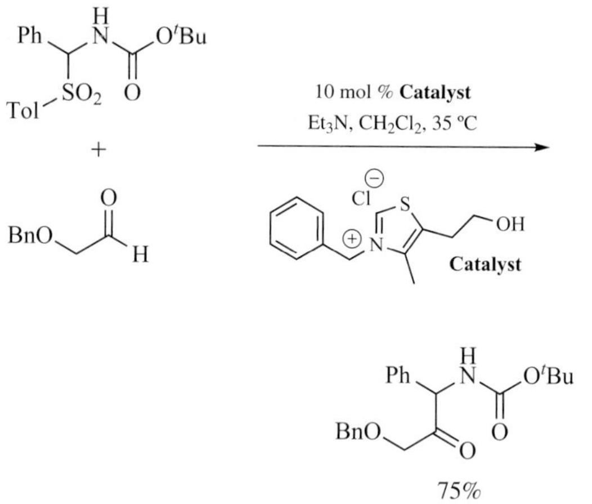

(9)

Asymmetric Synthesis. 2-(Phenylmethoxy)acetaldehyde has been extensively utilized in asymmetric transformations. This aldehyde can also form bidentate complexes with a variety of Lewis acids, and several families of chiral ligands or catalysts have been utilized. While not systematically highlighted below, asymmetric reactions using 2-(phenylmethoxy)acetaldehyde are routinely used to provide starting points for the assembly of complex natural products. Selected reactions using chiral reagents, chiral promoters, and chiral catalysts will be described below.

A variety of "allylmetal"-type reagents have been added to 2-(phenylmethoxy)acetaldehyde with high enantio- and stereocontrol. For example, (*E*)-crotylsilanes are able to undergo an asymmetric addition to aldehydes, including 2-(phenylmethoxy)acetaldehyde, and sequential silicon-directed heterocyclization that results in the formation of tetrahydrofurans. Upon treatment with antimony pentachloride, these cyclic compounds undergo ring opening through an E2-type elimination to yield either *anti* (eq 10) or *syn* (eq 11) homoallylic alcohols with excellent stereocontrol.[26–28] The relative stereochemistry in the intermediates and products is dictated by the choice of the Lewis acid.

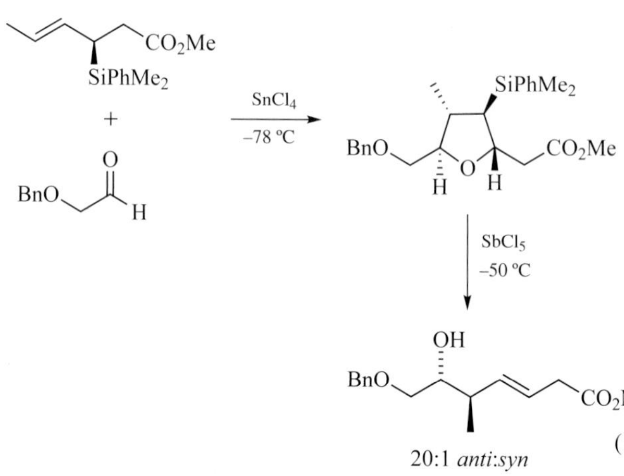

(10)

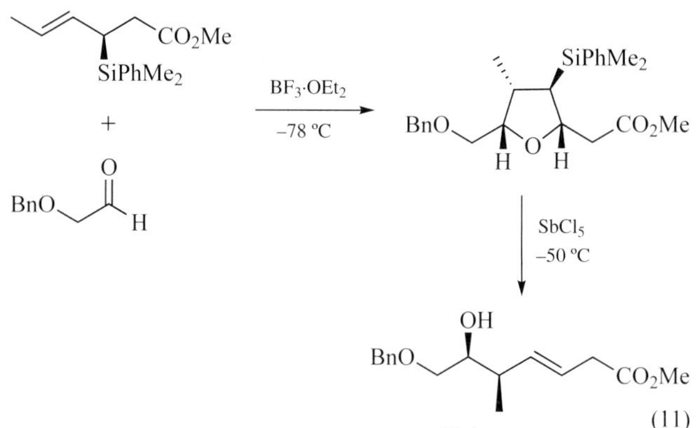

20:1 *syn*:*anti* (11)

Asymmetric Additions. Additions of chiral silanes to 2-(phenylmethoxy)acetaldehyde have been employed to yield highly enantioenriched homoallylic alcohols without the use of added Lewis acids. This system provides access to both *anti* (eq 12) and *syn* (eq 13) derivatives with excellent stereocontrol.[30–32]

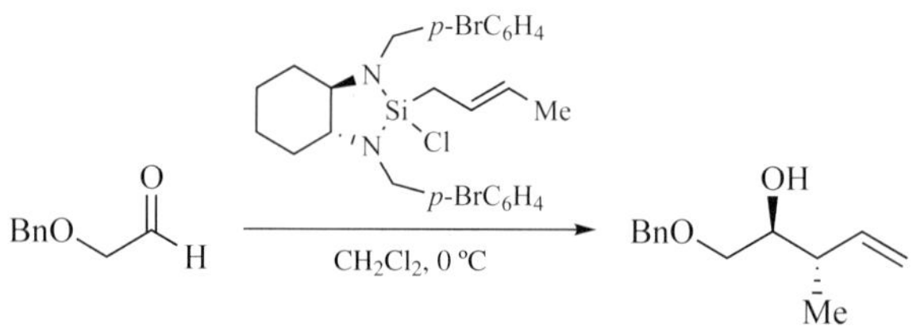

(12) 83%, 99% ee

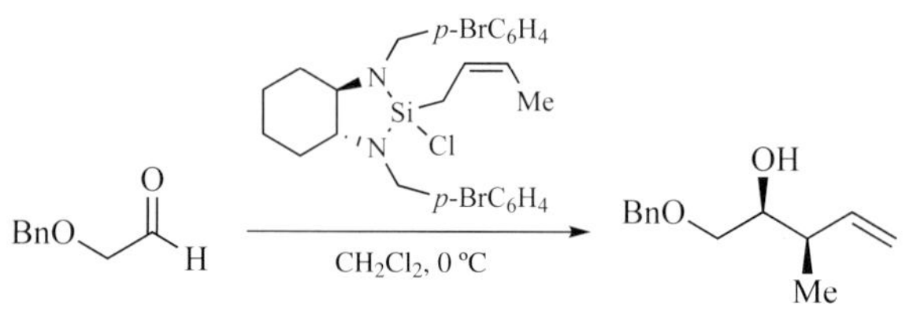

(13) 82%, 96% ee

Asymmetric Catalysis. Many chiral catalysts employ the propensity of 2-(phenylmethoxy)acetaldehyde to form bidentate complexes. Representative examples of the use of 2-(phenylmethoxy)acetaldehyde are provided below.

Asymmetric Mukaiyama Aldol Reactions. 2-(Phenylmethoxy)acetaldehyde can be reacted with silyl ketene acetals in copper-catalyzed asymmetric aldol reactions.[33,34] Both bisoxazolines (box, **1**) and pyridine bisoxazoline (pybox, **2**) ligands allow highly enantioselective reactions (eqs 14 and 15). (*Z*)-Propionate-derived silyl ketene acetals provided high preference for the formation of *syn*-aldol products (eq 16).

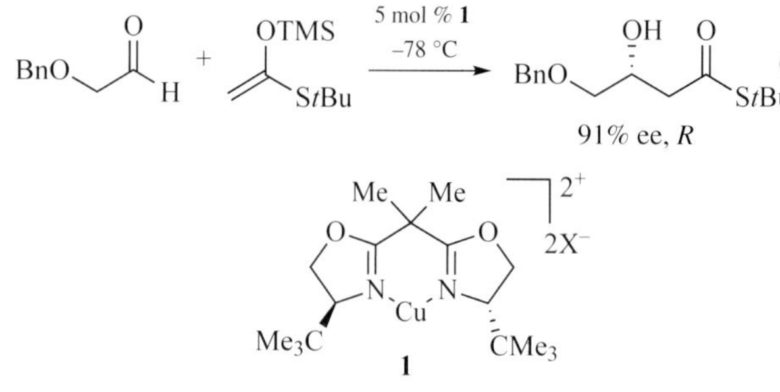

(14) 91% ee, *R*

1

X = OTf

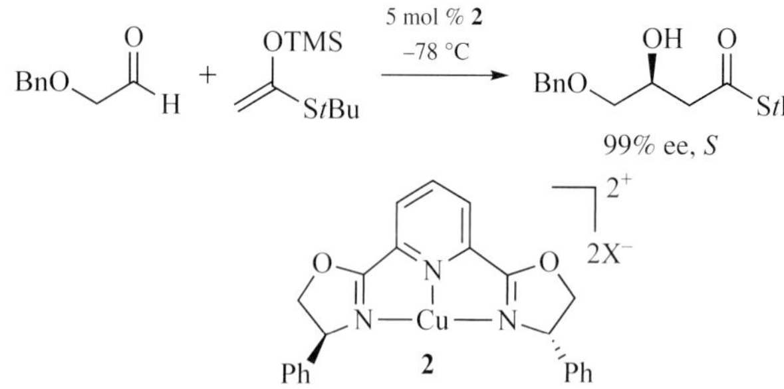

(15) 99% ee, *S*

2

X = SbF_6

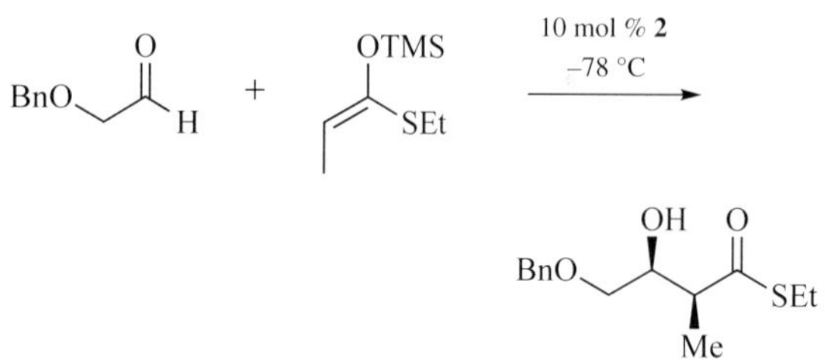

(16) 97% ee, 97:3 *syn*:*anti*

The title compound was applied in the preparation of highly enantioenriched monoprotected *anti*-1,2-diol motifs through a direct, proline-catalyzed, enantioselectivedimerization (eq 17). Cross-coupling of aldehyde substrates can also be achieved by taking advantage of the increased electrophilicity and acidity of 2-(phenylmethoxy)acetaldehyde to form the enamine intermediate (eq 18).[35–37] A two-step hexose synthesis can be accomplished utilizing 2-(phenylmethoxy)acetaldehyde and L-prolinecatalysis (eq 19). This methodology provides a synthesis of carbohydrate analogs that is inexpensive and relatively simple.[38]

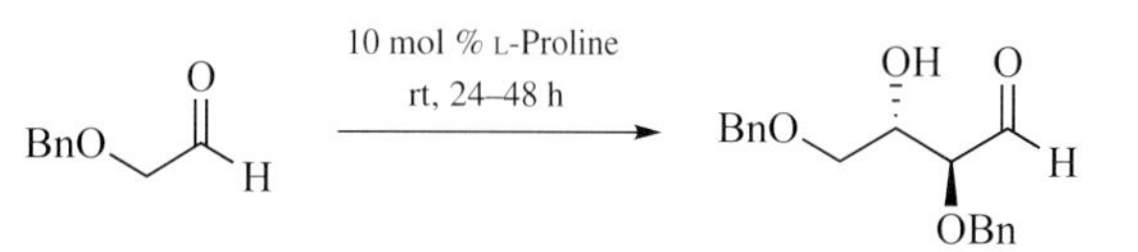

(17) 78%, 4:1 *anti*:*syn* 98% ee

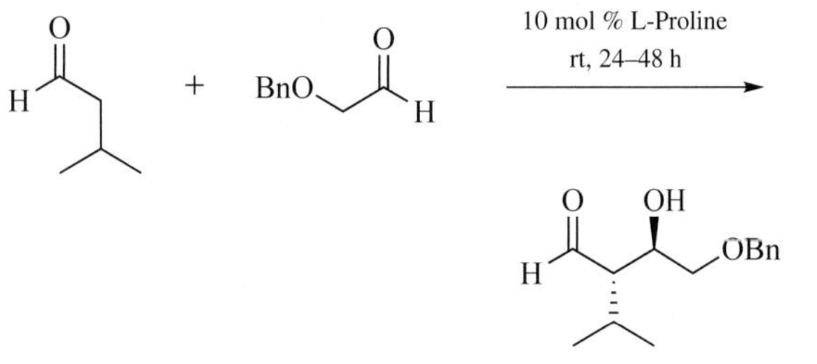

(18)

64%, 4:1 *anti:syn*
94% ee

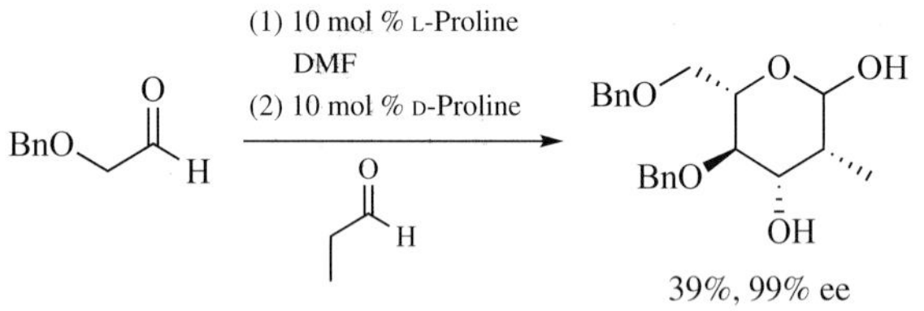

(19)

39%, 99% ee

[2+2] Cycloadditions. C_2-Symmetric Cu(II) complexes have been used as chiral Lewis acids to catalyze the cycloaddition of 2-(phenylmethoxy)acetaldehyde and a silyl ketene in excellent yield and with high enantioselectivity (87% ee, eq 20). Due to chelation, the reaction operates through an organized substrate–catalyst complex.[39] 2-(Phenylmethoxy)acetaldehyde was also reported undergoing a similar reaction with a Lewis acid–Lewis base bifunctional catalyst with near perfect enantioselectivity (eq 21). There are several related reports with Al(III) and a cinchona alkaloid–Lewis acid-catalyzed system.[40–44]

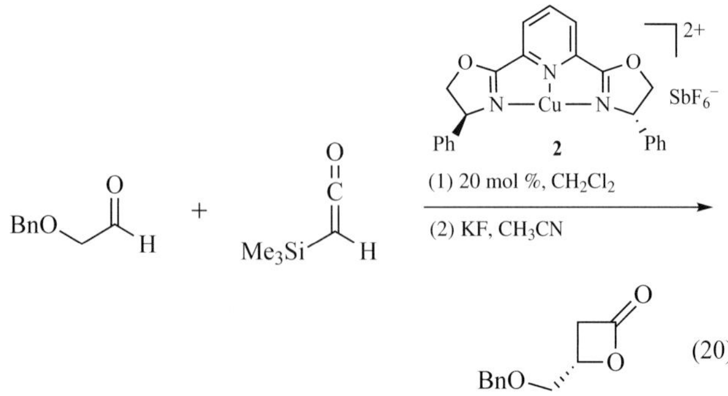

(20)

92%, 87% ee

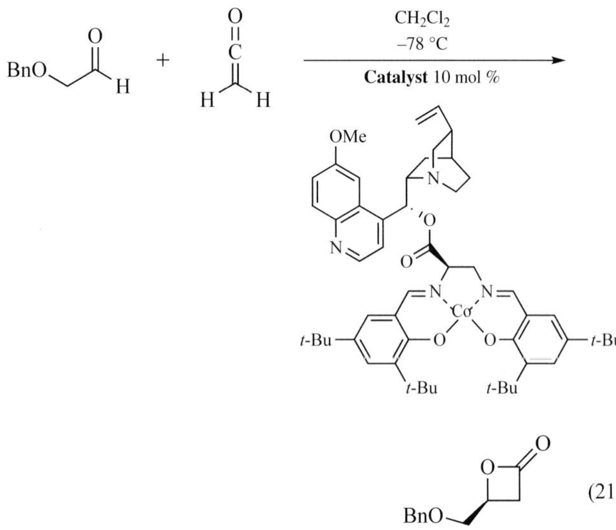

(21)

83%, 99% ee

Hetero-Diels–Alder Reactions. 2-(Phenylmethoxy)-acetaldehyde has been used to synthesize dihydropyran derivatives through a hetero-Diels–Alder reaction. Utilizing a chiral tridentate chromium(III) catalyst, three stereocenters are set with high enantioselectivity (94% ee, eq 22). This effective methodology was one of the first hetero-Diels–Alder approaches to enable dienes bearing less than two oxygen substituents to react with unactivated carbonyl compounds.[45]

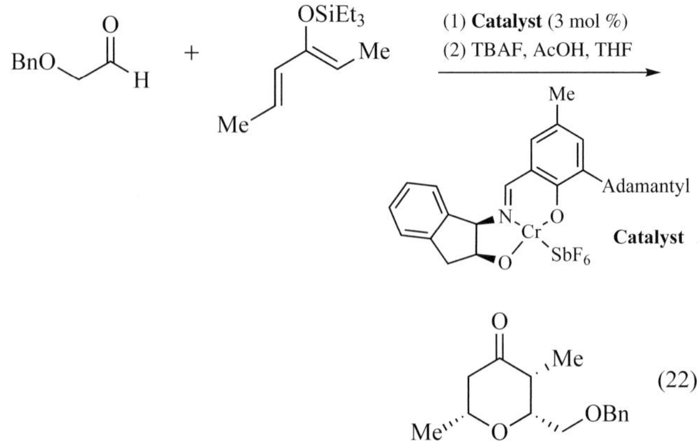

(22)

89%, 94% ee

A catalytic asymmetric hetero-Diels–Alder reaction was reported between 2-aza-3-silyloxy-1,3-butadienes and 2-(phenylmethoxy)acetaldehyde. The system uses a chiral dirhodium(II) Lewis acid catalyst, Rh_2(*S*-BPTPI)$_4$, and gives access to exclusively *cis*-substituted 1,3-oxazin-4-ones in high yields and with excellent enantioselectivities (eq 23). Monooxygenated dienes also react with high selectivity.[46,47]

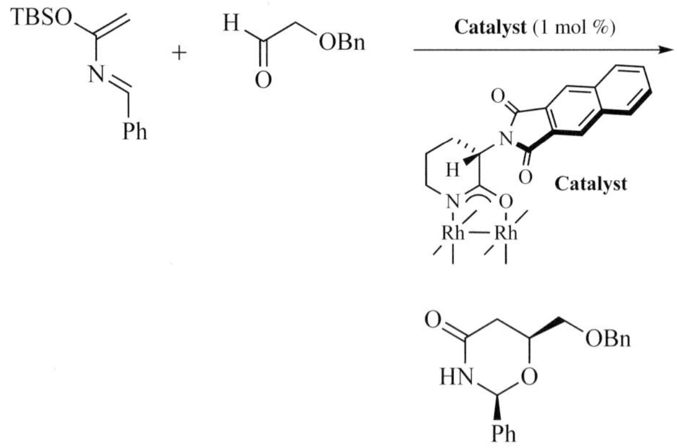

(23)

86%, 97% ee

Multicomponent reactions have been developed via the in situ formation of imines derived from 2-(phenylmethoxy)acetaldehyde. The synthesis of highly selective polysubstituted tetrahydroquinolines has been accomplished using the title aldehyde as the aldehyde component in a highly diastereoselective and enantioselective (96% ee) Povarov reaction (inverse electron-demand aza-Diels–Alder, eq 24).[48,49] In this example, high enantiocontrol is achieved upon activation of the imine by the chiral phosphoric acid catalyst.[50–52]

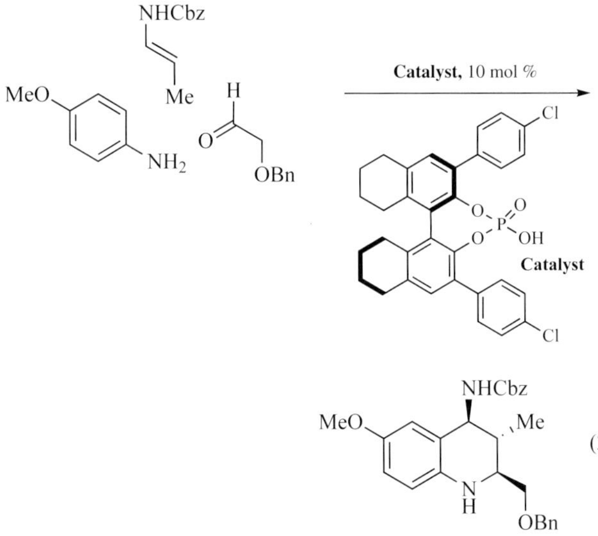

(24)

72%, 96% ee

Aldehyde Catalysis. 2-(Phenylmethoxy)acetaldehyde has been used to catalyze the intermolecular Cope-type hydroamination of *N*-benzylhydroxylamines and *N*-allylamines in moderate to good yields at room temperature (eq 25). This approach uses the electrophilicity and inductive destabilization of 2-(phenylmethoxy)acetaldehyde to favor the preassociation between the two reagents and thus catalyze intermolecular hydroaminations via the "temporary intramolecularity" present in the aminal below. This work provides a rare example of a catalytic tethering approach, and the room temperature reactivity demonstrates the efficiency of such organocatalysis to achieve difficult intermolecular reactions.[54–56]

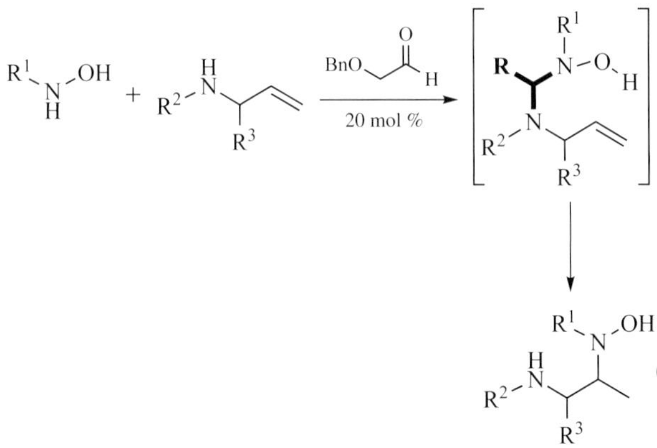

(25)

10 examples, up to 83%

Other Reaction Conditions. 2-(Phenylmethoxy)-acetaldehyde has been applied in a variety of reaction conditions. To complement the illustrative examples above, it is worth noting that this aldehyde is also compatible with a variety of transition metal catalysts, including ruthenium,[56] rhodium,[57] palladium,[58] chromium,[59] and silver[60] based systems. 2-(Phenylmethoxy)acetaldehyde has also been used as a substrate in enzymatic synthesis.[61]

1. Nishizono, N.; Akama, Y.; Agata, M.; Sugo, M.; Yamaguchi, Y.; Oda, K., *Tetrahedron* **2011**, *67*, 358.
2. Pollex, A.; Millet, A.; Müller, J.; Hiersemann, M.; Abraham, L., *J. Org. Chem.* **2005**, *70*, 5579.
3. Zhong, Y.-L.; Shing, T. K. M., *J. Org. Chem.* **1997**, *62*, 2622.
4. Vanier, S. F.; Larouche, G.; Wurz, R. P.; Charette, A. B., *Org. Lett.* **2010**, *12*, 672.
5. Collins, I.; Caldwell, J.; Fonseca, T.; Donald, A.; Bavetsias, V.; Hunter, L.-J. K.; Garrett, M. D.; Rowlands, M. G.; Aherne, G. W.; Davies, T. G.; Berdini, V.; Woodhead, S. J.; Davis, D.; Seavers, L. C. A.; Wyatt, P. G.; Workman, P.; McDonald, E., *Bioorg. Med. Chem.* **2006**, *14*, 1255.
6. Nicolai, S.; Erard, S.; González, D. F.; Waser, J., *Org. Lett.* **2010**, *12*, 384.
7. Müller, D.; Pitsch, S.; Kittaka, A.; Wagner, E.; Wintner, C. E.; Eschenmoser, A.; Ohlofjgewidmet, G. N., *Helv. Chim. Acta* **1990**, *73*, 1410.
8. Iturrate, L.; Sánchez-Moreno, I.; Oroz-Guinea, I.; Pérez-Gil, J.; García-Junceda, E., *Chem. Eur. J.* **2010**, *16*, 4018.
9. Aggarwal, V. K.; Drabowicz, J.; Grainger, R. S.; Gultekin, Z.; Lightowler, M.; Spargo, P. L., *J. Org. Chem.* **1995**, *60*, 4962.
10. Chen, C.-T.; Lin, Y.-D.; Liu, C.-Y., *Tetrahedron* **2009**, *65*, 10470.
11. Panek, J. S.; Yang, M., *J. Org. Chem.* **1991**, *56*, 5755.
12. Brown, M. J.; Harrison, T.; Herrinton, P. M.; Hopkins, M. H.; Hutchinson, K. D.; Overman, L. E.; Mishra, P., *J. Am. Chem. Soc.* **1991**, *113*, 5365.
13. Boger, D. L.; Robarge, K. D., *J. Org. Chem.* **1988**, *53*, 5793.
14. Boger, D. L.; Curran, T. T., *J. Org. Chem.* **1990**, *55*, 5439.
15. Boger, D. L.; Corbett, W. L.; Wiggins, J. M., *J. Org. Chem.* **1990**, *55*, 2999.
16. Boger, D. L.; Corbett, W. L.; Curran, T. T.; Kasper, A. M., *J. Am. Chem. Soc.* **1991**, *113*, 1713.
17. Denmark, S. E.; Ghosh, S. K., *Tetrahedron* **2007**, *63*, 8636.
18. Denmark, S. E.; Chung, W.-J., *J. Org. Chem.* **2008**, *73*, 4582.
19. Goodman, S. N.; Jacobsen, E. N., *Adv. Synth. Catal.* **2002**, *344*, 953.
20. Sibi, M. P.; Rutherford, D.; Renhowe, P. A.; Li, B., *J. Am. Chem. Soc.* **1999**, *121*, 7509.
21. Billard, F.; Robiette, R.; Pospíšil, J., *J. Org. Chem.* **2012**, *77*, 6358.
22. Borg, T.; Tuzina, P.; Somfai, P., *J. Org. Chem.* **2011**, *76*, 8070.
23. Bartrum, H. E.; Blakemore, D. C.; Moody, C. J.; Hayes, C. J., *J. Org. Chem.* **2010**, *75*, 8674.
24. Murry, J. A.; Frantz, D. E.; Soheili, A.; Tillyer, R.; Grabowski, E. J. J.; Reider, P. J., *J. Am. Chem. Soc.* **2001**, *123*, 9696.
25. Toh, Q. Y.; McNally, A.; Vera, S.; Erdmann, N.; Gaunt, M. J., *J. Am. Chem. Soc.* **2013**, *135*, 3772.
26. Wrona, I. E.; Lowe, J. T.; Turbyville, T. J.; Johnson, T. R.; Beignet, J.; Beutler, J. A.; Panek, J. S., *J. Org. Chem.* **2009**, *74*, 1897.
27. Panek, J. S.; Cirillo, P. F., *J. Org. Chem.* **1993**, *58*, 999.
28. Panek, J. S.; Yang, M., *J. Am. Chem. Soc.* **1991**, *113*, 9868.
29. Panek, J. S.; Yang, M.; Xu, F., *J. Org. Chem.* **1992**, *57*, 5790.
30. Hackman, B. M.; Lombardi, P. J.; Leighton, J. L., *Org. Lett.* **2004**, *6*, 4375.
31. Kubota, K.; Leighton, J. L., *Angew. Chem., Int. Ed.* **2003**, *42*, 946.
32. Kinnaird, J. W. A.; Ng, P. Y.; Kubota, K.; Wang, X.; Leighton, J. L., *J. Am. Chem. Soc.* **2002**, *124*, 7920.
33. Evans, D. A.; Kozlowski, M. C.; Murry, J. A.; Burgey, C. S.; Campos, K. R.; Connell, B. T.; Staples, R. J., *J. Am. Chem. Soc.* **1999**, *121*, 669.
34. Evans, D. A.; Murry, J. A.; Kozlowski, M. C., *J. Am. Chem. Soc.* **1996**, *118*, 5814.
35. Northrup, A. B.; Mangion, I. K.; Hettche, F.; MacMillan, D. W. C., *Angew. Chem., Int. Ed.* **2004**, *43*, 2152.
36. Mangion, I. K.; Northrup, A. B.; MacMillan, D. W. C., *Angew. Chem., Int. Ed.* **2004**, *43*, 6722.

37. Northrup, A. B.; MacMillan, D. W. C., *Science* **2004**, *305*, 1752.
38. Cordova, A.; Ibrahem, I.; Casas, J.; Sundén, H.; Engqvist, M.; Reyes, E., *Chem. Eur. J.* **2005**, *11*, 4772.
39. Evans, D. A.; Janey, J. M., *Org. Lett.* **2001**, *3*, 2125.
40. Lin, Y.-M.; Boucau, J.; Li, Z.; Casarotto, V.; Lin, J.; Nguyen, A. N.; Ehrmantraut, J., *Org. Lett.* **2007**, *9*, 567.
41. Nelson, S. G.; Peelen, T. J.; Wan, Z., *J. Am. Chem. Soc.* **1999**, *121*, 9742.
42. Zhu, C.; Shen, X.; Nelson, S. G., *J. Am. Chem. Soc.* **2004**, *126*, 5352.
43. Nelson, S. G.; Zhu, C.; Shen, X., *J. Am. Chem. Soc.* **2004**, *126*, 14.
44. Nelson, S. G.; Spencer, K. L., *Angew. Chem., Int. Ed.* **2000**, *39*, 1323.
45. Dossetter, A. G.; Jamison, T. F.; Jacobsen, E. N., *Angew. Chem., Int. Ed.* **1999**, *38*, 2398.
46. Watanabe, Y.; Washio, T.; Krishnamurthi, J.; Anada, M.; Hashimoto, S., *Chem. Commun.* **2012**, *48*, 6969.
47. Anada, M.; Washio, T.; Shimada, N.; Kitagaki, S.; Nakajima, M.; Shiro, M.; Hashimoto, S., *Angew. Chem., Int. Ed.* **2004**, *43*, 2665.
48. Dagousset, G.; Drouet, F.; Masson, G.; Zhu, J., *Org. Lett.* **2009**, *11*, 5546.
49. Dagousset, G.; Zhu, J.; Masson, G., *J. Am. Chem. Soc.* **2011**, *133*, 14804.
50. Akiyama, T.; Takaya, J.; Kagoshima, H., *Adv. Synth. Catal.* **2002**, *344*, 338.
51. Akiyama, T.; Morita, H.; Fuchibe, K., *J. Am. Chem. Soc.* **2006**, *128*, 13070.
52. Sorimachi, K.; Terada, M., *J. Am. Chem. Soc.* **2008**, *130*, 14452.
53. MacDonald, M. J.; Schipper, D. J.; Ng, P. J.; Moran, J.; Beauchemin, A. M., *J. Am. Chem. Soc.* **2011**, *133*, 20100.
54. Guimond, N.; MacDonald, M. J.; Lemieux, V.; Beauchemin, A. M., *J. Am. Chem. Soc.* **2012**, *134*, 16571.
55. Pascal, R., *Liebigs Ann. Chem.* **2003**, *2003*, 1813.
56. Leung, J. C.; Patman, R. L.; Sam, B.; Krische, M. J., *Chem. Eur. J.* **2011**, *17*, 12437.
57. Ishida, M.; Shibata, Y.; Noguchi, K.; Tanaka, K., *Chem. Eur. J.* **2011**, *17*, 12578.
58. Zhu, S.-F.; Qiao, X.-C.; Zhang, Y.-Z.; Wang, L.-X.; Zhou, Q.-L., *Chem. Sci.* **2011**, *2*, 1135.
59. Xia, G.; Yamamoto, H., *J. Am. Chem. Soc.* **2006**, *128*, 2554.
60. Mandai, H.; Mandai, K.; Snapper, M. L.; Hoveyda, A. H., *J. Am. Chem. Soc.* **2008**, *130*, 17961.
61. Kimura, T.; Vassilev, V. P.; Shen, G.-J.; Wong, C. H., *J. Am. Chem. Soc.* **1997**, *119*, 11734.

Melissa J. MacDonald & André M. Beauchemin
University of Ottawa, Ontario, Canada

(S)-Proline[1]

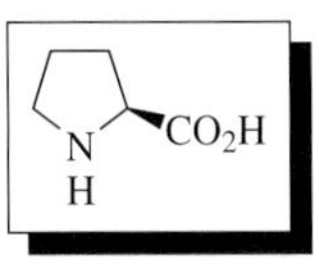

[147-85-3] $C_5H_9NO_2$ (MW 115.15)

InChI = 1S/C5H9NO2/c7-5(8)4-2-1-3-6-4/h4,6H,1-3H2,(H,7,8)/t4-/m0/s1

InChIKey = ONIBWKKTOPOVIA-BYPYZUCNSA-N

(chiral auxiliary[1] in asymmetric synthesis)

Physical Data: mp 228–233 °C (dec.); $[\alpha]_D^{20} = -84°$ ($c = 4$, H_2O); ninhydrin yellow in color.
Solubility: sol H_2O, alcohol; insol ether.
Form Supplied in: white solid; widely available; inexpensive.
Analysis of Reagent Purity: measurement of optical rotation; mp.
Handling, Storage, and Precautions: cold and dry storage.

Original Commentary

Sabine Wallbaum & Jürgen Martens
Universität Oldenburg, Oldenburg, Germany

General Considerations. In addition to its use in peptide chemistry, (*S*)-proline is often applied as a chiral precursor in the total syntheses of natural products, e.g. odorin,[2] pumiliotoxin,[3] petasinecine,[4] or threonine.[5] Some highly effective pharmaceuticals, such as optically pure ACE inhibitors, are prepared from L-proline.[6] In the last two decades, (*S*)-proline has attracted much attention as an optically active auxiliary in asymmetric synthesis.

Asymmetric Aldolization. Proline mediates the asymmetric aldol cyclization of the prochiral triketone (**1a**) to the optically active bicyclic enedione (*S*)-(**3a**) (eq 1).[7,8] Optical yields up to 94% are realized, depending on the solvent used. The enedione (**3**) is prepared directly in the presence of an acid such as $HClO_4$. In the absence of acid, the aldol (**2**) is frequently isolated. Among the amino acids tested, (*S*)-proline gives the best results in almost every case. With (*S*)-proline the (*S*) configured products (**3**) are usually obtained. Enediones such as (**3**) are important building blocks for the synthesis of steroids or alkaloids[9] because natural steroids have the same configuration at C-13. Analogously, the Wieland–Miescher ketone (*S*)-(**3e**) is prepared with 70% ee from the cyclic prochiral triketone (**1e**) in the presence of (*S*)-proline.[10] Optically pure enedione (*S*)-(**3e**) is obtained by a single crystallization of the product mixture having an enantiomeric excess over 50%.

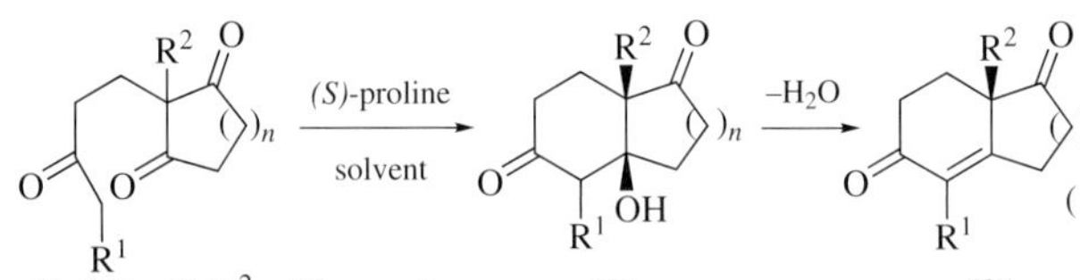

(1)

(**1a**) R = H, R^2 = Me, $n = 1$ (**2**) (**3**)
(**1b**) R = H, R^2 = Et, $n = 1$
(**1c**) R = $(CH_2)_2CH{=}CH_2$, R^2 = Me, $n = 1$
(**1d**) R = $(CH_2)_2$-2-(6-Me-pyridyl), R^2 = Me, $n = 1$
(**1e**) R = H, R^2 = Me, $n = 2$
(**1f**) R = Me, R^2 = Me, $n = 2$

An asymmetric aldolization was successfully applied to the preparation of gibbane.[11] The total synthesis of the macrolide antibiotic erythromycin was developed involving an asymmetric aldolization step catalyzed by proline.[12] Since the mid-1970s, a flood of papers has appeared dealing with the asymmetric aldolization of various triketones. Some results are listed for comparison in Table 1.

Table 1 Asymmetric aldolization of prochiral triketones (**1**)

Educt	(*S*)-Proline (mol %)	Solvent	Product	Chemical Yield (%)	Optical Yield (%)
(**1a**)	4	DMF	(**2a**)[13]	97	94
(**1a**)	47/$HClO_4$	MeCN	(**3a**)[13]	90	82
(**1a**)	100	MeCN	(**2a**)[8]	97	97
(**1a**)	50/$HClO_4$	MeCN	(**3a**)[7]	87	84
(**1a**)	3	DMF	(**2a**)[8]	100	93
(**1b**)	30	DMF	(**2b**)[8]	71	100
(**1b**)	50/HCl	DMF	(**3b**)[7]	76	80
(**1c**)	100/$HClO_4$	MeCN	(**3c**)[14]	22	20
(**1d**)	120/$HClO_4$	MeCN	(**3d**)[15]	67	26
(**1e**)	5	DMSO	(**3e**)[10]	71	70
(**1e**)	50/$HClO_4$	MeCN	(**3e**)[7]	83	71
(**1f**)	13.7	DMSO	(**3f**)[16]	43	38
(**1f**)	50/$HClO_4$	MeCN	(**3f**)[17]	74	15

The intramolecular asymmetric cyclization of open chain symmetrical triketone (**4**) leading to (*R*)-(**5**) proceeds with 16% ee (eq 2).[18]

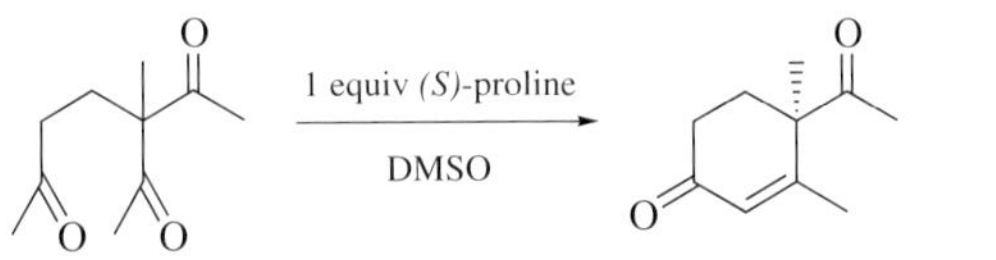

(**4**) (*R*)-(**5**) (2)

Even acyclic 1,5-diketones (**6**) are cyclized enantioselectivitely in the presence of (*S*)-proline.[19] Depending on the structure of the cyclic α,β-unsaturated ketone, (R)-(**7**) is obtained in up to 43% (R = Me) optical yield (eq 3).

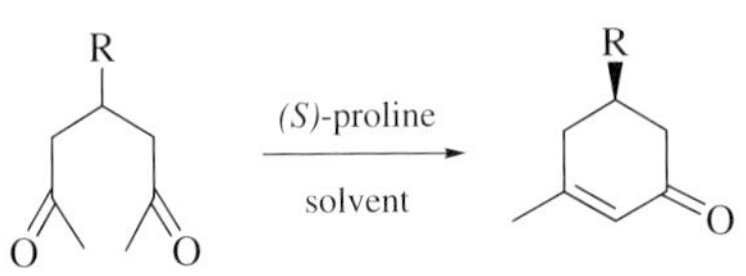

(**6**) (*R*)-(**7**) (3)

The mechanism of the proline-catalyzed enantioselective aldol reaction has been studied.[20] An extension of the asymmetric aldolization deals with the cyclization of diketones.[21] Also investigated was the dehydration of racemic β-ketols in the presence of (*S*)-proline and a kinetic resolution was observed.[22]

Asymmetric Michael Addition. An intramolecular Michael reaction catalyzed by (*S*)-proline leads to the chiral thiadecalin (**9**) and thiahydrindan (**11**) and (**12**).[23] Enone (**8**) undergoes cyclization in the presence of (*S*)-proline to give exclusively the *trans* isomer (**9**) (eq 4). The thiahydrindandions (**11**) and (**12**) are obtained from (**10**) as a 1:1 mixture of the *cis* and *trans* isomers (eq 5).

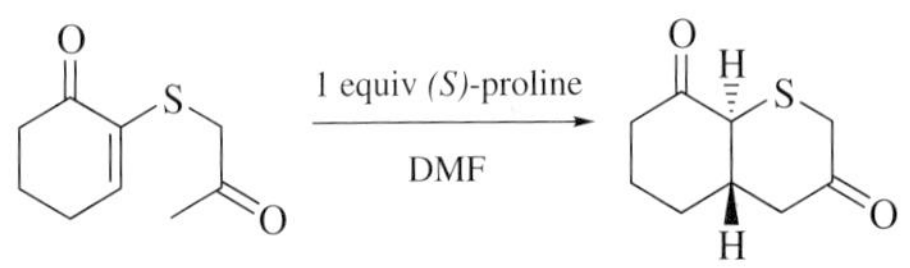

(**8**) (**9**) (4)

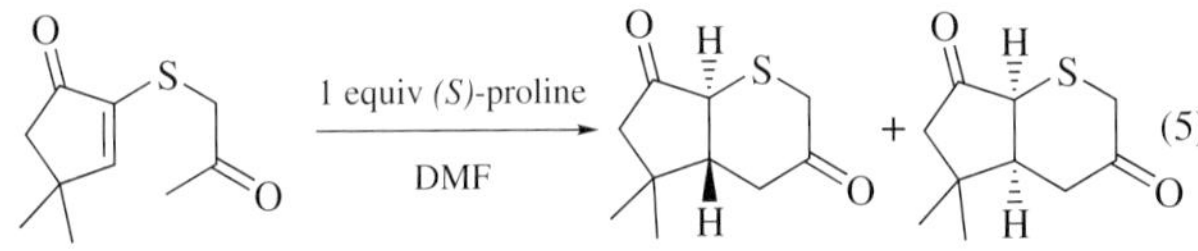

(**10**) (**11**) 1:1 (**12**) (5)

The intramolecular asymmetric Michael reaction of acyclic compounds obtained from chiral alkaloid building blocks using amines and (*S*)-proline has been investigated.[24] The Michael addition of dimethyl malonate to α,β-unsaturated aldehydes proceeds smoothly with a catalytic amount of (*S*)-proline lithium salt.[25] However, no asymmetric induction was observed.

Asymmetric Halolactonization. An asymmetric halolactonization reaction using proline as a chiral auxiliary has been reported.[26] Optically active α-hydroxy acids (**16**) are prepared from α,β-unsaturated acids via the corresponding (*S*)-proline amide (**13**) involving an asymmetric bromolactonization step (eq 6).[26a]

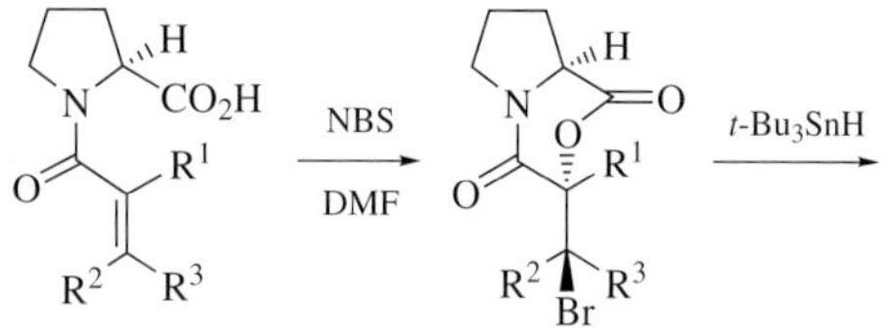

(13) (14)

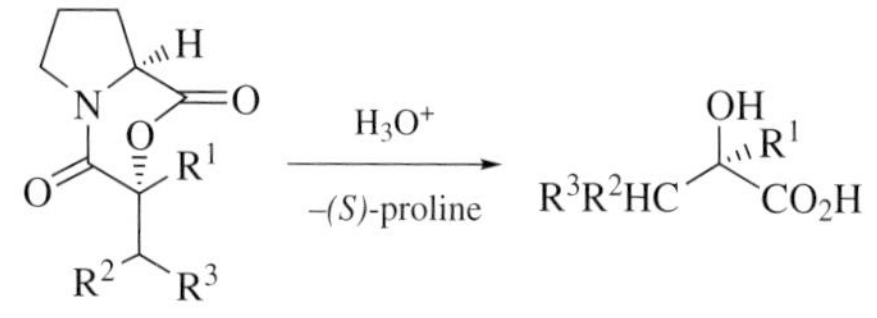

(15) (16) (6)

The unsaturated carboxylic acid (**13**) undergoes an asymmetric bromolactonization when treated with **N-Bromosuccinimide** in DMF. The bromolactone (**14**) and its diastereomer are obtained in a 94.5:5.5 ratio. Reduction and hydrolysis yields the α-hydroxy acid (**16**) in an overall optical yield of 90%. The same procedure gives chiral α-hydroxy ketones.[26c]

A modification of the asymmetric bromolactonization leads to optically active α,β-epoxy aldehydes (**18**).[26d,e] Treatment of the bromolactone (**14**) with **Sodium Methoxide** results in the formation of the epimeric epoxy ester (**17**) in a ratio of 2:1 (eq 7).

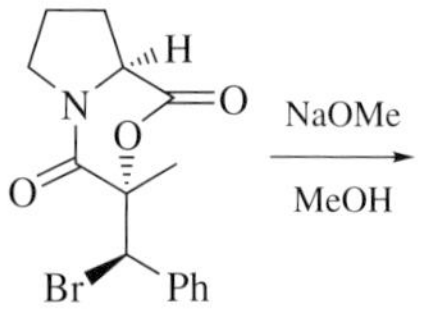

(14)

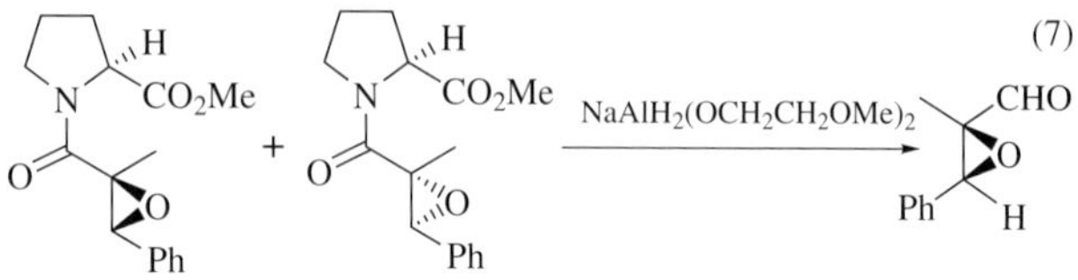

(7)

(17a) 2:1 (17b) (2*R*,3*S*)-(18)

The reductive cleavage of the proline derivative yields the chiral α,β-epoxy aldehyde (2*R*,3*S*)-(**18**) in 98% ee. Even natural product syntheses can be realized utilizing the bromolactonization procedure.[26h,i]

Reduction of C=O and C=N Bonds. Asymmetric reductions of prochiral ketones (**19**) to the corresponding chiral alcohols (**20**) using (*S*)-proline-modified borohydride reagents as the reductant have been published. The borane reductions of ketones (**19**) employing (*S*)-proline as chiral mediator proceeds with enantiomeric excesses up to >95%. It is proposed that the in situ produced (*S*)-prolinol reacts with borane to form the oxazaborolidine (*S*)-(**21**) as the reducing catalyst (eq 8).[27]

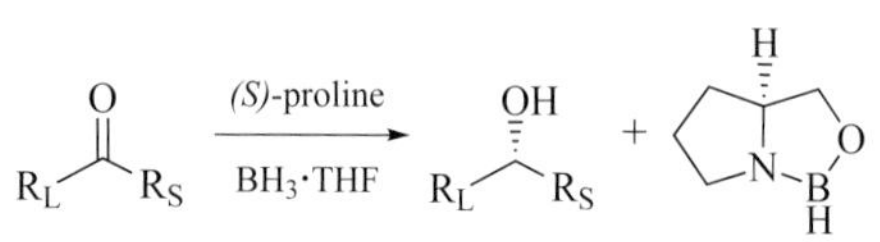

(8)

(19) (*R*)-(20) (*S*)-(21)

The (*S*)-prolinate–borane complex (*S*)-**22** reduces ketones to the corresponding alcohols with optical yields up to 50%.[28] The asymmetric reduction of cyclic imines (**24**) with chiral sodium triacyloxyborohydride (*S*)-**23** was utilized to prepare optically active alkaloids (**25**) with optical yields up to 86% (eq 9).[29]

(*S*)-(22) (*S*)-(23)

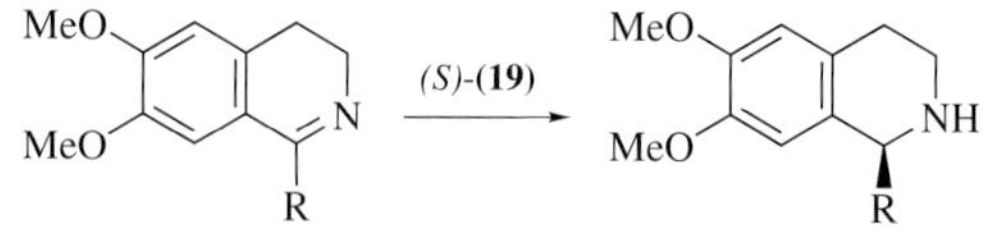

(9)

(24) (25)

The hydrogenation of various ketones with heterogeneous ***Palladium on Carbon*** or ***Raney Nickel*** catalysts in the presence of (*S*)-proline proceeds to produce the corresponding optically active alcohols with low optical yields (up to 23%).[30]

Reduction of C=C Bonds. The reduction of the C=C double bond of isophorone (**26**) with Pd/C in the presence of (*S*)-proline yields the saturated ketone (**27**) with 60% optical purity (eq 10).[30a,31] With (*S*)-proline ester/Pd (or Pt) systems the hydrogenation of ethyl pyruvate, an α-keto ester, was investigated, but only insignificant enantioselectivities were reached.[32]

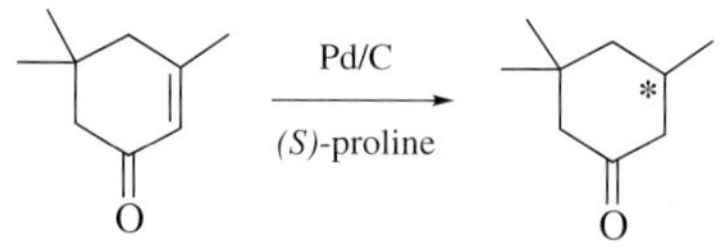

(10)

(26) (27)

An efficient synthesis of (*S*)-amino acids from α-keto acids via a diastereoselective hydrogenation step with (*S*)-proline as the chiral inducer was reported (eq 11).[33] Optical yields up to 90% were reached.

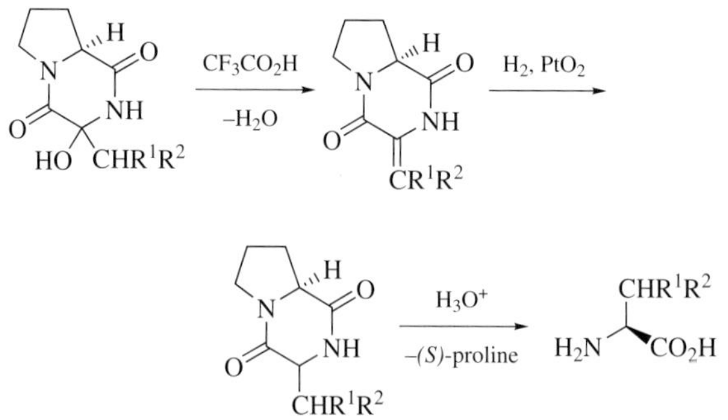

(11)

Racemization of Amino Acids. The synthesis of (*R*)-alanine was achieved starting from (*S*)-alanine via formation of the imidazoline with (*S*)-proline.[34] This result can be explained in terms of epimerization and stereoselective protonation with asymmetric induction by the chiral center originating from (*S*)-proline.

Resolution of Amino Acids. For the optical resolution of racemic threonine via replacing resolution, (*S*)-proline was utilized as an optically active cosolute although the structure of the imino acid is different from that of threonine.[35] The same procedure was applied less sucessfully to the resolution of (*R,S*)-thiazolidine carboxylic acid.[36]

Synthesis of Unnatural (S)-Proline Derivatives. The condensation of pivaladehyde with (*S*)-proline yields stereoselectively, after lithiation and reaction with an electrophile, the bicyclic compound (**28**), which is a versatile educt for the synthesis of many α-substituted proline analogs (**29**) (eq 12).[37] The reactions proceed via the formation of a chiral lithium enolate without the use of a chiral auxiliary (self-reproduction of chirality). The reaction with a variety of electrophiles *cis* to the *t*-Bu group yields a plethora of α-substituted (*S*)-proline derivatives (**29**). A limitation of this strategy is the acetal cleavage of some substituted products (**28**).[38]

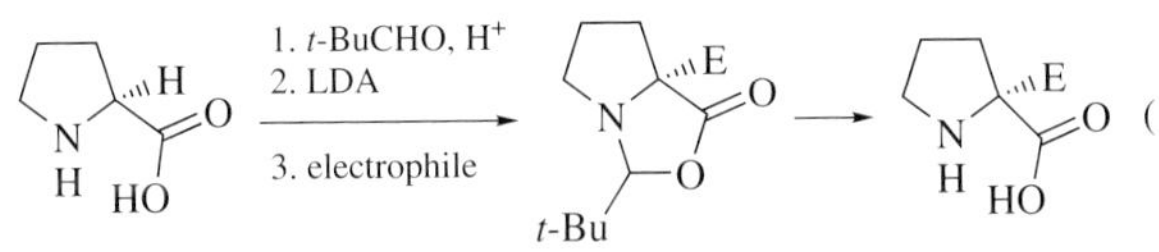

(**28**) (**29**) (12)

The β-amino acid homoproline can be synthesized via an Arndt–Eistert reaction from (*S*)-proline.[39]

Synthesis of Optically Active Phophorus Compounds. A series of chiral organophosphorus compounds (**33**) have been prepared in which the phophorus atom is the stereogenic center (eq 13).[40] The best stereoselectivity is reached with (*S*)-proline esters (**30**) as the chiral auxiliary. The reaction of phosphonic acid chloride (**31**) with (*S*)-proline ethyl ester affords a mixture of diastereomeric amides (**32**) in high stereoselectivity. The diastereomers can easily be purified by chromatography. The chiral organophosphorus compounds (**33**) are obtained from hydrolysis of (**32**).

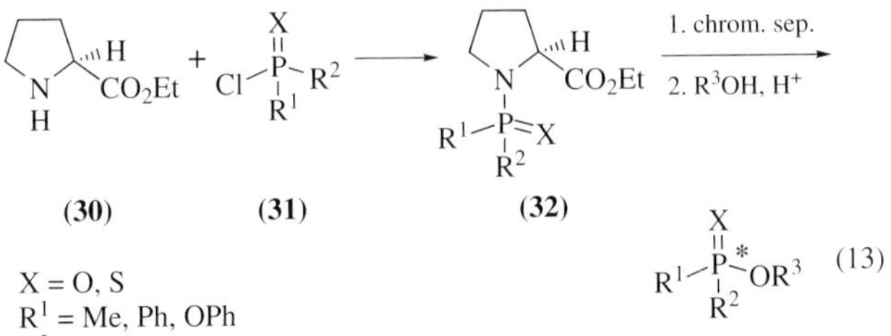

(**30**) (**31**) (**32**) (**33**) (13)

X = O, S
R^1 = Me, Ph, OPh
R^2 = Et, Bu, Ph, OPh
R^3 = Me, Et, Pr, *i*-Pr

Alkylations and Allylations. The asymmetric alkylation of chiral enamines derived from (*S*)-proline esters has been disclosed.[41] The α-alkylation of cyclohexanone proceeds with an optical purity of 59%. (*S*)-Proline catalyzes the alkylation of xanthopurpurin (**34**) by 2-hydroxytetrahydropyran[42] yielding (**35**), which was later used in the synthesis of the racemate of the pigment averufanin (eq 14).[43] With phenylacetaldehyde, two molecules react to build up an anomeric mixture of lactols with a new pyran ring.[44] In each case, no enantioselectivity is detected.

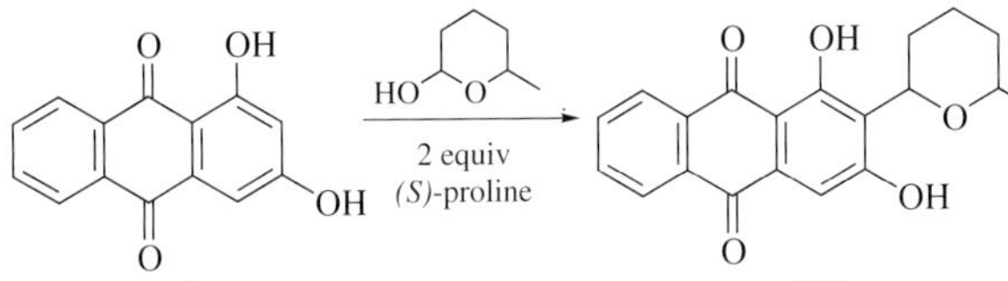

(**34**) (**35**) (14)

The stereoselective allylation of aldehydes was reported to proceed with allyltrifluorosilanes in the presence of (*S*)-proline.[45] The reaction involves pentacoordinate silicate intermediates. Optical yields up to 30% are achieved in the copper-catalyzed allylic acetoxylation of cyclohexene with (*S*)-proline as a chiral ligand.[46]

The intramolecular asymmetric palladium-catalyzed allylation of aldehydes, including allylating functionality in the molecules, via chiral enamines prepared from (*S*)-proline esters has been reported (eq 15).[47] The most promising result was reached with the (*S*)-proline allyl ester derivative (**36**). Upon treatment with **Tetrakis(triphenylphosphine)palladium(0)** and PPh_3 in THF, the chiral enamine (**36**) undergoes an intramolecular allylation to afford an α-allyl hemiacetal (**37**). After an oxidation step the optically active lactones (**38**) with up to 84% ee were isolated in high chemical yields. The same authors have also reported sucessful palladium-catalyzed asymmetric allylations of chiral allylic (*S*)-proline ester enamines[48] and amides[49] with enantiomeric excesses up to 100%.

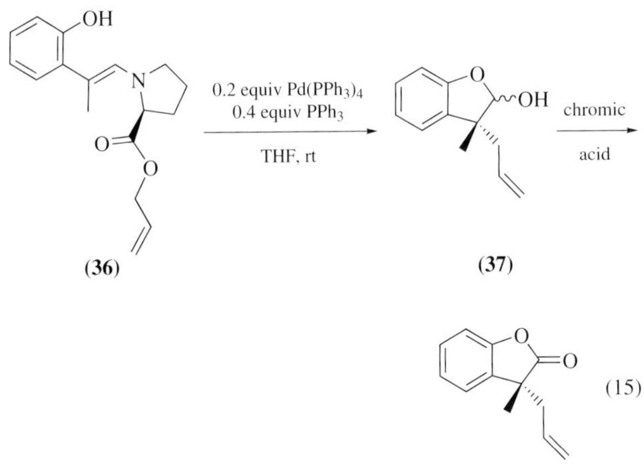

(**36**) (**37**) (**38**) (15)

First Update

Frederic Minassian
Université Joseph Fourier, Grenoble, France

General Considerations. In the last few years, the concept of asymmetric organocatalysis[50] has gained importance in synthesis, and the use of proline, considered as an enzyme mimic,[51] has been widely described in this context.[52] Proline catalyzes several reactions according to an enamine mechanism.[53–55]

Asymmetric Intramolecular Aldol Reaction. Proline is a useful catalyst for the Hajos-Parrish-Eder-Sauer-Wiechert asymmetric ketol cyclization. Mechanism of this reaction has been studied through theoretical[56] and spectroscopic[57] studies considering triketones (**1a** and **1e**) (eq 1) as starting materials. A one-pot sequence has been studied starting from methyl vinyl ketone and 2-methylcyclohexane-1,3-dione producing the Wieland-Mischer ketone (**3e**) with 76% ee and 49% yield.[58]

Compound **3e** was then used in the total synthesis of sesquiterpenoids.[59] Reaction has been performed using proline without the use of any solvent, but isolated yields are moderate and ee's of the (*S*)-diones (**3a** and **3e**) are in a 33–63% range upon the experimental conditions.[60] This was applied in one-pot procedure for the synthesis of spiroenediones, but yields and ee's are still moderate.[61]

This annulation procedure was extended to other polyketones for the syntheses of taxol,[62] steroid analogs,[63] antibiotic,[64] and antimalarial agents.[65] Chiral triketoacetonides can be converted to the corresponding bicyclic enediones with good levels of diastereoselectivity, albeit in moderate conversion ratios.[66] 1,5-Dicarbonyl compounds react via an Enolexo process in the presence of a catalytic amount of (*S*)-proline in dichloromethane, providing access to β-hydroxy cyclohexane carbonyl derivatives with good to excellent yields, good diastereoselectivities, and nearly quantitative ee's.[67] Total synthesis of cocaine has been performed via an intramolecular aldol key-step with *meso*-dialdehyde (**39**) in toluene, leading to an inseparable 1:1 mixture of axial/equatorial epimers (**40**) in good yield (eq 16).[68]

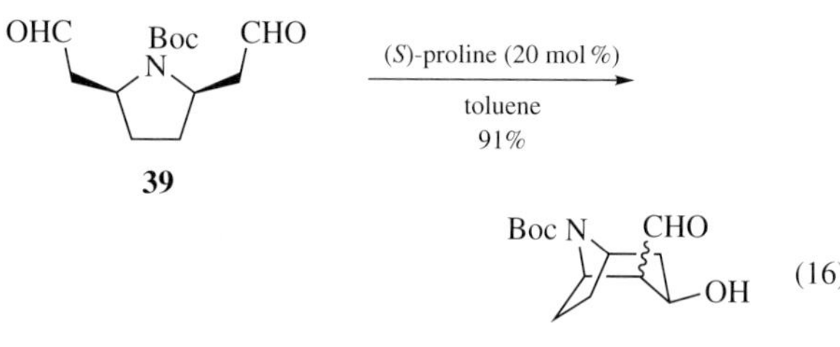

39 **40** (16)

Elaboration of a pyran ring has also been performed using this method. In the total synthesis of (−)-arisugacin A, formal [3+3] cycloaddition involves a proline-iminium ion (**41**) and pyrone (**42**), and leads to the formation of one single diastereomer (**43**) (eq 17).[69]

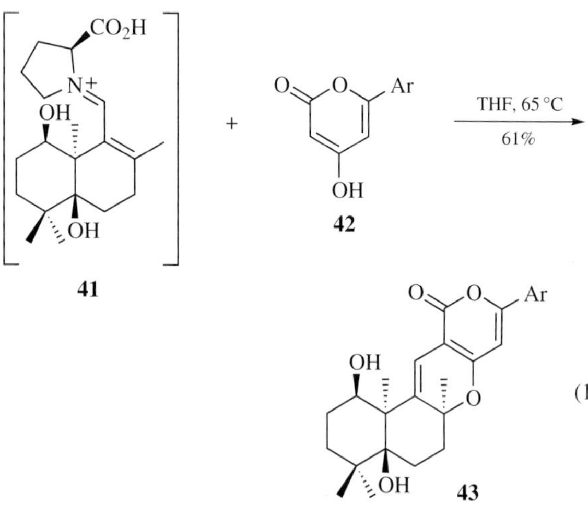

41 **42** **43** (17)

Recent studies on the subject have dealt with the use of glycal-derived enal partners (this process exhibits high degrees of diastereoselectivities), the impact of the experimental conditions as well as mechanistic considerations.[70]

Asymmetric Intermolecular Aldol Reaction. In the last decade, much attention has been given to the proline-catalyzed asymmetric intermolecular aldol reaction.[71,72] Direct intermolecular aldol process is observed by reacting acetone with a wide variety of aldehydes.[73] Reaction of *iso*-butyraldehyde with acetone in the presence of a catalytic amount of (*S*)-proline leads to the corresponding product **44** in excellent yield and ee (eq 18).[73]

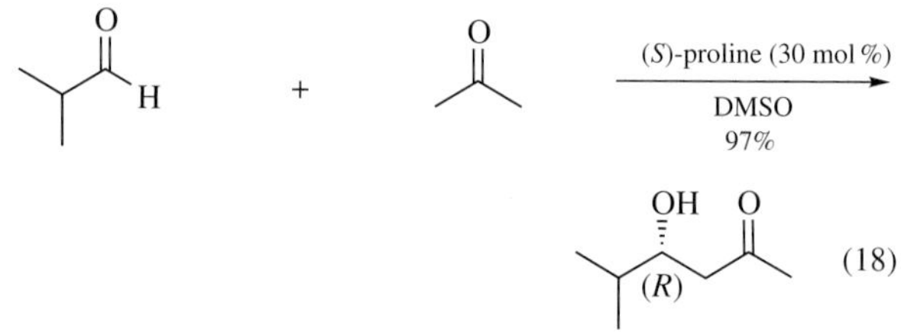

(18)

44
96% ee

The mechanism of this reaction has been studied,[74] and influence of experimental conditions have been described, such as the use of additive water,[75] buffered aqueous media,[76] or aqueous micelles.[77] By performing the procedure under high pressure, yields and enantioselectivities could be improved.[78] Recycling of the (*S*)-proline catalyst is possible by performing aldol reactions in poly(ethylene glycol),[79] or in ionic liquids.[80] Combination of the proline-catalyzed aldol reaction with enzyme-catalyzed kinetic resolutions produces enantiopure β-hydroxy ketones in moderate yields.[81]

Using DMSO as the solvent, reaction of α-unsubstituted aldehydes with acetone does not lead to cross-aldol products but mainly aldol condensation and aldehyde self-aldolization. When DMSO is replaced by acetone, reaction of α-unsubstituted aldehydes produces cross-aldol products in good ee's (around 70%) albeit in moderate yields.[82] Cyclohexanone reacts with *iso*-valeraldehyde in chloroform with good levels of enantio- and diastereoselectivity (eq 19).[73]

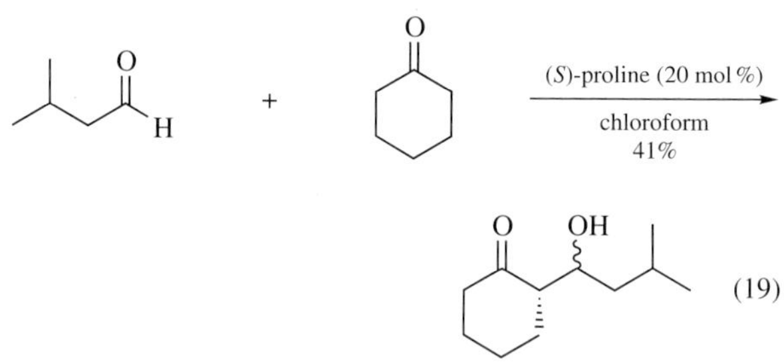

(19)

syn/*anti* = 1:7
ee *syn* = 89%
ee *anti* = 86%

(*S*)-proline-catalyzed cross-aldol reaction involving cyclopentanone has been applied to the total synthesis of a natural mosquito attractant pheromone,[83] as well as the dynamic kinetic resolution of atropoisomeric amides.[84] Reaction with protected diacetone hexulose aldehydes leads to nearly optically pure Claisen-Schmidt aldol products.[85]

Proline is also an efficient organocatalyst for transfer aldol reactions between diacetone alcohol and aldehydes.[86]

When hydroxyacetone is reacted with various aldehydes, the regioselective formation of *anti*-1,2-diols is observed with good yields and diastereoselectivities, as well as with almost quantitative ee's.[87] Cross-aldol reaction of hydroxy acetone with various α-aminoaldehydes has been studied and applied to the preparation of valuable synthetic intermediates.[88] Reaction of *O-tert*-butyldimethylsilyl-protected hydroxy acetone is also possible, with less diastereo- and enantioselectivity but with good yields and regioselectivity.[89] Both aldol donors have been applied to the stereoselective synthesis of a steroid side chain.[90]

Dihydroxy acetone derivatives react with aldehydes to give the corresponding polyols.[91] Compound **45** reacts in DMF with cyclopentane carboxaldehyde to give polyol (**46**) in good yield and excellent levels of stereoselectivity (eq 20).

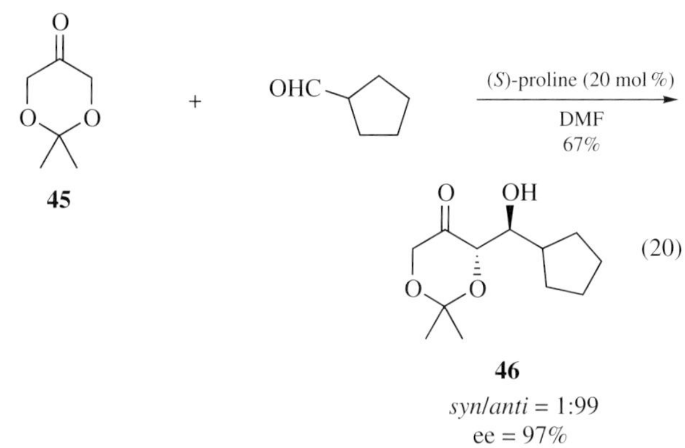

(20)

The use of the easily obtained zinc-proline complex $Zn(Pro)_2$ has been studied in the aldolization of glycolaldehyde giving tetroses and hexoses.[92]

An astonishing two-step synthesis of carbohydrates in the glucose, allose, and mannose series has been described involving first a proline-catalyzed dimerization of α-oxyaldehydes followed by tandem Mukaiyama aldol addition-cyclization step.[93]

Proline-catalyzed cross-aldol reaction of tetrahydro-4*H*-thiopyranone and some of its aldehyde derivatives is a useful route to polypropionate skeletons.[94]

It is also possible to replace the ketone donor by an aldehyde in the presence of 10 mol % of (*S*)-proline, which is known as cross-aldol reaction of aldehydes and yields almost optically pure products.[95] It is possible to perform this reaction in ionic liquid media with excellent results.[96] When only one aldehyde is used in excess, self-aldol products are obtained with good ee's albeit in low yields.[97] This has been applied to the syntheses of polyketides.[98,99] As in the case of a ketone donor, α-substituted aldehydes could easily be used. The use of α-oxyaldehydes in dimerization or in cross-aldol procedures gives rise to highly optically active polyols.[100] With some glycine-derived α-aminoaldehydes, use of catalytic (*S*)-proline in NMP provides an efficient access to enantiopure β-hydroxy-α-amino acid derivatives.[101] Cross-aldol reaction using α-dithioacetal aldehydes is also an efficient and valuable synthetic method because of the latent carbonyl functionality remaining at the end of the process.[102]

If the donor is diethyl oxomalonate or an ethyl trifluoropyruvate, reaction with various aldehydes and a catalytic amount of (*S*)-proline in dichloromethane gives an efficient approach to chiral β-hydroxy carboxylic acid derivatives considering the high observed yields and ee's.[103] If nitrones are used instead of aldehydes, optically active β-hydroxy nitrones are produced in good yield.[104] Theoretical mechanistic studies have been performed on this particular reaction.[105]

Dimethyl malonate and ethyl cyanoacetate react similarly with aldehydes in the presence of 10 mol % of (*S*)-proline in a Knoevenagel process to afford alkylidene and arylidene malonates in good yields.[106]

Asymmetric Mannich Addition. Direct three-component Mannich condensation using (*S*)-proline as the catalyst is a valuable synthetic method, owing to the facts that this small chiral amine is able to form enamine with unmodified ketones together with activation of the subsequent nucleophilic addition.[107] Yields are moderate to excellent. For example, reaction of *p*-nitrobenzaldehyde, *p*-anisidine, and excess hydroxy acetone in the presence of 20 mol % of (*S*)-proline leads to the nearly exclusive formation of enantiopure *syn*-1,2-amino alcohol (**47**) (eq 21).

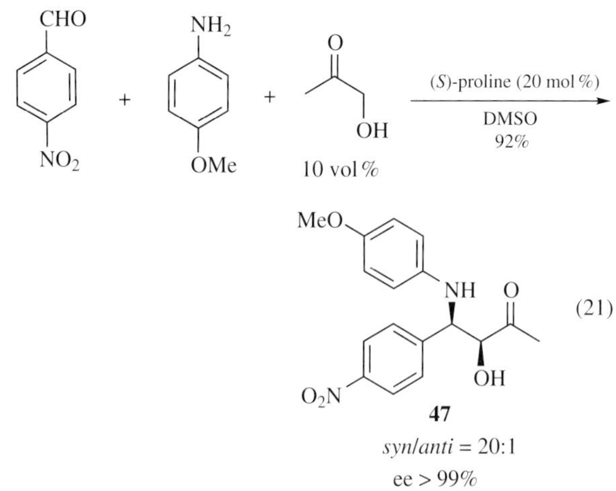

(21)

Origins of the opposite stereoselectivities observed in proline-catalyzed aldol and Mannich reactions have been studied.[108] Organocatalytic effect of (*S*)-proline has also been envisaged in [bmim]BF_4 ionic liquid,[109] Mannich products being obtained in nearly optically pure forms after only 30 min. The resulting *syn*-1,2-amino alcohols could be easily converted to 5-acetoxy oxazolidin-2-ones in a three-step sequence including Baeyer-Villiger oxidation.[110] Asymmetric Mannich reaction could also be performed with β-carbolines[111] or with preformed polyfluoroalkylated aldimines[112] with moderate yields and ee's up to 98%. Mannich reaction involving unmodified aldehydes as donors is usually a *syn*-selective process too, affording 3-amino aldehydes with excellent diastereomeric ratios and ee's.[113,114] Subsequent in situ reduction using sodium borohydride affords γ-amino alcohols in good yields and ee's.[115]

The Mannich reaction could also be performed under high pressure by freezing water with good to excellent enantioselectivities.[116]

When ethyl glyoxylate-based imines are involved, use of (*S*)-proline as the catalyst leads concisely to highly diastereo- and enantioselective synthesis of functionalized α- and β-amino acids, β-lactams, and amino alcohols.[117] For example, Mannich condensation product (**48**) represents a convenient precursor for the elaboration of spiro β-lactam (**49**) (eq 22).[118]

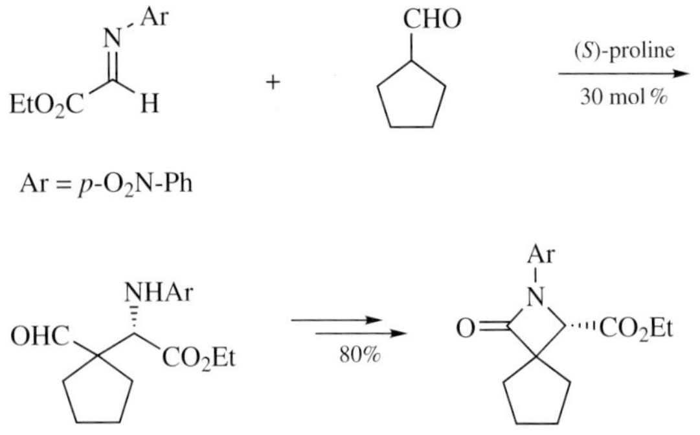

(22)

48 98% ee

49

In the case of *iso*-valeraldehyde, combination of Mannich reaction with indium-promoted allylation and subsequent lactonization affords γ-allyl substituted lactone (**50**) (eq 23).[119]

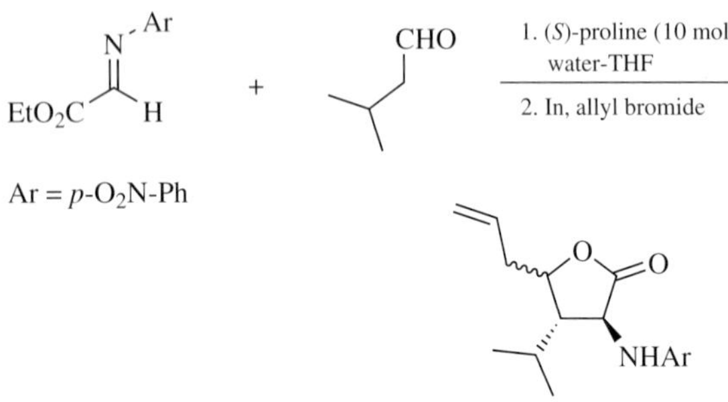

(23)

50
dr = 1:1
> 99% ee

Combination with a cyanation reaction is also possible to yield β-cyano α-amino acid derivatives with excellent ee's.[120] Starting imines can also be generated in situ from the corresponding *N*-protected amines by a catalytic aerobic oxidation performed before the Mannich step.[121]

The (*S*)-proline-catalyzed Biginelli reaction shows mechanistic similarities with Mannich condensation and allows access to 3,4-dihydropyrimidin-2(1*H*)-ones under solvent-free conditions.[122]

Asymmetric Michael Addition. Catalytic asymmetric conjugate addition of nitroalkanes[123] to cycloalkenones has been investigated in detail. Rubidium salt of proline catalyzes the Michael addition of nitrocyclohexane to cycloheptenone giving rise to the major dextrorotatory compound **51** in good yield (eq 24).[124]

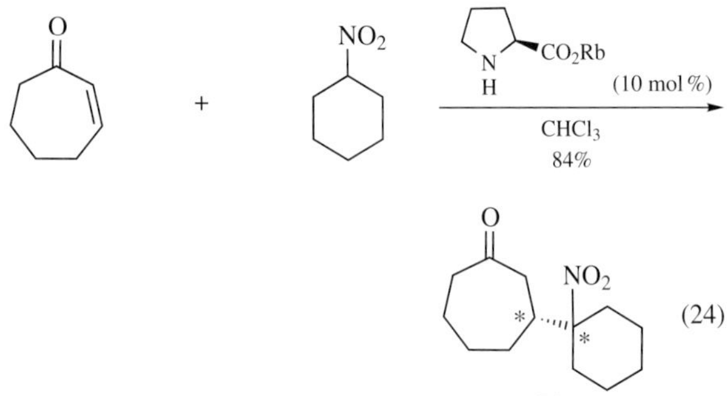

(24)

51
84% ee (+)

With only 5–7 mol % of (*S*)-proline and *trans*-2,5-dimethylpiperazine as an additive in chloroform, a Michael addition involving the same reactants produces **51** in 89% ee.[125] In this second procedure, an important nonlinear effect is observed, the ee of the product remaining almost constant when the ee of proline varies in the 20–80% range.

Ketones add efficiently to nitro olefins via proline-enamine catalysis. Butanone reacts with (*E*)-β-nitrostyrene in the presence of catalytic amount of (*S*)-proline to afford **52** in good yield, regio- and diastereoselectivity, and low enantioselectivity, (eq 25).[126]

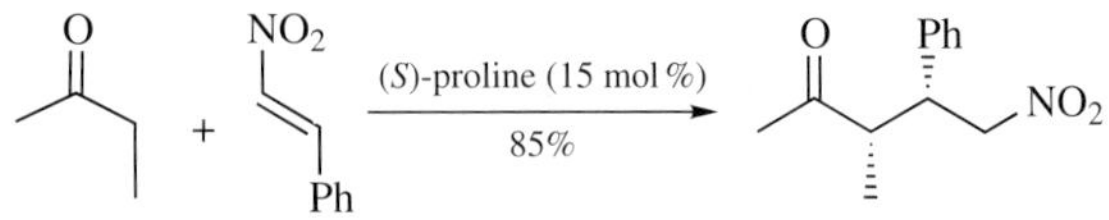

(25)

52
rr > 20:1
dr = 3:1
ee = 10%

This reaction could be also performed in THF with various aldehydes,[127] or in alcoholic solvents,[128] as well as in an ionic liquid.[129]

Intramolecular enamine-Michael addition to enal (**53**) leads mainly to *cis*-product (**54**) and has been applied successfully to the total synthesis of brasoside and littoralisone (eq 26).[130]

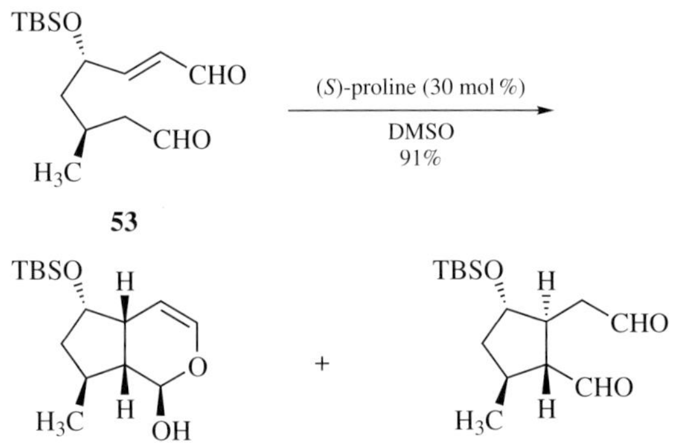

(26)

53

54 10:1 **55**

Malonate anions react with prochiral acceptors in the presence of (*S*)-proline Rb(I) salt and cesium fluoride as an additive with good yields and ee's.[131] Addition to (*E*)-enones is observed at the *si*(α)/*si*(β) face, while it takes place at the *si*(α)/*re*(β) face with (*Z*)-enones. This strategy has been applied to the synthesis of (*S*)-(+)-muscone.

Methyl vinyl ketone reacts with 1,3-diketones in a domino proline-catalyzed Michael-aldol process with good yields and enantioselectivities.[132]

Enantioselective hydride addition to α,β-unsaturated aldehydes involving a dihydropyridine as the hydride donor has been tested under (*S*)-proline-catalysis conditions, but conversion ratio and ee are low compared to those obtained with other catalysts.[133]

Sugar-based Michael acceptor could also be prepared by a proline-catalyzed dehydration reaction.[134]

Asymmetric α-Carbonyl Transformations. Proline-based enamine organocatalysis has been applied to various α-carbonyl functionalization reactions.

Asymmetric α-oxidation of carbonyls could be accomplished by an aminoxylation procedure involving nitrosobenzene with great efficiency.[135,136] It allows access to α-oxyamino-β-(3-indolyl)aldehyde (**56**) in high yield and ee (eq 27).

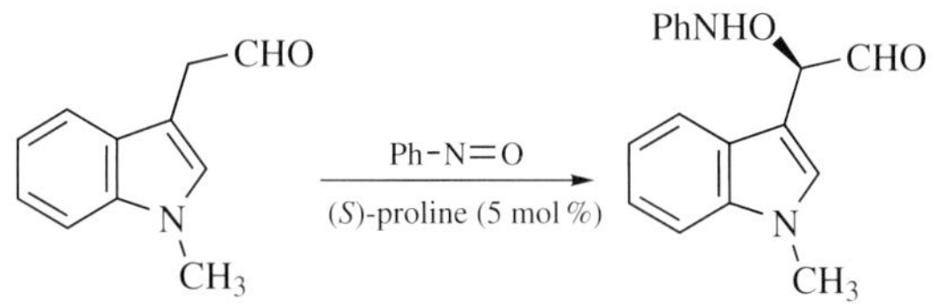

(27)

56
ee = 98%

Mechanism of this proline-mediated reaction has been theoretically[137] and experimentally[138] studied, showing an amplification of the enantiomeric excess.

Tandem aminoxylation/N–O bond heterolysis could be performed for desymmetrization of highly functionalized *meso*-compounds such as ketone (**57**) with an excellent level of enantioselectivity (eq 28).

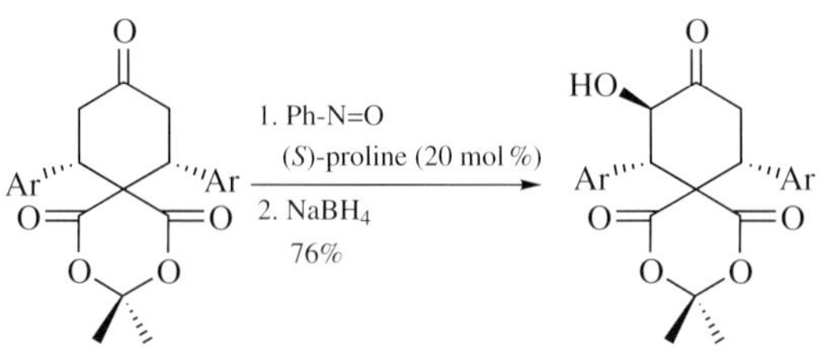

(28)

57 **58**
ee = 99%

Ar =

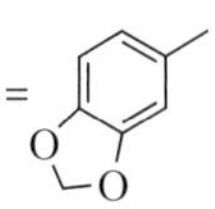

Tandem sequences involving aminoxylation and either Michael addition,[139] or olefination reaction,[140] or allylation reaction[141] have been described.

The proline-catalyzed asymmetric incorporation of molecular dioxygen to the α-position of aldehydes has been demonstrated to proceed via an addition of singlet dioxygen on a proline-based enamine.[142] Intermediate α-hydroperoxides are directly reduced to the corresponding 1,2-diols in excellent overall yields but low enantioselectivities.

By using a catalytic amount of (*S*)-proline in the presence of iodosylbenzene or in the presence of *N*-sulfonyloxaziridines, α-hydroxy ketones could be directly obtained in moderate yields and ee's.[143] However, these two procedures are complementary because they afford products **59** and **60** having opposite optical activities from cyclohexanone (eq 29).

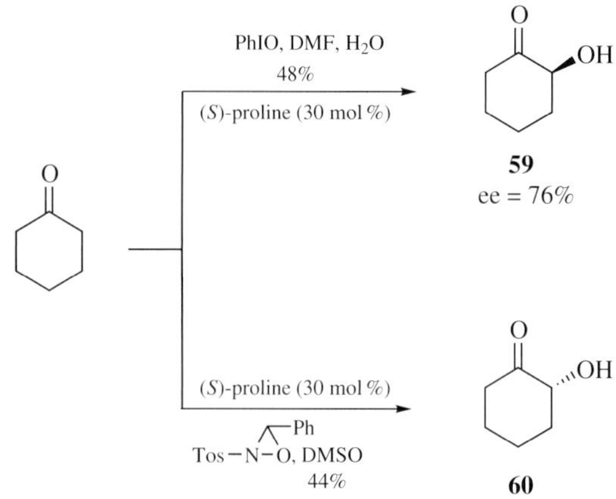

(29)

59
ee = 76%

60
ee = 29%

On another side, asymmetric α-amination of unmodified aldehydes has been described by the same method nearly at the same time by two independent teams.[144,145] It consists in the addition of dialkyl azodicarboxylates with (*S*)-proline as the catalyst which affords *N*-protected α-hydrazino aldehydes such as **61** in high yields and ee's (eq 30). The obtained products are valuable synthetic precursors for both α-amino acids and α-amino alcohols.

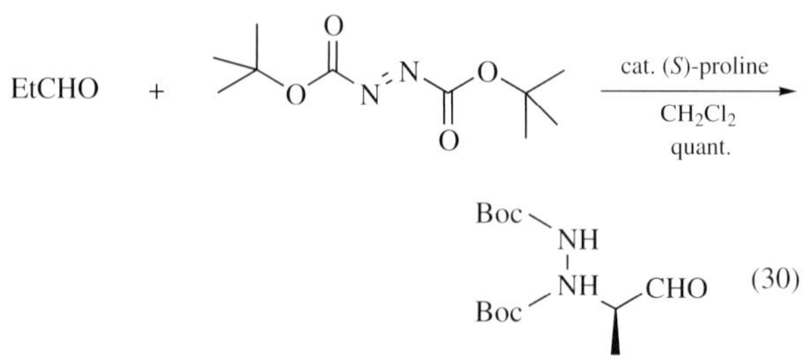

(30)

61
ee = 89%

This transformation was then applied to ketones,[146] its mechanism was studied,[147] and its application to the preparation of both enantiomers of piperazic acid was described.[148] An assembly reaction involving together aldol and α-amination processes has been studied with success and applied to a short synthesis of a renin inhibitor.[149]

Asymmetric α-halogenation reactions using proline-enamine catalysis have received much attention recently. Direct α-chlorination of aldehydes is possible by the use of a slight excess of either *N*-chlorosuccinimide or perchlorinated quinone (**62**) as chlorinating agents (eq 31).[150,151] In the case of octanal, α-chlorinated product (**63**) is obtained in moderate to excellent yields but in very low ee (eq 31).

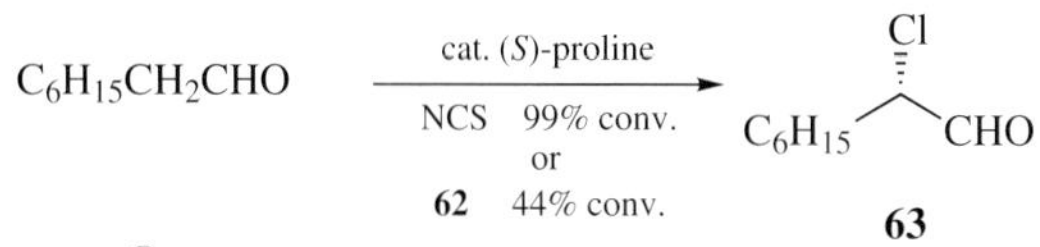

(31)

63

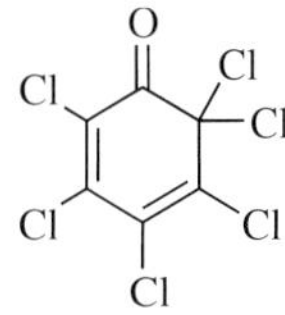

62

ee = 2% for both methods

Proline-promoted direct α-fluorination of aldehydes and ketones has also been studied using Selectfluor™ in acetonitrile.[152] Yields are quite good while enantiomeric excesses are rather low.

Asymmetric Cycloaddition Reactions. (*S*)-Proline catalyzes Diels-Alder reactions of α,β-unsaturated ketones to nitro olefins by an in situ generation of a 2-amino-1,3-diene moiety.[153]

In the absence of an external dienophile, self-Diels-Alder reaction of α,β-unsaturated ketones is observed.[154] For example, self-reaction of compound **64** affords the highly functionalized cyclohexanone (**65**) as a mixture of diastereoisomers (eq 32).

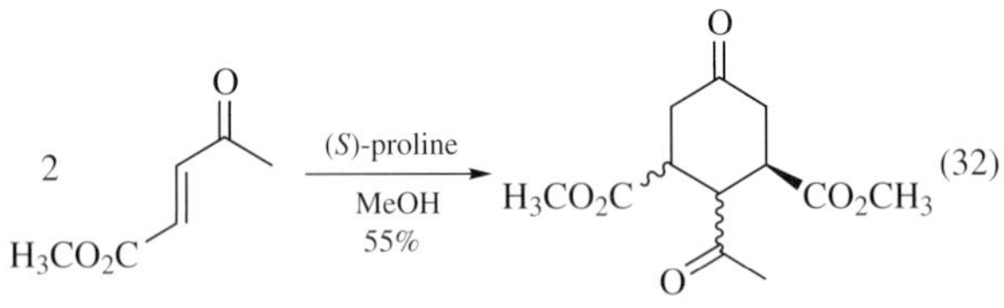

64 **65**

A similar strategy has been applied to an organocatalytic asymmetric domino Knoevenagel/Diels-Alder pathway in various solvents.[155] Further epimerization gives access to useful synthons of benzoannelated centropolyquinanes with excellent degrees of diastereoselectivity.[156] By analogy, a domino Knoevenagel-hetero-Diels-Alder mechanism was postulated for a three-component reaction involving two carbonyl compounds together with 2,2-dimethyl-1,3-dioxane-4,6-dione (Meldrum's acid).[157]

(*S*)-proline hydrochloride promotes a formal intramolecular aza-[3+3] cycloaddition in moderate yield and enantioselectivity.[158]

Baylis-Hillman Reaction. Various aldehydes react with methyl vinyl ketone in a Baylis-Hillman fashion in the presence of small amounts of (*S*)-proline and imidazole as synergistic cocatalysts, although (*S*)-proline is not efficient as sole catalyst.[159] However, the observed ee's are very low. Further screening of a variety of peptides as cocatalysts allows an increase of the ee's up to 80%.[160]

Asymmetric Alkylation. Intramolecular α-alkylation of ε-iodo aldehyde (**66**) in the presence of a catalytic amount of (*S*)-proline leads to the cyclopentane carboxaldehyde analog (**67**) in good yield and moderate ee (eq 33).[161]

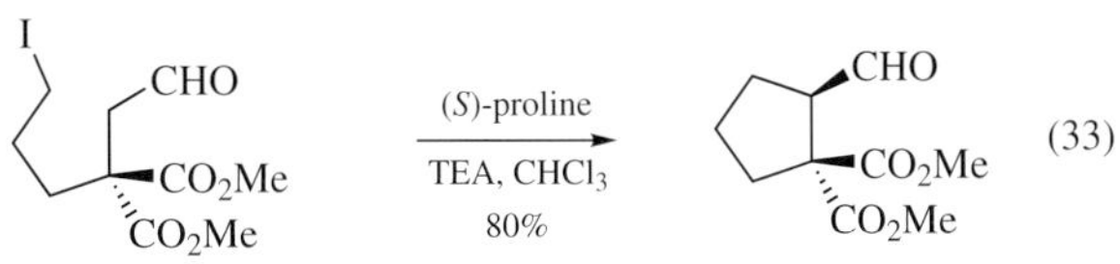

66 **67**
68% ee

Asymmetric Allylation. Allylation of aldehydes has been performed using (*S*)-proline as a chiral promoter with allylstannanes. The desired allylic alcohols are obtained in moderate yields and low ee's.[162]

Asymmetric Allylic Oxidation. An enantioselective version of the Kharash-Sosnovsky allylic benzoyloxylation catalyzed by a combination of (*S*)-proline and copper(I) oxide has been investigated.[163] With perbenzoic acid as the oxidant, cyclopentene leads under these conditions to the corresponding allylic benzoate in 39% yield and 54% ee. Reaction was investigated in water as well, but still ee's remain low.[164]

Friedländer Condensation. Preparation of 1,8-naphthyridine (**68**) has been achieved by a proline-catalyzed Friedländer condensation from a methyl ketone and 2-amino-3-formylpyridine in moderate yield (eq 34).[165]

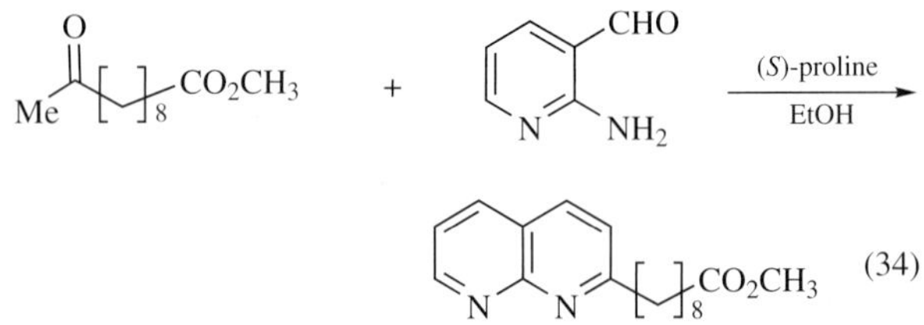

68

Aryl Coupling Reactions. A variety of proline-mediated aryl coupling reactions has been recently described. Ullmann coupling reaction between 1,4-diiodobenzene and a 1.5-fold excess of *n*-hexylamine in the presence of (*S*)-proline, copper(I) iodide, and potassium carbonate affords arylamine (**69**) in good yield (eq 35).[166]

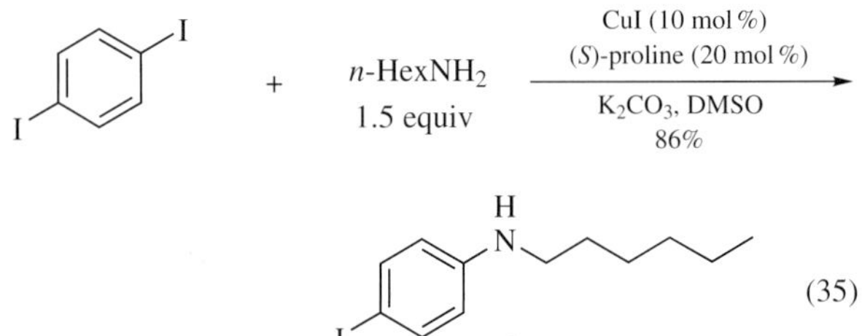

69

This method has also been applied successfully to protected phenols[167] and nitrogen heterocycles.[168] Its extension to the preparation of aryl and vinyl azides has been described, replacing amines by sodium azide and turning on sodium hydroxide as a base.[169] This safe procedure has been applied to the preparation of vinyl azide (**70**) in good yield (eq 36).

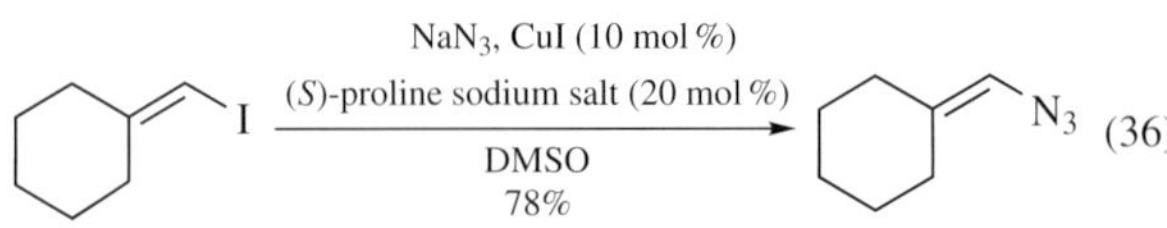

70

When sodium benzenesulfinate is used instead of sodium azide, aryl sulfonylation is observed and has been applied to the synthesis of the *p*-benzenesulfonyl analog of *N*-Boc-L-phenylalanine methyl ester (**71**) (eq 37).[170]

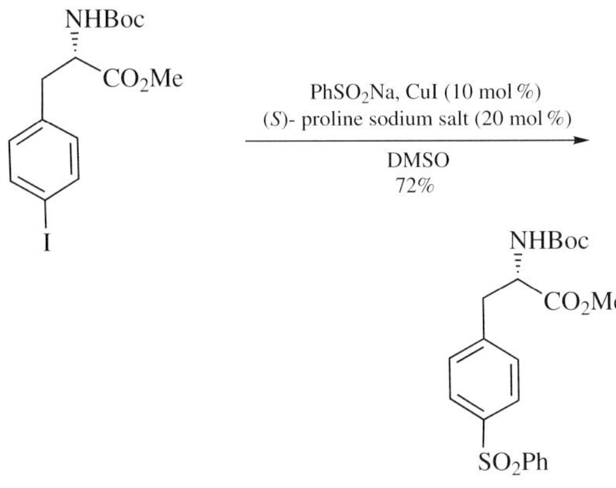

(37)

71

Asymmetric Alkyne Addition Reactions. Addition of phenylacetylene to benzaldehyde in the presence of diethylzinc could be performed with moderate yields and ee's using (*S*)-proline and some of its derivatives as ligands.[171] A combination of copper(I) iodide and (*S*)-proline mediates the Kinugasa reaction of nitrones with propargylic alcohols.[172] For example, propargylic alcohol (**72**) reacts with *N*-phenyl-furfuraldonitrone (**73**) to afford a mixture of *cis*- and 3-*exo*-methylene β-lactams (**74** and **75**) (eq 38).

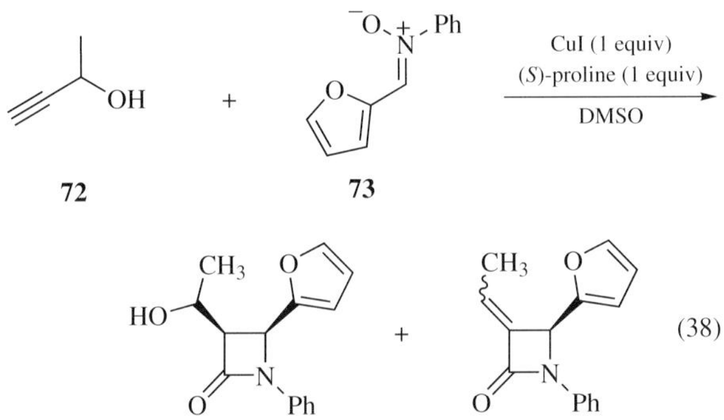

(38)

72 **73** **74** 10% **75** 75%

(*S*)-Proline Derivatives. Numerous examples of proline structurally based catalysts or auxiliaries are described in literature.[173]

Proline and some of its derivatives form carboxylate anions which represent useful ligands for Rhodium(II) complexes. Among them, tetrakis(*N*-arylsulfonyl-(*S*)-prolinato)dirhodium complexes[174] are widely applicable catalysts in numerous synthetic domains such as C–H activation involving carbenoid species.[175]

Esters derived from (*S*)-proline are useful as chiral auxiliaries in a three-component reaction leading to propargylamines,[176] in conjuguate addition[177] and in oxa-Diels-Alder reactions.[178] Proline methyl ester is easily homologated to the corresponding β-amino ester by Kowalski procedure with high yield and clean preservation of the initial optical purity.[179] Derivatization of alcohols to esters by an esterification reaction with *N*-benzenesulfonyl-(*S*)-proline has been performed for some crystallographic purposes in the total synthesis of palmarumycin and preussomerin natural products.[180]

Prolinamides and proline-derived peptides are used as organocatalysts for various reactions,[181] such as α-selenylation of aldehydes[182] or aldol condensation.[183] They are also employed as chiral deactivators.[184] The *N*-acryloyl analogs such as compound **76** are useful chiral auxiliaries in the asymmetric epoxidation reaction catalyzed by polyaniline-supported cobalt(II) salen (PASCOS) (eq 39).[185]

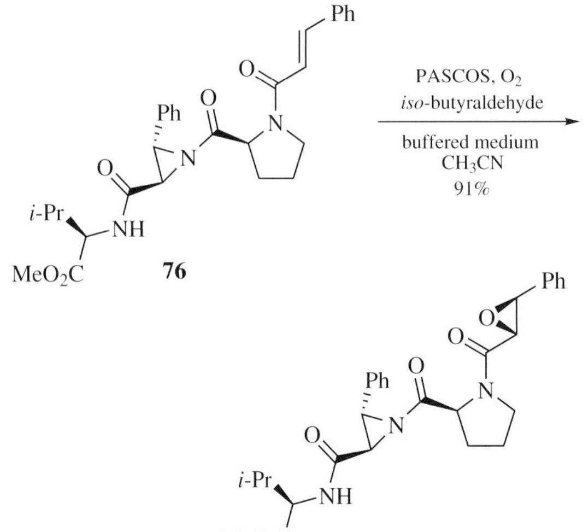

(39)

76 **77** de = 95%

Prolinol-based ligands are involved in catalytic asymmetric reactions such as organozinc additions[186] and cycloadditions.[187] SAMP-hydrazones and related compounds derived from prolinol are good chiral auxiliaries for the addition of organometallic reagents to C=N bonds.[188] Bisprolinol-biszinc(II) complexes are efficient for the catalysis of aldol reactions.[189]

Although prolinol-derived oxazaborolidines are well-known in the area of asymmetric reduction of ketones,[190] their use as catalysts has been extended to some intermolecular Diels-Alder reactions in combination with trifluoromethanesulfonic acid affording high yields and selectivities.[191] It has to be noted that resolution of racemic prolinols is successfully achieved using a combination of 1,1′-bi-2-naphthol and boric acid.[192]

Replacement of the prolinol's alcohol moiety by a fluorine atom has been carried out and applied to an organocatalytic epoxidation process producing epoxides with up to 60% ee.[193]

Proline-derived diamines are employed as ligands in the Pd(II)-catalyzed (complex (**78**)) asymmetric rearrangement of allylic imidate (**79**) to the corresponding allylic amide (**80**) (eq 40).[194]

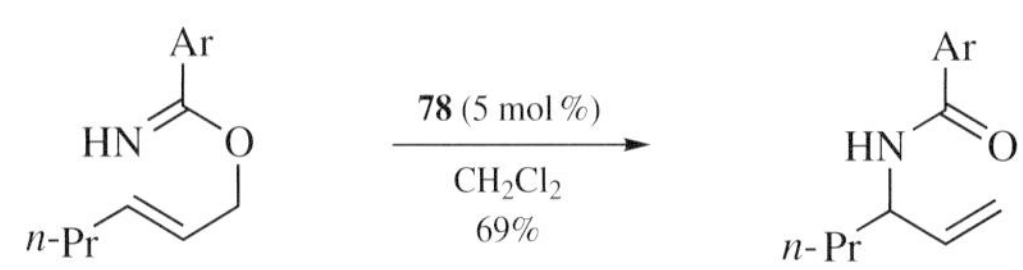

(40)

79 Ar = *p*-F$_3$C-Ph- **80** ee = 55%

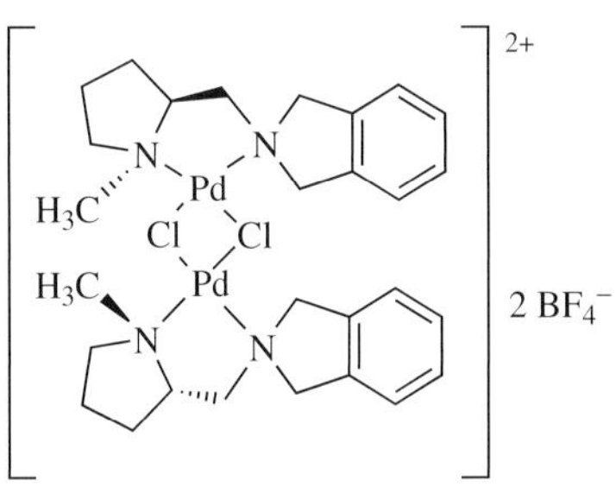

78

Asymmetric oxidative coupling leading to binaphthol derivatives is catalyzed by a combination of copper(I) chloride and proline-based diamine.[195] 1,2-Diamines of this type are also useful catalysts for the kinetic resolution of glycerol derivatives by selective acylation.[196]

Proline-based tetrazole (**81**) is a recent powerful organocatalyst for enamine-based reactions.[197] It has been used during a key-step construction of a stereogenic quaternary center of compound **82** by an efficient aminoxylation reaction. The enantioselective synthesis of this chiral center was necessary for the total synthesis of BIRT-377, a cell adhesion inhibitor (eq 41).[198]

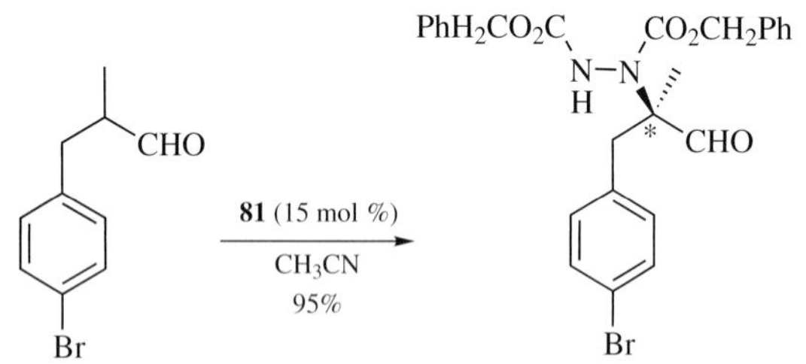

82
ee = 80%
ee = 99% after 1 recryst.

(41)

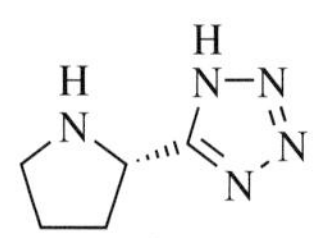

81

Proline-based oxazoline ligands bearing phosphine on the nitrogen atom are efficient ligands for asymmetric catalysis, mainly in allylic alkylation or Heck reactions.[199] A solid-phase version of such systems has also been developed.[200]

A wide variety of phosphorus-containing (*S*)-proline derivatives are easily accessible via phosphorylation reactions.[201] Among them, phosphoramides have been studied, for example, as catalysts in allylation reaction of aldehydes.[202]

Resolution of Chiral Compounds using (*S*)-Proline. Resolution of racemic C_2-symmetric carboxylic acids has been successfully observed via the formation of salts with (*S*)-proline.[203]

Resolution of racemic 1,1′-bi-2-naphthol is achieved via formation of spirocyclic 1,1′-bi-2-naphtholboric (*S*)-proline anhydrides.[204] The two diastereomeric compounds having quite different solubilities in THF, this process affords finally both enantiomers of 1,1′-bi-2-naphthol in enantiopure forms and with good yields.

Second Update

Nicholas G. Jentsch
University of Southern Mississippi, Hattiesburg, MS, USA

General Considerations. The use of proline for asymmetric organocatalysis has remained a significant tool in the synthetic community. Following the previous update, a wide variety of applications utilizing both asymmetric synthesis and carbon–carbon bond formation have been reported.[205]

Asymmetric Intramolecular Aldol Reaction. Construction of chiral bicyclic enediones resulted in inversion of enantioselectivity as a function of ring size when synthesizing six-membered or seven-membered rings.[206] Transannular aldolizations were utilized toward the total synthesis of (+)-hirsutene for construction of fused bicycles (**83**) with a minimum of two new stereogenic centers formed with moderate enantioselectivity of 54% ee (eq 42).[207]

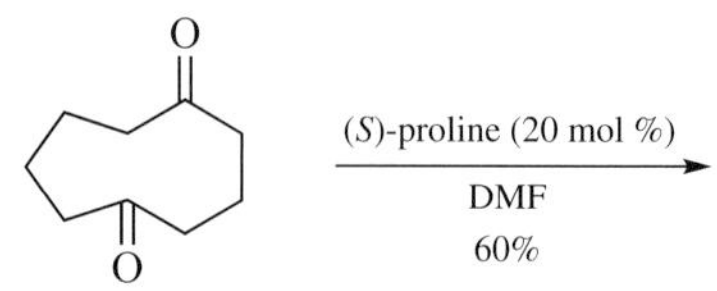

83
54% ee

(42)

Synthesis of the nucleoside phosphoramidite was achieved via diastereoselective aldol cyclization of **84** to afford the tricyclic core (**85**) in good yields (eq 43).[208]

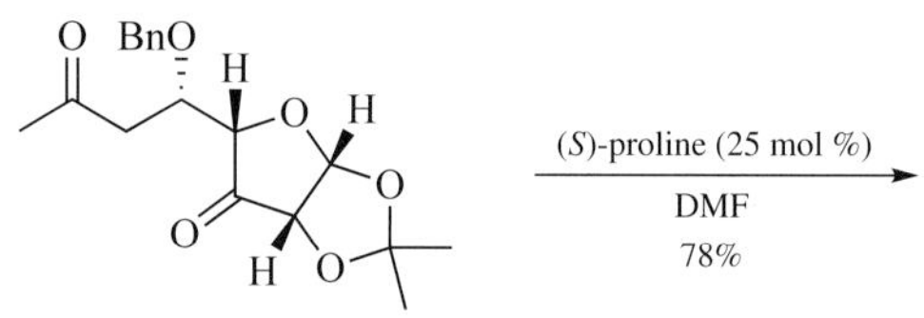

84

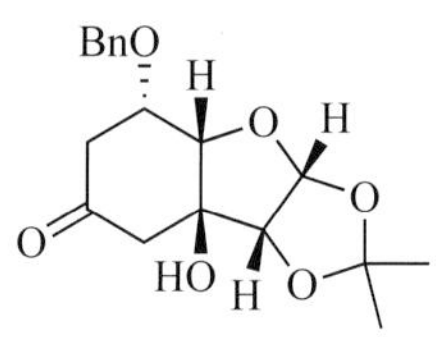

(43)

85

Asymmetric aldol cyclization of triketone (**86**) afforded fused bicyclic diketone (**87**) in moderate yields and high enantioselectivity (eq 44).[209]

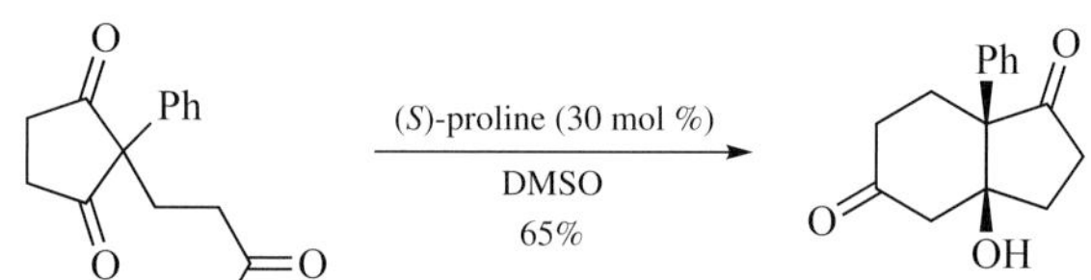

(44)

86

87
94% ee

Optically pure ionic liquids have been implemented for improvement of enantioselectivity affording products in 80% yield and ee's up to 77%.[210] A computational study of 1,7-dialdehydes confirmed the enamine-based mechanism for intramolecular aldol reactions and confirmed the importance of water to influence stereoselectivity.[211]

Asymmetric Intermolecular Aldol Reaction. Protected α-amino aldehydes (**88**) have been reacted with acetone to yield β-lactam aldols (**89**) in 94% yield and diastereoselectivity >99%. This reaction was successfully extended to include cyclopentanone with good yields and diastereoselectivity (eq 45).[212]

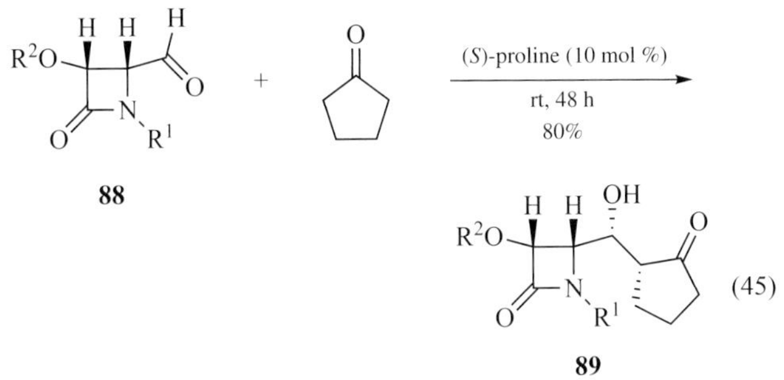

The previous update discussed the use of added water for improved enantioselectivity and recent work has attempted the reaction of cyclohexanone with *p*-nitrobenzaldehyde in only water to afford no conversion. A diamine analog (**90**) with trifluoroacetic acid as an additive was necessary to yield **91** in high conversion (98%) and 94% ee (eq 46).[213]

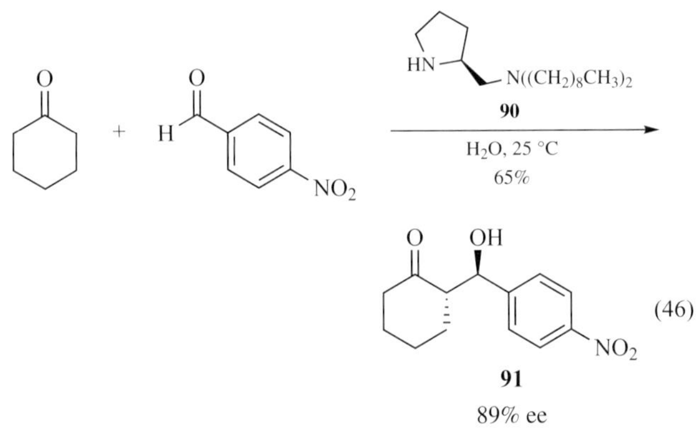

Aldol chemistry has been extended to α-keto phosphonates (**92**) and ketones to afford α-hydroxyphosphonates (**93** and **94**) in a range of 50–95% yield and 71–99% ee (eq 47).[214]

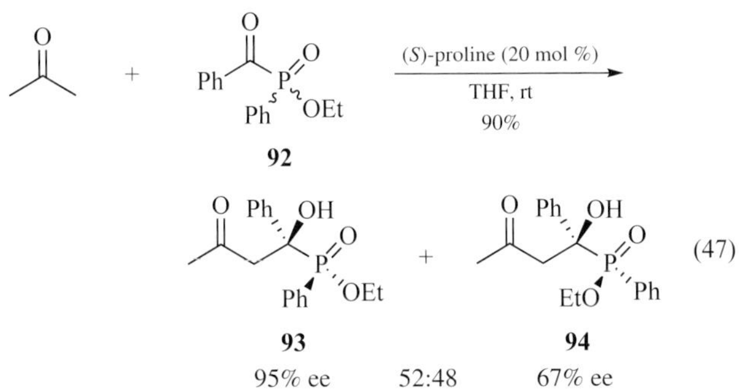

The use of proline has been extended to the synthesis of iminosugars. The β-azidoaldehyde (**95**) and protected 1,3-dihydroxyacetone (**96**) afforded iminosugar (**97**) in 90% de and good yield (eq 48).[215]

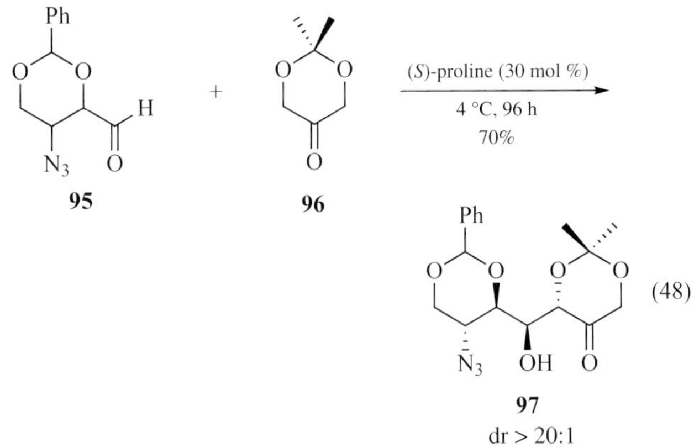

Studies of protected dihydroxyacetone derivatives toward stereoselective synthesis of carbohydrates with >80% yield and 80–99% ee have been reported.[216] Progress toward the synthesis of L-1-deoxynojirimycin (**98**) was achieved using 2-*tert*-butyl-2-methyldioxanone (**99**) with (*S*)-isoserinal hydrate derivative (**100**) to produce chiral intermediate (**101**) in 69% yield and 96% de (eq 49).[217] Syntheses of higher sugars via sequential aldol reactions[218] and sulfur-based sugars L-*erythro*-thioketoses[219] have been reported.

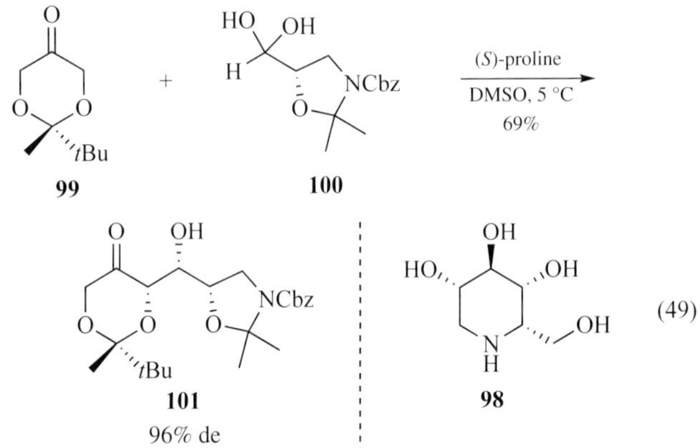

The total synthesis of (−)-genioheptolide A derivatives has been achieved with de >95% and ee >99%.[220] Preparation of azasugars with a key step involving the reaction of isoserinal hydrate with hydroxyacetone has been performed, but results produced low yields and poor diastereoselectivity when catalyzed by proline.[221]

In progress toward the synthesis of chiral spirocyclic tetrahydrofuran heterocycles, a reaction of (*Z*/*E*)-sulfanylcycloalkanecarbaldehydes (**102**) with acetone was achieved with high yields and excellent enantioselectivity to afford aldol products (**103**) (eq 50).[222]

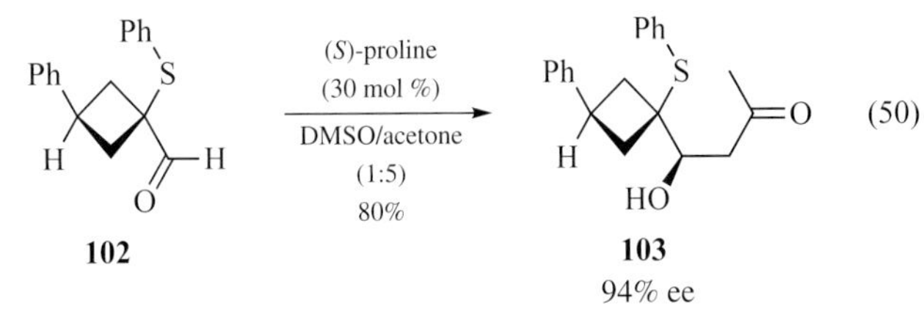

Cross-aldol Reaction. β-(Hydroxyalkyl)-γ-butyrolactones (**104**) have been synthesized in a one-pot method via cross-aldol reaction of methyl-4-oxobutyrate with a variety of aldehydes followed by sodium borohydride reduction with moderate yields and excellent stereoselectivity of up to 99% ee and 92% de.[223] This work was extended to the total synthesis of (−)-enterolactone (**105**) with the first chiral inducing step involving reaction between methyl-4-oxobutanoate and *m*-methoxybenzaldehyde to afford intermediate (**104**) in 55% yield and excellent stereoselectivities of 97% ee and 92% de (eq 51).[224]

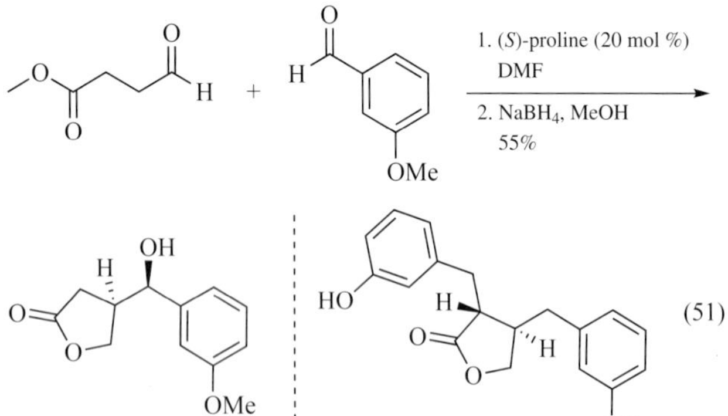

104 ee 97% de 92% **105** (51)

α-Halo aldehydes have also been utilized to afford *anti*-selective cross-aldol products in 50% yield and 97% ee.[225] The asymmetric total synthesis of (-)-amphidinolide V has been achieved using a cross-aldol reaction of **106** and **107**, which afforded an advanced intermediate (**108**) in 66% yield and good regioselectivity (*E*/*Z* = 12.5:1) (eq 52).[226]

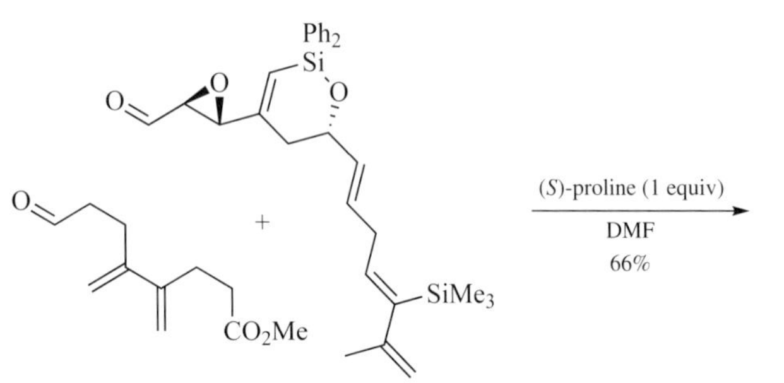

106 **107**

108
E/*Z* = 12.5:1 (52)

Cocatalysts. The use of cocatalysts is an emerging trend for the improvement of stereoselectivity. Crossover of diastereoselectivity based on the use of varied borate-derived guanidinium salts as a cocatalyst has been reported. Tetrafluoroborate guanidinium salts afforded *anti*-aldols, while tetraphenylborate guanidinium salts (**109**) afforded *syn*-aldols (**110**) (eq 53).[227]

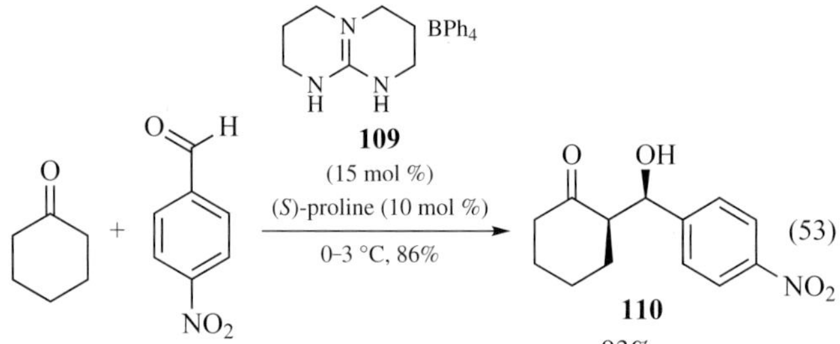

109 **110** 93% ee 30% de (53)

Chiral diols including *1,1′-Bi-2-naphthol* have also been shown to improve enantioselectivity.[228]

Dihydrocupreidine (**113**) (25 mol %) has been used as a cocatalyst for the construction of bicyclic intermediates toward the synthesis of bicyclo[2.2.2]octadiene ligands **111** and **112** with 83% yield and 48–66% ee (eq 54).[229]

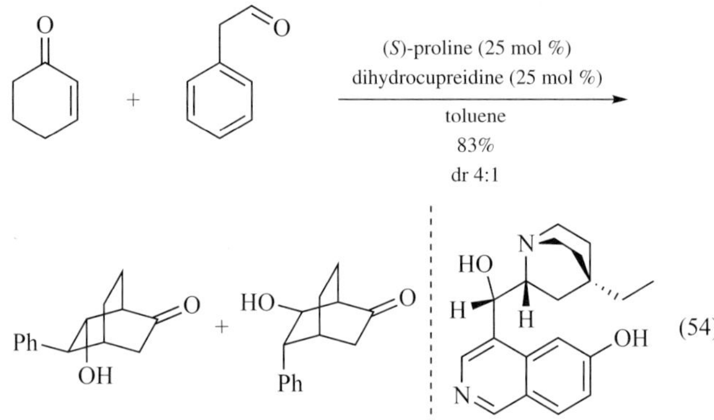

111 48% ee **112** 66% ee **113** (54)

Mechanistic Studies. Several attempts to describe the transition states and mechanistic aspects involved in proline-catalyzed aldol reactions have been made recently. Computational methods have been invoked to describe intramolecular aldol reactions,[230] stereoselectivity of proline,[231] and the origin of double asymmetric induction.[232] NMR[233] and kinetic isotope effects[234] have also been utilized for mechanistic studies.

Solid-state/Polymer Networks. A variety of polymer-based and solid-state supports have also been utilized for improved catalysis with proline. A recoverable catalytic system of (*S*)-proline supported on poly(diallyldimethylammonium) salts performed aldol reactions in yields ranging from 83 from 96% and 66–69% ee.[235] Merrifield resin-supported ionic liquids have also been used as efficient and recyclable catalyst systems for asymmetric aldol reactions.[236] Heterogenized (*S*)-proline on mesoporous supports catalyzes aldol reactions in a wide range of solvents, which would otherwise afford poor yields due to proline insolubility.[237] Proline-modified DNA has been utilized for aldol catalysis between an oligonucleotide template bearing a conjugated aldehyde at the 3′ terminus and acetone to afford the product in 92% yield.[238]

Reactions in water have been facilitated via 2-propanol-modified hyperbranched polyethyleneimine to afford **114** (eq 55).[239]

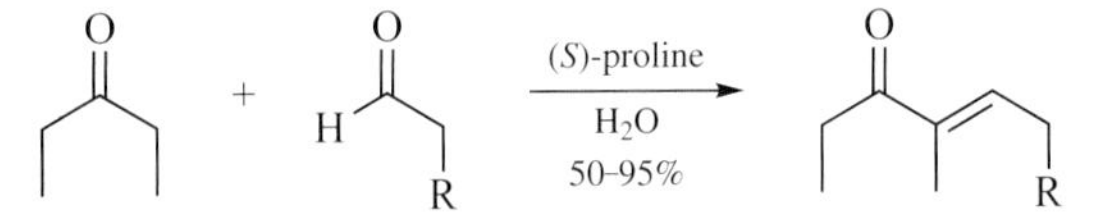

114 (55)

Mannich Reaction. The proline-catalyzed Mannich reaction has remained a useful tool in the synthetic community due to the one-pot direct condensation and a variety of novel applications have been recently reported. A facile one-pot synthesis of hexahydropyrimidines (**115**) with moderate to good yields has been achieved and optimum yields were realized when utilizing the displayed stoichiometry of the three starting materials (eq 56).[240]

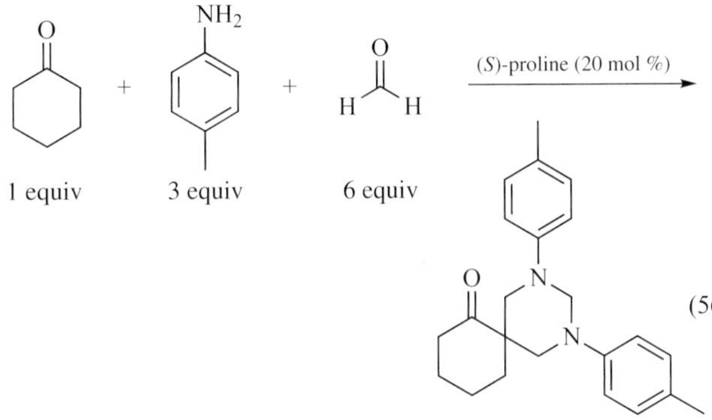

115
73% (56)

Additionally, one-pot double Mannich reactions of acetaldehyde and various enamines (**116**) have afforded β,β'-diamino aldehydes (**117**) containing three adjacent stereogenic centers (eq 57).[241] *syn*-Mannich reactions in a sequential, one-pot synthesis have been utilized toward the synthesis of densely functionalized pyrrolidines resulting in 60–80% yield and 92–99% ee.[242]

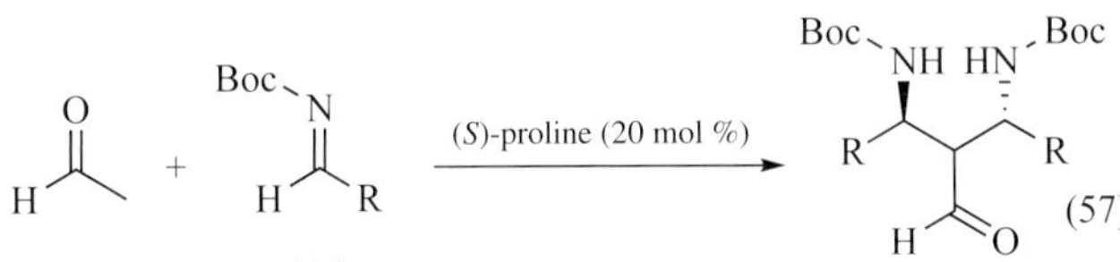

116 (3 equiv)

117 de 98% ee 99% (57)

anti-Selective Mannich-type reactions of unmodified ketones with high diastereo- and enantioselectivity have been reported. This study further determined that the position of the acid group on the pyrrolidine ring will direct either *syn*- or *anti*-products.[243] Thermal effects from microwave radiation afforded the Mannich product (**118**) in excellent yields and enantioselectivity (eq 58).[244]

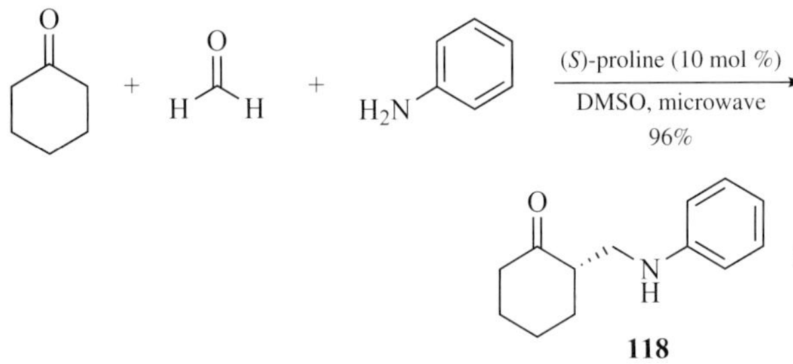

118
98% ee (58)

There has also been continued use of the Mannich reaction for the synthesis of natural products and synthetic derivatives. The synthesis of quinolizidine skeletons (**120**) has been achieved from acetone, *p*-methoxyphenylamine, and (**119**) in excellent enantiomeric excess (eq 59).[245]

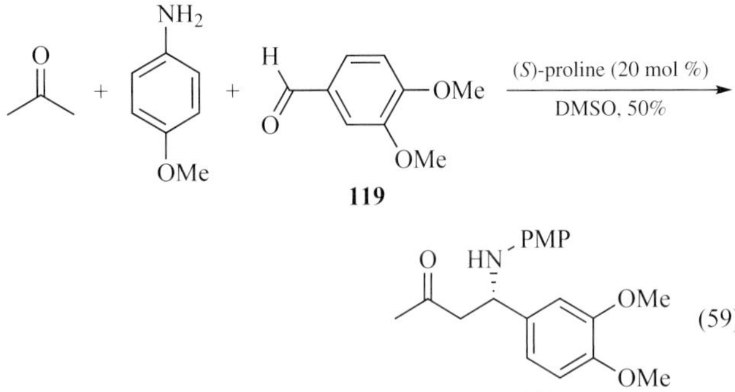

120
99% ee (59)

A γ-keto-α-amino ester precursor (**121**) toward the total synthesis of (2*S*,3*R*,4*S*)-4-hydroxyisoleucine (**122**), a sugar analog for diabetes treatment, was prepared in 72% yield and 99% ee (eq 60).[246]

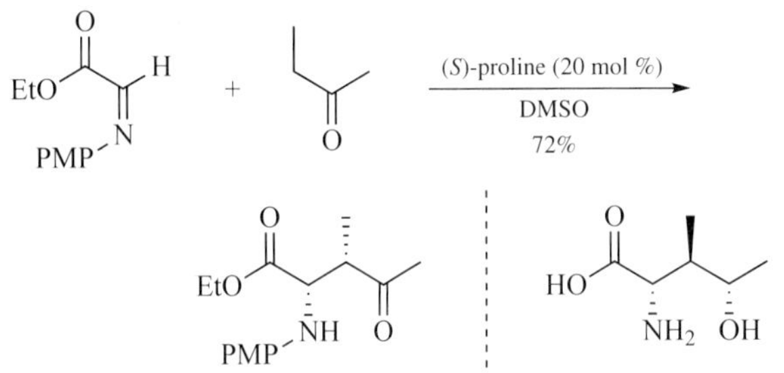

121 99% ee **122** (60)

The reaction of glyoxalate-derived *p*-methoxyphenyl-protected imine and phenylacetaldehyde followed by immediate reduction afforded *syn*-β-phenyl-substituted homoserine in 72% yield with excellent stereoselectivity (*syn*/*anti* 92:8, >90% ee).[247]

Synthesis of β^2-amino acids has been achieved with 20 mol % of L-proline with excellent yields of >90% and diastereoselectivity of 72% de.[248] Sequential asymmetric Mannich–aza-Michael reactions toward the synthesis of highly substituted pyrrolidines have afforded the product in a range of 44–97% yield and 99% ee.[249] β-Lactam Mannich bases (**124**) have been synthesized from a carbapenem core (**123**) in moderate yields of 52–55% and diastereoselectivities of >98% (eq 61).[250]

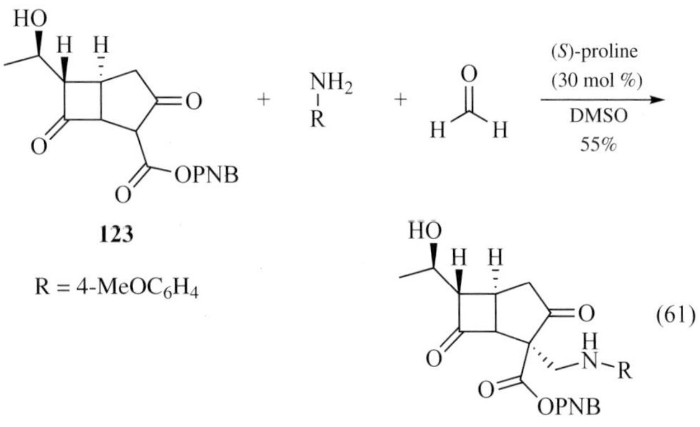

123

R = 4-MeOC$_6$H$_4$

124
98% de

(61)

A key step in the total synthesis of (−)-agelastatin A (**128**) involved a *syn*-Mannich reaction between **125** and **126**, which produced vicinal diamines (**127**) in yields ranging from 63 to 83% and 94% ee (eq 62).[251] Computational studies using density functional theory confirmed the proposed transition state for the stereoselective step between cyclohexanone, formaldehyde, and *p*-methoxyaniline.[252]

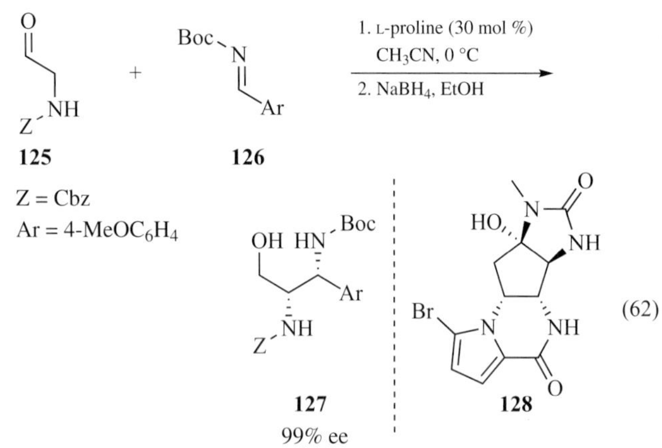

125 126

Z = Cbz
Ar = 4-MeOC$_6$H$_4$

127
99% ee

128

(62)

Michael Additions. Asymmetric Michael addition of dioxanone (**129**) to nitroalkenes (**130**) to afford highly functionalized 1,4-adducts (**131**) was achieved, though use of a trifluorosulfonamide proline derivative showed increased ee and yields (eq 63).[253]

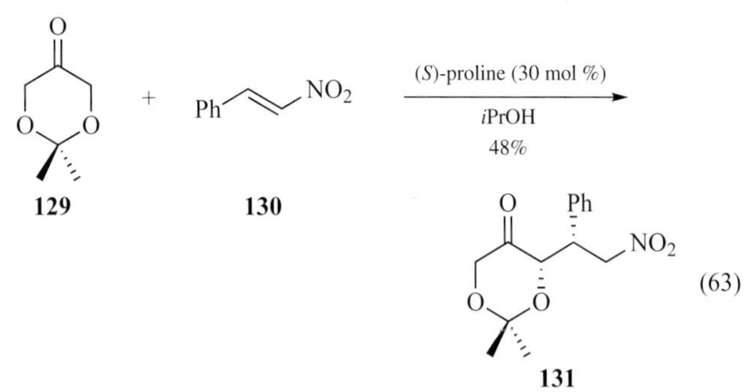

129 130

131
43% ee
98% de

(63)

An intramolecular heteroatom Michael addition of **132** for the synthesis of enantioenriched pyrrolidine, indoline, and piperidine rings (**133**) has been reported, but poor ee was observed (eq 64). TMS diphenylprolinol derivative (**134**) was utilized for improved results.[254]

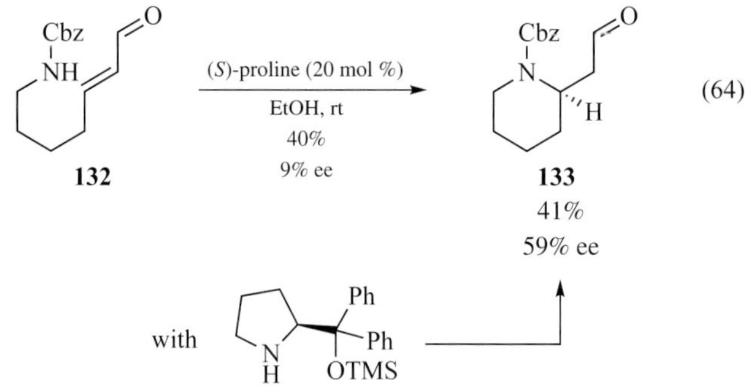

132

133
41%
59% ee

134

(64)

The proline-derived amino alcohol (**134**) was utilized in Michael additions as an improvement to attempted proline catalysis for a variety of publications, including the tandem Michael additions of **135** and **136** in the formal synthesis of (−)-α- and (−)-β-lycorane (**137**) (eq 65),[255] the reaction of acetaldehyde with β-nitrostyrene,[256] tandem Michael addition–hemiacetalization between aliphatic aldehydes (**138**) and nitrovinyl phenols (**139**) to construct benzopyran backbones (**140**) (eq 66),[257] and the synthesis of α-cyclopropylphosphonates.[258]

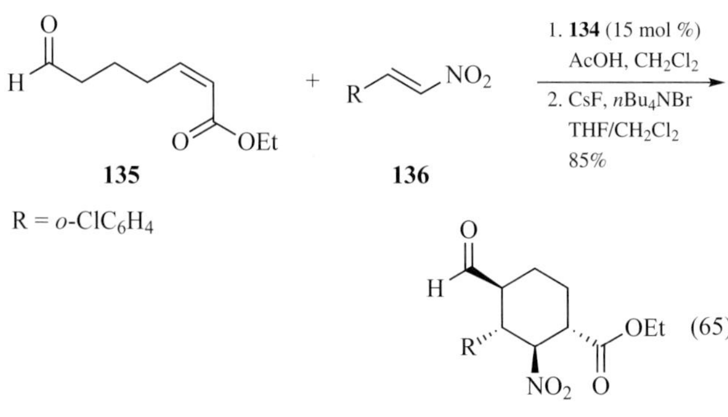

135 136

R = *o*-ClC$_6$H$_4$

137
95% ee

(65)

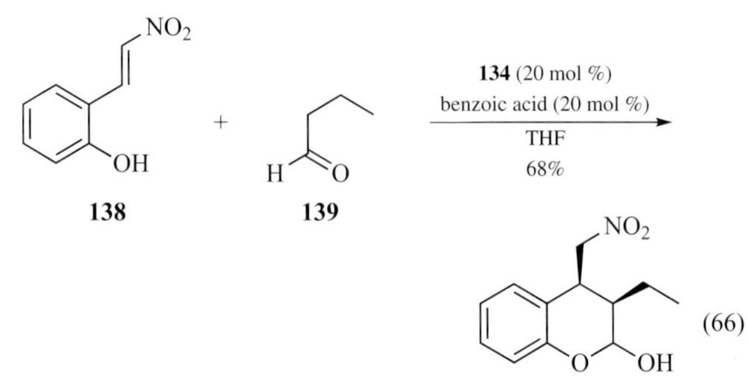

138 139

140
18% ee

(66)

A double Michael addition performed between 1,3-cyclohexanone and methyl propiolate has produced Michael products (**141**) in good yield (eq 67).[259]

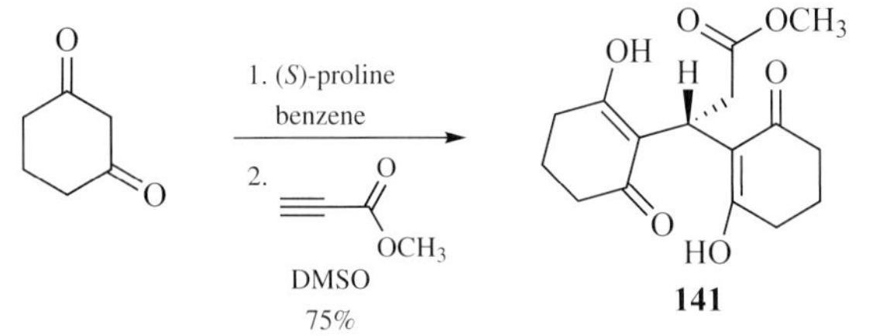

(67)

An intramolecular Michael addition has afforded 3-aryl-4-nitrocyclohexanones via the reaction of Morita–Baylis–Hillman adducts with excellent de of >99% but poor ee of 20%.[260] A report has shown success in aqueous media with a $Yb(OTf)_3$ cocatalyst in 5 mol % and L-proline in 12 mol % to afford Michael products in high yields and good enantioselectivity for Michael acceptors containing a 1,2-disubstituted alkene.[261] Reaction of α,β-unsaturated ketones with aromatic nitroso compounds afforded 3-oxa-2-aza bicycloketone (**142**) in poor to moderate yields but with 99% ee (eq 68).[262] Computational modeling has been used for transition states of proline-catalyzed asymmetric Michael reactions in basic and base-free conditions.[263] NMR studies have also been utilized to examine reaction profiles of proline-catalyzed aldol condensations and Mannich-type reactions.[264]

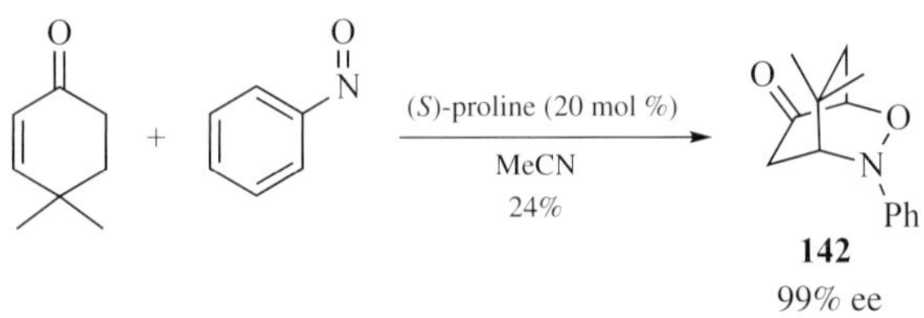

(68)

Ullmann Coupling. Arylation via Ullmann coupling with a Cu(I) catalyst and L-proline ligand has received increased attention in the synthesis of a variety of heterocycles. For example, coupling of 2-halotrifluoroacetanilides with β-keto esters and amides was utilized toward the synthesis of 2,3-substituted indoles.[265] *N*-Functionalized indoles have also been constructed via intramolecular cyclization of acyclic acrylonitriles and primary amines.[266]

A variety of other heterocycles have also been constructed via the Ullmann method. For instance, arylation of 2*H*-isoindole-1-carbaldehydes (**143**) and 2-iodoanilines (**144**) has afforded isoindoloquinoxalines (**145**) in yields of 70–80% (eq 69).[267]

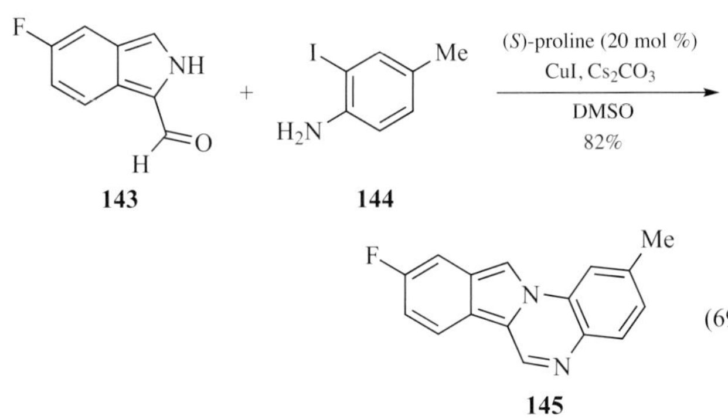

(69)

Catalyzed coupling of iodobenzenes (**146**) with pyrrole-2-carboxylate esters (**147**) has afforded intermediates toward the synthesis of pyrrolo[1,2-α]quinoxalines (**148**) with 70–96% yield (eq 70).[268]

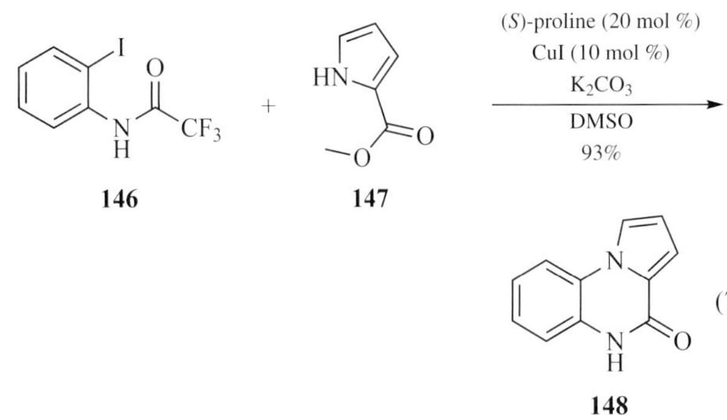

(70)

Synthesis of tertiary aryl squaramides has been achieved through C–N bond formation between aryl bromides and secondary squaramides in moderate yields.[269] *N*-Arylation of benzimidazole derivatives (**149**) with *o*-bromobenzaldehyde to synthesize benzimidazo[1,2-α]quinazolines (**150**) has been achieved with good yields of 66–82% (eq 71).[270] Phenothiazines have also been constructed in high yields via coupling of 2-iodoanilines and 2-bromobenzenethiols.[271]

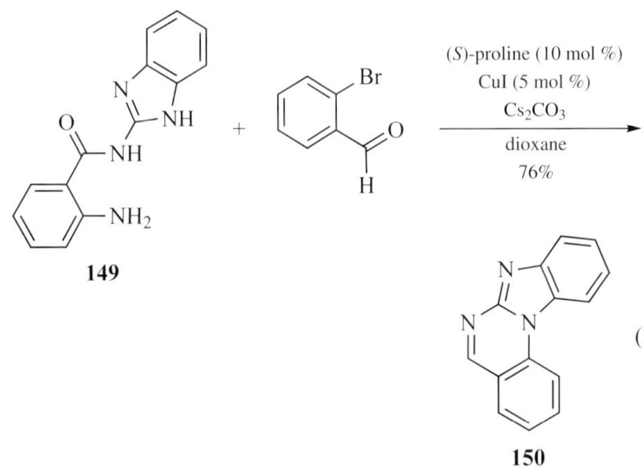

(71)

Direct C–H arylations of unactivated arenes via the Ullmann method have been achieved with addition of 3 equiv of potassium *t*-butoxide, which afforded the product in 87% yield.[272] Ionic liquids containing an imidazolium–proline ion pair have been utilized to perform CuI-catalyzed *N*-arylation of nitrogen-containing heterocycles.[273] A modified Ullmann reaction to achieve direct guanidinylation with *p*-methoxybenzyl guanidine and iodobenzene in a range of 43–96% yields has been reported.[274]

Ring Closing/Cyclization. A variety of cyclizations have been achieved via the same CuI/L-proline catalysis as previously discussed with the Ullmann coupling. A Sonogashira cross-coupling of vinyl iodides and 1-alkynes was extended to a domino coupling/cyclization process of aryl acetylenes and 2-bromotrifluoroacetanilides (**151**) to afford indoles (**152**) substituted at the 2,5-, 2,6-, or 2,7-positions. Most reactions succeeded at high yields (eq 72).[275]

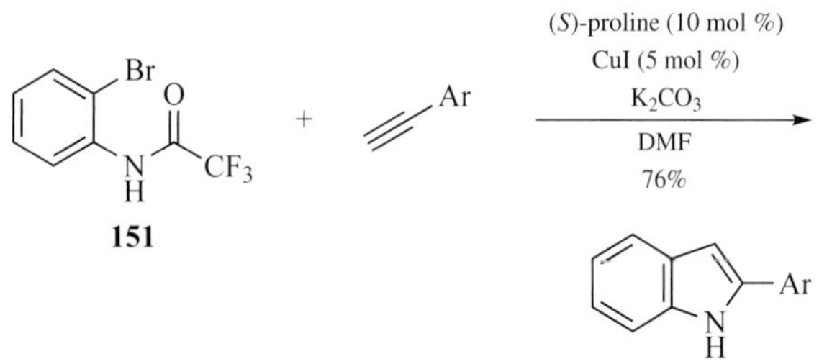

151 152 (72)

A three-step one-pot synthesis of dihydropyrazolo indoles has also been reported, which included C–N coupling/intramolecular cyclization reaction.[276] A versatile one-pot synthesis of various *N*-substituted 2-heterobenzimidazoles has been achieved via catalyzed intermolecular cyclization of *o*-haloarylcarbodiimides and *N*-/*O*-nucleophiles.[277] Coupling of aqueous ammonia with 2-iodoacetanilides (**153**) afforded 1*H*-benzimidazoles (**154**) (eq 73).[278] A CuI/L-proline-catalyzed intramolecular C–N/C–C bond formation has resulted in the construction of seven-membered heterocycles.[279]

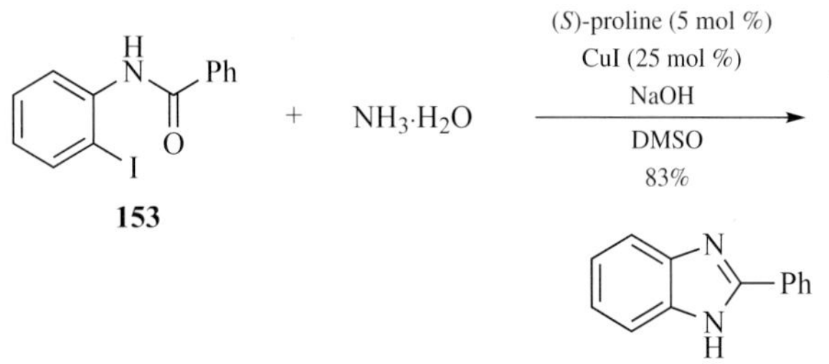

153 154 (73)

A three-component aldol/Michael/5-*exo-trig* cyclization between 5-aryltetrahydro-3-thiophenone (**155**), sulfonyl acetates (**156**), and aromatic aldehydes has afforded thiopyran derivatives (**157**) in yields of 49–79% (eq 74).[280]

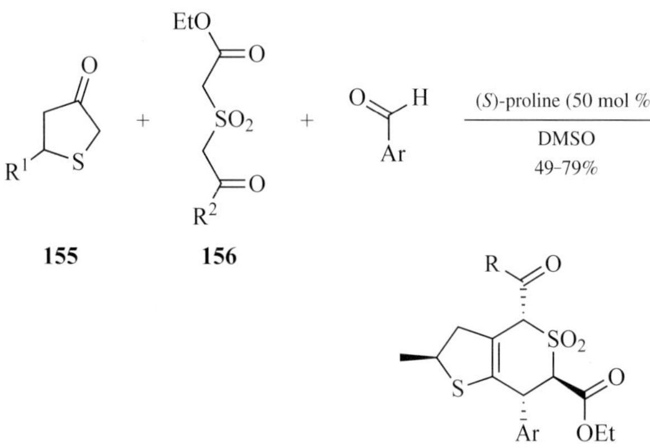

155 156 157 (74)

$AgNO_3$/L-proline-catalyzed reactions of quinolone carbaldehydes with aniline and acetone afforded naphthyridines in good yields (55–79%).[281] Aziridine aldehyde dimers, proline, and isocyanides have been shown to undergo sequential aziridine ring opening/intramolecular cyclization to afford piperazinones in moderate yields of 40–84% and good regioselectivity (*cis*:*trans* 93:7).[282] A recent use of proline-catalyzed Friedlander annulations has afforded 4-trifluoromethyl quinolones in high yields.[283]

Cycloadditions. L-Proline has been utilized to perform self-condensation reactions of β-methyl-substituted α,β-unsaturated aldehydes to synthesize ring-fused homodimers (**158**) (eq 75).[284]

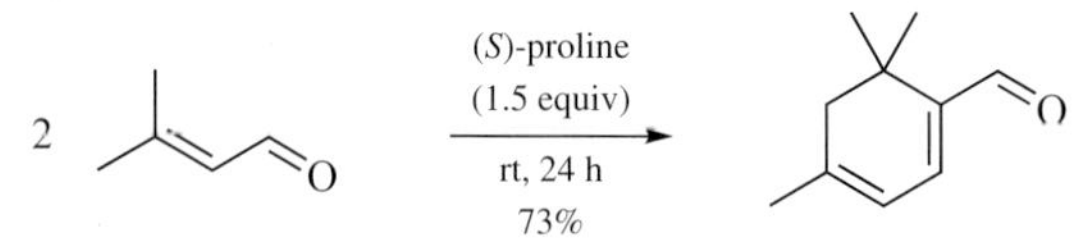

158 (75)

The total synthesis of (+)-palitantin (**160**) was achieved utilizing a key Robinson annulation between a variety of α,β-unsaturated aldehydes to afford cyclic intermediates (**159**) in 70–80% yields and 80–90% ee (eq 76).[285] A CuI/L-proline-catalyzed azide–alkyne dipolar cycloaddition has been reported for the synthesis of 1,2,3-triazole-3-alanine azapeptides.[286]

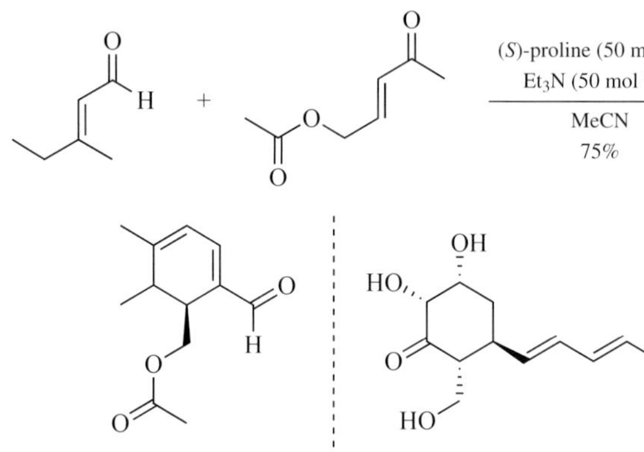

159 82% ee 160 (76)

α-Carbonyl Transformations. A variety of α-aminations have been reported recently. For instance, α-amination of aldehydes with dibenzyl azodicarboxylate has afforded the product in 80% yield and high regioselectivity (13:1 *anti*:*syn*).[287] A stereocontrolled synthesis of *syn*- and *anti*-1,3-diamines using sequential α-amination and Horner–Wadsworth–Emmons olefination has been developed.[288] α-Amination of **161** to afford **162** has been utilized as a chiral inducing step toward the total synthesis of clavaminol H (**163**) in 75% yield (eq 77).[289] A mechanistic study has revealed that addition of various tertiary amine bases can result in the reversal of proline-mediated α-amination of aldehydes.[290]

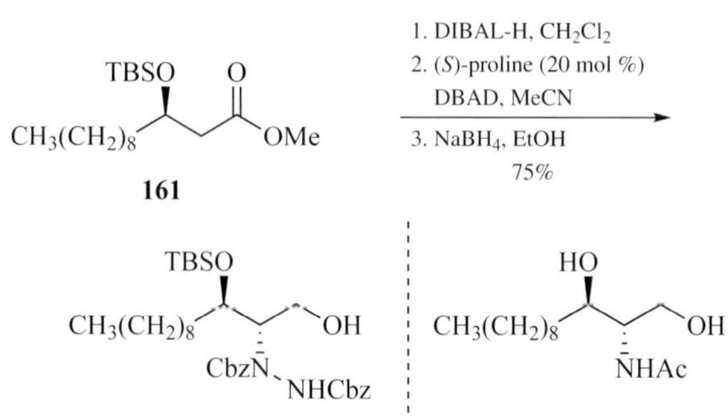

161 162 163 (77)

α-Aminoxylations have also been utilized in a recent work. The key chiral inducing reaction toward the total synthesis of

(+)-α-conhydrine and (−)-sedamine utilized an L-proline α-aminoxylation.[291] α-Aminoxylation of an aldehyde intermediate with nitrosobenzene has been utilized in progress toward the total synthesis of (*R*)-antofine and (*R*)-cryptopleurine in moderate yield.[292] Room-temperature ionic liquids have been implemented as reaction media to perform direct asymmetric α-aminoxylation of propanal or cyclohexanone with high yields and enantioselectivites.[293]

Asymmetric α-alkylation of aldehydes has been achieved when utilizing activated iminium ion intermediates generated from arylsulfonyl indoles (**164**) to afford 3-indolyl derivatives (**165**) in 74–92% yield and 75–90% ee (eq 78).[294]

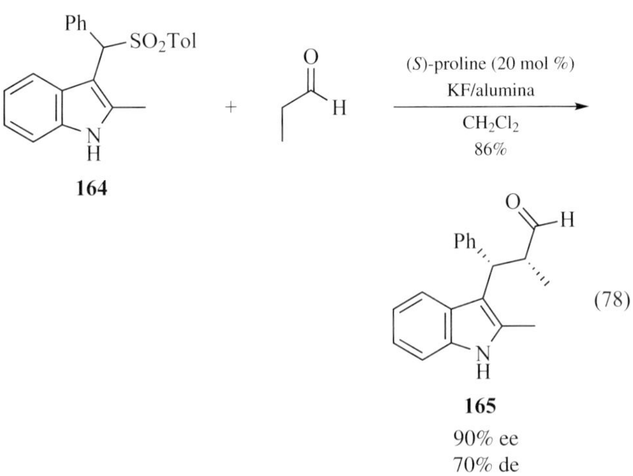

(78)

α-Sulfamidations of α-branched aldehydes with arylsulfonyl azides have afforded the product in moderate yields and enantioselectivity.[295] The scope and limitations of this sulfamidation have been sufficiently examined.[296]

Baylis–Hillman Reaction. Insight into the mechanism of the Baylis–Hillman reaction was gained through observation of the reaction between methyl or ethyl vinyl ketone and aryl aldehydes. It was determined that the amine catalyst attacks the vinyl ketone, forming a zwitterionic iminium species that undergoes intramolecular cyclization to form a bicyclic enaminolactone species. The C–C bond formation is then assisted via hydrogen bond assistance from the hydrogen carbonate ion.[297]

An asymmetric Baylis–Hillman reaction between aromatic aldehydes and methyl vinyl ketone has been achieved using cyclic tertiary amine/L-proline cocatalysis with good yields and enantioselectivity up to 83% ee.[298]

Knoevenagel Condensation. An organocatalytic Knoevenagel condensation reaction between methyl ketones or active methylene compounds with dimethylformamide dimethyl acetal has afforded dimethylamino-2-propenones in high yields.[299] A multicatalyst cascade reaction has been developed with the initial step involving a proline-catalyzed Knoevenagel reaction between benzaldehyde (**166**) and ethyl cyanomalonate (**167**) in the presence of Hantzsch ester (**170**) to afford **168** in 95% yield (eq 79).[300]

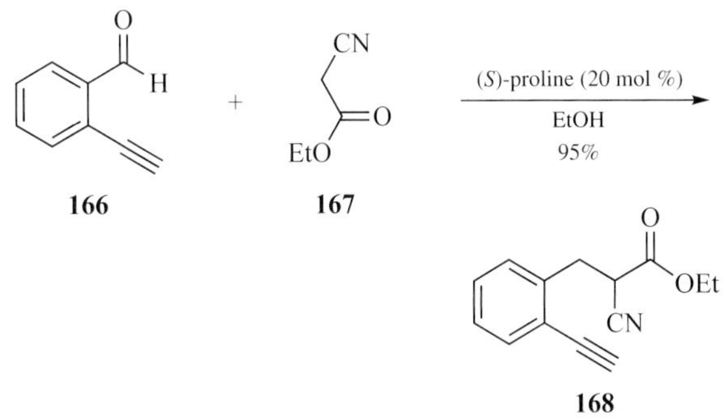

(79)

Reductive alkylations have been performed on (*R*)-glyceraldehyde (**169**) with CH-acids in the presence of the Hantzsch ester (**170**) to produce chiral building block (**171**) for natural product synthesis (eq 80).[301] A proline-catalyzed Knoevenagel/hydrogenation/Robinson annulation cascade for the synthesis of Wieland–Miescher ketone analogs has been demonstrated.[302]

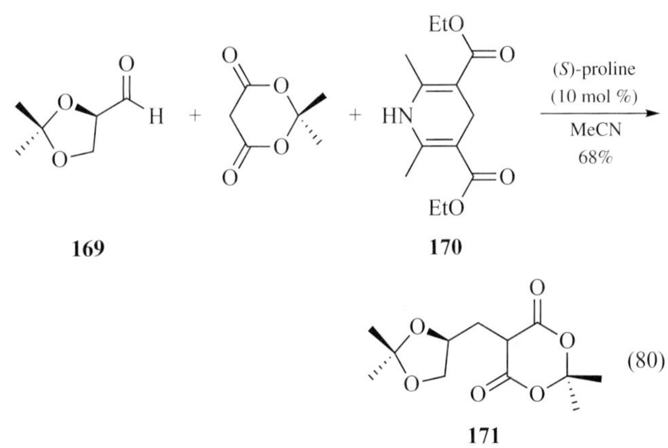

(80)

Alkyne Addition. Propargylic amine derivatives have been synthesized in 95% yield via CuI-catalyzed aldehyde- and ketone-induced intermolecular tandem decarboxylation–coupling of proline with alkynes.[303]

Microwave-assisted Reaction. Microwave irradiation has been utilized to examine specific or nonthermal microwave effects for standard aldol and Mannich-type reactions, though no evidence of such effects was observed.[304] Anomerization of α-*C*-glycosylmethyl aldehydes and ketones has been observed under microwave conditions.[305] Reaction rates between aldehydes and azodicarboxylates have been increased, which resulted in increased yields and enantioselctivity.[306] The cyclization of proline to tricyclic 2,5-diketopiperazine (**173**) via microwave irradiation in the presence of methyl dichlorophosphite, trimethylamine, and ionic liquid (**172**) has been achieved (eq 81). This chemistry can easily be expanded to achieve the synthesis of highly functionalized symmetrical and unsymmetrical 2,5-diketopiperazines in a single-step, one-pot reaction.[307]

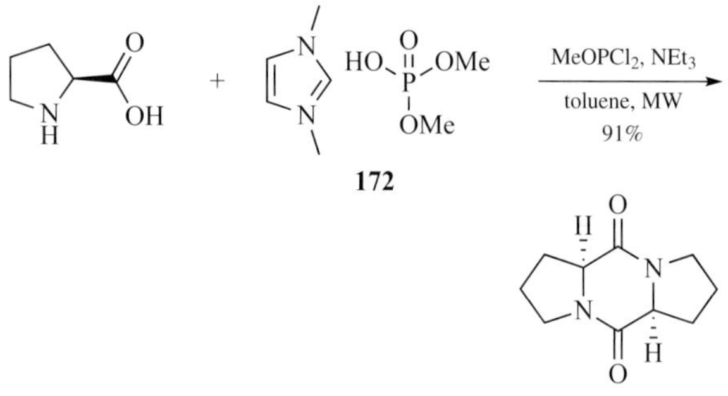

172

173

(81)

Proline Derivatives. Pyridine *N*-oxide catalysts derived from proline have been utilized for the allylation of aldehydes with allyl(trichloro)silane to afford homallylic alcohols in good yield and enantioselectivity.[308] Proline-oxazoline ligands have been applied in the asymmetric Nozaki–Hiyama–Kishi reaction that also afforded allylic alcohols in high yields and moderate enantioselectivities.[309] The reaction of a broad range of aldehydes and oxazolones has produced intermediates for transformation into highly functionalized optically active γ-quaternary proline derivatives.[310] C_1-Symmetric salan ligands derived from proline have been applied for highly enantioselective epoxidation of styrene derivatives.[311] An (*S*)-proline-oxazolidinone derivative has been synthesized to perform sequential Seebach stereoselective alkylation and hydrolysis toward total synthesis of hypoestestatin 1.[312]

Unique Addition Reactions. A one-pot synthesis of 1-phenyl-4-vinyl-1,2,3-triazoles has been achieved through azidation of iodobenzene with L-proline in the presence of CuI.[313] Phenylhydrazine and trialkyl phosphites in the presence of CuO (10 mol %) and L-proline (10 mol %) have been reacted to afford the product, but modest yield (35%) was achieved.[314] A study of enantioselective hydrogenation of isophorone reported stereoselective reduction when using proline-directed, Pd-catalyzed hydrogenation.[315] A reductive coupling of aldehydes with activated methylene compounds has also been reported in progress toward the total synthesis of chiloglottone analogs.[316]

---

1. (a) Coppola, G. M.; Schuster, H. F. *Asymmetric Synthesis: Construction of Chiral Molecules Using Amino Acids*; Wiley: New York, 1987; pp 267–345. (b) Drauz, K.; Kleemann, A.; Martens, J., *Angew. Chem., Int. Ed. Engl.* **1982**, *21*, 584. (c) Martens, J., *Top. Curr. Chem.* **1984**, *125*, 165.
2. Babidge, P. J.; Massy-Westropp, R. A.; Pyne, S. G.; Shiengthong, D.; Ungphakorn, A.; Veerachat, G., *Angew. Chem., Int. Ed. Engl.* **1980**, *33*, 1841.
3. (a) Ito, A.; Takahashi, R.; Baba, Y., *Chem. Pharm. Bull.* **1975**, *23*, 3081. (b) Overman, L. E.; Bell, K. L., *J. Am. Chem. Soc.* **1981**, *103*, 1851. (c) Overman, L. E.; Bell, K. L., Ito, F., *J. Am. Chem. Soc.* **1984**, *106*, 4192.
4. Mulzer, J.; Shanyoor, M., *Tetrahedron Lett.* **1993**, *34*, 6545.
5. (a) Berlokon, Y. N.; Zeltzer, I. E.; Ryzhov, M. G.; Saporovskaya, M. B.; Bakhmutov, V. I.; Belikov, V. M., *Chem. Commun./J. Chem. Soc., Chem. Commun.* **1982**, 180. (b) Belokon, Y. N.; Bulychev, A. G.; Vitt, S. V.; Struchkov, Y. T.; Batsanov, A. S.; Timofeeva, T. V.; Tsyryapkin, V. A.; Ryzhov, M. G.; Lysova, L. A.; Bakhmutov, V. I.; Belikov, V. M., *J. Am. Chem. Soc.* **1985**, *107*, 4252.
6. (a) Kim, D. H., *J. Heterocycl. Chem.* **1980**, *17*, 1647. (b) Cushman, D. W.; Cheung, H. S.; Sabo, E. F.; Ondetti, M. A., *Biochemistry* **1977**, *16*, 5484. (c) Suh, J. T.; Skiles, J. W.; Williams, B. E.; Youssefyeh, R. D.; Jones, H.; Loev, B.; Neiss, E. S.; Schwab, A.; Mann, W. S.; Khandwala, A.; Wolf, P. S.; Weinryb, I., *J. Med. Chem.* **1985**, *28*, 57.
7. Eder, U.; Sauer, G.; Wiechert, R., *Angew. Chem., Int. Ed. Engl.* **1971**, *10*, 496.
8. Hajos, Z. G.; Parrish, D. R., *J. Org. Chem.* **1974**, *39*, 1615.
9. (a) Nagasawa, K.; Hiroi, K.; Yamada, S., *Yakugaku Zasshi* **1975**, *95*, 46. (b) Micheli, R. A.; Hajos, Z. G.; Cohen, N.; Parrish, D. R.; Portland, L. A.; Sciamanna, W.; Scott, M. A.; Wehrli, P. A., *J. Org. Chem.* **1975**, *40*, 675. (c) Cohen, N.; Banner, B. L.; Eichel, W. F.; Parrish, D. R.; Saucy, G.; Cassal, J.-M.; Meier, W.; Fürst, A., *J. Org. Chem.* **1975**, *40*, 681. (d) Eder, U.; Sauer, G.; Haffer, G.; Ruppert, J.; Wiechert, R.; Fürst, A.; Meier, W., *Helv. Chim. Acta* **1976**, *59*, 999. (e) Eder, U.; Gibian, H.; Haffer, G.; Neef, G.; Sauer, G.; Wiechert, R., *Ber. Dtsch. Chem. Ges./Chem. Ber.* **1976**, *109*, 2948.
10. Gutzwiller, J.; Buchschacher, P.; Fürst, A., *Synthesis* **1977**, 167.
11. Takano, S.; Kasahara, C.; Ogasawara, *Chem. Commun./J. Chem. Soc., Chem. Commun.* **1981**, 635.
12. Woodward, R. B.; Logusch, E.; Nambiar, K. P.; Sakan, K.; Ward, D. E.; Au-Yeung, B.-W.; Balaram, P.; Browne, L. J.; Card, P. J.; Chen, C. H.; Chênevert, R. B.; Fliri, A.; Frobel, K.; Gais, H.-J.; Garratt, D. G.; Hayakawa, K.; Heggie, W.; Hesson, D. P.; Hoppe, D.; Hoppe, I.; Hyatt, J. A.; Ikeda, D.; Jacobi, P. A.; Kim, K. S.; Kokube, Y.; Kojima, K.; Krowicki, K.; Lee, V. J.; Leutert, T.; Malchenko, S.; Martens, J.; Matthews, R. S.; Ong, B. S.; Press, J. B.; Rajan Babu, T. V.; Rousseau, G.; Sauter, H. M.; Suzuki, M.; Tatsuta, K.; Tolbert, L. M.; Truesdale, E. A.; Uchida, I.; Ueda, Y.; Uyehara, T.; Vasella, A. T.; Vladuchick, W. C.; Wade, P. A.; Williams, R. M.; Wong, H. N.-C., *J. Am. Chem. Soc.* **1981**, *103*, 3210, 3213, 3215.
13. Buchschacher, P.; Cassal, J.-M.; Fürst, A.; Meier, W., *Helv. Chim. Acta* **1977**, *60*, 2747.
14. Shimizu, I.; Naito, Y.; Tsuji, J., *Tetrahedron Lett.* **1980**, *21*, 487.
15. Danishefsky, S.; Cain, P., *J. Am. Chem. Soc.* **1976**, *98*, 4975.
16. Coisne, J.-M.; Pecher, J.; Declercq, J.-P.; Germain, G.; Van Meerssche, M., *Bull. Soc. Chim. Belg.* **1981**, *90*, 481.
17. Uma, R.; Rajagopalan, K.; Swaminathan, S., *Tetrahedron* **1986**, *42*, 2757.
18. Terashima, S.; Sato, S.; Koga, K., *Tetrahedron Lett.* **1979**, 3469.
19. Agami, C.; Platzer, N.; Sevestre, H., *Bull. Soc. Claim. Fr.* **1987**, 358.
20. (a) Agami, C., *Bull. Soc. Claim. Fr.* **1988** 499. (b) Agami, C.; Puchot, C., *J. Mol. Catal.* **1986**, *38*, 341. (c) Puchot, C.; Samuel, O.; Duñach, E.; Zhao, S.; Agami, C.; Kagan, H. B., *J. Am. Chem. Soc.* **1986**, *108*, 2353. (d) Agami, C.; Levisalles, J.; Puchot, C., *Chem. Commun./J. Chem. Soc., Chem. Commun.* **1985**, 441. (e) Agami, C.; Sevestre, H., *Chem. Commun./J. Chem. Soc., Chem. Commun.* **1984**, 1385. (f) Agami, C.; Puchot, C.; Sevestre, H., *Tetrahedron Lett.* **1986**, *27*, 1501.
21. Agami, C.; Levisalles, J.; Sevestre, H., *Chem. Commun./J. Chem. Soc., Chem. Commun.* **1984**, 418.
22. Agami, C.; Puchot, C., *Tetrahedron* **1986**, *42*, 2037.
23. Kozikowski, A. P.; Mugrage, B. B., *J. Org. Chem.* **1989**, *54*, 2274.
24. Hirai, Y.; Terada, T.; Yamazaki, T.; Momose, T., *J. Chem. Soc., Perkin Trans. 1* **1992**, 509.
25. Yamaguchi, M.; Yokota, N.; Minami, T., *Chem. Commun./J. Chem. Soc., Chem. Commun.* **1991**, 1088.
26. (a) Terashima, S.; Jew, S., *Tetrahedron Lett.* **1977**, 1005. (b) Jew, S.-S.; Terashima, S.; Koga, K., *Tetrahedron* **1979**, *35*, 2337. (c) Jew, S.-S.; Terashima, S.; Koga, K., *Tetrahedron* **1979**, *35*, 2345. (d) Terashima, S.; Hayashi, M.; Koga, K., *Tetrahedron Lett.* **1980**, 2733. (e) Hayashi, M.; Terashima, S.; Koga, K., *Tetrahedron* **1981**, *37*, 2797. (f) Terashima, S.; Jew, S.; Koga, K., *Chem. Lett.* **1977**, 1109. (g) Terashima, S.; Jew, S.; Koga, K., *Tetrahedron Lett.* **1977**, 4507. (h) Jew, S.; Terashima, S.; Koga, K., *Chem. Pharm. Bull.* **1979**, *27*, 2351. (i) Terashima, S.; Jew, S.; Koga, K., *Tetrahedron Lett.* **1978**, 4937. (j) Rüeger, H.; Benn, M., *Heterocycles* **1982**, *19*, 23.

27. (a) Wallbaum, S.; Martens, J., *Tetrahedron: Asymmetry* **1992**, *3*, 1475. (b) Mehler, T.; Behnen, W.; Wilken, J.; Martens, J., *Tetrahedron: Asymmetry* **1994**, *5*, 185.
28. Umino, N.; Iwakuma, T.; Itoh, N., *Chem. Pharm. Bull.* **1979**, *27*, 1479.
29. (a) Yamada, K.; Takeda, M.; Iwakuma, T., *Tetrahedron Lett.* **1981**, *22*, 3869. (b) Yamada, K.; Takeda, M.; Ohtsuka, H.; Iwakuma, T., *Chem. Pharm. Bull.* **1983**, *31*, 70. (c) Yamada, K.; Takeda, M.; Iwakuma, T., *J. Chem. Soc., Perkin Trans. 1* **1983**, 265.
30. (a) Tungler, A.; Kajtar, M.; Mathé, T.; Toth, G.; Fogassy, E.; Petró, J., *Catal. Today* **1989**, *5*, 159. (b) Tungler, A.; Tarnai, T.; Máthé, T.; Petró, J., *J. Mol. Catal.* **1991**, *67*, 277.
31. Tungler, A.; Máthé, T.; Petró, J.; Tarnai, T., *J. Mol. Catal.* **1990**, *61*, 259.
32. Tungler, A.; Tarnai, T.; Máthé, T.; Petró, J., *J. Mol. Catal.* **1991**, *70*, L5.
33. (a) Bycroft, B. W.; Lee, G. R., *Chem. Commun./J. Chem. Soc., Chem. Commun.* **1975**, 988. (b) Poisel, H.; Schmidt, U., *Ber. Dtsch. Chem. Ges./Chem. Ber.* **1973**, *106*, 3408.
34. (a) Shibata, S.; Matsushita, H.; Noguchi, M.; Saburi, M.; Yoshikawa, S., *Chem. Lett.* **1978**, 1305. (b) Shibata, S.; Matsushita, H.; Kato, K.; Noguchi, M.; Saburi, M.; Yoshikawa, S., *Bull. Chem. Soc. Jpn.* **1979**, *52*, 2938.
35. Shiraiwa, T.; Yamauchi, M.; Yamamoto, Y.; Kurokawa, H., *Bull. Chem. Soc. Jpn.* **1990**, *63*, 3296.
36. Shiraiwa, T.; Yamauchi, M.; Tatsumi, T.; Kurokawa, H., *Bull. Chem. Soc. Jpn.* **1992**, *65*, 267.
37. Seebach, D.; Naef, R., *Helv. Chim. Acta* **1981**, *64*, 2704.
38. Seebach, D.; Boes, M.; Naef, R.; Schweizer, W. B., *J. Am. Chem. Soc.* **1983**, *105*, 5390.
39. Rüeger, H.; Benn, M., *Heterocycles* **1982**, *19*, 1677.
40. (a) Koizumi, T.; Kobayashi, Y.; Amitani, H.; Yosshii, E., *J. Org. Chem.* **1977**, *42*, 3459. (b) Koizumi, T.; Amitani, H.; Yoshii, E., *Tetrahedron Lett.* **1978**, 3741. (c) Koizumi, T.; Amitani, H.; Yoshii, E., *Synthesis* **1979**, 110. (d) Koizumi, T.; Takagi, H.; Yoshii, E., *Chem. Lett.* **1980**, 1403. (e) Koizumi, T.; Yanada, R.; Takagi, H.; Hirai, H.; Yoshii, E., *Tetrahedron Lett.* **1981**, 477 and 571.
41. (a) Yamada, S.; Hiroi, K.; Achiwa, K., *Tetrahedron Lett.* **1969**, 4233. (b) Hiroi, K.; Achiwa, K.; Yamada, S.-I., *Chem. Pharm. Bull.* **1972**, *20*, 246. (c) Hiroi, K.; Yamada, S.-I., *Chem. Pharm. Bull.* **1973**, *21*, 47. (d) Hiroi, K.; Yamada, S.-I., *Chem. Pharm. Bull.* **1973**, *21*, 54.
42. Castonguay, A.; Berger Y., *Chem. Commun./J. Chem. Soc., Chem. Commun.* **1978**, 951.
43. Castonguay, A.; Berger Y., *Tetrahedron* **1979**, *35*, 1557.
44. Castonguay, A.; Berger Y., *Angew. Chem., Int. Ed. Engl.* **1979**, *32*, 2681.
45. Kira, M.; Sato, K.; Sakurai H., *J. Am. Chem. Soc.* **1990**, *112*, 257.
46. Muzart, J., *J. Mol. Catal.* **1991**, *64*, 381.
47. Hiroi, K. Abe J., *Heterocycles* **1990**, *30*, 283.
48. (a) Hiroi, K.; Suya, K.; Sato, S., *Chem. Commun./J. Chem. Soc., Chem. Commun.* **1986**, 469. (b) Hiroi, K.; Abe, J.; Suya, K.; Sato, S., *Tetrahedron Lett.* **1989**, *30*, 1543.
49. Hiroi, K.; Maezuru, K.; Kimura, M.; Ito, N., *Chem. Lett.* **1989**, 1751.
50. Seayad, J.; List, B., *Org. Biomol. Chem.* **2005**, *3*, 719.
51. Gröger, H.; Wilken, J., *Angew. Chem., Int. Ed. Engl.* **2001**, *40*, 529.
52. List, B., *Tetrahedron* **2002**, *58*, 5573.
53. List, B., *Acc. Chem. Res.* **2004**, *37*, 548.
54. Allemann, C.; Gordillo, R.; Clemente, F. R.; Cheong, P. H.; Houk, K. N., *Acc. Chem. Res.* **2004**, *37*, 558.
55. Notz, W.; Tanaka, F.; Barbas III, C. F., *Acc. Chem. Res.* **2004**, *37*, 580.
56. Hoang, L.; Bahmanyar, S.; Houk, K. N.; List, B., *J. Am. Chem. Soc.* **2003**, *125*, 16.
57. Rajagopal, D.; Moni, M. S.; Subramanian, S.; Swaminathan, S., *Tetrahedron: Asymmetry* **1999**, *10*, 1631.
58. Bui, T.; Barbas, III, C. F., *Tetrahedron Lett.* **2000**, *41*, 6951.
59. Díaz, S.; González, A.; Bradshaw, B.; Cuesta, J.; Bonjoch, J., *J. Org. Chem.* **2005**, *70*, 3749.
60. Rajagopal, D.; Rajagopalan, K.; Swaminathan, S., *Tetrahedron: Asymmetry* **1996**, *7*, 2189.
61. Rajagopal, D.; Narayanan, K.; Swaminathan, S., *Tetrahedron Lett.* **2001**, *42*, 4887.
62. Danishefsky, S. J.; Masters, J. J.; Young, W. B.; Link, J. T.; Snyder, L. B.; Magee, T. V.; Jung, D. K.; Isaacs, R. C. A.; Bornmann, W. G.; Alaimo, C. A.; Coburn, C. A.; Di Grandi, M. J., *J. Am. Chem. Soc.* **1996**, *118*, 2843.
63. Przezdźiecka, A.; Stepanenko, W.; Wicha, J., *Tetrahedron: Asymmetry* **1999**, *10*, 1589.
64. Li, F.; Warshakoon, N. C.; Miller, M. J., *J. Org. Chem.* **2004**, *69*, 8836.
65. Avery, M. A.; Fan, P.; Karle, J. M.; Bonk, J. D.; Miller, R.; Goins, D. K., *J. Med. Chem.* **1996**, *39*, 1885.
66. Inomata, K.; Barragué, M.; Paquette, L. A., *J. Org. Chem.* **2005**, *70*, 533.
67. Pidathala, C.; Hoang, L.; Vignola, N.; List, B, *Angew. Chem., Int. Ed. Engl.* **2003**, *42*, 2785.
68. Mans, D. M.; Pearson, W. H., *Org. Lett.* **2004**, *6*, 3305.
69. Sunazuka, T.; Handa, M.; Nagai, K.; Shirahata, T.; Harigaya, Y.; Otoguro, K.; Kuwajima, I.; Omura, S., *Tetrahedron* **2004**, *60*, 7845.
70. Sagar, R.; Singh, P.; Kumar, R.; Maulik, P. R.; Shaw, A. K., *Carbohydr. Res.* **2005**, *340*, 1287.
71. Bahmanyar, S.; Houk, K. N.; Martin, H. J.; List, B., *J. Am. Chem. Soc.* **2003**, *125*, 2475.
72. Saito, S.; Yamamoto, H., *Acc. Chem. Res.* **2004**, *37*, 570.
73. List, B.; Lerner, R. A.; Barbas, III, C. F., *J. Am. Chem. Soc.* **2000**, *122*, 2395.
74. Sakthivel, K.; Notz, W.; Bui, T.; Barbas III, C. F., *J. Am. Chem. Soc.* **2001**, *123*, 5260.
75. Nyberg, A. I.; Usano, A.; Pihko, P. M., *Synlett* **2004**, 1891.
76. Córdova, A.; Notz, W.; Barbas III, C. F., *Chem. Commun./J. Chem. Soc., Chem. Commun.* **2002**, 3024.
77. Peng, Y.-Y.; Ding, Q.-P.; Li, Z.; Wang, P. G.; Cheng, J.-P., *Tetrahedron Lett.* **2003**, *44*, 3871.
78. Hayashi, Y.; Tsuboi, W.; Shoji, M.; Suzuki, N., *Tetrahedron Lett.* **2004**, *45*, 4353.
79. Chandrasekhar, S.; Narsihmulu, C.; Reddy, N. R.; Sultana, S. S., *Tetrahedron Lett.* **2004**, *45*, 4581.
80. Gruttadauria, M.; Riela, S.; Lo Meo, P.; D'Anna, F.; Noto, R., *Tetrahedron Lett.* **2004**, *45*, 6113.
81. Edin, M.; Bäckvall, J.-E.; Córdova, A., *Tetrahedron Lett.* **2004**, *45*, 7697.
82. List, B.; Pojarliev, P.; Castello, C., *Org. Lett.* **2001**, *3*, 573.
83. Sun, B.; Peng, L.; Chen, X.; Li, Y.; Li, Y.; Yamasaki, K., *Tetrahedron: Asymmetry* **2005**, *16*, 1305.
84. Chan, V.; Kim, J. G.; Jimeno, C.; Carroll, P. J.; Walsh, P. J., *Org. Lett.* **2004**, *6*, 2051.
85. Izquierdo, I.; Plaza, M. T.; Robles, R.; Mota, A. J.; Franco, F., *Tetrahedron: Asymmetry* **2001**, *12*, 2749.
86. Chandrasekhar, S.; Narsihmulu, C.; Reddy, N. R.; Sultana, S. S., *Chem. Commun./J. Chem. Soc., Chem. Commun.* **2004**, 2450.
87. Notz, W.; List, B., *J. Am. Chem. Soc.* **2000**, *122*, 7386.
88. Pan, Q.; Zou, B.; Wang, Y.; Ma, D., *Org. Lett.* **2004**, *6*, 1009.
89. Liu, H.; Peng, L.; Zhang, T.; Li, Y., *New. J. Chim.* **2003**, *27*, 1159.
90. Peng, L.; Liu, H.; Zhang, T.; Zhang, F.; Mei, T.; Li, Y.; Li, Y., *Tetrahedron Lett.* **2003**, *44*, 5107.
91. Suri, J. T.; Ramachary, D. B.; Barbas, III, C. F., *Org. Lett.* **2005**, *7*, 1383.
92. Kofoed, J.; Reymond, J.-L.; Darbre, T., *Org. Biomol. Chem.* **2005**, *3*, 1850.
93. Northrup, A. B.; MacMillan, D. W. C., *Science* **2004**, *305*, 1752.

94. Ward, D. E.; Jheengut, V.; Akinnusi, O. T., *Org. Lett.* **2005**, *7*, 1181.
95. Northrup, A. B.; MacMillan, D. W. C., *J. Am. Chem. Soc.* **2002**, *124*, 6798.
96. Córdova, A., *Tetrahedron Lett.* **2004**, *45*, 3949.
97. Córdova, A.; Notz, W.; Barbas III, C. F., *J. Org. Chem.* **2002**, 301.
98. Chowdari, N. S.; Ramachary, D. B.; Córdova, A.; Barbas III, C. F., *Tetrahedron Lett.* **2002**, *43*, 9591.
99. Pihko, P. M.; Erkkilä, A., *Tetrahedron Lett.* **2003**, *44*, 7607.
100. Northrup, A. B.; Mangion, I. K.; Hettche, F.; MacMillan, D. W. C., *Angew. Chem., Int. Ed. Engl.* **2004**, *43*, 2152.
101. Thayumanavan, R.; Tanaka, F.; Barbas III, C. F., *Org. Lett.* **2004**, *6*, 3541.
102. Storer, R. I.; MacMillan, D. W. C., *Tetrahedron* **2004**, *60*, 7705.
103. Bøgevig, A.; Kumaragurubaran, N.; Jorgensen, K. A., *Chem. Commun./J. Chem. Soc., Chem. Commun.* **2002**, 620.
104. Bøgevig, A.; Poulsen, T. B.; Zhuang, W.; Jørgensen, K. A., *Synlett* **2003**, 1915.
105. Arnó, M.; Zaragóza,, R. J.; Domingo, L. R., *Tetrahedron: Asymmetry* **2004**, *15*, 1541.
106. Cardillo, G.; Fabbroni, S.; Gentilucci, L.; Gianotti, M.; Tolomelli, A., *Synth. Commun.* **2003**, *33*, 1587.
107. List, B.; Pojarliev, P.; Biller, W. T.; Martin, H. J., *J. Am. Chem. Soc.* **2002**, *124*, 827.
108. Bahmanyar, S.; Houk, K. N., *Org. Lett.* **2003**, *5*, 1249.
109. Chowdari, N. S.; Ramachary, D. B.; Barbas, III, C. F., *Synlett* **2003**, 1906.
110. Pojarliev, P.; Biller, W. T.; Martin, H. J.; List, B., *Synlett* **2003**, 1903.
111. Itoh, T.; Yokoya, M.; Miyauchi, K.; Nagata, K.; Ohsawa, A., *Org. Lett.* **2003**, *5*, 4301.
112. Funabiki, K.; Nagamori, M.; Goushi, S.; Matsui, M., *Chem. Commun./J. Chem. Soc., Chem. Commun.* **2004**, 1928.
113. Córdova, A., *Chem. Eur. J.* **2004**, *10*, 1987.
114. Hayashi, Y.; Tsuboi, W.; Ashimine, I.; Urushima, T.; Shoji, M.; Sakai, K., *Angew. Chem., Int. Ed. Engl.* **2003**, *42*, 3677.
115. Córdova, A., *Synlett* **2003**, 1652.
116. Hayashi, Y.; Tsuboi, W.; Shoji, M.; Suzuki, N., *J. Am. Chem. Soc.* **2003**, *125*, 11208.
117. Notz, W.; Tanaka, F.; Watanabe, S.-I.; Chowdari, N. S.; Turner, J. M.; Thayumanavan, R.; Barbas III, C. F., *J. Org. Chem.* **2003**, *68*, 9624.
118. Chowdari, N. S.; Suri, J. T.; Barbas III, C. F., *Org. Lett.* **2004**, *6*, 2507.
119. Córdova, A.; Barbas III, C. F., *Tetrahedron Lett.* **2003**, *44*, 1923.
120. Watanabe, S.-I.; Córdova,, A.; Tanaka, F.; Barbas III, C. F., *Org. Lett.* **2002**, *4*, 4519.
121. Ibrahem, I.; Samec, J. S.; Bäckvall, J. E.; Córdova, A., *Tetrahedron Lett.* **2005**, *46*, 3965.
122. Yadav, J. S.; Kumar, S. P.; Kondaji, G.; Rao, R. S.; Nagaiah, K., *CL* **2004**, *33*, 1168.
123. Ballini, R.; Bosica, G.; Fiorini, D.; Palmieri, A.; Petrini, M., *Chem. Rev.* **2005**, *105*, 933.
124. Yamaguchi, M.; Igarashi, Y.; Reddy, R. S.; Shiraishi, T.; Hirama, M., *Tetrahedron* **1997**, *53*, 11223.
125. Hannessian, S.; Pham, V., *Org. Lett.* **2000**, *2*, 2975.
126. List, B.; Pojarliev, P.; Martin, H. J.; *Org. Lett.* **2001**, *3*, 2423.
127. Betancort, J. M.; Barbas III, C. F., *Org. Lett.* **2001**, 3737.
128. Enders, D.; Seki, A., *Synlett* **2002**, 26.
129. Kotrusz, P.; Toma, S.; Schmalz, H.-G.; Adler, A., *EJOC* **2004**, 1577.
130. Mangion, I. K.; MacMillan, D. W. C., *J. Am. Chem. Soc.* **2005**, *127*, 3696.
131. Yamaguchi, M.; Shiraishi, T.; Hirama, M., *J. Org. Chem.* **1996**, *61*, 3520.
132. Gryko, D., *Tetrahedron: Asymmetry* **2005**, *16*, 2377.
133. Ouellet, S. G.; Tuttle, J. B.; MacMillan, D. W. C., *J. Am. Chem. Soc.* **2005**, *127*, 32.
134. Schoevaart, R.; Kieboom, T., *Tetrahedron Lett.* **2002**, *43*, 3399.
135. Brown, S. P.; Brochu, M. P.; Sinz, C. J.; MacMillan, D. W. C., *J. Am. Chem. Soc.* **2003**, *125*, 10808.
136. Zhong, G., *Angew. Chem., Int. Ed. Engl.* **2003**, *42*, 4247.
137. Cheong, P. H.-Y.; Houk, K. N., *J. Am. Chem. Soc.* **2004**, *126*, 13912.
138. Mathew, S. P.; Iwamura, H.; Blackmond, D. G., *Angew. Chem., Int. Ed. Engl.* **2004**, *43*, 3317.
139. Sundén, H.; Dahlin, N.; Ibrahem, I.; Adolfsson, H.; Córdova, A., *Tetrahedron Lett.* **2005**, *46*, 3385.
140. Zhong, G.; Yu, Y., *Org. Lett.* **2004**, *6*, 1637.
141. Zhong, G, *Chem. Commun./J. Chem. Soc., Chem. Commun.* **2004**, 606.
142. Córdova, A.; Sundén, H.; Engqvist, M.; Ibrahem, I.; Casas, J., *J. Am. Chem. Soc.* **2004**, *126*, 8914.
143. Engqvist, M.; Casas, J.; Sundén, H.; Ibrahem, I.; Córdova,, A., *Tetrahedron Lett.* **2005**, *46*, 2053.
144. Bøgevig, A.; Juhl, K.; Kumaragurubaran, N.; Zhuang, W.; Jorgensen, K. A., *Angew. Chem., Int. Ed. Engl.* **2002**, *41*, 1790.
145. List, B., *J. Am. Chem. Soc.* **2002**, *124*, 5656.
146. Kumaragurubaran, N.; Juhl, K.; Zhuang, W.; Bøgevig, A.; Jørgensen, K. A., *J. Am. Chem. Soc.* **2002**, *124*, 6254.
147. Iwamura, H.; Wells, D. H.; Mathew, S. P.; Klussmann, M.; Armstrong, A.; Blackmond, D. G., *J. Am. Chem. Soc.* **2004**, *126*, 16312.
148. Henmi, Y.; Makino, K.; Yoshitomi, Y.; Hara, O.; Hamada, Y., *Tetrahedron: Asymmetry* **2004**, *15*, 3477.
149. Chowdari, N. S.; Ramachary, D. B.; Barbas III, C. F., *Org. Lett.* **2003**, *5*, 1685.
150. Brochu, M. P.; Brown, S. P.; MacMillan, D. W. C., *J. Am. Chem. Soc.* **2004**, *126*, 4108.
151. Halland, N.; Braunton, A.; Bachmann, S.; Marigo, M.; Jorgensen, K. A., *J. Am. Chem. Soc.* **2004**, *126*, 4790.
152. Enders, D.; Hüttl, R. M., *Synlett* **2005**, 991.
153. Thayumanavan, R.; Dhevalapally, B.; Sakthivel, K.; Tanaka, F.; Barbas III, C. F., *Tetrahedron Lett.* **2002**, *43*, 3817.
154. Ramachary, D. B.; Chowdari, N. S.; Barbas III, C. F., *Tetrahedron Lett.* **2002**, *43*, 6743.
155. Ramachary, D. B.; Chowdari, N. S.; Barbas III, C. F., *Angew. Chem., Int. Ed. Engl.* **2003**, *42*, 4233.
156. Ramachary, D. B.; Anebouselvy, K.; Chowdari, N. S.; Barbas III, C. F., *J. Org. Chem.* **2004**, *69*, 5838.
157. List, B.; Castello, C., *Synlett* **2001**, 1687.
158. Gerasyuto, A.; Hsung, R. P.; Sydorenko, N.; Slafer, B., *J. Org. Chem.* **2005**, *70*, 4248.
159. Shi, M.; Jiang, J.-K.; Li, C.-Q., *Tetrahedron Lett.* **2002**, *43*, 127.
160. Imbriglio, J. E.; Vasbinder, M. M.; Miller, S. J.; *Org. Lett.* **2003**, *5*, 3741.
161. Vignola, N.; List, B., *J. Am. Chem. Soc.* **2004**, *126*, 450.
162. Yanagisawa, A.; Nakamura, Y.; Arai, T., *Tetrahedron: Asymmetry* **2004**, *15*, 1909.
163. Levina, A.; Muzart, J., *Tetrahedron: Asymmetry* **1995**, *6*, 147.
164. Le Bras, J.; Muzart, J., *Tetrahedron: Asymmetry* **2003**, *14*, 1911.
165. Whitman, D. B.; Askew, B. C.; Duong, L. T.; Fernandez-Metzler, C.; Halczenko, W.; Hartman, G. D.; Hutchinson, J. H.; Leu, C.-T.; Prueksaritanont, T.; Rodan, G. E.; Rodan, S. B.; Duggan, M. E., *Bioorg. Med. Chem. Lett.* **2004**, *14*, 4411.
166. Ma, D.; Cai, Q.; Zhang, H., *Org. Lett.* **2003**, *5*, 2453.
167. Fletcher, S. P.; Clive, D. L. J.; Peng, J.; Wingert, D. A., *Org. Lett.* **2005**, *7*, 23.
168. Ma, D.; Cai, Q., *Synlett* **2004**, 128.
169. Zhu, W.; Ma, D., *Chem. Commun./J. Chem. Soc., Chem. Commun.* **2004**, 888.

170. Zhu, W.; Ma, D., *J. Org. Chem.* **2005**, *70*, 2696.
171. Han, Z.-J.; Wang, R.; Zhou, Y.-F.; Liu, L., *Eur. J. Org. Chem.* **2005**, 934.
172. Basak, A.; Ghosh, S. C., *Synlett* **2004**, 1637.
173. Fache, F.; Schulz, E.; Tommasino, M. L.; Lemaire, M., *Chem. Rev.* **2000**, *100*, 2159.
174. Davies, H. M. L., *EJOC* **1999**, 2459.
175. Davies, H. M. L.; Beckwith, R. E. J., *Chem. Rev.* **2003**, *103*, 2861.
176. Shi, L.; Tu, Y.-Q.; Wang, M.; Zhang, F.-M.; Fan, C.-A., *Org. Lett.* **2004**, *6*, 1001.
177. Kim, B. H.; Lee, H. B.; Hwang, J. K.; Kim, Y. G., *Tetrahedron: Asymmetry* **2005**, *16*, 1215.
178. Bläser,, E.; Kolar, P.; Fenske, D.; Goesmann, H.; Waldmann, H., *EJOC* **1999**, 329.
179. Gray, D.; Concellón,, C.; Gallagher, T., *J. Org. Chem.* **2004**, *69*, 4849.
180. Barrett, A. G. M.; Blaney, F.; Campbell, A. D.; Hamprecht, D.; Meyer, T.; White, A. J.; P.; Witty, D.; Williams, D. J., *J. Org. Chem.* **2002**, *67*, 2735.
181. Miller, S. J., *Acc. Chem. Res.* **2004**, *37*, 601.
182. Wang, W.; Wang, J.; Li, H., *Org. Lett.* **2004**, *6*, 2817.
183. Tang, Z.; Yang, Z.-H.; Cun, L.-F.; Gong, L.-Z.; Mi, A.-Q.; Jiang, Y.-Z., *Org. Lett.* **2004**, *6*, 2285.
184. Mikami, K.; Yamanaka, M., *Chem. Rev.* **2003**, *103*, 3369.
185. Nandy, J. P.; Prabhakaran, E. N.; Kumar, S. K.; Kunwar, A. C.; Iqbal, J., *J. Org. Chem.* **2003**, *68*, 1679.
186. France, S.; Guerin, D. J.; Miller, S. J.; Lectka, T., *Chem. Rev.* **2003**, *103*, 2985.
187. Xie, L.; Jones, G. B., *Tetrahedron Lett.* **2005**, *46*, 3579.
188. Bloch, R., *Chem. Rev.* **1998**, *98*, 1407.
189. Trost, B. M.; Silcoff, E. R.; Ito, H., *Org. Lett.* **2001**, *3*, 2497.
190. Xavier, L. C.; Mohan, J. J.; Mathre, D. J.; Thompson, A. S.; Carroll, J. D.; Corley, E. G.; Desmond, R., *Org. Synth.* **1997**, *74*, 50.
191. Corey, E. J.; Shibata, T.; Lee, T. W., *J. Am. Chem. Soc.* **2002**, *124*, 3808.
192. Periasamy, M.; Kumar, N. S.; Sivakumar, S.; Rao, V. D.; Ramanathan, C. R.; Venkatraman, L., *J. Org. Chem.* **2001**, *66*, 3828.
193. Ho, C.-Y.; Chen, Y.-C.; Wong, M.-K.; Yang, D., *J. Org. Chem.* **2005**, *70*, 898.
194. Calter, M.; Hollis, T. K.; Overman, L. E.; Ziller, J.; Zipp, G. G., *J. Org. Chem.* **1997**, *62*, 1449.
195. Nakajima, M.; Miyoshi, I.; Kanayama, K.; Hashimoto, S.-I.; Noji, M.; Koga, K., *J. Org. Chem.* **1999**, *64*, 2264.
196. Terakado, D.; Koutaka, H.; Oriyama, T., *Tetrahedron: Asymmetry* **2005**, *16*, 1157.
197. Cobb, A. J. A.; Shaw, D. M.; Longbottom, D. A.; Gold, J. B.; Ley, S. V., *Org. Biomol. Chem.* **2005**, *3*, 84.
198. Chowdari, N. S.; Barbas III, C. F., *Org. Lett.* **2005**, *7*, 867.
199. McManus, H. A.; Guiry, P. J., *Chem. Rev.* **2004**, *104*, 4151.
200. Benito, J. M.; Christensen, C. A.; Meldal, M. *Org. Lett.* **2005**, *7*, 581.
201. Moonen, K.; Laureyn, I.; Stevens, C. V., *Chem. Rev.* **2004**, *104*, 6177.
202. Iseki, K.; Kuroki, Y.; Takahashi, M.; Kishimoto, S.; Kobayashi, Y., *Tetrahedron* **1997**, *53*, 3513.
203. Ramanathan, C. R.; Periasamy, M., *Tetrahedron: Asymmetry* **1998**, *9*, 2651.
204. Shan, Z.; Xiong, Y.; Zhao, D., *Tetrahedron* **1999**, *55*, 3893.
205. Enders, D.; Narine, A. A., *J. Org. Chem.* **2008**, *73*, 7857.
206. Nagamine, T.; Inomata, K.; Endo, Y.; Paquette, L. A., *J. Org. Chem.* **2007**, *72*, 123.
207. Chandler, C. L.; List, B., *J. Am. Chem. Soc.* **2008**, *130*, 6737.
208. Hanessian, S.; Schroeder, B. R.; Merner, B. L.; Chen, B.; Swayze, E. E.; Seth, P. P., *J. Org. Chem.* **2013**, *78*, 9051.
209. Kennedy, J. W. J.; Vietrich, S.; Weinmann, H.; Brittain, D. E. A., *J. Org. Chem.* **2008**, *73*, 5151.
210. Gonzalez, L.; Escorihuela, J.; Altava, B.; Burguete, M. I.; Luis, S. V., *Eur. J. Org. Chem.* **2014**, 5356.
211. Duarte, F. J. S.; Cabrita, E. J.; Frenking, G.; Santos, A. G., *Eur. J. Org. Chem.* **2008**, 3397.
212. Alcaide, B.; Almendros, P.; Luna, A.; Torres, M. R., *J. Org. Chem.* **2006**, *71*, 4818.
213. Mase, N.; Nakai, Y.; Ohara, N.; Yoda, H.; Takabe, K.; Tanaka, F.; Barbas, C. F., III, *J. Am. Chem. Soc.* **2006**, *128*, 734.
214. Samanta, S.; Zhao, C.-G., *J. Am. Chem. Soc.* **2006**, *128*, 7442.
215. Calderon, F.; Doyagueez, E. G.; Fernandez-Mayoralas, A., *J. Org. Chem.* **2006**, *71*, 6258.
216. Suri, J. T.; Mitsumori, S.; Albertshofer, K.; Tanaka, F.; Barbas, C. F., III, *J. Org. Chem.* **2006**, *71*, 3822.
217. Palyam, N.; Majewski, M., *J. Org. Chem.* **2009**, *74*, 4390.
218. Niewczas, I.; Majewski, M., *Eur. J. Org. Chem.* **2009**, 33.
219. Fanton, J.; Camps, F.; Castillo, J. A.; Guerard-Helaine, C.; Lemaire, M.; Charmantray, F.; Hecquet, L., *Eur. J. Org. Chem.* **2012**, 203.
220. Enders, D.; David, S.; Deckers, K.; Greb, A.; Raabe, G., *Synthesis* **2012**, *44*, 3483.
221. Nicolas, C.; Pluta, R.; Pasternak-Suder, M.; Martin, O. R.; Mlynarski, J., *Eur. J. Org. Chem.* **2013**, 1296.
222. Secci, F.; Frongia, A.; Rubanu, M. G.; Sechi, M. L.; Sarais, G.; Arca, M.; Piras, P. P., *Eur. J. Org. Chem.* **2014**, 6659.
223. Hajra, S.; Giri, A. K., *J. Org. Chem.* **2008**, *73*, 3935.
224. Hajra, S.; Giri, A. K.; Hazra, S., *J. Org. Chem.* **2009**, *74*, 7978.
225. Kano, T.; Sugimoto, H.; Maruoka, K., *J. Am. Chem. Soc.* **2011**, *133*, 18130.
226. Volchkov, I.; Lee, D., *J. Am. Chem. Soc.* **2013**, *135*, 5324.
227. Martinez- Castaneda, A.; Rodriguez-Solla, H.; Concellon, C.; del Amo, V., *J. Org. Chem.* **2012**, *77*, 10375.
228. Zhou, Y.; Shan, Z., *J. Org. Chem.* **2006**, *71*, 9510.
229. Abele, S.; Inauen, R.; Spielvogel, D.; Moessner, C., *J. Org. Chem.* **2012**, *77*, 4765.
230. Duarte, F. J. S.; Cabrita, E. J.; Frenking, G.; Santos, A. G., *J. Org. Chem.* **2010**, *75*, 2546.
231. (a) Sharma, A. K.; Sunoj, R. B., *Angew. Chem., Int. Ed.* **2010**, *49*, 6373. (b) Fu, A.; Li, H.; Yuan, S.; Si, H.; Duan, Y., *J. Org. Chem.* **2008**, *73*, 5264.
232. Calderon, F.; Doyaguez, E. G.; Cheong, P. H.-Y.; Fernandez-Mayoralas, A.; Houk, K. N., *J. Org. Chem.* **2008**, *73*, 7916.
233. Schmid, M. B.; Zeitler, K.; Gschwind, R. M., *Angew. Chem., Int. Ed.* **2010**, *49*, 4997.
234. Zhu, H.; Clemente, F. R.; Houk, K. N.; Meyer, M. P., *J. Am. Chem. Soc.* **2009**, *131*, 1632.
235. Kucherenko, A. S.; Struchkova, M. I.; Zlotin, S. G., *Eur. J. Org. Chem.* **2006**, 2000.
236. Wang, Z.; Yan, J.; Zhang, X.; Wang, L., *Synthesis* **2009**, 3744.
237. Doyagueez, E. G.; Calderon, F.; Sanchez, F.; Fernandez-Mayoralas, A., *J. Org. Chem.* **2007**, *72*, 9353.
238. Tang, Z.; Marx, A., *Angew. Chem., Int. Ed.* **2007**, *46*, 7297.
239. Chi, Y.; Scroggins, S. T.; Boz, E.; Frechet, J. M. J., *J. Am. Chem. Soc.* **2008**, *130*, 17287.
240. Wei, H.-L.; Yan, Z.-Y.; Niu, Y.-N.; Li, G.-Q.; Liang, Y.-M., *J. Org. Chem.* **2007**, *72*, 8600.
241. Chandler, C.; Galzerano, P.; Michrowska, A.; List, B., *Angew. Chem., Int. Ed.* **2009**, *48*, 1978.
242. Aher, R. D.; Kumar, B. S.; Sudalai, A., *J. Org. Chem.* **2015**, *80*, 2024.

243. Zhang, H.; Mifsud, M.; Tanaka, F.; Barbas, C. F., III, *J. Am. Chem. Soc.* **2006**, *128*, 9630.
244. Rodriguez, B.; Bolm, C., *J. Org. Chem.* **2006**, *71*, 2888.
245. Verkade, J. M. M.; van der Pijl, F.; Willems, M. M. J. H. P.; Quaedflieg, P. J. L. M.; van Delft, F. L.; Rutjes, F. P. J. T., *J. Org. Chem.* **2009**, *74*, 3207.
246. De Lamo Marin, S.; Catala, C.; Kumar, S. R.; Valleix, A.; Wagner, A.; Mioskowski, C., *Eur. J. Org. Chem.* **2010**, 3985.
247. Janey, J. M.; Hsiao, Y.; Armstrong, J. D., III, *J. Org. Chem.* **2006**, *71*, 390.
248. Chi, Y.; English, E. P.; Pomerantz, W. C.; Horne, W. S.; Joyce, L. A.; Alexander, L. R.; Fleming, W. S.; Hopkins, E. A.; Gellman, S. H., *J. Am. Chem. Soc.* **2007**, *129*, 6050.
249. Enkisch, C.; Schneider, C., *Eur. J. Org. Chem.* **2009**, 5549.
250. Cele, Z. E. D.; Pawar, S. A.; Naicker, T.; Maguire, G. E. M.; Arvidsson, P. I.; Kruger, H. G.; Govender, T., *Eur. J. Org. Chem.* **2014**, 2253.
251. Kano, T.; Sakamoto, R.; Akakura, M.; Maruoka, K., *J. Am. Chem. Soc.* **2012**, *134*, 7516.
252. Parasuk, W.; Parasuk, V., *J. Org. Chem.* **2008**, *73*, 9388.
253. Enders, D.; Chow, S., *Eur. J. Org. Chem.* **2006**, 4578.
254. Carlson, E. C.; Rathbone, L. K.; Yang, H.; Collett, N. D.; Carter, R. G., *J. Org. Chem.* **2008**, *73*, 5155.
255. Bor-Cherng, H.; Nimje, R. Y.; Wu, M.-F.; Sadani, A. A., *Eur. J. Org. Chem.* **2008**, 1449.
256. Garcia-Garcia, P.; Ladepeche, A.; Halder, R.; List, B., *Angew. Chem., Int. Ed.* **2008**, *47*, 4719.
257. Lu, D.; Li, Y.; Gong, Y., *J. Org. Chem.* **2010**, *75*, 6900.
258. Faisca Phillips, A. M.; Barros, M. T., *Eur. J. Org. Chem.* **2014**, 152.
259. Luna, L. E.; Seoane, G.; Cravero, R. M., *Eur. J. Org. Chem.* **2008**, 1271.
260. Reddy, C. R.; Reddy, M. D.; Haribabu, K., *Eur. J. Org. Chem.* **2012**, 6414.
261. Aplander, K.; Ding, R.; Krasavin, M.; Lindstroem, U. M.; Wennerberg, J., *Eur. J. Org. Chem.* **2009**, 810.
262. Momiyama, N.; Yamamoto, Y.; Yamamoto, H., *J. Am. Chem. Soc.* **2007**, *129*, 1190.
263. Sharma, A. K.; Sunoj, R. B., *J. Org. Chem.* **2012**, *77*, 10516.
264. Schmid, M. B.; Zeitler, K.; Gschwind, R. M., *J. Org. Chem.* **2011**, *76*, 3005.
265. Chen, Y.; Xie, X.; Ma, D., *J. Org. Chem.* **2007**, *72*, 9329.
266. Vijay Kumar, S.; Saraiah, B.; Parameshwarappa, G.; Ila, H.; Verma, G. K., *J. Org. Chem.* **2014**, *79*, 7970.
267. Biswas, S.; Batra, S., *Eur. J. Org. Chem.* **2013**, 4895.
268. Yuan, Q.; Ma, D., *J. Org. Chem.* **2008**, *73*, 5159.
269. Ramalingam, V.; Bhagirath, N.; Muthyala, R. S., *J. Org. Chem.* **2007**, *72*, 3976.
270. Li, C.; Zhang, W.-T.; Wang, X.-S., *J. Org. Chem.* **2014**, *79*, 5847.
271. Ma, D.; Geng, Q.; Zhang, H.; Jiang, Y., *Angew. Chem., Int. Ed.* **2010**, *49*, 1291.
272. Tanimoro, K.; Ueno, M.; Takeda, K.; Kirihata, M.; Tanimori, S., *J. Org. Chem.* **2012**, *77*, 7844.
273. Chen, W.; Zhang, Y.; Zhu, L.; Lan, J.; Xie, R.; You, J., *J. Am. Chem. Soc.* **2007**, *129*, 13879.
274. Hammoud, H.; Schmitt, M.; Bihel, F.; Antheaume, C.; Bourguignon, J.-J., *J. Org. Chem.* **2012**, *77*, 417.
275. Liu, F.; Ma, D., *J. Org. Chem.* **2007**, *72*, 4844.
276. Liu, H.; Zhang, L.; Zhao, F.; Liu, H., *Eur. J. Org. Chem.* **2014**, 1047.
277. Lv, X.; Bao, W., *J. Org. Chem.* **2009**, *74*, 5618.
278. Diao, X.; Wang, Y.; Jiang, Y.; Ma, D., *J. Org. Chem.* **2009**, *74*, 7974.
279. Ghorai, M. K.; Sahoo, A. K.; Bhattacharyya, A., *J. Org. Chem.* **2014**, *79*, 6468.
280. Indumathi, S.; Perumal, S.; Menendez, J. C., *J. Org. Chem.* **2010**, *75*, 472.
281. Verma, A. K.; Kotla, S. K. R.; Choudhary, D.; Patel, M.; Tiwari, R. K., *J. Org. Chem.* **2013**, *78*, 4386.
282. Zaretsky, S.; Adachi, S.; Rotstein, B. H.; Hickey, J. L.; Scully, C. C. G.; St. Denis, J. D.; Courtemanche, R.; Yu, J. C. Y.; Chung, B. K. W.; Yudin, A. K., *J. Org. Chem.* **2014**, *79*, 9948.
283. Jiang, B.; Dong, J.-J.; Jin, Y.; Du, X.-L.; Xu, M., *Eur. J. Org. Chem.* **2008**, 2693.
284. Bench, B. J.; Liu, C.; Evett, C. R.; Watanabe, C. M. H., *J. Org. Chem.* **2006**, *71*, 9458.
285. Hong, B.-C.; Wu, M.-F.; Tseng, H.-C.; Huang, G.-F.; Su, C.-F.; Liao, J.-H., *J. Org. Chem.* **2007**, *72*, 8459.
286. Proulx, C.; Lubell, W. D., *J. Org. Chem.* **2010**, *75*, 5385.
287. Khumsubdee, S.; Zhou, H.; Burgess, K., *J. Org. Chem.* **2013**, *78*, 11948.
288. Kumar, P.; Jha, V.; Gonnade, R., *J. Org. Chem.* **2013**, *78*, 11756.
289. Ait-Youcef, R.; Moreau, X.; Greck, C., *J. Org. Chem.* **2010**, *75*, 5312.
290. Blackmond, D. G.; Moran, A.; Hughes, M.; Armstrong, A., *J. Am. Chem. Soc.* **2010**, *132*, 7598.
291. Shaikh, T. M.; Sudalai, A., *Eur. J. Org. Chem.* **2010**, 3437.
292. Cui, M.; Song, H.; Feng, A.; Wang, Z.; Wang, Q., *J. Org. Chem.* **2010**, *75*, 7018.
293. Huang, K.; Huang, Z.-Z.; Li, X.-L., *J. Org. Chem.* **2006**, *71*, 8320.
294. Shaikh, R. R.; Mazzanti, A.; Petrini, M.; Bartoli, G.; Melchiorre, P., *Angew. Chem., Int. Ed.* **2008**, *47*, 8707.
295. Vogt, H.; Baumann, T.; Nieger, M.; Brase, S., *Eur. J. Org. Chem.* **2006**, 5315.
296. Baumann, T.; Vogt, H.; Brase, S., *Eur. J. Org. Chem.* **2007**, 266.
297. Gruttadauria, M.; Giacalone, F.; Meo, P. L.; Marculescu, A. M.; Riela, S.; Noto, R., *Eur. J. Org. Chem.* **2008**, 1589.
298. Tang, H.; Zhao, G.; Zhou, Z.; Gao, P.; He, L.; Tang, C., *Eur. J. Org. Chem.* **2008**, 126.
299. Kumar, D.; Kommi, D. N.; Chopra, P.; Ansari, M. I.; Chakraborti, A. K., *Eur. J. Org. Chem.* **2012**, 6407.
300. Ramachary, D. B.; Mondal, R.; Venkaiah, C., *Eur. J. Org. Chem.* **2010**, 3205.
301. Ramachary, D. B.; Vijayendar Reddy, Y., *J. Org. Chem.* **2010**, *75*, 74.
302. Ramachary, D. B.; Kishor, M., *J. Org. Chem.* **2007**, *72*, 5056.
303. Bi, H.-P.; Teng, Q.; Guan, M.; Chen, W.-W.; Liang, Y.-M.; Yao, X.; Li, C.-J., *J. Org. Chem.* **2010**, *75*, 783.
304. Hosseini, M.; Stiasni, N.; Barbieri, V.; Kappe, C. O., *J. Org. Chem.* **2007**, *72*, 1417.
305. Massi, A.; Nuzzi, A.; Dondoni, A., *J. Org. Chem.* **2007**, *72*, 10279.
306. Baumann, T.; Baechle, M.; Hartmann, C.; Brase, S., *Eur. J. Org. Chem.* **2008**, 2207.
307. Jainta, M.; Nieger, M.; Brase, S., *Eur. J. Org. Chem.* **2008**, 5418.
308. Pignataro, L.; Benaglia, M.; Annunziata, R.; Cinquini, M.; Cozzi, F., *J. Org. Chem.* **2006**, *71*, 1458.
309. Hargaden, G. C.; Muller-Bunz, H.; Guiry, P. J., *Eur. J. Org. Chem.* **2007**, 4235.
310. Cabrera, S.; Reyes, E.; Aleman, J.; Milelli, A.; Kobbelgaard, S.; Joergensen, K. A., *J. Am. Chem. Soc.* **2008**, *130*, 12031.
311. Matsumoto, K.; Oguma, T.; Katsuki, T., *Angew. Chem., Int. Ed.* **2009**, *48*, 7432.
312. Su, B.; Cai, C.; Wang, Q., *J. Org. Chem.* **2012**, *77*, 7981.
313. Thibault, R. J.; Takizawa, K.; Lowenheilm, P.; Helms, B.; Mynar, J. L.; Frechet, J. M. J.; Hawker, C. J., *J. Am. Chem. Soc.* **2006**, *128*, 12084.
314. Chen, S.-Y.; Zeng, R.-S.; Zou, J.-P.; Asekun, O. T., *J. Org. Chem.* **2014**, *79*, 1449.
315. McIntosh, A. I.; Watson, D. J.; Burton, J. W.; Lambert, R. M., *J. Am. Chem. Soc.* **2006**, *128*, 7329.
316. Madhavachary, R.; Ramachary, D. B., *Eur. J. Org. Chem.* **2014**, 7317.

(2S)-2-Pyrrolidinemethanol-α,α-bis(3,5-dimethylphenyl) and (2R)-2-Pyrrolidinemethanol-α, α-bis(3,5-dimethylphenyl)

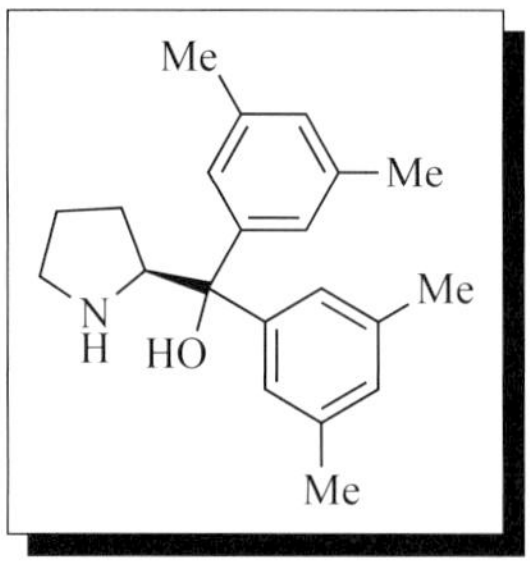

[131180-63-7] $C_{21}H_{27}NO$ (MW 309.45)
InChI = 1S/C21H27NO/c1-14-8-15(2)11-18(10-14)21(23,20-6-5-7-22-20)19-12-16(3)9-17(4)13-19/h8-13,20,22-23H,5-7H2,1-4H3/t20-/m0/s1
InChIKey = MXIOBZPVLNQGIU-FQEVSTJZSA-N

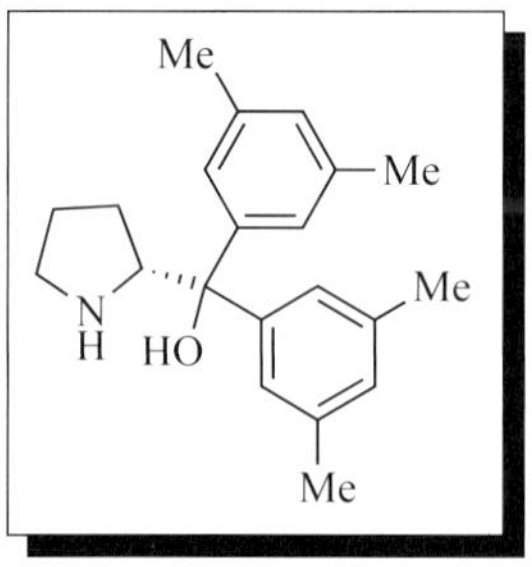

[948595-01-5] $C_{21}H_{27}NO$ (MW 309.45)
InChI = 1S/C21H27NO/c1-14-8-15(2)11-18(10-14)21(23,20-6-5-7-22-20)19-12-16(3)9-17(4)13-19/h8-13,20,22-23H,5-7H2,1-4H3/t20-/m1/s1
InChIKey = MXIOBZPVLNQGIU-HXUWFJFHSA-N

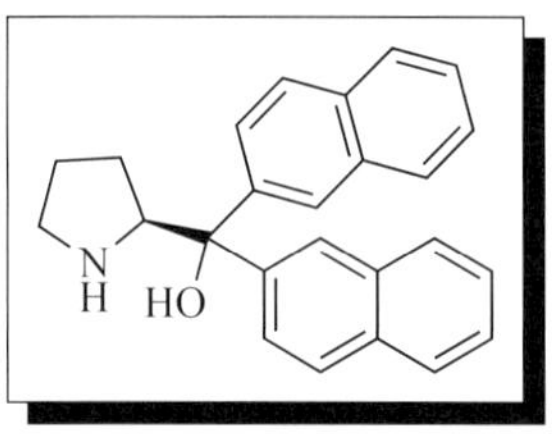

[127986-84-9] $C_{25}H_{23}NO$ (MW 353.46)
InChI = 1S/C25H23NO/c27-25(24-10-5-15-26-24,22-13-11-18-6-1-3-8-20(18)16-22)23-14-12-19-7-2-4-9-21(19)17-23/h1-4,6-9,11-14,16-17,24,26-27H,5,10,15H2/t24-/m0/s1
InChIKey = LNUDNNFEXRHHGY-DEOSSOPVSA-N

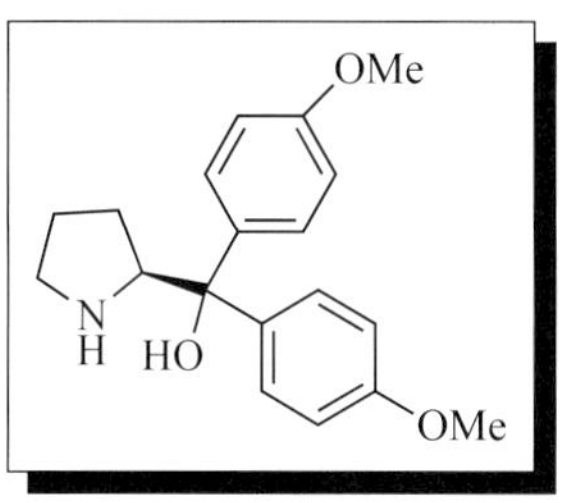

[131180-57-9] $C_{19}H_{23}NO_3$ (MW 313.39)
InChI = 1S/C19H23NO3/c1-22-16-9-5-14(6-10-16)19(21,18-4-3-13-20-18)15-7-11-17(23-2)12-8-15/h5-12,18,20-21H,3-4,13H2,1-2H3/t18-/m0/s1
InChIKey = DIHSOOPDCJLLHA-SFHVURJKSA-N

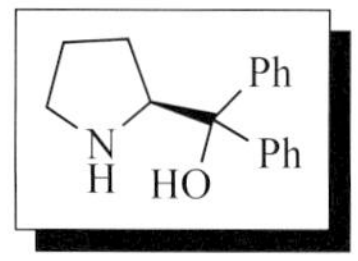

[112068-01-6] $C_{17}H_{19}NO$ (MW 253.34)
InChI = 1S/C17H19NO/c19-17(16-12-7-13-18-16,14-8-3-1-4-9-14)15-10-5-2-6-11-15/h1-6,8-11,16,18-19H,7,12-13H2/t16-/m0/s1
InChIKey = OGCGXUGBDJGFFY-INIZCTEOSA-N

(catalyst in asymmetric epoxidation of α,β-unsaturated carbonyl compounds and β-peroxidation of nitroalkenes with *tert*-butyl hydroperoxide as oxidant)

Alternative Names: (*S*)-(−)-2-[bis(3,5-dimethylphenyl)hydroxymethyl]pyrrolidine, α,α-L-bis-(3,5-dimethylphenyl)prolinol.
Physical Data: white powder, melting point 96–98 °C.
Solubility: in most organic solvents, for example, toluene, $CHCl_3$, Et_2O, insoluble in water.
Preparative Methods: it is commercially available as is the enantiomeric 2-pyrrolidinemethanol-α,α-bis(3,5-dimethylphenyl)-(2*R*) [948595-01-5].
Handling, Storage, and Precautions: compatible with acids and bases and mild oxidants, such as commercially available alkyl hydroperoxides (*tert*-butyl hydroperoxide (TBHP) and cumyl hydroperoxide (CHP)). Incompatible with strong oxidizing agents. Recommended storage temperature range is 2–8 °C. Skin/eye contact and inhalation should be avoided.

Diastereo- and Enantioselective Epoxidation of α,β-Unsaturated Carbonyl Compounds. α,α-Diaryl prolinols are well-established ligands in the enantioselective Corey–Shibata reduction of ketones to chiral nonracemic secondary alcohols.[1] More recently, new applications of these compounds have been disclosed in the context of asymmetric oxidations. The asymmetric epoxidation of electron-poor alkenes is an important area of organic synthesis to access functionalized chiral nonracemic epoxides.[2] These compounds are highly valuable intermediates amenable to further useful elaborations either through epoxide ring opening or selective manipulation of the carbonyl group.[3] A large number of methodologies, based on the use of a chiral–metal complex coupled with TBHP or CHP as oxidants, are currently available to effect the asymmetric epoxidation of α,β-unsaturated carbonyl compounds.[2] Among them, the lanthanides-(*R*)-BINOL/TBHP-based systems developed by Shibasaki and coworkers can be considered the most effective from different viewpoints: (i) wide substrate scope, (ii) low catalyst loading, and (iii) level of asymmetric induction (ee >90%).[4] More recently, systems based on the use of optically pure organic molecules as catalysts have been developed.[2b,5] Chiral nonracemic secondary amines derived from L-proline and primary amines derived from *Cinchona* alkaloids are able to catalyze, at generally 20 mol % loading, the asymmetric epoxidation of challenging alkenes such as *trans*-enals[6] and cyclic enones[7] achieving good to high level of diastereo- and enantioselectivities. Interestingly, it has been shown that hydrogen peroxide is a suitable and more effective terminal oxidant when using some of these organocatalysts in the process, which represents an advantage for large-scale applications in terms of cost and by-product disposal.[6a–8]

In 2006, it was demonstrated that 2-pyrrolidinemethanol-α,α-bis(3,5-dimethylphenyl)-(2*S*) is an effective organocatalyst in the asymmetric epoxidation of *trans*-α,β-unsaturated ketones when used at 20 mol % loading with TBHP as oxidant in hexane at 0 °C.[9] Differently substituted *trans*-chalcones ($R^1 = R^2$ = aryl) are epoxidized with complete control of the diastereoselectivity for the *trans*-epoxide in 70–98% yield and with 82–92% ee (eq 1). Aryl, alkyl, or dialkyl (R^1 = Me, Ph; R^2 = Me, Ph, $(CH_2)_2Ph$, C_5H_{11}) enones are obtained with the same diastereocontrol in 70–87% yield, although with a somewhat decreased level of enantiocontrol (34–79% ee). The asymmetric epoxidation of dialkyl enones is generally achieved at best with good level of enantiocontrol by other methods.

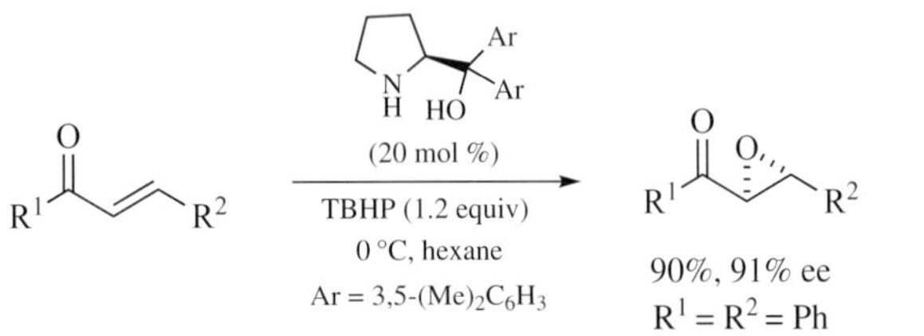

(1)

Moreover, enantiomerically enriched epoxides of opposite absolute configuration are accessible by using 2-pyrrolidinemethanol-α,α-bis(3,5-dimethylphenyl)-(2*R*) as catalyst.

Zhao and coworkers have applied the same system in the first enantioselective epoxidation of more electrophilic aromatic *trans*-β,γ-unsaturated α-keto esters (R^1 = aryl; R^2 = Me, Et, allyl, *i*-Pr, *t*-Bu, Bn) (eq 2). In this case, 10 mol % loading of the catalyst is required to obtain the epoxides in good to high yield (52–96%) with complete control of the diastereoselectivity and high enantioselectivity (92–98% ee).[10]

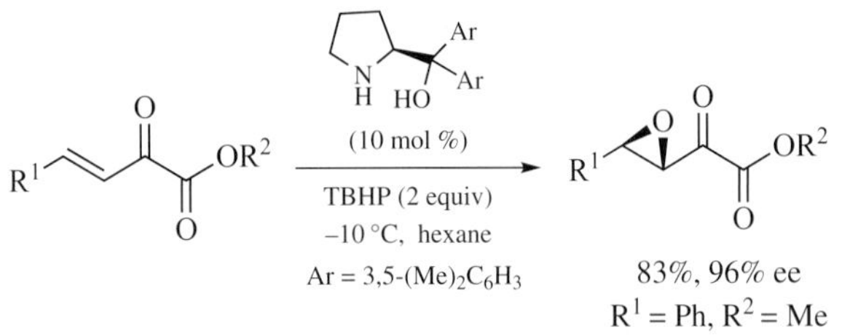

(2)

It has been shown that α,β-unsaturated trichloromethyl ketones can also be epoxidized at room temperature (R^1 = aryl, Me, CH=CHPh) (eq 3). β-Aromatic substituted derivatives are transformed into the epoxides in good yield (36–93%) and with moderate to high enantioselectivity (56–99% ee).

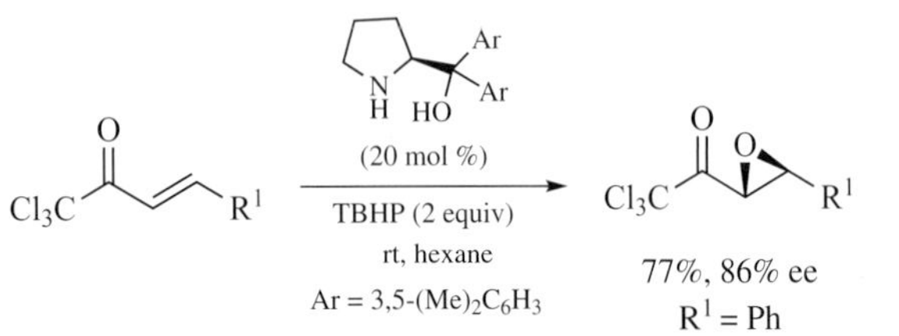

(3)

Enantiomerically enriched epoxides illustrated in eqs 2 and 3 can be synthetically elaborated to access useful intermediates and biologically interesting compounds.[10]

The methodologies reported to date for the asymmetric epoxidation of electron-poor alkenes are mainly focused on *trans*-1,2-disubstituted alkenes and most of them include enones as substrates. Although efforts to develop efficient methods by introducing new catalytic systems are clearly important, likewise expanding the scope of the nucleophilic epoxidation in terms of substrate scope should be pursued. From this point of view, simple and easily accessible catalytic systems are expected to be particularly appealing to check for their ability in promoting the epoxidation of structurally different alkenes.

A new finding features the first asymmetric epoxidation of *trans*-2-aroyl-3-arylacrylonitriles (R^1 = aryl, Me; R^2 = aryl, *trans*-cinnamyl, cyclohexyl) by using 10 mol % 2-pyrrolidinemethanol-α,α-bis(3,5-dimethylphenyl)-(2*S*) and TBHP in *m*-xylene at −18 °C (eq 4).[11]

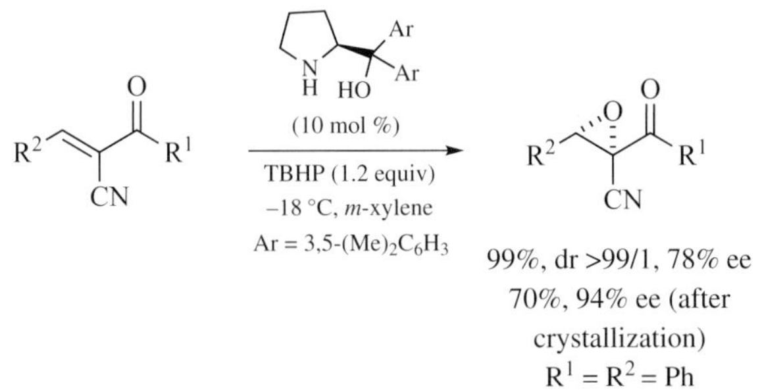

(4)

Complete control of the diastereoselectivity is observed in favor of *trans*-epoxides, which are recovered in high yield (77–99%) and generally with good enantioselectivity (43–84% ee). Aromatic and heteroaromatic residues at the β-position and methyl ketones in the starting alkenes are tolerated. It is worth noting that in the epoxidation, two vicinal quaternary and tertiary stereogenic centers are installed in stereocontrolled manner.

The asymmetric epoxidation reactions catalyzed by the prolinol/TBHP system are advantageous in terms of (i) easy commercial accessibility of the organocatalysts and (ii) mild reaction conditions and performance under air in undistilled solvents. The main drawbacks of this system are that (i) the relatively high catalyst loading is required compared to metal-catalyzed nucleophilic epoxidations and (ii) its low reaction rate (few hours to several days).

Asymmetric β-Peroxidation of Nitroalkenes. A limited number of stereoselective oxa-Michael reactions have been reported using nitroalkenes as acceptors, although the corresponding products are very useful synthetic intermediates.[12] The strong basicity and poor nucleophilicity of the oxygen nucleophiles disfavor the addition process, whereas polymerization becomes a highly competitive side reaction.

A single procedure is available to obtain enantiomerically enriched nitroepoxides starting from simple nitroalkenes. Enders et al. disclosed a highly diastereoselective (de >98%) and moderately enantioselective (36–82% ee) formation of nitroepoxides by treating aliphatic nitroalkenes with a stoichiometric reagent formed by $Zn(Et)_2$/(*R*,*R*)-*N*-methyl pseudoephedrine/O_2.[13] A highly diastereoselective approach to nitroepoxide has been recently developed by reacting enantiopure *O*-protected 3-hydroxy-1-nitroalkenes with potassium *tert*-butyl peroxide at low temperature (*anti*/*syn* up to 10/1).[14]

The potential of the diaryl prolinols/TBHP system as an epoxidizing system has also been studied using highly reactive nitroalkenes as starting compounds. Interestingly, under these reaction conditions, an unprecedented formation of enantioenriched β-peroxy adducts is observed. 2-Pyrrolidinemethanol-α,α-bis(3,5-dimethylphenyl)-(2*S*) and TBHP in cyclohexane at $-18\,°C$ enable the synthesis of aromatic nitroperoxides (R^1 = aryl) in moderate to high yield (20–83%) and with good enantioselectivity (43–84% ee) (eq 5).[15] The β-peroxidation of less reactive aliphatic nitroalkenes proceeds sluggishly and with lower enantiocontrol.

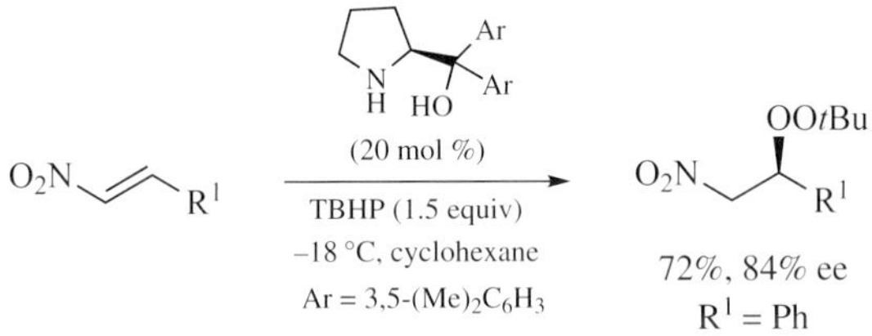

(5)

Parent α,α-Diaryl Prolinols. The asymmetric epoxidation of *trans*-α,β-unsaturated ketones was originally developed using (*S*)-α,α-diphenyl-2-pyrrolidinemethanol/TBHP system achieving good yield (27–87%), complete control of diastereoselectivity and moderate to good enantioselectivity (63–80% ee) for the epoxides (eq 6).[16]

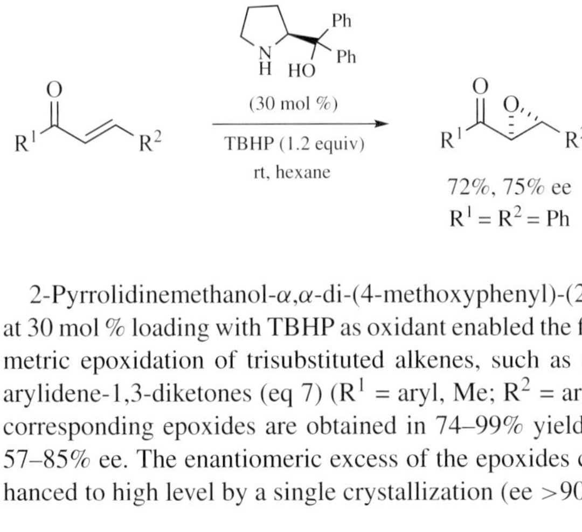

(6)

2-Pyrrolidinemethanol-α,α-di-(4-methoxyphenyl)-(2*S*)[17] used at 30 mol % loading with TBHP as oxidant enabled the first asymmetric epoxidation of trisubstituted alkenes, such as acyclic 2-arylidene-1,3-diketones (eq 7) (R^1 = aryl, Me; R^2 = aryl).[18] The corresponding epoxides are obtained in 74–99% yield and with 57–85% ee. The enantiomeric excess of the epoxides can be enhanced to high level by a single crystallization (ee >90%).

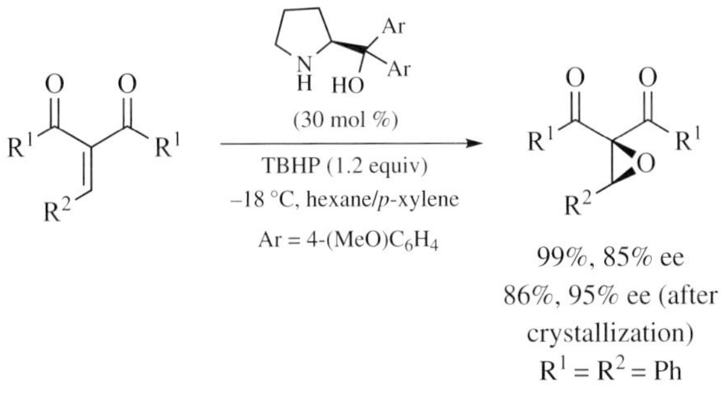

(7)

Commercially available 2-pyrrolidinemethanol-α,α-di-2-naphthalenyl-(2*S*) is the best catalyst to use in the asymmetric epoxidation of less sterically demanding cyclic alkenes derived from 1,3-indandione (R^1 = aryl) in the presence of TBHP or CHP as the oxygen source (eq 8).

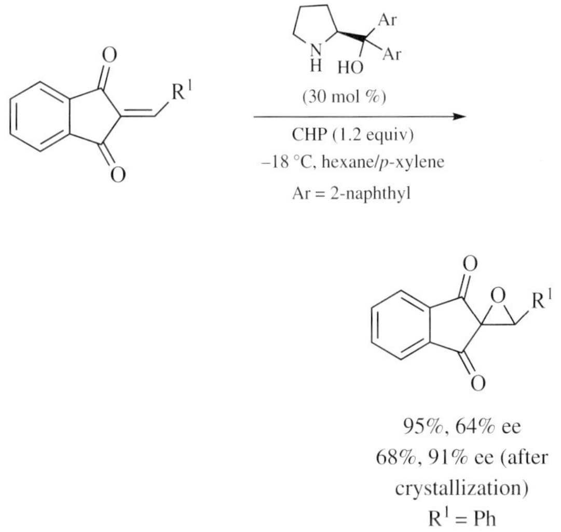

(8)

The epoxides, whose absolute configurations were not determined, are obtained in high yield (95–99%) and with moderate to good level of enantioselectivity (64–80% ee), which is improved by a single crystallization. These enantiomerically enriched epoxides are synthetically useful as starting material to produce the first potent inhibitors of human papillomavirus HPV-11 E1–E2 protein–protein interaction.[19]

Gasperi and coworkers recently employed the (*S*)-α,α-diphenyl-2-pyrrolidinemethanol/TBHP system in the first asymmetric epoxidation of *trans*-α-ylideneoxindoles (R^1 = aryl, Me, Bn; R^2 = H, Cl, F, OMe, OCF_3, *i*-Pr; R^3 = H, F, Cl, Br) employing 30 mol % of the organocatalyst in hexane at room temperature (eq 9).[20]

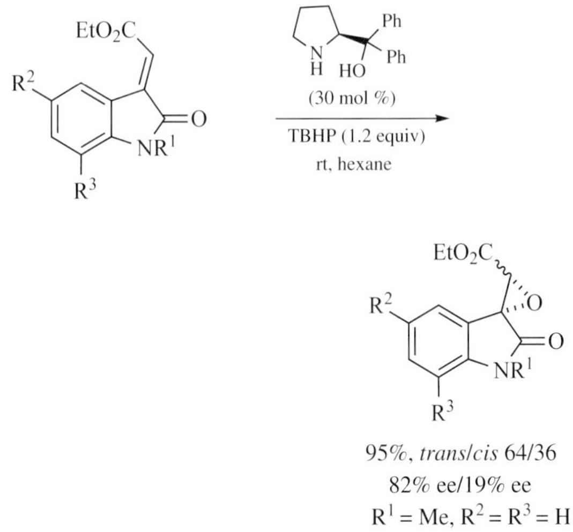

(9)

The products are obtained in good to high yield (71–95%), with low diastereocontrol in favor of the *trans*-epoxide (dr 54/46 to 75/25) and with moderate to good enantioselectivity for the *trans*-epoxide (52–88% ee). The spirocyclic oxindolic oxiranes represent attractive building blocks for further synthetic elaboration.

1. Corey, E. J.; Bakshi, R. K.; Shibata, S., *J. Am. Chem. Soc.* **1987**, *109*, 5551.
2. (a) Colonna, S.; Perdicchia, D. In *Stereoselective Synthesis 1*; de Vries, J. G., Ed.; Thieme-Verlag: Weinheim, 2011, Chapter 1.3, p 123. (b) Lattanzi, A. In *Catalytic Asymmetric Conjugate Reactions*; Córdova, A., Ed.; Wiley-VCH: Weinheim, Germany, 2010, p 351.
3. Lauret, C., *Tetrahedron: Asymmetry* **2001**, *12*, 2359.
4. (a) Bougauchi, M.; Watanabe, S.; Arai, T.; Sasai, H.; Shibasaki, M., *J. Am. Chem. Soc.* **1997**, *119*, 2329. (b) Nemoto, T.; Ohshima, T.; Shibasaki, M., *J. Am. Chem. Soc.* **2001**, *123*, 9474. (c) Nemoto, T.; Kakei, H.; Gnanadesikan, V.; Tosaki, S.; Ohshima, T.; Shibasaki, M., *J. Am. Chem. Soc.* **2002**, *124*, 14544. (d) Kinoshita, T.; Okada, S.; Park, S.-R.; Matsunaga, S.; Shibasaki, M., *Angew. Chem., Int. Ed.* **2003**, *42*, 4680.
5. Weib, M. K.; Tsogoeva, S. B., *Chem. Rec.* **2011**, *11*, 18.
6. (a) Marigo, M.; Franzén, G.; Poulsen, T. B.; Zhuang, W.; Jørgensen, K. A., *J. Am. Chem. Soc.* **2005**, *127*, 6964. (b) Wang, X.; List, B., *Angew. Chem., Int. Ed.* **2008**, *47*, 1119.
7. Wang, X.; Reisinger, C. M.; List, B., *J. Am. Chem. Soc.* **2008**, *130*, 6070.
8. For a recent review on catalytic asymmetric epoxidation using hydrogen peroxide, see: De Faveri, G.; Ilyashenko, G.; Watkinson, M., *Chem. Soc. Rev.* **2011**, *40*, 1722.
9. Lattanzi, A., *Adv. Synth. Catal.* **2006**, *348*, 339.
10. Zheng, C.; Li, Y.; Yang, Y.; Wang, H. F.; Cui, H.; Zhang, J.; Zhao, G., *Adv. Synth. Catal.* **2009**, *351*, 1685.
11. De Fusco, C.; Tedesco, C.; Lattanzi, A., *J. Org. Chem.* **2011**, *76*, 676.
12. (a) Berner, O. M.; Tedeschi, L.; Enders, D., *Eur. J. Org. Chem.* **2002**, 1877. (b) Nising, C. F.; Bräse, S., *Chem. Soc. Rev.* **2008**, *37*, 1218. (c) Enders, D.; Heartwig, A.; Raabe, G.; Runsink, J., *Angew. Chem., Int. Ed.* **1996**, *35*, 2388. (d) Adderley, N. J.; Buchanan, D. J.; Dixon, D. J.; Lainé, D. I., *Angew. Chem., Int. Ed.* **2003**, *42*, 4241. (e) Dinér, P.; Nielsen, M.; Bertelsen, S.; Niess, B.; Jørgensen, K. A., *Chem. Commun.* **2007**, 3646.
13. Enders, D.; Kramps, L.; Zhu, J., *Tetrahedron: Asymmetry* **1998**, *9*, 3959.
14. Jain, A.; Rodríguez, S.; López, I.; González, F. V., *Tetrahedron* **2009**, *65*, 8362.
15. Russo, A.; Lattanzi, A., *Adv. Synth. Catal.* **2008**, *350*, 1991.
16. Lattanzi, A., *Org. Lett.* **2005**, *7*, 2579.
17. Mathre, D. J.; Jones, T. K.; Xavier, L. C.; Blacklock, T. J.; Reamer, R. A.; Mohan, J. J.; Turner Jones, E. T.; Hoogsteen, K.; Baum, M. W.; Grabowski, E. J. J., *J. Org. Chem.* **1991**, *56*, 751.
18. Russo, A.; Lattanzi, A., *Org. Biomol. Chem.* **2010**, *8*, 2633.
19. Goudreau, N.; Cameron, D. R.; Déziel, R.; Haché, B.; Jakalian, A.; Malenfant, E.; Naud, J.; Ogilvie, W. W.; O'Meara, J.; White, P. W.; Yoakim, C., *Bioorg. Med. Chem.* **2007**, *15*, 2690.
20. Palumbo, C.; Mazzeo, G.; Mazziotta, A.; Gambacorta, A.; Loreto, M. A.; Migliorini, A.; Superchi, S.; Tofani, D.; Gasperi, T., *Org. Lett.* **2011**, *13*, 6248.

Alessandra Lattanzi
Università di Salerno, Fisciano, Italy

1-(2-Pyrrolidinylmethyl)pyrrolidine

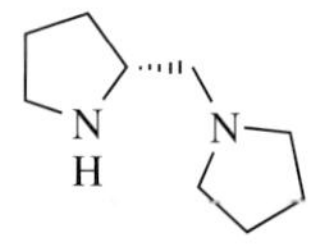

(*R*)-(−)-1-(2-Pyrrolidinylmethyl)pyrrolidine

[60419-23-0] $C_9H_{18}N_2$ (MW 154.25)
InChI = 1S/C9H18N2/c1-2-7-11(6-1)8-9-4-3-5-10-9/h9-10H,1-8H2/t9-/m1/s1
InChIKey = YLBWRMSQRFEIEB-SECBINFHBO

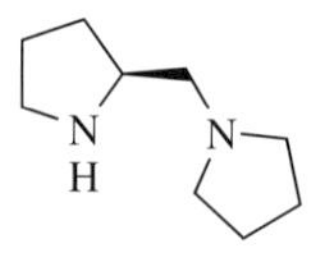

(*S*)-(+)-1-(2-Pyrrolidinylmethyl)pyrrolidine

[51207-66-0] $C_9H_{18}N_2$ (MW 154.25)
InChI = 1S/C9H18N2/c1-2-7-11(6-1)8-9-4-3-5-10-9/h9-10H,1-8H2/t9-/m0/s1
InChIKey = YLBWRMSQRFEIEB-VIFPVBQESA-N

Physical Data: fp 190 °C; bp 99–101 °C (2 mm Hg); d 0.946 g/mL at 25 °C, $\alpha_D^{19} = +7.0°$ (c 2.4, EtOH), $n_D^{20} = 1.487$.
Solubility: H_2O, alcohol, ether, THF, toluene. Li-**1** is only soluble in THF.
Form Supplied in: neat liquid.
Purification: dried over CaH_2 and distilled under reduced pressure.
Handling, Storage, and Precautions: to prevent moisture, it should be deoxygenated and treated with 4 Å molecular sieves. 1-(2-Pyrrolidinylmethyl)pyrrolidine is irritating to the eyes, the respiratory system, and the skin, and suitable protective clothing, gloves, and eye/face protection should be used.

Introduction. The lithium salt of the chiral amine (*S*)-(+)-1-(2-pyrrolidinylmethyl)pyrrolidine was introduced in organic chemistry as a strong chiral base for stereoselective deprotonation reactions. Later, this chiral amine has been explored in various areas of asymmetric synthesis. Notably, it has been used as a chiral ligand in the addition of Et_2Zn to aldehydes and ketones, and in recent years it has also found application as a chiral amine in organocatalysis. Furthermore, it has also been explored in asymmetric epoxidation reactions and in heterogeneous asymmetric hydrogenation reactions.

Asymmetric Deprotonation of Epoxides. In 1958, Cope and co-workers reported that lithium amides, which are strong bases, can deprotonate epoxides to obtain allylic alcohols.[1] In 1980, Whitesell and Felman showed that a chiral lithium amide could be used in asymmetric deprotonation and rearrangement of cyclohexene oxide.[2] In their study, cyclohexene oxide (a meso-epoxide) was converted into optically active (*R*)-2-cyclohexenol in 36% enantiomeric excess (Figure 1).

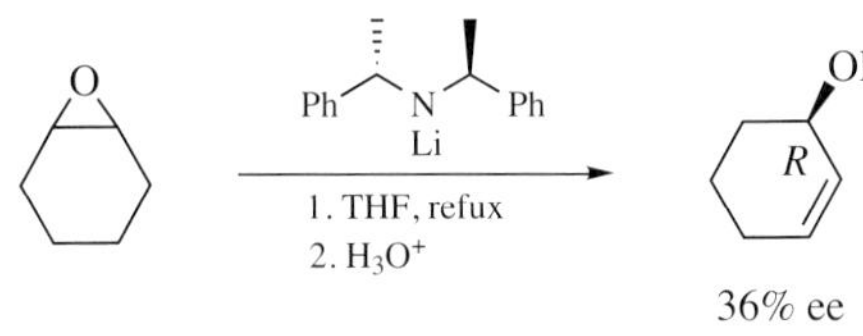

Figure 1 Stereoselective deprotonation of cyclohexene oxide by a chiral lithium amide.

In the mid 1980s, Asami and co-workers reported that the lithium salt of (*S*)-(+)-1-(2-pyrrolidinylmethyl)pyrrolidine (**1**), readily prepared from (*S*)-proline, mediates enantioselective rearrangement of cylohexene oxide to 1-cyclohexenol with 80% ee (Figure 2).[3]

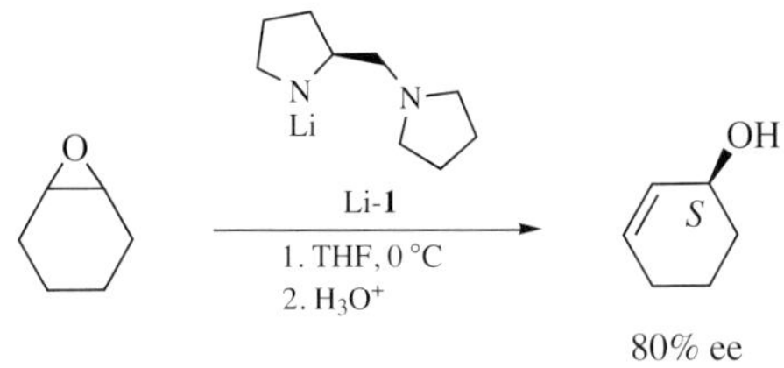

Figure 2 Stereoselective deprotonation of cyclohexene oxide by the chiral lithium amide Li-**1**.

Asami proposed that the chiral lithium amide base Li-**1** enantioselectively deprotonates one of the pseudo axial protons of cyclohexene oxide. According to this simple model, the predominate formation of the (S)-isomer is due to the unfavorable interaction in the expected TS-1 (Figure 3).

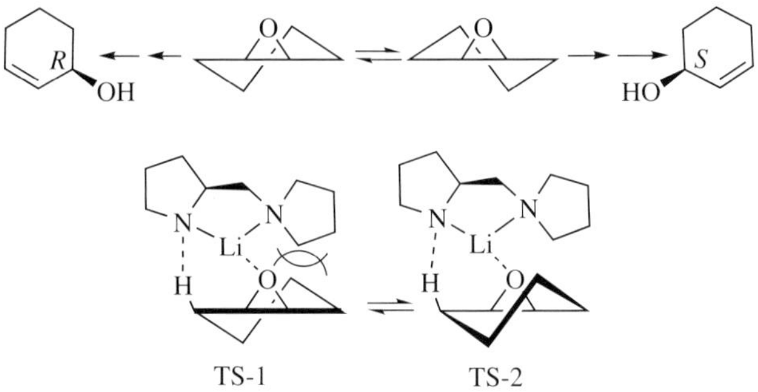

Figure 3 A simple model that describes the unfavorable steric interactions responsible for the observed stereoselectivity in the deprotonation of cyclohexene oxide with Li-**1**.

Later work by the Asami group showed that the chiral amide deprotonates other *meso*-epoxides to allylic alcohols with high enantioselectivities. Cyclopentenol derivatives are important key intermediates for the synthesis of prostaglandins and cyclopentanoids, thus there is much interest in asymmetric synthesis of such compounds. Asami reported in 1985 that Li-**1** mediates high enantioselectivity in the rearrangement reaction in coordinating or noncoordinating solvents, Figure 4.

The *cis-meso*-cyclopentene oxide in Figure 5 was converted to the corresponding alcohol in 92% yield and 90% ee.[4] The total synthesis of (−)-Carbovir, an anti-HIV carbocyclic nucleoside, by Asami and co-workers utilized a chiral alcohol obtained from

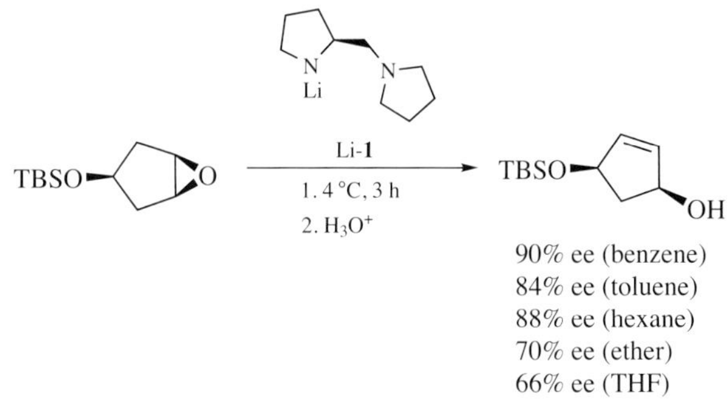

Figure 4 Solvent dependency in the Li-**1** mediated stereoselective deprotonation of TBS protected *meso*-cyclopentenol oxide.

an asymmetric reaction with Li-**1**. Later Asami also showed that the ee in the reactions varied with solvent. In THF and ether, they obtained only 66 and 70% ee, respectively, while in the noncoordinating solvents benzene and hexane, they observed ees of 90 and 88%, respectively.[5]

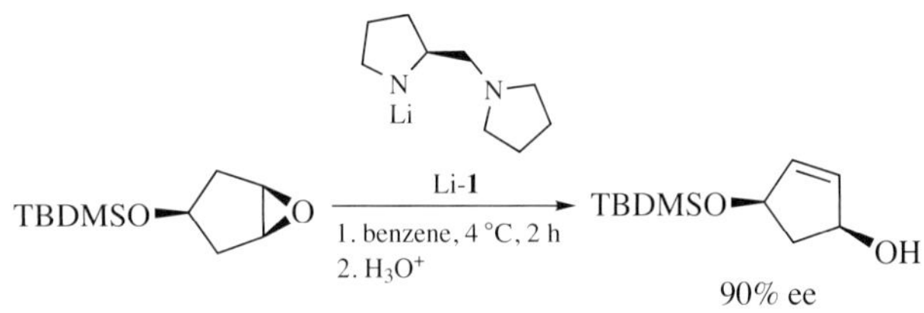

Figure 5 Stereoselective deprotonation of TBDMS-protected *meso*-cyclopentenol oxide mediated by Li-**1**.

The asymmetric synthesis of chiral allylic alcohols was also employed in the synthesis of the natural product (−)-Untenone A.[6] The chiral lithium amide Li-**1** was also used in the stereoselective deprotonation of a *meso*-epoxide (Figure 6). The chiral allylic alcohol was obtained in 89% ee when the reaction was conducted at 0 °C in toluene for 5 h.

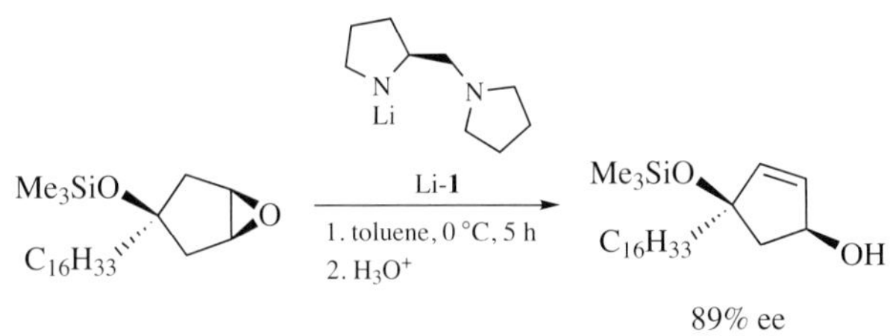

Figure 6 Stereoselective deprotonation of a *meso*-epoxide by Li-**1** gives access to a key intermediate for the synthesis of (−)-Untenone A.

Several years later, in 1997, Asami showed that the chiral lithium amide Li-**1** was more reactive toward epoxides than lithium diisopropylamide (LDA) or lithium diethylamide.[7] From this observation they developed a catalytic system in which the more reactive chiral amine could be used in substoichiometric amounts in the presence of a less reactive lithium amide, e.g., LDA.[8] By employing the achiral lithium amide LDA in the reaction, the catalytic cycle was achieved, and the chiral lithium amide was regenerated during the progress of the reaction, Figure 7.

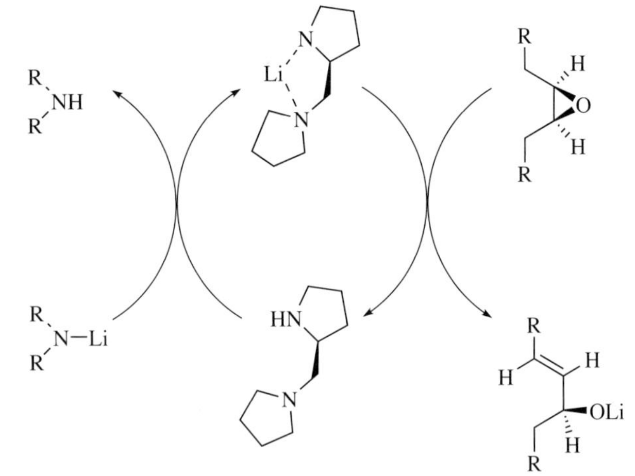

Figure 7 Deprotonation with a catalytic amount of Li-**1**.

Various additives were tested and the addition of 1.65 equiv of 1,8-diazabicyclo[5.4.0]undec-7-ene (DBU) was observed to result in a significant increase in chemical yield and enantiomeric excess.[8] Later work by Ahlberg et al. have shown that DBU undergoes rapid lithiation by, for example, LDA, to yield a weak base incapable of deprotonating cyclohexene oxide.[9] On the basis of the results by the Ahlberg group, it is likely that a mixed complex is formed between Li-**1** and Li-DBU (Figure 8).

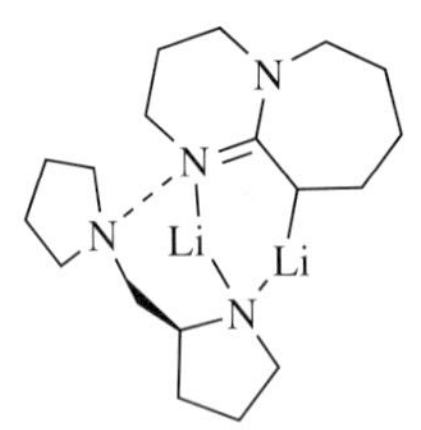

Figure 8 Proposed mixed complex formed between Li-**1** and Li-DBU.

The first kinetic resolution method applied in the resolution of a racemic epoxide was published by Asami et al. in 2000.[10] Using a chiral lithium amide, they could perform the kinetic resolution of *cis*-3-alkenecyclohexene oxides. However, a stochiometric amount of the chiral base was necessary for the reaction to produce both the alcohol and the unreacted epoxide in good enantiomeric excess.

Seki and Asami have also explored the catalytic use of Li-**1** in the presence of excess cross-linked polymer-bound lithium amides.[11] The use of the polymer-bound lithium amide was particularly successful because the reactivity is much lower than that for the free lithium amide Li-**1**. The stereoselectivity of the reaction of *meso*-epoxides was also enhanced in comparison with that of the stoichiometric reactions.

The structure of the lithium amide Li-**1** remained unknown until Hilmersson and Håkansson succeeded in crystallizing it from toluene solution, obtaining the X-ray structure in 2003.[12] In the solid state, it forms a ladder-type tetramer with the two pyrrolidine nitrogens solvating the two lithiums at the end of the ladder, that is, (Li-**1**)$_4$ (Figure 9).

The ^{6}Li NMR spectra of the lithium amide Li-**1** in the presence of excess **1** or the diamine TMEDA shows two signals in a 1:1

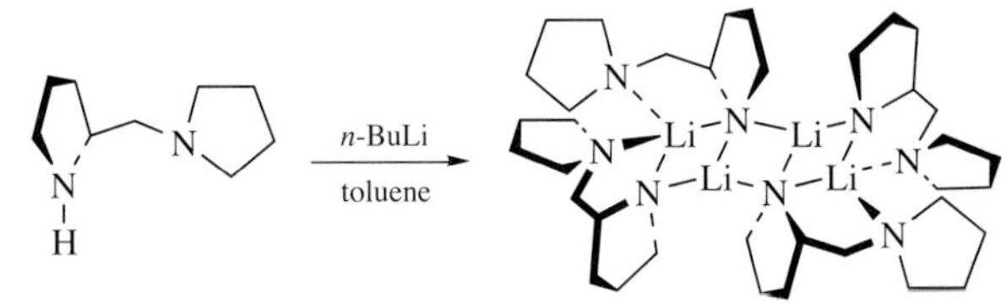

1 (Li-**1**)$_4$

Figure 9 Li-**1** forms tetramers in toluene as determined by X-ray diffraction studies.

ratio. On the basis of extensive NMR experiments Hilmersson and coworkers could unravel the solution structure of this complex to arise from a lithium pyrrolidide amide dimer solvated by a pyrrolidine or TMEDA, that is, diamine, (Li-**1**)$_2$·**1**, or (Li-**1**)$_2$·TMEDA, Figure 10.[13]

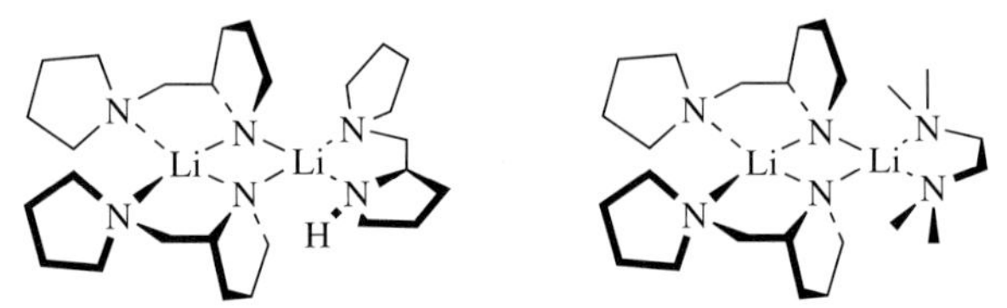

(Li-**1**)$_2$ · **1** (Li-**1**)$_2$ · TMEDA

Figure 10 Detailed NMR studies of Li-**1** have revealed that it forms dimers solvated by one amine (**1**) or one TMEDA ligand in solution.

Detailed kinetic studies by Olsson and Ahlberg on the reaction between cyclohexene oxide and the chiral amide Li-**1** in the presence of the diamine (**1**) was reported in 1997. On the basis of the NMR data of the initial state in solution and the determined reaction orders from the kinetic studies, the transition state for the reaction was proposed to be composed of one molecule of epoxide and a dimer of the amide Li-**1**.

An interesting application of Li-**1** as a synthetic key intermediate was reported by Anders and co-workers in 2004.[14] The chiral amine (**1**) was used in the synthesis of chiral carbamates, using CO_2 as opposed to toxic compounds as isocyanates and phosgene, giving a white precipitate in 79% yield, Figure 11. Chiral carbamates are important in biotin-mediated reactions and they are important intermediates in the synthesis of several drugs.

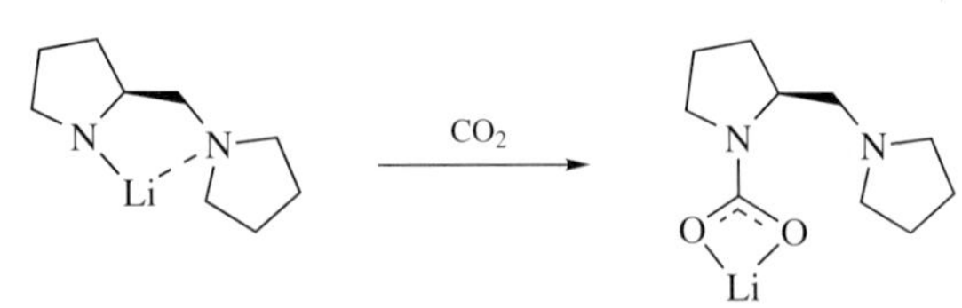

Figure 11 Addition of CO_2 to the chiral amide Li-**1**.

Organocatalytic Reactions.

Nitrocyclopropanation Reaction. Hansen and co-workers have reported that when the chiral amine **1** (15 mol %) was employed with *trans*-2,5-dimethylpiperazine, 2-cyclohexene-1-one, and bromonitromethane in CH_2Cl_2, an enantioselective nitrocyclopropanation reaction was accomplished with moderate yield (69%) and ee (64%).[15]

Asymmetric Michael Addition of Ketones and Aldehydes to Nitroolefins. In conjugate addition reactions of α-heterosubstituted ketones to nitrostyrene, Andrey and co-workers tested various organocatalysts. With the chiral diamine **1** in CH_2Cl_2, the 1, 4-adducts were obtained in moderate yield (59%) and low ee (10%).[16] In addition, when aldehydes were used with nitrostyrene in THF at room temperature the yield (80%) and ee (79%) increased.[17]

Direct organocatalytic Aldol Reactions in Buffered Aqueous Media. In another report by Córdova and co-workers, buffered aqueous media was used as solvent in cross-aldol reactions, catalyzed by the diamine **1**.[18] By using various ketones as donors with *p*-nitrobenzaldehyde in phosphate buffer, the yield increased dramatically (5-90%) with increasing polarity of the ketone, where the dihydroxyacetone received the best yield (90%) and a de of 1:1. The latter reaction was not as beneficial in pure water (40%), but in phosphate buffer/DMSO the yield increased to 87% and the diastereomeric excess increased to 2:1 (Figure 12).

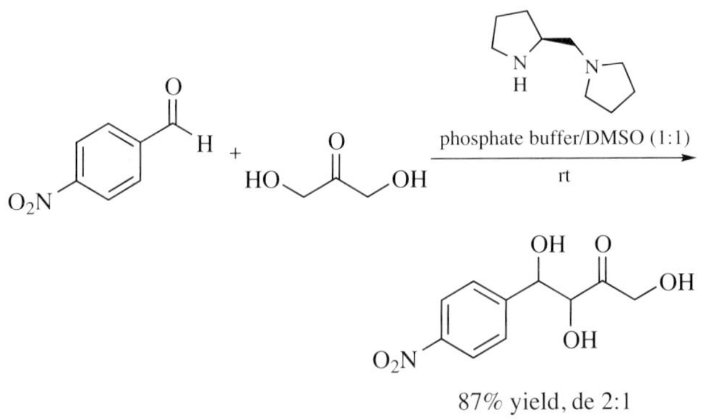

Figure 12 Organocatalytic aldol reaction catalyzed by the chiral amine **1**.

Furthermore, when varying the aldehydes in the described reaction above, the yields decreased to moderate yields (35–55%). A diastereomeric ratio of >20:1 was achieved, but decreased when only phosphate buffer was used as solvent.

Asymmetric Addition Reactions. The chiral amine **1** has been used as chiral inducer in the addition of Et_2Zn to imines. Gao et al. have found that there is a strong coordination between the *N*-tosylimine and Et_2Zn in nonpolar solvents, resulting in quantitative reduction of *N*-tosylimines. However, in polar solvents and/or chelating amines this coordination is lost and instead a 1,2-addition of the Et group to the imine group was observed. When the chiral amine **1** was added to Et_2Zn (1:1) in toluene solution at −78 °C they obtained the addition product in 98% yield and 77% enantioselectivity (Figure 13).[19]

In the aza-Henry reaction (nitro-Mannich reaction), the chiral amine **1** was used both as a catalyst and in stochiometric amounts with *n*-BuLi as a base in THF to give an enantioselectivity of 2% at rt and 56% at −78 °C.[20] Trost and Ito have proposed a mechanism for the catalytic aza-Henry reaction.[21]

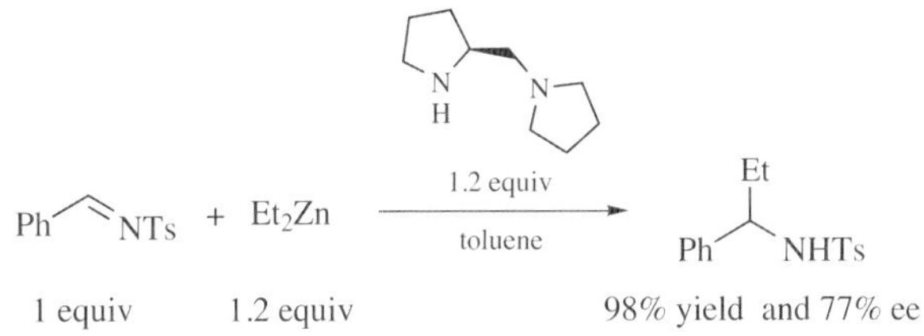

Figure 13 Asymmetric addition of Et_2Zn to *N*-tosylimine catalyzed by the chrial amine **1**.

Asymmetric Epoxidation. Direct asymmetric epoxidation of α,β-unsaturated aldehydes is a relatively new frontier in asymmetric catalysis. Sundén et al. screened **1** and other chiral pyrrolidine derivatives for their ability to mediate the asymmetric epoxidation of cinnamic aldehyde with hydrogen peroxide in chloroform.[22] In this reaction, **1** turned out to be a poor catalyst, giving an enantioselectivity of only 7% and low diastereoselectivity (45:54 dr). However, when the diamine **1** was employed with aliphatic α,β-unsaturated aldehydes and hydrogen peroxide, the diastereoselectivity increased to 84:16, but the enantioselectivity remained low (6%) (Figure 14).

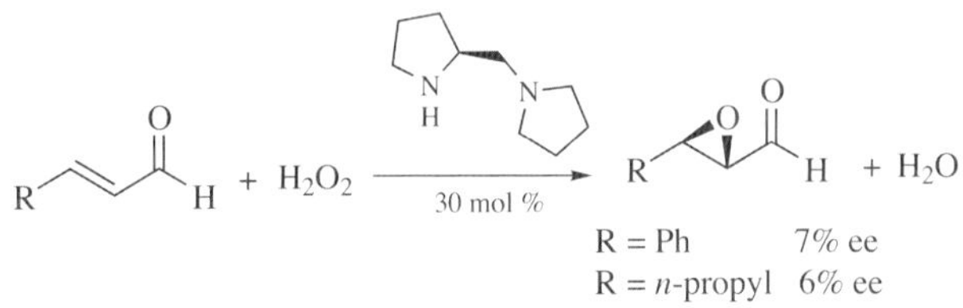

Figure 14 Asymmetric epoxidation of α,β-unsaturated aldehydes.

The amine **1** was also used as a catalyst in the epoxidation of *trans*-stilbene, with oxone and sodium hydrogen carbonate in CH_3CN/H_2O (10:1); however, they observed a product that was nearly racemic (5% ee).[23]

Heterogeneous Hydrogenation. Raja, Thomas, and co-workers have reported that the chiral amine **1** in a complex with Rh^+ and cyclooctadiene (COD) induces asymmetry in the hydrogenation of methyl benzoylformate.[24] They observed a modest stereoselectivity with CF_3SO_2 as the counteranion. However, when the same reaction was carried out within Mesoporous supports (Davisson 923, silica of 38 Å) the reaction was 95% complete in 2 h and with an enantioselectivity of 94% (Figure 15).

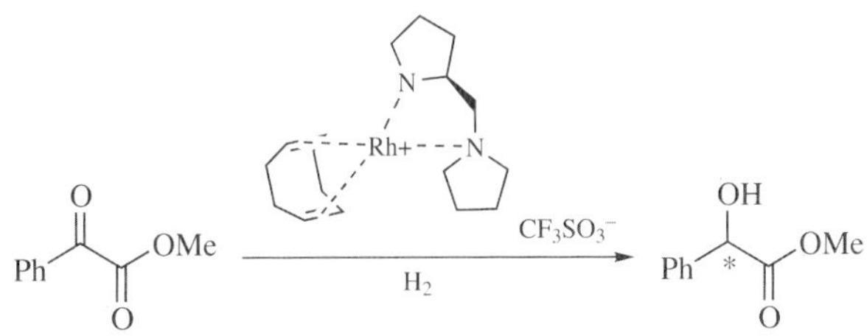

Figure 15 Catalytic heterogeneous asymmetric reduction.

The complex between **1**, Rh, and COD was also verified by X-ray analysis. Replacing Rh–COD with Pd–allyl resulted in significantly lower stereoselectivity, 66% ee.

Polymerization. The chiral amine **1** can also act as a ligand in polymerization reactions.[25] In the presence of CuCl-(+)PMP as a catalyst and 2,3-dihydroxynaphthalene in CH_2Cl_2 a binaphthyl is formed. With pentaethylene glycol di(*p*-toluenesulfonate) and KO*t*-Bu in THF the binaphthyl is converted to binaphthyl crown ether, followed by hydrogenolysis to form poly(1,4-naphthylene) via asymmetric oxidative coupling polymerization. The yields are low (33%) but it gives rigid compounds with a chiral backbone that can be used in chiral recognition toward amino acids (Figure 16).

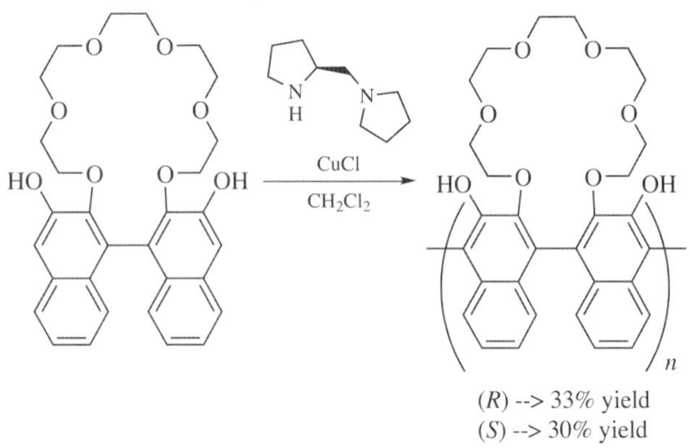

Figure 16 Polymerization reaction catalyzed by the chiral amine **1** and CuCl.

Poly(1,4-naphthylene) prepared from the (*R*)- and (*S*)-isomers of binaphthyl had a stronger ability to complex to the (*S*)- and (*R*)-isomers, respectively, of the amino acids. In addition, the amount of extracted guests and the ees are low with 2-phenylglycine methyl ester hydrochloride and 2-phenylglycine hydrochloride as guest compounds in the chiral recognition, Table 1.

Table 1 Chiral recognition of poly (1,4-naphthylene) toward amino acids with CuCl-(+)PMP as catalyst

Guest	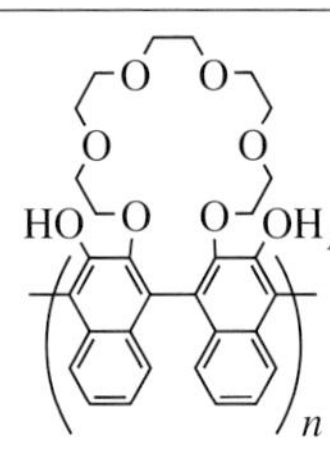	Amount extracted guest (%)	ee of extracted guest (%)
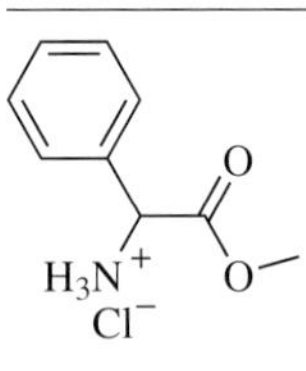			
	(*R*)	20.1	5.4 (*S*)
	(*S*)	21.4	5.7 (*R*)
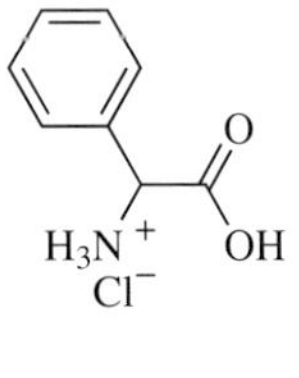			
	(*R*)	2.2	48.1 (*S*)
	(*S*)	3.5	47.7 (*R*)

Availability. The diamine has been commercially available at a relatively moderate cost since about 1997. Mohammed and Ahlberg have reported a short and efficient synthesis of the diamine from an oxazolidinone and reduction with $LiAlH_4$ (Figure 17).[26]

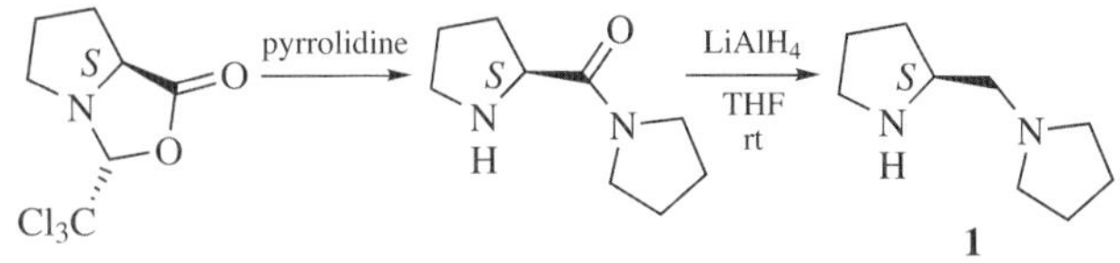

Figure 17 A simple synthetic route to the chiral amine **1**.

In addition, (*R*)-(+)-1-(2-pyrrolidinylmethyl)pyrrolidine **2** can be synthesized from commercially available *N*-Boc-D-proline with pyrrolidine. The Boc-group is deprotected with TFA, followed by reduction with $LiAlH_4$ (Figure 18).[27]

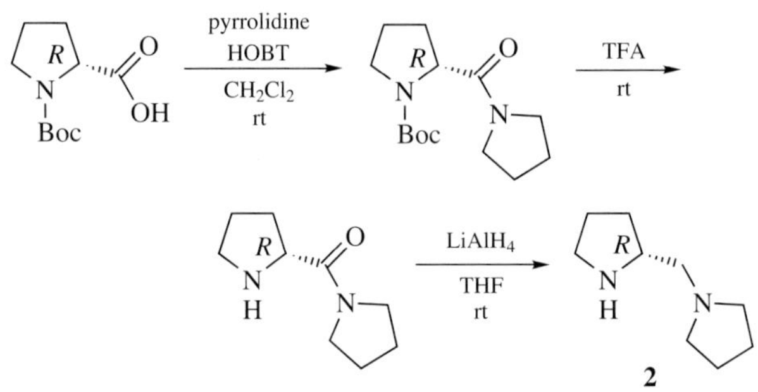

Figure 18 Synthesis of the enantiomer of **1**, that is, the chiral amine **2**.

The X-ray crystallographic characterization of **1**•HCl has been reported by Davidsson and co-workers. The diamine readily crystallizes from an HCl saturated CH_2Cl_2 solution.[28]

---

1. Cope, A. C.; Lee, H.-H.; Petree, H. E., *J. Am. Chem. Soc.* **1958**, *80*. 2849.
2. Whitesell, J. K.. Felman, S. W., *J. Org. Chem.*, **1980**, *45*, 755.
3. Asami, M., *Chem. Lett.* **1984**, 829.
4. Asami, M.; Takahashi, J.; Inoue, S., *Tetrahedron: Asymmetry* **1994**, *5*, 1649.
5. Asami, M.; Inoue, S., *Tetrahedron* **1995**, *51*, 11775.
6. Asami, M.; Ishizaki, T.; Inoue, S., *Tetrahedron* **1995**, *51*, 1893.
7. Asami, M.; Suga, T.; Honda, K.; Inoue, S., *Tetrahedron Lett.* **1997**, *37*, 6425.
8. Asami, M.; Ishizaki, T.; Inoue, S., *Tetrahedron: Asymmetry* **1994**, *5*, 793.
9. Pettersen, D.; Amedjkouh, M.; Nilsson, L. S. O.; Ahlberg, P., *J. Chem. Soc., Perkin Trans. 2* **2002**, 1397.
10. Asami, M.; Sato, S.; Honda, K.; Inoue, S., *Heterocycles* **2000**, *52*, 1029.
11. Asami, M., Seki, A., *Tetrahedron* **2002**, *58*, 4655.
12. Vestergren, M.; Eriksson, J.; Hilmersson, G.; Håkansson, M., *J. Organomet. Chem.* **2003**, *682*, 172.
13. Arvidsson, P. I.; Hilmersson, G.; Ahlberg, P., *J. Am. Chem. Soc.* **1999**, *121*, 1883.
14. (a) Nilsson, L. S. O.; Köhn, U., Anders, E., *Eur. J. Org. Chem.* **2004**, 2868.
(b) Köhn, U., Anders, E., *Tetrahedron Lett.* **2002**, *43*, 9585.

15. Hansen, H. M.; Longbottom, D. A.; Ley, S., *Chem. Commun.* **2006**, 4838.
16. Andrey, O.; Alexakis, A.; Tomassini, A.; Bernardinelli, G., *Adv. Synth. Catal.* **2004**, *346*, 1147.
17. Betancort, J. M.; Barbas, C. F., *Org. Lett.* **2001**, *3*, 3737.
18. Córdova, A., Notz, W.; Barbas, C. F., *Chem. Commun.* **2002**, 3024.
19. Gao, F.; Deng, M.; Qian, C., *Tetrahedron* **2005**, *61*, 12238.
20. Gao, F.; Zhu, J.; Tang, Y.; Deng, M.; Qian, C., *Chirality* **2006**, *18*, 741.
21. Trost, B. M.; Pinkerton, A. B.; Kremzow, D., *J. Am. Chem. Soc.* **2000**, *122* (48): 12007.
22. Sundén, H., Ibrahem, I.; Cordóva, A., *Tetrahedron Lett.* **2006**, *47*, 99.
23. Ho, C.; Chen, Y.; Wong, M.; Yang, D., *J. Org. Chem.* **2005**, *70*, 898.
24. Raja, R.; Thomas, J. M.; Jones, M. D.; Johnson, B. F. G.; Vaughan, D. E. W., *J. Am. Chem. Soc.* **2003**, *125*, 14982.
25. Habaue, S.; Seko, T.; Okamoto, Y., *Polymer* **2003**, *44*, 7377.
26. Amedjkouh, M.; Ahlberg, P., *Tetrahedron: Asymmetry* **2002**, *13*, 2229.
27. de Costa, B. R.; Dominguez, C.; He, X.; Williams, W.; Radesca, L.; Bowen, W., *J. Med. Chem.* **1992**, *35* (23): 4334.
28. Arvidsson, P. I.; Khan, A. Z.-Q.; Hilmersson, G.; Davidsson, Ö.; Håkansson, M., *Enantiomer* **1999**, *4*, 445.

Göran Hilmersson & Petra Rönnholm
Göteborg University, Göteborg, Sweden

4-(1-Pyrrolidinyl)pyridine

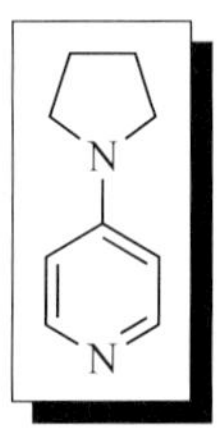

[2456-81-7] $C_9H_{12}N_2$ (MW 148.20)
InChI = 1S/C9H12N2/c1-2-8-11(7-1)9-3-5-10-6-4-9/h3-6H,1-2,7-8H2
InChIKey = RGUKYNXWOWSRET-UHFFFAOYSA-N

(nucleophilic organocatalyst)

Alternate Names: PPY; PP.
Physical Data: mp 54–58 °C.
Solubility: soluble in most organic solvents.
Form Supplied in: commercially available as a light yellow solid.
Preparative Methods: prepared by reaction of 4-chloropyridine with pyrrolidine. This S_NAr reaction requires strong activation and can be carried out efficiently under high pressure,[1] microwave irradiation,[2] or organometallic catalysis.[3]
Handling, Storage, and Precautions: skin irritant; corrosive, toxic solid.

4-(1-Pyrrolidinyl)pyridine: A Powerful Nucleophilic Organocatalyst. The use of 4-(1-pyrrolidinyl)pyridine (PPY) as an organocatalyst was first reported shortly following the discovery of the application of (4-(dimethylamino)pyridine) (DMAP) in a similar role.[4–6] 4-(1-Pyrrolidinyl)pyridine is more efficient than DMAP in some cases, presumably due to its slightly higher nucleophilicity (*N* values of 15.8 and 15.9 for DMAP and PPY, respectively).[7] The title compound was also shown to be significantly more basic than DMAP, as was quantified through the values of methyl cation affinities (MCA) that increase from 581.2 kJ mol^{-1} for DMAP to 590.1 kJ mol^{-1} for PPY.[8] As the use of PPY in place of DMAP can improve the efficiency of reactions, 4-(1-pyrrolidinyl)pyridine is an ideal organocatalyst in "difficult cases" for which DMAP reaches its limits. Such scenarios include the acetylation of hindered alcohols, as well as the direct esterification of acids with poorly nucleophilic alcohols in the presence of DCC.

Catalyst for Acylation of Alcohols in the Presence of Anhydrides. In the case of acylation of hindered alcohols, PPY proved to be a superior catalyst than DMAP. An acceleration (×1.6) in the acetylation of 1,1-diphenylethanol (neat) was reported when PPY was used in place of DMAP as a catalyst (eq 1).[9]

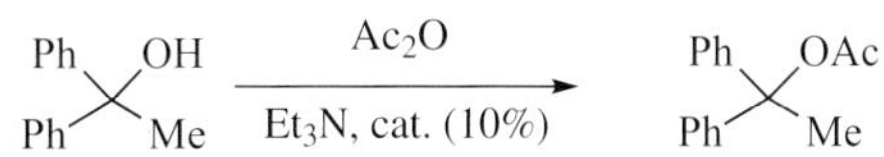

(1)

cat. = PPY, relative rate = 1
cat. = DMAP, relative rate = 0.69

The half-life reaction for the acetylation of 1-ethynyl cyclohexanol in $CDCl_3$ was studied and a 2.25-fold increase of rate was reported with PPY acting as catalyst compared to DMAP (eq 2).[10]

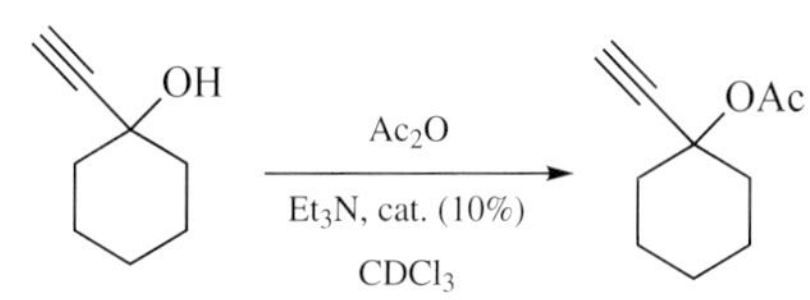

(2)

cat. = PPY, $t_{1/2}$ = 67 ± 0.1 min
cat. = DMAP, $t_{1/2}$ = 151 ± 2 min

The use of PPY for the acetylation of other tertiary alcohols has also been demonstrated (Table 1).[9]

Table 1 Acetylation of hindered tertiary alcohols catalyzed by PPY

Alcohol	Reaction Time	Yield (%)
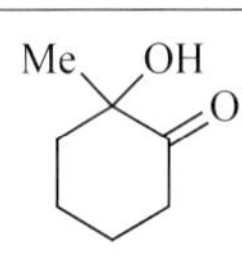	45 min	87
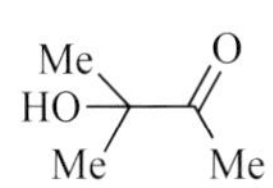	30 min	97
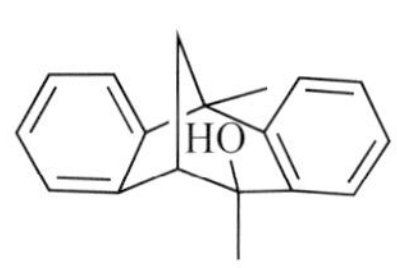	24 h	80

4-(1-Pyrrolidinyl)pyridine is an efficient catalyst for benzoylation reactions and can be much more efficient than DMAP in the case of hindered hydroxyl groups (eq 3).[11]

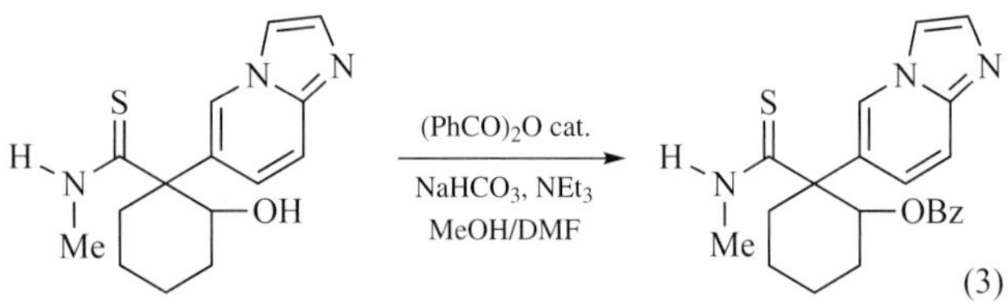

(3)

cat. = DMAP, yield = 6%
cat. = PPY, yield = 98%

Catalyst for the Direct Esterification of Alcohols and Carboxylic Acids. The title compound is a catalyst of choice for the direct esterification of poorly nucleophilic alcohols in the presence of carboxylic acids and carbodiimide derivatives: 1,3-dicyclohexylcarbodiimide (DCC) or 1-ethyl-3-(3-dimethylaminopropyl)carbodiimide (EDC). In these reactions PPY increases the rate of reaction by reacting with the intermediate *O*-acylurea, generating an acyl pyridinium that reacts further with the alcohol. Thus, carboxylic acids can be directly activated as pentafluorophenyl esters,[12] as *N*-hydroxysuccinimide esters,[13] or as trichlorophenyl esters in order to perform Yamaguchi macrolactonization (eq 4).[14]

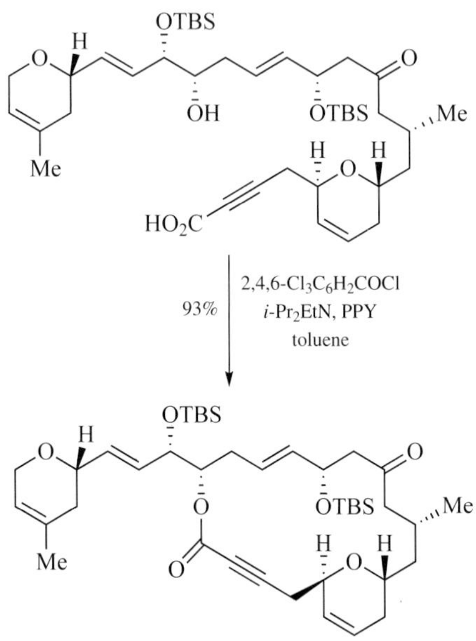

(4)

Macrolactonization of seco acids can also be conducted via activation of the acid through a mixed carbonate, first prepared by reacting the acid with Boc_2O (di-*tert*-butyl dicarbonate) with NEt_3. The corresponding intermediate is efficiently lactonized under high dilution, by reaction with DMAP or PPY. The title compound proved to be more efficient than DMAP (eq 5).[15]

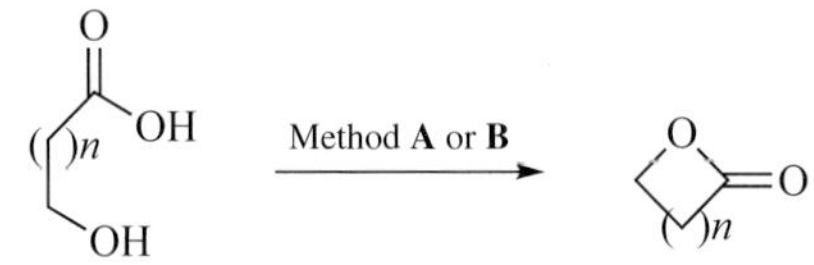

Method **A**: Boc_2O, Et_3N then DMAP, toluene, 90 °C, 0.002 M
Method **B**: Boc_2O, *i*-Pr_2NEt then PPY, toluene, 90 °C, 0.002 M

n	Yield (%) **A**	Yield (%) **B**
9	39	45
10	44	67
13	71	87

(5)

Carboxylic acids can be transformed into *tert*-butyl esters by reaction with Boc_2O in the presence of PPY. In this example, PPY demonstrated modest improvement of efficiency relative to DMAP.[16] Phenols can be acylated efficiently,[17] and direct esterifications involving elaborate carboxylic acids and alcohols are catalyzed with PPY (eq 6)[18]

Catalyst for *N*-Acylation of Sulfonamides. 4-(1-Pyrrolidinyl)pyridine catalyzes the formation of *N*-acyl[19] or *N*-carboxyl sulfonamides.[20] Upon reaction of DCC and PPY, *N*-tosyl β-lactams can be efficiently prepared (eq 7). Sulfonamides are carboxylated by reaction with chloroformates in pyridine, in the presence of PPY (eq 8).

Miscellaneous Reactions. Transesterification processes can be conducted in the presence of PPY and *N*,*N*(-[3,5-bis(trifluoromethyl)phenyl]thiourea.[21] In this example, PPY proved more efficient than DMAP (eq 9).

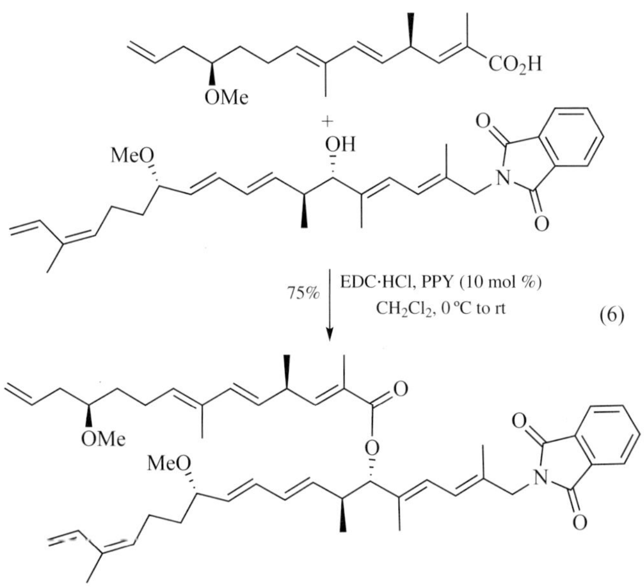

(6)

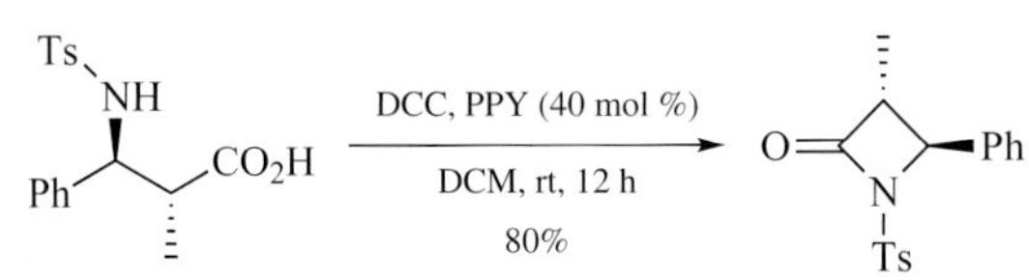

(7)

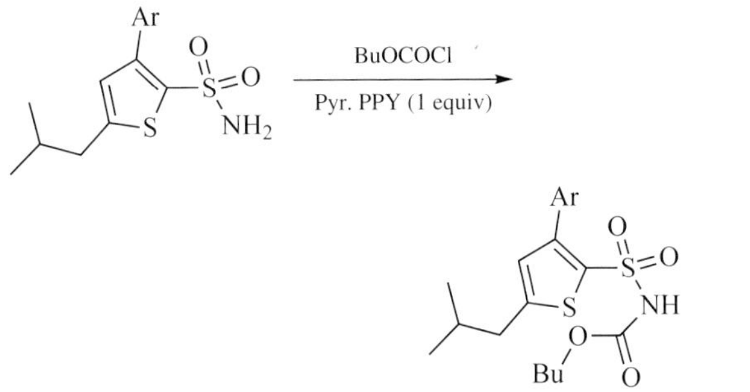

(8)

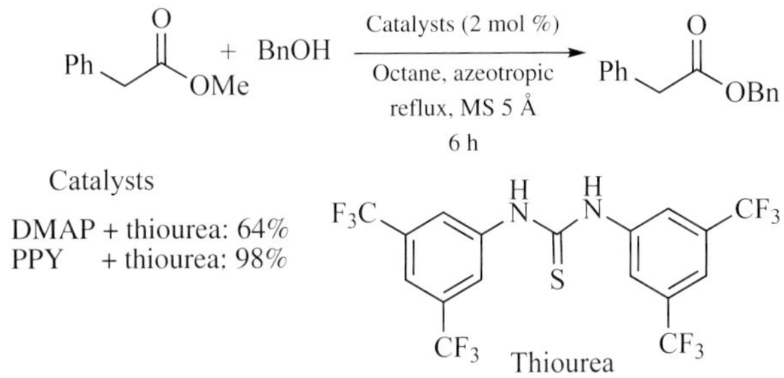

(9)

4-(1-Pyrrolidinyl)pyridine can add reversibly to conjugated alkenes and catalyze reactions involving this elementary step in their mechanism. Relative to DMAP, the title compound almost doubles the rate of *aza*-Morita-Baylis-Hillman reactions (eq 10).[10b]

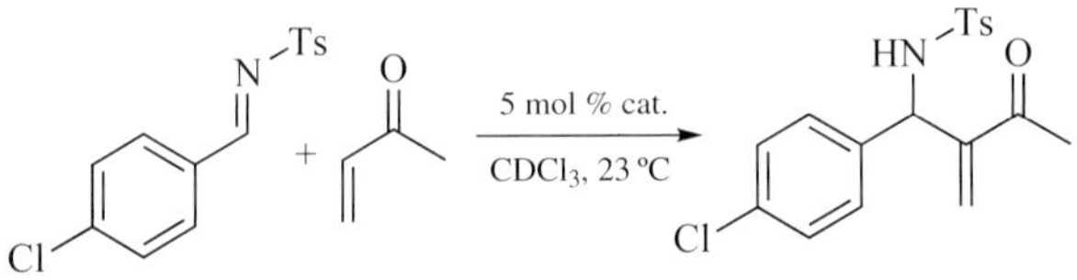

cat.
DMAP, $t_{1/2} = 745 \pm 2$ min
PPY, $t_{1/2} = 355 \pm 2$ min

(10)

In certain cases, 4-(1-pyrrolidinyl)pyridine improves the conjugate addition of alcohols on 2-nitroglycals[22] (eq 11). It catalyzes the enantioselective addition of malonates to enones in the presence of chiral thiourea.[23] Carboxylic acids can be activated to ketenes, leading to pyridinium enolates upon reaction with PPY and methanesulfonyl chloride, which was used as a key step for the synthesis of (−)-salinosporamide.[24]

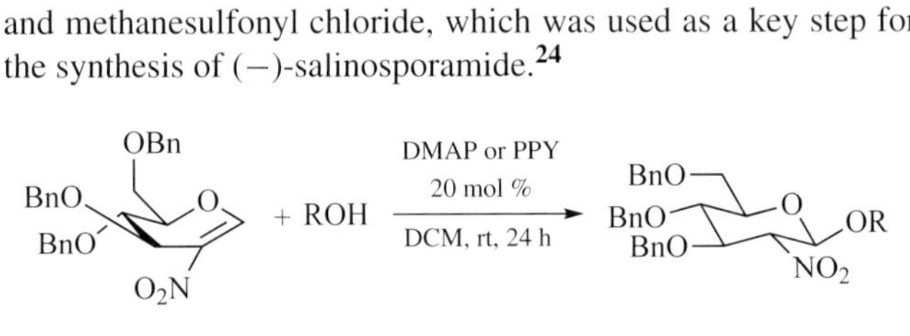

ROH	Yield with DMAP (%)	Yield with PPY (%)
HO⁀TMS	50	95
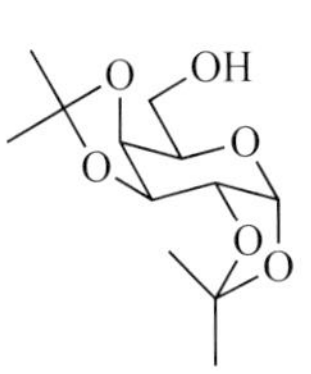	17	71

(11)

Other Transformations of Alcohols Catalyzed by PPY. Similarly to DMAP, PPY can catalyze the transformation of alcohols into tosylates[25] of silyl ethers[26] or carbonates.[27] When DMAP denotes poor reactivity or is nonselective, it may be advantageous to replace it by PPY. For example, selective carbonate protection of the diol shown in eq 12 by di-*tert*-butyl dicarbonate catalyzed by PPY is highly regioselective while DMAP gives a mixture of regioisomers.[27]

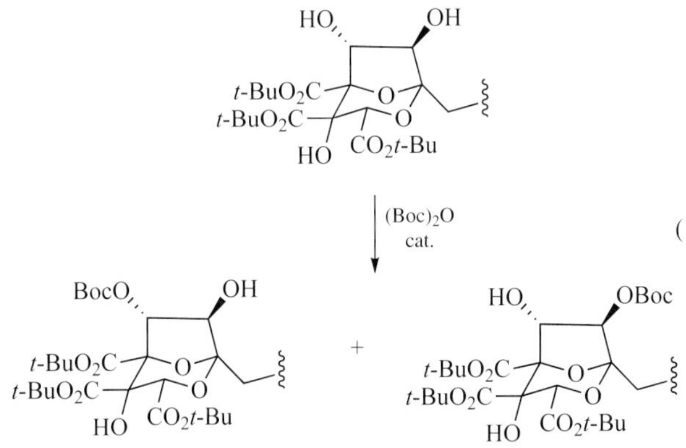

(12)

cat. = DMAP:	1	:	3 (Yield = 54%)
cat. = PPY:	1	:	>20 (Yield = 80–85%)

PPY can catalyze the elimination of hindered tertiary alcohols upon reaction with trifluoroacetic anhydride (TFAA) and triethylamine.[28] Chiral versions of PPY have been developed in order to catalyze enantioselective transformations such as acyl transfer[29] or kinetic resolutions of racemic alcohols through acetylation.[30]

Related Reagents. DMAP (4-(dimethylamino)pyridine); tributylphosphine; DABCO (1,4-diazabicyclo[2.2.2]octane).

1. Kotsuki, H.; Sakai, H.; Shinoura, T., *Synlett* **2000**, *1*, 116.
2. Narayan, S.; Seelhammer, T.; Gawley, R. E., *Tetrahedron Lett.* **2004**, *45*, 757.
3. Gradel, B.; Brenner, E.; Schneider, R.; Fort, Y., *Tetrahedron Lett.* **2001**, *42*, 5689.
4. Steglich, W.; Höfle, G., *Tetrahedron Lett.* **1970**, *54*, 4727.
5. Litvinenko, L. M.; Kirichenko, A. I., *Dokl. Akad. Nauk SSSR, Ser. Khim.* **1967**, *176*, 97.
6. Steglich, W.; Höfle, G., *Angew. Chem., Int. Ed.* **1969**, *8*, 981.
7. Brotzel, F.; Kempf, B.; Singer, T.; Zipse, H.; Mayr, H., *Chem. Eur. J.* **2007**, *13*, 336.
8. Lindner, C.; Tandon, R.; Maryasin, B.; Larionov, E.; Zipse, H., *Beilstein J. Org. Chem.* **2012**, *8*, 1406.
9. Hassner, A.; Krepski, L. R.; Alexanian, V., *Tetrahedron*, **1978**, *34*, 2069.
10. (a) Held, I.; Xu, S.; Zipse, H., *Synthesis*, **2007**, 1185. (b) D'Elia, V.; Liu, Y.; Zipse, H., *Eur. J. Org. Chem.* **2011**, 1527.
11. (a) Eisai Co., Ltd, U.S. Pat. 5,444,066A1 (1995). (b) Hart, T.; Guillochon, D.; Perrier, G.; Sharp, B. W.; Toft, M. P.; Vacher, B.; Walsh, R. J. A., *Terahedron Lett.* **1992**, *33*, 7211.
12. (a) Yamada, H.; Imahori, H.; Nishimura, Y.; Yamazaki, I.; Ahn, T. K.; Kim, S. K.; Kim, D.; Fukuzumi, S., *J. Am. Chem. Soc.* **2003**, *125*, 9129. (b) Schmidt, U.; Lieberknecht, A.; Griesser, H.; Talbiersky, J., *J. Org. Chem.* **1982**, *47*, 3261.

13. Vávrová, K.; Hrabálek, A.; Doležal, P.; Šalamová, L.; Palát, K.; Zbytovská, J.; Holas, T.; Klimentová, J., *Bioorg. Med. Chem. Lett.* **2003**, *11*, 5381.
14. Nelson, S. G.; Cheung, W. S.; Kassiks, A. J.; Hilfiker, M. A., *J. Am. Chem. Soc.* **2002**, *124*, 13654.
15. Nagarajan, M.; Kumar, V. S.; Rao, V., *Tetrahedron* **1999**, *55*, 12349.
16. Held, I.; von den Hoff, P.; Stephenson, D. S.; Zipse, H., *Adv. Synth. Catal.* **2008**, *350*, 1891.
17. Curini, M.; Epifano, F.; Maltese, F.; Marcotullio, M. C.; Tubaro, A.; Altinier, G.; Prieto Gonzales, S.; Rodriguez, J. C., *Bioorg. Med. Chem. Lett.* **2004**, *14*, 2241.
18. (a) Gagnepain, J.; Moulin, E.; Nevado, C.; Waser, M.; Maier, A.; Kelter, G.; Fiebig, H.-H.; Fürstner, A., *Chem. Eur. J.* **2011**, *17*, 6973. (b) Pettit, G. R.; Flahive, E. J., US Pat. 6,686,445B1 (2004). (c) Xiao, X.; Wua, J.; Trigili, C.; Chen, H.; Chu, J. W. K.; Zhao, Y.; Lu, P.; Sheng, L.; Li, Y.; Sharom, F. J.; Barasoain, I.; Diaz, F. J.; Fang, W.-S., *Bioorg. Med. Chem. Lett.* **2011**, *20*, 4852.
19. Davis, F. A.; Minsoo, H. Q.; Gaddiraju, N. V., *J. Org. Chem.* **2009**, *74*, 2798.
20. Mahalingam, A. K.; Wan, Y.; Murugaiah, A. M. S.; Wallinder, C.; Wu, X.; Plouffe, B.; Botros, M.; Nyberg, F.; Hallberg, A.; Gallo-Payet, N.; Alterman, M., *Bioorg. Med. Chem.* **2010**, *18*, 4570.
21. Ishiara, K.; Niwa, M.; Kosugi, Y., *Org. Lett.* **2008**, *10*, 2187.
22. Xue, W.; Sun, J.; Yu, B., *J. Org. Chem.* **2009**, *74*, 5079.
23. Moritaka, M.; Miyamae, N.; Nakano, K.; Ichikawa, Y.; Kotsuki, H., *Synlett* **2012**, 2554.
24. Nguyen, H.; Ma, G.; Gladisheva, T.; Fremgen, T.; Romo, D., *J. Org. Chem.* **2011**, *76*, 2.
25. Sokolov, V. V.; Kozhushkov, S. I.; Nikolskaya, S.; Belov, V.; Es-Sayed, M.; de Meijere, A., *Eur. J. Org. Chem.* **1998**, 777.
26. Danishevsky, S.; Morris, J.; Mullen, G.; Gammill, R., *J. Am. Chem. Soc.* **1982**, *104*, 7591.
27. Carreyra, E.; Du Bois, J., *J. Am. Chem. Soc.* **1995**, *117*, 8106.
28. Asaoka, M.; Yanagida, M.; Ishibashi, K.; Takei, H., *Tetrahedron Lett.* **1981**, *22*, 4269.
29. (a) Nguyen, H. V.; Butler, D. C. D.; Richards, C. J., *Org. Lett.* **2006**, *8*, 769. (b) Ruble, C. J.; Fu, G. C., *J. Am. Chem. Soc.* **1998**, *120*, 11532.
30. (a) Dàlaigh, C. O.; Hynes, S. J.; Maher, D. J.; Connon, S. J., *Org. Biomol. Chem.* **2005**, *3*, 981. (b) Spivey, A. C.; Maddaford, A.; Fekner, T.; Redgrave, A. J.; Frampton, C. S., *J. Chem. Soc., Perkin Trans. 1*, **2000**, 3460.

François Couty & Olivier David
Institut Lavoisier de Versailles, Versailles, France

5H-Pyrrolo[2,1-*c*]-1,2,4-triazolium, 7-Fluoro-6,7-dihydro-5-(1-methylethyl)-2-(2,3,4,5,6-pentafluorophenyl)-, (5R,7*R*)-, Tetrafluoroborate(1-)

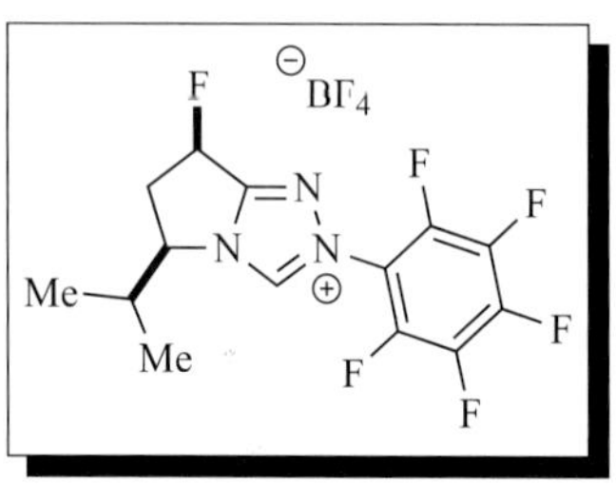

[1311991-39-5] $C_{12}H_6BF_{10}N_3$ (MW 392.99)

InChI = 1S/C14H12F6N3.BF4/c1-5(2)7-3-6(15)14-21-23(4-22(7)14)13-11(19) 9(17)8(16)10(18)12(13)20;2-1(3,4)5/h4-7H,3H2,1-2H3;/q+1;-1/t6-,7-;/m1./s1

InChIKey = LJRYBUKHSUIFRP-ZJLYAJKPSA-N

(used to generate chiral acyl anion and homoenolate equivalents from aldehydes)

Physical Data: mp 154–155 °C; $[\alpha]D^{21}$ = +22.8 (c = 0.010 g ml^{-1}, MeOH).

Solubility: soluble in most organic solvents.

Form Supplied in: colorless solid; not widely available.

Preparative Methods: chiral pool synthesis starting from L-valine.[1]

Purification: crystallized from EtOAc.

Handling, Storage, and Precautions: indefinitely stable solid under ambient conditions.

Asymmetric Stetter Reactions. 5*H*-Pyrrolo[2,1-*c*]-1,2,4-triazolium, 7-fluoro-6,7-dihydro-5-(1-methylethyl)-2-(2,3,4,5,6-pentafluorophenyl)-, (5*R*,7*R*)-, tetrafluoroborate(1-) (**1**) catalyzes the formal enantioselective 1,4-addition of aldehydes to β-substituted nitroalkenes[1] and ketoesters[2] in the presence of base. Hindered amine bases and polar protic solvents are generally preferred to obtain the best selectivities and highest reactivity. Hetaryl aldehydes react with β-substituted nitroalkenes in high yield and with excellent enantioselectivity. Enals are also competent nucleophiles in the Stetter reaction with nitroalkenes; however, catechol must be used as an additive to enhance catalytic turnover due to a markedly slow proton transfer event.[3] Primary alkyl substitution of the nitroalkene leads to diminished enantioselectivities. Furthermore, α-substitution of the enal is not tolerated (eqs 1 and 2).

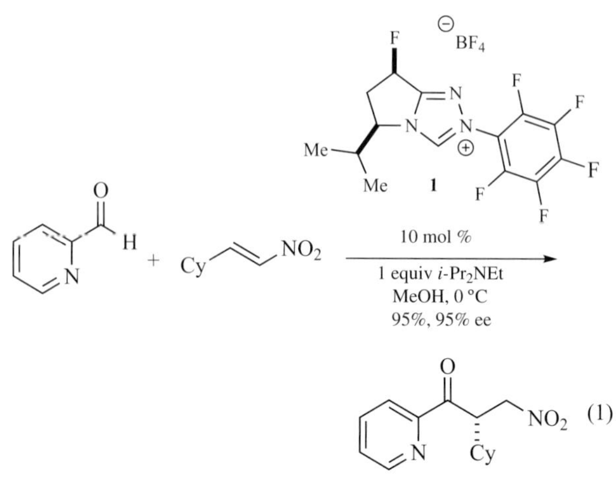

(1)

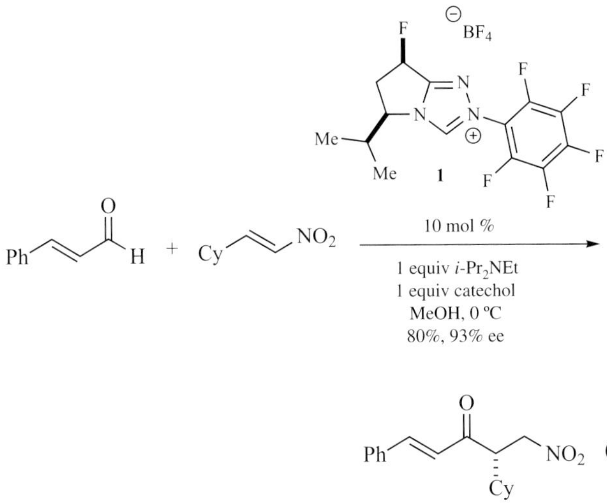

(2)

β,γ-Unsaturated-α-ketoesters can be coupled with hetaryl aldehydes in high yield and with excellent enantioselectivity under modified conditions.[2] Although mainly limited to hetaryl aldehydes, an example of a modestly selective reaction was reported with the use of an activated benzaldehyde derivative (eq 3).

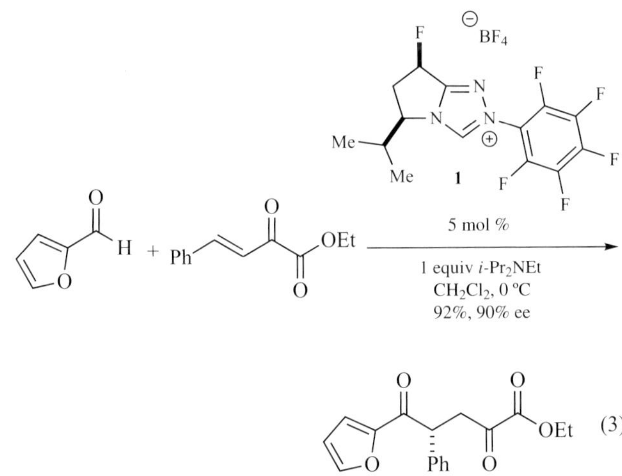

(3)

Asymmetric Homoenolate Processes. 5*H*-Pyrrolo[2,1-*c*]-1,2,4-triazolium, 7-fluoro-6,7-dihydro-5-(1-methylethyl)-2-(2,3,4,5,6-pentafluorophenyl)-, (5*R*,7*R*)-, tetrafluoroborate(1-) catalyzes the enantioselective homo- and heterodimerization of enals in the presence of 4-nitro-pyridine *N*-oxide, providing aryl-substituted cyclopentanones in high yield and with good enantioselectivity.[4] While heterodimerizations are shown to proceed in good yield and with high enantioselectivity, an excess of one substrate must be used to prevent homodimerization (eq 4).

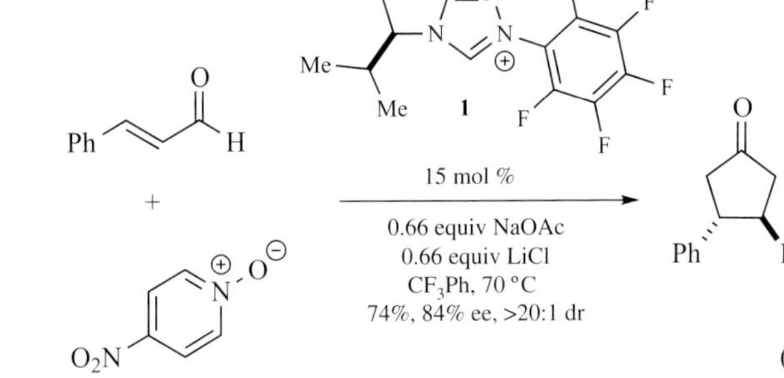

(4)

5*H*-Pyrrolo[2,1-*c*]-1,2,4-triazolium, 7-fluoro-6,7-dihydro-5-(1-methylethyl)-2-(2,3,4,5,6-pentafluorophenyl)-, (5*R*,7*R*)-, tetrafluoroborate(1-) catalyzes the enantio- and diastereoselective homoenolate addition of enals to β-substituted nitroalkenes, exemplified in a two-step synthesis of (−)-paroxetine (eq 5).[5]

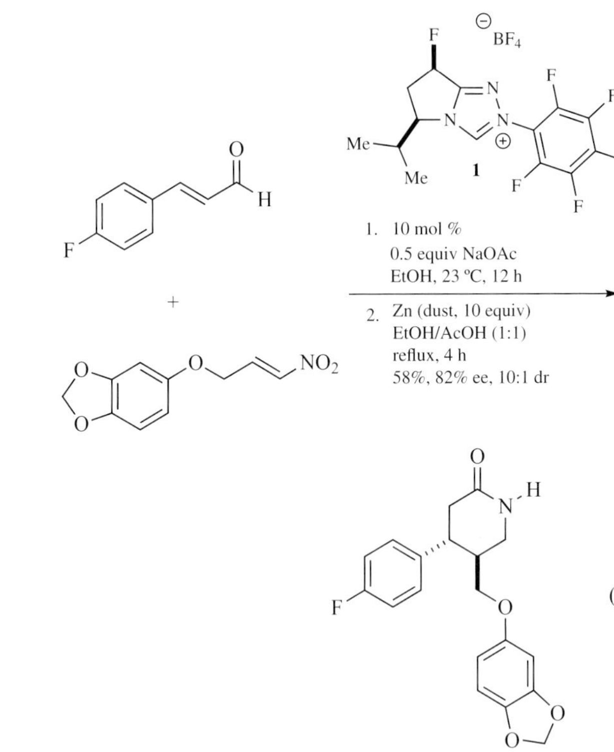

(5)

Related Reagents. (5*R*,7*S*)-5-(*tert*-Butyl)-7-fluoro-2-(perfluorophenyl)-6,7-dihydro-5*H*-pyrrolo[2,1-*c*][1,2,4]triazol-2-ium tetrafluoroborate (**2**) [1355451-60-3] has been shown to catalyze a highly enantioselective Stetter reaction between aliphatic aldehydes and β-nitrostyrenes under mild conditions, employing CsOAc as base. The scope of the linear aliphatic aldehydes is good; however, α-branching is not well tolerated. DFT calculations provide evidence to suggest the unique effectiveness of the catalyst is due to unfavorable electrostatic interactions between the nitro group and the F atom in the transition state (eq 6).[6]

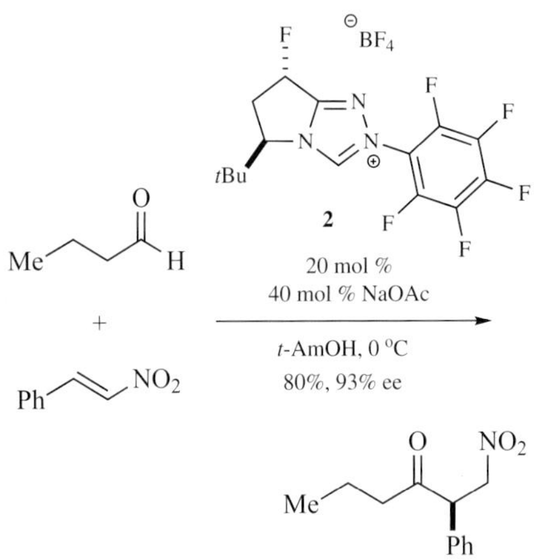

(6)

1. DiRocco, D. A.; Oberg, K. M.; Dalton, D. M.; Rovis, T., *J. Am. Chem. Soc.* **2009**, *131*, 10872.
2. Sánchez-Larios, E.; Thai, K.; Bilodeau, F.; Gravel, M., *J. Org. Chem.* **2011**, *13*, 4942.
3. DiRocco, D. A.; Rovis, T., *J. Am. Chem. Soc.* **2011**, *133*, 10402.
4. White, N. A.; Rovis, T., *J. Am. Chem. Soc.* **2015**, *137*, 10112.
5. White, N. A.; Ozboya, K. E.; Flanigan, D. M.; Rovis, T., *Asian J. Org. Chem.* **2014**, *3*, 442.
6. DiRocco, D. A.; Noey, E. L.; Houk, K. N.; Rovis, T., *Angew. Chem., Int. Ed.* **2012**, *51*, 2391.

Daniel A. DiRocco
Merck & Co Inc, Kenilworth, NJ, USA

Q

Quinine

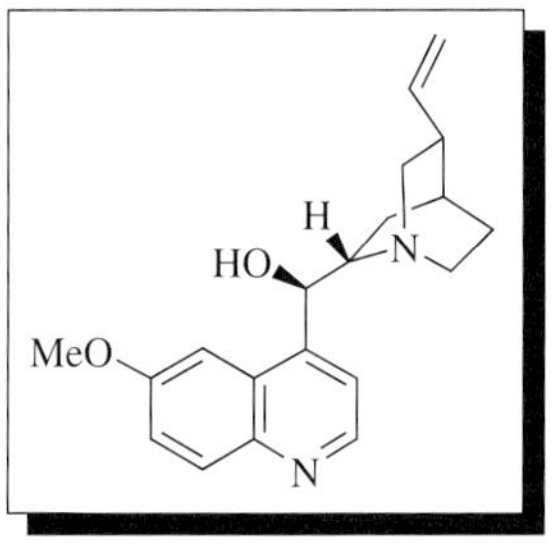

[130-95-0] $C_{20}H_{23}N_2O_2$ (MW 324.45)
InChI = 1/C20H24N2O2/c1-3-13-12-22-9-7-14(13)10-19(22)20(23)16-6-8-21-18-5-4-15(24-2)11-17(16)18/h3-6,8,11,13-14,19-20,23H,1,7,9-10,12H2,2H3/t13-,14-,19-,20-/m0/s1
InChIKey = LOUPRKONTZGTKE-FEBSWUBLSA-N

(chiral catalyst[1–14])

Physical Data: mp 173–175 °C.
Solubility: sol hot water, methanol, benzene, chloroform, ether, glycerol; insol pet ether.
Form Supplied in: crystalline solid; 90% purity.
Analysis of Reagent Purity: NMR, mp.
Preparative Methods: commercially available from several sources.
Purification: recrystallize from absolute ethanol.
Handling, Storage, and Precautions: toxic; irritant.

Original Commentary

Ellen M. Leahy
Affymax Research Institute, Palo Alto, CA, USA

Asymmetric Diels–Alder Reactions. Chiral bases, including quinine, have been used as catalysts in Diels–Alder reactions (eq 1).[1] The reactions take place at room temperature or below and require 1–10% equiv of the alkaloid. The asymmetric induction that is observed can be attributed to complex formation between the achiral dienolate and the chiral amine.[1]

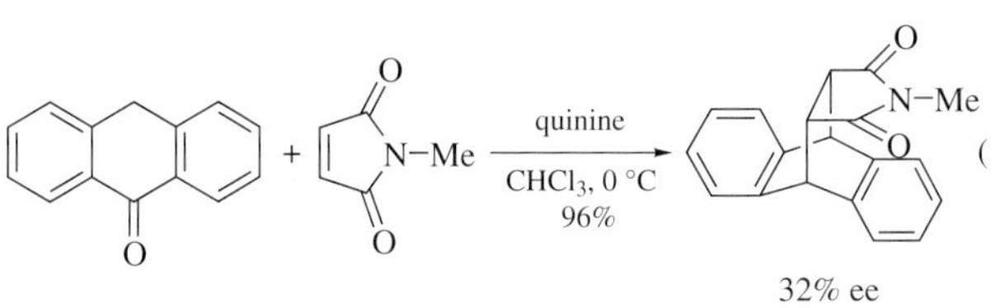
(1)

Preparation of Chiral Sulfinates. Optically active sulfinates can be prepared by reaction of a symmetrical sulfite with ***tert-Butylmagnesium Chloride*** in the presence of an optically active amino alcohol. The best enantioselectivity has been observed using quinine as the optically active amine (eq 2).[2] An alternative approach to this new enantioselective asymmetric synthesis of alkyl *t*-butylsulfinates would be reaction of a racemic sulfinate with *t*-butylmagnesium chloride complexed by optically active alkaloids (eq 3).[2] In this case, kinetic resolution of the racemic sulfinate leads to an optically active sulfinate and an optically active sulfoxide.

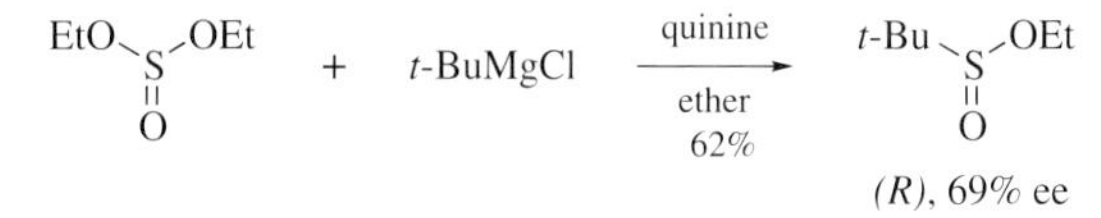
(2)

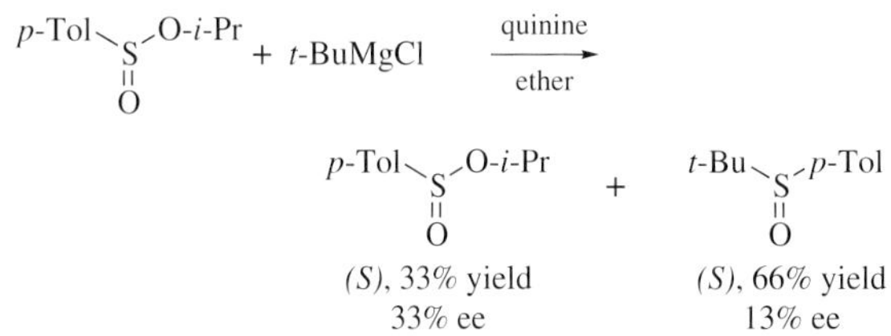
(3)

Stereoselective Addition of Diethylzinc to Aldehydes. Wynberg has found that the cinchona alkaloids catalyze the reaction of ***Diethylzinc*** and aldehydes to form optically active alcohols (eq 4).[3] The highest enantiomeric excess obtained was from reactions which used quinine as the catalyst. Results show that the hydroxyl group of the catalyst hydrogen bonds with the aldehyde and that the diethylzinc interacts with the vinyl group of the catalyst as well, but it has not been determined if one or two catalyst molecules are involved in the transition state. Similar results have been obtained using a furan aldehyde.[4]

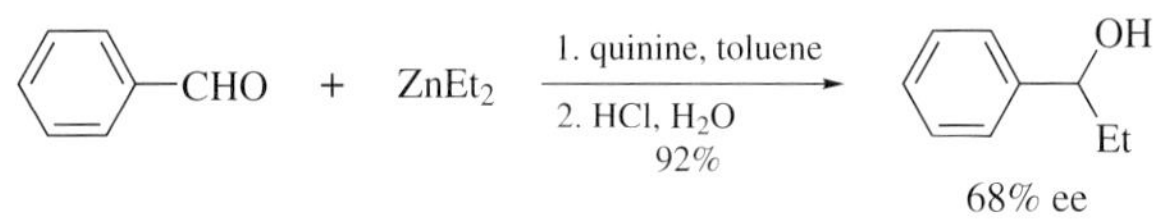
(4)

Synthesis of Optically Active Epoxides. Alkaloids and alkaloid salts have been successfully used as catalysts for the asymmetric synthesis of epoxides. The use of chiral catalysts such as quinine or quinium benzylchloride (QUIBEC) have allowed access to optically active epoxides through a variety of reaction conditions, including oxidation using ***Hydrogen Peroxide*** (eq 5),[5] Darzens condensations (eq 6),[6] epoxidation of ketones by ***Sodium Hypochlorite*** (eq 7),[6] halohydrin ring closure (eq 8),[6] and cyanide addition to α-halo ketones (eq 9).[6] Although the relative stereochemistry of most of the products has not been determined, enantiomerically enriched materials have been isolated. A more recent example has been published in which optically active 2,3-epoxycyclohexanone has been synthesized by oxidation with **tert-Butyl Hydroperoxide** in the presence of QUIBEC and the absolute stereochemistry of the product established (eq 10).[7]

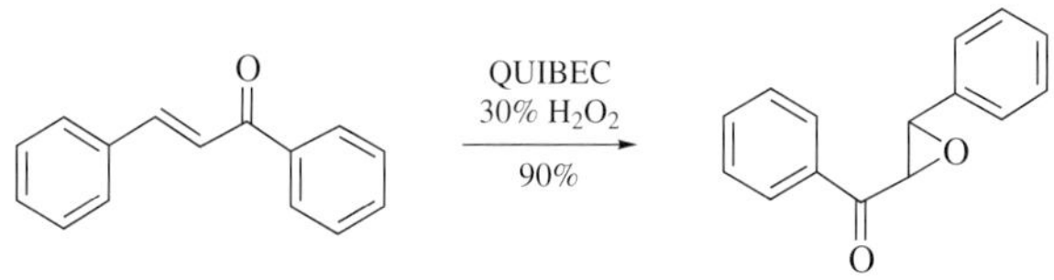
(5)

(optical purity not determined)

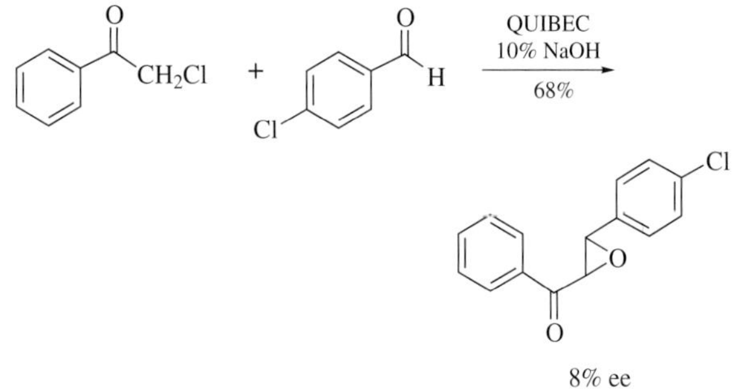

(6)

(7)

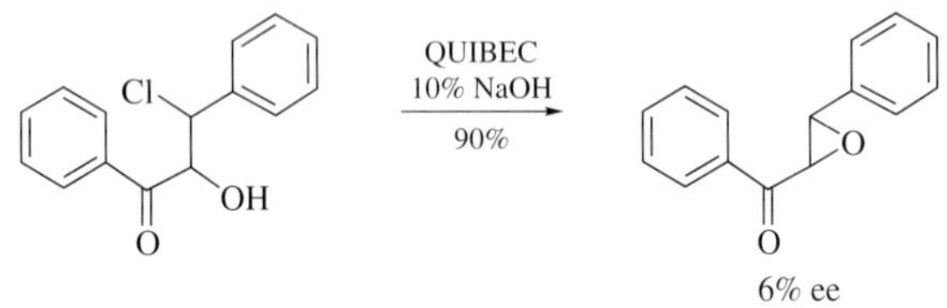

(8)

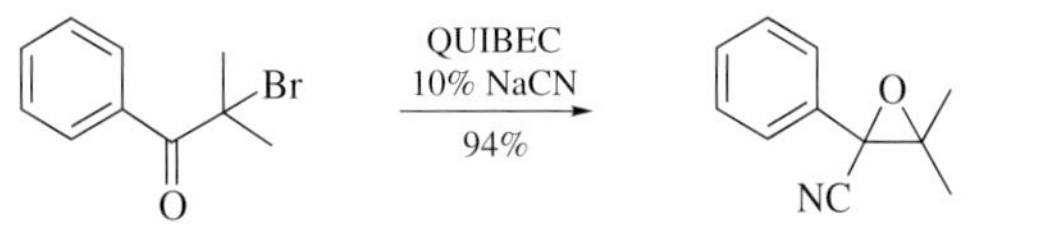

(9)

(optical purity not determined)

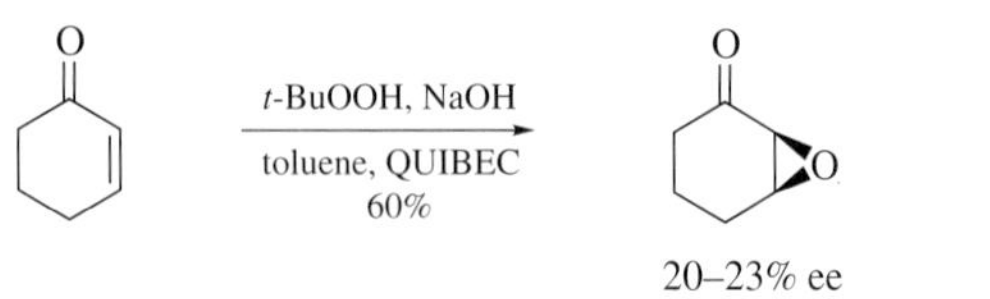

(10)

Asymmetric Michael Reactions. Asymmetric induction has been observed in Michael-type addition reactions that are catalyzed by chiral amines.[8] The *N*-benzyl fluoride salt of quinine has been particularly successful since the fluoride ion serves as a base and the aminium ion as a source of chirality.[9] Drastic improvements in optical purity (1–23%) have resulted by changing from quinine to the *N*-benzyl fluoride salt (eq 11).[9]

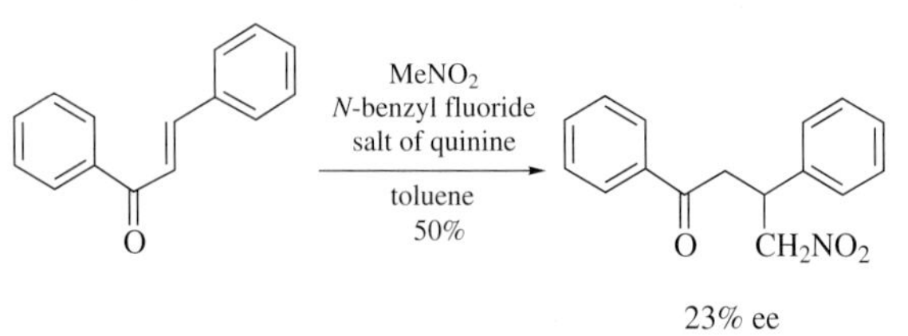

(11)

Asymmetric Synthesis of β-Keto Sulfides. Quinine can be used to catalyze asymmetrically the addition of thiols to cyclohexenone, thus forming β-keto sulfides (eq 12).[10] The absolute stereochemistry of the products has not been determined.

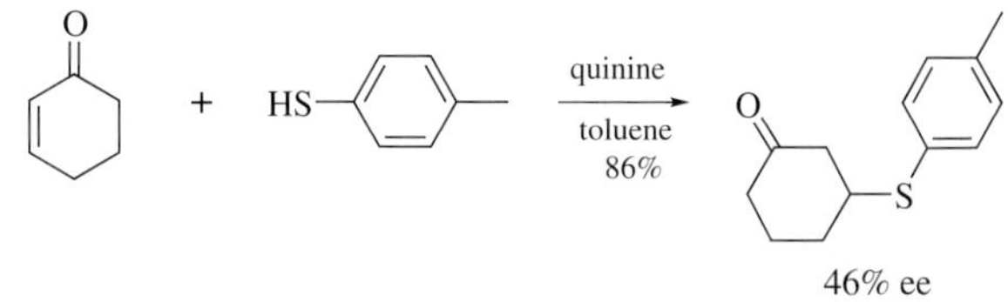

(12)

Asymmetric Reduction of Ketones. Alkyl phenyl ketones can be asymmetrically reduced to the corresponding alcohol using ***Sodium Borohydride*** under phase-transfer conditions in the presence of a catalytic amount of QUIBEC (eq 13).[11] The results indicate that the asymmetric reduction is due to the rigidity of the catalyst as well as the β-position of the hydroxyl group on the quinine molecule. The asymmetric induction is much lower with a γ-hydroxyl group.[11]

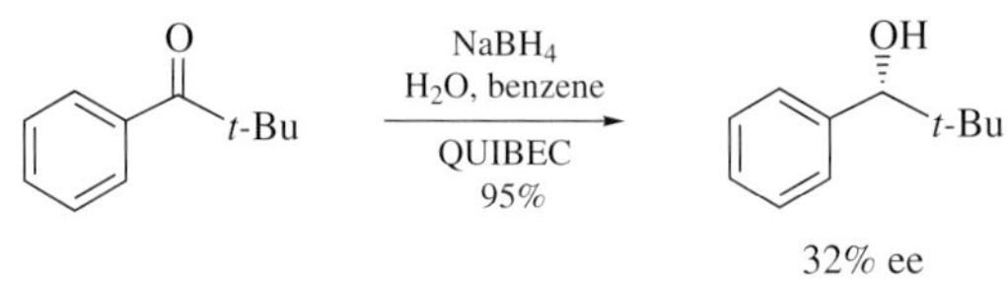

(13)

Synthesis of Optically Active β-Hydroxy Esters. Chiral amino alcohols such as quinine have been used in the enantioselective synthesis of β-hydroxy esters via an indium-induced Reformatsky reaction (eq 14).[12] Although the enantioselectivities are not particularly high, aromatic aldehydes have produced the best results to date. The absolute stereochemistry of the products has not yet been assigned.

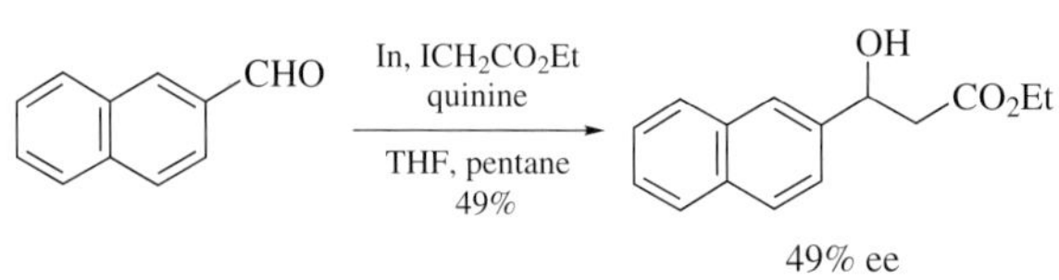

(14)

Preparation of Polymeric Catalyst. A quinine/***Acrylonitrile*** copolymer has been successfully synthesized via radical polymerization using ***Azobisisobutyronitrile*** (AIBN) as initiator (eq 15).[13] The polymer can be prepared such that the vinyl group is the connecting site and the amino alcohol portion can either be free or protected. These copolymers are thermally stable and are soluble in polar aprotic solvents such as DMF and DMSO, but insoluble in common organic solvents. Preliminary experiments have shown that these copolymers can be used as asymmetric catalysts.[13]

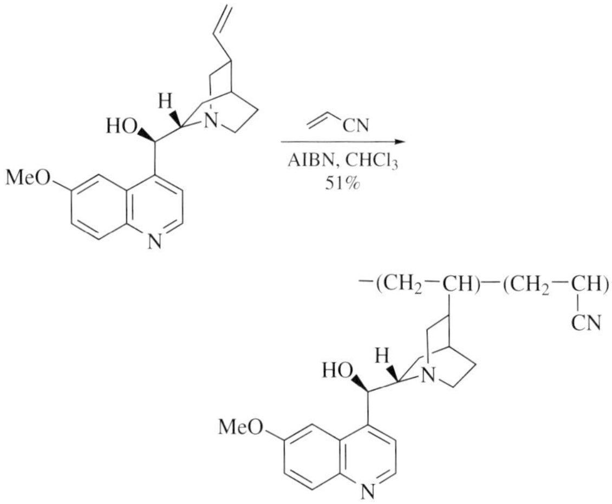

(15)

Asymmetric Addition of Thioglycolic Acid to Nitro Alkenes. Quinine has been used to catalyze the addition of thioglycolic acid to nitro alkenes (eq 16).[14] Enantiomerically enriched materials have been isolated, although the absolute stereochemistry of the products has not been assigned. The direction and extent of asymmetric induction seems to be dependent on the catalyst/acid ratio, thereby pointing to interaction between the carbonyl of the acid and the alkaloid nitrogen as being responsible for the asymmetric induction.[14]

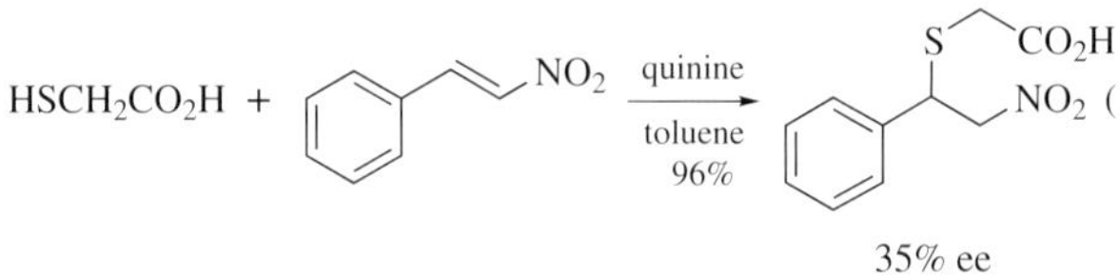

(16)

First Update

Yongwei Wu & Li Deng
Brandeis University, Waltham, MA, USA

Asymmetric reactions with cinchona alkaloids experienced an explosive growth during the last three decades. Nowadays it is well known to an organic chemist that cinchona alkaloids are one of the most "privileged" scaffolds for catalyst development in asymmetric catalysis, promoting numerous reactions of fundamentally distinct mechanisms with excellent reactivity and selectivity.[15] Cinchona alkaloids include two pairs of "pseudoenantiomers": quinine and quinidine, cinchonidine and cinchonine, all of which can be separated from the bark of cinchona trees and thus are readily available and inexpensive. This allows similarly facile access to both enantiomers of chiral products of particular interest generated by cinchona alkaloid-catalyzed asymmetric reactions. In most cases, quinidine- or cinchonine-derived catalysts afford reactivity and selectivity similar to those provided by quinine or cinchonidine derivatives. The easy accessibility of cinchona alkaloids and the presence of several sites for various derivatizations render cinchona alkaloids one of the most favored chiral backbones for catalyst discovery and development in asymmetric synthesis.

Cinchona Alkaloids as Ligands in Metal-Catalyzed Reactions. The basic quinuclidine nitrogen renders cinchona alkaloids effective ligands for chiral metal catalysts that proved to be remarkably efficient for the promotion of asymmetric transformations of synthetic importance.

Dihydroxylation of Olefins. The Sharpless asymmetric dihydroxylation of olefins represents one of the most important milestones in the field of asymmetric catalysis.[16] Consistently high enantioselectivity over an extremely broad range of simple olefins is achieved through chiral Os–metal complexes employing quinine-derived cinchona alkaloids such as $(DHQ)_2PHAL$ or their pseudoenantiomeric quinidine-derived counterparts. Prepackaged stable mixtures of $(DHQ)_2PHAL$, $K_2OsO_4{\cdot}2H_2O$, K_2CO_3, and $K_3Fe(CN)_6$ are made commercially available as AD-mix-α. Optically active diols could be reliably obtained in good yields and with high enantioselectivity from substituted terminal olefins and *trans*-disubstituted and trisubstituted olefins (eq 17).

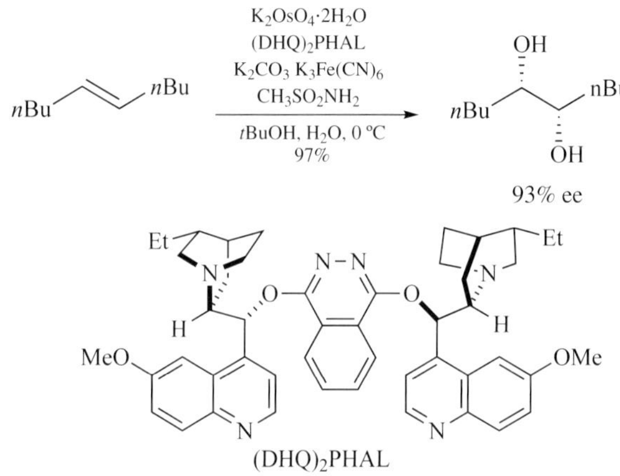

(17)

Aminohydroxylation of Olefins. Asymmetric aminohydroxylation of olefins, a distinct variant of the Sharpless asymmetric dihydroxylation, establishes a broadly useful catalytic approach for the preparation of optically active amino alcohols. However, regioselectivity was low unless the substituents of the olefins differ significantly from each other, such as cinnamates.[17] With Os complex derived from $(DHQ)_2PHAL$ and $MeSO_2N$-ClNa as the nitrogen source, the Sharpless aminohydroxylation of methyl cinnamate proceeded with excellent enantioselectivity and high regioselectivity (eq 18).[18]

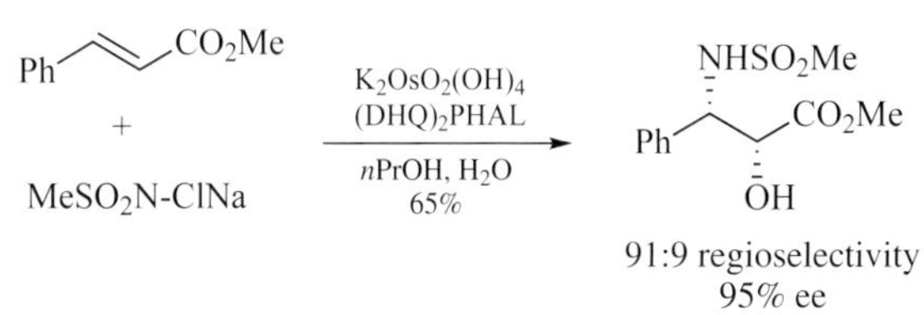

(18)

Reactions with Cinchona Alkaloids as Phase Transfer Catalysts. Cinchona alkaloid-derived quaternary ammonium salts constitute the most explored class of phase transfer catalysts. These ammonium salts serve as chiral phase transfer catalysts (PTCs) by forming ion pair with the anionic nucleophiles and then directing them to react with electrophiles to form optically active products. Because of their availability and demonstrated ability to promote highly enantioselective reactions under mild reaction conditions, cinchona alkaloid-derived phase transfer catalysts have been successfully applied to the promotion of a broad range of asymmetric reactions.

α-Alkylation of Carbonyls. Cinchonine (C)-derived ammonium salt [C-*N*-CF_3Bn]Br was applied in the first highly enantioselective reaction documented with a chiral phase transfer catalyst. Specifically, catalytic asymmetric methylation of indanone with methyl chloride was achieved in 95% yield and 92% ee (eq 19).[19] The optically active product was transformed to (*β*)-indacrinone in three steps (eq 19). The high enantioselectivity was attributed to the possible formation of a highly ordered chiral ion pair via electrostatic, hydrogen bonding and *π*–*π* stacking interactions between the chiral ammonium salt and the enolate.

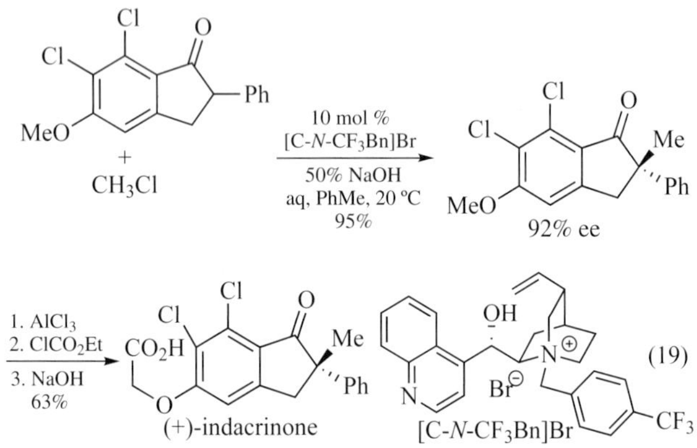

(19)

The catalytic asymmetric alkylation of benzophenone imines of glycine esters is achieved by several cinchona alkaloid-derived PTCs. O'Donnell et al. first reported that cinchonium salt [C-*N*-Bn]Br afforded the alkylation in moderate yield and enantioselectivity (eq 20).[20] Subsequently, O'Donnell et al. reported that the active catalyst was identified to be the *O*-alkylated cinchonium salt.[21] The next breakthrough in catalyst development was made by Corey and Noe by replacing the *N*-Bn group with the more bulky 9-anthracenylmethyl (*N*-ANTM) group. With cinchonidium salts [CD-*N*-ANTM]Cl[22] and [CD-9-OAllyl-*N*-ANTM]Br,[23] the alkylation proceeded in higher yield and significantly improved enantioselectivity (eq 20). Later, it was found that cinchonidium salt bearing a *N*-fluorinated benzyl group [DHCD-*N*-2,3,4-FBn]Br) could also give improved yields and enantioselectivity.[24] The fluorine atoms were postulated to form a hydrogen bond with a water molecule that led the catalyst to adopt a more optimal conformation for asymmetric induction.[25]

The preparation of optically active *α*,*α*-dialkyl amino esters was realized through phase transfer-catalyzed asymmetric alkylation of *α*-alkyl amino esters-derived aldimines. Cinchonium [C-*N*-Bn]Br afforded good yield but only 44% ee (eq 21).[26] Higher yields and significantly improved enantioselectivity were obtained with either [DHCD-*N*-ANTM]Cl (eq 21)[27] or [DHCD-*N*-2,3,4-FBn]Br (eq 22)[28] as the catalyst.

The highly enantioselective conjugate additions of glycine Schiff base to several Michael acceptors, including enones (eq 23),[29] acrylates (eq 24), and acrylonitrile (eq 25)[30], were successfully promoted by cinchonidine-derived PTC [CD-9-OAllyl-*N*-ANTM]Br. These reactions enabled facile enantioselective preparations of various functionalized *α*-alkyl amino acids.

Highly efficient asymmetric induction was achieved in the conjugate additions of silyl enol ethers to chalcones with cinchonine-derived PTC [C-9-OBn-*N*-ANTM]Br (eq 26),[31] affording chiral 1,5-diketones in good yields and with exceedingly high enantioselectivity.

α-Hydroxylation of Carbonyls. The enantioselective *α*-hydroxylation of indanones and tetralones with oxygen as the terminal oxidant was promoted by cinchonine-derived PTC [C-*N*-*p*-CF_3Bn]Br (eq 27).[32] The corresponding *α*-hydroxy carbonyls were obtained in good yields but with moderate enantioselectivity.

Aza-Henry Reaction. Quinium benzylchloride (QUIBEC) as a phase transfer catalyst in the presence of KOH promoted the in situ formation of *N*-carbamoyl imines from *α*-amido sulfones while activating nitromethane for asymmetric aza-Henry reactions (eq 28).[33] Notably, *N*-carbamoyl *β*-nitroamines bearing aliphatic substituents were obtained in good yields and with excellent enantioselectivity. The enantioselectivity remained high with aromatic *α*-amido sulfones as precursors for aryl *N*-carbamoyl imines. The scope of this asymmetric aza-Henry reaction could be extended to nitroethane by applying CsOH as the base (eq 29),[34] which afforded the *syn*-adduct with excellent diastereoselectivity and enantioselectivity. The presence of the free hydroxyl group in QUIBEC was shown to be critical for the observed catalytic activity.

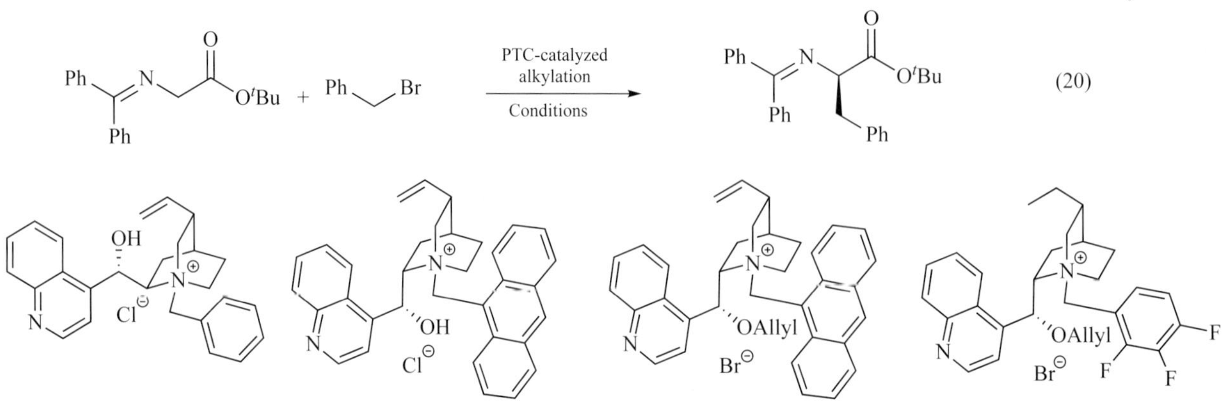

(20)

10 mol % [C-N-Bn]Cl	10 mol % [CD-*N*-ANTM]Cl	10 mol % [CD-9-Oallyl-*N*-ANTM]Br	10 mol % [DHCD-*N*-2,3,4-FBn]Br
50% NaOH aq. CH_2Cl_2, rt	50% KOH aq. PhMe, rt	$CsOH.H_2O$, CH_2Cl_2, –78 °C	50% KOH aq, PhMe/$CHCl_3$, –20 °C
75% yield, 66% ee	**68% yield, 91% ee**	**84% yield, 94% ee**	**92% yield, 98% ee**

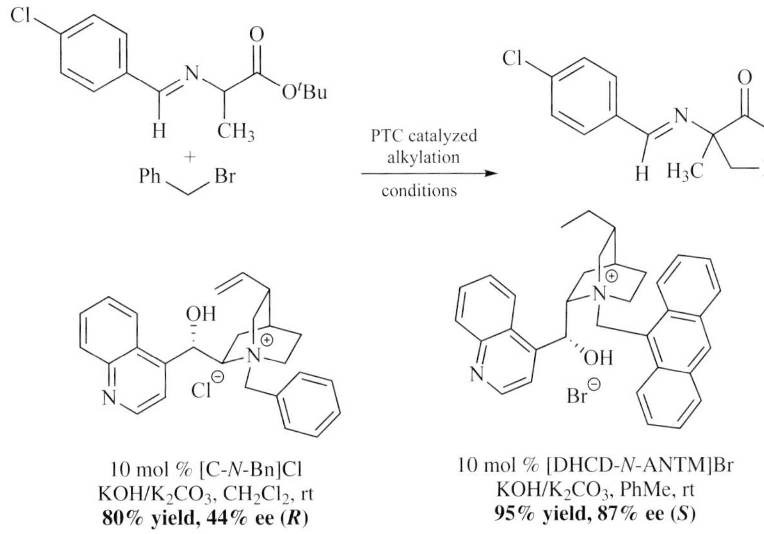
(21)

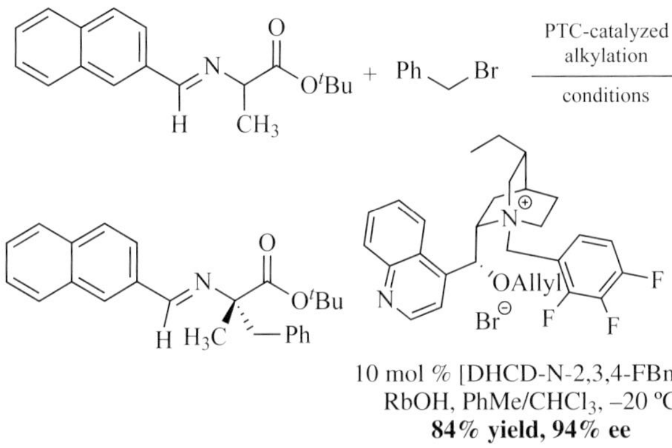
(22)

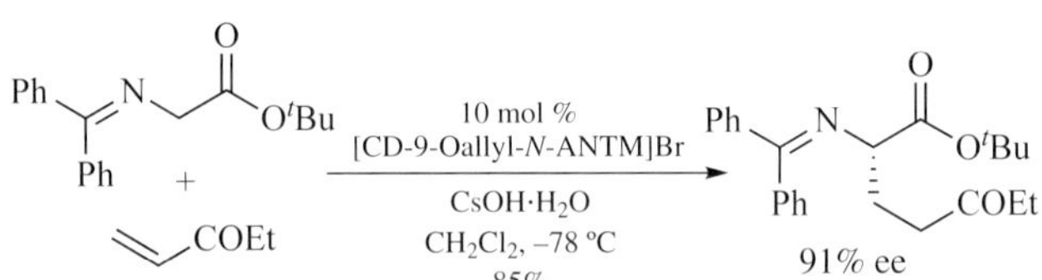
(23)

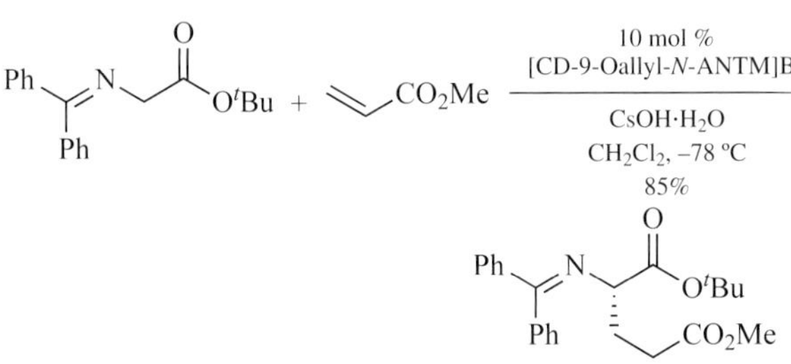
(24)

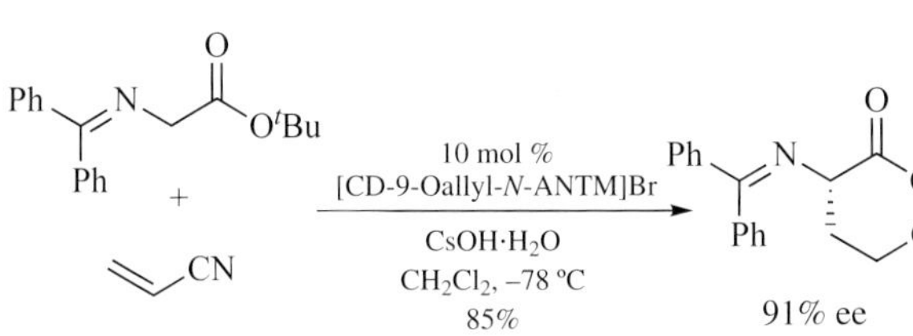
(25)

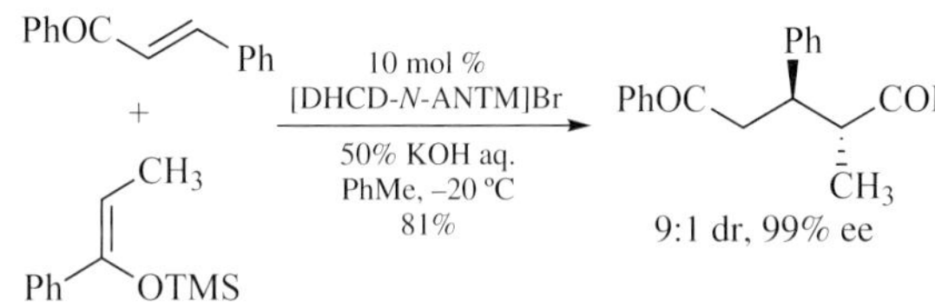
(26)

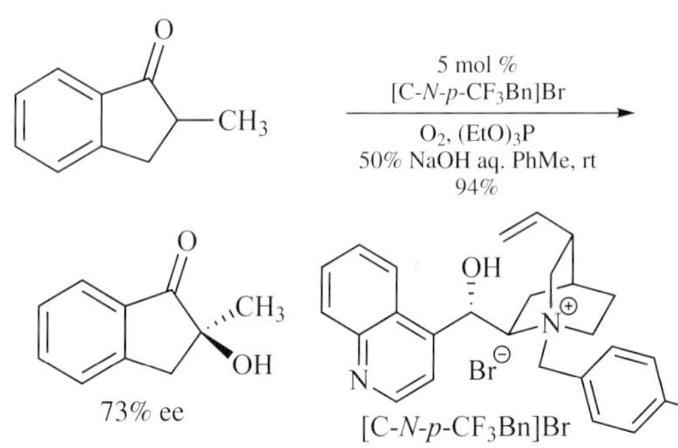
(27)

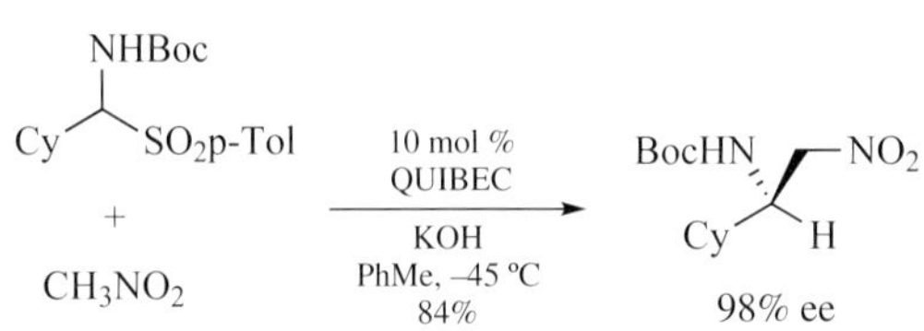
(28)

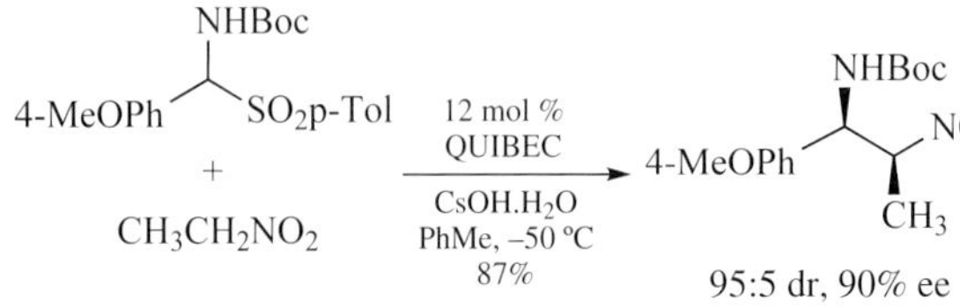
(29)

α-Alkylation of Sulfones. The highly enantioselective conjugate additions of 1-fluorobis(phenylsulfonyl)methane to enones were promoted by quinidine-derived PTC [QD-*N*-terphenyl]Br (eq 30).[35] The highly enantiomerically enriched bis-sulfones were obtained in moderate to good yield, and subsequently converted into monofluoromethylated compounds via a three-step sequence.

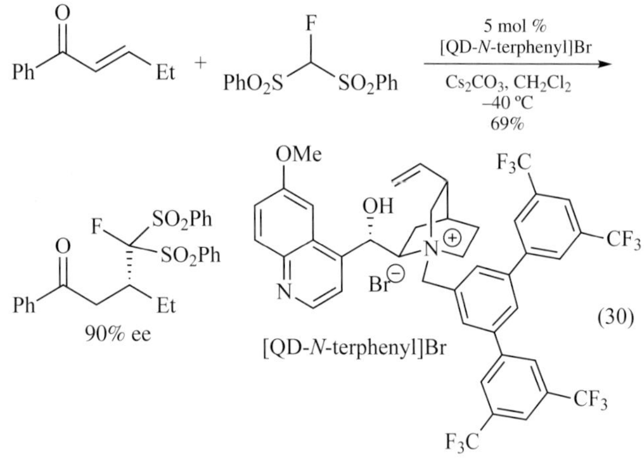

(30)

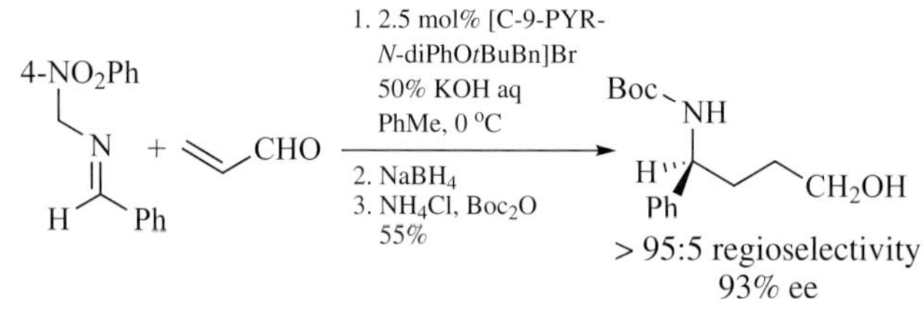

(32)

Umpolung Reactions of Imines. Asymmetric umpolung reactions of imines, in which imines reacted as nucleophiles with enals, were successfully developed, which was resulted from the catalyst design and development studies leading to the discovery of cinchonine-derived chiral PTC [C-9-PYR-*N*-diPhBn]Br.[36] This new chiral phase transfer catalyst promoted first the deprotonation of the *N*-benzyl trifluoromethylated imines and then directed the 2-azaallyanion thus formed to react with the enals with excellent chemo-, regio-, diastereo-, and enantioselectivity (eq 31). Optically active α,α-disubstituted trifluoromethyl amines, which were not easily accessible, could be synthesized in good yields. A gram-scale reaction in good yield and with excellent selectivity was successfully achieved with only 0.01 mol % catalyst. The reaction went to completion in 24 h, thereby showing the catalyst turnover rate of the chiral phase transfer catalyst approached those of biocatalysts. The substrate scope was extended to aryl and alkenyl aldimines, allowing the umpolung reaction to become useful for the preparation of chiral aryl and alkenyl amines in a highly enantioselective manner (eq 32). A π–π interaction between the aryl moiety of the quinuclidine *N*-substituent on the catalyst and the *N*-*p*-nitro-benzyl group of the imine is believed to play a key role in how the phase transfer catalyst affords the extraordinary reactivity and selectivity.

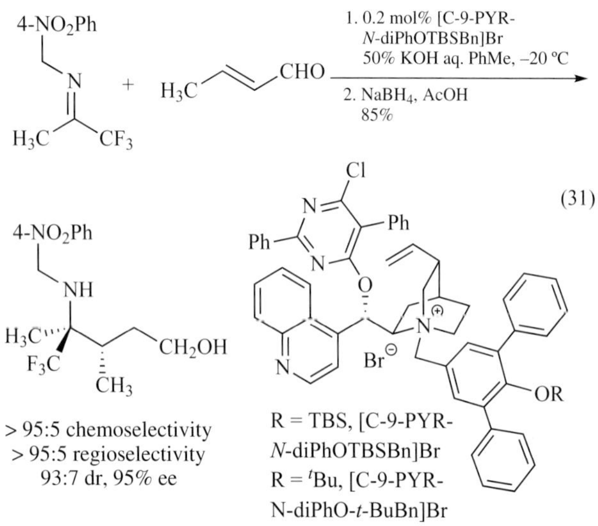

(31)

Reactions with Cinchona Alkaloids as Lewis Base Catalysts. Cinchona alkaloids have been employed as Lewis base catalysts for a broad spectrum of asymmetric transformations. A major mode of action involved cinchona alkaloids as a nucleophilic catalyst, when a nucleophilic attack by its quinuclidine nitrogen to a reactant led to the formation of a reactive intermediate covalently linked to the activated substrate, the cinchona alkaloid directed it to react enantioselectively with another reactant to form the optically active product.

Ketene–Aldehyde Cycloadditions. The reaction of ketene with chloral was promoted by natural cinchona alkaloids at −50 °C in toluene.[37] The β-(trichloromethyl)-β-propiolactone was generated in 76% ee when the reaction was mediated by quinine (eq 33). Excellent enantioselectivity (98% ee) was obtained with quinidine. Later, the active catalyst was believed to be the 9-OAc quinidine. The catalysis was initiated by the addition of quinidine to ketene. The resulted zwitterionic acyl ammonium enolate reacted with the chloral to form the corresponding aldolate, which proceeded to form the chiral β-lactone product while regenerating the catalyst via an acyl transfer reaction with the *N*-acyl ammonium salt.

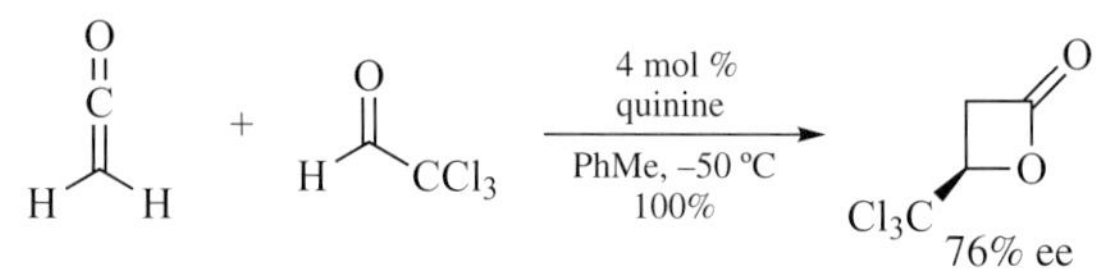

(33)

The substrate scope of ketene-aldehyde cycloadditions was greatly extended by combining Q-9-OTMS-mediated nucleophilic activation of ketenes and $LiClO_4$-mediated Lewis acid activation of aldehydes.[38] With this catalyst system, a wide range of alkyl and aryl aldehydes reacted with acid chlorides with high stereoselectivity to form chiral β-lactones in high yields (eq 34).

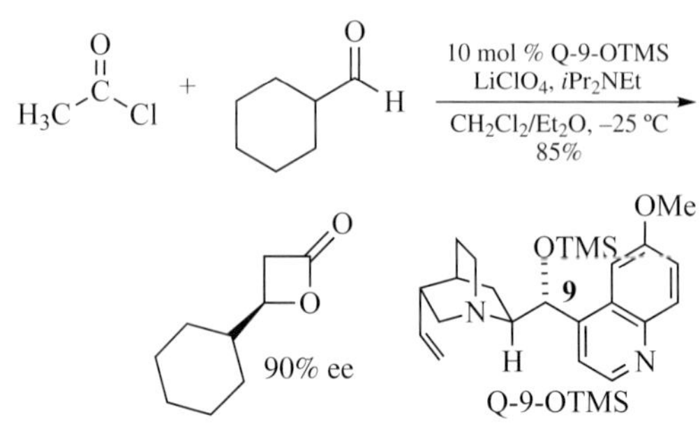

(34)

Nucleophilic Addition of Ketenes to Imines. Optically active β-lactams were prepared from acid chlorides and imines in the presence of a quinine derivative Q-9-OBz.[39] A variety of acid chlorides were in situ converted into ketenes, which reacted with imines to form chiral β-lactams with high diastereoselectivity and enantioselectivity (eq 35). Q-9-OBz played two roles: as a dehydrohalogenation agent and a chiral nucleophilic catalyst. The moderate yield initially obtained could be improved upon the use of indium Lewis acid as a cocatalyst.[40]

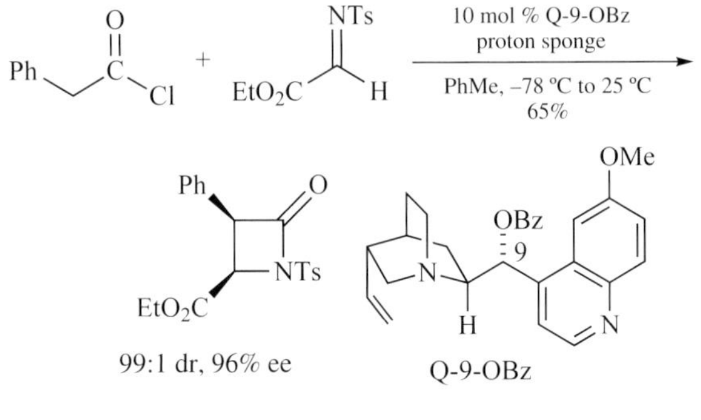

(35)

Hetero-Diels–Alder Reactions. A formal inverse electron demand Diels–Alder reaction of ketene enolates and *o*-benzoquinone diimides was enabled by applying a bifunctional catalytic system that activated the *o*-benzoquinone diimides with an achiral Lewis acid $Zn(OTf)_2$ in concert with the nucleophilic activation of the in situ-generated ketenes by the quinine derivative Q-9-OBz.[41] Chiral tetrahydroquinoxalines are prepared in moderate to good yields and with excellent enantioselectivity (eq 36).

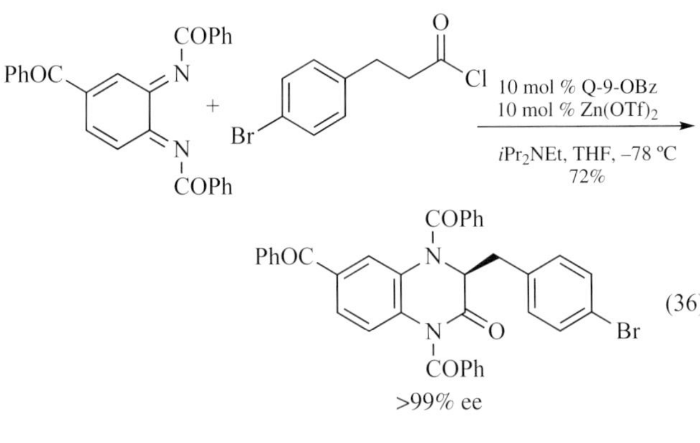

(36)

cis-4,5-Disubstituted 1,3-thiazin-6-one heterocycles were prepared from a formal cycloaddition of ketene and *N*-thioacyl imine mediated by Q-9-OTMS and $LiClO_4$, which proceeded in moderate to good yields and with high diastereoselectivity and enantioselectivity (eq 37).[42] The thiazinone product could function as an activated ester surrogate for the synthesis of α,β-dipeptides via coupling with an α-amino acid.

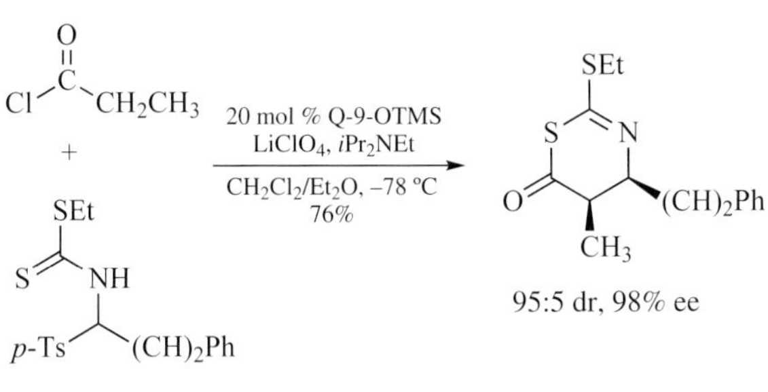

(37)

Cyanation of Ketones. Chiral tertiary cyanohydrin carbonates could be prepared by $(DHQ)_2AQN$-catalyzed asymmetric cyanation of ketones with ethyl cyanoformate (eq 38).[43] A variety of acyclic and cyclic dialkyl ketones were transformed to cyanohydrin carbonates with good to excellent enantioselectivity and in synthetically useful yield. The catalyst was postulated to react with ethyl cyanoformate to release cyanide and form a chiral acyl ammonium salt. The cyanide then reacted with the ketone to form a cyanoalkoxide with low or low enantioselectivity. The ee- and rate-determining step was found to be the acyl transfer step between the chiral acyl ammonium salt and the cyanoalkoxide.[44]

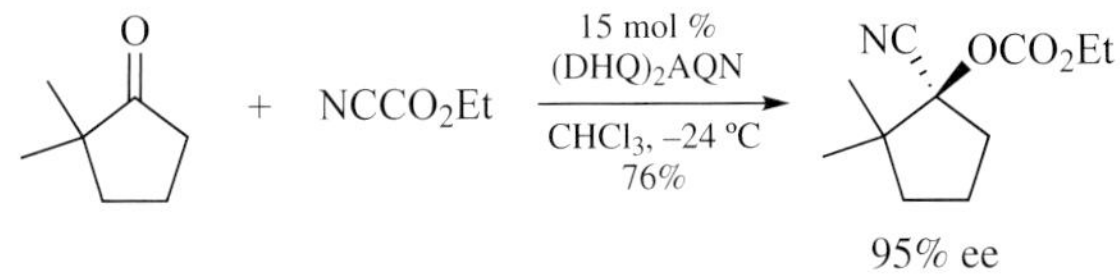

(38)

The highly enantioselective cyanosilylation of α-acetal ketones catalyzed by $(DHQ)_2AQN$ was reported shortly after (eq 39).[45] This asymmetric cyanation consistently afforded high enantioselectivity for alkyl, alkenyl, alkynyl, and aryl α-acetal ketones. Such a general scope in combination with the versatility of the acetal and cyano functionalities rendered this reaction useful for the enantioselective preparation of chiral tertiary alcohols, which was demonstrated in its application to the enantioselective total syntheses of bisorbicillinoids family of natural products.[46]

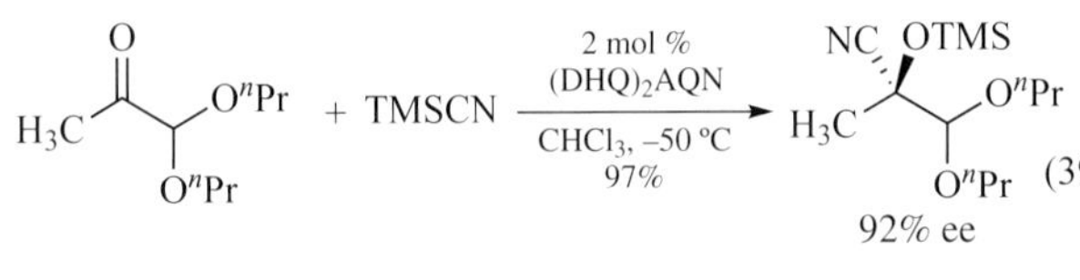

(39)

α-Fluorination of Carbonyls. Stoichiometric mediator *N*-fluoro-cinchona alkaloids have been used to facilitate asymmetric uorination of α-cyanoester and trimethylsilyl enol ethers. With in situ generated [DHQD-9-OAc-*N*-F]BF_4 or [DHQ-9-O-*p*-ClBz-*N*-F]BF_4, the corresponding α-uorinated compounds were obtained in good yield and high ee (eq 40).[47] These fluorinated cinchona alkaloids were also used in the fluorination of α-amino phenyl acetonitrile (eq 41).[48]

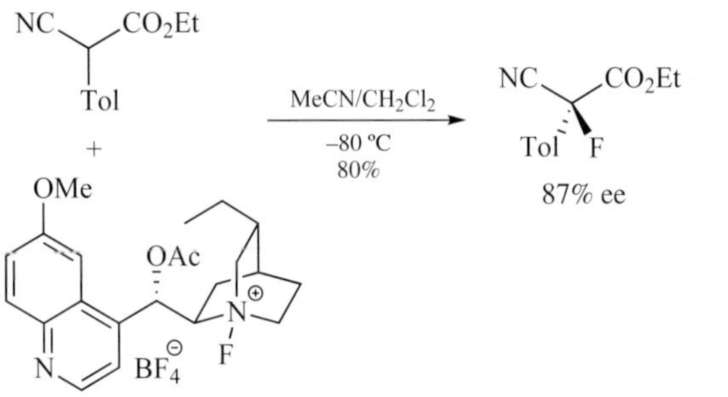

(40)

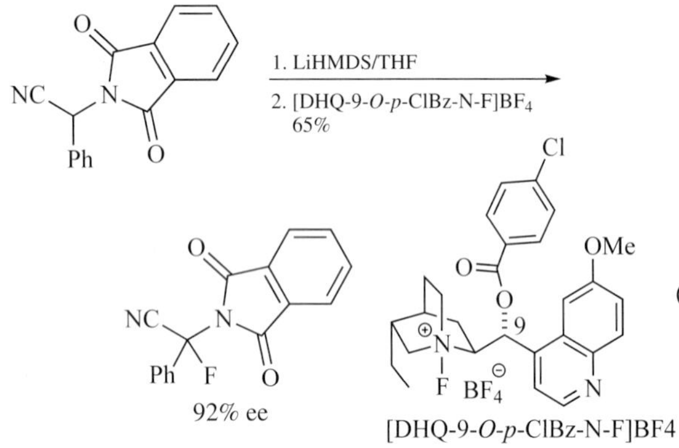

(41)

Morita–Baylis–Hillman* (MBH) *Reaction. Quinidine-derived β-isocupreidine (β-ICD) was reported by Hatakeyama and coworkers to achieve the first highly enantioselective catalytic MBH reaction (eq 42).[49] The reactions of hexafluoroisopropyl acrylate with a wide range of aromatic and aliphatic aldehydes provided the optically active adducts in moderate yields and with high ee. Both the rigidity of the catalyst and the presence of the 6′-OH group were crucial for the catalytic reactivity and enantioselectivity.

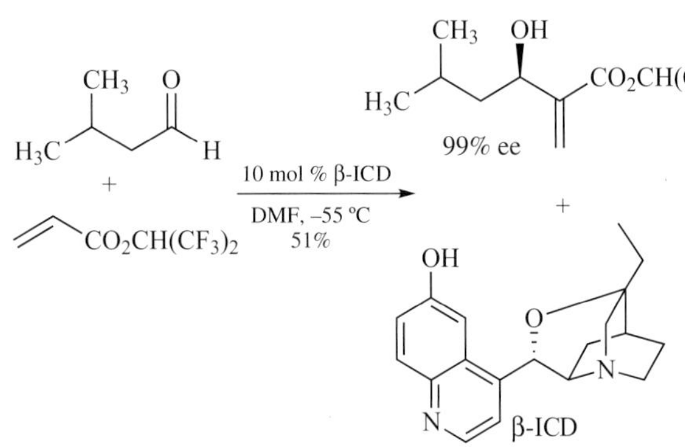

(42)

β-ICD was also shown to be an efficient catalyst for the asymmetric aza-MBH reactions of aryl imines with hexafluoroisopropyl and simple acrylates, enones, and enals (eq 43).[50]

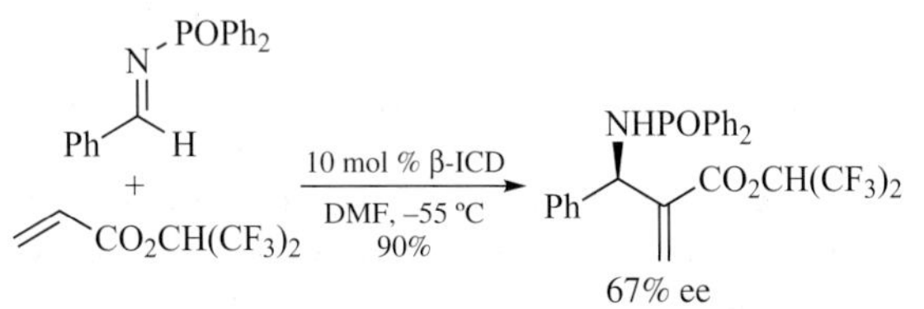

(43)

Reactions with Cinchona Alkaloids as General Base Catalysts. Cinchona alkaloids can serve as chiral bases to promote a broad range of asymmetric transformations by hydrogen bonding-mediated activation of nucleophiles.

Alcoholysis of Cyclic Anhydrides. Natural cinchona alkaloids such as quinine were explored by Oda as catalysts for the asymmetric alcoholysis of *meso*- and prochiral acid anhydrides (eq 44).[51] These asymmetric alcoholyses, normally carried out with 10 mol % catalyst loading, afforded the corresponding hemiesters in low to moderate ee.

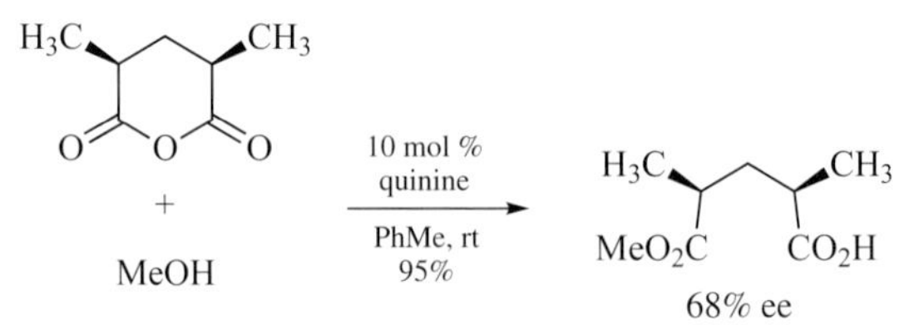

(44)

Highly enantioselective alcoholysis of cyclic anhydrides with alcohols was achieved with modified quinine derivatives.[52] With 5–30 mol % $(DHQ)_2AQN$, the reactions afforded chiral hemiesters in good yields and with high enantioselectivities (eq 45). The same catalyst when applied to the racemic monosubstituted succinic anhydrides led to an unusual parallel kinetic resolution, in which the two enantiomers of the starting anhydrides were converted into two highly enantiomerically enriched regioisomers of hemiesters (eq 46).[53] The scope of the cyclic anhydrides for the asymmetric alcoholysis was further extended to *N*- and *O*-carboxyanhydrides, allowing highly efficient kinetic resolutions of these readily available cyclic anhydrides to form highly enantiomerically enriched α-amino[54] and α-hydroxyl[55] acids (eqs 47 and eq 48). Mechanistic studies[54,55] revealed that the cinchona alkaloid catalysts mediated these asymmetric alcoholyses by highly enantioselective general base catalysis.

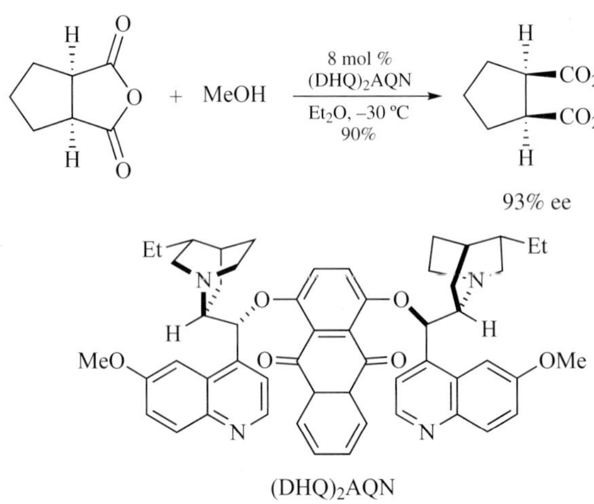

(45)

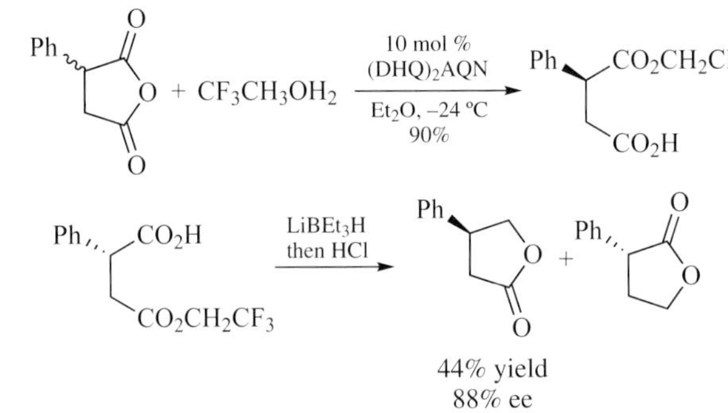

(46)

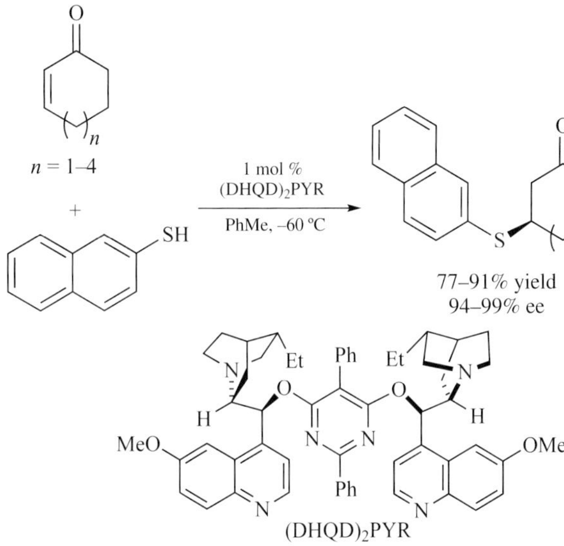

(49)

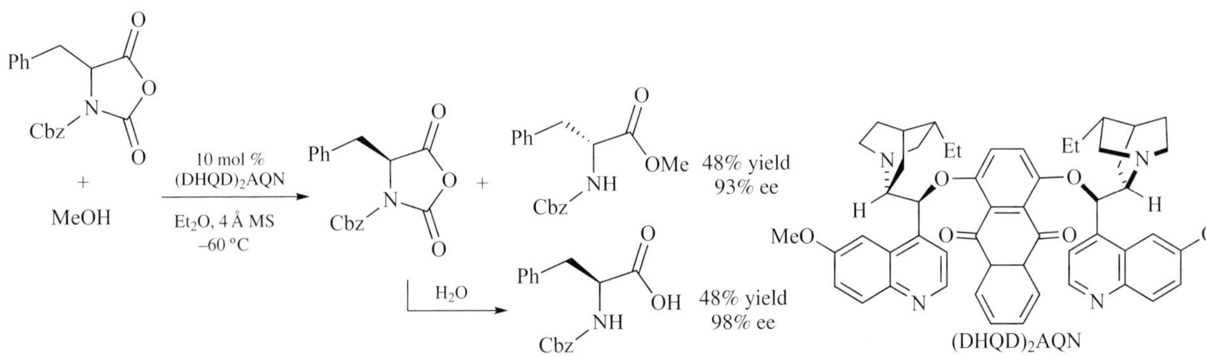

(47)

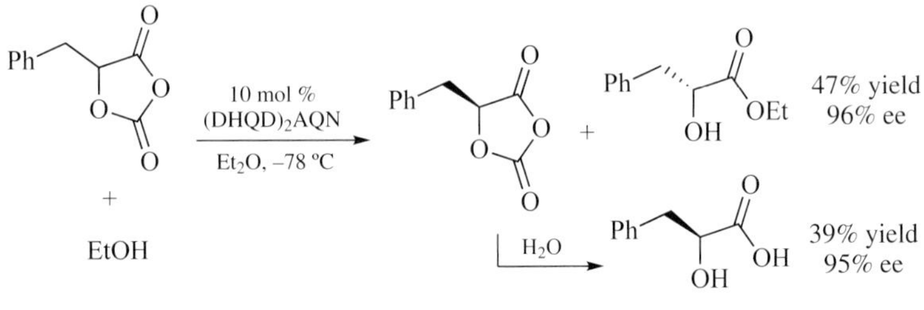

(48)

Synthesis of Chiral Sulfide. Quinine-catalyzed conjugate additions of aryl thiols to cyclic enones afforded up to 46% ee as reported by Wynberg and coworkers in 1977.[10] In 2002, McDaid et al. reported that modified cinchona alkaloids such as $(DHQD)_2PYR$ were highly effective catalysts for the activation of thiols for asymmetric reactions, promoting the additions of aryl thiols to cyclic enones of various ring size to afford the corresponding sulfides in good yields and with excellent enantioselectivity (eq 49).[57]

Reactions with Cinchona Alkaloids as Acid/Base Bifunctional Catalysts. Based on results from his earlier studies, Wynberg proposed that natural cinchona alkaloids such as quinine and cinchonidine were superior catalysts than modified cinchona alkaloids, because the former could act as acid–base bifunctional catalysts with the 9-OH as the hydrogen bond donor.[10] This perception in combination with the generally inadequate enantioselectivity of the broad range of transformation promoted by natural cinchona alkaloids discouraged the exploration of cinchona alkaloids for the promotion of acid–base bifunctional chiral catalysts. The more recent discovery of modified cinchona alkaloids as broadly effective chiral base catalysts encouraged a revisit to the topic of cinchona alkaloids as acid–base bifunctional catalysts. The advent of highly enantioselective yet broadly applicable acid–base bifunctional catalysts, such as 6′-OH-modified cinchona alkaloids,[58] marked the beginning of an extraordinary progress in both new catalyst and new reaction development. It has been demonstrated that these new generation of bifunctional cinchona alkaloid catalysts not only efficiently induce enantioselectivity via cooperative catalysis but also as dual-functional catalysts control stereochemistry in different stereoselective steps of a complex asymmetric reaction cascade.[59]

Conjugate Additions. The conjugate addition of malonates to nitroalkenes was promoted by 6′-OH-modified quinine (Q-6′-OH) in good yields and with excellent enantioselectivity.[56] This reaction was carried out under mild conditions that tolerated moisture and air (eq 50). With 10 mol % Q-6′-OH, the reaction could readily convert β-aryl and alkyl nitroalkenes into the corresponding 1,4-adducts in as high as 99% ee. The high catalytic enantioselectivity could be extended to the addition of ethyl acetoacetate to nitroalkenes, from which a 1:1 diastereomeric mixture with both diastereomers in 91% ee was isolated (eq 51). Subsequently, it was demonstrated that the tunable nature of the 6′-OH cinchona alkaloids could be explored for the development of efficient catalysts for achieving the conjugate additions of trisubstituted carbon Michael donors to nitroalkenes in good yields and with high diastereoselectivity and excellent enantioselectivity (eq 52).[60] This transformation established a versatile route for the stereoselective construction of adjacent quaternary and tertiary carbon centers. The 6′-OH cinchona alkaloids was designed to act as a bifunctional catalysis via simultaneous base activation of the malonates by the quinuclidine nitrogen and the hydrogen bond donor activation of the nitroalkene by the 6′-OH. Results from control experiments, kinetic studies of the reaction, and conformational studies of the catalyst indicated this design concept was operational.[56,58]

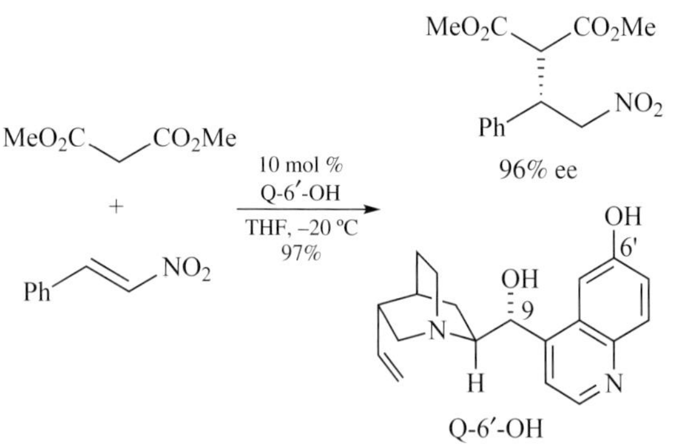

(50)

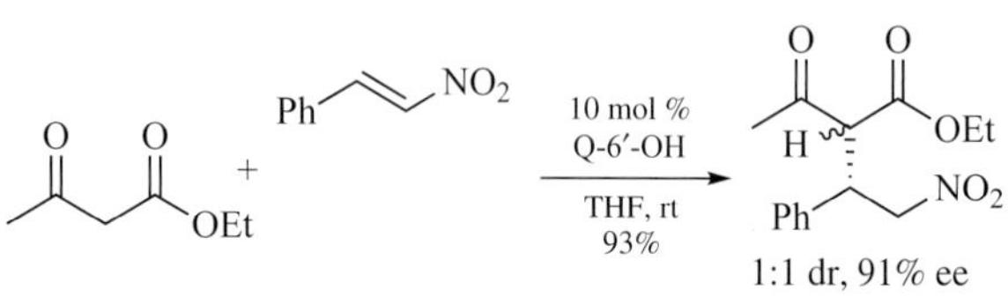

(51)

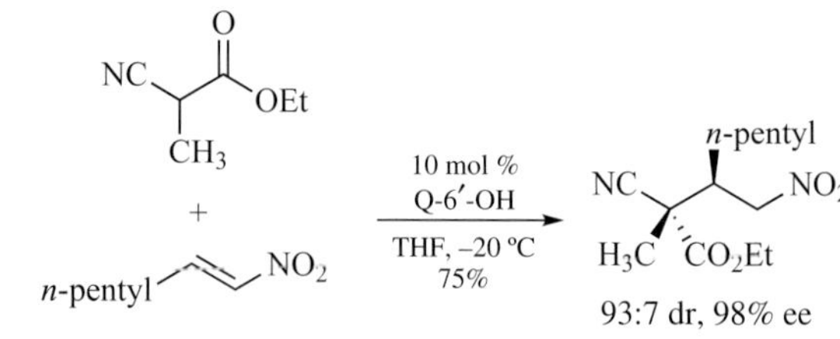

(52)

The conjugate addition of nitromethane to chalcones was catalyzed by Q- or QD-9-TU in high ee and chemical yields (eq 53).[61] This catalyst was designed to couple base activation of the nitromethane by the quinuclidine nitrogen with hydrogen bond donor activation of the chalcones by the thiourea moiety.

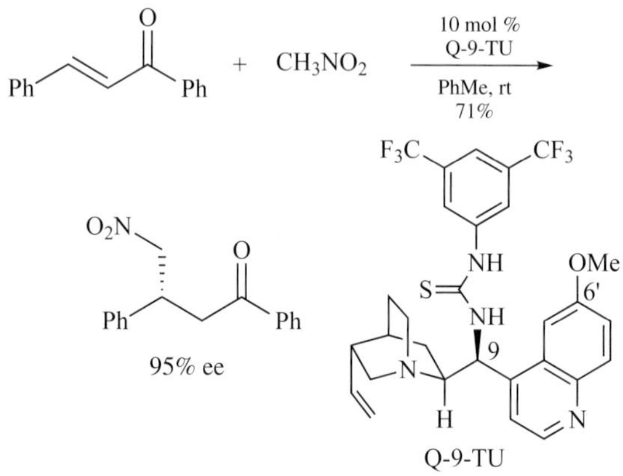

(53)

Subsequently, cinchonine-derived chiral squaramide C-*epi*-9-SQ was designed to be another bifunctional catalyst. It was demonstrated to effectively mediate the conjugate additions of 1,3-diketones and β-keto esters to β-aryl nitroalkenes (eq 54).[62]

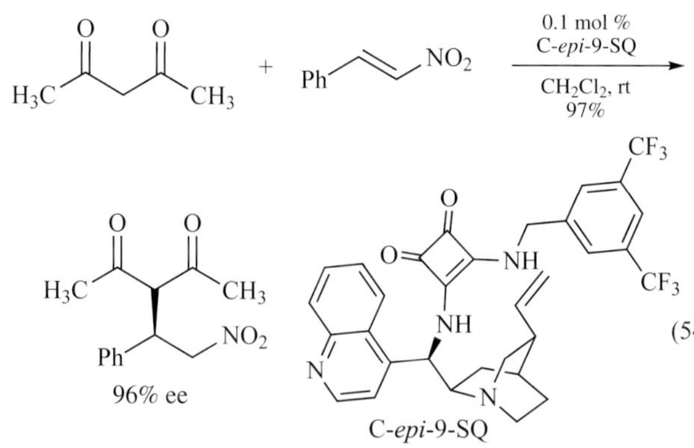

(54)

A highly enantioselective conjugate addition to vinyl sulfone was documented with 6′-OH-modified quinine Q-9-OPHN-6′-OH (eq 55). High asymmetric induction could be achieved in the additions of α-cyanoacetate to vinyl sulfones.[63] Aryl, allyl, and alkyl substituents on the α-cyanoacetates were all tolerated. The reaction employed an operationally simple protocol and practical conditions.

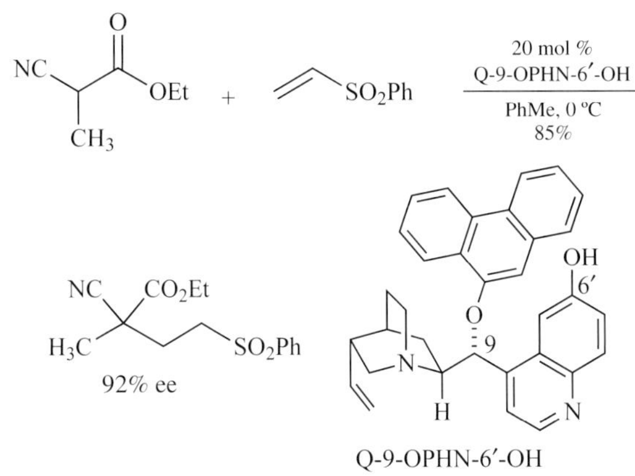
(55)

Asymmetric conjugate additions of β-ketoesters to enals were reported with 6′-OH-modified quinine derivatives (eq 56).[64] Nearly complete chemoselectivity and excellent diastereoselectivity and enantioselectivity were attained with only 1.0 mol % catalyst, affording the corresponding adduct in virtually quantitative yield. The reactions, performed at ambient temperature, went to completion in a few minutes. It was also reported that 6′-OH-modified quinine derivatives could catalyze highly enantioselective conjugate additions of various enolizable Michael donors to enones.[65]

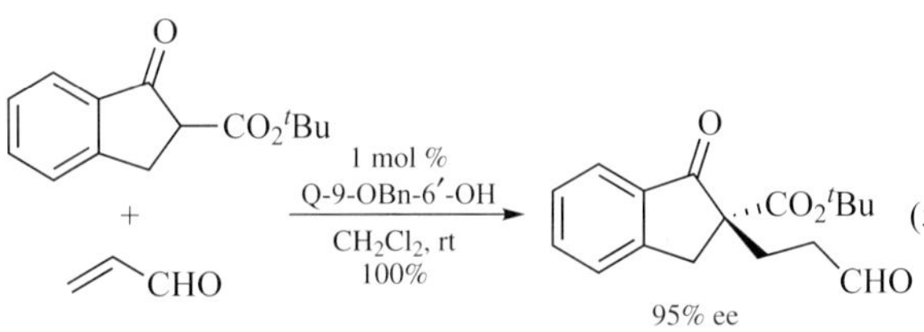
(56)

Conjugate Addition–Protonation Reaction Cascades. The direct creation of nonadjacent stereocenters was enabled by an asymmetric conjugate addition–protonation reaction cascade catalyzed by Q-OPHN-6′-OH (eq 57).[57] The asymmetric additions of enolizable Michael donors such as β-ketoesters and α-cyanoketones to 2-chloroacrylonitrile provided a versatile catalytic approach for the one-step construction of 1,3-tertiary quaternary stereocenters in good yields and with high diastereoselectivity and enantioselectivity. The catalyst played dual roles, mediating the enantioselective addition of the Michael donor to the 2-chloroacrylonitrile and promoting the asymmetric protonation of the carbanion intermediate generated from the addition step. Later, the same asymmetric reaction cascade but with a complementary sense of diastereoselectivity related to that with Q-OPHN-6′-OH was obtained by applying another quinine-derived catalyst Q-9-TU (eq 58).[66] These studies illustrated the power of bifunctional cinchona alkaloid catalysts in exercising control over both enantioselectivity and diastereoselectivity. The ability of cinchona alkaloids as efficient catalysts for asymmetric protonation of prochiral carbon centers was also revealed in these studies.

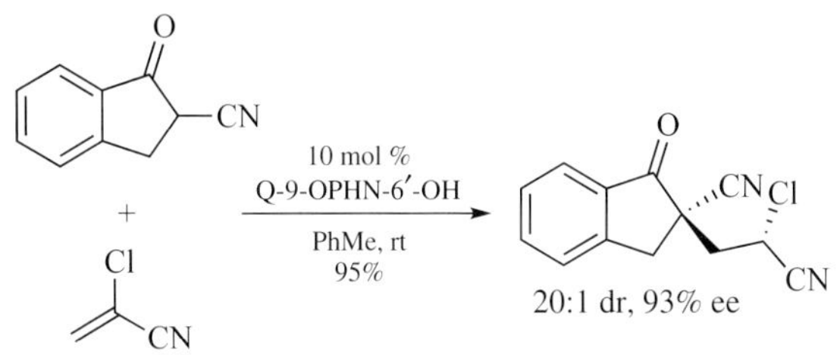
(57)

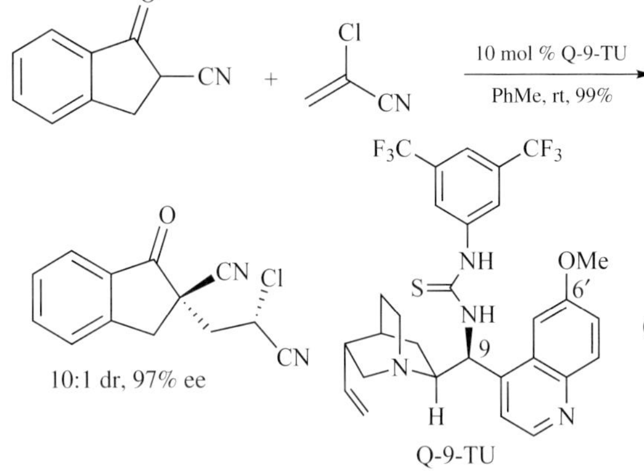
(58)

Henry Reaction. 6′-OH-modified quinine Q-9-OBz-6′-OH has been successfully used to promote enantioselective Henry reaction of α-ketoesters (eq 59).[67] With 5 mol % catalyst loading, optically active tertiary carbinols were obtained in high yields and with excellent enantioselectivity. Alkyl, aryl, and alkenyl α-ketoesters with varying electronic and steric properties were tolerated by the catalyst.

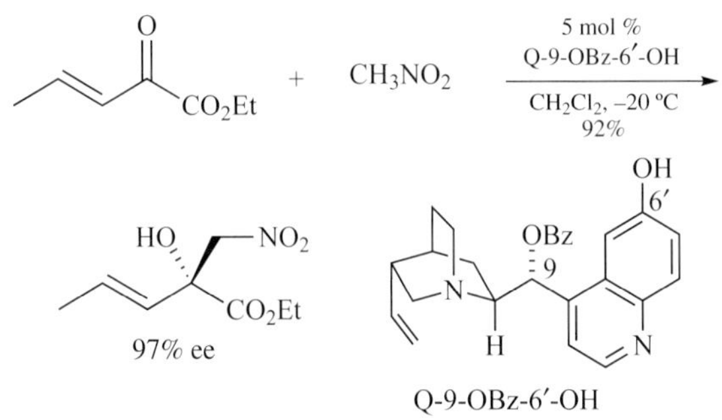
(59)

The highly enantioselective Henry reaction of aryl aldehydes was developed with quinine derivative Q-9-OBn-6′-TU, which featured a 6′-thiourea. This reaction worked well with aromatic aldehydes, affording the corresponding adducts in high yields and with good enantioselectivity (eq 60).[68]

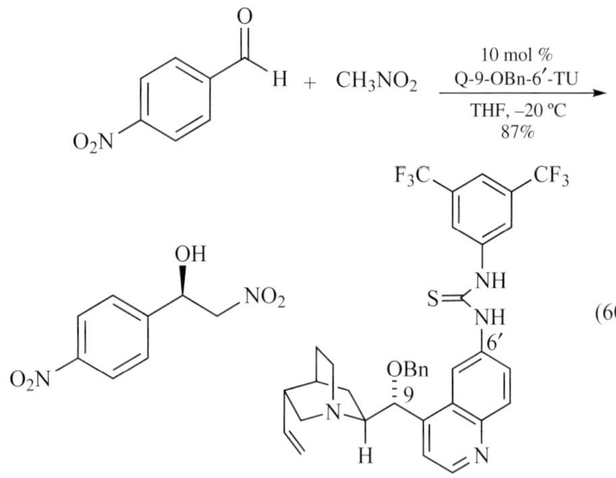

Q-9-OBn-6′-TU

(60)

The asymmetric addition of nitromethane to fluoromethyl ketones was catalyzed efficiently by a 6′-OH quinine derivative Q-9-OCF_3Bz-6′-OH (eq 61).[69] This reaction is applicable to aryl and aliphatic trifluoromethyl ketones, providing the corresponding fluorinated nitroalcohols in high yields and with good enantioselectivity. The Henry reaction of difluoromethylketones could also be promoted by the same catalyst with high enantioselectivity.

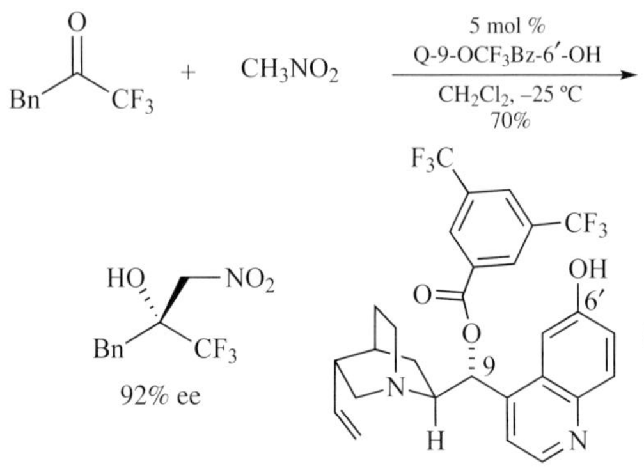

Q-9-OCF_3Bz-6′-OH

(61)

Aldol Reaction. Asymmetric vinylogous aldol reactions of silyloxy furans and aldehydes were enabled by the catalysis of a chiral carboxylate–ammonium salt, which was formed simply by mixing a quinine derivative Q-9-TU and a carboxylic acid (eq 62).[70] The reaction tolerated variations of both the silyloxy furans and aldehydes and afforded the corresponding optically active α,β-butenolides with high enantioselectivity and high *anti*-selective diastereoselectivity.

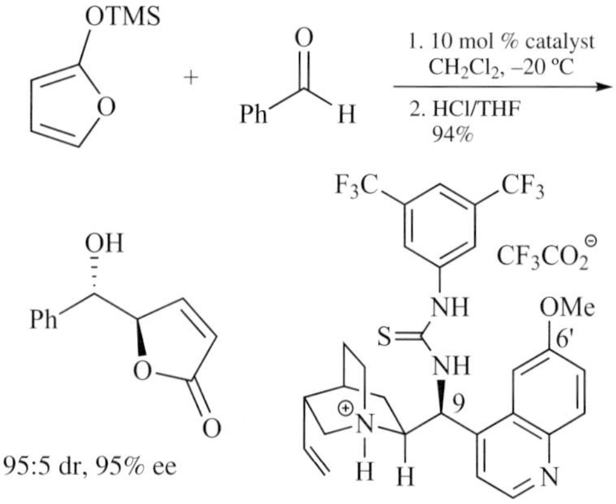

catalyst

(62)

The highly diastereoselective and enantioselective aldol reaction of α-alkyl azlactones and aliphatic aldehydes was achieved with the catalysis of 6′-OH quinine derivative DHQ-OPYR-6′-OH (eq 63).[71] This reaction established a useful approach toward β-hydroxy-α-amino acids bearing alkyl substituents at both stereogenic centers, a chiral motif presented in a range of natural products.

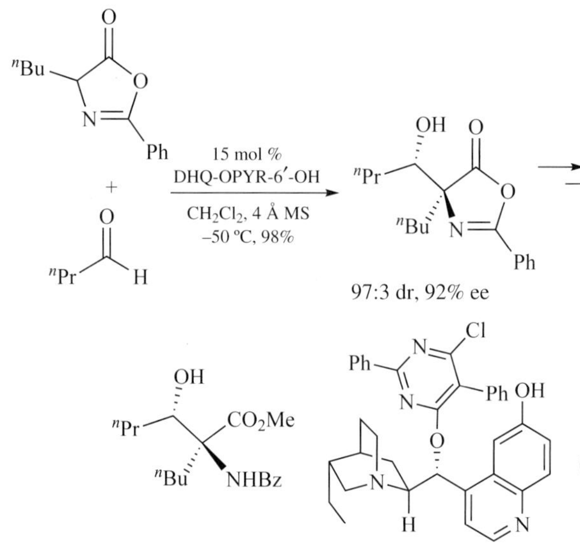

DHQ-OPYR-6′-OH

(63)

Mannich Reactions. Enantioselective Mannich reactions of malonates and β-ketoesters with aldimines were promoted by Q-9-TU[72] or C-9-TU[73] under mild conditions in good yields and with high enantioselectivity (eq 64). Both aryl and aliphatic imines are well tolerated in this transformation, although the unstable nature of the latter required a higher catalyst loading to furnish a reaction of useful yield. Later, an improved protocol involving the in situ generation of aliphatic imines from α-amido sulfones was developed to overcome this drawback.[74]

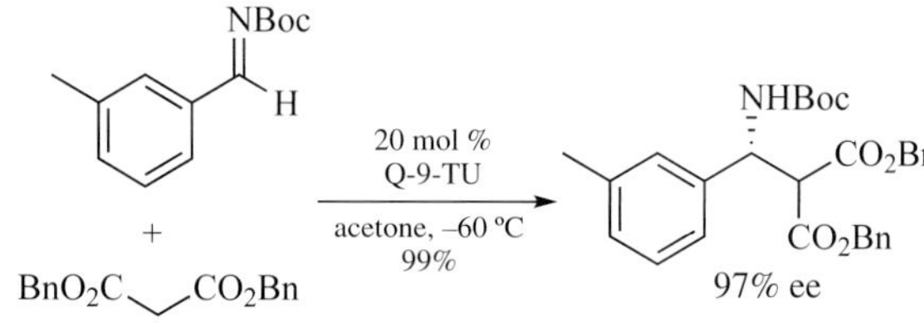

(64)

Hydrophosphonylation. A quinine-catalyzed enantioselective hydrophosphonylation of *ortho*-nitrobenzaldehyde was reported to produce the chiral α-hydroxyphosphonate ester in 85% ee (eq 65).[75] The high enantioselectivity was extended to the hydrophosphonylation of aryl *N*-Boc-imines, which afforded the corresponding α-aminophosphonates in 94% ee and moderate to good yields (eq 66).[76]

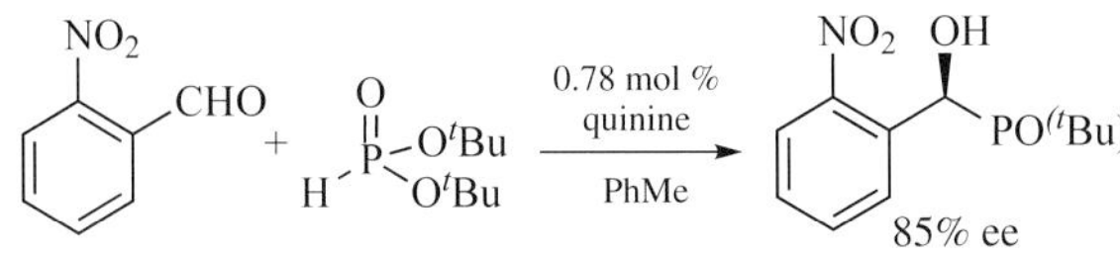

(65)

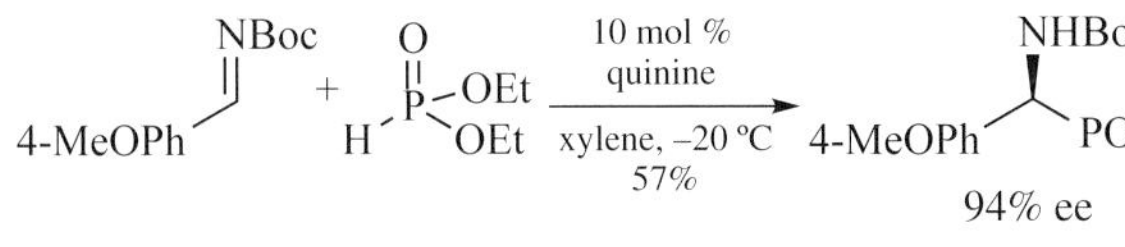

(66)

Friedel–Crafts Reactions. Highly enantioselective Friedel–Crafts reactions of indoles with carbonyls were achieved with the catalysis of 6′-OH quinine derivative Q-9-OPHN-6′-OH (eq 67).[77] In the presence of 10 mol % catalyst, a wide range of alkynyl and aryl α-ketoesters as well as aryl aldehydes reacted smoothly with indoles to afford the desired adduct in good to excellent yields and with high enantioselectivity.

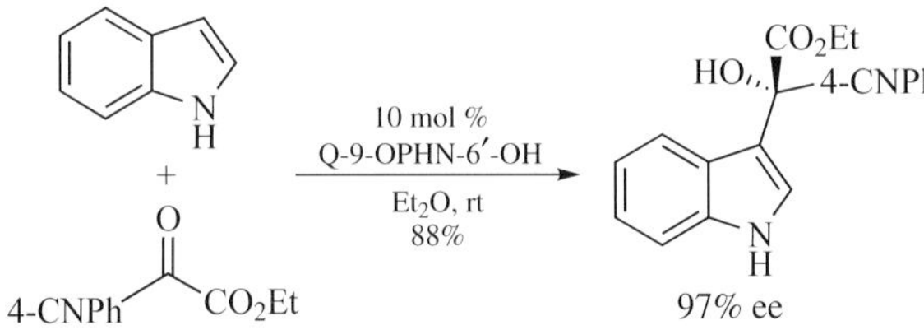

(67)

Bifunctional catalyst Q-9-TU was reported to promote highly enantioselective Fridel–Crafts reactions of indoles with imines at 50 °C in ethyl acetate (eq 68).[78] A broad range of indoles reacted with aryl and alkyl imines to form the corresponding adducts in good yields and with high enantioselectivity.

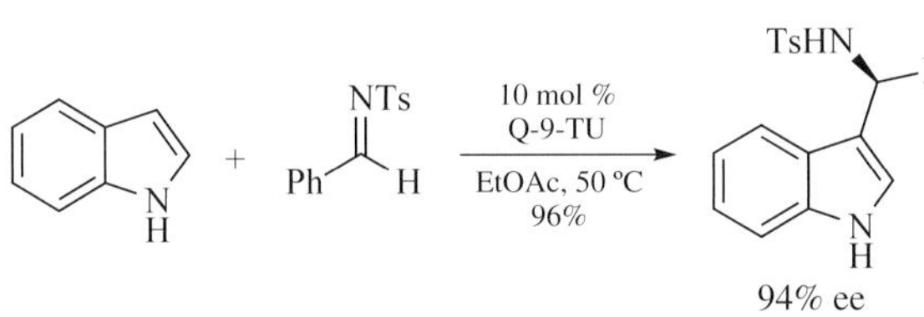

(68)

Diels–Alder Reactions. Highly enantioselective and diastereoselective catalytic Diels–Alder reaction with pyrones was developed by employing bifunctional quinine derivative DHQ-9-PHN-6′-OH (eq 69).[79] The reaction was performed with 5 mol % catalyst at ambient temperature. Substitutions on the pyrones were tolerated. By applying Q-9-TU as the catalyst for the reaction of pyrone and α-chloro acrylonitrile, a complementary sense of diastereoselectivity to that by catalyst DHQ-9-PHN-6′-OH was obtained (eq 70). Consequently, with 6′-OH and 9-Tu catalysts derived respectively from quinine and quinidine, each of the four possible stereoisomers of the D–A adducts could be obtained in a stereoselective fashion.

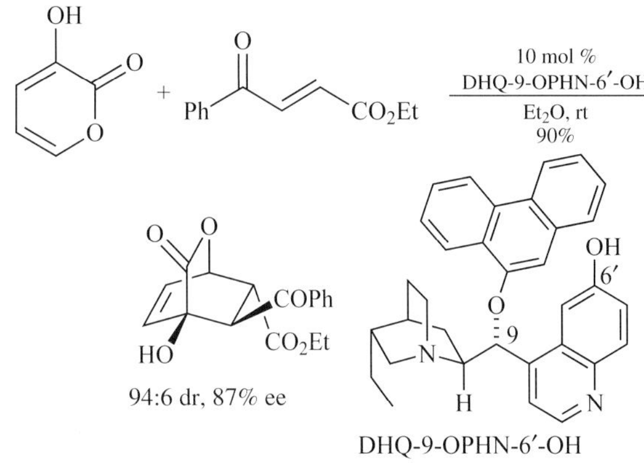

(69)

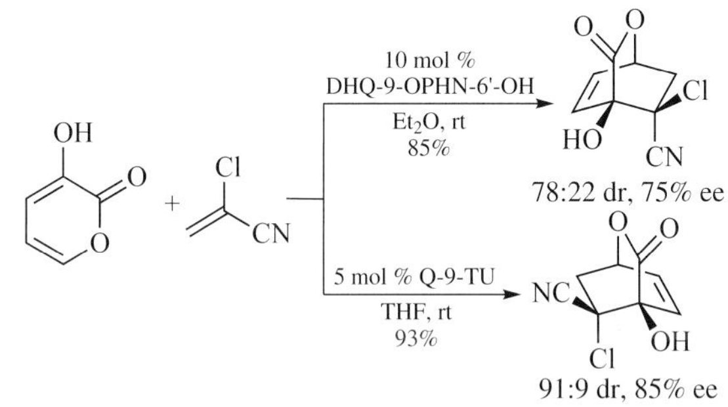

(70)

Kornblum–DeLaMare Rearrangement. Enantiomerically enriched cyclic 4-hydroxyenones were prepared from dienes by photochemical dioxygenation followed by a highly enantioselective desymmetrization of the resulting *meso*-endoperoxides with a 6′-OH cinchona alkaloid (eq 71).[80] The reaction was applicable to a variety of cyclic *meso*-endoperoxides of various ring sizes.

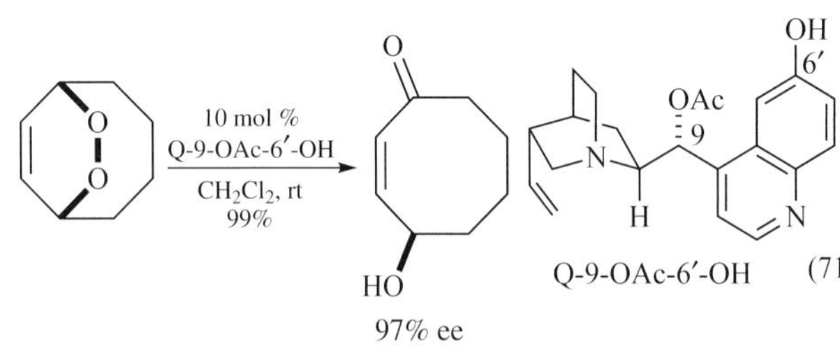

(71)

Asymmetric Protonation. In the presence of 10 mol % of quinine derivative Q-9-OBz, the nucleophilic addition of thiophenol to silylketenes followed by the enantioselective protonation of the resulting enolates furnished chiral α-silylthioesters in good yields and with moderate to high enantioselectivities (eq 72).[81]

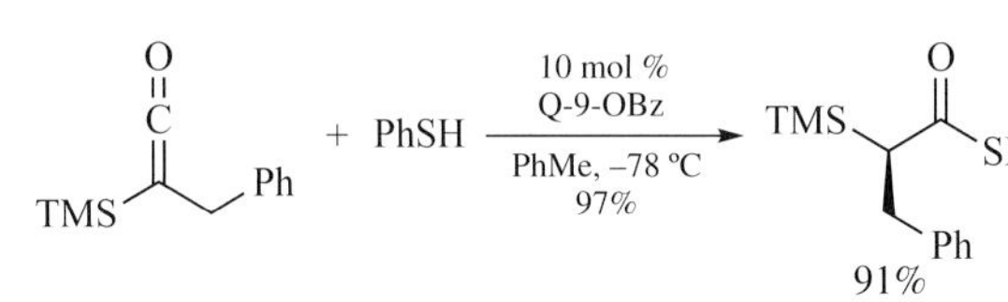

(72)

Enantioselective protonation of cyclic silyl enol ether was attempted with a dimeric cinchona alkaloid catalyst $(DHQ)_2AQN$ and a latent source of HF (eq 73).[82] Presumably, the cinchona alkaloid-hydrogen fluoride salt $(DHQ)_2AQN$-HF was responsible for the promotion of this enantioselective protonation. This enantioselective protonation proceeded in good yield and with moderate to high enantioselectivity.

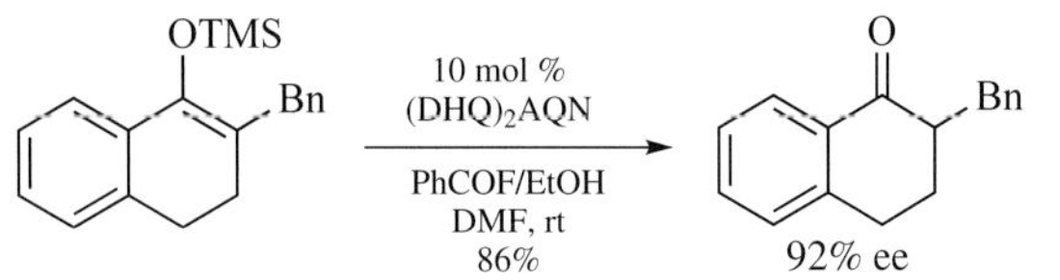

(73)

Enantioselective proton transfer catalysis involving the formation and protonation of a non-enolate carbanion was realized by the development of a highly enantioselective isomerization of trifluoromethyl imines (eq 76). This isomerization, mediated by a quinine derivative 2′-Cl DHQ, proceeded in a highly enantioselective fashion with either the aryl or alkyl trifluoromethyl imines. Consequently, it provided a highly general route toward optically active trifluoromethylated amines.[85]

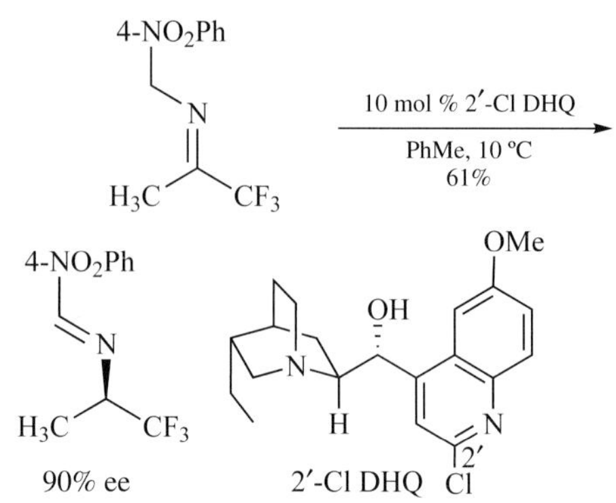

(76)

Asymmetric Olefin Isomerization. A broad range of mono- and disubstituted β,γ-unsaturated butenolides were transformed into the corresponding chiral α,β-unsaturated butenolides with good enantioselectivity and in high yield via biomimetic proton transfer catalysis by N'-O-DHQD-6′-OH (eq 74).[83] The oxidation of the quinoline ring into the corresponding N-oxide analog in the bifunctional catalyst enhanced the acidity of the 6′-OH group, which resulted in significantly improved reactivity and selectivity for the asymmetric isomerization. The reaction was performed at −20 °C or room temperature with as low as 0.5 mol % catalyst loading and tolerated functional groups such as unprotected alcohols. Kinetic isotope effect studies revealed that the protonation was the rate-determining step.

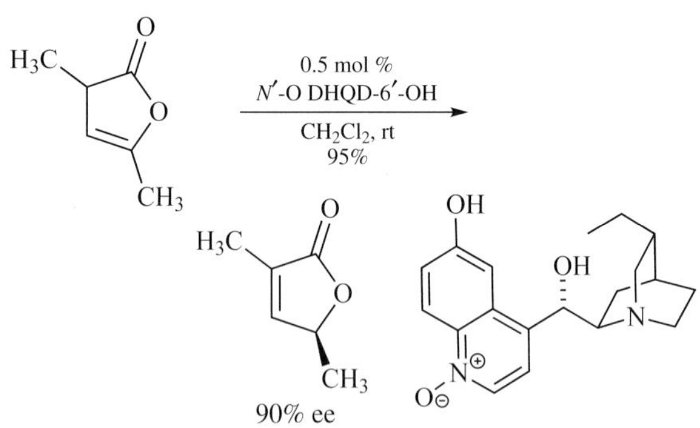

(74)

Transamination. Biomimetic transamination of α-imino esters generated in situ from *ortho*-chlorobenzyl amine and pyruvates was reported with the bifunctional catalysis of quinine-derived Q-9-^{n}Bu-6′-OH (eq 75).[84] The proton transfer catalysis-mediated isomerization of ketimine was the crucial asymmetric induction step. A wide variety of alkyl α-amino esters containing various functional groups were synthesized with high enantioselectivity and in useful yield upon acidic hydrolysis of the isomerization products.

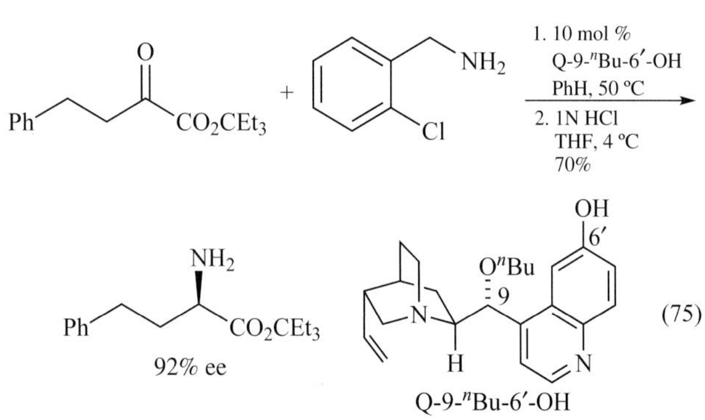

(75)

Reactions with Cinchona Alkaloids as Iminium or Enamine Catalysts. The *epi*-9-NH_2 cinchona alkaloids such as Q-*epi*-9-NH_2 have been extensively explored as chiral catalysts for asymmetric reactions with carbonyl nucleophiles and electrophiles.

Addition of α,α-Dicyanoalkenes to Enones Quinine-derived Q-*epi*-9-NH_2 was reported to catalyze the highly asymmetric conjugate addition of α,α-dicyanoalkenes to enones.[86] An acidic additive such as CF_3COOH was required as a cocatalyst. Presumably, the activation of the enone by the catalyst was accomplished by the acid-mediated formation of an iminium salt between the 9-amino cinchona alkaloid and the enone. The catalyst afforded complete *anti*-selectivity and high enantioselectivity and accepted both β-aryl and -alkyl enones (eq 77).

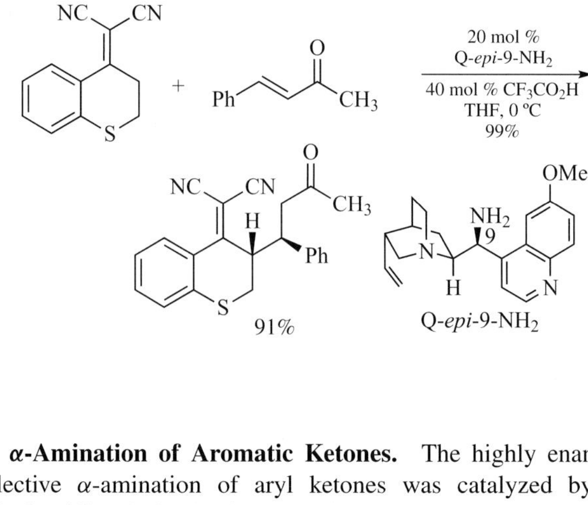

(77)

α-Amination of Aromatic Ketones. The highly enantioselective α-amination of aryl ketones was catalyzed by the cinchonidine-derived primary amine C-*epi*-9-NH_2 (eq 78).[87] A broad range of chiral α-aminoketones were obtained in good yields and with excellent enantioselectivity.

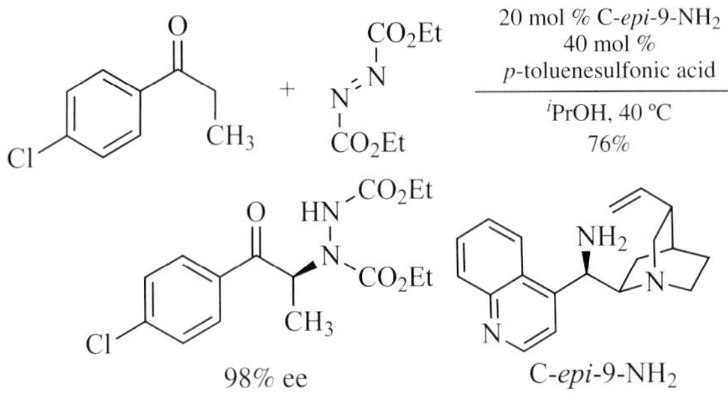

(78)

Epoxidation of Cyclic Enones. An enantioselective epoxidation of cyclic enones with hydrogen peroxide was developed through the catalysis of quinine derivative Q-*epi*-9-NH_2 (eq 79).[88] Epoxy cyclic ketones are prepared in good yields and with high enantioselectivity.

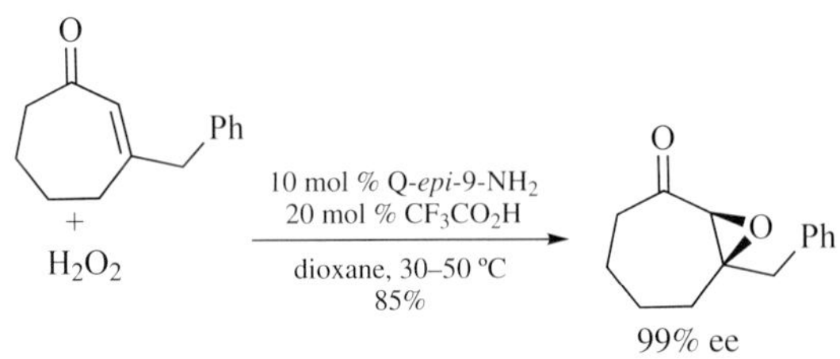

(79)

Peroxidation and Epoxidation of α,β-Unsaturated Ketones. An enantioselective route toward chiral peroxides was established by the development of an asymmetric peroxidation of hydroperoxides to α,β-unsaturated ketones with the quinine derivative Q-*epi*-9-NH_2(eq 80).[89] The reaction is applicable to a wide range of acyclic enones and various hydroperoxides. The employment of α-methoxy isopropyl hydroperoxide enabled the generation of chiral peroxides that could be readily converted into the corresponding chiral hydroperoxides suitable for further elaborations. The amine catalyst Q-*epi*-9-NH_2 not only directed the enantioselective addition of the hydroperoxide to the alkenyl iminium intermediate but also facilitated a rapid intermolecular protonation of the resulting peroxyenamine intermediate to override the intramolecular epoxide-forming step, thereby turning a conventional epoxidation pathway into a peroxidation pathway. Interestingly, it is found that, by simply elevating the temperature, the epoxidation pathway could be restored to form the corresponding chiral epoxides in a highly enantioselective manner (eq 81). The reduction of the chiral peroxides[87] readily afforded the corresponding β-hydroxy ketones without erosion in optical purity. When hydrogen peroxide was used as an oxidant, the same reaction furnished cyclic peroxyhemiketals (eq 82).[90]

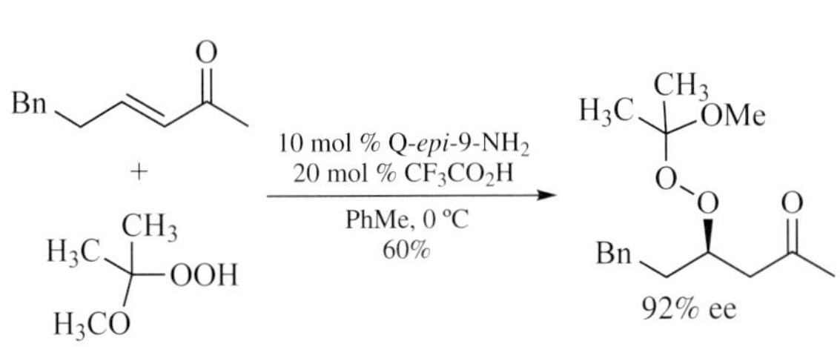

(80)

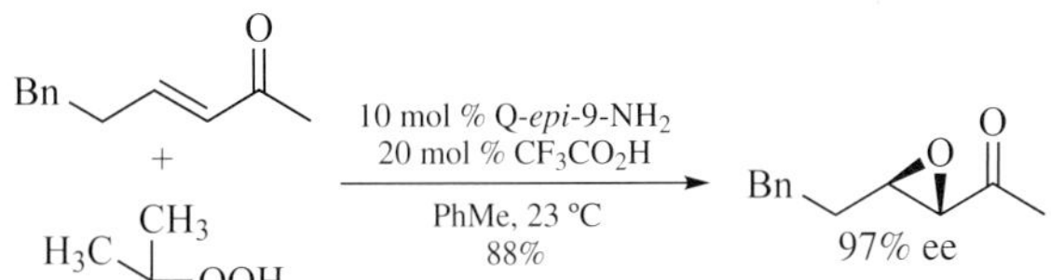

(81)

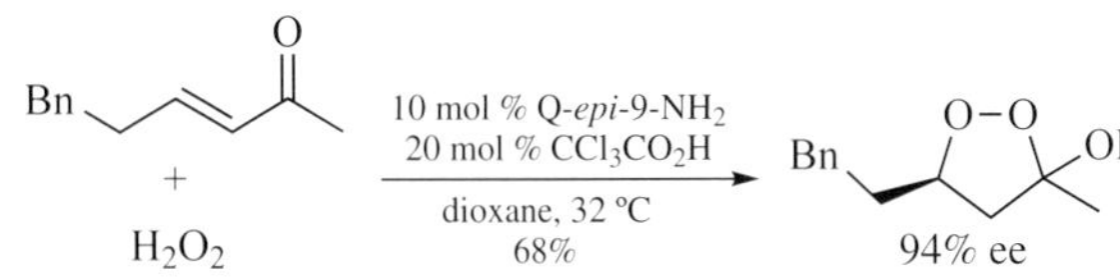

(82)

Aza-Michael Addition to Enones. The first highly enantioselective organocatalytic aza-Michael reactions with α,β-unsaturated ketones were enabled with the catalysis of bifunctional catalyst Q-*epi*-9-NH_2 (eq 83).[91] The commercially available reagent BocNH-OBn was applied as the nitrogen nucleophile. This new catalytic asymmetric aza-Michael reaction was effective for a broad range of alkyl vinyl ketones bearing both aryl and alkyl β-substituents, affording the corresponding β-amino ketones in moderate to good yields and with high enantioselectivity.

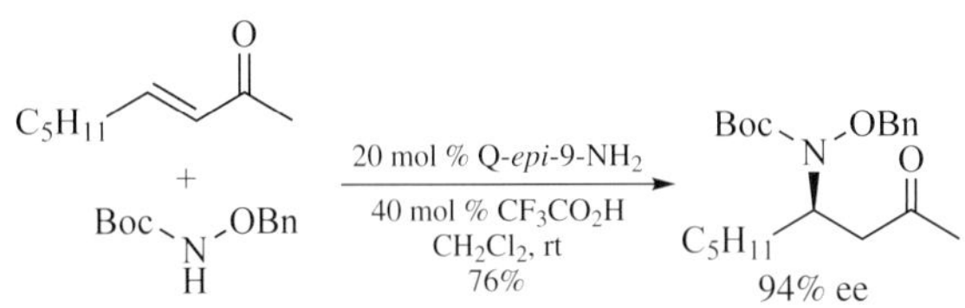

(83)

Aziridination of Enones. Cinchona alkaloid-derived phase transfer catalysts have been successfully used as catalysts for the asymmetric aziridination of enones.[92] However, chiral aziridines are obtained with low to moderate enantioselectivity under phase transfer conditions. By using a quinine-derived catalyst salt generated by combining DHQ-*epi*-9-NH_2 with D-*N*-Boc-phenylglycine, high enantioselectivity was achieved in the aziridination of various α,β-unsaturated ketones with *N*-protected hydroxyl amines (eq 84).[93] These reactions proceeded through iminium–enamine catalysis; the conjugate addition of hydroxyl amine to the iminium generated from the condensation of catalyst and enone afforded an enamine intermediate, which underwent intramolecular cyclization to form the desired aziridine in high yield and with good enantioselectivity.

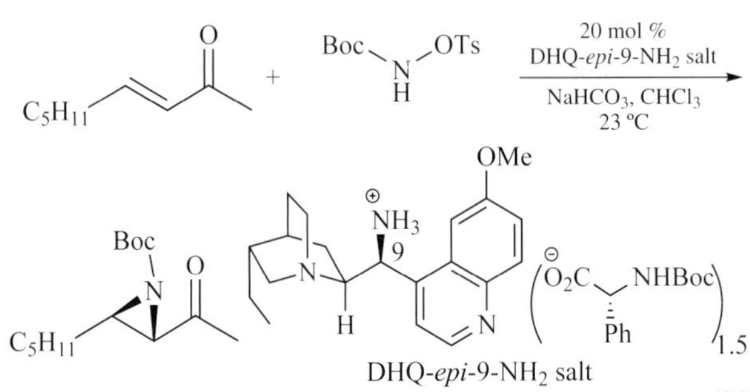

(84)

Olefin Isomerization. An enantioselective preparation of optically active chiral cyclohex-2-enones was realized through asymmetric olefin isomerization of prochiral β,γ-unsaturated cyclohex-3-enones catalyzed by a quinine-derived Q-*epi*-9-AN catalyst (eq 85).[94] Q-*epi*-9-AN was designed to utilize an electronically tunable aniline moiety to activate carbonyls by iminium or enamine catalysis. This catalyst accepted a wide range of cyclohex-3-enones, which were easily accessible in two steps from the readily available anisoles. The efficient cooperative iminium–base catalysis was proposed to be responsible for the high reactivity and selectivity of this asymmetric isomerization.

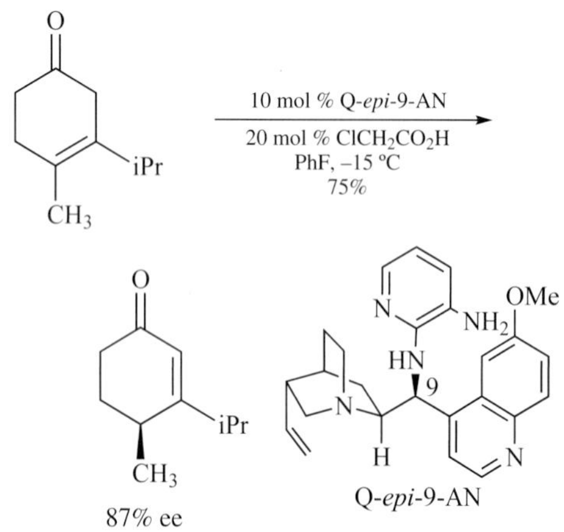

(85)

Aldol Condensation. Enantiomerically enriched 5-substituted-3-methyl-2-cyclohexene-1-ones were prepared through intramolecular condensation of 4-substituted-2,6-heptanediones in the presence of 20 mol % Q-*epi*-9-NH_2 and 60 mol % AcOH (eq 86).[95] The acidity of the cocatalyst was found to be crucial for achieving good catalytic activity and enantioselectivity.

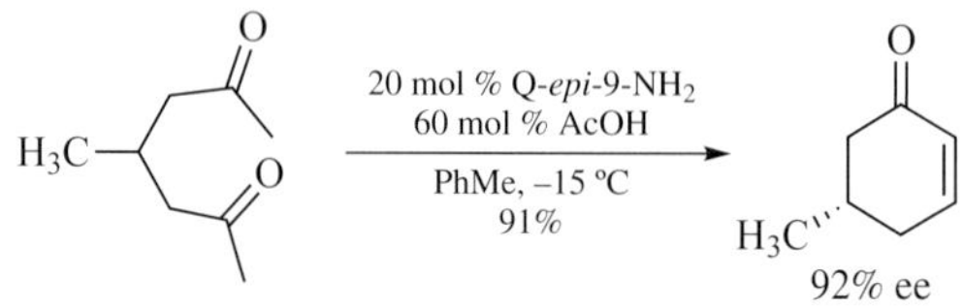

(86)

1. Riant, O.; Kagan, H. B. *Tetrahedron Lett.* **1989**, *30*, 7403.
2. Drabowicz, J.; Legedź, S.; Mikolajczyk, M., *Tetrahedron* **1988**, *44*, 5243.
3. Smaardijk, A. A.; Wynberg, H., *J. Org. Chem.* **1987**, *52*, 135.
4. Van Oeveren, A.; Menge, W.; Feringa, B. L., *Tetrahedron Lett.* **1989**, *30*, 6427.
5. Helder, R.; Hummelen, J. C.; Laane, R. W. P. M.; Wiering, J. S.; Wynberg, H., *Tetrahedron Lett.* **1976**, 1831.
6. Hummelen, J. C.; Wynberg, H., *Tetrahedron Lett.* **1978**, 1089.
7. Wynberg, H.; Marsman, B., *J. Org. Chem.* **1980**, *45*, 158.
8. Wynberg, H.; Helder, R., *Tetrahedron Lett.* **1975**, 4057.
9. Colonna, S.; Hiemstra, H.; Wynberg, H., *J. Chem. Soc., Chem. Commun.* **1978**, 238.
10. Helder, R.; Arends, R.; Bolt, W.; Hiemstra, H.; Wynberg, H., *Tetrahedron Lett.* **1977**, 2181.
11. Colonna, S.; Fornasier, R., *J. Chem. Soc., Perkin Trans. 1* **1978**, 371.
12. Johar, P. S.; Araki, S.; Butsugan, Y., *J. Chem. Soc., Perkin Trans. 1* **1992**, 711.
13. Kobayashi, N.; Iwai, K., *J. Am. Chem. Soc.* **1978**, *100*, 7071.
14. Kobayashi, N.; Iwai, K., *J. Org. Chem.* **1981**, *46*, 1823.
15. Song, C. E. *Cinchona Alkaloids in Synthesis and Catalysis*; Wiley-VCH Verlag GmbH & Co. KGaA, **2009**.
16. Kolb, H. C.; VanNieuwenhze, M. S.; Sharpless, K. B., *Chem. Rev.* **1994**, *94*, 2483.
17. Li, G.; Chang, H.-T.; Sharpless, K. B., *Angew. Chem., Int. Ed.* **1996**, *35*, 451.
18. Rudolph, J.; Sennhenn, P. C.; Vlaar, C. P.; Sharpless, K. B., *Angew. Chem., Int. Ed.* **1996**, *35*, 2810.
19. Dolling, U. H.; Davis, P.; Grabowski, E. J. J., *J. Am. Chem. Soc.* **1984**, *106*, 446.
20. O'Donnell, M. J.; Bennett, W. D.; Wu, S., *J. Am. Chem. Soc.* **1989**, *111*, 2353.
21. O'Donnell, M. J.; Wu, S.; Huffman, J. C., *Tetrahedro.* **1994**, *50*, 4507.
22. Lygo, B.; Wainwright, P. G., *Tetrahedron Lett.* **1997**, *38*, 8595.
23. Corey, E. J.; Xu, F.; Noe, M. C., *J. Am. Chem. Soc.* **1997**, *119*, 12414.
24. Jew, S.-S.; Yoo, M.-S.; Jeong, B.-S.; Park, I. Y.; Park, H.-G., *Org. Lett.* **2002**, *4*, 4245.
25. Yoo, M.-S.; Jeong, B.-S.; Lee, J.-H.; Park, H.-G.; Jew, S.-S. *Org. Lett.* **2005**, *7*, 1129.
26. O'Donnell, M. J.; Wu, S., *Tetrahedron: Asymmetry* **1992**, *3*, 591.
27. Lygo, B.; Crosby, J.; Peterson, J. A., *Tetrahedron Lett.* **1999**, *40*, 8671.
28. Jew, S.-S.; Jeong, B.-S.; Lee, J.-H.; Yoo, M.-S.; Lee, Y.-J.; Park, B.-S.; Kim, M. G.; Park, H.-G., *J. Org. Chem.* **2003**, *68*, 4514.
29. Corey, E. J.; Noe, M. C.; Xu, F., *Tetrahedron Lett.* **1998**, *39*, 5347.
30. Zhang, F.-Y.; Corey, E. J., *Org. Lett.* **2000**, *2*, 1097.
31. Zhang, F.-Y.; Corey, E. J., *Org. Lett.* **2001**, *3*, 639.
32. Masui, M.; Ando, A.; Shioiri, T., *Tetrahedron Lett.* **1988**, *29*, 2835.
33. Fini, F.; Sgarzani, V.; Pettersen, D.; Herrera, R. P.; Bernardi, L.; Ricci, A., *Angew. Chem., Int. Ed.* **2005**, *44*, 7975.
34. Palomo, C.; Oiarbide, M.; Laso, A.; López, R., *J. Am. Chem. Soc.* **2005**, *127*, 17622.
35. Furukawa, T.; Shibata, N.; Mizuta, S.; Nakamura, S.; Toru, T.; Shiro, M., *Angew. Chem., Int. Ed.* **2008**, *47*, 8051.
36. Wu, Y.; Hu, L.; Li, Z.; Deng, L., *Nature* **2015**, *523*, 445.
37. Wynberg, H.; Staring, E. G. J., *J. Am. Chem. Soc.* **1982**, *104*, 166.
38. Zhu, C.; Shen, X.; Nelson, S. G., *J. Am. Chem. Soc.* **2004**, *126*, 5352.
39. Taggi, A. E.; Hafez, A. M.; Wack, H.; Young, B.; Drury, W. J.; Lectka, T., *J. Am. Chem. Soc.* **2000**, *122*, 7831.
40. France, S.; Wack, H.; Hafez, A. M.; Taggi, A. E.; Witsil, D. R.; Lectka, T., *Org. Lett.* **2002**, *4*, 1603.
41. Abraham, C. J.; Paull, D. H.; Scerba, M. T.; Grebinski, J. W.; Lectka, T., *J. Am. Chem. Soc.* **2006**, *128*, 13370.
42. Xu, X.; Wang, K.; Nelson, S. G., *J. Am. Chem. Soc.* **2007**, *129*, 11690.
43. Tian, S.-K.; Deng, L., *J. Am. Chem. Soc.* **2001**, *123*, 6195.
44. Tian, S.-K.; Deng, L., *Tetrahedron* **2006**, *62*, 11320.
45. Tian, S. K.; Hong, R.; Deng, L., *J. Am. Chem. Soc.* **2003**, *125*, 9900.
46. Hong, R.; Chen, Y.; Deng, L., *Angew. Chem., Int. Ed.* **2005**, *44*, 3478.
47. Shibata, N.; Suzuki, E.; Takeuchi, Y., *J. Am. Chem. Soc.* **2000**, *122*, 10728.
48. Mohar, B.; Baudoux, J.; Plaquevent, J.-C.; Cahard, D., *Angew. Chem., Int. Ed.* **2001**, *40*, 4214.
49. Iwabuchi, Y.; Nakatani, M.; Yokoyama, N.; Hatakeyama, S., *J. Am. Chem. Soc.* **1999**, *121*, 10219.
50. Kawahara, S.; Nakano, A.; Esumi, T.; Iwabuchi, Y.; Hatakeyama, S., *Org. Lett.* **2003**, *5*, 3103.
51. Hiratake, J.; Yamamoto, Y.; Oda, J. I., *J. Chem. Soc., Chem. Commun.* **1985**, 1717.
52. Chen, Y.; Tian, S.-K.; Deng, L., *J. Am. Chem. Soc.* **2000**, *122*, 9542.
53. Chen, Y.; Deng, L., *J. Am. Chem. Soc.* **2001**, *123*, 11302.

54. Hang, J.; Tian, S. K.; Tang, L.; Deng, L., *J. Am. Chem. Soc.* **2001**, *123*, 12696.
55. Tang, L.; Deng, L., *J. Am. Chem. Soc.* **2002**, *124*, 2870.
56. Li, H.; Liu, X.; Wu, F.; Tang, L.; Deng, L., *Proc. Natl. Acad. Sci. USA.* **2010**, *107*, 20625.
57. McDaid, P.; Chen, Y.; Deng, L., *Angew. Chem., Int. Ed.* **2002**, *41*, 338.
58. Li, H.; Wang, Y.; Tang, L.; Deng, L., *J. Am. Chem. Soc.* **2004**, *126*, 9906.
59. Wang, Y.; Liu, X.; Deng, L., *J. Am. Chem. Soc.* **2006**, *128*, 3928.
60. Li, H.; Wang, Y.; Tang, L.; Wu, F.; Liu, X.; Guo, C.; Foxman, B. M.; Deng, L., *Angew. Chem., Int. Ed.* **2005**, *44*, 105.
61. Vakulya, B.; Varga, S.; Csámpai, A.; Soós, T., *Org. Lett.* **2005**, *7*, 1967.
62. Malerich, J. P.; Hagihara, K.; Rawal, V. H., *J. Am. Chem. Soc.* **2008**, *130*, 14416.
63. Li, H.; Song, J.; Liu, X.; Deng, L., *J. Am. Chem. Soc.* **2005**, *127*, 8948.
64. Wu, F.; Hong, R.; Khan, J.; Liu, X.; Deng, L., *Angew. Chem., Int. Ed.* **2006**, *45*, 4301.
65. Wu, F.; Li, H.; Hong, R.; Deng, L., *Angew. Chem., Int. Ed.* **2006**, *45*, 947.
66. Wang, B.; Wu, F.; Wang, Y.; Liu, X.; Deng, L., *J. Am. Chem. Soc.* **2007**, *129*, 768.
67. Li, H.; Wang, B.; Deng, L., *J. Am. Chem. Soc.* **2006**, *128*, 732.
68. Marcelli, T.; van der Haas, R. N. S.; van Maarseveen, J. H.; Hiemstra, H., *Angew. Chem., Int. Ed.* **2006**, *45*, 929.
69. Bandini, M.; Sinisi, R.; Umani-Ronchi, A., *Chem. Commun.* **2008**, 4360.
70. Singh, R. P.; Foxman, B. M.; Deng, L., *J. Am. Chem. Soc.* **2010**, *132*, 9558.
71. Zheng, Y.; Deng, L., *Chem. Sci.* **2015**, *6*, 6510.
72. Song, J.; Wang, Y.; Deng, L., *J. Am. Chem. Soc.* **2006**, *128*, 6048.
73. Ye, J.; Dixon, D. J.; Hynes, P. S., *Chem. Commun.* **2005**, 4481.
74. Song, J.; Shih, H. W.; Deng, L., *Org. Lett.* **2007**, *9*, 603.
75. Wynberg, H.; Smaardijk, A. A., *Tetrahedron Lett.* **1983**, *24*, 5899.
76. Pettersen, D.; Marcolini, M.; Bernardi, L.; Fini, F.; Herrera, R. P.; Sgarzani, V.; Ricci, A., *J. Org. Chem.* **2006**, *71*, 6269.
77. Li, H.; Wang, Y.-Q.; Deng, L., *Org. Lett.* **2006**, *8*, 4063.
78. Wang, Y. Q.; Song, J.; Hong, R.; Li, H.; Deng, L., *J. Am. Chem. Soc.* **2006**, *128*, 8156.
79. Wang, Y.; Li, H.; Wang, Y. Q.; Liu, Y.; Foxman, B. M.; Deng, L., *J. Am. Chem. Soc.* **2007**, *129*, 6364.
80. Staben, S. T.; Linghu, X.; Toste, F. D., *J. Am. Chem. Soc.* **2006**, *128*, 12658.
81. Blake, A. J.; Friend, C. L.; Outram, R. J.; Simpkins, N. S.; Whitehead, A. J., *Tetrahedron Lett.* **2001**, *42*, 2877.
82. Poisson, T.; Dalla, V.; Marsais, F.; Dupas, G.; Oudeyer, S.; Levacher, V., *Angew. Chem., Int. Ed.* **2007**, *46*, 7090.
83. Wu, Y.; Singh, R. P.; Deng, L., *J. Am. Chem. Soc.* **2011**, *133*, 12458.
84. Xiao, X.; Xie, Y.; Su, C.; Liu, M.; Shi, Y., *J. Am. Chem. Soc.* **2011**, *133*, 12914.
85. Wu, Y.; Deng, L., *J. Am. Chem. Soc.* **2012**, *134*, 14334.
86. Xie, J.-W.; Chen, W.; Li, R.; Zeng, M.; Du, W.; Yue, L.; Chen, Y.-C.; Wu, Y.; Zhu, J.; Deng, J.-G., *Angew. Chem., Int. Ed.* **2007**, *46*, 389.
87. Liu, T.-Y.; Cui, H.-L.; Zhang, Y.; Jiang, K.; Du, W.; He, Z.-Q.; Chen, Y.-C., *Org. Lett.* **2007**, *9*, 3671.
88. Wang, X.; Reisinger, C. M.; List, B., *J. Am. Chem. Soc.* **2008**, *130*, 6070.
89. Lu, X.; Liu, Y.; Sun, B.; Cindric, B.; Deng, L., *J. Am. Chem. Soc.* **2008**, *130*, 8134.
90. Reisinger, C. M.; Wang, X.; List, B., *Angew. Chem., Int. Ed.* **2008**, *47*, 8112.
91. Lu, X.; Deng, L., *Angew. Chem., Int. Ed.* **2008**, *47*, 7710.
92. Aires-de-Sousa, J.; Lobo, A. M.; Prabhakar, S., *Tetrahedron Lett.* **1996**, *37*, 3183.
93. Pesciaioli, F.; De Vincentiis, F.; Galzerano, P.; Bencivenni, G.; Bartoli, G.; Mazzanti, A.; Melchiorre, P., *Angew. Chem., Int. Ed.* **2008**, *47*, 8703.
94. Lee, J. H.; Deng, L., *J. Am. Chem. Soc.* **2012**, *134*, 18209.
95. Zhou, J.; Wakchaure, V.; Kraft, P.; List, B., *Angew. Chem., Int. Ed.* **2008**, *47*, 7656.

3-Quinuclidinol

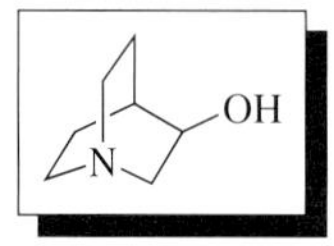

[1619-34-7] $C_7H_{13}NO$ (MW 127.21))

InChI = 1/C7H13NO/c9-7-5-8-3-1-6(7)2-4-8/h6-7,9H, 1-5H2/t7-/s2

InChIKey = IVLICPVPXWEGCA-JLDDOWRYBP

(cleavage of β-keto esters;[1] catalyst for Baylis–Hillman reaction;[2–6] synthesis of monoesters;[7] oxygenation of branched aldehydes[8]; catalyst for α-iodination of acrylate esters[9]; catalyst for aza-Baylis–Hillman reactions,[10–13] catalyst for cyanation of 2-chloropyrimidines[14])

Physical Data: mp 223–224 °C.
Solubility: sol water, ethanol, methanol.
Form Supplied in: crystalline solid; >98% purity.
Analysis of Reagent Purity: NMR, mp.
Preparative Methods: commercially available from several sources.
Purification: recrystallize from benzene or acetone.
Handling, Storage, and Precautions: corrosive.

Original Commentary

Ellen M. Leahy
Affymax Research Institute, Palo Alto, CA, USA

Cleavage of β-Keto Esters. 3-Quinuclidinol selectively cleaves β-keto esters to ketones (eq 1).[1] Typical reaction conditions involve heating the ester with a fivefold excess of quinuclidinol in *o*-xylene. The reaction is also successful for cleavage of vinylogous β-keto esters (eq 2).[1]

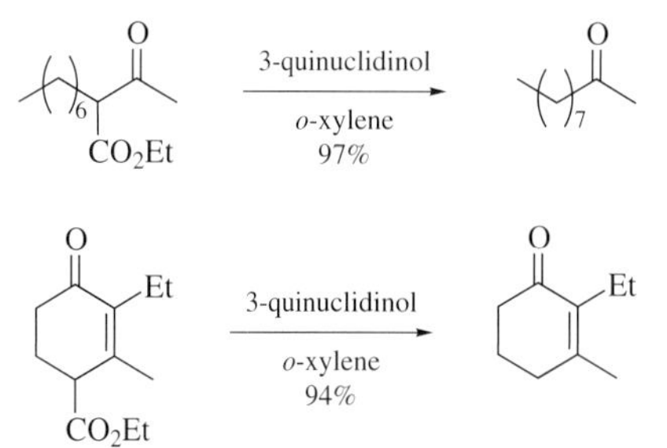

(1)

(2)

Baylis–Hillman Reaction. 3-Quinuclidinol is a much more efficient catalyst than ***1,4-Diazabicyclo[2.2.2]octane*** (DABCO) for the Baylis–Hillman reaction, yielding 2-hydroxyalkenoate esters (eqs 3 and 4).[2] Reaction times are much shorter, typically less than 24 h as compared to 4–7 days.[3] The same results have been observed for the synthesis of hydroxy enones (eq 5).[4] It has been speculated that the rate enhancement can be due to hydrogen-bonding stabilization of the catalyst–acrylate adduct[5] by the hydroxyl group of quinuclidinol, which does not occur with DABCO. This methodology has also been extended to include the use of aromatic aldehydes (eq 6).[6]

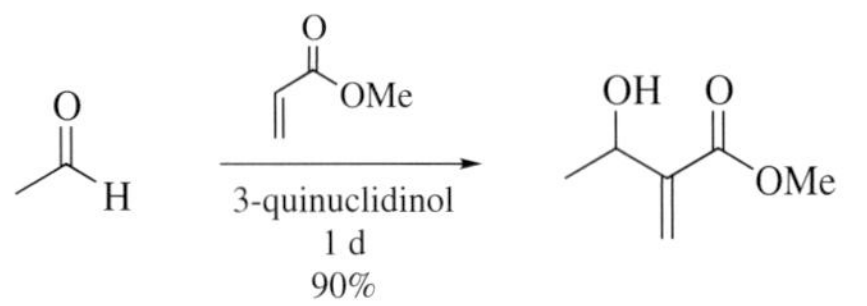
(3)

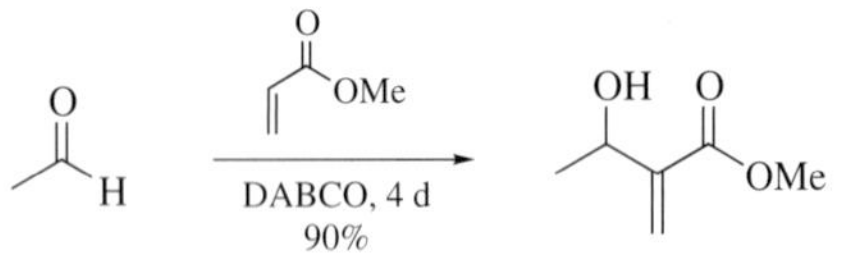
(4)

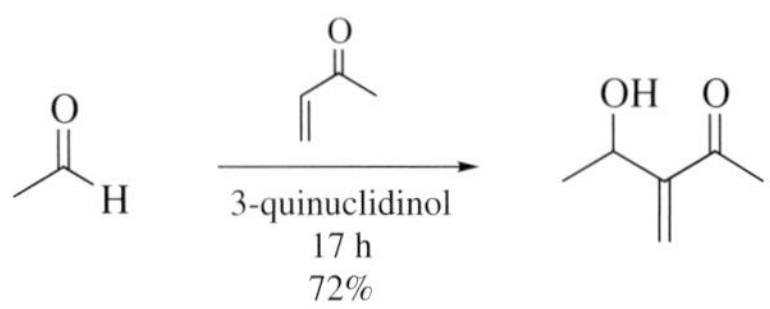
(5)

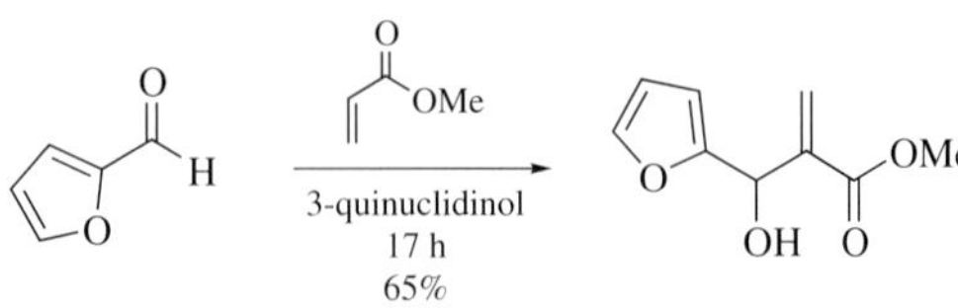
(6)

Synthesis of Monoesters. Geminal diesters are converted to the corresponding monoester with quinuclidinol (eq 7).[7] Other bicyclic amines such as DABCO and ***1,5-Diazabicyclo[4.3.0]non-5-ene*** (DBN) can also be used.[7] These cleavage conditions are advantageous in that they involve a one-step process compared with two steps involving formation of the monoacid followed by esterification. More importantly, these reaction conditions are perfectly suited to cases where acid or base labile groups are present or where the compounds are not soluble in aqueous solutions.

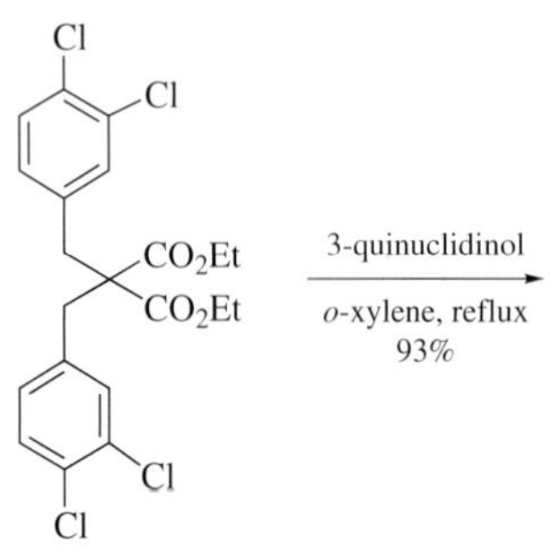
(7)

Oxygenation of Branched Aldehydes to Ketones. Copper-catalyzed oxygenation of α-branched aldehydes in weakly basic media yields ketones having one less carbon (eq 8).[8] Van Rheenen found that the optimum reaction conditions use ***Copper(II) Acetate*** complexed with 2,2′-bipyridyl or 1,10-phenanthroline as catalyst and bicyclic amines such as DABCO or 3-quinuclidinol in polar aprotic solvents rather than in alcohols.[8]

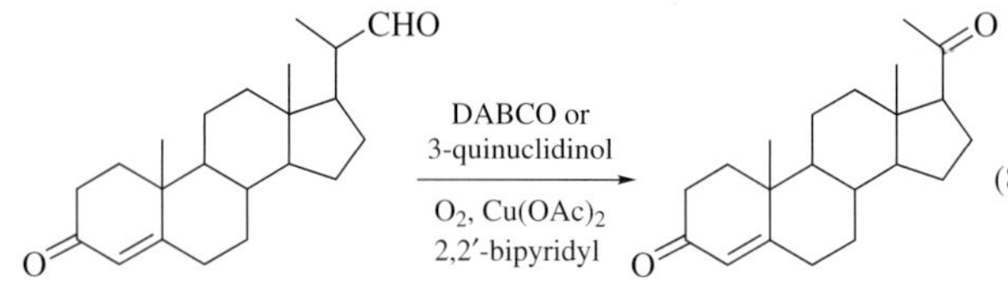
(8)

First Update

Yu Chen & Pauline Chiu
The University of Hong Kong, Pokfulam Road, Hong Kong

Chemoselective α-Iodination of Acrylate Esters. 3-Quinuclidinol is a nucleophilic catalyst for the chemoselective α-iodination of acrylate esters. The reaction proceeds via a Morita–Baylis–Hillman type of mechanism in which the enolate, generated in situ, reacts with an electrophilic iodinating agent.[9] Acrylate esters react chemoselectively over cinnamates or crotonates (eqs 9,10).

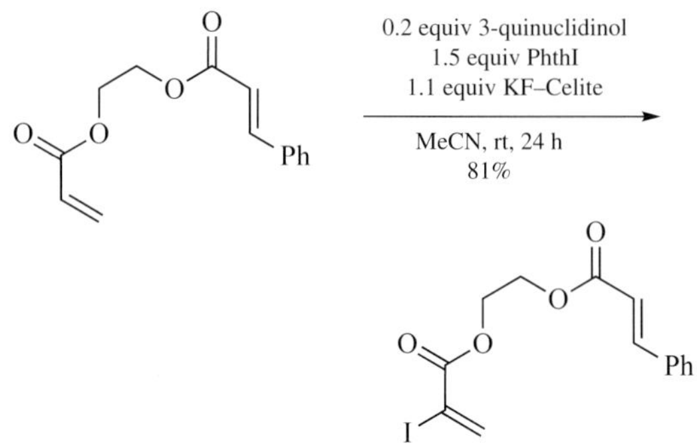
(9)

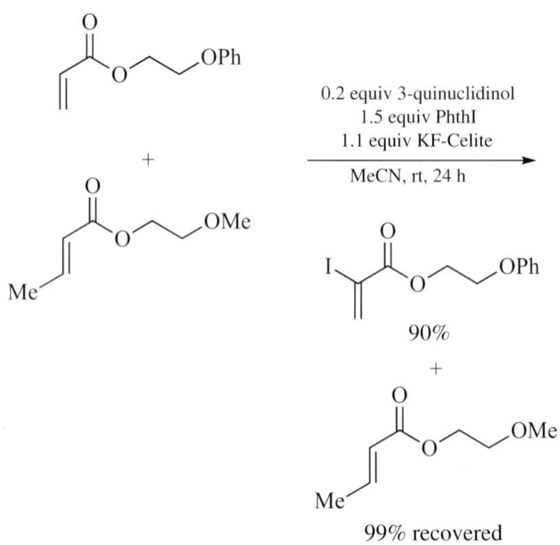
(10)

3-Quinuclidinol is a better nucleophilic catalyst compared with DABCO or quinuclidine, because it can stabilize the intermediate zwitterion by intramolecular hydrogen bonding. In this reaction, the use of alternative iodinating reagents, including iodine, results in inferior yields, as does the use of bases other than KF–Celite, which tends to induce further dehydroiodination to generate propiolate esters.

Aza-Baylis–Hillman Reaction. The aza-Baylis–Hillman reaction of methyl acrylate with aldimines is more efficiently catalyzed by 3-quinuclidinol than DABCO (eq 11).[10] The yields of the reaction are highest for electron-deficient imines (Ar = electron-withdrawing groups). However, the corresponding reaction of acrylonitrile was reported to proceed more efficiently under catalysis by DABCO.

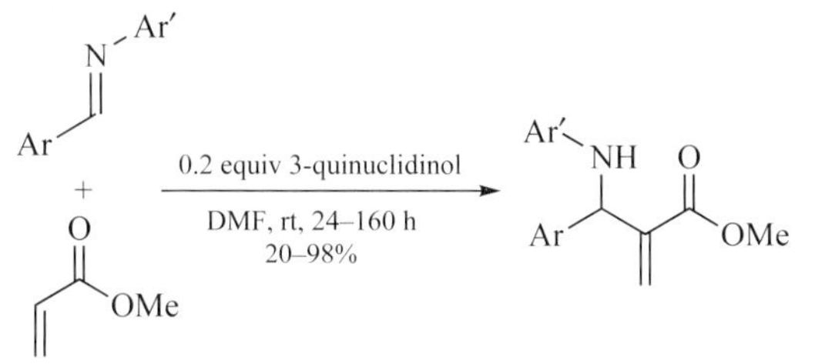

(11)

Ar = C_6H_5, C_6F_5, *p*-NO_2-C_6H_4, *o*-NO_2-C_6H_4
o-Br-C_6H_4, *o*-Cl-C_6H_4, *p*-Me-C_6H_4, *p*-MeO-C_6H_4

Ar′ = C_6F_5, 4-Cl-C_6F_4

Conjugated dienones, dienoate, and dienyl sulfone react with aldimines in the presence of 3-quinuclidinol to generate aza-Baylis–Hillman products in moderate to good yields (eq 12).[11] Generally, there is a tendency to form the (*E*)-product isomer in the catalyst regeneration step.

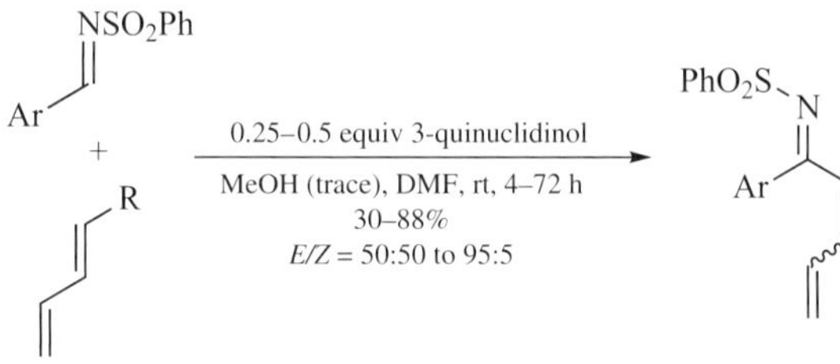

(12)

Ar = Ph, *p*-Cl-C_6H_4, *m*-Cl-C_6H_4, *o*-Cl-C_6H_4,
p-MeO-C_6H_4, *p*-NO_2-C_6H_4, *p*-MeO_2C-C_6H_4,
p-NC-C_6H_4, α-naphthyl, 3-pyridyl, mesityl, cyclohexyl

R = CO_2Me, COMe, COPh, SO_2Ph

Compared with both DABCO and PPh_3, 3-quinuclidinol is a more efficient catalyst for the three-component aza-Baylis–Hillman reaction. Allylic amines are obtained in a one-pot reaction of aldehydes, sulfonamides, and methyl acrylate (eq 13).[12] A variety of aromatic aldehydes and substituted furfurals have been used, as well as other Michael acceptors including *t*-butyl acrylate, acrylonitrile, and vinylsulfone. The success of the reaction is contingent upon the reaction of the sulfonamide with aldehyde, being faster than that between the quinuclidinol-activated nucleophile and the aldehyde, which would give rise to the typical Baylis–Hillman product. Molecular sieves increase the rate of aldimine formation by removal of water, resulting in an overall reaction rate enhancement and an increase in selectivity for the aza-Baylis–Hillman reaction.[12] In addition, catalytic amounts of Lewis acids such as $La(OTf)_3$ or $Ti(Oi\text{-}Pr)_4$ have been observed to further accelerate the reaction.[13]

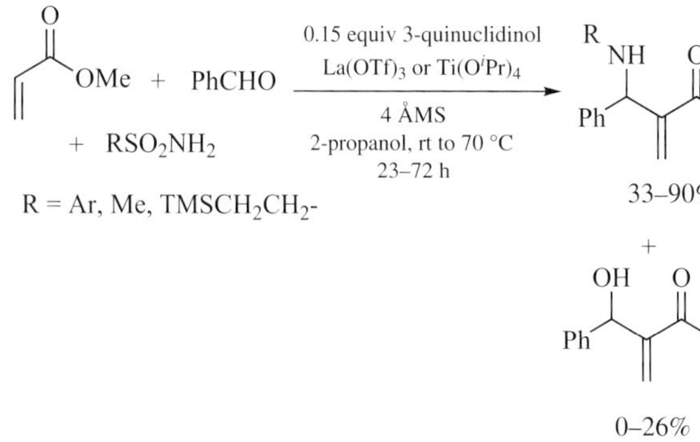

(13)

Cyanation of 2-Chloropyrimidines. 3-Quinuclidinol has been found to be superior to DABCO as a nucleophilic catalyst for the cyanation of 2-chloropyrimidines, which induced the reactions in better yields and less time, and also using less potassium cyanide (eq 14).[14] This was proposed to be due to the ability of 3-quinuclidinol to stabilize the pyrimidine intermediate by hydrogen bonding (eq 15).[14]

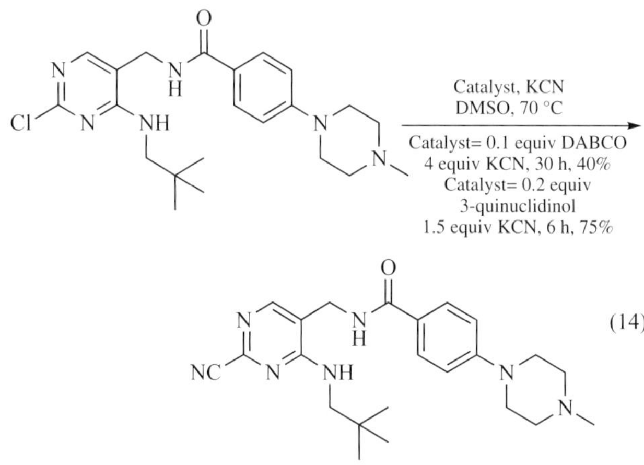

(14)

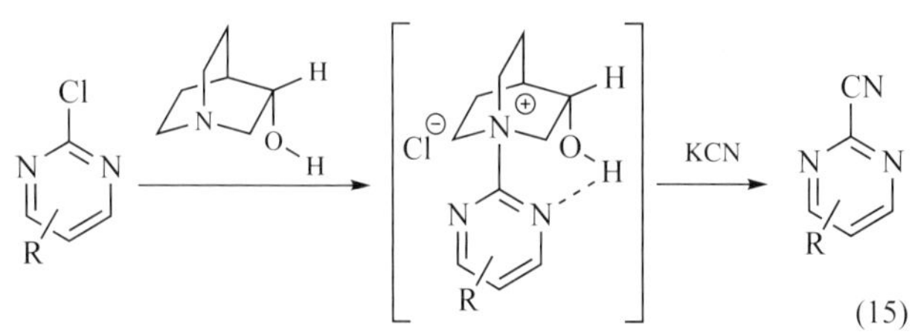

(15)

1. Parish, E. J.; Huang, B.-S.; Miles, D. H. *Synth. Commun.* **1975**, *5*, 341.
2. Drewes, S. E.; Freese, S. D.; Emslie, N. D.; Roos, G. H. P. *Synth. Commun.* **1988**, *18*, 1565.
3. Bode, M. L.; Kaye, P. T. *Tetrahedron Lett.* **1991**, *32*, 5611.
4. Bailey, M.; Markó, I. E.; Ollis, W. D.; Rasmussen, P. R. *Tetrahedron Lett.* **1990**, *31*, 4509.
5. Ameer, F.; Drewes, S. E.; Freese, S.; Kaye, P. T. *Synth. Commun.* **1988**, *18*, 495.
6. Drewes, S. E.; Emslie, N. D.; Khan, A. A.; Roos, G. H. P. *Synth. Commun.* **1989**, *19*, 959.
7. Miles, D. H.; Huang, B.-S. *J. Org. Chem.* **1976**, *41*, 208.

8. Van Rheenen, V. *Tetrahedron Lett.* **1969**, 985.
9. Pace, V.; Vilkauskaite, G.; Sackus, A.; Holzer, W., *Org. Biomol. Chem.* **2013**, *11*, 1085.
10. Liu, X.; Chai, Z.; Zhao, G.; Zhu, S., *J. Fluorine Chem.* **2005**, *126*, 1215.
11. (a) Sorbetti, J. M.; Clary, K. N.; Rankic, D. A.; Wulff, J. E.; Parvez, M.; Back, T. G., *J. Org. Chem.* **2007**, *72*, 3326.
    (b) Kristen, N.; Clary, T. G. B., *Synlett.* **2010**, 2802.
12. (a) Balan, D.; Adolfsson, H., *J. Org. Chem.* **2001**, *66*, 6498.
    (b) Balan, D.; Adolfsson, H., *J. Org. Chem.* **2002**, *67*, 2329.
    (c) Declerck, V.; Ribiére, P.; Martinez, J.; Lamaty, F., *J. Org. Chem.* **2004**, *69*, 8372.
13. Aggarwal, V. K.; Mereu, A.; Tarver, G. J.; McCague, R., *J. Org. Chem.* **1998**, *63*, 7183.
14. Kim, H. Y.; Shieh, W. C.; Prashad, M., *Tetrahedron Lett.* **2014**, *55*, 5055.

R

Ruthenium (II), Tris(2,2′-bipyrazine-κN^1, $\kappa N^{1'}$),[1] (OC-6-11)-

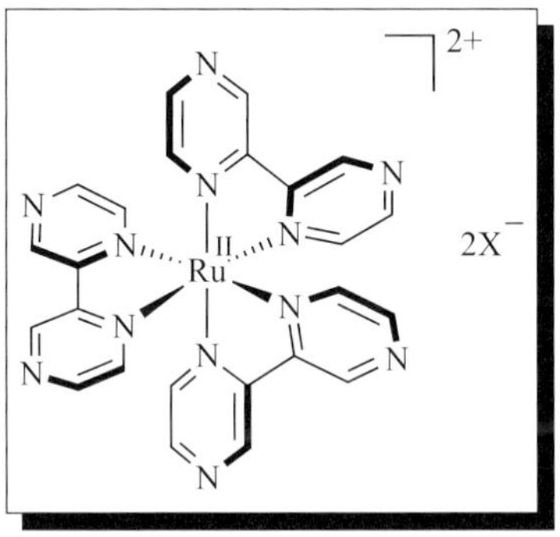

[80925-50-4] $C_{24}H_{18}N_{12}RuCl_2$ (MW 646.463)
(Cl complex)
InChI = 1S/3C8H6N4.2ClH.Ru/c3*1-3-11-7(5-9-1)8-6-10-2-4-12-8;;;/h3*1-6H;2*1H;/q;;;;;+2/p-2
InChIKey = XWSLUKLOHQAJCE-UHFFFAOYSA-L
[75523-96-5] $C_{24}H_{18}N_{12}Ru$
(cation)

InChI = 1S/3C8H6N4.Ru/c3*1-3-11-7(5-9-1)8-6-10-2-4-12-8;/h3*1-6H;/q;;;+2
InChIKey = LGAFSIQNRRRLLT-UHFFFAOYSA-N
[80907-56-8] $C_{24}H_{18}F_{12}N_{12}P_2Ru$
(PF_6 complex)
InChI = 1S/3C8H6N4.2F6P.Ru/c3*1-3-11-7(5-9-1)8-6-10-2-4-12-8;2*1-7(2,3,4,5)6;/h3*1-6H;;;/q;;;2*-1;+2
InChIKey = ;RRBHRYNFGGGCST-UHFFFAOYSA-N
[1350432-81-3] $C_{88}H_{42}B_2F_{48}N_{12}Ru$
(BArF complex)
InChI = 1S/2C32H12BF24.3C8H6N4.Ru/c2*34-25(35,36)13-1-14(26(37,38)39) 6-21(5-13)33(22-7-15(27(40,41)42)2-16(8-22)28(43,44)45,23-9-17(29(46,47) 48)3-18(10-23)-30(49,50)51)24-11-19(31(52,53)54)4-20(12-24)32(55,56)57;3*1-3-11-7 (5-9-1)8-6-10-2-4-12-8;/h2*1-12H;3*1-6H;/q2*-1;;;;+2
InChIKey = FCGIEHHPNNLZRC-UHFFFAOYSA-N

Alternate Names: $[Ru(bpz)_3]^{2+}$, ruthenium-tris(2,2′-bipyrazinyl).

Physical Data: λ_{max} = 441 nm, redox potentials (vs. saturated calomel electrode (SCE), as 0 V in CH_3CN); −0.26 V (Ru^{III}/Ru^{II*}), 1.86 V (Ru^{III}/Ru^{II}), 1.45 V (Ru^{II*}/Ru^{I}), and −0.80 V (Ru^{II}/Ru^{I}).[1]

Solubility: for X = PF_6, soluble in CH_3CN, acetone, DMF, CH_3NO_2, and DMSO; insoluble in H_2O and nonpolar, aprotic solvents.

Form Supplied in: red to orange crystals or powder; Commercially available from Sigma-Aldrich, USA.

Preparative Methods: for X = PF_6, the catalyst is prepared from $RuCl_3$, 2,2(-bipyrazine), and KPF_6 in ethylene glycol.[2]

Purification: purified by column chromatography using neutral alumina and CH_3CN as eluent.

Handling: containers should be stored under nitrogen atmosphere in the dark.

Similar to $[Ru(bpy)_3]^{2+}$ (bpy = 2,2′-bipyridine), upon exposure to visible light, $[Ru(bpz)_3]^{2+}$ (bpz = 2,2′-bipyrazine) enters into a photoexcited state that results from metal to ligand charge transfer (MLCT) and is fairly long-lived (740 ns in CH_3CN[1b]). The dual ability of the photoexcited state to accept or donate an electron is the bedrock of photoredox catalysis[1c]. $[Ru(bpz)_3]^{2+*}$(Ru^{2+*}/Ru^{1+}, 1.45 V vs. SCE) is a much stronger oxidant than $[Ru(bpy)_3]^{2+*}$(Ru^{2+*}/Ru^{1+}, 0.77 V vs. SCE[1b]).

Radical CationDiels–Alder Reactions. Radical cation Diels–Alder reactions allow electronically mismatched dienes and dienophiles to proceed.[3] Because of its high redox potential, the excited state of $Ru(bpz)_3(BArF)_2$ can readily oxidize electron-rich anethole to its corresponding radical cation without the aid of an oxidative quencher such as MV^{2+}.[4] The radical cation undergoes a facile cycloaddition with electron-rich isoprene to produce the cycloadduct with complete regio- and diastereocontrol (eq 1). BArF greatly increases the solubility of $Ru(bpz)_3^{2+}$, thus enabling the reaction to be performed in less polar solvents such as CH_2Cl_2. The radical cation Diels–Alder reaction has been applied to the synthesis of heitziamide A[5] (eq 2), a natural product isolated from the medicinal plant *Fagara heitzii*.[4]

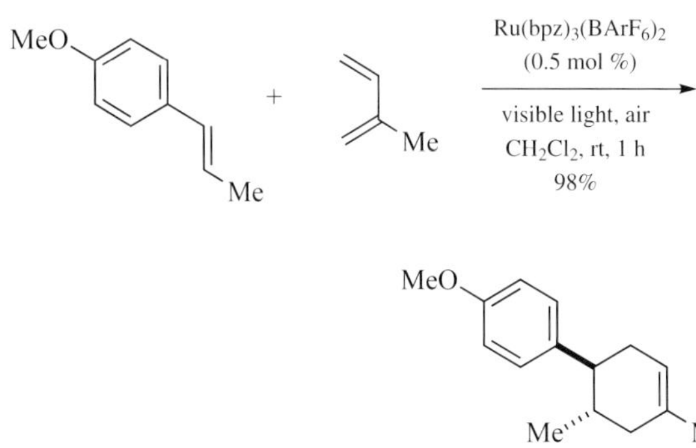

(1)

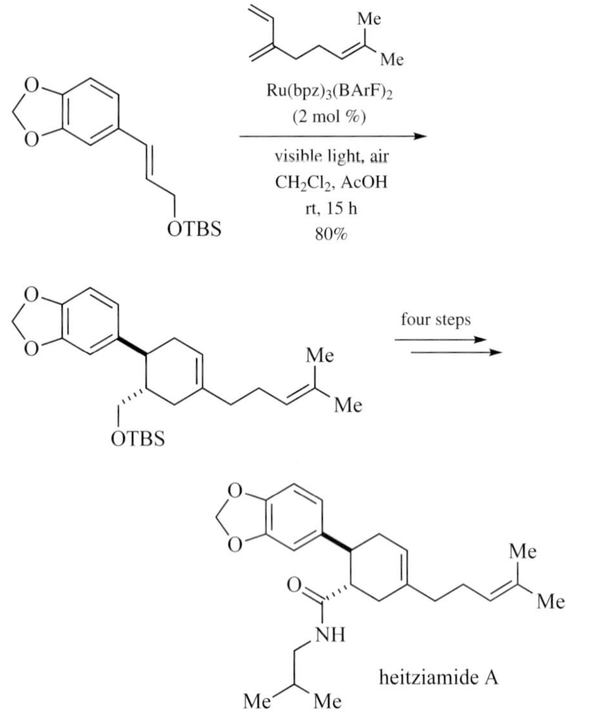

(2)

Aerobic [2+2+2] Cycloadditions. Electron-rich alkenes are known to undergo [2+2] cycloaddition to form cyclobutanes upon one-electron oxidation.[6] The primary intermediate involved in the cycloaddition is the corresponding radical cation obtained by the one-electron oxidation of one of the alkenes. This oxidation can be realized by chemical oxidants or photoinduced electron transfer with organic sensitizers. Recently, $[Ru(bpy)_3]^{3+}$, generated via oxidative quenching of $[Ru(bpy)_3]^{2+*}$ by MV^{2+}, has been shown to catalyze the same oxidation as well.[7] When $Ru(bpz)_3(PF_6)_2$ is used as the photocatalyst in conjunction with low reaction concentration, low temperature, and increased oxygen concentration, a different type of cycloaddition [2+2+2] aerobic photooxygenation, becomes the dominant pathway leading to endoperoxides (eq 3).[7b] It is worth noting that $Ru(bpy)_3(PF_6)_2$ and common organic sensitizers such as tetraphenylporphyrin, 9,10-dicyanoanthracene, and triphenylpyrylium·BF_4 are ineffective in this reaction.

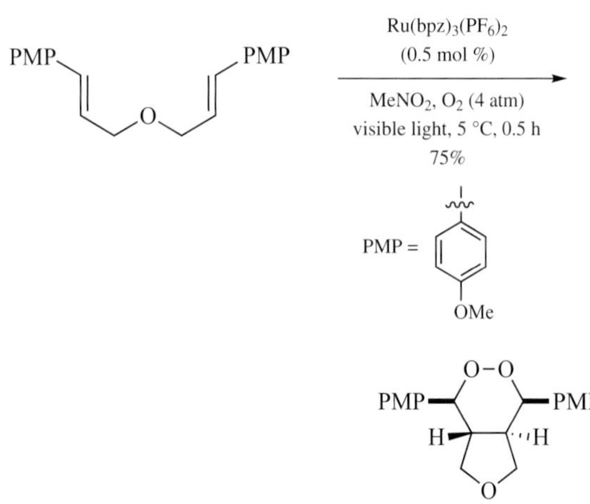

(3)

[3+2] Annulation of Cyclopropylamines with Olefins. One-electron oxidation of *N*-cyclopropylamines triggers opening of the cyclopropyl ring to produce β-carbon radical iminium ions, which can be added to alkenes intra- or intermolecularly to form the [3+2] annulation products. This oxidation has been performed using electrochemistry,[8a] UV light with a photosensitizer,[8b,d] strong chemical oxidants (e.g., CAN[8b,c]), or more recently the photoexcited state of $Ru(bpz)_3(PF_6)_2$.[9] Using $Ru(bpz)_3(PF_6)_2$ and light in degassed CH_3NO_2, monocyclic cyclopropylamines produce good to excellent yields of the annulation products but with little to poor diastereoselectivity (eq 4). Good yields and modest diastereoselectivity are achieved with bicyclic cyclopropylamines (eq 5).

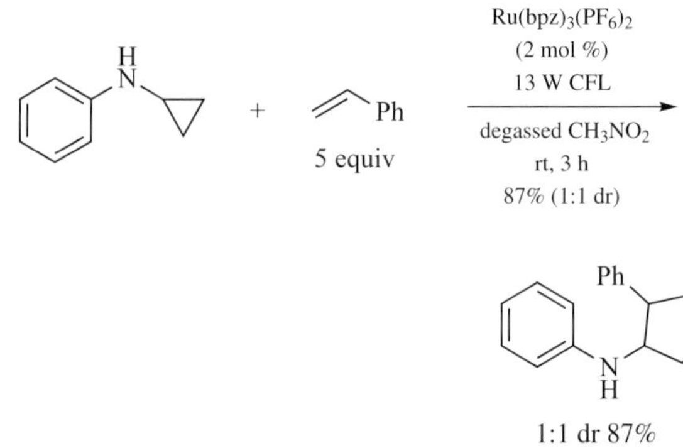

(4)

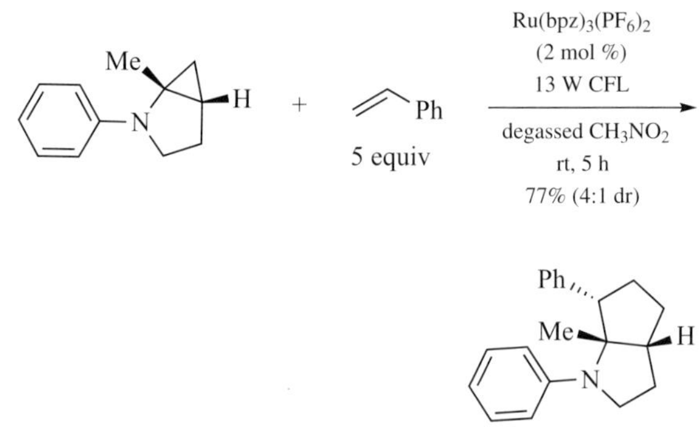

(5)

Radical Thiol–Ene Reactions. Thiol radical cations are produced via a one-electron oxidation of thiols by the photoexcited state of $Ru(bpz)_3(PF_6)_2$.[10] Subsequent deprotonation of the acidified S–H bond generates thiyl radicals, which are added across alkenes in an anti-Markovnikov fashion to afford thioethers in excellent yields (eq 6). For comparison, $Ru(bpy)_3Cl_2$ produces poor yields of thioethers.

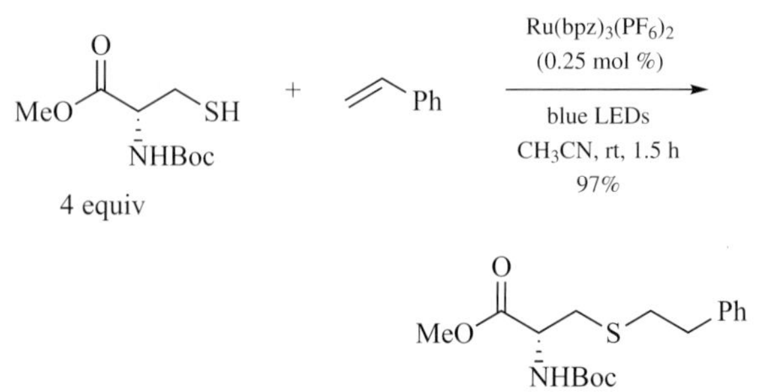

(6)

Indole Synthesis. The nitrogen radical cation of styryl anilines, which is formed by exposure to the excited state of $Ru(bpz)_3(PF_6)_2$, has been shown to undergo a cascade reaction including ctrophilic addition to the tethered alkene followed by aromatization. *N*-Arylindoles are obtained in good to excellent yields (eq 7).[11] When the β-carbon of styrene is disubstituted, a 1,2-carbon shift occurs prior to the aromatization, furnishing 2,3-disubstituted *N*-arylindoles (eq 8). Having the *p*-methoxyphenyl group is essential to realize the cascade reaction. Silica gel is found to greatly accelerate the reaction.[8b] $Ru(bpz)_3(PF_6)_2$ is a far superior catalyst to $Ru(bpy)_3(PF_6)_2$ in this specific reaction.

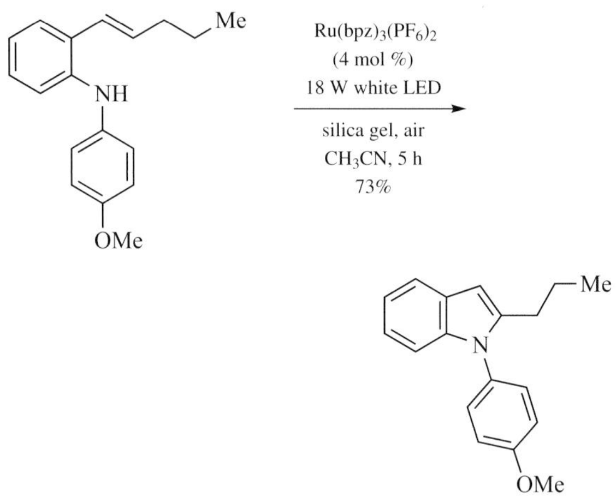

(7)

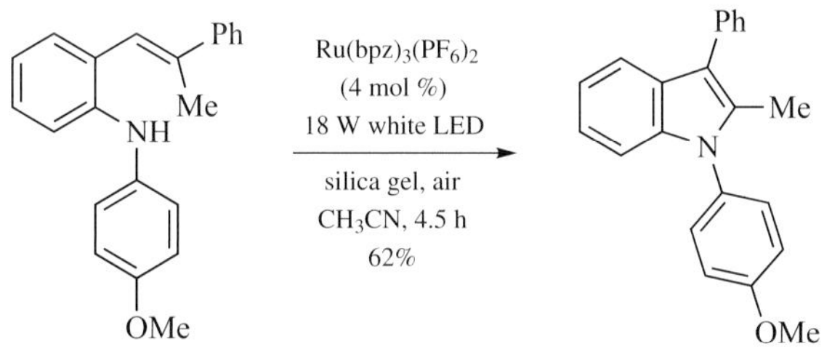

(8)

Cleavage of N–N Bonds. Cleavage of the N–N bonds of hydrazines and hydrazides has been achieved using $Ru(bpz)_3(PF_6)_2$ and light with air as the terminal oxidant.[12] The photoexcited state of $Ru(bpz)_3(PF_6)_2$ accepts an electron from hydrazines or hydrazides to produce the nitrogen radical cations, which trigger the cleavage of the N–N bonds. *N*- methyl-*N*-phenylhydrazine is converted to *N*-methylaniline in CH_3CN (eq 9), while electron poorer *N*-phenyl-*N*-benzoylhydrazine requires the use of a cosolvent CH_3OH with CH_3CN (eq 10).

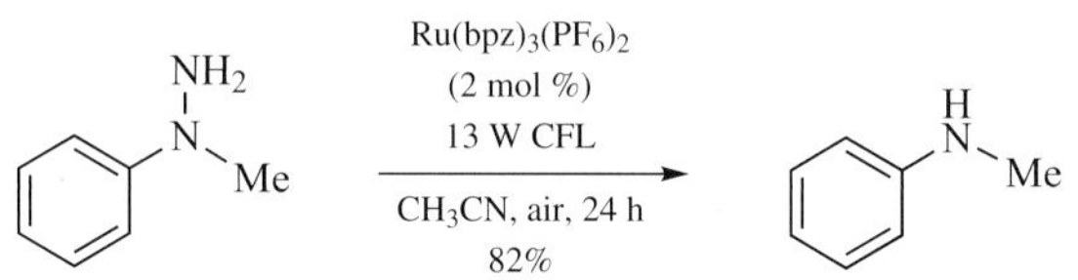

(9)

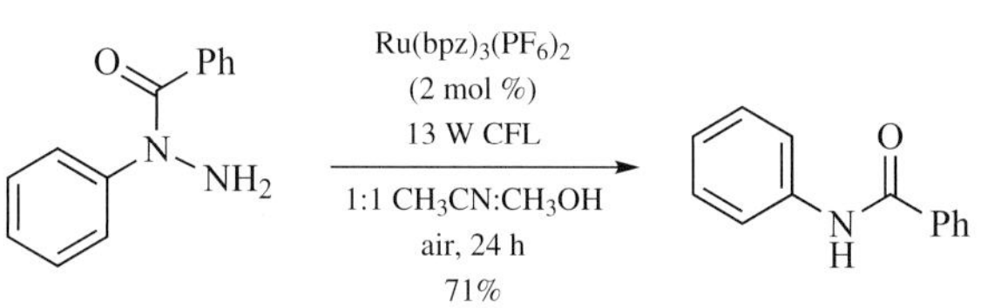

(10)

Oxidative Esterification of Aldehydes. Acylazolium ions are efficient acyl donors in *N*-heterocyclic carbene (NHC)-catalyzed reactions. They can be formed by oxidation of the Breslow intermediate.[13] Although a number of chemical oxidants have been reported to mediate this oxidation, the use of cheap and abundant terminal oxidants such as air is preferable.[14] $Ru(bpz)_3(PF_6)_2$ is far superior to $Ru(bpy)_3(PF_6)_2$ in catalyzing the oxidation with air as the terminal oxidant.[15] It is interesting to note that the ground state of $Ru(bpz)_3(PF_6)_2$ is strong enough to oxidize the Breslow intermediate and thus irradiation with light is not necessary. Aromatic and heteroaromatic aldehydes are converted to esters under these conditions (eq 11).

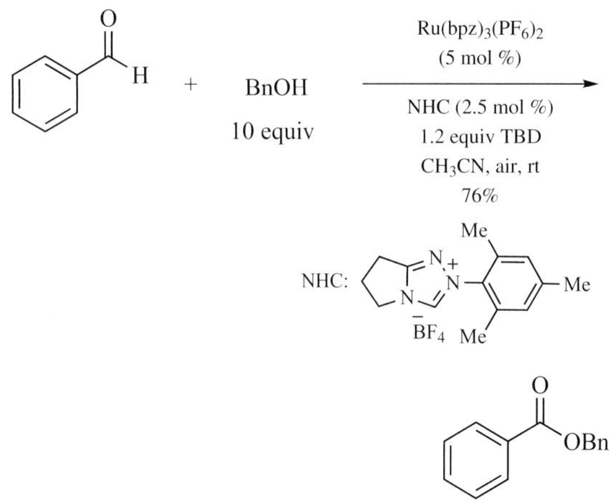

(11)

Related Reagents. [Tris(2,2′-bipyridine)ruthenium(II)]-[bis(2-phenylpyridinehexafluorophosphate, $[Ru(bpy)_3]PF_6$; [bis(2-phenylpyridine)(4,4′-*tert*-butyl-2,2′-bipyridine)iridium(III)] hexafluorophosphate, $[Ir(ppy)_2(dtbbpy)]PF_6$; 4,4′-bis(1,1-dimethyl-ethyl)-2,2′-bipyridine]bis[3,5-difluoro-2-[5-(trifluoromethyl)-2-pyridinyl-**κ***N*]phenyl]iridium(III)hexafluorophosphate, $[Ir(dF(CF_3)ppy)_2(dtbbpy)]PF_6$; [tris(2-phenylpyridinato-C^2, *N*)-iridium(III)], $Ir(ppy)_3$.

---

1. (a) Crutchley, R. J.; Lever, A. B. P., *J. Am. Chem. Soc.* **1980**, *102*, 7129. (b) Haga, M.-A.; Dodsworth, E. S.; Eryavec, G.; Seymour, P.; Lever, A. B. P., *Inorg. Chem.* **1985**, *24*, 1901. (c) Prier, C. K.; Rankic, D. A.; MacMillan, W. D. C., *Chem. Rev.* **2013**, *113*, 5322.
2. Rillema, D. P.; Allen, G.; Meyer, T. J.; Conrad, D., *Inorg. Chem.* **1983**, *22*, 1617.
3. (a) Bauld, N. L.; Bellville, D. J.; Pabon, R.; Chelsky, R.; Green, G., *J. Am. Chem. Soc.* **1983**, *105*, 2378. (b) Bellville, D. J.; Bauld, N. L.; Pabon, R.; Gardner, S. A., *J. Am. Chem. Soc.* **1983**, *105*, 3584. (c) Pabon, R. A.; Bauld, N. L., *J. Am. Chem. Soc.* **1984**, *106*, 1145. (d) Wiest, O.; Steckhan, E.; Grein, F., *J. Org. Chem.* **1992**, *57*, 4034. (e) Bauld, N. L.; Yang, J., *Org. Lett.* **1999**, *1*, 773. (f) Valley, N. A.; Wiest, O., *J. Org. Chem.* **2007**, *72*, 559. (g) Sevov, C. S.; Wiest, O., *J. Org. Chem.* **2008**, *73*, 7909.
4. Lin, S. S.; Ischay, M. A.; Fry, C. G.; Yoon, T. P., *J. Am. Chem. Soc.* **2011**, *133*, 19350.
5. Mbaze, L. M.; Lado, J. A.; Wansi, J. D.; Shiao, T. C.; Chiozem, D. D.; Mesaik, M. A.; Choudhary, M. I.; Lacaille-Dubois, M.-A.; Wandji, J.; Roy, R.; Sewald, N., *Phytochemistry*, **2009**, *70*, 1442.
6. For selected reviews, see: (a) Ledwith, A., *Acc. Chem. Res.* **1972**, *5*, 133. (b) Bauld, N. L.; Bellville, D. J.; Harirchian, B.; Lorenz, K. T.; Pabon, R. A., Jr.; Reynolds, D. W.; Wirth, D. D.; Chiou, H. S.; Marsh, B. K., *Acc. Chem. Res.* **1987**, *20*, 371. (c) Bauld, N. L., *Tetrahedron* **1989**, *45*, 5307.
7. (a) Ischay, M. A.; Lu, Z.; Yoon, T. P., *J. Am. Chem. Soc.* **2010**, *132*, 8572. (b) Parrish, J. D.; Ischay, M. A.; Lu, Z.; Guo, S.; Peters, N. R.; Yoon, T. P., *Org. Lett.* **2012**, *14*, 1640.
8. (a) Madelaine, C.; Six, Y.; Buriez, O., *Angew. Chem., Int. Ed.* **2007**, *46*, 8046. (b) Lee, H. B.; Sung, M. J.; Blackstock, S. C.; Cha, J. K., *J. Am. Chem. Soc.* **2001**, *123*, 11322. (c) Takemoto, Y.; Yamagata, S.; Furuse, S.; Hayase, H.; Echigo, T.; Iwata, C., *Chem. Commun.* **1998**, 651. (d) Lee, J.; Sun, J.; Blackstock, S. C.; Cha, J. K., *J. Am. Chem. Soc.* **1997**, *119*, 10241.

9. Maity, S.; Zhu, M. Z.; Shinabery, R. S.; Zheng, N., *Angew. Chem., Int. Ed.* **2012**, *51*, 222.

10. Tyson, E. L.; Ament, M. S.; Yoon, T. P., *J. Org. Chem.* **2013**, *78*, 2046.

11. Maity, S.; Zheng, N., *Angew. Chem., Int. Ed.* **2012**, *51*, 9562.

12. Zhu, M. Z.; Zheng, N., *Synthesis.* **2011**, *14*, 2223.

13. For recent reviews, see: (a) Wheeler, P.; Rovis, T.; Vora, H. U., *Adv. Synth. Catal.* **2012**, *354*, 1617. (b) Imami, A.; Wangelin, A.; Knappke, C. E. I., *ChemCatChem.* **2012**, *4*, 937. (c) Sarkar, S. D.; Biswas, A.; Samanta, R. C.; Studer, A., *Chem. Eur. J.* **2013**, *19*, 4664.

14. For a recent review, see: Shi, Z. Z.; Zhang, C., Tang, C.; Jiao, N., *Chem. Soc. Rev.* **2012**, *41*, 3381.

15. Zhao, J. F.; Mück-Lichtenfeld, C.; Studer, A., *Adv. Synth. Catal.* **2013**, *355*, 1098.

Jiang Wang, Theresa H. Nguyen, Jie Hu, & Nan Zheng
University of Arkansas, Fayetteville, AR, USA

Ruthenium(II), Tris(2,2′-bipyridine-κN1,κN1′)-, (OC-6-11)-

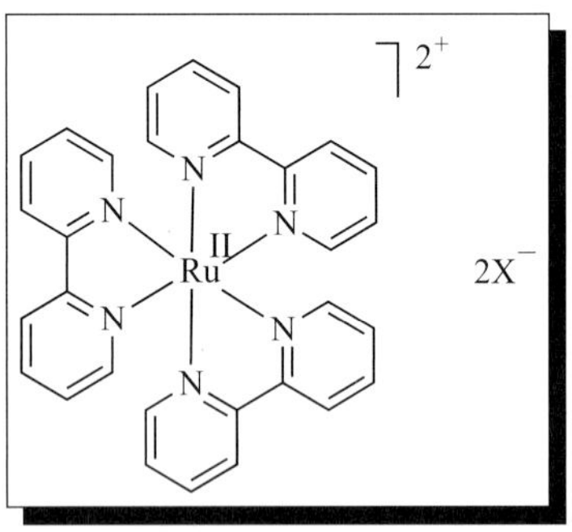

[14323-06-9]
Ruthenium(II), tris(2,2′-bipyridine-κN1, κN1′),[1] chloride(1:2), (OC-6-11)-
$C_{30}H_{24}N_6RuCl_2$ (MW 640.53)
InChI = 1S/3C10H8N2.2ClH.Ru/c3*1-3-7-11-9(5-1)10-6-2-4-8-12-10;;;/h3*1-8H;2*1H;/q;;;;;+2/p-2
InChIKey = SJFYGUKHUNLZTK-UHFFFAOYSA-L
(cation)
[15158-62-0]
InChI = 1S/3C10H8N2.Ru/c3*1-3-7-11-9(5-1)10-6-2-4-8-12-10;/h3*1-8H;/q;;;+2
InChIKey = HNVRWFFXWFXICS-UHFFFAOYSA-N
(PF_6 complex)
[60804-74-2]
InChI = S/3C10H8N2.2F6P.Ru/c3*1-3-7-11-9(5-1)10-6-2-4-8-12-10;2*1-7(2,3,4,5)6;/h3*1-8H;;;/q;;;2*-1;+2
InChIKey = KLDYQWXVZLHTKT-UHFFFAOYSA-N
(chloride hydrate)
[50525-27-4]
InChI = 1S/3C10H8N2.2ClH.6H2O.Ru/c3*1-3-7-11-9(5-1)10-6-2-4-8-12-10;;;;;;;;;;/h3*1-8H;2*1H;6*1H2;/q;;;;;;;;;;;+2/p-2
InChIKey = WHELTKFSBJNBMQ-UHFFFAOYSA-L
(perchlorate)
[15635-95-7]
InChI = S/3C10H8N2.2ClHO4.Ru/c3*1-3-7-11-9(5-1)10-6-2-4-8-12-10;2*2-1(3,4)5;/h3*1-8H;2*(H,2,3,4,5);/q;;;;;+2/p-2
InChIKey = BXKPAPTYLLPPEO-UHFFFAOYSA-L
(BF_4 complex)
[63950-81-2]
InChI = 1S/3C10H8N2.2BF4.Ru/c3*1-3-7-11-9(5-1)10-6-2-4-8-12-10;2*2-1(3,4)5;/h3*1-8H;;;/q;;;2*-1;+2
InChIKey = GWPUNVSVEJLECO-UHFFFAOYSA-N

(visible light-active catalyst used in single-electron transfer reductions and oxidations[1])

Alternate Names: $[Ru(bpy)_3]^{2+}$, Ruthenium-tris(2,2′-bipyridyl).
Physical Data: λ_{max} = 452 nm, redox potentials[2] (vs. SCE, saturated calomel electrode, as 0 V in CH_3CN); −0.86 V (Ru^{III}/Ru^{II}*), 1.29 V (Ru^{III}/Ru^{II}), 0.84 V (Ru^{II}*/Ru^{I}), and −1.33 (Ru^{II}/Ru^{I}).
Solubility: for X = Cl, sol in H_2O, CH_3OH, CH_3CN, DMF, and DMSO; insoluble in nonpolar, aprotic solvents; for X = PF_6, sol in CH_3CN, acetone, DMF, and DMSO; insoluble in H_2O.
Form Supplied in: the dichloride is commercially available from Strem Chemicals, Inc. and Sigma-Aldrich as the hexahydrate. $[Ru(bpy)_3]^{2+}$ complexes appear as red to orange crystals or powder.
Preparative Methods: for X = PF_6, the catalyst is prepared from $RuCl_3$,2,2′-bipyridine , and NH_4PF_6 in EtOH.[3]
Purification: recrystallized in boiling H_2O.
Handling, Storage, and Precautions: containers should be stored under nitrogen atmosphere in the dark. Keep away from strong oxidizing agents.

Upon visible light excitation of $[Ru(bpy)_3]^{2+}$ (bpy = 2,2′-bipyridine), the resulting metal-to-ligand charge transfer (MLCT) excited state species, *$[Ru(bpy)_3]^{2+}$, can be reductively or oxidatively quenched with the appropriate single-electron donor or acceptor, respectively. Reductive quenching of the excited state forms $[Ru(bpy)_3]^{+}$, which then performs a single-electron reduction to regenerate $[Ru(bpy)_3]^{2+}$. Alternatively, oxidative quenching forms $[Ru(bpy)_3]^{3+}$, which then oxidizes a suitable donor to regenerate ground state $[Ru(bpy)_3]^{2+}$.

Reductions of Carbon–Sulfur, Carbon–Phosphorus, and Carbon–Nitrogen Bonds. Suitably activated carbon–heteroatom bonds may be reduced by *$[Ru(bpy)_3]^{2+}$ resulting in C–H bond formation.[4–7] Phenylacylsulfonium salts have been reduced in the presence of $[Ru(bpy)_3]$ Cl_2 and a 1,4-dihydropyridine derivative under visible light irradiation to produce the corresponding acetophenone (eq 1).[8] Similarly, ammonium and phosphonium salts undergo efficient reduction under these conditions.[4] Reduction of aryl diazonium salts has historically found usage in radical aromatic substitution reactions via decomposition into N_2 and aryl radical species, a transformation classically promoted by copper(I) salts. A milder version of this reductive decomposition pathway has been demonstrated in a photocatalytic Pschorr reaction that uses *$[Ru(bpy)_3]^{2+}$ to reduce the C–N bond (eq 2).[5] Excellent yields are reported and no additional oxidants are necessary. Recently, reductive ring openings of chalcone-derived tosyl aziridines have been disclosed using Hantzsch ester (HE) to produce β-amino ketones (eq 3).[6]

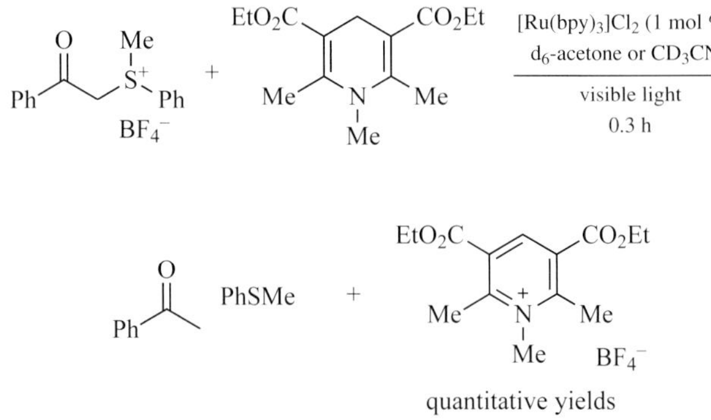

(1)

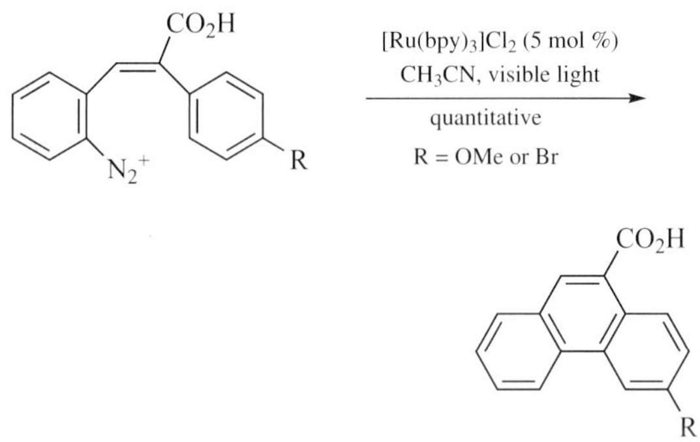

(2)

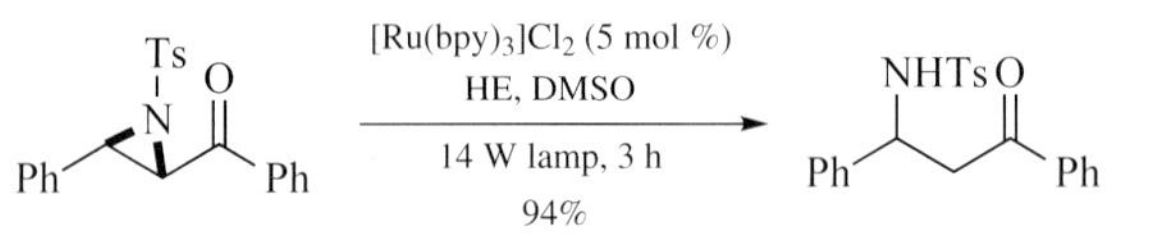

(3)

Reductions of Carbon–Oxygen Bonds. Aryl carbonyls can be transformed into alcohols through single-electron reduction of the carbon–oxygen double bond by $[Ru(bpy)_3]^+$, which is generated in situ using 1-benzyl-1,4-dihydronicotinimide (BNAH)[7] or Et_3N[9] as the reductive quencher of excited state $^*[Ru(bpy)_3]^{2+}$. The radical HO–C•R_2 intermediate formed in the $^*[Ru(bpy)_3]^{2+}$/BNAH photocatalytic cycle can be intercepted by the BNA• species to form secondary benzylic alcohols in good yields (eq 4).[10] $[Ru(bpy)_3](PF_6)_2$ has been used in combination with catalytic viologen species under oxidative quenching conditions of $^*[Ru(bpy)_3]^{2+}$ to promote Meerwein–Ponndorf–Verley-type reductions of aryl ketones.[11] Under reductive quenching conditions, radical decarboxylation of *N*-(acyloxy)phthalimides proceeds in the presence of $[Ru(bpy)_3]Cl_2$, BNAH, and light to generate alkyl radicals that react with α,β-unsaturated ketones to form β-alkylated products (eq 5).[12] Cleavage of carbon–oxygen single bonds can also be achieved via the photoreductive ring opening of β-epoxy ketones to form β-hydroxy ketones[6] using $[Ru(bpy)_3]Cl_2$ and a dimethylbenzimidazoline (eq 6).[13]

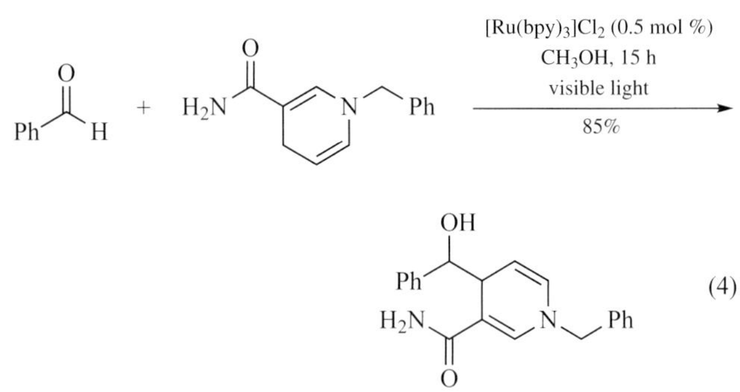

(4)

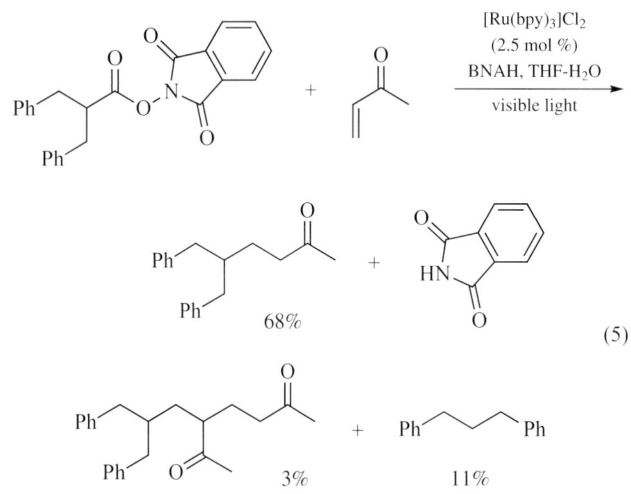

(5)

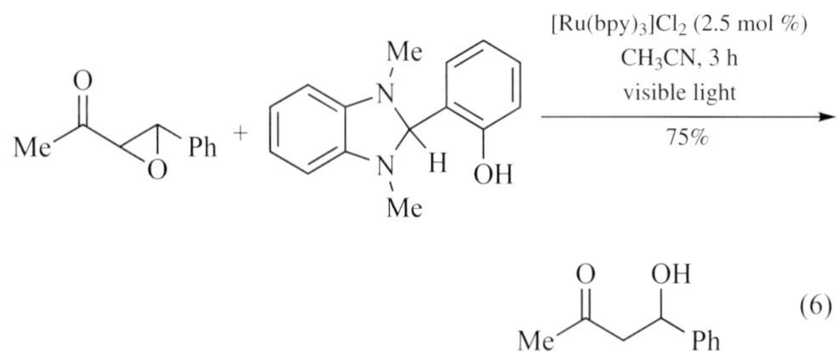

(6)

Reductions of Carbon–Halogen Bonds. Reductive dehalogenation of substrates bearing an α-electron withdrawing group have been well studied by several groups in the field of photocatalysis. $[Ru(bpy)_3]Cl_2$ has been used with 3-methyl-2,3-dihydrobenzothiazole and a fluorescent lamp to carry out the reduction of various organohalides such as α-chloride ketones and α-chloride esters, α-bromo esters, and bromomalonates.[14] Both the reductive and oxidative quenching pathways of $^*[Ru(bpy)_3]^{2+}$ have been utilized for the reduction of α-bromoacetophenone in the absence and presence of perchloric acid, respectively, using $[Ru(bpy)_3]Cl_2$, *N*-methyldihydroacridine and light (eq 7).[15] Modern development of this transformation has led to an expansion of the substrate scope to include bromopyrroloindolines and α-bromo- and α-chloroamides using $[Ru(bpy)_3]Cl_2$ in combination with $i Pr_2NEt$ and either HE or formic acid using a fluorescent light bulb as a light source (eq 8).[16] This study also demonstrated the chemoselective conversion of activated halides over aryl halides as well as high functional group tolerance. Dehalogenations of vicinal dibromides is known to produce stilbenes under oxidative quenching conditions utilizing light, $[Ru(bpy)_3]Cl_2$, substoichiometric viologen, and $(NH_4)_3EDTA$ in a biphasic EtOAc-H_2O system (eq 9).[17] Reductive quenching conditions for similar transformations can be accessed using a $[Ru(bpy)_3]Cl_2/Et_3N$[9] or $[Ru(bpy)_3]Cl_2$/1,5-dimethoxynaphthalene/ascorbic acid[18] catalytic cycle.

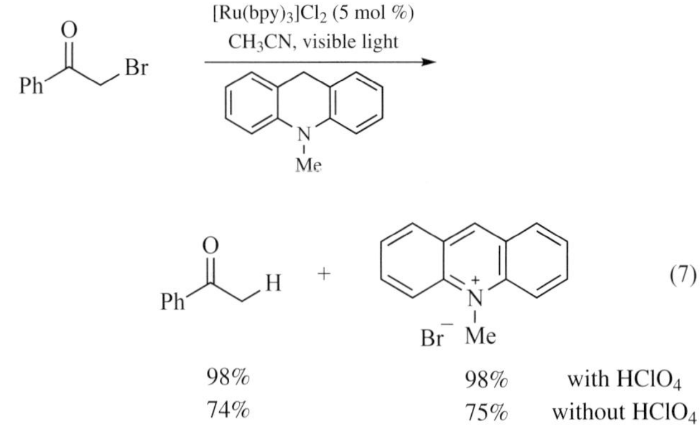

(7)

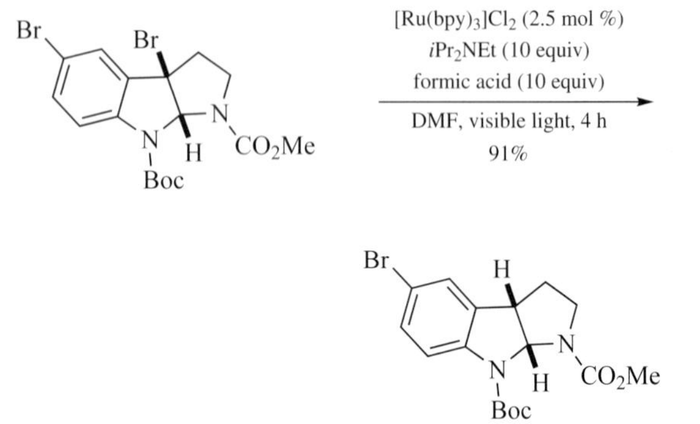

(8)

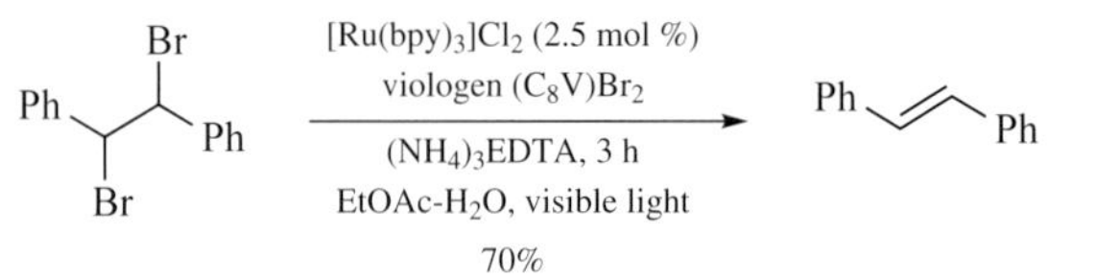

(9)

Oxidation of Carbon–Hydrogen Bonds. Alkyl radicals produced from halide reductions can be exploited for additions onto various π-systems through modification of the $[Ru(bpy)_3]Cl_2$ cycle. Enantioselective alkylations of aldehydes have been carried out using dual photocatalytic/organocatalytic cycles (eq 10).[19] Excellent enantioselectivities and yields are obtained using this system for the asymmetric α-trifluoromethylation[20] and α-benzylation[21] of aldehydes. Additions of TEMPO onto morpholine-derived enamines have been described using the reductive quenching cycle of $[Ru(bpy)_3](PF_6)_2$.[22]Silyl enol ethers of ketones, amides, and aldehydes can undergo α-perfluoroalkylations in the presence of $[Ru(bpy)_3]Cl_2$ and light in a one-pot process from the corresponding carbonyl compounds (eq 11).[23] The $[Ru(bpy)_3]Cl_2/Et_3N$ reductive quenching cycle has also been shown to mediate the intramolecular addition of activated bromides onto heterocycles[24] and electron-rich olefins (eq 12).[25] The analogous intermolecular bromomalonate additions onto heterocycles require the use of a non-hydrogen atom donor triarylamine as the reductive quencher.[26] Bromopyrroloindolines can be coupled with indoles under similar reductive quenching conditions to obtain complex bisindole structures, which have been recently demonstrated in the synthesis of gliocladin C (eq 13).[27] Related iridium-based photocatalysts are also competent in many of these reactions, including atom transfer reactions.[28] Glycosyl halides can be reduced using the reductive quenching cycle of $[Ru(bpy)_3](BF_4)_2$ and the resultant radical can add onto electron-deficient alkenes to obtain exclusive α-substituted products in high yields (eq 14).[29] In addition to C–H oxidation via reduction of organohalides, $[Ru(bpy)_3]Cl_2$ is known to perform direct oxidations of phenols to generate cross-linked proteins. Nearby tyrosine residues oxidize in the presence of $[Ru(bpy)_3]^{3+}$ through oxidative quenching of $^*[Ru(bpy)_3]^{2+}$ with persulfate ion and combine rapidly (~0.5 s) to form the bis(phenol)-linked protein in good yield (eq 15).[30]

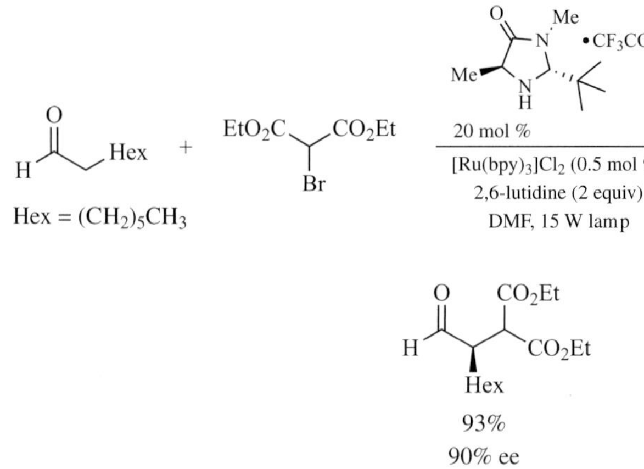

(10)

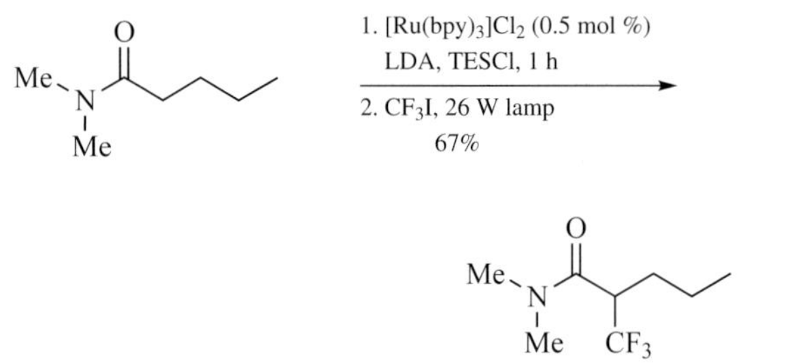

(11)

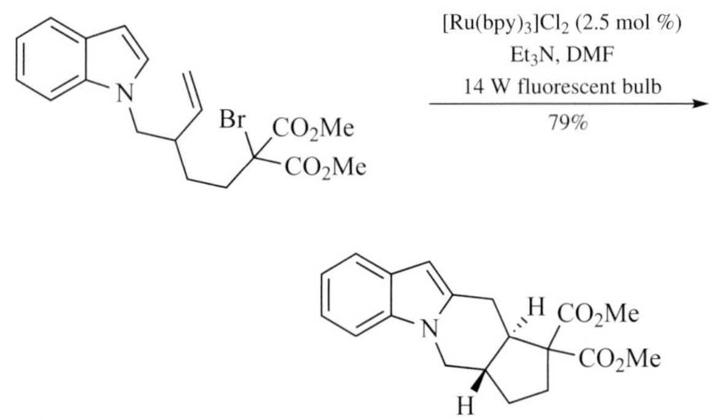

(12)

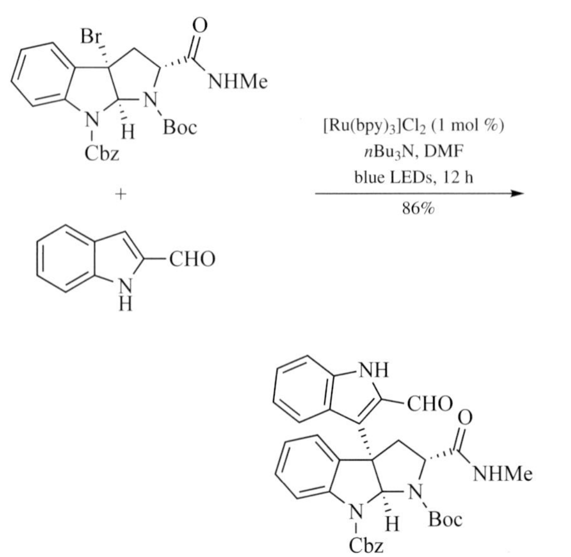

(13)

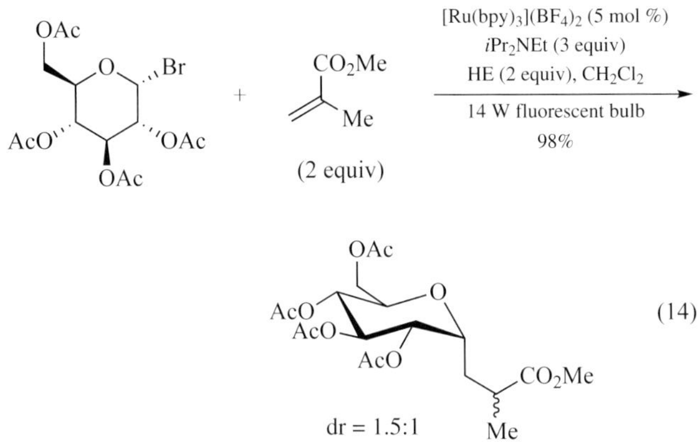

(14)

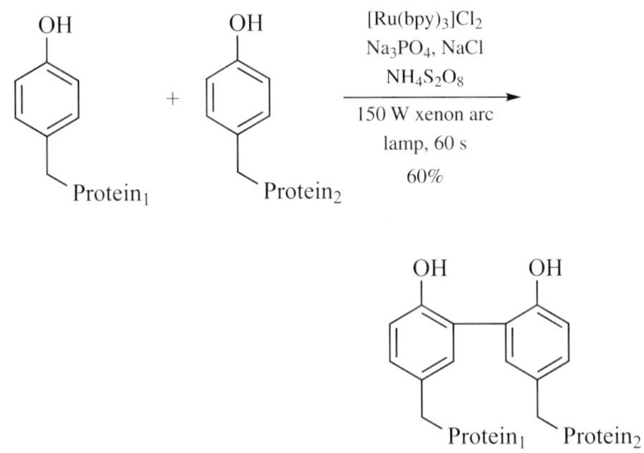

(15)

Oxidations of Carbon–Heteroatom Bonds. Iminium ions are produced by the single-electron oxidation of tertiary amines by *$[Ru(bpy)_3]^{2+}$, followed by hydrogen atom abstraction and can be trapped by nucleophiles to obtain diverse scaffolds. *N*-Aryl tetrahydroisoquinolines have been oxidized in the presence of $[Ru(bpy)_3]Cl_2$, visible light, and nitroalkanes to promote aza-Henry reactions under aerobic conditions in high yields (eq 16).[31] Azomethine ylides can be formed under similar conditions to undergo efficient [3 + 2]cycloadditions with various dipolarophiles, generating highly functionalized pyrrolo[2,1-*a*]isoquinolines.[32] Mannich reactions between acetone and *N*-aryl tetrahydroisoquinolines have been afforded using $[Ru(bpy)_3](PF_6)_2$, a fluorescent lamp, and a proline catalyst.[33] For transformations involving tetrahydroisoquinolines, the iridium-based photocatalysts have also been successful. The use of $[Ru(bpy)_3]Cl_2$, visible light, and base in the presence of oxygen has been shown to promote intramolecular cyclizations of amines onto in situ oxidized benzylamines to form tetrahydroimidazoles.[34] In addition to amine oxidations, when $[Ru(bpy)_3]Cl_2$ is combined with CBr_4 or CHI_3 in the presence of DMF, an aminal radical intermediate is formed, which is then oxidized by $[Ru(bpy)_3]^{3+}$ to form a Vilsmeier–Haack reagent that activates primary or secondary alcohols toward displacement by bromide or iodide to give high yields of halogenated products (eq 17).[35]

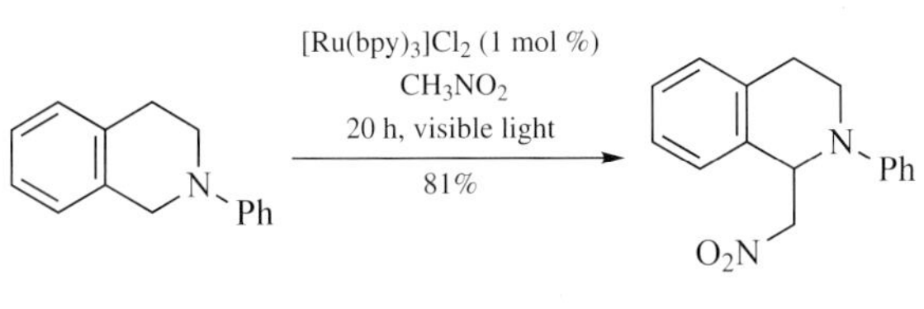

(16)

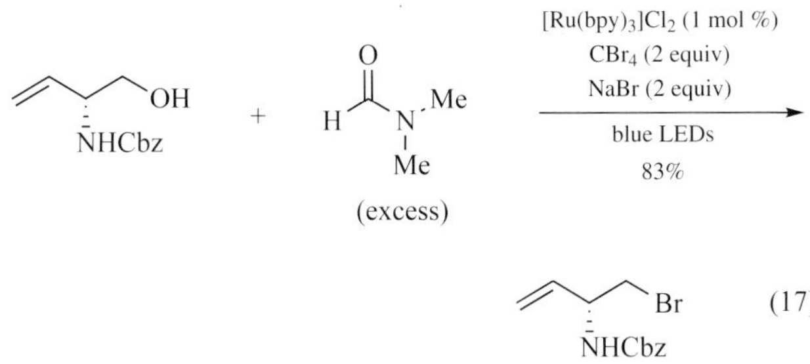

(17)

Cycloadditions. Bis-enones are well known to undergo single-electron transfer-mediated cyclizations under a variety of conditions.[36] The reduction of aryl enones to the radical anion species by $[Ru(bpy)_3]^+$ has been reported under reductive quenching conditions of *$[Ru(bpy)_3]^{2+}$ with $[Ru(bpy)_3]Cl_2$,*i*Pr_2NEt, and LiBr as a Lewis acid using a flood lamp or under ambient sunlight (eq 18).[37] This process was shown to initiate both intra- and intermolecular formal [2+ 2]cycloadditions of symmetrical and unsymmetrical[38] enones in high diastereomeric ratio (dr) and yield. The intramolecular formal [3 + 2]cycloaddition of aryl cyclopropyl ketones and alkenes likewise forms fused cyclopentane structures with good dr and in high yield in the presence of $[Ru(bpy)_3]Cl_2$, TMEDA, $La(OTf)_3$, and light (eq 19).[39] These conditions have considerable advantages over other metal-mediated or electrochemical-promoted processes with respect to yield and chemoselectivity. One of the most common types of cycloadditions is the Diels–Alder [4+ 2]reaction, which can establish six-membered cycles with excellent regio- and stereocontrol. Accordingly, single-electron reduction of bis-enones initiates intramolecular hetero [4+ 2] cycloadditions in the presence of $[Ru(bpy)_3]Cl_2$, *i*Pr_2NEt, LiBr, and light in H_2O–CH_3CN mixtures to give products in >10:1 dr (eq 20).[40] Notably, both reactive partners are electron deficient and the observed regioselectivity is opposite to that obtained using methods involving a two-electron pathway. The oxidative quenching cycle of *$[Ru(bpy)_3]^{2+}$ can alternatively be accessed for [2+ 2]cycloadditions of electron-rich styrenes using a $[Ru(bpy)_3](PF_6)_2$/methyl viologen (MV) system under sunlight via single-electron oxidation of the cyclopropyl ketone (eq 21).[41]

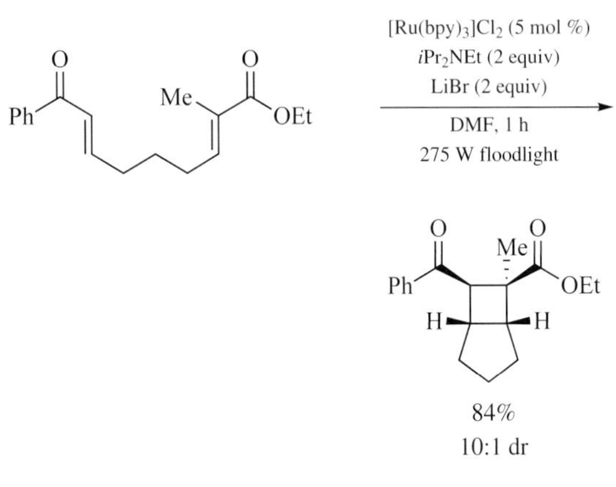

(18)

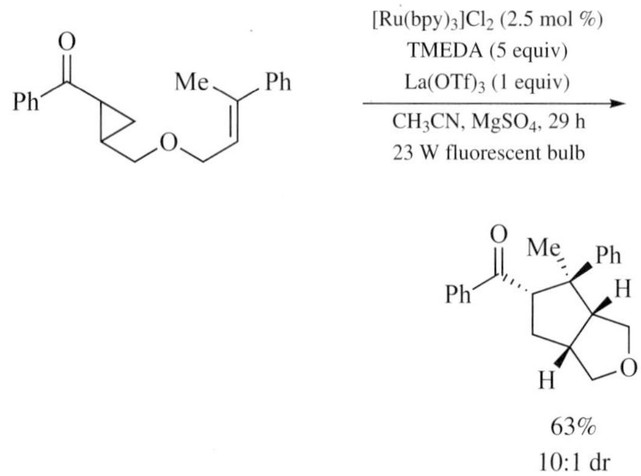

(19)

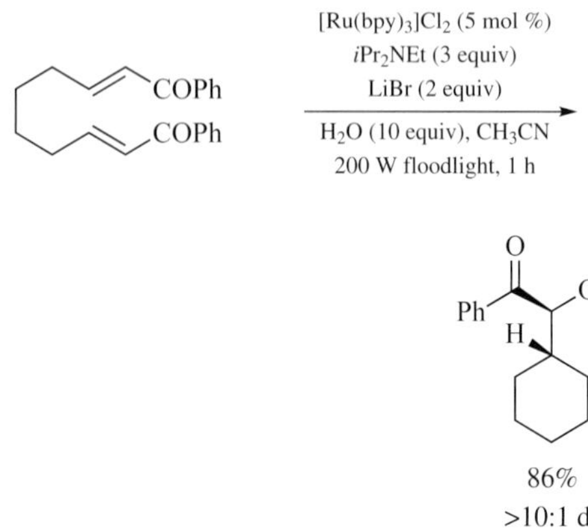

(20)

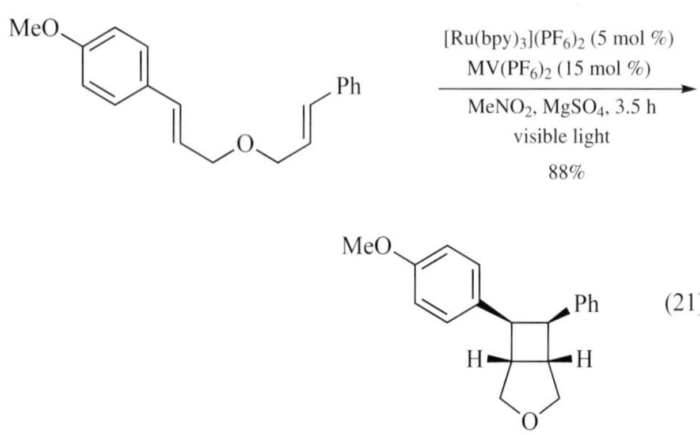

(21)

Miscellaneous. The reduction of alkyl azides and biomolecule-containing azides to the corresponding primary amine has been recently accomplished with $[Ru(bpy)_3]Cl_2$and visible light using an appropriate reductive quencher in organic as well as aqueous solvents (eq 22).[42] This transformation exhibits high levels of chemoselectivity and functional group tolerance, with alkyl azides reduced in the presence of sensitive moieties such as disulfides, aldehydes, alcohols, and alkyl halides. Aryl radicals have also been generated from aryl iodonium salts using $[Ru(bpy)_3]Cl_2$ and silane additives in the presence of visible light to promote the ring-opening radical polymerization of epoxides.[43] Irradiation with a green fluorescent light bulb was shown to improve the efficiency of the transformation.

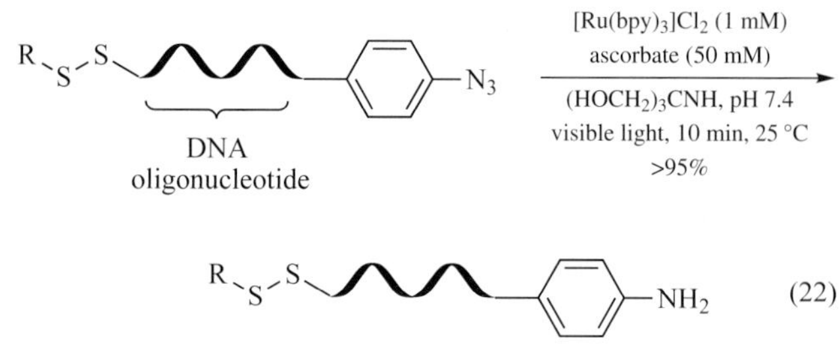

(22)

Related Reagents. [Bis(2-phenylpyridine)(4,4′-*tert*-butyl-2,2′-bipyridine)iridium(III)]hexafluorophosphate, $[Ir(ppy)_2(dtbbpy)]PF_6$; [4,4′-bis(1,1-dimethylethyl)-2,2′-bipyridine]-bis[3,5-difluoro-2-[5-(trifluoromethyl)-2-pyridinyl-κN]phenyl]-iridium(III)hexafluorophosphate, $[Ir(dF(CF_3)ppy)_2(dtbbpy)]PF_6$.

1. (a) Yoon, T. P.; Ischay, M. A.; Du, J., *Nat. Chem.* **2010**, *2*, 527. (b) Narayanam, J. M. R.; Stephenson, C. R. J.; *Chem. Soc. Rev.* **2011**, *40*,102. (c) Teplý, F. *Collect. Czech. Chem. Commun.* **2011**, *76*, 859.
2. (a) Kalyanasundaram, K., *Coord. Chem. Rev.* **1982**, *46*, 159. (b) Juris, A.; Balzani, V.; Barigelletti, F.; Campagna, S.; Belser, P.; von Zelewsky, A., *Coord. Chem. Rev.* **1988**, *84*, 85. (c) Balzani, V.; Bergamini, G.; Marchioni, F.; Ceroni, P., *Coord. Chem. Rev.* **2006**, *250*, 1254.
3. Mabrouk, P. A.; Wrighton, M. S., *Inorg. Chem.* **1986**, *25*, 526.
4. Van Bergen, T. J.; Hedstrand, D. M.; Kruizinga, W. H.; Kellog, R. M., *J. Org. Chem.* **1979**, *44*, 4953.
5. Cano-Yelo, H.; Deronzier, A., *J. Chem. Soc., Perkin Trans. 2.* **1984**, 1093.
6. Larraufie, M. H.; Pellet, R.; Fensterbank1, L., Goddard, J. P.; Lacôte, E.; Malacria, M.; Ollivier, C., *Angew. Chem., Int. Ed.* **2011**, *50*, 4463.
7. Ishitani, O.; Pac, C.; Sakurai, H., *J. Org. Chem.* **1983**, *48*, 2941.
8. Hedstrand, D. M.; Kruizinga, W. H.; Kellog, R. M., *Tetrahedron Lett.* **1978**, *19*, 1255.
9. Willner, I.; Tsfania, T.; Eichen, Y., *J. Org. Chem.* **1990**, *55*, 2656.
10. Ishitani, O.; Yanagida, S.; Takamuku, S.; Pac, C., *J. Org. Chem.* **1987**, *52*, 2790.
11. Herance, J. R.; Ferrer, B.; Bourdelande, J. L.; Marquet, J.; Garcia, H., *Chem. Eur. J.* **2006**, *12*, 3890.
12. Okada, K.; Okamoto, K.; Morita, N.; Okubo, K.; Oda, M., *J. Am. Chem. Soc.* **1991**, *113*, 9401.
13. Hasegawa, E.; Takizawa, S.; Seida, T.; Yamaguchi, A.; Yamaguchi, N.; Chiba, N.; Takahashi, T.; Ikeda, H.; Akiyama, K., *Tetrahedron* **2006**, *62*, 6581.
14. Mashraqui, S. H.; Kellog, R. M., *Tetrahedron Lett.* **1985**, *26*, 1453.
15. Fukuzumi, S.; Mochizuki, S.; Tanaka, T., *J. Phys. Chem.* **1990**, *94*, 722.
16. Narayanam, J. M. R.; Tucker, J. W.; Stephenson, C. R. J., *J. Am. Chem. Soc.* **2009**, *131*, 8756.
17. (a) Maidan, R.; Goren, Z.; Becker, J. Y.; Willner, I., *J. Am. Chem. Soc.* **1984**, *106*, 6217.

    (b) Goren, Z.; Willner, I., *J. Am. Chem. Soc.* **1983**, *105*, 7764.
18. Maji, T.; Karmakar, A.; Reiser, O., *J. Org. Chem.* **2011**, *76*, 736.
19. Nicewicz, D. A.; MacMillan, D. W. C., *Science* **2008**, *322*, 77.
20. Nagib, D. A.; Scott, M. E.; MacMillan, D. W. C., *J. Am. Chem. Soc.* **2009**, *131*, 10875.
21. Shih, H.-W.; Vander Wal, M. N.; Grange, R. L.; MacMillan, D. W. C., *J. Am. Chem. Soc.* **2010**, *132*, 13600.
22. Koike, T.; Akita, M., *Chem. Lett.* **2009**, *38*, 166.
23. Pham, P. V.; Nagib, D. A.; MacMillan, D. W. C., *Angew. Chem., Int. Ed.* **2011**, *50*, 6119.

24. Tucker, J. W.; Narayanam, J. M. R.; Krabbe, S. W.; Stephenson, C. R. J., *Org. Lett.* **2010**, *12*, 368.
25. Tucker, J. W.; Nguyen, J. D.; Narayanam, J. M. R.; Krabbe, S. W.; Stephenson, C. R. J., *Chem., Commun.* **2010**, *46*, 4985.
26. Furst, L.; Matsuura, B. S.; Narayanam, J. M. R.; Tucker, J. W.; Stephenson, C. R. J., *Org. Lett.* **2010**, *12*, 3104.
27. Furst, L.; Narayanam, J. M. R.; Stephenson, C. R. J., *Angew. Chem., Int. Ed.* **2011**, DOI: 10.1002/anie.201103145.
28. Nguyen, J. D.; Tucker, J. W.; Konieczynska, M.; Stephenson, C. R. J., *J. Am. Chem. Soc.* **2011**, *133*, 4160.
29. Andrews, R. S.; Becker, J. J.; Gagné, M. R., *Angew. Chem., Int. Ed.* **2010**, *49*, 7274.
30. Fancy, D. A.; Kodadek, T., *Proc. Natl. Acad. Sci. USA* **1999**, *96*, 6020.
31. Condie, A. G.; González-Gómez, J.-C., Stephenson, C. R. J., *J. Am. Chem. Soc.* **2010**, *132*, 1464.
32. Zou, Y.-Q.; Lu, L.-Q.; Fu, L.; Chang, N.-J.; Rong, J.; Chen, J.-R.; Xiao, W.-J., *Angew. Chem., Int. Ed.* **2011**, *50*, 7171.
33. Rueping, M.; Vila, C.; Koenigs, R. M.; Poscharny, K.; Fabry, D. C., *Chem. Commun.* **2011**, *47*, 2360.
34. Xuan, J.; Cheng, Y.; An, J.; Lu, L.-Q.; Zhang, X.-X.; Xiao, W.-J., *Chem. Commun.* **2011**, *47*, 8337.
35. Dai, C.; Narayanam, J. M. R.; Stephenson, C. R. J., *Nature Chem.* **2011**, *3*, 140.
36. (a) Yang, J.; Felton, G. A. N.; Bauld, N. L.; Krische, M. J., *J. Am. Chem. Soc.* **2004**, *126*, 1634.
(b) Wang, L.-C.; Jang, H.-Y.; Lynch, V.; Krische, M. J., *J. Am. Chem. Soc.* **2002**, *124*, 9448.
(c) Baik, T. G.; Wang, L.-C.; Luiz, A.-L., Krische, M. J., *J. Am. Chem. Soc.* **2001**, *123*, 6716.
37. Ischay, M. A.; Anzovino, M. E.; Du, J.; Yoon, T. P., *J. Am. Chem. Soc.* **2008**, *130*, 12886
38. Du, J.; Yoon, T. P., *J Am. Chem. Soc.* **2009**, *131*, 14604.
39. Lu, Z.; Shen, M.; Yoon, T. P., *J. Am. Chem. Soc.* **2011**, *133*, 1162.
40. Hurtley, A. E.; Cismesia, M. A.; Ischay, M. A.; Yoon, T. P., *Tetrahedron* **2011**, *67*, 4442.
41. Ischay, M. A.; Lu, Z.; Yoon, T. P., *J. Am. Chem. Soc.* **2010**, *132*, 8572.
42. Chen, Y.; Kamlet, A. S.; Steinman, J. B.; Liu, D. R., *Nat. Chem.* **2011**, *3*, 146.
43. Lalevee, J.; Blanchard, N.; Tehfe, M. A.; Morlet-Savary, F.; Fouassier, J. P., *Macromolecules* **2010**, *43*, 10191.

Laura Furst & Corey R. J. Stephenson
Boston University, Boston, MA, USA

Ruthenium(II), Tris(1,10-phenanthroline-κN1,κN10)-, (OC-6-11)-

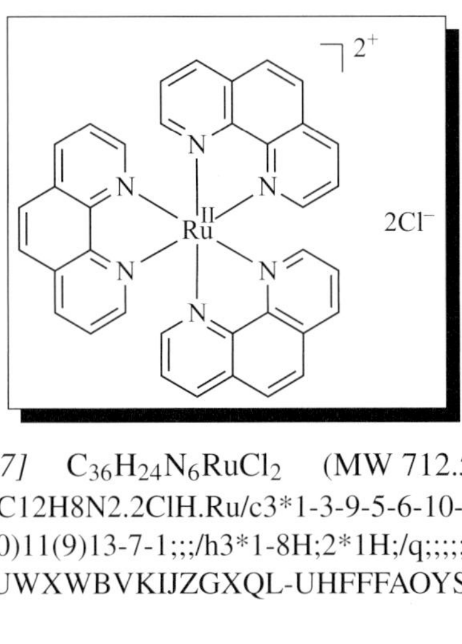

[207802-45-7] $C_{36}H_{24}N_6RuCl_2$ (MW 712.59 anhydrous)
InChI = 1S/3C12H8N2.2ClH.Ru/c3*1-3-9-5-6-10-4-2-8-14-12(10)11(9)13-7-1;;;/h3*1-8H;2*1H;/q;;;;;+2/p-2
InChIKey = UWXWBVKIJZGXQL-UHFFFAOYSA-L

(visible light-absorbing photocatalyst used for single-electron transfer reactions)[1]

Alternative Names: $[Ru(phen)_3]Cl_2$, ruthenium-tris(1,10-phenanthroline), tris(1,10-phenanthroline)ruthenium(II) chloride, dichlorotris(1,10-phenanthroline)ruthenium(II).
Physical Data: λ_{max} = 447 nm,[2,3] redox potentials[1] (vs. saturated calomel electrode (SCE), as 0 V in CH_3CN); −0.87 V (Ru^{III}/Ru^{II*}), +1.26 V (Ru^{III}/Ru^{II}), +0.82 V (Ru^{II*}/Ru^{I}), and −1.36 V (Ru^{II}/Ru^{I}).
Solubility: soluble in H_2O, CH_3CN, CH_2Cl_2, CH_3OH, CH_3CH_2OH.
Form Supplied in: $[Ru(phen)_3]Cl_2$ is a red to dark red or red-orange crystalline powder. The hydrate is commercially available from Sigma-Aldrich.
Preparative Method: the complex can be prepared from $Ru(DMSO)_4Cl_2$ and 1,10-phenanthroline,[4] or alternatively can be obtained directly from $RuCl_3$, 1,10-phenanthroline, and $NaPO_2H_2$ in 0.1 N aqueous HCl.[5]
Purification: recrystallized from distilled water.
Handling, Storage, and Precautions: the catalyst should be stored in the dark under an inert atmosphere.

Trifluoromethylation of Alkenes, Alkynes, and Arenes. Polypyridyl metal photocatalysts are capable of reducing a variety of carbon–halogen bonds to form carbon-centered radical intermediates.[6,7] $[Ru(phen)_3]Cl_2$ has been used in this context for the generation of a trifluoromethyl radical through the single-electron reduction of trifluoromethyl iodide (CF_3I). The trifluoromethyl radical formed through this process is capable of efficient addition to terminal alkenes and is followed by selective elimination to provide the corresponding trifluoromethylated (*E*)-olefin (eq 1).[8] $[Ru(phen)_3]Cl_2$ outperformed the more ubiquitous $[Ru(bpy)_3]Cl_2$ in this transformation by providing a cleaner reaction profile and higher yields. This reactivity has further been realized in the iodotrifluoromethylation of terminal alkynes (eq 2).[9] $[Ru(phen)_3]Cl_2$ has also been shown to efficiently form the trifluoromethyl radical through single-electron reduction of trifluoromethanesulfonyl chloride, the radical anion of which is capable of undergoing an entropically favored fragmentation to yield a trifluoromethyl radical, a chloride anion, and SO_2 gas.[10] In this way, $[Ru(phen)_3]Cl_2$ was used to trifluoromethylate a wide variety of aromatics and heterocycles in good yield at room temperature (eq 3).

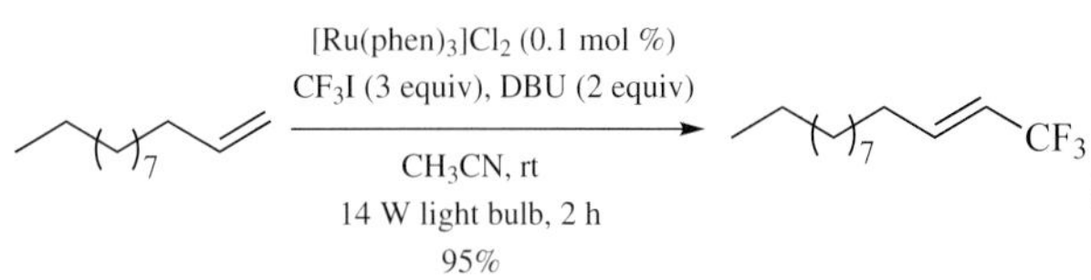

(1)

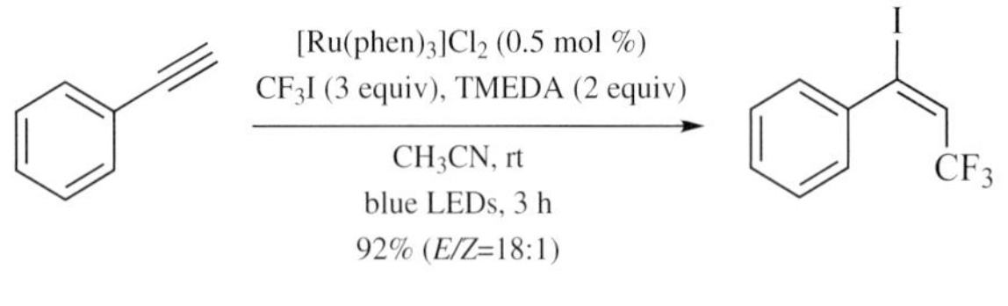

(2)

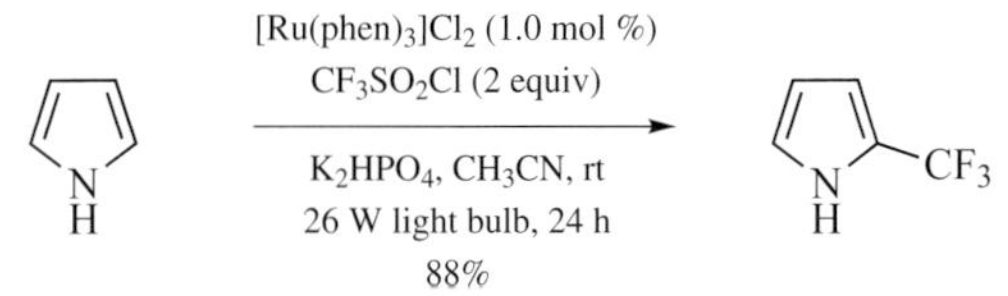

(3)

Decarboxylation and Oxidative Amidation of α-Keto Acids. $[Ru(phen)_3]Cl_2$ has also been shown to be effective in promoting the oxidative decarboxylation and amidation of α-keto acids (eq 4).[11] This process is thought to occur through the formation of a peroxyl radical via reduction of molecular oxygen, which in turn is proposed to perform a single-electron oxidation of the starting α-keto acid. Upon decarboxylation, the benzoyl radical undergoes nucleophilic attack by an amine, and subsequent deprotonation and oxidation of the intermediate provide the product amide in good yield. This transformation is not limited to aryl α-keto acids, but product yield decreased when an aliphatic substrate was used. The reaction was found to be more tolerant in the scope of amines used, but again aromatic reactants provided the most efficient reactivity.

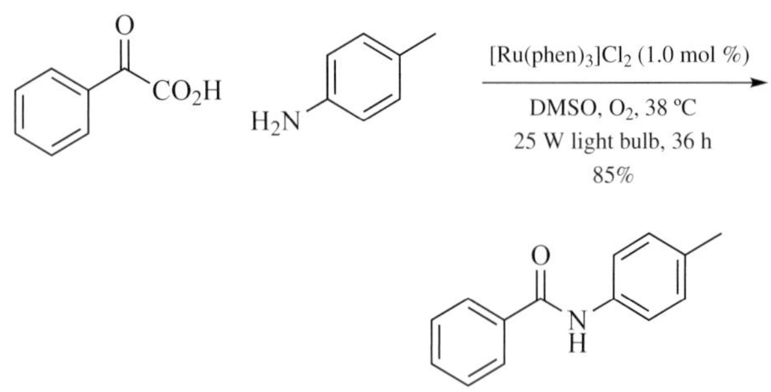

(4)

Asymmetric Oxidation of Sulfides to Sulfoxides. Resolved $Ru(phen)_3^{2+}$ enantiomers have been used to provide chiral induction during photocatalytic reactions. The catalyst was shown to be incapable of enantioinduction in the homogeneous oxidation of ±-phenyl(1-phenylpropyl) sulfane to the corresponding sulfoxide; however, when immobilized on montmorillonite clay, mild enantioselectivities were observed (eq 5).[12] Δ-$[Ru(phen)_3](ClO_4)_2$ was capable of inducing 30% ee for both sets of diastereomers, outperforming the 9% ee achieved by Δ-$[Ru(bpy)_3](ClO_4)_2$. Importantly, for both $[Ru(phen)_3](ClO_4)$ and $[Ru(bpy)_3](ClO_4)$, the Δ- and Λ-enantiomers gave opposite selectivities, supporting the role of the catalyst in the observed chiral induction. The sulfoxides were recovered in quantitative yield.

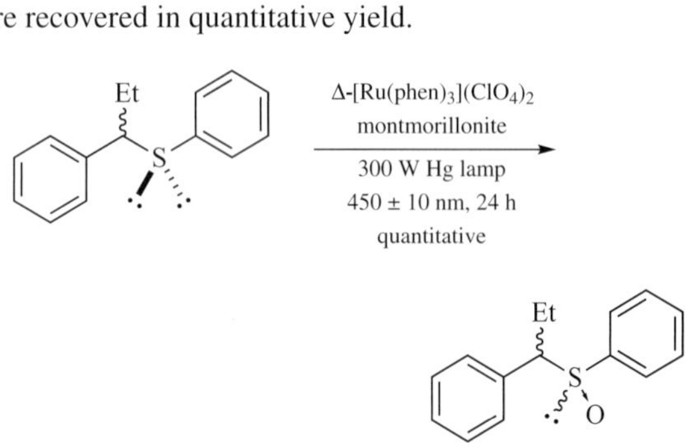

(5)

30% ee (*RS:SR*)
29% ee (*SS:RR*)

$[Ru(phen)_3]^{2+}$ as a Chiral Mediator of 1,2-Addition. Chiral $[Ru(phen)_3]^{2+}$ has also been used as a chiral initiator for the 1,2-addition of diisopropylzinc to a pyrimidine-5-carboxaldehyde (eq 6).[13] In contrast to the asymmetric oxidation of sulfides previously mentioned, this reaction produced good levels of enantioselectivity in homogeneous conditions. The reaction was achieved with both enantiomers of the catalyst, with Λ- and Δ-$[Ru(phen)_3](ClO_4)_2$ providing the (*R*)- and (*S*)-enantiomers, respectively, in good yield.

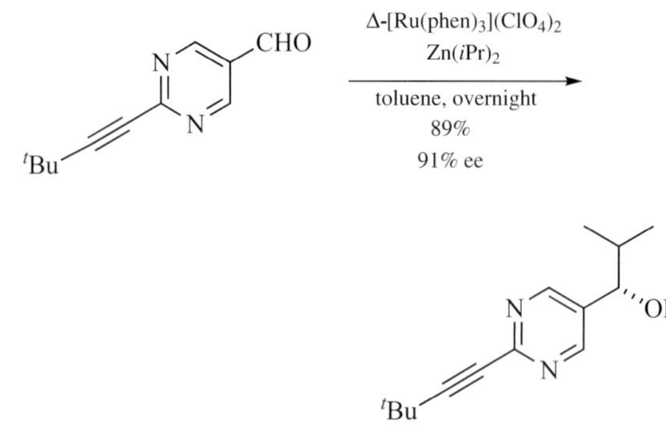

(6)

Related Reagents. [Tris(2,2′-bipyridine)ruthenium(II)]-chloride, $[Ru(bpy)_3]Cl_2$; [bis(2-phenylpyridine) (4,4′-*tert*-butyl-2,2′-bipyridine)iridium(III)] hexafluorophosphate, $[Ir(ppy)_2(dtbbpy)]PF_6$; [4,4′-bis(1,1-dimethylethyl)-2,2′-bipyridine]-bis[3,5-difluoro-2-[5-(trifluoromethyl)-2-pyridinyl-**κ**N]phenyl]-iridium(III)hexafluorophosphate, $[Ir(dF(CF_3)ppy)_2(dtbbpy)]PF_6$.

1. Prier, C. K.; Rankic, D. A.; MacMillan, D. W. C., *Chem. Rev.* **2013**, *113*, 5322.
2. Lin, C. T.; Boettcher, W.; Chou, M.; Creutz, C.; Sutin, N., *J. Am. Chem. Soc.* **1976**, *98*, 6536.
3. Young, R. C.; Meyer, T. J.; Whitten, D. G., *J. Am. Chem. Soc.* **1976**, *98*, 286.
4. Kawanishi, Y.; Kitamura, N.; Tazuke, S., *Inorg. Chem.* **1989**, *28*, 2968.
5. Kratochvil, B.; Zatko, D. A., *Anal. Chem.* **1964**, *36*, 527.
6. Nicewicz, D. A.; Macmillan, D. W. C., *Science* **2008**, *322*, 77.
7. Nguyen, J. D.; D'Amato, E. M.; Narayanam, J. M. R.; Stephenson, C. R. J., *Nat. Chem.* **2012**, *4*, 854.
8. Iqbal, N.; Choi, S.; Kim, E.; Cho, E. J., *J. Org. Chem.* **2012**, *77*, 11383.
9. Iqbal, N.; Jung, J.; Park, S.; Cho, E. J., *Angew. Chem., Int. Ed.* **2013**, doi: 10.1002/anie.201308735. *53*, 539.
10. Nagib, D. A.; MacMillan, D. W. C., *Nature* **2011**, *480*, 224.
11. Liu, J.; Liu, Q.; Yi, H.; Qin, C.; Bai, R.; Qi, X.; Lan, Y.; Lei, A., *Angew. Chem., Int. Ed.* **2013**, doi: 10.1002/anie.201308614. *53*, 502.
12. Fujita, S.; Sato, H.; Kakegawa, N.; Yamagishi, A., *J. Phys. Chem. B.* **2006**, *110*, 2533.
13. Kawasaki, T.; Omine, T.; Suzuki, K.; Sato, H.; Yamagishi, A.; Soai, K., *Org. Biomol. Chem.* **2009**, *7*, 1073.

Joel W. Beatty & Corey R. J. Stephenson
University of Michigan, Ann Arbor, MI, USA

S

4,4′-Spirobi[4*H*-dinaphth[2,1-*c*:1′,2′-*e*]-azepinium],3,3′,5,5′-tetrahydro- 2,6-bis(3,4,5-trifluorophenyl), bromide

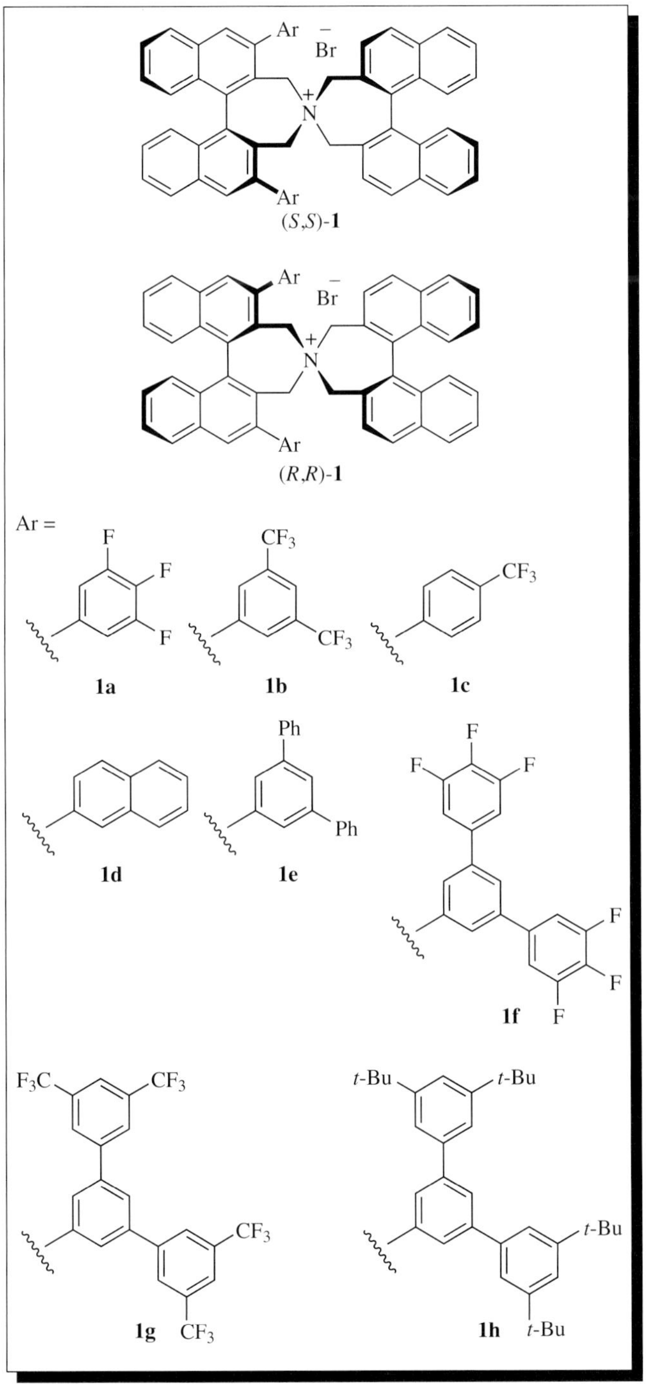

(*S,S*-**1a**)
[287384-12-7] $C_{56}H_{34}BrF_6N$ (MW 914.77)
InChI = 1/C56H34F6N.BrH/c57-47-23-37(24-48(58)55(47)61)
43-21-33-11-3-7-15-41(33)53-45(43)29-63(30-46-44
(38-25-49(59)56(62)50(60)26-38)22-34-12-4-8-16-42
(34)54(46)53)27-35-19-17-31-9-1-5-13-39(31)51(35)
52-36(28-63)20-18-32-10-2-6-14-40(32)52;/h1-26H,
27-30H2;1H/q+1;/p-1/fC56H34F6N.Br/h;1h/qm;-1
InChIKey = NYWUJLJJIDDKII-CEYVMBDECL

(*R,R*-**1a**)
[534570-50-8] $C_{56}H_{34}BrF_6N$
InChI = 1/C56H34F6N.BrH/c57-47-23-37(24-48(58)55(47)61)
43-21-33-11-3-7-15-41(33)53-45(43)29-63(30-46-44
(38-25-49(59)56(62)50(60)26-38)22-34-12-4-8-16-42
(34)54(46)53)27-35-19-17-31-9-1-5-13-39(31)51(35)
52-36(28-63)20-18-32-10-2-6-14-40(32)52;/h1-26H,
27-30H2;1H/q+1;/p-1/fC56H34F6N.Br/h;1h/qm;-1
InChIKey = NYWUJLJJIDDKII-CEYVMBDECL

(*S,S*-**1b**)
[438002-03-0] $C_{80}H_{56}BrN$
InChI = 1/C80H56N.BrH/c1-5-21-53(22-6-1)63-41-64(54-23-7-
2-8-24-54)44-67(43-63)73-47-59-31-15-19-35-71(59)
79-75(73)51-81(49-61-39-37-57-29-13-17-33-69(57)77
(61)78-62(50-81)40-38-58-30-14-18-34-70(58)78)52-
76-74(48-60-32-16-20-36-72(60)80(76)79)68-45-65(55-
25-9-3-10-26-55)42-66(46-68)56-27-11-4-12-28-56;/h1-
48H,49-52H2;1H/q+1;/p-1/fC80H56N.Br/h;1h/qm;-1
InChIKey = LRSGDSUVBVXQMY-FNTALQFFCW

(*R,R*-**1b**)
[515137-97-0] $C_{60}H_{36}BrF_{12}N$
InChI = 1/C60H36F12N.BrH/c61-57(62,63)41-21-39(22-42(27-
41)58(64,65)66)49-25-35-11-3-7-15-47(35)55-51(49)
31-73(29-37-19-17-33-9-1-5-13-45(33)53(37)54-38
(30-73)20-18-34-10-2-6-14-46(34)54)32-52-50(26-36-
12-4-8-16-48(36)56(52)55)40-23-43(59(67,68)69)28-
44(24-40)60(70,71)72;/h1-28H,29-32H2;1H/q+1;/p-1/
fC60H36F12N.Br/h;1h/qm;-1
InChIKey = GSWZUYLKIPVEBY-DCOIOKSICH

(*S,S*-**1c**)
[446017-36-3] $C_{58}H_{38}BrF_6N$
InChI = 1/C58H38F6N.BrH/c59-57(60,61)43-25-21-37(22-26-
43)49-29-39-11-3-7-15-47(39)55-51(49)33-65(34-52-
50(30-40-12-4-8-16-48(40)56(52)55)38-23-27-44(28-
24-38)58(62,63)64)31-41-19-17-35-9-1-5-13-45(35)53
(41)54-42(32-65)20-18-36-10-2-6-14-46(36)54;/h1-30
H,31-34H2;1H/q+1;/p-1/fC58H38F6N.Br/h;1h/qm;-1
InChIKey = XAPCXFDRXQSGJN-UPGIYUHZCJ

(*S,S*-**1d**)
[237762-42-4] $C_{64}H_{44}BrN$
InChI = 1/C64H44N.BrH/c1-3-17-45-33-49(29-25-41(45)13-1)
57-35-47-19-7-11-23-55(47)63-59(57)39-65(40-60-58
(36-48-20-8-12-24-56(48)64(60)63)50-30-26-42-14-2-
4-18-46(42)34-50)37-51-31-27-43-15-5-9-21-53(43)61
(51)62-52(38-65)32-28-44-16-6-10-22-54(44)62;/h1-
36H,37-40H2;1H/q+1;/p-1/fC64H44N.Br/h;1h/qm;-1
InChIKey = JUYKOKMBQKFTTB-AFYRJNNJCN

(*S,S*-**1e**)
[534570-44-0] $C_{80}H_{56}BrN$

InChI = 1/C80H56N.BrH/c1-5-21-53(22-6-1)63-41-64(54-23-7-2-8-24-54)44-67(43-63)73-47-59-31-15-19-35-71(59)79-75(73)51-81(49-61-39-37-57-29-13-17-33-69(57)77(61)78-62(50-81)40-38-58-30-14-18-34-70(58)78)52-76-74(48-60-32-16-20-36-72(60)80(76)79)68-45-65(55-25-9-3-10-26-55)42-66(46-68)56-27-11-4-12-28-56;/h1-48H,49-52H2;1H/q+1;/p-1/fC80H56N.Br/h;1h/qm;-1
InChIKey = LRSGDSUVBVXQMY-FNTALQFFCW

(*R*,*R*-**1e**)
[466679-93-6] $C_{80}H_{56}BrN$
InChI = 1/C80H56N.BrH/c1-5-21-53(22-6-1)63-41-64(54-23-7-2-8-24-54)44-67(43-63)73-47-59-31-15-19-35-71(59)79-75(73)51-81(49-61-39-37-57-29-13-17-33-69(57)77(61)78-62(50-81)40-38-58-30-14-18-34-70(58)78)52-76-74(48-60-32-16-20-36-72(60)80(76)79)68-45-65(55-25-9-3-10-26-55)42-66(46-68)56-27-11-4-12-28-56;/h1-48H,49-52H2;1H/q+1;/p-1/fC80H56N.Br/h;1h/qm;-1
InChIKey = LRSGDSUVBVXQMY-FNTALQFFCW

(*S*,*S*-**1f**)
[871130-09-5] $C_{80}H_{44}BrF_{12}N$
InChI = 1/C80H44F12N.BrH/c81-65-29-51(30-66(82)77(65)89)47-21-48(52-31-67(83)78(90)68(84)32-52)24-55(23-47)61-27-43-11-3-7-15-59(43)75-63(61)39-93(37-45-19-17-41-9-1-5-13-57(41)73(45)74-46(38-93)20-18-42-10-2-6-14-58(42)74)40-64-62(28-44-12-4-8-16-60(44)76(64)75)56-25-49(53-33-69(85)79(91)70(86)34-53)22-50(26-56)54-35-71(87)80(92)72(88)36-54;/h1-36H,37-40H2;1H/q+1;/p-1/fC80H44F12N.Br/h;1h/qm;-1
InChIKey = RFMFXPCGQXJVNJ-HZTQMOCFCN

(*S*,*S*-**1g**)
[503538-60-1] $C_{88}H_{48}BrF_{24}N$
InChI = 1/C88H48F24N.BrH/c89-81(90,91)61-27-55(28-62(37-61)82(92,93)94)51-21-52(56-29-63(83(95,96)97)38-64(30-56)84(98,99)100)24-59(23-51)73-35-47-11-3-7-15-71(47)79-75(73)43-113(41-49-19-17-45-9-1-5-13-69(45)77(49)78-50(42-113)20-18-46-10-2-6-14-70(46)78)44-76-74(36-48-12-4-8-16-72(48)80(76)79)60-25-53(57-31-65(85(101,102)103)39-66(32-57)86(104,105)106)22-54(26-60)58-33-67(87(107,108)109)40-68(34-58)88(110,111)112;/h1-40H,41-44H2;1H/q+1;/p-1/fC88H48F24N.Br/h;1h/qm;-1
InChIKey = MUSHOMOSUQEONW-PHTUWOSACO

(*R*,*R*-**1g**)
[515137-98-1] $C_{88}H_{48}BrF_{24}N$
InChI = 1/C88H48F24N.BrH/c89-81(90,91)61-27-55(28-62(37-61)82(92,93)94)51-21-52(56-29-63(83(95,96)97)38-64(30-56)84(98,99)100)24-59(23-51)73-35-47-11-3-7-15-71(47)79-75(73)43-113(41-49-19-17-45-9-1-5-13-69(45)77(49)78-50(42-113)20-18-46-10-2-6-14-70(46)78)44-76-74(36-48-12-4-8-16-72(48)80(76)79)60-25-53(57-31-65(85(101,102)103)39-66(32-57)86(104,105)106)22-54(26-60)58-33-67(87(107,108)109)40-68(34-58)88(110,111)112;/h1-40H,41-44H2;1H/q+1;/p-1/fC88H48F24N.Br/h;1h/qm;-1
InChIKey = MUSHOMOSUQEONW-PHTUWOSACO

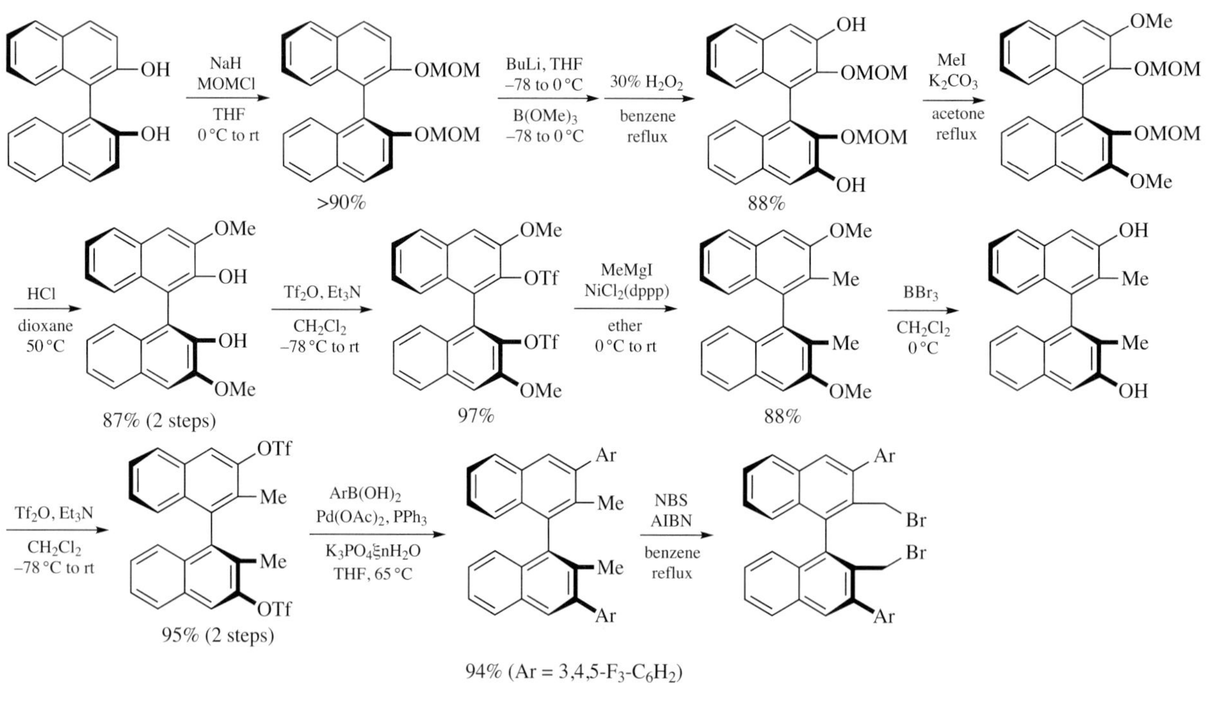

(*S,S*-**1h**)
[501934-21-0] $C_{112}H_{120}BrN$
InChI = 1/C112H120N.BrH/c1-105(2,3)85-51-79(52-86(61-85)106(4,5)6)75-45-76(80-53-87(107(7,8)9)62-88(54-80)108(10,11)12)48-83(47-75)97-59-71-35-27-31-39-95(71)103-99(97)67-113(65-73-43-41-69-33-25-29-37-93(69)101(73)102-74(66-113)44-42-70-34-26-30-38-94(70)102)68-100-98(60-72-36-28-32-40-96(72)104(100)103)84-49-77(81-55-89(109(13,14)15)63-90(56-81)110(16,17)18)46-78(50-84)82-57-91(111(19,20)21)64-92(58-82)112(22,23)24;/h25-64H,65-68H2,1-24H3;1H/q+1;/p-1/fC112H120N.Br/h;1h/qm;-1
InChIKey = IYDFYUZGKSMTQH-WRHYBRNJCK

Preparation. The parent compound 4,4′-spirobi[4*H*-dinaphth[2,1-c:1′,2′-e]azepinium],3,3′,5,5′-tetrahydro-2,6-bis(3,4,5-trifluorophenyl) bromide (abbreviated as 3,4,5-trifluorophenyl-NAS bromide) **1a** is available from Wako Pure Chemicals. This compound and its analogs **1b**–**1h** containing different aryl groups at the 3,3′-positions of the binaphthyl unit can be synthesized starting from either (*S*)- or (*R*)-1,1′-bi-2-naphthol following the procedure summarized below. These ammonium salts are bench stable, and can be stored without special caution for a extended periods of time.

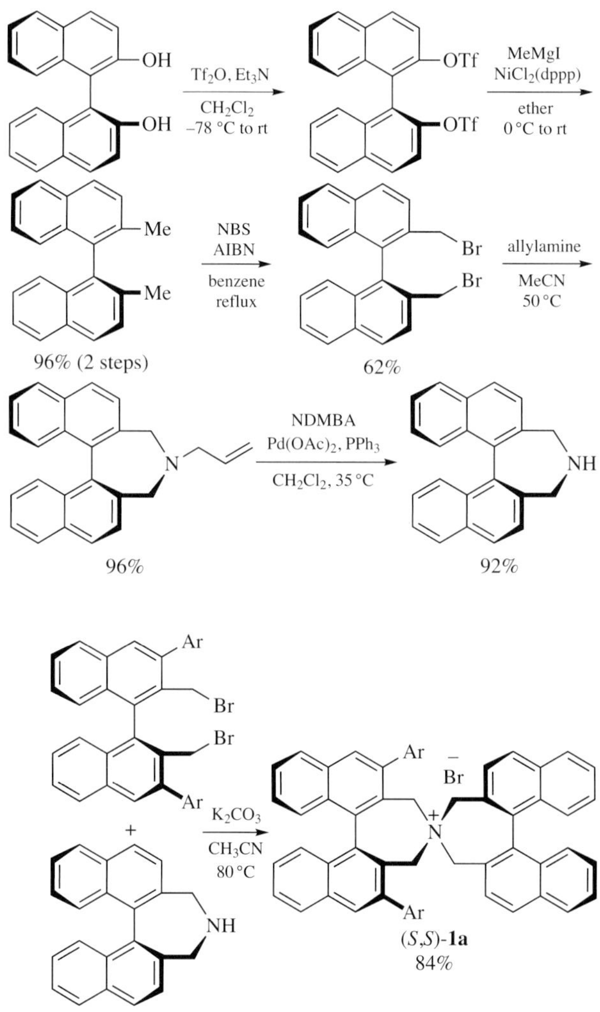

Asymmetric Alkylation for the Synthesis of α-Amino Acids Derivatives. The use of chiral *N*-spiro quaternary ammonium salts of type **1** as phase-transfer catalysts[1] was first reported in the highly efficient, asymmetric alkylation of benzophenone imine glycine Schiff base **2a** (eq 1).[2] The reaction generally provides α-amino acids derivatives in almost perfect enantioselectivity (96–99% ee). Since both enantiomers of the catalyst can be readily assembled, a wide variety of natural and unnatural α-amino acids can be synthesized in enantiomerically pure form. Compared to other phase-transfer catalysts, such as cinchonidinium salts, NAS bromide **1** generally requires lower catalyst loading and provides alkylated compounds in higher enantiomeric excesses. At the same time, the requisite reaction conditions using toluene as solvent at mild reaction temperatures is considered to be the additional advantage of NAS bromide **1**.

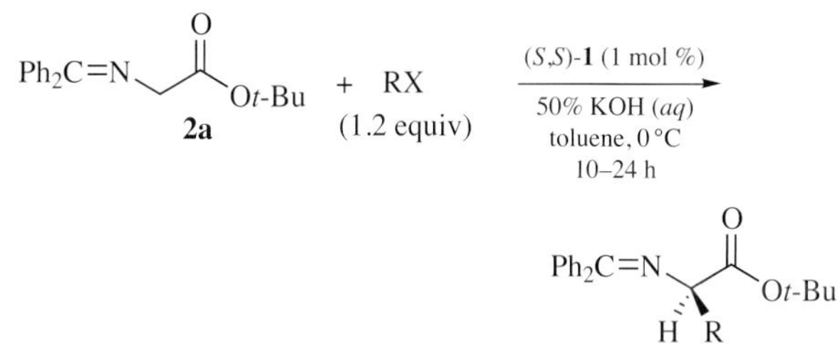

(1)

Furthermore, the catalyst loading can be as low as 0.05 mol % by the use of cocatalysts such as 18-crown-6 (eq 2).[3]

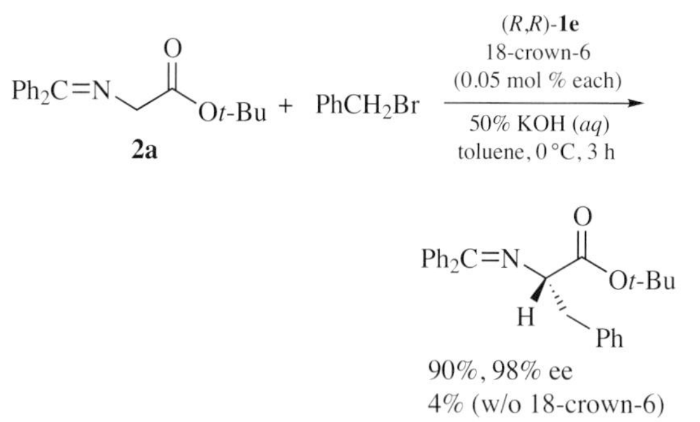

(2)

The salient feature of **1a** as a chiral phase-transfer catalyst is its ability to catalyze the asymmetric alkylation of glycine methyl and ethyl ester derivatives **2b** and **2c** with excellent enantioselectivities (eq 3).[4]

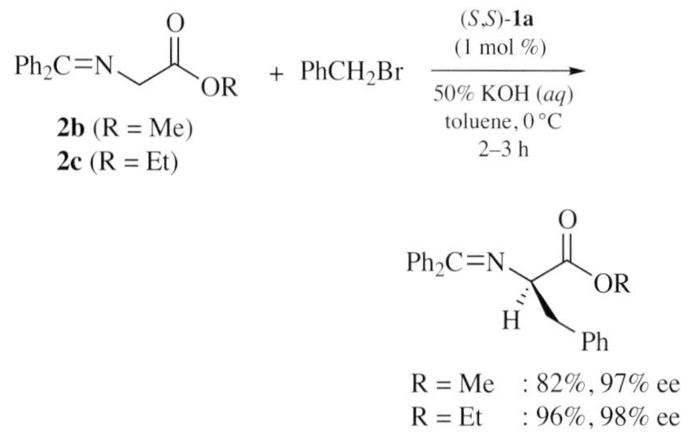

(3)

Not only the benzophenone imine Schiff bases **2a**–**c** but the aldimine Schiff base **2d** can be utilized for highly enantioselective synthesis of α-amino acids using (*R,R*)-**1a** as catalyst (eq 4).[5]

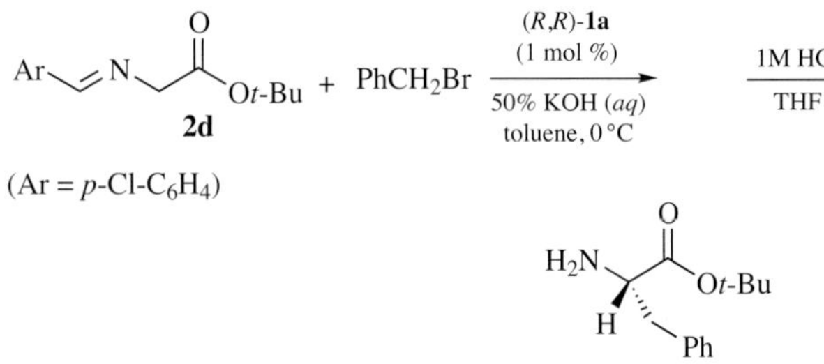

(4)

99%, 98% ee

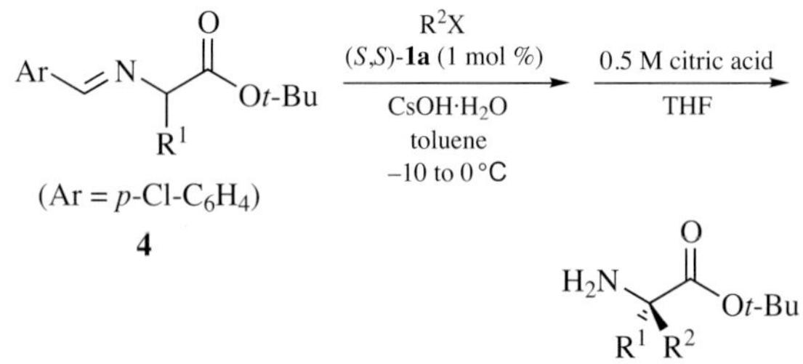

(7)

R^1 = Me, R^2X = EtI

71%, 99% ee

R^1 = $PhCH_2$, R^2X = CH_2=$CHCH_2Br$

71%, 97% ee

Asymmetric alkylation of **2a** catalyzed by **1g** offers an approach to the direct construction of β-branched-α-amino acid derivatives by the use of racemic secondary alkyl halides as alkylation partners, which involves the kinetic resolution of alkyl halides (eq 5).[6]

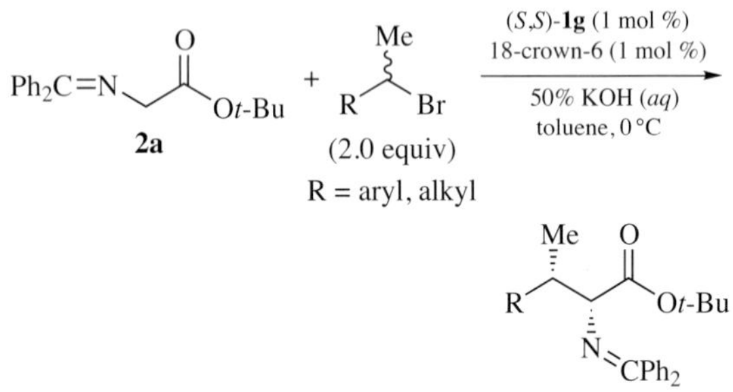

(5)

syn:anti up to >95:5
86–99% ee

The asymmetric alkylation of glycine amide **3** can also be achieved with high enantioselectivity by the use of chiral ammonium salt **1h**. The reduction of the amide moiety of the alkylated compound furnishes a wide range of vicinal diamine derivatives in optically pure form. Additionally, this substrate has increased nucleophilicity compared to **2**, and even undergoes smooth alkylation with less reactive secondary alkyl halides (eq 6).[7]

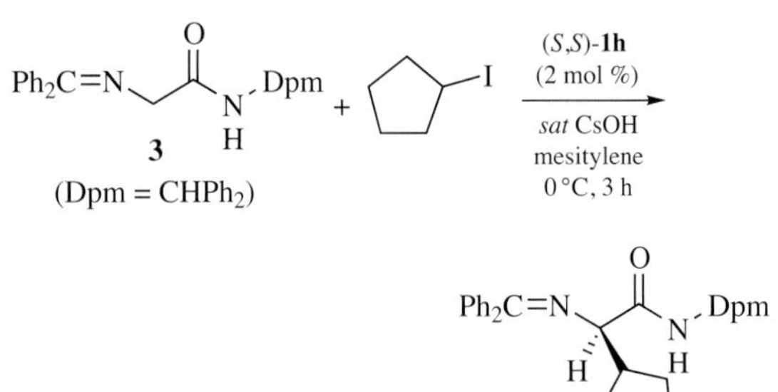

(6)

91%, 96% ee

Asymmetric Alkylation for the Synthesis of α,α-Dialkyl-α-amino Acids. Phase-transfer catalysis using **1** can be further extended to the asymmetric alkylation of aldimine Schiff bases for the synthesis of chiral α,α-dialkyl-α-amino acids (eq 7). Irrespective of the substituent R^1 of aldimine Schiff base **4**, α,α-dialkyl-α-amino acids can be obtained in almost perfect enantioselectivity.[8]

Utilization of (*S*,*S*)-**1a** catalyzed asymmetric alkylation of phenyl oxazoline derivative **5a** is beneficial for the stereoselective synthesis of α-alkyl serines. The reaction is general and provides practical access to a variety of optically active α-alkyl serines through acidic hydrolysis.[9] This method can also be applied to similar asymmetric alkylation of 2-phenyl-2-thiazoline-4-carboxylic acid ester **5b** to give optically enriched α-alkyl cystein derivatives (eq 8).[10]

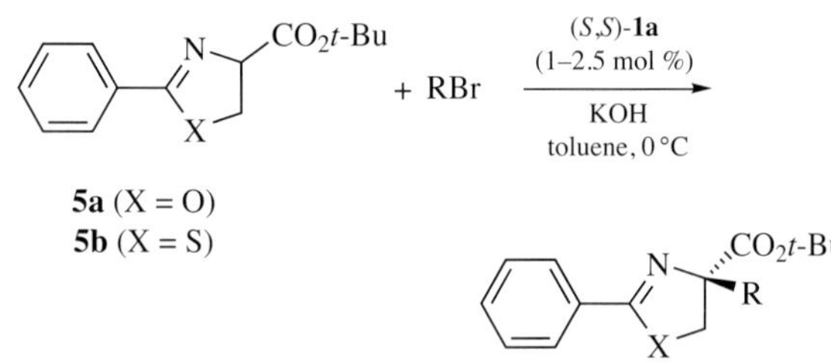

(8)

X = O : 93–99% ee
= S : 67–99% ee

Asymmetric Alkylation of Schiff Base-activated Peptides. The introduction of side chains directly onto a peptide backbone represents a powerful method for the preparation of unnatural peptides. The alkylation of the peptide activated by the formation of Schiff base constructs α-alkyl and α,α-dialkyl α-amino acid residues at the peptide terminus with high stereoselectivity (eqs 9 and 10).[11]

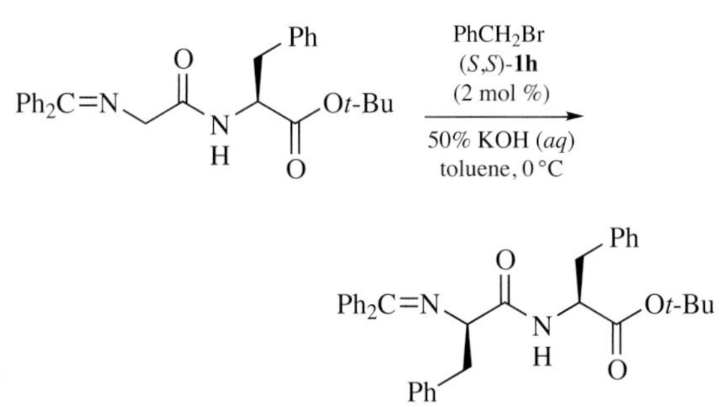

(9)

97%, 97% de

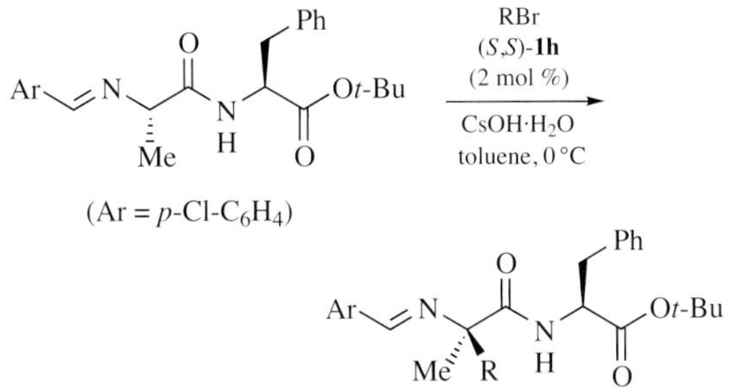

(10)

R = $PhCH_2$: 87%, 92% de
CH_2=$CHCH_2$: 91%, 92% de

Other Asymmetric Alkylations. The ammonium salt **1b** catalyzes the asymmetric alkylation of β-keto esters to give the alkylated compound containing an all-carbon quaternary center with excellent enantioselectivity (eq 11).[12] The resulting alkylation products can be easily converted into the corresponding β-hydroxy and β-amino esters.

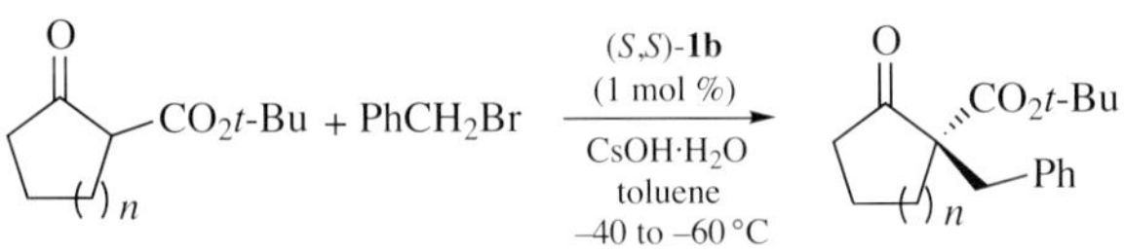

(11)

n = 1: 94%, 97% ee
n = 2: 88%, 92% ee

Another approach for the construction of all-carbon quaternary stereocenters on 1,3-dicarbonyl compounds can be realized by the asymmetric alkylation of α-acyl-γ-butyrolactones catalyzed by **1b** under mild phase-transfer conditions (eq 12).[13]

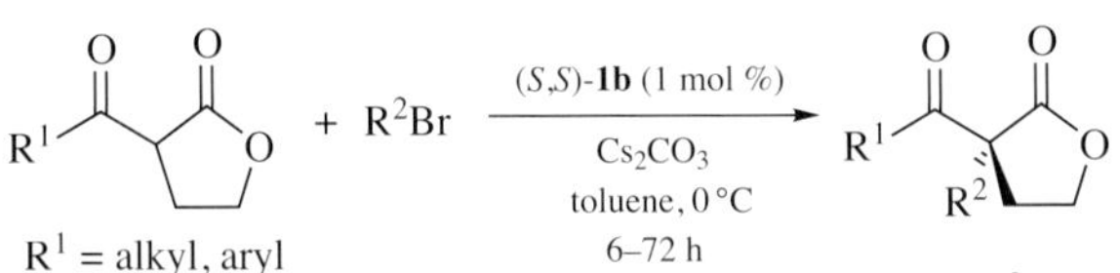

(12)

R^1 = Ph, R^2 = $PhCH_2$
85%, 88% ee

Highly enantioselective alkylation of cyclic α-amino β-keto esters **6** with chiral quaternary ammonium salt **1b** as catalyst gives rise to a unique class of unnatural α-amino acids (eq 13).[14]

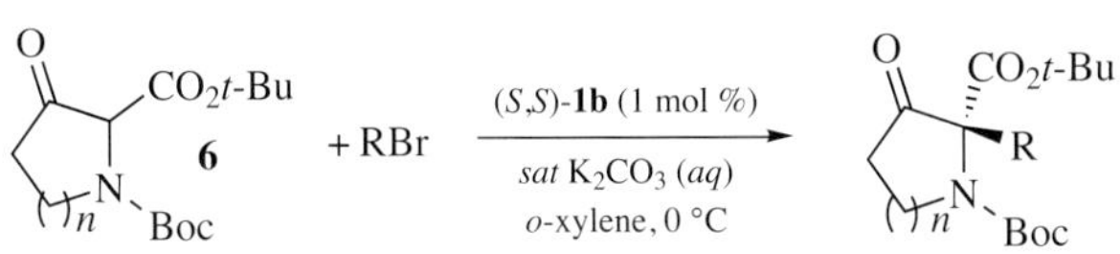

(13)

n = 1, 2

87–95% ee (R)

The designed ammonium salt **7** (CAS registry number [905708-30-7]) catalyzes the asymmetric alkylation of 3,5-diaryloxazolidin-2,4-diones **8** to give tertiary α-aryl-α-hydroxy carboxylic acid derivatives in high enantiomeric excesses (eq 14).[15]

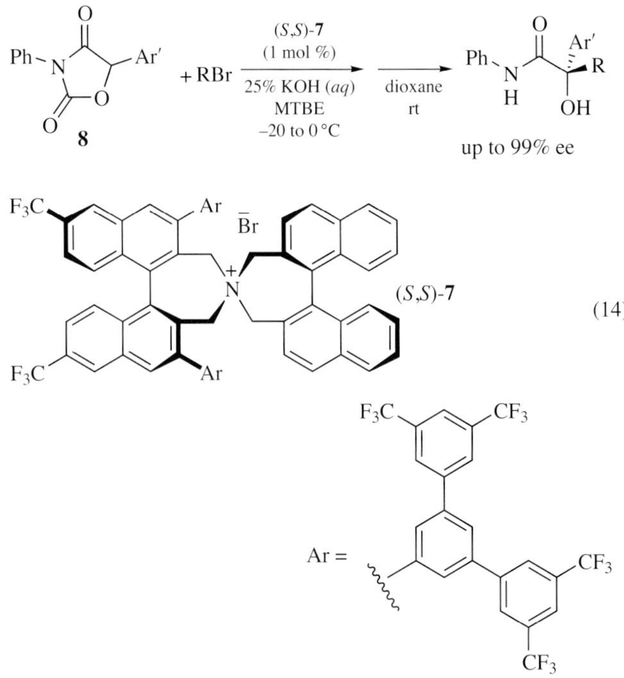

(14)

Asymmetric alkylation of α-cyanoacetates catalyzed by **1b** offers an attractive method for the synthesis of useful chiral building blocks bearing all-carbon quaternary centers (eq 15).[16]

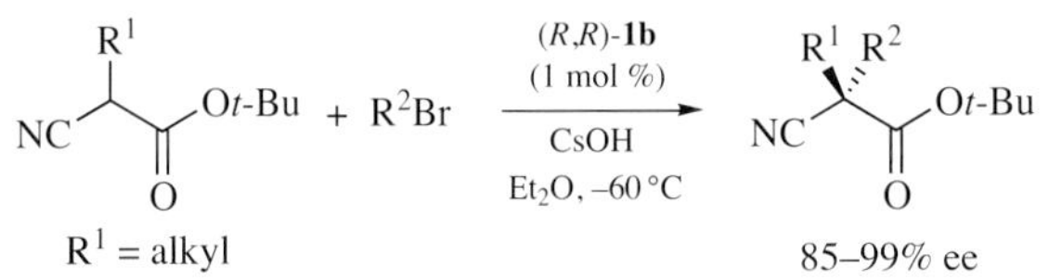

(15)

Asymmetric Michael Addition. The enantioselective construction of all-carbon quaternary centers has been achieved by quaternary ammonium salt **1b**-catalyzed Michael addition of cyclic β-keto esters with α,β-unsaturated aldehydes and ketones under phase-transfer conditions (eqs 16 and 17).[12]

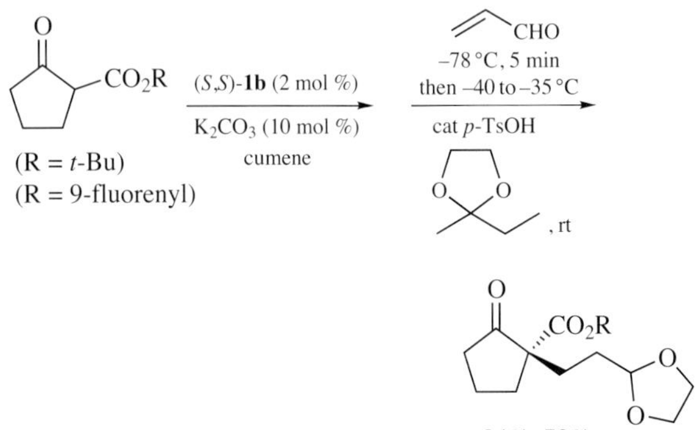

(16)

84%, 79% ee
(R = t-Bu)
92%, 90% ee
(R = 9-fluorenyl)

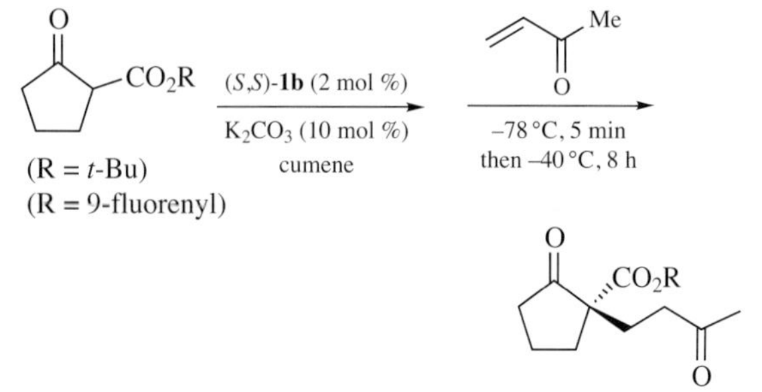

(17)

quant, 97% ee
(R = 9-fluorenyl)

The efficient asymmetric phase-transfer catalyzed Michael addition of 2-naphthalen-1-yl-2-oxazoline-4-carboxylic acid *tert*-butyl ester **9** was reported by Jew, Park, and co-workers (eq 18).[17] This method is a useful tool for the enantioselective syntheses of (2*S*)-α-(hydroxymethyl)glutamic acid.

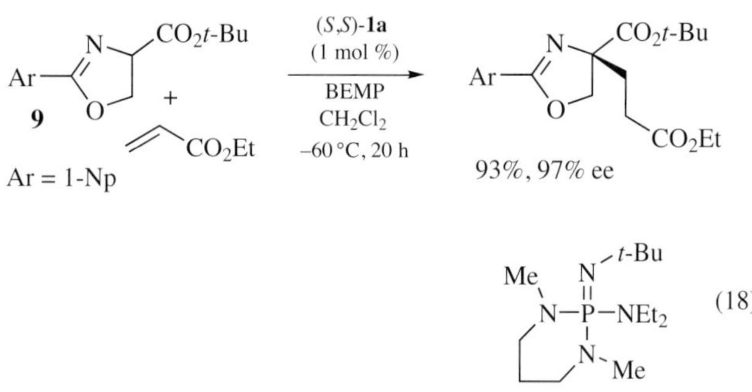

(18)

BEMP

The diastereo- and enantioselective conjugate addition of nitroalkanes to alkylidenemalonates can be achieved under phase-transfer conditions employing *N*-spiro chiral quaternary ammonium salt **1g**. This method provides a new and practical entry to optically active γ-amino acid derivatives (eq 19).[18]

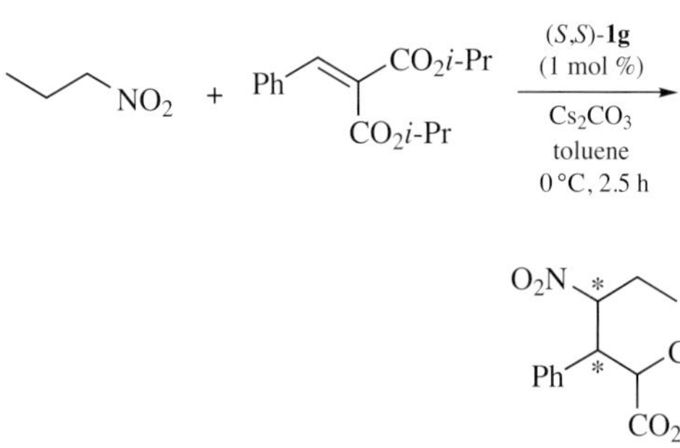

(19)

anti:syn = 86:14
97% ee (anti isomer)

Diastereo-and enantioselective Michael addition of nitroalkanes to cyclic α,β-unsaturated ketones can be implemented by the use of catalyst **1f** (eq 20).[19]

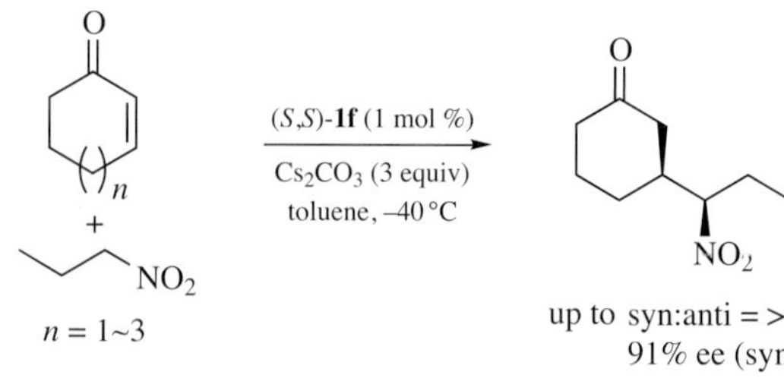

(20)

up to syn:anti = >95:5
91% ee (syn isomer)

Asymmetric Aldol Reaction. The aldol reaction between benzophenone imine glycine Schiff base and aldehydes is one of the powerful methods for the direct construction of the primary structure of β-hydroxy-α-amino acids. The diastereo- and enantioselective direct aldol reaction of these substances can be realized by using *N*-spiro chiral quaternary ammonium bromide **1g** as catalyst (eq 21).[20]

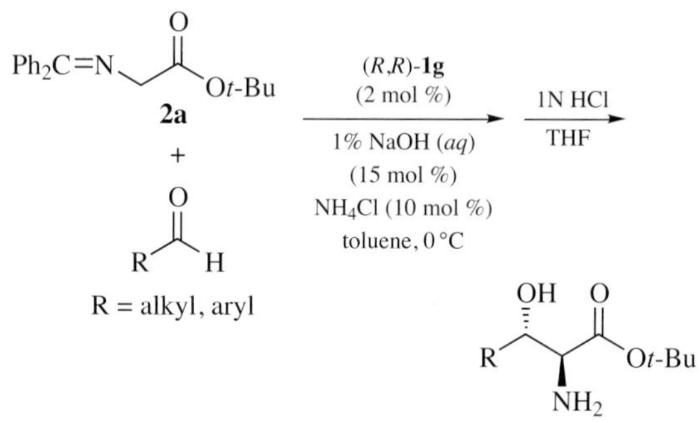

(21)

anti:syn up to >96:4
up to 99% ee (anti isomer)

Asymmetric Mannich Reaction. The asymmetric Mannich reaction of benzophenone imine glycine Shiff base **2a** with α-imino ester under phase-transfer conditions is accomplished with high enantioselectivity employing chiral quaternary ammonium bromide **1a** as catalyst (eq 22).[21]

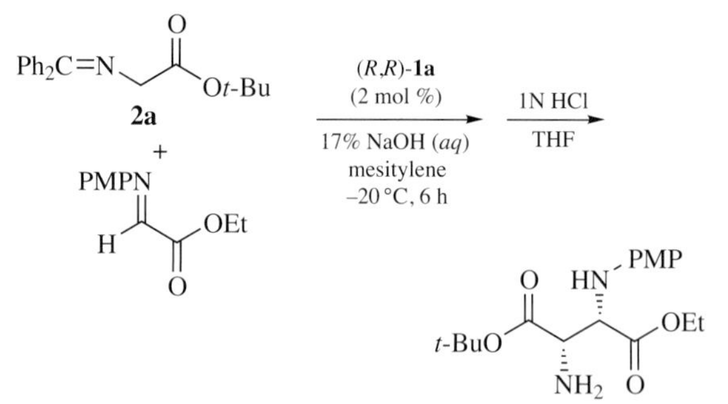

(22)

syn:anti = 82:18
91% ee (syn isomer)

Asymmetric Neber Rearrangement. The Neber rearrangement of oxime sulfonates has been recognized as a distinctive method for the preparation of a variety of α-amino ketones. The asymmetric phase-transfer catalyzed Neber rearrangement of oxime sulfonates, generated in situ from the parent oxime **10**, provides the protected α-amino ketones in high yield with moderate enantioselectivity (eq 23).[22]

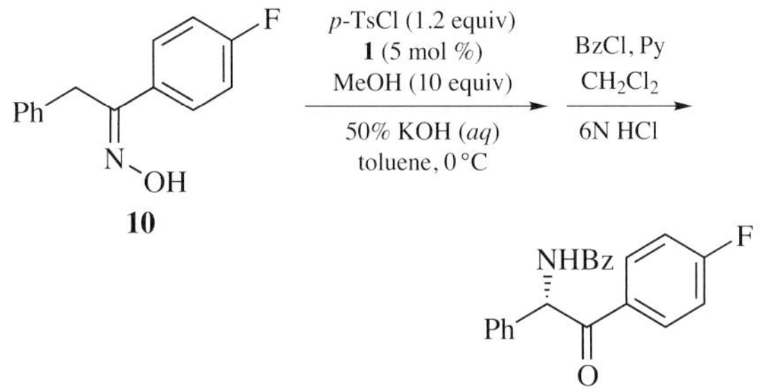

(*S*,*S*)-**1c** : 90%, 63% ee (23)

1. (a) Maruoka, K.; Ooi, T., *Chem. Rev.* **2003**, *103*, 3013.
   (b) Hashimoto, T.; Maruoka, K., *Chem. Rev.* **2007**, *107*, 5656.
2. (a) Ooi, T.; Kameda, M.; Maruoka, K., *J. Am. Chem. Soc.* **1999**, *121*, 6519.
   (b) Ooi, T.; Kameda, M.; Maruoka, K., *J. Am. Chem. Soc.* **2003**, *125*, 5139.
3. Shirakawa, S.; Yamamoto, K.; Kitamura, M.; Ooi, T.; Maruoka, K., *Angew. Chem., Int. Ed.* **2005**, *44*, 625.
4. Ooi, T.; Uematsu, Y.; Maruoka, K., *Tetrahedron Lett.* **2004**, *45*, 1675.
5. Ooi, T.; Arimura, Y.; Hiraiwa, Y.; Yuan, L. M.; Kano, T.; Inoue, T.; Matsumoto, J.; Maruoka, K., *Tetrahedron: Asymmetry* **2006**, *17*, 603.
6. Ooi, T.; Kato, D.; Inamura, K.; Ohmatsu, K.; Maruoka, K., *Org. Lett.* **2007**, *9*, 3945.
7. Ooi, T.; Sakai, D.; Takeuchi, M.; Tayama, E.; Maruoka, K., *Angew. Chem., Int. Ed.* **2003**, *42*, 5868.
8. Ooi, T.; Takeuchi, M.; Kameda, M.; Maruoka, K., *J. Am. Chem. Soc.* **2000**, *122*, 5228.
9. Jew, S.-S.; Lee, Y.-J.; Lee, J.; Kang, M.; J.; Jeong, B.-S.; Lee, J.-H.; Yoo, M.-S.; Kim, M.-J.; Choi, S.-H.; Ku, J.-M.; Park, H.-G., *Angew. Chem., Int. Ed.* **2004**, *43*, 2382.
10. Kim, T.-S.; Lee, Y.-J.; Jeong, B.-S.; Park, H.-G.; Jew, S.-S., *J. Org. Chem.* **2006**, *71*, 8276.
11. (a) Ooi, T.; Tayama, E.; Maruoka, K., *Angew. Chem., Int. Ed.* **2003**, *42*, 579.
    (b) Maruoka, K.; Tayama, E.; Ooi, T., *Proc. Natl. Acad. Sci. USA* **2004**, *101*, 5824.
12. Ooi, T.; Miki, T.; Taniguchi, M.; Shiraishi, M.; Takeuchi, M.; Maruoka, K., *Angew. Chem., Int. Ed.* **2003**, *42*, 3796.
13. Ooi, T.; Miki, T.; Fukumoto, K.; Maruoka, K., *Adv. Synth. Catal.* **2006**, *348*, 1539.
14. Ooi, T.; Miki, T.; Maruoka, K., *Org. Lett.* **2005**, *7*, 191.
15. Ooi, T.; Fukumoto, K.; Maruoka, K., *Angew. Chem., Int. Ed.* **2006**, *45*, 3839.
16. Nagata, K.; Sano, D.; Itoh, T., *Synlett* **2007**, 547.
17. Lee, Y.-J.; Lee, J.; Kim, M.-J.; Jeong, B.-S.; Lee, J.-H.; Kim, T.-S.; Lee, J.; Ku, J.-M.; Jew, S.-S.; Park, H..-G., *Org. Lett.* **2005**, *7*, 3207.
18. Ooi, T.; Fujioka, S.; Maruoka, K., *J. Am. Chem. Soc.* **2004**, *126*, 11790.
19. Ooi, T.; Takada, S.; Fujioka, S.; Maruoka, K., *Org. Lett.* **2005**, *7*, 5143.
20. (a) Ooi, T.; Taniguchi, M.; Kameda, M.; Maruoka, K., *Angew. Chem., Int. Ed.* **2002**, *41*, 4542.
    (b) Ooi, T.; Kameda, M.; Taniguchi, M.; Maruoka, K., *J. Am. Chem. Soc.* **2004**, *126*, 9685.
21. Ooi, T.; Kameda, M.; Fujii, J.-I.; Maruoka, K., *Org. Lett.* **2004**, *6*, 2397.
22. Ooi, T.; Takahashi, M.; Doda, K.; Maruoka, K., *J. Am. Chem. Soc.* **2002**, *124*, 7640.

Takuya Hashimoto, Kazuhiro Fukumoto & Keiji Maruoka
Kyoto University, Kyoto, Japan

Spiro[6*H*-1,3-dioxolo[4,5-*c*]pyran-6,5′-oxazolidine]-3′-carboxylic acid, tetrahydro-2,2-dimethyl-2′,7-dioxo-, 2-methyl-2-propyl ester, (3a*R*, 5′*S*, 7a*R*)-

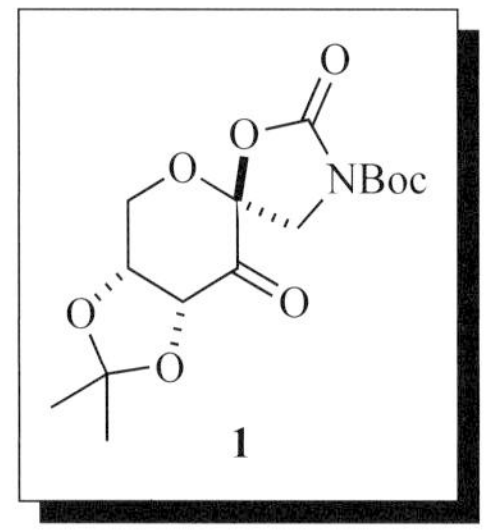

[319491-89-9] $C_{15}H_{21}NO_8$ (MW 343.33)

InChI = 1S/C15H21NO8/c1-13(2,3)23-11(18)16-7-15(24-12(16)19)10(17)9-8(6-20-15)21-14(4,5)22-9/h8-9H,6-7H2,1-5H3/t8-,9-,15+/m1/s1

InChIKey = RZRLSDWFGSKEQD-XWRVAKJBSA-N

(ketone catalyst used in the epoxidation of conjugated *cis*, terminal, *trans*, trisubstituted, and amphiphilic olefins and in the C–H activation of alcohols)

Alternate Names: (3a*R*,5′*S*,7a*R*)-*tert*-butyl 2,2-dimethyl-2′,7-dioxotetrahydrospiro [[1,3]dioxolo[4,5-*c*]pyran-6,5′-oxazolidine]-3′-carboxylate.

Physical Data: white solid; melting point = 139–140 °C; $[\alpha]^{20}_D = -47.9$ (*c* 0.83, $CHCl_3$) molecular weight = 343.33 g mol^{-1}.[1,2]

Preparative Methods: two literature methods have been reported for the synthesis of ketone **1** from D-glucose using Amadori rearrangement and a series of functional group manipulations.[1,2]

Handling, Storage, and Precautions: avoid inhalation and contact with eyes, skin, and clothing. Store in a tightly sealed container. Avoid contact with strong oxidizing agents. Thorough toxicity studies have not been performed.

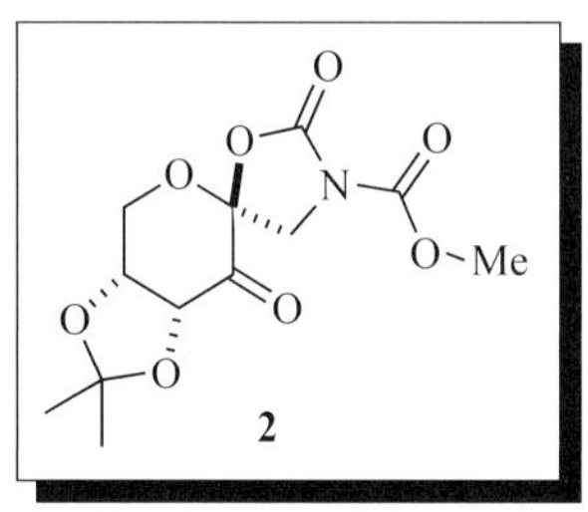

Spiro[6*H*-1,3-dioxolo[4,5-*c*]pyran-6,5′-oxazolidine]-3′-carboxylic acid, tetrahydro-2,2-dimethyl-2′,7-dioxo-, methyl ester, (3a*R*, 5′*S*, 7a*R*)-

[906371-65-1] $C_{12}H_{15}NO_8$ (MW 301.25)

InChI = 1S/C12H15NO8/c1-11(2)19-6-4-18-12(8(14)7(6)20-11)5-13(9(15)17-3)10(16)21-12/h6-7H,4-5H2,1-3H3/t6-,7-,12+/m1/s1

InChIKey = JQXROEASUDOFLI-QHMRUWFYSA-N

(ketone catalyst used in stereoselective C–H activation of vicinal diols)

Alternate Names: (3a*R*,5′*S*,7a*R*)-methyl 2,2-dimethyl-2′,7-dioxotetrahydrospiro [[1,3] dioxolo[4,5-*c*]pyran-6,5′-oxazolidine]-3′-carboxylate.

Physical Data: colorless crystals.

Handling, Storage, and Precautions: avoid inhalation and contact with eyes, skin, and clothing. Store in a tightly sealed container. Avoid contact with strong oxidizing agents. Thorough toxicity studies have not been performed.

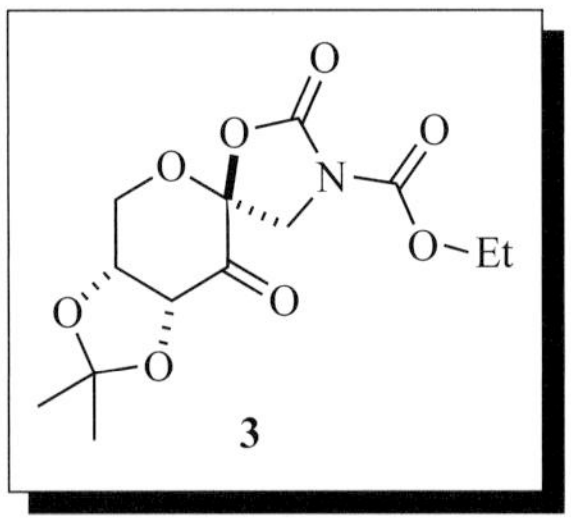

3

Spiro[6*H*-1,3-dioxolo[4,5-*c*]pyran-6,5′-oxazolidine]-3′-carboxylic acid, tetrahydro-2,2-dimethyl-2′,7-dioxo-, ethyl ester, (3a*R*, 5′*S*, 7a*R*)-,

[906371-64-0] $C_{13}H_{17}NO_8$ (MW 315.28)

InChI = 1S/C13H17NO8/c1-4-18-10(16)14-6-13(22-11(14)17) 9(15)8-7(5-19-13)20-12(2,3)21-8/h7-8H,4-6H2,1-3H3/ t7-,8-,13+/m1/s1

InChIKey = KYIZCUPXRAOQPJ-RBDZCENOSA-N

(ketone catalyst used in stereoselective C–H activation of vicinal diols)

Alternate Names: (3a*R*,5′S,7a*R*)-ethyl 2,2-dimethyl-2′,7-dioxotetrahydrospiro[[1,3]dioxolo[4,5-c]pyran-6,5′-oxazolidine]-3′-carboxylate.

Physical Data: colorless crystals.

Handling, Storage, and Precautions: avoid inhalation and contact with eyes, skin, and clothing. Store in a tightly sealed container. Avoid contact with strong oxidizing agents. Thorough toxicity studies have not been performed.

General. The *N*-Boc-protected ketone **1**[1,2] is an easily handled asymmetric epoxidation catalyst[3] that can be readily prepared from D-glucose. Upon treatment with a stoichiometric oxidant, a dioxirane forms that then behaves as a chiral oxidant. This ketone has shown efficacy in the asymmetric epoxidation of conjugated *cis*-, terminal, *trans*-, and trisubstituted olefins as well as nonconjugated amphiphilic olefins, providing a complement to the previously reported ketone catalysts (see also 1,2:4,5-di-*O*-isopropylidene-*β*-D-*erythro*-hexo-2,3-diluo-2,6-pyranose).[4–7] Moreover, understanding of this ketone has led to the dynamic development of other ketone catalysts.[3c–e,g–l,8,9] The ketone and its two derivatives (**2** and **3**) have also been shown to promote desymmetrization of *meso* vicinal diols and kinetic resolution of vicinal diols.[10,11]

Epoxidation. Following up on their studies of the asymmetric epoxidation of *trans*- and trisubstituted olefins,[4–6] vinyl silanes,[12] hydroxyalkenes,[13] dienes,[14] enynes,[15] enol ethers, and enol esters[16,17] using a fructose-derived catalyst, Shi and coworkers reported the synthesis and preliminary substrate scope studies of ketone **1** in 2000.[1] The ketone promoted good to high yields and induced high enantioselectivities in the epoxidation of conjugated *cis*-, terminal-, *trans*-, and trisubstituted olefins (eq 1) (Table 1).[1,8a,b,18] Based upon the absolute configurations of the resulting epoxides, a beneficial interaction between the conjugating group and the oxazolidinone of the catalyst was proposed (eq 2).[1,3c–e,g–l,8,9,18–22] The epoxidation was also stereospecific; no *trans*-epoxide was formed from the reaction of a *cis*-olefin. In addition, a few nonconjugated substrates were successfully epoxidized (Table 1—entry 18—and Table 2).[1,8a,b,21]

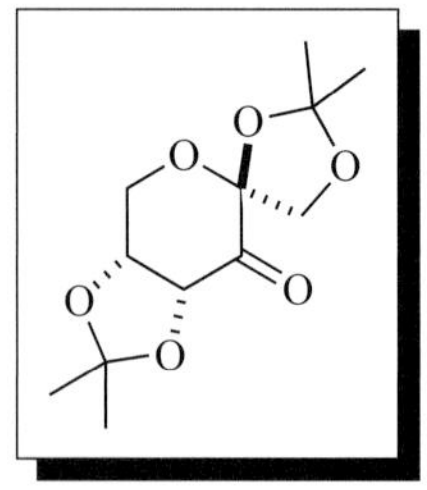

1,2:4,5-Di-*O*-isopropylidene-*β*-D-*erythro*-hexo-2,3-diluo-2,6-pyranose

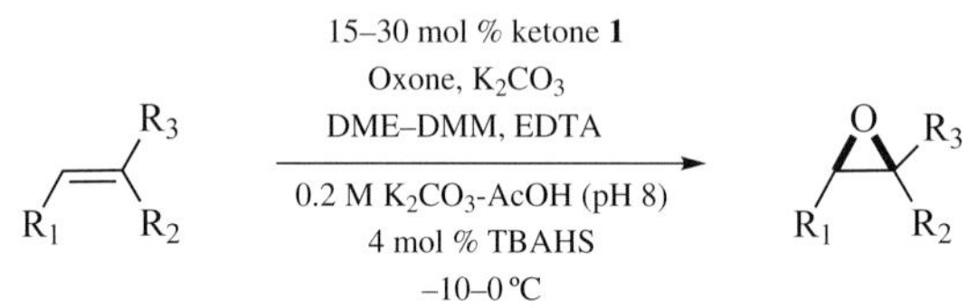

(1)

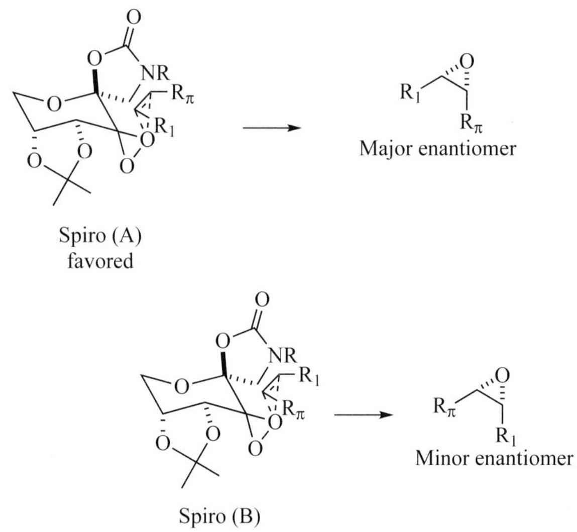

(2)

Shi and coworkers later reported that various *N*-substituted derivatives (eq 3) could provide a similar directing effect as ketone **1** and offer ease of synthesis.[8e,22] Ketones **b** and **c** are now commercially available.

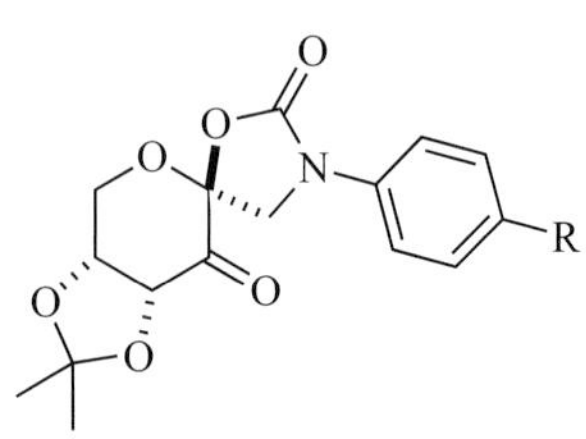

a: R = SO_2Me
b: R = Me
c: R = Et
d: R = Ph
e: R = *n*Bu (3)

Table 1 Representative examples of ketone **1**-catalyzed epoxidation

Entry	Product	Yield (%)	ee (%)
1	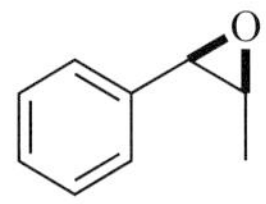	87	91
2	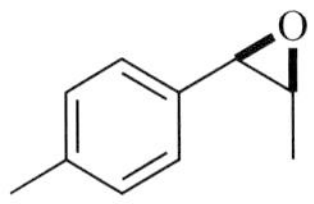	76	92
3	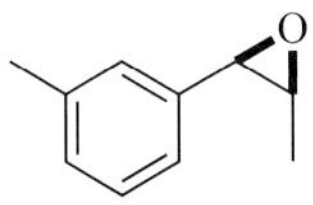	79	88
4	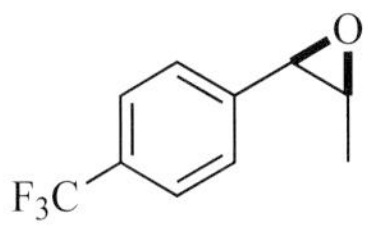	63	90
5	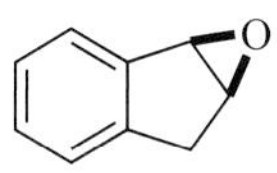	88	83
6	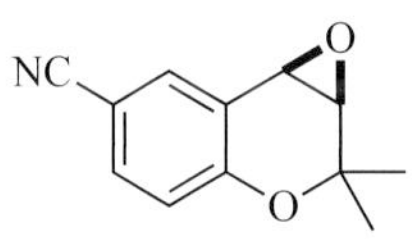	61	91
7	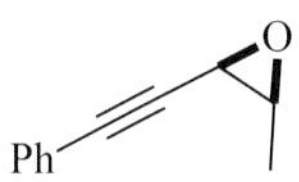	82	91
8	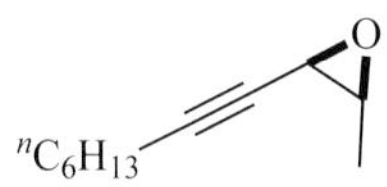	77	87
9	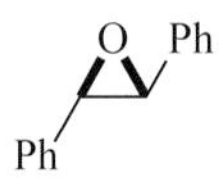	65	94
10	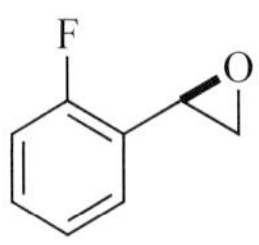	87	82
11	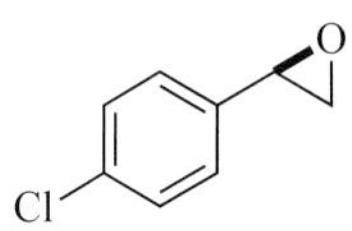	90	85
12	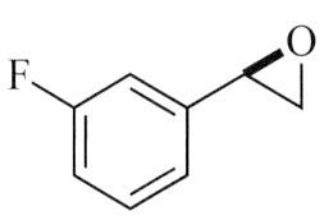	93	81
13	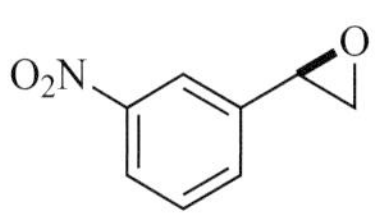	88	74
14	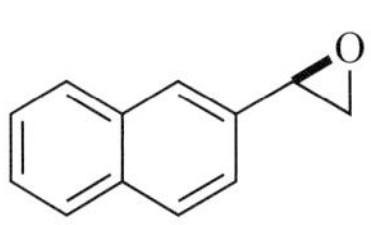	86	84
15	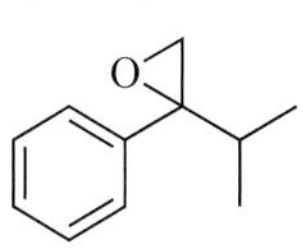	87	58

Table 1 Continued

Entry	Product	Yield (%)	ee (%)
16	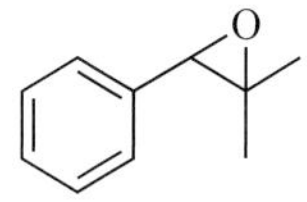	78	95
17	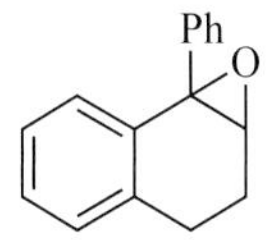	55	80
18	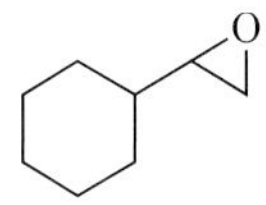	93	71
19	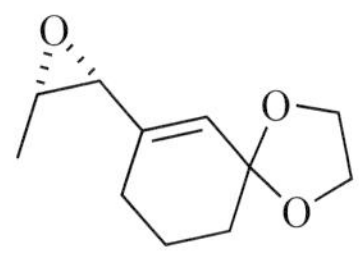	61	93

Table 2 Representative examples of ketone **1**-catalyzed epoxidation of nonconjugated substrates

Entry	Product	Yield (%)	ee (%)
1	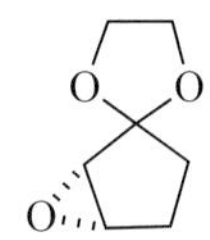	47	96
2	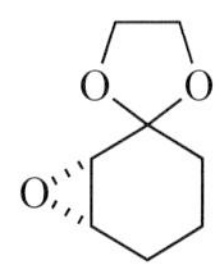	61	97
3	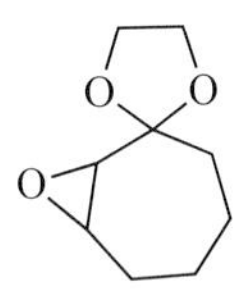	88	94
4	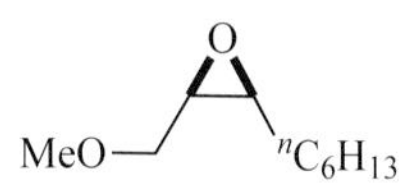	79	61
5	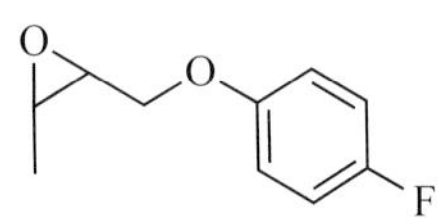	73	86
6	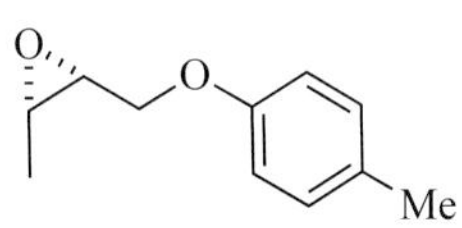	71	85
7	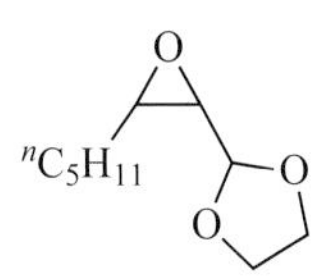	76	92
8	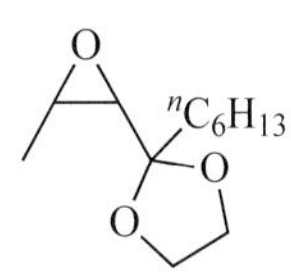	39	51

In 2009, Shi and coworkers extended the substrate scope of the epoxidation to that of nonconjugated amphiphilic *cis*-olefins (eq 4).[21] Allylic ethers and allylic acetals favor aligning themselves proximal to the oxazolidinone presumably for electronic reasons (Table 2). In studies of the epoxidation of allylic alcohols and carboxylic acids with ketone **e**, it was shown that the increased hydrophilicity of the polar side chain overrides the oxygen/oxazolidinone attractive interaction, thus further favoring the hydrophobic interaction between the nonpolar substituent of the olefin and the oxazolidinone of the catalyst. The *n*-butyl group of ketone **e** can further increase the nonpolar substituent/oxazolidinone interaction.[21]

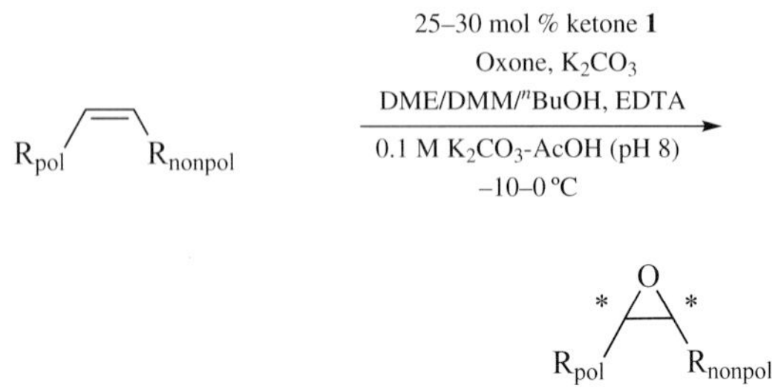

(4)

Desymmetrization/Kinetic Resolution of Vicinal Diols. Zhao, Adam, and coworkers have studied the dioxirane-mediated desymmetrizing C–H activation of *meso*-hydrobenzoins and the kinetic resolution of racemic hydrobenzoins.[10,11] In 2006, Jakka and Zhao reported that ketone **1** could catalyze the desired processes.[11] Some representative examples are shown (eq 5). In both cases, the dioxirane prefers to activate the C–H bond of *S*-configured chiral centers. Ketone analogs **2** and **3** could also promote the desymmetrization of *meso*-hydrobenzoin (eq 6).[11]

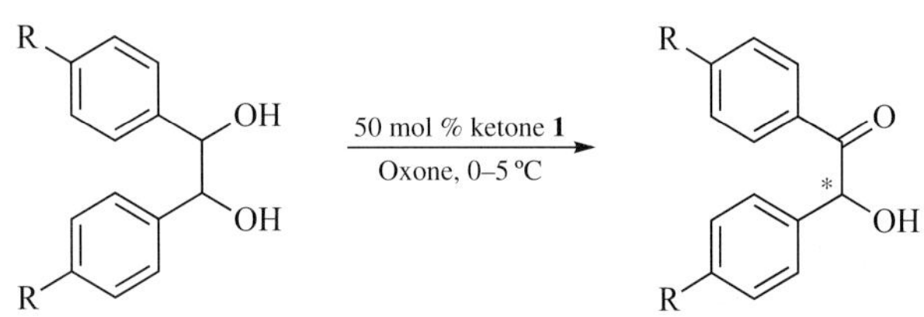

meso vic diols

R	Yield (%)	ee (%)
H	80	87
Me	85	77
F	65	87
Cl	80	77
OMe	60	76
Br	74	76

racemic *vic* diols

R	Yield (%)	ee (%)
H	48	87
Me	42	85
F	45	90
Cl	38	84
Br	40	84

(5)

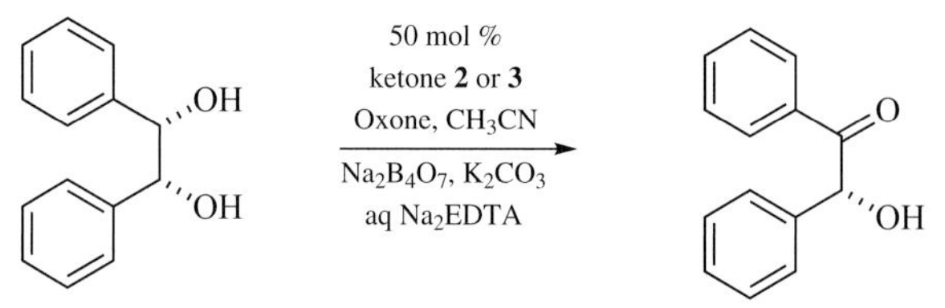

(6)

Ketone	Yield (%)	ee (%)
2	85	80
3	90	80

1. Tian, H.; She, X.; Shu, L.; Yu, H.; Shi, Y., *J. Am. Chem. Soc.* **2000**, *122*, 11551.
2. Shu, L.; Shen, Y.-M.; Burke, C.; Goeddel, D.; Shi, Y., *J. Org. Chem.* **2003**, *68*, 4963.
3. For leading reviews on asymmetric epoxidation mediated by chiral ketones, see: (a) Denmark, S. E.; Wu, Z., *Synlett* **1999**, 847. (b) Frohn, M.; Shi, Y., *Synthesis* **2000**, 1979. (c) Shi, Y., *J. Syn. Org. Chem.* **2002**, *60*, 342. (d) Shi, Y., In *Modern Oxidation Methods*; Bäckvall, J.-E. Ed.; Wiley-VCH: Weinheim, 2004; Chapter 3. (e) Shi, Y., *Acc. Chem. Res.* **2004**, *37*, 488. (f) Yang, D., *Acc. Chem. Res.* **2004**, *37*, 497. (g) Shi, Y. In *Handbook of Chiral Chemicals*; Ager, D., Ed.; CRC Press, Taylor & Francis Group: Boca Raton, 2006; Chapter 10. (h) Goeddel, D.; Shi, Y. In *Catalysts for Fine Chemical Synthesis*; Roberts, S. M.; Whittall, J., Eds.; Wiley: Chichester, 2007, vol. 5, p. 215. (i) Wong, O. A.; Shi, Y., *Chem. Rev.* **2008**, *108*, 3958. (j) Goeddel, D.; Shi, Y. In *Science of Synthesis*; Forsyth, C. J., Ed., Georg Thieme Verlag KG: Stuttgart, Germany, 2008, vol. 37, p. 277. (k) Wong, O. A.; Shi, Y., *Top. Curr. Chem.* **2009**, *291*, 201. (l) Shi, Y. In *Modern Oxidation Methods*; Bäckvall, J.-E., Ed.; Wiley-VCH: Weinheim, 2010; 2nd., Chapter 3.
4. Wong, O. A.; Shi, Y., In *Encyclopedia of Reagents for Organic Synthesis*; Wiley: Chichester, 2008.
5. (a) Tu, Y.; Wang, Z.-X.; Shi, Y., *J. Am. Chem. Soc.* **1996**, *118*, 9806. (b) Wang, Z.-X.; Tu, Y.; Frohn, M.; Zhang, J.-R.; Shi, Y., *J. Am. Chem. Soc.* **1997**, *119*, 11224.
6. Zhao, M.-X.; Shi, Y., *J. Org. Chem.* **2006**, *71*, 5377.
7. (a) Wu, X.-Y.; She, X.; Shi, Y., *J. Am. Chem. Soc.* **2002**, *124*, 8792. (b) Wang, B.; Wu, X.-Y.; Wong, O. A.; Nettles, B.; Zhao, M.-X.; Chen, D.; Shi, Y., *J. Org. Chem.* **2009**, *74*, 3986.
8. (a) Tian, H.; She, X.; Xu, J.; Shi, Y., *Org. Lett.* **2001**, *3*, 1929. (b) Tian, H.; She, X.; Yu, H.; Shu, L.; Shi, Y., *J. Org. Chem.* **2002**, *67*, 2435. (c) Hickey, M.; Goeddel, D.; Crane, Z.; Shi, Y., *Proc. Natl. Acad. Sci. USA* **2004**, *101*, 5794. (d) Crane, Z.; Goeddel, D.; Gan, Y.; Shi, Y., *Tetrahedron* **2005**, *61*, 6409. (e) Goeddel, D.; Shu, L.; Yuan, Y.; Wong, O. A.; Wang, B.; Shi, Y., *J. Org. Chem.* **2006**, *71*, 1715.
9. (a) Wang, B.; Wong, O. A.; Zhao, M.-X.; Shi, Y., *J. Org. Chem.* **2008**, *73*, 9539. (b) Wong, O. A.; Wang, B.; Zhao, M.-X.; Shi, Y., *J. Org. Chem.* **2009**, *74*, 6335.
10. (a) Adam, W.; Saha-Möller, C. R.; Zhao, C.-G., *Tetrahedron: Asymmetry* **1998**, *9*, 4117. (b) Adam, W.; Saha-Möller, C. R.; Zhao, C.-G., *J. Org. Chem.* **1999**, *64*, 7492. (c) Adam, W.; Zhao, C.-G. In *Handbook of CH Transformations: Applications in Organic Synthesis*; Dyker, G., Ed.; Wiley-VCH: Weinheim, **2005**; Vol. 2, p 507.
11. Jakka, K.; Zhao, C.-G., *Org. Lett.* **2006**, *8*, 3013.
12. Warren, J. D.; Shi, Y., *J. Org. Chem.* **1999**, *64*, 7675.
13. Wang, Z.-X.; Shi, Y., *J. Org. Chem.* **1998**, *63*, 3099.
14. Frohn, M.; Dalkiewicz, M.; Tu, Y.; Wang, Z-X.; Shi, Y., *J. Org. Chem.* **1998**, *63*, 2948.
15. (a) Cao, G.-A.; Wang, Z.-X.; Tu, Y.; Shi, Y., *Tetrahedron Lett.* **1998**, *39*, 4425. (b) Wang, Z.-X.; Cao, G.-A.; Shi, Y., *J. Org. Chem.* **1999**, *64*, 7646.

16. (a) Zhu, Y.; Tu, Y.; Yu, H.; Shi, Y., *Tetrahedron Lett.* **1998**, *39*, 7819. (b) Zhu, Y.; Manske, K. J.; Shi, Y., *J. Am. Chem. Soc.* **1999**, *121*, 4080. (c) Feng, X.; Shu, L.; Shi, Y., *J. Am. Chem. Soc.* **1999**, *121*, 11002. (d) Zhu, Y.; Shu, L.; Tu, Y.; Shi, Y., *J. Org. Chem.* **2001**, *66*, 1818.

17. Adam, W.; Fell, R. T.; Saha-Möller, C. R.; Zhao, C.-G., *Tetrahedron: Asymmetry* **1998**, *9*, 397.

18. Burke, C. P.; Shi, Y., *Angew. Chem., Int. Ed.* **2006**, *45*, 4475.

19. For studies on the transition state of epoxidation with dimethyldioxirane, see: (a) Baumstark, A. L.; McCloskey, C. J., *Tetrahedron Lett.* **1987**, *28*, 3311. (b) Baumstark, A. L.; Vasquez, P. C., *J. Org. Chem.* **1988**, *53*, 3437.

20. For computational studies on the transition state of epoxidation with dioxiranes, see: (a) Bach, R. D.; Andrés, J. L.; Owensby, A. L.; Schlegel, H. B.; McDouall, J. J. W., *J. Am. Chem. Soc.* **1992**, *114*, 7207. (b) Houk, K. N.; Liu, J.; DeMello, N. C.; Condroski, K. R., *J. Am. Chem. Soc.* **1997**, *119*, 10147. (c) Jenson, C.; Liu, J.; Houk, K. N.; Jorgensen, W. L., *J. Am. Chem. Soc.* **1997**, *119*, 12982. (d) Deubel, D. V., *J. Org. Chem.* **2001**, *66*, 3790. (e) Singleton, D. A.; Wang, Z., *J. Am. Chem. Soc.* **2005**, *127*, 6679. (f) Hirschi, J. S.; Takeya, T.; Hang, C.; Singleton, D. A., *J. Am. Chem. Soc.* **2009**, *131*, 2397.

21. Burke, C. P.; Shi, Y., *Org. Lett.* **2009**, *11*, 5150.

22. (a) Shu, L.; Wang, P.; Gan, Y.; Shi, Y., *Org. Lett.* **2003**, *5*, 293. (b) Zhao, M.-X.; Goeddel, D.; Li, K.; Shi, Y., *Tetrahedron* **2006**, *62*, 8064.

Thomas A. Ramirez & Yian Shi
Colorado State University, Fort Collins, CO, USA

Squaramide

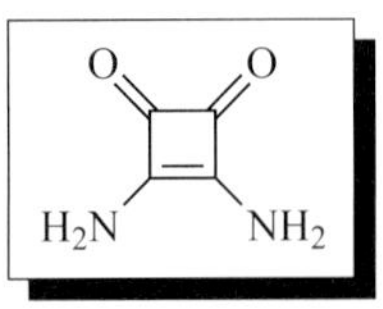

[5231-89-0] $C_4H_4N_2O_2$ (MW 112.08)
InChI = 1S/C4H4N2O2/c5-1-2(6)4(8)3(1)7/h5-6H2
InChIKey = WUACDRFRFTWMHE-UHFFFAOYSA-N

(reagent used as a building block for the construction of *N*-substituted squaramides)

Alternate Name: 3,4-Diaminocyclobut-3-ene-1,2-dione
Physical Data: mp 338–340°C.
Solubility: soluble in water and many organic solvents.
Form Supplied in: white solid, available from various suppliers.
Analysis of Reagent Purity: IR, NMR.
Preparative Methods: see below for preparation routes.
Handling, Storage, and Precautions: harmful, irritant, target organs: respiratory system, eyes, skin. Avoid breathing dust or vapor. Avoid contact with skin and eyes. Avoid ingestion and inhalation. Store in a cool, dry place. Store in a tightly closed container. Stable under normal temperatures and pressures. *Incompatibilities with Other Materials:* strong oxidizing agents, strong acids, strong bases.

Introduction and Physicochemistry. Squaramide is a member of the diketocyclobutene class of reagents, which also includes squaric acid, squaric acid dichloride, and squarate esters (eq 1). The syntheses and applications of these reagents have been the subject of numerous reviews.[1–3]

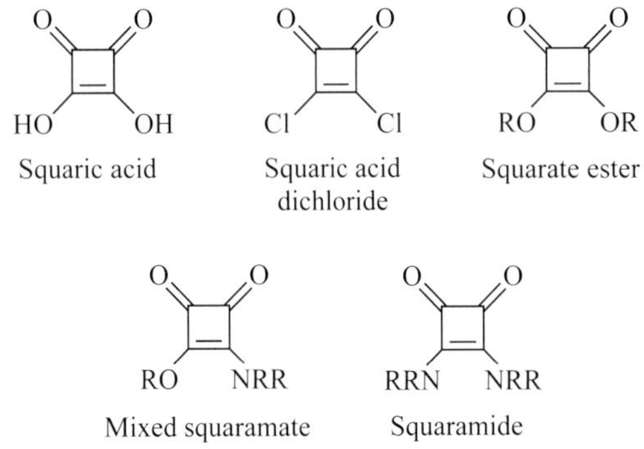

(1)

This relatively unusual ring system was first synthesized as squaric acid in the 1960s.[4] However, the full extent of the relatively special physical properties that underpin many of the applications of this ring system have not been fully established until recently. Squaric acid was initially characterized and observed to have an unusually high double acidity ($pK_{a_1} = 0.54$; $pK_{a_2} = 3.58$), which was later shown by numerous methods to be largely driven by an increased aromatic stabilization on formation of the dianion relative to the protonated molecule in accordance with the Hückel rule $4n+2$; $n = 0$ (eq 2).[5–9]

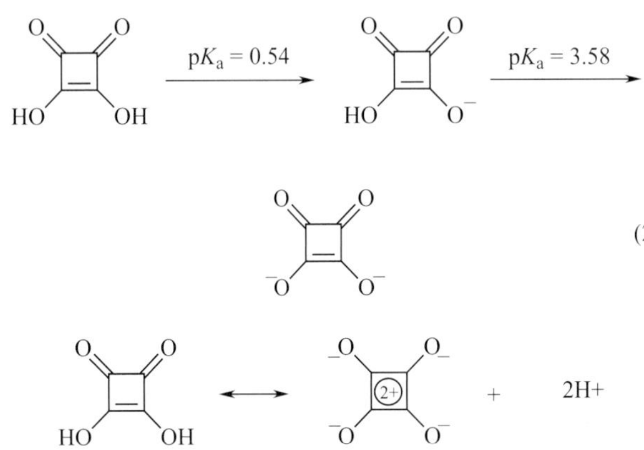

(2)

Although several squarate derivatives have been incorporated into a variety of chemical reactions, the majority of product applications have come from squaramide derivatives owing to their chemical stability coupled with an ability to form strong hydrogen-bonding intermolecular interactions. Applications of squaramides to date cover bioconjugate chemistry, material science, medicinal chemistry, molecular recognition, organometallic chemistry, and organocatalysis.[1]

The relative strength and nature of hydrogen bonds to squaramides were studied via the design of squaramide-based artificial receptors.[10–14] The interactions with a series of anions and cations were studied, measuring markers for square ring system aromaticity, including analyses of bond lengths and ring magnetism via Nucleus-independent chemical shifts (NICS—more negative implies more aromatic).[15] These studies concluded that squaramide systems become more aromatic on formation of hydrogen-bonding interactions (eq 3). Furthermore, it was concluded that this motif favors the formation of multiple simultaneous cooperative hydrogen-bonding interactions with both donors and acceptors.

Preparation of Squaramide. Although squaramide can now be purchased directly it can also be readily synthesized from other available squarate ester reagents. A recent example from Tidwell and coworkers illustrates that stirring diethyl squarate with aqueous ammonium hydroxide for 10 min at room temperature provides squaramide in quantitative yield (eq 5).[19]

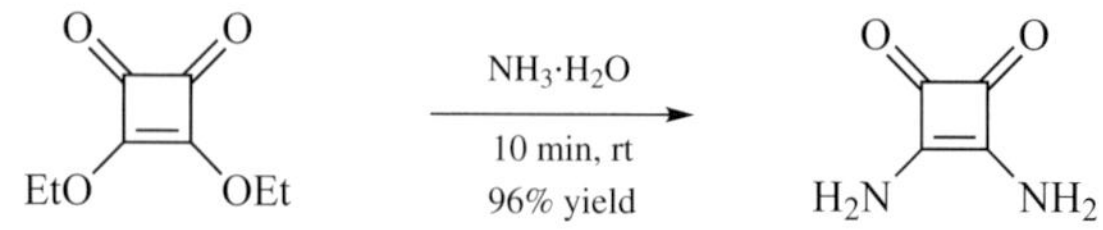

(5)

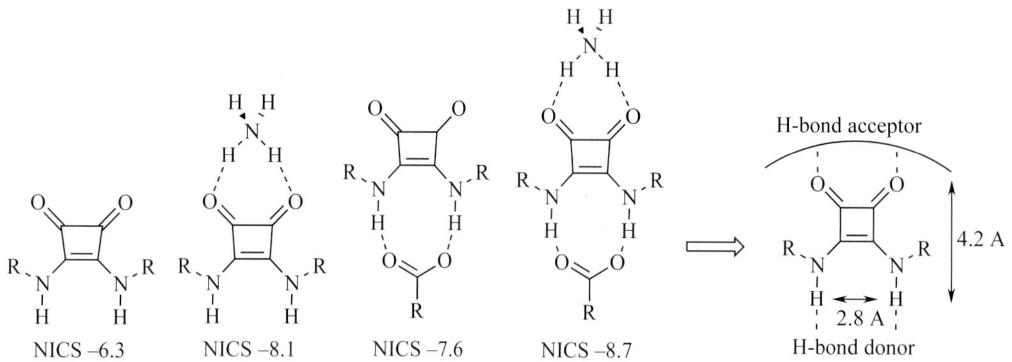

(3)

Conformational Analysis. Squaramides prefer to adopt a rigid, planar structural arrangement, stabilized by the nitrogens, which are essentially sp^2-hybridised, having lone pairs available for conjugation from *N*(p-orbital) into the π-system orthogonal to the plane.[16] For substituted squaramides there is the additional consideration of *anti* vs. *syn* substituent conformations, where amide-like restricted rotation about the C–N bond of a squaramide might, in principle, provide mixtures of *anti/syn* conformers (eq 4).[12] However, numerous solution NMR studies and solid state X-ray structures have confirmed that the majority of bis-secondary squaramides prefer to adopt an extended *anti/anti* conformation.[17,18]

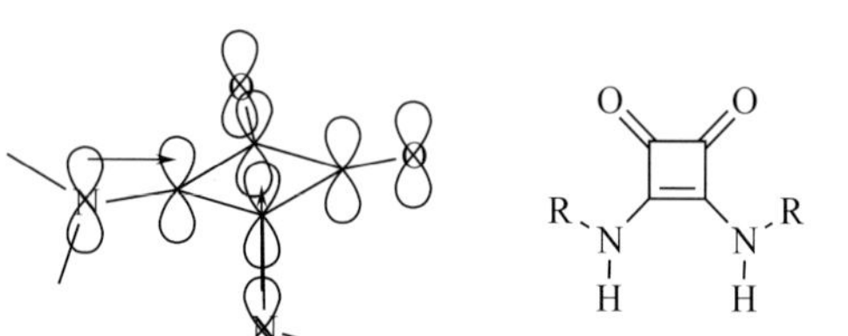

Lone pair conjugation stabilises planar conformation

anti/anti conformation preferred

(4)

Chen et al. have shown that squaramide can also be formed in high yield from a range of other squarate diesters by treatment with ammonium hydroxide. These include more hindered and less reactive ester substituents some of which require heating up to 70°C to facilitate conversion.[20]

Squaramide as a Nucleophilic Squarate Reagent. The majority of synthetic work to build functionalized squaramides has so far exploited electrophilic squarate reagents in combination with nucleophilic amine sources. However, recently there has been demand to synthesize systems with electron-poor secondary amino substituents in order to access acidic bis-sulfonamide squaramides. Yamamoto and coworkers initially explored the use of diethyl squarate as an electrophilic with a deprotonated perfluorinated sulfonamide as the nucleophile (eq 6). However, this reaction only proceeded to give a monosubstituted product as opposed to the desired bis-substituted derivative.[21,22]

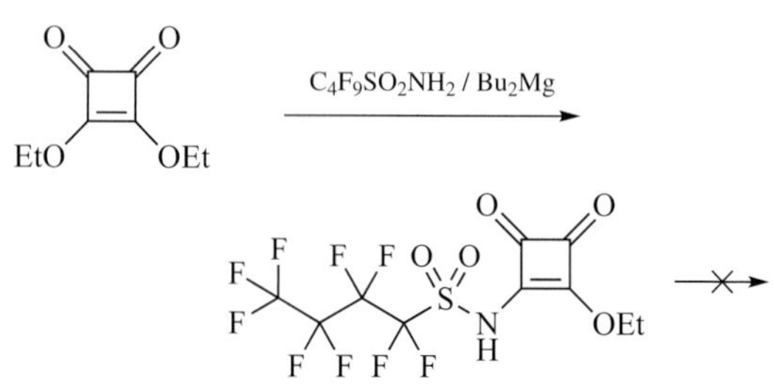

(6)

Yamamoto and coworkers then reversed the reactivity by application of squaramide as the nucleophile in combination with a perfluorinated sulfonyl fluoride electrophile to successfully provide the desired bis-substituted products in high yield (eq 7). This method was further exploited to provide multiple sulfonamide derivatives bearing longer perfluoroalkyl chains, which were subsequently applied as Brønsted acid catalysts to facilitate Mukaiyama aldol and Mukaiyama–Michael addition reactions in good yields. Further mechanistic work revealed that these reactions were catalyzed by the desired Brønsted acid pathway via activation of the carbonyl as opposed to being catalyzed by the Lewis acidity of the silylated acid.[21]

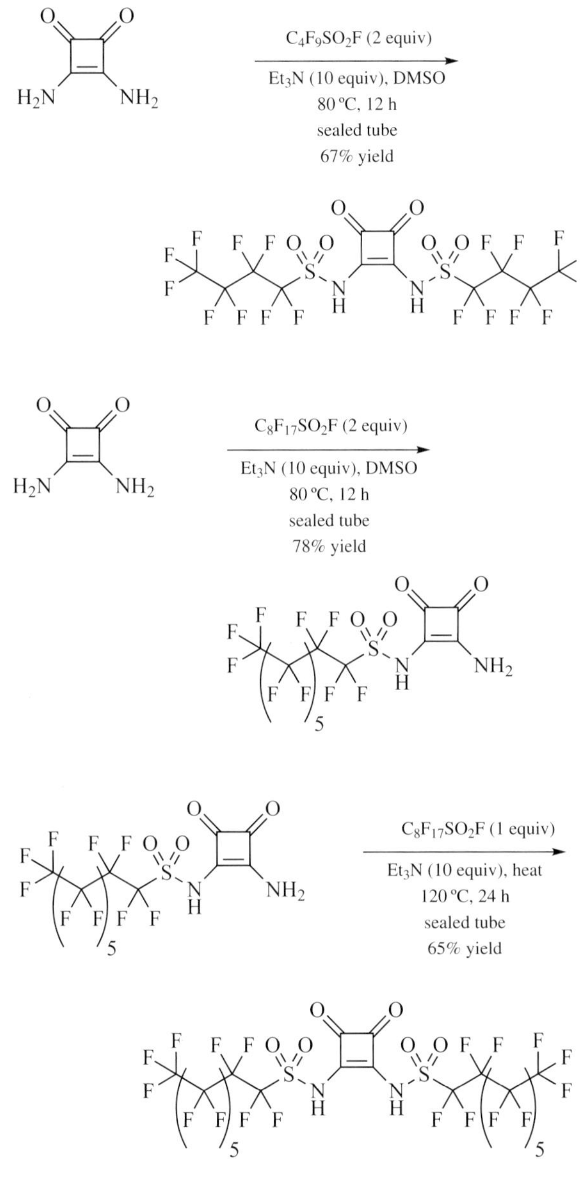

(7)

Related Reagents. Although there have recently been applications of squaramide as a nucleophilic source of the squarate unit for the synthesis of acidic sulfonamide derivatives, to date squaramide has not received much application as either an intermediate or reagent within organic synthesis. This is largely due to the greater availability, precedent, and versatility of several electrophilic squaramide reagents that already enable direct access to the many functionalized squaramide derivatives (eq 8).[1] In particular, reactions with squarate esters and squaric acid dichloride have been developed to offer stepwise, controlled reactions with nucleophiles including amines for the synthesis of functionalized squaramides. Examples of these reagents and their corresponding CAS numbers are highlighted below (LG = leaving group, Nuc = nucleophile) (eq 8).

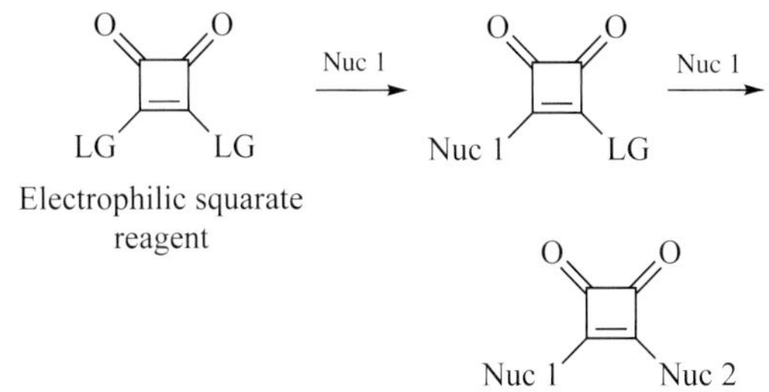

(8)

squaric acid (3,4-dihydroxycyclobut-3-ene-1,2-dione) [2892-51-5]; dimethyl squarate (3,4-dimethoxy-3-cyclobutene-1,2-dione) [5222-73-1]; diethylsquarate (3,4-diethoxycyclobut-3-ene-1,2-dione) [5231-87-8]; isopropyl squarate (3,4-diisopropoxy-3-cyclobutene-1,2-dione) [61699-62-5]; squaric acid dichloride (3,4-dichlorocyclobutene-1,2-dione) [2892-63-9].

---

1. Storer, R. I.; Aciro, C.; Jones Lyn, H., *Chem. Soc. Rev.* **2011**, *40*, 2330.
2. Schmidt, A. H., *Synthesis* **1980**, 961.
3. Mukkanti, A.; Periasamy, M., *ARKIVOC* **2005**, *11*, 48.
4. Cohen, S.; Cohen, S. G., *J. Am. Chem. Soc.* **1966**, *88*, 1533.
5. Schwartz, L. M.; Howard, L. O., *J. Phys. Chem.* **1971**, *75*, 1798.
6. West, R.; Niu, J. *In Chemistry of the Carbonyl Group*; Zabicky, J., Ed.; Wiley–Interscience: New York, **1970**; Vol. **2**, pp 241–275.
7. West, R.; Niu, H. Y., *J. Am. Chem. Soc.* **1963**, *85*, 2580.
8. Schleyer, P. v. R.; Najafian, K.; Kiran, B.; Jiao, H., *J. Org. Chem.* **2000**, *65*, 426.
9. Quinonero, D.; Frontera, A.; Ballester, P.; Deya, P. M., *Tetrahedron Lett.* **2000**, *41*, 2001.
10. Quinonero, D.; Prohens, R.; Garau, C.; Frontera, A.; Ballester, P.; Costa, A.; Deya, P. M., *Chem. Phys. Lett.* **2002**, *351*, 115.
11. Prohens, R.; Tomas, S.; Morey, J.; Deya, P. M.; Ballester, P.; Costa, A., *Tetrahedron Lett.* **1998**, *39*, 1063.
12. Tomas, S.; Prohens, R.; Vega, M.; Rotger, M. C.; Deya, P. M.; Ballester, P.; Costa, A., *J. Org. Chem.* **1996**, *61*, 9394.
13. Quinonero, D.; Frontera, A.; Suner, G. A.; Morey, J.; Costa, A.; Ballester, P.; Deya, P. M., *Chem. Phys. Lett.* **2000**, *326*, 247.
14. Rotger, C.; Soberats, B.; Quinonero, D.; Frontera, A.; Ballester, P.; Benet-Buchholz, J.; Deya, P. M.; Costa, A., *Eur. J. Org. Chem.* **2008**, 1864.
15. Schleyer, P. v. R.; Maerker, C.; Dransfeld, A.; Jiao, H.; van Eikema Hommes, N. J. R., *J. Am. Chem. Soc.* **1996**, *118*, 6317.
16. Rotger, M. C.; Pina, M. N.; Frontera, A.; Martorell, G.; Ballester, P.; Deya, P. M.; Costa, A., *J. Org. Chem.* **2004**, *69*, 2302.
17. Portell, A.; Barbas, R.; Braga, D.; Polito, M.; Puigjaner, C.; Prohens, R., *CrystEngComm* **2009**, *11*, 52.
18. Malerich, J. P.; Hagihara, K.; Rawal, V. H., *J. Am. Chem. Soc.* **2008**, *130*, 14416.

19. Fu, N.; Allen, A. D.; Kobayashi, S.; Tidwell, T. T.; Vukovic, S.; Arumugam, S.; Popik, V. V.; Mishima, M., *J. Org. Chem.* **2007**, *72*, 1951.

20. Chen, Y.; Li, J.; Li, W.; Yu, L., *Sichuan Da xue Xue bao, Ziran Kexueban* **1995**, *32*, 612.

21. Cheon, C. H.; Yamamoto, H., *Tetrahedron* **2010**, *66*, 4257.

22. Cheon, C. H.; Yamamoto, H., *Bull. Korean Chem. Soc.* **2010**, *31*, 539.

R. Ian Storer
Pfizer Neusentis, Cambridge, UK

T

Tetrachlorosilane

$SiCl_4$

[10026-04-7] Cl_4Si (MW 169.89)
InChI = 1S/Cl4Si/c1-5(2,3)4
InChIKey = FDNAPBUWERUEDA-UHFFFAOYSA-N

(precursor of chlorotrialkylsilanes; Lewis acid catalyst)

Physical Data: mp −70 °C; bp 57.6 °C; *d* 1.483 g cm −3.
Solubility: sol benzene, ether, chloroform, petroleum ether, dichloromethane.
Form Supplied in: neat as clear colorless liquid; as 1.0 M solution in dichloromethane.
Handling, Storage, and Precautions: the neat liquid and solutions react violently with water and protic solvents to produce HCl gas. The neat reagent reacts violently and/or explosively with DMSO and DMF. Avoid contact with metals, strong bases, strong acids, and glass. Corrosive; stench; handle and store under N_2; store cool. Use in a fume hood.

Original Commentary

David P. Sebesta
University of Colorado, Boulder, CO, USA

Chlorosilane Precursor. Tetrachlorosilane serves as a useful precursor to tri-, di-, and monosilyl chlorides.[1] Substitution by metal alkoxides and phenoxides,[2] as well as alkyl, vinyl,[3] and aryl organometal nucleophiles, have been reported (eqs 1 and 2). An extremely hindered chlorosilane has been prepared from tri-*t*-butylphenoxide and tetrachlorosilane.[4]

$$2\,ROH + 2\,MeLi \xrightarrow{SiCl_4} (RO)_2SiCl_2 \quad (1)$$

R = alkyl, aryl

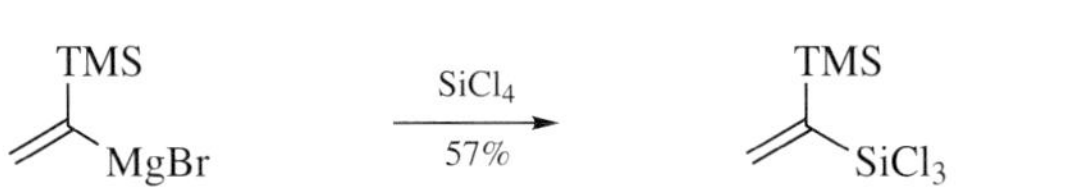
(2)

Treatment of tetrachlorosilane with 4 equiv of a propargyl Grignard reagent afforded, after zirconocene-promoted cyclization and sulfurization, a spiro-fused tetracyclic compound which serves as a precursor to orthogonally fused conducting polymers (eq 3).[5]

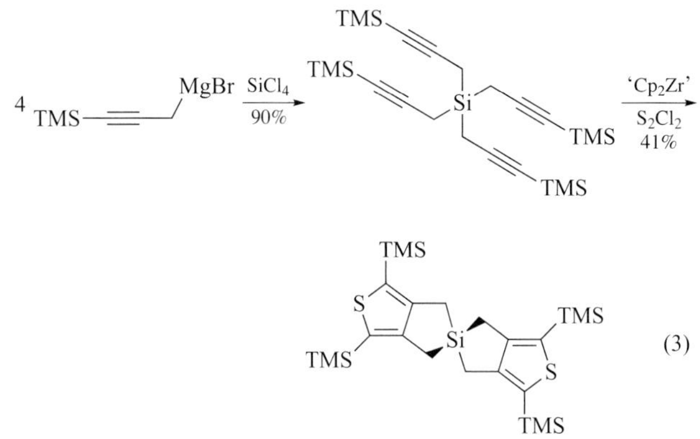

Coupling Reagent for Amide Formation. Tetrachlorosilane in pyridine has been shown to facilitate amide formation from carboxylic acids and alkyl and aryl amines (eq 4).[6]

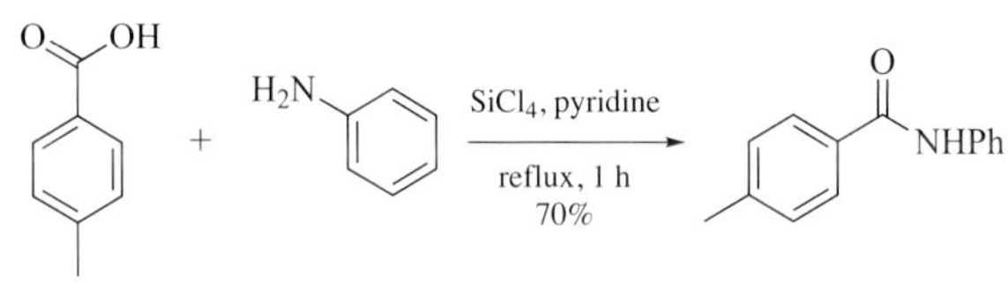
(4)

Acetal and Thioacetal Formation. Tetrachlorosilane has been shown to convert α,β-unsaturated ketones, in the presence of ethylene glycol, to β-chloro acetals (eq 5).[7] Similarly, a propargyl ketone was transformed into the corresponding β,β-dichloro acetal under the same conditions (eq 6).

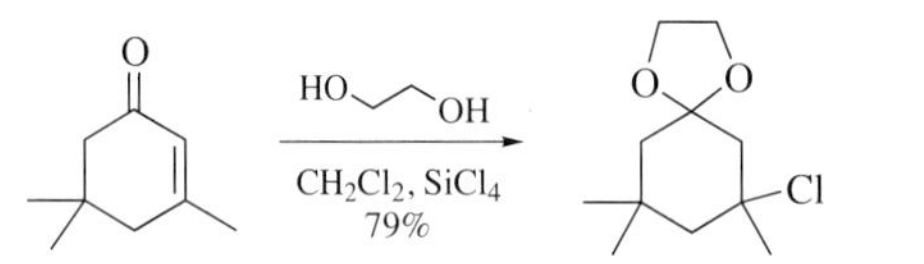
(5)

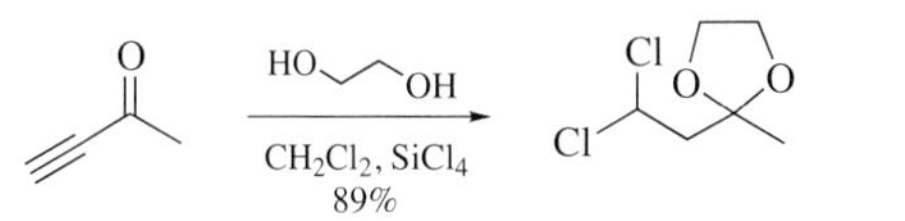
(6)

Dithioacetals of aryl aldehydes are efficiently formed upon treatment with 2 equiv of thiol in the presence of tetrachlorosilane (eq 7).[8] Aryl ketones (e.g. acetophenone or benzophenone) were not converted under the reaction conditions, while treatment of aliphatic ketones with tetrachlorosilane and dimercaptoethane resulted in some vinyl sulfide byproduct, as well as the desired dithioacetal (eq 8).

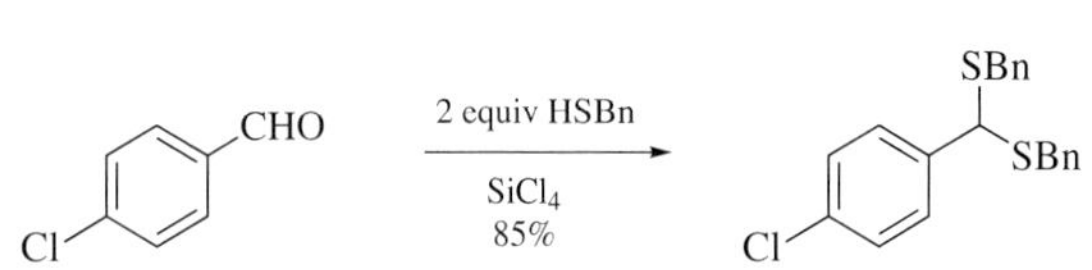
(7)

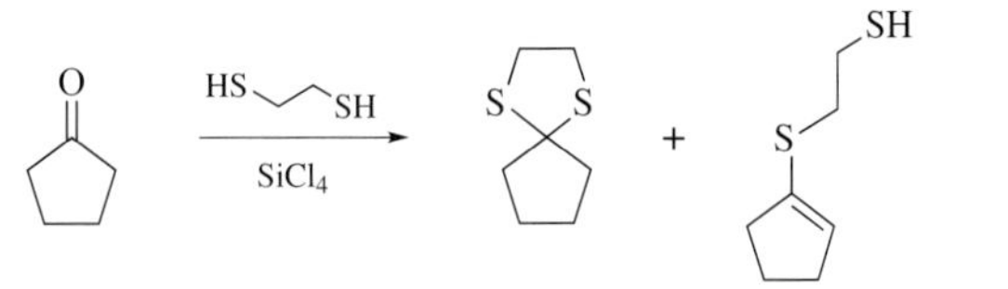

(8)

Cyclization Promoter. Tetrachlorosilane was the Lewis acid promoter of choice for a critical cyclization step towards the synthesis of the alkaloids julandine and ipalbidine (eq 9).[9] Other Lewis acids evaluated were $TiCl_4$, $SnCl_4$, $BF_3{\cdot}Et_2O$, $ZrCl_4$, $AlCl_3$, and $FeCl_3$. Tetrachlorosilane, and several other Lewis acids, were evaluated as promoters for Fischer indole synthesis reactions.[10]

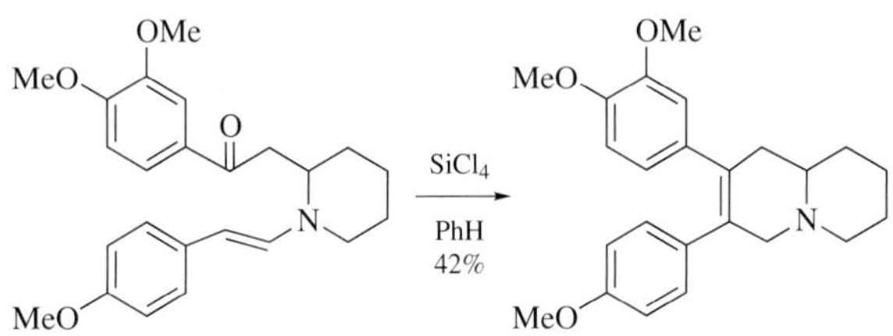

(9)

Disulfide Bond Formation. In conjunction with diphenyl sulfoxide, tetrachlorosilane in trifluoroacetic acid (TFA) has been shown to effectively facilitate disulfide bond formation in the synthesis of cystine-containing peptides.[11,12] As shown in eq 10, the tetrachlorosilane–diphenyl sulfoxide–TFA system serves to activate the protected SH moiety of cysteine residue side-chains in peptides to afford, after nucleophilic attack of a second protected cysteine thiol, a disulfide bridge. A number of S-protecting groups were employed in the study, including acetamidomethyl (Acm), trimethylacetamidomethyl (Tacm), benzamidomethyl (Bam), *t*-butyl, isopropyl, 4-methoxybenzyl, 4-methylbenzyl, and 4-nitrobenzyl. Similarly, methyltrichlorosilane can be used in place of tetrachlorosilane to effect this transformation and, due to its ease of handling relative to tetrachlorosilane, is recommended by the authors.

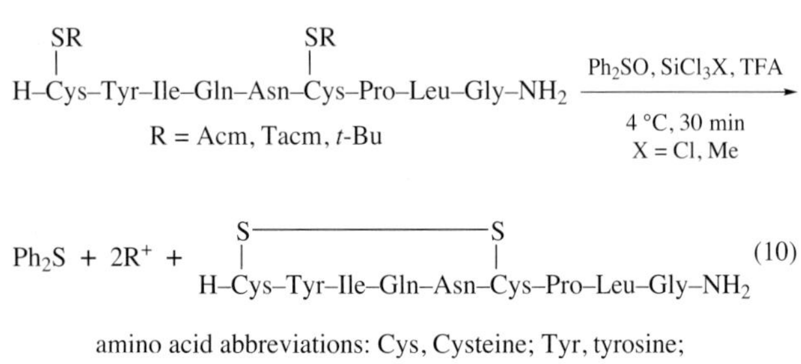

(10)

amino acid abbreviations: Cys, Cysteine; Tyr, tyrosine; Ile, isoleucine; Gln, glutamine; Asn, asparagine; Pro, proline; Leu, leucine; Gly, glycine

The chlorosilane–diphenyl sulfoxide oxidation system is operationally superior to both air and iodine, which are also used to facilitate peptide disulfide bond formation, due to shorter reaction times and improved regio- and chemoselectivity.

Friedel–Crafts Acylation Catalyst. Tetrachlorosilane–***Silver(I) Perchlorate*** has been employed as a catalyst to effect the Friedel–Crafts acylation of anisole by hexanoic acid anhydride (eq 11).[13] Other Lewis acids used in combination with $AgClO_4$ which were evaluated in the study and shown to be as effective or better than $SiCl_4$ were $GaCl_4$, $GeCl_4$, $SnCl_4$, BCl_3, and $AlCl_3$. Lewis acids which were not as effective as $SiCl_4$ for this application were $InCl_3$, $SbCl_5$, $TiCl_4$, $ZrCl_4$, and $HfCl_4$.

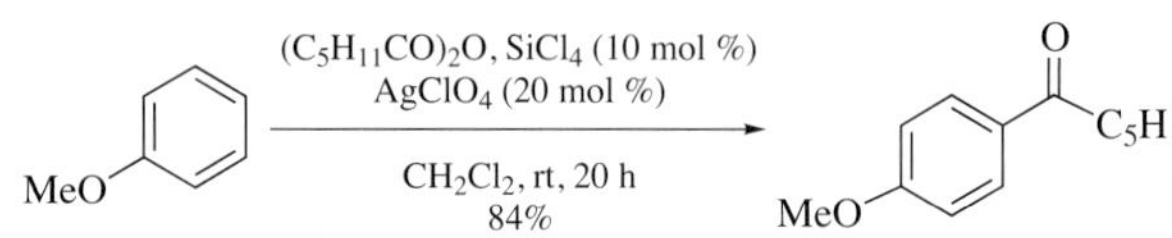

(11)

First Update

D. David Hennings
Array BioPharma, Boulder, CO, USA

Introduction. This update is arranged by reagent use and/or reaction type: $SiCl_4$ used as a silane precursor, as a reagent for amide coupling, tetrazole formation, acetal formation, cyclization, chlorohydrin formation from epoxides, and aminonitrile synthesis, followed by chlorination, reduction, and deprotection.

Silane Precursor. Tetrachlorosilane can be used with various nucleophiles to prepare substituted tri-, di-, and monosilyl chlorides as well as aryl-, vinyl-, alkynyl-, alkyl-, amino-, and alkoxysilanes. Dichlorosilacyclopentadienes (siloles) have been prepared by addition of dilithiobutadienes to tetrachlorosilane (eq 12). [14] Silicon containing polymers derived from siloles have been identified as promising candidates for use in electroluminescent devices because they are highly luminescent in their aggregation state. Siloles have also been useful in studying the chemistry of silole dianions[15] and organosilicon diradical dianions.[16] Spiro-fused structural motifs have also been prepared by addition of 4 equiv of aryl lithium or Grignard species to tetrachlorosilane.[17,18] Ferrocenyllithium species have also been utilized to prepare tetra-arylated silanes that potentially could possess nonlinear optical properties.[19] Two examples of interest owing to their unusual physical and/or optical properties are shown below.[17,19] Both were prepared from tetrachlorosilane and the appropriate organometallic nucleophiles.

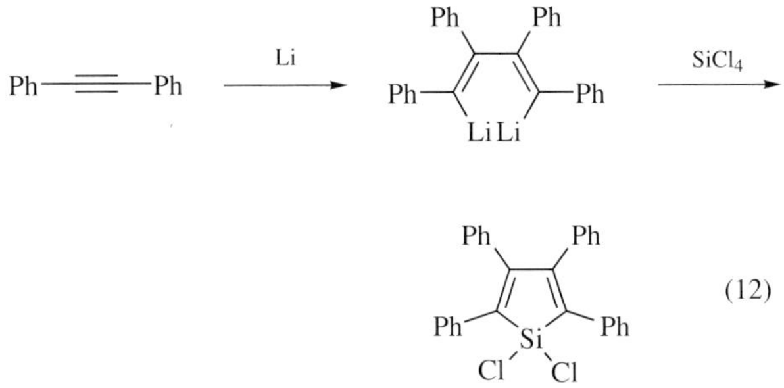

(12)

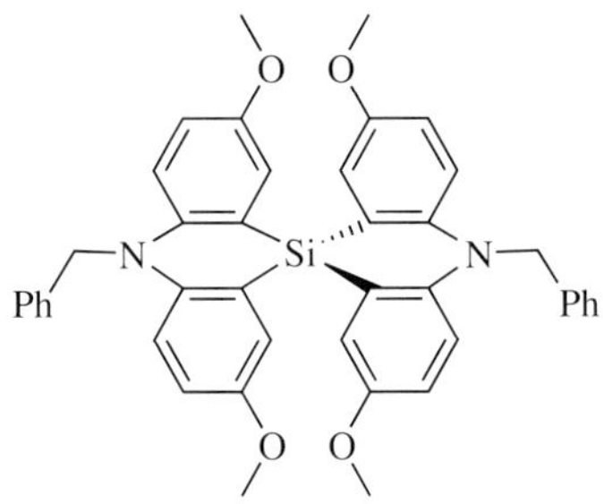

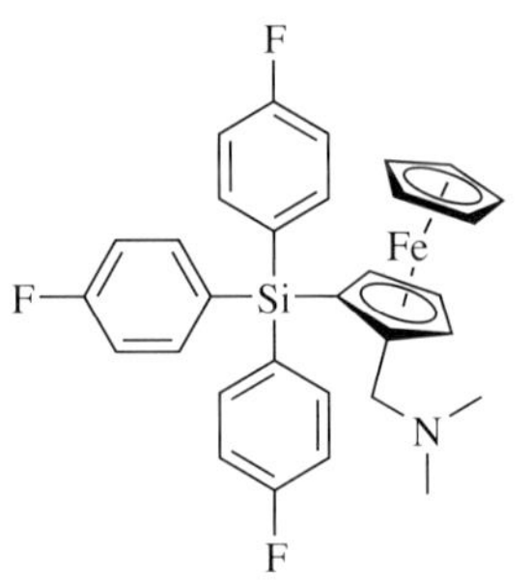

Amines and alcohols react readily with tetrachlorosilane, and depending on the stoichiometry, can give chlorosilanes or fully substituted amino or alkoxysilanes. Tetrachlorosilane has been used to tether two molecules of binaphthol together during radical copolymerization with methyl methacrylate. The latter reagent was required for the preparation of polymer-bound binaphthol.[20] Tetrachlorosilane has also been used in combinatorial chemistry to prepare the hydrochloride salts of amino acid tri-*tert*-butoxysilyl ester building blocks directly from the amino acid zwitterions (eq 13).[21]

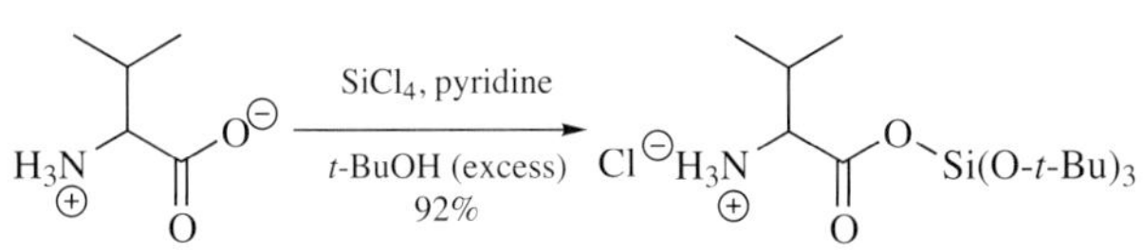

(13)

Trichlorosilyl enolates have proven to be synthetically useful substrates. Using catalytic amounts of a chiral Lewis base, it is possible to achieve highly diastereo- and enantioselective aldol reactions by the addition of geometrically defined trichlorosilyl enolates to aldehydes,[22a] esters,[22b] and ketones.[22c–e] The base that has been exploited in these aldol reactions is a chiral phosphoramide. There are three methods that have been used to generate trichlorosilyl enolates. One method uses α-chloro ketones with trichlorosilane and a trialkylamine.[23] The other two methods use the corresponding trimethylsilyl (TMS) enol ether as the starting material. Cleavage of the TMS enol ether using MeLi occurs with retention of the double bond geometry.[22a,24] Trapping the resulting enolate with tetrachlorosilane gives the trichlorosilyl enolate (eq 14). The second method utilizes metathesis with either a stannyl enolate[22b] or a catalytic mercury-catalyzed metathesis reaction to generate the trichlorosilyl enolate from tetrachlorosilane (eq 15).[22c]

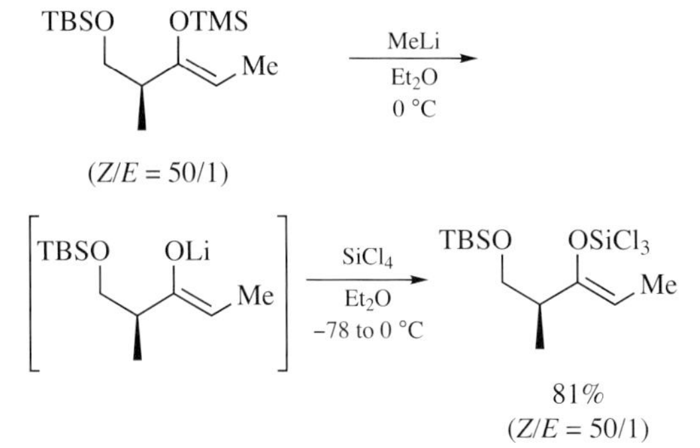

(14)

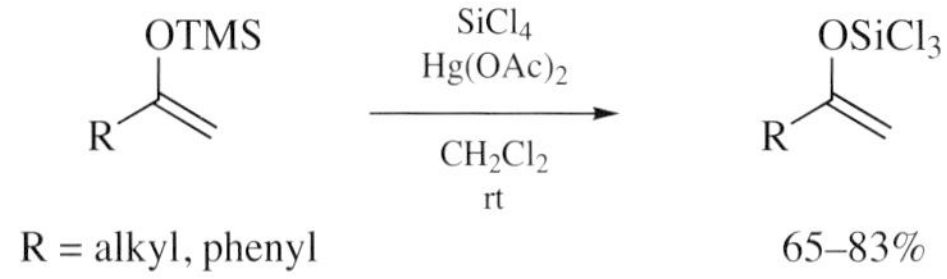

(15)

Coupling Reagent for Amide Formation. Tetrachlorosilane was used to promote a novel cyclodehydration.[25] The initial step involved intramolecular amide formation followed by activation of a tertiary amide to dehydration (eq 16).

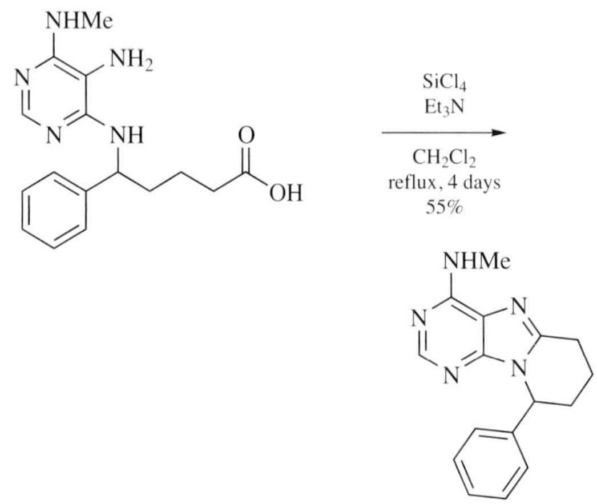

(16)

Tetrazole Formation. Amides can be converted to tetrazoles using tetrachlorosilane with sodium azide (eq 17).[26] This method has proven useful in several medicinal chemistry programs, including the preparation of Endothelin-A receptor antagonists[27a] and Histamine H_3 receptor antagonists.[27b]

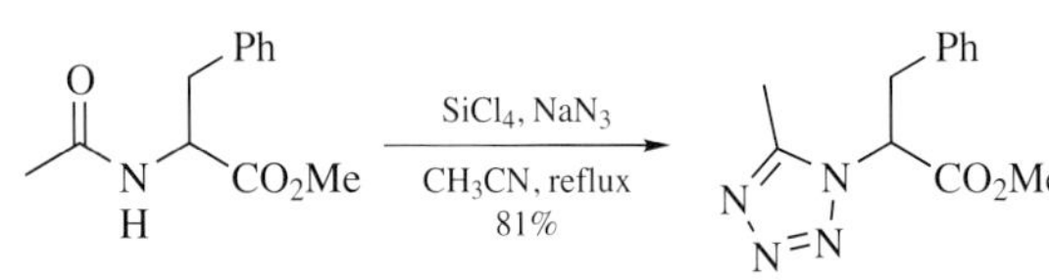

(17)

Acetal Formation. Protection of a tetracyclic dione was accomplished using tetrachlorosilane in the presence of ethanol to provide a caged polycylic (eq 18).[28] Interestingly, the dione was liberated using tetrachlorosilane in combination with sodium iodide.

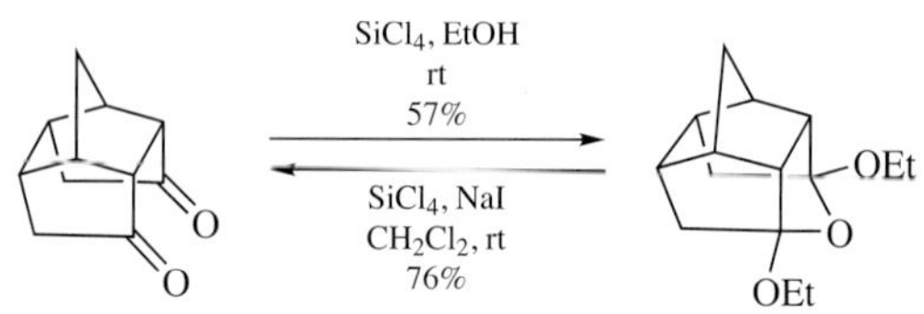

(18)

Cyclization Promoter. The cyclotrimerization of aryl methyl ketones can be accomplished using tetrachlorosilane in refluxing ethanol to give 1,3,5-triarylbenzenes (eq 19).[29] Using cyclohexanone or cyclopentanone under the same conditions gives triannulated benzenes.

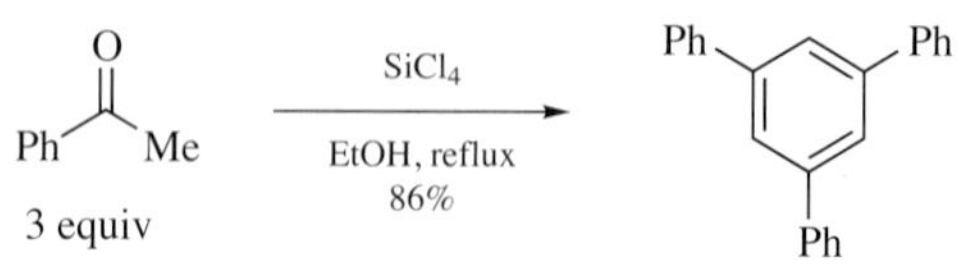

(19)

Chlorohydrin Preparation from Epoxides. Chlorohydrins can be prepared from epoxides using tetrachlorosilane in combination with Lewis bases such as phosphoramides[30] (eq 20) or pyridine *N*-oxides.[31] The use of a chiral *N*-oxide catalyst has proven useful in the enantioselective ring opening of *meso*-epoxides.[31a, 31b]

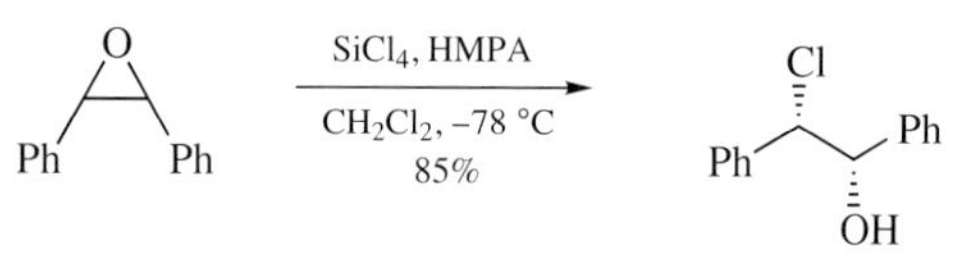

(20)

Aminonitrile Synthesis. Using tetrachlorosilane with potassium cyanide is useful in activating imines and oximes to addition of cyanide.[32] This provides an alternative to using hydrogen cyanide to generate α-aminonitriles (eqs 21 and 22).

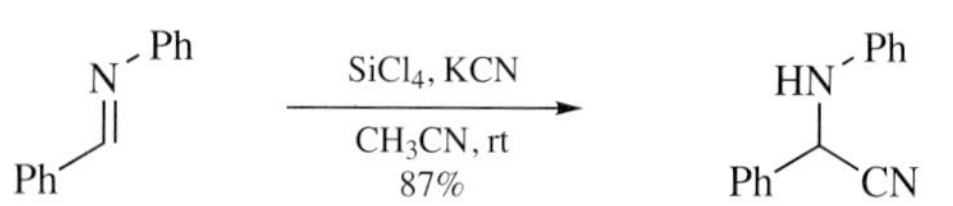

(21)

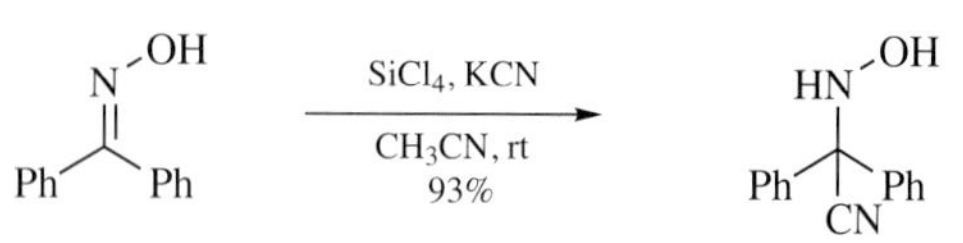

(22)

Chlorination. Tetrachlorosilane can be used to prepare organodichloroboranes directly from organotrifluoroborates (eq 23).[33] The latter can be readily accessed from boronic esters. Formation of the dichloro species is accompanied by evolution of gaseous tetrafluorosilane.

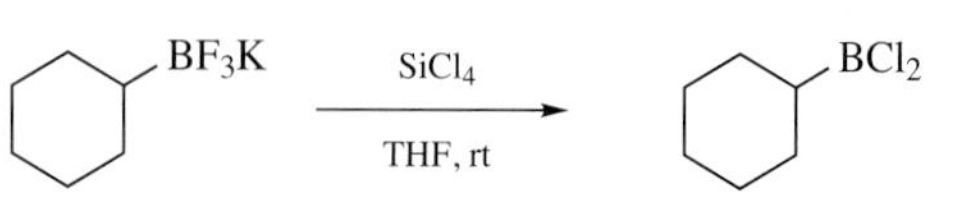

(23)

Deprotection and Reduction. Tetrachlorosilane has been used to deprotect various groups, typically in conjunction with other reagents or catalysts. Oximes, hydrazones, and semicarbazones can be cleaved to liberate the corresponding carbonyl compounds using tetrachlorosilane with wet silica gel.[34] Iodotrichlorosilane (ITCS) can be prepared from sodium or potassium iodide and tetrachlorosilane. ITCS was used to deprotect the benzyloxycarbonyl (*Z*) group in the solid phase synthesis of dermorphin and oxytocin.[35] Chemoselective reduction of α,β-unsaturated ketones and nitriles has been accomplished using ITCS generated from tetrachlorosilane (eq 24).[36]

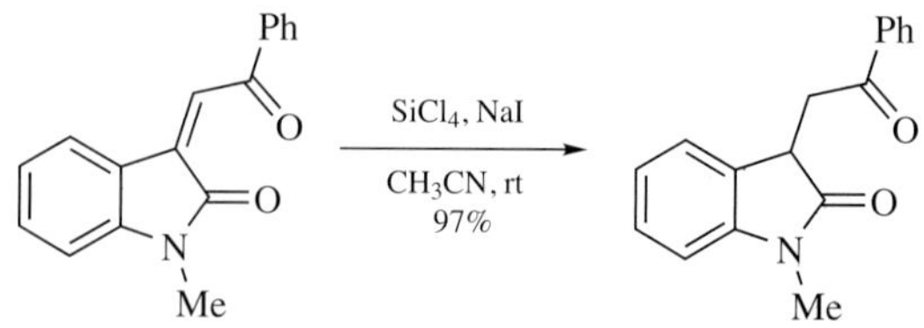

(24)

A combination of tetrachlorosilane and lithium chloride with catalytic BF_3 was utilized for the *O*-debenzylation of an estrogen receptor (eq 25).[37] This system provided an economic advantage and a cleaner reaction profile over trimethylsilyl iodide.

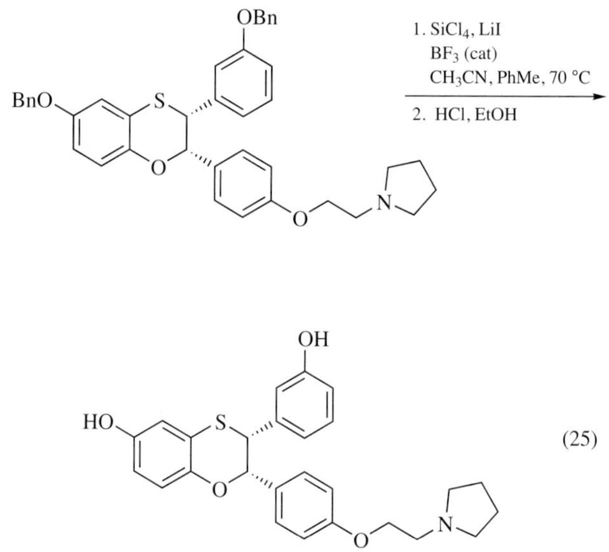

(25)

Defluoridation of organotrifluoroborates in the presence of organic azides using tetrachlorosilane provides secondary amines.[38]

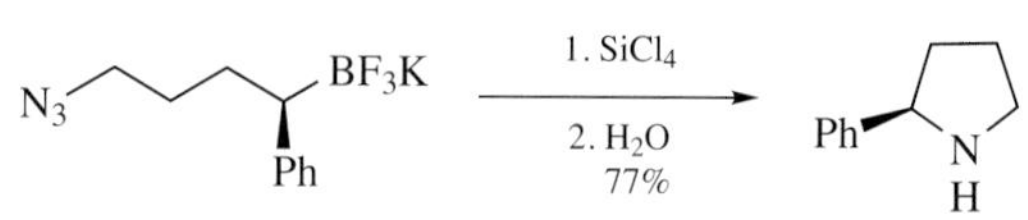

(26)

Second Update

Dr Thomas Wynn
Biogen Idec, Inc., Cambridge, MA, USA

Introduction. This update follows the original organization for reagent use and/or reaction type for $SiCl_4$. The major expansion of this update covers the use of $SiCl_4$ in Lewis base-activated Lewis acid-promoted reactions.

Cyclization Promoter. Tetrachlorosilane has seen some success promoting Biginelli's dihydropyrimidone three- component cyclization reaction. Using a catalytic amount of tetrachlorosilane, a variety of aromatic aldehydes can be coupled with urea and a variety of β-ketoesters (eq 27).[39]

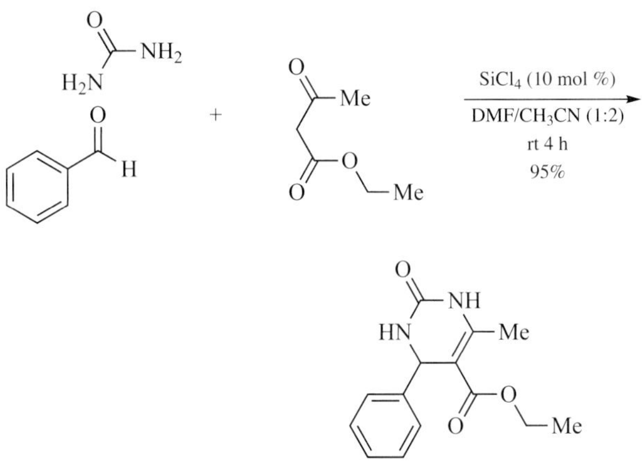

(27)

Lewis Base Activation of Lewis Acids. The Denmark lab has over the past 15 years championed a variant of organocatalysis where Lewis bases are used to greatly increase the Lewis acidity of weak Lewis acids. To date, the weak Lewis acid that has seen the greatest success in these studies is tetrachlorosilane.

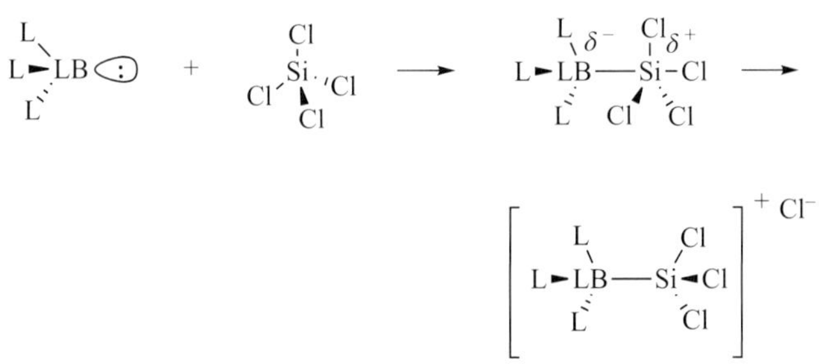

The principle uncovered in these studies was that strong Lewis bases can cause the redistribution of electrons between the acid/base pair wherein a partial negative charge forms on the central atom of the Lewis base and a partial positive charge forms on the silicon atom. When the Lewis base is of sufficient strength, this polarization can cause one of the chloride atoms to dissociate from the silicon, thus forming a full formal positive charge on the complex. This cationic Lewis base–$SiCl_3$ complex has been found to be quite competent in promoting a variety of Lewis acid-catalyzed reactions. One benefit of generating a strong Lewis acid in this manner is that if the Lewis base is chiral, the cationic LB/$SiCl_3$ complex will also be chiral, thus, a strong chiral Lewis acid is formed. Another benefit of this method is that the weakly Lewis acidic $SiCl_4$ can be used in stoichiometric amounts without reacting with the substrate and only when the chiral Lewis base is present, the catalytically competent species is generated. This method eliminates achiral background reactions often observed with in situ generated Lewis acid catalysts. This process has been applied to several different reaction classes with several different Lewis base promoters.[40]

Chlorohydrin Preparation from Epoxides. Several catalyst systems have been developed that can smoothly open *meso* epoxides to the corresponding chlorohydrins. The initial studies used chiral phosphoramides[41] and later chiral *N*-oxides,[42] and chiral phosphines[43] were also shown to be competent (eq 28).

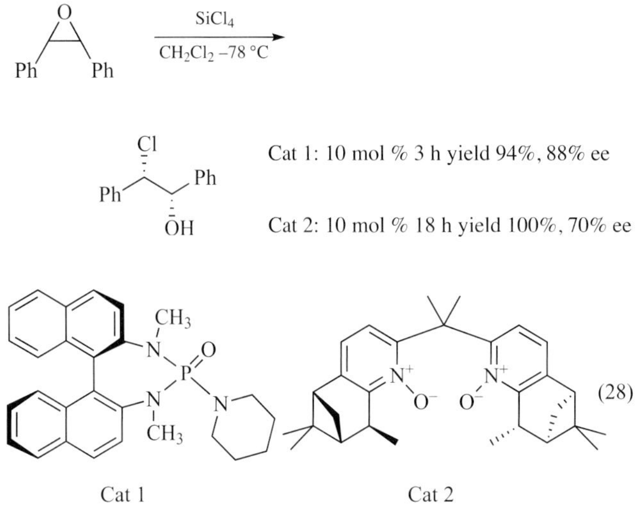

(28)

Chiral Homoallylic Alcohols from Aldehydes and Allyltributylstannane. The synthetic utility of Lewis base activation of $SiCl_4$ was further expanded with the discovery that chiral bisphosphoramides can catalyze the allylation of aromatic aldehydes by allyltributylstannane in the presence of a stoichiometric amount of $SiCl_4$ (eq 29).[44] Both the high degree of stereoinduction in the reaction and the observation that $SiCl_4$ by itself does not promote the allylation lead credence to the idea that the Lewis acid is generated by the interaction of the Lewis base (phosphoramide) with the $SiCl_4$. The facility of the reaction depends somewhat on the strength of the nucleophile where allyltributylstannanes react at reasonable rates at low temperatures, but allyltrimethyl silane does not. This finding has led to investigation of a variety of other nucleophiles in these reactions.

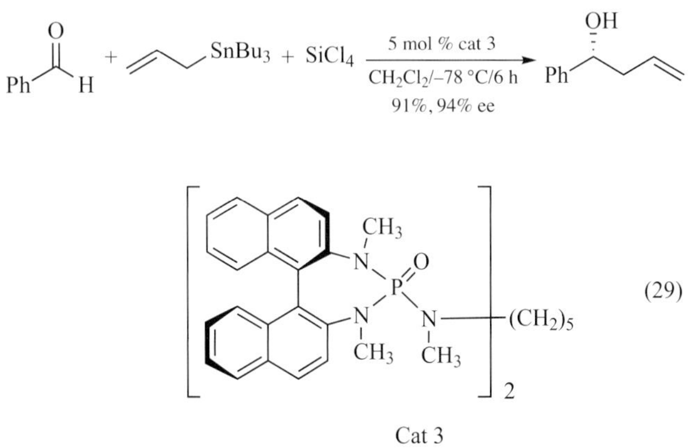

(29)

Aldol Reaction of Silyl Ketene Acetals and Aldehydes. The true utility of Lewis base activation of $SiCl_4$ was seen when these studies were extended to the addition of silyl ketene acetals to aldehydes (eq 30).[45] The transformation also occurs with vinylogous aldols (eq 31)[46] and using isocyanates[47] (eq 32), enol ethers,[48] *O,N* silyl ketene acetals,[49] silyl ketene imines,[50] and malonates[51] as nucleophiles.

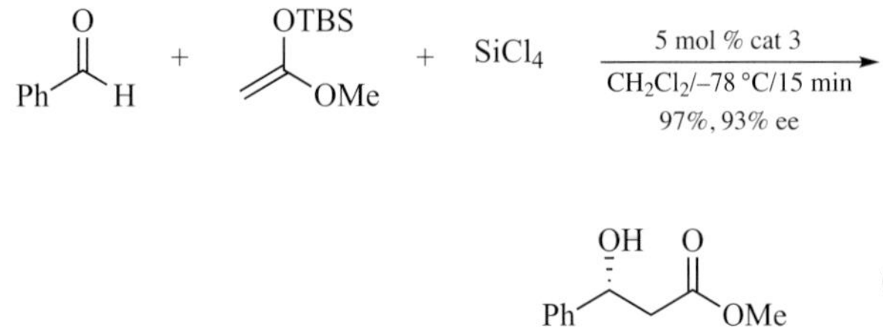

(30)

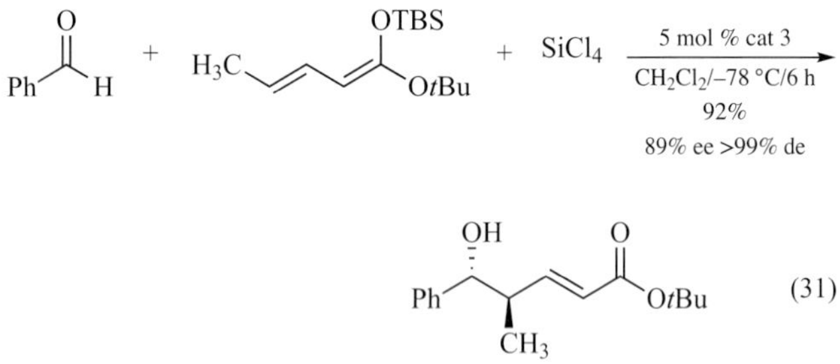

(31)

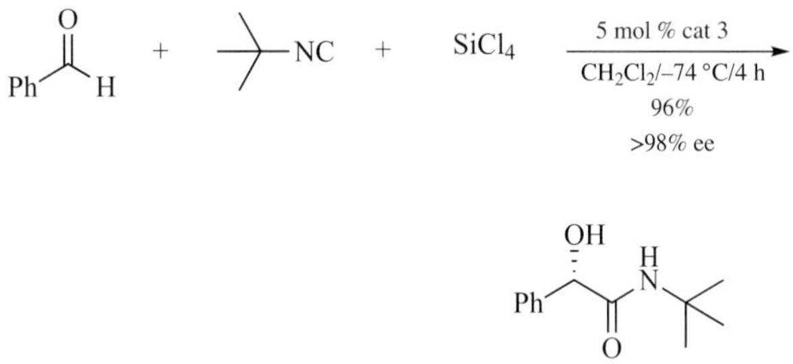

(32)

These reactions have shown the best results using aromatic aldehydes, but there has been some success with aliphatic aldehydes.[49a]

β-Amino Ketone Synthesis Using $SiCl_4/ZnCl_2$. Tetrachlorosilane has also been used in conjunction with $ZnCl_2$ to affect the three-component, one-pot coupling to generate functionalized β-amino ketones in reasonable yields (eq 33).[52] In these reactions, a large excess of $SiCl_4$ is required.

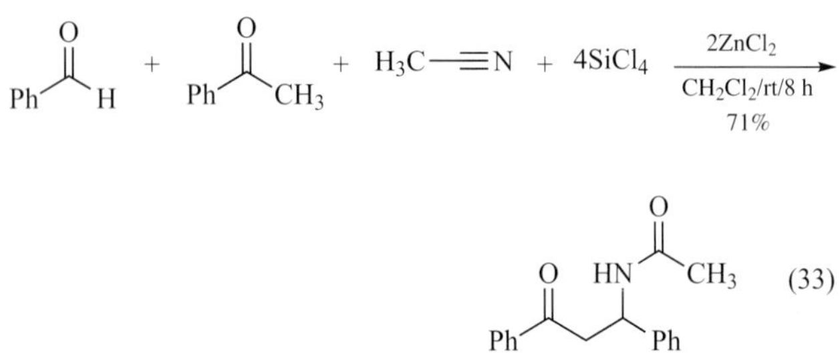

(33)

Abramov Phosphonylations. The use of chiral Lewis bases to activate $SiCl_4$ has also been extended to non carbon nucleophiles, notably addition of trialkyl phosphites to aromatic aldehydes (eq 34).[53] These reactions proceed in moderate to good yields and with moderate enantioselectivities with a series of chiral phosphine oxides.

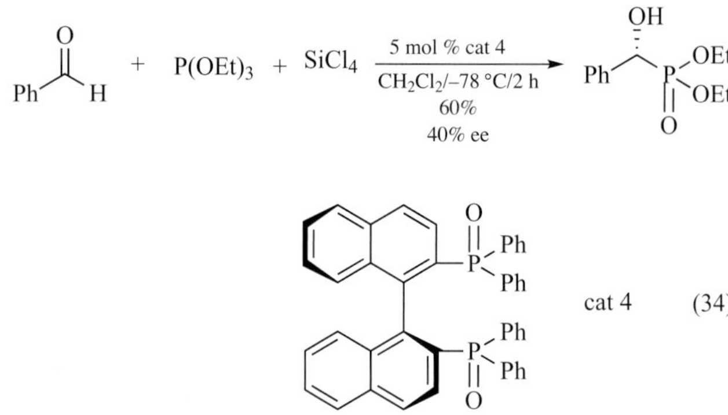

cat 4 (34)

Cyanohydrin Formation From Ketones. Tetrachlorosilane has also been used to promote the addition of potassium cyanide to methyl ketones to form cyanohydrins (eq 35).[54] Of note is that these reactions do not require any other activators in order to cause the transformation.

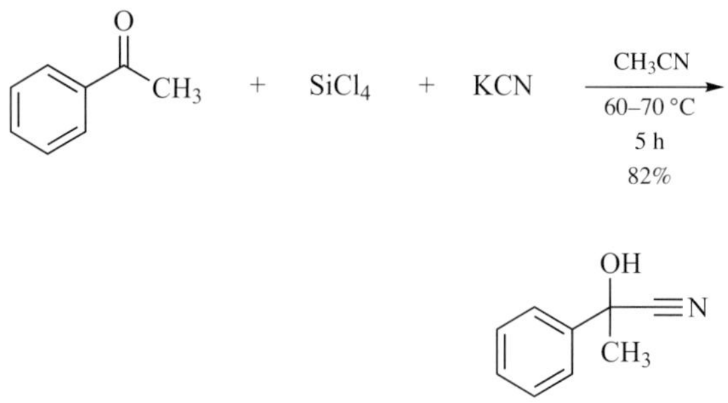

(35)

Aza-Diels–Alder Reactions. Despite being a weak Lewis Acid tetrachlorosilane has been shown to act as a Lewis acid catalyst in aza-Diels–Alder reactions using Danishefsky's diene (eq 36).[55]

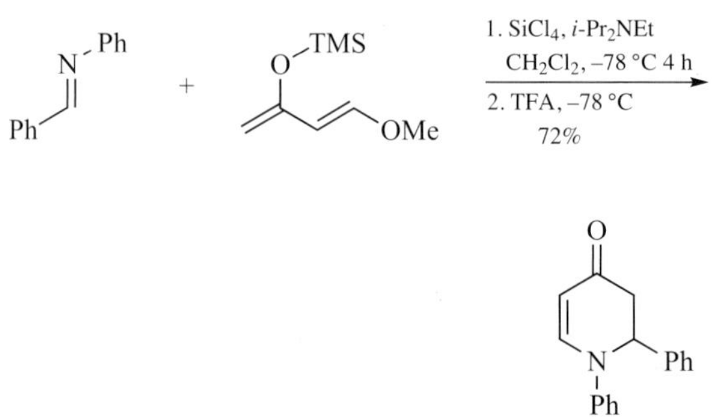

(36)

The authors were also able to generate the aromatic imine in situ and couple more sterically hindered aldehydes when a Lewis base was added to the reaction to activate the $SiCl_4$ (eq 37). These reactions afforded good yields over the three steps and provided moderate *cis/trans* selectivities.

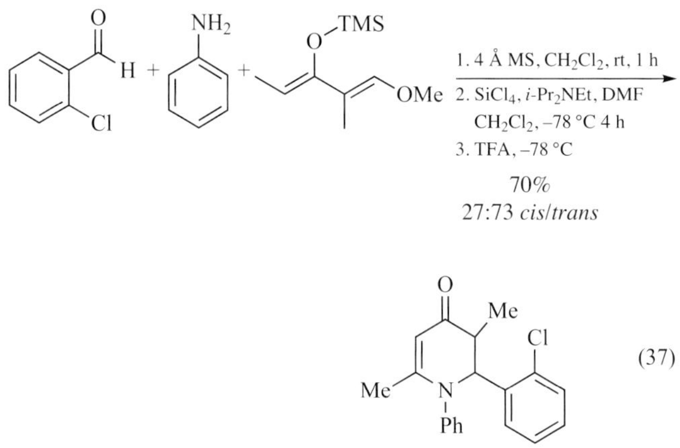

(37)

Related Reagents. Iodotrichlorosilane; trimethylsilyl iodide.

1. (a) *Silicon Chemistry*; Corey, J. Y.; Corey, E. R.; Gaspar, P. P., Eds.; Horwood: Chichester, **1987**. (b) *Organosilicon Compounds*; Eaborn, C., Ed.; Butterworths: London, **1960**.
2. Leis, C.; Wilkinson, D. L.; Handwerker, H.; Zybill, C.; Müller, G., *Organometallics* **1992**, *11*, 514.
3. Ziche, W.; Auner, N.; Behm, J., *Organometallics* **1992**, *11*, 2494.
4. Schafer, A.; Weidenbruch, M.; Pohl, S.; Saak, W., *Z. Naturforsch., Teil B* **1990**, *45b*, 1363.
5. Tour, J. M.; Wu, R.; Schumm, J. S., *J. Am. Chem. Soc.* **1990**, *112*, 5662.
6. Chan, T. H.; Wong, L. T. L., *J. Org. Chem.* **1969**, *34*, 2766.
7. Gil, G., *Tetrahedron Lett.* **1984**, 3805.
8. Oh, D. Y.; Ku, B., *Synth. Commun.* **1989**, 433.
9. Cragg, J. E.; Hedges, S. H.; Herbert, R. B., *Tetrahedron Lett.* **1981**, 2127.
10. Prochazka, M. P.; Carlson, R., *Acta Chem. Scand.* **1989**, *43*, 651.
11. Akaji, K.; Tatsumi, T.; Yosida, M.; Kimura, T.; Fujiwara, Y.; Kiso, Y., *J. Am. Chem. Soc.* **1992**, *114*, 4137.
12. Akaji, K.; Tatsumi, T.; Yosida, M.; Kimura, T.; Fujiwara, Y.; Kiso, Y., *J. Chem. Soc., Chem. Commun.* **1991**, 167.
13. Harada, T.; Ohno, T.; Kobayashi, S.; Mukaiyama, T., *Synthesis* **1991**, 1216.
14. Chen, J.; Law, C. C. W.; Lam, J. W. Y.; Dong, Y.; Lo, S. M. F.; Williams, I. D.; Zhu, D.; Tang, B. Z., *Chem. Mater.* **2003**, *15*, 1535.
15. Choi, S. B.; Boudjouk, P.; Wei, P., *J. Am. Chem. Soc.* **1998**, *120*, 5814.
16. Liu, Y.; Stringfellow, T. C.; Ballweg, D.; Guzei, I. A.; West, R., *J. Am. Chem. Soc.* **1998**, *120*, 5814.
17. Ito, A.; Urabe, M.; Tanaka, K., *Angew. Chem. Int. Ed.* **2003**, *42* 921.
18. Russell, A. G.; Spencer, N. S.; Philp, D.; Kariuki, B. M.; Snaith, J. S., *Organometallics* **2003**, *22*, 5589.
19. Beyer, C.; Bohme, U.; Pietzsch, C.; Roewer, G., *J. Organomet. Chem.* **2002**, *654*, 187.
20. Arai, T.; Sekiguti, T.; Otsuki, K.; Takizawa, S.; Sasai, H., *Angew. Chem. Int. Ed.* **2003**, *42*, 2144.
21. Johansson, A.; Akerblom, E.; Ersmark, K.; Lindeberg, G.; Hallberg, A., *J. Comb. Chem.* **2000**, *2*, 496.
22. (a) Denmark, S. E.; Ghosh, S. K., *Angew. Chem. Int. Ed.* **2001**, *40*, 4759. (b) Denmark, S. E.; Winter, S. B. D.; Su, X.; Wong, K. T., *J. Am. Chem. Soc.* **1996**, *118*, 7404. (c) Denmark, S. E.; Stavenger, R. A.; Wong, K. T., *J. Org. Chem.* **1998**, *63*, 918. (d) Denmark, S. E.; Stavenger, R. A., *J. Org. Chem.* **1998**, *63*, 9524. (e) Denmark, S. E.; Stavenger, R. A., *J. Am. Chem. Soc.* **2000**, *122*, 8837.
23. Benkeser, R. A.; Smith, W. E., *J. Am. Chem. Soc.* **1968**, *90*, 5307.
24. Denmark, S. E.; Fujimori, S., *Org. Lett.* **2002**, *4*, 3473.
25. Desaubry, L.; Wermuth, C. G.; Bourguignon, J. J., *Tetrahedron Lett.* **1995**, *36*, 4249.
26. Esikov, K. A.; Morozova, S. E.; Malin, A. A.; Ostrovskii, V. A., *Russian Journal of Organic Chemistry* **2002**, *38*, 1370.
27. (a) Ishizuka, N.; Matsumura, K.; Sakai, K.; Fujimoto, M.; Mihara, S.; Yamamori, T., *J. Med. Chem.* **2002**, *45*, 2041. (b) Grabmann, S.; Sadek, B.; Ligneau, X.; Elz, S.; Ganellin, C. R.; Arrang, J. M.; Schwartz, J. C.; Stark, H.; Schunack, W., *Eur. J. Med. Chem.* **2002**, *15*, 367.
28. Kotha, S.; Chakraborty, K., *Indian, J., Chem.* **2000**, *39B*, 382.
29. Elmorsy, S. S.; Pelter, A.; Smith, K., *Tetrahedron Lett.* **1991**, *32*, 4175.
30. Denmark, S. E.; Barsanti, P. A.; Wong, K. T.; Stavenger, R. A., *J. Org. Chem.* **1998**, *63*, 2428.
31. (a) Tao, B.; Lo, M. M. C.; Fu, G. C., *J. Am. Chem. Soc.* **2001**, *123*, 353. (b) Nakajima, M.; Saito, M.; Uemura, M.; Hashimoto, S., *Tetrahedron Lett.* **2002**, *43*, 8827.
32. El-Ahl, A. A. S., *Synth. Commun.* **2003**, *33*, 989.
33. Kim, B. J.; Matteson, D. S., *Angew. Chem. Int. Ed.* **2004**, *43*, 3056.
34. Shirini, F.; Zolfigol, M. A.; Khaleghi, M.; Mohammadpour-Baltork, I., *Synth. Commun.* **2003**, *33*, 1839.
35. Sivanandaiah, K. M.; Babu, V. V. S.; Renukeshwar, H. C.; Gangadhar, B. P., *Indian, J., Chem.* **1995**, *34B*, 487.
36. Elmorsy, S. S.; El-Ahl, A. A. S.; Soliman, H.; Amer, F. A., *Tetrahedron Lett.* **1996**, *37*, 2297.
37. Zewge, D.; King, A.; Weissman, S.; Tschaen, D., *Tetrahedron Lett.* **2004**, *45*, 3729.
38. Matteson, D. S.; Kim, G. Y., *Org. Lett.* **2002**, *4*, 2153.
39. Ramalingan, C.; Kwak, Y.-W., *Tetrahedron* **2008**, *64*, 5023.
40. A review on the use of Lewis base activation of weak Lewis acids: Denmark, S. E.; Beutner, G. L., *Angew. Chem.* International Edition **2008**, *47*, 1560.
41. (a) Denmark, S. E.; Barsanti, P. A.; Wong, K.-T.; Stavenger, R. A., *J. Org. Chem.* **1998**, *63*, 2428. (b) Mechanistic studies: Denmark, S. E.; Barsanti, P. A.; Beutner, G. L.; Wilson, T. W., *Adv. Synth. Cat.* **2007**, *349*, 567.
42. (a) Chelucci, G.; Baldino, S.; Pinna, G. A.; Benaglia, M.; Buffa, L.; Guizzetti, S., *Tetrahedron*, **2008**, *64*, 7574. (b) Malkov, A. V.; Gordon, M. R.; Stoncius, S.; Hussain, J.; Kocovsky, P., *Org. Lett.* **2009**, *11*, 5390.
43. Tokuoka, E.; Kotani, S.; Matsunaga, H.; Ishizuka, T.; Hashimoto, S.; Nakajima, M., *Tetrahedron: Asymmetry* **2005**, *16*, 2391.
44. Denmark, S. E.; Wynn, T., *J. Am. Chem. Soc.* **2001**, *123*, 6199.
45. (a) Denmark, S. E.; Beutner, G. L.; Wynn, T.; Eastgate, M. D., *J. Am. Chem. Soc.* **2005**, *127*, 3774. (b) Mechanistic studies: Denmark, S. E.; Eklov, B. M.; Yao, P. J.; Eastgate, M. D., *J. Am. Chem. Soc.* **2009**, *131*, 11770.
46. (a) Denmark, S. E.; Beutner, G. L., *J. Am. Chem. Soc.* **2003**, *125*, 7800. (b) Acocella, M. R.; De Rosa, M.; Massa, A.; Palombi, L.; Villano, R.; Scettri, A., *Tetrahedron* **2005**, *61*, 4091. (c) Villano, R.; Acocella, M. R.; Massa, A.; Palombi, L.; Scettri, A., *Tetrahedron: Asymmetry* **2006**, *17*, 3332. (c) Shimoda, Y.; Tando, T.; Kotani, S.; Sugiura, M.; Nakajima, M., *Tetrahedron: Asymmetry* **2009**, *20*, 1369. (d) Denmark, S. E.; Bui, T., *J. Org. Chem* **2005**, *70*, 10190.
47. Denmark, S. E.; Fan, Yu., *J. Am. Chem. Soc.* **2003**, *125*, 7825.
48. Denmark, S. E.; Heemstra, J. R., Jr., *Synlett* **2004**, 2411.
49. (a) Denmark, S. E.; Heemstra, J. R., Jr., *J. Am. Chem. Soc.* **2006**, *128*, 1038. (b) Curti, C.; Sartori, A.; Battistini, L.; Rassu, G.; Zanardi, F.; Casiraghi, G., *Tetrahedron Lett.* **2009**, *50*, 3428.
50. Denmark, S. E.; Wilson, T. W.; Burk, M. T.; Heemstra, J. R., Jr., *J. Am. Chem. Soc.* **2007**, *129*, 14864.
51. Massa, A.; Scettri, A.; Filosa, R.; Capozzolo, L., *Tetrahedron Lett.* **2009**, *50*, 7318.
52. Salama, T. A.; Elmorsy, S. S.; Khalil, A.-G. M.; Ismail, M. A., *Tetrahedron Lett.* **2007**, *48*, 6199.
53. Nakanishi, K.; Kotani, S.; Sugiura, M.; Nakajima, M., *Tetrahedron* **2008**, *64*, 6415.
54. Salama, T. A.; Elmorsy, S. S.; Khalil, A.-G. M.; Girges, M. M.; El-Ahl, A.-A. S., *Synth. Commun.* **2007**, *37*, 1313.

55. Massa, A.; de Sio, V.; Villano, R.; Acocella, M. R.; Palombi, L.; Sellitto, G.; Peduto, A.; Filosa, R.; De Capraris, P.; Scettri, A., *Synthesis* **2009**, 643.

2,3,5,6-Tetrahydro-2,6-bis(phenylmethyl)-1*H*-imidazo[1,2-*a*] imidazole[1]

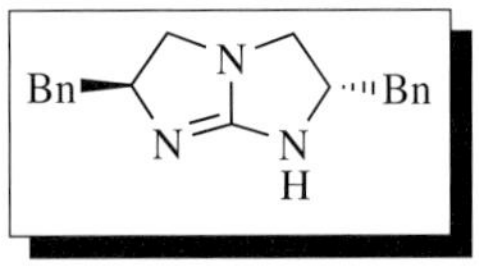

[877773-30-3] $C_{19}H_{21}N_3$ (MW 291.39)

InChI = 1/C19H21N3/c1-3-7-15(8-4-1)11-17-13-22-14-18(21-19(22)20-17)12-16-9-5-2-6-10-16/h1-10,17-18H,11-14H2,(H,20,21)/t17-,18-/m0/s1/f/h20H

InChIKey = WPGKFCUFRDEVPU-CQUJRPSSDE

(highly enantioselective catalyst for the base-catalyzed reactions of anthrones)

Physical Data: (pale yellow oil) $[\alpha]_D^{25} = -22.3$ (*c* 1.0, CHCl3).

Solubility: soluble in organic solvents such as CH_2Cl_2, toluene, ethyl acetate, and THF.

Analysis of Reagent Purity: ^1H NMR in $CDCl_3$, ^{13}C NMR in $CDCl_3$,[2] and IR.[2]

Form Supplied in: not commercially available.

Preparative Methods: prepared from (*S*)-phenylalaninol in five steps (eq 1).[2] This method was also used to prepare several other chiral bicyclic guanidines. The catalyst was prepared as the guanidinium iodide. Before use the catalyst is basified with K_2CO_3: the guanidinium salt was dissolved in CH_2Cl_2 and loaded onto a column packed with solid K_2CO_3 from which the free base was eluted with CH_2Cl_2.

Purification: silica gel chromatography.

Handling, Storage, and Precautions: the reagent is stable as both the free base or the guanidinium salt. Both forms are typically stored under nitrogen at 0°C.

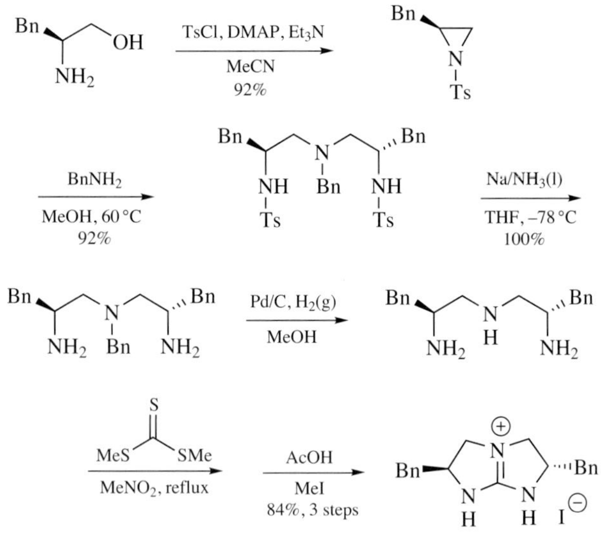

(1)

Diels–Alder Reaction.[1] This chiral guanidine is an excellent catalyst for the Diels–Alder reaction of anthrones and maleimides. The reaction allows the formation of adducts in a highly enantio– and diastereoselective manner (up to 99% ee) under mild reaction conditions (eq 2). Only 10 mol % of catalyst is sufficient for efficient conversion.

Excellent regioselectivities were also observed when 1,5-dichloro-9-anthrone and 4-(*N*-methylamino)-9-anthrone were used as the dienes, leading to the products illustrated below, whose regiochemistry was determined by X-ray crystallographic analysis.

Conjugate Addition.[1,2] The guanidine catalyst has also been employed in conjugate additions of 1,8-dihydroxy-9-anthrone (dithranol) with maleimides (eq 3) and other activated olefins such as methyl *trans*-4-oxo-2-pentenoate and ethyl *trans*-3-benzoylacrylate (eq 4). The reaction of dithranol always leads to the exclusive formation of the Michael adducts. It was found that high enantiomeric excesses and yields can be obtained for the Michael adducts (up to 99% ee) under mild reaction conditions.

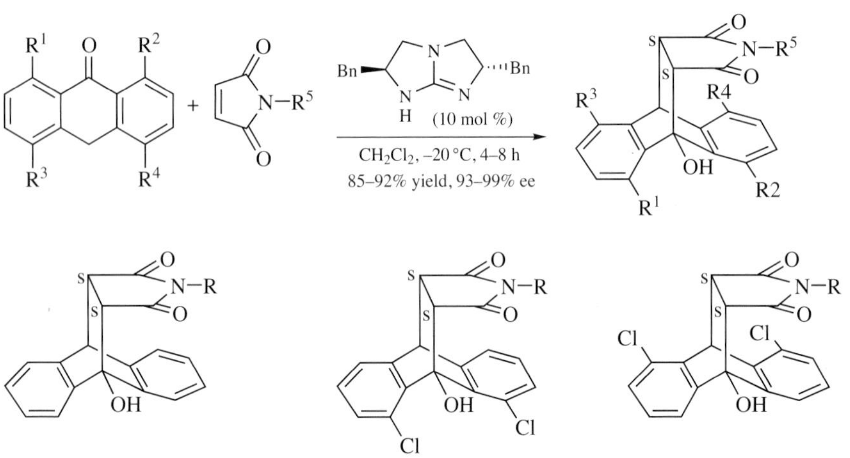

(2)

R = 2-$NO_2C_6H_4$: 87%, 98% ee
R = 2,5-$Cl_2C_6H_3$: 88%, 95% ee

R = CH_2Ph: 92%, 95% ee
R = *c*-C_6H_{11}: 88%, 98% ee
R = CMe_3: 87%, 93% ee
R = 4-$ClC_6H_4CH_2$: 85%, 98% ee

R = Ph: 92%, 99% ee
R = 2,6-$F_2C_6H_3$: 92%, 99% ee
R = 2-$MeOC_6H_4$: 90%, 98% ee

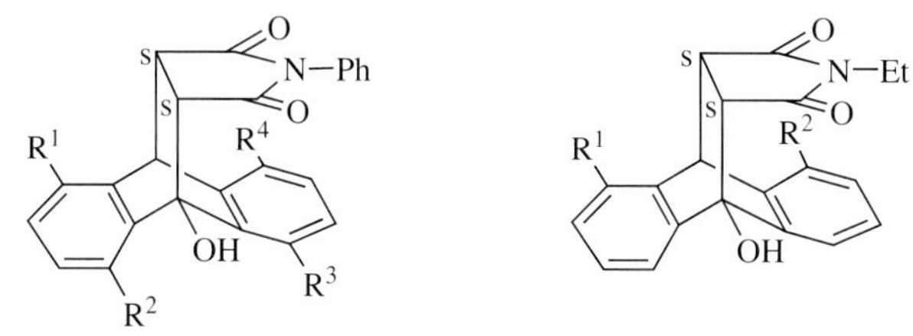

ratio = 8:1

$R^1 = R^3 = Cl$; $R^2 = R^4 = H$, major
87%, 99% ee

$R^1 = R^3 = H$; $R^2 = R^4 = Cl$, minor

$R^1 = NHMe$, $R^2 = H$, major
95%, 98% ee

$R^1 = H$, $R^2 = NHMe$, negligible

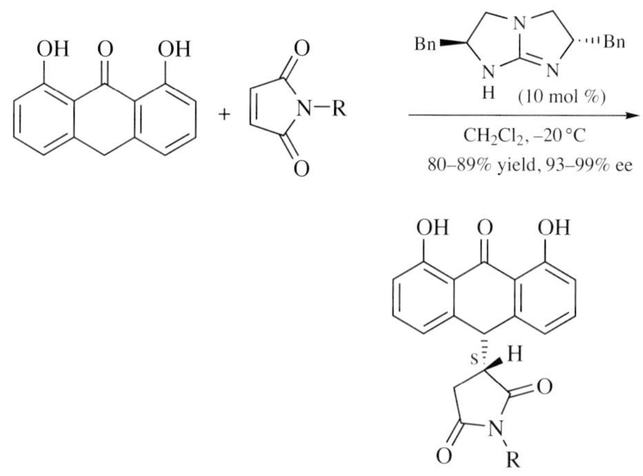

(3)

R = Ph: 80%, 99% ee
$R = 2\text{-}NO_2C_6H_4$: 87%, 97% ee
$R = 3,4\text{-}Cl_2C_6H_3$: 89%, 98% ee
$R = CH_2Ph$: 86%, 93% ee

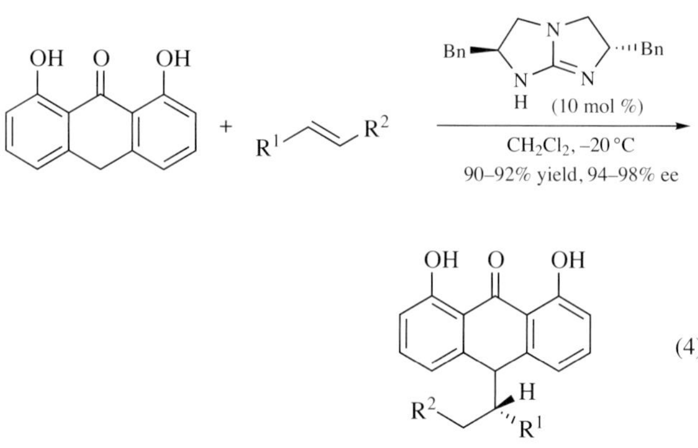

(4)

$R^1 = CO_2Me$, $R^2 = CH_3CO$: 92%, 98% ee
$R^1 = CO_2Et$, $R^2 = PhCO$: 92%, 95% ee
$R^1 = R^2 = CN$: 90%, 94% ee

Related Reagents. 2,3,5,6-Tetrahydro-2,6-bis(phenyl)-1*H*-imidazo[1,2-*a*]imidazole;[3] 2,3,5,6-Tetrahydro-2,6-bis(*tert*-butyl)-1*H*-imidazo[1,2-*a*]imidazole.[4]

---

1. Shen, J.; Nguyen, T. T.; Goh, Y. P.; Ye, W.; Fu, X.; Xu, J.; Tan, C. H., *J. Am. Chem. Soc.* **2006**, *128*, 13692.
2. Ye, W.; Leow, D.; Goh, S. L. M.; Tan, C. T.; Chian, C. H.; Tan, C. H., *Tetrahedron Lett.* **2006**, *47*, 1007.
3. Corey, E. J.; Grogan, M. J., *Org. Lett.* **1999**, *1*, 157.
4. (a) Fu, X.; Jiang, Z.; Tan, C. H., *Chem. Commun.* **2007**, 5058.
(b) Ye, W.; Jiang, Z.; Zhao, Y.; Goh, S. L. M.; Leow, D.; Soh, Y. T.; Tan, C.H., *Adv. Synth. Catal.* **2007**, *349*, 2454.

Choon-Hong Tan & Juan Shen
National University of Singapore, Singapore

2,2,6,6-Tetramethylpiperidin-1-oxyl[1]

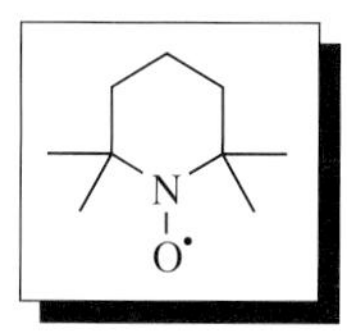

[2564-83-2] $C_9H_{18}NO$ (MW 156.28)
InChI = 1S/C9H18NO/c1-8(2)6-5-7-9(3,4)10(8)11/h5-7H2, 1-4H3
InChIKey = QYTDEUPAUMOIOP-UHFFFAOYSA-N

(oxidizing agent for the conversion of primary alcohols to aldehydes[2,3] or carboxylic acids,[2] of secondary alcohols to ketones,[2] and of diols to lactones or hydroxy aldehydes[4])

Alternate Name: TEMPO.
Physical Data: mp 40.0 °C; fp 67 °C.
Solubility: soluble in all organic solvents and 0.03 mol L^{-1} in H_2O at 25 °C.
Form Supplied in: red orange solid; commercially available.
Analysis of Reagent Purity: IR (Nujol) 1330, 1180, 1060, 975, 950 cm^{-1};[5] UV–Vis, λ (ε) = heptane, 470 nm (10.5), dioxane, 465 nm (10.4), 240–242 nm (2030).[5]
Purification: sublimation.[6]
Handling, Storage, and Precautions: TEMPO is a toxic substance, and a severe irritant which is readily absorbed through the skin.[7] Its toxicity is probably related to the formation of hydroxylamine metabolites.[8]

Original Commentary

Fernando Montanari & Silvio Quici
Universit'a di Milano, Milan, Italy

Oxammonium Salts. Oxammonium salts (**2**) can be prepared in situ by oxidation of the nitroxide (**1**).[1e] There is evidence that the oxoammonium ion (**2**) is actually the product of an acid-catalyzed disproportionation between two nitroxide molecules (**1**), which produces one oxoammonium ion (**2**) and one hydroxylamine molecule (**3**).[1e] Peracids, for instance, do not oxidize nitroxides to oxoammonium ions, yet do act as the secondary oxidant. The chemical stability of (**2**) depends on the counterion, and this can be easily exchanged; the chloride salt is not very stable. The oxammonium salt is the active species in the oxidation of primary and secondary alcohols to carbonyl derivatives and can be used in stoichiometric or catalytic fashion.[1e] Catalytic procedures require a co-oxidant such as Copper(II) Chloride–O_2,[9] peroxy acids,[10] electrooxidation,[11] or Sodium Hypochlorite[2–4] Oxammonium salts can be titrated iodometrically.[1e]

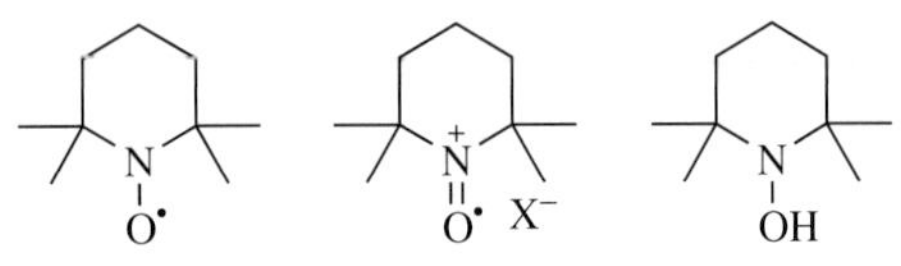

(1) (2) (3)

It has been proposed that (**2**) reacts with alcohols, forming carbonyl derivatives and hydroxylamine (**3**).

Oxidation of Alcohols.[2,3,7] Primary alcohols are converted into aldehydes by the following catalytic system: 0.01–0.002 equiv of TEMPO (or its 4-methoxy derivative), 0.05 equiv of KBr as co-catalyst, and aqueous NaOCl buffered at pH 8.5–9.5 in a CH_2Cl_2/H_2O two-phase system. At this pH, HOCl is the co-oxidant and in the presence of KBr it is likely transformed into HOBr which is more efficient in the oxidation of (**1**) to (**2**). The reaction is exothermic. For reactions carried out on a 1–10 mmol scale the temperature can be easily maintained at 0 °C and conversion is complete in few minutes. On a larger scale, a very efficient cooling system is required. A compromise is to maintain the temperature in the range 10–15 °C with an ice bath; note that higher temperatures lead to fast decomposition of the catalyst. The pH is buffered by adding the appropriate amount of $NaHCO_3$ to the aqueous NaOCl. $NaH_2PO_4{\cdot}2H_2O$ and $Na_2HPO_4{\cdot}2H_2O$ have been used as alternatives.[12]

Oxidation of secondary alcohols similarly affords ketones. The oxidation can be applied to saturated alkyl and aryl–alkyl substrates; relatively unstable protecting groups such as acetonide derivatives of diols are not affected. Side reactions occur with substrates with isolated and conjugated double bonds, leading to lower yields. The reaction rates are markedly decreased by the presence of electron-donor groups in the aromatic ring of benzyl alcohol, but they can be speeded up by addition of catalytic amounts of quaternary ammonium salts (Q^+X^-). At the pH of commercial bleach (12.7), reactions are very slow. At lower pH, HOCl is distributed between the aqueous and the organic phase, thus making the phase transfer catalyst unnecessary. Optically active alcohols afford the corresponding aldehydes in good yields and high enantiomeric purity (eq 1).[3,13]

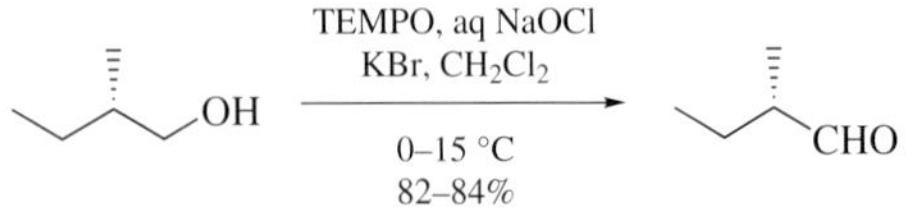

(1)

Oxidation of Diols. Diols are easily oxidized with this reagent; the nature of the products depends on the amount of oxidant used, the difference in oxidation rates of primary vs. secondary alcohols, the presence of Q^+X^-, and on the relative distance between the OH groups in the aliphatic chain. Under the reaction conditions reported above, primary–secondary diols afford hydroxy or keto aldehydes. In the presence of a catalytic amount of Q^+X^-, primary–secondary diols lead to ketocarboxylic acids. Selectivity in these oxidations also depends on the substituents on the 2,2,6,6-tetramethylpiperidine ring.[14]

Lactones are obtained from 1,4- and 1,5-diols, whereas α,ω-diols give unresolvable mixtures of polymeric products.[4] Oxidation of hydrophilic 1,4- and 1,5-diols to γ- and δ-lactones and of hydrophilic alcohols to aldehydes is best conducted using CH_2Cl_2/solid LiOCl in the presence of solid $NaHCO_3$. Commercial LiOCl contains 7% of H_2O so that reaction conditions are those of a pseudo solid–liquid system.[4]

Oxidation of Primary Alcohols and Aldehydes to Carboxylic Acids.[2] The addition of catalytic amounts of quaternary salt to the oxidizing system leads to the fast and direct formation of carboxylic acids. This oxidation requires the presence, in the organic phase, of ClO^- and/or BrO^- anions which behave as strong bases. Again, electron-donor groups in benzyl alcohol or aromatic aldehydes strongly lower the reaction rates.

First Update

Huda Henry-Riyad & Thomas T. Tidwell
University of Toronto, Toronto, Ontario, Canada

Oxidation of Alcohols. The general procedure, according to the protocol of Anelli et al.,[3,4] goes through an oxoammonium cation that acts as the active oxidant (eq 2). Typically these transformations employ 1 mol % of TEMPO and a stoichiometric amount of a secondary oxidant, e.g., sodium hypochlorite,[15] trichloroisocyanuric acid,[2,16] *m*-chlorobenzoic acid,[10a,17] sodium bromite,[18] sodium chlorite,[19] and oxone.[20] Recently, iodine as a secondary oxidant proved to be superior in cases of electron-rich and heteroaromatic rings.[21] The secondary oxidant (hypochlorite) oxidizes TEMPO to the oxoammonium salt, which converts the alcohol to a ketone, aldehyde, or an acid. The product hydroxylamine is converted back to oxoammonium salt by hypochlorite.

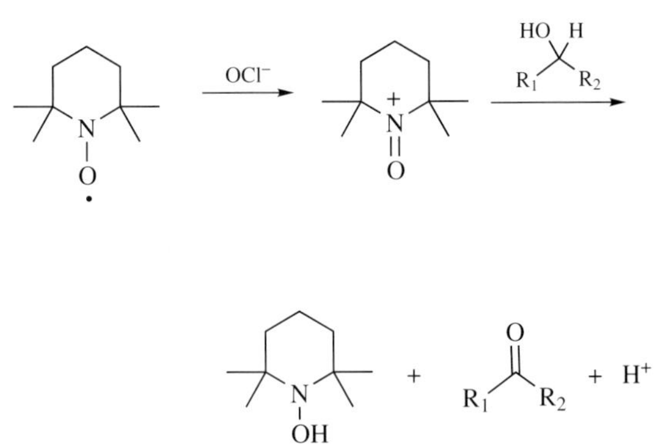

(2)

Recent Advances in TEMPO Recovery After Oxidation. To provide better regeneration procedures for TEMPO after oxidation, Fey and Bolm reported silica-supported TEMPO catalyst systems,[22] and more recently a polymer-supported TEMPO has been developed.[23] 'Ring-opening metathesis polymerization' (ROMP) of various strained norbornene systems is used to anchor TEMPO units. (TEMPO catalyst, Figure 1) to generate the homogeneous polymeric system, and 1 mol % of the catalyst was employed, setting the number of TEMPO units per molecule (*n*) at about 50. The performance of the catalyst in oxidations of various primary alcohols was investigated under the conditions shown in eq 3.

$$\underset{(0.8\ \text{mmol})}{RCH_2OH} \xrightarrow[CH_2Cl_2,\ 0\,^\circ C]{\substack{\text{TEMPO catalyst (1 mol \%)}\\ \text{NaOCl (2.7 mL, 0.37 M)}\\ \text{KBr (0.16 mL, 0.5 M)}}} \underset{72\text{–}85\%}{RCHO} \quad (3)$$

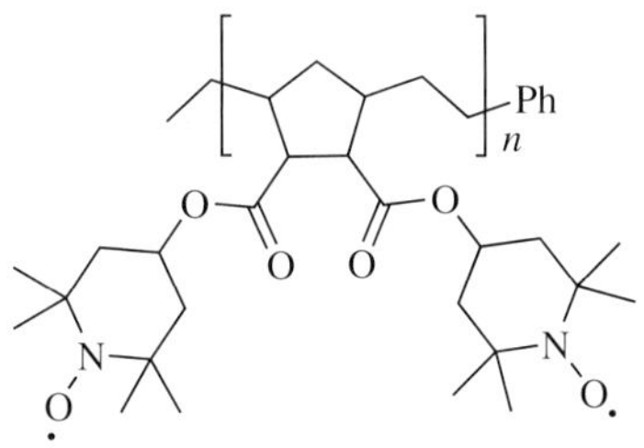

Figure 1 Polymeric TEMPO catalyst

Both the alcohols and the TEMPO containing polymers were soluble in two-phase systems (H_2O:CH_2Cl_2). After oxidation, the catalyst could be recovered from the organic phase by decreasing the polarity of the medium and filtration, leaving the resultant carbonyl compounds. For up to three runs of the recovered catalyst, no drastic decrease was observed in its catalytic activity.

Living Free Radical Polymerization.[24] TEMPO has attracted great interest in living free radical polymerization, where it reversibly combines with growing radical chains (**4**) in styrene polymerization.[25] The formation of a dissociating complex **5**, between the growing chain **4** and the additive TEMPO, is the key feature in stable free radical addition polymerization (SFRP) (eq 4).

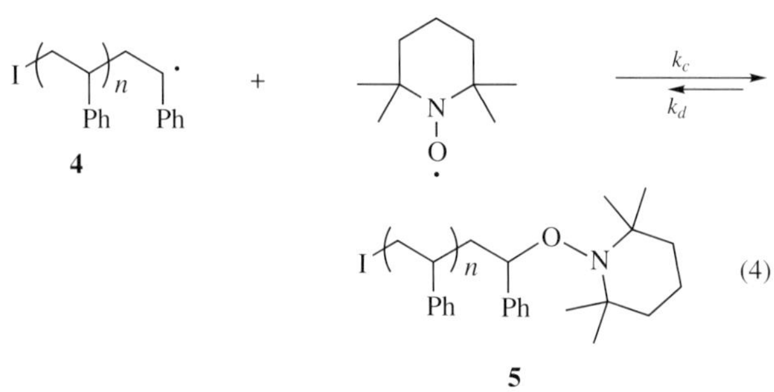

(4)

The low concentration of the growing chains (**4**) in the system governs the propagation rate. Chains start growing after generation of some initiating radicals in the system through thermal decomposition of added benzoyl peroxide. Covalent complex **5** removes most of the potential polystyrene chains from active propagation. Complex **5** is involved in an equilibrium with rate constants for complex formation (k_c) and dissociation (k_d) such that the equilibrium favors the covalent adduct. The basis of the ';living' description is the low probability of growing chains (**4**) coming together causing mutual termination, which is typically at least an order of magnitude smaller than it is in the conventional radical addition system.[26] Typical conditions for this process involve heating a mixture of monomer or monomers, free radical initiator (benzoyl peroxide, BPO), and the stable free radical (TEMPO), under argon at 95 °C for 3.5 h, followed by heating at 123 °C for 69 h yielding polystyrene with a polydispersity of 1.26.

Free Radical Hydroxylation Reactions of Alkylboronates.[27] Alkylboronic ester derivatives afford, in the presence of TEMPO, the corresponding oxygenated products. TEMPO reacts cleanly with *B*-alkylcatecholboranes to afford alkyl radicals that can be trapped by a second equivalent of TEMPO to give alkoxyamines. Reduction of the alkoxyamines with zinc in acetic acid affords the desired alcohols. The reaction is based on the exceptional reactivity of *B*-alkylcatecholborane derivatives toward TEMPO and the efficient trapping of the generated carbon-centered radicals by TEMPO (eq 5).

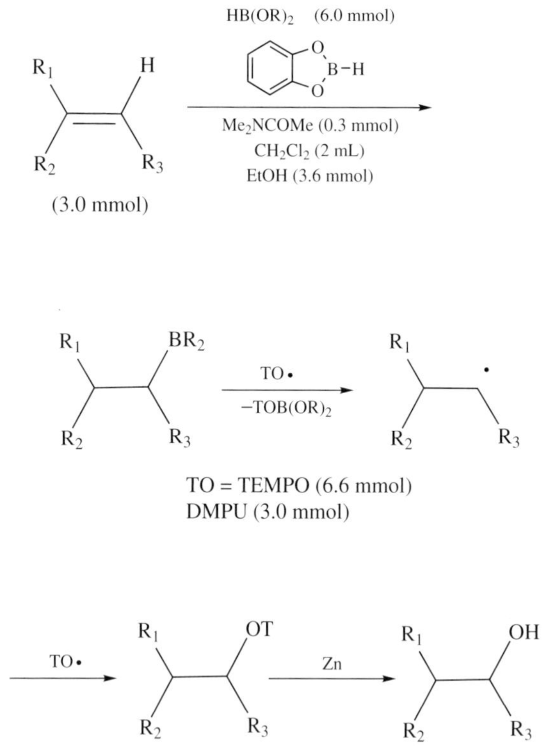

(5)

Despite the radical-radical recombination nature of these processes, a fair level of stereoselectivity and yield were reached. The radical nature of the TEMPO-mediated hydroxylation reaction was probed by a radical-clock experiment with (+)-2-carene (eq 6).

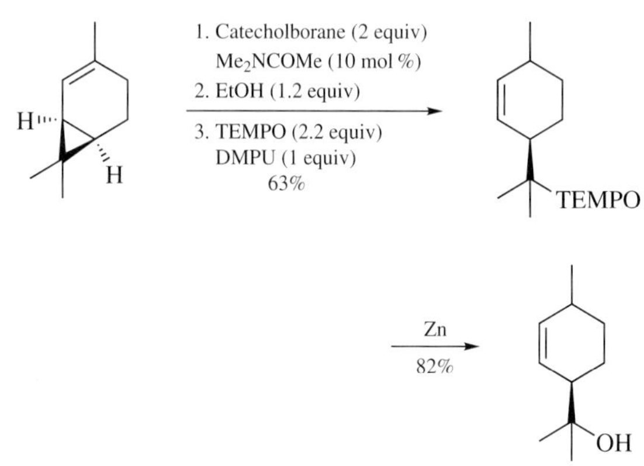

(6)

Recent Developments. The combination of $RuCl_2(PPh_3)_3$ and TEMPO affords an efficient catalytic system for the aerobic oxidation of a variety of primary and secondary alcohols, giving the corresponding aldehydes and ketones, in >99% selectivity in all cases.[28] The authors postulate a hydridometal mechanism, involving a '$RuH_2(PPh_3)_3$' species as the active catalyst. TEMPO acts as a hydrogen transfer mediator and is either regenerated by oxygen, under catalytic aerobic conditions, or converted to hydroxylamine under stoichiometric anaerobic conditions (Figure 2).

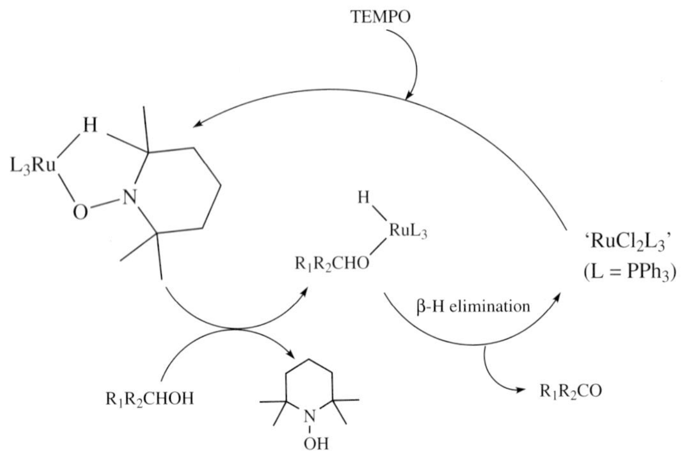

Reaction Conditions:
15 mmol of substrate, Ru:TEMPO = 1:3, 30 mL of chlorobenzene, 10 mL/min O_2/N_2 (8/92; v/v), p (total pressure) = 10 bars, T = 100 °C

Figure 2 Alcohol oxidation using TEMPO/Ru as a catalyst system

The selective oxidative conversion of a variety of highly functionalized 1°, 2°-1,5-diols into the corresponding δ-lactones has been effected simply and efficiently using a reagent system comprising catalytic TEMPO and excess bis-acetoxyiodobenzene (BAIB).[29] The generality and functional group compatibility of this process has been highlighted through the formation of several highly substituted systems as shown in eq 7.

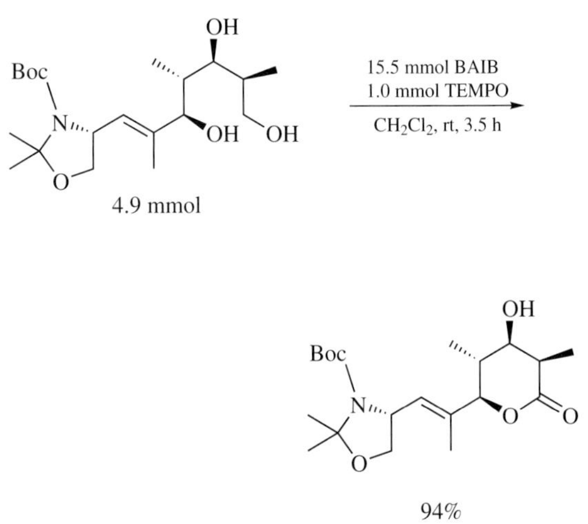

(7)

Many structural analogs of TEMPO have been utilized, including optically active compounds[30] and novel structures to optimize living free radical polymerization.[31]

TEMPO as Trapping Agent. TEMPO is finding utility as a trapping agent for free radicals and reactive intermediates:

1. In multistep synthesis, Boger and Boyce have used TEMPO to trap radicals generated by 5-*exo-trig* aryl radical cyclization onto an unactivated alkene in 1,2,9,9a-tetrahydrocyclopropa[*c*]pyrido[3,2-*e*]indol-4-one-7-carboxylate (CPyI) synthesis.[32]

2. TEMPO is used to trap benzyl radicals derived from di(benzyloxy)carbene generated through the thermolysis of 2,2-dibenzyloxy-5,5-dimethyl-Δ^3-1,3,4-oxadiazoline in benzene at 110 °C.[33]

3. Tidwell and co-workers have examined the reactivity of TEMPO with different ketenes.[34] The preferred attack at C_α, predicted by calculations,[34d] has been confirmed experimentally for TEMPO's reaction with PhCH=C=O.[34] α-Ester radical **6** was first generated and then trapped by a second TEMPO to form 1,2-bisadduct **7** (eq 8). This sequence of addition was demonstrated by the strong dependence of the rate of TEMPO addition on the ability of the substituent R to stabilize the forming radical **6**.[34a] Typical conditions are 2.1 equiv of TEMPO added to the preformed ketenes in hydrocarbon solvent at room temperature.

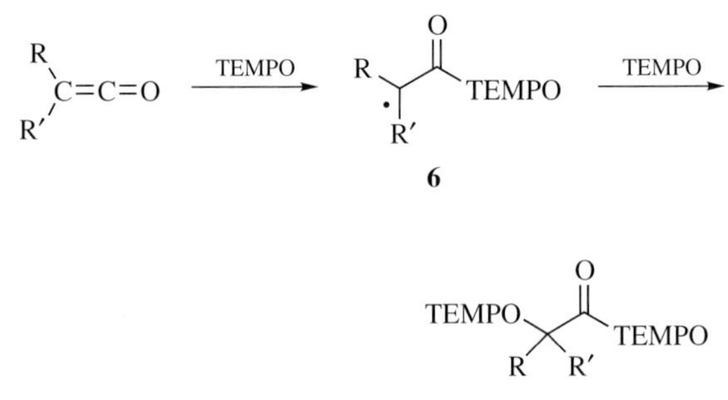

(8)

7

Reduction of TEMPO. TEMPO is reduced to its corresponding hydroxylamine, 1-hydroxy-2,2,6,6-tetramethylpiperidine (TEMPOH), by ascorbic acid (eq 9),[35a] Grignard reagents,[35b] or phenylhydrazine.[35c,d,e]

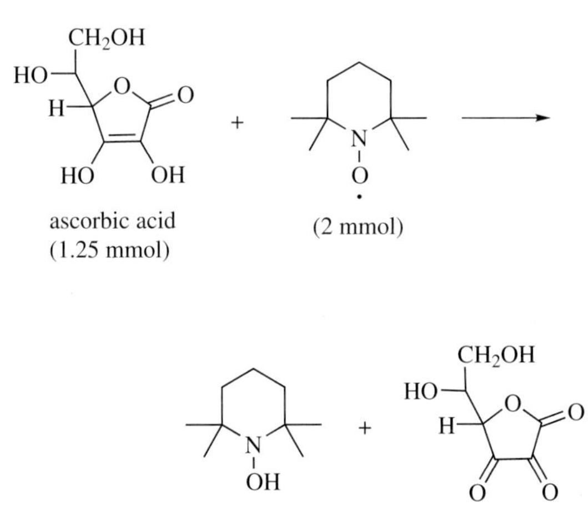

(9)

Second Update

Armido Studer & Thomas Vogler
Westfälische Wilhelms-Universität Münster, Münster, Germany

Oxidation of Alcohols. In the oxidation of alcohols according to eq. (2), TEMPO is in most cases used in catalytic amounts in combination with a stoichiometric oxidant. Meanwhile many stoichiometric co-oxidants have successfully been applied as already described above.[15–20] Recently, laccase, a Cu-containing enzyme, was used to reoxidize hydroxylamine (**3**) to *N*-oxoammonium salt **2** under aerobic conditions, which eventually oxidizes alcohols to the corresponding aldehydes.[36] To simplify the purification procedure, polymer bound co-oxidants such as bisacetoxybromate(I) anion **6**[37] and poly[4-(diacetoxy)iodo]styrene (**7**)[38] were introduced for the oxidation of primary and secondary alcohols in combination with catalytic amounts of TEMPO (Figure 3). Work-up is achieved by simple filtration and subsequent removal of the solvent.

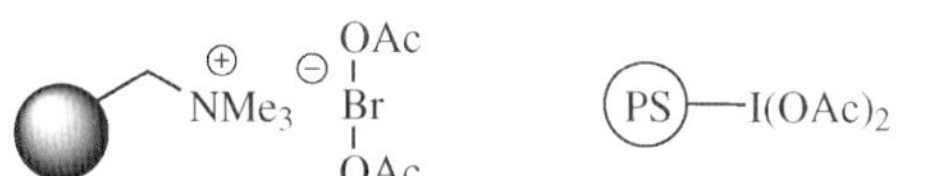

6 derived from IRA-900 anion exchange resin

7 (PS = polystyrene)

Figure 3 Polymeric co-oxidants

During the last years, research on aerobic metal-mediated TEMPO oxidations of alcohols has been intensified. Along with Ru- and Cu-catalysts reported above,[9,28] a $Mn(NO_3)_2/Cu(NO_3)_2$ catalyst couple in combination with TEMPO and oxygen as stoichiometric oxidant was shown to oxidize various primary and secondary alcohols efficiently.[39] Even air works as terminal oxidant for $CuBr_2$(2,2′-bipyridine)-TEMPO-catalyzed alcohol oxidations (eq 10).[40] $FeCl_3$-TEMPO-$NaNO_3$ catalyzes the mild aerobic oxidation of a broad range of alcohols to the corresponding aldehydes and ketones.[41] Environmentally benign transition-metal free alcohol oxidation using TEMPO-Br_2-$NaNO_3$ in CH_2Cl_2 with air as the stoichiometric oxidant was reported.[42] Aerobic transition-metal free oxidations of alcohols can also be achieved using catalytic amounts of TEMPO, polymeric [bis(acetoxy)iodo]styrene (**7**), and KNO_2.[43]

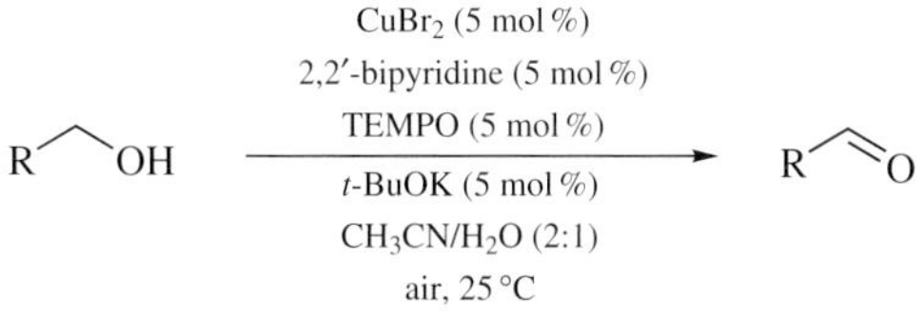

(10)

Recovery of TEMPO-derivatives. TEMPO-catalyzed oxidations were successfully conducted in ionic liquids. Acetamido-TEMPO[44] and, even more elegantly, an ionic-liquid-based derivative 8[45] in combination with the ionic-liquid [bmim][PF_6] (bmim = butylmethylimidazolium) were used as catalytic oxidants in biphasic systems (Figure 4). The ionic liquid containing the catalytically active TEMPO derivatives can be readily separated from the oxidation products and reused without loss of activity. Oxidation works also under aerobic conditions in [bmim][PF_6] using TEMPO in combination with CuCl.[46] Organically modified silicas doped with TEMPO prepared via the sol–gel process can be used as recyclable catalysts for the selective alcohol oxidation.[47] Reactions are best conducted in CH_2Cl_2 with aq. NaOCl as stoichiometric oxidant. Transformation of alcohols into the corresponding aldehydes or ketones was also achieved under organic-aqueous biphasic conditions with catalytic amounts of TEMPO-polystyrene resin **9**.[48] As terminal oxidant, aqueous $K_4[Fe(CN)_6]$ was added. 1,4- and 1,5-Diols were readily oxidized to the corresponding γ- and δ-lactones under these conditions. Resins of type **9** can also be oxidized to the corresponding oxoammonium resins which can eventually be applied as polymeric stoichiometric alcohol oxidizing reagents.[49]

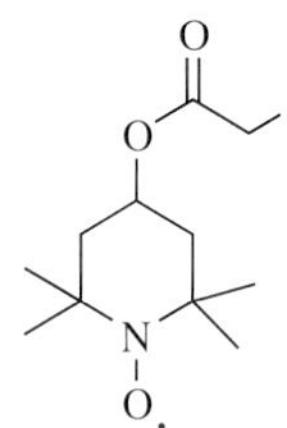

8

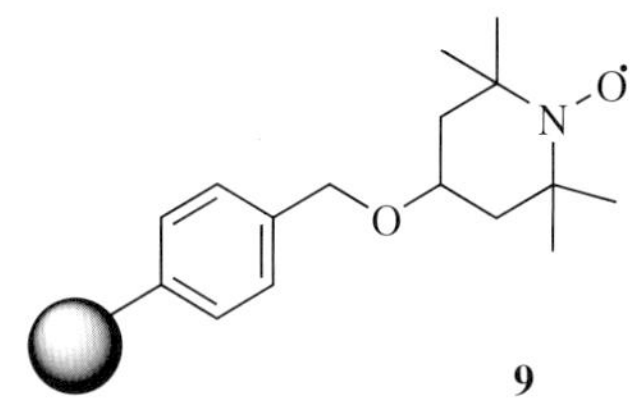

9

1% divinylpolystyrene resin

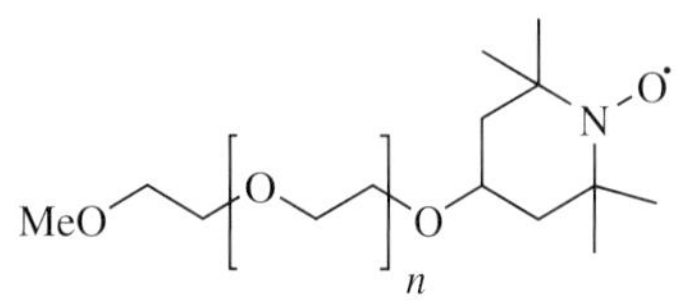

10 (M_w = 5000)

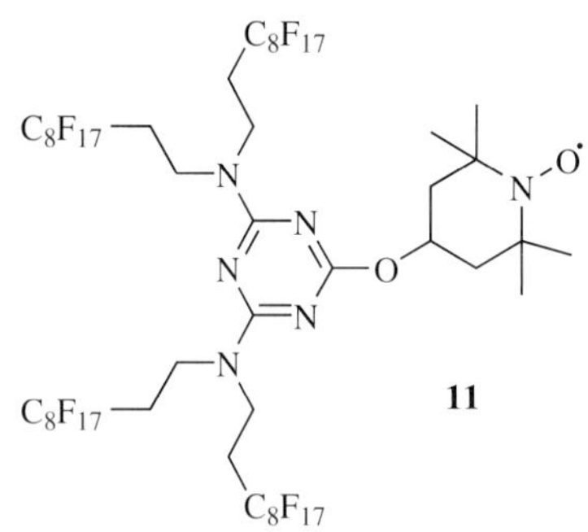

11

Figure 4 Readily recoverable TEMPO derivatives

Poly(ethylene glycol)-TEMPO conjugate **10** was prepared and used for the oxidation of alcohols with stoichiometric amounts of organic ([bis(acetoxy)iodo]benzene, trichloroisocyanuric acid), or inorganic (NaOCl) reagents as terminal oxidants.[50] The polymer-supported TEMPO can be reused, conserving the high activity. Aerobic oxidations can also be conducted using TEMPO-conjugate **10** in combination with $Mn(NO_3)_2/Cu(NO_3)_2$ as cocatalysts.[51] "Fluorous-tagged" TEMPO derivative **11** was successfully used in alcohol oxidations with either [bis(acetoxy)iodo]benzene or aqueous NaOCl as terminal oxidant.[52] The fluorous-TEMPO derivative was readily isolated from the reaction products by liquid–liquid or solid-phase extraction.

Miscellaneous Oxidations. A Cu(II)-salen catalyst with H_2O_2 as oxidant efficiently transforms various sulfides to the corresponding sulfoxides.[53] Addition of TEMPO enhances the conversion and suppresses the overoxidation to the corresponding sulfones (eq 11). The role of TEMPO in these processes is not yet understood. Styrene derivatives can be directly transferred to the corresponding styryl-TEMPO derivatives as described in eq 12.[54] A Mn(salen)Cl-catalyst and $NaBH_4$ as stoichiometric reducing reagent in the presence of TEMPO were used for these reactions. The alkoxyamines obtained can be applied as regulators/initiators for the controlled living radical polymerization (eq. 4).[24–26] TEMPO has been reported to oxidize allylic or benzylic compounds to the corresponding alkoxyamines under mild conditions. As an example, the oxidation of cyclohexene with hydroxy-TEMPO is presented (eq 13).[55] For benzylic oxidations, these reactions proceed via H-abstraction with TEMPO followed by TEMPO trapping of the intermediately formed benzylic radical. For allylic oxidations. However, the major pathway is TEMPO addition onto the olefin with subsequent β-hydrogen abstraction on the intermediate TEMPO-adduct radical.[55b] Oxidations of activated CH-bonds with TEMPO can also be achieved upon photochemical activation of TEMPO.[56]

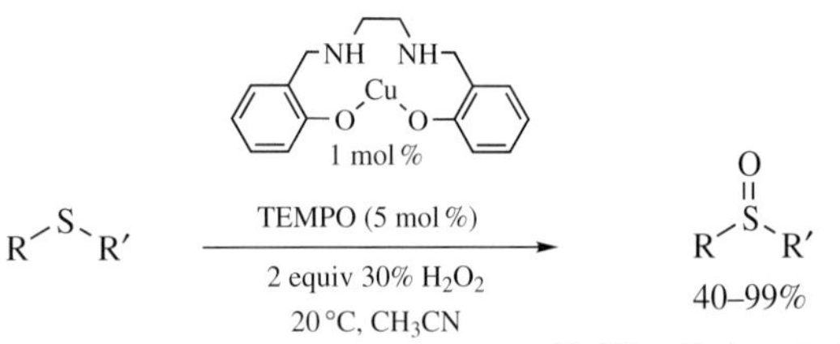

(11)

R, R′ = alkyl, aryl, allyl

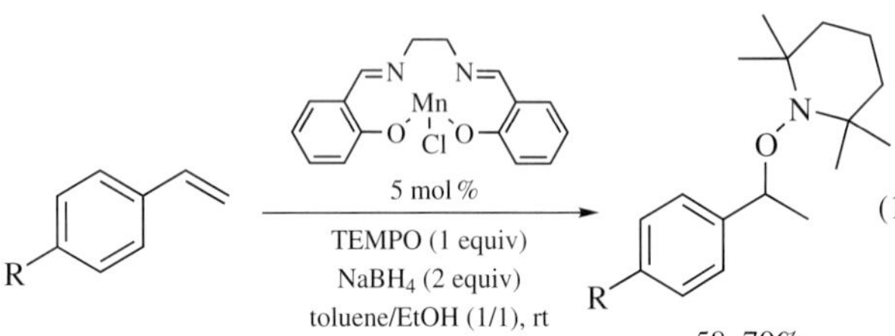

(12)

58–70%

R = CH_2Cl, CH_2OH, CO_2Me, vinyl, CH_2OR'

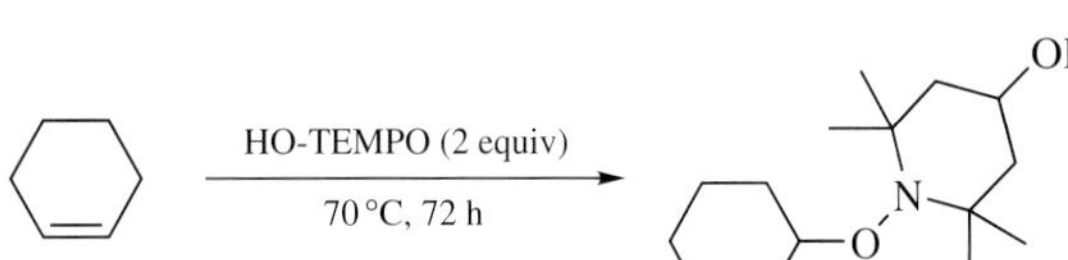

(13)

solvent

Oxidation of organometallic compounds can be realized using TEMPO. One equivalent of TEMPO is used to oxidize the metalated carbanion to the corresponding C-centered radical, which is then trapped with the second equivalent of TEMPO to give the alkoxyamine. R-Li,[57–59] R-MgBr[58], R-Ti(O*i*-Pr)$_3$,[58] R-Zr(Cl)Cp$_2$,[58] R-CuCN,[58] and R-SmI$_2$[58] were successfully transformed to the corresponding TEMPO-R alkoxyamines using this approach (eq 14). Oxoammonium salts of type **2** can directly react with activated electron rich olefins to form alkoxyamines in an electrophilic addition reaction (eq 15).[60,61] Moreover, oxoammonium salts of TEMPO undergo efficient reaction with various enolates to form the corresponding alkoxyamines.[62]

$$\text{R–M} \xrightarrow{\text{TEMPO}} \text{TEMPO}^{\ominus}\text{M}^{\oplus} + \text{R}^{\bullet} \xrightarrow{\text{TEMPO}} \text{TEMPO–R} \quad (14)$$

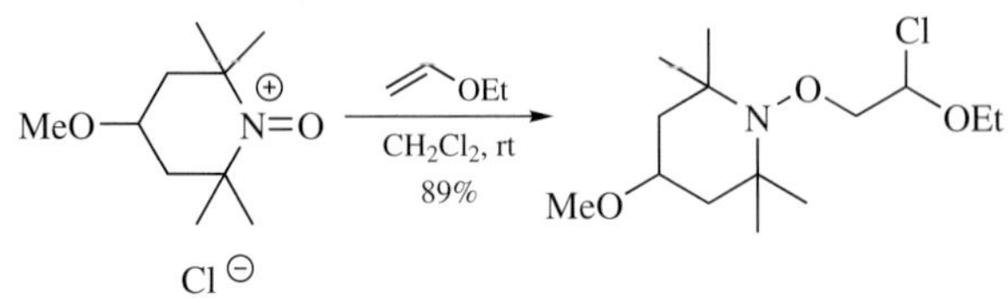

(15)

TEMPO Mediated Radical Cyclization and Addition Reactions. TEMPO-derived alkoxyamines undergo clean thermal homolysis to the corresponding C-centered radicals and TEMPO if stabilized C-radicals are formed.[63] Thermal C–O bond homolysis processes are reversible and can be used for the generation of C radicals to conduct radical cyclization and intermolecular addition reactions. Heating of alkoxyamine **12** in *t*-BuOH at 132 °C for 24 h provides the isomerized alkoxyamine **13** in good yield (eq 16).[64] Intermolecular alkoxyamine addition can also be preformed using this concept.[65] Importantly, these thermal alkoxyamine addition reactions can be conducted highly efficiently using microwave irradiation (eq 17).

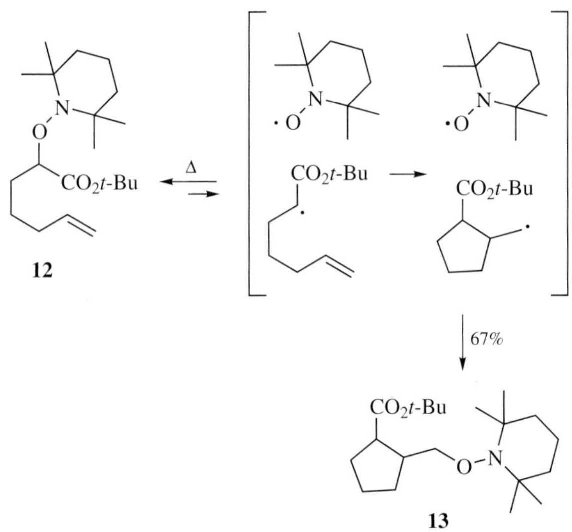

12 (16) **13**

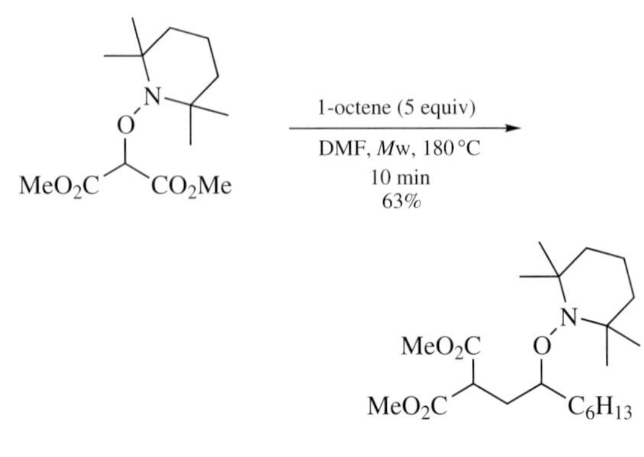

(17)

Third Update

Yu Rao, Chao Zhang, & Yu Zong
Tsinghua University, Beijing, China

Oxidation of Alcohols. TEMPO can be converted into TEMPO oxoammonium salts that are effectively providing an efficient method for oxidative rearrangement of tertiary allylic alcohols into β-substituted α,β-unsaturated carbonyl compounds. This reaction can be applied to acyclic substrates, medium-sized rings, and macrocyclic substrates (eq 18).[66]

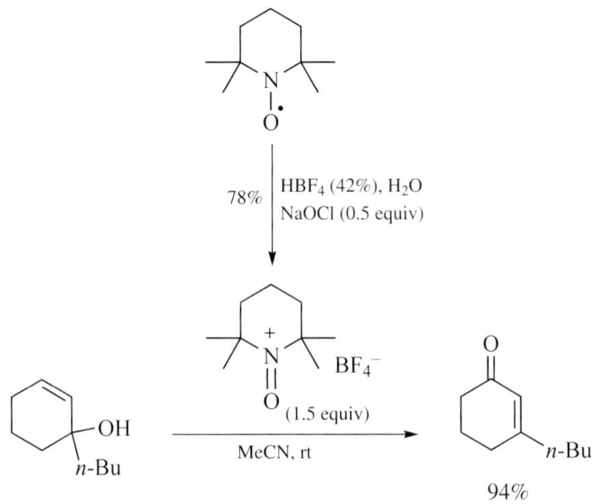

(18)

It is possible to transform 1,6- and 1,7-diols into the corresponding lactones under the aforementioned TEMPO/PhI(OAc)$_2$ oxidative system. This process is economical, easy to handle, and may proceed without the need for protection/deprotection sequences (eq 19).[67]

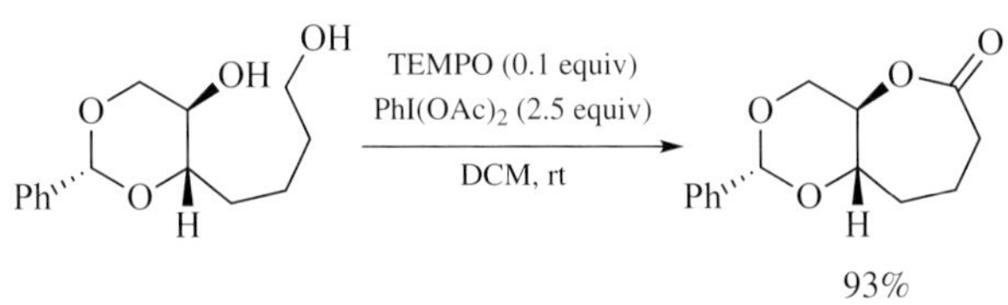

(19)

Positively charged TEMPO, generated by the oxidation of TEMPO, is an effective and selective oxidant. Alcohols may be oxidized to the corresponding aldehydes or ketones in presence of the title compound and TiO_2 under visible-light irradiation through a two-electron transfer mechanism (eq 20).[68] Cu–NHC–TEMPO complexes, derived from imidazolium salts with copper powder, are efficient catalysts in the oxidation of primary alcohols (eq 21).[69] Later a (bpy) CuI/TEMPO catalyst system was reported to also enable the efficient and selective aerobic oxidation of alcohols to the corresponding aldehydes with ambient air as the oxidant. This catalytic system is compatible with a wide range of functional groups mitigating the need for protecting groups (eq 22).[70] In presence of an oxygen atmosphere, a catalytic ternary system ($CuCl_2$–$NaNO_2$–TEMPO) oxidizes alcohols to aldehydes that may be subjected to standard Passerini reaction conditions to react with isonitriles and carboxylic acids to furnish the α-acyloxy amides (eq 23).[71]

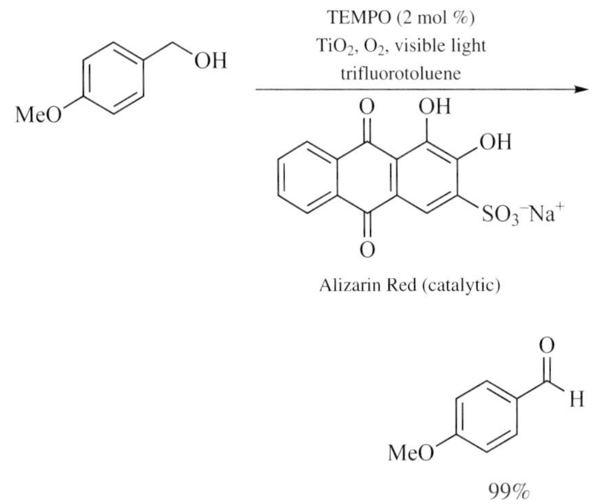

(20)

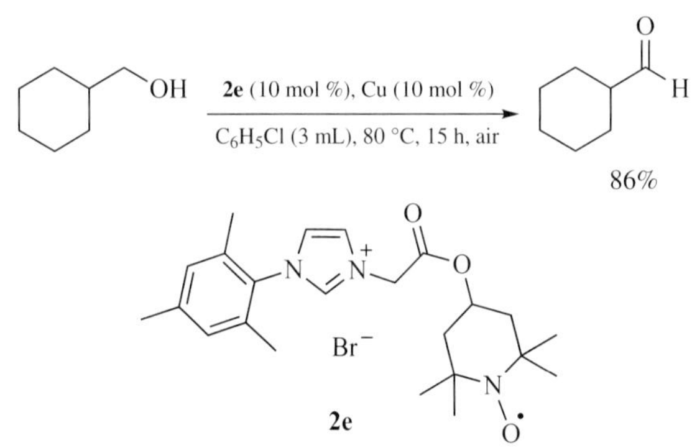

(21)

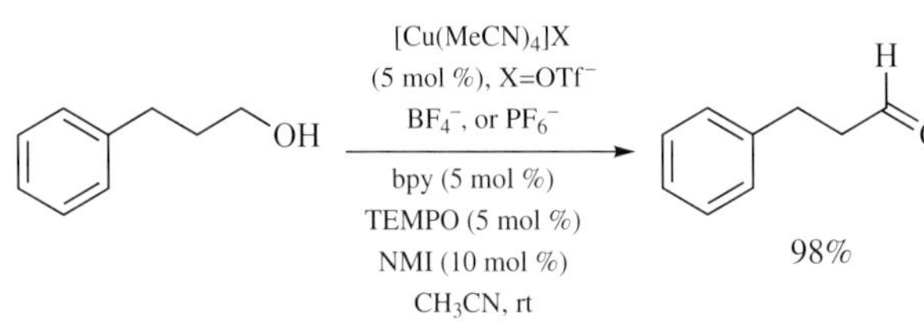

(22)

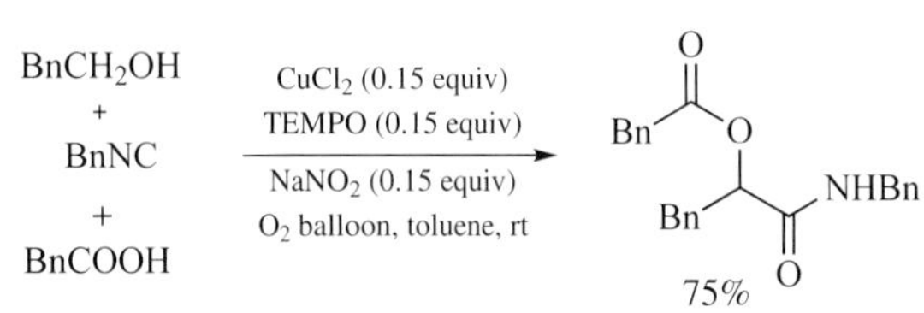

(23)

Other Oxidative Applications of TEMPO. Using a combination of an NHC catalyst and TEMPO, aldehydes were oxidized by TEMPO to the corresponding esters through a single-electron transfer (SET) process (eq 24).[72] A TEMPO-catalyzed aerobic oxygenation and nitrogenation of hydrocarbons via C=C double bond cleavage has also been disclosed. Here oxygen was used as the terminal oxidant, supplying an effective method for preparation of industrially and pharmaceutically important *N*- and *O*-containing motifs (eq 25).[73] Similarly, a copper · bipyridine complex/TEMPO system was applied in the preparation of nitriles from aldehydes or alcohols (eq 26).[74] Although the reaction was equally effective with several TEMPO derivatives, the use of 4-hydroxy TEMPO facilitated purification of the product.

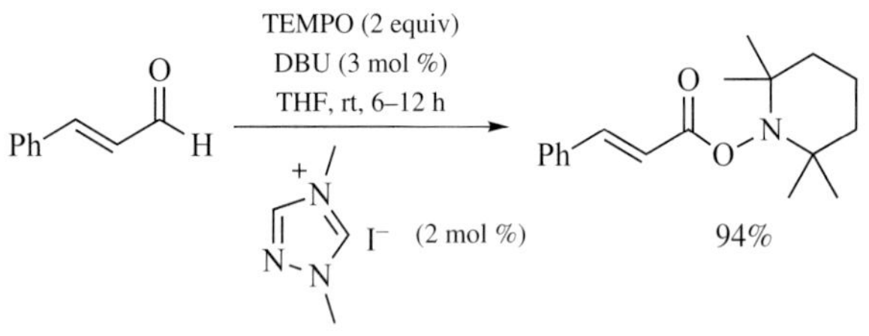

(24)

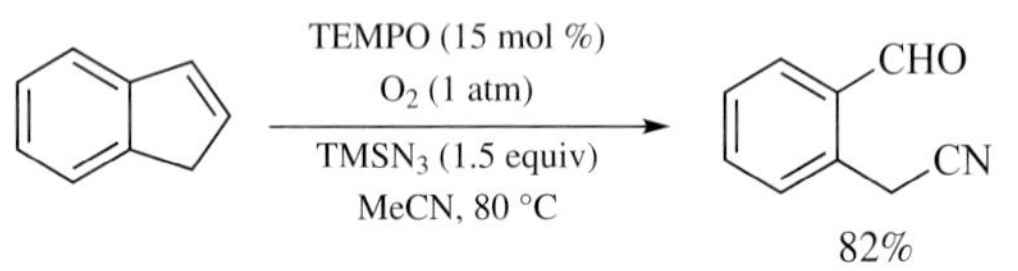

(25)

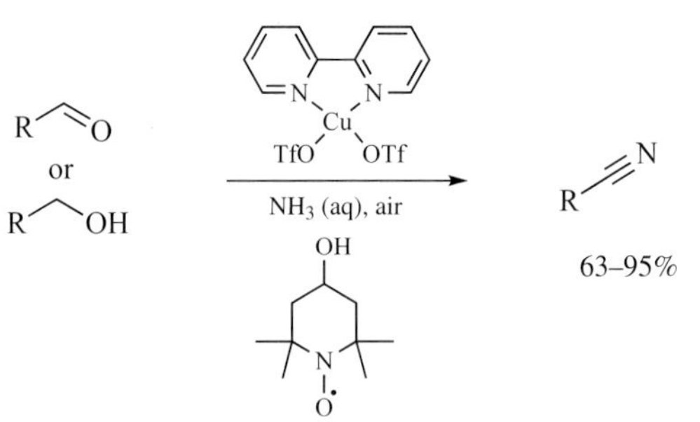

(26)

TEMPO Used in C–C Bond Formation. A TEMPO–O_2 catalyzed alkylation of sp^3 C–H bonds using molecular oxygen as the oxidant has been described, providing a new strategy for the formation of C–C bonds (eq 27).[75]

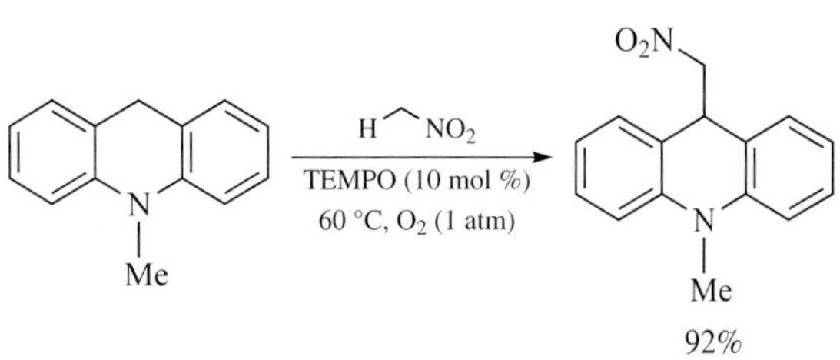

(27)

The Role of TEMPO in C–H Activation. TEMPO has been employed as a stoichiometric oxidant in various C–H activation processes. Rhodium (eq 28)[76] and palladium (eq 29)[77] catalyzed direct C–H arylation of different arenes with various arylboronic acids have been delineated. *N*-Protected indoles are directly arylated under standard conditions with high levels of diastereoselectivity (dr >99:1; HPLC analysis).

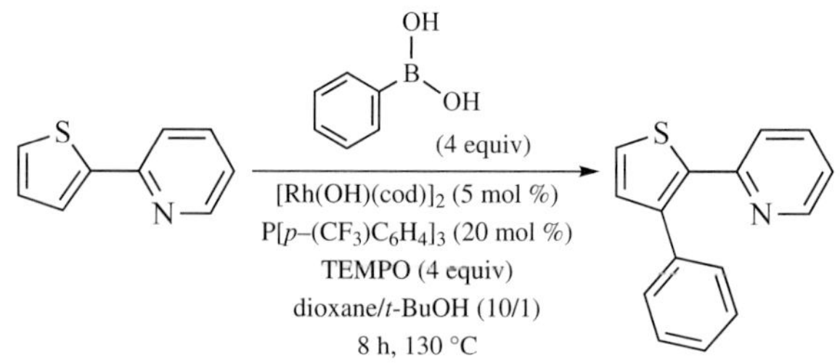

(28)

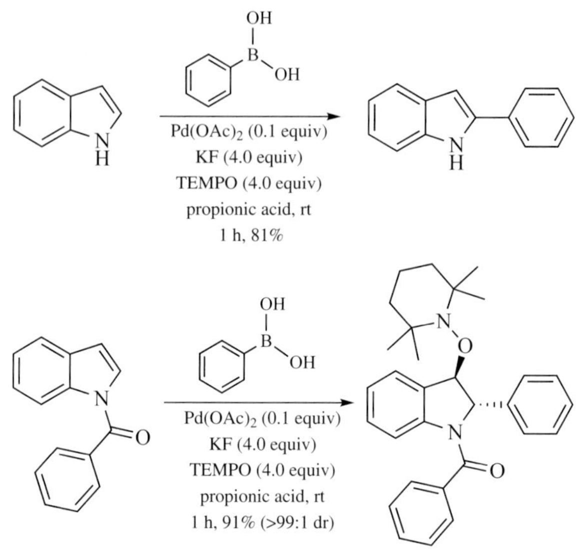

(29)

TEMPO-Oxidized Homo- and Heterocoupling. The oxidative homocoupling of aryl, alkenyl, and alkynyl Grignard reagents with TEMPO and dioxygen as the oxidant has been reported. The reaction required 15 mol % of TEMPO, using dioxygen as the terminal oxidant (eq 30).[78] A later report described the preparation of conjugated polyfluorenes. Bisorganomagnesium compounds are oxidized by TEMPO to furnish the target homocoupled product. This method provides a facile means to form copolymers containing the bisalkyne or the bisethene moiety alternating with a fluorene unit in the polymer back bone (eq 31).[79] In addition to the polymerization of bisorgano magnesium compounds, the title compound also mediates cross-coupling reactions between aryl, alkenyl, and alkynyl magnesium compounds. This constitutes is a new metal-free Sonogashira-type reaction that tolerates many functional groups, such as esters, amides, and cyanides, under the reaction conditions (eq 32).[80]

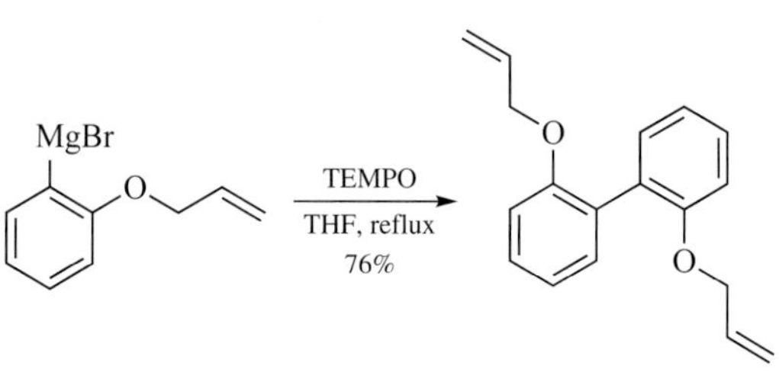

(30)

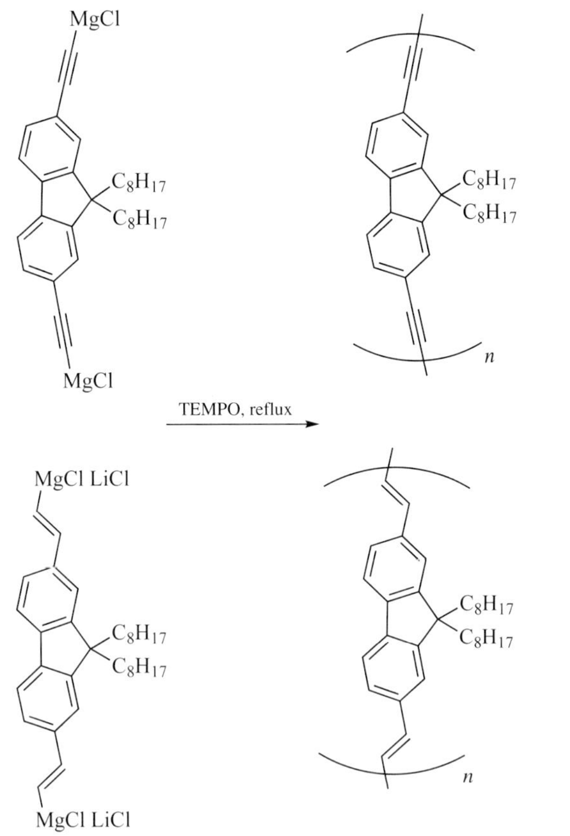

(31)

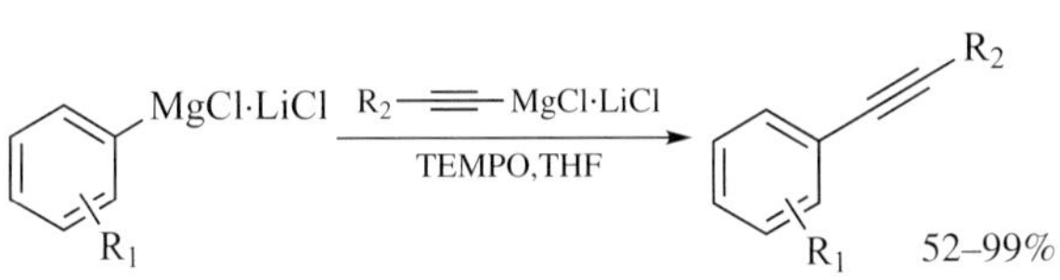

(32)

R_1 = COO_2t-Bu, CON(i-Pr)$_2$, CN, halide, alkyl, etc.
R_2 = aryl, allkyl, alkenyl, heteroaryl

TEMPO Used in C–N Bond Formation. A $Mn(OAc)_3$/TEMPO-mediated tandem azidation-1,2-carbon migration provides α-aryl-alkyl β-azido ketones with a quaternary stereocenter. With this system, allylic silyl ethers can be easily transformed into the corresponding ketone derivatives (eq 33).[81] Another example involved a TEMPO oxidation of styrene into the nitro-olefins using silver nitrite ($AgNO_2$) as the nitration reagent. Relative to traditional synthetic methods, this novel transformation shows good regio- and stereoselectivity (eq 34).[82]

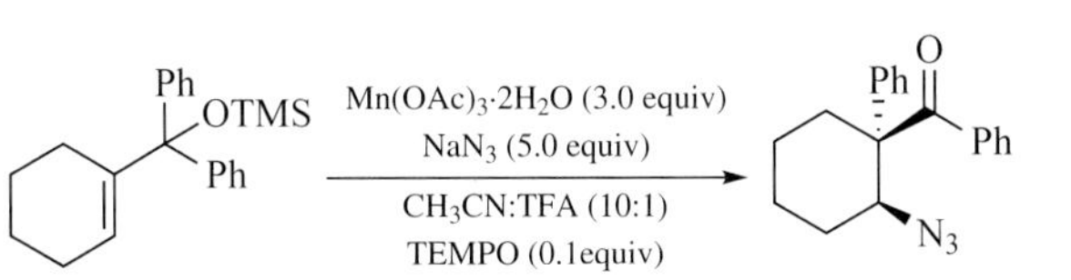

(33)

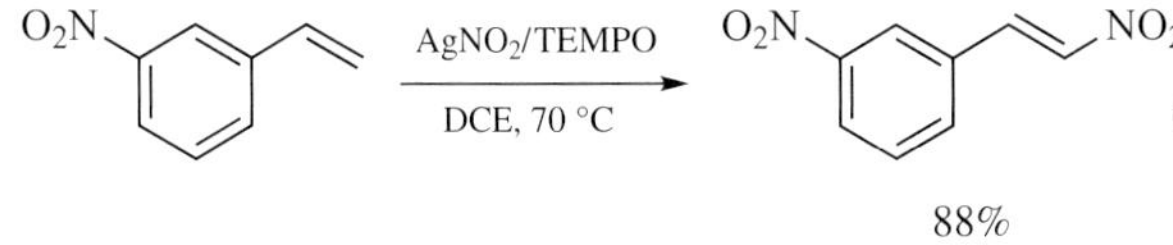

(34)

C–O Bond Formation. Several cases have been reported using TEMPO as an aminoxylation regent. Aryl radicals prepared from $FeSO_4$ react with alkenes to generate a radical that can be trapped with TEMPO (eq 35).[83] Second, intramolecular aminooxygenation of alkenes has been accomplished with $Cu(OTf)_2$ as the catalyst. In this reaction, the TEMPO is an oxidant as well as the oxygen source. Furthermore, with a chiral ligand, this method can readily afford a series of chiral pyrrolidines and indolines (eq 36).[84]

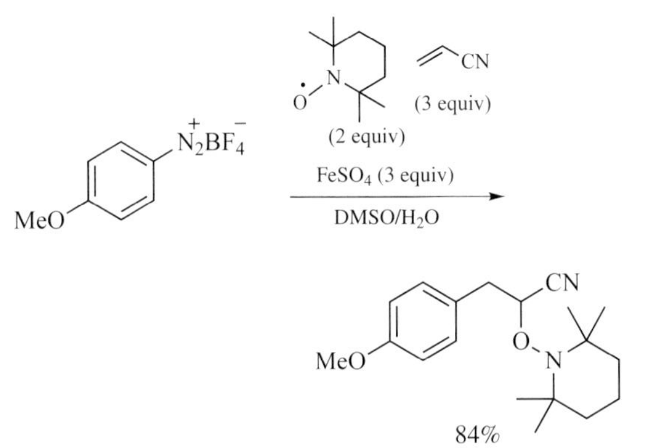

(35)

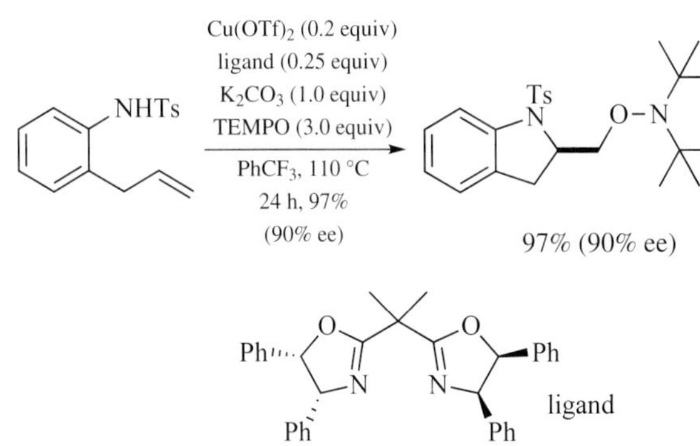

(36)

TEMPO promotes the oxidation of catecholboron enolates to provide α-alkoxyamine (eq 37).[85] The reaction presumably proceeds via a stabilized enolyl radical that is trapped with TEMPO. Aldehydes react with TEMPO, in presence of a resin-supported peptide to undergo asymmetric α-oxyamination (eq 38).[86] An iron-catalyzed amino-oxygenation of ketones, esters, amides, acids, or nitriles has also been reported (eq 39).[87] The reactions involve an initial deprotonation with a strong base. The N–O bond of reaction products can be reductively cleaved upon treatment with zinc in acetic acid to generate the α-hydroxy derivatives.

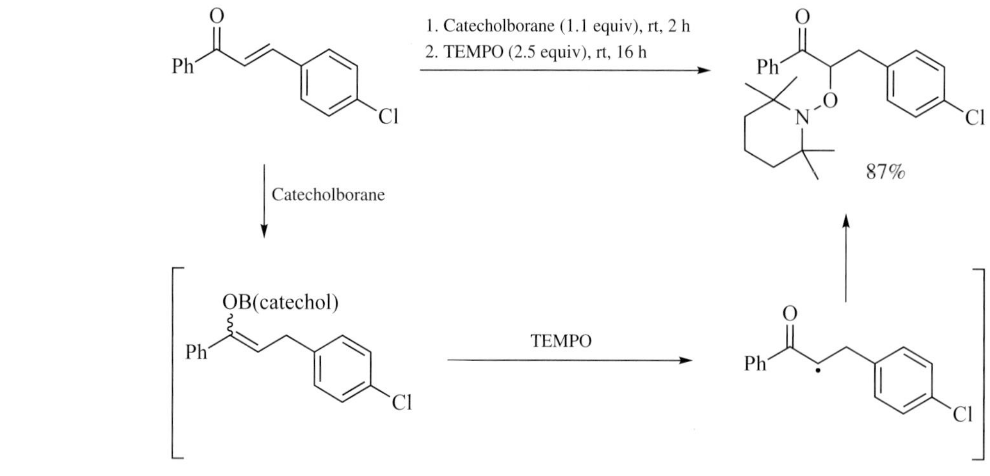

(37)

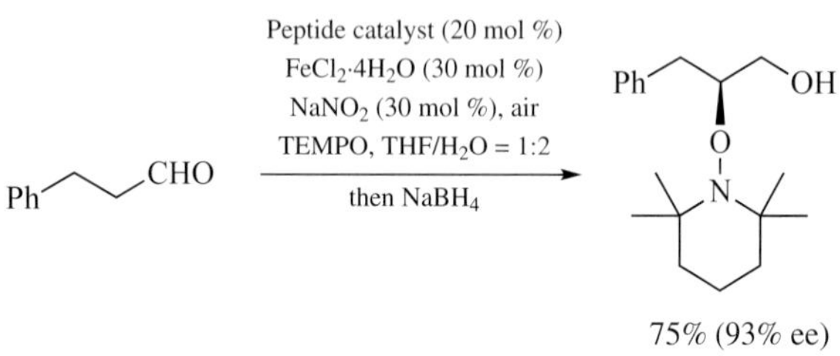

(38)

Peptide catalyst = Pro–D-Pro–Aib–Trp–Trp–$(Leu)_{25.4}$
–HN–CH_2–CH_2–PEG-PS

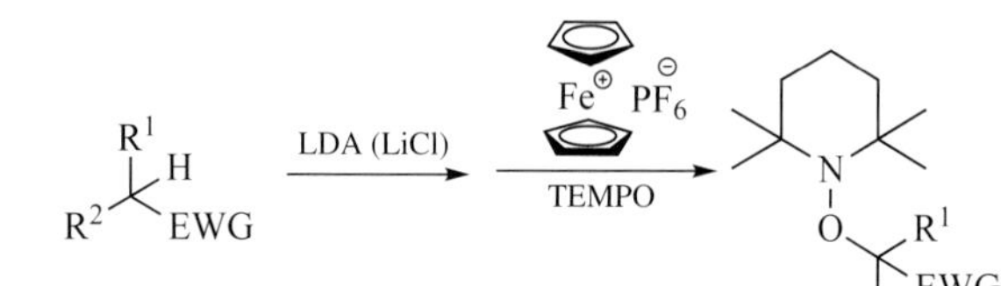

(39)

EWG = CO_2R, $CONR_2$, CO_2H, COR^3

TEMPO-Mediated Cyclization. A novel and efficient protocol for synthesis of 2-aryl quinazolines via benzyl C–H bond amination by a one-pot reaction of arylmethanamines with 2-aminobenzoketones and 2-aminobenzaldehydes has been carried out using the 4-hydroxy-TEMPO radical as the catalyst (eq 40).[88] For the preparation of carbonylated benzofurans, it occurs a radical 5-*exo-dig* cyclization of phenol-linked 1,6-enynes with O_2, TEMPO, and *t*-BuONO (eq 41).[89]

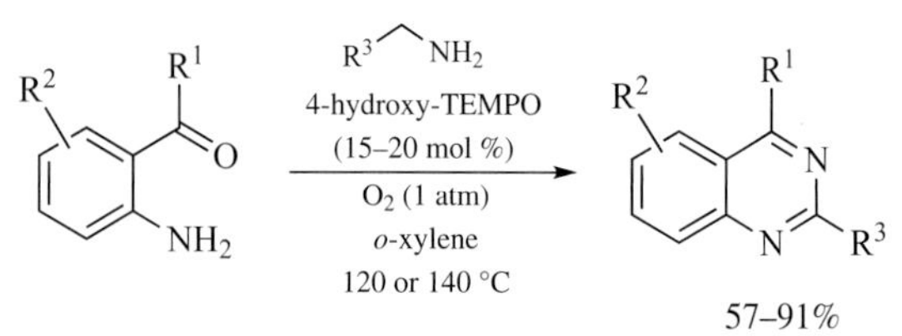

(40)

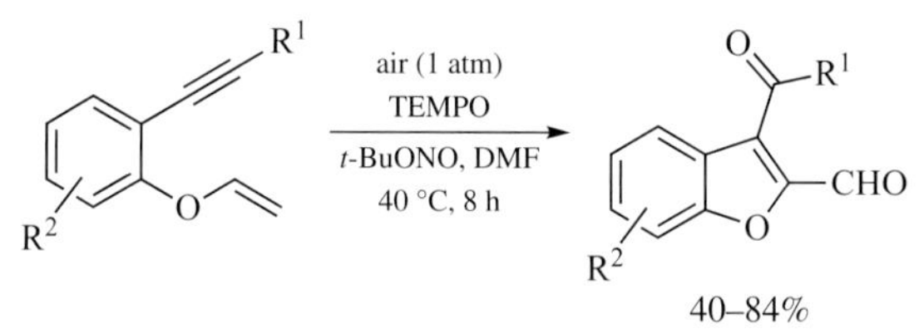

(41)

TEMPO also enables the concise assembly of substituted isoxazole and pyrazole skeletons via aliphatic C–H bond oxidation of oximes and hydrazones (eq 42).[90] It offers an effective method to convert hydrazones to 1*H*-azaindazoles and 1*H*-indazoles via TEMPO, a basic additive, and oxygen that was used as the terminal oxidant (eq 43).[91] In the synthesis of the natural product 15-E_2-IsoP, the key step for the assembly of the cyclopentane core consisted of an oxidative dianion cyclization derived from a β-hydroxyester and subsequent trapping of the allylic radical by TEMPO (eq 44).[92]

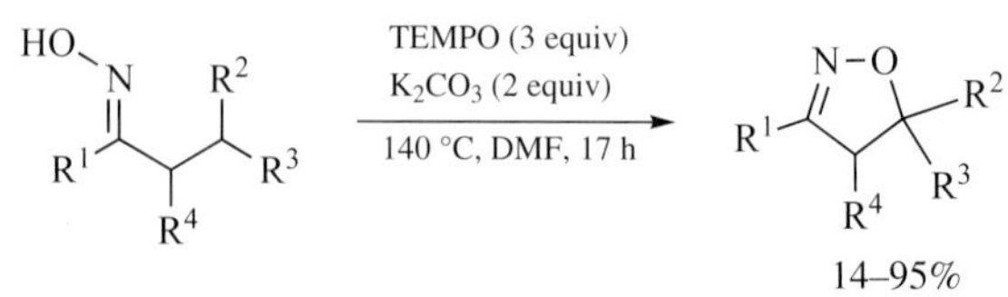

(42)

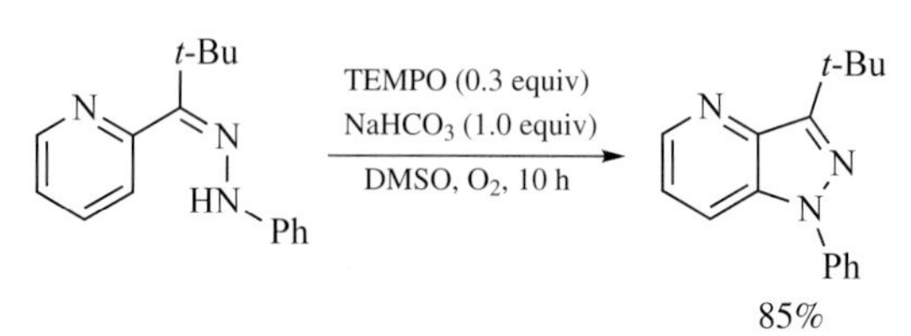

(43)

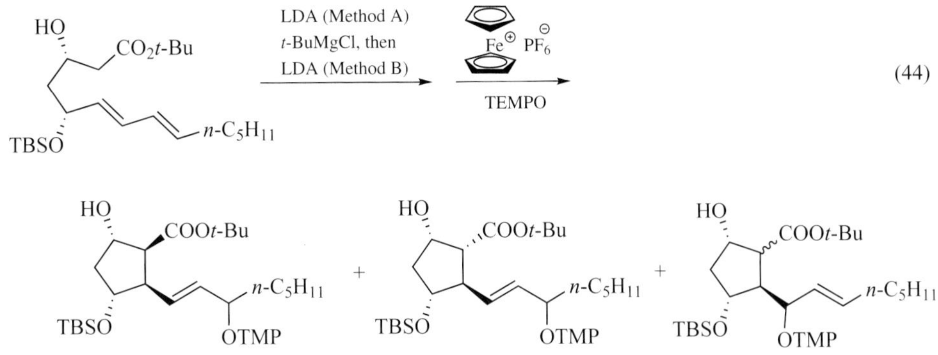

(44)

Method A: 4 : 2.8 : 1 (70%)
Method B: 1 : 6.0 : 1.2 (70%)

TEMPO Used as an Auxiliary Group. TEMPO-derived esters have been used in Knoevenagel condensation (eq 45),[93] aldol reactions (eq 46),[94] and enolate alkylations (eq 47). The selectivity may be caused by the steric hindrance between the TEMPO and the R group (eq 45). Another case, treated with LDA, the *O*-alkanoyl TEMPOs produce the corresponding anionic species that are potent nucleophiles for alkylations, Michael additions, aldol reactions, and so on. Furthermore, for *O*-propionoyl derivatives, with TEMPO as an auxiliary, a good *anti*- or *syn*-selectivities is attained in aldol reactions (eq 46).

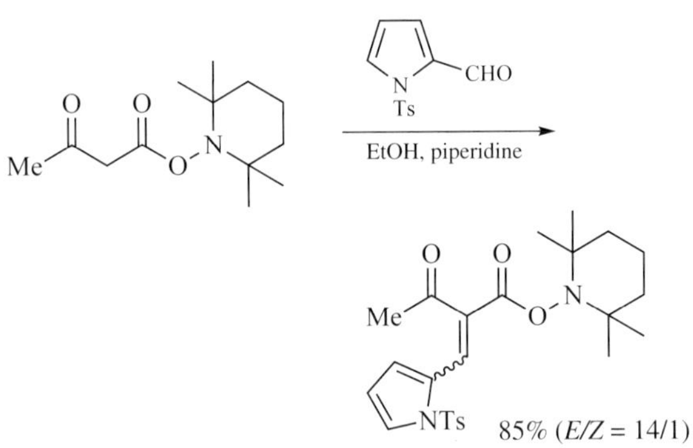

(45)

85% (*E/Z* = 14/1)

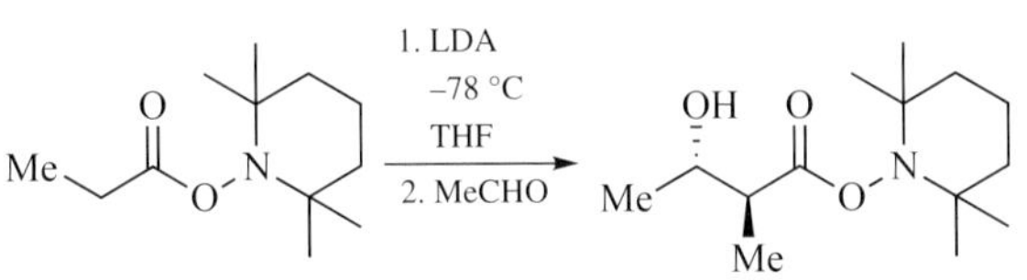

(46)

86% (*anti/syn* = 2.5/1)

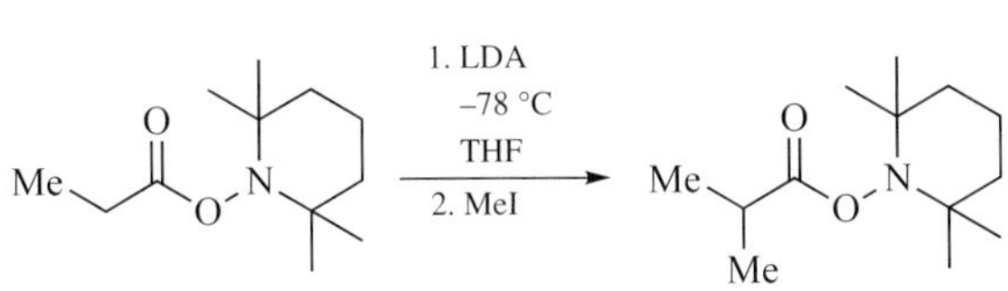

(47)

74%

Other Application of TEMPO. TEMPO is also known to oxidize Rh(II) complexes into Rh(III) complexes.[95] The latter is a particularly useful catalyst for C–C and C–H bond activation. A similar oxidation of a Ru(II) complex to Ru(III) is also known.[96]

1. (a) Rozantsev, E. G.; Sholle, V. D., *Synthesis* **1971**, 190. (b) *Synthesis* **1971**, 401. (c) Keana, J. F. W., *Chem. Rev.* **1978**, *78*, 37. (d) Yamaguchi, M.; Miyazawa, T.; Takata, T.; Endo, T., *Pure Appl. Chem.* **1990**, *62*, 217. (e) Bobbitt, J. M.; Flores, M. C. L., *Heterocycles* **1988**, *27*, 509. (f) Ma, Z.; Bobbitt, J. M., *J. Org. Chem.* **1991**, *56*, 6110.
2. Anelli, P. L.; Biffi, C.; Montanari, F.; Quici, S., *J. Org. Chem.* **1987**, *52*, 2559.
3. Anelli, P. L.; Montanari, F.; Quici, S., *Org. Synth.* **1990**, *69*, 212.
4. Anelli, P. L.; Banfi, S.; Montanari, F.; Quici, S., *J. Org. Chem.* **1989**, *54*, 2970.
5. Brière, R.; Lemaire, H.; Rassat, A., *Bull. Soc. Chem. Fr.* **1965**, 3273.
6. Mahoney, L. R.; Mendenhall, G. D.; Ingold, K. U., *J. Am. Chem. Soc.* **1973**, *95*, 8610.
7. Straub, T. S., *J. Chem. Educ.* **1991**, *68*, 1048.
8. Luzhkov, V. B., *Dokl. Akad. Nauk SSSR* **1983**, *268*, 126 (*Chem. Abstr.* **1983**, *98*, 155 990v).
9. Semmelhack, M. F.; Schmid, C. R.; Cortés, D. A.; Chou, C. S., *J. Am. Chem. Soc.* **1984**, *106*, 3374.
10. (a) Cella, J. A.; Kelley, J. A.; Kenehan, E. F., *J. Org. Chem.* **1975**, *40*, 1860. (b) Cella, J. A.; McGrath, J. P.; Kelley, J. A.; El Soukkary, O.; Hilpert, L., *J. Org. Chem.* **1977**, *42*, 2077.
11. (a) Semmelhack, M. F.; Chou, C. S.; Cortés, D. A., *J. Am. Chem. Soc.* **1983**, *105*, 4492. (b) Inokuchi, T.; Matsumoto, S.; Torii, S., *J. Org. Chem.* **1991**, *56*, 2416.
12. BASF A.-G. Ger. Offen. 4 007 923, **1990** (*Chem. Abstr.* **1991**, *114*, 163 728e).
13. Leanna, M. R.; Sowin, T. J.; Morton, H. E., *Tetrahedron Lett.* **1992**, *33*, 5029.
14. Siedlecka, R.; Skarzewski, J.; Mlochowski, J., *Tetrahedron Lett.* **1990**, *31*, 2177.
15. (a) de Nooy, A. E. J.; Besemer, A. C.; van Bekkum, H., *Synthesis* **1996**, 1153 and references therein. (b) Bobbitt, J. M.; Flores, M. C. L., *Heterocycles* **1988**, *27*, 509.
16. Dijksman, A.; Arends, I. W. C. E.; Sheldon, R. A., *Chem. Commun.* **2000**, 271.
17. Rychnovsky, S. D.; Vaidyanathan, R. J., *J. Org. Chem.* **1999**, *64*, 310.
18. Inokudi, T.; Matsumoto, S.; Nishiyama, T.; Torii, S., *J. Org. Chem.* **1990**, *55*, 462.
19. Zhao, M.; Li, J.; Mano, E.; Song, Z.; Tschaen, D. M.; Grabowski, E. J. J.; Reider, P. J., *J. Org. Chem.* **1999**, *64*, 2564.
20. Bolm, C.; Magnus, A. S.; Hildebrand, J. P., *Org. Lett.* **2000**, *2*, 1173.
21. Miller, R. A.; Hoerrner, R. S., *Org. Lett.* **2003**, *5*, 285.

22. Fey, T.; Fischer, H.; Bachmann, S.; Albert, K.; Bolm, C., *J. Org. Chem.* **2001**, *66*, 8154.
23. Tanyeli, C.; Gümüs, A.; *Tetrahedron Lett.* **2003**, *44*, 1639.
24. (a) U.S. Pat. 4,581,429 (to Solomon, D. H.; Waverley, G.; Rizzardo, E.; Hill, W.; Cacioli, P. (*Chem. Abstr.*, **1985**, *102*, 221335q). (b) Georges, M. K.; Veregin, R. P. N.; Kazmaier, P. M.; Hamer, G. K., *Macromolecules* **1993**, *26*, 2987. (c) Connolly, T. J.; Scaiano, J. C., *Tetrahedron Lett.* **1997**, *38*, 1133.
25. (a) Hawker, C. J., *Acc. Chem. Res.* **1997**, *30*, 373. (b) Kazmaier, P. M.; Daimon, K.; Georges, M. K.; Hamer, G. K.; Veregin, R. P. N., *Macromolecules* **1997**, *30*, 2228. (c) Nakamura, T.; Busfield, W. K.; Jenkins, I. D.; Rizzardo, E.; Thang, S. H.; Suyama, S., *J. Am. Chem. Soc.* **1997**, *119*, 10987. (d) Nakamura, T.; Busfield, W. K., Jenkins, I. D.; Rizzardo, E.; Thang, S. H.; Suyama, S., *J. Org. Chem.* **1997**, *62*, 5578. (e) Georges, M. K.; Veregin, R. P. N.; Kazmaier, P. K.; Hamer, G. K., *Macromol. Symp.* **1994**, *88*, 89.
26. Campbell, I. M., *Introduction to Synthetic Polymers*, 2nd ed.; Oxford University Press London, **2000**; p 119.
27. Cadot, C.; Dalko, P. I.; Cossy, J.; Ollivier, C.; Chuard, R.; Renaud, P., *J. Org. Chem.* **2002**, *67*, 7193.
28. Dijksman, A.; Marino-González, A.; Papyeras, A. M. i.; Arends, I. W. C. E.; Sheldon, R. A., *J. Am. Chem. Soc.* **2001**, *123*, 6826.
29. Benoit, D.; Chaplinski, V.; Braslau, R.; Hawker, C. J., *J. Am. Chem. Soc.* **1999**, *121*, 3904.
30. Hansen, T. M.; Florence, G. J.; Lugo-Mas, P.; Chen, J.; Abrams, J. N.; Forsyth, C. J., *Tetrahedron Lett.* **2003**, *44*, 57.
31. Naik, N.; Braslau, R., *Tetrahedron* **1998**, *54*, 667.
32. Boger, D. L.; Boyce, C. W., *J. Org. Chem.* **2000**, *65*, 4088.
33. Merkley, N.; Warkentin, J., *Can. J. Chem.* **2000**, *78*, 942.
34. (a) Allen, A. D.; Cheng, B.; Fenwick, M. H.; Huang, W.; Missiha, S.; Tahmassebi, D.; Tidwell, T. T., *Org. Lett.* **1999**, *1*, 693. (b) Carter, J.; Fenwick, M. H.; Huang, W.; Popic, V. V.; Tidwell, T. T., *Can. J. Chem.* **1999**, *77*, 806. (c) Allen, A. D.; Fenwick, M. F.; Henry-Riyad, H.; Tidwell, T. T., *J. Org. Chem.* **2001**, *66*, 5759. (d) Huang, W.; Henry-Riyad, H.; Tidwell, T. T., *J. Am. Chem. Soc.* **1999**, *121*, 3939. (e) Allen, A. D.; Cheng, B.; Fenwick, M. H.; Henry-Riyad, H.; Nikolaev, V. A.; Shikova, E. A.; Tahmassebi, D.; Tidwell, T., *J. Org. Chem.* **2001**, *66*, 2611.
35. (a) Paleos, C. M.; Dais, P., *J. Chem. Soc., Chem. Commun.* **1977**, 345. (b) Root, K. S.; Hill, C. L.; Lawrence, L. M.; Whitesides, G. M., *J. Am. Chem. Soc.* **1989**, *111*, 5405. (c) Rozantesv, E. G.; Golubev, V. A., *Bull. Acad. Sci., U.S.S.R* **1966**, 852. (d) Muller, R. N.; Haverbeke, Y. V.; Bonnet, P. A.; Chapat, J.-P.; Vallet, P., *Anal. Chem.* **1990**, *62*, 1671. (e) Lee, T. D.; Keana, J. F. W., *J. Org. Chem.* **1975**, *40*, 3145.
36. (a) Fabbrini, M.; Galli, C.; Gentili, P.; Macchitella, D., *Tetrahedron Lett.* **2001**, *42*, 7551. (b) Fabbrini, M.; Galli, C.; Gentili, P., *J. Mol. Catal. B: Enzymatic* **2002**, *16*, 231.
37. (a) Brünjes, M.; Sourkouni-Argirusi, G.; Kirschning, A., *Adv. Synth. Catal.* **2003**, *345*, 635. (b) Kloth, K.; Brünjes, M.; Kunst, E.; Jöge, T.; Gallier, F.; Adibekian, A.; Kirschning, A., *Adv. Synth. Catal.* **2005**, *347*, 1423.
38. Sakuratani, K.; Togo, H., *Synthesis* **2003**, 21.
39. Cecchetto, A.; Fontana, F.; Minisci, F.; Recupero, F., *Tetrahedron Lett.* **2001**, *42*, 6651.
40. Gamez, P.; Arends, I.W.C.E.; Sheldon, R. A.; Reedijk, J., *Adv. Synth. Catal.* **2004**, *346*, 805.
41. Wang, N.; Liu, R.; Chen, J.; Liang, X., *Chem. Commun.* **2005**, 5322.
42. Liu, R.; Liang, X.; Dong, C.; Hu, X., *J. Am. Chem. Soc.* **2004**, *126*, 4112.
43. Herrerías, C. I.; Zhang, T. Y.; Li., C.-J., *Tetrahedron Lett.* **2006**, *47*, 13.
44. Jiang, N.; Ragauskas, A. J., *Tetrahedron Lett.* **2005**, *46*, 3323.
45. Wu, X.-E.; Ma, L.; Ding, M.-X.; Gao, L.-X., *Synlett* **2005**, 607.
46. Ansari, I. A.; Gree, R., *Org. Lett.* **2002**, *4*, 1507.
47. (a) Ciriminna, R.; Bolm, C.; Fey, T.; Pagliaro, M., *Adv. Synth. Catal.* **2002**, *344*, 159. (b) Testa, M. L.; Ciriminna, R.; Hajji, C.; Zaballos Garcia, E.; Ciclosi, M.; Sepulveda Arques, J.; Pagliaro, M., *Adv. Synth. Catal.* **2004**, *346*, 655.
48. Kashiwagi, Y.; Ikezoe, H.; Ono, T., *Synlett* **2006**, 69.
49. Weik, S.; Nicholson, G.; Jung, G.; Rademann, J., *Angew. Chem., Int. Ed.* **2001**, *40*, 1436.
50. Pozzi, G.; Cavazzini, M.; Quici, S.; Benaglia, M.; Dell'Anna, G., *Org. Lett.* **2004**, *6*, 441.
51. Benaglia, M.; Puglisi, A.; Holczknecht, O.; Quici, S.; Pozzi, G., *Tetrahedron* **2005**, *61*, 12058.
52. Holczknecht, O.; Cavazzini, M.; Quici, S.; Shepperson, I.; Pozzi, G., *Adv. Synth. Catal.* **2005**, *347*, 677.
53. Velusamy, S.; Kumar, A. V.; Saini, R.; Punniyamurthy, T., *Tetrahedron Lett.* **2005**, *46*, 3819.
54. Dao, J.; Benoit, D.; Hawker, C. J., *J. Polym. Sci. Part A: Polym. Chem.* **1998**, *36*, 2161.
55. (a) Babiarz, J. E.; Cunkle, G. T.; DeBellis, A. D.; Eveland, D.; Pastor, S. D.; Shum, S. P., *J. Org. Chem.* **2002**, *67*, 6831. (b) Coseri, S.; Ingold, K. U., *Org. Lett.* **2004**, *6*, 1641.
56. Johnston, L. J.; Tencer, M.; Scaiano, J. C., *J. Org. Chem.* **1986**, *51*, 2806.
57. Whitesides, G. M.; Newirth, T. L., *J. Org. Chem.* **1975**, *40*, 3448.
58. Nagashima, T.; Curran, D. P., *Synlett* **1996**, 330.
59. Dalko, P. I., *Tetrahedron Lett.* **1999**, *40*, 4035.
60. Takata, T.; Tsujino, Y.; Nakanishi, S.; Nakamura, K.; Yoshida, E.; Endo, T., *Chem. Lett.* **1999**, *28*, 937.
61. Breton, T.; Liaigre, D.; Belgsir, E. M., *Tetrahedron Lett.* **2005**, *46*, 2487.
62. Schämann, M.; Schäfer, H. J., *Synlett* **2004**, 1601.
63. Marque, S.; Fischer, H.; Baier, E.; Studer, A., *J. Org. Chem.* **2001**, *66*, 1146.
64. (a) Studer, A., *Angew. Chem., Int. Ed.* **2000**, *39*, 1108. (b) Molawi, K.; Schulte, T.; Siegenthaler, K. O.; Wetter, C.; Studer, A., *Chem. Eur. J.* **2005**, *11*, 2335.
65. (a) Wetter, C.; Jantos, K.; Woithe, K.; Studer, A., *Org. Lett.* **2003**, *5*, 2899. (b) Wetter, C.; Studer, A., *Chem. Commun.* **2004**, 174.
66. Shibuya, M.; Tomizawa, M.; Iwabuchi, Y., *J. Org. Chem.* **2008**, *73*, 4750.
67. Ebine, M.; Suga, Y.; Fuwa, H.; Sasaki, M., *Org. Biomol. Chem.* **2010**, *8*, 39.
68. Zhang, M.; Chen, C. C.; Ma, W. H.; Zhao, J. C., *Angew. Chem., Int. Ed.* **2008**, *47*, 9730.
69. Liu, X. L.; Xia, Q. Q.; Zhang, Y. J.; Chen, C. Y.; Chen, W. Z., *J. Org. Chem.* **2013**, *78*, 8531.
70. Hoover, J. M.; Stahl, S. S., *J. Am. Chem. Soc.* **2011**, *133*, 16901.
71. Brioche, J.; Masson, G.; Zhu, J. P., *Org. Lett.* **2010**, *12*, 1432.
72. Guin, J.; Sarkar, S. D.; Grimme, S.; Studer, A., *Angew. Chem., Int. Ed.* **2008**, *47*, 8727.
73. Wang, T.; Jiao, N., *J. Am. Chem. Soc.* **2013**, *135*, 11692.
74. Dornan, L. M.; Cao, Q.; James, C. A. F.; Crawford, J. J.; Cook, M. J.; Muldoon, M. J., *Chem. Commun.* **2013**, *49*, 6030.
75. Zhang, B.; Cui, Y. X.; Jiao, N., *Chem. Commun.* **2012**, *48*, 4498.
76. Vogler, T.; Studer, A., *Org. Lett.* **2007**, *10*, 129.
77. Kirchberg, S.; Frohlich, R.; Studer, A., *Angew. Chem., Int. Ed.* **2009**, *48*, 4235.
78. Maji, M. S.; Pfeifer, T.; Studer, A., *Angew. Chem., Int. Ed.* **2008**, *47*, 9547.
79. Maji, M. S.; Pfeifer, T.; Studer, A., *Chem. Eur. J.* **2010**, *16*, 5872.
80. Maji, M. S.; Murarka, S.; Studer, A., *Org. Lett.* **2010**, *12*, 3878.
81. Chen, Z. M.; Zhang, Z.; Tu, Y. Q.; Xu, M. H.; Zhang, F. M.; Li, C. C.; Wang, S. H., *Chem. Commun.* **2014**, *50*, 10805.
82. Maity, S.; Manna, S.; Rana, S.; Naveen, T.; Mallick, A.; Maiti, D., *J. Am. Chem. Soc.* **2013**, *135*, 3355.

83. Heinrich, M. R.; Wetzel, A.; Kirschstein, M., *Org. Lett.* **2007**, *9*, 3833.
84. Fuller, P. H.; Kim, J.-W.; Chemler, S. R., *J. Am. Chem. Soc.* **2008**, *130*, 17638.
85. Pouliot, M.; Renaud, P.; Schenk, K.; Studer, A.; Vogler, T., *Angew. Chem., Int. Ed.* **2009**, *48*, 6037.
86. Akagawa, K.; Fujiwara, T.; Sakamoto, S.; Kudo, K., *Org. Lett.* **2010**, *12*, 1804.
87. Dinca, E.; Hartmann, P.; Smrcek, J.; Dix, I.; Jones, P. G.; Jahn, U., *Eur. J. Org. Chem.* **2012**, 4461.
88. Han, B.; Wang, C.; Han, R. F.; Yu, W.; Duan, X. Y.; Fang, R.; Yang, X. L., *Chem. Commun.* **2011**, *47*, 7818.
89. Hu, M.; Song, R. J.; Li, J. H., *Angew. Chem., Int. Ed.* **2015,** *54*, 608.
90. Zhu, X.; Wang, Y. F.; Ren, W.; Zhang, F. L.; Chiba, S., *Org. Lett.* **2013**, *15*, 3214.
91. Hu, J. T.; Xu, H. C.; Nie, P. J.; Xie, X. B.; Nie, Z. X.; Rao, Y., *Chem. Eur. J.* **2014**, *20*, 3932.
92. Jahn, U.; Dinca, E., *J. Org. Chem.* **2010**, *75*, 4480.
93. Inokuchi, T.; Kawafuchi, H., *J. Org. Chem.* **2006**, *71*, 947.
94. Inokuchi, T.; Kawafuchi, H., *J. Org. Chem.* **2007**, *72*, 1472.
95. Chan, K. S.; Li, X. Z.; Dzik, W. I.; Bruin, B., *J. Am. Chem. Soc.* **2011**, *133*, 18022.
96. Wu, A.; Mayer, J. M., *J. Am. Chem. Soc.* **2008**, *130*, 14745.

(*S*)-2-(5-1*H*-Tetrazolyl)pyrrolidine and (*R*)-2-(5-1*H*-Tetrazolyl)pyrrolidine[1]

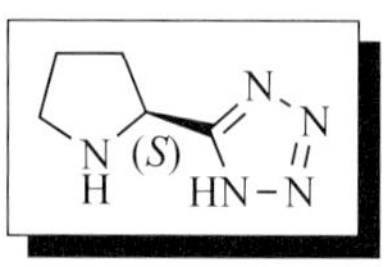

[33878-70-5] $C_5H_9N_5$ (MW 139.086)
InChI = 1S/C5H9N5/c1-2-4(6-3-1)5-7-9-10-8-5/h4,6H,1-3H2,(H,7,8,9,10)/t4-/m0/s1
InChIKey = XUHYQIQIENDJER-BYPYZUCNSA-N
(*S*)-isomer
[702700-79-6] $C_5H_9N_5$ (MW 139.086)
InChI = 1S/C5H9N5/c1-2-4(6-3-1)5-7-9-10-8-5/h4,6H,1-3H2,(H,7,8,9,10)/t4-/m1/s1
InChIKey = XUHYQIQIENDJER-BYPYZUCNSA-N

(asymmetric organocatalyst; used in the asymmetric Mannich;[2,3] Aldol;[4,5] α-oxidation;[6,7] α-amination;[8] and nitro-Michael[3,9] reactions.)

Alternative Names: L-pyrrolidin-2-yl-1*H*-tetrazole, (*S*)-5-pyrrolidine-2-yl-1*H*-tetrazole, (*S*)-2-tetrazolyl pyrrolidine.
Physical Data: mp 272–274 °C.
Solubility: soluble in most protic solvents. Partially soluble in aprotic solvents such as dichloromethane, tetrahydrofuran, and acetonitrile.
Form Supplied in: amorphous off-white solid powder. Purity >95%.
Analysis of Reagent Purity: ^{1}H NMR (400 MHz, CD_3COOD): δ = 4.95 (1H, *t*, *J* = 8.1 Hz, NHC*H*CN), 3.45 (2H, m, NHCH_2), 2.60–2.16 (4H, m, CH_2CH_2). [α]25D = +1.2 (*c*, 0.5, H_2O).
Preparative Methods: it has been noted that the published synthesis of this compound may be hazardous due to the formation of ammonium azide, a shock sensitive, explosive material. This problem has been circumvented through the use of triethylammonium acetate in place of ammonium chloride and by using *N*-methylpyrrolidinone as solvent.[11]
Purification: reagent can be used as supplied. See references 3, 6, and 10 for more details.
Handling, Storage, and Precautions: store in a cool, dry place. Avoid heating to extreme temperatures. Use a plastic spatula and weigh into a vessel which has been pre-treated with an electrostatic gun to avoid difficulties in compound containment.

Original Commentary

Alexander J. André Cobb, Steven V. Ley, Deborah A. Longbottom, David M. Shaw & Robbie J. Mutton
University of Cambridge, Cambridge, UK

Asymmetric Organocatalytic Reactions. The reagent behaves as an alternative to L-proline in various asymmetric processes. It has the advantage over proline of being more soluble in a broad range of solvents. The tetrazole's increased solubility also provides increased reactivity leading to greater turnover numbers than can be achieved with proline in many catalytic reactions. Hence the reagent can be used in lower concentrations in a wider range of solvents. In the presence of an enolizable ketone or aldehyde, a chiral enamine is formed which will react with electrophiles asymmetrically in a number of processes.

The Asymmetric Mannich Reaction.[2,3] The reaction between an enolizable carbonyl compound and *N*-*p*-methoxyphenyl-protected ethyl glyoxalate using a catalytic amount of this reagent has been described (eq 1).[2,3] Symmetrical ketones have been shown to be ideal substrates for this reaction, which proceeds with excellent diastereoselectivities and enantioselectivities. The reagent can be used at levels of 1 mol % in dichloromethane, demonstrating its increased reactivity over proline. Under the same conditions, proline does not give any reaction and even in more polar solvents such as DMSO is normally used at far greater loadings.[1]

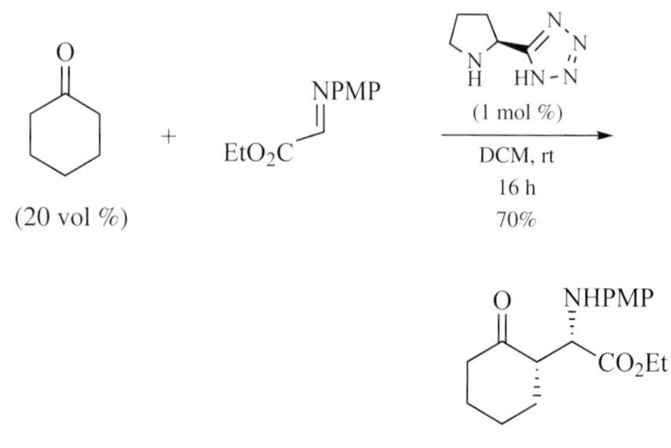

(1)

The Asymmetric Aldol Reaction.[4,5] The reaction between an enolizable carbonyl compound and a range of non-enolizable aldehydes using a catalytic amount of this reagent has been described (eqs 2 and 3).

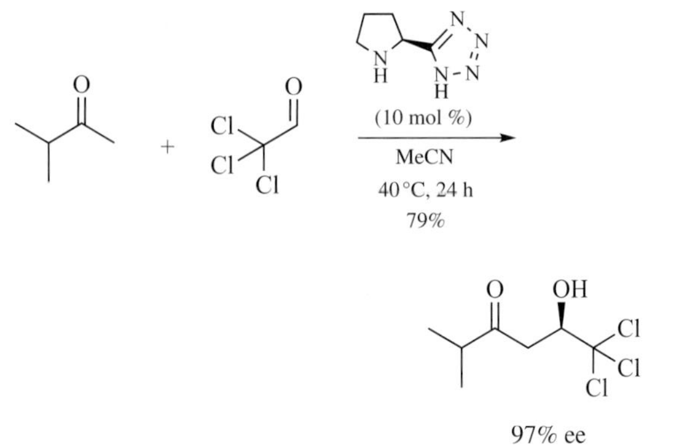

(2)

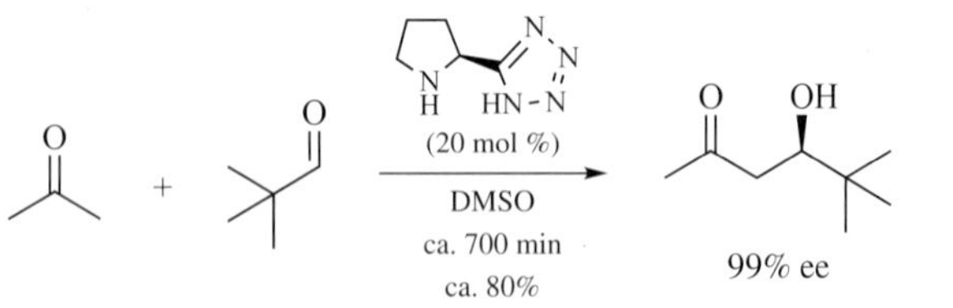

(3)

In the case of eq 3, the use of a less reactive aldehyde shows the advantage of employing a more reactive organocatalyst. In this example, using the tetrazole catalyst, about 80% of the aldehyde was consumed after 12 h, whereas only 20% conversion was observed in the same time with L-proline as catalyst.

The Asymmetric α-Oxidation.[6,7] The reaction between an enolizable carbonyl compound and nitrosobenzene, using a catalytic amount of this reagent, is described (eq 4).

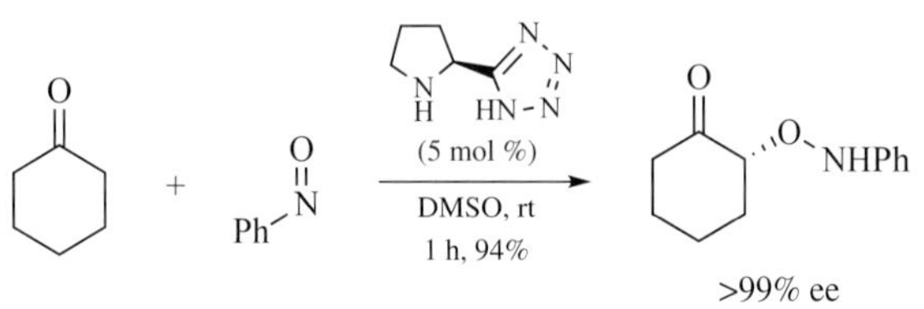

(4)

A range of aldehydes and ketones were tested, all of which gave good yields and excellent enantioselectivities. The resulting nitrogen-oxygen bond can be cleaved using copper (II) sulfate in methanol at 0 °C. The increased reactivity over proline was demonstrated by performing the same reaction in the presence of equimolar amounts of this reagent and D-proline, when the (*R*)-enantiomer was formed preferentially (eq 5).

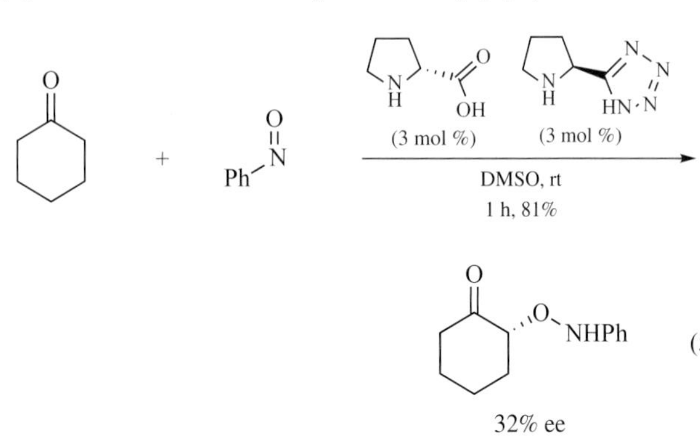

(5)

The Asymmetric α-Amination.[8] The reaction between an enolizable aldehyde and dibenzyl azodicarboxylate has been achieved in high enantioselectivity using this reagent in the synthesis of LFA-1 antagonist BIRT-377 (eq 6). The resulting aminoaldehyde was obtained in 80% ee, which can be increased to >99% on recrystallization from ethyl acetate/hexane. This synthesis is particularly noteworthy for its construction of a quaternary stereocenter.

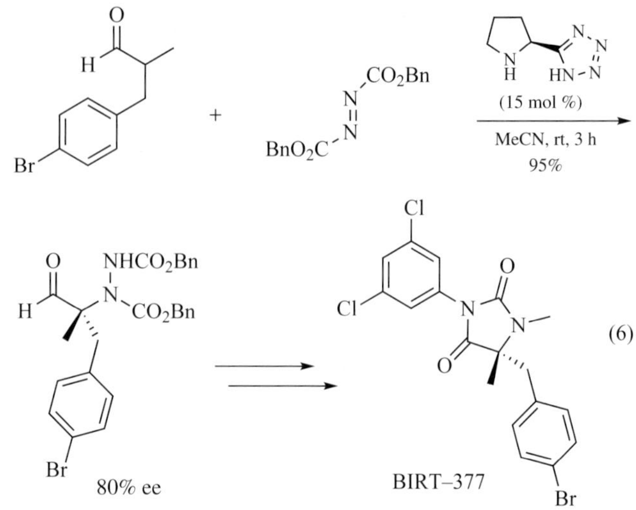

(6)

The Asymmetric Nitro-Michael Reaction.[3,9] The reaction between an enolizable carbonyl compound and a range of nitroolefins has been described. It was again shown that this reagent out performs proline in a range of solvents (eq 7). Alcoholic solvents were found to be best for this reaction, although it also proceeded in dichloromethane, in which proline was unreactive.[9]

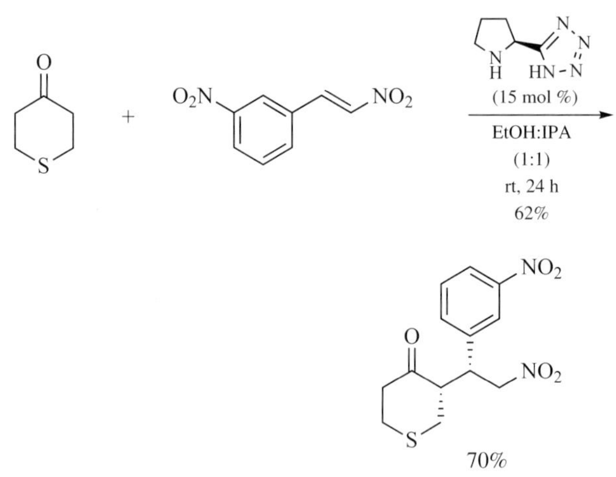

(7)

First Update

Alexander J. André Cobb, Steven V. Ley, Deborah A. Longbottom, David M. Shaw & Robbie J. Mutton
University of Cambridge, Cambridge, UK

(*S*)-2-(5-1*H*-Tetrazolyl)pyrrolidine has been successfully applied as an organocatalyst to a range of processes, including aldol, Mannich, Michael, and other α-functionalization reactions, some of which are highlighted below.

Aldol Processes. Perhaps the most well studied asymmetric reaction when it comes to asymmetric enamine catalysis, is the aldol reaction. Unsurprisingly, the proline tetrazole catalyst has been successfully applied to a range of these processes. Wang and coworkers utilized it in the addition of a variety of branched aldehydes with isatin **1** to generate the corresponding 3-substituted 3-hydroxyindol-2-ones in good to excellent enantioselectivities and reasonable yields (eq 8).[13]

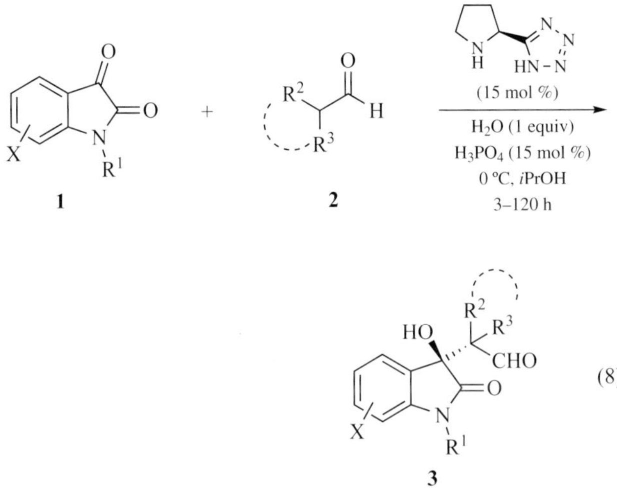

(8)

Yield : 46–92%
ee : 49–93%

Ward et al. have synthesized the polypropionate serricornin and membrenone B using a (*S*)-2-(5-1*H*-tetrazolyl)pyrrolidine induced organocatalytic aldol process involving a dynamic kinetic resolution (eq 9).[14] Interestingly, Ward first reported this process using L-proline,[14d] but later identified that the increased solubility of (*S*)-2-(5-1*H*-tetrazolyl)pyrrolidine was advantageous to the reaction.

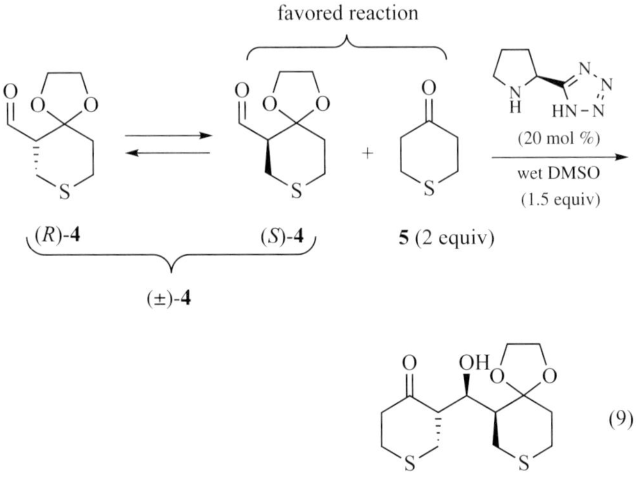

(9)

Yield : 75%
ee : >98%

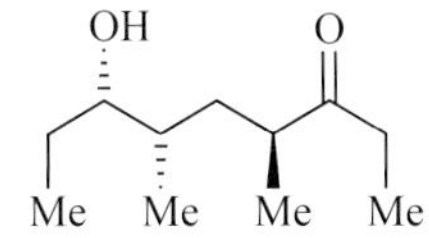

serricornin, **7**

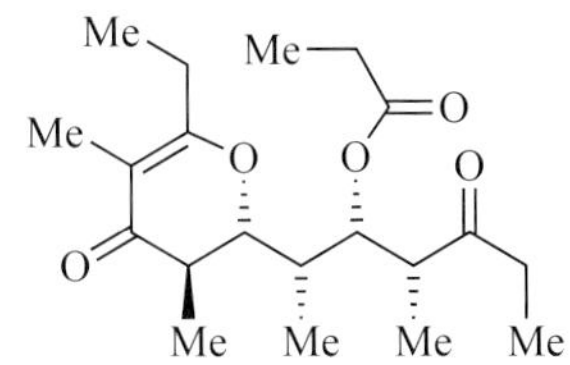

(+)-membrenone B, **8**

Funabiki et al. have utilized the tetrazole catalyst in the direct aldol reaction of a CF_3CHO hemiacetal with a range of aromatic methyl ketones (eq 10). The advantage of this approach, is that there is no need to generate the aldehyde itself, which would unhelpfully be a gas at room temperature.[15]

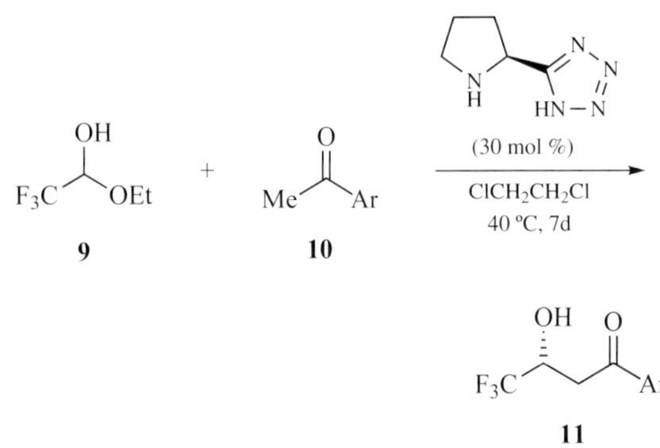

(10)

Yield : 60–89%
ee : 44->99%

Mannich Reactions. Barbas and coworkers have developed an organocatalytic Mannich reaction catalyzed by (*S*)-2-(5-1*H*-tetrazolyl)pyrrolidine, where the regioselectivity of the reaction is dependent on the nitrogen protecting group used (eq 11). If azide is used as the protecting group, then the more substituted enamine forms, as deprotonation of the α-carbon is facilitated by the enhanced acidity provided by azide substitution. This enamine is thermodynamically more stable than the enamine generated by deprotonation of the other α-carbon, due to resonance stabilization. In the case of the phthalimidoketone, the enamine formed at the methyl position is more stable than the one formed at the methylene position owing to steric hindrance in the transition state of the latter.[16]

Melchiorre and coworkers have developed a (*S*)-2-(5-1*H*-tetrazolyl)pyrrolidine catalyzed reaction whereby the imine is generated *in situ* from the corresponding α-amidosulfone **18** (eq 12).[17] The report focuses on the control of *syn* and *anti* selectivity, where the tetrazole catalyst gives the *syn*-Mannich product.

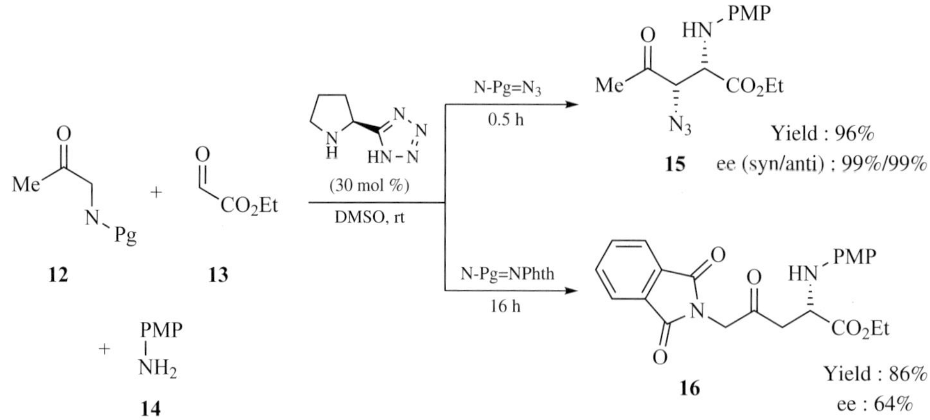

(11)

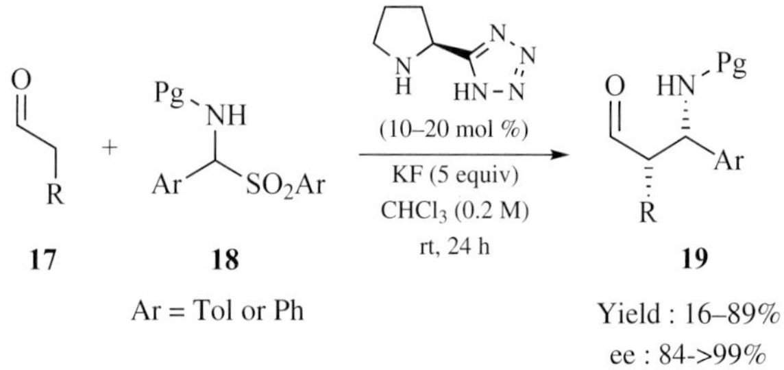

(12)

Michael Additions. (*S*)-2-(5-1*H*-tetrazolyl)pyrrolidine has also been used by Ley and coworkers for the addition of a variety of nitronates to various enones, presumably through iminium activation (eq 13).[18]

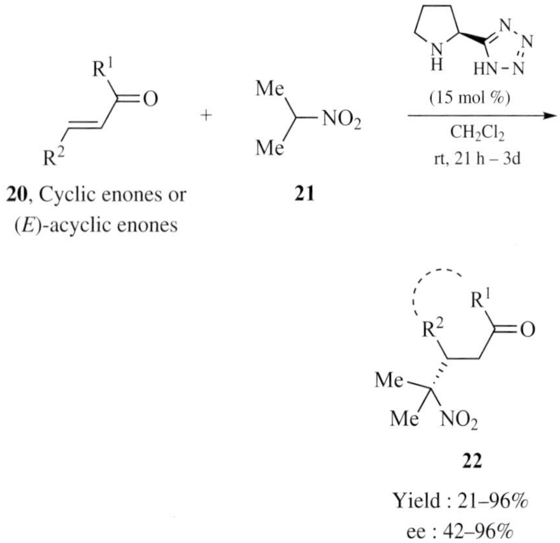

(13)

Similarly, Ley and coworkers have utilized the tetrazole catalyst in the addition of malonates to various enones.[19] In this investigation, addition of an external base was required to deprotonate the malonate. A typical substrate scope from this investigation is shown (eq 14).

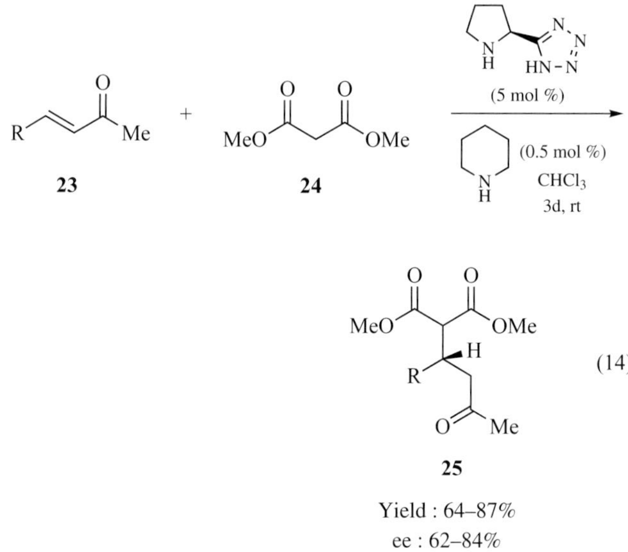

(14)

α-Aminoxylation. (*S*)-2-(5-1*H*-tetrazolyl)pyrrolidine has been used to good effect in various α-aminoxylation processes. Barbas and coworkers demonstrated its application towards an asymmetric desymmetrization (eq 15).[20] Under the reaction conditions, O–N bond heterolysis also occurred to give the free α hydroxy compound.

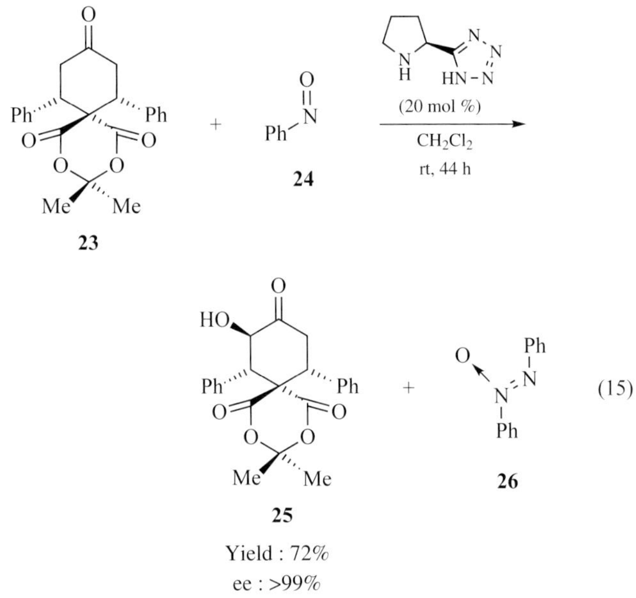

(15)

Ley and coworkers used the α-aminoxylation to gain access to a range of chiral dihydro-1,2-oxazines.[21] This was achieved in two steps; the first being the catalytic reaction, and the second being nucleophilic attack by the resulting nitrogen on a vinyl phosphonium salt to generate a substrate that cyclized via an intramolecular Wittig process (eq 16).

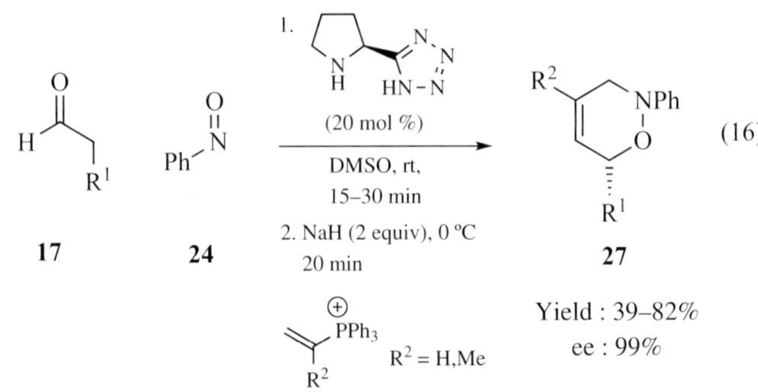

(16)

Yamamoto also used a two-stage process where the asymmetric α-aminoxylation was followed by Grignard addition to the carbonyl to generate a range of chiral 1,2-diols (eq 17).[22]

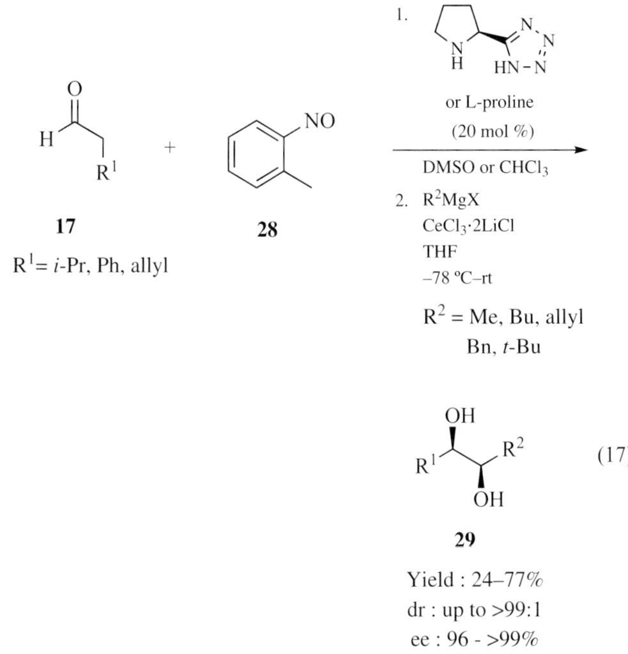

(17)

Finally, one other interesting application of the α-aminoxylation was reported by Frongia and coworkers, who discovered a fascinating ring expansion of an achiral cyclobutanone upon exposure to nitrosobenzene **29** and (S)-2-(5-1H-tetrazolyl)pyrrolidine (eq 18).[23]

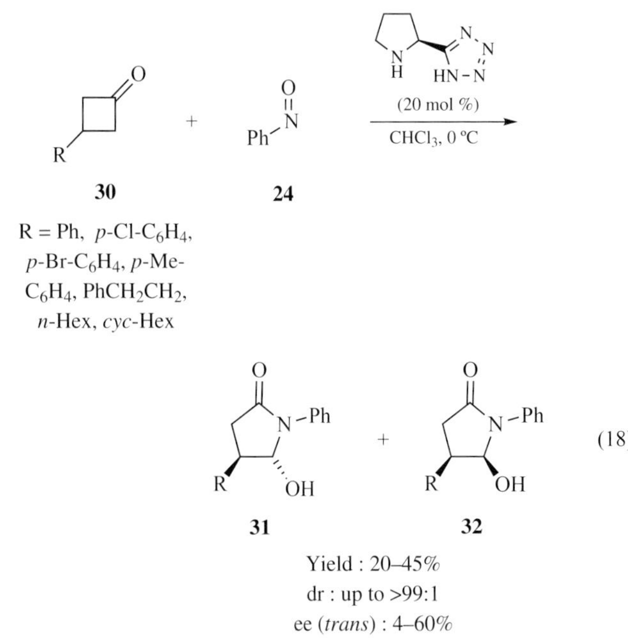

(18)

***α*-Aminations.** The procedure described in section 4 for synthesis of chiral dihydro-1,2-oxazines has also been applied to the synthesis of pyridazines (eq 19)[21c,24] by Ley and coworkers, who subsequently applied this methodology to the synthesis of the natural product chloptosin **35**.[25]

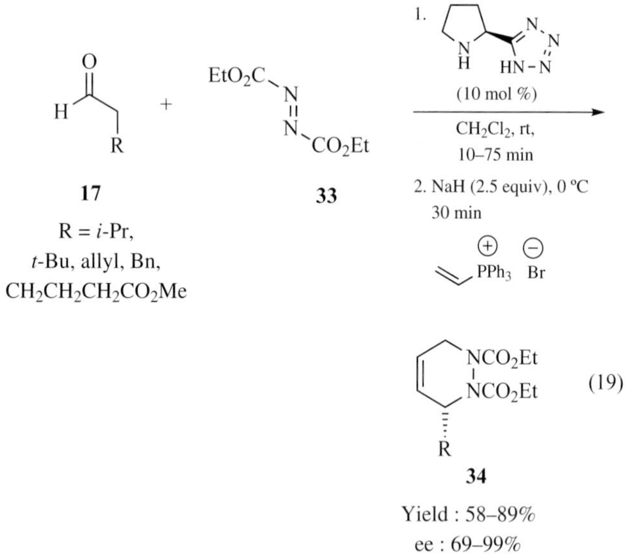

In the same paper, the authors modify catalyst structure to optimize the performance.

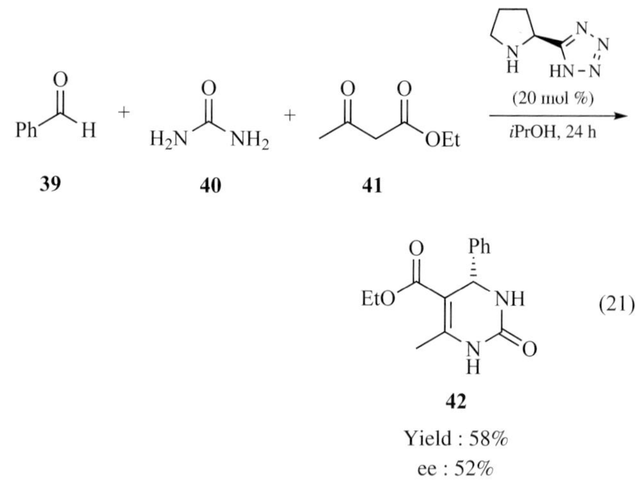

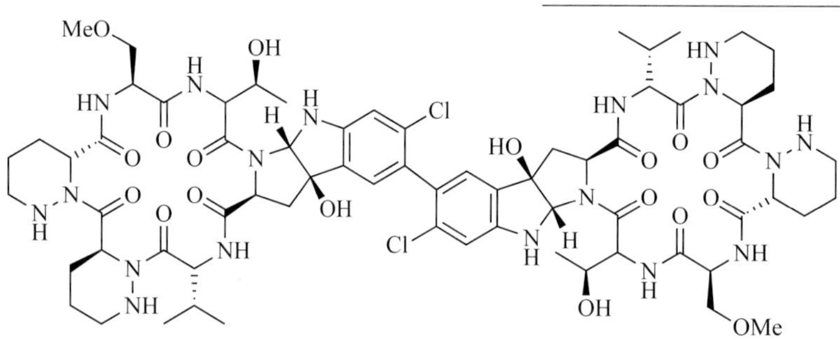

chloptosin **35**

Finally, Shigenaga et al. have utilized the tetrazole catalyzed α-amination (using (*R*)-2-(5-1*H*-tetrazolyl)pyrrolidine in this instance), in the synthesis of a stimulus-responsive amino acid **36** for the control of peptidyl function (eq 20).[26]

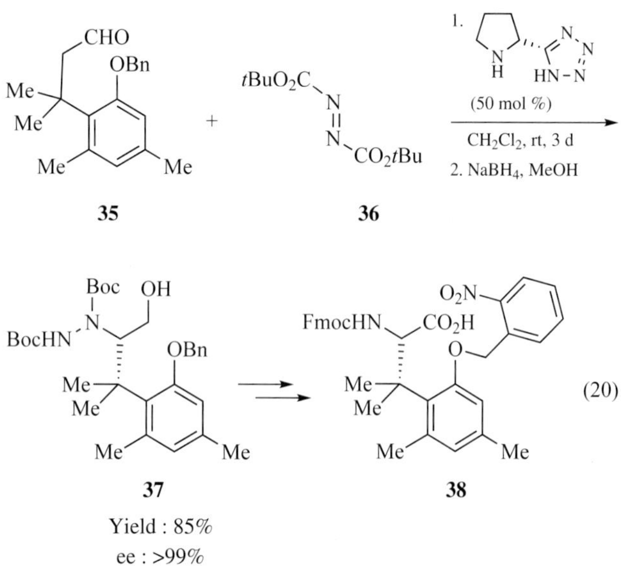

Multicomponent Reactions. The tetrazole catalyst has been successfully applied to the asymmetric Biginelli reaction. Zhao and coworkers achieved the coupling of benzaldehyde **39**, urea **40**, and ethyl acetoacetate **41** in moderate enantioselectivity (eq 21).[27]

In a similar vein, Su et al. used the tetrazole catalyst in a one-pot Hantzsch synthesis, using ammonium bicarbonate as a nitrogen source (eq 22).[28]

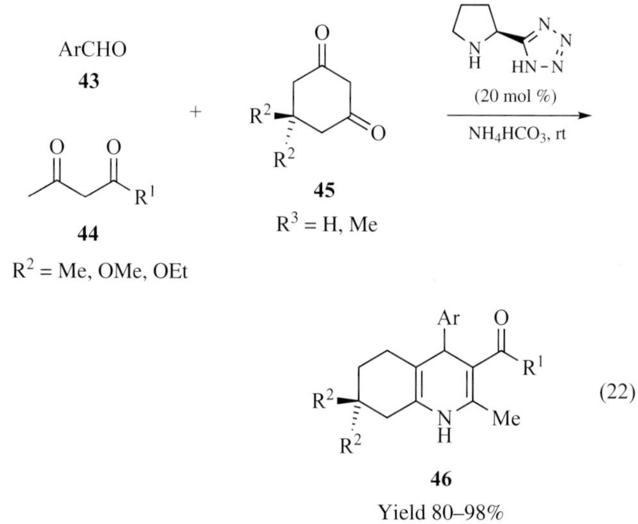

Conclusions. This update has highlighted some of the applications of (*S*)-2-(5-1*H*-tetrazolyl)pyrrolidine and (*R*)-2-(5-1*H*-tetrazolyl)pyrrolidine between 2005 and 2011. This tetrazole organocatalyst has proven itself to be a dynamic and useful addition to the synthetic chemist's toolbox.

Related Reagents. D-Proline and L-Proline; (*R*)2-(5-1*H*-Tetrazolylmethyl)Pyrrolidine.[12]

Second Update

Alexander J. André Cobb, Steven V. Ley, Deborah A. Longbottom, David M. Shaw & Robbie J. Mutton
University of Cambridge, Cambridge, UK

Aldol Reactions. An interesting application of proline-based asymmetric catalysts was applied to the atroposelective aldol condensation to form binaphthalene carbaldehyde.[29] (*S*)-2-(5-1*H*-Tetrazolyl)pyrrolidine as the catalyst was shown to produce excellent selectivity (98% ee) (eq 23) and after optimization the scope of the reaction was demonstrated by employing a series of substituted ζ-ketoaldehydes. Generally, yields and enantiomeric excesses of the product remained high in the majority of reactions studied.

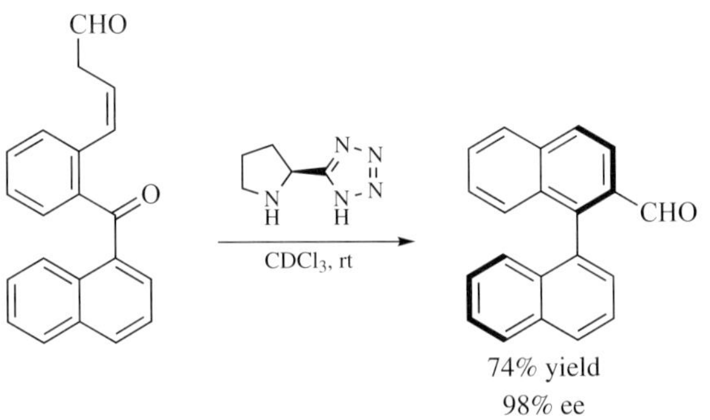

(23)

Both the (*S*)- and (*R*)-catalysts were utilized in the synthesis of various polyketide precursors.[30] Yamamoto and coworkers, in an attempt to identify a catalyst that would best favor an *anti*-Felkin outcome, found that the (*S*)-pyrrolidinyl tetrazole was shown to give a 91:9 dr of the desired product. Additionally, the authors found that by using the (*R*)-catalyst the Felkin isomer was the preferred outcome at 95:5 dr (eq 24).

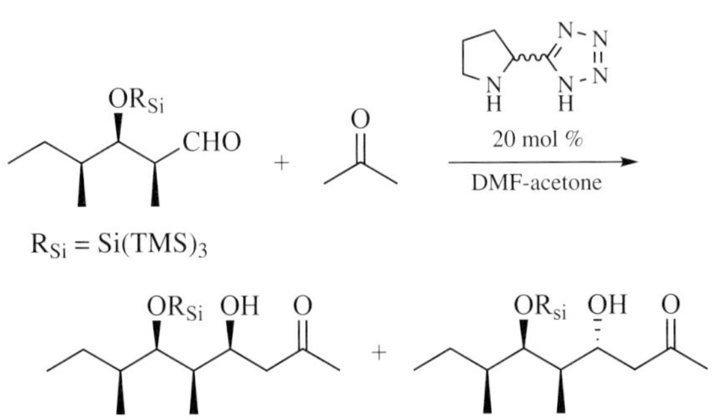

(24)

(*R*)-catalyst: Felkin isomer, 65% yield, 95:5 dr
(*S*)-catalyst: *anti*-Felkin isomer, 71% yield, 91:9 dr

Secci et al. reported their efforts for introducing chirality into benzothiophene derivatives.[31] Benzothiophenes and selected derivatives were reacted with acetoxyacetone in the presence of (*S*)-2-(5-1*H*-tetrazolyl)pyrrolidine. While the main product was obtained (65–80% yield, 90–91% ee), other minor components were also produced (eq 25).

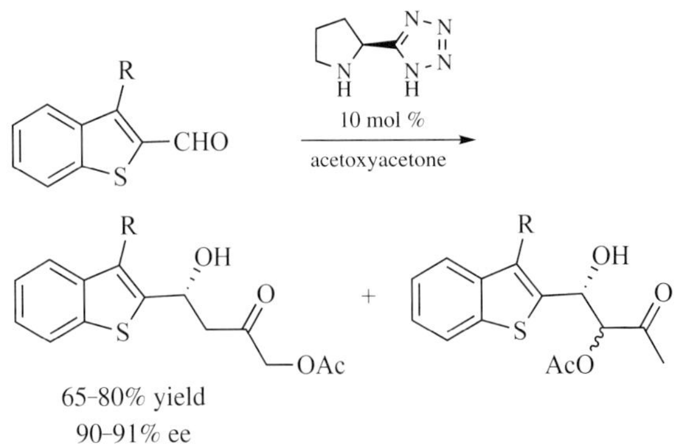

(25)

Michael Additions. While asymmetric catalyzed Michael additions using pyrrolidinyl tetrazole in a batch mode have been reviewed by Ley and coworkers[3] previously, no studies on Michael additions using (*S*)-2-(5-1*H*-tetrazolyl)pyrrolidine under flow mode conditions have been reported. Described here is the optimized asymmetric Michael addition of cyclohexanone and β-nitrostyrene in a microflow reactor affording a 94% yield after a residence time of just 100 min at 80 °C (eq 26).[32] Diastereomeric ratio of the products was found to be high at 90:10, with the enantiomeric excess of the major diastereomer at 40%. The scope of reaction was studied by subjecting a range of substituted cyclohexanones and β-nitrostyrenes under optimized conditions. Yield and diastereomeric ratio were not dramatically affected by the change in substitution pattern of the reactants; however, there was a slight decrease (5–10%) in the enantiomeric excess of the major diastereoisomer.

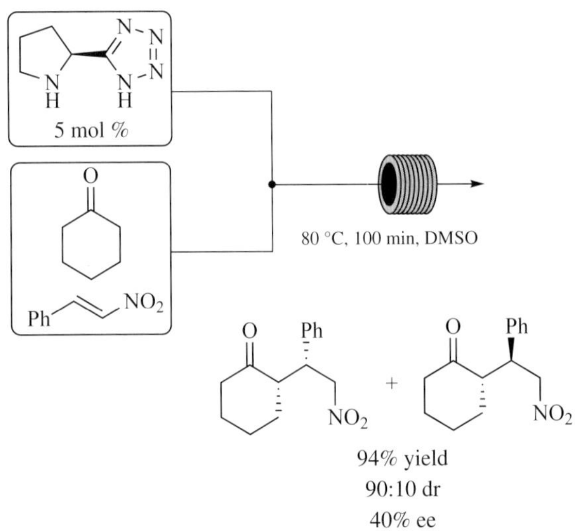

(26)

A synthetic pathway for the natural product atropurpuran utilizes (*S*)-2-(5-1*H*-tetrazolyl)pyrrolidine in an intramolecular Michael addition.[33] Initial screening gave an overall yield of 92% for the four diastereoisomers with a ratio of 24:22:12:42. Optimization of the reaction reduced the reaction time from 4 to 3.5 days, while maintaining the yield by switching solvents from DCM to MeOH, with only minor changes in the diastereomeric ratio (24:24:11:41) (eq 27).

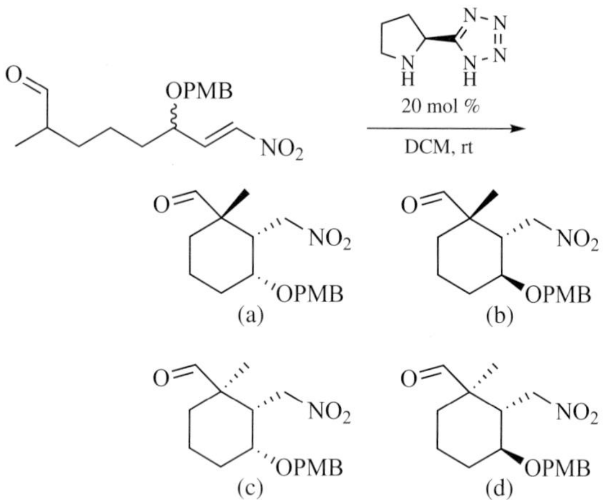

(27)

91% yield
24(a):24(b):11(c):41(d) dr

α-Aminoxylation. There continues to be further development in the use of the (S)-2-(5-1H-tetrazolyl)pyrrolidine catalyst in α-aminoxylation reactions. β-Amino alcohols are structurally important motifs due to their prevalence in natural products and pharmaceutical compounds; therefore, developing enantioselective methodologies represents significant opportunities. Yamamoto utilized the catalyst for the enantioselective formation of β-amino alcohols from aldehydes and an in situ generated nitrosocarbonyl species from the corresponding hydroxylamine.[34] In the reaction of 3-phenylpropanal and N-BocNHOH and in the presence of catechol as the Brønsted acid, an enantiomeric excess of 98% of the product was eventually obtained (eq 28). The scope of the reaction was studied by subjecting a range of substrates to the optimized protocol (51–69% yield, 92–98% ee). Yields decreased to a low of 41% and enantiomeric excesses to 91% by studying the effects of different protecting groups on the hydroxylamine.

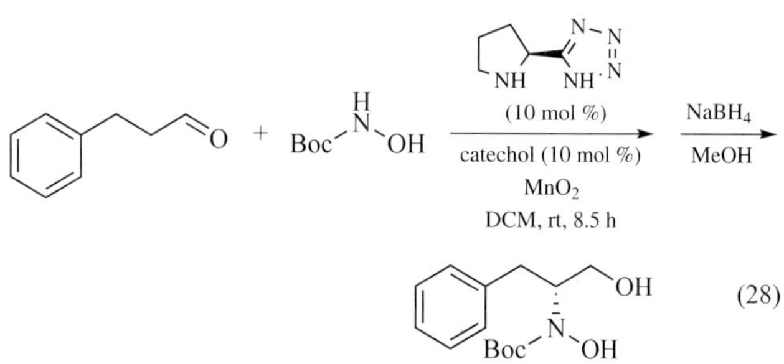

(28)

65% yield
98% ee

Copper-catalyzed Oxidation. While almost all the examples of (S)- and (R)-2-(5-1H-tetrazolyl)pyrrolidine utilization are in the asymmetric synthesis of various organic compounds, the (S)-catalyst has been used in the oxidation of alcohols to the corresponding carbonyl compound (eq 29).[35] In this work, the author proposes that the pyrrolidinyl tetrazole chelates to copper(I) triflate. Following oxidation to the corresponding Cu(II) superoxide derivative, the complex is stabilized by hydrogen atom transfer with TEMPO-H forming a peroxide. Upon reacting with water, the resulting complex undergoes a condensation reaction with the alcohol and eventually breaks down to the corresponding carbonyl compound, leaving the pyrrolidinyl tetrazole copper(I) triflate complex. The reaction works most efficiently with a range of primary and secondary benzylic alcohols, where yields have been report in excess of 95%. Reaction times tend to be between 0.5 and 1 h; occasionally, longer times are required when bulkier groups are present on the starting material.

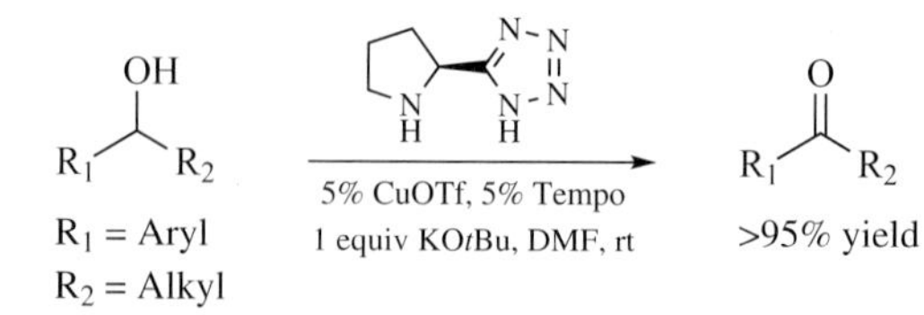

(29)

Dynamic Kinetic Resolution. Dynamic kinetic resolution (DKR) allows a racemic mixture of a particular starting material to be converted to an enantiomerically enriched product, in the presence of a chiral catalyst. Using this concept, Pandey et al. utilized (S)-2-(5-1H-tetrazolyl)pyrrolidine to initiate DKR during the preparation of *trans*-2,5-diallylcylohexanones (eq 30).[36] Optimized conditions gave a high diastereomeric ratio and enantiomeric excess (87:13 and 88%).

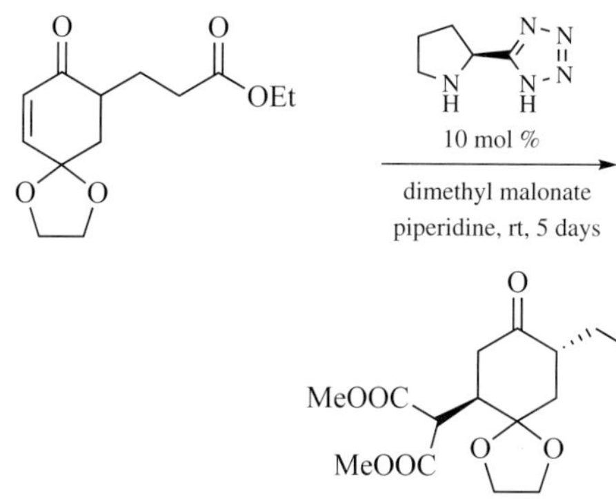

(30)

70% conversion
87:13 dr
88% ee

1. Dalko, P. I.; Moisan, L., *Angew. Chem. Int. Ed.* **2004**, *43*, 5138.
2. Cobb, A. J. A.; Shaw, D. M.; Ley, S. V., *Synlett* **2004**, *3*, 558.
3. Cobb, A. J. A.; Shaw, D. M.; Longbottom, D. A.; Gold, J. B.; Ley, S. V., *Org. Biomol. Chem.* **2005**, *3*, 84.
4. (a) Torii, H.; Nakadai, M.; Ishihara, K.; Saito, S.; Yamamoto, H., *Angew. Chem. Int. Ed.* **2004**, *43*, 1983. (b) Saito, S.; Yamamoto, H., *Acc. Chem. Res.* **2004**, *37*, 570. (c) Saito, S. Japanese Pat. JP 2004-262826 (CAN 141:277248). (d) Suri, J. T.; Ramachary, D. B.; Barbas, C. F., III., *Org. Lett.* **2005**, DOI: ol0502533.
5. Hartikka, A.; Arvidsson, P. I., *Tetrahedron: Asymmetry* **2004**, *15*, 1831.
6. Momiyama, N.; Torii, H.; Saito, S.; Yamamoto, H., *Proc. Natl. Acad. Sci. USA* **2004**, *101*, 5374. (b) Ramachary, D. B.; Barbas, C. F., III., *Org. Lett.* **2005**, DOI: ol050246e.
7. Yamamoto, Y.; Momiyama, N.; Yamamoto, H., *J. Am. Chem. Soc.* **2004**, *126*, 5962.
8. Chowdari, N. S.; Barbas, C. F., III., *Org. Lett.* **2005**, *7*, 867.
9. Cobb, A. J. A.; Longbottom, D. A.; Shaw, D. M.; Ley, S. V., *Chem. Commun.* **2004**, 1808.

10. Almquist, R. G.; Chao, W.-R.; Jennings-White, C., *J. Med. Chem.* **1986**, *28*, 1067.
11. (a) Bridge, A. W.; Jones, R. H.; Kabir, H.; Kee, A. A.; Lythgoe, D. J.; Nakach, M.; Pemberton, C.; Wrightman, J. A., *Org. Proc. Res. Dev.* **2001**, *5*, 9. (b) Koguro, K.; Oga, T.; Mitsui, S.; Orita, R., *Synthesis* **1998**, 910.
12. Mitchell, C. E. T.; Cobb, A. J. A.; Ley, S. V., *Synlett* **2005**, *4*, 611.
13. Xue, F.; Zhang, S.; Liu, L.; Duan, W.; Wang, W., *Chem. Asian J.* **2009**, *4*, 1664.
14. (a) Ward, D. E.; Jheengut, V.; Beye, G. E., *J. Org. Chem.* **2006**, *71*, 8989. (b) Jheengut, V.; Ward, D. E., *J. Org. Chem.* **2007**, *72*, 7805 (c) Ward, D. E.; Jheengut, V.; Beye, G. E.; Gillis, H. M.; Karagiannis, A., *Synlett* **2011**, 508 (d) Ward, D. E.; Jheengut, V.; Akinnusi, O. T., *Org. Lett.* **2005**, *7*, 1181.
15. Funabiki, K.; Itoh, Y.; Kubota, Y.; Matsui, M., *J. Org. Chem.* **2011**, *76*, 3545.
16. Chowdari, N. S.; Ahmad, M.; Albertshofer, K.; Tanaka, F.; Barbas, C. F., III, *Org. Lett.* **2006**, *8*, 2839.
17. Galzerano, P.; Agostino, D.; Bencivenni, G.; Sambri, L.; Bartoli, G.; Melchiorre, P., *Chem. Eur. J.* **2010**, *16*, 6069.
18. (a) Mitchell, C. E. T.; Brenner, S. E.; Ley, S. V., *Chem. Commun.* **2005**, 5346 (b) Mitchell, C. E. T.; Brenner, S. E.; García-Fortanet, J., *Org. Biomol.Chem.* **2006**, *4*, 2039.
19. (a) Knudsen, K. R.; Mitchell, C. E. T.; Ley, S. V., *Chem. Commun.* **2006**, 66. (b) Wascholowski, V.; Knudsen, K. R.; Mitchell, C. E. T.; Ley, S. V., *Chem. Eur. J.* **2008**, *14*, 6155.
20. Ramachary, D. B.; Barbas, C. F., III, *Org. Lett.* **2005**, *7*, 1577.
21. (a) Kumarn, S.; Shaw, D. M.; Longbottom, D. A.; Ley, S. V., *Org. Lett.* **2005**, *7*, 4189. (b) Kumarn, S.; Shaw, D. M.; Ley, S. V., *Chem. Commun.* **2006**, 3211. (c) Kumarn, S.; Oelke, A. J.; Shaw, D. M.; Longbottom, D. A.; Ley, S. V., *Org. Biomol. Chem.* **2007**, *5*, 2678.
22. Jiao, P.; Kawasaki, M.; Yamamoto, H., *Angew. Chem., Int. Ed.* **2009**, *48*, 3333.
23. Capitta, F.; Frongia, A.; Ollivier, J.; Piras, P. P.; Secci, F., *Synlett* **2011**, 89.
24. Oelke, A. J.; Kumarn, S.; Longbottom, D. A.; Ley, S. V., *Synlett* **2006**, 2548.
25. (a) Oelke, A. J.; France, D. J.; Hofmann, T.; Wuitschik, G.; Ley, S. V., *Angew. Chem., Int. Ed.* **2010**, *49*, 6139. (b) Oelke, A. J.; Antonietti, F.; Bertone, L.; Cranwell, P. B.; France, D. J.; Goss, R. J. M.; Hofmann, T.; Knauer, S.; Moss, S. J.; Skelton, P. C.; Turner, R. M.; Wuitschik, G.; Ley, S. V., *Chem. Eur. J.* **2011**, *17*, 4183.
26. Shigenaga, A.; Yamamoto, J.; Nishioka, N.; Otaka, A., *Tetrahedron* **2010**, *66*, 7367.
27. Wu, Y.-Y.; Chai, Z.; Liu, X.-Y.; Zhao, G.; Wang, S.-W., *Eur. J. Org. Chem.* **2009**, 904.
28. Su, W.; Li, J.; Jianjun, L., *Aust. J. Chem.* **2008**, *61*, 860.
29. Link, A.; Sparr, C., *Angew. Chem., Int. Ed.* **2014**, *53*, 5458.
30. Brandy, P. B.; Albert, B. J.; Akakura, M.; Yamamoto, H., *Chem. Sci.* **2013**, *4*, 3223.
31. Secci, F.; Cadoni, E.; Fattuoni, C.; Frongia, A.; Bruno, G.; Nicolò, F., *Tetrahedron* **2012**, *68*, 4773.
32. Carroccia, L.; Musio, B.; Degennaro, L.; Romanazzi, G.; Luisi, R., *J. Flow Chem.* **2013**, *3*, 29.
33. Chen, H.; Zhang, D.; Xue, F.; Qin, Y., *Tetrahedron* **2013**, *69*, 3141.
34. Maji, B.; Yamamoto, H., *Angew. Chem., Int. Ed.* **2014**, *53*, 8714.
35. Liu, Y.; Xie, A.; Li, J.; Xu, X.; Dong, W.; Wang, B., *Tetrahedron* **2014**, *70*, 9791.
36. Pandey, G.; Adate, P. A.; Puranik, V. G., *Org. Biomol. Chem.* **2012**, *10*, 8260.

1,2,4-Triazolo[4,3-α]pyridinium, 5,6,7,8-tetrahydro-2-(2,3,4,5,6-pentafluorophenyl)-, Tetrafluoroborate

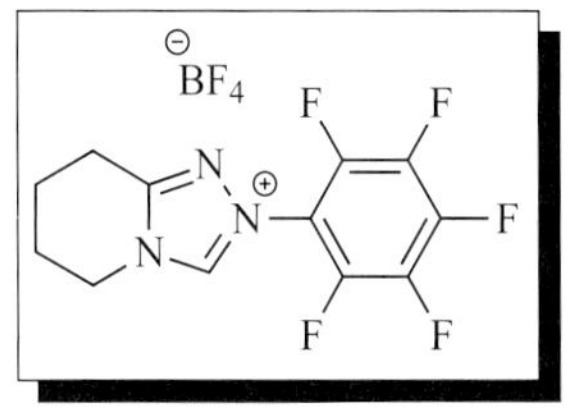

[1609525-88-3] $C_{12}H_9BF_9N_3$ (MW 377.02)

InChI = 1S/C12H9F5N3.BF4/c13-7-8(14)10(16)12(11(17)9(7)15)20-5-19-4-2-1-3-6(19)18-20;2-1(3,4)5/h5H,1-4H2;/q+1;-1

InChIKey = OQGPAXJABLJIMD-UHFFFAOYSA-N

(in the presence of base, catalyst for the preparation of benzoins from aldehydes, particularly cross-benzoin products from aromatic and aliphatic aldehydes[1])

Physical Data: amorphous white solid; mp 222-223 °C.

Solubility: soluble in H_2O, CH_2Cl_2, $CHCl_3$, EtOAc, MeOH, EtOH, CH_3CN; slightly soluble in toluene; insoluble in hexanes, Et_2O.

Analysis of Reagent Purity: ^{1}H NMR; ^{13}C NMR; ^{18}F NMR (may contain residual $C_6F_5NHNH_2$).

Preparative Method: can be prepared from δ-valerolactam in a one-pot procedure similar to that for the preparation of 6,7-dihydro-2-pentafluorophenyl-5*H*-pyrrolo[2,1-*c*]-1,2,4-triazolium tetrafluoroborate [862095-91-8].[1,2]

Handling, Storage, and Precautions: store dry.

Formation of Benzoins from Aldehydes. Many azolium salts catalyze the benzoin reaction in the presence of a base.[3] Homocoupling reactions proceed with relative ease. Cross-benzoin reactions between aromatic and aliphatic aldehydes typically afford a mixture of all four benzoin products or rely on substrate control to achieve chemoselectivity.[4] The title reagent (1) may be used to achieve a high degree of chemoselectivity in such cross-benzoin reactions without the need for substrate-specific control.[1] This selectivity surpasses that of the more common pyrrolidine-derived triazolium (**2**, [86209-91-8]) under similar conditions (eq 1).

Amine (i-Pr_2NEt) and acetate (NaOAc) bases, as well as DBU in catalytic amounts, are equally compatible with **1**. Carbonate bases are effective but result in slightly diminished yields. Common aprotic solvents (THF, CH_2Cl_2, and PhMe) are uniformly suitable. Elevated temperatures (up to 70 °C) do not negatively impact chemoselectivity and are advised to reduce reaction times. The use of **1** provides chemoselectivity in couplings between a variety of aliphatic and aromatic aldehydes (eq 2).

Branched aliphatic aldehydes require larger excess and extended reaction times for comparable yields. The use of **1** does not result in chemoselective cross-benzoin reactions between heteroaromatic and aliphatic aldehydes. Computational and experimental results suggest that chemoselective cross-benzoin reactions using **1** are kinetically controlled.[5]

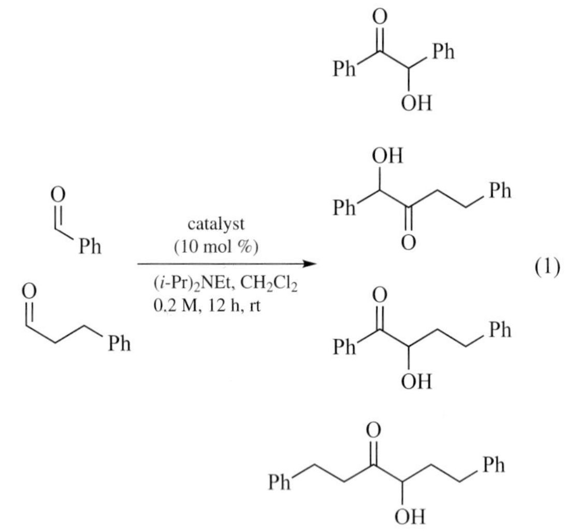

(1)

catalyst (**1**) = 2 : 94 : 0 : 4
catalyst (**2**) = 2 : 59 : 25 : 14
(top to bottom)

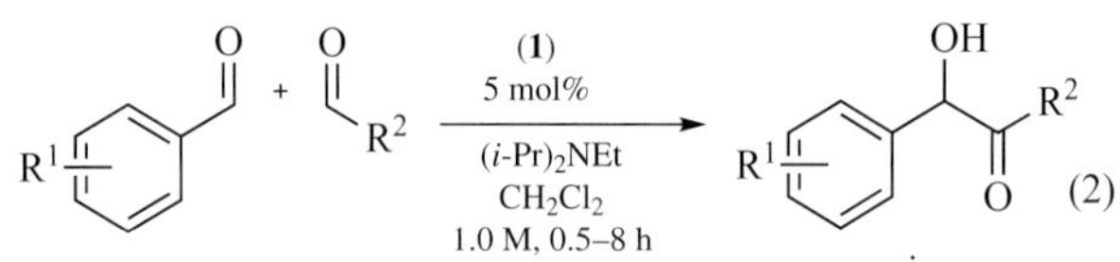

(2)

major
61–99%

R^1 = 2-Br, 2-OMe, 3-OMe, 4-OMe, 4-Me, 4-CO_2Me; R^2 = CH_2CH_2Ph
R^1 = Ph; R^2 = Me, Et, $(CH_2)_6CH_3$, CH_2*i*-Pr, *i*-Pr, CH_2CH_2Ph

Related Reagents. Pyrrolidine- and caprolactam-derived triazolium salts (**2** [86209-91-8] and **3** [1609525-91-8], respectively) can be used under similar conditions.[5]

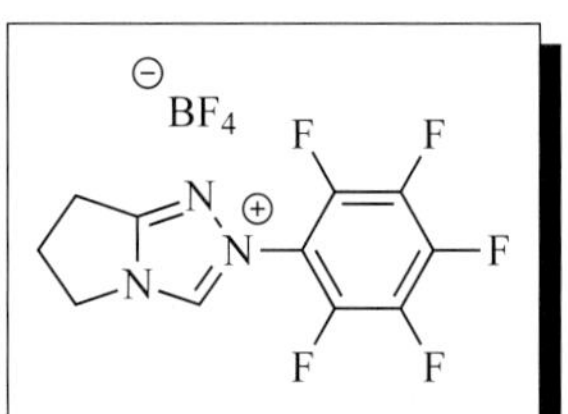

(2)

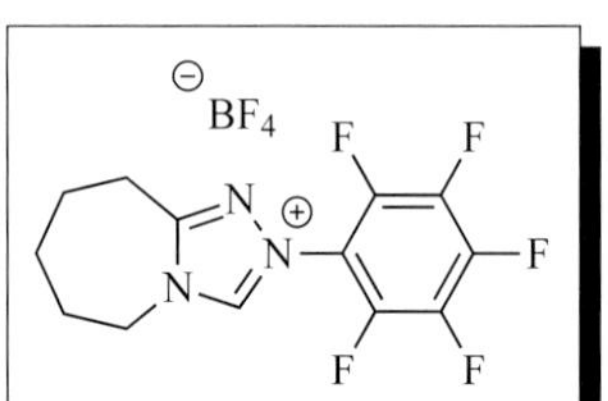

(3)

With elevated temperatures, the use of **2** may provide thermodynamically controlled chemoselectivity in situations where an energetic bias favors one of the cross-benzoin products. Using catalyst **3** offers chemoselectivity under kinetic control but results are highly variable; trace moisture or oxygen has a significant deleterious effect on the reactivity of **3**. The differing controlling factors are a function of the increased energies of activation (and proportional rate decrease) with increasing ring size; catalyst **1** catalyzes the cross-benzoin reaction approximately five times slower than **2** and five times faster than **3**.[5]

1. Langdon, S. M.; Wilde, M. M. D.; Thai, K.; Gravel, M., *J. Am. Chem. Soc.* **2014**, *136*, 7539.
2. Vora, H. U.; Lathrop, S. P.; Reynolds, N. T.; Kerr, M. S.; Read de Alaniz, J.; Rovis, T., *Org. Synth.* **2010**, *87*, 350.
3. (a) Bugaut, X. In *Comprehensive Organic Synthesis II*; Marek, I., Knochel, P., Ed.; Elsevier: Amsterdam, 2014; Vol. *1*, pp 424-470; (b) O'Bryan, E. A.; Scheidt, K. A., Acyloin Coupling Reactions. In *Comprehensive Organic Synthesis II*, 2 ed.; Marek, I., Knochel, P., Ed.; Elsevier: Amsterdam, 2014; Vol. *3*, pp 621-655; (c) Flanigan, D. M.; Romanov-Michailidis, F.; White, N. A.; Rovis, T., *Chem. Rev.* **2015**, *115*, 9307.
4. (a) Stetter, H.; Dambkes, G., *Synthesis* **1977**, 403; (b) Rose, C. A.; Gundala, S.; Connon, S. J.; Zeitler, K., *Synthesis* **2011**, 190; (c) O'Toole, S. E.; Rose, C. A.; Gundala, S.; Zeitler, K.; Connon, S. J., *J. Org. Chem.* **2011**, *76*, 347.
5. Langdon, S. M.; Legault, C. Y.; Gravel, M., *J. Org. Chem.* **2015**, *80*, 3597.

Steven M. Langdon & Michel Gravel
University of Saskatchewan, Saskatoon, Saskatchewan, Canada

Triflimide

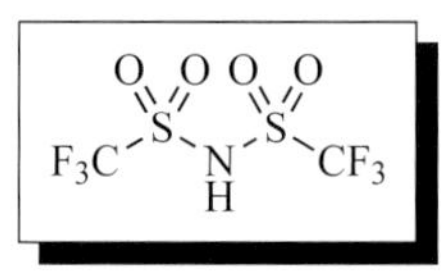

[82113-65-3] $C_2HF_6NO_4S_2$ (MW 281.15)
InChI = 1S/C2HF6NO4S2/c3-1(4,5)14(10,11)9-15(12,13)2(6,7)8/h9H
InChIKey = ZXMGHDIOOHOAAE-UHFFFAOYSA-N

(reagent used as a strong Brønsted acid)

Alternate Names: bis(trifluoromethanesulfonyl)imide; trifluoromethanesulfonimide.
Physical Data: mp 46–57° C; bp 90–91° C; *d* 1.360 g cm^{-3}.[1]
Solubility: sol in H_2O and most organic solvents.
Form Supplied in: white crystalline solid; solution in methylene chloride.
Purification: sublimation at 60° C followed by recrystallization from $CFCl_3$ at −50° C.[1,2]
Handling, Storage, and Precautions: triflimide fumes in air and dissolves exothermally in H_2O. It should be kept in a tightly closed container in a dry and well-ventilated place. It is extremely destructive to tissue of the mucous membranes and upper respiratory tract, eyes, and skin. It should be used in a well-ventilated fume hood.

Original Commentary

Jianwei Sun
The Hong Kong University of Science and Technology, Kowloon, Hong Kong

Preparation of Lewis Acids. As a strong acid,[3] triflimide (Tf_2NH) can participate in a range of acid–base reactions, most of which represent the preparation of useful Lewis acids.[2–6] For example, treatment of triflimide with silver carbonate results in silver triflimide ($AgNTf_2$), which is a versatile π-acid (eq 1).[2,4] Trimethylsilyl triflimide can be readily prepared from trimethylallylsilane and triflimide (eq 2).[5] Protonation of chiral proline-derived oxazaborolidine **1** with triflimide generates chiral cationic boron Lewis acid **2**, which proved to be an efficient catalyst for a range of enantioselective Diels–Alder reactions (eq 3).[6]

$$Tf_2NH + Ag_2CO_3 \longrightarrow AgNTf_2 \quad (1)$$

(2)

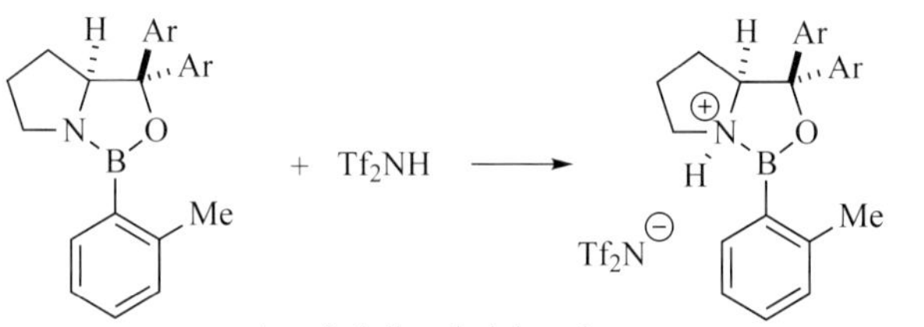

Ar = 3,5-dimethylphenyl

(**1**) (**2**) (3)

Cycloadditions.[7] Tf_2NH can serve as an efficient catalyst (or precatalyst) for a variety of cycloaddition reactions, such as [2 + 2],[8–10] [3 + 2],[11,12] and [4 + 2][13] cycloadditions. For example, in the presence of 0.1 mol % of Tf_2NH, the [2 + 2] cycloaddition reaction between silyl enol ether **3** and methyl acrylate **4** proceeds to afford cyclobutane **5** with excellent efficiency and diastereoselectivity (eq 4).[8a] In contrast, no reaction was observed when TfOH was used as the catalyst. When a propiolate is employed, the corresponding cyclobutene product can be generated in good yield. In the presence of a catalytic amount of Tf_2NH, allylsilanes can participate in [2 + 2] cycloaddition reactions with a range of activated alkenes and alkynes, furnishing cyclobutanes (or cyclobutenes) **6** regio-and/or stereoselectively (eq 5).[9] It is believed that in these reactions Tf_2NH acts as a precatalyst to produce the real catalyst silyl triflimide. Chiral organoammonium salts, prepared from Tf_2NH and chiral amines, have also been shown to catalyze the enantioselective [2 + 2] cycloaddition of unactivated alkenes with α-acyloxyacroleins.[10]

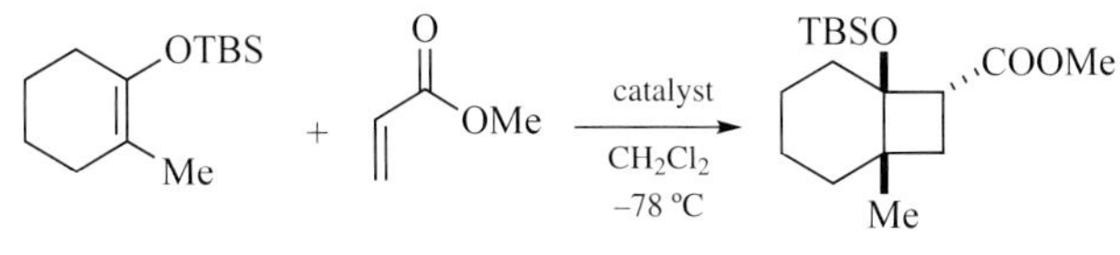

(**3**) (**4**) (**5**) (4)

Catalyst	Yield (*trans:cis*)
Tf_2NH (0.1 mol %)	98% (~100:trace)
TfOH	0%

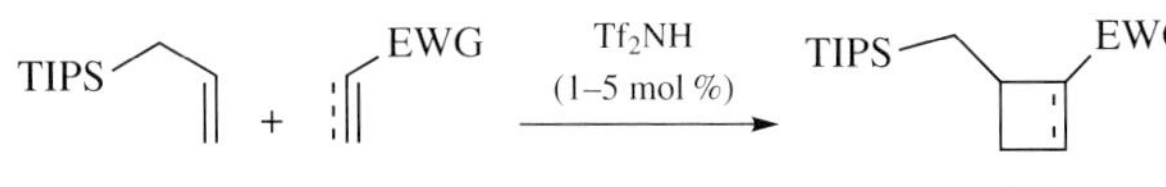

(5)

EWG = CO_2Me, CN, COMe, CHO

In the presence of Tf_2NH (6 mol %), the [3 + 2] cycloaddition of silyl enol ether **7** with donor–acceptor cyclopropane **8** proceeds to afford highly substituted cyclopentane **9** in 84% yield, but with low diastereoselectivity (eq 6).[11a] Vinylidenecyclopropanes are also efficient [3 + 2] cycloaddition partners with electron-deficient alkenes when activated by Tf_2NH.[11b]

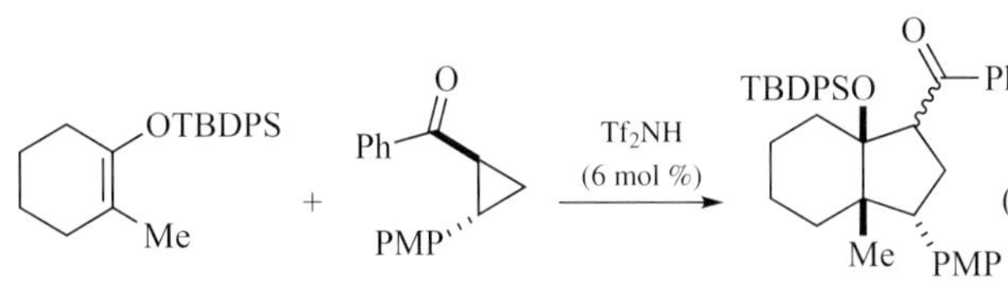

(6)

(**7**) (**8**) PMP = *p*-methoxyphenyl (**9**)

84% dr = 1:1

Tf_2NH has also been demonstrated to catalyze [4 + 2] cycloadditions.[13] With Lewis acid catalysis, some of these reactions proceed less efficiently or behave differently. For example, in a competition Diels–Alder reaction with cyclopentadiene, 2-hexen-4-one is more favorable than crotonaldehyde in the presence of Tf_2NH, while the opposite chemoselectivity is observed when a bulky Lewis acid is employed (eq 7).[13a] This unusual chemoselectivity may arise partially from the relative insensitivity of the Brønsted acid to steric hindrance around the carbonyl group.

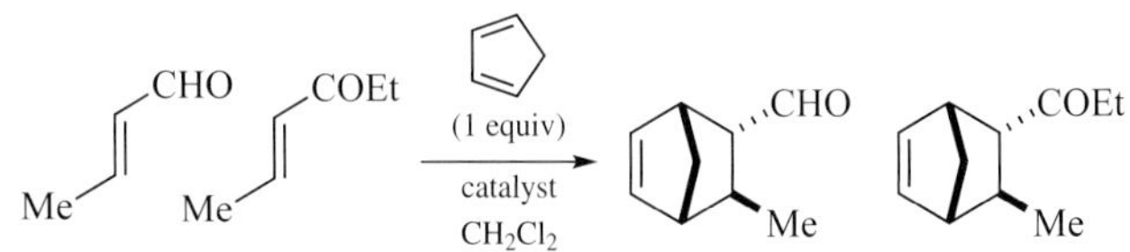

(1:1) (**10**) (**11**)

Catalyst	Yield (10:11)
Tf_2NH (5 mol %, –40 °C)	75% (10:90)
$B(C_6F_5)_3$ (30 mol %, –20 °C)	84% (92:8)

(7)

Takasu and coworkers have developed several triflimide-catalyzed aza-[4 + 2] cycloaddition reactions.[13d,e] In particular, with a catalytic amount of Tf_2NH, substituted quinolines can be formed in good yields by cycloaddition of aryl almidines and allylsilanes followed by oxidation (eq 8).[13e] Interestingly, the cycloaddition proceeds in a [3 + 2] fashion if α,α-dimethylallylsilane is employed.[12]

Tf_2NH-catalyzed cascade reactions and multicomponent cycloadditions have also been explored. These processes can provide rapid access to a number of complex skeletons.[7,11,14]

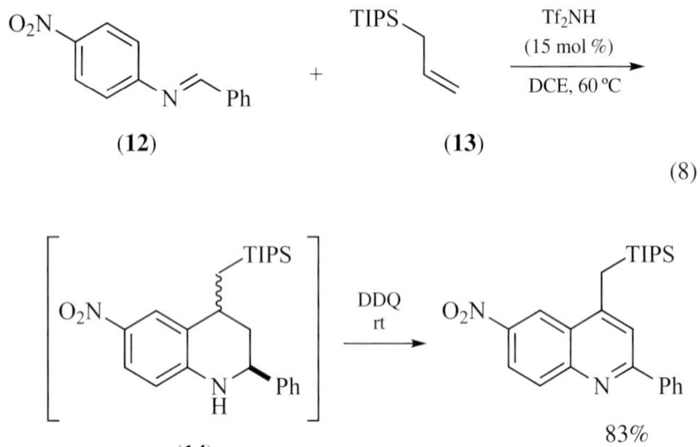

(12) (13) (14) (15) (8)

Aldol Reactions. Yamamoto and coworkers have demonstrated that triflimide can serve as very efficient precatalyst for Mukaiyama aldol reactions.[15] Specifically, in the presence of a small amount of Tf_2NH (0.1 mol %), the reaction of super silyl enol ether **16** with aldehyde **17** afforded *syn* aldol **18** in 90% yield and with excellent stereocontrol (eq 9).[15e] Several one-pot cascade reactions have been developed for stereoselective preparation of other synthetically useful molecules, such as β,δ,γ-tris(hydroxyl)aldehydes, β,δ,γ-tris(hydroxyl)aldehydes, and tertiary carbinols.[15] It is believed that tris(trimethylsilyl)silyl (TTMSS) group plays an important role in achieving the amazing efficiency in such transformations, presumably due to its unusual electrophilic and steric nature.

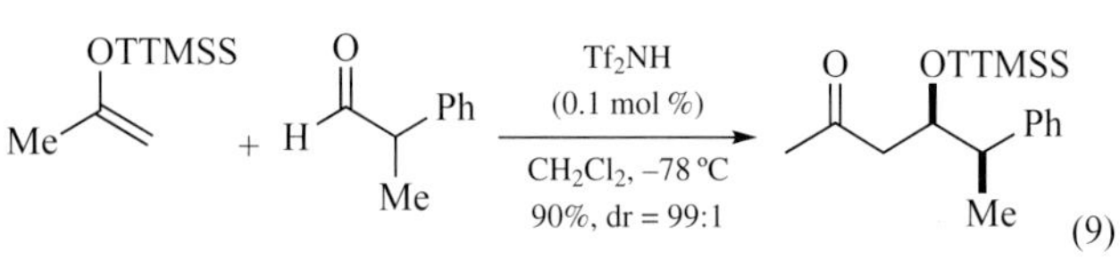

(16) (17) (18) (9)

Allylations. A range of allylations of aldehydes and α,β-unsaturated compounds with allylsilanes or allylboronates can also be catalyzed by triflimide.[15a,16] For example, under mild conditions, efficient allylation of enone **19** occurs to form highly substituted cyclopentanone **20** with high diastereoselectivity (eq 10).[16a] In contrast, the reaction does not proceed with $TiCl_4$ as the catalyst. The use of copper-catalyzed addition of allylmagnesium chloride gives the desired product in only 25% yield.[16a]

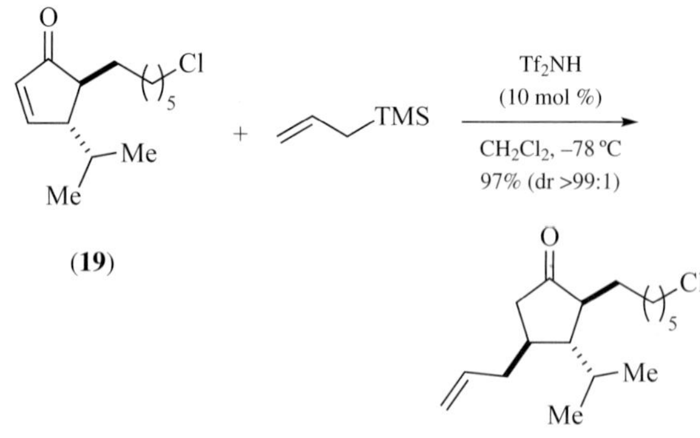

(19) (20) (10)

Friedel–Crafts Reactions.[16b,17] A number of Friedel–Crafts reactions can be catalyzed by triflimide under a variety of conditions, such as the use of microwave and ionic liquids.[17a,b] Under refluxing conditions, triflimide catalyzes the acylation of anisole with heptanoic acid, affording the corresponding ketone in 89% yield (eq 11).[17c]

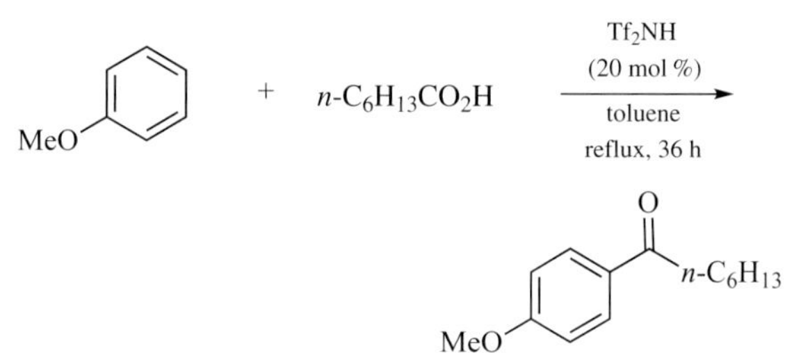

(11)

Nazarov Cyclizations. With a substoichiometric amount of triflimide, the microwave-assisted Nazarov cyclization of pyrrole derivative **21** proceeds efficiently to form bicyclic product **22** as a single diastereomer in 95% yield (eq 12).[18]

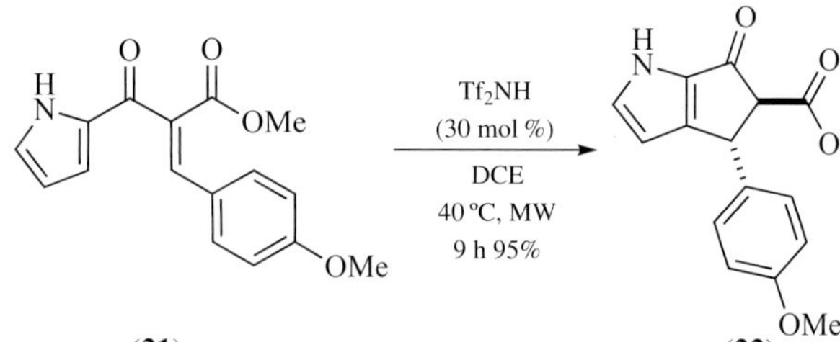

(21) (22) (12)

Michael Additions. Triflimide also catalyzes Michael additions of carbamates, alcohols, and thiols to α,β-unsaturated ketones, alkylidene malonates, and acrylimides.[19] For example, under very mild conditions, the additions of carbamate **24**, benzyl alcohol **26**, and benzyl mercaptan **28** to enone **23** proceed smoothly to form the corresponding 1,4-addition products in excellent yields (eqs 13–15). Triflimide shows the best catalytic ability among all the Brønsted acids evaluated. In addition, the reaction scope, rates, and yields are superior to those corresponding processes catalyzed by Lewis acids.

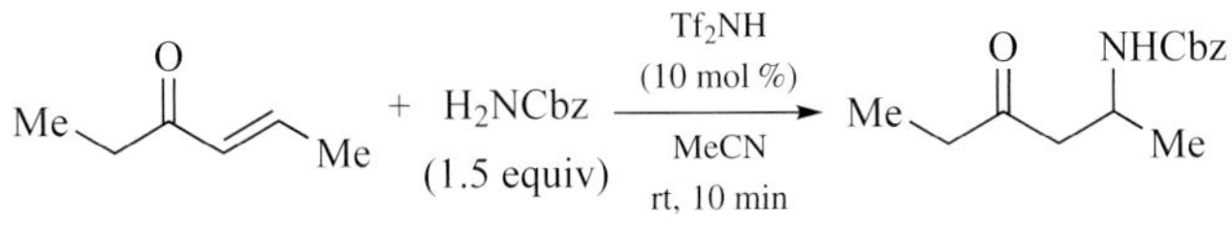

(23) (24) (25) 94%

(13)

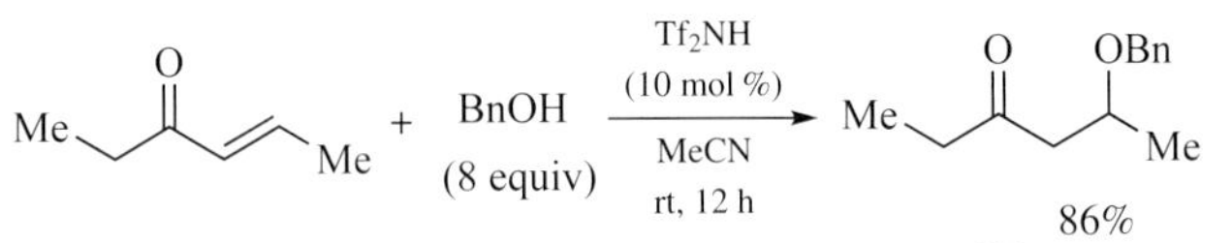

(23) (26) (27) 86%

(14)

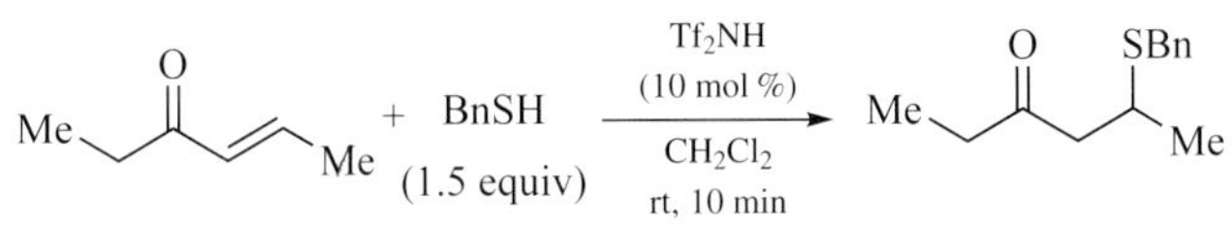

(23) (28) (29) 93%

(15)

Mannich Reactions. Triflimide is also competent in promoting efficient carbon–carbon bond formation between in situ generated *N*-acyliminium ions and silyl enol ethers. Specifically, in the presence of 5 mol % of triflimide, the Mannich reaction between 4,5-diacetoxy lactam **30** and silyl enol ether **31** results in the formation of α-amido alkylation product **32** in good yield and with good diastereoselectivity (eq 16).[20]

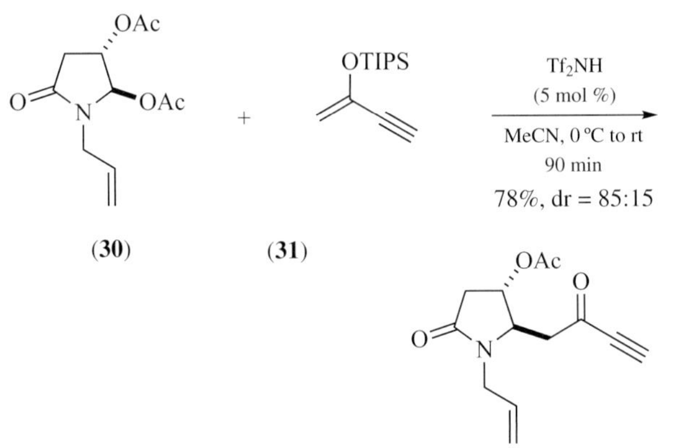

(30) (31) (32)

(16)

Other Carbon–Carbon and Carbon-Heteroatom Bond Formations. In addition to the above types of reactions, triflimide has also been sporadically shown to promote a variety of other carbon–carbon and carbon–heteroatom bond formation reactions, including novel cyclizations and rearrangements.[21–25] For example, in the presence of a stoichiometric amount of triflimide, an efficient *5-endo-dig* cyclization of 1,5-siloxydiyne **33**, followed by an unusual abstraction of a chlorine atom from methylene chloride, affords enone **34** in 70% yield (eq 17).[22b] It is believed that the strong acidity of triflimide combined with the weak nucleophilicity of its counterion is responsible for the observed high efficiency.

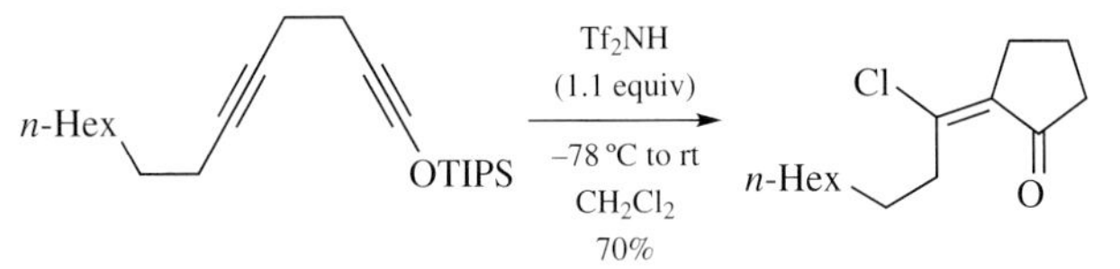

(33) (34)

(17)

Related Reagents. Trifluoromethanesulfonic acid (TfOH); tris[(trifluoromethyl)sulfonyl]methane (Tf_3CH); tetrafluoroboric acid (HBF_4); fluoroantimonic acid ($HSbF_6$); perchloric acid ($HClO_4$); trifluoromethanesulfonamide ($TfNH_2$); trifluoroacetic acid (TFA).

First Update

Jianwei Sun
The Hong Kong University of Science and Technology, Kowloon, Hong Kong

In the past few years, increasing attention has been paid to triflimide, including the studies on its structure, acidity, and other physical properties.[26] In organic synthesis, triflimide has also been more and more employed as a catalyst/cocatalyst, a promoter, or an additive in a variety of organic reactions.[7,27] The recent representative progress has been summarized herein.

Cycloadditions. Takasu and Takemoto and coworkers pioneered various triflimide-catalyzed cycloaddition reactions.[7] Recently, they have extended these cycloaddition reaction to a range of catalytic domino reactions based on their established [2+2], [3+2], and [4+2] cycloadditions.[28] They also utilized one of these reactions in natural product synthesis.[29] Furthermore, the same group demonstrated that flow microreactor system could be used to enhance the efficiency of these reactions.[30]

In 2008, Hsung and coworkers reported that triflimide is an excellent promoter of the intramolecular Gassman [2+2] cycloaddition.[31a] In the presence of 1 equiv of triflimide, the acetal **35** undergoes smooth cycloaddition to form **36** in 50% yield at $-78\,^{\circ}C$ (eq 18). Activation of the acetal moiety to form an oxonium intermediate is probably the key driving force of the process.[31a] Later, the same group reported a similar [2+2] reaction employing a different cleavable tether between the two double bonds, which can be considered an intermolecular equivalent.[31b]

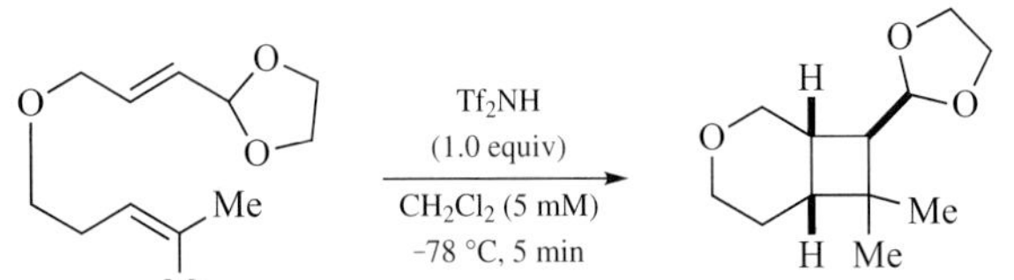

(35) 50% (36)

(18)

Saito et al. discovered that triflimide can be employed to promote the [2+2+1] annulation between alkynes, nitriles, and iodosobenzene for the efficient synthesis of oxazoles at room temperature (eq 19). Triflic acid (TfOH) can also be used to achieve comparable efficiency in this reaction, but typically requiring much higher temperature. The acid promoter is probably required to activate the iodosobenzene to enhance its electrophilicity.[32]

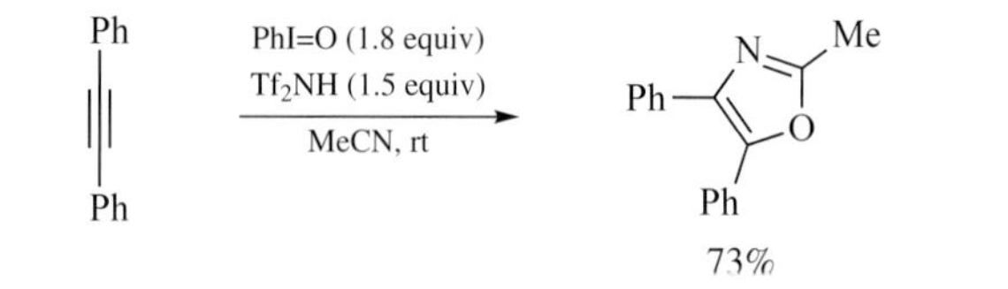

(19)

Li and Shi developed several formal cycloaddition reactions of vinylidenecyclopropanes with different reaction partners.[11b,33] For example, in the presence of a catalytic amount of triflimide, allene **37** and methyl vinyl ketone undergo [3+2] cycloaddition to form polysubstituted cyclopentane **38** with excellent efficiency at room temperature (eq 20).[11b] Other Brønsted acids (e.g., TfOH and *p*-TsOH) exhibit much lower catalytic activity. The use of a 5,5-diarylpenta-2,3,4-trienoate as the reaction partner forms a tricyclic molecule.[33]

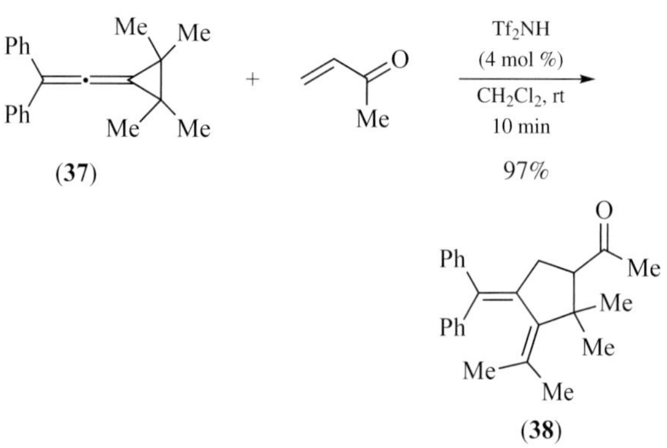

(20)

In 2006, Hsung and coworkers reported that triflimide is capable of catalyzing the intramolecular [2+2] cycloaddition reactions between ynamide and aldehyde. The oxetene adduct formed in situ undergoes facile ring opening to generate the observed yne-carbonyl metathesis product (eq 21).[23b] In 2011, Takemoto and coworkers reported an intermolecular variation using an imine electrophile. It is noteworthy that Tf_2NH triggers the highly torquoselective outward opening of the azetine adduct **43** to form the (*E*)-amidine **44** with high efficiency (eq 22).[34a] In contrast, the use of a weak acid catalyst, such as methanesulfonic acid (MsOH) and 10-camphorsulfonic acid (CSA), results in mainly inward opening to form the (*Z*)-isomer. Later, the same group reported that when a ketimine (instead of aldimine) is used, the [2+2] cycloaddition process is truncated, forming an acyclic *N,N*-divinylamine as the product.[34b] Further investigations indicated that the electron-withdrawing group on the ynamides can also affect the reaction outcome. The reactions of ynamides bearing a sulfonamide moiety instead of carbamate give formal [4+2] cycloaddition products.[34c]

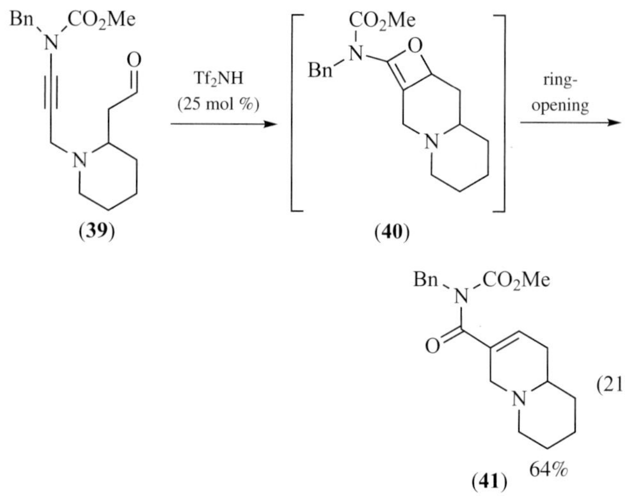

(21)

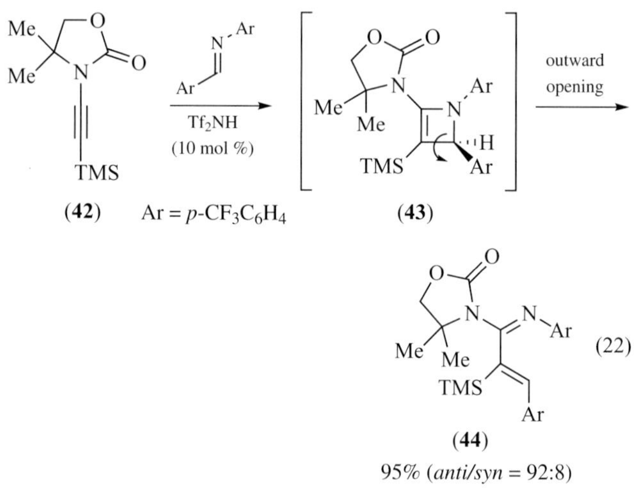

(22)

Triflimide is also known to activate alkyl electrophiles to generate carbocations for subsequent bond formation. For example, upon activation by triflimide, alcohol **45** forms 2-(silyloxy)allyl cation **46**, which undergoes [4+3] cycloaddition with pyrrole **47** to form tropinone product **48** in 85% yield (eq 23).[35] Triflic acid promotes the same reaction with lower efficiency due to the formation of a Friedel-Crafts alkylation product.

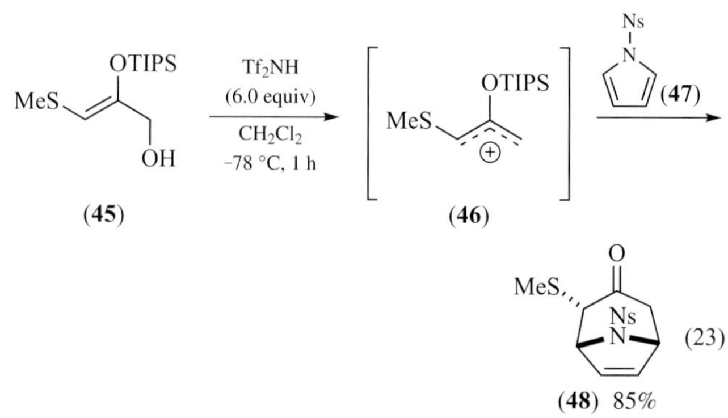

(23)

Recently, Sun and coworkers reported a triflimide-catalyzed formal [6+2] cycloaddition process for eight-membered lactone synthesis. The reaction is enabled by the design of a new family of (1,6)-amphoteric molecules (e.g., **49**) containing oxetane and aldehyde moieties.[36a] The reaction may involve the formation of a series of intermediates, such as oxonium, ketenium, and silyl ketene acetal, together with sequential bond formation and cleavage events. A limitation of the process is the requirement of an arene tether between the oxetane and aldehyde moieties. To address this limitation, the same group has recently reported a new ring expansion strategy using cyclic acetals as the starting material.[36b] More recently, this medium ring lactone synthesis has been successfully extended to medium lactam synthesis (eq 24).[36c] In the presence of a catalytic amount of triflimide, the cyclic hemiaminal **51** generates the corresponding iminium in situ and then undergoes [2+2] cycloaddition with siloxy alkyne **50** to form azetenium **52**. Further ring expansion and desilylation form the medium ring lactam **53** (eq 25).

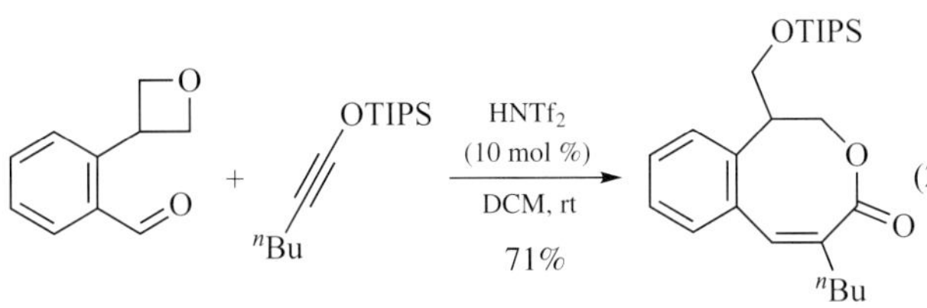

(49) **(50)** **(51)** (24)

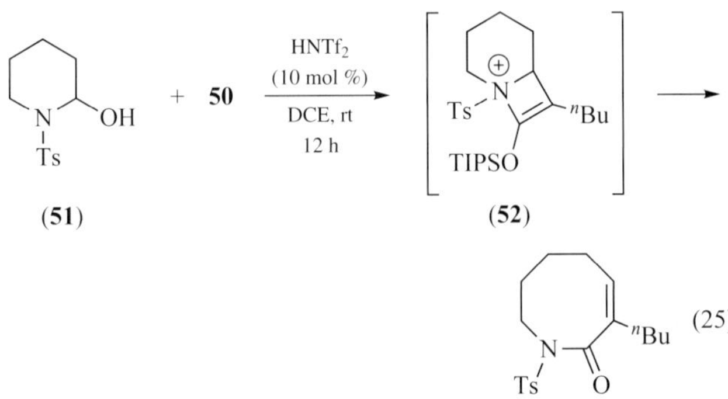

(51) **(52)** **(53)** (25)

Aldol Reactions. Yamamoto and coworkers pioneered Mukaiyama aldol reactions using silyl enol ethers based on the "super-silyl" group [tris(trimethylsilyl)silyl, TMMSS], which is unique in taming the reactivity and selectivity of these transformations. They have identified that triflimide is a superior precatalyst in these reactions (eq 9).[15] The same group also extended the catalytic systems to 1,5-*anti* aldol reactions of β-siloxy methyl ketones with aldehydes. In contrast to *anti*-selectivity with silyl enol ethers by triflimide catalysis, the reaction with lithium enolates exhibits excellent *syn*-selectivity.[37] In 2010, the same group reported an amazingly efficient triple cascade Mukaiyama aldol reaction (eq 26). With only 0.1 mol % loading of triflimide, cyclohexylaldehyde **54** and silyl enol ether **55** react efficiently to afford the triple-aldol product **56** in 87% yield and with good diastereoselectivity.[38] The cocatalyst organoiodide might react with the in situ-generated silyl triflimide to form a cationic and sterically less demanding complex, which proved able to promote the same reaction. More recently, the Yamamoto group has extended the cascade protocol to an intermolecular/intramolecular sequential aldol reaction of disilyl enol ethers to form five-, six-, and seven-membered ring products bearing four or more contiguous stereogenic centers with good stereocontrol.[39]

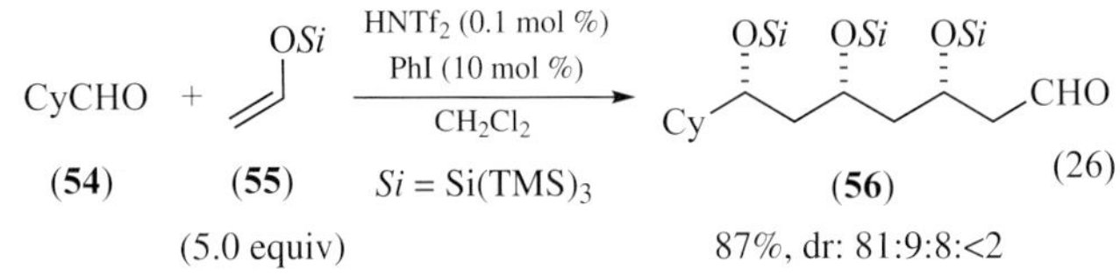

(26)

In 2011, Yamamoto and coworkers reported that triflimide can catalyze the efficient *anti*-selective Mukaiyama aldol reaction between chloro silyl enol ether **55** and benzaldehyde **44**. However, the extension of this highly selective process to aliphatic aldehydes requires a modified catalytic system (eq 27).[40] The *anti*-β-siloxy-α-haloaldehyde products are useful precursors to other stereodefined building blocks. The reaction with methyl silyl enol ethers can also be catalyzed by triflimide.[41]

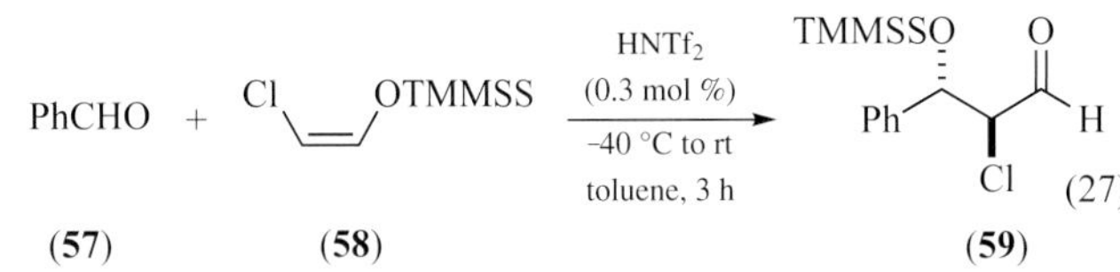

(57) **(58)** **(59)** (27)

99%, *anti/syn* > 99:1

These highly efficient Mukaiyama aldol reactions have also been applied in the synthesis of useful fragments of polyketides.[42]

Allylations. Carbocations generated via activation of alkyl electrophiles by triflimide can undergo allylation reactions with allylsilanes. As shown in eqs 28 and 29, the electrophiles can be allylic and benzylic acetates and sulfonamides, and the reaction efficiency is generally excellent.[43] Triflimide might be the precatalyst for in situ generation of the actual Lewis acid silyl triflimide.

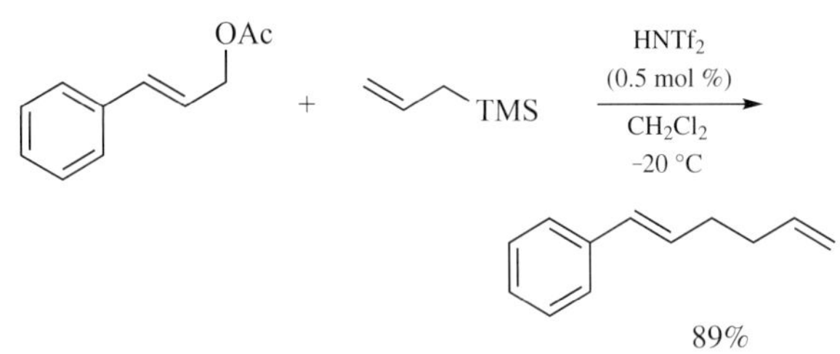

(28)

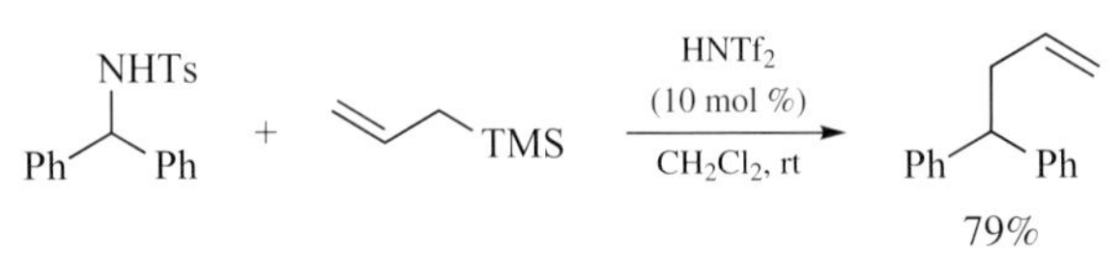

(29)

Friedel-Crafts and Related Reactions. Triflimide can activate a range of functional groups, such as carbon-carbon double and triple bonds to generate carbon-based electrophilic species for subsequent inter- and intramolecular reactions with arenes.[44,45] In 2004, Zhang and Kozmin reported that siloxy alkynes can be activated by triflimide to generate electrophilic silyl ketenium species followed by trapping with a properly tethered arene unit to form tetralone derivatives (eq 30).[22a] The extension of this reaction to ynamides was later reported by Hsung and coworkers.[23a,45] More recently, Thibaudeau[46] and coworkers reported a novel and efficient cascade polycyclization of ynamide **60** to form polycycle **61** catalyzed by triflimide (eq 31). The initially generated ketenimium ion triggers an internal hydride shift from the benzylic position, which is followed by sequential Nazarov and Friedel-Crafts cyclizations to form the observed product.

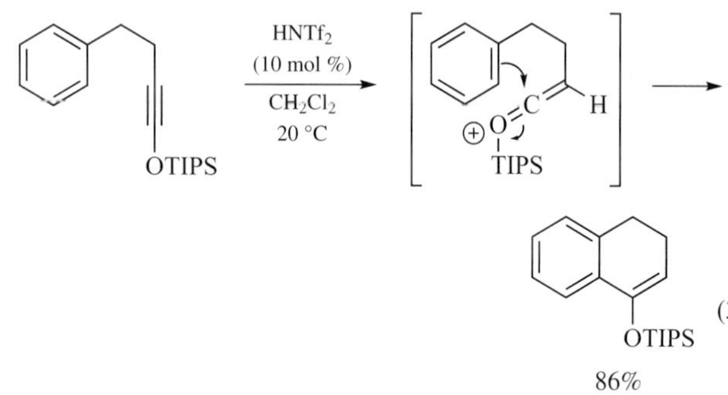

(30)

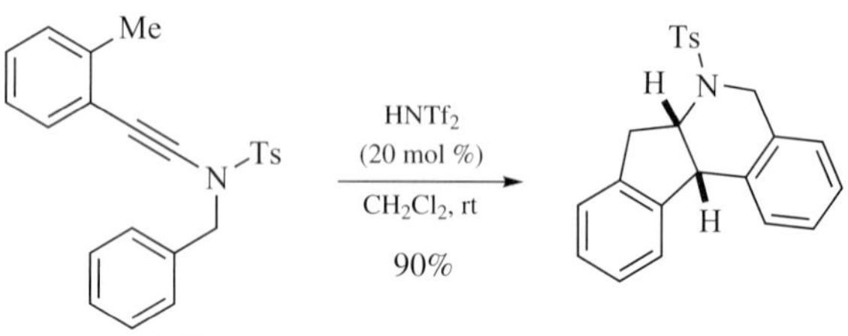

(31)

(60) **(61)**

In 2014, Nomiyama and Tsuchimoto reported that triflimide can activate ketones for subsequent alkylation of pyrroles in the presence of a silane reductant. The alkylation takes place selectively at the β-position of the pyrrole (eq 32).[47]

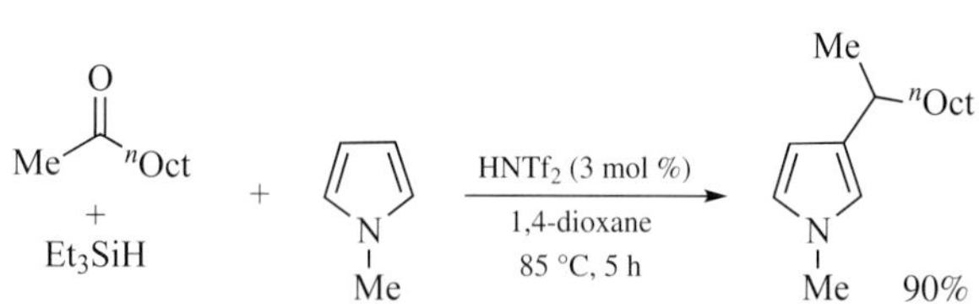

(32)

In addition to using arene as the trapping reagent for the activated intermediate, some other internal and external carbon-carbon double and triple bonds can also trap the intermediate to trigger cycloisomerization or formal cycloaddition reactions.[48]

Nazarov Cyclizations. In 2013, Tius and coworkers reported a highly diastereospecific Nazarov cyclization of fully substituted dienones to generate vicinal all-carbon quaternary stereocenters. With 20 mol % loading of triflimide, the strongly polarized "push-pull" vinylogous carbonate **62** smoothly cyclizes to form the highly substituted cyclopentenone **63** in 80% yield (eq 33). The SEM group in the starting dienone is also important because it rapidly collapses to suppress erosion of the stereochemical integrity.[49]

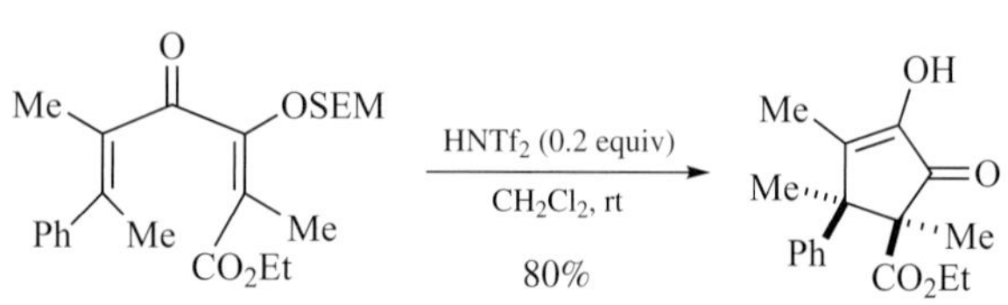

(33)

(62) **(63)**

Other Carbon-Carbon and Carbon-Heteroatom Bond Formations. In the past few years, a number of other new processes catalyzed or mediated by triflimide have been reported. In 2001, Bolte and Gagosz reported an efficient synthesis of spiro tetrahydrofurans and tetrahydropyrans featuring a hydride shift and cyclization sequence (eq 34).[50] Interestingly, a gold(I) complex catalyzes the cyclization to form a completely different bicyclic product, thereby demonstrating the complementary feature between triflimide and gold catalysis.

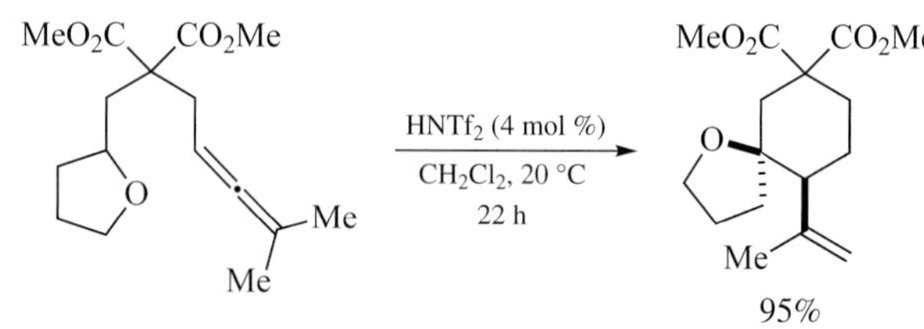

(34)

In 2007, Maruoka and coworkers reported a stereodefined construction of 1,1,2,2-tetrasubstituted cyclopropanes from aryldiazoacetates and α-substituted acroleins.[51] The reaction proceeds via initial Michael addition followed by intramolecular substitution. Among all the Brønsted and Lewis acids evaluated, triflimide provides the best efficiency and diastereoselectivity. With 20 mol % of triflimide, the reaction between diazoacetate **64** and substituted acrolein **65** affords the cyclopropane product **66** in 64% yield with high diastereoselectivity (eq 35).[51a]

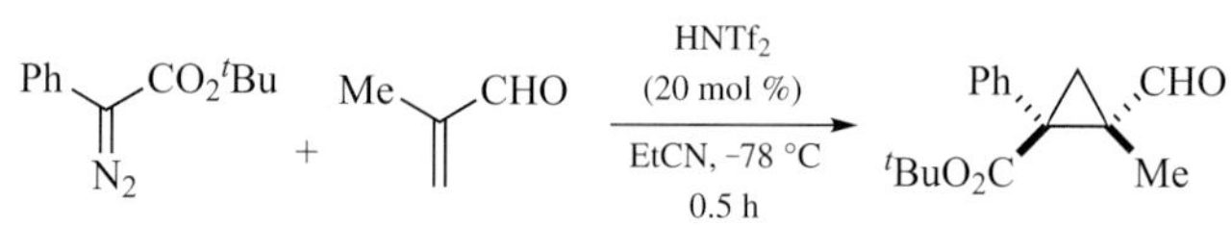

(64) **(65)** **(66)**

(35)

The [3,3] sigmatropic rearrangements are powerful reactions in organic synthesis. Thomson and coworkers realized a triflimide-catalyzed variant of the Stevens [3,3] sigmatropic rearrangement of *N*-allylhydrazones. The reaction generates a *s*-bond between two unfunctionalized sp^3 carbons (eq 36).[52a] The method represents a "traceless" bond formation strategy that may have applications in complex molecule synthesis. Further investigations on this reaction, including reaction mechanism and scope expansion, have also been reported.[52b,c]

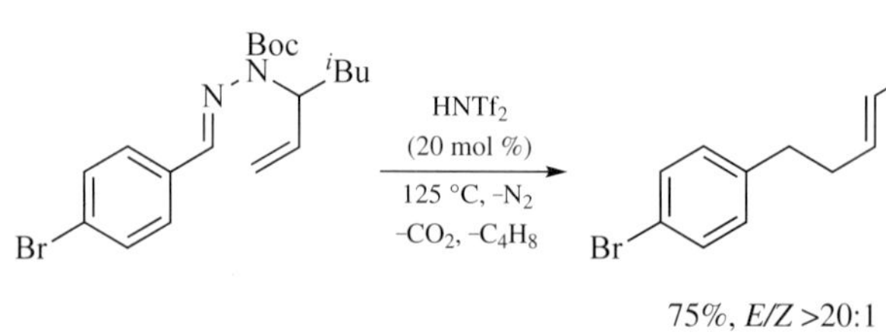

(36)

Recently, Togni and coworkers discovered that triflimide can activate the trifluoromethylation reagent **68**.[53a] This activation further enables the transfer of the electrophilic trifluoromethyl unit to acetonitrile, which then installs the *N*-(trifluoromethyl)imidoyl group to benzotriazole **67** to form the observed product **69** (eq 37). The same research group further extended this reaction to direct *N*-trifluoromethylation of azoles.[53b]

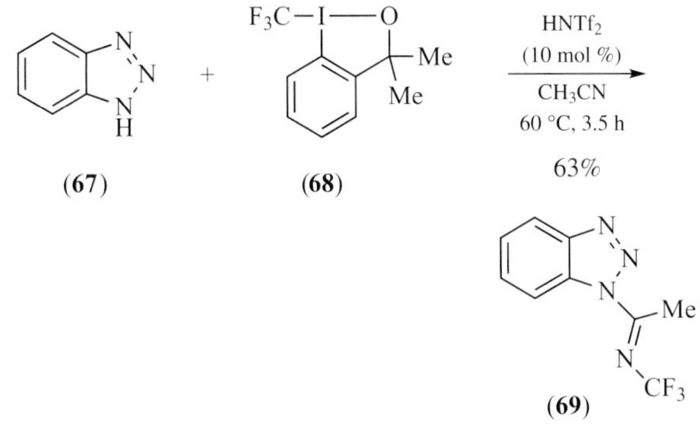

(67) (68) (69) (37)

Triflimide has also shown superior activity in promoting a range of glycosidation reactions.[54] For example, in 2005, Hashimoto and coworkers reported a direct and stereoselective glycosidation of 2-acetamido-2-deoxyglycopyranosyl diethyl phosphites with various alcohols to form 1,2-*trans*-β-linked glycosides with high efficiency (eq 38). Among all the Brønsted and Lewis acids, triflimide provides the best reaction efficiency and selectivity.[54a]

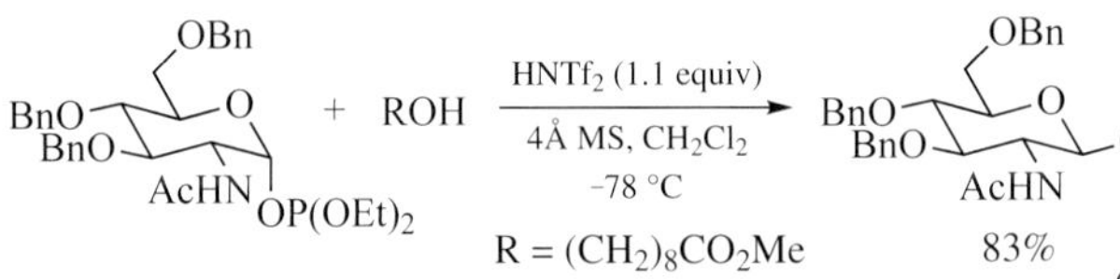

(38)

Owning to its strong Brønsted acidity, triflimide is also known for catalyzing alkyne hydration and related hydrofunctionalization reactions.[55] In 2000, Shirakawa and coworkers reported that triflimide itself can catalyze the hydration of a range of alkynes with good efficiency (eq 39).[55a] Later on, this reaction was utilized and modified to other catalytic hydrofunctionalization reactions. Naka and coworkers have recently reported that, in combination with a cobalt porphyrin complex, triflimide can be used in low loading to cocatalyze the hydration of a range of terminal alkynes.[55d]

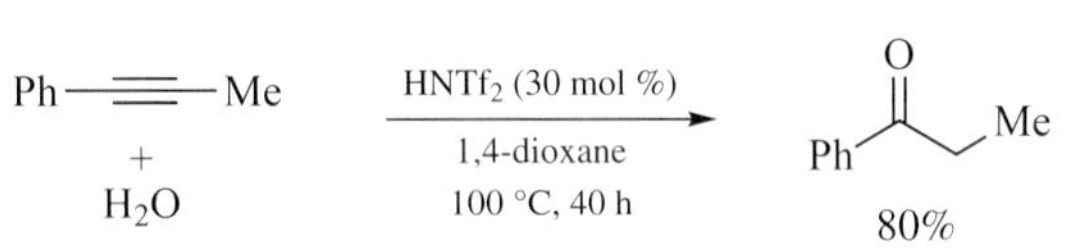

(39)

Finally, triflimide has been used as a cocatalyst in a wide range of reactions. In many of these reactions, triflimide serves as a precatalyst to form cationic metal triflimide complexes for better catalytic activity. However, its role can go beyond that. For example, in 2006, Arisawa and Yamaguchi reported a palladium-catalyzed synthesis of 1-alkylphosphonium salts from 1-alkenes. In this reaction, triflimide is not only a proton source but also a uniquely effective activator of the palladium catalyst (eq 40).[56]

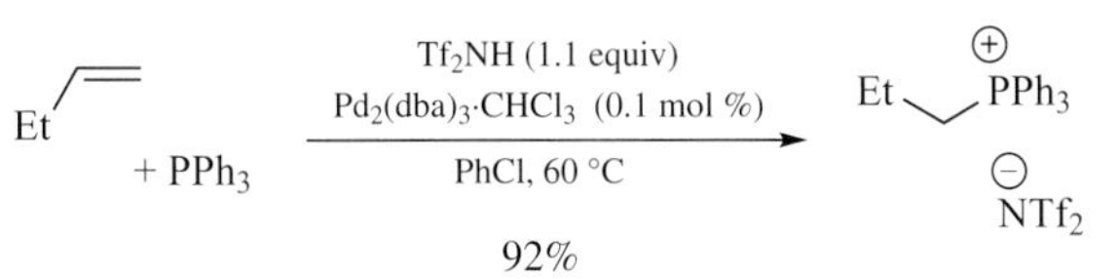

(40)

1. Foropoulos, J.; DesMarteau, D. D., *Inorg. Chem.* **1984**, *23*, 3720.
2. Vij, A.; Zheng, Y. Y.; Kirchmeier, R. L.; Shreeve, J. M., *Inorg. Chem.* **1994**, *33*, 3281.
3. For references on its acidity, see: (a) Ref. 1. (b) Koppel, I. A.; Taft, R. W.; Anvia, F.; Zhu, S.-Z.; Hu, L.-Q.; Sung, K.-S.; DesMarteau, D. D.; Yagupolskii, L. M.; Yagupolskii, Y. L.; Vlasov, V. M.; Notario, R.; Maria, P.-C., *J. Am. Chem. Soc.* **1994**, *116*, 3047. (c) Thomazeau, C.; Olivier-Bourbigou, H. O.; Magna, L.; Luts, S.; Gilbert, B., *J. Am. Chem. Soc.* **2003**, *125*, 5264. (d) Chipanina, N. N.; Sterkhova, I. V.; Aksamentova, T. N.; Sherstyannikova, L. V.; Kukhareva, V. A.; Shainyan, B. A., *Russ. J. Gen. Chem.* **2008**, *78*, 2363. (e) Leito, I.; Raamat, E.; Kütt, A.; Saame, J.; Kipper, K.; Koppel, I. A.; Koppel, I.; Zhang, M.; Mishima, M.; Yagupolskii, L. M.; Garlyauskayte, R. Y.; Filatov, A. A., *J. Phys. Chem. A* **2009**, *113*, 8421.
4. For selected examples of $AgNTf_2$-catalyzed reactions, see: (a) Sweis, R. F.; Schramm, M. P.; Kozmin, S. A., *J. Am. Chem. Soc.* **2004**, *126*, 7442. (b) Sun, J.; Kozmin, S. A., *Angew. Chem., Int. Ed.* **2006**, *45*, 4991. (c) Sun, J.; Keller, V. A.; Meyer, S. T.; Kozmin, S. A., *Adv. Synth. Catal.* **2010**, *352*, 839.
5. Mathieu, B.; Ghosez, L., *Tetrahedron Lett.* **1997**, *38*, 5497.
6. Ryu, D. H.; Corey, E. J., *J. Am. Chem. Soc.* **2003**, *125*, 6388.
7. For a review on Tf_2NH-catalyzed cycloaddition reactions, see: Takasu, K., *Synlett* **2009**, 1905.
8. (a) Inanaga, K.; Takasu, K.; Ihara, M., *J. Am. Chem. Soc.* **2005**, *127*, 3668. (b) Takasu, K.; Miyakawa, Y.; Ihara, M.; Tokuyama, H., *Chem. Pharm. Bull.* **2008**, *56*, 1205.
9. Takasu, K.; Hosokawa, N.; Inanaga, K.; Ihara, M., *Tetrahedron Lett.* **2006**, *47*, 6053.
10. Ishihara, K.; Nakano, K., *J. Am. Chem. Soc.* **2007**, *129*, 8930.
11. (a) Takasu, K.; Nagao, S.; Ihara, M., *Adv. Synth. Catal.* **2006**, *348*, 2376. (b) Li, W.; Shi, M., *J. Org. Chem.* **2009**, *74*, 856.
12. Shindoh, N.; Tokuyama, H.; Takemoto, Y.; Takasu, K., *Heterocycles* **2009**, *77*, 187.
13. (a) Nakashima, D.; Yamamoto, H., *Org. Lett.* **2005**, *7*, 1251. (b) Sakakura, A.; Suzuki, K.; Nakano, K.; Ishihara, K., *Org. Lett.* **2006**, *8*, 2229. (c) Jung, M. E.; Ho, D. G., *Org. Lett.* **2007**, *9*, 375. (d) Takasu, K.; Shindoh, N.; Tokuyama, H.; Ihara, M., *Tetrahedron* **2006**, *62*, 11900. (e) Shindoh, N.; Tokuyama, H.; Takasu, K., *Tetrahedron Lett.* **2007**, *48*, 4749.
14. Takasu, K.; Inanaga, K.; Ihara, M., *Tetrahedron Lett.* **2008**, *49*, 4220.
15. (a) Ishihara, K.; Hiraiwa, Y.; Yamamoto, H., *Synlett* **2001**, 1851. (b) Boxer, M. B.; Yamamoto, H., *J. Am. Chem. Soc.* **2006**, *128*, 48. (c) Boxer, M. B.; Yamamoto, H., *Nat. Protoc.* **2006**, *1*, 2434. (d) Boxer, M. B.; Yamamoto, H., *J. Am. Chem. Soc.* **2007**, *129*, 2762. (e) Boxer, M. B.; Akakura, M.; Yamamoto, H., *J. Am. Chem. Soc.* **2008**, *130*, 1580. (f) Boxer, M. B.; Yamamoto, H., *Org. Lett.* **2008**, *10*, 453.
16. (a) Kuhnert, N.; Peverley, J.; Robertson, J., *Tetrahedron Lett.* **1998**, *39*, 3215. (b) Cossy, J.; Lutz, F.; Alauze, V.; Meyer, C., *Synlett* **2002**, 45. (c) Yu, S. H.; Ferguson, M. J.; McDonald, R.; Hall, D. G., *J. Am. Chem. Soc.* **2005**, *127*, 12808.
17. (a) Hakala, U.; Wähälä, K., *Tetrahedron Lett.* **2006**, *47*, 8375. (b) Yamada, C.; Sasaki, K.; Matsumura, S.; Toshima, K., *Tetrahedron Lett.* **2007**, *48*, 4223. (c) Kawamura, M.; Cui, D.-M.; Shimada, S., *Tetrahedron* **2006**, *62*, 9201. (d) Terada, M.; Sorimachi, K., *J. Am. Chem. Soc.* **2008**, *130*, 14452.
18. Bachu, P.; Akiyama, T., *Biorg. Med. Chem. Lett.* **2009**, *19*, 3764.
19. (a) Wabnitz, T. C.; Spencer, J. B., *Org. Lett.* **2003**, *5*, 2141. (b) Wabnitz, T. C.; Yu, J.-Q., Spencer, J. B., *Chem. Eur. J.* **2004**, *10*, 484. (c) Chen, H.; Wang, Y.; Wei, S.; Sun, J., *Tetrahedron: Asymmetry* **2007**, *18*, 1308.
20. Othman, R. B.; Bousquet, T.; Othman, M.; Dalla, V., *Org. Lett.* **2005**, *7*, 5335.
21. Bouchez, L. C.; Craita, C.; Vogel, P., *Org. Lett.* **2005**, *7*, 897.
22. For examples of cyclizations of siloxy alkyne, see: (a) Zhang, L.; Kozmin, S. A., *J. Am. Chem. Soc.* **2004**, *126*, 10204. (b) Sun, J.; Kozmin, S. A., *J. Am. Chem. Soc.* **2005**, *127*, 13512.
23. For examples of ynamide reactions, see: (a) Zhang, Y.; Hsung, R. P.; Zhang, X.; Huang, J.; Slafer, B. W.; Davis, A., *Org. Lett.* **2005**, *7*, 1047. (b) Kurtz, K. C. M.; Hsung, R. P.; Zhang, Y., *Org. Lett.* **2006**, *8*, 231.

24. Rowland, G. B.; Zhang, H.; Rowland, E. B.; Chennamadhavuni, S.; Wang, Y.; Antilla, J. C., *J. Am. Chem. Soc.* **2005**, *127*, 15696.
25. Ko, C.; Hsung, R. P., *Org. Biomol. Chem.* **2007**, *5*, 431.
26. (a) Kögel, J. F.; Linder, T.; Schröder, F. G.; Sundermeyer, J.; Goll, S. K.; Himmel, D.; Krossing, I.; Kütt, K.; Saame, J.; Leito, I., *Chem. Eur. J.* **2015**, *21*, 5769. (b) Haas, A.; Klare, C.; Betz, P.; Bruckmann, J.; Kruger, C.; Tsay, Y.-H.; Aubke, F., *Inorg. Chem.* **1996**, *35*, 1918. (c) Burk, P.; Koppel, I. A.; Koppel, I.; Yagupolskii, L. M.; Taft, R. W., *J. Comp. Chem.* **1996**, *17*, 30. (d) Stoyanov, E. S.; Kim, K.-C.; Reed, C. A., *J. Phys. Chem. A* **2004**, *108*, 9310. (e) Munson, K. T.; Vergara, J.; Yu, L.; Vaden, T. D., *J. Phys. Chem. B* **2015**, *119*, 6304. (f) Zhang, M.; Sonoda, T.; Mishima, M.; Honda, T.; Leito, I.; Koppel, I. A.; Bonrath, W.; Netscher, T., *J. Phys. Org. Chem.* **2014**, *27*, 676. (g) Shainyan, B. A.; Chipanina, N. N.; Oznobikhina, L. P., *J. Phys. Org. Chem.* **2012**, *25*, 738. (h) McCune, J. A.; He, P.; Petkovic, M.; Coleman, F.; Estager, J.; Holbrey, J. D.; Seddon, K. R.; Swadzba-Kwasny, M., *Phys. Chem. Chem. Phys.* **2014**, *16*, 23233.
27. Vij, A.; Kirchmeier, R. L.; Shreeve, J. M.; Verman, R. D., *Coord. Chem. Rev.* **1997**, *158*, 413.
28. (a) Takasu, K.; Tanaka, T.; Azuma, T.; Takemoto Y., *Chem. Commun.* **2010**, *46*, 8246. (b) Azuma, T.; Takemoto, Y.; Takasu, K., *Chem. Pharm. Bull.* **2011**, *59*, 1190. (c) Shindoh, N.; Tokuyama, H.; Takemoto, Y.; Takasu, K., *J. Org. Chem.* **2008**, *73*, 7451.
29. Yoshii, Y.; Otsu, T.; Hosokawa, N.; Takasu, K.; Okano, K.; Tokuyama, H., *Chem. Commun.* **2015**, *51*, 1070.
30. Kurahashi, K.; Takemoto, Y.; Takasu, K., *ChemSusChem* **2012**, *5*, 270.
31. (a) Ko, C.; Feltenberger, J. B.; Ghosh, S. K.; Hsung, R. P., *Org. Lett.* **2008**, *10*, 1971. (b) Deng, J.; Hsung, R. P.; Ko, C., *Org. Lett.* **2012**, *14*, 5562.
32. Saito, A.; Taniguchi, A.; Kambara, Y.; Hanzawa, Y., *Org. Lett.* **2013**, *15*, 2672.
33. Li, W.; Shi, M., *Org. Biomol. Chem.* **2009**, *7*, 1775.
34. (a) Shindoh, N.; Kitaura, K.; Takemoto, Y.; Takasu, K., *J. Am. Chem. Soc.* **2011**, *133*, 8470. (b) Shindoh, N.; Takemoto, Y.; Takasu, K., *Heterocycles* **2011**, *82*, 1133. (c) Kuroda, Y.; Shindoh, N.; Takemoto, Y.; Takasu, K., *Synthesis* **2013**, *45*, 2328.
35. Fuchigami, R.; Namba, K.; Tanino, K., *Tetrahedron Lett.* **2012**, *53*, 5725.
36. (a) Zhao, W.; Wang, Z.; Sun, J., *Angew. Chem., Int. Ed.* **2012**, *51*, 6209. (b) Zhao, W.; Li, Z.; Sun, J., *J. Am. Chem. Soc.* **2013**, *135*, 4680. (c) Zhao, W.; Qian, H.; Li, Z.; Sun, J., *Angew. Chem., Int. Ed.* **2015**, *54*, doi: 10.1002/anie.201504926.
37. Yamaoka, Y.; Yamamoto, H., *J. Am. Chem. Soc.* **2010**, *132*, 5354.
38. Albert, B.; Yamamoto, H., *Angew. Chem., Int. Ed.* **2010**, *49*, 2747.
39. Izumiseki, A.; Yamamoto, H., *J. Am. Chem. Soc.* **2014**, *136*, 1308.
40. Saadi, J.; Akakura, M.; Yamamoto, H., *J. Am. Chem. Soc.* **2011**, *133*, 14248.
41. Brady, P.; Yamamoto, H., *Angew. Chem., Int. Ed.* **2012**, *51*, 1942.
42. (a) Albert, B. J.; Yamaoka, Y.; Yamamoto, H., *Angew. Chem., Int. Ed.* **2011**, *50*, 2610. (b) Brady, P. B.; Albert, B. J.; Akakura, M.; Yamamoto, H., *Chem. Sci.* **2013**, *4*, 3223.
43. (a) Yang, B.-L.; Tian, S.-K., *Chem. Commun.* **2010**, *46*, 6180. (b) Mendoza, O.; Rossey, G.; Ghosez, L., *Tetrahedron Lett.* **2010**, *51*, 2571. (c) Ding, F.; William, R.; Wang, F.; Liu, X.-W., *Chem. Commun.* **2012**, *48*, 8709.
44. (a) Earle, M. J.; Hakala, U.; McAuley, B. J.; Nieuwenhuyzen, M.; Ramani, A.; Seddon, K. R., *Chem. Commun.* **2004**, 1368. (b) Mendoza, O.; Rossey, G.; Ghosez, L., *Tetrahedron Lett.* **2011**, *52*, 2235.
45. (a) Zhang, Y., *Tetrahedron Lett.* **2005**, *46*, 6483. (b) Zhang, Y., *Tetrahedron* **2006**, *62*, 3917.
46. Theunissen, C.; Métayer, B.; Henry, N.; Compain, G.; Marrot, J.; Martin-Mingot, A.; Thibaudeau, S.; Evano, G., *J. Am. Chem. Soc.* **2014**, *136*, 12528.
47. Nomiyama, S.; Tsuchimoto, T., *Adv. Synth. Catal.* **2014**, *356*, 3881.
48. (a) Li, H.-H., *Chin. Chem. Lett.* **2015**, *26*, 320. (b) Kondo, M.; Sakamoto, R.; Uchikawawa, M.; Zhang, W.-W.; Namiki, K.; Kume, S.; Murata, M.; Kobayashi, Y.; Nishihara, H., *Angew. Chem., Int. Ed.* **2007**, *46*, 6271. (c) Nagahora, N.; Wasano, T.; Nozaki, K.; Ogawa, T.; Nishijima, S.; Motomatsu, D.; Shioji, K.; Okuma, K., *Eur. J. Org. Chem.* **2014**, 1423. (d) Ponra, S.; Vitale, M. R.; Michelet, V.; Ratovelomanana-Vidal, V., *J. Org. Chem.* **2015**, *80*, 3250. (e) Huo, Z.; Gridnev, I. D.; Yamamoto, Y., *J. Org. Chem.* **2010**, *75*, 1266. (f) Jin, T.; Himuro, M.; Yamamoto, Y., *J. Am. Chem. Soc.* **2010**, *132*, 5590.
49. Jolit, A.; Vazquez-Rodriguez, S.; Yap, G. P. A.; Tius, M. A., *Angew. Chem., Int. Ed.* **2013**, *52*, 11102.
50. Bolte, B.; Gagosz, F., *J. Am. Chem. Soc.* **2011**, *133*, 7696.
51. (a) Hashimoto, T.; Naganawa, Y.; Kano, T.; Maruoka, K., *Chem. Commun.* **2007**, 5143. (b) Hashimoto, T.; Nakatsu, H.; Yamamoto, K.; Watanabe, S.; Maruoka, K., *Chem. Asian J.* **2011**, *6*, 607.
52. (a) Mundal, D. A.; Avetta, C. T., Jr.; Thomson, R. J., *Nat. Chem.* **2010**, *2*, 294. (b) Guitierrez, O.; Strick, B. F.; Thomson, R. J.; Tantillo, D. J., *Chem. Sci.* **2013**, *4*, 3997. (c) Dittrich, S.; Bracher, F., *Tetrahedron* **2015**, *71*, 2530.
53. (a) Niedermann, K.; Früh, N.; Vinogradova, E.; Wiehn, M. S.; Moreno, A.; Togni, A., *Angew. Chem., Int. Ed.* **2011**, *50*, 1059. (b) Niedermann, K.; Füh, N.; Senn, R.; Czarniecki, B.; Verel, R.; Togni, A., *Angew. Chem., Int. Ed.* **2012**, *51*, 6511.
54. (a) Arihara, R.; Nakamura, S.; Hashimoto, S., *Angew. Chem., Int. Ed.* **2005**, *44*, 2245. (b) Yamanoi, T.; Matsuda, S.; Yamazaki, I.; Inoue, R.; Hamasaki, K.; Watanabe, M., *Heterocycles* **2006**, *68*, 673. (c) Arihara, R.; Kakita, K.; Suzuki, N.; Nakamura, S.; Hashimoto, S., *J. Org. Chem.* **2015**, *80*, 4259. (d) Yamamoi, T.; Inoue, R.; Matsuda, S.; Katsuraya, K.; Hamasaki, K., *Tetrahedron: Asymmetry* **2006**, *17*, 2914. (e) Sasaki, K.; Nagai, H.; Matsumura, S.; Toshima, K., *Tetrahedron Lett.* **2003**, *44*, 5605.
55. (a) Tsuchimoto, T.; Joya, T.; Shirakawa, E.; Kawakami, Y., *Synlett* **2000**, 1777. (b) Tachinami, T.; Nishimura, T.; Ushimaru, R.; Noyori, R.; Naka, H., *J. Am. Chem. Soc.* **2013**, *135*, 50. (c) Leyva-Pérez, A.; Cabrero-Antonino, J. R.; Rubio-Marqués, P.; Al-Resayes, S. I.; Corma, A., *ACS Catal.* **2014**, *4*, 722. (d) Cabrero-Antonino, J. R.; Levya-Pérez, A.; Corma, A., *Chem. Eur. J.* **2013**, *19*, 8627.
56. Arisawa, M.; Yamaguchi, M., *J. Am. Chem. Soc.* **2006**, *128*, 50.

N-[4-(Trifluoromethyl)benzyl]-cinchoninium Bromide[1]

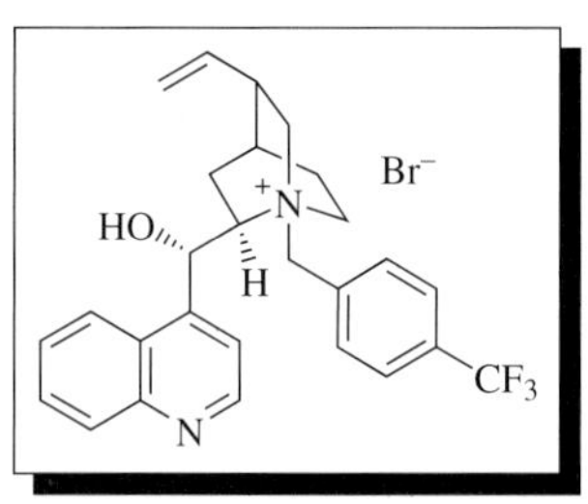

[95088-20-3] $C_{27}H_{28}BrF_3N_2O$ (MW 533.47)

InChI = 1/C27H28F3N2O.BrH/c1-2-19-17-32(16-18-7-9-21(10-8-18)27(28,29)30)14-12-20(19)15-25(32)26(33)23-11-13-31-24-6-4-3-5-22(23)24;/h2-11,13,19-20,25-26,33H,1,12,14-17H2;1H/q+1;/p-1/t19-,20-,25+,26-,32+;/m0./s1/fC27H28F3N2O.Br/h;1h/qm;-1

InChIKey = LOCWWLLFHKDFLF-AZXAVISJDN

(chiral phase-transfer catalyst for asymmetric alkylations,[2a] amino acid synthesis,[7] hydroxylations,[6] Michael additions,[3] and Robinson annulations[2b])

Physical Data: mp 245 °C (dec).

Solubility: < 10^{-5} M in toluene; 20 mM in toluene as dimer.[2a]

Form Supplied in: Supplied in: crystalline salt; commercially available. Can contain anywhere from 0 to 25 mol% of the dihydro analog, usually 15 mol %.

Handling, Storage, and Precautions: do not breath dust; avoid contact with skin and eyes.

Asymmetric Alkylation.[1,2] *N*-[4-(Trifluoromethyl)-benzyl]cinchoninium bromide (**1**) has been used as chiral phase-transfer catalyst[1,2] in the alkylation of indanones (eq 1).[2a] For the alkylation of α-aryl-substituted carbonyl compounds the diastereomeric *N*-[4-(trifluoromethyl)benzyl]cinchonidinium bromide (**2**) was used to obtain the opposite stereochemistry (eqs 2 and 3).[5] The asymmetric alkylation of oxindoles was used as the key step in an asymmetric synthesis of (−)-physostigmine (eq 4).[4]

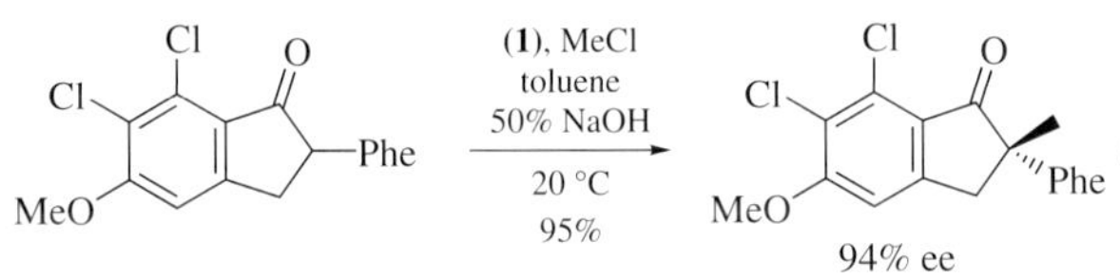
(1)

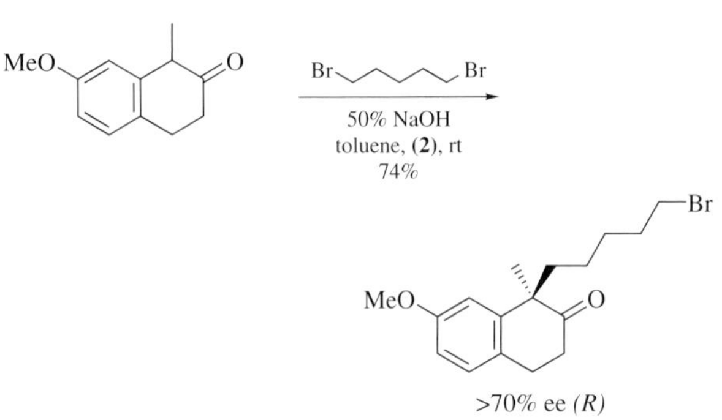
(2)

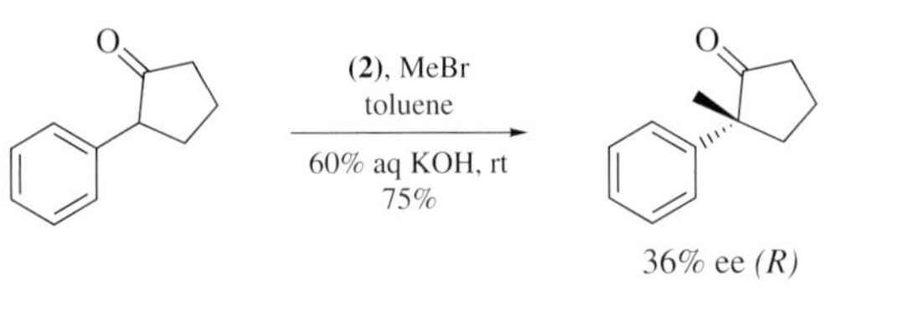
(3)

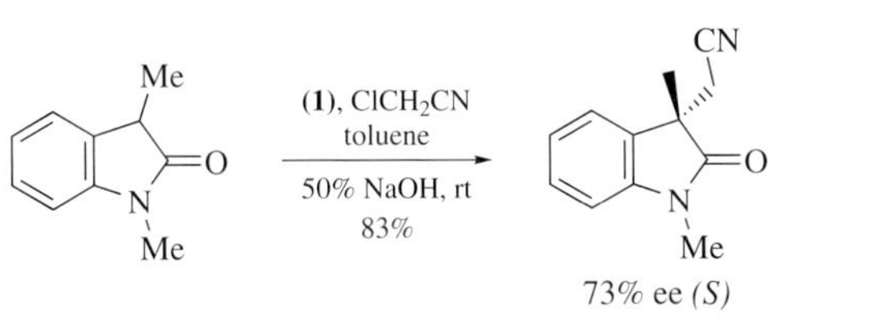
(4)

In all cases it was reported that the trifluoromethyl group enhances the interaction in the prochiral ion pair, resulting in higher ee. The exception appears to be the asymmetric synthesis of α-amino acids via alkylation of the benzophenone Schiff base of glycine alkyl esters with allyl bromide, which produced a 56% ee with the trifluoromethyl-substituted catalyst compared to 66% with the unsubstituted catalysts *N*-benzylcinchoninium chloride (**3**) or *N*-benzylcinchonidinium chloride (**4**) (eq 5).[7]

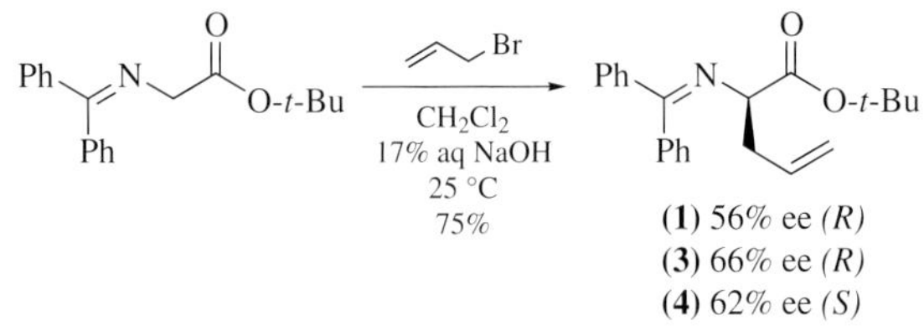
(5)

The unsubstituted catalyst (**3**) was also used in an asymmetric Gabriel synthesis of α-amino acids via solid–liquid chiral phase-transfer alkylation of potassium phthalimide with 2-bromocarboxylates.[10]

In general, nonpolar solvents and less reactive alkyl halides (Cl > Br > I) give higher ee values. Aqueous 50% NaOH is the preferred base, as it acts as a dehydrating agent and keeps the organic solvent dry, thus promoting a tight ion pair. During the reaction the catalyst is extracted into the organic layer as a dimer of ammonium bromide and its zwitterionic oxide,[2a] but it reacts as a monomer resulting in an order of 0.5 for the catalyst in the alkylation reaction. The catalyst degrades during the reaction via Hofmann elimination to tertiary amines, which are readily removed by acid extractions during the workup.[2a] Therefore the catalyst usually cannot be recovered. An economical alternative catalyst is *N*-(3,4-dichlorobenzyl)cinchoninium chloride, which may give equivalent results.[2a,3,4a,9] Reduction in catalyst concentration and reaction times may be achieved by addition of a very small amount of a PEG cocatalyst (PEG 400 or PEG *p*-isooctylphenyl ether).[2e,9] The ee will depend on the interaction of the catalyst with the substrate in the ion pair and in certain cases an electron-donating substituent on the benzyl group may provide optimal interaction, as was found for an asymmetric Michael reaction using *N*-methyl-*N*-benzyl ephedrinium bromide as the phase-transfer catalyst.[8]

Asymmetric Michael Addition. A chiral catalytic addition of methyl vinyl ketone to 2-propylindanone in 93% yield and 80% ee (*S* enantiomer) has been reported (eq 6).[3]

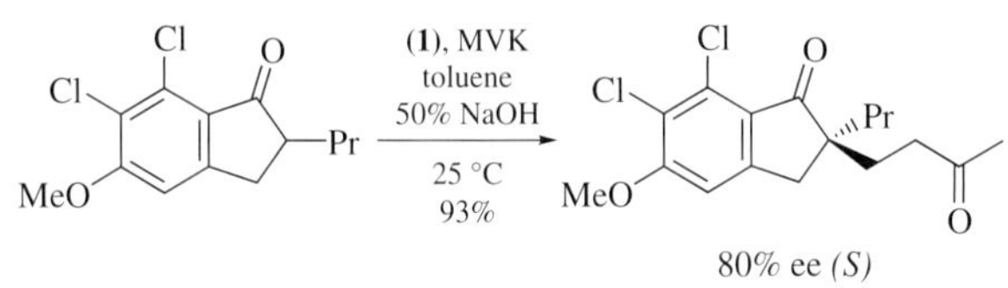
(6)

The (*R*) enantiomer was prepared with 40% ee using the cinchonidine catalyst (**2**). When the vinyl group of the catalyst was hydrogenated to the ethyl group (*N*-[4-(trifluoromethyl)benzyl]dihydrocinchonidinium bromide, **5**), the ee improved to 52%. Equally good results were obtained using the basic catalyst dimer[2a] in a homogeneous system which allowed the use of Michael acceptors not compatible with hydroxide bases.

Asymmetric Hydroxylation. The catalyst has been used for asymmetric α-hydroxylations of indanones and α-tetralones using the standard conditions in combination with oxygen and Triethyl Phosphite (eq 7).[6] Substituents on the aromatic ring of the substrates will influence the π–π interaction in the ion pair and affect the ee. Similarly, (*E*)-2-ethylidene-1-tetralone was oxidized to the α-hydroxy ketone (eq 8).

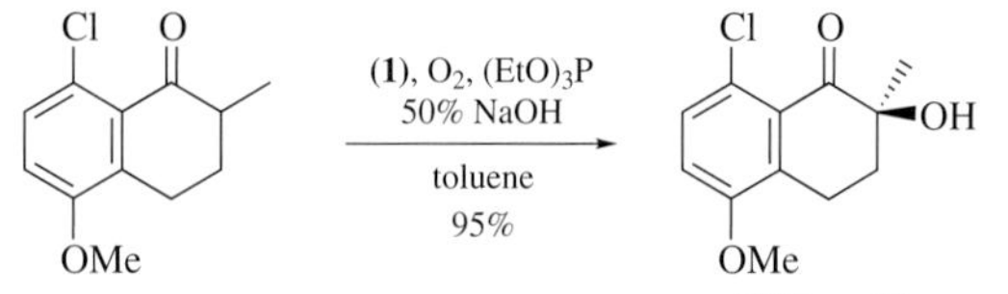

79% ee (S) (7)

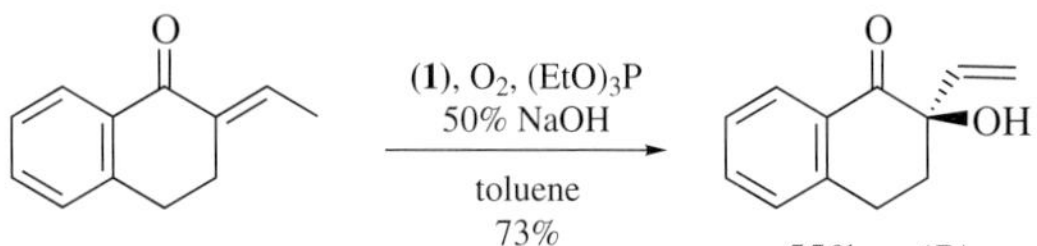

55% ee (R) (8)

Asymmetric Robinson Annulation. 2-Propyl-1-indanone undergoes Robinson annulation with the catalyst and methyl vinyl ketone (eq 6).[3] Higher ee values were achieved using 1,3-dichloro-2-butene (Wichterle Reagent) as an MVK surrogate for the Michael addition and overall Robinson annulation (eq 9).[2b,d,e]

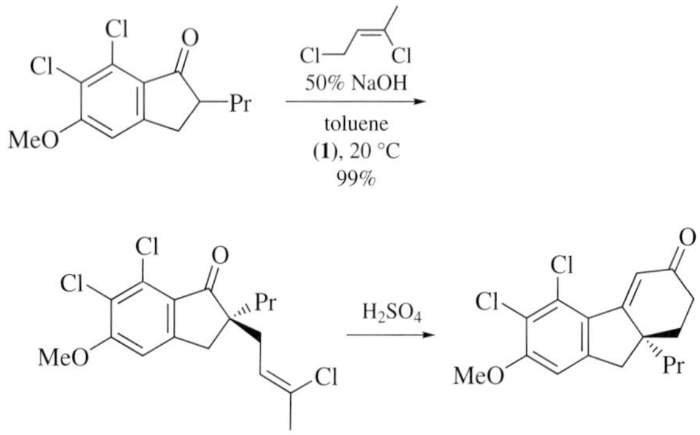

92% ee (S) 90% overall (9)

Using (**2**) as catalyst provided the (*R*) enantiomer in 99% yield, 78% ee. The key introduction of asymmetry during the synthesis of (+)-podocarp-8(14)-en-13-one was the phase-transfer-catalyzed Robinson annulation of 6-methoxy-1-methyl-2-tetralone with ethyl vinyl ketone. The authors carried out a comparative study of the *N*-(4-trifluoromethyl)benzyl derivatives of cinchonine, cinchonidine, dihydrocinchonine, and dihydrocinchonidine and found that (**5**) produced the highest ee of the desired (*S*) enantiomer at −45 °C using toluene and 60% aq KOH (eq 10).[5]

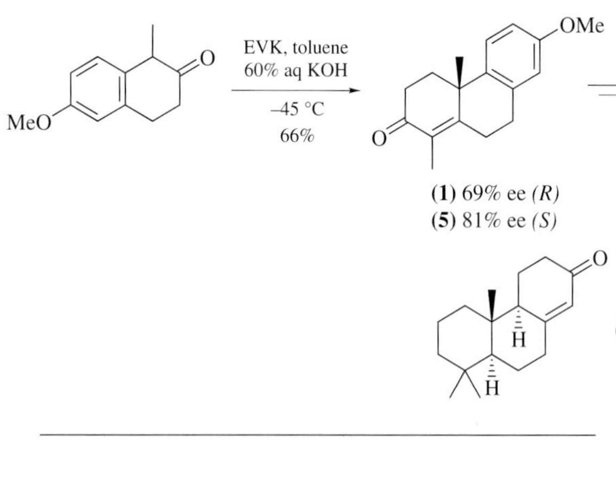

(**1**) 69% ee (*R*)
(**5**) 81% ee (*S*)

(10)

1. Dehmlow, E. V.; Dehmlow, S. S. *Phase Transfer Catalysis*; VCH: Weinheim, **1993**; p 80.
2. (a) Hughes, D. L.; Dolling, U.-H.; Ryan, K. M.; Schoenewaldt, E. F.; Grabowski, E. J. J., *J. Org. Chem.* **1987**, *52*, 4745. (b) Bhattacharya, A.; Dolling, U.-H.; Grabowski, E. J. J.; Karady, S.; Ryan, K. M.; Weinstock, L. M. *Angew. Chem., Int. Ed. Engl.* **1986**, *25*, 476. (c) Dolling, U.-H.; Davis, P.; Grabowski, E. J. J. *J. Am. Chem. Soc.* **1984**, *106*, 446. (d) Dolling, U.-H.; Hughes, D. L.; Bhattacharya, A.; Ryan, K. M.; Karady, S.; Weinstock, L. M.; Grabowski, E. J. J. In *Phase-Transfer Catalysis*; Starks, C. M., Ed.; American Chemical Society: Washington, **1987**; p 67. (e) Dolling, U.-H.; Hughes, D. L.; Bhattacharya, A.; Ryan, K. M.; Karady, S.; Weinstock, L. M.; Grenda, V. J.; Grabowski, E. J. J. In *Catalysis of Organic Reactions*; Rylander, P. N.; Greenfield, H.; Augustine, R. L.; Eds.; Dekker: New York, **1988**; p 65.
3. Conn, R. S. E.; Lovell, A. V.; Karady, S.; Weinstock, L. M. *J. Org. Chem.* **1986**, *51*, 4710.
4. (a) Lee, T. B. K.; Wong, G. S. K., *J. Org. Chem.* **1991**, *56*, 872. (b) Chen, B. H.; Ji, Q. E. *Acta Chim. Sinica* **1989**, *47*, 350 (*Chem. Abstr.* **1989**, *111*, 194 508).
5. Nerinckx, W.; Vandewalle, M., *Tetrahedron: Asymmetry* **1990**, *1*, 265.
6. Masui, M.; Ando, A.; Shioiri, T., *Tetrahedron Lett.* **1988**, *29*, 2835.
7. (a) O'Donnell, M. J.; Bennett, W. D.; Wu, S., *J. Am. Chem. Soc.* **1989**, *111*, 2353. (b) O'Donnell, M. J.; Wu, S., *Tetrahedron: Asymmetry* **1992**, *3*, 591. (c) Imperiali, B.; Prins, T. J.; Fisher, S. L., *J. Org. Chem.* **1993**, *58*, 1613.
8. Loupy, A.; Zaparucha, A., *Tetrahedron Lett.* **1993**, *34*, 473.
9. Dolling, U.-H. U. S. Patent 4 605 761, **1986** (*Chem. Abstr.* **1987**, *106*, 4697).
10. Guifa, S.; Lingchong, Y., *Synth. Commun.* **1993**, *23*, 1229.

Ulf-H. Dolling
Merck Research Laboratories, Rahway, NJ, USA

3-(2,4,6-Trimethylphenyl)-1-methyl-1*H*-benzo[*d*]imidazol-3-ium Iodide

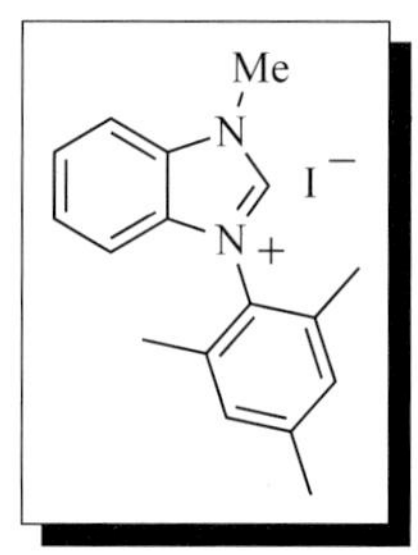

[937807-73-3] $C_{17}H_{19}IN_2$ (MW 378.25)

InChI = 1S/C17H19N2.HI/c1-12-9-13(2)17(14(3)10-12)19-11-18(4)15-7-5-6-8-16(15)19;/h5-11H,1-4H3;1H/q+1;/p-1

InChIKey = GZUMUMKCXFZBLQ-UHFFFAOYSA-M

(*N*-heterocyclic carbene, umpolung, homoenolate, organocatalysis)

Physical Data: mp 266–269 °C.

Solubility: sol DCM, chloroform; insol ether, alkanes.

Form Supplied in: colorless to off-white solid.

Analysis of Reagent Purity: IR (film) 3018.1, 2947.8, 1610.2, 1560.3, 1260.7, 753.8, 731.9 cm^{-1}; ^1H NMR (400 MHz, $CDCl_3$) δ 10.75 (s, 1H), 7.90 (d, J = 8.2 Hz, 1H), 7.72 (t, J = 7.6 Hz, 1H); 7.62 (t, J = 7.6,1H), 7.23 (t, J = 8.2 Hz, 1H), 7.06 (s, 2H), 4.54 (s, 3H), 2.37 (s, 3H), 2.02 (s, 6H).

Preparative Methods: the original synthesis is a five-step procedure utilizing two consecutive Buchwald–Hartwig amination coupling reactions (eq 1).[1]

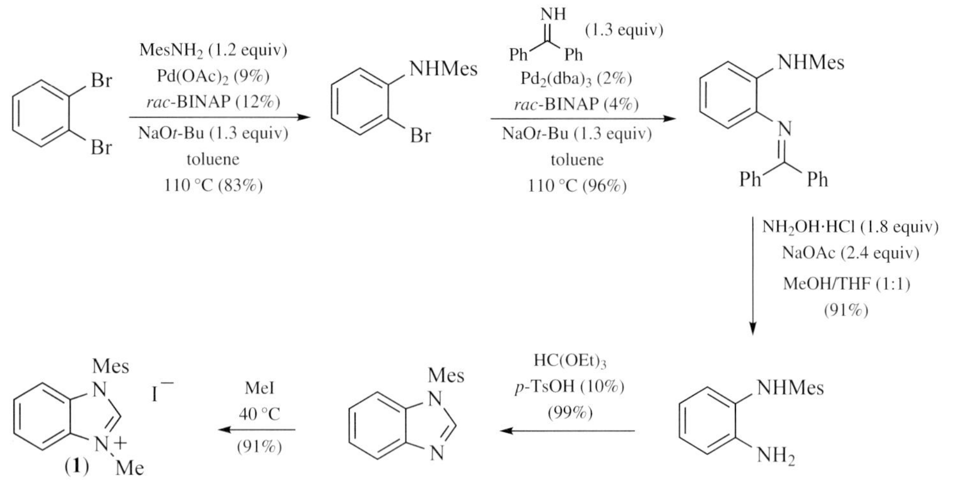

(1)

An alternate synthesis was reported, which is shorter and relies upon a copper-catalyzed amination reaction (eq 2).[2]

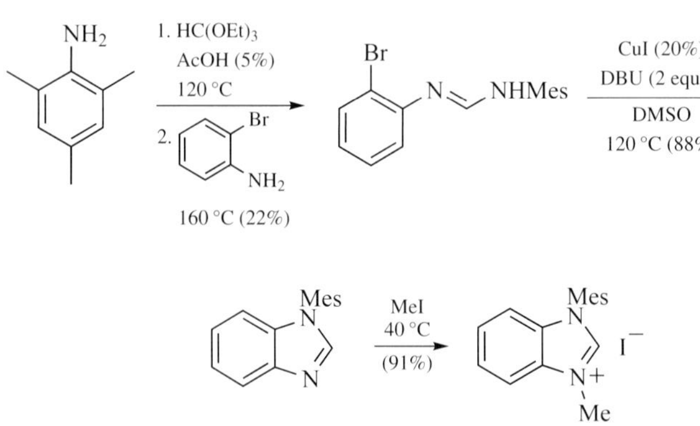

(2)

Mes = 2,4,6-trimethylphenyl

Nucleophilic Addition to Aldehydes. The carbene generated from deprotonation of the benzimidazolium salt **1** can act as a nucleophile in the presence of cinnamaldehyde derivatives to generate the Breslow intermediate **2**, enabling homoenolate umpolung reactions. As such, formal [3 + 3]-cycloadditions between the conjugated Breslow intermediates and azomethine-imines are possible to generate six-membered heterocycles. The benzimidazolium salt **1** gives the best yield for the transformation compared to other commercially available carbenes (eq 3).[3]

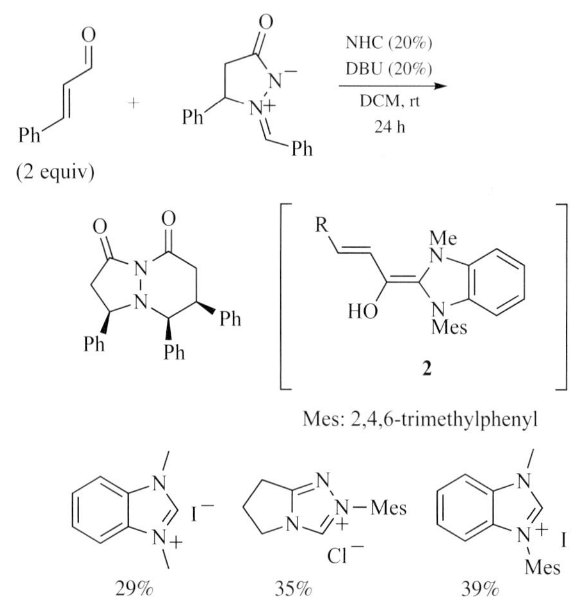

(3)

The formal [3 + 3]-cycloaddition reaction has a wide substrate scope with regards to the aldehyde; however, the imine must be substituted by an aromatic group to achieve sufficient reactivity (eq 4).

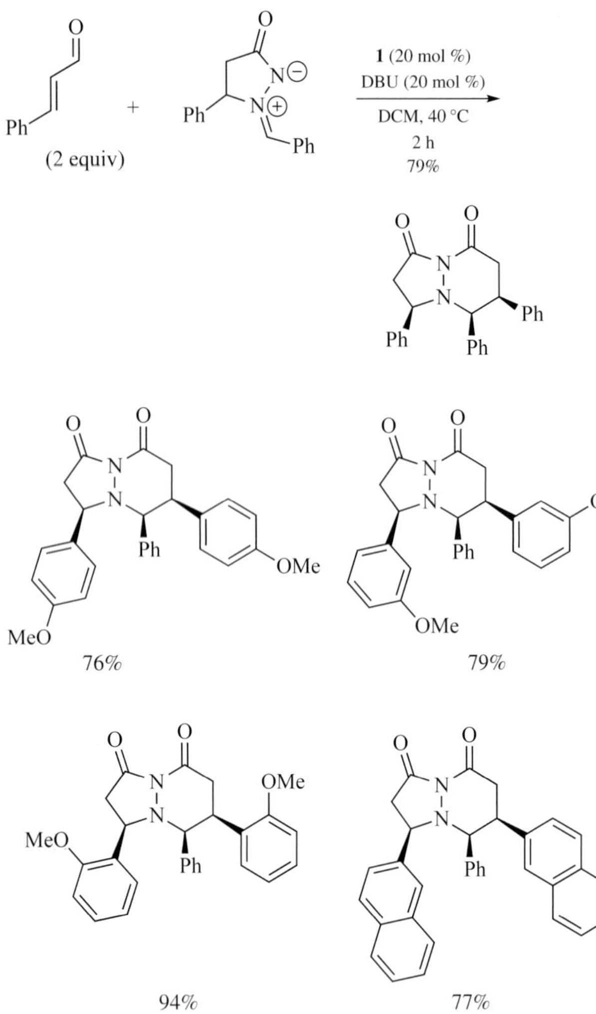

(4)

The Breslow intermediates formed from the reaction of **1** and cinnamaldehyde can also react with *N*-amidoimines with high diastereoselectivity (eq 5).[4]

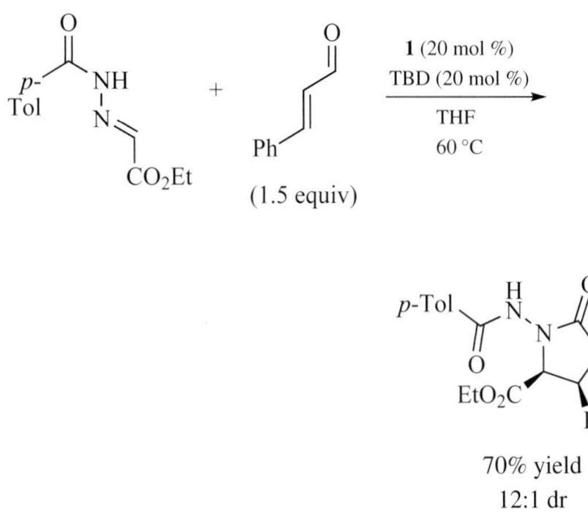

(5)

The benzimidazolium salt **1** is effective at selectively oxidizing cinnamaldehyde to the corresponding ester product (eq 6).[4] A comparison with other commercial reagents is shown below.

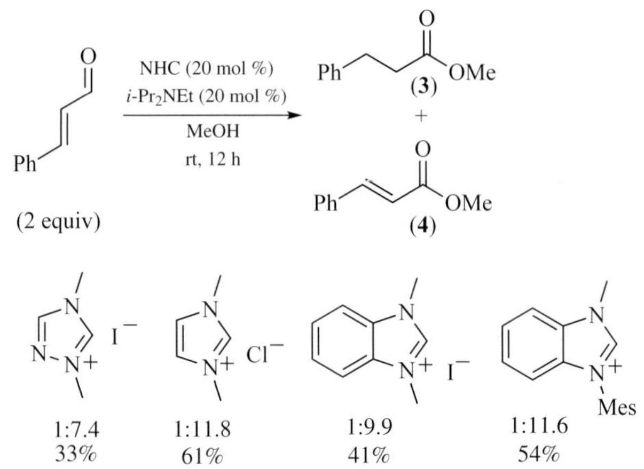

(6)

Mes: 2,4,6-trimethylphenyl

---

1. Chan, A.; Scheidt, K. A., *J. Am. Chem. Soc.* **2007**, *129*, 5334.
2. Hirano, K.; Biju, A. T.; Glorius, F., *J. Org. Chem.* **2009**, *74*, 9570.
3. Raup, D. E. A.; Cardinal-David, B.; Holte, D.; Scheidt, K. A., *Nat. Chem.* **2010**, *2*, 766.
4. Maki, B. E.; Patterson, E. V.; Cramer, C. J.; Scheidt, K. A., *Org. Lett.* **2009**, *11*, 3942.

Michael Holtz-Mulholland & Shawn K. Collins
Université de Montréal, Montréal, Québec, Canada

(5*S*)-2,2,3-Trimethyl-5-(phenylmethyl)-4-imidazolidinone

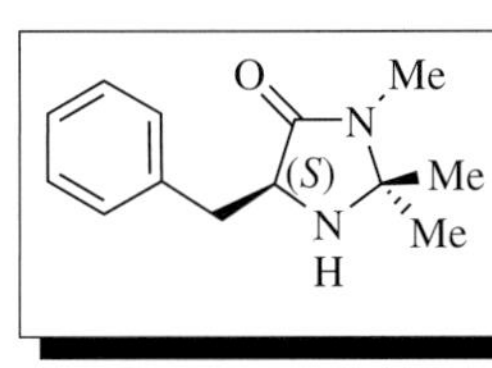

1

[132278-23-2] $C_{13}H_{18}N_2O$ (MW 218.29)

InChI = 1/C13H18N2O/c1-13(2)14-11(12(16)15(13)3)9-10-7-5-4-6-8-10/h4-8,11,14H,9H2,1-3H3/t11-/m0/s1

InChIKey = UACYWOJLWBDSHG-NSHDSACABQ

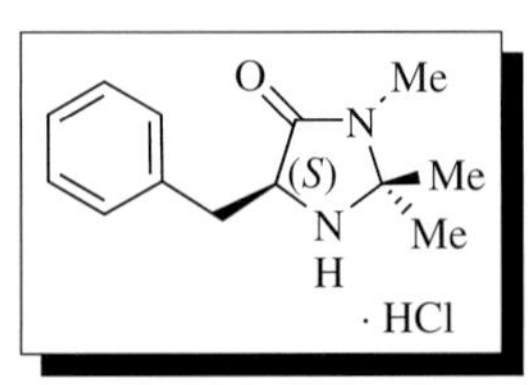

[278173-23-2] $C_{13}H_{18}N_2O{\cdot}HCl$ (MW 254.76)

InChI = 1/C13H18N2O.ClH/c1-13(2)14-11(12(16)15(13)3)9-10-7-5-4-6-8-10;/h4-8,11,14H,9H2,1-3H3;1H/t11-; /m0./s1

InChIKey = YIYFEXGDFJLJGM-MERQFXBCBZ

(highly enantioselective organocatalyst for a wide array of asymmetric reactions)

Physical Data: imid·HCl·colorless crystals; mp 157–161 °C; $[\alpha]_D^{20}$ −64°; c = 1 in H_2O.

Solubility: imid: ether, MeCN, EtOAc, most organic solvents; imid·HCl: sol. H_2O, alcohol/H_2O mixtures.

Form Supplied in: colorless crystals, synthesized from the addition of HCl-dioxane (4 M) to (5S)-2,2,3-trimethyl-5-(phenylmethyl)-4-Imidazolidinone in Et_2O. The precipitate was recrystallized from *iso*-propanaol. Available from Aldrich.

Handling, Storage, and Precautions: Store under inert gas in a cool dry place. The salt is hygroscopic, so exposure to moisture should be avoided. Should not be exposed to strong oxidizing agents. Its hazardous decomposition products are carbon monoxide, carbon dioxide, nitrogen oxides, and hydrogen chloride gas.

Original Commentary

Barbara J. Morgan & Marisa C. Kozlowski
University of Pennsylvania, Philadelphia, PA, USA

MacMillan and co-workers developed **1** based upon design features utilized in Lewis acid catalysis. It was reasoned that LUMO lowering activation and the kinetic lability toward ligand substitution that enables Lewis acid catalyst turnover (eq 1) might also be emulated by chiral Lewis bases, thus enabling reversible formation of iminium ions from α, β-unsaturated aldehydes and amines. Catalytic systems based on this premise provide a new platform for the design of organocatalytic processes.[1] Significantly, the chiral amines are effective as enantioselective catalysts for a range of transformations that have traditionally employed metal salts.

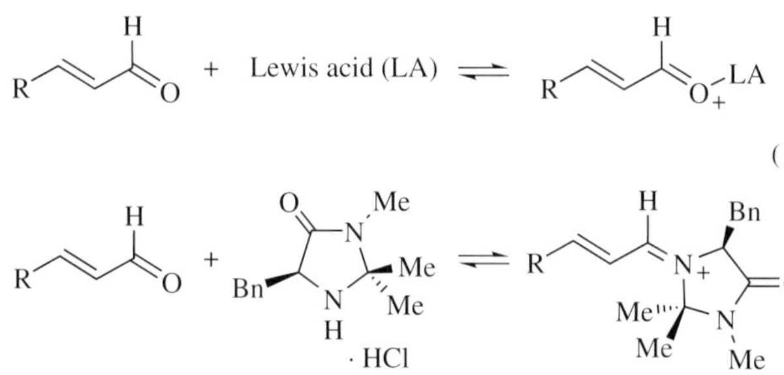

(1)

Diels–Alder Reactions. The utility of **1** was first shown in an organocatalytic Diels–Alder reaction. Uniformly high enantioselectivity, yield, and turnover were observed upon the addition of various dienophiles to cyclopentadiene (eq 2).[1] The synthetic utility of this protocol has been demonstrated in the highly enantioselective (93 % ee) preparation of the key intermediate in the total synthesis of (+)-hapalindole Q.[2] The reaction illustrated in eq 3 establishes the proper arrangement of the required four contiguous stereocenters, and represents the first successful application of an enantioselective organocatalyzed Diels–Alder reaction in total synthesis.

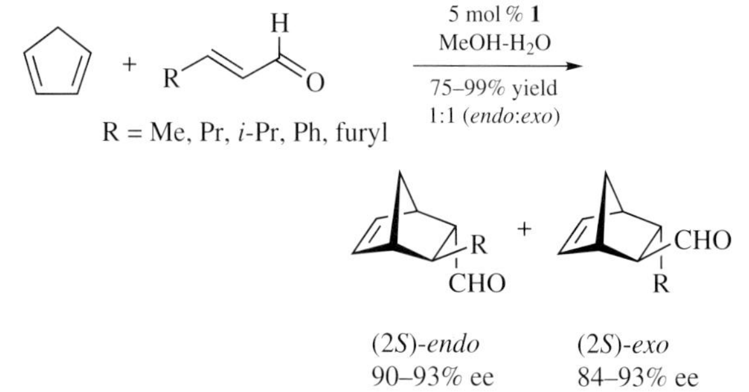

(2)

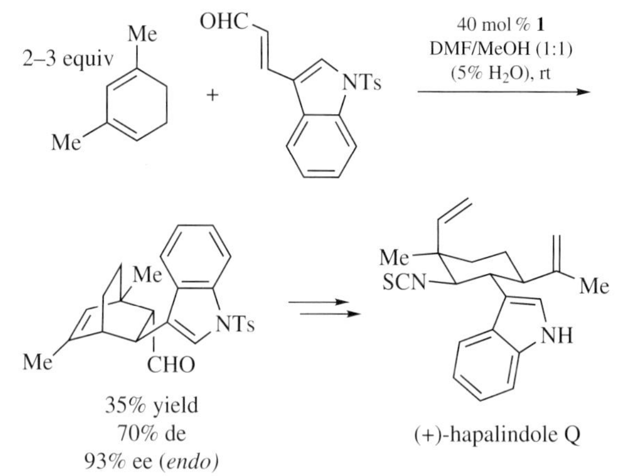

(3)

Imidazolidinone **1** has also found utility in enantioselective *intramolecular* Diels-Alder (IMDA) reactions. Although there are numerous reports of diastereoselective IMDA reactions, examples of high levels of enantiocontrol in the catalytic asymmetric IMDA reactions are not as plentiful. The LUMO-lowering ability of **1** has proven to be effective for the enantioselective cycloisomerization of a range of trienal aldehydes (eq 4).[3]

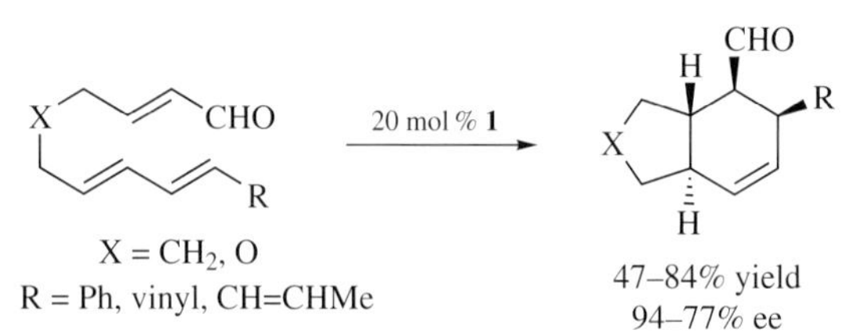

(4)

Further work in enantioselective Diels-Alder catalysis with **1** centered on ionic liquids, and microwave activation. Rate acceleration was found both when hydrophobic ionic liquids such as $[Bmim]PF_6$ and $[Bmim]SbF_6$ (Bmim = 1-*n*-butyl-3-methylimidazolium), were used as the solvent system and also via microwave activation.[4]

1,3-Dipolar Additions. The catalyst system is also useful for the [3 + 2] cycloadditions between nitrones and α,β-unsaturated aldehydes to provide isoxazolidines (eq 5).[5] Enantioselective formation of the *endo*-(4*S*)-isoxazolidine adduct was observed in all cases; the sense of asymmetric induction and diastereoselectivity are consistent with the (*E*)-iminium isomer, which is in direct analogy with the Diels-Alder reports.

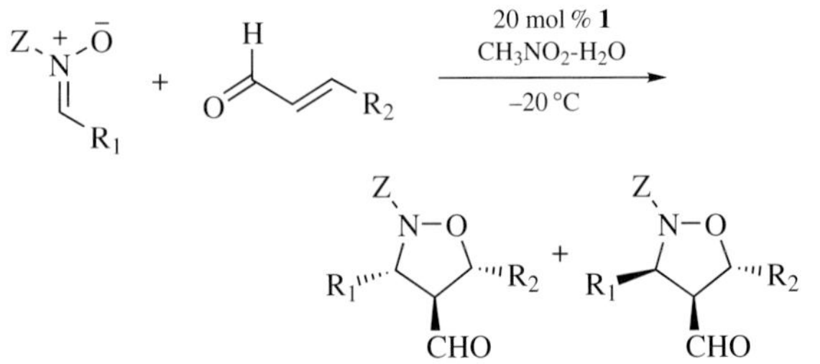

(5)

endo:*exo* = 81:19–99:1
yield = 66–98%
%ee (*endo*) = 90–99%

Z = Bn, Allyl, Me
R_1 = Ph, C_6H_3Cl-4, C_6H_3OMe-4, C_6H_3Me-4, 2-naph, *c*-Hx
R_2 = Me, H

Conjugate Additions. MacMillan further demonstrated the utility of **1** in the first enantioselective conjugate pyrrole additions.[6] The asymmetric catalyzed alkylation of pyrroles with α,β-unsaturated aldehydes generates β-pyrrolyl carbonyls in high yield and enantioselectivity (eq 6). In addition, **1** catalyzed indole alkylations in high yields and moderate enantioselectivities.[7]

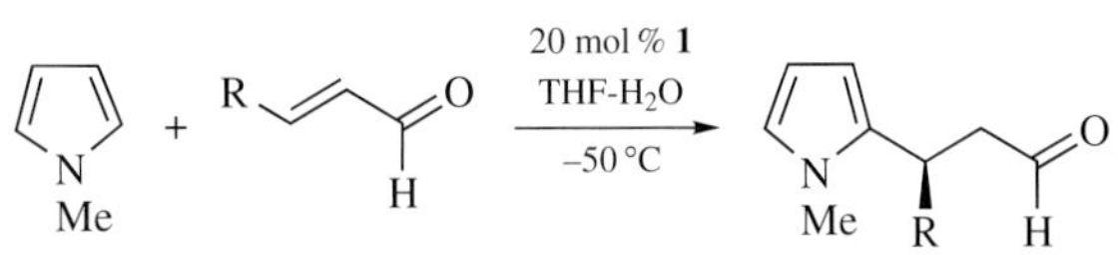

(6)

R = Me, Pr, *i*-Pr, Ph, 4-MeOPh, CH_2OBn, CO_2Me — 72–87% yield, 87–93% ee

The intermolecular aldehyde-enone Michael addition reactions can be catalyzed by **1**, provided that an appropriate hydrogen-bond-donating cocatalyst is employed (eq 7).[8] The imidazolidinone-derived enamine has been isolated and shown to be a competent nucleophile. This is a departure from the prior reports of imidazolidinone-catalyzed conjugate additions that involved electrophilic activation of α,β-unsaturated aldehydes via iminium ion formation.[6,7]

List reported an efficient enantioselective cyclization of formyl enones, constituting the first catalytic asymmetric intramolecular Michael reaction of aldehydes (eq 8). Presumably, an imidazolidinone derived nucleophilic enamine is also involved in this related process. The reaction furnishes cyclic ketoaldehydes in high yields and enantioselectivities under mild and convenient reaction conditions.[9]

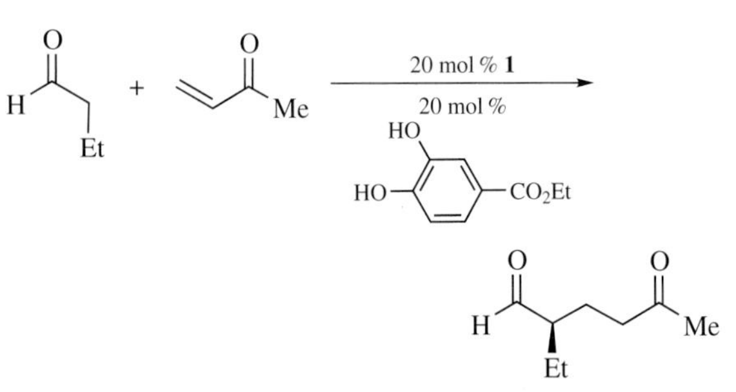

(7)

78% yield, 81% ee

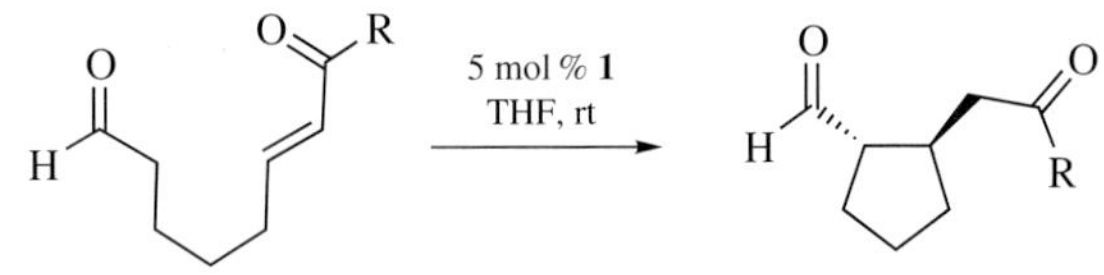

(8)

R = Ph, naph, Me, Et, H — 85–99% yield, 80–97% ee, 17:1–49:1 (*anti/syn*)

Related Reagents. MacMillan has shown that modifications to reduce the steric demand of catalyst system **1** can be beneficial. In particular, the related reagent, **2**, has been found to catalyze a wide variety of reactions with high yield and enantioselectivity, including Diels-Alder reactions, conjugate additions, hydride reductions, and cascade reactions.[3,7,10]

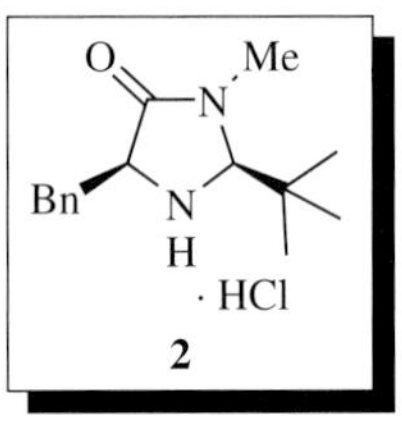

First Update

Aiguo Song & Wei Wang
University of New Mexico, Albuquerque, NM, USA

Diels–Alder Reactions. MacMillan catalyst **1** has been extensively employed as efficient promoter in enantioselective Diels–Alder reactions for asymmetric synthesis of important cores of natural products via iminium activation. For example, the core of eunicellin was generated as a single diastereomer in good yield by the treatment of triene **3** with catalyst **1** (eq 9).[11]

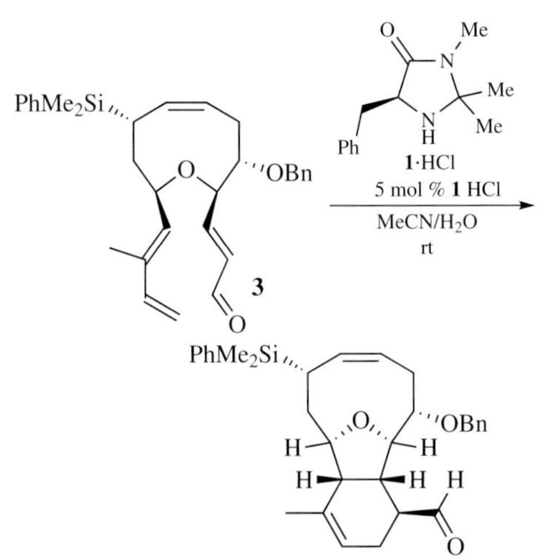

(9)

67% yield, single diastereomer

Moreover, catalyst **1** mediated Diels–Alder reaction and intramolecular cyclization cascade sequence were utilized to construct the tetracyclic alkaloid core architecture in one step for the total synthesis of (−)-vincorine from tryptamine **4** and enal **5**.[12] The core structure **6** bearing multiple stereogenic centers including one all-carbon quaternary center with the requisite relative and absolute configuration was obtained in 70% yield and with 95% ee (eq 10).

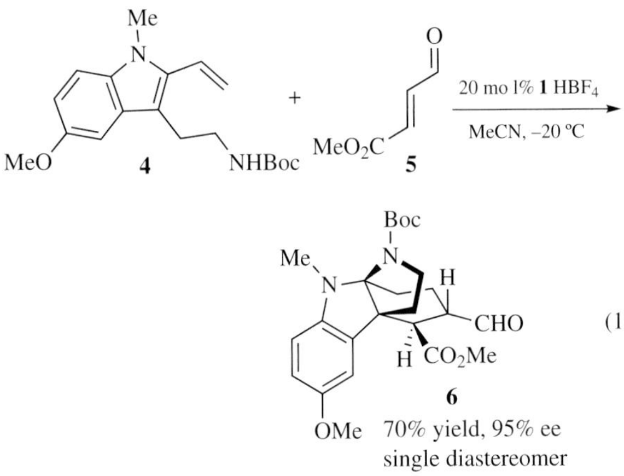

(10)

Based on the fully developed secondary amine-catalyzed Diels–Alder reactions via iminium activation, the first organocatalytic double Diels–Alder reaction was disclosed (eq 11).[13] The complex cycloaddition products was produced from π-bond-rich hydrocarbons and enals with amplified enantiopurity (er >99.5:0.5) as predicted by the Horeau principle. The chiral double cycloadducts were transformed into highly functionalized hydrophenanthrenes by a third Diels–Alder cycloaddition.

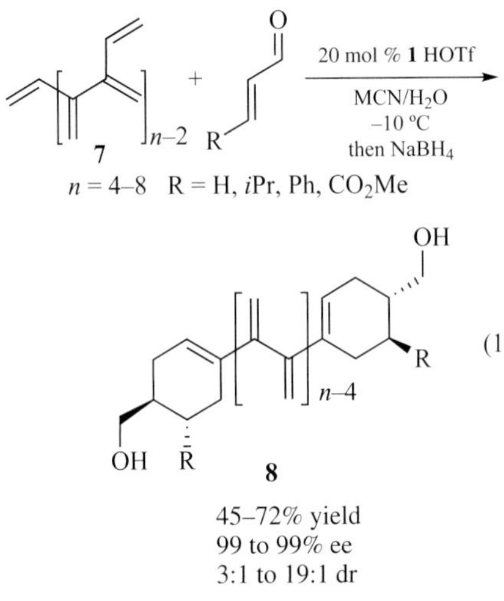

(11)

Moreover, organocatalyst **1** was found to be capable of promoting inverse-electron-demanding aza-Diels–Alder reaction between the β, γ-bond of crotonaldehyde and *N*-tosyl-1-aza-1,3-butadiene **9** via dienamine catalysis to provide the desired cycloadduct **10** with improved enantioselectivity compared with the product obtained by using (*S*)-2-(diphenyl((trimethylsilyl)oxy)methyl)pyrrolidine catalyst (eq 12).[14]

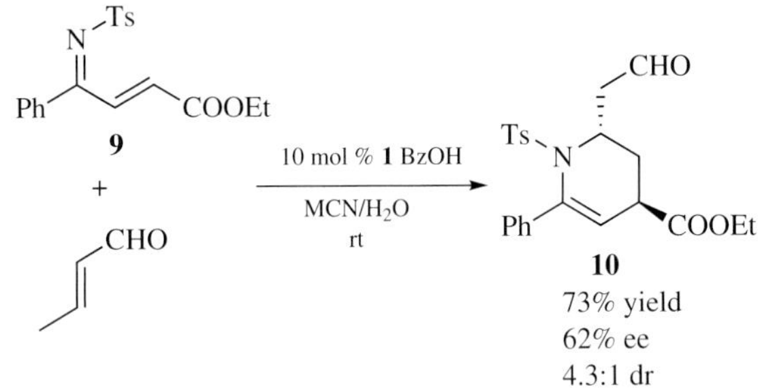

(12)

α-Chlorination. In addition to the iminium activation catalytic systems through LUMO-lowering strategy, enamine activation catalytic protocols via HOMO-raising mode employing catalyst **1** can also be achieved to furnish α-functionalized products of aldehydes. The first highly enantioselective direct α-chlorination of aldehydes using chlorinated quinone **11** as chlorination source was developed to provide chiral α-chloroaldehydes **12** with good yields (eq 13).[15] Application of this efficient catalytic method to the production of key intermediate for the synthesis of chiral β-lactams was also reported.[16]

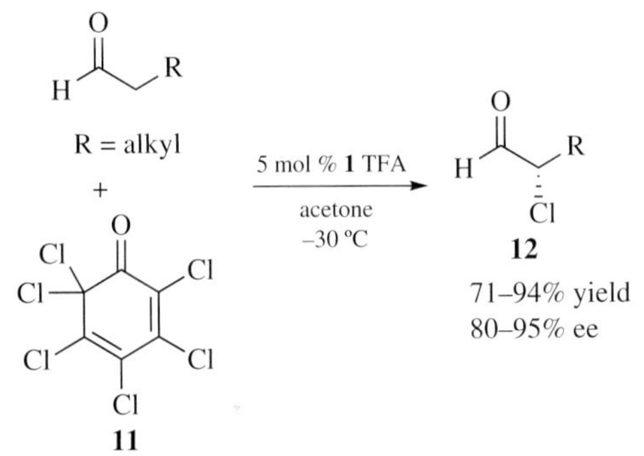

(13)

α-Fluorination. Naturally, chiral α-fluoroaldehydes can be generated in a similar way by replacing chlorinated quinone with a fluorinating reagent. Organocatalyst **1** catalyzed α-fluorination of aldehydes using *N*-fluorobenzenesulfonimide was reported by Barbas' and MacMillan's groups independently at the same time (eq 14).[17]

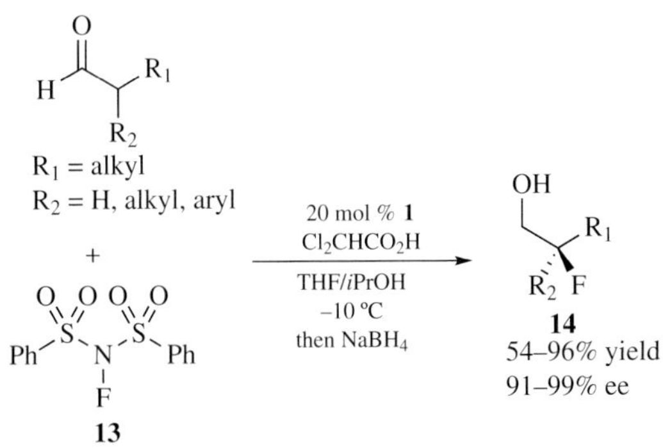

(14)

α-Oxyamination. Enantioselective α-oxyamination of aldehydes was also achieved by the combination of secondary amine **1** promoted organocatalysis and radical-mediated C-O bond-forming process.[18] It was proposed that enamines formed from catalyst **1** and aldehydes could be oxidized by a single-electron transfer reagent to yield radicals that could be trapped stereoselectively by TEMPO providing access to α-oxyaminated aldehydes. The important chiral α-oxyaminated aldehydes were obtained from the proposed processes in moderate to good yields with $FeCl_3$ single-electron transfer reagent and $NaNO_2/O_2$ cooxidant (eq 15).

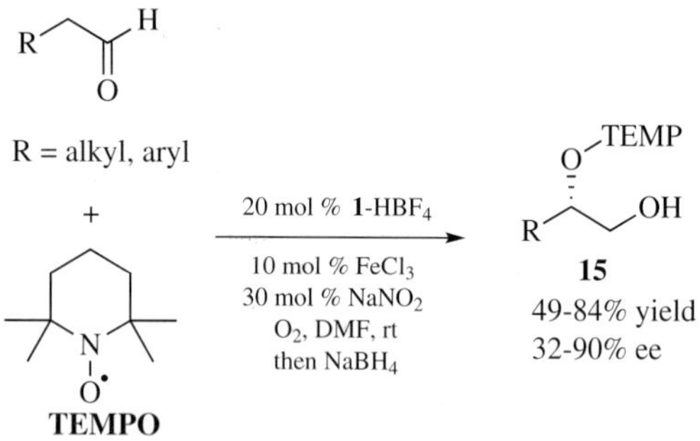

(15)

However, it was reasoned that this reaction proceeded via a traditional enamine catalysis pathway rather than the combination of enamine catalysis and radical-mediated process sequence.[19] The electrophilic complex formed from TEMPO and $FeCl_3$ is subjected to the nucleophilic enamine attack to generate the α-oxyaminated product. It was also noted that the real cooxidant was likely to be TEMPO regarding the fact that identical levels of conversion were obtained without $NaNO_2$ and O_2.

***α*-Alkylation.** Although various chiral α-alkylated aldehydes can be easily obtained from asymmetric Michael addition reactions using α, β-unsaturated electrophiles, the enantioselective installation of special groups, such as methyl group, to the α-position of aldehydes is still challenging. It is problematic to employ chiral secondary amine-mediated enamine catalytic process to produce α-alkylated aldehydes due to the side reactions between the catalyst and alkylating reagents, such as MeI, to give tertiary amine or quaternary ammonium salt and the side reactions beween alkylating reagents and water resulting from enamine formation. However, it was proved that enantioselective α-alkylation of aldehydes could be successfully achieved by the choice of stable benzodithiolylium tetrafluoroborate alkylating reagent **16** (eq 16).[20] This is because even if the catalyst can be alkylated by **16**, the catalyst alkylation complex is so unstable that the catalyst can be released easily. The presence of the base NaH_2PO_4 was necessary to capture the HBF_4 liberated by the reaction of the carbenium ion **16**.

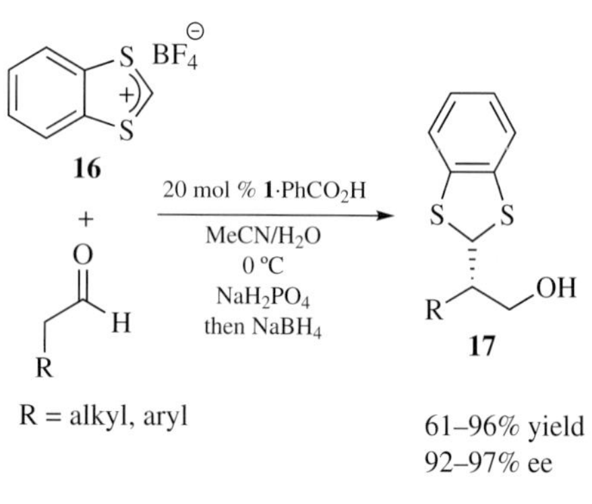

(16)

This organocatalytic α-alkylation provides an easy access to chiral α-methylated aldehydes since the 1,3-benzodithiol group can be easily removed by Raney Ni in the presence of hydrogen (eq 17).[20]

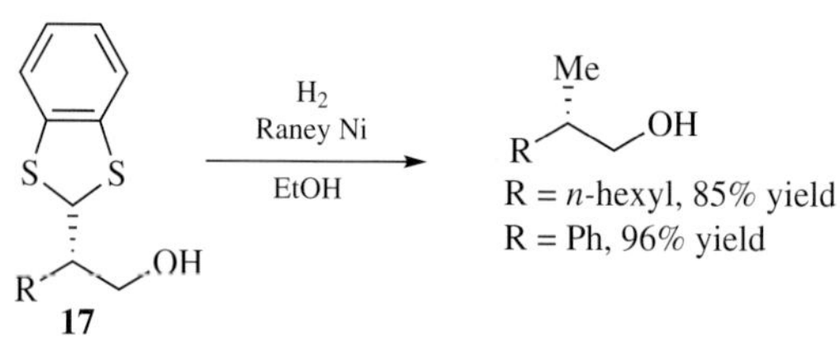

(17)

Oxocarbenium ions formed from acetals through Lewis acid catalysis were found to be alternative suitable electrophiles for aldehyde α-alkylation via enamine catalysis, since oxocarbenium ions will form stable products with carbon nucleophilies and unstable complexes with nitrogen nucleophilic sites.[21] The optically active α-alkylation aldehydes **19** were generated with good enantioselectivities but low diastereoselectivities (eq 18). A similar asymmetric organocatalytic Mannich cyclization employing catalyst **1** for the construction of bicyclic skeleton of izidine alkaloids was disclosed and applied to the total syntheses of (−)-epilupinine, (−)-tashiromine, and (−)-trachelanthamidine.[22]

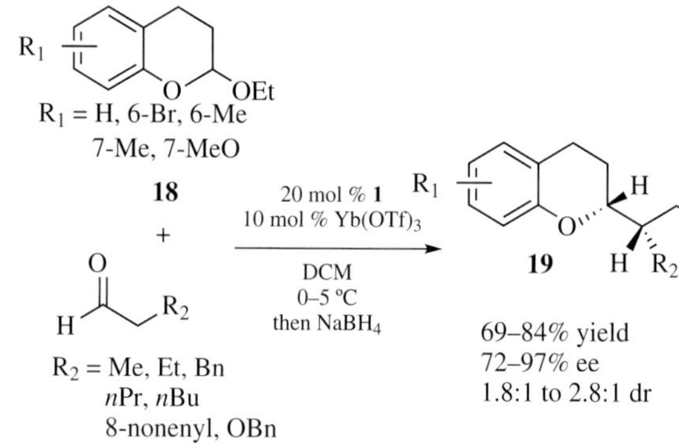

(18)

The stabilized carbocation generated from oxidation of benzylic C-H bond was also subjected to the enamine catalysis process using secondary amine **1** to furnish chiral α-alkylated aldehydes **21** via cross-dehydrogenative coupling process (eq 19).[23]

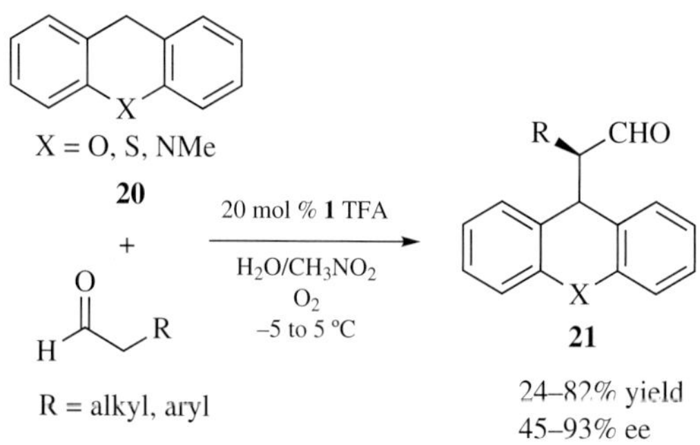

(19)

***β*,*α*-Cascade Functionalization.** MacMillan and coworkers proposed that the aldehyde intermediates resulting from β-conjugation of enals via iminium catalysis could be subjected to α-functionalization via enamine catalysis using the same secondary amine catalyst in a one-pot fashion for the rapid access to structural complexity.[24] This proposed cascade iminium–enamine catalysis sequence was achieved by consecutive addition of suitable nucleophilic and electrophilic reagents to the enal and secondary amine catalyst mixture. Amazingly, they also envisioned that the iminium and enamine steps might be discretely controlled by cycle-specific catalysts to produce building block **23** with various controlled diastereo-induction via judicious selection of the catalyst enantiomers (eq 20).

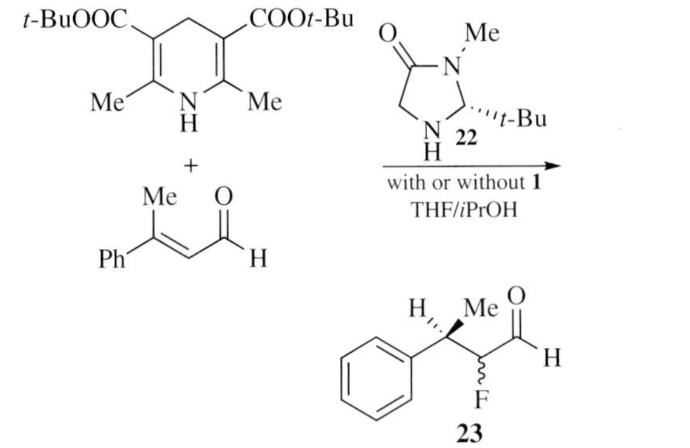

(20)

22 only: 60% yield, 99% ee, *syn:anti* = 3:1
22 and (5*R*)-**1**: 62% yield, 99% ee, *syn:anti* = 9:1
22 and (5*S*)-**1**: 81% yield, 99% ee, *anti:syn* = 16:1

γ,α-Cascade Functionalization. Significantly, optically active γ,α-functionalized aldehydes were generated by a similar pathway via secondary amine catalyst mediated activation of cyclopropane carbaldehydes.[25] Nucleophilic addition to the transient cyclopropyl iminium species followed by the capture of the transient enamine with an electrophile will give 1,3-difunctionalized products in a single operation. Chiral 1,3-dichlorides were produced in good yields with good enantio- and diastereoselectivities (eq 21).

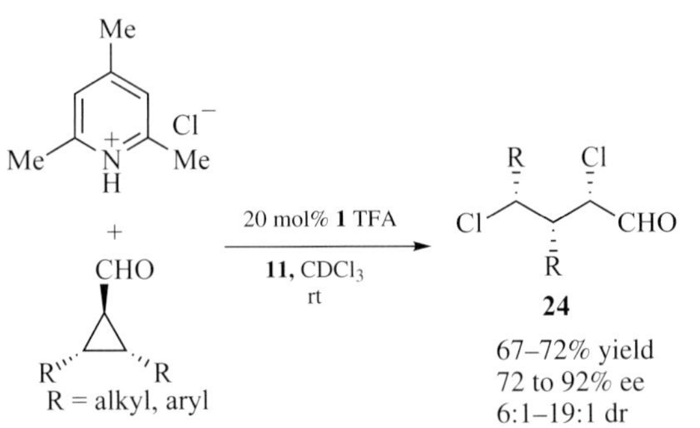

(21)

1. MacMillan, D. W. C.; Ahrendt, K. A.; Borths, C. J., *J. Am. Chem. Soc.* **2000**, *122*, 4243.
2. Kinsman, A. C.; Kerr, M. A., *J. Am. Chem. Soc.* **2003**, *125*, 14120.
3. MacMillan, D. W. C.; Wilson, R. M.; Jen, W. S., *J. Am. Chem. Soc.* **2005**, *127*, 11616.
4. Park, J. K.; Sreekanth, P.; Kim, B. M., *Adv. Synth. Catal.* **2004**, *346*, 49; Mosse, S.; Alexakis, A., *Org. Lett.* **2006**, *8*, 3577.
5. MacMillan, D. W. C.; Jen, W. S.; Wiener, J. J. M., *J. Am. Chem. Soc.* **2000**, *122*, 9874.
6. MacMillan, D. W. C.; Paras, N. A., *J. Am. Chem. Soc.* **2001**, *123*, 4370.
7. MacMillan, D. W. C.; Austin, J. F., *J. Am. Chem. Soc.* **2002**, *124*, 1172.
8. Peelen, T. J.; Chi, Y.; Gellman, S. H., *J. Am. Chem. Soc.* **2005**, *127*, 11598.
9. Fonseca Hechavarria, M. T.; List, B., *Angew. Chem. Int. Ed.* **2004**, *43*, 3958.
10. MacMillan, D. W. C.; Ouellet, S. G.; Tuttle, J. B., *J. Am. Chem. Soc.* **2005**, *127*, 32; MacMillan, D. W. C.; Huang, Y.; Walji, A. M.; Larsen, C. H., *J. Am. Chem. Soc.* **2005**, *127*, 15051.
11. Gilmour, R.; Prior, T. J.; Burton, J. W.; Holmes, A. B., *Chem. Commun.* **2007**, 3954.
12. Horning, B. D.; MacMillan, D. W. C., *J. Am. Chem. Soc.* **2013**, *135*, 6442.
13. Green, N. J.; Lawrence, A. L.; Bojase, G.; Willis, A. C.; Paddon-Row, M. N.; Sherburn, M. S., *Angew. Chem., Int. Ed.* **2013**, *52*, 8333.
14. Han, B.; He, Z. Q.; Li, J. L.; Li, R.; Jiang, K.; Liu, T. Y.; Chen, Y. C., *Angew. Chem., Int. Ed.* **2009**, *48*, 5474.
15. Brochu, M. P.; Brown, S. P.; MacMillan, D. W. C., *J. Am. Chem. Soc.* **2004**, *126*, 4108.
16. D'hooghe, M.; Brabandt, W. V.; Dekeukeleire, S.; Dejaegher, Y.; Kimpe, N. D., *Chem. Eur. J.* **2008**, *14*, 6336.
17. (a) Steiner, D. D.; Mase, N.; III Barbas, C. F., *Angew. Chem., Int. Ed.* **2005**, *44*, 3706. (b) Beeson, T. D.; MacMillan, D. W. C., *J. Am. Chem. Soc.* **2005**, *127*, 8826.
18. Sibi, M. P.; Hasegawa, M., *J. Am. Chem. Soc.* **2007**, *129*, 4124.
19. Humbeck, J. F. V.; Simonovich, S. P.; Knowles, R. R.; MacMillan, D. W. C., *J. Am. Chem. Soc.* **2010**, *132*, 10012.
20. Gualandi, A.; Emer, E.; Capdevila, M. G.; Cozzi, P. G., *Angew. Chem., Int. Ed.* **2011**, *50*, 7842.
21. Rueping, M.; Volla, C. M. R.; Atodiresei, I., *Org. Lett.* **2012**, *14*, 4642.
22. Koley, D.; Krishna, Y.; Srinivas, K.; Khan, A. A.; Kant, R., *Angew. Chem., Int. Ed.* **2014**, *53*, 13196.
23. Zhang, B.; Xiang, S. K.; Zhang, L. H.; Cui, Y. X.; Jiao, N., *Org. Lett.* **2011**, *13*, 5212.
24. Huang, Y.; Walji, A. M.; Larsen, C. H.; MacMillan, D. W. C., *J. Am. Chem. Soc.* **2005**, *127*, 15051.
25. Sparr, C.; Gilmour, R., *Angew. Chem., Int. Ed.* **2011**, *50*, 8391.

1,5,7-Trimethyl-7-(5,6,7,8-tetrahydronaphth[2,3-*d*]oxazol-2-yl)-3-azabicyclo[3.3.1]nonan-2-one[1]

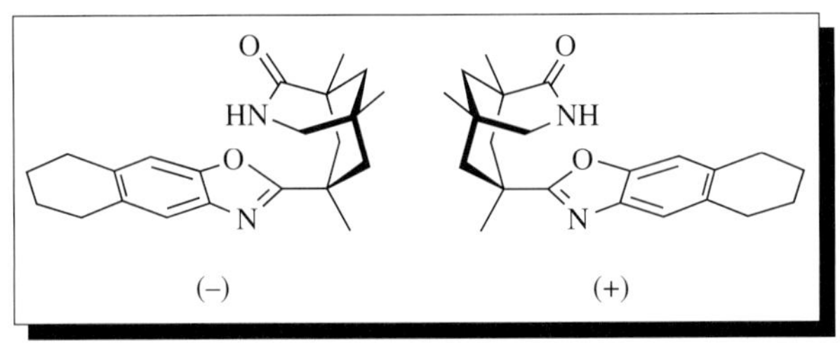

Racemate:
[299162-26-8] $C_{22}H_{28}N_2O_2$ (MW 352.47)
InChI = 1S/C22H28N2O2/c1-20-10-21(2,18(25)23-13-20)12-22(3,11-20)19-24-16-8-14-6-4-5-7-15(14)9-17(16)26-19/h8-9H,4-7,10-13H2,1-3H3,(H,23,25)/t20-,21+22-s3
InChIKey = OPOYSJNBRWSBCV-BDTNDASRSA-N
(+)-(1*R*,5*R*,7*S*)-1,5,7-Trimethyl-7-(5,6,7,8-tetrahydronaphth-[2,3-*d*]oxazol-2-yl)-3-azabicyclo[3.3.1]nonan-2-one
[299162-45-1] $C_{22}H_{28}N_2O_2$ (MW 352.47)
InChI = 1S/C22H28N2O2/c1-20-10-21(2,18(25)23-13-20)12-22(3,11-20)19-24-16-8-14-6-4-5-7-15(14)9-17(16)26-19/h8-9H,4-7,10-13H2,1-3H3,(H,23,25)/t20-,21+,22+/m0/s1
InChIKey = OPOYSJNBRWSBCV-BDTNDASRSA-N
(−)-(1*S*,5*S*,7*R*)-1,5,7-Trimethyl-7-(5,6,7,8-tetrahydronaphth-[2,3-*d*]oxazol-2-yl)-3-azabicyclo[3.3.1]nonan-2-one

[299162-27-9] $C_{22}H_{28}N_2O_2$ (MW 352.47)

InChI = 1S/C22H28N2O2/c1-20-10-21(2,18(25)23-13-20)12-22(3,11-20)19-24-16-8-14-6-4-5-7-15(14)9-17(16)26-19/h8-9H,4-7,10-13H2,1-3H3,(H,23,25)/t20-,21+,22-/m1/s1

InChIKey = OPOYSJNBRWSBCV-BHIFYINESA-N

(reagent used as chiral template in enantioselective reactions of prochiral amides)

Physical Data: mp 205 °C (enantiopure compound); (a) $[\alpha]_D^{20} = +20.5(c = 0.90, CH_2Cl_2)$; (b) $\alpha_D^{20} = -21.9(c = 0.90, CH_2Cl_2)$).

Solubility: sol in most organic solvents; insol in water.

Form Supplied in: colorless solid.

Analysis of Reagent Purity: ^{1}H NMR and/or HPLC (for enantiomeric purity chiral stationary phase: Chiralpak AD or AD-H).

Preparative Method: the racemate is prepared in high yield from commercially available Kemp's triacid (cis,cis-1,3,5-Trimethyl-1,3,5-cyclohexanetricarboxylic Acid *[79410-20-1]*) in five steps.[2]

Purification: chromatographic purification (silica, pentane: diethyl ether = 1:5); separation of the enantiomers either by chemical resolution (derivatization with (−)-menthyl chloroformate)[2] or by HPLC resolution on chiral stationary phase (Chiralpak AD).

Handling, Storage, and Precautions: the crystalline compound can be stored at room temperature without specific precautions. Under prolonged irradiation with UV-B light (λ = 300–350 nm), the compound decomposes very slowly, upon irradiation with UV-C light (e.g., λ = 254 nm), decomposition is slightly faster.

Enantioselective Photochemical Reactions. The bicyclic core structure of 1,5,7-trimethyl-7-(5,6,7,8-tetrahydronaphth-[2,3-*d*]oxazol-2-yl)-3-azabicyclo[3.3.1]nonan-2-one, which is derived from *Kemp's* triacid,[2] serves as source of chirality in a variety of enantioselective photochemical reactions. The chiral induction stems from the complexation of a prochiral amide substrate to the amide bond of this chiral complexing agent by means of hydrogen bonding (eq 1). A molecule bound by two hydrogen bonds to this template is placed in a chiral environment that allows the differentiation of the two enantiotopic faces of the substrate. Upon irradiation, the bound substrates react to give enantiomerically enriched products. The quality of chiral induction depends on the degree of complex formation between the chiral template and the substrate. The addition of typically 2.5 equiv of the template reagent in a nonpolar solvent (toluene or α,α,α-trifluorotoluene) at low temperatures (−60 or −25 °C, respectively) leads in most cases to almost quantitative complex formation and thereby to optical purities of typically more than 90% ee.

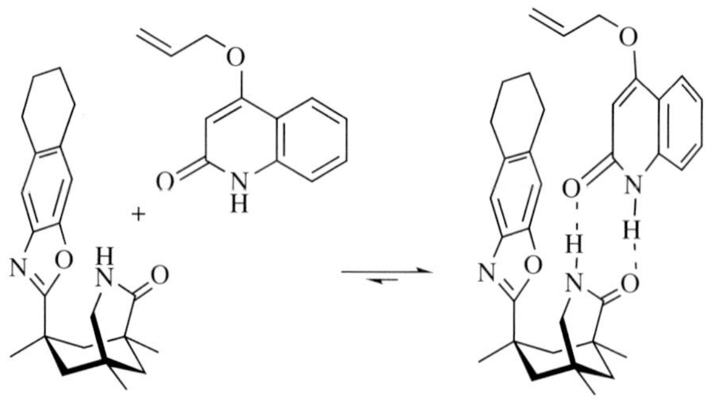

(1)

[2 + 2]-Photocycloaddition reactions were among the first reactions that were performed in the presence of this template. Both intra-[3,4] and intermolecular[4,5] reactions between 2-quinolones and alkenes gave the corresponding photocycloaddition products in yields typically between 75% and 90% and enantiomeric excesses of 81–98% ee (eq 2).

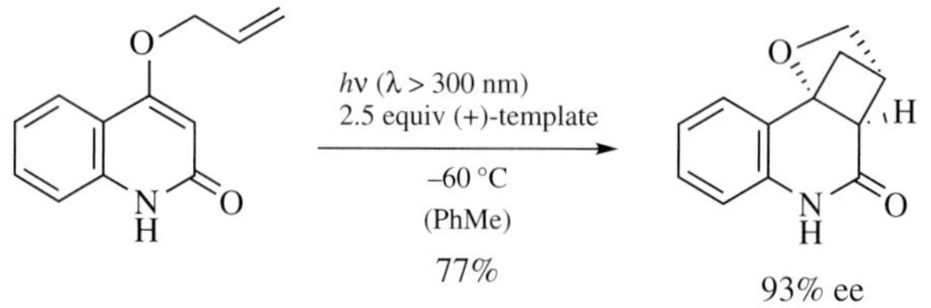

(2)

The reaction shown above has also been performed in the presence of modified template reagents, which allow for a simplified workup procedure either by immobilization of the template on a Wang resin or by precipitation of a MPEG-2000 derivatized template, which is soluble under the reaction conditions.[6] The results with these modified templates were comparable to those with the parent compound.

The template reagent was applied in the enantioselective total synthesis of the Melodinus alkaloid (+)-meloscine.[7,8] The enantioselective [2 + 2]-photocycloaddition of a quinolone precursor and a silyl enol ether derived from methyl pyruvate (see also Methyl-2-(trimethylsilyloxy)-acrylate [18269-31-3]) gave access to all four stereocenters of this complex polycyclic natural compound (eq 3).

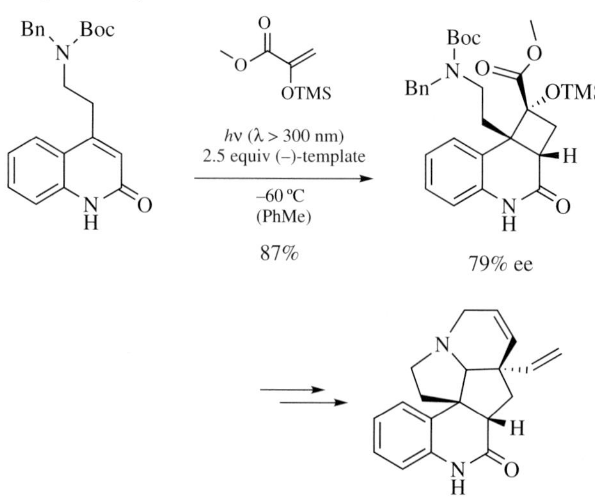

(3)

(+)-meloscine

[2 + 2]-Photocycloaddition reactions of 3- and 4-substituted dihydropyridones[9,10] proved the applicability of the template in photoreactions involving UV-C irradiation (λ = 254 nm). Although the template slowly decomposes under these conditions, moderate to good enantioselectivities could be obtained, as long as the reaction was sufficiently fast. In the case of a 4-substituted pyridone, this boundary could not be met, as the reaction was very slow.[9] Consequently, the enantiomeric excess decreased from 75% ee at 6% conversion after 3 h to as low as 59% ee after 29 h of irradiation. After this time, only small amounts of the template could be recovered (<50%).

The photocycloaddition reactions of 3-substituted pyridones turned out to be much faster.[10] Times to completion of the reaction were always below 6 h, resulting in recoveries of the template of typically 90% (eq 4). The good enantiomeric excesses obtained in these reactions are in line with radical reactions of related compounds (see below). The stereoselectivity could be increased in almost all cases when a modified template with a 6-azabicyclo[3.2.1]octane scaffold (m = 0) was used (eq 4).

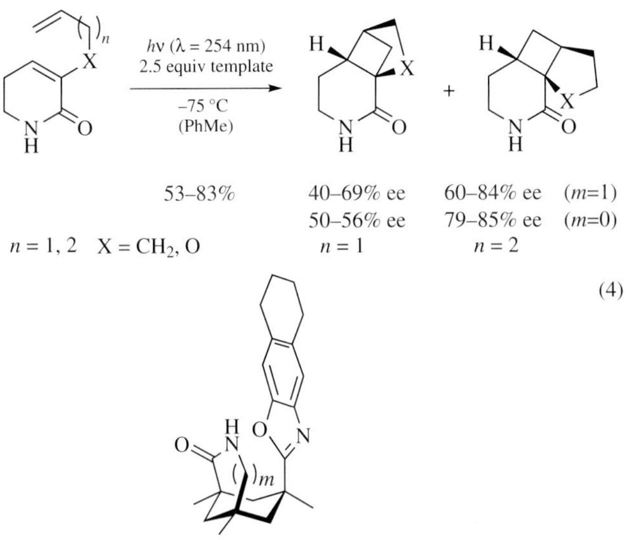

(4)

Besides the [2 + 2]-photocycloaddition chemistry, a variety of other photochemical reactions have been proven to be compatible with this methodology. Electrocyclization and [4 + 4]-cycloaddition reactions of pyridones[11] as well as electrocyclization reactions of enamides[12] have been shown to give the respective products in moderate to good enantiomeric excess. Norrish–Yang cyclization reactions[13] were the first reactions that proved the ability of the template to induce stereoselectivity in free radical reactions (see below). The cyclization after photochemically induced hydrogen abstraction occurred in the presence of the template (2.5 equiv) in 70% yield (dr 80:20) with an enantiomeric excess of 60% ee for the major diastereomer (eq 5).

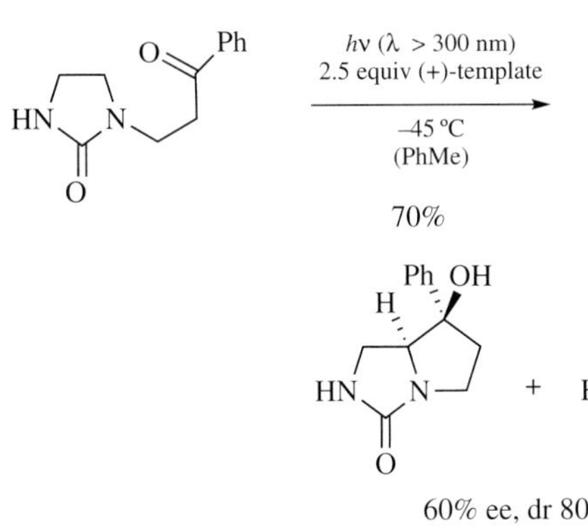

(5)

Another example for a photochemically induced thermal reaction is the Diels–Alder reaction of a photochemically generated *ortho*-quinodimethane[14] with electron-poor olefins. In this reaction, up to four stereocenters were established in 78% yield and optical purities as high as 96% ee (eq 6).

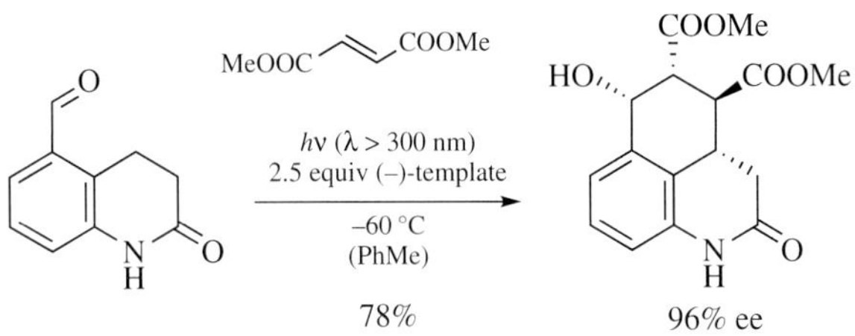

(6)

Enantioselective Radical Reactions. Reductive cyclization reactions of α,β-unsaturated ω-iodoalkenyl-piperidinones[15] were investigated in the presence of the template. As stated above, reactions with the template are to be conducted in nonpolar solvents at low temperature. The former is known to be applicable in radical chemistry whereas the latter required some optimization. The ideal choice for low-temperature radical initiation turned out to be the oxygen-induced decomposition of triethylborane. As expected, 5-*exo-trig* cyclizations gave the best results, with respect to both yield (81%) and stereoselectivity (84% ee) (eq 7).

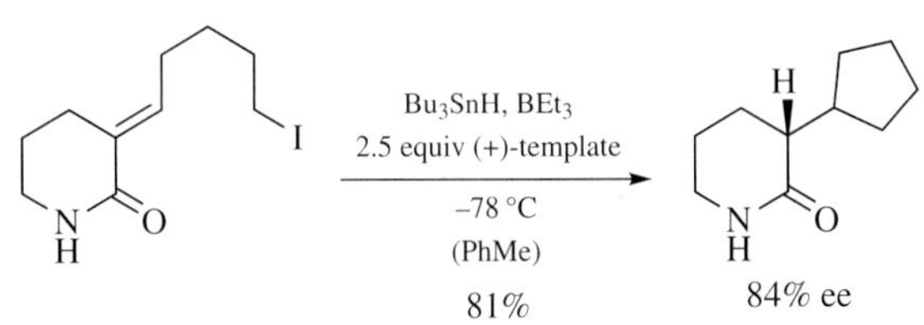

(7)

A series of 4-ω-iodoalkyl-substituted 2-quinolones were tested in radical cyclization reactions.[16] The results exceeded both the results of previous radical reactions and the results of the photochemical reactions with 2-quinolones. The reaction of the iodide shown in (eq 8) resulted in the predominant formation of the *endo*-cyclization product in 79% yield and an enantiomeric purity of the predominant *trans*-product of 99% ee using trifluorotoluene as the solvent.

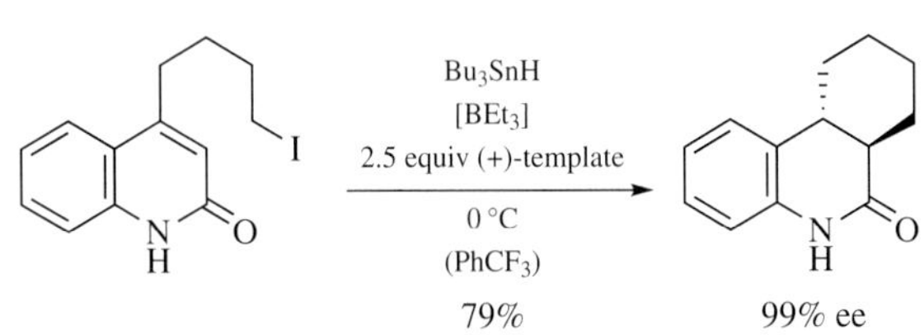

(8)

In addition, it was shown in this study that the template reagent has, under certain circumstances, a catalytic effect with respect to the enantiomeric excess that can be achieved. Possibly due to phase-transfer catalysis, the complexing agent showed the ability to induce a chirality multiplication, by solubilizing the substrate, which thus reacts faster than in the absence of the chiral reagent. The selectivities have been explained by the ideal interplay of substrate association and rigidity that allows for a nearly perfect differentiation of the enantiotopic faces of the prochiral substrates.

Other binding motifs such as primary amides, carboxylic acids, ureas, and hydrazides are in principle also suited for this kind of reactions,[17] although enantioselectivities remained low with these substrates and only 59% ee could be obtained in the best case. The reasons for this are either insufficient coordination to the template or – in the case of urea-binding motifs – the increased distance between binding motif and prostereogenic reaction center.

NMR Shift Reagent. The formation of hydrogen bonds can be detected by NMR spectroscopy. The enantiomers of chiral lactams differ significantly in their complexation behavior in the presence of the enantiomerically pure template reagent. This fact can be used for the determination of the enantiomeric ratio of enantiomerically enriched samples.[18] The most significant difference in association behavior was observed for compounds that are closely related in structure to the template reagent. Bicyclic amides derived from *Kemp*'s triacid cannot dimerize in a homochiral complex, as the substituents in the 7-position of the core structure would come very close in space.

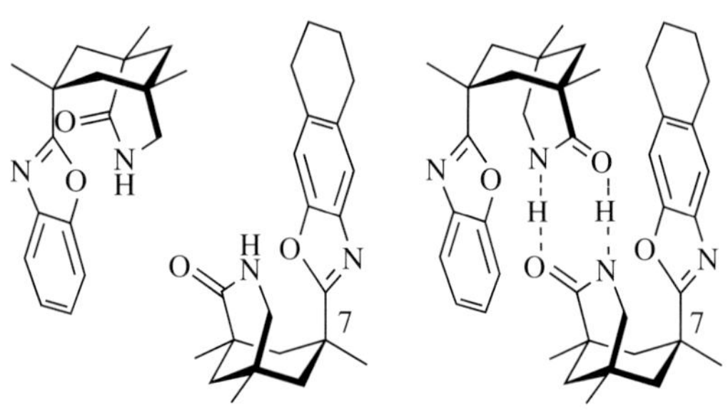

no homochiral complex heterochiral complex

Dimerization of compounds with opposite absolute configuration at the amide bridge is feasible, however. This difference in association behavior can also be monitored by NMR spectroscopy, which allows absolute configuration determination by means of simple NMR titration experiments.[19]

Related Reagents. As stated above, the template can be immobilized on a solid support to facilitate reisolation.[6] The modification of the parent scaffold gave better enantiomeric excesses in some cases.[10] The replacement of the photochemically inert tetrahydronaphthalene part of the molecule by a sensitizing residue such as benzophenone[20] or xanthone[21] allows the application of substoichiometric amounts of these now catalytically active molecules.

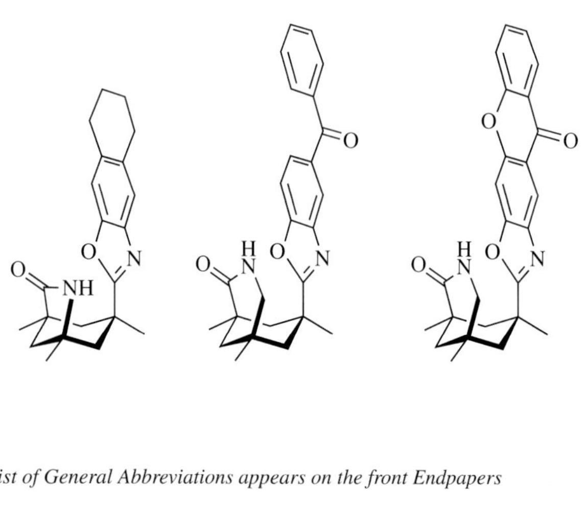

In a radical cyclization reaction induced by catalytic photoelectron transfer,[20] up to 70% ee could be obtained with 30 mol % of the benzophenone catalyst. The [2 + 2]-photocycloaddition reaction of 4-butenyloxy-2-quinolone with the xanthone as sensitizer resulted in 90% ee with only 5 mol % of catalyst.

---

1. Grosch, B.; Bach, T.; In *CRC Handbook of Organic Photochemistry and Photobiology*, 2nd ed. (Eds.: Horspool, W.; Lenci, F.;), CRC Press, Boca Raton, FL, **2004**, pp 61/1–61/14.
2. Bach, T.; Bergmann, H.; Grosch, B.; Harms, K.; Herdtweck, E., *Synthesis* **2001**, 1395.
3. Bach, T.; Bergmann, H.; Harms, K., *Angew. Chem., Int. Ed.* **2000**, *39*, 2302.
4. Bach, T.; Bergmann, H.; Grosch, B.; Harms, K., *J. Am. Chem. Soc.* **2002**, *124*, 7982.
5. Bach, T.; Bergmann, H., *J. Am. Chem. Soc.* **2000**, *122*, 11525.
6. Breitenlechner, S.; Bach, T., *Angew. Chem. Int. Ed.* **2008**, *47*, 7957.
7. Selig, P.; Herdtweck, E.; Bach, T., *Chem. Eur. J.* **2009**, *15*, 3509.
8. Selig, P.; Bach, T., *Angew. Chem., Int. Ed.* **2008**, *47*, 5082.
9. Albrecht, D.; Basler, B.; Bach, T., *J. Org. Chem.* **2008**, *73*, 2345.
10. Albrecht, D.; Vogt, F.; Bach, T., *Chem. Eur. J.* **2010**, *16*, 4284.
11. Bach, T.; Bergmann, H.; Harms, K., *Org. Lett.* **2001**, *3*, 601.
12. Bach, T.; Grosch, B.; Strassner, T.; Herdtweck, E., *J. Org. Chem.* **2003**, *68*, 1107.
13. Bach, T.; Aechtner, T.; Neumüller, B., *Chem. Eur. J.* **2002**, *8*, 2464.
14. Grosch, B.; Orlebar, C. N.; Herdtweck, E.; Massa, W.; Bach, T., *Angew. Chem., Int. Ed.* **2003**, *42*, 3693.
15. Aechtner, T.; Dressel, M.; Bach, T., *Angew. Chem., Int. Ed.* **2004**, *43*, 5849.
16. Dressel, M.; Bach, T., *Org. Lett.* **2006**, *8*, 3145.
17. Kapitan, P.; Bach, T., *Synthesis* **2008**, 1559.
18. Bergmann, H.; Grosch, B.; Sitterberg, S.; Bach, T., *J. Org. Chem.* **2004**, *69*, 970.
19. Bauer, A.; Bach, T., *Tetrahedron: Asymmetry* **2004**, *15*, 3799.
20. Bauer, A.; Westkämper, F.; Grimme, S.; Bach, T., *Nature* **2005**, *436*, 1139.
21. Müller, C.; Bauer, A.; Bach, T., *Angew. Chem., Int. Ed.* **2009**, *48*, 6640.

Andreas Bauer & Thorsten Bach
Technische Universität München, Munich, Germany

2-Triphenylmethylpyrrolidine

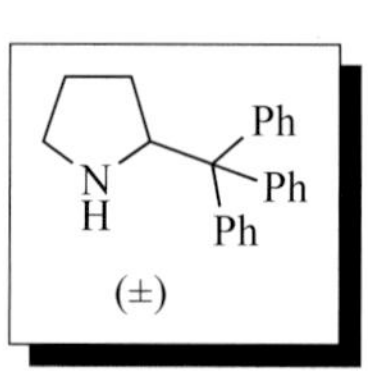

(±)

[1131010-26-8] $C_{23}H_{23}N$ (MW 313.44)

InChI = 1S/C23H23N/c1-4-11-19(12-5-1)23(22-17-10-18-24-22,20-13-6-2-7-14-20)21-15-8-3-9-16-21/h1-9,11-16,22,24H,10,17-18H2

InChIKey = JHNAIVKYDKSQEA-UHFFFAOYSA-N

(*S*)
[362610-72-8] $C_{23}H_{23}N$ (MW 313.44)
InChI = 1S/C23H23N/c1-4-11-19(12-5-1)23(22-17-10-18-24-22,20-13-6-2-7-14-20)21-15-8-3-9-16-21/h1-9,11-16,22,24H,10,17-18H2/t22-/m0/s1
InChIKey = JHNAIVKYDKSQEA-QFIPXVFZSA-N

(catalyst for the α-benzoyloxylation of aldehydes)[1]

Physical Data: $[\alpha]_D^{25}$ +15 (c = 1.0, $CHCl_3$) for (*S*)-2-tritylpyrrolidine.[1]
Analysis of Reagent Purity: by NMR and optical rotation or chiral HPLC for optically active samples (Daicel Chiracel OD-H, hexane/2-propanol/diethylamine = 100:1:0.1; retention time: 6.0 min (*R*), 6.4 min (*S*)).
Preparative Methods: prepared from diphenyl(pyrrolidin-2-yl)-methanol and benzene in the presence of trifluoromethanesulfonic acid.[2] An alternative synthetic approach involves addition of lithiated triphenylmethane to pyrrolidine nitrone and hydrogenation (Pd/C) of the resulting 2-tritylpyrrolidin-1-ol. Enantiopure samples of 2-tritylpyrrolidine are obtained by fractional crystallization with (*S*)-malic acid.[1]
Purification: chromatography on silica gel column using a hexane/EtOAc gradient from 4/1 to 0/10.

Organocatalytic Asymmetric Benzoyloxylation of Aldehydes. A catalytic amount of (*S*)-2-tritylpyrroline promotes the direct benzoyloxylation of aldehydes by benzoyl peroxide (BPO), a newly discovered reaction for the synthesis of α-oxygenated carbonyl compounds. These reactions proceed smoothly to give α-benzoyloxyaldehydes in good yields and enantiomeric excesses of over 90% (7 examples, see eq 1). Under routine conditions, these reactions are carried out in the presence of 10 mol % catalyst; however, a catalyst loading of 5 mol % gave similar results, with only a slightly longer reaction time.

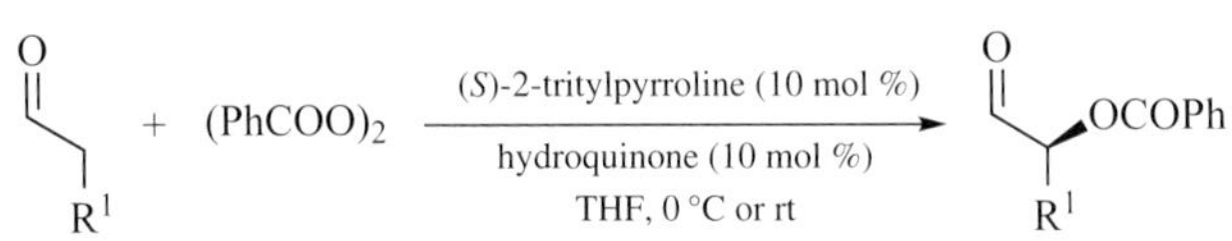

(1)

R^1	Temp. (°C)	Time (h)	Yield (%)	ee (%)
Me	0	2	62	94
Et	0	4	73	93
Bu	0	5	70	94
Bn	rt	2	72	93
Allyl	rt	1.5	62	92
i-Pr	rt	5	64	92
C_6H_{11}	rt	6	65	94

The absolute configuration of the final product, for R = Et, was determined to be (*S*), by comparison with literature data.[3] When the benzoyloxylation reactions were performed in the presence of 10 mol % hydroquinone (HQ), an increase in yield was observed. The HQ additive was postulated to behave as a radical scavenger inhibiting radical side reactions, while the desired benzoyloxylation continues to operate. Therefore, it was speculated that the benzoyloxylation reaction follows an ionic pathway via nucleophilic addition of enamine intermediates to dibenzoyl peroxide.

The method represents a straightforward approach to enantiomerically enriched α-oxygenated aldehydes. An alternative organocatalyst affording comparable levels of enantioselectivity is the (*S*)-diphenylprolinol *tert*-butyldimethyl silyl ether.[4] The synthetic potential of the method has been demonstrated by the selective conversion of the final products into the corresponding optically active diols and monoprotected diols.

---

1. Kano, T.; Mii, H.; Maruoka, K., *J. Am. Chem. Soc.* **2009**, *131*, 3450.
2. Klumpp, D. A.; Aguirre, S. L.; Sanchez, G. V. Jr.; de Leon, S. J., *Org. Lett.* **2001**, *3*, 2781.
3. Abate, A.; Brenna, E.; Fuganti, C.; Malpezzi, L.; Serra, S., *Tetrahedron: Asymmetry* **2007**, *18*, 1145.
4. Gotoh, H.; Hayashi, Y., *Chem. Commun.* **2009**, 3083.

Angela Marinetti
Institut de Chimie des Substances Naturelles, Gif-sur-Yvette, France

Tris[(trifluoromethyl)sulfonyl]methane and Related Salts

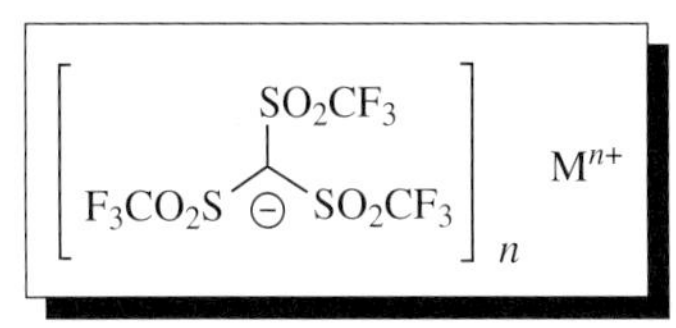

M=H, n=1
[60805-12-1] Tf_3CH (MW 412.2)
InChI = 1/C4HF9O6S3/c5-2(6,7)20(14,15)1(21(16,17)3(8,9)10)22(18,19)4(11,12)13/h1H
InChIKey = MYIAPBDBTMDUDP-UHFFFAOYAR
M=Yb, n=3
[224317-58-2] $(Tf_3C)_3Yb)$ (MW 1406.7)
InChI = 1/3C4F9O6S3.Yb/c3*5-2(6,7)20(14,15)1(21(16,17)3(8,9)10)22(18,19)4(11,12)13;/q3*-1;+3
InChIKey = GNGMSTGPEJMQKM-UHFFFAOYAF
(anhydrous
M=Sc, n=3
[224317-59-3] $(Tf_3C)_3Sc)$ (MW 1278.6)
InChI = 1/3C4F9O6S3.Sc/c3*5-2(6,7)20(14,15)1(21(16,17)3(8,9)10)22(18,19)4(11,12)13;/q3*-1;+3
InChIKey = LRDKRKSPTYEXCA-UHFFFAOYAO
(anhydrous
M=Cu, n=2
[261728-13-6] $(Tf_3C)_2Cu)$ (MW 886.0)
InChI = 1/2C4F9O6S3.Cu/c2*5-2(6,7)20(14,15)1(21(16,17)3(8,9)10)22(18,19)4(11,12)13;/q2*-1;+2
InChIKey = DERWDIPDZXUVNB-UHFFFAOYAT

M=Cs, *n*=1
[114395-68-5] Tf_3CCs (MW 544.1)
InChI = 1/C4F9O6S3.Cs/c5-2(6,7)20(14,15)1(21(16,17)3(8,9)10)22(18,19)4(11,12)13;/q-1;+1
InChIKey = ZVPDKZBKUGXYRF-UHFFFAOYAY
M=Ag, *n*=1
[114395-71-0] Tf_3CAg) (MW 519.1)
InChI = 1/C4F9O6S3.Ag/c5-2(6,7)20(14,15)1(21(16,17)3(8,9)10)22(18,19)4(11,12)13;/q-1;+1
InChIKey = LNHRIDDGXXLCSC-UHFFFAOYAC

(M=H, *n*=1: a Bronsted superacid, which can be transformed into recyclable, highly catalytically active metal salts)

Structure: the core framework of the tris[(trifluoromethyl)sulfonyl] methide anion comprises three C-S bonds, with the negative charge formally on the central carbon atom. However, the bulk negative charge is almost certainly on the (six) sulfone oxygens by resonance delocalization. The X-ray crystal structures of scandium and ytterbium triflide have seven and eight water aqua ligands respectively, (and one included water molecule each, for a total of eight and nine H_2O/M), the assemblage is reinforced by hydrogen bonds between the aqua ligands, the included water molecule, and the sulfone oxygen atoms. The central carbon—which shows sp^2 hybridization— is not associated with the water ensemble. For the free acid, presumably there is much dissociation and the main interaction with the proton is through oxygen atoms. The ^{19}F and ^{13}C NMR spectra show only one and two resonances respectively, demostrating rapid exchange. The same is true for the lanthanide triflides in aqueous solution where the triflide ligands are effectively dissociated.

Alternate Names: M=H, *n*=1: tris(triflyl)methane, triflidic acid; M=Yb, *n*=3: ytterbium(III) tris[tris(trifluoromethylsulfonyl) methide], ytterbium(III) triflide; M=Sc, *n*=3: scandium(III) tris[tris(trifluoromethylsulfonyl)methide], scandium(III) triflide; M=Cu, *n*=2: copper(II) bis[tris(trifluoromethylsulfonyl) methide], copper(II) triflide; M=Cs, *n*=1: cesium(I) tris(trifluoromethylsulfonyl) methide, cesium(I) triflide; M=Ag, *n*=1: silver(I) tris(trifluoromethylsulfonyl)methide, silver(I) triflide.

Solubility: M=H, *n*=1: soluble in water and diethyl ether.

Analysis of Reagent Purity: M=H, *n*=1: titrated against a standard NaOH solution (phenolphthalein as indicator) inwater immediately after sublimation.

Preparative Methods: tris(trifluoromethanesulfonyl)methane ('triflidic acid'), $HCTf_3$, was first prepared by Turowsky and Seppelt using gaseous CF_3SO_2F as the electrophilic triflyl group source.[1] More recently, Barrett et al. have developed a convenient synthesis from triflic anhydride.[2,3] Preparation of salts (methides or 'triflides') of triflidic acid have been performed by three general methods: by reaction of the free acid with a metal carbonate (K, Rb, Cs, Ag) in water (eq 1),[1] the reaction of $AgC(SO_2CF_3)_3$ with metal chlorides in dichloromethane (Yb, Sc, Cu) (eq 2),[4] or by the action of the free acid on metal oxides in water (Yb, Sc) (eq 3).[2,3] A bismuth(III) salt has also been prepared by the action of triflidic acid on $Bi(OEt)_3$ in diethyl ether.[5]

$$M_2CO_3 + 2\,HCTf_3 \xrightarrow{H_2O} 2\,MCTf_3 + CO_2 + H_2O \quad (1)$$
(M = K, Rb, Cs, Ag)

$$x\,AgCTf_3 + MCl_x \xrightarrow{CH_2Cl_2} M(CTf_3)_x + x\,AgCl\downarrow \quad (2)$$
[M = Yb, Sc, Cu(II)]

$$M_2O_3 + 6\,HCTf_3 \xrightarrow{H_2O} 2\,M(CTf_3)_3 + 3\,H_2O \quad (3)$$
(M = Yb, Sc)

^{13}C & ^{19}F NMR Data for $(Tf_3C)_nM$

M	*n*	$^{19}F(\delta/ppm)$	Ref.	$^{13}F(\delta/ppm)$	Ref.
H	1	−76.8[a]	1		
		−78.4[b]	5	122.6(q, *J*=326 Hz)[2,3]	5
Cs	1	−76.6[a]	2	121.7(q, *J*=326 Hz), 83.8[a]	2
Yb	3	−76.5[a]	2	122.7(q, *J*=326 Hz), 84.1[b]	2
Sc	3	−75.9[b]	2	122.6(q, *J*=326 Hz), 84.0[b]	2
Cu	2	−12.59[2,4]	4		
Ag	1	No data available	1	No data available	

[a] CD_3CN.
[b] D_2O.
[c] Central $C(SO_2CF_3)_3$ carbon not recorded.
[d] Referenced against $CH_3C_6H_5$— all other ^{19}F shifts against $CFCl_3$.

Purity: M=H, *n*=1: sublimation.

Handling, Storage, and Precautions: M=H, *n*=1: triflidic acid is a Bronsted superacid. Care should be taken in its handling. It is conveniently stored as the cesium salt (M=Cs, *n*=1) which readily precipitates from water. Sublimation from concentrated sulfuric acid (CARE) liberates triflidic acid as required.

General. Triflidic acid is a Bronsted superacid, with an estimated *p*Ka of ca. −18,[6] which compares with a *p*Ka of ca. −12 for trifluoromethanesulfonic (triflic) acid. On the basis of its obvious structural similarity to triflic acid, but its much-increased acidity, triflide salts should show interesting activities when compared to their corresponding triflate salts. In particular, lanthanide triflide salts could theoretically be used for a myriad of catalytic processes in place of their corresponding related counterparts, the lanthanide trifluoromethanesulfonates (triflates), which have already found use as recyclable catalysts in important carbon-carbon, carbon-nitrogen, and carbon-oxygen bond forming reactions such as Friedel–Crafts acylations, Mukaiyama aldol reactions, Diels–Alder cycloadditions, electrophilic aromatic nitrations, and esterifications (see related reagents). To date only isolated reports on the catalytic use of lanthanide triflides have appeared, viz., aromatic nitration[2,7] and debenzylation of esters, alcohols, and amides.[4] Importantly, these early reports suggest that the triflide salts are up to several orders of magnitude more active than the corresponding triflate salts under comparable conditions.

Aromatic Nitration. Scandium(III) and ytterbium(III) triflides function as catalysts (10 mol %) for the nitration of electron deficient aromatic compounds using 1 equiv of 69% nitric acid where the only side product is water, and the catalyst can be recycled and reused without any loss in activity.[2] For example, *o*-nitrotoluene was nitrated to >90% conversion under standard conditions in 24 h (eq 1). The corresponding ytterbium(III) triflate salt was essentially ineffective for carrying out this transformation under the same conditions, and scandium(III) triflate gave 50% conversion.

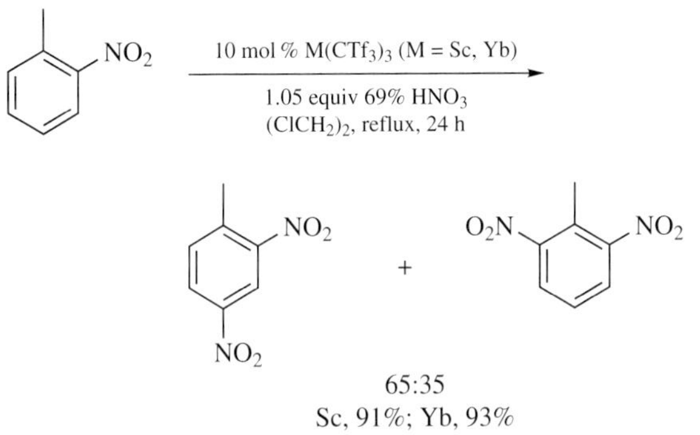

(4)

Scandium(III) and ytterbium(III) triflides have also been demonstrated to be effective catalysts for the nitration of fluoroarenes under the same conditions as above (eq 2).[7] Good-to-excellent yields were maintained in the mononitration of monofluoroarenes, difluoroarenes, and trifluoroarenes, but ytterbium(III) triflate became progressively less effective with increasing fluorine content in the substrate. The common problem of HF liberation in the nitration of fluoroarenes, through S_NAr displacement of fluoride from the newly created nitroarenes with the (increasing quantities of) water produced acting as a nucleophile, did not arise. Additionally, the catalysts were recovered by simple evaporation of the aqueous phase after the work-up procedure, and could be recycled with little or no loss in activity. Attempted catalyzed nitration of tetrafluorobenzene or pentafluorobenzene, however, was unsuccessful.

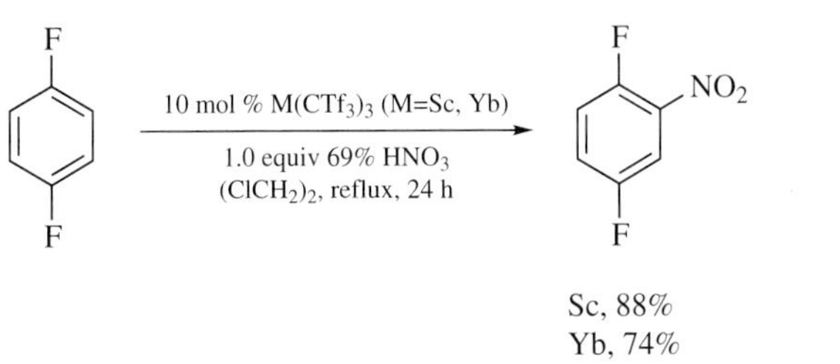

(5)

Debenzylation. Both triflidic acid and scandium(III) triflide were shown to be highly effective homogeneous catalysts for the debenzylation of benzyl esters in anisole, and in the case of benzyl phenylacetate >99% yield was reported at catalyst loadings of 0.1 mol % (eq 3).[4] Under these conditions only 6% yield was obtained with scandium(III) triflate as catalyst.

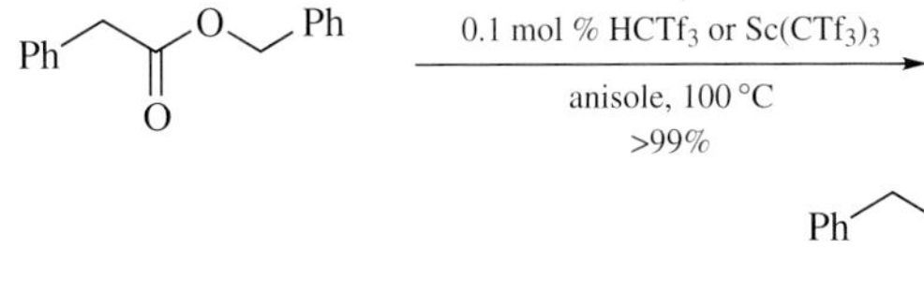

(6)

Benzyl ethers of phenols and primary alcohols were cleaved in anisole in the presence of 1 mol % scandium(III) triflide in excellent yields (eq 4).[4]

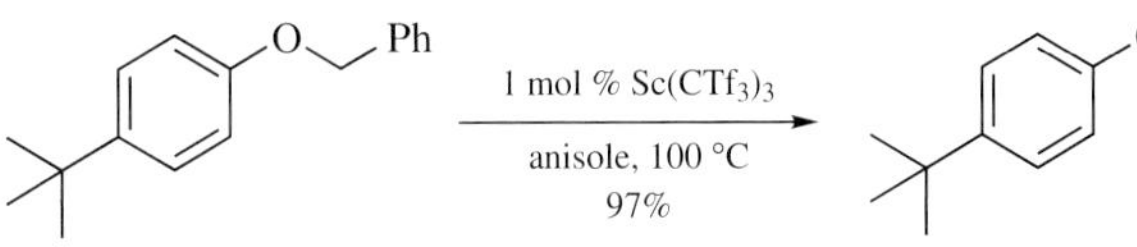

(7)

The ability of triflidic acid to mediate debenzylation of the less reactive secondary *N*-benzylamides in anisole was examined, and only a trace amount of product was observed in the case of *N*-benzyl-3-phenylpropionamide. However, triflidic acid, scandium(III) triflide, and copper(II) triflide catalyzed the deprotection of *N*-(*p*-methoxybenzyl) amides derived from saturated alkyl, unsaturated alkyl, and aromatic carboxylic acids in typically quantitative yields (eq 5).[4]

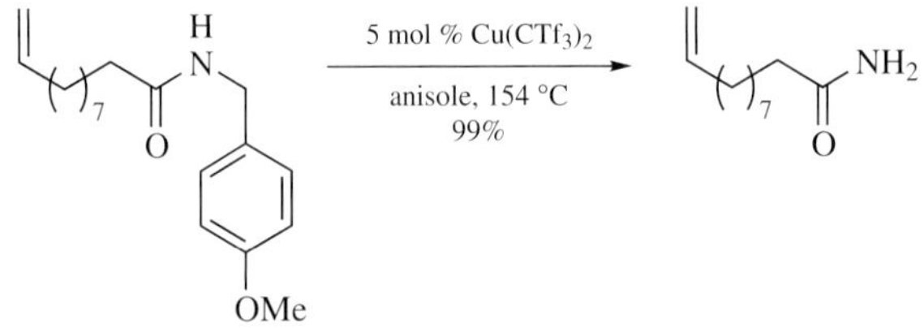

(8)

Related Reagents. Ytterbium(III) trifluoromethanesulfonate (triflate); scandium(III) trifluoromethanesulfonate (triflate); ytterbium(III) bis(trifluoromethanesulfonyl)imide (triflimide); scandium(III) bis(trifluoromethanesulfonyl)imide (triflimide); tris[(perfluoroalkyl)sulfonyl]methanes and related salts.

1. Turowsky, L.; Seppelt, K., *Inorg. Chem.* **1988**, *27*, 2135–2137.
2. Waller, F. J.; Barrett, A. G. M.; Braddock, D. C.; Ramprasad, D.; McKinnell, R. M.; White, A. J. P.; Williams, D. J.; Ducray, R., *J. Org. Chem.* **1999**, *64*, 2910–2193.
3. Barrett, A. G. M.; Braddock, D. C.; Peyralans, J. J.-P., In *Handbook of Fluorous Chemistry*; Gladysz, J. A.; Curran, D. P.; Horvàth, I. T., Eds.; Wiley-VCH: Weinheim, **2004**, in press.
4. Ishihara, K.; Hiraiwa, Y.; Yamamoto, H., *Synlett* **2000**, 80–82.
5. McKinnell, R. M., Ph.D. Thesis, Imperial College London, **2000**.
6. Koppel, I. A.; Taft, R. W.; Anvia, F.; Zhu, S.-Z.; Hu, L.-Q.; Sung, K.-S.; DesMarteau, D. D.; Yagupolskii, L. M.; Yagupolskii, Y. L.; Ignat'ev, N. V.; Kondratenko, N. V.; Volkonskii, A. Y.; Vlasov, V. M.; Notario, R.; Maria, P.-C., *J. Am. Chem. Soc.* **1994**, *116*, 3047–3057.
7. Barrett, A. G. M.; Braddock, D. C.; Ducray, R.; McKinnell, R. M.; Waller, F. J., *Synlett* **2000**, 57–58.

Anthony G. M. Barrett, D. Christopher Braddock & Gurubaran S. Raju
Imperial College London, London, UK

L-Tryptophan, *N,N′*-[[(2*S*,5*S*)-1-(4-pyridinyl)-2,5-pyrrolidinediyl]-dicarbonyl]bis-, 3,3′-dioctyl Ester

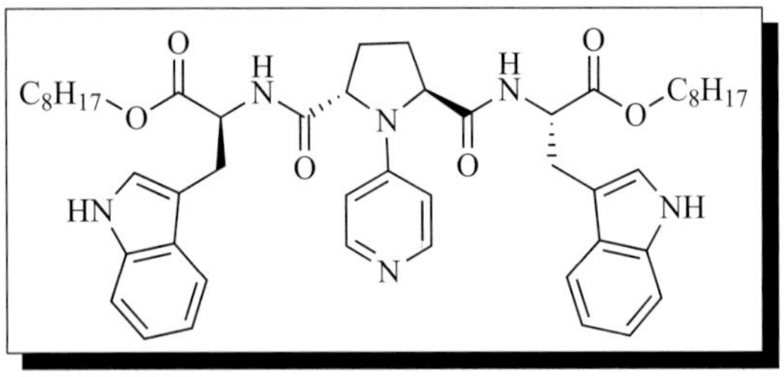

L-Tryptophan, *N,N′*-[[(2*S*,5*S*)-1-(4-pyridinyl)-2,5-pyrrolidinediyl]dicarbonyl]bis-, 3,3′-dioctyl ester

[958229-27-1] $C_{49}H_{64}N_6O_6$ (MW 833.09)

InChI = 1S/C49H64N6O6/c1-3-5-7-9-11-17-29-60-48(58)42(31-35-33-51-40-21-15-13-19-38(35)40)53-46(56)44-23-24-45(55(44)37-25-27-50-28-26-37)47(57)54-43(49(59)61-30-18-12-10-8-6-4-2)32-36-34-52-41-22-16-14-20-39(36)41/h13-16,19-22,25-28,33-34,42-45,51-52H,3-12,17-18,23-24,29-32H2,1-2H3,(H,53,56)(H,54,57)/t42-,43-,44-,45-/m0/s1

InChIKey = HGTRVHNYZINJPH-HRBFTXTFSA-N

D-Tryptophan, *N,N′*-[[(2*S*,5*S*)-1-(4-pyridinyl)-2,5-pyrrolidinediyl]dicarbonyl]bis-, 3,3′-dioctyl ester

[958229-28-2] $C_{49}H_{64}N_6O_6$ (MW 833.09)

(reagent used for chemo- and regioselective acylation of alcohols and carbohydrates)

Physical Data: colorless crystals with mp: 191–193 °C and $[\alpha]_D^{20} = -60$ (c 1.0, CH_3OH).

Solubility: soluble in common organic solvents.

Form Supplied in: white solid (Wako Pure Chemicals).

Preparative Method: prepared in six steps from commercially available L-pyroglutamic acid (eq 1).[1]

Handling, Storage, and Precautions: toxicity unknown, handle under fume hood with proper safety procedures.

Original Commentary

Vikas Sikervar
Arekere Mico Layout, Bangalore, India

Chemo- and Regioselective Acylation of Carbohydrates. Octyl β-D-glucopyranoside can be regioselectively acylated using 10 mol % catalyst **1** at the 4-OH in high yields and selectivity with only minor amounts of diacylated product (eq 2).[1] In contrast using DMAP as catalyst gives 66% yield for monoacylated product with poor regioselectivity (6-*O*/4-*O*/3-*O*: 33/24/43) along with 21% diacylated product. The reaction is postulated to proceed through transition state **2** in which the more reactive primary 6–OH is hydrogen bonded to the amide carbonyl of catalyst **1** placing the 4–OH in proximity to the acyl pyridinium moiety, thus resulting in the observed high regioselectivity. When 6–OH is protected as a methyl ether, the product ratio is 4-*O*/3-*O*/2-*O*: 31:67:2, indicating the importance of free 6–OH in achieving high regioselectivity (eq 3). Competitive acylation reaction between octyl β-D-glucopyranoside and 2-phenylethanol using 10 mol % catalyst1gives exclusively monoacylation of octyl β-D-glucopyranoside in high regioselectivity and yield (eq 4).

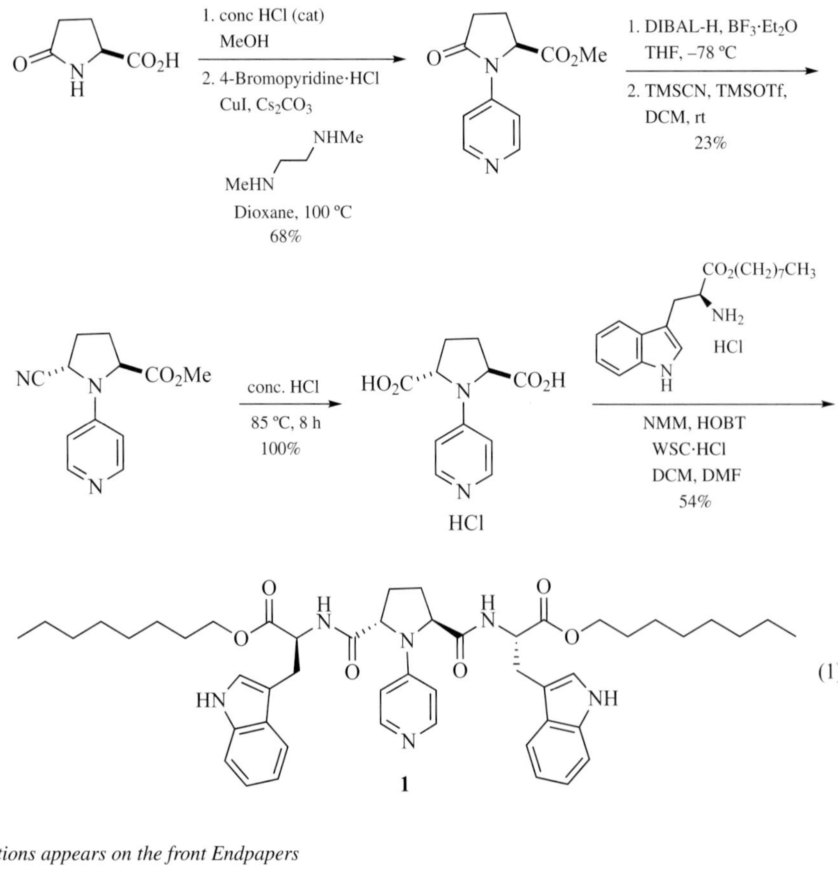

(1)

1

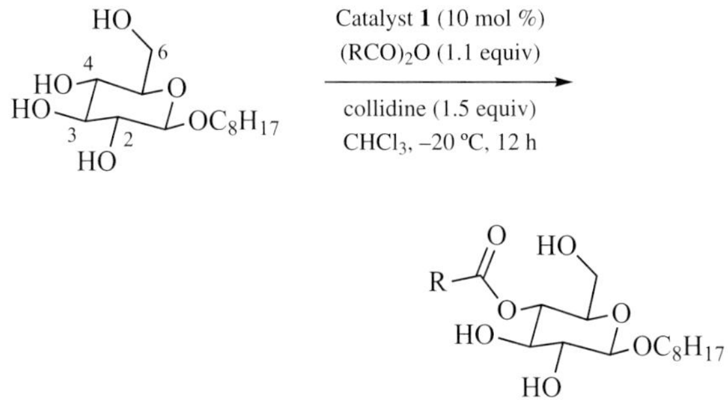

(2)

R = *i*-Pr, yield: 98%, regioselectivity 4-*O*/3-*O*: 99/1.
R = Ac, yield: 96%, regioselectivity 4-*O*/3-*O*: 96/4
diacylate yield: 4%

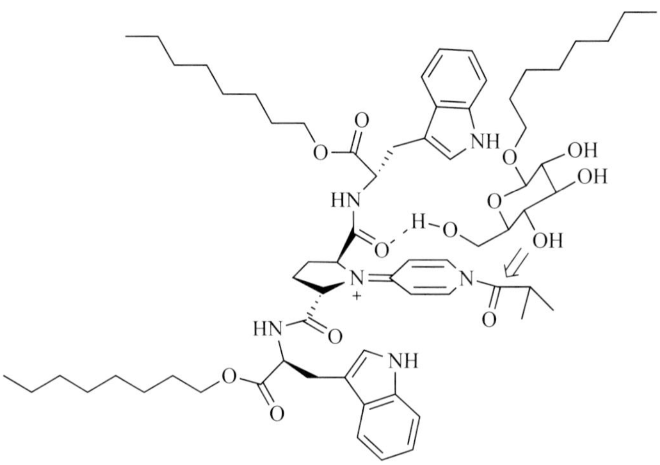

2

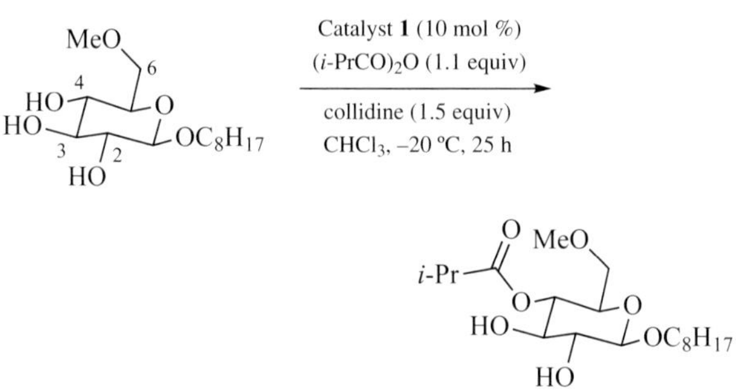

(3)

monoacylation yield: 95%,
regioselectivity 4-*O*/3-*O*/2-*O*: 31/67/2

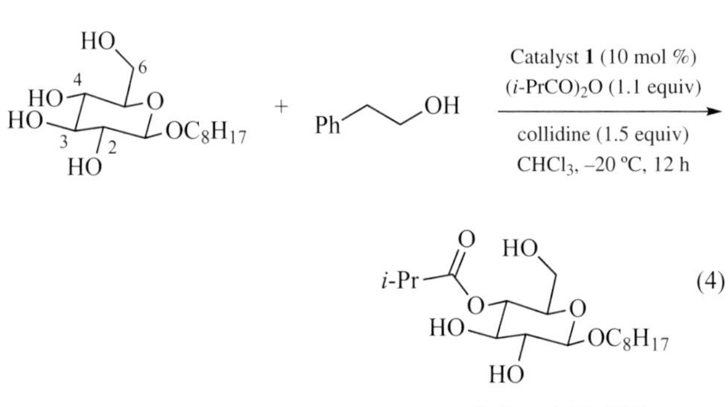

(4)

monoacylation yield: 98%,
regioselectivity 4-*O*/3-*O* = 99:1

Compound **1** also catalyzes the acylation of disaccharide in high yield and regioselectivity on reaction with isobutyric anhydride (eq 5).[2]

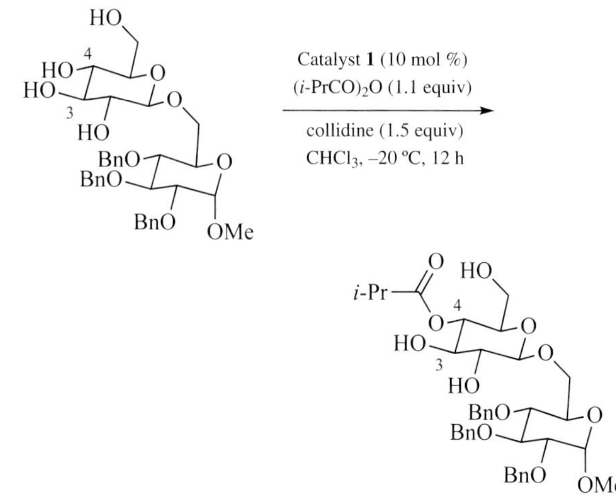

(5)

monoacylation yield: 94%.
94% regioselectivity

The catalyst **1** is used for the regioselective acylation of digitoxin, which is a clinically used cardiac glycoside. High regioselectivity and yields are obtained in exclusively forming the 4‴-acetate with no 3‴-,4‴-diacetate product and very little recovery of starting material (eq 6).[3] The reaction proceeds via formation of transition state **3** in which C-3‴ O–H is hydrogen bonded to the amide carbonyl of the catalyst **1**, placing equatorial C-4‴ O–H in proximity to the reactive acyl pyridinium ion. A control reaction using DMAP as catalyst and acetic anhydride as the acylating agent gives a 66% yield of 4‴-acetate along with 18% of the 3‴-, 4‴-diacetate product and 24% recovered starting material.

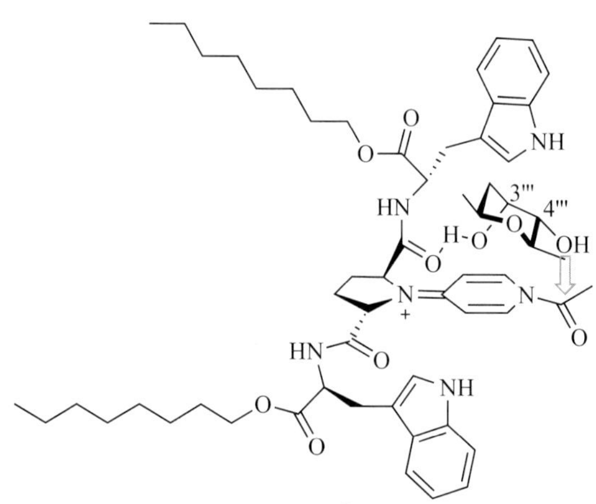

3

Catalyst **1** is also utilized for the regioselective acylation of lanatoside C at 4‴–OH. High regioselectivities are obtained for 4‴–OH compared with 3‴–OH on reaction with several mixed anhydrides generated in situ by reacting the corresponding carboxylic acid with *t*-butyl acyl chloride (eq 7).[4] Importantly, catalyst **1** generates the 4‴–OH product in contrast to the enzyme-catalyzed reactions that facilitate regioselective acylation at 6‴–OH.

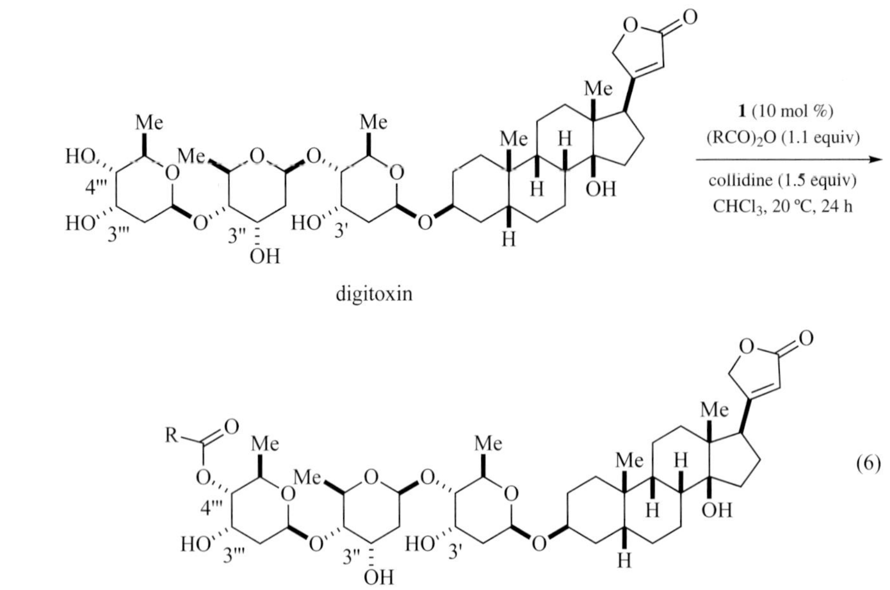

(6)

R = Me, regioselectivity >99%, yield: 98%
R = *i*-Pr, regioselectivity >99%, yield: 98%
R = Ph, regioselectivity >99%, yield: 90%

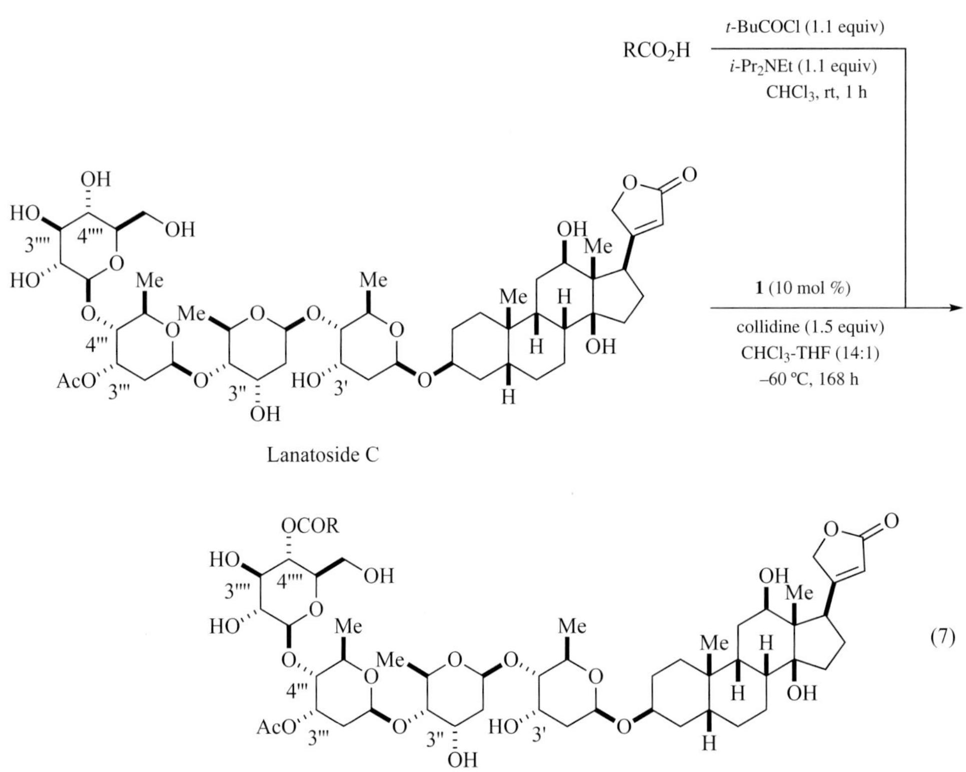

(7)

R = $C_{15}H_{31}$-, regioselectivity 4''''-*O*:3''''-*O*: 82:18, monoacylate yield: 86%

R = 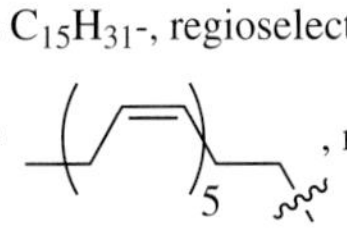 , regioselectivity 4''''-*O*:3''''-*O*: 82:18, monoacylate yield: 86%

Chemoselective Acylation of Linear Diols. Catalyst **1** has been used as an organocatalyst for the chemoselective monoacylation of linear diols. 1,5-Pentanediol on reaction with isobutyric anhydride and 10 mol % catalyst **1** gives monoacylated product in high yield with only 3% recovery of starting material (eq 8).[5] By comparison, using DMAP as catalyst leads to the formation of 26% diacylated product, 45% monoacylated product, and 23% recovery of starting material. Solvent plays a dramatic role in controlling the mono/diacylation selectivity with DMF and toluene giving low selectivity of 1.7 and 8.2, respectively, while chloroform gives high selectivity of 71.0. The mono/diacylation selectivity is also dependent on the carbon chain length of linear diols. For linear diols with carbon chain length of greater than 5, mono/diacylation selectivity is observed to be <7. Competing acylation reaction between 1,5-pentandiol and several mono-alcohols using isobutyric anhydride and 10 mol % catalyst **1** provides monoacylation of 1,5-pentandiol in high yield with only minor amount of acylation on added mono-alcohols (eq 9). Under similar conditions, use of DMAP as catalyst gives 38% monoacylated product **4**, 13% diacylated product, and 52% acylation on the mono-alcohol.

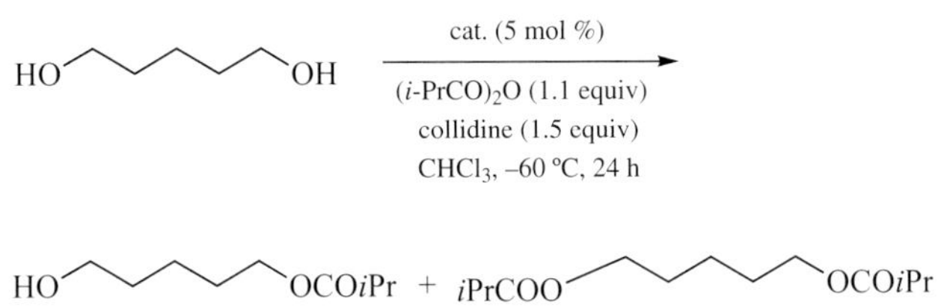

4 **5**

cat.: DMAP, yield **4**: 45%, yield **5**: 26%, recovery of SM: 23%
cat.: **1**, yield **4**: 92%, yield **5**: 3%, recovery of SM: 3% (8)

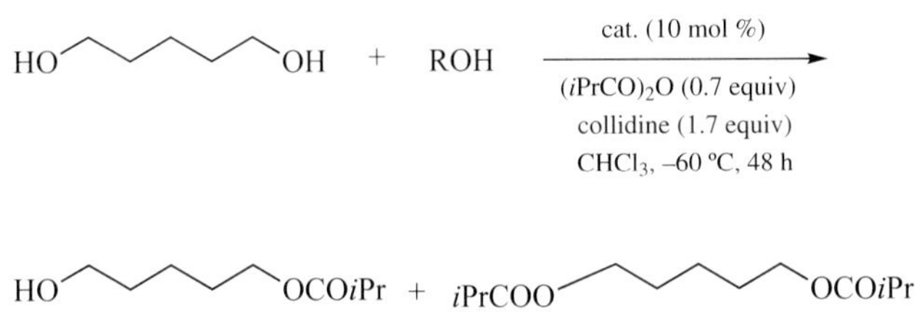

4 **5**

+ R-OCO*i*Pr (9)

6

cat.: **1**, R = AcO(CH_2)$_5$-, yield **4**: 91%, yield **5**: 8%, yield **6**: 4%
cat.: DMAP, R = AcO(CH_2)$_5$-, yield **4**: 38%, yield **5**: 13%, yield **6**: 52%
cat.: **1**, R = MeO(CH_2)$_5$-, yield **4**: 92%, yield **5**: 6%, yield **6**: 7%
cat.: **1**, R = Ph(CH_2)$_5$-, yield **4**: 90%, yield **5**: 7%, yield **6**: 6%

First Update

Yoshihiro Ueda & Takeo Kawabata
Kyoto University, Uji, Japan

Application of Regioselective Acylation of Glucopyranosides to Total Synthesis of Natural Glycosides, Ellagitannins. Short-step total syntheses of a family of ellagitannins can be achieved by sequential and regioselective acylation of unprotected glucose.[6] The key reaction is regioselective introduction of a galloyl group into the inherently less reactive secondary hydroxyl group of 1-*O*-gallate (**7**) catalyzed by catalyst **1**. Glycoside **7** is readily prepared in one step from unprotected D-glucose by the reaction with a gallic acid derivative under Mitsunobu conditions. Galloylation of **7** catalyzed by **1** takes place at C(4)-OH in $CHCl_3$/collidine (9:1) at $-40\,°C$ to give desired 4-*O*-acylate (**9**) in 91% yield (eq 10). Furthermore, a combination of catalyst- and substrate-controlled regioselective acylation provides a key intermediate (**10**) in 51% yield in one pot, which enables straightforward total syntheses of strictinin and tellimagrandin II in five and six steps, respectively, from unprotected D-glucose (eq 11).

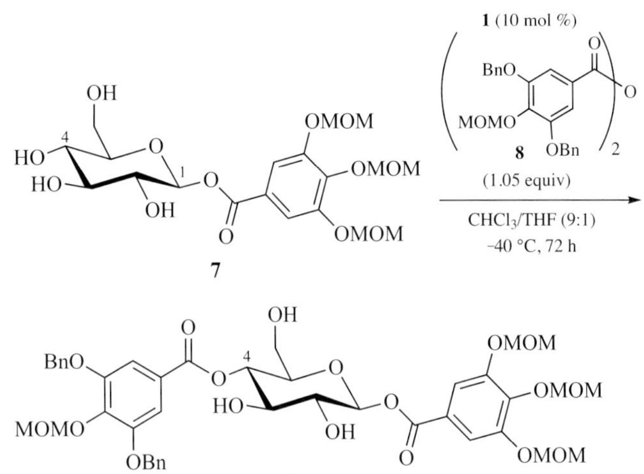

7

9

91% yield (10)

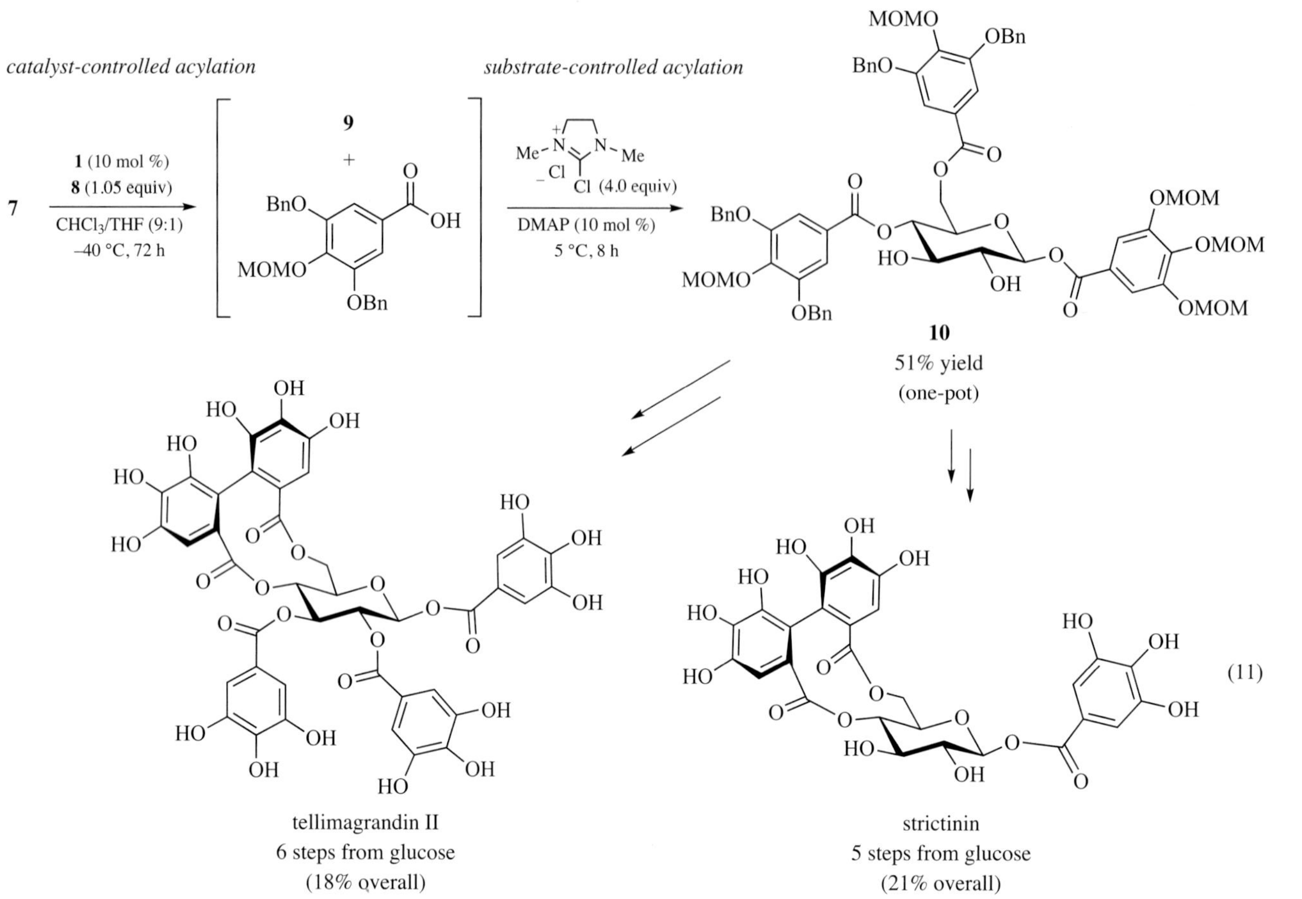

1. Kawabata, T.; Muramatsu, W.; Nishio, T.; Shibata, T.; Schedel, H., *J. Am. Chem. Soc.* **2007**, *129*, 12890.
2. Ueda, Y.; Muramatsu, W.; Mishiro, K.; Furuta, T.; Kawabata, T., *J. Org. Chem.* **2009**, *74*, 8802.
3. Yoshida, K.; Furuta, T.; Kawabata, T., *Tetrahedron Lett.* **2010**, *51*, 4830.
4. Ueda, Y.; Mishiro, K.; Yoshida, K.; Furuta, T.; Kawabata, T., *J. Org. Chem.* **2012**, *77*, 7850.
5. Yoshida, K.; Furuta, T.; Kawabata, T., *Angew Chem., Int. Ed.* **2011**, *50*, 4888.
6. Takeuchi, H.; Mishiro, K.; Ueda, Y.; Fujimori, Y.; Furuta, T.; Kawabata, T., *Angew. Chem., Int. Ed.* **2015**, *54*, 6177.

U

Urea, *N*-[3,5-Bis(trifluoromethyl)phenyl]-*N′*-[(8a,9*S*)-6′-methoxycinchonan-9-yl]-; Urea, *N*-[3,5-Bis(trifluoromethyl)phenyl]-*N′*-(9*R*)-cinchonan-9-yl-

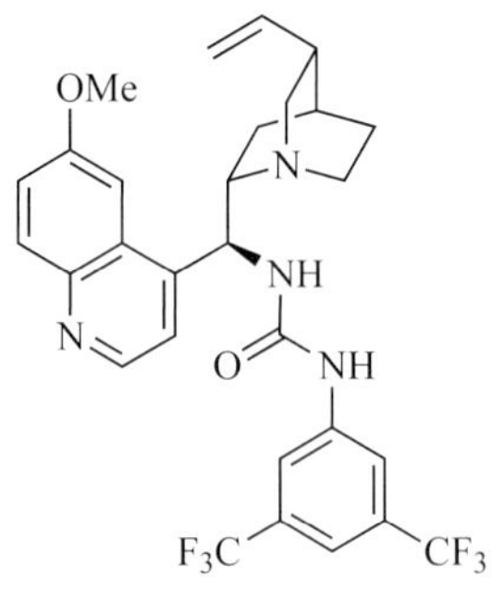

(1)

N-[3,5-Bis(trifluoromethyl)phenyl]-*N′*-(8a,9*S*)-6′-methoxycinchonan-9-yl-urea (**1**)
[957770-66-0] $C_{29}H_{28}F_6N_4O_2$ (MW 578.55)

InChI = 1S/C29H28F6N4O2/c1-3-16-15-39-9-7-17(16)10-25(39)26(22-6-8-36-24-5-4-21(41-2)14-23(22)24)38-27(40)37-20-12-18(28(30,31)32)11-19(13-20)29(33,34)35/h3-6,8,11-14,16-17,25-26H,1,7,9-10,15H2,2H3,(H2,37,38,40)/t16?,17?,25?,26-/m0/s1

InChIKey = QGLQHRHYJCQTQI-STGLZICKSA-N

(quinidine/quinine/cinchonidine/cinchonine-derived bifunctional organocatalysts for stereoinduction in a range of reactions)

Physical Data:[1] mp 134–135 °C;[2] $[\alpha]_{rt}^{D}$ −129.8 (c = 1.04, $CHCl_3$); IR;[2] ^{1}H NMR; ^{13}C NMR.

Solubility: toluene, benzene, mesitylene, chlorobenzene, *m*-xylene, *n*-hexane, 1,4-dioxane, CH_2Cl_2, CHClCHCl, $CHCl_3$, CCl_4, DME (1,2-dimethoxyethane), EVE (ethyl vinyl ether), CPME (cyclopentyl methyl ether), ^{t}BuOMe, Et_2O, THF, CH_3CN, EtOAc, pyridine, DMSO, MeOH, EtOH, ^{i}PrOH, ^{t}BuOH.

Form Supplied in: solid.

Analysis of Reagent Purity: NMR.

Preparative Methods: a solution of a 9-amino(9-deoxy) epicinchona alkaloid derivative in anhydrous THF was added slowly to a solution of 3,5-bis(trifluoromethyl)phenyl isocyanate (1.1 equiv) in anhydrous THF at 0 °C. The reaction mixture was allowed to warm to rt and stirred overnight. The resulting mixture was concentrated in vacuo affording ureas **1–4** usually in yields of over 90%.

Purification: Column chromatography (elution gradient: EtOAc:CH_3OH 95:5 to 97.5:2.5 or EtOAc.

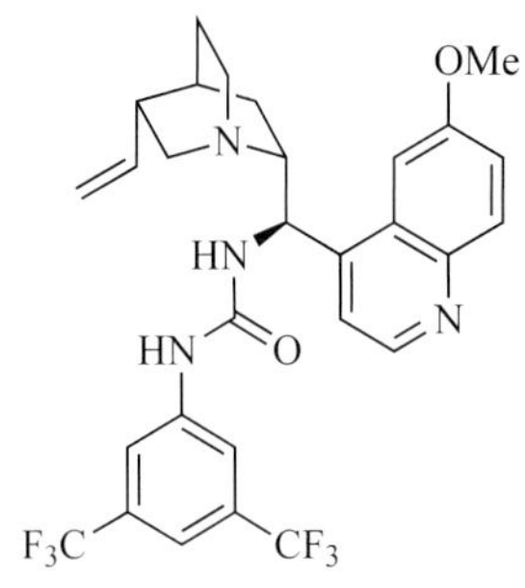

(2)

N-[3,5-Bis(trifluoromethyl)phenyl]-*N′*-(8a,9*R*)-6′-methoxycinchonan-9-yl-urea (**2**)
[1326239-56-8] $C_{29}H_{28}F_6N_4O_2$ (MW 578.55)

InChI = 1S/C29H28F6N4O2/c1-3-16-15-39-9-7-17(16)10-25(39)26(22-6-8-36-24-5-4-21(41-2)14-23(22)24)38-27(40)37-20-12-18(28(30,31)32)11-19(13-20)29(33,34)35/h3-6,8,11-14,16-17,25-26H,1,7,9-10,15H2,2H3,(H2,37,38,40)/t16?,17?,25?,26-/m1/s1

InChIKey = QGLQHRHYJCQTQI-COIZRTFASA-N

Physical Data:[3] mp 133–133.5 °C[2]; $[\alpha]_{rt}^{D}$ +113.1 (c = 1.58, $CHCl_3$); IR; ^{1}H NMR; ^{13}C NMR.

Solubility: toluene, benzene, mesitylene, chlorobenzene, *m*-xylene, *n*-hexane, 1,4-dioxane, CH_2Cl_2, CHClCHCl, $CHCl_3$, CCl_4, DME, (1,2-dimethoxyethane), EVE (ethyl vinyl ether), CPME (cyclopentyl methyl ether), TBME, Et_2O, THF, CH_3CN, EtOAc, pyridine, DMSO, MeOH, EtOH, ^{i}PrOH, ^{t}BuOH.

Form Supplied in: solid.

Analysis of Reagent Purity: NMR.

Purification: column chromatography (elution gradient CH_2Cl_2:MeOH:Et_3N 100:2:3).

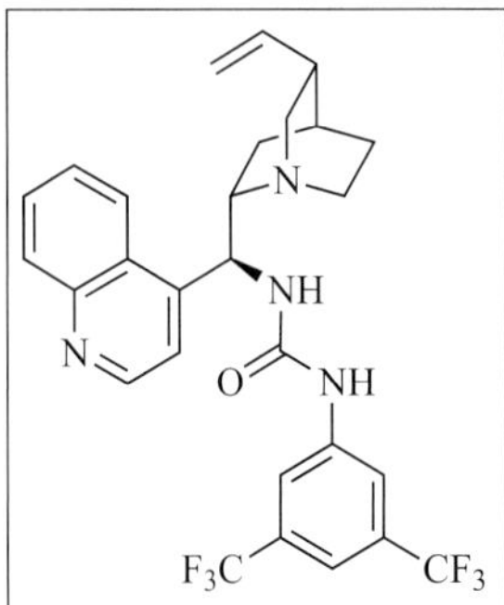

(3)

N-[3,5-Bis(trifluoromethyl)phenyl]-*N′*-(8a, 9*S*)-cinchonan-9-yl-urea (**3**)
[957770-67-1] $C_{28}H_{26}F_6N_4O$ (MW 548.52)

InChI = 1S/C28H26F6N4O/c1-2-16-15-38-10-8-17(16)11-24(38)25(22-7-9-35-23-6-4-3-5-21(22)23)37-26(39)36-20-13-18(27(29,30)31)12-19(14-20)28(32,33)34/h2-7,9,12-14,16-17,24-25H,1,8,10-11,15H2,(H2,36,37,39)/t16?,17?,24?,25-/m0/s1

InChIKey = VNRLCKBQNVNWFW-ZWNNSKJJSA-N

Physical Data:[4] mp 134–136 °C; $[\alpha]_{rt}^{D}$ −6.0 (c = 1.00, MeOH); IR; ^{1}H NMR; ^{13}C NMR.
Solubility: benzene, toluene, *o*-, *m*-, *p*-xylene, mesitylene, CH_2Cl_2, CHClCHCl, $CHCl_3$, CCl_4, THF, CH_3CN, Et_2O, 1,4-dioxane, ^{t}BuOMe, acetone, EtOH.
Form Supplied in: solid.
Analysis of Reagent Purity: NMR.
Purification: column chromatography (elution gradient: EtOAc:MeOH:Et_3N 100:2:0 to 100:10:3).

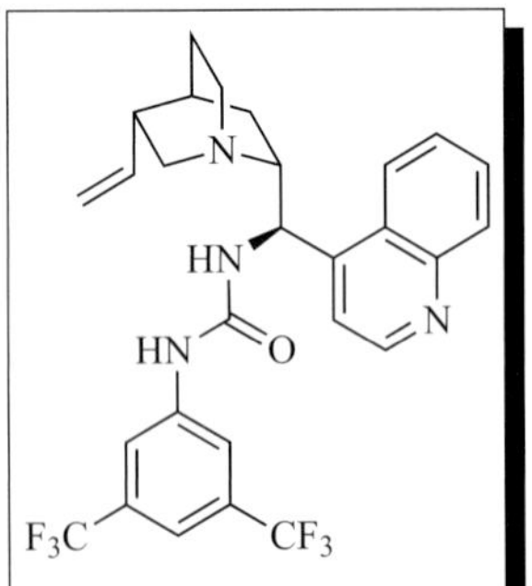

(4)

N-[3,5-Bis(trifluoromethyl)phenyl]-*N′*-(8a, 9*R*)-cinchonan-9-yl-urea (**4**)
[945985-98-8] $C_{28}H_{26}F_6N_4O$ (MW 548.52)

InChI = 1S/C28H26F6N4O/c1-2-16-15-38-10-8-17(16)11-24(38)25(22-7-9-35-23-6-4-3-5-21(22)23)37-26(39)36-20-13-18(27(29,30)31)12-19(14-20)28(32,33)34/h2-7,9,12-14,16-17,24-25H,1,8,10-11,15H2,(H2,36,37,39)/t16?,17?,24?,25-/m1/s1
InChIKey = VNRLCKBQNVNWFW-SPBIILDVSA-N

Physical Data:[5] mp 200–204 °C[6]; $[\alpha]_{rt}^{D}$ +130.0 (c = 0.97, $CHCl_3$); IR; ^{1}H NMR; ^{13}C NMR.
Solubility: benzene, toluene, *o*-, *m*-, *p*-xylene, mesitylene, CH_2Cl_2, CHClCHCl, $CHCl_3$, CCl_4, THF, CH_3CN, Et_2O, 1,4-dioxane, ^{t}BuOMe, acetone, EtOH.
Form Supplied in: solid.
Analysis of Reagent Purity: NMR.
Purification: column chromatography (elution gradient: CH_2Cl_2:MeOH:Et_3N 97:3:0 to 89:10:1).
Handling, Storage, and Precautions: bench stable, structurally similar commercial quinidine is toxic.

Like the other cinchona thioureas, ureas **1**–**4** could be used as an organocatalyst in a variety of reactions from Michael to Mannich additions, cycloadditions, or anhydride desymmetrization. According to the reaction, sometimes ureas show better yields and diastereo/enantioselectivity than thioureas. These organocatalysts have also been immobilized on magnetic nanoparticles[7] and orthosilicates,[8] showing good yields and ee in Michael additions, or have been alkylated in the preparation of useful phase-transfer catalysts for enantio- and diastereoselective nitro-Mannich reaction of amidosulfones.[9] Usually both enantiomers of the product are obtained by changing cinchonine for pseudoenantiomeric cinchonidine (or quinine for quinidine) as starting material in the preparation of the catalyst.

Michael Additions

Intermolecular Sulfa-Michael. Cinchona alkaloid-derived urea **1** (0.1 mol %) has shown good yields (up to >99%) and enantiomeric excesses (up to >99%) in the addition of thiophenols to α,β-unsaturated ketones (eq 1).[10] The reaction was performed in toluene at room temperature for 5–10 h and tolerates well the steric hindrance and electronic nature of the substituents on the aromatic rings. The reaction works well with both acyclic and cyclic ketones. Surprisingly, enantioselectivities are lower with cinchona alkaloid-derived thioureas. Thioacids also add to α,β-unsaturated ketones in the presence of **1** with yields up to 99% and 83% ee (eq 1).[11] 3,5-Bis(trifluoromethyl)phenyl group of the catalyst was found to be crucial to obtain high enantioselectivity. The reaction was shown to work well with cyclic ketones, but the enantioselectivities were lower with acyclic ketones.

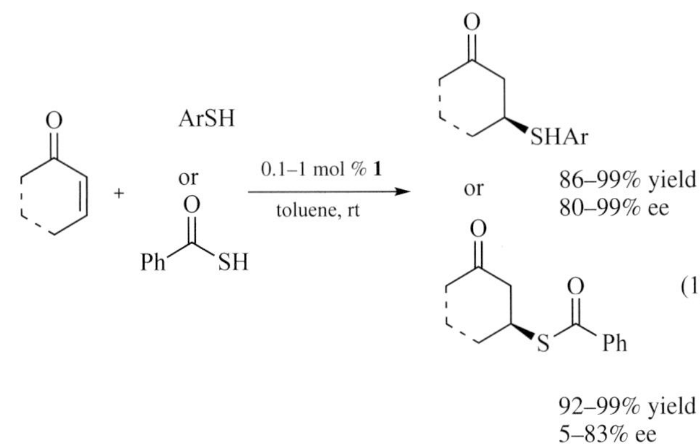

(1)

Intramolecular Aza-Michael. Urea **2** has been shown to catalyze the intramolecular aza-Michael addition for the synthesis of 2-substitued indolines (eq 2).[12] The reaction was found to be applicable to a wide range of substrates. Low-polarity aromatic solvents were necessary to achieve high enantioselectivity and acceptable yields. Substrates possessing the *N*-Cbz protecting group gave the best results. Interestingly, while urea **2** proved to be optimal with electron-poor enones, thiourea analogs were more suitable for electron-rich ones.

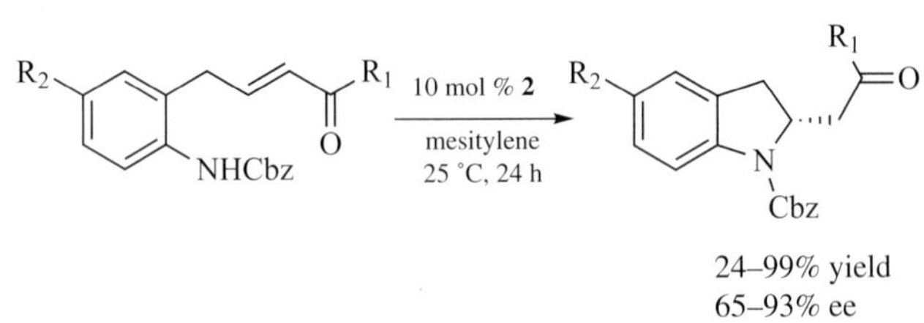

(2)

Intramolecular Oxy-Michael. Ureas **1**–**4** are more efficient than the corresponding thiourea derivatives in the preparation of 2-substituted chromans in high yield (up to 99%) and ee (up to 84%) (eq 3).[13] 4-Fluorophenyl and 2-naphthyl groups on R_1 produced slightly lower enantioselectivity. Whereas an α,β-unsaturated methyl ketone (R_1 = Me) proved to be an unsuccessful substrate, the reaction worked well for α,β-unsaturated thioesters.

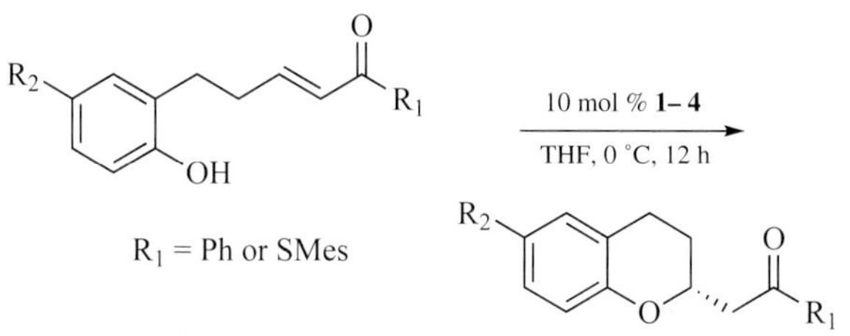

38–99% yield (3)
36–84% ee

Intermolecular Michael Addition of 1,3-Dicarbonyl Compounds. Urea **2** catalyzes the decarboxylative 1,4-addition of malonic acid half thioesters to nitroolefins with good yields (up to 97%) and ee (up to 90%) (eq 4).[14] Urea **2** was found to generate better enantioselectivity than ureas **1** and **3**, showing the importance of the quinuclidine configuration as well as the methoxyquinoline moiety. Urea catalysts possessing the 3,5-bis(trifluoromethyl)phenyl moiety proved to be superior to the analogous thiourea catalysts. The reaction could be carried out in THF or ethyl vinyl ether (which affords lower yields and better enantioselectivity). Electron-deficient aromatic nitroolefins gave better yields than electron-rich ones. In a similar way, ureas **3** and **4** have been used in the addition of α-methyl monothiomalonate esters to nitroolefins (eq 5).[15] Regardless of the protecting group on the phenolic hydroxyl group (R_2 = PMB, allyl, MOM, or Ts), the desired conjugate addition products were obtained in excellent yields (68–84%) and enantioselectivities (95–99%) and high diastereoselectivities (up to 13:1 dr). Taking advantage of the difference in reactivity of thioester and oxoester moieties, the addition products have been used to generate dihydrocoumarins, dihydroquinolinones, or indolin-3-yl acetates (eq 5)[16] with very high control over the relative and absolute configuration using urea **3** or **4**.

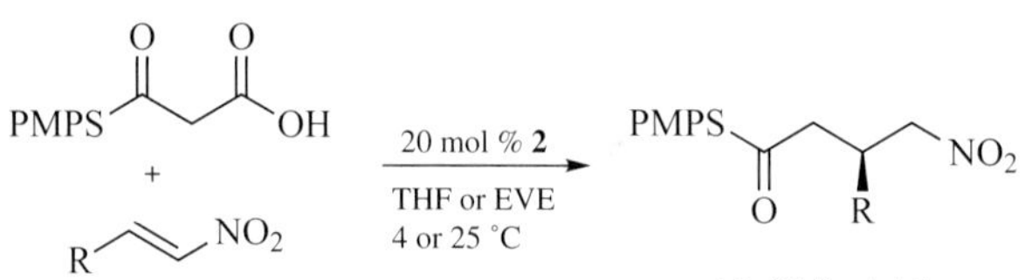

13–99% yield (4)
55–90% ee

Ureas **1–4** also catalyze the conjugate addition of cyclic 1,3-dicarbonyl compounds to β-substituted 2-enoylpyridines with yields up to 96% and 98% ee (eq 5).[17] The reaction is compatible with most general solvents with the exception of acetonitrile and both electron-releasing and electron-withdrawing groups were found to be well tolerated in the 2-enoylpyridine moiety. However, the enantioselectivity was found to be slightly lower with aliphatic Michael acceptors (R_1 = alkyl). Similar results were obtained with malononitrile as nucleophile (eq 6).[18]

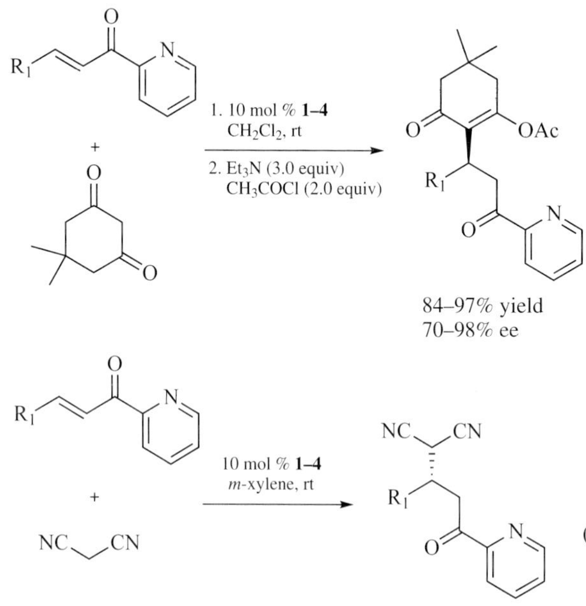

84–97% yield
70–98% ee

78–95% yield
94–97% ee

(6)

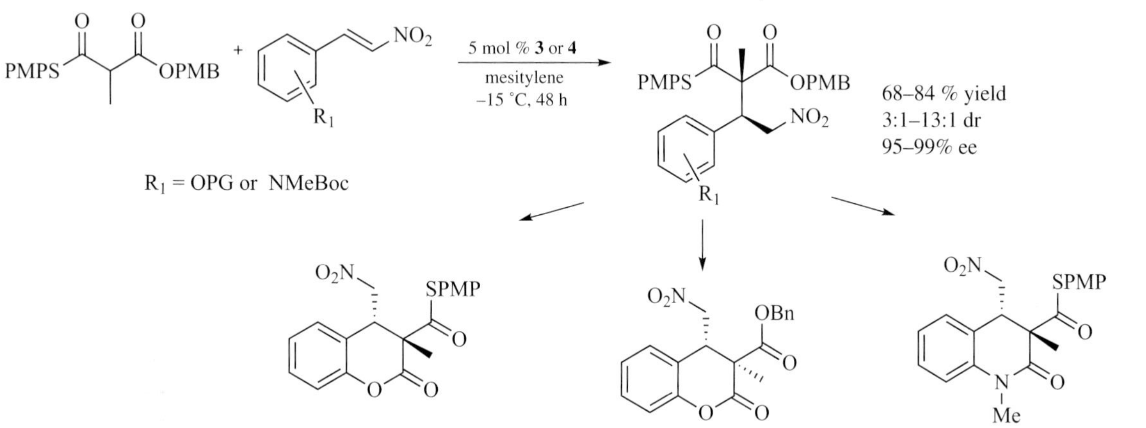

68–84 % yield
3:1–13:1 dr
95–99% ee

(5)

Other C-nucleophiles. Ureas **1** and **2** have been used to carry out a *Z/E*-selective and enantioselective conjugate addition of oxindoles to butyl but-2-ynedioate derivatives (eq 7).[19] Surprisingly, neither quinine nor quinine-derived thiourea catalysts afforded the desired product. However, urea **1** in CCl_4 at 25 °C under a N_2 atmosphere in the presence of 4 Å molecular sieves afforded yields up to 95%, *Z/E* selectivities of 82:18, and up to 96% ee for the *Z*-isomer. The substituent on the nitrogen atom of the oxindole was found to be critical for the success of the reaction. The enantioselectivity was found to be highly dependent on the size of the acetylenedicarboxylate ester moiety with less sterically hindered groups giving rise to lower enantioselectivity.

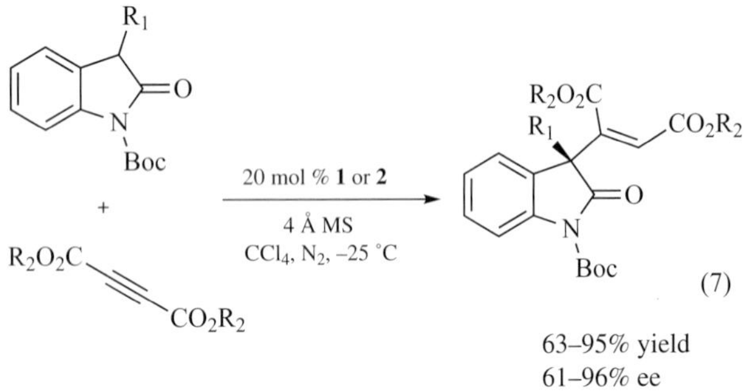

(7)

Urea **2** catalyzes the addition of nitroalkenes to oxindolylideneindolenines providing access to 3,3-disubstitued oxindole derivatives in high yields (77–97%) and ee (78–99%) (eq 8).[20] The transformation has been applied to the preparation of a key intermediate in a formal total synthesis of (+)-gliocladin C. The urea-containing catalysts were found to be more enantioselective than the analogous thiourea derivatives. The N-substituents of the oxidole moiety were found to have a significant effect on both the yield and enantioselectivity. The reaction could be scaled up with good yield and no loss in enantioselectivity. *α*-Branched nitroalkanes were also able to participate in the reaction with good yield and ee.

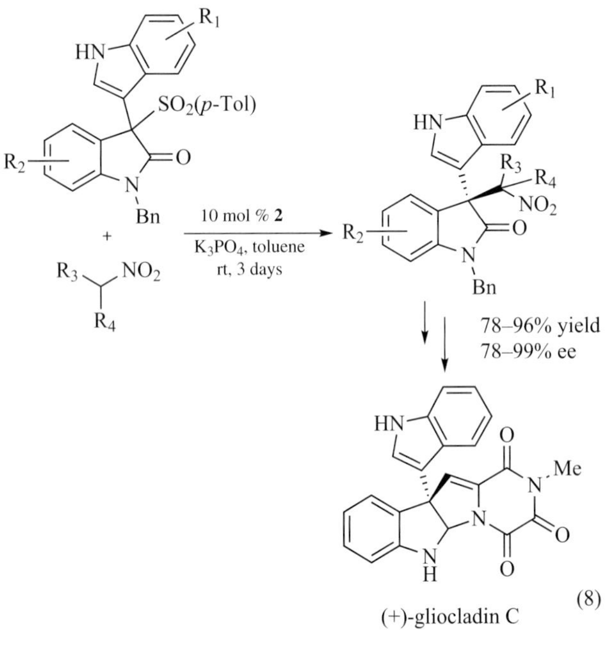

(8)

Cyclohexadienones bearing a bisphenylsulfonyl methylene group were desymmetrized in the presence of urea **4** with good yields (up to 98% and ee of 90%) (eq 9). The reaction was found to work well in $CHCl_3$, CCl_4, toluene, or DCE but not in THF or ether.[21]

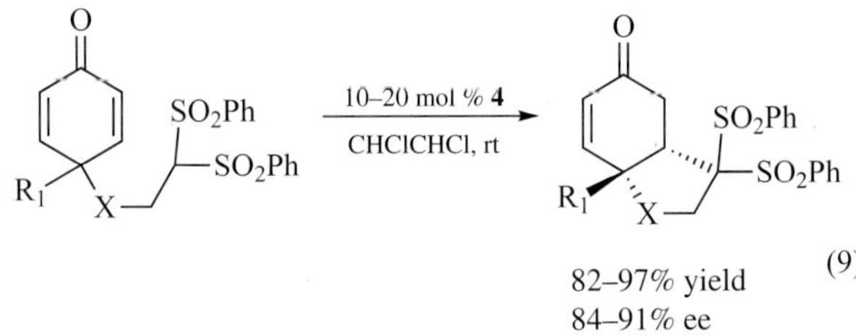

(9)

Mannich. Urea **1** has been used in the addition of monothiomalonate esters to Cbz-protected imines (from aryl and alkyl aldehydes) with yields up to 99%, high diastereoselectivity (>20:1), and excellent enantioselectivity (up to 99%) (eq 10).[22] This strategy provides direct and highly stereoselective access to acyclic β^2- and $\beta^{2,3,3}$-aminothioesters with adjacent tertiary and quaternary stereocenters. The substrate scope proved to be broad with respect to variations in both the imine and the *α*-substituent of the nucleophiles. In comparison with the previously reported methods for the synthesis of such *β*-aminothioesters, the low catalyst loading, the short reaction times, and the excellent enantioselectivities for a broad range of substrates were noteworthy.

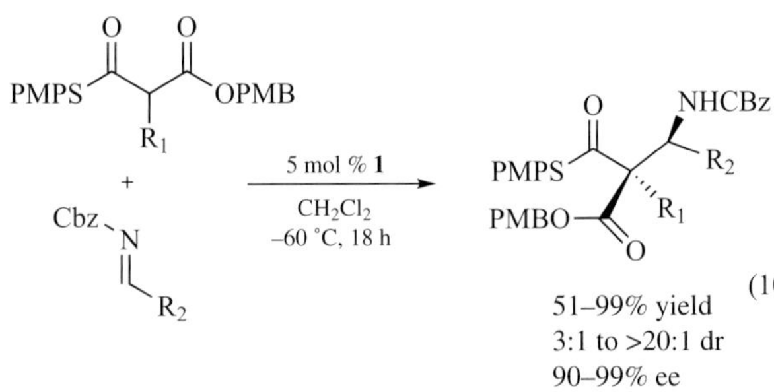

(10)

Both **3** and **4** are able to catalyze the Mannich reaction of 5*H*-oxazol-4-ones onto several sulfonimides derived from aryl and alkyl aldehydes with excellent yields (69–90%) as well as enantio- (90–99%) and diastereoselectivities (>19:1) (eq 11).[23] From the Mannich adducts, important *α*-methyl-*α*-hydroxy-*β*-amino acid derivatives (present in Taxol and Taxotere structures) can be readily accessed. The best results are obtained in Et_2O at −10 °C (−20 °C for sulfonimides derived from aliphatic aldehydes) and the reaction can be performed on gram scale without compromising the enantioselectivity.

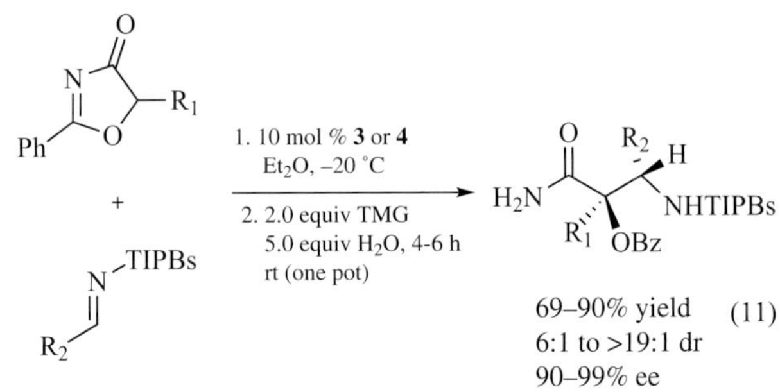

(11)

Other Reactions. Urea **1** catalyzes the solvent-free addition of nitroalkanes to alkylideneindolenines (prepared in situ from arylsulfonylindoles with K_3PO_4) with good yields (60–98%) and ee (up to 99%) although with low diastereoselectivity (ca. 60:40) (eq 12).[24]

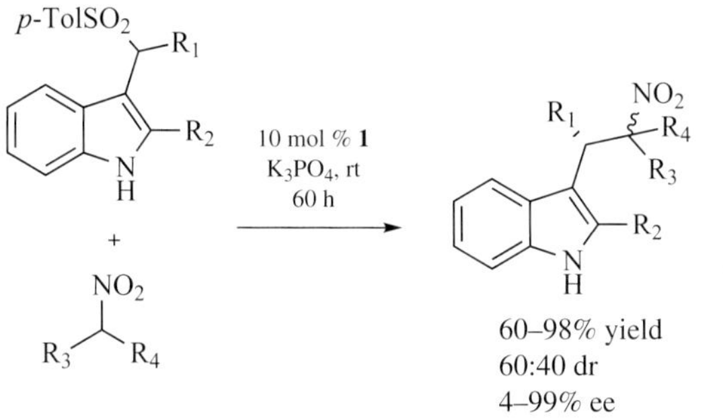

(12)

Ureas **1**–**4** have been used in the enantioselective oxidation of diaryl-substituted *meso*-1,2-diols with *N*-bromophthalimide (NBP) as oxidant at room temperature, affording a wide range of α-hydroxyketones in yields up to 94% and 95% ee (eq 13).[25] *Para*-, *meta*-, and *ortho*-substituted aryl moieties were well tolerated, although dialkyl-substitued diols afforded worse yields and enantioselectivities. The system has also been used for the resolution of racemic *syn*-diols.

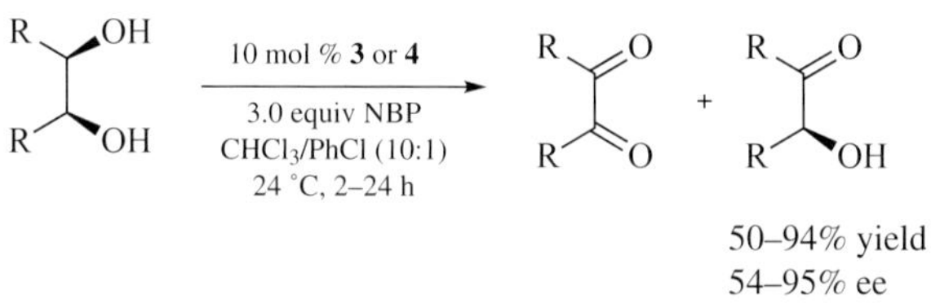

(13)

Ureas **3** and **4** in combination with Cu(I) are able to catalyze the Conia-ene reaction of different β-ketoesters with excellent yields (up to 98%) and enantioselectivities (up to 93%) (eq 14). The best yields were obtained with a molar ratio urea/Cu(I) = 4/1. The reaction is not sensitive to air or moisture.[26]

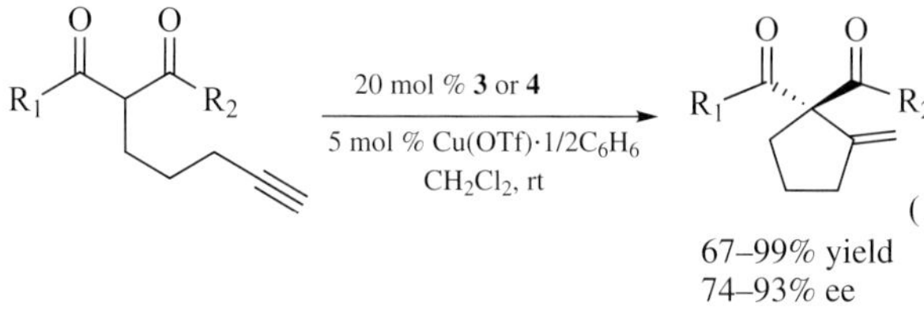

(14)

The aldol reaction between unactivated ketones and several isatin derivatives catalyzed by ureas **1** and **4** afforded functionalized 3-alkyl-3-hydroxyindolin-2-ones in high yields (up to 98%) and enantioselectivities (up to 92%) (eq 15).[27] The reaction works well with acetophenones and acetone, but lower yields and enantioselectivities are obtained with acetaldehyde.

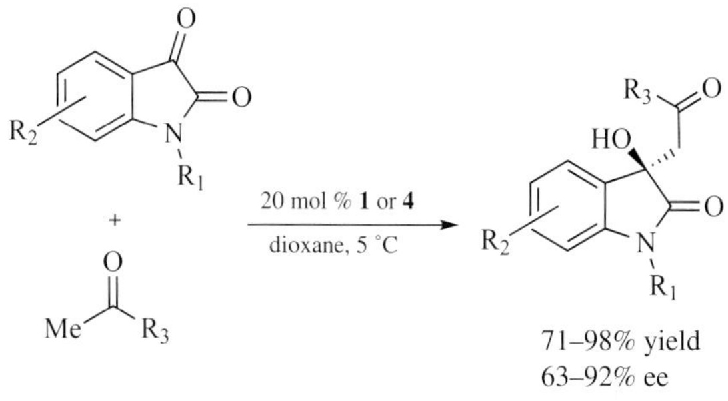

(15)

The aryloxylation of cyclic β-keto esters with *o*-quinones was catalyzed by ureas **2** and **4** yielding cyclic dioxins with good yields (48–92%) and enantioselectivities (65–87%) (eq 16).[3]

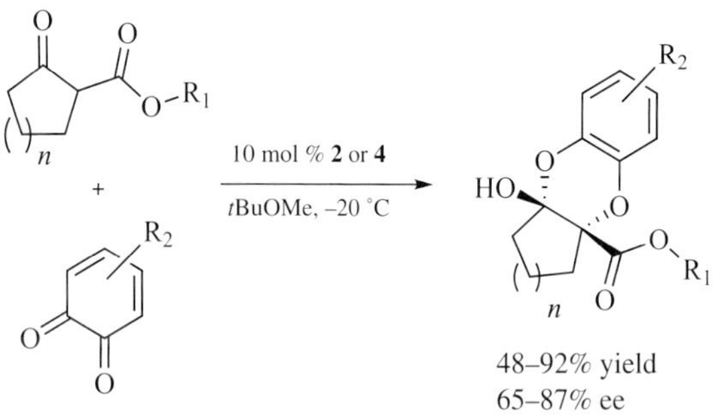

(16)

Urea **2** has also been used in a novel, highly enantioselective aromatic electrophilic bromination of isoquinoline *N*-oxides, being able to translate its molecular torsion into axial chirality in the product (eq 17).[28] Best results (92–99% yield and 93–99% ee) were obtained at 0 °C, with NBA as bromination agent and THF as solvent. This methodology could also be applied to the synthesis of axially chiral benzamides.

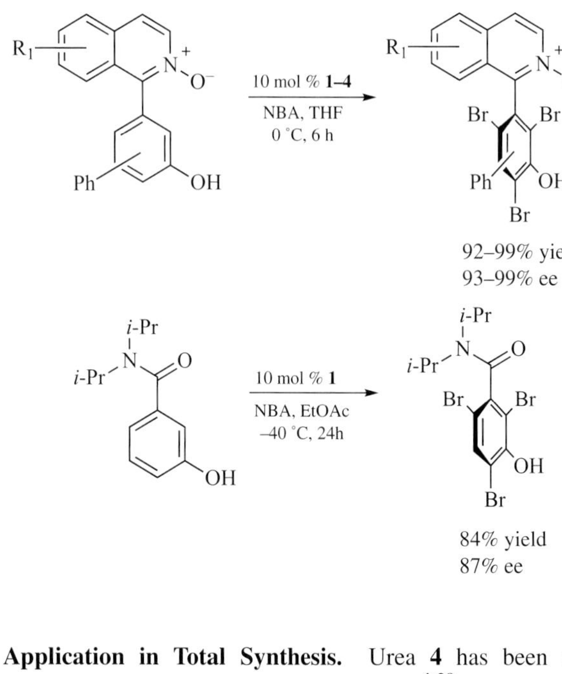

(17)

Application in Total Synthesis. Urea **4** has been used in the total synthesis of (−)-nakadomarin A[4,29] for generating key intermediate building blocks with excellent yield and good diastereoselectivities (eq 18).

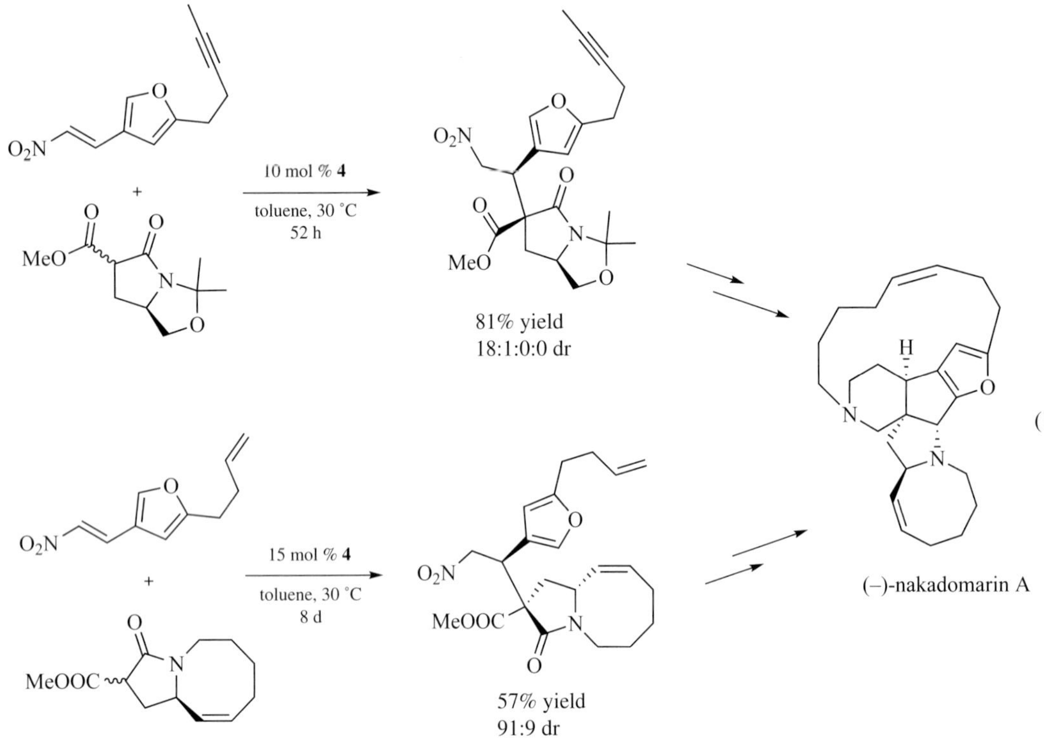

(18)

1. Wu, W.; Min, L.; Zhu, L.; Lee, C.-S., *Adv. Synth. Catal.* **2011**, *353*, 1135.
2. Miyaji, R.; Asano, K.; Matsubara, S., *Org. Lett.* **2013**, *15*, 3658.
3. Greenaway, K.; Dambruoso, P.; Ferrali, A.; Hazelwood, A. J.; Sladojevich, F.; Dixon, D. J., *Synthesis* **2011**, 1880.
4. Kyle, A. F.; Jakubec, P.; Cockfield, D. M.; Cleator, E.; Skidmore, J.; Dixon, D. J., *Chem. Commun.* **2011**, *47*, 10037.
5. Amere, M.; Lasne, M.-C.; Rouden, J., *Org. Lett.* **2007**, *9*, 2621.
6. Jakubec, P.; Cockfield, D. M.; Hynes, P. S.; Cleator, E.; Dixon, D. J., *Tetrahedron: Asymmetry* **2011**, *22*, 1147.
7. Gleeson, O.; Davies, G.-M.; Peschiulli, A.; Tekoriute, R.; Gun'ko, Y.-K.; Connon, S. J., *Org. Biomol. Chem.* **2011**, *9*, 7929.
8. García-García, P.; Zagdoun, A.; Copéret, C.; Lesage, A.; Díaz, U.; Corma, A., *Chem. Sci.* **2013**, *4*, 2006.
9. Johnson, K. M.; Rattley, M. S.; Sladojevich, F.; Barber, D. M.; Nuñez, M. G.; Goldys, A. M.; Dixon, D. J., *Org. Lett.* **2012**, *14*, 2492.
10. Rana, N. K.; Selvakumar, S.; Singh, V. K., *J. Org. Chem.* **2010**, *75*, 2089.
11. Rana, N. K.; Unhale, R.; Singh, V. K., *Tetrahedron Lett.* **2012**, *53*, 2121.
12. Miyaji, R.; Asano, K.; Matsubara, S., *Org. Lett.* **2013**, *15*, 3658.
13. Miyaji, R.; Asano, K.; Matsubara, S., *Org. Biomol. Chem.* **2014**, *12*, 119.
14. Lubkoll, J.; Wennemers, H., *Angew. Chem., Int. Ed.* **2007**, *46*, 6841.
15. Engl, O. D.; Fritz, S. P.; Käslin, A.; Wennemers, H., *Org. Lett.* **2014**, *16*, 5454.
16. Kolarovic, A.; Käslin, A.; Wennemers, H., *Org. Lett.* **2014**, *16*, 4236.
17. Molleti, N.; Allu, S.; Ray, S. K.; Singh, V. K., *Tetrahedron Lett.* **2013**, *54*, 3241.
18. Molleti, N.; Rana, N. K.; Singh, V. K., *Org. Lett.* **2012**, *14*, 4322.
19. Kang, G.; Wu, Q.; Liu, M.; Xu, Q.; Chen, Z.; Chen, W.; Luo, Y.; Ye, W.; Jiang, J.; Wua, H., *Adv. Synth. Catal.* **2013**, *355*, 315.
20. Huang, J.-Z.; Wu, X.; Gong, L.-Z., *Adv. Synth. Catal.* **2013**, *355*, 2531.
21. Gu, Q.; You, S.-L., *Org. Lett.* **2011**, *13*, 5192.
22. Bahlinger, A.; Fritz, S. P.; Wennemers, H., *Angew. Chem., Int. Ed.* **2014**, *53*, 8779.
23. Han, Z.; Yang, W.; Tan, C.-H.; Jiang, Z., *Adv. Synth. Catal.* **2013**, *355*, 1505.
24. Fochi, M.; Gramigna, L.; Mazzanti, A.; Duce, S.; Fantini, S.; Palmieri, A.; Petrini, M.; Bernardia, L., *Adv. Synth. Catal.* **2012**, *354*, 1373.
25. Rong, Z.-Q.; Pan, H.-J.; Yan, H.-L.; Zhao, Y., *Org. Lett.* **2014**, *16*, 208.
26. Yang, T.; Ferrali, A.; Sladojevich, F.; Campbell, L.; Dixon, D. J., *J. Am. Chem. Soc.* **2009**, *131*, 9140.
27. Allu, S.; Molleti, N.; Panem, R.; Singh, V. K., *Tetrahedron Lett.* **2011**, *52*, 4080.
28. Miyaji, R.; Asano, K.; Matsubara, S., *J. Am. Chem. Soc.* **2015**, *137*, 6766.
29. Jakubec, P.; Cockfield, D. M.; Dixon, D. J., *J. Am. Chem. Soc.* **2009**, *131*, 16632.

Darren Dixon & Angel L. Fuentes de Arriba
University of Oxford, Oxford, UK

List of Contributors

Varinder K. Aggarwal — *University of Bristol, Bristol, UK*
- Bicyclo[2.2.1]heptan-2-one, 7,7-Dimethyl-1-(1*S*,3*S*,4*R*)-2-thiabicyclo-[2.2.1]hept-3-yl-, (1*S*,4*R*) 88

Takahiko Akiyama — *Gakushuin University, Tokyo, Japan*
- Dinaphtho[2,1-*d*:1′,2′-*f*][1,3,2]dioxaphos- phepin, 4-Hydroxy-2,6-diphenyl-, 4-Oxide, (11b*R*)- (Family of Reagents) 345

Andrés R. Alcántara — *Complutense University, Madrid, Spain*
- 2-Methyltetrahydrofuran 422

Alexander J. André Cobb — *University of Cambridge, Cambridge, UK*
- (*S*)-2-(5-1*H*-Tetrazolyl)pyrrolidine and (*R*)-2-(5-1*H*-Tetrazolyl)pyrrolidine 595

Merritt B. Andrus — *Brigham Young University, Provo, UT, USA*
- Cinchonanium, 1-(9-Anthracenylmethyl)-9-(propen-1-yloxy)-, Bromide (1:1), (8α,9*R*) 227

Cosimo Annese — *University of Bari, Bari, Italy*
- Methyl(trifluoromethyl)dioxirane 428

Stellios Arseniyadis — *ESPCI ParisTech, Paris, France*
- *N*-[(1*S*)-1-(Hydroxydiphenylmethyl)-3-methylbutyl]-2-pyrrolidinecarboxamide 393

Dev P. Arya — *Northeastern University, Boston, MA, USA*
- Benzyl 1,3-Butadiene-1-carbamate 66

Thorsten Bach — *Technische Universität München, Munich, Germany*
- 1,5,7-Trimethyl-7-(5,6,7,8-tetrahydro-naphth[2,3-*d*]oxazol-2-yl)-3-azabicyclo-[3.3.1]nonan-2-one 621

Katrina Badiola — *University of Oxford, Oxford, UK*
- *N*-Benzylquininium Chloride 78

Edward I. Balmond — *University of Bristol, Bristol, UK*
- *N,N′*-Bis[3,5-bis(trifluoromethyl)-phenyl]thiourea 133

Anthony G. M. Barrett — *Imperial College London, London, UK*
- Tris[(trifluoromethyl)sulfonyl]methane and Related Salts 625

Maria Teresa Barros — *Universidade Nova de Lisboa, Caparica, Portugal*
- (*S*)-2-(Anilinomethyl)pyrrolidine 51

Andreas Bauer — *Technische Universität München, Munich, Germany*
- 1,5,7-Trimethyl-7-(5,6,7,8-tetrahydro-naphth[2,3-*d*]oxazol-2-yl)-3-azabicyclo-[3.3.1]nonan-2-one 621

Joel W. Beatty — *University of Michigan, Ann Arbor, MI, USA*
- Ruthenium(II), Tris(1,10- phenanthroline-$\kappa N1$,$\kappa N10$)-, (OC-6-11)- 559

André M. Beauchemin — *University of Ottawa, Ontario, Canada*
- 2-(Phenylmethoxy)acetaldehyde 489

William S. Bechara — *Université de Montréal, Montréal, Québec, Canada*
- Diethyl 1,4-Dihydro-2,6-dimethyl-3,5-pyridinedicarboxylate 254

Vladimir B. Birman — *Washington University, St. Louis, MO, USA*
- 2,3-Dihydro-2-phenylimidazo[2,1-*b*]benzothiazole 280

James M. Bobbitt — *University of Connecticut, Storrs, CT, USA*
- 4-Acetylamino-2,2,6,6-tetramethylpiperidine-1-oxoammonium Tetrafluoroborate 2
- 4-Acetamido-2,2,6,6-tetramethyl-1-piperidinyloxyl 1

Jeffrey W. Bode — *Eidgenössische Technische Hochschule, Zürich, Switzerland*
- (4a*R*,9a*S*)-4-Hydroxy-4,4a,9,9a-tetrahydroindeno[2,1-*b*][1,4]oxazin-3(2*H*)-one 399

Mary Ellen Bos — *RWTH, Aachen, Germany*
- *N*-Benzylquininium Chloride 78

Gregory R. Boyce — *University of North Carolina at Chapel Hill, Chapel Hill, NC, USA*
- *tert*-Butyl 2-(*tert*-butyldimethylsilyl)-2-oxoacetate 210

D. Christopher Braddock — *Imperial College London, London, UK*
- Tris[(trifluoromethyl)sulfonyl]methane and Related Salts 625

Cyril Bressy — *Université Paul Cezanne, Marseille, France*
- *N*-[(1*S*)-1-(Hydroxydiphenylmethyl)-3-methylbutyl]-2-pyrrolidinecarboxamide 393

Stefanie Breukers — *RWTH Aachen University, Aachen, Germany*
- 2,2-Dimethyl-1,3-dioxan-5-one 315

Jean Michel Brunel — *Aix-Marseille Université, Marseille, France*
- (*R*)-1,1′-Bi-2,2′-naphthol 101

Paul T. Buonora — *Lamar University, Beaumont, TX, USA*
- (*R*)-1,1′-Bi-2,2′-naphthol 101
- Methyl Vinyl Ketone 438

Rich G. Carter — *Oregon State University, Corvallis, OR, USA*
- *N*-[(4-Dodecylphenyl)sulfonyl]-2-pyrrolidinecarboxamide (Hua Cat) 373
- Dodecyl 4-[[[(2*S*)-2-pyrrolidinyl-carbonyl]- amino]sulfonyl]benzoate (Hua Cat II) 376

André B. Charette — *Université de Montréal, Montréal, Québec, Canada*
- Diethyl 1,4-Dihydro-2,6-dimethyl-3,5-pyridinedicarboxylate 254

Dian-Feng Chen — *Colorado State University, Fort Collins, CO, USA*
- 2-Phenyl-6,10b-dihydro-4*H*,5a*H*-5-oxa-3,10c-diaza-2-azonia-cyclopenta[*c*] fluorene tetrafluoroborate; 2-Pentafluorophenyl-6,10b-dihydro-4*H*, 5a*H*-5-oxa-3,10c-diaza-2-azoniacyclopenta[*c*]fluorene tetrafluoroborate; 6,7-Dihydro-2-phenyl-5-(phenylmethyl)-5*H*-pyrrolo[2,1-*c*]-1,2,4-triazolium chloride 474

Ying-Chun Chen — *Sichuan University, Chengdu, China*
- (9*R*)-9-Amino-9-deoxyquinidine 16
- (9*S*)-Amino-9-deoxyquinine 24

Yu Chen — *The University of Hong Kong, Pokfulam Road, Hong Kong*
- 3-Quinuclidinol 547

Yuzhong Chen — *DuPont Crop Protection, Newark, DE, USA*
- Dispiro[2*H*-pyran-2,4′-[4*H*-5,6,8b]-triazaacenaphthylene-7′(5′*H*),2″-[2*H*]-pyran], 1′,2′,2′a,3,3′,3″,4,4″,5,5″,6,6″,8′,8′a-tetradecahydro-1′,2′-dimethoxy-6,6″-dimethyl-, monohydrochloride, (1′*S*,2*R*,2′*S*,2″*R*,2′a*S*,6*S*,6″S,8′a*S*)- 372

Pauline Chiu — *The University of Hong Kong, Pokfulam Road, Hong Kong*
- 3-Quinuclidinol 547

J. Michael Chong — *University of Waterloo, Waterloo, Ontario, Canada*
- [1,1′-Binaphthalene]-2,2′-diol, 3,3′-Bis[3,5-bis(trifluoromethyl)phenyl]-, (1*R*)- 95

Hervé Clavier — *Institute of Chemical Research of Catalonia (ICIQ), Tarragona, Spain*
- 1,3-Dimesityl-4,5-dihydroimidazol-2-ylidene 304
- 1,3-Dimesityl-1*H*-imidazolium Salts and 1,3-Dimesityl-imidazolylidene 306

Daniel T. Cohen — *Northwestern University, Evanston, IL, USA*
- (*S*)-5-Benzyl-2-mesityl-6,6-dimethyl-6,8-dihydro-5*H*-[1,2,4]triazolo[3,4-*C*][1,4]-oxazin-2-ium Tetrafluoroborate 72

John A. Colapret — *Lamar University, Beaumont, TX, USA*
- Methyl Vinyl Ketone 438

Shawn K. Collins — *Université de Montréal, Montréal, Québec, Canada*
- 1-Mesityl-3,4-dimethyl-4*H*-1,2,4-triazol-1-ium Tetrafluoroborate 411
- 3-(2,4,6-Trimethylphenyl)-1-methyl-1*H*-benzo[*d*]imidazol-3-ium Iodide 614
- 1,3-Dihydro-1,3-bis(tricyclo[3.3.1.13,7] dec-1-yl)-2*H*-imidazol-2-yliden 270

Université de Montréal, Montréal, Québec, Canada
- 1,4-Dimethyl-1*H*-1,2,4-triazol-4-ium odide 335

Daniel J. Coughlin — *Johnson Matthey Pharma Services, Devens, MA, USA*
- Diethyl Oxomalonate 261

François Couty — *Institut Lavoisier de Versailles, Versailles, France*
- 4-(1-Pyrrolidinyl)pyridine 525

Pier Giorgio Cozzi — *Universitá degli Studi di Bologna, Bologna, Italy*
- Iron(2+), tris(2,2′-bipyridine-*N,N*′)-, Dibromide, (OC-6-11) 403

Jack K. Crandall — *Indiana University, Bloomington, IN, USA*
- Dimethyldioxirane 320
- Methyl(trifluoromethyl)dioxirane 428

Carina I. C. Crucho — *Universidade Nova de Lisboa, Caparica, Portugal*
- (*S*)-2-(Anilinomethyl)pyrrolidine 51

Ruggero Curci — *Universitá di Bari, Bari, Italy*
- Methyl(trifluoromethyl)dioxirane 428
- Dimethyldioxirane 320

Lucia D'Accolti — *University of Bari, Bari, Italy*
- Methyl(trifluoromethyl)dioxirane 428
- Dimethyldioxirane 320

Stephen G. Davey — *Imperial College London, UK*
- 1-[3,5-Bis(*tert*-butyl)phenyl]octahydro-3,6,6-trimethylcyclopenta[*b*]phosphol 137

Olivier David — *Institut Lavoisier de Versailles, Versailles, France*
- 4-(1-Pyrrolidinyl)pyridine 525

Tyler A. Davis — *Merck Research Laboratories, Rahway, NJ, USA*
- 1,2-Cyclohexanediamine, N^1,N^2-Bis[4-(1-pyrrolidinyl)-2-quinolinyl]-, (1*R*,2*R*)- 235

Li Deng — *Brandeis University, Waltham, MA, USA*
- Quinine 531

Scott E. Denmark — *University of Illinois, Urbana, IL, USA*
- 1,5-Pentanediamine, *N,N*″-Bis[(11b*R*)-3,5-dihydro-3,5-dimethyl-4-oxido-4*H*-dinaphtho[2,1-*d*:1″,2″-*f*]-[1,3,2]diazaphosphepin-4-yl]-*N,N*″-dimethyl 466

University of Bari, Bari, Italy
- [4*S*-(4α,5α)]-1-(1,3-Dimethyl-2-oxido-4,5-diphenyl-1,3,2-diazaphospholidine-2-yl)piperidine 330

Daniel A. DiRocco — *Merck & Co Inc, Kenilworth, NJ, USA*
- 5H-Pyrrolo[2,1-*c*]-1,2,4-triazolium, 7-Fluoro-6,7-dihydro-5-(1-methylethyl)-2-(2,3,4,5,6-pentafluorophenyl)-, (5*R*,7*R*)-, Tetrafluoroborate(1-) 528

Darren Dixon	*University of Oxford, Oxford, UK*	
	• Urea, *N*-[3,5-Bis(trifluoromethyl)phenyl]-*N*′-[(8a,9*S*)-6′–methoxycinchonan-9-yl]-; Urea, *N*-[3,5-Bis(trifluoromethyl)phenyl]-*N*′-(9*R*)-cinchonan-9-yl-	633
	• 1-[3,5-Bis(trifluoromethyl)phenyl]-3-{(2*S*)-3,3-dimethyl-1-[(triphenyl phosphoranylidene)amino] butan-2-yl}thiourea	189
	• *N*-[3,5-Bis(trifluoromethyl)phenyl]-*N*′-(9*R*)-cinchonan-9-yl-thiourea	183
Ulf-H. Dolling	*Merck Research Laboratories, Rahway, NJ, USA*	
	• *N*-[4-(Trifluoromethyl)benzyl]-cinchoninium Bromide	612
Wei Du	*Université de Caen, Caen, France*	
	• (9*R*)-9-Amino-9-deoxyquinidine	16
	Sichuan University, Chengdu, China	
	• (9*S*)-Amino-9-deoxyquinine	24
Richard Eaves	*University of Warwick, Coventry, UK*	
	• (1*R*,2*S*)-1-Amino-2,3-dihydro-1*H*-inden-2-ol	35
Chad C. Eichman	*Northwestern University, Evanston, IL, USA*	
	• (*S*)-5-Benzyl-2-mesityl-6,6-dimethyl-6,8-dihydro-5*H*-[1,2,4]triazolo[3,4-*C*][1,4]-oxazin-2-ium Tetrafluoroborate	72
Dieter Enders	*RWTH Aachen University, Aachen, Germany*	
	• 2,2-Dimethyl-1,3-dioxan-5-one	315
Daniel R. Fandrick	*Boehringer Ingelheim Pharmaceuticals, Inc., Ridgefield, CT, USA*	
	• 1,3-Di-*tert*-butylimidazol-2-ylidene	247
Xin Fang	*College of Chemistry and Molecular Sciences, Wuhan University, Wuhan, China*	
	• 3-[[3,5-Bis(trifluoromethyl)phenyl]-amino]-4-[[(1*R*,2*R*)-2-(dimethylamino)-cyclohexyl]amino]-3-cyclobutene-1,2-dione	180
Alistair J. M. Farley	*University of Oxford, Oxford, UK*	
	• 1-[3,5-Bis(trifluoromethyl)phenyl]-3-{(2*S*)-3,3-dimethyl-1-[(triphenylphosphoranylidene) amino] butan-2-yl} thiourea	189
Stephanie Felten	*University of Utah, Salt Lake City, UT, USA*	
	• 1,3-bis(2,6-diisopropylphenyl)-4,5-dihydroimidazol-2-ylidene	139
Nicolas Fleury-Bregeot	*University of Pennsylvania, Philadelphia, PA, USA*	
	• 1,4-Diazabicyclo [2.2.2]octane	241
George C. Fortman	*University of St. Andrews, St. Andrews, UK*	
	• 1,3-Bis(2,6-diisopropylphenyl)imidazolidinium Chloride	146
Andrea Lindsay Frederick	*Eli Lilly and Company, Indianapolis, IN, USA*	
	• (*S*)-2-(Anilinomethyl)pyrrolidine	51
Antonio Simón Fuentes	*Universidad de Valencia, Burjassot, Spain*	
	• (–)-(4*R*,5*R*)-4,5-Bis[hydroxy(diphenyl)methyl]-2,2-dimethyl-1,3-dioxolane	152
Angel L. Fuentes de Arriba	*University of Oxford, Oxford, UK*	
	• Urea, *N*-[3,5-Bis(trifluoromethyl)phenyl]-*N*′-[(8a,9*S*)-6′-methoxycinchonan-9-yl]-; Urea, *N*-[3,5-Bis(trifluoromethyl)phenyl]-*N*′-(9*R*)-cinchonan-9-yl-	633
Kazuhiro Fukumoto	*Kyoto University, Kyoto, Japan*	
	• 4,4′-Spirobi[4*H*-dinaphth[2,1-*c*:1′,2′-*e*]-azepinium],3,3′,5,5′-tetrahydro-2,6-bis(3,4,5-trifluorophenyl), bromide	561
Shunichi Fukuzumi	*Osaka University, Osaka, Japan*	
	• 9-Mesityl-10-methylacridinium Perchlorate	413

Laura Furst — *Boston University, Boston, MA, USA*
- Ruthenium(II), Tris(2,2′-bipyridine-*κN*1,*κN*1′)-, (OC-6-11)- 551

Caterina Fusco — *University of Bari, Bari, Italy*
- Dimethyldioxirane 320
- Methyl(trifluoromethyl)dioxirane 428

M. Carmen Galan — *University of Bristol, Bristol, UK*
- *N,N*′-Bis[3,5-bis(trifluoromethyl)-phenyl]thiourea 133

Jean-Pierre Genêt — *Ecole Nationale Supérieure de Chimie, Paris, France*
- 4,5-Dihydro-4-phenyl-3*H*-dinaphtho[2,1-*c*:1′,2′-*e*]phosphepin 277

Serafino Gladiali — *Università di Sassari, Sassari, Italy*
- 4,5-Dihydro-4-phenyl-3*H*-dinaphtho[2,1-*c*:1′,2′-*e*]phosphepin 277

R. Richard Goehring — *Scios Nova, Baltimore, MD, USA*
- Chloramine-T 219

Liu-Zhu Gong — *University of Science and Technology of China, Hefei, China*
- (11b*R*)-Dinaphtho[2,1-*d*:1′,2′-*f*][1,3,2] dioxaphosphepin, 2,6-Bis[3,5-bis (trifluoromethyl)phenyl]-4-hydroxy-, 4-Oxide 339
- (11b*R*)-Dinaphtho[2,1-*d*:1′,2′-*f*][1,3,2]dioxaphosphepin, 4-Hydroxy-2,6-di-2-naphthalenyl-, 4-Oxide 343

Kurt V. Gothelf — *Aarhus University, Aarhus, Denmark*
- Nitroethane 451

Gabriela A. Grasa — *University of New Orleans, LA, USA*
- 1,3-Bis(2,6-diisopropylphenyl)imidazol-2-ylidene 149

Michel Gravel — *University of Saskatchewan, Saskatoon, Saskatchewan, Canada*
- 3-Benzyl-5-(2-hydroxyethyl)-4-methyl-1,3-thiazolium Chloride 69
- 1,2,4-Triazolo[4,3-*a*]pyridinium, 5,6,7,8-tetrahydro-2-(2,3,4,5,6-pentafluorophenyl)-,Tetrafluoroborate 603

Andrea Gualandi — *Universitá degli Studi di Bologna, Bologna, Italy*
- Iron(2+), tris(2,2′-bipyridine-*N,N*′)-, Dibromide, (OC-6-11) 403

Gabriela Guillena — *University de Alicante, Alicante, Spain*
- *N,N*′-(1*S*)-[1,1′-Binaphthalene]-2,2′-diylbis-(2*S*,2′*S*)-pyrrolidine-2-carboxamide 98

Pouyan Haghshenas — *University of Saskatchewan, Saskatoon, Saskatchewan, Canada*
- 3-Benzyl-5-(2-hydroxyethyl)-4-methyl-1,3-thiazolium Chloride 69

Alison Hart — *University of Southern Mississippi, Hattiesburg, MS, USA*
- 4-Dimethylaminopyridine 309

Aiko Hasegawa — *Shionogi & Co., Ltd., Hyogo, Japan*
- 1-[Bis(trifluoromethanesulfonyl)methyl]-2,3,4,5,6-pentafluorobenzene 174

Takuya Hashimoto — *Kyoto University, Kyoto, Japan*
- *N*-Benzylquininium Chloride 78
- 4,4′-Spirobi[4*H*-dinaphth[2,1-*c*:1′,2′-*e*]-azepinium],3,3′,5,5′-tetrahydro-2,6-bis(3,4,5-trifluorophenyl), bromide 561

Alfred Hassner — *Bar-Ilan University, Ramat Gan, Israel*
- 4-Dimethylaminopyridine 309

Susumi Hatakeyama — *Nagasaki University, Nagasaki, Japan*
- *β*-Isocupreidine (*β*-ICD) 406

Manabu Hatano — *Nagoya University, Nagoya, Japan*
- 1,1′-Binaphthyl-2,2′-disulfonic Acid 124

Harry Heaney — *Loughborough University, Loughborough, UK*
- Chloramine-T 219

Hui Jiang
Washington University, St. Louis, MO, USA
- 2,3-Dihydro-2-phenylimidazo[2,1-*b*]benzothiazole 280

Peng Jiao
Beijing Normal University, Beijing, China
- Benzyl 1,3-Butadiene-1-carbamate 66

Jeffrey S. Johnson
University of North Carolina at Chapel Hill, Chapel Hill, NC, USA
- *tert*-Butyl 2-(*tert*-butyldimethylsilyl)-2-oxoacetate 210

Jeffrey N. Johnston
Vanderbilt University, Nashville, TN, USA
- Bromonitromethane 205

Ahmed Kamal
CSIR—Indian Institute of Chemical Technology Hyderabad, India
- *trans*-4-Hydroxy-L-proline 395

Pavan K. Kancharla
Wayne State University, Michigan, USA
- (2*S*)-*N*-(2,3,4,5,6-Pentafluorophenyl)-2-pyrrolidine carboxamide 465

Tai-Ran Kang
Sichuan University, Chengdu, China
- (9*S*)-Amino-9-deoxyquinine 24

Takeo Kawabata
Kyoto University, Uji, Japan
- L-Tryptophan, *N*,*N*′-[[(2*S*,5*S*)-1-(4-pyridinyl)-2,5-pyrrolidinediyl]-dicarbonyl]bis-, 3,3′-dioctyl Ester 628
- D-2-Naphthylalanine, *N*,*N*′-[[(2*S*,5*S*)-1-(4-Pyridinyl)-2,5-pyrrolidinediyl] dicarbonyl]bis-, 3,3′-Dioctyl Ester[1] and L-2-Naphthylalanine, *N*,*N*′-[[(2*S*,5*S*)-1-(4-Pyridinyl)-2,5-pyrrolidinediyl]dicarbonyl]bis-, 3,3′-Dioctyl Ester 449

Roy A. Kelly
University of New Orleans, LA, USA
- 1,3-Bis(2,6-diisopropylphenyl)imidazol-2-ylidene 149

Nessan J. Kerrigan
Oakland University, Rochester, MI, USA
- 6′-Methoxy-(8*S*,9*R*)-cinchonan-9-yl *N*-*tert*-butoxycarbonyl-2-pyrrolidine-carboxylate 417

Pavel Kočovský
University of Glasgow, Glasgow, UK
- 2′-Amino-[1,1′-binaphthalen]-2-ol 13

Marcin Kowalczyk
Monash University, Victoria, Australia
- 1,3-Dihydro-4,5-dimethyl-1,3-bis(1-methylethyl)-2*H*-imidazol-2-ylidene 273

Marisa C. Kozlowski
University of Pennsylvania, Philadelphia, PA, USA
- (5*S*)-2,2,3-Trimethyl-5-(phenylmethyl)-4-imidazolidinone 616

Heinrich Kuhlmann
RWTH Aachen, Aachen, Germany
- 3-Benzyl-5-(2-hydroxyethyl)-4-methyl-1,3-thiazolium Chloride 69
- 3-Benzyl-4-methyl-1,3-thiazolium Chloride 75

Amol A. Kulkarni
Howard University, Washington, DC, USA
- Benzoyl trimethylsilane 61

Steven M. Langdon
University of Saskatchewan, Saskatoon, Saskatchewan, Canada
- 1,2,4-Triazolo[4,3-*a*]pyridinium, 5,6,7,8-tetrahydro-2-(2,3,4,5,6-pentafluorophenyl)-, Tetrafluoroborate 603

Marie-Claire Lasne
Université de Caen, Caen, France
- (9*R*)-9-Amino-9-deoxyquinidine 16

Stephen P. Lathrop
Colorado State University, Fort Collins, CO, USA
- 2-Phenyl-6,10b-dihydro-4*H*,5a*H*-5-oxa-3,10c-diaza-2-azonia-cyclopenta[*c*]fluorene tetrafluoroborate; 2-Pentafluorophenyl-6,10b-dihydro-4*H*M,5a*H*-5-oxa-3,10c-diaza-2-azoniacyclopenta[*c*]fluorene tetrafluoroborate; 6,7-Dihydro-2-phenyl-5-(phenylmethyl)-5*H*-pyrrolo[2,1-*c*]-1,2,4-triazolium chloride 474

Alessandra Lattanzi — *Università di Salerno, Fisciano, Italy*
- 2-Pyrrolidinemethanol-α,α-bis(3,5-dimethylphenyl)-(2*S*); 2-Pyrrolidinemethanol-α,α-bis(3,5-dimethylphenyl)-(2*R*) 517

Tatiana Le Gall-Diop — *Université de Montréal, Montréal, Québec, Canada*
- 1-Mesityl-3,4-dimethyl-4*H*-1,2,4-triazol-1-ium Tetrafluoroborate 411

Ellen M. Leahy — *Affymax Research Institute, Palo Alto, CA, USA*
- Quinine 531
- 3-Quinuclidinol 547

Alison Levens — *Monash University, Victoria, Australia*
- 1,3-Dihydro-4,5-dimethyl-1,3-bis(1-methylethyl)-2*H*-imidazol-2-ylidene 273

Steven V. Ley — *University of Cambridge, Cambridge, UK*
- (*S*)-2-(5-1*H*-Tetrazolyl)pyrrolidine and (*R*)-2-(5-1*H*-Tetrazolyl)pyrrolidine 595

Ximin Li — *Washington University, St. Louis, MO, USA*
- 2,3-Dihydro-2-phenylimidazo[2,1-*b*]benzothiazole 280

Rajinikanth Lingampally — *Brigham Young University, Provo, UT, USA*
- *N*-[(4-Dodecylphenyl)sulfonyl]-2-pyrrolidinecarboxamide (Hua Cat) 373
- Dodecyl 4-[[[(2*S*)-2-pyrrolidinyl-carbonyl]- amino]sulfonyl]benzoate (Hua Cat II)

Guy C. Lloyd-Jones — *University of Bristol, Bristol, UK*
- 3,3′-Bis(trifluoromethanesulfonyl)-2,2′-dihydroxy-1,1′-binaphthalene 172

Deborah A. Longbottom — *University of Cambridge, Cambridge, UK*
- (*S*)-2-(5-1*H*-Tetrazolyl)pyrrolidine and (*R*)-2-(5-1*H*-Tetrazolyl)pyrrolidine 595

Carlos del Pozo Losada — *Universidad de Valencia, Burjassot, Spain*
- (−)-(4*R*,5*R*)-4,5-Bis[hydroxy(diphenyl)methyl]-2,2-dimethyl-1,3-dioxolane 152

Janis Louie — *University of Utah, Salt Lake City, UT, USA*
- 1,3-bis(2,6-diisopropylphenyl)-4,5-dihydroimidazol-2-ylidene 139

Yi Luan — *Boston University, Boston, MA, USA*
- (*R*)-2,2′-Dihydroxy-3,3′diphenyl-1,1′-binaphthyl 301

Ottorino De Lucchi — *Università di Venezia, Venezia, Italy*
- 1,2-Bis(phenylsulfonyl)ethylene 163

David W. Lupton — *Monash University, Victoria, Australia*
- 1,3-Dihydro-4,5-dimethyl-1,3-bis(1-methylethyl)-2*H*-imidazol-2-ylidene 273

Li-Jian Ma — *Okayama University, Okayama, Japan*
- 4-Acetylamino-2,2,6,6-tetramethylpiperidine-1-oxoammonium Tetrafluoroborate 2

Zhenkun Ma — *University of Connecticut, Storrs, CT, USA*
- 4-Acetamido-2,2,6,6-tetramethyl-1-piperidinyloxyl 1

Melissa J. MacDonald — *University of Ottawa, ON, Canada*
- 2-(Phenylmethoxy)acetaldehyde 489

Uppuluri V. Mallavadhani — *Regional Research Laboratory, Bhubaneswer, India*
- 1,4-Diazabicyclo [2.2.2]octane 241

Elisabetta Manoni — *Universitá degli Studi di Bologna, Bologna, Italy*
- Iron(2+), tris(2,2′-bipyridine-*N*,*N*′)-, Dibromide, (OC-6-11) 403

Angela Marinetti — *Institut de Chimie des Substances Naturelles, Gif-sur-Yvette, France*
- 2-Triphenylmethylpyrrolidine 624

Jürgen Martens — *Universität Oldenburg, Oldenburg, Germany*
- (*S*)-Proline 494

Keiji Maruoka
Kyoto University, Kyoto, Japan
- (1*R*)-3,3′-Bis(2,4,6-trimethylphenyl)-[1,1′-binaphthalene]-2,2′-dicarboxylic acid 199
- *N*-Benzylquininium Chloride 78
- 4,4′-Spirobi[4*H*-dinaphth[2,1-*c*:1′,2′-*e*]-azepinium],3,3′,5,5′-tetrahydro-2,6-bis(3,4,5-trifluorophenyl), bromide 561

David J. Mathre
Merck Research Laboratories, Rahway, NJ, USA
- α,α-Diphenyl-2-pyrrolidinemethanol 365

Eoghan M. McGarrigle
University of Bristol, Bristol, UK
- Bicyclo[2.2.1]heptan-2-one,7,7-Dimethyl-1-(1*S*,3*S*,4*R*)-2-thiabicyclo[2.2.1]hept-3-yl-,(1*S*,4*R*) 88

University College Dublin, Belfield, Ireland
- *N*,*N*′-Bis[3,5-bis(trifluoromethyl)-phenyl]thiourea 133

Luca Mengozzi
Universitá degli Studi di Bologna, Bologna, Italy
- Iron(2+), tris(2,2′-bipyridine-*N*,*N*′)-, Dibromide, (OC-6-11) 403

Nabyl Merbouh
University of Connecticut, Storrs, CT, USA
- 4-Acetylamino-2,2,6,6-tetramethylpiperidine-1-oxoammonium Tetrafluoroborate 2

David J. Meyers
Johns Hopkins University School of Medicine, Baltimore, MD, USA
- Dinaphtho[2,1-*d*:1′,2′-*f*][1,3,2]dioxaphos-phepin, 4-Hydroxy-2,6-diphenyl-, 4-Oxide, (11b*R*)-(Family of Reagents) 345

Véronique Michelet
Ecole Nationale Supérieure de Chimie, Paris, France
- 4,5-Dihydro-4-phenyl-3*H*-dinaphtho[2,1-*c*:1′,2′-*e*]phosphepin 277

Koichi Mikami
Tokyo Institute of Technology, Tokyo, Japan
- (*R*)-1,1′-Bi-2,2′-naphthol 101

Frederic Minassian
Université Joseph Fourier, Grenoble, France
- (*S*)-Proline 494

Xavier Mollat du Jourdin
BASF SE, Ludwigshafen, Am Rhein, Germany
- (*R*)-1,1′-Bi-2,2′-naphthol 101

Fernando Montanari
Universitá di Milano, Milan, Italy
- 2,2,6,6-Tetramethylpiperidin-1-oxyl 583

Barbara J. Morgan
University of Pennsylvania, Philadelphia, PA, USA
- (5*S*)-2,2,3-Trimethyl-5-(phenylmethyl)-4-imidazolidinone 616

Shin A. Moteki
Kyoto University, Kyoto, Japan
- (1*R*)-3,3′-Bis(2,4,6-trimethylphenyl)-[1,1′-binaphthalene]-2,2′-dicarboxylic acid 199

Yukihiro Motoyama
Tokyo Institute of Technology, Tokyo, Japan
- (*R*)-1,1/-Bi-2,2/-naphthol 101

Alberto Munoz
Colorado State University, Fort Collins, CO, USA
- 6,7-Dihydro-2-phenyl-5-(phenylmethyl)-5*H*-pyrrolo[2,1-*c*]-1,2,4-triazolium Chloride 286

Robbie J. Mutton
University of Cambridge, Cambridge, UK
- (*S*)-2-(5-1*H*-Tetrazolyl)pyrrolidine and (*R*)-2-(5-1*H*-Tetrazolyl)pyrrolidine 595

Carmen Nájera
University de Alicante, Alicante, Spain
- *N*,*N*′-(1*S*)-[1,1′-Binaphthalene]-2,2′-diylbis-(2*S*,2′*S*)-pyrrolidine-2-carboxamide 98

David J. Nelson
University of Strathclyde, Glasgow, UK
- 1,3-Bis(2,6-diisopropylphenyl)imidazol-2-ylidene 149

SonBinh T. Nguyen
Northwestern University, Evanston, IL, USA
- 1,3-bis(2,6-diisopropylphenyl)-4,5-dihydroimidazol-2-ylidene 139

Theresa H. Nguyen
University of Arkansas, Fayetteville, AR, USA
- Ruthenium(II), Tris(2,2′-bipyrazine-κN^{1}, $\kappa N^{1'}$),[1] (OC-6-11)- 551

David A. Nicewicz
University of North Carolina at Chapel Hill, Chapel Hill, NC, USA
- 9-Mesityl-10-methylacridinium Perchlorate 413

Steven P. Nolan
University of St. Andrews, St. Andrews, UK
- 1,3-Bis(2,6-diisopropylphenyl)imidazolidinium Chloride 146

University of New Orleans, LA, USA
- 1,3-Bis(2,6-diisopropylphenyl)imidazol-2-ylidene 149

Institute of Chemical Research of Catalonia (ICIQ), Tarragona, Spain
- 1,3-Dimesityl-1*H*-imidazolium Salts and 1,3-Dimesityl-imidazolylidene 306
- 1,3-Dimesityl-4,5-dihydroimidazol-2-ylidene 304

Ryoji Noyori
Nagoya University, Nagoya, Japan
- 1,1′-Binaphthyl-2,2′-diyl Hydrogen Phosphate 129

Kei Ohkubo
Osaka University, Osaka, Japan
- 9-Mesityl-10-methylacridinium Perchlorate 413

Takashi Ooi
Nagoya University, Nagoya, Japan
- (3*R*,5*R*,8*R*)-3,8-Bis(1-methylethyl)-2,2,7,7-tetrakis(4-methylphenyl)-1,4,6,9-tetraaza-5λ5-phosphaspiro[4,4]non-5-ene, Hydrochloride salt (1:1) 162

Gareth J. J. Owen-Smith
University of Bristol, Bristol, UK
- 3,3′-Bis(trifluoromethanesulfonyl)-2,2′-dihydroxy-1,1′-binaphthalene 172

Vittorio Pace
University of Vienna, Vienna, Austria
- 2-Methyltetrahydrofuran 422

Jon R. Parquette
The Ohio State University, Columbus, OH, USA
- (1*R*,2*S*,4*R*,5*S*)-2,5-Dimethyl-7-phenyl-7-phosphabicyclo[2.2.1]heptane 334

Lucia Pasquato
CNR-CMRO, Padova, Padova, Italy
- 1,2-Bis(phenylsulfonyl)ethylene 163

Aniruddha P. Patwardhan
Michigan State University, East Lansing, MI, USA
- 2,2′-Diphenyl-[3,3′-biphenanthrene]-4,4′-diol (VAPOL) and 3,3′-Diphenyl-[2,2′-binaphthrene]-1,1′-diol (VANOL) 353

Nicholas J. Peraino
Oakland University, Rochester, MI, USA
- 6′-Methoxy-(8*S*,9*R*)-cinchonan-9-yl *N*-*tert*-butoxycarbonyl-2-pyrrolidine-carboxylate 417

Son M. Pham
University of illinais, Urban, IL, USA
- [4*S*-(4a,5a)]-1-(1,3-Dimethyl-2-oxido-4,5-diphenyl-1,3,2-diazaphospholidine-2-yl)piperidine 330

Julie A. Pigza
University of Southern Mississippi, Hattiesburg, MS, USA
- 4-Dimethylaminopyridine 309

Petri M. Pihko
University of Jyväskylä, Jyvaskyla, Finland
- (2*S*,5*S*)-2-(Bis(4-methoxyphenyl)((trimethylsilyl)oxy)methyl)-5-(4-methoxy-3,5-dimethylphenyl)pyrrolidine 161

Andrey Yu. Platonov
Saint Petersburg, Russia
- (8α,9*R*)-(8″α,9″*R*)-1,1″-[1,3-Phenylene-bis(methylene)]bis[9-hydroxy-cinchonanium] Dibromide and (9*S*)-(9″*S*)-1,1″-[1,3-Phenylene-bis(methylene)] bis[9-hydroxy-cinchonanium] Dibromide 486

Puttaswamy
Bangalore University, Bangalore, India
- Bromamine-T 202

Bangalore University, Central College Campus, Bangalore, India
- Chloramine-T 219

Silvio Quici
Universitá di Milano, Milan, Italy
- 2,2,6,6-Tetramethylpiperidin-1-oxyl 583

Gurubaran S. Raju — *Imperial College London, London, UK*
- Tris[(trifluoromethyl)sulfonyl]methane and Related Salts 625

Thomas A. Ramirez — *Colorado State University, Fort Collins, CO, USA*
- Spiro[6*H*-1,3-dioxolo[4,5-*c*]pyran-6,5′-oxazolidine]-3′-carboxylic acid, tetrahydro-2,2-dimethyl-2′,7-dioxo-,2-methyl-2-propyl ester, (3a*R*, 5′*S*, 7a*R*) 567

Yu Rao — *Tsinghua University, Beijing, China*
- 2,2,6,6-Tetramethylpiperidin-1-oxyl 583

Jonathan T. Reeves — *Boehringer Ingelheim Pharmaceuticals, Inc., Ridgefield, CT, USA*
- 1,3-Di-*tert*-butylimidazol-2-ylidene 247

Linus F. Reichenbach — *RWTH Aachen University, Aachen, Germany*
- 2,2-Dimethyl-1,3-dioxan-5-one 315

Jason M. Rohde — *The Scripps Research Institute, La Jolla, CA, USA*
- (*S*)-1-Methyl-2-[(dihydroisoindol-2-yl)methyl]pyrrolidine 421

Patrick Rollin — *Université d'Orléans, Orléans, France*
- 1,2-Bis(phenylsulfonyl)ethylene 163

Fedor Romanov-Michailidis — *Colorado State University, Fort Collins, CO, USA*
- 1,2-Bis(phenylsulfonyl)ethylene 163

Petra Rönnholm — *Góteborg University, Göteborg, Sweden*
- 1-(2-Pyrrolidinylmethyl)pyrrolidine 520

Jacques Rouden — *Université de Caen, Caen, France*
- (9*R*)-9-Amino-9-deoxyquinidine 16

Tomislav Rovis — *Colorado State University, Fort Collins, CO, USA*
- 2-Phenyl-6,10b-dihydro-4*H*,5a*H*-5-oxa-3,10c-diaza-2-azonia-cyclopenta[*c*]fluorene tetrafluoroborate; 2-Pentafluorophenyl-6,10b-dihydro-4*H*M,5a*H*-5-oxa-3,10c-diaza-2-azoniacyclopenta[*c*]fluorene tetrafluoroborate; 6,7-Dihydro-2-phenyl-5-(phenylmethyl)-5*H*-pyrrolo[2,1-*c*]-1,2,4-triazolium chloride 474
- 6,7-Dihydro-2-phenyl-5-(phenylmethyl)-5*H*-pyrrolo[2,1-*c*]-1,2,4-triazolium Chloride 286

University of Cambridge, Cambridge, UK
- 6,7-Dihydro-2-phenyl-5-(phenylmethyl)-5*H*-pyrrolo[2,1-*c*]-1,2,4-triazolium Chloride 286

Marie-Noelle Roy — *Université de Montréal, Montreal, Québec, Canada*
- Nitroethane 451

Rebecca T. Ruck — *Merck & Co., Inc., Rahway, NJ, USA*
- (1R,2S)-1-Amino-2,3-dihydro-1H-inden-2-ol 35

Kazuhiko Sato — *Nagoya University, Nagoya, Japan*
- 1,1′-Binaphthyl-2,2′-diyl Hydrogen Phosphate 129

Paul B. Savage — *The Ohio State University, Columbus, OH, USA*
- Benzyltriethylammonium Chloride 86

Scott E. Schaus — *Boston University, Boston, MA, USA*
- (*R*)-2,2′-Dihydroxy-3,3′diphenyl-1,1′-binaphthyl 301

Karl A. Scheidt — *Northwestern University, Evanston, IL, USA*
- (*S*)-5-Benzyl-2-mesityl-6,6-dimethyl-6,8-dihydro-5*H*-[1,2,4]triazolo[3,4-*C*][1,4]-oxazin-2-iumTetrafluoroborate 72

Christoph Schneider — *Georg-August-Universität zu Göttingen, Germany*
- Diethyl Oxomalonate 261

Mitchell J. Schultz — *Archer Daniels Midland Company, Decatur, IL, USA*
- 1*H*-Imidazolium, 1,3-Bis[2,6-bis(1-methylethyl)phenyl], tetrafluoroborate(1-) 400

David P. Sebesta	*University of Colorado, Boulder, CO, USA*	
	• Tetrachlorosilane	575
Hirofumi Seike	*The Scripps Research Institute, La Jolla, CA, USA*	
	• (*S*)-1-Methyl-2-[(dihydroisoindol-2-yl)methyl]pyrrolidine	421
Chris H. Senanayake	*Boehringer Ingelheim Pharmaceuticals, Inc., Ridgefield, CT, USA*	
	• 1,3-Di-*tert*-butylimidazol-2-ylidene	247
David M. Shaw	*University of Cambridge, Cambridge, UK*	
	• (*S*)-2-(5-1*H*-Tetrazolyl)pyrrolidine and (*R*)-2-(5-1*H*-Tetrazolyl)pyrrolidine	595
Juan Shen	*National University of Singapore, Singapore*	
	• 2,3,5,6-Tetrahydro-2,6-bis(phenylmethyl)-1*H*-imidazo[1,2-*a*]imidazole	582
Yian Shi	*Colorado State University, Fort Collins, CO, USA*	
	• Spiro[6*H*-1,3-dioxolo[4,5-*c*]pyran-6,5′-oxazolidine]-3′-carboxylic acid, tetrahydro-2,2-dimethyl-2′,7-dioxo-, 2-methyl-2-propyl ester, (3a*R*, 5′*S*, 7a*R*)-	567
Ichiro Shinkai	*Merck Research Laboratories, Rahway, NJ, USA*	
	• α, α-Diphenyl-2-pyrrolidinemethanol	365
Seiji Shirakawa	*Kyoto University, Kyoto, Japan*	
	• (1*R*)-3,3′-Bis(2,4,6-trimethylphenyl)-[1,1′-binaphthalene]-2,2′-dicarboxylic acid	199
Vikas Sikervar	*Arekere Mico Layout, Bangalore, India*	
	• L-Tryptophan, *N,N*′-[[(2*S*,5*S*)-1-(4-pyridinyl)-2,5-pyrrolidinediyl]-dicarbonyl]bis-, 3,3′-dioctyl Ester	628
Martin D. Smith	*University of Oxford, Oxford, UK*	
	• *N*-Benzylquininium Chloride	78
Kenso Soai	*Science University of Tokyo, Tokyo, Japan*	
	• (2*S*,4*S*)-2-(Anilinomethyl)-1-ethyl-4-hydroxypyrrolidine	50
	• (S)-2-(Anilinomethyl)pyrrolidine	51
Aiguo Song	*University of New Mexico, Albuquerque, NM, USA*	
	• α,α-Diphenyl-2-pyrrolidinemethanol	365
	• (5*S*)-2,2,3-Trimethyl-5-(phenylmethyl)-4-imidazolidinone	616
Jinhua J. Song	*Boehringer Ingelheim Pharmaceuticals, Inc., Ridgefield, CT, USA*	
	• 1,3-Di-*tert*-butylimidazol-2-ylidene	247
Erik J. Sorensen	*The Scripps Research Institute, La Jolla, CA, USA*	
	• (*S*)-1-Methyl-2-[(dihydroisoindol-2-yl)methyl]pyrrolidine	421
Alan C. Spivey	*Imperial College London, UK*	
	• 1-[3,5-Bis(*tert*-butyl)phenyl]octahydro-3,6,6-trimethylcyclopenta[*b*]phosphole	137
Linus Stegbauer	*University of Oxford, Oxford, UK*	
	• *N*-[3,5-Bis(trifluoromethyl)phenyl]-*N*′-(9*R*)-cinchonan-9-yl-thiourea	183
Corey R. J. Stephenson	*Boston University, Boston, MA, USA*	
	• Ruthenium(II), Tris(2,2′-bipyridine-$\kappa N1$,$\kappa N1'$)-, (OC-6-11)-	554
	University of Michigan, Ann Arbor, MI, USA	
	• Ruthenium(II), Tris(1,10-phenanthroline-$\kappa N1$,$\kappa N10$)-, (OC-6-11)-	559
R. Ian Storer	*Pfizer Neusentis, Cambridge, UK*	
	• Squaramide	571
Armido Studer	*Westfälische Wilhelms-Universität Münster, Münster, Germany*	
	• 2,2,6,6-Tetramethylpiperidin-1-oxyl	583
Farheen Sultana	*CSIR—Indian Institute of Chemical Technology Hyderabad, India*	
	• *trans*-4-Hydroxy-L-proline	395
Jianwei Sun	*The Hong Kong University of Science and Technology, Kowloon, Hong Kong*	
	• Triflimide	604

Keisuke Takahashi — *Nagasaki University, Nagasaki, Japan*
- β-Isocupreidine (β-ICD) 406

Kiyosei Takasu — *Kyoto University, Kyoto, Japan*
- *N*-[3,5-Bis(trifluoromethyl)phenyl]-*N'*-[(1*R*,2*R*)-2-(dimethylamino)cyclohexyl]-thiourea 186

Yoshiji Takemoto — *Kyoto University, Kyoto, Japan*
- *N*-[3,5-Bis(trifluoromethyl)phenyl]-*N'*-[(1*R*,2*R*)-2-(dimethylamino)cyclohexyl]-thiourea 186
- Bromonitromethane 205

Choon-Hong Tan — *National University of Singapore, Singapore*
- 2,3,5,6-Tetrahydro-2,6-bis(phenylmethyl)-1*H*-imidazo[1,2-*a*]imidazole 582

Zhulin Tan — *Boehringer Ingelheim Pharmaceuticals, Inc., Ridgefield, CT, USA*
- 1,3-Di-*tert*-butylimidazol-2-ylidene 247

Arnaud Tatibouët — *Université d'Orléans, Orléans, France*
- 1,2-Bis(phenylsulfonyl)ethylene 163

Katherine L. Thomas — *Ohio Wesleyan University, Delaware, OH, USA*
- 2-[Bis-(3,5-bistrifluoromethyl)-phenyl]-[(trimethylsilyl)oxy]-methyl]-pyrrolidine 192

Thomas T. Tidwell — *University of Toronto, Toronto, Ontario, Canada*
- 2,2,6,6-Tetramethylpiperidin-1-oxyl 583

Lutz F. Tietze — *Georg-August-Universität zu Göttingen, Germany*
- Diethyl Oxomalonate 261

Nicholas C. O. Tomkinson — *University of Strathclyde, Glasgow, UK*
- 5,6-Dioxaspiro[2.4]heptane-4,7-dione 351

Kurt B. G. Torssell — *Aarhus University, Aarhus, Denmark*
- Nitroethane 451

Patrick Yves Toullec — *Ecole Nationale Supérieure de Chimie, Paris, France*
- 4,5-Dihydro-4-phenyl-3*H*-dinaphtho[2,1-*c*:1',2'-*e*]phosphepin 277

David M. Tschaen — *Merck & Co., Inc., Rahway, NJ, USA*
- (1R,2S)-1-Amino-2,3-dihydro-1H-inden-2-ol 35

Sergey V. Tsukanov — *Vanderbilt University, Nashville, TN, USA*
- Bromonitromethane 205

Yoshihiro Ueda — *Kyoto University, Uji, Japan*
- L-Tryptophan, *N,N'*-[[(2*S*,5*S*)-1-(4-pyridinyl)-2,5-pyrrolidinediyl]-dicarbonyl]bis-, 3,3'-dioctyl Ester 628
- D-2-Naphthylalanine, *N,N'*-[[(2*S*,5*S*)-1-(4-Pyridinyl)-2,5-pyrrolidinediyl] dicarbonyl]bis-, 3,3'-Dioctyl Ester and L-2-Naphthylalanine, *N,N'*-[[(2*S*,5*S*)-1-(4-Pyridinyl)-2,5-pyrrolidinediyl]dicarbonyl]bis-, 3,3'-Dioctyl Ester 449

Daisuke Uraguchi — *Nagoya University, Nagoya, Japan*
- (3*R*,5*R*,8*R*)- 3,8-Bis(1-methylethyl)-2,2,7,7-tetrakis(4-methylphenyl)-1,4,6,9-tetraaza-5λ5-phosphaspiro[4,4]non-5-ene, Hydrochloride Salt (1:1) 162

Thomas Vogler — *Westfälische Wilhelms-Universität Münster, Münster, Germany*
- 2,2,6,6-Tetramethylpiperidin-1-oxyl 583

Harit U. Vora — *Colorado State University, Fort Collins, CO, USA*
- 2-Phenyl-6,10b-dihydro-4*H*,5a*H*-5-oxa-3,10c-diaza-2-azonia-cyclopenta[*c*]fluorene tetrafluoroborate; 2-Pentafluorophenyl-6,10b-dihydro-4*H*M,5a*H*-5-oxa-3,10c-diaza-2-azoniacyclopenta[*c*]fluorene Tetrafluoroborate; 6,7-Dihydro-2-phenyl-5-(phenylmethyl)-5*H*-pyrrolo[2,1-*c*]-1,2,4-triazolium Chloride 474

	Merck Research Laboratories, West Point, PA	
	• (–)-(4*R*,5*R*)-4,5-Bis[hydroxy(diphenyl)methyl]-2,2-dimethyl-1,3-dioxolane (TADDOL)	152
Sabine Wallbaum	*Universität Oldenburg, Oldenburg, Germany*	
	• (*S*)-Proline	494
Chun-Jiang Wang	*College of Chemistry and Molecular Sciences, Wuhan University, Wuhan, China*	
	• 3-[[3,5-Bis(trifluoromethyl)phenyl]-amino]-4-[[(1*R*,2*R*)-2-(dimethylamino)-cyclohexyl]amino]-3-cyclobutene-1,2-dione	180
Jiang Wang	*University of Arkansas, Fayetteville, AR, USA*	
	• Ruthenium(II), Tris(2,2′-bipyrazine-κN^1, $\kappa N^{1'}$), (OC-6-11)-	551
Wei Wang	*University of New Mexico, Albuquerque, NM, USA*	
	• α, α-Diphenyl-2-pyrrolidinemethanol	365
	• (5*S*)-2,2,3-Trimethyl-5-(phenylmethyl)-4-imidazolidinone	616
Nicholas A. White	*Colorado State University, Fort Collins, CO, USA*	
	• 4-Nitropyridine *N*-Oxide	460
Martin Wills	*University of Warwick, Coventry, UK*	
	• (1*R*,2*S*)-1-Amino-2,3-dihydro-1*H*-inden-2-ol	35
Yongwei Wu	*Brandeis University, Waltham, MA, USA*	
	• Quinine	531
William D. Wulff	*Michigan State University, East Lansing, MI, USA*	
	• 2,2′-Diphenyl-[3,3′-biphenanthrene]-4,4′-diol (VAPOL) and 3,3′-Diphenyl-[2,2′-binaphthrene]-1,1′-diol (VANOL)	353
Thomas Wynn	*Biogen Idec, Inc., Cambridge, MA, USA*	
	• Tetrachlorosilane	575
Ming Xian	*Washington State University, Pullman, WA, USA*	
	• 2-[Bis-(3,5-bistrifluoromethyl)-phenyl]-[(trimethylsilyl)oxy]-methyl]-pyrrolidine	192
Hua Yang	*Central South University, Changsha, Hunan, China*	
	• Dodecyl 4-[[[(2*S*)-2-pyrrolidinyl-carbonyl]-amino]sulfonyl]benzoate (Hua Cat II)	376
	• *N*-[(4-Dodecylphenyl)sulfonyl]-2-pyrrolidinecarboxamide (Hua Cat)	373
Qingwei Yao	*Northern Illinois University, DeKalb, IL, USA*	
	• 9-Azabicyclo[3.3.1]nonane-*N*-oxyl	59
	• (11b*R*)-4-(1,1-Dimethylethyl)-4,5-dihydro-3*H*-dinaphtho[2,1-*c*:1′,2′-*e*]phosphepin	327
	• 1-Methyl-2-azaadamantane *N*-Oxyl	420
	• 1,1′-Binaphthyl-2,2′-diyl Hydrogen Phosphate	129
Song Ye	*Chinese Academy of Sciences, Beijing, China*	
	• (*S*)-5-((*tert*-Butyldimethylsilyloxy)diphenylmethyl)-2-phenyl-6,7-dihydro-5*H*-pyrrolo[2,1-*c*][1,2,4]triazol-2-ium Tetrafluoroborate 1a (and Related Reagents)	214
Zheng-Qing Ye	*Shanghai Institute of Organic Chemistry, Chinese Academy of Sciences, Peoples' Republic of China*	
	• *cis*-4-Benzyloxy-α,α-bis-(3,5-dimethylphenyl)-L-prolinol (2*S*,4*R*)	76
Nathan K. Yee	*Boehringer Ingelheim Pharmaceuticals, Inc., Ridgefield, CT, USA*	
	• 1,3-Di-*tert*-butylimidazol-2-ylidene	247
Gu Zhan	*Sichuan University, Chengdu, China*	
	• (9*S*)-Amino-9-deoxyquinine	24
Chao Zhang	*Tsinghua University, Beijing, China*	
	• 2,2,6,6-Tetramethylpiperidin-1-oxyl	583
Gang Zhao	*Shanghai Institute of Organic Chemistry, Chinese Academy of Sciences, Peoples' Republic of China*	
	• *cis*-4-Benzyloxy-α,α-bis-(3,5-dimethylphenyl)-L-prolinol (2*S*,4*R*)	76

Reagent Formula Index

Subject Index

Index notes: **Bold headings** indicate main reagent titles, *see* indicates alternative names and acronyms, *see also* indicates related reagents and concepts.